The chemistry of
organic silicon compounds
Volume 2

THE CHEMISTRY OF FUNCTIONAL GROUPS

*A series of advanced treatises under the general editorship of
Professors Saul Patai and Zvi Rappoport*

The chemistry of alkenes (2 volumes)
The chemistry of the carbonyl group (2 volumes)
The chemistry of the ether linkage
The chemistry of the amino group
The chemistry of the nitro and nitroso groups (2 parts)
The chemistry of carboxylic acids and esters
The chemistry of the carbon–nitrogen double bond
The chemistry of amides
The chemistry of the cyano group
The chemistry of the hydroxyl group (2 parts)
The chemistry of the azido group
The chemistry of acyl halides
The chemistry of the carbon–halogen bond (2 parts)
The chemistry of the quinonoid compounds (2 volumes, 4 parts)
The chemistry of the thiol group (2 parts)
The chemistry of the hydrazo, azo and azoxy groups (2 volumes, 3 parts)
The chemistry of amidines and imidates (2 volumes)
The chemistry of cyanates and their thio derivatives (2 parts)
The chemistry of diazonium and diazo groups (2 parts)
The chemistry of the carbon–carbon triple bond (2 parts)
The chemistry of ketenes, allenes and related compounds (2 parts)
The chemistry of the sulphonium group (2 parts)
Supplement A: The chemistry of double-bonded functional groups (3 volumes, 6 parts)
Supplement B: The chemistry of acid derivatives (2 volumes, 4 parts)
Supplement C: The chemistry of triple-bonded functional groups (2 volumes, 3 parts)
Supplement D: The chemistry of halides, pseudo-halides and azides (2 volumes, 4 parts)
Supplement E: The chemistry of ethers, crown ethers, hydroxyl groups
and their sulphur analogues (2 volumes, 3 parts)
Supplement F: The chemistry of amino, nitroso and nitro compounds and their derivatives
(2 volumes, 4 parts)
The chemistry of the metal–carbon bond (5 volumes)
The chemistry of peroxides
The chemistry of organic selenium and tellurium compounds (2 volumes)
The chemistry of the cyclopropyl group (2 volumes, 3 parts)
The chemistry of sulphones and sulphoxides
The chemistry of organic silicon compounds (2 volumes, 5 parts)
The chemistry of enones (2 parts)
The chemistry of sulphinic acids, esters and their derivatives
The chemistry of sulphenic acids and their derivatives
The chemistry of enols
The chemistry of organophosphorus compounds (4 volumes)
The chemistry of sulphonic acids, esters and their derivatives
The chemistry of alkanes and cycloalkanes
Supplement S: The chemistry of sulphur-containing functional groups
The chemistry of organic arsenic, antimony and bismuth compounds
The chemistry of enamines (2 parts)
The chemistry of organic germanium, tin and lead compounds

UPDATES

The chemistry of α-haloketones, α-haloaldehydes and α-haloimines
Nitrones, nitronates and nitroxides
Crown ethers and analogs
Cyclopropane derived reactive intermediates
Synthesis of carboxylic acids, esters and their derivatives
The silicon–heteroatom bond
Synthesis of lactones and lactams
Syntheses of sulphones, sulphoxides and cyclic sulphides
Patai's 1992 guide to the chemistry of functional groups — *Saul Patai*

The chemistry of
organic silicon compounds

Volume 2

Part 3

Edited by

ZVI RAPPOPORT

The Hebrew University, Jerusalem

and

YITZHAK APELOIG

Technion–Israel Institute of Technology, Haifa

1998

JOHN WILEY & SONS
CHICHESTER–NEW YORK–WEINHEIM–BRISBANE–SINGAPORE–TORONTO
An Interscience® Publication

Copyright © 1998 John Wiley & Sons Ltd,
Baffins Lane, Chichester,
West Sussex PO19 1UD, England

National 01243 779777
International (+44) 1243 779777
e-mail (for orders and customer service enquiries): cs-books@wiley.co.uk
Visit our Home Page on http://www.wiley.co.uk
or http://www.wiley.com

All Rights Reserved. No part of this publication may be reproduced, stored in a retrieval system, or transmitted, in any form or by any means, electronic, mechanical, photocopying, recording, scanning or otherwise, except under the terms of the Copyright Designs and Patents Act 1988 or under the terms of a licence issued by the Copyright Licensing Agency, 90 Tottenham Court Road, London W1P 9HE, UK, without the permission in writing of the Publisher

Other Wiley Editorial Offices

John Wiley & Sons, Inc., 605 Third Avenue,
New York, NY 10158-0012, USA

WILEY-VCH Verlag GmbH, Pappellallee 3,
D-69469 Weinheim, Germany

Jacaranda Wiley Ltd, 33 Park Road, Milton,
Queensland 4064, Australia

John Wiley & Sons (Asia) Pte Ltd, Clementi Loop #02-01,
Jin Xing Distripark, Singapore 129809

John Wiley & Sons (Canada) Ltd, 22 Worcester Road,
Rexdale, Ontario M9W 1L1, Canada

British Library Cataloguing in Publication Data

A catalogue record for this book is available from the British Library

ISBN 0 471 96757 2

Typeset in 9/10pt Times by Laser Words, Madras, India
Printed and bound in Great Britain by Biddles Ltd, Guildford, Surrey
This book is printed on acid-free paper responsibly manufactured from sustainable forestry, in which at least two trees are planted for each one used for paper production.

Dedicated to

Saul Patai

teacher, editor and friend

Contributing authors

Wataru Ando	Department of Chemistry, University of Tsukuba, Tsukuba, Ibaraki 305, Japan
Yitzhak Apeloig	Department of Chemistry and the Lise-Meitner Minerva Center for Computational Quantum Chemistry, Technion–Israel Institute of Technology, Haifa 32000, Israel
D. A. ('Fred') Armitage	Department of Chemistry, King's College London, Strand, London, WC2R 2LS, UK
Norbert Auner	Fachinstitut für Anorganische und Allgemeine Chemie, Humboldt-Universität zu Berlin, Hessische Str. 1–2, D-10115 Berlin, Germany
David Avnir	Institute of Chemistry, The Hebrew University of Jerusalem, Jerusalem 91904, Israel
Alan R. Bassindale	Department of Chemistry, The Open University, Milton Keynes, MK7 6AA, UK
Rosa Becerra	Instituto de Quimica Fisica 'Rocasolano', C/Serrano 119, 28006 Madrid, Spain
Johannes Belzner	Institut für Organische Chemie der Georg-August-Universität Göttingen, Tammannstrasse 2, D-37077 Göttingen, Germany
M. B. Boisen Jr	Department of Materials Science and Engineering, Virginia Tech, Blacksburg, VA 24061, USA
Mark Botoshansky	Department of Chemistry, Technion–Israel Institute of Technology, Haifa 32000, Israel
A. G. Brook	Lash Miller Chemical Laboratories, University of Toronto, Toronto, Ontario M5S 3H6, Canada
C. Chatgilialoglu	I. Co. C. E. A., Consiglio Nazionale delle Ricerche, Via P. Gobetti 101, 40129 Bologna, Italy
Buh-Luen Cheng	Department of Chemistry, Tsing Hua University, Hsinchu, Taiwan 30043, Republic of China
Nami Choi	Department of Chemistry, University of Tsukuba, Tsukuba, Ibaraki 305, Japan
Ernest W. Colvin	Department of Chemistry, University of Glasgow, Glasgow, G12 8QQ, UK

Contributing authors

Uwe Dehnert	Institut für Organische Chemie der Georg-August-Universität Göttingen, Tammannstrasse 2, D-37077 Göttingen, Germany
Robert Drake	Dow Corning Ltd, Cardiff Road, Barry, South Glamorgan, CF63 2YL, UK
Jacques Dubac	Hétérochimie Fondamentale et Appliqué, ESA-CNRS 5069, Université Paul-Sabatier, 118 route de Narbonne, 31062 Toulouse Cedex, France
Moris S. Eisen	Department of Chemistry, Technion–Israel Institute of Technology, Kiryat Hatechnion, Haifa 32000, Israel
C. Ferreri	Departimento di Chimica Organica e Biologica, Università di Napoli 'Federico II', Via Mezzocannone 16, 80134 Napoli, Italy
Toshio Fuchigami	Department of Electrochemistry, Tokyo Institute of Technology, 4259 Nagatsuta, Midori-ku, Yokohama 226, Japan
Peter P. Gaspar	Department of Chemistry, Washington University, St Louis, Missouri 63130-4899, USA
Christian Guérin	Chimie Moléculaire et Organisation du Solide, UMR-CNRS 5637, Université Montpellier II, Place E. Bataillon, 34095 Montpellier Cedex 5, France
G. V. Gibbs	Department of Materials Science and Engineering, Virginia Tech, Blacksburg, VA 24061, USA
T. Gimisis	I. Co. C. E. A., Consiglio Nazionale delle Ricerche, Via P. Gobetti 101, 40129 Bologna, Italy
Simon J. Glynn	Department of Chemistry, The Open University, Milton Keynes, MK7 6AA, UK
Norman Goldberg	Technische Universität Braunschweig, Institut für Organische Chemie, Hagenring 30, D-38106 Braunschweig, Germany
Edwin Hengge	(Deceased)
Reuben Jih-Ru Hwu	Department of Chemistry, Tsing Hua University, Hsinchu, Taiwan 30043, Republic of China
Jörg Jung	Institut für Organische Chemie der Justus-Liebig Universität Giessen, Heinrich-Buff-Ring 58, D-35392 Giessen, Germany
Peter Jutzi	Faculty of Chemistry, University of Bielefeld, Universitätsstr. 25, D-33615 Bielefeld, Germany
Yoshio Kabe	Department of Chemistry, University of Tsukuba, Tsukuba, Ibaraki 305, Japan
Menahem Kaftory	Department of Chemistry, Technion–Israel Institute of Technology, Haifa 32000, Israel
Inna Kalikhman	Department of Chemistry, Ben-Gurion University of the Negev, Beer Sheva 84105, Israel
Moshe Kapon	Department of Chemistry, Technion-Israel Institute of Technology, Haifa 32000, Israel

Miriam Karni	Department of Chemistry and the Lise-Meitner Minerva Center for Computational Quantum Chemistry, Technion–Israel Institute of Technology, Haifa 32000, Israel
Mitsuo Kira	Department of Chemistry, Graduate School of Science, Tohoku University, Aoba-ku, Sendai 980-77, Japan
Sukhbinder S. Klair	Department of Chemistry, Loughborough University, Loughborough, Leicestershire, LE11 3TU, UK
Lisa C. Klein	Ceramics Department, Rutgers–The State University of New Jersey, Piscataway, New Jersey 08855-0909, USA
Daniel Kost	Department of Chemistry, Ben-Gurion University of the Negev, Beer Sheva 84105, Israel
Takahiro Kusukawa	Department of Chemistry, University of Tsukuba, Tsukuba, Ibaraki 305, Japan
R. M. Laine	Department of Chemistry, University of Michigan, Ann Arbor, Michigan 48109-2136, USA
David Levy	Instituto de Ciencia de Materiales de Madrid, C.S.I.C., Cantoblanco, 28049 Madrid, Spain
Larry N. Lewis	GE Corporate Research and Development Center, Schenectady, NY 12309, USA
Zhaoyang Li	Department of Chemistry, State University of New York at Stony Brook, Stony Brook, New York 11794-3400, USA
Paul D. Lickiss	Department of Chemistry, Imperial College of Science, Technology and Medicine, London, SW7 2AY, UK
Shiuh-Tzung Liu	Department of Chemistry, National Taiwan University, Taipei, Taiwan 106
Tien-Yau Luh	Department of Chemistry, National Taiwan University, Taipei, Taiwan 106
Gerhard Maas	Abteilung Organische Chemie I, Universität Ulm, Albert-Einstein-Allee 11, D-89081 Ulm, Germany
Iain MacKinnon	Dow Corning Ltd, Cardiff Road, Barry, South Glamorgan, CF63 2YL, UK
Svetlana Kirpichenko	Irkutsk Institute of Chemistry, Siberian Branch of the Russian Academy of Sciences, 1 Favorsky St, 664033 Irkutsk, Russia
Christoph Maerker	Laboratoire de Chimie Biophysique, Institut Le Bel, Université Louis Pasteur, 4 rue Blaise Pascal, F-67000 Strasbourg, France
Günther Maier	Institut für Organische Chemie der Justus-Liebig Universität Giessen, Heinrich-Buff-Ring 58, D-35392 Giessen, Germany
Michael J. McKenzie	Department of Chemistry, Loughborough University, Loughborough, Leicestershire, LE11 3TU, UK
Andreas Meudt	Institut für Organische Chemie der Justus-Liebig Universität Giessen, Heinrich-Buff-Ring 58, D-35392 Giessen, Germany

Contributing authors

Philippe Meunier	Synthèse et Electrosynthèse Organométalliques, UMR-CNRS 5632, Universitë de Bourgogne, 6 Boulevard Gabriel, 21004 Dijon Cedex, France
Takashi Miyazawa	Photodynamics Research Center, The Institute of Physical and Chemical Research, 19-1399, Koeji, Nagamachi, Aoba-ku, Sendai 980, Japan
Thomas Müller	Fachinstitut für Anorganische und Allgemeine Chemie, Humboldt-Universität zu Berlin, Hessische Str. 1–2, D-10115 Berlin, Germany
Shigeru Nagase	Department of Chemistry, Faculty of Science, Tokyo Metropolitan University, Hachioji, Tokyo 192-03, Japan
Iwao Ojima	Department of Chemistry, State University of New York at Stony Brook, Stony Brook, New York 11794-3400, USA
Renji Okazaki	Department of Chemistry, School of Science, The University of Tokyo, Bunkyo-ku, Tokyo 113, Japan
Harald Pacl	Institut für Organische Chemie der Justus-Liebig Universität Giessen, Heinrich-Buff-Ring 58, D-35392 Giessen, Germany
Philip C. Bulman Page	Department of Chemistry, Loughborough University, Loughborough, Leicestershire, LE11 3TU, UK
Vadim Pestunovich	Irkutsk Institute of Chemistry, Siberian Branch of the Russian Academy of Sciences, 1 Favorsky St, 664033 Irkutsk, Russia
Stephen Rosenthal	Department of Chemistry, Loughborough University, Loughborough, Leicestershire, LE11 3TU, UK
Hideki Sakurai	Department of Industrial Chemistry, Faculty of Science and Technology, Science University of Tokyo, Yamazaki 2641, Noda, Chiba 278, Japan
Paul von Ragué Schleyer	Center for Computational Quantum Chemistry, The University of Georgia, Athens, Georgia 30602, USA
Ulrich Schubert	Institute for Inorganic Chemistry, The Technical University of Vienna, A-1060 Vienna, Austria
Helmut Schwarz	Institut für Organische Chemie der Technischen Universität Berlin, Straße des 17 Juni 135, D-10623 Berlin, Germany
Akira Sekiguchi	Department of Chemistry, Graduate School of Science, University of Tsukuba, Tsukuba, Ibaraki 305, Japan
A. Sellinger	Sandia National Laboratory, Advanced Materials Laboratory, 1001 University Blvd, University of New Mexico, Albuquerque, New Mexico 87106, USA
Hans-Ullrich Siehl	Abteilung für Organische Chemie I der Universität Ulm, D-86069 Ulm, Germany
Harald Stüger	Institut für Anorganische Chemie, Erzherzog-Johann-Universität Graz, Stremayrgasse 16, A-8010 Graz, Austria
Reinhold Tacke	Institut für Anorganische Chemie, Universität Würzburg, Am Hubland, D-97074 Würzburg, Germany

Toshio Takayama	Department of Applied Chemistry, Faculty of Engineering, Kanagawa University, 3-27-1 Rokkakubashi, Yokohama, Japan 221
Yoshito Takeuchi	Department of Chemistry, Faculty of Science, Kanagawa University, 2946 Tsuchiya, Hiratsuka, Japan 259-12
Peter G. Taylor	Department of Chemistry, The Open University, Milton Keynes, MK7 6AA, UK
Richard Taylor	Dow Corning Ltd, Cardiff Road, Barry, South Glamorgan, CF63 2YL, UK
Norihiro Tokitoh	Department of Chemistry, School of Science, The University of Tokyo, Bunkyo-ku, Tokyo 113, Japan
Shwu-Chen Tsay	Department of Chemistry, Tsing Hua University, Hsinchu, Taiwan 30043, Republic of China
Mikhail Voronkov	Irkutsk Institute of Chemistry, Siberian Branch of the Russian Academy of Sciences, 1 Favorsky St, 664033 Irkutsk, Russia
Stephan A. Wagner	Institut für Anorganische Chemie, Universität Würzburg, Am Hubland, D-97074 Würzburg, Germany
Robin Walsh	The Department of Chemistry, The University of Reading, P O Box 224, Whiteknights, Reading, RG6 6AD, UK
Robert West	Department of Chemistry, University of Wisconsin at Madison, Madison, Wisconsin 53706, USA
Anna B. Wojcik	Ceramics Department, Rutgers–The State University of New Jersey, Piscataway, New Jersey 08855-0909, USA
Jiawang Zhu	Department of Chemistry, State University of New York at Stony Brook, Stony Brook, New York 11794-3400, USA
Wolfgang Ziche	Fachinstitut für Anorganische und Allgemeine Chemie, Humboldt-Universität zu Berlin, Hessische Str. 1–2, D-10115 Berlin, Germany

Foreword

The preceding volume in 'The Chemistry of Functional Groups' series, *The chemistry of organic silicon compounds* (S. Patai and Z. Rappoport, Eds), appeared a decade ago and was followed in 1991 by an update volume, *The silicon–heteroatom bond*. Since then the chemistry of organic silicon compounds has continued its rapid growth, with many important contributions in the synthesis of new and novel types of compounds, in industrial applications, in theory and in understanding the chemical bonds of silicon, as well as in many other directions. The extremely rapid growth of the field and the continued fascination with the chemistry of this unique element, a higher congener of carbon — yet so dramatically different in its chemistry — convinced us that a new authoritative book in the field is highly desired.

Many of the recent developments, as well as topics not covered in the previous volume are reviewed in the present volume, which is the largest in 'The Chemistry of Functional Groups' series. The 43 chapters, written by leading silicon chemists from 12 countries, deal with a wide variety of topics in organosilicon chemistry, including theoretical aspects of several classes of compounds, their structural and spectral properties, their thermochemistry, photochemistry and electrochemistry and the effect of silicon as a substituent. Several chapters review the chemistry of various classes of reactive intermediates, such as silicenium ions, silyl anions, silylenes, and of hypervalent silicon compounds. Multiple-bonded silicon compounds, which have attracted much interest and activity over the last decade, are reviewed in three chapters: one on silicon–carbon and silicon–nitrogen multiple bonds, one on silicon–silicon multiple bonds and one on silicon–hereroatom multiple bonds. Other chapters review the synthesis of several classes of organosilicon compounds and their applications as synthons in organic synthesis. Several chapters deal with practical and industrial aspects of silicon chemistry in which important advances have recently been made, such as silicon polymers, silicon-containing ceramic precursors and the rapidly growing field of organosilica sol–gel chemistry.

The literature covered in the book is mostly up to mid-1997.

Several of the originally planned chapters, on comparison of silicon compounds with their higher group 14 congeners, interplay between theory and experiment in organisilicon chemistry, silyl radicals, recent advances in the chemistry of silicon–phosphorous,-arsenic,-antimony and –bismuth compounds, and the chemistry of polysilanes, regrettably did not materialize. We hope to include these important chapters in a future complementary volume. The current pace of research in silicon chemistry will certainly soon require the publication of an additional updated volume.

We are grateful to the authors for the immense effort they have invested in the 43 chapters and we hope that this book will serve as a major reference in the field of silicon chemistry for years to come.

We will be grateful to readers who will draw our attention to mistakes and who will point out to us topics which should be included in a future volume of this series.

Jerusalem and Haifa
March, 1998

ZVI RAPPOPORT
YITZHAK APELOIG

The Chemistry of Functional Groups
Preface to the series

The series 'The Chemistry of Functional Groups' was originally planned to cover in each volume all aspects of the chemistry of one of the important functional groups in organic chemistry. The emphasis is laid on the preparation, properties and reactions of the functional group treated and on the effects which it exerts both in the immediate vicinity of the group in question and in the whole molecule.

A voluntary restriction on the treatment of the various functional groups in these volumes is that material included in easily and generally available secondary or tertiary sources, such as Chemical Reviews, Quarterly Reviews, Organic Reactions, various 'Advances' and 'Progress' series and in textbooks (i.e. in books which are usually found in the chemical libraries of most universities and research institutes), should not, as a rule, be repeated in detail, unless it is necessary for the balanced treatment of the topic. Therefore each of the authors is asked not to give an encyclopaedic coverage of his subject, but to concentrate on the most important recent developments and mainly on material that has not been adequately covered by reviews or other secondary sources by the time of writing of the chapter, and to address himself to a reader who is assumed to be at a fairly advanced postgraduate level.

It is realized that no plan can be devised for a volume that would give a complete coverage of the field with no overlap between chapters, while at the same time preserving the readability of the text. The Editors set themselves the goal of attaining reasonable coverage with moderate overlap, with a minimum of cross-references between the chapters. In this manner, sufficient freedom is given to the authors to produce readable quasi-monographic chapters.

The general plan of each volume includes the following main sections:

(a) An introductory chapter deals with the general and theoretical aspects of the group.

(b) Chapters discuss the characterization and characteristics of the functional groups, i.e. qualitative and quantitative methods of determination including chemical and physical methods, MS, UV, IR, NMR, ESR and PES — as well as activating and directive effects exerted by the group, and its basicity, acidity and complex-forming ability.

(c) One or more chapters deal with the formation of the functional group in question, either from other groups already present in the molecule or by introducing the new group directly or indirectly. This is usually followed by a description of the synthetic uses of the group, including its reactions, transformations and rearrangements.

(d) Additional chapters deal with special topics such as electrochemistry, photochemistry, radiation chemistry, thermochemistry, syntheses and uses of isotopically labelled compounds, as well as with biochemistry, pharmacology and toxicology. Whenever applicable, unique chapters relevant only to single functional groups are also included (e.g. 'Polyethers', 'Tetraaminoethylenes' or 'Siloxanes').

This plan entails that the breadth, depth and thought-provoking nature of each chapter will differ with the views and inclinations of the authors and the presentation will necessarily be somewhat uneven. Moreover, a serious problem is caused by authors who deliver their manuscript late or not at all. In order to overcome this problem at least to some extent, some volumes may be published without giving consideration to the originally planned logical order of the chapters.

Since the beginning of the Series in 1964, two main developments have occurred. The first of these is the publication of supplementary volumes which contain material relating to several kindred functional groups (Supplements A, B, C, D, E, F and S). The second ramification is the publication of a series of 'Updates', which contain in each volume selected and related chapters, reprinted in the original form in which they were published, together with an extensive updating of the subjects, if possible, by the authors of the original chapters. A complete list of all above mentioned volumes published to date will be found on the page opposite the inner title page of this book. Unfortunately, the publication of the 'Updates' has been discontinued for economic reasons.

Advice or criticism regarding the plan and execution of this series will be welcomed by the Editors.

The publication of this series would never have been started, let alone continued, without the support of many persons in Israel and overseas, including colleagues, friends and family. The efficient and patient co-operation of staff-members of the publisher also rendered us invaluable aid. Our sincere thanks are due to all of them.

The Hebrew University	SAUL PATAI
Jerusalem, Israel	ZVI RAPPOPORT

Contents

1. Theoretical aspects and quantum mechanical calculations of silaaromatic compounds 1
   Yitzhak Apeloig and Miriam Karni

2. A molecular modeling of the bonded interactions of crystalline silica 103
   G. V. Gibbs and M. B. Boisen

3. Polyhedral silicon compounds 119
   Akira Sekiguchi and Shigeru Nagase

4. Thermochemistry 153
   Rosa Becerra and Robin Walsh

5. The structural chemistry of organosilicon compounds 181
   Menahem Kaftory, Moshe Kapon and Mark Botoshansky

6. ^{29}Si NMR spectroscopy of organosilicon compounds 267
   Yoshito Takeuchi and Toshio Takayama

7. Activating and directive effects of silicon 355
   Alan R. Bassindale, Simon J. Glynn and Peter G. Taylor

8. Steric effects of silyl groups 431
   Jih Ru Hwu, Shwu-Chen Tsay and Buh-Luen Cheng

9. Reaction mechanisms of nucleophilic attack at silicon 495
   Alan R. Bassindale, Simon J. Glynn and Peter G. Taylor

10. Silicenium ions: Quantum chemical computations 513
    Christoph Maerker and Paul von Ragué Schleyer

11. Silicenium ions — experimental aspects 557
    Paul D. Lickiss

12. Silyl-substituted carbocations 595
    Hans-Ullrich Siehl and Thomas Müller

13. Silicon-substituted carbenes 703
    Gerhard Maas

Contents

14 Alkaline and alkaline earth silyl compounds — preparation and structure   779
Johannes Belzner and Uwe Dehnert

15 Mechanism and structures in alcohol addition reactions of disilenes and silenes   827
Hideki Sakurai

16 Silicon-carbon and silicon-nitrogen multiply bonded compounds   857
Thomas Müller, Wolfgang Ziche and Norbert Auner

17 Recent advances in the chemistry of silicon-heteroatom multiple bonds   1063
Norihiro Tokitoh and Renji Okazaki

18 Gas-phase ion chemistry of silicon-containing molecules   1105
Norman Goldberg and Helmut Schwarz

19 Matrix isolation studies of silicon compounds   1143
Günther Maier, Andreas Meudt, Jörg Jung and Harald Pacl

20 Electrochemistry of organosilicon compounds   1187
Toshio Fuchigami

21 The photochemistry of organosilicon compounds   1233
A. G. Brook

22 Mechanistic aspects of the photochemistry of organosilicon compounds   1311
Mitsuo Kira and Takashi Miyazawa

23 Hypervalent silicon compounds   1339
Daniel Kost and Inna Kalikhman

24 Silatranes and their tricyclic analogs   1447
Vadim Pestunovich, Svetlana Kirpichenko and Mikhail Voronkov

25 Tris(trimethylsilyl)silane in organic synthesis   1539
C. Chatgilialoglu, C. Ferreri and T. Gimisis

26 Recent advances in the direct process   1581
Larry N. Lewis

27 Acyl silanes   1599
Philip C. Bulman Page, Michael J. McKenzie, Sukhbinder S. Klair and Stephen Rosenthal

28 Recent synthetic applications of organosilicon reagents   1667
Ernest W. Colvin

29 Recent advances in the hydrosilylation and related reactions   1687
Iwao Ojima, Zhaoyang Li and Jiawang Zhu

Contents

30	Synthetic applications of allylsilanes and vinylsilanes Tien-Yau Luh and Shiuh-Tzung Liu	1793
31	Chemistry of compounds with silicon-sulphur, silicon-selenium and silicon-tellurium bonds D. A. ('Fred') Armitage	1869
32	Cyclic polychalcogenide compounds with silicon Nami Choi and Wataru Ando	1895
33	Organosilicon derivatives of fullerenes Wataru Ando and Takahiro Kusukawa	1929
34	Group 14 metalloles, ionic species and coordination compounds Jacques Dubac, Christian Guérin and Philippe Meunier	1961
35	Transition-metal silyl complexes Moris S. Eisen	2037
36	Cyclopentadienyl silicon compounds Peter Jutzi	2129
37	Recent advances in the chemistry of cyclopolysilanes Edwin Hengge and Harald Stüger	2177
38	Recent advances in the chemistry of siloxane polymers and copolymers Robert Drake, Iain MacKinnon and Richard Taylor	2217
39	Si-containing ceramic precursors R. M. Laine and A. Sellinger	2245
40	Organo-silica sol–gel materials David Avnir, Lisa C. Klein, David Levy, Ulrich Schubert and Anna B. Wojcik	2317
41	Chirality in bioorganosilicon chemistry Reinhold Tacke and Stephan A. Wagner	2363
42	Highly reactive small-ring monosilacycles and medium-ring oligosilacycles Wataru Ando and Yoshio Kabe	2401
43	Silylenes Peter P. Gaspar and Robert West	2463
Author index		2569
Subject index		2721

List of abbreviations used

Ac	acetyl (MeCO)
acac	acetylacetone
Ad	adamantyl
AIBN	azoisobutyronitrile
Alk	alkyl
All	allyl
An	anisyl
Ar	aryl
Bn	benzyl
Bz	benzoyl (C_6H_5CO)
Bu	butyl (also t-Bu or Bu^t)
CD	circular dichroism
CI	chemical ionization
CIDNP	chemically induced dynamic nuclear polarization
CNDO	complete neglect of differential overlap
Cp	η^5-cyclopentadienyl
Cp*	η^5-pentamethylcyclopentadienyl
DABCO	1,4-diazabicyclo[2.2.2]octane
DBN	1,5-diazabicyclo[4.3.0]non-5-ene
DBU	1,8-diazabicyclo[5.4.0]undec-7-ene
DIBAH	diisobutylaluminium hydride
DME	1,2-dimethoxyethane
DMF	N,N-dimethylformamide
DMSO	dimethyl sulphoxide
ee	enantiomeric excess
EI	electron impact
ESCA	electron spectroscopy for chemical analysis
ESR	electron spin resonance
Et	ethyl
eV	electron volt

Fc	ferrocenyl	
FD	field desorption	
FI	field ionization	
FT	Fourier transform	
Fu	furyl(OC_4H_3)	
GLC	gas liquid chromatography	
Hex	hexyl(C_6H_{13})	
c-Hex	cyclohexyl(C_6H_{11})	
HMPA	hexamethylphosphortriamide	
HOMO	highest occupied molecular orbital	
HPLC	high performance liquid chromatography	
i-	iso	
Ip	ionization potential	
IR	infrared	
ICR	ion cyclotron resonance	
LAH	lithium aluminium hydride	
LCAO	linear combination of atomic orbitals	
LDA	lithium diisopropylamide	
LUMO	lowest unoccupied molecular orbital	
M	metal	
M	parent molecule	
MCPBA	*m*-chloroperbenzoic acid	
Me	methyl	
MNDO	modified neglect of diatomic overlap	
MS	mass spectrum	
n	normal	
Naph	naphthyl	
NBS	*N*-bromosuccinimide	
NCS	*N*-chlorosuccinimide	
NMR	nuclear magnetic resonance	
Pc	phthalocyanine	
Pen	pentyl(C_5H_{11})	
Pip	piperidyl($C_5H_{10}N$)	
Ph	phenyl	
ppm	parts per million	
Pr	propyl (also *i*-Pr or Pr^i)	
PTC	phase transfer catalysis or phase transfer conditions	
Pyr	pyridyl (C_5H_4N)	

R	any radical
RT	room temperature
s-	secondary
SET	single electron transfer
SOMO	singly occupied molecular orbital
t-	tertiary
TCNE	tetracyanoethylene
TFA	trifluoroacetic acid
THF	tetrahydrofuran
Thi	thienyl(SC_4H_3)
TLC	thin layer chromatography
TMEDA	tetramethylethylene diamine
TMS	trimethylsilyl or tetramethylsilane
Tol	tolyl(MeC_6H_4)
Tos or Ts	tosyl(*p*-toluenesulphonyl)
Trityl	triphenylmethyl(Ph_3C)
Xyl	xylyl($Me_2C_6H_3$)

In addition, entries in the 'List of Radical Names' in *IUPAC Nomenclature of Organic Chemistry*, 1979 Edition, Pergamon Press, Oxford, 1979, p. 305-322, will also be used in their unabbreviated forms, both in the text and in formulae instead of explicitly drawn structures.

CHAPTER 30

Synthetic applications of allylsilanes and vinylsilanes

TIEN-YAU LUH and SHIUH-TZUNG LIU

Department of Chemistry, National Taiwan University, Taipei, Taiwan 106, Republic of China
Fax: +886-2-364-4971;   e-mail: tyluh@ccms.ntu.edu.tw

I. INTRODUCTION	1794
II. ELECTROPHILIC SUBSTITUTION OF C−Si BONDS	1794
A. Substitution of C−Si Bond by an Electrophile Other than Carbon . . .	1795
1. Protodesilylation of allyl- and vinylsilanes	1795
2. Halodesilylation of allyl- and vinylsilanes	1796
3. Other heteroatom electrophiles	1797
B. Substitution of C−Si Bond by a Carbon Electrophile	1798
1. Reactions with acetals	1798
2. Reactions with aldehydes and ketones	1801
3. Reactions with enones	1811
4. Reactions with iminium ions	1814
5. Reactions with acid chlorides	1818
6. Reactions with alkyl halides and related electrophiles	1820
7. Reactions with alkenes	1822
III. ORGANOMETALLIC REAGENTS PROMOTED C−Si CLEAVAGE REACTIONS	1824
A. Coupling Reactions Involving the Cleavage of the Carbon–Silicon Bond	1824
1. Heck reactions	1824
a. Silyl group as a halide-like leaving group	1824
b. Vinylsilanes as organometallic acceptors	1825
2. Homocoupling reactions	1828
3. Other coupling reactions	1828
B. Transmetallation	1828
IV. MISCELLANEOUS C−Si CLEAVAGE REACTIONS	1829
V. REACTIONS OF ALLYLSILANES AND VINYLSILANES WITHOUT CLEAVAGE OF THE C−Si BOND	1832

The chemistry of organic silicon compounds, Vol. 2
Edited by Z. Rappoport and Y. Apeloig   © 1998 John Wiley & Sons Ltd

A. Lewis Acid Catalyzed Addition of an Electrophile to the Double Bond	1832
B. Radical Reactions	1834
C. Organometallic Reagents Promoted Reactions	1836
1. Cross-coupling reactions	1836
2. Heck reaction	1837
3. Murai reaction	1838
4. Carbonylation	1839
5. Metathesis	1840
D. Oxidation Reactions	1840
1. Epoxidation	1840
2. OsO_4 oxidation	1844
3. Other oxidation reactions	1846
E. Cycloaddition Reactions	1846
1. [2 + 1] Cycloaddition	1846
2. [2 + 2] Cycloaddition	1847
3. [3 + 2] Cycloaddition	1850
4. [4 + 2] Cycloaddition	1853
a. Silyl-substituted alkenes as dienophiles	1854
b. Silyl-substituted alkenes as dienes	1854
5. [5 + 2] Cycloaddition	1855
F. Rearrangements	1856
1. Ene reaction	1856
2. Sigmatropic rearrangements	1858
3. Miscellaneous rearrangements	1860
G. Other Reactions	1860
VI. REFERENCES	1862

I. INTRODUCTION

During the last three decades, an ever increasing number of silicon-based reagents and reactions has been discovered to provoke an enormous variety of fascinating transformations under mild conditions. It is the unique properties of the silicon element that renders these synthetic applications[1-6]. The use of allylsilanes and vinylsilanes in organic synthesis has been particularly attractive and the chemistry has been covered extensively in several excellent monographs[1-6] and numerous review articles[7-32]. This chapter summarizes recent developments of the synthetic applications of these silyl functionalities. Material covered in earlier reviews will not be repeated here. As in-depth discussions regarding the syntheses of allyl- and vinylsilanes can easily be found in these reviews[1-32], the syntheses of these organosilicon compounds are, in general, not included, unless the allyl- or vinylsilanes are used for the synthesis of other analogs. Literature dates up to early 1996, but comprehensive coverage has not been attempted. Personal prejudice and ignorance become factors dictating the selection or omission of various important works for which the authors assume full responsibility.

II. ELECTROPHILIC SUBSTITUTION OF C—Si BONDS

Substitution of the carbon–silicon bond in allyl- and vinylsilanes by an electrophile has served as a powerful tool in organic synthesis. Electrophiles ranging from proton, carbon and main group heteroatoms to certain transition metal species have been employed. A comprehensive review covers literature up to early 1989[19]. The directive effect of the silicon moiety and the mechanistic aspects of these substitution reactions have been discussed in detail[19,33]. Scheme 1 summarizes the typical reaction patterns for the displacement of C—Si bonds in allyl- and vinylsilanes by electrophiles E^+.

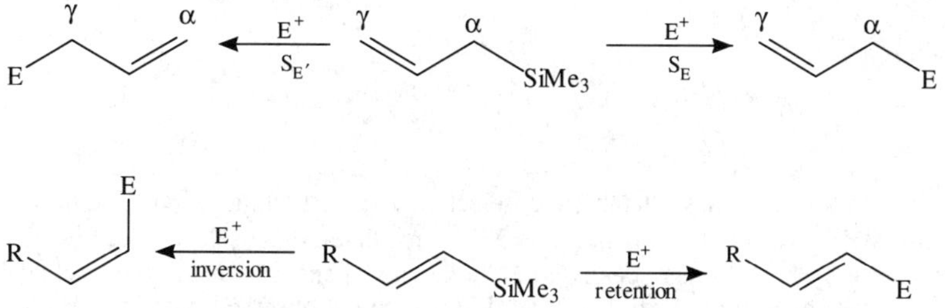

SCHEME 1

In the case of allylsilanes, most reactions occur via an $S_{E'}$ manner. In other words electrophile attacks selectively at the γ-position. This process is applied particularly when the silyl group has strongly electronegative substituent(s) and carbonyl electrophiles or the like are employed. A possible six-membered ring transition state has been suggested. The reaction is, in general, stereoselective. As can be seen in Section II.B.6, electrophiles derived from sterically hindered alkyl halides may attack preferentially at the α-position.

In the case of vinylsilanes, retention of configuration is usually observed, although inversion of configuration is also known. Depending on the relative reactivity of the electrophile, nucleophilic or Lewis acid catalysts are occasionally employed to facilitate the displacement reactions. The applications of these reactions in the synthesis of natural products have recently been reviewed[32]. Selected recent examples are presented here for illustration.

A. Substitution of C—Si Bond by an Electrophile Other than Carbon

1. Protodesilylation of allyl- and vinylsilanes

The proton serves as the simplest electrophile to displace the silyl group stereo- and regioselectively. Numerous conditions have been used for the electrophilic protodesilylation of allyl- and vinylsilanes[19]. (E)-vinyl sulfones **2** are prepared from the silylallylic sulfones **1** in the presence of protic acids in high yield (equation 1)[34]. Diastereoselective protodesilylation of allylsilane **3** gives **4** with excellent control of the geometry of the double bond exocyclic to the ring (equation 2)[35].

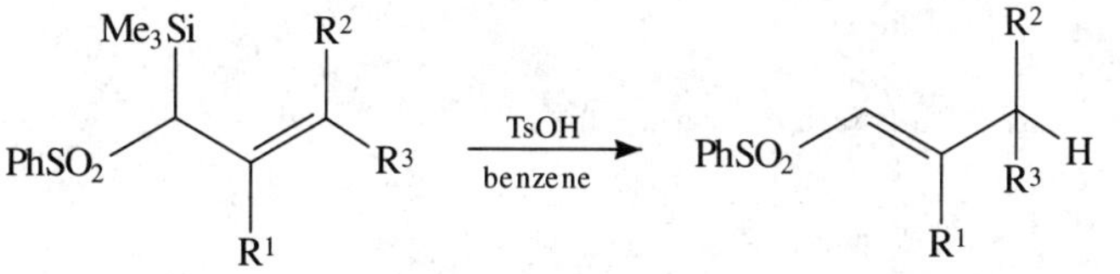

(1)

R^1 = H, Me; R^2, R^3 = H, Me; R^2R^3 = $(CH_2)_5$

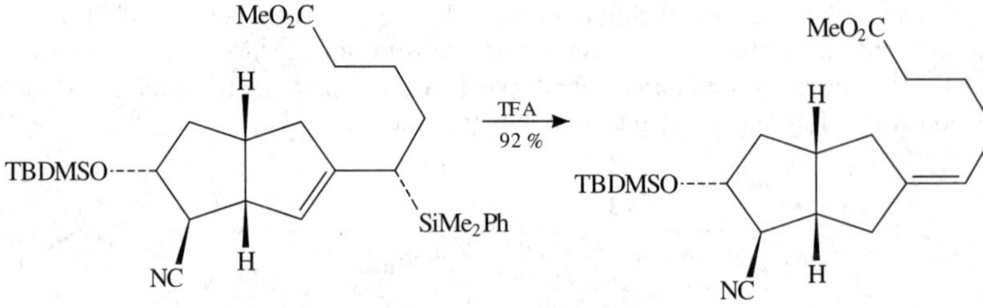

[Equation 2 scheme: compound (3) → TFA, 92% → compound (4)]

Vinylsilanes can be desilylated stereoselectively upon treatment with iodine in water (equation 3)[36,37]. Under basic conditions, TBAF-mediated stereoselective desilylation proceeds smoothly[38]. Although desilylation of a vinylsilane can readily be achieved by treatment with hydroiodic acid, selective destannation is observed when a substrate contains both stannyl and silyl groups at the olefinic carbons (equation 4)[39].

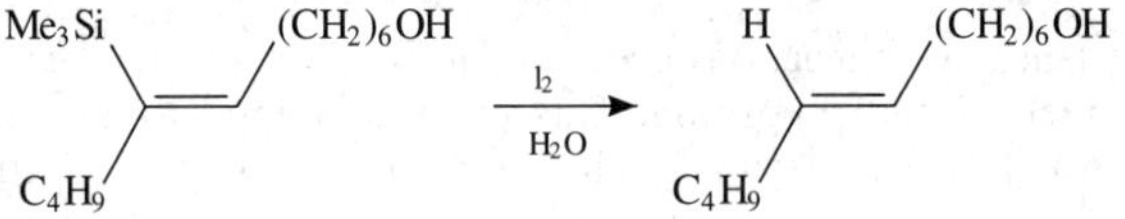

[Equation 3 scheme: vinylsilane + I_2/H_2O → desilylated alkene]

[Equation 4 scheme: stannyl/silyl substrate + HI, $Bu_4N^+I^-$, 0 °C → vinylsilane product]

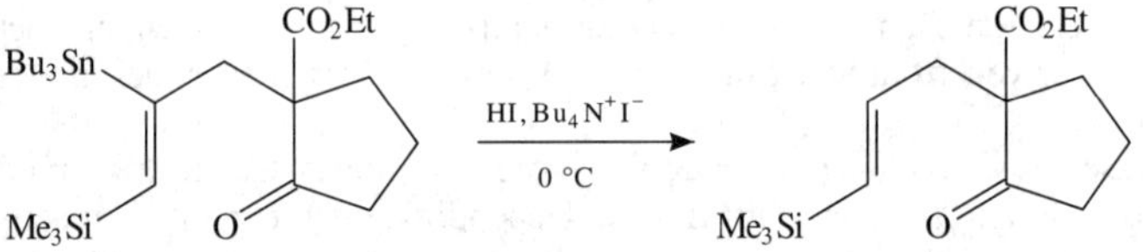

2. Halodesilylation of allyl- and vinylsilanes

Whereas an allylsilane can serve as an allyl anion synthon, the reaction of 1,3-bis(silyl)propene with electrophiles can afford the 1,3-disubstituted propene. Thus, treatment of a mixture of (*E*)- and (*Z*)-2-aryl-1,3-bis(trimethylsilyl)propenes **5** with 2 equivalents of NBS at −78 °C stereoselectively yields the corresponding (*Z*)-2-aryl-1,3-dibromopropene **6**. When 1 equivalent of NBS is employed, the monobromo product **7** is obtained (equation 5). The reactions apparently proceed via the pattern of sequential displacement of allylsilane moieties[40,41].

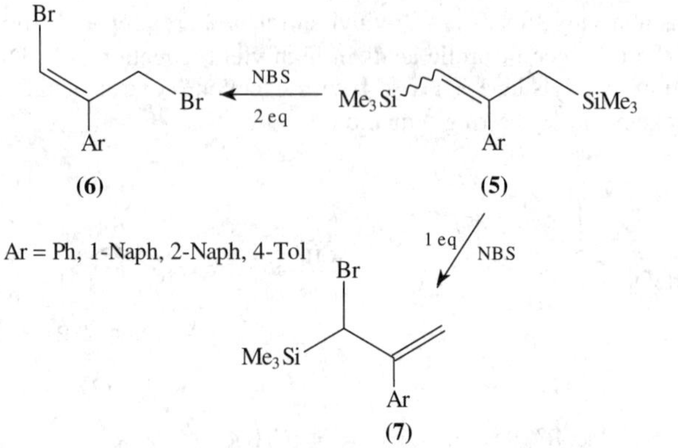

[Equation 5 scheme: compound (5) ⇌ NBS 2 eq → (6); NBS 1 eq → (7); Ar = Ph, 1-Naph, 2-Naph, 4-Tol]

Reaction of **8** with chlorine followed by a brief exposure to KF in DMSO at 25 °C gives mycorrhizin in good yield with an inversion of stereochemistry (equation 6)[42–44].

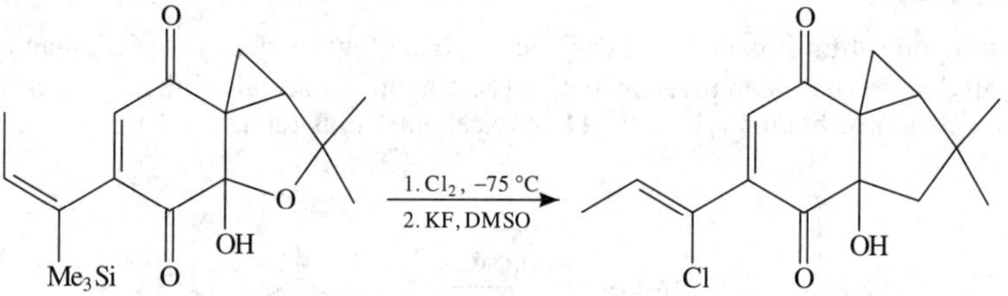

$$\text{(6)}$$

Iododesilylation of vinylsilanes gives regio- and stereoselectively the corresponding vinyl iodides. Depending on the nature of the iodination reagent and the reaction conditions, both retention and inversion of configuration take place[19,33,45]. When the reaction is carried out in DMF, HMPA or NMP solvent, the products predominantly, if not exclusively, retain their configurations[46]. The substituent on the silicon atom and its coordination state also determine the stereoselectivity of the halogenolysis of vinylsilanes. Retention of stereochemistry for vinylpentafluorosilicates is commonly found[46]. Electrophilic attack on hypervalent silicon species is suggested. More recently, IPy_2BF_4 has been found to be a useful reagent for the stereospecific iodine–silicon exchange in vinylsilanes[47]. Labile (Z)-vinyl(phenyl)iodonium salts **9** are synthesized from (Z)-vinylsilanes by the reaction with (diacetoxyiodo)benzene in the presence of $BF_3 \cdot Et_2O$[48].

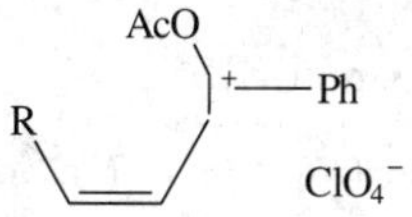

3. Other heteroatom electrophiles

Sulfonation of allylsilanes with SO_3 gives the corresponding trimethylsilyl allylsulfonates in excellent yields. The sultone intermediate **10** has been detected (equation 7)[49].

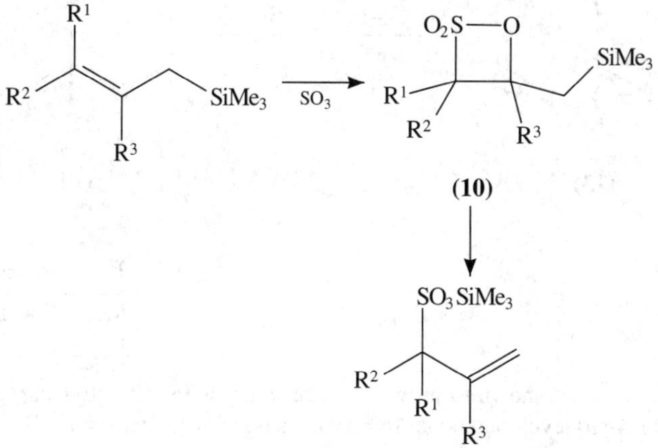

$$\text{(7)}$$

B. Substitution of C—Si Bond by a Carbon Electrophile

1. Reactions with acetals

Using liquid sulfur dioxide as a Lewis-acid solvent for the alkoxyalkylation and alkylation of allylsilanes has been investigated[50]. The reaction of acetals with allylsilanes results in the formation of homoallyl ethers **11** in excellent yields (equation 8).

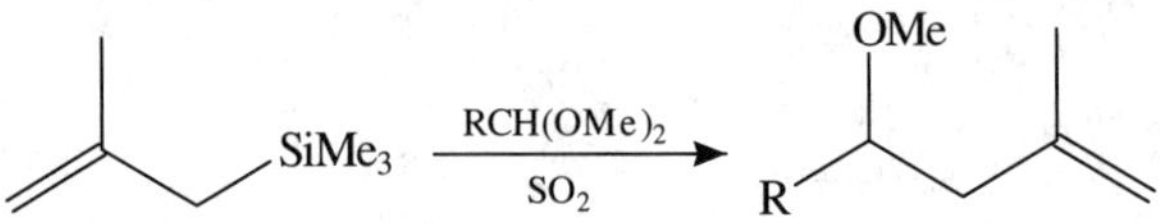

$$\text{(8)}$$

R = 1° Alk, Ph

The coupling reaction of allylsilane with the α-thiomethoxyacetal is catalyzed by TMSOTf[51]. TiCl$_4$-mediated reaction of α-bromoallylsilane **7** with 1,1-diethoxyethane leads to homoallylic ether **12** stereoselectively in excellent yield (equation 9)[40]. Under similar reaction conditions, double substitution of allylsilane to diketals **13** affords **14** in high diastereoselectivity (equation 10)[52].

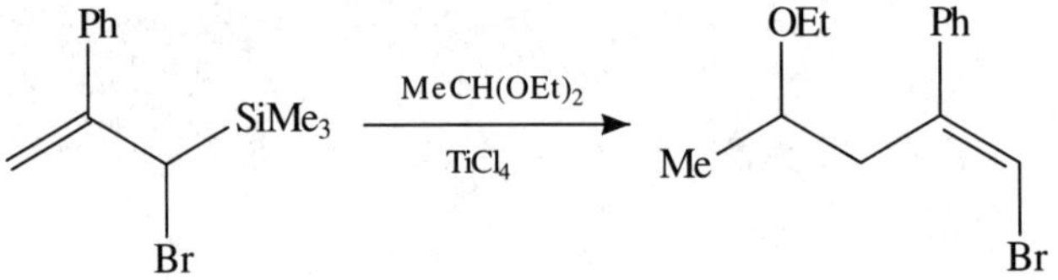

$$\text{(9)}$$

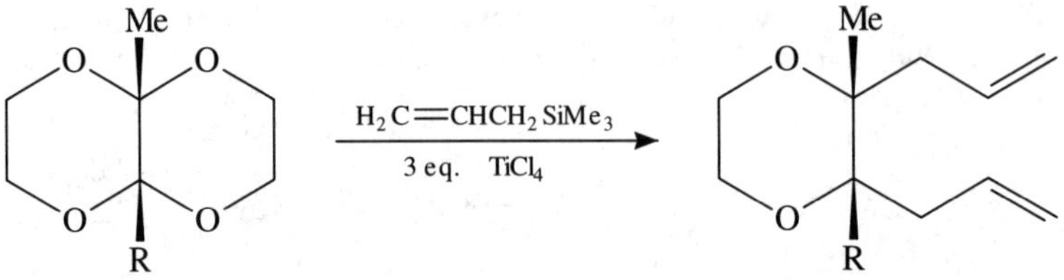

$$\text{(10)}$$

R	cis:trans
Me	85 : 15
Et	90 : 10

BF$_3$ · OEt$_2$ has been shown to promote the reaction of **15** with allylsilanes to yield the corresponding 4-allylcyclobutenone **16**[53], which undergoes stereoselective thermal ring opening and subsequent intramolecular [2+2] cycloaddition[54] leading to **17** (equation 11).

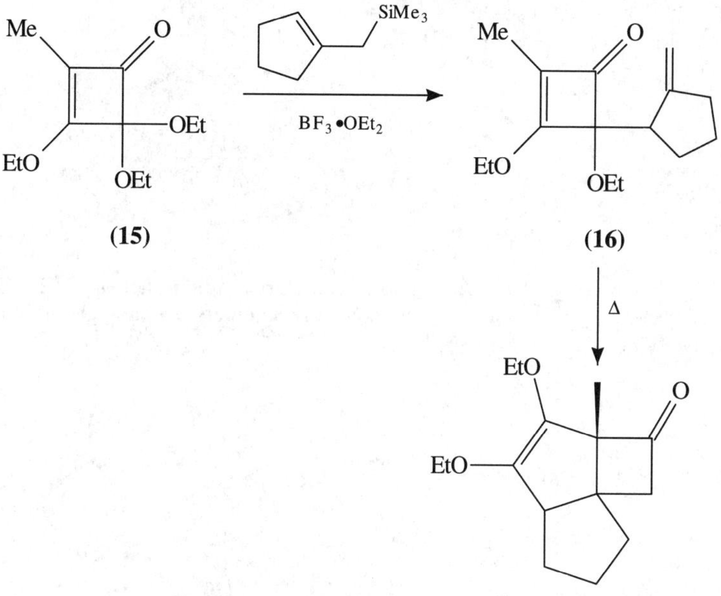

(11)

Neighboring methoxy group assisted asymmetric induction is demonstrated in the TMSOTf-catalyzed reaction of γ-alkoxy acetal **18** with allylsilane. A possible cyclic oxocarbenium ion intermediate **19** is proposed (equation 12)[55]. Siladioxanes **20** react with allylsilane in the presence of a catalytic amount of the Brønsted superacid $TfOH_2^+ B(OTf)_4^-$ to afford the allylated products **21** diastereoselectively (equation 13)[56,57].

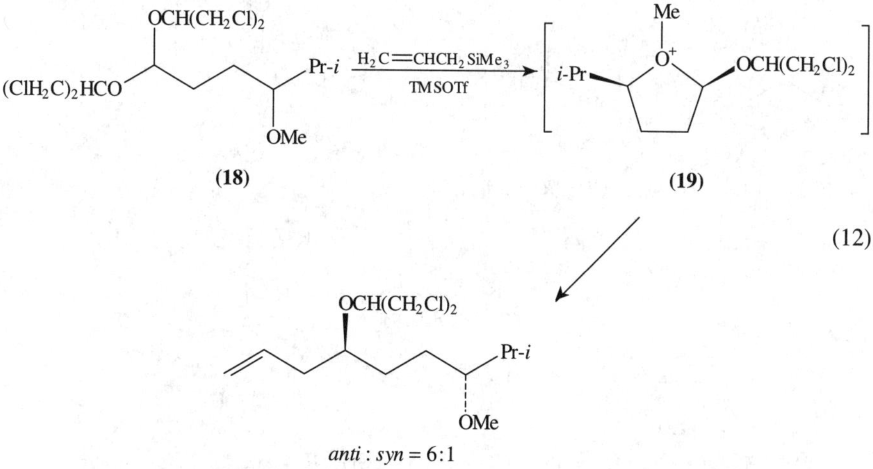

(12)

anti : syn = 6:1

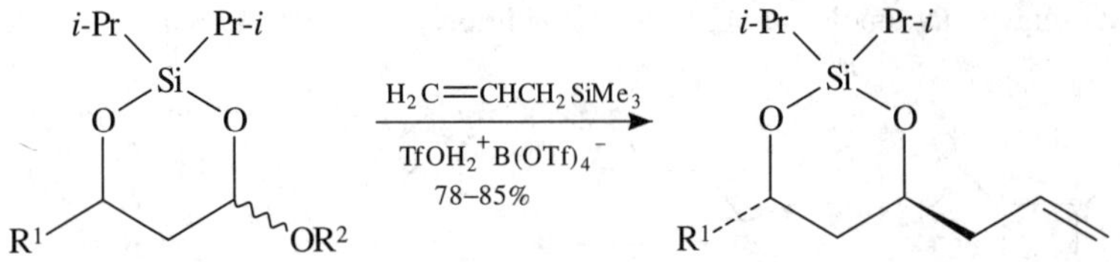

(13)

(20)

R^1 = Pr, Bu, Ph
R^2 = Me, Et

(21)

anti : syn = 20–23 : 1

Optically active crotylsilane **22** functions as a chiral carbon nucleophile in TMSOTf-catalyzed reactions with acetals giving homoallylic ethers **23** in high diastereo- and enantioselectivities (equation 14)[58,59].

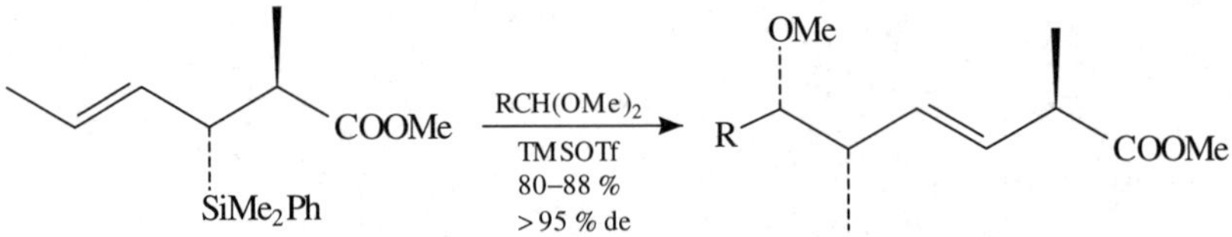

(14)

(22)   (23)

R = BnOCH$_2$, BnOCH$_2$CH$_2$, Ar

(2-Siloxyallyl)silane **24**, acting as a synthetic equivalent of acetone α,α'-dianion, readily undergoes double alkylation (equation 15)[60]. An acid-catalyzed transacetalization-ring closure reaction occurs in the reaction of 6-hydroxy substituted allylsilanes **25** with acetals to afford the corresponding trisubstituted tetrahydropyrans in moderate to good yield with high diastereoselectivity (equation 16)[61].

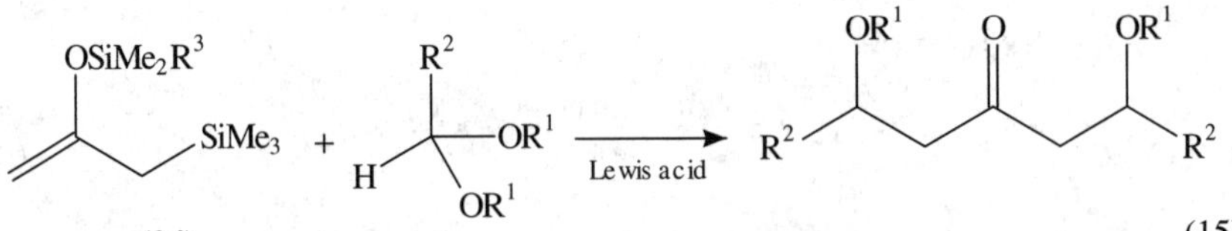

(15)

(24)

R^1 = Me; R^2 = Alk, Ar, BrCH$_2$; R^3 = Me, t-Bu

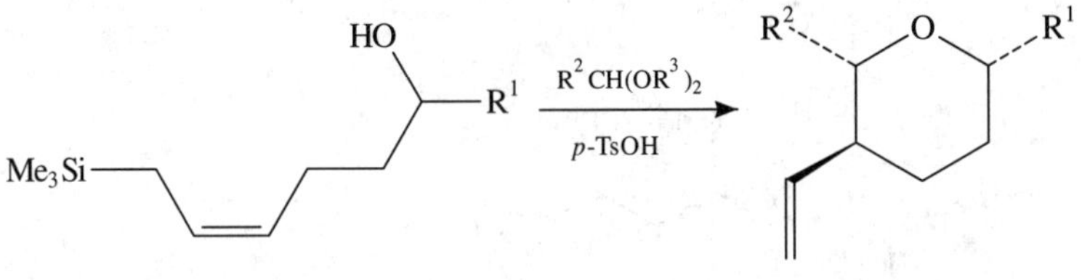

(16)

(25)

R^1 = Alk, Ph, EtOOCH$_2$; R^2 = Alk, HC(CH$_2$)$_2$, MeOCH$_2$; R^3 = Me, Et

Selenoacetals are found to be good electrophiles towards allylsilanes. Under $SnCl_4$ catalyzed conditions, selenoketals couple with allylsilane to produce the corresponding homoallyl selenides in moderate to good yield[62]. Reaction of allylsilanes with tris(phenylseleneno)methane or tris(phenylthio)methane in the presence of a Lewis acid furnishes the corresponding homoallylchalcogenoacetals **26** in moderate to good yields (equation 17)[63]. Thionium ion generated *in situ* from **27** leads to the corresponding coupling product (equation 18)[64].

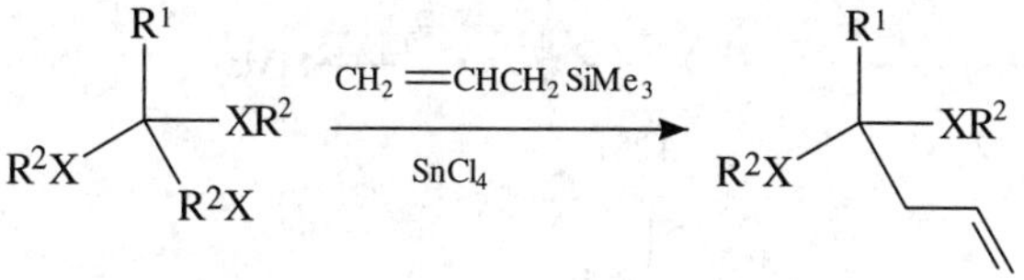

R^1 = Substituted allyl; X = S, Se   (**26**)

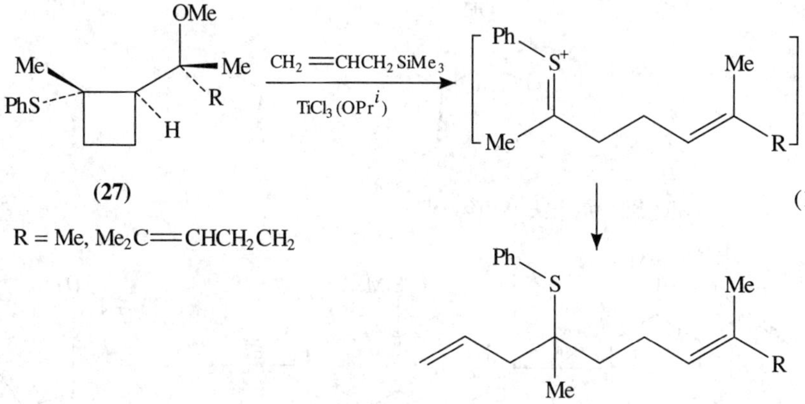

(**27**)
R = Me, Me_2C=$CHCH_2CH_2$

$Cp_2Ti(OTf)_2$ has been found to be an effective catalyst for the Sakurai–Hosomi reaction of allylsilanes with a variety of electrophiles[65].

2. Reactions with aldehydes and ketones

The reaction of allylsilanes with a carbonyl functionality has been continuously useful for the formation of carbon–carbon bonds. In the presence of a Lewis acid allylsilane, in general, undergoes $S_{E'}$ reaction with a carbonyl compound to afford the corresponding homoallylic alcohol[19,24,32]. The addition of allylsilane to carbonyl compounds catalyzed by the Brønsted acid $HN(SO_2F)_2$ gives the corresponding homoallylic alcohols in high yields[66]. It is interesting to note that cyclopentanone and acetophenone do not undergo addition reaction, whereas cyclohexanone and cyclododecanone give the corresponding homoallylic alcohols in excellent yields. Selective addition to aldehydes in the presence of α-tetralone or cyclopentanone can be achieved. Occasionally, ene reaction product has been found to be the major side product in these coupling reactions (*cf* Section V.F.1)[67]. $TiCl_4$-mediated reactions of **28** with aliphatic aldehydes afford 2,5-divinylhexane-1,6-diols in good yield with very high diastereoselectivity (equation 19)[68]. Fluoride ion promoted intramolecular cyclization of the allylsilanes **30** produced from anionic oxy-Cope rearrangement of 1,2-divinylcyclohexanols **29** leads to hydroazulenols with the *cis* ring fusion (equation 20)[69]. It is noteworthy that the Lewis acid catalyzed cyclization proceeds with high selectivity via an *anti* $S_{E'}$ pathway, which suggests that the silicon electrofuge is located away from the approaching electrophile regardless of the nature of Lewis acid or of the orientation of the double bond[70]. BF_3-catalyzed reactions of siloxy-substituted allylsilanes **31** with aldehydes give the corresponding unsaturated heterocycles **32** (equation 21)[71].

(28)

(19)

R = Ph, 4-$O_2NC_6H_4$, Alk

(29) **(30)**

(20)

(31) **(32)**

(21)

R = Me, Et, *i*-Pr, Ph

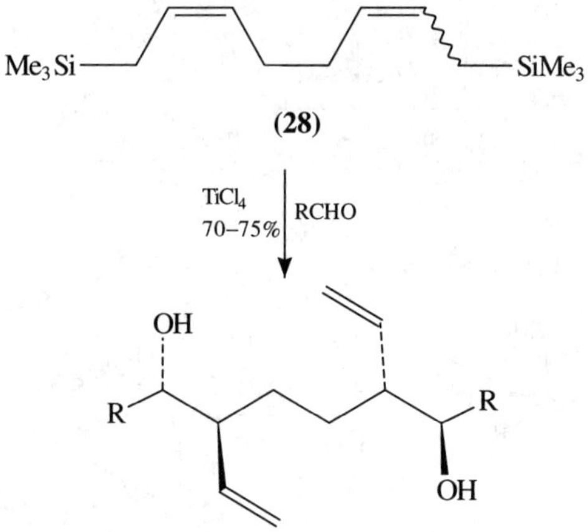

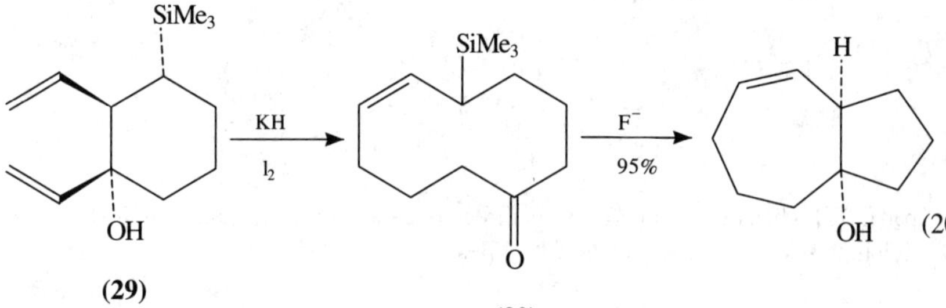

The nature of the Lewis acid catalyst has been found to play a pivotal role in the chemo- and stereoselectivities of the Sakurai–Hosomi reaction. For example, the reaction of **33**, which contains both allylsilane and allylstannane moieties, with benzaldehyde leads to the silyl-substituted homoallylic alcohol **34** in moderate yield when a mild Lewis acid such as Et_3Al is used. On the other hand, when $BF_3 \cdot OEt_2$ or fluoride ion is employed,

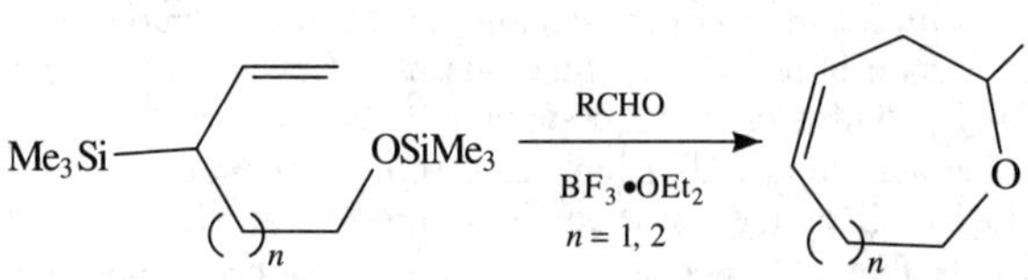

30. Synthetic applications of allylsilanes and vinylsilanes

the silyl group is displaced to give **35** (equation 22)[72]. Other strong Lewis acids such as $TiCl_4$, $SnCl_4$, $FeCl_3$, $MgBr_2$, $ZnCl_2$ and TMSOTf, however, give unsatisfactory results. As expected, the trimethylsilyl group is a much better leaving group than the diphenylmethylsilyl moiety in these $S_{E'}$ reactions.

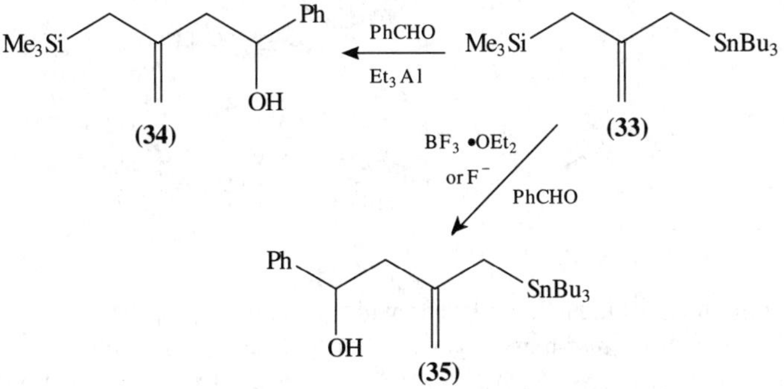

Carbamate and amide groups have been found to be stable under these coupling conditions[73]. In the presence of $TiCl_4$ or $SnCl_4$, chiral α-keto amides **36** react with allylsilane to produce, after hydrolysis, optically active tertiary alcohols **37** with extremely high optical selectivity (equation 23)[74]. The addition reaction appears to occur from the *Si* face of the carbonyl group. In a similar manner, a high degree of stereoselectivity is obtained from the reactions of *N*-Boc-α-amino aldehydes **38** with 2-substituted allylsilanes (equation 24)[75].

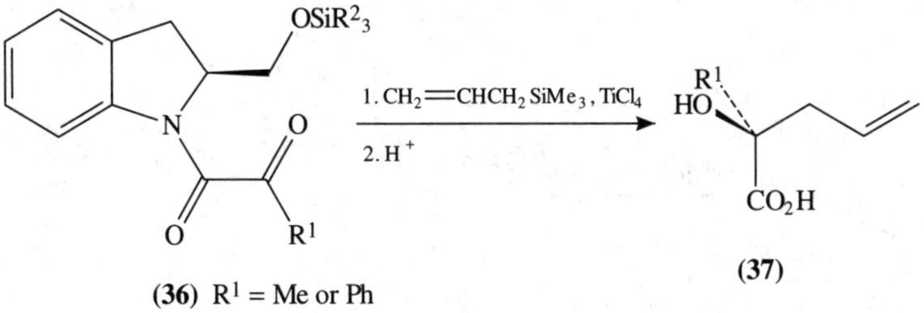

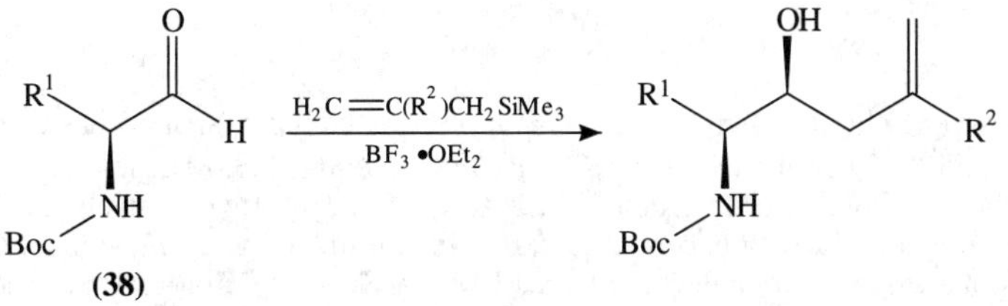

R^1 = 1° and 2° Alk; R^2 = *i*-Pr, $ClCH_2$, $PhCH=CH$

Parallel to an earlier work on the highly diastereoselective reactions of aliphatic aldehydes with allylsilane in the presence of **39**[76], treatment of methyl ketones under the same conditions yields the corresponding tertiary homoallylic ether with a diastereomeric excess of up to 90% (equation 25)[77].

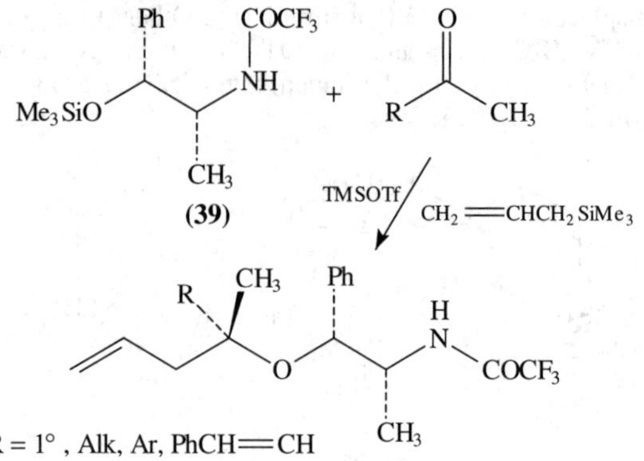

The reaction of (Z)-β-methylcrotylsilane with 2-benzyloxypropanal in the presence of chelative SnCl$_4$ gives the *anti*-homoallylic alcohol **40** diastereoselectively (equation 26)[78]. The use of chelative TiCl$_4$ versus nonchelative BF$_3 \cdot$OEt$_2$ Lewis acids also gives different stereoselectivity in the coupling of allylsilane with the aldehydes **41** (equation 27)[79].

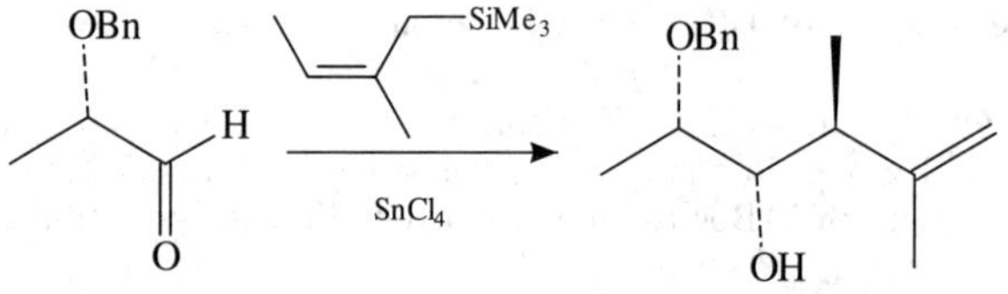

The addition of chiral (*E*)-crotylsilanes **42** to (*S*)-2-alkoxypropanal **43** has been investigated in detail[80]. When the (*S*)-silane **42** and **43a** are employed, nonchelating BF$_3 \cdot$OEt$_2$ promotes the addition reaction from the *Re* face of the aldehyde moiety to give *syn* homoallylic alcohol **44a** with excellent levels of Felkin induction. *Anti* homoallylic alcohol **45** is obtained with high levels of *anti*-Felkin selectivity from the *Si* face of the aldehyde group when a chelative Lewis acid such as TiCl$_4$ is employed (equation 28). Interestingly, both TiCl$_4$ and BF$_3 \cdot$OEt$_2$ mediated condensations of (*R*)-silane **42** with **43a** yield the same *syn* disposed crotylsilation product **46a** having *anti*-Felkin induction (equation 29). Reactions of sterically hindered **43b** with (*S*)-**42** yield **44b** predominantly (>30 : 1) in the presence of TiCl$_4$ or BF$_3 \cdot$OEt$_2$. In these cases, the bulky silicon group prevents the Lewis acid from chelate formation with the aldehyde. In a similar manner, treatment of (*R*)-**42** with **43b** affords **46b** as the major product no matter which

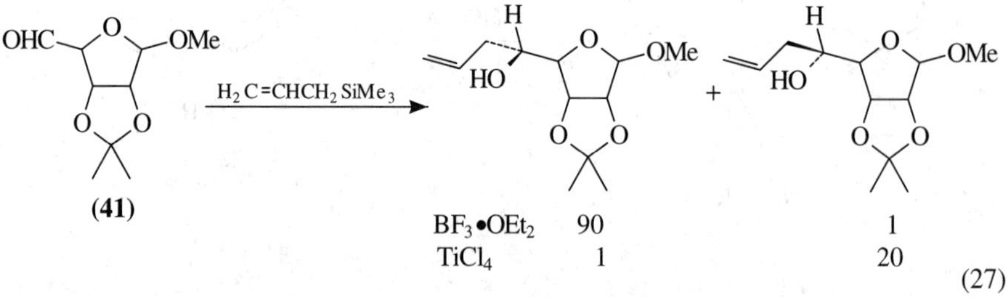

30. Synthetic applications of allylsilanes and vinylsilanes

catalyst (TiCl$_4$ or BF$_3$·OEt$_2$) is used (equation 29). An open transition state model has been employed to rationalize the stereoselectivity (Scheme 2).

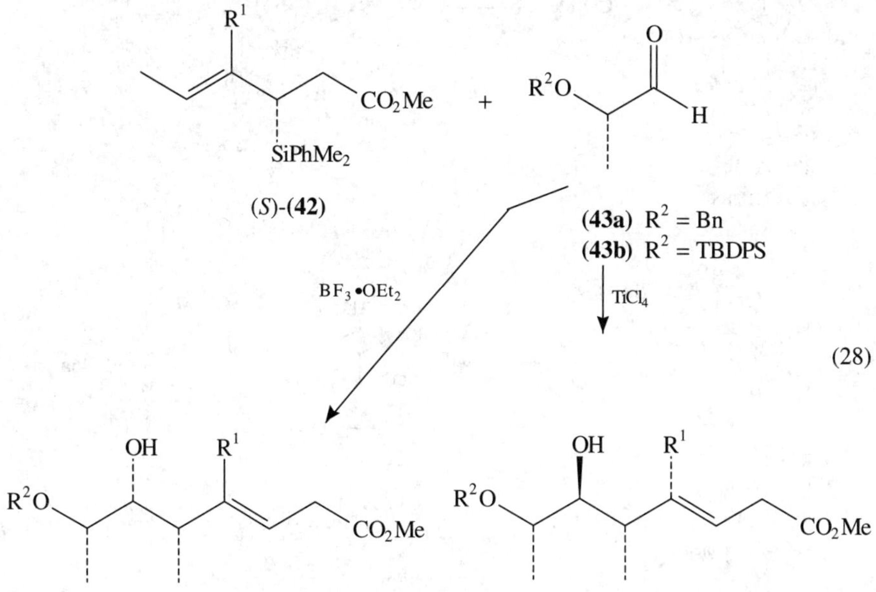

(28)

(S)-(42) + (43a) R^2 = Bn
(43b) R^2 = TBDPS

(44a) R^2 = Bn
(44b) R^2 = TBDPS
R^1 = Me, Et

(45) R^2 = Bn

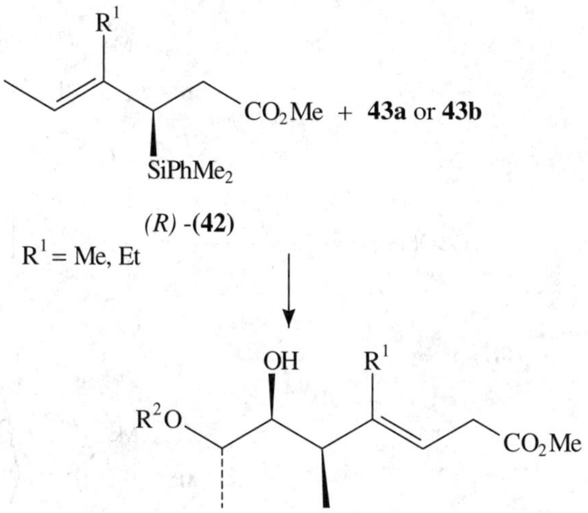

(R)-(42) + 43a or 43b
R^1 = Me, Et

(29)

(46a) R^2 = Bn
(46b) R^2 = TBDPS

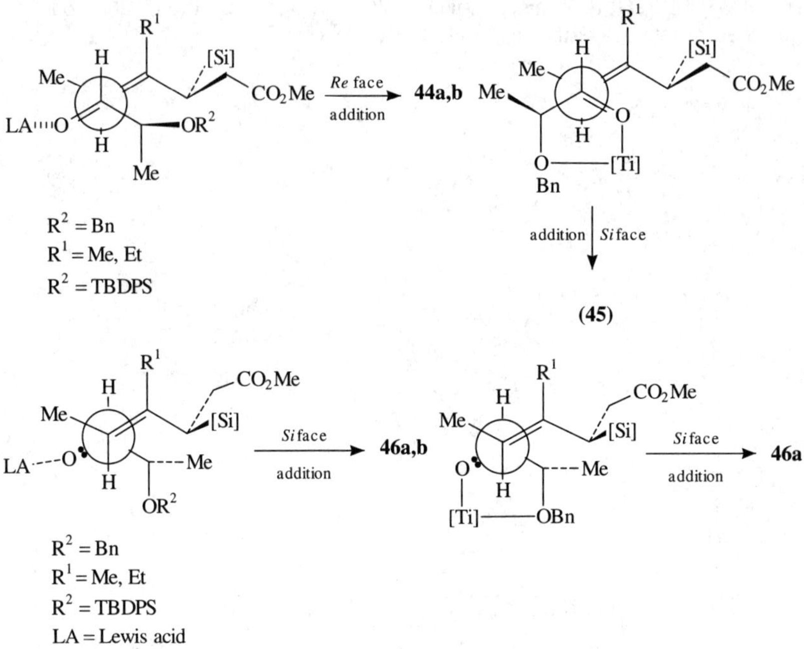

SCHEME 2

Treatment of chiral (E)-crotylsilanes **47** with **43a** in the presence of $BF_3 \cdot OEt_2$ gives tetrahydrofuran derivatives in good yield with 96% de. Interestingly, 1,2-silyl group migration competes favorably with elimination of the silyl group after condensation with **43a** (equation 30)[81].

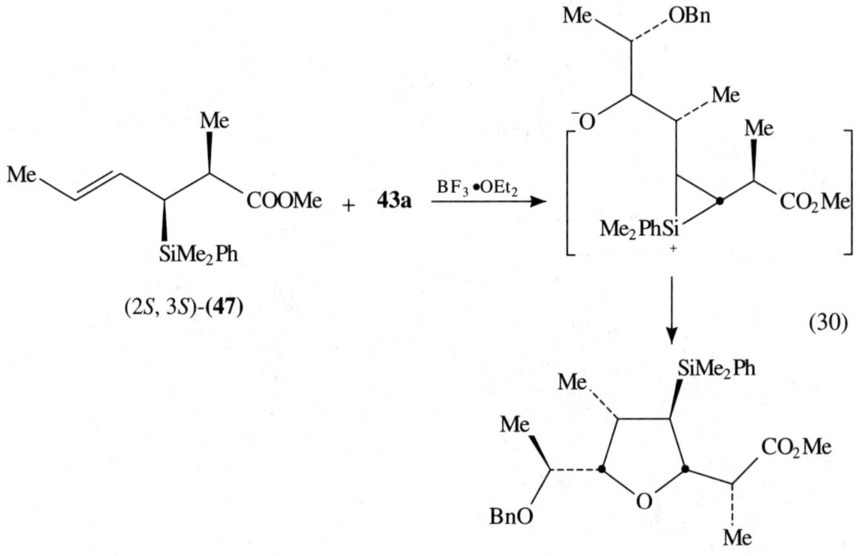

30. Synthetic applications of allylsilanes and vinylsilanes

The TiCl$_4$-induced three-component coupling reaction of an α-haloacylsilane, allylsilane and another carbonyl compound gives **48** in good yield. A silyl enol ether intermediate is suggested (equation 31)[82]. The reaction of a cyclopropyl ketone with allylsilane yields a mixture of skeletal rearranged products[83].

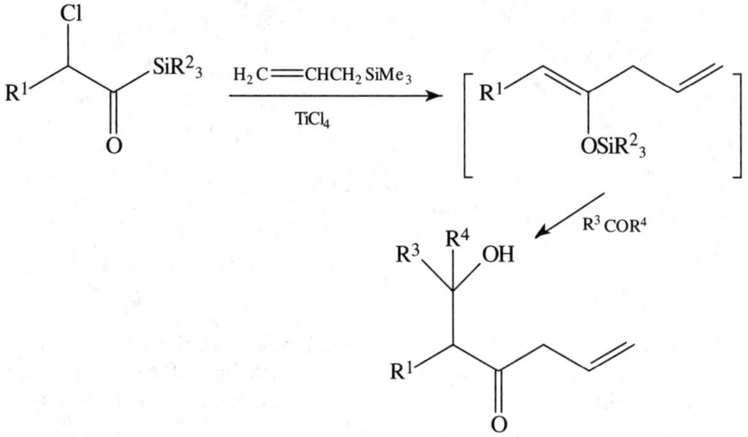

$$R^1 = n\text{-}C_3H_7, n\text{-}C_6H_{13}; R^2 = Me, t\text{-}Bu, Ph; R^3 = Me, Et, Ph; R^4 = H, Me$$

The TMSOTf-catalyzed reaction of allylsilane, a carbonyl compound and a trimethylsilyl ether gives the corresponding homoallylic ether (equation 32)[84]. Tetrahydropyran derivatives **49** are thus obtained conveniently (equation 33)[85–87]. The extension of this intramolecular reaction to vinylsilane derivatives also yields dihydropyran derivatives **50** (equation 34)[88]. Interestingly, two equivalents of the carbonyl compounds react with **51** in the presence of BF$_3$ · OEt$_2$ catalyst to yield tetrahydropyran derivatives (equation 35). Presumably, an ene reaction may occur first[89]. The methodology consists of the coupling between a carbonyl compound and a silyl ether, containing a judiciously positioned allylsilane moiety. The oxonium cation initially formed is the key intermediate leading to these products.

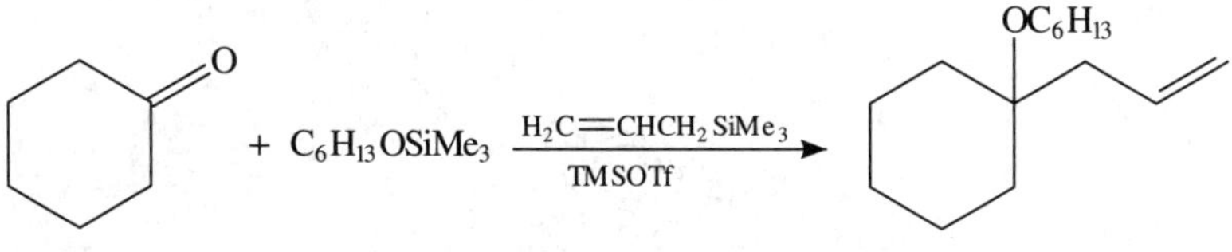

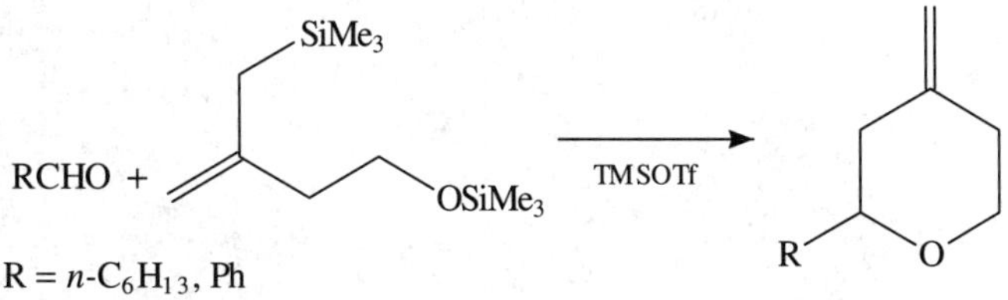

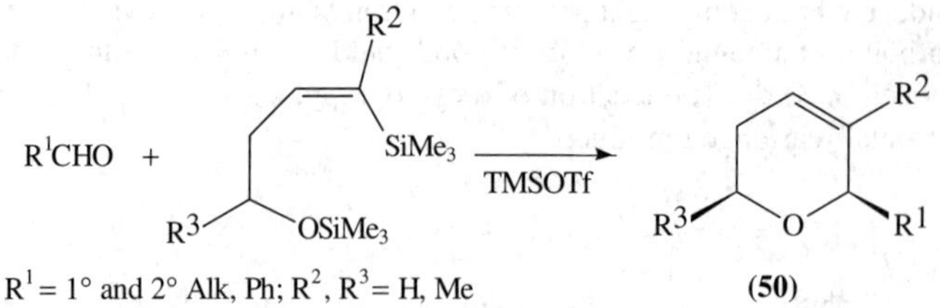

R^1CHO + [allylsilane with R^2, R^3, SiMe$_3$, OSiMe$_3$] $\xrightarrow{\text{TMSOTf}}$ (**50**) (34)

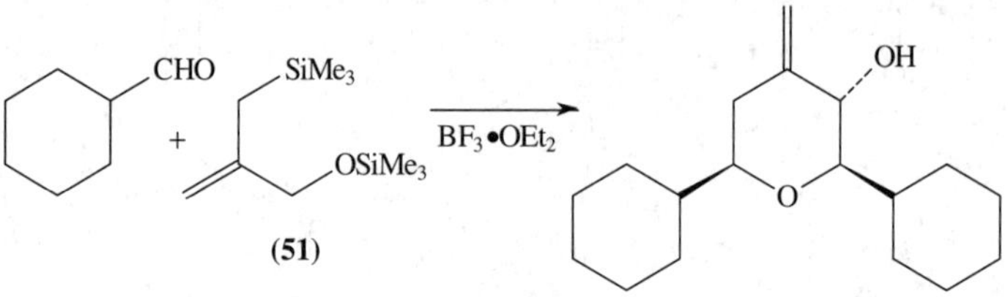

R^1 = 1° and 2° Alk, Ph; R^2, R^3 = H, Me

[cyclohexyl-CHO] + [allylsilane (**51**)] $\xrightarrow{\text{BF}_3 \cdot \text{OEt}_2}$ [tetrahydropyran product] (35)

Structural variation at the silicon atom changes the reactivity and the selectivity. The reactions of pentacoordinated allylic silicates are remarkably different from their tetracoordinated allylic silane counterparts[90–95]. Extremely high regio- and stereospecificites in reactions with carbonyl compounds are well documented[20]. Thus, allylation of aldehydes with allyltrifluorosilane in the presence of catechol and triethylamine gives the corresponding homoallylic alcohols in a regiospecific manner. The pentacoordinated allylsilicate **52** is suggested as the key intermediate. Unlike reactions with tetracoordinated allylsilanes which occur via the acyclic $S_{E'}$ mechanism, the reactions of an aldehyde with **52** proceeds via a cyclic transition state **53** (equation 36)[90,91]. Similar reactions with α-hydroxyketones[92] and with β-functionalized α,β-unsaturated enones[93] give the corresponding homoallylic alcohols with excellent diastereoselectivity. $BF_3 \cdot OEt_2$-mediated condensation of allylalkoxysilane with aldehydes gives 2,4,6-trisubstituted tetrahydropyrans **54** (equation 37)[95].

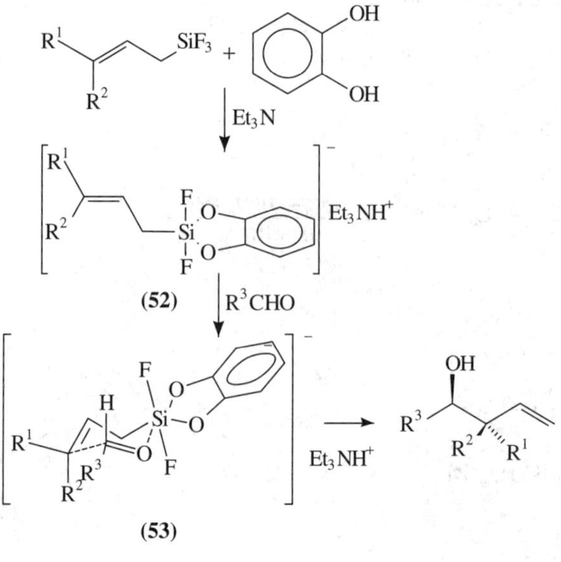

(36)

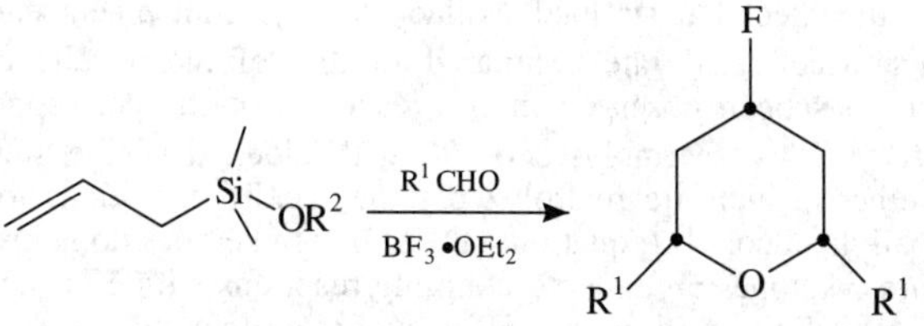

[Equation (37) shows an allylsilane with OR² group reacting with R¹CHO in the presence of BF₃·OEt₂ to give tetrahydropyran product (54) with F substituent.]

$R^1 = n\text{-}C_6H_{13}, n\text{-}C_8H_{17}$; $R^2O = (-)$-menthoxy

Allyltrichlorosilanes react regioselectively with aldehydes in the presence of an amide additive[96] to afford the corresponding homoallylic alcohols in high yields. The reaction can also be carried out in DMF without a catalyst (equation 38)[97,98]. It is noted that *syn* and *anti* homoallylic alcohols are obtained stereospecifically from (Z)- and (E)-crotyltrichlorosilanes, respectively. Again, a cyclic transition state similar to **53** is proposed. A one-pot synthesis of homoallylic alcohols from the reactions of allyl halides with trichlorosilane in the presence of triethylamine and copper(I) chloride followed by the addition of an aldehyde is achieved (equation 39)[99]. Allyltrichlorosilane is believed to be generated *in situ* under these conditions. The reactions of 1-chloro-2,4-pentadiene with aldehydes under the same conditions give **55** exclusively, which suggests that the incoming electrophile attacks at the γ position of the diene system (equation 40). This result further supports the early suggestions that the reaction may proceed via a cyclic transition state like **53**. It is noteworthy that 2,4-pentadienyltrimethylsilane reacts with aldehydes at the terminal position of the diene[100–102].

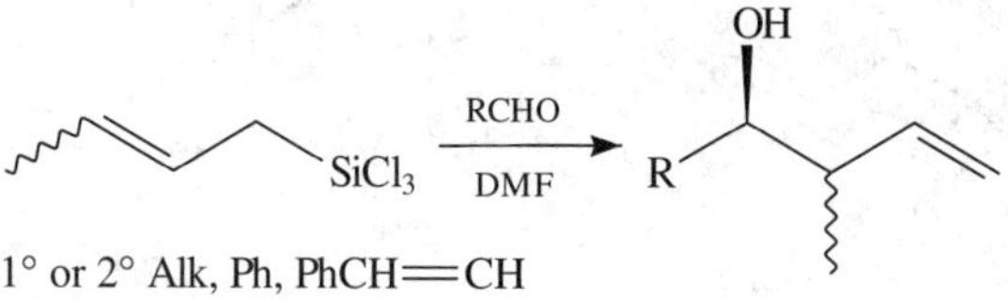

[Equation (38): allyl-SiCl₃ + RCHO in DMF gives homoallylic alcohol.]

R = 1° or 2° Alk, Ph, PhCH=CH

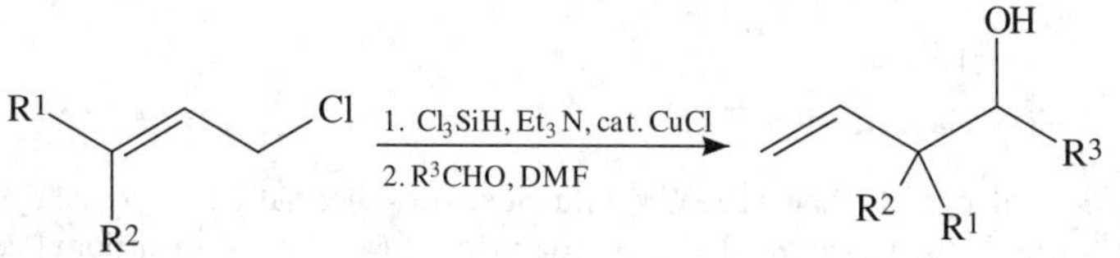

[Equation (39): allyl chloride with R¹, R² substituents reacts with 1. Cl₃SiH, Et₃N, cat. CuCl; 2. R³CHO, DMF to give homoallylic alcohol.]

R^1 = H, Me, Me₂C=CHCH₂
R^2 = H, Me

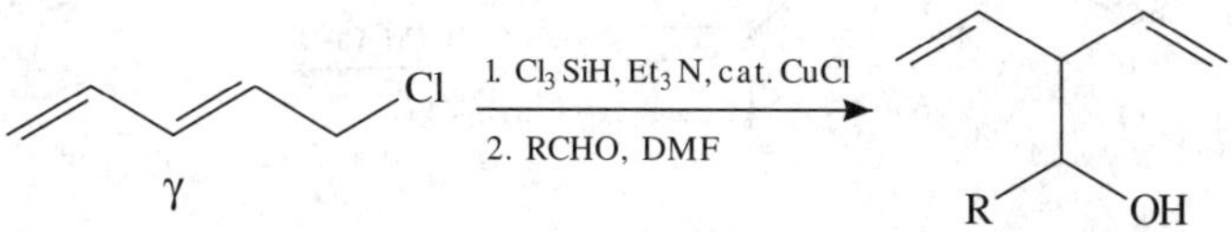

[Equation (40): 1-chloro-2,4-pentadiene (γ position labeled) reacts with 1. Cl₃SiH, Et₃N, cat. CuCl; 2. RCHO, DMF to give product (55).]

R = Ph, PhCH₂CH₂, PhCH=CH

It is well documented that strained 1-silacyclobutyl enol ethers undergo aldol reactions at a much accelerated rate compared to that of the acyclic counterpart[103,104]. Similar behavior has been observed in the reactions of the corresponding allylsilanes with aldehydes[105]. Thus, thermolyses of **56** with aldehydes give stereospecifically 1-silacyclobutyl ethers which are hydrolyzed under acidic conditions to yield the corresponding homoallylic alcohols (equation 41). Although ketones do not react under similar conditions, hydroxyketones **58** undergo coupling reactions with **57** to give diols **60** stereoselectively. The cyclic transition state **59** is suggested (equation 42).

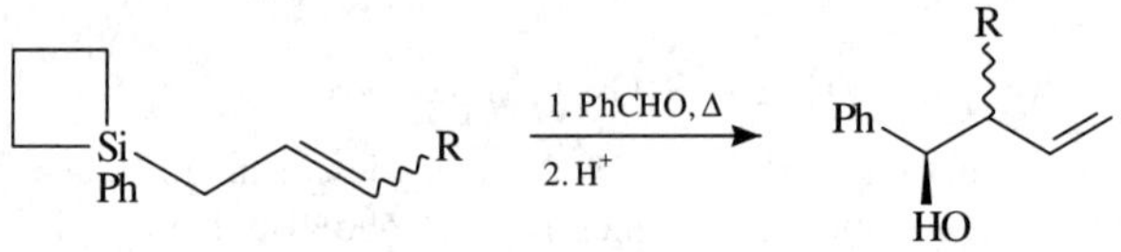

(56)
R = Ph, n-Pr

Z-(56) syn : anti = 5 : 95
E-(56) syn : anti = 95 : 5

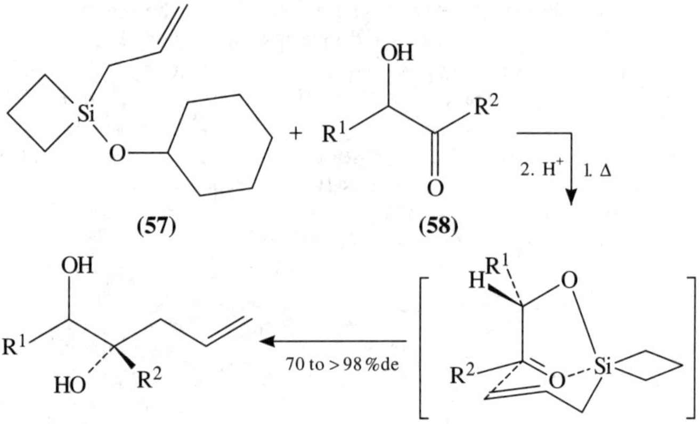

(41)

(57) + (58)

(60)
(59)

R^1 = Ph, n-Bu
R^2 = Ph, Me

R^1 = Ph, n-Bu; R^2 = Ph, Me

(42)

70 to > 98 %de

3-Methylenesilacyclobutane **61** reacts with aldehydes thermally to give the cyclic siloxy adduct **62**[105]. In the presence of $BF_3 \cdot OEt_2$, treatment of **61** with 1,4-dicarbonyl compounds yields the corresponding 8-oxabicyclo[3.2.1]octane skeleton **63** (equation 43)[106].

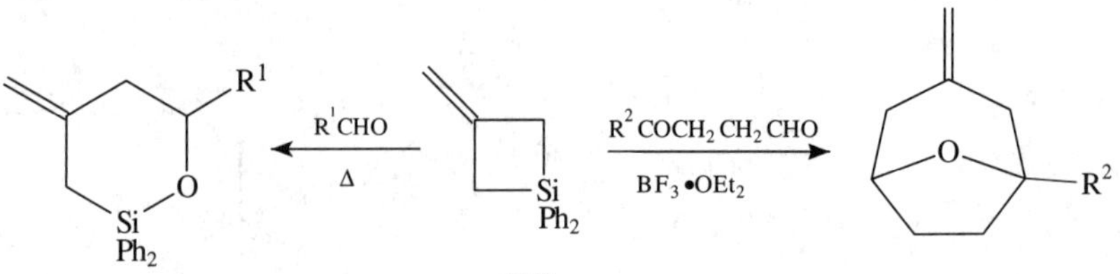

(62)
(61)
(63)

R^1 = n-C_6H_{13}, Ph; R^2 = Me, Ph

(43)

Cathodic coupling of ketones with **64** takes place regioselectively at the β-position to the silyl substituent to give **65** in good yields (equation 44). A radical anionic intermediate derived from the carbonyl compound is proposed[107].

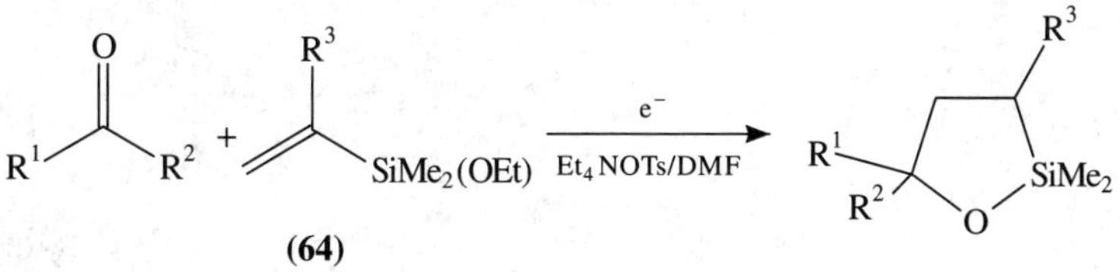

(44)

$R^1, R^2 = \text{Alk} ; R^3 = H, n\text{-Pr}$

3. Reactions with enones

Since allylsilane can be considered as a very soft nucleophile because of the involvement of σ–π conjugation between the π-electrons of the double bond and the σ-electrons in the carbon–silicon bond, the addition of allylsilanes to α,β-unsaturated enone moiety occurs exclusively via 1,4-addition[19,20]. A typical Sakurai–Hosomi reaction is illustrated in the reaction of allylsilane **66** with enone **67** in the presence of TiCl$_4$ to give **68** which is used for the synthesis of a cyclic enediyne (equation 45)[108]. A similar reaction has been used for the synthesis of *ent*-herbasolide **70** from enone **69** (equation 46)[109]. Prenylsilane undergoes 1,4-addition with squaric acid chloride **71** followed by dechlorosilylation to give **72** as the predominant product (equation 47). Interestingly, other simple allylsilanes react in a 1,2-addition fashion to yield **73**[110].

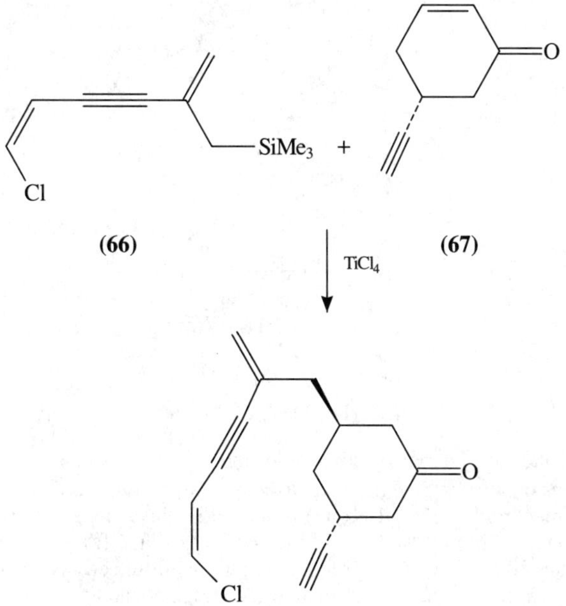

(45)

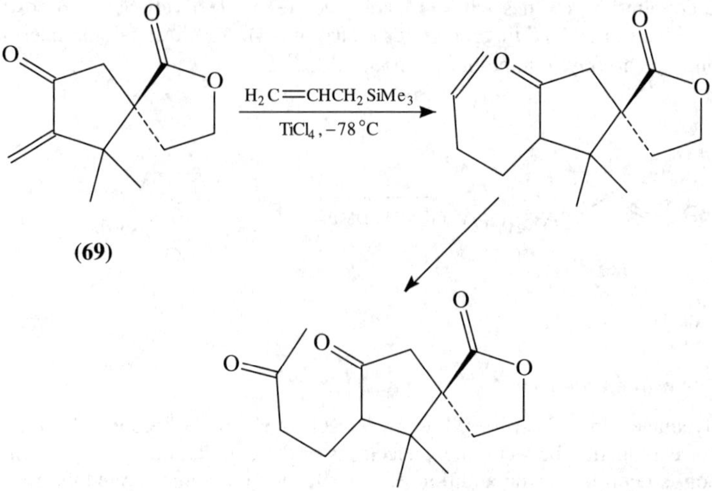

(69) → H₂C=CHCH₂SiMe₃, TiCl₄, −78 °C → (46)

↓

(70)

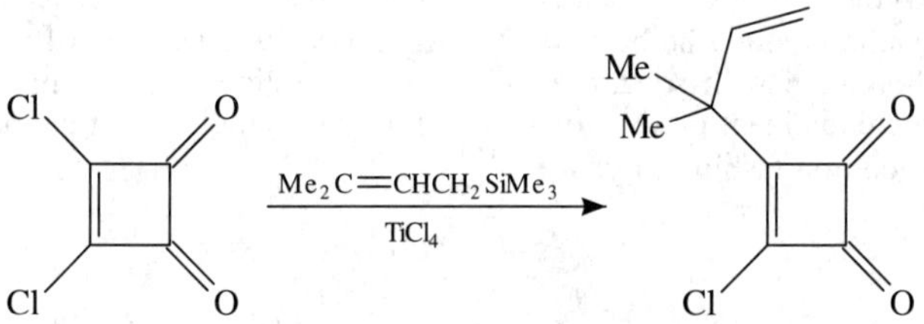

(71) → Me₂C=CHCH₂SiMe₃, TiCl₄ → (72)   (47)

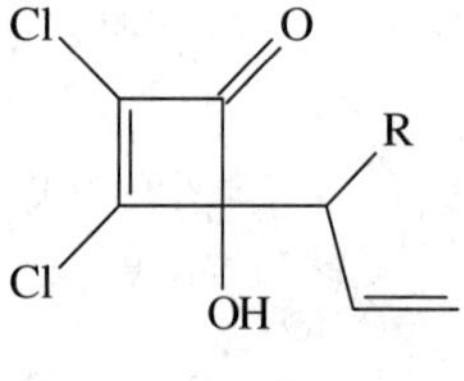

(73)

R = H or Me

Lewis acid-catalyzed stereoselective addition of crotylsilanes to chiral **74** has been studied in detail[111,112]. The presence of the chiral auxiliary at C_2 (e.g. *p*-tolylsulfinyl or menthoxy carbonyl group) induces the diastereofacially selective addition of cyclopentenones with crotylsilanes. Thus, (*E*)-crotylsilane favors the *erythro* product, whilst (*Z*)-isomer favors the *threo* product. High enantioselectivity is observed in both reactions (equation 48). In a similar manner, conjugated addition of allylsilane to **75** proceeds with high efficiency (equation 49)[113]. Interestingly, the yield and enantiomeric excess of the product is dependent on the amount of TiCl₄ used and the best selectivity

30. Synthetic applications of allylsilanes and vinylsilanes

(94–96 %ee) is obtained with 0.8–0.95 equivalent of the Lewis acid catalyst.

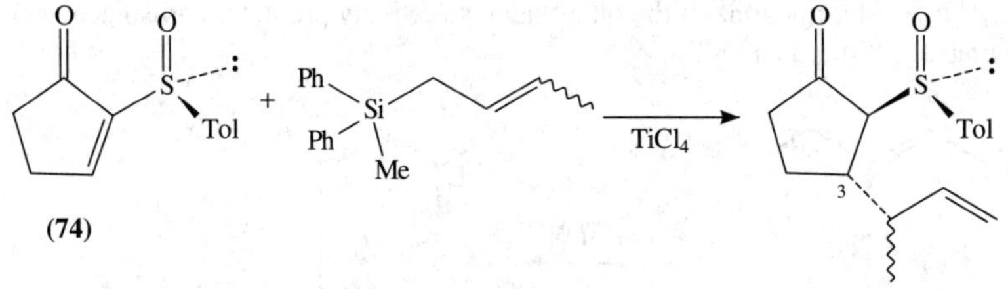

(74)

$$\text{(48)}$$

E erythro : thero = 9 : 1   3S : 3R = 99 : 1
Z erythro : thero = 1 : 14 3S : 3R = 99 : 1

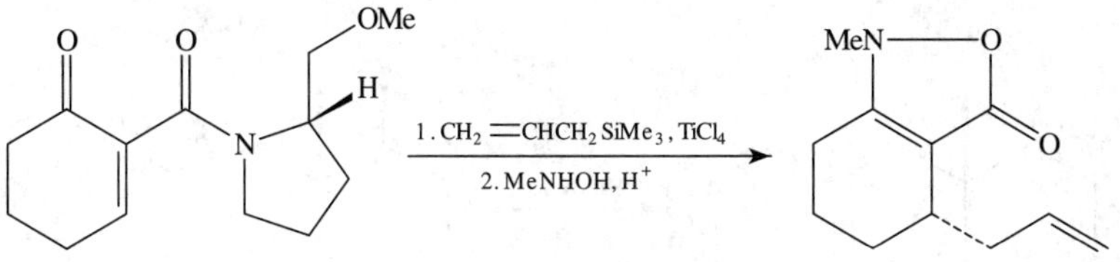

(75)

$$\text{(49)}$$

TMSOTf-mediated intramolecular cyclization of allylsilanes with alkylidene 1,3-dioxo moiety gives almost exclusively the *trans*-1,2-disubstituted cyclopentanes **76** in high yield (equation 50). Other Lewis acids or fluoride ion can also promote similar reactions[114]. In the presence of the chiral auxiliary, such intramolecular cyclization gives the corresponding *trans*-1,2-disubstituted cyclopentanes and cyclohexanes **77** (equation 51) in excellent enantioselectivity[115].

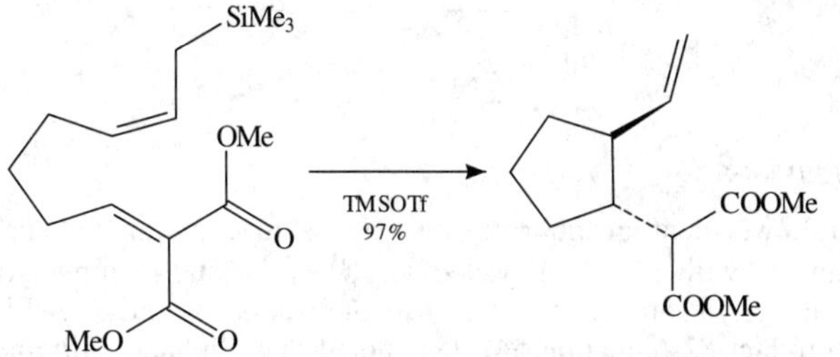

(76)

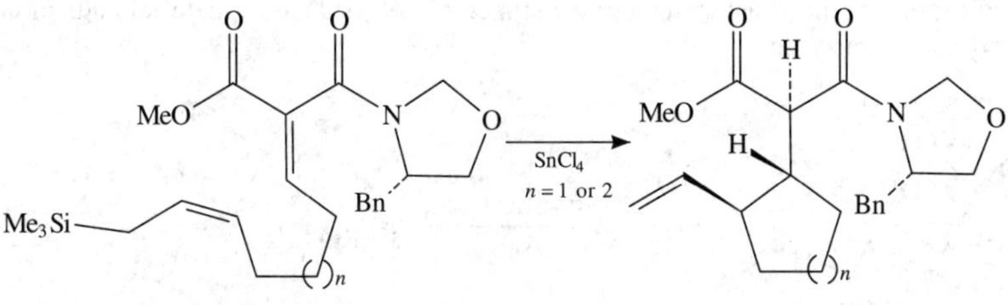

$$\text{(50)}$$

(77)

$$\text{(51)}$$

Intramolecular cyclization of allylsilane moiety with conjugated dienone groups occurs readily. Both S_E and $S_{E'}$ reactions are observed depending on the nature of the catalyst.

Amberlyst 15 and $EtAlCl_2$ catalysts give γ-substitution products **78** (equation 52) and **79** (equation 53)[116,117]. Fluoride ion, on the other hand, selectively promotes α-displacement reaction (giving e.g. **80**) (equation 53)[118].

(78) (52)

(79) (53)

(80)

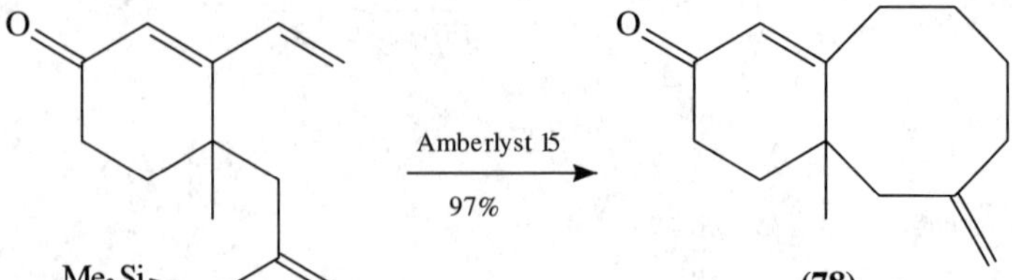

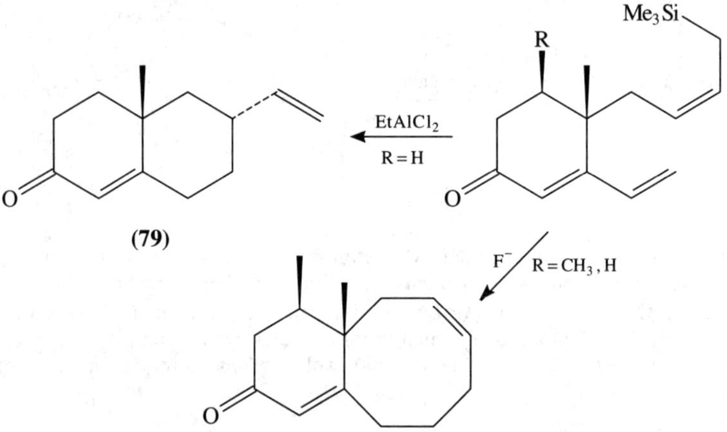

4. Reactions with iminium ions

In the presence of a Lewis acid or fluoride ion, imines react with allylsilane to yield the homoallylic amines with high stereoselectivity[119,120]. Thus, treatment of *N*-galactosylaldimine **81** with allylsilane in the presence of excess of $SnCl_4$ yields the corresponding allylated product **82** (equation 54). It is noted that aliphatic aldimines do not react under these conditions. Fluoride ion promoted crotylation of aldimines proceeds in a regiospecific and diasteroselective manner[121]. A pentacoordinate silicate moiety is involved in this reaction.

(81) (82) (54)

R = *t*-BuCO

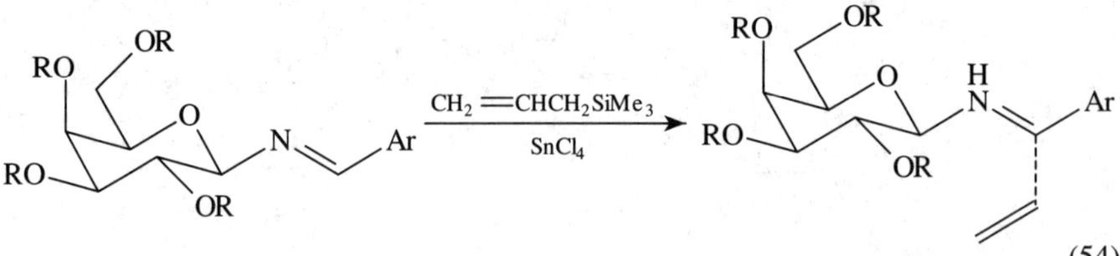

30. Synthetic applications of allylsilanes and vinylsilanes

Treatment of an α-acylaminal derivative **83** with allylsilanes in the presence of a Lewis acid or a Brønsted acid gives the corresponding condensation products (equation 55)[122-125]. The reaction can also proceed intramolecularly, e.g. in the formation of **84** and **85** (equations 56 and 57). A piperidine skeleton **86** can also be formed by such intramolecular cyclization (equation 58)[126].

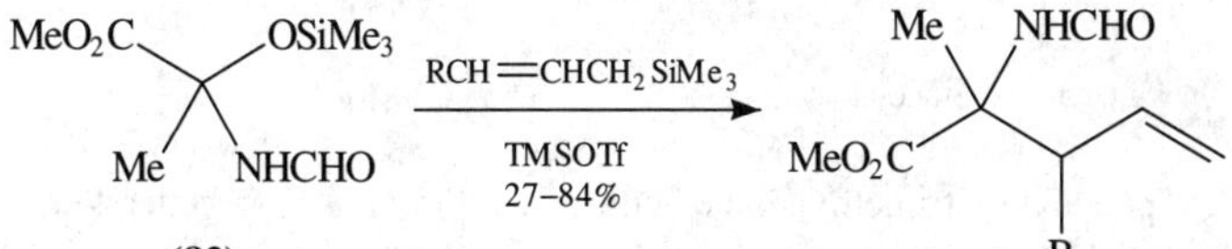

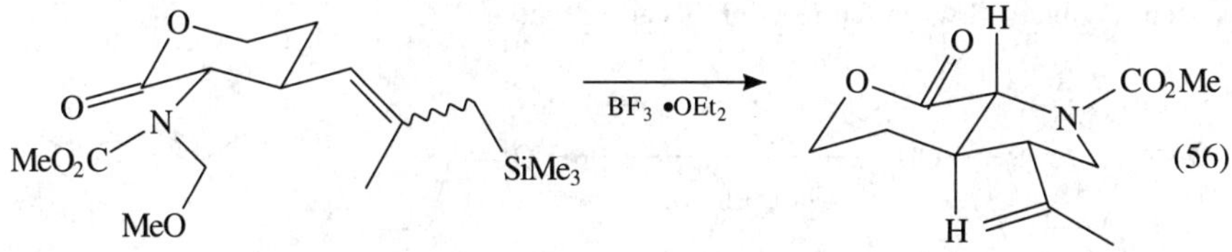

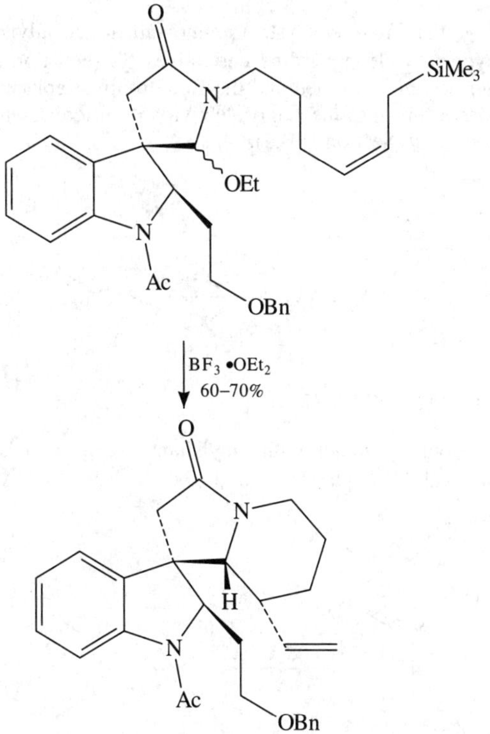

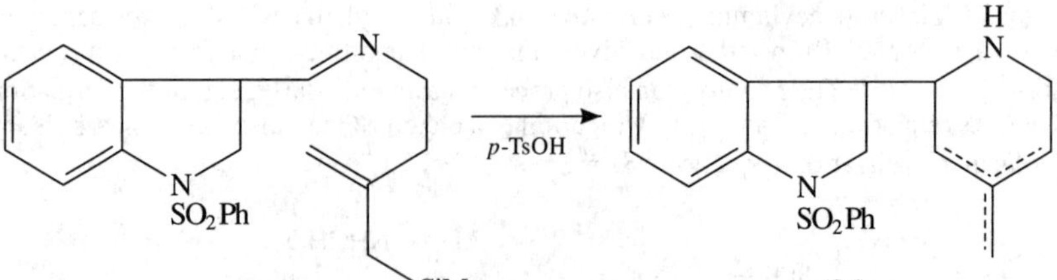

The reaction of 2,4-pentadienyltrimethylsilane with N-acyliminium ion generated in situ at −78 °C results in the formation of pentadienyl-substituted lactams **87** in good yield (equation 59). It is noted that the reaction temperature is an important factor for controlling the ε-regioselectivity. A mixture of ε- and γ-substitution products at the dienyl system is obtained when the reaction is carried out at 0 °C[127].

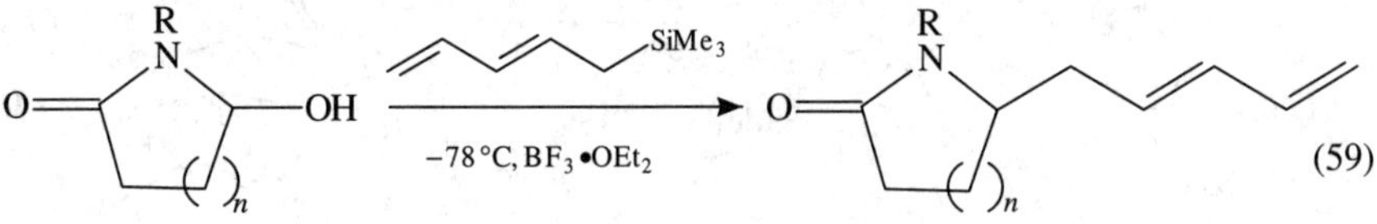

R = $PhCH_2$, $PhCH_2OCH_2$ $CH=CHCH_2$

In the presence of the $TiCl_4$ catalyst, δ-amino-substituted allylsilanes readily react with isobutyraldehyde to yield piperidine derivatives **88** (equation 60)[128]. Presumably, an iminium ion intermediate is formed and the Boc group is replaced under the reaction conditions. It is noteworthy that the regioselectivity is different when acid chloride in used as the electrophile (cf Section II.B.5).

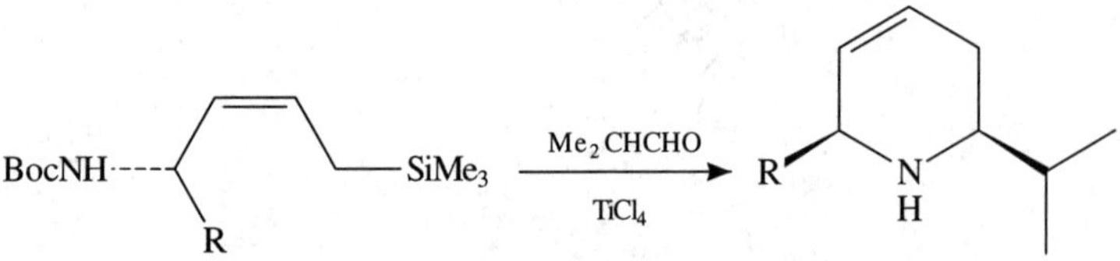

R = Me, Me_2CHCH_2, Et(Me)CH

The coupling reactions of imines with vinylsilanes serve as a key protocol for the synthesis of various alkaloids (equation 61 and 62)[129–132].

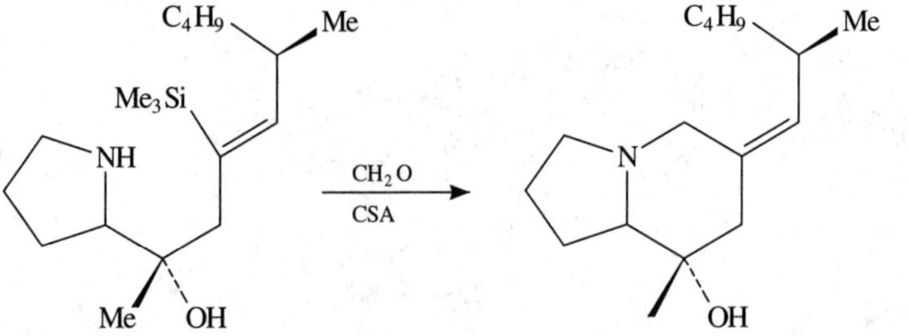

(62)

CSA = Camphor-10-sulfonic acid

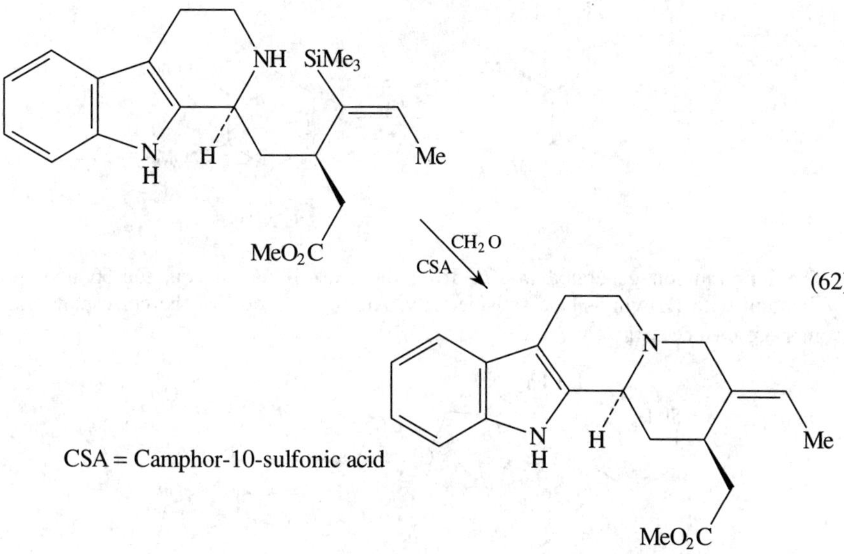

Protonation of an imine also leads to the corresponding intramolecular cyclization product[133]. Acyliminium ion behaves similarly[134,135]. Representative examples are shown in equations 63–65.

(63)

(64)

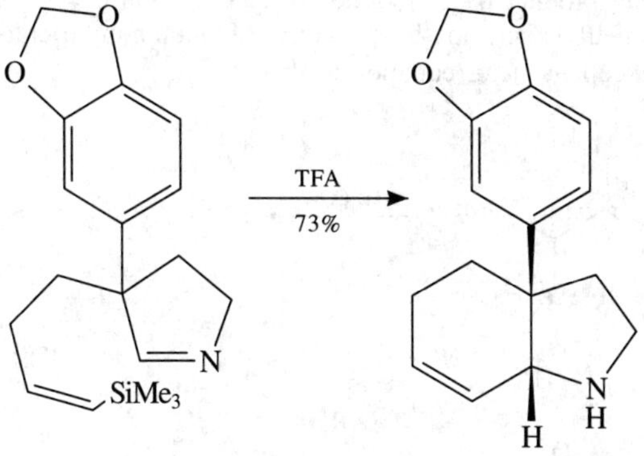

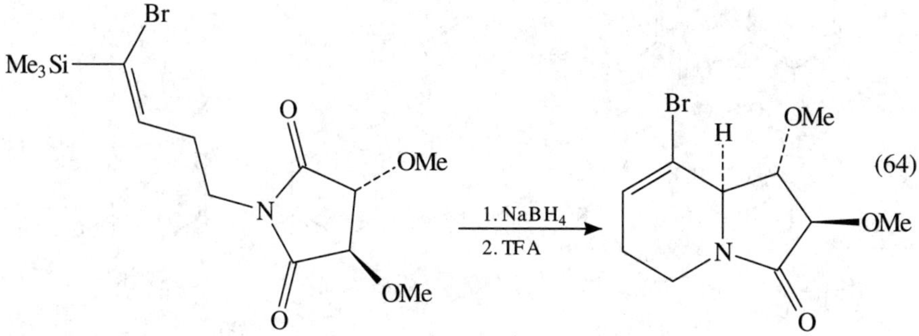

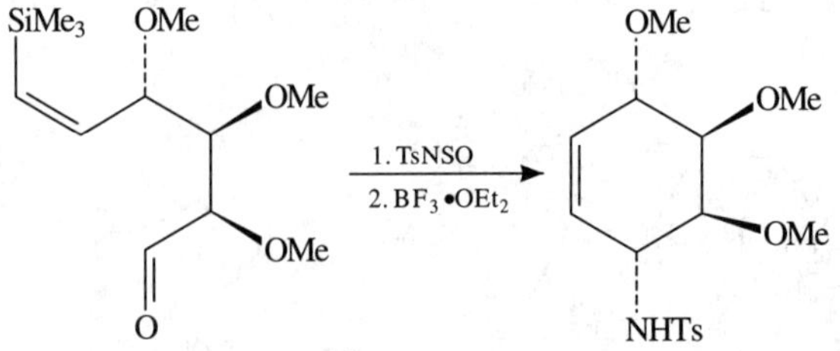

$$\text{(65)}$$

The iminium ion generated *in situ* from the oxazolidine in equation 66 undergoes cyclization with (Z)-vinylsilane satisfactorily. Attempts to cyclize the corresponding E-isomer are unsuccessful[136].

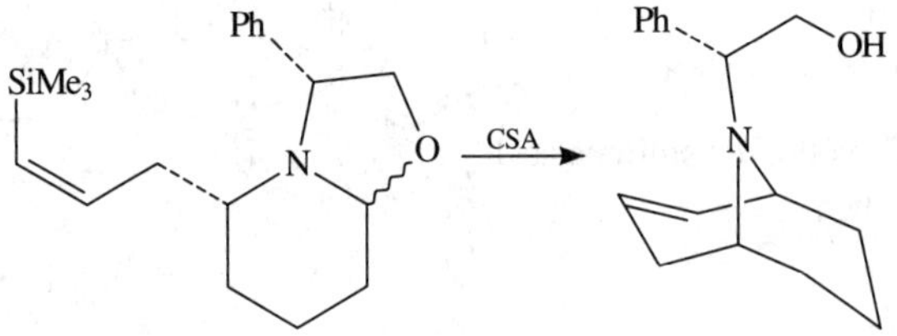

$$\text{(66)}$$

5. Reactions with acid chlorides

Treatment of **7** with acid chloride at $-60\,°C$ in the presence of $TiCl_4$ gives the corresponding coupling product **89** stereoselectively (equation 67)[40]. The reaction of aryl chloroformate with allylsilane in the presence of aluminum trichloride gives aryl 4-alkenoates **90** in excellent yield (equation 68)[137].

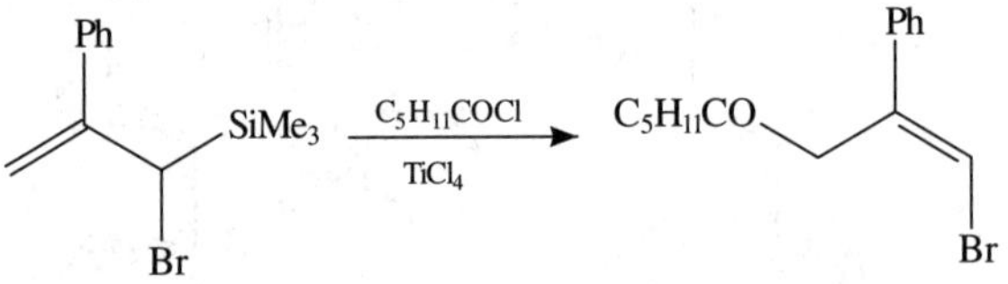

$$\text{(67)}$$

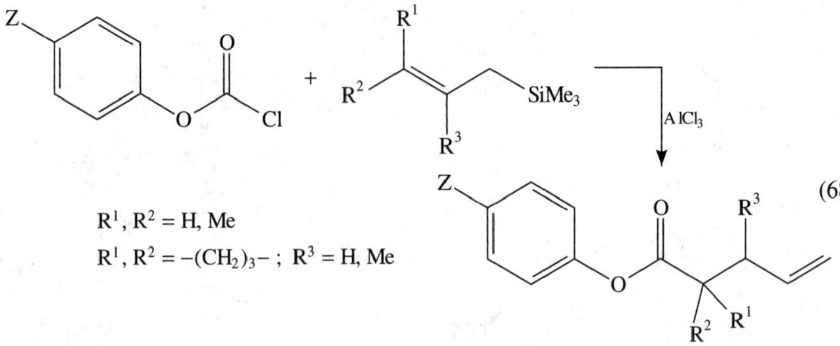

$R^1, R^2 = H, Me$
$R^1, R^2 = -(CH_2)_3-$; $R^3 = H, Me$

$$\text{(68)}$$

Depending on the nature of the substrate, treatment of acid chloride with δ-aminoallylsilane derivatives **91** gives different regioisomers. When R^1 is methyl or a primary alkyl group, a normal S_E' reaction takes place. When R^1 is the sterically hindered isopropyl group, direct displacement of the silyl group occurs (equation 69)[128]. Similar regioselectivity is observed when an aldehyde is employed as the electrophile (see Section II.B.6).

(69)

(91)

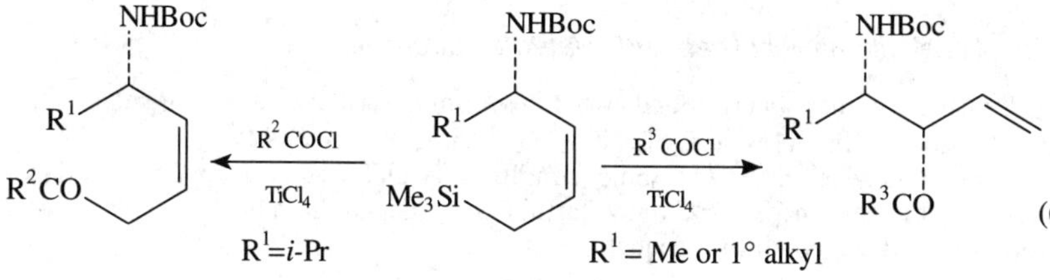

R^2 = Me, $PhCH_2$
R^3 = Me, $PhCH_2$, *n*-C_6H_{13}

Reactions of 2-stannyl-3-silylpropene with acid chlorides in the presence of Lewis acid afford vinylsilanes **92** via 1,2-silyl migration followed by destannylation (equation 70)[138].

(70)

(92)

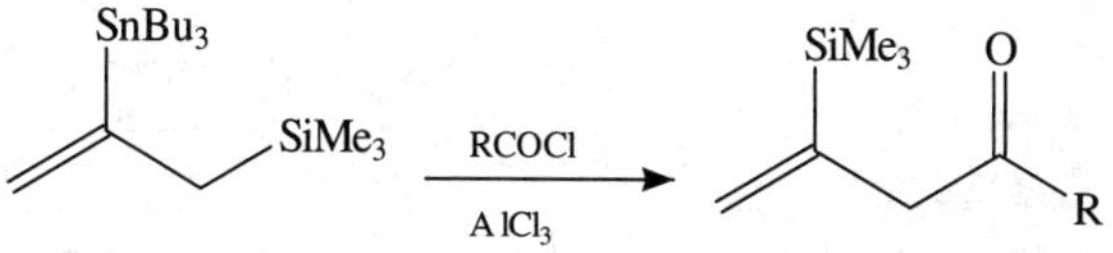

Lewis acid-mediated reactions of acid chlorides with vinylsilanes have demonstrated numerous fascinating transformations leading to carbon–carbon bond formation. The reactions can occur either intermolecularly or intramolecularly. Various key intermediates in the total syntheses of natural products are prepared by these coupling processes[28]. In the presence of Lewis acids, the reaction of vinylsilanes with chlorothiocarbonate gives α,β-unsaturated thiocarboxylates **93** in good yield (equation 71)[139]. The reactivity of the silyl substituent as an electrofugal group following electrophilic attack depends on the substituent on the silicon. Whereas trimethylsilyl is highly reactive, *tert*-butyldiphenylsilyl is much less reactive (equation 72)[140]. Sequential replacement of silyl groups with different acyl groups has been reported (equation 73)[141,142].

$$RCH{=}CHSiMe_3 \xrightarrow[AlCl_3,\ -60\,°C]{ClCOSEt} RCH{=}CHCOSEt \qquad (71)$$

(93)

(72)

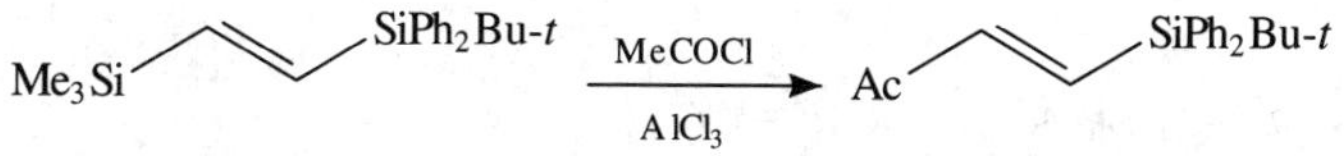

R^1, R^2 = alkyl, aryl (73)

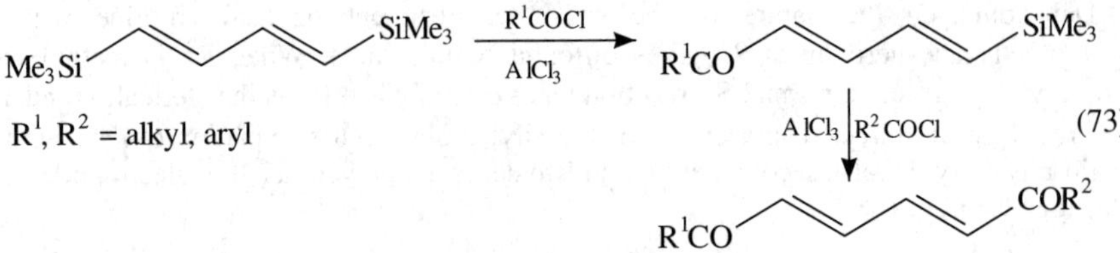

6. Reactions with alkyl halides and related electrophiles

S_E-Substitution has been found when α-bromoallylsilane **7** is treated with *tert*-alkyl chloride in the presence of $TiCl_4$ at −78 °C to give **94**[40]. The reaction with α-bromoethylbenzene under the same conditions yields the corresponding indane **96**. Apparently, the intermediate **95** may further undergo an intramolecular Friedel-Crafts reaction under these conditions (equation 74)

(94) **(7)** Ar = Ph **(95)** **(96)**

(74)

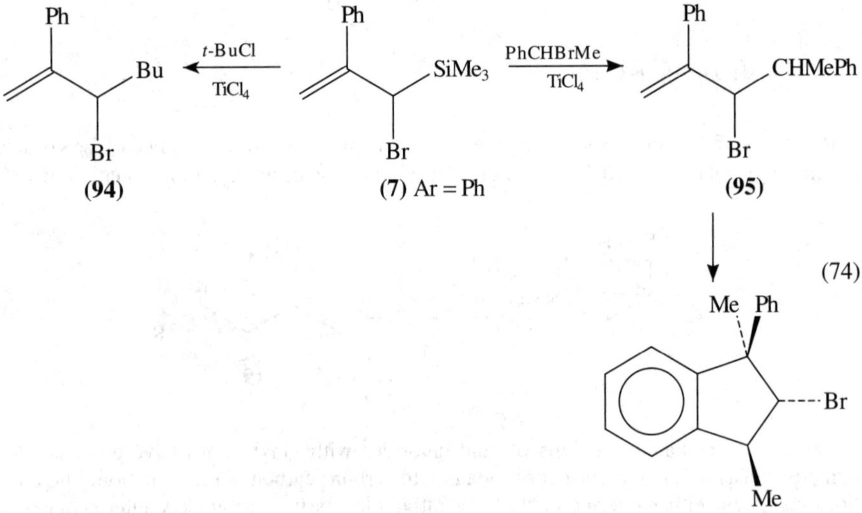

Without a catalyst, tropylium ion reacts with allylsilanes regioselectively to give 7-alkylated cycloheptatriene **97** (equation 75)[143].

(97) (75)

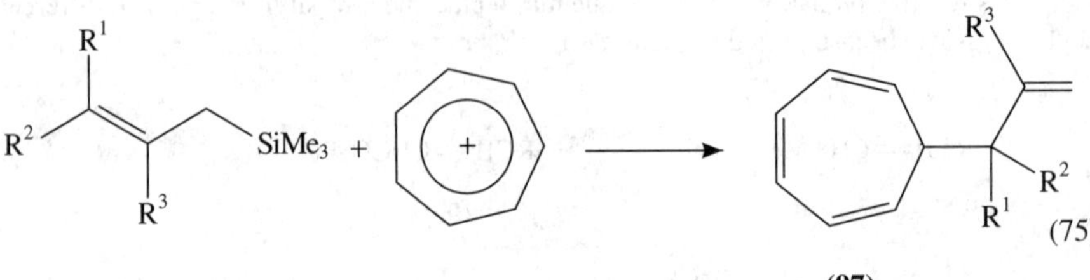

R^1 = H, Me; R^2 = H, Me; R^3 = H; $R^2R^3 = -(CH_2)_3-$

The intramolecular reaction of allylsilane with the tosylate moiety induced by fluoride ion is designed in the last step of the synthesis of (±)-norartemeseol **98** (equation 76)[144].

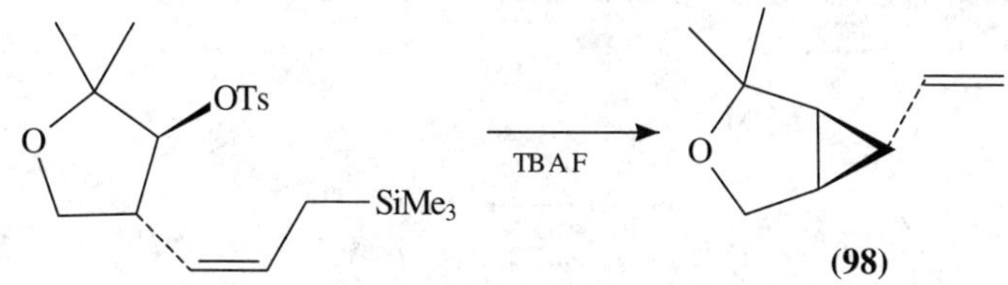
OTs
O
SiMe$_3$
TBAF
O

(76)

(98)

Benzylic silyl ethers couple with allylsilanes in the presence of trityl tetrakis [3,5-bis(trifluoromethyl)phenyl]borate (TFPB) catalyst leading to carbon–carbon bond formation (equation 77). The corresponding $ZnCl_2$-catalyzed reactions with an allyl silyl ether lead to a mixture of regioisomers[145,146].

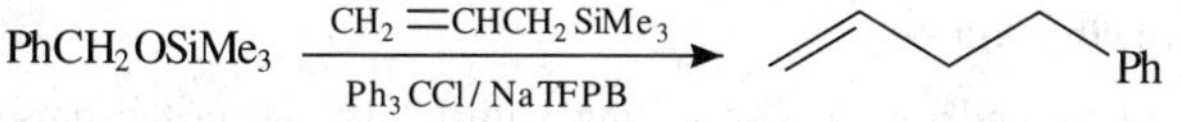
PhCH$_2$OSiMe$_3$
CH$_2$=CHCH$_2$SiMe$_3$
Ph$_3$CCl/NaTFPB
Ph

(77)

Difficulties have been found in the intermolecular coupling reactions of allylsilanes with epoxides other than the simplest ethylene oxide[19]. For example, reactions of allylsilane **99** with epichlorohydrins in the presence of $TiCl_4$ or $EtAlCl_2$ give chlorohydrins in moderate to good yields. However, treatment of epichlorohydrin with **99** furnishes the expected allylated chlorohydrin product **100** (equation 78)[147]. In the intramolecular addition of the allylsilane moiety to 2,3-epoxyether moiety of **101**, the ratio of 6- and 7-membered ring products is affected by the nature of the Lewis acid. When $BF_3 \cdot OEt_2$ is used, a mixture of **102** and **103** is obtained (equation 79). Interestingly, **102** is the exclusive product in the $TiCl_4$-catalyzed reaction[148].

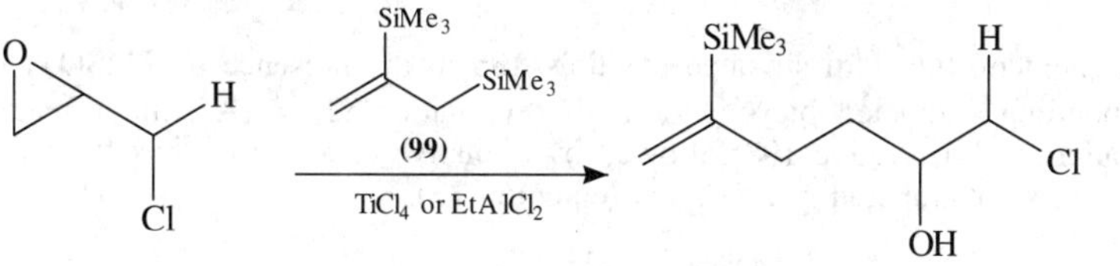
O
H
Cl
SiMe$_3$
SiMe$_3$
(99)
TiCl$_4$ or EtAlCl$_2$
SiMe$_3$
H
Cl
OH

(78)

(100)

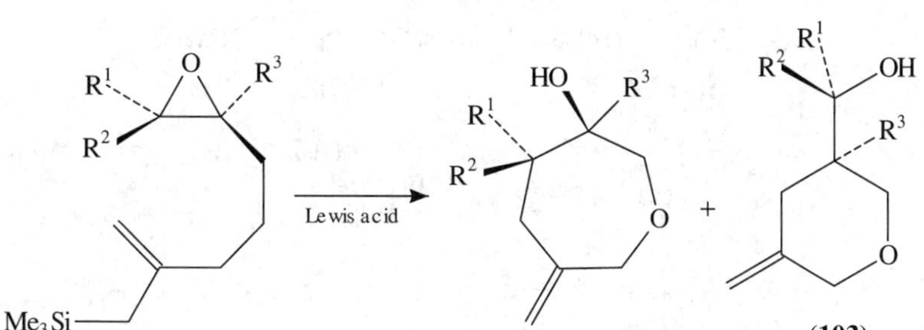
R^1
O
R^3
R^2
Me$_3$Si
Lewis acid
HO
R^3
R^1
R^2
O
+
R^1
R^2
OH
R^3
O

(101) **(102)** **(103)** (79)

$R^1, R^2 = H, Me, n\text{-}Pr, Me_2C{=}CHCH_2CH_2;\ R^3 = H, Me$

Lewis acid catalyzed intramolecular coupling of allylsilanes with aziridines serves as a useful route for the synthesis of carbocycles having the γ-amino olefin unit **104** (equation 80)[149]. However, this reaction does not occur in an intermolecular manner.

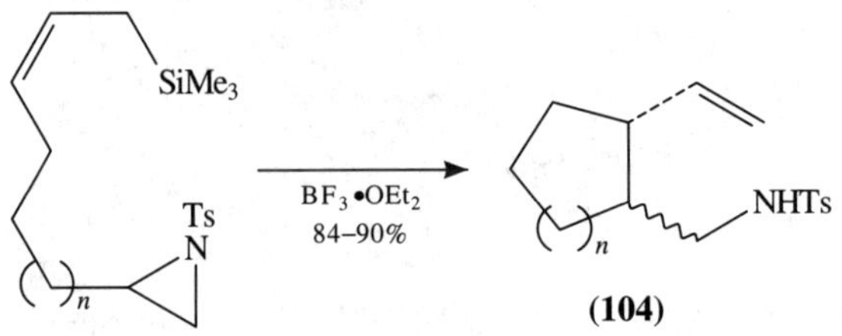

$$ \text{(80)} $$

$n = 1,2$

$trans : cis = 2.7 : 1$

7. Reactions with alkenes

Silyl enol ethers undergo cross-coupling with allylsilane in the presence of a Lewis acid. For example, an oxovanadium(V) complex can induce such condensation reactions (equation 81)[150].

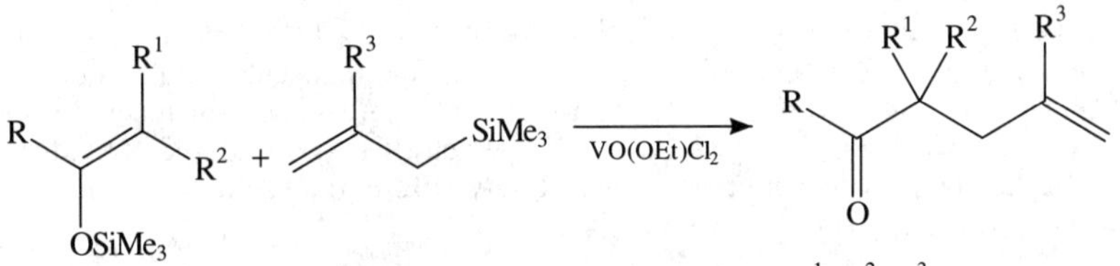

$$ \text{(81)} $$

R, R^1, R^2, R^3 = H, alkyl, aryl

Treatment of **106** with silylated enediols **105** in the presence of TMSOTf gives the corresponding annulated bicyclic diols **107** (equation 82)[151]. A similar reaction with cyclobutenediol derivative **108** followed by oxidative cleavage of thus formed diol **109** yields a seven-membered ring product (equation 83).

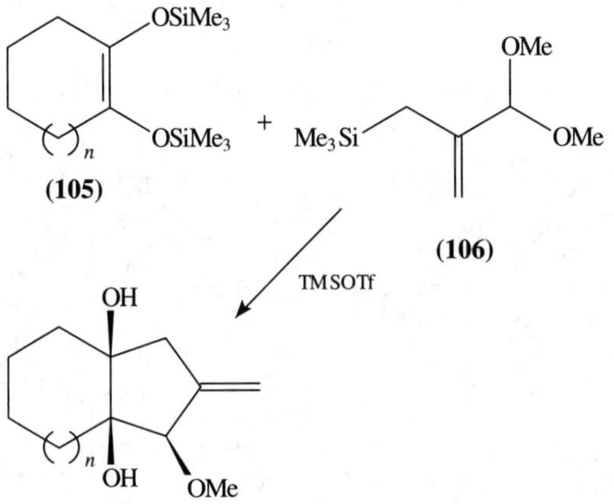

$$ \text{(82)} $$

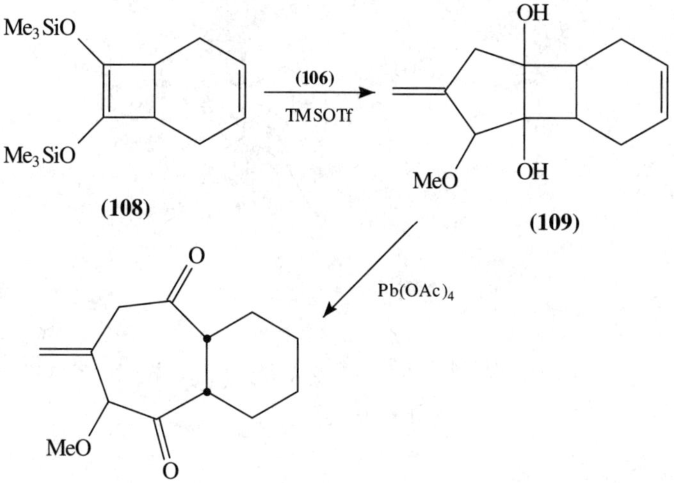

Intramolecular anodic oxidation of allylsilane moiety with enol ether moiety gives the corresponding cyclized product **110** (equation 84)[152].

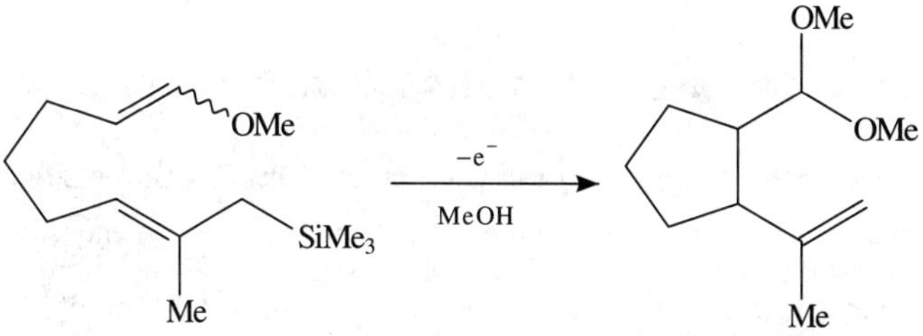

In the presence of the Lewis acid, the carbocation generated *in situ* can serve as an electrophile to replace, intramolecularly, the silyl substituent at the olefinic carbon[153–155]. The stereochemical requirements at the silyl substituent appear not to be important (equations 85–87).

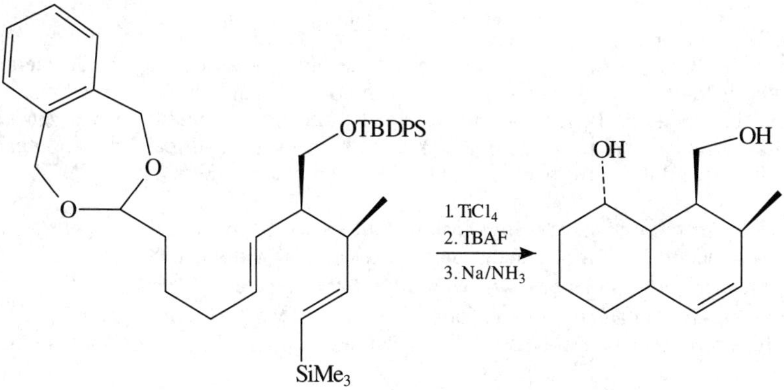

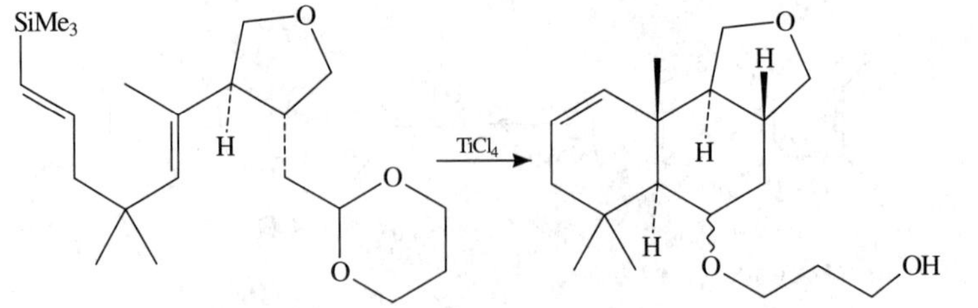

(86)

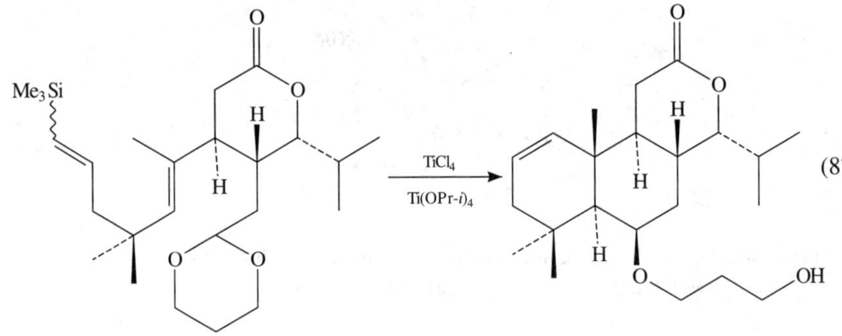

(87)

III. ORGANOMETALLIC REAGENTS PROMOTED C—Si CLEAVAGE REACTIONS

A. Coupling Reactions Involving the Cleavage of the Carbon—Silicon Bond

Transition-metal catalyzed cross-coupling reactions have been demonstrated to be an extremely powerful tool for the construction of carbon–carbon bonds in regio- and stereoselective manners. These metal catalysts enable the transmetallation of carbon–silicon bonds into other carbon–metal bonds from which further reactions can proceed, leading to carbon–carbon bond formation. The use of palladium catalyst[156] in the activation of the carbon–silicon bond is particularly noteworthy[31]; various important transformations have thus been developed. The presence of fluoride ions will make the reaction more facile, pentacoordinate organofluorosilicates being presumably formed under these conditions[157,158].

1. Heck reactions

Organosilicon compounds are versatile in the Heck reaction. The silyl moiety can serve as a halide-like leaving group and transfer the vinyl or aryl group to the olefin. Alternatively, it can serve as an organometallic acceptor and couple with the vinyl or aryl group from corresponding halides or triflates. In the latter case, the silyl group can be eliminated stereoselectively under the reaction conditions or a β-hydride elimination can occur readily to yield the corresponding silyl-substituted alkenes (Section V.C.2).

a. Silyl group as a halide-like leaving group. Both styryl and 1-alkenyl pentafluorosilicates **111** can undergo the Heck reaction in the presence of a stoichiometric quantity of palladium acetate[159]. A similar reaction with allyl chloride gives the corresponding 1,4-diene (equation 88). Apparently, the reaction may proceed via β-chloride elimination[159]. The Heck-type reaction can also be used for the coupling of (*E*)-β-(trimethylsilyl)styrene

with methyl acrylate or related alkenes[160]. Under the carbon monoxide atmosphere, an
α,β-unsaturated ester is obtained (equation 88)[159].

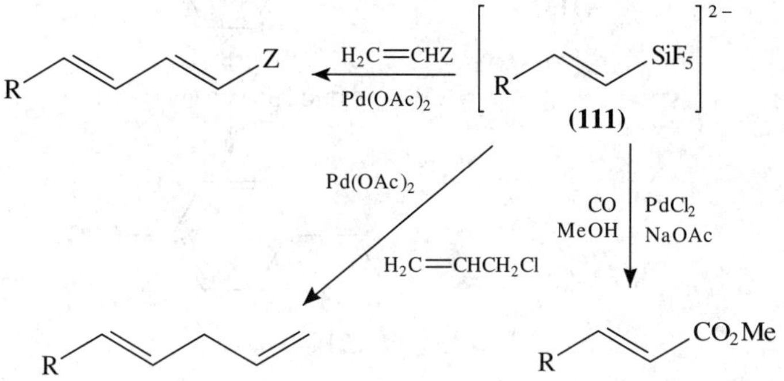

$$R = 1° \text{ or } 3° \text{ Alk, Ph}$$

(88)

b. Vinylsilanes as organometallic acceptors. Vinylsilanes have been used extensively as the acceptor for the arylpalladium intermediate generated *in situ* in the Heck reaction[161–168]. In the absence of a promoter, the reaction is generally nonselective. The desilylation product is a major side product if not the predominant product (Section V.C.2). Moreover, the addition reaction occasionally gives a mixture of regioisomers[163–165]. Nevertheless, when the silyl group at the olefinic carbon is displaced, an inversion of configuration is noted (equation 89)[163,168].

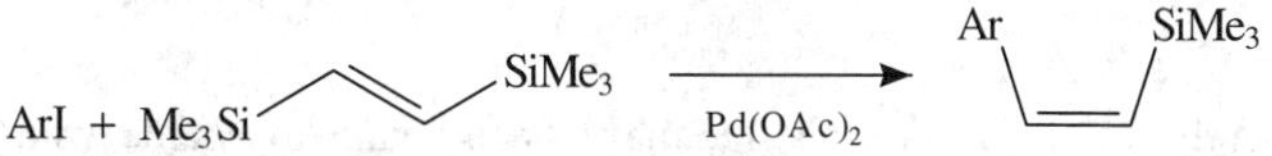

$$\text{ArI} + \text{Me}_3\text{Si} \diagup\!\!\!\diagdown \text{SiMe}_3 \xrightarrow{\text{Pd(OAc)}_2} \text{Ar} \diagup\!\!\!\diagdown \text{SiMe}_3 \quad (89)$$

Ar = 1-Naph, 4-XC$_6$H$_4$(X = H, NO$_2$, MeO)

$(\text{Et}_2\text{N})_3\text{S}^+$ SiMe$_3$F$_2^-$ (TASF) has been known to generate petacoordinate silicates. Thus, treatment of vinyl or aryl iodides with vinylsilanes in the presence of TASF and a catalytic amount of $(\eta^3\text{-C}_3\text{H}_5\text{PdCl})_2$ affords the corresponding coupling products in good to excellent yield[169,170]. Retention of configuration is generally observed. This version of a silicon-based coupling reaction has provided a general and highly stereospecific route to 1,3-dienes **112** (equation 90) as well as alkenylarenes. The triflate leaving group can couple efficiently under similar conditions[171]. It is noteworthy that alkylfluorosilanes **113** also undergo this coupling reaction in satisfactory yield with retention of configuration (equation 91). Unlike the reactions of organotin compounds, lithium halide is not necessary for the reactions with silanes. Under a carbon monoxide atmosphere, carbonylative cross-coupling reactions have been found when an aryl halide is treated with an aryl- or an alkenylfluorosilane[172].

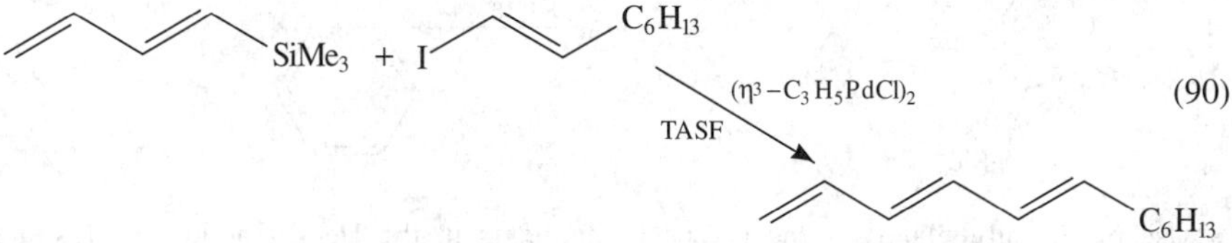

(90)

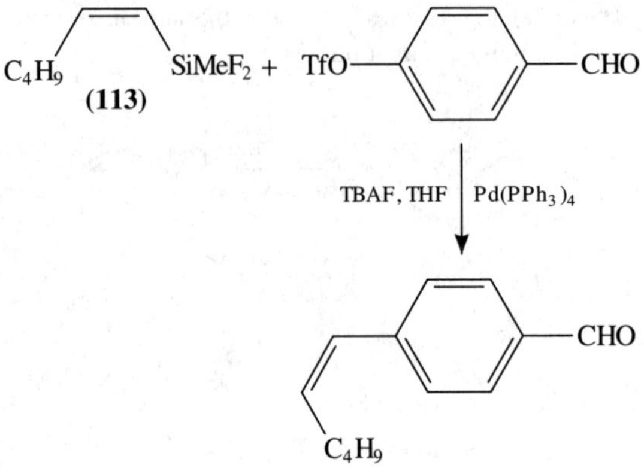

$$\text{C}_4\text{H}_9\text{-CH=CH-SiMeF}_2 \;+\; \text{TfO-C}_6\text{H}_4\text{-CHO} \quad \textbf{(113)}$$

$$\xrightarrow[\text{TBAF, THF}]{\text{Pd(PPh}_3)_4} \quad \text{C}_4\text{H}_9\text{-CH=CH-C}_6\text{H}_4\text{-CHO} \tag{91}$$

Cross-coupling reactions of alkenylsilicates **114** with aryl or vinyl halides or triflates are catalyzed by a palladium complex to give the corresponding alkenes (equation 92)[173].

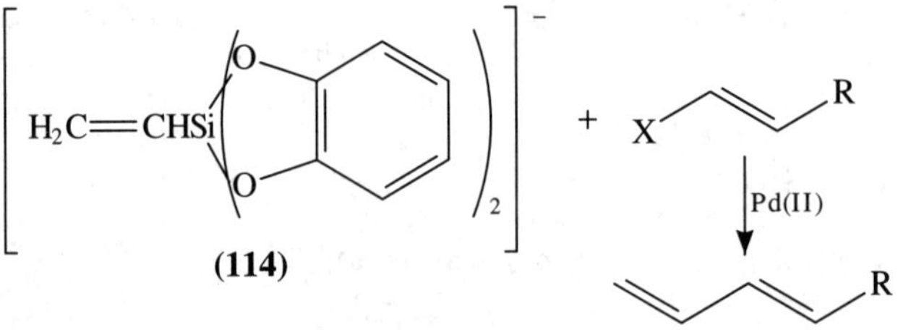

$$\left[\text{H}_2\text{C=CHSi(O-C}_6\text{H}_4\text{-O)}_2\right]^{-} \;+\; \text{X-CH=CH-R} \xrightarrow{\text{Pd(II)}} \text{CH}_2\text{=CH-CH=CH-R} \tag{92}$$

(114)

More recently, allylic carbonates **115** have been found to couple with alkenylfluorosilanes in the presence of a palladium catalyst to yield the corresponding 1,4-dienes (equation 93)[174]. The corresponding palladium-catalyzed reaction of allylsilane with allyl acetate **116** has been explored (equation 94)[175].

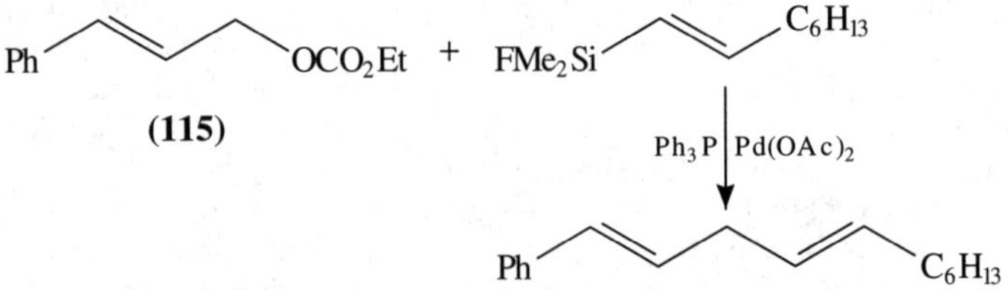

$$\text{Ph-CH=CH-CH}_2\text{-OCO}_2\text{Et} \;+\; \text{FMe}_2\text{Si-CH}_2\text{-CH=CH-C}_6\text{H}_{13} \quad \textbf{(115)}$$

$$\xrightarrow[\text{Ph}_3\text{P}]{\text{Pd(OAc)}_2} \quad \text{Ph-CH=CH-CH}_2\text{-CH}_2\text{-CH=CH-C}_6\text{H}_{13} \tag{93}$$

$$\text{Indene-SiMe}_3 \;+\; \text{MeO}_2\text{C-cyclohexenyl-OAc} \;\textbf{(116)}\; \xrightarrow{\text{Pd(PPh}_3)_4} \text{coupled product with CO}_2\text{Me} \tag{94}$$

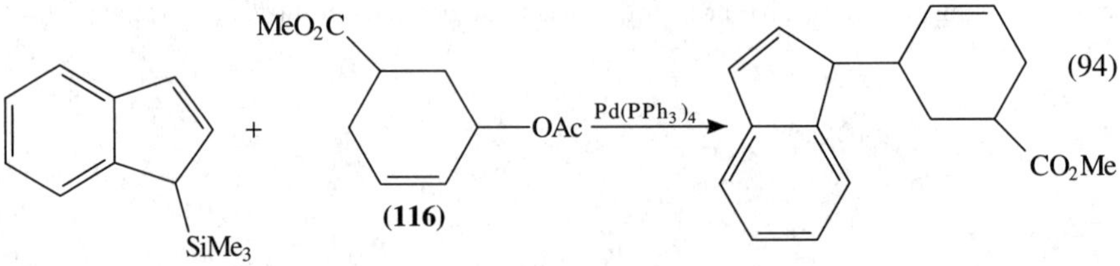

Use of the allylsilanes as the alkene components in the Heck reaction is illustrated in the palladium catalyzed γ-allylation of organic halides or triflate or acetate[176]. The reaction has been used for the synthesis of ibuprofen **117** (equation 95). Interestingly, the regioselectivity of such a coupling reaction is quite dependent on the phosphine ligands associated with the metal center. α-Selectivity is obtained by using good chelating bidentate ligands, whereas γ-selectivity is favored by mondentate ligands, as shown in the reaction of **118** (equation 96)[177].

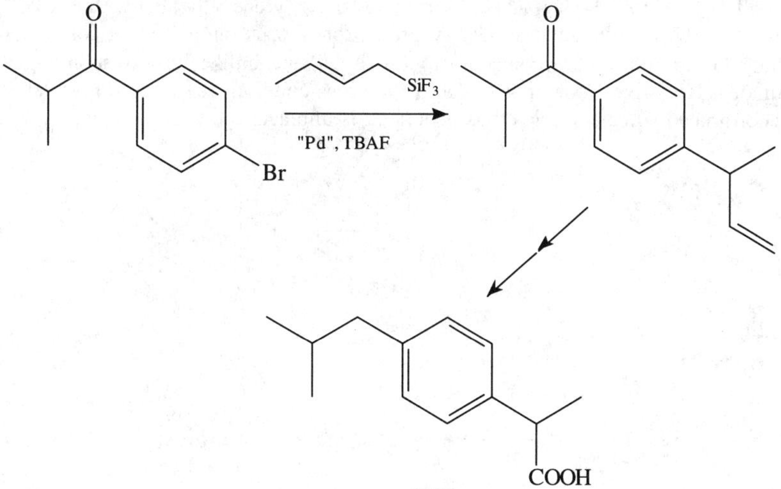

(95)

(117)

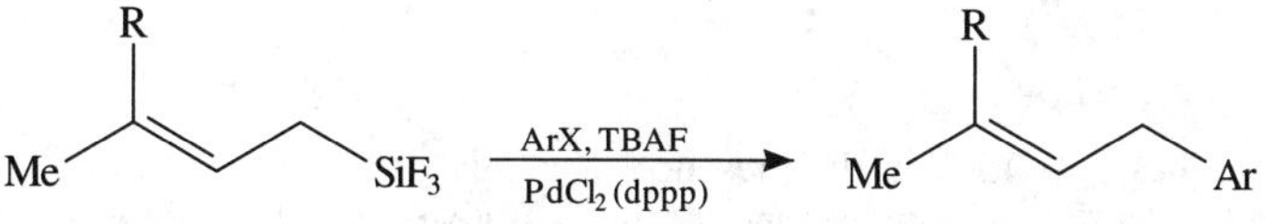

(96)

(118)

R = H, Me, 4-YC$_6$H$_4$ (Y = MeCO, CN, MeO, CO$_2$Me); X = Br, I, OTf

Chiral benzyl or allyl silanes **119** couple with aryl triflates to give the corresponding products having 58–99% enantioselectivities (equation 97)[176,178,179]. The selectivity appears to be very much dependent on the reaction conditions.

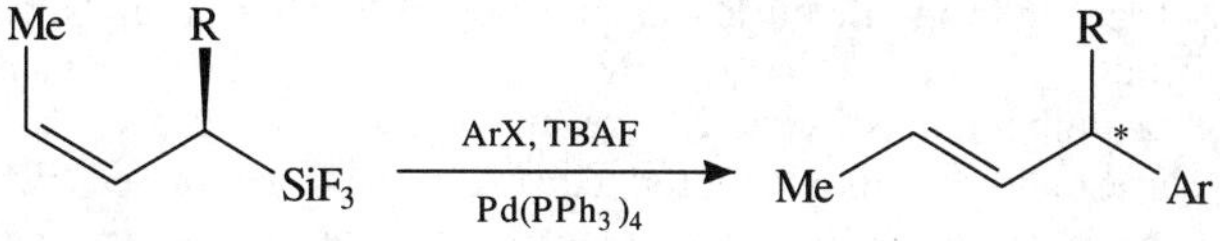

(97)

(119)

R = Me, Et; Ar = 1- and 2-Naph

Use of the allylsilanes as the alkene components in the Heck reaction is illustrated in the regioselective intramolecular cyclization of **120** (equation 98)[180]. In addition, a high enantioselectivity of 90 %ee is obtained when the chiral phosphine (*S*)-BINAP is employed. The propargylsilane analog undergoes a similar reaction[181].

2. Homocoupling reactions

In the presence of a stoichiometric amount of $PdCl_2$, styrylpentafluorosilicate (**111**, R = Ph) dimerizes to give (E,E)-1,4-diphenyl-1,3-butadiene in moderate yield (equation 99)[159]. Interestingly, the corresponding 1-alkenyl analogs give only trace amount of the dimers under the same conditions. A similar catalytic homocoupling reaction has been found when $(E)-\beta$-(trimethylsilyl)styrene is treated with $PdCl_2$[160,182]. Heteroatom substituents such as alkoxy groups or fluorine atoms on silicon in general facilitate the reaction[183]. Surprisingly, trifluorosilylethene, unlike the corresponding mono- or difluoro derivatives, does not undergo the cross-coupling reactions. Presumably, the hexacoordinated silicate species thus generated is unreactive toward coupling.

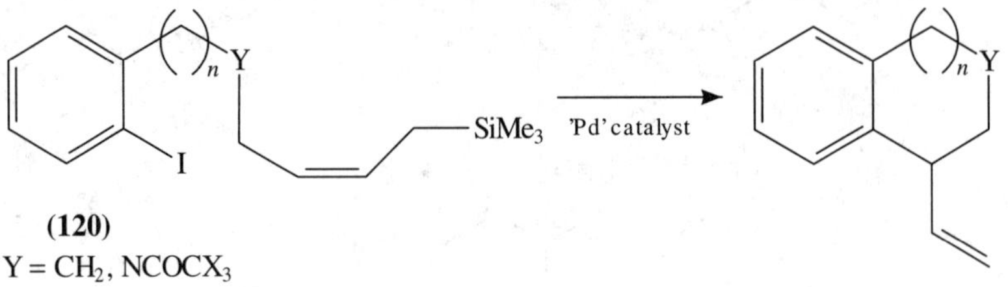

(98)

(**120**)
Y = CH_2, $NCOCX_3$

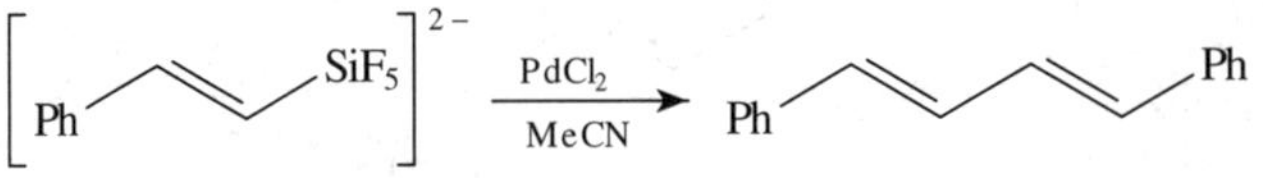

(99)

(**111**) R = Ph

3. Other coupling reactions

The reaction of **121** with LiCl in the presence of a catalytic amount of $LiPdCl_4$ and p-benzoquinone (p-$C_6H_4O_2$) and in acetone–acetic acid gives a mixture of two isomeric allylic chlorides **122a** and **122b** in a ratio of 3:1 (equation 100). Employment of the corresponding (Z)-allylsilane under the same conditions provides the same products but with opposite selectivity (**122a : 122b** = 1:3). This reaction demonstrates a Pd(II)-catalyzed intramolecular C–C bond formation by addition of an allylsilane to the coordinated diene leading to the highly stereoselective 1,4-*syn*-adduct[184].

B. Transmetallation

Allyl group transfer from silicon to bismuth takes place when allylsilanes are treated with triarylbismuth difluorides in the presence of $BF_3 \cdot OEt_2$ (equation 101). The resulting allylbismuthonium species **123** reacts readily with arenes and nucleophiles to give the corresponding allylated products[185]. Transformation of allyl group from silanes to other elements has been performed using $(NH_4)_2Ce(NO_3)_6$[186], $Tl(O_2CCF_3)_3$[187], $PhTe(O)O_2CCF_3$[188], $PhIO/BF_3 \cdot OEt_2$[189,190] and $Ti(OPr^i)$[191]

Vinylsilanes react with boron trichloride to give the corresponding borodesilylation products in good yield which, in turn, can be transformed into boronic esters **124** by alcoholysis (equation 102). The initial dichloroorganoborane products can be used directly in the Suzuki–Miyaura cross-coupling reaction[192]. Replacement of a carbon–silicon bond by a carbon–tin bond in fluorinated alkenes (e.g. **125**) can be achieved by the reaction of silanes with Bu_3SnCl and KF in DMF under mild conditions (equation 103)[193]. It is noteworthy that a vinylic iodine remains intact under these conditions.

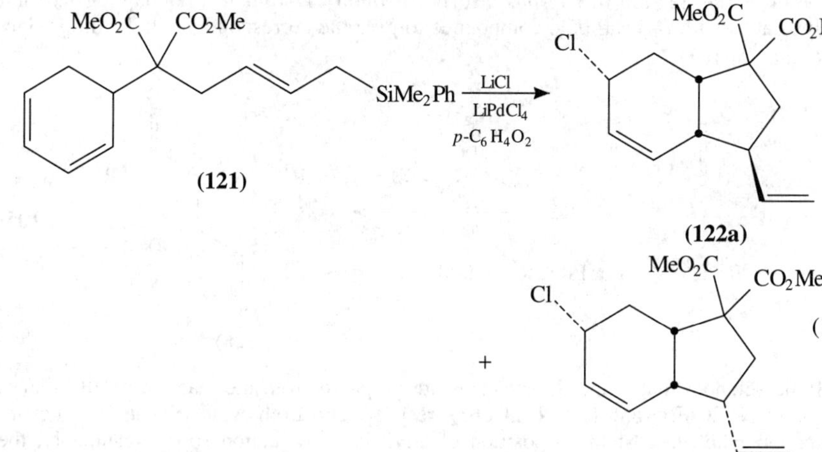

(121) → (122a) + (122b) [LiCl, LiPdCl$_4$, p-C$_6$H$_4$O$_2$] (100)

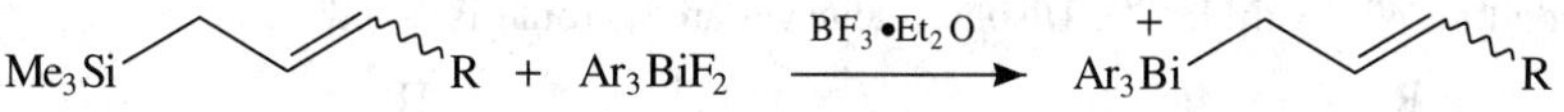

Me$_3$Si~R + Ar$_3$BiF$_2$ →[BF$_3$·Et$_2$O] Ar$_3$Bi + ~R (101)

(123)

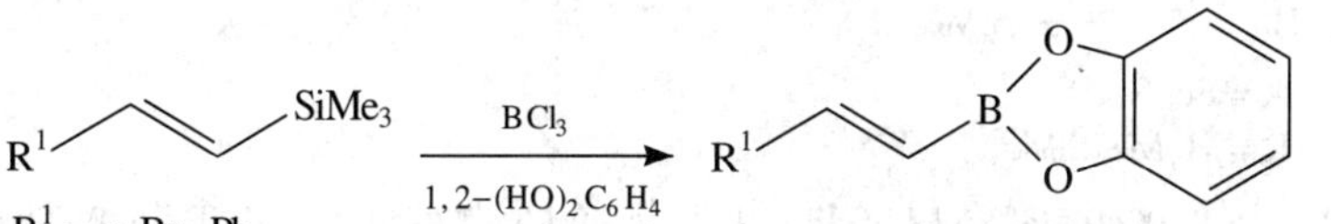

R^1~SiMe$_3$ →[BCl$_3$, 1,2-(HO)$_2$C$_6$H$_4$] R^1~B(catechol) (102)

R^1 = n-Bu, Ph

(124)

R^{1}FC=CFSiR$^2_3$ →[KF, DMF, Bu$_3$SnCl; rt to 80 °C] R^{1}FC=CFSnBu$_3$ (103)

R^1 = alkyl, aryl, F or I; R^2 = Me, Et

(125)

Disproportionation between vinylsilanes and monosubstituted alkenes catalyzed by ruthenium complex (equation 104) has been suggested to occur via a β-silyl group elimination. The intermediate silylruthenium complex, RuCl(CO)(PPh$_3$)$_2$(SiMe$_2$R^1), has been characterized by spectroscopic means[194].

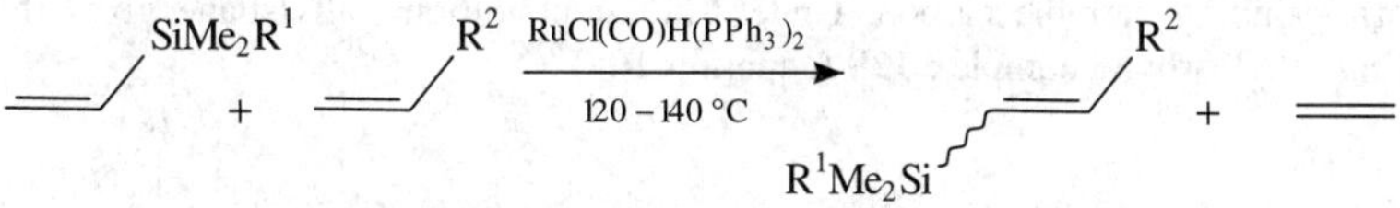

=/SiMe$_2$R^1 + =/R^2 →[RuCl(CO)H(PPh$_3$)$_2$, 120–140 °C] R^2/=/SiMe$_2$R^1 + = (104)

R^1 = Me, OEt; R^2 = Me$_3$Si, Me, Ph, CO$_2$Me, OBu-n

IV. MISCELLANEOUS C–Si CLEAVAGE REACTIONS

As discussed in Section II, allylsilane normally serves as a nucleophile which reacts with an electrophile. However, pentafluorosilicates occasionally behave like an electrophile in certain palladium-catalyzed reactions (Section III.A.1). It is interesting to note that $VO(EtO)Cl_2$ (*cf* Section II.B.7) has also been found to promote the coupling reaction of allylsilanes with β-dicarbonyl compounds to give the corresponding allylation product **126** (equation 105)[195].

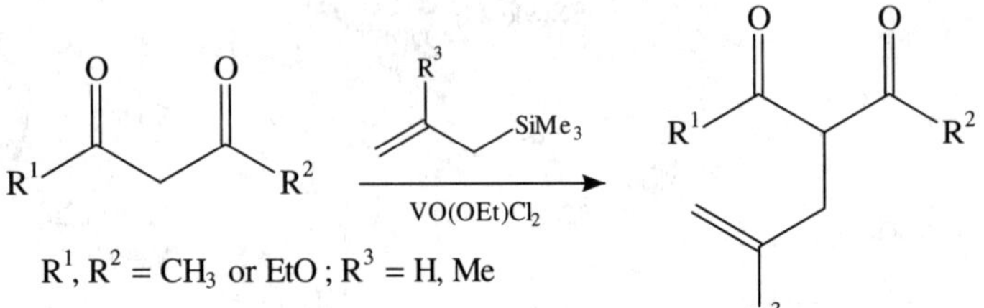

$$R^1, R^2 = CH_3 \text{ or EtO}; R^3 = H, Me$$

(105)

(**126**)

Allylic selenocyanates and thiocyanates are prepared from the reaction of allylsilanes with $KSeCN \cdot 2CuBr_2$ and $KSCN \cdot 2CuBr_2$, respectively. Both cyanoselenation and cyanothianation take place at the α-position of allylsilanes (equation 106). Presumably, the electrophile 'XCN' generated *in situ* is first added to the γ-position, followed by a [1,3]-migration[196]. Fluoride induced reactions of allylsilanes with thioketones yield the corresponding allyl sulfides[197]. Displacement via an S_E route is noted.

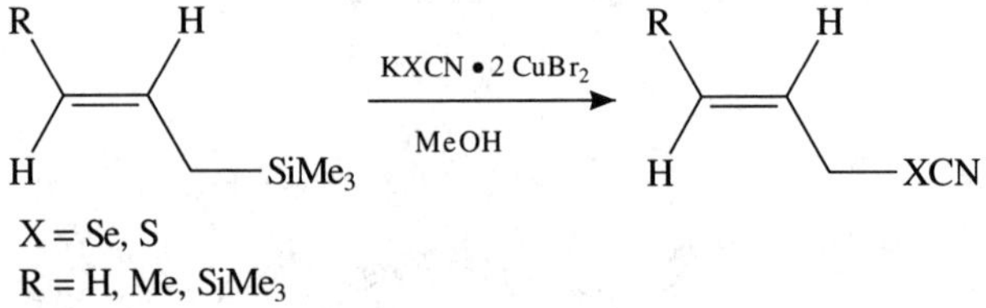

$$X = Se, S$$
$$R = H, Me, SiMe_3$$

(106)

Insertion of a bromocarbene unit into the Si—C bond has been observed in the reaction of **61** with CH_2Br_2 in the presence of LDA (equation 107)[105,106].

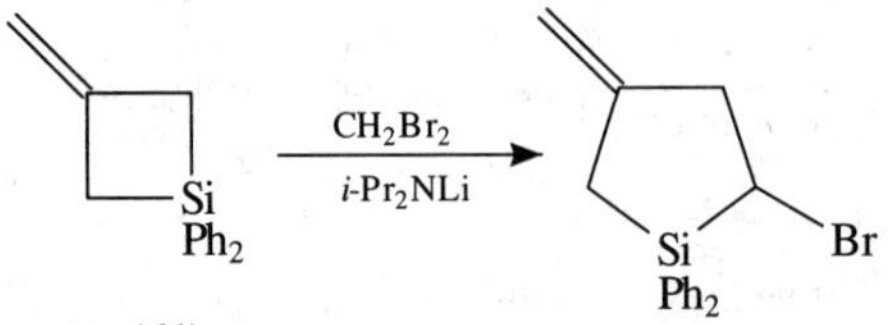

(**61**)

(107)

Upon treatment with $WCl_2(PMe_3)_4$ at 65 °C, vinylsilane can serve as a precursor of methylcarbyne complex **127**[198]. It is noteworthy that the C1 carbon is converted into the corresponding carbyne carbon. Under similar conditions, allylsilane gives the corresponding ethylcarbyne complex **128** (equation 108)[198].

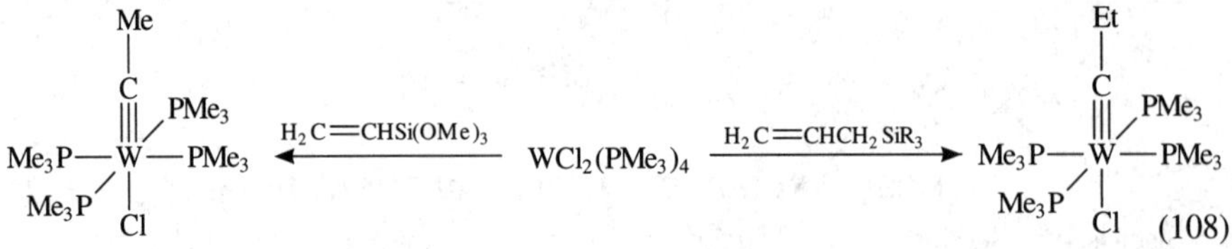

(**127**) (**128**)

(108)

30. Synthetic applications of allylsilanes and vinylsilanes

Electrochemical oxidation of 1-(trimethylsilyl)dienes in MeOH affords **129** in moderate to satisfactory yield. It is worth noting that the reaction shows some diastereoselectivity (equation 109)[199].

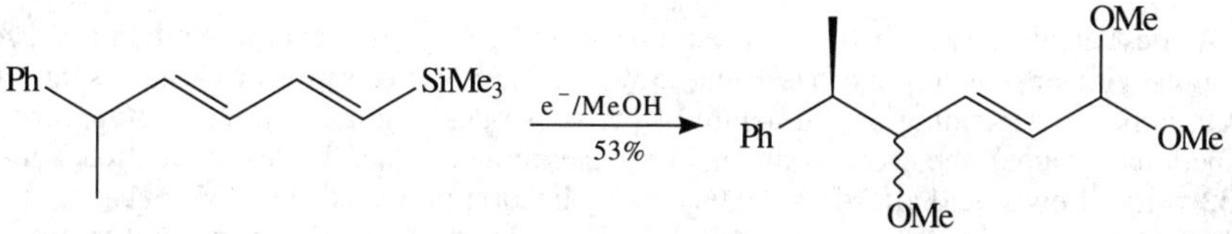

$$\text{Ph} \diagup\!\!\diagdown\!\!\diagup\!\!\diagdown\!\!\diagup\text{SiMe}_3 \xrightarrow[53\%]{e^-/\text{MeOH}} \text{Ph} \diagup\!\!\diagdown\!\!\diagup\!\!\diagdown\!\!\diagup \text{OMe} \quad (109)$$

(129)

syn : anti = 4 : 1

Although the elimination of saturated β-hydroxysilane giving olefin is well-documented, a similar reaction involving β-silylallylic alcohol normally gives a mixture of products. When NaH is employed in refluxing octane, allene **130** is obtained as the major product (equation 110)[200]. However, when the reaction is carried out in the presence of a catalytic amount of NaH in HMPA at room temperature, a 1,3-silyl group shift from carbon to oxygen occurs readily (e.g. to give **131**) (equation 111)[201].

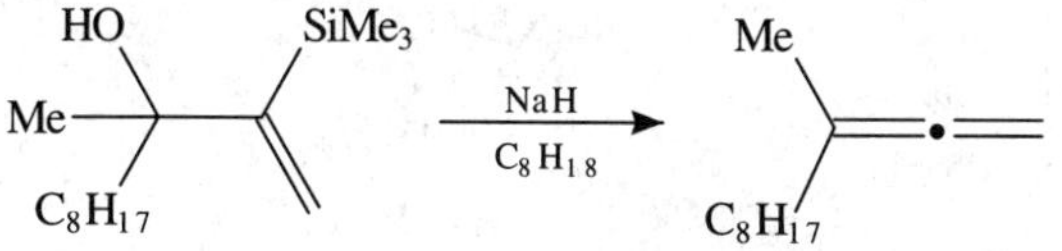

$$\quad (110)$$

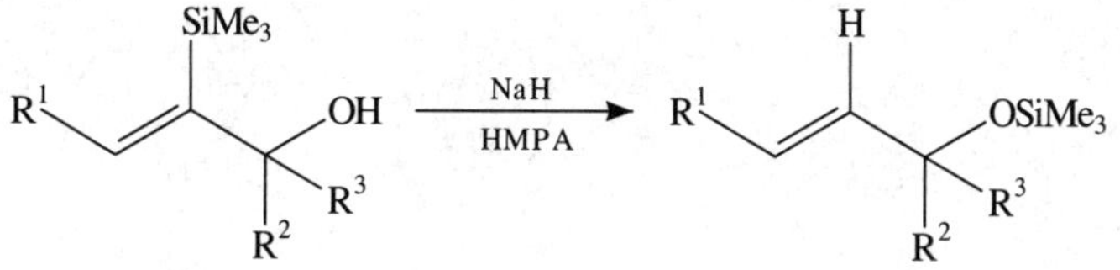

(130)

$$\quad (111)$$

(131)

$R^1 = H, n\text{-Bu}; R^2 = H, \text{Me}; R^3 = H, 1° \text{Alk, Ph, CH}_2={=}\text{CH}$

Tris(trimethylsilyl)silyl radical is relatively stable and can therefore serve as a radical leaving group. This reaction has been extended to the radical-initiated allylation of organic halides[202,203]. Thus, thermolyses of bromides α to a carbonyl substituent **144** or of simple iodides with allyltris(silyl)silane in the presence of a radical initiator gives the corresponding allylation products (equation 112).

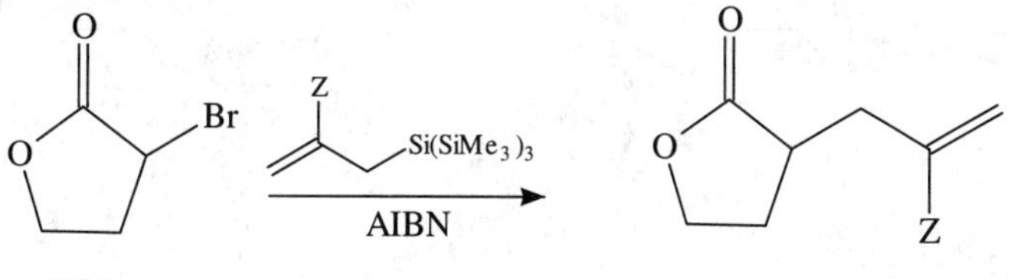

$$\quad (112)$$

(144)

Z = H, Me, CN, CO$_2$Et

V. REACTIONS OF ALLYLSILANES AND VINYLSILANES WITHOUT CLEAVAGE OF THE C—Si BOND

A. Lewis Acid Catalyzed Addition of an Electrophile to the Double Bond

As described in Section II, Lewis acid catalyzed desilylative carbon–carbon bond formation with an electrophile has been shown to be very versatile in organic synthesis. Occasionally, depending on the nature of the substrates (e.g. the presence of appropriate functional groups), the carbon–silicon bond may remain intact. For example, treatment of **132** with a Lewis acid affords a mixture of cyclization products **133–135** (equation 113). The isolation of **133** indicates that the carbocation intermediate thus formed is trapped by the oxygen nucleophile before elimination of the silyl moiety occurs[204].

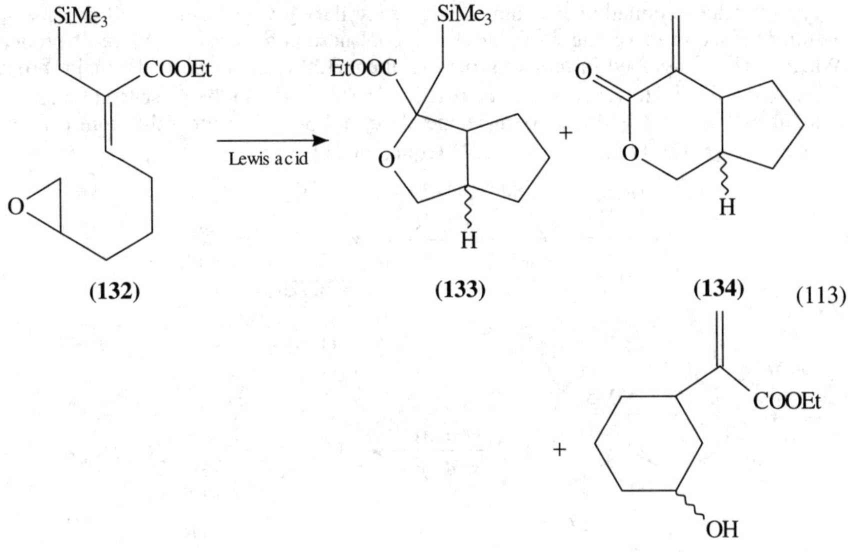

SnCl$_4$-mediated intramolecular cyclization of vinylsilane **136** with the acetal moiety leads to the corresponding cyclization product **137** (equation 114). The electrophile apparently attacks at the β-carbon to the silyl substituent[205]. It is noteworthy that desilylation has been observed in the similar case of **138** when BCl$_3$ is employed as the Lewis acid catalyst (equation 115)[206].

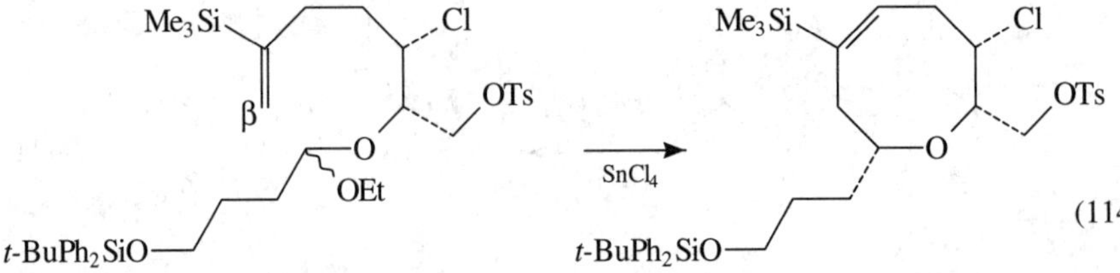

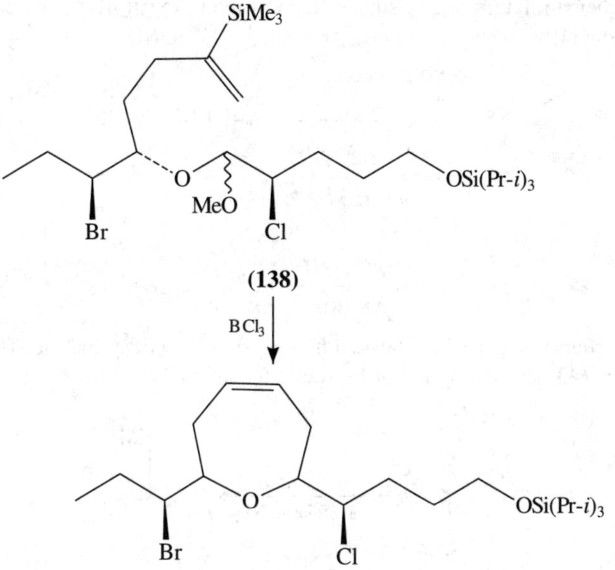

(138) (115)

↓ BCl₃

Silicon-directed stereoselective *syn* addition of the hydroxy group to the olefinic double bond occurs intramolecularly in a TiCl₄- or TsOH-catalyzed cyclization of vinylsilanes **139** bearing a hydroxy group[207]. The reaction gives mainly the *trans*-isomer (*trans* : *cis* = 80–90:20–10) (equation 116)[208]. It is noteworthy that no cyclization is observed in the absence of the silyl substituent under the same conditions.

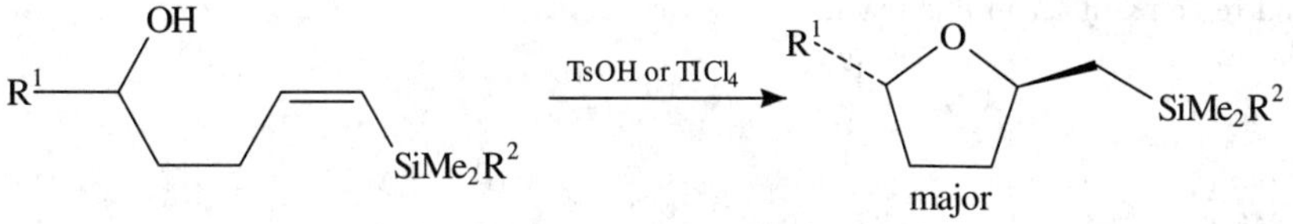

(139) (116)

R^1 = *i*-Pr, *n*-C₆H₁₃, Ph; R^2 = H, Me, *t*-Bu, Ph

1-Trimethylsilylbutadiene **140** is treated with acetyl chloride at −80 °C to give the adducts, which are dehydrochlorinated to give **141** (equation 117)[209]. No carbon–silicon bond cleavage is observed.

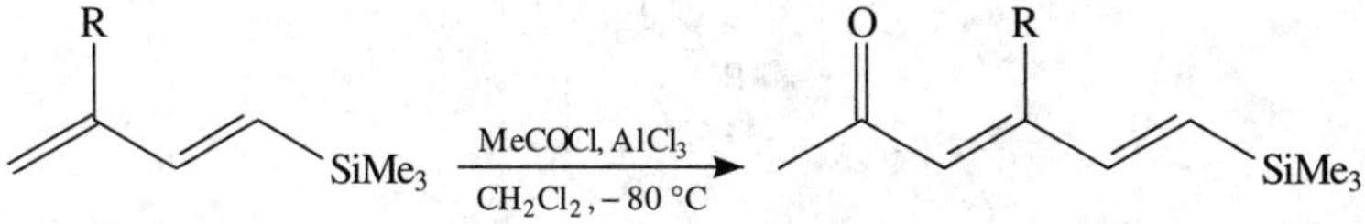

(140) (141) (117)

R = Me, Et, *i*-Pr

Convenient procedures for the synthesis of β-nitrovinylsilanes **142** from vinylsilanes have been developed (equation 118)[210]. Similarly, nitration of allylsilane at the γ-position is achieved when a mixture of allylsilane, NaNO$_2$ and CAN in HOAc is treated ultrasonically (equation 119)[211].

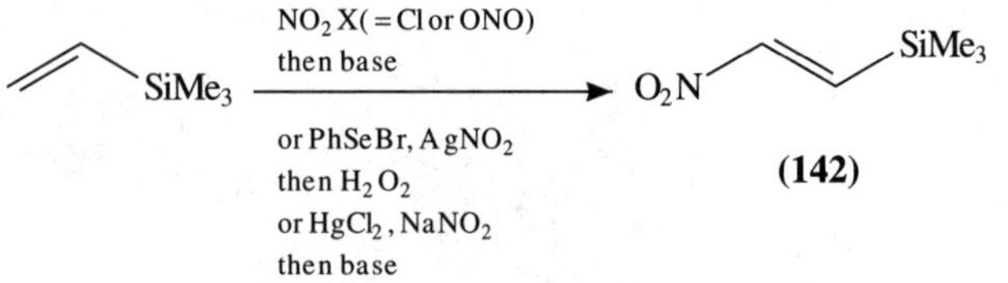

$$\text{CH}_2=\text{CHCH}_2\text{SiMe}_3 \xrightarrow[\substack{\text{or PhSeBr, AgNO}_2 \\ \text{then H}_2\text{O}_2 \\ \text{or HgCl}_2,\text{NaNO}_2 \\ \text{then base}}]{\substack{\text{NO}_2\text{X}(=\text{Cl or ONO}) \\ \text{then base}}} \text{O}_2\text{N}-\text{CH=CH}-\text{CH}_2\text{SiMe}_3 \quad (118)$$

$$\textbf{(142)}$$

$$\text{CH}_2=\text{CHCH}_2\text{SiEt}_3 \xrightarrow[\text{CAN, HOAc}]{\text{NaNO}_2\;)))} \text{O}_2\text{N}-\text{CH=CH}-\text{CH}_2\text{SiEt}_3 \quad (119)$$

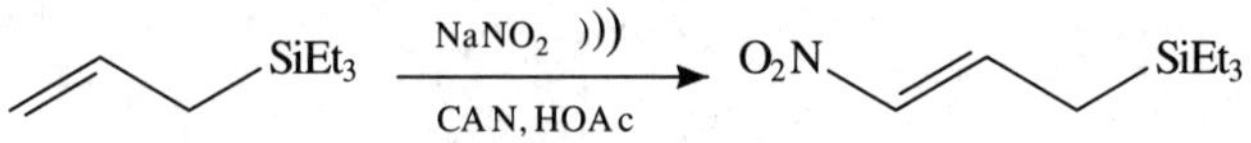

During the course of the acid-catalyzed hydrolysis of allylic acetal moiety, the vinylsilane moiety in **143** has been found to be stable (equation 120)[212].

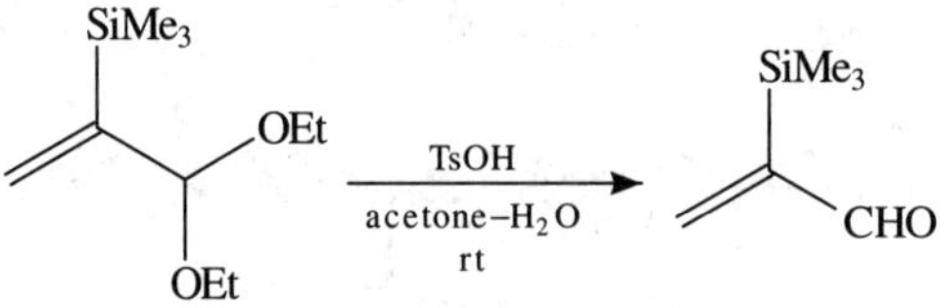

$$\textbf{(143)} \xrightarrow[\substack{\text{acetone}-\text{H}_2\text{O} \\ \text{rt}}]{\text{TsOH}} \quad \text{SiMe}_3/\text{CHO product} \quad (120)$$

(143)

B. Radical Reactions

A three-component radical coupling reaction involving allylsilanes has been employed leading to **145** (equation 121)[213]. It is noteworthy that the corresponding carbon–silicon bond remains intact in this reaction.

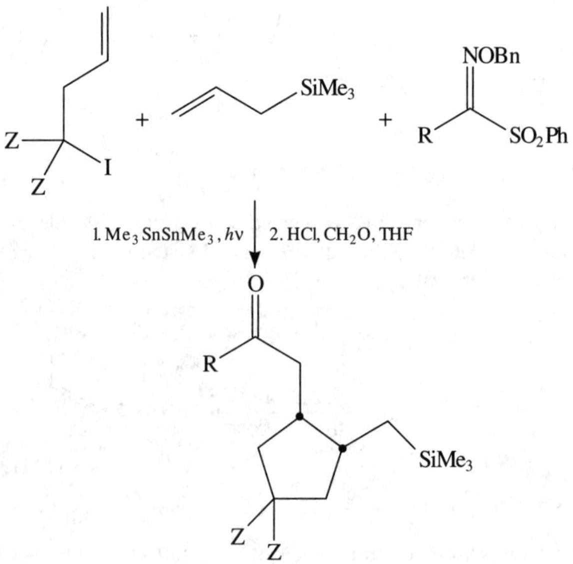

$$\text{(reagents)} \xrightarrow[\text{2. HCl, CH}_2\text{O, THF}]{\text{1. Me}_3\text{SnSnMe}_3, h\nu} \textbf{(145)} \quad (121)$$

30. Synthetic applications of allylsilanes and vinylsilanes

Intermolecular free-radical addition of iodoalkyl sulfones to vinylsilanes yields regioselectively the adduct **146** which can be further transformed to cyclopropene derivatives (equation 122)[214].

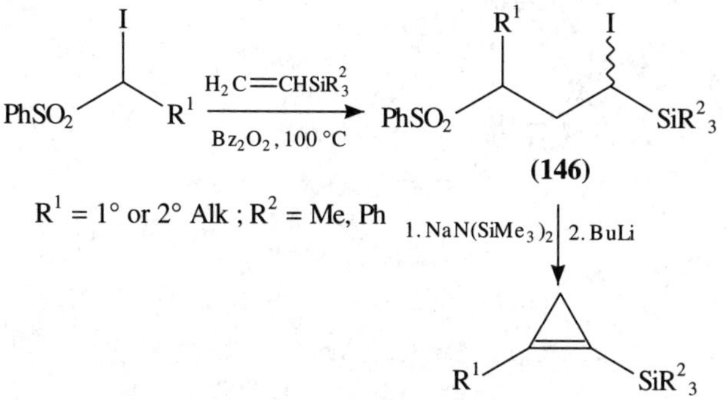

$$R^1 = 1° \text{ or } 2° \text{ Alk}; R^2 = Me, Ph \qquad (122)$$

Treatment of a vinylsilane with ketones in the presence of 2 equivalents of MnO_2 and a drop of acetic acid at 140 °C in a sealed tube produces the corresponding adduct **147** in moderate to good yields (equation 123)[215].

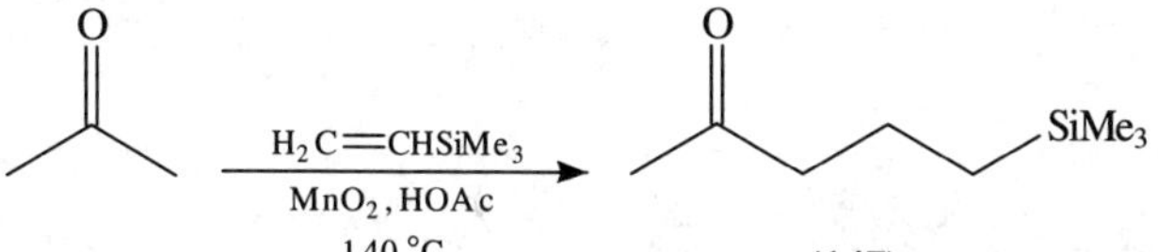

$$(123)$$

Intramolecular radical addition to vinylsilane moiety has been used for the synthesis of statine **148** (equation 124)[216] and the pyrrolizidine skeleton[217]. Tandem [2 + 1] radical cyclization of bromides **149** with Bu_3SnH is extremely sensitive to the nature of the substituent on the diene moiety. Bicyclo[3.1.0] skeleton **150** is obtained in the presence of the silyl substituent (equation 125). When the silyl substituent in **149** is replaced by a *t*-butyl group, the six-membered ring product **151** is obtained exclusively[218].

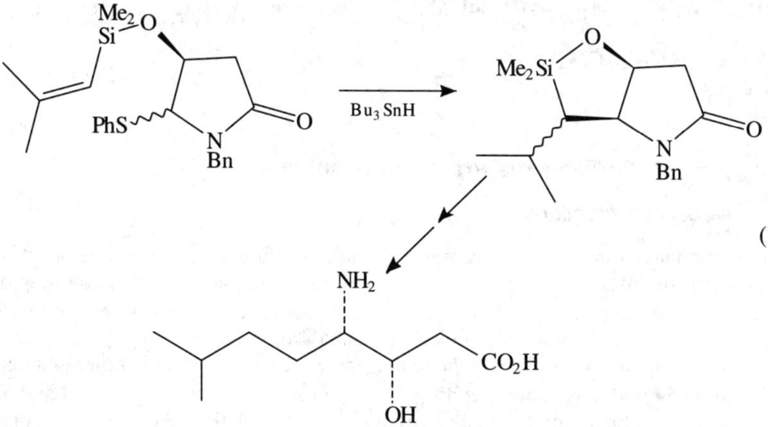

$$(124)$$

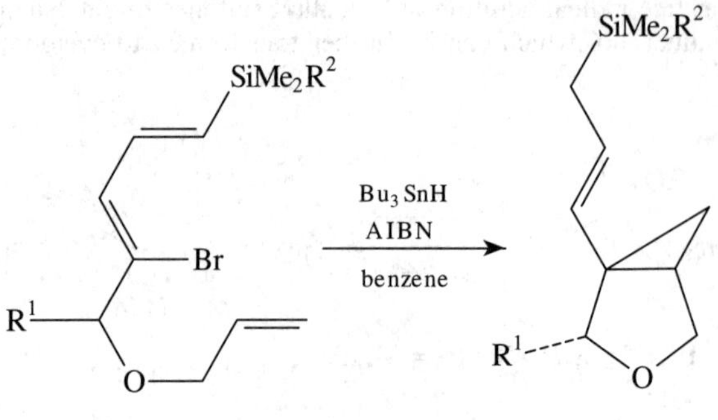

[Structure **(149)** → **(150)** via Bu$_3$SnH, AIBN, benzene] (125)

R^1 = Me, n-C$_6$H$_{13}$, Ph; R^2 = Me, Ph

(151)

Decomposition of perfluoroalkyl peroxide in the presence of trimethoxyvinylsilane gives the corresponding oligomers **152** (equation 126)[219]

$$(R_FCO)_2O_2 + H_2C=CHSi(OMe)_3 \xrightarrow{30\,°C} R_F[CH_2-CHSi(OMe)_3]_nR_F$$

R$_F$ = C$_3$F$_7$C$_3$F$_7$OC(Me)F
n = 2,3

(152) (126)

C. Organometallic Reagents Promoted Reactions

1. Cross-coupling reactions

Allyl- and vinylsilanes having bromide or iodide substituents at the double bond undergo various kinds of cross-coupling reactions. For example, Stephens–Castro coupling of such halides gives polyunsaturated silanes **153** in good to excellent yields (equation 127)[220]. A hydroxy group can be tolerated under these conditions. The cross-coupling reactions of vinylsilane phosphate with organocuprate reagents give a single stereoisomer of substituted vinylsilanes **154** with retention of configuration (equation 128)[221]. Aryl, vinyl and alkynyl bromides undergo the efficient Suzuki–Miyaura coupling with **155** under basic conditions to produce the corresponding *trans*-styryl-, dienyl- (equation 129) and enynylsilanes in

30. Synthetic applications of allylsilanes and vinylsilanes

good to excellent yields[222,223]. It is also noteworthy that the carbon–silicon bond in allylsilanes remains intact under Stille coupling conditions[224].

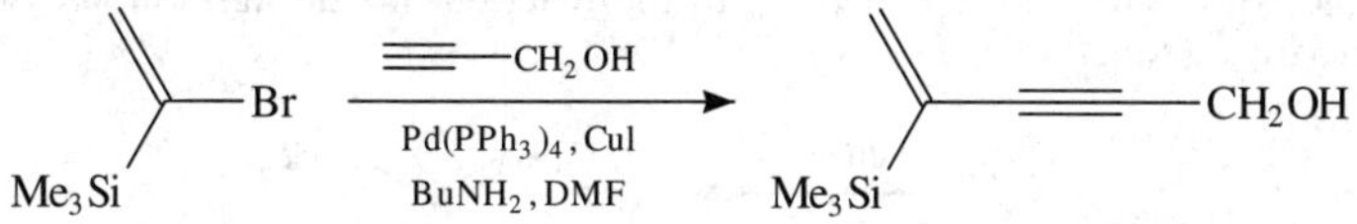

(127)

(153)

(128)

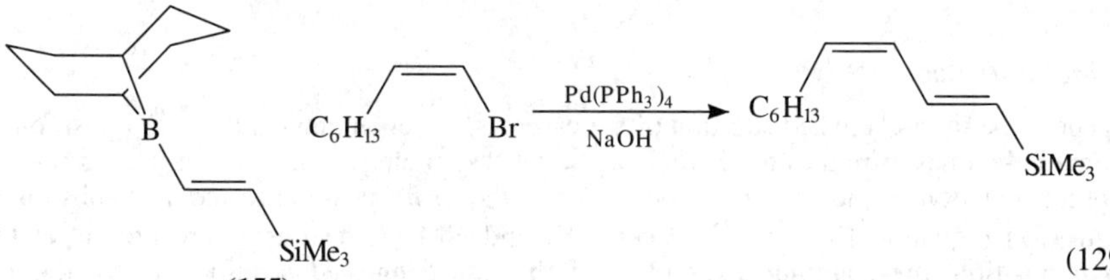

(154)

(129)

(155)

Organozinc reagent prepared from the 2-(trimethylsilyl)allyl bromide **156** and zinc react with alkynes or carbonyl compounds as well as nitriles to give the corresponding coupling products in moderate to good yields (equation 130)[225].

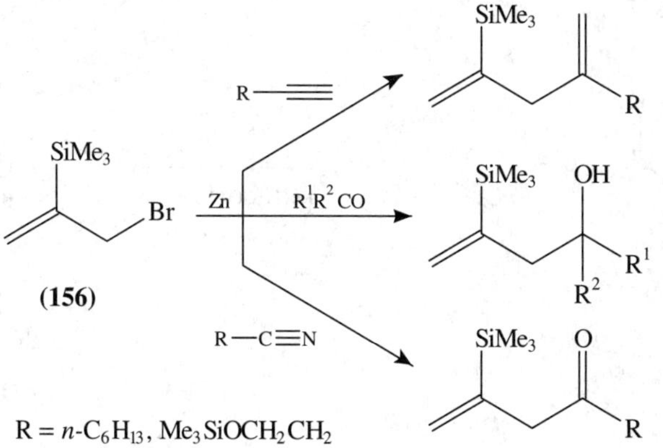

(130)

(156)

$R = n\text{-}C_6H_{13}$, $Me_3SiOCH_2CH_2$
$R^1 = H$, Me ; $R^2 = Ph$, $PhCH(Me)$, $n\text{-}C_6H_{13}$
$R = 1°Alk$, $c\text{-}C_6H_{11}$, Ph

2. Heck reaction

As described in Section III.A.1, the silyl group in vinylsilanes is replaced under the Heck reaction conditions. Several procedures are known to enable the carbon–silicon bond to remain intact. Thus, under phase transfer conditions, arylation of vinylsilanes in the presence of Pd(OAc)$_2$ catalyst gives (E)-β-(trialkylsilyl)styrenes in excellent yield[226]. The addition of the Ag(I) salt[227,228] or Ph$_3$P[229] to the reaction mixture also suppresses the desilylation products (equation 131).

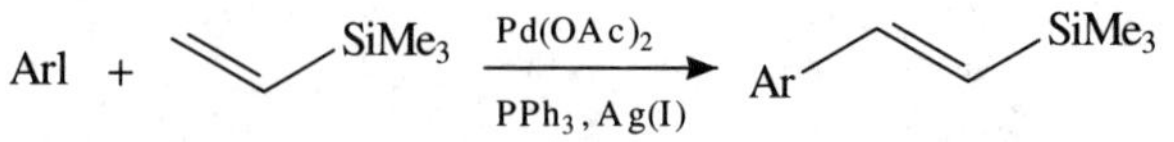

$$\text{ArI} + \diagup\!\!\!\diagdown\text{SiMe}_3 \xrightarrow[\text{PPh}_3,\text{Ag(I)}]{\text{Pd(OAc)}_2} \text{Ar}\diagup\!\!\!\diagdown\text{SiMe}_3 \quad (131)$$

Aryl bromides or iodides react with di- or monochloro(vinyl)silanes or triethoxy(vinyl)silane in the presence of Et$_3$N and palladium catalyst to give β-arylvinylsilanes in moderate to excellent yields. In contrast to the simple silyl substituent, the presence of the electron-withdrawing group on silicon is essential to avoid the elimination of the silyl group. No silver ion is necessary in these examples[230]. The palladium-catalyzed reaction of aryldiazonium tetrafluoroborates with vinylsilanes gives a mixture of terminal and internal vinylsilanes together with styrenes[231].

3. Murai reaction

The ruthenium-catalyzed addition of the carbon–hydrogen bond at the *ortho* position of aromatic ketones or imines to vinylsilanes serves as an elegant method for the regioselective introduction of the carbon–carbon bond at the *ortho* position to the acyl substituent (e.g. **157**) (equation 132)[232–235]. When CF$_3$ and NMe$_2$ substituents are present at the *meta* position, the coupling takes place at the less congested position to give the 1:1 adduct as the sole product[236,237]. The reaction can also be extended to the activation of the olefinic C—H bond. α, β-Unsaturated carbonyl compounds behave similarly and give the corresponding coupling products **158** (equation 133)[238,239]. Alkynylsilanes can also be used to give the corresponding (E)-vinylsilanes **159** (equation 134)[240]. Aromatic imines behave similarly[241].

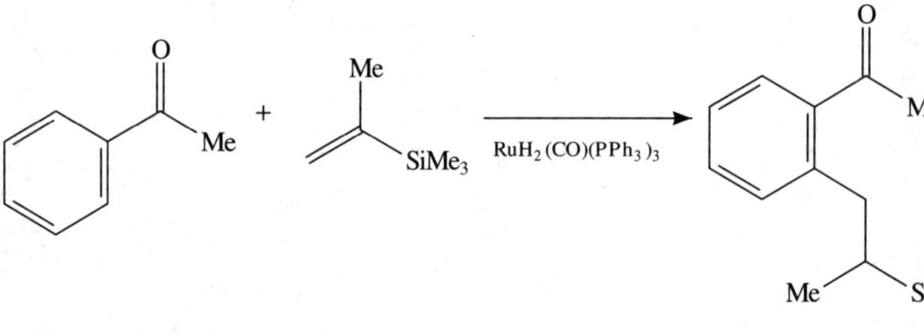

(132)

(**157**)

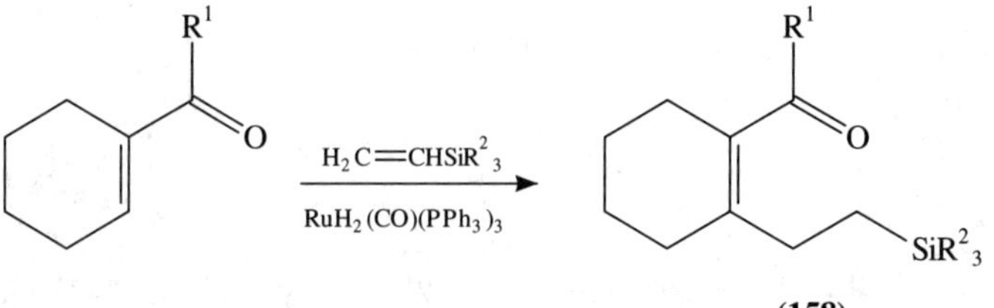

(133)

(**158**)

R^1 = Me or OR^3

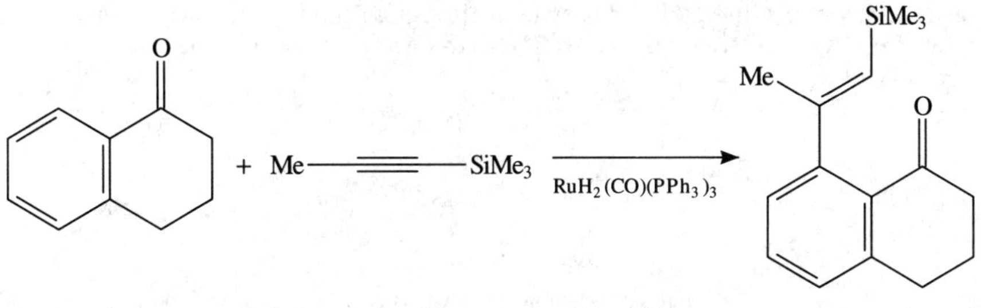

This reaction can also be applied to achieve step-growth copolymerization of aromatic ketone with α,ω-dienes and to synthesize high molecular weight copolymers (equation 135)[242,243].

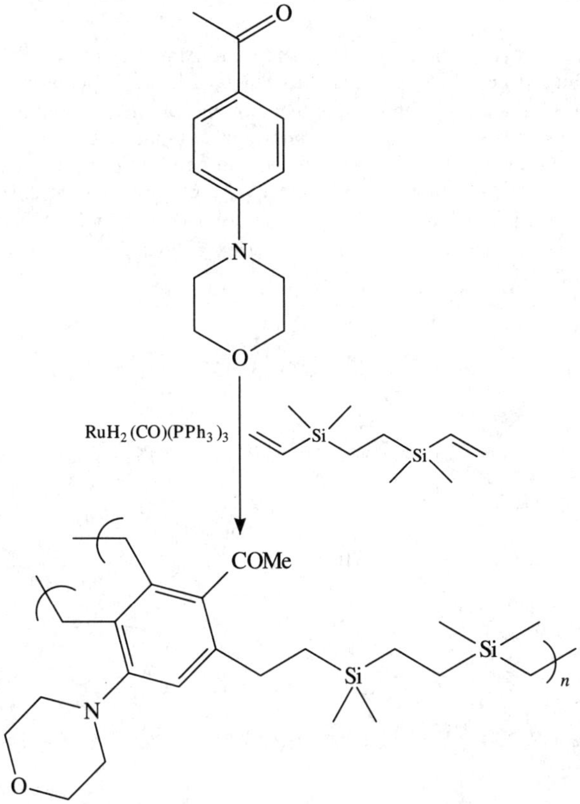

(135)

4. Carbonylation

The regioselectivity of carbonylation of vinylsilanes appears to depend on the nature of the catalysts. Thus, the $PdCl_2(PPh_3)_2$-catalyzed hydroesterification of vinylsilanes yields β-silylesters in high yields with high regioselectivity[244]. On the other hand, when $Co_2(CO)_8$ is used as the catalyst, α-silylesters are obtained predominantly (equation 136)[245].

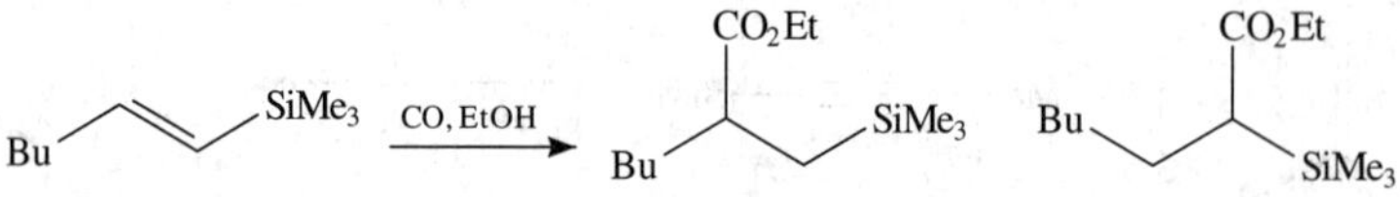
SiMe3
Bu
CO, EtOH
CO2Et
Bu
SiMe3
CO2Et
Bu
SiMe3

$PdCl_2(PPh_3)_2$	96	4
$Co_2(CO)_8$	0	100

(136)

It is interesting to note that hydroformylation of vinylsilanes catalyzed by a Co or a Pt catalyst gives β-silylaldehydes exclusively. A rhodium catalyst exhibits poor regioselectivity. However, in the presence of a large excess of Ph_3P, the regioselectivity improves to that of a normal aldehyde[246–248].

5. Metathesis

Metathesis of silacyclopent-3-ene **(160)** and diallylsilane using the Schrock molybdenum catalyst [Mo(NC_6H_3 *i*-Pr-2,6) ($CHCMe_2Ph$){$OCMe(CF_3)_2$}$_2$] produces a mixture of disilacyclodeca-3,8-dienes **161** and silicon-containing polymers (equation 137)[249]. Under similar conditions, cross-metathesis reactions of allylsilane with π-substituted terminal olefins such as styrene, 1-phenyl-1,3-butadiene and acrylonitrile give substituted allylsilanes in excellent yields with very high selectivity. Employment of simple α-olefins in these reaction gives lower but still useful selectivities (equation 138)[250]. Ruthenium and tungsten chlorides also catalyze metathesis between vinylsilanes and α-olefins[251,252].

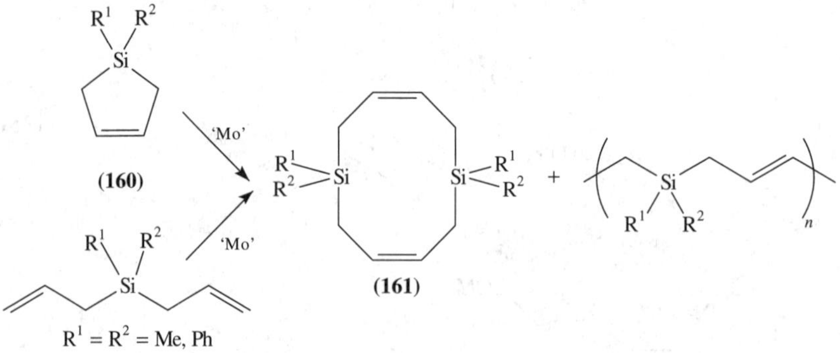
R1 R2
Si
'Mo'
R1
R2
Si
Si
R1
R2
+
Si
R1 R2
n
R1 R2
Si
'Mo'
R1 = R2 = Me, Ph

(160)

(161)

(137)

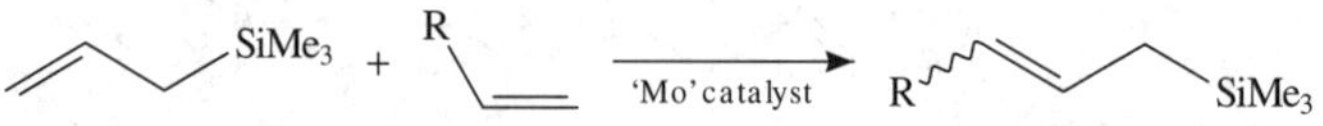
SiMe3
+
R
'Mo' catalyst
R
SiMe3

(138)

R = Ar, CN, $(CH_2)_2Br$, $(CH_2)_nOTBS$, $(CH_2)_nCN$

D. Oxidation Reactions

1. Epoxidation

Epoxidation of ω-hydroxy allylsilanes **162** with *t*-BuOOH in the presence of a catalytic amount of $VO(acac)_2$ in toluene gives the corresponding epoxide **163** with high *erythro*-

selectivity[253]. The resulting epoxide is treated with fluoride to yield the 1,3-diol species (equation 139). Epoxidation of allylsilane **164** with MCPBA at −20 °C gives **165** in extremely high diastereoselectivity, which is then converted into an allylic alcohol under acidic conditions (equation 140)[254].

(162) **(163)** (139)

(164) **(165)** (140)

TBS = Bu_3Si

TBS = Bu_3Si

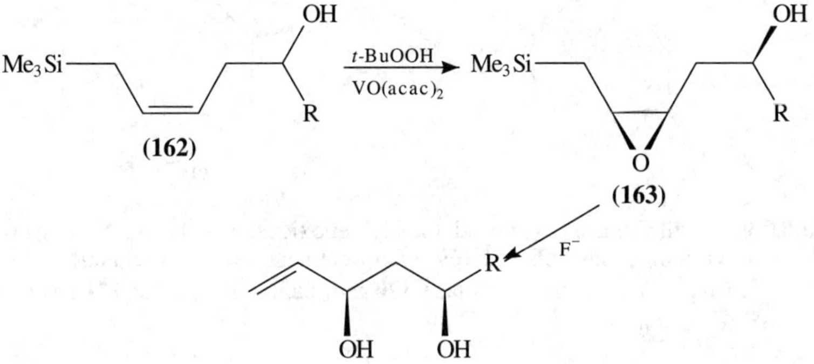

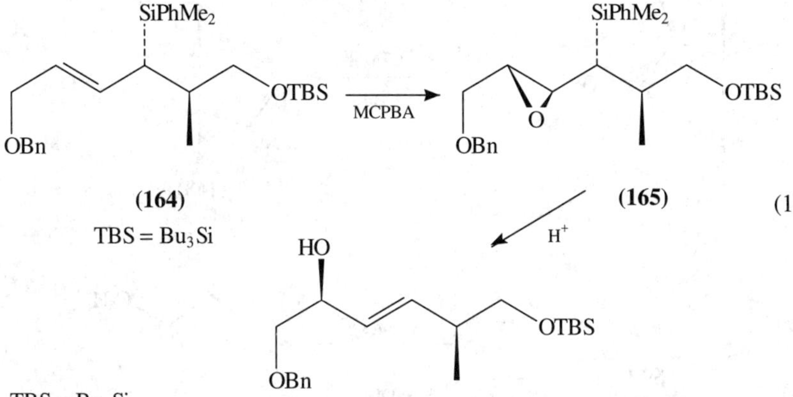

Dimethyldioxirane oxidizes acyclic vinylsilanes at room temperature to the corresponding epoxides **166** in excellent yield (equation 141)[255]. Allylic oxidation is found in appreciable amounts when cyclic vinylsilanes are used. It is interesting to note that simple alkenes react faster with dioxirane than vinylsilanes. The trend appears to be reversed when MCPBA is employed as the oxidant.

(141)

(166)

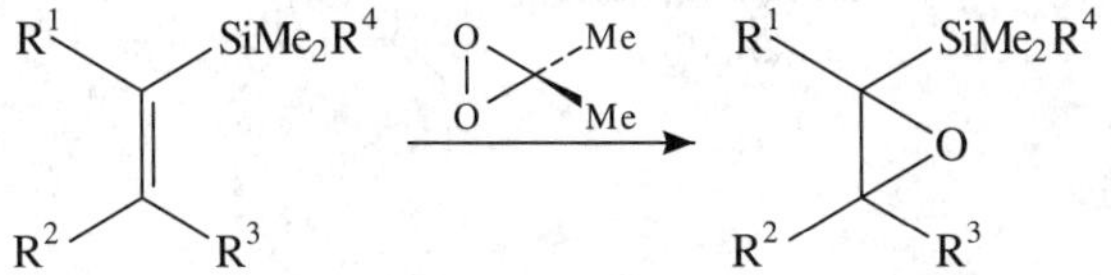

R^1 = H, Me, Et; R^2 = H, 1°Alk; R^3 = H, Me; R^4 = Me, Ph

Photooxygenation of vinylsilanes in the presence of $Ti(Pr\text{-}i)_4$ affords, regio- and diastereoselectively, the epoxy alcohols **167** in satisfactory yield (equation 142).

Cycloalkenones and α-silylcycloalkenones are obtained as by-products from these reactions[256].

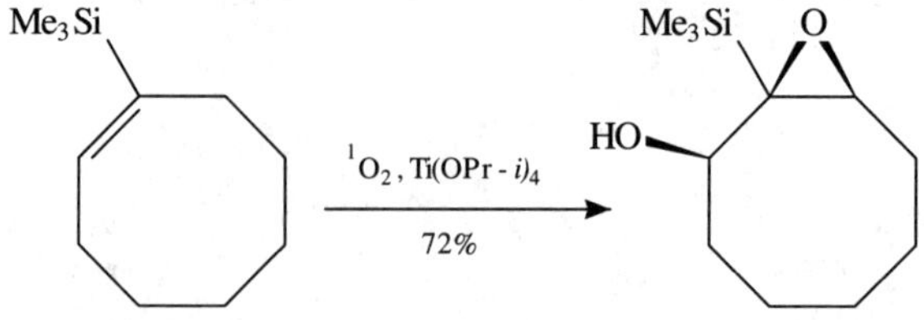

$$\text{(142)}$$

(167)

Epoxidation of vinylsilanes will lead to silyl epoxides, which can be transformed into the carbonyl compounds **168** and **169** upon treatment with acid (equations 143 and 144)[36,257,258]. Conversion of the vinylsilane **170** into methyl enol ether **171** has recently been reported (equation 145)[259].

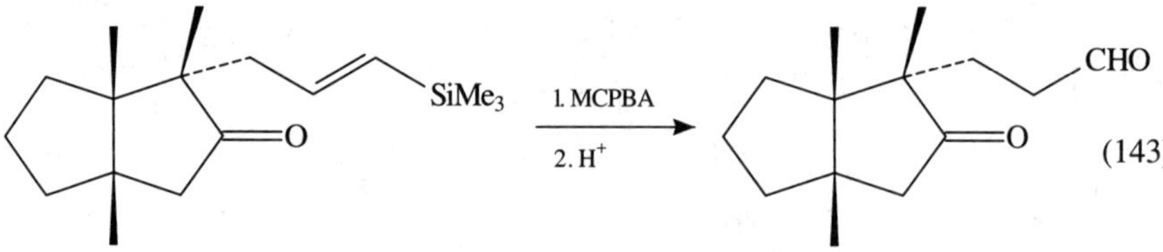

$$\text{(143)}$$

(168)

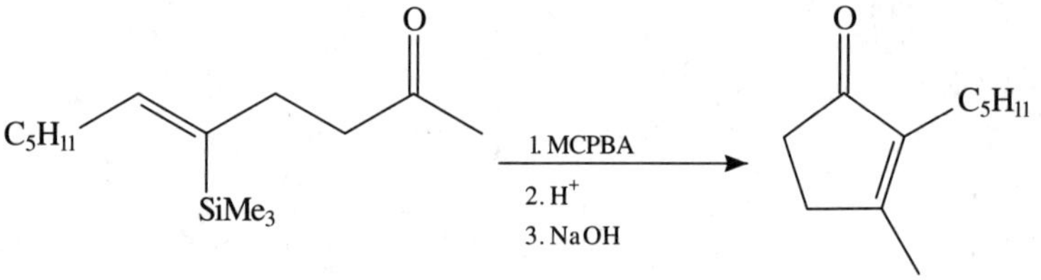

$$\text{(144)}$$

(169)

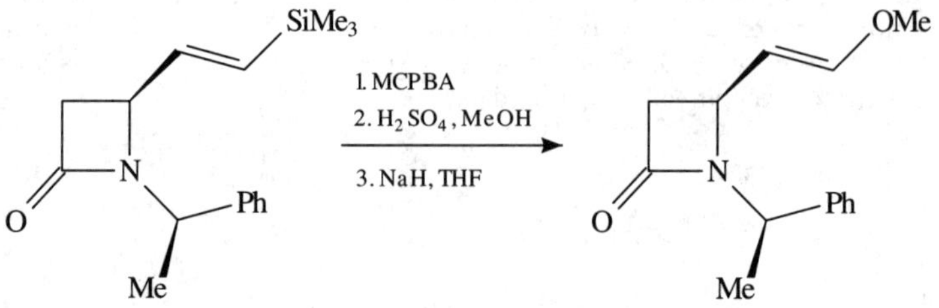

$$\text{(145)}$$

(170)   **(171)**

As mentioned in Section II.B.5, the *tert*-butyldiphenylsilyl group attached to the olefinic carbon is a poor electrofugal group and hence does not leave easily in electrophilic substitution reactions[140]. On the other hand, the epoxide **172**, obtained from the corresponding vinylsilane having such a bulky silyl substituent, gives the carbonyl compound upon

30. Synthetic applications of allylsilanes and vinylsilanes

treatment with acid, with the silyl group remaining intact (equation 146)[260].

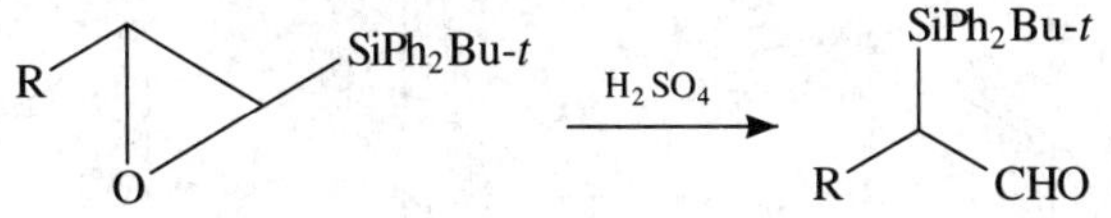

$$R\underset{O}{\triangle}SiPh_2Bu\text{-}t \xrightarrow{H_2SO_4} R\underset{CHO}{\overset{SiPh_2Bu\text{-}t}{\wedge}} \quad (146)$$

(172)
R = 1° Alk

Electro-initiated oxygenation of alkenylsilanes is found to proceed in the presence of thiophenol to give the corresponding α-(phenylthio)carbonyl compounds **173** regiospecifically (equation 147)[261]. Hydroxy and carbonyl functionalities remain intact under these conditions.

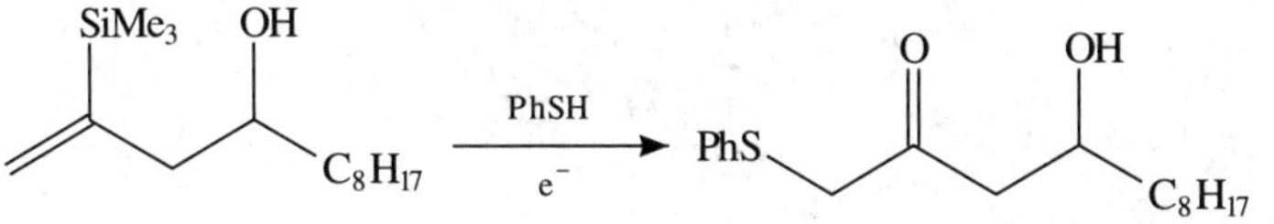

$$\text{SiMe}_3\ \text{OH} \qquad \xrightarrow[e^-]{\text{PhSH}} \qquad \text{PhS}\underset{\text{(173)}}{\overset{O\ \ OH}{\wedge\wedge}}C_8H_{17} \quad (147)$$

Treatment of vinylsilane epoxides with nucleophiles followed by acid promoted *anti*-elimination or base-promoted *syn* elimination gives the corresponding Z- or E-alkenes **174**, respectively (equation 148)[262,263]. A similar approach has been employed for the synthesis of *exo*-brevicomin **175** (equation 149)[264]. Allylsilanes are obtained stereoselectively under similar conditions when $Li_2Cu(CN)(CH_2SiMe_3)_2$ is employed as the nucleophile[265].

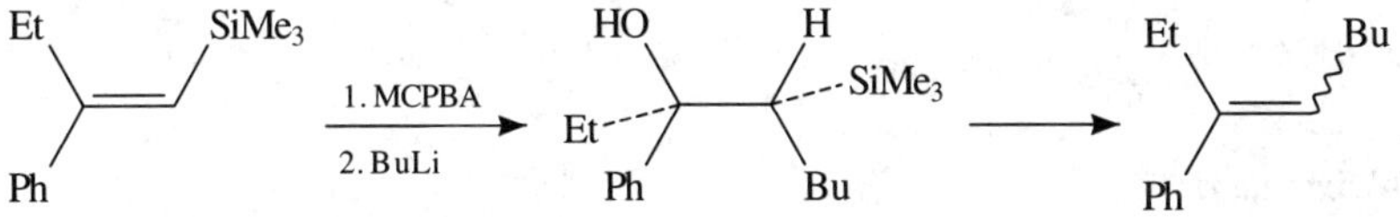

$$\underset{Ph}{\overset{Et}{\diagdown}}=\underset{}{\overset{SiMe_3}{\diagup}} \xrightarrow[\text{2. BuLi}]{\text{1. MCPBA}} \underset{Ph}{\overset{HO\ \ H}{Et\cdots\diagup\diagdown\cdots SiMe_3}}_{Bu} \longrightarrow \underset{Ph}{\overset{Et}{\diagdown}}=\underset{}{\overset{Bu}{\diagup}} \quad (148)$$

(174)

BF₃   (Z) 65%
KH   (E) 50%

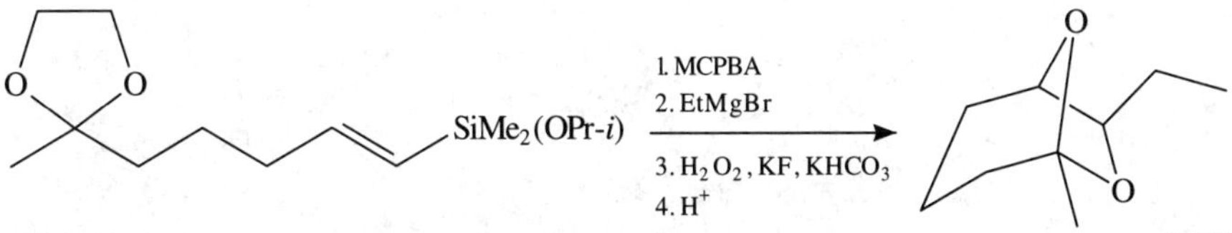

$$\xrightarrow{\text{SiMe}_2(\text{OPr-}i)} \xrightarrow[\substack{\text{1. MCPBA}\\\text{2. EtMgBr}\\\text{3. H}_2\text{O}_2,\text{KF, KHCO}_3\\\text{4. H}^+}]{} \quad (149)$$

(175)

Addition of nitrene to allylsilane gives the corresponding aziridine intermediate **176** which undergoes a ring-opening process to give allylamine **177** as the major product (equation 150)[266,267]. Aziridination of β-trimethylsilylstyrene with the chiral 3-acetoxyaminoquinazolinone reagent[268] lead to aziridine **178** in high diastereoselectivity (equation 151)[269].

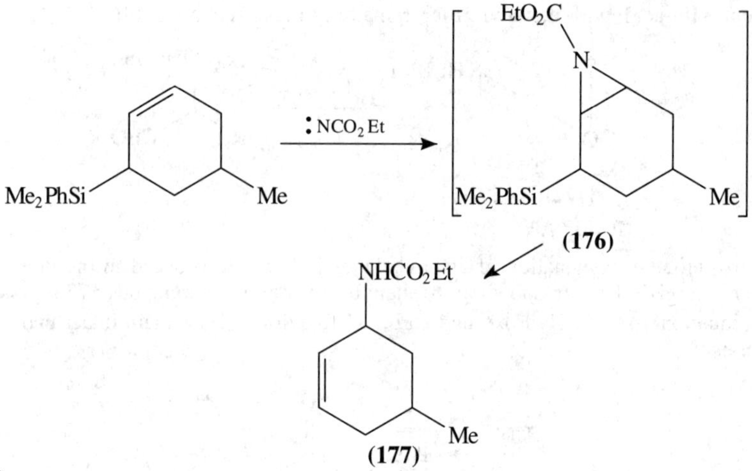

(150)

Q*NHOAc

(178) diastereomer ratio: 11 : 1

(151)

2. OsO_4 oxidation

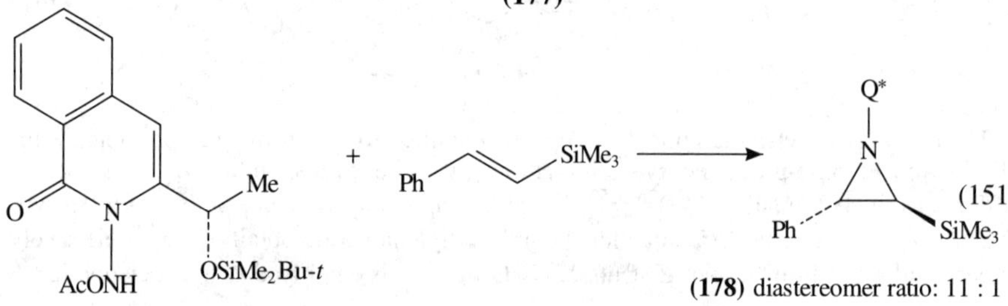

Dihydroxylation of allylsilanes with OsO_4 has been investigated extensively. The reaction has been used for the synthesis of β-damascone **179** (equation 152)[270].

(152)

(179)

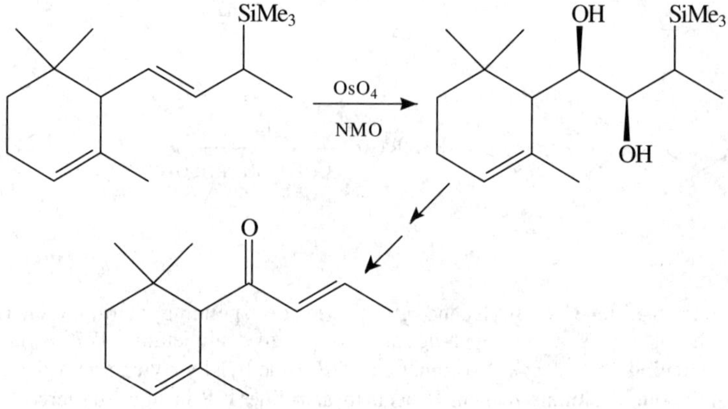

30. Synthetic applications of allylsilanes and vinylsilanes

α-Hydroxycrotylsilane **180** undergoes diastereoselective bishydroxylation with OsO_4 to form 1,2-*anti*-1,2,3-triol **181** (equation 153)[271,272]. It is noted that by increasing the size of the silyl group, the *anti* selectivity improves.

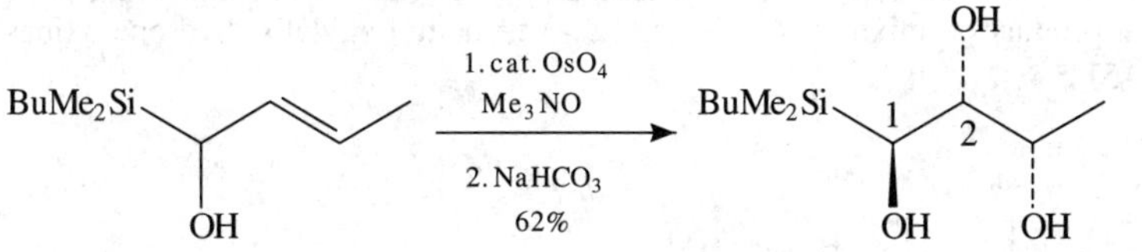

(153)

(**180**)   (**181**) *anti* : *syn* = 120–147 : 1

Sharpless asymmetric dihydroxylation of simple allylsilanes yields the corresponding diols **182** with moderate enantioselectivity (equation 154)[273]. However, when **183** is treated under similar conditions, substituted γ-lactones **184a** and **184b** are obtained in high diastereo- and enantioselectivities (equation 155)[274–277].

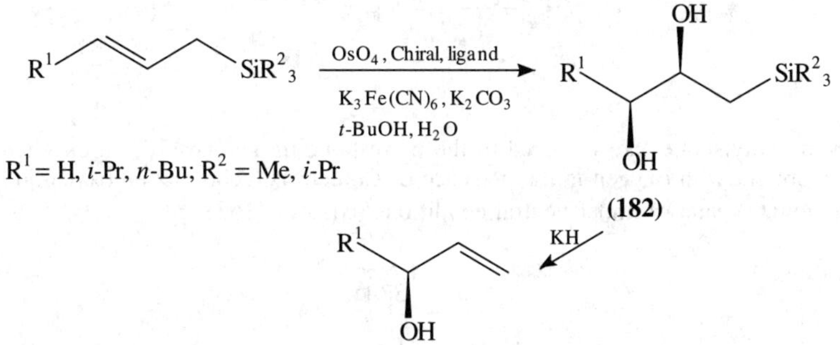

R^1 = H, *i*-Pr, *n*-Bu; R^2 = Me, *i*-Pr

(154)

(**182**)

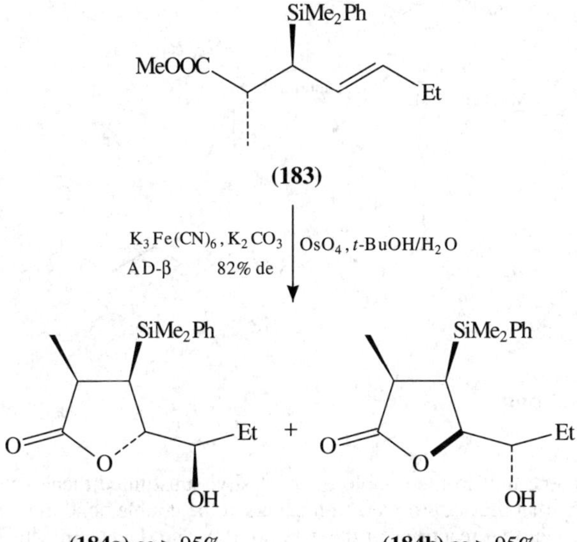

(**183**)

(155)

(**184a**) ee > 95%   (**184b**) ee > 95%

3. Other oxidation reactions

Photooxygenation of vinylsilane in the presence of acetic anhydride and pyridine furnishes a convenient procedure for the synthesis of α-trimethylsilylenones **185** in moderate to good yields (equation 156)[278]. Dye-sensitized photooxidations of allylsilane with singlet oxygen produces a mixture of (E)- and (Z)-3-trimethyl-silylallyl hydroperoxides **186** (equation 157)[279].

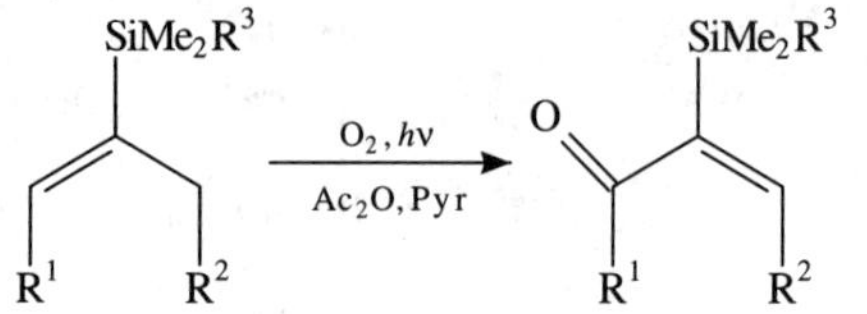

$$R^1 = Me, n\text{-Bu}; R^2 = H$$
$$R^1R^2 = (CH_2)_n, n = 2, 3; R^3 = Me, Ph$$

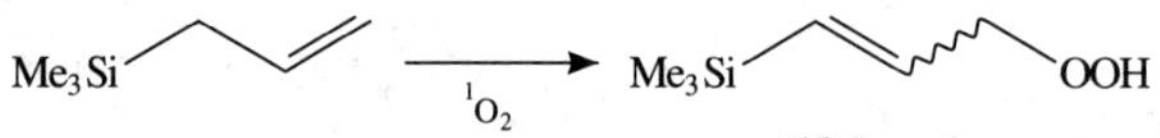

(156)

E : Z = 22 : 78

(157)

Various vinylsilanes are oxidized to the corresponding ketones **187** in excellent yield upon treatment with oxygen in the presence of Co(ecbo)$_2$ (ecbo = 2-ethoxycarbonyl-1,3-butanedionato) catalyst under neutral conditions (equation 158)[280].

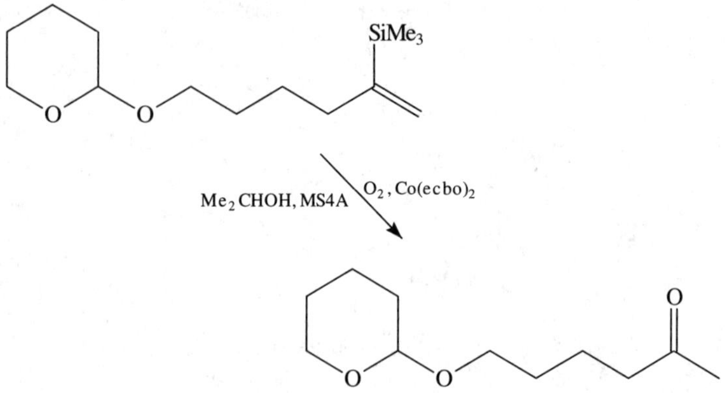

(158)

E. Cycloaddition Reactions

1. [2 + 1] Cycloaddition

The addition of a carbene unit to the double bond of silyl-substituted dienes has been executed. It is noteworthy that the addition reaction occurs at the double bond more remote from the silyl substituent. Further reactions of these cyclopropyl products **188** with AgBF$_4$

give the functionalized silyl-substituted dienes **189** in moderate yield (equation 159)[281]. In a similar manner, dichlorocyclopropane **190** is converted into the corresponding diene **191** by desilylation (equation 160)[282].

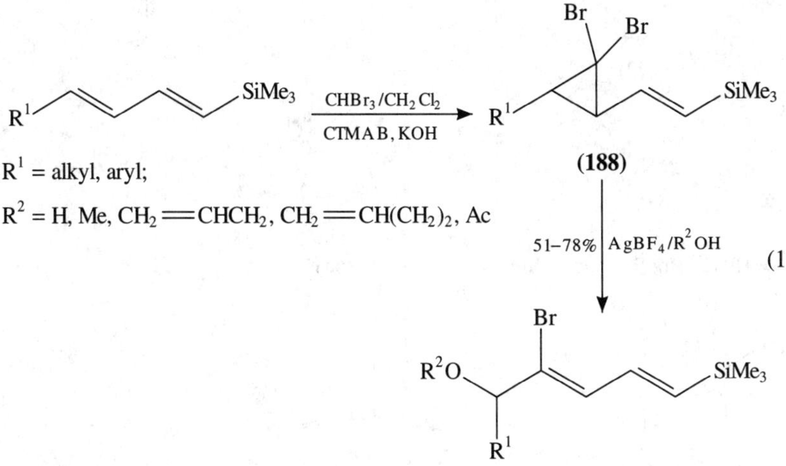

SiMe3
R1
CHBr3/CH2Cl2
CTMAB, KOH
Br
Br
R1
SiMe3
(188)
R1 = alkyl, aryl;
R2 = H, Me, CH2=CHCH2, CH2=CH(CH2)2, Ac
51–78%
AgBF4/R2OH
Br
R2O
SiMe3
R1

(189)

(159)

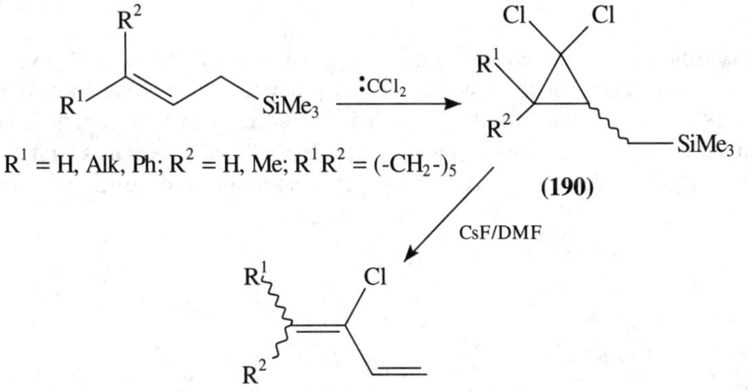

R2
R1
SiMe3
:CCl2
Cl
Cl
R1
R2
SiMe3
R1 = H, Alk, Ph; R2 = H, Me; R1R2 = (-CH2-)5
(190)
CsF/DMF
R1
Cl
R2

(191)

(160)

The Simmons–Smith reaction of chiral (*E*)-crotylsilane **192** produces **193**, which subsequently undergoes a ring-opening reaction followed by cyclization under acidic conditions to give **194** (equation 161)[283].

2. [2 + 2] Cycloaddition

The reaction of vinylsilane with dichloroketene gives **195** as the major product in 37% yield (equation 162). The minor product **196** decomposes during the course of chromatography on silica gel[284]. The regioselectivity of this reaction is in accord with frontier molecular orbital predictions.

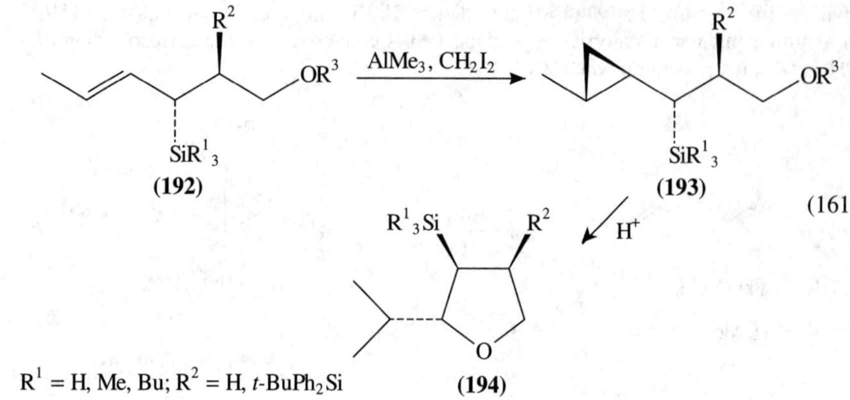

In the cycloaddition of triisopropylallylsilane to α,β-unsaturated lactams **197**, cyclobutane adducts **198** have been found to be the kinetic products whereas the formation of cyclopentanes **199** is thermodynamically controlled[285]. Reactions of allenylmethylsilanes with activated unsaturated esters and nitriles (equation 163)[286] and allylsilanes with unsaturated esters[287] are other examples of using [2+2] cycloaddition to construct cyclobutane derivatives.

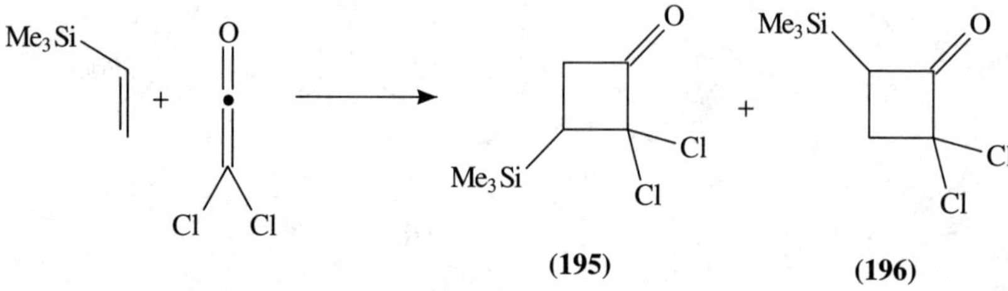

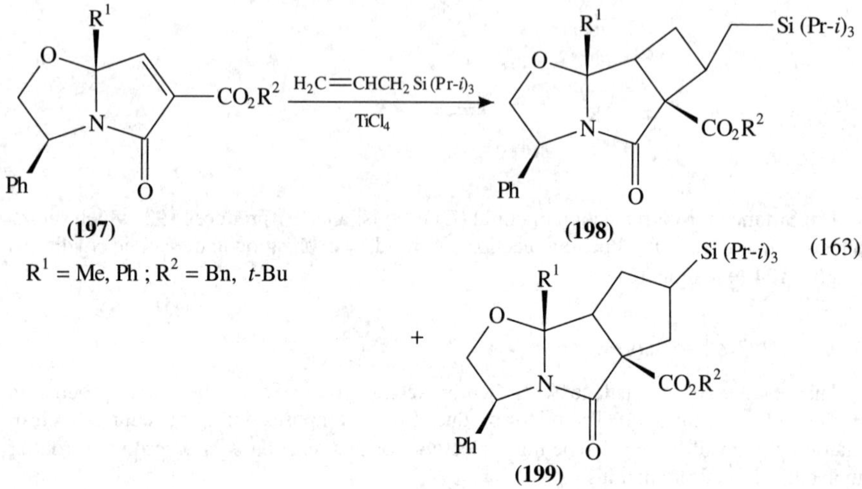

Lewis acids play a dominant role on the chemoselectivity of the cycloaddition of allylsilanes and 3-butyn-2-ones. $AlCl_3$ and $EtAlCl_2$ promote [2+2] cycloaddition predominantly, whereas $TiCl_4$-mediated reaction gives significant amount of the [3+2] adduct[288,289]. The reaction of 2,2-bis(trifluoromethyl)ethylene-1,1-dicarbonitrile with allylsilane without any Lewis acid catalysts yields the cyclobutane derivative via a nonconcerted fashion[290].

Photocycloaddition proceeds between allylsilane and N-methylphthalimide to yield a mixture of [2 + 2] and [4 + 2] adducts along with the allylated product[291]. Intramolecular cycloadditions of the vinylsilanes with the cyclopentenone moieties in **200** furnish good yields of cyclic products stereoselectively (equation 164)[292]. In the presence of 1,4-dicyanonaphthalene, diallylsilane **201** undergoes an intramolecular photocycloaddition reaction in an aromatic solvent to give a four-membered ring product (equation 165)[293].

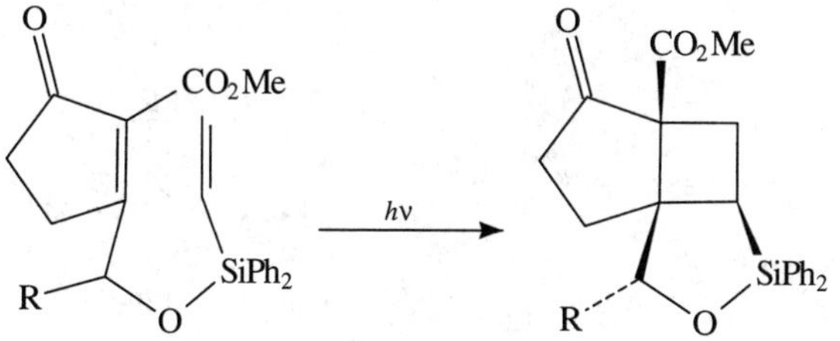

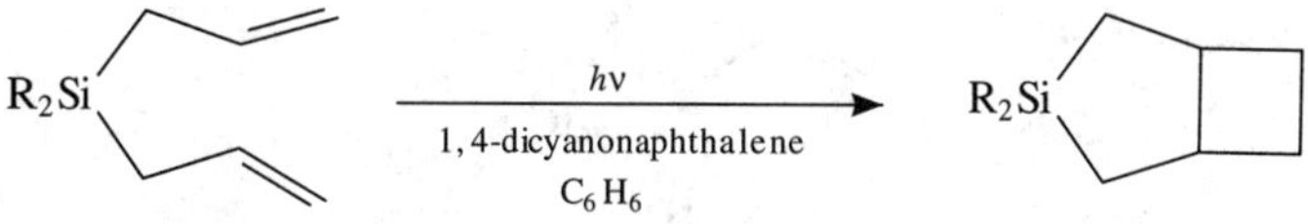

$TiCl_4^-$ or $BF_3 \cdot OEt_2$-catalyzed addition of N-acylated aromatic aldimine to triisopropylallylsilane gives azetidine derivatives **202** (equation 166) accompanied by some allylated products[294]. Cycloaddition of allylsilanes and chlorosulfonyl isocyanate yields the silyl-substituted β-lactam **203**[295,296]. The presence of an electronegative substituent on the silicon will induce ring opening and fragmentation to yield allyl cyanides **204** (equation 167)[296].

Oxetanes **205** are produced with a high degree of stereoselectivity when an allylsilane with bulky substituents at the silicon center reacts with α-keto ester in the presence of $TiCl_4$ (equation 168). It is interesting to note that tetrahydrofuran derivatives are obtained at high temperature[297].

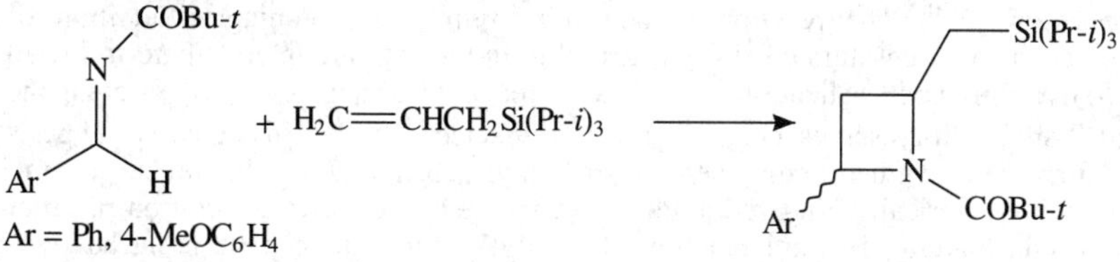

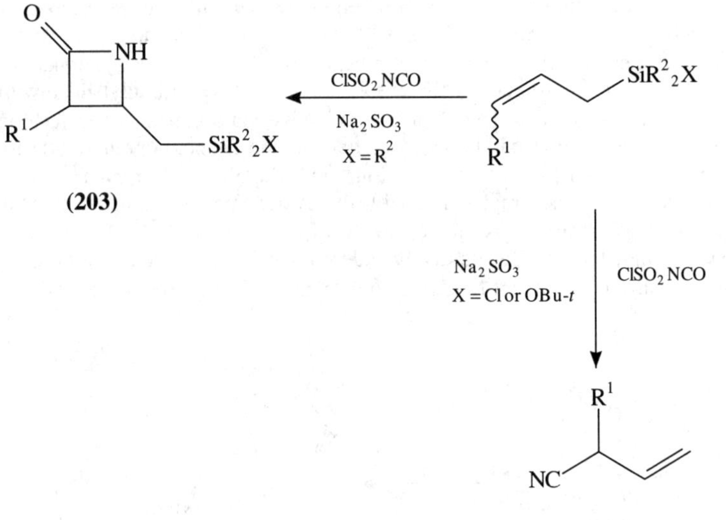

3. [3 + 2] Cycloaddition

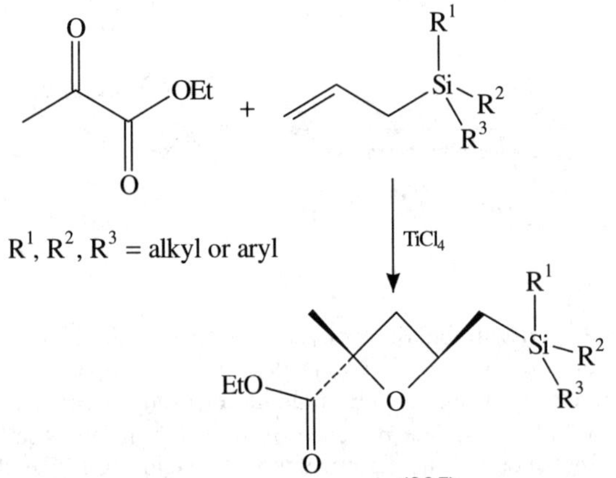

In the presence of a Lewis acid, allylsilane can act as a three-carbon component and react with α,β-unsaturated ketones to yield cyclopentane derivatives **206** (equation 169)[298–310]. The reaction is believed to involve a conjugate addition of allylsilane to the α,β-unsaturated system, resulting in the cationic intermediate followed by a 1,2-silyl shift and cyclization. The new stereogenic center generated through the cationic 1,2-silyl shift preserves a high degree of *anti* selectivity[309]. This outcome suggests that allylsilane is a synthetic equivalent for a 2-silyl substituted 1,3-dipole. In general, allylsilane with a sterically hindered silicon center is required for cycloaddition in order to suppress the Sakurai–Hosomi reaction. The silyl group can then be replaced by a Tamao oxidation[309]. High diastereoselectivities are obtained when chiral allylsilanes are employed in the [3 + 2] annulation with α, β-unsaturated carbonyl compounds as in the formation of **207** (equation 170)[303,304,310].

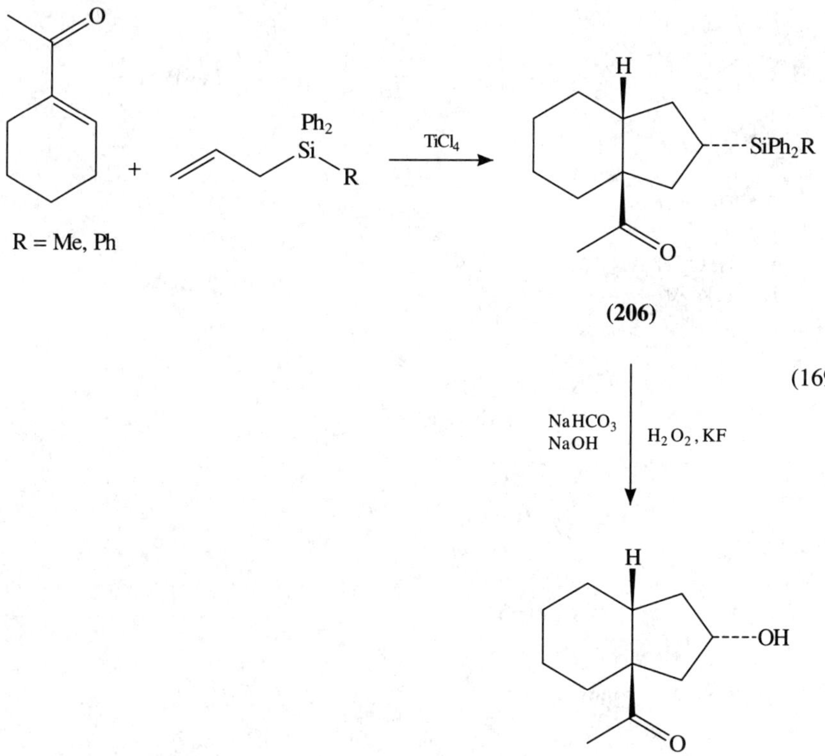

R = Me, Ph

(**206**)

(169)

NaHCO$_3$ / NaOH | H$_2$O$_2$, KF

Y = H, OCH$_3$
R^1 = R^3 = H, Me; R^2 = H, Me, Et

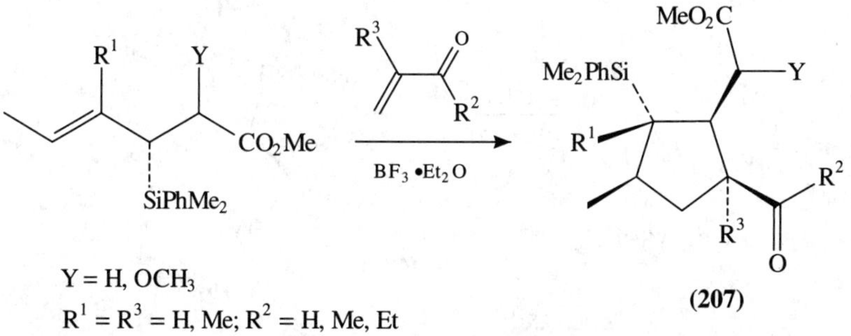

(**207**)   (170)

The [3 + 2] annulation reaction can be extended to the synthesis of heterocycles. Thus, the reaction of crotylsilanes with *in situ*-generated imines produces, diastereoselectively, [3 + 2] adducts **208** at low temperature or homoallylic *N*-acylamines **209** at elevated temperature (equation 171)[311,312].

Another approach leading to pyrrolidine **210** via [3 + 2] cycloaddition is the reaction of *N*-protected α-amino aldehydes with allylsilanes in the presence of a catalytic amount of BF$_3$ · OEt$_2$ (equation 172). No 1,2-silyl group migration occurs in these annulation processes[313].

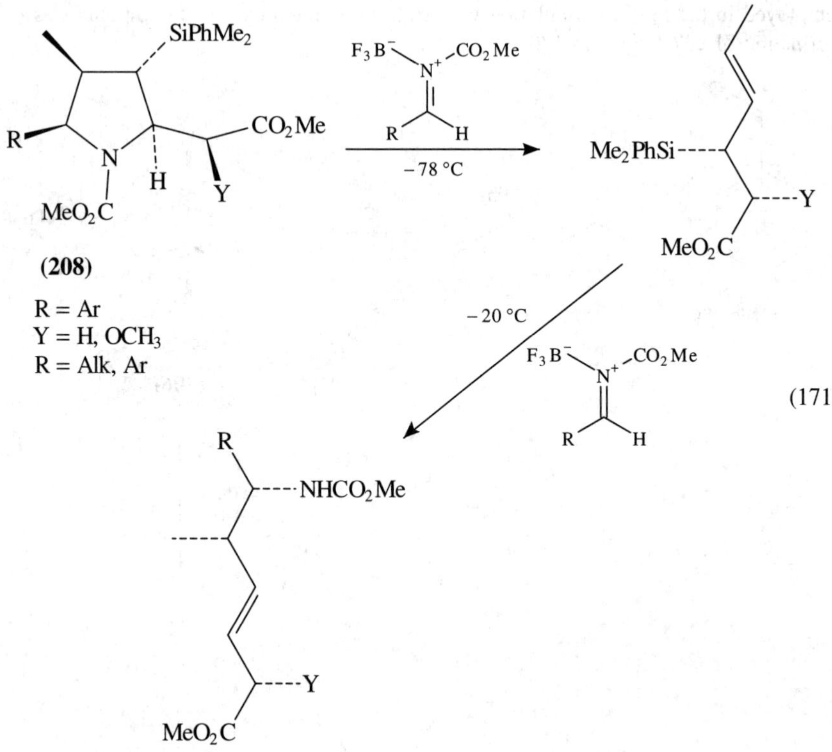

(208)

R = Ar
Y = H, OCH$_3$
R = Alk, Ar

(209)

(171)

R = Me, *i*-Pr, *s*-Bu, PhCH$_2$

(210)

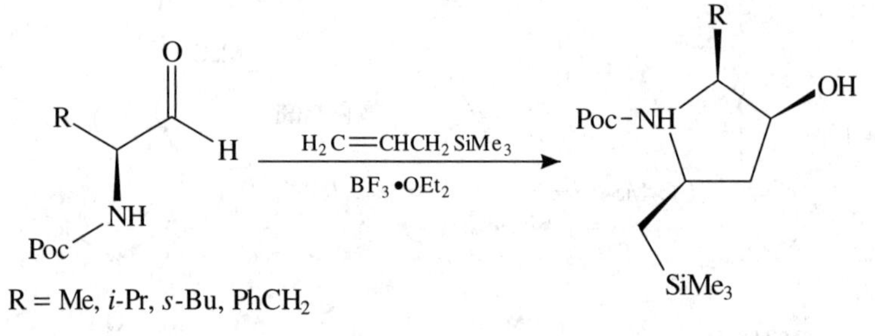

(172)

As for oxygen heterocycles, several reports involving the diastereoselective synthesis of highly substituted tetrahydrofuran derivatives have appeared. SnCl$_4$-mediated [3 + 2] cycloaddition of allylsilane with optically active α-ketoesters affords **211** with excellent diastereoselectivity (equation 173)[314–317].

When chiral allylsilanes are used for coupling with aldehydes, the diastereoselectivity of the reaction depends on the nature of the Lewis acid employed. Chelative SnCl$_4$-promoted reaction gives **212** whereas BF$_3$ · OEt$_2$ catalyzes the annulation reactions leading to **213** (equation 174)[315–317].

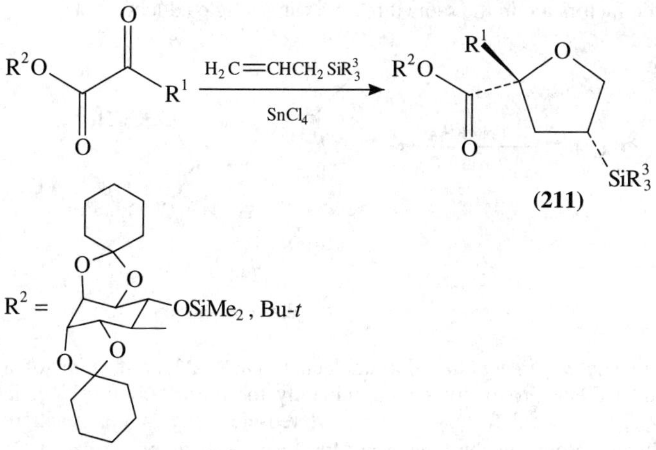

$R^2 =$ [sugar structure with $OSiMe_2$, Bu-t]

$R^1 = Me, Ph; R_3^3 = Me_3, Me_2Ph, t\text{-}BuMe_2, t\text{-}BuPh_2$

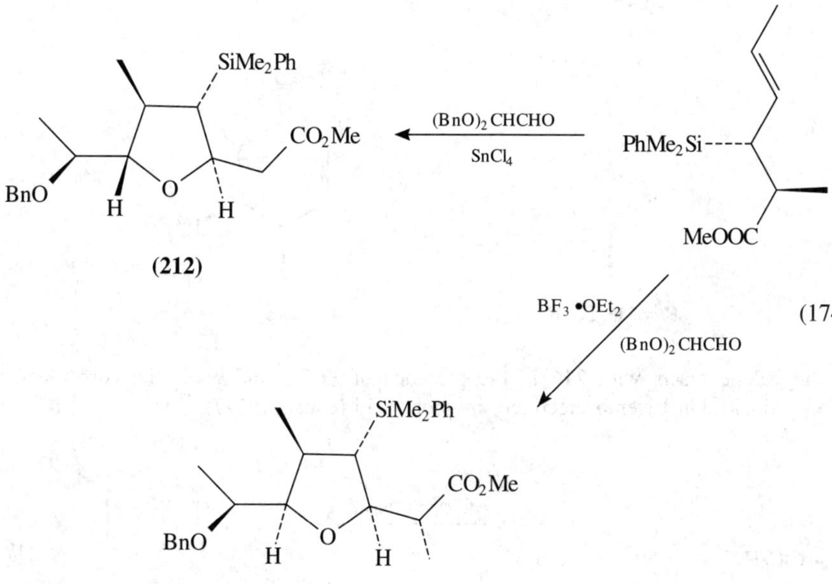

(173)

(174)

Formation of Δ^2-isoxazolines **214** can be achieved by the cycloaddition of allylsilanes to NOBF$_4$ in excellent diastereoselectivity (equation 175)[318].

4. [4 + 2] Cycloaddition

Silyl-substituted alkenes or dienes can undergo the Diels–Alder reaction[28]. The reactivity of these alkenes, however, is generally lower than those without silyl substituent. In addition, the degree of regioselectivity of the silyl group is evidently small. Only when both steric and electronic factors are in the same direction can a single adduct be obtained.

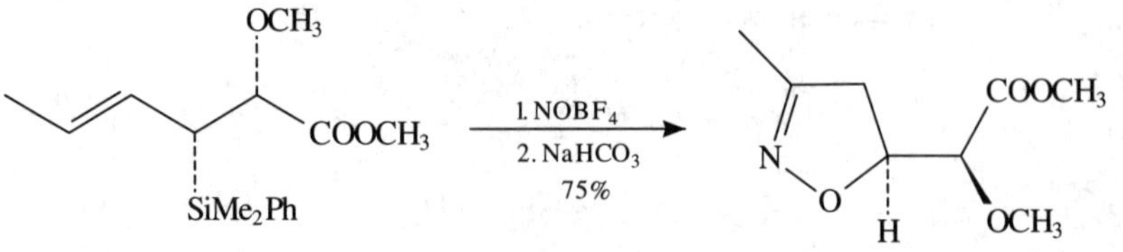

$$\underset{\text{SiMe}_2\text{Ph}}{\text{CH}_2=\text{CH-CH}_2\text{-CH(OCH}_3\text{)-CH(COOCH}_3\text{)}} \xrightarrow[\text{2. NaHCO}_3]{\text{1. NOBF}_4} \underset{(214)}{\text{isoxazoline-CH(COOCH}_3\text{)(OCH}_3\text{)}} \quad (175)$$

(214)

diastereomer ratio 40 : 1

a. Silyl-substituted alkenes as dienophiles. Intramolecular Diels–Alder reaction of a vinylsilane moiety with a diene group gives predominantly the corresponding bicyclic product **215** (equation 176). Several factors affect the stereoselectivity of this reaction. When $n = 1$, the bulky R^1 group on the silicon and the *trans* substituent group at C_3 appear to favor a *trans*-fused bicyclic skeleton. Poor selectivity, however, is observed when n is larger than 1[319–321].

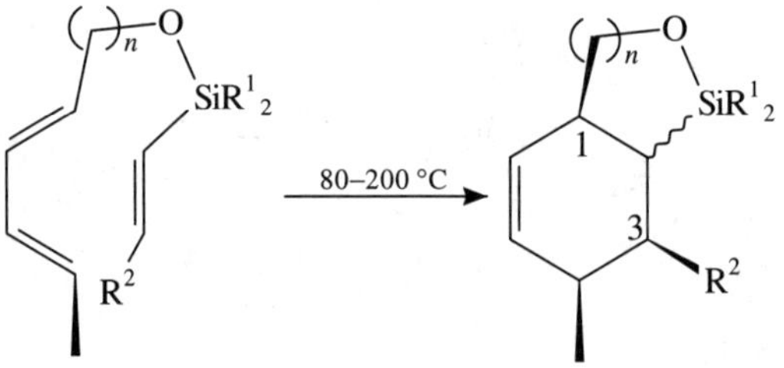

$$\text{(diene-CH}_2\text{-O-SiR}^1_2\text{-CH=CH-R}^2\text{)} \xrightarrow{80-200\,°C} \text{(215)} \quad (176)$$

$R = \text{Me, Ph}; R^2 = \text{H, Ph}$ **(215)**

Vinylsilane reacts with **216** in the presence of ZnCl_2 and gives the corresponding Diels–Alder adduct stereoselectively in good yield (equation 177)[322].

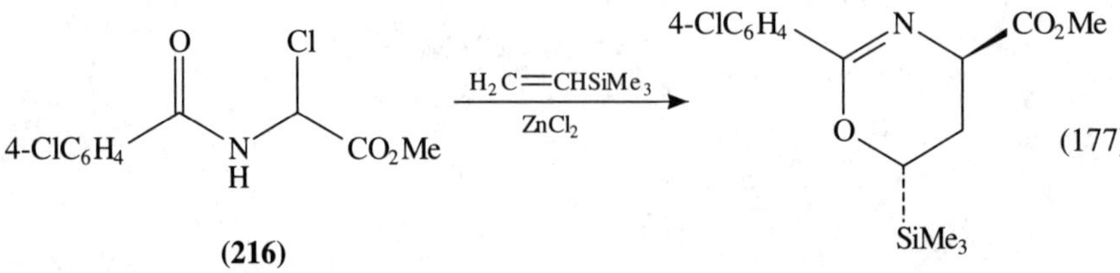

$$\underset{(216)}{4\text{-ClC}_6\text{H}_4\text{-C(O)-N(H)-CH(Cl)-CO}_2\text{Me}} \xrightarrow[\text{ZnCl}_2]{\text{H}_2\text{C}=\text{CHSiMe}_3} \underset{\text{SiMe}_3}{4\text{-ClC}_6\text{H}_4\text{-N=C-O-CH(SiMe}_3\text{)-CH}_2\text{-CH(CO}_2\text{Me)}} \quad (177)$$

b. Silyl-substituted alkenes as dienes. The use of silyl-substituted dienes as dienes in the Diels–Alder reaction has been reviewed[26,27]. Only recent representative examples are summarized here. As the carbon–silicon bonds are readily converted into a number of other functionalities, silicon-tethered intramolecular Diels–Alder reactions have paved the way for the synthesis of functionalized cyclohexane derivatives **217–219** (equation 178–180)[28]. When the silicon atom is directly attached to the diene moiety, stereoselective cycloaddition has been achieved[323–325].

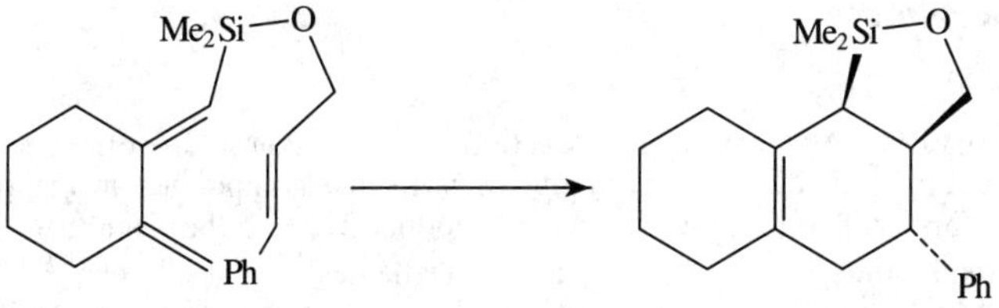

30. Synthetic applications of allylsilanes and vinylsilanes

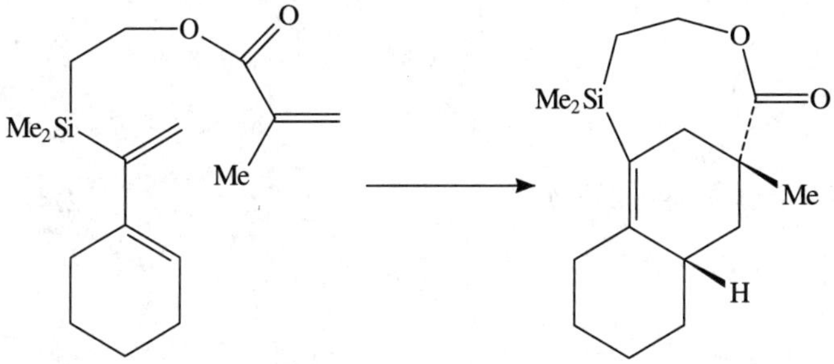

(178)

(217)

(179)

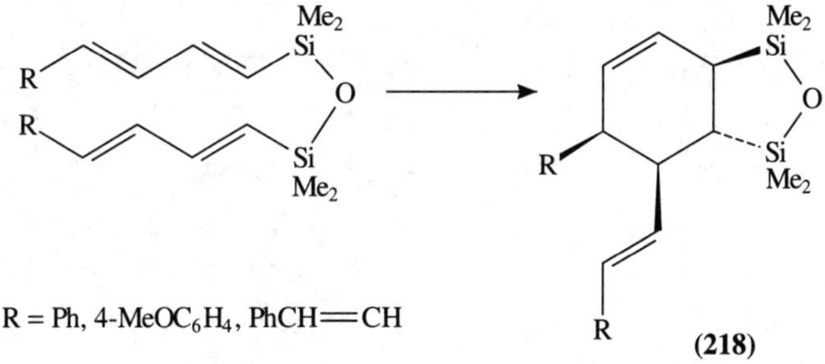

R = Ph, 4-MeOC$_6$H$_4$, PhCH=CH

(218)

(180)

(219)

5. [5 + 2] Cycloaddition

A silicon-tethered [5+2] intramolecular cycloaddition of vinylsilane moiety to a pyrone ring to **220** has been executed in regio- and stereoselective manners (equation 181)[326].

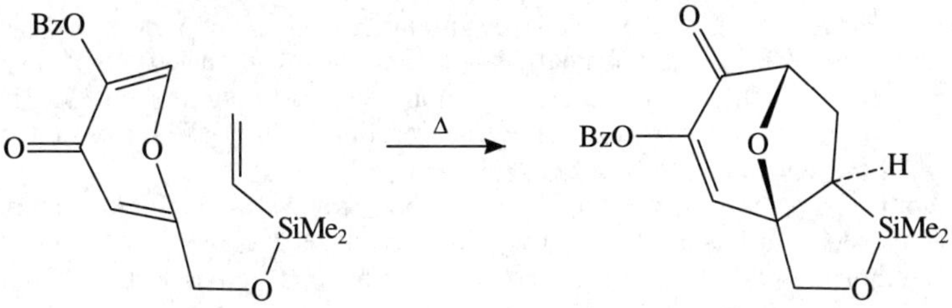

(181)

(220)

F. Rearrangements

1. Ene reaction

The ene reaction is the major side reaction in the Sakurai–Hosomi coupling reaction. Thus, treatment of **221** with carbonyl or with azo compounds in the presence of TiCl$_2$(OPr-i)$_2$ furnishes a mixture of ene-type product **222** and the Sakurai–Hosomi-type product **223** (equation 182)[327]. Allylsilylation of alkenes[328] and alkynes[329,330] proceeds regioselectively to give **224** (equation 183) and **225** (equation 184), respectively, when aluminum catalysts are used.

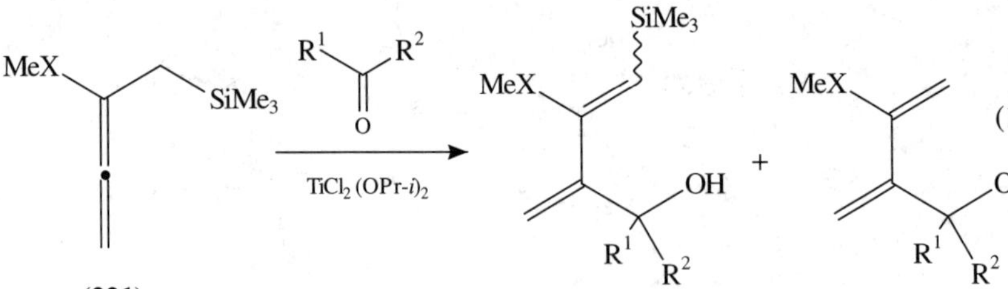

(182)

(221)

X = O, S
R^1 = Ar, CO$_2$Et ; R^2 = H, CO$_2$Et

(222) (223)

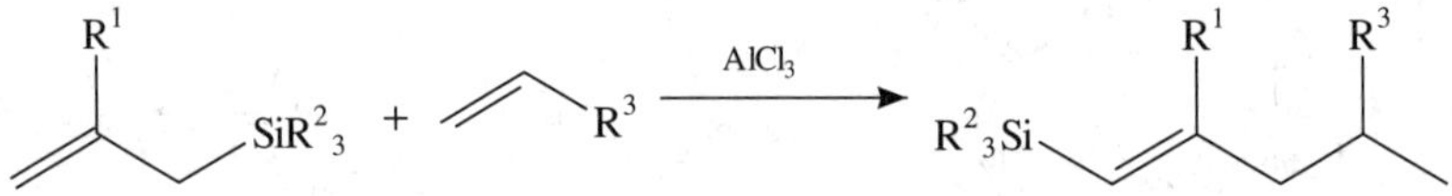

R^1 = H, Me; R^2 = Me, Ph; R^3 = n-Bu, n-C$_6$H$_{13}$, PhCH$_2$ (224) (183)

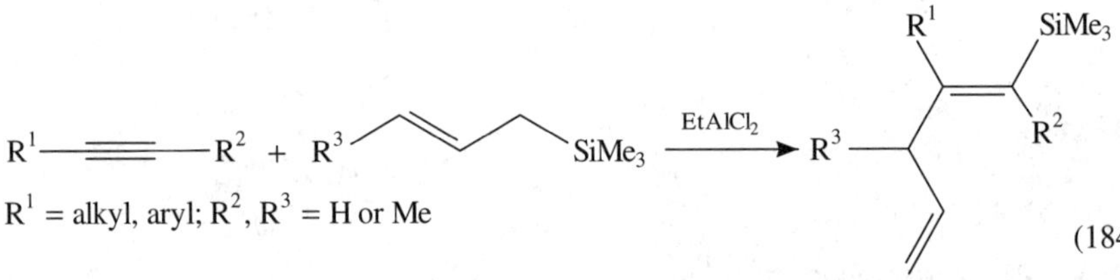

R^1 = alkyl, aryl; R^2, R^3 = H or Me

(225) (184)

It is noteworthy that the ene product **227** is obtained exclusively when allylsilane **226** is treated with butynone in the presence of ZnI$_2$ catalyst and molecular sieve (ms) 4Å (equation 185)[331]. Methyl vinyl ketone behaves similarly. High enantioselectivity has been observed in product **228** when triphenylallylsilane is coupled with methyl glyoxylate in the presence of (R)-(BINOL)TiCl$_2$ catalyst (equation 186)[332]. In addition to [2 + 2] cycloaddition, TCNE undergoes a regiospecific ene reaction with γ-alkyl substituted allylsilanes to yield the substituted olefins **229** (equation 187)[333].

The presence of the silyl substituent at the olefinic carbon provides an important feature in the regio- and stereochemical control of the Lewis acid-catalyzed carbonyl-ene reactions which gives **230** (equation 188)[334]. The changeover of the olefinic stereoselectivity from '*trans*' to '*cis*' is observed (*cf* **231a**) (equation 189). when R is a silyl group[334]. Without a silyl group (R = H) '*trans*' product **231b** is obtained predominantly[334]. It is noteworthy

30. Synthetic applications of allylsilanes and vinylsilanes

that a *trans* selectivity is widely recognized for the ene reaction with alkenes without silyl group[335–339].

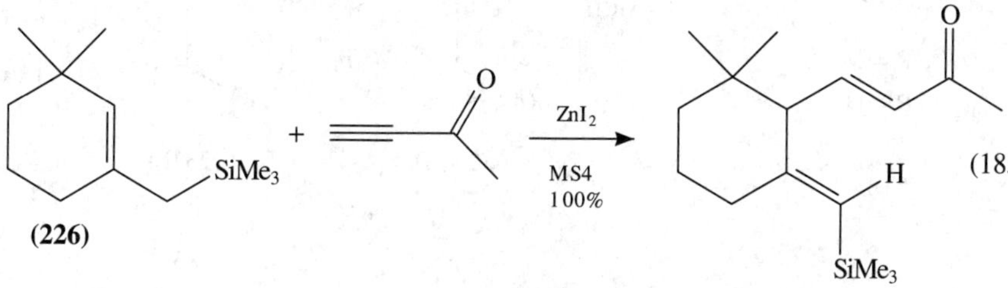

(226) + propynone $\xrightarrow[\text{MS4, 100\%}]{\text{ZnI}_2}$ (227)   (185)

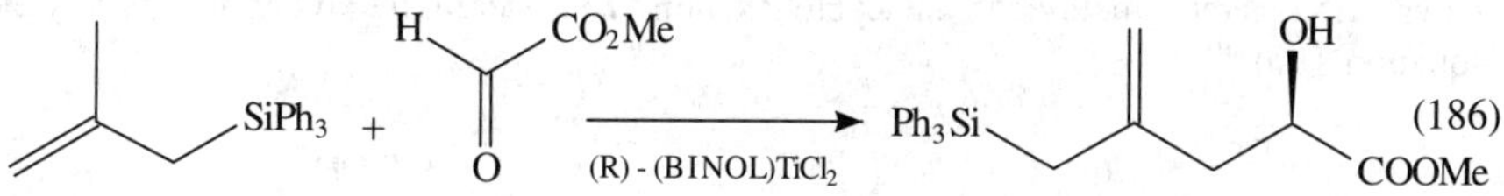

methallyl-SiPh$_3$ + methyl glyoxylate $\xrightarrow{\text{(R)-(BINOL)TiCl}_2}$ (228)   (186)

BINOL = 1,1'-Binaphthyl-2,2'-diol

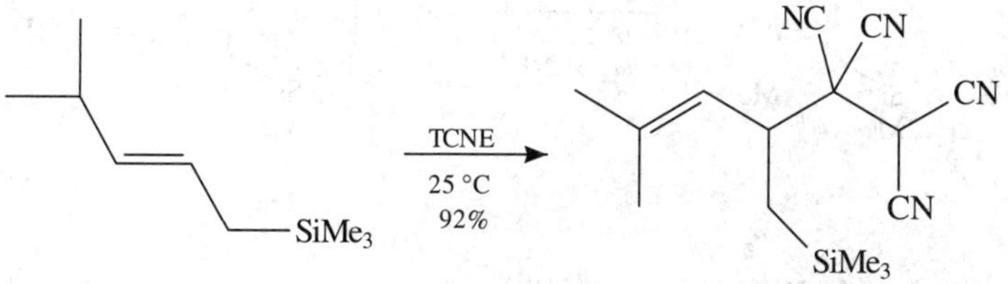

allylsilane $\xrightarrow[\text{25 °C, 92\%}]{\text{TCNE}}$ (229)   (187)

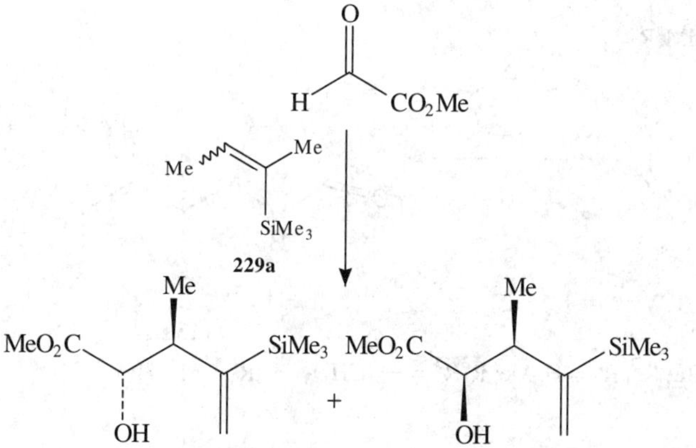

229a + methyl glyoxylate → (230a) + (230b)   (188)

From *E* – **229a** : 7   93
From *Z* – **229a** : 98   2

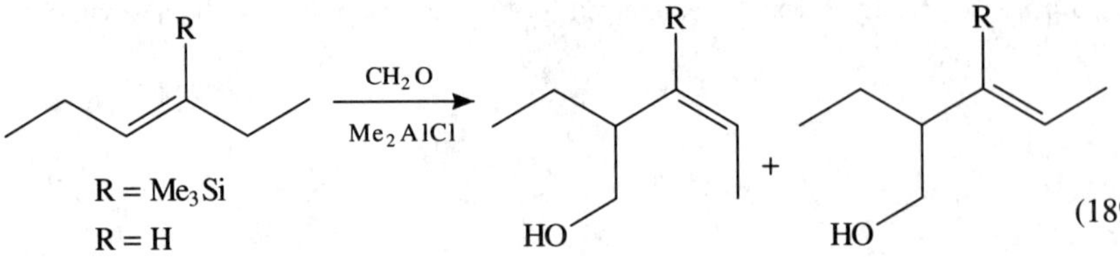

$$\begin{array}{c} R = Me_3Si \\ R = H \end{array} \quad \begin{array}{ccc} (231a) & & (231b) \\ 98 & : & 2 \\ 15 & : & 85 \end{array} \tag{189}$$

Silicon-tethered intramolecular carbonyl-ene reactions of vinylsilanes have been shown to generate methylene silacyclohexanols **232** stereoselectively in good yield (equation 190)[340].

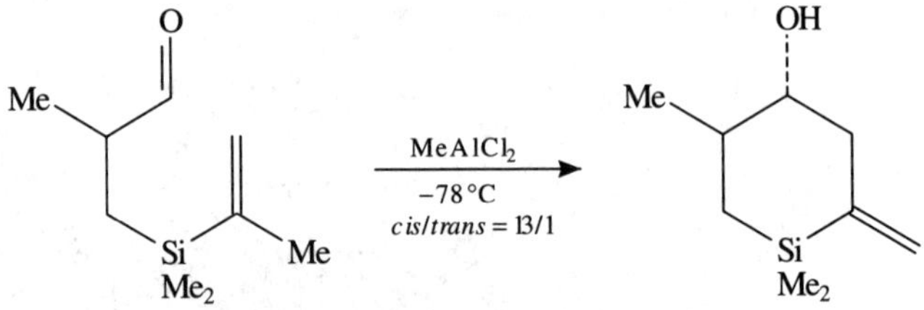

(190) **(232)**

The ene reaction of *N*-methyltriazolinedione with vinylsilanes yields **233** (equation 191). Only hydrogen at the carbon geminal to the silyl group will be abstracted[340–342].

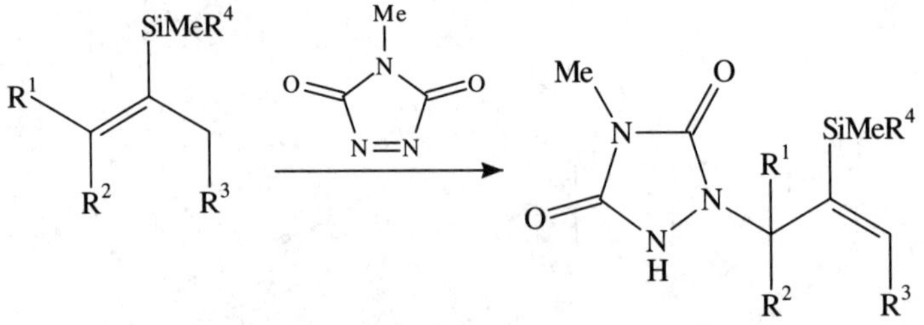

(191) **(233)**

R^1 = H, Me, *n*-Bu; R^2, R^3 = H, Me; R^2R^3 = —(CH$_2$)$_3$—; R^4 = Me, Ph

2. Sigmatropic rearrangements

β-(Silyl)allylic ethers **234** readily undergo 2,3-Wittig rearrangement upon treatment with an amide base to give, after methylation, **235** (equation 192)[37,343]. Tandem Claisen-2,3-Wittig-oxy-Cope rearrangement has been observed when **236** is treated under similar

30. Synthetic applications of allylsilanes and vinylsilanes

conditions to give **238** (equation 193)[343].

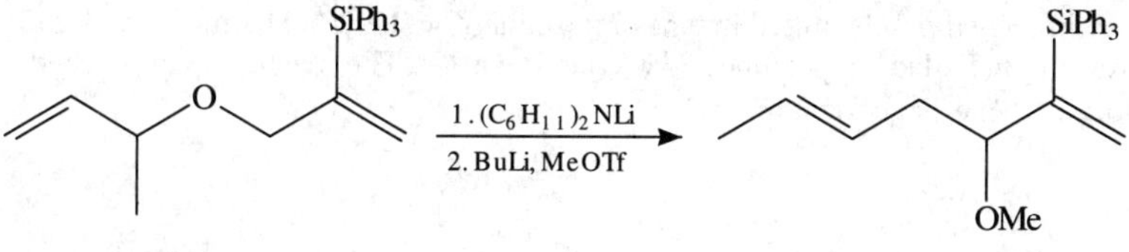

(234) → (235)　(192)

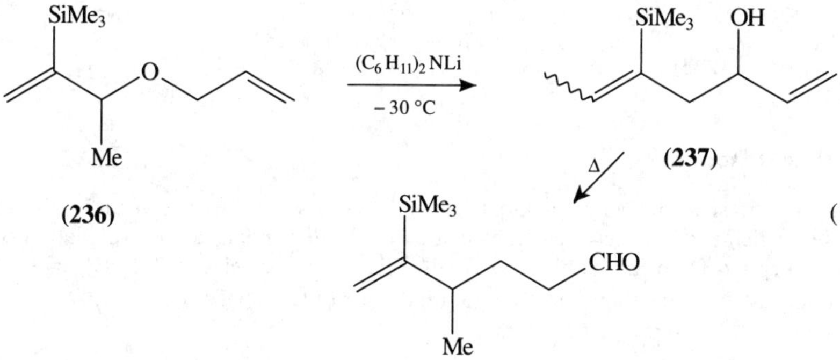

(236) → (237) → (238)　(193)

Claisen–Ireland rearrangement of **239** followed by DIBALH reduction gives a mixture of *syn* and *anti* isomers **240** (equation 194)[34–347]. In a similar manner, orthoester Claisen rearrangements of **241** give stereoselectively the corresponding chiral allylsilanes **242** (equation 195)[348].

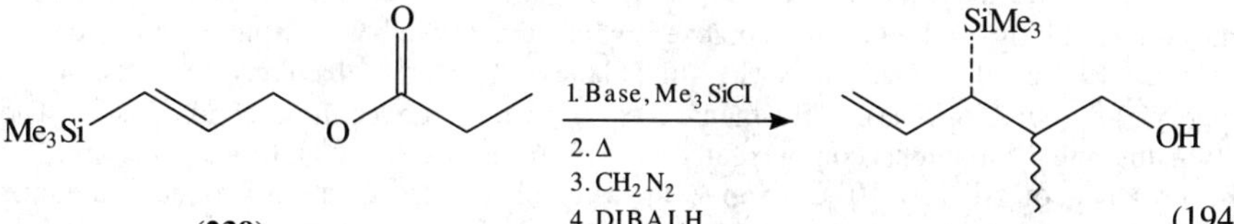

(239) → (240)　(194)

Base	syn/anti
LDA	1/2
LiH(SiMe$_3$)$_2$/HMPA	4/1

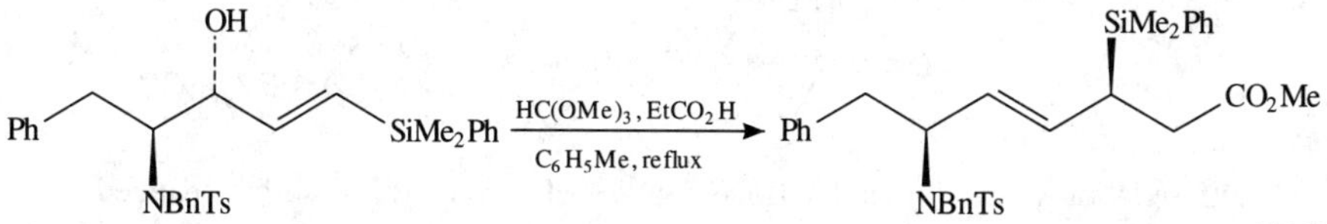

(241) → (242)　(195)

3. Miscellaneous rearrangements

β-Silylallylic alcohols undergo rhodium-catalyzed double bond migration to give the corresponding α-silyl-substituted ketones in excellent yields[349]. Thermolysis of **243** gives the corresponding silacyclopentenes **244** (equation 196). The reaction may proceed via an olefin to a carbene isomerization[325,350,351].

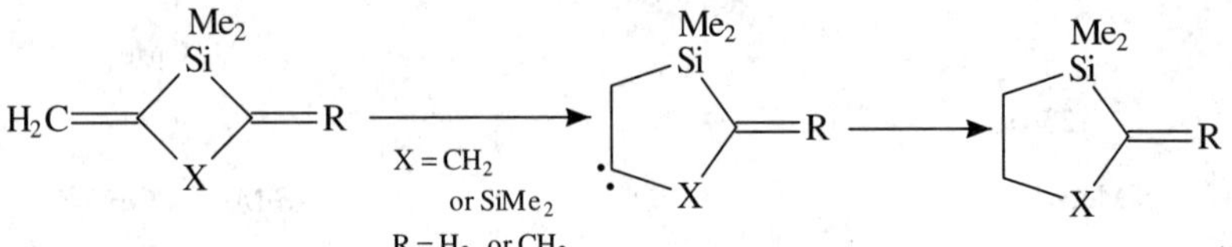

$$X = CH_2 \text{ or } SiMe_2$$
$$R = H_2, \text{ or } CH_2$$

(196)

G. Other Reactions

The use of silylallyl anion in organic synthesis has been extensive[24,29]. The regio- and stereochemistry of these reactions can be controlled. For example, alkylation of the anion generated from the corresponding allylsilane with an electrophile E^+ takes place selectively at the γ-position due to steric hindrance (equation 197)[352].

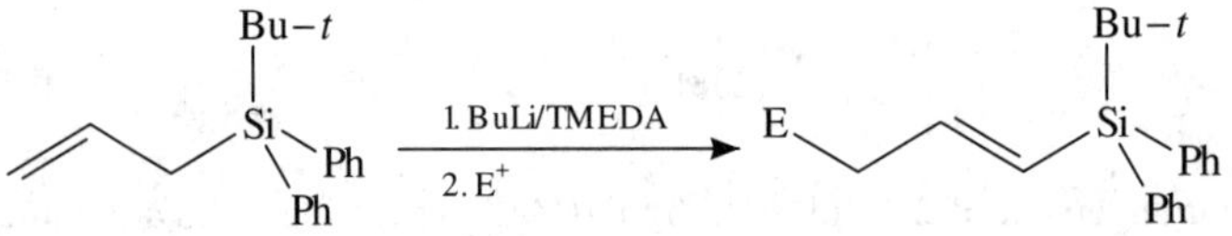

(197)

E = H, Me, 1° or 2° Alk, MeCH(OH), PhCH(OH)

3-Silyl-substituted allyl acetates react with $Fe_2(CO)_9$ under a CO atmosphere to give the corresponding η^2-$Fe(CO)_4$ complexes which are treated with a silyl enol ether in the presence of $BF_3 \cdot OEt_2$ to yield vinylsilanes **245** stereoselectively in moderate to good yields (equation 198). Allylstannane can also react under these conditions[353]. The silyl-substituted butadiene complex of tricarbonyliron is treated with acetyl chloride in the presence of $AlCl_3$ at 0 °C to give the Friedel–Crafts acylation products, dienone complexes, without desilylation[354,355].

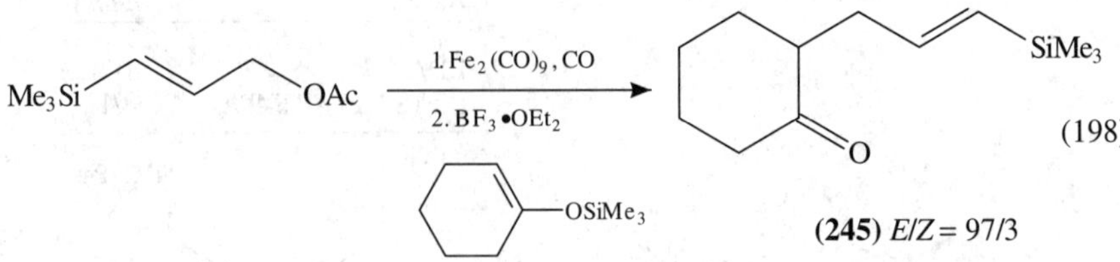

(198)

(245) E/Z = 97/3

Desulfonylation of β-silylvinyl sulfone **246** with Bu_3SnLi followed by treatment with aldehydes affords γ-hydroxyvinylsilanes **247** in good yields (equation 199)[356]. Reactions of trimethyl(2-nitrovinyl)silane with organometallic reagents give regioselectively the corresponding Michael adducts in excellent yields[357].

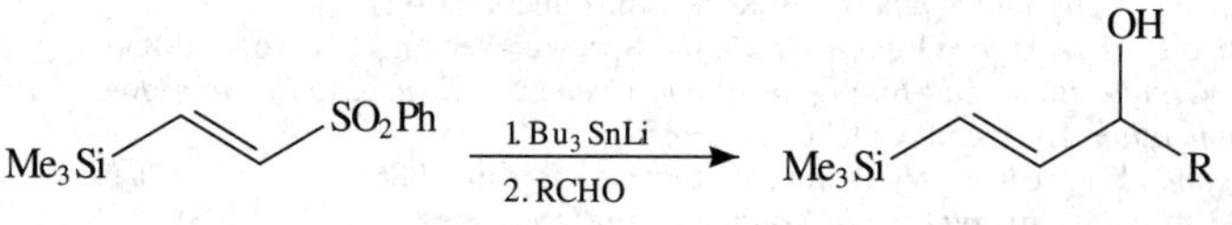
OH
SO_2Ph
1. Bu_3SnLi
2. RCHO
Me_3Si
Me_3Si
R

(199)

(246) **(247)**

R = 1° or 2° Alk

Hydroboration of **248** takes place at the terminal position which provides an approach to prepare ε-(trimethylsilyl)homogeraniol (equation 200)[358].

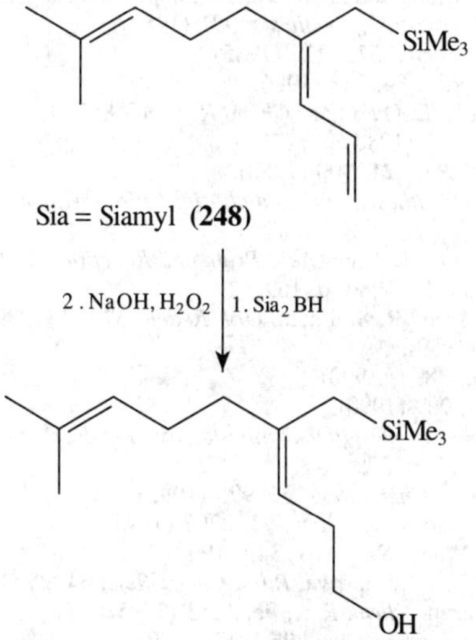
$SiMe_3$
Sia = Siamyl **(248)**
2. NaOH, H_2O_2
1. Sia_2BH
$SiMe_3$
OH

(200)

The double bond in vinylsilanes can be hydrogenated under various catalytic conditions. Ni/Et_3SiH[359] and rhodium[360] catalysts have been used recently (equation 201).

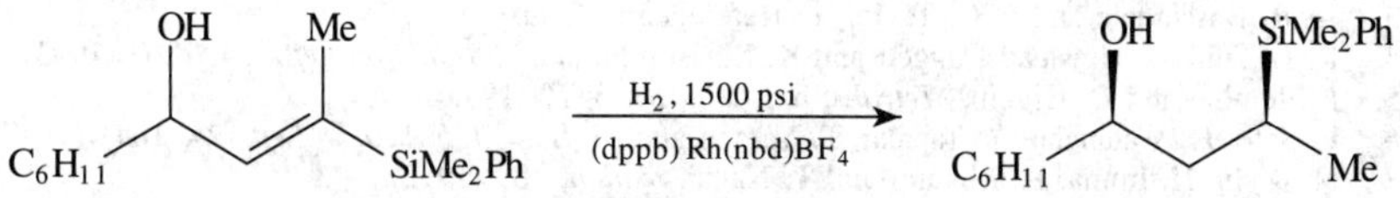
OH
Me
C_6H_{11}
$SiMe_2Ph$
H_2, 1500 psi
(dppb) Rh(nbd)BF_4
OH
$SiMe_2Ph$
C_6H_{11}
Me

dppb = $Ph_2PCH_2(CH_2)_2CH_2PPh_2$ (>500 : 1) (201)

nbd = 2,5-Norbornadiene

Vinylsilane works as an effective chain transfer agent in the Mo- or W-catalyzed polymerization of various substituted acetylenes. The molecular weight of the polymer can be controlled[361]. Perfluorovinylsilanes undergo fluoride ion promoted polymerization reaction[362]. The polymeric product which is insoluble in most of the solvents but HMPA at elevated temperature (200–230 °C) shows infrared absorption at 1630 cm^{-1} which indicates the presence of conjugated fluoroolefin moiety. Apparently some of the silyl groups have been eliminated during the polymerization process. The polymer forms metallic luster on the glass wall and shows a conductivity of $8.5 \times 10^{-9}\ \Omega^{-1}\ cm^{-1}$.

VI. REFERENCES

1. E. W. Colvin, *Silicon in Organic Synthesis*, Butterworth, London, 1981.
2. W. P. Weber, *Silicon Reagents for Organic Synthesis*, Springer-Verlag, New York, 1983.
3. H. Sakurai (Ed.), *Organosilicon and Bioorganosilicon Chemistry, Structure, Bonding, Reactivity and Synthetic Application*, Horwood, Chichester, 1985.
4. S. Pawlenko, *Organosilicon Chemistry*, Walter de Gruyter, Berlin, 1986.
5. E. W. Colvin, *Silicon Reagents in Organic Synthesis*, Academic Press, London, 1988.
6. S. Patai and Z. Rappoport (Eds.), *The Chemistry of Organic Silicon Compounds*, Parts 1 and 2, Wiley, Chichester, 1989.
7. T. H. Chan and I. Fleming, *Synthesis*, 761 (1979).
8. I. Fleming, in *Comprehensive Organic Chemistry*, Vol. 3 (Eds. D. H. R. Barton and W. D. Ollis), Chap. 13, Pergamon Press, Oxford, 1979.
9. H. Sakurai, *Pure Appl. Chem.*, **54**, 1 (1982).
10. P. D. Magnus, T. Sarkar and S. Djuric, in *Comprehensive Organometallic Chemistry*, Vol. 8 (Eds. G. Wilkinson, F. G. A. Stone and E. W. Abel), Chap. 48.4, Pergamon Press, New York, 1983.
11. Z. N. Parnes and G. I. Bolestova, *Synthesis*, 991 (1984).
12. H. Sakurai, *Pure Appl. Chem.*, **57**, 1759 (1985).
13. L. A. Paquette, *Chem. Rev.*, **86**, 733 (1986).
14. T. A. Blumenkopf and L. E. Overman, *Chem. Rev.*, **86**, 857 (1986).
15. D. Schinzer, *Synthesis*, 263 (1988).
16. A. Hosomi, *Acc. Chem. Res.*, **21**, 200 (1988).
17. G. Majetich, in *Organic Synthesis, Theory and Applications*, (Ed. T. Hudlicky), JAI Press, Greenwich, CT, 1989.
18. G. Majetich, in *Selectivities in Lewis Acid Promoted Reactions*, (Ed. D. Schinzer), Kluwer Academic Publishers, Dordrecht, 1989, p. 169.
19. I. Fleming, J. Dunoguès and R. Smithers, *Org. React.*, **37**, 57 (1989).
20. H. Sakurai, *Synlett*, 1 (1989).
21. D. J. Ager, *Org. React.*, **38**, 1 (1990).
22. T. K. Sarkar, *Synthesis*, 969 (1990).
23. J. S. Panek, in *Comprehensive Organic Synthesis*, Vol. 1 (Eds. B. M. Trost and I. Fleming), Pergamon Press, Oxford, 1991, p. 579.
24. T. H. Chan and D. Wang, *Chem. Rev.*, **92**, 995 (1992).
25. Y. Yamamoto and N. Asao, *Chem. Rev.*, **93**, 2207 (1993).
26. T.-Y. Luh and K.-T. Wong, *Synthesis*, 349 (1993).
27. M. D. Stadnichuk and T. I. Voropaeva, *Russ. Chem. Rev.*, **64**, 25 (1995).
28. M. Bols and T. Skrydstrup, *Chem. Rev.*, **95**, 1253 (1995).
29. T. H. Chan and D. Wang, *Chem. Rev.*, **95**, 1279 (1995).
30. C. E. Masse and J. S. Panek, *Chem. Rev.*, **95**, 1293 (1995).
31. K. A. Horn, *Chem. Rev.*, **95**, 1317 (1995).
32. E. Langkopf and D. Schinzer, *Chem. Rev.*, **95**, 1375 (1995).
33. A. R. Bassindale and P. G. Taylor, in Reference 6, Chap. 14.
34. R. L. Funk, J. Umstead-Daggett and K. M. Brummond, *Tetrahedron Lett.*, **34**, 2867 (1993).
35. I. Fleming and D. Higgins, *Tetrahedron Lett.*, **30**, 5777 (1989).
36. F. Sato, H. Watanabe, Y. Tanaka, T. Yamaji and M. Sato, *Tetrahedron Lett.*, **24**, 1041 (1983).
37. N. Kishi, H. Imma, K. Mikami and T. Nakai, *Synlett*, 189 (1992).
38. A. Degl'Innocenti, E. Stucchi, A. Capperucci, A. Mordini, G. Reginato and A. Ricci, *Synlett*, 329 (1992).
39. M. Mori, N. Watanabe, N. Kaneta and M. Shibasaki, *Chem. Lett.*, 1615 (1991).
40. W.-W. Weng and T.-Y. Luh, *J. Org. Chem.*, **57**, 2760 (1992).
41. E. Schaumann, A. Kirschning and F. Narjes, *J. Org. Chem.*, **56**, 717 (1991).
42. E. R. Koft and A. B. Smith, III, *J. Am. Chem. Soc.*, **104**, 2659 (1982).
43. A. B. Smith, III, Y. Yokoyama, D. M. Huryn and N. K. Dunlap, *Tetrahedron Lett.*, **28**, 3659 (1987).
44. A. B. Smith, III, Y. Yokoyama and N. K. Dunlap, *Tetrahedron Lett.*, **28**, 3663 (1987).
45. T. H. Chan and K. Koumaglo, *J. Organomet. Chem.*, **285**, 109 (1985).
46. K. Tamao, M. Akita, K. Maeda and M. Kumada, *J. Org. Chem.*, **52**, 1100 (1987).
47. J. Barluenga, L. J. Alverez-García and J. M. González, *Tetrahedron Lett.*, **36**, 2153 (1995).
48. M. Ochiai, K. Oshima and Y. Musaki, *J. Chem. Soc., Chem. Commun.*, 869 (1991).

49. H. Cerfontain, J. B. Kramer, R. M. Schonk and R. H. Bakker, *Recl. Trav. Chim. Pays-Bas*, **114**, 410 (1995).
50. H. Mayr, G. Gorath and B. Bauer, *Angew. Chem., Int. Ed. Engl.*, **33**, 788 (1994).
51. K. Saigo, K. Kudo, Y. Hashimoto, H. Kimoto and M. Hasegawa, *Chem. Lett.*, 941 (1990).
52. H. Pellissier and M. Santelli, *J. Chem. Soc., Chem. Commun.*, 607 (1995).
53. Y. Yamamoto, M. Ohno and S. Eguchi, *Chem. Lett.*, 525 (1995).
54. S. L. Xu, H. Xia and H. W. Moore, *J. Org. Chem.*, **56**, 6094 (1991).
55. G. A. Molander and J. P. Haar, Jr., *J. Am. Chem. Soc.*, **115**, 40 (1993).
56. A. P. Davis and S. C. Hegarty, *J. Am. Chem. Soc.*, **114**, 2745 (1992).
57. A. P. Davis and M. Jaspars, *J. Chem. Soc., Chem. Commun.*, 1176 (1990).
58. J. S. Panek and M. Yang, *J. Org. Chem.*, **56**, 5755 (1991).
59. J. S. Panek and M. Yang, *J. Am. Chem. Soc.*, **113**, 6594 (1991).
60. A. Hosomi, H. Hayashida and Y. Tominaga, *J. Org. Chem.*, **54**, 3254 (1989).
61. P. Mohr, *Tetrahedron Lett.*, **36**, 2453 (1995).
62. B. Hermans and L. Hevesi, *Tetrahedron Lett.*, **31**, 4363 (1990).
63. C. C. Silveira, G. L. Fiorin and A. L. Braga, *Tetrahedron Lett.*, **37**, 6085 (1996).
64. T. Fujiwara, T. Iwasaki, J. Miyagawa and T. Takeda, *Chem. Lett.*, 343 (1994).
65. T. K. Hollis, N. P. Robinson, J. Whelan and B. Bosnich, *Tetrahedron Lett.*, **34**, 4309 (1993).
66. G. Kaur, K. Manju and S. Trehan, *J. Chem. Soc., Chem. Commun.*, 581 (1996).
67. H. Monti and M. Féraud, *Synth. Commun.*, **26**, 1721 (1996).
68. H. Pellissier, L. Toupet and M. Santelli, *J. Org. Chem.*, **59**, 1709 (1994).
69. J. Li, T. Gallardo and J. B. White, *J. Org. Chem.*, **55**, 5426 (1990).
70. S. E. Denmark and N. G. Almstead, *J. Org. Chem.*, **59**, 5130 (1994).
71. B. Guyot, J. Pornet and L. Miginiac, *J. Organomet. Chem.*, **373**, 279 (1989).
72. G. Majetich, H. Nishidie and Y. Zhang, *J. Chem. Soc., Perkin Trans. 1*, 453 (1995).
73. M. Tsukazaki and V. Snieckus, *Tetrahedron Lett.*, **34**, 411 (1993).
74. Y. H. Kim and S. H. Kim, *Tetrahedron Lett.*, **36**, 6895 (1995).
75. F. D'Aniello, A. Mann, D. Mattii and M. Taddei, *J. Org. Chem.*, **59**, 3762 (1994).
76. L. F. Tietze, A. Dölle and K. Schiemann, *Angew. Chem., Int. Ed. Engl.*, **31**, 1372 (1992).
77. L. F. Tietze, K. Schiemann and C. Wegner, *J. Am. Chem. Soc.*, **117**, 5851 (1995).
78. K. Mikami, K. Kawamoto, T. -P. Loh and T. Nakai, *J. Chem. Soc., Chem. Commun.*, 1161 (1990).
79. S. Danishefsky and M. DeNinno, *Tetrahedron Lett.*, **26**, 823 (1985).
80. N. F. Jain, P. F. Cirillo, R. Pelletier and J. S. Panek, *Tetrahedron Lett.*, **36**, 8727 (1995).
81. J. S. Panek and M. Yang, *J. Am. Chem. Soc.*, **113**, 9868 (1991).
82. Y. Horiuchi, K. Oshima and K. Utimoto, *J. Org. Chem.*, **61**, 4483 (1996).
83. H. Monti, M. Afshari and G. Léandri, *J. Organomet. Chem.*, **486**, 69 (1995).
84. A. Mekhalfia and I. E. Markó, *Tetrahedron Lett.*, **32**, 4779 (1991).
85. A. Mekhalfia, I. E. Markó and H. Adams, *Tetrahedron Lett.*, **32**, 4783 (1991).
86. I. E. Markó and A. Mekhalfia, *Tetrahedron Lett.*, **33**, 1799 (1992).
87. I. E. Markó, A. Mekhalfia, D. J. Bayston and H. Adams, *J. Org. Chem.*, **57**, 2211 (1992).
88. I. E. Markó and D. J. Bayston, *Tetrahedron*, **50**, 7141 (1994).
89. I. E. Markó and D. J. Bayston, *Tetrahedron Lett.*, **34**, 6595 (1993).
90. M. Kira, K. Sato and H. Sakurai, *J. Am. Chem. Soc.*, **112**, 257 (1990).
91. M. Kira, K. Sato, H. Sakurai, M. Hada, M. Izawa and J. Ushio, *Chem. Lett.*, 387 (1991).
92. K. Sato, M. Kira and H. Sakurai, *J. Am. Chem. Soc.*, **111**, 6429 (1989).
93. M. Kira, K. Sato, K. Sekimoto, R. Gewald and H. Sakurai, *Chem. Lett.*, 281 (1995).
94. A. Hosomi, S. Kohra, K. Ogata, T. Yanagi and T. Tominaga, *J. Org. Chem.*, **55**, 2415 (1990).
95. Z. Y. Wei, D. Wang, J. S. Li and T. H. Chan, *J. Org. Chem.*, **54**, 5768 (1989).
96. S. E. Denmark, D. M. Coe, N. E. Pratt and B. D. Griedel, *J. Org. Chem.*, **59**, 6161 (1994).
97. S. Kobayashi and K. Nishio, *Tetrahedron Lett.*, **34**, 3453 (1993).
98. S. Kobayashi and K. Nishio, *J. Org. Chem.*, **59**, 6620 (1994).
99. S. Kobayashi and K. Nishio, *Chem. Lett.*, 1773 (1994).
100. D. Seyferth, J. Pornet and R. M. Weinstein, *Organometallics*, **1**, 1651 (1982).
101. D. Seyferth and J. Pornet, *J. Org. Chem.*, **45**, 1722 (1980).
102. A. Hosomi, M. Saito and H. Sakurai, *Tetrahedron Lett.*, **21**, 3783 (1980).
103. A. G. Myers, S. E. Kephart and H. Chen, *J. Am. Chem. Soc.*, **114**, 7922 (1992).
104. S. E. Denmark, B. D. Griedel, D. M. Coe and M. E. Schnute, *J. Am. Chem. Soc.* **116**, 7026 (1994).

105. K. Matsumoto, K. Oshima and K. Utimoto, *J. Synth. Org. Chem. Jpn.*, **54**, 289 (1996).
106. K. Okada, K. Matsumoto, K. Oshima and K. Utimoto, *Tetrahedron Lett.*, **36**, 8067 (1995).
107. S. Kashimura, M. Ishifune, Y. Murai and T. Shono, *Chem. Lett.*, 309 (1996).
108. D. Schinzer and J. Kabbara, *Synlett*, 766 (1992).
109. T. -L. Ho and F. -S. Liang, *Chem. Commun.*, 1887 (1996).
110. M. Ohno, Y. Yamamoto and S. Eguchi, *J. Chem. Soc., Perkin Trans. 1*, 2272 (1991).
111. L. -R. Pan and T. Tokoroyama, *Chem. Lett.*, 1999 (1990).
112. L. -R. Pan and T. Tokoroyama, *Tetrahedron Lett.*, **33**, 1469 (1992).
113. A. G. Schultz and H. Lee, *Tetrahedron Lett.*, **33**, 4397 (1992).
114. L. F. Tietze and M. Ruther, *Chem. Ber.*, **123**, 1387 (1990).
115. L. F. Tietze and C. Schünke, *Angew. Chem., Int. Ed. Engl.*, **34**, 1731 (1995).
116. G. Majetich, K. Hull and R. Desmond, *Tetrahedron Lett.*, **26**, 2751 (1985).
117. D. Schinzer and K. Ringe, *Synlett*, 463 (1994).
118. G. Majetich, D. Lowery and V. Khetani, *Tetrahedron Lett.*, **31**, 51 (1990).
119. S. Laschat and H. Kunz, *J. Org. Chem.*, **56**, 5883 (1991).
120. S. Laschat and H. Kunz, *Synlett*, 51 (1990).
121. M. Kira, T. Hino and H. Sakurai, *Chem. Lett.*, 277 (1991).
122. E. C. Roos, H. Hiemstra, W. N. Speckamp, B. Kaptein, J. Kamphuis and H. E. Schoemaker, *Synlett*, 451 (1992).
123. H. H. Mooiweer, H. Hiemstra and W. N. Speckamp, *Tetrahedron*, **45**, 4627 (1989).
124. H. H. Mooiweer, H. Hiemstra and W. N. Speckamp, *Tetrahedron*, **47**, 3451 (1991).
125. J. Mittendorf, H. Hiemstra and W. N. Speckamp, *Tetrahedron*, **46**, 4049 (1990).
126. M. Rubiralta, A. Diez and D. Miguel, *Synth. Commun.*, **22**, 359 (1992).
127. J. M. Takacs and J. J. Weidner, *J. Org. Chem.*, **59**, 6480 (1994).
128. M. Franciotti, A. Mann, A. Mordini and M. Taddei, *Tetrahedron Lett.*, **34**, 1355 (1993).
129. L. E. Overman and K. L. Bell, *J. Am. Chem. Soc.*, **103**, 1851 (1981).
130. L. E. Overman and N. -H. Lin, *J. Org. Chem.*, **50**, 3670 (1985).
131. L. E. Overman, K. L. Bell and F. Ito, *J. Am. Chem. Soc.*, **106**, 4192 (1984).
132. L. E. Overman and A. L. Robichaud, *J. Am. Chem. Soc.*, **111**, 300 (1989).
133. L. E. Overman and R. M. Burk, *Tetrahedron Lett.*, **25**, 5739 (1984).
134. C. J. Flann and L. E. Overman, *J. Am. Chem. Soc.*, **109**, 6115 (1987).
135. M. C. McIntosh and S. M. Weinreb, *J. Org. Chem.*, **56**, 5010, (1991).
136. L. Vidal, J. Royer and H. -P. Husson, *Tetrahedron Lett.*, **36**, 2991 (1995).
137. G. A. Olah, D. S. Van Vliet, Q. Wang and G. K. Surya Prakash, *Synthesis*, 159 (1995).
138. K. -T. Kang, J. C. Lee and J. S. U, *Tetrahedron Lett.*, **33**, 4953 (1992).
139. E. Schaumann and B. Mergardt, *J. Chem. Soc., Perkin Trans. 1*, 1361 (1989).
140. A. Barbero, P. Cuadrodo, I. Fleming, A. M. González, F. J. Pulido and A. Sánchez, *J. Chem. Soc., Perkin Trans. 1*, 1525 (1995).
141. F. Babudri, V. Fiandanese, G. Marchese and F. Naso, *J. Chem. Soc., Chem. Commun.*, 237 (1991).
142. F. Babudri, V. Fiandanese and F. Naso, *J. Org. Chem.*, **56**, 6245 (1991).
143. G. Picotin and P. Miginiac, *Tetrahedron Lett.*, **29**, 5897 (1988).
144. A. Kirschning, F. Narjes and E. Schaumann, *Justus Liebigs Ann. Chem.*, 933 (1991).
145. M. Kira, T. Hino and H. Sakurai, *Chem. Lett.*, 555 (1992).
146. T. Yokozawa, K. Furuhashi and H. Natsume, *Tetrahedron Lett.*, **36**, 5243 (1995).
147. L. E. Overman and P. A. Renhowe, *J. Org. Chem.*, **59**, 4138 (1994).
148. G. A. Molander and S. W. Andrews, *J. Org. Chem.*, **54**, 3114 (1989).
149. S. C. Bergmeier and P. P. Seth, *Tetrahedron Lett.*, **36**, 3793 (1995).
150. T. Hirao, T. Fujii and Y. Ohshiro, *Tetrahedron*, **50**, 10207 (1994).
151. T. V. Lee, F. S. Roden and J. R. Porter, *J. Chem. Soc., Perkin Trans 1*, 2139 (1989).
152. C. M. Hudson, M. R. Marzabadi, K. D. Moeller and D. G. New, *J. Am. Chem. Soc.*, **113**, 7372 (1991).
153. S. D. Burke and D. N. Deaton, *Tetrahedron Lett.*, **32**, 4651 (1991).
154. S. D. Burke, K. Shankaran and M. J. Helber, *Tetrahedron Lett.*, **32**, 4655 (1991).
155. S. D. Burke, M. E. Kort, M. S. S. Strickland, H. M. Organ and L. A. Silks, III, *Tetrahedron Lett.*, **35**, 1503 (1994).
156. J. Tsuji, *Palladium Reagents and Catalysts*, Wiley, Chichester, 1995
157. R. J. P. Corriu, R. Perz and C. Reys, *Tetrahedron*, **39**, 999 (1983).

158. G. G. Furin, O. A. Vyazankina, B. A. Gostevsky and N. S. Vyazankin, *Tetrahedron*, **44**, 2675 (1988).
159. J. Yoshida, K. Tamao, H. Yamamoto, T. Kakui, T. Uchida and M. Kumada, *Organometallics*, **1**, 542 (1982).
160. W. P. Weber, R. A. Felix, A. K. Willard and K. E. Koenig, *Tetrahedron Lett.*, 4701 (1971).
161. A. Hallberg and C. Westerlund, *Chem. Lett.*, 1993 (1982).
162. N. M. Chistovalova, I. S. Akhrem, E. V. Reshetova and M. E. Vol'pin, *Izvest. Akad. Nauk SSR, Ser. Khim.*, 2342 (1984); *Chem. Abstr.*, **102**, 78961 (1985).
163. K. Kikukawa, K. Ikenaga, F. Wada and T. Matsuda, *Tetrahedron Lett.*, **25**, 5789 (1984).
164. K. Ikenaga, K. Kikukawa and T. Matsuda, *J. Org. Chem.*, **52**, 1276 (1987).
165. K. Kikukawa, K. Ikenaga, F. Wada and T. Matsuda, *Chem. Lett.*, 1337 (1983).
166. K. Ikenaga, K. Kikukawa and T. Matsuda, *J. Chem. Soc., Perkin Trans. 1*, 1959 (1986).
167. M. E. Garst and B. J. McBride, *J. Org. Chem.*, **54**, 249 (1989).
168. K. Karabelas and A. Hallberg, *J. Org. Chem.*, **54**, 1773 (1989).
169. Y. Hatanaka and T. Hiyama, *J. Org. Chem.*, **53**, 918 (1988).
170. Y. Hatanaka and T. Hiyama, *J. Org. Chem.*, **54**, 268 (1989).
171. Y. Hatanaka and T. Hiyama, *Tetrahedron Lett.*, **31**, 2719 (1990).
172. Y. Hatanaka and T. Hiyama, *Chem. Lett.*, 2049 (1989).
173. M. Hojo, C. Murakami, H. Aihara, E. -i. Komori, S. Kohra, Y. Tominaga and A. Hosomi, *Bull. Soc. Chem. Fr.*, **132**, 499 (1995).
174. H. Matsuhashi, Y. Hatanaka, M. Kuroboshi and T. Hiyama, *Tetrahedron Lett.*, **36**, 1539 (1995).
175. B. M. Trost and E. Keinan, *Tetrahedron Lett.*, **21**, 2595 (1980).
176. Y. Hatanaka, Y. Ebina and T. Hiyama, *J. Am. Chem. Soc.*, **113**, 7075 (1991).
177. Y. Hatanaka, K. -i. Goda and T. Hiyama, *Tetrahedron Lett.*, **35**, 6511 (1994).
178. Y. Hatanaka and T. Hiyama, *J. Am. Chem. Soc.*, **112**, 7793 (1990).
179. Y. Hatanaka, K. -i. Goda and T. Hiyama, *Tetrahedron Lett.*, **35**, 1279 (1994).
180. L. F. Tietze and R. Schimpf, *Angew. Chem., Int. Ed. Engl.*, **33**, 1089 (1994).
181. L. F. Tietze and R. Schimpf, *Chem. Ber.*, **127**, 2235 (1994).
182. G. D. Daves, Jr. and A. Hallberg, *Chem. Rev.*, **89**, 1433 (1989).
183. K. Tamao, K. Kobayashi and Y. Ito, *Tetrahedron Lett.*, **30**, 6051 (1989).
184. A. M. Castaño and J. -E. Bäckvall, *J. Am. Chem. Soc.*, **117**, 560 (1995).
185. Y. Matano, M. Yoshimune and H. Suzuki, *Tetrahedron Lett.*, **36**, 7475 (1995).
186. S. R. Wilson and C. E. Augelli-Szafran, *Tetrahedron*, **44**, 3983 (1988).
187. M. Ochiai, M. Arimoto and E. Fujita, *Tetrahedron Lett.*, **22**, 4491 (1981).
188. M. Ochiai, E. Fujita, M. Arimoto and H. Yamaguchi, *Chem. Pharm. Bull.*, **33**, 41 (1985).
189. K. Lee, D. Y. Kim and D. Y. Oh, *Tetrahedron Lett.*, **29**, 667 (1988).
190. N. X. Hu, Y. Aso, T. Otsubo and F. Ogura, *Tetrahedron Lett.*, **29**, 4949 (1988).
191. Y. Ikeda and H. Yamamoto, *Bull. Chem. Soc. Jpn.*, **59**, 657 (1986).
192. G. M. Farinola, V. Fiandanese, L. Mazzone and F. Naso, *J. Chem. Soc., Chem. Commun.*, 2523 (1995).
193. L. Xue, L. Lu, S. Pederson, Q. Liu, R. Narske and D. J. Burton, *Tetrahedron Lett.*, **37**, 1921 (1996).
194. Y. Wakatsuki,, H. Yamazaki, M. Nakano and Y. Yamamoto, *J. Chem. Soc., Chem. Commun.*, 703 (1991).
195. T. Hirao, M. Sakaguchi, T. Ishikawa and I. Ikeda, *Synth. Commun.*, **25**, 2579 (1995).
196. A. S. Guram, *Synlett*, 259 (1993).
197. A. Capperucci, M. C. Ferrara, A. Degl'Innocenti, B. F. Bonini, G. Mazzanti, P. Zani and A. Ricci, *Synlett*, 880 (1992).
198. L. M. Atagi, S. C. Critchlow and J. M. Mayer, *J. Am. Chem. Soc.*, **114**, 1483 (1992).
199. J.-i. Yoshida, T. Murata and S. Isoe, *Tetrahedron Lett.*, **28**, 211 (1987).
200. P. F. Hurdrlik, A. M. Kassim, E. L. O. Agwaramgbo, K. A. Doonquah, R. R. Roberts and A. M. Hudrlik, *Organometallics*, **12**, 2367 (1993).
201. F. Sato, Y. Tanaka and M. Sato, *J. Chem. Soc., Chem. Commun.*, 165 (1983).
202. M. Kosugi, H. Kurata, K. -i. Kawata and T. Migita, *Chem. Lett.*, 1327 (1991).
203. C. Chatgilialoglu, C. Ferreri, M. Ballestri and D. P. Curran, *Tetrahedron Lett.*, **37**, 6387 (1996).
204. K. Nishitani, Y. Harada, Y. Nakamura, K. Yokoo and K. Yamakawa, *Tetrahedron Lett.*, **35**, 7809 (1994).
205. L. E. Overman and A. S. Thompson, *J. Am. Chem. Soc.*, **110**, 2248 (1988).

206. D. Berger, L. E. Overman and P. A. Renhowe, *J. Am. Chem. Soc.*, **115**, 9305 (1993).
207. K. Miura, S. Okajima, T. Hondo and A. Hosomi, *Tetrahedron Lett.*, **36**, 1483 (1995).
208. K. Miura, T. Hondo, S. Okajima and A. Hosomi, *Tetrahedron Lett.*, **37**, 487 (1996).
209. J. -P. Pillot, J. Dunoguès and R. Calas, *J. Chem. Res. (S)*, 268 (1977).
210. R. F. Cunico, *Synth. Commun.*, **18**, 917 (1988).
211. J. R. Hwu, K. -L. Chen, S. Ananthan and H. V. Patel, *Organometallics*, **15**, 499 (1996).
212. H. Okumoto and J. Tsuji, *Synth. Commun.*, **12**, 1015 (1982).
213. S. Kim, I. Y. Lee, J. -Y. Yoon and D. H. Oh, *J. Am. Chem. Soc.*, **118**, 5138 (1996).
214. P. Jankowski, M. Masnuk and J. Wicha, *Synlett*, 866 (1995).
215. J. R. Hwu, B. -L. Chen and S. -S. Shiao, *J. Org. Chem.*, **60**, 2448 (1995).
216. W. -J. Koot, R. van Ginkel, M. Kranenburg, H. Hiemstra, S. Louwrier, M. J. Moolenaar and W. N. Speckamp, *Tetrahedron Lett.*, **32**, 401 (1991).
217. J. M. Dener and D. J. Hart, *Tetrahedron*, **44**, 7037 (1988).
218. W. -W. Weng and T. -Y. Luh, *J. Org. Chem.*, **58**, 5574 (1993).
219. H. Sawada and M. Nakayama, *J. Chem. Soc., Chem. Commun.*, 677 (1991).
220. J. Kabbara, C. Hoffmann and D. Schinzer, *Synthesis*, 299, (1995).
221. F. L. Koerwitz, G. B. Hammond and D. F. Wiemer, *J. Org. Chem.*, **54**, 743 (1989).
222. J. A. Soderquist and J. C. Colberg, *Tetrahedron Lett.*, **35**, 27 (1994).
223. J. A. Soderquist and G. León-Colón, *Tetrahedron Lett.*, **32**, 43 (1991).
224. K. -T. Kang, S. S. Kim and J. C. Lee, *Tetrahedron Lett.*, **32**, 4341 (1991).
225. P. Knochel and J. F. Normant, *Tetrahedron Lett.*, **25**, 4383 (1984).
226. N. M. Chistovalova, I. S. Akhrem, V. F. Sizoi, V. V. Bardin and M. E. Vol'pin, *Izvest. Akad. Nauk SSR, Ser. Khim.*, 1180 (1988); *Chem. Abstr.*, **110**, 95337 (1989).
227. K. Karabelas and A. Hallberg, *Tetrahedron Lett.*, **26**, 3131 (1985).
228. K. Karabelas and A. Hallberg, *J. Org. Chem.*, **51**, 5286 (1986).
229. K. Karabelas and A. Hallberg, *J. Org. Chem.*, **53**, 4909 (1988).
230. H. Yamashita, B. L. Roan and M. Tanaka, *Chem. Lett.*, 2175 (1990).
231. K. Kikukawa, K. Ikenaga, K. Kono, K. Toritani, F. Wada and T. Matsuda, *J. Organomet. Chem.*, **270**, 277 (1984).
232. S. Murai, F. Kakiuchi, S. Sekine, Y. Tanaka, A. Kamatani, M. Sonoda and N. Chatani, *Nature*, **369**, 529 (1993).
233. S. Murai, F. Kakiuchi, S. Sekine, Y. Tanaka, A. Kamatani, M. Sonoda and N. Chatani, *Pure Appl. Chem.*, **66**, 1527 (1994).
234. S. Murai, *J. Synth. Org. Chem. Jpn.*, **52**, 993 (1994).
235. F. Kakiuchi, S. Sekine, Y. Tanaka, A. Kamatani, M. Sonoda, N. Chatani and S. Murai, *Bull. Chem. Soc., Jpn.*, **68**, 62 (1995).
236. M. Sonoda, F. Kakiuchi, N. Chatani and S. Murai, *J. Organomet. Chem.*, **504**, 151 (1995).
237. M. Sonoda, F. Kakiuchi, A. Kamatani, N. Chatani and S. Murai, *Chem. Lett.*, 109 (1996).
238. B. M. Trost, K. Imi and I. W. Davies, *J. Am. Chem. Soc.*, **117**, 5371 (1995).
239. F. Kakiuchi, Y. Tanaka, T. Sato, N. Chatani and S. Murai, *Chem. Lett.*, 679 (1995).
240. F. Kakiuchi, Y. Yamamoto, N. Chatani and S. Murai, *Chem. Lett.*, 681 (1995).
241. F. Kakiuchi, M. Yamauchi, N. Chatani and S. Murai, *Chem. Lett.*, 111 (1996).
242. H. Guo, M. A. Tapsak and W. P. Weber, *Macromolecules*, **28**, 4714 (1995).
243. H. Guo, G. Wang, M. A. Tapsak and W. P. Weber, *Macromolecules*, **28**, 5686 (1995).
244. R. Takeuchi, N. Ishii and N. Sato, *J. Chem. Soc., Chem. Commun.*, 1247 (1991).
245. R. Takeuchi, N. Ishii, M. Sugiura and N. Sato, *J. Org. Chem.*, **57**, 4189 (1992).
246. R. Takeuchi and N. Sato, *J. Organomet. Chem.*, **393**, 1 (1990).
247. M. M. Doyle, W. R. Jackson and P. Perlmutter, *Tetrahedron Lett.*, **30**, 233 (1989).
248. C. M. Crudden and H. Alper, *J. Org. Chem.*, **59**, 3091 (1994).
249. J. A. Anhaus, W. Clegg, S. P. Collingwood and V. C. Gibson, *J. Chem. Soc., Chem. Commun.*, 1720 (1991).
250. W. E. Crowe, D. R. Goldberg and Z. J. Zhang, *Tetrahedron Lett.*, **37**, 2117 (1996).
251. Z. Foltynowicz and B. Marciniec, *J. Organomet. Chem.*, **376**, 15 (1989).
252. M. Berglund, C. Andersson and R. Larsson, *J. Organomet. Chem.*, **292**, C15 (1985).
253. P. Mohr, *Tetrahedron Lett.*, **33**, 2455 (1992).
254. P. J. Murphy and G. Procter, *Tetrahedron Lett.*, **31**, 1059 (1990).
255. W. Adam, F. Prichtl, M. J. Richter and A. K. Smerz, *Tetrahedron Lett.*, **36**, 4991 (1995).
256. W. Adam and M. Richter, *Tetrahedron Lett.*, **33**, 3461 (1992).

257. Y. -K. Han and L. A. Paquette, *J. Org. Chem.*, **44**, 3731 (1979).
258. L. A. Paquette, R. A. Galemmo, Jr., J. -C. Caille and R. S. Valpey, *J. Org. Chem.*, **51**, 686 (1986).
259. S. G. Davies, C. J. R. Hedgecock and J. M. McKenna, *Tetrahedron: Asymmetry*, **6**, 2507 (1995).
260. J. M. Muchowski, R. Naef and M. L. Maddox, *Tetrahedron Lett.*, **26**, 5375 (1985).
261. J. -i. Yoshida, S. Nakatani and S. Isoe, *J. Org. Chem.*, **54**, 5655 (1989).
262. S. -S. P. Chou, H. -L. Kuo, C. -J. Wang, C. -Y. Tsai and C. -M. Sun, *J. Org. Chem.*, **54**, 868 (1989).
263. Y. Ukaji, A. Yoshida and T. Fujisawa, *Chem. Lett.*, 157 (1990).
264. K. Tamao, E. Nakajo and Y. Ito, *J. Org. Chem.*, **52**, 4412 (1987).
265. J. A. Soderquist and B. Santiago, *Tetrahedron Lett.*, **30**, 5693 (1989).
266. M. A. Loreto, P. A. Tardella and D. Tofani, *Tetrahedron Lett.*, **36**, 8295 (1995).
267. S. Fioravanti, M. A. Loreto, L. Pellacani, S. Raimondi and P. A. Tardella, *Tetrahedron Lett.*, **34**, 4101 (1993).
268. R. S. Atkinson, M. J. Grimshire and B. J. Kelly, *Tetrahedron*, **45**, 2875 (1989).
269. R. S. Atkinson, M. P. Coogan and I. S. T. Lochrie, *Tetrahedron Lett.*, **37**, 5179 (1996).
270. E. Azzari, C. Faggi, N. Gelsomini and M. Taddei, *J. Org. Chem.*, **55**, 1106 (1990).
271. J. S. Panek and P. F. Cirillo, *J. Am. Chem. Soc.*, **112**, 4873 (1990).
272. P. F. Cirillo and J. S. Panek, *J. Org. Chem.*, **59**, 3055 (1994).
273. J. A. Soderquist, A. M. Rane and C. J. López, *Tetrahedron Lett.*, **34**, 1893 (1993).
274. R. A. Ward and G. Procter, *Tetrahedron*, **51**, 12821 (1995).
275. M. J. Daly, R. A. Ward, D. F. Thompson and G. Procter, *Tetrahedron*, **51**, 7545 (1995).
276. M. J. Daly and G. Procter, *Tetrahedron*, **51**, 7549 (1995).
277. R. W. Ward and G. Procter, *Tetrahedron Lett.*, **33**, 3363 (1992).
278. W. Adam and M. J. Richter, *Synthesis*, 176 (1994).
279. N. Shimizu, F. Shibata, S. Imazu and Y. Tsuno, *Chem. Lett.*, 1071 (1987).
280. K. Kato and T. Mukaiyama, *Chem. Lett.*, 2233 (1989).
281. W. -W. Weng and T. -Y. Luh, *J. Chem. Soc., Perkin Trans. 1*, 2687 (1993).
282. M. Mitani, Y. Kobayashi and K. Koyama, *J. Chem. Soc., Perkin Trans. 1*, 653 (1995).
283. J. S. Panek, R. M. Garbaccio and N. F. Jain, *Tetrahedron Lett.*, **35**, 6453 (1994).
284. R. L. Danheiser and H. Sard, *Tetrahedron Lett.*, **24**, 23 (1983).
285. G. P. Brengel, C. Rithner and A. I. Meyers, *J. Org. Chem.*, **59**, 5144 (1994).
286. M. Hojo, K. Tomita, Y. Hirohara and A. Hosomi, *Tetrahedron Lett.*, **34**, 8123 (1993).
287. H. -J. Knölker, G. Baum and R. Graf, *Angew. Chem., Int. Ed. Engl.*, **33**, 1612 (1994).
288. H. Monti, G. Audran, J. -P. Monti and G. Léandri, *Synlett*, 403 (1994).
289. H. Monti, G. Audran, G. Léandri and J. -P. Monti, *Tetrahedron Lett.*, **35**, 3073 (1994).
290. R. Huisgen and R. Brückner, *Tetrahedron Lett.*, **31**, 2553 (1990).
291. Y. Kubo, E. Taniguchi and T. Araki, *Heterocycles*, **29**, 1857 (1989).
292. M. T. Crimmins and L. E. Guise, *Tetrahedron Lett.*, **35**, 1657 (1994).
293. K. Nakanishi, K. Mizuno and Y. Otsuji, *J. Chem. Soc., Chem. Commun.*, 90 (1991).
294. T. Uyehara, M. Yuuki, H. Masaki, M. Matsumoto, M. Ueno and T. Sato, *Chem. Lett.*, 789 (1995).
295. E. W. Colvin and M. Monteith, *J. Chem. Soc., Chem. Commun.*, 1230 (1990).
296. C. Nativi, E. Perrotta, A. Ricci and M. Taddei, *Tetrahedron Lett.*, **32**, 2265 (1991).
297. T. Akiyama and M. Kirino, *Chem. Lett.*, 723 (1995).
298. H. -J. Knölker, P. G. Jones and J. -B. Pannek, *Synlett*, 429 (1990).
299. K. Ohkata, K. Ishimaru, Y. -g. Lee and K. -y. Akiba, *Chem. Lett.*, 1725 (1990).
300. Y. -G. Lee, K. Ishimaru, H. Iwasaki, K. Ohkata and K. -y. Akiba, *J. Org. Chem.*, **56**, 2058 (1991).
301. H. -J. Knölker, P. G. Jones, J. -B. Pannek and A. Weinkauf, *Synlett*, 147 (1991).
302. B. B. Snider and Q. Zhang, *J. Org. Chem.*, **56**, 4908 (1991).
303. R. L. Danheiser, B. R. Dixon and R. W. Gleason, *J. Org. Chem.*, **57**, 6094 (1992).
304. R. L. Danheiser, T. Takahashi, B. Bertók and B. R. Dixon, *Tetrahedron Lett.*, **34**, 3845 (1993).
305. H. -J. Knölker and R. Graf, *Tetrahedron Lett.*, **34**, 4765 (1993).
306. H. -J. Knölker, N. Foitzik, H. Goesmann and R. Graf, *Angew. Chem., Int. Ed. Engl.*, **32**, 1081 (1993).
307. M. -J. Wu and J. -Y. Yeh, *Tetrahedron*, **50**, 1073 (1994).
308. H. -J. Knölker and R. Graf, *Synlett*, 131 (1994).

309. H. -J. Knölker and G. Wanzl, *Synlett*, 378 (1995) and references cited therein.
310. J. S. Panek and N. F. Jain, *J. Org. Chem.*, **58**, 2345 (1993).
311. J. S. Panek and N. F. Jain, *J. Org. Chem.*, **59**, 2674 (1994).
312. A. Stahl, E. Steckhan and M. Nieger, *Tetrahedron Lett.*, **35**, 7371 (1994).
313. S. -i. Kiyooka, Y. Shiomi, H. Kira, Y. Kaneko and S. Tanimori, *J. Org. Chem.*, **59**, 1958 (1994).
314. T. Akiyama, T. Yasusa, K. Ishikawa and S. Ozaki, *Tetrahedron Lett.*, **35**, 8401 (1994).
315. T. Akiyama, K. Ishikawa and S. Ozaki, *Chem. Lett.*, 627 (1994).
316. J. S. Panek and R. Beresis, *J. Org. Chem.*, **58**, 809 (1993).
317. K. Shanmuganthan, L. G. French and B. L. Jensen, *Tetrahedron: Asymmetry*, **5**, 797 (1994).
318. J. S. Panek and R. T. Beresis, *J. Am. Chem. Soc.*, **115**, 7898 (1993).
319. G. Stork, T. -Y. Chan and G. A. Breault, *J. Am. Chem. Soc.*, **114**, 7578 (1992).
320. S. M. Sieburth and L. Fensterbank, *J. Org. Chem.*, **57**, 5279 (1992).
321. D. F. Taber, R. S. Bhamidipati and L. Yet, *J. Org. Chem.*, **60**, 5537 (1995).
322. S. Ebeling, D. Matthies and D. McCarthy, *Phosphorus, Sulfur and Silicon*, **60**, 265 (1991).
323. K. Tamao, K. Kobayashi and Y. Ito, *J. Am. Chem. Soc.*, **111**, 6478 (1989).
324. R. -M. Chen, W. W. Weng and T. -Y. Luh, *J. Org. Chem.*, **60**, 3272 (1995).
325. K. J. Shea, A. J. Staab and K. S. Zandi, *Tetrahedron Lett.*, **32**, 2715 (1991).
326. A. Rumbo, L. Castedo, Mouriño and J. L. Mascareñas, *J. Org. Chem.*, **58**, 5585 (1993).
327. M. Hojo, C. Murakami, H. Aihara, K. Tomita, K. Miura and A. Hosomi, *J. Organomet. Chem.*, **499**, 155 (1995).
328. S. H. Yeon, B. W. Lee, B. R. Yoo, M. -Y. Suk and I. N. Jung, *Organometallics*, **14**, 2361 (1995).
329. S. H. Yeon, J. S. Han, E. Hong, Y. Do and I. N. Jung, *J. Organomet. Chem.*, **499**, 159 (1995).
330. N. Asao, E. Yoshikawa and Y. Yamamoto, *J. Org. Chem.*, **61**, 4874 (1996).
331. G. Audran, H. Monti, G. Léandri and J. -P. Monti, *Tetrahedron Lett.*, **34**, 3417 (1993).
332. K. Mikami and S. Matsukawa, *Tetrahedron Lett.*, **35**, 3133 (1994).
333. S. Imazu, N. Shimizu and Y. Tsuno, *Chem. Lett.*, 1845 (1990).
334. K. Mikami, T. -P. Loh and T. Nakai, *J. Am. Chem. Soc.*, **112**, 6737 (1990).
335. K. Mikami, M. Tarada, M. Shimizu and T. Nakai, *J. Synth. Org. Chem. Jpn.*, **48**, 292 (1990).
336. B. B. Snider, *Acc. Chem. Res.*, **13**, 426 (1980).
337. H. M. R. Hoffmann, *Angew. Chem., Int. Ed. Engl.*, **8**, 556 (1969).
338. J. K. Whitesell, *Acc. Chem. Res.*, **18**, 280 (1985).
339. J. Dubac and A. Laporterie, *Chem. Rev*, **87**, 319 (1987).
340. J. Robertson, G. O'Connor and D. S. Middleton, *Tetrahedron Lett.*, **37**, 3411 (1996).
341. W. Adam and M. Richter, *Chem. Ber.*, **125**, 243 (1992).
342. M. Orfanopoulos, Y. Elemes and M. Stratakis, *Tetrahedron Lett.*, **31**, 5775 (1978).
343. K. Mikami, N. Kishi and T. Nakai, *Chem. Lett*, 1643 (1982).
344. S. Marumoto and I. Kuwajima, *J. Am. Chem. Soc.*, **115**, 9021 (1993).
345. S. Marumoto and I. Kuwajima, *Chem. Lett.*, 1421 (1992).
346. M. A. Sparks and J. S. Panek, *J. Org. Chem.*, **56**, 3431 (1991).
347. M. A. Sparks and J. S. Panek, *Tetrahedron Lett.*, **32**, 4085 (1991).
348. M. Heneghan and G. Procter, *Synlett*, 489 (1992).
349. R. Takeuchi, S. Nitta and D. Watanabe, *J. Org. Chem.*, **60**, 3045 (1995).
350. T. J. Barton, J. Lin, S. Ijadi-Maghsoodi, M. D. Power, X. Zhang, Z. Ma, H. Shimizu and M. S. Gordon, *J. Am. Chem. Soc.*, **117**, 11695 (1995).
351. R. H. Conlin, H. B. Huffaker and Y. -W. Kwak, *J. Am. Chem. Soc.*, **107**, 731 (1985).
352. F. J. Blanco, P. Cuadrado, A. M. González and F. J. Pulido, *Synthesis*, 42 (1996).
353. C. Gajda and J. R. Green, *Synlett*, 973 (1992).
354. M. Franck-Neumann, M. Sedrati and M. Mokhi, *J. Organomet. Chem.* **326**, 389 (1987).
355. M. Franck-Neumann, M. Sedrati and A. Abdali, *J. Organomet. Chem.* **339**, C9 (1988).
356. M. Ochiai, T. Ukita and E. Fujita, *Tetrahedron Lett.*, **24**, 4025 (1983).
357. T. Hayama, S. Tomoda, Y. Takeuchi and Y. Nomura, *Tetrahedron Lett.*, **24**, 2795 (1983).
358. P. V. Fish, *Synth. Commun.*, **26**, 433 (1996).
359. J. -P. Picard, J. Dunogues and A. Elyusufi, *Synth. Commun.*, **14**, 95 (1984).
360. M. Lautens, C. H. Zhang and C. M. Crudden, *Angew. Chem., Int. Ed. Engl.*, **31**, 232 (1992).
361. T. Masuda, H. Kouzai and T. Higashimura, *J. Chem. Soc., Chem. Commun.*, 252 (1991).
362. T. Hiyama, K. Nishide and M. Obayashi, *Chem. Lett.*, 1765 (1984).

CHAPTER 31

Chemistry of compounds with silicon–sulphur, silicon–selenium and silicon–tellurium bonds

D. A. ('FRED') ARMITAGE

Department of Chemistry, King's College London, Strand, London WC2R 2LS, UK

I. INTRODUCTION	1869
II. THIOSILANES	1870
III. SILANE THIOLS AND POLYSULPHIDES	1874
IV. LINEAR DISILTHIANES	1875
V. SILICON–SULPHUR DOUBLE BOND	1878
VI. CYCLIC SILTHIANES	1879
VII. MISCELLANEOUS SILICON–SULPHUR RINGS	1881
VIII. SELENOSILANES	1884
IX. BIS(SILYL)SELENIDES	1885
X. CYCLIC SELENIDES	1887
XI. TRIS(TRIMETHYLSILYL)SILYL SELENIDE DERIVATIVES	1888
XII. TELLUROSILANES	1889
XIII. TRIS(TRIMETHYLSILYL)SILYL TELLUROL DERIVATIVES	1890
XIV. REFERENCES	1891

I. INTRODUCTION

Since 1990, several reviews have appeared covering silicon–chalcogen chemistry. These include the spectroscopic properties of silicon–sulphur compounds along with structural data[1]. The synthesis of metal–chalcogen clusters using $(Me_3Si)_2X$ (X = S, Se, Te), Me_3SiX^-, $(Me_3Si)_3SiX^-$ are covered[3]. The only compound of this chapter to be included recently in *Inorganic Syntheses* is hexamethyldisilthiane, prepared by a method improving the yield of previous reports[4]. A report of the structures of $(PhSe)_4Si$ with

---

The chemistry of organic silicon compounds, Vol. 2
Edited by Z. Rappoport and Y. Apeloig © 1998 John Wiley & Sons Ltd

Si–S bond lengths of 212.3 and 212.9 pm) and (PhSe)$_4$Si (Si–Se bond lengths are 227.2 and 227.4 pm) includes a comprehensive list of the structures of silicon–sulphur and silicon–selenium compounds. These suggest a decrease in the Si–S bond length from 216 pm to 212 pm as the number of Si–S bonds in the molecule increases[5].

II. THIOSILANES

Though there are no new routes to supercede the most convenient synthetic route to thiosilanes — thiolation of a halosilane — two new classes of thiosilanes have been prepared. Dialkyl (S-trimethylsilyl) tetrathiophosphates result from P$_4$S$_{10}$ and alkylthiosilanes (equation 1). They react with (EtO)$_2$CHR to give (1-alkoxy)alkyltetrathiophosphates and with (EtS)$_2$CH$_2$ and Et$_2$S$_2$ at 150 °C to give the alkylthiosilane (equation 2)[6].

$$P_4S_{10} + 8\ Me_3SiSR \longrightarrow 4\ Me_3SiSP(S)(SR)_2 + 2(Me_3Si)_2S \quad (1)$$

$$R(EtO)HCSP(S)(SR)_2 \xleftarrow[-EtOSiMe_3]{(EtO)_2CHR} Me_3SiSP(S)(SR)_2 \xrightarrow[-Me_3SiSEt]{Et_2S_2} EtSSP(S)(SR)_2 \quad (2)$$

Aliphatic thioketones result tautomerically pure from the reaction of H$_2$S with ketals in the presence of ZnCl$_2$. Deprotonation with LDA gives the enethiolate which can be silylated to form the silyl vinyl sulphide. These are fairly stable to traces of water but react with MeOH to give the enethiol (equation 3)[7].

$$R^1R^2HC-C(S)R^3 \xrightarrow[\text{ii. Me}_3\text{SiCl}]{\text{i. LDA/THF}} R^1R^2C{=}C(R^3)SSiMe_3$$

$$\Big\downarrow MeOH \quad (3)$$

$$R^1R^2C{=}C(R^3)SH$$

In the case of R^3 = MeS, the reaction with imines in the presence of Lewis acid catalysts give the *syn* and *anti* β-amino dithioesters with *anti* selectivity (equation 4)[8].

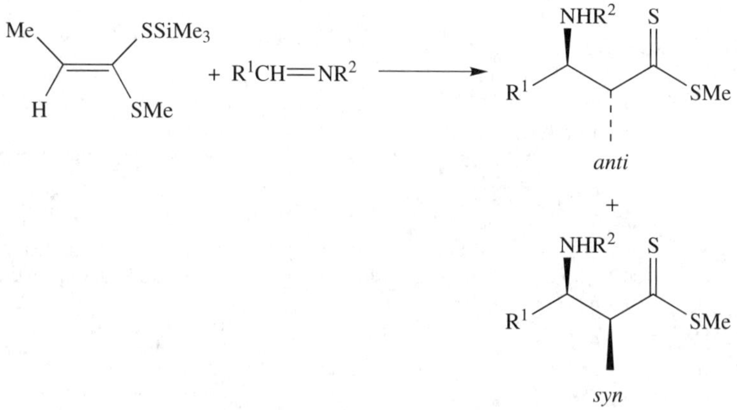

Monothioacetals have been prepared from alkyl and arylthiosilanes generated *in situ* from the thiol and silylimidazole in the presence of Me$_3$SiOTf as catalyst[9]. Yields are greater than 80%, and somewhat better than those obtained from the reaction of PhSSiMe$_3$

with aldehydes using silyl triflate as catalyst[10], or with acetals using dicyanoketene acetals as catalyst[11]. Monothioacetals can be reduced by Et_3SiH using $Me_3SiCl/InCl_3$ to give the monosulphide (**1**). The monosulphide also results from the reaction of a ketone and thiosilane (equation 5)[12].

$$R^1R^2C(OSiMe_3)SR^3 + Et_3SiH \xrightarrow{Me_3SiCl/InCl_3} R^1R^2CHSR^3 \quad (\mathbf{1})$$

$$\uparrow Me_3SiCl/InCl_3 \mid Et_3SiH$$

$$R^1R^2CO + R^3SiSiMe_3$$

(5)

With (*E*)-chalcone dimethyl acetal, in the presence of trityl perchlorate, $PhSSiMe_3$ adds to give the 1,3-disubstituted-1-propene as the major product (equation 6)[13].

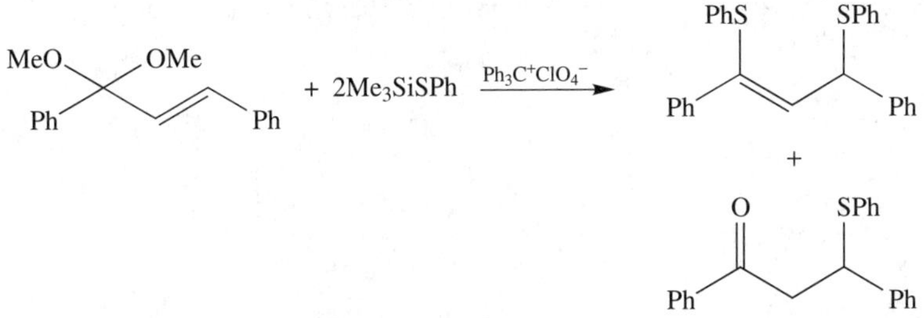

α-Silyl vinyl ketones react with $PhSSiMe_3$ in the presence of $Et_2O \cdot BF_3$ to give silylated bis(phenylthio)propene which serves as a silyl β-acyl carbanion equivalent, and readily substitutes position 3 with electrophiles (E^+) (equation 7)[14].

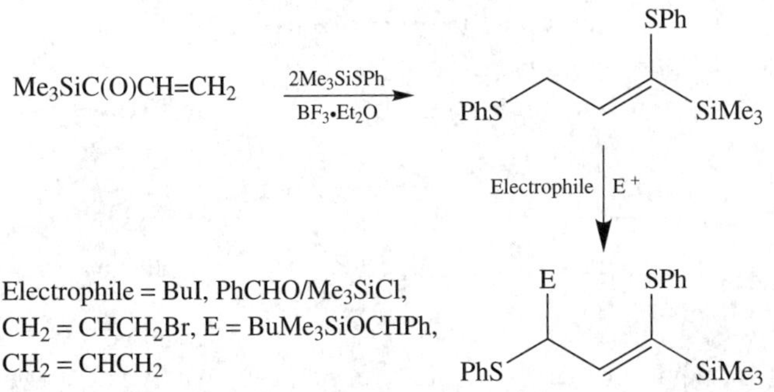

Electrophile = BuI, $PhCHO/Me_3SiCl$, $CH_2=CHCH_2Br$, E = $BuMe_3SiOCHPh$, $CH_2=CHCH_2$

i-$PrSSiMe_3$ catalyses the asymmetric allylation of aldehydes using (*S*)-BINOL–Ti(IV) complex (equation 8)[15].

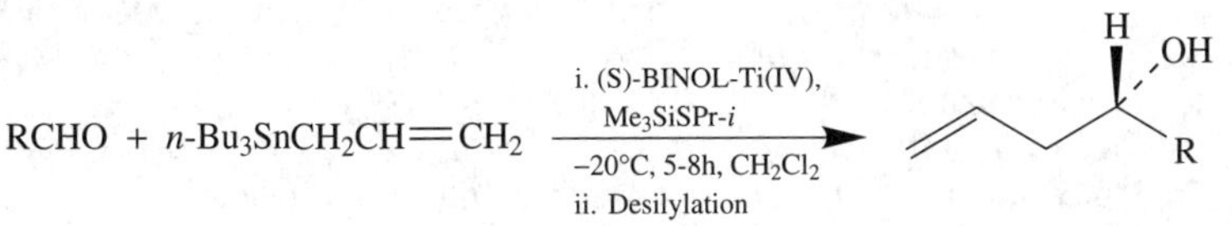

$$RCHO + n\text{-}Bu_3SnCH_2CH=CH_2 \xrightarrow[\text{ii. Desilylation}]{\substack{\text{i. (S)-BINOL-Ti(IV),} \\ Me_3SiPr\text{-}i \\ -20°C, 5\text{-}8h, CH_2Cl_2}} \quad (8)$$

Thiosilanes provide the first step in the conversion of amino sugars aryl and alkylthio substituted 2-azido-2-deoxyhexopyranosyl building blocks for oligosaccharide synthesis[16]. 2-(2-Phenylthiocyclobutyl)oxirane and oxetane (**2**) undergo acid-catalyzed ring opening with PhSSiMe$_3$ to give the *gem*-phenylthio substituted allyl and homoallyl alcohol (equation 9)[17].

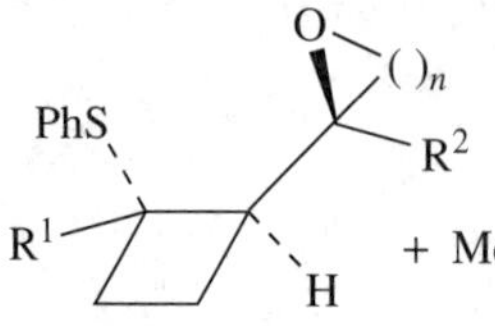

$$\text{(2)} + Me_3SiSPh \longrightarrow R^1(PhS)_2C(CH_2)_2CH=C(R^2)(CH_2)_nOH \quad (9)$$

$n = 1, 2$

Silicon is most important in the cycloaddition of alkynyl silyl sulphides to ynamines, the silyl shift giving the 4-silylcyclobut-2-enethione which, on S-methylation and desilylation, yields the cyclobutadiene (equation 10)[18].

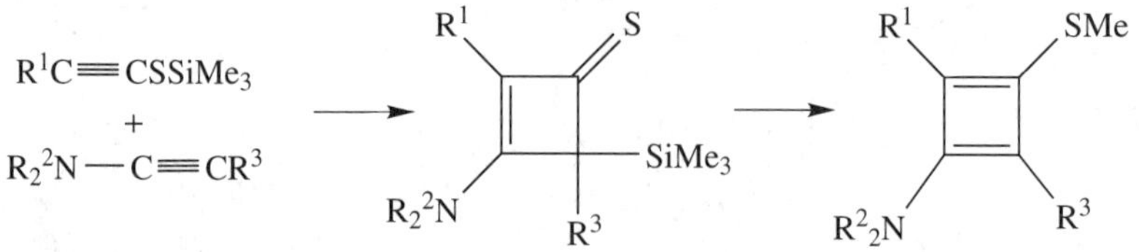

$$R^1C\equiv CSSiMe_3 + R_2^2N-C\equiv CR^3 \longrightarrow \cdots \longrightarrow \cdots \quad (10)$$

Reacting Ph$_2$SiH$_2$ with Rh(SAr)(PMe$_3$)$_3$ gives the complex *mer*-RhH(SiHPh$_2$)SAr (PMe$_3$)$_3$ (**3**) which isomerizes through SR transfer to silicon to give *fac*-RhH$_2$(SiPh$_2$SAr)(PMe$_3$)$_3$ with a Si—S bond length of 222.8 pm. Since the silyl group in **3** is *trans* to the thio group and the isomerization is intramolecular, it is suggested that *cis–trans* isomerization occurs prior to transfer (equation 11)[19].

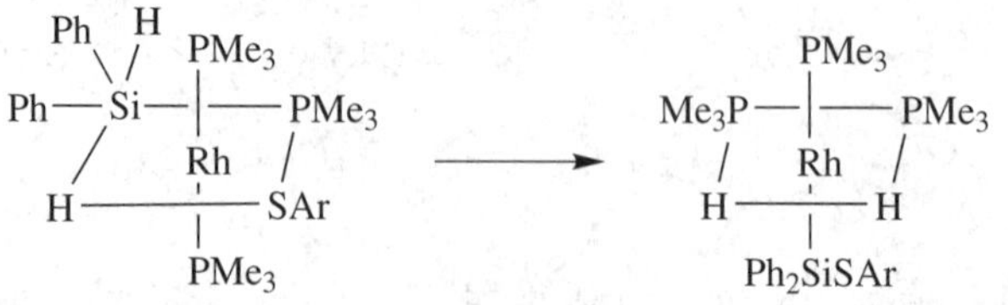

$$\text{(3)} \longrightarrow \cdots \quad (11)$$

Rhodium(I) catalyses the dehydrogenative coupling of silanes with arene thiols. Thus Ar$_2$SiH$_2$ reacts with 1,3- and 1,4-benzene dithiols to give polymers with molecular weights in the range 1.5 to 7×10^3. In the presence of alkynes, arylthiolation of PhSiH$_3$ results in the formation of arylthio substituted styryl silanes PhCH=CHSi(SAr)$_2$H in excellent yield[20]. The intermediacy of metal–silyl intermediates in such a reaction has been established through the reaction of (EtS)$_3$SiH with Rh(I) and Ir(I)[21], while Me$_3$SiSNa substitutes [Me$_3$Ru(Br)(N)]$^-$ to give [Ph$_4$P][Me$_3$Ru(N)SSiMe$_3$] which can be readily desilylated to give the monomeric sulphide and, with [(Me$_3$SiCH$_2$)$_2$Os(N)Cl]$_2$, couples to give the mixed sulphide with Ru—S—Os bridge with chlorosilane loss[22].

Bis(benzylthio)dimethylsilane readily undergoes a double rearrangement in the presence of 2 equivalents of t-BuLi in THF to give a mixture of the meso- and d,l-forms of the dithiol [PhCH(SH)]$_2$SiMe$_2$. It models dithiol bimetallic action of nitrogeneous substrates by the Mo-containing enzyme nitrogenase. The meso isomer reacts with MoCl$_5$ to give Mo$_2$Cl$_4$[(PhCHS)$_2$SiMe$_2$]$_3$ which complexes Me$_2$NNHPh[23].

The silyl substituted dithiol (Me$_3$SiSCF$_2$)$_2$, prepared from the sulphenyl chloride ClSC$_2$F$_4$SCl (4) and (Me$_3$Si)$_2$Hg, reacts with the dichlorides RECl$_2$ (E = P, As) to give the heterocycle, the phosphorus derivative undergoing a Michaelis–Arbuzov rearrangement, but not the arsenic one (equation 12). These heterocycles also result directly from (4) and the bis-silyl phosphine or arsine[24].

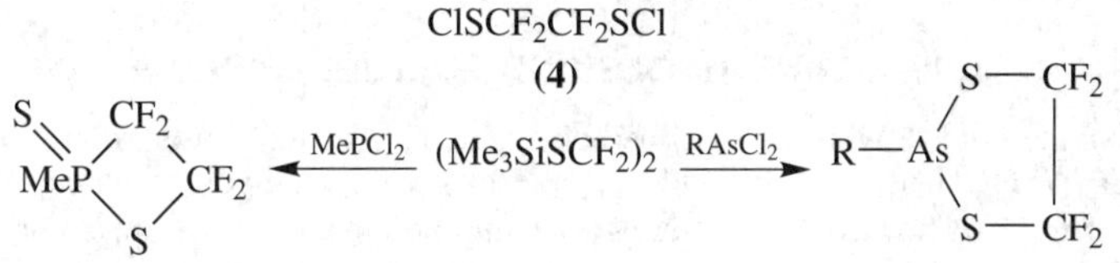

$$\text{MeP}\underset{S}{\overset{S\diagdown CF_2}{\diagup}}\!\!\!\underset{}{\overset{}{\diagdown}}\!\!\!\underset{}{\overset{CF_2}{\diagup}} \xleftarrow{\text{MePCl}_2} (\text{Me}_3\text{SiSCF}_2)_2 \xrightarrow{\text{RAsCl}_2} \text{R}\!-\!\text{As}\underset{S-CF_2}{\overset{S-CF_2}{\diagup\diagdown}} \quad (12)$$

$$\text{ClSCF}_2\text{CF}_2\text{SCl} \quad (4)$$

The hindered silyl chlorides (Ar*X)$_3$SiCl (X = S, Se) react with AgClO$_4$ to give a range of sulphides and selenides which are thought to result through the intermediacy of the silicenium ion which loses Ar*X$^+$ and (Ar*X)$_2$Si: (X = S) or Ar*X$^\bullet$ (X = Se). With the thio derivative, the products are the disulphide (5), sulphide (6) and arene (equation 13), while for the selenium chloride, diselenides and selenides dominate (equation 14)[25].

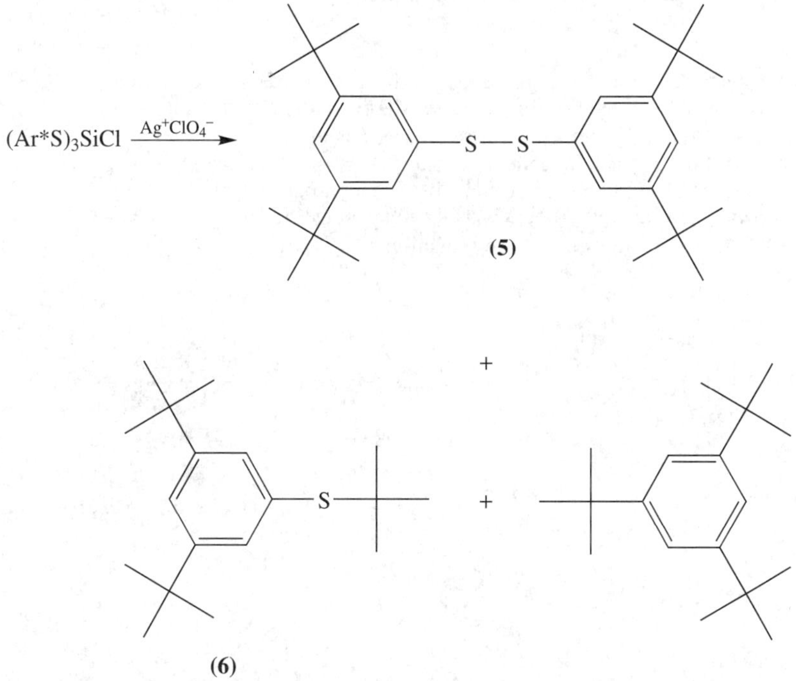

$$(\text{Ar*S})_3\text{SiCl} \xrightarrow{\text{Ag}^+\text{ClO}_4^-}$$

(5)

(6)

(13)

Ar* = 2,4,6-t-Bu$_3$C$_6$H$_2$

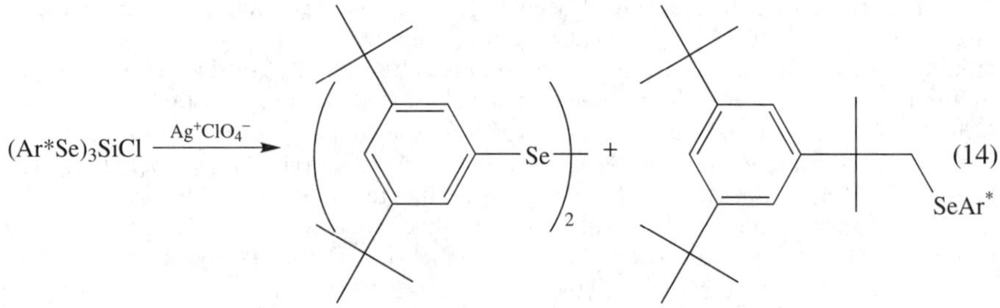

$(Ar*Se)_3SiCl \xrightarrow{Ag^+ClO_4^-}$ [diaryl selenide]$_2$ + [aryl selenide with SeAr*] (14)

III. SILANE THIOLS AND POLYSULPHIDES

Triphenylsilane thiol, a white crystalline solid, attacks epoxides on warming in the presence of Et$_3$N/MeOH on the less hindered side to give the β-hydroxymercaptans. In the presence of CsCO$_3$, however, the β-dihydroxysulphide results (equation 15 and 16)[26].

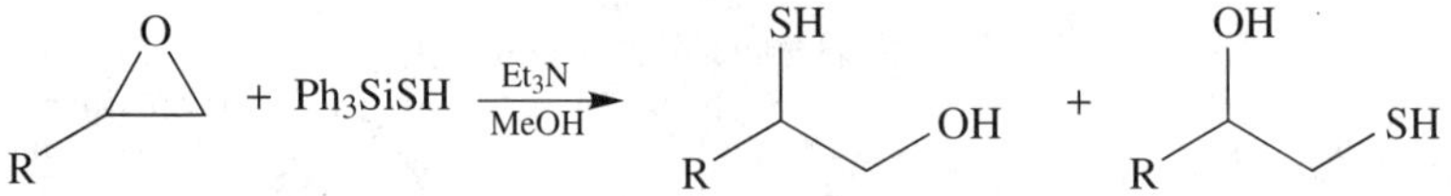

R-epoxide + Ph$_3$SiSH $\xrightarrow{Et_3N, MeOH}$ R-CH(SH)-CH$_2$OH + R-CH(OH)-CH$_2$SH (15)

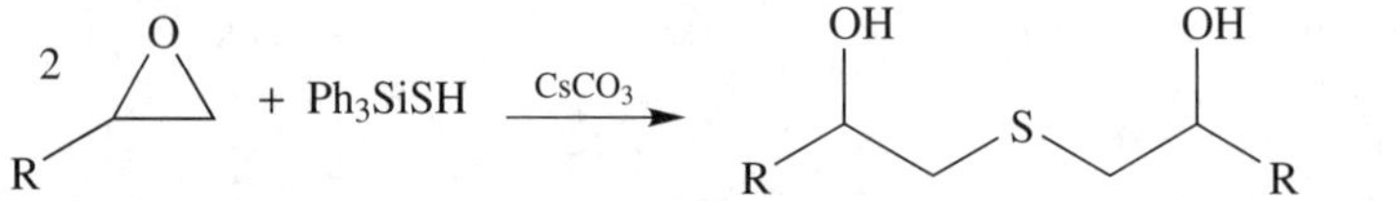

2 R-epoxide + Ph$_3$SiSH $\xrightarrow{CsCO_3}$ R-CH(OH)-CH$_2$-S-CH$_2$-CH(OH)-R (16)

The triphenylsilylthiyl radical, generated from Ph$_3$SiSH in the presence of AIBN or photolytically, adds to olefins through anti-Markovnikov addition to give the H$_2$S adduct RCH$_2$CH$_2$SH on desilylation with trifluoroacetic acid[27].

The silanethiyl radical (**7**) generated from the silane thiol, which is prepared from the silane using first CuCl$_2$ and then LiSH, abstracts hydrogen enantioselectively from silicon in (±)-*trans*-2,5-dimethyl-1-phenyl-1-silacyclopentane, to bring about kinetic resolution of the latter. Such reactions are rare (equation 17)[28].

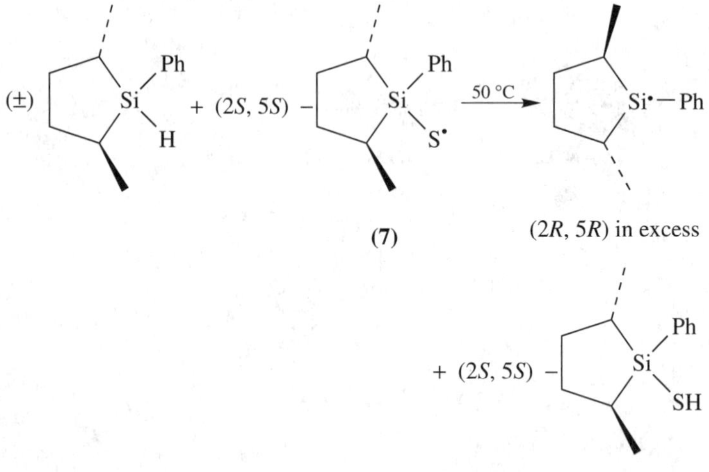

(±) [silacyclopentane with Si-H, Ph] + (2S, 5S) – [silacyclopentane with Si-S•, Ph] (**7**) $\xrightarrow{50\ °C}$ [Si•–Ph silacyclopentane] (2R, 5R) in excess

+ (2S, 5S) – [silacyclopentane with Si-SH, Ph] (17)

Ph$_3$SiSH can be S-halogenated using the N-halosuccinimide. The bromo and iodo derivatives decompose to release sulphur in polar solvents, but a structure determination of the bromo derivative shows the Si—S bond to be 216 pm, compared with 215 pm in the parent thiol. Surprisingly, the S—Br bond is 217 pm long[29].

Reacting Ph$_3$SiSNa with SCl$_2$ or S$_2$Cl$_2$ gives the bis-silyl tri- and tetra-sulphides. Both are thermally stable. The structure of the latter indicates a range of conformations present within the molecule involving *cis–cis*, *cis–trans* and *trans–trans* arrangements[30].

(Me$_3$Si)$_3$SiSH results from the thiolation of (Me$_3$Si)$_3$SiLi(THF)$_3$ followed by protolysis[31]. It also results from the silane (Me$_3$Si)$_3$SiH through chlorination with CCl$_4$, amination with NH$_3$ and then reaction with H$_2$S. It isomerizes in toluene at 80 °C in the presence of AIBN to give the silyl substituted silthiane (equation 18) through a 1,2-shift in 74% yield. At 85 °C, both compounds reduce terminal alkenyl bromides to the alkene or cycloalkane. With (Me$_3$Si)$_2$Si(Me)SH, the thiyl radical reduces terminal alkyl bromides to the alkene and likewise isomerizes to a silthiane (equation 19), while allylSSi(SiMe$_3$)$_3$ is less efficient than tin-substituted propenes in radical allylations[32].

$$(Me_3Si)_3SiSH \xrightarrow[\text{AIBN}]{80\,°C,\,\text{toluene}} (Me_3Si)_2Si(H)SSiMe_3 \quad (18)$$

$$(Me_3Si)_2Si(Me)SH \xrightarrow[\text{initiator}]{80\,°C,\,\text{benzene}} (Me_3Si)(Me)Si(H)SSiMe_3 \quad (19)$$

Reacting Me$_2$SiCl$_2$ with H$_2$S at −78 °C gives the silane dithiol as an unstable oil which shows ν_{S-H} stretching at 2500 cm^{-1}. The alkali metal derivatives of this dithiol result from the sulphide M$_2$S (M = Li, Na, K) and react with RI (R = Me, Et) to give the bis (alkylthio)silane and show ν_{Si-S} stretching frequencies at 400 and 447 cm^{-1}[33].

IV. LINEAR DISILTHIANES

Hexamethyldisilthiane can be conveniently prepared from Na$_2$S, freshly synthesized from the elements in THF with naphthalene, silylation with Me$_3$SiCl giving the silthiane in 90–95% yield[4]. 1,3-Diphenyldisilthiane results from H$_2$S, PhH$_2$SiCl and Et$_3$N in 83% yield, in contrast to the time-consuming *trans*-silylation of (Me$_3$Si)$_2$S with PhSiH$_2$Cl when only 40% yield could be obtained[34].

Electron diffraction studies of the three methylsilyl sulphides (Me$_n$H$_{3-n}$Si)$_2$S (n = 1–3) show Si—S bond lengths of 214.1, 214.6 and 215.4 pm respectively, with SiSSi angles of 97.9°, 100.8° and 105.8°, hindrance indicating a steady increase without undue lengthening of the Si—S bonds[35].

The photoelectron spectra of (Et$_3$Si)$_2$S and the silylalkyl sulphides (Et$_3$Si(CH$_2$)$_n$)$_2$S ($n = 0$–3) support a greater silyl donor influence for $n = 0, 2$ than for $n = 1, 3$, while the charge transfer spectra show a first ionization potential greatest for $n = 0$[36].

While organic sulphides can be readily synthesized from silthiane and organic halides, diallyl sulphides result in good yield from the allyl alcohols in the presence of 1 equivalent of Et$_2$O·BF$_3$. As the sodium salt, generated from NaH/(Me$_3$Si)$_2$S, it induces bis-O-demethylation of aryl methyl ethers through initial demethylation with Me$_3$SiSNa or NaH. The anion so formed can then react with (Me$_3$Si)$_2$S to give the silylated phenol and Me$_3$SiS$^-$ (Scheme 1)[37].

(Me$_3$Si)$_2$S in the presence of Me$_3$SiOTf induces the rearrangement of bis[(Mes)methanoyl] bicyclo[2.2.1]hept-5-enes to 2-oxabicyclo[3.3.0]octa-3,7-dienes (equation 20)[38].

3-Oxo dithioic acids react with a combination of (Me$_3$Si)$_2$S and N-chlorosuccinimide in the presence of imidazole to give oxidative ring closure to the 3H-1,2-dithiole-3-thiones (equation 21)[39].

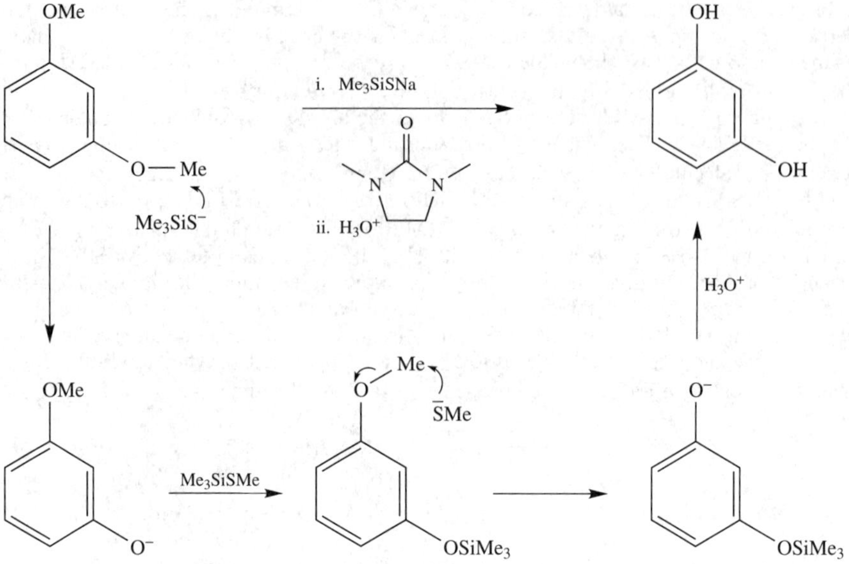

SCHEME 1

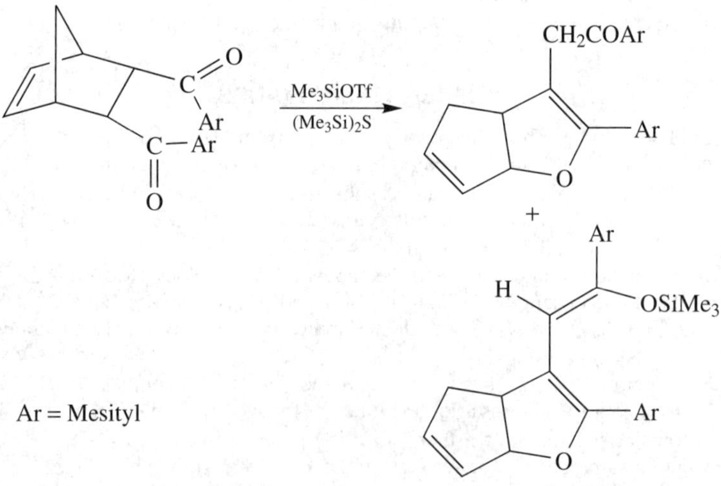

Ar = Mesityl

(20)

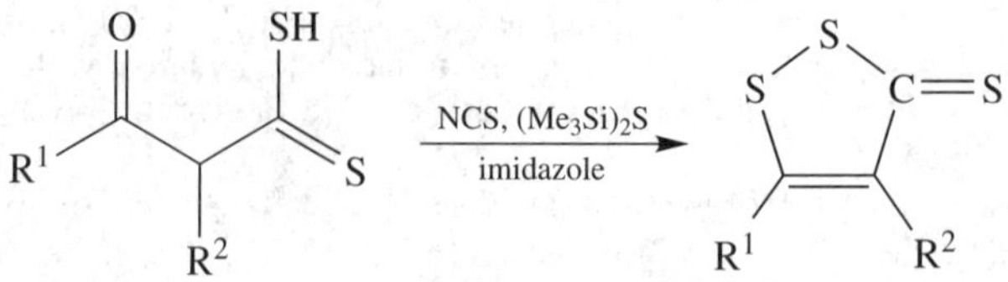

(21)

Hexamethyldisilthiane, despite its disgusting smell, has been extensively used in the conversion of carbonyl compounds to thiocarbonyls.

In the presence of CoCl$_2$·6H$_2$O, thioaldehydes can be generated from (Me$_3$Si)$_2$S and RCHO *in situ*, and form Diels–Alder adducts with cyclohexa-1,3-diene. Ketones react similarly in the presence of Me$_3$SiOTf to give the thioketone in good yield after 2 days. *o*-Azidohetarenecarboxaldehydes react with (Me$_3$Si)$_2$S to give the thioaldehyde in the presence of Co(II), and likewise adds dienes. In the presence of hydrochloric acid, however, intramolecular cyclization occurs to give the fused isothiazole, while with MeOH/HCl, the stable furan and thiophene *o*-aminothioaldehydes result (equation 22)[40].

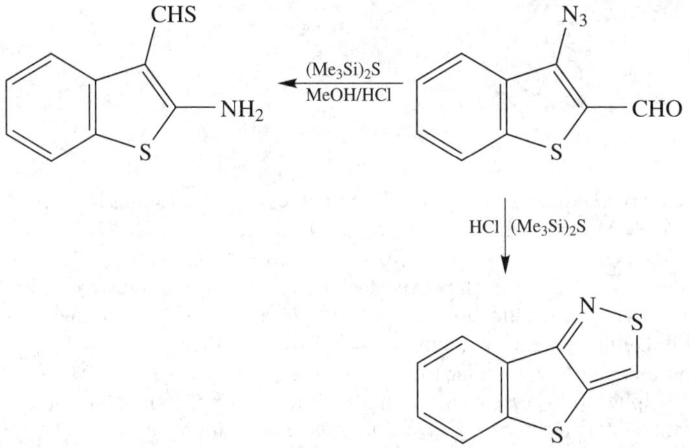

(22)

Electrophilic aromatic and heteroaromatic azides react with (Me$_3$Si)$_2$S to provide a route on methanolysis to the amine through N$_2$ and S loss and the intermediacy of the silylthio triazane ArNH−N=N−NSSiMe$_3$. Reaction times range from 1/2 to 6 h, and yields from 45 to 92%[41]. Peptides coupled to Kaiser's oxime ester resin can be cleaved using (Me$_3$Si)$_2$S/Bu$_4$NF to give the peptide C-terminal thioacids, which are useful for peptide fragment coupling[42]. Reductively condensing primary nitro compounds using KH/(Me$_3$Si)$_2$S provides a route to thiohydroxamic acids, which give nitriles on acidification (equation 23)[43].

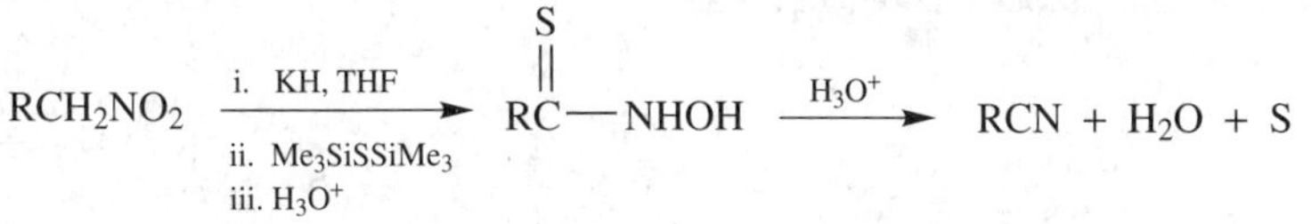

$$RCH_2NO_2 \xrightarrow[\substack{\text{ii. Me}_3\text{SiSSiMe}_3 \\ \text{iii. H}_3\text{O}^+}]{\text{i. KH, THF}} \underset{\|}{RC}-NHOH \xrightarrow{H_3O^+} RCN + H_2O + S \quad (23)$$

Aryl nitriles with (Me$_3$Si)$_2$S in the presence of MeONa/1,3-dimethyl-2-imidazolidinone results in the conversion to primary thioamides, while amides and lactams yields the thio derivative using (Me$_3$Si)$_2$S and an oxophilic promoter such as POCl$_3$, triphosgene or oxalyl chloride. Thus the intermediate [ClRC=NR'R'']$^+$Cl$^-$ reacts with the silthiane to give the thioamide (equation 24)[44].

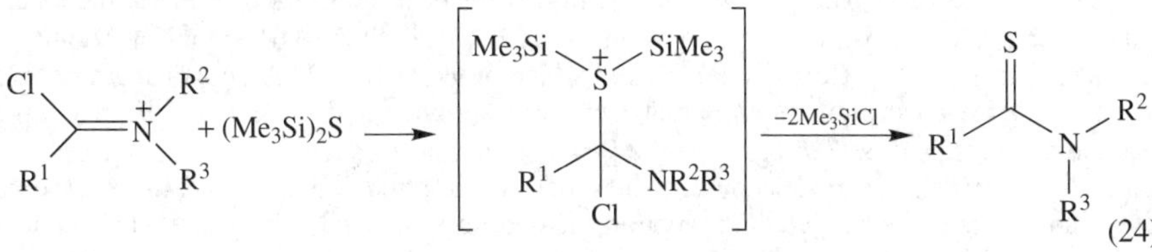

(24)

Gallium-chalcogen clusters $(Cp^*GaE)_4$ (E = S, Se) result from the reaction of $(Me_3Si)_2E$ with Cp^*GaCl_2, while the P–N heterocycle (8) can be thiolated at P(V) using $(Me_3Si)_2S$ (equation 25)[45].

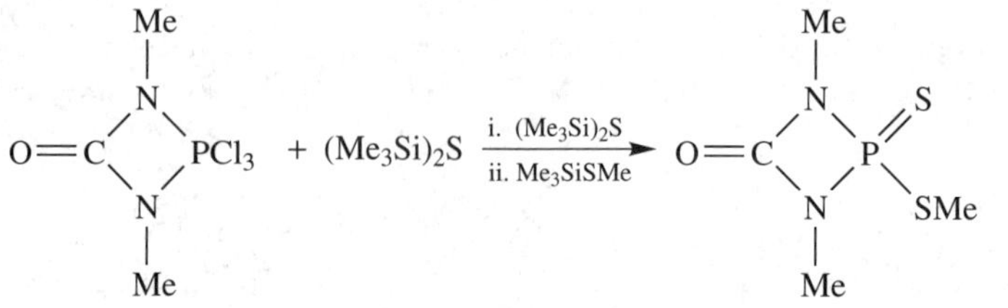

$$O=C\underset{\underset{Me}{N}}{\overset{\overset{Me}{N}}{\diagup\hspace{-0.5em}\diagdown}}PCl_3 + (Me_3Si)_2S \xrightarrow[\text{ii. Me}_3\text{SiSMe}]{\text{i. (Me}_3\text{Si)}_2\text{S}} O=C\underset{\underset{Me}{N}}{\overset{\overset{Me}{N}}{\diagup\hspace{-0.5em}\diagdown}}P\overset{S}{\underset{SMe}{\diagup\hspace{-0.5em}\diagdown}} \qquad (25)$$

(8)

Molybdenum and tungsten chlorides react to give the thiochlorides $MoSCl_3$ and $WSCl_4$, $M(O)Cl_4$ (M = Mo, W) reacting to give the mixed derivatives $M(O)(S)Cl_2$[46]. Tungsten hexafluoride reacts similarly to give WSF_4 at room temperature, and MCl_5 (M = Nb, Ta) yield $MSCl_3$[47]. With $CpZrCl_3$, however, the zirconium sulphide cluster $Cp_6Zr_6S_9$ results, in which the Zr_6 octahedron surrounds a μ_6-S atom with μ_3-S covering each triangular face[48]. With the nickel and palladium salts $(R_3P)_2MCl_2$, the clusters $M_3S_2Cl_2(PPh_3)_4$ are formed and contain the triangular M_3 core with capping μ_3-S ligands[49].

Copper sulphide clusters result from the reaction of Cu(I) acetate and $(Me_3Si)_2S$, with Ph_2EtP and Et_3P giving $Cu_{12}S_6(phosphine)_8$, while with Ph_3P, $Cu_{20}S_{10}(PPh_3)_8$ results. With the more hindered phosphines i-Pr_2MeP and t-Bu_2MeP, the larger clusters $Cu_{24}S_{12}(PMePr$-$i_2)_{12}$, $Cu_{28}S_{14}(PMeBu$-$t_2)_{12}$ and $Cu_{50}S_{25}(PMeBu$-$t_2)_{16}$ are formed[50].

V. SILICON–SULPHUR DOUBLE BOND

Silane thiones have been shown to be formed as intermediates in a range of reactions, and either oligomerize or react with other suitable acceptors. Thus, pyrolysing the propargylthiosilane 9 at 600 °C gives the silanethione which dimerises, while in the presence of ketene at 800 °C the 4-membered ring 10 results. The latter decomposes at 900 °C to give silanone and thioketene (equation 26)[51].

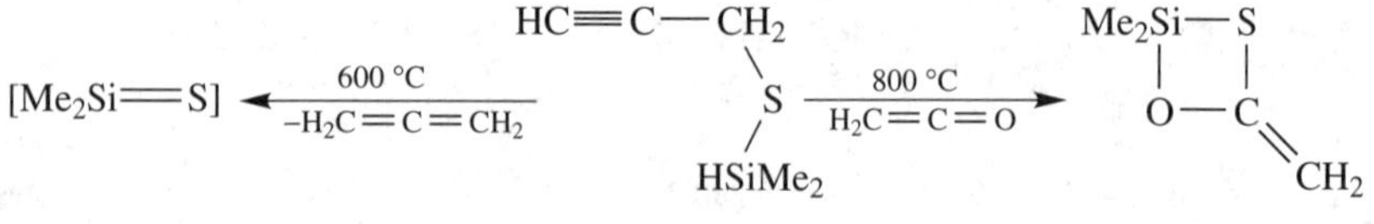

$$[Me_2Si\!=\!S] \xleftarrow[-H_2C=C=CH_2]{600\ °C} \underset{HSiMe_2}{\overset{HC\equiv C-CH_2}{\diagdown\hspace{-0.5em}\diagup}}S \xrightarrow[H_2C=C=O]{800\ °C} \begin{array}{c}Me_2Si-S\\|\quad\ \ |\\O-C\diagdown\\ \qquad CH_2\end{array} \qquad (26)$$

(9) \qquad\qquad\qquad\qquad (10)

The highly hindered dibromosilane $[2,4,6$-$\{(Me_3Si)_2CH\}_3C_6H_2](2,4,6$-$(i$-$Pr)_3C_6H_2)SiBr_2$ [Tbt(Tip)SiBr$_2$] on reduction and reaction with elemental sulphur gives the tetrathiasilolane 11 as pale yellow crystals. In the presence of 3 moles of Ph_3P, the silane thione 12 results as yellow crystals (equation 27). The ^{29}Si NMR signal at 166.56 is downfield from that of Corriu's complexed silanethione (13) at 22.3, supporting a double bond, while the absorption maximum at 396 nm is assigned to the n–π^* transition and the Raman spectral peak at 724 cm^{-1} to the Si=S stretching frequency. The silathiacarbonyl unit shows trigonal planar geometry and the Si–S bond length of 194.8 pm, some 9% shorter than the single bond, supports a double bond and is shorter than Corriu's complexed silathione (201.3 pm)[52].

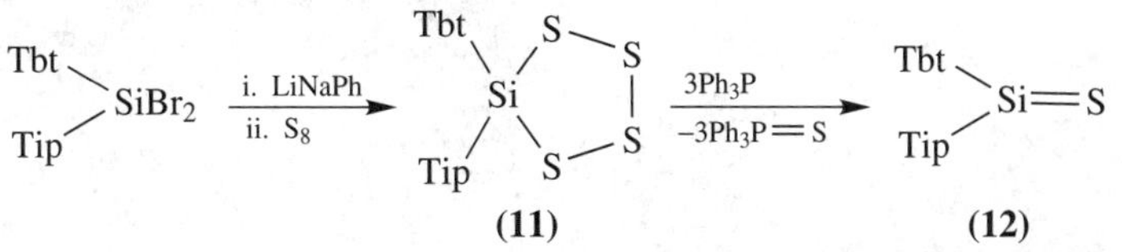

$$\text{Tbt}\diagdown_{\text{Tip}}\diagup\text{SiBr}_2 \xrightarrow[\text{ii. S}_8]{\text{i. LiNaPh}} \begin{array}{c}\text{Tbt}\diagdown\diagup\text{S}-\text{S}\\ \text{Si}|\\ \text{Tip}\diagup\diagdown\text{S}-\text{S}\\(\mathbf{11})\end{array} \xrightarrow[-3\text{Ph}_3\text{P}=\text{S}]{3\text{Ph}_3\text{P}} \begin{array}{c}\text{Tbt}\diagdown\\\text{Si}=\text{S}\\ \text{Tip}\diagup\\(\mathbf{12})\end{array} \quad (27)$$

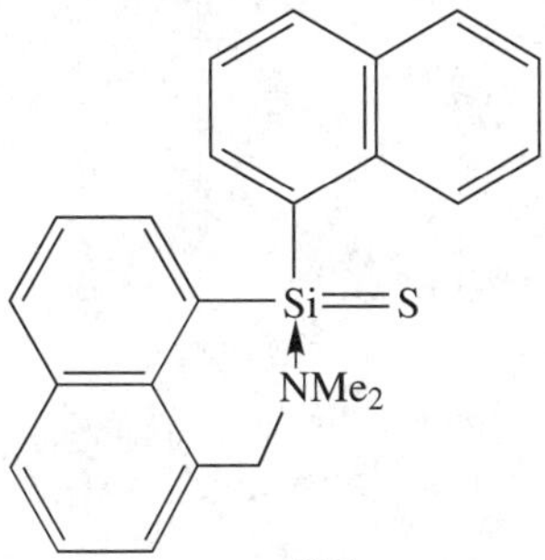

(13) — a naphthyl-substituted Si=S compound with NMe$_2$ donor.

The structures of a range of organothio substituted silylene and silyl complexes of transition metals support a degree of Si—S multiple bonding. The Si—S bond lengths in [Cp*Ru(PMe$_3$)$_2$Si(o-phen)STol-p]$^{2+}$ and Cp*Ru(PMe$_3$)$_2$Si[Os(CO)$_4$]STol-p are 217.9 and 217.2 pm respectively, somewhat shorter than in Cp*Ru(PMe$_3$)$_2$RuSi(STol-p)$_3$, implying some degree of π-bonding. This is also suggested for the short Si—S bonds of 209.2 and 207.4 pm in (Cy$_3$P)$_2$(H)Pt=Si(SEt)$_2$]$^+$, and therefore supporting some silylsulphonium resonance[53].

VI. CYCLIC SILTHIANES

Cyclotrisilthianes (R$_2$SiS)$_3$ have been prepared from Li$_2$S and R$_2$SiCl$_2$ (R = alkyl) in 70% yield with a view to using them as solid electrolytes for rechargeable batteries, and they also result in very good yield from the silyl triflate and H$_2$S in the presence of Et$_3$N[54]. The methyl derivative reacts with RLi to give the thiolate, which couples with a range of organic bromides to give the sulphide in good yield[55]. Reacting (C$_6$F$_5$)$_2$SiH$_2$ with sulphur or (Me$_3$Si)$_2$S gives the cyclodisilthiane [(C$_6$F$_5$)$_2$SiS]$_2$[56]. The adamantane-like tricyclo[3.3.1.1^{3,7}] tetrasilathiones (RSi)$_4$S$_6$ (R = Me, Et) result from reaction of sodium sulphide and RSiCl$_3$ in THF at room temperature for 1 day. The analogous selenium compounds result similarly[57].

The silicocene Cp*$_2$Si: reacts with a range of heterocumulenes and it is pertinent to include CO$_2$ along with the sulphur ones COS, CS$_2$ and RNCS (R = Me, Ph) for completeness. CO$_2$ gives an adduct which loses CO to form the silanone, which then adds more CO$_2$ to give the [2 + 2] cyclic adduct, the carbonate **14**. This dimerizes in pyridine through ring opening to give **15** and adds silanone Cp*$_2$Si=O in toluene through a further [2 + 2] cycloaddition to give the orthocarbonate **16** (Scheme 2)[58].

With COS, CO loss on addition gives the silathione which dimerizes to the disiladithietane (cyclodisilthiane). With CS$_2$ however, the cyclodisilthiane **17** also results through carbocation rearrangement of the adduct without CS loss (Scheme 3).

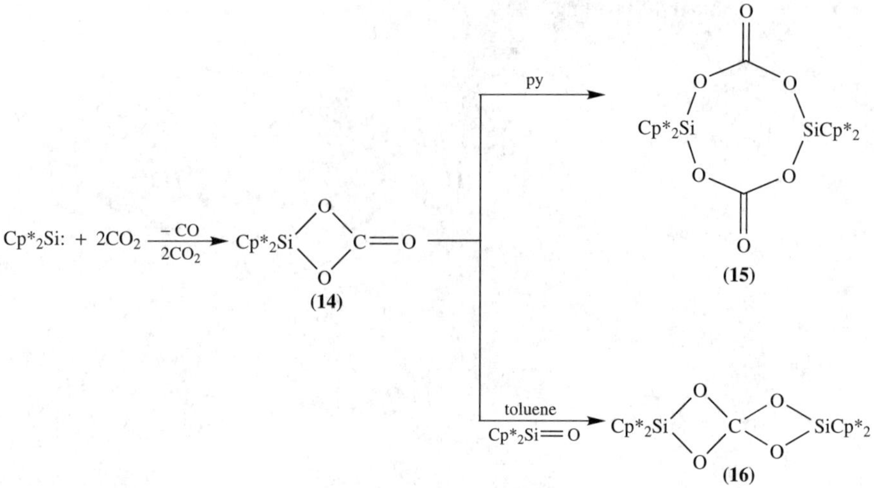

SCHEME 2

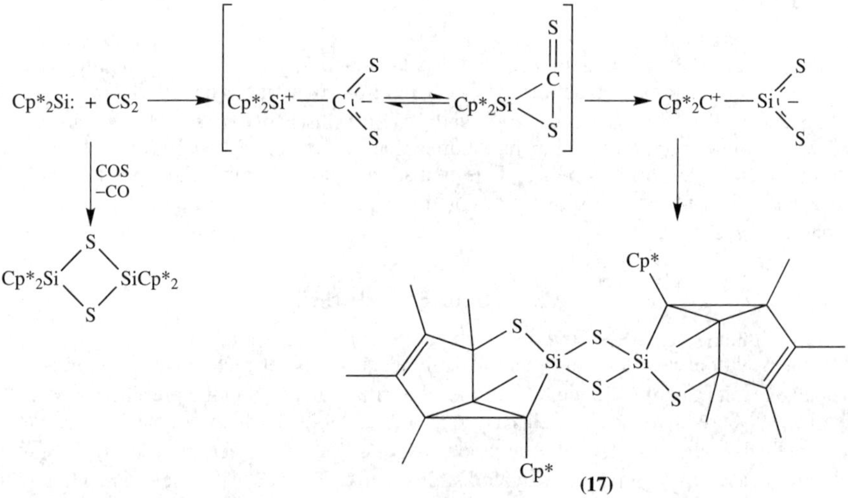

SCHEME 3

With RNCS, $Cp^*_2Si=S$ gives, together with isocyanide loss, further [2 + 2] addition of isothiocyanate leading to the dithiasilatane, while for R = Ph, further insertion of Cp^*_2Si: results in the formation of **18** (equation 28)[58].

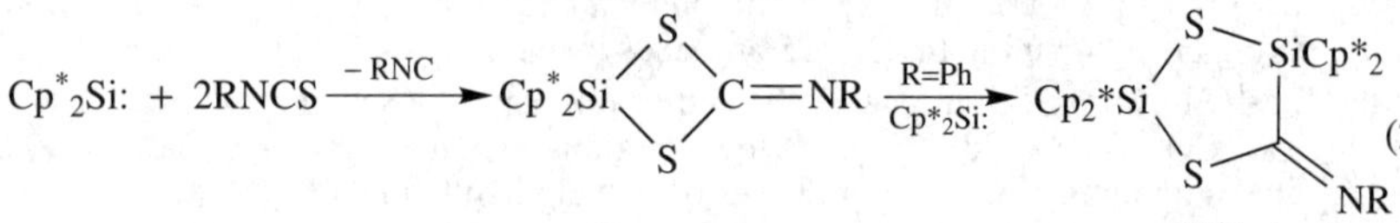

Reacting the hindered silane $(Me_3Si)_3CSiH_3$ ($TsiSiH_3$) with sulphur in Ph_2O at 200–210 °C gives the 2,3,5,6-tetrathia-1,4-disilabicyclo[2.1.1]hexane (**19**) in 6.1% yield as a crystalline solid, mp ⩽330°. In decalin, however, the [3.2.1] and [2.2.1] homologues also result, while the [2.1.1] derivative **19** can be readily desulphurized photolytically in the presence of Ph_3P to give **20**, as can the higher homologues Scheme 4.

VII. MISCELLANEOUS SILICON–SULPHUR RINGS

Thiadisilacyclopropanes result from the reaction of disilenes with sulphur and episulphides. From the mechanistic standpoint, the reaction of $RR'Si=SiRR'$ is of great interest and the separation of the (E)- and (Z)-isomers (*trans* and *cis*) [R = 2,4,6-(i-Pr)$_3C_6H_2$, R′ = t-Bu] has led to the separate addition of sulphur to each. This occurs within a minute to give *trans* and *cis*-isomers, respectively, the latter with slightly different Si–S bond lengths[60]. Propylene sulphide reacted similarly, and the reaction of $(t-Bu)_2Si=Si(Bu-t)_2$ with thiophene leads to sulphur abstraction with the formation of the thiadisilacyclopropane, with Si–S bonds of 217.1 pm, along with the 1,2-disilacyclohexa-3,5-diene and 2,6-disilabicyclo[3.1.0]hex-3-ene (equation 29)[61].

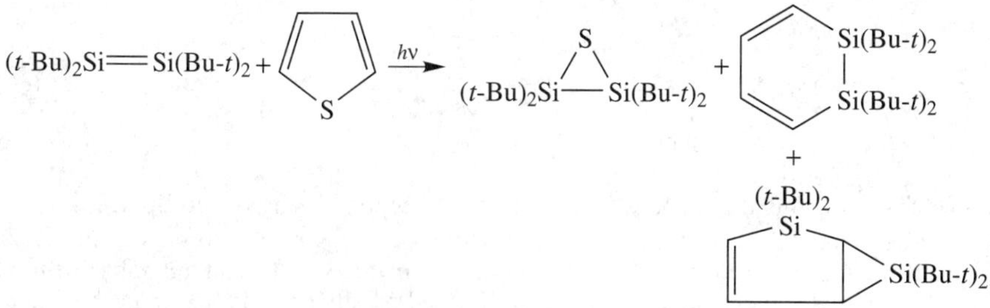

(29)

The stable silathiirane **21** results from the silene and elemental sulphur on photolysis. It has an Si–S bond of 212.9 pm, slightly longer than that found in the silathiirane formed from $Mes_2Si:$ addition to tetramethyl-2-indanethione[62]. In addition, the cyclodisilthiane results and can also be made from the disilabicyclo[1.1.1] pentane (**20**) on hydrolysis. It possesses Si–S bonds of length 217–219 pm (Scheme 4)[59].

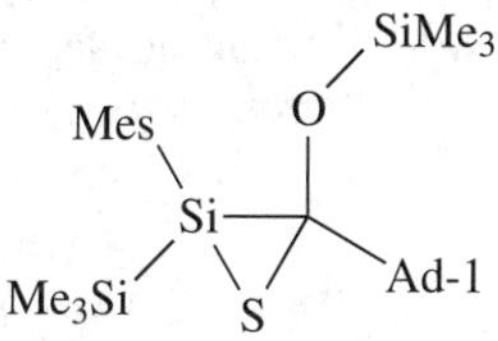

(**21**)

Mes = Mesityl, Ad = Adamntyl

Reacting tetramesityldisilene with PhHC=S gives the air-stable 1,2,3-disilathietane with a long Si–Si bond of 244.3 pm, longer than that in the oxetane. On photolysis in EtOH,

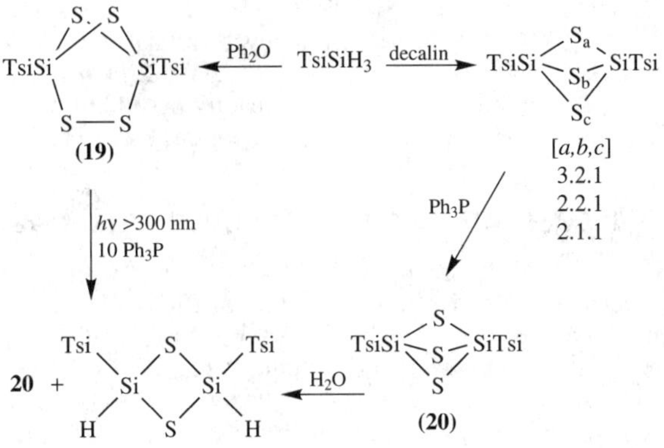

SCHEME 4

the ethoxysilane and silanethiol are formed (equation 30)[63].

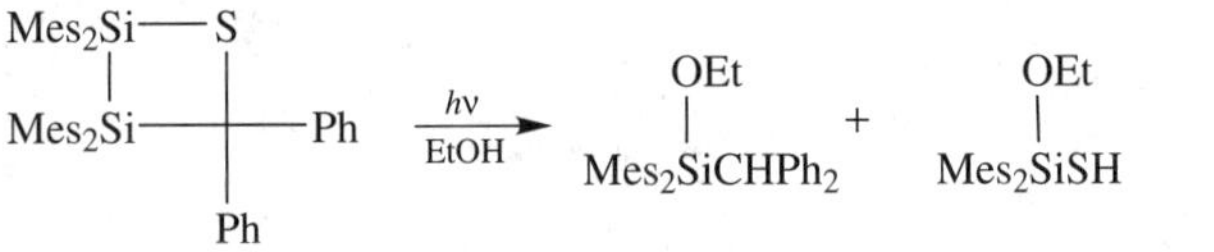

$$\text{Mes}_2\text{Si-S} \atop \text{Mes}_2\text{Si}{-}\text{Ph} \atop \text{Ph} \quad \xrightarrow{h\nu}_{\text{EtOH}} \quad \text{Mes}_2\text{SiCHPh}_2^{\text{OEt}} + \text{Mes}_2\text{SiSH}^{\text{OEt}} \qquad (30)$$

The first 1,2,4-thiadisiletane results from the reaction of carbon disulphide and the hindered silylene $[2,4,6-\{(\text{Me}_3\text{Si})_2\text{CH}\}_3\text{C}_6\text{H}_2]$MesSi: (TbtMesSi:) formed from the Z-disilene precursor. The mechanism is thought to involve a skeletal rearrangement of the 3,3'-spirobi(1,2-thiasilirane) intermediate formed by silylene addition to each carbon-sulphur double bond (equation 31)[64].

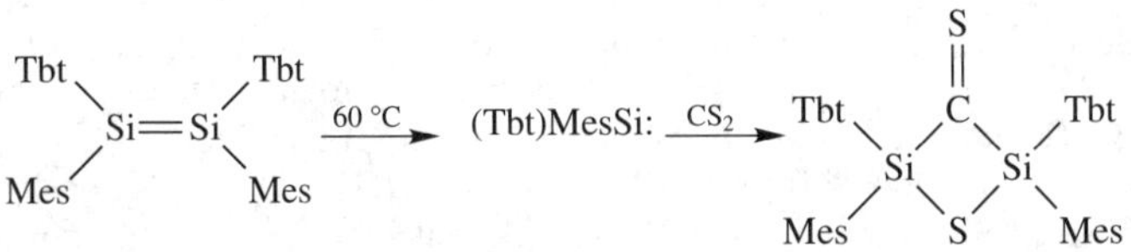

$$\text{Tbt, Tbt} \atop \text{Si=Si} \atop \text{Mes Mes} \quad \xrightarrow{60\,°\text{C}} \quad (\text{Tbt})\text{MesSi:} \xrightarrow{\text{CS}_2} \text{Tbt-Si-C(=S)-Si-Tbt (ring with S)} \qquad (31)$$

Coupling Na_2S with $\text{ClMe}_2\text{SiCH}_2\text{GeMe}_2\text{Cl}$ gives the 1,3,2-germasilathietane which slowly dimerizes on heating, with subsequent decomposition through silene and germathione formation to give 6-membered heterocycles containing the CSiSGe unit (equation 32)[65].

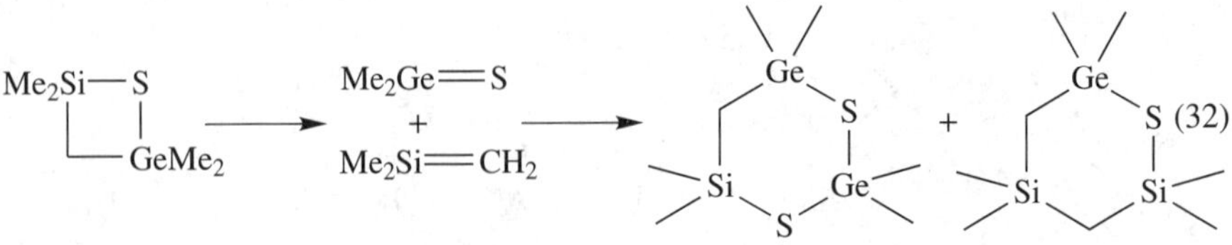

$$\text{Me}_2\text{Si-S} \atop \text{GeMe}_2 \quad \longrightarrow \quad \text{Me}_2\text{Ge=S} + \text{Me}_2\text{Si=CH}_2 \quad \longrightarrow \quad \text{(6-membered rings)} \qquad (32)$$

Stepwise insertion of sulphur into the silirane ring gives the 1,2-silathietane and siladithietane in 60% and 19% yield, respectively (equation 33). Pyrolysis of the

1,2-silathietane gives the cyclodisilthiane and the 2,4-dithia-1,3-disilacyclohexane[66].

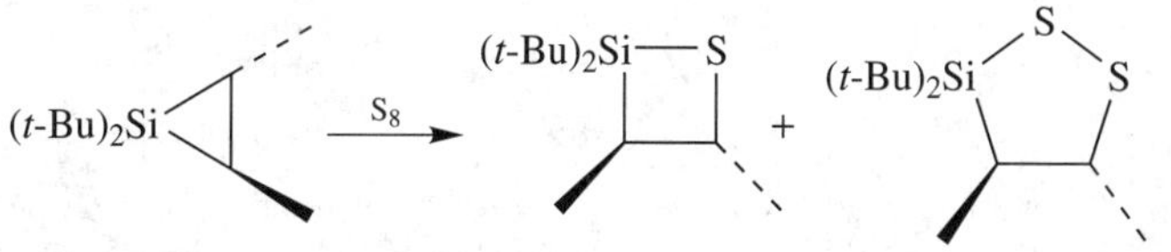

$$(33)$$

Similarly, ring expansion of fused ring siliranes also occurs stepwise with silabicyclo[3.1.0]hexane and silabicyclo[4.1.0]heptane (equation 34). With 1,1-dimethyl-1-silacyclobutane, sulphur insertion occurs to give the 1-thia-2-silacyclopentane in 72% yield[67].

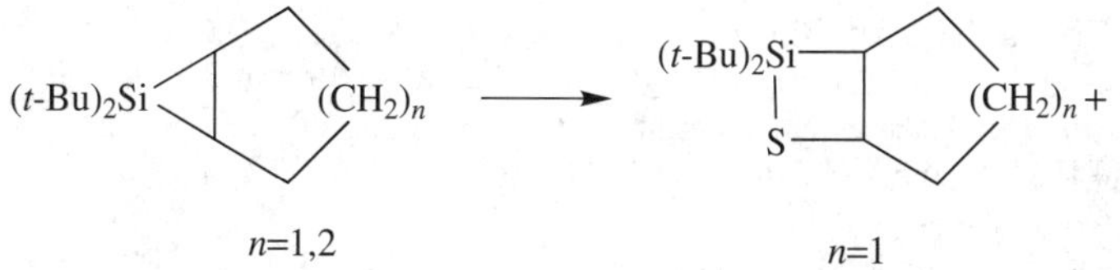

$$(34)$$

The structure of the 2,3-dithia-4-silabicyclo[4.3.0]nonane shows a Si−S bond length of 217.3 pm. The analogous selenium compound has a Si−Se bond length of 232.4 pm.

The dephenylation of $(t\text{-BuPh}_2\text{Si})_2$ using CF_3SO_3H, followed by thiolysis with H_2S gives the *meso*-dithiol in 67% yield. On lithiation and cyclization with a metallocene dichloride, the metallocycles react with a range of dichlorides to give chalcogeno substituted rings (equation 35)[68].

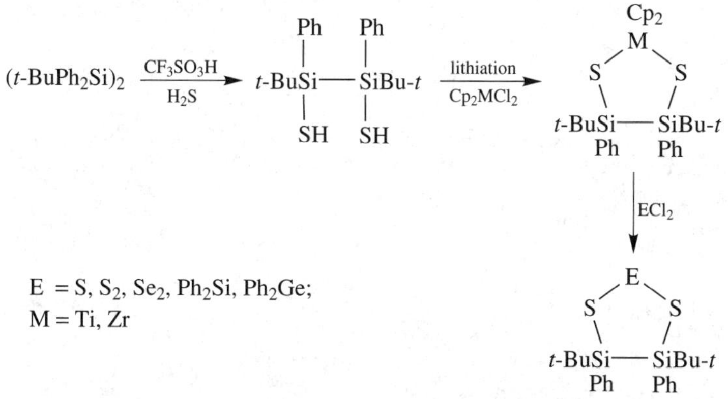

$$(35)$$

$E = S, S_2, Se_2, Ph_2Si, Ph_2Ge;$
$M = Ti, Zr$

Phosphines attack the disilatrithiane ring to give the cyclodisilthiane predominantly as the *trans* isomer, with Si−S bonds of 214.9 pm, a little shorter than in the disilatrithiane (217–218 pm), while in the titanocene derivatives, they are 210–213 pm. Photolysing the 6-membered selenium heterocycle eliminates one atom of Se to give a 5-membered mixed chalcogeno derivative (equation 36). $(Me_2SiOTf)_2$ in the presence of H_2S/Et_3N, gives the 1,2,4,5-tetrasila-3,6-dithiacyclohexane in excellent yield[54b].

The 4-membered titanocycle **22** reacts with S_2Cl_2 to eliminate Cp_2TiCl_2 and give the sulphur-rich silatetrathialane as a colourless wax which decomposes above −20 °C

(equation 37)[69].

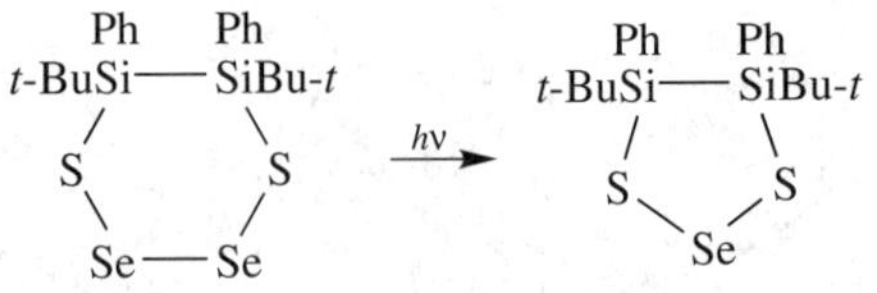

$$\text{(36)}$$

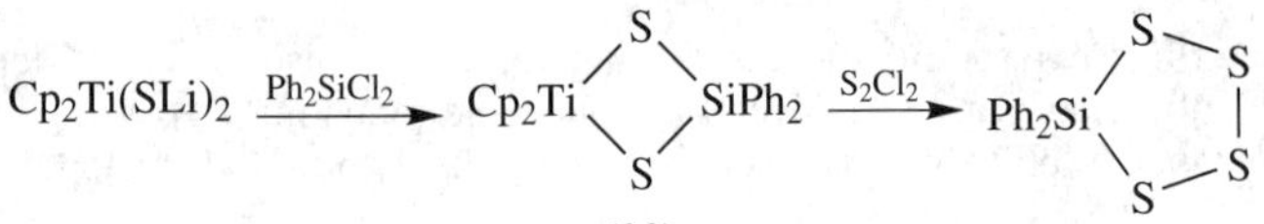

$$\text{(37)}$$

(22)

Heating TbtPhSiH$_2$ with sulphur at 280 °C gives **23** while the mesityl derivative gives TbtMesSi(SH)OH. However, the silatetrathialane **24** does form at 230 °C and the crystals have a distorted half-chair conformation and Si—S bonds of 216 and 222 pm[70]. It also results from the silylene and sulphur at 70 °C[71].

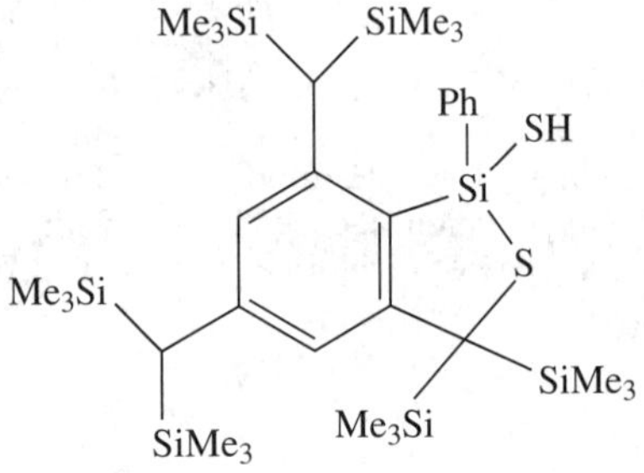

(23) 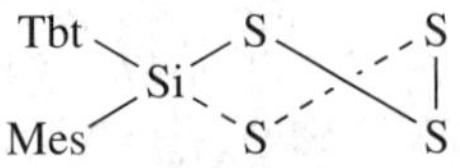 **(24)**

Condensing EtB(Cl)C(Et)=C(Me)SiMe$_2$Cl with LiSXSLi [X = (CH$_2$)$_n$, n = 2–4, o-phenylene] gives the heterocycles **25**. The structure of the o-phenylene derivative shows a Si—S bond length of 221.9 pm and a transannular donor S → B bond some 10 pm longer than that within the ring (equation 38)[72].

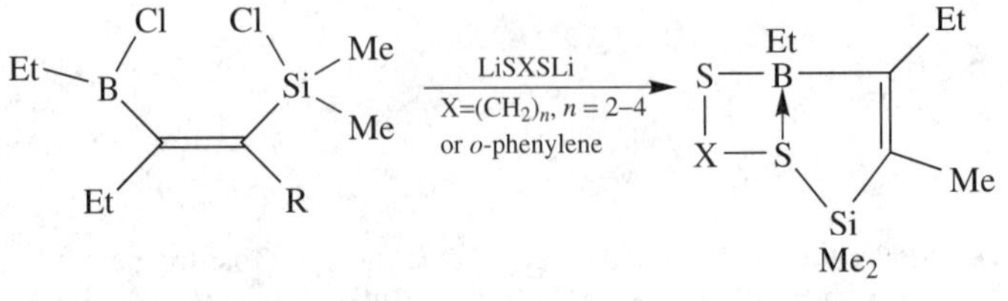

$$\text{(38)}$$

R=Me, H$_2$C=C(Me)   **(25)**

VIII. SELENOSILANES

There have been no recent, nor more efficient, general methods to prepare organoselenasilanes. However, Mes*SeSiMe$_3$ results from Mes*SeLi in (Mes* = 2,4,6-(t-Bu)$_3$C$_6$H$_2$) in

94% yield, while i-$C_3F_7SeSiMe_3$ is formed from Me_3SiI and $(i$-$C_3F_7Se)_2Hg$ in 83% yield[73].

Monoselenoacetals can be conveniently synthesized from acetals using $Me_3SiSePh$ in the presence of a catalytic amount of Me_3SiOTf with yields in most cases in excess of 80%. With 2-methoxytetrahydropyrans, selenation is highly chemoselective. It gives no ring-opened product, in contrast to the selenation with i-$Bu_2AlSePh$ or n-$Bu_3SnSePh$. In addition, the reaction is tolerant to ether, halogen and ester groups[74]. Radical cyclization of ω-halogeno-1-alkenes and $PhSeSiPh_2Bu$-t leads to the $PhSeCH_2$-substituted cyclized product in yields of between 55 and 87% (equation 39)[75].

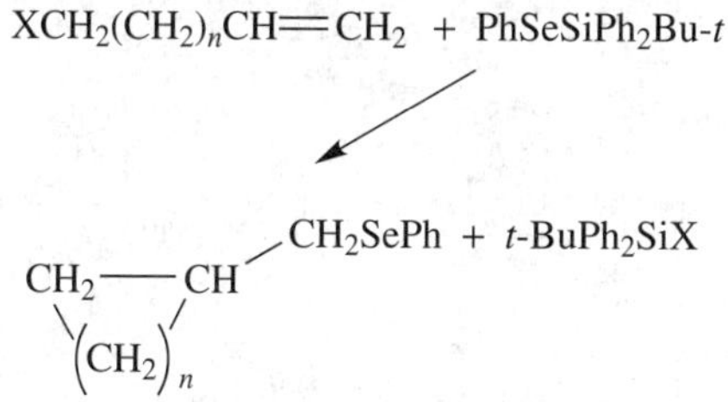

$$XCH_2(CH_2)_nCH = CH_2 + PhSeSiPh_2Bu\text{-}t$$

$$\begin{array}{c} CH_2SePh + t\text{-}BuPh_2SiX \end{array} \quad (39)$$

$$CH_2 \underset{(CH_2)_n}{\overset{}{\diagdown}} CH$$

In the presence of XeF_2, $PhSeSiMeR_2$ readily adds to acetylenes to give the vicinal (E)-fluoroalkenyl selenides (equation 40), but only diphenyl ditelluride results using $PhTeSiMe_3$[76].

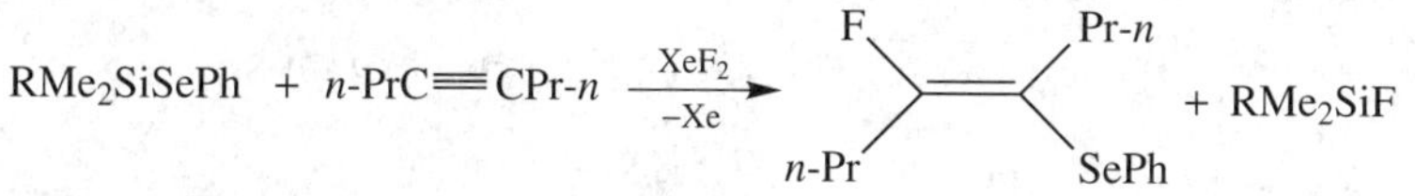

$$RMe_2SiSePh + n\text{-}PrC \equiv CPr\text{-}n \xrightarrow[-Xe]{XeF_2} \underset{n\text{-}Pr}{\overset{F}{\diagup}} \!\!\!\! = \!\!\!\! \underset{SePh}{\overset{Pr\text{-}n}{\diagdown}} + RMe_2SiF \quad (40)$$

$Me_3SiSeMe$ reacts with both $(MeCN)WCl_4(NCl)$ and $[Ph_3PMe]^+[Cl_5W=NCl]^-$ to give N-selenenation without cleavage of the W−Cl bonds. However, condensing $(MePh_2P)_4WCl_2$ with $Me_3SiSePh$ in THF results in both Si−Se and C−Se cleavage by W−Cl to give $W_2Se_2(PPh_2Me)_2(SePh)_4$ with a W−W bond bridged by the selenide ions with each tungsten coordinated to one phosphine ligand as well as two selenolate ligands[77].

IX. BIS(SILYL)SELENIDES

Bis(phenylsilyl)selenide results from freshly prepared K_2Se in 23% yield[33].

In the presence of $Et_2O\cdot BF_3$, $(Me_3Si)_2Se$ reacts with nitriles to give selenoamides while cyanates give the selenourea. Amides and tetramethylurea behave similarly, but the reaction with benzoates gives benzoin and 2,3,5,6-tetraphenyl-1,4-diselenin via selenoesters. These selenoesters can be trapped as conjugated diene cycloadducts (Scheme 5)[78].

A series of O-triorganosilyl selenocarboxylates $RC(Se)OSiR'_3$ result from the isomeric Se-triorganosilyl selenocarboxylates formed initially from $RC(O)Se^-M^+$ (M = Na, K) and R'_3SiCl. These O-silyl esters are thermally stable but isomerize in the mass spectrometer to Se-silyl derivatives. The carboxylates t-$BuMe_2SiOC(Se)Ar$ react readily with acyl halides to give selenoanhydrides, while a range of organometalloid chlorides cleave the Si−O bond to give the Se-metalloido selenocarboxylates (equation 41)[79].

An extensive range of transition metal selenide clusters result from $(Me_3Si)_2Se$ and a range of transition metal salts. $Cp'TiCl_3$ ($Cp' = MeC_5H_4$) reacts with $(Me_3Si)_2Se$ to give the cluster $Cp'_5Ti_5Se_6$, which possesses a distorted trigonal bipyramidal structure with

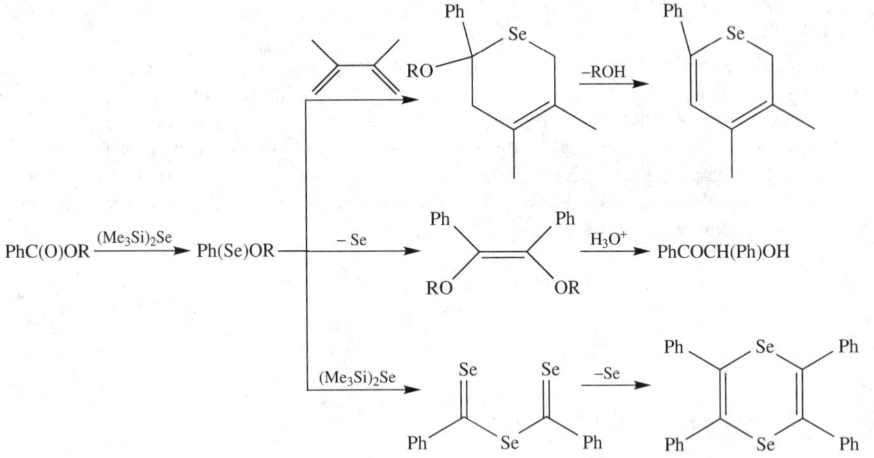

SCHEME 5

μ_3-Se ligands[80]. CpVCl$_2$(PMe$_3$) reacts to give V$_6$Se$_8$O(PMe$_3$)$_6$ with oxygen within the octahedral V$_6$ cluster[48]. With (Ph$_2$PCl)Cr(CO)$_5$, Si–Se bond cleavage by P–Cl leads to three products [(CO)$_5$CrPPh$_2$]$_2$Se, (CO)$_5$Cr[Ph$_2$P)$_2$Se] and Se(PPh$_2$)$_2$Cr(CO)$_4$, involving coordination of the (Ph$_2$P)$_2$Se ligand, in each case through phosphorus[81].

$$t\text{-BuMe}_2\text{SiOC(Se)Ar} + \text{ArCOCl} \longrightarrow (\text{ArCO})_2\text{Se} \qquad (41)$$

Ar = Ph, *p*-Ar

Phosphine substituted nickel(II) chloride gives a range of clusters, the cluster size being determined by the subtle steric and electronic effects of the chelating phosphine. Thus [(PhCH$_2$CH$_2$CH$_2$)$_3$P]$_2$NiCl$_2$ (L$_2$NiCl$_2$) gives Ni$_4$Se$_3$L$_5$, while (Et$_2$MeP)$_2$NiCl$_2$(L$_2'$NiCl$_2$) yields Ni$_5$Se$_4$Cl$_2$L$_6''$. With (*i*-Pr$_3$P)$_2$NiCl$_2$(L$_2''$NiCl$_2$), however, two larger clusters Ni$_7$Se$_5$L$_6''$ and Ni$_8$Se$_6$L$_4''$ result, while with (PhEt$_2$P)$_2$NiCl$_2$(L$_2'''$NiCl$_2$), Ni$_8$Se$_6$L$_6'''$ and Ni$_{21}$Se$_{14}$L$_{12}'''$ are formed. In the smaller clusters, the selenium atoms bridge but three metal atoms while, in the larger clusters, coordination to four metal atoms tends to dominate[82].

With CpNi(PPh$_3$)$_2$SnCl$_3$, the mixed Ni–Sn cluster Cp$_4$(Ph$_3$P)$_3$Ni$_5$Sn$_6$Se$_9$ results, in which a polycyclic Sn$_6$Se$_9$ cage encloses a central Cp$_2$Ni unit and coordinates three tin atoms to three (Ph$_3$P)CpNi units. Each tin atom interacts with 3 selenium atoms[83].

With [(PhCH$_2$CH$_2$)$_3$P]$_2$NiCl$_2$, (Me$_3$Si)$_2$Se gives the Ni$_3$Se$_2$(SeSiMe$_3$)$_2$[P(CH$_2$CH$_2$Ph)$_3$]$_4$ in which the μ_3-Se ligands cap the Ni$_3$ unit[84]. A similar Pd complex **26**, Pd$_3$Se$_2$(SeSiMe$_3$)$_2$(PPh$_3$)$_4$, results from (Ph$_3$P)$_2$PdCl$_2$, together with [Ph$_3$PPdSe]$_n$ (*n* = 5, 8). **26** reacts further with CpCrCl$_2$(THF), and MCl$_2$(PPh$_3$)$_2$ (M = Ni, Pd) to give a range of chloro substituted Pd$_n$ clusters with *n* = 5, 6, 7, 8[85].

Interest in copper selenide clusters is more recent, possibly because the clusters are even larger. Reacting copper(I) acetate with (Me$_3$Si)$_2$Se and Ph$_2$PEt yields the solitary cluster Cu$_{12}$Se$_6$(Ph$_2$EtP)$_8$ as red crystals[50]. However, relatively small clusters tend to be the exception in the products produced by this route. Thus *t*-Bu$_2$MeP leads to the two clusters Cu$_{31}$Se$_{15}$(SeSiMe$_3$)(PMe(Bu-*t*)$_2$)$_{12}$ and Cu$_{70}$Se$_{35}$(PMe(Bu-*t*)$_2$)$_{21}$ while Me$_2$PPh gives Cu$_{48}$Se$_{24}$(PMe$_2$Ph)$_{20}$. With copper(I) chloride, various size clusters result depending on the phosphine used. *i*-Pr$_3$P gives Cu$_n$Se$_{15}$(P(Pr-*i*)$_3$)$_{12}$ (*n* = 29, 30), *t*-Bu$_3$P gives Cu$_{36}$Se$_{18}$(P(Bu-*t*)$_3$)$_{12}$, while with PhPEt$_2$ and *n*-BuP(Bu-*t*)$_2$, Cu$_{44}$Se$_{22}$(PEt$_2$Ph)$_{18}$ and

$Cu_{44}Se_{22}(P(Bu-n)(Bu-t)_2)_{12}$ are formed, with each containing a deltahedron of 20 selenium atoms[86].

With Et_3P, a little $Cu_{20}Se_{13}(PEt_3)_{12}$ results but the major product is $Cu_{70}Se_{35}(PEt_3)_{22}$ and both are believed to be formed via a Cu_9 intermediate[87]. An even larger cluster results from copper(I) chloride and Ph_3P in THF, with $Cu_{146}Se_{73}(PPh_3)_{30}$ as the product with the selenium atoms arranged in layers of 21, 31 and 21 atoms, respectively, a similar layer structure being found in β-Cu_2Se[88].

The mononuclear mercury(II) derivatives result from 2,4,6-t-$Bu_3C_6H_2SeSiMe_3$ with $HgCl_2$[89], while with $PhSeSiMe_3$ and $Fe(CO)_4(HgCl)_2$, $Hg_{32}Se_{14}(SePh)_{36}$ results with a cubic space group. However, with $HgCl_2$, the trigonal polymorph results. With $CdCl_2(PPh_3)_2$, the analogous Cd compound results with a $Cd_{32}Se_{50}$ cage[90].

X. CYCLIC SELENIDES

The cyclic selenides $(R_2SiSe)_2$ or $(R_2SiSe)_3$ (R = Me, Et) result from Na_2Se and R_2SiCl_2 as thermally unstable oils, but $[Me_3Si)_2SiSe]_3$, a green crystalline solid, is more stable and, on heating with hexamethylcyclotrisiloxane, gives the silaseleninone insertion derivative[91]. There is no evidence for silylene extrusion from the Si—Se ring. In DMP, M_2E (E = Se, Te; M = Na, K) reacts with R_2SiX_2 to give the silicon—selenium rings[92], while with $RSiCl_3$, the adamantane-like cage derivatives **27** result[57].

Heating $TsiSiH_3$ with selenium at 150–160 °C in decalin containing DBU gives 8.4% yield of the 2,3,5,6-tetraselena-1,4-bicyclodisila[2.1.1] hexane **28**, which on irradiation loses selenium to give the bicyclo[1.1.1]pentane **29**[59]. This shows Si—Se bonds of 232 pm, a little longer than the normal length of 227 pm[93].

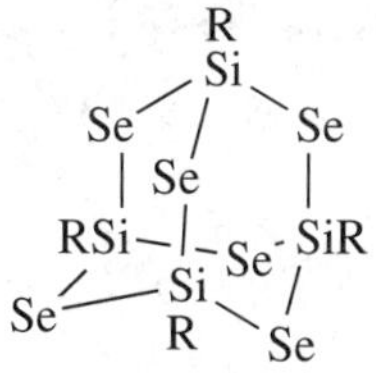

(27), 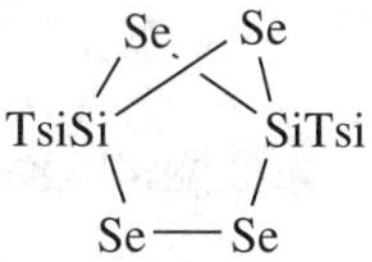 **(28)**,  **(29)**

Tsi = $(Me_3Si)_3C$

Addition of selenium and tellurium to tetramesityldisilene gives the selenirane and tellurirane (equation 42), which continue the trend with the chalcogen derivatives of a steady increase of the Si—Si bond length in going from the oxirane to the tellurirane[94].

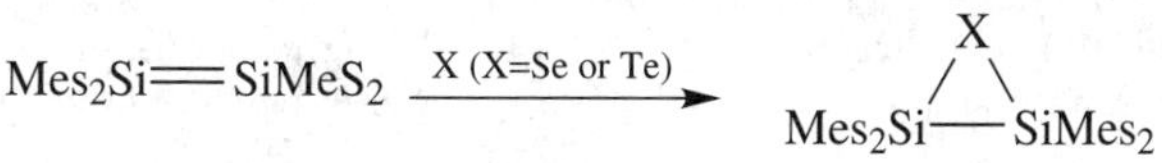

$$Mes_2Si=SiMes_2 \xrightarrow{X \ (X=Se \ or \ Te)} Mes_2Si\overset{X}{\underset{}{\triangle}}SiMes_2 \quad (42)$$

Reacting the silene $Mes(Me_3Si)Si=C(OSiMe_3)Ad$-1 with selenium gives the silaselenirane with a 3-membered CSiSe ring, which shows no tendency to dimerize[62].

The silylene Tbt(Ph)Si: reacts with selenium to give the 6-membered heterocycle **30** through insertion of silylene, once the 5-membered derivative **31** has been formed

(equation 43)[71].

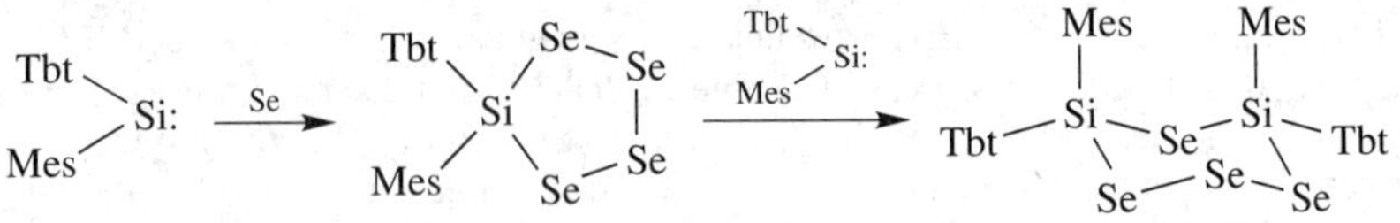

$$\text{Tbt}_2\text{Si: (Mes)} \xrightarrow{\text{Se}} \text{[Tbt(Mes)Si(Se)}_2\text{(Se)]} \longrightarrow \text{[Tbt(Mes)Si-Se-Si(Mes)Tbt with Se bridges]} \quad (43)$$

(31)   (30)

XI. TRIS(TRIMETHYLSILYL)SILYL SELENIDE DERIVATIVES

Reacting $(Me_3Si)_3SiLi(THF)_3$ with selenium results in the insertion of selenium into the Si—Li bond to give the dimeric $(Me_3Si)_3SiSeLi(THF)_2$[31]. The DME (1,2-dimethoxyethane) derivative is similarly dimeric with Si—Se bonds of 227.5 pm and Si—Si bonds of 234 pm[95]. Protonation gives the selenenol, pK_a 8.3, which reacts with a range of amides and organo derivatives of p-block, d-block and f-block elements. In addition, $(Me_3Si)_3SiSeLi(THF)_2$ reacts with a range of chlorides of these elements to give the seleno derivatives.

Thus for the p-block elements, the range of products are summarized in equations 44–46 and Scheme 6[96].

$$M(NR_2)_3 + 3(Me_3Si)_3SiSeH \longrightarrow [(Me_3Si)_3SiSe]_3M \quad (44)$$

M = Al, Bi; R = Me$_3$Si; M = As, Sb; R = Me

$$Cp_3In + 3(Me_3Si)_3SiSeH \longrightarrow [(Me_3Si)_3SiSe]_3In \quad (45)$$

$$CpM + (Me_3Si)_3SiSeH \longrightarrow (Me_3Si)_3SiSeM \quad (46)$$

M = In, Te

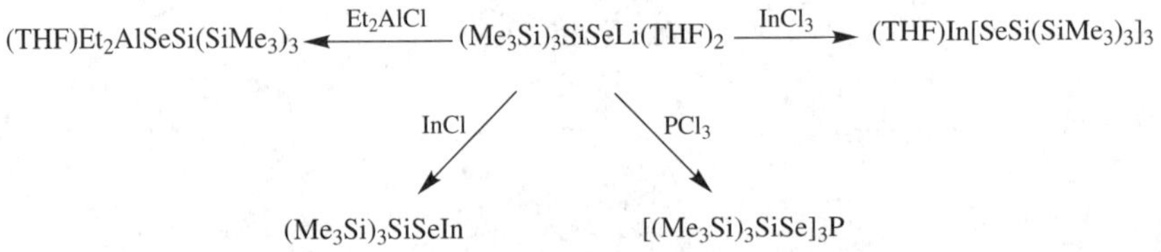

$$(THF)Et_2AlSeSi(SiMe_3)_3 \xleftarrow{Et_2AlCl} (Me_3Si)_3SiSeLi(THF)_2 \xrightarrow{InCl_3} (THF)In[SeSi(SiMe_3)_3]_3$$

$$\downarrow InCl \qquad \qquad \downarrow PCl_3$$

$$(Me_3Si)_3SiSeIn \qquad [(Me_3Si)_3SiSe]_3P$$

SCHEME 6

A similar range of derivatives of Ti, Zr and Hf result as indicated in Scheme 7, and of particular interest are the reactions of **32** with DMPM to give the 'A-frame'-like complexes **33** with disilyl selenide elimination[97].

The vanadium(III) complex **34** has a Si—Se bond length of 230.3 and, like the Ph$_3$Si derivative, reacts with styrene oxide, propylene sulphide and selenium or Ph$_3$P=Se, to give the vanadium(V) derivative (equation 47)[98].

Lanthanide(II) selenides result similarly from their THF amide complexes $[Me_3Si)_2N]_2Ln(THF)_2$ (Ln = Yb, Eu, Sm) with $(Me_3Si)_3SiSeH$ in Et$_2$O with excess TMEDA or DMPE. Yb(SeSi(SiMe$_3$)$_3$)$_2$·(TMEDA)$_2$ shows the selenide ligands in a *trans* relationship while $[Eu\{SeSi(SiMe_3)_3\}_2(DMPE)_2]_2\{\mu\text{-DMPE}\}$ crystallizes as ligand bridged dimers. The Yb derivative pyrolyses at 200 °C to give $[(Me_3Si)_3Si]_2Se$ and

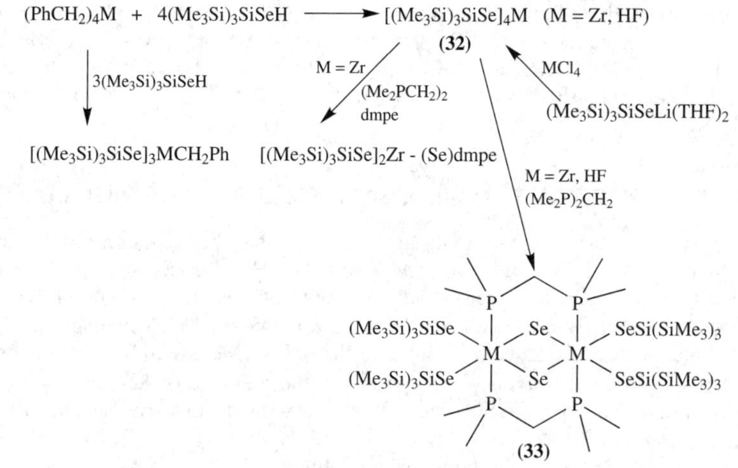

31. Chemistry of compounds with Si–S, Si–Se and Si–Te bonds

$(PhCH_2)_4M + 4(Me_3Si)_3SiSeH \longrightarrow [(Me_3Si)_3SiSe]_4M$ (M = Zr, Hf)

(32)

$\downarrow 3(Me_3Si)_3SiSeH$

M = Zr, (Me$_2$PCH$_2$)$_2$ dmpe

MCl$_4$

$[(Me_3Si)_3SiSe]_3MCH_2Ph$ $[(Me_3Si)_3SiSe]_2Zr$ - (Se)dmpe $(Me_3Si)_3SiSeLi(THF)_2$

M = Zr, Hf
(Me$_2$P)$_2$CH$_2$

(Me$_3$Si)$_3$SiSe — M — Se — M — SeSi(SiMe$_3$)$_3$
(Me$_3$Si)$_3$SiSe   Se   SeSi(SiMe$_3$)$_3$

(33)

SCHEME 7

YbSe[99]. With (Me$_3$Si)$_2$N]$_3$M (M = La, Ce) and (Me$_3$Si)SiSeH, 3-coordinate monomers result in toluene but the Y derivative occurs in solution as a selenium bridged dimer. The La, Sm and Yb(III) derivatives complex with 2 moles of THF[100].

$$[(Me_3Si)_2N]_2VSeSi(SiMe_3)_3 \xrightarrow[-Ph_3P]{Ph_3P=Se} [(Me_3Si)_2N]_2V\begin{matrix}Se\\ SeSi(SiMe_3)_3\end{matrix} \quad (47)$$

(34)

XII. TELLUROSILANES

Coupling sodium tellurocarboxylates with R′$_3$SiCl gives the unstable silyl derivative which isomerizes to the more thermodynamically stable RC(Te)OSiR′$_3$ through a facile rearrangement that can be readily followed by ^{13}C NMR spectroscopy. This contrasts with the reaction of alkali metal thiocarboxylates which give the O-silyl ester directly. There is no evidence for the S-silyl intermediate[101].

Ph$_3$SiTeLi(THF)$_3$ results from the Ph$_3$SiLi(THF)$_3$ and tellurium and reacts with CF$_3$CO$_2$H to give the tellurol, the structure of which indicates a Si–Te bond of 251.1 pm. It couples with Ph$_3$SiCl to give the disilyl telluride and gives Te derivatives of the group 4 metalocenes. (t-BuC$_5$H$_4$)$_2$Zr(TeSiPh$_3$)$_2$ reacts with t-butylpyridine to give the disilyl telluride and [(t-BuC$_5$H$_4$)$_2$ZrTe]$_2$, its formation supporting an intramolecular elimination of the disilyl telluride[102]. With Cp′TiCl$_3$ and (Me$_3$Si)$_2$Te, Cp′$_5$Ti$_5$Te$_6$ results while Cp′$_2$TiCl$_2$ gives Cp′$_4$Ti$_2$Te$_4$, the structure of which comprises a symmetrical chair conformation[80].

An extensive range of copper telluride clusters results from (Me$_3$Si)$_2$Te and copper(I) chloride in Et$_2$O with a range of phosphines. With i-Pr$_3$P(L) the small cluster Cu$_4$Te$_4$[P(Pr-i)$_3$]$_4$ results along with Cu$_{23}$Te$_{13}$L$_{10}$ and Cu$_{29}$Te$_{16}$L$_{12}$, while with Et$_3$P(L′), Cu$_{16}$Te$_9$L′$_8$ results. With Et$_2$PPh (L″), Cu$_{16}$Te$_9$L″$_8$ and Cu$_{28}$Te$_{17}$L″$_{12}$ form and, for (t-Bu)$_3$P (L^t) in Et$_2$O/THF, Cu$_{26}$Te$_{16}$L$^t_{10}$ is formed[103].

Both alkyl and aryl tellurosilanes act as a good source of RTe$^-$ and Te^{2-} ligands, forming transition metal–tellurium clusters with cobalt(II) and copper(I) salts. The nuclearity of the cluster is determined primarily by the phosphine present, but also by the metal. Cobalt gives but one derivative, $Co_6Te_8(PPh_2(Pr-n))_6$, with each triangular face of the Co_6 octahedron bridged by Te ions, thereby supporting C−Te cleavage. With copper, however, the 4 derivatives $Cu_{11}(\mu_3\text{-Te(Bu-}n))_7(\mu_4\text{-Te(Bu-}n))_2(\mu_7\text{-Te})(PPh_3)_5$, $Cu_{18}(\mu_3\text{-Te(Bu-}n))_6Te_6(P(Pr-n)_3)_8$, $Cu_{58}Te_{32}(PPh_3)_{16}$ and $Cu_{23}Te_{13}(PEt_3)_{12}$ result, and again C−Te cleavage occurs[104].

XIII. TRIS(TRIMETHYLSILYL)SILYL TELLUROL DERIVATIVES

For the past few years, silicon–tellurium chemistry has been dominated by the use of $(Me_3Si)_3SiTe^-$ and its derivatives. $(Me_3Si)_3SiLi(THF)_3$ reacts at room temperature with tellurium in THF to give the lithium telluride as a THF complexed dimer $[(Me_3Si)_3SiTeLi(THF)]_2$ (**35**) with Si−Te bonds of 248 pm[31]. DME displaces THF to give the dimer with a planar Li_2Te_2 ring and the bulky $(Me_3Si)_3SiTe$ substituents *trans* to one another and Si−Te bonds of 250 pm[105]. In the presence of 12-crown-4, however, the ionic $[Li(12\text{-crown-}4)_2]^+[TeSi(SiMe_3)_3]^-$ results with the ions separated and an Si−Te distance of 246.8 pm[106].

Acidification of (**35**) gives the highly stable tellurol $(Me_3Si)_3SiTeH$ as an air-sensitive wax, with pK_a 7.3, which readily oxidizes to the ditelluride and gives base-free derivatives of the alkali metal with $(Me_3Si)_2NM$ (M = Li, Na) or KOBu-t[31].

Alkaline earth metal derivatives can also be prepared from the tellurol and $[Me_3Si)_2N]_2M(THF)_2$ (M = Ca, Sr, Ba) in hexane and crystallize from THF with 4 ligands (M = Ca, Sr) or 5 (M = Ba). The magnesium compound is prepared from Bu_2Mg and the tellurol as a THF complex of tetrahedral structure. The Mg derivative substitutes Cp_2MCl_2 to give the group IV tellurides[107].

A range of group 13 derivatives results from the tellurol and $[Me_3Si)_2N]_3Al$ and Cp_3In, the Ga derivative from the lithium telluride and $GaCl_3$. They readily give complexes as does the In(I) derivative[96].

The gallium derivative $[(Me_3Si)_2CH]_2GaTeSi(SiMe_3)_3$ results from either metathetical exchange between the Li–telluride and gallium chloride, or from the digallane and ditelluride. It is monomeric and, with the C_2Ga−TeSi unit lying in the mirror plane of the molecule, a short Ga−Te bond and high barrier to rotation suggests a π-interaction between Te and Ga[108].

The tin(II) and lead(II) tellurides result from the amide $M[N(SiMe_3)_2]_2$ and tellurol. The tin derivative is dimeric with the Sn_2Te_2 unit adopting a butterfly-like structure with the terminal tellurol groups *cis*[109]. The antimony and bismuth derivatives result from amides and the tellurol[96].

The zinc, cadmium and mercury derivatives result from the silyl amide and tellurol and, while the zinc and cadmium derivatives are dimeric in solution, the mercury derivative is a monomer. Vapour pressure molecular weight measurements indicate a decreasing tendency to dimerization with increasing atomic weight of the metal. The zinc derivative crystallizes with a planar telluride bridged Zn_2Te_2 unit, and give monomeric 4-coordinate complexes with pyridine and bipyridyl. The cadmium results in a 6-coordinate derivative $Cd[TeSi(SiMe_3)_3]_2(bipy)_2$ but 4-coordinate with dmpe[110].

$(Me_3Si)_3SiTeH$ demethylates Cp_2ZrMe_2 in a stepwise manner to give $Cp_2Zr[TeSi(SiMe_3)_3]_2$[111] while this and the Ti and Hf derivatives also result from the metallocene dichloride and the lithium telluride. The methyl tellurate $Cp_2Zr(Me)[TeSi(SiMe_3)_3]$ reacts with CO to give the η_2-acyl derivative. Addition of base to the titanium derivative $Cp_2Ti[TeSi(SiMe_3)_3]L$ (L = phosphine or isocyanide), which can be prepared directly from $(Cp_2TiCl)_2 \cdot Cp_2Ti[TeSi(SiMe_3)_3]_2$, eliminates the ditelluride $[Me_3Si)_3SiTe]_2$ on coordination to CO, CO_2 or CS_2[112].

The lithium telluride substitutes $TiCl_3(THF)_3$ and $(PhCH_2)_4M$ react with the tellurol, the $[(Me_3Si)_3SiTe]_4M$ derivatives complexing with isocyanides or dmpe, the latter inducing decomposition to the disilyl telluride and metal telluride[97] (equation 48).

$$[(Me_3Si)_3SiTe]_4Mdmpe \xrightarrow[-[(Me_3Si)_3Si]_2Te]{dmpe \atop M=Zr, Hf} [(Me_3Si)_3SiTe]_2(dmpe)_2M=Te \quad (48)$$

The vanadium(III) derivative $[(Me_3Si)_2N]_2VTeSi(SiMe_3)_3$ shows a Si−Te bond length of 252.7 pm, slightly longer than that in $[(Me_3Si)_2N]_2VTeSiPh_3$ (250.7 pm), and readily reacts with styrene oxide, propylene sulphide and elemental selenium or $Ph_3P=Se$ to give the vanadium(V) derivative. The sulphur derivative decomposes on heating to eliminate ditelluride and form the vanadium(III) sulphide bridged dimer[98].

Reacting $(Me_3Si)_3SiTeLi(THF)_2$ with $MCl_2(dmpe)$ (M = Cr, Mn, Fe) gives $M[TeSi(SiMe_3)_3]_2(dmpe)_2$, but $CoBr_2(PMe_3)_3$ gives $Co[TeSi(SiMe_3)_3](PMe_3)_3$ through disproportionation giving the ditelluride $[(Me_3Si)_3SiTe)_2$[113].

Tellurolysis of $[(Me_3Si)_2N]_2M$ with $(Me_3Si)_3SiTeH$ in the presence of base gives the 4-coordinate ditellurides, while the structures of the dmpe complexes of Mn(II) and Fe(II) show Si−Te bond lengths of 250 −252 pm, shorter than those in Fe(II) complexes (with values 253.7 −254.6 pm), while that in the Co(I) derivative was shorter than both at 249.4 pm.

Tellurolysis of $[(Me_3Si)_2N]_3M$ (M = La, Ce) gives the tritellurols, which complex dmpe but decompose at 20 °C to give the clusters $M_5Te_3[TeSi(SiMe_3)_3]_9$ through elimination of the disilyl telluride. The structures of these clusters indicate a Ce_3Te_3 ring capped by $Ce[TeSi(SiMe_3)_3]_3$ units, with terminal $(Me_3Si)_3SiTe$ units on the three Ce atoms. Pyrolysis at 600 °C gives M_2Te_3. These tellurols, like that of yttrium, complex with dmpe, while the europium derivative gives a 7-coordinate complex with dmpe bridging two $[(Me_3Si)_3SiTe]_2Eu(dmpe)_2$ units. The ytterbium complex $[(Me_3Si)_3SiTe]_2Yb$ (TMEDA) pyrolyses to the disilyl telluride and YbTe. The selenide behaves similarly [114].

XIV. REFERENCES

1. H.-G. Horn, *J. prakt. Chem.*, **334**, 201 (1992).
2. I. Dance and K. Fisher, *Prog. Inorg. Chem.*, **41**, 637 (1994).
3. J. Arnold, *Prog. Inorg. Chem.*, **43**, 353 (1995).
4. *Inorg. Synth.*, **29**, 30 (1992).
5. R. K. Sibao, N. L. Keder and H. Eckert, *Inorg. Chem.*, **29**, 4163 (1990).
6. I. S. Nizamov, V. A. Kuznetzov, E. S. Batyeva, V. A. Al'fonsov and A. N. Pudovik, *Phosphorus, Sulphur, Silicon*, **93**, 179 (1993); I. S. Nizamov, V. A. Kuznetzov and E. S. Batyeva, *Phosphorus, Sulphur, Silicon*, **90**, 249 (1994).
7. A.-M. Le Nocher and P. Metzner, *Tetrahedron Lett.*, **33**, 6151 (1992).
8. P. Beslin and P. Marion, *Tetrahedron Lett.*, **33**, 5339 (1992).
9. M. B. Sassaman, G. K. S. Prakash and G. A. Olah, *Synthesis*, 104 (1990).
10. A. Kusche, R. Hoffmann, I. Munster, P. Keiner and R. Bruckner, *Tetrahedron Lett.*, **32**, 467 (1991).
11. T. Miura and Y. Masaki, *Tetrahedron*, **51**, 10477 (1995).
12. T. Mukaiyama, T. Ohno, T. Nishimura, J. S. Han and S. Kobayashi, *Bull. Chem. Soc. Jpn.*, **64**, 2524 (1991).
13. T. Soga, H. Takenoshita, M. Yamada, J. S. Han and T. Mukaiyama, *Bull. Chem. Soc. Jpn.*, **64**, 1108 (1991).
14. A. Degl'Innocenti, P. Ulivi, A. Capperucci, G. Reginato, A. Mordini and A. Ricci, *Synlett*, 883 (1992).
15. C.-M. Yu, H.-S. Choi, W.-H. Jung and S.-S. Lee, *Tetrahedron Lett.*, **37**, 7095 (1996).

16. T. Buskas, P. J. Garegg, P. Konradsson and J.-L. Maloisel, *Tetrahedron: Asymmetry*, **5**, 2187 (1994).
17. Y. Lim and W. K. Lee, *Tetrahedron Lett.*, **36**, 8431 (1995).
18. M. Muller, W. R. Forster, A. Holst, A. J. Kingma, W. Schaumann and G. Adiwidjaja, *Chem. Eur. J.*, **2**, 949 (1996).
19. K. Osakada, K. Hataya and T. Yamamoto, *J. Chem. Soc., Chem. Commun.*, 2315 (1995).
20. J. B. Baruah, K. Osakada and T. Yamamoto, *Organometallics*, **15**, 456 (1996); J. B. Baruah, *Polyhedron*, **15**, 3709 (1996).
21. M. Aizenberg, R. Goikhman and D. Milstein, *Organometallics*, **15**, 1075 (1996).
22. H.-C. Liang and P. A. Shapley, *Organometallics*, **15**, 1331 (1996).
23. J. Zubieta, E. Block, G. Ofori-Okai and K. Tang, *Inorg. Chem.*, **29**, 4595 (1990).
24. H. W. Roesky and U. Otten, *J. Fluorine Chem.*, **46**, 433 (1990).
25. N. Tokitoh, T. Imakubo and R. Okazaki, *Tetrahedron Lett.*, **33**, 5819 (1992).
26. J. Brittain and Y. Gareau, *Tetrahedron Lett.*, **34**, 3363 (1993).
27. B. Hache and Y. Gareau, *Tetrahedron Lett.*, **35**, 1837 (1994).
28. H.-S. Dang and B. P. Roberts, *Tetrahedron Lett.*, **36**, 3731 (1995).
29. R. Minkowitz, A. Kornath and H. Preut, *Z. Anorg. Allg. Chem.*, **619**, 877 (1993).
30. R. Minkowitz, A. Kornath and H. Preut, *Z. Anorg. Allg. Chem.*, **620**, 981 (1994).
31. P. J. Bonasai, V. Christou and J. Arnold, *J. Am. Chem. Soc.*, **115**, 6777 (1993).
32. M. Ballestri, C. Chatgilialoglu and G. Seconi, *J. Organomet. Chem.*, **408**, C1 (1991); J. Daroszewski, J. Lusztyk, M. Degueil, C. Navarro and B. Maillard, *J. Chem. Soc., Chem. Commun.*, 586 (1991); D. P. Curran and B. Yoo, *Tetrahedron Lett.*, **33**, 6931 (1992).
33. G. Gattow and H.-P. Dewald, *Z. Anorg. Allg. Chem.*, **604**, 63 (1991).
34. N. W. Mirzel, A. Schier, H. Baruda and H. Schmidbaur, *Chem. Ber.*, **125**, 1053 (1992).
35. D. G. Anderson, G. A. Forsyth and D. W. H. Rankin, *J. Mol. Struct.*, **221**, 45 (1990); D. G. Anderson, V. A. Campbell, G. A. Forsyth and D. W. H. Rankin, *J. Chem. Soc., Dalton Trans.*, 2125 (1990).
36. D. N. Dolenko, M. G. Voronkov, V. P. Elin, E. V. Dolenko, T. O. Pavlova, M.Yu.Maroshina, V. V. Belova and T. I. Zhidkova, *J. Mol. Struct.*, **326**, 221 (1994); M.Yu.Maroshina, N. N. Vlasova and M. G. Voronkov, *J. Organomet. Chem.*, **406**, 279 (1991).
37. S.-C. Tsay, G. L. Yep, B.-L. Chen, L. C. Lin and J. R. Hwu, *Tetrahedron*, **49**, 8969 (1993); J. R. Hwu and S.-C. Tsay, *J. Org. Chem.*, **55**, 5987 (1990).
38. F. Freeman, J. D. Kim, M. Y. Lee and X. Wang, *Tetrahedron*, **52**, 5699 (1996).
39. T. J. Curphey and H. H. Joyner, *Tetrahedron Lett.*, **34**, 7231 (1993).
40. A. Capperucci, A. Degl'Innocenti, A. Ricci, A. Mordini and G. Reginato, *J. Org. Chem.*, **56**, 7323 (1991); A. Degl'Innocenti, A. Capperucci, A. Mordini, G. Reginato, A. Ricci and F. Cerreta, *Tetrahedron Lett.*, **34**, 873 (1993); A. Degl'Innocenti, M. Funicello, P. Scafato and P. Spagnolo, *Chem. Lett.*, 1873 (1994) and 147 (1995); A. Capperucci, A. Degl'Innocenti, M. Funicello, P. Scafato and P. Spagnolo, *Synthesis*, 1185 (1996).
41. A. Capperucci, A. Degl'Innocenti, M. Funicella, G. Mauriello, P. Scafato and P. Spagnolo, *J. Org. Chem.*, **60**, 2254 (1995).
42. A. W. Schabacher and T. L. Maynard, *Tetrahedron Lett.*, **34**, 1269 (1993).
43. J. R. Hwu and S.-C. Tsay, *Tetrahedron* **46**, 7413 (1990); S.-C. Tsay, P. Gani and J. R. Hwu, *J. Chem. Soc., Perkin Trans. 1*, 1493 (1991).
44. P.-Y. Lin, W.-S. Ku and M.-J. Shaio, *Synthesis*, 1219 (1992); D. C. Smith, S. W. Lee and P. L. Fuchs, *J. Org. Chem.*, **59**, 348 (1994).
45. S. Schulz, E. G. Gillan, J. L. Ross, L. M. Rogers, R. D. Rogers and A. R. Barron, *Organometallics*, **15**, 4880 (1996); J. Breker, U. Wermuth and R. Schmutzler, *Z. Naturforsch., Teil B*, **45**, 1398 (1990).
46. V. C. Gibson, T. P. Kee and A. Shaw, *Polyhedron*, **9**, 2293 (1990).
47. K. K. Banger, C. S. Blackman and A. K. Brisdon, *J. Chem. Soc., Dalton Trans.*, 2975 (1996); B. Siewek, G. Koellner, K. Ruhlandt-Senge, F. Schmock and U. Muller, *Z. Anorg. Allg. Chem.*, **593**, 160 (1991).
48. D. Fenske, A. Grissinger, M. Loos and J. Magull, *Z. Anorg. Allg. Chem.*, **598/9**, 121 (1991).
49. D. Fenske, H. Fleischer, H. Krautscheid and J. Magull, *Z. Naturforsch., Teil B*, **45**, 127 (1990).
50. S. Dehnen, A. Schafer, D. Fenske and R. Ahlrichs, *Angew. Chem., Int. Ed. Engl.*, **33**, 746 (1994); S. Dehnen and D. Fenske, *Chem. Eur. J.*, **2**, 1407 (1996).
51. V. Lefevre and J.-L. Ripoll, *Tetrahedron Lett.*, **37**, 7017 (1996).

52. H. Suzuki, N. Tokitoh, S. Nagase and R. Okazaki, *J. Am. Chem. Soc.*, **116**, 11578 (1994).
53. S. D. Grumbine, R. K. Chadha and T. D. Tilley, *J. Am. Chem. Soc.*, **114**, 1518 (1992); S. D. Grumbine, T. D. Tilley and A. L. Rheingold, *J. Am. Chem. Soc.*, **115**, 358 (1993); S. D. Grumbine, T. D. Tilley, F. P. Arnold and A. L. Rheingold, *J. Am. Chem. Soc.*, **115**, 7884 (1993).
54. (a) E. I. Band and S. T. Eberhart, US Patent 4,885,378; *Chem. Abstr.*, **112**, 179472m (1990). (b) W. Uhlig, *Z. Anorg. Allg. Chem.*, **588**, 133 (1990).
55. G. A. Kraus and B. Andersh, *Tetrahedron Lett.*, **32**, 2189 (1991).
56. H.-G. Horn and M. Probst, *Monatsh. Chem.*, **126**, 1169 (1995).
57. S. R. Bahr and P. Boudjouk, *Inorg. Chem.*, **31**, 712 (1992).
58. P. Jutzi, D. Eikenberg, A. Mohrke, B. Neumann and H.-G. Stammler, *Organometallics*, **15**, 753 (1996).
59. H. Yoshida, Y. Kabe and W. Ando, *Organometallics*, **10**, 27 (1991); H. Yoshida and W. Ando, *Phosphorus, Sulphur, Silicon*, **67**, 45 (1992); N. Choi, K. Asano and W. Ando, *Organometallics*, **14**, 3146 (1995); N. Choi, K. Asano, N. Sato and W. Ando, *J. Organomet. Chem.*, **516**, 155 (1996).
60. J. E. Magnette, D. R. Powell and R. West, *Organometallics*, **14**, 3551 (1995).
61. E. Kroke, M. Weidenbruch, W. Saak, S. Pohl and H. Marsmann, *Organometallics*, **14**, 5695 (1995).
62. A. G. Brook, R. Kumarathasan and A. J. Lough, *Organometallics*, **13**, 424 (1994).
63. K. Kabeta, D. R. Powell, J. Hanson and R. West, *Organometallics*, **10**, 827 (1991).
64. N. Tokitoh, H. Suzuki and R. Okazaki, *J. Chem. Soc., Chem. Commun.*, 125 (1996).
65. J. Barrau, N. Ben Hamida and J. Stage, *Synth. React. Inorg. Met.-Org. Chem.*, **20**, 1373 (1990).
66. P. Boudjouk and U. Samaraweera, *Organometallics*, **9**, 2205 (1990).
67. P. Boudjouk, E. Black, R. Kumarathasan, U. Samaraweera, S. Castellino, J. P. Oliver and J. W. Kampf, *Organometallics*, **13**, 3715 (1994).
68. N. Choi, S. Morino, S. Sugi and W. Ando, *Bull. Chem. Soc. Jpn.*, **69**, 1613 (1996).
69. J. Albertsen and R. Steudel, *J. Organomet. Chem.*, **424**, 153 (1992).
70. N. Tokitoh, M. Takahashi, T. Matsumoto, H. Suzuki, Y. Matsuhashi and R. Okazaki, *Phosphorus, Sulphur, Silicon*, **59**, 455 (1991); N. Tokitoh, H. Suzuki, T. Matsumoto, Y. Matsuhashi, R. Okazaki and M. Goto, *J. Am. Chem. Soc.*, **113**, 7047 (1991).
71. H. Suzuki, N. Tokitoh and R. Okazaki, *Bull. Chem. Soc. Jpn.*, **68**, 2481 (1995).
72. R. Koster, G. Seidel and R. Boese, *Chem. Ber.*, **123**, 2109 (1990).
73. W.-W. du Mont, S. Kubiniok, L. Lange, S. Pohl, W. Saak and I. Wagner, *Chem. Ber.*, **124**, 1315 (1991); A. Haas, C. Limberg and M. Spehr, *Chem. Ber*, **124**, 423 (1991).
74. M. Sakakibara, K. Katsumata, Y. Watanabe, T. Toru and Y. Ueno, *Synlett*, 965 (1992).
75. G. Pandey and K. S. S. P. Rao, *Angew. Chem., Int. Ed. Engl.*, **34**, 2669 (1995).
76. H. Poleschner, M. Heydenreich, K. Spindler and G. Haufe, *Synthesis*, 1043 (1994).
77. D. Fenske, A. Frankenau and K. Dehnicke, *Z. Naturforsch., Teil B*, **45**, 427 (1990); P. M. Boorman, H.-B. Kraatz and M. Parvez, *J. Chem. Soc., Dalton Trans.*, 3281 (1992).
78. K. Shimada, S. Hikage, Y. Takeishi and Y. Takikawa, *Chem. Lett.*, 1403 (1990); Y. Takikawa, H. Watanabe, R. Sasaki and K. Shimada, *Bull. Chem. Soc. Jpn.*, **67**, 876 (1994).
79. S. Kato, M. H. Kageyama, Y. Kawahara, T. Murai and H. Ishihara, *Chem. Ber.*, **125**, 417 (1992); H. Kageyama, K. Kido, S. Kato and T. Murai, *J. Chem. Soc., Perkin Trans. 1*, 1083 (1994).
80. D. Fenske and A. Gressinger, *Z. Naturforsch., Teil B*, **45**, 1309 (1990).
81. K. Merzweiler and H.-J. Kersten, *Z. Naturforsch., Teil B*, **46**, 1025 (1991).
82. D. Fenske, H. Krautscheid and M. Muller, *Angew. Chem., Int. Ed. Engl.*, **31**, 321 (1992).
83. K. Merzweiler and L. Weisse, *Z. Naturforsch., Teil B*, **46**, 695 (1991).
84. D. Fenske, H. Fleischer, H. Krautscheid and J. Magull, *Z. Naturforsch., Teil B*, **45**, 127 (1990).
85. D. Fenske, H. Fleischer, H. Krautscheid, J. Magull, C. Oliver and S. Weisgerber, *Z. Naturforsch., Teil B*, **46**, 1384 (1991).
86. D. Fenske, H. Krautscheid and S. Balter, *Angew. Chem., Int. Ed. Engl.*, **29**, 796 (1990); S. Dehnen and D. Fenske, *Angew. Chem., Int. Ed. Engl.*, **33**, 2287 (1994).
87. D. Fenske and H. Krautscheid, *Angew. Chem., Int. Ed. Engl.*, **29**, 1452 (1990).
88. H. Krautscheid, D. Fenske, G. Baum and M. Semmelmann, *Angew. Chem., Int. Ed. Engl.*, **32**, 1303 (1993).
89. I. Wagner and W.-W. du Mont, *J. Organomet. Chem.*, **395**, C23 (1990).
90. S. Behrens, M. Bettenhausen, A. C. Deveson, A. Eichhofer, D. Fenske, A. Lohde and U. Woggon, *Angew. Chem. Int., Ed. Engl.*, **35**, 2215 (1996).

91. P. Boudjouk, S. R. Bahr and D. P. Thompson, *Organometallics*, **10**, 778 (1991).
92. P. Boudjouk, *Polyhedron*, **10**, 1231 (1991).
93. H. Yoshida, Y. Takahara, T. Erata and W. Ando, *J. Am. Chem. Soc.*, **114**, 1098 (1992).
94. R. P.-K. Tan, G. R. Gillette, D. R. Powell and R. West, *Organometallics*, **10**, 546 (1991).
95. K. E. Flick, P. J. Bonasia, D. E. Gindelberger, J. E. B. Katari and D. Schwartz, *Acta Crystallogr.*, **C50**, 674 (1994).
96. S. P. Wuller, A. L. Seligson, G. P. Mitchell and J. Arnold, *Inorg. Chem.*, **34**, 4861 (1995).
97. C. P. Gerlach, V. Christou and J. Arnold, *Inorg. Chem.*, **35**, 2758 (1996).
98. C. P. Gerlach and J. Arnold, *Inorg. Chem.*, **35**, 5770 (1996).
99. D. R. Cary and J. Arnold, *Inorg. Chem.*, **33**, 1791 (1994).
100. D. R. Cary, G. E. Ball and J. Arnold, *J. Am. Chem. Soc.*, **117**, 3492 (1995).
101. S. Kato, H. Kageyama, T. Kanda, T. Murai and T. Kawamura, *Tetrahedron Lett.*, **31**, 3587 (1990).
102. D. E. Gindelberger and J. Arnold, *Organometallics*, **13**, 4462 (1994).
103. D. Fenske and J.-C. Steck, *Angew. Chem., Int. Ed. Engl.*, **32**, 238 (1993).
104. J. F. Corrigan, S. Balter and D. Fenske, *J. Chem. Soc., Dalton Trans.*, 729 (1996).
105. G. Becker, K. W. Klinkhammer, S. Lartiges, P. Bottcher and W. Poll, *Z. Anorg. Allg. Chem.*, **613**, 7 (1992).
106. P. J. Bonasia, D. E. Gindelberger, B. O. Dabbousi and J. Arnold, *J. Am. Chem. Soc.*, **114**, 5209 (1992).
107. G. Becker, K. W. Klinkhammer, W. Schwarz, M. Westerhausen and T. Hildenbrand, *Z. Naturforsch., Teil B*, **47**, 1225 (1992); D. E. Gindelberger and J. Arnold, *J. Am. Chem. Soc.*, **114**, 6242 (1992).
108. W. Uhl, M. Layh, G. Becker, K. W. Klinkhammer and T. Hildenbrand, *Chem. Ber.*, **125**, 1547 (1992).
109. A. L. Seligson and J. Arnold, *J. Am. Chem. Soc.*, **115**, 8214 (1993).
110. P. J. Bonasia and J. Arnold, *Inorg. Chem.*, **31**, 2508 (1992).
111. B. O. Dabbousi, P. J. Bonasai and J. Arnold, *J. Am. Chem. Soc.*, **113**, 3186 (1991).
112. V. Christou, S. P. Wuller and J. Arnold, *J. Am. Chem. Soc.*, **115**, 10545 (1993).
113. D. E. Gindelberger and J. Arnold, *Inorg. Chem.*, **32**, 5813 (1993).
114. D. R. Cary and J. Arnold, *J. Am. Chem. Soc.*, **115**, 2520 (1993).

CHAPTER 32

Cyclic polychalcogenide compounds with silicon

NAMI CHOI and WATARU ANDO

Department of Chemistry, University of Tsukuba, Tsukuba, Ibaraki 305, Japan
Fax: +81 298 53 6503

I. INTRODUCTION	1896
II. MONOCYCLOSILACHALCOGENIDES	1897
A. Introduction	1897
B. Cyclodisiloxanes and Cyclotrisiloxanes	1897
C. Oxidation of Disiliranes with Molecular Oxygen	1899
D. Stereochemistry and Azetidinium Imide Intermediate	1901
E. Cyclic Silathianes and Silaselenanes	1903
1. Synthesis of cyclic silathianes and silaselenanes	1903
2. Syntheses of disilathianes via silametallacycles	1904
3. Syntheses of 1,4-dithia-2,3-disilametallacyclopentanes (**32a** and **32b**)	1906
4. Reaction of disilametallacycle (**32a** and **32b**) with various electrophiles	1906
F. Reaction of Silathianes and Silaselenanes	1907
1. Thermolysis or photolysis of cyclosilathianes and cyclosilaselenanes	1907
2. Dechalcogenation	1907
G. Structures of Cyclic Silathianes	1908
1. Structures of dithiadisilatitanacycle **32a**, trithiadisilacyclopentane **33a** and cyclodisilathiane **37**	1908
III. BICYCLOSILACHALCOGENIDES	1910
A. Introduction	1910
B. Synthesis and Reactions	1911
C. Structure	1913
1. X-ray analysis of polythiadisilabicyclo[$k.l.m$]alkanes and polyselenadisilabicyclo[$k.l.m$]alkanes	1913
D. Spectroscopic Data	1915
1. NMR spectra of trithia- and triselenadisilabicyclo[1.1.1]pentanes	1915
2. UV spectra of polythiadisilabicyclo[$k.l.m$]alkanes	1915

The chemistry of organic silicon compounds, Vol. 2
Edited by Z. Rappoport and Y. Apeloig © 1998 John Wiley & Sons Ltd

E. Reactions 1916
  1. Hydrolysis of trithiadisilabicyclo[1.1.1]pentane 1916
  2. Oxidation of trithiadisilabicyclo[1.1.1]pentane 1916
IV. TRICYCLOSILACHALCOGENIDES 1916
  A. Introduction 1916
  B. Synthesis and Reactions 1917
    1. Adamantane and double-decker type 1917
    2. Nor- and bis-nor-adamantane type sesquichalcogenides 1921
  C. Structure 1923
    1. Adamantane and double-decker type sesquichalcogenides 1923
V. POLYCYCLOSILOXANES 1923
  A. Introduction 1923
  B. Synthesis and Reactions 1923
    1. Prismane and cubane type silsesquioxanes 1923
    2. Transformation of functionalized spherosilicate 1925
VI. REFERENCES 1925

I. INTRODUCTION

The chemistry of silachalcogenides has been studied widely and reviewed extensively[1,2]. A variety of polycyclic silachalcogenides has been synthesized during the past decade. The structures of these species are strongly dependent upon the substituents. Recently, some facile syntheses of cyclic silachalcogenides have been reported[3–5]. In this chapter, the chemistry of cyclic silachalcogenides will be reviewed, mainly with respect to disilapolychalcogenides and silasquichalcogenides, although some reference will be made to cyclic siloxanes.

Another class of compounds that has attracted much interest are the sesquioxides and sesquichalcogenides of the type $(RE)_{2n}Y_{3n}$ [$(M_2Y_3)_n$] (E = Si, Ge, Sn; Y = O, S, Se), due to their unique structural properties. These molecules have polyhedral cage-like structures analogous to those of bicyclo[1.1.1]pentane ($n = 1$) **(1)**, adamantane ($n = 2$)

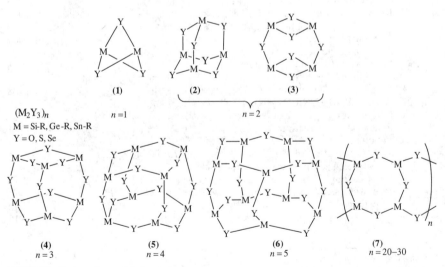

FIGURE 1. Series of sesquichalcogenides

32. Cyclic polychalcogenide compounds with silicon

(2) or double-decker ($n = 2$) (3), prismane ($n = 3$) (4), cubane ($n = 4$) (5), pentagonal prismane ($n = 5$) (6), ladder ($n = 20-30$) (7) etc. (Figure 1). The isolated sesquisulfide and selenide (Y = S, Se) have been restricted to the adamantane-type structure (2) or the double-decker structure (3) while higher homologous molecules are known for the sesquioxide (4, M = Ge, Y = O; M = Si, Y = O)[6]. Although many silsesquioxanes have been reported, there is no example of an adamantane-type structure.

At an early stage in silicon chemistry, Kipping and coworkers proposed a propellane-type sesquioxide (1, M = Si, Y = O) as an intermediate in the hydrolysis of trichlorosilanes[7]. However, the synthesis of this attractive molecule has not been achieved despite much interest from both theoretical and practical chemists. Under the appropriate conditions, trihydro- and trihalosilanes react with elemental chalcogenides or chalcogen transfer reagents to give adamantane-type silachalcogenanes[8-11] and/or bicyclopolychalcogenasilanes[12], dependent upon the bulkiness of the substituents. The smallest silasesquichalcogenide has a 2,4,5-chacogena-1,3-disilabicyclo[1.1.1]pentane skeleton. The synthesis of the sulfur and selenium analogues has been achieved by kinetic stabilization[13-15].

II. MONOCYCLOSILACHALCOGENIDES

A. Introduction

Cyclic siloxanes are important precursors in the silicon industry, being formally dimers or trimers of silanone ($R_2Si=O$), a known intermediate. Cyclic siloxanes have been synthesized by four routes, the conventional methods being the condensation of silanediol or the hydrolysis of species such as halosilanes or aminosilanes (Scheme 1)[16-20]. Alternatively, oxidation of disilene by triplet oxygen (equation 1)[21-27] or oxidation of oxadisiliranes by singlet oxygen (equation 2)[28-31] may be utilized.

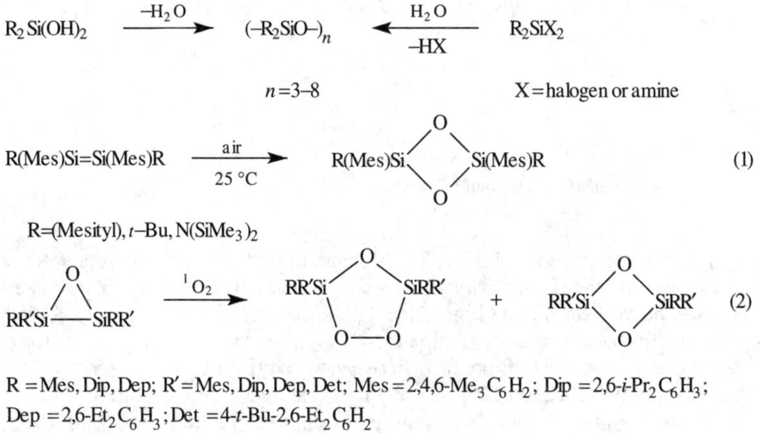

$$R_2Si(OH)_2 \xrightarrow{-H_2O} (-R_2SiO-)_n \xleftarrow[-HX]{H_2O} R_2SiX_2$$

$$n = 3-8 \qquad\qquad X = \text{halogen or amine}$$

$$R(Mes)Si=Si(Mes)R \xrightarrow[25\,°C]{air} R(Mes)Si\underset{O}{\overset{O}{\diamond}}Si(Mes)R \qquad (1)$$

R = (Mesityl), t–Bu, N(SiMe$_3$)$_2$

$$RR'Si\underset{}{\overset{O}{\triangle}}SiRR' \xrightarrow{^1O_2} RR'Si\underset{O-O}{\overset{O}{\diamond}}SiRR' + RR'Si\underset{O}{\overset{O}{\diamond}}SiRR' \qquad (2)$$

R = Mes, Dip, Dep; R' = Mes, Dip, Dep, Det; Mes = 2,4,6-Me$_3$C$_6$H$_2$; Dip = 2,6-i-Pr$_2$C$_6$H$_3$;
Dep = 2,6-Et$_2$C$_6$H$_3$; Det = 4-t-Bu-2,6-Et$_2$C$_6$H$_2$

SCHEME 1

B. Cyclodisiloxanes and Cyclotrisiloxanes

Tetramesityldisiloxane [(Mes$_2$SiO)$_2$] (8) has been prepared by the reaction of tetramesityldisilene with atmospheric oxygen[22,24]. It has been reported that 1,2-cyclodisiloxane is

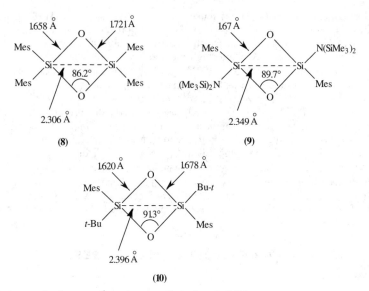

FIGURE 2. Atomic distances (Å) and angles (deg) of cyclodisiloxanes

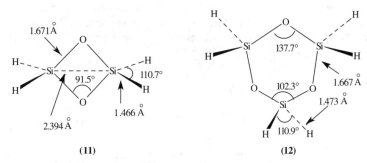

FIGURE 3. Geometries of $(H_2SiO)_2$ and $(H_2SiO)_3$

stereospecifically isomerized to 1,3-cyclodisiloxane in both solution and the solid state[32]. Crystal structures of the 1,3-cyclodisiloxanes **8, 9** and **10**, differing only in their substituents, have been determined (Figure 2). The structure of tetramesitylcyclodisiloxane consists of roughly a square cyclodisiloxane ring orthogonal to a slightly distorted planar skeleton due to the steric hindrance between mesityl groups. The mesityl rings are diposed in a roughly helical fashion about this skeleton. Therefore, the cyclodisiloxane ring has two independent silicon—oxygen bond lengths. 1.658(4) and 1.721(4) Å. The Si—Si distance in **8** of 2.306 Å is shorter than the theoretical value (*vide infra*); this may be explained by the fact that the four-membered ring is not planar, but distorted with a O—Si—Si—O angle of 6°[25]. The Si—Si distance in **9** is slightly shorter than the calculated distance of $(H_2SiO)_2$ (**11**) 2.394 Å[33], and the HF/6-31G* value of 2.469 Å[34] in the cyclic dimer $(SiO)_2$. By comparison of the calculated distances in $(F_2SiO)_2$ (2.341 Å) and $(H_2SiO)_2$ (2.394 Å), it appears that the greater the charge separation (or ionic character) the shorter the Si—Si bond distance. Moreover, the HF/3-21G optimized structure of

32. Cyclic polychalcogenide compounds with silicon 1899

cyclic trimer $(H_2SiO)_3$ (12) (Figure 3) has D_{3h} symmetry with alternative silicon and oxygen atoms arranged in a planar six-membered ring[33]. The planarity of the Si_3O_3 ring has also been observed in the electron diffraction study of hexamethylcyclotrisiloxane $(Me_2SiO)_3$ (13)[35,36]. The Si—O—Si and O—Si—O angles of 137.7° and 102.3° in the trimer $(H_2SiO)_3$ are much larger than those in the dimer $(H_2SiO)_2$. The disproportionation energies for $3(H_2SiO)_2 \rightarrow 2(H_2SiO)_3$ are calculated to be -162.3 kcal mol^{-1} at the HF/3-21G level and -110.4 kcal mol^{-1} at the HF/6-31G*//3-21G level. These large negative values clearly favor the trimer over the dimer[33]. In contrast, the disproportionation energy for $3(H_2SiS)_2 \rightarrow 2(H_2SiS)_3$ is calculated to be much less negative $(-30.3$ kcal mol$^{-1})$[33].

C. Oxidation of Disiliranes with Molecular Oxygen

Photochemical reactions of silicon–silicon σ-bonds of disilirane (14) with molecular oxygen have been studied in three modes: charge transfer (CT), electron transfer (ET) and singlet oxygen. An oxygen-saturated acetonitrile solution of 1,1,2,2-tetramesityl-1,2-disilirane (14a) was irradiated to give 3,3,5,5-tetramesityl-1,2-dioxa-3,5-disilolane (15a) in 36% yield (equation 3). Similar results were obtained with 2,2,3,3-tetrakis(2,6-diisopropylphenyl)oxadisilirane (14b) and 2,2,3,3-tetrakis(2,6-diethylphenyl)oxadisilirane (14c) affording the corresponding 1,2,4-trioxa-3,5-disilolanes (15b and 15c) (equation 3). The structure of 15c is given in Figure 4[37]. The UV absorption spectrum of the disilirane 14a in oxygen-saturated solvents such as methylene chloride or acetonitrile reveals a weak broad contact CT band with a maximum at 300 nm. Photolysis of 14a in an oxygen matrix at 16 K resulted in smooth oxygenation, giving a new species with an intense IR band at 1078 cm^{-1}. When $^{18}O_2$ is employed, the IR band shifts by 39 cm^{-1} to a lower wave number. Upon subsequent warming of the matrix to temperatures higher than 50 K, the intense band at 1078 cm^{-1} disappeared and the oxygen matrix became cloudy[38]. Product analysis at room temperature showed the formation of 15a and 1,2,4-oxadisiletane 16a. The intense band at 1078 cm^{-1} is ascribable to the characteristic O—O stretching mode of the disilirane–oxygen adduct 17 in the triplet state. Accordingly, the silicon–silicon bond with its low ionization energy easily donates σ-electrons to the oxygen molecule, thereby forming a CT complex. A calculation of vibrational frequencies predicts that the O—O stretching frequency of 17 should be near 1083 cm^{-1} and should shift to 1026 cm^{-1} upon double ^{18}O substitution. Thus, the labile intermediate which is formed in CT photooxygenation of 14 can be represented by structure 17.

$$R_2Si\overset{X}{\underset{}{\triangle}}SiR_2 \xrightarrow[h\nu < 370\ nm]{CT, ET\ and\ ^1O_2} R_2Si\overset{X}{\underset{O-O}{\diagup\diagdown}}SiR_2 + R_2Si\overset{X}{\underset{O}{\diagup\diagdown}}SiR_2 \quad (3)$$

(14) (15) (16)

(a) R = Mes; X = CH$_2$
(b) R = Dip; X = O
(c) R = Dep; X = O
(d) R = Mes; X = O
(e) R = Xyl; X = O
(f) R = Mes; X = N-Ph

CT = $h\nu$ (<370 nm)/O$_2$/C$_6$H$_6$
ET = $h\nu$ (>400 nm)/O$_2$/DCA/CH$_2$Cl$_2$
or $h\nu$ / MB/O$_2$/–78 °C/CH$_2$Cl$_2$
$^1O_2$ = $h\nu$/TPP/O$_2$/C$_6$H$_6$

(Dip = 2,6-i-Pr$_2$C$_6$H$_3$; Det:4-t-Bu-2,6-Et$_2$C$_6$H$_2$; Dep = 2,6-Et$_2$C$_6$H$_3$; Xyl = 2,6-Me$_2$C$_6$H$_3$)

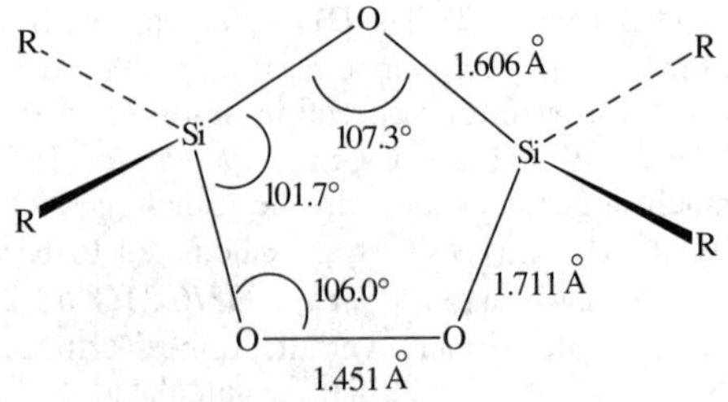

(15c)

FIGURE 4. X-ray structure analysis of **15c**

Irradiation of oxadisilirane **14c** in an acetonitrile–methylene chloride solvent mixture in the presence of 9,10-dicyanoanthracene (DCA) as the sensitizer with tungsten–halogen lamps under an oxygen flow resulted in the formation of **15c** in 69% yield[31]. In the absence of the sensitizer, no reaction occurred and **14c** was recovered quantitatively. A similar result was also obtained in the reaction of **14c** with $^3O_2$ in the presence of 10 mol% of tris(p-bromophenyl)aminium hexachloroantimonate [$(p\text{-BrC}_6\text{H}_4)_3\text{N}^+\text{SbCl}_6^-$] as the single ET reagent, giving **15c** in 58% yield. The free-energy changes (ΔG) of -26.7 kcal mol^{-1} for electron transfer from **14a** to DCA and -23.0 kcal mol^{-1} in the case of **14c** are indicative of an exothermic ET. The DCA fluorescence was efficiently quenched by **14c** [$k_q = 1.12 \times 10^{10}$ (M s^{-1})]. The proposed mechanism for ET oxygenation of **14** is shown in Scheme 2. The initially formed radical cation **14**$^{+\bullet}$ is attacked by either $O_2^{-\bullet}$ or $^3O_2$, directly giving **15** or [[**14**] · O_2]$^{+\bullet}$ as the intermediate (Scheme 2). In the

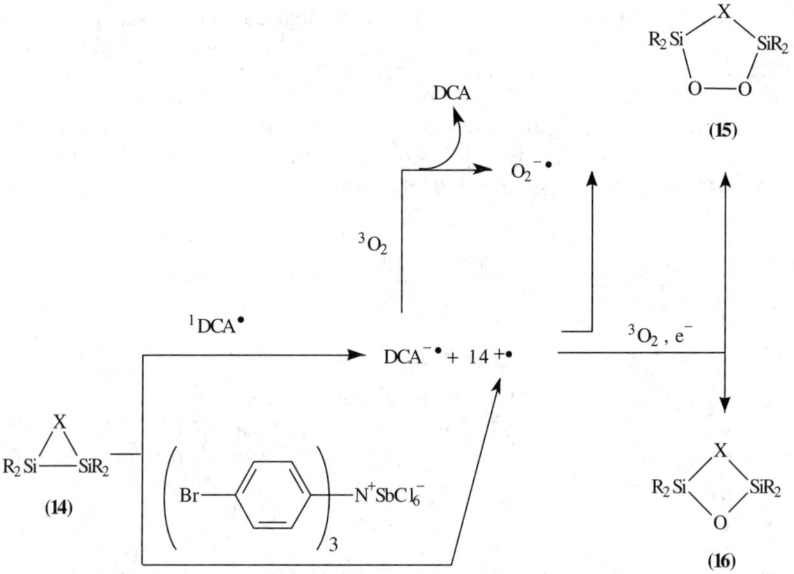

SCHEME 2

latter case, ET to [[**14**] • O$_2$]$^{+•}$ affords **15**. It should be noted that participation of O$_2^{-•}$ is excluded when (p-BrC$_6$H$_4$)$_3$N$^+$SbCl$_6^-$ is present.

Although singlet oxygen reacts with various electron-rich olefins and heteroatomic molecules, so far there have been no reports concerning direct reactions with σ-bonds. The photooxygenations of disiliranes **14b–f**, carried out in benzene with tetraphenylporphine (TPP) as the sensitizer, gave cyclic peroxides **15b–f** in 40–71% isolated yield with a small amount of monooxygenated products **16**[29–31,39]. The reactions were inhibited by addition of DABCO, a known ^{1}O$_2$ quencher, whereas the addition of triphenylmethane, a free-radical scavenger, did not have any influence on the course of the reactions.

D. Stereochemistry and Azetidinium Imide Intermediate

DCA-sensitized photooxygenations of *cis*- and *trans*-**18** gave 68:32 and 41:59 mixtures of *cis*- and *trans*-**19**, respectively[31]. These results are in agreement with the formation of the open intermediate **17**, and the fact that rotation around the Si−O bonds is fairly competitive with the collapse to **19**. Similar results, i.e. nonstereospecific formation of *cis*- and *trans*-**19** with almost identical selectivity, were also observed in the contact CT oxygenation of *cis*- and *trans*-**19** with (p-BrC$_6$H$_4$)$_3$N$^+$SbCl$_6^-$ (Scheme 3 and Table 1). Moreover, stereochemical studies of singlet oxygenation were carried out. The TPP-sensitized photooxygenation of *trans*-**18** gave *trans*-**19** as the exclusive product in 84% yield. Similarly, photooxygenation of *cis*-**18** afforded only *cis*-**19** in 79% yield[31]. Clearly, the TPP-sensitized photooxygenations of *cis*- and *trans*-**18** proceed in a stereospecific manner, which is consistent with a peroxonium ion mechanism with **20** as the intermediate. A similar mechanism has been proposed for 1,2-dioxetane formation by ^{1}O$_2$ (Scheme 4).

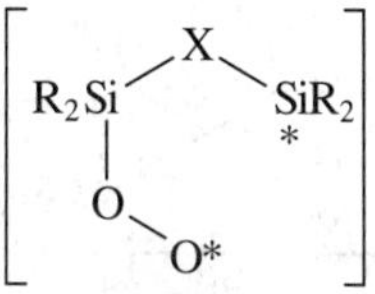

$$\left[\begin{array}{c} R_2Si\diagup^{X}\diagdown SiR_2 \\ * \\ O\diagdown_{O^*} \end{array} \right]$$

(17)

* = • or +/−

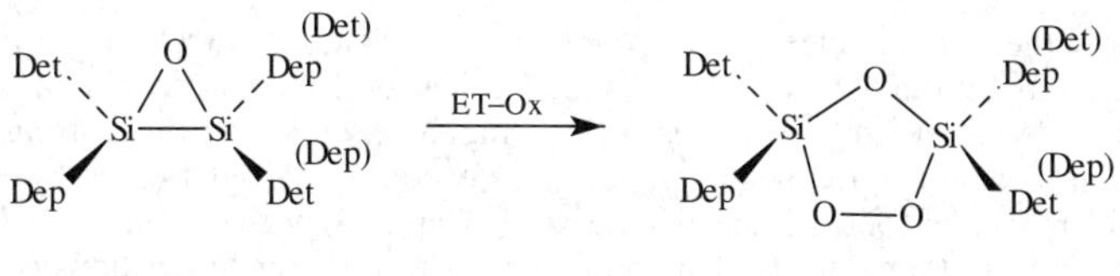

(18) →[ET–Ox] **(19)**

SCHEME 3. (For definition of Det and Dep, see equation 3)

4-Phenyl-1,2,4-triazoline-3,5-dione (PTAD) is a reagent which mimics the behavior of singlet oxygen. Thus it is capable of undergoing [2 + 2] and [2 + 4] cycloadditions, as well as ene reactions, and exhibits a similar reactivity toward azines and sulfides. Moreover, in most cases, similar mechanistic pathways have been proposed for both reagents, while in singlet-oxygen chemistry, the lifetime of intermediates is often too

TABLE 1

Substrate	Condition	Products and yields	
		19 (%)	cis/trans
cis-18	hν(λ > 400 nm)/O_2/DCA[a]/CH_2Cl_2	51	68/32
trans-18	hν(λ > 400 nm)/O_2/DCA[a]/CH_2Cl_2	35	41/59
cis-18	M.B.[b]/O_2/−78 °C/CH_2Cl_2	88	66/34
trans-18	M.B.[b]/O_2/−78 °C/CH_2Cl_2	95	42/58

[a] 9,10-Dicyanoanthracene.
[b] $(p\text{-BrC}_6H_4)_3N^+SbCl_6^-$; 10 mol%

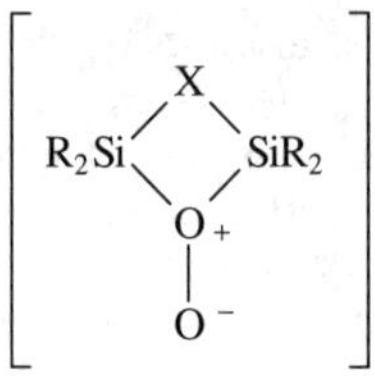

$$\left[\begin{array}{c} R_2Si \overset{X}{\underset{O_+}{\diamond}} SiR_2 \\ | \\ O^- \end{array} \right]$$

(20)

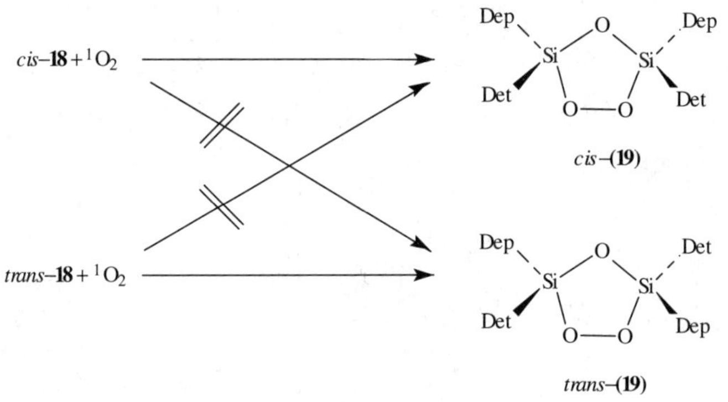

SCHEME 4

short to allow their direct detection. PTAD intermediates have a much longer lifetime and can be detected. Consequently, PTAD might be a useful reagent to obtain insight into the structure of reactive intermediates in the singlet oxygenation of disiliranes (**14** and **18**). As mentioned above, the results of addition reactions of singlet oxygen with strained silicon–silicon σ-bonds of disilirane derivatives (**14** and **18**) provide considerable evidence of peroxonium ion intermediates. This prompted us to investigate reactions of disiliranes with PTAD instead of singlet oxygen. Addition of PTAD to a $CHCl_3$ solution of 1,1,2,2-tetramesityldisilirane **14a** led to rapid disappearance of the red color of PTAD, and gave adduct **21** as colorless crystals in 38% isolated yield (69% yield in the crude reaction mixture as determined by ^1H NMR) (equation 4). The course of the reaction of PTAD with **14a** was monitored by ^1H, ^{13}C and ^{29}Si NMR spectroscopy at −78 °C. A series of new resonances in the ^{13}C NMR spectrum appeared at δ 157.29(s), 150.70(s), 143.73(s), 140.68(s), 141.39(s), 129.81(d), 129.62(d) 127.58(d), 126.48(d), 130.91(d), 129.32(d), 24.30(q), 23.61(q), 21.10(q) and 5.81(t) which gradually decreased at −55 °C, while those of **21** increased. The presence of two different carbonyl carbon resonances at δ 157.29 and 150.70 showed that the mirror plane orthogonal to the plane of the PTAD moiety is lost; similarly, the appearance of two signals for the diastereomeric methylene protons in the ^{1}H NMR spectra indicates a loss of symmetry. The 2D NMR and C–H COSY spectra imply the presence of a four-membered ring, as indicated by a comparison of the observed ^{13}C-^{1}H coupling constants with those of other cyclic disiletanes, thus the structure of the intermediate is best represented by the azetidinium structure **22**. The rate of disappearance of intermediate **22** has been measured by the integration of characteristic ^{1}H NMR signals at several temperatures; from these values an activation energy of 20.6 ± 2.0 kcal mol^{-1} required for the breakdown of **21** is calculated. A plausible rationale for these observations is that in the reaction of **14a** with PTAD, **22** is initially formed, which collapses to **21**, similar to the singlet oxygenation of **14a**. Meanwhile, comparable cycloadditions of cyclopropanes with PTAD have been reported. Roth and Martin[40] postulated the formation of a diradical intermediate in the reaction of PTAD with the strained central σ-bond of a bicyclo[2.1.0]pentane derivative. Nevertheless, although [$_\pi 2 + {}_\sigma 2$] additions have been a subject of continual interest, no direct observation of such intermediates has been reported so far. The above results provide not only proof of a peroxonium intermediate (**20**) in the singlet oxygenation of disiliranes, but also represent the first example of a direct observation of an intermediate in [$_\pi 2 + {}_\sigma 2$] addition.

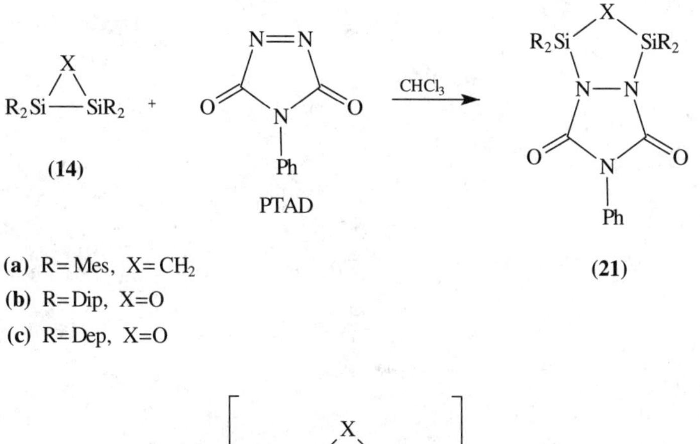

(14) + PTAD $\xrightarrow{\text{CHCl}_3}$ (21)   (4)

(a) R= Mes, X= CH$_2$
(b) R= Dip, X= O
(c) R= Dep, X= O

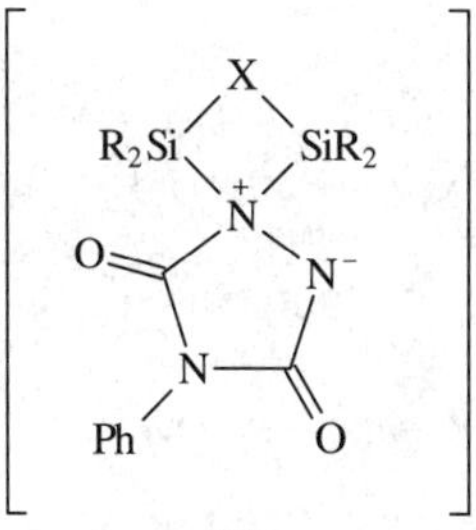

(22)

E. Cyclic Silathianes and Silaselenanes

1. Synthesis of cyclic silathianes and silaselenanes

Heterocycles containing group 14 elements[1,41–44] have received much attention in terms of both their chemistry and their use in material science. Cyclic silathianes were synthesized by the reaction of dihalosilanes or trihalosilanes with alkali sulfides or dihydrosilanes or trihydrosilanes with elemental sulfur (equation 5)[1,41b]. The facile syntheses of cyclosilachalcogenides by the reaction of halosilanes with alkali sulfides or selenides were reported by Thompson and Boudjouk (equation 6)[45]. Rauchfuss and coworkers[46,47] and Albertsen and Steudel[3] have reported the synthesis of silapolychalcogenides via silatitanacycles. Disilachalcogenides via disilatitanacycles and disilazirconacycles were reported[4,5]. Okazaki, Tokitoh and coworkers have devised a method to prepare silachalcogenanes bearing bulky groups (Scheme 5)[48–50].

$$M_2E + R_2SiCl_2 \xrightarrow{E=S,Se} (R_2SiE)_n \quad (5)$$
$$M = Li, Na \qquad\qquad n = 2, 3$$

$$Na_2Se + R_2SiCl_2 \xrightarrow{THF, rt} \begin{array}{c} R_2Si\text{---}Se \\ | \quad\quad | \\ Se\text{---}SiR_2 \end{array} + \begin{array}{c} R_2\\ Si \\ Se \quad Se \\ | \quad\quad | \\ R_2Si \quad SiR_2 \\ Se \end{array} \quad (6)$$

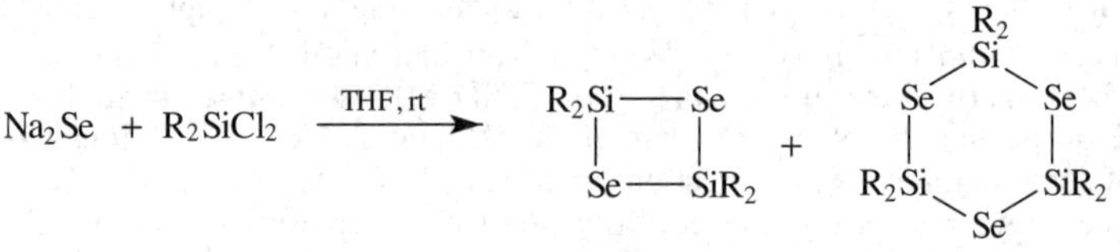

R = Me (6%)   R = Me (35%)
R = Et (40%)   R = Et (30%)
R = Me$_3$Si (35%)   R = Ph (40%)

SCHEME 5

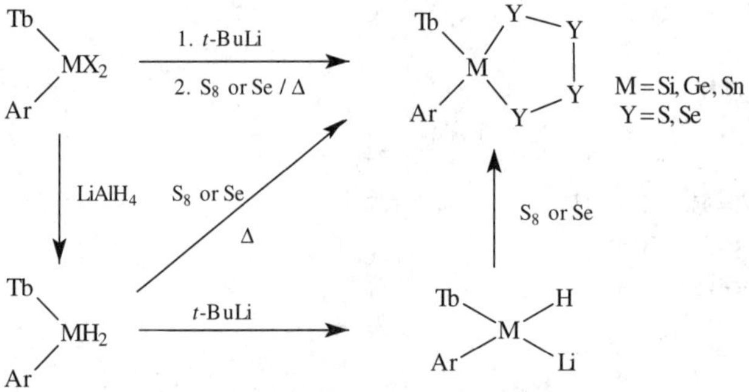

Tb = 2,4,6-Tris[bis(trimethylsilyl)methyl]phenyl
Ar = Mesityl or 2,4,6-Triisopropylphenyl

M = Si, Ge, Sn
Y = S, Se

2. Syntheses of disilathianes via silametallacycles

The halides Ph$_2$SiCl$_2$, Ph$_2$GeBr$_2$ and Ph$_2$SnCl$_2$ react with Li$_2$S and (η^5-C$_5$H$_4$R)$_2$TiCl$_2$ (R = H, Me) to give the corresponding four-membered titanacycles Cp$_2$TiS$_2$MPh$_2$ (M = Si, Ge, Sn) (**23**) (equation 7). Cp$_2$TiS$_2$SiPh$_2$ (**23**) reacts with S$_2$Cl$_2$ to afford Ph$_2$SiS$_4$ (**24**) quantitatively (equation 8)[3]. Dimesityldichlorogermane reacted with 2 eq. of Li$_2$S and did not yield lithium germanedithiolate (**26**) but rather tetramesityl dithiadigermetane (**25**) (Scheme 6)[3]. On the other hand, dilithium dimesitylgermanedithiolate **26** was readily obtained by the treatment of a THF solution of Mes$_2$Ge(SH)$_2$ (**27**) with a THF solution of LiBEt$_3$H at room temperature (equation 9). Successive addition of a THF solution of

32. Cyclic polychalcogenide compounds with silicon

dichlorobis(η^5-cyclopentadienyl)titanium to Mes$_2$Ge(SLi)$_2$ (**26**) gave a germatitanacycle **28a** as a green crystalline solid in 70% yield (equations 9 and 10). Germazirconacycle **28b** was also obtained as a yellow solid in 90% yield in a similar manner. Germatitanacycle **28a** underwent ready substitution reaction at the titanocene moiety with S$_2$Cl$_2$ in CS$_2$ to give tetrathiagermolane **29** as a yellow oil in 53% yield (equation 11). When a toluene solution of **28b** and Ph$_2$GeCl$_2$ was refluxed for 20 hours, unsymmetric 1,3-dithia-2,4-digermetane **30** was prepared in 52% yield (equation 12).

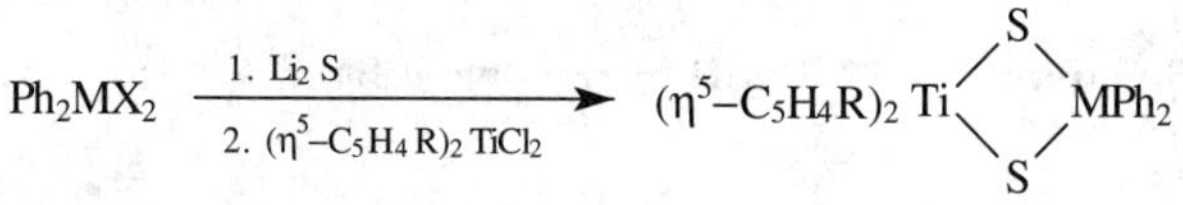

$$\text{Ph}_2\text{MX}_2 \xrightarrow[\text{2. } (\eta^5\text{-C}_5\text{H}_4\text{R})_2\text{TiCl}_2]{\text{1. Li}_2\text{S}} (\eta^5\text{-C}_5\text{H}_4\text{R})_2\text{Ti} \begin{matrix} \text{S} \\ \diagup \diagdown \\ \diagdown \diagup \\ \text{S} \end{matrix} \text{MPh}_2 \qquad (7)$$

M = Si, Ge, Sn   R = H, Me   (**23**)
X = Cl or Br

$$\textbf{23}, \text{M = Si} \xrightarrow{\text{S}_2\text{Cl}_2} \text{Ph}_2\text{Si} \begin{matrix} \text{S---S} \\ | \quad | \\ \text{S---S} \end{matrix} \qquad (8)$$

(**24**)

$$\text{Mes}_2\text{Ge} \begin{matrix} \text{S} \\ \diagup \diagdown \\ \diagdown \diagup \\ \text{S} \end{matrix} \text{GeMes}_2 \longleftarrow \text{Mes}_2\text{GeCl}_2 + 2\text{Li}_2\text{S} \not\longrightarrow \text{Mes}_2\text{Ge(SLi)}_2$$

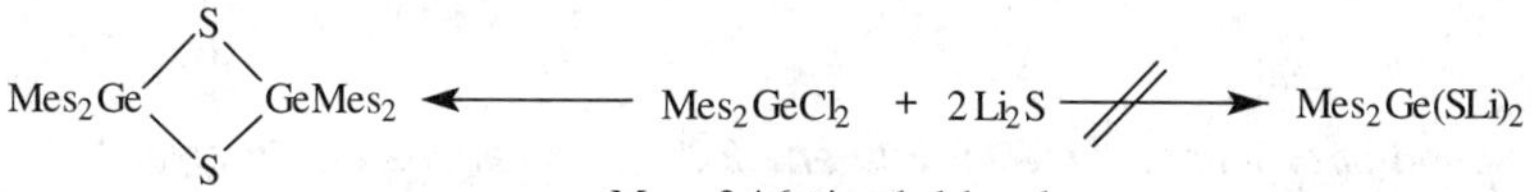

Mes = 2,4,6-trimethylphenyl

(**25**)   (**26**)

SCHEME 6

$$\text{Mes}_2\text{Ge(SH)}_2 + 2\text{ LiBEt}_3\text{H} \xrightarrow{\text{THF, rt}} \textbf{26} \qquad (9)$$
(**27**)

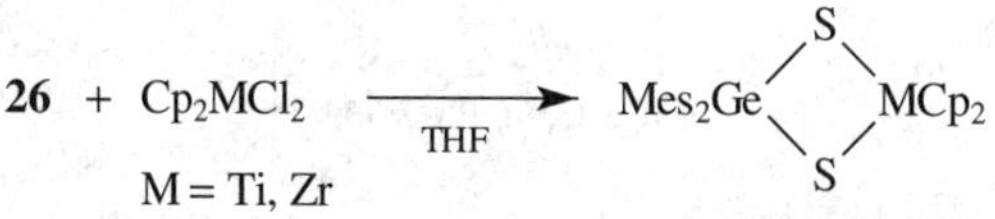

$$\textbf{26} + \text{Cp}_2\text{MCl}_2 \xrightarrow{\text{THF}} \text{Mes}_2\text{Ge} \begin{matrix} \text{S} \\ \diagup \diagdown \\ \diagdown \diagup \\ \text{S} \end{matrix} \text{MCp}_2 \qquad (10)$$

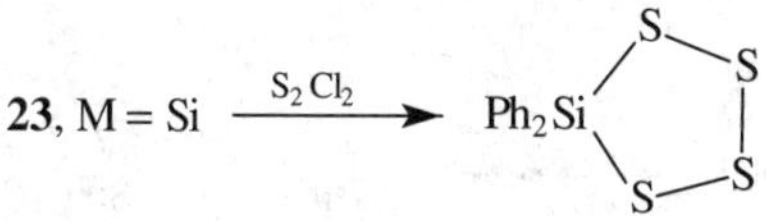

M = Ti, Zr
(**28a**) M = Ti
(**28b**) M = Zr

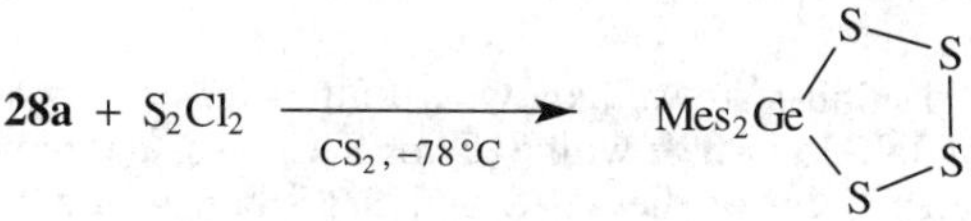

$$\textbf{28a} + \text{S}_2\text{Cl}_2 \xrightarrow{\text{CS}_2, -78\,°\text{C}} \text{Mes}_2\text{Ge} \begin{matrix} \text{S---S} \\ \diagup \quad | \\ \diagdown \quad \text{S} \\ \text{S---S} \end{matrix} \qquad (11)$$

(**29**)

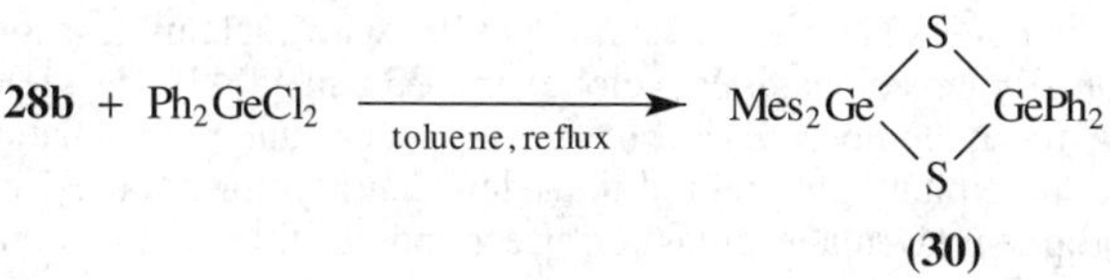

$$\textbf{28b} + \text{Ph}_2\text{GeCl}_2 \xrightarrow{\text{toluene, reflux}} \text{Mes}_2\text{Ge} \begin{matrix} \text{S} \\ \diagup \diagdown \\ \diagdown \diagup \\ \text{S} \end{matrix} \text{GePh}_2 \qquad (12)$$

(**30**)

3. Syntheses of 1,4-dithia-2,3-disilametallacyclopentanes (32a and 32b)

Disilametallacyclopentanes, **32a** and **32b**, can be prepared from *meso*-1,2-di-*t*-butyl-1,2-diphenyldisilane-1,2-dithiol (**31**) in high yield (equations 13 and 14)[4,5]. Disilatitanacycle **32a** can be isolated as green crystals in 80% yield, and zirconacycle **32b** as a yellow crystalline solid in 90% yield. Disilatitanacycle **32a** is relatively stable toward moisture and air, but zirconacycle **32b** is unstable toward moisture and is air-sensitive.

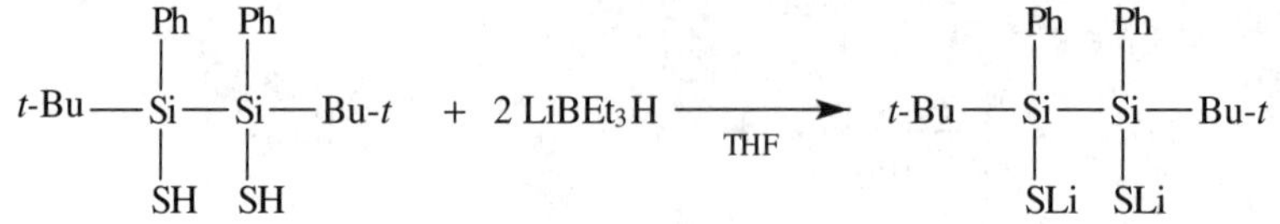

$$t\text{-Bu}-\underset{\underset{SH}{|}}{\overset{\overset{Ph}{|}}{Si}}-\underset{\underset{SH}{|}}{\overset{\overset{Ph}{|}}{Si}}-Bu\text{-}t + 2\,LiBEt_3H \xrightarrow{THF} t\text{-Bu}-\underset{\underset{SLi}{|}}{\overset{\overset{Ph}{|}}{Si}}-\underset{\underset{SLi}{|}}{\overset{\overset{Ph}{|}}{Si}}-Bu\text{-}t \quad (13)$$

(**31**) *meso*

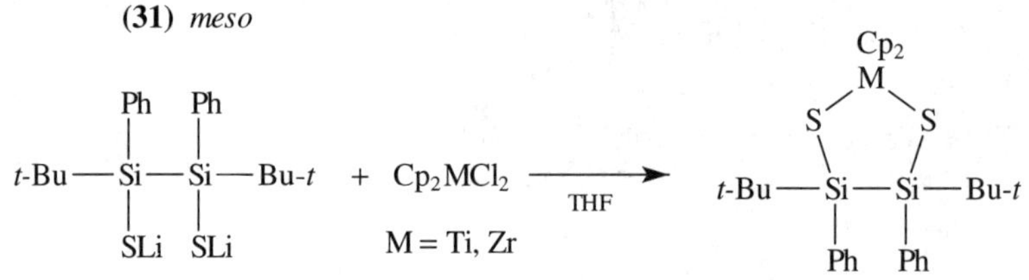

$$t\text{-Bu}-\underset{\underset{SLi}{|}}{\overset{\overset{Ph}{|}}{Si}}-\underset{\underset{SLi}{|}}{\overset{\overset{Ph}{|}}{Si}}-Bu\text{-}t + Cp_2MCl_2 \xrightarrow{THF} \quad (14)$$

M = Ti, Zr

(**32a**) M = Ti
(**32b**) M = Zr

4. Reaction of disilametallacycle (32a and 32b) with various electrophiles

By using the disilametallacycles **32a** and **32b**, various disilaheterocycles are readily prepared under appropriate conditions, as shown in Scheme 7 and Table 2[4,5].

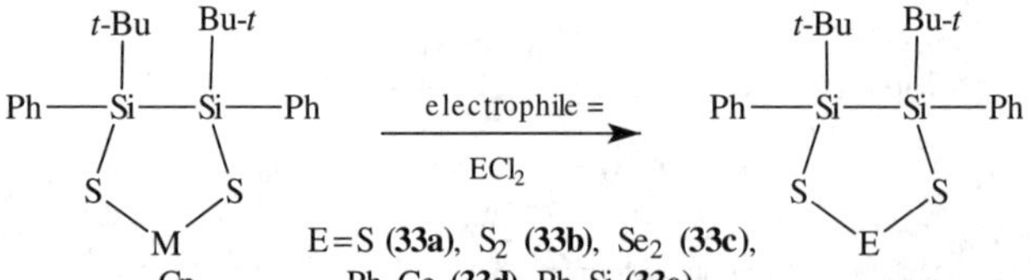

electrophile = ECl_2

E = S (**33a**), S_2 (**33b**), Se_2 (**33c**), Ph_2Ge (**33d**), Ph_2Si (**33e**)

(**32a**) M = Ti
(**32b**) M = Zr

(**33**)

SCHEME 7

Although silyl-substituted sulfides are well known as sulfur-transfer reagents[51] by Si—S bond cleavage, the reaction of **32a** or **32b** with SCl_2 or S_2Cl_2 was performed on the site of Ti or Zr. The relative reactivity of the Ti—S and Si—S bonds in a four-membered TiS_2Si ring is explained by Rauchfuss and coworkers[47] as an example of frontier orbital control.

Treatment of disilametallacycles **32a** and **32b** with selenium monochloride at $-78\,°C$ affords the dithiadiselenadisilacyclohexane **33c** in 90–100% yield. Compound **33c** is stable at room temperature, but decomposes under irradiation, liberating selenium as a red precipitate to yield dithiaselenadisilacyclopentane. The reaction of **32a** with dichlorodiphenylgermane gives a corresponding dithiadisilagermacyclopentane

32. Cyclic polychalcogenide compounds with silicon

TABLE 2. Reactions of **32a** or **32b** with various electrophiles

Entry	Electrophile	Product	Solvent	Time (h)	Temp. (°C)	Yield (%) 32a	Yield (%) 32b
1	SCl_2	**33a**	CS_2	0.5	−78	90	93
2	S_2Cl_2	**33b**	CS_2	0.5	−78	92	66
3	Se_2Cl_2	**33c**	CS_2	0.5	−78	90	quant.
4	Ph_2GeCl_2	**33d**	toluene	24	110	55	40
5	Ph_2SiCl_2	**33e**	xylene[a]	48	140	—	—

[a]Isomeric mixture.

derivative **33d**. On the other hand, the reaction of **32a** with halosilane to give dithiatrisilacyclopentane **33e** was unsuccessful, even under xylene reflux for 72 hours.

F. Reaction of Silathianes and Silaselenanes

1. Thermolysis or photolysis of cyclosilathianes and cyclosilaselenanes

Pyrolysis of cyclic silachalcogenides is a well-known route to reactive intermediates of the $R_2Si=X$ (X = O, S, Se) species (equation 15)[52]. $(Me_2SiSe)_3$ and $(Et_2SiSe)_3$ undergo ring transformation, photochemically and thermally, to form $(Me_2SiSe)_2$ and $(Et_2SiSe)_2$, respectively (equation 16)[53]. Photolysis or thermolysis of $(Me_2SiSe)_3$ or $(Me_2SiSe)_2$ in the presence of $(Me_2SiO)_3$ gives trioxatetrasilaselenocane derivative (**36**), which is produced by trapping of silaneselone ($R_2Si=Se$, **35**) with $(Me_2SiO)_3$ (Schemes 8 and 9).

$$(R_2E - X)_n \xrightarrow{\Delta'} nR_2E = X$$

$$R = \text{alkyl, H, halogen}; E = Si, Ge; X = O, S, Se \qquad (15)$$

$$2(R_2SiSe)_3 \longrightarrow 3(R_2SiSe)_2 \qquad (16)$$

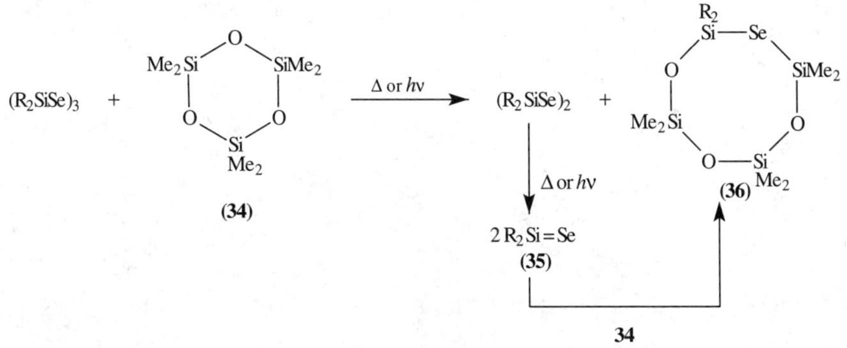

SCHEME 8

2. Dechalcogenation

Polychalcogenasilane is a good precursor of silanechalcogenone (equation 17)[48−51]. Desulfurization of polychalcogenadisilane **33a** and **33b** gives 1,3-cyclodisilathiane (**37**).

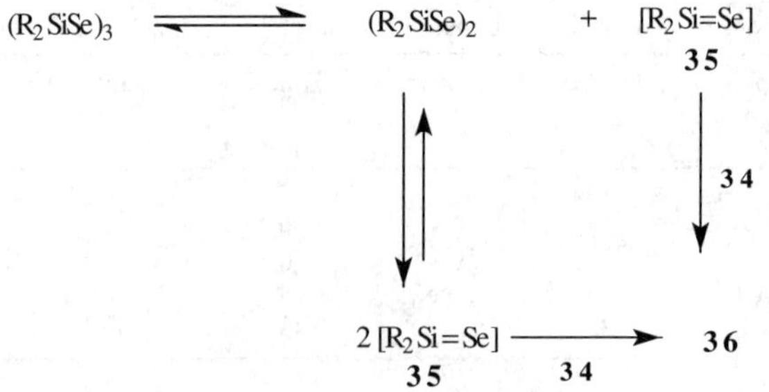

$(R_2SiSe)_3 \rightleftharpoons (R_2SiSe)_2 + [R_2Si=Se]$
                                                35

$\Big\updownarrow$ $\Big\downarrow$ **34**

$2 [R_2Si=Se] \xrightarrow{} \mathbf{36}$
35   **34**

SCHEME 9

Desulfurization of **33a** by PPh$_3$ (1.1 eq.) gives 2,4-di-*t*-butyl-2,4-diphenyl-1,3-dithia-2,4-disilacyclobutane **37** containing *cis*- and *trans*-isomers (ratio: $cis/trans = 1/9$) without 1,2-dithia-3,4-disilacyclobutane **36**, as shown in Scheme 10. The desulfurization of **33a** by hexamethylphosphorous triamide at room temperature also gave similar results. A mixture of *cis*- and *trans*-isomers of **37** was also obtained from desulfurization of **33b** (*cis*). The formation of **37** suggests two plausible routes, as shown in Scheme 11. One possible route is an intramolecular rearrangement via an intermediate **36**. The stereospecific rearrangements of 1,2-dioxa-3,4-disilacyclobutanes to 1,3-dioxa-2,4-disilacyclobutanes are reported by West and coworkers[26]. Since this reaction is not stereospecific, the first route (a) is not likely. Another, more likely route is (b), which is the dimerization of thioxosilane derivative **39**.

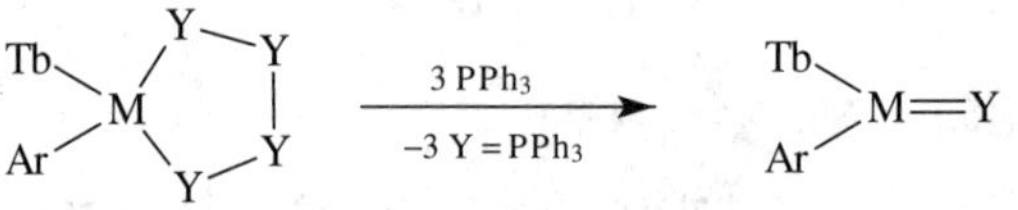

$$\begin{array}{c} Tb \diagdown \quad Y \diagup Y \\ M \\ Ar \diagup \quad Y \diagdown Y \end{array} \xrightarrow[-3Y=PPh_3]{3\,PPh_3} \begin{array}{c} Tb \diagdown \\ M=Y \\ Ar \diagup \end{array} \tag{17}$$

M = Si, Ge, Sn
Y = S, Se, Te
Tb = 2,4,6-{(Me$_3$Si)$_2$CH}$_3$C$_6$H$_2$
Ar = 2,4,6-R$_3$C$_6$H$_2$ (R=Me, *i*-Pr)

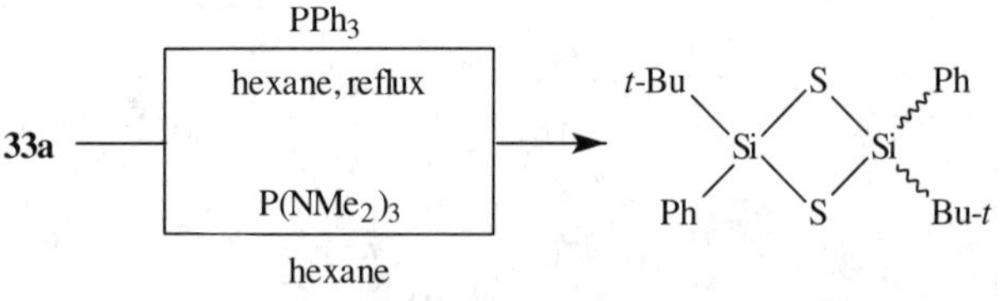

33a → [PPh$_3$, hexane, reflux / P(NMe$_2$)$_3$ hexane] → *t*-Bu, S, Ph, Si, Si, Ph, S, Bu-*t*

(**37**)
cis : *trans* = 1 : 9

SCHEME 10

G. Structures of Cyclic Silathianes

1. Structures of dithiadisilatitanacycle **32a**, trithiadisilacyclopentane **33a** and cyclodisilathiane **37**

The titanacycle **32a** exhibits a half-chair conformation as depicted in Figure 5. The Ti–S bond lengths (2.430 Å and 2.437 Å) are within the normal range (2.42–2.45

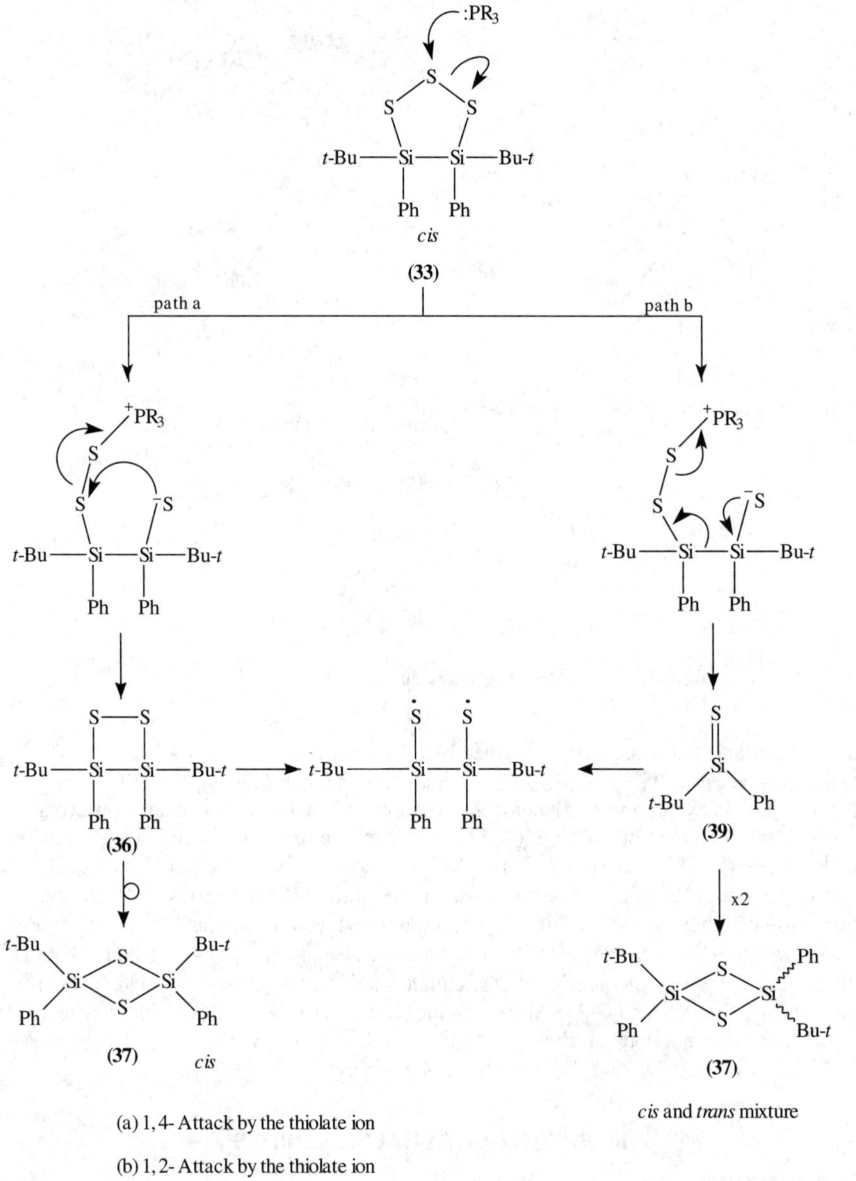

SCHEME 11

(a) 1,4-Attack by the thiolate ion
(b) 1,2-Attack by the thiolate ion

Å[46]), the Si—S bonds (2.13 Å) are slightly shortened (usually 2.16–2.17 Å)[46]. Interestingly, the Ti—S—Si bond angles (113.0° and 108.4°) are larger than those of other five-membered dithiatitanacycles [e.g. o-(—SC$_6$H$_4$S—)TiCp$_2$: 95.7–97.0°[54]; (—SCH=CHS—)TiCp$_2$: 94.2–95.0°[55]] and resemble those of acyclic derivatives [e.g. Cp$_2$Ti(SPh)$_2$: 113.6–115.5°[56]]. On the other hand, the S—Ti—S bond angle (93.9°) takes

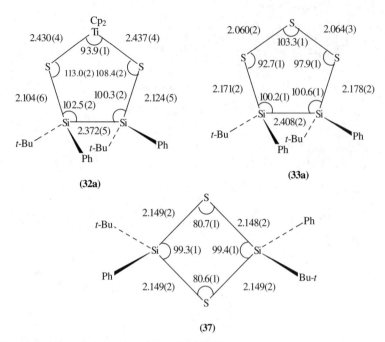

FIGURE 5. Bond distances and angles of **32a**, **33a** and **37**.

an intermediate value between five-membered titanacycles [e.g. o-($-SC_6H_4S-$)TiCp$_2$: 82.2°[54]; ($-SCH=CHS-$)TiCp$_2$: 83.2°[55]] and acyclic ones [e.g. Cp$_2$Ti(SPh)$_2$: 99.3°[56]]. The trithiadisilacyclopentane skeleton in compound **33a** has a half-chair conformation. The central four-membered ring of **37** was found to be completely planar (dihedral angle; S1—Si1—Si2—S2 = 0.7°). The Si—S bond lengths [2.148(2)–2.149(2) Å] are within the normal range[46]. The Si—S—Si angles [80.64(6) and 80.66(6)°] are smaller than those of other cyclodisilathianes [e.g. tetramethylcyclodisilathiane 82.46(6)°[57]; tetra-t-butoxycyclodisilathiane 82.2(1)°[58]] and the S—Si—S angles [99.40(7) and 99.30(7)°] are larger than those of other cyclodisilathiane derivatives [97.54(7)[57] and 97.8(1)°[58]]. Nonbonding Si---Si and S---S distances are 2.78 and 3.28 Å, respectively, which are within the normal range of those of cyclodisilathianes (R$_2$Si)$_2$S$_2$ [Si---Si; R = Me, 2.837(2)[57]; R = Cl, 2.725(1)[59]; R = Br, 2.741(6)[59] Å].

III. BICYCLOSILACHALCOGENIDES

A. Introduction

There has been considerable interest in the chemistry of group 4 propellanes (**40a**) and bicyclo[1.1.1]pentanes (**41b**) in view of the nature of the bridgehead bonds[60]. Pentasila[1.1.1]propellane has not been isolated as yet, although a derivative of bicyclo[1.1.1]pentasilane[61] has been synthesized. According to both calculations and experiments, M—M distances in [1.1.1]propellanes (M = C, Si, Ge, Sn; Y = CH$_2$, SiH$_2$, GeH$_2$, SnH$_2$) are much shorter than those in bicyclo[1.1.1]pentanes except for the

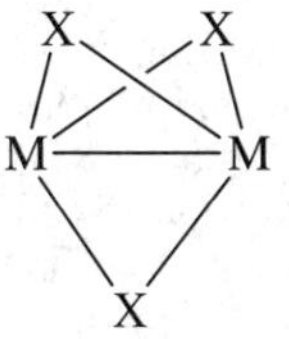

32. Cyclic polychalcogenide compounds with silicon

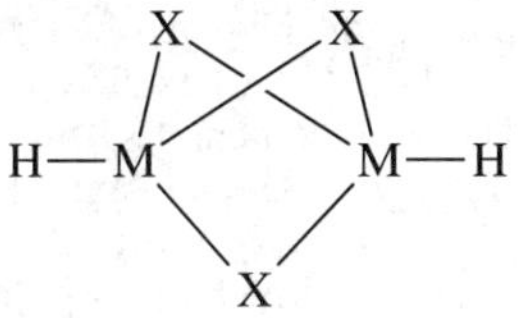

M = C, Si, Ge
X = CH$_2$, SiH$_2$, GeH$_2$, O, S, Se

(40)   (41)

TABLE 3. Calculated atomic distances (Å) and angles (deg) of disilabicyclo[1.1.1]pentane derivatives.^{a}

X	Atomic distance or angle	40 (M = Si)	41 (M = Si)
(a) SiH$_2$	Si$_b$---Si$_b$	2.719	2.915
	Si$_p$---Si$_p$	2.347	2.364
	Si$_b$—Si$_p$—Si$_b$	70.8	76.2
(b) O	Si---Si	2.089	2.060
	Si—O	1.714	1.698
	Si—O—Si	75.1	74.7
(c) S	Si---Si	2.347	2.363
	Si—S	2.171	2.176
	Si—S—Si	65.4	65.8
(d) Se	Si---Si	2.457	2.484
	Si—Se	2.340	2.339
	Si—Se—Si	63.3	64.1

^{a}Si$_b$ denotes bridgehead Si, Si$_p$ denotes peripheral Si.

tin analogue[60-74]. Theoretically, substitution of electronegative groups relative to silicon (e.g. O, CH$_2$) at the peripheral positions should stabilize the central M—M interaction for M = Si[72]. Therefore, the structures of [1.1.1]propellanes of the type M$_2$X$_3$ [M = Si, Ge, Sn; Y = O, S, Se] (40) and the corresponding bicyclo[1.1.1]pentanes H$_2$M$_2$X$_3$ (41) are very similar. The calculated geometries of silicon analogues are shown in Table 3[70,73,74b]. According to theoretical studies, trioxadisilabicyclo[1.1.1]pentane (41b) is predicted to have a short distance between the bridgehead silicon atoms of only 2.060 Å[74a]; this is shorter than that of a silicon–silicon double bond. The corresponding distance in the sulfur analogue, trithiadisilabicyclo[1.1.1]pentane (41c), has been predicted to be 2.363 Å[74a], which corresponds to a Si—Si single bond. Trithiadisilabicyclo[1.1.1]pentane derivative and its selenium analogue were isolated by the introduction of the bulky substituent, tris(trimethylsilyl)methyl group, on the silicon atom[14,15]. The next section will describe the synthesis and reactivities of polychalcogenadisilabicyclo[k.l.m]alkanes.

B. Synthesis and Reactions

Chalcogenation of conventional substituted trihydrosilanes gives silachalcogenides having adamantane skeletons (see Section IV.B)[12-15,75]. Tris(trimethylsilyl)methylsilane (TsiSiH$_3$) (42, M = Si) reacts with elemental sulfur to give 2,4,5,6-tetrathiadisilabicyclo[2.1.1]hexane derivative (43a, 22% yield) along with (TsiSi)$_2$S$_5$ (44a, 12%) and (TsiSi)$_2$S$_6$ (45a 12%) as shown in equation 18 and Table 4. Bicyclosilathianes 43-45 (M=Si) are stable toward air and moisture. The structures of polythiadisilabicyclo[k.l.m]alkanes 43a, 44a and 45a are confirmed by X-ray structure analysis to be 2,4,5,6-tetrathia-1,3-disilabicyclo[2.1.1]hexane, 2,4,5,6,7-pentathia-1,3-disilabicyclo[2.2.1]heptane and 2,4,5,6,7,8-hexathia-1,3-disilabicyclo[3.2.1]octane, respectively. The reaction of TsiGeH$_3$ with elemental sulfur gave 2,4,5,6,7,8,9-heptathia-1,3-disilabicyclo[3.3.1]nonane **46a** along with **43a, 44a** and **45a** (equation 18 and Table 4). Trithiadisilabicyclo[1.1.1]pentanes **47a** can be synthesized by photolysis of tetrathiadisilabicyclo[2.1.1]hexane in the presence of trimethylphosphine or triphenylphosphine in 100% or 64% yield, respectively. Tetraselenadisilabicyclo[2.1.1]hexane can be prepared by the reaction of TsiSiH$_3$ with selenium/1,8-diazabicyclo[5.4.0]undec-7-ene. Triselena-1,3-disilabicyclo[1.1.1]pentane and its germanium analogue can be prepared in a similar manner (Scheme 12).

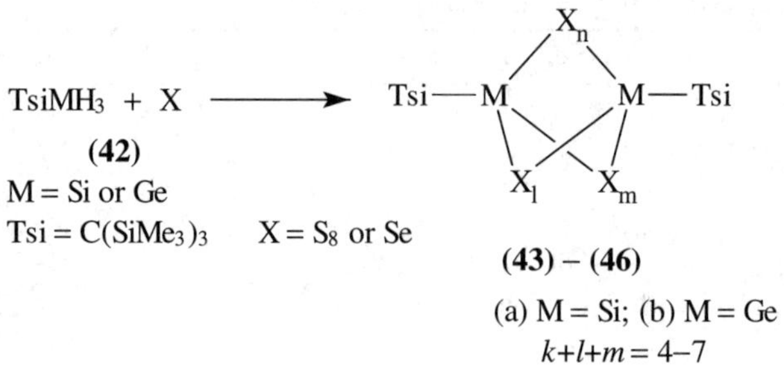

$$\text{TsiMH}_3 + X \longrightarrow \text{Tsi—M} \begin{smallmatrix} X_n \\ \\ X_l \ X_m \end{smallmatrix} \text{M—Tsi} \tag{18}$$

(42)

M = Si or Ge

Tsi = C(SiMe$_3$)$_3$ X = S$_8$ or Se

(43) – (46)

(a) M = Si; (b) M = Ge

$k+l+m = 4-7$

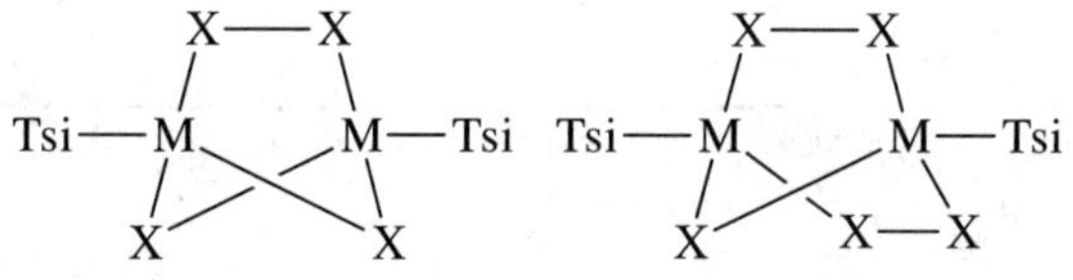

(43)   (44)

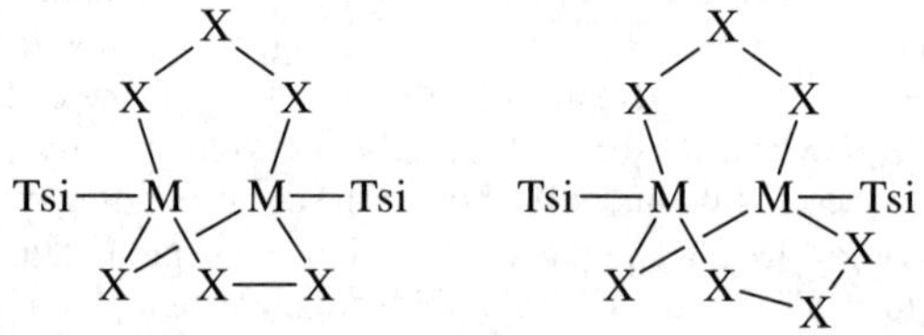

(45)   (46)

TABLE 4. Yield and reaction conditions in equation 18

M	X	Reaction conditions		Yield (%)			
				43	44	45	46
			[k.l.m]	[2.1.1]	[2.2.1]	[3.2.1]	[3.3.1]
Si	S$_8$	30 eq. decalin, 190–200 °C		22	12	12	—
	Se	10 eq. decalin, DBU 160 °C		19	—	—	—
Ge	S$_8$	30 eq. Ph$_2$O, 150–160 °C		8	8	26	9
	Se	10 eq. decalin, DBU 160 °C		19	12	27	—

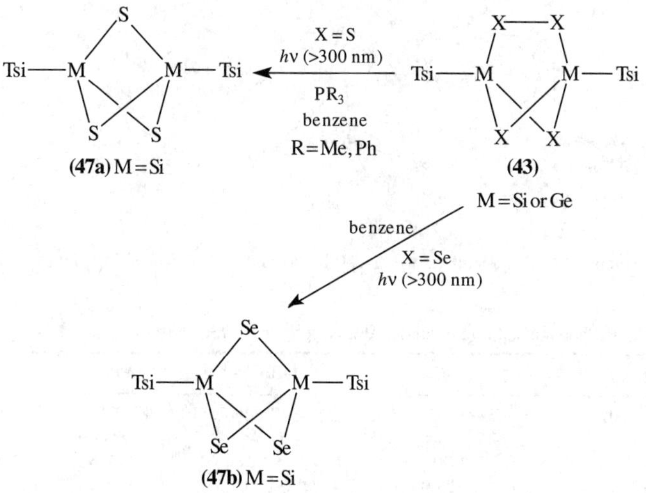

32. Cyclic polychalcogenide compounds with silicon   1913

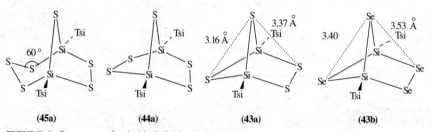

SCHEME 12

C. Structure

1. X-ray analysis of polythiadisilabicyclo[k.l.m]alkanes and polyselenadisilabicyclo[k.l.m]alkanes

The results of single-crystal X-ray diffraction analyses of polychalcogenadisilabicyclo-[k.l.m]alkanes are summarized in Figures 6 and 7 and Tables 5 and 6.

FIGURE 6. Structures of polythiadisilabicyclo[k.l.m]alkanes

The five-membered ring in bicyclo[2.1.1]hexane skeleton is an envelope conformation, and torsion angles of Si−S−S−Si = 0.05–0.08°. The structure of **47a**[13] is in good agreement with the results of the quantum chemical calculation[73,74], which is reported by Gordon and coworkers as shown in Table 3. The bicyclo[1.1.1]pentane skeleton is constructed from three cyclobutanes (dihedral angles of each of the cyclobutanes are 120.40°, 119.75° and 118.85°, av. 120.0°). The nonbonding Si- - -Si distances of **44a, 43a** and **47a** are 3.00, 2.66 and 2.407 Å. The Si−S−Si angles of **44a, 43a** and **47a** are 89.2°, 75.9° (av.) and 67.08° (av.). Considering the common Si−S−Si angles (ca 82°)[76–79] in

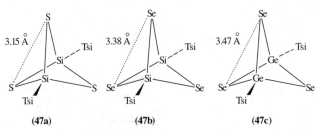

FIGURE 7. Structures of 2,4,5-trichalcogena-1,3-disila- and digerma-bicyclo[1.1.1]pentanes

TABLE 5. Atomic distances, angles and UV absorptions of **43a**, **43b**, **44a** and **45a**

X = S or Se	43a	43b	44a	45a
Si−X bond distance	2.160−2.180 Å	2.234−2.325 Å	2.128−2.168 Å	2.122−2.183 Å
Si---Si distance	2.66 Å	2.82 Å	3.00 Å	3.21 Å
S−S bond distance	2.096 Å	2.353 Å	2.035 Å	2.028, 2.073 Å
Si−X−Si angle	75.9°	74.8, 77.9°	89.2°	108.7°
X−Si−X angle	94.1, 102.2°	95.9, 100.9°	101.5−106.4°	102.5−109.3°
Si−X−X−Si torsion angle	0°	0°	11−19°	22°
UV λ_{max} (nm)	280 ($\varepsilon = 320$), 366 ($\varepsilon = 41$)	290 ($\varepsilon = 470$), 440 ($\varepsilon = 48$), 458 ($\varepsilon = 48$), 505 ($\varepsilon = 25$)	287 ($\varepsilon = 365$), 375 ($\varepsilon = 200$)	265 ($\varepsilon = 2200$), 350 ($\varepsilon = 340$)

TABLE 6. Atomic distances, angles and UV absorptions of **47a**, **47b** and **47c**

M = Si, Ge; X = S, Se	47a	47b	47c
M−X bond distance	2.171−2.186 Å	2.316−2.327 Å	2.398−2.412 Å
M---M distance	2.407 Å	2.515 Å	2.672 Å
M−X−M angle	67.08 nm	65.54 nm	67.45 nm
X−M−X angle	91.76−92.87°	93.61°	92.15°
UV λ_{max} nm	257 ($\varepsilon = 210$), 290 ($\varepsilon = 38$)	300 ($\varepsilon = 300$), 366 ($\varepsilon = 33$)	265 ($\varepsilon = 4220$), 365 ($\varepsilon = 100$)

cyclodisilathianes, these Si−S−Si values of **43a** and **47a** are extremely sharp. Especially, in **47a** the acute Si−S−Si angles [66.88(3)°, 66.98(3)° and 67.27(4)°] can be regarded as angles of three-membered ring compounds rather than those of normal four-membered ring compounds. One of the features of **47a** is a very short S---S distance, 3.15 Å, which is within the sum of van der Waals radii of sulfur (3.70 Å). Interestingly, the bridgehead Si---Si distance of 2.407 Å lies within the range of common Si−Si single bond lengths (2.23−2.70 Å)[80], and is ca 0.1 Å shorter than that of its selenium analogue (2.515 Å)[15]. This value is very short relative to that of the cyclodisilathianes (2.78−2.83 Å)[76−79]. This unusual shortening of the Si---Si distance is caused by the difference in electronegativity between the peripheral sulfur atom and the bridgehead silicon atom. Actually, no such shortening is observed in the corresponding silicon analogue, bicyclo[1.1.1]pentasilane (Si---Si, 2.92 Å)[81]. The repulsion between sulfur atoms affects the shortening of the Si---Si distance of **47a**. Then the unusual short Si---Si distance is due to the geometrical factor, thus a tug-of-war between the three Si−S−Si and six S−Si−S angles occurs.

32. Cyclic polychalcogenide compounds with silicon

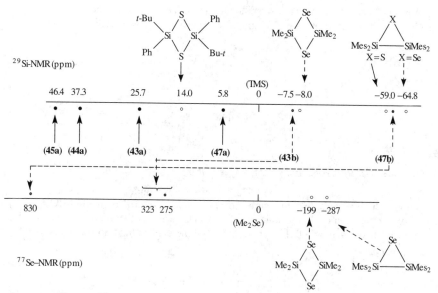

FIGURE 8. ^{77}Se and ^{29}Si-NMR data of cyclosilthianes and cyclosilselenane

D. Spectroscopic Data

1. NMR spectra of trithia- and triselena-disilabicyclo[1.1.1]pentanes

^{29}Si NMR chemical shift of the bridgehead silicons in **45a**, **44a**, **43a** and **47a** are 46.4, 37.3, 25.7 and 5.8 ppm, which are in the order of the strain of silathiane rings (Figure 8). On the other hand, the shifts of **43b**[14] and **47b**[15] are −7.5 and −59.0 ppm, respectively. The difference of the bridgehead silicons chemical shift between **47a** and **47b** is derived from the stereo-compression effect between the peripheral atoms[75].

2. UV spectra of polythiadisilabicyclo[k.l.m]alkanes

Although the absorption maximum of polysulfides in the UV-VIS spectrum is shifted to longer wavelengths with the number of sulfur atoms[82], that of cyclic polysulfides is shifted to shorter wavelengths with the number of ring atoms[83]. The smaller dihedral angle of cyclic polysulfides is responsible for the increase of the ring strain. Typical cyclic disulfides, tetramethylene disulfide and trimethylene disulfide have dihedral angles (C−S−S−C) of 60° and 27° and absorption maxima at 286 and 330 nm, respectively[83]. Interestingly, **43a** has a dihedral (Si−S−S−Si) angle of 0° and the absorption maximum is at 366 nm. Compound **44a** has also small dihedral angles of Si−S−S−Si, 11° and 19°, and absorption maximum at 375 nm. The longer wavelength of 375 nm in **44a** than in **43a** is due to the through-space interaction between the two intramolecular disulfides. A remarkable S---S interaction in **47a** was observed by the 290 nm maximum (ε 38) in the electronic spectrum. The following two reasons were invoked that the weak but definite absorption at 290 nm was assigned as originating from the S---S interaction in **47a**. Firstly, cyclodisilathiane **48** does not have a definite absorption around 290 nm. It suggests that the absorption of 290 nm in **47a** is not derived from the Si−S bond. Secondly, the selenium analogue **47b** displays an absorption band at 365 nm (ε 33)[14].

Moreover, **47c** also exhibits an absorption maximum in the same region (365 nm). It was concluded that the absorption at 365 nm was derived from Se---Se interaction[75].

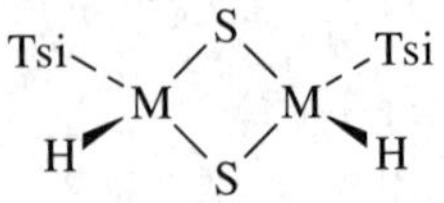

(48)

E. Reactions

1. Hydrolysis of trithiadisilabicyclo[1.1.1]pentane

Trithiadisilabicyclo[1.1.1]pentane **47a** is thermally quite stable (>300 °C) but is slowly hydrolyzed by moisture to provide 1,3-bis{tris(trimethylsilyl)methyl}-1-hydroxy-3-mercaptocyclodisilathiane **49**, which was further hydrolyzed to 1,3-bis{tris(trimethylsilyl)methyl}-1,3-dihydroxy-1,3-dimercaptosulfide **50** (equation 19)[13,14].

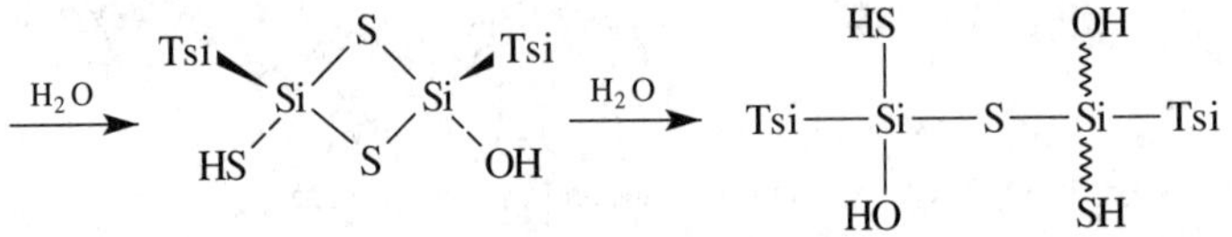

(19)

(49) (50) $dl : meso = 1:1$

2. Oxidation of trithiadisilabicyclo[1.1.1]pentane

Oxidation of trithiadisilabicyclo[1.1.1]pentane **47a** by 1 eq. of *m*-chloroperbenzoic acid (*m*-CPBA) gives 27% of the ring expansion product **51** and recovered **47a** in 37% yields, respectively (equation 20)[84]. By using 2 eq. of *m*-CPBA, **51** is obtained in 62% yield.

(20)

(51)

IV. TRICYCLOSILACHALCOGENIDES

A. Introduction

In the series of sesquichalcogenides containing group 14 elements, the structures of sesquisulfides and sesquiselenides are known to have adamantane (**2**) and double-decker (**3**) skeletons. On the other hand, sesquioxides have prismane (**4**) or cubane (**5**) skeletons. Neither adamantane nor double-decker type silsesquioxanes have been synthesized. These results are due to the M−X−M angles, i.e. the M−O−M angle is much larger than the M−S−M or M−Se−M angles. There are nor-adamantane

(52), bis-nor-adamantane (53), nor-double-decker (54) and bis-nor-double-decker (55) type cage compounds, which are derived from the removal of one or two chalcogen atoms from adamantane or double-decker type sesquichalcogenides (Scheme 13). In this section, adamantane and double-decker type sesquichalcogenides and their nor- and bis-nor derivatives are reviewed.

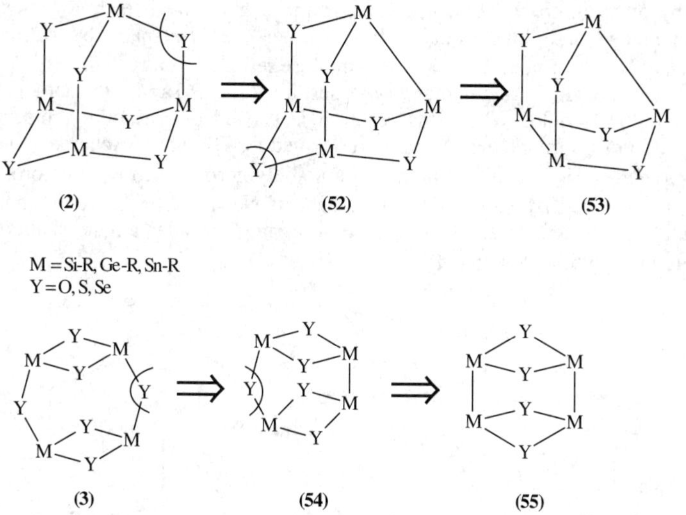

SCHEME 13

B. Synthesis and Reactions

1. Adamantane and double-decker type

Adamantane-type silathianes can be made by heating organosilanes with hydrogen sulfide or elemental sulfur. The tricyclic silathianes are also synthesized by the reaction of halosilanes with sulfurization reagents, e.g. Li_2S, Na_2S, $(Me_3Si)_2S$ and H_2S (Scheme 14). Phenyl or t-butyltrichlorosilane reacts with Na_2S to afford adamantane skeleton silsesquisulfides (56). These bulky substituted trichlorosilanes do not react with H_2S/pyridine or $(NH_4)_2S_5$ (equation 21). On the contrary, halogermane or halostannane easily reacts with H_2S/pyridine or $(NH_4)_2S_5$ to afford 57 (equation 22). Silaselenides

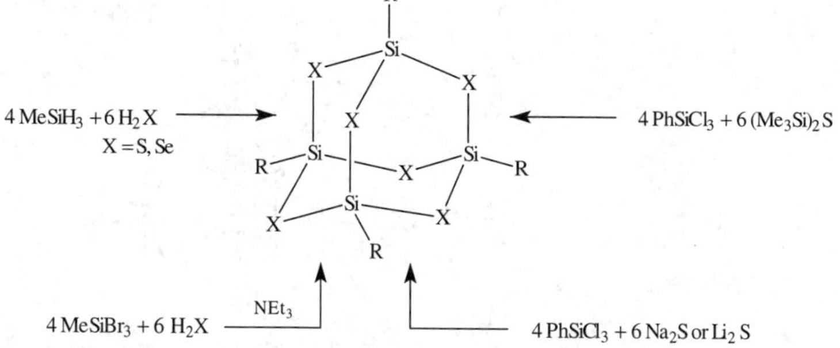

SCHEME 14

with adamantane skeleton (**58**) are also prepared by the reaction of methyl- or ethylsilane or halosilanes with H_2Se[8,85]. Adamantane-like structure silaselenanes are prepared by reaction of organochlorosilanes with Na_2Se in improved yields (equation 23)[86]. The double-decker structure silathiane is one of two reasonable isomers from these reactions. In 1996, the first example of double-decker type silsesquisulfide (**59**) was reported by Matsumoto and coworkers using the conventional method (equation 24)[87]. Both adamantane and double-decker type isomers can be made by sulfurization of halogermanes[88]. Treatment of *tert*-butyltrichlorogermane (**60a**) with H_2S/pyridine in refluxing benzene yields adamantane-type sesquisulfide (**61a**) of the stoichiometry (*t*-BuGe)$_4$S$_6$ in 67% yield. When this reaction is carried out at room temperature, two diastereomeric bis(germanethiols) (**62a**) are formed in 43% yield (Scheme 15). The latter thermally releases H_2S to afford **61a**[6]. Tetra(*t*-butylgermanium)sesquichlorosulfide and selenide are prepared by using 5/4 equivalents of Na_2Y (Y = S, Se) (equation 25)[89]. By using 1/2 equivalent of Li_2S, sesquisulfide having nor-adamantane skeleton (**64a**) is yielded along with **63a** (Scheme 15)[88].

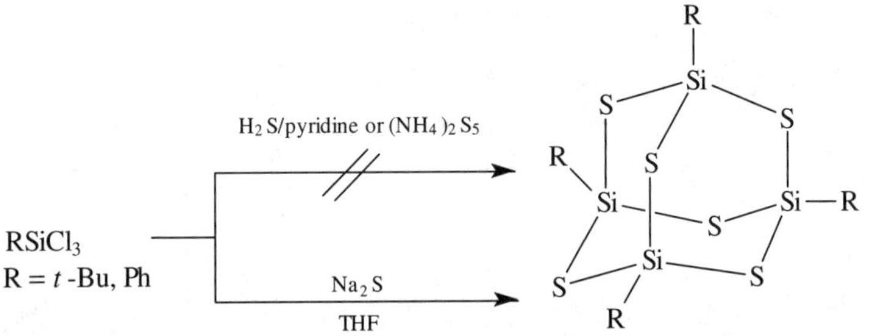

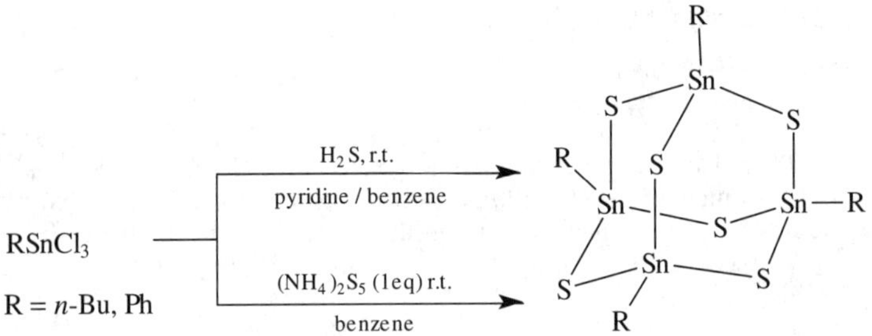

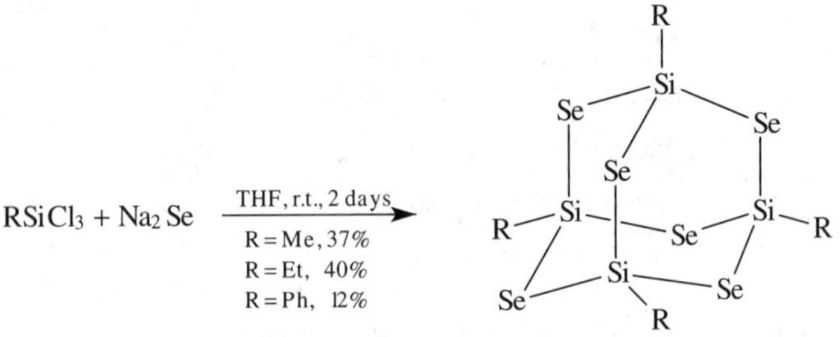

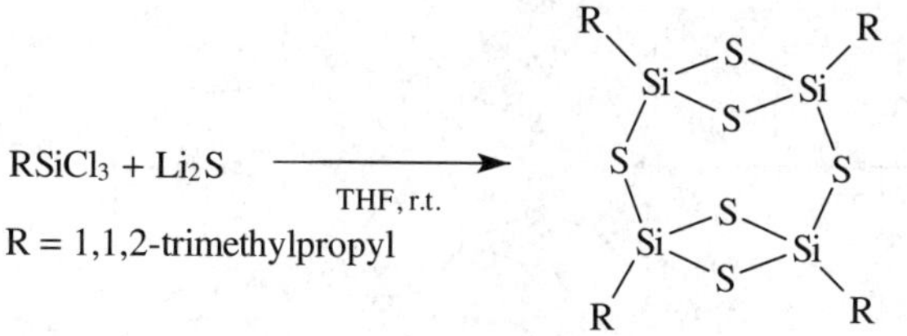

32. Cyclic polychalcogenide compounds with silicon

$$\text{RSiCl}_3 + \text{Li}_2\text{S} \xrightarrow{\text{THF, r.t.}} \textbf{(59)} \quad (24)$$

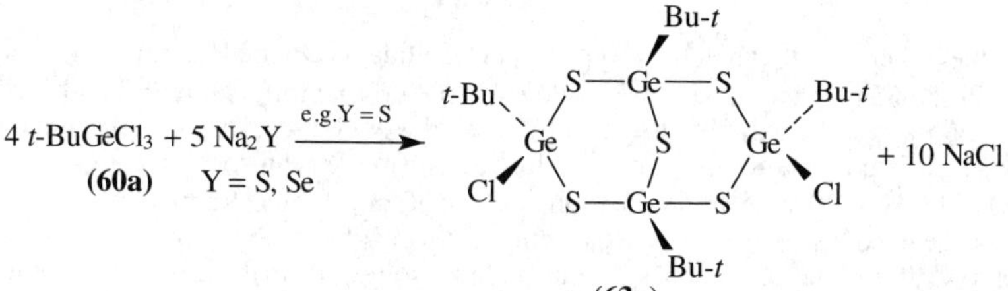

R = 1,1,2-trimethylpropyl

$$4\ t\text{-BuGeCl}_3 + 5\ \text{Na}_2\text{Y} \xrightarrow{\text{e.g. Y = S}} \textbf{(63a)} + 10\ \text{NaCl} \quad (25)$$

(60a)   Y = S, Se

t-BuGeCl$_3$

(60a)

- H$_2$S, Δ, pyridine/benzene → **(61a)** ⇌ (−H$_2$S, Δ)
- H$_2$S, r.t., pyridine/benzene → (cis,cis-**62a**) + (cis,trans-**62a**)
- Li$_2$S (0.5 eq), −78 °C → r.t., THF → (trans,trans-**63a**) + **(64a)**
- (Me$_3$Si)$_2$S, benzene, 80 °C → **(61a)**

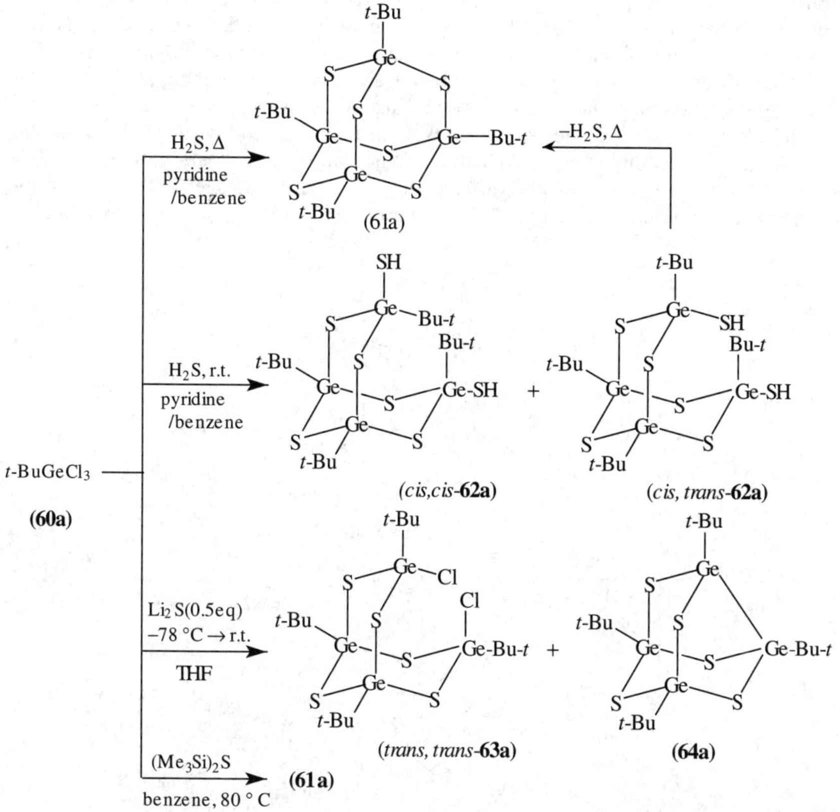

SCHEME 15

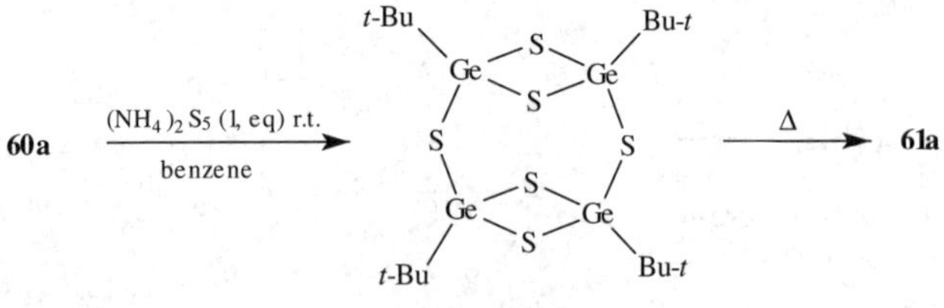

SCHEME 16

On the other hand, when ammonium pentasulfide is applied as the sulfur source, sesquisulfide **65a** is obtained in 30% yield (Scheme 16). Unfortunately, ^{1}H and ^{13}C NMR data of **65a** and **61a** cannot distinguish between the double-decker and adamantane-type structures. It is possible to distinguish both structures by Raman spectroscopy. Thus in the range of the Ge−S framework vibrations (200−650 cm^{-1}), the Raman spectrum of **65a** exhibits six bands, whereas that of **61a** only gives two bands. This observation does not only allow differentiation of the two isomers, but also reveals that the molecular symmetry of **65a** (D_{2h}) is lower than that of **61a** (T_d). Furthermore, the rearrangement of **65a** to **61a** was followed by Raman spectroscopy, i.e. five bands of **65a** vanished and simultaneously a strong band of **61a** appeared. According to X-ray analysis of the double-decker structure type **65a**, the molecule crystallographically possesses three orthogonal 2-fold axes. The Ge atoms and S atoms form two four-membered and two eight-membered rings which are mutually perpendicular; the four-membered rings are almost planar.

This unique formation of the double-decker structure is highly dependent on the correct choice of the substituents. When the reaction is performed with MesGeCl$_3$ (Mes = 2,4,6-trimethylphenyl) **(60b)** instead of *tert*-butyltrichlorogermane **(60a)**, only the adamantane-like sesquisulfide **(61b)** is formed (Scheme 17).

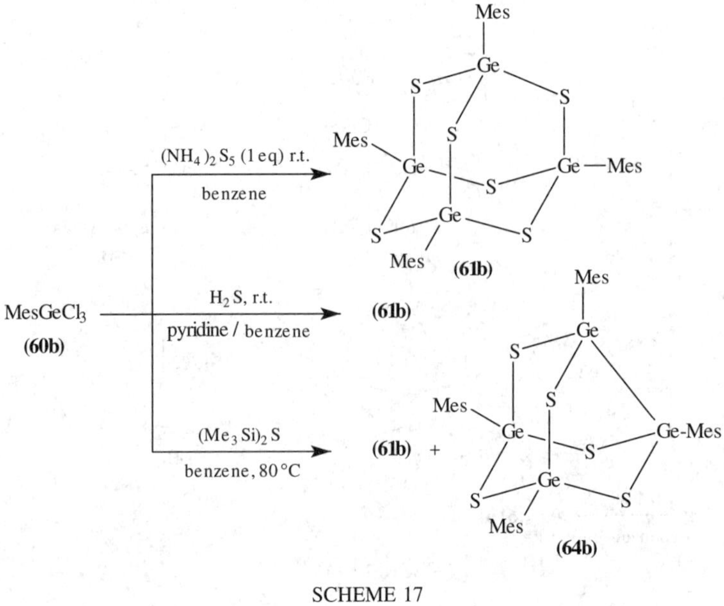

SCHEME 17

2. Nor- and bis-nor-adamantane type sesquichalcogenides

The reaction of tetrachlorodisilane with chalcogenation reagents is one of the synthetic methods to bis-nor-adamantane or double-decker type compounds. When a solution of di-*tert*-butyltetrachlorodisilane **66** in THF is refluxed together with one equivalent of lithium sulfide or stirred with lithium selenide at room temperature, the tetra(*tert*-butylsilicon)pentachalcogenides **67** and **68**, respectively, are formed (Scheme 18)[6]. Each pentachalcogenide exhibits one resonance of a *tert*-butyl group in the ^{1}H and ^{13}C NMR spectra. Most likely, the bis-nor-adamantane derivatives **69** and **70** are initially formed. Insertion of a sulfur or selenium atom into one of the two strained Si—Si bonds would then lead to the observed products.

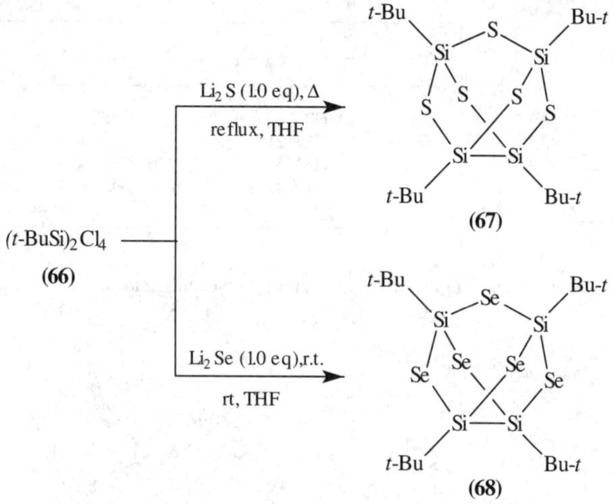

SCHEME 18

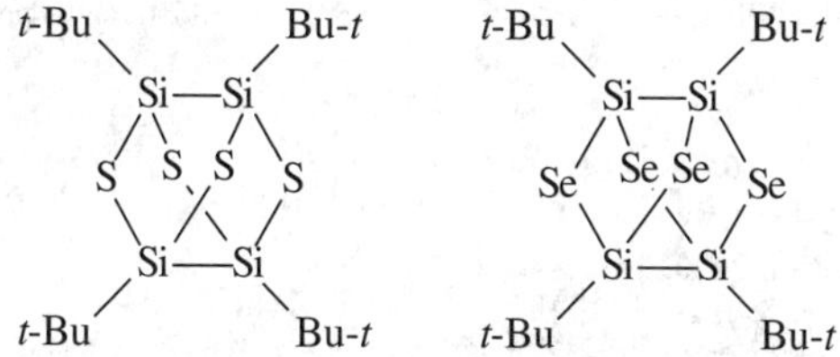

On the other hand, when di-*tert*-butyltetrachlorodigermane (**71**) is reacted with lithium sulfide at −78 °C, the corresponding bis-nor-adamantane **72** is formed along with pentasulfide (**64a**) (Scheme 19). Presumably, due to the long Ge—Ge bond distance, **72** is less strained than the silicon analogues **69** and **70**, and thus sulfur insertion is less favored.

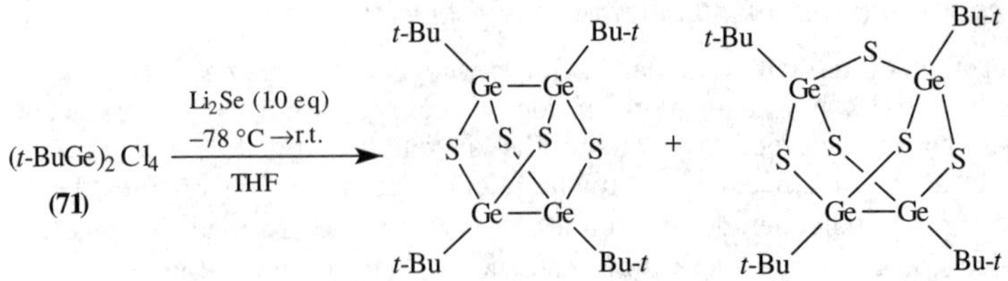

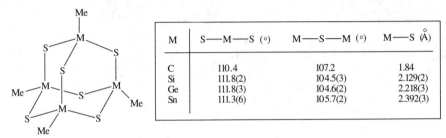

(72) **(64a)**

SCHEME 19

M	S—M—S (°)	M—S—M (°)	M—S (Å)
C	110.4	107.2	1.84
Si	111.8(2)	104.5(3)	2.129(2)
Ge	111.8(3)	104.6(2)	2.218(3)
Sn	111.3(6)	105.7(2)	2.392(3)

FIGURE 9. Bond distances and angles of adamantane-type sesquisulfides.

(56), R = *t*-Bu

(59) R = 1,1,2-trimethylpropyl

(67)

(65a)

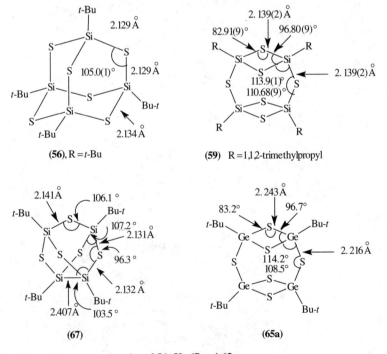

FIGURE 10. Bond distances and angles of **56**, **59**, **67** and **65a**.

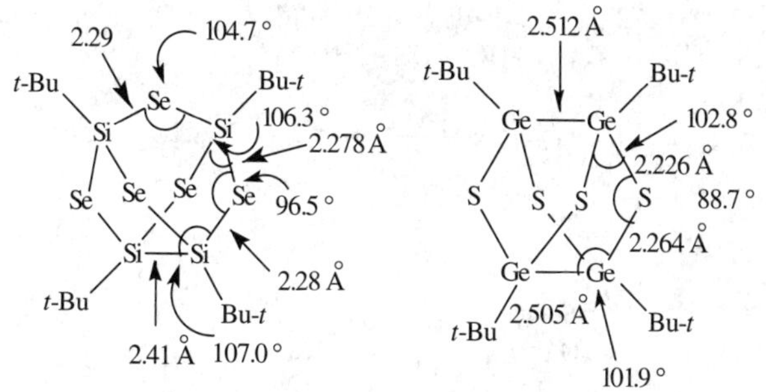

32. Cyclic polychalcogenide compounds with silicon 1923

(68)   (72)

FIGURE 10. Bond distances and angles of **68** and **72**

C. Structure

1. Adamantane and double-decker type sesquichalcogenides

Structures of adamantane and nor-adamantane type silsesquichalcogenides are shown in Figure 9[90] and Figure 10 along with those of adamantane, nor-adamantane, bis-nor-adamantane and double-decker type germasesquichalcogenides[6,86,88,91].

V. POLYCYCLOSILOXANES
A. Introduction

Organosilsesquioxanes $(RSiO_{1.5})_n$[92–94] are organometallic analogues of silicate anions $[Si_nO_{2.5}]^{n-}$. The silsesquioxanes were first prepared by hydrolysis of organotrichlorosilanes[95,96]. Recently, the yields of silsesquioxanes were much improved by a new synthetic method. In the past decade, the structures of organosilsesquioxanes were determined by X-ray structure analysis.

B. Synthesis and Reactions
1. Prismane and cubane type silsesquioxanes

Silsesquioxanes are reported as $(RSiO_{1.5})_n$ ($n \geq 4$) where R is either H, Cl or any of a large number of organic groups[97–100]. The first silsesquioxane $(O_h\text{-}H_8Si_8O_{12}; T_8)$ **(73)** was synthesized in 1959 by Müller, Köhne and Silwinski (<1% yield from $HSiCl_3$) and shown to be **73**[101]. In 1970 an improved synthesis of this compound (T_8) was reported by Frye and Collins [ca 13% yield from $HSi(OCH_3)_3$][98]. They also reported an unusual preparative procedure that yielded a mixture of hydridospherosiloxanes in highly variable yields (T10, 4%; T12, 43%; T14, 39%; T16, 14% by the treatment of $HSiCl_3$ with H_2SO_4 and successive hydrolysis). Day, Klemperer and coworkers reported the synthesis of $O_h\text{-}(CH_3O)_8Si_8O_{12}$ by the reaction of $O_h\text{-}Cl_8Si_8O_{12}$ with methyl nitrite in 45% yield (equation 26)[99]. Agaskar, Day and Klemperer reported in 1987 that a modification of Frye and Collins' procedure gave improved yields (>30%) and also described a purification protocol that yielded gram quantities of four pure hydridospherosiloxanes[102]. The structures of the compounds $O_h\text{-}H_8Si_8O_{12}$ **(73)**, $D_{5h}\text{-}H_{10}Si_{10}O_{15}$ **(74)**, $D_{2d}\text{-}H_{12}Si_{12}O_{18}$ and $D_{3h}\text{-}H_{14}Si_{14}O_{21}$, which were obtained in ca 0.5, 3.5, 3.5 and 0.5% yields, respectively, have been determined by X-ray analysis. They have reported a new synthetic procedure to give O_h-$H_8Si_8O_{12}$ (**73**) and D_{5h}-$H_{10}Si_{10}O_{15}$ (**74**) (*ca* 3.5 : 1) in *ca* 27.2% yield by treatment of HSiCl$_3$ with FeCl$_3$–HCl–CH$_3$OH[103,104].

$$[Si_8O_{12}]Cl_8 + 8CH_3ONO \xrightarrow{25\,°C} [Si_8O_{12}](OCH_3)_8 + 8\,NOCl \qquad (26)$$

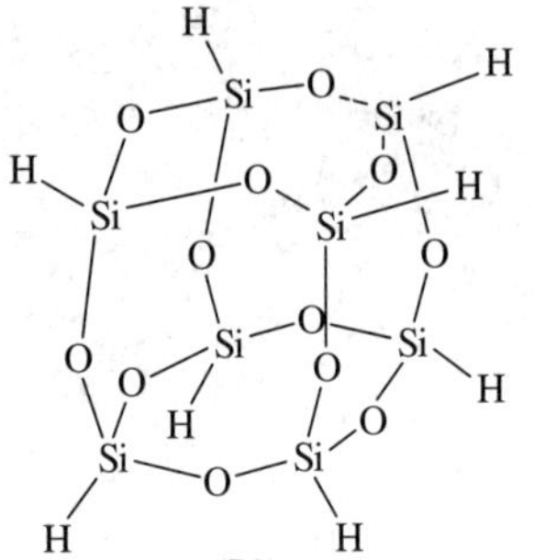

(**73**) 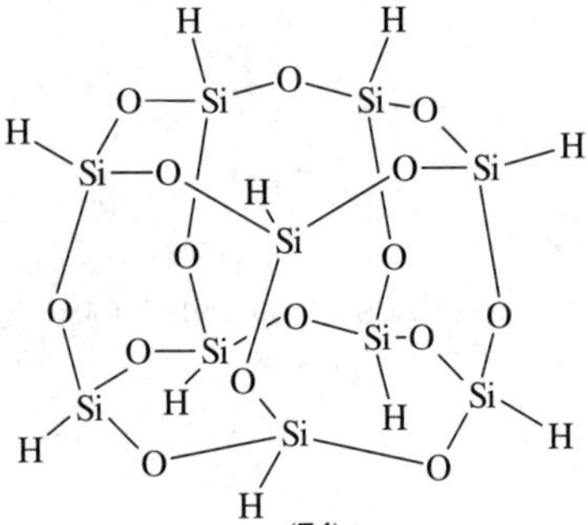   (**74**)

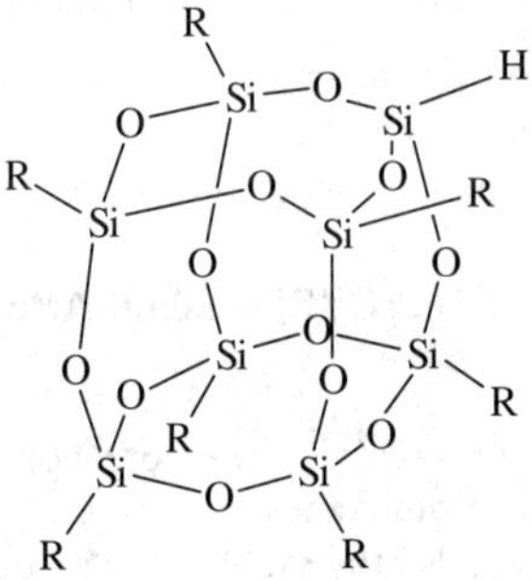

(**75**)   R = PhCH$_2$

Silsesquioxane (**75**, R = PhCH$_2$) can be synthesized by the hydrolytic condensation of benzyltrichlorosilane in 95% ethanol in 24% yield[105]. Feher and coworkers reported the facile synthesis of O_h-H(c-C$_6$H$_{11}$)$_7$Si$_8$O$_{12}$ (**77**) by the reaction of **76** with HSiCl$_3$ in Et$_3$N/Et$_2$O (equation 27)[106]. Matsumoto and coworkers reported the facile synthesis of octakis(silsesquioxane) [(RSi)$_8$O$_{12}$, R = t-BuMe$_2$Si] by the oxidation of octasilacubane[107].

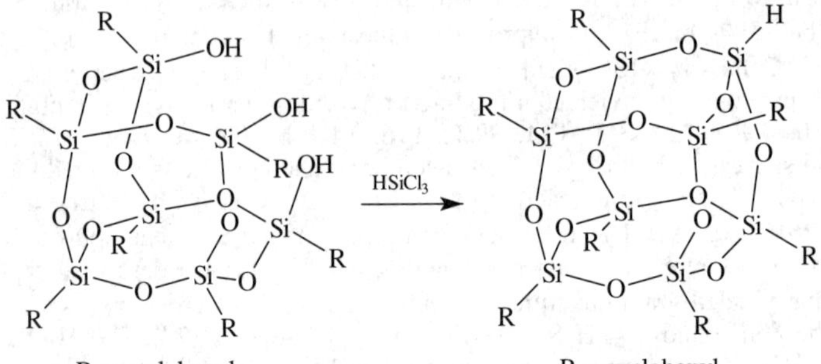

R = cyclohexyl   R = cyclohexyl

(**76**)   (**77**)

(27)

32. Cyclic polychalcogenide compounds with silicon

The synthesis of hexakis(cyclohexylsilsesquioxane) in 10% yield by the treatment of cyclohexyltrichlorosilane with water was first reported in 1965[108]. The X-ray structure of hexakis(cyclohexylsilsesquioxane) was determined by Molloy and coworkers in 1994[109]. The hexasilsesquioxanes (80) substituted by t-butyl or 1,1,2-trimethylpropyl groups are prepared by condensation of the corresponding silanetriol (78) or tetrahydroxysiloxane (79) using dicyclohexylcarbodiimide (DCC) as a dehydrating reagent in DMSO or DMF (Scheme 20)[110].

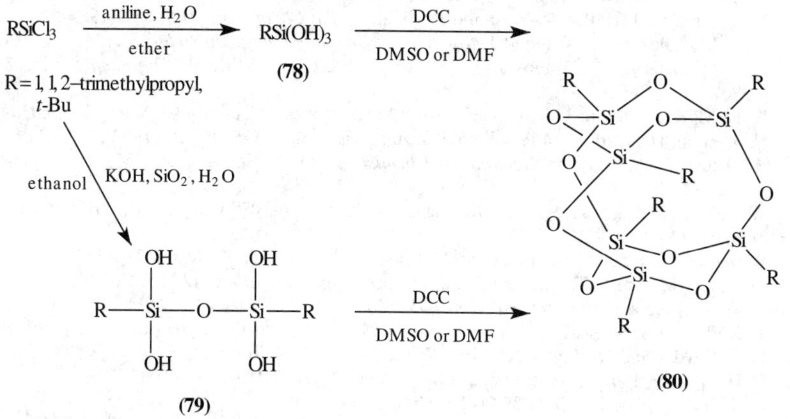

SCHEME 20

2. Transformation of functionalized spherosilicate

A hydridespherosiloxane, $H_{10}Si_{10}O_{15}$ (74), can be easily converted to $[Si_{10}O_{25}]$ $(Si(CH_3)_2(CH=CH_2))_{10}$ by using the reagent $(CH_3)_3NO·ClSi(CH_3)_2(CH=CH_2)$[111].

VI. REFERENCES

1. D. A. ('Fred') Armitage, in *The Chemistry of Organic Silicon Compounds*, Part 2 (Eds. S. Patai and Z. Rappoport); Chap. 23, Wiley, New York, 1989, pp. 1395–1414.
2. D. A. ('Fred') Armitage, in *The Silicon–Heteroatom Bond* (Eds. S. Patai and Z. Rappoport), Wiley, New York, 1991, Chap. 7, pp. 213–231; Chap. 8, pp. 233–243.
3. J. Albertsen and R. Steudel, *J. Organomet. Chem.*, **424**, 153 (1992).
4. N. Choi, S. Sugi and W. Ando, *Chem. Lett.*, 1395 (1994).
5. N. Choi, S. Morino, S. Sugi and W. Ando, *Bull. Chem. Soc. Jpn.*, **69**, 1613 (1996).
6. W. Ando, T. Kadowaki, A. Watanabe, N. Choi, Y. Kabe, T. Erata and T. Ishii, *Nippon Kagaku Kaishi*, 214 (1994); *Chem. Abstr.*, **121**, 205539w (1994).
7. (a) F. S. Kipping and R. Robinson, *J. Chem. Soc.*, **105**, 484 (1914).
   (b) F. S. Kipping, *J. Chem. Soc.*, **101**, 2108 (1912).
   (c) F. S. Kipping, *J. Chem. Soc.*, **101**, 2125 (1912).
8. J. A. Forstner and E. L. Muetterties, *Inorg. Chem.*, **5**, 552 (1966).
9. S. R. Bahr and Boudjouk, *Inorg. Chem.*, **30**, 712 (1991).
10. R. H. Benno and C. J. Fritchie, *J. Chem. Soc., Dalton Trans.*, 543 (1973).
11. H. Berwe and A. Haas, *Chem. Ber.*, **120**, 1175 (1987).
12. H. Yoshida, Y. Kabe and W. Ando, *Organometallics*, **10**, 27 (1991).
13. N. Choi, K. Asano and W. Ando, *Organometallics*, **14**, 3146 (1995).
14. N. Choi, K. Asano, N. Sato and W. Ando, *J. Organomet. Chem.*, **516**, 155 (1996).
15. H. Yoshida, Y. Takahara, T. Erata and W. Ando, *J. Am. Chem. Soc.*, **114**, 1098 (1992).

16. L. Birkofer and O. Stuhl, in *The Chemistry of Organic Silicon Compounds*, Part 1 (Eds. S. Patai and Z. Rappoport), Chap. 10, Wiley, New York, 1989, pp. 655–762.
17. N. G. Bokii, G. N. Zakharova and Yu. T. Z. Struchkov, *Z. Strukt. Khim.*, **13**, 291 (1972); *Chem. Abstr.*, **77**, 25794q (1972).
18. M. A. Hossain, M. D. Hursthouse and K. M. A. Malik, *Struct. Crystallogr. Cryst. Chem.*, **B35**, 522 (1979).
19. R. K. Harris, B. J. Kimber, M. D. Wood and A. Holt, *J. Organomet. Chem.*, **116**, 291 (1976).
20. G. Engelhardt, M. Magi and E. Lippmaa, *J. Organomet. Chem.*, **54**, 115 (1973).
21. R. West, M. J. Fink and J. Michl, *Science (Washington, D.C.)*, **214**, 1343 (1981).
22. M. J. Fink, D. J. De Young, R. West and J. Michl, *J. Am. Chem. Soc.*, **105**, 1070 (1983).
23. M. J. Fink, K. J. Haller, R. West and J. Michl, *J. Am. Chem. Soc.*, **106**, 823 (1984).
24. M. J. Michalczyk, R. West and J. Michl, *J. Chem. Soc., Chem. Commun.*, **22**, 1525, (1984).
25. M. J. Michalczyk, M. J. Fink, K. J. Haller, R. West and J. Michl, *Organometallics*, **5**, 531 (1986).
26. K. L. McKillop, G. R. Gillette, D. R. Powell and R. West, *J. Am. Chem. Soc.*, **114**, 5203 (1992).
27. N. Tokitoh, H. Suzuki and R. Okazaki, *J. Am. Chem. Soc.*, **115**, 10428 (1993).
28. W. Ando, in *Reviews on Heteroatom Chemistry*, Vol. 2 (Ed. S. Oae), MYU, Tokyo, 1994, pp. 121–142.
29. W. Ando, M. Kako, T. Akasaka, S. Nagase, T. Kawai, Y. Nagai and T. Sato, *Tetrahedron Lett.*, **30**, 6705 (1989).
30. W. Ando, M. Kako, T. Akasaka and Y. Kabe, *Tetrahedron Lett.*, **31**, 4117 (1990).
31. W. Ando, M. Kako, T. Akasaka and S. Nagase, *Organometallics*, **12**, 1514 (1993).
32. R. West, *Angew. Chem., Int. Ed. Engl.*, **26**, 1201 (1987).
33. T. Kudo and S. Nagase, *J. Am. Chem. Soc.*, **107**, 2589 (1985).
34. L. C. Snyder and K. Raghavachari, *J. Chem. Phys.*, **80**, 5076 (1984).
35. E. H. Aggarwal and S. H. Bauer, *J. Chem. Phys.*, **18**, 42 (1950).
36. (a) G. Peyronel, *Chim. Ind.*, **36**, 441 (1954).
    (b) G. Peyronel, *Atti Accad. Naz. Lincei, Cl. Sci. Fis., Mat. Nat. Rend.*, **16**, 231 (1954).
37. W. Ando, M. Kako and T. Akasaka, *Chem. Lett.*, 1679 (1993).
38. The annealing temperature, melting point and boiling point of Oxygen are 26, 54 and 90 K, respectively. M. Moskovits and G. A. Ozin, in *Cryochemistry* (Eds. M. Moskovits and G. A. Ozin), Wiley, New York, 1976, p. 24.
39. W. Ando (Ed.), *Singlet O_2*, CRC, Boca Raton, 1985.
40. (a) W. R. Roth and M. Martin, *Tetrahedron Lett.*, 4695 (1967).
    (b) W. R. Roth, F. G. Klarner, W. Grimme, H. G. Roser, R. Busch, B. Muskulus, R. Breuckmann, B. P. Scholz and H. W. Lennartz, *Chem. Ber.*, **116**, 2717 (1983).
41. (a) L. Birkofer and O. Stuhl, in *The Chemistry of Organic Silicon Compounds* (Eds. S. Patai and Z. Rappoport), Chap. 10, Wiley, New York, 1989, pp. 706–721.
    (b) D. A. Armitage ('Fred'), in *The Chemistry of Organic Silicon Compounds* (Eds. S. Patai and Z. Rappoport), Chap. 22, Wiley, New York, 1989, pp. 1363–1394.
42. J. X. McDermott, M. E. Wilson and G. M. Whitesides, *J. Am. Chem. Soc.*, **98**, 6529 (1976).
43. M. A. Chaudhari and F. G. A. Stone, *J. Chem. Soc. (A)*, 838 (1966).
44. D. Sen and U. N. Kantak, *Indian J. Chem.*, **9**, 254 (1971).
45. D. P. Thompson and P. Boudjouk, *J. Chem. Soc., Chem. Commun.*, 1466 (1987).
46. D. M. Giolando, T. B. Rauchfuss, A. L. Rheingold and S. R. Wilson, *Organometallics*, **6**, 667 (1987).
47. D. M. Giolando, T. B. Rauchfuss and G. M. Clark, *Inorg. Chem.*, **26**, 3080 (1987).
48. N. Tokitoh, H. Suzuki, T. Matsumoto, Y. Matsuhashi, R. Okazaki and M. Goto, *J. Am. Chem. Soc.*, **113**, 7047 (1991).
49. N. Tokitoh, M. Takahashi, T. Matsumoto, H. Suzuki, Y. Matsuhashi and R. Okazaki, *Phosphorous, Sulfur and Silicon*, **59**, 161 (1991).
50. N. Tokitoh, T. Matsumoto and R. Okazaki, *Tetrahedron Lett.*, **33**, 2531 (1992).
51. N. Yamazaki, S. Nakahama, K. Yamaguchi and T. Yamaguchi, *Chem. Lett.*, 1355 (1980).
52. For reviews on heavy atom analogues of ketones see:
    (a) J. C. Guziec, in *Organoselenium Chemistry* (Ed. D. Liotta), Wiley, New York, 1987, p. 237.
    (b) C. Raabe and J. Michl, *Chem. Rev.*, **85**, 419 (1985).
    (c) J. Stage, *Pure Appl. Chem.*, **56**, 137 (1984).
53. P. Boudjouk, *Polyhedron*, **10**, 1231 (1991).

32. Cyclic polychalcogenide compounds with silicon

54. V. A. Kutoglu, Z. Anorg. Allg. Chem., **390**, 195 (1972).
55. V. A. Kutoglu, Acta Crystallogr., **B29**, 2891 (1973).
56. E. G. Müller, J. L. Petersen and L. F. Dahl, J. Organomet. Chem., **111**, 91 (1976).
57. W. E. v. Schklower, Y. Yu. T. Strutschkow, L. E. Guselnikow, W. W. Wolkowa and W. G. Awakyam, Z. Anorg. Allg. Chem., **501**, 153 (1983).
58. K. Peters, D. Weber and H. G. von Schnering, Z. Anorg. Allg. Chem., **519**, 134 (1984).
59. J. Peters, J. Mandt, M. Meyring and B. Krebs, Z. Kristallogr., **156**, 90 (1981).
60. [1.1.1]propellanes, M=C.
    (a) K. B. Wiberg and F. H. Walker, J. Am. Chem. Soc., **104**, 5239 (1982).
    (b) K. B. Wiberg, W. P. Dailey, F. H. Walker, S. T. Waddell, L. S. Crocker and M. D. Newton, J. Am. Chem. Soc., **107**, 7247 (1985).
    (c) L. Hedberg and K. Hedberg, J. Am. Chem. Soc., **107**, 7257 (1985).
    (d) E. Honegger, H. Huber, E. Heilbronner, W. P. Dailey and K. B. Wiberg, J. Am. Chem. Soc., **107**, 7172 (1985).
    (e) J. Belzner, U. Bunz, K. Semmler, G. Szeimies, K. Opitz and A. D. Schlüter, Chem. Ber., **122**, 397 (1989).
    (f) P. Seiler, Helv. Chim. Acta, **73**, 1574 (1990).
    M = Sn.
    (g) L. R. Sita and I. Kinoshita, J. Am. Chem. Soc., **112**, 8839 (1990).
    (h) L. R. Sita and I. Kinoshita, J. Am. Chem. Soc., **114**, 7024 (1992).
61. Y. Kabe, T. Kawase and J. Okada, Angew. Chem., Int. Ed. Engl., **29**, 794 (1990).
62. W. D. Stohrer and R. Hoffmann, J. Am. Chem. Soc., **94**, 779 (1972).
63. M. D. Newton and J. M. Schulman, J. Am. Chem. Soc., **94**, 773 (1972).
64. J. E. Jackson and L. C. Allen, J. Am. Chem. Soc., **106**, 591 (1984).
65. D. B. Kitchen, J. E. Jackson and L. C. Allen, J. Am. Chem. Soc., **112**, 3408 (1990).
66. N. D. Epiotis, J. Am. Chem. Soc., **106**, 3170 (1984).
67. R. F. W. Bader, T. T. Nguyen-Dang and Y. Tal, Rep. Prog. Phys., **44**, 893 (1981).
68. P. v. R. Schleyer and R. Janoschek, Angew. Chem., Int. Ed. Engl., **26**, 1267 (1987).
69. S. Nagase and T. Kudo, Organometallics, **6**, 2456 (1987).
70. S. Nagase, T. Kudo and T. Kurakake, J. Chem. Soc., Chem. Commun., 1063 (1988).
71. S. Nagase and T. Kudo, Organometallics, **7**, 2534 (1988).
72. A. Streitwieser, J. Chem. Soc., Chem. Commun., 1261 (1989).
73. M. S. Gordon, K. A. Nguyen and M. T. Carroll, Polyhedron, **10**, 1247 (1991).
74. (a) K. A. Nguyen, M. T. Carroll and M. S. Gordon, J. Am. Chem. Soc., **113**, 7924 (1991).
    (b) S. Nagase, Polyhedron, **10**, 1299 (1991).
75. W. Ando, S. Watanabe and N. Choi, J. Chem. Soc., Chem. Commun., 1683 (1995).
76. W. Wojnowski, K. Peters, D. Weber and H. G. v. Schnering, Z. Anorg. Allg. Chem., **519**, 134 (1984).
77. W. E. v. Schklower, Y. T. Strutschkow, L. E. Guselnikow, W. W. Wolkowa and W. G. Awakyan, Z. Anorg. Allg. Chem., **501**, 153 (1983).
78. J. Peters and B. Krebs, Acta Crystallogr., Sect. B, **B38**, 1270 (1982).
79. J. Peters, J. Mandt, M. Meyring and B. Krebs, Z. Kristallogr., **156**, 90 (1981).
80. W. S. Sheldrick, in The Chemistry of Organic Silicon Compounds, Part 1 (Eds. S. Patai and Z. Rappoport), Chap. 3, Wiley, Chichester 1989, pp. 277–303.
81. The distances between the bridgehead silicon atoms were investigated by ab initio calculations and found to be 2.69–2.89 Å[60,68].
82. F. Serrano, Rev. Fac.ciênc., Lisboa, 2a Sér., **B7**, 105 (1959–60); Chem. Abstr., **55**, 20978h (1961).
83. M. S. Kharasch, W. Nundnberg and G. Mantell, J. Org. Chem., **16**, 524 (1951).
84. N. Choi, K. Asano, S. Watanabe and W. Ando, Phosphorous, Sulfur and Silicon (1997), in press.
85. A. Hass, R. Hitze, C. Kruger and K. Angermund, Z. Naturforsch., **39B**, 890 (1984).
86. P. Boudjouk, Polyhedron, **10**, 1231 (1991).
87. M. Unno, H. Shioyama and H. Matsumoto, Phosphorous, Sulfur and Silicon, (1997), in press.
88. W. Ando, T. Kadowaki and Y. Kabe, Angew. Chem., Int. Ed. Engl., **31**, 59 (1992).
89. H. Puff, K. Braun, S. Franken, T. R. Kök and W. Schuh, J. Organomet. Chem., **335**, 167 (1987).
90. J. C. J. Bart and J. J. Daly, J. Chem. Soc., Dalton Trans., 2063 (1975).
91. W. Ando, N. Choi, S. Watanabe, K. Asano, T. Kadowaki, Y. Kabe and H. Yoshida, Phosphorous, Sulfur and Silicon, **93–94**, 51 (1994).

92. U. Dittmar, B. J. Hendan, U. Floerke and H. C. Marsmann, *J. Organomet. Chem.*, **489**, 185 (1995).
93. R. H. Baney, M. Itoh, A. Sakakibara and T. Suzuki, *Chem. Rev.*, **95**, 1409 (1995).
94. F. J. Feher, K. Rahimian, T. A. Budzichowski and Z. W. Ziller, *Organometallics*, **14**, 3920 (1995).
95. D. W. Scott, *J. Am. Chem. Soc.*, **68**, 356 (1946).
96. J. F. Brown, L. H. Vogt and P. I. Prescott, *J. Am. Chem. Soc.*, **86**, 1120 (1964).
97. M. G. Voronkov and V. I. Lavrent'yev, *Top. Curr. Chem.*, **102**, 199 (1982).
98. C. L. Frye and W. T. Collins, *J. Am. Chem. Soc.*, **92**, 5586 (1970).
99. V. W. Day, W. G. Klemperer, V. V. Mainz and D. M. Millar, *J. Am. Chem. Soc.*, **107**, 8262 (1985).
100. F. J. Feher and T. A. Budzichowski, *J. Organomet. Chem.*, **379**, 33 (1989).
101. R. Müller, F. Köhne and S. Silwinski, *J. Prakt. Chem.*, **9**, 71 (1959).
102. P. A. Agaskar, V. W. Day and W. G. Klemperer, *J. Am. Chem. Soc.*, **109**, 5554 (1987).
103. P. A. Agaskar, *Inorg. Chem.*, **29**, 1603 (1990).
104. P. A. Agaskar, *Inorg. Chem.*, **30**, 2707 (1991).
105. F. J. Feher and T. A. Budzichowski, *J. Organomet. Chem.*, **373**, 153 (1989).
106. F. J. Feher and K. J. Weller, *Organometallics*, **9**, 2638 (1990).
107. H. Matsumoto, K. Higuchi and M. Goto, *31st Symposium of Organometallic Chemistry*, Osaka, Japan, abstract (1986).
108. J. F. Brown and L. H. Vogt, *J. Am. Chem. Soc.*, **87**, 4313 (1965).
109. H. Behbehani, B. J. Brisdon, M. F. Mahon and K. C. Molloy, *J. Organomet. Chem.*, **469**, 19 (1994).
110. M. Unno, S. B. Alias, H. Saito and H. Matsumoto, *Organometallics*, **15**, 413 (1996).
111. P. A. Agaskar, *J. Am. Chem. Soc.*, **111**, 6858 (1989).

CHAPTER 33

Organosilicon derivatives of fullerenes

WATARU ANDO and TAKAHIRO KUSUKAWA

Department of Chemistry, University of Tsukuba, Tsukuba, Ibaraki 305, Japan

I. INTRODUCTION	1929
II. PHOTOCONDUCTIVITY OF FULLERENE-DOPED POLYSILANES	1930
III. PHOTOCHEMICAL REACTIONS OF 2,2-DIARYLTRISILANE	1931
A. Reaction of 2,2-Diaryltrisilane with C_{60}	1931
B. Reaction of 2,2-Diaryltrisilane with C_{70}	1934
IV. PHOTOCHEMICAL AND THERMAL ADDITIONS OF CYCLO-SILANES TO FULLERENES	1936
A. Additions of Disilirane to C_{60}	1936
B. Additions of Oxadisilirane to C_{60}	1939
C. Additions of Disilirane to C_{70}	1940
D. Photochemical and Thermal Additions of Disilirane to Metallofullerene	1940
V. PHOTOCHEMICAL ADDITIONS OF POLYSILACYCLOBUTANES AND -BUTENE TO C_{60}	1944
A. Additions of Benzodisilacyclobutene	1944
B. Additions of Bis(alkylidene)disilacyclobutanes	1945
C. Additions of Cyclotetrasilanes and Cyclotetragermane	1947
VI. REACTIONS OF SILYLLITHIUM REAGENTS WITH C_{60}	1951
VII. REACTIONS OF SILICON-SUBSTITUTED NUCLEOPHILES	1957
A. Reactions of Silylmethyl Grignard Reagents	1957
B. Reactions of Silyl-acetylene Nucleophiles	1958
VIII. REACTIONS OF SILYL DIAZONIUM COMPOUNDS	1959
IX. REFERENCES	1960

I. INTRODUCTION

The fullerenes (C_{60}, C_{70} and higher molecular allotropes)[1a], discovered by Kroto, Smalley and coworkers in 1985[1b], represent the third form of carbon, following graphite and diamond. However, it was not until 1990, when Kräschmer, Huffman and coworkers[1c,1d]

The chemistry of organic silicon compounds, Vol. 2
Edited by Z. Rappoport and Y. Apeloig © 1998 John Wiley & Sons Ltd

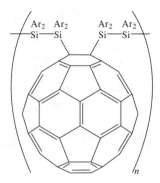

FIGURE 1. Fullerene-bonded polysilane

discovered the method of arc vaporization of graphite affording these marvelous new materials in macroscopic amounts, that the fullerene emerged from the realm of being a scientific curiosity. Since the isolation of C_{60} in preparatively useful quantities, the chemical functionalization of this new allotropic form of carbon has attracted much interest, and has led to fascinating results[1a,2–5]. In spite of a number of studies with different types of substrates, relatively few examples with silicon are known and this prompted recent investigations. One particular target which is attractive in view of its possible photoconductivity is a fullerene-bonded polysilane (Figure 1).

II. PHOTOCONDUCTIVITY OF FULLERENE-DOPED POLYSILANES

Polysilanes are a unique class of polymers with the backbone consisting entirely of tetrahedrally coordinated silicon atoms[6,7]. Because of the σ-conjugation, carrier transport along the silicon backbone is very efficient. The hole mobility of polysilanes, ca 10^{-4} cm^2 V^{-1} s^{-1}, is among the highest observed for polymers. Fullerenes are known to be good electron acceptors. In the presence of electron donors such as aromatic amines, weakly bonded charge-transfer complexes can be formed. Poly(methylphenylsilylene) (PMPS, **1** Scheme 1) can be doped with 1.6% by weight with C_{60}. As a result, its photoinduced discharge rate (a tungsten lamp, 50 mW cm^{-2}) is enhanced by an order of magnitude[8]. The same result is obtained when the samples are irradiated with monochromatic irradiation at 340 nm. However, in the shorter wavelength region (<350 nm), PMPS is a light absorber and the charge is generated by an electron transfer from an excited state of the polysilane to the fullerene. Extension of this approach to other polysilanes such as cyclohexylmethylpolysilane and dimethylpolysilane has been tried. However, no significant photoinduced discharge can be observed from these polysilanes. These results suggest that specific interaction exists between the fullerene and the phenyl group in PMPS which leads to the formation of a weakly bonded complex between them and enhances the transfer of an electron from PMPS to fullerene.

The photoinduced electron transfer between C_{60} and polysilane was investigated by laser flash photolysis[9]. The transient absorption spectra of the C_{60} triplet ($^3C_{60}{}^*$) (**2**), the C_{60} radical anion (**3**) and the poly(methylphenylsilylene) radical cation (**4**, M = Si) (Scheme 2) were observed in the region 600–1600 nm in a polar solvent.

Electron transfer from the germanium analog of **1**, (PMePhGe)$_n$, to photoexcited C_{60} in benzene–acetonitrile solution has also been investigated by 532 nm laser flash photolysis[10]. The transient absorption band at 730 nm which appears immediately after laser exposure is attributed to $^3C_{60}{}^*$ (**2**). With the decay of $^3C_{60}{}^*$ (**2**), the intensities of the

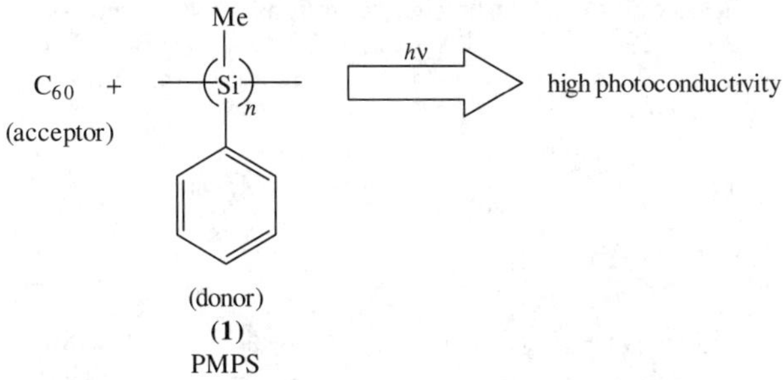

33. Organosilicon derivatives of fullerenes

SCHEME 1. A Fullerene-doped polysilane photoconductor

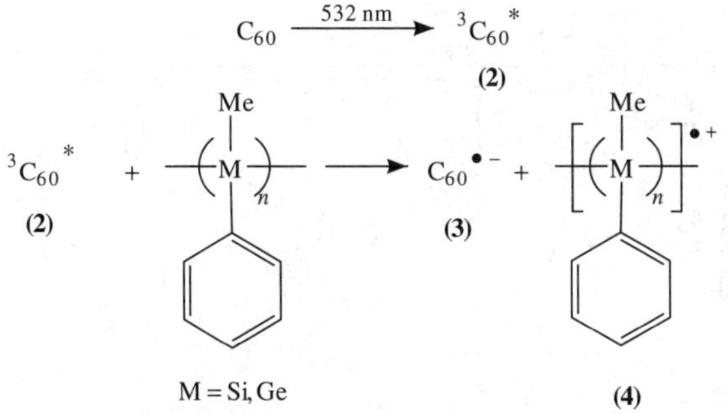

SCHEME 2. Photoinduced electron transfers from polysilane and polygermane to C_{60}

absorption bands increase in the region 900–1600 nm. The absorption band at 1030 nm is a characteristic band of the radical anion of C_{60} ($C_{60}^{\bullet -}$) (**3**). The broad absorption band extending from 1200 to 1600 nm may be attributed to the radical cation of $(PMePhGe)_n$, $(PPhMeGe^{\bullet +})_n$ (**4**, M = Ge). Such radical ions generated by the electron transfer were not observed in neat benzene. $^3C_{60}^*$ is the only transient species observed by nanosecond laser flash photolysis in the nonpolar solvent benzene.

III. PHOTOCHEMICAL REACTIONS OF 2,2-DIARYLTRISILANE

A. Reaction of 2,2-Diaryltrisilane with C_{60}

Ando, Akasaka and coworkers have studied the reaction of 2,2-diaryltrisilane with C_{60}[11]. When 2,2-bis(2,6-diisopropylphenyl)hexamethyltrisilane (**5**) the silylene (**6**) precursor, was photolyzed with a low-pressure mercury lamp in a toluene solution of C_{60}, the color of the solution changed from purple to dark brown. Flash chromatography on silica gel furnished thermally stable **7** and **8** in 58% and 27% yields, respectively. Small amounts of **9** and **10** were also obtained (equation 1). It has been observed that the product composition of the silylene-addition reaction varies with the amount of the trisilane used. The fast atom bombardment (FAB) mass spectrometry of **7** displays a peak for adduct **7** at 1074–1070 as well as for C_{60} at 723–720 which arises from loss of diarylsilylene (**6**). The FAB mass spectra of the $C_{60}(Dip_2Si)_2$ (**8**), $C_{60}(Dip_2Si)_3$ (**9**) and $C_{60}(Dip_2Si)_4$ (**10**) adducts of C_{60} were reasonably analyzed. The UV-vis absorption spectra of **7** is virtually identical to that of C_{60} except for subtle differences in the 400–700 nm region (Figure 2). Adduct **7** exhibits a new band at 421 nm but lacks the C_{60} band at 406 nm.

Relative to C_{60}, adduct **7** shows stronger absorptions at 463 and 508 nm and weaker absorptions at 539 and 599 nm. The FAB mass, UV-vis spectra of **7** contain a number of unique features, but also suggest that this new fullerene retains the essential electronic and structural character of C_{60}.

$$\text{Dip}_2\text{Si}(\text{SiMe}_3)_2 \xrightarrow[-\text{Me}_3\text{SiSiMe}_3]{h\nu} [\text{Dip}_2\text{Si:}]$$
$$(5) \qquad\qquad\qquad\qquad (6)$$

$$\xrightarrow{C_{60}}$$

$$C_{60}\text{Dip}_2\text{Si} + C_{60}(\text{Dip}_2\text{Si})_2 + C_{60}(\text{Dip}_2\text{Si})_3 + C_{60}(\text{Dip}_2\text{Si})_4$$
$$(7) \qquad\qquad (8) \qquad\qquad\quad (9) \qquad\qquad\quad (10)$$

(1)

$\text{Dip} = 2,6\text{-}i\text{-Pr}_2\text{C}_6\text{H}_3$

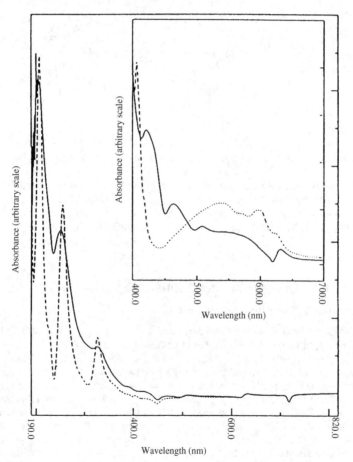

FIGURE 2. UV-vis spectra of C_{60} (- - - -) and **7** (———) from 190 to 829 nm in hexane Insert: Spectra of C_{60} (- - - -) and **7** (———) from 400 to 700 nm in toluene. Reprinted with permission from Ref. 11. Copyright 1993 American Chemical Society

33. Organosilicon derivatives of fullerenes

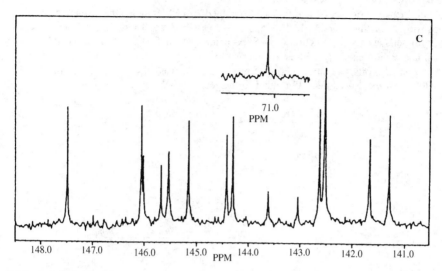

FIGURE 3. ^{13}C NMR spectrum of **7**. Reprinted with permission from Ref. 11. Copyright 1993 American Chemical Society

The possible structures of **7** are **7a** and **7b**. The silirane structure **7a**, of C_{2v} symmetry, is derived from the addition of the diarylsilylene across the reactive 1,2-junction. Silamethano[10]annulene **7b**, analogous to the structure proposed for Ph_2C_{61}, could arise via isomerization of **7a**. The ^{1}H NMR of **7** is consistent with the Dip_2Si adduct of C_{60}. The ^{13}C-NMR spectrum of **7** shows 17 signals for the C_{60} skeleton, of which four correspond to two carbons each and thirteen correspond to four carbon atoms each; one signal is at 71.12 ppm and the remainder between 140 and 150 ppm (Figure 3). This is the appropriate number and ratio of peak intensities for a C_{60} adduct of C_{2v} symmetry. The ^{13}C-NMR signal at 71.12 ppm strongly supports structure **7a** rather than **7b** (Scheme 3). The ^{29}Si-NMR spectrum of **7** shows a peak at -72.74 ppm which is also assigned to the silicon atom of **7a**. These data are fully consistent with the structural representation of **7a**.

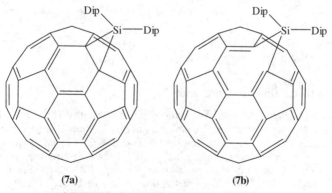

SCHEME 3. Two possible structures of adduct **7**

Further support for structure **7a** is obtained by AM1 molecular orbital calculations performed on the reaction of C_{60} and silylenes (Ph_2Si:). These calculations show that Ph_2Si adds across the junction of two six-membered rings in C_{60} to give the 1,2-addition silirane analogous to **7a** with an exothermicity of 61.3 kcal mol^{-1}. The analog of **7b** was not located on the potential energy surface. Also, **7a** was 19.4 and 10.7 kcal mol^{-1} more stable than the 1,6-adducts **11a** and **11b**, respectively (Scheme 4).

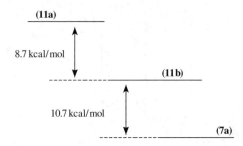

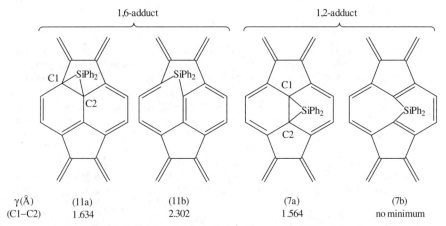

SCHEME 4. Relative energies and bond distances (C1–C2) of four isomers

Thus, we have shown that the addition of a diarylsilylene with C_{60} occurs in a facile manner analogous to the carbene addition[3] reported recently.

B. Reaction of 2,2-Diaryltrisilane with C_{70}

The silylene precursor, 2,2-bis(2,6-diisopropylphenyl)hexamethyltrisilane (**5**), was photolyzed with a low-pressure mercury lamp in a toluene solution of C_{70}[12]. The adduct **12** obtained (equation 2) contains two isomers of $(Dip)_2SiC_{70}$ (**12a** and **12b**) which were separated by flash chromatography on silica gel. FAB mass spectrometry of **12** displays a peak for adduct **12** at 1190–1194 as well as for C_{70} at 840–843 which arises from the loss of diarylsilylene. The UV-vis absorption spectra of **12** are virtually identical to those of C_{70} with bands at 333, 381 and 471 nm. AM1 molecular orbital calculations on the reaction of C_{70} with dihydrosilylene show that the 1,2-adduct is more stable than the 1,6-adduct. In the two adducts the silyl group is positioned differently with respect to the mirror planes of C_{70} such that they each correspond to a characteristic number of symmetry-independent carbons and protons (Table 1 and Figure 4).

$$Dip_2Si(SiMe_3)_2 \xrightarrow[-Me_3SiSiMe_3]{h\nu} [Dip_2Si:] \xrightarrow{C_{70}} C_{70}Dip_2Si \qquad (2)$$
$$(5) \qquad\qquad\qquad (6) \qquad\qquad (12)$$
$$Dip = 2,6\text{-}i\text{-}Pr_2C_6H_3$$

The ^{1}H-NMR spectrum of **12a** displays four methyl signals at 1.59, 1.35, 1.21 and 0.93, and two methine signals at 3.81 and 3.60 ppm. Similarly, four methyl signals at 1.75, 1.47, 1.39 and 1.32, and two methine signals at 4.09 and 3.83 ppm are observed in the ^{1}H-NMR spectrum of **12b**. The ^{13}C-NMR spectrum of **12a** displays two signals at 78.51 and 66.77 ppm for the C_{70} skeleton which are attributed to the sp^3 fullerene carbons. From Table 1, **12a** and **12b** correspond to a=b and c=c addition products, respectively. The kinetically controlled regioselectivity observed for the addition of silylene **6** to C_{70} agrees qualitatively with the AM1 calculations on C_{70}. Products **12a** and **12b** are calculated to

TABLE 1. Number of independent protons and carbons in the 1,2-isomeric adducts of $(Dip)_2SiC_{70}$

Isomer	Symmetry	CH	CH$_3$	sp^3 Carbon on C_{70}
12a	C_s	2	4	2
12b	C_s	2	4	1
a,b-	C_s	2	4	2
c,c-	C_s	2	4	1
d,e-	C_1	4	8	2
e,e-	C_{2v}	1	1	1

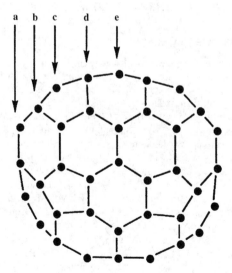

FIGURE 4. Structure of C_{70}

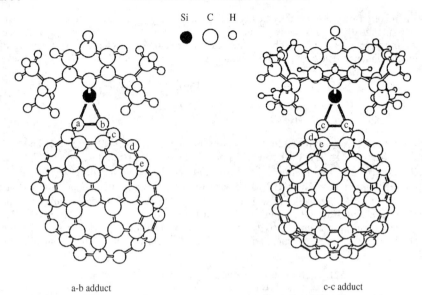

a-b adduct   c-c adduct

FIGURE 5. Optimized structures of **12a** (left) and **12b** (right) calculated by the AM1 method. Reproduced by permission of the Royal Society of Chemistry from Ref. 12

have almost the same energy; **12a** and **12b** are formed with an exothermicity of 46.3 and 46.4 kcal mol^{-1}. The LUMO electron densities fit very well with the product ratios of the adducts. The C_s structures of **12a** and **12b** obtained by full geometry optimization at the AM1 level are shown in Figure 5.

IV. PHOTOCHEMICAL AND THERMAL ADDITIONS OF CYCLOSILANES TO FULLERENES

A. Additions of Disilirane to C_{60}

C_{60} is a strong electron acceptor capable of taking on as many as six electrons and the photoexcited C_{60} is an even stronger acceptor than the C_{60} in the ground state. On the other hand, strained Si—Si bonds can act as the source of electrons which shows low oxidation potentials and high ionization potentials (Table 2).

Photochemical reaction of C_{60} with disilirane (**13**) was reported by Ando and coworkers[13]. Irradiation of a toluene solution of 1,1,2,2-tetramesityl-1,2-disilirane (**13**) and C_{60} with a high-pressure mercury lamp resulted in the formation of 1,1,3,3-tetramesityl-1,3-disilorane (**14**) in 82% yield with the complete consumption of C_{60} (equation 3). The UV-vis spectrum of **14** and the progress of its generation by irradiation are shown in Figure 6. FAB mass spectroscopy of **14** displays a peak at m/z 1270–1266 besides showing loss of **13** in the fragmentation. The ^{1}H-NMR spectrum of **14** displays six methyls and four *meta*-proton signals of mesityl groups. An AB quartet ($J = 13.0$ Hz) is displayed by two methylene protons at 3.61 and 2.38 ppm supporting a C_s symmetry for the molecules. The ^{13}C-NMR spectrum of **14** shows 32 signals for the C_{60} skeleton. Of these, 28 signals have a relative intensity of 2 and four signals have a relative intensity of 1: one of these signals appears at 73.36 ppm and the remainder between 130 and 150 ppm. These spectral data suggest a C_s symmetry for **14**.

33. Organosilicon derivatives of fullerenes

TABLE 2. Oxidation potential and ionization potential of disilirane and digermirane

Substrate	E_{ox} (V) vs SCE[a]	I_p (eV)[b]
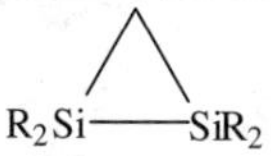 R_2Si—SiR_2 (three-membered ring)	0.81 (R = Mes)	8.9 (R = H)
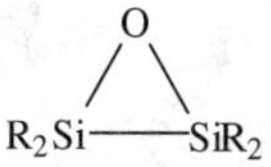 R_2Si—SiR_2 with O bridge	0.97 (R = Dep) 0.79 (R = Dip)	9.4 (R = H)
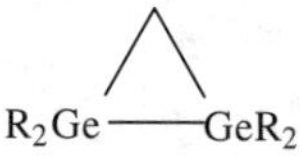 R_2Ge—GeR_2 (three-membered ring)	0.72 (R = Dep)	

[a]Measured by cyclic voltammetry at a platinum electrode in CH_2Cl_2 with 0.1 M n-Bu_4 $NClO_4$ as a supporting electrolyte.
[b]Calculated at the HF/6-31G* level.

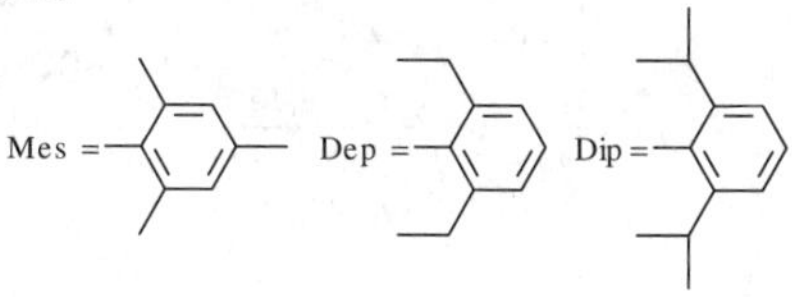

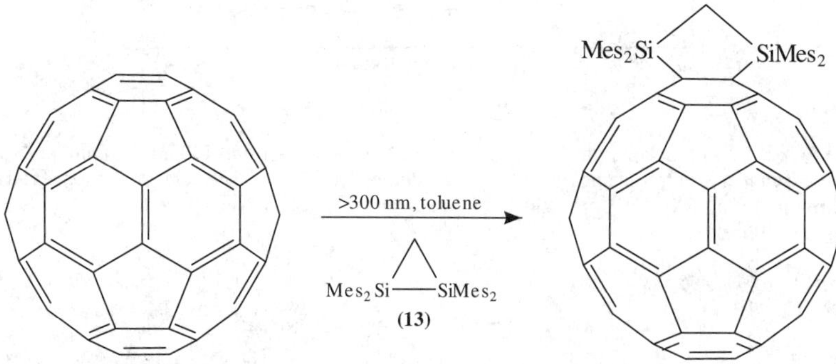

$$Mes = \text{(mesityl)} \quad Dep = \text{(2,6-diethylphenyl)} \quad Dip = \text{(2,6-diisopropylphenyl)}$$

$$C_{60} \xrightarrow[\text{Mes}_2Si\text{—}SiMes_2 \; (13)]{>300 \text{ nm, toluene}} \text{(14)} \quad (3)$$

Symmetry arguments support the following possibilities for the structure of **14**: (i) addition at the 1,6-junction on the fullerene without free rotation of mesityl groups at 30 °C resulting in a 'frozen' (no ring inversion) single conformer in the envelope conformation, (ii) a 1,6-junction addition on the fullerene without free rotation of mesityl groups at 30 °C and (iii) a 1,2-ring junction addition on the C_{60} without free rotation of mesityl groups at 30 °C and with a 'frozen' conformer (no ring inversion) (Scheme 5).

To obtain further information on the structure of **14**, a variable-temperature ^{1}H-NMR measurement was carried out. Coalescence of the *para*-methyl signals at 2.06 and 1.96 ppm at 44 °C, reflecting a conformational change of the molecule, was observed with an activation free energy of $\Delta G^{\neq} = 17.0$ kcal mol^{-1}. The two pairs of four resonances of *meta*-protons between 6.7 and 6.3 ppm and mesityl *ortho*-methyl protons between 3.3 and 2.3 ppm also coalesce at 60 °C and 80 °C, respectively (Figure 7). $\Delta G^{\neq}$ for the coalescence of both the *meta*-protons and *ortho*-methyl groups at the coalescence

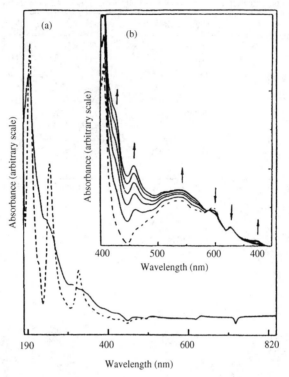

FIGURE 6. (a) UV-vis spectra of C_{60} (- - - -) and **14** (———) from 190 to 820 nm in hexane. (b) Time-dependent spectral changes of a toluene solution of C_{60} and **13** upon irradiation. Reprinted with permission from Ref. 13. Copyright 1993 American Chemical Society

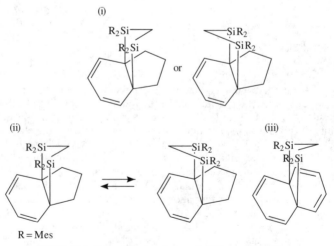

SCHEME 5. Schematic description of disilirane-C_{60} adduct

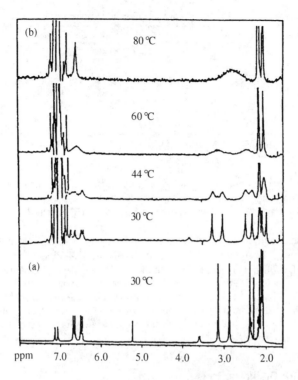

FIGURE 7. ^{1}H-NMR spectra of **14** (a) in CS_2/CD_2Cl_2 and (b) in $CD_3C_6D_5$ at different temperatures. Reprinted with permission from Ref. 13. Copyright 1993 American Chemical Society

temperature is 16.2 kcal mol^{-1}. These results reveal the equivalency of the methylene, methyl and aromatic protons at the coalescence temperature and support possibility (iii) in Scheme 5 as the structure of **14**.

B. Additions of Oxadisilirane to C_{60}

In contrast to the reaction of **13**, the oxadisilirane (**15**) reacts with C_{60} to give mixtures of the mono adducts, 1,2-envelope form (**16**) and the 1,4-twist form (**17**) in 17% yield, and bis-adducts in 39% yield (equation 4). The stabilities of the products were calculated

TABLE 3. Relative energies (kcal mol^{-1}) of C$_{60}$ (Mes$_2$Si)$_2$O isomers

1,2-Addition		1,4-Addition	
envelope form	twist form	envelope form	twist form
0.0(Cs)	8.7(C$_2$)	1.8(Cs)	2.7(Cs)	−2.9(C$_1$)

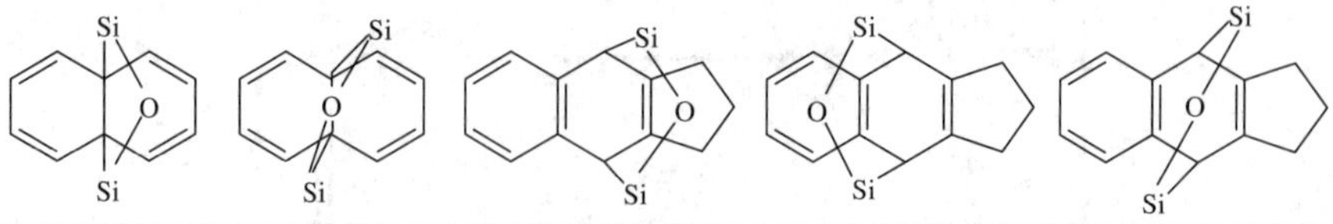

by semiempirical calculation and found to be fully consistent with the experimental results (Table 3). A photochemical and thermal rearrangement of the 1,2-adduct (**16**) to the 1,4-adduct (**17**) was also observed, as shown by the ^{1}H-NMR spectral changes with time (Figure 8)[14].

On the other hand, the digermirane (**18**) reacts with C$_{60}$ to afford the 1,4-adduct (**19**, twist form) selectively[15]. The structure of the 1,4-adduct was confirmed by ^{13}C–^{13}C INADUQUATE (Incredible Natural Abundance Double Quantum Experiment) spectroscopy (equation 5). The rate of disappearance of C$_{60}$ was suppressed by addition of diazabicyclo[2.2.2]octane and 1,2,4,5-tetramethoxybenzene. Furthermore, the reaction was completely inhibited by addition of rubrene as triplet quencher. One plausible rationale for these observations is that an exciplex intermediate may be responsible for formation of the products (Scheme 6).

C. Additions of Disilirane to C$_{70}$

Irradiation of a toluene solution of 1,1,2,2-tetramesityl-1,2-disilirane (**13**) and C$_{70}$ with a high-pressure mercury lamp resulted in the formation of 1,1,3,3-tetramesityl-1,3-disilorane (**20**) in 85% yield with complete consumption of C$_{70}$ (equation 6)[16]. FAB mass spectroscopy of **20** displays a peak at m/z 1390–1386. The ^{1}H-NMR spectrum of **20** displays six methyls and four *meta*-proton signals of the mesityl groups and a singlet for the methylene protons at 1.83 ppm. The ^{13}C-NMR spectrum of **20** shows 35 signals for the C$_{70}$ skeleton: one at 60.61 ppm and the remainder between 128 and 160 ppm. The ^{29}Si-NMR spectrum of **20** shows a peak at −15.88 ppm. In the variable-temperature ^{1}H NMR, coalescence of the *meta*-protons was observed, with an activation free energy of $\Delta G^{\neq} = 18.1$ kcal mol^{-1}. This result reveals the equivalency of aromatic protons above the coalescence temperature, indicative of (a) 1,2-addition across a 6,6-ring junction at the equatorial belt (e–e bond) or (b) 1,4-addition at the same position. The equatorial bonding structure for **20**, in which either 1,2-adduct (**20a**) or 1,4-adduct (**20b**) structures are conceivable, was confirmed by AM1 molecular orbital calculation (Figure 9).

D. Photochemical and Thermal Additions of Disilirane to Metallofullerene

A toluene solution of La@C$_{82}$ and 1,1,2,2-tetramesityl-1,2-disilirane (**13**) was photoirradiated with a tungsten halogen lamp (cutoff <400 nm) (equation 7)[17]. The MALDITOF mass spectrum of the product shows the presence of La@C$_{82}$(Mes$_2$Si)$_2$CH$_2$ (**21**). The EPR spectrum of the adduct (**21**) in 1,2,4-trichlorobenzene shows the presence of at least two species, presumably two positionally isomeric forms of the disilirane

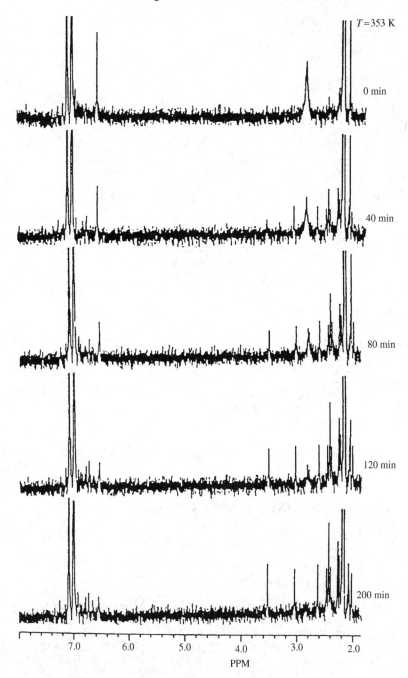

FIGURE 8. Thermal isomerization of 1,2-adduct (16) to 1,4-adduct (17)

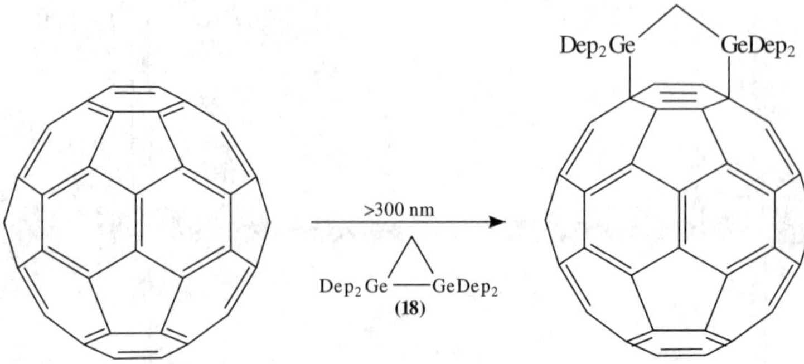

Dep = 2,6-$Et_2C_6H_3$

twist form

(19)

(5)

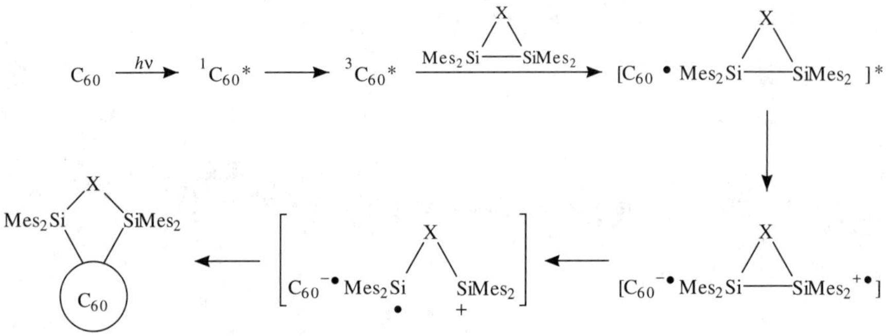

$X = CH_2, O$

SCHEME 6. A possible mechanism for the formation of bis-silylated fullerene

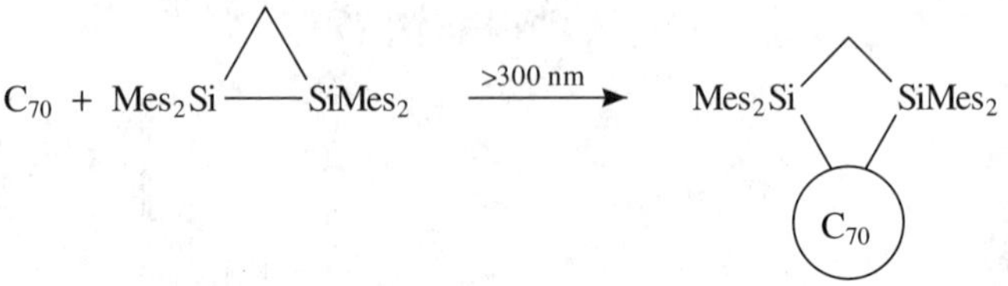

(13) **(20)**

(6)

derivative. La@C_{82} is also thermally reactive towards **13**; in a reaction at 80 °C for 10 h, formation of the adduct **21** was verified by means of mass and EPR spectroscopic studies. The high thermal reactivity of La@C_{82} towards **13** can be rationalized on the basis of its stronger electron donor and acceptor properties relative to C_{60} and C_{70}: *ab initio* calculations predict that while the ionization potential of La@C_{82} (6.19 eV) is much smaller than that of C_{60} (7.78 eV) or C_{70} (7.64 eV), its electron affinity is larger than either that of C_{60} (2.57 eV) or of C_{70} (2.69 eV). Gd@C_{82}[18], La_2@C_{80}[19] and Sc_2@C_{82}[19] are also thermally reactive with disilirane **13**. This can be rationalized on the basis of

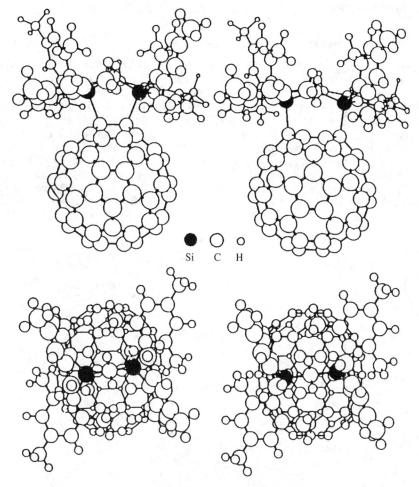

FIGURE 9. Two views of the optimized C_2 structures of **20a** (left) and **20b** (right) calculated with the AM1 method. Reprinted with permission from Ref. 16. Copyright 1994 American Chemical Society

their stronger electron donor acceptor reactivities.

$$\text{La@C}_{82} + \text{Mes}_2\text{Si}\triangle\text{SiMes}_2 \xrightarrow{h\nu} \text{Mes}_2\text{Si}\diamond\text{SiMes}_2\text{-La@C}_{82} \quad (7)$$

(13)   (21)

V. PHOTOCHEMICAL ADDITIONS OF POLYSILACYCLOBUTANES AND -BUTENE TO C_{60}

A. Additions of Benzodisilacyclobutene

The reaction of 3,4-benzo-1,2-disilacyclobutene (**22**) with C_{60} yields the corresponding disilacyclohexane derivative (**23**)[20]. Irradiation of a solution of disilacyclobutene **22** and C_{60} in toluene with a low-pressure mercury lamp (254 nm) afforded the brown adduct **23** in 14% yield (based on unreacted C_{60}) (equation 8). The FAB mass spectrum of **23** exhibits one peak at m/z 1024–1027 ($C_{78}H_{32}Si_2$, M$^+$, molecular cluster ion), as well as one for C_{60} at m/z 720–723. The ^{1}H-NMR spectrum of **23** showed a symmetrical spectrum with two diastereotopic isopropyl methyl protons, one isopropyl methine proton and a AA'BB' pattern assigned to phenyl protons. The ^{13}C-NMR spectrum of **23** shows 17 signals for the C_{60} skeleton, of which four correspond to two carbon atoms each and 13 correspond to four carbon atoms each: one signal appears at 63.93 ppm and the remainder between 130 and 160 ppm (Scheme 7, Figure 10). This pattern is consistent

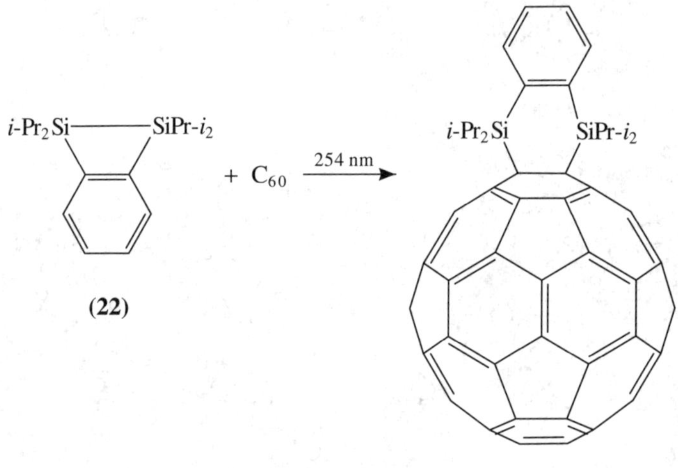

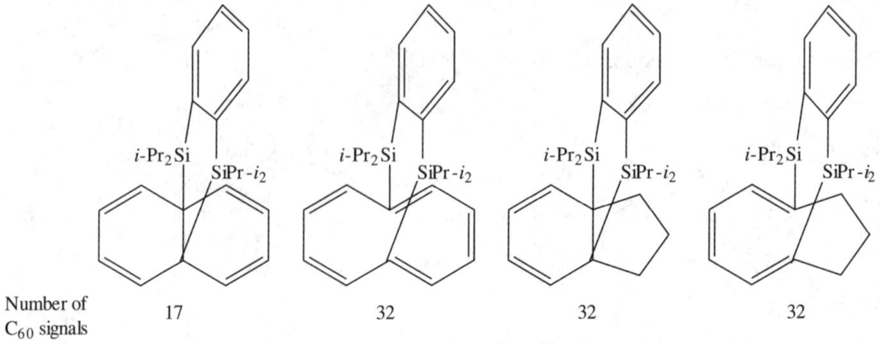

SCHEME 7. Schematic description of benzodisilacyclobutene-C_{60} adduct

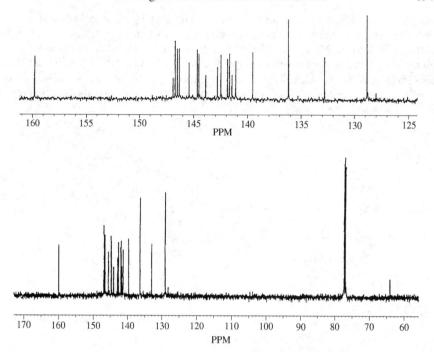

FIGURE 10. ^{13}C NMR spectrum (125 MHz, 1 : 1 CS_2-$CDCl_3$) of **23**. Reprinted by permission of Elsevier Science Ltd from Ref. 20

for a C_{60} adduct of **23**. Furthermore, the ^{13}C-NMR signal at 63.93 ppm strongly supports 1,2-addition. Adduct **23** did not form under the irradiation of a high-pressure mercury lamp (>300 nm), which indicates that the diradical formed by the initial Si−Si bond cleavage is trapped by C_{60} to afford adduct **23**.

B. Additions of Bis(alkylidene)disilacyclobutanes

The photochemical reaction of bis(alkylidene)disilacyclobutane (**24a** and **24b**) with C_{60} by a high-pressure mercury lamp (cutoff <300 nm) proceeds to afford the adducts **26a** (61%) and **26b** (52%) instead of **25**[21]. The formation of these products is a result of an unexpected rearrangement of the disilacyclobutane moiety (equation 9). The FAB mass spectrum of **26a** exhibits one peak at m/z 1056–1059 ($C_{80}H_{40}Si_2$, $M^+ + 1$ cluster), as well as one for C_{60} at m/z 720–723.

The C_{60} unit of **26a** displays 52 resonances in the ^{13}C-NMR spectrum which indicates the absence of any symmetry element in this molecule. One signal has a relative intensity of four, two signals have each a relative intensity of three and one signal has a relative intensity of two; thus the number of carbon atoms sums up to 60. While two fullerene carbon atoms resonate at 63.99 (C^i) and 77.50 (C^h) ppm, the other 50 signals appear in the region between 130–165 ppm. In the ^{1}H-NMR spectrum of **26a**, the four alkylidene methyl protons gave rise to new resonances; two isopropyl methyl protons H^a and $H^{a'}$, isopropyl methine proton H^b, one methyl proton H^g and methylene protons H^f and $H^{f'}$. These observations support the migration of an alkylidene methyl proton to an alkylidene quaternary carbon to form an isopropyl and a 1-propanyl-2-ylidene group. Analysis of the $^{13}C-^{1}H$ three-bond coupling as shown in structure **26** from the $^{13}C-^{1}H$ COLOC and HMBC (^{1}H-Detected Multiple-bond Heteronuclear Multiple Quantum Coherence Spectrum) NMR of **26a** allowed one to show that the isopropyl group (H^a, H^b) is connected to the olefinic carbon (C^c), whereas methyl (H^g) and methylene (H^f, $H^{f'}$) groups are bonded to another olefinic carbon (C^d) and fullerene carbon (C^h) through quaternary carbon (C^e) shown in **26a** and **26b** (Scheme 8).

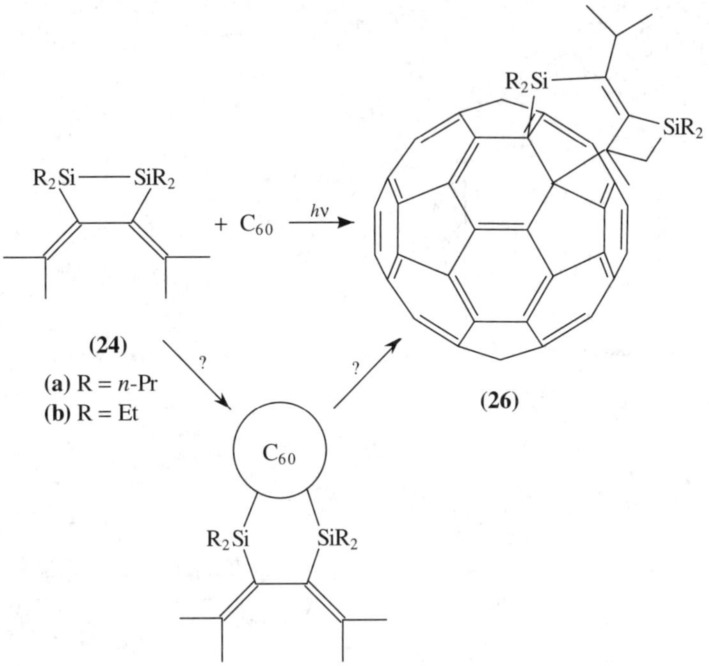

(24)

(a) R = *n*-Pr

(b) R = Et

(26)

(25)

(9)

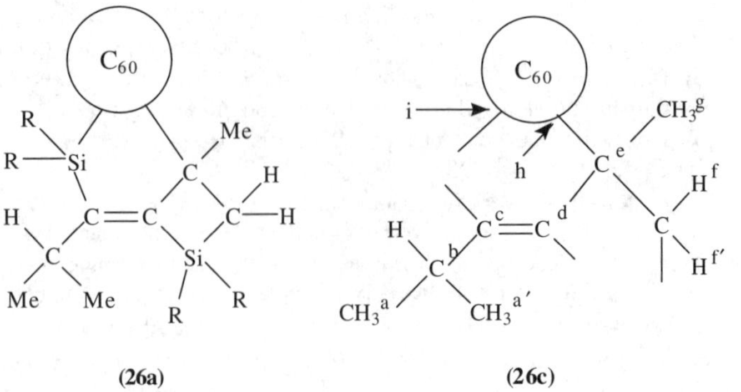

(26a) **(26c)**

SCHEME 8. Connectivities derived from $^{1}H-^{1}H$ and $^{1}H-^{13}C$ shift correlation experiments

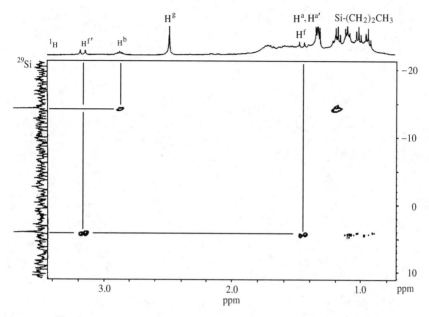

FIGURE 11. ^{29}Si-^{1}H HMBC NMR spectrum of **26a**. Reprinted with permission from Ref. 21. Copyright 1994 American Chemical Society

In order to narrow down the possibilities, Ando and coworkers[21] measured ^{29}Si-^{1}H HMBC spectra. The ^{29}Si-NMR spectrum of **26a** showed two peaks at 4.20 and −14.30 ppm. From the ^{29}Si-^{1}H HMBC spectrum of **26a** the silicon resonance at −14.30 ppm correlates with isopropyl methine proton (H^b) and the silicon resonance at 4.20 ppm correlates with the methylene protons (H^f, H$^{f'}$) (Figure 11). Furthermore, NOE enhancement of **26a** was observed for signals between isopropyl methyl and methine protons and the four Si-n-propyl methyl groups. Clearly this is only possible for structure **26a**. With regard to the addition pattern of the fullerene moiety, a 1,2-junction of the silabutene fragment is most probable. In this arrangement, two diastereomeric adducts are expected.

C. Additions of Cyclotetrasilanes and Cyclotetragermane

Photochemical reactions of cyclotetrasilane **27a** with C_{60} by irradiation of a solution of the reactants in toluene with a high-pressure mercury lamp (>300 nm) for 6 h yielded **28a** and **29a** in 13% and 46% yields, respectively (based on unreacted C_{60}) (equation 10)[22]. Similarly, the adducts **29b** and **29c** were obtained from the reaction of **27b** and **27c** in 87% and 41% yields, respectively, but **28b** and **28c** were not obtained.

The FAB mass spectrum of **28a** exhibits one peak at m/z 1560–1563 ($C_{116}H_{56}Si_4$, M$^+$+1, molecular cluster ion), as well as one for C_{60} at m/z 720–723. The ^{13}C-NMR spectrum of **28a** displays 38 signals for all quaternary carbons. The ^{1}H-NMR spectrum of **28a** displays 4 methyl signals and 4 pairs of AB quartets, supporting a C_s symmetry for the molecule. The ^{29}Si-NMR spectrum of **28a** shows two peaks at −22.25 and −11.34 ppm which are assigned to the silicon atoms of **28a**. Symmetry arguments support the following

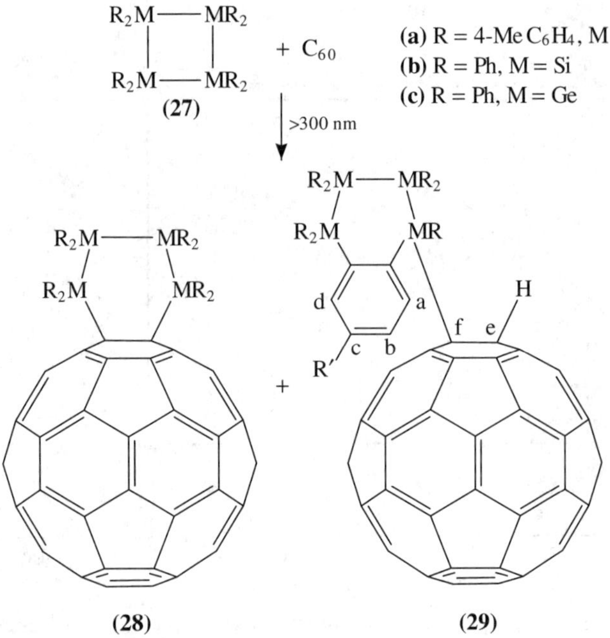

$R_2M\text{———}MR_2$
$||$
$R_2M\text{———}MR_2$
(27)

$+ C_{60}$

(a) R = 4-MeC_6H_4, M = Si
(b) R = Ph, M = Si
(c) R = Ph, M = Ge

↓ >300 nm

(28) + (29)

(10)

(a) R = 4-MeC_6H_4, R′ = Me, M = Si
(b) R = Ph, R′ = H, M = Si
(c) R = Ph, R′ = H, M = Ge

possibilities: (i) a 1,6-junction or a 1,4-addition to the C_{60} with ring inversion (a frozen conformer with this addition mode would give 60 signals in the ^{13}C NMR for the C_{60} moiety) and (ii) a 1,2-junction on the C_{60} with a frozen conformer (no ring inversion) (Scheme 9). A chemical shift change of the p-tolyl groups, reflecting a conformational change of the molecule, was observed by variable-temperature ^1H-NMR measurement. Therefore, a 1,2-junction on the C_{60} moiety with a frozen conformer is the most probable structure for **28a**. The FAB mass spectrum of **29a** exhibits one peak at m/z 1560–1563 ($C_{116}H_{56}Si_4$, M$^+$+1, molecular cluster ion), as well as one for C_{60} at m/z 720–723. The ^{13}C-NMR spectrum of **29a** displays 64 signals for all quaternary carbon which indicates the absence of any symmetry element in this structures. One 4-methyl-o-phenylene group and other seven 4-MeC_6H_4 groups appear in the ^1H, ^{13}C, ^1H-^1H COSY ^1H-^{13}C COSY experiments. The presence of one hydrogen connected to C_{60} is deduced from the appearance of one doublet at 60.35 (C^e) ppm in the ^{13}C NMR and one singlet at δ 6.91 (H^e) in the ^1H-NMR spectrum, respectively. The ^{29}Si-NMR spectrum shows four peaks at −44.83, −41.26, −24.30 and −12.66 ppm which are assigned to the silicon atoms of **29a**. The connectivities between these structural elements were determined by ^{29}Si-^1H HMBC experiment. It was shown that the silicon resonance at −12.66 ppm correlated with both H^a and H^e methine proton signals (Figures 12). The mechanistic pathway suggested for the formation of **28** and **29** was supported by control experiments; photochemical reaction (>300 nm) of **27b** in the presence of CCl_4 affords 1,4-dichloro-1,1,2,2,3,3,4,4-octaphenyltetrasilane (**30b**) and 1,3-dichloro-1,1,2,2,3,3-hexaphenyltrisilane (**31b**) in 37% and 31% yield, respectively. Under identical photolytic conditions, adduct **28a** did not convert to **29a** in a control experiment. These findings indicate that biradical **32** might be

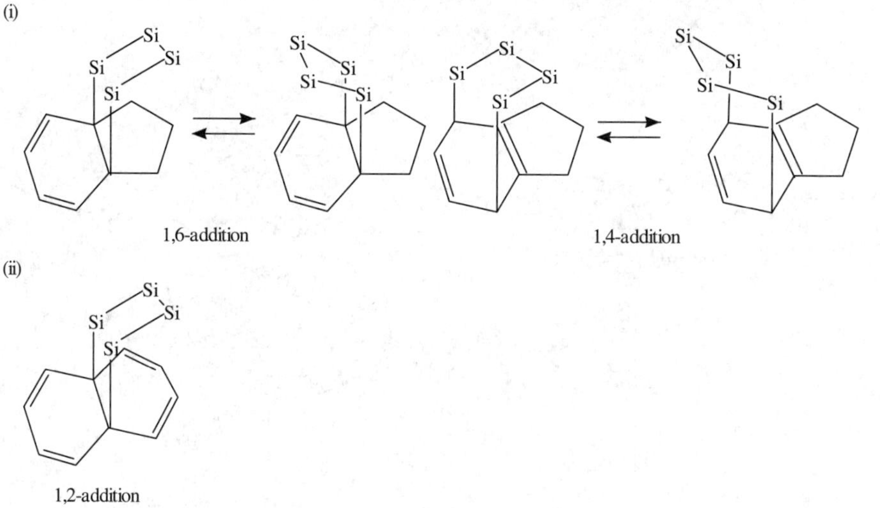

33. Organosilicon derivatives of fullerenes

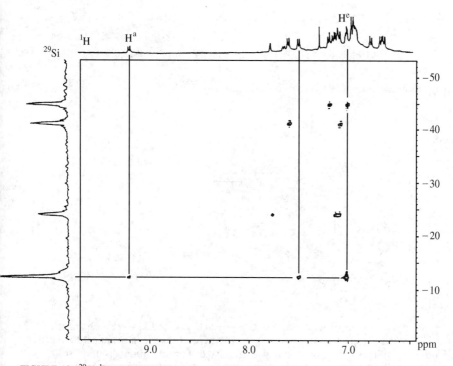

SCHEME 9. Schematic description of cyclotetrasilane-C_{60} adduct

FIGURE 12. ^{29}Si-^{1}H HMBC NMR spectrum of **29a**. Reprinted with permission from Ref. 22. Copyright 1995 American Chemical Society

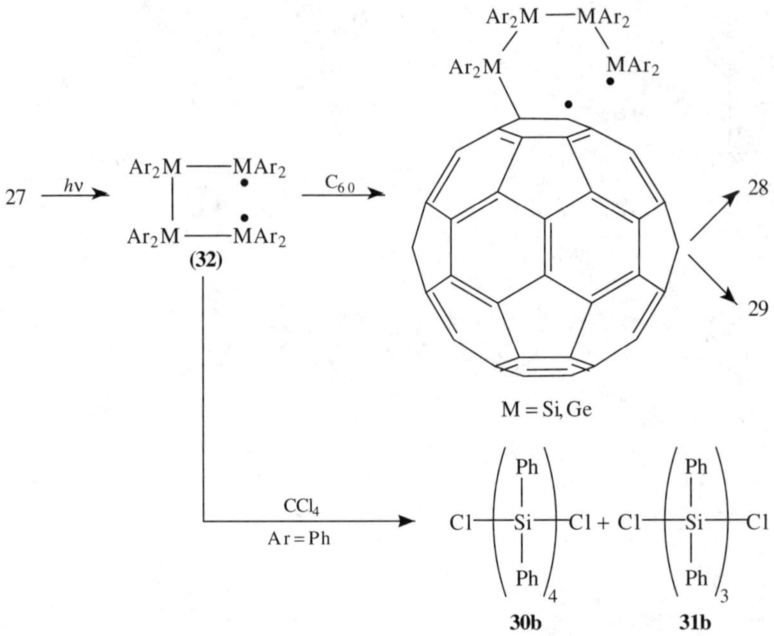

SCHEME 10. A possible mechanism for the formation of **28** and **29**

involved in the course of the reaction (Scheme 10).

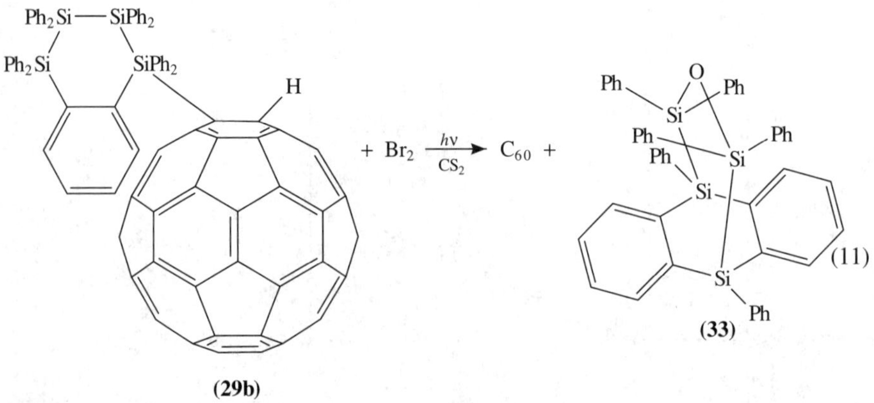

In the reaction of cyclotetragermane **27c**, products **28c** and **29c** were obtained in 43% and 37% yields, respectively, based on unreacted C_{60}. In order to examine the reactivity of fullerene-silicon derivatives, **29b** was irradiated with a carbon disulfide solution of bromine. Purification of the product by means of gel-permeation chromatography afforded C_{60} and **33** in 93% and 45% yields, respectively (equation 11)[23]. The structure of **33** was determined by X-ray crystallographic analysis (Figure 13). Adduct **33b** did not react with bromine in the dark, and therefore a bromine radical seems to participate in this reaction.

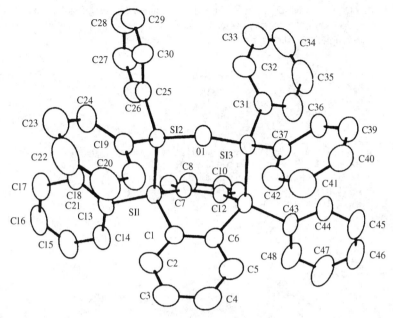

FIGURE 13. X-ray structure of **33**

VI. REACTIONS OF SILYLLITHIUM REAGENTS WITH C_{60}

Nucleophilic addition to C_{60} is among the most common reaction types in fullerene chemistry. The reactions of silyllithium reagents with C_{60} have been reported by Kusukawa and Ando (equation 12)[24]. Reactions of C_{60} with Ph_3SiLi have been unsuccessful and formation of a brown precipitate was observed at $-78\,°C$; when the suspension was quenched with EtOH, only C_{60} was recovered. The analogous reaction with other silyllithium derivatives ($Me_2PhSiLi$, $MePh_2SiLi$, $Et_2PhSiLi$, $EtPh_2SiLi$) leads to similar results. On the other hand, diisopropylphenylsilyllithium **34a** reacted with C_{60} under these conditions to produce **35a** in 78% yield (based on unreacted C_{60}) upon quenching with EtOH (Table 4). The ^{13}C-NMR spectrum of **35a** displays 30 signals for the C_{60} carbons, supporting a C_s symmetry for the molecule. Similarly, diphenylisopropyl- and diphenyl-*t*-butyl-silyllithiums, **34b** and **34c**, react with C_{60} to give monoadducts **35b** and **35c**, respectively.

On the other hand, the reaction of tris(trimethylsilyl)silyllithium **34d** with equivalent amounts of C_{60} at $-78\,°C$ give bisadduct **36d** in 69% yield (based on unreacted C_{60}), and the monoadduct **35d** could not be detected. The FAB mass spectra of **36d** exhibit one peak at m/z 1215–1218 ($C_{78}H_{54}Si_8$, $M^+ +1$, molecular cluster ion), as well as one for C_{60} at m/z 720–723. The ^{13}C-NMR spectrum of **36d** displays 29 signals for all quaternary carbons. Symmetry arguments support a 1,6-addition or 1,4-addition to the C_{60}; however, a 1,6-addition of the bulky substituent [$(TMS)_3Si$] is unfavorable. Modification **36d** crystallized in the triclinic space group $P-1$, with two enantiomorphic molecules of **36d** and two CS_2 molecules in the unit cell, and the crystal structure of **36d** and selected bond angles are shown in Figure 14. The C—C bond lengths C(2)—C(12), C(9)—C(10), C(11)—C(28) and C(13)—C(30) are located within 1.363(7)–1.390(7) Å, which are close to the C—C double-bond length. The bond angles between fullerene C atoms, excluding the bonds in rings A and B, are 107.3(4)–109.2(4) deg in the pentagons and 118.3(4)–121.5(5) deg in the hexagons, values which are within the normal range. The lengths of the bond in the C_{60} skeleton, excluding the bonds in rings A and B, range from 1.371(7) to 1.474(8)Å (Figure 15). The structure of **36d** is a new type of addition mode to the C_{60}.

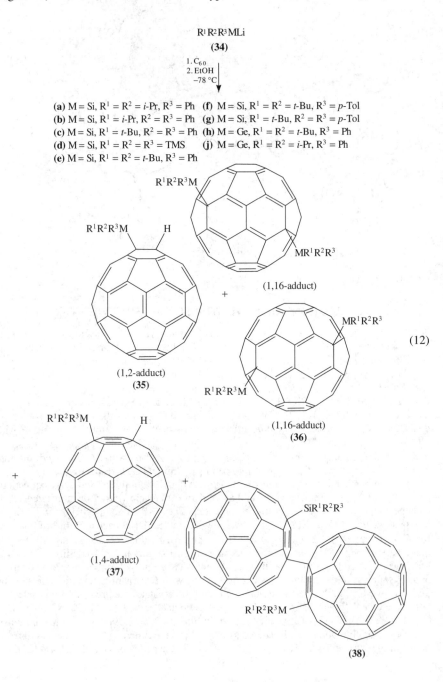

(a) M = Si, $R^1 = R^2 = i$-Pr, $R^3 = $ Ph (f) M = Si, $R^1 = R^2 = t$-Bu, $R^3 = p$-Tol
(b) M = Si, $R^1 = i$-Pr, $R^2 = R^3 = $ Ph (g) M = Si, $R^1 = t$-Bu, $R^2 = R^3 = p$-Tol
(c) M = Si, $R^1 = t$-Bu, $R^2 = R^3 = $ Ph (h) M = Ge, $R^1 = R^2 = t$-Bu, $R^3 = $ Ph
(d) M = Si, $R^1 = R^2 = R^3 = $ TMS (j) M = Ge, $R^1 = R^2 = i$-Pr, $R^3 = $ Ph
(e) M = Si, $R^1 = R^2 = t$-Bu, $R^3 = $ Ph

33. Organosilicon derivatives of fullerenes

TABLE 4. Reactions of silyllithium reagents with C_{60}

Silyllithium	Product and isolated yield (%)			
	35	36	37	38
34a	78			
34b	80			
34c	50	22		
34d		69		
34e		72		
34f	11	42	11	6
34g		29	19	7
34h		87		
34i	76			

Relative energies of the bis adducts ([(TMS)$_3$Si]$_2$C$_{60}$) calculated by the AM1 method (Scheme 11) are: 1,6 (1041.6 kcal mol^{-1}) >1,2 (1024.8 kcal mol^{-1}) >1,4 (987.5 kcal mol^{-1}) >1,16 or 1,29 (977.9 kcal mol^{-1}). Additions of two (TMS)$_3$Si groups to the 1,2-and 1,4-positions are unfavourable because of steric hindrance. Therefore, the (TMS)$_3$Si groups might be bonded to the 1,16 (or 1,29) positions. The UV-vis absorption spectrum of **35a–35c** shows absorption at 416–417 nm and 444–445 nm. On the other hand, bisadducts **36d** and **36e** show broad absorption at 520–600 nm.

Interestingly, silyllithium reagents substituted with aromatic electron-releasing groups (**34f**) and **34g**) gave **35** (1,2-adduct), **36** (1,16- and 1,29-adduct), **37** (1,4-adduct) and **38** (dimer at 1,4-position) which can be separated by gel permeation chromatography (GPC). The ^{13}C-NMR spectrum of **37f** displays 60 signals for all quaternary carbons.

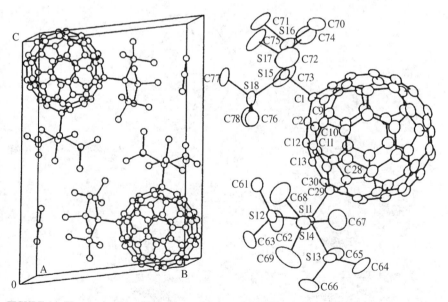

FIGURE 14. X-ray crystallographic characterization of **36d**·CS$_2$. Left: arrangement in the unit cell. Right: crystal structure of one of the two enantiomorphic molecules in the unit cell. The ellipsoids represent 50% probability. From Ref. 26

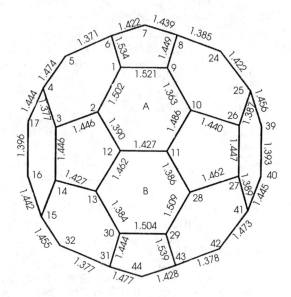

FIGURE 15. Details of bond lengths (Å) and selected bond angles (°) in **36d**: C(2)−C(1)−C(6) 121.9, C(2)−C(1)−C(10) 109.1, C(6)−C(1)−C(10) 117.3, C(1)−C(2)−C(3) 123.2, C(2)−C(3)−C(4) 110.2, C(3)−C(4)−C(5) 124.0, C(4)−C(5)−C(6) 120.6, C(1)−C(6)−C(5) 117.2, C(1)−C(6)−C(7) 122.0, C(5)−C(6)−C(7) 108.9, C(6)−C(7)−C(8) 123.3, C(7)−C(8)−C(9) 109.7, C(8)−C(9)−C(10) 124.3, C(1)−C(10)−C(9) 120.4. Reproduced by permission of VCH Verlagsgesellschaft mbH from Ref. 24

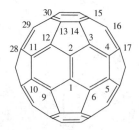

AM1 calculations of [(TMS)$_3$Si]$_2$C$_{60}$ PM3/MNDO calculations of C$_{60}$H$_2$

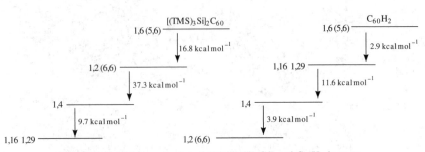

SCHEME 11. Relative energies of [(TMS)$_3$Si]$_2$C$_{60}$ and C$_{60}$H$_2$ isomers

TABLE 5. HOMO energies of hydrosilanes calculated by semiempirical methods

Hydrosilane	HOMO (eV)		Product formed
	PM3	AM1	
(TMS)$_3$SiH	−8.60	−9.24	
t-Bu$_2$(p-Tol)SiH	−9.01	−9.22	Bis-adduct
t-Bu(p-Tol)$_2$SiH	−9.02	−9.23	
t-Bu$_2$PhSiH	−9.16	−9.45	
t-BuPh$_2$SiH	−9.20	−9.49	
i-Pr$_2$PhSiH	−9.22	−9.50	Mono-adduct
i-PrPh$_2$SiH	−9.29	−9.52	
Et$_2$PhSiH	−9.30	−9.53	
MePh$_2$SiH	−9.34	−9.59	
Ph$_3$SiH	−9.36	−9.52	No reaction
Me$_2$PhSiH	−9.39	−9.62	
EtPh$_2$SiH	−9.41	−9.59	

One fullerene carbon resonates at 62.37 ppm whereas the signals of all other carbons appear in the region between δ 130 and 160 ppm. The partial structure of the fragment annulated to the C$_{60}$ moiety can be derived from the NMR spectroscopic properties for **37f**. The presence of one hydrogen connected to C$_{60}$ is deduced from the appearance of one doublet at 49.54 ppm in the ^{13}C-NMR spectrum and one singlet at 5.44 ppm in the ^{1}H-NMR spectrum, respectively. The ^{29}Si-NMR spectrum of **37f** shows one signal at 3.59 ppm. From these findings, a 1,4-ring junction is most probable for **37f**. The MALDI-TOF mass spectrum of **38f** exhibits one peak at m/z 1907 (C$_{150}$H$_{50}$Si$_2$, M$^+$, molecular cluster ion), as well as one peak for C$_{60}$ at m/z 720. The ^{13}C-NMR spectrum of **38f** displays 60 signals for all quaternary carbons; two fullerene carbons resonate at 75.49 and 62.19 ppm and the signals of all other carbons appear in the region between 135 and 160 ppm. The ^{29}Si-NMR spectrum of **38f** shows one signal at 6.17 ppm, which is assigned to the silicon atoms of **38f**. Additionally, germyllithium compounds **34h** and **34i** also gave mono- and bis-adducts, selectively.

The reactivity difference of the silyllithium reagents raises some questions concerning the mechanism. To clarify the mechanistic pathway, it was examined by semiempirical calculations using hydrosilanes instead of silyllithium compounds (Table 5)[25]. As deduced from the HOMO energies of Table 5, electron-releasing substituents favor the formation of adducts **35** and **36**, which indicate that their formation should involve a radical reaction proceeding via electron transfer from the silyllithium to C$_{60}$.

In order to clarify the mechanistic pathway, the photochemical reactions of disilanes **39** with C$_{60}$ were carried out and gave 1,16- and 1,29-adducts (equation 13, Table 6)

TABLE 6. Photochemical reactions of disilanes with C$_{60}$

Disilane	Product and isolated yield (%)	
	36	40
39c	56	
39e	62	
39g	54	
39j	43	24

$$R^1R^2R^3Si\text{—}SiR^1R^2R^3 + C_{60}$$
(39)

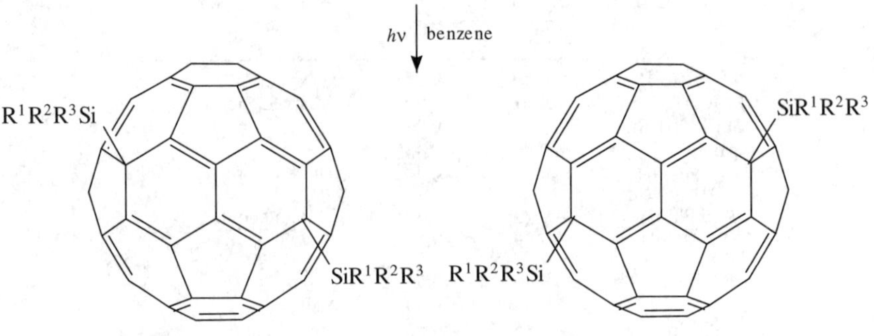

$h\nu$ | benzene

(1,16-adduct)   (1,29-adduct)
(36)

(13)

(c) $R^1 = t\text{-Bu}, R^2 = R^3 = Ph$
(e) $R^1 = R^2 = t\text{-Bu}, R^3 = Ph$
(g) $R^1 = t\text{-Bu}, R^2 = R^3 = p\text{-Tol}$
(k) $R^1 = TMS, R^2 = R^3 = Ph$

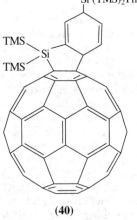

(40)

$$t\text{-BuPh}_2Si\text{—}Si(Bu\text{-}t)(p\text{-Tol})_2 + C_{60}$$
(41)

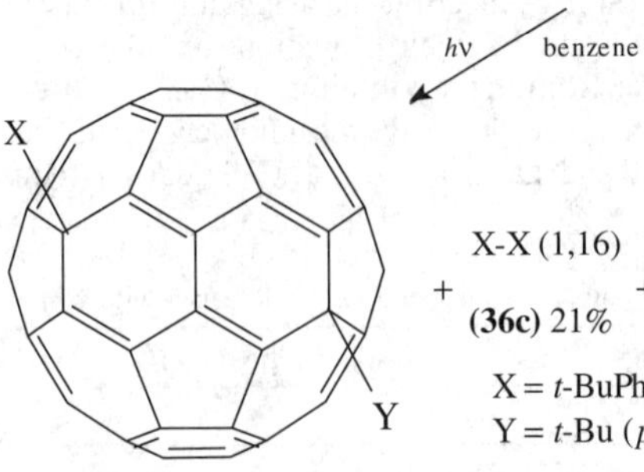

$h\nu$ / benzene

(1,16-adduct)
(42) 17%

+

X-X (1,16)   Y-Y (1,16)
+
(36c) 21%   **(36g)** 10%

$X = t\text{-BuPh}_2Si$
$Y = t\text{-Bu}(p\text{-Tol})_2Si$

(14)

in moderate yields; in the case of **39k**, rearranged product **40** was also obtained as a by-product, probably via a radical reaction. When unsymmetrical disilane **41** was reacted with C_{60}, bis-adducts **36c**, **36g** and **42** which should involve the additions of silyl radicals, were formed (equation 14).

VII. REACTIONS OF SILICON-SUBSTITUTED NUCLEOPHILES

A. Reactions of Silylmethyl Grignard Reagents

Addition reactions of Me_2YSiCH_2MgCl (**43**) to C_{60} provided two products, $HC_{60}CH_2SiYMe_2$ (**44**) and $[Me_2YSiCH_2]_2C_{60}$ (**45**) (Y = Me, H, CH=CH$_2$, Ph, OPr-i)[26] (equation 15). Selective preparation of either **44** or **45** was accomplished by a selection of the solvent. **44** is formed in THF, whereas **45** was produced in toluene. ^{13}C-NMR resonances of the C_{60} moiety of **44** consist of 30 carbons in the aromatic region and two carbons around 62 ppm (one CH and one quaternary carbon). However, ^{13}C-NMR resonances of **45** are compatible with 31 carbons in the aromatic region and one carbon at 54.90 ppm.

$$C_{60} + Me_2SiYCH_2MgCl \quad (43)$$

(a) Y = Me (d) Y = CH=CH$_2$
(b) Y = H   (e) Y = OPr-i
(c) Y = Ph

(15)

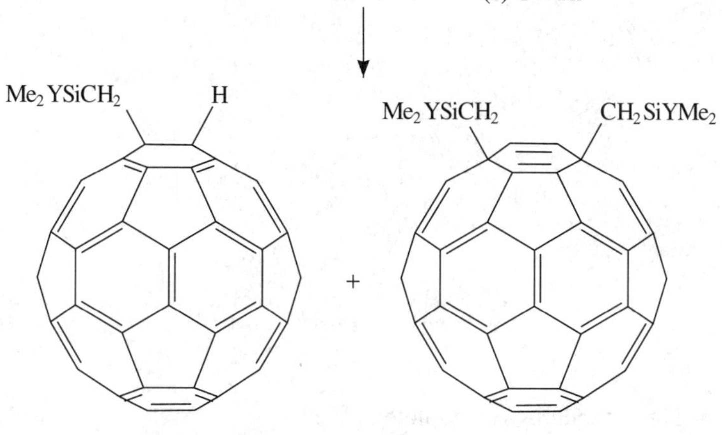

(**44**)   (**45**)

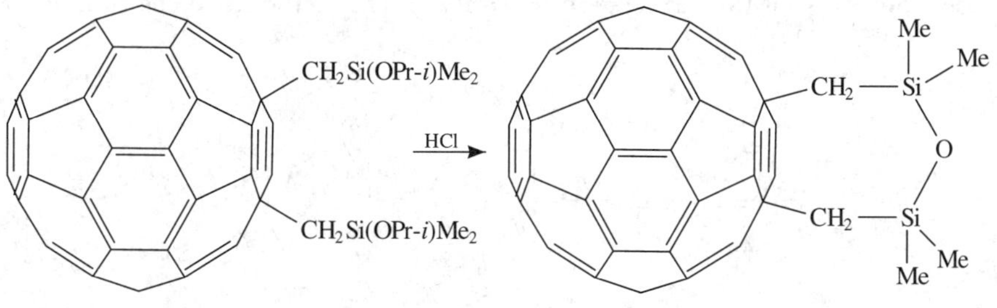

(**45e**)   (**46**)   (16)

Nagashima and coworkers prepared cyclic siloxane **46** in 76% yield by acidic hydrolysis of adduct **45e**[27]. This is the first well-characterized siloxane containing C_{60} (equation 16).

Treatment of **44e** with 3 equivulents of $AlCl_3$ gave chlorosilane **47**. Treatment of silica with **47** and pyridine in refluxing toluene gave modified silica (**48**) (equation 17). Its carbon content was 7.18%, suggesting that 0.21 molecule of C_{60} is present per 100 $Å^2$ of silica surface. A mixture of C_{60} and C_{70} was well separated by means of modified silica (**48**). No separation occurred using unmodified silica as a stationary phase.

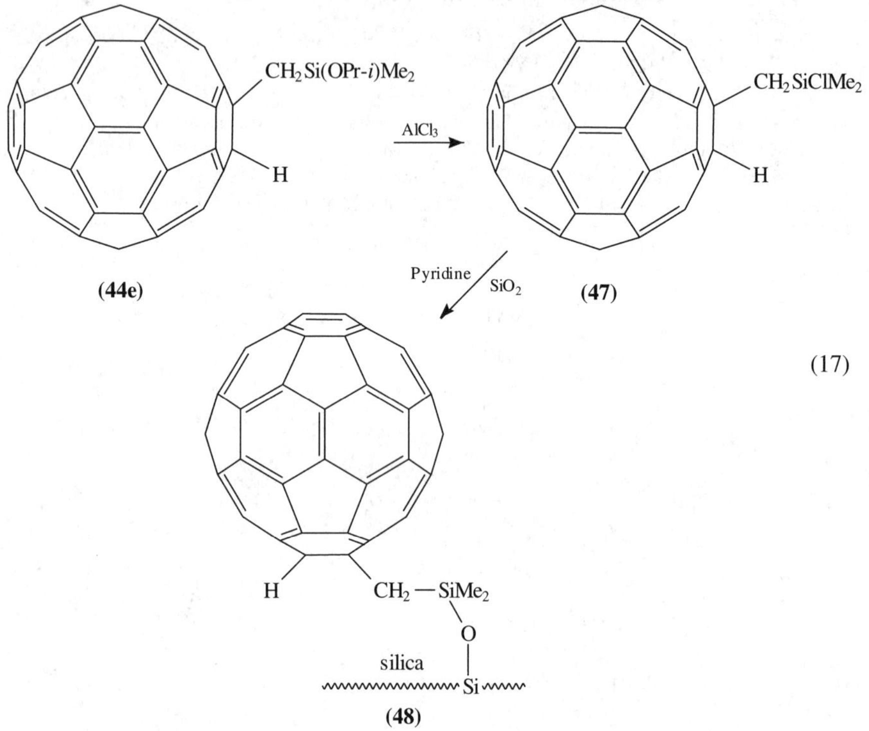

(17)

B. Reactions of Silyl-acetylene Nucleophiles

Nucleophilic addition of lithium trimethylsilylacetylide (**49**) to C_{60} to produce the adduct **50** (equation 18) was reported by Diederich and coworkers[28]. The structure of

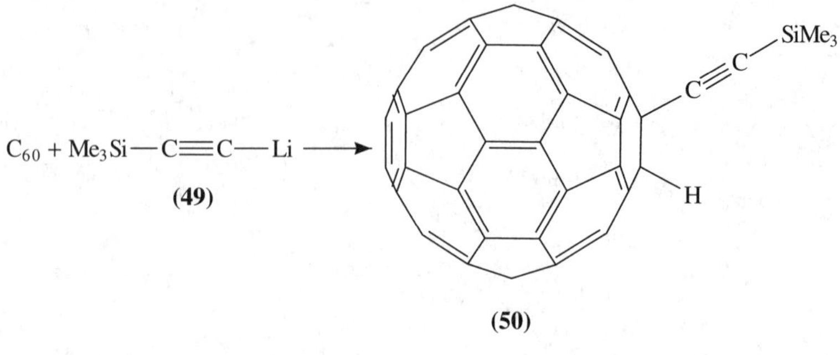

$$C_{60} + Me_3Si-C{\equiv}C-Li \longrightarrow$$
(**49**)

(18)

(**50**)

50 is evident from its spectral data. Its ^{1}H-NMR spectrum shows one singlet corresponding to the fullerene proton. The addition proceeds in a 1,2 fashion, as judged by the 32 fullerene resonances in the ^{13}C-NMR spectrum of **50**. At the same time, Komatsu and coworkers also reported the addition of **49** to the C_{60}[29].

VIII. REACTIONS OF SILYL DIAZONIUM COMPOUNDS

A broad variety of methano-bridged fullerenes are accessible by the reaction of C_{60} with different diazomethanes. This chemical transformation of C_{60} was discovered by Wudl and is based on the findings that C_{60} behaves as an 1,3-dipolarphile[3].

Reactions of the silyldiazonium compounds with C_{60} were examined[30]. To the toluene solution of C_{60} was slowly added the silyldiazonium compounds at room temperature and the products obtained were separated by gel permeation chromatography (equation 19). The yields of the mono-, bis- and tris-adducts are given in Table 7. The ^{1}H-NMR chemical shift of the methine proton and the $^{2,3}J_{C-H}$ coupling constants with the C_{60} carbon support the methanoannulene structure (**51**).

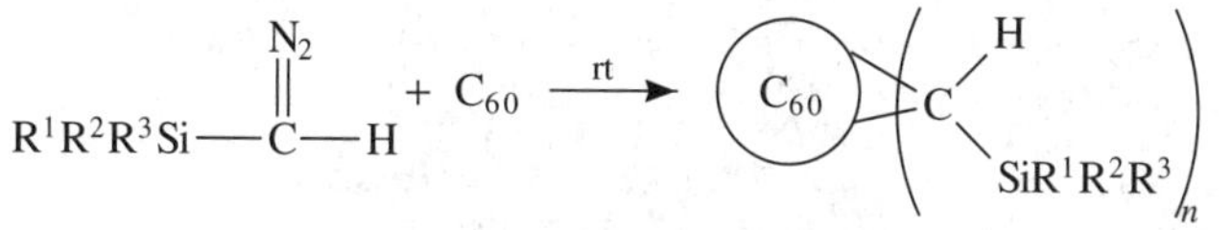

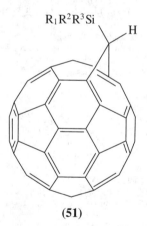

(19)

(**51**)

Under similar conditions, bis(silyl)diazonium compounds (**52**) do not react with C_{60} even under reflux (equation 20).

TABLE 7. Yields of methanoannulene products from reaction 19

	$n = 1$	$n = 2$	$n = 3$
$R^1 = R^2 = R^3 = Me$	34%	24%	22%
$R^1 = Ph, R^2 = R^3 = Me$	41%	29%	8%
$R^1 = 3,4\text{-}(MeO)_2C_6H_3, R^2 = R^3 = Me$	41%	29%	8%

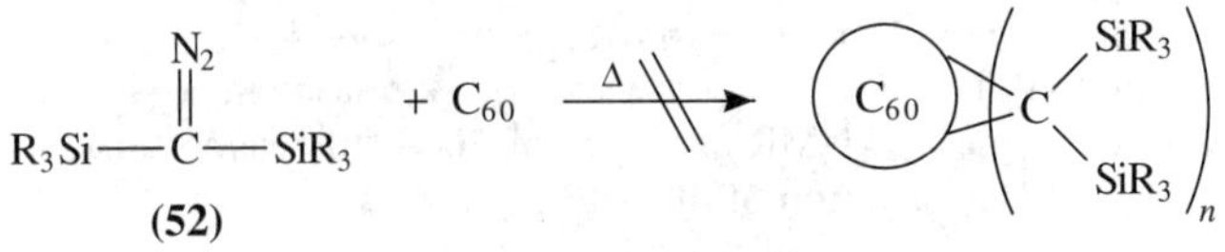

$$R_3Si-\overset{N_2}{\underset{\|}{C}}-SiR_3 \quad + \quad C_{60} \quad \xrightarrow{\Delta} \quad \left(C_{60} \diagdown C \diagup \overset{SiR_3}{\underset{SiR_3}{}} \right)_n \quad (20)$$

(52)

IX. REFERENCES

1. (a) H. W. Kroto, A. W. Allaf and S. P. Balm, *Chem. Rev.*, **91**, 1213 (1991).
   (b) H. W. Kroto, J. R. Heath, S. C. O'Brien, R. F. Curl and R. E. Smalley, *Nature*, **318**, 162 (1985).
   (c) W. Kräschmer, L. Lamb, K. Fostiropoulos and D. R. Huffman, *Nature*, **347**, 354 (1990).
   (d) W. Kräschmer, K. Fostiropoulos and D. R. Huffman, *Chem. Phys. Lett.*, **170**, 167 (1990).
2. See the issue of *Acc. Chem. Res.*, **25**, 98–175 (1992).
3. F. Diederich, L. Isaacs and D. Philip, *Chem. Soc. Rev.*, 243 (1994).
4. A. Hirsch, *Angew. Chem., Int. Ed. Engl.*, **32**, 1138 (1993).
5. G. Hammond and V. J. Kuck (Eds.), *Fullerenes: Synthesis, Properties, and Chemistry of Large Carbon Clusters*, ACS Symposium Series 481, American Chemical Society, Washington, DC, 1992.
6. R. West, In *The Chemistry of Organic Silicon Compounds* (Eds. S. Patai and Z. Rappoport), Chap. 19, Wiley, Chichester, 1989, pp. 1207–1240.
7. R. D. Miller and J. Michl, *Chem. Rev.*, **89**, 1359 (1989).
8. Y. Wang, R. West and C.-H. Yuan, *J. Am. Chem. Soc.*, **115**, 3844 (1993).
9. A. Watanabe and O. Ito, *J. Phys. Chem.*, **98**, 7736 (1994).
10. A. Watanabe, O. Ito and K. Mochida, *Organometallics*, **14**, 4281 (1995).
11. T. Akasaka, W. Ando, K. Kobayashi and S. Nagase, *J. Am. Chem. Soc.*, **115**, 1605 (1993).
12. T. Akasaka, E. Mitsuhida, W. Ando, K. Kobayashi and S. Nagase, *J. Chem. Soc., Chem. Commun.*, 1529 (1995).
13. T. Akasaka, W. Ando, K. Kobayashi and S. Nagase, *J. Am. Chem. Soc.*, **115**, 10366 (1993).
14. T. Akasaka, E. Mitsuhida, W. Ando, K. Kobayashi and S. Nagase, unpublished results.
15. T. Akasaka, T. Mizushima, W. Ando, K. Kobayashi and S. Nagase, unpublished results.
16. T. Akasaka, E. Mitsuhida, W. Ando, K. Kobayashi and S. Nagase, *J. Am. Chem. Soc.*, **116**, 2627 (1994).
17. T. Akasaka, T. Kato, K. Kobayashi, S. Nagase, K. Yamamoto, H. Funasaka and T. Takahashi, *Nature*, **374**, 600 (1995).
18. T. Akasaka, S. Nagase, K. Kobayashi, T. Suzuki, T. Kato, K. Yamamoto, H. Funasaka and T. Takahashi, *J. Chem. Soc., Chem. Commun.*, 1343 (1995).
19. T. Akasaka, S. Nagase, K. Kobayashi, T. Suzuki, T. Kato, K. Kikuchi, Y. Achiba, K. Yamamoto, H. Funasaka and T. Takahashi, *Angew. Chem., Int. Ed. Engl.*, **34**, 2139 (1995).
20. T. Kusukawa, A. Shike and W. Ando, *Tetrahedron*, **52**, 4995 (1996).
21. T. Kusukawa, Y. Kabe, T. Erata, B. Nestler and W. Ando, *Organometallics*, **13**, 4186 (1994).
22. T. Kusukawa, Y. Kabe and W. Ando, *Organometallics*, **14**, 2142 (1995).
23. T. Kusukawa, K. Ohkubo and W. Ando, *Organometallics*, in press.
24. T. Kusukawa and W. Ando, *Angew. Chem., Int. Ed. Engl.*, **35**, 1315 (1996).
25. T. Kusukawa and W. Ando, unpublished results.
26. H. Nagashima, H. Terasaki, E. Kimura, K. Nakajima and K. Itoh, *J. Org. Chem.*, **59**, 1246 (1994).
27. H. Nagashima, H. Terasaki, E. Kimura, K. Nakajima and K. Itoh, *J. Org. Chem.*, **60**, 4966 (1995).
28. H. L. Anderson, R. Faust, Y. Rubin and F. Diederich, *Angew. Chem., Int. Ed. Engl.*, **33**, 1366 (1994).
29. K. Komatsu, Y. Murata, N. Takimoto, S. Mori, N. Sugita and T. S. M. Wan, *J. Org. Chem.*, **59**, 6101 (1994).
30. Y. Kabe, K. Suzuki, Y. Yakushigawa and W. Ando, unpublished results.

CHAPTER 34

Group 14 metalloles, ionic species and coordination compounds

JACQUES DUBAC

Hétérochimie Fondamentale et Appliquée (ESA-CNRS 5069), Université Paul-Sabatier, 118, Route de Narbonne, 31062 Toulouse Cedex, France
Fax: (33)5 61 55 82 04;   e-mail: dubac@irit.ups-tlse.fr

CHRISTIAN GUÉRIN

Chimie Moléculaire et Organisation du Solide (UMR-CNRS 5637), Université Montpellier II, Place E. Bataillon, 34095 Montpellier Cedex 5, France
Fax: (33)4 67 14 38 88;   e-mail: chguerin@crit.univ-montp2.fr

and

PHILIPPE MEUNIER

Synthèse et Electrosynthèse Organométalliques (UMR-CNRS 5632), Université de Bourgogne, 6 Boulevard Gabriel, 21004 Dijon Cedex, France
Fax: (33)3 80 39 61 00;   e-mail: philippe.meunier@u.bourgogne.fr

I. INTRODUCTION	1963
II. SYNTHESIS	1964
A. C-unsubstituted Metalloles	1964
1. Siloles	1964
2. Germoles	1968
3. Stannoles	1970
B. C-substituted Metalloles	1971
1. Synthetic methods involving direct formation of the dienic ring	1971
a. Cyclization of 1,4-dilithio-1,3-butadienes by a polyfunctional compound $R_n EX_{4-n}$ $(n = 0-2)$	1971
b. Cyclization of two acetylenic molecules with a $\Sigma_2 E$ fragment	1972

The chemistry of organic silicon compounds, Vol. 2
Edited by Z. Rappoport and Y. Apeloig   © 1998 John Wiley & Sons Ltd

      i. Reaction between an alkyne and a disilane, a silirene
         (1-silacyclopropene) or a cyclotrisilane 1973
      ii. Reaction between an alkyne and a
         7-metallanorbornadiene 1976
      iii. Reactions involving 1,n-diynes 1977
   c. Cyclization of two acetylenic groups bonded to a group 14
     element 1978
      i. Reaction of trialkylboranes with dialkynyl derivatives of a
         group 14 element 1978
      ii. Intramolecular reductive cyclization of diethynylsilanes ... 1979
   d. Cyclization of a 1,3-diyne with a group 14 dihydride 1979
 2. Synthetic methods involving a group 14 heterocycle as
    precursor 1980
   a. Dehydrogenation of metallacyclopentanes or
     metallacyclopentenes 1980
   b. Dehydrohalogenation of halogenometallacyclopentanes 1980
   c. Dehydration of 1-metallacyclopent-4-en-3-ols 1981
   d. Thermolysis of esters of 1-metallacyclopent-4-en-3-ols 1982
   e. Gas-phase pyrolysis of silacyclopent-3-enes 1984
 3. Exchange reactions (transmetallation) with other
    heterocyclopentadienes 1986
C. Dibenzometalloles or 9-Metallafluorenes 1988
D. Benzometalloles or 1-Metallaindenes 1991
 1. Cyclization of 1-lithio-2-(2'-lithiophenyl)ethylene 1991
 2. Transmetallation reactions 1992
 3. Nucleophilic substitution on the metal centre 1993
 4. Thermolytic reactions 1994
 5. Other reaction 1995
III. ORGANIC CHEMICAL PROPERTIES OF GROUP 14
     METALLOLES 1996
A. Stability and Isomerizations 1996
 1. Kinetic stability toward Diels–Alder dimerization 1996
 2. Geometric isomerization into transoid dienes 1996
 3. Tautomerism. 2H- and 3H-metalloles 1997
 4. Stability of functional and spiro derivatives 1998
B. Cycloaddition Reactions 1998
 1. Diels–Alder [4 + 2] cycloadditions 1998
   a. With ethylenic dienophiles 1998
   b. With acetylenic dienophiles 1999
   c. Group 14 metalloles as dienophiles 2002
 2. [2 + 2] Cycloadditions 2002
C. Reactions with Halogens 2003
 1. Silole series 2003
 2. Germole and stannole series 2004
D. Reactions with Acids 2005
E. Reactions with Bases 2006
F. Reactions with Organometallic Reagents 2007
G. Oxidation 2009
H. Reduction 2010
 1. Alkali metal reduction 2010
 2. Electrode reactions 2011
 3. Reaction with hydrides 2011

34. Group 14 metalloles, ionic species and coordination compounds   1963

  I. Transmetallation Reactions 2011
  J. Ring Expansion 2012
IV. POLYMERIC SILOLE-CONTAINING π-CONJUGATED
    SYSTEMS 2013
  A. Synthesis, Structure and UV-Visible Absorption Data of
     Oligosiloles 2013
  B. Silole–Thiophene and -Pyrrole Cooligomers and Copolymers 2015
  C. $\sigma-\pi$ Conjugated Polymers with Alternating Arrangement of a
     Disilanylene Unit and 3,4-Diethynylene-Substituted Siloles 2016
  D. Polymers Involving only the Silicon Atom of the Silole Ring in the
     Main Chain 2017
 V. IONIC SPECIES AND COORDINATION COMPOUNDS 2018
  A. Group 14 Metallole Anionic Species 2019
     1. Silacyclopentadienide and silafluorenide anions 2019
     2. Germacyclopentadienide anions 2021
     3. Germa- and stannaindenide anions 2022
     4. Silole dianions 2022
     5. Germole dianions 2024
  B. Stable η^5-Sila- and η^5-Germacyclopentadienyl Transition Metal
     Complexes 2026
VI. ADDENDUM 2027
VII. CONCLUSION 2029
VIII. ACKNOWLEDGEMENTS 2030
IX. REFERENCES 2030

I. INTRODUCTION

Five-membered heterocyclic dienes (**1**) are a very important class of organic heterocyclic compounds. Furan, thiophene and pyrrole derivatives are the best known, but numerous compounds have been described in which E is a main group or transition element.

(**1**)

The rings are generally named heterocyclopentadienes, *heteroles* or *metalloles* according to the non-metallic or metallic character of E. The heterocycle can be 6 π aromatic (E = O, S, NR, PR), 4 π simple diene (E = CR_2, SiR_2) or 4 π antiaromatic (E = BR, AlR). Their stability obviously decreases in the same order, and boroles, which are particularly unstable, have been isolated only in the Diels–Alder dimer form[1]. However, in the case of a trivalent or tetravalent E atom, [1,5]-sigmatropic shifts can give rise to isomerization of **1** (1H-heterole) into tautomeric forms (2H or 3H-heterole), in particular for the phospholes[2].

With regard to group 14 metalloles, two reviews were published in 1990, the first on their synthesis and their properties[3], the second on their ionic species and their coordination compounds[4]. Here, we present a simplified review by concentrating particularly on the last five years' work.

II. SYNTHESIS

A. C-unsubstituted Metalloles

1. Siloles

Much sought-after by several laboratories, C-unsubstituted group 14 metalloles and parent metalloles, such as the silole (**2**), have been the subject of one of the most interesting challenges in organometallic chemistry over more than three decades.

After a fairly large number of unfruitful attempts between 1960 and 1980[3], the synthesis of the first C-unsubstituted silole, 1-methylsilole (**3**), was described by Barton and Burns[5] (equation 1).

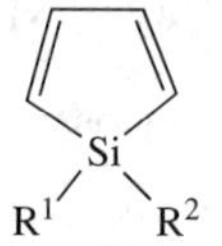

$R^1 = R^2 = H$ (**2**)
$R^1 = Me, R^2 = H$ (**3**)
$R^1 = R^2 = Me$ (**4**)

The flash vacuum pyrolysis (FVP) of 1-allyl-1-methyl-1-silacyclopent-3-ene gives rise to elimination of propene and formation of silole **3** identified as its dimer **5**, and by its Diels–Alder adducts with maleic anhydride and hexafluorobutyne. However, the silole **3** has been detected as a monomer by the MS-FVP technique[6].

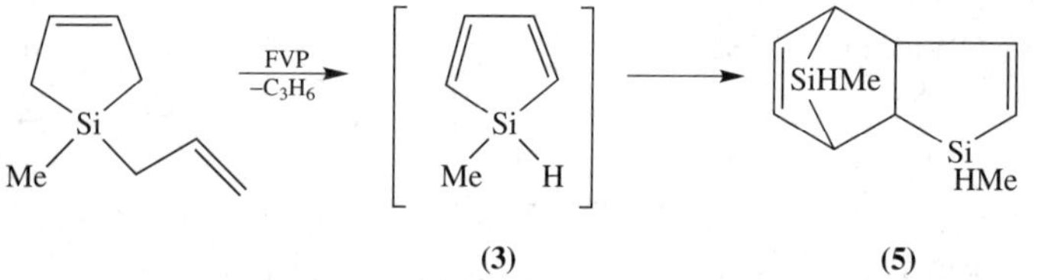

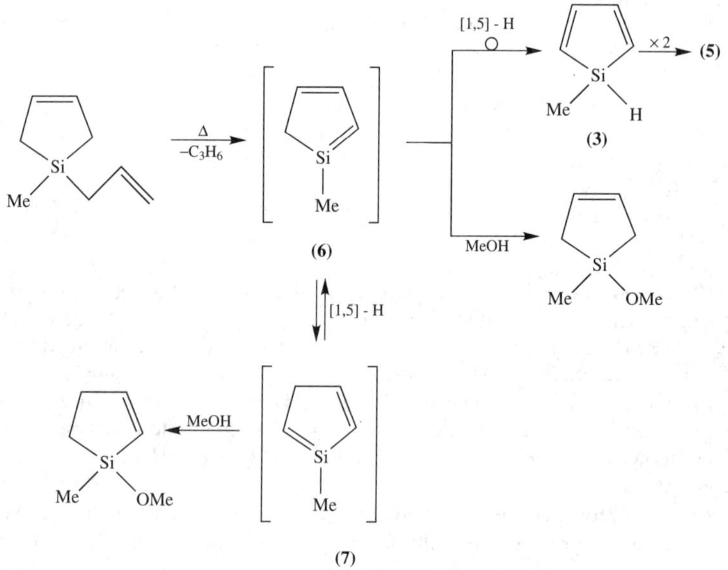

SCHEME 1

Thermolysis of allysilanes has been studied by several authors[7,8], but when carried out under low pressure (FVP), the main process is a retroene elimination of the corresponding alkene, for example propene[9]. In a kinetic study of the FVP of 1-allylsilacyclopent-3-enes, Davidson, Dubac and coworkers[6] have shown that the Arrhenius factor ($\log A = 11.0-11.6$) is the same as for acyclic allylsilanes, but the activation energy is significantly lower, $E_a = 179$ kJ mol^{-1} instead of 230 kJ mol^{-1} (for allyltrimethylsilane). Consequently, the loss of propene involving the methyl group of 1-allyl-1-methylsilacyclopent-3-ene giving a Si=C exocyclic bond is insignificant compared with the endocyclic elimination process (Scheme 1). The latter involves the formation of a 2H-silole, 1-methyl-1-silacyclopenta-1,3-diene (**6**), which is in equilibrium with its tautomers, 1-methyl-1-silacyclopenta-1,4-diene (**7**), a 3H-silole, and 1-methyl-1-silacyclopenta-2,4-diene (**3**). Indeed, the intermediate isosiloles **6** and **7** have been trapped by MeOH to give a 80 : 20 mixture of 1-methoxy-1-methylsilacyclopent-3-ene and 1-methoxy-1-methylsilacyclopent-2-ene (Scheme 1). Additionally, the exocyclic retroene process has also been excluded by means of MeOD trapping experiments.

The mechanistic study of this reaction has also shown that in the case of a substituted allyl group an exocyclic [1,3]-silatropic rearrangement is in competition ($E_a = 173-176$ kJ mol^{-1}) with the endocyclic retroene reaction[6]. Thus the yield of silole reaches at most about 40%. However, this method can be used for the synthesis of C-methylated siloles having a Si—H bond (see Section II.B.2.e).

The first C-unsusbstituted silole to be isolated, 1,1-dimethylsilole (**4**), was reported simultaneously by Dubac and coworkers[10], and Burns and Barton[11] in 1981. This silole was prepared by dehydration of 1,1-dimethyl-1-silacyclopent-4-en-3-ol in the gas phase on alumina or thoria[10,12,13] (Scheme 2). In the presence of acidic reagents, the reaction results in β C—Si rather than β C—H bond cleavage, giving the dienic siloxane (Me$_2$SiC$_4$H$_5$)$_2$O. The Barton method consists of the flash vacuum pyrolysis of 3-(benzoyloxy)-1,1-dimethyl-1-silacyclopent-4-ene at 540°C (equation 2).

The thermolysis of the corresponding N-phenylcarbamate takes place at a lower temperature (310°C) (Scheme 2)[12,14].

The silole **4** can be identified in the monomeric state (NMR), but it dimerizes within a few minutes at RT. It can be trapped by a dienophile or in the form of a η^4-coordination complex (Scheme 2). Thus, the above methods[10-14] are successful because they take place in the gas phase, silole **4** being trapped at low temperature.

The decomposition of the S-methylxanthate ester of 1,1-dimethyl-1-silacyclopent-4-en-3-ol takes place in refluxing ether, but this ester cannot give the silole **4** which dimerizes *in situ* (61% yield), together with siloxane formation (Me$_2$SiC$_4$H$_5$)$_2$O (31%) due to a β C—Si elimination[14].

However, in the case of a stable silole having hindered groups at silicon, this mild method allowed Nakadaira and coworkers to prepare 1,1-dimesitylsilole[15].

As regards the silole **2** itself, this has been the subject of several studies some of which could not be reproduced or have been refuted[3]. As we shall show, this was probable owing to the great instability of this compond, which has been identified only recently.

In 1974, Gaspar and coworkers suggested that **2** is formed in a gas-phase reaction of silicon atoms, resulting from the ^{31}P (n,p) ^{31}Si nuclear transformation, with 1,3-butadiene[16]. They proposed a mechanism in which the silylene **9**, 1-silacyclopent-3-ene-1,1-diyl, rearranges to silole **2**.

Other mechanisms have been proposed, and the formation of **2** was verified by high-temperature catalytic hydrogenation[17]. Although the identification of **2** in these experiments seems questionable[5,10], Gaspar and coworkers, in 1986, reported that gas-phase pyrolysis of 1,1,1,3,3,3-hexamethyltrisilane with butadiene gives the silole [2 + 4]

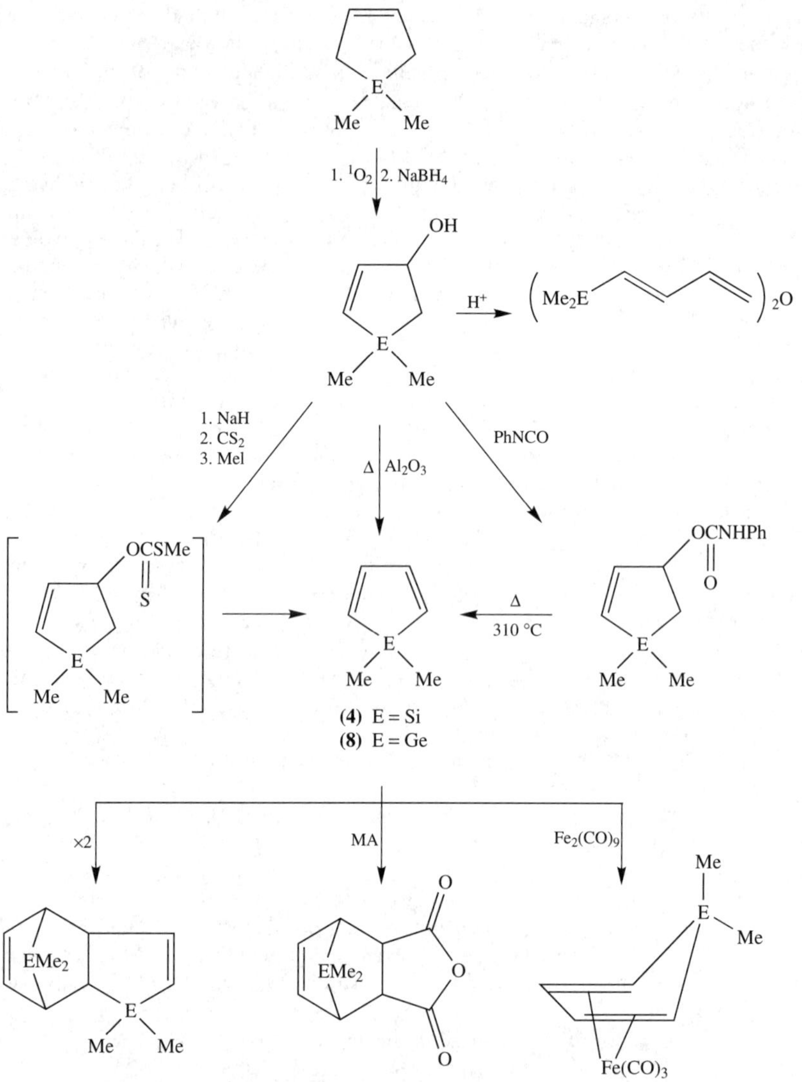

SCHEME 2

dimer in low yield[18]. More recently, the same research group isolated and identified definitively this dimer **10**[19] (Scheme 3).

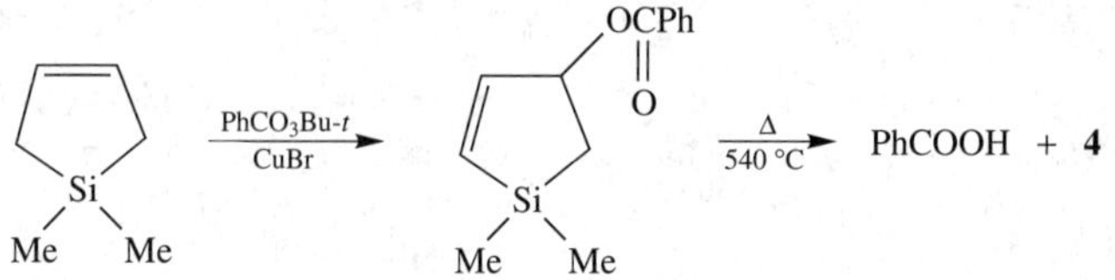

$$\text{(2)}$$

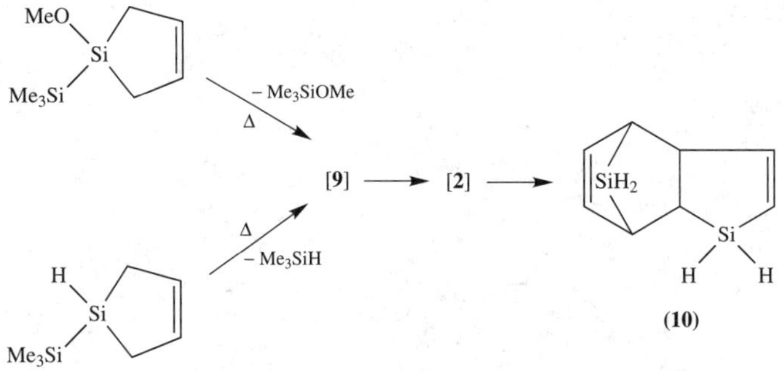

34. Group 14 metalloles, ionic species and coordination compounds 1967

SCHEME 3

Attempts to isolate the silole monomer **2** or to trap it by reaction with maleic anhydride or perfluoro-2-butyne have failed. This is surprising given the successful trapping of 1-methylsilole[5] and 1,1-dimethylsilole[10]. The authors conclude that silole **2** is not very reactive in Diels–Alder cycloadditions, except toward self-reaction[19].

The mechanisms proposed by Gaspar and coworkers in their approach to the problem have a common point, the formation of the intermediate **9**. This silylene then undergoes two prototropic rearrangements: initial isomerization to a silene, the 2H-silole (**11**), followed by a second to give the 1H-silole (**2**)[19] (equation 3).

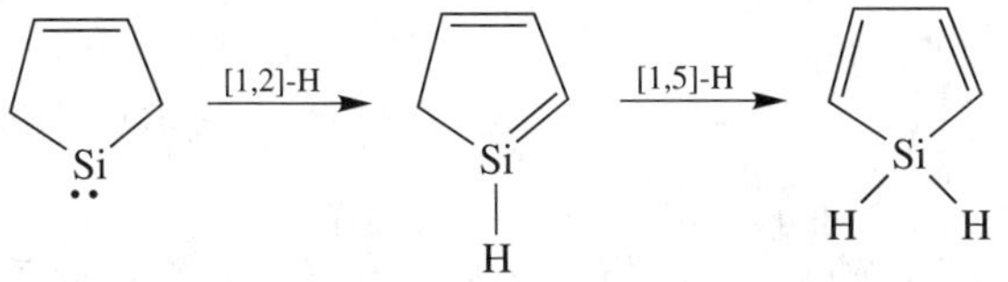

$$\underset{(9)}{\text{Si:}} \xrightarrow{[1,2]\text{-H}} \underset{(11)}{\underset{\text{H}}{\text{Si}}} \xrightarrow{[1,5]\text{-H}} \underset{(2)}{\underset{\text{H}\quad\text{H}}{\text{Si}}} \quad (3)$$

This proposition is in agreement with the previous work of Davidson, Dubac and coworkers[6] (see above and Scheme 1).

Also in 1992, Michl, Nefedov and coworkers[20a,b] reported that vacuum pyrolysis (800 °C, 10^{-3}–10^{-4} Torr) of 5-silaspiro[4.4]nona-2,7-diene (**12**) resulted in the formation of silole (**2**), that was isolated in an Ar matrix, and characterized by its IR and UV-visible spectra. Matrix photolysis and vacuum pyrolysis of 1,1-diazido-1-silacyclopent-3-ene (**13**) also produced the parent silole. After warming the pyrolysate to RT, the [2 + 4] dimer **10** of silole **2** was identified by GC-MS.

A very interesting photoreversible interconversion of **2** with **9, 11** and **14**, its three tautomeric forms, has been observed upon irradiation at selected wavelengths[20a,b]. Thus, Scheme 4 summarizes the mechanistic aspects of both the pyrolysis of **12**, and the pyrolysis or photolysis of **13** which proceed via common intermediates **9** and **11**. These results are in perfect agreement with the mechanistic study of the FVP of 1-allyl-1-methylsilacyclopent-3-ene[6] in which the silicon-methylated analogues of **11** and **14** have also been identified (Scheme 1). However, we think that **14** must be regarded as a possible intermediate not only in photolysis but also in pyrolysis reactions, because the methylated analogue **7** (Scheme 1) can be readily trapped.

A similar study, which supports previous claims involving the 3,4-dimethylsilole and its corresponding isomeric intermediates, has been also carried out on the 3,4-dimethyl derivatives of **12** and **13**[20a,b].

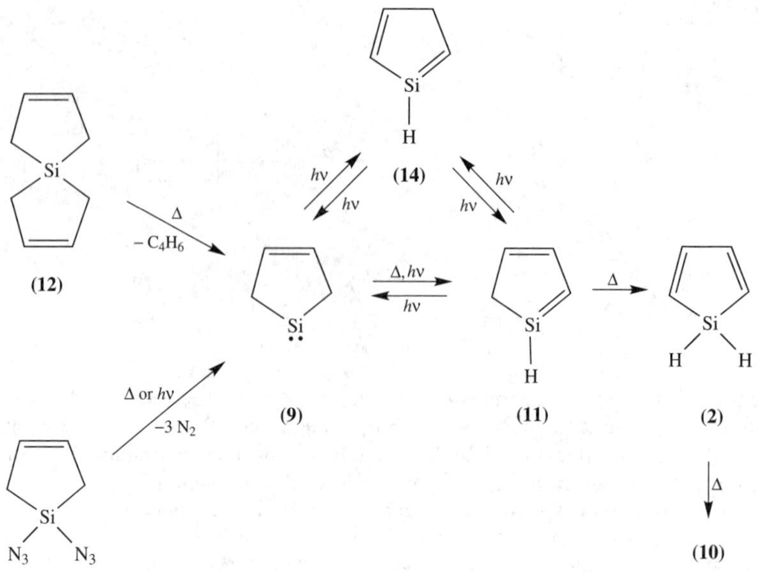

SCHEME 4

Vacuum pyrolysis of **12** has been recently studied by high-resolution mass spectrometry[20c]. The ionization potential (IP = 9.17±0.10 eV) for silole **2** was found to be in excellent agreement with the IP (9.26 eV) calculated by the semiempirical PM3 method. Thermodynamic calculations on possible decomposition mechanisms of **12** have shown that **2** is probably formed from the primary intermediate silylene **9** via two consecutive hydrogen shifts (equation 3).

2. *Germoles*

The first C-unsubstituted germole, 1,1-dimethylgermole (**8**), was prepared in 1981 by Dubac and coworkers[21] by dehydration in the gas phase of 1,1-dimethyl-1-germacyclopent-4-en-3-ol (Scheme 2). At room temperature, as in the case of 1,1-dimethylsilole, the [2+4] dimer is formed. At low temperature, the monomer gives a Diels–Alder adduct with maleic anhydride, and forms a η^4-tricarbonyl iron complex[12,21]. The indirect dehydration of the same alcohol, via its *N*-phenylcarbamate, also produces the germole **8**[13,22] (Scheme 2).

Among synthetic methods of group 14 metalloles involving direct formation of the dienic ring[23a] are reactions of two alkyne molecules with a Σ_2E fragment[23b]. These reactions give C-substituted metalloles (see Section II.B.1.b). However, from ethyne, no C-unsubstituted derivative was obtained until Ando and coworkers described in 1990 the first Pd-catalysed reactions of germylenes with ethyne giving 1,1-dialkylmetalloles[24]. These reactions are not chemospecific. For example, in the presence of a catalytic amount of Pd(PPh$_3$)$_4$, hexamesitylcyclotrigermane and ethyne, at 80°C in toluene, react to give two products, 1,1-dimesitylgermole (**15**) and 1,1,4,4-tetramesityl-1,4-digermacyclohexa-2,5-diene (**16**) (1.25 : 1) in 76% overall yield (equation 4).

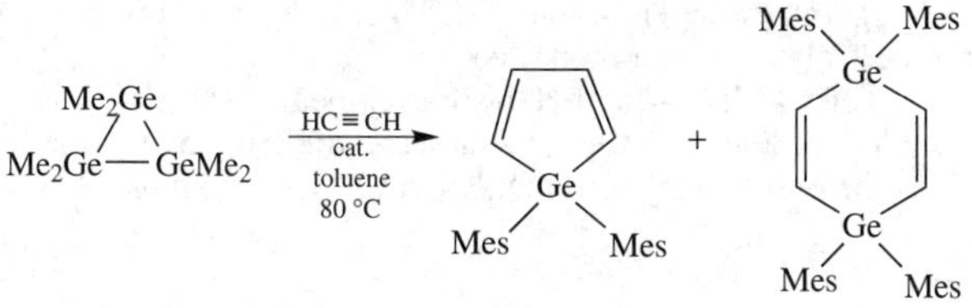

34. Group 14 metalloles, ionic species and coordination compounds 1969

$$\text{Me}_2\text{Ge} \overset{\text{Me}_2\text{Ge}\text{—}\text{GeMe}_2}{\diagdown} \xrightarrow[\substack{\text{cat.} \\ \text{toluene} \\ 80\,°\text{C}}]{\text{HC}\equiv\text{CH}} \quad \underset{\text{Mes}\quad\text{Mes}}{\overset{\text{Ge}}{\bigcirc}} \quad + \quad \underset{\substack{\text{Ge} \\ \text{Mes}\quad\text{Mes}}}{\overset{\substack{\text{Mes}\quad\text{Mes} \\ \text{Ge}}}{\bigcirc}} \qquad (4)$$

(15)   (16)

In contrast to **8**, germole **15** is stable in the monomeric state, thanks to steric shielding by the bulky mesityl groups (see Section III.A.1). It reacts, nevertheless, with maleic anhydride and diiron nonacarbonyl.

The proposed mechanism for reaction 4 involves the thermal decomposition of cyclotrigermane into a germylene ($Mes_2Ge:$) and a digermene ($Mes_2Ge=GeMes_2$) which give products **15** and **16**, respectively[24] (see Section II.B.1.b and Scheme 11).

The germylene generated by thermal or photolytic cycloreversion from a 7-germanorbornadiene gives with alkynes many different products[25]. However, in the presence of a Pd-catalyst a germole can be obtained together with a digermacyclohexadiene[26]. From a 7-germanorbornadiene prepared from an easily accessible C-phenylated germole, the reaction with ethyne allows a C-unsubstituted germole **(17)** to be obtained specifically without formation of the corresponding digermacyclohexadiene[24a] (Scheme 5). In the absence of the Pd-catalyst the reaction

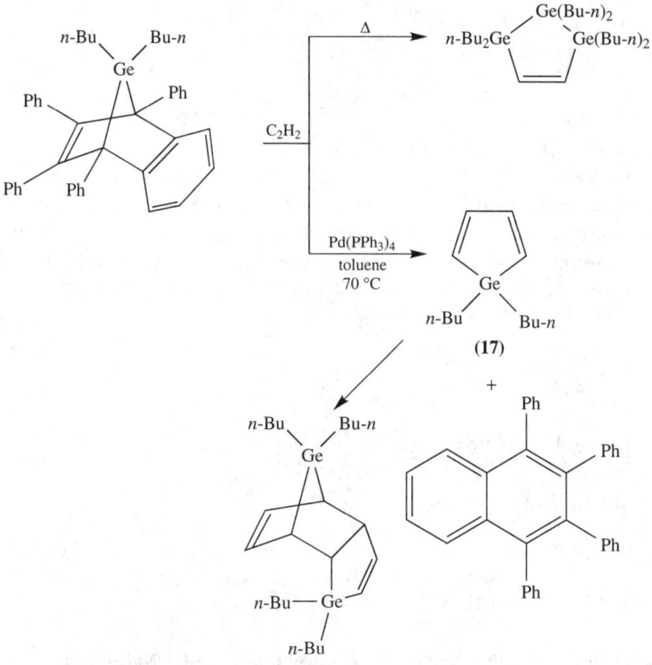

SCHEME 5

1970   Jacques Dubac, Christian Guérin and Philippe Meunier

afforded a 1,2,3-trigermacyclopent-4-ene. The germole **17** dimerizes to **18**, but by distillation it can be isolated as the monomer.

Although FVP of 1-allyl-1-methyl-1-silacyclopent-3-ene affords 1-methylsilole (Section II.A.1), FVP of the germanium analogue did not give 1-methylgermole, but rather butadiene as the only volatile product, together with an amorphous solid containing germanium[27].

3. Stannoles

In 1988, Ashe and Mahmoud prepared the first C-unsubstituted stannole, 1,1-dibutylstannole **(19)**, by cyclization of 1,4-dilithio-1,3-butadiene with dibutyltin dichloride[28] (equation 5). This stannole, stable in the monomeric state, has been isolated by distillation (28% yield).

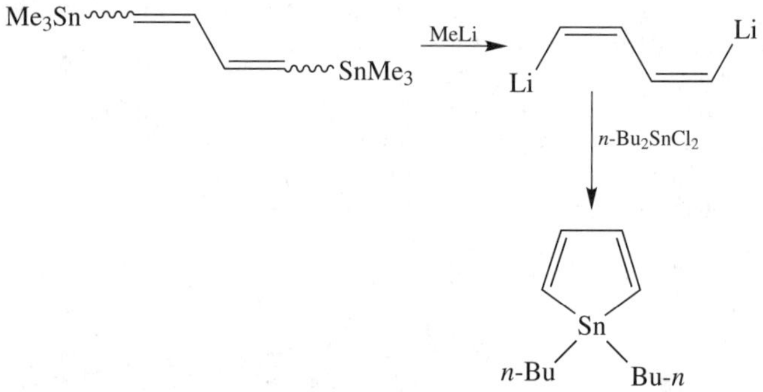

$$Me_3Sn\text{\textasciitilde}\!\!=\!\!=\!\!\text{\textasciitilde}SnMe_3 \xrightarrow{MeLi} Li\!-\!\!=\!\!=\!\!-Li \xrightarrow{n\text{-}Bu_2SnCl_2} \text{(19)} \tag{5}$$

As with the germylene (see above), an alkyne can react with a stannylene to give a stannole. Here also, the result is dependent on experimental conditions[25]. With ethyne, SnR_2 [R = $CH(SiMe_3)_2$] reacts at room temperature to form a linear chain coupling product[29a], while metal-coordinated ethyne reacts differently[29b]. In particular, complexes **20** or **21** are converted by ethyne at room temperature into the Pd^0 ethyne complex **22** and the C-unsubstituted stannole $R_2Sn(C_4H_4)$ **(23)**[29b]. It must be pointed out that complexes **21** and **22** are analogues to the proposed intermediates for the Pd-catalysed reaction between a germylene and ethyne[24].

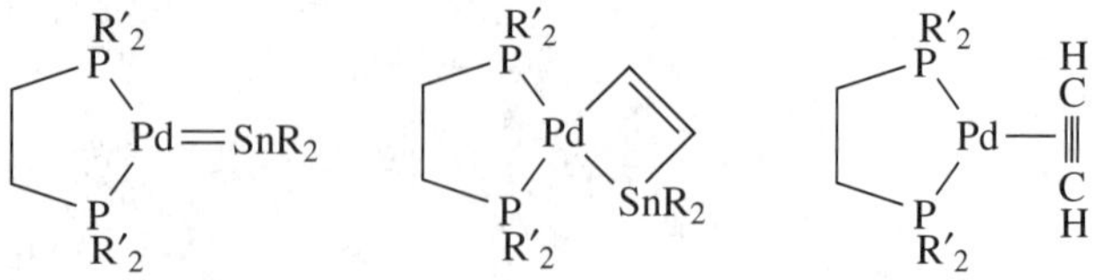

(20)   (21)   (22)

R = $CH(SiMe_3)_2$; R′ = i-Pr, t-Bu

The Pd-catalysed reaction is possible with a stannylene on condition that it is carried out at low temperature, in order to avoid the competing uncatalysed formation of a polymeric product, and in the presence of a labile complex as catalyst (equation 6). The stannole **23** was obtained in 87% yield.

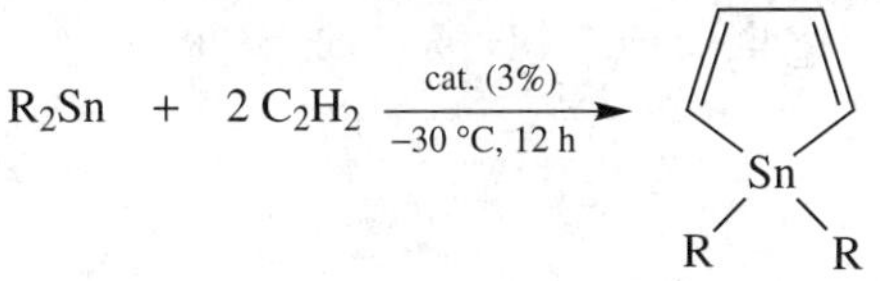

$$R_2Sn + 2\,C_2H_2 \xrightarrow[-30\,°C,\,12\,h]{cat.\,(3\%)}$$

(23)

R = CH(SiMe$_3$)$_2$; catalyst

(6)

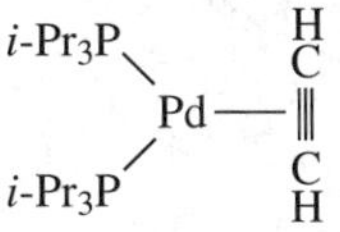

B. C-substituted Metalloles

These compounds form the largest class of group 14 metalloles known today[30]. We will review again the more outstanding results concerning the long-established synthetic methods for these heterocycles, as well as the more recent ones.

1. Synthetic methods involving direct formation of the dienic ring

a. Cyclization of 1,4-dilithio-1,3-butadienes by a polyfunctional compound R_nEX_{4-n} (n = 0 − 2). Leavitt, Manuel and Johnson[31] prepared the first group 14 metalloles containing germanium or tin, from 1,4-dilithio-1,2,3,4-tetraphenyl-1,3-butadiene (DTB) (equation 7). The silicon[32] and lead[33] analogues have been prepared in the same way.

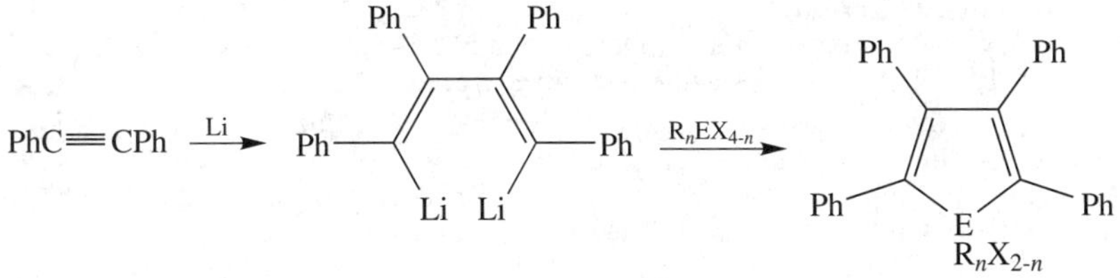

$$PhC\equiv CPh \xrightarrow{Li} \text{(intermediate)} \xrightarrow{R_nEX_{4-n}} \text{(metallole)}\quad(7)$$

(24)

Since one or two E−X bonds (e.g. X = Cl) can survive the cyclization reaction, numerous derivatives bearing a variety of groups at the heteroatom can be prepared by substitution reactions[30]. Structures of the spirobimetallole type (25) are also known.

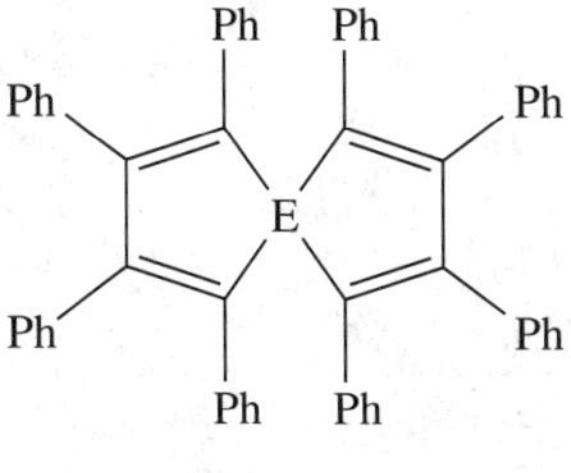

(25)

In 1967, an original synthesis of 1,4-dilithio-1,4 diphenyl-1,3-butadiene (DDB) from 1,1-dimethyl-2,5-diphenylsilole (see equation 17 below) was proposed by Gilman and coworkers[34]. This dilithium reagent reacts with polyhalides $R_n EX_{4-n}$ to yield 2,5-diphenylmetalloles (**26**, E = Si[34,35], Ge[36], Sn[34]) (Scheme 6).

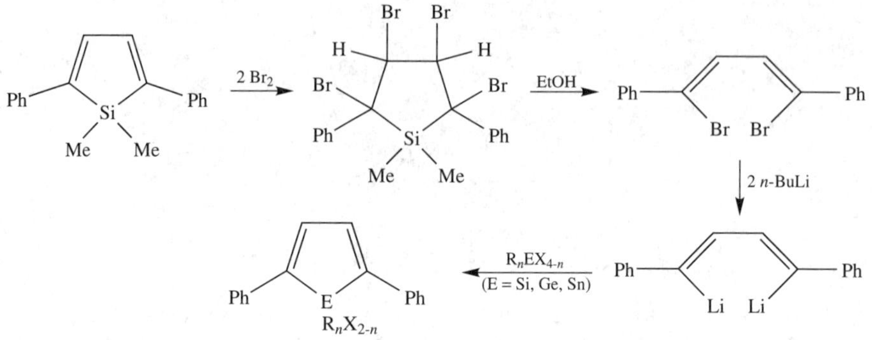

SCHEME 6

Corriu and coworkers[37] showed that alkoxysilanes $R_n Si(OR')_{4-n}$ [in particular $RSi(OMe)_3$] give better yields for the DDB cyclization than do chlorosilanes. The reaction of anionic pentacoordinated silicon complexes $[RSi(O_2C_6H_4-O)_2]^-$ Na^+ with DDB and subsequent $LiAlH_4$ reduction give 1-R-diphenylsiloles (R = Me, Ph)[38]. Bis(silacyclopentadien-1-yl)alkanes were formed from DDB and α, ω-bis(dihalomethylsilyl)alkanes[39].

In analogy to the preparation of DDB from a 2,5-diphenylated silole (Scheme 6), 1,4-dilithio-1,2,3,4-tetramethylbutadiene has been recently prepared from a C-methylated zirconacyclopentadiene. Thus, West and coworkers[40] were successful in preparing 1,1-dichloro-2,3,4,5-tetramethylsilole (**27**), the first C-methylated silole having Si—Cl bonds, from bis(cyclopentadienyl)-2,3,4,5-tetramethylzirconacyclopenta-2,4-diene via (1Z, 3Z)-1,4-diiodo-1,2,3,4-tetramethylbuta-1,3-diene (equation 8).

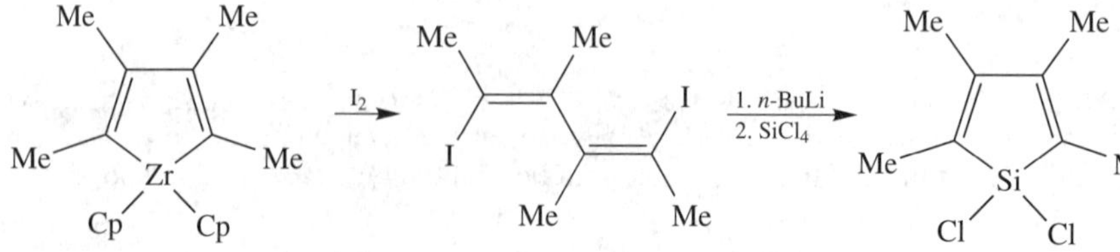

b. Cyclization of two acetylenic molecules with a Σ_2E fragment. This method can be carried out in two ways: the direct cyclization (Scheme 7, path a) of the two acetylenic molecules by a convenient reagent, e.g. Σ_2E: or Σ_2EX_2, or their cyclization through another element M (Scheme 7, path b), e.g. a transition element. This latter possibility will be developed below (see Section II.B.3).

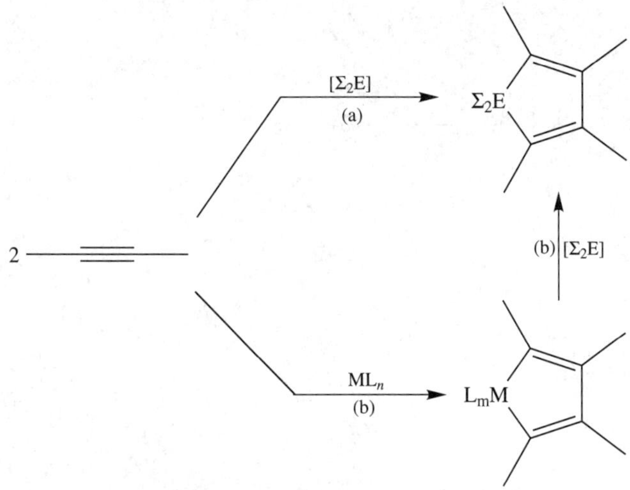

34. Group 14 metalloles, ionic species and coordination compounds

SCHEME 7

Many substrates have been used to prepare C-substituted group 14 metalloles by this method, in particular those which generate analogues of carbenes Σ_2E:[23b].

i. Reaction between an alkyne and a disilane, a silirene (1-silacyclopropene) or a cyclotrisilane.. The substrates generally used for the title reaction are disilanes[41–43] (or stannylsilanes[44]) and silirenes[45–50]. The reactions are usually catalytic.

Treatment of 1,1,2,2-tetramethyldisilane with an alkyne in the presence of a catalyst [NiCl$_2$(PEt$_3$)$_2$ or PdCl$_2$(PEt$_3$)$_2$] gives C-substituted siloles **28**[41,42] (equation 9).

$$HMe_2SiSiMe_2H + 2RC\equiv CR' \xrightarrow{cat.} Me_2SiH_2 + \text{(28)} \quad (9)$$

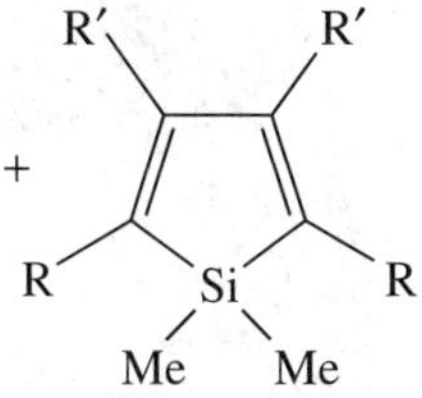

(R, R' = alkyl, aryl or silyl groups)

(28)

Depending on the alkynes, the yields are variable, but they are very high in the case of disubstituted symmetrical alkynes (R = R' = Me, Et, n-Bu). The reaction is sometimes not specific, giving also a 1,4-disilacyclohexa-2,5-diene. It was recently used by Ishikawa and coworkers to obtain a 3,4-diethynylsilole **(29)** (Scheme 8) which is a precursor of polymers containing silole rings[43] (see Section IV.C).

From (trialkylstannyl)dimethylsilane and terminal alkynes this method gives 3,4-disubstituted siloles (**28**, R = H, R' = alkyl, aryl or alkoxy groups) in moderate to good yields[44]. Catalysis by Pd is better than by Ni or Rh. The proposed mechanism involves a palladium–silylene species **30** as an intermediate. This reacts with an alkyne giving successively a palladasilacyclobutene **(31)** and a palladacyclohexadiene **(32)**. In the final

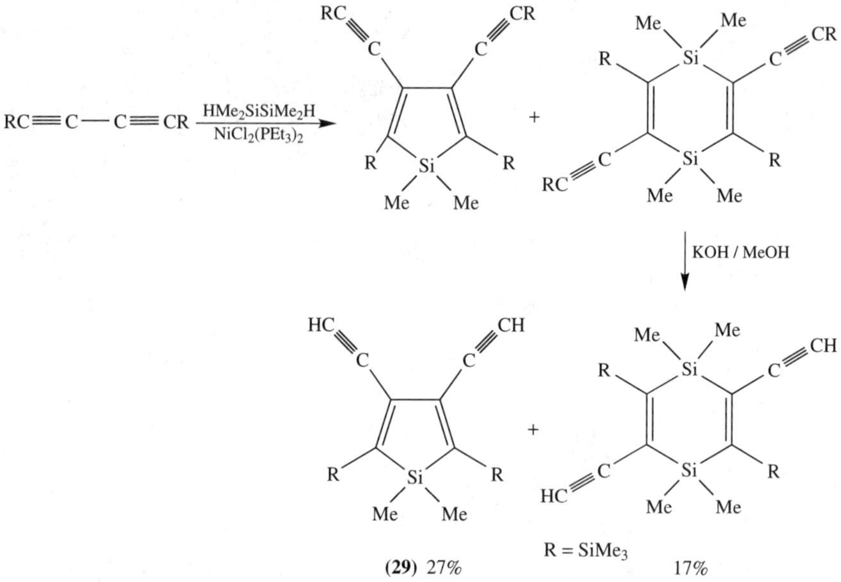

SCHEME 8

step, reductive elimination of silole **28** regenerates the Pd^0-complex[44]. This mechanism is slightly different from that proposed for the reaction of a stannylene with ethyne[29b] (see Section II.A.3).

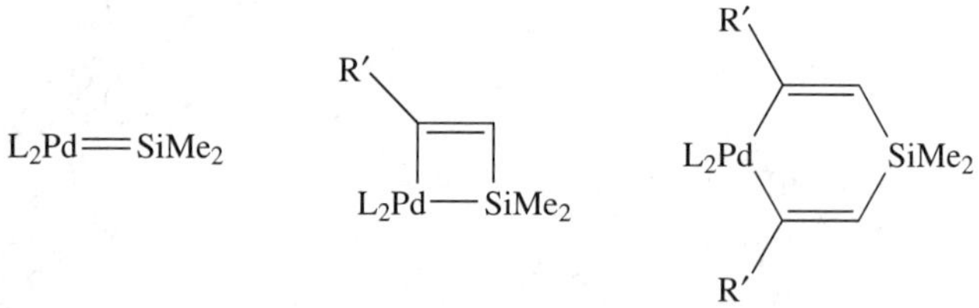

(30)   (31)   (32)

A silirene may also generate a silylene. In the absence of catalyst, thermolysis of a silirene can give various products resulting from its decomposition into a silylene and an alkyne[50]. Thus siloles are formed together with 1,2-disilacyclobutanes, 1,4-disilacyclohexadienes and other products. With silirenes bearing bulky substituents on the ring carbon atoms, siloles become the major products (Scheme 9).

The reaction of a silirene with an alkyne in the presence of a palladium catalyst allows cyclization of two molecules of the alkyne with the silylene, as in equation 9 above. For example, Seyferth and coworkers have prepared the silole **33** in 80% yield from 1,1-dimethyl-2,3-bis (trimethylsilyl) silirene and phenylacetylene (equation 10)[45]. Without catalyst, this reaction yielded the silole **34** and the ene-yne **35**, resulting respectively from ring expansion and cleavage by PhC≡CH of the silirene. Under UV irradiation, **35** alone was formed.

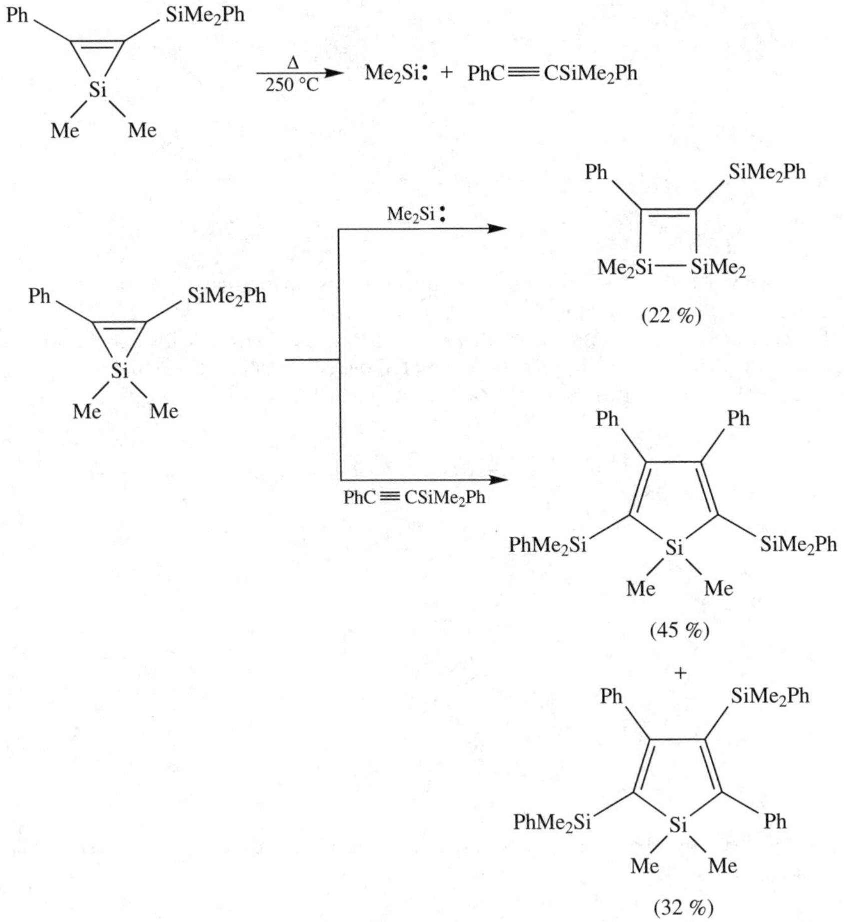

SCHEME 9

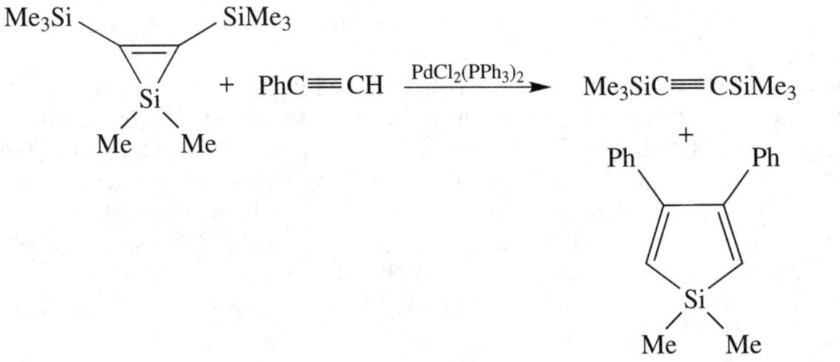

(10)

(33)

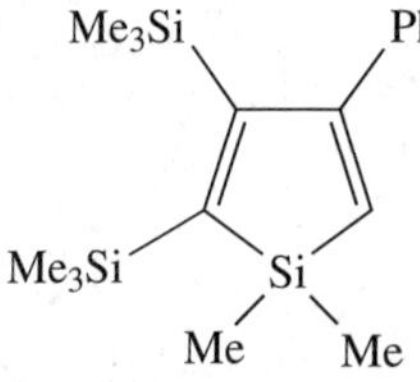

(34) 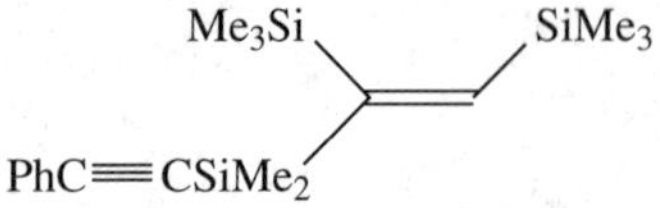 **(35)**

This method did not afford C-unsubstituted siloles, in particular in the case of the reaction of a silirene with ethyne[45d].

The catalysed cleavage of cyclotrisilanes (or trisiliranes) gives a disilene and a silylene. The latter can be trapped by various multiple-bonded compounds, in particular by 1-phenylacetylene leading to the silole **36** (equation 11)[51].

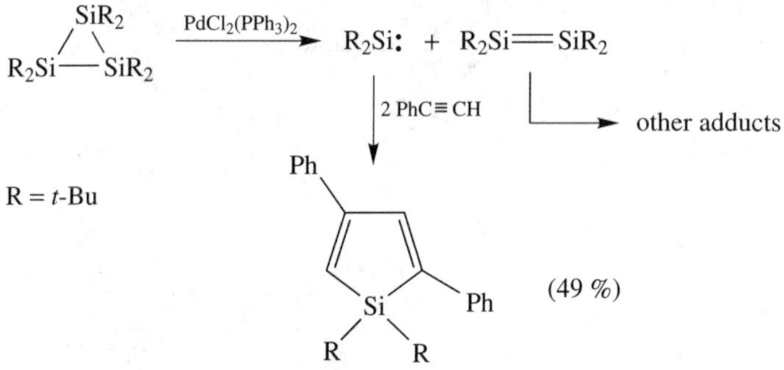

$$R_2Si\text{—}SiR_2 \text{ (cyclotrisilane)} \xrightarrow{PdCl_2(PPh_3)_2} R_2Si: + R_2Si=SiR_2$$

$R = t\text{-Bu}$

$\downarrow 2\ PhC\equiv CH$ → other adducts

(36) (49 %) (11)

The same reaction between a cyclotrigermane and ethyne resulted in a C-unsubstituted germole (equation 4)[24].

ii. Reaction between an alkyne and a 7-metallanorbornadiene. In 1964, Gilman and coworkers[52] showed that a 7-silanorbornadiene thermally decomposes with formation of a silylene. Since then, many studies have developed the chemistry of group 14 metallanorbornadienes, demonstrating that their stability depends on both the metal and the substituents[53].

Since a 7-metallanorbornadiene is easily obtained in the C-phenylated metallole series, it appeared possible, by decomposition in the presence of an alkyne, to generate differently substituted metalloles (**37, 38**, Scheme 10). However, this reaction is only known for the germoles (Neumann and coworkers)[25,26], probably because decomposition temperature of germanorbornadiene is in the same range as that of the trapping of germylene. Nevertheless, the reaction can result in numerous products, and the selective formation of germoles, including that of regioisomer **37**, is limited to some alkynes in the presence of Pd-catalysts (e.g. $R^3/R^4 = n-Bu/H$, 2-methylphenyl/H, CO_2Me/Me, CO_2Et/Ph, Ph/Ph)[26].

The Pd-catalysed reaction between ethyne and the germanorbornadiene derived from 1,1-dibutyl-2,3,4,5-tetraphenylgermole and benzyne gives a C-unsubstituted germole (**17**) (Scheme 5)[24a].

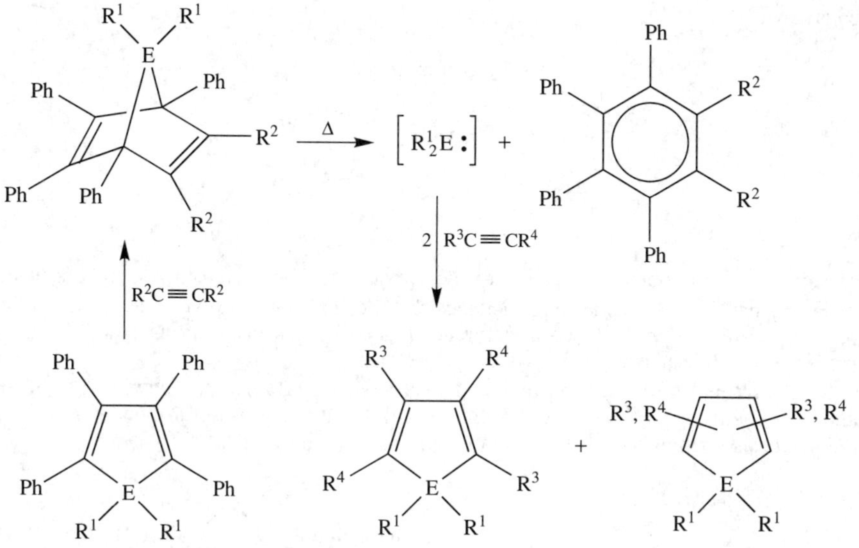

34. Group 14 metalloles, ionic species and coordination compounds

SCHEME 10

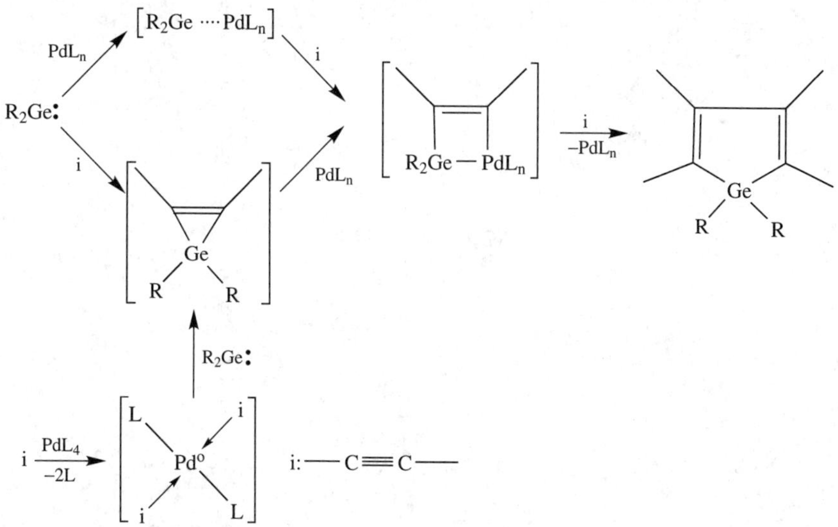

SCHEME 11

Mechanisms for the formation of germoles in these reactions have been discussed in terms of Scheme 11[24a,26].

iii. Reactions involving 1,n-diynes. If the two acetylenic groups are borne by the same carbon chain, cyclization on a Σ_2E fragment can give a bicyclic metallole as shown in equation 12[54a].

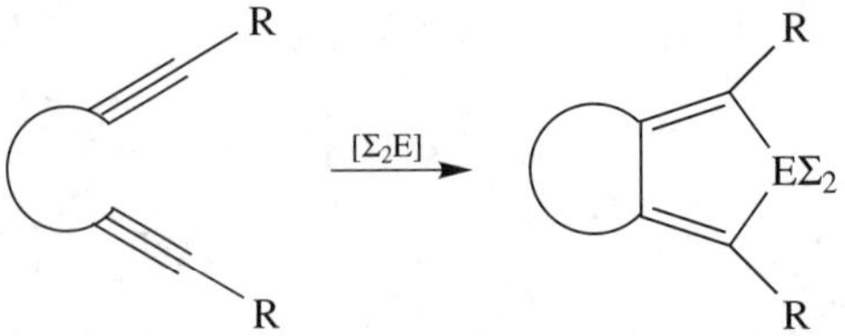

$$\text{(diyne structure)} \xrightarrow{[\Sigma_2 E]} \text{(silole structure with } E\Sigma_2\text{)} \qquad (12)$$

This is the intramolecular version of the previous reactions, and the precursors of the $\Sigma_2 E$ fragment are the same (e.g. disilanes)[54]. The reaction, however, has been applied only to 1,6-diynes bearing two aryl (or heteroaryl) groups in positions 1 and 7, in the presence of a Ni-catalyst.

Siloles **39** and **40** have been prepared in this way by Tamao and coworkers[54] in 64 and 40% yield, respectively, the 2,5-dithienylsilole **40** being a useful monomer unit for the synthesis of thiophene–silole copolymers (see Section IV.B).

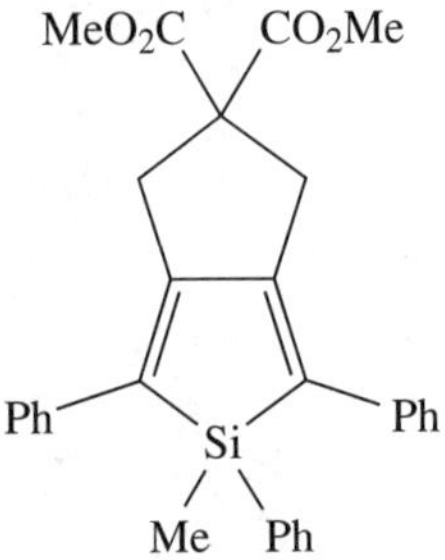

(**39**) 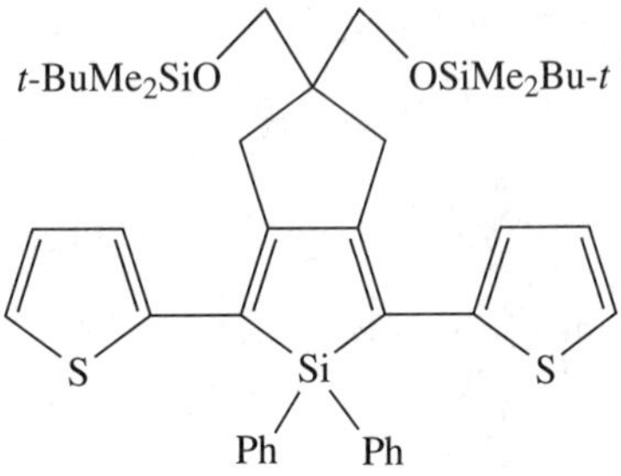   (**40**)

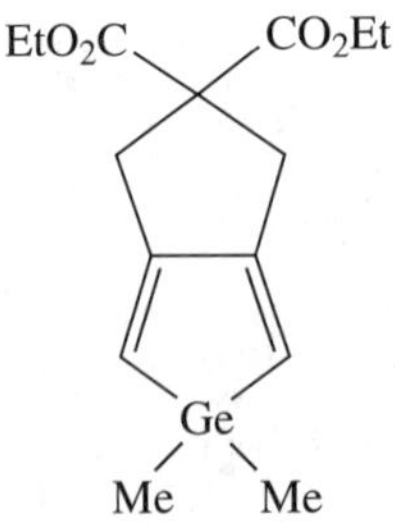

(**41**)

The reaction of diethyl dipropargylmalonate with cyanotrimethylgermane in refluxing toluene, with $PdCl_2$ as catalyst, gives the germole **41** in 77% yield[55].

c. Cyclization of two acetylenic groups bonded to a group 14 element. i. Reaction of trialkylboranes with dialkynyl derivatives of a group 14 element. The cyclization of two acetylenic groups bonded to a group 14 element to yield a metallole was achieved by

34. Group 14 metalloles, ionic species and coordination compounds

Wrackmeyer and coworkers using trialkylboranes[56−58] (equation 13).

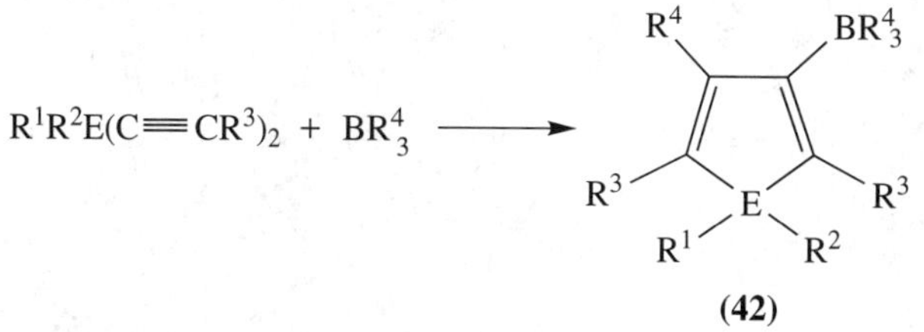

$$R^1R^2E(C≡CR^3)_2 + BR^4_3 \longrightarrow (42)$$

(13)

$E = Si^{56}, Ge^{56a,b}, Sn^{57}, Pb^{58}$

R^1, R^2 = Me, Et or t-BuN(CH$_2$)$_3$NBu-t; R^3 = alkyl, aryl, SiMe$_3$, SnMe$_3$, PPh$_2$ or organic group containing N, O, S; R^4 = Me, Et, i-Pr or n-Bu

Zwitterionic intermediates **43**, in which a tin centre is stabilized by intramolecular side-on coordination to the C≡C bond of an alkynylborate moiety, have been detected in solution by multinuclear NMR[57]. Moreover, the structures of some of these compounds **43** (E = Sn[57c,e,g], Pb[58b]) have been established by X-ray analysis.

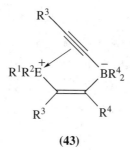

(**43**)

These reactions are not chemospecific, but the formation of other heterocycles can be avoided, and the yields of metallole are generally high. Some heteroatomic groups attached to the ring carbons can be substituted by hydrogen (e.g. in protodeborylation or protodestannylation reactions)[56c,d].

ii. Intramolecular reductive cyclization of diethynylsilanes. Diethynylsilanes undergo intramolecular reductive cyclization in an *endo-endo* mode upon treatment with lithium naphthalenide (NpLi) to form 2,5-dilithiosiloles **44** (equation 14)[59]. By treatment with electrophiles, these dilithium derivatives are converted into 2,5-difunctional siloles (**45**) which are precursors of oligosiloles (see Section IV.A).

Though this method is still limited to 3,4-diphenylsiloles, it represents a distinct advance in group 14 metallole synthesis.

d. Cyclization of a 1,3-diyne with a group 14 dihydride. An attractive and simple route to group 14 metalloles is the cyclization of a 1,3-diyne with a dihydride, e.g. R$_2$EH$_2$. However, only a single group 14 metallole, 1,1-dibutyl-2,5-dimethylstannole (**46**), has been synthesized in this way in low yield (15%)[60] (equation 15).

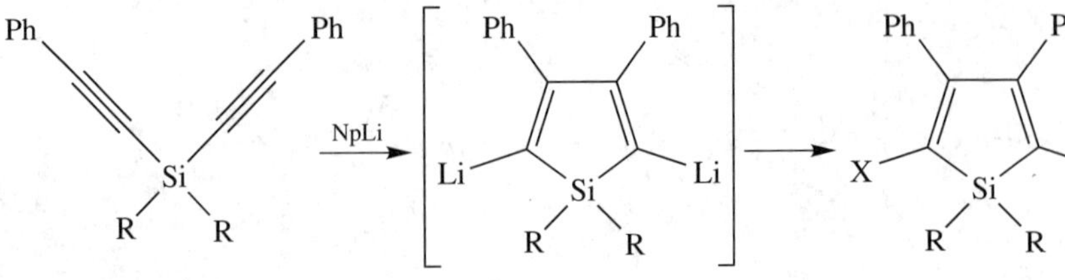

$$\text{(44)} \xrightarrow{\text{NpLi}} \text{(45)} \longrightarrow \quad (14)$$

X, Y = H, Br, SiMe$_3$, SnMe$_3$, SePh, C≡CPh, (thienyl)

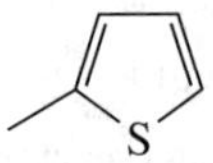

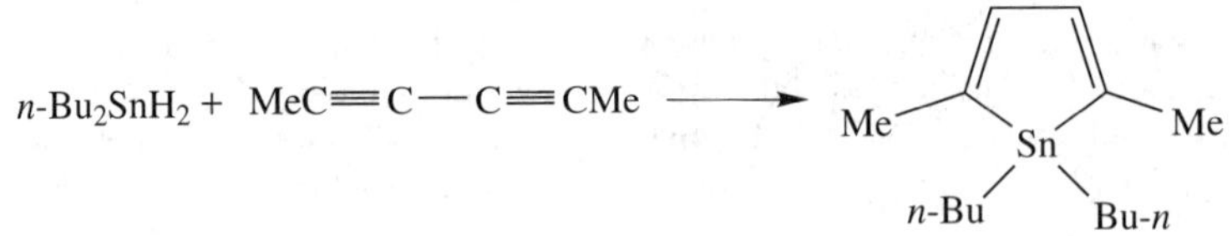

$$n\text{-Bu}_2\text{SnH}_2 + \text{MeC}{\equiv}\text{C}-\text{C}{\equiv}\text{CMe} \longrightarrow \text{(46)} \quad (15)$$

2. Synthetic methods involving a group 14 heterocycle as precursor

a. Dehydrogenation of metallacyclopentanes or metallacyclopentenes. Attempts to dehydrogenate C-unsubstituted metallacyclopentanes or -pentenes in order to obtain C-unsubstituted metalloles have failed[3]. The reaction is only successful for 2,5-diphenylated derivatives in the presence of DDQ (equation 16)[61].

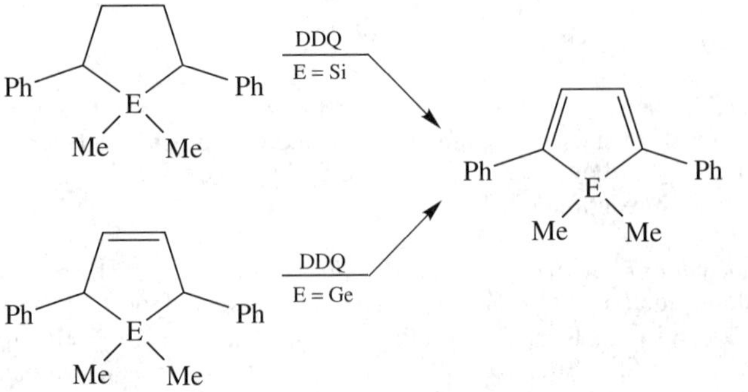

$$\quad (16)$$

In the case of 3-methyl or 3,4-dimethylsiloles, the same reaction gives the transoid isomer of the corresponding silole[62].

b. Dehydrohalogenation of halogenometallacyclopentanes. One of the most noteworthy group 14 metallole syntheses is that of Gilman and coworkers[34], reported in 1967. This involves the synthesis of 1,1-dimethyl-2,5-diphenylsilole (47) by bromination/dehydrobromination of 1,1-dimethyl-2,5-diphenylsilacyclopentane which is

34. Group 14 metalloles, ionic species and coordination compounds 1981

easily accessible from styrene (equation 17).

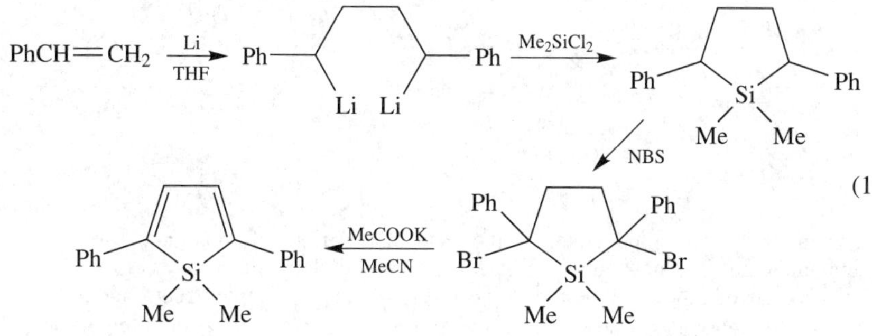

(17)

(**47**)

This method has been extended successfully to the germanium analogue[63], and to other 2,5-diphenylsiloles[39,64]. However, its main interest lies in the possibility of preparing 1,4-dilithio-1,4-diphenyl-1,3-butadiene (DDB) from **47** (see Scheme 6) giving access to other 2,5-diphenylmetalloles, especially those functionalized at the heteroatom[37].

Since the dehydrohalogenation method could not be extended to other group 14 metalloles and in particular to the simple metalloles[65], except for 1,1,3,4-tetramethylsilole[66], it seems clear today that Gilman's synthesis is specific to 2,5-diphenylsiloles owing to the benzylic positions involved.

c. Dehydration of 1-metallacyclopent-4-en-3-ols. The synthesis of sila- or germacyclopent-4-en-3-ols by photooxidation ($^1O_2$)/reduction of metallacyclopent-3-enes[10,21] (see Section II.A.1, Scheme 2) has been extended successfully to the C-methylated (equations 18 and 19)[12,67] and C-phenylated (equation 20)[68] derivatives.

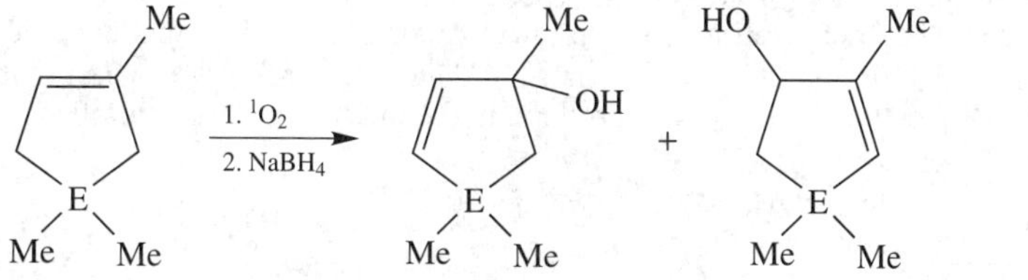

(18)

E = Si, Ge   (**48**)   (**49**)

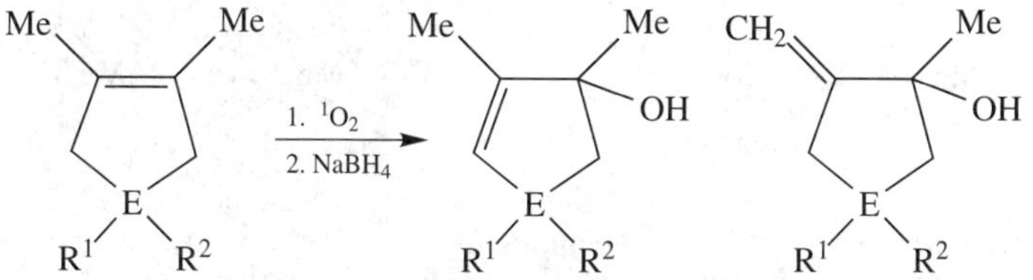

(19)

E = Si, Ge   (**50**)   (**51**)
R^1, R^2 = alkyl, alkenyl or aryl groups

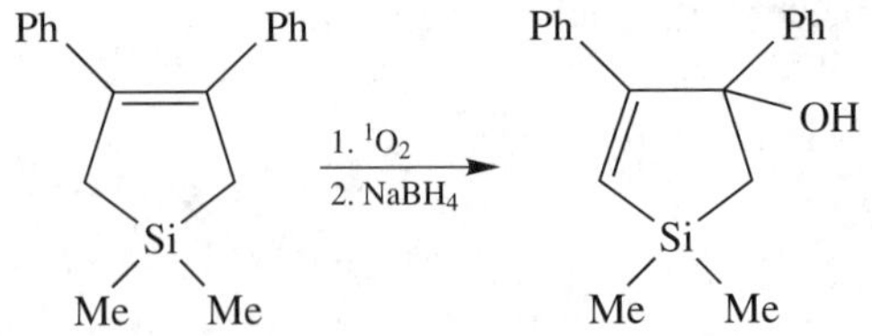

[Structures for equation (20): compound (52) with Ph, Ph, Si(Me)(Me) reacting with 1. ¹O₂, 2. NaBH₄ to give the corresponding alcohol]

(20)

(52)

The method gives good yields in the case of a variety of substituents at silicon (or germanium), but it is not regiospecific for the C-methylated derivatives.

In the case of 3,4-dimethylated derivatives, the base-promoted rearrangement of the corresponding epoxides is a preferable method because it is regiospecific (equation 21)[69]. However, it does not take place in the case of derivatives having an alkenyl group at the heteroatom[12,67c].

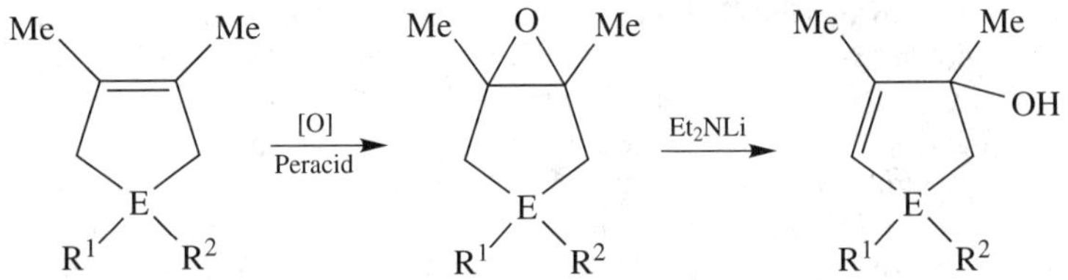

[Structures for equation (21): dimethyl metallacyclopentene → epoxide (50) via peracid → alcohol via Et_2NLi]

(21)

(50)

E = Si, Ge

R^1, R^2 = alkyl or aryl groups

The catalytic dehydration of metallacyclopentenols is not a convenient method for the synthesis of C-methylated metalloles since:

(a) catalysis by a protonic acid leads to a β C−E elimination, with formation of a dienic open siloxane (or germoxane) (Scheme 2);

(b) gas phase dehydration with alumina or thoria results in a mixture of C-methylated metalloles **53**, or **54**, and isomeric transoid dienes **55**, or **56**[67a,b,70]. The isomer ratio, in which the metallole is the minor product, is thermodynamically controlled.

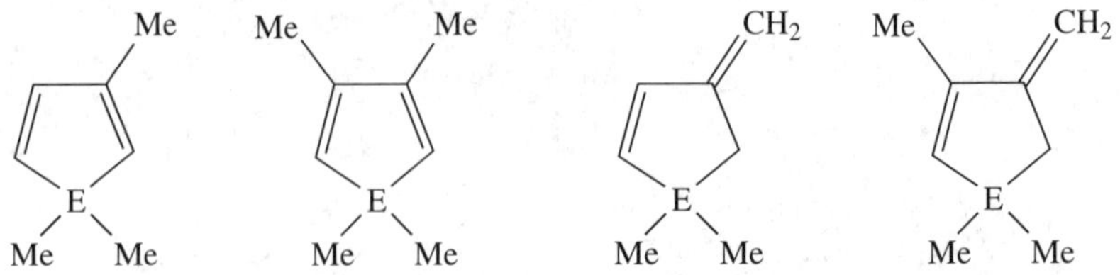

[Structures (53), (54), (55), (56)]

(53)   (54)   (55)   (56)

d. Thermolysis of esters of 1-metallacyclopent-4-en-3-ols. The synthesis of C-methylated metalloles **53** and **54** which avoids their isomerization requires milder conditions than those reported above. With this in mind, Dubac and coworkers investigated the thermolysis of esters of metallacyclopentenols, known to give elimination reactions at relatively low temperatures, such as carbamates or xanthates[12,14,22,67c,71].

In the case of 3,4-dimethylmetalloles **58**, which are stable in their monomeric form (see Section III.A.1), the method proposed was the thermolysis of *N*-phenylcarbamates **57**, the decomposition of which is regioselective (equation 22). Intracyclic β C—H elimination giving the metallole **58** is the main process (80–90%), and the transoid isomer **59** is always the minor product. Tertiary *N*-phenylcarbamates **57** decompose at about 70 °C in common solvents. In carbon tetrachloride, one can follow easily the progress of the reaction by NMR spectroscopy. An interesting one-pot synthesis is possible: with two equivalents of phenyl isocyanate, the side products are gaseous carbon dioxide and diphenylurea which precipitates (equation 23)[14,22,67c].

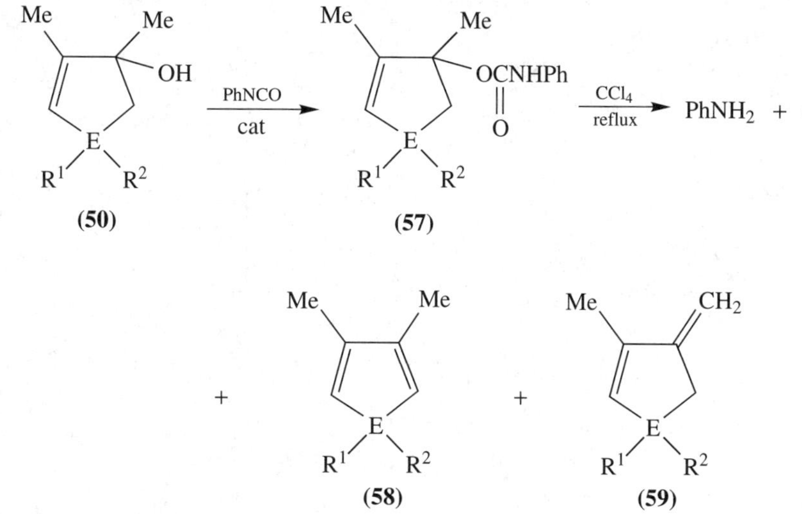

$E = Si, Ge$; R^1, R^2 = alkenyl or aryl groups   4:1 to 9:1

Cat. = $[Me(CH_2)_6COO]_2 Sn$

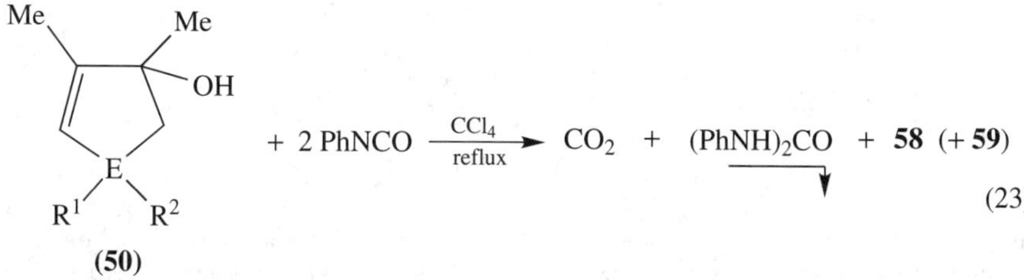

This route has allowed the synthesis of spirometalloles (**60**[72], **61**[73]) and of 1,1-dimethyl-3,4-diphenylsilole[68] in good yields.

S-Methylxanthates **62** derived from the same alcohols **50** are thermally less stable than the corresponding carbamates **57**. They decompose during their synthesis according to two competitive elimination pathways, β C—H and β C—Si (3/2) eliminations[14] (Scheme 12), giving rise to a lower metallole yield than from carbamates **57**.

The synthesis of 3-methylmetalloles is more difficult because they are unstable as monomers. Thermolysis of the tertiary carbamate corresponding to the alcohol **48** (under the experimental conditions of equations 22 or 23) gives the Diels–Alder dimer of metallole **53**. Thermolysis at 310 °C of the secondary carbamate corresponding to the alcohol **49** affords 1,1,3-trimethylmetalloles (**53**) which can be trapped in monomeric state at low temperature[14,22].

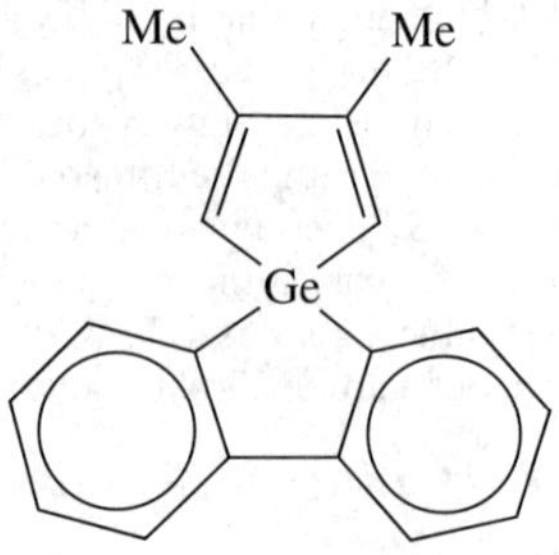

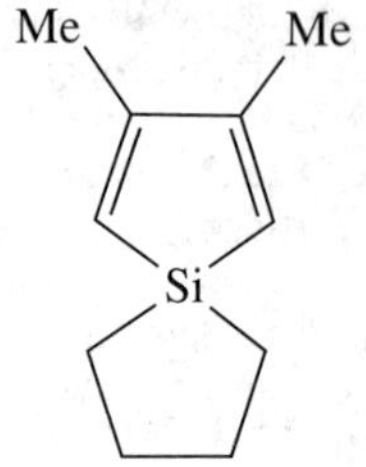

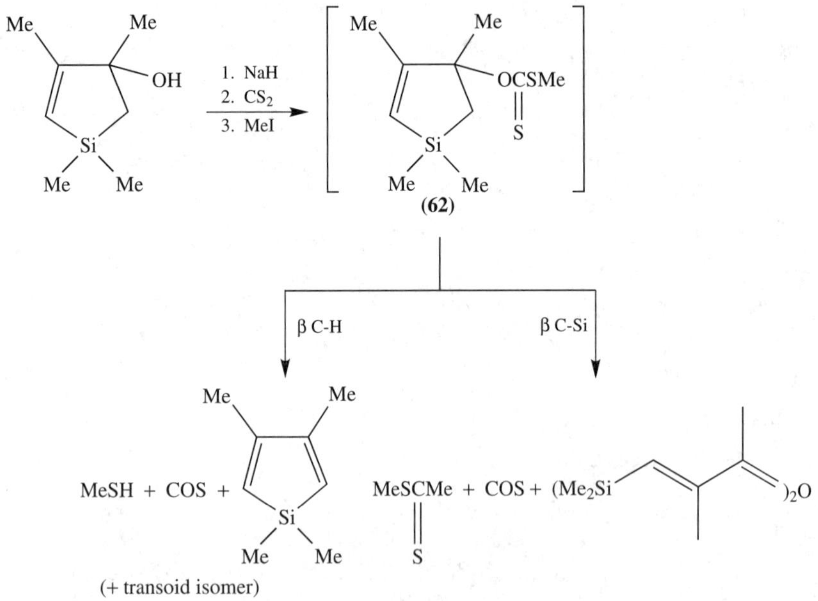

e. Gas-phase pyrolysis of silacyclopent-3-enes. Flash vacuum pyrolysis of 1-allyl-1-methylsilacyclopent-3-ene gives 1-methylsilole (**3**) (equation 1), the first C-unsubstituted silole, which is unstable in the monomeric state[5]. 3,4-Dimethylsiloles having one or two silicon–hydrogen bonds (**63, 64**[6], **65**[74]), which are stable in the monomeric state, were obtained in the same way. However, an exocyclic [1,3]-silyl shift in competition with the endocyclic retroene reaction[6] limits the silole yield which does not exceed 40%. Moreover, the transoid isomer of the silole is also formed.

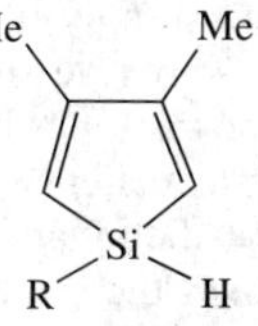

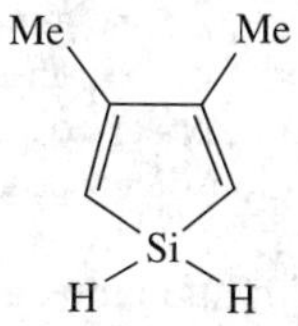

34. Group 14 metalloles, ionic species and coordination compounds

(63) R = Me
(64) R = Ph

(65)

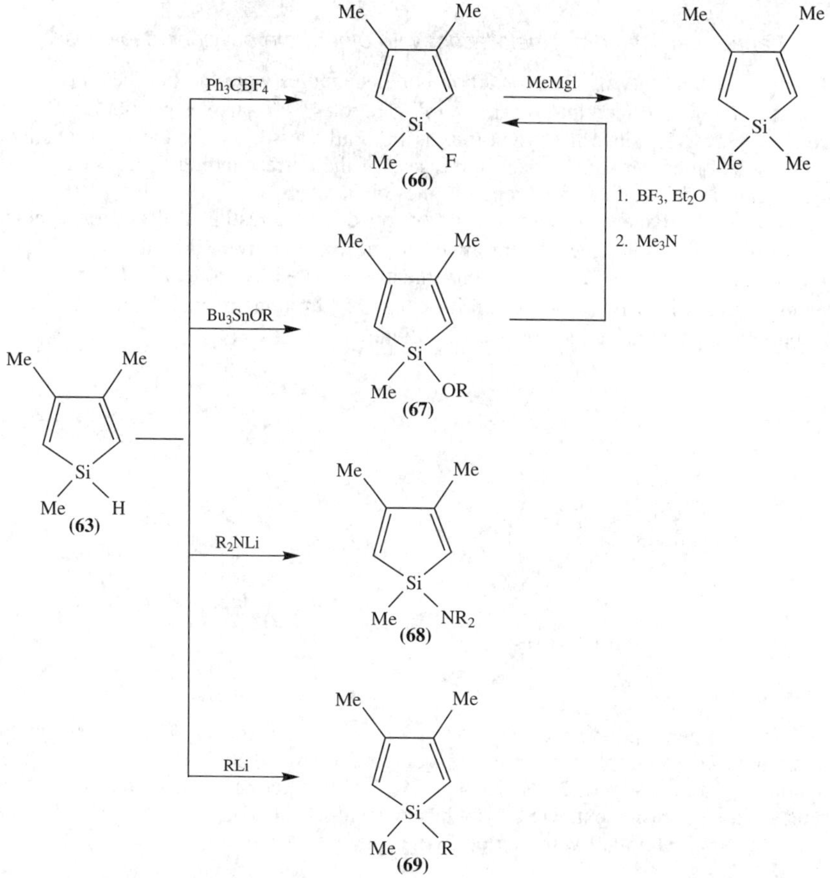

SCHEME 13

These three metalloles are the lower C-alkyl substituted Si—H bond-containing siloles which were isolated. The possible substitution of hydrogen bonded to silicon by another atom or a functional group appeared very attractive. Scheme 13 shows the functionalization reactions of 1,3,4-trimethylsilole (**63**) described by Dubac and coworkers[75]. 1-Fluoro-1,3,4-trimethylsilole (**66**) has been identified spectroscopically and chemically, but it is not stable enough to be isolated. Astonishingly, the chlorinated analogue was not obtained by chlorination of **63**; the chlorosilole seems to be very unstable (see Section III.A.4). This would appear to be unknown for a chlorosilane. In contrast, the alkoxylation and the amination of **63** proceed in excellent yield to give stable 1-alkoxy-1,3,4-trimethylsiloles (**67**, R = Me, *i*-Pr) and 1-diethylamino-1,3,4-trimethylsilole (**68**, R = Et). Reaction of **63** with lithium reagents gives the same siloles **69** as those obtained by the preceding method (equation 22).

Vacuum pyrolysis of the 2,3,7,8-tetramethyl derivative of spiro compound **12** gives the silole **65**, isolated in an Ar matrix[20]. A study of phototransformations of this silole showed isomeric forms analogous to those found for the parent silole (Scheme 4).

3. Exchange reactions (transmetallation) with other heterocyclopentadienes

Owing to the lability of the tin–carbon bond, stannoles were the first heterocyclopentadienes to be transformed into other heteroles: boroles[56a,76], arsoles or stiboles[60a,77] and auroles[78]. However, since their synthesis is presently less easy for the C-unsubstituted or alkyl substituted derivatives than in the case of the corresponding siloles or germoles, they are not useful reagents for preparing these heteroles.

More recently, Fagan and coworkers developed an interesting and general synthetic route to main group heterocycles by metallacycle transfer from zirconium[1,79]. The zirconacyclopentadienes **70–72** can be conveniently prepared by reduction of commercially available Cp_2ZrCl_2 ($Cp = \eta^5\text{-}C_5H_5$) with either Mg turnings or *n*-BuLi in the presence of 2-butyne, diphenylacetylene or deca-2,8-diyne[80], respectively.

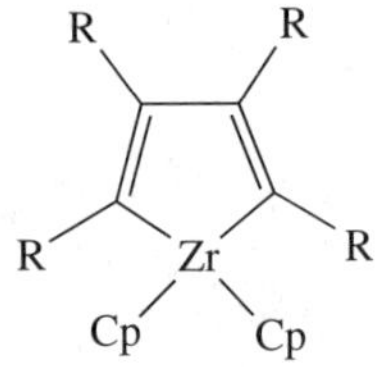

(**70**) R = Me
(**71**) R = Ph

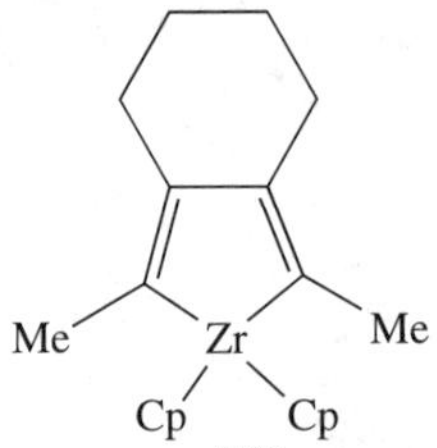

(**72**)

In THF solution the zirconacycles **70** and **72** reacted immediately at room temperature with main group electrophilic reagents affording main group heterocycles and zirconocene dihalide (equations 24 and 25) in about 80% yield. After removal of the solvent, the products can be easily extracted with hexane and isolated in the pure state by flash chromatography (alumina) with elution by hexane.

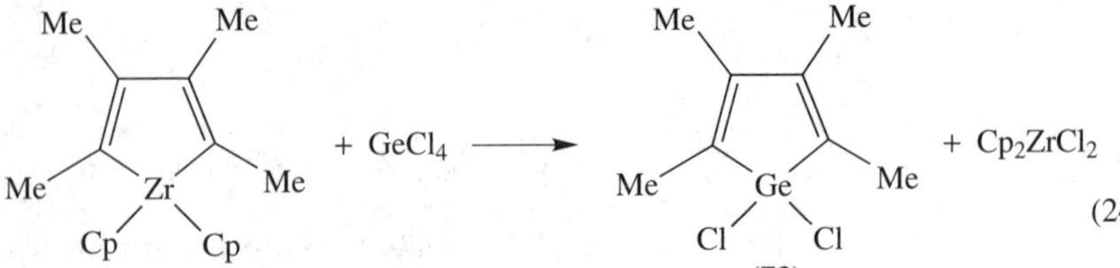

(**70**) + $GeCl_4$ ⟶ (**73**) + Cp_2ZrCl_2 (24)

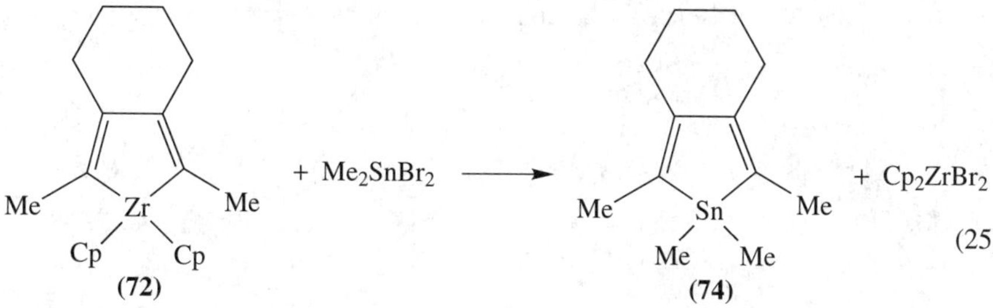

34. Group 14 metalloles, ionic species and coordination compounds

$$\text{(72)} + Me_2SnBr_2 \longrightarrow \text{(74)} + Cp_2ZrBr_2 \quad (25)$$

where (72) is the zirconacycle with Me, Zr(Cp)(Cp), Me substituents on a fused bicyclic system, and (74) is the corresponding stannole with Me, Sn(Me)(Me), Me.

By using the same method, many other group 13 (P, As, Sb, Bi) or 15 (S, Se) heterocycles can be readily obtained in good yields (50–85%).

Depending on the synthetic target, the zirconium metallacycle does not need to be isolated, and 'one pot' syntheses can be carried out. Addition of n-butyllithium to a THF slurry of Cp_2ZrCl_2 at $-78\,°C$ in the presence of the relevant alkyne, followed by warming to room temperature, generates zirconacycles. Dropwise addition of an hexane solution of a dichloride derivative of main group elements to this solution leads to a variety of heterocycles.

This methodology was extended by Dubac and coworkers[81] and Meier-Brocks and Weiss[82]. Titanacycles, zirconacycles and hafnacycles were treated with various halogenogermanes (equation 26).

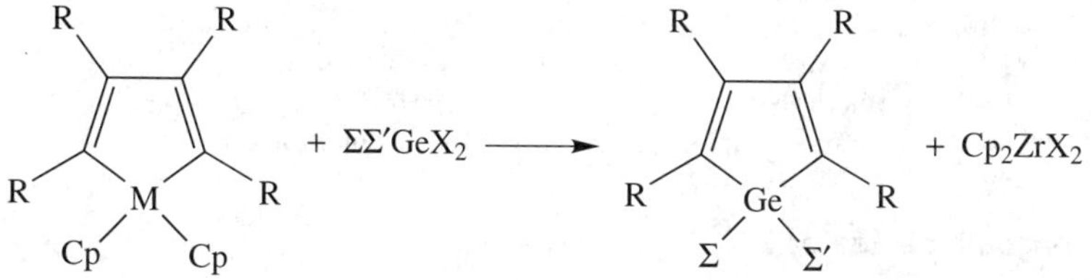

$$\text{M-cycle} + \Sigma\Sigma'GeX_2 \longrightarrow \text{Ge-cycle} + Cp_2ZrX_2 \quad (26)$$

(70) M = Zr, R = Me $\Sigma = \Sigma' =$ Me, Ph, Cl   X = Cl, Br
(71) M = Zr, R = Ph
(75) M = Ti, R = Ph
(76) M = Hf, R = Me

Starting from the titanium, zirconium or hafnium complex, germanium tetrachloride or tetrabromide undergoes transmetallation reactions under mild conditions (room temperature, THF or benzene solution) in good yields (70–90%). Dimethyldichlorogermane appeared less reactive than the tetrahalides in zirconium–germanium or hafnium–germanium exchange. Hafnium complexes are less efficient than the zirconium one. Methyltrichlorogermane readily reacts with zirconacycle **70** but the authors did not succeed in characterizing or isolating 1-chloropentamethylgermole, probably because it is unstable.

In the same way, the direct synthesis of 1-chloro-1-phenyltetramethylgermole **77** by Zr/Ge transmetallation starting from **70** and phenyltrichlorogermane was carried out in moderate yield owing to the low reactivity of $PhGeCl_3$. The desired chlorogermole **77** could be also obtained by chlorination of 1-phenyltetramethylgermole[81] (equation 27).

In the particular case of electrophilic silicon derivatives, the exchange reaction is very difficult to achieve. Transfer from Zr to Si is not possible in THF solution at room temperature and the zirconium complex could be recovered unchanged even after heating in neat $SiCl_4$. However, when the zirconacycle **70** was treated with neat $SiBr_4$ for 2 days at 150 °C, silole **78** was isolated as an impure liquid in 28% yield[1b] (equation 28).

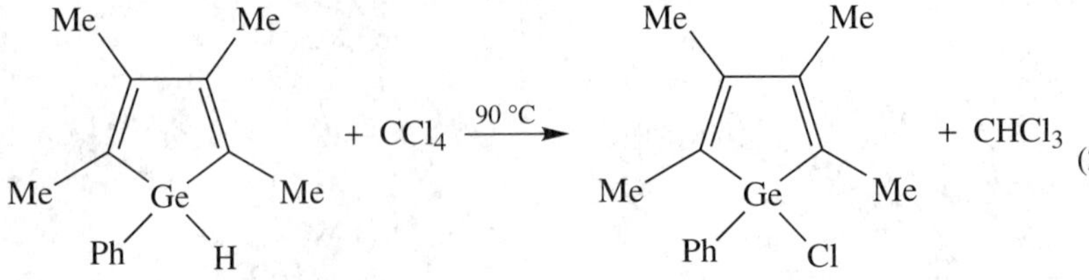

$$\text{(77)} + CCl_4 \xrightarrow{90\ °C} \text{(77-Cl)} + CHCl_3 \quad (27)$$

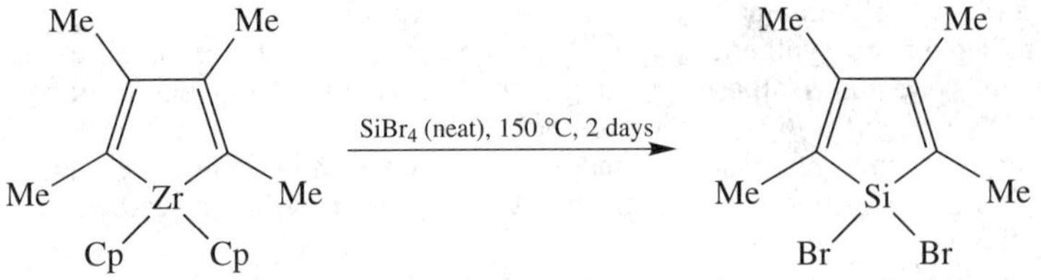

$$\text{(70)} \xrightarrow{SiBr_4\ (neat),\ 150\ °C,\ 2\ days} \text{(78)} \quad (28)$$

The Zr-based route to main group heteroles is a convenient alternative to the method using 1,4-dilithio-2,3-butadienes (see Section II.B.1.a), the more so as these dilithium reagents are easily available from zirconacyclopentadienes. Thus, the synthesis of 1,1-dichloro-2,3,4,5-tetramethylsilole **(27)** was readily performed from **70** by use of the I_2/n-BuLi/SiCl$_4$ sequence of reagents[40] (see Section II.B.1, equation 8)

C. Dibenzometalloles or 9-Metallafluorenes

The main method for the synthesis of 9-metallafluorenes was first described by Gilman and coworkers[83]. A one-step reaction of 2,2′-dilithiobiphenyl with various organometallic halides provided a great number of sila-, germa-, stanna- and plumbafluorenes with various groups on the heteroatom (equation 29).

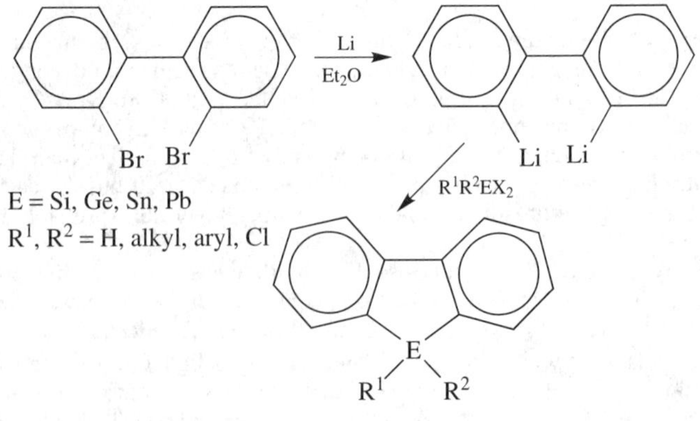

E = Si, Ge, Sn, Pb
R^1, R^2 = H, alkyl, aryl, Cl

(29)

2,2′-Dilithiobiphenyl also reacts with silicon, germanium or tin tetrahalides or hexachlorodisilane to give 5,5′-spirobi[dibenzometalloles] **(79)**[83b,d,84,85].

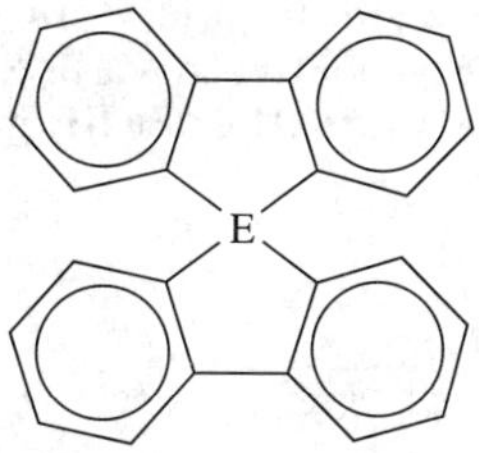

(79)

Reaction of 2-chloro-2′-bromobiphenyl with magnesium powder followed by addition of Me$_2$SiCl$_2$ gives in good yield (87%) 9,9-dimethyl-9-silafluorene **(80)**[86] (equation 30).

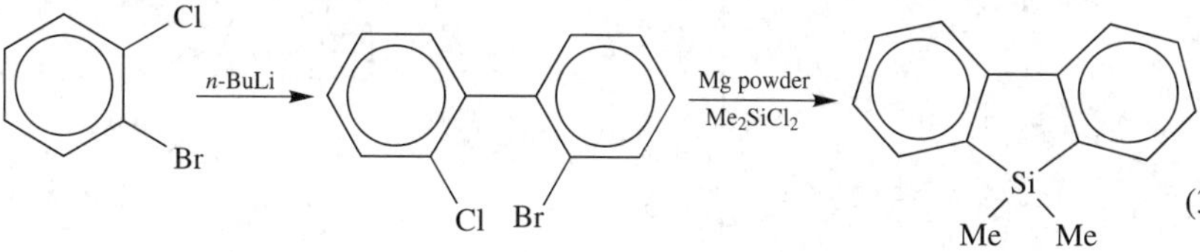

(30)

(80)

Metallaspirofluorenes **81**[87] and 9,9-dichloro-9-silafluorene **(82)**[88] were prepared by high temperature reactions of 2,2′-dihalogenobiphenyl with Ge or Sn powder or organochlorosilanes. The dichloride **82** was obtained in high yield by thermal reaction of (o-chlorophenyl)phenyldichlorosilane with various organochlorosilanes[89,90]. Substitution reactions on the metal also led to other functional metallafluorenes[91].

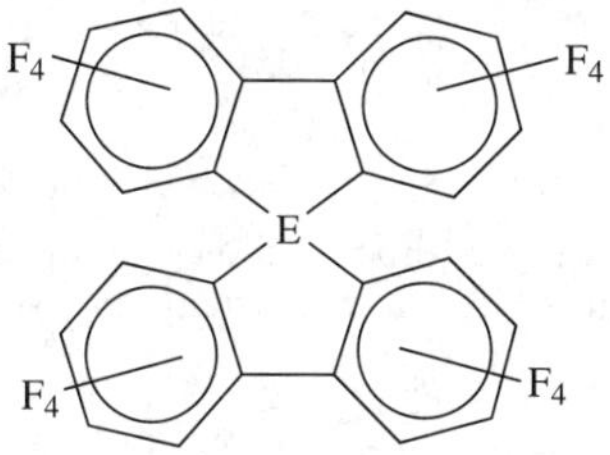

(81) 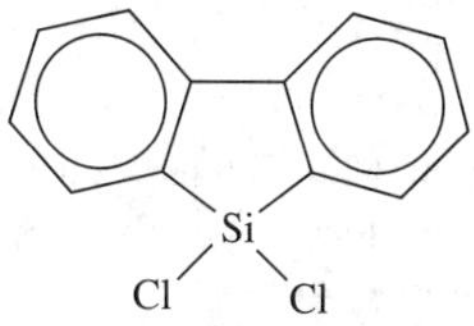   (82)

Irradiation of 1,1,2,2-tetramethyl-1,2-disila (or germa) dibenzocyclohexa-3,5-diene gave rise to 9,9-dimethyl-9-sila (or germa) fluorene **(83)** by a diradical process[92] (equation 31). This silafluorene could also be produced by pyrolysis of dimethyldiphenylsilane[93].

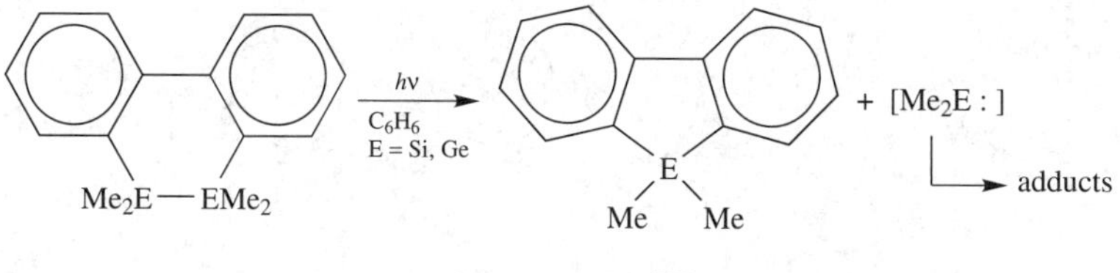

(31)

(83)

The photolytic cleavage of the Si—Si bond in 9-(trimethylsilyl)-9-silafluorenes in the presence of acetone afforded a Si—O(C) bond via a Si=C double bond[94]. The same reactions performed in neat alcohol (MeOH or EtOH) give rise to direct formation of a Si—H bond (Scheme 14).

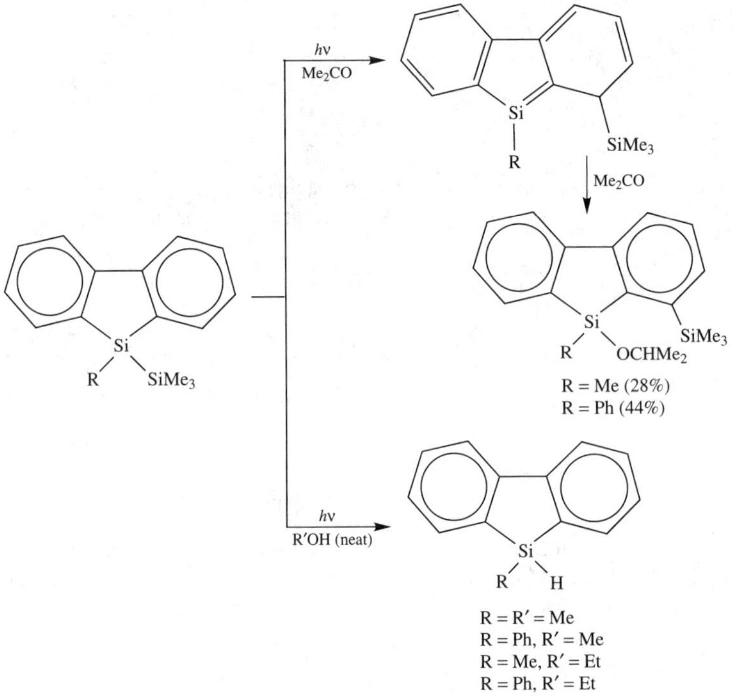

R = Me (28%)
R = Ph (44%)

R = R' = Me
R = Ph, R' = Me
R = Me, R' = Et
R = Ph, R' = Et

SCHEME 14

The reaction of MeLi and n-BuLi with various dibenzosiloles produces 1,1-dialkyl-dibenzosiloles in quantitative yields, while PhLi, and particularly t-BuLi, are less reactive[95] (equation 32).

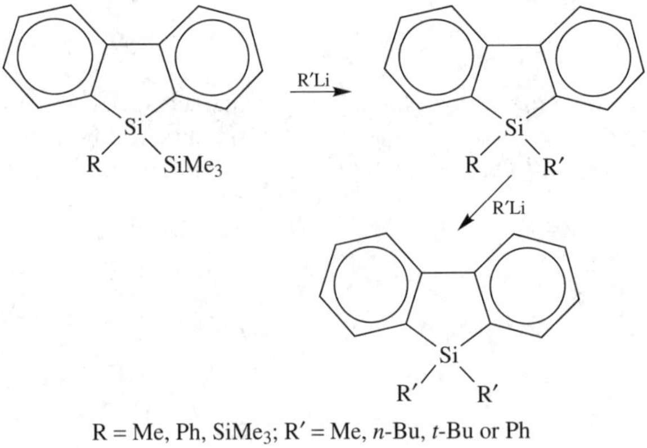

(32)

R = Me, Ph, SiMe$_3$; R' = Me, n-Bu, t-Bu or Ph

D. Benzometalloles or 1-Metallaindenes

Several methods have been recently developed for synthesizing various metallaindenes.

1. Cyclization of 1-lithio-2-(2'-lithiophenyl)ethylene

Reaction of organometallic dihalides with the title dilithio reagent afforded silicon and tin metallaindenes (equation 33)[96].

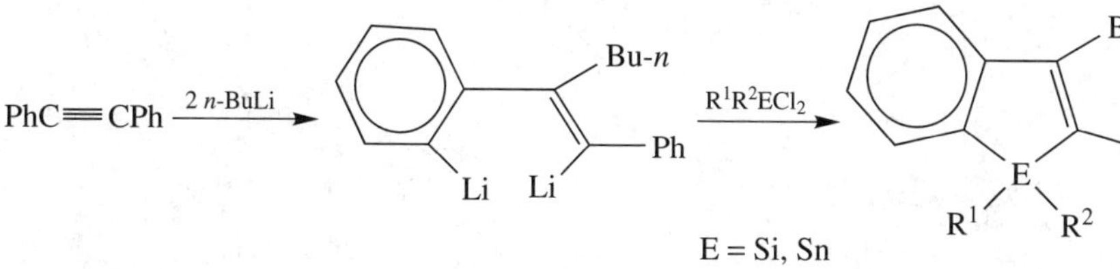

$$PhC \equiv CPh \xrightarrow{2\ n\text{-BuLi}} \text{[aryl-vinyl dilithio intermediate]} \xrightarrow{R^1R^2ECl_2} \text{[benzometallole]} \quad E = Si, Sn \quad (33)$$

An extension of this method was developed by Kurita and coworkers[97]. 2-Trimethylsilyl-1-benzometalloles are prepared in four steps from phenylacetylene in about 60% yield. The advantage of the trimethylsilyl substituent is that it can be easily removed, leading to C-unsubstituted 1,1-dimethyl-1-benzometalloles (**84**) (Scheme 15)

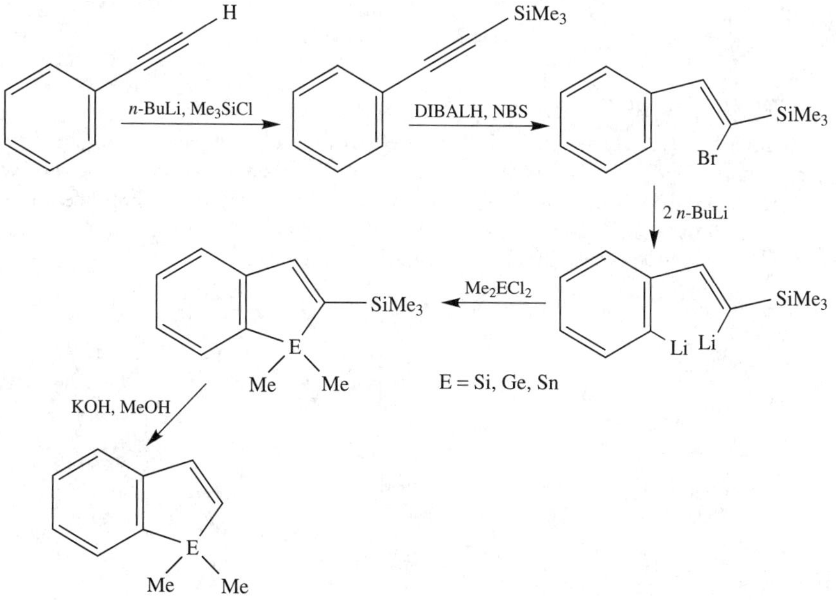

SCHEME 15

1,1-Dibutyl-1-benzostannole has been prepared in two steps from benzo[b]tellurophene by successive reactions of n-butyllithium and di-n-butyldichlorostannane (**85**) (equation 34)[98]. When Si, Ge and Sn tetrahalides are reacted with dilithio reagent **86**, spirometallaindenes **87–89** are obtained (equation 35)[96].

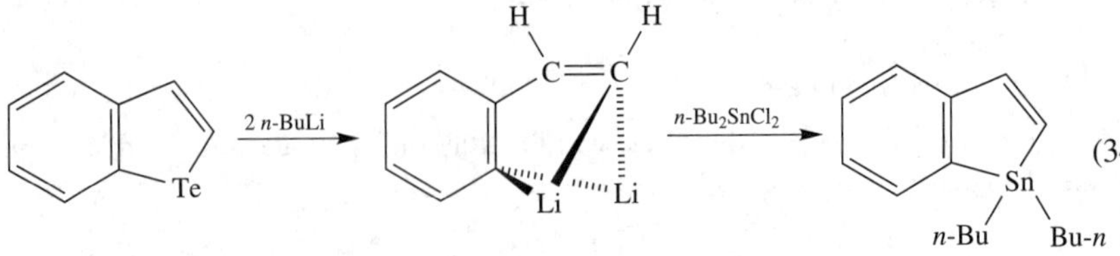

(34)

(85)

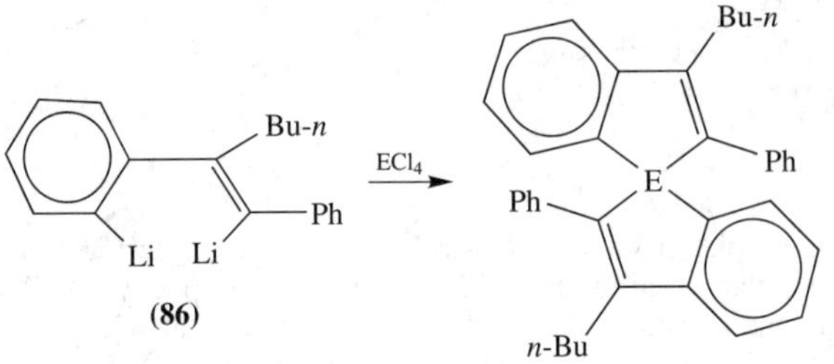

(35)

(**87**) E = Si (22%)
(**88**) E = Ge (28%)
(**89**) E = Sn (28%)

2. Transmetallation reactions

1-Zirconaindenes, which are easily obtained by insertion of various alkynes or an ynamine into the Zr—C bond of *in situ* generated[99,100] benzyne-zirconocene, are very good precursors of 1,1-dichloro-1-germaindenes (**90**) (Scheme 16)[101]. This method has been successfully extended to tin chemistry (Scheme 17)[102]. In this case, a convenient reaction was obtained even with alkyl or aryltrichlorostannane.

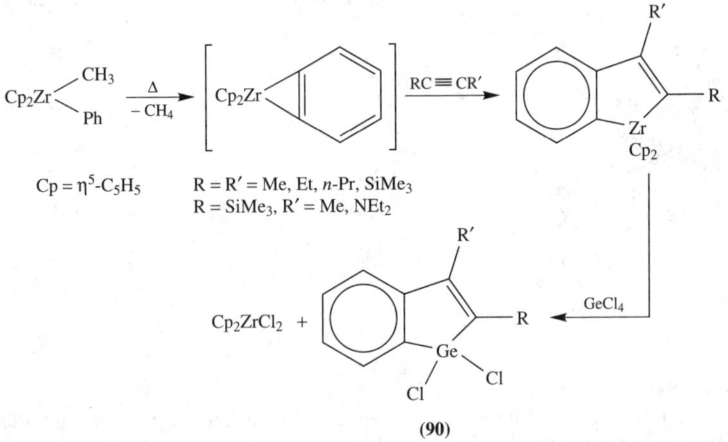

Cp = η^5-C$_5$H$_5$
R = R' = Me, Et, *n*-Pr, SiMe$_3$
R = SiMe$_3$, R' = Me, NEt$_2$

(**90**)

SCHEME 16

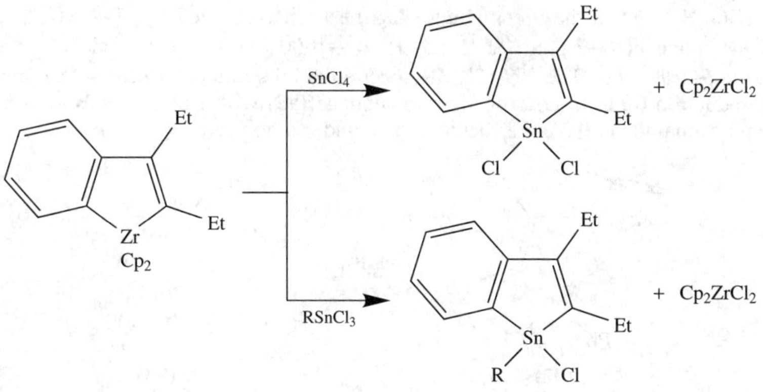

SCHEME 17

3. Nucleophilic substitution on the metal centre

Contrary to the tin series, transmetallation reaction with alkyl- or arylgermanium trichloride starting from zirconaindenes does not give monochloro 1-germaindene in good yield[103]. However, the required compounds were synthesized from 1,1-dichloro-1-germaindenes, by reaction of phenyl- or silyl-tris(trimethylsilyl)lithium (Scheme 18).

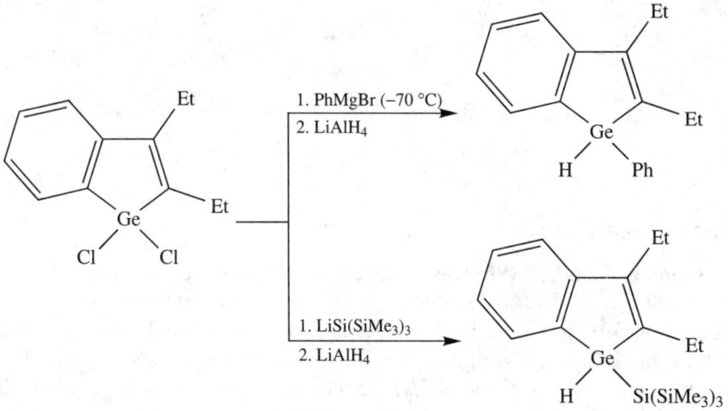

SCHEME 18

Sn−H bond-containing stannaindenes (**91**) are accessible by lithium aluminium hydride reduction of the monochloro derivative (equation 36)[103].

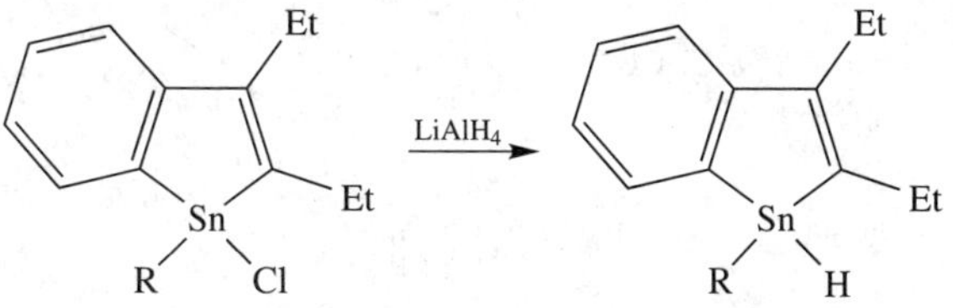

(36)

(**91**)

The Sn—H bond in these compounds has been characterized by ^{1}H NMR as a quartet in the range of 6–7 ppm ($^1J^{117}$ Sn—H = −1800 to −1900 Hz and $^1J^{119}$ Sn—H = −1880 to −1980 Hz). The ^{119}Sn NMR spectra exhibits a doublet near −150 ppm. Similarly, reduction of 1-chloro-1-phenylstannaindene (**92**) with LiAlD$_4$ leads to 1-deuterio-1-phenylstannaindene (**93**) after deuteriolysis under argon (equation 37).

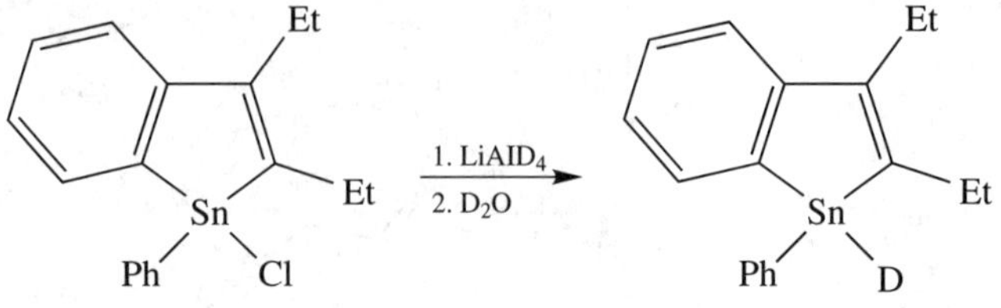

$$\text{(92)} \xrightarrow[\text{2. D}_2\text{O}]{\text{1. LiAlD}_4} \text{(93)} \quad (37)$$

In this case, the ^{119}Sn NMR spectrum showed a triplet at −153 ppm with a coupling constant $^1J^{119}$ Sn—D = −306 Hz.

It is well known[104] that organotin hydrides are very efficient reagents for organic halide reduction. Not surprisingly, therefore, 1-stannaindene (**94**) in CDCl$_3$ solution gives rise to 1-chloro-1-phenyl-1-stannaindene (**95**) (equation 38).

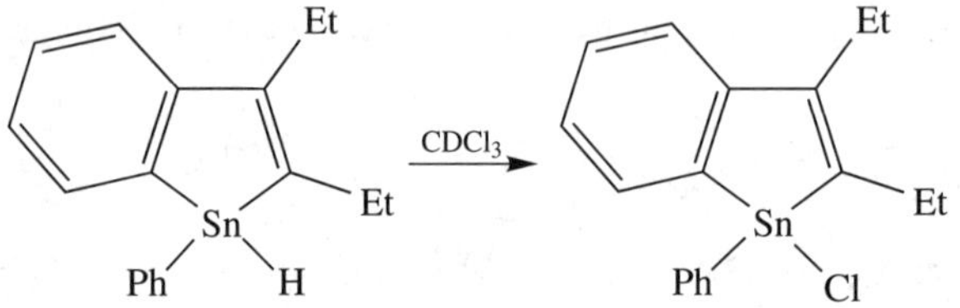

$$\text{(94)} \xrightarrow{\text{CDCl}_3} \text{(95)} \quad (38)$$

4. Thermolytic reactions

1,1-Dichloro-1-silaindene (**96**) was obtained in moderate yield by reaction of dichlorosilylene (generated from hexachlorodisilane) and chlorinated phenylvinylchlorosilanes at 500 °C (equation 39)[89a]. Similarly, 1,1-dimethyl-1-silaindene (**97**) was generated in high yield by FVP of (o-dimethylsilylphenyl)acetylene at 800 °C (equation 40)[105a]. The same reaction, starting from (o-dimethylsilylphenyl)dimethylsilylacetylene, also affords the silaindene **97**[105a]. The formation of this product is likewise observed on addition of dimethylsilylene to cyclooctatetraene[105b].

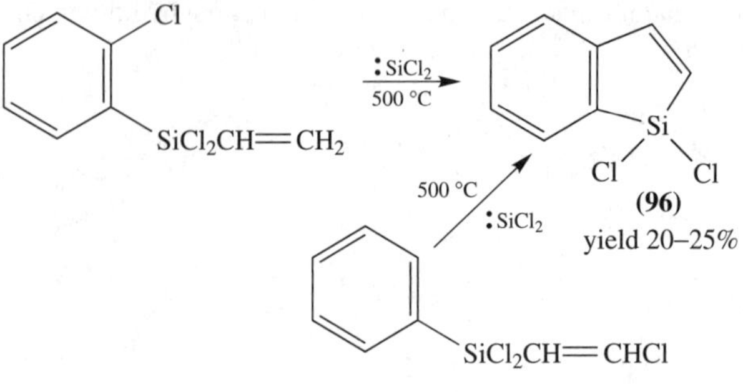

$$(39)$$

yield 20–25%

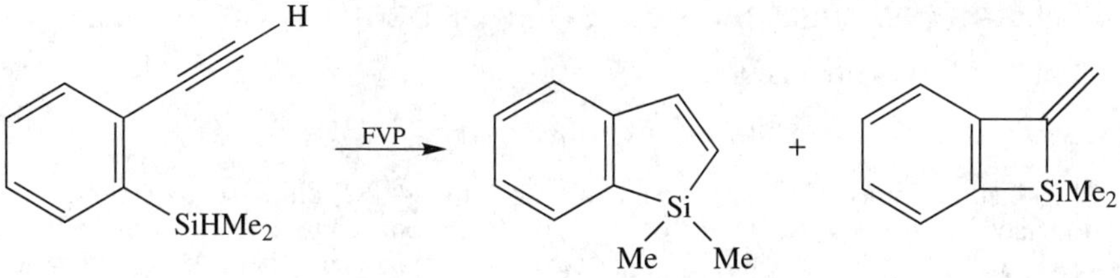

(40)

(97)

800 °C	84%	0%
650 °C	47%	34%

Barton and Burns[106] obtained 1-methyl-1*H*-1-silaindene **(98)** in moderate yield by FVP of 2-allyl-2-methyl-2-silaindane. The formation of this product was explained by the rearrangement of an initially formed 1*H*-2-silaindene via a retroene reaction (Scheme 19).

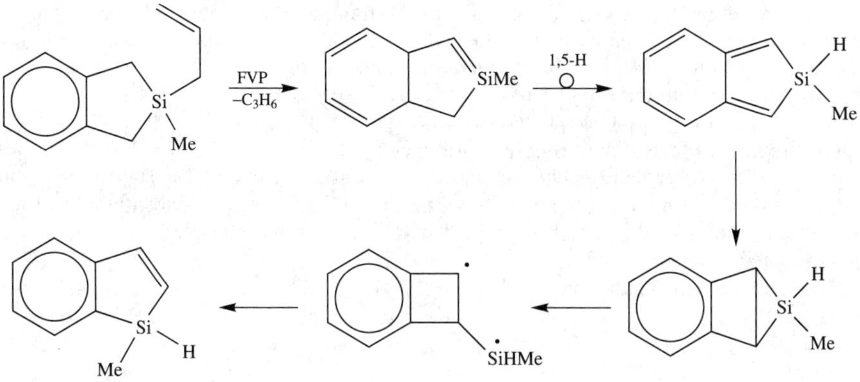

(98)

SCHEME 19

5. Other reaction

Seyferth and coworkers observed an interesting insertion reaction of a silirene into the C≡C bond of benzyne affording a 1,1-dimethyl-1-silaindene derivative **(99)** (equation 41)[45c].

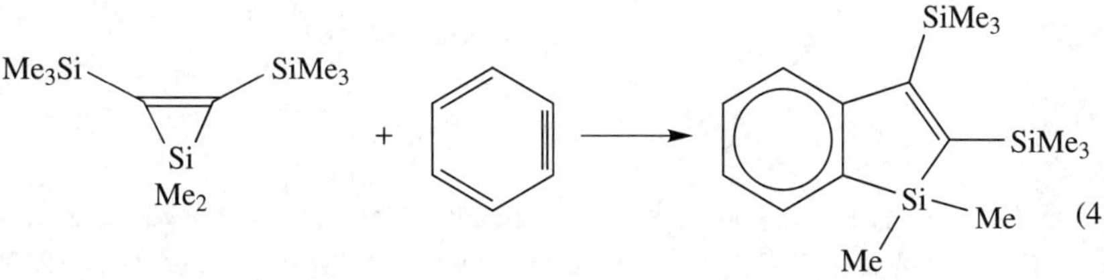

(41)

(99)

III. ORGANIC CHEMICAL PROPERTIES OF GROUP 14 METALLOLES

A. Stability and Isomerizations

1. Kinetic stability toward Diels–Alder dimerization

The kinetic stability of group 14 metalloles towards [4 + 2] dimerization is dependent on the nature of the heteroatom and on the substituents bonded to it and the ring carbons. Like cyclopentadiene, the silole **2** is unstable as a monomer. The silicon methylated derivatives (**3** and **4**) are also unstable, but two mesityl groups provide stabilization by steric hindrance of the [4 + 2] dimerization transition state[15]. The same effect has been observed for 1,1-dimesitylgermole[24].

Some observations seem to us interesting:

(a) 1,1-dimethylgermole (**18**) dimerizes more slowly than 1,1-dimethylsilole (**4**)[21];

(b) 1,1-di-n-butylstannole (**19**) is stable as a monomer[28], and 1,1-di-n-butylgermole (**17**), which can be distilled, dimerizes only very slowly[24a];

(c) a kinetic stability comparison between 1,1-di-n-butylsilole and **17** [prepared by dehydration of the corresponding metallacyclopentenols (see Section II.A.1)] suggests that **17** dimerizes much slower than the silicon analogue[107].

Hence, kinetic stability towards [4+2] dimerization increases from siloles to stannoles. Intracyclic bond lengthening (Si−C < Ge−C < Sn−C) must give rise to a decrease in the ring angular strain and to HOMO stabilization. This inhibits dimerization by an increase in the frontier orbital (HOMO/LUMO) energy-level difference.

A similar effect brought about by electronic factors can also be obtained by polymethylation of the dienic ring, as with cyclopentadienes[108]. The 3,4-dimethylsiloles and -germoles are generally stable in the monomeric state, except for some functional derivatives[20,75]. The close analogy between the stability of C-unsubstituted and C-methylated siloles on the one hand and that of isoelectronic phospholium ions on the other has been emphasized[14].

Aromatic groups on the ring carbons stabilize strongly the C-phenylated group 14 metalloles, as in the case with phospholes[109], but the reactivity of the dienic system (i.e. Diels–Alder cycloadditions, formation of transition metal complexes) decreases[3,4].

2. Geometric isomerization into transoid dienes

C-alkylated metalloles, particularly C-methylated ones (**53, 55, 58, 60, 61, 63–69**), isomerize readily into transoid dienes (**100**), their thermodynamically more stable isomers.

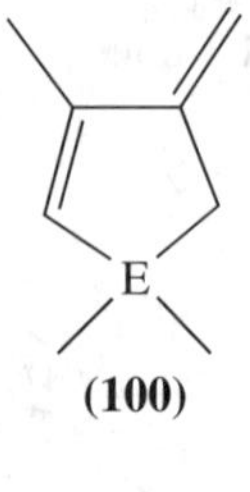

(**100**) 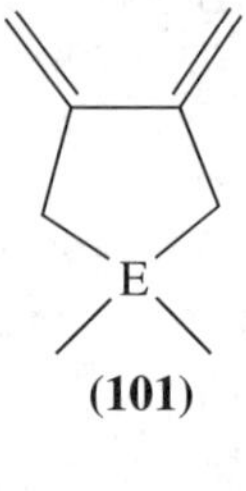   (**101**)

This disadvantage is overcome only with difficulty. Many catalysts or reagents induce this isomerization, e.g. alumina, thoria[67] and lithium reagents[110]. The best synthetic conditions which minimize production of the transoid diene are non-catalytic (see Section II.B.2.d.). In the pure state, the extent of isomerization is small at room temperature. However, storage at $-20\,°C$ or below is desirable. This isomerization under non-catalytic conditions implies a thermally allowed [1,3]-H antarafacial shift[111].

The presence of exocyclic cisoid isomers **101** has never been recorded. These derivatives have been prepared in other ways for silicon and germanium[112].

3. Tautomerism. 2H- and 3H-metalloles

The tautomerism phenomenon resulting from sigmatropic migrations of hydrogen atom or a substituent in cyclopentadiene and heterocyclopentadienes (pyrroles, phospholes) is well known[2d,113,114]. In pioneering work, Barton and coworkers[35c] observed during silole **102** thermolysis the formation of a [4 + 2] dimer **(104)** in which the $2H$-tautomer **103** is the diene component and the $1H$-silole **102** the dienophile (equation 42). **103** results from **102** by a [1,5]-Si shift. Other trapping products providing evidence for the formation of the reactive silole **103** were described (Scheme 20).

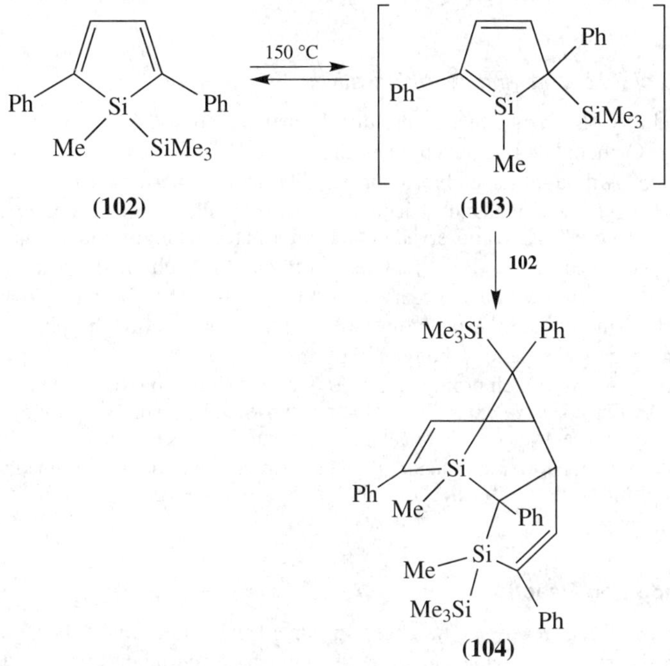

$$(42)$$

$2H$-Siloles and $3H$-siloles were suggested as reaction intermediates in the FVP of 1-allylsilacyclopentenes[5,6] (see Section II.A.1, Scheme 1), on the basis of trapping experiments. They were spectrometrically detected during vacuum pyrolysis of other silacyclopentenes[19,20] (see Section II.A.1, Scheme 4). A $2H$-benzosilole has also been postulated in the case of FVP of a 2-allylsilaindane[106].

Comparing the [1,5]-sigmatropic hydrogen shifts in pyrrole and phosphole, at the MP2/6-31G* level, Bachrach has shown that the more stable tautomeric form is, respectively, $1H$-pyrrole and $2H$-phosphole[2d]. The activation enthalpies were computed to be 186 kJ mol^{-1} for $1H$-pyrrole to $2H$-pyrrole transformation and only 67 kJ mol^{-1}

for 1H-phosphole to 2H-phosphole rearrangement. However, in substituted phospholes, substitution is preferred at P rather than C[115]. Hence the 1H-isomer is stabilized, probably for steric reasons[2d,115] and/or orbital overlap between the π-dienic system and the σ orbital of the P—R exocyclic bond[2a].

An extensive theoretical investigation does not exist for the siloles, but PM3 calculations of formation enthalpies of **2** and its tautomers have indicated that the 1H-silole is the most thermodynamically stable species[20c]. The activation barrier for **11 → 2** isomerization was calculated to be 96 kJ mol^{-1}, comparable to that for cyclopentadiene[2d,116]. The (1H+1H) dimer **10**[19] is isolated rather than the (2H + 1H) dimer as in the case of phosphole. This directly confirms the thermodynamic stability and the Diels–Alder kinetic instability of **2**. The marked difference in the stability of the parent silole and phosphole was explained[3] by the relative stabilities of the σ bonds in silanes and phosphines (Si > P) and of the π bonds in silenes and phosphenes (P > Si)[117].

No tautomeric forms of 1H-germoles or stannoles are as yet known. In these metalloles, it is likely that the carbene analogue isomer (like **9**) should be more stable than the 2H- or 3H-forms.

4. Stability of functional and spiro derivatives

The instability of some halogenated group 14 metallole derivatives has been previously reported[118]. C-phenylated halometalloles are stable, but C-methylated derivatives are much less so. Although 1,1-dichloro- or 1,1-dibromo-2,3,4,5-tetramethylsilole[1b,40] and -germole[1,81] have been isolated, 1-halo-1,3,4-trimethylsiloles are unstable[75]: 1-fluoro-1,3,4-trimethylsilole has been observed but the chlorinated analogue could not be prepared. The latter has been isolated in complex form as (η^4-1-chloro-1,3,4-trimethylsilole) carbonyliron[119]. It has been also observed that 1-chloro-2,3,4,5-tetramethylphosphole[120], -stibole and -bismole[79] are thermally unstable. 1-Halo-3,4-dimethylphospholes have been stabilized as σ-tungsten pentacarbonyl complexes[121].

These observations, which prompted a theoretical investigation, have been explained[118] by an effect similar to *spiroconjugation* which is known to be responsible for modifications in the electronic spectra and chemical reactivity of halocyclopentadienes, cyclopentadiene ketals and 1,1-dioxothiophene[122]. This spiroconjugation phenomenon could be also involved in the destabilization of spirobisiloles with respect to spirobigermoles and spirobistannoles[118].

B. Cycloaddition Reactions

The cycloaddition reactions involving group 14 metalloles have been extensively reported in the previous review[123]. Here we summarize the main points, and give new results.

1. Diels–Alder [4 + 2] cycloadditions

a. With ethylenic dienophiles. Ethylenic dienophiles give relatively stable adducts (7-metallanorbornenes) with siloles and germoles[123] (equation 43). C-phenylated metalloles (R^3-R^6 = Ar) are less reactive. For example, with maleic anhydride (MA), C-unsubstituted[10,15] and C-methylated siloles[67a,c,74,110] react at room temperature while C-phenylated siloles react at higher temperatures[52,64,68,124,125]. Indeed, for the reaction between 1,1-dimethyl-2,3,4,5-tetraphenylsilole and MA, the activation energy is higher

(60.3 kJ mol^{-1}) than for 1,1-dimethyl-2,5 (or 3,4)-diphenylsilole (39.3 and 34.7 kJ mol^{-1}, respectively), and the large negative activation entropy values (-30 to -40 e.u.) are consistent with those previously found for Diels–Alder reactions.

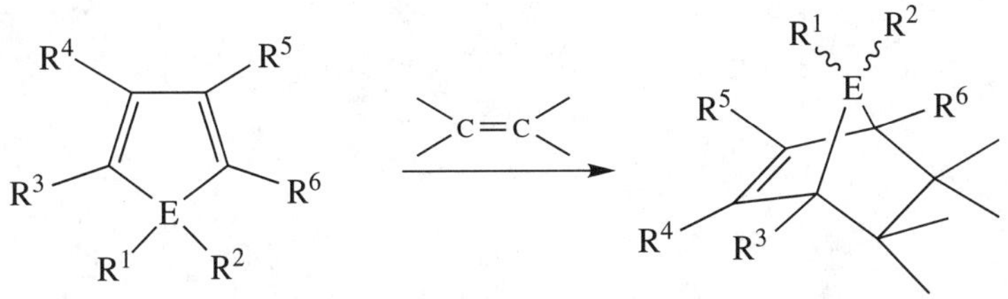

$$\begin{array}{c}\text{(structure with } R^1, R^2, R^3, R^4, R^5, R^6, E) + \text{C=C} \longrightarrow \text{adduct}\end{array} \qquad (43)$$

MA cycloaddition to metalloles seems to be always stereospecific and, according to Alder's rule, the *endo* configuration has been attributed to the adduct[125]. When two different groups R^1 and R^2 are present (equation 43), the major adduct results from the sterically more suitable transition state, i.e. the more bulky group is preferentially *syn* with respect to the C=C bond of the norbornene[6,67c].

Concerning the Diels–Alder dimers of unstable C-unsubstituted metalloles, only one ($1H + 1H$) dimer has been isolated, in a single configuration[10,12,19,21]. In the absence of crystallographic analysis of a solid dimer, an NMR study of hydrogen coupling constants was of help in determining the structure. The ^{1}H NMR of silole dimer (**10**)[19] did not allow measurement of the two hydrogen coupling constants which could give the required information. Dubac and coworkers[107] have calculated $J_{1,2}$ and $J_{6,7}$, which are related to H–C$_1$–C$_2$–H and H–C$_6$–C$_7$–H dihedral angles in the dimer of 1,1-dimethylsilole, from the ^{1}H NMR spectrum, decoupling experiments and the 2D COSY spectrum. The value obtained for these coupling constants (2 Hz) agrees with the corresponding dihedral angles of 60° and an *endo* configuration for the adduct.

The 7-silanorbornene-type adducts are relatively stable products, whose thermolysis or photolysis gives rise to the retro-Diels–Alder reaction or decomposition with extrusion of silylene[123]. With 7-stannanorbornenes this latter decomposition takes place at low temperatures[126]. Nucleophilic reagents provoke either silylene elimination[124] or cleavage of a single bridge Si–C bond[127].

From a 2*H*-silole and benzophenone or stilbene, Diels–Alder adducts **105** and **106** are formed from the conjugated diene rather than the expected [2 + 2] adducts with the silicon–carbon double bond[35c] (Scheme 20).

The reaction between a disilene (*cis*- and *trans*-MePhSi=SiPhMe) and a silole (1,1-dimethyl-2,5-diphenylsilole) is a stereospecific Diels–Alder addition, as in the case of an ethylenic dienophile[128]. A digermene reacts in a similar way[129].

b. With acetylenic dienophiles. The [4 + 2] cycloaddition of an alkyne with a 1*H*-metallole leads to a 7-metallanorbornadiene[123] (equation 44). These products are much less stable than 7-metallanorbornenes: they decompose by thermolysis or photolysis by elimination of the bridge [R^{1}R^{2}E:] and formation of a substituted aromatic compound. The instability increases from silicon to tin derivatives. This type of decomposition occurs when the substituents bonded to the norbornadiene are alkyl or aryl groups, including condensed rings. For example, 2,3-benzo-7,7-dimethyl-1,4,5,6-tetraphenyl-7-sila(or germa)norbornadiene decomposes at 300 °C^{52} (or 70 °C for the germanium derivative)^{61b}

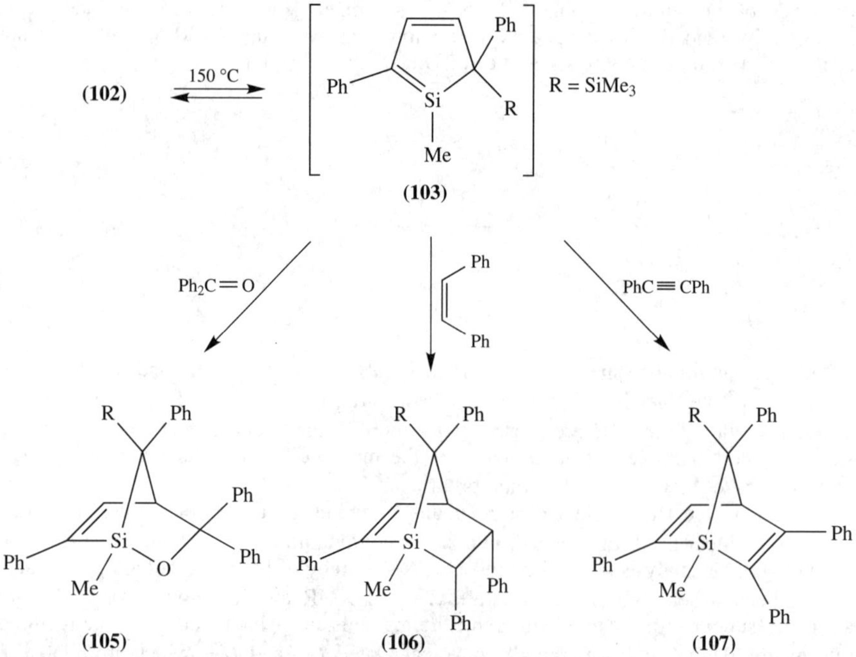

SCHEME 20

to give 1,2,3,4-tetraphenylnaphthalene and dimethylsilylene (or germylene) which can be trapped by reaction with diphenylacetylene (equation 45).

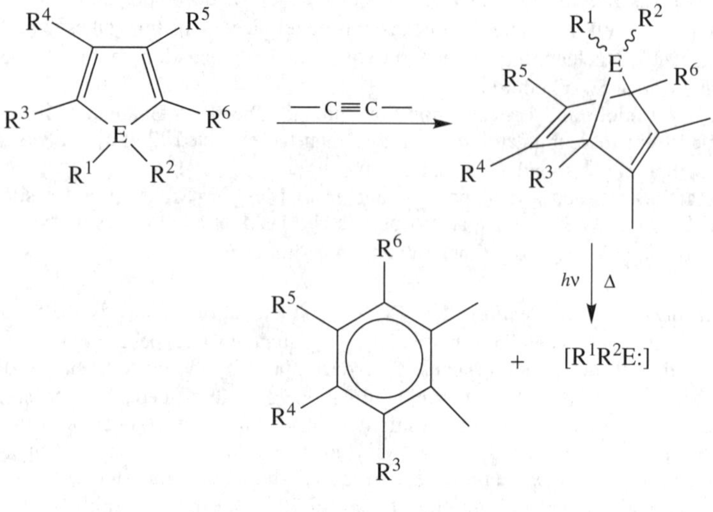

(44)

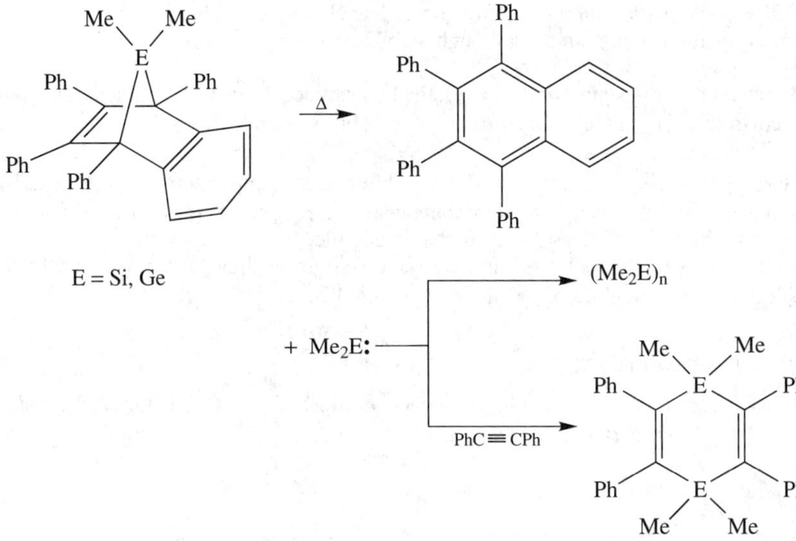

34. Group 14 metalloles, ionic species and coordination compounds

E = Si, Ge

(45)

The main interest of this reaction is the possibility of generating silylenes, germylenes or other highly reactive group 14 species. One of the applications is the synthesis of germoles when the decomposition of germanorbornadiene is carried out in presence of an alkyne and a catalyst (see Section II.A.2, Scheme 5, and Section II.B.1.b.ii, Scheme 10). Other noteworthy applications are the generation of species with multiple bonding to silicon[7,123], for example disilenes[130] or a disilyne[131].

Another type of decomposition of metallanorbornadienes can arise in the case of derivatives incorporating certain substituents, in particular C=O or CF$_3$. They decompose by rearrangement with formation of stable Si−O or Si−F bonds[123]. These rearrangement reactions give several products depending on the solvent used.

Rearrangements of [4 + 2] adducts of siloles and dimethyl acetylenedicarboxylate have been extensively studied for C-phenylated siloles[52,64,124b,132,133] and, more recently, for C-methylated siloles[73]. However, the previously described decomposition pathway with silylene extrusion and formation of a substituted dimethyl phthalate is also observed[73,132]. An iron-substituted silylene [Cp(CO)$_2$FeSiMe] has been generated by using this method[134].

Only dimethyl acetylenedicarboxylate adducts with 2,3,4,5-tetraphenylsiloles are isolable but not those with 2,5-diphenyl- or 3,4-dimethylsiloles. The corresponding adducts with C-phenylated germoles[64,135] or a C-tetraphenylstannole[136] are not isolable either.

Reactions of 7-sila(or germa)norbornadienes with nucleophiles resulting in the cleavage of one or two endocyclic Si (or Ge)−C bonds have been described[137].

Ab initio SCF MO studies of the electronic structure of 7-sila(germa or stanna)norbornadienes[138] have shown that these molecules possess an inverted sequence of π levels [that is, the π_- (b_1) orbital lies energetically below the π_+ (a_1) orbital] in contrast to that found for the parent norbornadiene. The degree of level inversion increases along the series: Si(−0.17 eV) < Ge(−0.26 eV) < Sn(−0.87 eV). The main cause of this trend is probably the increase in the strength of through-bond interactions between the π-orbital and the bridge σ-molecular orbital. Replacement of C-7 in norbornadiene by the larger and more polarizable Si, Ge or Sn atom raises the energy of this σ-orbital, and also polarizes it. Moreover, the crystal structure determination of a 7-germanobornadiene showed a very small value (78.5°) for the endocyclic C–Ge–C angle[139]. All these data explain why these compounds, although stabilized by E or C-substitution, nevertheless remain fragile.

In contrast to the 7-silanorbornadienes, the less strained 1-silanorbornadiene ring system (derived from a 2H-silole) proved to be remarkably stable[35c].

c. Group 14 metalloles as dienophiles. At elevated temperatures, 1,1-dimethyl-2,5-diphenylsilole and dienes (2,3-dimethylbutadiene[140], tropone[141]) form [4+2] Diels-Alder adducts in which the silole behaves as the dienophile.

A 1H-silole can thermally isomerize to a 2H-silole leading to a dimer in which the 1H-silole is the dienophile fragment[35c] (equation 42).

2. [2 + 2] Cycloadditions

It has been reported that the low-energy irradiation of 1,1-dimethyl-2,5-diphenylsilole[142,143] and germole[142] yields [2 + 2] photodimers (equation 46).

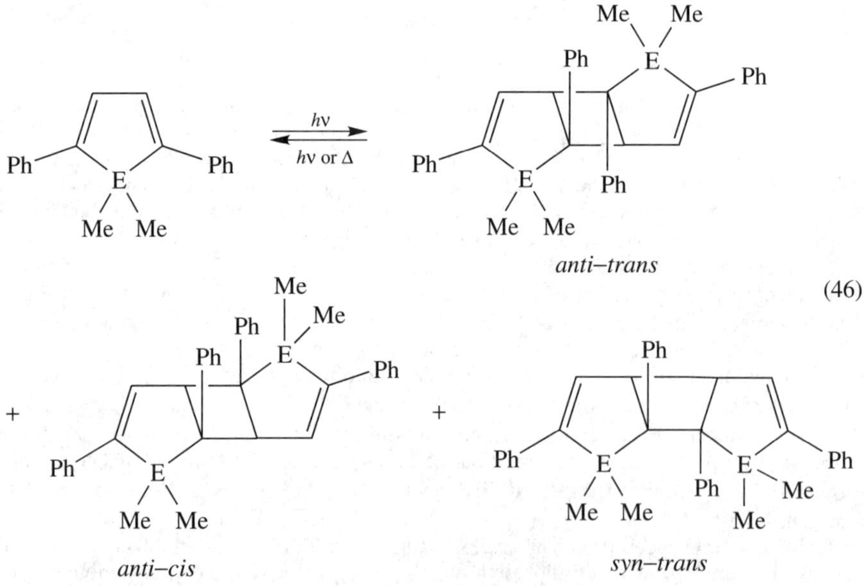

(46)

anti–trans

anti–cis

syn–trans

The major (almost exclusive) isomer formed is the *anti–trans* dimer, but two other dimers (*anti–cis* and *syn–trans*) are also obtained under a variety of conditions. The analogous intramolecular dimerization of α, ω-bis(1-methyl-2,5-diphenylsilacyclopentadienyl)alkanes has been also described[39].

A mechanistic study suggested that this photodimerization involves the excited single state of the silole[39]. It is thermally and photochemically reversible[39,143].

1,1-Dimethyl-2,3,4,5-tetraphenylsilole[144] is reported to be photochemically inactive whereas 1,1-dimethyl-2,5-diphenylstannole undergoes Sn–C cleavage leading to polymeric products[142].

The [2 + 2] photoadduct between 1,1-dimethyl-2,5-diphenylsilole and 1,1-dimethoxyethene was an intermediate in the synthesis of the first silacyclohepta-2,4,6-triene (silepin)[145]. Another, similar adduct is also formed when this silole reacts with N-chlorosulphonyl isocyanate[146].

C. Reactions with Halogens

Depending on the nature of both the halogen and the metal, the group 14 metalloles can react with halogens in three different ways: (i) addition of 1 or 2 equivalents of halogen to the diene system, with the cyclic structure preserved, (ii) cleavage of one or two endocyclic E-C bonds, with destruction of the cyclic structure, and (iii) substitution of one or two exocyclic substituents, with preservation of the metallole structure.

1. Silole series

Bromine follows reaction (i) with 1,1-dimethyl-2,5-diphenylsilole. If one equivalent of bromine is used, a mixture of dibromocyclopentenes in which the two halogen atoms are in the *trans* position is obtained (equation 47).

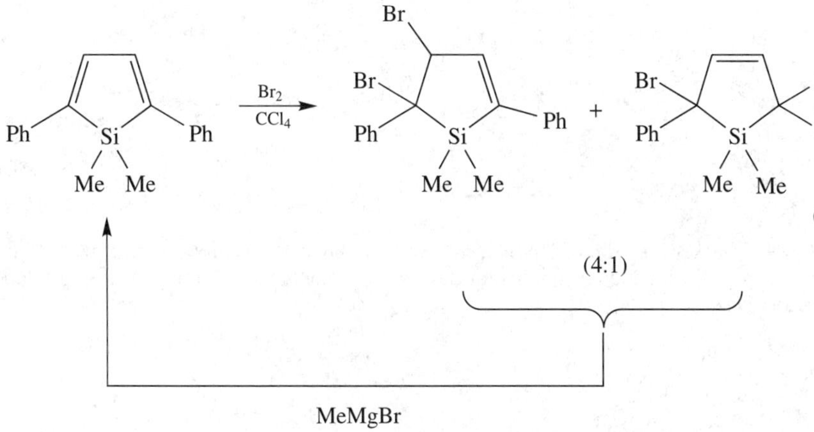

Treatment of the mixture with methylmagnesium bromide regenerates the starting silole[34]. Reaction of two equivalents of bromine with 1,1-dimethyl-2,5-diphenylsilole led to tetrabromosilacyclopentane from which a double β-elimination afforded (E,E)-1,4-dibromo-1,4-diphenylbutadiene (equation 48).

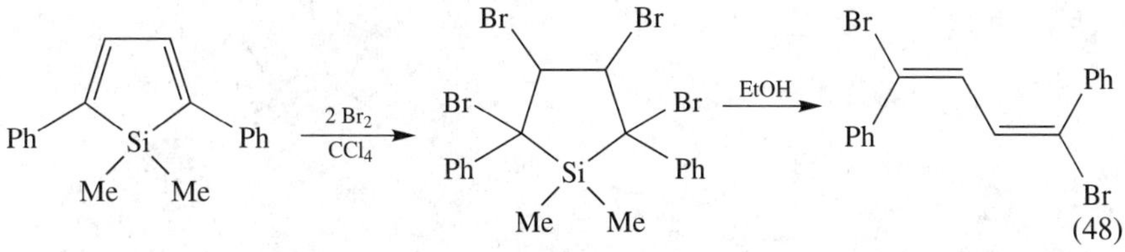

Iodine reacts differently with the C-methylated silole ring. Cleavage of the ring and formation of a complex mixture instead of the expected 1,4-diiodo-2,3-dimethylbutadiene is observed[12].

2. Germole and stannole series

Chlorine, bromine and iodine generally cleave germole and stannole rings[85c,147−150]. From 1,1-dimethyltetraphenylgermole or -stannole, [(Z,Z)-butadienyl]metal halides are formed in a quantitative yield (equation 49). The methyl–metal bonds remain intact. In the tin series, further chlorination or bromination gives rise to dimethyltin dihalide and dihalobutadiene (equation 50).

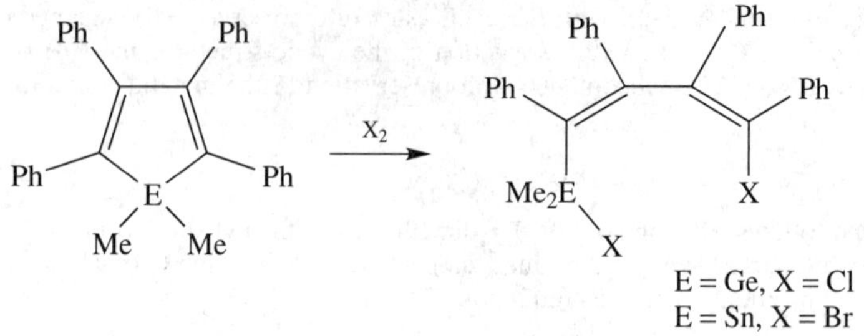

$$E = Ge, X = Cl$$
$$E = Sn, X = Br$$

(49)

$$E = Ge, Sn$$

$$+ Me_2EX_2$$

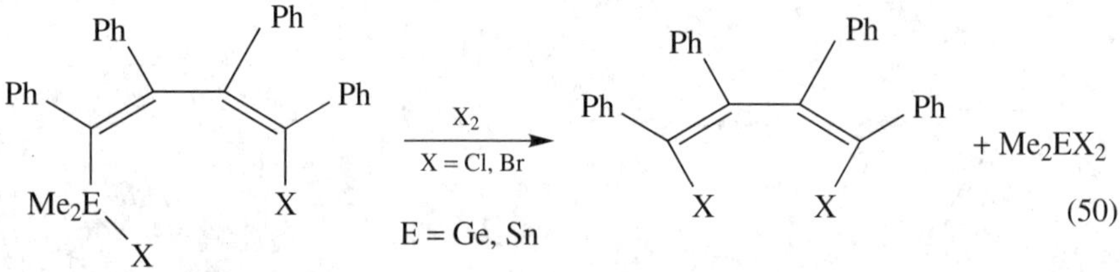

(50)

With the weaker electrophile iodine, mixed dihalobutadienes can be synthesized. Similarly, treatment with iodine monochloride converts the stannole into the same product (equation 51).

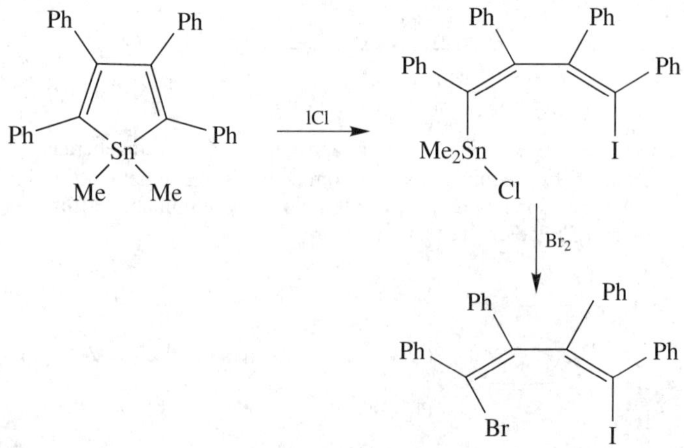

(51)

The reaction of iodine with 1,1-di-n-butyl-2,5-dimethylstannole leads to (Z,Z)-2,5-diiodohexa-2,4-diene, which is an interesting precursor for stibole and bismole synthesis[60b] (equation 52). By contrast, substitution of exocyclic substituents with conservation of the metallole structure has been discovered by Zuckerman and coworkers in the controlled bromination or iodination of hexaphenylstannole[151,152] (equation 53).

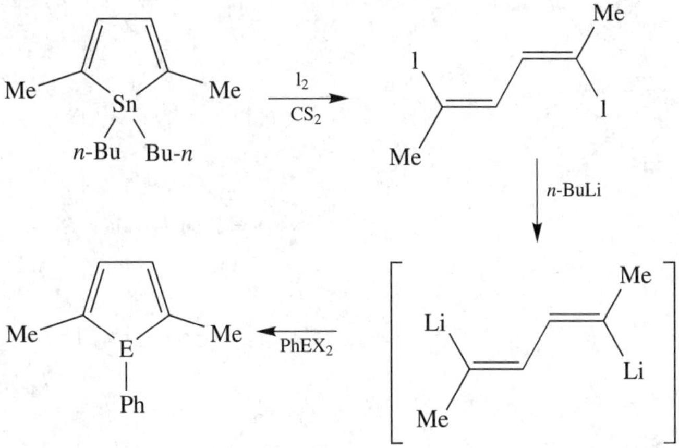

$$E = Sb, X = Cl$$
$$E = Bi, X = I$$

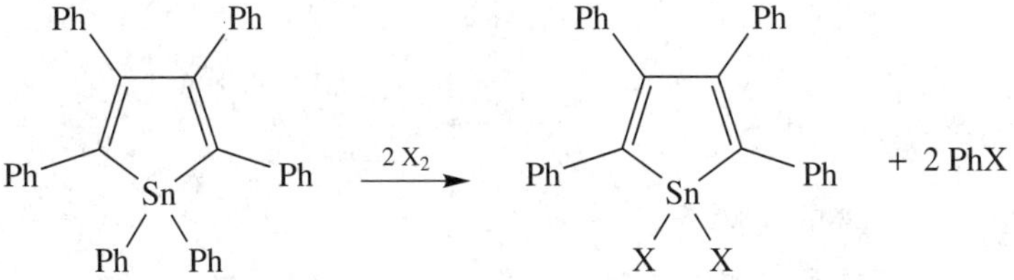

(52)

$$X = Br, I$$

(53)

Dihalostannoles are very useful starting materials for the synthesis of many substituted stannoles[153]. They also give rise to the only known group 14 metalloles having a five- or six-coordinated heteroatom in the form of anionic or cationic species[152,153].

The particular reactivity of halogens toward stannoles was demonstrated by Sandel and coworkers[154]. Treatment of 1,1-dimethyltetraphenylstannole with iodine trichloride led to cleavage products and the Hückel aromatic 2,3,4,5-tetraphenyliodonium ion[154] (equation 54).

D. Reactions with Acids

The action of acids on siloles results in the cleavage of the two endocyclic Si—C bonds. Boiling concentrated hydrochloric acid, hydrogen bromide or glacial acetic acid promote the reaction, and the substituted butadienes, in which the geometry of the parent metallole is retained, are produced in high yield[34,155] (equation 55).

Stannoles are more easily cleaved by acids than are siloles. At room temperature, brief exposure of 1,1-dimethyltetraphenylstannole to a dilute solution of acetic acid in alcohol resulted in rapid cleavage of the two tin–carbon bonds, affording (E,E)-1,2,3,4-tetraphenylbutadiene.

The mechanism of the cleavage is probably the same for both silole and stannole, consisting of a two-step protodemetallation reaction (equation 56). The product of the first cleavage has been isolated in the case of stannole[156].

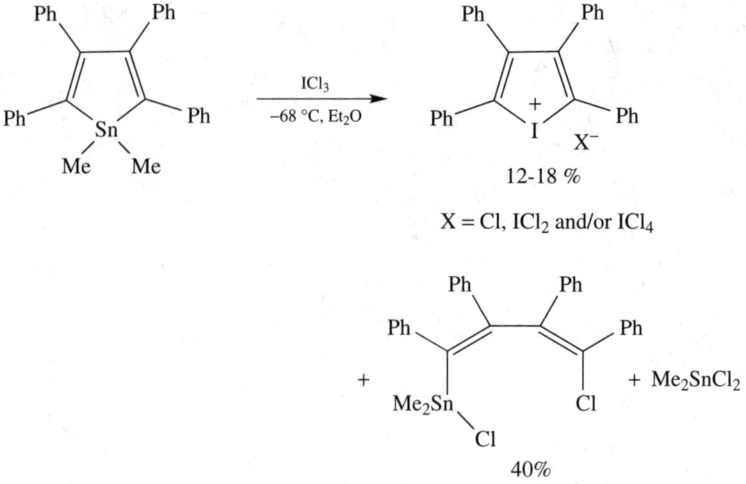

(54)

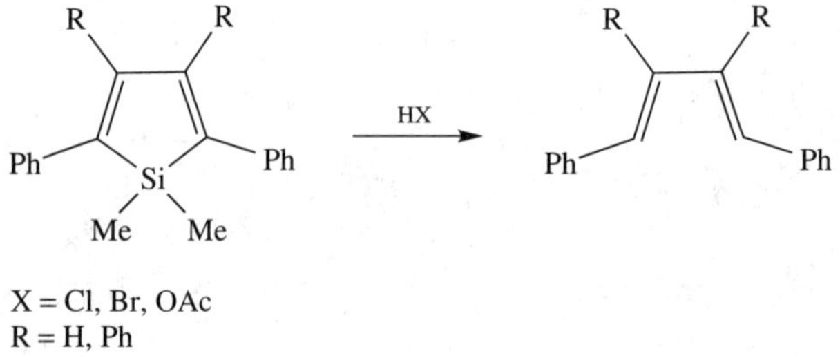

(55)

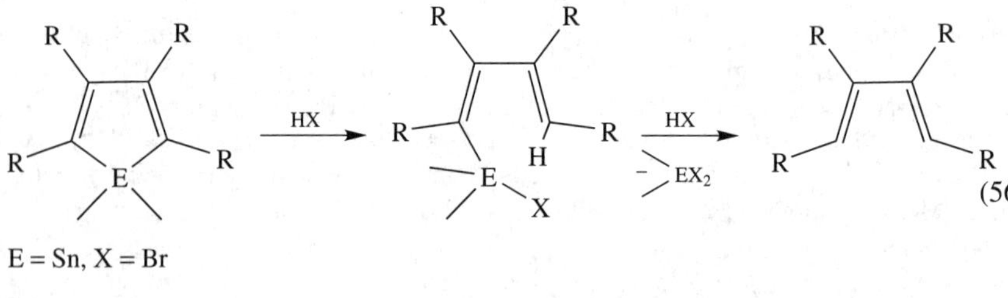

X = Cl, Br, OAc
R = H, Ph

(56)

E = Sn, X = Br

E. Reactions with Bases

The alkaline cleavage of siloles (by ammonia[157] or sodium hydroxide[155,157]) leads to the same substituted butadiene as that obtained with the acid reaction (equation 57).

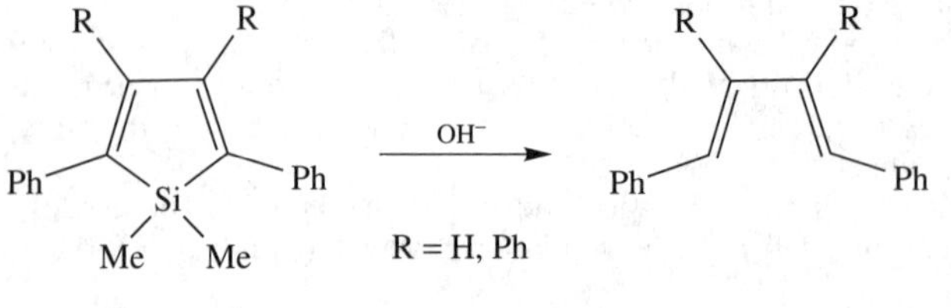

(57)

R = H, Ph

34. Group 14 metalloles, ionic species and coordination compounds 2007

The reaction probably proceeds by nucleophilic attack of the hydroxide ion on silicon and subsequent ring opening with retention of the geometry of the starting silole[155]. In the case of 1,1,3,4-tetramethylsilole, the reaction stops after the first Si−C cleavage giving a dienyl siloxane[12]. Furthermore, the reaction of dilute NH$_4$OH with a methylene chloride solution of 1-chlorotetraphenylsilole or germole derivatives produces the corresponding 1-hydroxysilole and -germole[157] (equation 58).

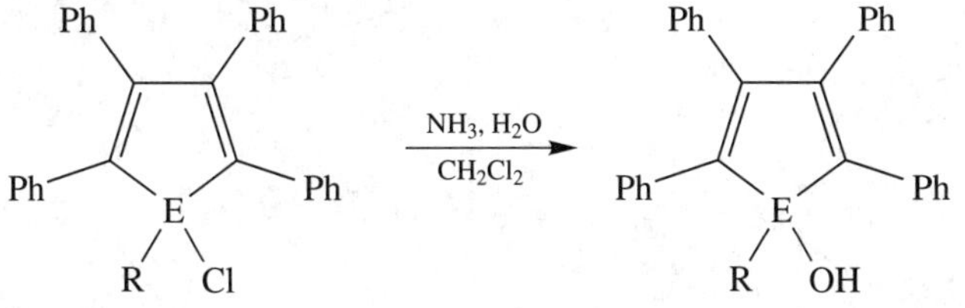

$$\text{(58)}$$

E = Si, R = Me
E = Ge, R = Ph

F. Reactions with Organometallic Reagents

These reagents are among the most important in group 14 chemistry for the promotion of substitution reactions (S$_N$2−M)[158] and metallacyclopentadienide anion generation (see Section V.A). Thus, 1,3,4-trimethylsilole (63) undergoes a classical S$_N$2−Si substitution[75,159] when treated with lithium reagents (Scheme 13). By contrast, in the case of pentaphenylsilole a secondary reaction occurs; a silacyclopentene ring is formed by 1,4-addition of the lithium hydride produced during the S$_N$2−Si reaction[160] (Scheme 21). This secondary reaction can be avoided by scavenging LiH *in situ* with Me$_3$SiCl[95a].

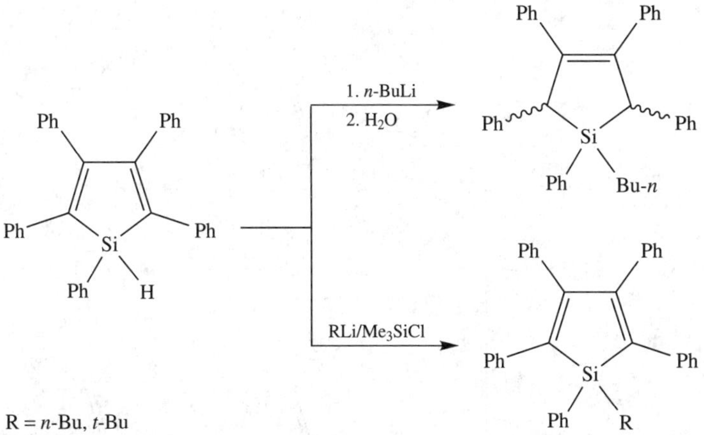

R = *n*-Bu, *t*-Bu

SCHEME 21

The hydrogermoles show different behaviour towards the lithium reagents. Metallation at the germanium atom is observed and germacyclopentadienide anions[157,161] are produced (see Section V.A).

The reactivity of lithium reagents with metalloles devoid of leaving groups on the heteroatom is very different. Methyllithium reacts with 1-silylsiloles by 1,4-addition to the π-system[160], leading essentially to the rearranged addition product together with a small amount of the 'normal' trimethylsilyl group substitution product[95a] (equation 59).

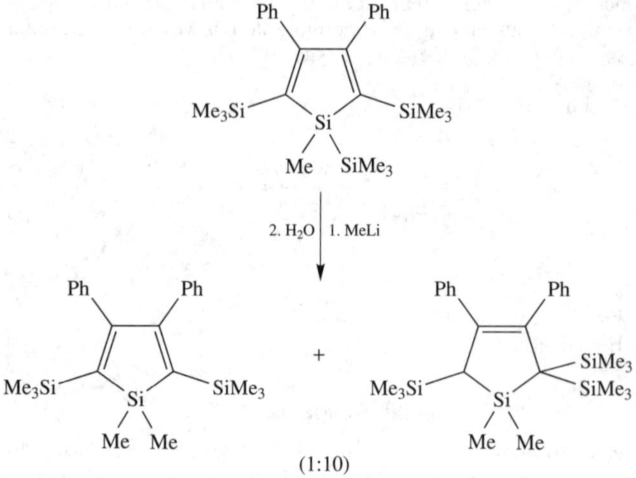

(59)

On the other hand, the reaction of RLi (R = n-Bu, Ph) at $-70\,°C$ in THF with 1,1,3,4-tetramethylsilole or -germole gives rise to a mono- or disubstitution reaction at the metal centre, depending on the proportion of the organolithium reagent[34,110] (Scheme 22). Phenyllithium is less reactive than n-butyllithium (9% and 70% yield, respectively, with tetramethylsilole) and the phenyl group is displaced in preference to the methyl group by

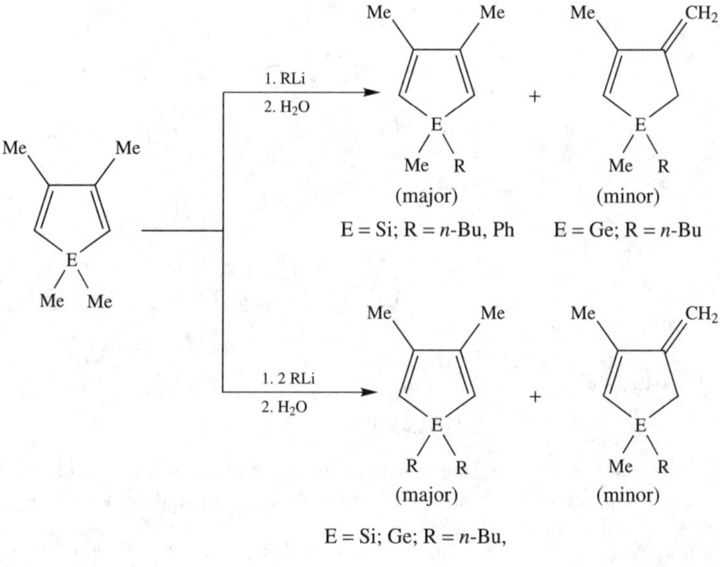

SCHEME 22

n-BuLi in 1-phenyl-1,3,4-trimethylsilole[34] (equation 60).

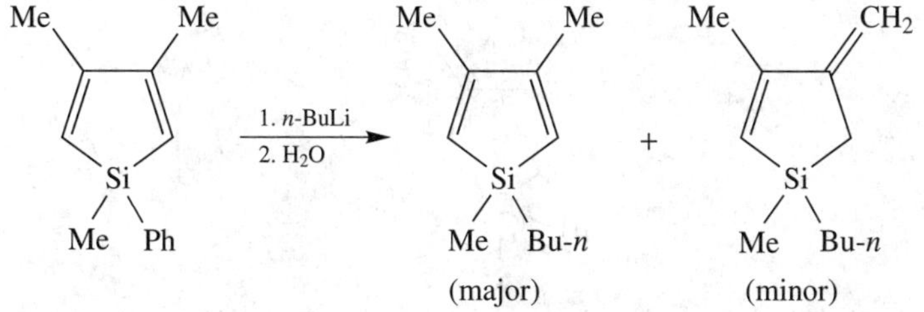

$$(60)$$

A predominant isomerization of metalloles to transoidal dienes occurs when t-BuLi is used[34]. This isomerization involves the formation of an allylic carbanion, which is protonated by water at the position α to Si (or Ge) and is silylated at the exocyclic carbon by Me$_3$SiCl[95a,110] (Scheme 23).

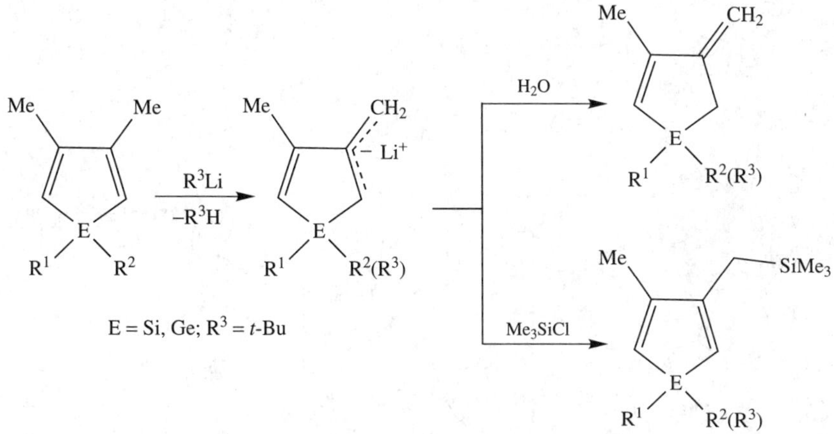

SCHEME 23

Substitution reactions[110] as well as certain rearrangements[95a] of exocyclic groups bonded to the heteroatom can occur via a five-coordinate anionic complex (**108**).

In the stannole series, the reactions could be explained by the extraordinary reactivity of the exocyclic tin–carbon bonds with regard to lithium reagents. Thus, the reaction of 1,4-dilithio-1,2,3,4-tetraphenyl-1,3-butadiene with 1,1-dialkyl-2,3,4,5-tetraphenylstannole leads to alkyl–tin bond cleavage–cyclization by the dilithium reagent[33] (equation 61).

G. Oxidation

Photooxygenation of 1,1-dimethyl-2,3,4,5-tetraphenylsilole was studied by Sato and coworkers[144], and Sakurai and coworkers[162]. The intermediate endoperoxide was produced by 1,4-addition of singlet oxygen to the cyclic diene, leading to cis- and trans-dibenzoylstilbenes in addition to the major product (1 : 1.45), 3,3-dimethyl-1,5,6,7-tetraphenyl-2,4-dioxa-3-silabicyclo[3.2.0]heptene (Scheme 24). Under the same conditions, the photooxygenation of 1-methyl-1-vinyl-2,3,4,5-tetraphenylsilole leads to analogous products[163].

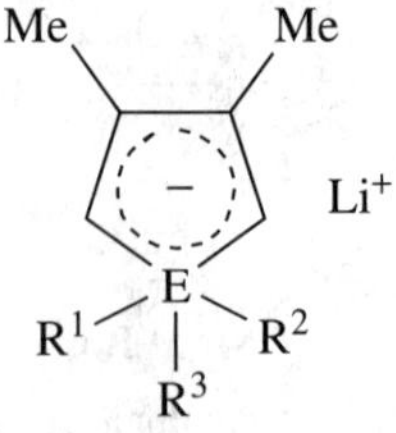

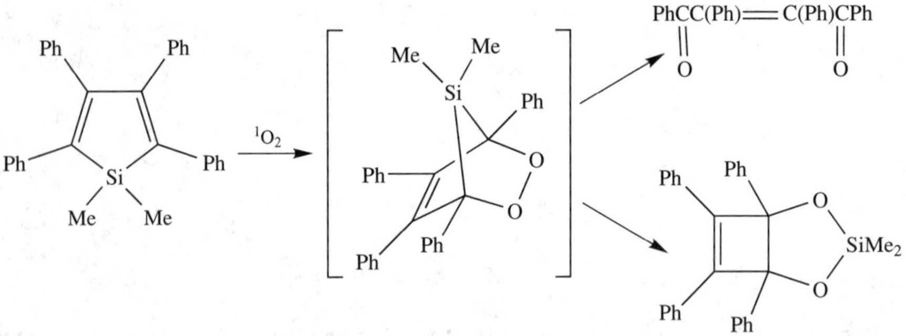

SCHEME 24

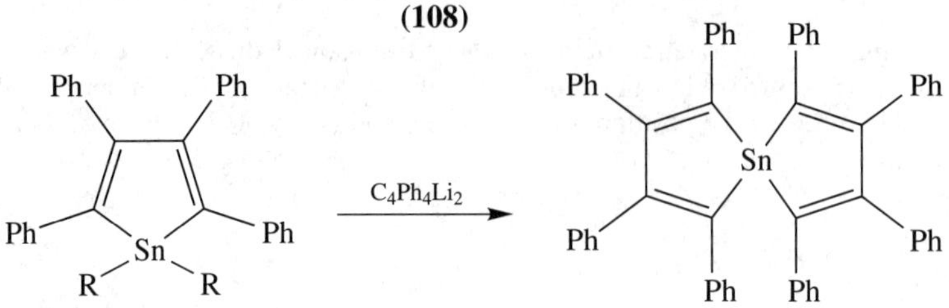

When treated with perbenzoic acid, 1-methyl-1,2,3,4,5-pentaphenylsilole yields a mixture of products, resulting from the oxidation of the diene system, various cleavages and rearrangements[163].

H. Reduction

1. Alkali metal reduction (see also Section V.A)

Formation of radical anions **109** and **110** from 1,1-dimethyl-2,5-diphenylsilole and 1,1-dimethyl-2,3,4,5-tetraphenylsilole[164], which are further reduced to the dianions **111** and **112**, respectively, has been shown by electron spin resonance and UV-visible spectrometry (equation 62).

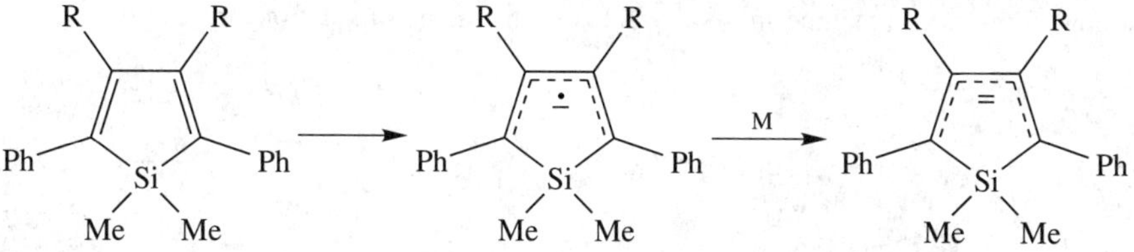

34. Group 14 metalloles, ionic species and coordination compounds

(109) R = H
(110) R = Ph

(111) R = H
(112) R = Ph

(62)

M = Li, Na, K / THF or DME

The presence of these dianions was also shown by aqueous quenching and the isolation of *cis*-dibenzylstilbene[155]. The dianions are remarkably stable even in the presence of alkali metal. Addition of metal to a solution of the dianion **112** (R = Ph) results in the formation of a new anionic species. ^{13}C NMR chemical shifts provide strong evidence that this is the tetraanion **113**.

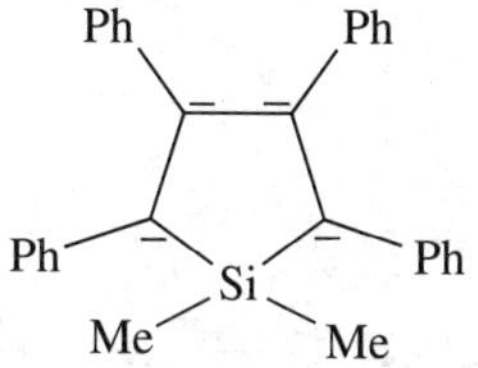

(113)

Reaction with an alkali metal gives substitution instead of reduction if the metallole possesses an exocyclic Si—Cl or Ge—Cl bond[160,161].

2. Electrode reactions

Two polarographic waves were observed in the electrochemical reduction of the same metalloles[165,166]. This means that they are reduced to a radical anion at the potential of the first wave and to a dianion at the potential of the second. The radical anions are stable enough to show ESR spectra.

3. Reaction with hydrides

LiAlH$_4$ is widely used for the synthesis of metalloles or benzometalloles with M—H bonds[167]. This hydride does not react by attacking the diene ring. Lithium hydride formed *in situ* may, however, give an addition reaction to the diene system[160].

Reaction of KH with C$_4$Ph$_4$SiMeH gives rise to a pentavalent silicate through simple addition of H$^-$ to the silicon centre (see Section V.A.1).

I. Transmetallation Reactions

Stannoles differ significantly from other group 14 metalloles in their ability to undergo exchange reactions with boron, arsenic or antimony halides. Boroles[56a,76], arsoles and stiboles[60a,77] were obtained in good yields by this method (equations 63 and 64).

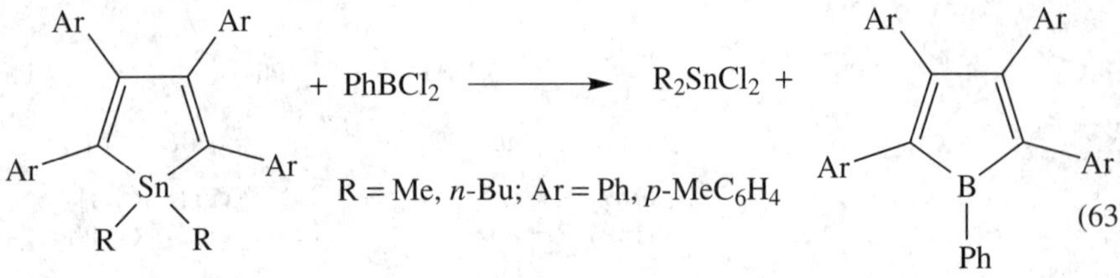

$$\text{Ar-[Sn(R)(R)]ring(Ar,Ar,Ar)} + PhBCl_2 \longrightarrow R_2SnCl_2 + \text{Ar-[B(Ph)]ring(Ar,Ar,Ar,Ar)} \quad (63)$$

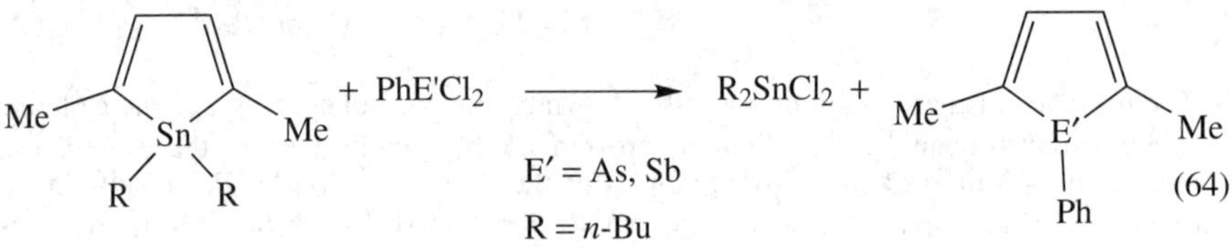

R = Me, n-Bu; Ar = Ph, p-MeC$_6$H$_4$

$$\text{Me-[Sn(R)(R)]ring(Me)} + PhE'Cl_2 \longrightarrow R_2SnCl_2 + \text{Me-[E'(Ph)]ring(Me)} \quad (64)$$

E' = As, Sb
R = n-Bu

2,3,4,5-Tetraphenystannoles were also used in obtaining auroles by reaction with AuCl$_3$, THT[78] (equation 65).

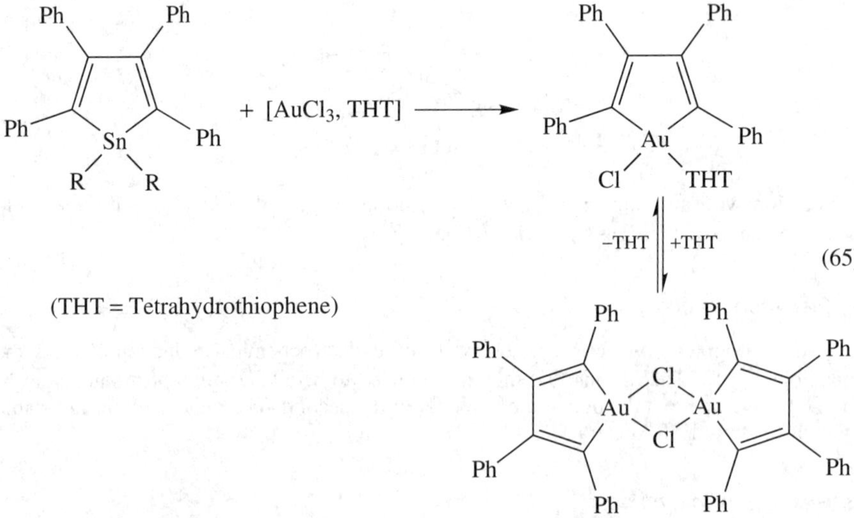

$$\text{Ph-[Sn(R)(R)]ring(Ph,Ph,Ph)} + [AuCl_3, \text{THT}] \longrightarrow \text{Ph-[Au(Cl)(THT)]ring(Ph,Ph,Ph)} \rightleftharpoons_{+\text{THT}}^{-\text{THT}} \text{[Au-Cl-Au dimer]} \quad (65)$$

(THT = Tetrahydrothiophene)

J. Ring Expansion

Ring expansion of a silacyclopentadienylcarbene generated from thermolysis or photolysis of (1-methyl-2,3,4,5-tetraphenylsilacyclopentadienyl)-diazomethane or -diazirine to silabenzene is known[168] (Scheme 25). With AlCl$_3$, 1-(chloromethyl)-1-methyl-2,3,4,5-tetraphenylsilole undergoes ring expansion leading to 1-chloro-1-methyl-2,3,4,5-tetraphenylsilacyclohexa-2,4-diene[169].

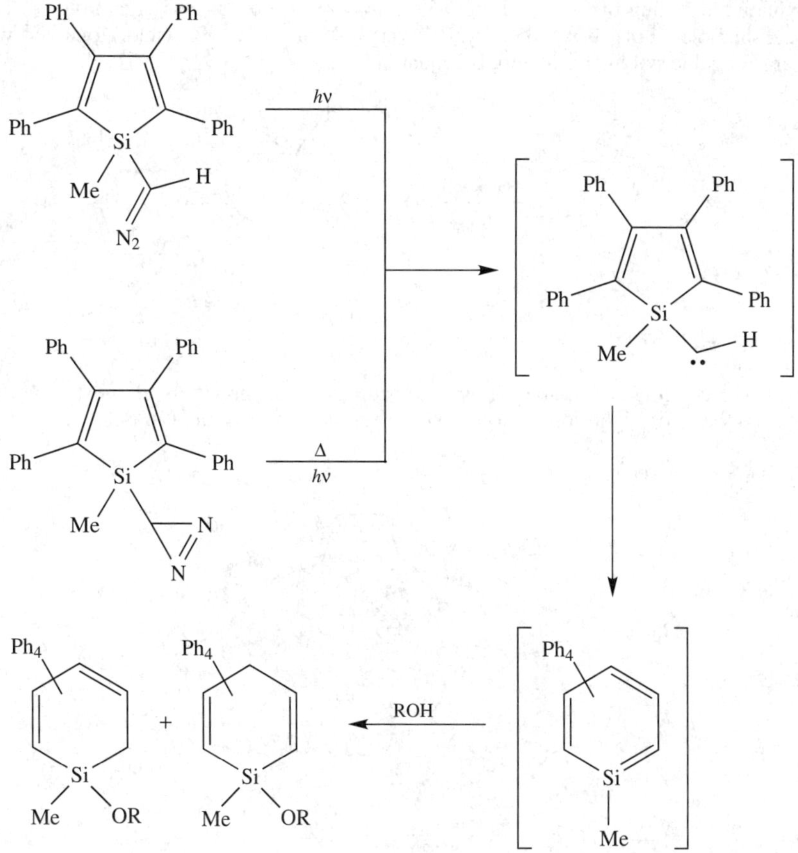

SCHEME 25

IV. POLYMERIC SILOLE-CONTAINING π-CONJUGATED SYSTEMS

π-Conjugated polymers containing silole rings have recently been highlighted as promising candidates for novel π-electronic materials, because of their potential properties such as conductivity, thermochromism and non-linear optical properties. Substituted siloles can behave as electron-accepting components in such polymeric derivatives[164]. Furthermore, *ab initio* molecular orbital calculations have demonstrated the unusual electronic structure of the silole ring which has relatively lower LUMO levels, compared with either cyclopentadiene or thiophene[170–172].

A. Synthesis, Structure and UV-Visible Absorption Data of Oligosiloles

Oligosiloles, from bisiloles to quatersiloles, have been prepared from functional siloles[59], the synthesis of which is discussed earlier (*vide supra*) (cf Section II.B.1.c).

Oxidative coupling of the 2,5-dilithiosilole (**44**) by use of an Fe(III) complex affords the 2,2′-bisilole (**114**) isolated as yellow crystals in only 9% yield, together with uncharacterizable polymeric materials (equation 66).

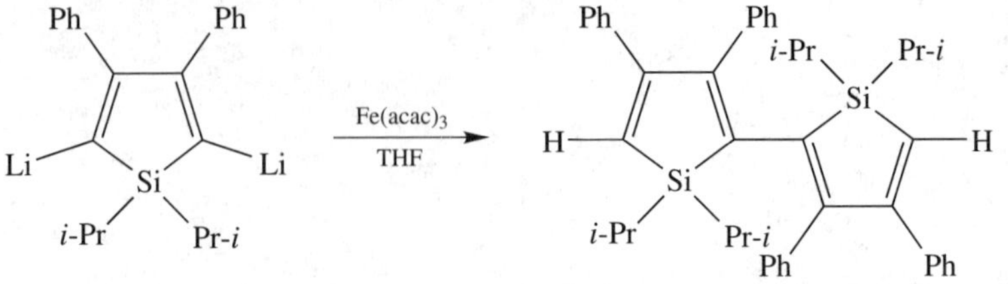

$$\text{(44, R} = i\text{-Pr)} \xrightarrow[\text{THF}]{\text{Fe(acac)}_3} \text{(114)} \quad 9\% \tag{66}$$

Oxidative coupling via higher order cyanocuprates was shown to be most promising for the synthesis of higher oligosiloles and polysiloles, as illustrated in Scheme 26.

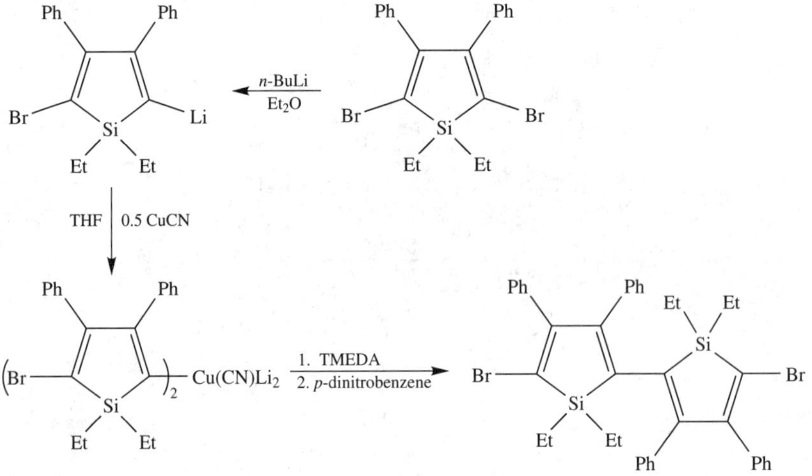

SCHEME 26

Under essentially the same optimum conditions, dibromoquatersilole (**116**) was also obtained in 16% yield as an orange powder by coupling 5-bromo-5′-lithio-2,2′-bisilole, prepared *in situ* from **115** by monolithiation.

X-ray crystal structures of the bisiloles **114** and **115** show highly twisted arrangements between the two silole rings with torsion angles of 62–64°. ^{1}H NMR studies indicate a rapid equilibration between non-coplanar conformers in solution. In the UV-visible spectra, nevertheless, all of the oligosiloles have unusually long absorption maxima, i.e. λ_{max} 398 and 417 nm for **114** and **115**, respectively. Noticeably, there is a large bathochromic shift, more than 90 nm, upon changing from monosilole **117** to bisilole **115**, suggesting the development of a unique π-electronic structure by combination of two silole rings and a crucial role of the silicon atom therein. As pointed out by Tamao and coworkers[59a], the introduction of the bisilole component into a π-conjugated chain would be one of the promising routes to novel π-electronic materials.

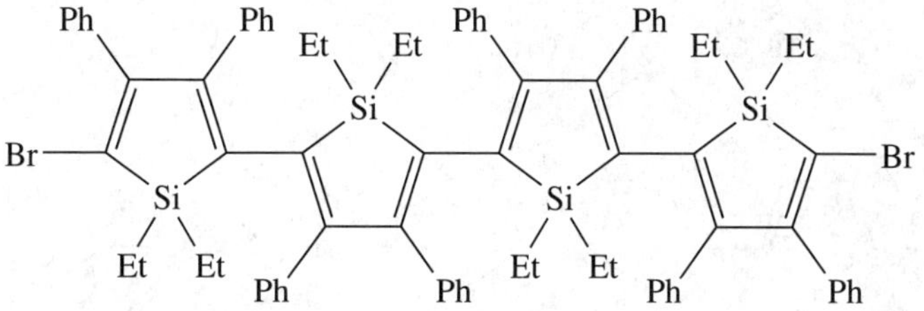

(116)

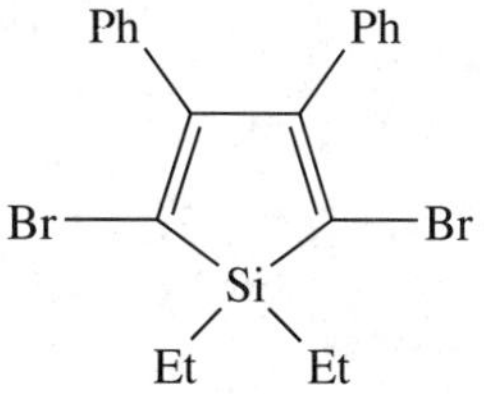

(117)

B. Silole–Thiophene and –Pyrrole Cooligomers and Copolymers

Air-stable silole–thiophene alternating 1 : 1 cooligomers (up to a seven-ring system) have been prepared by nickel(0)-promoted intramolecular cyclization of the corresponding thiophene–(1,6-heptadiyne) alternating cooligomers with hydrodisilanes[54b,173], as illustrated in Scheme 27.

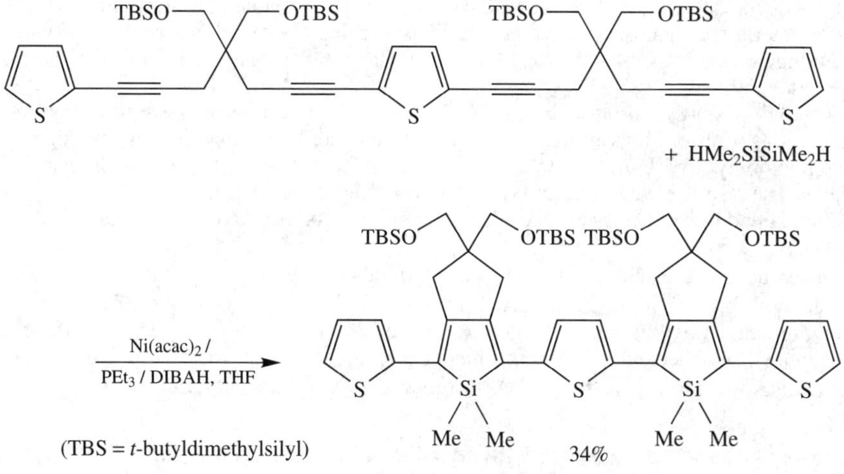

SCHEME 27

Silole–thiophene copolymers, with varying silole:thiophene ratios from 1 : 2 to 1 : 4, have been obtained by a palladium cross-coupling reaction[173], as outlined in Scheme 28.

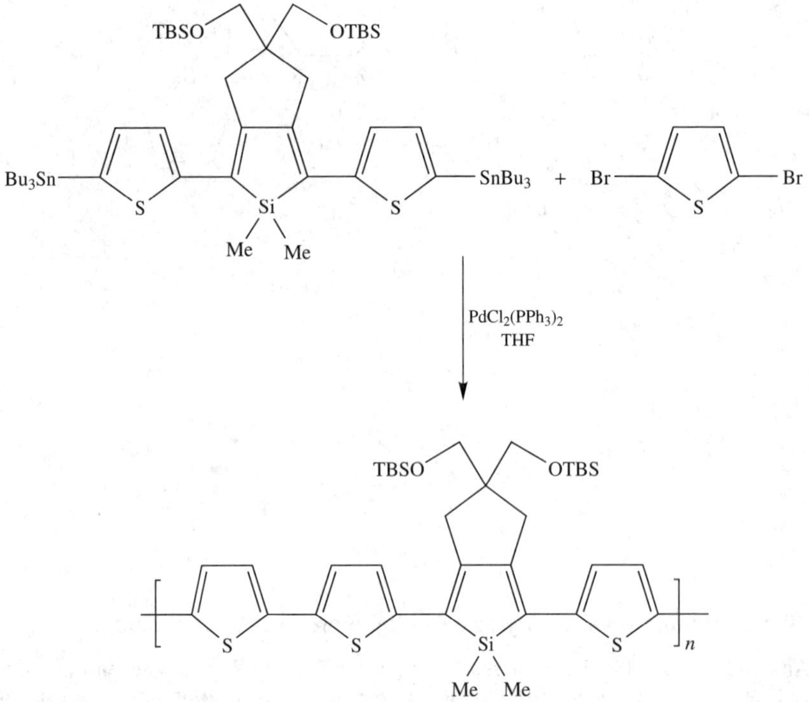

SCHEME 28

All of the series of the silole–thiophene copolymers are air-stable and soluble in common organic solvents such as THF, CH_2Cl_2 and $CHCl_3$. Molecular weight distributions were determined against polystyrene standards; the degrees of polymerization (DP) are in the range of about 20–40; the total number of rings in the main chain varies from *ca* 70 to 200.

The UV-visible absorption data in $CHCl_3$ of the silole–thiophene mixed systems show much longer absorptions in the visible region compared with thiophene homooligomers and homopolymers. Furthermore, significant bathochromic shifts accompany higher silole ratios. The results have been ascribed to the lowering of the LUMO level in the π-systems accompanied by an increase in the silole content, as demonstrated by *ab initio* calculation studies[173].

In contrast, the electrical conductivity tends to become higher with a lower silole content, reaching up to 2.4 S cm^{-1} upon doping with iodine. The bulky (*t*-butyldimethylsiloxy)methyl side groups in the silole unit are oriented out of the π-conjugated polymer plane and hence hinder the interaction between the polymer chains, which is known to be one of the key factors for high conductivity values[174].

C. σ–π Conjugated Polymers with an Alternating Arrangement of a Disilanylene Unit and 3,4-Diethynylene-Substituted Siloles

Polymers in which the regular alternating arrangement of a disilanyl moiety, $-SiR_2-SiR_2-$, and a π-electron system such as phenylene[175], ethynylene[176] and a butenyne group[177] is found in the backbone, are photoreactive and show conducting properties upon doping by exposure to SbF_5 vapour.

Treatment of 3,4-diethynylsilole **29** with 2 equivalents of lithium diisopropylamide, followed by the reaction of the resulting dilithio compound with 1,2-dichlorotetramethyl-, 1,2-dichlorotetraethyl- or 1,2-dibutyl-1,2-dichlorodimethyldisilane produces the cyclic silole derivatives **118**[43] (Scheme 29).

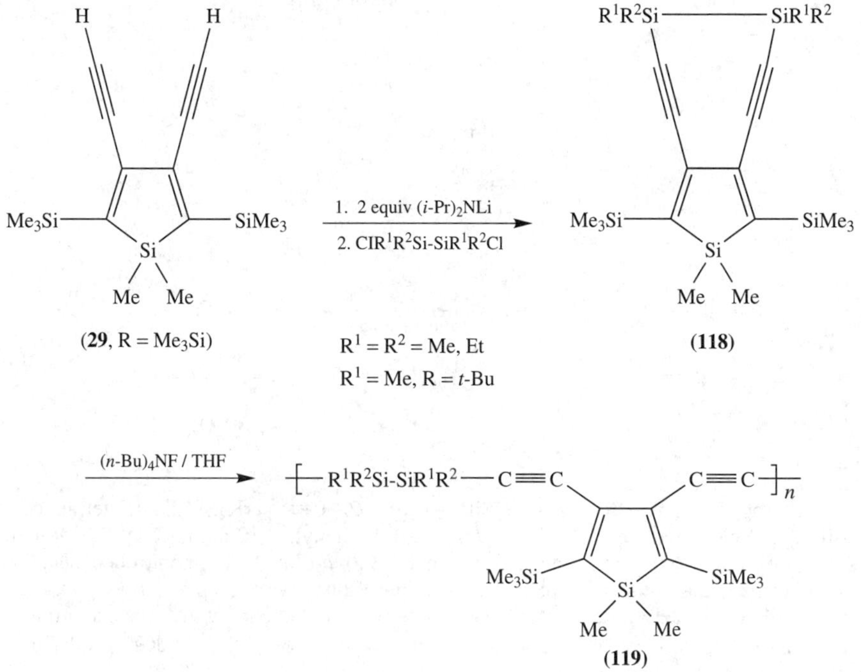

SCHEME 29

The anionic ring-opening polymerization of **118**[178] was carried out with the use of tetrabutylammonium fluoride (2 mol%) as catalyst in THF to give rise to the poly[(disilanylene)ethynylenes] **119** with high molecular weights.

D. Polymers Involving only the Silicon Atom of the Silole Ring in the Main Chain

The reaction of dilithiobutadiyne with 1-chloro-2,5-diphenylsilacyclopentadiene results in the cleavage of both exocyclic bonds, giving rise to the desired polymer[179] (equation 67).

In order to prepare polycarbosilanes containing organometallic fragments which could be precursors of transition-metal-containing ceramics[180], **120** was reacted with dicobalt octacarbonyl to give **121** with approximatively half of the triple bonds complexed with $Co_2(CO)_6$ moieties[181] (equation 68). η^4-Complexation of the silole ring was performed with Fe$_2$(CO)$_9$ and Mo(CO)$_4$(COD)[181].

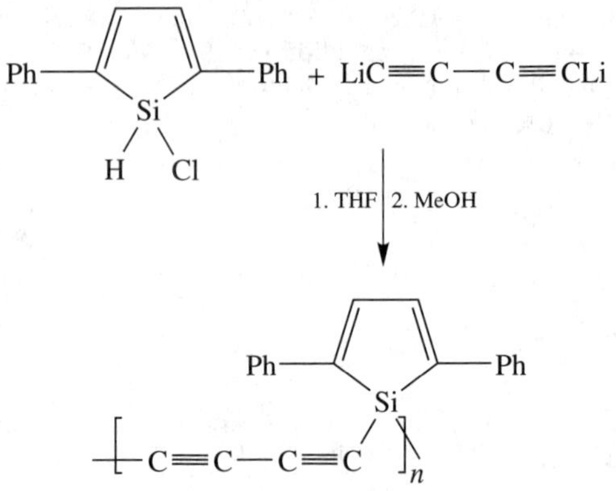

(67)

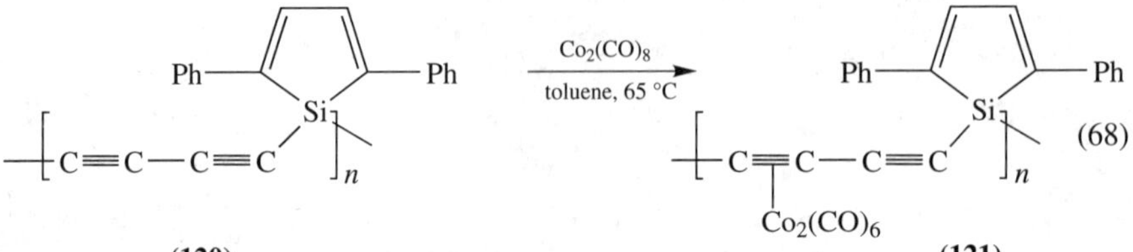

(68)

(120)   (121)

Polycarbosilanes with the structure [SiR$_2$–C≡C–Z–C≡C]$_n$ (R$_2$Si : 2,3,4,5-tetraphenyl-1-sila-2,4-cyclopentadiene; Z: 1,4-benzene, 4,4′-biphenyl, 9,10-anthracene, 2,7-fluorene and 2,6-pyridine, 6,6′-bipyridine, 2,5-thiophene, 2,6-p-dimethylaminonitrobenzene, 2,6-p-nitroaniline, 2,6-p-nitrophenol and 2,7-fluoren-9-one) were prepared by the reaction of 1,1-diethynyl-2,3,4,5-tetraphenyl-1-sila-2,4-cyclopentadiene with the appropriate (hetero)aromatic dibromide or diiodide in the presence of [(PPh$_3$)$_2$PdCl$_2$] and CuI[182] (equation 69).

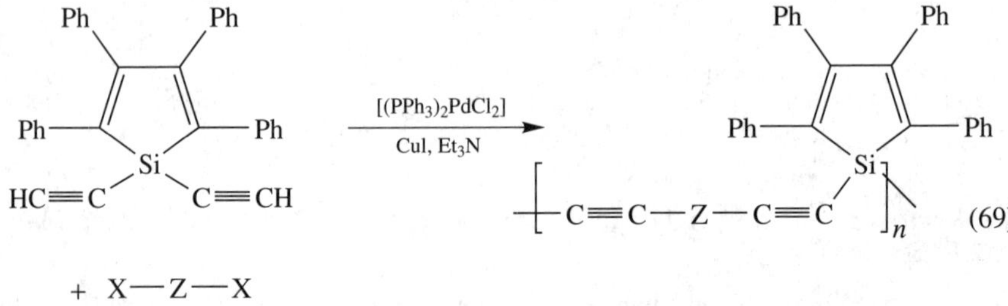

(69)

The oligomer where Z = p-C$_6$H$_4$ reacted with Fe(CO)$_5$ under UV irradiation to give an oligomer containing iron carbonyl units attached to ca 30% of the silole groups.

V. IONIC SPECIES AND COORDINATION COMPOUNDS

Recent years have seen renewed interest in the problem of chemical bonding in metalloles and metallole anions. After the initial report by Joo and coworkers[183] on the generation of the silole dianion $C_4Ph_4Si^{2-}$, a number of investigations concerned with characterizing the structural and chemical properties of π-electron heterole anionic systems, which may possess some degree of aromaticity, have been reported. η^5-Sila- or η^5-germacyclopentadienyl complexes have been vigorously pursued as potentially accessible sila-or germaaromatic derivatives, since transition metal fragments are known to stabilize many reactive species by coordination. Theoretical calculations support these views and provide some insight into the determination of the factors leading to aromaticity in these substances[184].

The earlier work, up to 1990, was the subject of a review by Corriu and coworkers[4]; the intent of this section is to focus upon the latest developments.

A. Group 14 Metallole Anionic Species

1. Silacyclopentadienide and silafluorenide anions

The reaction of 1-chlorosiloles with alkali metals leads to unstable species that react with methyl iodide, ethyl bromide or chlorotrimethylsilane in the manner expected for the 1-silacyclopentadienide anion. Interestingly, X-ray quality crystals were isolated from the reaction with Li in THF and the molecular structure was shown to be that of a [2 + 2] head-to-tail dimer, which is formed by the 1,5-rearrangement of the anion in the silole ring[185] (equation 70).

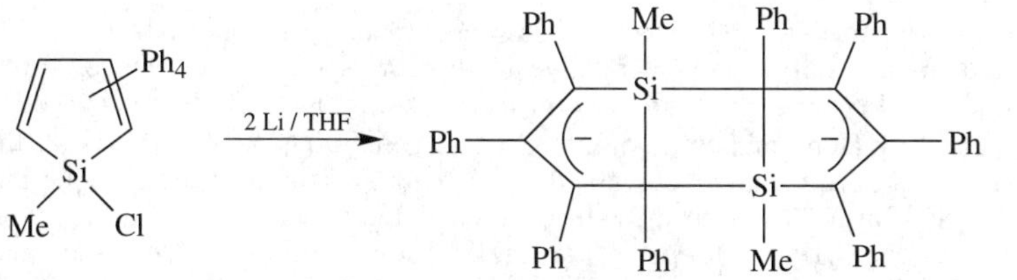

Recently, changing the methyl group bonded to silicon into the sterically demanding *t*-butyl substituent allowed Hong and Boudjouk to generate unambiguously the sodium and lithium derivatives of $[Ph_4C_4Si(Bu-t)]^-$ in THF via reductive cleavage of the Si−Si bond in $C_4Ph_4(Bu-t)Si-Si(t-Bu)C_4Ph_4$ with sodium or lithium, respectively[186] (equation 71).

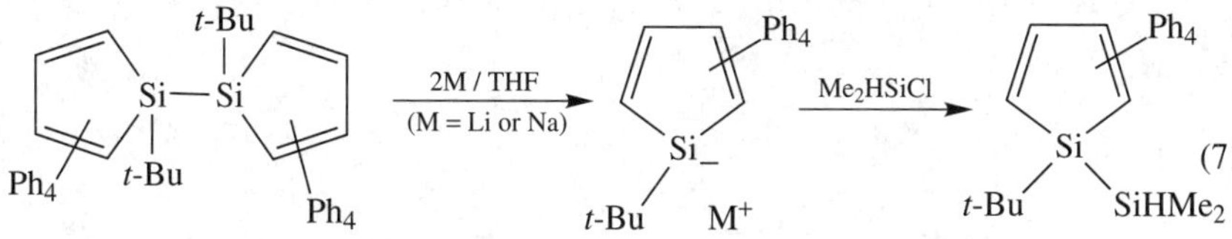

(122)   (123)

Evidence for the formation of **123** was obtained by subsequent quenching with an excess of dimethylchlorosilane to produce the expected $C_4Ph_4Si(Bu-t)$ (SiHMe) in 74% yield. Analysis of the NMR parameters suggests some delocalization of the negative charge into the butadiene moiety. Consistent with sp^2 character of the ring silicon is the downfield shift of ^{29}Si NMR resonances on going from the starting disilane **122** (δ 3.62 ppm) to $[C_4Ph_4Si(Bu-t)]^-$ (δ 26.12 ppm; $\Delta\delta$ 22.50 ppm).

Nucleophilic cleavage of a Si−Si bond with benzylpotassium in the presence of 18-crown-6 provides an alternative approach[187]. This method works for the synthesis of 'free' silolyl anions **125** and **127** from **124** and **126**, respectively (equations 72 and 73).

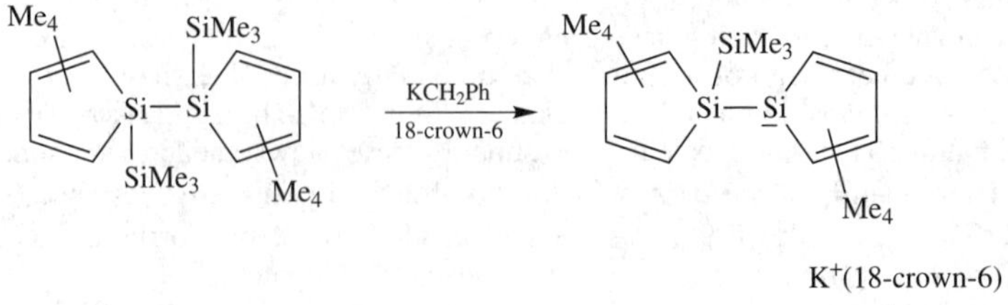

$$(72)$$

(124)   **(125)**

$$(73)$$

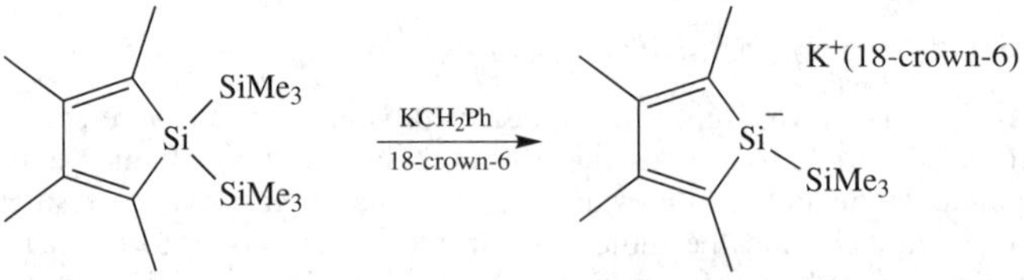

(126)   **(127)**

Interestingly, the ^{29}Si NMR resonances for **125** ($\delta = -53.43$ ppm) and **127** ($\delta = -41.52$ ppm) are shifted upfield relative to the corresponding resonances for the parent siloles (**124**: $\delta = -34.71$ ppm; **126**: $\delta = -34.26$ ppm). This trend is opposite to that reported by Hong and Boudjouk for M[C$_4$Ph$_4$Si(Bu-t)] (M = Li, Na)[186]; their ^{29}Si NMR resonances are shifted downfield relative to that of the starting material C$_4$Ph$_4$(Bu-t)Si—Si(t-Bu)C$_4$Ph$_4$. The difference between these two systems may be attributed to phenyl vs methyl substitution in the rings, and/or to different degrees of interaction between the alkali metal ion and the silolyl ring.

Solutions of 1-lithio-1-methyl-1-silafluorenide anion (**128**) were obtained by sonication of bis(1-methyl-1-silafluorenyl) and lithium in THF[188] (equation 74).

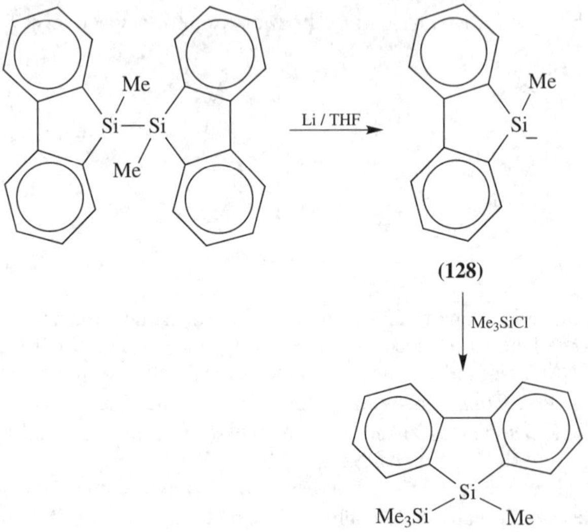

$$(74)$$

(128)

The resulting dark green solution was converted into the expected 1-methyl-1-(trimethylsilyl)-1-silafluorene by reaction with an excess of Me$_3$SiCl. The ^{29}Si NMR chemical shift for **128** ($\delta = -22.09$ ppm) is in the range of aryl-substituted silyllithium compounds, e.g. Ph$_2$MeSi$^-$Li$^+$, which have no delocalization of the negative charge on silicon to the phenyl substituents. Downfield shifts of the ring carbons are also consistent with a localized silyl anion.

Treatment of hydrosilanes with potassium hydride is a known route to silyl anions and has been used to generate silacyclopentadienide anions. Reinvestigation of the reaction of KH with C$_4$Ph$_4$SiMeH has shown that the dominant pathway is simple addition of the hydride anion to the silicon centre to give the pentavalent silicate [Ph$_4$C$_4$SiMeH$_2$]$^-$ K$^+$ (**129**)[189] (equation 75). The Si–H coupling constants in **129** ($J_{Si-H} = 192.5$ and 179.6 Hz) are very close to $J_{Si-H} = 192$–225 Hz in [HSi(OR)$_4$]$^-$ K^{+}[190] and are consistent with a trigonal–bipyramidal structure in which one hydrogen and two carbons of the butadiene moiety occupy the equatorial positions.

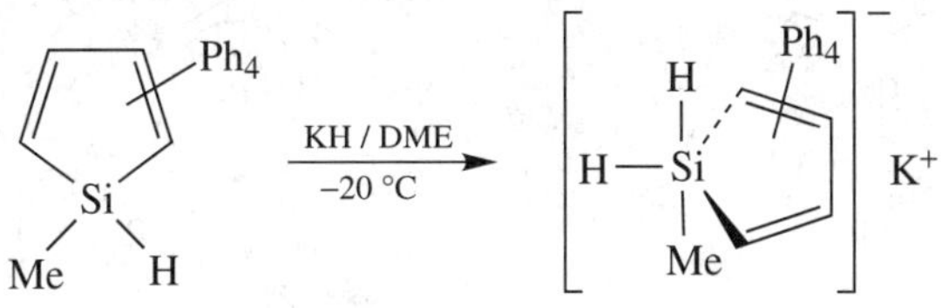

$$\text{(75)}$$

(129)

2. Germacyclopentadienide anions

The C-methylated germacyclopentadiene anion, [C$_4$Me$_4$GePh]$^-$, was reported in 1990 by Dubac and coworkers[191] (equation 76). ^{13}C NMR data were consistent with a substantial localization of negative charge on germanium.

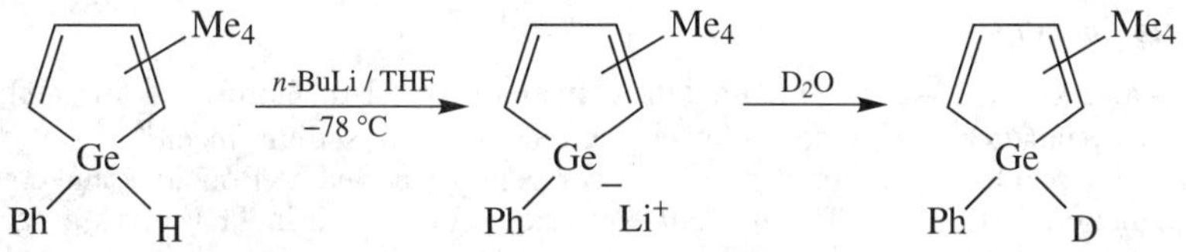

$$\text{(76)}$$

Tilley and coworkers have recently succeeded in the synthesis of isolated germacyclopentadienide derivatives, [C$_4$Me$_4$GeSi(SiMe$_3$)$_3$]$^-$M$^+$, as the alkaline metal salts (M = Li, K) *via* a deprotonation reaction[192], as outlined in equation 77.

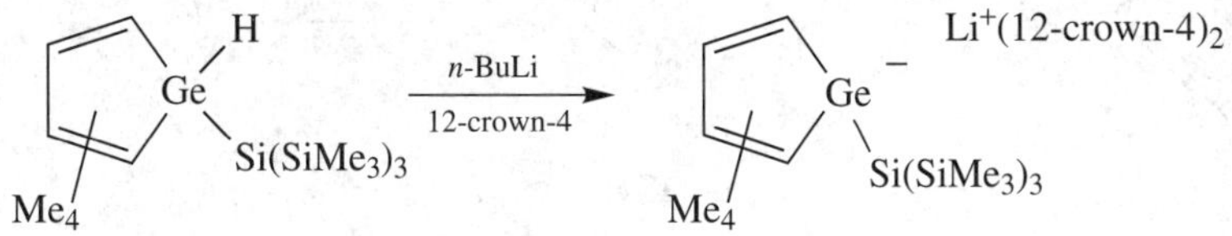

$$\text{(77)}$$

(130)   **(131)**

Use of KN(SiMe$_3$)$_2$ in the presence of 18-crown-6 affords the corresponding potassium salt; the latter was also prepared from C$_4$Me$_4$Ge[Si(SiMe$_3$)$_3$]$_2$ by heterolytic cleavage of one Ge–Si bond with benzylpotassium.

The solid-state structure of [Li(12-crown-4)$_2$] [C$_4$Me$_4$GeSi(SiMe$_3$)$_3$] was determined by X-ray diffraction and shows well-separated cations and anions. A pyramidalization at the germanium centre is evidenced by the angle between the C$_4$Ge plane and the Ge−Si bond of only 100.1° versus 131.0° for **130**. The carbon portion of the ring contains inequivalent C−C distances that reflect isolated single [1.46(6) Å] and double [1.36(6), 1.35(5) Å] bonds and thus has a considerable diene character. The data clearly suggest that **131** gains very little stabilization by π-delocalization of the negative charge.

3. Germa- and stannaindenide anions

1-Lithio-1-metallaindenide anions are readily available from the parent hydrogermanes and stannane[193] (equation 78).

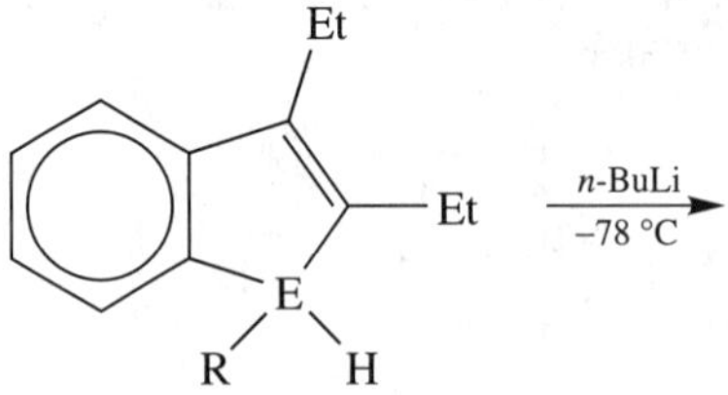

$$\underset{\underset{R\quad H}{E}}{\text{indene-Et}} \xrightarrow[-78\,°C]{n\text{-BuLi}} \underset{\underset{R\quad Li^+}{E^-}}{\text{indenide-Et}} \quad (78)$$

(**132**) E = Ge
(**133**) E = Sn

Chemical characterization of **132** and **133** was obtained by subsequent treatment with PhBr and CH$_3$I or deuteriolysis, respectively. As observed previously for [C$_4$Me$_4$GePh]$^-$ (*vide supra*), ^{13}C NMR data give no significant evidence of aromatic delocalization in both cases.

4. Silole dianions

In 1990, Joo and coworkers reported the conversion of 1,1-dichloro-2,3,4,5-tetraphenyl-1-silacyclopentadiene to the disodium salt by reaction with sodium metal[183].

The corresponding dilithium derivative was recently reported by Boudjouk and coworkers. Reduction of C$_4$Ph$_4$SiCl$_2$ with four equivalents of lithium in THF gives a dark red brown solution of dianion **134** as determined by subsequent quenching with Me$_3$SiCl[194] (equation 79).

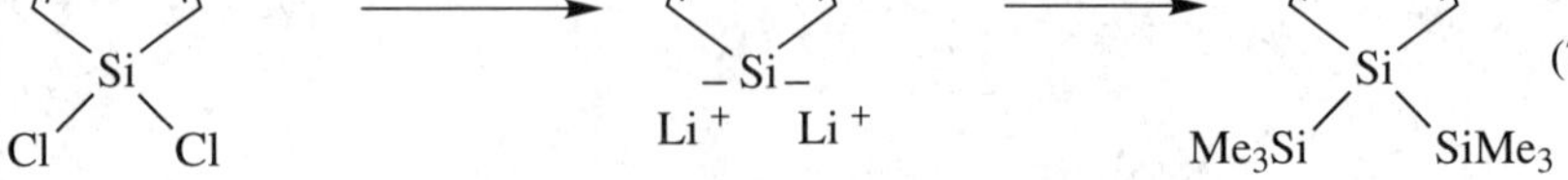

$$\underset{Cl\quad Cl}{\overset{Ph_4}{Si}} \xrightarrow{4\text{Li}\,/\,\text{THF}} \underset{Li^+\quad Li^+}{\overset{Ph_4}{\underline{Si}}} \xrightarrow{Me_3SiCl} \underset{Me_3Si\quad SiMe_3}{\overset{Ph_4}{Si}} \quad (79)$$

(**134**)

Interestingly, the ^{29}Si NMR resonance for this dianion ($\delta = 68.54$ ppm) is shifted downfield relative to the resonance for the parent C$_4$Ph$_4$SiCl$_2$ ($\Delta\delta = 61.74$ ppm). It is further downfield than the ^{29}Si shift of [C$_4$Ph$_4$Si(Bu-t)]$^-$ anion but in the range of the silenes ($\delta = 40$–140 ppm). In addition, Boudjouk and coworkers observed upfield shifts of the C$_\alpha$ and C$_\beta$ atoms in the ring. These ^{13}C and ^{29}Si NMR observations were interpreted in terms of a silole dianion with considerable aromatic delocalization and strong participation of a resonance form with silylene character, **135**[194]. The latter is isolobal with the stable π-electron ring silylene **136**[195].

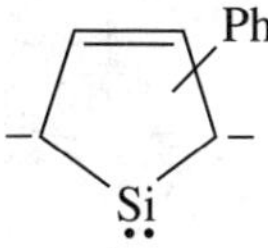

(135) 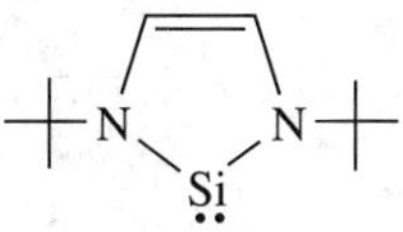   (136)

The dianion **134** was isolated in a crystalline form from THF[196]. The structure, **134a**, contains two different lithium atoms. One lithium is η^5-bonded to the silole ring and also coordinated to two THF molecules; the other is η^1-bonded to the silicon atom and coordinated to three THF molecules. The silole ring is almost planar; the C−C distances within the ring are nearly equal, ranging from 142.6 to 144.8 pm. The long Si−Li$_{(1)}$ distance suggests that the in-plane Si−Li interaction is predominantly ionic. The ^{13}C and ^{29}Si NMR spectra for solid **134a** are also consistent with the X-ray structure.

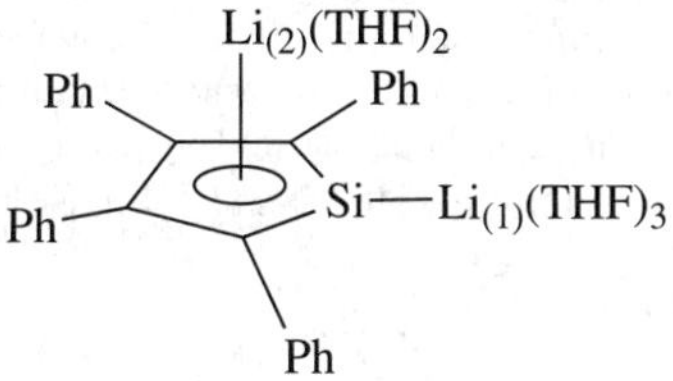

(134a)

When carrying out the sonication of $C_4Ph_4SiCl_2$ with lithium in a 1 : 3 ratio, the 1,1′-disila-2,2′,3,3′,4,4′,5,5′-octaphenylfulvalene dianion **137** is formed as the major product in addition to **134**[194]. Further reaction with lithium produces the silolyl dianion **134** exclusively (equation 80).

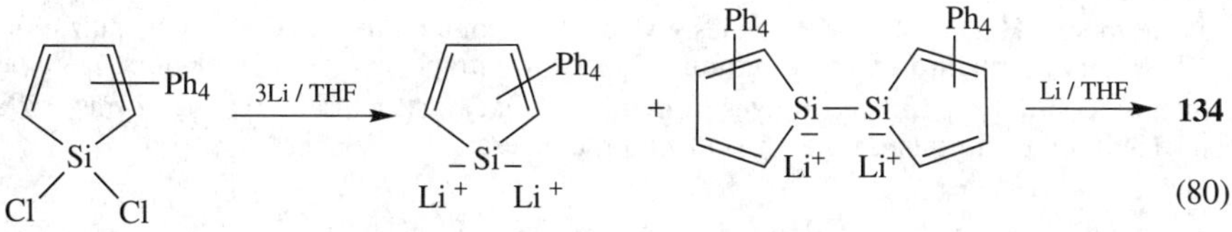

(134)   (137)   (80)

Reaction of $C_4Me_4SiCl_2$ with four equivalents of lithium in THF gives rise to the analogous dianion[40]. As observed for **134**, ^{1}H, ^{13}C and ^{29}Si NMR shifts suggest delocalization of the negative charge into the silole ring.

The solid-state structure of the silolyl dianion is substantially affected by the nature of the substituents attached to the silole ring and of the metal alkali counterions. Stirring $C_4Me_4SiBr_2$ with four equivalents of potassium in the presence of 18-crown-6 gives complex **138** which was isolated in crystalline form[187] (equation 81).

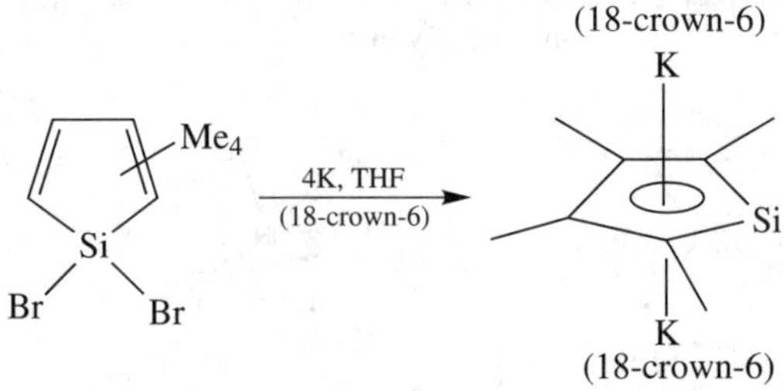

$$\text{Br}_2\text{SiMe}_4\text{(ring)} \xrightarrow[\text{(18-crown-6)}]{\text{4K, THF}} \text{[K(18-crown-6)]}_2[\text{C}_4\text{Me}_4\text{Si}] \quad (138) \quad (81)$$

In the structure of **138**, the electrons within the five-membered ring are highly delocalized, which leads to nearly equal C–C bond lengths. Furthermore, **138** has a $\eta^5-\eta^5$ reverse-sandwich structure. The two potassium cations, each coordinated to a 18-crown-6 molecule, lie above and below the C_4Si ring within bonding distance of all five ring atoms.

5. Germole dianions

The germacyclopentadienide anions present structures consistent with substantial localization of the negative charge on germanium unlike silicon (*vide supra*); germole dianions would be reasonable candidates for stable aromatic germanium derivatives.

Tetraphenylgermole dianions have been studied by Hong and Boudjouk. Sonication of $C_4Ph_4GeCl_2$ with excess of lithium in THF gives $[C_4Ph_4Ge]^{2-}$ $(Li^+)_2$ (**139**) as the major product[197] (equation 82).

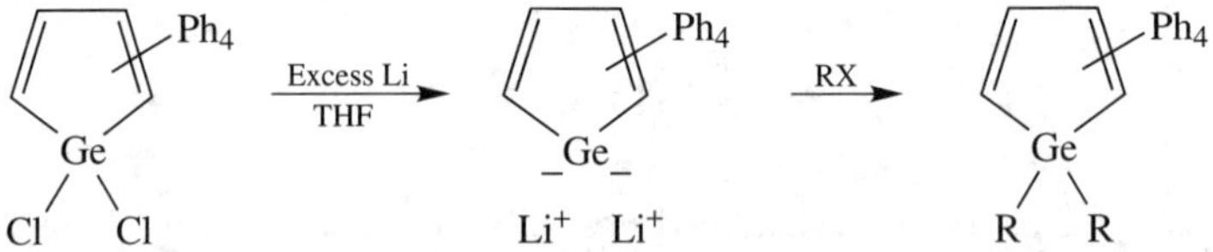

$$\text{C}_4\text{Ph}_4\text{GeCl}_2 \xrightarrow[\text{THF}]{\text{Excess Li}} [\text{C}_4\text{Ph}_4\text{Ge}]^{2-}(\text{Li}^+)_2 \xrightarrow{\text{RX}} \text{C}_4\text{Ph}_4\text{GeR}_2 \quad (82)$$
$$\qquad\qquad\qquad\qquad\qquad\qquad (139)$$

Derivatization of this salt with a series of organic halides, chlorodimethylsilane and chlorotrimethylsilane gives the 1,1-disubstituted-1-germacyclopentadienes in high yields. Furthermore, ^1H and ^{13}C NMR studies give evidence of a high degree of π-delocalization.

The X-ray structure of the dilithium salt of the tetraphenylgermole dianion **139** was recently reported[198]. **139** crystallizes from dioxane in two structurally distinct forms, **139a** and **139b**, depending upon the crystallization temperature.

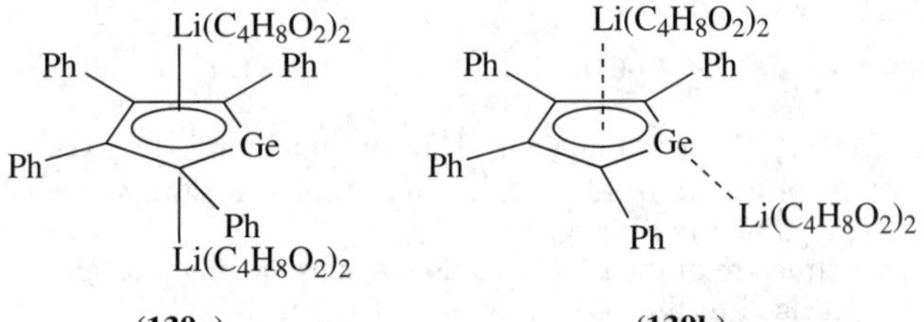

(**139a**)   (**139b**)

The crystals obtained from dioxane at $-20\,°C$, **139a**, have a reverse-sandwich structure with two η^5-coordinated lithium ions lying above and below the C_4Ge ring. Crystals of **139b** obtained at $25\,°C$ show a $\eta^1-\eta^5$ dilithium structure; one lithium atom is η^1-coordinated to the germanium and the other is η^5-coordinated to the ring atoms. In both structures, the electrons are highly delocalized within the ring. Finally, the latter structure **139b** can be related to that published for $Li_2[C_4Ph_4Si]\cdot 5THF$ which also contains η^1- and η^5-coordinated lithium centres (*vide supra*). A significant difference between the structures is that in **134a** the arrangement at silicon is nearly planar, whereas in **139b** the η^1-coordinated lithium is shifted to the hemisphere *anti* to the η^5-coordinated lithium. The angle between the Li—Ge vector and the C—Ge—C plane is $42.9°$.

In addition to the problem of the chemical bonding in metallole anions, these species may offer the promise of rich chemistry and lead to unusual structures. One example is provided by the synthesis and characterization of a novel trisgermole complex of lithium **140**[199] (equation 83).

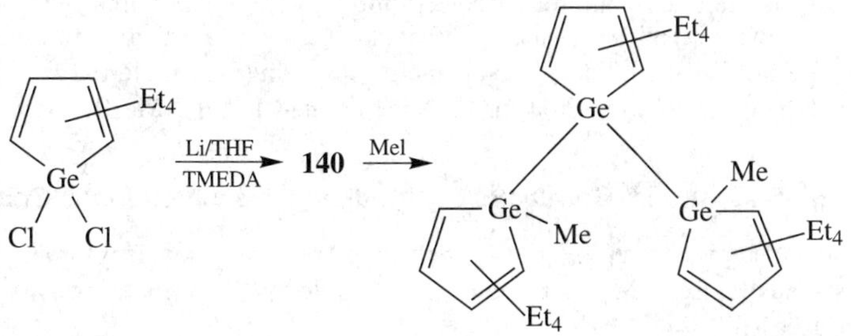

$$\text{(83)}$$

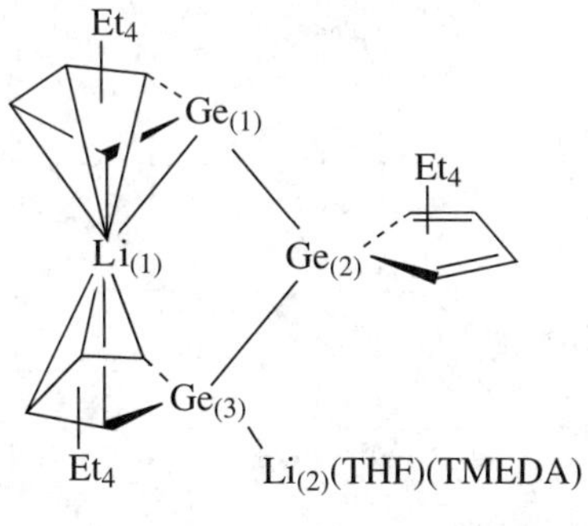

(140)

X-ray analysis of a single crystal of **140** reveals that this species has a sandwich structure in the solid state.

The essential structural features may be analyzed as follows: (1) two differently π-complexed rings, one with only π-bonding to the Li centre, the other ring with π-bonding to the $Li_{(1)}$ centre plus a σ $Ge_{(3)}-Li_{(2)}$ bond [$Ge_{(3)}$ is sp^3 hybridized]; (2) apparent η^5- and η^4-bonding to $Li_{(1)}$; (3) the least-squares planes for the two terminal rings form an angle of $10.7°$; (4) $Li_{(1)}$ is positioned close to the centre of the two planes defined by $C_4Ge_{(1)}$ and $C_4Ge_{(3)}$, similar to the situation in lithocene or ferrocenophane derivatives.

In summary, the anions of group 14 metalloles (C_4E rings) have either localized non-aromatic or delocalized aromatic structures, depending of the metal E, Si vs Ge and Sn, and on the substituents, methyl vs phenyl group. As pointed out by West and coworkers[198], structural studies of metallole anions and dianions of this kind with different substituents will be of value. Additionally, it should be noted that experimentally observed structures in the solid state for silole and germole dianions are greatly influenced by the nature of the alkali metal counterion.

Recent theoretical calculations support these views and provide new insights into the determination factors leading to aromaticity in these species[184a,b]. For example, an approach based on structural, energetic and magnetic criteria demonstrates that lithium silolide, C_4H_4SiHLi, exhibits significant aromatic character and 80% of the stabilization energy of $Li^+C_5H_5^-$; in addition, Li^+ η^5-coordination in the ground state increases the aromaticity over that in free $[C_4H_4SiH]^-$ strongly. The high degree of aromatic character in the dianionic siloles $C_4H_4SiM_2$ (M = Li, Na, K) and $[C_4H_4SiLi]^-$ is also evident from structural, energetic and magnetic aspects. These criteria for aromaticity reveal that $[\eta^5\text{-}CHSiLi]^-$ is more aromatic than isoelectronic phosphole and thiophene systems and even approach the aromaticity of the cyclopentadiene anion. Furthermore, inverse sandwich structure with η^5-coordinated alkali metal ions would be preferred by $C_4H_4SiM_2$ species; such a structure is reported for $[K\,([18]\text{ crown-}6)^+]_2\,[C_4Me_4Si^{2-}]^{187}$.

B. Stable η^5-Sila- and η^5-Germacyclopentadienyl Transition Metal Complexes

Numerous attempts to prepare η^5-sila- and η^5-germacyclopentadienyl transition metal complexes, usually from the corresponding η^4-silole or η^4-germole derivatives, have proven unsuccessful[4].

In 1993, Tilley and coworkers reported the first isolation and complete characterization of a stable η^5-germacyclopentadienyl ruthenium complex (**141**)[200], as indicated in equation 84.

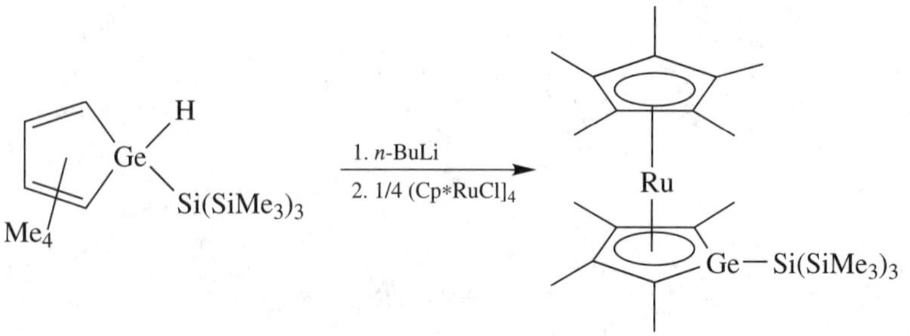

(84)

(**141**)

The molecular structure of **141** consists of two planar five-membered rings bound in a sandwich fashion to the ruthenium atom. The least-squares planes for the two five-membered rings form an angle of 8.6°, and the ruthenium atom is symmetrically positioned between the two planes. The germanium atom deviates by only 0.02 Å from the C_4Ge ring's least-squares plane; the summation of the angles about the Ge atom (358.1°) further reflects sp^2 hybridization. Thus, it appears that the η^5-$C_4Me_4GeSi(SiMe_3)_3$ and Cp^* ligands are bonded similarly to Ru.

An irreversible oxidation wave was observed for **141** at $E_{1/2} = 0.250$ V (vs SCE) by cyclic voltametry. For comparison, decamethylruthenocene displays a reversible oxidation wave at 0.42 V (vs SCE)[201]. These data therefore suggest that the η^5-C$_4$Me$_4$SiSi(SiMe$_3$)$_3$ ligand is more electron-donating than η^5-C$_5$Me$_5$.

The preparation of the analogous transition metal η^5-silacyclopentadienyl complex requires a different synthetic approach[202], as shown in Scheme 30. The intermediate cationic species **143** has the general appearance of a protonated metallocene. The five-membered C$_4$Si ring of the silacyclopentadienyl ligand is planar; the summation of angles at the ring silicon atom (355.1°) reflect the considerable sp^2 character for that atom. An additional indication of electron delocalization in the silacyclopentadienyl ligand is the small difference between the C−C bond distances. The hydride ligand is nearer to the silacyclopentadienyl ring. The Ru−H−Si interaction is similar to Mn−H−Si arrangements in Cp(CO)(L)Mn(η^2-HSiR) complexes[203]. The silyl group bends away from the ruthenium centre; the angle between the Si−Si bond and the C$_4$Si least-squares plane is 19°.

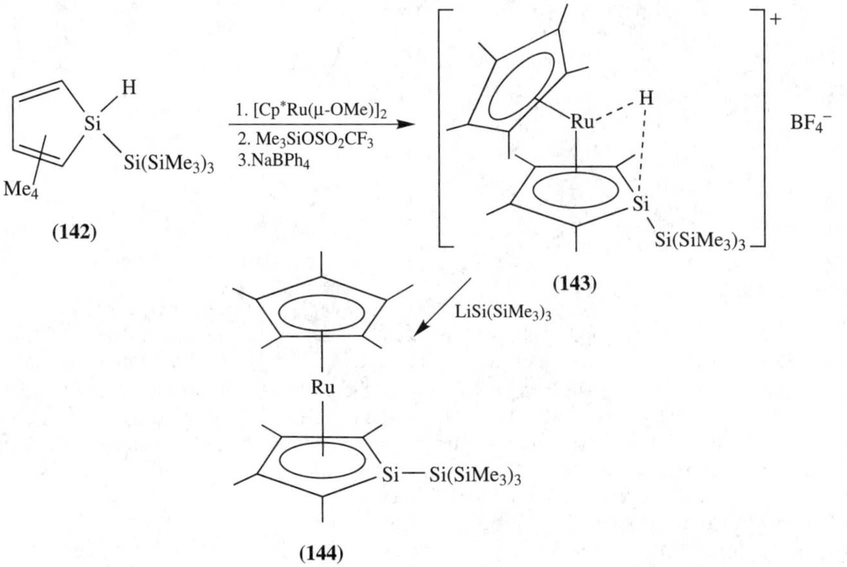

SCHEME 30

The deprotonation of **143** was only successful with (THF)$_3$LiSi(SiMe$_3$)$_3$ and gave rise to **144** in 50% yield. The ^{1}H and ^{13}C NMR spectra of **144** resemble those of **141**. Also consistent with increased sp^2 character for the ring silicon is the downfield progression of ^{29}Si NMR shifts on going from **142** ($\delta = -32.68$ ppm, $J_{Si-H} = 181$ Hz) to **143** ($\delta = -27.14$ ppm, $J_{Si-H} = 41$ Hz) to **144** ($\delta = -7.35$ ppm). The latter physical data suggest some delocalization of electron density in the SiC$_4$ ring of **143**, and considerably more for the related ring in **144**.

VI. ADDENDUM

Section II.A.2

Michl, Nefedov and coworkers[204] reported that UV irradiation of matrix-isolated 1,1-diazido-1-germacyclopent-3-ene (the germanium analog of **13**) produced 1H-germole and

1-germacyclopent-3-ene-1,1-diyl (the germanium analogs of **2** and **9**) as major products along with 2H-germole and 3H-germole (the germanium analogs of **11** and **14**) as minor products. As in the case of silicon species (Scheme 4), reversible photoconversion of these tautomeric germoles has been observed. Similar experiments were carried out on deuterated analogs. IR and UV-VIS spectra of these germoles are interpreted and compared with those of silole isomers.

Section III.B.1.b

Stannoles bearing dialkylboryl groups in the 3-position (**145**) react with phosphaalkynes $R^4C\equiv P$ ($R^4 = t$-Bu, CH_2Bu-t) by [4 + 2] cycloaddition (isomeric adducts were not detected) and elimination of stannylene to give phosphabenzenes (**146** and **147**) in high yield[205].

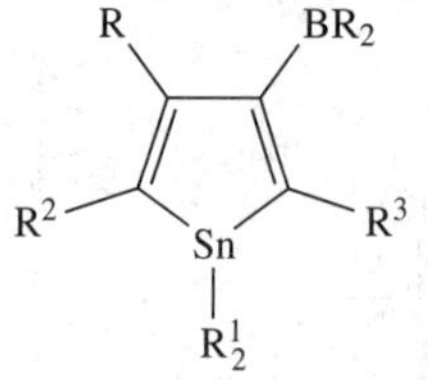

(145) 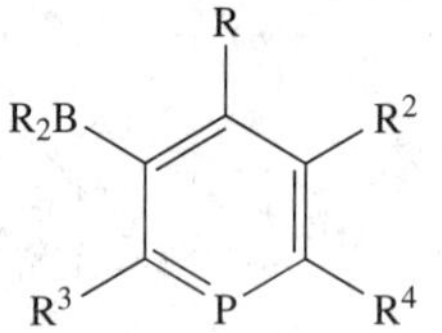 (146) 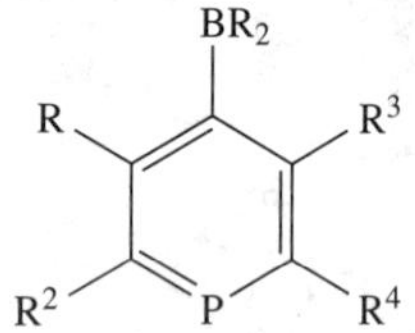 (147)

Section IV.A

A recent report from Yamaguchi and Tamao[206] describes semiempirical and *ab initio* calculations to elucidate the electronic structure of bisilole in comparison with that of bicyclopentadiene[207]. The silole ring, as well as the extended π-conjugated bisilole system, has an unusually low-lying LUMO level arising from $\sigma^*-\pi^*$ conjugation which is unexpectedly enhanced by molecular distortions in the case of bisilole. As a support of the calculation data, redox potentials of some silole derivatives and their carbon analogs have been determined by cyclic voltammetry. Siloles and polysiloles would thus have quite different electronic structures and properties from those of the conventional thiophene or pyrrole analogs. Noteworthy is the performance of the silole ring as a core component for efficient electron transporting materials[208].

Section IV.B

Silole–pyrrole cooligomers (containing up to nine rings) have been synthesized[209]. As expected, the combination of a π-electron-deficient silole[206] with a π-electron-rich pyrrole causes long UV-VIS absorption maxima in spite of twisted structures. As indicated by *ab initio* calculations, the LUMO of the parent pyrrole–silole–pyrrole is almost localized on the silole ring and lies at a much lower level than that of the parent terpyrrole; as a consequence, the HOMO–LUMO energy gap is about 1.3 eV smaller, strongly suggesting promising development in the area of novel low band π-conjugated polymers.

Section V.A.

A significant contribution by Tilley and coworkers, in addition to material cited earlier[187,192], describes the synthesis, structure and electronic properties of various silole and germole anions and dianions, in the $[C_4(alkyl)_4E-R]^-$ and $[C_4(alkyl)_4E]^{2-}$ series (E = Si, Ge; alkyl = Me, Et), respectively[210]. A brief discussion follows.

Alkali metal reduction of dihalide precursors is shown to be a valuable route to various silole and germole dianions in solution. Crystal structure determinations for $[K(18\text{-crown-}6)^+]_2(C_4Me_4Si^{2-})$ and the dimer $[K_4(18\text{-crown-}6)_3][C_4Me_4Ge]_2$ are consistent with the presence of delocalized π-systems and with η^5-bonding modes for all of the potassium cations.

Deprotonation of hydrogermoles, reductive cleavage of the Ge—Ge bond or nucleophilic cleavage of a E—SiMe$_3$ bond (E = Si or Ge) can be employed for the generation of silolyl and germolyl anions. They exhibit ^{13}C and ^{29}Si NMR chemical shifts consistent with significant localization of the negative charge on the group 14 element and a non-aromatic structure; the countercation seems to have little effect on the electronic structure. Crystallographic studies reveal highly pyramidalized Ge and Si centres and pronounced double bond localization in the carbon portion of the ring. Nevertheless, the low barriers to inversion observed by variable-temperature ^{1}H NMR spectroscopy for K[C$_4$Et$_4$E(SiMe$_3$)] and [Li(12-crown-4)$_2$][C$_4$Et$_4$E(SiMe$_3$)] ion pairs (E = Si, Ge), i.e. approximatively 41.8 kJ mol^{-1} for germolyl anions and 29–33 kJ mol^{-1} for silolyl anions, suggest a stabilization of the transition state by delocalization of π-electron density in the ring. The difference in the barriers apparently reflects a slightly greater degree of delocalization in the silicon compounds; they also seem to be related to the nature of the countercation.

VII. CONCLUSION

Studies in the chemistry of group 14 metalloles over the past few years have been numerous. Mention should be made of the synthesis of the C-unsubstituted 1*H*-silole; the structures of the product itself, as well as of its tautomeric forms and dimer, have been established. Progress has also been made in developing new synthetic routes. By way of a transmetallation reaction from zirconacyclopentadiene, the synthesis of group 14 heterocyclopentadiene derivatives was considerably facilitated. On the other hand, though the method is limited to 3,4-diphenylsiloles, a further noticeable contribution was a general and versatile synthesis of the corresponding 2,5-difunctional derivatives.

Functional siloles have found applications in the elaboration of silole-containing π-conjugated systems. Because of its unique electronic properties, the introduction of a silole component or a silole cooligomer into an unsaturated chain should be a promising route to novel π-electronic materials and, in this connection, the preparation of new heterocyclopentadiene monomeric precursors remains a current challenge.

Stable group 14 heterole anionic or dianionic species are no longer curiosities. Recent work has focused on the structural behaviour, spectroscopic features and bonding patterns of heterole anionic or dianionic species so that we can use the data obtained so far as guidelines for further work. As far as the reactivity of the latter derivatives is concerned, much effort will be directed to the structure of the heterolyl salts and its dependence on the counterion.

Finally, heterocyclopentadienes or -dienides have been shown to be suitable as complex ligands. They can bind in a π-fashion to a metal complex fragment like the classical cyclopentadienyl ligand. The synthesis of **141** and **144** deserves particular attention in the light of vain attempts, until recently, to detect unambiguously a π-coordinated germa- and silacyclopentadienide in a complex. It is to be hoped that the remaining open questions pertaining to the chemical reactivity of such species will be answered by investigations in the near future.

VIII. ACKNOWLEDGEMENTS

We are grateful to Mrs S. Julia, Mrs M. Maris, Prof. M. A. Brook (McMaster University, Hamilton), Dr W. Douglas (Université de Montpellier) and Prof. B. Gautheron (Université de Bourgogne, Dijon) for their assistance in the production of the manuscript.

IX. REFERENCES

1. (a) P. J. Fagan, E. G. Burns and J. C. Calabrese, *J. Am. Chem. Soc.*, **110**, 2979 (1988).
   (b) P. J. Fagan, W. A. Nugent and J. C. Calabrese, *J. Am. Chem. Soc.*, **116**, 1880 (1994).
2. (a) F. Mathey, *Chem. Rev.*, **88**, 429 (1988).
   (b) F. Zurmühlen and M. Regitz, *J. Organomet. Chem.*, **332**, C1 (1987).
   (c) F. Laporte, F. Mercier, L. Ricard and F. Mathey, *Bull. Soc. Chim. Fr.*, **130**, 843 (1993).
   (d) S. M. Bachrach, *J. Org. Chem.*, **58**, 5414 (1993).
3. J. Dubac, A. Laporterie and G. Manuel, *Chem. Rev.*, **90**, 215 (1990).
4. E. Colomer, R. J. P. Corriu and M. Lheureux, *Chem. Rev.*, **90**, 265 (1990).
5. T. J. Barton and G. T. Burns, *J. Organomet. Chem.*, **179**, C17 (1979).
6. J. P. Béteille, M. P. Clarke, I. M. T. Davidson and J. Dubac, *Organometallics*, **8**, 1292 (1989).
7. G. Raabe and J. Michl, *Chem. Rev.*, **85**, 419 (1985).
8. J. Dubac and A. Laporterie, *Chem. Rev.*, **87**, 319 (1987).
9. (a) T. J. Barton, S. A. Burns, I. M. T. Davidson, S. Ijadi-Maghsoodi and I. T. Wood, *J. Am. Chem. Soc.*, **106**, 6367 (1984).
   (b) N. Auner, I. M. T. Davidson and S. Ijadi-Maghsoodi, *Organometallics*, **4**, 2210 (1985).
10. (a) A. Laporterie, P. Mazerolles, J. Dubac and H. Iloughmane, *J. Organomet. Chem.*, **206**, C25 (1981).
    (b) A. Laporterie, J. Dubac, P. Mazerolles and H. Iloughmane, *J. Organomet. Chem.*, **216**, 321 (1981).
11. G. T. Burns and T. J. Barton, *J. Organomet. Chem.*, **209**, C25 (1981).
12. H. Iloughmane, *Thesis*, Université Paul Sabatier, Toulouse, No. 1247 (1986).
13. A. Laporterie, H. Iloughmane and J. Dubac, *J. Organomet. Chem.*, **244**, C12 (1983).
14. J. Dubac, A. Laporterie and H. Iloughmane, *J. Organomet. Chem.*, **293**, 295 (1985).
15. M. Kako, S. Oba and Y. Nakadaira, *J. Organomet. Chem.*, **461**, 173 (1993).
16. (a) P. P. Gaspar, R.-J. Hwang and W. C. Eckelman, *J. Chem. Soc., Chem. Commun.*, 242 (1974).
    (b) R.-J. Hwang and P. P. Gaspar, *J. Am. Chem. Soc.*, **100**, 6626 (1978).
    (c) P. P. Gaspar, Y.-S. Chen, A. P. Helfer, S. Konieczny, E. C.-L. Ma and S.-H. Mo, *J. Am. Chem. Soc.*, **103**, 7344 (1981).
17. E. E. Siefert, K. L. Loh, R. A. Ferrieri and Y. N. Tang, *J. Am. Chem. Soc.*, **102**, 2285 (1980).
18. B. H. Boo and P. P. Gaspar, *Organometallics*, **5**, 698 (1986).
19. D. Lei, Y.-S. Chen, B. H. Boo, J. Frueh, D. L. Svoboda and P. P. Gaspar, *Organometallics*, **11**, 559 (1992).
20. (a) V. N. Khabashesku, V. Balaji, S. E. Boganov, S. A. Bashkirova, P. M. Matveichev, E. A. Chernyshev, O. M. Nefedov and J. Michl, *Mendeleev Commun.*, 38 (1992).
    (b) V. N. Khabashesku, V. Balaji, S. E. Boganov, O. M. Nefedov and J. Michl, *J. Am. Chem. Soc.*, **116**, 320 (1994).
    (c) V. N. Khabashesku, S. E. Boganov, V. I. Faustov, A. Gömöry, I. Besenyei, J. Tamas and O. M. Nefedov, *High Temp. Mater. Sci.*, **33**, 125 (1995).
21. A. Laporterie, G. Manuel, J. Dubac, P. Mazerolles and H. Iloughmane, *J. Organomet. Chem.*, **210**, C33 (1981).
22. C. Guimon, C. Pfister-Guillouzo, J. Dubac, A. Laporterie, G. Manuel and H. Iloughmane, *Organometallics*, **4**, 636 (1985).
23. (a) Reference 3, pp. 218–224 and references cited therein.
    (b) Reference 3, pp. 221–224 and references cited therein.
24. (a) T. Tsumuraya and W. Ando, *Organometallics*, **9**, 869 (1990).
    (b) T. Tsumuraya, Y. Kabe and W. Ando, *J. Organomet. Chem.*, **482**, 131 (1994).
25. W. P. Neumann, *Chem. Rev.*, **91**, 311 (1991).
26. (a) H. Brauer and W. P. Neumann, *Synlett*, 431 (1991).
    (b) G. Billeh, H. Brauer, W. P. Neumann and M. Weisbeck, *Organometallics*, **11**, 2069 (1992).
27. J. P. Béteille, A. Laporterie and J. Dubac, unpublished work.
28. A. J. Ashe III and S. Mahmoud, *Organometallics*, **7**, 1878 (1988).

29. (a) C. Pluta and K. R. Pörschke, *J. Organomet. Chem.*, **453**, C11 (1993).
    (b) J. Krause, C. Pluta, K. R. Pörschke and R. Goddard, *J. Chem. Soc., Chem. Commun.*, 1254 (1993).
30. Reference 3, pp. 218–231 and pp. 235–236 and references cited therein.
31. F. C. Leavitt, T. A. Manuel and F. Johnson, *J. Am. Chem. Soc.*, **81**, 3163 (1959).
32. K. W. Hübel and E. H. Braye, U. S. Patent 3 426 052 (1969); *Chem. Abstr.*, **70**, 106663 (1969).
33. J. G. Zavistoski and J. J. Zuckerman, *J. Org. Chem.*, **34**, 4197 (1969).
34. W. H. Atwell, D. R. Weyenberg and H. Gilman, *J. Org. Chem.*, **32**, 885 (1967).
35. (a) V. Hagen and K. Rühlman, *Z. Chem.*, **8**, 114 (1968).
    (b) R. Müller, *Z. Chem.*, **8**, 262 (1968).
    (c) T. J. Barton, W. D. Wulff, E. V. Arnold and J. Clardy, *J. Am. Chem. Soc.*, **101**, 2733 (1979).
36. T. J. Barton, A. J. Nelson and J. Clardy, *J. Org. Chem.*, **37**, 895 (1972).
37. (a) J. Y. Corey, C. Guérin, B. Henner, B. Kolani, W. W. C. Wong Chi Man and R. J. P. Corriu, *C. R. Séances Acad. Sci., Sér. 2*, **300**, 331 (1985).
    (b) F. Carré, E. Colomer, J. Y. Corey, R. J. P. Corriu, C. Guérin, B. J. L. Henner, B. Kolani and W. W. C. Wong Chi Man, *Organometallics*, **5**, 910 (1986).
38. A. Boudin, G. Cerveau, C. Chuit, R. J. P. Corriu and C. Reye, *Bull. Chem. Soc. Jpn.*, **61**, 101 (1988).
39. H. Sakurai, A. Nakamura and Y. Nakadaira, *Organometallics*, **2**, 1814 (1983).
40. U. Bankwitz, H. Sohn, D. R. Powell and R. West, *J. Organomet. Chem.*, **499**, C7 (1995).
41. (a) H. Okinoshima, K. Yamamoto and M. Kumada, *J. Am. Chem. Soc.*, **94**, 9263 (1972).
    (b) H. Okinoshima, K. Yamamoto and M. Kumada, *J. Organomet. Chem.*, **86**, C27 (1975).
42. M. Ishikawa, *Organomet. Synth.*, **4**, 527 (1988).
43. E. Toyoda, A. Kunai and M. Ishikawa, *Organometallics*, **14**, 1089 (1995).
44. K. Ikenaga, K. Hiramatsu, N. Nasaka and S. Matsumoto, *J. Org. Chem.*, **58**, 5045 (1993).
45. (a) D. Seyferth, D. P. Duncan and S. C. Vick, *J. Organomet. Chem.*, **125**, C5 (1977).
    (b) D. Seyferth, S. C. Vick, M. L. Shannon, T. F. O. Lim and D. P. Duncan, *J. Organomet. Chem.*, **135**, C37 (1977).
    (c) D. Seyferth, S. C. Vick and M. L. Shannon, *Organometallics*, **3**, 1897 (1984).
    (d) D. Seyferth, S. C. Vick and M. L. Shannon and T. F. O. Lim, *Organometallics*, **4**, 57 (1985).
46. H. Sakurai, Y. Kamiyama and Y. Nakadaira, *J. Am. Chem. Soc.*, **99**, 3879 (1977).
47. J. Belznerand and H. Ihmels, *Tetrahedron Lett.*, **34**, 6541 (1993).
48. (a) M. Ishikawa, H. Sugisawa, O. Harata and M. Kumada, *J. Organomet. Chem.*, **217**, 43 (1981).
    (b) M. Ishikawa, S. Matsuzawa, T. Higuchi, S. Kamitori and K. Hirotsu, *Organometallics*, **4**, 2040 (1985).
    (c) M. Ishikawa, S. Matsuzawa, T. Higuchi, S. Kamitori and K. Hirotsu, *Organometallics*, **3**, 1930 (1984).
49. M. Ishikawa, J. Ohshita, Y. Ito and J. Iyoda, *J. Am. Chem. Soc.*, **108**, 7471 (1986).
50. (a) M. Ishikawa, T. Fuchikami and M. Kumada, *J. Organomet. Chem.*, **142**, C45 (1977).
    (b) M. Ishikawa, H. Sugisawa, M. Kumada, H. Kawakami and T. Yamabe, *Organometallics*, **2**, 974 (1983).
51. A. Shäfer, M. Weidenbruch and S. Pohl, *J. Organomet. Chem.*, **282**, 305 (1985).
52. H. Gilman, S. G. Cottis and W. H. Atwell, *J. Am. Chem. Soc.*, **86**, 1596 and 5584 (1964).
53. Reference 3, pp. 240–245 and references cited therein.
54. (a) K. Tamao, K. Kobayashi and Y. Ito, *Synlett*, 539 (1992).
    (b) K. Tamao, S. Yamaguchi, M. Shiozaki, Y. Nakagawa and Y. Ito, *J. Am. Chem. Soc.* **114**, 5867 (1992).
55. N. Chatani, T. Morimoto, T. Muto and S. Murai, *J. Organomet. Chem.*, **473**, 335 (1974).
56. (a) B. Wrackmeyer, *J. Chem. Soc., Chem. Commun.*, 397 (1986).
    (b) B. Wrackmeyer, *J. Organomet. Chem.*, **310**, 151 (1986).
    (c) R. Köster, G. Seidel, J. Süb and B. Wrackmeyer, *Chem. Ber.*, **126**, 1107 (1993).
    (d) B. Wrackmeyer, G. Kehr and J. Süb, *Chem. Ber.*, **126**, 2221 (1993).
57. (a) L. Killian and B. Wrackmeyer, *J. Organomet. Chem.*, **132**, 213 (1977).
    (b) B. Wrackmeyer, G. Kehr and R. Boese, *Chem. Ber.*, **125**, 643 (1992).
    (c) B. Wrackmeyer, S. Kundler and R. Boese, *Chem. Ber.*, **126**, 1361 (1993).
    (d) B. Wrackmeyer, S. Kundler and A. Ariza-Castolo, *Phosporus, Sulfur and Silicon*, **91**, 229 (1994).
    (e) B. Wrackmeyer, S. Kundler, W. Milius and R. Boese, *Chem. Ber.*, **127**, 333 (1994).

(f) B. Wrackmeyer, G. Kehr and S. Ali, *Inorg. Chim. Acta*, **216**, 51 (1994).
(g) B. Wrackmeyer, G. Kehr and R. Boese, *Angew. Chem., Int. Ed. Engl.*, **30**, 1370 (1991) and references cited therein.
58. (a) B. Wrackmeyer and K. Horchler, *J. Organomet. Chem.*, **399**, 1 (1990).
(b) B. Wrackmeyer, K. Horchler and R. Boese, *Angew. Chem., Int. Ed. Engl.*, **28**, 1500 (1989).
59. (a) K. Tamao, S. Yamaguchi and M. Shiro, *J. Am. Chem. Soc.*, **116**, 11715 (1994).
(b) K. Tamao and S. Yamaguchi, *Jpn. Kokai Tokkyo Koho* JP 07,179,477 (1995); *Chem. Abstr.*, **124**, 56295 (1995).
60. (a) A. J. Ashe III and T. R. Diephouse, *J. Organomet. Chem.*, **202**, C95 (1980).
(b) A. J. Ashe III and F. J. Drone, *Organometallics*, **3**, 495 (1984).
61. (a) T. J. Barton and E. E. Gottsman, *Synth. React. Inorg. Met.-Org. Chem.*, **3**, 201 (1973).
(b) M. Schriewer and W. P. Neumann, *J. Am. Chem. Soc.*, **105**, 897 (1983).
62. (a) T. J. Barton, in *Comprehensive Organometallic Chemistry* (Eds. G. Wilkinson, F. G. A. Stone and E. W. Abel), Vol. 2, Pergamon Press, Oxford, 1982, p. 252.
(b) J. Dubac and A. Laporterie unpublished results.
63. A. Laporterie, *Thesis*, Université Paul Sabatier, Toulouse, n° 703 (1976).
64. J. C. Brunet and N. Demey, *Ann. Chim.*, **8**, 123 (1973).
65. Reference 3, pp. 217, 225.
66. W. C. Joo, H. S. Hwang and J. H. Hong, *Bull. Korean Chem. Soc.*, **6**, 348 (1985).
67. (a) A. Laporterie, G. Manuel, J. Dubac and P. Mazerolles, *Nouv. J. Chim.*, **6**, 67 (1982).
(b) A. Laporterie, G. Manuel, H. Iloughmane and J. Dubac, *Nouv. J. Chim.*, **8**, 437 (1984).
(c) J. Dubac, A. Laporterie, G. Manuel, H. Iloughmane, J. P. Béteille and P. Dufour, *Synth. React. Inorg. Met.-Org. Chem.*, **17**, 783 (1987).
68. G. K. Henry, R. Shinimoto, Q. Zhou and W. P. Weber, *J. Organomet. Chem.*, **350**, 3 (1988).
69. G. Manuel, G. Bertrand and F. El Anba, *Organometallics*, **2**, 391 (1983).
70. J. Dubac, A. Laporterie, G. Manuel and H. Iloughmane, *Phosphorus and Sulfur*, **27**, 191 (1986).
71. A. Laporterie, H. Iloughmane and J. Dubac, *Tetrahedron Lett.*, **24**, 3521 (1983).
72. J. D. Andriamisaka, C. Couret, J. Escudié, A. Laporterie, G. Manuel and M. Regitz, *J. Organomet. Chem.*, **419**, 57 (1991).
73. D. Terunuma, M. Hirose, Y. Motoyama and K. Kumano, *Bull. Chem. Soc. Jpn.*, **66**, 2682 (1993).
74. J. P. Béteille, A. Laporterie and J. Dubac, *J. Organomet. Chem.*, **426**, C1 (1992).
75. J. P. Béteille, A. Laporterie and J. Dubac, *Organometallics*, **8**, 1799 (1989).
76. (a) J. J. Eisch, N. K. Hota and S. Kozima, *J. Am. Chem. Soc.*, **91**, 4575 (1969).
(b) G. E. Herberich, B. Buller, B. Hessner and W. Oschmann, *J. Organomet. Chem.*, **195**, 253 (1980).
(c) L. Killian and B. Wrackmeyer, *J. Organomet. Chem.*, **148**, 137 (1978).
77. A. J. Ashe, III, W. M. Butler and T. R. Diephouse, *Organometallics*, **2**, 1005 (1983).
78. (a) R. Uson, J. Vicente and M. T. Chicote, *Inorg. Chim. Acta*, **35**, L 205 (1979).
(b) R. Uson, J. Vicente and M. T. Chicote, *J. Organomet. Chem.*, **209**, 271 (1981).
(c) R. Uson, J. Vicente, M. T. Chicote, P. G. Jones and G. M. Sheldrick, *J. Chem. Soc., Dalton Trans.*, 1131 (1983).
79. P. J. Fagan and W. A. Nugent, *J. Am. Chem. Soc.*, **110**, 2310 (1988).
80. (a) W. A. Nugent, D. L. Thorn and R. L. Harlow, *J. Am. Chem. Soc.*, **109**, 2788 (1987).
(b) A. Famili, M. F. Farona and S. Thanedar, *J. Chem. Soc., Chem. Commun.*, 435 (1983).
(c) E. Negishi, F. E. Cederbaum and T. Tamatsu, *Tetrahedron Lett.*, **27**, 2829 (1986).
81. P. Dufour, M. Dartiguenave, Y. Dartiguenave and J. Dubac, *J. Organomet. Chem.*, **384**, 61 (1990).
82. F. Meier-Brocks and E. Weiss, *J. Organomet. Chem.*, **453**, 33 (1993).
83. (a) H. Gilman and R. D. Gorsich, *J. Am. Chem. Soc.*, **77**, 6380 (1955).
(b) H. Gilman and R. D. Gorsich, *J. Am. Chem. Soc.*, **80**, 1883 (1958).
(c) D. Wittenberg and H. Gilman, *J. Am. Chem. Soc.*, **80**, 2677 (1958).
(d) H. Gilman and R. D. Gorsich, *J. Am. Chem. Soc.*, **80**, 3243 (1958).
84. R. Gelius, *Chem. Ber.*, **93**, 1759 (1960).
85. (a) S. C. Cohen and A. G. Massey, *J. Chem. Soc., Chem. Commun.*, 457 (1966).
(b) S. C. Cohen and A. G. Massey, *Tetrahedron Lett.*, 4393 (1966).
(c) S. C. Cohen and A. G. Massey, *J. Organomet. Chem.*, **10**, 471 (1967).
86. J. Y. Corey and L. S. Chang, *J. Organomet. Chem.*, **307**, 7 (1986).
87. S. C. Cohen, M. L. N. Reddy and A. G. Massey, *J. Chem. Soc., Chem. Commun.*, 451 (1967).

88. E. A. Chernyshev, S. A. Shehepinov, T. L. Krasnova, N. P. Filimonova and E. I. Petrova, *Izobret. Prom. Obraztsy, Tovarnye Znaki*, **45**, 21 (1968); *Chem. Abstr.*, **69**, 7250 (1968).
89. (a) E. A. Chernyshev, N. G. Komalenkova and S. A. Bashkirova, *J. Organomet. Chem.*, **271**, 129 (1984).
    (b) E. A. Chernyshev, N. G. Komalenkova, O. V. Elagina, V. L. Rogachevskii, S. A. Bashkirova and L. V. Dunaeva, *Zh. Obshch. Khim.*, **55**, 2314 (1985); *Chem. Abstr.*, **105**, 172554m (1986).
90. (a) E. A. Chernyshev, N. G. Komalenkova, L. N. Shamshin and S. A. Shchepinov, *Zh. Obshch. Khim.*, **41**, 843 (1971); *Chem. Abstr.*, **75**, 49195e (1971).
    (b) E. A. Chernyshev, N. G. Komalenkova and L. N. Shamshin, *Otkrytiya, Izobret., Prom. Obraztsy, Tovarnye Znaki*, **48**, 65 (1971); *Chem. Abstr.*, **75**, 390 (1971).
91. B. Becker, R. J. P. Corriu, B. J. L. Henner, W. Wojnowski, K. Peters and H. G. von Schnering, *J. Organomet. Chem.*, **312**, 305 (1986).
92. M. Kira, K. Sakamoto and H. Sakurai, *J. Am. Chem. Soc.*, **105**, 7469 (1983).
93. R. W. Coutant and A. Levy, *J. Organomet. Chem.*, **10**, 175 (1967).
94. M. Ishikawa, T. Tabohashi, M. Kumada and J. Iyoda, *J. Organomet. Chem.*, **264**, 79 (1984).
95. (a) M. Ishikawa, T. Tabohashi, H. Sugisawa, K. Nishimura and M. Kumada, *J. Organomet. Chem.*, **250**, 109 (1983).
    (b) M. Ishikawa, K. Nishimura, H. Sugisawa and M. Kumada, *J. Organomet. Chem.*, **218**, C21 (1981).
96. (a) M. D. Rausch and L. P. Klemann, *J. Am. Chem. Soc.*, **89**, 5732 (1967).
    (b) M. D. Rausch, L. P. Klemann and W. H. Boon, *Synth. React. Inorg. Met.-Org. Chem.*, **15**, 923 (1985).
97. J. Kurita, M. Ishuii, S. Yasuike and T. Tsuchiya, *J. Chem. Soc., Chem. Commun.*, 1309 (1993).
98. A. Maercker, H. Bodenstedt and L. Brandsma, *Angew. Chem., Int. Ed. Engl.*, **31**, 1339 (1992).
99. S. L. Buchwald, B. T. Watson and J. C. Huffman, *J. Am. Chem. Soc.*, **108**, 7411 (1986).
100. S. L. Buchwald and R. B. Nielsen, *Chem. Rev.*, **88**, 1047 (1988).
101. A. Kanj, P. Meunier, B. Gautheron, J. Dubac and J. C. Daran, *J. Organomet. Chem.*, **454**, 51 (1993).
102. A. Kanj, P. Meunier, B. Hanquet, B. Gautheron, J. Dubac and J. C. Daran, *Bull. Soc. Chim. Fr.*, **131**, 715 (1994).
103. A. Kanj, P. Meunier, C. Legrand, B. Gautheron, J. Dubac and P. Dufour, to appear.
104. M. Pereyre, J. P. Quintard and A. Rahm, *Tin in Organic Synthesis*, Butterworths, London, 1987.
105. (a) T. J. Barton and B. L. Groh, *Organometallics*, **4**, 575 (1985).
    (b) T. J. Barton and M. Juvet, *Tetrahedron Lett.*, 3893 (1975).
106. T. J. Barton and G. T. Burns, *Organometallics*, **1**, 1455 (1982).
107. J. Dubac, A. Laporterie and G. Manuel, unpublished work.
108. (a) O. A. Arefev, N. S. Vorob'era, V. I. Epishev and A. Petrov, *Neftekhimiya*, **12**, 171 (1972); *Chem. Abstr.*, **77**, 61348f (1977).
    (b) V. A. Mironov, A. P. Ivanov and A. A. Akhrem, *Izv. Akad. Nauk. SSSR, Ser. Khim.*, 363 (1973); *Chem. Abstr.*, **78**, 159013p (1973).
    (c) S. D. Mekhtiev, M. R. Musaev, M. A. Mardanov, S. M. Sharifova, G. T. Badirova and M. A. Aliev, *Dokl. Akad. Nauk SSSR*, **30**, 49 (1974); *Chem. Abstr.*, **82**, 15952d (1975).
    (d) S. Cradock, R. H. Findlay and M. H. Palmer, *J. Chem. Soc., Dalton Trans.*, 1650 (1974).
    (e) N. T. Anh, E. Canadell and O. Eisenstein, *Tetrahedron*, **34**, 2283 (1978).
109. F. Mathey, *Top. Phosphorus Chem.*, **10**, 1 (1980).
110. J. Dubac, H. Iloughmane, A. Laporterie and C. Roques, *Tetrahedron Lett.*, **26**, 1315 (1985).
111. I. Fleming, *Frontier Orbitals and Organic Chemical Reactions*, Wiley, New York, 1976.
112. (a) R. B. Bates, B. Gordon III, T. K. Highsmith and J. J. White, *J. Org. Chem.*, **49**, 2981 (1984).
    (b) P. Mazerolles and C. Laurent, *J. Organomet. Chem.*, **406**, 119 (1991).
    (c) C. Laurent and P. Mazerolles, *Synth. React. Inorg. Met.-Org. Chem.*, **22**, 1183 (1992).
113. (a) R. B. Woodward and R. Hoffmann, *Angew. Chem., Int. Ed. Engl.*, **8**, 781 (1969).
    (b) C. W. Spangler, *Chem. Rev.*, **76**, 187 (1976).
    (c) P. Jutzi, *Chem. Rev.*, **86**, 983 (1986).
114. (a) P. K. Chiu and M. P. Sammes, *Tetrahedron*, **46**, 3439 (1990).
    (b) F. Mathey, *Acc. Chem. Res.*, **25**, 90 (1992) and references cited therein.
115. S. M. Bachrach and L. Perriot, *J. Org. Chem.*, **59**, 3394 (1994).
116. W. R. Roth, *Tetrahedron Lett.*, 1009 (1964).

117. H. Sun, D. A. Hrovat and W. T. Borden, *J. Am. Chem. Soc.*, **109**, 5275 (1987).
118. Reference 3, p. 239
119. G. T. Burns, E. Colomer, R. J. P. Corriu, M. Lheureux, J. Dubac, A. Laporterie and H. Iloughmane, *Organometallics*, **6**, 1398 (1987).
120. T. Douglas and K. H. Theopold, *Angew. Chem., Int. Ed. Engl.*, **28**, 1367 (1989).
121. J. M. Alcaraz, J. Svara and F. Mathey, *Nouv. J. Chim.*, **10**, 321 (1986).
122. (a) H. E. Simmons and T. Fukunaga, *J. Am. Chem. Soc.*, **89**, 5208 (1967).
    (b) R. Breslow, J. M. Hoffmann Jr. and C. Perchonock, *Tetrahedron Lett.*, 3723 (1973).
    (c) E. W. Garbisch Jr. and R. F. Sprecher, *J. Am. Chem. Soc.*, **88**, 3433, 3434 (1966).
123. Reference 3, pp. 239–246 and referneces cited therein.
124. (a) B. Résibois, J. C. Brunet and J. Bertrand, *Bull. Soc. Chim. Fr.*, **2**, 681 (1968).
    (b) B. Résibois and J. C. Brunet, *Ann. Chim.*, **5**, 199 (1970).
125. R. Balasubramanian and M. V. George, *Tetrahedron*, **29**, 2395 (1973).
126. C. Grugel, W. P. Neumann and M. Schriewer, *Angew. Chem., Int. Ed. Engl.*, **18**, 543 (1979).
127. D. Terunuma, K. Kumano and Y. Motoyama, *Bull. Chem. Soc. Jpn.*, **67**, 2763 (1994).
128. H. Sakurai, Y. Nakadaira and T. Kobayashi, *J. Am. Chem. Soc.*, **101**, 487 (1979).
129. H. Sakurai, Y. Nakadaira and H. Tobita, *Chem. Lett.*, 1855 (1982).
130. (a) Y. Nakadaira, T. Kobayashi, T. Otsuda and H. Sakurai, *J. Am. Chem. Soc.*, **101**, 486 (1979).
    (b) H. Sakurai, H. Sakaba and Y. Nakadaira, *J. Am. Chem. Soc.*, **104**, 6156 (1982).
    (c) H. Sakurai, Y. Nakadaira and H. Sabaka, *Organometallics*, **2**, 1484 (1983).
131. (a) A. Sekiguchi, S. S. Zigler, R. West and J. Michl, *J. Am. Chem. Soc.*, **108**, 4241 (1986).
    (b) A. Sekiguchi, G. R. Gillette and R. West, *Organometallics*, **7**, 1226 (1988).
132. T. J. Barton, W. F. Goure, J. L. Witiak and W. D. Wulff, *J. Organomet. Chem.*, **225**, 87 (1982).
133. H. Appler, L. W. Gross, B. Mayer and W. P. Neumann, *J. Organomet. Chem.*, **291**, 9 (1985).
134. A. Marinetti-Mignani and R. West, *Organometallics*, **6**, 141 (1987).
135. N. K. Hota and C. J. Willis, *J. Organomet. Chem.*, **15**, 89 (1968).
136. K. Kunô, K. Kobayashi, M. Kawasini, S. Kosima and T. Hitomi, *J. Organomet. Chem.*, **137**, 349 (1977).
137. M. P. Egorov, A. M. Galminas, M. B. Ezhova, S. P. Kolesnikov and O. M. Nefedov, *Metalloorg. Khim.*, **6**, 15 (1993).
138. (a) M. N. Paddon-Row and K. D. Jordan, *J. Chem. Soc., Chem. Commun.*, 1508 (1988).
    (b) M. N. Paddon-Row, S. S. Wong and K. D. Jordan, *J. Chem. Soc., Perkin Trans. 2*, 417 (1990).
139. M. P. Egorov, M. B. Ezhova, M. Yu. Antipin and Yu. T. Struchkov, *Main Group Met. Chem.*, **14**, 19 (1991).
140. R. Maruca, *J. Org. Chem.*, **36**, 1626 (1971).
141. Y. Fujise, Y. Chonan, H. Sakurai and S. Ito, *Tetrahedron Lett.*, 1585 (1974).
142. T. J. Barton and J. Nelson, *Tetrahedron Lett.*, 5037 (1969).
143. Y. Nakadaira and H. Sakurai, *Tetrahedron Lett.*, 1183 (1971).
144. T. Sato, N. Moritami and M. Matsuyama, *Tetrahedron Lett.*, 5113 (1969).
145. T. J. Barton, R. C. Kippenhan and A. J. Nelson, *J. Am. Chem. Soc.*, **96**, 2272 (1974).
146. T. J. Barton and R. J. Rogido, *J. Org. Chem.*, **40**, 582 (1975).
147. H. H. Freedman, US Patent 3090 797 (1963); *Chem. Abstr.*, **59**, 11560c (1963).
148. H. H. Freedman, *J. Org. Chem.*, **27**, 2298 (1962).
149. H. H. Freedman, *J. Am. Chem. Soc.*, **83**, 2194 (1961).
150. V. R. Sandel and H. H. Freedman, *J. Am. Chem. Soc.*, **90**, 2059 (1968).
151. W. Z. Rhee and J. J. Zuckerman, *J. Am. Chem. Soc.*, **97**, 2291 (1975).
152. W. A. Gustavson, L. M. Principe, W. Z. Rhee and J. J. Zuckerman, *J. Am. Chem. Soc.*, **103**, 4126 (1981).
153. W. A. Gustavson, L. M. Principe, W. Z. Rhee and J. J. Zuckerman, *Inorg. Chem.*, **20**, 3460 (1981).
154. V. R. Sandel, G. R. Buske, S. G. Maroldo, D. K. Bates, D. Whitman and G. Sypniewski, *J. Org. Chem.*, **46**, 4069 (1981).
155. R. Balasubramanian and M. V. George, *J. Organomet. Chem.*, **85**, 311 (1975).
156. J. J. Eisch, J. E. Galle and S. Kozima, *J. Am. Chem. Soc.*, **108**, 379 (1986).
157. M. D. Curtis, *J. Am. Chem. Soc.*, **91**, 6011 (1969).
158. (a) Si chemistry: R. J. P. Corriu, C. Guérin and J. J. E. Moreau, *Top. Stereochem.*, **15**, 158 (1984).
    (b) Ge chemistry: J. Dubac, J. Cavezzan, A. Laporterie and P. Mazerolles, *J. Organomet. Chem.*, **209**, 25 (1981) and references cited therein.

159. J. P. Béteille, *Thesis*, Université Paul Sabatier, Toulouse, n° 335 (1988).
160. P. Jutzi and A. Karl, *J. Organomet. Chem.*, **214**, 289 (1981).
161. P. Jutzi and A. Karl, *J. Organomet. Chem.*, **215**, 19 (1981).
162. Y. Nakadaira, T. Nomura, S. Kanouchi, R. Sato, C. Kabuto and H. Sakurai, *Chem. Lett.*, 209 (1983).
163. M. P. Mahayan, R. Balasubramanian and M. V. George, *Tetrahedron*, **32**, 1549 (1976).
164. E. G. Janzen, J. B. Pickett and W. H. Atwell, *J. Organomet. Chem.*, **10**, P6 (1967).
165. R. E. Dessy and R. L. Pohl, *J. Am. Chem. Soc.*, **90**, 1995 (1968).
166. N. Tamaka, T. Ogata, Y. Uratani, Y. Nakadaira and H. Sakurai, *Inorg. Nucl. Chem. Lett.*, **8**, 1041 (1972).
167. Reference 3, pp. 235, 236 and references cited therein.
168. (a) W. Ando, H. Tanikawa and A. Sekiguchi, *Tetrahedron Lett.*, **24**, 4245 (1983).
    (b) A. Sekiguchi, H. Tanikawa and W. Ando, *Organometallics*, **4**, 584 (1985).
169. V. Hagen and K. Rühlmann, *Z. Chem.*, **9**, 309 (1969).
170. K. Tamao and S. Yamaguchi, *Pure Appl. Chem.*, **68**, 139 (1996).
171. Y. Yamaguchi and J. Shioya, *Mol. Eng.*, **2**, 339 (1993).
172. (a) V. N. Khabashesku, V. Balaji, S. E. Boganov, O. M. Nefedov and J. Michl, *J. Am. Chem. Soc.*, **116**, 320 (1994).
    (b) B. Goldfuss and P. v. R. Schleyer, *Organometallics*, **14**, 1553 (1995) and references cited therein.
173. K. Tamao, S. Yamaguchi, Y. Ito, Y. Matsuzaki, T. Yamabe, M. Fukushima and S. Mori, *Macromolecules*, **28**, 8668 (1995).
174. M. G. Hill, K. R. Mann, L. L. Miller and J.-F. Penneau, *J. Am. Chem. Soc.*, **114**, 2728 (1992).
175. (a) M. Ishikawa and K. Nate, in *Inorganic and Organometallic Polymers* (Eds. M. Zeldin, K. J. Wynne and H. R. Allcock), ACS Symposium Series 360, Chap 16, American Chemical Society, Washington DC, 1988.
    (b) J. Ohshita, D. Kanaya and T. Yamanaka, *J. Organomet. Chem.*, **369**, C18 (1989).
    (c) T. Imori, H-G. Woo, J. F. Walzer and T. D. Tilley, *Chem. Mater.*, **5**, 1487 (1993).
    (d) A. Kunai, E. Toyoda, K. Horata and M. Ishikawa, *Organometallics*, **14**, 714 (1995).
176. (a) S. Ijadi-Maghsoodi and T. J. Barton, *Macromolecules*, **24**, 1257 (1991).
    (b) T. Iwahara, S. Hayase and R. West, *Macromolecules*, **23**, 4485 (1990).
    (c) See also Reference 179.
177. (a) J. Ohshita, K. Furumori, M. Ishikawa and T. Yamanaka, *Organometallics*, **8**, 2084 (1989).
    (b) M. Ishikawa, *Pure Appl. Chem.*, **63**, 851 (1991).
    (c) J. Ohshita, A. Matsuguchi, K. Furumori, R.-F. Hong, M. Ishikawa, T. Yamanaka, T. Koike and J. Shioya, *Macromolecules*, **25**, 2134 (1992).
178. M. Ishikawa, T. Hatano, Y. Hasegawa, T. Horio, A. Kunai, A. Miyai, T. Ishida, T. Tsukihara, T. Yamanaka, T. Koike and J. Shioya, *Organometallics*, **11**, 1604 (1992) and references cited therein.
179. J. L. Brefort, R. Corriu, P. Gerbier, C. Guerin, B. Henner, A. Jean, T. Kuhlmann, F. Garnier and A. Yassar, *Organometallics*, **11**, 2500 (1992).
180. R. Corriu, N. Devylder, C. Guérin, B. Henner and A. Jean, *J. Organomet. Chem.*, **509**, 249 (1996).
181. R. Corriu, N. Devylder, C. Guérin, B. Henner and A. Jean, *Organometallics*, **13**, 3194 (1994).
182. R. Corriu, W. Douglas and Z.-X. Yang, *J. Organomet. Chem.*, **456**, 35 (1993).
183. W.-C. Joo, J.-H. Hong, S.-B. Choi and H.-E. Son, *J. Organomet. Chem.*, **391**, 27 (1990).
184. (a) H. Grützmacher, *Angew. Chem., Int. Ed. Engl.*, **34**, 295 (1995).
    (b) B. Goldfuss and P. v. R. Schleyer, *Organometallics*, **14**, 1553 (1995).
    (c) B. Goldfuss, P. v. R. Schleyer and F. Hampel, *Organometallics*, **15**, 1755 (1996).
185. W.-C. Joo, J.-H. Hong, H.-E. Son and J.-H. Kim, *Proceedings of the 9th International Symposium on Organosilicon Chemistry*, University of Edinburgh, UK, July 16–20, 1990, B 8.
186. J.-H. Hong and P. Boudjouk, *J. Am. Chem. Soc.*, **115**, 5883 (1993).
187. W. P. Freeman, T. D. Tilley, G. P. A. Yap and A. L. Rheingold, *Angew. Chem., Int. Ed. Engl.*, **35**, 882 (1996).
188. J.-H. Hong, P. Boudjouk and I. Stoenescu, *Organometallics*, **15**, 2179 (1996).
189. J.-H. Hong and P. Boudjouk, *Organometallics*, **14**, 574 (1995).
190. R. Corriu, C. Guérin, B. Henner and Q. Wang, *Organometallics*, **10**, 2297 (1991).
191. P. Dufour, J. Dubac, M. Dartiguenave and Y. Dartiguenave, *Organometallics*, **9**, 3001 (1990).

192. W. P. Freeman, T. D. Tilley, F. P. Arnold, A. L. Rheingold and P. K. Gantzel, *Angew. Chem., Int. Ed. Engl.*, **34**, 1887 (1995).
193. P. Meunier and J. Dubac, to appear.
194. J.-H. Hong, P. Boudjouk and S. Castellino, *Organometallics*, **13**, 3387 (1994).
195. M. Denk, R. Lennon, R. Hayashi, R. West, A. V. Belyakov, H. R. Verne, M. Wagner and N. Metzler, *J. Am. Chem. Soc.*, **116**, 2691 (1994).
196. R. West, H. Sohn, V. Bankwitz, J. Calabrese, Y. Apeloig and T. Mueller, *J. Am. Chem. Soc.*, **117**, 11608 (1995).
197. J.-H. Hong and P. Boudjouk, *Bull. Soc. Chim. Fr.*, **132**, 495 (1995).
198. R. West, H. Sohn, D. R. Powelll, T. Mueller and Y. Apeloig, *Angew. Chem., Int. Ed. Engl.*, **35**, 1002 (1996).
199. J.-H. Hong, Y. Pan and P. Boudjouk, *Angew. Chem., Int. Ed. Engl.*, **35**, 186 (1996) and references cited therein.
200. W. P. Freeman, T. D. Tilley, A. L. Rheingold and R. L. Ostrander, *Angew. Chem., Int. Ed. Engl.*, **32**, 1744 (1993).
201. P. G. Gassman and C. H. Winter, *J. Am. Chem. Soc.*, **110**, 6130 (1988).
202. W. P. Freeman, T. D. Tilley and A. L. Rheingold, *J. Am. Chem. Soc.*, **116**, 8428 (1994).
203. U. Schubert, G. Scholz, J. Müller, K. Ackermann, B. Wörle and R. F. D. Stansfield, *J. Organomet. Chem.*, **306**, 303 (1986).
204. V. N. Khabashesku, S. E. Boganov, D. Antic, O. M. Nefedov and J. Michl, *Organometallics*, **15**, 4714 (1996).
205. B. Wrackmeyer and U. Klaus, *J. Organomet. Chem.*, **520**, 211 (1996).
206. S. Yamaguchi and K. Tamao, *Bull. Chem. Soc. Jpn.*, **69**, 2327 (1996).
207. S. Yamaguchi and K. Tamao, *Tetrahedron Lett.*, **37**, 2983 (1996).
208. K. Tamao, M. Uchida, T. Izumizawa, K. Furukawa and S. Yamaguchi, *J. Am. Chem. Soc.*, in press.
209. K. Tamao, S. Ohno and S. Yamaguchi, *J. Chem. Soc., Chem. Commun.*, 1873 (1996).
210. W. P. Freeman, T. D. Tilley, L. M. Liable-Sands and A. L. Rheingold, *J. Am. Chem. Soc.*, **118**, 10457 (1996).

CHAPTER 35

Transition-metal silyl complexes

MORIS S. EISEN

Department of Chemistry, Technion— Israel Institute of Technology, Kiryat Hatechnion, Haifa 32000, Israel

I.	ABBREVIATIONS	2038
II.	INTRODUCTION	2038
III.	GROUP-3 SILYL COMPLEXES	2038
IV.	LANTHANIDE SILYL DERIVATIVES	2031
V.	GROUP-4 SILYL COMPLEXES	2042
VI.	ACTINIDE SILYL COMPLEXES	2053
VII.	GROUP-5 SILYL COMPLEXES	2054
VIII.	GENERAL INTRODUCTION TO LATE-TRANSITION-METAL SILICON-CONTAINING COMPLEXES	2057
IX.	GROUP-6 SILYLENE COMPLEXES	2058
X.	GROUP-6 SILYL COMPLEXES	2061
	A. η^2-Silyl Complexes	2064
XI.	GROUP-7 SILYLENE COMPLEXES	2069
XII.	GROUP-7 SILYL DERIVATIVES	2069
XIII.	GROUP-8 SILYL DERIVATIVES	2071
	A. Bimetallic Silyl Complexes	2072
	B. Clusters Containing Silyl Groups	2075
	C. Silyl Complexes	2075
	D. Bis(silyl) Complexes	2079
	E. Silacyclopentadienyl Complexes	2082
	F. Silene and Silylene Complexes	2083
	G. η^4-Silatrimethylenemethane Complexes	2089
XIV.	GROUP-9 SILICON-CONTAINING COMPLEXES	2095
	A. Silylene Complexes	2095
	B. Silyl and Bis(silyl) Complexes	2095
	C. Silanol Complexes	2107
XV.	GROUP-10 SILICON-CONTAINING COMPLEXES	2110
	A. Silyl, Bis(silyl) and Silanol Complexes	2110
	B. Silylene Complexes	2117
XVI.	COPPER AND MERCURY SILYL COMPLEXES	2121
XVII.	ACKNOWLEDGMENTS	2122
XVIII.	REFERENCES	2122

The chemistry of organic silicon compounds, Vol. 2
Edited by Z. Rappoport and Y. Apeloig © 1998 John Wiley & Sons Ltd

I. ABBREVIATIONS

acac	acetylacetonate	dppb	bis(1,4-diphenyl-phosphino) butane
Ad	adamantyl	dppe	bis(1,2-diphenyl-phosphino) ethane
Cbb	cyclobuta-η^6-benzene	dppm	bis(diphenyl-phosphino) methane
cod	cyclooctadiene	HMPA	hexamethyl-phosphoric triamide
coe	cyclooctene	Mes	mesityl (2,4,6-Me$_3$ C$_6$H$_2$)
cot	cyclooctatetraene	Np	neopentyl
Cp	η^5-C$_5$H$_5$	OTf	triflate
Cp*	η^5-C$_5$(CH$_3$)$_5$	Ph$_2$Pyr	2-(diphenylphosphino) pyridine
Cy	cyclohexyl	Pyr	pyridine
dcpe	bis(1,2-dicyclohexylphosphino) ethane	TFB	tetrafluorobenzobarrelene
dmpe	bis(1,2-dimethylphosphino) ethane	Trityl	triphenylmethyl
dmpm	bis(dimethylphosphino) methane	vdpp	vinylidenebis(diphenyl-phosphine)
dmpp	bis(1,3-dimethylphosphino) propane		

II. INTRODUCTION

Transition-metal chemistry is currently one of the most rapidly developing research areas. From Hein's discovery in 1941 of the first metal–silicon complex, followed by Wilkinson's Cp(CO)$_2$FeSiMe$_3$ complex synthesis in 1956 and Speier's catalytic hydrosilylation discovery in 1977, a new area of organometallic chemistry was entered in 1987 with the discovery of complexes with silicon multiply bonded to transition metals. In the last decade, coordination complexes of silylenes, silylynes, silenes, disilenes, silanimines, silaallenes and metallasilaallenes have been prepared with some of the compounds being the key intermediates in important technological processes. Numerous developments in this field in the last few years have provided evidence of the interest in the stoichiometric and catalytic chemistry of metal–silicon bonds. This chapter is an update of an earlier chapter[1] giving information on some of the newer developments, covering the literature from 1992 to approximately the middle of 1996. This review is organized according to the order of the groups in the Periodic Table and at the end of each group a compilation of metal–silicon bond distances and ^{29}Si NMR chemical shifts of the aforementioned complexes are given. Lanthanide and actinide silyl complexes are reviewed after group 3 and group 4, respectively, due to their isolobal relationship.

III. GROUP-3 SILYL COMPLEXES

Early-transition-metal complexes containing silicon ligands are becoming increasingly important in organometallic chemistry[1]. The first group-3 silyl complex

$Cp_2Sc[Si(SiMe_3)_3](THF)$ was prepared by Tilley and coworkers in 1990 by reacting the dimeric scandocene chloride $[Cp_2ScCl]_2$ with $(Me_3Si)_3SiLi·3THF$. The complex obtained undergoes insertion with carbon monoxide or xylyl isocyanides yielding CO—CO coupling and ketene-derived products or the monoinsertion η^2-iminosilaacyl complex, respectively, whereas the analogous complex $Cp_2Sc(SiPh_2Bu-t)(THF)$ insert CO_2 yielding the dimeric silane carboxylate complex $[Cp_2Sc(\mu-O_2-CSiR_3)]_2$[1]. Recently, increasing the small number of group-3 silyl derivatives, Tilley and coworkers prepared a series of scandium complexes via the addition of the appropriate lithium silyl reagent to the dimeric scandocene chloride (equation 1).

$$1/2[Cp_2ScCl]_2 + (THF)_3LiSiR_3 \xrightarrow[-LiCl]{-2\,THF} Cp_2Sc\begin{matrix}SiR_3\\O\end{matrix}\bigcirc \quad (1)$$

$SiR_3 = Si(SiMe_3)_3, Si(SiMe_3)_2Ph, SiPh_2(Bu-t), SiPh_3$

Solid state studies on the complex $Cp_2Sc[Si(SiMe_3)_3](THF)$ reveal a tetrahedral environment for the metal with a Sc—Si distance of 2.863(2) Å. This bond length is shorter than the isolobal Sm—Si bond [3.052(8) Å] (*vide infra*) found in the dimeric complex $[Cp_2*SmH(SiMe_3)_2]_2$ by a distance of 0.19 Å, reflecting the difference in the anionic radii between Sc^{3+} and Sm^{3+2}.

These unusually reactive d^0 silyl complexes have been found to catalyze the polymerization of ethylene to polyethylene with polydispersity ranges from 3.6 to 5.5 and with molecular weights from 27,000 to 410,000. It is noteworthy that the reaction of $Cp_2Sc[Si(SiMe_3)_3](THF)$ with propene yielded the organic compound $HSi(SiMe_3)_3$; however the allylic activation complex, $Cp_2Sc(\eta^3-C_3H_5)$, was not formed. Presumably, the formation of the $HSi(SiMe_3)_3$ is through a vinylic activation as found for lanthanides[3]. In addition, no reactivity was found for the silyl complex $Cp_2Sc[Si(SiMe_3)_3](THF)$ with butadiene or cyclohexene. However, addition of $PhC\equiv CH$ yielded $HSi(SiMe_3)_3$ and the expected dimeric complex $[Cp_2Sc(\mu-C\equiv CPh)]_2$[2].

IV. LANTHANIDE SILYL DERIVATIVES

Organolanthanide complexes have been found to be active catalysts for the dehydrogenative coupling of silanes, arguing that a lanthanide–silicon bond should be involved in such a process (*vide infra*). The reactivity of lanthanide silyl complexes is based on the σ-bond metathesis reaction, which appears to be highly sensitive to steric effects. For example, the isolobal group III yttrium complex $Cp_2*YMe(THF)$ converts $PhSiH_3$ to $PhMeSiH_2$, indicating that the primary σ-bond metathesis process involves a four-center transition state that transfers the silyl group to the carbon rather than to the yttrium atom. Recently, based on this mechanism, Molander and Nichols have shown the use of $Cp_2*YMe(THF)$ for the cyclization/silylation of 1,5- and 1,6-dienes[4]. In addition, Marks and coworkers have shown that silanes can be used as terminating chain transfer agents with a similar mechanism in the lanthanide-mediated ethylene polymerization with $(Cp_2*SmH)_2$ and $PhSiH_3$[5].

The more hindered lanthanide alkyl complexes, as in $Cp_2*LnCH(SiMe_3)_2$ (Ln = Nd, Sm), were found not to react with bulky silanes such as t-Bu_2SiH_2 or Ph_2MeSiH, but to react with unhindered silanes to produce $CH_2(SiMe_3)_2$, the corresponding dimeric hydride $(Cp_2*LnH)_2$ complex, and a disilane resulting from the dehydrocoupling reaction (*vide infra*)[6]. Until 1992, the only lanthanide silyl derivatives reported were those corresponding

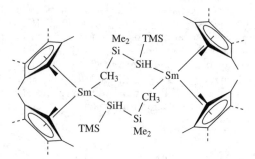

FIGURE 1. Dimeric structure of complex **1**

to the salt of type $[Li(DME)_3]^+[Cp_2Ln(SiMe_3)_2]^-$ (DME = dimethoxyethane)[7]. The first neutral lanthanide silyl complexes $Cp_2^*LnSiH(SiMe_3)_2$ [Ln = Sm (**1**), Nd] were obtained via a σ-bond metathesis of an excess of $SiH_2(SiMe_3)_2$ with $Cp_2^*LnCH(SiMe_3)_2$ or $Cp''_2LnCH(SiMe_3)_2$ (Ln = Nd, Sm; Cp'' = η^5-C_5Me_4Et)[8,9].

In the solid state, these lanthanide silyl compounds are dimeric (Figure 1) formed via intermolecular Sm···CH$_3$−Si interactions ensuring a Sm−C(Me) distance of 2.97(2) Å. The Sm−Si bond length of 3.052(8) Å in **1** is longer than the Lu−Si bond of 2.888(2) Å in the anionic $[Cp_2Lu(SiMe_3)_2]^-$ complex (Table 1)[7]. This difference of 0.16 Å is somewhat larger than the one attributed to the lanthanide contraction (*ca* 0.10 Å). These new silyl complexes are monomeric in solution, displaying in the ^{1}H NMR chemical shifts a Curie–Weiss behavior up to −80 °C (Table 2). The Cp* resonances for the Nd compound coalesce into two singlets at 10 °C, which corresponds to an activation barrier for rotation about the Nd−Si bonds of 13.3 ± 1.0 kcal mol^{-1}[9].

Interestingly, Marks and coworkers have shown that lanthanide complexes of the type Cp_2^*LnR [Ln = La, Nd, Sm, Y, Lu; R = H, CH(TMS)$_2$] in hydrocarbon solvents react with R'SiH$_3$ by eliminating H$_2$ (R = H) or CH$_2$(TMS)$_2$ [R = CH(TMS)$_2$], respectively, forming the corresponding lanthanocene silyl derivatives and the subsequent formation of the dimeric hydrides $[Cp_2^*LnH]_2$ and polysilanes. This result argues that the difference in reactivity found for the similar lanthanide complexes may be 'tuned' by either the silane group (R'), the presence of an olefin, as in the cyclization/silylation of dienes (*vide supra*), or specifically in the yttrium complex, by the solvent (*vide infra*). The relative rates of the dehydrogenative coupling of silanes for the different lanthanide complexes were found to follow the trend Lu > Y > Sm > Nd > La. This trend reveals no parallel with lanthanide III/II redox characteristics but correlates well with the metal ionic radius[10]. Lately, Marks and coworkers have proposed, based on kinetic measurements, mechanistic pathways for the dehydrogenative coupling of silanes (Scheme 1) and the hydrosilylation of olefins (Scheme 2) catalyzed by organolanthanides[5].

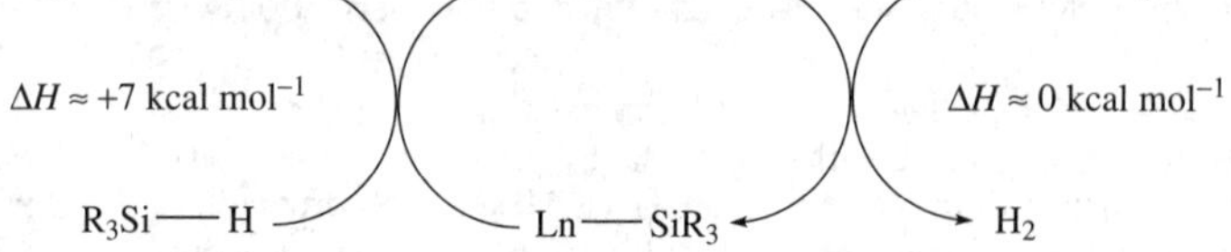

SCHEME 1

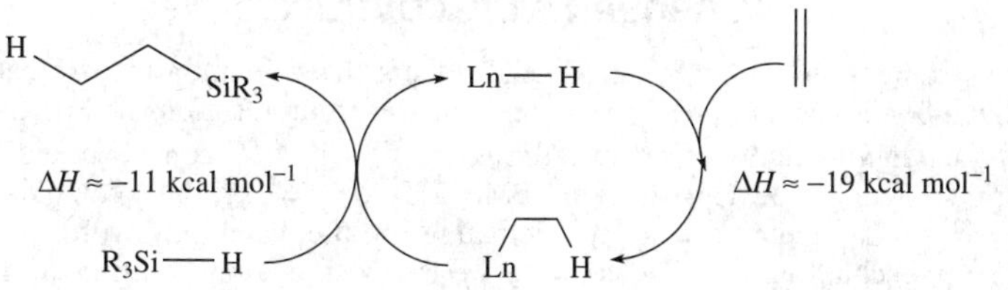

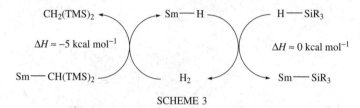

SCHEME 2

A close look at the hydrogenolysis reaction of the $Cp_2^*SmCH(TMS)_2$ complex with $H_2Si(SiMe_3)_2$ shows that the two compounds *do not react directly*. The kinetic profile shows an S-shape form with an induction time. This shape is consistent with a second-order autocatalytic mechanism in which the reaction is catalyzed by a product or an intermediate. Indeed, the induction period was completely eliminated by addition of catalytic amounts of H_2, $[Cp_2^*SmH]_2$, or of the complex $[Cp_2^*Sm\,(\mu\text{-}H)\,(\mu\text{-}CH_2C_5Me_4)SmCp^*]$, which is a decomposition product of the $[Cp_2^*SmH]_2$. Thus, two possible mechanistic pathways (Schemes 3 and 4) have been proposed to explain the hydrogenolysis reaction.

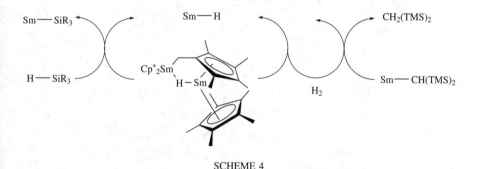

SCHEME 3

SCHEME 4

Kinetic studies by Radu and Tilley have shown that both Schemes 3 and 4 are observed, with a large preference for Scheme 3[11]. A similar S-shaped reaction profile has been observed recently for chiral Cp_2^*SmR complexes in the presence of $PhSiH_3$[5].

The first neutral lanthanide(II) silyl complex $(Ph_3Si)_2Yb(THF)_4$ was synthesized by the reaction of Ph_3SiCl with metallic ytterbium in THF. The solid complex has a centrosymmetrical octahedral structure with a Yb atom bonded to four oxygen atoms of the THF molecules in the equatorial positions and to two Si atoms of the $SiPh_3$ fragment in the axial positions. The Yb−Si distance of 3.158(2) Å is 0.1 Å larger than the Sm−Si distance in $Cp_2^*SmSiH(SiMe_3)_2$ which is in agreement with the difference in ionic radius between Ln(II) and Ln(III) complexes[12].

V. GROUP-4 SILYL COMPLEXES

Pioneering studies by Tilley, Harrod and others have demonstrated that group-4 metal–silicon bonds are quite reactive, undergoing insertion reactions and participating in silane polymerizations and σ-bond metathesis reactions[1,13]. Zirconocene and hafnocene silyl hydride of the type $Cp_2M(SiPh_3)(PMe_3)(H)$ (M = Zr, Hf) complexes, which are believed to be active species participating in the catalytic hydrosilylation and dehydrogenative coupling of silanes, were synthesized by the addition of $HSiPh_3$ to solutions of the corresponding $Cp_2M(olefin)(PMe_3)$ (M = group-4). For Hf, intermediate **2** was observed, arguing in favor of an addition–elimination pathway instead of an oxidative addition mechanism as being responsible for the silyl hydride formation (equation 2). Additional evidence supporting the proposed mechanism was provided by deuterium labeling and the observation of complex **4** resulting from the reversible elimination of $HSiPh_3$ or $DSiPh_3$ from the proposed intermediate complex **3** (equation 3).

The hydride zirconium compound $Cp_2Zr(H)(SiPh_3)(PMe_3)$ reacts with t-BuNC, t-BuCN and acetone via insertion into the Zr–H bond, and undergoes σ-bond metathesis with Ph_2SiH_2 producing $HSiPh_3$ and two isomers of $Cp_2Zr(H)(SiHPh_2)(PMe_3)$. In addition, the reaction of $Cp_2Zr(H)(SiPh_3)(PMe_3)$ with the internal alkyne $CH_3C\equiv CCH_3$ produces the metallacycle complex **5** by elimination of the silane $HSiPh_3$ with the concomitant formation of the alkyne complex followed by two subsequent oxidative couplings (equation 4).

The X-ray crystal structure of the silyl-hydrido zirconium derivative $Cp_2Zr(H)(SiPh_3)(PMe_3)$ showed that the silyl and phosphine ligands are in the outward position in the CpZr wedge, with the hydride ligand in between[14].

$$\underset{Me_3P}{\overset{SiPh_3}{Cp_2Zr}}\diagdown H \quad \xrightarrow[\substack{-HSiPh_3 \\ -PMe_3}]{2\ H_3CC\equiv CCH_3} \quad Cp_2Zr\diagdown\diagup\overset{Me}{\underset{Me}{\diagdown\diagup}}\overset{Me}{\underset{Me}{}} \quad (4)$$

(5)

Similar silyl-hydrido metallocene derivatives, $Cp_2Zr(H)(SiHPh_2)(PR_2Me)$ (R = Me, Ph), have been found to be efficient catalysts for the high regioselective hydrosilylation of olefins. These complexes are obtained by trapping the organometallic hydride intermediate $Cp_2Zr(H)(SiHPh_2)$ with phosphines. In the absence of phosphine, the hydride dimerizes producing the surprisingly stable complex $[(\mu\text{-H})Cp_2Zr(SiHPh_2)]_2$ (6), which is inactive in the hydrosilylation of olefins. This dimer was also found to be inert to acetone, PMe_3 or 3N HCl. X-ray analysis of the trapped $Cp_2Zr(H)(SiHPh_2)(PMe_3)$ complex shows similar geometry features to those of the complex $Cp_2Zr(H)(SiPh_3)(PMe_3)$, although the Zr—Si bond is shorter by 0.014 Å[15].

For titanium silyl complexes of the type $Cp_2Ti(SiHRPh)(PMe_3)$ (R = H, Me) the X-ray analysis shows similar geometries with close bond lengths of 2.652(1) Å (R = H) and 2.646(2) Å (R = Me), respectively[16]. Titanium complexes of the type $Cp_2M(SiHRR')L$ [L = NH_3, M = Ti(III); L = H, M = Ti(IV); R = aryl or alkyl; R' = H, $RSiH_2$ or $(RSiH)_n$] have been found to induce the dehydrocoupling of ammonia and silanes as shown in Scheme 5[17].

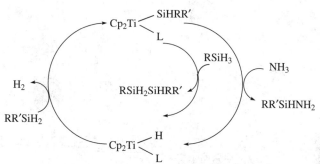

For Ti(IV), L = H; For Ti (III), L = NH_3, R = aryl, R' = H, $RSiH_2$, $(RSiH)_n$

SCHEME 5

Although transition-metal silyl hydrides have been proposed as intermediates in the catalytic hydrosilylation of olefins and the polymerization of organosilanes, the complexes have been described classically as having η^1 or η^2 σ-bonds. The oxidative addition of Ph_2SiH_2 to $Cp_2Ti(PMe_3)_2$ produces the unstable complex $Cp_2Ti(Ph_2SiH_2)(PMe_3)$ (7) with an agostic hydrogen atom bridging the silyl–metal σ-bond. The complex decomposes to the binuclear species $[(\mu\text{-}(\eta^1:\eta^5\text{-}C_5H_4))(Cp)Ti(PMe_3)]_2$ and Ph_2SiH_2. The silane moiety in complex 7 is weakly bound and is readily displaced by nitrogen according to equation 5, leading to formation of the dinitrogen complex 8. X-ray analysis of complex 7 shows a Ti—Si bond length of 2.597(2) Å, which is normal as compared to other titanocenes

(Table 1), although it displays a very interesting distorted trigonal bipyramid geometry at the silicon atom. The two hydrogens occupy the apical positions while the Ti and *ipso* carbons define the equatorial plane.

$$\text{Cp}_2\text{Ti}(\text{PMe}_3)(\text{H})(\text{SiPh}_2\text{H}) + N_2 \underset{\text{vacuum}}{\rightleftarrows} \text{Cp}_2\text{Ti}(\text{PMe}_3)-\text{N}\equiv\text{N}-\text{TiCp}_2(\text{PMe}_3) + \text{Ph}_2\text{SiH}_2 \quad (5)$$

$$(7) \qquad\qquad\qquad (8)$$

^{29}Si NMR of **7** is consistent with a bridging hydrogen showing a doublet of doublets, indicating a coupling of two different hydrogens ($J_{\text{SiH}_1} = 28$ Hz, $J_{\text{SiH}_2} = 161$ Hz). The smaller value is slightly larger than the typical values observed for silyl-hydride complexes in which there is no interaction between Si and H[18].

Non-π-bound titanium amido silyl complex (Me$_2$N)$_3$TiSi(SiMe$_3$)$_3$ (Figure 2) was prepared from the titanium chloride complex (Me$_2$N)$_3$TiCl with the corresponding lithium salt (Me$_3$Si)$_3$SiLi·3THF. X-ray analysis shows a normal Ti−Si bond length of 2.635(2) Å[19].

The unsymmetrical silyl-hydrido zirconium dimer, Cp$_2$Zr(SiPhMeH) (μ-H)$_2$(SiPhMe$_2$)ZrCp$_2$, similar to the above-mentioned complex **6**, was isolated from the dehydrogenative coupling reaction of phenylmethylsilane with dimethylzirconocene. The X-ray analysis of the dimer shows a disorder which is produced by the ordered superposition of two

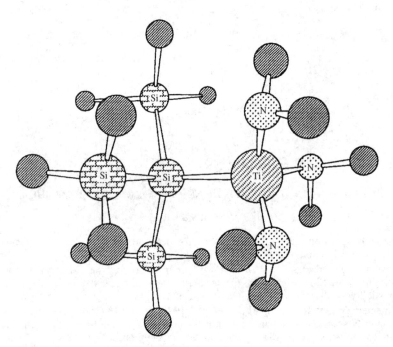

FIGURE 2. Structure view of (Me$_2$N)$_3$TiSi(SiMe$_3$)$_3$

enantiomers as shown in equation 6[13].

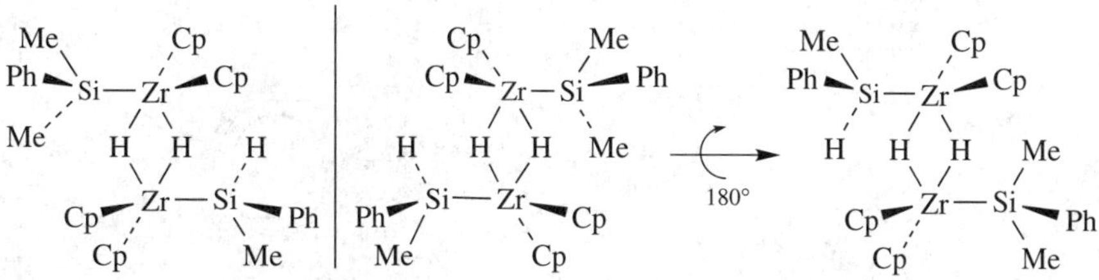

$$
\begin{array}{c}
\text{(structures of Zr silyl enantiomers, related by 180° rotation)}
\end{array}
\quad (6)
$$

Group-4 monosilyl complexes of the types CpCp*M(SiPh$_3$)Cl and Cp$_2$*M(SiPh$_3$)Cl (M = Zr, Hf) were synthesized by the reaction of the corresponding metallocenes with Ph$_3$SiLi·3THF. These silyl complexes decompose slowly with a concomitant elimination of HSiPh$_3$. The decompositions in solution are accelerated in the light according to the following trends: CpCp*M(SiPh$_3$)Cl > Cp$_2$*M(SiPh$_3$)Cl; Hf > Zr. The reaction of complexes Cp$_2$*M(SiPh$_3$)Cl (M = Zr, Hf) with MeMgBr affords the corresponding methyl derivatives Cp$_2$*M(SiPh$_3$)Me (M = Zr, Hf) and hydrogenolysis of the triphenyl silyl complexes rapidly yield the corresponding hydrides with the elimination of HSiPh$_3$. Reaction of the mixed cyclopentadienyl complex CpCp*Hf(SiPh$_3$)Cl with one equivalent of PhSiH$_3$ under ambient fluorescent light afforded the σ-bond metathesis product CpCp*Hf(SiH$_2$Ph)Cl and the same reaction under ambient laboratory conditions produced the corresponding hydride and silane dehydrocoupling products. The reaction of a biscyclopentadienyl silyl complexes of Zr with CO and of a mixed cyclopentadienyl Hf complex with nitriles is shown in equations 7 and 8, respectively.

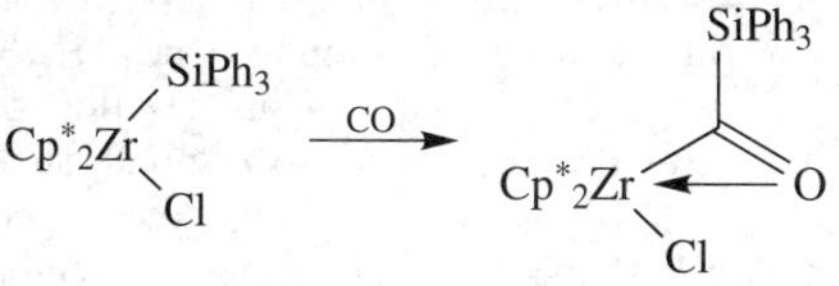

$$
\text{Cp*}_2\text{Zr(SiPh}_3\text{)Cl} \xrightarrow{\text{CO}} \text{Cp*}_2\text{Zr(Cl)}[\eta^2\text{-C(SiPh}_3\text{)=O}] \quad (7)
$$

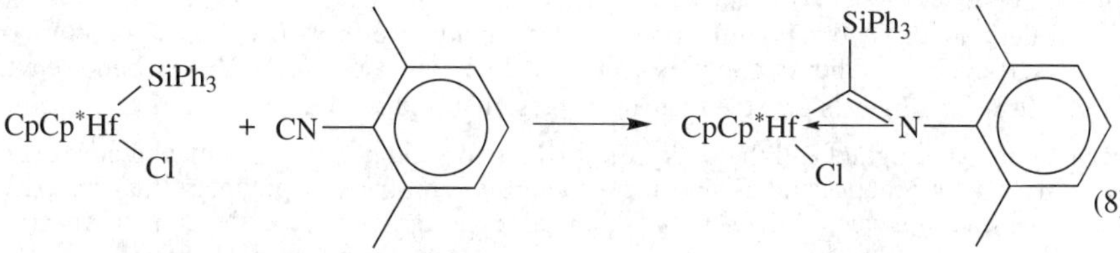

$$
\text{CpCp*Hf(SiPh}_3\text{)Cl} + \text{CN-Ar} \longrightarrow \text{CpCp*Hf(Cl)}[\eta^2\text{-C(SiPh}_3\text{)=N-Ar}] \quad (8)
$$

Interestingly, the Zr and Hf complexes, Cp$_2$*M(SiPh$_3$)Cl and CpCp*M(SiPh$_3$)Cl, were found not to react with acetylenes, ethylene, CS or phenyl isocyante[20]. σ-Bond metathesis reactions of Cp$_2$M(SiR$_3$)Cl (M = Zr, Hf; R = Me, SiMe$_3$) with PhSiH$_3$ yield the exchange products R$_3$SiH and Cp$_2$M(SiH$_2$Ph)Cl (M = Zr, Hf). The zirconium complexes rapidly combine with PhSiH$_3$ producing the polymeric hydride zirconium [CpZrHCl]$_n$, PhH$_2$Si—SiH$_2$Ph and PhH$_2$Si—SiHPh—SiH$_2$Ph. In contrast, similar reactions of the mixed cyclopentadienyl hafnium complex CpCp*Hf(Si(SiMe$_3$)$_3$)Cl (9) with hydrosilanes RR'SiH$_2$ under fluorescent room lighting cleanly gave the corresponding silyl derivatives CpCp*Hf(SiHRR')Cl [SiHRR' = SiH$_2$Ph, SiH$_2$(C$_6$H$_4$Me-p), SiH$_2$(C$_6$H$_4$OMe-p), SiH$_2$(C$_6$H$_4$F-p), SiH$_2$Mes, SiH$_2$CH$_2$Ph, SiH$_2$CH$_2$Cy, SiHPh$_2$, SiHPhMe and SiHPhSiH$_2$Ph]. The last two complexes are obtained as a 7:5 and 1:1

mixture of diastereomers, respectively. Reaction of complex **9** with 0.5 equivalent of the 1,4-bis(silyl)benzene affords the bimetallic silyl complex **10** through the monohafnium intermediate (equation 9).

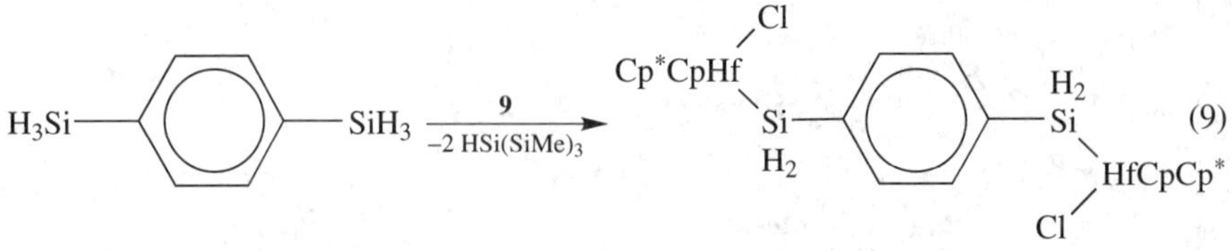

$$H_3Si-\langle C_6H_4 \rangle-SiH_3 \xrightarrow[-2\ HSi(SiMe_3)_3]{9} Cp^*CpHf(Cl)(H_2)Si-\langle C_6H_4 \rangle-Si(H_2)HfCpCp^*(Cl) \quad (9)$$

(10)

Following the course of equation 9, a bimetallic thiophene complex 2,5-[CpCp*Hf(SiH$_2$)Cl]$_2$C$_4$H$_2$S can be similarly obtained. The Hf−Cl bonds in the bimetallic complex are *trans* to one another, producing one set of diastereomeric SiH$_2$ protons as observed in the ^{1}H NMR spectrum (Table 2).

In the dark, the mixed cyclopentadienyl complex CpCp*Hf(SiH$_2$Ph)Cl reacts with another mixed complex, CpCp*Hf(SiH$_2$(C$_6$H$_4$Me-*p*))Br, yielding an equilibrating mixture of the exchanged halide complexes. Similarly, the σ-bond metathesis of CpCp*Hf(SiH$_2$Ph)Cl with hydrosilanes leads to an equilibrium mixture of the silyl-exchanged complexes. The reactions in the dark are slower although well behaved. The σ-bond metathesis reaction of **9** with PhSiH$_3$ follows a second-order rate law (rate = k[**9**][PhSiH$_3$]), showing a kinetic isotope effect {k(PhSiH$_3$)/k(PhSiD$_3$)} of 2.5, arguing in favor of a four-center transition state. Under fluorescent room lighting, the reactions are very sensitive to the bulk of the silane and to low concentration of Lewis acids, indicating that in these photochemical σ-bond metathesis reactions, a reactive coordinatively unsaturated intermediate is involved. Reactions of complex **9** with alkoxyhydrosilanes HSi(OMe)$_2$R″ provides a route to the new complex CpCp*Hf(SiH$_2$Me)Cl when R″ = Me and CpCp*Hf(SiH$_3$)Cl when R″ = OMe. This reaction proceeds through initial metal-catalyzed redistribution of the alkoxyhydrosilanes, 3HSi(OMe)$_2$R″ $\rightleftharpoons$ R″SiH$_3$ + 2R″Si(OMe)$_3$, followed by trapping of the new hydrosilane by complex **9**[21].

X-ray analysis of complex **9** shows two independent molecules in the unit cell. The Hf−Si bonds of 2.881(4) Å and 2.888(4) Å are longer than those found in other d^0 Zr−Si derivatives (Table 1), reflecting the steric hindrance in **9**. The crystal structure of the mixed cyclopentadienyl complex CpCp*Hf(SiH$_2$Ph)Cl reveals a Hf−Si bond length of 2.729(3) Å, which is shorter than in **9**, presumably due to a lower steric interaction between the silyl ligand and the bulky Cp*[21]. In a very elegant way, Tilley and coworkers have shown by kinetic studies and the structure of the derived polymers that polysilyl and hydride complexes of early-transition metals are intermediates in the dehydrogenative coupling of silanes by the σ-bond metathesis mechanism. Thus, CpCp*Hf[Si(SiMe$_3$)$_3$]Me, which is prepared by the reaction of CpCp*Hf[Si(SiMe$_3$)$_3$]Cl with MeMgBr, reacts with silanes to yield the corresponding hydride complexes and the dehydrogenative coupling product of silanes as shown in equation 10[22]. The σ-bond metathesis mechanism proposed by Tilley is shown in Scheme 6[23].

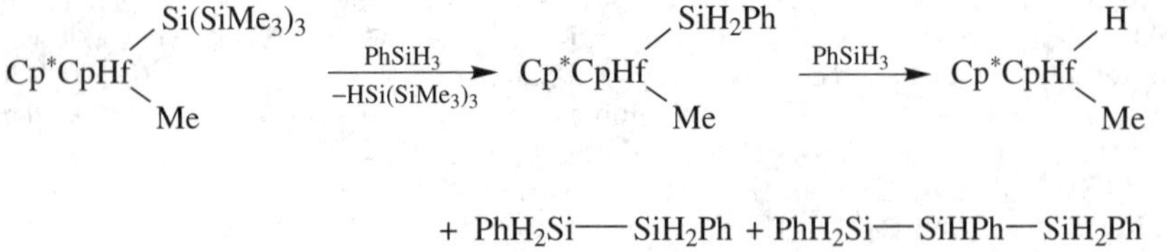

$$Cp^*CpHf(Si(SiMe_3)_3)(Me) \xrightarrow[-HSi(SiMe_3)_3]{PhSiH_3} Cp^*CpHf(SiH_2Ph)(Me) \xrightarrow{PhSiH_3} Cp^*CpHf(H)(Me) \quad (10)$$

$$+ \ PhH_2Si-SiH_2Ph + PhH_2Si-SiHPh-SiH_2Ph$$

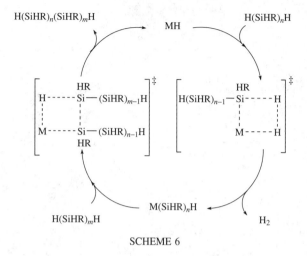

SCHEME 6

Interestingly, the complexes CpCp*Hf(SiHPhSiHPhSiH$_2$Ph)Cl and CpCp*Hf(SiHPh SiH$_2$Ph)Cl are observed to form concurrently with the small polysilanes. It seems that these complexes are not necessarily the intermediates in the dehydrogenative process as suggested in Scheme 6, but perhaps only side products formed via non-productive σ-bond metathesis reactions (equation 11)[22].

$$2\ Cp^*CpHf\begin{matrix}SiH_2Ph\\Cl\end{matrix} \xrightarrow{PhSiH_3} Cp^*CpHf\begin{matrix}SiHPhSiH_2Ph\\Cl\end{matrix} + Cp^*CpHfHCl$$

$$\xrightarrow{PhSiH_3}$$

$$Cp^*CpHf\begin{matrix}SiHPhSiHPhSiH_2Ph\\Cl\end{matrix} + PhH_2Si{-\!\!-}SiH_2Ph + PhH_2Si{-\!\!-}SiHPh{-\!\!-}SiH_2Ph$$

(11)

Cationic silicenium complexes of zirconium, **11** and **12**, have also been observed as intermediates in the high molecular weight dehydrogenative coupling of phenylsilanes. The cation-like structure is formed by the reaction of the corresponding zirconocene dichloride complexes with BuLi, (C$_6$F$_5$)$_3$B and PhSiH$_3$. The incorporation of (C$_6$F$_5$)$_3$B into the catalytic system reduces the overall activity of the catalyst, and strongly suppresses the ring formation. Special features of the complexes **11** and **12** are the silicenium-like ligand (^{29}Si δ = 106–108 ppm) and the diastereotopic hydride presence in **12** suggesting that silicon is chiral[24].

Besides the σ-bond metathesis mechanism proposed by Tilley[23] for the dehydrogenative coupling of silanes, a Zr(II) pathway[25] and a silylene mechanism[26] have been proposed based on the nature of the products. The dehydrogenative polymerization of 1,2,3-trimethyltrisilane or of a mixture of diastereomers of 1,2,3,4-tetramethyltetrasilane showed evidence that, besides Tilley's mechanism, a further mechanism is present. The product formation can be explained by a silylene mechanism where the silylenes are formed by α-elimination from the silyl complexes by a new type of β-elimination which involves Si—Si bond cleavage (β*-bond elimination) as described in Scheme 7[27].

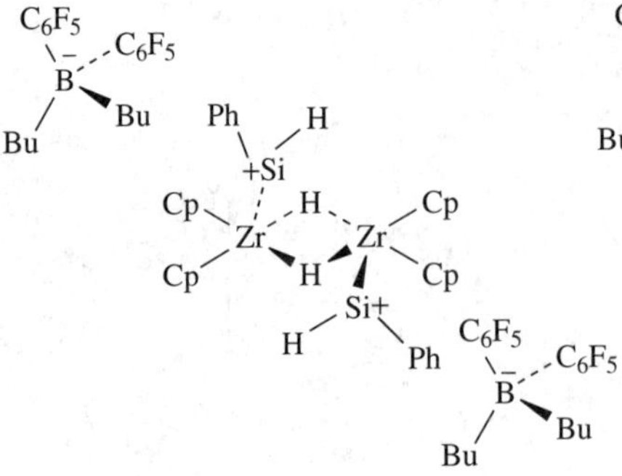

(11) 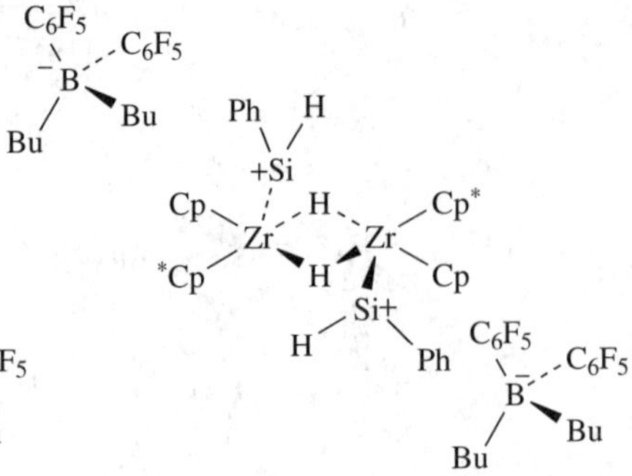 (12)

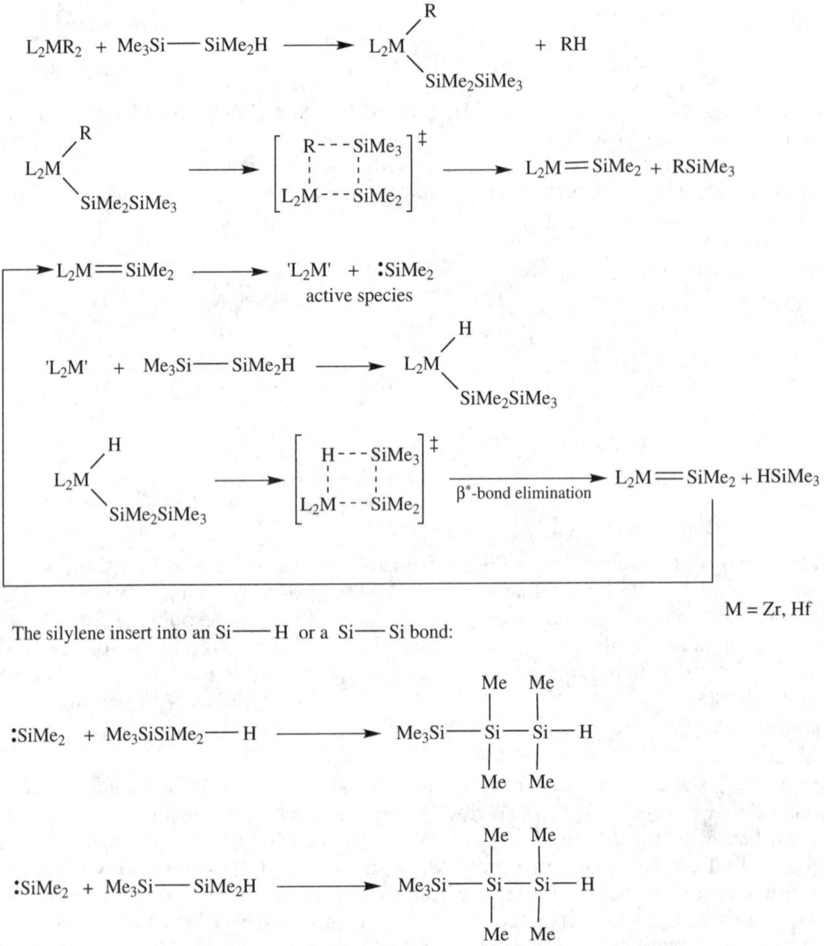

$L_2MR_2 + Me_3Si\text{—}SiMe_2H \longrightarrow L_2M\overset{R}{\underset{SiMe_2SiMe_3}{}} + RH$

$L_2M\overset{R}{\underset{SiMe_2SiMe_3}{}} \longrightarrow \left[\begin{array}{c} R\text{---}SiMe_3 \\ L_2M\text{---}SiMe_2 \end{array}\right]^{\ddagger} \longrightarrow L_2M=SiMe_2 + RSiMe_3$

$\longrightarrow L_2M=SiMe_2 \longrightarrow \text{'}L_2M\text{'} + :SiMe_2$
$ \text{active species}$

$\text{'}L_2M\text{'} + Me_3Si\text{—}SiMe_2H \longrightarrow L_2M\overset{H}{\underset{SiMe_2SiMe_3}{}}$

$L_2M\overset{H}{\underset{SiMe_2SiMe_3}{}} \longrightarrow \left[\begin{array}{c} H\text{---}SiMe_3 \\ L_2M\text{---}SiMe_2 \end{array}\right]^{\ddagger} \xrightarrow{\beta^*\text{-bond elimination}} L_2M=SiMe_2 + HSiMe_3$

M = Zr, Hf

The silylene insert into an Si—H or a Si—Si bond:

$:SiMe_2 + Me_3SiSiMe_2\text{—}H \longrightarrow Me_3Si\text{—}\underset{Me}{\overset{Me}{Si}}\text{—}\underset{Me}{\overset{Me}{Si}}\text{—}H$

$:SiMe_2 + Me_3Si\text{—}SiMe_2H \longrightarrow Me_3Si\text{—}\underset{Me}{\overset{Me}{Si}}\text{—}\underset{Me}{\overset{Me}{Si}}\text{—}H$

SCHEME 7

The silyl chloride complex (η^5-C$_5$H$_4$SiMe$_3$)$_2$Zr[Si(SiMe$_3$)$_3$]Cl and the silyl–tin chloride complexes Cp$_2$M[Si(SnMe$_3$)$_3$]Cl (M = Ti, Zr, Hf) were prepared from the corresponding metallocene dichlorides with the corresponding silyl lithium reagent. Concomitant methylation to the methyl complexes is obtained by use of the methyl Grignard reagent. The molecular structure for the two Zr chloride complexes shows Zr–Si distances of 2.833(3) Å and an average of 2.768(4) Å for the Zr–Si(SnMe$_3$)$_3$ derivatives, respectively. The shorter Zr–Si in the latter compound presumably reflects a lower steric crowding. The complex (η^5 - C$_5$H$_4$SiMe$_3$)$_2$ Zr[Si(SiMe$_3$)$_3$]Cl has been found to be an effective catalyst for the dehydropolymerization of PhSiH$_3$ and n-Bu$_2$SnH$_2$ to low molecular weight polymers. In as much as the complexes Cp$_2$M[Si(SnMe$_3$)$_3$]Cl (M = Zr, Hf) do not react with CO, the Ti analog complex undergoes a CO-induced reductive elimination to Cp$_2$Ti(CO)$_2$ and ClSi(SnMe$_3$)$_3$[28]. Interestingly, theoretical studies on these Schrock-type silylidene complexes shows that, if formed, these complexes consist of contributions of the σ- and π-covalent, dative and back-bonding resonance structures based on the corresponding metal[29].

η^2-Silanimine complex Cp$_2$Zr(η^2-Me$_2$Si=NBu-t)(PMe$_3$) (**13**) was obtained by addition of LiCH$_2$SiMe$_3$ to Cp$_2$Zr(N(Bu-t)SiMe$_2$H)I in the presence of PMe$_3$. The η^2-bond ligation results from the intermolecular elimination of SiMe$_4$ from Cp$_2$Zr(CH$_2$SiMe$_3$)(N(Bu-t)SiMe$_2$H). The structure of Cp$_2$Zr(η^2-Me$_2$Si=NBu-t)(PMe$_3$) reveals short Zr–Si [2.654(1) Å] and long Si–N [1.687(3) Å] distances, suggesting the importance of the metallacyclic, Zr(IV) resonance form[30]. The preparation of complex **13** and its reaction with carbon monoxide, hydrogen, ethylene and ketene are shown in Scheme 8[31–33]. Reaction of η^2-silanimine complexes Cp$_2$Zr(η^2-Me$_2$Si=NBu-t)(L) (L = CO, PMe$_3$) with CO$_2$ leads to a net oxygen insertion into the Zr–Si bond. The products are CO, and the dimeric oxazirconacycle [cyclo-Cp$_2$Zr(OSiMe$_2$NBu-t)]$_2$. Low-temperature ^{1}H and ^{13}C {H} NMR studies *in situ* allowed the observation of the *cyclo*-Cp$_2$Zr[OC(=O)SiMe$_2$NBu-t] as the initial intermediate formed by the insertion of one of the C=O double bonds into the Zr–Si bond of Cp$_2$Zr(η^2-Me$_2$Si=NBu-t)(L). Concomitant decarboxylation of the *cyclo*-Cp$_2$Zr[OC(=O)SiMe$_2$NBu-t] at room temperature and dimerization leads to the oxazirconacycle complex [*cyclo*-Cp$_2$Zr(OSiMe$_2$NBu-t)]$_2$. Carbon disulfide was found to react with **13**, yielding the monomeric four-membered metallacycle complex *cyclo*-Cp$_2$Zr(SSiMe$_2$NBu-t). Interestingly, the analog carbonyl monosulfide leads to a mixture of the oxygen and sulfur atom insertion products, as shown in equation 12[32].

$$\begin{array}{c}
\text{CP}_2\text{Zr}\overset{\text{PMe}_3}{\underset{\text{N}}{\diagup\!\!\!\diagdown}}\text{SiMe}_2 + \text{O}=\text{C}=\text{S} \xrightarrow{-\text{PMe}_3} \text{CP}_2\text{Zr}\overset{\text{CO}}{\underset{\text{N}}{\diagup\!\!\!\diagdown}}\text{SiMe}_2 \\
\text{Bu-}t \text{Bu-}t
\end{array}$$

$$+ \quad \begin{array}{c}\text{Me}_2\text{Si} - \text{N} \\ \diagdown \quad \diagdown \\ \text{O} - \text{ZrCp}_2 \\ \diagup \quad \diagup \\ \text{Cp}_2\text{Zr} - \text{O} \\ \diagdown \quad \diagdown \\ \text{N} - \text{SiMe}_2 \\ \diagup \\ \text{Bu-}t\end{array} \quad + \quad \text{Cp}_2\text{Zr}\overset{\text{S}}{\underset{\text{N}}{\diagup\!\!\!\diagdown}}\text{SiMe}_2 \text{Bu-}t \quad (12)$$

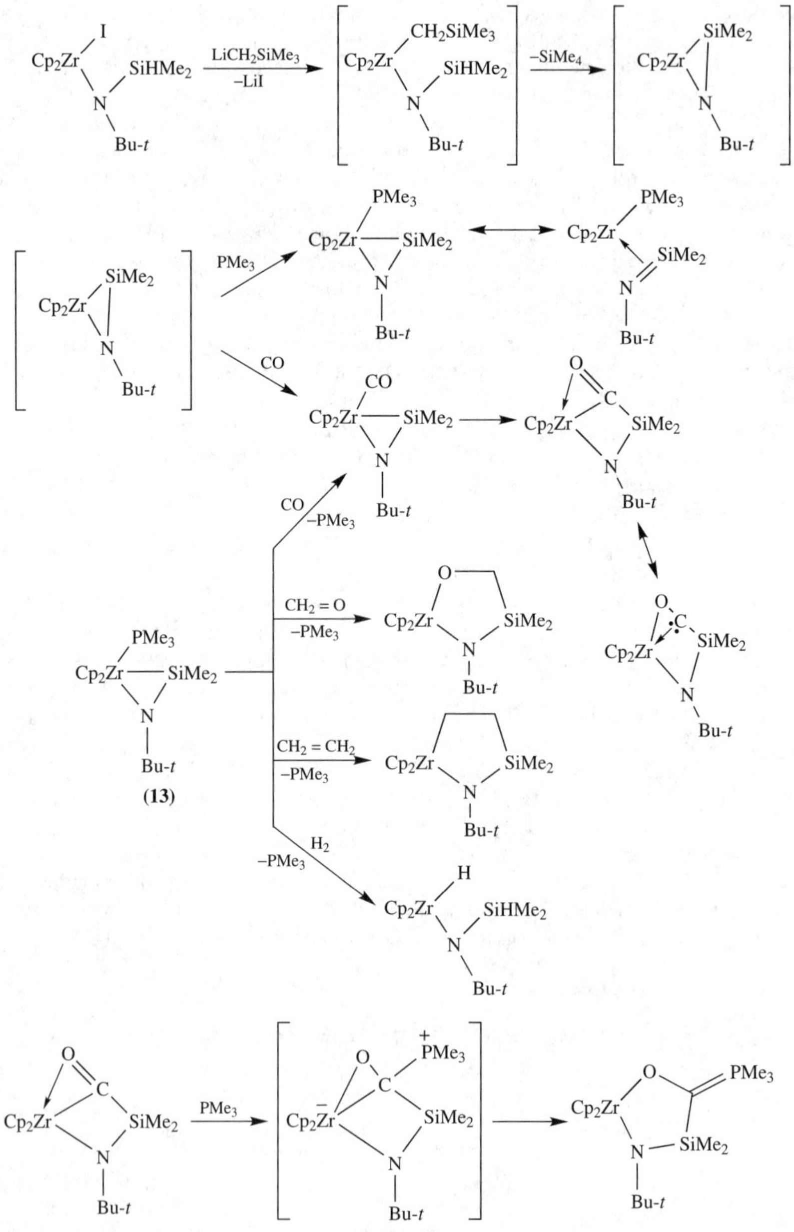

SCHEME 8

Reaction of $Cp_2Ti(\eta^2\text{-}Me_3SiC\equiv CSiMe_3)$ with $t\text{-}BuC\equiv CSiMe_2H$ in hexane leads to the alkyne substitution. The titanocene complex shows a temperature-dependent Si−H−Ti interaction as demonstrated by IR spectroscopy. This agostic interaction is ascertained by the ^{29}Si NMR (193 K) spectrum showing a signal at $\delta = 17.6$ ppm ($\delta = -0.5$ ppm at 303 K) with a coupling constant J_{SiH} of 93 Hz ($J_{SiH} = 123$ Hz at 303 K) compared with the J_{SiH} of 199 Hz of the free alkyne, indicating that the Ti−H bond formation is not complete and that the Si−H bond is not fully broken. The solid structure of the titanium complex reveals normal distances for the Ti−H [1.82(5) Å], Si−H [1.42(6) Å] and Ti−Si [2.655(2) Å] bonds. The symmetrical Cp$_2$Ti(η^2-HMe$_2$SiC≡CSiMe$_2$H) which was obtained in a similar way carries two Si−H functions, exhibiting a dynamical behavior showing a 'flip-flop' coordination[34].

Free of anionic π-ligands, the bulky monomeric silyl complex Np$_3$ZrSi(SiMe$_3$)$_3$ (Np = neopentyl) has been prepared by the metathesis of Np$_3$SiCl with (Me$_3$Si)$_3$SiLi·3THF. Interestingly, ^{29}Si {H} NMR of the complex exhibits a large upfield shift of the Zr−Si signal to −7.64 ppm (Table 2)[35].

TABLE 1. Selected M−Si distances in group-4 transition-metal silyl complexes

Compound	d^n Configuration	M−Si (Å)	Reference
Cp$_2$Zr(H)(SiPh$_3$)(PMe$_3$)	d^0	2.721(2)	14
Cp$_2$Zr(H)(SiHPh$_2$)(PMe$_3$)	d^0	2.707(5)	15
[Cp$_2$Zr(μ-H)(SiHPh$_2$)]$_2$	d^0	2.759(8)	15
Cp$_2$Zr(SiPhMeH)(μ-H)$_2$(SiPhMe$_2$)ZrCp$_2$	d^0	2.806(4)	13
CpCp*Hf[Si(SiMe$_3$)$_3$]Cl	d^0	2.881(4), 2.888(4)	21
CpCp*Hf(SiH$_2$Ph)Cl	d^0	2.729(3)	21
Cp$_2$Ti(SiH$_2$Ph)PMe$_3$	d^1	2.650(1)	16
Cp$_2$Ti(SiHMePh)PMe$_3$	d^1	2.646(2)	16
[η_5-C$_5$H$_4$SiMe$_3$)$_2$Zr[Si(SiMe$_3$)$_3$Cl]	d^0	2.833(4)	28
Cp$_2$Zr[Si(SnMe$_3$)$_3$Cl]	d^0	2.772(4), 2.765(4)	28
Cp$_2$Ti(SiH$_2$Ph)PMe$_3$	d^1	2.597(2)	18
(Me$_2$N)$_3$TiSi(SiMe$_3$)$_3$	d^0	2.635(2)	19
Cp$_2$Zr(η^2-Me$_2$Si=NBu-t)(PMe$_3$)	d^0	2.654(1)	31
Cp$_2$Zr(η^2-Me$_2$Si=NBu-t)(CO)	d^0	2.706(1)	30
Cp$_2$Ti(η^2-(t-Bu)C≡CSiMe$_2$H)	d^0	2.655(2)	34

TABLE 2. Selected ^{29}Si NMR chemical shifts for group-4 transition-metal silyl complexes^a

Compound	^{29}Si NMR shift (ppm)	Reference
Cp$_2$Zr(H)(SiPh$_3$)(PMe$_3$)	39.48 ($^2J_{SiP} = 15$ Hz)	14
Cp$_2$Zr(H)(SiHPh$_2$)(PMe$_3$)	50.46	14
CpCp*Zr(SiPh$_3$)Cl	42.42	20
CpCp*Hf(SiPh$_3$)Cl	39.96	20
Cp$_2$*Zr(SiPh$_3$)Cl	47.34	20
Cp$_2$*Hf(SiPh$_3$)Cl	42.86	20
CpCp*Zr[Si$_a$(Si$_b$Me$_3$)$_3$]Cl	−87.30 (Si$_a$); −6.03 (Si$_b$)	21
CpCp*Hf[Si$_a$(Si$_b$Me$_3$)$_3$]Cl	−77.87 (Si$_a$); −4.85 (Si$_b$)	21
CpCp*Zr(SiH$_2$Ph)Cl	−14.33 ($^1J_{SiH} = 144$ Hz)	21

(*continued overleaf*)

TABLE 2. (continued)

Compound	^{29}Si NMR shift (ppm)	Reference
CpCp*Hf(SiH$_2$Ph)Cl	1.49 ($^1J_{SiH} = 155$ Hz)	21
CpCp*Hf[SiH$_2$(Tol-p)]Cl	1.46 ($^1J_{SiH} = 157$ Hz)	21
CpCp*Hf[SiH$_2$(Tol-p)]Br	7.74 ($^1J_{SiH} = 153$ Hz)	21
CpCp*Hf[SiH$_2$(C$_6$H$_4$OMe-p)]Cl	1.56 ($^1J_{SiH} = 169$ Hz)	21
CpCp*Hf[SiH$_2$(C$_6$H$_4$F-p)]Cl	1.17 ($^1J_{SiH} = 160$ Hz)	21
CpCp*Hf[SiH$_2$(Mes)]Cl	1.50 ($^1J_{SiH} = 157$ Hz)	21
CpCp*Hf(SiH$_2$CH$_2$Ph)Cl	9.69 ($^1J_{SiH} = 153$ Hz)	21
CpCp*Hf(SiH$_2$Cy)Cl	14.83 ($^1J_{SiH} = 145$ Hz)	21
Cp$_2$Hf(SiH$_2$Cy)Cl	11.07 ($^1J_{SiH} = 149$ Hz)	21
CpCp*Hf(SiHPh$_2$)Cl	32.25 ($^1J_{SiH} = 158$ Hz)	21
Cp$_2$Hf(SiHPh$_2$)Cl	31.88 ($^1J_{SiH} = 148$ Hz)	21
CpCp*Zr(SiPh$_3$)Cl	42.42	21
CpCp*Hf(SiPh$_3$)Cl	39.96	21
CpCp*Hf(SiHPhMe)Cl (diastereomers)	21.75 ($^1J_{SiH} = 153$ Hz) 25.10 ($^1J_{SiH} = 153$ Hz)	21
CpCp*Hf(Si$_a$HPhSi$_b$H$_2$Ph)Cl (diastereomers)	−9.05 (Si$_a$; $^1J_{SiH} = 152$ Hz) −9.86 (Si$_a$; $^1J_{SiH} = 152$ Hz) −43.91 (Si$_b$; $^1J_{SiH} = 183$ Hz) −50.43 (Si$_b$; $^1J_{SiH} = 183$ Hz)	21
CpCp*Hf (SiH$_2$C$_6$H$_4$SiH$_3$-p)Cl	1.16 (Si$_a$; $^1J_{SiH} = 162$ Hz) −63.10 (Si$_b$; $^1J_{SiH} = 199$ Hz)	21
2,5-[CpCp*Hf (Cl) SiH$_2$]$_2$ C$_4$H$_2$S	−14.09 ($^1J_{SiH} = 158$ Hz)	21
CpCp*Hf (SiH$_2$CH$_3$)Cl	−7.36 ($^1J_{SiH} = 149$ Hz)	21
CpCp*Hf (SiH$_3$)Cl	−46.52 ($^1J_{SiH} = 156$ Hz)	21
Cp$_2$Hf [Si(SnMe$_3$)$_3$Cl]	7.61 ($^1J_{SiSn} = 130.7$ Hz)	28
Cp$_2$Zr[Si(SnMe$_3$)$_3$Cl]	1.88 ($^1J_{SiSn} = 119.6$ Hz)	28
Cp$_2$Ti(SiH$_2$Ph)PMe$_3$	49.90 ($^1J_{SiH} = 161; 28$ Hz)	18
[Cp$_2$Zr(SiHPh)(μ-H)]$_2^{2+}$ [Bu$_2$(C$_6$F$_5$)$_2$B]$_2^{2−}$	108.04	24
[CpCp*Zr(SiHPh)(μ-H)]$_2^{2+}$ [Bu$_2$(C$_6$F$_5$)$_2$B]$_2^{2−}$	106.0	24
Cp$_2$Zr(η^2-Me$_2$Si=NBu-t)(CO)	69.9 ($^1J_{SiC} = 24$ Hz)	30
Cp$_2$Ti(η^2-(t-Bu)C≡CSiMe$_2$H)	−0.5 ($^1J_{SiH} = 123$ Hz at 303 K) 17.6 ($^1J_{SiH} = 93$ Hz at 193 K)	34
Cp$_2$Ti(η^2-HMe$_2$SiC≡CSiMe$_2$H)	0.46 ($^1J_{SiH} = 147$ Hz at 303 K)	34
Np$_3$ZrSi$_a$(Si$_b$Me$_3$)$_3$[b]	−7.64 (Si$_a$); −85.77 (Si$_b$)	35

[a]Reporting ^{29}Si NMR data for transition-metal silyl complexes is now a routine. Although some trends have emerged, it is sometimes difficult to interpret satisfactorily the meaning of chemical shift data. However, ^{29}Si NMR spectroscopy can be very useful in cases where coupling constants can be obtained, or where comparison among closely related systems is possible.
[b]Np = neopentyl.

VI. ACTINIDE SILYL COMPLEXES

For reasons that are currently unclear, actinide silyl complexes have proven very difficult to isolate, with the first reported success being Cp_3USiPh_3[36]. Similarly, $Cp_3USi(TMS)_3$ has been prepared by the metathesis reaction of $(Me_3Si)_3SiLi\cdot 3THF$ with the corresponding Cp_3UCl. The uranium–silicon bond dissociation enthalpies have been measured to be as low as ca 37 ± 3 kcal mol^{-1}, arguing presumably for the high reactivity of the actinide–silyl bonds[37]. Recently, use of the benzamidinate ancillary ligands, which are regarded as cyclopentadienyl equivalents, allows the preparation of thorium and uranium silyl complexes as described in equation 13. These benzamidinate complexes were found to be active for the dehydrogenative coupling of $PhSiH_3$[38].

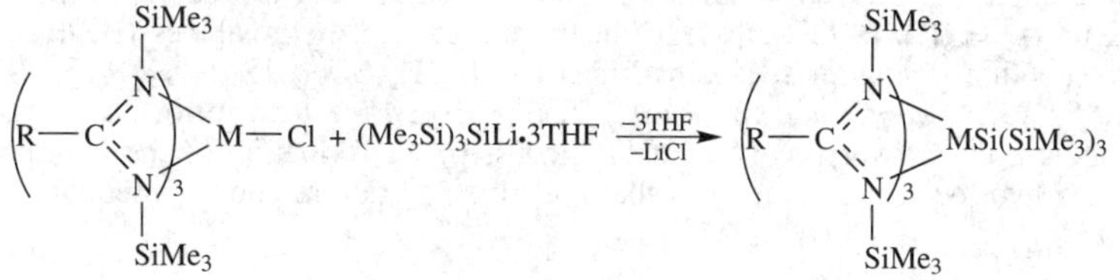

$$\left(R-C\underset{\underset{SiMe_3}{|}}{\overset{\overset{SiMe_3}{|}}{\underset{N}{\overset{N}{\diagup}}}}\right)_3 M-Cl + (Me_3Si)_3SiLi\cdot 3THF \xrightarrow[-LiCl]{-3THF} \left(R-C\underset{\underset{SiMe_3}{|}}{\overset{\overset{SiMe_3}{|}}{\underset{N}{\overset{N}{\diagup}}}}\right)_3 MSi(SiMe_3)_3 \qquad (13)$$

R = 4-MeC$_6$H$_4$
R = Ph
M = Th, U

Recently, double insertion of CO into the reactive thorium–silicon bond of $Cp_2{}^*ThSiR_3(Cl)$ allowed the isolation of the first metalloxy ketene complex **14**. Reactivity studies of complex **14** are described in Scheme 9[39].

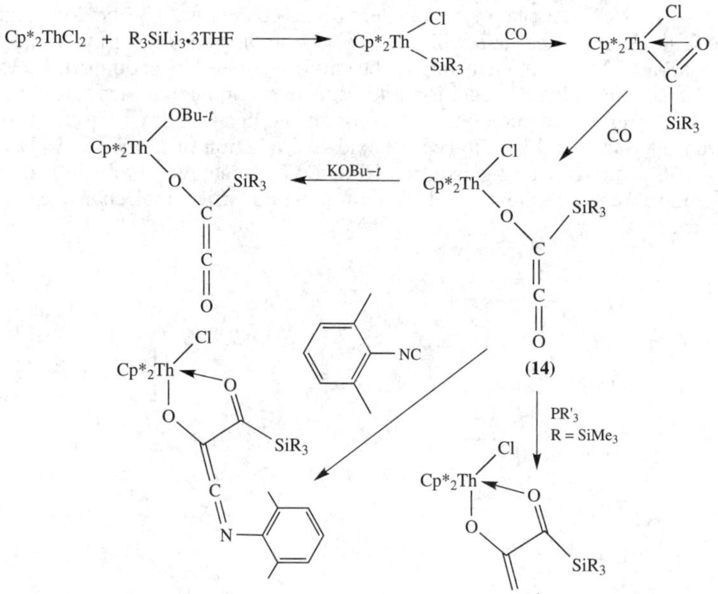

R=Si(Me$_3$)$_3$, Si(Bu-t)Ph$_2$
PR'$_3$=PMe$_3$, PMePh$_2$

SCHEME 9

VII. GROUP-5 SILYL COMPLEXES

Organosilicon complexes of group-5 metals are among the most recently explored in the transition-metal series. The first vanadium silyl compound was reported in 1976[40], whereas the first niobium and tantalum silyl complexes were reported only in 1985[41]. The two most common methods for preparing metal–silicon bonds are: (1) Salt elimination reactions, starting either from halosilanes and an anionic metal complex or from a metal halide and the corresponding anionic silyl derivative. (2) Oxidative addition of hydrosilanes to coordinatively unsaturated transition-metal complexes. The latter route has been widely applied for the synthesis of late-transition-metal silyl complexes and was recently developed to prepare group-5 silyl complexes by the direct reaction of hydrosilanes with tantalum complexes. Thus, thermolysis (>100 °C) of Cp_2TaH_3 or photolysis (350 nm) of $Cp_2Ta(L)H$ (L = CO, PMe_3, ethylene) in the presence of hydrosilanes ($HSiR_3$) affords the corresponding tantalum silyl dihydrides $Cp_2Ta(H)_2(SiR_3)$ [$SiR_3 = SiH_3$, $SiMeH_2$, $SiMe_2H$, $SiMe_3$, $SiMe_2Cl$ and $Si(OMe)_3$]. The proposed mechanism for the formation of these tantalum complexes involves, at the first step, the formation of the coordinatively unsaturated hydride which undergoes the oxidative addition as shown in equation 14.

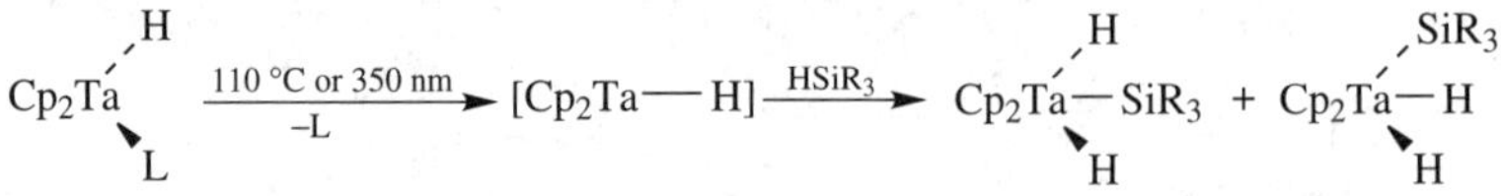

$$Cp_2Ta\begin{smallmatrix}H\\L\end{smallmatrix} \xrightarrow[-L]{110\,°C\ or\ 350\ nm} [Cp_2Ta-H] \xrightarrow{HSiR_3} Cp_2Ta\begin{smallmatrix}H\\-SiR_3\\H\end{smallmatrix} + Cp_2Ta\begin{smallmatrix}SiR_3\\-H\\H\end{smallmatrix} \quad (14)$$

^{1}H NMR spectroscopy shows that both isomers interconvert in solution with an isomerization rate for SiR_3 which follows the trend $SiH_3 \approx SiMe_2Cl < Si(OMe)_3 \approx SiMeH_2 < SiMe_2H < SiMe_3$[42]. Bis(silyl) complexes of Ta(V) $Cp_2Ta(H)(SiR_3)_2$ [$SiR_3 = SiMeH_2$, $SiMe_2H$, $SiMe_3$, $SiMe_2Cl$ and $Si(OMe)_3$] were prepared by the reaction of the Ta(III) complex $Cp_2Ta(L)(CH_3)$ (L = PMe_3, ethylene) with an excess of the corresponding hydrosilane $HSiR_3$ under photolytic or thermolytic conditions. The predominant isomer observed is the symmetrical one with the two silyl groups occupying the two equivalent lateral positions. The mechanism for the bis(silyl) complex formation can be explained in terms of simple oxidative addition and reductive elimination steps as described in Scheme 10. A thermal or photolytic activation of the ligand L from $Cp_2Ta(L)Me$ to the 16-electron intermediate **15**, followed by oxidative addition of $HSiR_3$ to **16**, could give the two rapidly interconverting isomers **16** and **17**. Reductive elimination of methane from **17** and subsequent oxidative addition of a second $HSiR_3$ molecule would yield the bis(silyl) product.

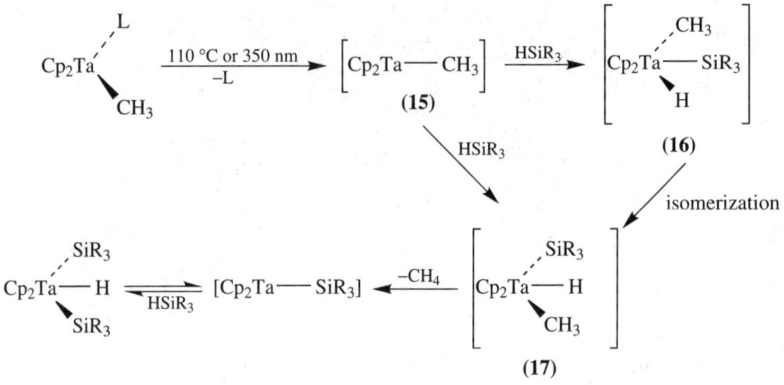

SCHEME 10

Tantalum(III) silyl complexes of the type $Cp_2Ta(PMe_3)(SiR_3)$ [$SiR_3 = SiMeH_2$, $SiMe_3$, $SiMe_2Cl$ and $Si(OMe)_3$] have been prepared from the corresponding bis(silyl) compounds by thermolysis in the presence of a phosphine (equation 15). In addition, the $Cp_2Ta(PMe_3)$ [$SiH(Bu-t)_2$] complex has been prepared in a one-pot synthesis from $Cp_2Ta(PMe_3)Me$ and large excess of $H_2Si(Bu-t)_2$ under 350-nm radiation. The reaction should be performed in non-aromatic solvents since the silyl complex induces a C−H activation of arenes[42].

$$Cp_2Ta\overset{SiR_3}{\underset{SiR_3}{-}}H + PMe_3 \xrightarrow{110\ °C} Cp_2Ta\overset{SiR_3}{\underset{PMe_3}{-}} + HSiR_3 \quad (15)$$

Interestingly, thermodynamic studies on the bond dissociation energies of $Cp_2Ta(PMe_3)(SiH(Bu-t)_2)$ and $Cp_2Ta(PMe_3)(SiMe_3)$ which react in neat arenes (ArH) to produce the C−H activation complex $Cp_2Ta(PMe_3)Ar$ have shown that the Ta−Si bonds in these two complexes are respectively 5.4 and 7.9 kcal mol^{-1} weaker than the Ta−Ph bond in $Cp_2Ta(PMe_3)Ph$. However, metal–phenyl bonds are generally much stronger than metal–alkyl bonds and the Ta−Si bond dissociation energies (BDE) are probably similar to or greater than those of the tantalum–alkyl bonds. For the tantalum–silyl complexes, the bulkier silyl exhibits the lower Ta−Si BDE, but surprisingly the lower stability of $Cp_2Ta(PMe_3)(SiH(Bu-t)_2)$ with respect to the phenyl complex $Cp_2Ta(PMe_3)Ph$ is primarily due to a large and favorable entropy change (34 ± 3 eu) resulting from the reduction of steric congestion upon converting the silyl into the phenyl complex[43]. The phosphine ligand in the Ta(III) complexes $Cp_2Ta(PMe_3)(SiR_3)$ [$SiR_3 = SiMe_3$, $SiH(Bu-t)_2$] can be replaced by CO, either by thermolysis or by photolytic cleavage, yielding the corresponding $Cp_2Ta(CO)(SiR_3)$ [$SiR_3 = SiMe_3$, $SiH(Bu-t)_2$] complexes.

Solid state structures of $Cp_2TaH(SiMe_2H)_2$ and $Cp_2Ta(PMe_3)SiMe_3$ show normal Ta−Si bond lengths of 2.633(2) Å and 2.624(2) Å for the former complex and 2.639(4) Å for the latter complex (Table 3)[42]. In addition, for the carbon monoxide analog, $Cp_2Ta(CO)(SiMe_3)$, a similar bond length of 2.631(2) Å was found[32]. Regarding the structural similarities and ignoring the hydride, all the molecules exhibit tetrahedral geometries which are typical for bent metallocene complexes. Solid structures for the sterically hindered silyl tantalum complexes $Cp_2Ta(PMe_3)(SiH(Bu-t)_2)$ and $Cp_2Ta(Co)(SiH(Bu-t)_2)$ exhibit bond lengths of 2.740(1) Å and 2.684(1) Å, respectively. It is noteworthy that the Ta−Si bond lengths in the two di-*tert*-butyl silyl complexes are quite different and that the more electron-rich phosphine complex exhibits the longer distance whereas the ^{29}Si NMR chemical shifts of both complexes are similar (Table 4).

Monosilyl complexes of niobium of the type $Nb(\eta^5-C_5H_4SiMe_3)_2H_2(SiR_3)$ ($SiR_3 = SiMe_2Ph$, $SiMePh_2$, $SiPhH_2$, $SiPh_2H$, $SiPh_3$) have been obtained by the thermolytic reaction of $Nb(\eta^5-C_5H_4SiMe_3)_2H_3$ with the corresponding organosilicon hydride. In all cases only the isomer with the silyl group in the central equatorial position was obtained. The Nb−Si bond length of 2.616(3) Å in the molecular structure of $Nb(\eta^5-C_5H_4SiMe_3)_2)_2H_2$ ($SiPh_2H$) is *ca* 0.016 Å shorter than in similar Ta(V) complexes (Table 3), and in agreement with the corresponding ionic radius[44].

Niobium bis(silyl) complexes of the type $Cp_2Nb(SiMe_2X)_2H$ (X = Cl, Ph) have been prepared by the oxidative addition of $HSiMe_2X$ to the corresponding monoalkyl-niobocene $Cp_2NbCH_2CH_2Ph$. The structures of the bis(silyl)niobocene complexes are isostructural and exhibit the typical trisubstituted geometry. In $Cp_2Nb(SiMe_2Cl)_2H$ the Nb−Si bond lengths are 2.584(5) Å and 2.611(5) Å. Those distances were found to be shorter than in the $Cp_2Nb(SiMe_2Ph)_2H$ derivative, with Nb−Si bond lengths of 2.646(5) Å and 2.665(5) Å, respectively. The short Nb−Si and the long Si−Cl distances found in the former complex have been explained as due to interaction of 5c–6e hypervalent ligand Cl···Si···H···Si···Cl[45].

Reaction of the tantalum(III) complex $Cp_2Ta(PMe_3)(Si(Bu-t)_2H)$ with 2 equivalents of an amine–borane adduct, L·BH$_3$ (L = Et$_3$N, N-methylmorpholine), results in the formation of the silyl substituted η^2-borohydride complex $Cp_2Ta\{\eta^2\text{-BH}_3(Si(Bu-t)_2H)\}$ (**18**). Complex **18** reacts with PMe$_3$ forming the $Cp_2Ta\{\eta^1\text{-BH}_3(Si(Bu-t)_2H)\}PMe_3$ (**19**) complex and with a large excess of the phosphine, yielding the silylborane adduct and the $Cp_2Ta(PMe_3)(H)$ complex, as shown in equation 16[46].

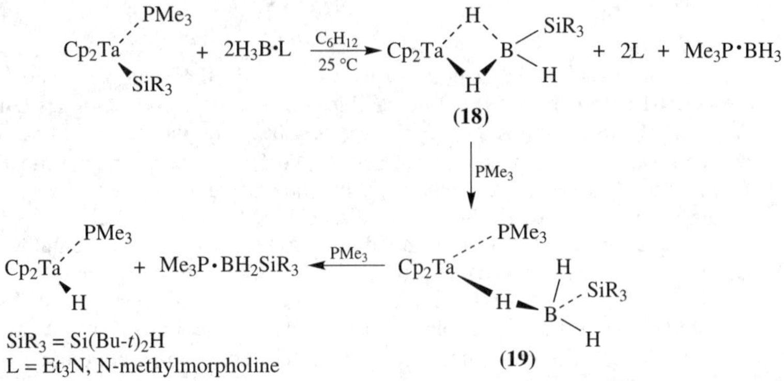

$$Cp_2Ta\genfrac{}{}{0pt}{}{PMe_3}{SiR_3} + 2H_3B\cdot L \xrightarrow[25\,°C]{C_6H_{12}} Cp_2Ta\genfrac{}{}{0pt}{}{H\cdots SiR_3}{H\cdots B\cdots H} + 2L + Me_3P\cdot BH_3$$

(**18**)

↓ PMe$_3$ (16)

$$Cp_2Ta\genfrac{}{}{0pt}{}{PMe_3}{H} + Me_3P\cdot BH_2SiR_3 \xleftarrow{PMe_3} Cp_2Ta\genfrac{}{}{0pt}{}{PMe_3}{H\cdots B\cdots SiR_3\;H}$$

SiR$_3$ = Si(Bu-t)$_2$H
L = Et$_3$N, N-methylmorpholine   (**19**)

Berry and coworkers have observed the thermally labile silyl alkylidene complex $Cp_2Ta(=CH_2)(Si(Bu-t)_2H)$ *in situ* at $-70\,°C$ in the photolysis of $Cp_2Ta(PMe_3)(Si(Bu-t)_2H)$ in the presence of $Me_3P=CH_2$[46]. In contrast, and besides the theoretical studies in Schrock-type metal silylidene complexes which predict a strong Nb=Si bond[47], the alkylidene silyl complex $Np_2Ta(Si(SiMe_3)_3)(=CHBu-t)$ was obtained by the metathesis reaction of Np_3TaCl_2 with 2 equivalents of $(Me_3Si)_3SiLi\cdot 3THF$ and the concomitant elimination of one equivalent of $HSi(SiMe_3)_3$. The molecular structure of $Cp_2Ta(=CH_2)(Si(Bu-t)_2H)$ imposed a 3-fold rotation axis containing the Ta–Si bond, giving a disorder among the *t*-BuCH= moiety and the two neopentyl groups. The Ta–Si bond length of 2.680(15) Å is comparable to those in other Ta–silyl complexes (Table 3)[35].

TABLE 3. Selected M–Si distances in group-5 transition-metal silyl complexes

Compound	d^n Configuration	M–Si (Å)	Reference
$Cp_2Ta(PMe_3)(SiMe_3)$	d^2	2.639(4)	42
$Cp_2TaH(SiMe_2H)_2$	d^0	2.633(2), 2.624(2)	42
$Cp_2Ta(CO)(SiMe_3)$	d^2	2.631(2)	43
$Cp_2Ta(PMe_3)(Si(Bu-t)_2H)$	d^2	2.740(1)	43
$Cp_2Ta(CO)(Si(Bu-t)_2H)$	d^2	2.684(1)	43
$Np_2Ta(Si(SiMe_3)_3)(=CHBu-t)$	d^0	2.680(15)	35
$Cp_2Nb(SiMe_2Cl)_2H$	d^0	2.584(5), 2.611(5)	45
$Cp_2Nb(SiMe_2Ph)_2H$	d^0	2.646(5), 2.665(5)	45
$Nb(\eta^5\text{-}C_5H_4SiMe_3)_2H_2(SiPh_2H)$	d^0	2.616(3)	44

TABLE 4. Selected ^{29}Si NMR chemical shifts for group-5 transition-metal silyl complexes

Compound	^{29}Si NMR shift (ppm)	Reference
Cp$_2$TaH$_2$(SiH$_3$)	−74.1 symmetrical isomer	42
Cp$_2$TaH(SiH$_3$)H	−40.1 unsymmetrical isomer	42
Cp$_2$TaH$_2$(SiMeH$_2$)	−33.5 symmetrical isomer	42
Cp$_2$TaH(SiMeH$_2$)H	−7.0 unsymmetrical isomer	42
Cp$_2$TaH$_2$(SiMe$_2$H)$_2$	−5.2 symmetrical isomer	42
Cp$_2$TaH(SiMe$_2$H)H	15.0 unsymmetrical isomer	42
Cp$_2$TaH$_2$(SiMe$_3$)	10.4 symmetrical isomer	42
Cp$_2$TaH(SiMe$_3$)H	27.3 unsymmetrical isomer	42
Cp$_2$TaH$_2$(SiMe$_2$Cl)	76.2 symmetrical isomer	42
Cp$_2$TaH(SiMe$_2$Cl)H	83.0 unsymmetrical isomer	42
Cp$_2$TaH$_2$(Si(OMe)$_3$)	40.9 symmetrical isomer	42
Cp$_2$TaH(Si(OMe)$_3$)H	40.7 unsymmetrical isomer	42
Cp$_2$TaH(SiMeH$_2$)$_2$	−11.7 symmetrical isomer	42
Cp$_2$TaH(SiMeH$_2$)$_2$	−5.9; −23.4 unsymmetrical isomer	42
Cp$_2$TaH(SiMe$_2$H)$_2$	11.5 symmetrical isomer	42
Cp$_2$TaH(SiMe$_3$)$_2$	24.5 symmetrical isomer	42
Cp$_2$TaH(SiMe$_2$Cl)$_2$	81.0 symmetrical isomer	42
Cp$_2$TaH(Si(OMe)$_3$)$_2$	36.6 symmetrical isomer	42
Cp$_2$Ta(PMe$_3$)(SiMeH$_2$)	−21.3 ($^2J_{\text{PSi}}$ = 15.1 Hz)	42
Cp$_2$Ta(PMe$_3$)(SiMe$_3$)	9.7 ($^2J_{\text{PSi}}$ = 11.5 Hz)	42
Cp$_2$Ta(PMe$_3$)(SiMe$_2$Cl)	89.7 ($^2J_{\text{PSi}}$ = 8.6 Hz)	42
Cp$_2$Ta(PMe$_3$)(Si(OMe)$_3$)	30.4 ($^2J_{\text{PSi}}$ = 22.9 Hz)	42
Cp$_2$Ta(PMe$_3$)(Si(Bu-t)$_2$H)	46.4 ($^2J_{\text{PSi}}$ = 8.9 Hz)	42
Cp$_2$Ta(CO)SiMe$_3$	7.5	42
Cp$_2$Ta((CO)(Si(Bu-t)$_2$H)	51.1	42
Np$_2$Ta(Si(SiMe$_3$)$_3$)(=CHBu-t)	−53.47 (SiSiMe$_3$), −5.3 (SiMe$_3$)	35
Nb(η^5-C$_5$H$_4$SiMe$_3$)$_2$H$_2$(SiMePh$_2$)	26.7 (SiMePh$_2$), −3.0 (SiC$_5$H$_4$)	44
Nb(η^5-C$_5$H$_4$SiMe$_3$)$_2$H$_2$(SiPh$_2$H)	23.8 ($^2J_{\text{SiH}}$ = 177 Hz, SiPh$_2$H) −0.3 (SiC$_5$H$_4$)	44

VIII. GENERAL INTRODUCTION TO LATE-TRANSITION-METAL SILICON-CONTAINING COMPLEXES

Major advances in organometallic chemistry during the last years have been achieved in the area of silicon–metal multiple bonding and silicon with low coordination numbers. For late transition metals, new complexes have been synthesized such as silanediyl (**A**), silene (**B**), silaimine (**C**), disilene (**D**), silatrimethylenemethane (**E**), silacarbynes (**F**), cyclic silylenes (**G**), silacyclopentadiene (**H**) and metalla-sila-allenes (**I**) (Figure 3).

The existence of such a diverse silaorganometallic chemistry made it clear that coordination compounds of subvalent silicon ligands are common species playing an important role in many reactions and having important applications in industrial processes. The links made by silaorganometallic chemistry among organic and inorganic chemistry and the exploration of this highly interdisciplinary field raise a challenge for the future[48–50].

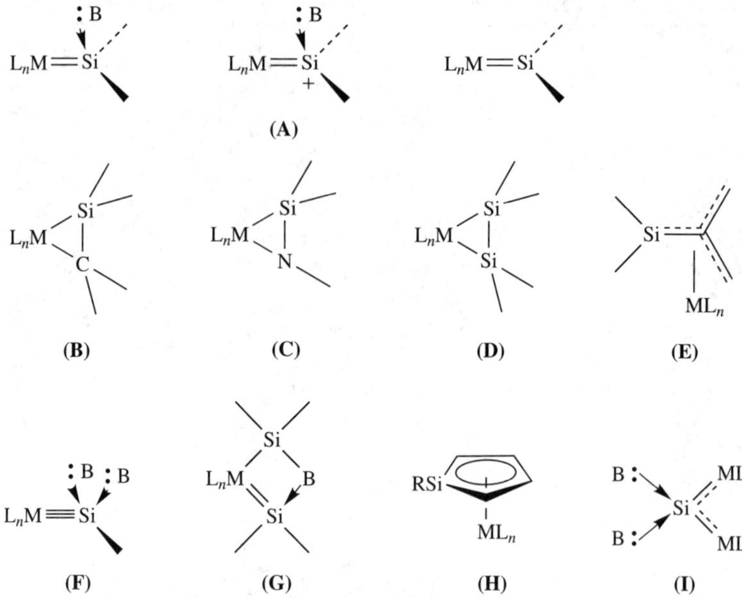

FIGURE 3. Some late-transition-metal silyl derivatives

IX. GROUP-6 SILYLENE COMPLEXES

The formation of a metal–silicon double bond is most effective by the 'salt elimination route'. It is accomplished in a one-step procedure reacting the supernucleophilic metallate dianions with dihalosilanes in polar solvents as HMPA, as shown for chromium in equation 17.

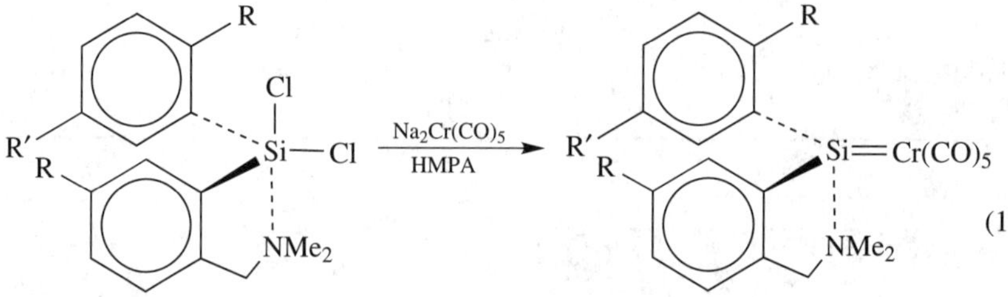

(20) R = R' = H
(21) R = H, R' = CH$_2$NMe$_2$
(22) R = CH$_2$NMe$_2$, R' = t-Bu

(17)

The chromium complexes are proved to be silanediyl complexes, as shown by the silicon–transition metal bond lengths (Table 5) and by the extreme low field shift of the ^{29}Si NMR signals (124.9 and 121.2) at 22 °C for R = H and CH$_2$NMe$_2$, respectively (Table 6). The ^{29}Si NMR shifts of these complexes are temperature-dependent due to the hindered rotation of the phenyl ring and dynamic coordination of the nitrogens to the Si atom.

Thus in **21**, below 21 °C (coalescence barrier $\Delta G^{\ddagger} = 12.9 \pm 1$ kcal mol^{-1}) one nitrogen is attached to the Si atom, at 58 °C the two substituents on the Si atom are magnetically equivalent and above 58 °C the dynamic behavior disappears and no N−Si bonds are observed as shown by the ^{29}Si NMR signal at 138.8 ppm which is in line with a three-coordinative silicon atom. For **20**, the dimethylamino substituent is rigidly coordinated to the silicon atom and coalescence was obtained only at 95 °C ($\Delta G^{\ddagger} = 19.1 \pm 1$ kcal mol^{-1}). The solid structure of **21** shows the silanediyl ligand coordinated to the octahedral metal fragment with a short Cr−Si bond length of 2.408(1) Å[51,52]. Photochemical activation of CO in **21** or **22** results in a 1,2-shift of the amine donor-substituent from the silicon to the metal, yielding the complex (2-Me$_2$NCH$_2$C$_6$H$_4$)(2-Me$_2$NCH$_2$C$_6$H$_4$)Si=Cr(CO)$_4$. This shift causes a shortening in the Cr=Si bond to 2.3610(4) Å, reflecting the stronger Cr=Si bond obtained by the electron transfer from the amine to the metal atom. Similar tandem photolytic 1,2-shifts have been observed in the phosphane-stabilized silylene complex as depicted in Scheme 11. The starting phosphane exhibits, in analogy to what was presented above, the flip-flop coordination and has a Cr=Si bond length of 2.414(1) Å, which is comparable to other chromium silylene complexes (Table 5)[53]. A similar strategy has been used to prepare a large number of base-stabilized group 6 and 7 (*vide infra*) silylene complexes as shown in equation 18.

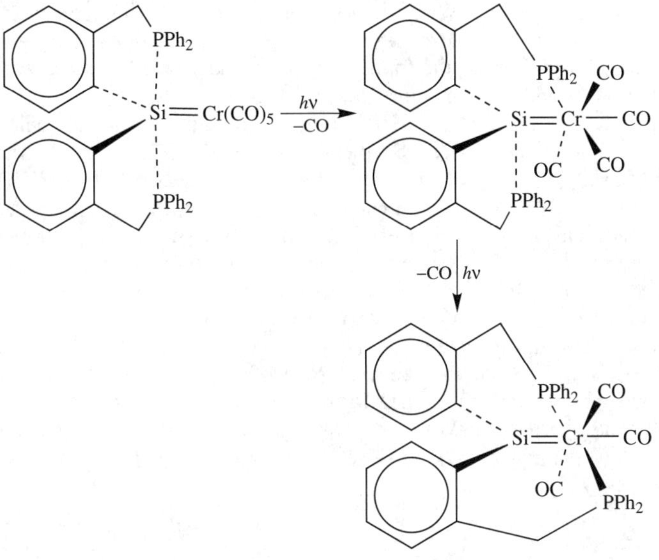

SCHEME 11

$$\text{Na}_2\text{Cr(CO)}_5 + \text{R}_2\text{SiCl}_2 \xrightarrow[-2\text{ NaCl}]{\text{HMPA}} \text{R}_2\text{Si=Cr(CO)}_4 \cdot \text{R}' \tag{18}$$

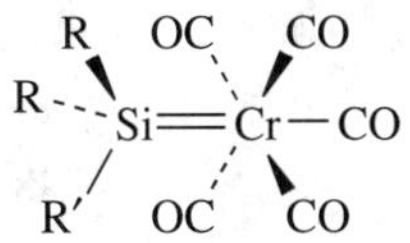

R = *t*-BuO, *t*-BuS, CH$_3$, Cl, 1-AdO, 2−AdO, NpO, Ph$_3$CO; R′ = HMPA

In addition, on reacting Na$_2$Cr(CO)$_5$ with *t*-Bu$_2$Si(OTf)$_2$ in THF, similarly to equation 18, the salt adduct (CO)$_5$Cr=Si(Bu-*t*)$_2$·Na(OTf)·2THF was obtained with an oxygen atom of the triflate anion coordinated to the subvalent silicon atom. The solid structure of the salt adduct complex exhibits a Cr=Si bond length of 2.475(1) Å. The salt adduct is readily exchange by HMPA, yielding the respective (CO)$_5$Cr=Si(Bu-t)$_2$(HMPA) complex[54]. For these chromium complexes, the ^{29}Si NMR chemical shift data cover a range of 100 ppm with no linear correlation with the electronegativity of the particular substituent at the silicon atom. Instead, a quadratic dependency between the overall electronegativity of the silicon substituents and the ^{29}Si NMR chemical shifts was observed. This result argues for a superposition of diamagnetic and paramagnetic shift influences on silicon. The observed trend on the ^{29}Si NMR shifts is analogous to that known for regular tetravalent silicon. The coupling constants P–Si for the HMPA complexes were found to be invariant with changes of temperature in the range of −90 to +110 °C, indicating the rigid coordination to the HMPA base.

Solid state structures of the silylene complexes (CO)$_5$Cr=Si(HMPA)Me$_2$, (CO)$_5$Cr=Si (HMPA)Cl$_2$, (CO)$_5$Cr=Si(Bu-t)$_2$•Na(OTf)•2THF and (CO)$_5$Cr=Si(HMPA)(Bu-t)$_2$ have been shown to be isostructural with the Si ligand in the apical position of the octahedral metal environment and to have bond lengths of 2.410(1) Å, 2.342(1) Å, 2.475(1) Å and 2.527(1) Å, respectively. The latter complex can be described better as a σ-donor Si → Cr complex[53,55]. Photolysis of the complex (CO)$_5$Cr=Si(HMPA)(OBu-t)$_2$ in the presence of triphenylphosphine yields the *trans* diphosphine complex (CO)$_5$Cr(PPh$_3$)$_2$ together with HMPA and a polysilane. When the reaction, as described in equation 18, is carried out in the presence of dimethyl carbonate, a sila-Wittig reaction (metathesis of M=Si and C=O) takes place forming the (dimethoxycarbene)chromium pentacarbonyl complex (CO)$_5$Cr=C(OMe)$_2$ and hexamethyltrisiloxane[55].

An alternative synthetic pathway to silylene complexes can be achieved by the photolytic reaction of the 16-electron metal complex Cr(CO)$_6$ with the trihydrosilane 2-Me$_2$NCH$_2$C$_6$H$_4$SiH$_3$, yielding complex **23** after elimination of H$_2$ and CO, as shown in equation 19[52,56]. Complex **23** eliminates two CO molecules in a tandem fashion under photolytic conditions in the presence of phosphine, phosphite or diphosphines. In the latter case the *cis* isomer is obtained. At room temperature, complex **23** was found to react with either Ph$_3$CBF$_4$, CH$_3$COBr, Ph$_3$COSO$_2$CF$_3$ or with CH$_3$COCl exchanging the hydrosilane hydrogen, forming respectively the F, Br, OSO$_2$CF$_3$ or Cl derivatives of complex **23**[57]. An unexpected reactivity was found for complex **23** and for its halide substituted complexes, when they reacted with alkyl lithium reagents. Instead of the expected nucleophilic attack at the CO ligand, the alkyl substituted complex is obtained as shown for the hydride complex **23** in equation 20. In general, the reactivity trend of **23** and analogous complexes was found to follow the order of F > H ≫ Br ⩾ Cl[57].

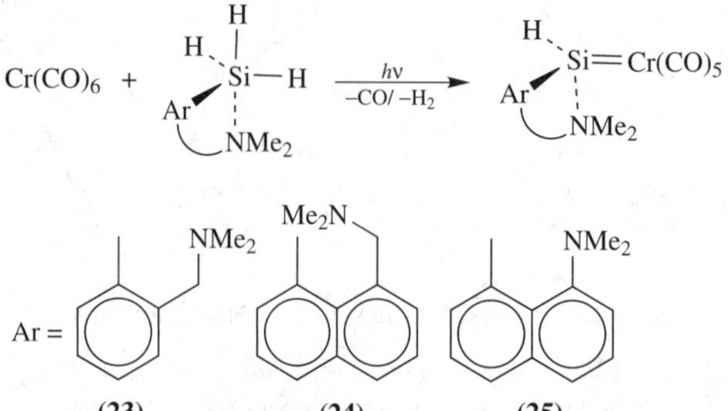

$$Cr(CO)_6 + \underset{\underset{NMe_2}{Ar}}{\overset{H}{\underset{|}{Si}}}-H \xrightarrow[-CO/-H_2]{h\nu} \underset{\underset{NMe_2}{Ar}}{\overset{H}{\underset{|}{Si}}}=Cr(CO)_5$$

(19)

Ar = (23) , (24) , (25)

(with NMe$_2$, Me$_2$N / NMe$_2$, NMe$_2$ substituents on aryl groups)

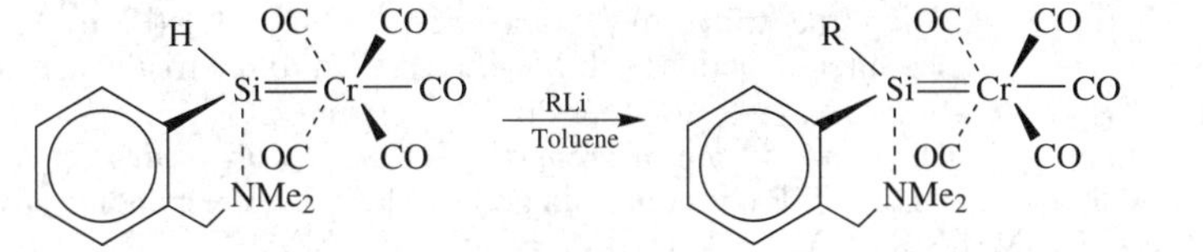

(20)

R = Ph, Me, t-Bu, Me$_3$SiC≡C, PhC≡C

X. GROUP-6 SILYL COMPLEXES

Reaction of the lithium or the tetramethylphosphonium metallates M′[M(PMe$_3$)(CO)$_2$C$_5$R$_5$] (M′ = Li, Me$_4$P; M = Mo, W; R = H, Me) with the trichlorosilanes R′SiCl$_3$ (R′ = H, SiCl$_3$) leads to the formation of monosilyl complexes of the type C$_5$R$_5$(CO)$_2$(Me$_3$P)M−SiCl$_2$R′. These complexes react with LiAlH$_4$ replacing all the chlorine atoms, forming the corresponding hydrosilane complexes as shown in Scheme 12[58]. When R′ = SiCl$_3$, the hydrido disilene complex can be reconverted back into the pentachloro(metallo) disilane by reacting it with tetrachloromethane[59].

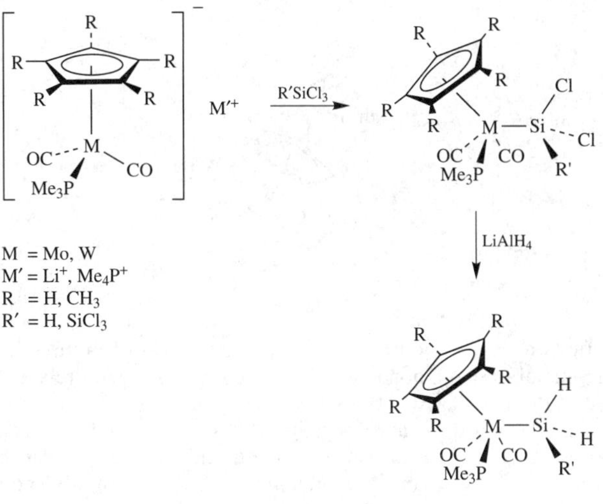

M = Mo, W
M′ = Li$^+$, Me$_4$P$^+$
R = H, CH$_3$
R′ = H, SiCl$_3$

SCHEME 12

In a similar fashion the monosilyl complexes C$_5$Me$_5$(CO)$_2$(PMe$_3$)W−SiR$_3$ (SiR$_3$ = SiHMe$_2$, SiHClMe and SiHCl$_2$) are prepared from the lithium metallate C$_5$Me$_5$(CO)$_2$(PMe$_3$)WLi with the corresponding chlorosilane. The latter two chlorine complexes have been reacted with LiAlH$_4$ yielding the hydrosilyl complexes C$_5$Me$_5$(CO)$_2$(PMe$_3$)W−SiH$_3$ and C$_5$Me$_5$(CO)$_2$(PMe$_3$)W−SiH$_2$Me, respectively. Crystal data for the complexes C$_5$Me$_5$(CO)$_2$(PMe$_3$)W−SiCl$_2$SiCl$_3$ and C$_5$Me$_5$(CO)$_2$(PMe$_3$)W−SiH$_2$R (R = H, Me) reveal a pseudo-tetrahedral arrangement of the ligands around the metal with the cyclopentadienyl ligand in the axial position. The 2.469(2) Å W−Si bond length for the former is ca 0.1 Å shorter than the bond lengths of 2.533(3) Å and 2.559(2) Å for the latter two complexes, respectively, reflecting the difference in the inductive −I effect. This effect can also be recognized in the slightly enhanced coupling constant $^1J_{SiW}$ of the last two complexes (Table 6)[60]. The monosilyl complexes C$_5$Me$_5$(CO)$_2$(PMe$_3$)M−SiH$_3$ (M = Mo, W) were found to react with dimethyloxirane, yielding the trihydroxy complexes, $C_5Me_5(CO)_2(PMe_3)M-Si(OH)_3$ (M = Mo, W). A concomitant reaction with $Me_2Si(H)Cl$ led to the trisiloxane complexes $C_5Me_5(CO)_2(PMe_3)MSi(OSiMe_2H)_3$ (M = Mo, W), respectively[61,62].

Photochemical reactions at 350 nm of group-6 metallocene dihydrides Cp_2MH_2 (M = Mo, W) with hydrosilanes, $HSiR_3$, produced in good yields the corresponding silyl hydride complexes, $Cp_2MH(SiR_3)$ [M = Mo; SiR_3 = $SiMe_2H$, $SiMe_2Cl$, $SiMe_3$, $Si(Bu-t)_2H$, $Si(Bu-t)_2Cl$; M = W; SiR_3 = $SiMe_2Cl$, $SiMe_3$]. The pentamethylcyclopentadienyl derivatives $Cp_2*M(H)(SiR_3)$ [M = Mo; SiR_3 = $SiMe_3$, $Si(Bu-t)_2H$] were prepared in a similar fashion, whereas bis(silyl) complexes of the type $Cp_2W(SiMe_3)(SiR_3)$ were synthesized from the silene complex $Cp_2W(\eta^2- SiMe_2=CH_2)$ upon reaction with the corresponding hydrosilane as described, together with the proposed mechanism, in Scheme 13[63].

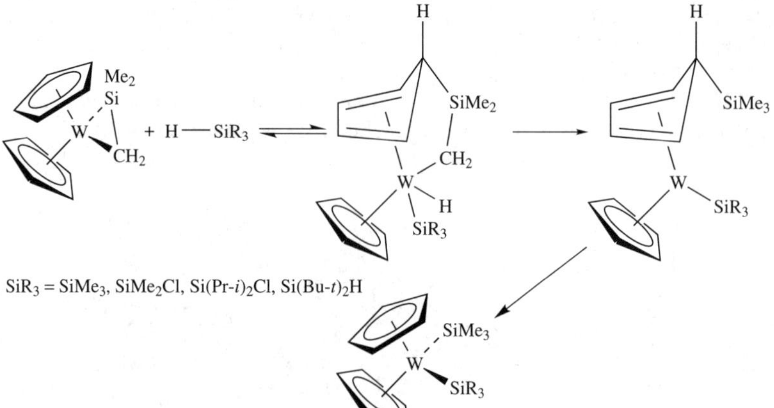

SCHEME 13

Crystal structures for some mono and bis(silyl) complexes of group 6 have been determined showing metal–silicon bond lengths which generally correlate with the steric congestion around the metal. Exceptions have been found for the chlorosilyl derivatives which exhibit short Group 6–Si and long Si–Cl distances (Table 5). This peculiar feature of the chlorosilyl complexes is ascribed to both inductive and π-backbonding effects from the metal to the Si–Cl σ^* orbital, inducing some degree of silylene character. This silylene character is evidenced by the ^{29}Si NMR ($^1J_{WSi}$) (Table 6) coupling constants[63]. The bis(silyl)tungsten complexes $Cp_2W(SiMe_3)(SiR_2X)$ (R = i-Pr, CH_3, CD_3; X = Cl, OTf) have been found to undergo thermal isomerization of the methyl groups through a silyl–silylene intermediate forming the mixed $Cp_2W(SiMe_2X)(SiR_2Me)$ complexes. The proposed mechanism is described in Scheme 14 in which an electrophilic cationic silyl(silylene) tungstenocene is formed by dissociation of X^-, followed by migration of a methyl group to the electrophilic silylene center, and reassociation of X^- to the second silicon to complete the process. As can be expected, the rates are strongly dependent on the nature of the R and X moieties, solvent polarity and the presence of strong Lewis acid catalysts, such as $B(C_6F_5)_3$[64].

Using the same strategy Berry and coworkers recently described the synthesis of mixed tungsten germyl silyl complexes $Cp_2W(SiMe_3)(GeR_3)$ [GeR_3 = $GeMe_3$, $GeMe_2H$, $Ge(Bu-t)_2H$, $GePh_2H$, $GeMe_2Cl$, $Ge(Bu-t)_2Cl$, $GePh_2Cl$, $GeMe_2OTf$, $Ge(Bu-t)_2Me$], which display the same methyl rearrangement described in Scheme 14 for the bis(silyl) complexes[65].

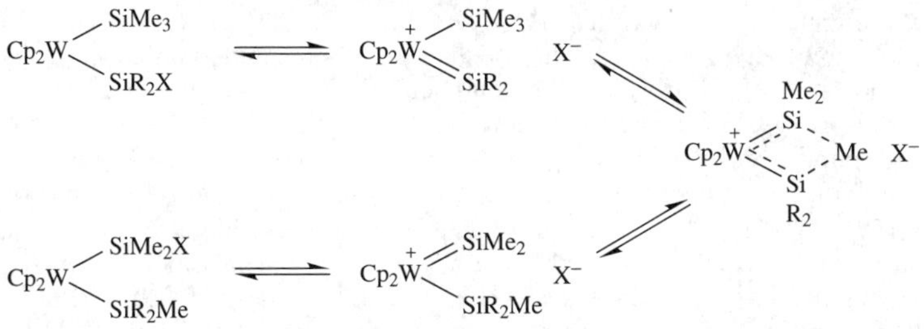

SCHEME 14

Arene complexes of the type (η^6-arene)Cr(CO)$_2$(H)(SiCl$_3$) (η^6-arene = C$_6$H$_6$, 1,3,5-Me$_3$C$_6$H$_3$, C$_6$Me$_6$, C$_6$H$_5$Me, Cbb) have been obtained by the photolytic CO activation of the corresponding chromium tricarbonyl complex after oxidative addition of the trichlorosilane. Further photolysis of the complexes causes the elimination of H$_2$ and formation of the (η^6-arene)Cr(CO)$_2$(SiCl$_3$)$_2$ complexes, respectively. During the last step, the arene ligand becomes labilized and can be exchanged with other arene ligands[66,67]. Recently, the intermediate complexes, (η^6-arene)Cr(CO)(H)$_2$(SiCl$_3$)$_2$ (η^6-arene = C$_6$H$_5$F, 1,3,5-Me$_3$C$_6$H$_3$), in the formation of (η^6-arene)Cr(CO)$_2$(SiCl$_3$)$_2$, were isolated and characterized. The former crystallized as a 1:1 cocrystallite with the product (η^6-arene)Cr(CO)$_2$(SiCl$_3$)$_2$. The structure is described as a distorted three-legged piano stool where two legs are trichlorosilyl ligands and the third leg is the carbonyl. The two hydrides bridge the Si−Cr bonds. In the latter molecule, a similar three-legged piano-stool structure is observed, except that one hydrogen is considerably further and almost equidistant from the two silicon atoms. Thus, the best formulation of these compounds is as the dihydrides of formally Cr(IV). Noteworthy is the fluxional behavior of the two hydrogen atoms in the intermediate complex, which exchange in solution according to the mechanism proposed in Scheme 15[68].

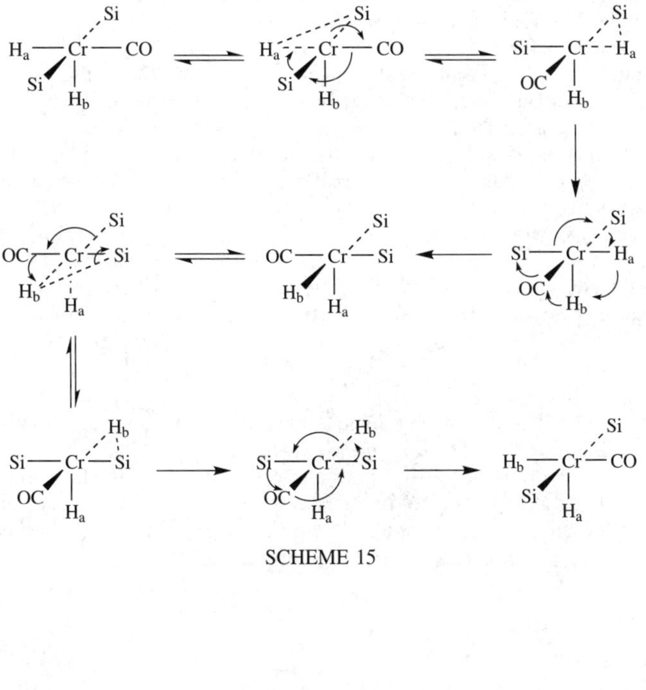

SCHEME 15

Chelate complexes of tungsten $(CO)_4\overline{W(PPh_2PCH_2CH_2SiHR_2)}$ (R = Me, Ph) containing three-center two-electron bonds were prepared photochemically from $W(CO)_6$ and $Ph_2PCH_2CH_2SiR_2H$. The non-chelated phosphino silyl tungsten complexes $(CO)_5W(PPh_2PCH_2CH_2SiR_2H)$ have been found to be the intermediates in this reaction. The NMR coupling constants $^1J_{SiWH}$ in the chelating complexes (98.1 and 95.2 Hz for R = Me and Ph, respectively) are lower than the coupling constant for the uncoordinated complex $(CO)_5W(PH_2PCH_2CH_2SiR_2H)$ ($^1J_{SiH}$ = 196.5 Hz) indicating that, due to the chelation, an earlier stage of the oxidative addition of the silane bond is frozen in the corresponding non-chelated complexes. The stabilization obtained by the chelating ligand is slightly larger than the electronic effect caused by substituting another CO by a PR_3 ligand or by replacing SiR_3 (R = alkyl, aryl) by $SiCl_3$ in the non-chelating complexes[69,70].

The first monosilyl alkylidyne complex of tungsten without anionic π-bonding ligands $Np_2W(\equiv CBu\text{-}t)(Si(SiMe_3)_3)$ was prepared recently upon metathesis of a neopentyl tungstenyl chloride derivative with the salt $(Me_3Si)_3SiLi$ (equation 21)[35].

$$Np_3W\equiv CBu\text{-}t \xrightarrow[Et_2O]{HCl} Np_2\overset{\overset{\displaystyle Cl}{|}}{W}\equiv CBu\text{-}t \xrightarrow{(Me_3Si)_3SiLi} Np_2\overset{\overset{\displaystyle Si(SiMe_3)_3}{|}}{W}\equiv CBu\text{-}t \quad (21)$$

Np = neopentyl

A. η^2-Silyl Complexes

Berry and coworkers have shown that $Cp_2Mo(\eta^2\text{-}Me_2Si=SiMe_2)$ reacts readily with elemental sulfur to yield the symmetrical insertion product $cyclo$-$Cp_2Mo(Me_2SiSSiMe_2)$[71]. Recently, Berry and coworkers have shown that the isolobal tungsten disilene complex $Cp_2W(\eta^2\text{-}Me_2Si=SiMe_2)$ (**26**) reacts with the heavier elemental chalcogens to yield the symmetrical insertion products $cyclo$-$Cp_2W(Me_2SiESiMe_2)$ (E = S, Se, Te)[72]. Although complex **26** does not react cleanly with oxygen, the insertion product $cyclo$-$Cp_2W(Me_2SiOSiMe_2)$ can be prepared by the reaction of **26** with trimethylamine N-oxide. The four-membered ring compounds $cyclo$-$Cp_2W(Me_2SiESiMe_2)$ (E = S, Se) can likewise be formed in the reaction of **26** with phosphine chalcogenides, although varying amounts of the unsymmetrical isomers $cyclo$-$Cp_2W(EMe_2SiSiMe_2)$ are produced as well. The ratio of unsymmetrical to symmetrical isomers formed have been found to be strongly dependent on the size of the phosphine chalcogenide. The proposed mechanism for the formation of both sulfur isomers (Scheme 16) involves an initial nucleophilic attack by the phosphine sulfide at either a silicon atom or at the tungsten atom. Cleavage of the silicon–silicon bond would generate a zwitterionic intermediate, **J**, consisting of silyl anion and phosphonium cation moieties. Intramolecular nucleophilic displacement would then lead to the symmetrical four-membered ring complex. In the second case, breaking the metal–silicon bond would form a different zwitterionic complex, **K**, and intramolecular nucleophilic displacement of the phosphine will generate the unsymmetrical complex. As the size of the phosphine sulfide is increased, the greater is the hindrance at the metal center, disfavoring the metal–silicon cleavage and hence the formation of the unsymmetrical isomer[72].

The structures of the oxygen and selenium metallacycle complexes $cyclo$-$Cp_2W(Me_2SiESiMe_2)$ (E = O, Se) were determined by single-crystal X-ray diffraction methods. When E = O, the metal lies on the crystallographic plane bisecting the two Cp rings and the planar WSi_2O ring. When E = S, two unique molecules are present in the unit cell with the only difference among them being the planarity of the WSi_2Se rings. The unique W–Si distance in the former complex [2.551(2) Å] is essentially identical

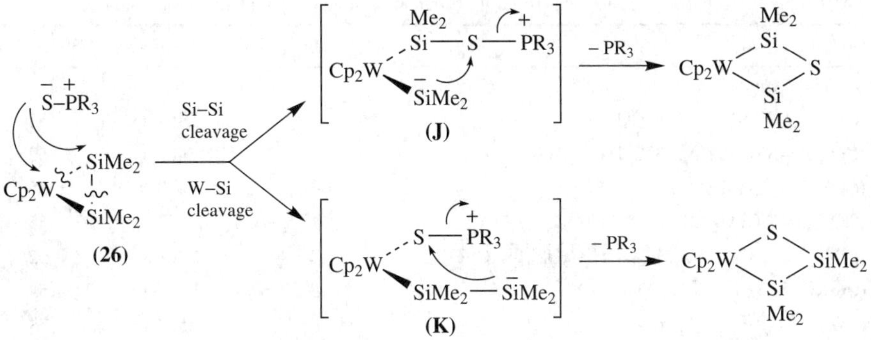

SCHEME 16

to the average (2.554 Å) of the four W−Si distances in the latter which range from 2.544(4) to 2.565(4) Å. The ^{29}Si NMR chemical shifts for the symmetrical complexes cyclo-Cp$_2$W(Me$_2$SiESiMe$_2$) are 20.6 ($^1J_{WSi}$ = 85.9 Hz), −17.8 ($^1J_{WSi}$ = 96.4 Hz), −30.5 ($^1J_{WSi}$ = 99.2 Hz) and −68.1 ppm ($^1J_{WSi}$ = 103.1 Hz) for E = O, S, Se and Te, respectively. Hence, two trends are apparent from the ^{29}Si NMR data. The first is the upfield shift obtained by going down the series as found for the disilyl chalcogenides (Me$_3$Si)$_2$E, and has been ascribed to more effective shielding of silicon by the larger chalcogens. The second is the increase in the one-bond ^{29}Si−^{183}W coupling constant down the series indicating greater s-orbital character in the Si−W bonds of the heavier chalcogenide derivatives[72].

η^2-Silane complexes of molybdenum have been prepared by the reaction of the formally 16-electron complex Mo(CO)(R$_2$PCH$_2$CH$_2$PR$_2$)$_2$ with silanes (equation 22.)

Mo(CO)(R$_2$PCH$_2$CH$_2$PR$_2$)$_2$ + R″R′SiH$_2$ ⟶ [Mo complex product] (22)

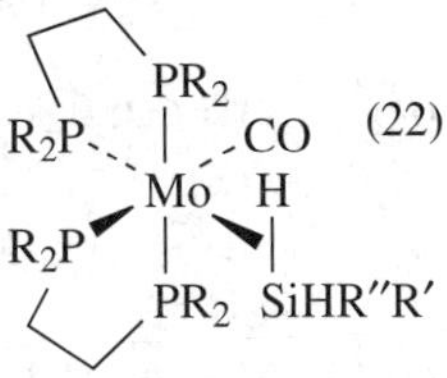

R = Et, R′ = R″ = Ph, n−C$_6$H$_{13}$
R = Et, R′ = H, R″ = Ph
R = CH$_2$Ph, R′ = H, R″ = Ph, n−C$_6$H$_{13}$
R = Ph, i-Bu, R′ = R″ = H

The η^2-silane coordination was established by the observation of the $^1J_{SiH}$ coupling constants for the bound η^2-Si−H bonds which are in the range of 20−70 Hz. Consequently, these complexes are better described as six-coordinate η^2-silane complexes than seven-coordinate hydride complex[73,74]. Remarkably, in solution the η^2-SiH$_4$ complex cis-Mo(η^2-SiH$_4$)(CO)(Et$_2$PCH$_2$CH$_2$PEt$_2$)$_2$ was found to be in equilibrium with its seven - coordinate hydrosilyl tautomer Mo(CO)(H)(SiH$_3$)(Et$_2$PCH$_2$CH$_2$PEt$_2$)$_2$. Crystal structures for the complexes cis-Mo(η^2-SiH$_2$Ph−H)(CO)(Et$_2$PCH$_2$CH$_2$PEt$_2$)$_2$ and cis-Mo(η^2-SiH$_3$−H)(CO)(Et$_2$PCH$_2$CH$_2$PEt$_2$)$_2$ evince an octahedral environment with the η^2-silane ligand cis to the CO, and similar Mo−Si bond lengths of 2.501(2) Å and 2.556(4) Å (Table 5), respectively[74].

TABLE 5. Selected M–Si distances in group-6 transition-metal silylene and silyl complexes

Compound	d^n Configuration	M–Si (Å)	Reference
Silylene Complexes			
$(CO)_5Cr=Si(o-Me_2NCH_2C_6H_4)_2$	d^6	2.408(1)	51
$(CO)_5Cr=Si(o-Me_2NCH_2C_6H_4)(C_6H_5)$	d^6	2.409(1)	52
$(CO)_5Cr=Si(HMPA)Me_2$	d^6	2.410(1)	55
$(CO)_5Cr=Si(HMPA)Cl_2$	d^6	2.342(1)	55
$[2-Me_2NCH_2C_6H_4][2-Me_2\overline{NCH_2C_6H_4]\,Si=C}r(CO)_4$	d^6	2.3610(4)	52
$(CO)_5Cr=Si(Bu\text{-}t)_2 \cdot NaOTf \cdot 2THF$	d^6	2.475(1)	53
$(CO)_5Cr=Si(Bu\text{-}t)_2(HMPA)$	d^6	2.527(1)	53
Silyl Complexes			
$(C_5Me_5)(CO)_2(Me_3P)W-SiH_3$	d^6	2.533(3)	60
$(C_5Me_5)(CO)_2(Me_3P)W-SiH_2Me$	d^6	2.559(2)	60
$(C_5Me_5)(CO)_2(Me_3P)W-SiCl_2SiCl_3$	d^6	2.469(2)	59
$(C_5H_5)_2Mo(H)SiMe_2H$	d^2	2.538(2), 2.541(2)	63
$(C_5H_5)_2Mo(H)SiMe_2Cl$	d^2	2.513(1)	63
$(C_5Me_5)_2Mo(H)SiMe_3$	d^2	2.560(1)	63
$(C_5H_5)_2Mo(H)Si(Bu\text{-}t)_2H$	d^2	2.604(1)	63
$(C_5H_5)_2W(H)SiMe_3$	d^2	2.560(1)	63
$(C_5H_5)_2W(SiMe_3)(Si(Pr\text{-}i)_2Cl)$	d^2	2.602(1)	63
$(C_5Me_5)_2W(SiMe_3)(Si(Bu\text{-}t)_2H)$	d^2	2.599(1)	63
$(C_5H_5)_2W(SiMe_3)(GeMe_2Cl)$	d^2	2.591(3)	65
$(\eta^6\text{-}C_6H_5F)Cr(CO)_2(SiCl_3)_2$	d^4	2.376(2) 2.377(2)	68
$(\eta^6\text{-}C_6H_5F)Cr(CO)(H)_2(SiCl_3)_2$	d^2	2.361(2) 2.368(2)	68
$(\eta^6\text{-}1,3,5\text{-}Me_3C_6H_3)Cr(CO)(H)_2(SiCl_3)_2$	d^4	2.365(1) 2.373(1)	68
$(\eta^6\text{-}1,3,5\text{-}Me_3C_6H_3)Cr(CO)_2(SiCl_3)_2$	d^4	2.383(3) 2.382(3)	66
cyclo-$Cp_2W(Me_2SiOSiMe_2)$	d^4	2.551(2)	72
cyclo-$Cp_2W(Me_2SiSeSiMe_2)$	d^4	2.547(4) 2.559(4) 2.544(4) 2.565(4)	72
cis-$Mo(\eta^2\text{-}SiH_2Ph\text{-}H)(CO)(Et_2PCH_2CH_2PEt_2)_2$	d^6	2.501(2)	73
cis-$Mo(\eta^2\text{-}SiH_3\text{-}H)(CO)(Et_2PCH_2CH_2PEt_2)_2$	d^6	2.556(4)	74

TABLE 6. Selected ^{29}Si NMR chemical shifts for group-6 transition-metal silylene and silyl complexes

Compound	^{29}Si NMR shift (ppm)	Reference
Silylene Complexes		
(CO)$_5$Cr=Si(*o*-Me$_2$NCH$_2$C$_6$H$_4$)$_2$	124.9 (58 °C)	51
(CO)$_5$Cr=Si(*o*-Me$_2$NCH$_2$C$_6$H$_4$)(C$_6$H$_5$)	121.2 (22 °C)	51
(CO)$_5$Cr=Si(HMPA)(OBu-*t*)$_2$	12.7 ($^2J_{PSi} = 37.2$ Hz)	55
(CO)$_5$Cr=Si(HMPA)Me$_2$	101.4 ($^2J_{PSi} = 31.3$ Hz)	55
(CO)$_5$Cr=Si(HMPA)Cl$_2$	55.0 ($^2J_{PSi} = 41.4$ Hz)	55
(CO)$_5$Cr=Si(HMPA)(OAd-1)$_2$	11.9 ($^2J_{PSi} = 30.1$ Hz)	55
(CO)$_5$Cr=Si(HMPA)(OAd-2)$_2$	11.7 ($^2J_{PSi} = 30.2$ Hz)	55
(CO)$_5$Cr=Si(HMPA)(ONp)$_2$	12.5 ($^2J_{PSi} = 31.1$ Hz)	55
(CO)$_5$Cr=Si(HMPA)(OCPh$_3$)$_2$	10.9 ($^2J_{PSi} = 32.0$ Hz)	55
(CO)$_5$Cr=Si(HMPA)(SBu-*t*)$_2$	83.2 ($^2J_{PSi} = 31.0$ Hz)	55
[*o*-Me$_2$NCH$_2$C$_6$H$_4$]-[*o*-Me$_2$$\overline{\text{NCH}_2\text{C}_6\text{H}_4}$]Si=Cr(CO)$_4$	143.2	52
(CO)$_5$Cr=Si(HMPA)(Bu-*t*)$_2$	133.1 ($^2J_{PSi} = 37.2$ Hz)	53
(CO)$_5$Cr=Si(THF)(Bu-*t*)$_2$	149.7	53
(CO)$_5$Cr=Si(Bu-*t*)$_2$• NaOTf•2THF	150.7	53
(CO)$_5$Cr=Si(H)(*o*-Me$_2$NCH$_2$C$_6$H$_4$)	110.9 ($^1J_{HSi} = 162.3$ Hz)	56
(CO)$_5$Cr=Si(H)[8-(Me$_2$NCH$_2$)C$_{10}$H$_6$]	102.1 ($^1J_{HSi} = 165.3$ Hz)	57
(CO)$_5$Cr=Si(H)[8-(Me$_2$N)C$_{10}$H$_6$]	120.4 ($^1J_{HSi} = 164.9$ Hz)	57
(CO)$_5$Cr=Si(F)(*o*-Me$_2$NCH$_2$C$_6$H$_4$)	117.2 ($^1J_{FSi} = 398.5$ Hz)	57
(CO)$_5$Cr=Si(Br)(*o*-Me$_2$NCH$_2$C$_6$H$_4$)	119.4	57
(CO)$_5$Cr=Si(OSO$_2$CF$_3$)(*o*-Me$_2$NCH$_2$C$_6$H$_4$)	123.5	57
(CO)$_5$Cr=Si(Cl)(*o*-Me$_2$NCH$_2$C$_6$H$_4$)	122.9	57
(CO)$_5$Cr=Si(Ph)(*o*-Me$_2$NCH$_2$C$_6$H$_4$)	122.0	57
(CO)$_5$Cr=Si(Me)(*o*-Me$_2$NCH$_2$C$_6$H$_4$)	126.1	57
(CO)$_5$Cr=Si(Bu-*t*)(*o*-Me$_2$NCH$_2$C$_6$H$_4$)	138.5	57
(CO)$_5$Cr=Si(Me$_3$SiC≡C)(*o*-Me$_2$NCH$_2$C$_6$H$_4$)	92.3 (SiC≡), −18.1 (SiMe$_3$)	57
(CO)$_5$Cr=Si(PhC≡C)(*o*-Me$_2$NCH$_2$C$_6$H$_4$)	92.2	57
Silyl Complexes		
(C$_5$Me$_5$)(CO)$_2$(Me$_3$P)WSiH$_3$	−43.2 ($^2J_{SiP} = 13.2$ Hz, $^1J_{SiW} = 49.8$ Hz)	60
(C$_5$Me$_5$)(CO)$_2$(Me$_3$P)WSiH$_2$Me	10.3 ($^2J_{SiP} = 12.5$ Hz, $^1J_{SiW} = 45.8$ Hz)	60
(C$_5$Me$_5$)(CO)$_2$(Me$_3$P)WSiHMe$_2$	7.4 ($^2J_{SiP} = 11.7$ Hz, $^1J_{SiW} = 41.8$ Hz)	60

(*continued overleaf*)

TABLE 6. (*continued*)

Compound	^{29}Si NMR shift (ppm)	Reference
$(C_5Me_5)(CO)_2(Me_3P)WSiH(Cl)Me$	58.3 ($^2J_{SiP}$ = 15.4 Hz, $^1J_{SiW}$ = 64.1 Hz)	60
$(C_5Me_5)(CO)_2(Me_3P)WSiHCl_2$	60.87 ($^2J_{SiP}$ = 14.7 Hz)	60
$(C_5Me_5)(CO)_2(Me_3P)WSiCl_2SiCl_3$	52.87 (W−Si) ($^2J_{SiP}$ = 18.1 Hz) 4.41 (Si−Si) ($^3J_{SiP}$ = 4.4 Hz)	59
$(C_5H_5)(CO)_2(Me_3P)WSiCl_2SiCl_3$	52.91 (W−Si) ($^2J_{SiP}$ = 17.0 Hz) 3.80 (Si−Si)	59
$(C_5Me_5)(CO)_2(Me_3P)WSiH_2SiH_3$	−61.2 ($^2J_{SiP}$ = 13.9 Hz, $^1J_{WSi}$ = 52.0 Hz) −94.4 (SiH$_3$) ($^3J_{PSi}$ = 1.5 Hz)	59
$(C_5H_5)(CO)_2(Me_3P)WSiH_2SiH_3$	−76.84 ($^2J_{SiP}$ = 14.6 Hz, $^1J_{WSi}$ = 45.4 Hz) −96.5 (SiH$_3$)	59
$(C_5Me_5)(CO)_2(Me_3P)WSiH_3$	−56.87 ($^2J_{SiP}$ = 15.1 Hz, $^1J_{WSi}$ = 44.0 Hz)	58
$(C_5H_5)_2Mo(H)SiMe_2H$	19.9	63
$(C_5H_5)_2Mo(H)SiMe_2Cl$	86.4	63
$(C_5H_5)_2Mo(Cl)SiMe_2Cl$	70.6	63
$(C_5H_5)_2Mo(H)SiMe_3$	27.0	63
$(C_5Me_5)_2Mo(H)SiMe_3$	27.8	63
$(C_5H_5)_2Mo(H)Si(Bu\text{-}t)_2H$	66.8	63
$(C_5Me_5)_2Mo(H)Si(Bu\text{-}t)_2H$	68.2	63
$(C_5H_5)_2Mo(H)Si(Bu\text{-}t)_2Cl$	108.3	63
$(C_5H_5)_2W(H)SiMe_2Cl$	55.8 ($^1J_{WSi}$ = 118 Hz)	63
$(C_5H_5)_2W(H)SiMe_3$	0.5 ($^1J_{WSi}$ = 84 Hz)	63
$(C_5H_5)_2W(SiMe_3)_2$	0.3 ($^1J_{WSi}$ = 106 Hz)	63
$(C_5H_5)_2W(SiMe_3)(SiMe_2Cl)$	70.0 (SiMe$_2$Cl), ($^1J_{WSi}$ = 141 Hz) 4.5 (SiMe$_3$), ($^1J_{WSi}$ = 100 Hz)	63
$(C_5H_5)_2W(SiMe_3)(Si(Pr\text{-}i)_2Cl)$	70.7 (Si(Pr-i)$_2$Cl), ($^1J_{WSi}$ = 144 Hz), 2.7 (SiMe$_3$) ($^1J_{WSi}$ = 102 Hz)	63
$(C_5Me_5)_2W(SiMe_3)(Si(Bu\text{-}t)_2H)$	38.8 (Si(Bu-t)$_2$H), ($^1J_{WSi}$ = 118 Hz) −3.3 (SiMe$_3$), ($^1J_{WSi}$ = 111 Hz)	63
$(C_5H_5)_2W(SiMe_3)(GeMe_3)$	2.1 ($^1J_{WSi}$ = 90 Hz)	65
$(C_5H_5)_2W(SiMe_3)(GeMe_2H)$	1.3 ($^1J_{WSi}$ = 94 Hz)	65
$(C_5H_5)_2W(SiMe_3)(Ge(Bu\text{-}t)_2H)$	−0.6 ($^1J_{WSi}$ = 107 Hz)	65
$(C_5H_5)_2W(SiMe_3)(GePh_2H)$	1.6 ($^1J_{WSi}$ = 90 Hz)	65
$(C_5H_5)_2W(SiMe_3)(GeMe_2Cl)$	2.7 ($^1J_{WSi}$ = 95 Hz)	65
$(C_5H_5)_2W(SiMe_3)(Ge(Bu\text{-}t)_2Cl)$	−1.5 ($^1J_{WSi}$ = 103 Hz)	65
$(C_5H_5)_2W(SiMe_3)(GePh_2Cl)$	3.5 ($^1J_{WSi}$ = 90 Hz)	65

TABLE 6. (continued)

Compound	^{29}Si NMR shift (ppm)	Reference
$(C_5H_5)_2W(SiMe_3)(GeMe_2OTf)$	2.4 ($^1J_{WSi} = 88$ Hz)	65
$(C_5H_5)_2W(SiMe_3)(Ge(Bu\text{-}t)_2Me)$	113.6 ($^1J_{WSi} = 154$ Hz)	65
cyclo-$Cp_2W(Me_2SiOSiMe_2)$	20.6 ($^1J_{WSi} = 85.9$ Hz)	72
cyclo-$Cp_2W(Me_2SiSSiMe_2)$	-30.5 ($^1J_{WSi} = 96.4$ Hz)	72
cyclo-$Cp_2W(Me_2SiSeSiMe_2)$	-30.5 ($^1J_{WSi} = 99.2$ Hz)	72
cyclo-$Cp_2W(Me_2SiSeSiMe_2)$	-61.3 ($^1J_{WSi} = 103.1$ Hz)	72

XI. GROUP-7 SILYLENE COMPLEXES

The first donor-stabilized manganese bis(silylene) complex $(CO)_4\overline{MnSiMe_2(\mu\text{-OMe})Si}$ Me_2 has been prepared by the photolytic decarbonylation of $(CO)_5Mn(SiMe_2SiMe_2OMe)$ with a concomitant 1,2-shift of the silane, allowing cyclization/stabilization by the oxygen atom[75]. Analogous to complexes **23–25**, silylene complexes of manganese of the type $Ar(L)Si(H)=Mn(CO)_2(C_5H_4R)$ (L = H, R = H, Me[57]; L = Ar, R = H, Me[76]) have been prepared following an analogous synthetic procedure to that of equation 19. Cationic base-free silylene rhenium complexes of the form $[CpRe(NO)(PPh_3)(=SiMe_2)]^+X^-$ (X = $AlCl_4^-$, $Al_2Cl_7^-$) were found to be in equilibrium with the silyl-adduct derivative $CpRe(NO)(PPh_3)(SiMe_2Cl)-AlCl_3$. Interestingly, other rhenium silyl and Lewis acid ECl_3 (E = B, Al) derivatives have shown the complexed $Re-ECl_3$ and $ReNO-ECl_3$ structures in different ratios[77]. Structural data for group VII–silylene complexes are given in Table 7 and the corresponding ^{29}Si NMR data are presented in Table 8.

XII. GROUP-7 SILYL DERIVATIVES

The reaction of $CpRe(NO)(PPh_3)(SiMe_2H)$ and the halomethane substrates $CHCl_3$, CBr_4 and CHI_3 gave the halosilyl complexes $CpRe(NO)(PPh_3)(SiMe_2X)$ (X = Cl, Br, I), respectively. $CpRe(NO)(PPh_3)(SiMe_2Cl)$ was found to react with Me_3SiOTf, yielding the corresponding triflate which reacts either with pyridine yielding the base stabilized $[CpRe(NO)(PPh_3)\{SiMe_2(NC_5H_5)\}]^+OTf^-$ complex or with $(Me_2N)_3S^+[SiMe_3F_2]^-$ to give the fluoro derivative $CpRe(NO)(PPh_3)(SiMe_2F)$. Solution of the complex $CpRe(NO)(PPh_3)(SiMe_2Cl)$ was found to form with BCl_3 the $CpRe(NOBCl_3)(PPh_3)(SiMe_2Cl)$ adduct which has been structurally characterized (Table 7)[77]. Bridging siloxy hydride rhenium complexes has been obtained as described in equation 23, with comparable Re–Si bond lengths in the range of 2.477(3) Å –2.499(3) Å for all the three silyloxy complexes (Table 7)[78].

Photolysis in hydrocarbon solutions of $CpMn(CO)(P_2)$ [P_2 = dmpe, dmpp, dmpm, $(PMe_3)_2$] with the silanes $SiPh_nH_{4-n}$ ($n = 1\text{-}3$) yielded the corresponding manganese silyl hydrides complexes $CpMn(CO)(P_2)(H)(SiPh_nH_{3-n})$ [P_2 = dmpe, $n = 1\text{-}3$; P_2 = dmpp, $n = 1\text{-}3$; P_2 = dmpm, $n = 2$; P_2 = $(PMe_3)_2$; $n = 2$]. The crystal structure of the monosilyl derivative $CpMn(CO)(dmpe)(H)(SiPh_2H)$ features a distorted three-legged piano-stool geometry with a vacant coordination site presumably occupied by the (non-observed) hydride ligand. The silyl and hydride ligands exchange positions through a pseudorotation of a seven-coordinate Mn center[79].

High-valent rhenium silyl hydrides $ReH_5(PPh(Pr\text{-}i)_2)_2(SiHPh_2)_2$ (**27**) and $ReH_5(PCy_3)_2$ $(SiH_2Ph)_2$ have been characterized by neutron and X-ray diffraction, respectively. The complexes were prepared from the corresponding hydrides $ReH_7(PPh(Pr\text{-}i)_2)_2$ and

ReH$_7$(PCy$_3$)$_2$ with SiH$_2$Ph$_2$ and SiH$_3$Ph, respectively. The neutron diffraction study on **27** shows that the geometry coordination of the rhenium is better described as a monocapped twisted rhombic antiprism (Figure 4) with metal–silicon bond lengths of 2.510(9) Å and 2.501(2) Å for the ReH$_5$(PPh(Pr-i)$_2$)$_2$(SiHPh$_2$)$_2$ (**27**) and ReH$_5$(PCy$_3$)$_2$(SiH$_2$Ph)$_2$ complexes, respectively[80].

(23)

Similar rhenium complexes having a chelating disilyl ligand ReH$_5$(disilyl)(PPh$_3$)$_2$ [disilyl = 1,2-bis(dimethylsilyl)benzene (**28**) and 1,2-bis(dimethylsilyl)ethane] have been prepared by similar reaction of the disilyl derivatives with ReH$_7$(PPh$_3$)$_2$[81]. The structure of the aromatic silyl rhenium derivative is not the tricapped trigonal prism as found for other nine-coordinate rhenium polyhydrides or the monocapped twisted rhombic prism as in complex **27**, but rather a dodecahedral structure typical for eight-coordination is adopted. Thus, the compound is described as a ReH$_3$(H$_2$)(disilyl)(PPh$_3$)$_2$ containing a stretched η^2-H$_2$ ligand (Figure 4). These complexes were found to be active catalysts for the dehydrogenation of *tert*-butylethylene although with low turnover frequencies[82].

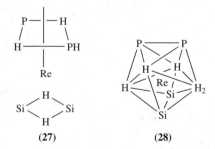

FIGURE 4. Coordination geometries for ReH$_5$(PPh(Pr-i)$_2$)$_2$(SiHPh$_2$)$_2$) (**27**, monocapped antiprism) and ReH$_5$(disilyl)(PPh$_3$)$_2$ (**28**, eight-coordinate octahedron)

35. Transition-metal silyl complexes

TABLE 7. Selected M—Si distances in group-7 transition-metal silylene and silyl complexes

Compound	d^n/configuration	M—Si (Å)	Reference
$(CO)_4\overline{MnSiMe_2}(\mu\text{-OMe})\dot{S}iMe_2$	d^4	2.336(2)	75
		2.344(2)	
$CpRe(NOBCl_3)(PPh_3)(SiMe_2Cl)$	d^6	2.476(3)	77
$CpMn(CO)(dmpe)(H)(SiPh_2H)$	d^3	2.319(4)	79
$Re_2(CO)_8[Si(\mu\text{-}\eta^2\text{-OMe})(OMe)_2](\mu\text{-H})$	d^6	2.477(3)	78
$Re_2(CO)_7(PMe_2Ph)[Si(\mu\text{-}\eta^2\text{-OMe})(OMe)_2](\mu\text{-H})$	d^6	2.477(4)	78
$Re_2(CO)_7(PMe_2Ph)[Si(OMe)_3](H_2O)(\mu\text{-H})$	d^6	2.499(3)	78
$ReH_5(PPh(Pr\text{-}i)_2)_2((SiHPh_2)_2)$	d^0	2.510(9)	80
$ReH_5(PCy_3)_2((SiH_2Ph)_2)$	d^0	2.501(2)	80
$ReH_3(H_2)[o\text{-}C_6H_4(Me_2SiH)_2](PPh_3)_2$	d^0	2.495(2)	82
		2.489(3)	

TABLE 8. Selected ^{29}Si NMR chemical shifts for group-7 transition-metal silylene complexes

Compound	^{29}Si NMR shift (ppm)	Reference
$(CO)_4\overline{MnSiMe_2}(\mu\text{-OMe})\dot{S}iMe_2$	115.4	75
$Cp(CO)_2Mn=SiH$ ($o\text{-Me}_2NCH_2C_6H_4$)	130.89 ($^1J_{HSi}$ = 156.8 Hz)	57
$MeCp(CO)_2Mn=SiH$ ($o\text{-Me}_2NCH_2C_6H_4$)	131 ($^1J_{HSi}$ = 149 Hz)	57
$Cp(CO)_2Mn=SiH$ [$8\text{-}(Me_2NCH_2)C_{10}H_6$]	119.82 ($^1J_{HSi}$ = 165.46 Hz)	57
$MeCp(CO)_2Mn=SiH$ [$8\text{-}(Me_2NCH_2)C_{10}H_6$]	124.49 ($^1J_{HSi}$ = 165.96 Hz)	57
$MeCp(CO)_2Mn=SiH$ [$8\text{-}(Me_2N)C_{10}H_6$]	141.85 ($^1J_{HSi}$ = 163.97 Hz)	57
$MeCp(CO)_2Mn=Si(Ph)$ ($o\text{-Me}_2NCH_2C_6H_4$)	145.2	76
$MeCp(CO)_2Mn=Si(Ph)$ [$8\text{-}(Me_2N)C_{10}H_6$]	124.3	76
$Cp(CO)_2Mn=Si(Ph)$ [$8\text{-}(Me_2N)C_{10}H_6$]	125.2	76
$Cp(CO)_2Mn=Si(Np)$ [$8\text{-}(Me_2N)C_{10}H_6$]	127.1	76
$Cp(CO)_2Mn=Si$ ($o\text{-Me}_2NCH_2C_6H_4$)$_2$	147.4	76

XIII. GROUP-8 SILYL DERIVATIVES

During the last five years, a vast amount of research has been focused on the chemistry of group-8 silicon complexes. This section takes into account only those complexes containing well-characterized metal–silicon bonds but does not take into account intermediary compounds in stoichiometric or catalytic reactions in which such species may be formed.

A. Bimetallic Silyl Complexes

Phosphine substituted hydrido silyl complexes of the type mer-FeH(SiR$_3$)(CO)$_3$L [R = OMe, L = dppm, Ph$_2$PCH$_2$C(=O)Ph, PPh$_2$H; R = OEt, L = dppm, R = Ph, L = dppb] in which only one phosphorous atom is bonded to the metal center, or the chelating complex cis-FeH{Si(OMe)$_3$}(CO)$_2$L [L = vinylidene(diphenylphosphine) = vdpp] have been easily prepared by carbonyl substitution in the complex cis-[FeH(SiR$_3$)(CO)$_4$] with the corresponding phosphine. The anionic potassium silyl complexes K[Fe(SiR$_3$)(CO)$_3$L] [R = OMe, L = dppm, Ph$_2$PCH$_2$C(=O)Ph; R = OEt, L = dppm; R = Ph, L = dppb] or the sodium complexes Na[Fe(CO)$_3$(PR$_3'$)(SiR$_3$)] (PR$_3'$ = PMe$_3$, SiR$_3$ = SiMePh$_2$, SiMe$_3$; PR$_3'$ = PBu$_3$, SiR$_3$ = SiMePh$_2$; PR$_3'$ = PPh$_2$H, SiR$_3$ = SiMe$_2$Ph, SiMePh$_2$)[83] were prepared from the corresponding hydrides in polar solvents with KH or NaH, respectively. These anionic silyl complexes have been used widely to prepare metal–metal heterobinuclear silicon-containing complexes[83]. Thus, for example, mer-[(CO)$_3$(Ph$_2$MeSi)Fe–ML] (ML = AgPPh$_2$CH$_2$C$_6$H$_5$, AuPPh$_2$CH$_2$C$_6$H$_5$, HgBr)[83]; mer-[LM(μ-dppm)FeSi(OMe)$_3$(CO)$_3$] (M = Cu, Ag, Au, L = PPh$_3$; M = Cu, L = MeCN, M = Ag, L = AsPh$_3$)[84] and (M = Pd, L = η^3-2-Me-allyl, η^3-allyl; M = Pt, L = η^3-allyl)[85] were prepared from their anionic silyl complexes with the corresponding counterpart in the form of the metal halide complexes. Enticingly the arsine ligand, AsPh$_3$, in the complex mer-[Ph$_3$AsAg(μ-dppm)FeSi(OMe)$_3$(CO)$_3$] dissociates in solution forming the corresponding mer-[$\overline{\text{Ag}(\mu\text{-dppm})\text{Fe}\{(\text{OMe})}$Si(OMe)$_2$}(CO)$_3$ with a η^2-μ_2-SiO bridge among the two metal centers[84]. A similar bridged complex with the η^2-μ_2-SiO moiety $\overline{\text{Fe}\{\mu\text{-Si(OMe)}_2(\text{OMe})\}(\text{CO})_3(\mu\text{-dppm})\text{Pd}}$(SnPh$_3$) was obtained by the replacement of the allylic moiety and elimination of hydrogen in [η^3-2-Me-allyl)Pd(μ-dppm)FeSi(OMe)$_3$(CO)$_3$] with an excess of HSnPh$_3$[85]. The reaction of the anionic complex, [Fe(SiR$_3$)(CO)$_3$PR$_3'$]$^-$, with CdX$_2$ (X = Cl, Br) results in the formation of three different types of products, depending on the nature of the phosphine and X ligands. The PMe$_3$ substituted complexes [Fe(SiR$_3$)(CO)$_3$(PMe$_3$)]$^-$ (SiR$_3$ = SiPh$_3$, SiMePh$_2$) exclusively gave the 2:1 (Fe:Cd) ratio complexes [fac-(Me$_3$P)(R$_3$Si)(CO)$_3$Fe]$_2$Cd (29a). The 2:1 ratio complexes [{(MeO)$_3$P}(Ph$_2$MeSi)(CO)$_3$Fe]$_2$ (29b) and [{(MeO)$_3$Si}(CO)$_3$Fe(μ-Ph$_2$PPyr)]$_2$Cd can be also obtained similarly; however, the arrangement of the CO ligands was found to be meridional.

$$\begin{array}{c}
\text{CO} \quad\quad \text{PMe}_3 \\
| \ \nearrow\text{SiR}_3 \ \text{OC} \ | \\
\text{OC}-\text{Fe}-\text{Cd}-\text{Fe}-\text{CO} \\
| \quad\searrow\text{CO R}_3\text{Si}\nearrow | \\
\text{Me}_3\text{P} \quad\quad\quad\quad \text{CO}
\end{array} \quad\quad \begin{array}{c}
\text{MePh}_2\text{Si} \quad\quad \text{P(OMe)}_3 \\
| \ \nearrow\text{CO} \ \text{OC} \ | \\
\text{OC}-\text{Fe}-\text{Cd}-\text{Fe}-\text{CO} \\
| \searrow\text{CO} \quad \text{OC}\nearrow | \\
\text{(MeO)}_3\text{P} \quad\quad\quad \text{SiMePh}_2
\end{array}$$

(29a)   (29b)

In addition to the 2:1 metal ratio complexes, the 1:1 ratio compounds mer-[(R$_3'$P)(R$_3$Si)(CO)$_3$FeCd(μ-X)]$_2$ were also obtained by the reaction of the ferrosilyl anion [Fe{Si(OMe)$_3$}(CO)$_3$(Ph$_2$PPyr)]$^-$ with CdBr$_2$, the reaction of [Fe(SiPh$_2$Me)(CO)$_3$P(OMe)$_3$]$^-$ with CdCl$_2$, and the reaction of [Fe(SiR$_3$)(CO)$_3$PPh$_2$H]$^-$ (SiR$_3$ = SiPh$_3$, SiMePh$_2$) with CdCl$_2$ and CdBr$_2$[86]. Similar bimetallic Fe–late transition metal complexes, main group elements and trimetallic systems have been studied comprehensively by Braunstein and coworkers. Structural Fe–Si data for the bimetallic and trimetallic complexes are collected in Table 9[87–95].

When the salt [Bu$_4$N][AuX$_2$] (X = Cl, Br) was reacted with 2 equiv of the dppm-substituted iron silyl metallate K[Fe{Si(OMe)$_3$}(CO)$_3$(dppm)], instead of the expected [Fe—Au—Fe]$^-$ silyl derivative anion chain, the complex [AuFe{Si(OMe)$_3$}(CO)$_3$(μ-dppm)]$_2$·CH$_2$Cl$_2$ (**30**) was obtained with a Au—Au and two Au—Fe bonds. It is interesting to point out that complex **30** forms a 10-membered-ring structure in which the Au—Au interaction is reminiscent of the transannular interactions that occur in organic cycles of middle size (8–12). The similar reaction with HgCl$_2$ yielded the expected Fe—Hg—Fe complex (**31**)[96].

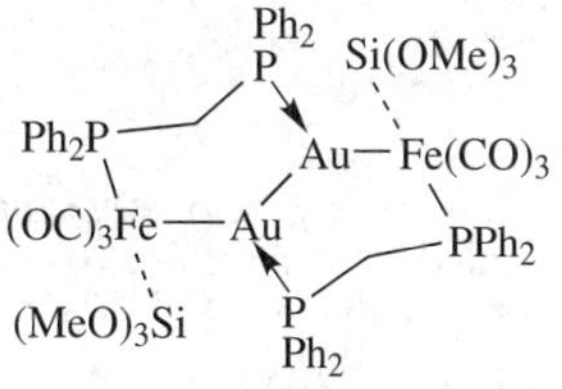

(30) 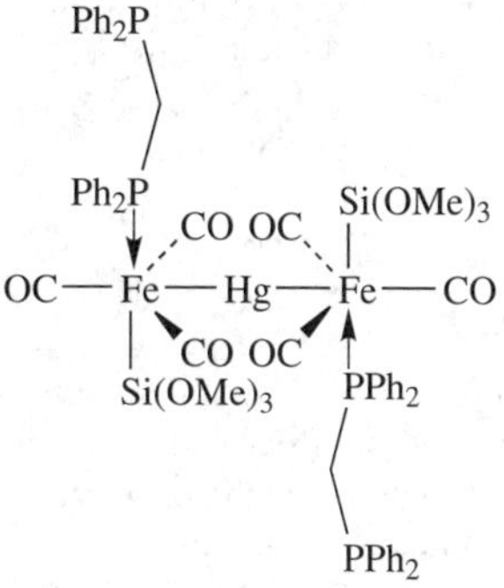　　　　　(31)

Photochemical decarbonylation reaction of (CO)$_2$(dppe)Fe(H)(SiR$_3$) with HSiR$_3$ [SiR$_3$ = Si(OMe)$_3$, Si(OEt)$_3$, SiMe$_3$, SiMe$_2$Ph, SiPh$_3$] yields the trihydrido silyl complexes (CO)(dppe)FeH$_3$(SiR$_3$). The analogous silyl triphosphine complexes (PR'Ph$_2$)$_3$FeH$_3$(SiR$_3$) can also be prepared by the reaction of the H_2-complexes, (PR'Ph$_2$)$_3$FeH$_2$(H$_2$), with the corresponding silanes HSiR$_3$. Additional silyl derivatives of the above-mentioned carbonyl and phosphine trihydrido complexes were obtained by simple silane exchange starting from the corresponding (CO)(dppe)FeH$_3$(SiR$_3$) or (PR'Ph$_2$)$_3$FeH$_3$(SiR$_3$), respectively. Interestingly, the mono and trihydrido silyl complexes (CO)$_2$(dppe)Fe(H)(SiR$_3$) and (PR'Ph$_2$)$_3$FeH$_3$(SiR$_3$) were found to show fluxional behavior in solutions with an exchange mechanism involving a η^2-HSiR$_3$ ligation[97]. The trihydrido silyl complexes (PEtPh$_2$)$_3$FeH$_3$(SiPh$_n$Me$_{3-n}$)(n = 1–3) were found to react selectively with HBF$_4$ to give the corresponding fluorinated complexes (PEtPh$_2$)$_3$FeH$_3$(SiF$_n$Me$_{3-n}$) by cleavage of the Si—Ph bonds[98]. Trihydrido silyl complexes of the type Fe(CO)(dppe)H$_3$(Si(OEt)$_3$) were deprotonated in the presence of KH/18-crown-6 to the anionic silyl complex [K(18-crown-6)][Fe(CO)(dppe)H$_2$(Si(OEt)$_3$)]. This complex can react with Me$_3$ECl (E = Sn, Pb) affording the capped octahedral complex Fe(CO)(dppe)H$_2$(Si(OEt)$_3$(EMe$_3$) with the dppe, ER$_3$, and CO ligands in the plane, one hydride ligand in the axial position and the other hydride capping the triangular face[99].

Dinuclear silyl-substituted complexes of the type (CO)$_3$(R$_3$Si)Fe(μ-PR'R'')Pt(PPh$_3$)$_2$ [PHR'R'' = PHPh$_2$, PH$_2$Ph, PH$_2$Cy; SiR$_3$ = SiPh$_3$, SiPh$_2$Me, SiPhMe$_2$, Si(OMe)$_3$] were prepared by an oxidative addition across the P—H and Fe—H bonds of the corresponding hydro-silyl phosphine complexes with the platinum (0) complex, Pt(CH$_2$=CH$_2$)(PPh$_3$)$_2$. For the bridged (μ-PPh$_2$) complexes, the PPh$_3$ ligand *trans* to the bridge can be replaced by CO (Scheme 17)[100] or by a number of phosphine and diphosphine ligands[101]. Braunstein and coworkers have found the unprecedented CO-promoted transfer of a silyl group from a metal to a metal in (CO)$_3$(SiR$_3$)Fe(μ-PPh$_2$)Pt(PPh$_3$)CO(Fe—Pt) as described in Scheme 17.

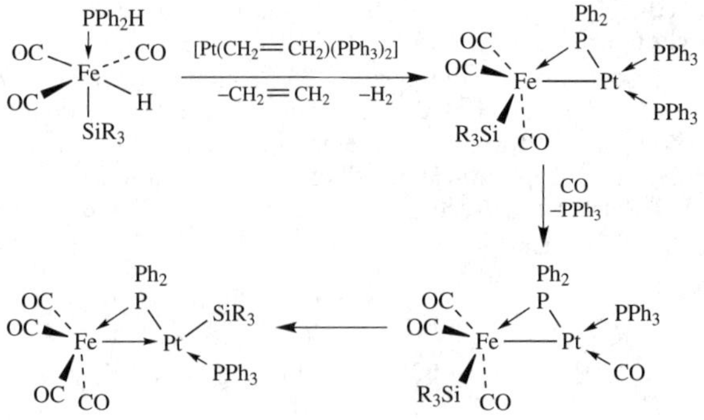

SCHEME 17

The migration was found to be faster for the methoxide silyl derivative than for the dimethylphenyl silyl derivative and it could be prevented by using $SiPh_3$ or stronger σ-donor / π-acceptor ligands on Pt such as t-BuCN and dppm[102]. In related dppm-bridged Fe—Pt complexes containing a Pt—Alkyl bond *trans* with respect to the metal—metal bond, facile CO insertion has been found, affording the corresponding bimetallic acyl complexes of the type η^2-μ_2-SiO[$(CO)_3$Fe{μ-Si(OMe)$_2$(OMe)}(μ-dppm)Pt{C(=O)R}] (R = Me, Et, norbornyl) with the acyl group *trans* with respect to the metal—metal bond[103]. The addition of phosphine or phosphite ligands induced silyl migration from the iron to the acyl group leading to the μ-siloxycarbene complex[104]. Similar silyl migration and μ-siloxycarbene bridged formation has been observed for the complex [$(CO)_3$Fe{μ-Si(OSiMe$_3$)$_2$(OSiMe$_3$)}(μ-dppm)Pd(Me)] upon reaction with CO[105]. Likewise, μ_2-η^2-SiO—Fe—Pd and Fe—Pt silyloxy complexes undergo methoxy exchange with fluoro atoms by the reaction of the corresponding complexes with $BF_3 \cdot Et_2O$ forming μ_2-η^2-SiF—Fe—Pd and μ-SiF—Fe—Pt—H interactions as outlined in Scheme 18[87].

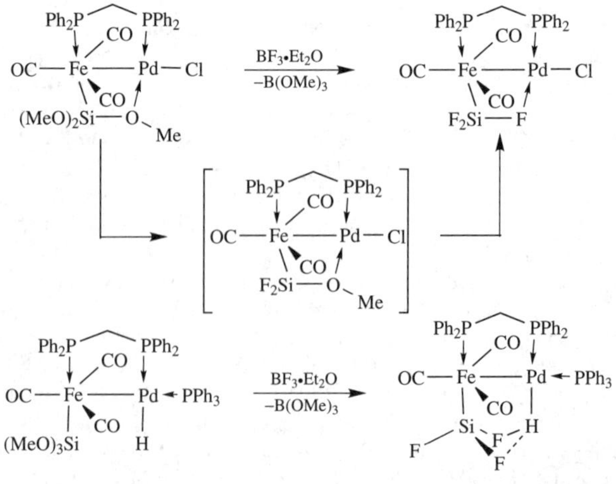

SCHEME 18

B. Clusters Containing Silyl Groups

The oxidative addition of $Os_3(CO)_{10}(NCMe)_2$ with $HSi(OEt)_3$ yielded the bridging hydride complex $Os_3(CO)_{10}(NCMe)[Si(OEt)_3](\mu\text{-H})$ (**32**), which was transformed to the first $\mu_3\text{-}\eta^3\text{-SiO}$ complex, $Os_3(CO)_9(NCMe)[\mu_3\text{-}\eta^3\text{-Si(OEt)}_3](\mu\text{-H})$ (**33**), by reflux in nonpolar solvents (equation 24). The molecular structure of **33** consists of a triangular triosmium cluster with a $\eta^3\text{-Si(OEt)}_3$ ligand bridging one face of the cluster. Complex **33** was found to add 2 equivalents of CO forming the complex $Os_3(CO)_{11}[Si(OEt)_3](\mu\text{-H})$, which is believed to be geometrically similar to complex **32** with a CO ligand replacing the NCMe ligand[106,107].

$$\begin{array}{c}
(EtO)_3Si \quad H \\
(CO)_3Os\text{------}Os(CO)_3 \\
\diagdown \quad \diagup \quad NCMe \\
Os \\
(CO)_4 \\
(32)
\end{array} \quad \xrightarrow[\text{reflux}]{\text{heptane}} \quad \begin{array}{c}
(CO)_3 \\
Os \\
H \diagup \quad | \diagdown OEt \\
| \quad OEt\text{-}Si \diagdown \\
| \quad \diagup \quad \quad OEt \\
(CO)_3Os\text{------}Os(CO)_3 \\
(33)
\end{array} \quad (24)$$

The comparable methoxy hydride cluster $Os_3(CO)_{10}(NCMe)[Si(OMe)_3](\mu\text{-H})$ (**34**) was found to react with terminal alkynes ($t\text{-BuC}{\equiv}CH$) forming a *tert*-butylvinyl η^2-bridge among the metal centers by insertion of the hydride into the triple bond. Concomitant addition of CO produced the hydrosilylation silyl olefin product *trans-t*-BuCH=CHSi(OMe)_3 [107]. Reaction of complex **34** with the phosphine ligand, dppm, replaced the NCMe ligand affording the complex $Os_3(CO)_{10}(\eta^1\text{-dppm})[Si(OMe)_3](\mu\text{-H})$, and reaction of the latter with $Pt(COD)_2$ (COD = 1, 5-cyclooctadiene) yields the cluster $(\mu\text{-H})PtOs_3(CO)_{10}(\mu\text{-}\eta^2\text{-dppm})[Si(OMe)_3]$ (**35**). The structure of **35** consist of a planar butterfly cluster of four metal atoms, with the platinum atom located in one of the 'hinge' sites. Irradiation of **35** produced the isomeric complex $PtOs_3(\mu\text{-H})(CO)_{10}(\mu\text{-}\eta^2\text{-dppm})[Si(OMe)_3]$ (**36**) which slowly converts back thermally to **35**. The transformation of **35** to **36** was proposed to occur via a tetrahedral-like cluster complex **37** through the formation of a bond between the wing tip metal atoms (Os) in **35** (equation 25)[108].

Bridging between two osmium centers has been observed through the oxidative addition of a sila-phosphine ligand and $Os_3(CO)_{10}(MeCN)_2$ as observed by the NMR coupling constants (Table 10)[109], whereas the oxidative addition reaction of an isolobal sila-amine (e.g. $2\text{-Me}_2SiHC_5H_4N$) ligand with the cluster $[Os_3(CO)_{10}(\mu\text{-H})_2]_2$ produced the disilyl triosmium cluster $Os_3(CO)_{10}(\mu\text{-H})_2\{2\text{-Me}_2SiHC_5H_4N\}_2$ with no bonding of the nitrogen to the metal center[110]. Similar diaza-bridged Ru carbonyl clusters have been prepared by the oxidative addition of the silanes $HSiR_3$, with the succeeding series of $[Ru_3(\mu\text{-H})(\mu_3,\eta^2\text{-ampy})(PPh_3)(CO)_{9-n}]$ (n = 0-2; Hampy = 2-amino-6-methylpyridine), $[Ru_3(\mu\text{-C}_4H_4N_2)(\mu\text{-CO})_3(CO)_7]$ ($C_4H_4N_2$ = pyridazine) and $[Ru_3(\mu\text{-dmpz})(\mu\text{-CO})_3(CO)_7]$ (Hdmpz = 3, 5-dimethylpyrazole) complexes[111-114]. An interesting symmetric triruthenium dihydrido cluster bridged by 1,1-ferrocenediyl bis(dimethylsilylene) was obtained by the reaction of 1,1'-bis(dimethylsilyl)ferrocene with $Ru_3(CO)_{12}$[115].

C. Silyl Complexes

Girolami and coworkers have shown that the methylene/silyl complex $Cp^*Ru(\mu\text{-}CH_2)(\mu\text{-Cl})Cp^*Ru(SiMe_3)$ undergoes low-and high-energy dynamic processes in which the silyl group migrates between the two metal centers. The low-energy dynamic process has $\Delta H^{\ddagger} = 8.9 \pm 0.1$ kcal mol^{-1} which involves a migration through a symmetric Ru(μ- SiMe$_3$)Ru intermediate whereas the consecutive high-energy dynamic process has $\Delta H^{\ddagger} = 12.0 \pm 0.1$ kcal mol^{-1} involving a remarkable reversible C−Si bond cleavage (equation 26)^{116,117}. Treatment of the silyl/methylene complex with acetylene gave the first unsubstituted metallabenzene complex [Ru$_2$(η^5-C$_5$Me$_5$)$_2$(η^2,η^5-C$_5$H$_5$)(SiMe$_3$)] (equation 27)^{116}.

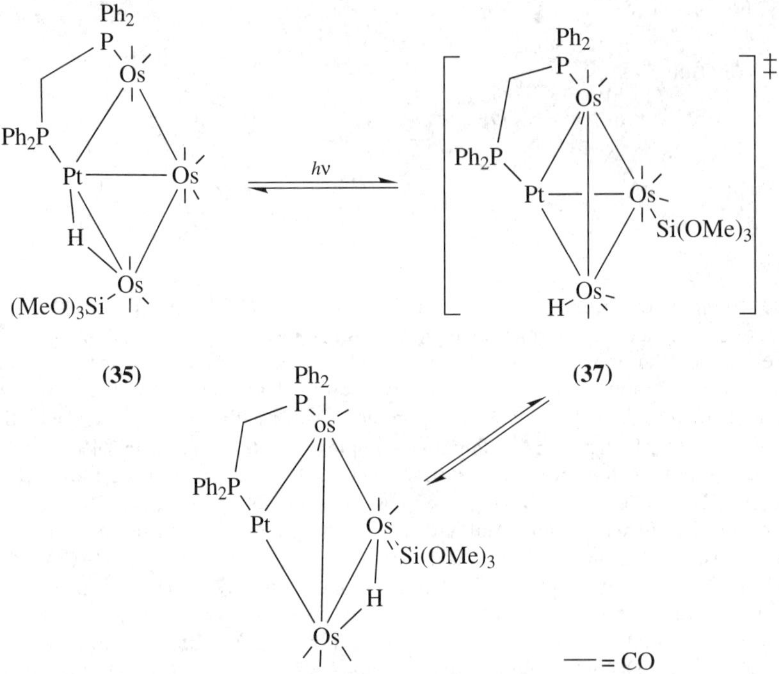

(25)

Oxidative addition of the penta-coordinative Os(CO)$_2$(PPh$_3$)$_3$ with the hydrosilanes HSiR$_3$ (SiR$_3$ = SiMe$_3$, SiEt$_3$, SiPh$_2$H) affords the corresponding octahedral monosilyl hydride complexes, OsH(SiR$_3$)(CO)$_2$(PPh$_3$)$_2$^{118}, whereas the reaction of the penta-coordinative phenyl complex Ru(Ph)(CO)Cl(PPh$_3$)$_2$ with the appropriate silanes HSiR$_3$ yields the corresponding five-coordinate silyl complexes Ru(SiR$_3$)Cl(CO)(PPh$_3$)$_2$ upon the reductive elimination of benzene. These complexes have been found to be effective catalysts for the hydrosilylation of alkynes to vinylsilanes^{119}.

Reaction of silyl–late transition metal complexes with amines normally induces the removal of the silicon ligand to obtain the Si−N bonded product. Interestingly, the reaction of Ru(CO)$_4$Si(Me$_2$Ph)I with benzylamine yields the octahedral complex, Ru(CO)$_2$(H$_2$NCH$_2$Ph)$_2$(SiMe$_2$Ph)I, with the carbonyl and the nitrogens of the benzylamine ligands in the *trans* position of the equatorial plane, and a short Ru−Si bond length compared with other Ru(II)–silyl complexes (Table 9)^{120}. Monomeric and dimeric silanetriols of similar osmium complexes were obtained by the basic hydrolysis of the corresponding trichlorosilanes as outlined in equation 28^{119}.

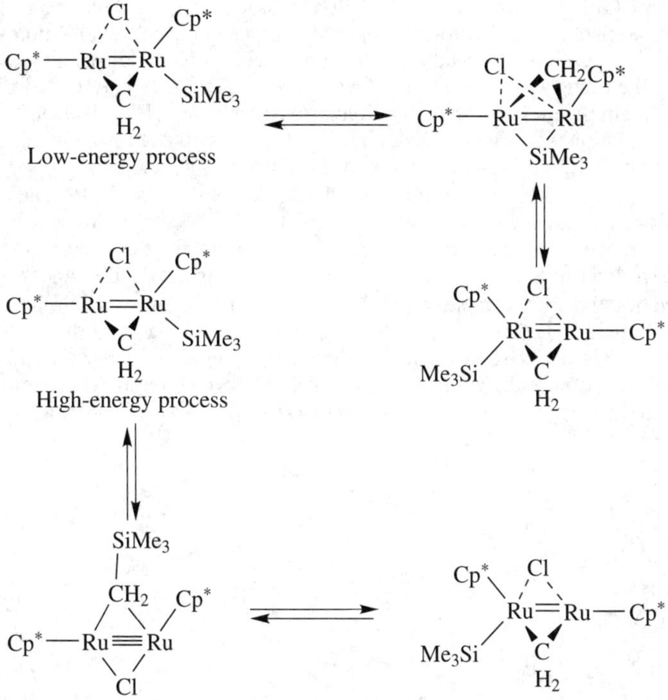

35. Transition-metal silyl complexes

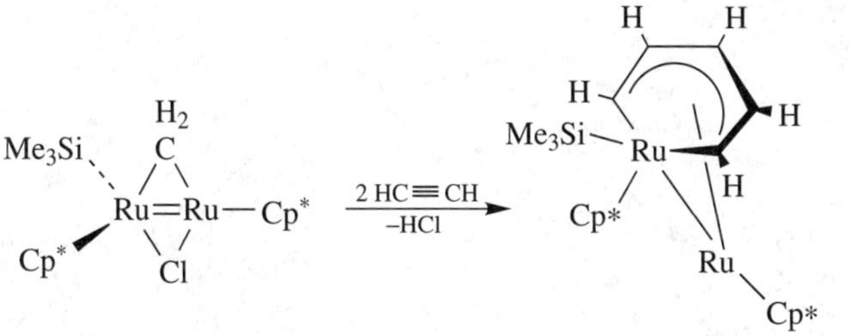

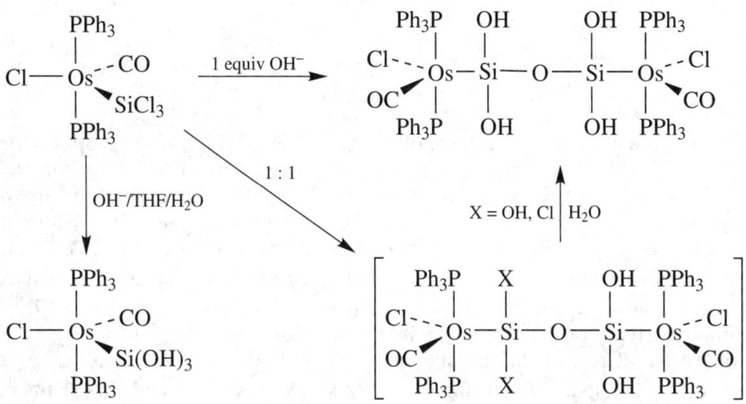

(26)

(27)

(28)

Other Cp(CO)(L)Fe (L = CO, PPh$_3$) silaneols, silanediols and silanetriols have been prepared via hydrolysis of the corresponding metallochlorosilanes or via oxyfunctionalization of metallo-hydrosilanes with dimethyldioxirane as described above for the tungsten complexes *(vide supra)*. Bifunctional metallosilylamines have been synthesized from the metallohydridosilanes by H/Cl exchange and upon dehydrohalogenation with Me$_3$P=CH$_2$, giving access to the corresponding silanimine complexes. The latter complexes add amines, acetone and alcohols in a 1,2 fashion yielding the corresponding disubstituted metallasilanes (Scheme 19)[62]. Recently, the ferro- and ruthenio-trihydridosilanes (C$_5$R$_5$)(CO)$_2$MSiH$_3$ (R = H, Me; M = Fe, Ru) were prepared via reaction of the ferro- and ruthenio-dichlorosilanes (C$_5$R$_5$)(CO)$_2$MSiCl$_2$H with LiAlH$_4$. Photo-induced substitution of the CO ligands with other ligands can be achieved for the iron complex with Me$_3$P or Ph$_3$P[121]. The ruthenio-hydridosilanes (C$_5$Me$_5$)(CO)$_2$RuSiR$_2$H (R = Me, *o*-MeC$_6$H$_4$) were accessible from the reaction of the metallating salt K[Ru(CO)$_2$C$_5$Me$_5$] and the corresponding chlorosilanes R$_2$SiHCl with a concomitant reduction by LiAlH$_4$. Oxofunctionalization with dimethyldioxirane afforded the corresponding ruthenio-silanol (C$_5$Me$_5$)(CO)$_2$RuSiR$_2$OH (R = Me, *o*-MeC$_6$H$_4$) complexes[122].

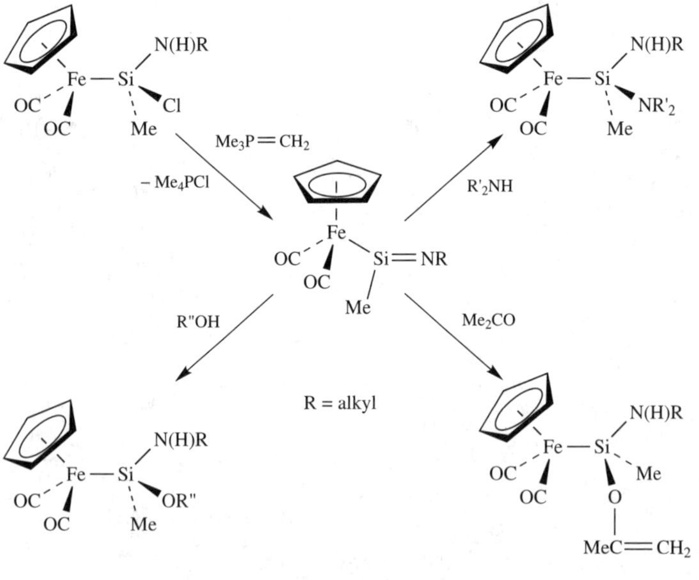

SCHEME 19

A series of ruthenium silyl complexes of the type CpRu(PMe$_3$)$_2$(SiR$_3$) (SiR$_3$ = SiHCl$_2$, SiCl$_3$, SiMeHCl, SiMeCl$_2$, SiMe$_2$Cl) have been obtained by HCl elimination from the reaction of the Cp(Me$_3$P)$_2$RuH with the corresponding chlorosilanes, presumably through a nucleophilic exchange mechanism involving a cationic ruthenium silyl hydride derivative exhibiting an agostic Ru−Si−H interaction. Interestingly, a trend of decreasing the $^2J_{SiP}$ coupling constants is observed when less electronegative substituents replaced the electronegative atoms at the silicon center, in line with Bent's rule (Table 10)[123]. The first reported example of a cationic metal silyl hydride was prepared by the protonation of Cp(Me$_3$P)$_2$RuSiCl$_3$ with the borane salt [H(Et$_2$O)$_2$][B(C$_6$H$_3$(CF$_3$)$_2$-3,5)$_4$] as shown in equation 29. The small $^2J_{PH}$ value of 29 Hz suggests a weakened Ru−H interaction, arguing for an agostic interaction which is also consistent with the large $^2J_{SiH}$ value of 48 Hz[123].

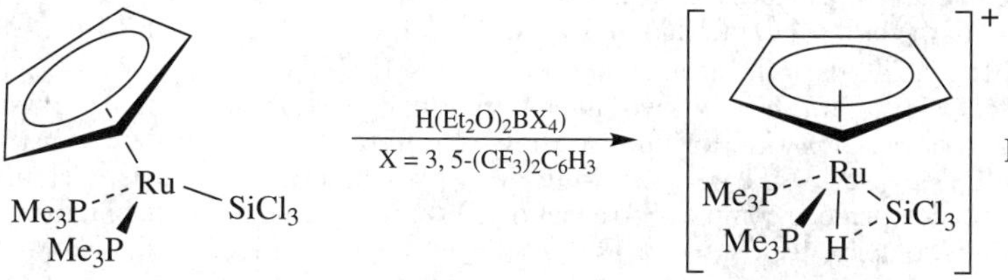

$$\text{Cp(PMe}_3)_2\text{Ru(SiCl}_3) \xrightarrow{\text{H(Et}_2\text{O)}_2\text{BX}_4} [\text{Cp(PMe}_3)_2\text{Ru(H)(SiCl}_3)]^+ \text{BX}_4^- \quad (29)$$

$$X = 3,5\text{-(CF}_3)_2\text{C}_6\text{H}_3$$

Recently, Lemke and coworkers prepared the Cp(PMe$_3$)$_2$RuSiCl$_2$(η^1-Cp*) complex by reacting Cp(PMe$_3$)$_2$RuCH$_2$SiMe$_3$ with (η^1-Cp*) SiHCl$_2$. The Cp* in the complex is σ-bound to the silicon and behaves as 'static' on the NMR time scale at room temperature. The η^1-Cp* configuration was confirmed by single-crystal X-ray diffraction showing a three-legged piano-stool geometry with a Ru−Si bond length of 2.335(1) Å, consistent with a single bond within the range of 2.27−2.51 Å (Table 9) as observed for other d^6 ruthenium silyl complexes[124]. Similar Ru−Si−H agostic interaction was obtained in the 1-sila-allene complex of Ru which was prepared using Tilley and coworkers'[125] synthesis for encumbered silenes as shown in equation 30. Here also, in contrast with similar systems prepared by Tilley and coworkers[125], the values $^2J_{SiH} = 66$ Hz and $^2J_{SiP} = 34.7$ Hz show that the hydrogen is not totally transferred to the metal center. This observation is verified by the hydrogen position at the silicon atom, which is disposed in the crystal structure almost symmetrically between the Ru and Si atoms [Ru−H = 1.58(5) Å and H−Si = 1.70(3) Å][126]. Caulton and coworkers have shown that no agostic Ru−H−Si interactions are observed in the similar systems Ru(CO)H(P(Bu-t)$_2$Me)$_2$SiHPh$_2$ and Cp*RuH(P(Pr-i)$_2$ Ph)HSiHPh$_2$ prepared by the reactions of Ru(CO)(P(Bu-t)$_2$Me)$_2$ and Cp*Ru(P(Pr-i)$_2$ Ph)(OCH$_2$CF$_3$); respectively, with H$_2$SiPh$_2$ (Table 10)[127,128].

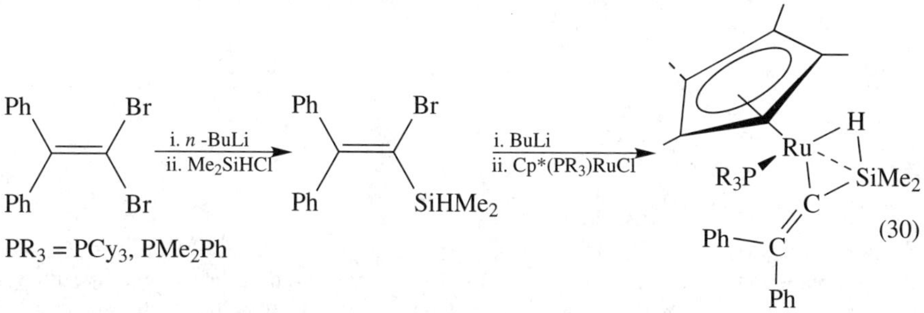

$$\text{Ph}_2\text{C=CBr}_2 \xrightarrow[\text{ii. Me}_2\text{SiHCl}]{\text{i. }n\text{-BuLi}} \text{Ph}_2\text{C=C(Br)(SiHMe}_2) \xrightarrow[\text{ii. Cp*(PR}_3)\text{RuCl}]{\text{i. BuLi}} \quad (30)$$

PR$_3$ = PCy$_3$, PMe$_2$Ph

D. Bis(silyl) Complexes

Four-legged 'piano stool' bis(silyl) complexes of the arrangement (η^6-arene)RuH$_2$ (SiMe$_3$)$_2$ (arene = C$_6$Me$_6$, C$_6$H$_6$, p-cymene) have been prepared by the oxidative addition of an excess of Me$_3$SiH with the dimeric [(η^6-arene)Ru(Cl)$_2$]$_2$ complex. The solid structure of the complex (arene = C$_6$Me$_6$) revealed a transoid disposition of the silyl and hydride ligands. In general, these complexes show a remarkable reactivity when heated at 150 °C with deuterated benzene which results in a complete H/D exchange in *all* the positions of the molecule, presumably through a η^2-Si=C intermediate, with concurrent arene exchange[129]. Similar (η^6-arene)Fe(H)$_2$(SiX$_3$)$_2$(X = Cl, F) complexes have been prepared from the arene solvated iron atoms in the presence of the trihalosilane and their solid state crystal structures have been determined[130–132].

Reaction of the bridged dinuclear tetrahydride complex [(C$_5$Me$_5$)Ru(μ-H)$_4$Ru(C$_5$Me$_5$)] with R$_2$SiH$_2$ yields the dinuclear μ-silyl complexes [{(C$_5$Me$_5$)Ru(μ-η^2-HSiR$_2$)}$_2$ (μ-H)(H)] (R = Et, Ph), having two three-center two-electron interactions. Pyrolysis of the latter complex in toluene at 150 °C afforded the dinuclear μ-silylene complex [(C$_5$Me$_5$)Ru(μ-SiPh$_2$)$_2$(μ-H)$_2$Ru(C$_5$Me$_5$)] which inserts acetylene, yielding the disilaruthenacyclopentene complex **38** (equation 31) or reacts with CO under high pressure to obtain the complex [(C$_5$Me$_5$)(CO)Ru(μ-SiPh$_2$)$_2$Ru(CO)(C$_5$Me$_5$)] with the two Cp* rings in the transoid position with respect to the Ru–Ru vector[133,134]. Starting from the coordinatively unsaturated dimer [Cp*Ru(μ-OMe)]$_2$ with the silane Ph$_2$SiH$_2$, the triple-bridged complex {[Cp*Ru]$_2$(μ-SiPhOMe)(μ-OMe)(μ-H)} was obtained[135]. Recently, similar complexes with one μ-silyl bridged group have been obtained in a similar manner, by the reaction of the bridged dinuclear tetrahydride complex [(C$_5$Me$_5$)Ru(μ-H)$_4$Ru(C$_5$Me$_5$)] with bulky t-Bu$_2$SiH$_2$. This complex can react with an excess of a less crowded silane to obtain mixed μ-silyl bridged derivatives (Table 10)[136].

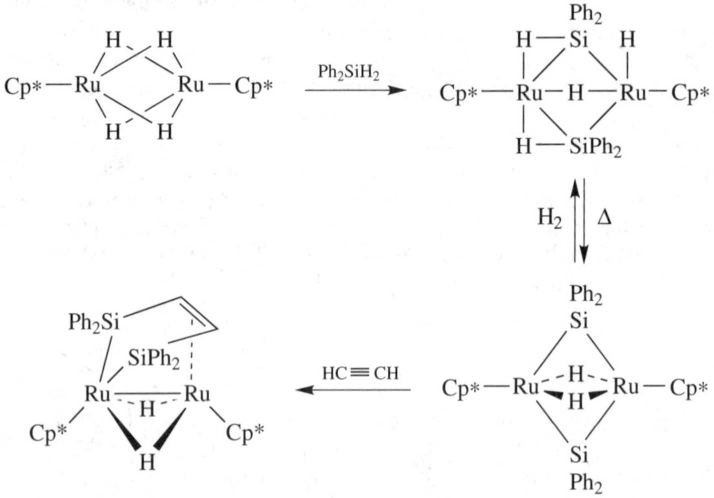

(31)

Akita and coworkers have shown the use of silanes as an H$_2$ equivalent for the reduction of organometallic ruthenium complexes. Thus, the reaction of the ruthenium carbonyl complex Cp$_2$Ru(CO)$_4$ with di- or trihydrosilanes at 150 °C resulted in the deoxygenative reduction of the carbonyl ligand, yielding a mixture of mono- and di-μ-methylene complexes. Further treatment of the mono-μ-methylene complex with monohydrosilanes produced methane (Scheme 20). The reaction mechanism, which involves an initial dissociation of CO, was investigated by using the labile μ-methylene acetonitrile complex which afforded instantaneously the hydrido-silyl-μ-CH$_2$ and disilyl-μ-CH$_2$ complexes, and by concomitant reaction with HSiR$_3$ yielded methane[137–141].

Reactions of the silene hydride complex [Cp*Ru(P(Pr-i)$_3$)(H)(η^2-CH$_2$=SiPh$_2$)] with hydrosilanes proceed via an initial migration of the hydride to the silene ligand, affording as the final products either the disilyl hydride or the mono silyl dihydrido ruthenium(IV) complexes as described in Scheme 21. Similar reductive elimination and

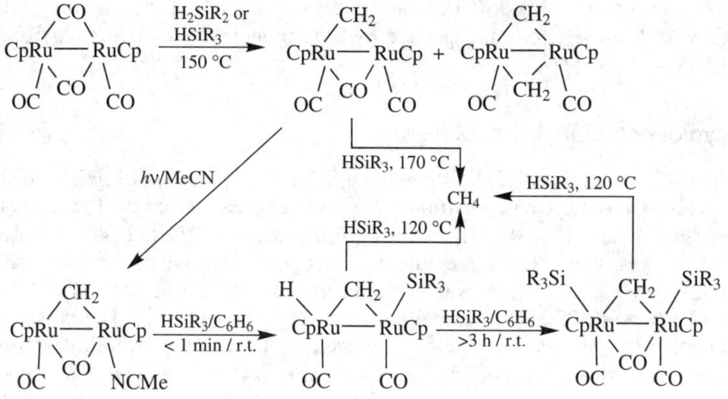

35. Transition-metal silyl complexes

SCHEME 20

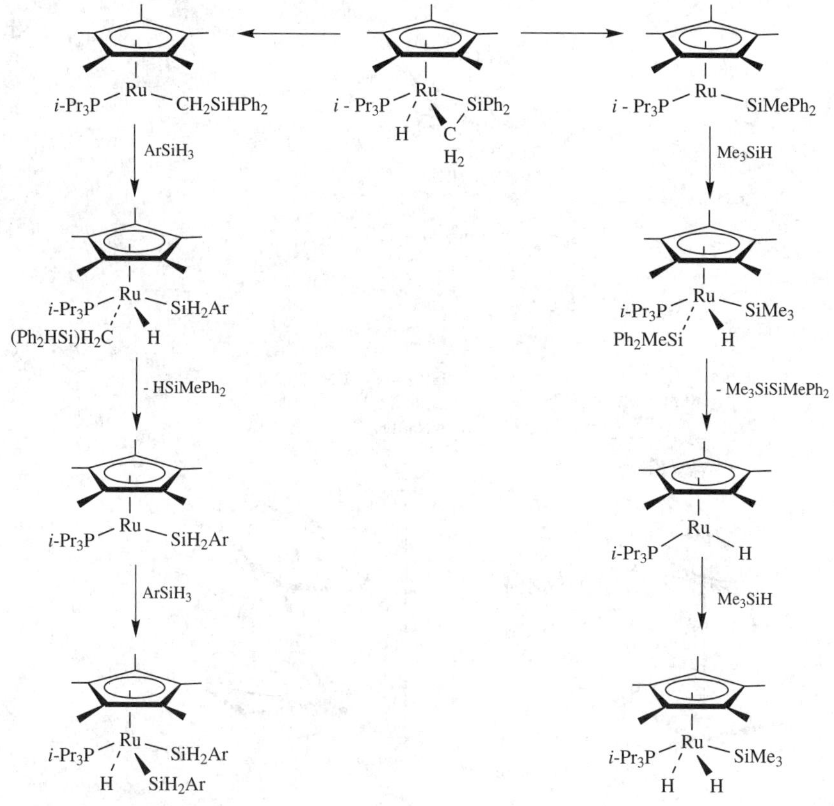

SCHEME 21

oxidative addition of a MesSi(Cl)H ligand was applied in the reaction of Cp*Ru(P(Pr-i)$_3$)Cl with MesSiH$_3$, producing the dihydrido monosilane complex Cp*Ru(P(Pr-i)$_3$)(H)$_2$(SiHClMes)[125,135,142].

E. Silacyclopentadienyl Complexes

The first silacyclopentadienyl ruthenium complex [Cp*Ru{η^5-Me$_4$C$_4$SiSi(SiMe$_3$)$_3$}] was prepared by Tilley and coworkers by the deprotonation of the corresponding tetraphenylborate salt, [Cp*Ru(H){η^5-Me$_4$C$_4$SiSi(SiMe$_3$)$_3$}][BPh$_4$] (**40**), with (Me$_3$Si)$_3$SiLi·3THF in 50% yield. The ruthenium salt was generated by the oxidative addition of the silole **39** to Cp*Ru$^+$, which was prepared *in situ* by reacting the dimer [Cp*Ru(μ-OMe)]$_2$ with Me$_3$SiOTf, after the addition of NaBH$_4$ (Scheme 22). The crystal structure of the complex salt shows the general protonated metallocene appearance with identical Ru–ring centroid distances within the experimental error (1.84 Å). The Ru–Si bond length of 2.441(3) Å is representative of a single bond (Table 9) and the $^1J_{SiH}$ coupling constant of 41 Hz is in the lower end for η^2-Si–H interactions[143].

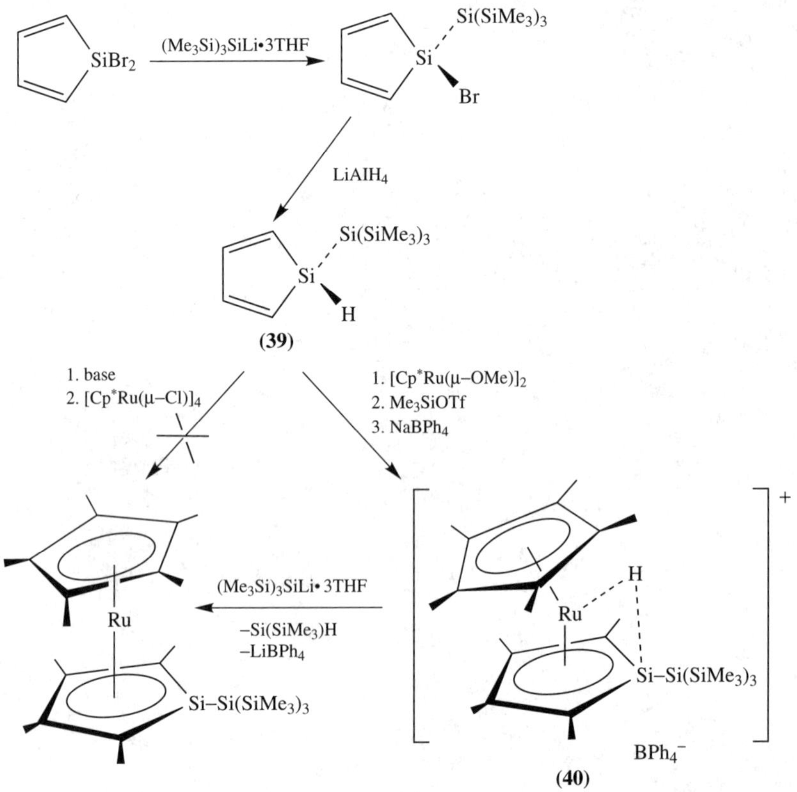

SCHEME 22

η^1-Tetraphenylsilacyclopentadienyl complexes of Fe and Ru have been prepared from the corresponding tetraphenylalkyl silyl chloride with NaMCp(CO)$_2$ (M = Ru, Fe). The complexes were found to react as dienes in spite of the presence of the electronegative phenyl substituents. Thus, the reaction of the substituted silacyclopentadiene complex with dimethyl acetylenedicarboxylate was found to yield the 7-silanorbornadiene complex with a Ru−Si bond length of 2.372(1) Å. The reaction of the latter with 2,3-dimethylbutadiene or 2,3-butanedione formed the corresponding 2,3-silacyclopentene complex and the dimethyl tetraphenylphthalate (Scheme 23)[144].

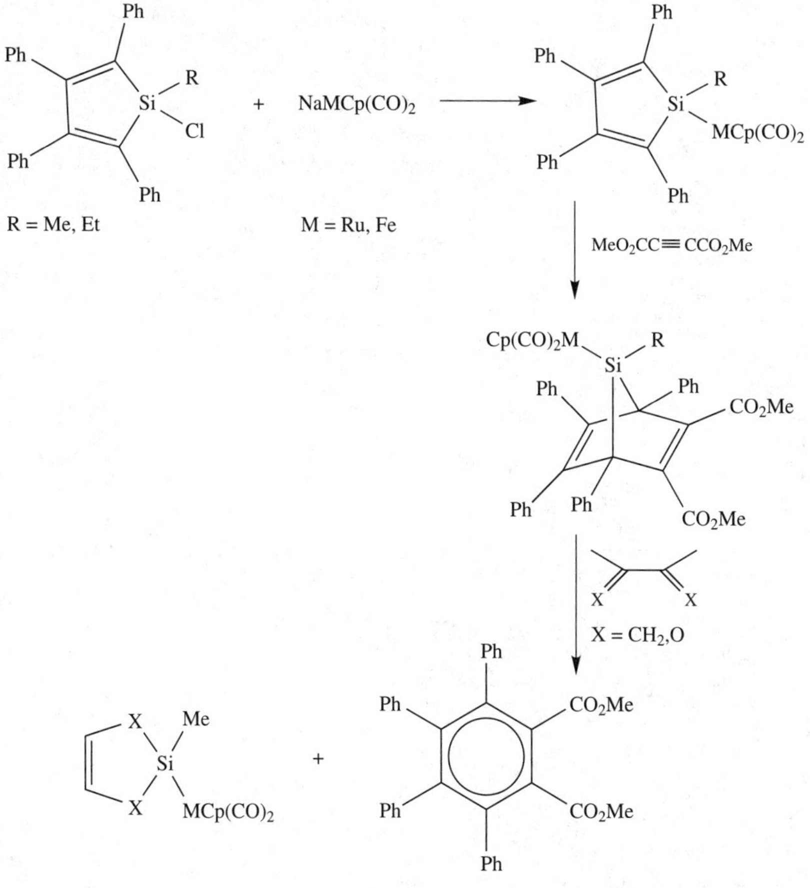

SCHEME 23

Many cyclic and linear transition metal–silicon compounds have been obtained by the elimination of alkali halides with the corresponding transition metal salts[145,146]. The synthesis and reactivity of the Fe–oligosilane systems have been studied in detail by Pannell[147–154] and this area has been recently reviewed[155].

F. Silene and Silylene Complexes

Lewis-base-stabilized iron–silylene complexes have been prepared by Corriu (equation 19) and Zybill (equation 17 and 18) in a similar fashion to that described for groups 6 and 7 silylene complexes (Tables 5–8)[55–57,76,156]. The reactivity of the Fe=Si bond with dienes, acetylenes and alcohols can be summarized for the silylene complex **41** as shown in Scheme 24.

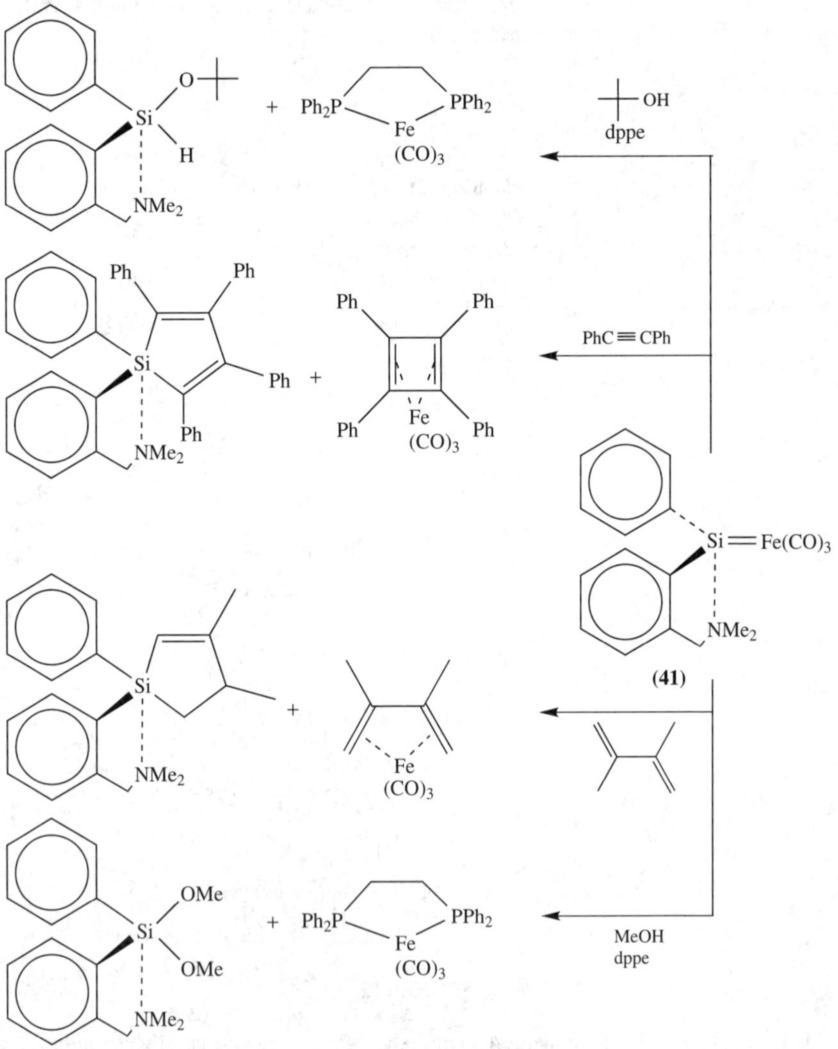

SCHEME 24

The photochemical reaction of Fe(CO)$_5$ with HSi(NMe$_2$)$_3$ yields the amine-stabilized silylene complex (CO)$_4$Fe=Si(NMe$_2$)$_2$(NHMe$_2$) (**42**). Reaction of complex **42** with Pt(C$_2$H$_4$)(PR$_3$)$_2$ (R = Ph, p-MeC$_6$H$_4$) resulted in the displacement of the dimethylamino and the ethylene ligands yielding the η^2-Pt–disilene complex **43**. The ^{29}Si NMR chemical shift of $\delta = 119$ found in complex **43** is consistent with a silylene complex and the $^2J_{\text{SiPt}} = 136.4$ Hz indicates that the Pt(PR$_3$)$_2$ fragment does not rotate freely. Similar photochemical reaction of the complex Fe(CO)$_4${P(OEt)$_3$} with HSi(NMe$_2$)$_3$ proceeds with an unprecedented ligand rearrangement to produce complex **44**. Two ethoxy groups of the phosphite ligand were exchanged with two NMe$_2$ substituents groups of the silicon atom forming a HNMe$_2$-stabilized diethoxysilylene complex having one of the shortest Fe−Si bond lengths [2.218(2) Å, 2.216(2) Å] (Table 9)[157].

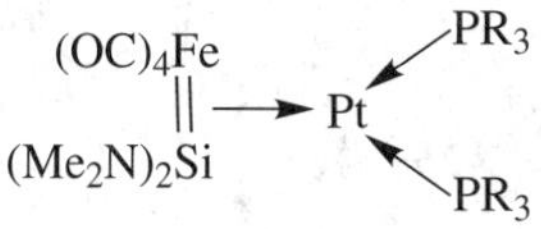

(**43**) 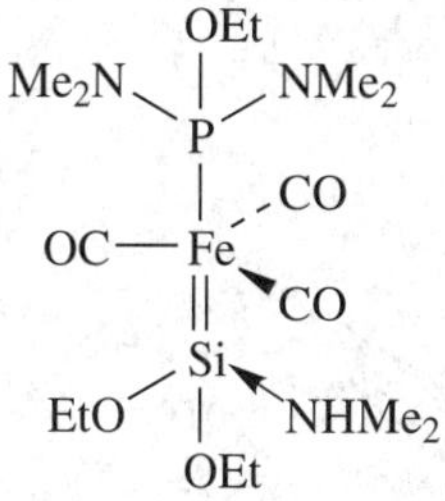   (**44**)

Recently, additional silacyclobutenes **45** and **46**[158] and isopropoxide[159] Fe=Si stabilized complexes were prepared from Collman's salt (Na$_2$Fe(CO)$_4$) with the corresponding dichlorosilane as described in equations 32 and 33, respectively. In complexes **45** and **46** a mixture of the E/Z diastereomers was obtained because of the possibility for the HMPA solvent to coordinate to the silicon center either E or Z relative to the neopentyl group. Thus, in the ^{29}Si NMR spectra of complexes **45** and **46**, the coordination to HMPA leads to a line splitting into doublets at 75.58 ($^2J_{SiP}$ = 36.4 Hz) for E-**45** and 77.90 ($^2J_{SiP}$ = 47.0 Hz) for Z-**45**, or at 72.03 ($^2J_{SiP}$ = 32.3 Hz) for E-**46** and 76.25 ($^2J_{SiP}$ = 29.4 Hz) for Z-**46**. The signals in **45** and **46** are shifted to higher fields (Table 10) due to the strong shielding at the silicon atom produced by the more electron-rich silacyclobutene rings. The two isopropoxy methyl groups are diastereotopic due to the pyramidal configuration of the silicon atom which is maintained even at high temperatures (100 °C). Similar base-stabilized ruthenium(0) silanediyl complexes have been prepared by Zybill following the approach described in equation 16 (Table 10)[160].

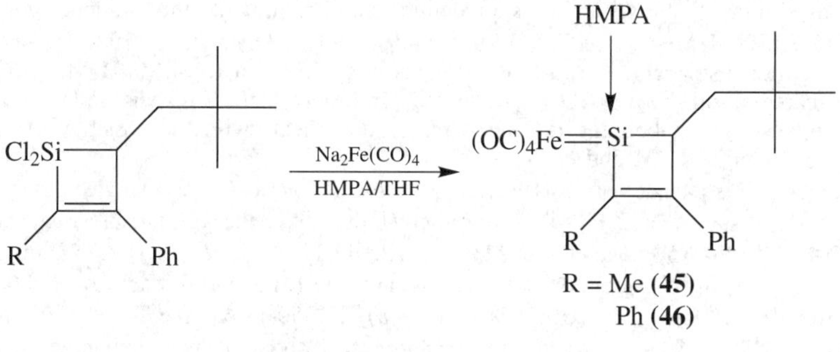

R = Me (**45**)
Ph (**46**)

(32)

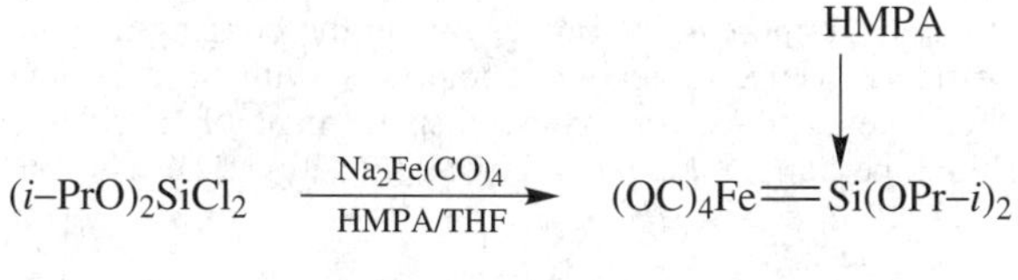

(33)

Base-stabilized ruthenium silylyne complex [Cp*(PMe$_3$)$_2$Ru≡Si(SC$_6$H$_4$Me-p)(phen)] [(OTf)$_2$] (**47**) has been prepared by Tilley and coworkers by the displacement of two triflate groups from Cp*(PMe$_3$)$_2$RuSi(SC$_6$H$_4$Me-p)(OTf)$_2$ in the presence of 1,10-phenantroline (Phen) (Scheme 25). Reaction of the silylyne complex with sodium amalgam produced complex **48**, which resulted from a reductive coupling of **47** leading to a carbon–carbon bond formation. The Ru–Si bond length in **47** [2.269(5) Å] is quite short and the rather long Si–N distances [1.95(1) Å and 1.91(1) Å] reflect the coordinate dative bond distances[161].

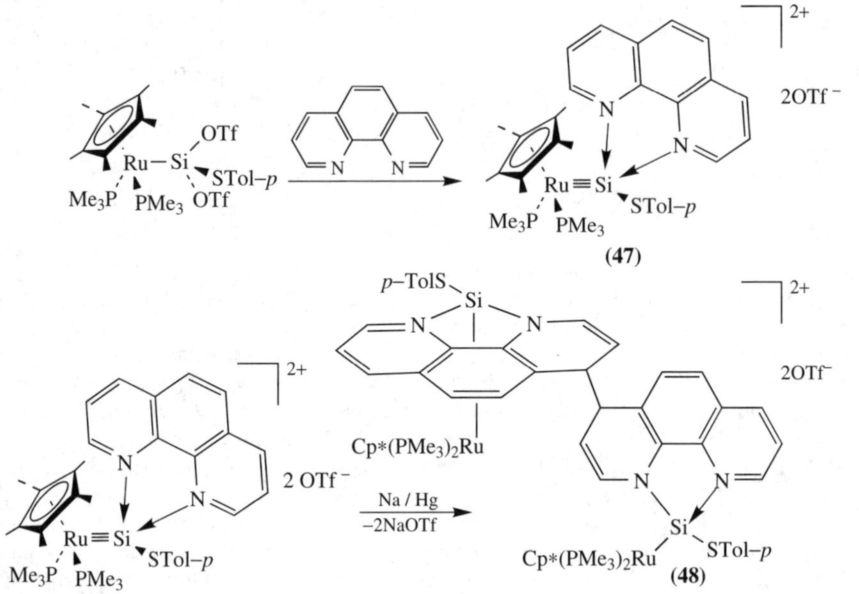

SCHEME 25

In a similar fashion, an organometallic stabilization of the silylene complex (OC)$_4$OsSi(SC$_6$H$_4$Me-p) [Ru(Cp*)(PMe$_3$)$_2$] (**49**), predicted by Aylett in 1980[162], has been obtained after displacement of the two triflates in Cp*(PMe$_3$)$_2$RuSi(SC$_6$H$_4$Me-p)(OTf)$_2$ in the reaction with Na$_2$Os(CO)$_4$. The molecular structure of **49** reveals a planar silicon atom suggesting a π-bonding to one or more substituents with the possible resonance structures given in **L**, **M** and **N**.

The Os–Si distance of 2.419(2) Å is typical for a Os–Si single bond. The Si–S bond distance of 2.172(4) Å is rather shorter than the comparable distances in Cp*(PMe$_3$)$_2$RuSi(SC$_6$H$_4$Me-p)$_3$ [2.223(1) Å, 2.195(1) Å and 2.196(1) Å[163]], implying some degree of Si–S π-bonding interactions, and the Ru–Si bond length of 2.286(2) Å is shorter than in Cp*(PMe$_3$)$_2$RuSi(SC$_6$H$_4$Me-p)$_3$ [2.350(1) Å] (see Table 9 for similar complexes)[163]. Thus, it seems that the structure corresponding to a ruthenium silylene complex (**N**) is the most important hybrid describing the bonding in complex **49**.

The first base-free silylene ruthenium complexes without a π-donor stabilization were prepared by Tilley and coworkers by displacement of a triflate ion by a non-coordinating lithium perfluorotetraphenyl borate (equation 34). The base-free silylene

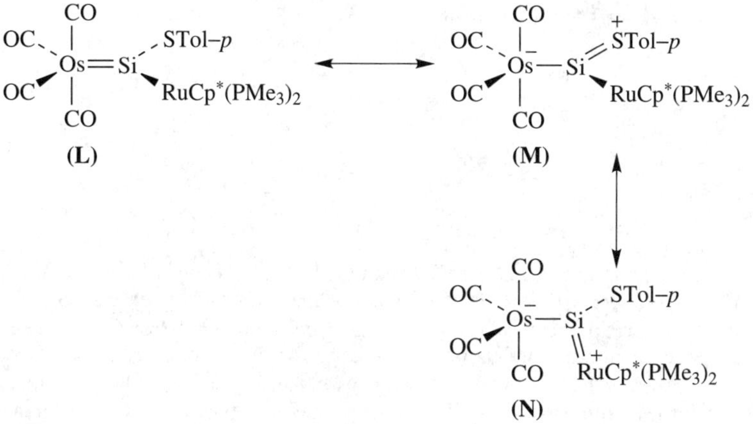

35. Transition-metal silyl complexes

[Structures L, M, N showing Os-Si complexes with RuCp*(PMe₃)₂ groups and STol-p substituents in resonance]

complexes were found to be unstable in solutions and decomposed with half-lives up to 7 hours. The structure of the silylene complex (R = Me) consists of a planar silicon with a Ru—Si bond length of 2.238(2) Å, which is the shortest yet reported distance for a ruthenium silylene complex (Table 9)[164]. A similar abstraction route to a cationic base-stabilized silylene complex $Cp(CO)_2Fe=SiMe(o-Me_2NCH_2C_6H_4)^+PF_6^-$ has been recently developed by the reaction of $Cp(CO)_2FeSi\{HMe(o-Me_2NCH_2C_6H_4)\}$ with Ph_3CPF_6. The cationic silylene complex reacts with MeOH affording the methoxy silyl complex $FpSiMe(OMe)\{o-Me_2N(H)CH_2C_6H_4\}$ [Fp = $CpFe(CO)_2$] with a Fe—Si distance of 2.305(2) Å and a ^{29}Si NMR signal at $\delta = 7.40$ ppm. These values can be compared with a Fe—Si distance of 2.266(1) Å and $\delta = 118.3$ ppm for the parent cationic complex[165].

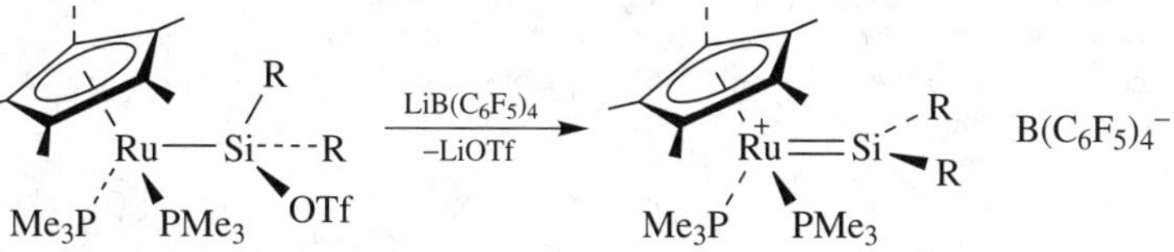

[Equation 34: Ru complex with Si-R, OTf converting via LiB(C₆F₅)₄/-LiOTf to cationic Ru=Si with B(C₆F₅)₄⁻]

R = Me, Ph

Photolysis of the complex $FpSiMe_2SiMe_2$ [Fp = $CpFe(CO)_2$] has been studied extensively by Pannell and Sharma[150,155] and Ogino and coworkers[166]. At early stages of the photolysis a mixture of all the geometric isomers of $[CpFe(CO)]_2(\mu\text{-}CO)\{\mu\text{-}SiMe(SiMe_3)\}$ and small amounts of $[CpFe(CO)]_2(\mu\text{-}SiMe_2)_2$ were obtained. The isomer ratio is 44:29:27 for the cis-SiMe₃ : cis-Me : trans isomers, respectively[167], whereas a prolonged photolysis gave only the cis and trans $[CpFe(CO)]_2(\mu\text{-}SiMe_2)_2$ complexes quantitatively[166]. Exchanging the iodine atom on the silanediyl bridge of the μ-silanediyl complex **50** by a strong Lewis base, N-methylimidazole (NMI) or 4-dimethylaminopyridine (DAMP), resulted in the formation of a donor-stabilized cationic silanetriyl diiron complex (equation 35). The Fe—Si bond lengths of 2.262(2) Å and 2.266(3) Å for the NMI and DAMP complexes respectively are shorter than those for other neutral silanediyl-bridged diiron complexes, although larger than those for the mononuclear iron silyl complexes (Table 9)[168].

$$\begin{array}{c}
\text{t-Bu}\diagdown\diagup\text{I} \\
\text{OC}\diagdown\text{Si}\diagup\text{CO} \\
\text{Fe}\text{---}\text{Fe} \\
\text{Cp}\diagup\text{C}\diagdown\text{Cp} \\
\text{O} \\
\text{(50)}
\end{array}
\xrightarrow{\text{NMI}}
\left[\begin{array}{c}
\text{t-Bu}\diagdown\diagup\text{NMI} \\
\text{OC}\diagdown\text{Si}\diagup\text{CO} \\
\text{Fe}\text{=}\text{Fe} \\
\text{Cp}\diagup\text{C}\diagdown\text{Cp} \\
\text{O}
\end{array}\right]^{+}
+ \text{I}^{-} \quad (35)$$

The cationic complexes can be easily reduced by $NaBH_4$ to the corresponding silane [CpFe(CO)]$_2(\mu$-CO)(μ-Si(Bu-t)H) **(51)** whereas the iodine complex **50** does not react under similar conditions[169]. μ-Halosilanes have been prepared from complex **51** with the corresponding di-, tri- and tetra-halomethane for the iodine, bromine and chlorine complexes, respectively[168]. Complex **51** has also been prepared by the photoreaction of CpFe(CO)$_2$SiMe$_3$ with t-BuSiH$_3$. Its X-ray crystal structure revealed that it adopts a geometry in which the two Cp rings and a Si–H bond are located on the same side with respect to the SiFe$_2$C four-membered ring. The ^{29}Si NMR spectra of the complex shows a signal at a remarkably low field ($\delta = 254.4$) which is characteristic for silylene bridging complexes[170]. Thermal or photochemical isomerization processes were found to take place among the similar cis and $trans$-Cp$'_2$Fe$_2$(CO)$_3(\mu$-SiHCH$_2$C$_6$H$_5$) complexes. However, the composition obtained by thermal equilibration ($cis:trans = 2:98$) differs considerably from the photostationary composition ($cis:trans = 70:30$)[170].

Upon irradiation of the alkoxy-substituted disilairon complexes FpSiMe$_2$SiMe(OMe)$_2$ [Fp = Cp$'$Fe(CO)$_2$, Cp$'$ = Cp or Cp*] an interconverting mixture of bis(silylene) iron complexes with syn and $anti$ configurations was obtained (equation 36)[171,172]. Both complexes have been observed as a mixture in the solid state crystal structure [$anti$ isomer Fe–Si 2.222(3) Å and 2.207(3) Å], ^{29}Si NMR ($\delta = 121.1, 101.9$). Each exist as a combination of two resonance structures where the silylene group is base stabilized by the methoxy group[173]. Similarly, mixed silylene–germylene methoxy stabilized iron complexes or the amino-bridged bis(silylene) iron complexes were obtained starting from the corresponding complexes, FpGeMe$_2$SiMe$_2$OMe or FpSiMe$_2$SiMe$_2$NEt$_2$, respectively[174–176].

$$\begin{array}{c}
\text{Cp}' \\
\diagdown \\
\text{OC}\diagup\text{Fe}\text{---}\text{SiMe}_2\text{SiMe(OMe)}_2 \\
\text{OC}
\end{array}
\xrightarrow[-\text{CO}]{h\nu}
\begin{array}{c}
\text{MeO}\diagdown\diagup\text{Me} \\
\text{Cp}'\diagdown\text{Si} \\
\text{OC}\diagup\text{Fe}\diagdown\text{OMe} \\
\text{OC}\diagdown\text{Si}\diagup \\
\text{Me}\diagup\diagdown\text{Me}
\end{array}
+
\begin{array}{c}
\text{Me}\diagdown\diagup\text{OMe} \\
\text{Cp}'\diagdown\text{Si} \\
\text{OC}\diagup\text{Fe}\diagdown\text{OMe} \\
\text{OC}\diagdown\text{Si}\diagup \\
\text{Me}\diagup\diagdown\text{Me}
\end{array}
\quad (36)$$

The first donor stabilized bis(silylene)ruthenium complex of the type Cp(PPh$_3$)$_2$Ru(SiMe$_2$–O(Me)–SiMe$_2$) was synthesized by the thermal reaction of Cp(PPh$_3$)$_2$RuMe with HSiMe$_2$SiMe$_2$OMe. The X-ray crystal structure reveals a distorted piano-stool structure, containing a RuSi$_2$O four-membered chelate ring with two equal short Ru–Si distances of 2.333(5) Å and two long Si–O bond distances [1.79(1) Å and 1.85(1) Å]. The ^{29}Si NMR chemical shift of 108.7 ppm ($^2J = 24$ Hz) is comparable to those for the isolobal alkoxy-bridged iron complexes (Table 10)[177]. In the gas phase, cationic iron silylene and silenes complexes have been generated by either an electron impact of Fe(CO)$_5$ which produces FeCO$^+$, or by Fe(CH$_2$CH$_2$)$^+$ produced by the reaction of Fe$^+$ formed by laser desorption with n-butane, with silanes Me$_x$SiH$_{4-x}$[178–180].

G. η^4-Silatrimethylenemethane Complexes

η^4-Silatrimethylenemethane complexes of iron and ruthenium have been obtained by the reaction of an alkylidenesilirane with the corresponding metal carbonyl as described in Scheme 26. Complexes (Z/E)-**52** and (Z/E)-**53** are obtained regioselectively from the corresponding Z/E-alkylidenesilirane. Of particular interest are the ^{29}Si chemical shifts of 43.5 (Z-**52**), 23.1 (E-**52**), 40.1 (Z-**53**) and 18.9 (E-**53**) and 19.1 ppm for **54**, respectively. These shifts are at considerably lower field than the shifts for other silene complexes (Table 10). The X-ray diffraction of these complexes exhibited an umbrella-shape ligation for the η^4-silatrimethylenemethane ligand with the three substituents at the central carbon bent away from the metal center and with Fe−Si bond lengths of 2.422(2) Å and 2.395(1) Å for complexes Z-**52** and **54**, respectively (Table 9). The formation of Z-**53** was accompanied by a competitive formation of an alkylidenesilirane containing triruthenium complex, thus reducing the yield of **53** to 17%[181,182].

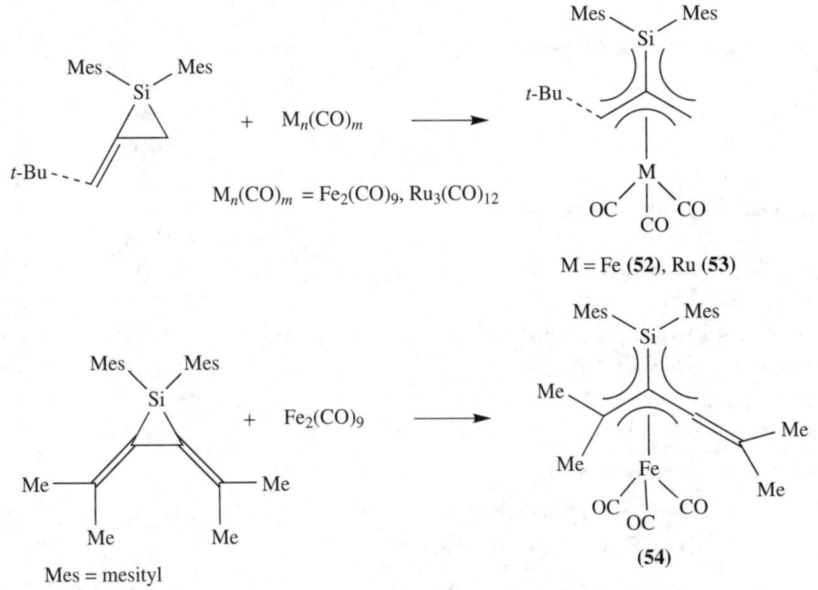

M = Fe (**52**), Ru (**53**)

Mes = mesityl

SCHEME 26

TABLE 9. Selected M−Si distances in group-8 transition-metal silicon-containing complexes

Compound	d^n Configuration	M−Si (Å)	Reference
Silylene Complexes			
(CO)$_4$Fe=Si(C$_6$H$_5$)(o-Me$_2$NCH$_2$C$_6$H$_4$)	d^8	2.259(1)	76
[{(Me$_2$N)$_2$OEt}P(CO)$_3$Fe=Si{NHMe$_2$}(OEt)$_2$]	d^8	2.218(2), 2.216(2)	157
(CO)$_4$Fe=Si(HMPA)Me$_2$	d^8	2.280(1), 2.294(1)	55
(CO)$_4$Fe=Si(HMPA)Cl$_2$	d^8	2.214(5), 2.221(1)	55

(*continued overleaf*)

TABLE 9. (continued)

Compound	d^n Configuration	M–Si (Å)	Reference
$(CO)_4Fe=Si(HMPA)(SBu$-$t)_2$	d^8	2.278(1)	156
$(CO)_4Fe=Si(HMPA)(OPr$-$i)_2$	d^8	2.261(2)	159
E-45	d^8	2.277(1)	158
E-46	d^8	2.272(1)	158
$(CO)_4Ru=Si(HMPA)(OBu$-$t)_2$	d^8	2.414(1)	160
$[Cp^*(PMe_3)_2Ru\equiv Si(STol$-$p)(phen)][(OTf)_2]^a$	d^6	2.269(5)	161
48	d^6	2.281(5)	161
$(CO)_4OsSi(STol$-$p)[Ru(Cp^*)(PMe_3)_2]$	d^6	2.419(2) (Os–Si) 2.286(2) (Ru–Si)	183
$[(Cp^*)(PMe_3)_2Ru=Si(Me)_2][B(C_6F_5)_4]$	d^6	2.238(2)	164
$Cp^*(PMe_3)_2RuSi(STol$-$p)_3$	d^6	2.350(1)	163
$Cp^*(PMe_3)_2RuSi(STol$-$p)_2OTf$	d^6	2.306(2)	163
$Cp^*(PMe_3)_2RuSi(STol$-$p)(OTf)_2$	d^6	2.269(3)	163
$[(Cp^*)(PMe_3)_2Ru=Si(STol$-$p)_2(NCMe)][BPh_4]$	d^6	2.284(3)	163
$[(Cp^*)(PMe_3)_2Ru=SiMe_2(NCMe)][BPh_4]$	d^6	2.258(4), 2.190(14)	163
$[CpFe(CO)]_2(\mu$-$CO)\{\mu$-$SiMe(SiMe_3)\}$	d^8	2.294(1)	166
$[\{CpFe(CO)\}_2\{\mu$-$Si(Bu$-$t)(NMI)\}]I^b$	d^8	2.262(1)	169
$[\{CpFe(CO)\}_2\{\mu$-$Si(Bu$-$t)(DAMP)\}]I^c$	d^8	2.266(3)	168
$Cp(CO)_2\overline{Fe=SiMe(o\text{-}Me_2NCH_2C_6H_4)}\ PF_6$	d^6	2.266(1)	165
Silyl Complexes			
$[PPh_3Cu(\mu$-$dppm)\overline{Fe\{Si(OMe)_3\}(CO)_3}$	d^8	2.271(4)	84
$[(CO)_3(SiF_3)\overline{Fe(\mu\text{-}dppm)Pt}(H)(PPh_3)]$	d^8	2.249(3)	87
$(\eta^3$-2-Me-allyl$)\overline{Pd(\mu\text{-}dppm)Fe\{Si(OMe)_3\}(CO)_3}$	d^8	2.303(1)	85
$\overline{Fe\{\mu\text{-}Si(OMe)_2(OMe)\}(CO)_3(\mu\text{-}dppm)Pd}(SnPh_3)$	d^8	2.258(1)	85
$[(CO)_3\overline{Fe\{\mu\text{-}Si(OSiMe_3)_2(OSiMe_3)\}(\mu\text{-}dppm)Pd}(Cl)]$	d^8	2.277(3)	105
$[(PPh_3)(CO)_3\overline{Fe\{\mu\text{-}Si(OMe)_2(OMe)\}}InCl_2-$ (O=PPh_3)$]$	d^8	2.311(2)	88
$[\{(MeO)_3Si\}(CO)_3\overline{Fe(\mu\text{-}dppm)In}Cl\{Mo(C_5H_4Me)(CO)_3\}$	d^8	2.307(1)	88
$[(CO)_3\overline{Fe\{\mu\text{-}Si(OMe)_2(OMe)\}(\mu\text{-}dppm)Pt}\{C\equiv N$-(xylyl-2,6)\}][PF_6]$.	d^8	2.263(2)	91
$[\{[(MeO)_3Si\}(CO)_3\overline{Fe(\mu\text{-}dppm)Cd}(\mu\text{-}Cl)\}_2]\cdot C_6H_{14}$	d^8	2.286(5)	90
$[(CO)_3\{Si(OMe)_3\}\overline{Fe(\mu\text{-}dppm)Hg}(C_6Cl_5)]$	d^8	2.311(3)	94
$[(CO)_3\{Si(OMe)_3\}\overline{Fe(\mu\text{-}dppm)HgPt}(C_6Cl_5)(t$-$BuNC)(PPh_3)]\cdot CH_2Cl_2$	d^8	2.316(13)	94

TABLE 9. (continued)

Compound	d^n Configuration	M–Si (Å)	Reference
[(CO)$_3${Si(OMe)$_3$}Fe(μ-dppm)(μ-SnCl$_2$)PtCl(PEt$_3$)]·CH$_2$Cl$_2$	d^8	2.326(3)	94
$\overline{\text{Fe}\{\mu\text{-Si(OMe)}_2(\text{OMe})\}(\text{CO})_3(\mu\text{-dppm})\text{R}}$h(CO	d^8	2.249(1)	89
[(CO)$_3$(Ph$_2$MeSi)Fe–AgPh$_2$CH$_2$C$_6$H$_5$]	d^8	2.327(3)	83
[(CO)$_3$(Ph$_2$MeSi)Fe–AuPh$_2$CH$_2$C$_6$H$_5$]	d^8	2.330(10)	83
[AuFe{Si(OMe)$_3$}(CO)$_3$(μ-dppm)]$_2$·CH$_2$Cl$_2$	d^8	2.268(8), 2.282(7)	96
Hg[Fe{Si(OMe)$_3$}(CO)$_3$(dppm)]$_2$	d^8	2.290(3), 2.285(3)	96
[(CO)$_3$(Ph$_2$MeSi)Fe–HgBr]	d^8	2.399(6)	83
mer-[{(MeO)$_3$Si}(CO)$_3$Fe(μ-Ph$_2$PPyr)]$_2$Cd	d^8	2.286(1)	86
mer-[(Ph$_2$HP)(Ph$_3$Si)(CO)$_3$FeCd(μ-Br)]$_2$	d^8	2.364(5)	86
(CO)(dppe)FeH$_3$(Si(OEt)$_3$)	d^6	2.250(4)	97
(CO)$_3$(R$_3$Si)Fe(μ-PR'R'')Pt(PPh$_3$)$_2$(Fe–Pt)	d^8	2.339(2)	100
(CO)$_4$Fe(μ-PPh$_2$)Pt{Si(OMe)$_3$}(PPh$_3$)(Fe–Pt)	d^8	2.288(1)	102
Os$_3$(CO)$_{10}$(NCMe)[Si(OEt)$_3$](μ-H)	d^6	2.39(1)	106
Os$_3$(CO)$_9$[μ_3-η^3-Si(OEt)$_3$](μ-H)	d^6	2.32(1)	106
Os$_3$(CO)$_{10}$[Si(OMe)$_3$][μ-CH=CHBu-t]	d^6	2.427(4)	107
Os$_3$(CO)$_{10}$(μ-H)$_2${o-Me$_2$SiHC$_5$H$_4$N}$_2$	d^6	2.434(4), 2.431(4)	110
Os$_3$(CO)$_8$[μ-PMe$_2$(o-C$_6$H$_4$)][Si(OMe)$_3$](μ-H)$_2$	d^6	2.369(5)	184
Os$_3$(CO)$_{10}$(μ-H)(MeCN){o-BrC$_6$H$_4$SiH$_2$Me$_2$}	d^6	2.452(5)	110
Os$_3$(CO)$_{10}$(μ-H)(PPh$_3$){o-BrC$_6$H$_4$SiH$_2$Me$_2$}	d^6	2.463(9)	110
Os$_3$(CO)$_{10}$(μ-H)(o-Ph$_2$PC$_6$H$_4$CH$_2$SiMe$_2$)	d^6	2.444(4)	109
Os$_3$(CO)$_{10}$(μ-H)(o-Me$_2$SiC$_6$H$_4$CH$_2$PPh$_2$)	d^6	2.453(5)	109
Os$_3$(CO)$_{10}$(μ-H)(o-MeHSiC$_6$H$_4$CH$_2$PPh$_2$)	d^6	2.425(7)	109
Os$_3$(CO)$_8$(μ-H)$_3$(o-Me$_2$SiC$_6$H$_4$CH$_2$PPh$_2$)	d^6	2.432(6)	109
Ru$_3$(μ-H)$_2$(μ-SiMe$_2$C$_5$H$_4$FeC$_5$H$_4$SiMe$_2$)(CO)$_{10}$	d^6	2.459(4)	115
Ru$_3$(μ-H)$_2$(SiEt$_3$)$_2$(μ-C$_4$H$_4$N$_2$)(CO)$_8^d$	d^6	2.455(5)	112
Ru$_3$(μ-H)$_2$(SiEt$_3$)(μ_3, η^2-ampy)(CO)$_8^e$	d^6	2.435(4)	111
Os(SiEt$_3$)H(PPh$_3$)$_2$(CO)$_2$	d^6	2.493(2)	118
Ru(CO)$_2$(NH$_2$CH$_2$CH$_2$Ph)$_2$(SiMe$_2$Ph)I	d^6	2.412(2)	120
Cp$_2^*$Ru$_2$(μ-CH$_2$)(μ-Cl)(SiMe$_3$)	d^4	2.387(2)	117
OsCl(CO)(PPh$_3$)$_2$(Si(OH)$_3$)	d^6	2.319(2)	119
[OsCl(CO)(PPh$_3$)$_2$Si(OH)$_2$]$_2$O	d^6	2.318(5), 2.337(5)	119
Cp(CO)$_2$Fe{Si(Cl)(Tol-o)(NMes)}	d^6	2.284(2)	62
(C$_5$Mes$_5$)(CO)$_2$FeSiH$_3$	d^6	2.287(2)	121
(C$_5$Mes$_5$)(CO)$_2$RuSi(Tol-o)$_2$OH	d^6	2.411(2)	122

(*continued overleaf*)

TABLE 9. (continued)

Compound	d^n Configuration	M–Si (Å)	Reference
Cp*(PCy$_3$)$\overline{\text{Ru}-\text{C}(=\text{CPh}_2)\text{SiMe}_2\text{H}}$	d^6	2.507(2)	126
(η^6-C$_6$H$_6$)RuH$_2$(SiMe$_3$)$_2$	d^4	2.396(1), 2.391(1)	129
(η^6-C$_6$H$_5$Me)FeH$_2$(SiCl$_3$)$_2$	d^4	2.222(2), 2.218(2)	130
(η^6-p-Me$_2$C$_6$H$_4$)FeH$_2$(SiCl$_3$)$_2$	d^4	2.226(2), 2.222(2)	132
(η^6-C$_6$H$_6$)FeH$_2$(SiCl$_3$)$_2$	d^4	2.210(3), 2.207(3)	132
(η^6-C$_6$H$_5$Me)FeH$_2$(SiF$_3$)$_2$	d^4	2.251(5), 2.261(5)	131
{Cp*Ru(μ-η^2-HSiEt$_2$)}$_2$(μ-H)H	d^4	2.544(2), 2.551(2) 2.338(1), 2.335(1)	133
[{Cp*Ru(μ-SiPh$_2$)(μ-H)}$_2$]	d^4	2.364(1), 2.360(1)	133
{Cp*Ru(μ-SiPh$_2$CH=CHSiPh$_2$)(μ-H)$_2$RuCp*}	d^4	2.410(2), 2.421(2)	134
[Cp*Ru(CO)$_2$]$_2$(μ-η^2:η^2-H$_2$Si(Bu-t)$_2$)	d^6	2.447(1), 2.457(1)	136
[Cp*Ru]$_2$(μ-η^2-HSi(Bu-t)$_2$H)(μ-η^2-HSiPhH)(μ-H)H	d^4	2.300(1), 2.375(1) 2.438(1), 2.675(1)	136
[Cp*Ru]$_2$(μ-η^2-HSi(Bu-t)$_2$H)(μ-η^2-HSiEtH)(μ-H)H	d^4	2.322(3), 2.365(3) 2.414(4), 2.680(3)	136
Cp*Ru(P(Pr-i)$_3$)H$_2$(SiHClMes)	d^4	2.303(3)	135
{[Cp*Ru]$_2$(μ-SiPh(OMe))(μ-OMe)(μ-H)}	d^4	2.309(10), 2.288(11)	135
[Cp*Ru{C$_4$Me$_4$Si(μ-H)Si(SiMe$_3$)$_3$}][BPh$_4$]	d^6	2.441(3)	143
Cp*Ru(P(Pr-i)$_3$)H(η^2-CH$_2$=SiPh$_2$)	d^6	2.382(4)	125
FpSiMe(OMe){o-Me$_2$N(H)CH$_2$C$_6$H$_4$}	d^6	2.305(2)	165
Z-Fe(CO)$_3$[η^4-Mes$_2$SiC(CH$_2$)CHBu-t)]	d^6	2.422(2)	181
Fe(CO)$_3$[η^4-Mes$_2$SiC(CMe$_2$)$_2$(C=CMe$_2$)]	d^6	2.395(1)	182

[a] Phen = 1,10-phenanthroline.
[b] NMI = N-methylimidazole.
[c] DAMP = dimethylaminopyridine.
[d] $C_2H_4N_2$ = pyridazine.
[e] ampy = 2-amino-6-methylpyridinate.

TABLE 10. Selected ^{29}Si NMR chemical shifts for group-8 transition-metal silicon containing complexes

Compound	^{29}Si NMR shift (ppm)	Reference
Cp(Me$_3$P)$_2$RuSiHCl$_2$	66.79 ($^1J_{SiH}$ = 199.6 Hz, $^2J_{SiP}$ = 36.2 Hz)	123
Cp(Me$_3$P)$_2$RuSiCl$_3$	42.11 ($^2J_{SiP}$ = 43.6 Hz)	123
Cp(Me$_3$P)$_2$RuSiHMeCl	57.17 ($^1J_{SiH}$ = 163.6 Hz, $^2J_{SiP}$ = 30.5 Hz)	123
Cp(Me$_3$P)$_2$RuSiCl$_2$Me	92.16 ($^2J_{SiP}$ = 35.4 Hz)	123
Cp(Me$_3$P)$_2$RuSiClMe$_2$	87.29 ($^2J_{SiP}$ = 30.2 Hz)	123
Cp(PMe$_3$)$_2$RuSiCl$_2$(η^1-Cp*)	82.71 ($^2J_{SiP}$ = 33.0 Hz)	124
[Cp(Me$_2$P)$_2$RuHSiCl$_3$] [B(3,5-(CF$_3$)$_2$C$_6$H$_3$)$_4$]	30.60 ($^2J_{SiH}$ = 48.8 Hz)	123
Ru(SiMe$_3$)Cl(CO)(PPh$_3$)$_2$	55.74 (t, $^2J_{SiP}$ = 10.9 Hz)	185
Ru(SiEt$_3$)Cl(CO)(PPh$_3$)$_2$	71.70 (t, $^2J_{SiP}$ = 11.7 Hz)	185
Ru(SiPh$_3$)Cl(CO)(PPh$_3$)$_2$	40.3 (t, $^2J_{SiP}$ = 13.1 Hz)	185
Cp*(PCy$_3$)Ru—C≡CPh$_2$)SiMe$_2$H	−81.30 ($^1J_{SiH}$ = 66 Hz, $^2J_{SiP}$ = 34.7 Hz)	126
Cp*Ru(P(Pr-i)$_2$Ph)H$_2$Si(OMe)$_3$	−55.2	127
Cp*Ru(P(Pr-i)$_2$Ph)H$_2$SiPh$_3$	−8.5	127
Cp*Ru(P(Pr-i)$_2$Ph)H$_2$SiMe$_3$	10.7 ($^2J_{SiP}$ = 8 Hz)	127
Cp*Ru(P(Pr-i)$_2$Ph)H$_2$(Si(H)Ph$_2$)	5.86 ($^1J_{SiH}$ = 182 Hz)	127
{Cp*Ru(μ-SiPh$_2$CH=CHSiPh$_2$)(μ-H)$_2$RuCp*}	4.6 (broad)	134
[Cp*Ru(μ-H)]$_2$(μ-η^2:η^2-H$_2$Si(Bu-t)$_2$)	75.5 (quintet, $^1J_{SiH}$ = 34.2 Hz)	136
[Cp*Ru(CO)$_2$]$_2$(μ-η^2:η^2-H$_2$Si(Bu-t)$_2$)	186.2 (t, $^1J_{SiH}$ = 22.4 Hz)	136
[Cp*Ru(CO)$_2$]$_2$ (μ-η^2-H$_2$Si(Bu-t)$_2$)H	168.7 (dd, $^1J_{SiH}$ = 31.6, 7.9 Hz)	136
[Cp*Ru{C$_4$Me$_4$Si (μ-H)Si(SiMe$_3$)$_3$}] [BPh$_4$]	−27.14 ($^1J_{SiH}$ = 41)	143
Cp*Ru {C$_4$Me$_4$Si-Si(SiMe$_3$)$_3$}	7.35	143
{[Cp*Ru]$_2$ (μ-SiPh (OMe))(μ-OMe)(μ-H)}	211.12	135
Cp*Ru(P(Pr-i)$_3$)(H)(η^2-CH$_2$=SiPh$_2$)	6.14 (m, ^2J = 21 Hz)	125
Cp*(PMe$_3$)$_2$RuSi(STol-p)$_3$	49.03 (t, $^2J_{SiP}$ = 34 Hz)	163
Cp*(PMe$_3$)$_2$RuSi(OTol-p)$_3$	−1.22 (t, $^2J_{SiP}$ = 42 Hz)	163
Cp*(PMe$_3$)$_2$RuSiMe$_2$(STol-p)	50.19 (t, $^2J_{SiP}$ = 42 Hz)	163
Cp*(PMe$_3$)$_2$RuSi(STol-p)$_2$OTf	77.14 (t, $^2J_{SiP}$ = 36 Hz)	163
Cp*(PMe$_3$)$_2$RuSi(OTol-p)$_2$OTf	−0.01 (t, $^2J_{SiP}$ = 45 Hz)	163
Cp*(PMe$_3$)$_2$RuSiMe$_2$OTf	133.29 (t, $^2J_{SiP}$ = 33 Hz)	163
Cp*(PMe$_3$)$_2$RuSi(STol-p)(OTf)$_2$	37.10 (t, $^2J_{SiP}$ = 39 Hz)	163
FpSiMe(OMe){o-Me$_2$ N(H)CH$_2$C$_6$H$_4$}	7.40	165
(CO)$_4$Fe=SiH(C$_6$H$_5$)•DMI[a]	83.52 ($^1J_{HSi}$ = 199)	57
(CO)$_4$Fe=SiH(1-C$_{10}$H$_7$)•DMI[a]	81.51 ($^1J_{HSi}$ = 218.61)	57
(CO)$_4$Fe=SiMe(o-Me$_2$NCH$_2$C$_6$H$_4$)	123.6	76
Cp(CO)$_2$Fe=SiMe(o-Me$_2$NCH$_2$C$_6$H$_4$)	118.3	165

(*continued overleaf*)

TABLE 10. (continued)

Compound	^{29}Si NMR shift (ppm)	Reference
$(CO)_4Fe=SiPh(o-Me_2NCH_2C_6H_4)$	118.1	76
$(CO)_4Fe=SiPh(8-(Me_2NCH_2)C_{10}H_6)$	101.1	76
$(CO)_4Fe=SiPh(8-(Me_2N)C_{10}H_6)$	125.7	76
$(CO)_4Fe=Si(o-Me_2NCH_2C_6H_4)_2$	115	76
$(CO)_4Fe=Si(NMe_2)_2(NHMe_2)$	68.2	157
$[(Ph_3P)_2Pt(\eta^2-\{(CO)_4Fe=Si(NMe_2)_2\})]$	113.9 ($^2J_{SiPt} = 136.4$ Hz)	157
$[\{(Me_2N)_2EtO\}P(CO)_3Fe=Si\{NHMe_2\}(OEt)_2]$	57.4 (d, $^2J_{SiP} = 34$ Hz)	157
$(CO)_4Fe=Si(HMPA)(OBu-t)_2$	7.1 (d, $^2J_{SiP} = 26.4$ Hz)	55
$(CO)_4Fe=Si(HMPA)(SBu-t)_2$	74.7 (d, $^2J_{SiP} = 25.3$ Hz)	156
$(CO)_4Fe=Si(HMPA)Me_2$	92.3 (d, $^2J_{SiP} = 17.5$ Hz)	55
$(CO)_4Fe=Si(HMPA)Cl_2$	59.7 (d, $^2J_{SiP} = 31.2$ Hz)	55
$(CO)_4Fe=Si(HMPA)(OPr-i)_2$	20.9 (d, $^2J_{SiP} = 25.6$ Hz)	159
E-**45**	75.58 ($^2J_{SiP} = 36.4$ Hz)	158
Z-**45**	77.90 ($^2J_{SiP} = 47.0$ Hz)	158
E-**46**	72.03 ($^2J_{SiP} = 32.3$ Hz)	158
Z-**46**	76.25 ($^2J_{SiP} = 29.4$ Hz)	158
$(CO)_4Ru=Si(HMPA)Me_2$	79.0 ($^2J_{SiP} = 23$ Hz)	160
$(CO)_4Ru=Si(HMPA)Cl_2$	33.2 ($^2J_{SiP} = 32.2$ Hz)	160
$(CO)_4Ru=Si(HMPA)(OBu-t)_2$	−5.3 ($^2J_{SiP} = 25.7$ Hz)	160
$(CO)_4Ru=Si(HMPA)(Ph)_2$	73.9 ($^2J_{SiP} = 21.1$ Hz)	160
$(CO)_4OsSi(STol-p)[Ru(Cp^*)(PMe_3)_2]$	19.43 ($^2J_{SiP} = 23$ Hz)	183
$[(Cp^*)(PMe_3)_2Ru=SiMe_2][B(C_6F_5)_4]$	311.4	164
$[(Cp^*)(PMe_3)_2Ru=SiPh_2][B(C_6F_5)_4]$	299 (t, $^2J_{SiP} = 32$ Hz)	164
$[(Cp^*)(PMe_3)_2Ru=Si(STol-p)_2(NCMe)][BPh_4]$	58.30 (t, $^2J_{SiP} = 39$ Hz)	163
$[(Cp^*)(PMe_3)_2Ru=Si(OTol-p)_2(NCMe)][BPh_4]$	14.33 (t, $^2J_{SiP} = 47$ Hz)	163
$[(Cp^*)(PMe_3)_2Ru=SiMe_2(NCMe)][BPh_4]$	110.03 (t, $^2J_{SiP} = 30$ Hz)	163
$[CpFe(CO)]_2(\mu-CO)\{\mu-SiMe(SiMe_3)\}$	232.1 (μ-Si), −3.7 (SiMe$_3$)	166
$[CpFe(CO)]_2(\mu-SiMe_2)_2$	243.8, 229.5	166
$[\{CpFe(CO)\}_2\{\mu-Si(Bu-t)(NMI)\}]I^b$	251.5	169
$[\{CpFe(CO)\}_2\{\mu-Si(Bu-t)(DMAP)\}]I^c$	264.6	168
$[\{CpFe(CO)\}_2(\mu-CO)(\mu-Si(Bu-t)H)]$	254.4	168
$[\{CpFe(CO)\}_2(\mu-CO)(\mu-Si(Bu-t)Cl)]$	276.3	168
$[\{CpFe(CO)\}_2(\mu-CO)(\mu-Si(Bu-t)Br)]$	284.8	168
$[\{CpFe(CO)\}_2(\mu-CO)(\mu-Si(Bu-t)I)]$	289.1	168
$[\{CpFe(CO)\}_2(\mu-CO)(\mu-Si(Bu-t)Me)]$	267.4	168
(Z)-Fe(CO)$_3[\eta^4$-Mes$_2$SiC(CH$_2$)CHBu-t]	43.6	181
(E)-Fe(CO)$_3[\eta^4$-Mes$_2$SiC(CH$_2$)CHBu-t]	23.1	182
(Z)-Ru(CO)$_3[\eta^4$-Mes$_2$SiC(CH$_2$)CHBu-t]	40.1	181
(E)-Ru(CO)$_3[\eta^4$-Mes$_2$SiC(CH$_2$)CHBu-t]	18.9	182
Fe(CO)$_3[\eta^4$-Mes$_2$SiC(CMe$_2$)$_2$(C=CMe$_2$)]^d	19.1	182

TABLE 10. (continued)

Compound	^{29}SiNMR shift (ppm)	Reference
$(C_5H_5)(CO)_2FeSiCl_2H$	71.7	121
$(C_5H_5)(CO)_2RuSiCl_2H$	52.6	121
$(C_5Me_5)(CO)_2FeSiCl_2H$	79.9	121
$(C_5H_5)(CO)_2FeSiH_3$	−42.2	121
$(C_5H_5)(CO)_2RuSiH_3$	−64.6	121
$(C_5Me_5)(CO)_2FeSiH_3$	−31.24	121
$(C_5Me_5)(CO)_2RuSiH_3$	−49.44	121
$(C_5H_5)(CO)(Ph_3P)FeSiH_3$	−35.68	121
$(C_5H_5)(Ph_3P)_2FeSiH_3$	−36.4	121
$(C_5Me_5)(CO)_2RuSiMe_2H$	13.70	122
$(C_5Me_5)(CO)_2RuSi(Tol\text{-}o)_2H$	10.18	122
$(C_5Me_5)(CO)_2RuSiMe_2OH$	52.41	122
$(C_5Me_5)(CO)_2RuSi(Tol\text{-}o)_2OH$	48.24	122

^{a}DMI = 1,3-dimethyl-2-imidazolidinone.
^{b}NMI = N-methylimidazole.
^{c}DAMP = 4-dimethylaminopyridine.
^{d} η^4-silatrimethylenemethane framework.

XIV. GROUP-9 SILICON-CONTAINING COMPLEXES

A. Silylene Complexes

Recently, the first base-stabilized iridium silylene complexes, $IrH_2\{Si(OTf)Ph_2\}$ (TFB)(PR$_3$) (TFB = tetrafluorobenzobarrelene), were prepared and characterized following the reaction sequence of Scheme 27. As expected, in the complex with PR$_3$ = P(Pr-i)$_3$, the Ir−Si distance of 2.337(2) Å is significantly shorter than those determined previously for the six-coordinate silyliridium(III) complexes (Table 11). The silylene character was also supported by the relatively long Si−O bond of 1.790(5) Å compared with the normal range of 1.63–1.66 Å[186].

B. Silyl and Bis(silyl) Complexes

The reaction of the complex [Ir(μ-OMe)(TFB)]$_2$ with acetic acid in polar solvents leads to the formation of the dimeric carboxylate compound (Ir(μ, η^2-O$_2$ CMe$_3$)(TFB)]$_2$. This dimeric acetate complex reacted with monodentate phosphine ligands yielding the square-planar monomeric derivatives, Ir(OCOMe)(TFB)L [L = Pyr, PPh$_3$, PCy$_3$, P(Pr-i)$_3$], which upon reaction with trisubstituted silanes of the type HSiEt$_3$ or HSiPh$_3$ afforded the corresponding octahedral dihydride silyl complexes IrH$_2$ (SiR$_3$)(TFB)L (R = Et, Ph) and with the dihydrosilane, H$_2$SiPh$_2$, the complex IrH$_2$(Ph$_2$SiOCOMe)(TFB)L [L = PPh$_3$, PCy$_3$, P(Pr-i)$_3$] was formed. IrH$_2$(SiEt$_3$)(TFB)L [L = PPh$_3$, PCy$_3$, P(Pr-i)$_3$] were found to catalyze the hydrosilylation and the dehydrogenative silylation of phenylacetylene with triethylsilane, yielding in all cases a mixture of PhCH=CH$_2$, cis- and trans-PhCH=CHSiEt$_3$ and PhC(SiEt$_3$)=CH$_2$. Interestingly, the reaction of these dihydrosilane complexes with CO causes the reductive elimination of the silane and the formation of Ir (η^1 : η^2-C$_{12}$F$_4$H$_7$)(CO)$_2$L (equation 37)[187].

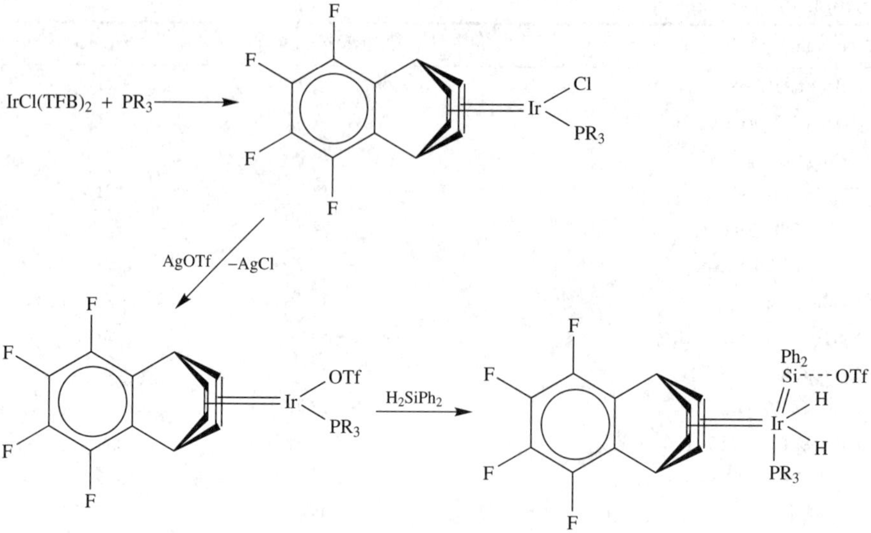

$PR_3 = PCy_3, P(Pr\text{-}i)_3$

SCHEME 27

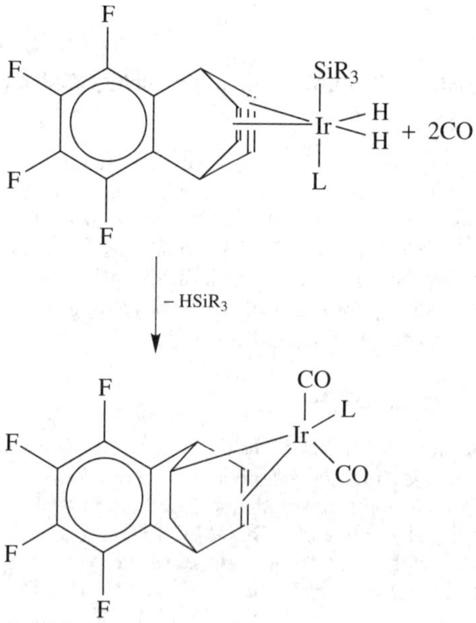

(37)

$L = PPh_3, PCy_3, P(Pr\text{-}i)_3$

The complex Ir(OCOMe)(TFB)(PCy_3) was found to react with CO, yielding the *cis*-dicarbonyl compound Ir(η^1-OCOMe)$(CO)_2$(PCy_3) **(55)**. Treatment of **55** with 1 equivalent each of $HSiPh_3$ and CO in toluene led to the air-stable tricarbonyl complex $Ir(CO)_3(SiPh_3)$

(PCy$_3$) (**56**) exhibiting a P−Si coupling of 66 Hz, which suggests a *trans* disposition of the silane and phosphine ligands. Complex **56** reacts with molecular hydrogen to give the dihydrido compound **57**. Reaction of complex **55** with 3 equivalents of the HSiR$_3$ gives **57** or its triethylsilyl analogue (equation 38). The reaction of **55** with 1 equivalent of H$_2$SiPh$_2$ yields the analogous dihydride IrH$_2$(Ph$_2$SiOCOMe)(CO)$_2$(PCy$_3$) and with 2 equivalents of H$_{x+1}$SiPh$_{3-x}$ affords the bis(silyl) derivatives **58** and **59** (equation 39). The bis(silyl) complex **58** reacts with acetic acid yielding IrH$_2$(Ph$_2$SiOCOMe)(CO)$_2$(PCy$_3$). **58** and **59** react with alcohols, yielding the alkoxysilyl and dialkoxysilyl derivatives, respectively[188].

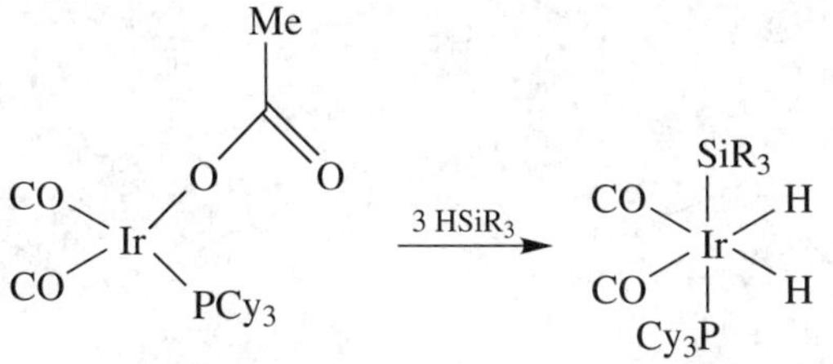

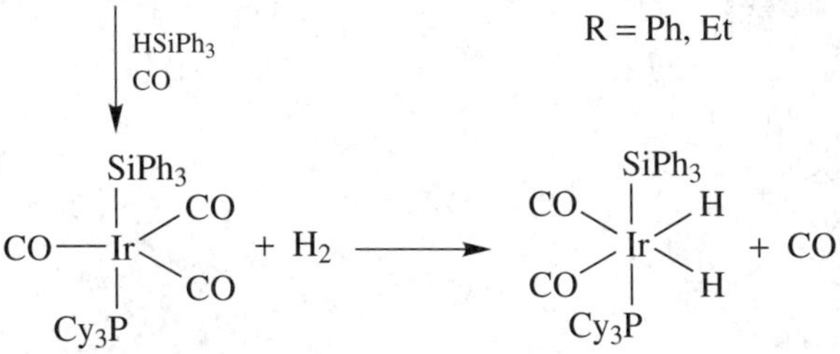

The square-planar *cis*-dicarbonyl acetylide complex Ir(C≡CPh)(CO)$_2$(PCy$_3$) was found to react with HSiR$_3$ to give the corresponding alkynyl-hydridosilyl derivatives HIr(C≡CPh)(CO)$_2$(SiR$_3$) (SiR$_3$ = SiEt$_3$, SiPh$_3$, SiHPh$_2$, SiH$_2$Ph). The oxidative addition of the silane to the metal center is generally viewed as a diastereoselective concerted *cis* addition process with a specific substrate orientation. The exclusive formation of the complex with *trans* P−Si orientation argues that the addition of the silane occurs along the OC−Ir−P axis with the silicon atom occupying a position above the carbonyl group. The reaction of the related Ir(C≡CPh)(TFB)(PCy$_3$) complexes with HSiR$_3$ (R = Et, Ph) and H$_2$SiPh$_2$ is described in Scheme 28[189].

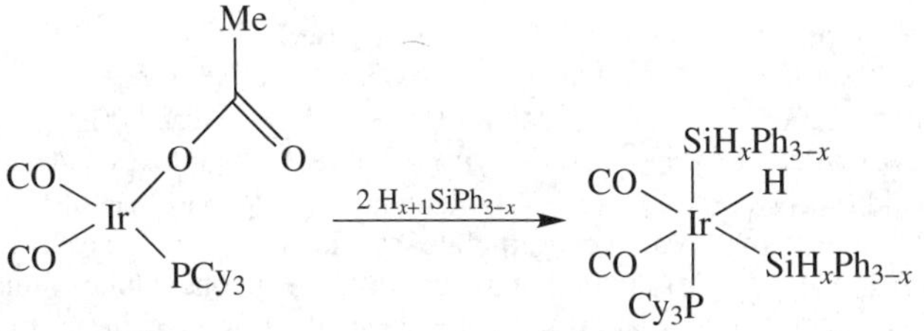

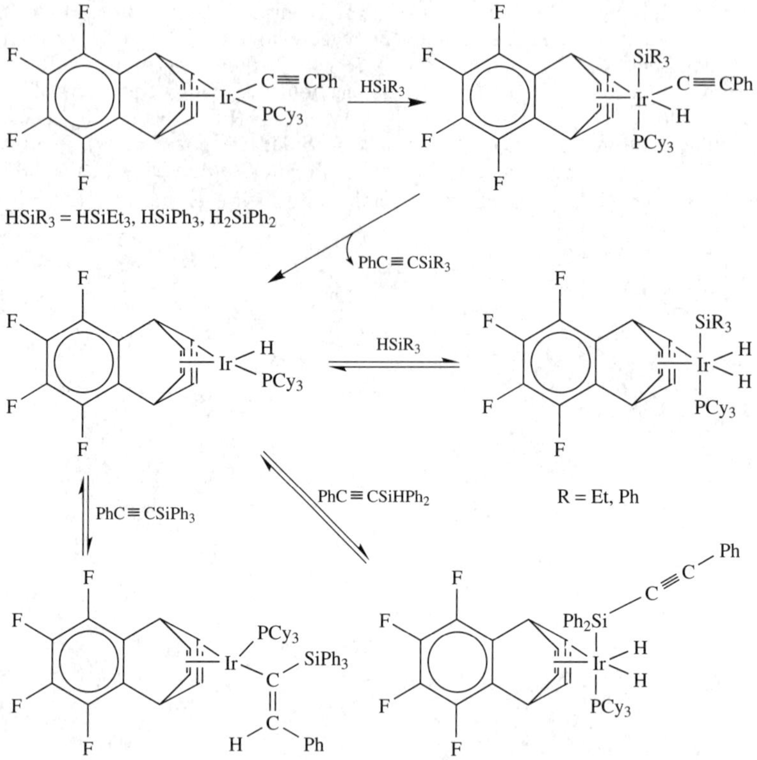

SCHEME 28

Complexes of the type Ir(acac)(coe)L [L = PCy$_3$ (**60**), P(Pr-i)$_3$ (**61**); coe = cyclooctene] reacted with dimethyl acetylenedicarboxylate to give the corresponding complexes Ir(acac)(η^2-MeO$_2$C−C≡C−CO$_2$Me)L [L = PCy$_3$ (**62**), P(Pr-i)$_3$ (**63**)]. Consecutive reactions with H$_2$SiPh$_2$ afforded the silavinyl complexes **64** and **65** (Scheme 29)[190]. In addition, complex **60** reacts with silanes HSiR$_3$ (SiR$_3$ = SiEt$_3$, SiPh$_3$, SiHPh$_2$) yielding the square-pyramid five-coordinate hydridosilyl complexes Ir(acac)H(SiR$_3$)(PCy$_3$) (**65a**). These complexes do not react with tricyclohexylphosphine, but the addition of 1 equivalent of the silane in the presence of the phosphine affords the six-coordinate hydridosilyl derivatives with the phosphine ligands in *trans* disposition (Scheme 29)[191]. The formation of complexes **64** and **65** can be regarded as a result of a net transformation involving addition of one Si−H bond across the C=O bond and an addition of an Ir−H across the alkyne triple bond. The X-ray diffraction of **65a** indicates that the bonding situation in the Ir−Si−O sequence can be described as an intermediate state between a metal silylene stabilized by an oxygen base and a tetrahedral silicon. The Ir−Si distance of 2.264(2) Å in **65a** (R = Et) is shorter than in Ir(acac)H(SiEt$_3$)(PCy$_3$) [Ir−Si 2.307(1) Å] and other Ir(III) complexes (Table 11)[190,191].

The pentahydride complex IrH$_5$(PPh$_3$)$_2$ was found to react with chelating silanes with loss of hydrogen to form the classical seven-coordinate, distorted pentagonal bipyramidal silyl polyhydride complexes, IrH$_3$(disilyl)(PPh$_3$)$_2$ [disilyl = 1,2-dimethylsilylbenzene (dmsb) (**66**), 1,1,3,3-tetraisopropyldisiloxane (tids)], and with monodentate silanes to give

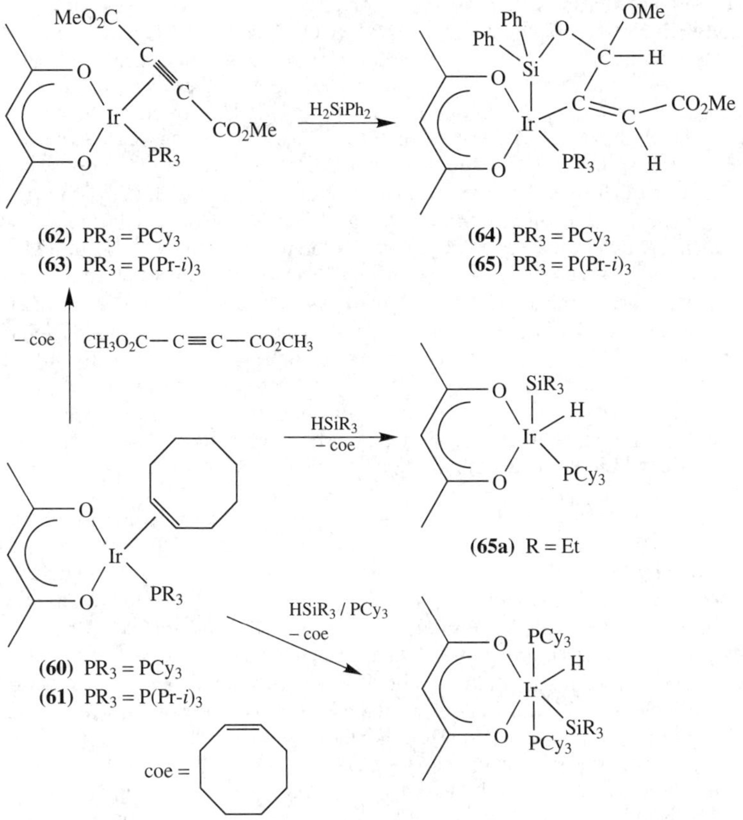

35. Transition-metal silyl complexes

SCHEME 29

the complexes $IrH_4(SiR_3)(PPh_3)_2$ (R = Et, Ph). Complex **66** displays coalescence of the hydrides on Ir at 70 °C which implies a barrier of 13 kcal mol^{-1} for the fluxional exchange. The X-ray diffraction studies on this complex reveal Ir—Si bond lengths of 2.437(1) Å and 2.430(1) Å, which are comparable to other Ir(III) silyl complexes (Table 11). The structure of complex **66** is unusual since the phosphine ligands are in *cis* configuration, although other examples of *cis* complexes are known. Complex **66** reacts with CO to yield the complex $Ir(dmsb)H(CO)_2PPh_3$ which has not been isolated yet in a pure condition, and reacts with $P(OMe)_3$ yielding $Ir(dmsb)\{P(OMe)_3\}_2(H)PPh_3$ (equation 40)[192].

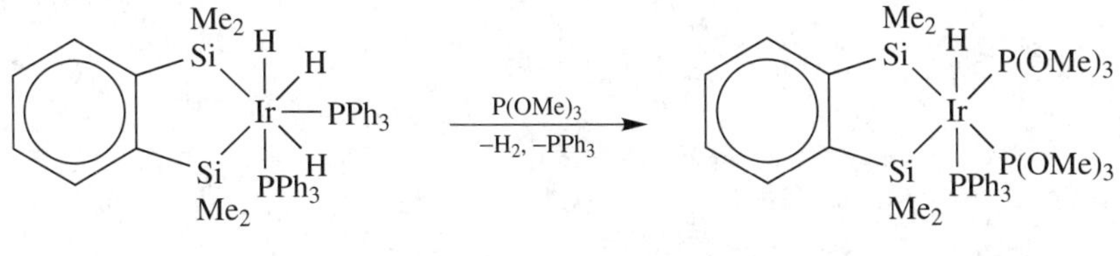

IrH$_3$(CO)(dppe) reacts with primary, secondary and tertiary silanes yielding mono- and bis(silyl) hydride complexes, IrH$_2$(SiRR$_2'$)(CO)(dppe) and IrH(SiRR$_2'$)$_2$(CO)(dppe), respectively. In the mono(silyl) dihydride complexes, the hydrides are disposed in a *cis* configuration with one hydride being *trans* to CO and the other *trans* to the phosphine. The silyl ligand is *trans* to the second phosphine ligand. The structures for the bis(silyl) complexes show equivalent silyl groups disposed *trans* to the dppe ligand. The solid state structures of IrH(SiClMe$_2$)$_2$(CO)(dppe) and IrH(SiClMe$_2$)(SiHMe$_2$)(CO)(dppe) have been characterized. In the former the Ir−Si bond lengths are 2.396(2) Å and 2.397(2) Å whereas for the latter the Ir−Si bond lengths are 2.394(3) Å and 2.418(3) Å in the Ir−SiClMe$_2$ and Ir−SiHMe$_2$ moieties, respectively. This difference is attributed to the chloro substituent on the silyl group which makes the Ir−Si bond of the former slightly stronger. This view was corroborated by the Ir−P and Ir−CO bond distances [2.368(2) Å and 1.93(1) Å, respectively] which are slightly longer relative to those in closely related systems. Interestingly, the reaction of the bis(silyl) complex IrH(SiH$_2$Et)$_2$(CO)(dppe) with BF$_3$·OEt$_2$ afforded the bis(fluorosilyl) complex IrH(SiF$_2$Et)$_2$(CO)(dppe) having similar Ir−Si distances [2.349(7) Å; 2.372(7) Å] to those in the bis(chlorosilyl) complex[193].

Wang and Eisenberg have studied the reaction between primary and secondary silanes with dppm-bridged dinuclear Rh dihydride complexes which gives bis μ-silylene bridged complexes for primary silanes (R' = H), and unusual P−Si bond containing products for secondary silanes. In both cases, the reaction proceeds through the intermediacy of the fluxional μ-SiRR' dihydride 67 (equation 41). For the μ-SiRH derivatives of the fluxional complex 67, the hydrogens at the Rh and Si atoms undergo exchange. The mechanism proposed for this process involves a reductive elimination/oxidative addition via the intermediacy of the unsaturated complex 68[194]. Complex 67 has also been found to promote the reaction between silanes and simple amines leading to silazane oligomers with a similar dihydride complex, 69, as an intermediate[195]. Exchange reactions between complex 67 and free alkyl silanes show that the complexes with μ-SiRR' bridges with R' = H are more stable than those with disubstituted bridges and that aryl substituents on the silicon increase the bridge stability. The relative reactivity of silanes with the dihydride complex Rh$_2$H$_2$(CO)$_2$(dppm)$_2$ was found to follow the order PhSiH$_3$ > n-HexSiH$_3$ > MePhSiH$_2$ > Ph$_2$SiH$_2$ > Et$_2$SiH$_2$. The primary silanes are more reactive than the secondary silanes, and the order or reactivity is correlated with the stability order[194].

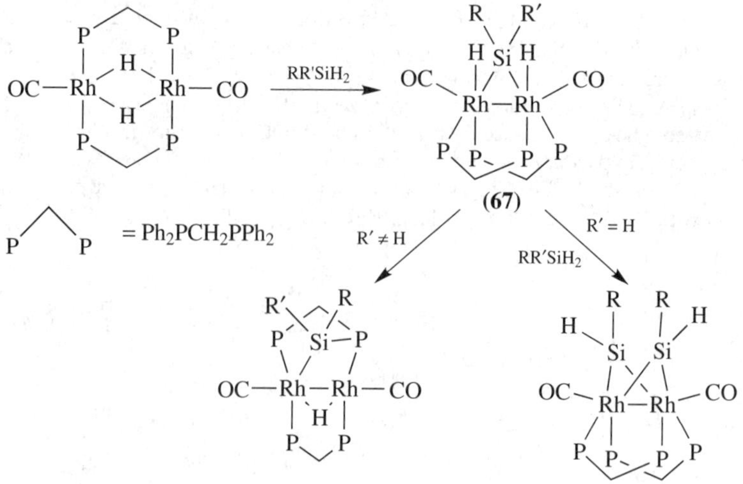

(41)

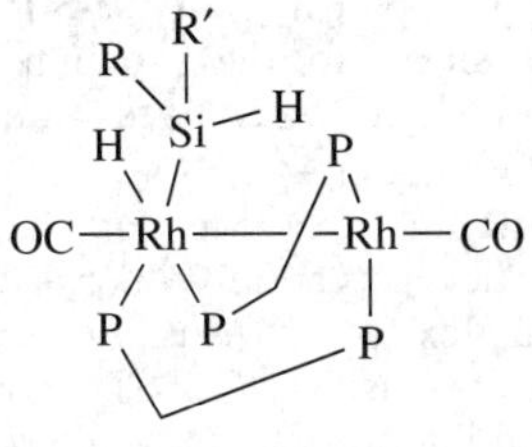

35. Transition-metal silyl complexes

(68) 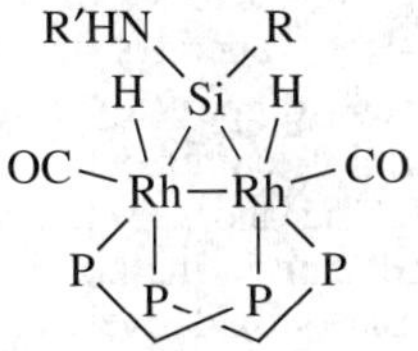 **(69)**

P⌒P = Ph$_2$PCH$_2$PPh$_2$

R' = Me, H
R = Et, Hex

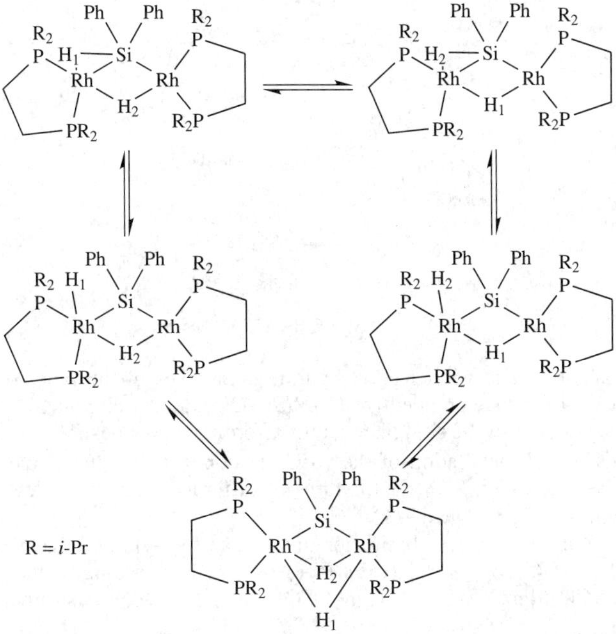

SCHEME 30

R = i-Pr

Stoichiometric reaction of Ph$_2$SiH$_2$ with [(dippe)Rh]$_2$(μ-H)$_2$ [dippe = 1,2-di(isopropylphosphino)ethane] produced the fluxional complex [(dippe)Rh]$_2$(μ-η^2-HSiPh$_2$)(μ-H) **(70)** which exchanges the silane hydride with the bridging hydride presumably through the mechanism described in Scheme 30[196]. The X-ray structure of **70** indicates the presence of a three-center, two-electron Rh—Si—H interaction similar to the interaction found for complex **67** with the bridging μ-SiPh$_2$[194]. Addition of a second equivalent of H$_2$SiPh$_2$ to complex **70** gives an equilibrium reaction to the symmetric complex [(dippe)Rh]$_2$(μ-SiPh$_2$)$_2$[196]. The reaction of the dinuclear rhodium complex [(dippe)Rh]$_2$(μ-H)$_2$ with 2 equivalents of primary silanes afforded the dinuclear bis(μ-silylene) complexes [(dippe)Rh]$_2$(μ-SiR$_2$)$_2$ (R = n-Bu, p-CH$_3$C$_6$H$_4$) which were reacted with hydrogen producing the corresponding [(dippe)Rh]$_2$(μ-η^2-H-SiR$_2$)$_2$ complexes. Addition of a large excess of the primary silane to the rhodium hydride dimer [(dippe)Rh]$_2$(μ-H)$_2$ generates the dinuclear rhodium derivatives with three silicon-containing ligands [(dippe)Rh]$_2$(μ-SiHR)(μ-η^2-H-SiHR)$_2$[197].

The rhodium silyl complexes, (Me$_3$P)$_3$RhSiR$_2$Ph (R = Me, Ph), have been shown to induce a catalytic and selective C−F activation of partially fluorinated and perfluorobenzenes, producing in the latter case the complex (Me$_3$P)$_3$RhC$_6$F$_5$ (Scheme 31). This complex adds silanes oxidatively and eliminates exclusively C$_6$F$_5$H, reforming the starting silyl complex. For (EtO)$_3$SiH, the octahedral intermediate in which perfluorobenzene and the silane are disposed *trans* to one another has been structurally characterized. This geometry is responsible for the selective C−H elimination of the C$_6$F$_5$H[198].

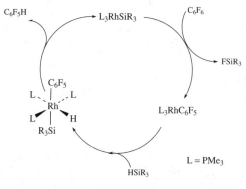

SCHEME 31

Similar octahedral facial silyl methyl hydride complexes of the type IrL$_3$H(SiR$_3$)Me have been shown to induce competitive C−H/Si−C reductive elimination depending on the electronic properties of the silyl ligand, thus affording a novel example of a metallation of silyl ligands or the metallation of the sp^{3}C−H bond of the ethyl moiety when R = OEt. For the complex with R = Et, mixtures of different complexes are formed by the thermolysis with benzene (Scheme 32)[198,199].

Similar competitive reductive elimination of C−H and Si−H bonds have been observed for the complex IrH(SiHPh$_2$)(Mes)(CO)(dppe) in benzene, leading to the iridium complexes Ir(Mes)(CO)(dppe) and Ir(SiHPh$_2$)(CO)(dppe) following first-order kinetics for both eliminations[200].

Duckett and Perutz have shown the stoichiometric reaction of the CpRh(C$_2$H$_4$)(SiR$_3$)H (R = Et, *i*-Pr) complexes (Scheme 33)[201]. These complexes have been found to act as precursors to the catalytically active species for the hydrosilylation of ethene with Et$_3$SiH but are not within the catalytic cycle. The mechanism proposed in Scheme 34 for the hydrosilylation of ethene was found to be equivalent to the Seitz–Wrighton hydrosilylation mechanism catalyzed by cobalt carbonyls complexes[202].

Silanes of the form HSiR$_3'$ (SiR$_3'$ = SiMe$_2$Ph, SiMeEt$_2$, SiEt$_3$, SiPh$_3$) have been found to cleave the cobalt acetyl complexes CH$_3$C(O)Co(CO)$_3$(PR$_3$) (PR$_3$ = PPh$_2$Me, PPh$_3$) at room temperature to give cobalt silyl compounds R$_3'$SiCo(CO)$_3$(PR$_3$) and ethoxysilanes. Interestingly, acetaldehyde or α-siloxyethyl complexes, CH$_3$CH(OSiR$_3'$)Co(CO)$_3$(PR$_3$), were not produced. The presence of CO, phosphine or benzaldehyde inhibits the reaction whereas the presence of RhCl(PPh$_3$)$_3$ has no effect on the reaction rate or on the outcome other than in the hydrosilylation of acetaldehyde with PhMe$_2$SiH to give EtOSiMe$_2$Ph[203].

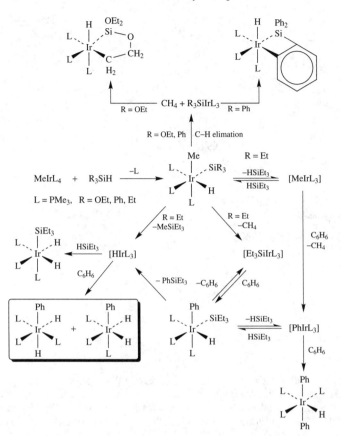

SCHEME 32

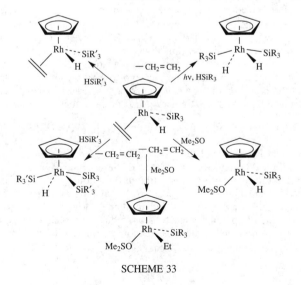

SCHEME 33

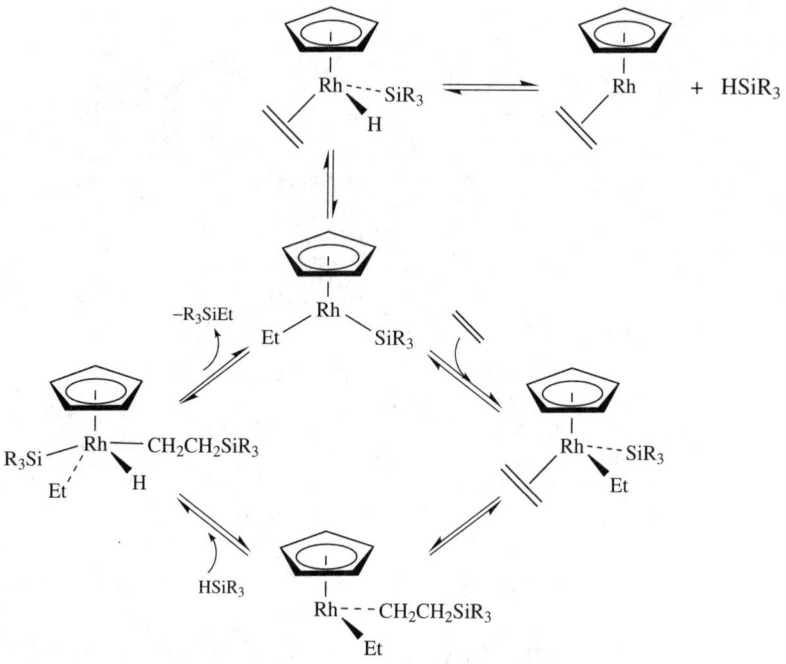

SCHEME 34

Using the metalloradical reactivity of the Rh(II)OEP (OEP = 2,3,7,8,12,13,17,18-octaethylphorphynato) dimer, the preparation of silyl rhodium complexes was achieved by the hydrogen elimination reaction with silanes $R_2R'SiH$ ($R = R' = $ Et, Ph; R = Me, $R' = $ Ph, OEt). The Rh–Si bond length of 2.32(1) Å, found when R = Et, is comparable to those in other Rh(III) complexes (Table 11). The crystal packing indicates that all the ethyl groups on the porphyrin periphery are directed toward the silyl group. Consequently, the aromatic part of one complex molecule is in contact with the aromatic part of the next molecule and the aliphatic part is in contact with the aliphatic part of the next molecule[204].

Complexes of the type $M(PPh_3)_3(CO)H$ (M = Rh, Ir) were reacted with an excess of dmsiqn-H [dmsiqn-H = (8-quinolyl)dimethylsilane] in toluene producing the tris N–Si chelating facial five-membered ring $M(dmsiqn)_3$ (equation 42). The fused aromatic quinoline ring system provides a low energy π acceptor orbital which, in conjunction with the silyl group and the easily oxidized metal center such as M(III), leads to complexes with low-energy metal-to-ligand charge transfer (MLCT) transitions.

$$M(PPh_3)_3(CO)H + 3dmsiqn\text{-}H \rightarrow M(dmsiqn)_3 + 3PPh_3 + 2H_2 + CO \qquad (42)$$

For M = Rh, the Rh–Si bond lengths [2.28–2.31 (Å)] are shorter than in comparable complexes (Table 11). An unusual feature of these complexes is the packing of the molecules in the crystal lattice. The two different enantiomers of the complex segregate into columns which are aligned in adjacent rows, creating alternating sheets of Δ and Λ isomers. These sheets are also arranged in such a way that the 3-fold axes of the complexes are oriented in parallel to each other so that the molecular dipoles point in the same direction. The alignment of the molecules impart the crystal with a net polarization which leads to the observation of non-linear optical behavior[205].

Reaction of $(Me_3P)_3RhCl$ with the lithium salt, $(Me_3Si)_3SiLi·3THF$, generated the thermally unstable complex $(Me_3P)_3RhSiMe_2SiMe(SiMe_3)_3$ presumably via a series of 1,2- and 1,3-migrations as shown for the Cp(CO)Fe fragment by Sharma and Pannell[155], by Ogino and coworkers[171–173] and by Berry and coworkers[64] for the Cp_2W moiety (*vide supra*). The analogous reaction of the iridium complex results in a similar rearrangement, yielding a mixture of diastereomers of the iridacycle complex $(Me_3P)_3(H)IrSiMe_2SiMe(SiMe_3)SiMe_2\dot{C}H_2$ (Scheme 35). The two isomers can be interconverted by catalytic amounts of hydrogen to their equilibrium mixture by reversing the processes cyclometallation, oxidative addition of hydrogen to the Ir(I) center, rotation and cyclometallation with elimination of hydrogen[206].

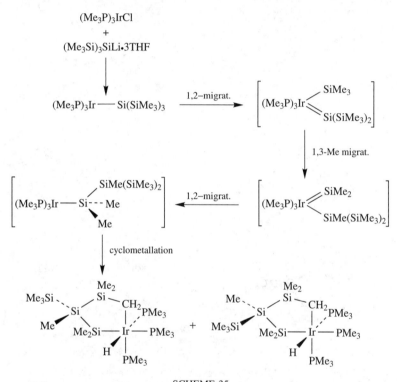

SCHEME 35

Fac-tris(trimethylphosphino)iridium(III) silane complexes of the form $IrHX(PMe_3)_3SiRR'_2$ (X = H, D; R = H, D, Cl, Ph; R' = Ph, *t*-Bu) have been prepared by the oxidative addition of $SiXRR'_2$ to $IrH(PMe_3)_4$ with the loss of PMe_3. The exclusive formation of the facial isomers is attributed to a preferential *cis* addition of the R_3Si-H moiety parallel to the P–Ir–P axis, yielding the *cis* hydride products in which the silicon is *trans* to a phosphine[207].

The methyl iridium triflate complex $Cp^*(Me_3P)IrMe(OTf)$ reacted efficiently with silanes leading to rearrangement products of the presumably initial Si–H products. Thus, trimethylsilane produced methane and, instead of the expected (trimethylsilyl)iridium complex, the methyliridium dimethylsilyl triflate $Cp^*(Me_3P)Ir(Me)(SiMe_2OTf)$ was obtained. Similarly, phenylsilane produced the analog $Cp^*(Me_3P)Ir(Ph)(SiPh_2OTf)$.

Phenyldimethylsilane and diphenylsilane gave the phenyl and the hydride migration products, Cp*(Me$_3$P)Ir(Ph)(SiMe$_2$OTf) and Cp*(Me$_3$P)IrH(SiPh$_2$OTf), respectively[208].

Direct alkyne insertion into a Rh—Si bond has been observed for the intermediate rhodium silyl complex (dtbpm) Rh[Si(OEt)$_3$] (PMe$_3$) [dtbpm = di(*tert*-butyl)phosphino methane] in the hydrosilylation of 2-butyne with triethoxysilane catalyzed by the rhodium alkyl complex (dtbpm)RhMe(PMe$_3$). The crystal structure of (dtbpm)Rh[Si(OEt)$_3$] (PMe$_3$) shows that the coordination around the Rh metal is planar with a Rh—Si bond length [2.325(2) Å] similar to that found for the complex (Me$_3$P)$_3$RhH(C$_6$F$_5$){Si(OEt)$_3$} (Table 11)[209]. The proposed mechanism for the hydrosilylation reaction of 2-butyne with HSi(OEt)$_3$ yielding mainly the *E*-isomer of MeCH=C(Me)Si(OEt)$_3$ is outlined in Scheme 36.

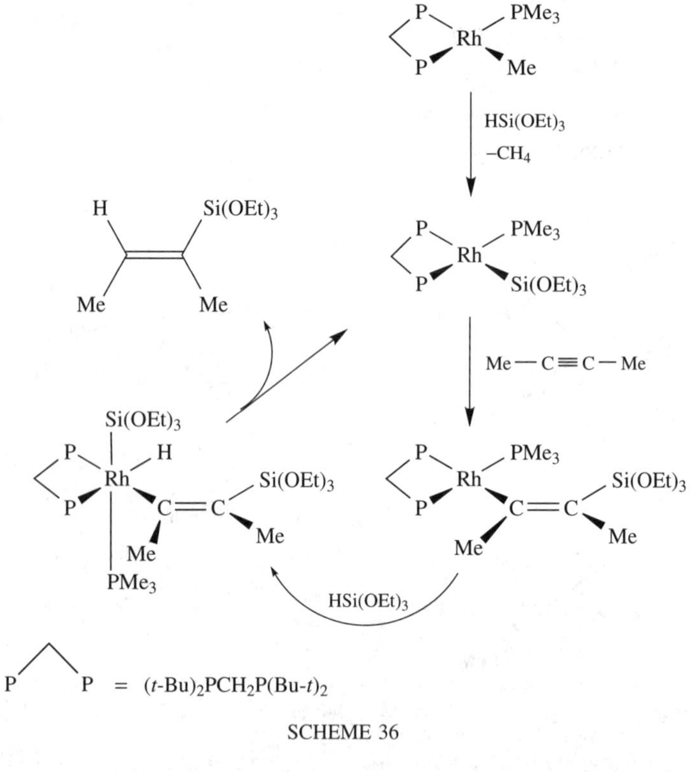

SCHEME 36

The behavior of the SiMe$_2$H-substituted alkyne in Me$_2$HSiC≡CSiMe$_2$H toward [RhCl(P(Pr-*i*)$_3$)$_2$]$_n$ has been shown to be different from that of the analog Me$_3$SiC≡CSiMe$_3$. Instead of the expected mononuclear vinylidene complex *trans*-[ClRh(C(SiMe$_2$H)=CH(SiMe$_2$H))(P(Pr-*i*)$_3$)$_2$], the dinuclear symmetrical hydrido(silyl) complex was obtained by oxidative addition of both terminal Si—H bonds to two metal centers (equation 43)[210]. Interestingly, reaction of the *in situ* prepared phenylthio rhodium complex Rh(SPh)(PMe$_3$)$_3$ with two equivalents of the more flexible disilane, HMe$_2$SiCH$_2$CH$_2$SiMe$_2$H, afforded a mixture of the complexes *cis, mer*-[RhH$_2$(SPh)(PMe$_3$)$_3$] and disilarhodacycle *fac*-[Rh(SiMe$_2$CH$_2$CH$_2$SiMe$_2$)H(PMe$_3$)$_3$] with a concomitant elimination of thiophenol. The Rh—Si distances [2.383(2) Å and

2.389(2) Å] observed for the latter complex are longer than for similar Rh complexes (Table 11)[211].

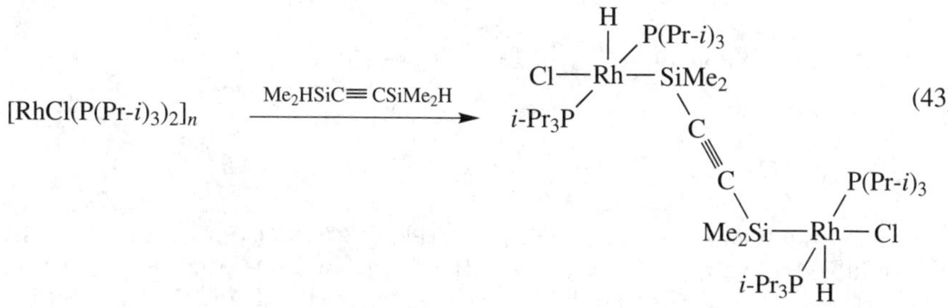

$$[RhCl(P(Pr\text{-}i)_3)_2]_n \xrightarrow{Me_2HSiC \equiv CSiMe_2H}$$

[structure of binuclear Rh complex with bridging bis(silyl)acetylene ligand, P(Pr-i)$_3$, Cl, H, SiMe$_2$ groups] (43)

μ-Cyclopentadienyldiene dicobalt complexes (**71**) reacted with trimethylsilylacetylenes by eliminating ethene and forming the unusual binuclear complexes (**72**) with a cobaltabicyclic ring system as a ligand (equation 44). The crystal structure of complex **72** with R$^1{}_5$ = Me$_4$Et, R^2 = SiMe$_3$ exhibit similar Co—Si bond distance [2.281(2) Å][212] to those found in the silasesquioxane complex Co(CO)$_4$(H$_7$Si$_8$O$_{12}$) [2.285(1) Å][213]. A number of interesting multinuclear cobalt silicon clusters have been obtained by the reactions of Co$_2$CO$_8$ with chloro and alkyl silanes[214−218]. In addition, a monosubstituted cobalt silyl octanuclear silasesquioxane cage exhibiting a metal–silicon bond of 2.285(1) Å was recently obtained from the reaction of Co$_2$CO$_8$ with octahydrosilasesquioxane H$_8$Si$_8$O$_{12}$[213].

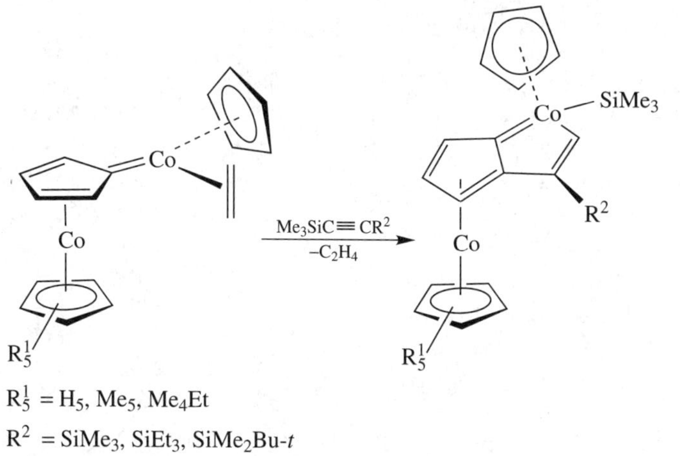

[structures of dicobalt complexes **71** and **72** connected by arrow labeled Me$_3$SiC≡CR^2, −C$_2$H$_4$] (44)

R$^1_5$ = H$_5$, Me$_5$, Me$_4$Et
R^2 = SiMe$_3$, SiEt$_3$, SiMe$_2$Bu-t

(**71**) (**72**)

C. Silanol Complexes

The iridium silanol complexes of the form (Et$_3$P)$_2$IrH(Cl)(SiR$_2$OH) [R = i-Pr (**73**), t-Bu] have been prepared from low-valent Ir(I) complexes and the corresponding dialkylsilanol. The 16-electron coordinatively unsaturated complex **73** with a short Ir—Si bond length [2.313(6) Å] adopts a trigonal bipyramidal arrangement rather than a square pyramidal configuration, due to the partial multiple bond present between the metal center and the π-donor chlorine atom. Reaction of **73** with a strong base (BuLi) selectively deprotonates the silanol, yielding a planar dihydride metallosilanolate **74** (equation 45)[219].

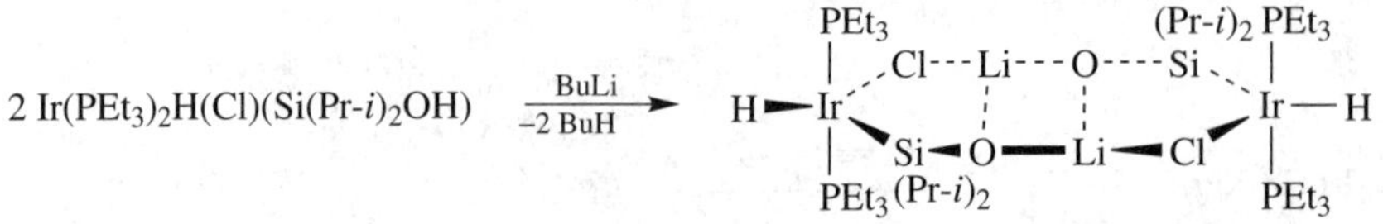

$$2\ \text{Ir(PEt}_3)_2\text{H(Cl)(Si(Pr-}i)_2\text{OH)} \xrightarrow[-2\ \text{BuH}]{\text{BuLi}} \text{structure } \mathbf{74}$$

(73) (74) (45)

Analogous iridium silanethiol complexes $(PPh_3)_2(CO)IrH_2(Si(SEt)_3)$, fac-$(PMe_3)_3$IrMeH(Si(SEt)$_3$) and mer-$(PMe_3)_3$Ir(C_6F_5)H(Si(SEt)$_3$) were prepared by the oxidative addition of HSi(SEt)$_3$ to HIr(CO) (PPh$_3$)$_3$, MeIr(PMe$_3$)$_4$ and C_6F_5Ir(PMe$_3$)$_3$, respectively. Unlike the extremely easily hydrolyzable parent silane, the iridium complexes are stable in neutral and basic H$_2$O/THF solutions. This stabilization is attributed to the electron-donating capacity of the metal center which reduces efficiently the electrophilicity of the silicon. Interestingly, the rhodium analog mer-$(Me_3P)_3$Rh (C_6F_5)H(Si(SEt)$_3$) reacts with 5 equivalents of H$_2$O, producing the thioethyl complex mer-$(Me_3P)_3$Rh(C_6F_5)H(SEt) due to its high propensity to reductively eliminate the Si−H bond[220].

TABLE 11. Selected M−Si distances in group-9 transition-metal silicon-containing complexes

Compound	d^n Configuration	M−Si (Å)	Reference
Ir(acac){C[CH(OCH$_3$)OSiPh$_2$]=CHCO$_2$CH$_3$}PCy$_3$	d^6	2.264(2)	190
In(acac)H(SiEt$_3$)(PCy$_3$)	d^6	2.307(1)	191
Ir(dmsb)H$_3$(PPh$_3$)$_2$[a]	d^4	2.437(1), 2.430(1)	192
IrH(SiClMe$_2$)$_2$(CO)(dppe)[b]	d^6	2.396(2), 2.397(2)	193
IrH(SiClMe$_2$)(SIHMe$_2$)(CO)(dppe)[b]	d^6	2.394(3), 2.418(3)	193
IrH(SiF$_2$Et)$_2$(CO)(dppe)[b]	d^6	2.349(7), 2.372(7)	193
[(dippe)Rh]$_2$(μ-H)($\mu - \eta^2$-HSiPh$_2$)][c]	d^8	2.298(2), 2.487(2)	196
[(dippe)Rh]$_2$(μ-Si(Bu-n)$_2$][c]	d^6	2.334(1), 2.335(1)	197
[(dippe)Rh]$_2$(μ-SiHTol-p)(μ-η^2-H-SiH Tol-p)$_2$][c]	d^7	2.336(2), 2.349(2) 2.444(2), 2.477(2) 2.350(2), 2.356(2)	197
(PMe$_3$)$_3$RhH(C$_6$F$_5$){Si(OEt)$_3$}	d^6	2.325(4)	198
fac-(Me$_3$P)$_3$IrMeH(SiPh$_3$)	d^6	2.381(3)	199
fac-(Me$_3$P)$_3$IrMeH(SiEt$_3$)	d^6	2.424(2)	199
IrH$_2${Si(OTf)Ph$_2$}(TFB)(P(Pr-i)$_3$)[d]	d^6	2.337(2)	186
(Me$_3$P)$_3$Ir{SiPh$_2$(o-C$_6$H$_4$)	d^6	2.404(3)	198
(Et$_3$P)$_2$IrH(Cl)(Si(Pr-i)$_2$OH)	d^6	2.313(6)	219
[(Et$_3$P)$_2$IrH(Cl)(Si(Pr-i)$_2$OLi)]$_2$	d^6	2.344(2)	219
(OEP)RhSiEt$_3$[e]	d^6	2.32(1)	204
Rh(dmsiqn)$_3$[f]	d^6	2.278(1), 2.290(1) 2.301(1)	205

TABLE 11. (continued))

Compound	d_n Configuration	M–Si (Å)	Reference
(Me$_3$P)$_3$HIrSiMe$_2$SiMe(SiMe$_3$)SiMe$_2$CH$_2$	d^6	2.441(3)	206
IrH$_2$(PMe$_3$)$_3$(SiHPPh$_2$)	d^6	2.361(3), 2.369(3)	207
IrH$_2$(PMe$_3$)$_3$(SiPPh$_3$)	d^6	2.382(4)	207
IrH$_2$(PMe$_3$)$_3$(Si(Bu-t)$_2$Cl	d^6	2.392(3)	207
(dtbpm)Rh[Si(OEt)$_2$](PMe$_3$)^g	d^6	2.325(2)	209
fac-[Rh(SiMe$_2$CH$_2$CH$_2$SiMe$_2$)H(PMe$_3$)$_3$]	d^6	2.383(2), 2.389(2)	211
72	d^6	2.281(2)	212
Co(CO)$_4$(H$_7$Si$_8$O$_{12}$)	d^8	2.285(1)	213

^{a}dmsb = 1,2-dimethylsilylbenzene.
^{b}dppe = diphenylphosphinoethane.
^{c}dippe = di(isopropyl)phosphinoethane.
^{d}TFB = tetrafluorobenzobarrelene.
^{e}OEP = 2,3,7,8,12,13,17,18-octaethylphorphynato.
^{f}dmsiqn = (8 − quinoyl)dimethylsilane.
^{g}dtbpm = di($tert$-butyl)phosphinomethane.

TABLE 12. Selected ^{29}Si NMR chemical shifts for group-9 transition-metal silicon-containing complexes

Compound	^{29}Si NMR shift (ppm)	Reference
[(dippe)Rh]$_2$(μ-Si(Bu-n)H)(μ-η^2-H-SiH(Bu-n))$_2$^{a}	125.5 (br t, 2Si, $^1J_{Si-H}$ = 180 Hz)	197
[(dippe)Rh]$_2$(μ-SiHTol-p)(μ-η^2-H-SiHTol-P)$_2$^{a}	100–142 br	197
CpRh(Si(Pr-i)$_3$)$_2$H$_2$	52.2 (d, $^1J_{SiRh}$ = 18.3 Hz)	202
PhMe$_2$SiCo(CO)$_3$(PPh$_2$Me)	32.3 (d, $^2J_{SiP}$ = 28.9 Hz)	203
EtMe$_2$SiCo(CO)$_3$(PPh$_2$Me)	42.2 (d, $^2J_{SiP}$ = 26.0Hz)	203
Et$_3$SiCo(CO)$_3$(PPh$_2$Me)	51.8 (d, $^2J_{SiP}$ = 25.0 Hz)	203
Ph$_3$SiCo(CO)$_3$(PPh$_2$Me)	29.8 (d, $^2J_{SiP}$ = 33.0 Hz)	203
PhMe$_2$SiCo(CO)$_3$(PPh$_3$)	33.1 (d, $^2J_{SiP}$ = 29.4 Hz)	203
Et$_3$SiCo(CO)$_3$(PPh$_3$)	52.9 (d, $^2J_{SiP}$ = 25.4 Hz)	203
(OEP)Rh(SiEt$_3$)^b	51.98 (d, $^1J_{SiRh}$ = 29.3 Hz)	204
(OEP)Rh(SiPh$_3$)^b	11.56 (d, $^1J_{SiRh}$ = 35.7 Hz)	204
(OEP)Rh(SiMe$_2$Ph)^b	28.3 (d, $^1J_{SiRh}$ = 30.2 Hz)	204
(OEP)Rh{SiMe$_2$(OEt)}^{b}	37.24 (d, $^1J_{SiRh}$ = 30.5 Hz)	204
Rh(dmsiqn)$_3$^{c}	23.60 (d, $^1J_{SiRh}$ = 41.7 Hz)	205
Ir(dmsiqn)$_3$^{c}	−10.14	205
IrH$_2$(PMe$_3$)$_3$(Si(Bu-t)$_2$H)	23.70 (dt, $^1J_{SiH}$ = 126 Hz, $^3J_{SIH}$ = 9 Hz)	207
IrH$_2$(PMe$_3$)$_3$(SiPPh$_3$)	0.094 (dt, $^1J_{SiH}$ = 134 Hz, $^3J_{SiH}$ = 8 Hz)	207
[Rh$_2$H$_2$Cl$_2$(P(Pr-i)$_3$)$_4$(μ-Me$_2$SiC≡CSiMe$_2$)]	14.52 (dt, $^1J_{SiRh}$ = 36.4 Hz, $^2J_{SiP}$ = 9.7 Hz)	210

^{a}dippe = di(isopropyl)phosphinoethane.
^{b}OEP = 2,3,7,8,12,13,17,18-octaethylphorphynato.
^{c}dmsiqn = (8 − quinoyl)dimethylsilane.

XV. GROUP-10 SILICON-CONTAINING COMPLEXES

A. Silyl, Bis(silyl) and Silanol Complexes

Reaction of halotrimethylsilanes with electron-rich Pt(II) complexes resulted in a facile oxidative addition of the Si−X bond (X = Cl, Br, I) (equation 46). The crystal structure of the complex [PtIMe$_2$(SiMe$_3$)(2,2'-bipyridyl)] shows normal Pt−Si bond distance of 2.313(6) Å (Table 13) and exceptionally long Pt−I bond distance which arises from the strong *trans* influence of the trimethylsilyl group[221].

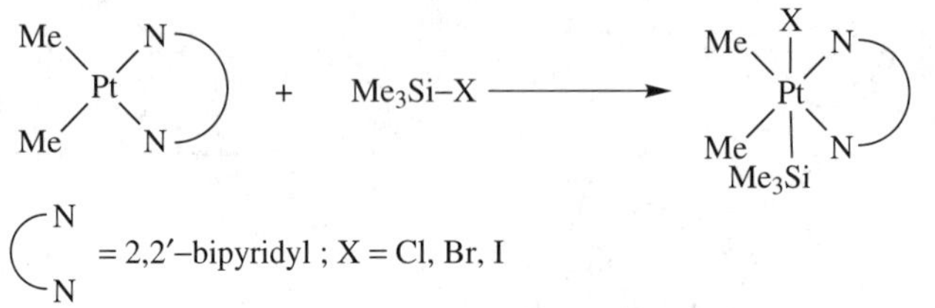

$$\text{Me}_2\text{Pt(N-N)} + \text{Me}_3\text{Si-X} \longrightarrow \text{Me}_2(\text{Me}_3\text{Si})(\text{X})\text{Pt(N-N)} \quad (46)$$

N⌒N = 2,2'-bipyridyl ; X = Cl, Br, I

Cis- and *trans*-PtMe(SiPh$_3$)(PMePh$_2$)$_2$ have been prepared selectively by reacting for both cases the complex *trans*-Pt(Cl)(SiPh$_3$)(PMePh$_2$)$_2$ either with an excess amount of MeLi followed by methanolysis, or by Me$_2$Mg in THF, respectively. Thermolysis of both isomers in benzene solutions yields quantitatively MeSiPh$_3$ as the reducing elimination product. The thermolysis of the *cis* isomer was initiated by dissociation of the PMePh$_2$ ligand and has been accelerated by the addition of alkynes and olefins[222]. Oxidative addition of Me$_3$SiH to Pt(C$_2$H$_4$)(PPh$_3$)$_2$ afforded the distorted square-planar hydrido *cis*-PtH(SiMe$_3$)(PPh$_3$)$_2$ complex. The Pt−P bond *trans* to the silicon is significantly longer than that of the *cis* Pt−P bond to the silicon, in accord with the *trans* influence of the trimethylsilyl group[223].

Reaction of 1-silyl dienol silyl ethers (**75**) with Li$_2$PdCl$_4$ in the presence of Li$_2$CO$_3$ in MeOH gave the (η^3-1-(silylcarbonyl)allyl)palladium chloride complexes (**76**) (equation 47). These complexes undergo catalytic decarbonylation to give the corresponding (η^3-1-(silylallyl))palladium chloride complexes. The mechanism proposed proceeds through the formation of an η^1-allyl complex (**77**) as the first step, followed by β-elimination of the TMS group affording the vinyl ketene intermediate complex (**78**). Subsequent addition of the silyl-palladium moiety to the ketene in the reverse regiochemistry yielded the acylpalladium complex (**79**) and its decarbonylation lead to the (η^1-allyl)palladium and (η^3-allyl)palladium species (Scheme 37)[224].

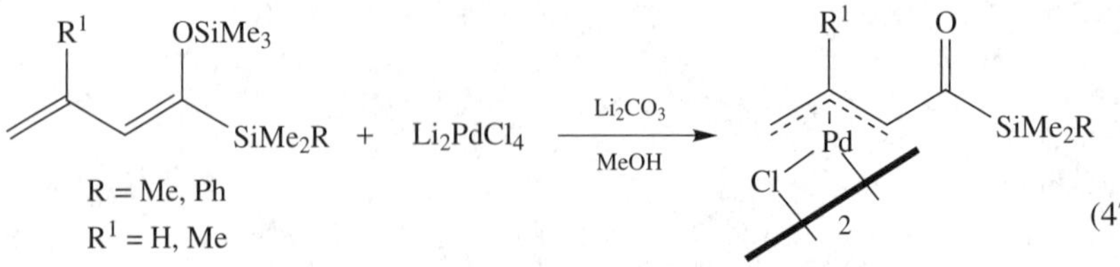

$$\text{(75)} + \text{Li}_2\text{PdCl}_4 \xrightarrow[\text{MeOH}]{\text{Li}_2\text{CO}_3} \text{(76)} \quad (47)$$

R = Me, Ph
R^1 = H, Me

Braunstein and coworkers have shown (*vide supra*) that in the bimetallic Fe−Pt complex (CO)$_3$(SiR$_3$)Fe(μ-PPh$_2$)Pt(PPh$_3$)CO(Fe−Pt), a silyl migration occurred from the Fe to the Pt center (Scheme 17)[102]. Remarkably, the first example of a well-characterized insertion of a transition-metal fragment into a strained silicon−carbon bond of a silicon-bridged [1]ferrocenophane was recently reported by Sheridan, Lough and Manners. The

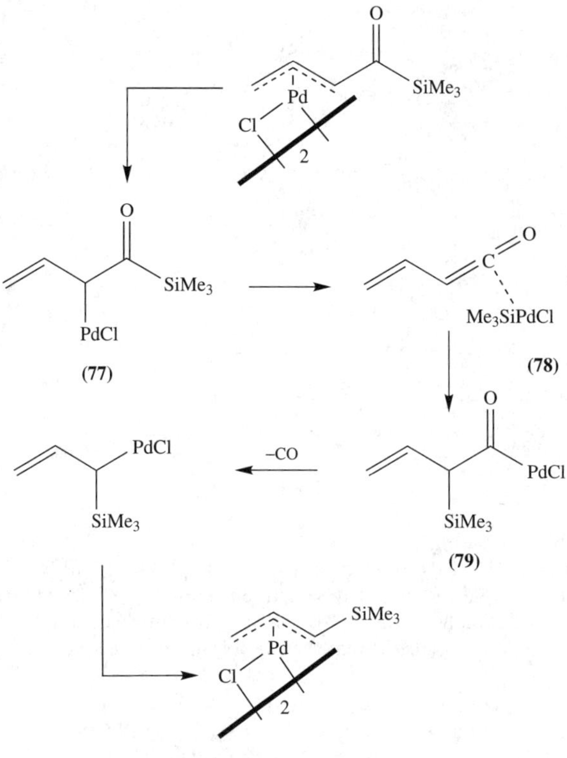

SCHEME 37

reaction of Fe(η^5-C$_5$H$_4$)$_2$SiMe$_2$ with Pt(PEt$_3$)$_3$ afforded the complex [2]platinasilaferrocenophane Fe(η^5-C$_5$H$_4$)$_2$Pt(PEt$_3$)$_2$SiMe$_2$ (equation 48). X-ray diffraction study of the complex reveals that the platinum metal is in a distorted square-planar environment with large P—Pt—P angles and compressed P—Pt—Cp and Cp—Pt—Si angles (Table 13)[225].

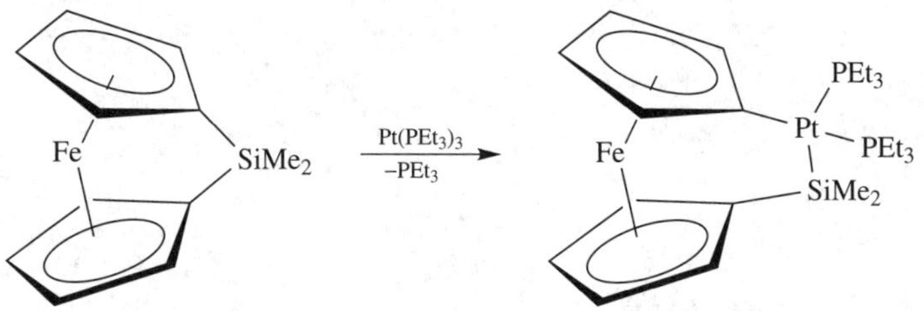

(48)

Thermolytic orthometallation of the polysilyl chloroplatinum complex **80** in a refluxing mixture of methanol in benzene (1:8) yielded the cyclometallated complex **81** which was cleaved by water to complex **82** having a platinum–phenyl bond and a silanol moiety on the polysilyl ligand. Interestingly, in the absence of methanol, under refluxing conditions, complex **80** undergoes an exchange reaction among the trimethylsilyl and chloride moieties across the platinum–silicon bond yielding complex **83**. Complex **83** reacted with methanol producing the bis(silyl) complex **84** (Scheme 38)[226].

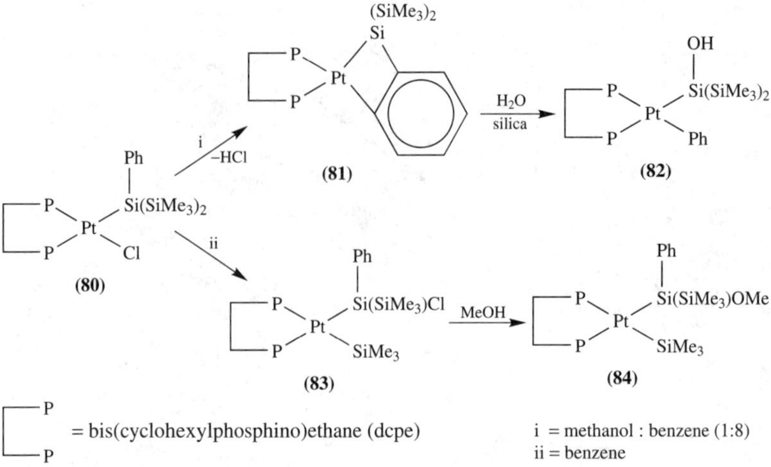

SCHEME 38

Similar bis(silyl) chelating platinum complexes have been obtained by the oxidative addition of disilanes H_3SiSiH_3 to the dihydride complex $(dcpe)PtH_2$. A rapid addition of the disilane to the hydride afforded selectively complex **85** whereas a slow addition afforded a mixture of complexes **85** and **87**, presumably through complex **86** via an α-silyl shift. X-ray determination indicates that the Pt_2Si_4 ring core adopts a chair conformation in **85** (Scheme 39)[227].

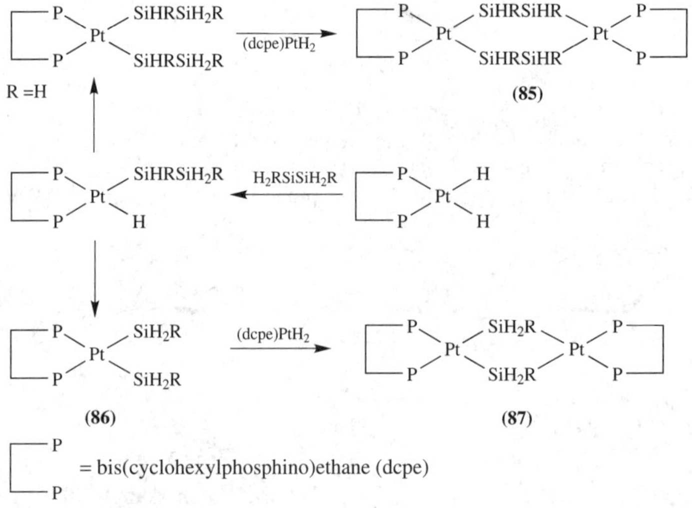

SCHEME 39

Similar reactivity has been found for the isolobal Pd hydrides upon reaction with either hydrosilanes or 1,2-dihydrodisilanes. Thus, the reaction of the dinuclear palladium hydride $[(dcpe)Pd]_2(\mu-H)_2$ with either H_2SiPhR (R = Ph, Me, H) or $HR'MeSiSiMeR'H$ (R' = H, Me) afforded the mononuclear $(dcpe)Pd(SiHRPh)_2$ and $(dcpe)Pd(SiHR'Me)_2$

complexes, respectively. The latter complex with R′ = Me has been characterized by X-ray diffraction. It exhibit Pd−Si bond lengths of 2.3563(9) Å and 2.359(1) Å, which are within the range for Pt−Si bond lengths in similar complexes (Table 13). The palladium bis(silyl) complex (dcpe)Pd(SiHMe$_2$)$_2$ was found to react with 2.5 equivalents of MeO$_2$CC≡CCO$_2$Me affording stereospecifically the double hydrosilylation compound, dimethyl bis(dimethylsilyl)maleate, and the palladium π-complex of the corresponding alkyne (equation 49)[228].

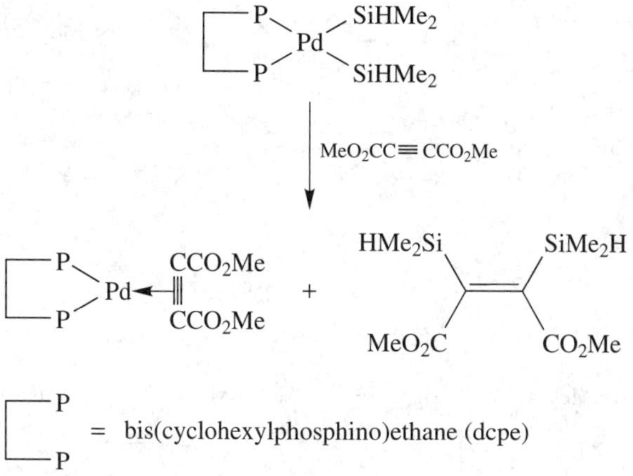

$$\begin{array}{c}
\left[\begin{array}{c} P \\ P \end{array}\right] Pd \begin{array}{c} SiHMe_2 \\ SiHMe_2 \end{array} \\
\downarrow MeO_2CC \equiv CCO_2Me \\
\left[\begin{array}{c} P \\ P \end{array}\right] Pd \Leftarrow \left|\left|\right| \begin{array}{c} CCO_2Me \\ CCO_2Me \end{array} \quad + \quad \begin{array}{c} HMe_2Si \\ \diagup \\ MeO_2C \end{array} = \begin{array}{c} SiMe_2H \\ \diagdown \\ CO_2Me \end{array}
\end{array} \quad (49)$$

$$\left[\begin{array}{c} P \\ P \end{array}\right] = \text{bis(cyclohexylphosphino)ethane (dcpe)}$$

Tilley and coworkers[229] have shown that the reactivity of Pt(0) complexes with arylsilanes is strongly dependent on the ratio of the reactants. The reaction of 1 equivalent of ArSiH$_3$ with 2 equivalents of Pt(PEt$_3$)$_3$ gave quantitatively the Tessier−Young dimers (Et$_3$P)$_2$Pt(η^2,η^2-ArHSiSiHAr)Pt(PEt$_3$)$_2$ (Ar = Ph, p-MeC$_6$H$_4$, Mes). Performing the same reaction with a 3:1 ratio of silane to the platinum complex gave the corresponding cis-bis(silyl) complexes, cis-(Et$_3$P)$_2$Pt(SiH$_2$Ar)$_2$, which isomerized to cis/trans mixtures in solution. The cis isomers decompose in solution to the bridged disilane dimers (equation 50).

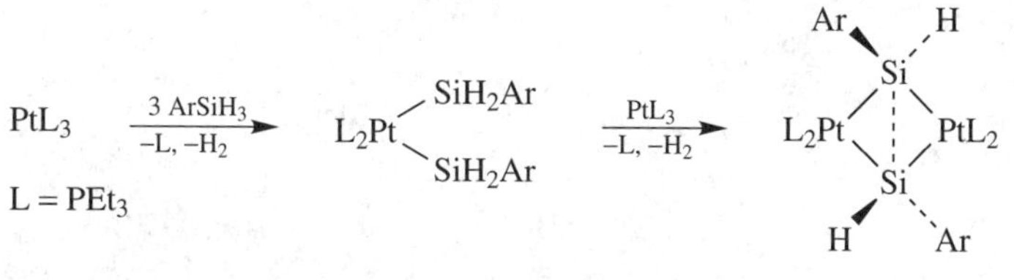

$$PtL_3 \xrightarrow[-L, -H_2]{3 \text{ ArSiH}_3} L_2Pt \begin{array}{c} SiH_2Ar \\ SiH_2Ar \end{array} \xrightarrow[-L, -H_2]{PtL_3} L_2Pt \begin{array}{c} Ar \diagdown Si \diagup H \\ \vdots \quad \vdots \\ \diagup Si \diagdown \\ H \quad Ar \end{array} PtL_2 \quad (50)$$

L = PEt$_3$

Reaction of the bis(silyl) complex (Et$_3$P)$_2$Pt(SiH$_2$Mes)$_2$ with the chelating phosphine dmpe afforded the expected ligand exchange producing the complex (dmpe)Pt(SiH$_2$Mes)$_2$. The reaction with less sterically demanding silyl ligands gave dimeric species [(dmpe)Pt(SiH$_2$Ar)$_2$]$_2$(μ-dmpe) (Ar = Ph, p-CH$_3$C$_6$H$_4$), which were found to be in equilibrium with the corresponding monomers. The thermolysis of these dimeric complexes produced the unusual compounds (dmpe)HPt(μ^2-SiHAr)$_2$[μ-η^1,η^1-ArHSiSiHAr]PtH(dmpe) (**88**) (Ar = Ph), which possess an η^1,η^1-disilene ligand and two bridging silylene ligands. Complex **88** was found to react with 4 equivalents of PhSiH$_3$ yielding the disilane PhH$_2$SiSiH$_2$Ph and the Pt(IV) silyl fac-(dmpe)PtH(SiH$_2$Ph)$_3$ (**89**), which thermally decomposes with loss of H$_2$ and PhSiH$_3$, regenerating complex

88 (equation 51)[229]. Similarly, Tanaka and coworkers have found that the reaction of an excess of the disilane $HMe_2SiSiMe_2H$ with $Pt(PEt_3)_3$ affords the complex cis-$(HMe_2Si)_2Pt(PEt_3)_2$ and oligosilanes of the type $H(SiMe_2)_nH$ ($n < 7$). The thermolysis of this platinum complex produced mainly Me_2SiH_2 while the reaction with 3 equivalents of phenylacetylene afforded the platinacycle complex **90** that, under heating, reductively eliminates 1,1-dimethyl-3,4-diphenyl-1-sila-2,4-cyclopentadiene (**91**) (equation 52)[230].

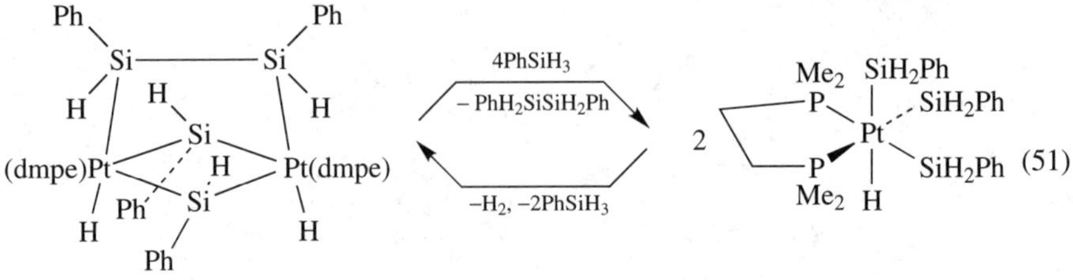

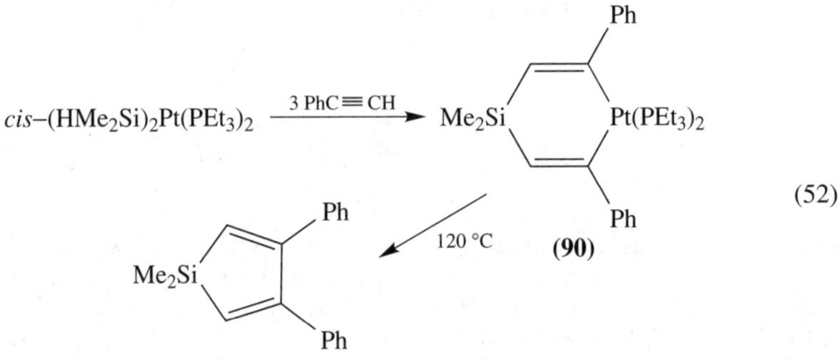

Oxidative addition of the Si−C bond of silacyclobutanes to $Pt(PEt_3)_3$ afforded selectively 1-platina-2 silacyclopentanes. Thus, complex **92** was obtained from 1,1-diphenyl-1-silacyclobutane. Complex **92** exhibits a slightly distorted square-planar coordination geometry, with a Pt−Si bond length of 2.354(1) Å. For the dichloro silacyclobutane compound, in addition to a complex analogous to complex **92**, 10% of the Si−Cl addition product trans-**93** was also observed, whereas from the dimethysilacyclobutane the reaction afforded, besides the analogous complex **92**, large amounts of its dimer **94**[231].

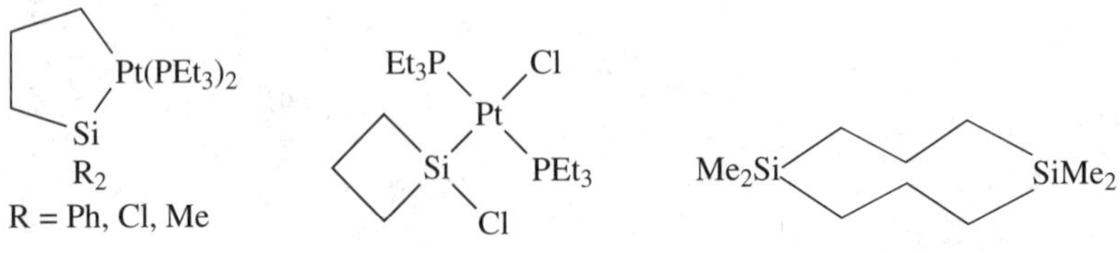

The reactivity of the Pt(0) complex, $Pt(PEt_3)_3$, with equivalent amounts of 1,2-disilanes at different temperatures afforded the first $Pt(IV)Si_4P_2$ (**95**) type complex which was characterized by X-ray diffraction and the dinuclear mixed valence $Pt(II)Pt(IV)Si_4P_4$ (**96**)

type complex (Scheme 40)[232]. Large excess of the disilanes allowed the formation of the well known cyclic Pt(II)Si$_2$P$_2$ complex reported in 1973 by Eaborn and coworkers[233].

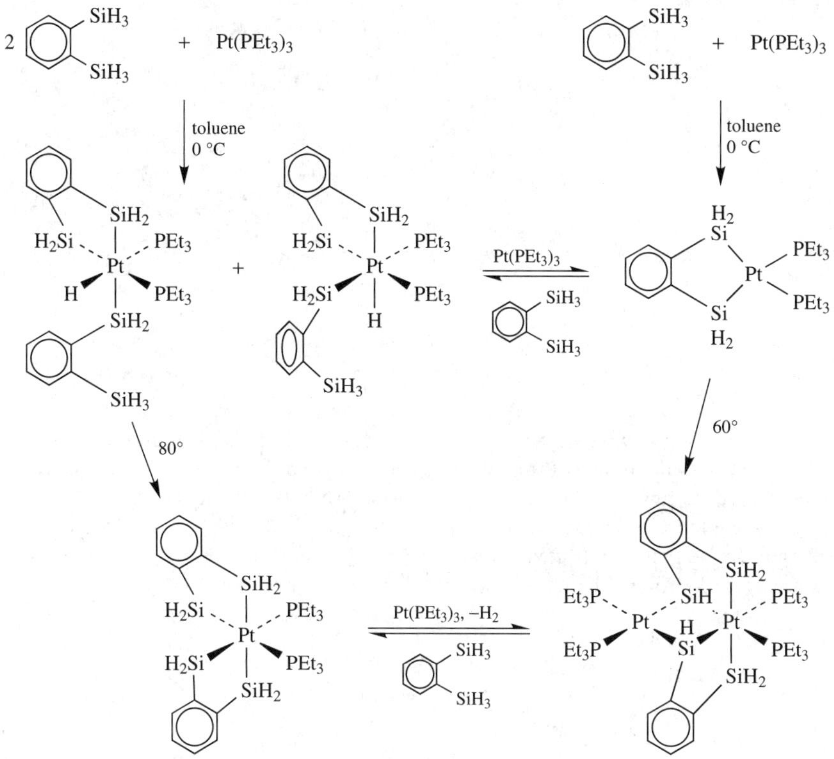

SCHEME 40

Secondary silanes RR'SiH$_2$ (R = R' = Me, Et, Ph; R = Me, R' = Ph) were found to react with the binuclear zero-valent platinum complex [Pt$_2$(μ-CO)(CO)$_2$(μ-dppm)$_2$] to give the corresponding μ-SiRR' complexes of the form [Pt$_2$(μ-SiRR')(CO)$_2$(μ-dppm)$_2$]. The ^{31}P{H} NMR data suggest that these compounds are better formulated as W frame complexes. When R = R' = Ph, low-temperature NMR data showed that the reaction proceeds via the intermediate [Pt$_2$H(SiHPh$_2$)(CO)(μ-dppm)$_2$] which is the result of the Si—H addition to one Pt atom with concomitant loss of two CO molecules[234]. Enticingly, the reaction of Cl(SiMe$_2$)$_3$Cl with Pt(PEt$_3$)$_3$ afforded the complex cis-(ClMe$_2$Si)(Me$_3$SiClMeSi)Pt(PEt$_3$)$_2$ (**97**). Complex **97** can also be obtained by the reaction of Pt(PEt$_3$)$_3$ with ClMe$_2$SiSiClMeSiMe$_3$. Thermolysis of **97** produced the complex cis-(ClMe$_2$Si)$_2$Pt(PEt$_3$)$_2$ and the dimethylsilylene [Me$_2$Si:], presumably via a silylene–platinum intermediate. This reaction has been used to explain the unexpected formation of 1,4-disilacyclohexa-2,5-dienes in the palladium-catalyzed reaction of Cl(SiMe$_2$)$_3$Cl with acetylenes[235].

The bis(silylation) of C—C triple bonds is a reaction in which two Si—C bonds are created in the same molecule. Palladium complexes are the most often used catalysts and the redox cycle of the palladium in the generally postulated mechanism consists of two major pathways: Oxidative addition of the Si—Si bond to the palladium(0), and transfer of the two organosilyl groups to the C—C triple bond regenerating the palladium(0) (Scheme 41).

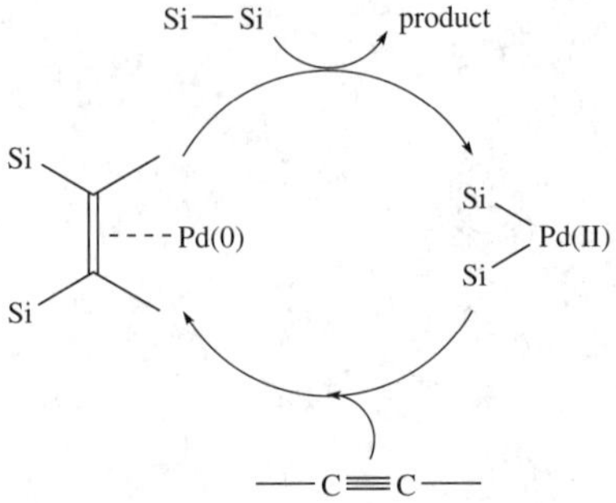

SCHEME 41

Recently Ito and coworkers have prepared, structurally determined and interconverted bis(silyl)alkane palladium(II) **(98)** and bis(silyl)palladium(0) **(99)** complexes, supporting the postulated mechanism for the bis(silylation) of triple bonds (equation 53)[236]. Similar supporting evidence has been recently obtained using mixed silyl–stannyl compounds[237], by the unsymmetrical substituted disilane Me_3SiSiF_2Ph[238], and by theoretical studies by Sakaki[239–241] Márquez[242,243] and others[244–247].

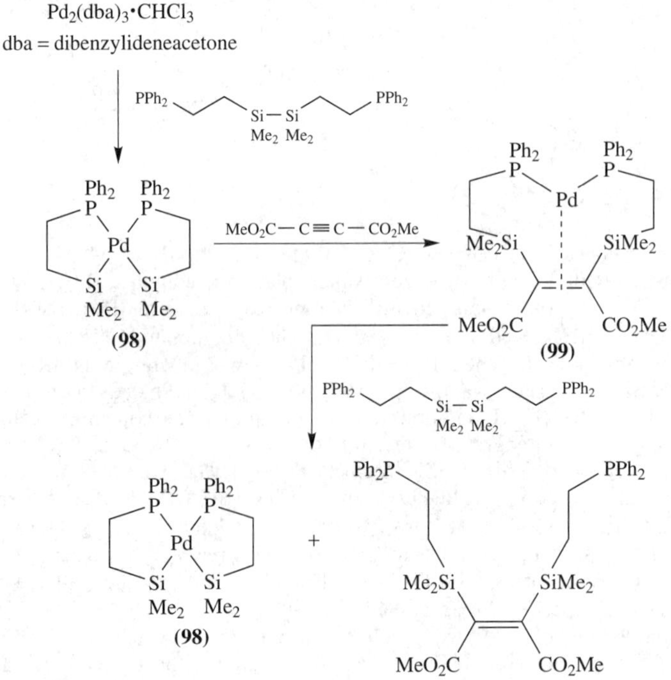

Activation of two Si—Si bonds in bis(disilanyl)alkanes with palladium(0) bis(*tert*-alkyl isocyanide) induced the formation of the cyclic bis(silyl)palladium(II) bis(*tert*-alkyl isocyanide) complexes (**100**) and disilanes described schematically in Scheme 42. These complexes were found to react with phenylacetylene, affording different amounts of five-membered cyclic products and acyclic products which are derived from the insertion of the alkyne into the general intermediate complex **101** (Scheme 42, equation 54). The bis(silanyl)dithiane palladium complex (**102**) was isolated and characterized in the solid state: the two silicon atoms, the two isocyano carbons and the palladium atom are nearly in a plane with a short cross-ring Si—Si distance of 2.613(2) Å, suggesting the possibility of covalently bonded two Si—Si atoms in the four-membered ring. Similar reaction with cyclic disilanes afforded oligomers, and cyclic 20-membered compounds have been prepared in the presence of nitriles[248,249].

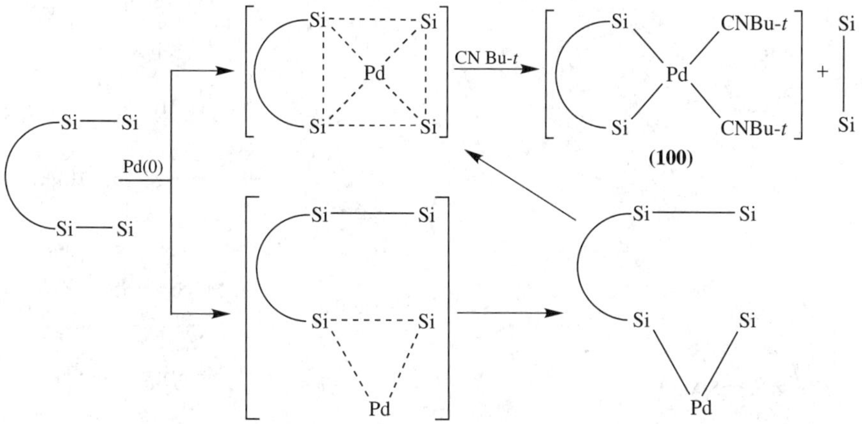

SCHEME 42

Synthetic application of group-10 complexes in catalytic and stoichiometric reactions with silanes has produced a large number of interesting compounds which have been recently reviewed elsewhere[231,250–253].

B. Silylene Complexes

Fischer-type cationic platinum silylene complexes of the type *cis*-[(Cy$_3$P)$_2$HPt=Si(DMAP)(SEt)$_2$] BPh$_4$ and *trans*-[(Cy$_3$P)$_2$HPt=Si(SEt)$_2$] BPh$_4$ were prepared by Tilley and coworkers. The former base-stabilized silylene complex was obtained by displacing a labile triflate group with the neutral two-electron donor (dimethylamino)pyridine (DMAP) from the complex *trans*-(Cy$_3$P)$_2$HPt{Si(SEt)$_2$OTf}. Isomerization to the *cis* complex took place upon recrystallization. The latter base-free silylene complex was formed in the reaction of *trans*-(Cy$_3$P)$_2$HPt{Si(SEt)$_2$OTf} with NaBPh$_4$ (Scheme 43). The structure of the cationic [(Cy$_3$P)$_2$HPt=Si(SEt)$_2$]$^+$ reveals a square-planar geometry for the platinum with no interionic interactions. The silylene fragment is planar, reflecting the sp^2 hybridization. The plane of the silylene ligand is rotated 76° out of the least-squares plane of platinum donor atoms, implying that π-donation from the d$_{xz}$ level contributes more to the molecular bonding. The Pt—Si bond length of 2.270(2) Å is short in comparison with related complexes (Table 13) and the large downfield ^{29}Si chemical shift of 308.65 ppm is in accord with other values for base-free silylene complexes (Table 14)[254].

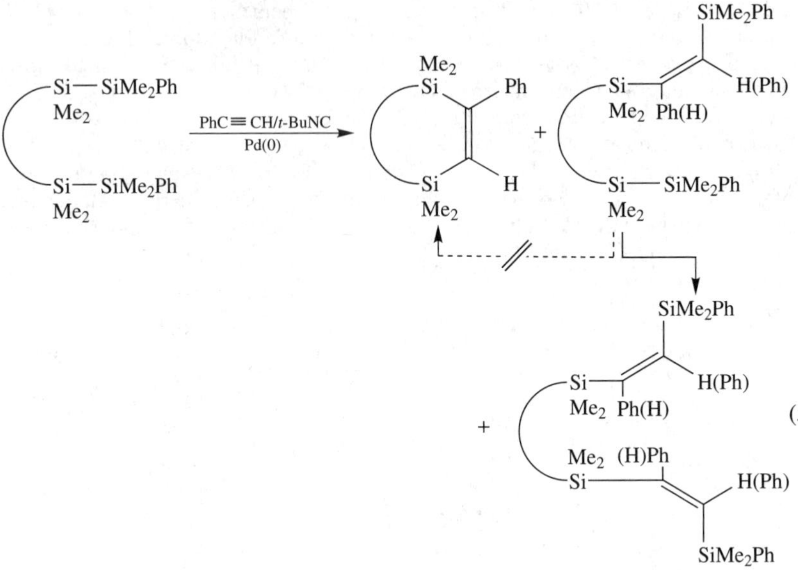

(54)

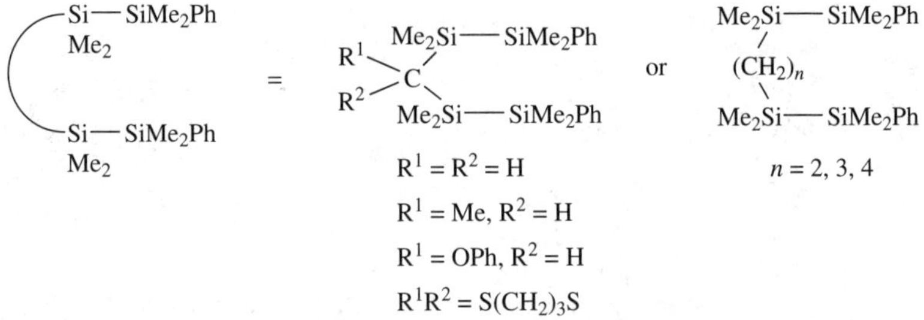

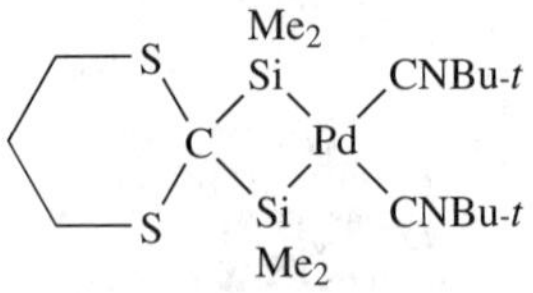

(102)

The first donor-free thermally stable (mp = 160 °C) bis(silylene)nickel complex **105** has been obtained from the reaction of the stable silylene 1,3-di-*tert*-butyl-2,3-dihydro-1*H*-1,3,2-diazasilol-2-ylidine **(103)** with nickel tetracarbonyl[255]. Remarkably, no monosubstitution complex **(104)** or comproportionation of complex **105** with Ni(CO)$_4$ was observed (equation 55). The structure of **105** shows a slightly distorted tetrahedral arrangement of two silylene and two carbonyl ligands around the metal. The small N−Si−N angle (90°) is close to that predicted for free silylenes and the Si−Ni bond lengths of 2.207(2) Å and 2.216(2) Å are close to the theoretical expectations[29]. Interestingly, the deshielding

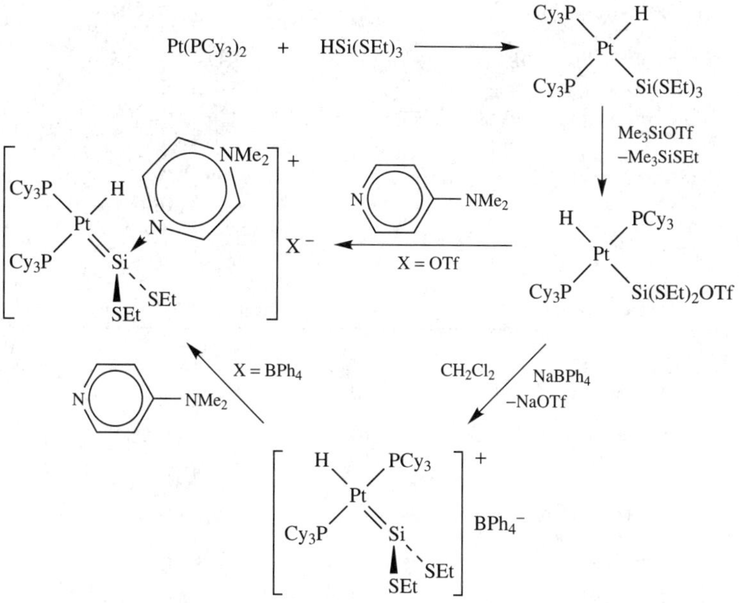

35. Transition-metal silyl complexes

SCHEME 43

expected for this base-free silylene is only moderate in comparison with the starting silylene (^{29}Si δ_{103} = 78.4 ppm; δ_{105} = 97.5 ppm)^{255}.

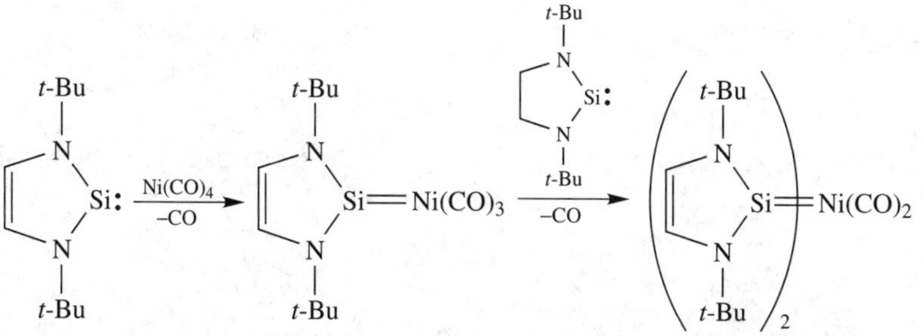

$$(55)$$

Recently, Tamao and coworkers have found that palladium catalyzed the skeletal rearrangement of alkoxy(oligosilanes) (equation 56), presumably via formation of a silyloxy base-stabilized silylene complex (**106**) (equation 57)^{256}.

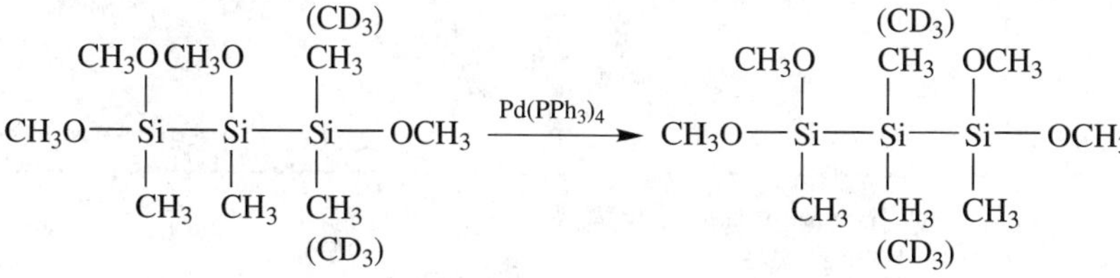

$$(56)$$

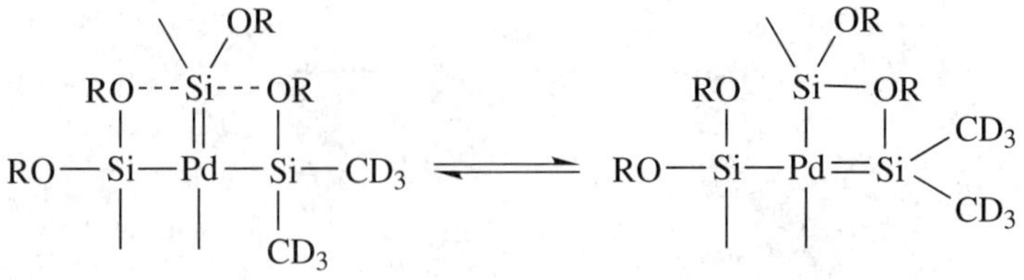

$$\begin{array}{c}\text{RO}\diagdown\text{Si}\diagup\text{OR} \\ \text{RO}-\text{Si}-\text{Pd}-\text{Si}-\text{CD}_3 \\ |\quad|\quad\text{CD}_3\end{array} \rightleftharpoons \begin{array}{c}\text{RO}\diagdown\text{Si}-\text{OR} \\ \text{RO}-\text{Si}-\text{Pd}=\text{Si}\diagup\text{CD}_3 \\ |\quad|\quad\diagdown\text{CD}_3\end{array} \quad (57)$$

(106)

TABLE 13. Selected M−Si distances in group-10 transition-metal silicon-containing complexes

Compound	d^n Configuration	M−Si (Å)	Reference
PtIMe$_2$(SiMe$_3$)(2,2′-bipyridyl)	d^6	2.313(6)	221
cis-PtMe(SiPh$_3$)(PMePh$_2$)$_2$•OEt$_2$	d^8	2.381(2)	222
cis-Pt(H)(SiMe$_3$)(PPh$_3$)$_2$•1/2OEt$_2$	d^8	2.357(3)	223
(dcpe)Pt(Ph){Si(SiMe$_3$)$_2$OH}^{a}	d^8	2.360(2)	226
Fe(η^5-C$_5$H$_4$)$_2$Pt(PEt$_3$)$_2$SiMe$_2$	d^8	2.385(1)	225
[(dcpe)Pt]$_2$(μ_2-SiH$_2$SiH$_2$)$_2$^{a}	d^8	2.378(1), 2.369(1)	227
(dcpe)Pd(SiHMe$_2$)$_2$^{a}	d^8	2.3563(9), 2.359(1)	228
(dmpe)HPt(μ-SiHPh)$_2$[μ-η^1,η^1-ArHSiSiHPh]PtH(dmpe)^b	d^6	2.382(4), 2.379(4) 2.426(4)	229
fac-(dmpe)Pt(H)(SiH$_2$Ph)$_3$^{b}	d^6	2.385(5), 2.406(5) 2.362(5)	229
(Et$_3$P)$_2$Pt{Si(Ph$_2$)CH$_2$CH$_2$CH$_2$}	d^8	2.354(1)	231
(Et$_3$P)$_2$Pt[1,2-(SiH$_2$)$_2$C$_6$H$_4$]$_2$	d^6	2.428(2), 2.383(1) 2.430(2), 2.376(2)	232
96	d^8, d^6	2.415(9), 2.44(1) 2.40(1), 2.41(1) 2.39(1), 2.352(9)	232
(η^2-SiMe$_2$CH$_2$CH$_2$PPh$_2$)$_2$Pd	d^8	2.368(1), 2.367(1)	236
(η^2-SiMe$_2$CH$_2$CH$_2$PPh$_2$)Pd(η^2-SnMe$_2$CH$_2$CH$_2$PPh$_2$)	d^8	2.59(1), 2.43(1)	237
102	d^8	2.336(1), 2.356(1)	248
Silylene Complexes			
[(Cy$_3$P)$_2$(H)Pt=Si(SEt)$_2$] BPh$_4$	d^8	2.270(2)	254
105	d^8	2.207(2), 2.216(2)	255

^{a}dcpe = bis(1,2-dicyclohexylphosphino)ethane.
^{b}dmpe = bis(1,2-dimethylphosphino)ethane.

TABLE 14. Selected ^{29}Si NMR chemical shifts for group-10 transition-metal silicon-containing complexes

Compound	^{29}Si NMR shift (ppm)	Reference
Fe(η^5-C$_5$H$_4$)$_2$Pt(PEt$_3$)$_2$SiMe$_2$	5.18 ($^1J_{PtSi}$ = 1312 Hz, $^2J_{PSi}$ = 181, 14.5 Hz)	225
(dcpe)Pt(Ph){Si(SiMe$_3$)$_2$OH}^{a}	30.88 ($^1J_{SiPt}$ = 1222 Hz, $^2J_{SiP}$ = 7, 179 Hz)	226
(dcpe)$\overline{\text{Pt}\{o\text{-C}_6\text{H}_4\text{Si}}$(SiMe$_3$)$_2$}^{a}	36.47 ($^1J_{SiPt}$ = 1246 Hz, $^2J_{SiP}$ = 5, 183 Hz)	226
(Et$_3$P)$_2$$\overline{\text{Pt}\{\text{Si(Ph}_2)\text{CH}_2\text{CH}_2\text{CH}_2\}}$	35.9 (dd, $^1J_{SiPt}$ = 1421 Hz, $^2J_{SiP}$ = 14.5, 172 Hz)	231
(Et$_3$P)$_2$$\overline{\text{Pt}\{\text{Si(Cl}_2)\text{CH}_2\text{CH}_2\text{CH}_2\}}$	79.4 (dd, $^1J_{SiPt}$ = 2061 Hz, $^2J_{SiP}$ = 19.3, 257 Hz)	231
(Et$_3$P)$_2$$\overline{\text{Pt}\{\text{Si(Me}_2)\text{CH}_2\text{CH}_2\text{CH}_2\}}$	34.6 (dd, $^1J_{SiPt}$ = 1281 Hz, $^2J_{SiP}$ = 14.0, 163 Hz)	231
trans-$\overline{(\text{CH}_2\text{CH}_2\text{CH}_2\text{(Cl)Si})}$PtCl(PEt$_3$)$_2$	34.4 (t, $^1J_{SiPt}$ = 1574 Hz, $^2J_{SiP}$ = 14.5 Hz)	231
(Et$_3$P)$_2$Pt[1,2-(SiH$_2$)$_2$C$_6$H$_4$]$_2$	−27.36 (t, $^1J_{SiPt}$ = 617 Hz, $^2J_{SiP}$ = 14.Hz)	232
	−12.89 (dd, $^1J_{SiPt}$ = 596 Hz, $^2J_{SiP}$ = 136, 21 Hz)	
96	−54.94 (ddd, $^1J_{SiPt}$ = 362, 771 Hz, $^2J_{SiP}$ = 99, 104, 8Hz)	232
	−26.86 (t, $^1J_{SiPt}$ = 733 Hz, $^2J_{SiP}$ = 17 Hz, $^3J_{PtSi}$ = 16 Hz)	
(ClMe$_2$Si)(Me$_3$Si(Cl)(Me)Si)Pt(PEt$_3$)$_2$	43.0 (t, $^1J_{Ptsi}$ = 1220 Hz, $^2J_{PSi}$ = 70 Hz)	235
	46.4 (t, $^1J_{PtSi}$ = 1361 Hz, $^2J_{PSi}$ = 76 Hz)	
Silylene Complexes		
[(Cy$_3$P)$_2$(H)Pt=Si(SEt)$_2$]BPh$_4$	308.65 ($^1J_{PtSi}$ = 1558)	254
105	97.5	255

^{a}dcpe = bis(1,2-dicyclohexylphosphino)ethane.

XVI. COPPER AND MERCURY SILYL COMPLEXES

Copper silyl complexes of the type (Ph$_3$P)$_3$CuSiR$_3$ were synthesized by reacting the hexameric complex, [Ph$_3$PCuH]$_6$, with equivalent amounts of the corresponding silane R$_3$SiH (equation 58)[257].

$$1/6[\text{Ph}_3\text{PCuH}]_6 + \text{HSiR}_3 \xrightarrow[\text{SiR}_3 = \text{SiPh}_3, \text{SiPh}_2\text{Cl}]{} (\text{Ph}_3\text{P})_3\text{Cu}-\text{SiR}_3 + \text{H}_2 \quad (58)$$

The first mercury(I) silyl complex [(Me$_3$SiMe$_2$Si)$_3$Si]$_2$Hg$_2$ was prepared by the reaction of an excess of (Me$_3$SiMe$_2$Si)$_3$SiH with (t-Bu)$_2$Hg. The solid structure of the complex displays linear Si−Hg−Hg−Si fragment with regular metal−silicon bond lengths

[Hg–Si = 2.485(2) Å] and large metal–metal bond distances [2.656(1) Å]. Irradiation of the complex leads to the coupling silyl product and metallic mercury (Scheme 44)[258].

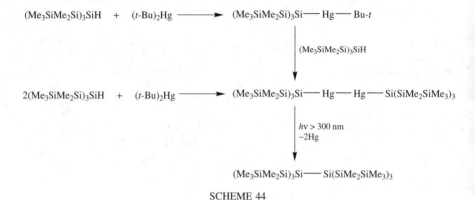

SCHEME 44

XVII. ACKNOWLEDGMENTS

The author thanks Profs. Y. Apeloig and M. Akita and Dr D. Bravo-Zhivotovskii who assisted in the preparation of this review by supplying manuscripts prior to publication.

XVIII. REFERENCES

1. (a) T. D. Tilley, in *The Silicon–Heteroatom Bond* (Eds. S. Patai and Z. Rappoport), Wiley, Chichester, 1991, pp. 245–359.
   (b) T. D. Tilley, in *The Chemistry of Organic Silicon Compounds* (Eds. S. Patai and Z. Rappoport), Chap. 24, Wiley, New York 1989.
2. B. K. Campion, R. H. Heyn and T. D. Tilley, *Organometallics*, **12**, 2584 (1993).
3. P. L. Watson and G. W. Parshall, *Acc. Chem. Res.*, **18**, 51 (1985).
4. G. A. Molander and P. J. Nichols, *J. Am. Chem. Soc.*, **117**, 4415 (1995).
5. F. Peng-Fei, L. Brard, L. Yanwu and T. J. Marks, *J. Am. Chem. Soc.*, **117**, 7157 (1995).
6. T. D. Tilley, in *Organosilicon Chemistry. From Molecules to Materials* (Eds. N. Auner and J. Weis), VCH, Weinheim, 1994, p. 225.
7. H. Schumann, A. Meese-Marktscheffel and F. E. Hahn, *J. Organomet. Chem.*, **390**, 301 (1990).
8. N. S. Radu, T. D. Tilley and A. L. Rheingold, *J. Organomet. Chem.*, **516**, 41 (1996).
9. N. S. Radu, T. D. Tilley and A. L. Rheingold, *J. Am. Chem. Soc.*, **114**, 8293 (1992).
10. C. M. Forsyth, S. P. Nolan and T. J. Marks, *Organometallics*, **10**, 2543 (1991).
11. N. S. Radu and T. D. Tilley, *J. Am. Chem. Soc.*, **117**, 5863 (1995).
12. L. N. Bochkarev, V. M. Makarov, Y. N. Hrzhanovskaya, L. N. Zakharov, G. K. Fukin, A. I. Yanovsky and Y. T. Struchkov, *J. Organometal. Chem.*, **467**, C3 (1994).
13. Y. Mu, C. Aitken, B. Cote, J. F. Harrod and E. Samuel, *Can. J. Chem.*, **69**, 264 (1991).
14. K. A. Kreutzer, R. A. Fisher, W. M. Davis, E. Spaltenstein and S. L. Buchwald, *Organometallics*, **10**, 4031 (1991).
15. T. Takahashi, M. Hasegawa, N. Suzuki, M. Saburi, C. J. Rousset, P. E. Fanwick and E. Negishi, *J. Am. Chem. Soc.*, **113**, 8564 (1991).
16. J. Britten, Y. Mu, J. F. Harrod, J. Polowin, M. C. Baird and E. Samuel, *Organometallics*, **12**, 2672 (1993).
17. H. Q. Liu and J. F. Harrod, *Organometallics*, **11**, 822 (1992).
18. E. Spaltenstein, P. Palma, K. A. Kreutzer, C. A. Willoughby, W. M. Davis and S. L. Buchwald, *J. Am. Chem. Soc.*, **116**, 10308 (1994).

19. Y. E. Ovechinikov, V. A. Igonin, T. V. Timofeeva, S. V. Lindeman, Y. T. Struchkov, M. V. Ustinov and D. A. Bravo-Zhivotovskii, *Metallorgan. Khim.*, **5**, 1155 (1992); *Chem. Abstr.*, **118**, 147614s (1992).
20. H. G. Woo, W. P. Freeman and T. D. Tilley, *Organometallics*, **11**, 2198 (1992).
21. H. G. Woo, R. H. Heyn and T. D. Tilley, *J. Am. Chem. Soc.*, **114**, 5698 (1992).
22. H. G. Woo, J. F. Walzer and T. D. Tilley, *J. Am. Chem. Soc.*, **114**, 7047 (1992).
23. T. D. Tilley, *Acc. Chem. Res.*, **26**, 22 (1993).
24. V. K. Dioumaev and J. F. Harrod, *Organometallics*, **13**, 1548 (1994).
25. J. Y. Corey, X. H. Zhu, T. C. Bedard and L.D. Lange, *Organometallics*, **10**, 924 (1991).
26. J. F. Harrod, C. T. Aitken and E. Samuel, *J. Am. Chem. Soc.*, **108**, 4059 (1986).
27. E. Hengge and M. Weinberger, *J. Organomet. Chem.*, **443**, 167 (1993).
28. T. Imori, R. H. Heyn, T. D. Tilley and A. L. Rheingold, *J. Organomet. Chem.*, **493**, 83 (1995).
29. T. R. Cundari and M. S. Gordon, *J. Phys. Chem.*, **96**, 631 (1992).
30. L. J. Procopio, P. J. Carroll and D. H. Berry, *Polyhedron*, **14**, 45 (1995).
31. L. J. Procopio, P. J. Caroll and D. H. Berry, *J. Am. Chem. Soc.*, **113**, 1870 (1991).
32. L. J. Procopio, P. J. Carroll and D. H. Berry, *Organometallics*, **12**, 3087 (1993).
33. L. J. Procopio, P. J. Carroll and D. H. Berry, *J. Am. Chem. Soc.*, **116**, 177 (1994).
34. A. Ohff, P. Kosse, W. Baumann, A. Tillack, R. Kempe, H. Görls, V. V. Burlakov and U. Rosenthal, *J. Am. Chem. Soc.*, **117**, 10399 (1995).
35. Z. Xue, L. Li, L. K. Hoyt, J. B. Diminnie and J. L. Pollitte, *J. Am. Chem. Soc.*, **116**, 2169 (1994).
36. M. Porchia, N. Brianse, U. Casellato, F. Ossola, G. Rossetto, P. Zanella and R. Graziani, *J. Chem. Soc., Dalton Trans.*, 677 (1989).
37. W. A. King and T. J. Marks, *Inorg. Chim. Acta*, **229**, 343 (1995).
38. A. Lisovski and M. S. Eisen, unpublished results.
39. N. S. Radu, M. P. Engeler, C. P. Gerlach, T. D. Tilley and A. L. Rheingold, *J. Am. Chem. Soc.*, **117**, 3261 (1995).
40. J. S. Allison, B. J. Aylett and H. M. Colquhoun, *J. Organomet. Chem.*, **112**, C7 (1976).
41. M. D. Curtis, L. G. Bell and W. M. Butler, *Organometallics*, **4**, 701 (1985).
42. Q. Jiang, P. J. Carroll and D. H. Berry, *Organometallics*, **10**, 3648 (1991).
43. Q. Jiang, D. C. Pestana, P. J. Carrol and D. H. Berry, *Organometallics*, **13**, 3679 (1994).
44. A. Antiñolo, F. Carrillo, M. Fajarado, A. Otero, M. Lanfranchi and M. A. Pellinghelli, *Organometallics*, **14**, 1518 (1995).
45. G. I. Nikonov, L. G. Kuzmina, D. A. Lemenovskii and V. V. Kotov, *J. Am. Chem. Soc.*, **117**, 10133 (1995).
46. Q. Jiang, P. J. Carroll and D. H. Berry, *Organometallics*, **12**, 177 (1993).
47. H. Nakatsuji, M. Hada and K. Kondo, *Chem. Phys. Lett.*, **196**, 404 (1992).
48. C. Zybill, H. Handwerker and H. Friedrich, *Adv. Organomet. Chem.*, **36**, 229 (1994).
49. C. E. Zybill and C. Liu, *Synlett*, **7**, 687 (1995).
50. P. Braunstein and M. Knorr, *J. Organomet. Chem.*, **500**, 21 (1995).
51. R. Probst, C. Leis, S. Gamper, E. Herdtweck, C. Zybill and N. Auner, *Angew. Chem., Int. Ed. Engl.*, **30**, 1132 (1991).
52. H. Handwerker, C. Leis, R. Probst, P. Bissinger, A. Grohmann, P. Kiprof, E. Herdtweck, J. Blümel, N. Auner and C. Zybill, *Organometallics*, **12**, 2162 (1993).
53. H. Handwerker, M. Paul, J. Blümel and C. Zybill, *Angew. Chem., Int. Ed. Engl.*, **32**, 1313 (1993).
54. H. Handwerker, M. Paul, J. Riede and C. Zybill, *J. Organomet. Chem.*, **459**, 151 (1993).
55. C. Leis, D. L. Wilkinson, H. Handwerker and C. Zybill, *Organometallics*, **11**, 514 (1992).
56. R. J. P. Corriu, G. F. Lanneau and B. P. S. Chauhan, *Organometallics*, **12**, 2001 (1993).
57. R. J. P. Corriu, B. P. S. Chauhan and G. F. Lanneau, *Organometallics*, **14**, 1646 (1995).
58. W. Malisch, R. Lankat, S. Schmitzer, R. Pikl, U. Posset and W. Kiefer, *Organometallics*, **14**, 5622 (1995).
59. W. Malisch, R. Lankat, W. Seelbach, J. Reising, M. Noltemeyer, R. Pikl, U. Posset and W. Kiefer, *Chem. Ber.*, **128**, 1109 (1995).
60. S. Schmitzer, U. Weis, H. Käb, W. Buchner, W. Malisch, T. Polzer, U. Posset and W. Kiefer, *Inorg. Chem.*, **32**, 303 (1993).
61. W. Malisch, R. Lankat, S. Schmitzer and J. Reising, *Inorg. Chem.*, **34**, 5701 (1995).
62. W. Malisch, S. Schmitzer, G. Kaupp, K. Hindahl, H. Käb and U. Wachtler, in *Organosilicon Chemistry. From Molecules to Materials* (Eds. N. Auner and J. Weis), VCH, Weinheim, 1994, p. 185.

63. T. S. Koloski, D. C. Pestana, P. J. Carrol and D. H. Berry, *Organometallics*, **13**, 489 (1994).
64. D. C. Pestana, T. S. Koloski and D. H. Berry, *Organometallics*, **13**, 4173 (1994).
65. L. K. Figge, P. J. Carroll and D. H. Berry, *Organometallics*, **15**, 209 (1996).
66. G. N. Glavee, B. R. Jagirdar, J. J. Schneider, K. J. Klabunde, L. J. Radonovich and K. Dodd, *Organometallics*, **11**, 1043 (1992).
67. B. R. Jagirdar and K. J. Klabunde, *J. Coord. Chem.*, **34**, 31 (1995).
68. B. R. Jagirdar, R. Palmer, K. J. Klabunde and L. J. Radonovich, *Inorg. Chem.*, **34**, 278 (1995).
69. U. Schubert and H. Gilges, *Organometallics*, **15**, 2373 (1996).
70. U. Schubert, *Progress in Organosilicon Chemistry*, Gordon and Breach, Basel, 1995.
71. D. H. Berry, J. C. Chey, H. S. Zipin and P. J. Carroll, *Polyhedron*, **10**, 1189 (1991).
72. P. Hong, N. H. Damrauer, P. J. Carrol and D. H. Berry, *Organometallics*, **12**, 3698 (1993).
73. X. -L. Lou, G. J. Kubas, J. C. Bryan, C. J. Burns and C. J. Unkefer, *J. Am. Chem. Soc.*, **116**, 10312 (1994).
74. X. Luo, G. J. Kubas, C. J. Burns, J. C. Bryan and C. J. Unkefer, *J. Am. Chem. Soc.*, **117**, 1159 (1995).
75. T. Takeuchi, H. Tobita and H. Ogino, *Organometallics*, **10**, 835 (1991).
76. B. P. S. Chauhan, R. J. P. Corriu, G. F. Lanneau, C. Priou, N. Auner, H. Handwerker and E. Herdtweck, *Organometallics*, **14**, 1657 (1995).
77. K. E. Lee, A. M. Arif and J. A. Gladysz, *Chem. Ber.*, **124**, 309 (1991).
78. R. D. Adams, J. E. Cortopassi and J. H. Yamamoto, *Organometallics*, **12**, 3036 (1993).
79. J. Sun, R. S. Lu, R. Bau and G. K. Yang, *Organometallics*, **13**, 1317 (1994).
80. J. A. K. Howard, P. A. Keller, T. Vogt, A. L. Taylor, N. D. Dix and J. L. Spencer, *Acta Crystallogr.*, **B48**, 438 (1992).
81. R. H. Crabtree, *Angew. Chem., Int. Ed. Engl.*, **32**, 789 (1993).
82. M. L. Loza, S. R. De Gala and R. H. Crabtree, *Inorg. Chem.*, **33**, 5073 (1994).
83. G. Reinhard, B. Hirle and U. Schubert, *J. Organomet. Chem.*, **427**, 173 (1992).
84. P. Braunstein, M. Knorr, U. Schubert, M. Lanfranchi and A. Tripicchio, *J. Chem. Soc., Dalton Trans.*, 1507 (1991).
85. P. Braunstein, M. Knorr, H. Piana and U. Schubert, *Organometallics*, **10**, 828 (1991).
86. G. Reinhard, B. Hirle, U. Schubert, M. Knorr, P. Braunstein, A. DeCian and J. Fischer, *Inorg. Chem.*, **32**, 1656 (1993).
87. P. Braunstein, E. Colomer, M. Knorr, A. Tiripicchio and M. Tiripicchio-Camellini, *J. Chem. Soc., Dalton Trans.*, 903 (1992).
88. P. Braunstein, M. Knorr, M. Strampfer, A. DeCian and J. Fischer, *J. Chem. Soc., Dalton Trans.*, 117 (1994).
89. P. Braunstein, M. Knorr, E. Villarroya, A. DeCian and J. Fischer, *Organometallics*, **10**, 3714 (1991).
90. P. Braunstein, L. Douce, M. Knorr, M. Strampfer, M. Lanfranchi and A. Tiripicchio, *J. Chem. Soc., Dalton Trans.*, 331 (1992).
91. P. Braunstein, T. Faure, M. Knorr, F. Balegroune and D. Grandjean, *J. Organomet. Chem.*, **462**, 271 (1993).
92. M. Strampfer, M. Knorr and P. Braunstein, in *Organosilicon Chemistry. From Molecules to Materials* (Eds. N. Auner and J. Weis), VCH, Weinheim, 1994, p. 199.
93. T. Faure, M. Knorr and P. Braunstein, in *Organosilicon Chemistry. From Molecules to Materials* (Eds. N. Auner and J. Weis), VCH, Weinheim, 1994, p. 201.
94. P. Braunstein, M. Knorr, M. Strampfer, A. Tiripicchio and F. Ugozzoli, *Organometallics*, **13**, 3038 (1994).
95. P. Braunstein, T. Faure, M. Knorr, T. Stährfeldt, A. DeCian and J. Fischer, *Gazz. Chim. Ital.*, **125**, 35 (1995).
96. P. Braunstein, M. Knorr, A. Tiripicchio and M. Tiripicchio-Camellini, *Inorg. Chem.*, **31**, 3685 (1992).
97. S. Gilbert, M. Knorr, S. Mock and U. Schubert, *J. Organomet. Chem.*, **480**, 241 (1994).
98. S. Gilbert and U. Schubert, *J. Organomet. Chem.*, **444**, C12 (1993).
99. U. Schubert, S. Gilbert and M. Knorr, *J. Organomet. Chem.*, **454**, 79 (1993).
100. G. Reinhard, M. Knorr, P. Braunstein, U. Schubert, S. Khan, C. E. Strouse, H. D. Kaesz and A. Zinn, *Chem. Ber.*, **126**, 17 (1993).
101. M. Knorr, T. Stährfeldt, P. Braunstein, G. Reinhard, P. Hauenstein, B. Mayer, U. Schubert, S. Khan and H. D. Kaesz, *Chem. Ber.*, **127**, 295 (1994).

102. P. Braunstein, M. Knorr, B. Hirle, G. Reinhard and U. Schubert, *Angew. Chem., Int. Ed. Engl.*, **31**, 1583 (1992).
103. P. Braunstein, M. Knorr and T. Stährfeldt, *J. Chem. Soc., Chem. Commun.*, 1913 (1994).
104. M. Knorr, P. Braunstein, A. DeCian and J. Fischer, *Organometallics*, **14**, 1302 (1995).
105. M. Knorr, P. Braunstein, A. Tiripicchio and F. Ugozzoli, *Organometallics*, **14**, 4910 (1995).
106. R. D. Adams, J. E. Cortopassi and M. P. Pompeo, *Inorg. Chem.*, **30**, 2960 (1991).
107. R. D. Adams, J. E. Cortopassi and M. P. Pompeo, *Organometallics*, **11**, 1 (1992).
108. R. D. Adams, J. E. Cortopassi, J. Aust and M. Myrick, *J. Am. Chem. Soc.*, **115**, 8877 (1993).
109. H. G. Ang, B. Chang and W. L. Kwik, *J. Chem. Soc., Dalton Trans.*, 2161 (1992).
110. H. G. Ang, B. Chang, W. L. Kwik and E. S. H. Sim, *J. Organomet. Chem.*, **474**, 153 (1994).
111. J. A. Cabeza, A. Llamazares, V. Riera, S. Triki and L. Ouahab, *Organometallics*, **11**, 3334 (1992).
112. J. A. Cabeza, R. J. Franco, A. Llamazares, V. Riera, C. Bois and Y. Jeannin, *Inorg. Chem.*, **32**, 4640 (1993).
113. J. A. Cabeza, S. García-Granda, A. Llamazares, V. Riera and J. F. Van der Maelen, *Organometallics*, **12**, 2973 (1993).
114. J. A. Cabeza, R. J. Franco, V. Riera, S. García-Granda and J. F. Van der Maelen, *Organometallics*, **14**, 3342 (1995).
115. S. Kotani, T. Tanizawa, T. Adaci, T. Yoshida and K. Sonogashira, *Chem. Lett.*, 1665 (1994).
116. W. Lin, S. R. Wilson and G. S. Girolami, *J. Chem. Soc., Chem. Commun.*, 284 (1993).
117. W. Lin, S. R. Wilson and G. S. Girolami, *Organometallics*, **13**, 2309 (1994).
118. G. R. Clark, K. R. Flower, C. E. F. Rickard, W. R. Roper, D. M. Salter and L. J. Wright, *J. Organomet. Chem.*, **462**, 331 (1993).
119. C. E. F. Rickard, W. R. Roper, D. M. Salter and L. J. Wright, *J. Am. Chem. Soc.*, **114**, 9682 (1992).
120. B. J. Rappoli, K. J. McGrath, C. F. George and J. C. Cooper, *J. Organomet. Chem.*, **450**, 85 (1993).
121. W. Malisch, S. Möller, O. Fey, H.-U. Wekel, R. Pikl, U. Posset and W. Kiefer, *J. Organomet. Chem.*, **507**, 117 (1996).
122. S. Möller, O. Fey, W. Malisch and W. Seelbach, *J. Organomet. Chem.*, **507**, 239 (1996).
123. F. R. Lemke, *J. Am. Chem. Soc.*, **116**, 11183 (1994).
124. F. R. Lemke, R. S. Simons and W. J. Youngs, *Organometallics*, **15**, 216 (1996).
125. B. K. Campion, R. H. Heyn, T. D. Tilley and A. L. Rheingold, *J. Am. Chem. Soc.*, **115**, 5527 (1993).
126. J. Yin, J. Klosin, K. A. Abboud and W. M. Jones, *J. Am. Chem. Soc.*, **117**, 3298 (1995).
127. T. J. Johnson, P. S. Coan and K. G. Caulton, *Inorg. Chem.*, **32**, 4594 (1993).
128. R. H. Heyn, J. C. Huffman and K. G. Caulton, *New. J. Chem.*, **17**, 797 (1993).
129. P. I. Djurovich, P. J. Carroll and D. H. Berry, *Organometallics*, **13**, 2551 (1994).
130. V. S. Asirvatham, Z. Yao and K. J. Klabunde, *J. Am. Chem. Soc.*, **116**, 5493 (1994).
131. Z. Yao and K. J. Klabunde, *Organometallics*, **14**, 5013 (1995).
132. Z. Yao, K. J. Klabunde and S. Asirvatham, *Inorg. Chem.*, **34**, 5289 (1995).
133. H. Suzuki, T. Takao, M. Tanaka and Y. Moro-oka, *J. Chem. Soc., Chem. Commun.*, 476 (1992).
134. T. Takao, H. Suzuki and M. Tanaka, *Organometallics*, **13**, 2554 (1994).
135. B. K. Campion, R. H. Heyn and T. D. Tilley, *J. Chem. Soc., Chem. Commun.*, 1201 (1992).
136. T. Takao, S. Yoshida, H. Suzuki and M. Tanaka, *Organometallics*, **14**, 3855 (1995).
137. M. Akita, R. Hua, T. Oku and Y. Moro-oka, *J. Chem. Soc., Chem. Commun.*, 1031 (1992).
138. M. Akita, R. Hua, T. Oku and Y. Moro-oka, *J. Chem. Soc., Chem. Commun.*, 1670 (1993).
139. M. Akita, R. Hua, T. Oku and Y. Moro-oka, *J. Chem. Soc., Chem. Commun.*, 541 (1996).
140. M. Akita, N. Kazumi, T. Yoshiaki and Y. Moro-oka, *Organometallics*, **14**, 5209 (1995).
141. M. Akita, personal communication.
142. T. D. Tilley, B. K. Campion, S. D. Grumbine, D. A. Straus and R. H. Heyn, *Transition-metal Complexes of Reactive Silicon Intermediates*, Royal Society of Chemistry, London, 1991.
143. W. P. Freeman, T. D. Tilley and A. L. Rheingold, *J. Am. Chem. Soc.*, **116**, 8428 (1994).
144. C. Paek, J. Ko, Y. Kong, C. H. Kim and M. E. Lee, *Bull. Korean Chem. Soc.*, **15**, 460 (1994).
145. B. Stadelmann, P. Lassacher, H. Stüger and E. Hengge, *J. Organomet. Chem.*, **482**, 201 (1994).
146. E. Hengge, M. Eibl, B. E. Stadelmann, A. Zechmann and H. Siegl, in *Organosilicon Chemistry. From Molecules to Materials* (Eds. N. Auner and J. Weis), VCH, Weinheim, 1994. p. 213.

147. S. Sharma, J. Cervantes, J. L. Mata-Mata, M. -C. Brun, F. Cervantes-Lee and K. H. Pannell, *Organometallics*, **14**, 4269 (1995).
148. K. H. Pannell, M. -C. Brun, H. Sharma, K. Jones and S. Sharma, *Organometallics*, **13**, 1075 (1994).
149. K. H. Pannell, T. Kobayashi and R. N. Kapoor, *Organometallics*, **11**, 2229 (1992).
150. K. H. Pannell and S. Sharma, *Organometallics*, **10**, 954 (1991).
151. K. L. Jones and K. H. Pannell, *J. Am. Chem. Soc.*, **115**, 11336 (1993).
152. C. Hernandez, H. K. Sharma and K. H. Pannell, *J. Organomet. Chem.*, **462**, 259 (1993).
153. H. Sharma and K. H. Pannell, *Organometallics*, **13**, 4946 (1994).
154. Z. Zhang, R. Sanchez and K. H. Pannell, *Organometallics*, **14**, 2605 (1995).
155. H. K. Sharma and K. H. Pannell, *Chem. Rev.*, **95**, 1351 (1995).
156. C. Leis, C. Zybill, J. Lachmann and G. Müller, *Polyhedron*, **10**, 163 (1991).
157. U. Bodensieck, P. Braunstein, W. Deck, T. Faure, M. Knorr and C. Stern, *Angew. Chem., Int. Ed. Engl.*, **33**, 2440 (1994).
158. N. Auner, C. Wagner, E. Herdtweck, M. Heckel and W. Hiller, *Bull. Soc. Chim. Fr.*, **132**, 599 (1995).
159. M. E. Lee, J. S. Han and C. H. Kim, *Bull. Korean Chem. Soc.*, **15**, 335 (1994).
160. H. Handwerker, C. Leis, S. Gamper and C. Zybill, *Inorg. Chim. Acta*, **198-200**, 763 (1992).
161. S. D. Grumbine, R. K. Chadha and T. D. Tilley, *J. Am. Chem. Soc.*, **114**, 1518 (1992).
162. B. J. Aylett, *J. Organomet. Chem.*, **327**, 9 (1980).
163. S. K. Grumbine, D. A. Straus, T. D. Tilley and A. L. Rheingold, *Polyhedron*, **14**, 127 (1995).
164. S. K. Grumbine, T. D. Tilley, F. P. Arnold and A. L. Rheingold, *J. Am. Chem. Soc.*, **116**, 5495 (1994).
165. H. Kobayashi, K. Ueno and H. Ogino, *Organometallics*, **14**, 5490 (1995).
166. K. Ueno, N. Hamashima, M. Shimoi and H. Ogino, *Organometallics*, **10**, 959 (1991).
167. K. Ueno, N. Hamashima and H. Ogino, *Organometallics*, **11**, 1435 (1992).
168. Y. Kawano, H. Tobita, M. Shimoi and H. Ogino, *J. Am. Chem. Soc.*, **116**, 8575 (1994).
169. Y. Kawano, H. Tobita and H. Ogino, *Angew. Chem., Int. Ed. Engl.*, **30**, 843 (1991).
170. Y. Kawano, H. Tobita and H. Ogino, *Organometallics*, **11**, 499 (1992).
171. K. Ueno and H. Ogino, *Bull. Chem. Soc. Jpn.*, **68**, 1955 (1995).
172. K. Ueno, S. Seki and H. Ogino, *Chem. Lett.*, 2159 (1993).
173. K. Ueno, H. Tobita and H. Ogino, *J. Organomet. Chem.*, **430**, 93 (1992).
174. J. R. Koe, H. Tobita and H. Ogino, *Organometallics*, **11**, 2479 (1992).
175. K. Ueno, S. Ito, K. -I. Endo, H. Tobita, S. Inomata and H. Ogino, *Organometallics*, **13**, 3309 (1994).
176. H. Tobita and H. Ogino, *J. Synth. Org. Chem. (Japan)*, **53**, 530 (1995).
177. H. Tobita, H. Wada, K. Ueno and H. Ogino, *Organometallics*, **13**, 2545 (1994).
178. R. Bakhtiar, C. M. Holzangel and D. B. Jacobson, *J. Am. Chem. Soc.*, **115**, 3038 (1993).
179. R. Bakhtiar and D. B. Jacobson, *Organometallics*, **12**, 2876 (1993).
180. D. B. Jacobson and R. Bakhtiar, *J. Am. Chem. Soc.*, **115**, 10830 (1993).
181. W. Ando, T. Yamamoto, H. Saso and Y. Kabe, *J. Am. Chem. Soc.*, **113**, 2791 (1991).
182. Y. Kabe, T. Yamamoto and W. Ando, *Organometallics*, **13**, 4606 (1994).
183. S. D. Grumbine, T. D. Tilley and A. L. Rheingold, *J. Am. Chem. Soc.*, **115**, 358 (1993).
184. R. D. Adams and J. E. Cortopassi, *J. Clus. Sci.*, **6**, 437 (1995).
185. S. M. Maddock, C. E. F. Rickard, W. R. Roper and L. J. Wright, *Organometallics*, **15**, 1793 (1996).
186. W. Chen, A. J. Edwards, M. A. Esteruelas, F. J. Lahoz, M. Oliván and L. A. Oro, *Organometallics*, **15**, 2185 (1996).
187. M. A. Esteruelas, O. Nürnberg, M. Oliván, L. A. Oro and H. Werner, *Organometallics*, **12**, 3264 (1993).
188. M. A. Esteruelas, F. J. Lahoz, M. Oliván, E. Oñate and L. A. Oro, *Organometallics*, **13**, 4246 (1994).
189. M. A. Esteruelas, M. Oliván and L. A. Oro, *Organometallics*, **15**, 814 (1996).
190. M. A. Esteruelas, F. J. Lahoz, E. Oñate, L. A. Oro and L. Rodríguez, *Organometallics*, **14**, 263 (1995).
191. M. A. Esteruelas, F. J. Lahoz, E. Oñate, L. A. Oro and L. Rodríguez, *Organometallics*, **15**, 823 (1996).
192. M. Loza, J. W. Faller and R. H. Crabtree, *Inorg. Chem.*, **34**, 2937 (1995).

193. M. K. Hays and R. Eisenberg, *Inorg. Chem.*, **30**, 2623 (1991).
194. W. -D. Wang and R. Eisenberg, *Organometallics*, **11**, 908 (1992).
195. W. -D. Wang and R. Eisenberg, *Organometallics*, **10**, 2222 (1991).
196. M. D. Fryzuk, L. Rosenberg and S. J. Rettig, *Organometallics*, **10**, 2537 (1991).
197. M. D. Fryzuk, L. Rosenberg and S. J. Rettig, *Inorg. Chim. Acta*, **222**, 345 (1994).
198. M. Aizenberg and D. Milstein, *Angew. Chem., Int. Ed. Engl.*, **33**, 317 (1994).
199. M. Aizenberg and D. Milstein, *J. Am. Chem. Soc.*, **117**, 6456 (1995).
200. B. P. Cleary, R. Mehta and R. Eisenberg, *Organometallics*, **14**, 2297 (1995).
201. S. B. Duckett and R. N. Perutz, *J. Chem. Soc., Chem. Commun.*, 28 (1991).
202. S. B. Duckett and R. N. Perutz, *Organometallics*, **11**, 90 (1992).
203. B. T. Gregg and A. R. Cutler, *Organometallics*, **11**, 4276 (1992).
204. T. Mizutani, T. Uesaka and H. Ogoshi, *Organometallics*, **14**, 341 (1995).
205. P. I. Djurovich, A. L. Safir, N. L. Keder and R. J. Watts, *Inorg. Chem.*, **31**, 3195 (1992).
206. G. P. Mitchell, T. D. Tilley, G. P. A. Yap and A. L. Rheingold, *Organometallics*, **14**, 5472 (1995).
207. E. A. Zarate, V. O. Kennedy, J. A. McCune, R. S. Simons and C. A. Tessier, *Organometallics*, **14**, 1802 (1995).
208. P. Burger and R. G. Bergman, *J. Am. Chem. Soc.*, **115**, 10462 (1993).
209. P. Hofmann, C. Meier, W. Hiller, M. Heckel, J. Riede and M. U. Schmidt, *J. Organomet. Chem.*, **490**, 51 (1995).
210. H. Werner, M. Baum, D. Schneider and B. Windmüller, *Organometallics*, **13**, 1089 (1994).
211. K. Osakada, K. Hataya, Y. Nakamura, M. Tanaka and T. Yamamoto, *J. Chem. Soc., Chem. Commun.*, 576 (1993).
212. H. Wadepohl, W. Galm, H. Pritzkow and A. Wolf, *Angew. Chem., Int. Ed. Engl.*, **31**, 1058 (1992).
213. G. Calzaferri, R. Inhof and K. W. Törnroos, *J. Chem. Soc., Dalton Trans.*, 3741 (1993).
214. J. Borgdorff, N. W. Duffy, B. H. Robinson and J. Simpson, *Inorg. Chim. Acta*, **224**, 73 (1994).
215. H. Lang, U. Lay and L. Zsolnai, *J. Organomet. Chem.*, **417**, 377 (1991).
216. M. Van Tiel, K. M. Mackay and B. K. Nicholson, *J. Organomet. Chem.*, **462**, 79 (1993).
217. S. G. Anema, S. K. Lee, K. M. Mackay and B. K. Nicholson, *J. Organomet. Chem.*, **444**, 211 (1993).
218. G. C. Barris, K. M. Mackay and B. K. Nicholson, *Acta Crystallogr.*, **C48**, 1204 (1992).
219. R. Goikhman, M. Aizenberg, H. B. Kraatz and D. Milstein, *J. Am. Chem. Soc.*, **117**, 5865 (1995).
220. M. Aizenberg, R. Goikhman and D. Milstein, *Organometallics*, **15**, 1075 (1996).
221. C. J. Levy, R. J. Puddephatt and J. J. Vittal, *Organometallics*, **13**, 1559 (1994).
222. F. Ozawa, T. Hikida and T. Hayashi, *J. Am. Chem. Soc.*, **116**, 2844 (1994).
223. L. Abdol Latif, C. Eaborn, A. P. Pidcock and N. S. Weng, *J. Organomet. Chem.*, **474**, 217 (1994).
224. S. Ogoshi, K. Ohe, N. Chatani, H. Kurosawa and S. Murai, *Organometallics*, **10**, 3813 (1991).
225. J. B. Sheridan, A. J. Lough and I. Manners, *Organometallics*, **15**, 2195 (1996).
226. L. S. Chang, M. P. Johnson and M. J. Fink, *Organometallics*, **10**, 1219 (1991).
227. M. J. Michalczyk, C. A. Recatto, J. C. Calabrese and M. J. Fink, *J. Am. Chem. Soc.*, **114**, 7955 (1992).
228. Y. Pan, J. T. Mague and M. J. Fink, *Organometallics*, **11**, 3495 (1992).
229. R. H. Heyn and T. D. Tilley, *J. Am. Chem. Soc.*, **114**, 1917 (1992).
230. H. Yamashita, M. Tanaka and M. Goto, *Organometallics*, **11**, 3227 (1992).
231. H. Yamashita and M. Tanaka, *Bull. Chem. Soc. Jpn.*, **68**, 403 (1995).
232. S. Shimada, M. Tanaka and A. Honda, *J. Am. Chem. Soc.*, **117**, 8289 (1995).
233. C. Eaborn, T. N. Metham and A. Pidcock, *J. Organomet. Chem.*, **63**, 107 (1973).
234. K. A. Brittingham, T. N. Gallaher and S. Schreiner, *Organometallics*, **14**, 1070 (1995).
235. Y. Tanaka, H. Yamashita and M. Tanaka, *Organometallics*, **14**, 530 (1995).
236. M. Murakami, T. Yoshida and Y. Ito, *Organometallics*, **13**, 2900 (1994).
237. M. Murakami, T. Yoshida, S. Kawanami and Y. Ito, *J. Am. Chem. Soc.*, **117**, 6408 (1995).
238. F. Ozawa, M. Sugawara and T. Hayashi, *Organometallics*, **13**, 3237 (1994).
239. S. Sakaki and M. Eiki, *Inorg. Chem.*, **30**, 4218 (1991).
240. S. Sakaki and M. Eiki, *J. Am. Chem. Soc.*, **115**, 2373 (1993).
241. S. Sakaki, M. Ogawa, Y. Musashi and T. Arai, *J. Am. Chem. Soc.*, **116**, 7258 (1994).
242. A. Márquez and J. F. Sanz, *J. Am. Chem. Soc.*, **114**, 2903 (1992).
243. A. Márquez and J. F. Sanz, *J. Am. Chem. Soc.*, **114**, 10019 (1992).
244. N. Koga and K. Morokuma, *J. Am. Chem. Soc.*, **115**, 6883 (1993).

245. D. G. Musaev, K. Morokuma and N. Koga, *J. Chem. Phys.*, **99**, 7859 (1993).
246. T. R. Cundari and M. S. Gordon, *J. Mol. Struct.*, **313**, 47 (1994).
247. H. Jacobsen and T. Ziegler, *Organometallics*, **14**, 224 (1995).
248. M. Sunginome, H. Oike and Y. Ito, *Organometallics*, **13**, 4148 (1994).
249. M. Sunginome, H. Oike and Y. Ito, *J. Am. Chem. Soc.*, **117**, 1665 (1995).
250. K. Horn, *Chem. Rev.*, **95**, 1317 (1995).
251. P. Hofmann, in *Organosilicon Chemistry. From Molecules to Materials* (Eds. N. Auner and J. Weis), VCH, Weinheim, 1994.
252. U. Schubert, *Angew. Chem., Int. Ed. Engl.*, **33**, 419 (1994).
253. C. J. Herzig, in *Organosilicon Chemistry. From Molecules to Materials* (Eds. N. Auner and J. Weis), VCH, Weinheim, 1994.
254. S. D. Grumbine, T. D. Tilley, F. P. Arnold and A. L. Rheingold, *J. Am. Chem. Soc.*, **115**, 7884 (1993).
255. M. Denk, R. K. Hayashi and R. West, *J. Chem Soc., Chem. Commun.*, 33 (1994).
256. K. Tamao, G. -R. Sun and A. Kawachi, *J. Am. Chem. Soc.*, **117**, 8043 (1995).
257. U. Schubert, B. Mayer and C. Rub, *Chem. Ber.*, **127**, 2189 (1994).
258. D. Bravo-Zhivotovskii, M. Yuzefovich, M. Bendikov, Y. Apeloig and K. Klinkhammer, personal communication.

CHAPTER 36

Cyclopentadienyl silicon compounds

PETER JUTZI

Faculty of Chemistry, University of Bielefeld, 33615 Bielefeld, Germany
Fax: +49 521 106 6026; e-mail: peter.jutzi@uni-bielefeld.de

I. INTRODUCTION	2130
II. COMPOUNDS CONTAINING CYCLOPENTADIENYL-SILICON	
σ-BONDS	2130
A. Basic Features of Structure and Bonding	2130
1. Silicon bound to an sp^3-hybridized carbon atom	2131
2. Silicon bound to an sp^2-hybridized carbon atom	2134
B. Synthesis of Compounds with Silicon–Cyclopentadienyl Bonds	2138
1. Silylated cyclopentadienes	2139
2. Silylcyclopentadienyl metal complexes	2140
C. Methods of Silicon–Cyclopentadienyl Bond Cleavage	2143
D. Multiple Silylation of Cyclopentadiene and Cyclopentadienyl	
Complexes	2146
1. Effects of multisilylation	2146
2. Synthetic methods	2147
E. Silicon-bridged Cyclopentadienyl Complexes	2149
1. Single-bridged cyclopentadienyl systems	2149
2. Double-bridged cyclopentadienyl systems	2155
3. Cooperative effects in silicon-bridged ferrocenes	2159
III. COMPOUNDS CONTAINING CYCLOPENTADIENYL–SILICON	
π-BONDS	2163
A. Introduction	2163
B. Decamethylsilicocene	2163
1. Synthesis, structure and bonding	2163
2. Chemistry of decamethylsilicocene	2166
C. Other π-Complexes	2169
IV. ACKNOWLEDGEMENTS	2170
V. REFERENCES	2170

The chemistry of organic silicon compounds, Vol. 2
Edited by Z. Rappoport and Y. Apeloig © 1998 John Wiley & Sons Ltd

I. INTRODUCTION

Cyclopentadienyl systems are the most common ligands in the molecular chemistry of s-, p-, d- and f-block elements. Their versatile application is based on the wide variety of possible bonding modes which allow for the most appropriate interaction with each acceptor. Furthermore, several kinds of substituents can be readily introduced to the C_5-perimeter, offering many variations in structure and chemistry. Finally, the potential of cyclopentadienyl ligands to behave as leaving groups gives them interesting utility in synthetic chemistry.

Not surprisingly, cyclopentadienyl ligands also play an important role in silicon chemistry, where novel bonding leads to fascinating variations in structure and properties. This account presents basic features of structure and bonding in the various classes of cyclopentadienyl silicon compounds and describes the chemistry with a special emphasis on organometallic aspects. Some directions of this chemistry have already been treated in earlier reports. A review by Abel and coworkers[1] gave the state-of-the-art in cyclopentadienyl silicon chemistry at the beginning of the 1970s. Discussion of structure, bonding and reactivity are spread over several chapters of part 1 and part 2 of this series[2] and in the series *Comprehensive Organometallic Chemistry*[3] and *Comprehensive Organic Chemistry*[4]. Finally, reviews dealing with the fluxionality in cyclopentadienyl compounds[5] and with the π-complex chemistry of main group elements[6] are also available.

II. COMPOUNDS CONTAINING CYCLOPENTADIENYL-SILICON σ-BONDS

A. Basic Features of Structure and Bonding

Compounds with a cyclopentadienyl-silicon σ-bond adopt a variety of bonding arrangements, which can be classified on the basis of the hybridization of the corresponding cyclopentadienyl carbon atom, as displayed in Scheme 1. In species of type **1**, the silicon atom is bound to an sp^3-hybridized carbon atom, i.e. to an allylic carbon within a cyclopentadiene unit. The silicon atom is bound to an sp^2-hybridized carbon atom in species of type **2–6**, including silylcyclopentadienyl radicals (**2**), ionic silylcyclopentadienide species (**3**), and silylcyclopentadienyl fragments bound in a η^5-fashion to a metal centre (**4**). Species **5** represents the ionic resonance structure of a silafulvene, and structure **6** stands for the vinylic isomers of a σ-cyclopentadienylsilane (type **1**), but these two types of compound are not examined in detail in this article.

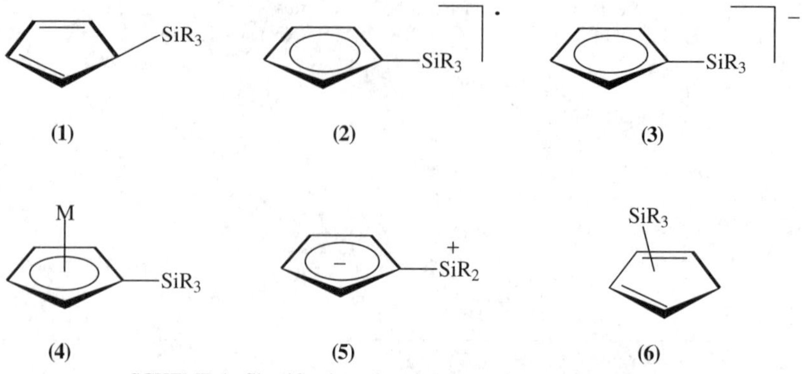

SCHEME 1. Classification of σ-cyclopentadienyl silicon compounds

TABLE 1. Structural parameters of **7** and **8** in Å and deg

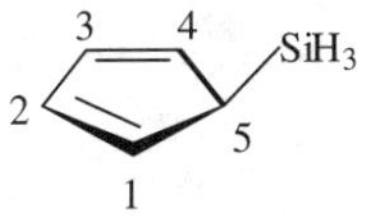

(7) 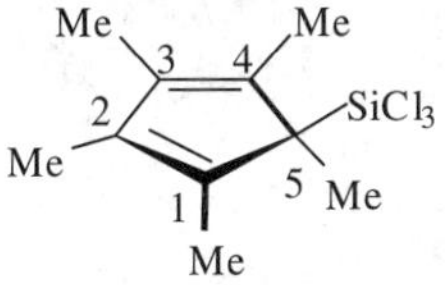 (8)

	7 (GED)	8 (X-ray)
C(1)−C(2)	1.389 ± 0.013	1.34
C(2)−C(3)	1.436 ± fixed	1.46
C(3)−C(4)	1.389 ± 0.013	1.34
C(4)−C(5)	1.500 ± 0.013	1.51
C(5)−C(1)	1.500 ± 0.013	1.51
Si−C(5)	1.881 ± 0.010	1.87
C(4)C(5)C(1)	100.3 ± 1.5	103.2
C(5)C(1)C(2)	112.0 ± 1.0	108.7
C(1)C(2)C(3)	107.9 ± 0.6	109.6

1. Silicon bound to an sp^3-hybridized carbon atom

a. Structural data. The molecular structure of the parent cyclopentadienylsilane, $H_5C_5SiH_3$ (**7**), in the gas phase was determined in the 1970s by a gas electron diffraction (GED) study[7]. The cyclopentadiene ring was found to be planar, consistent with theoretical predictions[8,9], and the carbon−carbon bond lengths and bond angles (Table 1) are close to those found in cyclopentadiene itself. The structure is typical of the whole class of cyclopentadienyl silicon compounds with the silicon atom in an allylic position.

The crystal structure parameters for pentamethylcyclopentadienyltrichlorosilane, $Me_5C_5SiCl_3$ (**8**)[10], obtained from X-ray data, are also presented in Table 1 for comparison. The C_5 ring is again essentially planar, and the Si−C bond length is typical despite the steric demand of the Me_5C_5 group, which is evident from the fact that the ClSiCl bond angles within the $SiCl_3$ moiety are 2−5° less than the tetrahedral angle. Even more pronounced steric effects are observed for silicon compounds with two pentamethylcyclopentadienyl ligands, e.g. for $(Me_5C_5)_2SiCl_2$[11] and for $(Me_5C_5)_2Si(OH)_2$[12].

b. Bonding and fluxionality. Experimental and theoretical studies together have led to an understanding of both the bonding and the dynamic behaviour of cyclopentadienylsilanes. Ab initio calculations[13] for cyclopentadiene derivatives with a silyl group in allylic position predict strong interactions between the σ-C−Si bond and the π-system. The drawings of the occupied π-molecular orbitals for a silylated cyclopentadiene in Scheme 2 show that silyl substitution will have a more pronounced effect on the $b_1(\pi_2)$-type orbital because of the nodal plane through the substitution centre in the $a_2(\pi_1)$-orbital.

This feature is illustrated experimentally by comparing the vertical π_1 and π_2 ionization energies of cyclopentadiene and some silyl derivatives, which have been obtained from the PE spectra[14,15], as shown in Table 2. Adjustments in the ionization energies due to silylation are more pronounced for the $b_1(\pi_2)$ MO than for the $a_2(\pi_1)$ MO. This is consistent with a more effective σ/π interaction in the b_1 MO made possible by the fixed orientation with the σC−Si bond collinear to the π-system. Generally, trialkylsilyl groups in a β-position to a π-system are powerful electron donors, which enormously

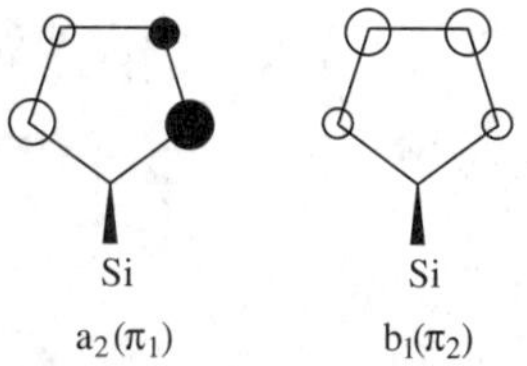

2132   Peter Jutzi

Si   Si
$a_2(\pi_1)$ $b_1(\pi_2)$

SCHEME 2. Representation of the π-MOs in a silylated cyclopentadiene

modify the properties of the parent π-system[15,16]. Thus, the data obtained for the silylated cyclopentadienes are typical for the introduction of silyl groups in a β-position to a π-system.

The fluxionality of η^1-cyclopentadienyl compounds of silicon and of Main Group elements in general is also well defined[5]. The 1,2 migration of silicon, germanium and tin was first discussed in 1965[17], and the important observation that competitive prototropic rearrangements might occur was presented soon after[18]. In cyclopentadienyl silicon compounds of the type $H_5C_5SiR_3$, degenerate silicon and non-degenerate hydrogen shifts take place, as illustrated in Figure 1. These dynamic processes are classified as symmetry-allowed, suprafacial 1,5-sigmatropic rearrangements[19]. Substituents at the migrating element have a surprisingly large influence on the rate of sigmatropic rearrangements, especially in boron and phosphorus chemistry. In silicon chemistry, these substituent effects are comparatively small due to restricted possibilities of influencing transition state energies[5].

TABLE 2. Vertical π-ionization energies (eV) of cyclopentadiene and some silyl derivatives

	IE (π_1)	IE(π_2)		IE(π_1)	IE (π_2)
H_6C_5	8.6	10.7	$H_5C_5SiMe_3$	8.3	9.1
$H_5C_5SiH_3$	8.7	10.2	$H_4C_5(SiMe_3)_2$	8.0	9.1

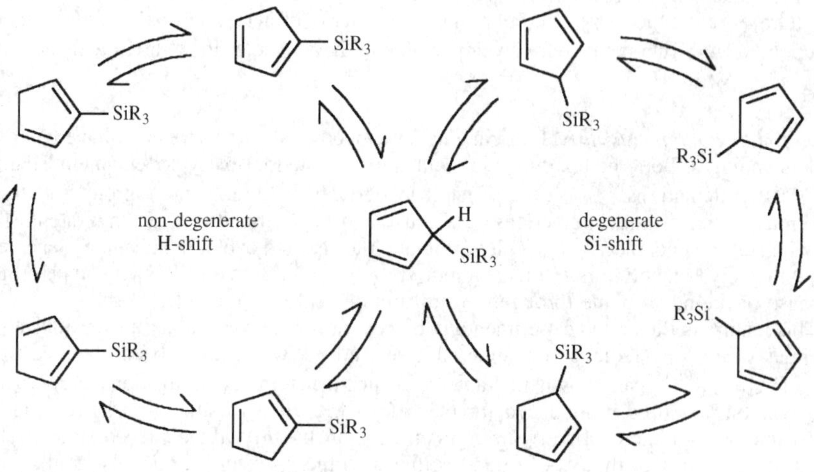

FIGURE 1. Sigmatropic rearrangements in $H_5C_5SiR_3$-type compounds

Activation energies for hydrogen shifts are higher than those for silicon shifts. As seen in Figure 1, hydrogen shifts allow the formation of isomers, in which the silicon group resides in a vinylic position of the cyclopentadiene system rather than allylic. This process may be desired or not desired depending on the synthetic target. The substitution of all hydrogen atoms is necessary to avoid hydrogen shifts. For this reason, several pentamethylcyclopentadienyl silicon compounds were synthesized and studied in more detail[5,11]. In the case of only partial hydrogen substitution, the silyl group prefers a ring position with a geminal hydrogen atom. For example, an overall 1,3 silicon shift caused by two successive 1,2 shifts is favoured in the rearrangement process of (1,2,4-trimethylcyclopentadienyl)trimethylsilane (**9**) to **9′**[20], as shown in equation 1. Also in 5,5-bis(trimethylsilyl)cyclopentadiene (**10**), an overall 1,3 silicon shift to **12** takes place[21,22], as depicted in equation 2. Due to steric constraints of the trimethylsilyl group, the intermediate 1,5-bis(trimethylsilyl)cyclopentadiene (**11**) is of higher energy and cannot be observed by NMR techniques or trapped by chemical reactions. On the other hand, even more pronounced steric constraints can prevent the overall 1,3 shift and enforce isomer formation via a 1,2 silicon shift. This is demonstrated with two typical examples in equation 3 [a degenerate rearrangement (**13** ⇌ **13′**)][23] and in equation 4 [a non-degenerate process (**14** ⇌ **15**)[24]].

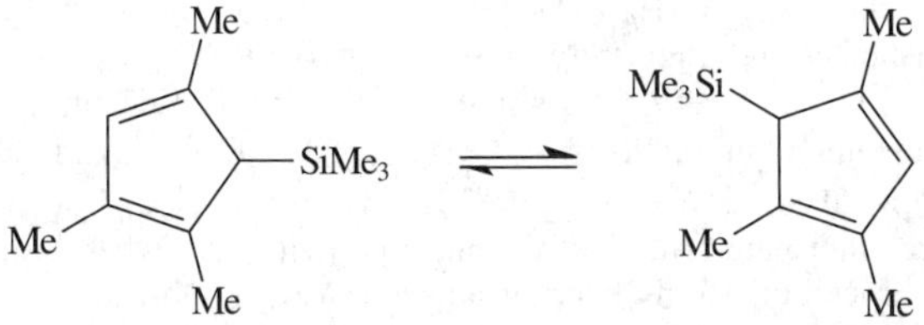

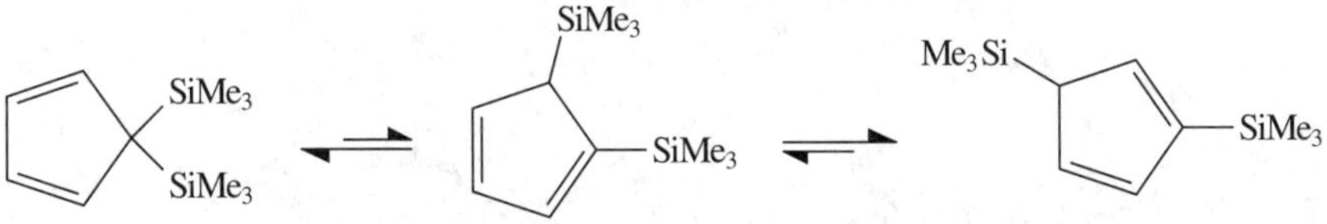

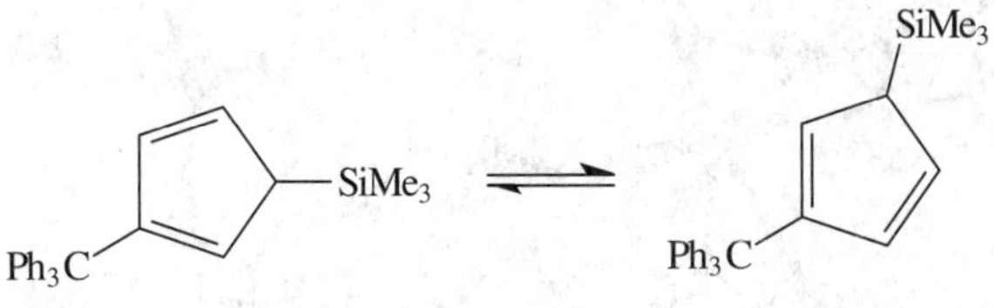

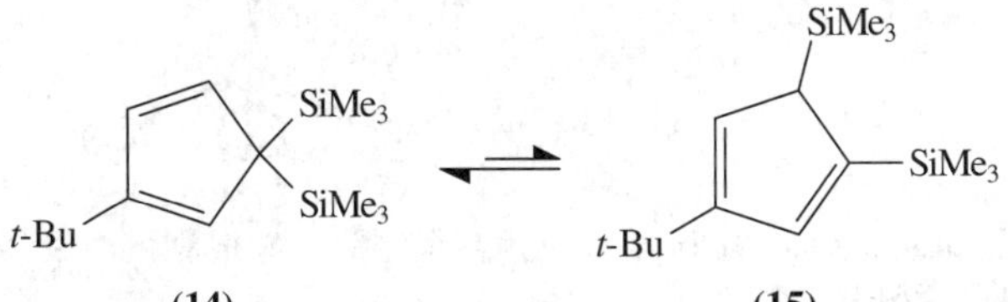

2134   Peter Jutzi

Dynamic processes, such as those in equations 1–4, are of great importance in synthetic cyclopentadienylsilicon chemistry. They allow design of multiply substituted cyclopentadienyl compounds using the regioselectivity in silicon–carbon and also in carbon–carbon bond formation (see Sections II.D and II.E).

2. Silicon bound to an sp^2-hybridized carbon atom

In this chapter, compounds of type **2–4** are described in which the respective silyl group is directly bonded to a cyclopentadienyl π-system, raising more questions concerning the consequences of electronic interaction. Answers will be given on the basis of X-ray crystal structure investigations and of physico-chemical measurements (X-PES, CV, ESR) performed on typical silylcyclopentadienyl species.

a. X-ray crystal structure data. An η^0-substituted cyclopentadienide species of type **3** in Scheme 1 is present in the compound [Li(12crown4)$_2$][1,2,4-(Me$_3$Si)$_3$C$_5$H$_2$] **(16)**[25] and is the most appropriate example to discuss the influence of trimethylsilyl groups on the structure of a cyclopentadienide anion. **16** is comprised of well-separated cations and anions, as depicted in Figure 2, and does not involve the cation contacts evident in compounds like $\{[\mu - \eta^5: \eta^5\text{-}C_5H_4(SiMe_3)]Li\}_n$[26] and $\{[\mu - \eta^5: \eta^5\text{-}C_5H_4(SiMe_3)]K\}_n$[27]. Selected bond distances and angles are presented in Table 3.

The average C–C bond length within the cyclopentadienide unit is 1.42 Å, which is significantly greater than in the free C$_5$H$_5$ ion (1.399 Å)[28] and the average C–C bond length within the η^5-H$_5$C$_5$ ring of (12crown4)Li(η^5-H$_5$C$_5$) (1.395 Å)[25]. Four of the C–C distances in **16** are remarkably similar [1.414(3) Å], whereas the C$_1$–C$_2$ bond, which involves the carbon atoms bearing adjacent Me$_3$Si substituents, is 1.446(3). The Si–C distances in **16** are in the range 1.835(2) Å to 1.842(2) Å, and the average Si–C bond length is 1.84 Å, slightly shorter than normally observed (Si–C$_{sp^2}$ 1.85 Å[29]). The

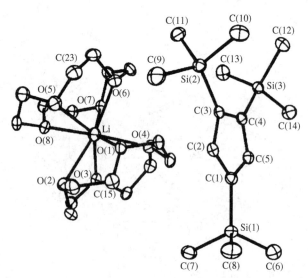

FIGURE 2. Solid-state structure of **16**. Reprinted with permission from Ref. 25. Copyright (1991) American Chemical Society

36. Cyclopentadienyl silicon compounds

TABLE 3. Selected bond distances (Å) and angles (deg) for the anion of (16)

	C(1)–C(2)	1.446(3)	C(1)–C(2)–C(3)	106.3(2)
	C(2)–C(3)	1.414(3)	C(2)–C(3)–C(4)	110.9(2)
	C(3)–C(4)	1.414(3)	C(3)–C(4)–C(5)	105.4(2)
	C(4)–C(5)	1.414(3)	C(4)–C(5)–C(1)	110.7(2)
	C(5)–C(1)	1.413(3)	C(5)–C(1)–C(2)	106.7(2)
	Si(1)–C(1)	1.836(2)	C(5)–C(4)–Si(3)	127.9(1)
	Si(2)–C(2)	1.842(2)	C(5)–C(1)–Si(1)	122.1(1)
	Si(3)–C(4)	1.835(2)	C(1)–C(2)–Si(2)	132.3(2)

relatively long C–C distances as well as the slightly shortened Si–C distances in **16** are probably a result of the electronic effects of the Me$_3$Si substituents, whose ability to stabilize negatively charged species such as carbanions is well documented[2]. The three Me$_3$Si groups provide stabilization by lowering the negative charge density through delocalization by π-σ^* interaction.

In η^5-silylcyclopentadienyl metal complexes of type **4** in Scheme 1, the π-orbitals of the cyclopentadienyl fragment are intimately involved in bonding with low-lying metal valence orbitals. The ferrocene derivatives are representative of this class of compounds, and illustrate the influence of silylation on structural and electronic parameters. A single-crystal X-ray diffraction study of 1,1′,3,3′-tetrakis(trimethylsilyl)ferrocene, Fe[η^5-(1.3-SiMe$_3$)$_2$H$_3$C$_5$]$_2$ (**17**)[30], has revealed a conformation of C$_2$ symmetry in which the trimethylsilyl groups are arranged in a staggered fashion, with the five-membered rings nearly eclipsed, as portrayed in Figure 3. Selected bond distances and angles are presented in Table 4.

The average Fe–C(ring) bond length of 2.059(2) Å is comparable to that in ferrocene, as are the C–C bond distances [average 1.432(3) Å] within the five-membered ring. The trimethylsilyl groups are bent away from the metal centre (average 8°) probably for steric reasons. The average Si–C distance is 1.854(2) Å. It is informative to compare C–C and Si–C bond lengths within the silylated cyclopentadienyl fragments in the ionic

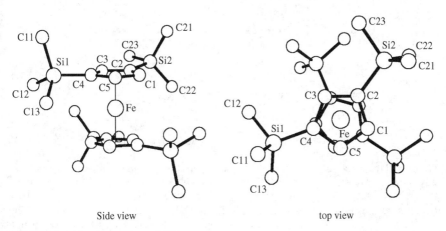

FIGURE 3. Molecular structure of (**17**)

TABLE 4. Selected bond distances (Å) and angles (deg) for **17** (for numbering scheme see Figure 3)

C(1)-C(2)	1.442(3)	C(1)-C(2)-C(3)	105.5(2)
C(2)-C(3)	1.424(3)	C(2)-C(3)-C(4)	112.0(2)
C(3)-C(4)	1.429(3)	C(3)-C(4)-C(5)	104.6(2)
C(4)-C(5)	1.441(3)	C(1)-C(5)-C(4)	109.5(2)
C(5)-C(1)	1.422(3)	C(2)-C(1)-C(5)	108.7(2)
Si(1)-C(4)	1.848(2)	Si(2)-C(2)-C(1)	126.0(2)
Si(2)-C(2)	1.861(2)	Si(2)-C(2)-C(3)	128.3(2)

cyclopentadienide species (**16**) with those in the covalent ferrocene derivative (**17**), which exhibits significantly longer C−C and Si−C distances. This is likely due to the transfer of electron density from the cyclopentadienyl ligands to the metal centre, which weakens the C−C as well as the Si−C bonds to some extent.

b. Cyclic voltammetry data. It is well known that ferrocene and its derivatives exhibit reversible one-electron oxidations, and cyclic voltammetry is a useful tool to determine the effect of different kind of substituents on the redox properties of the corresponding compounds. The trimethylsilylated ferrocenes **17–21** have been studied by cyclic voltammetry[31]. Their redox potentials, $E_{1/2}$, which are internally referenced to ferrocene (**22**), are presented in Table 5. For the sake of comparison, the $E_{1/2}$ values are also given for $(Me_5C_5)_2Fe$ (**23**)[33], $Me_5C_5(H_5C_5)Fe$ (**24**)[34] and $[1,3(t\text{-}Bu)_2H_3C_5]_2Fe$ (**25**)[31].

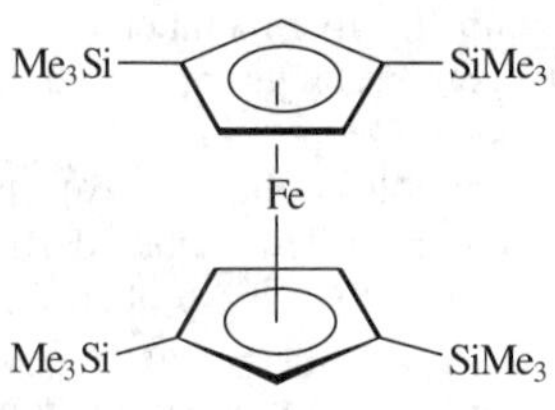

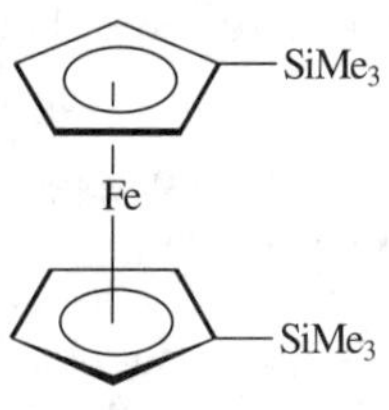

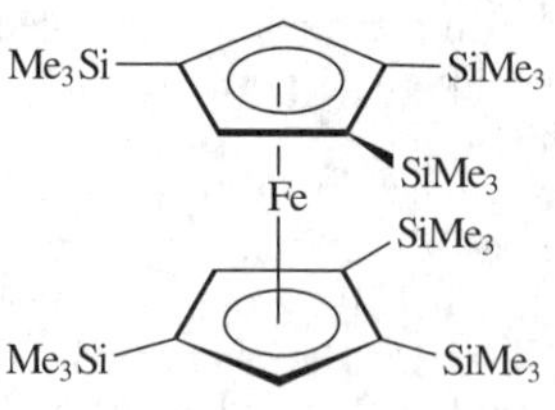

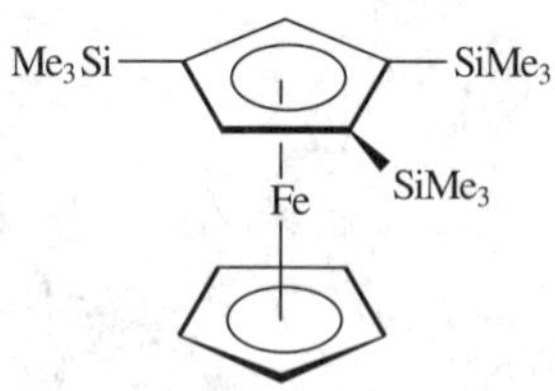

Table 5 shows that alkyl groups such as methyl or *t*-butyl at the ring periphery cathodically shift the oxidation potential of the substituted ferrocene by *ca* 50 mV per alkyl group relative to the parent compound. This reflects the steady increase of electron density at the iron centre by successive ring substitution with electron-donating groups[31]. On the other hand, anodic shifts of the oxidation potential are observed for the trimethylsilyl substituted ferrocenes **17**, **20** and **21** compared with those of the parent compounds **22**

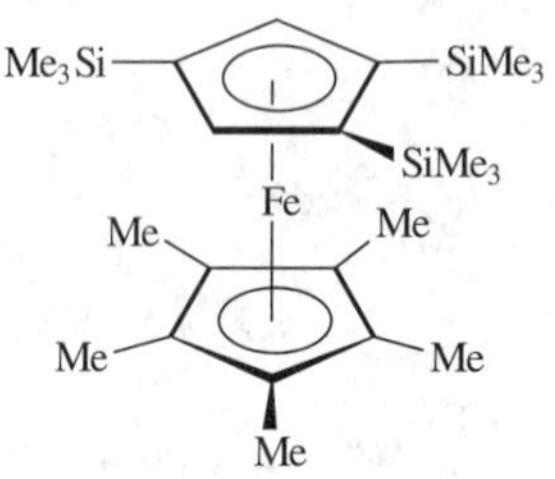

TABLE 5. Redox potential data for ferrocene derivatives[31]

		$E_{1/2}$ (mV)^a			$E_{1/2}$ (mV)^a
[1,3-(Me$_3$Si)$_2$H$_3$C$_5$]$_2$Fe	(17)	+33^b	(H$_5$C$_5$)$_2$Fe	(22)	0^b
[(Me$_3$Si)H$_4$C$_5$]$_2$Fe	(18)	0^{c,32}	(Me$_5$C$_5$)$_2$Fe	(23)	-550^{c,33}
[1,2,4-(Me$_3$Si)$_3$H$_2$C$_5$]$_2$Fe	(19)	-104^b	(Me$_5$C$_5$)(H$_5$C$_5$)Fe	(24)	-270^{b,34}
1,2,4-(Me$_3$Si)$_3$H$_2$C$_5$(H$_5$C$_5$)Fe	(20)	+17^b	[1,3-(Me$_3$C)$_2$H$_3$C$_5$]$_2$Fe	(25)	-250^b
1,2,4-(Me$_3$Si)$_3$H$_2$C$_5$(Me$_5$C$_5$)Fe	(21)	-359^b			

$^a E_{1/2} = 1/2 (E_p^{ox} + E_p^{red})$.
^{b}In CH$_2$Cl$_2$ solution; Ag/AgCl reference electrode; Bu$_4$N$^+$PF$_6^-$ as supporting electrolyte.
^{c}In CH$_3$CN solution.

and **23**. This reflects the decrease of electron density at the respective iron centre by successive ring substitution with electron-accepting trimethylsilyl groups. These observations further support the concept of π-σ^* interaction. The unexpectedly low $E_{1/2}$ for the hexakis-substituted ferrocene derivative **19** must be due to steric factors[31].

c. X-PES data. The electronic effects of trimethylsilyl substituents in cyclopentadienyl π-complexes have also been probed by X-ray photoelectron spectroscopy for a series of ferrocenes and Group 14 metallocenes[35]. Table 6 lists the binding energies of inner-shell electrons for several compounds with trimethylsilyl substituents on the cyclopentadienyl ring and for appropriate reference compounds. For each metal, the binding energy decreases by about 0.1 eV per trimethylsilyl group, indicating that the trimethylsilyl group is unexpectedly weakly electron-donating according to this physical method.

d. ESR data. Electron spin-resonance (ESR) spectroscopy has provided detailed structural information on the paramagnetic silylated cyclopentadienyl species of type **2** in Scheme 1. The orbital degeneracy in the cyclopentadienyl radical is removed by substitution so that the form of the ESR spectra depends critically on the symmetry properties of the orbital in which the unpaired electron resides[36,37]. As an unpaired electron must

TABLE 6. Binding energies of inner-shell electrons for ferrocenes and Group IV metallocenes

Compound	Electron	Binding energy ±0.1 eV
(H$_5$C$_5$)Fe (22)	2p$_{3/2}$	708.0
[(Me$_3$Si)$_3$H$_2$C$_5$]$_2$Fe (19)	2p$_{3/2}$	707.3
(Me$_5$C$_5$)$_2$Fe (23)	2p$_{3/2}$	707.1
(H$_5$C$_5$)$_2$ZrCl$_2$	3d$_{5/2}$	181.7
[(Me$_3$Si)H$_4$C$_5$]$_2$ZrCl$_2$	3d$_{5/2}$	181.5
[(Me$_3$Si)$_2$H$_3$C$_5$]$_2$ZrCl$_2$	3d$_{5/2}$	181.4
[(Me$_3$Si)$_3$H$_2$C$_2$]$_2$ZrCl$_2$	3d$_{5/2}$	181.2
(Me$_5$C$_5$)$_2$ZrCl$_2$	3d$_{5/2}$	181.0
(H$_5$C$_5$)$_2$HfCl$_2$	4f$_{7/2}$	17.1
[(Me$_3$Si)$_3$H$_2$C$_5$]$_2$HfCl$_2$	4f$_{7/2}$	16.5
(Me$_5$C$_5$)$_2$HfCl$_2$	4f$_{7/2}$	16.5

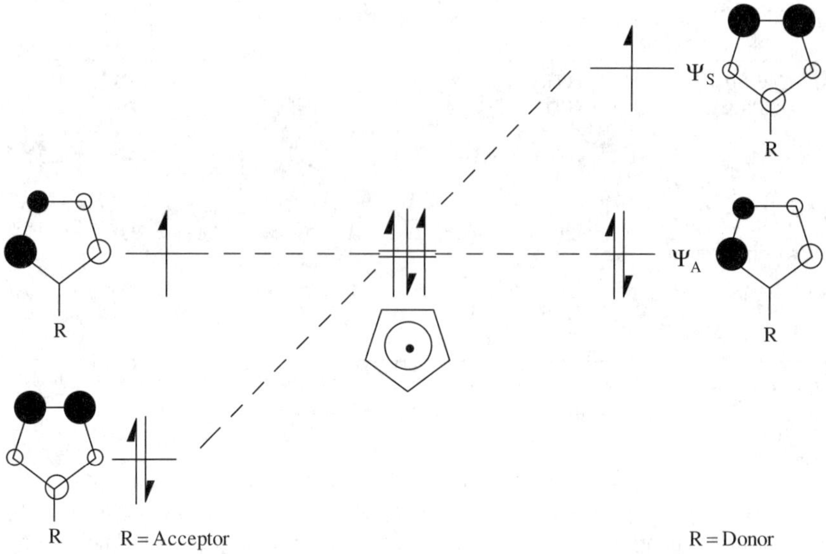

FIGURE 4. Breaking the degeneracy in the highest occupied molecular orbitals in a monosubstituted cyclopentadienyl radical[36,37]

occupy a bonding π-orbital, substituent effects are directly observed. Figure 4 is a simplified diagram illustrating the degeneracy of the highest occupied molecular orbitals in a cyclopentadienyl radical and how the degeneracy is split by electron-accepting or electron-donating substituents R. An electron-donating substituent raises the ψ_S level favouring the electronic configuration $\psi_A^2 \psi_S^1$, while an electron-accepting substituent lowers the ψ_S level favouring the $\psi_S^2 \psi_A^1$ configuration. Analyses of the temperature-dependent ESR spectra revealed a thermal equilibrium between radicals in the symmetric and antisymmetric orbitals.

Two experimental studies with silylated cyclopentadienyl radicals are reported in the literature which show how other substituents at the relevant silicon atom influence the donor/acceptor qualitites of a SiR$_3$ group. In one study the observed spin densities at the point of substituent attachment indicate that the Ph$_2$MeSi, PhMe$_2$Si, and Me$_3$Si fragments are electron-accepting and that the Me$_5$Si$_2$ and Me$_7$Si$_3$ fragments are electron-donating[36]. In the other study it was shown that the Me$_3$Si, Me$_2$HSi, Me$_2$ClSi, Me$_2$FSi, and Cl$_3$Si fragments are increasingly electron-accepting[37]. The electron attraction is due to π-σ^* conjugation, and the electron donation is due to π-σ conjugation.

From the different physical methods described in this chapter, the electronic effect of the most frequently used Me$_3$Si group can be summarized as weakly electron-accepting rather than weakly electron-donating.

B. Synthesis of Compounds with Silicon–Cyclopentadienyl Bonds

This section describes representative strategies for the synthesis of silylated cyclopentadienes of type **1**, and for silylcyclopentadienyl metal complexes of type **4**.

The basic synthetic principles (Schemes 3–9) will be illustrated by some typical examples. More comprehensive treatments can be found in standard textbooks for organic and organometallic synthesis[2–4] and recent review articles[6,38–40].

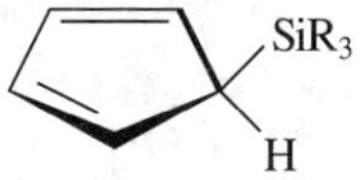

1. Silylated cyclopentadienes

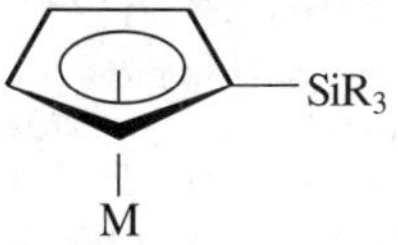

The reaction of metal cyclopentadienyl compounds with organosilicon electrophiles has been the most widely applied method for the synthesis of silylated cyclopentadienes (see Scheme 3) since the beginning of cyclopentadienylsilicon chemistry[1].

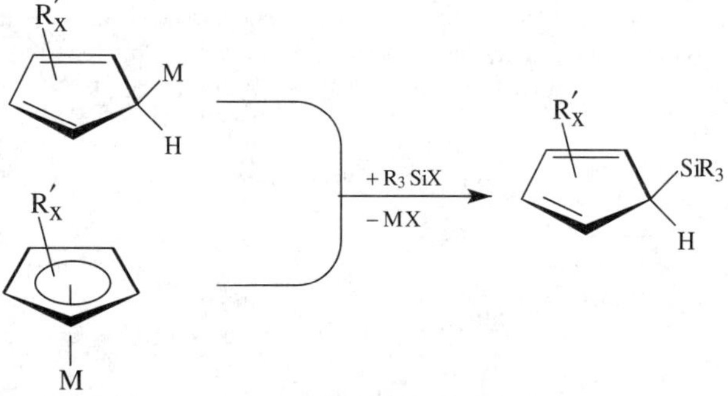

Cyclopentadienyl compounds of lithium, sodium, potassium, magnesium and thallium are most commonly used. The synthesis of these cyclopentadienyl-transfer agents is described in more detail in Section II.B.2. Organosilicon halides and alkoxides are the most convenient substrates. The synthesis of side-chain functionalized cyclopentadienyl silanes is described in equation 5[41,42] as a typical example of this strategy.

$$H_5C_5Li + Ph_2P(CH_2)_nSiMe_2Cl \longrightarrow H_5C_5SiMe_2(CH_2)_nPPh_2 \qquad (5)$$

The reaction sequence shown in Scheme 4 is exceptional for the synthesis of a multisilylated cyclopentadiene involving six-fold silylation in the reaction of hexabromocyclopentadiene with dimethylchlorosilane and magnesium[43]. Grignard-type compounds are presumably the intermediates in this process. More simple is the formation of the Grignard compound pentamethylcyclopentadienylmagnesium bromide from the reaction of bromopentamethylcyclopentadiene with magnesium[44].

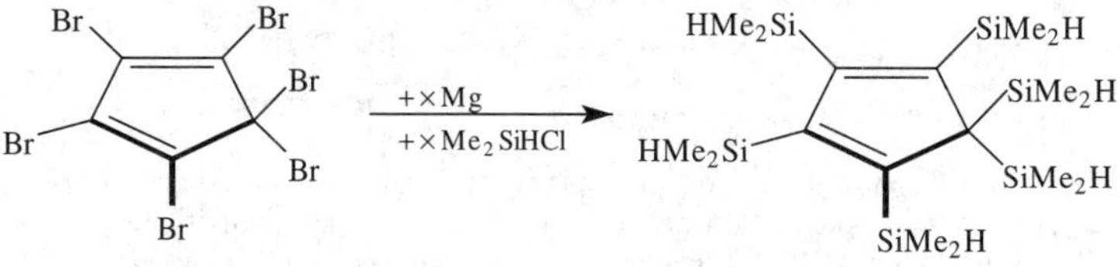

SCHEME 4

2. Silylcyclopentadienyl metal complexes

The most important synthetic methods for the preparation of silylcyclopentadienyl metal compounds are depicted in Schemes 5–9. Metallation of silylated cyclopentadienes, as depicted in Scheme 5, is a very often used and efficient strategy. Hydrogen abstraction and reduction can be easily performed with reactive metals like potassium (equation 6[45]), calcium, strontium or barium (equation 7[46]). An advantage of these reagents is their non-nucleophilic behaviour, which allows for the synthesis of compounds possessing functionalized silyl groups. Deprotonation takes place with butyllithium (equation 8[1]) and with dimethylmagnesium (equation 9[47]). Different kinds of metal amides can also be used, as shown for calcium amide (equation 10[48]) and for tetrakis(dimethylamino)zirconium (equation 11[49]). Under appropriate conditions metallation can be achieved with reagents possessing low nucleophilicity including thallium alkoxides (equation 12[50]) and cyclopentadienylindium (equation 13[51]). It can also be accomplished with transition-metal carbonyl complexes, as demonstrated in the reaction with pentacarbonyliron (equation 14[52]) and with octacarbonyldicobalt (equation 15[52]). In the corresponding reaction with decacarbonyldimanganese, competing silicon–carbon bond fission takes place[52].

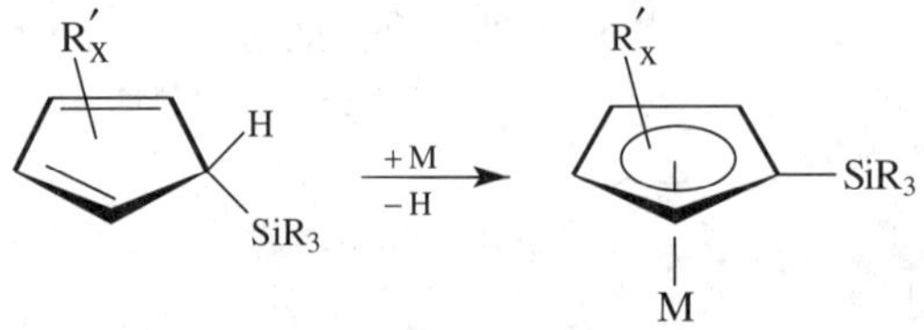

SCHEME 5

$$(Me_3Si)H_5C_5 + K \longrightarrow [(Me_3Si)H_4C_5]K + \tfrac{1}{2}H_2 \quad (6)$$

$$2[2,5\text{-}(Me_3Si)_2H_4C_5] + Ca(Sr,Ba) \longrightarrow [1,3\text{-}(Me_3Si)_2H_3C_5]_2Ca(Sr,Ba) + H_2 \quad (7)$$

$$(Me_3Si)H_5C_5 + LiBu \longrightarrow [(Me_3Si)H_4C_5]Li + BuH \quad (8)$$

$$2,4,5\text{-}(Me_3Si)_3H_3C_5 + MgMe_2 + TMEDA \longrightarrow [1,2,4\text{-}(Me_3Si)_3H_2C_5]$$
$$MgMe \cdot TMEDA + MeH \quad (9)$$

$$2(Me_3Si)H_5C_5 + Ca(NH_2)_2 \longrightarrow [(Me_3Si)H_4C_5]_2Ca + 2NH_3 \quad (10)$$

$$2,4,5\text{-}(Me_3Si)_3H_3C_5 + Zr(NMe_2)_4 \longrightarrow [1,2,4\text{-}(Me_3Si)_3H_2C_5]Zr(NMe_2)_3$$
$$+ Me_2NH \quad (11)$$

$$(Me_3Si)H_5C_5 + TlOEt \longrightarrow [(Me_3Si)H_4C_5]Tl + EtOH \quad (12)$$

$$(Me_3Si)H_5C_5 + InC_5H_5 \longrightarrow [(Me_3Si)H_4C_5]In + C_5H_6 \quad (13)$$

$$2(Me_3Si)H_5C_5 + 2Fe(CO)_5 \longrightarrow \{[Me_3Si)H_4C_5]Fe(CO)_2\}_2 + 6CO + H_2 \quad (14)$$

$$2(Me_3Si)H_5C_5 + Co_2(CO)_8 \longrightarrow 2[(Me_3Si)H_4C_5]Co(CO)_2 + 4CO + H_2 \quad (15)$$

Desilylation or destannylation of the corresponding cyclopentadiene derivatives can be easily performed in addition to deprotonation, as shown schematically in Scheme 6. The possibility for desilylation indicates the pronounced reactivity of the silicon–carbon bond, which is described in more detail in Section III. Nucleophiles as well as

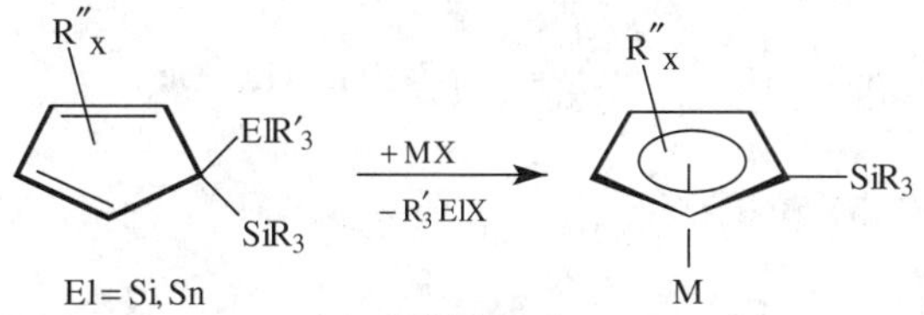

36. Cyclopentadienyl silicon compounds

SCHEME 6

electrophiles can initiate the desilylation (destannylation) process as illustrated in the following examples. A regioselective Si—C bond cleavage takes place in the reaction of hexakis(dimethylsilyl)cyclopentadiene with butyllithium (equation 16[43]) and also in the reaction of multiply trimethylsilylated cyclopentadienes with titanium or zirconium tetrachloride (equations 17[53] and 18[54]). An allylic silicon–carbon bond is cleaved in both cases. The comparatively more reactive trimethylstannylated cyclopentadienes are required in the reaction with less electrophilic transition-metal halides in the process of silylcyclopentadienyl metal complex formation (equations 19[52] and 20[52]).

$$(Me_2SiH)_6C_5 + BuLi \longrightarrow [(Me_2SiH)_5C_5]Li + Me_2Si(H)Bu \quad (16)$$

$$1,3,5\text{-}(Me_3Si)_3H_3C_5 + TiCl_4 \longrightarrow [1,3\text{-}(Me_3Si)_2H_3C_5]TiCl_3 + Me_3SiCl \quad (17)$$

$$2[2,5\text{-}(Me_3Si)_2H_4C_5] + ZrCl_4 \longrightarrow [(Me_3Si)H_4C_5]_2ZrCl_2 + 2Me_3SiCl \quad (18)$$

$$(Me_3Sn)(Me_3Si)H_4C_5 + (OC)_5MnBr \longrightarrow [(Me_3Si)H_4C_5]Mn(CO)_3$$
$$+ Me_3SnBr + 2CO \quad (19)$$

$$2(Me_3Sn)(Me_3Si)H_4C_5 + [Rh(CO)_2Cl]_2 \longrightarrow 2[(Me_3Si)H_4C_5]Rh(CO)_2$$
$$+ 2Me_3SnCl \quad (20)$$

An alternative approach to silylcyclopentadienyl metal complexes employs the deprotonation/silylation process in cyclopentadienyl metal complexes, as shown in Scheme 7. Deprotonation with an organolithium compound and subsequent reaction with an electrophilic organosilicon compound is shown in equations 21[55] and 22[56]. Halogen–lithium exchange is another possibility to produce a lithiated cyclopentadienyl complex which further reacts with an organosilicon halide to form the final product (equation 23[57]). Interestingly, a fivefold silylation can be achieved after a series of halogen–lithium exchange reactions (equation 24[58]). Furthermore, Friedel–Crafts type conditions can be chosen to introduce an electrophilic organosilyl group into a cyclopentadienyl fragment (equation 25[59,60]). Alternatively, trimethylsilyltrifluormethylsulphonate can be used as the electrophilic reagent (equation 26[61]).

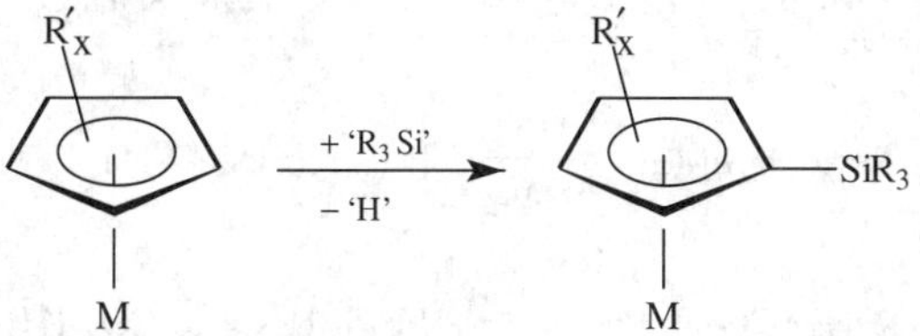

SCHEME 7

$$Fe(C_5H_5)_2 \xrightarrow[\text{2. Me}_3\text{SiCl}]{\text{1. LiBu}} Fe(C_5H_5)C_5H_4SiMe_3 \quad (21)$$

$$Sn(C_5H_5)_2 \xrightarrow[\text{2. 2Me}_3\text{SiCl}]{\text{1. 2LiBu}} Sn(C_5H_4SiMe_3)_2 \quad (22)$$

$$(OC)_3Mn(C_5Cl_4Br) \xrightarrow[\text{2. Me}_3\text{SiCl}]{\text{1. LiBu}} (OC)_3Mn(C_5Cl_4SiMe_3) \quad (23)$$

$$(OC)_3Mn(C_5Br_5) \xrightarrow[\text{2. 5 Me}_2\text{HSiCl}]{\text{1. 5 LiBu}} (OC)_3Mn[C_5(SiMe_2H)_5] \quad (24)$$

$$Fe(C_5H_5)_2 \xrightarrow[\text{AlCl}_3]{\text{Me}_3\text{SiCl}} Fe(C_5H_5)C_5H_4SiMe_3 \quad (25)$$

$$(Me_3P)_2CoC_5H_5 \xrightarrow[\text{2. NH}_4^+\text{ PF}_6^-]{\text{1. Me}_3\text{SiOSO}_2\text{CF}_3} (Me_3P)_2(H)CoC_5H_4SiMe_3^+PF_6^-$$

$$(Me_3P)_2(H)CoC_5H_4SiMe_3^+PF_6^- \xrightarrow[-H_2]{\text{NaH}} (Me_3P)_2CoC_5H_4SiMe_3 \quad (26)$$

Silyl group migration from a transition-metal centre to a carbanionic cyclopentadienyl carbon atom[62–70] is indicated in Scheme 8. In the first step, a cyclopentadienyl C—H unit is deprotonated and the resulting carbanionic carbon centre induces a silyl group migration from the transition metal to the carbon atom. Subsequently, an electrophile attacks the negatively charged transition-metal centre to form a neutral compound. Crossover experiments show the reactions to be intramolecular; the silyl group migration occurs with retention of configuration at silicon. A concerted mechanism for the migration involving front-side nucleophilic attack at silicon has been suggested[65]. Examples for this synthetic strategy from iron[62–67], ruthenium[66,67], rhenium[68,69] and molybdenum[70] chemistry are collected in equations 27–31. In the reaction shown in equation 30, a cyclization with LiCl elimination gives the final disilametallacycle[69].

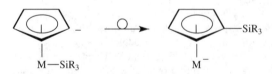

SCHEME 8

$$\begin{array}{l}H_5C_5Fe(CO)_2SiMe_2R \\ R = Me, Pr, Ph, NMe_2; R' = Me\end{array} \xrightarrow[\text{2. R'X}]{\text{1. LiBu or LiNMe}_2} (RMe_2Si)H_4C_5Fe(CO)_2R' \quad (27)$$

$$\begin{array}{l}H_5C_5M(CO)_2(SiMe_2)_nSiMe_3 \\ M = Fe, Ru\end{array} \xrightarrow[\text{2. MeI}]{\text{1. LiNMe}_2} [Me_3Si(SiMe_2)_n]H_4C_5M(CO)_2Me \quad (28)$$

$$\begin{array}{l}H_5C_5Re(H)(CO)_2SiPh_3 \\ N_p = \text{neopentyl}\end{array} \xrightarrow[\text{2. HX}]{\text{1. LiNp}} (Ph_3Si)H_4C_5Re(CO)_2(H)_2 \quad (29)$$

$$H_5C_5Re(NO)(PPh)_3SiMe_2SiMe_2Cl \xrightarrow{\text{LiBu}} (Me_2SiSiMe_2)H_4C_5Re(NO)(PPh)_3 \quad (30)$$

$$H_5C_5Mo(H)SiMe_2Ph \xrightarrow[\text{2. HX}]{\text{1. LiBu}} (PhMe_2Si)H_4C_5H_5C_5MoH_2 \quad (31)$$

A similar process to that shown in Scheme 8 also involves formation of a carbanionic cyclopentadienyl carbon centre. But in this case, silicon–carbon bond formation stems from the migration of a silylmethyl group, which proceeds with rearrangement, as shown in Scheme 9. These surprising rearrangement processes are observed with iron and with tungsten compounds[71,72], as shown in equations 32 and 33.

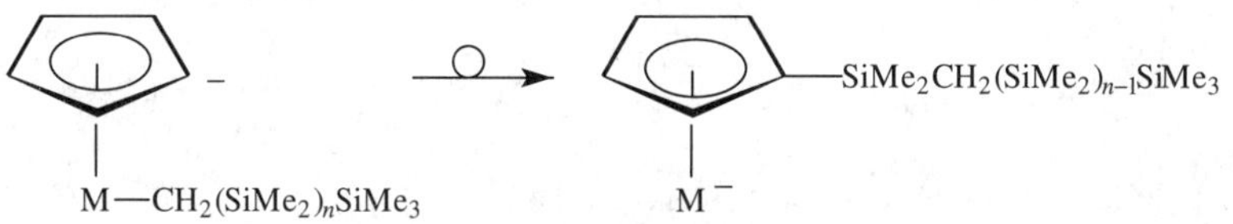

SCHEME 9

$$H_5C_5Fe(CO)_2CH_2(SiMe_2)_n\,SiMe_3 \xrightarrow[\text{2. MeI}]{\text{1. LiNMe}_2} \begin{array}{c}[Me_3Si(SiMe_2)_{n-1}CH_2SiMe_2]\\ H_4C_5Fe(CO)_2Me\end{array} \quad (32)$$

$$H_5C_5W(CO)_3CH_2(SiMe_2)_n SiMe_3 \xrightarrow[\text{2. MeI}]{\text{1. LiNMe}_2} \begin{array}{c}[Me_3Si(SiMe_2)_{n-1}CH_2SiMe_2]\\ H_4C_5W(CO)_3Me\end{array} \quad (33)$$

Presumably, a new kind of rearrangement process takes place in the reaction of dicyclopentadienyltungsten dichloride with tris(trimethylsilyl)silyllithium (equation 34[73]). Silyl complexes formed as intermediates spontaneously rearrange with formation of silylcyclopentadienyl complexes, but the mechanism of this process is not yet clear.

$$(H_5C_5)_2WCl_2 + 2LiSi(SiMe_3)_3 \longrightarrow \{[Me_3Si)_3Si]H_4C_5\}_2WH_2 + 2LiCl \quad (34)$$

C. Methods of Silicon–Cyclopentadienyl Bond Cleavage

The silicon–carbon bond in cyclopentadienyl silicon compounds of type **1** (Scheme 1) is comparatively weak, and is susceptible to sigmatropic rearrangements. Si–C bond fission might be a disadvantage in certain synthetic strategies or in some practical applications; however, it has been successfully used in the design of new silicon compounds as illustrated in the following section. Scheme 10 shows a collection of different kinds of Si–C(Cp) cleavage reactions involving cyclopentadienyltrimethylsilane.

Halide-promoted cleavage of C_{sp^3}–Si bonds is most extensively demonstrated with the fluoride ion, due to the high dissociation energy of the silicon–fluorine bond[74]. A novel chloride ion-induced desilylation process with trimethylsilylcyclopentadiene, bis(trimethylsilyl)cyclopentadiene or tris(trimethylsilyl)cyclopentadiene in alcoholic solvents has been described only recently[75]. Another nucleophilic substitution process with Si–C bond cleavage takes place in the reaction of trimethylsilylcyclopentadiene with secondary amines[76]. Electrophilic attack at the π-system is the basis for the cleavage reactions observed with several element halides, which clearly demonstrate that trimethylsilylcyclopentadiene is a mild and effective cyclopentadienyl transfer agent[52,77–81]. The reaction

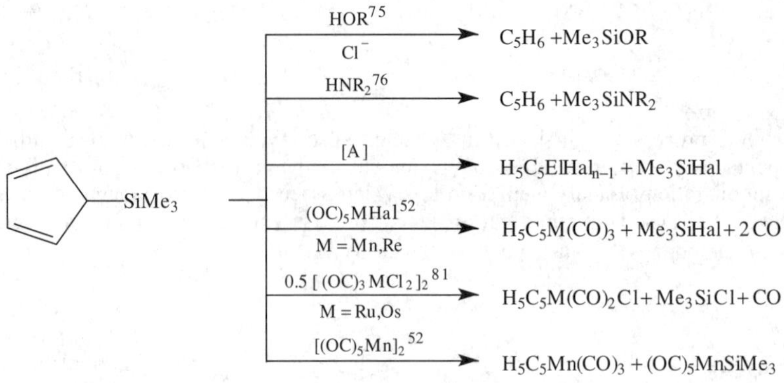

[Cp-SiMe3] reacts with various reagents:

- HOR [75] → $C_5H_6 + Me_3SiOR$
- Cl^-
- HNR_2 [76] → $C_5H_6 + Me_3SiNR_2$
- [A] → $H_5C_5ElHal_{n-1} + Me_3SiHal$
- $(OC)_5MHal$ [52], $M = Mn, Re$ → $H_5C_5M(CO)_3 + Me_3SiHal + 2CO$
- $0.5\,[(OC)_3MCl_2]_2$ [81], $M = Ru, Os$ → $H_5C_5M(CO)_2Cl + Me_3SiCl + CO$
- $[(OC)_5Mn]_2$ [52] → $H_5C_5Mn(CO)_3 + (OC)_5MnSiMe_3$

[A] = $BHal_3$ [77], $AsHal_3$ [78], R_2AsHal [78], $SbHal_3$ [79], $TiHal_3$ [80], $TiCl_4$ [80], $ZrCl_4$ [80], $HfCl_4$ [80], $NbHal_5$ [80], $TaHal_5$ [80], $H_5C_5ZrCl_3$ [55,80], $H_5C_5HfCl_3$ [80]

SCHEME 10. Si–C(Cp) bond cleavage in cyclopentadienyltrimethylsilane

of trimethylsilylcyclopentadiene with dimanganese decacarbonyl takes two paths[52]: One in which the silicon–ring bond remains intact, and the other in which silicon–ring fission takes place, as shown in Scheme 10.

Preparatively useful cleavage reactions with other cyclopentadienyl silicon substrates are shown in equations 35–40. A regioselective introduction of $BHal_2$ groups is possible according to the process in equation 35[77]. A novel chloride-catalysed cleavage reaction is the basis for the sequence shown in equation 36[82]. The transfer of BR_2^-, $ClMe_2Si$- and Me_3Si-substituted cyclopentadienyl fragments to Group IV transition metal centres is illustrated in equations 37–40[54,83–86].

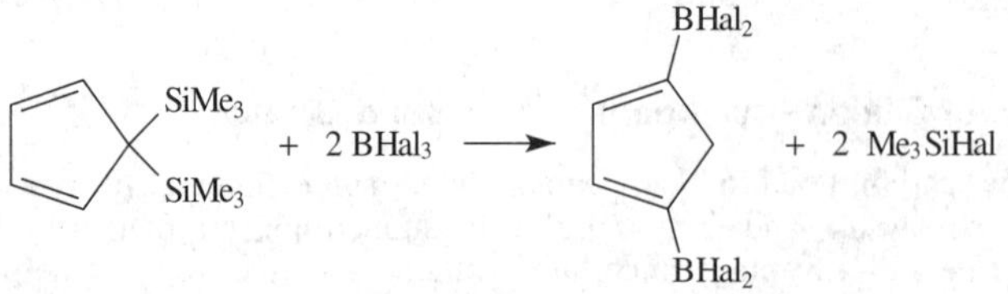

$$Cp(SiMe_3)_2 + 2\,BHal_3 \longrightarrow Cp(BHal_2)_2 + 2\,Me_3SiHal \quad (35)^{77}$$

$$Cp(SiMe_3)(SiMe_2Cl) + Me_2SiCl_2 \xrightarrow{Cl^-} Cp(SiMe_2Cl)_2 + Me_3SiCl \quad (36)^{82}$$

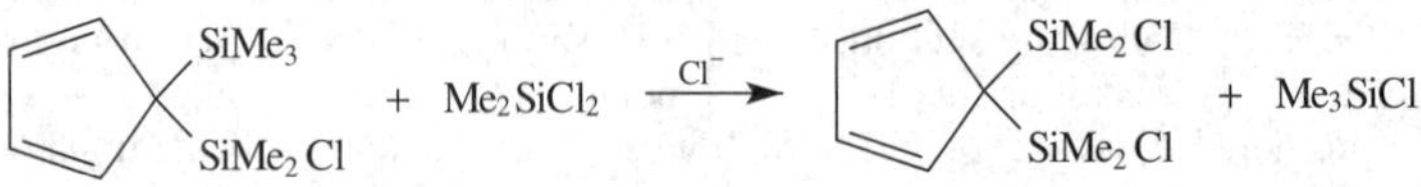

$$Cp(SiMe_3)(BR_2) + TiCl_4 \longrightarrow [(R_2B)H_4C_5]TiCl_3 + Me_3SiCl \quad (37)^{83}$$

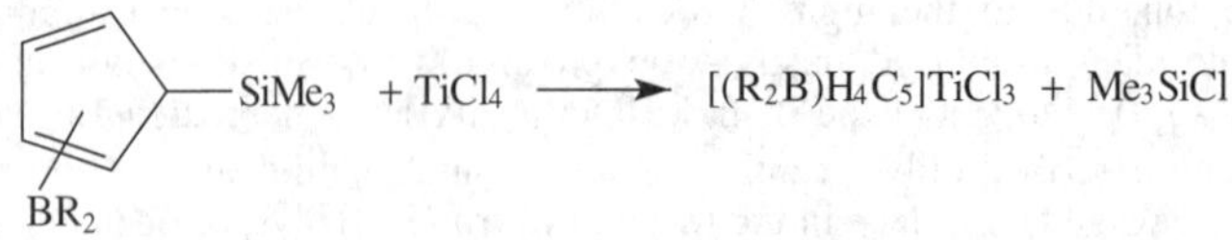

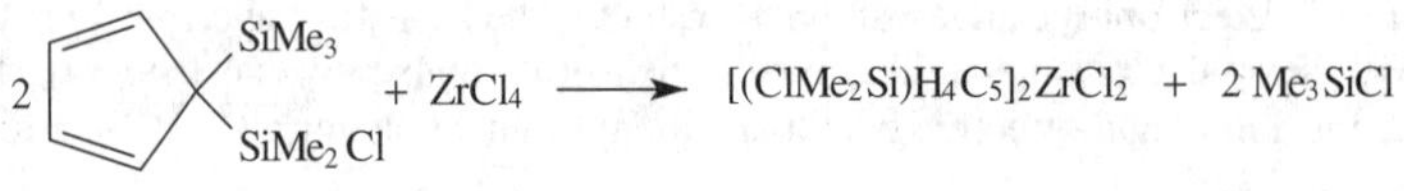

$$2\,Cp(SiMe_3)(SiMe_2Cl) + ZrCl_4 \longrightarrow [(ClMe_2Si)H_4C_5]_2ZrCl_2 + 2\,Me_3SiCl \quad (38)^{84}$$

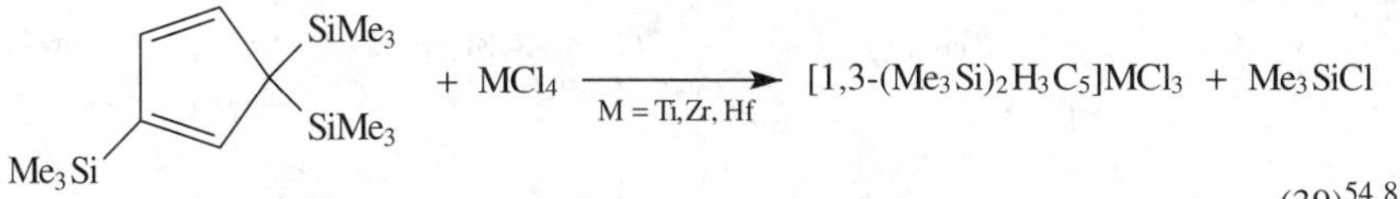

36. Cyclopentadienyl silicon compounds

[Structure: 1,3-bis(trimethylsilyl)cyclopentadiene] + MCl_4 $\xrightarrow{M = Ti, Zr, Hf}$ [1,3-$(Me_3Si)_2H_3C_5]MCl_3$ + Me_3SiCl

$(39)^{54,85}$

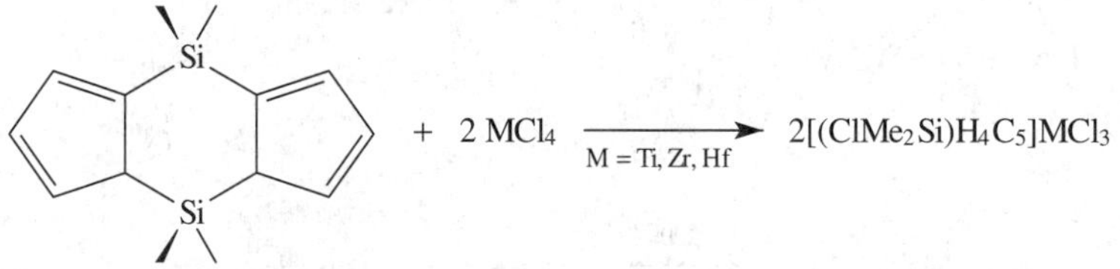

[Structure: bis(cyclopentadienyl)dimethyldisilane bridged compound] + 2 MCl_4 $\xrightarrow{M = Ti, Zr, Hf}$ 2[$(ClMe_2Si)H_4C_5]MCl_3$

$(40)^{86}$

Other examples of Si—C(Cp) bond cleavage reactions deal with pentamethylcyclopentadienyl (Cp*) substituted silicon compounds. Scheme 11 shows a collection of different kinds of reactions with (pentamethylcyclopentadienyl)trimethylsilane as the substrate. Nucleophilic as well as electrophilic attack is the basis for Si—C(Cp) bond fission[97]. Reactions with various element halides involved in pentamethylcyclopentadienyl transfer are very useful in synthetic chemistry[88–91] (see also Sections II.B and II.E).

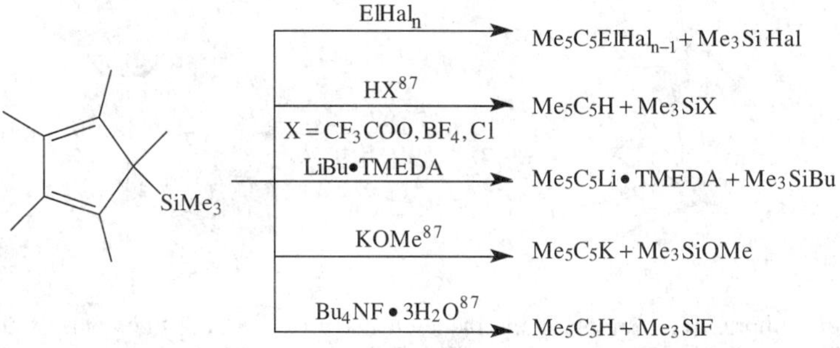

$ElHal_n$ → $Me_5C_5ElHal_{n-1}$ + Me_3SiHal

HX^{87}, X = CF_3COO, BF_4, Cl → Me_5C_5H + Me_3SiX

$LiBu•TMEDA$ → $Me_5C_5Li•TMEDA$ + Me_3SiBu

(Me$_5$C$_5$)SiMe$_3$

$KOMe^{87}$ → Me_5C_5K + Me_3SiOMe

$Bu_4NF•3H_2O^{87}$ → Me_5C_5H + Me_3SiF

$ElHal_n$ = $BHal_3^{88}$, $AlCl_3^{89}$, $TiHal_4^{90}$, $ZrCl_4^{90}$, $HfCl_4^{90}$, $NbCl_5^{91}$, $TaCl_5^{91}$

SCHEME 11. Si—C(Cp*) bond cleavage in (pentamethylcyclopentadienyl)trimethylsilane.

It is possible to selectively cleave a Si—C(Cp*)bond even in the presence of silicon–transition metal bonds as shown in reactions with nucleophiles, electrophiles and chlorinated hydrocarbons (equations 41 and 42)[92,93].

The thermal decomposition of Cp* substituted silicon compounds with the elimination of pentamethylcyclopentadiene (Cp*H) turns out to be synthetically very useful. As shown in equations 43–45, extreme conditions are necessary to selectively cleave the respective Si—C(Cp*) bond[94–96]. In this context, it is important to mention the use of cyclopentadienyl silicon compounds in the metalorganic chemical vapour deposition (MOCVD) process or in related techniques[95,96].

$(Me_5C_5)(Cl)_2Si-[M]$

$\xrightarrow{CDCl_3}$ $Cl_3Si-[M]$

$\xrightarrow{Py•HCl}$ $Cl_3Si-[M]$

$\xrightarrow{KOMe}$ $(MeO)_3Si-[M]$

$(41)^{92}$

$[M] = H_5C_5Ru(PPh_3)_2$

$$Me_5C_5(H)(R)Si-Fe(H_5C_5)(CO)_2 \xrightarrow{CCl_4} Cl(H)(R)Si-Fe(H_5C_5)(CO)_2 \quad (42)^{93}$$
$$R = Cl, Me$$

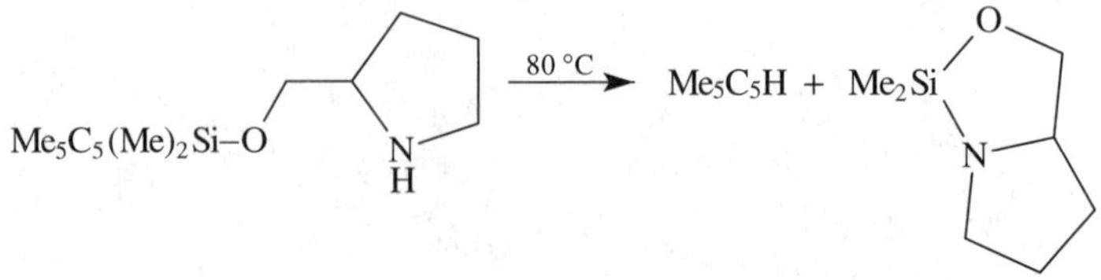

$$Me_5C_5(Me)_2Si-O\text{-pyrrolidine} \xrightarrow{80\,^\circ C} Me_5C_5H + Me_2Si(\text{oxazolidine}) \quad (43)^{94}$$

$$Me_5C_5SiH_3 \xrightarrow[\text{Plasma-MOCVD}]{200\,^\circ C} Si + Me_5C_5H + H_2 \quad (44)^{95}$$

$$[Me_5C_5(PH_2)SiPH]_2 \xrightarrow{700\,^\circ C} 2SiP_2 + 2Me_5C_5H + 2H_2 \quad (45)^{96}$$

There are some reports in the literature concerning the cleavage of bonds between silicon and an sp^2-hybridized carbon in silylcyclopentadienyl compounds of type **4** (Section II.B). For example, equation 46[97] shows hydrolytic fission in the reaction of a tris(trimethylsilyl) substituted ferrocene derivative with aqueous tetrabutylammonium fluoride.

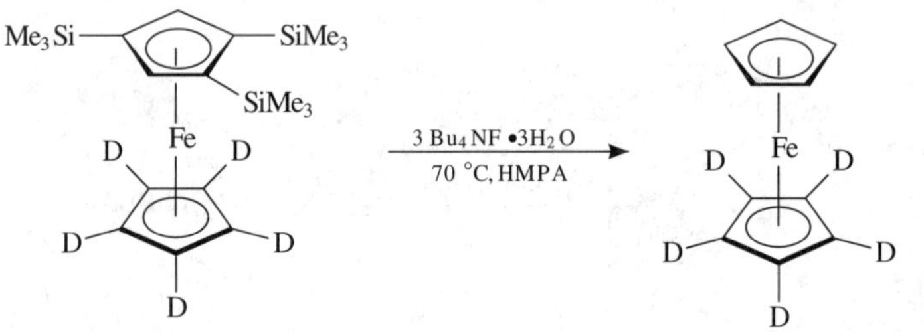

$$\text{tris(trimethylsilyl)ferrocene-}d_5 \xrightarrow[70\,^\circ C, HMPA]{3\,Bu_4NF \cdot 3H_2O} \text{ferrocene-}d_5 \quad (46)$$

Another impressive example is the facile cleavage of Si—C(Cp) bonds in strained [1]-ferrocenophanes with bridging silicon. This behaviour is the basis for the synthesis of high-molecular-weight poly(ferrocenylsilanes) (equation 47) and will be described in more detail in Section II.E.1.

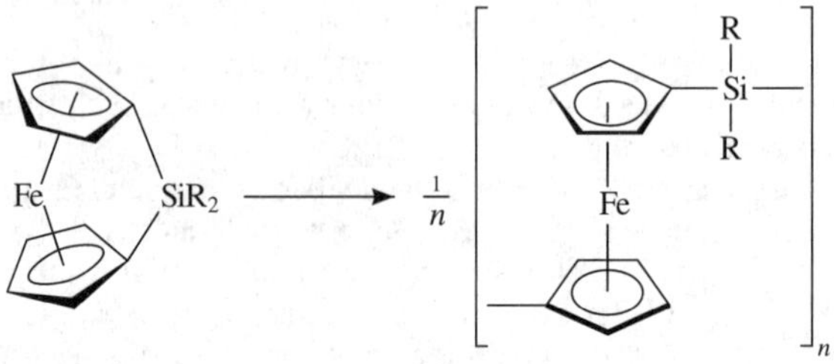

$$[\text{Fe}(C_5H_4)_2SiR_2] \longrightarrow \frac{1}{n}[\text{Fe}(C_5H_4)_2SiR_2]_n \quad (47)$$

D. Multiple Silylation of Cyclopentadiene and Cyclopentadienyl Complexes

1. Effects of multisilylation

Multisilylation of cyclopentadienyl compounds can lead to drastic changes in structure and properties due to steric interaction. For example, 1,1′,2,2′,4,4′-hexakis(trimethylsilyl)ferrocene **(19)** exhibits hindered rotation of the cyclopentadienyl rings, comparatively long Fe−C (ring) distances and an unexpected redox behaviour[31,98]. Drastic changes in solubility are also observed due to lipophilic wrapping. For example, the base-free 1,2,4-tris(trimethylsilyl)cyclopentadienyllithium (**26**) is soluble even in hydrocarbon solvents[25], whilst the parent compound, cyclopentadienyllithium, is essentially insoluble in most solvents. The base adduct (**26**)·THF is a monomer in solution and in the solid state[25], whereas cyclopentadienyllithium is an amorphous, polymeric species.

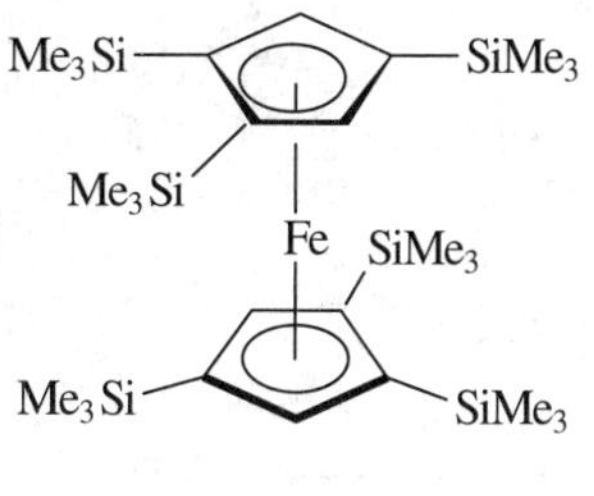

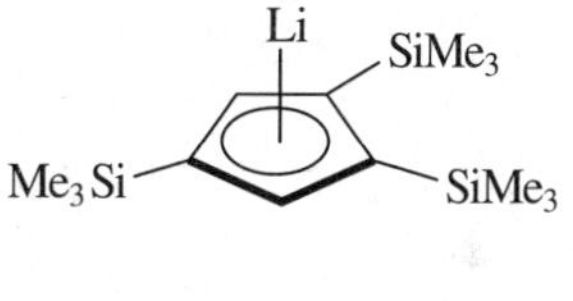

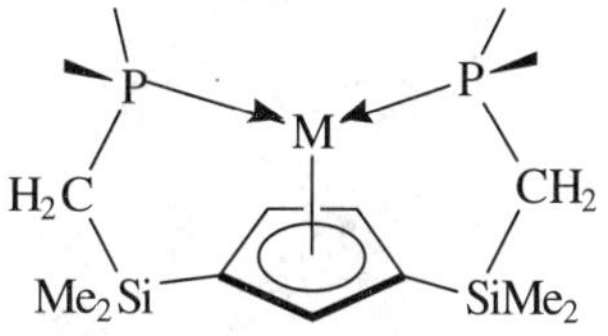

The introduction of functionalized silyl substituents into a cyclopentadienyl fragment allows for regioselective incorporation of side-chain functionalities and is therefore of high synthetic value. For example, two (di-isopropylphosphino)methyldimethylsilyl groups can be introduced in 1,3 position of a cyclopentadienyl system to create the potentially tridentate ligand, as shown in **27**[99]. Here, the chelate effect provides additional control of a reactive metal centre and generates a more stereochemically defined coordination sphere[100]. Finally, the five-fold dimethylsilylated cyclopentadienyl complexes of type **28**[43] (M = Li·THF) and **29**[58] [M = Mn(CO)$_3$] offer promising possibilities for functionalizations by hydrosilylation reactions.

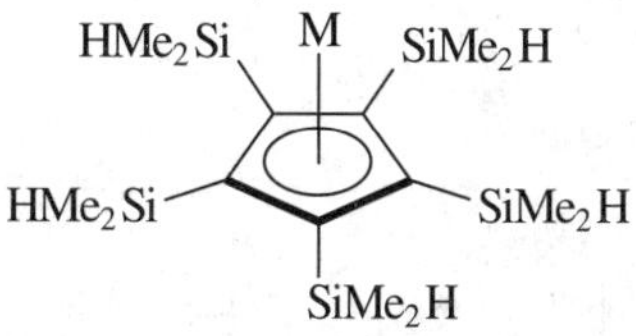

2. Synthetic methods

Two strategies can be followed for the synthesis of polysilylated cyclopentadienyl compounds. One employs sigmatropic rearrangements of silylated cyclopentadienes, and the other is based on substitution reactions in cyclopentadiene and in cyclopentadienyl metal complexes. As already pointed out in Section I, the fluxionality of silyl substituted cyclopentadienes involves sila- and prototropic rearrangement processes, while the former have comparatively lower activation energies. Silatropic shifts generate isomers regiospecifically with a hydrogen atom in allylic position by a formal 1,3 silicon shift. These isomers are sometimes present in negligible amounts in equilibrium, but deprotonation with stoichiometric amounts of organolithiums allow for quantitative isomerization. The underlying reaction sequences are shown in Scheme 12. Is is worth mentioning that further silatropic shifts are not observed in the tetrasilyl-substituted cyclopentadiene. It is possible to synthesize cyclopentadienyl metal complexes with silyl groups of different kinds in 1,3- and in 1,2,4-position of the cyclopentadienyl ring.

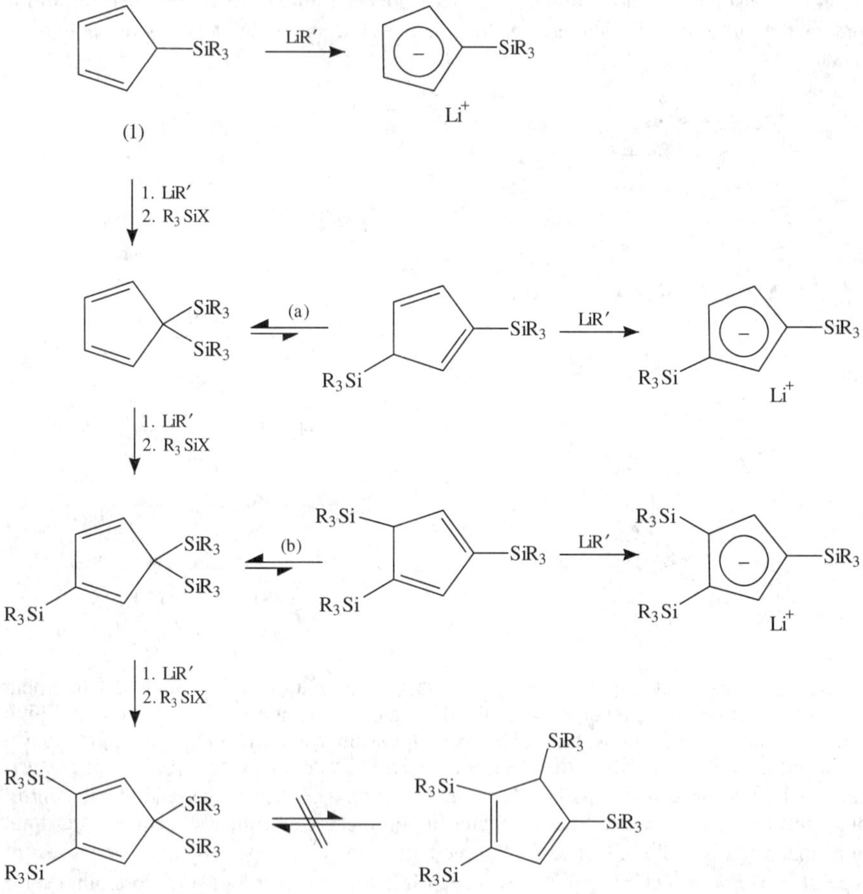

SiR$_3$: SiMe$_3$[22], Si$_2$Me$_5$[101]

SCHEME 12. Silatropic shifts and deprotonation/silylation reactions in silylated cyclopentadienes

As already illustrated in equations 3 and 4, steric interactions prevent the generally preferred 1,3-silicon shift in favour of a 1,2-shift allowing introduction of two silyl groups in neighbouring positions (equation 4). It should also be noted that the analogous 1,2-shift cannot be excluded in step (b) for the rearrangement processes shown in Scheme 12.

Other methods for the synthesis of polysilylated cyclopentadienyl compounds involve two different substitution reactions. Firstly, repeated halogen–metal exchange and subsequent reaction with dimethylchlorosilane allows the persilylation of a cyclopentadienyl metal complex and also of cyclopentadiene. (Pentabromocyclopentadienyl)tricarbonylmanganese can be converted to the corresponding pentasilylated complex by a series of halogen–lithium exchange reactions and silylations using dimethylchlorosilane[58] (equation 48). Similarly, a six-fold silylation takes place in the reaction of hexabromocyclopentadiene with dimethylchlorosilane and magnesium, as already pointed out in Section II, Scheme 4. Treatment of the reaction product with butyllithium in THF leads to the formation of the THF adduct of pentakis(dimethylsilyl)cyclopentadienyllithium (**28**)[43] (equation 49). Secondly, multifold silylation is possible by repeated deprotonation and silylation of cyclopentadienyl metal complexes. For example, the stepwise lithiation and trimethylsilylation of stannocene is shown in equation 50[56].

$$(OC)_3Mn(C_5Br_5) \xrightarrow[5\ Me_2SiHCl]{5\ LiBu} (OC)_3Mn[C_5(SiMe_2H)_5] \quad (48)$$

$$C_5(SiMe_2H)_6 \xrightarrow[THF]{LiBu} Li[C_5(SiMe_2H)_5]\cdot THF \quad (49)$$

$$Sn(C_5H_5)_2 \xrightarrow[2.\ 2\ Me_3SiCl]{1.\ 2\ LiBu} Sn[C_5H_4(SiMe_3)]_2 \xrightarrow[2.\ 2\ Me_3SiCl]{1.\ 2\ LiBu} Sn[C_5H_3(SiMe_3)_2\text{-}1,3]_2 \quad (50)$$

E. Silicon-bridged Cyclopentadienyl Complexes

In recent years, bridging of two or more cyclopentadienyl fragments by silicon-containing units has become a very important synthetic tool in the chemistry of s-, p-, d- and f-block elements. In mononuclear compounds, bridging limits the relative orientation of two cyclopentadienyl ligands enforcing conformations around the metal centre, which might lead to changes in structure and reactivity. In dinuclear compounds, bridging inherently retains the metal centres in close proximity, which might induce pronounced structural changes in comparison to the non-bridged species and might allow cooperative effects by direct metal–metal interaction or by propagation through the bridging silicon unit. Similar effects are expected in bridged polynuclear compounds. In view of the vast number of cyclopentadienyl compounds in organometallic chemistry and their huge synthetic potential, many applications for the strategy of bridging with silicon units can be envisaged.

The choice of silicon as the bridging atom results primarily from the ease with which these compounds can be prepared (*vide infra*). Nevertheless, certain silicon-containing bridging units have been chosen for electronic reasons; for example, it is well known that a disilanyl group qualitatively behaves like an ethylene unit.

Cyclopentadienyl systems bridged by a single silicon-containing unit are described in Section II.E.1. Cyclopentadienyl units which are held together by a double bridge are described in Section II.E.2. Section II.E.3. describes cooperative effects exerted by dimethylsilyl groups as bridging units, with special emphasis on polyferrocenyl compounds.

1. Single-bridged cyclopentadienyl systems

The most important single-bridged cyclopentadienyl systems described in the literature (**30–37**) are collected in Scheme 13. The silicon-containing bridging units range from the dimethylsilyl (type **30**) to a polysiloxane group (type **35**). Tetramethyldisilanyl groups can connect three (type **36**) or even four (type **37**) cyclopentadienyl fragments.

The strategies used to synthesize metal complexes with the cyclopentadienyl ligands portrayed in Scheme 13 are briefly described below. The most widely used procedure begins with the corresponding cyclopentadiene derivative, which can be transformed into a cyclopentadienyl metal complex by a variety of metallation methods as described in

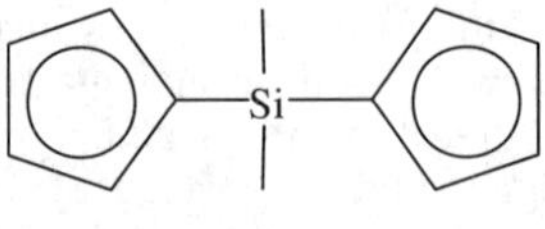

(30)[102]

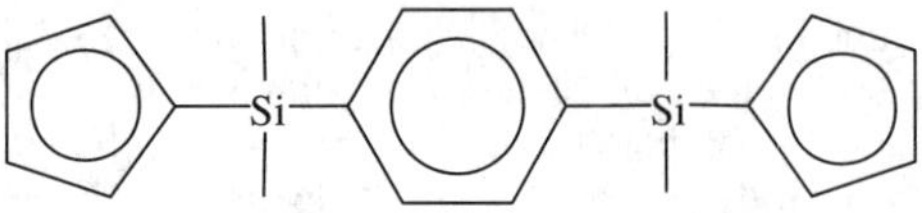

(31)[103]

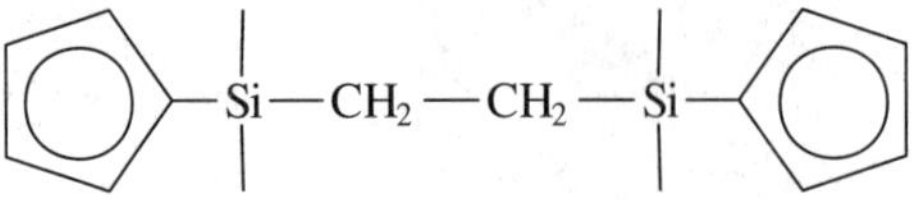

(32)[104]

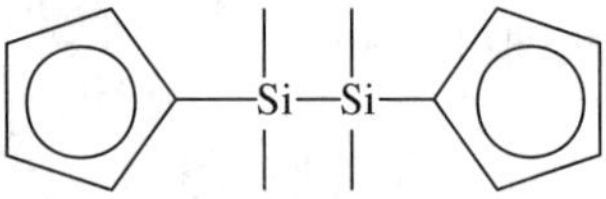

(33)[105]

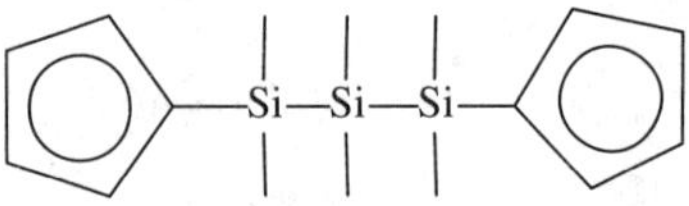

(34)[105]

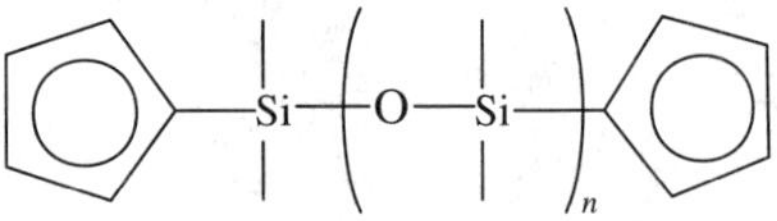

(35)[84,106–108]

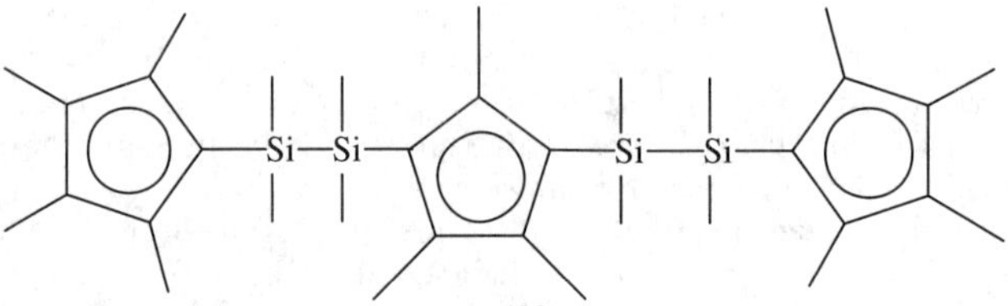

(36)[109]

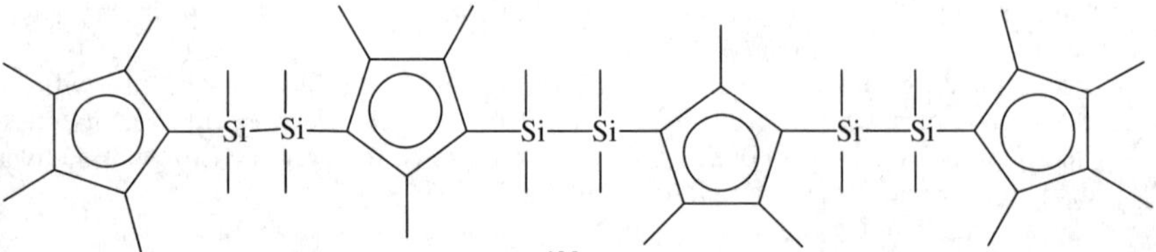

(37)[109]

SCHEME 13. Single-bridged cyclopentadienyl systems

Section II. A regioselective introduction of disilanyl bridges is necessary in the process to synthesize complexes with ligands of types **36** or **37**[109], and this has been made possible by methylation of the cyclopentadienyl fragments. Due to prototropic rearrangements, the protonated derivatives of **36** and **37** exist as a complicated mixture of isomers. The synthesis of the thallium derivative of **37**[109] is shown in equation 51.

$$Me_4HC_5(Si_2Me_4)Me_3HC_5(Si_2Me_4)Me_3HC_5(Si_2Me_4)C_5HMe_4$$
$$+4TlOEt \longrightarrow Tl_4(\mathbf{37}) + 4EtOH \qquad (51)$$

Another synthetic procedure is based on the reaction of dilithioferrocene with difunctional electrophilic silicon substrates, as shown in equation 52. For example, Me_2Si[110],

$Me_4Si_2^{105}$, and $Me_6Si_3^{105}$ groups have been introduced by this classical procedure.

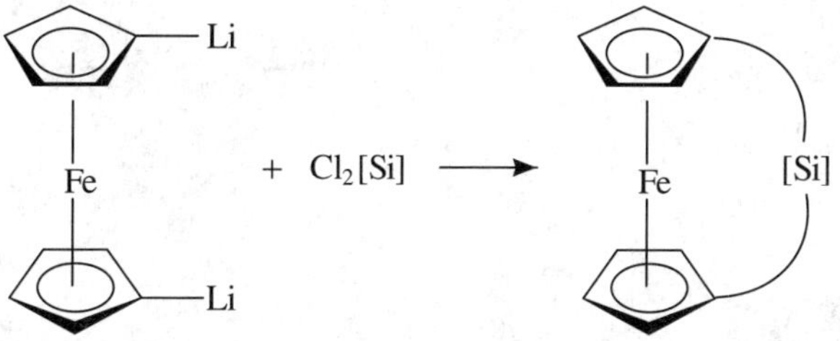

$$\text{Fe(Cp-Li)}_2 + Cl_2[Si] \longrightarrow \text{Fe(Cp)}_2[Si] \quad (52)$$

Bridges containing SiOSi units have been introduced by condensation reactions of dicyclopentadienyl complexes containing functionalized silyl groups (equation 53)[84,107] or by oxygen insertion into a preformed Si—Si bridge (equation 54)[108].

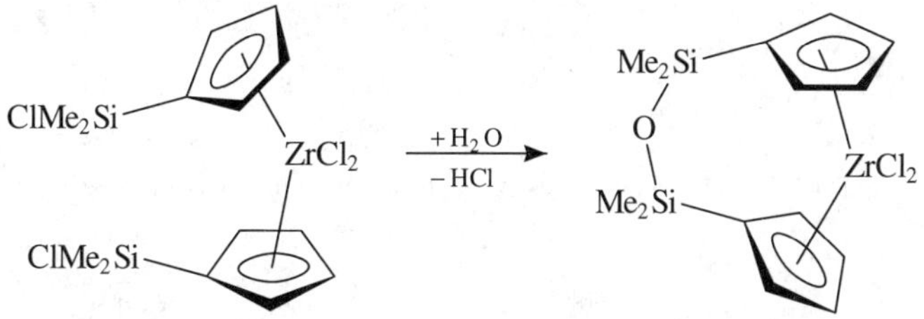

$$\text{(ClMe}_2\text{Si-Cp)}_2\text{ZrCl}_2 \xrightarrow[-\text{HCl}]{+H_2O} \text{(Me}_2\text{Si-O-SiMe}_2\text{)(Cp)}_2\text{ZrCl}_2 \quad (53)$$

$$\text{Fe(Cp-SiMe}_2\text{)}_2 \xrightarrow{\frac{1}{2}O_2} \text{Fe(Cp-SiMe}_2\text{-O-SiMe}_2\text{-Cp)} \quad (54)$$

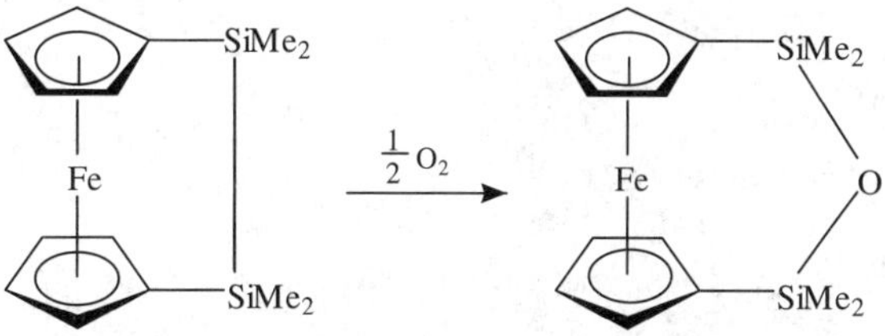

In the following section we will concentrate on the most important classes of compounds which have been studied with the dimethylsilyl-bis(cyclopentadienyl) ligand system (**30**). In Scheme 14, five different types of complexes are portrayed, showing two principally different coordination modes of type **30**: connection of two metal centres in binuclear units (type **38**, **39** and **40**) and chelation of one metal centre in mononuclear complexes (type **41** and **42**).

Complexes of type **38** have been realized with several metal centres. Representative examples are collected in Figure 5. Most of these complexes have been synthesized with the aim of studying the interaction between the two metal centres since the silicon retains these centres in close proximity. Limited rotational freedom in this class of compounds leads to preferred conformations in the solid and presumably also in solution. As portrayed in Figure 6, three conformations are possible: *exo/exo*, *exo/endo* and *endo/endo*.

X-ray crystal structure studies show only the *exo/exo* [M = Fe(CO)$_2$I[114]] and the *exo/endo* [M = Mn(CO)$_3$[113], TiCl$_3$[111]] conformations with some deviations from ideal geometry. These orientations allow the metals to achieve maximum separation. The activation energies for conformational changes depend on the steric bulk of the [M] unit. Reactions which result in the formation of metal–metal links by either M—M bonds or M—X—M bridges must involve transformation into the *endo/endo* conformation[111,117,118].

(38)

(39)

(40)

(41)

(42)

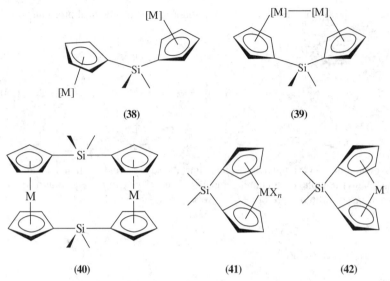

SCHEME 14. Structure types of complexes with the ligand system **30**

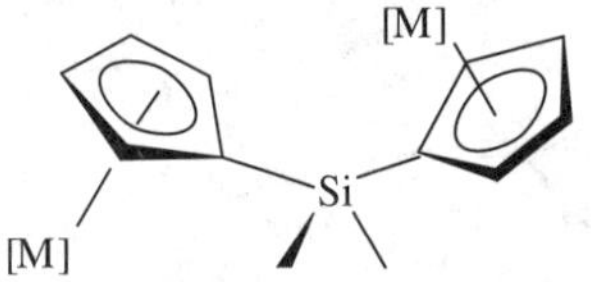

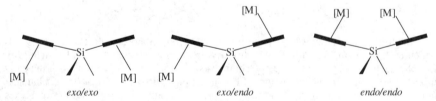

[M] = Li[102], Tl[111], $Co(CO)_2$[112], $Mn(CO)_3$[113], $Fe(CO)_2X$[114], $Mo(CO)_3X$[115], $TiCl_3$[111,116], $W(CO)_3X$[117], $ZrCl_3$[111]

FIGURE 5. Metal complexes of type **38**

exo/exo *exo/endo* *endo/endo*

FIGURE 6. Conformations in complexes of type **38**

In general, the induced geometrical restrictions in compounds of type **38** modify the chemical behaviour to some extent in relation to the corresponding unbridged species.

Representative examples of dinuclear Me_2Si-bridged metal complexes belonging to type **39** in Scheme 14 are listed in Figure 7. The additional metal–metal bridge enforces the *endo*/*endo* orientation of the metal fragments and induces a pronounced stereochemical rigidity. As a consequence, noticeable changes in structure and reactivity compared with the unbridged species are observed[112,119,121]. Furthermore, cooperative interactions of the two neighbouring metal centres have to be considered[120–123], but it is still an open question whether the observed electronic effects are mediated by the bridging silicon atom.

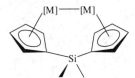

$L = $ 2-electron donor

$[M]-[M] = Fe_2(CO)_2L_2{}^{112,119,123}, Cr_2(CO)_4L_2{}^{120}, Mo_2(CO)_4L_2{}^{121}, W_2(CO)_4L_2{}^{121,122}$

FIGURE 7. Metal complexes of type **39**

Silicon-bridged [1,1]metallocenophanes of type **40** in Scheme 14 represent another class of dinuclear compounds with the ligand system (**30**). The synthesis of the 1,1,12,12-tetramethyl [1,1] silaferrocenophane was reported by two groups only recently. In one procedure, the dilithium salt of dicyclopentadienyldimethylsilane was reacted with ferrous chloride to give a mixture of the metallocenophane with poly(ferrocenyldimethylsilane) (equation 55)[124]. In the other procedure, this compound was prepared in a multistep synthesis starting from dilithio ferrocene, as shown in equation 56[125]

$$Me_2Si(C_5H_4Li)_2 + FeCl_2 \longrightarrow [Me_2Si(H_4C_5)_2Fe]_2 + [Me_2Si(H_4C_5)_2Fe]_n \quad (55)$$

$$(LiH_4C_5)_2Fe \xrightarrow{(a)} \{[(Et_2N)SiMe_2]H_4C_5\}_2Fe \xrightarrow{(b)} [(ClSiMe_2)H_4C_5]_2Fe$$

$$\xrightarrow{(c)} \{[(H_5C_5)SiMe_2]H_4C_5\}_2Fe \xrightarrow{(d)} [Me_2Si(H_4C_5)]_2Fe]_2 \quad (56)$$

(a) SiMe$_2$Cl(NEt$_2$); (b) MeCOCl; (c) LiC$_5$H$_5$; (d)1. BuLi; 2. FeCl$_2$

A single-crystal X-ray diffraction study revealed that the two ferrocene units are in *anti* conformation. Cyclic voltammetric results are in accord with a strong interaction between the iron centres. Electronic interactions through bonds are mediated by the bridging dimethylsilyl groups and this is discussed in more detail in Section II.E.3.

Bent metallocenes with bridging silicon of type **41** in Scheme 14 are known for many early d-block elements and also for many f-block elements (mainly actinides). Therefore, this class of compounds has been successfully utilized as catalysts or precatalysts in different types of reactions. The most prominent application is in the field of stereo- and enantio-selective polymerization of alkenes, which requires chiral catalysts. Chirality is introduced by a chiral auxiliary as cyclopentadienyl ring substituent, as bridging silicon substituent or by a chiral metal centre. The bridging silicon unit together with further substituents at the cyclopentadienyl rings introduces conformational constraints with respect to the coordination gap aperture angle and to the lateral coordination gap extension angle. Once again, the straightforward synthesis of silyl-substituted cyclopentadienyl compounds is the principal advantage. The interested reader is referred to recent original publications[126], review articles[41,127,128] and monographs[129].

There are very few examples of silicon-bridged bent-metallocenes of type **42** in Scheme 14 described in the literature. The class of silicon-bridged [1]-ferrocenophanes is most important due to the exceptional structure and reactivity, and representative examples **43–52** are listed in Scheme 15.

The first reported examples, Fe(C$_5$H$_4$)$_2$SiPh$_2$ (**47**) and the spiro compound Si(H$_4$C$_5$FeC$_5$H$_4$)$_2$ (**52**)[131], were synthesized by a classical procedure (see Section II) via the reaction of dilithioferrocene·TMEDA with diphenyldichlorosilane and tetrachlorosilane, respectively. The other members in this class were prepared analogously or via the reaction of ferrous chloride with the corresponding dilithiated bis(cyclopentadienyl)dimethylsilane[110,132–134].

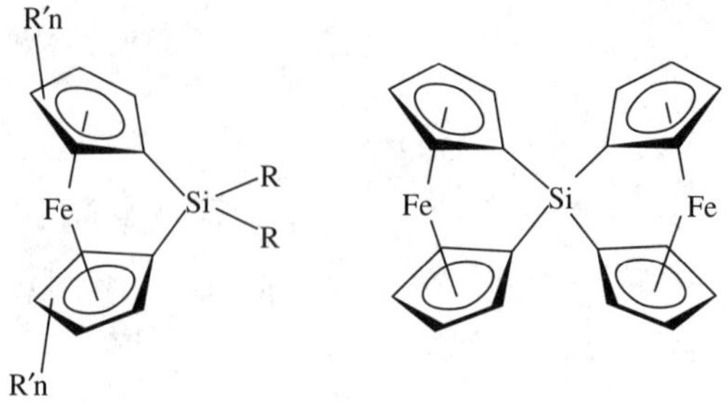

SCHEME 15. Iron complexes of type **42**

Compound	43[130]	44[110]	45[110]	46[110]	47[131]	48[132]	49[133]	50[132]	51[134]
$Cp^{R'}_n$	H_4C_5	H_3MeC_5	Me_4C_5	H_4C_5	H_4C_5	H_4C_5	H_4C_5	H_4C_5	H_4C_5
$Cp^{R'}_n$	H_4C_5	H_3MeC_5	Me_4C_5	Me_4C_5	H_4C_5	H_4C_5	RH_3C_5	H_4C_5	H_4C_5
R	Me	Me	Me	Me	Ph	Ph	Cl	Cl	Me
R	Me	Me	Me	Me	Ph	Cl	Cl	Cl	Cl

R = $Me_2N(Me)HC$

X-ray diffraction studies of the compounds **43–45** and **47–49** revealed highly strained, ring-tilted structures. In Table 7, the strain is expressed by the tilt angle α between the cyclopentadienyl ring planes and by the averaged angle β between the cyclopentadienyl ring planes and the *ipso* (Cp)–Si bonds. With increasing ring methylation, the tilt angle α decreases and the corresponding angle β increases. A similar effect is observed when a methyl group at silicon is replaced by a phenyl group or a chlorine atom.

The strain in silicon-bridged [1]-ferrocenophanes has drastic consequences for the chemistry of such compounds, which is dominated by an easy silicon–(Cp)carbon bond cleavage. This bond fission is used for the thermal or transition-metal catalysed or the anionic ring-opening polymerization, which leads to the interesting class of poly(ferrocenylsilanes)[134], as shown in equation 57. Reaction of **43** with Pt(PEt$_3$)$_3$ yields a novel [2]-platinasilaferrocenophane (equation 58), representing the first example of the insertion of a transition metal into the strained Si–C bond[138]. The compound is of interest as a model of the proposed intermediate during the transition-metal catalysed polymerization process.

Derivatization of surfaces via the reaction of strained Si–C bonds also results in ring-opening reactions. Thus, the reaction of silaferrocenophanes like **49** with an electrode surface is a useful means of attaching a ferrocenyl group, as shown in equation 59[133]. In

TABLE 7. Distortions in ferrocenophanes defining the angles α and β

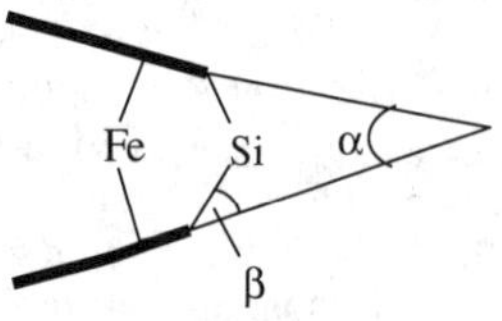

Compound	43[137]	44[110]	45[110]	47[136]	48[132]	49[133]
$\alpha(°)$	20.8(5)	18.6(3)	16.1(3)	19.1(10)	19.0(2)	19.0(2)
$\beta(°)$	37.0(6)	39.1(2)	40.3(2)	40.0(9)	40.96(2)	40.96(2)

a similar procedure, compound **50** has been shown to derivatize n-type semiconducting silicon photoelectrodes, thus representing the first example of a photoelectroactive surface-attached species[132].

$$43 \xrightarrow[\text{(b) [M]}]{\text{(a) heat}} \text{+Me}_2\text{Si}(\text{H}_4\text{C}_5)\text{Fe}(\text{C}_5\text{H}_4)\text{+}_n \tag{57}$$
(c) LiR

$$43 \xrightarrow[-\text{PEt}_3]{\text{Pt(PEt}_3)_3} \text{Fe} \begin{array}{c} \text{PEt}_3 \\ | \\ \text{Pt—PEt}_3 \\ | \\ \text{SiMe}_2 \end{array} \tag{58}$$

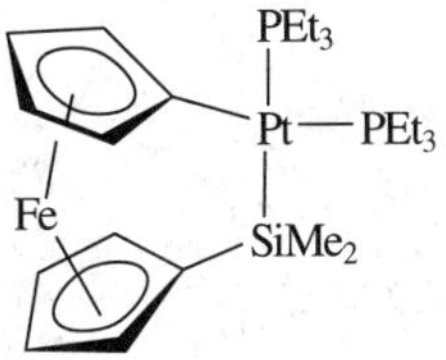

$$49 \xrightarrow{\text{Surface}} \text{[structure]} \quad \text{or} \quad \text{[structure]} \tag{59}$$

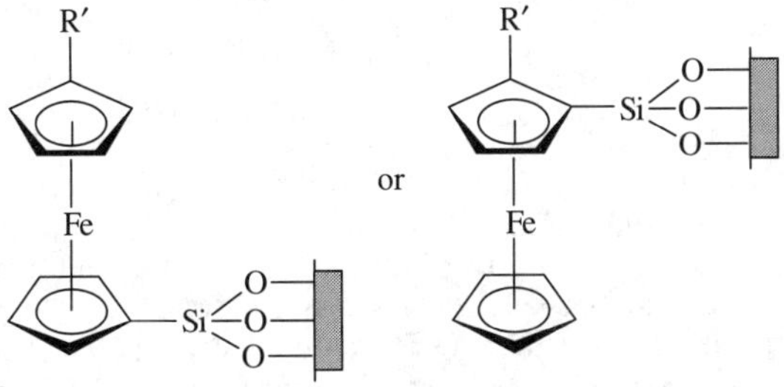

R' = H, CHMeNMe$_2$

The easy silicon–(Cp)carbon bond cleavage in compounds of type **42** is also apparent from electrochemical studies, which have been performed with compound **43** and with ferrocenophanes having tetramethyldisilanyl and hexamethyltrisilanyl bridges. Cyclic voltammetric studies showed a marked dependency upon the oligosilyl bridge[135]. Complex **43** undergoes decomposition upon oxidation, while the complex with the disilanyl bridge exhibits partial reversibility. Finally, the complex with the trisilanyl bridge shows a completely reversible oxidation process. This trend reflects the differing capacity of the three bridges to incorporate the increase in Fe—Cp ring distance that is established to occur upon oxidation to ferrocenium ions.

2. Double-bridged cyclopentadienyl systems

Double-bridging of two cyclopentadienyl units inevitably reduces the conformational flexibility and allows the formation of rather rigid systems. These are excellently suited for assembling metal centres in a stereochemically well-defined manner, and therefore are attractive building blocks for many purposes. The most important ligand systems of this type, which contain the dimethylsilyl(dimethylsilandiyl), the tetramethyldisilanyl and the tetramethyldisiloxanyl bridges, are listed in Scheme 16.

All of these ligand systems have only recently been introduced in metal complex chemistry, and strategies for their synthesis are described below.

Two cyclopentadienyl units are bridged in 1,2 position by two dimethylsilyl groups in the ligand system **53**. The isomeric mixture of 4,4,8,8-tetramethyltetrahydro-4,8-disila-*s*-indacene[139–144] is the starting material for metal complexes of a different kind. Deprotonation of this compound and metallation with alkali metal compounds have been investigated in detail[142,143] (equation 60).

In the ligand system **54**, two cyclopentadienyl units are bridged asymmetrically in 1,2 positions by a dimethylsilyl and by a teramethyldisilanyl group, which imposes chirality

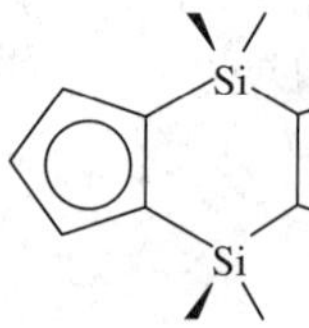

(53) 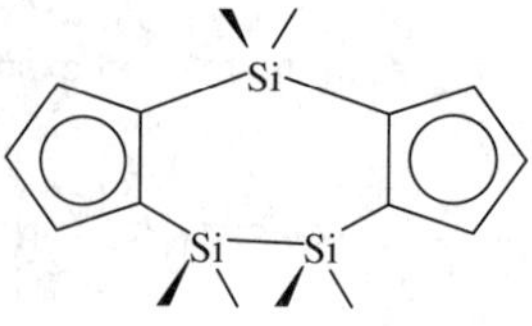 **(54)** 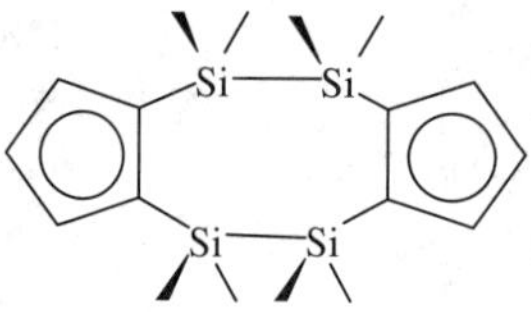 **(55)** 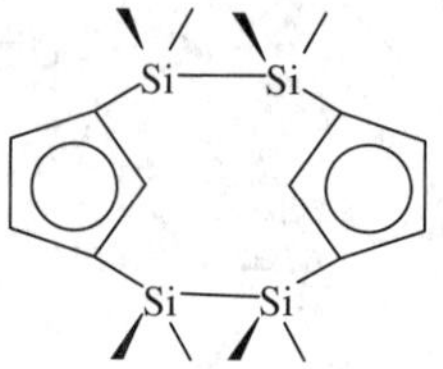 **(56)** 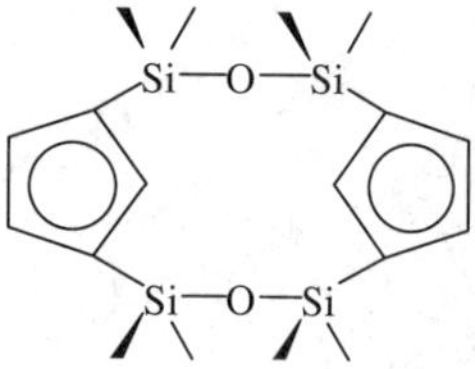 **(57)**

SCHEME 16. Double-bridged cyclopentadienyl systems

at the complexed metal centre[144]. As shown in equation 61, a silatropic shift in the protonated precursor is necessary to generate the desired ligand. Double-bridging with two tetramethyldisilanyl groups occurs in the ligand systems **55**, involving the 1,2 positions of the cyclopentadienyl fragments, and **56** involving the 1,3 positions. Interestingly, both ligand systems are generated from the same precursor molecule, as shown in equation 62[145]. Once more, silatropic shifts are necessary to create isomers which can be deprotonated. In the reaction with butyllithium, two different conformations of the ligand **55** can be generated; in the reaction with dicobalt octacarbonyl, the ligand **56** is formed[145]. Finally, a [3,3]-ferrocenophane containing the ligand system **57** in Scheme 16 is formed from a tetra(alkoxysilyl)ferrocene under hydrolytic conditions, as shown in equation 63[107].

In the following section, we will concentrate on the different kinds of metal complexes which can be derived from the ligand system **53**. The conformational freedom in **53** is restricted to a butterfly-like movement. Although the ligand is rather rigid, the fold angle between the two cyclopentadienyl planes may vary quite substantially, and therefore the ligand system can be adapted to the space-filling requirements of the respective metal fragments. Typical structures for complexes with ligand **53** are presented schematically in Scheme 17.

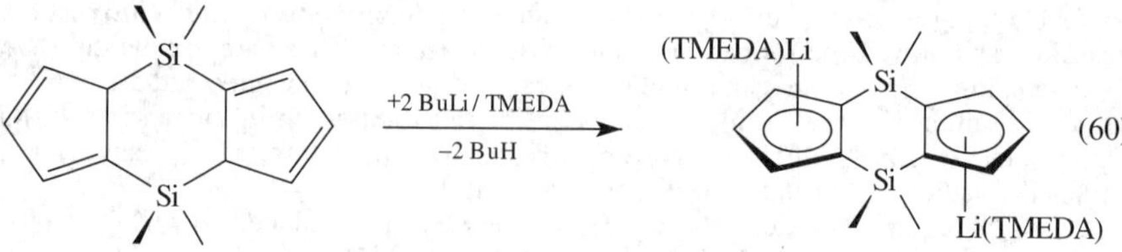

(60)

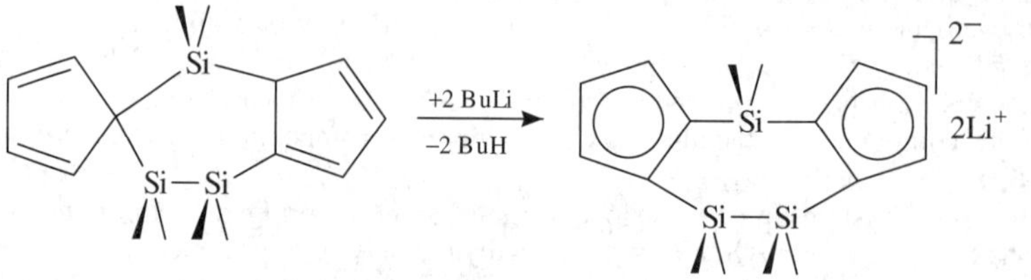

(61)

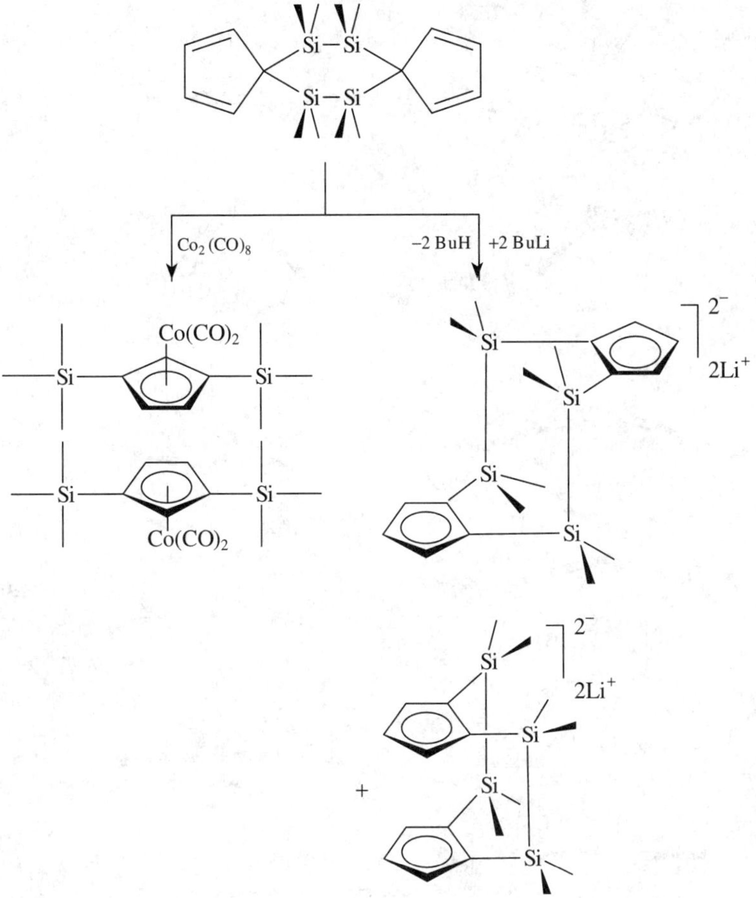

(62)

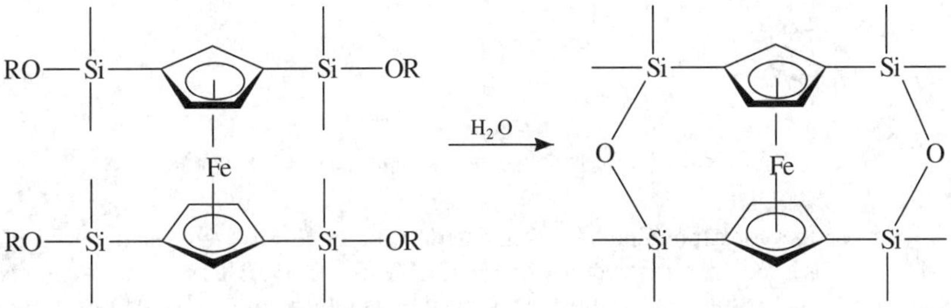

(63)

The comparatively smallest fold angles are found in the chelated bent-metallocenes of type **58**, which have been realized with the Group IV elements titanium[146], zirconium[146,147] and hafnium[146] as central atoms. In the X-ray crystal structure of [$H_3C_5(Me_2Si)_2C_5H_3$]ZrCl$_2$, the (Cp)C—Si—C(Cp) angles are nearly 92°, and the fold angle between the two cyclopentadienyl planes is 69.2°[147]. Comparable structural features

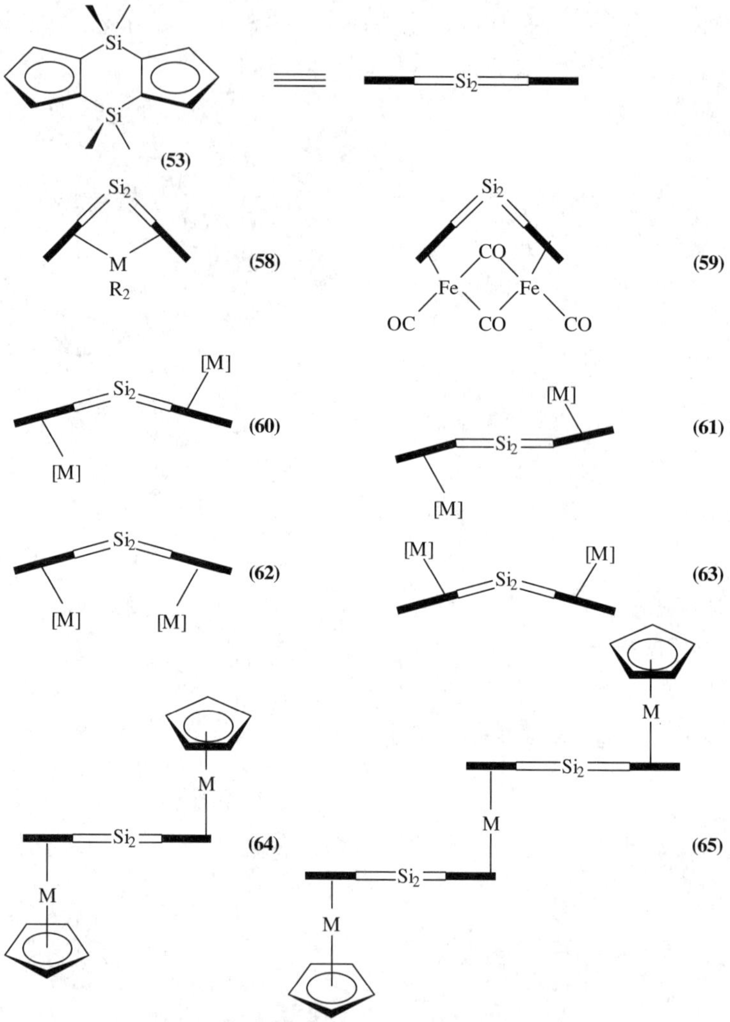

SCHEME 17. Structure types in complexes with ligand 53

are found for the corresponding Cp-methylated species[148]. An even smaller fold angle of 64.4° is observed in the structure of $[H_3C_5(Me_2Si)_2C_5H_3]TiCl_2$[146]. Compounds of this class are interesting model substances in the catalytic olefin polymerization. In the dinuclear complex **59**[143], the steric constraints induced by the $Fe_2(CO)_4$ fragment are likely to impose a small fold angle, but X-ray crystal structure data for this compound are not available.

Several dinuclear metal complexes with the ligand system **53** as bridging unit are now known, where steric constraints are much less pronounced and where fold angles are much greater or nearly 180°. These complexes belong to the structure types **60–63**, where the metal fragments reside in *trans*-configuration, i.e. on different sides of the π-ligand (types **60** and **61**), or in *cis*-configuration, i.e. on the same side of the ligand (types **62** and **63**). Differences in conformation are observed for the *trans*- as well as for the *cis*-configured compounds, at least in the solid state. In the *trans*-complexes, a conformation with a planar central Si_2C_4 ring (type **61**) is found beside the expected butterfly conformation (type **60**). In the *cis*-complexes, the metal fragments occupy either the convex (*exo/exo*; type **63**) or the concave surface (*endo/endo*; type **62**) of the bridging ligand. For steric grounds, the *endo/endo* fixation should be less favourable. Some homobimetallic complexes have been obtained as *cis* isomers, others as *trans* isomers, and others as *cis* and *trans* mixtures. The stereoselective formation of *cis* or *trans* complexes may depend on the synthetic route employed, as exemplified for the synthesis of the molybdenum complex $Cl(OC)_3MoC_5H_3(Me_2Si)_2H_3C_5Mo(CO)_3Cl$[149]: the *trans* complex belongs to (type **61**) with a planar central six-membered ring and a fold angle of 7.8° with the annelated cyclopentadienyl planes; the *cis*-complex belongs to type **62** with a fold angle of 21° (average value). In the titanium complex *trans*-$Cl_3TiC_5H_3(Me_2Si)_2H_3C_5TiCl_3$, the fold angle is $4(2)°$[146] (type **61**), so that the bridging ligand is almost exactly planar. A similar disposition has been found for the complex *trans* -$Li(tmeda)C_5H_3(Me_2Si)_2H_3C_5Li(tmeda)$[142].

Detailed investigations have been performed with metallocene-type species **64** and **65** using the ligand **53** as a stereochemically well-defined building block. In compounds of class **64**, the following metal combinations have been realized: FeFe[143,150,151], RuRu[143], CrCr[152], NiNi[152], CoCo[152], VV[152] and FeNi[152]. In the compounds of class **65**, the following MM'M combinations are found: FeFeFe[151,153], FeVFe[154], FeCrFe[154], FeNiFe[154], CoCrCo[154], NiVNi[154], NiCrNi[154,155], NiCoNi[154] and NiFeNi[154]. Due to the special redox properties of metallocenes[32], interesting electronic properties are envisaged for such molecules. In polymetallocenes with diamagnetic and paramagnetic metal centres, appreciable spin density can be transferred. The role of silicon as a spacer and as a mediator for electronic effects is of special interest and will be discussed in more detail in Section II.E.3.

3. Cooperative effects in silicon-bridged ferrocenes

The bridging of two or more metallocene units is of great interest with regard to the potential for electronic interaction between the metal centres. Cooperative effects may be observed as the result of electronic transmission through the bridge or by direct metal–metal interaction[156]. New physical and chemical properties can be envisaged. In partially oxidized or reduced polymetallocenes, different kinds of mixed-valence behaviour are expected. The properties of such mixed-valence compounds will depend on the amount of delocalization, i.e. on the extent of interaction between the metallocene centres involved (class I,II or III according to Robin and Day[157]).

The special interest in silicon-bridged metallocene units stems from the observation that silicon atoms possess electronic transmission characteristics of unsaturated alkene or alkyne fragments[158–161].

In the class of metallocenes, ferrocene systems fulfill several prerequisites for the investigation of electronic effects[32,162]. Therefore, it is not surprising that ferrocenyl-substituted silicon compounds have been studied most extensively. We will concentrate on this class of compounds in the following discussion.

The development of processable high-molecular-weight polymers with skeletal transition metal atoms, as present in ferrocenes, represents a synthetic challenge, and a variety of materials with novel electrical, optical and magnetic properties can be expected[163]. In this context, polymers containing repeating ferrocenyl units are of great interest. Some representative examples are listed in Scheme 18.

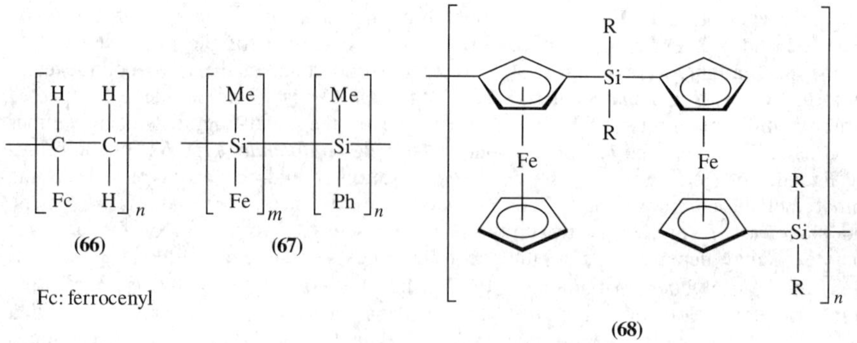

SCHEME 18. Polymers containing repeating ferrocenyl units

In polymers such as poly(vinylferrocene) (66), the iron centres are essentially non-interacting, and a single reversible oxidation wave is detected by cyclic voltammetry. Similarly, the ferrocenyl(methyl)silane–phenyl(methyl)silane copolymers of type 67 exhibit reversible redox properties with no evidence for electronic interactions between the ferrocenyl units or between the pendant ferrocenyl groups and the polysilane chain segments[164,165]. In contrast, poly(ferrocenyl)silanes of type 68 show two reversible oxidation waves in their cyclic voltammograms; this phenomenon has been interpreted in terms of cooperative interactions between the iron centres[166–169]. The peak-to-peak separation $\Delta E_{1/2}$ [$\Delta E_{1/2} = E_{pa}(1) - E_{pa}(2)$] in the cyclovoltammograms, which gives an indication of the degree of interaction, has been found to vary with the side groups R at silicon, which indicates that the electronic effects are tunable[166].

The copolymers of type 67 exhibit the characteristic photochemical depolymerization noted for polysilanes, a source of their potential as photoresist materials. However, the presence of the ferrocenyl substituents results in a significant retardation of this depolymerization, presumably due to the ability of ferrocene to quench the triplet state responsible for the polysilane photochemistry[164]. Polymers of type 67 can be regarded as polysilanes with reversible redox behaviour.

In Scheme 19, a series of silicon-bridged diferrocenyl compounds is collected, together with the corresponding $\Delta E_{1/2}$ values. The $(SiMe_2)_n$ groups ($n = 1-6$) in the compounds 72–75 serve as single bridging units; in compound 71, the $SiMe_2$-groups are double bridging (ligand system 53). The diferrocenyl compounds 69 and 70 with directly C—C connected cyclopentadienyl fragments are presented for comparison.

Most of these compounds exhibit two discrete reversible oxidations. The greatest separation between the oxidation potentials, and thus presumably the most pronounced interaction, is observed for the directly linked ferrocenyl units in compounds 69[169] and 70[170], with the highest degree of cooperation in the rigid and coplanar dicyclopentadienyl system in 69. The conformationally rather rigid double-bridged complex 71[151] offers the best possibilities for electronic interaction within the silicon-bridged species, which becomes progressively smaller in the sequence 72 > 73 > 74[171]. For compounds of type 75, only a single oxidation process is observed, which excludes any electronic interaction[171]. In comparison with the analogous carbon-bridged diferrocenyl compounds, $Fc(CH_2)_nFe$ (Fc = ferrocenyl), the silicon-bridged species clearly turn out to be superior for the mediation of electronic effects[171,172].

A series of [n,n]-ferrocenophanes is listed in Scheme 20 together with their $\Delta E_{1/2}$ values. The introduction of a second bridge enforces the *cis*-configuration and a comparatively shorter Fe—Fe distance. Thus, the geometrical restrictions enhance the possibility

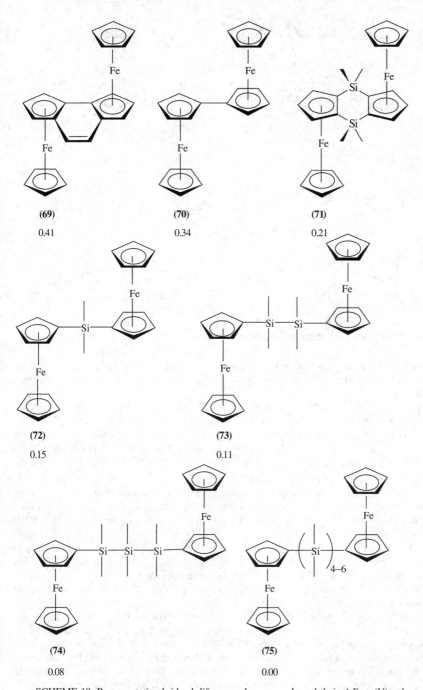

SCHEME 19. Representative bridged diferrocenyl compounds and their $\Delta E_{1/2}$ (V) values

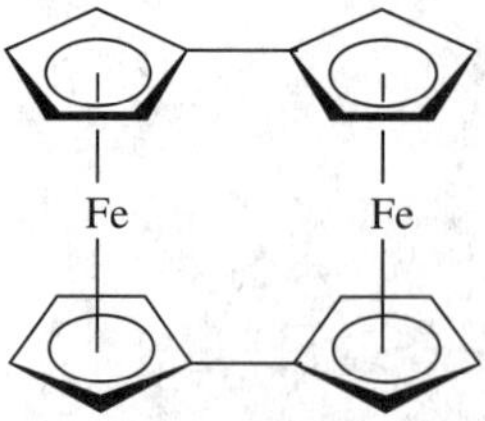

SCHEME 20. Representative [n,n]-ferrocenophanes and their $\Delta E_{1/2}$ (V) values

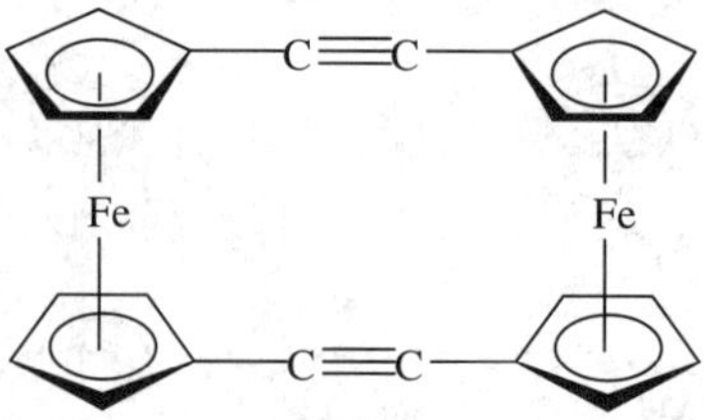

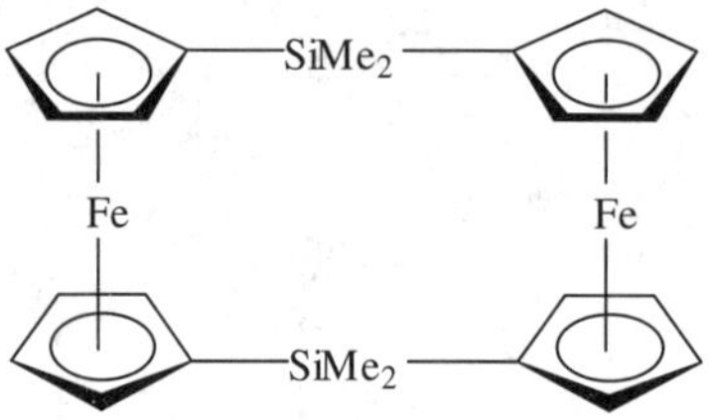

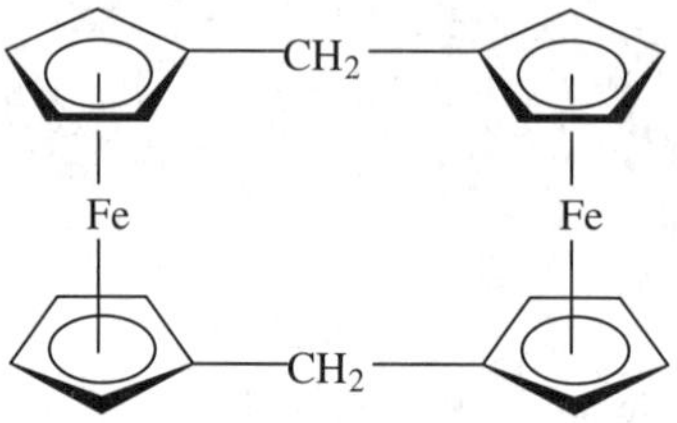

for cooperative effects by both through-bond and through-space interactions. It is interesting to compare the effect of different bridging situations on the extent of electronic interaction.

Very strong interactions have been observed for biferrocenylene (bisfulvalenediiron) (**76**) with directly linked cyclopentadienyl units and also for the bis(alkynyl)-bridged species (**77**); the monocations of these compounds are fully delocalized on all spectroscopic time scales[172,173] (class III according to Robin and Day[157]). On the basis of the measured $\Delta E_{1/2}$ values, the two dimethylsilyl bridges in **78**[124,125] mediate electron delocalization better than the two methylene bridges in **79**[174].

Substantial differences in the electrochemical behaviour have been observed for triferrocenes bridged by methylene (compound **80**) and by dimethylsilyl (compound **81**) units (Scheme 21). Cyclic voltammetry experiments reveal that **80** exhibits three reversible redox processes[175], whereas for **81** only two redox events are observed[176]. The three reversible waves in the carbon-bridged ferrocene (**80**) are in the ratio 1 : 1 : 1 and can be assigned to successive redox events at each iron center, as described in Scheme 21. The two reversible waves in the silicon-bridged ferrocene (**81**) are in the ratio 2 : 1 and are consistent with oxidation of alternate ferrocene units, with subsequent oxidation of the intervening iron site (see Scheme 21).

Comparable effects have been described for silicon-bridged species with up to eight ferrocenyl units[176] and also for poly(ferrocenylsilanes) of type **68**[163]. The difference in the electrochemical behaviour of carbon and silicon substituted multiferrocenes may be explained by the different electron donating or withdrawing effects exerted by these substituents on a ferrocene unit. The phenomena described show that a proper choice of the bridges allows one to determine whether a molecule with three or more ferrocenyl units is first oxidized in the center or at the periphery[151].

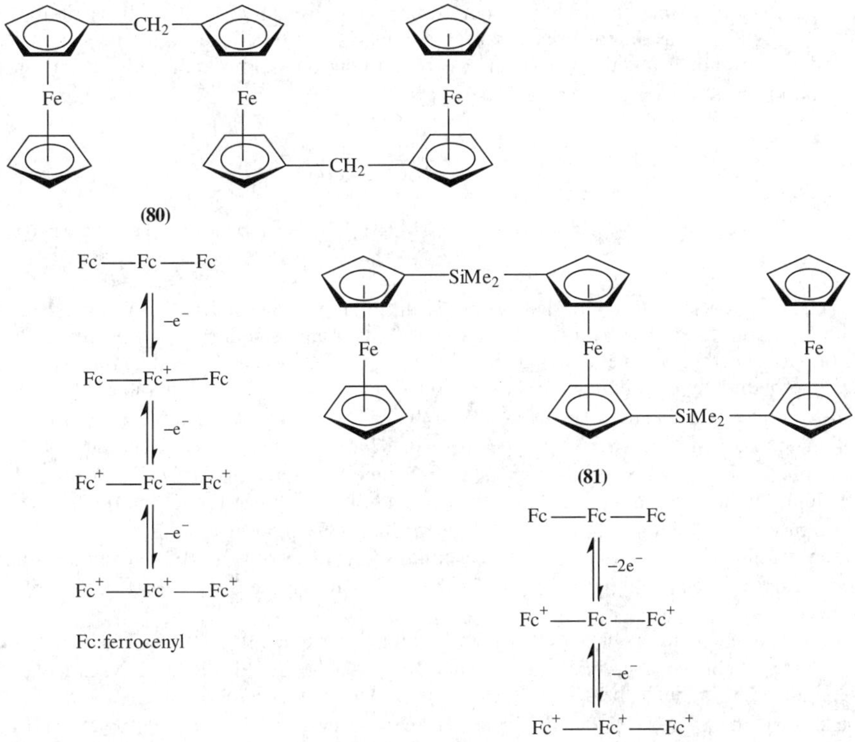

SCHEME 21. Carbon- or silicon-bridged triferrocenes and their redox behaviour

III. COMPOUNDS CONTAINING CYCLOPENTADIENYL–SILICON π-BONDS

A. Introduction

The area of metallocene chemistry started in 1952 with the successful synthesis of ferrocene[177], Fe(C$_5$H$_5$)$_2$, and with the description of a new type of bonding in this sandwich-like compound[178,179].

Only a short time after the discovery of ferrocene, plumbocene, Pb(C$_5$H$_5$)$_2$[180] was the first Group 14 sandwich compound to be synthesized. There are now many derivatives of Group 14 metallocenes reported[6], including decamethylsilicocene, Si(C$_5$Me$_5$)$_2$ (**82**), the first stable π-complex with divalent silicon as central atom[181,182]. The characterization of this compound and of other less stable silicon species with π-bonded cyclopentadienyl ligands is described below.

B. Decamethylsilicocene

1. Synthesis, structure and bonding

Reaction of dichloro(pentamethylcyclopentadienyl)silane with lithium, sodium or potassium naphthalenide gives a mixture of elemental silicon, the corresponding alkali metal pentamethylcyclopentadienide and decamethylsilicocene (**82**) (equation 64)[181]. Compound **82** is formed as the only product in the reduction of dibromobis(pentamethylcyclopentadienyl)silane with potassium anthracenide (equation 65)[182].

The complex is stable in air for a short time, but is sensitive to hydrolysis; it melts at 171 °C without decomposition. From equations 64 and 65 it can be assumed that **82** is reduced by alkali metal naphthalenides, but not by potassium anthracenide, and this assumption was proved in separate experiments[182].

$$(Me_5C_5)_2SiCl_2 \xrightarrow[-MCl, -MC_5Me_5, -Si, -C_{10}H_8]{+MC_{10}H_8} (Me_5C_5)_2Si \quad (64)$$
$$\qquad\qquad\qquad\qquad\qquad\qquad\qquad\qquad\qquad (\mathbf{82})$$

$$(Me_5C_5)_2SiBr_2 + 2KC_{14}H_{10} \longrightarrow (Me_5C_5)_2Si + 2KBr + 2C_{14}H_{10} \quad (65)$$
$$\qquad\qquad\qquad\qquad\qquad\qquad (\mathbf{82})$$

The X-ray crystal structure analysis of **82** surprisingly revealed two conformers, one with a parallel (**82a**) and the other with a bent (**82b**) arrangement (interplane angle: 25°) of the pentamethylcyclopentadienyl rings (see Figure 8). Molecular models indicate that the interplane angle in **82b** is the largest possible value. Both exhibit rather long silicon-carbon bond distances (**82a**, 2.42 Å; **82b** 2.32–2.54 Å) (see also Scheme 22). The solid-state (CP-MAS) ^{29}Si NMR spectrum of **82** reflects the gross structural features determinded in the X-ray analysis[183]. The ^{29}Si nuclear shielding in **82a** is found to be 20.2 ppm higher than that of the bent structure in **82b**, which is close to that found for **82** in solution[182]. The high-field chemical shift for **82** (−398 ppm in solution) is exceptional for an organosilicon compound, but corresponds to the observed shift for decamethylstannocene and for decamethylplumbocene in the ^{119}Sn- and in the ^{207}Pb-NMR spectrum[182], respectively.

Extensive calculations have been performed for the parent silicocene, $Si(C_5H_5)_2$, some of them probably at insufficient levels of theory. At the SCF DZP level[184], silicocene is predicted to adopt a bent C_s structure. However, the difference in energy between the low-symmetry (C_s, C_2) and C_{2v} conformers is only 2.4 kcal mol^{-1}, with the D_{5d} conformer lying 8.8 kcal mol^{-1} higher in energy. The d-polarization function indicates that d-orbitals play no significant role in the π-bonding[184]. It is evident from all calculations that the silicocene potential energy surface is flat with respect to the interconversion between several conformers and there is a low-energy barrier for the rotation of the

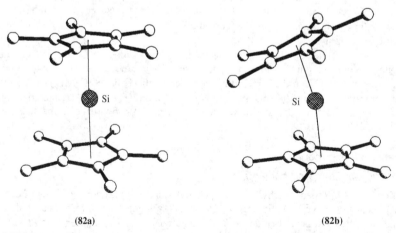

(82a)   (82b)

FIGURE 8. Molecular structure of decamethylsilicocene (**82**), reproduced by permission of VCH Verlagsgesellschaft, Weinheim, from Ref. 181

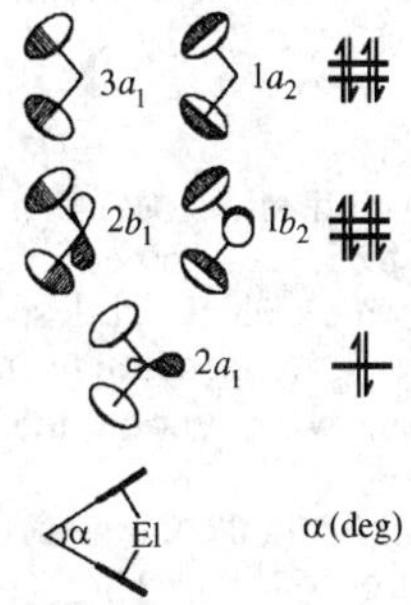

36. Cyclopentadienyl silicon compounds

			$(Me_5C_5)_2Si$	$(Me_5C_5)_2Ge$	$(Me_5C_5)_2Sn$	$(Me_5C_5)_2Pb$
$3a_1$	$1a_2$		6.70 6.96	6.60 6.75	6.60 6.60 7.64	6.88 6.88
$2b_1$	$1b_2$		8.06 8.30	7.91 8.05	7.64	7.38 7.38
$2a_1$			7.5	8.36	8.40	8.93
α El	α(deg)		0^a; 25.3^d	$22 + 2^b$	$36.4^{a,c}$ 35.4	37.1^a
Range of El–C Distances (Å)			2.42	2.32 - 2.54	2.58 – 2.77	2.69 – 2.90
Average of El–C Distances (Å)			2.42	2.42	2.52	2.64	2.79

SCHEME 22. MO-Scheme[e]), PES data (eV), and structural parameters of the Group 14-decamethyl-metallocenes [a]X-ray; [b]gED; [c]two independent molecules; [d]two conformers in the unit cell; [e]in orbital notation for C_{2v} symmetry

cyclopentadienyl rings. The total electron population of the silicon center is calculated to be 13.5, in agreement with weak π-interactions and explaining the rather long Si–C distances observed for **82**[16].

In Scheme 22, a qualitative molecular orbital scheme is presented for the highest occupied orbitals in the Group 14 (El) metallocenes, and the vertical ionization energies from the PE spectra of decamethylsilicocene[182], -germanocene[185], -stannocene[185], and -plumbocene[182] are presented and assigned in the context of Koopmans' theorem. In addition, some structural parameters of these metallocenes are given (interplane angle α, El–C distances)[6].

The following conclusions can be drawn from the MO scheme and the PE data:

(I) The HOMOs in all metallocenes are non-bonding with respect to cyclopentadi-enyl–metal interactions, as indicated by the first ionization energies which essentially remain constant.

(II) The $2b_1$ and $1b_2$ type orbitals provide the strongest contributions to cyclopentadi-enyl–metal bonding, and the respective ionization energies depend strongly on the metal involved.

(III) The orbital representing the 'lone-pair' at the Group 14 element is rather low in energy in the case of the germanium, tin and lead compound, but is higher in energy and in the region of the frontier orbitals in the case of the silicon compound. These differences in energy have implications concerning the chemistry of the Group 14 metallocenes.

(IV) The observed bathochromic shift in the electronic absorption spectra (**82**: colorless; Cp_2^*Ge: light yellow; Cp_2^*Sn: orange; Cp_2^*Pb: red) can be correlated with the lowering in energy of the respective LUMO, which can be described as the antibonding linear combination between the ligand π-MOs and the p orbitals at the relevant central atom[182].

Cyclovoltammetric data show that **82** cannot be reduced in the region available (up to −1.7 V versus SCE)[182], consistent with the chemical behaviour of **82** in reduction processes, as shown in equations 64 and 65. An irreversible oxidation process takes place at +0.4 V versus SCE. Presumably, the radical cation of **82** is unstable due to the easy loss of the pentamethylcyclopentadienyl radical[186,187]. The mass-spectrometric observations support this assumption: The molecular ion cannot be observed in EI as well as in CI studies, and the fragment with the highest mass corresponds to the $Me_5C_5Si^+$ ion. It is evident from these studies that the $(Me_5C_5)_2Si^+$ radical ion, which has its positive charge exclusively delocalized within the π-system of the pentamethylcyclopentadienyl ligands, is rather unstable in solution and in the gas phase, and easily loses a Me$_5$C$_5$ radical[182].

2. Chemistry of decamethylsilicocene

Decamethylsilicocene (**82**) can be regarded as an electron-rich silicon(II) compound containing a hypercoordinated silicon atom. The chemistry of **82** is determined by (a) the nucleophilicity of the silicon lone-pair (σ-donor function towards electrophiles, oxidative–addition processes) and (b) the weakness of the silicon–(cyclopentadienyl)carbon π-bond ($\eta^5-\eta^1$ rearrangement, Si−C bond cleavage). In the following section, the chemistry of **82** will be illustrated with some typical examples.

Reductive dehalogenation of substrates with geminal or vicinal halogen ligands can be performed under very mild conditions (equations 66–68[188]). Thus, organic dibromides with the halogen atoms in vicinal position are transferred to the corresponding alkene derivatives (equation 66), and the 1,2-bis(pentamethylcyclopenadienyl)-1,2-diiododiphosphane is converted to the corresponding diphosphene (equation 67). Finally, decamethylstannocene is formed from the tin(IV) compound bis(pentamethylcyclopentadienyl)dichlorostannane in the reaction with **82** (equation 68).

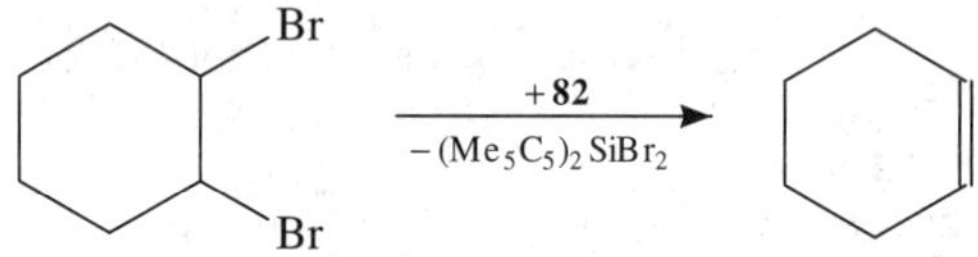

$$\text{cyclohexane-1,2-Br}_2 \xrightarrow[-(\text{Me}_5\text{C}_5)_2\text{SiBr}_2]{+\,\mathbf{82}} \text{cyclohexene} \tag{66}$$

$$\text{Me}_5\text{C}_5(\text{I})\text{P}-\text{P}(\text{I})\text{C}_5\text{Me}_5 \xrightarrow[-(\text{Me}_5\text{C}_5)_2\text{SiI}_2]{+\,\mathbf{82}} \text{Me}_5\text{C}_5\text{P}=\text{PC}_5\text{Me}_5 \tag{67}$$

$$(\text{Me}_5\text{C}_5)_2\text{SnCl}_2 \xrightarrow[-(\text{Me}_5\text{C}_5)_2\text{SiCl}_2]{+\,\mathbf{82}} (\text{Me}_5\text{C}_5)_2\text{Sn} \tag{68}$$

As expected for a nucleophilic silicon(II) compound, **82** does not react with triethylsilane, but with the electrophilic trichlorosilane to form the corresponding unsymmetrical disilane (equation 69)[189]. Other polar and non-polar substrates react in a similar fashion to give the corresponding oxidative addition products (equations 69[188] and 70)[190].

$$(\text{Me}_5\text{C}_5)_2\text{Si} + \text{RX} \longrightarrow (\text{Me}_5\text{C}_5)_2\text{Si}(\text{R})\text{X} \tag{69}$$
(**82**)

RX = HSiCl$_3$, MeBr, MeI, Me$_3$SiCN, BrCN

$$(\text{Me}_5\text{C}_5)_2\text{Si} + X_2 \longrightarrow (\text{Me}_5\text{C}_5)_2\text{SiX}_2 \tag{70}$$
(**82**)

X$_2$ = Br$_2$, I$_2$, RSSR

The reaction of **82** with protic substrates HX can follow different pathways, as depicted in Scheme 23. In most of the cases studied so far, simple oxidative addition takes place leading to compounds of the type (Me$_5$C$_5$)$_2$Si(H)X[190]. The reaction of **82** with tetrafluoroboric acid leads to elimination of pentamethylcyclopentadiene and to the formation of a cyclotetrasilane derivative, via several intermediates with divalent silicon from which the dimeric pentamethylcyclopentadienyl(fluoro)silylene was proved by Si NMR spectroscopy[191,192]. In contrast, an HF-oxidation product is formed in the reaction of **82** with the pyridinium salt of HBF_4[189]. The influence of the acidity of the relevant protic substrates on the final reaction products is further demonstrated by the reaction of **82** with two equivalents of trifluoromethanesulfonic acid giving an ionic species containing the bis(pentamethylcyclopentadienyl)hydridosilicon cation[193] and with its pyridinium salt giving the oxidative addition product[189].

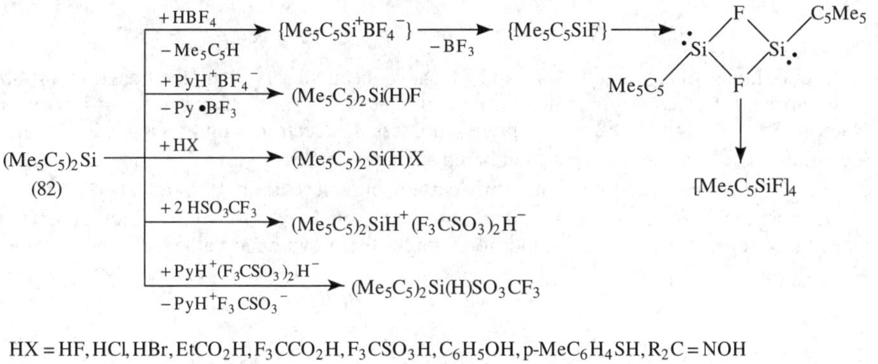

$HX = HF, HCl, HBr, EtCO_2H, F_3CCO_2H, F_3CSO_3H, C_6H_5OH, p\text{-}MeC_6H_4SH, R_2C=NOH$

SCHEME 23. Reaction of $(Me_5C_5)_2Si$ (**82**) with protic substrates

Differences in the reactivity of **82** and its heavier homologues towards protic substrates become evident also from the reaction with catechol: Whereas oxidation addition and subsequent H_2-elimination is observed with **82**, a substitution process takes place with decamethylstannocene (equations 71 and 72)[190]. **82** undergoes an oxidative addition with compounds formally delivering 6-electron species, giving intermediates of the type $(Me_5C_5)_2Si=X$ with a silicon element (p-p)π-bond. These species are highly reactive under the reaction conditions (equation 73). Substrates like trimethylamine oxide and carbon dioxide can be used as a source of oxygen[194,195], and elemental sulphur[194], organic isothiocyanates[195] and carbon oxosulphide[195] can function as a source of sulphur. Intermediates were not detected in the reaction with tri-n-butylphosphane telluride, and the final product is a tritelluradisilole derivative (equation 74)[194].

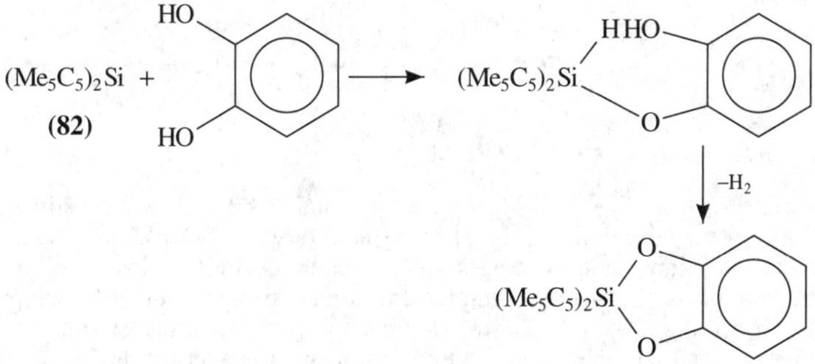

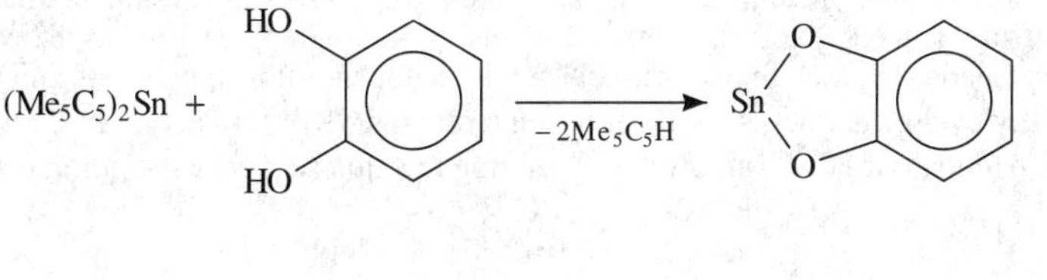

$$(Me_5C_5)_2Si + X \longrightarrow \{(Me_5C_5)_2Si = X\} \longrightarrow \longrightarrow \text{product} \quad (73)$$
(82)

X : O from CO_2, Me_3NO; S from S_8, COS, RNCS; Se from $n\text{-}Bu_3P=Se$

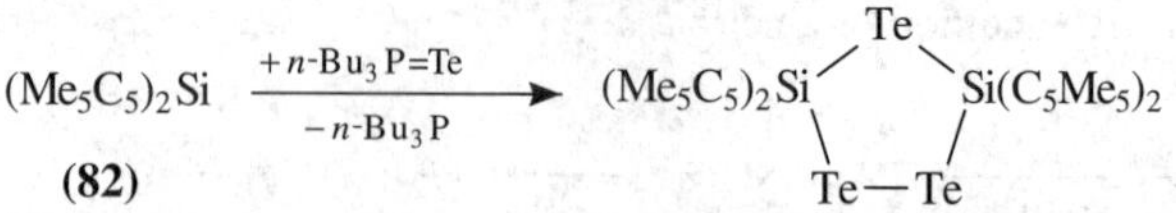

$$(Me_5C_5)_2Si \underset{-n\text{-}Bu_3P}{\overset{+n\text{-}Bu_3P=Te}{\longrightarrow}} (Me_5C_5)_2Si\begin{array}{c}Te\\ \diagup \diagdown\\ \vert \quad \vert\\ Te-Te\end{array}Si(C_5Me_5)_2 \quad (74)$$
(82)

Several kinds of cycloaddition products have been observed in the reaction of **82** with various activated double and triple bond species. A simple 2:1 cycloaddition takes place in the reaction of **82** with alkynes possessing electron-withdrawing substituents (equation 75)[199]. In contrast, five-membered silaheterocycles are formed regio- and stereo-specifically in multi-step reactions with certain organic carbonyl[196] (equation 76) and nitrile compounds (equation 77)[197]. The observed C−C bond formation, which generally takes place under mild reaction conditions, might be of synthetic value.

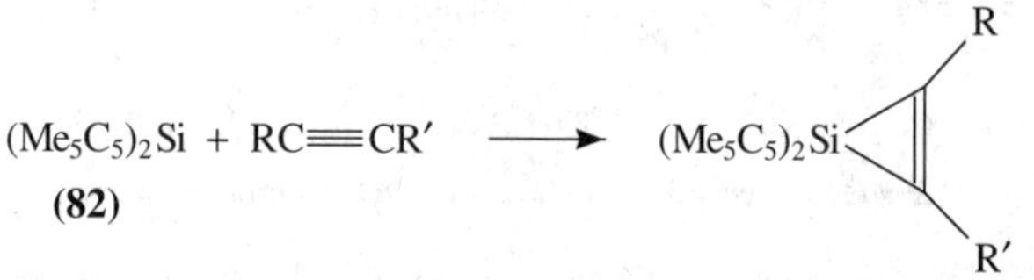

$$(Me_5C_5)_2Si + RC\equiv CR' \longrightarrow (Me_5C_5)_2Si\begin{array}{c}R\\ \diagup\\ \Vert\\ \diagdown\\ R'\end{array} \quad (75)$$
(82)

$RC\equiv CR'$: $MeO_2CC\equiv CCO_2Me$; $Me_3SiC\equiv CSO_2Ph$

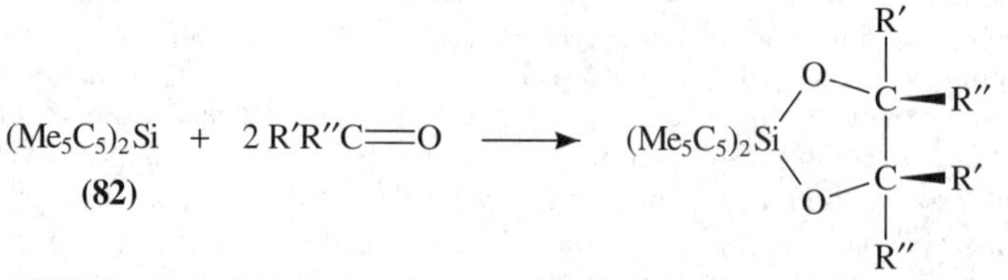

$$(Me_5C_5)_2Si + 2R'R''C=O \longrightarrow (Me_5C_5)_2Si\begin{array}{c}R'\\ O\diagdown \overset{\vert}{C}\text{-}R''\\ \vert\\ O\diagup \underset{\vert}{C}\text{-}R'\\ R''\end{array} \quad (76)$$
(82)

$R'R''CO$: $C_6H_5C(H)O$; $C_6H_5CH=CHC(H)O$; Me_2CO

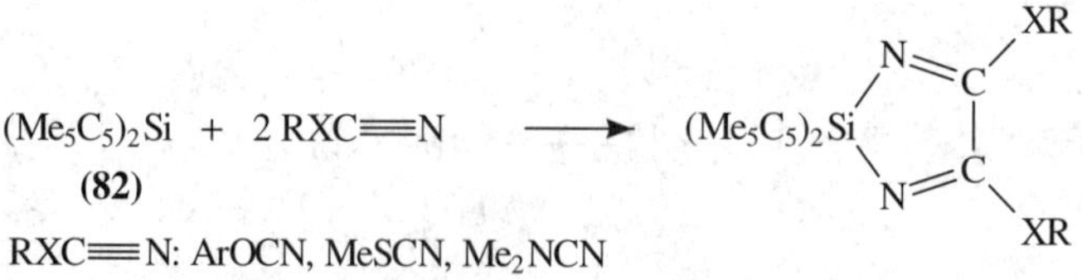

$$(Me_5C_5)_2Si + 2RXC\equiv N \longrightarrow (Me_5C_5)_2Si\begin{array}{c}XR\\ N=\overset{\vert}{C}\\ \vert\\ N=\underset{\vert}{C}\\ XR\end{array} \quad (77)$$
(82)

$RXC\equiv N$: ArOCN, MeSCN, Me_2NCN

The weakness of the Si−C(cyclopentadienyl) bond in **82** and its derivatives becomes apparent in several reactions. This phenomenon (see Section II.C) sometimes leads to complicated cyclopentadienyl migration processes as well as to the elimination of cyclopentadienyl fragments. For example, the formation of pentamethylcyclopentadiene has been observed in the stoichiometric reactions of **82** with tetrafluoroboric acid (see Scheme 23) and with pentacarbonyl(hydrido)manganese (equation 78)[189]. Dyotropic rearrangements are presumably the basis for the observed cyclopentadienyl group transfer from silicon to other elements. Thus, a multi-step reaction of **82** with boron trihalides leads to the final products with halogen substituents at silicon and a pentamethylcyclopentadienyl group at boron (equation 79)[188]. Reaction of **82** with carbon disulphide (equation 80)[195] involves a surprising reaction sequence including the exchange of two cyclopentadienyl and two sulphur ligands.

$$(Me_5C_5)_2Si + 2HMn(CO)_5 \longrightarrow Me_5C_5(H)Si[Mn(CO)_5]_2 + Me_5C_5H \quad (78)$$
(82)

$$(Me_5C_5)_2Si + BHal_3 \longrightarrow Me_5C_5(Hal)_2Si - B(Hal)C_5Me_5 \quad (79)$$
(82)
Hal : Cl, Br, I

$$2(Me_5C_5)_2Si + 2CS_2 \longrightarrow$$
(82)

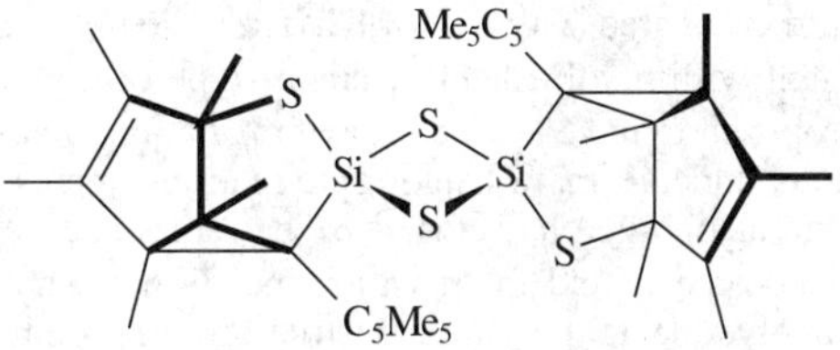

[structure showing two cyclopentadienyl groups bridged by Si-S-Si with sulfur bridges, bearing Me_5C_5 and C_5Me_5 substituents]

(80)

The formal oxidation state of silicon changes from +II to +IV in nearly all of the reactions with decamethylsilicocene **(82)** described so far, and the hapticity of the cyclopentadienyl ligand changes from η^5 to η^1. The latter phenomenon raises the question about the possibility of a haptotropic shift in the ground state molecule, which should lead to quite different silicon species in equilibrium, as shown in equation 81.

$$(\eta^5\text{-}Me_5C_5)_2Si \rightleftharpoons (\eta^1\text{-}Me_5C_5)_2Si \quad (81)$$
(82)   **(83)**

Variable-temperature NMR studies give no evidence for such an equilibrium[189], but cannot exclude the presence of small quantities of the classical silylene $(\eta^1\text{-}Me_5C_5)_2Si$ **(83)**. Recent experiments performed with decamethylsilicocene in cryosolutions and accompanying theoretical studies indicate the presence of small quantities of isomer **83** in equilibrium. In liquid xenon (LXe), the formation of the monocarbonyl complex $(Me_5C_5)_2Si(CO)$ and of the mono(dinitrogen) complex $(Me_5C_5)_2Si(N_2)$ has been proven by IR spectroscopic studies (equations 82 and 83)[198]. The reactions were found to be incomplete under a few bar of CO or N_2 and reversible when the pressure was released. From the spectroscopic studies it was concluded that less than one percent of the complex is formed under the reaction conditions. *Ab initio* calculations[199] show that an equilibrium as shown in equation 81 is the first step before CO or N_2 form adducts.

$$(Me_5C_5)_2Si \xrightarrow[\text{1.5 bar CO}]{\text{LXe/253 K}} (Me_5C_5)_2Si(CO) \quad (82)$$
(82)

$$(Me_5C_5)_2Si \xrightarrow[\text{20 bar } N_2]{\text{LXe/173 K}} (Me_5C_5)_2Si(N_2) \quad (83)$$
(82)

C. Other π-Complexes

During the course of the investigation of the chemistry of decamethylsilicocene **(82)**, few experiments gave evidence for the formation of compounds in which the π-bonding between silicon and the cyclopentadienyl ligand remained intact. These compounds will be presented shortly. As already described (see Section III.B.2; Scheme 23), reaction of **82** with one equivalent of a protic substrate HX does not lead to $(Me_5C_5)_2SiH^+X^-$ containing a silyl cation, in which silicon is bonded to hydrogen and to two π-pentamethylcyclopentadienyl ligands. This cation was, however, unexpectedly formed as one component in the multi-step reaction of **82** with catechol and with catechol derivatives[200], and is most likely also generated by the reaction of **82** with two equivalents of trifluoromethanesulphonic acid[200] (Scheme 23). The structure was inferred from the ^{29}Si NMR spectrum, showing a shift at δ -12.1, which represents a deshielding of 386 ppm compared with that of **82**. The observed $^{29}Si-H$ coupling constant of 302 Hz is consistent with involvement of an sp^2-type orbital at silicon. The cation $(Me_5C_5)_2SiH^+$ was proposed to have a structure with η^2/η^3-bound cyclopentadienyl ligands, whereby the π-ligands undergo rapid haptotropic rearrangements.

The nucleophilic character of **82** has stimulated experiments to use this compound as a silylene-type ligand in transition-metal chemistry but with only one useful result[189], presumably due to the steric restrictions imposed by the bulky pentamethylcyclopentadienyl ligands. Reaction of **82** with carbonylgold(I) chloride led to the formation of [bis(pentamethylcyclopentadienyl)silanediyl]gold chloride, $(Me_5C_5)_2SiAuCl$. The structure was established by temperature-dependent ^1H NMR studies, which indicate that one of the Cp ligands is bound in a σ-fashion and the other is bound in a π-fashion, but the exact hapticity of the π-bonded ligand could not be determined[201].

Half-sandwich complexes of the type $Me_5C_5El^+X^-$ with El = Ge, Sn and Pb are stable under standard conditions and have been well characterized. However, the corresponding silicon derivatives have only been identified by mass spectrometry. An environment of very low basicity and nucleophilicity will be necessary in order to further stabilize compounds of this type.

In general, the extensive series of metallocene derivatives that are known for the heavier elements of Group 14 have not been readily extrapolated to silicon chemistry. It is intriguing that in spite of a huge effort in this area over many years, decamethylsilicocene remains the only derivative that has been comprehensively characterized, and the challenges with this chemistry are still numerous.

IV. ACKNOWLEDGEMENTS

The indispensable and stimulating contributions of the coworkers mentioned in the references are very gratefully acknowledged. The author wishes to thank Mrs.B. Neumann for preparing the drawings, Mrs. H. Niermann for typing the manuscript and Prof. N. Burford (Dalhousie University, Halifax, Canada) for helpful comments and for improving the English. Finally, financial support of our work by the Deutsche Forschungsgemeinschaft, the Fonds der Chemischen Industrie and the University of Bielefeld is gratefully acknowledged.

V. REFERENCES

1. E. W. Abel, M. O. Dunster and A. Waters, *J.Organomet. Chem.*, **49**, 287 (1973).
2. S. Patai and Z. Rappoport (Eds.), *The Chemistry of Organic Silicon Compounds*, Wiley, New York, 1989.
3. D. A. Armitage, *'Organosilanes'*, in *Comprehensive Organometallic Chemistry*, Vol. 2 (Eds. G. Wilkinson, F. G. A. Stone and E. W. Abel), Pergamon Press, Oxford–New York, 1982.
4. I. Fleming, *'Organic Silicon Chemistry'*, in *Comprehensive Organic Chemistry*, Vol. 3 (Eds. D. Barton and W. D. Ollis), Pergamon Press, Oxford–New York, 1979.
5. P. Jutzi, *Chem. Rev.*, **86**, 983 (1986).

6. P. Jutzi, *Adv. Organomet. Chem.*, **26**, 217 (1986).
7. J. E. Bentham and D. W. H. Rankin, *J. Organomet. Chem.*, **30**, C54 (1971).
8. G. A. Shchembelov and Yu. A. Ustynyuk, *Dokl. Akad. Nauk SSSR*, **173**, 847 (1967); *Engl. Transl.*, **173**, 847 (1967).
9. A. F. Cuthbertson and C. Glidewell, *J. Organomet. Chem.*, **221**, 19 (1981).
10. A. H. Cowley, E. A. V. Ebsworth, S. K. Mehrotra, D. W. H. Rankin and M. D. Walkinshaw, *J. Chem. Soc., Chem. Commun.*, 1099 (1982).
11. P. Jutzi, D. Kanne, M. B. Hursthouse and A. J. Haves, *Chem. Ber.*, **121**, 1299 (1988).
12. S. S. Al-Inaid, C. Eaborn, P. P. Hitchcock. P. Lickiss, A. Möhrke and P. Jutzi, *J. Organomet. Chem.*, **384**, 33 (1990).
13. S. Cradock, R. H. Findlay and M. H. Palmer, *J. Chem. Soc., Dalton Trans.*, 1650 (1974).
14. S. Cradock, E. A. V. Ebsworth, H. Moretto and D. W. H. Rankin, *J. Chem. Soc., Dalton Trans.*, 390 (1975).
15. H. Bock and W. Kaim, *J. Am. Chem. Soc.*, **102**, 4429 (1980).
16. H. Bock, *Angew. Chem.*, **101**, 1659 (1989); *Angew. Chem., Int. Ed. Engl.*, **28**, 1627 (1989).
17. H. P. Fritz and C. G. Kreiter, *J. Organomet. Chem.*, **4**, 313 (1965).
18. Y. A. Ustynyuk, A. V. Kisin and D. W. Oksinoid, *Zh. Obshch Khim.*, **38**, 391 (1968); *Chem Abstr.*, **69**, 76368w (1969).
19. A. Bonny, R. D. Holmes-Smith, G. Hunter and S. R. Stobart, *J. Am. Chem. Soc.*, **104**, 1855 (1982).
20. Yu. A. Ustynyuk, A. V. Kisin, J. M. Pribytkova, A. A. Zarkin and N. D. Antonova, *J. Organometal. Chem*, **42**, 47 (1972).
21. R. Krallmann, Ph. D. Thesis, University of Bielefeld, 1991.
22. P. Jutzi and J. Sauer, *J. Organomet. Chem.*, **50**, C29 (1973).
23. W. Kläui and H. Werner, *Helv. Chim. Acta*, **59**, 844 (1978).
24. J. Okuda, *Chem. Ber.*, **122**, 1075 (1989).
25. H. Chen. P. Jutzi, W. Leffers, M. M. Olmstead and P. P. Power, *Organometallics*, **10**, 1282 (1991).
26. W. J. Evans, T. J. Boyle and J. W. Ziller, *Organometallics*, **11**, 3903 (1992).
27. P. Jutzi, W. Leffers, B. Hampel, S. Pohl and W. Saak, *Angew. Chem.*, **99**, 563 (1987); *Angew. Chem., Int. Ed. Engl.*, **26**, 583 (1987).
28. C. P. Casey, J. M. O'Connor and K. J. Haller, *J. Am. Chem. Soc.*, **107**, 1241 (1985).
29. W. S. Sheldrick, 'Structural Chemistry of Organic Silicon Compounds', in *The Chemistry of Organic Silicon Compounds* (Eds. S. Patai and Z. Rappoport), Wiley, New York, 1989.
30. J. Okuda and E. Herdtweck, *J. Organomet. Chem.*, **373**, 99 (1989).
31. J. Okuda, R. W. Albach, E. Herdtweck and F. E. Wagner, *Polyhedron*, **10**, 1741 (1991).
32. D. Astruc, (Ed.) *Electron Transfer and Radical Processes in Transition-Metal Chemistry*, VCH Publ., Weinheim, 1995.
33. U. Koelle and F. Khouzami, *Angew. Chem.*, **92**, 658 (1980); *Angew. Chem., Int. Ed. Engl.*, **19**, 640 (1980); J. L. Robbins, N. Edelstein, B. Spencer and J. C. Smart, *J. Am. Chem. Soc.*, **104**, 1882 (1982).
34. P. G. Gassman, D. W. Macomber and D. W. Hershberger, *Organometallics*, **2**, 1471 (1983).
35. P. G. Gassman, P. A. Deck, C. H. Winter, D. A. Dobbs and D. H. Cao, *Organometallics*, **11**, 959 (1992).
36. M. Kira, M. Watanabe and H. Sakurai, *J. Am. Chem. Soc.*, **99**, 7780 (1977).
37. P. J. Parker, A. G. Davies, R. Henriquez and J.-Y. Nedalec, *J. Chem. Soc., Perkin Trans. 2*, 745 (1982).
38. J. Okuda, *Top. Curr. Chem.*, **160**, 97 (1991).
39. C. Janiak and H. Schumann. *Adv. Organomet. Chem.*, **33**, 291 (1991).
40. R. L. Haltermann, *Chem. Rev.*, **92**, 965 (1992).
41. N. E. Schore, *J. Am. Chem. Soc.*, **101**, 7410 (1979).
42. N. E. Schore and S. Sundar, *J. Organomet. Chem.*, **184**, C44 (1980).
43. A. Sekegudi, Y. Sagai, K. Ebata, C. Kabuto and H. Sakurai, *J. Am. Chem. Soc.*, **115**, 1144 (1993).
44. P. Jutzi and K.-H. Schwartzen, *Chem. Ber.*, **122**, 287 (1989).
45. P. Jutzi, W. Leffers, B. Hampel, S. Pohl and W. Saak, *Angew. Chem.*, **99**, 563 (1987); *Angew. Chem., Int. Ed. Engl.*, **26**, 583 (1987).
46. L. M. Engelhardt, P. C. Junk, C. L. Rastah and A. H. White, *J. Chem Soc., Chem. Commun.*, 1500 (1988).

47. C. P. Morley, P. Jutzi, C. Krüger and J. M. Wallis, *Organometallics*, **6**, 1084 (1987).
48. P. Jutzi, W. Leffers, G. Müller and B. Huber, *Chem. Ber.*, **122**, 879 (1989).
49. A. K. Hughes, A. Meetsma and J. H. Teuben, *Organometallics*, **12**, 1936 (1993).
50. S. Harvey, C. L. Raston, B. W. Shelton, A. H. White, M. F. Lappert and G. Soistiva, *J. Organomet. Chem.*, **328**, C1 (1987).
51. P. Jutzi, W. Leffers and G. Müller, *J. Organomet. Chem.*, **334**, C24 (1987).
52. E. W. Abel and S. Moorhouse, *J. Organomet. Chem.*, **28**, 211 (1971).
53. P. Jutzi and M. Kuhn, *J. Organomet. Chem.*, **173**, 221 (1979).
54. C. H. Winter, X.-X. Zhau, D. A. Dobbs and M. J. Heeg, *Organometallics*, **10**, 210 (1991).
55. M. D. Rausch, M. Vogel and H. Rosenberg, *J. Organomet. Chem.*, **22**, 900 (1957).
56. A. H. Cowley, P. Jutzi, F. X. Kohl, J. G. Lasch, N. C. Norman and E. Schlüter, *Angew. Chem.*, **96**, 603 (1984).
57. K. Sünkel and D. Motz, *Chem. Ber.*, **121**, 799 (1988).
58. K. Sünkel and J. Hofmann, *Organometallics*, **11**, 3923 (1992).
59. G. P. Sollot and W. R. Peterson Jr., *J. Am. Chem. Soc.*, **89**, 5054 (1967).
60. G. A. Olah, T. Bach and G. K. S. Prakash, *New. J. Chem.*, **15**, 57 (1991).
61. H. Werner and W. Hofman, *Chem. Ber.*, **114**, 2681 (1981).
62. S. R. Berryhill and B. Sharmow, *J. Organomet. Chem.*, **221**, 143 (1981).
63. G. Thum, W. Ries, D. Greisinger and W. Malisch, *J. Organomet. Chem.*, **252**, C61 (1983).
64. S. R. Berryhill, G. L. Clevenger and F. Y. Burdulu, *Organometallics*, **4**, 1509 (1985).
65. S. R. Berryhill and R. J. P. Corriu, *J. Organomet. Chem.*, **370**, C1 (1989).
66. K. H. Pannell, J. M. Rozell and W. M. Tsai, *Organometallics*, **6**, 2085 (1987).
67. K. H. Pannell, J. Castillo-Ramirez and L. Cervantes, *Organometallics*, **11**, 3139 (1992).
68. P. Pasman and J. J. Suel, *J. Organomet. Chem.*, **301**, 329 (1986).
69. G. L. Crocco, C. S. Young, K. E. Lee and J. A. Gladysz, *Organometallics*, **7**, 2158 (1988).
70. S. Seebald, B. Mayer and U. Schubert, *J. Organomet. Chem.*, **462**, 225 (1993).
71. K. H. Pannell, S. R. Vincenti and R. S. Scott III, *Organometallics*, **6**, 1593 (1987).
72. S. Sharma, R. N. Kapoor, F. Cervantes-Lee and K. H. Pannell, *Polyhedron*, **10**, 1177 (1991).
73. U. Schubert and A. Schenkel, *Chem. Ber.*, **121**, 939 (1988).
74. G. G. Furin, O. A. Vyazankina, B. A. Gostevsky and N. S. Vyazankin, *Tetrahedron*, **44**, 2675 (1988).
75. C. H. Winter, S. Pirzard, D. D. Graf, D. H. Cao and M. J. Heeg, *Inorg. Chem.*, **32**, 3654 (1993).
76. A. P. Hagen and P. J. Russo, *Inorg. Nucl. Chem.*, **6**, 507 (1970).
77. P. Jutzi and A. Seufert, *Angew. Chem.*, **88**, 333 (1976).
78. P. Jutzi and M. Kuhn, *Chem. Ber.*, **107**, 1228 (1974).
79. P. Jutzi, M. Kuhn and F. Herzog, *Chem. Ber.*, **108**, 2439 (1975).
80. A. M. Cardoso, R. J. H. Clark and S. Moorhouse, *J. Chem. Soc., Dalton Trans.*, 1156 (1980).
81. S. Dev and J. P. Selegue, *J. Organomet. Chem.*, **469**, 107 (1994).
82. J. M. Rozell, Jr. and R. R. Janes, *Organometallics*, **4**, 2206 (1985).
83. P. Jutzi and A. Seufert, *J. Organomet. Chem.*, **169**, 373 (1979).
84. S. Ciruelos, J. Cuenca, P. Goméz-Sal, A. Manzanero and P. Royo, *Organometallics*, **14**, 177 (1995).
85. P. Jutzi and M. Kuhn, *J. Organomet. Chem.*, **173**, 221 (1979).
86. A. V. Churakov, *Khim.*, **35**, 483 (1994); *Chem Abstr.*, **122**, 187720 (1994)
87. P. Jutzi and M. Schneider, unpublished results.
88. P. Jutzi, B. Krato, M. Hursthouse and A. J. Howes, *Chem. Ber.*, **120**, 585 (1987).
89. H. W. Roesky, S. Schulz, H. J. Koch, G. M. Sheldrick, D. Stalke and A. Kuhn, *Angew. Chem.*, **105**, 1828 (1993).
90. G. H. Mena, F. Palacios, P. Royo and R. Serrano, *J. Organomet. Chem.*, **340**, 37 (1988).
91. T. Okamato, H. Yasuda, A. Nakamura, Y. Kai, N. Kanehisa and N. Kasai, *J. Am. Chem. Soc.*, **110**, 5008 (1988).
92. F. R. Lembke, R. S. Simons and W. J. Youngs, *Organometallics*, **15**, 216 (1996).
93. W. Malisch, G. Thum, D. Wilson, P. Lorz, U. Wachter and W. Seelbach, in *Silicon Chemistry* (Eds. J. Y. Corey, E. R. Corey and P. G. Gaspar), Ellis Horwood, Chichester, 1988.
94. U. The and N. E. Purs, *Polyhedron*, **14**, 1 (1995).
95. J. Dahlhaus, P. Jutzi, H. J. Frenck and W. Kulisch, *Adv. Mat.*, **5**, 321 (1993).
96. M. Waltz, M. Nieger, D. Gudat and E. Niecke, *Z. Anorg. Allg. Chem.*, **621**, 1951 (1995); M. Waltz, M. Nieger and E. Niecke, *Organosilicon Chemistry II, From Molecules to Materials* (Eds. N. Auner and J. Weis), VCH Publ., Weinheim, 1996.

97. A. F. Cummingham, Jr., *Organometallics*, **13**, 2480 (1994).
98. J. Okuda and E. Herdtweck, *Chem. Ber.*, **121**, 1899 (1988).
99. M. D. Fryzuk, S. S. M. Mao, P. B. Duval and S. J. Rettig, *Polyhedron*, **14**, 11 (1995).
100. J. Okuda, *Comments Inorg. Chem.*, **16**, 185 (1994).
101. P. Jutzi, J. Kleimeier, R. Krallmann, H.-G. Stammler and B. Neumann, *J. Organomet. Chem.*, **462**, 57 (1993).
102. H. Köpf and W. Kahl, *J. Organomet. Chem.*, **64**, C37 (1974).
103. H. Plenio, *J. Organomet. Chem.*, **435**, 21 (1992)
104. K. D. Janda, W. W. McConnell, G. O. Nelson and M. E. Wright, *J. Organomet. Chem.*, **259**, 139 (1983).
105. M. Kumada, T. Kondo, K. Mimura, H. Ishikawa, K. Yamamoto, S. Ikeda and M. Kondo, *J. Organomet. Chem.*, **43**, 293 (1972).
106. M. D. Curtis, J. J. D'Enrico, D. N. Duffi, P. S. Epstein and L. G. Bell, *Organometallics*, **2**, 1808 (1983).
107. U. Siemeling, B. Neumann and H. G. Stammler, *Chem. Ber.*, **126**, 1311 (1993).
108. U. Siemeling, R. Krallmann, P. Jutzi, B. Neumann and H. G. Stammler, *Monatsh. Chem.*, **125**, 579 (1994).
109. P. Jutzi and J. Dahlhaus, *Phosphorous, Sulfur, and Silicon*, **87**, 73 (1994).
110. J. K. Pudelski, D. A. Foucher, C. H. Honeyman, A. J. Lough, I. Manners, S. Barlow and D. O'Hare, *Organometallics*, **14**, 2470 (1995).
111. S. Ciruelos, T. Cuenca, J. C. Flores, R. Goméz, P. Goméz-Sol and P. Royo, *Organometallics*, **12**, 944 (1993).
112. P. A. Wegner, V. A. Uski, R. P. Kiesler, S. Dabestani and V. W. Day, *J. Am. Chem. Soc.*, **99**, 4847 (1977).
113. A. W. Cordes. B. Durham and E. Askew, *Acta Crystallogr.*, **C46**, 896 (1990).
114. V. W. Day, H. R. Thompson, G. O. Nelson and M. E. Wright, *Organometallics*, **2**, 494 (1983).
115. P. Goméz-Sal, E. de Jesús, A. J. Pérez and P. Royo, *Organometallics*, **12**, 4633 (1993).
116. J. E. Nifant'ev, K. A. Butakov, Z. G. Aliev and J. F. Urazovskii, *Metallorg. Khim.*, **4**, 1265 (1991); *Chem. Abstr.*, **116**, 59554a (1992).
117. F. Amor, P. Goméz-Sal, E. de Jesús, A. Martin, A. J. Pérez, P. Royo and A. Vázquez de Migel, *Organometallics*, **15**, 2103 (1996).
118. H. Werner, F. Lippert and T. Boley, *Z. Anorg. Allg. Chem.*, **620**, 2053 (1994).
119. G. O. Nelson and M. E. Wright, *J. Organomet. Chem.*, **206**, C21 (1981).
120. J. Heck, K. A. Kriebisch and H. Mellinghoff, *Chem. Ber.*, **121**, 1753 (1988).
121. W. Abriel, G. Baum, J. Heck and K. A. Kriebisch, *Chem. Ber.*, **123**, 1767 (1990).
122. W. Abriel, G. Baum, H. Burdorf and J. Heck, *Z. Naturforsch.*, **46b**, 841 (1991).
123. W. van den Berg, C. E. Boot, J. G. M. van der Linden, W. P. Bosman, J. M. M. Smits, P. T. Beurskens and J. Heck, *Inorg. Chim. Acta*, **216**, 1 (1994).
124. J. Park, Y. Seo, S. Cho, D. Wuang, K. Kim and I. Chang, *J. Organomet. Chem.*, **489**, 23 (1995).
125. D. L. Zechel, D. A. Foucher, J. K. Pudelski, G. P. A. Yap, A. L. Rheingold and I. Manners, *J. Chem. Soc., Dalton Trans.*, 1893 (1995).
126. M. A. Giardello, V. P. Conticello, L. Brard, M. R. Gagne and T. J. Marks, *J. Am. Chem. Soc.*, **116**, 12041 (1994).
127. H. Schumann, J. A. Meese-Marktschaffel and L. Esser, *Chem. Rev.*, **95**, 865 (1995).
128. H. H. Brintzinger, D. Fischer, R. Mühlhaupt, B. Rieger and R. Waymouth, *Angew. Chem.*, **107**, 1255 (1995); *Angew. Chem., Int. Ed. Engl.*, **34**, 1134 (1995).
129. G. Fink, R. Mühlhaupt and H. H. Brintzinger (Eds.), *Ziegler Catalysts*, Springer-Verlag, Berlin–Heidelberg, 1995.
130. A. B. Fischer, J. B. Kinney, R. H. Staley and M. S. Wrighton, *J. Am. Chem. Soc.*, **101**, 6501 (1979).
131. A. G. Osborne and R. H. Whiteley, *J. Organomet. Chem.*, **101**, C27 (1975).
132. M. S. Wrighton, M. C. Palazotto, A. B. Bocarsly, J. M. Bolts, A. B. Fischer and L. Nadjo, *J. Am. Chem. Soc.*, **100**, 7264 (1978).
133. J. R. Butler, W. R. Cullen and S. J. Rettig, *Can. J. Chem.*, **65**, 1452 (1987).
134. D. L. Zechel, K. C. Hultsch, R. Rulkens, D. Balaishis, Y. Ni, J. K. Pudelski, A. J. Lough, I. Manners and D. A. Foucher, *Organometallics*, **15**, 1972 (1996).
135. V. V. Dement'ev, F. Cervantes-Lee, L. Parkanyi, H. Sharma, K. H. Pannell, M. T. Nguyen and A. Diaz, *Organometallics*, **12**, 1983 (1993).

136. H. Evans-Stoeckli, A. G. Osborne and R. H. Whiteley, *Helv. Chim. Acta*, **59**, 2402 (1976).
137. W. Finckh, B. Z. Tang, D. A. Foucher, D. B. Zamble, R. Ziembinski, A. Lough and I. Manners, *Organometallics*, **12**, 823 (1993).
138. J. B. Sheridan, A. J. Lough and I. Manners, *Organometallics*, **15**, 2195 (1996).
139. P. R. Jones, J. M. Rozell, Jr. and B. M. Campbell, *Organometallics*, **4**, 133 (1985).
140. T. J. Barton, G. T. Burns, E. V. Arnold and J. Clardy, *Tetrahedron Lett.*, **22**, 7 (1981).
141. V. K. Belsky, N. N. Zemlyansky, J. V. Borisova, N. D. Kolosova and J. P. Beletskaya, *Cryst. Struct. Commun.*, **11**, 497 (1982).
142. J. Hiermeier, F. H. Köhler and G. Müller, *Organometallics*, **10**, 1787 (1991).
143. U. Siemeling, P. Jutzi, B. Neumann, H.-G. Stammler and M. B. Hursthouse, *Organometallics*, **11**, 1328 (1992).
144. P. Jutzi, I. Mieling, B. Neumann and H.-G. Stammler, *J. Organomet. Chem.*, in press.
145. P. Jutzi, R. Krallmann, G. Wolf, B. Neumann and H. G. Stammler, *Chem. Ber.*, **124**, 2391 (1991).
146. A. Cano, T. Cuenca, P. Goméz-Sal, B. Royo and P. Royo, *Organometallics*, **13**, 1688 (1994).
147. I. Mieling, Ph. D. Thesis, University of Bielefeld, 1993.
148. W. Mengele, J. Diebold, C. Troll, W. Röll and H. H. Brintzinger, *Organometallics*, **12**, 1931 (1993).
149. F. Amor, P. Goméz-Sal, E. de Jesús, P. Royo and A. V. de Miguel, *Organometallics*, **13**, 4322 (1994).
150. U. Siemeling, P. Jutzi, E. Bill and A. V. Trautwein, *J. Organomet. Chem.*, **463**, 151 (1993).
151. H. Atzkern, J. Hiermeier, F. H. Köhler and A. Steck, *J. Organomet. Chem.*, **408**, 281 (1991).
152. H. Atzkern, B. Bergerat, M. Fritz, J. Hiermeier, P. Hudeszek, O. Kahn, B. Kanellakopulus, F. H. Köhler and M. Ruhs, *Chem. Ber.*, **127**, 277 (1994).
153. U. Siemeling and P. Jutzi, *Chem. Ber.*, **125**, 31 (1992).
154. H. Atzkern, P. Bergerat, H. Beruda, M. Fritz, J. Hiermeier, P. Hudezek, O. Kahn, F. H. Köhler, M. Paul and B. Weber, *J. Am. Chem. Soc.*, **117**, 997 (1995).
155. P. Bergerat, J. Blümel, M. Fritz, J. Hiermeier, P. Hudeszek, O. Kahn and F. H. Köhler, *Angew. Chem.*, **104**, 1285 (1992).
156. D. O. Cowan, C. Le Vanda, J. Park and F. Kaufmann, *Acc. Chem. Res.*, **6**, 1 (1973).
157. M. V. Robin and P. Day, *Inorg. Chem. Radiochem.*, **10**, 247 (1967).
158. H. O. Findea and D. Hanshow, *J. Am. Chem. Soc.*, **114**, 3173 (1992).
159. C. Elschenbroich, A. Bretschneider-Hurley, J. Hurley, W. Massa, S. Wocadlo and J. Pebler, *Inorg. Chem.*, **32**, 5421 (1993).
160. H. Sakurai, S. Hoshi, A. Kamiga, A. Hosomi and C. Kabuto, *Chem. Lett.*, 1781 (1986).
161. J. Michl, 'Polysilanes', in *The Chemistry of Organic Silicon Compounds* (Eds. S. Patai and Z. Rappoport), Wiley, New York, 1989.
162. A. Togni and T. Hayashi, *Ferrocenes*, VCH Publ., Weinheim, 1995.
163. I. Manners, *Adv. Mater.*, **6**, 68 (1994).
164. A. F. Diaz, C. M- Seymour, K. H. Pannell and J. M. Rozell, *J. Electrochem. Soc.*, **137**, 503 (1990).
165. K. H. Pannell, J. M. Rozell and J. M. Ziegler, *Macromolecules*, **21**, 276 (1988).
166. D. A. Foucher, C. H. Honeyman, J. M. Nelson, B. Z. Tang and I. Manners, *Angew. Chem., Int. Ed. Engl.*, **32**, 1709 (1993).
167. I. Manners, *Adv. Organomet. Chem.*, **37**, 131 (1995).
168. M. T. Nguyen, A. F. Diaz, V. V. Dement'ev and K. H. Pannell, *Chem. Mat.*, **5**, 1389 (1993).
169. S. Iijima, J. Motoyama and H. Sano, *Bull. Chem. Soc. Jpn.*, **53**, 3180 (1980).
170. W. H. Morrison, S. Krogsrud and N. Hendrickson, *Inorg. Chem.*, **12**, 1998 (1973).
171. V. V. Dement'ev and K. H. Pannell, *Organometallics*, **12**, 1983 (1993).
172. C. Le Vanda, K. Bechgaard and D. O. Cowan, *J. Org. Chem.*, **41**, 2700 (1976).
173. C. Le Vanda, K. Bechgaard, D. O. Cowan, U. T. Mueller-Westerhoff, P. Eilbracht, G. A. Candela and R. L. Collins, *J. Am. Chem. Soc.*, **98**, 3181 (1976).
174. J. E. Gorton, H. L. Lentzner and W. E. Watts, *Tetrahedron*, **27**, 4353 (1971).
175. S. Barlow, V. J. Murphy, J. S. O. Evans and D. O'Hare, *Organometallics*, **14**, 3461 (1995).
176. R. Rulkens, A. J. Lough and I. Manners, *J. Am. Chem. Soc.*, **116**, 797 (1994).
177. T. J. Kealy and P. L. Pauson, *Nature (London)*, **168**, 1039 (1951).
178. E. O. Fischer and W. Pfab, *Z. Naturforsch.*, **B7**, 377 (1952).
179. G. Wilkinson, M. Rosenblum, M. C. Whiting and R. B. Woodward, *J. Am. Chem. Soc.*, **74**, 2125 (1952).

180. E. O. Fischer and H. Grubert, *Z. Anorg. Allg. Chem.*, **286**, 237 (1956).
181. P. Jutzi, D. Kanne and C. Krüger, *Angew. Chem.*, **98**, 163 (1986).
182. P. Jutzi, U. Holtmann, D. Kanne, C. Krüger, R. Blom, R. Gleiter and J. Hyla-Kryspin, *Chem. Ber.*, **122**, 1629 (1989).
183. B. Wrackmeyer, A. Sebald and L. H. Marwin, *Magn. Reson. Chem.*, **29**, 260 (1991).
184. T. J. Lee and J. E. Rice, *J. Am. Chem. Soc.*, **111**, 2011 (1989).
185. G. Bruno, E. Ciliberto, J. L. Fragala and P. Jutzi, *J. Organomet. Chem.*, **289**, 268 (1985).
186. A. G. Davies and J. Lusztyk, *J. Chem. Soc., Perkin Trans. 2*, 692 (1981).
187. P. Jutzi, *Comments Inorg. Chem.*, **6**, 123 (1987).
188. P. Jutzi, in *Frontiers of Organosilicon Chemistry* (Eds. A. R. Bassindale and P. P. Gaspar), The Royal Society of Chemistry, Cambridge, 1991.
189. P. Jutzi. U. Holtmann, A. Möhrke, E. A. Bunte and D. Eikenberg, unpublished results; P. Jutzi, in *Organosilicon Chemistry* (Eds. N. Auner and J. Weis), VCH-Publ., Weinheim, 1993.
190. P. Jutzi, E. A. Bunte, U. Holtmann, B. Neumann and H.-G. Stammler, *J. Organomet. Chem.*, **446**, 139 (1993).
191. P. Jutzi, U. Holtmann, H. Bögge and A. Müller, *J. Chem. Soc., Chem. Commun.*, 305 (1988).
192. J. Maxka and Y. Apeloig, *J. Chem. Soc., Chem. Commun.*, 737 (1990).
193. P. Jutzi and E. A. Bunte, *Angew. Chem.*, **104**, 1636 (1992); *Angew. Chem., Int. Ed. Engl.*, **31**, 1605 (1992).
194. P. Jutzi, A. Möhrke, A. Müller and H. Bögge, *Angew. Chem.*, **101**, 1527 (1989); *Angew. Chem., Int. Ed. Engl.*, **28**, 1518 (1989).
195. P. Jutzi, D. Eikenberg, A. Möhrke, B. Neumann and H. G. Stammler, *Organometallics*, **15**, 753 (1996).
196. P. Jutzi, D. Eikenberg, E. A. Bunte, A. Möhrke, B. Neumann and H.-G. Stammler, *Organometallics*, **15**, 1930 (1996).
197. P. Jutzi, D. Eikenberg, B. Neumann and H.-G. Stammler, *Organometallics*, **15**, 3659 (1996).
198. M. Tacke, C. Klein, D. J. Stufkens, A. Oskam, P. Jutzi and E. A. Bunte, *Z. Anorg. Allg. Chem.*, **619**, 865 (1993).
199. M. Tacke, in *Organosilicon Chemistry III* (Eds. N. Auner and J. Weis), VCH-Publ., Weinheim, in press.
200. P. Jutzi and E. A. Bunte, *Angew. Chem.*, **104**, 1636 (1992); *Angew. Chem., Int. Ed. Engl.*, **31**, 1605 (1992).
201. P. Jutzi and A. Möhrke, *Angew. Chem.*, **102**, 913 (1990); *Angew. Chem., Int. Ed. Engl.*, **29**, 893 (1990).

CHAPTER 37

Recent advances in the chemistry of cyclopolysilanes

EDWIN HENGGE[†] and HARALD STÜGER

Institute of Inorganic Chemistry, Technical University, Graz, Austria
Fax: +43 316 873 8701; e-mail: stueger@anorg.tu-graz.ac.at

I. INTRODUCTION	2177
II. GENERAL METHODS FOR CYCLOPOLYSILANE SYNTHESIS	2178
III. GENERAL PROPERTIES OF CYCLOPOLYSILANES	2180
IV. MONOCYCLIC SILANES	2182
A. Cyclotrisilanes	2182
B. Cyclotetrasilanes	2187
C. Cyclopenta- and Cyclohexasilanes	2188
V. POLYCYCLIC SILANES AND CAGES	2197
A. Linearly Connected Cyclosilanes	2197
B. Annelated Cyclosilanes	2204
C. Polysilane Cages	2205
VI. TRANSITION METAL CONTAINING CYCLOPOLYSILANES	2209
VII. ACKNOWLEDGMENTS	2213
VIII. REFERENCES	2213

I. INTRODUCTION

Cyclopolysilanes have been known since the fundamental work of Kipping in 1921, who synthesized the four-, five- and six-membered perphenylcyclopolysilanes from diphenyldichlorosilane and sodium.[1] After a long period of no activity in this field, Burkhard published the synthesis of dodecamethylcyclohexasilane ($Me_2Si)_6$, the first peralkylated polysilane ring, in 1949.[2] Since then numerous cyclopolysilane derivatives bearing organic substituents have been prepared mainly in the groups of Gilman, West and Kumada[3a–c]. The first solely inorganic substituted cyclopolysilanes $(SiX_2)_n$ ($n = 4$, 5 and 6; X = Cl, Br and I), from which the parent cyclopolysilanes Si_5H_{10} and Si_6H_{12} are readily accessible simply by $LiAlH_4$ reduction, have been found by Hengge and coworkers[4a–d].

[†] Deceased February 1997

The chemistry of organic silicon compounds, Vol. 2
Edited by Z. Rappoport and Y. Apeloig © 1998 John Wiley & Sons Ltd

The parent cyclotetrasilane Si_4H_8 and all the perfluorinated cyclopolysilanes are so far still unknown. More recently, the efforts in cyclopolysilane chemistry are mainly directed towards polycyclic and cage-like compounds and towards partially functionalized derivatives, which may be used as synthons for larger polysilane frameworks.

The chemistry and properties of cyclopolysilanes have been summarized in various earlier reviews. The review published in 1995 by Hengge and Janoschek provides the most up-to-date summary of the field of cyclopolysilane chemistry and covers the literature up to 1994[5a–d]. In the current chapter, therefore, recent results gained since 1994 will be emphasized and only a short survey of older investigations will be given. The interested reader is referred to the reviews cited above for further details of previous work. Cyclopolysilanes containing endocyclic heteroatoms are also beyond the scope of this article.

II. GENERAL METHODS FOR CYCLOPOLYSILANE SYNTHESIS

Organocyclopolysilanes are usually made by the dehalogenative coupling of dialkyl- or diaryldichlorosilanes with alkali metals. The cyclopolysilane yield and the preferred ring size obtained depend strongly on the kind of alkali metal, the solvent and the reaction conditions. Therefore, optimized procedures have to be worked out for the high-yield synthesis of specific compounds in most cases. The perphenylcyclopolysilanes $(Ph_2Si)_4$, $(Ph_2Si)_5$ and $(Ph_2Si)_6$, for instance, are usually made from Ph_2SiCl_2 using lithium metal in refluxing THF (equation 1).

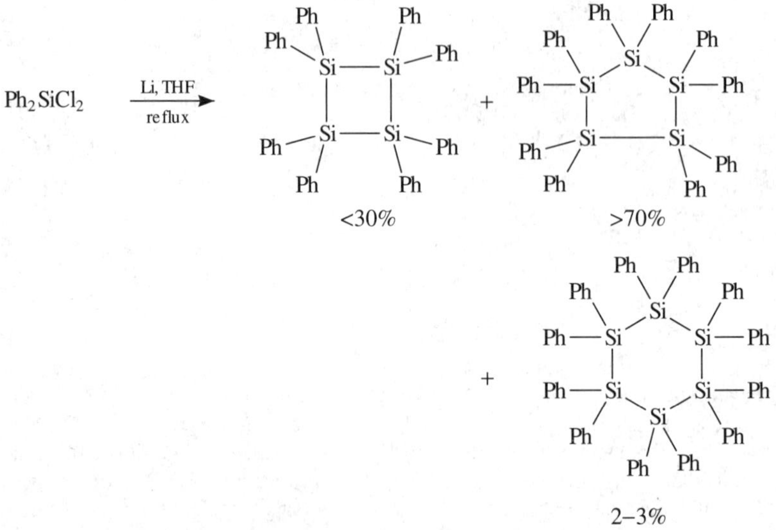

$$Ph_2SiCl_2 \xrightarrow{Li, THF, reflux} (Ph_2Si)_4 \;(<30\%) + (Ph_2Si)_5 \;(>70\%) + (Ph_2Si)_6 \;(2-3\%) \quad (1)$$

Carrying out the condensation reaction with 2.0 equivalents of lithium yields $(Ph_2Si)_4$ as the major product in yields up to 30% in a kinetically controlled reaction. Larger amounts of lithium and longer reaction times give rise to rearrangement reactions and lead finally to the thermodynamically most stable five-membered ring (optimum yield >70%). $(Ph_2Si)_6$ is only produced in yields of 2–3% as a by-product.

The six-membered permethylated analogue dodecamethylcyclohexasilane, however, can be prepared conveniently from Me_2SiCl_2 with sodium–potassium alloy in THF in the presence of an equilibrating catalyst, which causes depolymerization of the initially formed permethylpolysilane polymer[6,7] (equation 2).

$$\text{Me}_2\text{SiCl}_2 \xrightarrow[\text{naphthalene}]{\text{Na/K, THF}} \underset{>80\%}{(\text{Me}_2\text{Si})_6} + \underset{<10\%}{(\text{Me}_2\text{Si})_5} + \underset{\text{traces}}{(\text{Me}_2\text{Si})_7} \qquad (2)$$

The preferred ring size in the equilibrium mixture obtained from diorganodichlorosilanes and alkali metals, therefore, is additionally governed by the bulk of the organic groups attached to silicon. For small substituents like methyl, the six-membered ring is favored, while five- and four-membered rings are preferentially obtained in case of larger groups like phenyl or i-propyl. Even larger substituents like mesityl or t-butyl give rise to the formation of three-membered rings.

In order to quantify the dependence of the preferred ring size obtained in cyclopolysilane synthesis on the bulk of the substituent groups on the silicon Cartledge postulated a set of parameters derived from the rates of acid catalyzed hydrolysis of appropriate hydrosilanes[8]. The model, however, was applied only with limited success[9].

The electrochemical formation of Si–Si bonds by cathodic reduction of dichlorosilanes provides an alternative access to cyclopolysilanes[10,11]. Electrolyses are carried out in single compartment cells with THF/Bu$_4$NBF$_4$ as solvent/electrolyte system at current intensities of about 1 mA cm^{-2}. The cathode material is stainless steel, the anode consists of a sacrificial metal, e.g. magnesium or aluminum. A modified hydrogen anode especially designed for this type of electrolysis[12] can also be used alternatively, and this may be advantageous for large-scale or industrial processes because it avoids the formation of huge amounts of metal halides. Presumably, the reaction is neither thermodynamically nor kinetically controlled in all cases, the determinant factor obviously being the steric conditions on the cathode surface. Perarylated silanes like Ph$_2$SiCl$_2$ always give cyclotetrasilanes as the sole cyclic reaction products and polymeric material is formed only in negligible amounts. When the aryl substituents on the silicon are replaced by methyl groups, ring formation is observed only at low dichlorosilane concentrations. Thus, electrolyses carried out with silane concentrations below 0.5 mol l^{-1} yield (Me$_2$Si)$_6$ in case of Me$_2$SiCl$_2$ or (MePhSi)$_5$ in case of MePhSiCl$_2$, respectively, as the main cyclic reaction products. With increasing dichlorosilane concentration, intermolecular reaction favors the formation of linear polymers in both cases. Cyclic organohydropolysilanes (RHSi)$_n$ cannot be synthesized electrochemically at all. The electrolysis of organohydrodichlorosilane solutions under various conditions exclusively yields polymeric material[11]. In certain cases the electrochemical method can be used for the synthesis of cyclopolysilanes, which are not accessible by the usual Wurtz-type route. Bis(pentafluorophenyl)dichlorosilane, for instance, affords the corresponding four-membered ring only electrochemically but not under the influence of alkali metals. Further examples are given elsewhere[13,14].

Catalytic dehydrogenative coupling of organohydrosilanes, a common method for the formation of linear polysilanes, can also be applied to cyclopolysilane synthesis. Thus, benzylsilane reacts to give all-*trans*-hexabenzylcyclohexasilane in the presence of dimethyltitanocene (equation 3). However, extremely long reaction times are required and only moderate yields are obtained[15].

$$\text{PhCH}_2\text{SiH}_3 \xrightarrow[\text{6 weeks, r.t.}]{\text{Cp}_2\text{TiMe}_2, \text{ toluene}} \underset{35\%}{\left[\text{Si(H)CH}_2\text{Ph}\right]_6} + \text{oligomers} \qquad (3)$$

Cyclopolysilanes bearing several different substituents can be assembled in a step-by-step procedure from small appropriately substituted organosilane precursor molecules. An example originally presented by Uhlig[16] and modified later in order to obtain the reaction product free of polymers[17] is shown in Scheme 1, where

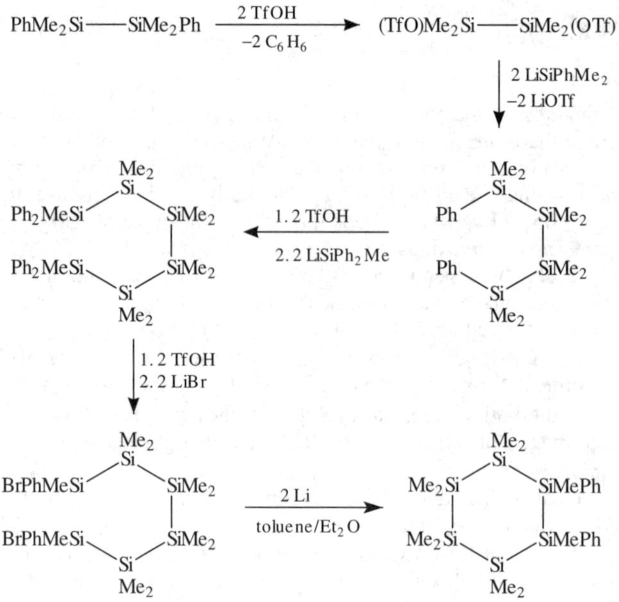

PhMe$_2$Si——SiMe$_2$Ph $\xrightarrow[-2\,C_6H_6]{2\,\text{TfOH}}$ (TfO)Me$_2$Si——SiMe$_2$(OTf)

$\downarrow$ 2 LiSiPhMe$_2$
-2 LiOTf

[Cyclic structure: Ph$_2$MeSi–SiMe$_2$/SiMe$_2$–Ph$_2$MeSi with Si(Me$_2$) bridges] $\xleftarrow[\text{2. 2 LiSiPh}_2\text{Me}]{\text{1. 2 TfOH}}$ [Cyclic structure: Ph–SiMe$_2$/SiMe$_2$–Ph with Si(Me$_2$) bridges]

1. 2 TfOH
2. 2 LiBr $\downarrow$

[Cyclic structure: BrPhMeSi–SiMe$_2$/SiMe$_2$–BrPhMeSi with Si(Me$_2$) bridges] $\xrightarrow[\text{toluene/Et}_2\text{O}]{\text{2 Li}}$ [Cyclic structure: Me$_2$Si–SiMePh/SiMePh–Me$_2$Si with Si(Me$_2$) bridges]

SCHEME 1

1,2-diphenyldecamethylcyclohexasilane is synthesized systematically starting from 1,2-diphenyltetramethyldisilane by several Si–Si bond formation and selective substituent exchange steps.

III. GENERAL PROPERTIES OF CYCLOPOLYSILANES

The chemical and thermal stability of cyclic polysilanes varies strongly with the ring size and the kind of substituents on silicon. Most of the peralkylated and perarylated cyclosilanes do not react with oxygen or atmospheric moisture and therefore can be handled and stored without any additional precautions. The perarylated cyclosilanes furthermore exhibit high melting points and are remarkably stable at elevated temperatures. With decreasing ring size or in the presence of reactive substituents, however, the reactivity of cyclopolysilanes is strongly enhanced. While (SiPh$_2$)$_6$, for instance, reacts with mild halogenating agents, hydrogen halides or lithium only under forcing conditions to give small fragments, (SiPh$_2$)$_4$ is easily cleaved to the corresponding 1,4-disubstituted products. The perhalo derivatives (SiX$_2$)$_n$ ($n = 4$, 5 and 6; X = Cl, Br and I) immediately react with traces of water due to the sensitivity of the Si–halogen bonds. (SiCl$_2$)$_4$, (SiBr$_2$)$_4$ and (SiI$_2$)$_4$, but also (SiMe$_2$)$_4$, additionally undergo thermal ring-opening polymerization even at ambient temperatures (compare Section IV.B). Cyclotrisilanes generally need to be stabilized by the presence of sterically demanding substituents because of their high ring strain. Hexakis(2,6-dimethylphenyl)cyclotrisilane, for instance, is totally stable against the atmosphere and melts above 200 °C without any decomposition, whereas cyclotrisilanes with smaller substituents only can be isolated in argon matrices at low temperatures.

Compared to hydrocarbon rings, cyclopolysilanes are much more flexible due to their rather flat potential energy surfaces. Five- and six-membered cyclopolysilanes usually adopt folded conformations. The energy differences between stable conformers, however, exhibit values of only 5–9.2 kJ mol^{-1} in Si$_5$H$_{10}$ and Si$_6$H$_{12}$. Additionally, the rotational barriers around the Si−Si bonds in cyclopolysilanes are low, and this generally facilitates the interconversion of different conformers[18,19]. The flexibility of the cyclopolysilane skeletons is also reflected by the variable structures of the four-membered rings. Depending on the substituents, dihedral angles between 39.39° in 1,1,2,2-tetraneopentyl-3,3,4,4-tetraisopropylcyclotetrasilane and 0° in [(Me$_3$Si)$_2$Si]$_4$ have been found. A compilation of the dihedral angles in cyclotetrasilanes measured so far can be found in Ref. 5d. (Me$_2$Si)$_4$ adopts a planar structure in the solid state according to X-ray crystallography[20]. Electron diffraction measurements, however, exhibit a puckered conformation with a folding angle of 29.4° for the same molecule in the gas phase[21]. An explanation for these marked differences in the structures of various cyclotetrasilane derivatives has not yet been given.

One of the most striking properties of cyclopolysilanes is their strong absorption of light in the near-UV/visible region; λ_{max} values for the lowest-energy UV absorption bands range from 250 up to nearly 420 nm, depending on the ring size and the substituents attached to the cyclopolysilane ring. Representative examples for low-energy UV absorption bands of cyclopolysilanes are summarized in Table 1.

In cyclic polysilanes, delocalization of σ-(Si−Si) electrons over the ring is regarded as responsible for the unusual electronic properties[22]. The UV absorption spectra of the cyclic compounds (R$_2$Si)$_n$ with $n = 3-8$ exhibit a bathochromic shift as the ring size decreases. Larger rings behave like linear polysilanes; the UV absorption maxima move to longer wavelengths with an increasing number of silicon atoms. Thus, the lowest electronic excitation energies are observed for cyclotri- and tetrasilanes, which at least can be partially explained by ring strain present in these small ring compounds causing an increase in the energy of the highest occupied molecular orbitals. The relative energies of the HOMOs in cyclopolysilanes can be easily derived from the UPS spectra[23]. As follows from comparison of the electronic excitation energies with the ionization potentials, the destabilization of the HOMOs cannot be responsible exclusively for the trends observed for the low-energy UV absorption bands of cyclopolysilanes. Progressive stabilization of the LUMOs with decreasing ring size also occurs and must be taken into account.

Substituent influences on UV excitation energies of cyclopolysilanes are also well pronounced. For cyclosilane derivatives (SiX$_2$)$_n$ of the same ring size, bathochromic shifts of the first absorption bands are observed in the order X = H < alkyl < MeO ∼ Cl < Br < I[24] (Table 1). Extraordinarily low energy first UV absorption bands are shown by the perhalocyclopolysilanes, so that some of these compounds even exhibit colors. UPS measurements and theoretical calculations of the HOMO−LUMO energies suggest that intramolecular charge transfer might be responsible for the unusual low energy of the corresponding electronic transitions[25].

The cyclic polysilanes can be reduced to radical anions either electrolytically or by reaction with alkali metals quite similar to aromatic rings. Numerous substituted

TABLE 1. Low-energy UV absorption bands of selected cyclopolysilanes

X	Si$_4$X$_8$		Si$_5$X$_{10}$		Si$_6$X$_{12}$		References
	λ_{max} (nm)	ε	λ_{max} (nm)	ε	λ_{max} (nm)	ε	
H	—	—	215	3000	192	6000	31
Me	302	250	275sh	700	258sh	1100	24,32
Et	304	180	266	1400	259sh	1900	33
MeO	—	—	295sh	1000	316	1000	24
Cl	394	50	305sh	1000	322	300	24
Br	402	50	320sh	900	344	300	24
I	424	75	380sh	500	400sh	450	24

cyclopolysilane radical anions have been made; however, attempts to reduce the parent compound Si_5H_{10} failed. Detailed ESR[26,27] and ENDOR[28] investigations of cyclosilane radical anions exhibit complete delocalization of the odd electron over the cyclosilane ring, consistent with the assumption of σ-(Si—Si) delocalization in larger Si—Si bonded systems. Important information about the nature of the unfilled orbitals in cyclopolysilanes can also be obtained from ESR studies of their radical anions. The corresponding results have been reviewed recently[22]. Radical cations of cyclopolysilanes have also been observed by ESR[29] after treatment of the solutions of cyclopolysilanes in CH_2Cl_2 with $AlCl_3$. The oxidation of cyclopolysilanes, however, has so far not been studied so extensively.

The magnitude of the Si—Si vibrational force constants in cyclopolysilane rings also depends on the ring size and on the substituents attached to silicon. Compared with other silanes like disilanes, the force constants are generally weaker. Detailed results are summarized elsewhere[5d].

Semiempirical calculations (AM1), which are in good agreement with *ab initio* calculations and experimental data, showed for cyclic polysilanes that the first ionization energy is higher with increasing number of silicon atoms. This stands in contrast to linear chains, where the ionization energy decreases with the number of silicon atoms[30].

IV. MONOCYCLIC SILANES

A. Cyclotrisilanes

Hexakis(2,6-dimethylphenyl)cyclotrisilane, the first known stable three-membered cyclosilane, was prepared in 1982 by Masamune and coworkers from dichlorobis(2,6-dimethylphenyl)silane by the reductive dechlorination with lithium naphthalenide in low yield[34]. Other cyclotrisilanes bearing substituents of comparable steric bulk have been synthesized analogously. The chemistry of the cyclotrisilanes presently known has recently been reviewed[35].

For the preparation of derivatives with sterically less demanding side groups, alternative routes have to be used. Thus, for instance, $[(Et_2CH)_2Si)]_3$ and $[(i-Pr)_2Si)]_3$ are accessible only by the reductive dechlorination of the corresponding 1,3-dichlorotrisilanes applying highly specific experimental conditions[36] (equation 4).

$$Cl(SiR_2)_3Cl \xrightarrow[\text{THF/pentane}]{t\text{-BuLi/Li/C}_{10}H_8} [R_2Si]_3 \quad (4)$$

$$R = Me_2CH, Et_2CH$$

In case of exceptionally strained cyclotrisilanes like $[(t\text{-Bu})_2Si)]_3$, it turned out to be advantageous to start from the diorganodibromo- or the diorganodiiodosilane instead of the dichlorosilane. Due to the lower Si—halogen bond energy in these cases, the reduction proceeds much smoother and the desired reaction product is obtained in good yield[37] (equation 5).

$$(t\text{-Bu})_2SiX_2 \xrightarrow{\text{Li/C}_{10}H_8} [(t\text{-Bu})_2Si]_3 \quad (5)$$

$$X = Br, I$$

The preparative application of photolytic ring contraction reactions of cyclotetrasilanes, which also lead to the formation of cyclotrisilanes, is largely impeded by the limited stability of the cyclotrisilane moiety. The photolysis of octaisopropylcyclotetrasilane, for instance, affords the corresponding cyclotrisilane only as an intermediate, which can be detected by UV spectroscopy. Further silylene extrusion gives rise to the formation of the unstable tetraisopropyldisilene and its successive decomposition products (equation 6)[38,39].

$$[(i\text{-Pr})_2\text{Si}]_4 \xrightarrow[-(i\text{-Pr})_2\text{Si:}]{h\nu} \underset{(i\text{-Pr})_2\text{Si}\text{———}\text{Si}(\text{Pr-}i)_2}{\overset{(i\text{-Pr})_2}{\underset{\text{Si}}{\triangle}}} \xrightarrow[-(i\text{-Pr})_2\text{Si:}]{h\nu} [(i\text{-Pr})_2\text{Si}\text{═}\text{Si}(\text{Pr-}i)_2] \downarrow \text{successive products} \quad (6)$$

Functionalized cyclotrisilane derivatives bearing oxygen- or nitrogen-containing substituents are also known. The synthesis of the aryl-alkoxy cyclotrisilane trimesityl-tris(2,6-diisopropylphenoxy)cyclotrisilane can be easily accomplished by the photolysis of 2-mesityl-2-(2,6-diisopropylphenoxy)-1,1,1,3,3,3-hexamethyltrisilane via several silylene extrusion and addition steps (equation 7)[40].

$$R^1(R^2O)\text{Si}(\text{SiMe}_3)_2 \xrightarrow{h\nu} R^1(R^2O)\text{Si:} \xrightarrow{\times 2} R^1(R^2O)\text{Si}\text{═}\text{Si}(OR^2)R^1$$

$R^1 = 2,4,6\text{-Me}_3\text{C}_6\text{H}_2$

$R^2 = 2,6\text{-}i\text{-Pr}_2\text{C}_6\text{H}_3$

$$R^1(R^2O)\text{Si:} \downarrow$$

$$[R^1(R^2O)\text{Si}]_3 \quad (7)$$

The aminofunctional cyclotrisilane derivative hexakis[2-(dimethylaminomethyl)phenyl]cyclotrisilane can be formed by the reductive dechlorination of the corresponding dichloride Ar_2SiCl_2 (Ar = 2-$\text{Me}_2\text{NCH}_2\text{C}_6\text{H}_4$) with magnesium[41].

Molecular structures and physical properties of cyclotrisilanes primarily reflect the inherent ring strain of the three-membered ring. The cyclotrisilanes typically show isosceles or equal-sided triangle structures. The Si—Si bond distances observed are generally longer compared with the normal value of about 234 pm. The longest Si—Si distance detected so far was found for $[(t\text{-Bu})_2\text{Si})]_3$ with a bond length of 251.1 pm[37]. Despite the elongated Si—Si bonds, the ^{29}Si-NMR spectra of the cyclotrisilanes exhibit considerable shielding of the ^{29}Si nuclei. Typical ^{29}Si-chemical shift values for the alkyl and arylcyclotrisilanes, for instance, fall into the range between −4 and −82 ppm. A more comprehensive discussion of structural and physicochemical aspects of cyclotrisilanes can be found in several recent reviews[35,42,43].

The chemical reactions of the three-membered polysilane rings are initiated quite generally by Si—Si bond cleavage in order to relieve the high ring strain. Three fundamental pathways for the degradation of the cyclotrisilane ring are possible: (1) scission of one Si—Si bond yielding linear trisilanes, (2) scission of two Si—Si bonds leading to the formation of a disilene and a silylene and (3) scission of all Si—Si bonds giving monosilanes (Scheme 2).

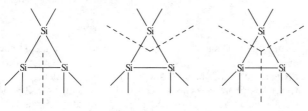

SCHEME 2

In the absence of light or a catalyst, cyclotrisilanes usually react to give acyclic trisilanes or products of ring expansion, depending on the reaction partner. Thus, [(t-Bu)$_2$Si)]$_3$ and the heavier halogens, for instance, afford the corresponding hexa-t-butyl-1,3-dihalotrisilanes. With elemental oxygen or DMSO, however, the oxatrisiletane and the 1,3-dioxatrisilolane are formed (Scheme 3)[35].

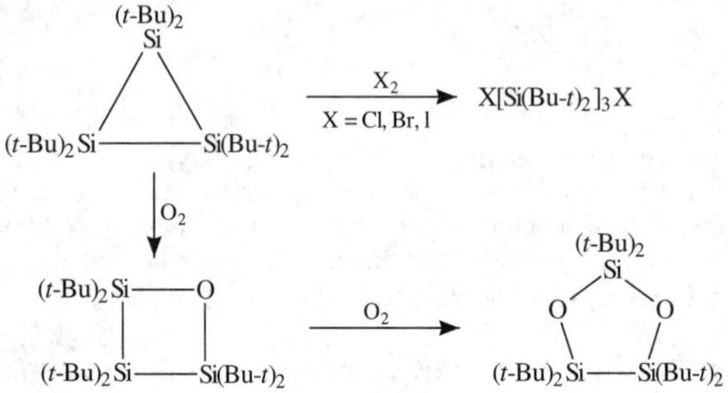

SCHEME 3

The photolysis of cyclotrisilanes bearing at least one aryl group on each silicon atom usually affords the corresponding disilene and a silylene, which dimerizes to give more disilene. Persilylated derivatives behave similarly; stable tetraalkyldisilenes, however, cannot be prepared using this route. Thus, when arylcyclotrisilanes are photolyzed in the presence of suitable trapping agents, disiliranes are produced selectively via [2 + 1]-cycloaddition reactions[35]. In the case of [(t-Bu)$_2$Si)]$_3$, which also gives the corresponding disilene and a short-lived silylene upon UV irradiation, the formation of trapping products of both reactive intermediates can be observed, depending on the nature of the trapping reagent a–b (Scheme 4)[44].

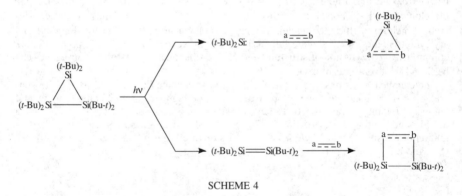

SCHEME 4

The photolysis and thermolysis of hexa-t-butylcyclotrisilane in the absence of trapping reagents was investigated recently[45]. Heating the compound in a sealed tube to 185 °C affords a 5-methylene-1,2,3,7,8,9-hexasilanonane derivative (Scheme 5).

When [(t-Bu)$_2$Si)]$_3$ is photolyzed in the UV with wavelengths < 300 nm, a complex mixture of products containing t-Bu−Si and H−Si bonds is obtained. The only isolated product was hepta-t-butylcyclotetrasilane (compare equation 8).

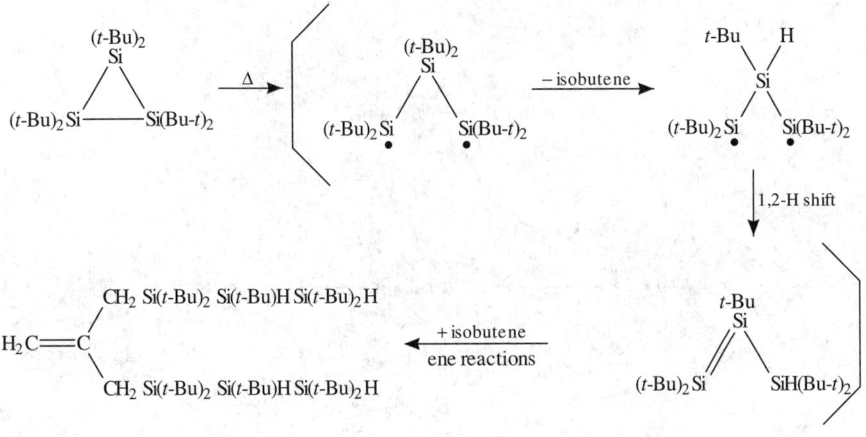

SCHEME 5

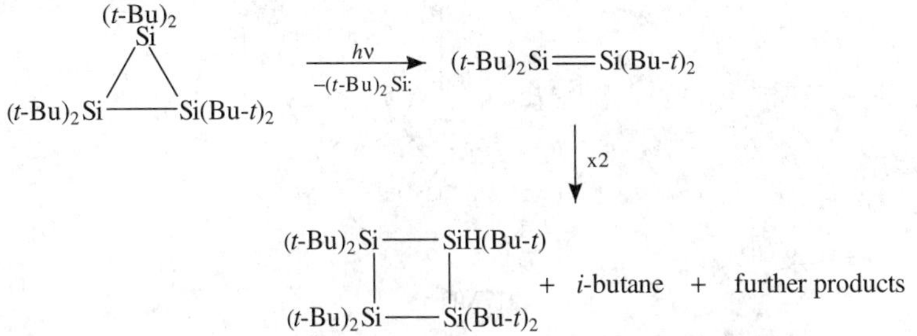

(8)

Further cycloaddition reactions of silylenes generated by the photolysis of cyclotrisilanes have been published since Weidenbruch and coworkers summarized these reactions in an excellent review. Different siliranes were prepared by [2+1]-cycloaddition of di-*t*-butylsilylene to various alkenes and dienes (Scheme 6)[46]. Quite interesting results are obtained from the photolysis of hexa-*t*-butylcyclotrisilane in the presence of unsaturated five-membered ring compounds[47] (Scheme 7). With cyclopentadiene and furane, [4 + 2]-cycloaddition of the photolytically generated disilene occurs only as a side reaction. Furthermore, [2 + 1]-cycloaddition of the intermediately formed silylene is highly favored and siliranes are primarily obtained. A totally different course is observed for the reaction in the presence of thiophene. The disilene abstracts the sulfur atom with the formation of the 1,2-disilathiirane as the major product with an extremely short Si–Si distance of 230.49 pm.

Addition of phenylacetylene or pent-1-yne to a solution of hexakis[2-(dimethylaminomethyl)phenyl]cyclotrisilane yields spectroscopically pure 1,2-disilacyclobutenes (equation 9)[41].

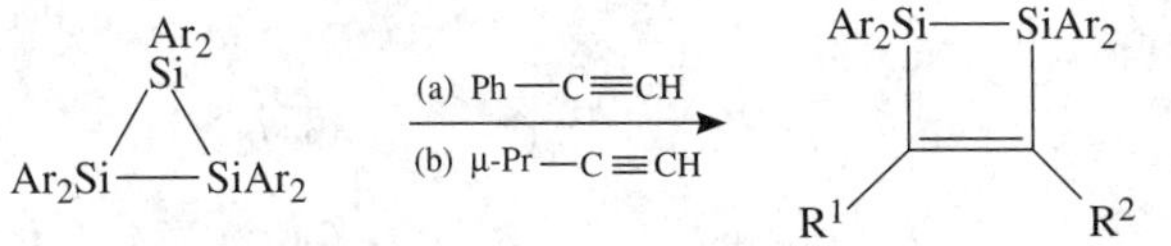

(9)

Ar = 2-(Me$_2$NCH$_2$)C$_6$H$_4$

(a) R^1 = Ph, R^2 = H
(b) R^1 = *n*-Pr, R^2 = H

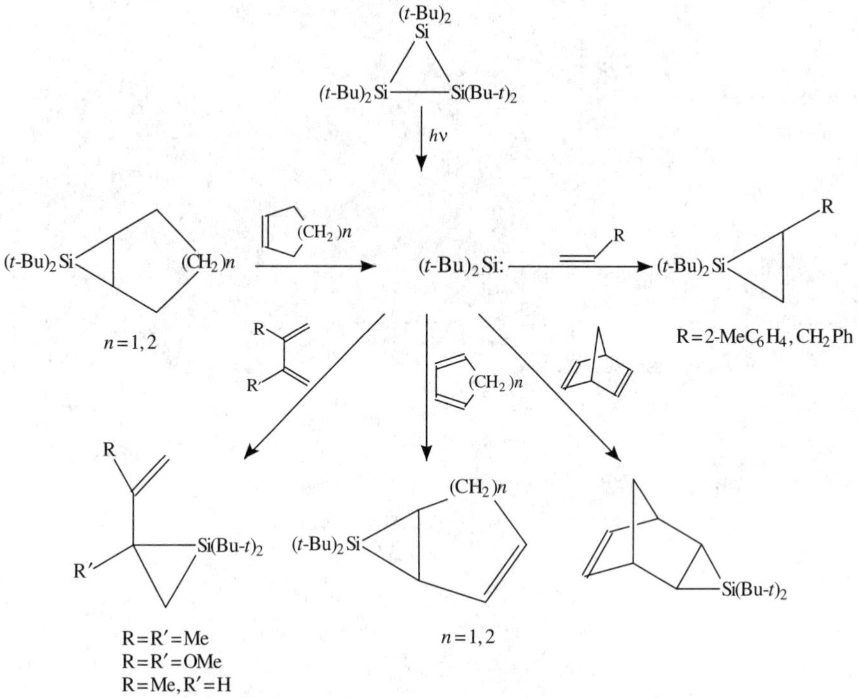

SCHEME 6

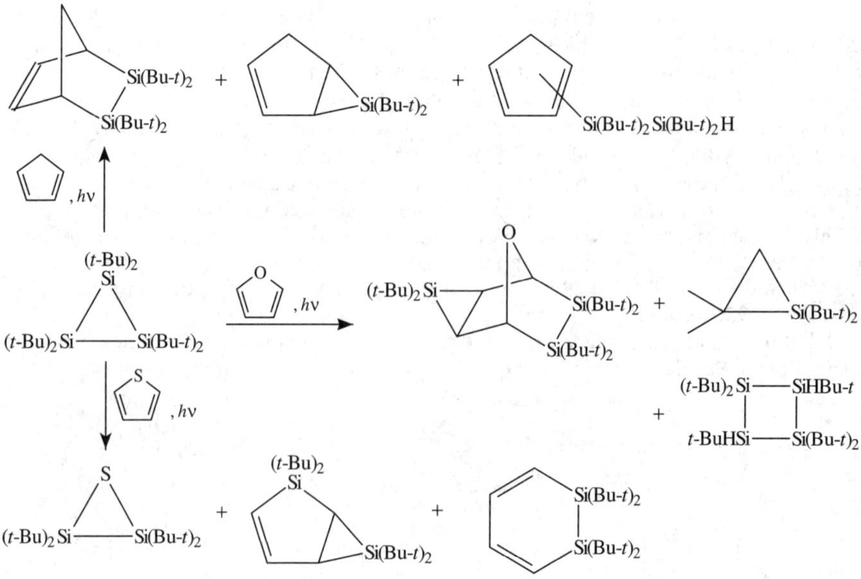

SCHEME 7

37. Recent advances in the chemistry of cyclopolysilanes   2187

Monitoring the reaction by ^{1}H-NMR spectroscopy clearly indicates the formation of the 1,2-disilacyclobutenes by a two-step addition–insertion process. Primarily, the silylene generated by thermal decomposition of the cyclotrisilane adds to the C—C triple bond yielding the silacyclopropene. Further insertion of a second silylene into the Si—C bond of the silacyclopropene finally affords the corresponding 1,2-disilacyclobutene.

Siliranes are also formed by the reaction of the cyclotrisilane [2-(Me$_2$NCH$_2$)C$_6$H$_4$]$_6$Si$_3$ with terminal and strained internal olefins under mild thermal conditions. The products obtained from the thermolysis of the siliranes thus prepared suggest a thermal equilibrium of the silirane with the cyclotrisilane and the corresponding alkene. This observation provides evidence for an equilibrium between the silylene and the cyclotrisilane and, moreover, proves that free silylenes are involved in the silylene transfer reaction[48].

B. Cyclotetrasilanes

As has already been mentioned in Section III for (SiPh$_2$)$_4$ cyclotetrasilanes are cleaved selectively by various nucleophiles and electrophiles like the halogens, hydrogen halides, Li, RLi, alkoxides etc. to the corresponding linear 1,4-disubstituted tetrasilanes. The reaction can be accomplished by most reagents known to cleave Si—Si bonds (for details consult Ref. 5d). Matyjaszewski and coworkers published a series of papers about the ring-opening polymerization of cyclotetrasilanes yielding highly polymeric polysilanes with molecular weights up to M = 100.000[49–52]. Compared with larger polysilanes, the strained four-membered ring turned out to be best suited for ring-opening polymerization. As for any chemical reaction, a decrease in the free energy is necessary for successful ring-opening polymerization. Polymerization, however, is usually accompanied by loss of entropy, because each monomer molecule loses three degrees of translational freedom on conversion to a polymer segment. Thus, a sufficient gain of enthalpy has to be provided by the ring cleavage reaction, and this can only be realized using strained monomers.

Anionic ring-opening polymerization of 1,2,3,4-tetramethyl-1,2,3,4-tetraphenylcyclotetrasilane is quite effectively initiated by butyllithium or silyl potassium initiators. The process resembles the anionic polymerization of other monomers where solvent effects play an important role. In THF, the reaction takes place very rapidly but mainly cyclic five- and six-membered oligomers are formed. Polymerization is very slow in nonpolar media (toluene, benzene); however, reactions are accelerated by the addition of small amounts of THF or crown ethers. The stereochemical control leading to the formation of syndiotactic, heterotactic or isotactic polymers is poor in all cases. In order to improve the stereoselectivity of the polymerization reaction, more sluggish initiators like silyl cuprates are very effective. A possible reaction mechanism is discussed elsewhere[49,52].

Strained and polymerizable cyclotetrasilanes Si$_4$Me$_n$Ph$_{8-n}$ (n = 3,4,5,6) are accessible by the partial dearylation of octaphenylcyclotetrasilane and subsequent displacement with MeMgBr[49]. 1,2,3,4-Tetramethyl-1,2,3,4-tetraphenylcyclotetrasilane, for instance, is formed by the reaction of (SiPh$_2$)$_4$ with 4 equivalents of triflic acid followed by treatment with 4 equivalents of methyl magnesium bromide. Three of the four possible isomers are obtained in a ratio of **1a** : **1b** : **1c** = 45 : 15 : 40, whereas the all-*cis*-isomer **1d** has not been observed (Scheme 8). Recrystallization from cold hexane yields isomer **1a** in 95% purity[53].

The introduction of the novel educt 1,2,3,4-tetraphenyl-1,2,3,4-tetra-*p*-tolylcyclotetrasilane enables the selective monofunctionalization of perarylated cyclotetrasilanes[54]. [Ph(*p*-Tol)Si$_4$] is easily prepared by the reaction of [Ph(*p*-Tol)SiCl$_2$] with lithium and shows significantly higher solubility and reactivity compared to (SiPh$_2$)$_4$. From [Ph(*p*-Tol)Si$_4$] the monofunctional cyclotetrasilane derivatives Si$_4$Ph$_4$(*p*-Tol)$_3$X with X = OSO$_2$CF$_3$, F, Cl, Br, I, H or *t*-Bu are accessible simply by dearylation with triflic acid and

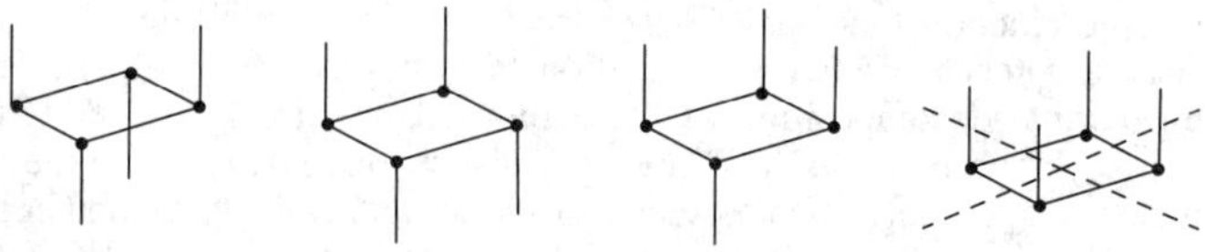

(1a)   (1b)   (1c)   (1d)

● = SiPh

SCHEME 8

subsequent reaction with lithium or potassium halides, $NaBH_4$ or t-BuLi (equations 10 and 11).

$$Ph(p\text{-Tol})SiCl_2 \xrightarrow[THF]{Li} [Ph(p\text{-Tol})Si]_4 \xrightarrow[TfOH = CF_3SO_3H]{TfOH} \quad (10)$$

$$Si_4Ph_4(p\text{-Tol})_3OTf \xrightarrow[X = F,Cl,Br,I]{LiX} Si_4Ph_4(p\text{-Tol})_3X$$

$$Si_4Ph_4(p\text{-Tol})_3H \xleftarrow{NaBH_4} Si_4Ph_4(p\text{-Tol})_3Br \xrightarrow{t\text{-BuLi}} Si_4Ph_4(p\text{-Tol})_3Bu\text{-}t \quad (11)$$

The reaction of $(t\text{-Bu})_2SiCl_2$ with lithium affords the highly hindered products hepta-t-butylcyclotetrasilane and *trans*-1,1,2,3,3,4-hexa-t-butylcyclotetrasilane[55] (equation 12).

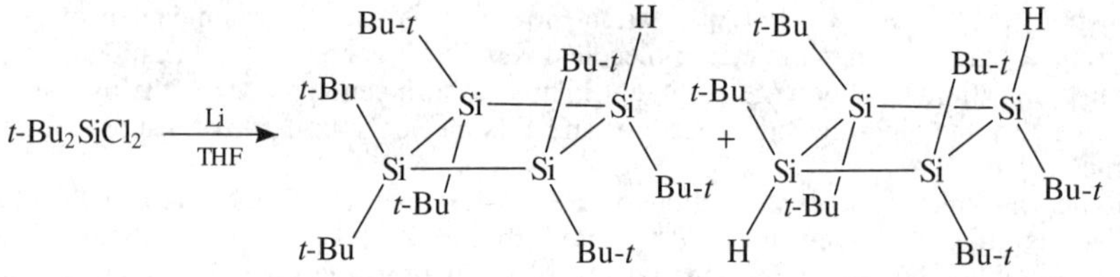

$$t\text{-}Bu_2SiCl_2 \xrightarrow[THF]{Li} \text{[hepta-}t\text{-butylcyclotetrasilane]} + \text{[hexa-}t\text{-butylcyclotetrasilane]} \quad (12)$$

$$+ \; H(t\text{-Bu})_2SiSi(Bu\text{-}t)_2H$$

Quite interesting structural features are found for both compounds. The Si—Si bond distance in hepta-t-butylcyclotetrasilane with 254.2 pm is the longest reported so far in cyclotetrasilanes, while the dihedral angles near 16° are relatively small compared with those of other t-butyl-substituted cyclotetrasilanes. The hexa-t-butyl derivative, however, adopts a planar conformation with unexceptional Si—Si bond lengths of 238.7 pm. In the UV spectrum of hepta-t-butylcyclotetrasilane the longest-wavelength absorption maximum appears at 315 nm. This is the lowest electron transition energy of all alkyl-substituted cyclotetrasilanes reported so far.

C. Cyclopenta- and Cyclohexasilanes

The Si—C bonds in the five- and six-membered peralkyl- and perarylcyclopolysilanes exhibit considerable stability under various conditions. This results in Si—Si bond scission prior to substituent exchange in the course of many reactions employing cyclopolysilane substrates. The introduction of reactive side groups into polysilane rings, therefore, is much desirable in order to allow the performance of chemical reactions under preservation of the polysilane ring skeleton. As a consequence, recent research in cyclopenta- and cyclohexasilane chemistry mainly focused on substituent exchange reactions giving rise to the formation of mono- or polyfunctional cyclosilanes and further derivatives thereof.

Perphenylcyclopolysilanes are easily dephenylated by anhydrous hydrogen halides[56]. The reactivity decreases sharply in the order HI > HBr > HCl. In the presence of catalytic amounts of the corresponding aluminum halide or under forcing conditions (excess HX, high pressure), all phenyl groups are replaced by halogen and the perhalocyclopolysilanes $(SiX_2)_n$ with X = Cl, Br and I and n = 4, 5 and 6 are obtained. Partial dephenylation, however, is sometimes also possible under proper reaction conditions. Generally, the reaction follows an electrophilic substitution mechanism of the aromatic ring. Alkyl-substituted phenyl groups like p-tolyl or mesityl exhibit enhanced reactivity, which can sometimes be utilized for the synthesis of partially functionalized cyclopolysilanes[54].

Triflic acid also readily cleaves the Si–C bonds in arylcyclopolysilanes. The method, which has been studied by Uhlig in detail[57,58], has been recently applied to the synthesis of previously unknown cyclopentasilane derivatives Si_5Ph_9X with X = CF_3SO_3, F, Cl, Br, I, H, NMe_2, PPh_2, i-PrO, Me, $SiPh_3$ and $SnPh_3$[59,60]. Dearylation of Si_5Ph_{10} using one equivalent of triflic acid selectively yields Si_5Ph_9OTf, from which the other derivatives are easily accessible by nucleophilic substitution of the triflate group (Scheme 9).

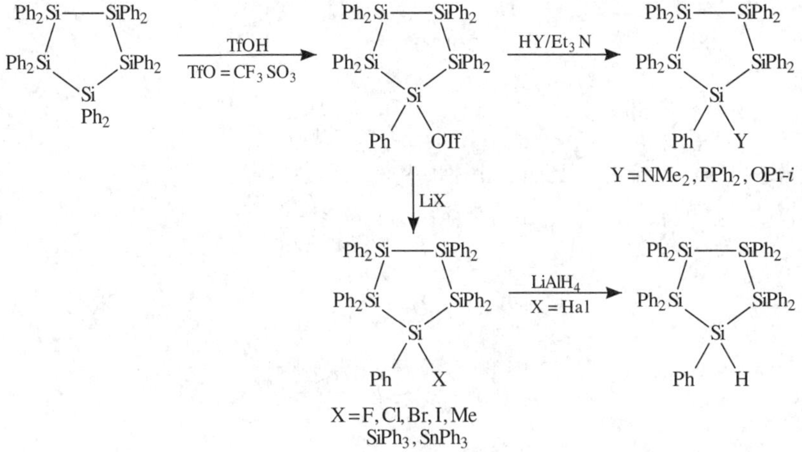

SCHEME 9

The scission of Si–aryl bonds with triflic acid takes place very easily under mild reaction conditions and shows high selectivity. Depending on the stoichiometric ratio, one or more phenyl groups can be removed step by step. When two phenyl groups are attached to the same silicon atom, the first one reacts much faster due to the high electronegativity of the triflate substituent[61]. Ditriflation of Si_5Ph_{10} by the addition of two equivalents of triflic acid, therefore, rather affords *trans*-1,3-$Si_5Ph_8(OTf)_2$ as the major product along with small amounts of the 1,2-isomers instead of the 1,1-disubstitution product[60] (Scheme 10).

The octa- and the nonahalocyclopentasilanes $H_2Si_5X_8$ and HSi_5X_9 with X = Cl, Br, I are accessible by dearylation of the corresponding phenylated cyclosilanes $H_2Si_5Ph_8$ and HSi_5Ph_9 with HX/AlX_3[62].

Methyl groups on cyclopolysilanes can also be substituted by halogens without destruction of the ring structure. With $SbCl_5$, for instance two or three methyl groups can be removed from $(Me_2Si)_6$ depending on the stoichiometric ratio of the reactants. In case of the dichlorodecamethylcyclohexasilanes, the 1,3- and 1,4-isomers are produced exclusively (Scheme 11). Formation of the 1,2 dichloride was never observed.

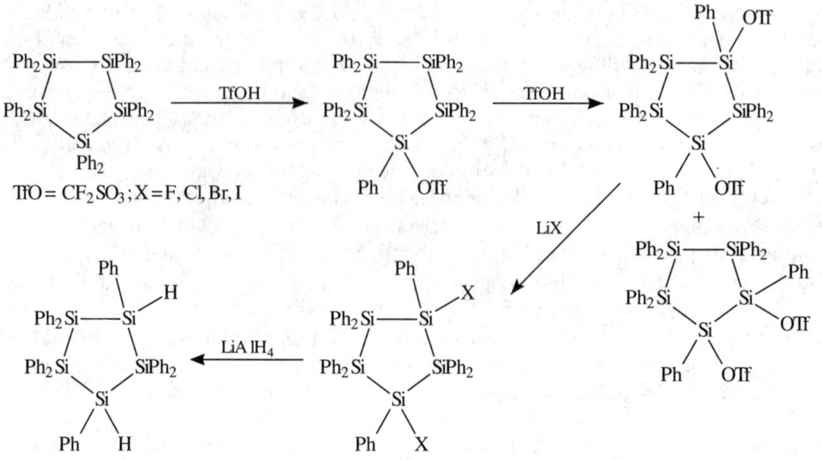

SCHEME 10

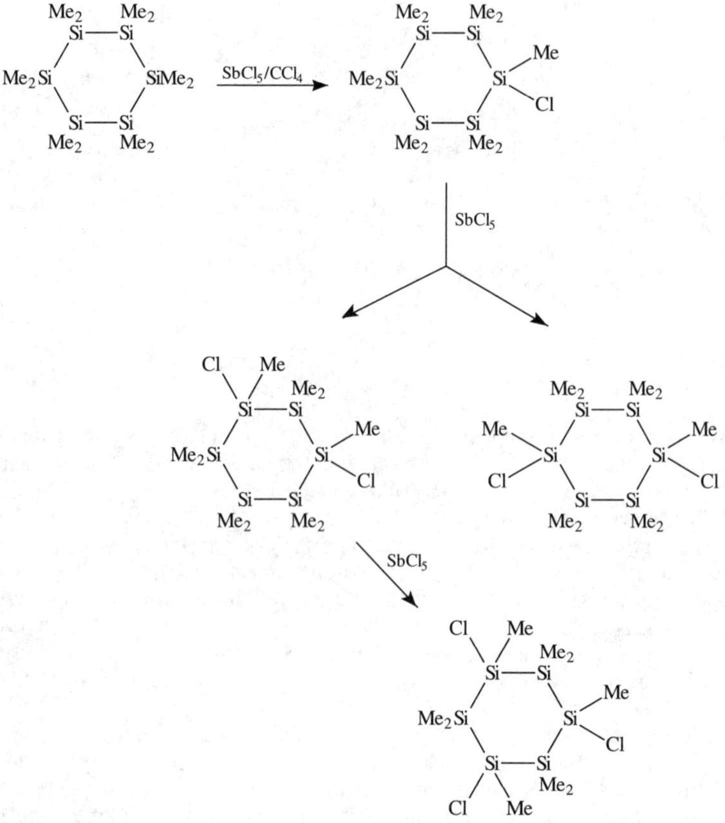

SCHEME 11

The course of the reaction is strongly influenced by solvent effects. Thus, a detailed study of the optimum reaction conditions has been performed using five-membered ring substrates. According to these investigations CCl_4 turned out to be the most suitable solvent[63].

Separation of 1,3- and 1,4-dichlorodecamethylcyclohexasilane can only be accomplished by derivatization. With Na[(CO)$_2$CpFe], for instance, both regioisomers are converted into the bis(iron) derivatives which show different solubilities[64]. Simple hydrolysis of mixtures of 1,3- and 1,4- dichlorodecamethylcyclohexasilane affords the corresponding dihydroxy derivatives, which exhibit remarkably different stabilities with respect to intramolecular condensation to heterocyclic siloxanes. While the 1,3-isomer is stable, the 1,4-dihydroxy compound readily loses water on heating and the oxygen-bridged heterocycle decamethyl-7-oxa-1,2,3,4,5,6-hexasilanorbornane is formed. The crystal structure of the condensation product depicted in Figure 1 clearly shows the Si_6 ring in boat conformation and the bridging oxygen[65]. Both products have different boiling points and may be purified by distillation. On prolonged heating, however, considerable amounts of siloxane polymers are obtained by intermolecular condensation. The Si—O bonds in 1,3-dihydroxydecamethylcyclohexasilane and in decamethyl-7-oxa-1,2,3,4,5,6-hexasilanorbornane can finally be chlorinated with acetyl chloride to give pure 1,3- or 1,4-dichlorodecamethylcyclohexasilane, respectively (Scheme 12)[66].

The partially chlorinated cyclosilanes are valuable starting materials for the synthesis of other derivatives like di- and polycyclic polysilanes (see Section V.A). The dehalogenative coupling with alkali metals, however, only gives low yields and many side products are formed. Salt elimination of cyclosilanyl halides and cyclosilanyl alkali metal compounds

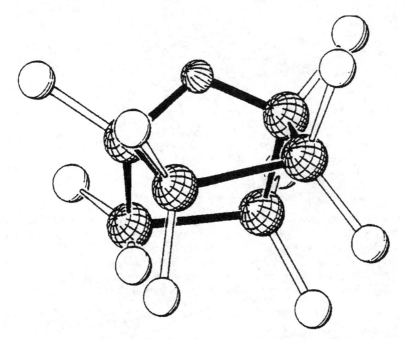

FIGURE 1. Crystal structure of decamethyl-7-oxahexasilanorbornane*(cocrystallized with 1,4-dihydroxydecamethylcyclohexasilane[66]. Reprinted from Hengge et al., J. Organomet. Chem., **499**, 241 (1995) with kind permission from Elsevier Science S.A., P.O. Box 564, 1001 Lausanne, Switzerland

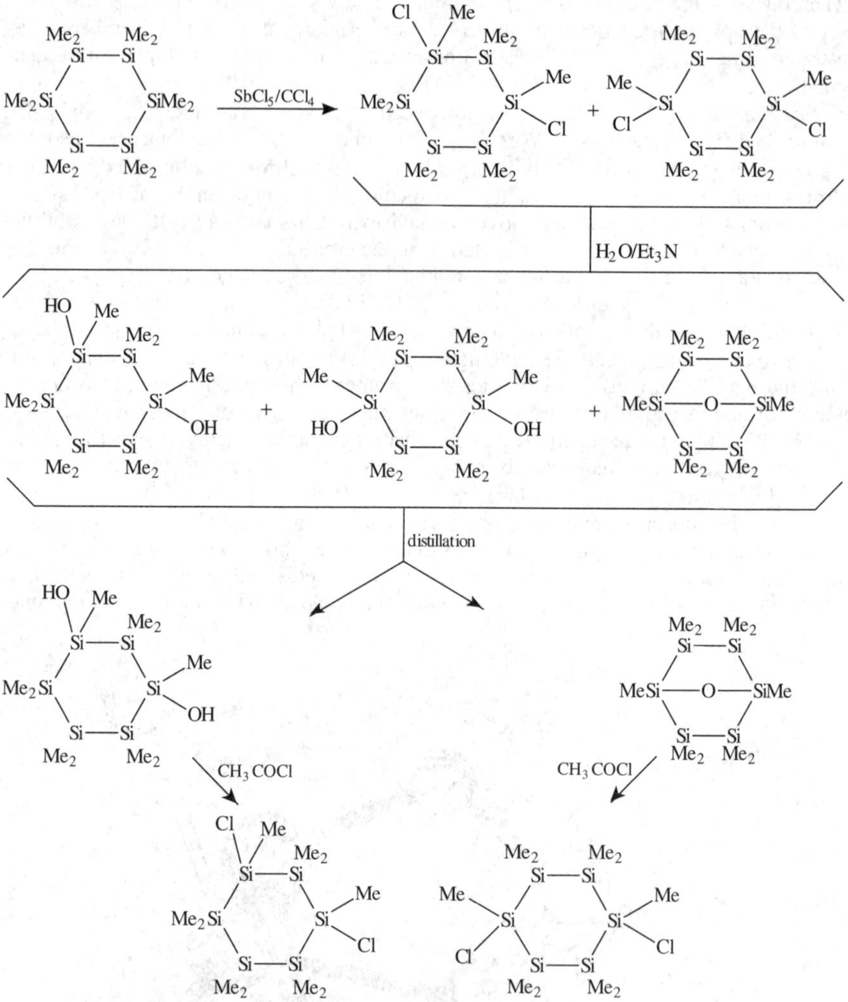

SCHEME 12

turned out to be much more effective, though alkali metal derivatives of cyclosilanes have only been available by a multistep synthesis via the corresponding mercury compounds until recently[67] (Scheme 13).

Lately, a more satisfactory synthesis was found[68]. When a diglyme solution of $(Me_2Si)_6$ with potassium t-butoxide is stirred at room temperature for 7–8 days, undecamethylcyclohexasilanylpotassium is obtained in 70% yield (equation 13).

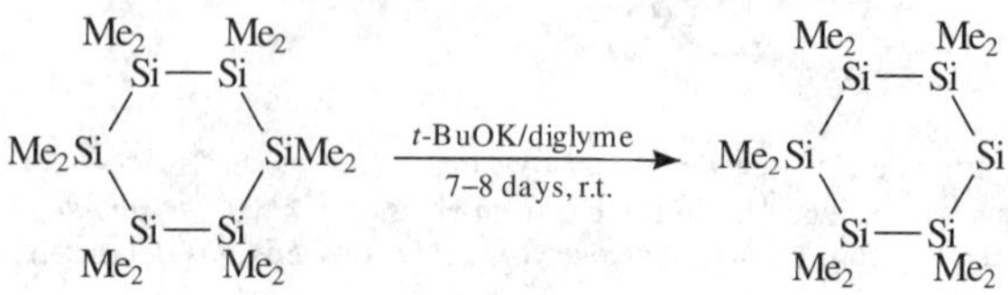

(13)

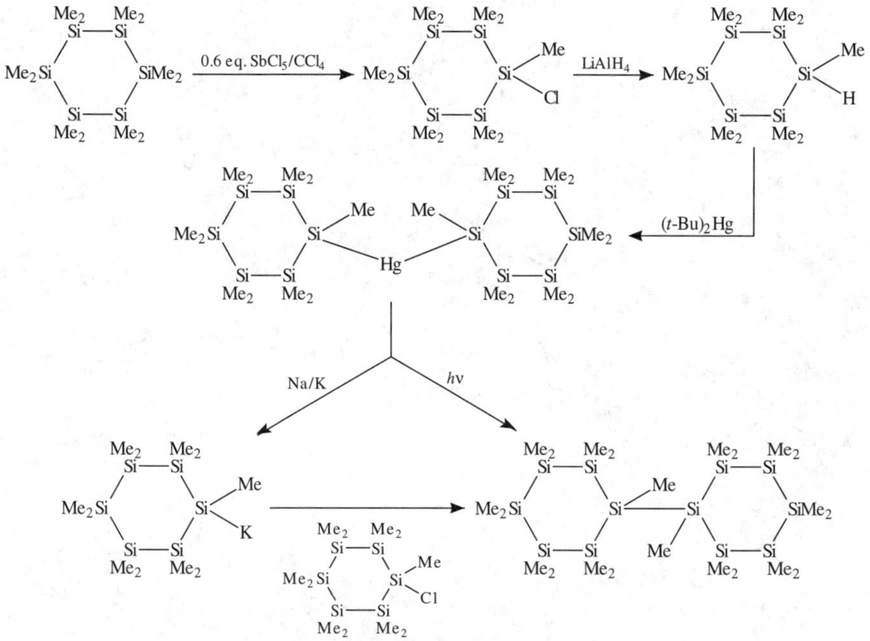

SCHEME 13

When the reaction is monitored by ^{29}Si-NMR, surprisingly, nonamethylcyclopentasilanylpotassium is found as an intermediate. This observation suggests the initial splitting of the six-membered ring by potassium t-butoxide followed by the formation of the kinetically preferred five-membered ring and trimethylsilyl t-butoxide, which can be isolated in the expected amounts. Subsequently, rearrangement to the thermodynamically stable six-membered ring may take place. Similar rearrangements have been described earlier[2,69-72].

Undecamethylcyclohexasilanylpotassium is a useful starting material for the synthesis of various cyclohexasilane derivatives, as shown in Scheme 14. Reactions affording bicyclic and polycyclic silanes are reviewed in Section V.A.

Treatment of 1-halo- or 1,4-dihalopermethylcyclohexasilane with alkali metal thiolates MSPh (M = Li, Na, K) yields the corresponding thiocyclohexasilanes which subsequently can be reacted with silanyl alkali metal compounds to give cyclohexasilanes containing exocyclic Si—Si bonds[73] (equations 14 and 15) without the formation of considerable amounts of by-products due to transmetallation reactions.

$$Si_6Me_{11}X + MSPh \longrightarrow Si_6Me_{11}SPh + MX \qquad (14)$$

$$X = Cl, Br; M = Li, Na, K$$

$$Si_6Me_{11}SPh + MSiR_3 \longrightarrow Si_6Me_{11}SiR_3 + MSPh \qquad (15)$$

$$R = Me, Ph; M = Li, K$$

Another route to cyclohexasilane rings with silanyl side chains begins with cyclosilanylpotassium and chlorosilanes $Ph_nMe_{3-n}SiCl$ (Scheme 15). Dephenylation of the resulting silyl-substituted cyclohexasilanes with triflic acid and subsequent reduction

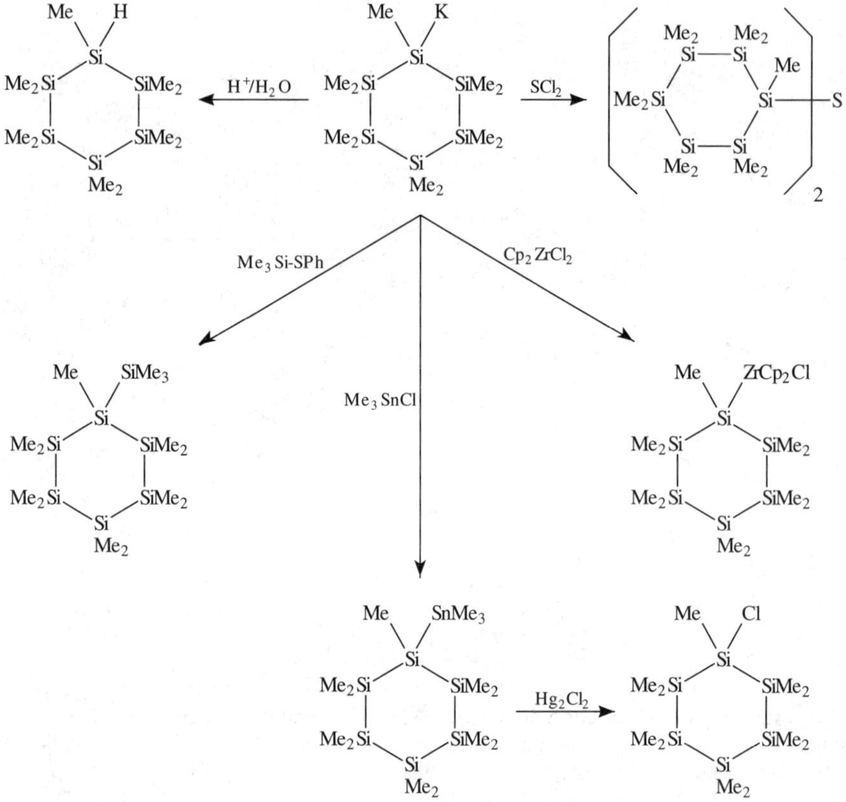

SCHEME 14

of the silyl triflates thus obtained affords cyclohexasilane derivatives with Si—H functional silanyl side chains[74], which can be used for the synthesis of new polycyclic silanes (see Section V.A).

A similar reaction sequence starting from 1,4-dichlorodecamethylcyclohexasilane and $Ph_nMe_{3-n}SiLi$ leads to 1,4-disilylcyclohexasilanes[74] (Scheme 16), which can also be converted to larger polycyclic frameworks (Section V.A).

Partially hydrogenated halocyclopolysilanes are not accessible by simple $LiAlH_4$ reduction of the corresponding perhalogenated precursor molecules. The reaction of Si_5Cl_{10}, for instance, with deficient amounts of $LiAlH_4$ only affords a mixture of Si_5Cl_{10} and Si_5H_{10}. With 5 equivalents of Me_3SnH or n-Bu_3SnH, which have been used successfully for the synthesis of hydrochlorodisilanes[75], one, two or three chlorine atoms in Si_5Cl_{10} are replaced by hydrogen and mixtures of partially hydrogenated cyclopentasilanes are obtained[76] (Scheme 17). Analogous reactions are possible using cyclohexasilane substrates.

Partially functionalized cyclopolysilanes recently attracted attention as model substances for siloxene and luminescent silicon. The yellow luminescent silicon is formed by the anodic oxidation of elemental silicon in HF-containing solutions and may be used for the development of silicon-based materials for light-emitting structures which could be integrated into optoelectronic devices[77]. Because the visible photoluminescence of

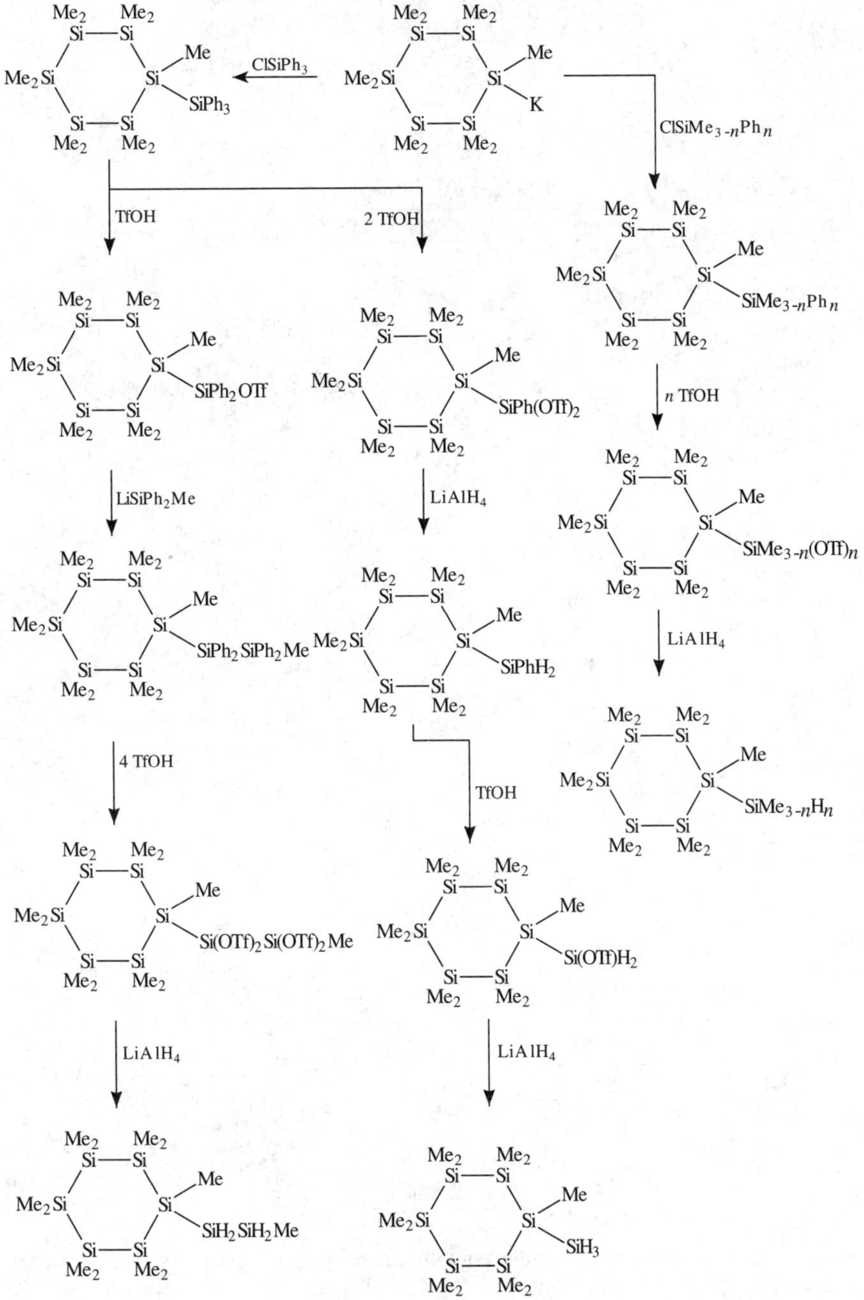

37. Recent advances in the chemistry of cyclopolysilanes

SCHEME 15

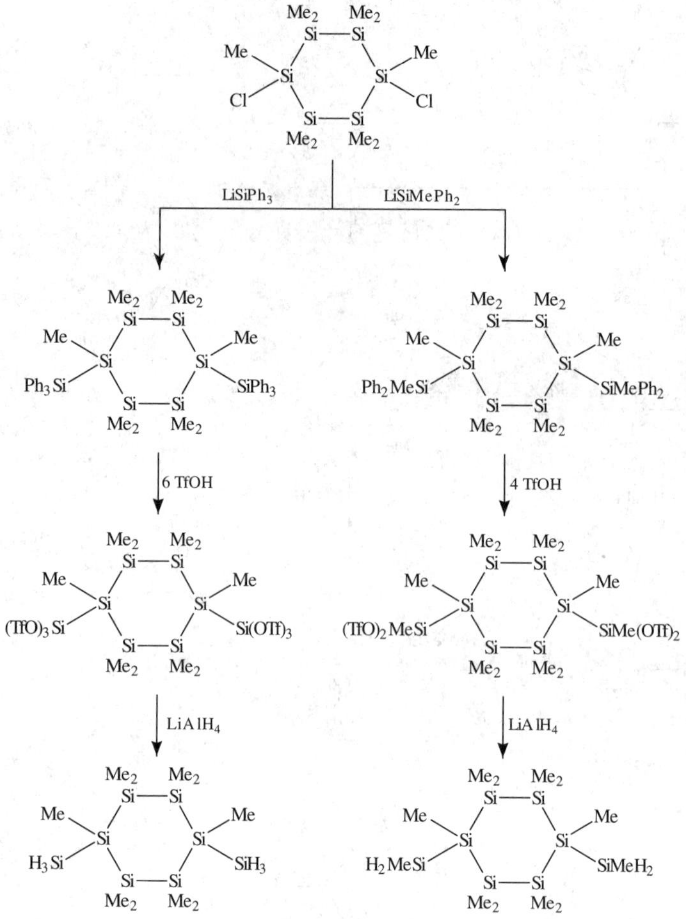

SCHEME 16

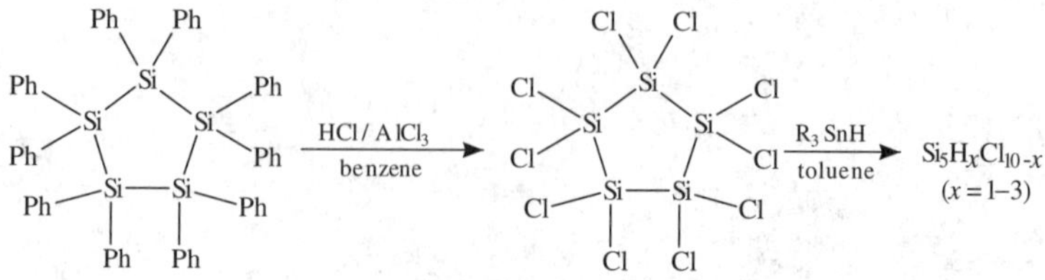

SCHEME 17

porous silicon is very similar to the photoluminescence of siloxene, some researchers have argued that layers of siloxene or siloxene-like structures are responsible for the optical properties of the material[78a]. Siloxene $(Si_6O_3H_6)_n$ was discovered by Wöhler as a product of the reaction of calcium disilicide with hydrochloric acid. The results of Wöhler were reinvestigated by Kautsky[78b] using milder reaction conditions resulting in a more

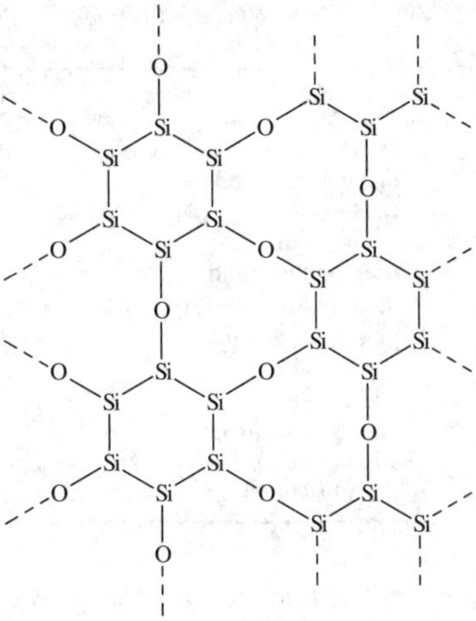

37. Recent advances in the chemistry of cyclopolysilanes

SCHEME 18

ClMe$_2$SiSiMe$_2$Cl
(1)

ClMe$_2$SiSiMeClSiMe$_2$Cl
(2)

ClMe$_2$Si(SiMeCl)$_2$SiMe$_2$Cl
(3)

Cl$_2$MeSi(SiMeCl)$_2$SiMeCl$_2$
(4)

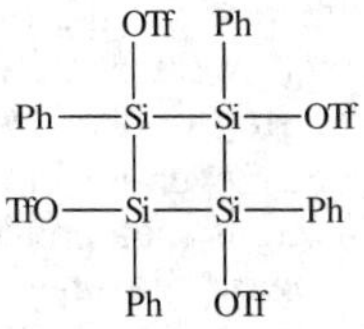

(5)

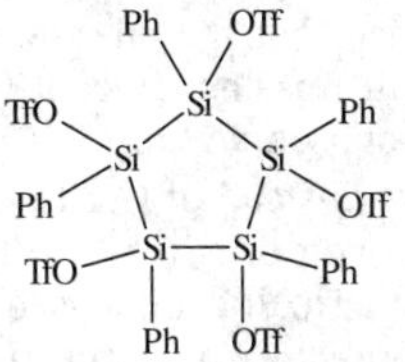

(6)

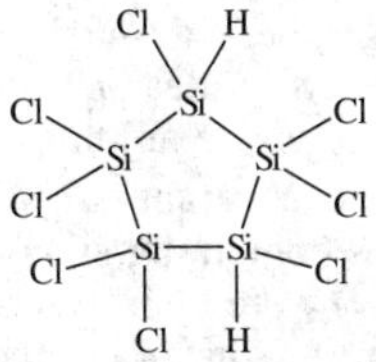

(7)

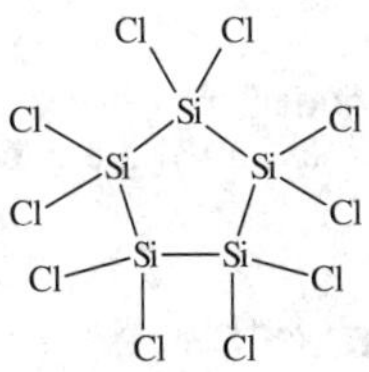

(8)

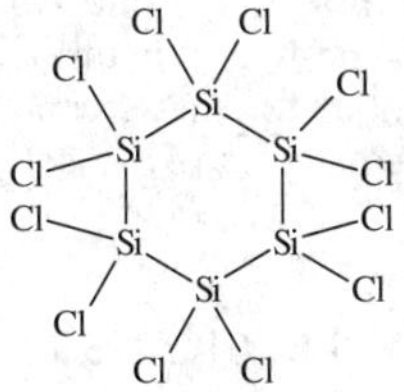

(9)

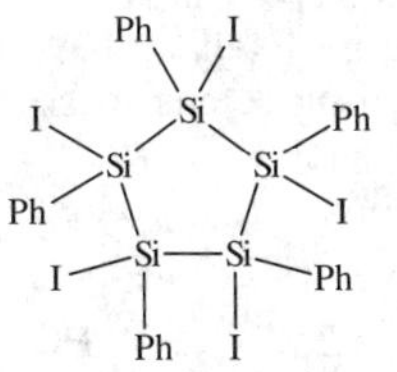

(10)

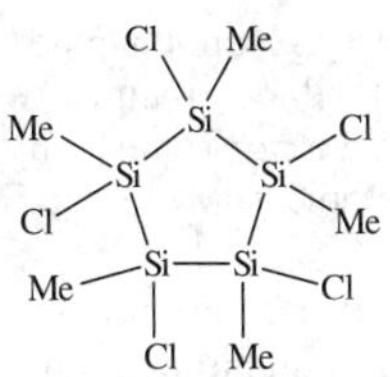

(11)

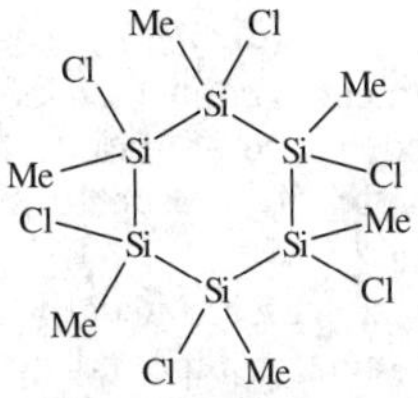

(12)

SCHEME 19

TABLE 2. Fluorescence maxima and appearance of siloxene-like polymers derived from compounds 1-12

Starting material		Product description	Fluorescence maximum (nm)
$Si_2Me_4Cl_2$	1	slightly yellowish, thin oil	none
$Si_3Me_5Cl_3$	2	opaque, thick oil	none
$Si_4Me_6Cl_4$	3	white solid	very weak (399)
$Si_4Me_4Cl_6$	4	white solid	weak (413)
$Si_4Ph_4(OTf)_4$	5	dark yellow solid	553
$Si_5Ph_5(OTf)_5$	6	dark yellow solid	522
$1,3-H_2Si_5Cl_8$	7	pale orange solid	486
Si_5Cl_{10}	8	white solid	400
Si_6Cl_{12}	9	white solid	432
$Si_5Ph_5I_5$	10	dark yellow-green solid	540
$Si_5Me_5Cl_5$	11	slightly yellow-green solid	505
$Si_6Me_6Cl_6$	12	slightly yellow-green solid	436

defined product. Kautsky proposed a structure for siloxene with cyclohexasilanyl rings in chair conformation connected by oxygen bridges to form a highly polymeric layer. The axial positions of the cyclohexasilane rings are occupied by hydrogen atoms (Scheme 18, hydrogens neglected).

The hydrogen atoms of siloxene can easily be replaced by stepwise bromination resulting in products of the general formula $(Si_6O_3)H_{6-n}Br_n$. The bromides can subsequently be converted to amines, hydroxides or alkoxy derivatives. All siloxene derivatives exhibit color and strong fluorescence. Detailed studies of the chemical and physicochemical properties, including substituent effects on the optical spectra of siloxene derivatives, have already been performed in the early days and have been thoroughly reviewed[79].

Although there is much evidence that the structure of siloxene fits the one depicted in Scheme 18, a clear proof is still missing. Recently, attempts were made to elucidate this question by rebuilding the proposed structure of siloxene starting from well-defined low molecular weight polysilanes[80] instead of $CaSi_2$. Therefore, several cyclic and linear polysilanes containing silicon–halogen or silicon–triflate functions (compounds 1-12 in Scheme 19) were hydrolyzed and thermally condensed to polymeric siloxanes. Depending on the starting materials, the fluorescence maxima of the products thus obtained range from 400 to 500 nm, as shown in Table 2.

Since only polymers obtained from the cyclic starting materials 5, 6, 7, 10, 11 and 12, which are likely to have siloxene-like structures, exhibit color and fluorescence, the polysilane ring seems to be essential for the exceptional optical properties of siloxene. This is in agreement with the original idea of Kautsky, who assumed the cycle to be the chromophore.

V. POLYCYCLIC SILANES AND CAGES

A. Linearly Connected Cyclosilanes

In contrast to annelated polycyclic polysilanes, where two cyclopolysilane units share two common silicon atoms, in linearly connected polycyclosilanes mainly two or three polysilane rings are linked by exocyclic Si–Si bonds. Selected compounds, which recently have been described in the literature, are presented in Scheme 20. More details and bibliography can be found elsewhere[5d,22].

Linearly connected polycyclic silanes are usually made by salt elimination reactions, whenever the cyclosilanyl alkali metal compound involved is stable enough. Otherwise, the photochemical cleavage of bis(cyclosilanyl)mercury compounds may be employed. An example is shown in Scheme 13, where bi(undecamethylcyclohexasilanyl) is synthesized either from XSi_6Me_{11} (X = Cl, Br) and KSi_6Me_{11} or by photolysis of $(Si_6Me_{11})_2Hg$. Both methods can also be used to make polycyclic systems containing cyclopentasilanyl units. When $BrSi_6Me_{11}$ or 1,4-$Br_2Si_6Me_{10}$ are reacted with KSi_5Me_9, polycyclic polysilanes are formed containing both five- and six-membered rings[81].

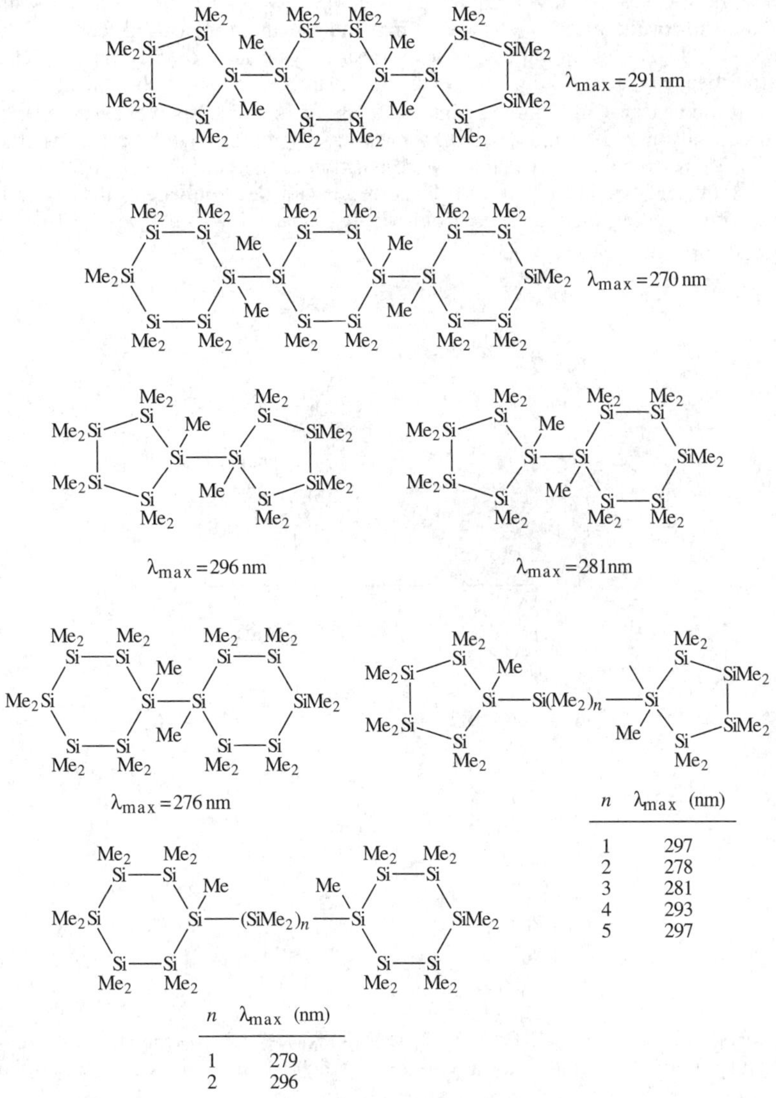

n	λ_{max} (nm)
1	297
2	278
3	281
4	293
5	297

n	λ_{max} (nm)
1	279
2	296

SCHEME 20

Interesting features are shown by the UV absorption spectra of the linearly connected polycyclic silanes (Scheme 20 also gives the longest-wavelength UV absorption maxima)[81]. A marked bathochromic shift of the first UV-maximum relative to the corresponding monocyclic compound results when two cyclosilanyl systems are bonded directly to each other, particularly in the case of two cyclopentasilane rings. When six-membered rings are involved, the effect becomes less pronounced; this strongly suggests the presence of distinct electronic interactions between two Si_5 moieties, which become less favorable with the six-membered analogs. Tricyclic systems behave similarly. Compounds with two cyclosilanyl rings linked by a $-(SiMe_2)_n-$ bridge exhibit a regular bathochromic shift of the first UV absorption maximum with an increasing length of the connecting $-(SiMe_2)_n-$ chain quite similar to linear polysilanes (compare Section III).

The partial dephenylation of diphenyldecamethylcyclohexasilanes with stoichiometric amounts of triflic acid provides an entry into the class of functional polycyclic silanes[82], which subsequently may be used as starting materials for further syntheses. Thus, 1,3- or 1,4-$Ph_2Si_6Me_{10}$, now easily available by the reaction of the pure isomers 1,3- or 1,4-$Cl_2Si_6Me_{10}$ (compare Section IV.C) with PhLi, react with one equivalent of CF_3SO_3H to give the corresponding monotriflates, from which 1,3- and 1,4-$XPhSi_6Me_{10}$ (X = Cl, Br) are accessible with LiX (Scheme 21).[†]

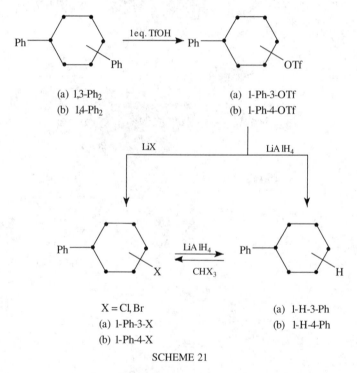

SCHEME 21

Further reaction of 1,3- or 1,4-$XPhSi_6Me_{10}$ with KSi_6Me_{11} affords the corresponding monophenylated permethylcyclohexasilanyl dimers, which again can be converted to the halo derivatives with CF_3SO_3H/LiX, as shown in equation 16.[†]

[†] In Scheme 21 and equation 16 a dot represents a silicon atom with methyl groups attached, sufficient to bring the total coordination number to four.

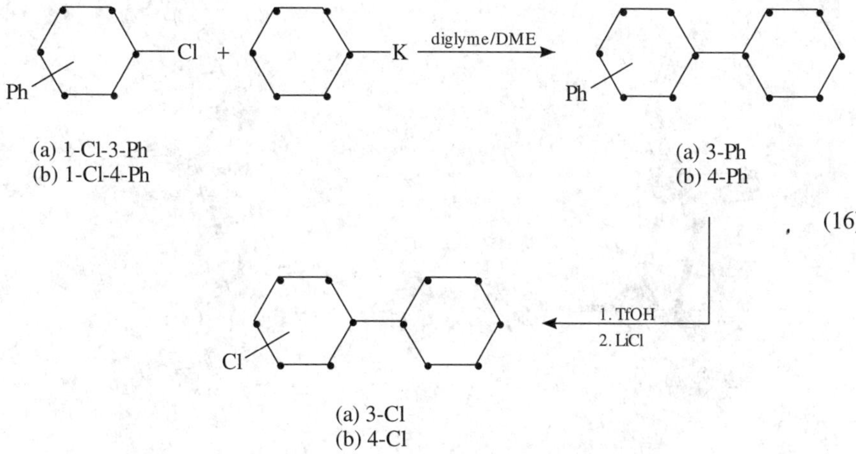

37. Recent advances in the chemistry of cyclopolysilanes

(a) 1-Cl-3-Ph
(b) 1-Cl-4-Ph

(a) 3-Ph
(b) 4-Ph

(a) 3-Cl
(b) 4-Cl

(16)

A system with two cyclohexasilanyl substituents on one phenyl ring is obtained from the phenyl-1,4-di-Grignard reagent and $ClSi_6Me_{11}$[83,84] (equation 17).

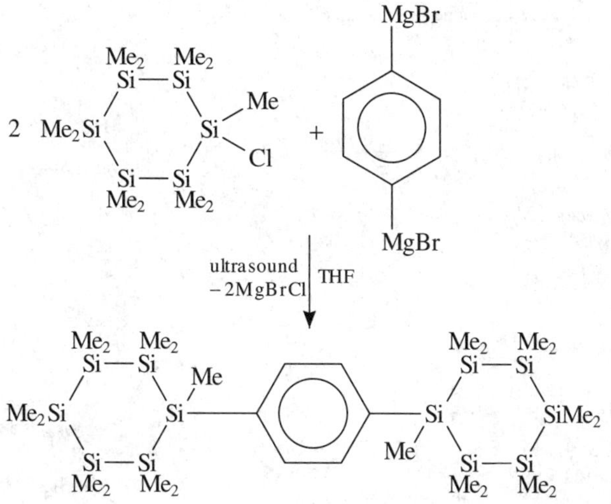

(17)

The X-ray structure of 1,4-bis(undecamethylcyclohexasilanyl)benzene (Figure 2) exhibits a triclinic unit cell with the benzene ring in axial positions of the two cyclohexasilanyl chairs, which both adopt *cis*-positions relative to the plane of the benzene ring.

Compounds containing two cyclohexasilanyl rings linked by a functional $-SiMeX-$ group have also been synthesized[85]. Starting materials are monocyclic silanes with $SiPh_nMe_{3-n}$ ($n = 2, 3$) side chains. A typical reaction sequence is shown in equation 18. $Ph_2MeSi-Si_6Me_{11}$ affords the corresponding dihalomethylsilyl derivative after treatment with two equivalents of $CF_3SO_3H/LiCl$, which subsequently can be converted to the Cl or H functional bicyclosilanes with KSi_6Me_{11} and $LiAlH_4$. At this point it needs to be emphasized that a phenyl group bonded to the central Si atom cannot be removed by the

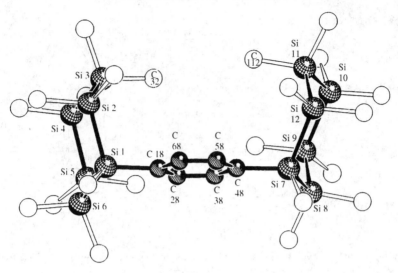

FIGURE 2. Crystal structure of 1,4-bis(undecamethylcyclohexasilanyl)benzene. Reprinted from Hengge et al., J. Organomet. Chem., **524**, 187, (1996) with kind permission from Elsevier Science S.A., P.O. Box. 564 1001 Lausanne, Switzerland

usual methods with triflic acid or with HX/AlX_3, very likely for steric reasons. Attempts to attach a third Si_6Me_{11} group to the central Si atom also failed. Three cyclohexasilanyl groups, however, can easily be linked to a larger central Sn atom as has been shown recently by Uhlig and coworkers[86].

Catalytic dehydrogenative coupling of cyclosilanes bearing Si—H functional side groups has also been used for the synthesis of polycyclic polysilanes[74]. Thus, $H_3SiSi_6Me_{11}$ reacts in the presence of catalytic amounts of Cp_2ZrMe_2 or Cp_2TiMe_2 to give the dimeric product, whereas $MeH_2SiH_2Si-Si_6Me_{11}$ or 1,4- $MeH_2Si-Si_6Me_{10}-SiH_2Me$ react to give oligomers containing 2 to 4 monomeric units. Under similar reaction conditions a soluble polymer, for which a linear structure is suggested by NMR investigations, is obtained from 1,4-$H_3Si-Si_6Me_{10}-SiH_3$ (equations 19-21). Gel permeation chromatography shows molecular weights of $M_w = 10500$ (with Cp_2ZrMe_2) and $M_w = 4200$ (with Cp_2TiMe_2), respectively, versus polystyrene.

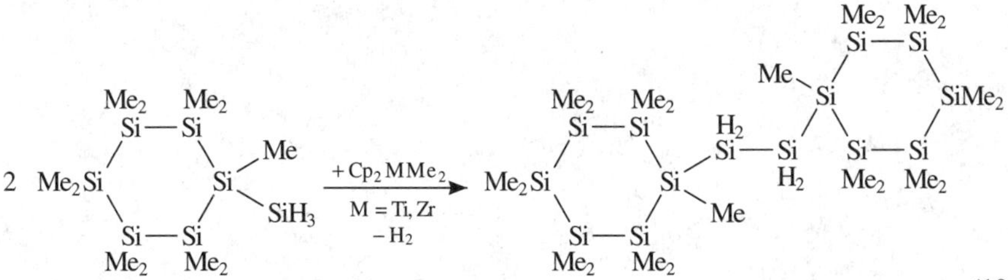

(19)

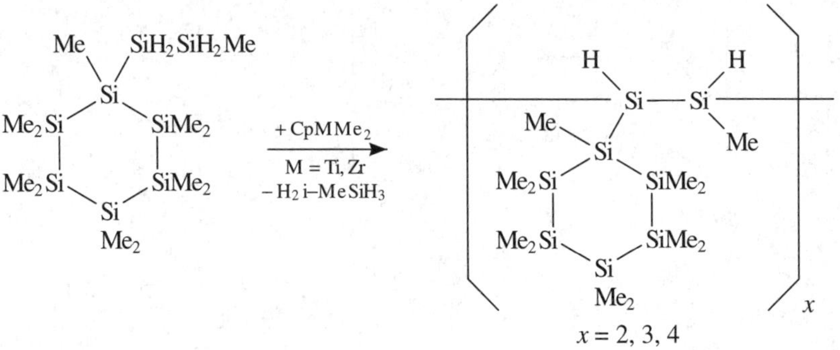

(20)

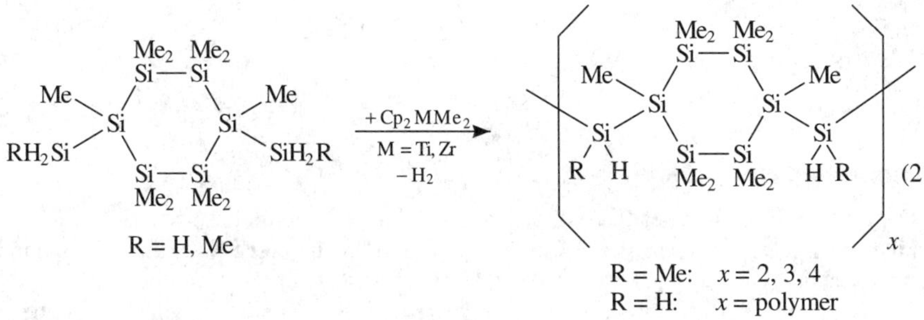

(21)

Cyclosilanes with hydrogen atoms directly attached to the ring do not react to give oligomers in the presence of the usual catalysts like Cp_2ZrMe_2 or Cp_2TiMe_2.

Recently, a first example of a purely inorganic cyclopolysilane dimer has been described[87]. As shown in Scheme 22, $Si_5Cl_9H^{62}$ can easily be converted to bi(nonachlorocyclopentasilanyl) upon treatment with $(t\text{-}Bu)_2Hg$. The reaction of $Si_5Ph_9Br^{60}$ with naphthyl lithium affords the phenyl derivative bi(nonaphenylcyclopentasilanyl). When the perhalogenated cyclopentasilanedimers are reacted with $LiAlH_4$, about 15% of Si_5H_{10} are obtained along with the expected product bi(cyclopentasilanyl), which demonstrates the remarkable sensitivity of the central Si—Si bond towards nucleophilic attack.

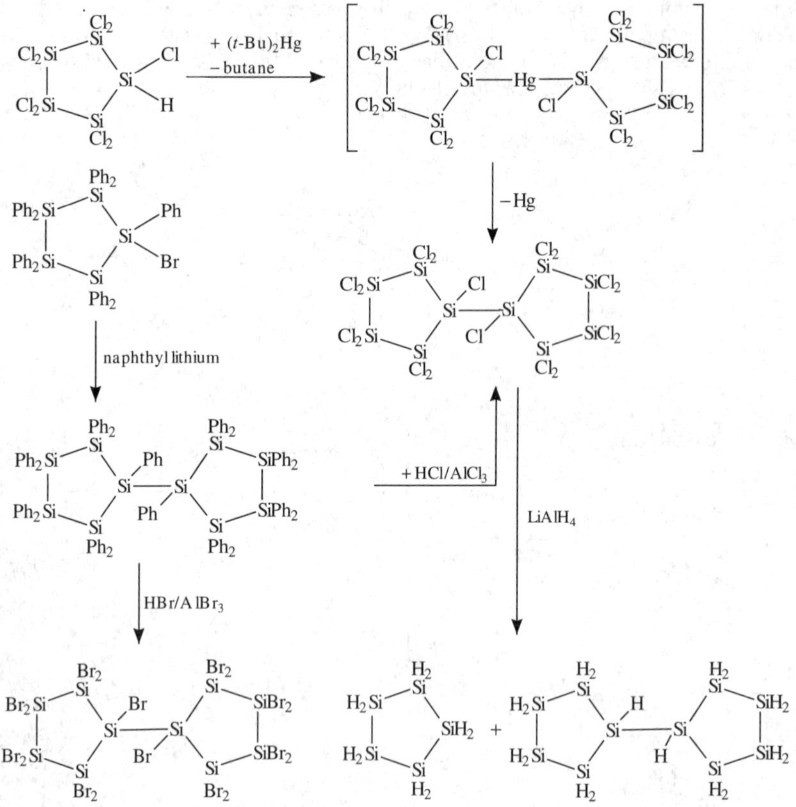

SCHEME 22

B. Annelated Cyclosilanes

Only a handful of annelated cyclosilanes are known. Nothing substantial has been published on the topic since the last review appeared[5d]. The best investigated compound is octadecamethylbicyclo[4.4.0]decasilane **13**. It is formed in the reaction of a mixture of Me_2SiCl_2 and $MeSiCl_3$ with Na/K alloy[88]. Recently, an improved synthesis has been published together with spectroscopic properties and an X-ray structure determination[89]. The radical anion of the compound, which can be generated by action of Na/K alloy at 130 K, is stable only at low temperatures. Further investigations will be necessary to understand the electronic structure of this bicyclic radical anion.

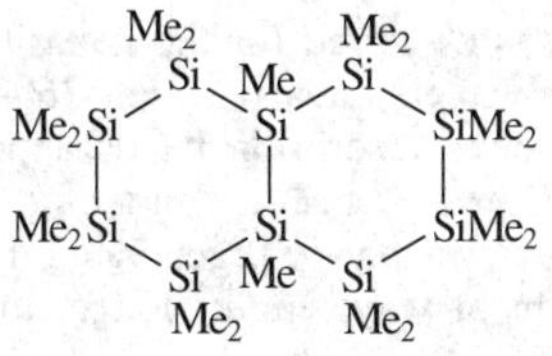

$$\begin{array}{c}
\text{Me}_2\text{Si}-\text{Si}(\text{Me}_2)-\text{Si}(\text{Me})-\text{Si}(\text{Me}_2)-\text{SiMe}_2 \\
| \qquad\qquad | \qquad\qquad | \\
\text{Me}_2\text{Si}-\text{Si}(\text{Me}_2)-\text{Si}(\text{Me})-\text{Si}(\text{Me}_2)-\text{SiMe}_2
\end{array}$$

(13)

C. Polysilane Cages

Recently, some attention has been focused on the synthesis and on the properties of strained cage compounds made up exclusively of Si atoms. As a consequence a number of derivatives of tetrasilatetrahedrane, hexasilaprismane and octasilacubane have been synthesized. The current knowledge in the field up to 1994 has been summarized elsewhere[5d,22,43].

Polysilane cages are usually prepared by the reductive coupling of bulky organohalosilane precursors mainly with Li, Na or Mg metal as shown in equations 22 and 23 for octakis(2,6-diethylphenyl)octasilacubane or hexakis(2,6-diisopropylphenyl)hexasilaprismane[90,91].

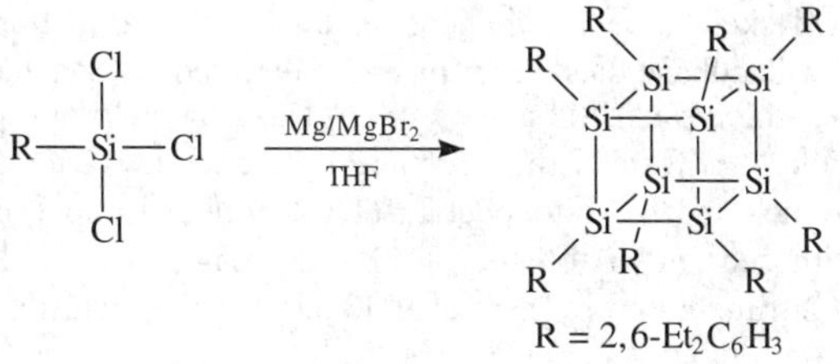

$$R-SiCl_3 \xrightarrow{\text{Mg/MgBr}_2, \text{THF}} \text{octasilacubane}(R)_8 \quad (22)$$

R = 2,6-Et$_2$C$_6$H$_3$

$$R-SiCl_3 \xrightarrow{\text{Mg/MgBr}_2 \text{ or Mg/THF}} \text{hexasilaprismane}(R)_6 \quad (23)$$

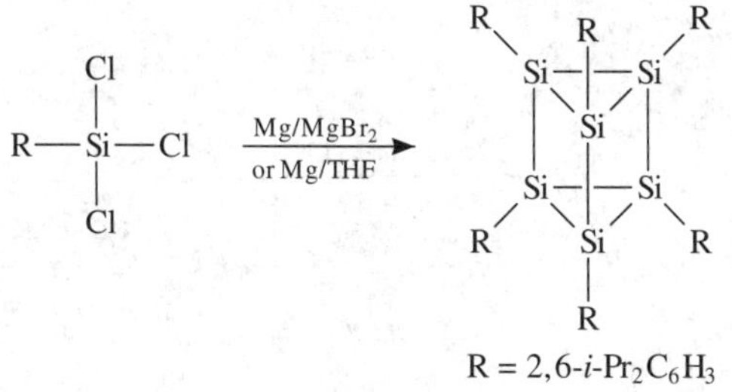

R = 2,6-i-Pr$_2$C$_6$H$_3$

The highly strained tetrasilatetrahedrane structure can only be stabilized using extremely bulky substituents like the tri-t-butylsilyl (= supersilyl) group. Thus the only tetrasilatetrahedrane derivative known so far has been synthesized by coupling $(t\text{-Bu})_3\text{SiSiBr}_2\text{SiBr}_2\text{Si}(\text{Bu-}t)_3$ with two equivalents of $(t\text{-Bu})_3\text{SiNa}$[92] (equation 24).

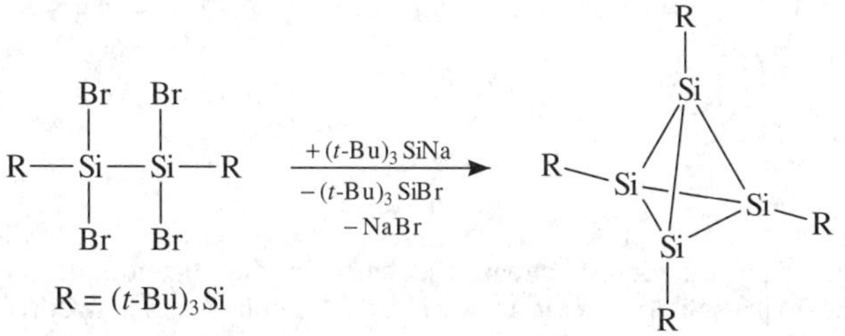

$$R-\underset{\underset{Br}{|}}{\overset{\overset{Br}{|}}{Si}}-\underset{\underset{Br}{|}}{\overset{\overset{Br}{|}}{Si}}-R \xrightarrow[\substack{-(t\text{-Bu})_3\text{SiBr} \\ -\text{NaBr}}]{+(t\text{-Bu})_3\text{SiNa}} \text{Si}_4R_4 \text{ tetrahedrane} \quad (24)$$

R = $(t\text{-Bu})_3$Si

Remarkable photosensitivity is observed for the hexasilaprismane mentioned above[90]. Upon irradiation with light of wavelengths between 360–380 nm in solution at 223 K or in a glass matrix at 77 K, new absorption bands appear which are very likely due to the formation of hexasila-Dewar benzene containing highly reactive Si=Si double bonds. The hexasilaprismane is immediately regenerated thermally or by excitation of the new absorption bands with light of wavelengths longer than 460 nm (equation 25). The activation energy for the reverse reaction has been calculated to be only 57.4 kJ mol^{-1}. This small E_a value is consistent with the high reactivity of Si=Si double bonds.

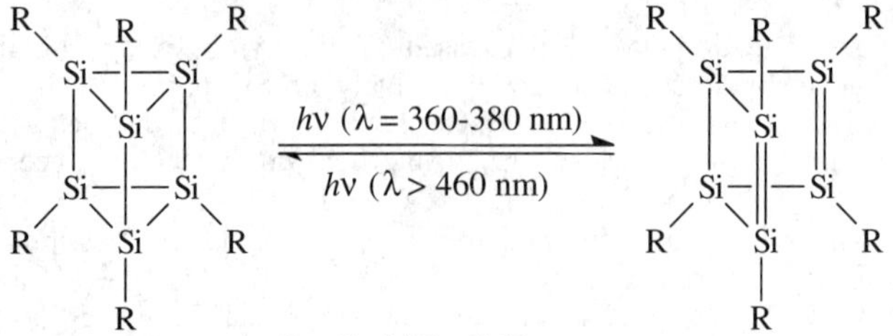

$$R = 2, 6\text{-}i\text{-}Pr_2C_6H_3 \tag{25}$$

An unprecedented skeletal rearrangement leading to the formation of octakis(1,1,2-trimethylpropyl)octasilacubane has been recently reported[93] (Scheme 23). The octasilacubane derivative, which is usually made from (1,1,2-trimethylpropyl)trichlorosilane and sodium in 2.6% yield[94], is formed in 70% yield by the reductive dehalogenation of 4,8-dichloro-, 4,8-dibromo- and 4,8-diiodooctakis(1,1,2-trimethylpropyl)tetracyclo[3.3.0.0^{2,7}0^{3,6}]octasilanes with sodium in toluene at 110 °C along with a reduced product 4,8-dihydrooctakis(1,1,2-trimethylpropyl)tetracyclo[3.3.0.0^{2,7}0^{3,6}]-octasilane.

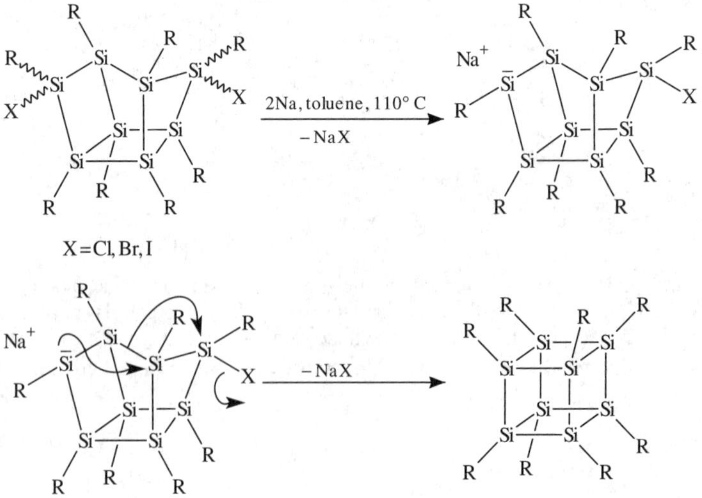

R = CMe$_2$CHMe$_2$

SCHEME 23

The formation of the octasilacubane skeleton in the reaction described above is quite remarkable in light of previous work by Masamune and coworkers, who found that treatment of 4,8-dichlorooocta-*t*-butyl-tetracyclo[3.3.0.0^{2,7}0^{3,6}]octasilane with lithium naphthalenide in toluene had provided only the reduced product 4,8-dihydroocta-*t*-butyl-tetracyclo[3.3.0.0^{2,7}0^{3,6}]octasilane[95].

The 4,8-dihalotetracyclooctasilane starting materials were synthesized by treating the octasilacubane with PCl$_5$, Br$_2$ or I$_2$, respectively (equation 26). Thus, not really a new route to octasilacubanes has been discovered, but a new mechanism, which might be able to provide an entry into new synthetic strategies.

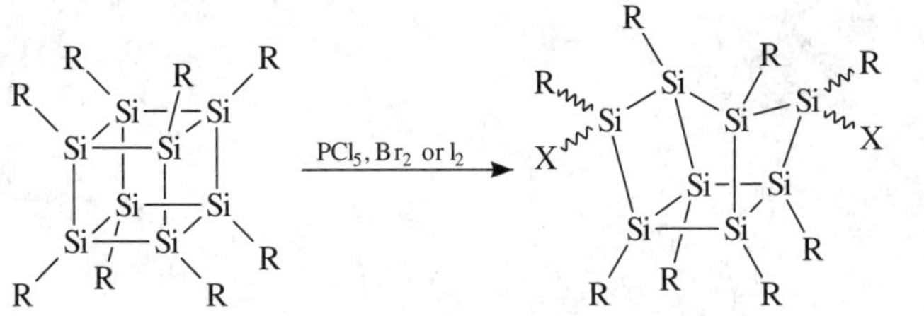

$$R = CMe_2CHMe_2$$
$$X = Cl, Br, I$$
(26)

The halogenation of octakis(1,1,2-trimethylpropyl)octasilacubane results in the formation of three possible stereoisomeric 4,8-dihalooctakis(1,1,2-trimethylpropyl)tetracyclo [3.3.0.0^{2,7}0^{3,6}] octasilanes (Scheme 24) which can be separated by preparative HPLC and characterized spectroscopically and by X-ray crystallography[93,96]. Compared to the structures of octasilacubanes, which are almost perfectly cubic, all three isomers adopt highly distorted conformations with folded cyclotetrasilane rings and Si—Si—Si bond angles varying from 81.2° to 93.9°. The Si—Si bond distances are considerably lengthened compared to the normal value of about 234 nm and vary from 235.1 to 247.1 pm. In agreement with the X-ray structures the ^{29}Si-NMR spectra of the *endo,exo* isomers show eight resonance lines, indicating that all silicon atoms are nonequivalent, whereas the other isomers only exhibit four signals because of their higher symmetry.

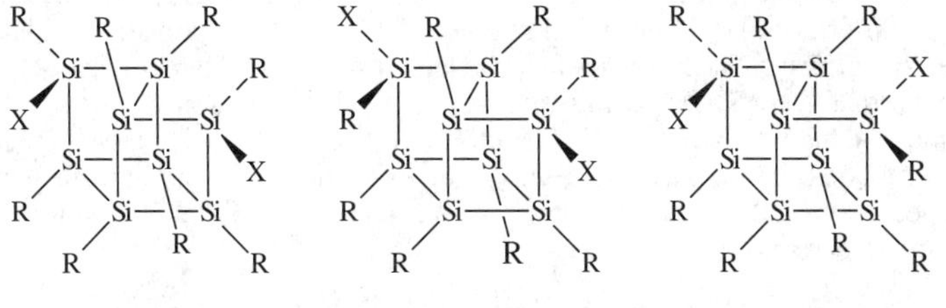

endo,exo   *exo,exo*   *endo,endo*

SCHEME 24

In the case of X = Cl or Br, the *endo,exo* isomer is generated preferentially in yields of 37 and 32%, respectively, whereas the *exo,exo* form of the diiodide is most abundant (39% yield). The *endo,endo* product is obtained in yields of about 10–15% in all cases.

A series of complex cage polysilanes has been synthesized starting from *t*-BuCl$_2$SiSiCl$_2$Bu-*t* by Masamune and coworkers[95]. When *t*-BuCl$_2$SiSiCl$_2$Bu-*t* is treated with lithium naphthalenide in DME, reductive oligomerization takes place and, somewhat surprisingly, tricyclo[2.2.0.0^{2,5}]hexasilane and tetracyclo[3.3.0.0^{2,7}.0^{3,6}]octasilane derivatives are obtained (Scheme 25). The unexpected structures of the products were established by double quantum coherence ^{29}Si-NMR spectroscopy[97] and X-ray crystallography.

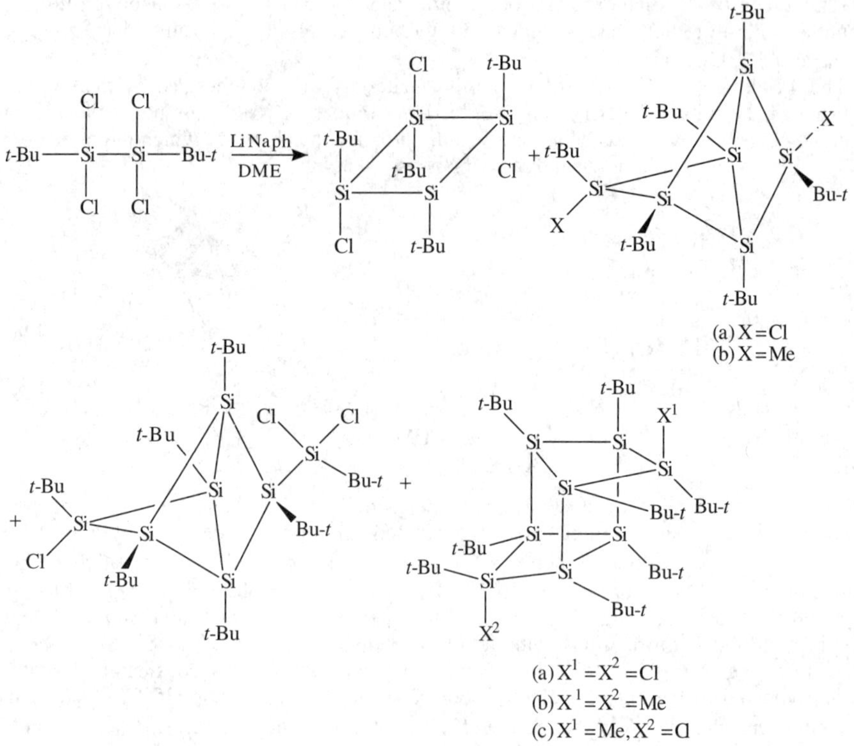

SCHEME 25

Another recent study[98] does not deal with the formation of homocyclic polysilane cages, but with the insertion of oxygen into a strained bicyclic ladder polysilane. The introduction of heteroatoms into ladder polysilane skeletons seems interesting because of possible perturbations of the σ-(Si—Si) conjugation system. The oxidation of decaisopropylbicyclo[2.2.0]hexasilane with a deficient amount of m-chloroperbenzoic acid (0.7 equivalents) affords the monooxidation products decaisopropyl-7-oxabicyclo[2.2.1]heptasilane in 32% yield and decaisopropyl-2-oxabicyclo[3.2.0]heptasilane in 18% yield (equation 27), which can be easily separated by HPLC.

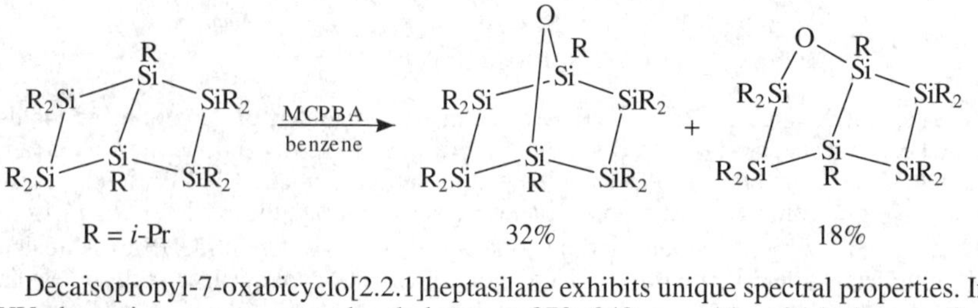

Decaisopropyl-7-oxabicyclo[2.2.1]heptasilane exhibits unique spectral properties. In the UV absorption spectrum two bands between 270–340 nm with a fairly large extinction coefficient of 5100 appear. The compound also shows considerably stronger fluorescence, with a maximum at 373 nm and a quantum yield of $\Phi = 0.014$, than the second oxidation product or the unoxidized ladder polysilane. *Ab initio* (STO-3G) calculations assign the unusual spectral properties of the compound to novel σ-n conjugation resulting from interaction of the oxygen lone pairs and the σ-(Si−Si) skeleton. An unusually small Si−O−Si bond angle of 120.2° and a long Si−O bond distance of 168.3 pm are also reported for decaisopropyl-7-oxabicyclo[2.2.1]heptasilane [compare $Me_3Si-O-SiMe_3$: ∠(Si−O−Si) = 148.8°; d(Si−O) = 162.6 pm].

VI. TRANSITION METAL CONTAINING CYCLOPOLYSILANES

Just a few reactions among the wide range of synthetic methods available for the formation of transition-metal silicon bonds turned out to be suitable for the synthesis of transition-metal derivatives of cyclopolysilanes. So far only some five- and six-membered polysilane rings bearing mostly iron or cobalt containing side groups have been made mainly by salt elimination reactions of cyclosilanyl halides and transition metal anions[5d]. Major problems associated with the cyclopolysilanyl substrates are ring cleavage and transmetallation reactions.

The first successful synthesis of a cyclosilanyl transition-metal compound was published by West and coworkers[99] in 1980, who prepared methylated cyclopentasilane derivatives containing one or two [Fe(CO)$_2$Cp] ligands from the cyclopentasilanyl chlorides and Na[Fe(CO)$_2$Cp] (equations 28 and 29).

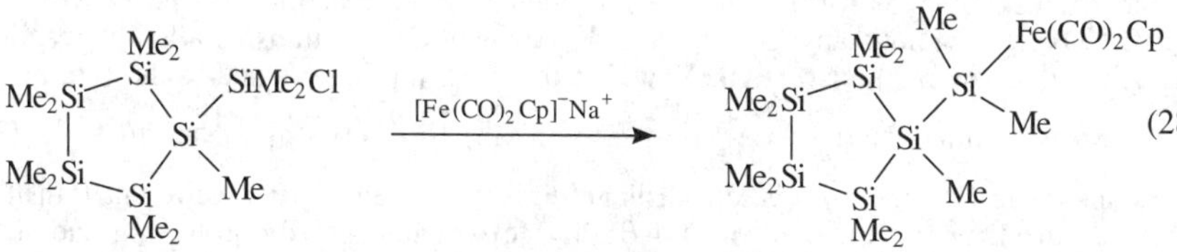

$$\text{(28)}$$

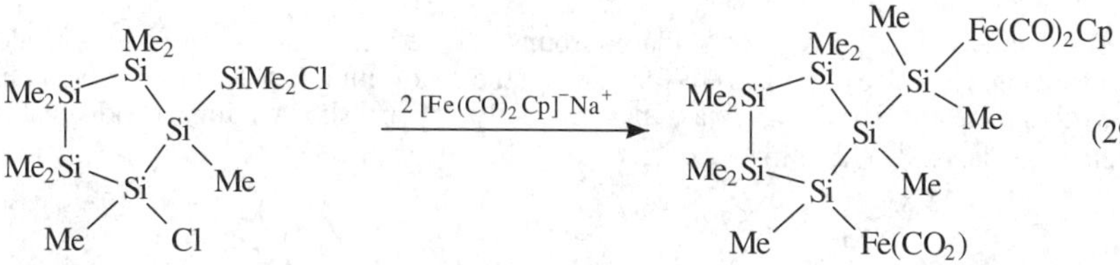

$$\text{(29)}$$

The cyclohexasilane derivatives [(CO)$_2$CpFe]−Si$_6$Me$_{11}$, [(CO)$_3$PPh$_3$Co]−Si$_6$Me$_{11}$, 1,3- or 1,4-[(CO)$_2$CpFe]$_2$−Si$_6$Me$_{10}$ and 1,4-[(CO)$_3$PPh$_3$Co]$_2$−Si$_6$Me$_{10}$ can be made quite similarly[100,101]. Manganese derivatives of cyclosilanes have also been prepared[102]. An 'inverse' salt elimination procedure, however, has to be applied in order to prevent extensive transmetallation reactions (equation 30).

$$Si_6Me_{11}K + [Mn(CO)_5]Br \longrightarrow Si_6Me_{11}[Mn(CO)_5] + KBr \qquad (30)$$

The synthesis of cobalt-substituted cyclosilanes is only possible in the presence of electron-donating ligands like PPh$_3$ at the transition metal center, increasing the nucleophilic character of the transition metal anion. Furthermore, use of cyclosilanyl triflates as the starting materials instead of the chloro derivatives turned out to be advantageous in order to minimize transmetallation reactions. Thus, a number of high-yield syntheses of cyclosilane derivatives containing [Fe(CO)$_2$Cp], [Co(CO)$_3$PPh$_3$], [W(CO)$_3$Cp′]

(Cp′ = methylcyclopentadienyl) or [Mo(CO)$_3$Cp] fragments can only be performed starting from the corresponding cyclosilanyl triflates[100,103,104] (equations 31–33).

$$Me_9Si_5-OTf \xrightarrow{Na[M]} Me_9Si_5-[M] \quad\quad (31)$$
(a) [M] = [Fe(CO)$_2$Cp]
(b) [M] = [Co(CO)$_3$PPh$_3$]

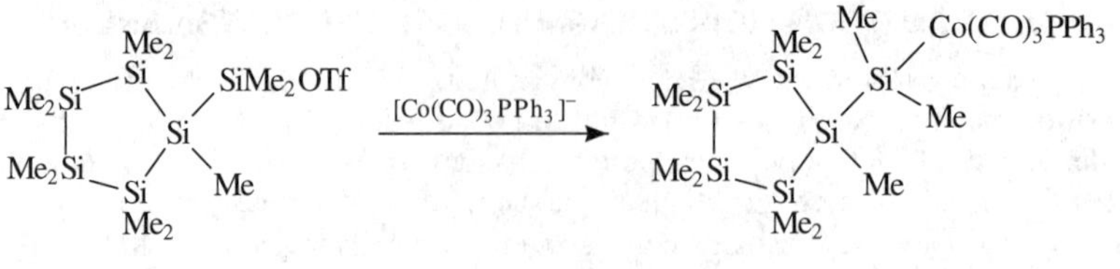

(32)

$$Me_{11}Si_6-OTf \xrightarrow{K[M]} Me_{11}Si_6-[M] \quad\quad (33)$$
(a) [M] = [W(CO)$_3$Cp′]
(b) [M] = [Mo(CO)$_3$Cp]

At least in one example the phenylthio substituent has also been used as a leaving group in the reaction of transition metal anions and cyclopolysilanes[73]. Thus, Si$_6$Me$_{11}$[Fe(CO)$_2$Cp] is formed in excellent yields by the reaction of Si$_6$Me$_{11}$SPh with Na[Fe(CO)$_2$Cp] without any detectable side-products due to transmetallation reactions (equation 34). The method very likely will gain increasing importance in the future.

$$Si_6Me_{11}SPh + Na[Fe(CO)_2Cp] \longrightarrow Si_6Me_{11}[Fe(CO)_2Cp] + NaSPh \quad\quad (34)$$

As shown in Scheme 26, cyclohexasilane derivatives bearing two different transition metal fragments are accessible from 1,4-Br$_2$Si$_6$Me$_{10}$ by temporarily protecting one of the Si–Br valences with hydrogen[101].

The increased reactivity of triflate groups bonded to silicon relative to halogen substituents can also be utilized for the synthesis of multifunctional cyclopolysilane transition metal derivatives, as has recently been shown for 1,3-disubstituted cyclohexasilanes[104] (equation 35).

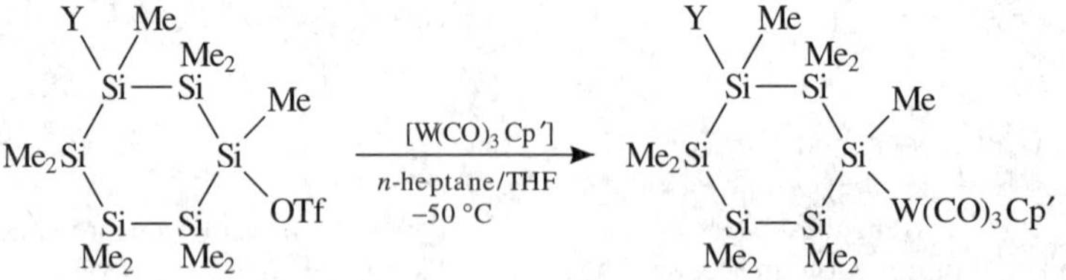

(35)

Y = Cl, Br
Cp′ = MeC$_5$H$_4$

The electron-rich late transition elements are best suited for the formation of stable bonds to silicon. Therefore, most of the cyclosilanyl transition metal compounds described in the literature so far contain transition metals belonging to the groups 7–10. A few examples with early transition elements, however, are also known[105]. Cp$_2$ZrCl$_2$, for instance, reacts with KSi$_6$Me$_{11}$ to give red crystals containing one cyclohexasilanyl and one chlorine substituent on the zirconium. The second chlorine atom can be easily replaced by treatment with MeLi. Addition of excess KSi$_6$Me$_{11}$ to the monochloride, however, does not afford the dicyclosilanyl derivative (Scheme 27).

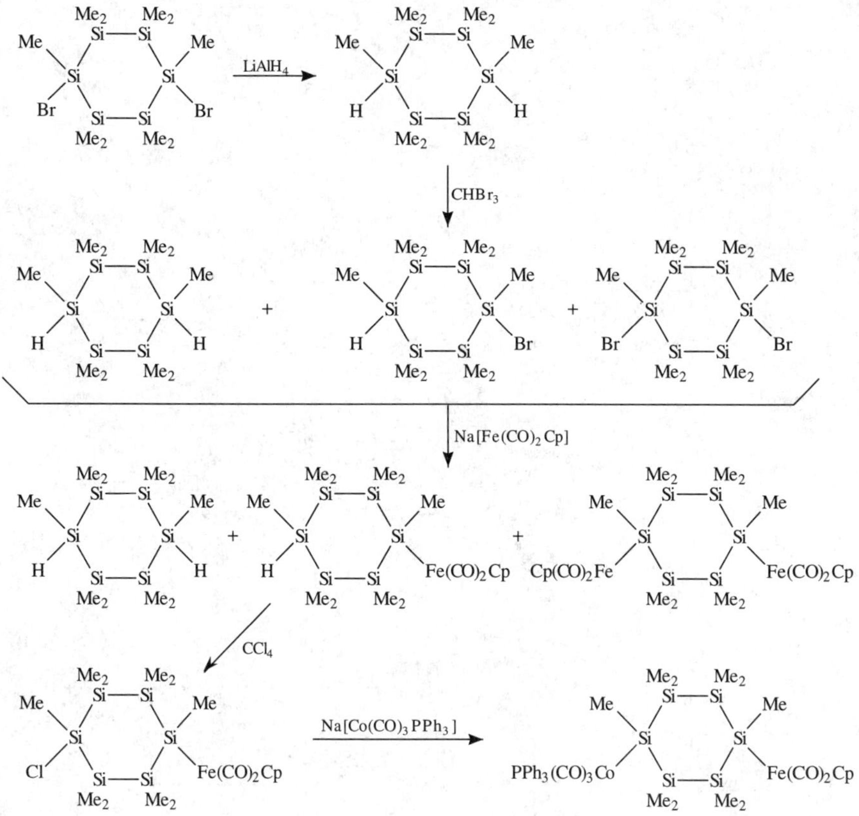

SCHEME 26

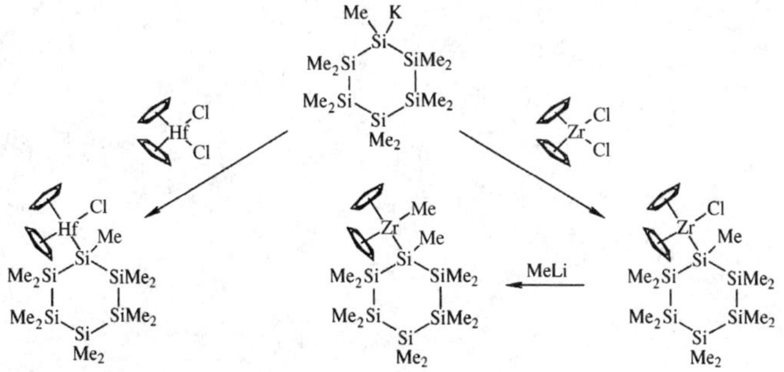

SCHEME 27

The corresponding hafnium compound behaves similarly. Attempts to make the titanium derivative using this method failed. Primarily titanium is reduced, as indicated by the color change of the solution from green to brown. Variation of the reaction conditions did not provide better results, in all cases only bi(undecamethylcyclohexasilanyl) could be identified among the reaction products.

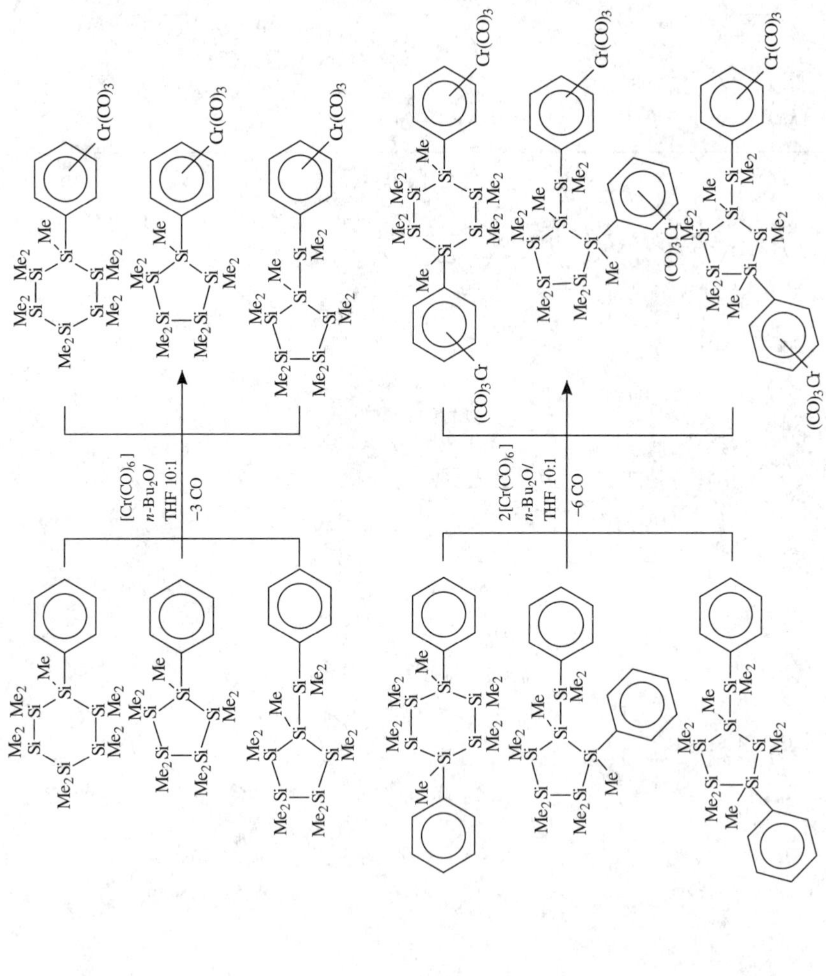

SCHEME 28

Compounds containing chromium directly linked to a cyclosilanyl ring are not known so far. From hexacarbonylchromium and mono- or diphenylcyclopenta- and hexasilanes, however, the corresponding tricarbonylchromium η^6-phenyl complexes have recently been synthesized[106] (Scheme 28).

The complexation of the phenyl groups causes a noticeable low-field shift of the ^{29}Si–NMR signal (ca 5 ppm) assigned to the silicon atom directly bonded to the aromatic ring, due to the presence of the electron-withdrawing tricarbonylchromium group. The different solubilities of the corresponding tricarbonylchromium complexes can be utilized for the isolation of pure 1-phenyl-2-(phenyldimethylsilyl)octamethylcyclopentasilane. This compound is usually obtained as a mixture with its 1,3-isomer, which cannot be separated by common techniques[107]. After separation of the isomeric chromium complexes by fractional crystallization the Cr(CO)$_3$ group can easily be removed by oxidation.

The chemical properties of cyclosilanyl transition metal compounds have hardly been investigated. The silicon transition metal bond is generally cleaved by acids, halogens or strong nucleophiles. However, it is remarkably stable against hydrogenation. Silicon–halogen bonds in linear oligosilanes, for instance, can be reduced with LiAlH$_4$ without any cleavage of adjacent Si–Fe bonds[108]; this also should be possible on cyclosilanyl substrates.

Some interesting rearrangement reactions of cyclosilanyl transition metal compounds were studied by Pannell and coworkers. Upon photolysis, Si$_6$Me$_{11}$[Fe(CO)$_2$Cp] undergoes ring contraction to dicarbonyl(η^5-cyclopentadienyl)[octamethyl(trimethylsilyl)cyclopentasilanyl]iron[109] (equation 36).

$$\begin{array}{c}\text{Me}_2\quad\text{Me}_2\\ \text{Si—Si}\\ \diagup\qquad\diagdown\text{Me}\\ \text{Me}_2\text{Si}\qquad\qquad\text{Si}\\ \diagdown\qquad\diagup\\ \text{Si—Si}\quad\text{Fe(CO)}_2\text{Cp}\\ \text{Me}_2\quad\text{Me}_2\end{array} \xrightarrow{h\nu} \begin{array}{c}\text{Me}_2\\ \text{Si}\\ \text{Me}_2\text{Si}\diagup\quad\diagdown\text{SiMe}_3\\ |\qquad\qquad\text{Si}\\ \text{Me}_2\text{Si}\diagdown_{\text{Si}}\diagup\text{Fe(CO)}_2\text{Cp}\\ \text{Me}_2\end{array} \qquad (36)$$

When dicarbonyl(η^5-indenyl)(undecamethylcyclohexasilanyl)iron is treated with (i-Pr)$_2$NLi and subsequently with MeI, migration of the cyclosilanyl group to the 2-position of the indenyl ligand takes place[110].

VII. ACKNOWLEDGMENTS

The authors would like to thank Dr Peter Gspaltl for drawing the formulae and his helpful comments, and Dr Christa Grogger for her assistance in the preparation of this manuscript.

VIII. REFERENCES

1. F. S. Kipping and J. E. Sands, *J. Chem. Soc.*, **119**, 830 (1921).
2. C. A. Burkhard, *J. Am. Chem. Soc.*, **71**, 963 (1949).
3. a) H. Gilman and G. L. Schwebke, *Adv. Organomet. Chem.*, **1**, 89 (1964). b) R. West and E. Carberry, *Science*, **184**, 179 (1975). c) M. Kumada and K. Tamao, *Adv. Organomet. Chem.*, **6**, 19 (1968).
4. a) E. Hengge and D. Kovar, *Angew. Chem.*, **93**, 698 (1981). b) E. Hengge and D. Kovar, *Z. Anorg. Allg. Chem.*, **458**, 163 (1979). c) E. Hengge and G. Bauer, *Monatsh. Chem.*, **106**, 503 (1975). d) E. Hengge and D. Kovar, *Angew. Chem.*, **89**, 417 (1977).
5. (a) R. West, in *Comprehensive Organometallic Chemistry*, Vol. 2 (Eds. G. Wilkinson, F. G. A. Stone and E. W. Abel), Chap. 9.4, Pergamon Press, Oxford, 1982, p. 365.
   (b) E. Hengge and K. Hassler, in *The Chemistry of Inorganic Homo- and Heterocycles* (Eds. I. Haiduc and D. B. Sowerby), Vol. 1, Academic Press, London, 1987, p. 191.

    (c) R. West, *Pure Appl. Chem.*, **54**, 1041 (1982).
    (d) E. Hengge and R. Janoschek, *Chem. Rev.*, **95**, 1495 (1995).
6. L. F. Brough and R. West, *J. Organomet. Chem.*, **145**, 139 (1980).
7. E. Carberry and R. West, *J. Am. Chem. Soc.*, **91**, 5440 (1969).
8. F. K. Cartledge, *Organometallics*, **2**, 425 (1983).
9. M. Weidenbruch, K.-L. Thom, S. Pohl and W. Saak, *J. Organomet. Chem.*, **329**, 151 (1987).
10. E. Hengge and G. Litscher, *Angew. Chem.*, **88**, 414 (1976).
11. S. Graschy, C. Grogger and E. Hengge, in *Organosilicon Chemistry III* (Eds. J. Weis and N. Auner), Verlag Chemie, Weinheim, 1997.
12. C. Jammegg, S. Graschy and E. Hengge, *Organometallics*, **13**, 2397 (1994).
13. E. Hengge and H. Firgo, *J. Organomet. Chem.*, **212**, 155 (1981).
14. E. Hengge and G. Litscher, *Monatsh. Chem.*, **109**, 1217 (1978).
15. H. Li, I. S. Butler and J. F. Harrod, *Organometallics*, **12**, 4553 (1993).
16. W. Uhlig, *J. Organomet. Chem.*, **452**, C6 (1993).
17. P. Gspaltl and E. Hengge, unpublished results.
18. Z. Smith, H. M. Seip, E. Hengge and G. Bauer, *Acta Chem. Scand.*, **A30**, 697 (1976).
19. Z. Smith, A. Almenningen, E. Hengge and D. Kovar, *J. Am. Chem. Soc.*, **104**, 4362 (1982).
20. C. Kratky, H. G. Schuster and E. Hengge, *J. Organomet. Chem.*, **247**, 253 (1983).
21. V. S. Mastryukov, S. A. Strelkov, L. V. Vilkov, M. Kolonits, B. Rozsondai, H. G. Schuster and E. Hengge, *J. Mol. Struct.*, **238**, 433 (1990).
22. R. West, in *Comprehensive Organometallic Chemistry II*, Vol. 2 (Eds. G. Wilkinson, F. G. A. Stone and E. W. Abel), Chap. 3.5, Pergamon Press, Oxford, 1995, p. 91–95 and references cited therein.
23. H. Bock and B. Solouki, in *The Chemistry of Organic Silicon Compounds* (Eds. S. Patai and Z. Rappoport), Wiley, Chichester, 1989, p. 203.
24. H. Stüger and E. Hengge, *Monatsh. Chem.*, **119**, 873 (1988).
25. H. Stüger and R. Janoschek, *Phosphorus, Sulfur, and Silicon*, **68**, 129 (1992).
26. C. L. Wadsworth and R. West, *Organometallics*, **4**, 1664 (1985).
27. C. L. Wadsworth, R. West, Y. Nagai, H. Watanabe and T. Muraoka, *Organometallics*, **4**, 1659 (1985).
28. B. Kirste, R. West and H. Kurreck, *J. Am. Chem. Soc.*, **107**, 3013 (1985).
29. H. Bock, W. Kaim, M. Kira and R. West, *J. Am. Chem. Soc.*, **101**, 7667 (1979).
30. Y. Apeloig and D. Danovich, *Organometallics*, **15**, 350 (1996).
31. H. Stüger, E. Hengge and R. Janoschek, *Phosphorus, Sulfur, and Silicon*, **48**, 189 (1990).
32. L. F. Brough and R. West, *J. Am. Chem. Soc.*, **103**, 3049 (1981).
33. C. W. Carlson and R. West, *Organometallics*, **2**, 1792 (1983).
34. S. Masamune, Y. Hanzawa, S. Murakami, T. Bally and J. F. Blount, *J. Am. Chem. Soc.*, **104**, 1150 (1982).
35. M. Weidenbruch, *Chem. Rev.*, **95**, 1479 (1995).
36. S. Masamune, H. Tobita and S. Murakami, *J. Am. Chem. Soc.*, **105**, 6524 (1983).
37. A. Schäfer, M. Weidenbruch, K. Peters and H. G. von Schnering, *Angew. Chem.*, **96**, 311 (1984).
38. H. Watanabe, Y. Kougo, M. Kato, H. Kuwabara, T. Okawa and Y. Nagai, *Bull. Chem. Soc. Jpn.*, **57**, 3019 (1984).
39. H. Watanabe, Y. Kougo and Y. Nagai, *J. Chem. Soc., Chem. Commun.*, 66 (1984).
40. G. Gillette, G. Nora and R. West, *Organometallics*, **9**, 2925 (1990).
41. J. Belzner, H. Ihmels, B. O. Kneisel and R. Herbst-Irmer, *J. Chem. Soc., Chem. Commun.*, 1989 (1994).
42. H. Watanabe, M. Kato, T. Okawa, Y. Kougo, Y. Nagai and M. Goto, *Appl. Organomet. Chem.*, **1**, 157 (1987).
43. T. Tsumuraya, S. A. Batcheller and S. Masamune, *Angew. Chem.*, **103**, 916 (1991).
44. M. Weidenbruch, *Front. Organosilicon Chem., [Proc. Int. Symp. Organosilicon Chem.]*, 9th, 122 (1990).
45. M. Weidenbruch, E. Kroke, S. Pohl, W. Saak and H. Marsmann, *J. Organomet. Chem.*, **499**, 229 (1995).
46. E. Kroke, S. Willms, M. Weidenbruch, W. Saak, S. Pohl and H. Marsmann, *Tetrahedron Lett.*, **37**, 3675 (1996).
47. E. Kroke, M. Weidenbruch, W. Saak, S. Pohl and H. Marsmann, *Organometallics*, **14**, 5695 (1995).

48. J. Belzner, H. Ihmels, B. O. Kneisel, R. O. Gould and R. Herbst-Irmer, *Organometallics*, **14**, 305 (1995).
49. E. Fossum, M. Mohan and K. Matyjaszewski, in: *Progress in Organosilicon Chemistry* (Eds. B. Marciniec, J. Chojnowski), Gordon and Breach, 1995, pp. 429–443.
50. E. Fossum and K. Matyjaszewski, *ACS Symp. Ser.*, **579**, 433 (1994).
51. K. Matyjaszewski, *Makromol. Chem., Macromol. Symp.*, **4243**, 269 (1991).
52. E. Fossum, J. Chrusciel and K. Matyjaszewski, *ACS Symp. Ser.*, **572**, 32 (1994).
53. E. Fossum, S. W. Gordon-Wylie and K. Matyjaszewski, *Organometallics*, **13**, 1695 (1994).
54. U. Pöschl and K. Hassler, *Organometallics*, **14**, 4948 (1995).
55. S. Kyushin, H. Sakurai and H. Matsumoto, *J. Organomet. Chem.*, **499**, 235 (1995).
56. Ref. 5d, pp. 195–203 and references therein.
57. W. Uhlig, *Organosilicon Chem.* (Eds. F. Weis and N. Auner), VCH, Weinheim, 1996, p. 21.
58. W. Uhlig, *Chem. Ber.*, **125**, 47 (1992).
59. W. Uhlig and C. Tretner, *J. Organomet. Chem.*, **436**, C1 (1992).
60. U. Pöschl, H. Siegl and K. Hassler, *J. Organomet. Chem.*, **506**, 93 (1996).
61. K. E. Ruehl and K. Matyjaszewski, *J. Organomet. Chem.*, **410**, 1 (1991).
62. U. Pöschl and K. Hassler, *Organometallics*, **15**, 3238 (1996).
63. E. Hengge and P. K. Jenkner, *Z. Anorg. Allg. Chem.*, **604**, 69 (1991).
64. E. Hengge and M. Eibl, *J. Organomet. Chem.*, **428**, 335 (1992).
65. A. Spielberger, P. Gspaltl and E. Hengge, *Phosphorus, Sulfur, and Silicon*, **93–94**, 355 (1994).
66. A. Spielberger, P. Gspaltl, H. Siegl, E. Hengge and K. Gruber, *J. Organomet. Chem.*, **499**, 241 (1995).
67. E. Hengge and F. K. Mitter, *Monatsh. Chem.*, **117**, 721 (1986).
68. F. Uhlig, P. Gspaltl, M. Trabi and E. Hengge, *J. Organomet. Chem.*, **493**, 33 (1995).
69. A. L. Allred, R. T. Smart and D. A. Van Beek Jr., *Organometallics*, **11**, 4225 (1992).
70. E. Carberry, R. West and G. E. Glass, *J. Am. Chem. Soc.*, **91**, 5446 (1969).
71. R. West and E. S. Kean, *J. Organomet. Chem.*, **96**, 323 (1975).
72. A. C. Buchanan III and R. West, *J. Organomet. Chem.*, **172**, 273 (1979).
73. F. Uhlig, B. Stadelmann, A. Zechmann, P. Lassacher, H. Stüger and E. Hengge, *Phosphorus, Sulfur, and Silicon*, **90**, 29 (1994).
74. E. Hengge P. Gspaltl and E. Pinter, *J. Organomet. Chem.*, **521**, 145 (1996).
75. U. Herzog, G. Roewer and U. Pätzold, *J. Organomet. Chem.*, **494**, 143 (1995).
76. E. Hengge and U. Pätzold, unpublished results.
77. M. Tischler, R. Collins, M. Thewalt and G. Abstreiter *Silicon-Based Optoelectronic Materials*, (Eds.), MRS Symposia Proceedings **290**, Materials Research Society, Pittsburg, 1993.
78. (a) M. S. Brandt, H. D. Fuchs, M. Stutzmann, J. Weber and M. Cardona, *Solid State Commun.*, **81**, 307 (1992); (b) H. Kautsky, *Z. Anorg. Allg. Chem.*, **117**, 209 (1921).
79. E. Hengge, *Fortschr. Chem. Forsch.*, **51**, 1 (1974).
80. A. Kleewein, U. Pätzold, E. Hengge, S. Tasch and G. Leising in *Organosilicon III* (Eds. J. Weis and N. Auner), VCH, Weinheim, 1997.
81. E. Hengge, P. Gspaltl and A. Spielberger, *J. Organomet. Chem.*, **479**, 165 (1994).
82. P. Gspaltl, A. Spielberger, A. Zechmann and E. Hengge, *J. Organomet. Chem.*, **503**, 129 (1995).
83. P. Gspaltl, A. Spielberger and E. Hengge, *Phosphorus, Sulfur, and Silicon*, **93–94**, 353 (1994).
84. P. Gspaltl, S. Graschy, H. Siegl, E. Hengge and K. Gruber *J. Organomet. Chem.*, **524**, 187 (1996).
85. E. Hengge, E. Pinter and F. Uhlig, to appear.
86. F. Uhlig, U. Hermann, K. Klinkhammer and E. Hengge, in *Organosilicon Chemistry III* (Eds. J. Weis and N. Auner), VCH, Weinheim, 1997.
87. H. Stüger, P. Lassacher and E. Hengge, *Z. Allg. Anorg. Chem.*, **621**, 1517 (1995).
88. R. West and A. Indriksons, *J. Am. Chem. Soc.*, **94**, 6110 (1972).
89. P. K. Jenkner, E. Hengge, R. Czaputa and C. Kratky, *J. Organomet. Chem.*, **446**, 83 (1993).
90. A. Sekiguchi, T. Yatabe, S. Do and H. Sakurai, *Phosphorus, Sulfur and Silicon*, **93–94**, 193 (1994).
91. A. Sekiguchi, T. Yatabe, H. Kamatani, C. Kabuto and H. Sakurai, *J. Am. Chem. Soc.*, **114**, 6260 (1992).
92. N. Wiberg, C. M. Finger and K. Polborn, *Angew. Chem.*, **105**, 1140 (1993).
93. M. Unno, H. Shioyama, M. Ida and H. Matsumoto, *Organometallics*, **14**, 4004 (1995).
94. H. Matsumoto, K. Higuchi, S. Kyushin and M. Goto, *Angew. Chem.*, **104**, 1410 (1992).

95. Y. Kabe, M. Kuroda, Y. Honda, O. Yamashita, T. Kawase and S. Masamune, *Angew. Chem.*, **100**, 1793 (1988).
96. M. Unno, K. Higuchi, M. Ida, H. Shioyama, S. Kyushin, H. Matsumoto and M. Goto, *Organometallics*, **13**, 4633 (1994).
97. M. Kuroda, Y. Kabe, M. Hashimoto and S. Masamune, *Angew. Chem.*, **100**, 1795 (1988).
98. S. Kyushin, H. Sakurai, H. Yamaguchi, M. Goto and H. Matsumoto, *Chem. Lett.* 815, (1995).
99. T. S. Drahnak, R. West and S. C. Calabrese, *J. Organomet. Chem.*, **198**, 55 (1980).
100. E. Hengge, M. Eibl and F. Schrank, *J. Organomet. Chem.*, **369**, C23 (1989).
101. E. Hengge, and M. Eibl, *Organometallics*, **10**, 3185 (1991).
102. E. Hengge, E. Pinter, M. Eibl and F. Uhlig, *Bull. Soc. Chim. France*, **132**, 509 (1995).
103. E. Hengge, H. Siegl and B. Stadelmann, *J. Organomet. Chem.*, **479**, 187 (1994).
104. E. Hengge and A. Zechmann, *J. Organomet. Chem.*, **508**, 227 (1996).
105. T. D. Tilley, *Acc. Chem. Res.*, **26**, 22 (1993).
106. B. Stadelmann, P. Gspaltl, A. Spielberger and E. Hengge, *Phosphorus, Sulfur and Silicon*, **93-94**, 357 (1994).
107. E. Hengge, P. K. Jenkner, A. Spielberger and P. Gspaltl, *Monatsh. Chem.*, **124**, 1005 (1993).
108. B. Stadelmann, P. Lassacher, H. Stüger and E. Hengge, *J. Organomet. Chem.*, **482**, 201 (1994).
109. K. H. Pannell, L. J. Wang and J. M. Rozell, *Organometallics*, **8**, 550 (1989).
110. K. Pannell, J. Castillo-Ramirez and F. Cervantes-Lee, *Organometallics*, **11**, 3139 (1992).

CHAPTER 38

Recent advances in the chemistry of siloxane polymers and copolymers

ROBERT DRAKE, IAIN MacKINNON and RICHARD TAYLOR

Dow Corning Ltd, Cardiff Road, Barry, South Glamorgan CF63 2YL, UK

I. THE POLYMERIZATION OF SILICONES	2218
A. Introduction	2218
B. Polymerization of Cyclosiloxanes	2218
1. Anionic polymerization of cyclosiloxanes	2218
2. Cationic polymerization of cyclosiloxanes	2220
C. Condensation Polymerization of Linear Siloxanes	2221
1. Cationic polycondensation of linear siloxanes	2221
2. Anionic polycondensation of linear siloxanes	2224
II. ORGANOFUNCTIONAL SILOXANES	2224
A. Synthesis of Organofunctional Siloxanes	2224
B. Applications of Organofunctional Siloxanes	2226
III. DEGRADATION OF POLYSILOXANES	2227
A. Introduction	2227
B. Linear Polydimethylsiloxanes	2228
1. Acid-catalysed rearrangement	2228
2. Thermal depolymerization of siloxanes	2228
3. Degradation of polydimethylsiloxane in soil	2229
4. Oxidation of polydimethylsiloxanes	2231
C. Thermal Degradation of Siloxane Resins	2231
IV. THE SURFACE ACTIVITY OF LINEAR SILOXANE POLYMERS AND COPOLYMERS	2234
A. Introduction	2234
B. Siloxane Copolymer Migration to Solid–Air Interfaces	2234
1. Block copolymers	2234
2. Graft copolymers	2236

The chemistry of organic silicon compounds, Vol. 2
Edited by Z. Rappoport and Y. Apeloig © 1998 John Wiley & Sons Ltd

C. Blend Compatibilization 2238
  1. Pre-formed copolymers 2238
  2. Reactive blending 2239
V. REFERENCES 2240

I. THE POLYMERIZATION OF SILICONES

A. Introduction

Several reviews have appeared between 1991 and 1996 concerned with the polymerization and properties of silicones. Chojnowski[1] reviews the kinetically controlled ring opening of hexamethylcyclotrisiloxane (D_3). Both anionic and cationic initiation mechanisms are discussed in connection with their use in polymer synthesis. The focus of the review is on association phenomena at the initiating site and oligomer formation processes. He also reviews the behaviour of the silanol group in polycondensation reactions[2]. A review by Penezek and Duda[3] on the thermodynamics, kinetics and mechanism of anionic ring-opening polymerization discusses both organic and silicone monomers. Methods for the synthesis, characterization and properties of cyclic polymers including siloxanes are reviewed by Semlyen, Wood and Hodge[4]. Kopylov and coworkers review the mechanism of hydrolytic polycondensation of organochlorosilanes[5]. Data on the reaction kinetics and the effect of reaction conditions, reagent ratio and various additives on the course of the reaction are analysed. Hetero and metallasiloxanes derived from silanediols, disilanols, silanetriols and trisilanols are reviewed by Roesky and coworkers[6]. Additional reviews on methods of catalysis of ring-opening polymerization of cyclosiloxanes[7], macrokinetics of hydrolytic polycondensation of organochlorosilanes[8], the synthesis of silicone polymers[9] and synthesis of new silicon-based condensation polymers[10] have also appeared.

B. Polymerization of Cyclosiloxanes

1. Anionic polymerization of cyclosiloxanes

Boileau and coworkers[11] have used a novel trimethylsilylmethyl lithium initiator Me_3SiCH_2Li (**1**), in combination with a cryptand [211], for the ring-opening polymerization of cyclosiloxanes. Initiation of hexamethylcyclotrisiloxane (D_3) and octamethylcyclotetrasiloxane (D_4) polymerization has been followed by ^1H, ^7Li, ^{13}C and ^{29}Si NMR.

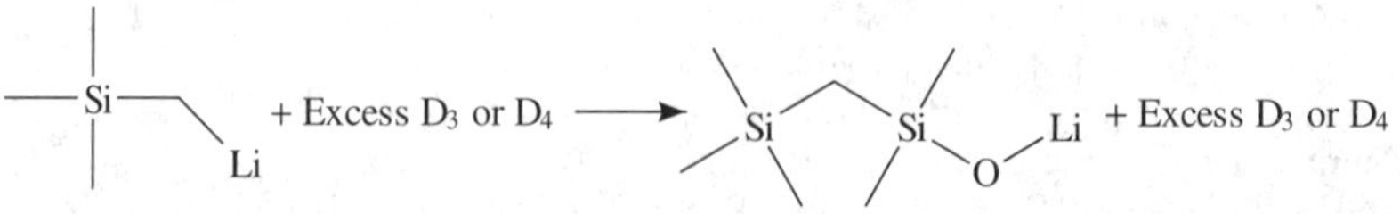

$$\text{—Si—Li} + \text{Excess } D_3 \text{ or } D_4 \longrightarrow \text{Si—Si—O—Li} + \text{Excess } D_3 \text{ or } D_4 \qquad (1)$$

(**1**)　　　　　　　　　　　　　　(**2**)

The complex $Me_3SiCH_2SiMe_2OLi$ (**2**) (^7Li shift 0.74 ppm) is the only product identified in the reaction of **1** (^7Li shift 2.5 ppm) with D_3 or D_4 at 20 °C in toluene even in the presence of excess siloxane, as shown in equation 1. Addition of the cryptand [211] shifted the Li resonance to -0.93 ppm in agreement with other lithium cryptand [211] complexes. This lithium silanolate was then shown to initiate polymerization of D_3, D_4, D_6 and functional cyclics such as $(SiMe(HC=CH_2)O)_4$ and $(SiMe(CH_2CH_2CF_3)O)_3$. Kinetic measurements using this initiator show a reactivity order of $D_3 \gg D_4 > D_5 > D_6$ and the results are in good agreement with those previously reported for anionic polymerization under similar conditions. Co-polymerization reaction involving vinyl dimethyl cyclics have been shown by ^{29}Si NMR to give, under certain conditions, polymers with a microblock structure where little intramolecular redistribution takes place.

Other novel initiators include the use of an aluminium–tetraphenylporphyrin complex Et–Al(TPP)[12]. Only D_3 is polymerized at 25 °C over 6 days to give a monodispersed ($M_n/M_w = 1.12$, $M_n = 1550$) polymer. D_4 and D_5 do not react under similar conditions and linear equilibration is suppressed. In addition, they report that this initiator will oligomerize functional cyclics such as $(SiMeHO)_4$ and $(SiPh_2O)_3$.

A fast catalyst for the ring-opening polymerization of D_4 in solution or in the bulk has been reported by Molenberg and Moller[13]. They report the use of a phosphazene base in combination with methanol as the initiator, compound **3**.

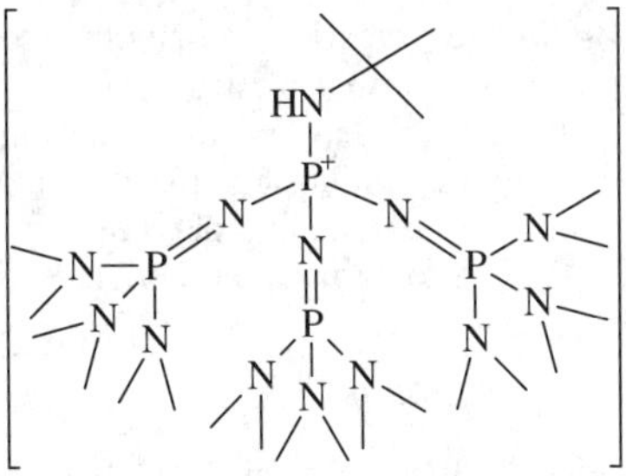

(3)

Very high polymerization and equilibration rates have been observed both in the bulk and in solution which are faster than those with KOH or CsOH. Some initial kinetic measurements have been reported. This catalyst system has recently been used by Van Dyke and Clarson[14] to initiate the ring-opening polymerization of tetraphenyltetramethyltetrasiloxane. Again very fast initiation is observed even at room temperature to give high molecular weight polymers within minutes. Using KOH under equivalent conditions gave no reaction.

Time-of-flight secondary ion mass spectroscopy (TOF-SIMS)[15] has been used for observing the initiating species in anionic ring-opening polymerizations of D_3. The PDMS mass spectrum shows peaks of varying intensity with local maxima every 3 silicon repeat units. This distribution is attributed to the relative concentrations of the three initiating species, I1, I2 and I3, shown in equation 2 which can be determined from the peak intensities.

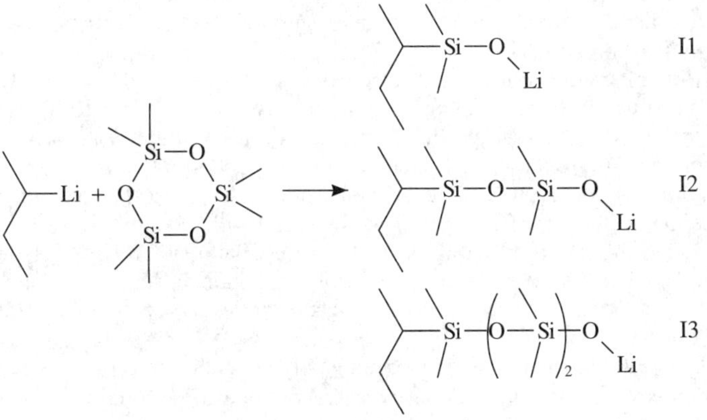

(2)

Hemery and coworkers[16] have investigated the polymerization of D_4 in aqueous emulsion. An emulsifying agent (benzyldimethyldodecylammonium hydroxide) was used as the initiator. Stable emulsions of α,ω-dihydroxypolydimethylsiloxanes with low polydispersities and controllable molecular weights up to 15,000 were obtained after polymerization. This method of polymerization was reported to give 70% conversion to polymer with only trace amounts of macrocycles, in contrast to bulk polymerizations which give large polydispersities and higher levels of macrocycles. Kinetic analysis and computer simulation were in good agreement with a mechanism involving a simultaneous polyaddition/polycondensation process. At high monomer conversion, polycondensation predominates leading to a broadening of the molecular weight distribution. Apparent rate constants for the initiation, propagation and polycondensation at different temperatures were determined. From the data activation parameters for these processes could be estimated.

There has been some interest in the preparation of monodispersed siloxanes which contain two different end groups on each polymer chain[17]. Typically, the process involves the ring opening of D_3 using a lithium reagent and then quenching the living polymer obtained with a chlorosilane. Two different groups can thus be incorporated, one from the lithium reagent used and one from the chlorosilane neutralizing agent. These materials can be used as macromonomers leading to siloxane/organic hybrid polymers.

2. Cationic polymerization of cyclosiloxanes

Cyclosiloxanes such as D_3 and D_4 have very different reactivities and generally do not give the same types of cyclic products when polymerization is initiated by cationic catalysts. Sigwalt and coworkers[18] have further examined the analogies and differences between these two monomers with respect to initiation by trifluoromethanesulphonic acid. The formation of water during these ring-opening reactions and the effect of water and other additives were examined. ^{1}H and ^{19}F NMR spectra were used to investigate the formation of triflate esters and hence the formation of water. Differences in the formation of water between D_3 and D_4 during the ring opening were not significant. The addition of water, however, has a large effect on the reactions. When water is premixed with the triflic acid prior to addition to the monomer, the ring opening initiation is effectively inhibited in both cases. If the water is added to the reactions *in situ*, then in the case of D_3 it acts as an effective cocatalyst whereas in the D_4 reaction it acts as an inhibitor. Extensions of this work[19] for D_3 ring opening initiated with triflic acid in CH_2Cl_2 have shown that the growth of linear high polymer as well as the formation of cyclics (mainly multiples of the monomer D_3, i.e. D_{3x}) involve a silyl triflate end group. These silyltriflic esters are inactive catalytic sites in the absence of triflic acid but are activated in the presence of the acid. The formation of small cyclics and macrocyclics is also discussed in depth. D_6 is mainly formed, sometimes in large amounts, which in some cases are greater than the amount of high polymer. The authors postulate that this occurs through a special type of back biting involving a transient tertiary oxonium ion PolyD-D_3^+ (**4**). This rearranges to form another oxonium ion polyD-D_6^+ (**5**) which leads to D_6 (Scheme 1). A smaller fraction of D_6 and the major fraction of D_9 are formed through cyclization of silanolesters HD_6OTf, HD_9OTf etc. resulting from addition of D_3 to the silanol ester HD_3OTf, as shown in Scheme 2. The acid reformed by this cyclization maintains a stationary concentration of silyl esters by reaction with D_3. The larger macrocyclics are also formed by cyclization of silanol esters.

The effect of ultrasound on the ring-opening polymerization of D_4 has been investigated by Price and coworkers[20]. Sonication increased the rate of polymerization over conventional stirring, giving polymers with higher molecular weights and lower polydispersities. This was explained in terms of more efficient mixing and dispersion of the acid catalyst by sonication.

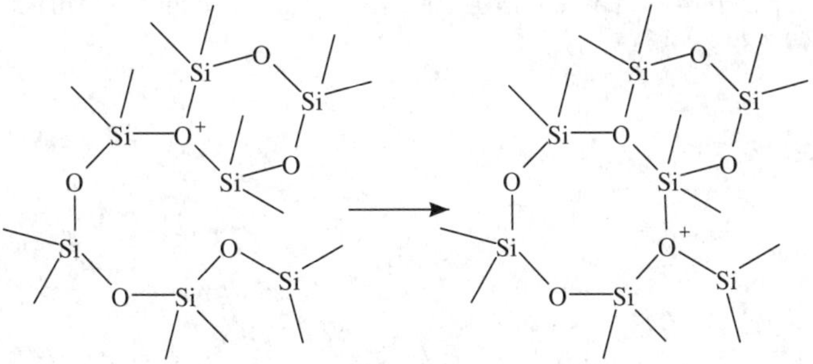

SCHEME 1

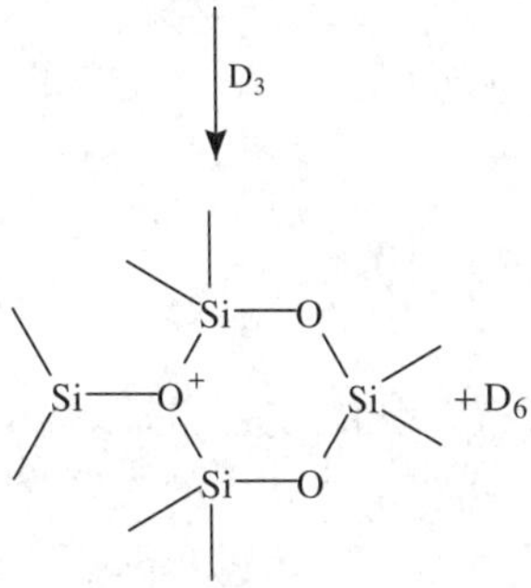

SCHEME 2

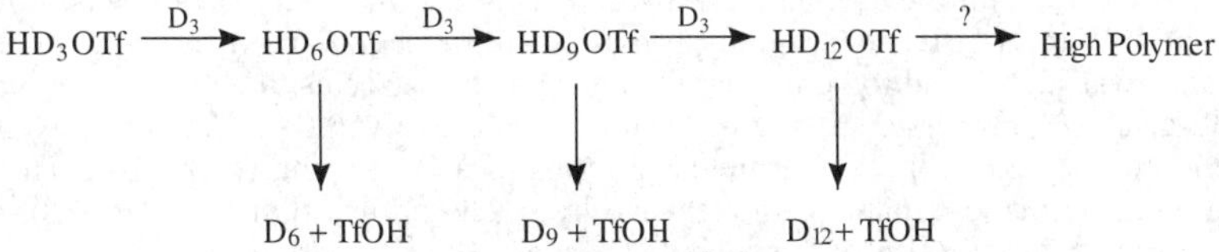

$$HD_3OTf \xrightarrow{D_3} HD_6OTf \xrightarrow{D_3} HD_9OTf \xrightarrow{D_3} HD_{12}OTf \xrightarrow{?} \text{High Polymer}$$

$$\downarrow \qquad\qquad \downarrow \qquad\qquad \downarrow$$

$$D_6 + TfOH \qquad D_9 + TfOH \qquad D_{12} + TfOH$$

C. Condensation Polymerization of Linear Siloxanes

1. Cationic polycondensation of linear siloxanes

There has been an increased interest in this area of siloxane polymerization in recent years. Many publications have appeared on the kinetics of acid-catalysed condensation and disproportionation. Chojnowski, Rubinsztajn and Cypryk[21] report the kinetics and disproportionation behaviour of oligo(dimethylsilanols), $HO(SiMe_2O)_nH$, $n = 2$ or 5 and $Me(SiMe_2O)_nH$, $n = 2$ or 5, in the presence of protic acids in both acid–base inert solvents such as methylene chloride and in the basic solvent dioxane. The kinetics in the two systems were very different. In methylene chloride they found that two molecules of $HO(SiMe_2O)_5H$ participate in the transition state and that water did not participate in this reaction. Siloxane bond cleavage was shown to be strongly affected by other silanol groups in the siloxane diols, $HO(SiMe_2O)_nH$, by comparison with the kinetics for the monosiloxanols, $Me(SiMe_2O)_nH$. The intra- (cf. **6**) and inter-molecular assistance (cf. **7**) of a third silanol group plays an important role in siloxane bond cleavage, shown in

Scheme 3. This third group acts as a base receiving a proton from the silanol group attacking the acid complex.

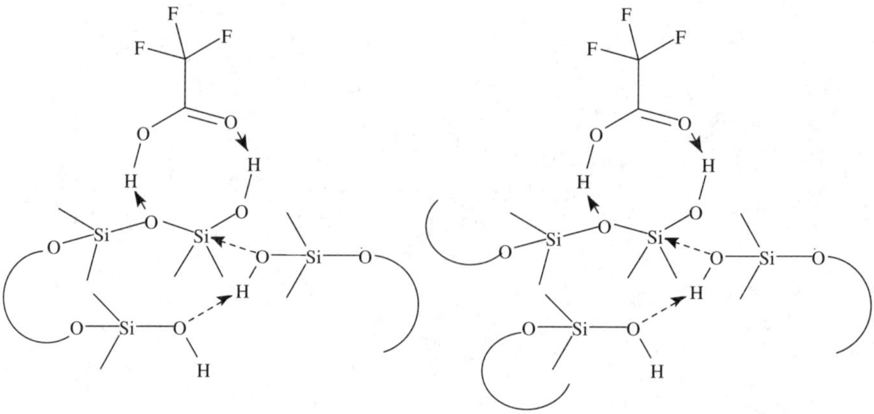

(6)   (7)

SCHEME 3

The kinetics of these reactions are strongly affected by the conversion of the protic acid catalyst to a silyl ester as shown in equation 3.

$$X(SiMe_2O)_nH + HOSO_2CF_3 \rightleftharpoons X(SiMe_2O)_nSO_2CF_3 + H_2O \qquad (3)$$

Because the rates of this reaction are very rapid, normal sampling techniques were not satisfactory and an infrared technique was used. This esterification reaction was shown to be about 100 times faster than the disproportionation reaction and inter–intra-molecular assistance was also found to be important. This assistance seems to be a common pattern in acid-catalysed processes of oligosiloxanols in inert solvents. In dioxane solvent the redistribution kinetics can be interpreted in terms of an unzipping mechanism. The rate-determining step is terminal silanol cleavage by water forming dimethylsilanediol which rapidly reacts with other substrate silanols (Scheme 4).

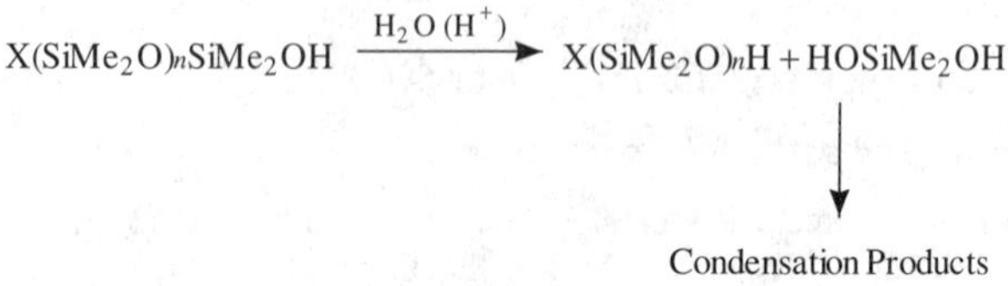

$$X(SiMe_2O)_nSiMe_2OH \xrightarrow{H_2O\;(H^+)} X(SiMe_2O)_nH + HOSiMe_2OH$$

↓

Condensation Products

SCHEME 4

Work by Sigwalt, Bischoff and Cypryk[22] have used this inter–and intramolecular catalysis to explain the condensation kinetics and cyclic formation processes in siloxane condensations. The kinetics show a very complex dependence on siloxane chain length, complicated by equilibria involving acid, silanol and water. They do indicate that the dominating reaction in the process is condensation and that chain disproportionation and chain scrambling are negligible! The kinetics of condensation are influenced by the involvement of triflic acid in several equilibria, i.e. the formation of triflate esters, shown in equation 3, the possible involvement of triflic acid in the reaction of these esters with silanol, shown in equation 4 and the hydration of the triflic acid itself, shown in equation 5.

$$X(SiMe_2O)_n SiMe_2OSO_2CF_3 + HOSiMe_2(OSiMe)_n X \rightleftharpoons HOSO_2CF_3$$
$$+ X(SiMe_2O)_n SiMe_2OSiMe_2(SiMe_2O)_n X \qquad (4)$$

$$HOSO_2CF_3 \cdot H_2O \xrightleftharpoons{H_2O} HOSO_2CF_3 \cdot H_2O \xrightleftharpoons{nH_2O} HOSO_2CF_3 \cdot (n+1)H_2O \qquad (5)$$

Alkoxysilane silanol condensation reactions play an important role in sol-gel technology, the manufacture of silicone resins, the vulcanization of silicones and in surface modification by alkoxysilanes. There have been recent investigations by Chojnowski and coworkers[23] into the kinetics of acid-catalysed heterofunctional condensation of model alkoxy and silanol functional siloxanes. The heterofunctional reaction involving SiOEt and SiOH competes with the homofunctional reaction of SiOH with SiOH. The rates of each process are similar, but are influenced by the medium and hence by the concentration of the reactants. Hydrolysis of the ethoxysiloxane as well as ethanolysis of the silanol groups leads to extensive interconversion of functional groups. These interconversion processes are two orders of magnitude faster than those of the condensation reactions.

Other publications have also appeared describing condensation kinetics with heterogeneous catalysts by a designed experimental approach to optimize reaction conditions[24], stannous octanoate condensation catalysis[25] and the role of bifunctional catalysis mechanisms[26]. A 'bifunctional transition state', shown for compound **8**, was postulated to explain the kinetics of heterofunctional condensation reactions catalysed by Cl_3CCOOH, Et_2NOH, $MeEtC=NOH$ or CH_3COOH (equation 6).

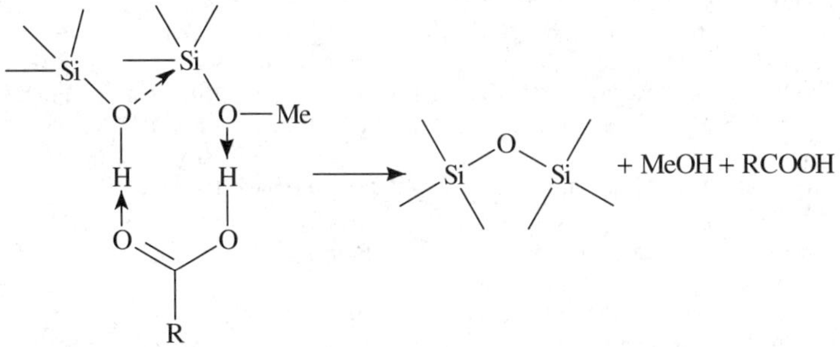

$$\text{(structure 8)} \longrightarrow \text{siloxane} + MeOH + RCOOH \qquad (6)$$

(8)

Most of the references cited in this section have been concerned with the kinetics and mechanism of the reactions. There have also been some relevant patents published on catalysts for silanol condensation. These catalysts are of the general family of chlorophosphazenes, $(-Cl_2P=N-)_n$. Wacker Chemie[27] have had several patents on the use of oxygen-containing phosphazenes for the condensation of hydroxy terminated siloxanes. Hagger[28] and Weis have also published their findings in journals and at conferences. Nunchritz[29] and General Electric[30] have also filed patents in this area although the Wacker patents seem to have priority. Dow Corning[31] has also patented the use of chlorophosphazenes and processes for their use. There has also been some discussion of the kinetics and mechanism of condensation using these unusual catalysts[32].

A very unusual initiator which has been discussed recently is the hotly disputed silicenium ion. A patent by Schulz Lambert, and Kania[33] shows that an initiator of the type R_3SiA, where A is a non-coordinating anion, can be used under anhydrous conditions to ring-open D_3 to give mixtures of cyclic and linear polymers. This initiator is also discussed by Olah and coworkers for the polymerization of organic and siloxane monomers[34].

2. Anionic polycondensation of linear siloxanes

An interesting approach to the synthesis of siloxanes has been patented by Kolaczkowski and Serbetcioglu[35]. Their process uses a K_3PO_4 heterogeneous catalyst dispersed on a monolithic ceramic support and can be used for condensation or ring-opening polymerizations. The advantages are that the catalyst is easily removed and high reaction rates and high throughput rates are obtained. Other anionic catalyst systems include the use of a quaternary ammonium borate, phosphate, carbonate or silicate by Westall[36].

The polymerization of silicones to give functional siloxane polymers has also been discussed. A commercially important class of polymers are the amine functional silicone fluids. McGrath and Spinu[37] discuss the use of D_4 and an aminofunctional cyclic siloxane in the presence of a tetramethylammonium silanolate catalyst. Sauvet and Helary[38] report condensation kinetic data for the condensation of an aminoalkylalkoxysilane ($H_2N(CH_2)_3SiMe(OMe)_2$) and a monosilanol ($PhMe_2SiOH$) as well as with oligosiloxanediols $HO(SiMe_2O)_nH$. This is a useful method for the synthesis of amine polymers of well-defined structure and proceeds without the need for a catalyst. Homofunctional condensation reactions and functional group exchange can be minimized by continuous removal of methanol and control of the reaction temperature. The reactivities of the two alkoxy groups in the silane are very different. The first reacts readily at room temperature, but higher temperatures are required to substitute the second. By using this large difference in reactivity of the two alkoxy groups, block copolymers of dimethyl and methylphenylsiloxanes were prepared. Polymers with arylamine functional groups have also been reported. Babu and coworkers[39] prepared a lithium complex of a protected aniline, $(Me_3Si)_2NC_6H_4Li$, to initiate the ring-opening polymerization of D_3 which was then terminated with a protected aniline functional chlorosilane, p-$(Me_3Si)_2NC_6H_4SiMe_2Cl$. The protected oligomer is then converted to the free amine with methanol.

An interesting variation of an ammonium silanolate catalyst was described by Hoffman and Leir[40]. They prepared tetramethylammonium 3-aminopropyldimethylsilanolate and used it to synthesize 3-aminopropyl terminated Polydimethylsiloxane (PDMS).

Another class of functional siloxane polymer that has received some attention are the fluorosiloxane materials, especially 3,3,3-trifluoropropylmethylsiloxanes. The use of conventional equilibration catalysts to produce these materials gives products which favour the cyclosiloxane in the bulk. Clarson and coworkers[41] report the use of specific condensation catalysts such as stannous octanoate, potassium carbonate and barium hydroxide to prepare hydroxy terminated fluorosilicone polymers.

Stannous octanoate has also been studied as a catalyst for the synthesis of polydimethyl- and poly(methylphenyl)siloxanes via polycondensation.

Other methods of polymerization that have appeared recently include plasma polymerization[42] and photopolymerization[43], but they give partially or highly crosslinked films.

II. ORGANOFUNCTIONAL SILOXANES

A. Synthesis of Organofunctional Siloxanes

As in previous years the major method used for attaching an organofunctional group to a siloxane backbone is hydrosilylation. Even in this much studied field there is still the potential for surprising and potentially very useful results. Thus, Crivello and Bi observed that the Rh-catalysed hydrosilylation of unsaturated epoxides with α, ω-SiH functional siloxanes proceeded in a discrete stepwise fashion[44]. Using $(Ph_3P)_3RhCl$, $(HMe_2Si)_2O$ added to the unsaturated epoxide, 3-vinyl-7-oxabicyclo[4.1.0]heptane, to yield almost exclusively the mono-adduct **9**. Rate studies attributed this to a large difference in the activation energies of the two SiH groups, $E_a =$ ca 109 kJ mol^{-1} for the first site and ca 243 kJ mol^{-1} for the second, so that with short reaction times and low temperatures, and even in the presence of excess unsaturated epoxide, the mono-adduct was formed in high yield. A further surprise was that under the same conditions, $HMe_2SiO(Me_2SiO)_2Me_2SiH$, with 6 bonds separating the two SiH groups, also formed the mono-adduct in ca 95% yield. This has already made possible the synthesis of a series of compounds with both SiH and epoxy functionality in the same molecule or of molecules with two dissimilar epoxy functional groups.

(9)

An alternative route to asymmetric α, ω-organofunctional siloxanes or to siloxanes bearing exclusively functionality at one end is by terminating living anionic polymerizations with appropriate functional-derivatives (see Section I.B.1)[45,46]. Alternatively, the asymmetric disiloxane, $H_2N(CH_2)_3Me_2SiOSiMe_2CH=CH_2$, was prepared by base-catalysed equilibration of the symmetrical disiloxanes, $[H_2N(CH_2)_3Me_2Si]_2O$ and $(CH_2=CHMe_2Si)_2O$, followed by fractional distillation of the statistical distribution of products. Although equilibration is probably not a route which is generally applicable for the preparation of pure higher homologues, a range of asymmetric α, ω-organofunctional siloxanes prepared by this method, $H_2N(CH_2)_3(Me_2SiO)_nMe_2SiCH=CH_2$, have been claimed as compatibilizers for polymer blends[47].

In some instances direct hydrosilylation of an olefin derivative of the intended functional group can lead to serious side-reactions, may proceed in poor yield or may simply call for unusual and difficult to prepare intermediates. Thus, Mitchell and coworkers were interested in preparing siloxanes bearing photoactive cinnamate moieties[48]. Hydrosilylation of a suitable substrate such as the allyl ester of 4-methoxycinnamic acid with poly(methylhydrogensiloxane) or poly(dimethyl-co-methylhydrogensiloxane) resulted in problems such as coloured material, crosslinking during work-up and reduction of some cinnamate double bonds. A far more successful method was to functionalize the siloxane backbone with allyl acetate, hydrolyse the acetate to the alcohol and then esterify with the acid chloride of the cinnamic acid derivative. Polymers so formed were nearly colourless, showed no signs of crosslinking during work-up and no reduction of the cinnamate double bond occurred. In a similar two-step route, workers at Wacker Chemie prepared anhydride functional siloxanes in good yield[49]. Hydrosilylation of an acetylenic alcohol gave a polysiloxane bearing alkenol groups, When heated in the presence of an acid catalyst these groups eliminate water to form a butadienyl-functional polysiloxane. When this dehydration is conducted in the presence of an unsaturated anhydride, e.g. maleic anhydride, the butadienyl-functional polysiloxane undergoes an immediate Diels–Alder reaction to form an anhydride functional polysiloxane (Scheme 5).

Some researchers continue to investigate alternatives to hydrosilylation or Grignard chemistry for the preparation of organofunctional siloxanes. One such example is the

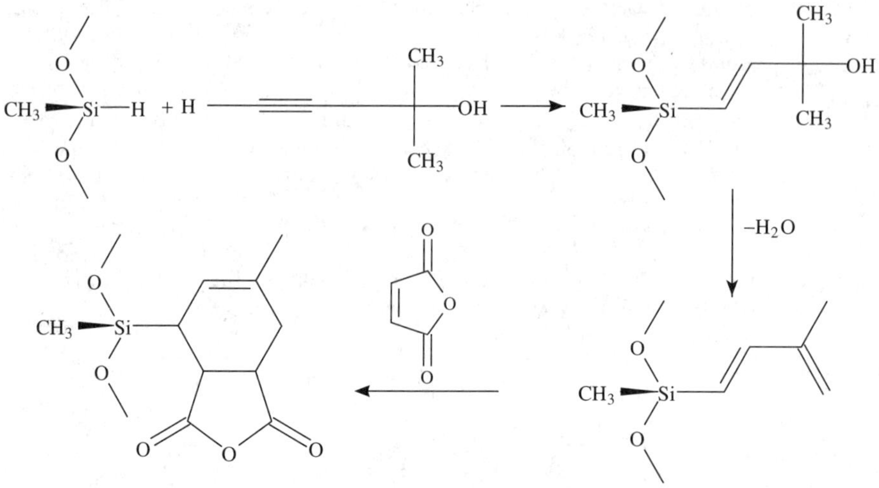

SCHEME 5

preparation by Bennetau, Dunoguès, Boileau and coworkers of precursors to highly fluorinated siloxanes, e.g. $(C_8F_{17}CH_2CH_2SiMeO)_4$[50]. Using ultrasound, with sodium dithionite as catalyst in a mixture of acetonitrile and water, the perfluoroalkyl iodide, $C_8F_{17}I$, added regioselectively to a vinylsiloxane to give an α-iodofunctional intermediate, $C_8F_{17}CH_2CH(I)Si\equiv$, which then underwent reductive cleavage with Bu_3SnH to give the desired product in 80–85% yield. Another interesting alternative strategy has been exploited by workers at Siemens AG who nitrated poly(methylphenyl)siloxane with a mixture of Ac_2O and concentrated HNO_3[51]. Other functionalities were introduced by reduction of the nitro group (preferably with Fe and AcOH in ethanol) to the corresponding amine which is then converted to, e.g., hydroxyl, halogen, cyano or maleimide.

B. Applications of Organofunctional Siloxanes

The crosslinking of siloxanes via UV-curable organofunctional groups is of ever increasing interest. Systems of this type generally have a low energy requirement (cf thermally cured systems) and produce no hazardous by-products (cf hydrolytic condensation). Several different approaches have been utilized. To date the most commonly studied method is the very rapid polymerization of (meth)acrylate-functional siloxanes and there remains a great deal of patent activity in this area[52]. Several research groups, most notably at GE, have also studied photopolymerization of epoxy-functional siloxanes[53] and another non-acrylate system has been much studied by workers at Loctite Corp. based on thiol-functional and norbornene-functional siloxanes (so-called thiol-ene polymerization)[54]. The thiol-ene systems were determined to have a higher photoresponse than comparable acrylate systems; this was partly ascribed to the low inhibition of the thiol-ene systems by ambient oxygen. Vinyl ethers and related cationically curable functionalities have also been the subject of much recent investigation[55]. Thus, Cazaux and Coqueret prepared telechelic vinyl ether-functional siloxanes by treating α,ω-SiH functional siloxanes with a twenty-fold excess of suitable commercial divinyl ethers, $CH_2=CHOROCH=CH_2$, R = $CH_2(CH_2OCH_2)_2CH_2$ or $CH_2(c\text{-Hex})CH_2$ and a Pt catalyst[56]. The large excess of divinyl ether prevented appreciable polycondensation. An alternative method, in this case to the related propenyl ether-functional siloxanes, was devised by Crivello and coworkers[57]. They used chemoselective hydrosilylation of 1-allyloxy-4-(1-propenoxy)butane [i.e. $CH_3CH=CHO(CH_2)_4OCH_2CH=CH_2$]. Hydrosilylation of a terminal double bond is so rapid when compared with hydrosilylation of an internal double bond that the propenyl ether-functional siloxane is effectively the sole product. Rather than just having a photopolymerizable siloxane, Coqueret and Pouliquen have developed siloxanes which bear the photoinitiators, e.g. a benzophenone or thioxanthone group in combination with a tertiary amine function[58]. Such polymeric photoinitiators may be more compatible with the monomeric system and give rise to reduced levels of volatiles and extractables after curing.

Interest in siloxanes bearing various mesogens is still an area of growing interest and patent activity. These liquid crystal polymers may be used in a variety of electro-optical, opto-optical, magneto-optical, mechanical or thermo-optical devices, with or without the addition of other polymeric or low molar mass mesogenic materials[59]. There is little else which is particularly novel in the area of applications. There has been a continuing growth in the number and variety of uses of functional siloxanes for hair care. Although 'traditional' functionality such as amine[60] and glycol[61] graft copolymers are well represented, other functionalities include acrylic (to replace peroxide)[62], thioglycolamide (for conditioning with good curl and low odour)[63], poly-N-acyl alkeneimine (gloss, smoothness)[64] or even protein-siloxane copolymers[65]. For skin care, glycerol functional polysiloxane have been shown to be useful as humectants, being durable and skin substantive[66]. Where the glycerol unit is linked to the siloxane via a Si—OC link, in use the compounds will hydrolyse on the skin, delivering the humectant glycerol and a silanol functional organo-Si compound which itself acts as a durable water-repellent and softening agent[67]. In the textile industry, the major use of siloxanes is as fabric softeners and many applications are based on amino-functional siloxanes. One of the major concerns is still imparting softness but avoiding the yellowing often seen with amino-functional siloxanes[68]. This is generally achieved by modifying the chemical structure of the amino residue in some manner, e.g. alkylation or acylation. One novel use of a polysiloxane in the textile industry is the incorporation of polysiloxane units into a copolymer, e.g. polyethylene terephthalate and polydimethylsiloxane(PDMS). A monofilament of this composition has been prepared, and claimed uses include the making of dense precision fabrics for filters, screens and screen-printing screens without flaking or attrition[69]. Besides their uses in the textile and hair-care areas, amino-functional siloxanes are used in a variety of other applications, e.g. they may be reacted with epoxy resins to give materials with increased fracture toughness while retaining the good mechanical and hot-wet properties of the original untoughened resins[70]. Amino-functional siloxanes have also been proposed for the reduction of atmospheric pollution through their use in the removal of CO_2, SO_x and NO_x from gaseous mixtures[71].

III. DEGRADATION OF POLYSILOXANES

A. Introduction

The degradation of linear siloxanes under conditions ranging from dry Michigan soils to outer space continues to be a topic of great interest, both academically and industrially. The previous two reviews in this series[72] have comprehensively covered the mechanisms and kinetics of thermolysis of linear polysiloxanes and little new work, on this aspect of siloxane degradation, appears to have been reported. The pyrolysis of resinous polysiloxane materials to give silicon-containing ceramics[73] is, however, an area of current interest and study.

B. Linear Polydimethylsiloxanes

1. Acid-catalysed rearrangement

Chojnowski and coworkers have examined the acid-catalysed disproportionation of oligomeric dimethylsiloxanols in dioxane in the presence of water[21b]. The results are interpreted in terms of a rate-determining cleavage of the terminal siloxane unit by water, followed by condensation of the resultant dimethylsilanediol (**10**) with the remaining dimethylsiloxanols (equations 7 and 8).

$$H_2O + H[OSiMe_2]_nOH \xrightarrow{slow} H[OSiMe_2]_{n-1}OH + Me_2Si(OH)_2 \qquad (7)$$
$$(\mathbf{10})$$

$$H[OSiMe_2]_nOH + Me_2Si(OH)_2 \xrightarrow{fast} H[OSiMe_2]_{n+1}OH \qquad (8)$$
$$(\mathbf{10})$$

This disproportionation offers an insight into the mechanism of hydrolytic degradation of PDMS, especially in the presence of traces of acid. Thus, the siloxane chain may undergo a slow scission in the presence of water to give a silanol ended chain, which then depolymerizes by loss of dimethylsiloxanediol.

2. Thermal depolymerization of siloxanes

One of the major commercial uses of polydimethylsiloxanes is the preparation of elastomeric materials. To obtain suitable material properties inorganic oxides, notably silica[74], are added to high molecular weight polydimethylsiloxanes, which are then crosslinked. The addition of an inorganic oxide can increase or decrease the thermal stability of polydimethylsiloxane depending on the acidic or basic nature of the filler and the specific interactions between polymer and filler[75]. Sohoni and Mark[75] showed that use of pyrogenic silica as a filler marginally increased the thermal stability of a crosslinked polydimethylsiloxane network. A much larger effect was observed, however[75], for silicas that were prepared by *in situ* hydrolysis of tetraethylorthosilicate within the polymer network. For example, the temperature for onset of degradation could be raised from around 410 °C to around 450 °C by incorporation of as little as 6% precipitated silica prepared *in situ*. The mechanism of this stabilization was postulated to result from deactivation of terminal silanol groups on the polymer by interaction with the silica[75]. A pyrogenic silica and an alumino silicate have been treated with a range of silanes and silicones and the thermal stability studied[76]. The results are shown in Table 1. The siloxanes grafted onto silica are stated to be slightly more stable than the corresponding polysiloxane polymers, though comparative data are not given[76]. Grafting of hexamethylcyclotrisiloxane to an aluminosilicate surface[76] produces a coating which has a decomposition temperature approximately 40 °C higher than that of the corresponding coating on silica. The order of stability depends on the organic substituent on silicon and decreases with the presence of hydrogen or higher alkyl substituents according to the following series:

$$PhMeSiO > Me_2SiO > Me_3SiO_{1/2} > CF_3CH_2CH_2MeSiO > MeHSiO > EtHSiO$$

Poly[oxybis(dimethylsilylene)][77] (**11**) starts to depolymerize at 250 °C predominatly by loss of octamethyl-1,4-dioxatetrasilacyclohexane (**12**).

The activation enthalpy for initiation of depolymerization was approximately 75 kJ mol^{-1}. This low value rules out spontaneous cleavage of Si—O or Si—Si bonds

38. Recent advances in the chemistry of siloxane polymers and copolymers

TABLE 1. Thermal breakdown data in air for organosiloxanes grafted to silica[76]

Treating Agent	T_0 (°C)[a]	T_{max} (°C)[b]	E_a (kJ mol^{-1})
(Me$_3$Si)$_2$NH	390	440	164
[Me$_2$SiO]$_3$	420	490 ± 10	176
[Me$_2$SiO]$_4$	420	480 ± 10	174
[MeHSiO]$_4$	380		
[EtHSiO]$_4$	250	280	126
[(H$_2$C=CH)MeSiO]$_4$	230	280	126
[CF$_3$CH$_2$CH$_2$MeSiO]$_4$	390	430	162
[PhMeSiO]$_4$	430	560	193

[a]Temperature of onset of degradation, as indicated by IR spectroscopy.
[b]Temperature of maximum rate of weight loss from TGA.

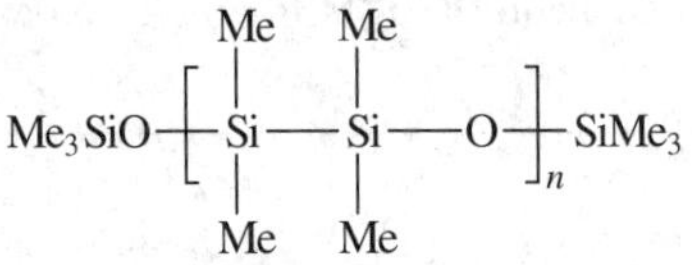

(11) 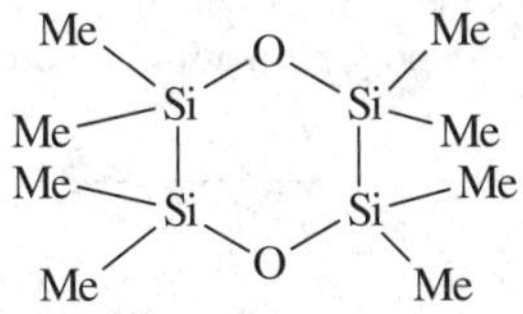   (12)

as the initiating reactions and suggests an ionic mechanism involving trace contaminants. Rearrangement of **11** to polysiloxane-b-polysilylene copolymers (**13**) (equation 9) was observed, as was ready oxidation of the polysilylene to polysiloxane[78].

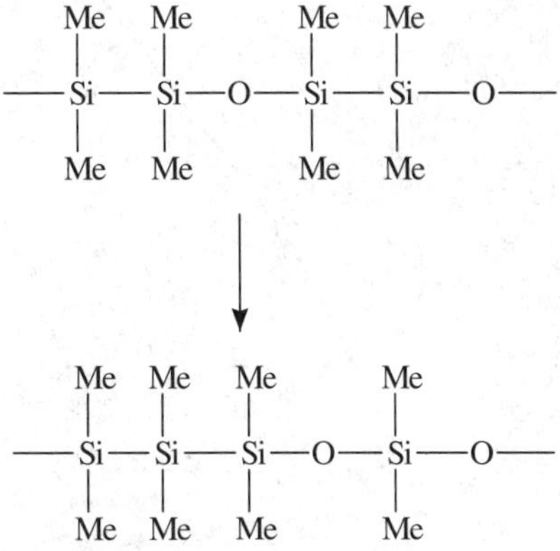

(9)

(13)

3. Degradation of polydimethylsiloxane in soil

The depolymerization of polydimethylsiloxane to dimethylsilanediol (**10**) on exposure to dry soil was reported in 1979 by Buch and Ingebridtson[79]. The presence of moisture markedly decreased the rate of depolymerization.

Several groups[80,81] have now further studied this depolymerization using a range of soils and conditions. Under moist soil conditions, very little depolymerization is observed with the majority of the polydimethylsiloxane remaining soluble in organic solvents[81a]. Once the soil moisture drops below about 4%, depolymerization of the siloxane occurs over a period of days to give siloxanol oligomers[80,81a] **(14)**, $n \leqslant 9$, which are water soluble. This was demonstrated (Figure 1) by using soil treated with ^{14}C-labelled polydimethylsiloxane (PDMS) and monitoring both the soil moisture level and the ^{14}C levels desorbed from the soil on water washing. As PDMS is insoluble in water, the presence of ^{14}C containing species was taken as being indicative of formation of siloxanol oligomers **(14)**.

$$HO(Me_2SiO)_nH (n = 2-9)$$

(14)

The mechanism of this depolymerization is unlikely to be biological but rather is thought to involve catalysis by clay minerals within the soil. At high water levels, access of the hydrophobic polydimethylsiloxane to the active sites on the clay may be limited, thus slowing the depolymerization. The dimethylsilanediol **(10)** is the major product species

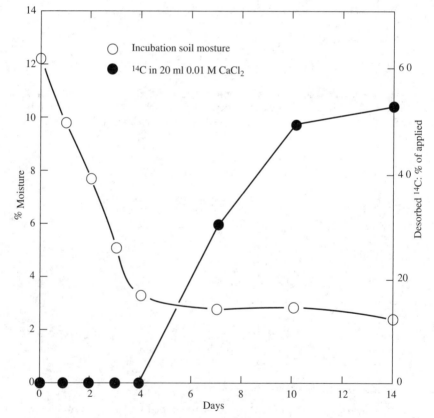

FIGURE 1. Breakdown of polydimethylsiloxane into water-soluble species in dry soil as evidenced by desorbable ^{14}C. Reproduced by permission of Pergamon Press from Reference 81a

found in all studies carried out to date. There is evidence[81b] that this species breaks down further with cleavage of the Si−C bond and release of CO_2. It is possible that this degradation may be due to microbes using **10** directly as a food source, but it is more likely that the known secretion of H_2O_2 and certain enzymes by some fungi results in a radical cleavage of the Si−C bond[81b]. Diemethylsilanediol (**10**) is a volatile substance and the major part of it is lost to the atmosphere by evaporation where it is thought to degrade by photo-oxidation to yield silica, carbon dioxide and water[81e].

4. Oxidation of polydimethylsiloxanes

A number of groups have looked at the photooxidation of siloxanes under irradiation at a range of wavelengths[82-86]. In all cases, loss of organic radicals from silicon as a range of oxidized species was observed with the residual material being silica. The rate of oxidation increases as the wavelength of UV light is decreased, as would be expected, due to the increasing energy of the photons. It is not clear, however, if the observed degradation is due to optically induced bond breaking or to local heating due to energy dissipation[84]. In the absence of oxygen no degradation is observed on irradiation at 193 nm[84] but loss of methane, ethane and phenyl groups was observed on irradiation at 147 nm and 123.6 nm[86]. Pelizzetti and coworkers[87] have shown that octaphenylcyclotetrasiloxane (**15**) and poly(methylphenylsiloxane) (**16**) can be photocatalytically degraded to CO_2 and inorganic silicates using TiO_2 and visible light. The mechanism of this degradation is postulated to involve an initial formation of radicals at the phenyl substituent either by hydrogen atom extraction or by •OH addition.

$$[Ph_2SiO]_4 \quad\quad [PhMeSiO]_n$$
$$(15) \quad\quad\quad (16)$$

Atomic oxygen is a highly reactive species that can pose serious problems for spacecraft in low Earth orbit due to degradation of materials of construction. In the case of siloxane elastomers and coatings[88,89], it has been shown that surface oxidation occurs to give a silica layer which then protects the bulk material from further oxidation, with the result that little loss of physical properties occurs.

C. Thermal Degradation of Siloxane Resins

Corriu and coworkers have studied the thermal decomposition of hydrosilsesquioxane resins[90] (**17**) and have observed the evolution of silane (**18**) between 350°C and 480°C (equation 10). Above this temperature hydrogen is observed.

$$4HSiO_{1.5} \longrightarrow SiH_4 + 3SiO_2 \quad\quad (10)$$
$$(17) \quad\quad\quad (18)$$

The loss of silane is due to redistribution reactions between Si−H and Si−O bonds which may be catalysed by residual silanol groups acting as basic catalysts[90]. The preparative method for the gel was shown to greatly affect the percentage weight loss due to silane evolution, depending on the presence or absence of catalytic species. In the presence of air, oxidation of the hydrosilsesquioxane to silica was observed[90]. The same workers extended this study to a series of resins containing D (Me_2SiO), D^H (MeHSiO), T ($MeSiO_{1.5}$) and T^H ($HSiO_{1.5}$) units. Use of TG/MS and solid state NMR clearly demonstrated that redistribution reactions could take place between Si−C and Si−O bonds (equation 11), as well as between Si−H and Si−O bonds; Si−O exchange with Si−O also occurs[91].

$$2MeSiO_{1.5} \longrightarrow Me_2SiO + SiO_2 \quad\quad (11)$$

Temperatures around 600 °C are required for exchange reactions involving silicon–carbon bonds whilst silicon–hydrogen bonds will redistribute at temperatures as low as 300 °C. The exchange of Si—C bonds has also been observed for mixed siloxane–titania gels derived from dimethyldiethoxysilane and titanium tetraisopropoxide[92]. At 600 °C, loss of methyl groups from silicon was observed with formation of Si—O bonds and Ti—C bonds, thus demonstrating an exchange of carbon between silicon and titanium.

Wachholz and coworkers[93] have used a combination of cryo-GC/FT-IR and GC/MS to study the siloxane pyrolysis products of a silicone resin prepared from a mixture of Me$_3$SiCl (0.34 mol), MeHSiCl$_2$ (0.56 mol), Me(CH$_2$=CH)SiCl$_2$ (0.56 mol) and PhSiCl$_3$ (1.56 mol). Heating at 500 °C gave an oil which contained over 30 compounds. FT-IR and MS analysis of three of the major signals allowed the identification of **19**, **20** and **21**. It is not clear if these species were present in the resin as made or if they result from siloxane bond redistribution during the pyrolysis.

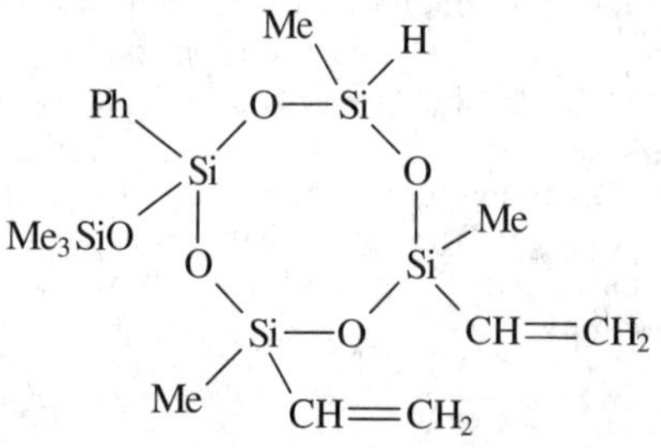

(19)

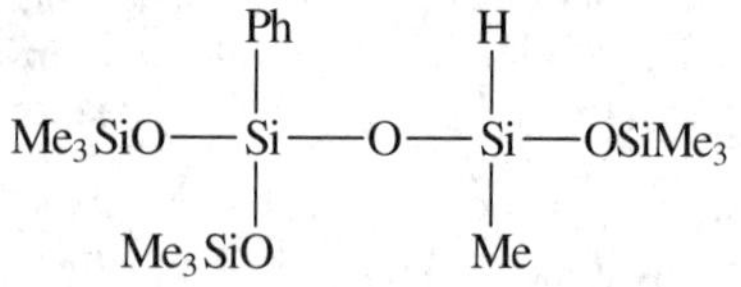

(20)

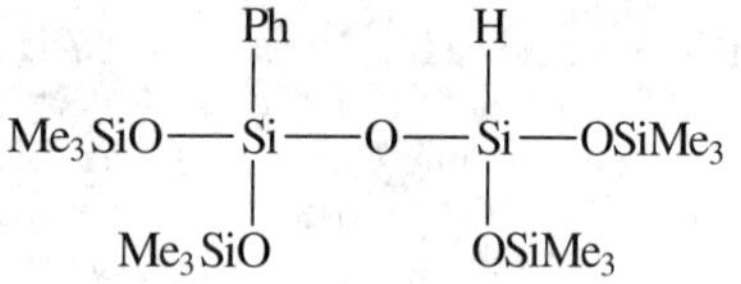

(21)

The pyrolysis of polyphenylsilsesquioxane above 500 °C proceeds with loss of benzene; some methane and hydrogen was observed above 600 °C. Approximately one-third of the phenyl rings were eliminated during pyrolysis[94] to give an amorphous silicon oxycarbide. Similar results were found for silsesquioxane copolymers containing mixtures of methyl and phenyl substituents on silicon[95]. Zank and coworkers have examined degradation pathways for a series of poly[methyl-*co*-phenylsilsesquioxane] resins containing vinyldimethylsiloxy units (**22**)[96].

$$(PhSiO_{1.5})_x(MeSiO_{1.5})_y[(H_2C=CH)Me_2SiO_{0.5}]_z$$

$$x = 0.25-0.75,\ y = 0.05-0.50,\ z = 0 \text{ or } 0.25$$

(22)

Benzene, ethylene and acetylene were the predominate observed volatiles at 550 °C whilst methane was evolved from 650 °C to 875 °C. An amorphous SiCO material was obtained at 1200 °C and bond redistribution and carbothermic reduction occurred up to 1800 °C to give a ceramic material composed of substantial amounts of crystalline β-silicon carbide. The preparation of bulk ceramic components from materials in the system

Si—Met—C—N—O (Met = Ti, Cr, V, Mo, Si, B, CrSi$_2$, MoSi$_2$ etc.) from organosilicon polymers has been well reviewed by Greil[73].

The thermolysis, under nitrogen, of polyhedral oligomeric cyclohexylsilsequioxane macromers and siloxane copolymers proceeds via a number of processes[97]. Fully condensed species such as c-Hex$_8$Si$_8$O$_{12}$ (**23**) undergo complete sublimation at around 380 °C with retention of the silicon–oxygen cages. For the partly condensed species **24**, silanol condensation between 390 °C and 450 °C prevents complete sublimation; loss of cyclohexane and cyclohexene occurs on further heating to 550 °C. The copolymer **25** exhibited loss of cyclic dimethylsiloxanes, D$_3$ and D$_4$ as well as loss of cyclohexane and cyclohexene between 400° and 500 °C.

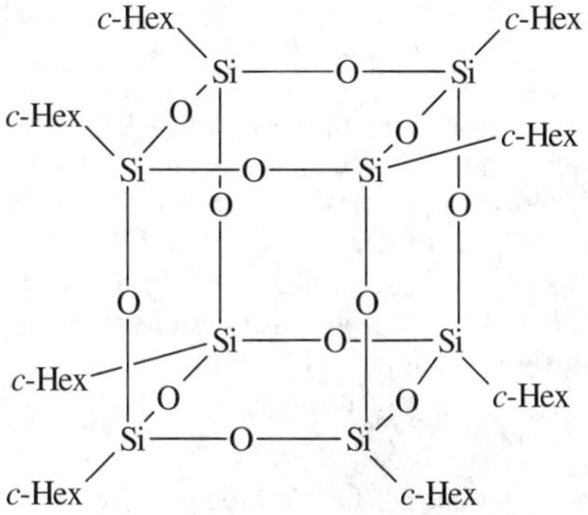

(**23**) 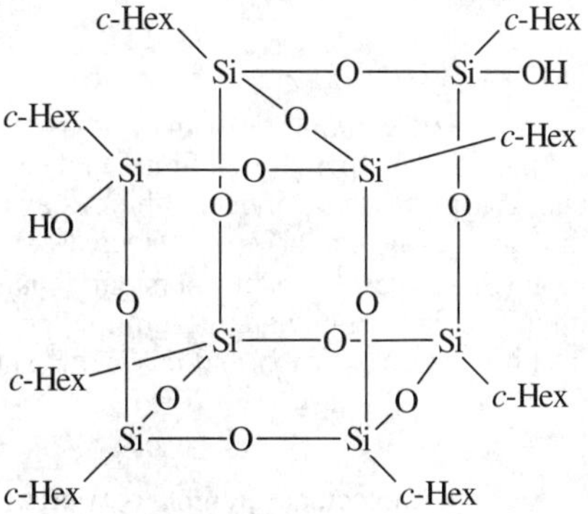   (**24**)

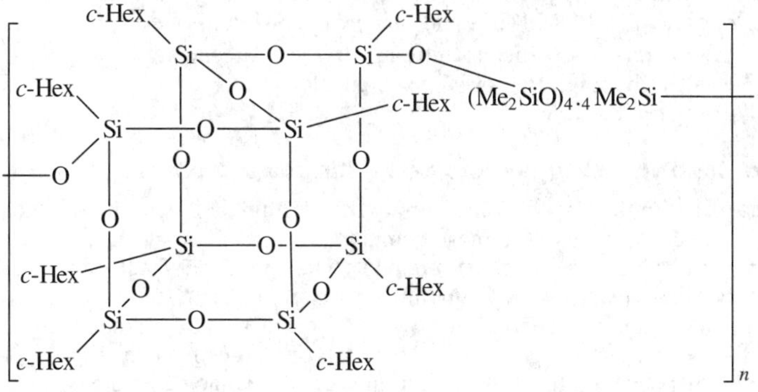

(**25**)

Silanes and silicones are extensively used as surface treating agents for silica, glass, aluminium and a range of other substrates[98] and the thermal stability of such coatings is an important factor in their performance. The polysiloxane resin coating obtaining on treating a glass fibre with an aqueous solution of 3-chloropropyltriethoxysilane[99] underwent an initial weight loss on heating to 130 °C due to loss of residual ethoxy groups and formation of siloxane bonds in the coating. Most of the organic functionality is lost at 360 °C, whilst the residue remaining after heating to 550 °C consists of SiO$_2$ with SiOH groups.

IV. THE SURFACE ACTIVITY OF LINEAR SILOXANE POLYMERS AND COPOLYMERS

A. Introduction

The novel surface properties of polydimethylsiloxanes (PDMS) derive from the combination of very surface active methyl groups attached to a very flexible and stable backbone, and they account for many of the industrial applications of PDMS. The surface chemistry and applications of polydimethylsiloxanes have recently been well reviewed by Owen[100] as have techniques for the surface analysis of silicones[101,102].

Many of the bulk properties of polymers are molecular weight dependent and can be expressed in the general form:

$$X = X_\infty - k_0/M_n$$

where X is the property of interest for a number-average molecular weight M_n and X_∞ is the same property at infinite molecular weight. This behaviour is attributed to the end groups of the polymer which may have differing properties from the repeat unit, thus as the molecular weight increases the influence of the end groups is diluted. The groups of Sauer[103] and Koberstein[104] have examined the dependence of surface tension on molecular weight and end-group type for PDMS and found a similar relationship when σ is the surface tension and σ_∞ is the surface tension at infinite dilution:

$$\sigma = \sigma_\infty - k_1/(M_n)^x$$

At high molecular weight x approaches unity, whilst at low molecular weight x approaches 2/3. The surface tension of the $-Me_2SiO-$ repeat unit was calculated to be 20.1 mN m^{-1} whilst the surface tensions of Me_3SiO-, $HOMe_2SiO-$ and $H_2N(CH_2)_3Me_2SiO-$ end groups were calculated to be 15.5, 26.9 and 32.8 mN m^{-1}, respectively[104a]. The surface of an amino terminated PDMS has been shown to be depleted in nitrogen as expected, given the high surface tension of this group[104b].

Whilst the migration of PDMS to liquid–air or liquid–liquid interfaces is well known and exploited commercially in silicone antifoams[100] and silicone surfactants[100,105], there is growing interest in the migration of siloxane copolymers to solid interfaces which will be the focus of this review.

B. Siloxane Copolymer Migration to Solid–Air Interfaces

The surface distribution of siloxane copolymers within polymer matrices has been studied by a range of techniques, most commonly X-ray Photoelectron Spectroscopy (XPS), combined with contact angle measurements and Attenuated Total Reflectance FTIR (ATR-FTIR) spectroscopy. By varying the incident angle of the X-ray beam with respect to the surface, the sampling depth can be varied and a profile of the various atomic constituents of the copolymers generated. There is good agreement in the literature that siloxanes copolymers can migrate to solid–air interfaces and that the nature of the resultant siloxane rich layer depends on a range of factors, including the length of the siloxane block and the architecture of the copolymer.

1. Block copolymers

Chen, Gardella and Kumler[106] have studied a series of polydimethylsiloxane-polystyrene block copolymers and examined the surface composition by ATR-FTIR and XPS. For AB-type PS-PDMS diblock copolymers (**26**) with siloxane block molecular weights of between 38,000 and 99,000, the surface was found to be exclusively polydimethylsiloxane down to a depth of 10 nm by XPS. ATR-FTIR, which samples

$-(CH_2CHPh)_m-(Me_2SiO)_n-$

(26)

at a depth of up to 2.4 μm[106], showed only PDMS in the case of a copolymer with a large siloxane block ($M_n = 99,000$), whereas in the case of a smaller siloxane block ($M_n = 38,000$) polystyrene was observed in the surface layer by ATR-FTIR but at a lower concentration than present in the bulk. Changing the copolymer morphology to an ABA-type PS-PDMS-PS triblock copolymer resulted in less efficient surface segregation of the siloxane. A siloxane segment with double the molecular weight of that used in the AB-type case was required to give a comparable siloxane surface layer. This result can be rationalized by assuming that the polystyrene segments are located within the bulk of the copolymer and that in the ABA-type copolymer case this will result in the siloxane segment being bent over, thus effectively halving its length (Figure 2). Annealing for the ABA-type copolymers was found to promote the surface segregation of the low-energy PDMS phase[106].

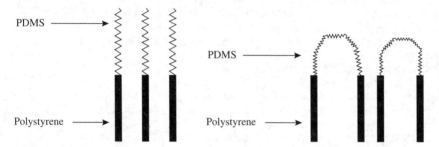

FIGURE 2. Schematic representations of a siloxane-polystyrene AB-type (left) and ABA-type (right) copolymer at a polystyrene surface

Jérôme and coworkers[107] have reported similar results with a PDMS-PS block copolymer in which the siloxane block was much smaller, $M_n = 3,300$. The copolymer was blended into polystyrene at up to 10% and films were prepared by either solvent casting from THF or by spin coating. Analysis by either XPS or Secondary Ion Mass Spectroscopy (SIMS) showed the air–polymer interface to be enriched in PDMS. However, polystyrene was also observable at even the shallowest sampling depths possible for either technique. Surface tension measurements showed that at above 0.5% added copolymer the surface properties of the blend closely match that of polydimethylsiloxane. With 10% of added copolymer no siloxane was observed beyond ca 27 nm depth. Interestingly, siloxane was also observed at the polymer–aluminium interface for spin cast films, but at only a quarter of the concentration observed at the air interface; on annealing the film this siloxane migrated to the air interface. A combination of Time of Flight Secondary Ion Mass Spectroscopy (TOF-SIMS) and XPS[108] has also shown surface layers with high polydimethylsiloxane character for blends of PDMS-PS block copolymers in polystyrene (2 wt% siloxane). Polydimethylsiloxane-urethane-ethylene oxide block copolymers of ABA **(27)** or BAB **(28)** architecture also showed differing efficiencies towards surface segregation, with the copolymers having siloxane as the terminal polymer **(27)** being the more efficient[109].

$[H(OSiMe_2)_{21}CH_2OCH_2CH_2OC(O)-[1,2,4]-NHC_6H_3(CH_3)NHC(O)(OCH_2CH_2)_n]_2O$

(27) $n = 4$ or 17

[H(OCH$_2$CH$_2$)$_m$OC(O)-[1,2,4]-NHC$_6$H$_3$(CH$_3$)NHC(O)OCH$_2$CH$_2$OCH$_2$(SiMe$_2$O)$_{10}$]$_2$SiMe$_2$

(28) $m = 8$ or 34

Interestingly, at very low additive levels both types of copolymer, 27 and 28, behave similarly which, it was suggested, may be due to a change in surface alignment of the siloxane segment of the copolymer from perpendicular to the surface to parallel. Pertsin and coworkers[110] have shown that, for blends of siloxane-sulphone or siloxane-carbonate multiblock copolymers with various homopolymers or copolymers, the nature of the surface layer depends on the amount of copolymer present. At low additive levels of the copolymer, the surface layer was thin and the copolymers were oriented parallel to the surface; at higher additive levels the surface layer was thicker and the copolymer tended to orient perpendicular to the surface. Gardella and coworkers[111] have demonstrated that siloxane block length, bulk concentration of siloxane and annealing of the sample all determine the thickness of the siloxane layer observed for a polydimethylsiloxane-polycarbonate multiblock copolymer.

Careful XPS analysis of a series of poly(dimethylsiloxane-urea-urethane) multiblock copolymers demonstrated that, as well as a surface layer of siloxane, there was a layer enriched in the hard block immediately beneath this. The thickness of both these layers depended on the molecular weights of the soft and hard block segments, respectively[112]. Annealing of these copolymers increased the thickness of both layers. The same authors have also shown that the thickness of these layers of hard and soft blocks could be modified by use of solvent mixtures which selectively precipitate the polar hard block during film formation by solvent casting[113].

TOF-SIMS analysis of the surface of poly(methylphenylsiloxane)-polystyrene copolymers gave signals due to both poly(methylphenylsiloxane) and polystyrene[114]. The higher surface energy of poly(methylphenylsiloxane) compared to polydimethylsiloxane would be expected to reduce the thermodynamic driving force for surface segregation of the siloxane in this case. The siloxane block of these copolymers was trimethylsilyl terminated and the strong signal observed for this group by TOF-SIMS is evidence that the terminal trimethylsilyl group was at the air interface, as might be expected considering both the low surface energy of the Me$_3$Si group and free volume considerations[114].

2. Graft copolymers

Poly(methyl methacrylate)-g-poly(dimethylsiloxane) copolymers have been prepared by free radical copolymerizations of acrylate-functional siloxane macromonomers (29) with methyl methacrylate. Siloxane macromonomers (29) of between 1,000 and 20,000 molecular weight were utilized to give a range of copolymers with between 4 and 17 wt% silicone[115].

Bu(Me$_2$SiO)$_n$Me$_2$Si(CH$_2$)$_3$OC(O)CMe=CH$_2$

(29)

Contact angle measurements and XPS demonstrated that the surface layer was silicone rich, especially for copolymers with high molecular weight siloxane arms. A layer of pure polydimethylsiloxane was observed to a depth of 1.5 nm for a copolymer prepared from 29 of 20,000 molecular weight. The corresponding copolymer derived from 29 of 10,000 molecular weight had a thinner surface layer of siloxane and only contained 95 wt% polydimethylsiloxane when probed to a depth of 1.5 nm. The percentage of siloxane in the surface layer continued to decrease as the molecular weight of 29 decreased[115]. Copolymers of methyl methacrylate and glycidyl methacrylate with acrylate functional siloxane macromonomers (29) of 5,000 molecular weight have been incorporated into epoxy resins[116]. Surface enrichment of the siloxane was observed, though XPS analysis

38. Recent advances in the chemistry of siloxane polymers and copolymers

suggested that the siloxane layer was less than 3 nm thick even if 10 wt% of these copolymers was added. The addition of these siloxane copolymers was shown to very effectively reduce the force needed to peel an adhesive tape from the surface of the epoxy resin[116], thus demonstrating the siloxane-like nature of the surface.

Ober and coworkers[117] have grafted siloxanes onto a styrene-isoprene block copolymer via hydrosilylation of the pendant vinyl groups of the diene block with pentamethyldisiloxane (30) (equation 12).

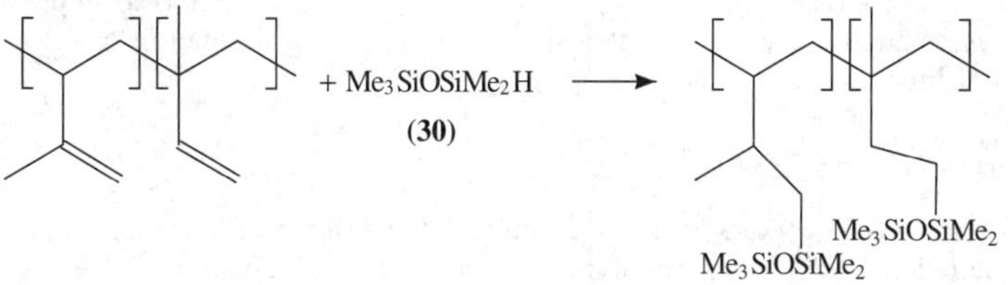

$$\text{[styrene-isoprene block copolymer]} + \text{Me}_3\text{SiOSiMe}_2\text{H} \longrightarrow \text{[grafted copolymer with Me}_3\text{SiOSiMe}_2\text{ side groups]} \qquad (12)$$

(30)

Blends of this copolymer with polystyrene were analysed by contact angle measurements and Rutherford backscattering (RBS) both of which showed that the surface was silicone rich.

Allcock and Smith[118] have prepared a series of poly(organophosphazenes) containing siloxane grafts and/or trifluoroethoxy side groups (31–34). Polymers 31 and 32 had critical surface tensions of 16–17 mN m^{-1} and surface layers which were enriched in fluorine. In the case of 32, silicone was not observed at the surface. Polymer 33 had a critical surface tension of 31 mN m^{-1}, which is higher than that for pure PDMS at 24 mN m^{-1} [119], and XPS analysis of this polymer showed that the surface was substantially enriched in siloxane. The surface modification of 34 via hydrosilylation of a swelled film was demonstrated with silicon being observed to a depth of 60 μm by energy-dispersive X-ray analysis of a cross section of the resultant film[118].

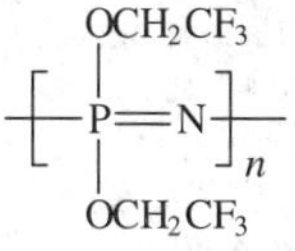

$$\begin{array}{c} \text{OCH}_2\text{CF}_3 \\ | \\ \text{[-P=N-]}_n \\ | \\ \text{OCH}_2\text{CF}_3 \end{array} \qquad \begin{array}{c} \text{OC}_6\text{H}_4\text{OCH}_2\text{CH}_2\text{CH}_2\text{Me}_2\text{SiOMe}_2\text{SiOSiMe}_3\text{-}p \\ | \\ \text{[-P=N-]}_n \\ | \\ \text{OCH}_2\text{CF}_3 \end{array}$$

(31) 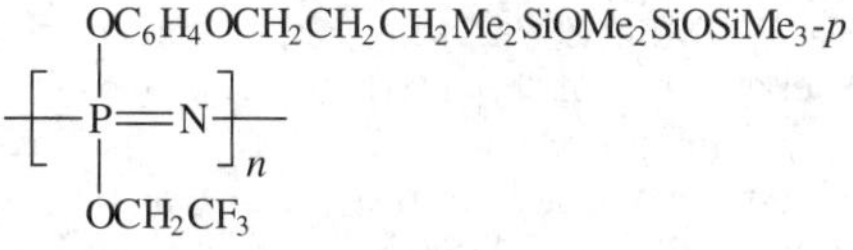 (32)

$$\begin{array}{c} \text{OC}_6\text{H}_4\text{OCH}_2\text{CH}_2\text{CH}_2\text{Me}_2\text{SiOMe}_2\text{SiOSiMe}_3\text{-}p \\ | \\ \text{[-P=N-]}_n \\ | \\ \text{OPh} \end{array} \qquad \begin{array}{c} \text{OCH}_2\text{CF}_3 \quad\quad \text{OC}_6\text{H}_4\text{OCH}_2\text{CH}=\text{CH}_2\text{-}p \\ | \quad\quad\quad\quad | \\ \text{[-P=N-]}_n\text{[-P=N-]}_m \\ | \quad\quad\quad\quad | \\ \text{OCH}_2\text{CF}_3 \quad\quad \text{OCH}_2\text{CF}_3 \end{array}$$

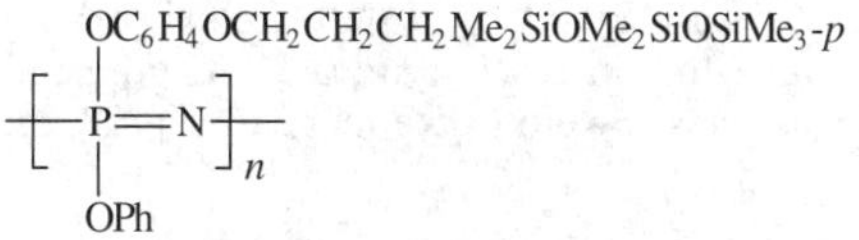

(33) 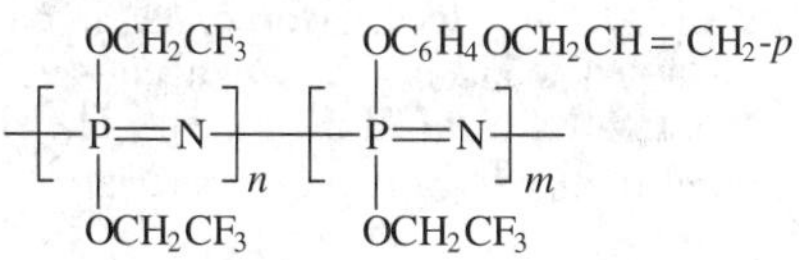 (34)

The combination of XPS with cryogenic sample-handling techniques has allowed the study of hydrated polymer surfaces and surface rearrangements on dehydration[120]. A polydimethylsiloxane grafted with poly(hydroxyethyl methacrylate) (HEMA) was found to contain 95 atomic% HEMA at the surface in the hydrated state, but on vaccum drying the surface became 90 atomic% PDMS. Unfortunately, equipment limitations prevented study of the kinetics of this rearrangement. Similar changes in surface composition for hydrated and dehydrated states were observed for a polyurethane containing a polydimethylsiloxane soft block[120].

C. Blend Compatibilization

The properties of immiscible polymers blends are strongly dependent on the morphology of the blend, with optimal mechanical properties only being obtained at a critical particle size for the dispersed phase. As the size of the dispersed phase is directly proportional to the interfacial tension between the components of the blend, there is much interest in interfacial tension modification. Copolymers, either preformed or formed *in situ*, can localize at the interface and effectively modify the interfacial tension of polymer blends. The incorporation of PDMS phases is desirable as a method to improve properties such as impact resistance, toughness, tensile strength, elongation at break, thermal stability and lubrication.

1. Pre-formed copolymers

Koberstein and coworkers[121] have examined the effects of a polydimethylsiloxane-polystyrene (PDMS-PS) block copolymer on the interfacial tensions of blends of PDMS and polystyrene. As little as 0.002 wt% of the copolymer, added to the siloxane phase, was sufficient to lower the interfacial tension by 82% in the case of a blend of polystyrene ($M_n = 4,000$) and PDMS ($M_n = 4,500$). No further reduction in interfacial tension was observed at higher copolymer levels due to micelle formation. Riess[122] has polymerized styrene in the presence of a silicon oil and a polydimethylsiloxane-polystyrene block copolymer to obtain a polystyrene in which 0.1–1 μm droplets of silicone oil are dispersed. This material displayed a lowered coefficient of kinetic friction on steel compared to pure polystyrene.

Addition of a polydimethylsiloxane-graft-polyethylene oxide copolymer to a blend of PDMS and polyethylene oxide (PEO) resulted in time-dependent changes in the interfacial tension over a period of hours[123]. The interfacial tension initially decreased to a minimum and then increased to a final equilibrium value. It is believed[123] that this represents adsorption of the copolymer at the interface followed by an ordering of the copolymer phase at the interface. The addition of triblock polydimethylsiloxane-polyethylene oxide copolymers, P(DMS$_n$-EO$_m$-DMS$_n$) (**35**) to blends of PDMS and polyethylene oxide demonstrated a dependence of the interfacial tension on the size of the siloxane segment and reached a minimum for **35** with $n \geq 16$[124]. Increasing the length of the ethylene oxide segment slightly increased the interfacial tension. Reversing the order of the polymer blocks to give a P(EO$_m$-DMS$_n$-EO$_m$) was found to give similar results to those obtained with **35** of comparable PDMS and PEO block lengths.

Branched copolymers of PDMS and poly(caprolactone) have been prepared as shown in equation 13. They were incorporated into epoxy resin via reaction at the terminal hydroxy groups and acted as toughening agents, with a three-fold increase in impact strength being observed with 5 wt% of **36** present. The siloxane phases were in the range of 10–20 nm, resulting in an optically clear material[125].

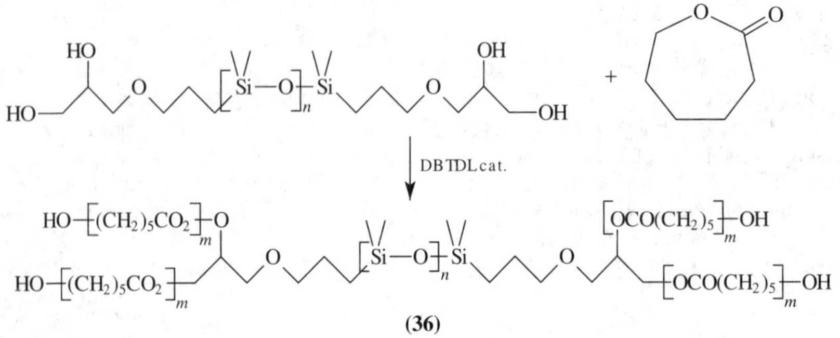

(13)

(**36**)

The end groups of a PDMS polymer have been shown to affect the interfacial tension of blends with poly(butadiene)[126]. Thus, substitution of an amine-terminated PDMS for a trimethylsilyl-terminated PDMS can reduce the interfacial tension by up to 30%. This effect is postulated to arise due to the amine end group having a surface energy closer to that of butadiene than does the trimethylsilyl group and thus being present at the interface.

2. Reactive blending

As an alternative to addition of pre-formed block copolymers to polymer blends, the copolymer compatibilizers can be formed *in situ* during processing; such reactive blending may have economic advantages. Nando and coworkers[127,128] have shown that a vinyl functional dimethylsiloxane (**37**) will react with an ethylene–methyl acrylate copolymer (EMA) (**38**) at temperatures over 150 °C. The mechanism of this reaction involves a homolytic cleavage of the hydrogen atom from the α-carbon adjacent to the ester group of **38**, followed by addition of the carbon radical to the double bond of **37** followed by hydrogen abstraction (Scheme 6).

$$-(Me_2SiO)_n-((H_2C=CH)MeSiO)_m- \qquad -(CH_2CH_2)_x-(CH_2CH(CO_2Me))_y-$$
$$(37) \qquad\qquad\qquad (38)$$

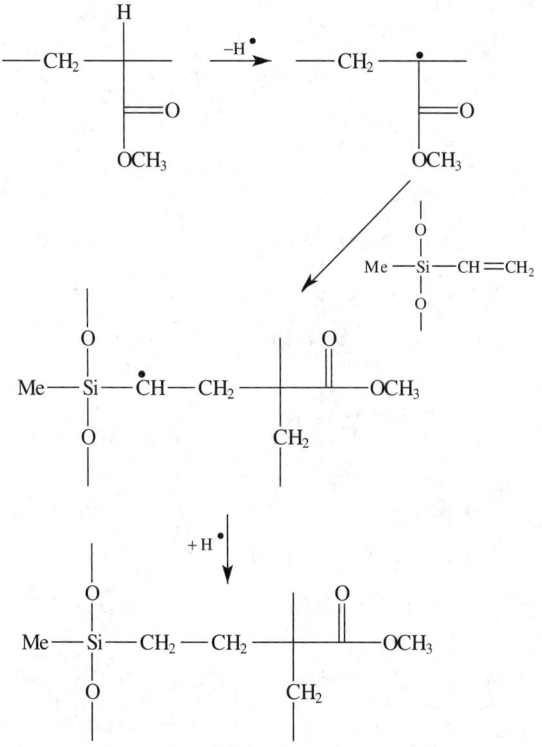

SCHEME 6. Mechanism of radical addition of EMA to vinyl functional PDMS

The resulting ethylene methyl acrylate-graft-polydimethylsiloxane (**39**) exhibited a single glass transition temperature, which was composition dependent, demonstrating polymer miscibility[127]. This reaction (Scheme 6) was carried out *in situ* during the blending of a 1:1 mixture of PDMS and polyethylene to give **39**, which acted as a blend compatibilizer. It was found that the presence of **39** improved adhesion between the blend components, reduced the dispersed domain size of PDMS in polyethylene and increased the degree of crystallinity within the blend[129]. **39** was also found to compatibilize blends of PDMS and an ethylene-propylene-diene rubber[130] and blends of PDMS and polyurethane[131], with improvements in tensile strength, elongation at break and thermal stability being observed. An alternative to reactive blending has been demonstrated[132] by addition of a carboxy-terminated polybutadiene and a difunctional amino-terminated PDMS to the respective homopolymers in blends of polybutadiene and PDMS. An ammonium carboxylate salt forms at the interface and this behaves like a block copolymer with reductions of up to 70% in the interfacial tension being observed.

It is clear that siloxane copolymers can migrate within solids and beneficially modify interface properties. The efficiency of this migration depends on a wide range of factors including the block lengths of the copolymer, the morphology of the copolymer and the method of sample preparation, and is an area of ongoing study.

V. REFERENCES

1. J. Chojnowski, *J. Inorg. Organomet. Polym.*, **1**, 299 (1991).
2. J. Chojnowski, *Spec. Publ. R. Soc. Chem.*, **166**, 59 (1995).
3. S. Penezek and A. Duda, *Makromol. Chem., Macromol. Symp.*, **67**, 15 (1993).
4. J. A. Semlyen, B. R. Wood and P. Hodge, *Polym. Adv. Technol.*, **5**, 473 (1994).
5. V. M. Kopylov, L. M. Khananashvili, O. V. Shkol'nik and A. G. Ivanov, *Vysokomol. Soedin., Ser. A*, **37**, 395 (1995); *Chem. Abstr.*, **123**, 170322 (1995).
6. R. Murugavel, A. Voight, M. G. Walawalker and H. W. Roesky, *Chem. Rev.*, **96**, 2205 (1996).
7. T. A. Kolomiets, N. N. Laskovenko and S. I. Omel'chaiko, *Kompoz. Polim. Mater.*, **50**, 9 (1991); *Chem. Abstr.*, **119**, 73139 (1991).
8. P. V. Ivanov, *Vysokomol. Soedin., Ser. A*, **37**, 417 (1995); *Chem. Abstr.*, **123**, 144672 (1995).
9. H. Inoue, *Nipon Gomu Kyokaishi*, **66**, 660 (1993); *Chem. Abstr.*, **120**, 325489 (1993).
10. Y. Imai, *J. Macromol. Sci. Chem., A*, **28**, 1115 (1991).
11. (a) L. Lestel and S. Boileau, *Makromol. Chem., Macromol. Symp.*, **47**, 293 (1991).
    (b) S. Boileau, *Makromol. Chem., Macromol. Symp.*, **73**, 177 (1993).
    (c) J. Ming Yu, D. Teyssie, R. B. Khalifa and S. Boileau, *Polym. Bull.*, **32**, 35 (1994).
    (d) T. Zundel, J. Ming Yu, L. Lestel, D. Teyssie and S. Boileau, *Makromol. Symp.*, **88**, 177 (1994).
12. M. K. Yoshinaga and Y. Iida, *Chem. Lett.*, 1057 (1991).
13. (a) A. Molenberg and M. Moller, *Macromol. Rapid Commun.*, **16**, 449 (1995).
    (b) A. Molenberg and M. Moller, *Macromol. Symp.*, **107**, 331 (1996).
14. M. E. Van Dyke and S. Clarson, *Poly. Prepr., Am. Chem. Soc.*, **37**, 668 (1996).
15. (a) M. O. Hunt, A. M. Belu, R. W. Linton and J. M. DeSimon, *Polym. Prepr., Am. Chem. Soc.*, **34**, 530 (1993).
    (b) M. O. Hunt, A. M. Belu, R. W. Linton and J. M. DeSimon, *Polym. Prepr., Am. Chem. Soc.*, **34**, 445 (1993).
    (c) M. O. Hunt, A. M. Belu, R. W. Linton and J. M. DeSimon, *Polym. Prepr., Am. Chem. Soc.*, **34**, 883 (1993).
16. A. DeGunzbourg, J-C. Favier and P. Hemery, *Polym. Int.*, **35**, 179 (1994).
17. (a) T. Okawa, Dow Corning Toray Silicone Co. Ltd., EP 679674, Nov. 1995.
    (b) I. Jansen, S. Kupfer and K. Ruehlmann, Huels Silicone GmbH, DE 4436076, Nov. 1996.
18. P. Sigwalt, P. Gobin, M. Moreau and M. Masure, *Makromol. Chem., Macromol. Symp.*, **42/43**, 229 (1991).
19. (a) G. Toskas, G. Besztercey, M. Moreau, M. Masure and P. Sigwalt, *Macromol. Chem. Phys.*, **196**, 2715 (1995).

(b) P. Nicol, M. Masure and P. Sigwalt, *Macromol. Chem. Phys.*, **195**, 2327 (1994).
(c) P. Sigwalt, M. Masure, M. Moreau and R. Bischoff, *Makromol. Chem., Macromol. Symp.*, **73**, 147 (1993).
20. G. J. Price, M. P. Hearn, E. N. K. Wallace and A. M. Patel, *Polymer*, **37**, 2303 (1996).
21. (a) M. Cypryk, S. Rubinsztajn and J. Chojnowski, *Macromolecules*, **26**, 5389 (1993).
(b) M. Cypryk, S. Rubinsztajn and J. Chojnowski, *J. Organomet. Chem.*, **446**, 91 (1993).
(c) M. Cypryk, *Polymery (Warsaw)*, **40**, 421 (1995).
22. (a) M. Cypryk and P. Sigwalt *Macromolecules*, **27**, 6245 (1994).
(b) R. Bischoff and P. Sigwalt, *Polym. Prepr., Am. Chem. Soc., Div. Polym. Chem.*, **37**, 349 (1996).
23. K. Kazmirski, J. Chojnowski and J. McVie, *Eur. Polym. J.*, **30**, 515 (1994).
24. M. Marcus, M. Simionescu, M. Cazacu and S. Lazarescu, *Iran. J. Polym. Sci. Technol.*, **3**, 95 (1994).
25. X. W. He and J. E. Herz, *Eur. Polym. J.*, **27**, 449 (1991).
26. H. K. Chu, R. P. Cross and D. I. Crossen, *J. Organomet. Chem.*, **425**, 9 (1992).
27. (a) R. Hagger and J. Weis, Wacker Chemie, EP0626414 May 1994.
(b) R. Hagger and J. Weis, Wacker Chemie, EP0626415 May 1994.
(c) R. Hagger R. Braun, O. Schneider and B. Deubzer, Wacker Chemie, DE 4422813, June 1996.
28. R. Hagger and J. Weis, *Z. Naturforsch., B: Chem. Sci.*, **49**, 1774 (1994).
29. (a) H. Schickmann, R. Lehnert, H-D. Wendt, H-G. Serbny and H. Rautaschek, Chemiewerk Nunchritz GmbH, DE 4344664, Dec. 1993.
(b) H. Schickmann, R. Lehnert, H-D. Wendt, H-G. Serbny and H. Rautaschek, Chemiewerk Nunchritz GmbH, DE 4323188, July 1993.
(c) H. Schickmann, R. Lehnert and H-D. Wendt, Chemiewerk Nunchritz GmbH, DE 4323185, July 1993.
(d) R. Lehnert, H-D. Wendt and H. Schickmann, Huels Silicone GmbH, DE 4446515, May 1996.
30. (a) S. Rubinsztajn, General Electric Company, GB 2279945, Jan 1995.
(b) J. S. Razzano, D. P. Thompson, P. P. Anderson and S. Rubinsztajn, General Electric Company, GB 2279959, January 1995.
31. (a) Dow Corning Ltd., EP 0522776, July 1991.
(b) Dow Corning Ltd., EP 0503825, March 1991.
(c) Dow Corning Ltd., GB 2252975, February 1991.
32. (a) J. Chojnowski, R. Taylor, J. Habimana, M. Cypryk, W. Fortuniak and K. Kazmierski, *J. Organomet. Chem.*, in press (1996).
(b) J. Chojnowski, R. Taylor, J. Habimana, M. Cypryk, W. Fortuniak and K. Kazmierski, *J. Organomet. Chem.*, in press (1996).
33. W. Schulz, J. Lambert and L. Kania, Dow Corning Corp., US 5196559, March 1993.
34. G. A. Olah, O. Wang, G. Rasol, G. K. S. Prakash, H. Zhang and T. E. Hogen-Esch, *Polym. Prepr., Am. Chem. Soc.*, **37**, 805 (1996).
35. (a) S. T. Kolaczkowski and S. Serbetcioglu, Dow Corning Ltd., EP 605143, July 1994.
(b) S. Serbetcioglu, PhD Thesis, University of Bath, 1995.
36. S. Westall, Dow Corning Ltd., EP 587343 A2, March 1994.
37. J. E. McGrath and M. Spinu, *J. Polym. Sci., Part A, Polym. Chem.*, **29**, 657 (1991).
38. G. Sauvet and G. Helary, *Eur. Polym. J.*, **28**, 37 (1992).
39. R. Babu, G. Sinai-Zingde and J. S. Riffle *J. Polym. Sci., Part A, Polym. Chem.*, **31**, 1645 (1993).
40. J. J. Hoffman and C. M. Leir, *Polym. Int.*, **24**, 131 (1991).
41. (a) L. E. Drechsler, A. van der Helm, C. M. Kuo and S. J. Clarson., *Polym. Prepr., Am. Chem. Soc.*, **33**, 986 (1992).
(b) C. M. Kuo and S. J. Clarson., *Polym. Prepr., Am. Chem. Soc.*, **32**, 183 (1991).
(c) S. J. Clarson and A. M. S. Al-Ghamdi., *Polym. Commun.*, **31**, 322 (1990).
42. (a) S. Cai, J. Fang and X. Yu, *J. Appl. Polym. Sci.*, **44**, 135 (1992).
(b) G. Akovali, Z. M. O. Rzaev and D. H. Mamedov, *Polym. Int.*, **37**, 119 (1995).
43. D. B. Yang, *J. Polym. Sci., Part A, Polym. Chem.*, **31**, 199 (1993).
44. J. V. Crivello and D. Bi, *J. Polym. Sci., Part A*, **31**, 2563, 2729, 3109, 3121 (1993); J. V. Crivello, D. Bi and Mingxin Fang, *J. Macromol. Sci., Pure Appl. Chem.*, **A31**, 1001 (1994).
45. T. Suzuki, S. Yamada and T. Okawa, *Polym. J. (Tokyo)*, **25**, 411 (1993).
46. M. A. Peters, A. M. Belu, R. W. Linton, L. Dupray, T. J. Meyer and J. M. DeSimone, *J. Am. Chem. Soc.*, **117**, 3380 (1995).

47. D. A. Williams, P. R. Willey and B. J. Ward, US 4992512, February 1991.
48. S. H. Barley, A. Gilbert and G. R. Mitchell, *Makromol. Chem.*, **192**, 2801 (1991).
49. C. Herzig and J. Esterbauer, US 5015700, May 1991.
50. E. Beyou, P. Babin, B. Bennetau, J. Dunoguès, D. Teyssié and S. Boileau, *Tetrahedron Lett.*, **36**, 1843 (1995).
51. H. Hacker, J. Huber, G. Kolodziej, G. Piecha and D. Wilhelm, DE 4014882, November 1991.
52. S. Aoki, Y. Hara, and T. Ohba, US 5190988, February 1992; W. V. Gentzkow, W. Rogler, F. Zapf and G. Zeidler, DE 4110654, October 1992; T. Saruyama and M. Yoshitake, EP 0656386, November 1995; T. E. Hohenwarter, RadTech 92 North Am. UV/EB Conf. Expo., Conference Proceedings, Vol. 1, RadTech Int. North Am., Northbrook, IL, 1992, pp. 108–111.
53. R. P. Eckberg, US 4987158, January 1991; R. F. Agars and R. P. Eckberg, EP 0599615, June 1994; R. P. Eckberg and R. Griswold, US 5397813, March 1995.
54. A. F. Jacobine, D. M. Glaser, S. T. Nakos, M. Masterson, P. J. Grabek, M. A. Rakas, D. Mancini and J. G. Woods, *Spec. Publ. R. Soc. Chem.*, **89** (Radiat. Curing Polym. 2), 342 (1991); M. A. Rakas and A. F. Jacobine, RadTech 92 North Am. UV/EB Conf. Expo., Conference Proceedings, Vol. 1, RadTech Int. North Am., Northbrook, IL, 1992, pp. 462–473; M. A. Rakas and A. F. Jacobine, *J. Adhes.*, **36**, 247 (1992).
55. S. V. Pertz and S. O. Glover, EP 0625533, May 1994; G. R. Homan and S. O. Glover, EP 0625534, May 1994.
56. F. Cazaux and X. Coqueret, *Eur. Polym. J.*, **31**, 521 (1995).
57. J. V. Crivello, Bo Yang and W. G. Kim, *J. Polym. Sci., Part A*, **33**, 2415 (1995).
58. X. Coqueret and L. Pouliquen, *Macromol. Symp.*, **87**, 17 (1994).
59. H. J. Coles, J. P. Hannington and D. R. Thomas, GB 2274649, August 1994; H. J. Coles, J. P. Hannington and D. R. Thomas, GB 2274652, August 1994.
60. S. A. Daunheimer and D. J. Halloran, EP 0574156, December 1993; D. J. Halloran and J. M. Vincent, US 5326483, July 1994.
61. G. S. Kohl, P. A. Giwa-Agbomeirele and J. M. Vincent, US 5063044, November 1991.
62. D. J. Halloran, EP 0437075, July 1991.
63. D. J. Halloran, EP 0437099, July 1991.
64. Y. Ito, EP 0524612, January 1993.
65. M. Humphries, *Cosmet. News*, **16**, 313 (1993).
66. A. M. Vincent, A. A. Wilson and A. Zombeck, US 5262155, March 1993.
67. A. J. Disapio, S. F. Rentsch and A. H. Ward, US 5208360, May 1993.
68. H. J. Lautenschlager, J. Bindl and K. G. Huhn, *Text. Chem. Color.*, **27**, 27 (1995).
69. H. Baris and E. Fleury, WO 9504847, February 1995.
70. G. T. Decker, K. Tobukuro and G. A. Gornowicz, EP 0475611, March 1992.
71. W. A. Starke and M. J. Ziemelis, EP 0674936, October 1995.
72. T. C. Kendrick, B. Parbhoo and J. W. White, in *The Silicon-Heteroatom Bond* (Eds S. Patai and Z. Rappoport), Chap. 3 and 4, Wiley, Chichester, 1991.
73. P. Greil, *J. Am. Ceram. Soc.*, **78**, 835 (1995).
74. E. L. Warrick, O. R. Pierce, K. E. Polmanteer and J. C. Saam, *Rubber Chem. Techol.*, **52**, 437 (1979).
75. G. B. Sohoni and J. E. Mark, *J. Appl. Polym. Sci.*, **45**, 1763 (1992).
76. G. Ya. Guba, V. I. Bogillo, M. I. Terets and A. A. Chuiko, *Ukr. Khim. Zh.*, **60**, 380 (1994); *Ukr. Chem. J.*, **60**, 4 (1994).
77. J. Chojnowski, J. Kurjata, S. Rubinsztajn, M. Scibiorek and M. Zeldin, *J. Inorg. Organomet. Polym.*, **2**, 387 (1992).
78. J. Chojnowski, J. Kurjata, S. Rubinsztajn and M. Scibiorek, in *Frontiers of Organosilicon Chemistry* (Eds. A. R. Bassindale and P. P. Gaspar), Royal Society of Chemistry, Cambridge, 1991, p. 70.
79. R. R. Buch and D. N. Ingebrigtson, *Environ. Sci. Technol.*, **13**, 676 (1979).
80. J. C. Carpenter, J. A. Cella and S. B. Dorn, *Environ. Sci. Technol.*, **29**, 864 (1995).
81. (a) R. G. Lehmann, S. Varaprath and C. L. Frye, *Environ. Toxicol. Chem.*, **13**, 1061 (1994).
    (b) R. G. Lehmann, S. Varaprath and C. L. Frye, *Environ. Toxicol. Chem.*, **13**, 1753 (1994).
    (c) R. G. Lehmann, S. Varaprath, R. B. Annelin and J. L. Arndt, *Environ. Toxicol. Chem.*, **14**, 1299 (1995).
    (d) R. G. Lehmann and T. R. Miller, *Environ. Toxicol. Chem.*, **15**, 1455 (1996).
    (e) R. Sommerlade, H. Parlar, D. Wrobel and P. Kocks, *Environ. Sci. Technol.*, **27**, 2435 (1993).

38. Recent advances in the chemistry of siloxane polymers and copolymers

82. Y. Israëli, J. Cavezzan and J. Lacoste, *Polym. Degrad. Stab.*, **37**, 201 (1992).
83. T. Imakoma, Y. Suzuki, O. Fujii and I. Nakajima, *Proc. Int. Conf. Prop. Appl. Dielectr. Mater.*, **1**, 306 (1994).
84. O. Joubert, G. Hollinger, C. Fiori, R. A. B. Devine, P. Paniez and R. Pantel, *J. Appl. Phys.*, **69**, 6647 (1991).
85. V. N. Vasilets, A. V. Kovalchuk and A. N. Ponomarev, *J. Photopolym. Sci. Technol.*, **7**, 165 (1994).
86. V. E. Skurat and Y. I. Dorofeev, *Angew. Makromol. Chem.*, **216**, 205 (1994).
87. C. Minero, V. Maurino and E. Pelizzetti, *Langmuir*, **11**, 4440 (1995).
88. A. F. Whitaker and B. Z. Jang, *SAMPE J.*, **30**, 30 (1994).
89. J. W. Connell, J. V. Crivello and D. Bi, *J. Appl. Polym. Sci.*, **57**, 1251 (1995).
90. (a) V. Belot, R. J. P. Corriu, A. M. Flank, D. Leclercq, P. H. Mutin and A. Vioux, *Eur. Mater. Res. Soc. Monogr. (Eurogel '91).*, 77 (1992).
    (b) V. Belot, R. J. P. Corriu, D. Leclercq, P. H. Mutin and A. Vioux, *Chem. Mater.*, **3**, 127 (1991).
91. V. Belot, R. J. P. Corriu, D. Leclercq, P. H. Mutin and A. Vioux, *J. Polym. Sci., Part A, Polym. Chem.*, **30**, 613 (1992).
92. S. Diré and F. Babonneau, *J. Sol-Gel Sci. Technol.*, **2**, 139 (1994).
93. S. Wachholz, U. Just. F. Keider, H. Geißler and K. Käppler, *Fresenius Z. Anal. Chem.*, **352**, 515 (1995).
94. A. M. Wilson, J. N. Riemers, E. W. Fuller and J. R. Dahn, *Solid State Ionics*, **74**, 249 (1994).
95. F. I. Hurwitz, P. Heimann, S. C. Farmer and D. M. Hembree Jr., *J. Mater. Sci.*, **28**, 6622 (1993).
96. G. T. Burns, R. B. Taylor, Y. Xu, A. Zangvil and G. A. Zank, *Chem. Mater.*, **4**, 1313 (1992).
97. J. D. Lichtenhan, R. A. Mantz, P. A. Jones, J. W. Gilman, K. P. Chaffe, I. M. K. Ismial and M. J. Burmiester, *Polym. Prepr., Am. Chem. Soc., Div. Polym. Chem.*, **36**, 334 (1995).
98. E. P. Pluddemann, *Silane Coupling Agents*, Plenum, New York, 1991
99. A. Lorena, M. C. Matias and J. Martinez Urreaga, *Spectrosc. Lett.*, **25**, 1121 (1992).
100. M. J. Owen, in *Siloxane Polymers* (Eds. S. J. Clarson and J. A. Semlyen), PTR Prentice Hall, New Jersey, 1993, pp. 309–372.
101. M. J. Owen in *The Analytical Chemistry of Silicones* Chap. V (Ed. A. L. Smith), Wiley, New York, 1991.
102. M. J. Owen, 'New Directions in Organosilicon Surface Science' in *Front. Polym. Adv. Mater., Proc. Int. Conf.*, Vol. 9 (Ed. P.N. Prasad), Plenum press New York, 1993, p. 677.
103. (a) B. B. Sauer and G. T. Dee, *Macromolecules*, **24**, 2124 (1991).
     (b) B. B. Sauer and G. T. Dee, *Mater. Res. Soc. Symp. Proc.*, **248**, 441 (1992).
     (c) G. T. Dee and B. B. Sauer, *Macromolecules*, **26**, 2771 (1993).
     (d) B. B. Sauer and G. T. Dee, *J. Colloid Interface Sci.*, **162**, 25 (1994).
     (e) G. T. Dee and B. B. Sauer, *Polymer*, **36**, 1673 (1995).
104. (a) C. Jalbert, J. T. Koberstein, I. Yilgor, P. Gallagher and V. Krukonis, *Macromolecules*, **26**, 3069 (1993).
     (b) C. Jalbert, J. T. Koberstein, R. Balaji, Q. Bhatia, L. Salvati and I. Yilgor, *Macromolecules*, **27**, 2409 (1994).
     (c) T. J. Lenk, D. H. T. Lee and J. T. Koberstein, *Langmuir*, **10**, 1857 (1994).
105. (a) M. He, R. M. Hill, H. A. Doumaux, F. S. Bates, H. T. Davis, D. F. Evans and L. E. Scriven, *ACS Symp. Ser.*, **578**, 192 (1994).
     (b) M. J. Owen and H. Kobayashi, *Surf. Coat. Int.*, **78**, 52 (1995).
106. X. Chen. J. A. Gardella and P. L. Kumler, *Macromolecules*, **25**, 6621 (1992).
107. S. Petitjean, G. Ghitti, R. Jérôme, Ph. Teyssié, J. Marzien, J. Riga and J. Verbist, *Macromolecules*, **27**, 4127 (1994).
108. C. E. Selby, J. O. Stuart, S. J. Clarson, S. M. Smith, A. Sabata, W. J. Van Ooij and N. G. Cave, *J. Inorg. Organomet. Polym.*, **4**, 85 (1994).
109. M. M. Gorelova, A. J. Pertsin, I. O. Volkov, L. V. Filimonova, L. I. Makarova and A. A. Zhdanov, *J. Appl. Polym. Sci.*, **57**, 227 (1995).
110. (a) A. J. Pertsin, M. M. Gorelova, V. Yu. Levin and L. I. Makarova, *J. Appl. Polym. Sci.*, **45**, 1195 (1992).
     (b) A. J. Pertsin, M. M. Gorelova, V. Yu. Levin and L. I. Makarova, *Makromol. Chem., Macromol. Symp.*, **44**, 317 (1991).
111. X. Chen, H. F. Lee and J. A. Gardella Jr., *Macromolecules*, **26**, 4601 (1993).

112. X. Chen, J. A. Gardella Jr., T. Ho and K. J. Wynne, *Macromolecules*, **28**, 1635 (1995).
113. J. A. Gardella Jr., T. Ho, K. J. Wynne and H.-Z. Zhong, *J. Colloid Interface Sci.*, **176**, 277 (1995).
114. S. J. Clarson, J. O. Stuart, C. E. Selby, A. Sabata, S. D. Smith and A. Ashraf, *Macromolecules*, **28**, 674 (1995).
115. S. D. Smith, J. M. Desimone, H. Huang, G. York, D. W. Dwight, G. L. Wilkes and J. E. McGrath, *Macromolecules*, **25**, 2575 (1992).
116. T. Kasemura, C. Komatu, H. Nishihara, S. Takahashi, Y. Oshibe, H. Onmura and T. Yamamoto, *J. Adhes.*, **47**, 17 (1994).
117. S. S. Hwang, C. K. Ober, S. Perutz, D. R. Iyengar, L. A. Schneggenburger and E. J. Kramer, *Polymer*, **36**, 1321 (1995).
118. H. A. Allcock and D. E. Smith, *Chem. Mater.*, **7**, 1469 (1995).
119. E. G. Shafrin in *Polymer Handbook*, 2nd ed. (Eds. J. Brandrup and E. H. Immurgut), Wiley, New York, 1975, pp. 221–228.
120. K. B. Lewis and B. D. Ratner, *J. Colloid Interface Sci.*, **159**, 77 (1993).
121. W. Hu, J. T. Koberstein, J. P. Lingelser and Y. Gallot, *Macromolecules*, **28**, 5209 (1995).
122. G. Riess, *Makromol. Chem., Macromol. Symp.*, **69**, 125 (1993).
123. G. Schreyeck and P. Marie, *C. R. Acad. Sci., Ser. II: Mec., Phys., Chim., Astron.*, **320**, 653 (1995); *Chem. Abstr.*, **123**, 113536 (1995).
124. M. Wagner and B. A. Wolf, *Polymer*, **34**, 1460 (1993).
125. R. Mülhaupt, U. Buchholz, J. Rösch and N. Steinhauser, *Angew. Makromol. Chem.*, **223**, 47 (1994).
126. C. A. Fleischer, J. T. Koberstein, V. Krukonis and P. A. Wetmore, *Macromolecules*, **26**, 4172 (1993).
127. R. A. Santra, S. Roy, A. K. Bhowmick and G. B. Nando, *Polym. Eng. Sci.*, **33**, 1352 (1993).
128. R. A. Santra, S. Roy and G. B. Nando, *Polym.-Plast. Technol. Eng.*, **33**, 23 (1994).
129. R. A. Santra, B. K. Samantaray, A. K. Bhowmick and G. B. Nando, *J. Polym. Sci.*, **49**, 1145 (1993).
130. S. Kole, R. Santra and A. K. Bhowmick, *Rubber Chem. Technol.*, **67**, 119 (1994).
131. R. A. Santra, S. Roy, V. K. Tikku and G. B. Nando, *Adv. Polym. Technol.*, **14**, 59 (1995).
132. C. A. Fleischer, A. R. Morales and J. T. Koberstein, *Macromolecules*, **27**, 379 (1994).

CHAPTER **39**

Si-containing ceramic precursors

R. M. LAINE

Departments of Materials Science and Engineering and Chemistry, University of Michigan, Ann Arbor, MI 48109-2136, USA
Fax: (313)764-6203; e-mail: talsdad@umich.edu

and

A. SELLINGER

Sandia National Laboratory, Advanced Materials Laboratory, 1001 University Blvd. S.E., University of New Mexico, Albuquerque, NM 87106, USA

I. INTRODUCTION	2246
II. CRITERIA THAT DEFINE A USEFUL PRECURSOR	2247
A. Rheology	2247
B. Latent Reactivity	2247
C. Pyrolytic Degradation	2247
D. High Ceramic Yield	2248
E. Selectivity to Phase and Chemically Pure Glasses or Ceramics	2249
F. Control of Microstructure and Densification	2249
III. PRECURSORS CONTAINING Si AND N	2251
A. Precursors to Si_3N_4	2252
B. Precursors to SiCN	2253
C. Precursors to Silicon Oxynitride (Si_2ON_2)	2261
D. Precursors to Si–Al–O–N Materials	2264
E. Precursors to Si–N–B–X Materials	2265
IV. PRECURSORS CONTAINING Si AND C	2272
A. Precursors to SiC	2272
1. PMS via dehalocoupling	2273
2. PMS by catalytic dehydrocoupling of $MeSiH_3$	2276
3. PMS by dehydrocoupling of $CH_3SiH_2SiH_2CH_3$	2279
4. PMS from the Bu_4PCl-catalyzed redistribution of chlorosilanes	2280
5. Polycarbosilanes as precursors to SiC	2281

The chemistry of organic silicon compounds, Vol. 2
Edited by Z. Rappoport and Y. Apeloig © 1998 John Wiley & Sons Ltd

6. Polysilaethylene (PSE) precursors 2283
7. Phase pure SiC via processing 2285
B. Precursors to Si−C−B Materials 2287
C. Precursors to Si−C−O Materials 2289
V. PRECURSORS TO Si−O CONTAINING MATERIALS 2294
A. SiO_2 Precursors 2295
B. Group I and II Silicate Precursors 2298
C. Precursors to Aluminosilicates 2300
D. Precursors to Transition Metal Silicates 2307
VI. REFERENCES 2310

I. INTRODUCTION

The purpose of this chapter is to provide an overview of how Si-containing compounds or polymers are used to form glass and ceramic shapes. In ceramics terminology, the fabrication of glass or ceramic materials using chemical compounds is called 'chemical processing'[1−7]. Chemical processing methods range in sophistication from simple precipitation of fine crystallites of common inorganic compounds, e.g. alum [$NH_4Al(SO_4)_2$], an alumina powder precursor], to spinning preceramic polymer fibers that will survive pyrolytic conversion to ceramic fibers. Chemical processing approaches can generally be divided into hydrolytic (sol-gel) and nonhydrolytic (precursor) methods. Sol-gel processing is a diverse area that has an immense literature, has been reviewed many times, is the subject of a well-written book[1a], is reviewed in this book[1b], and has its own journal[2]. Consequently, sol-gel processing will not be discussed here except as it relates to precursor processing methods.

Precursor processing can be subdivided into gas (e.g. chemical vapor deposition, CVD) and condensed phase methods. Gas phase methods, especially CVD (also plasma assisted CVD, etc.), because of microelectronics applications has spawned an entire field with an extensive literature that cannot be properly addressed here. Our focus will be on the use of silicon-containing organometallic and metalloorganic precursors as condensed phase sources of glass and ceramic materials. The term, organometallic, refers to compounds containing silicon−carbon bonds, e.g. polycarbosilanes, whereas the term metalloorganic refers to silicon compounds where Si−C bonds are absent, e.g. $Si(OR)_4$.

Although sol-gel processing is often used to generate materials for optical and electronic applications, with few exceptions, Si-containing precursors are used to produce glasses and/or ceramic materials of value because of their mechanical properties. Hence, most of the discussion that follows will be concerned with how the design, synthesis and processing of precursors can be used to realize better (optimal) mechanical properties. Despite these provisos, the literature available on precursor chemistry and precursor derived materials is still considerable. Consequently, even though we limit discussion to silicon-containing precursors to materials with structural applications, it is still not possible to cite all of the relevant literature. Because of the rapidity with which new precursor articles appear, and the extent of overlap of current publications (e.g. in SiC precursors), it was necessary to choose selected, exemplary articles in specific areas rather than attempt to review every publication.

We begin by providing a general definition of a precursor. We then discuss criteria that define 'useful' precursors in terms of processing shapes and then identify general types of materials properties and microstructures that are sought in the production of glass or ceramic materials. A combination of these criteria and desired materials properties/microstructures are used to identify precursors and precursor chemistries for both oxide and nonoxide glass and ceramic materials. Our overall goal is to delineate general principles that guide choices of chemistries, processing methods and materials properties, thereby permitting desired shapes with appropriate properties to be produced in a time and cost effective manner.

A precursor is any chemical compound that transforms, when decomposed in an energetic environment, to a glass or ceramic material. Thus alum [$NH_4Al(SO_4)_2$], which can be recrystallized to a purity of 99.995%, transforms on heating to high purity Al_2O_3 powder[8]. This type of chemical processing represents the simplest and most mature aspect of the field. The most exciting, newest and least mature aspect is the development of rheologically useful precursors that can be shaped[9]. This aspect represents the primary subject of this chapter.

To properly explore the subject area we begin by defining the concept 'useful'. Because processing ceramic fibers is the most demanding in terms of precursor properties, the criteria developed below emphasize 'useful' in terms of fiber precursors. To be 'useful' for fiber forming, a precursor must offer: (1) controllable rheology, (2) latent reactivity[6b], (3) controllable pyrolytic degradation, (4) high ceramic yield, (5) high selectivity to desired ceramic product and (6) controllable densification and microstructural development. The following paragraphs briefly define the specifics in each category.

II. CRITERIA THAT DEFINE A USEFUL PRECURSOR

A. Rheology

The type of fiber-forming procedure used (extrusion from solution or melt, with or without drawing) places some constraints on what is considered useful polymer rheology. In general, the precursor should exhibit thixotropic or non-newtonian viscoelastic behavior such that, during extrusion, it will flow readily without necking. The viscosity should be sufficiently high at zero shear so that the formed fiber retains its new shape and is self-supporting. Non-newtonian viscoelasticity normally arises in linear polymers with minimum molecular weights of 10–20 k Da as a result of chain entanglement. However, with a few notable exceptions (see below), all spinnable preceramics developed to date are low molecular weight, highly branched oligomers that exhibit non-newtonian behavior because the branches chain entangle. The correlation between rheology and 'spinnability' has been discussed in the literature[10,11].

B. Latent Reactivity

Not only must precursor fibers be self-supporting as extruded, they must also remain intact (e.g. not melt or creep) during pyrolytic transformation to ceramic fibers. Thus, precursor fibers (especially melt spun fibers) must retain some chemical reactivity so that the fibers can be rendered infusible before or during pyrolysis. Infusibility is commonly obtained through reactions that provide extensive crosslinking. These include free radical, condensation, oxidatively or thermally induced molecular rearrangements.

C. Pyrolytic Degradation

Most precursors contain extraneous organic ligands that are added to aid processability or provide latent reactivity. During pyrolysis these extraneous ligands must be eliminated as gaseous products. The rates and mechanisms of decomposition to gases require close monitoring to ensure conversion to the correct ceramic material, to prevent retention of impurities or creation of gas-generated flaws (e.g. pores). The processes involved can be likened to binder burnout in ceramic powder compacts[12,13]. In principle, this criterion is best satisfied if hydrogen is the only extraneous ligand required for stability and/or processability. Indeed, there is often a trade-off between precursor stability or processability and ceramic product purity that mandates processing with a less stable precursor to obtain higher quality ceramic products. In this instance, quality is equated with purity as detailed below. For example, Nicalon fibers derived from polycarbosilane, $-[\text{MeSiCH}_2]_n-$, are not stoichiometric SiC because the original precursor is only processable with a 2 : 1 C : Si ratio (see below). More recently, polymethylsilane (PMS), $-[\text{MeSiH}]_n-$, was found to provide access to phase and chemically pure SiC fibers. Unfortunately, in some forms this polymer is highly susceptible to air oxidation.

D. High Ceramic Yield

This criterion, which is product rather than precursor-property driven, is critical to the design and synthesis of new precursors. The need for high ceramic yields arises because of the excessive volume changes associated with pyrolytic conversion to ceramic materials. Scheme 1 illustrates these changes for a SiC precursor with an 80% ceramic yield of phase pure SiC (3.2 g ml^{-1}). Most precursors densities are close to 1 g ml^{-1}, whereas most Si ceramic densities range from 2.5 to 3.5 g ml^{-1}.

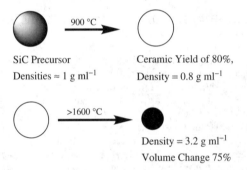

SCHEME 1. Volume changes during densification of an SiC precursor to phase pure SiC

If we have a ceramic yield of 100% (nothing is volatilized) and complete densification occurs, the total volume change will still be *ca* 70%. A 100% ceramic yield is unrealistic. Even PMS, which in principle only needs to lose 2H$_2$ molecules per monomer unit to form SiC, has a theoretical ceramic yield of 91 wt%. Thus, a precursor with a ceramic yield of *ca* 50 wt% (e.g. polycarbosilane)[7] will undergo volume changes of 85%. The possibility of achieving near net shape in the final ceramic product becomes very difficult. The result can be shape distortion. Furthermore, the 50 wt% that leaves as gases can cause pores, uneven densification and leave entrapped impurities. Only in processing thin films or fibers can a 50% ceramic yield still be viable because mass transport and shrinkage are minimal in at least one dimension (in fibers, the diametrical dimension) and shape integrity can be retained[7,9]. Because diffusion distances for mass transport are very short (in fibers, the diametrical direction), gaseous byproducts can leave easily, permitting ready densification at higher temperatures. Finally, gaseous byproducts represent potential pollution problems that must be dealt with in commercial processes.

Hence, for most applications, high ceramic yield precursors are essential. Consequently, it is important to formulate a preceramic polymer that contains minimal amounts of extraneous ligands that allow it to meet the processability criterion and yet provide high weight percent conversions to ceramic product. Thus, in many of the precursors discussed in this chapter, hydrogen and methyl are the ligands of choice. A good ceramic yield typically ranges 80–85 wt%, because most precursor syntheses produce some quantity of low molecular weight species that evaporate rather than decompose during pyrolysis.

Still another concern arises in reading or reporting ceramic yields. It is common to indicate ceramic yields at 1000 °C. For oxide ceramics, this temperature is usually acceptable. However, nonoxide ceramic precursors, especially SiC and Si_3N_4 systems, often retain 1–2 wt% hydrogen at temperatures up to 1400 °C. In 100 g of material this corresponds to one mole of H_2 per 2.5 moles of SiC, if the end product is phase pure SiC. A 1000 °C product with 2 wt% hydrogen can be thought of as a solid solution of SiC and hydrogen. These materials will not exhibit the properties of phase pure SiC, because the hydrogen, which is most likely present at grain boundaries, does not permit normal microstructural development to occur until it is eliminated. Furthermore, unanticipated outgassing at higher temperatures can cause cracking, compositional changes or pores in ceramic shapes. Thus, care must be taken in reading and reporting ceramic yields.

These last two points concern how precursor design and synthesis are influenced by the precursor-to-ceramic conversion process. The next set of criteria identify materials properties that drive precursor design and synthesis including product selectivity.

E. Selectivity to Phase and Chemically Pure Glasses or Ceramics

Chemical and phase purity are critical issues that drive precursor design because optimal mechanical properties are achieved only with high purity. For example, ceramics grade Nicalon fibers, with a chemical composition of ca $SiC_{1.45}O_{0.36}$ and densities of 2.3–2.5 g ml^{-1}[7,9], offer tensile strengths of 2.0–2.5 GPa and elastic moduli of ca 200 GPa. In the last five years, new precursors have been developed that provide sufficient chemical purity, following pyrolysis, to generate pure SiC. Stoichiometric SiC fibers were recently produced from these precursors, as discussed below. These fibers have densities >3.1 g ml^{-1}, tensile strengths of 3.0–3.5 GPa and elastic moduli of 400–470 GPa[7,9]. These values are equivalent to literature values for dense, pure SiC produced via standard ceramic processing methods.

Note: It is sometimes possible to achieve the same end product from off stoichiometry precursors or pyrolysis products by further processing steps after precursor pyrolysis. Thus, as will be discussed below, stoichiometrically correct precursors are not always the answer.

Chemical purity is sometimes insufficient to realize optimal strength properties. Phase purity is also quite important, although this issue is often ignored in precursor studies. Thus, it is relatively easy to obtain mullite ($2SiO_2 \cdot 3Al_2O_3$) composition gels using sol-gel processing[14–17]. However, these gels are amorphous and, despite much literature which suggests they crystallize to mullite at 1000 °C, they actually convert to phase pure, orthorhombic mullite only on heating to ca 1300 °C. Presumably true mullite properties are only realized at this temperature.

Chemical and phase purity are not always desirable. For example, H- and N-doped silicon carbide films behave as high temperature semiconductors, while silicon carbonitride glasses offer properties akin to glassy carbon with room temperature conductivities of 10^3 Ω cm^{-1}[18]. Additional reasons for targeting materials that are not chemically or phase pure stem from the desire to control microstructural properties.

F. Control of Microstructure and Densification

As noted above, densified products provide optimal mechanical properties. Unfortunately, heating a precursor to high temperatures to convert it to phase pure materials frequently does not lead to dense materials. For example, precursor-derived phase pure SiC will crystallize and undergo grain growth on heating to 1800 °C; however, grain growth occurs without coincidental sintering (densification) leading to porous materials[19]. This problem can be solved by adding small amounts of boron (0.3–0.5 wt%) which promotes densification without much grain growth. Thus, boron must be incorporated in the precursor synthesis or processing strategies to achieve the correct microstructure. In this instance, the desired microstructure drives precursor design.

In some instances, subtle changes in the precursor architecture can change the composition and microstructure of the final pyrolysis product. For example, pyrolysis of $-[\text{MeHSiNH}]_x-$ leads to amorphous, silicon carbide nitride (SiCN) solid solutions at >1000 °C (see SiCN section). At ca 1500 °C, these material transform to Si_3N_4/SiC nanocomposites, of interest because they undergo superplastic deformation[20]. In contrast, chemically identical but isostructural $-[H_2SiNMe]_x-$ transforms to Si_3N_4/carbon nanocomposites on heating, as discussed in more detail below[21].

Essentially all nonoxide and many oxide ceramic fibers currently produced commercially are amorphous or combinations of nanocrystallites in an amorphous phase. Indeed, until recently the general consensus in the ceramics community was that crystalline fibers would fail more readily than glassy fibers because of flaws at grain boundaries[6,22,23]. Thus, many studies discussed below are directed toward the development of high temperature, glassy fibers with good mechanical and thermal properties. As noted above, recent studies show that phase pure, microcrystalline SiC fibers provide properties and high temperature stability superior to current commercial fibers[7,9]. Thus, nano- or microcrystalline fibers may be better in some or many applications, especially where creep is a problem.

As a final comment on criteria for the design, synthesis and processing of precursors, if the precursor targeted is for coatings applications only, where the substrate provides most of the mechanical properties, a few additional criteria must be considered. First, the precursor must wet the substrate effectively to form uniform, adherent coatings. Some reaction with the substrate may or may not be desirable as a means of achieving either chemical or/and mechanical adhesion. Additionally, to process flaw (pore and crack) free ceramic coatings using dip, spin on or spray coating processes, it is generally necessary to limit coating thicknesses to <2 µm and more commonly to <1 µm. This is because mismatches in coefficients of thermal expansion and the overall densification process lead to compressive stresses in the films. These stresses can provide improved coating adhesion and abrasion resistance; however, at higher thicknesses the compressive stresses cause coatings to crack, unless a ceramic powder is used as a filler to offset dimensional changes.

These general criteria serve as a basis for the selection of candidate precursors potentially of use for processing both oxide and nonoxide ceramics. For specific materials, additional criteria can also play a role including ease of synthesis, purification and stability toward air and moisture. One final and critical criterion is cost. Costly syntheses can reduce the general utility of a given precursor. However, in ceramic fiber processing, the pyrolytic conversion and post-processing heat treatments designed to provide optimal fiber mechanical properties are often much more costly than the chemistry.

The above criteria provide guidelines for the design and synthesis of new precursors, and as a means of evaluating materials currently extant. The following sections attempt to provide an overview of recent work in nonoxide and oxide materials. We begin by examining precursors to Si–N containing materials beginning with phase pure Si_3N_4 and then moving to more complex Si–N–E (E = element) systems. We next focus on selected SiC precursors and follow with examples of Si–C–B and Si–C–O systems. We then move to precursors to Si–O, Si–Al–O precursors and end with transition metal silicate precursors.

III. PRECURSORS CONTAINING Si AND N

Precursors that fall in this category are generically called polysilazanes or polysilsesquiazanes. Synthetic routes to these materials have been reviewed[6a,24]; thus, we provide an overview of typical methods. The most common route to polysilazanes is via ammonolysis/aminolysis of chlorosilanes[25–27].

$$\text{MeHSiCl}_2 + 3\text{NH}_3 \xrightarrow{\text{Et}_2\text{O}/-78\,°\text{C}} 2\text{NH}_4\text{Cl} + -[\text{MeHSiNH}]_x- \quad (1)$$

$$\text{H}_2\text{SiCl}_2 + 3\text{MeNH}_2 \xrightarrow{\text{THF}/-78\,°\text{C}} 2\text{MeNH}_3\text{Cl} + -[\text{H}_2\text{SiNMe}]_x- \quad (2)$$

$$\text{MeSiCl}_3 + 4.5\text{NH}_3 \xrightarrow{\text{THF}/-78\,°\text{C}} 3\text{NH}_4\text{Cl} + -[\text{MeSi(NH)}_{0.5}]_x- \quad (3)$$

The products resulting from reactions 1 and 2 are short-chain oligomers mixed with 6- and 8-member cyclics, with M_n values ranging from 600–2000 Da. These molecular weights are generally too low to offer good rheological properties or ceramic yields unless modified (see below). Another drawback to these reactions is that two moles of ammonium chloride are produced per mole of polymer and it is difficult to separate the two products cleanly.

Silsesquiazanes such as formed in reaction 3 are often crosslinked and intractable, although they can afford good ceramic yields.

Polysilazanes can also be synthesized by catalytic dehydrocoupling (equation 4)[28,29]. The reaction works best with monosubstituted silanes (MeSiH$_3$ is flammable); however, as with the ammonolysis reactions, molecular weights are normally less than 2000 Da. Hence, as-formed oligomers are not useful precursors. However, combinations of ammonolysis followed by catalytic dehydrocoupling provide access to useful precursors as discussed in the SiCN section.

$$\text{RSiH}_3 + \text{R}'\text{NH}_2 \xrightarrow{\text{catalyst}} \text{H}_2 + \text{H}-[\text{RHSiNR}']_x-\text{H} \quad (4)$$

Recent work by Soum and coworkers provides a novel route to the first high molecular weight polysilazanes ever formed[30]. Although ring-opening polymerization (ROP) of cyclic silazanes has been attempted previously[24], Soum and coworkers report the first successful ionic ROP of cyclodisilazanes to form high molecular weight polysilazanes (equation 5). Molecular weights up to $M_n = 100$ kDa were observed. The key is to use low temperatures ($-40\,°$C) to inhibit depolymerization reactions. If vinyl groups are used instead of methyl groups, curing the resulting polymers with AIBN gives moderate to high ceramic yields of SiCN (exact ceramic yields were not provided). Without the latent reactivity offered by the vinyl groups, heating tends to depolymerize these linear polymers to form volatile cyclic species.

$$\begin{array}{c}\text{Me}\quad\quad\text{Me}\\ \diagdown\quad\diagup\\ \text{Si}-\text{N}\\ \text{Me}\quad|\quad|\quad\text{Me}\\ \text{N}-\text{Si}\\ \diagup\quad\diagdown\\ \text{Me}\quad\quad\text{Me}\end{array} \xrightarrow{\text{Na Naphthalene}} \left[\begin{array}{cc}\text{Me} & \text{Me}\\ | & |\\ \text{Si}-\text{N}\\ | \\ \text{Me}\end{array}\right]_n \quad (5)$$

Although related SiN precursor systems have been developed, the above sets of reactions provide launch points for synthesizing the majority of SiN-containing precursors studied to date. Various groups have learned to manipulate oligomers prepared as above to develop 'useful' precursors. We begin by discussing those that provide phase pure silicon nitride.

A. Precursors to Si_3N_4

Access to phase pure silicon nitride materials via processable precursors is limited to just three approaches. The first, shown in reaction 6, provides one of the first oligomers exploited as a preceramic polymer[24,25a]. This simple polysilazane, containing only Si, N and H, is known to be relatively unstable and will crosslink on its own to give intractable gels. Furthermore, it does not offer the 3Si:4N stoichiometry required for Si_3N_4. Nonetheless, it is useful as a binder and for fiber-reinforced ceramic matrix composites (CMCs)[31].

$$H_2SiCl_2 + 3NH_3 \xrightarrow{Et_2O/-78\,°C} 2NH_4Cl + -[H_2SiNH]_x- \quad (6)$$

Because H_2SiCl_2 is a flammable gas that is difficult to work with, researchers at Tonen devised a novel approach to ammonolysis wherein the silyl halide is modified by complexation to pyridine (py) prior to ammonolysis and subsequent heat treatment of the isolated oligomer[23,32]:

$$H_2SiCl_2 + 2py \xrightarrow{0\,°C} H_2SiCl_2 \cdot 2py \quad (7)$$

$$H_2SiCl_2 \cdot 2py + \text{excess } NH_3 \longrightarrow NH_4Cl(\text{or } py \cdot HCl) + -[H_2SiNH]_x- \quad (8)$$

$$-[H_2SiNH]_x- \xrightarrow{120\,°C/py/4\ h/NH_3} -[H_2SiNH]_x-(\text{white solid following solvent removal}) \quad (9)$$

The oligosilazanes formed via this approach give higher molecular weights than obtainable by direct ammonolysis. The product (M_n ca 1600 Da) is polydisperse and highly branched. This oligomer is also not stable and crosslinks to form an intractable product in a short time in the absence of solvent. Pyrolysis gives an 80% ceramic yield of silicon nitride mixed with 20 wt% Si as expected based on the 1:1 Si:N precursor stoichiometry. Silicon is found even when pyrolysis is conducted under an NH_3 atmosphere. A tentative structure for the oligomer can be proposed:

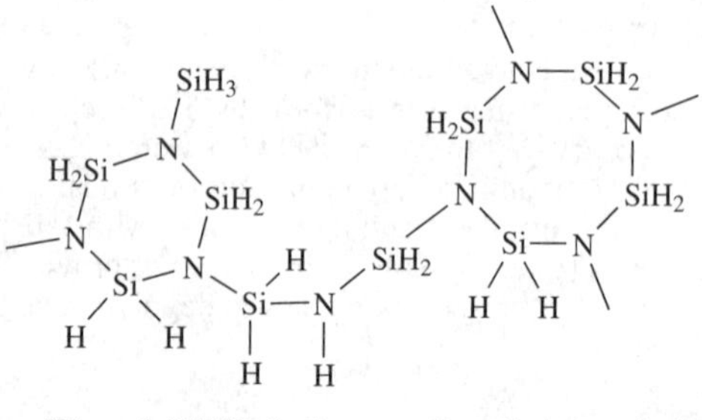

Characterization by IR and NMR indicates oligosilazane chains capped with SiH_3 groups. Analysis of the gases given off during synthesis reveals the presence of SiH_4 as a minor product. Both SiH_4 and the SiH_3 endcaps arise as a consequence of pyridine or ammonia catalyzed redistribution processes. The endcaps probably limit the degree to which higher molecular weight polymer can form.

In using the perhydropolysilazanes to process either composites or thin (10 μm diameter) fibers, heating to temperatures of 1100 °C, even in N_2, leads to crystallization of quantities of Si[31,32]. Blanchard and Schwab[31] indicate that up to 20 wt% Si appears at temperatures of 1200 to 1400 °C whether the processing atmosphere is N_2 or NH_3. This Si eventually reacts with the processing atmosphere, converting slowly to Si_3N_4 at higher temperatures. Conversion to Si_3N_4 is complete at ca 1700 °C in N_2 and 1580 °C in NH_3.

Work at Tonen indicates that fiber properties decrease with the appearance of Si metal, as it promotes Si_3N_4 crystallization. For example, the fiber tensile strengths, which are reported to be 2.5 GPa ($E = 250$ GPa) at room temperature, diminish by 10% after heating at 1200 °C for 10 h. On heating to 1300 °C for 10 h, the tensile strengths drop by 40%. Based on the observation that crystallization of Si_3N_4 is promoted by free Si and that crystallization leads to a loss in fiber strength, the fibers are processed to minimize Si content and up to 3 wt% oxygen is incorporated to aid in maintaining a glassy or amorphous nature.

As noted in the introduction, inexact stoichiometries and poor ceramic properties can be ameliorated during processing rather than by precursor design. The first example of this approach is the conversion of SiCN and SiC/C containing ceramics directly to Si_3N_4 by treating the precursor before, during or after initial pyrolysis processing with NH_3 at temperatures high enough to cleave N−H bonds (typically >500 °C).

Thus, a number of researchers have chosen ammonolysis as a route to phase and chemically pure Si_3N_4[33−37]. This process involves removal of excess free C and displacement of carbon in Si−C bonds with Si−N bonds. The process is assumed to involve free radical reactions. There is no apparent relationship between precursor structure or functionality and the extent of carbon removal[33]. *During the ammonolysis process, the carbon leaves as toxic HCN, thus care should be used in this approach*[38].

Schaible and coworkers find that ammonolysis of cyclotetra(dimethylsilazane) at 900 °C results in the formation of crystalline silicon diimide, Si_2N_2NH[37]. On further heating α-Si_3N_4 then crystallizes. It is important to recognize the potential for generating novel intermediate phases, as noted above, because these phases may be the key to developing the final microstructure.

The above studies indicate that phase pure Si_3N_4 can be obtained by careful choice of precursor, processing conditions or a combination of both. Several groups note that optimal mechanical properties for fibers are only obtained when the product is amorphous and often nonstoichiometric, because crystallization leads to grain grown and flaws at grain boundaries[6,22,23,32,36]. This then provides one motivation for the development of amorphous SiN-based ceramic systems. An additional driver is the fact that Si_3N_4 is not stable at >1700 °C, decomposing to Si and N_2. Thus, efforts to stabilize the system to nitrogen loss, while minimizing crystallization, have targeted modified SiN systems. These are discussed in the following sections. In closing, it is important to recognize that phase pure Si_3N_4 is still quite desirable for monolithic applications that do not involve fibers. In addition, pure Si_3N_4 remains a very valuable material in electronic applications.

B. Precursors to SiCN

Although there is considerable reason for developing modified SiN systems as just noted, most SiCN precursor systems were developed with the goal of processing Si_3N_4 fibers. In the early stages of this development, researchers explored methods that emphasized spinnability rather than product properties. However, the importance of minimizing C content was recognized, resulting in precursors with monomer units containing only −H and −CH_3 groups as shown below[21,24−28,39−48].

As noted above, for the most part, the first synthetic step resulted in oligomers with low ceramic yields, poor rheological properties and/or limited latent reactivity. As such, numerous approaches were explored with the idea of introducing latent reactivity and better viscoelastic properties. In the following, we discuss several of these synthetic approaches, some of the materials that result and a few applications.

Verbeek and coworkers[39] described the first processable SiCN precursors in the early 1970s. The approach, explored further recently[6,40], was to polymerize $MeSi(NHMe)_3$

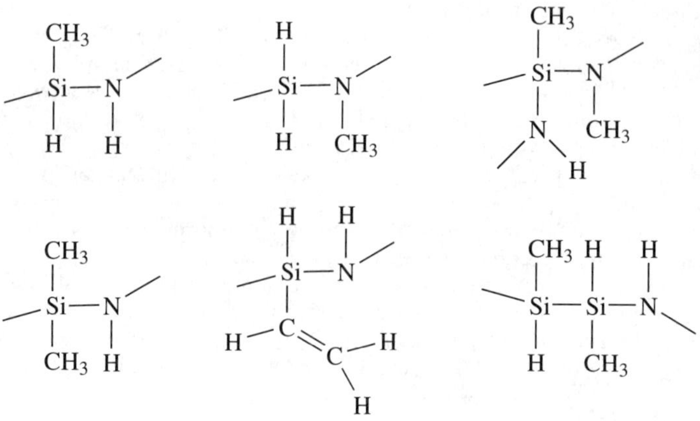

by self-condensation (equation 10) to obtain a polysilazane with $M_w \approx 4000$ Da. This polysilazane could be hand drawn to form 10–20 μm precursor fibers. In this instance, the precursor fibers were rendered infusible by exposure to humid air. Pyrolysis converts the crosslinked precursor fibers to ceramic fibers with ceramic yields of ca 55%. The ceramic products are mainly amorphous SiCN with some SiO_2 (a consequence of H_2O vapor used to crosslink the spun fibers)[6,40].

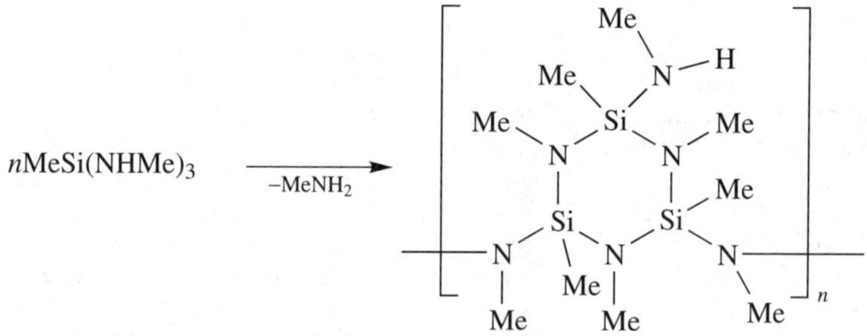

$$n\text{MeSi(NHMe)}_3 \xrightarrow{-\text{MeNH}_2} \quad (10)$$

Interrante and coworkers have explored the related condensation polymerization of $Si(NHEt)_4$. Unfortunately, polymerization leads to an insoluble, rather than a processable, polymer with a 60 wt% (1000 °C/Ar) ceramic yield. Based on solid state NMR data, the polymer structure given in equation 11 was proposed[41].

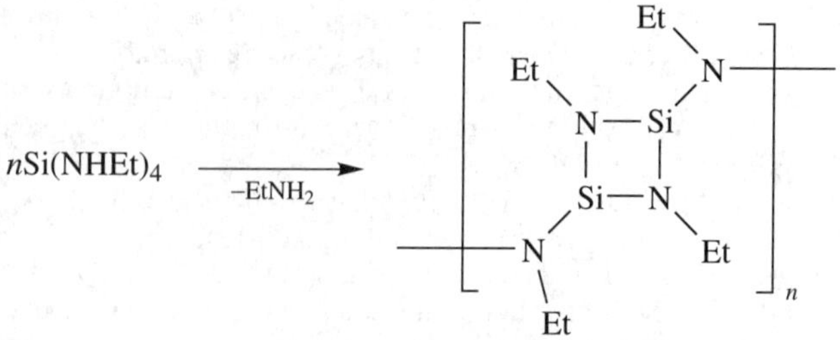

$$n\text{Si(NHEt)}_4 \xrightarrow{-\text{EtNH}_2} \quad (11)$$

^{29}Si NMRs taken of samples during the pyrolysis process indicate that the polymer first converts to a material in which all of the Si—N bonds are retained, and some Et groups are retained as free C. The conclusion is that amorphous Si_3N_4 forms coincident with free C. *Because the precursor does not contain Si—C bonds, SiCN is not formed as an intermediate product.* On heating to temperatures of 1500–1600 °C, the material transforms to either Si_3N_4/SiC (pyrolyzed in N_2) or SiC (pyrolyzed in Ar). Si_3N_4 begins to decompose at $\geqslant 1500$ °C (as noted above) and will react with free C to form SiC. A similar result is observed in pyrolysis studies of $-[H_2SiNMe]_x-$ (*N*-methylpolysilazane), see below[21,28].

A novel method of improving the processability of the cyclic $-[MeHSiNH]_x-$ oligosilazanes (from reaction 1) by reaction with urea was recently described by Seyferth and coworkers (equation 12)[42].

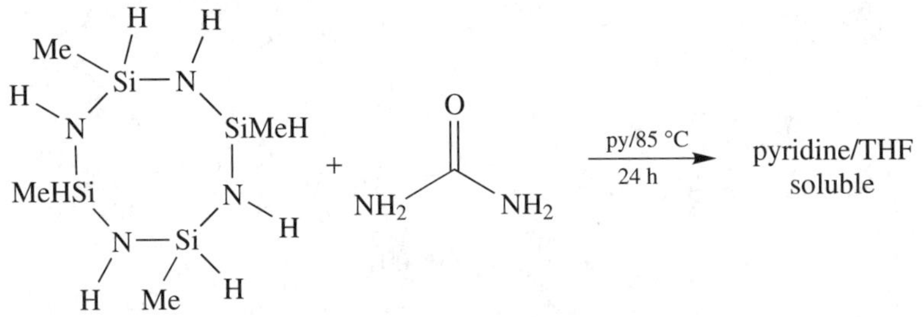

$$\text{cyclosilazane} + \underset{NH_2}{\overset{O}{\underset{\|}{C}}}{NH_2} \xrightarrow[24 \text{ h}]{py/85\,°C} \text{pyridine/THF soluble} \quad (12)$$

The original precursor materials give ceramic yields in the 15–20 wt% range, whereas the materials recovered from reaction 12 give ceramic yields (950 °C) of 80 + wt%. A typical sample gave a composition of $SiN_{1.06}C_{0.60}$. No precursor structure was offered; however, the presence of N—CO—N moieties was inferred from the gases evolved during pyrolysis and NMR data. Although some oxygen may be retained in the polymer, NMR studies do not indicate the presence of species with Si—O bonds in the final ceramic product. Because of the sensitivity of the polymer to hydrolysis, no molecular weight measurements were made. The processability of this polymer has not been described to date.

The above work follows earlier and more comprehensive efforts by the Seyferth group[25] on the use of base catalyzed crosslinking of cyclotetrasilazane, $-[MeHSiNH]_4-$, to increase its ceramic yield and processability. The currently proposed mechanism is suggested to involve silaimine formation followed by an addition reaction across the reactive double bond, as illustrated in equation 13[25,26].

The final oligomer has a M_n of 800–2000 Da depending on the reaction conditions and consists of two monomer types, $-[MeHSiNH]_x-$ and $-[MeSiN]_y-$. It is THF soluble and has ceramic yields of 80–85 wt% (1000 °C/N_2). A typical 1000 °C ceramic composition is $SiC_{0.50}N_{0.88}O_{0.04}$. Originally it was thought that this composition indicated an intimate mixture of SiC and Si_3N_4; however, recent solid state ^{29}Si NMR results suggest otherwise[21,43]. On heating to $\leqslant 1400$ °C, the material remains amorphous and shows the presence of numerous species of the type SiN_xC_{4-x}. Only on heating to $\geqslant 1500$ °C does the material transform to a true Si_3N_4/SiC mixture.

Transition-metal-catalyzed dehydrocoupling is an alternate catalytic method of turning lower molecular weight oligosilazanes into processable polysilazanes[21,28,44]. This process begins with the oligomers formed in reactions 1, 2 or 6. These oligomers usually contain internal Si—H bonds and $-NH_2$ or $-MeNH$ endcaps. These endcaps are more reactive toward dehydrocoupling with Si—H bonds (reaction 4) than internal N—H bonds. Consequently, it is possible to form highly branched, polydisperse, viscoelastic precursors (equation 14)[28,44]. The GPC trace of a polydisperse polymer prepared via reaction 14 is shown in Figure 1.

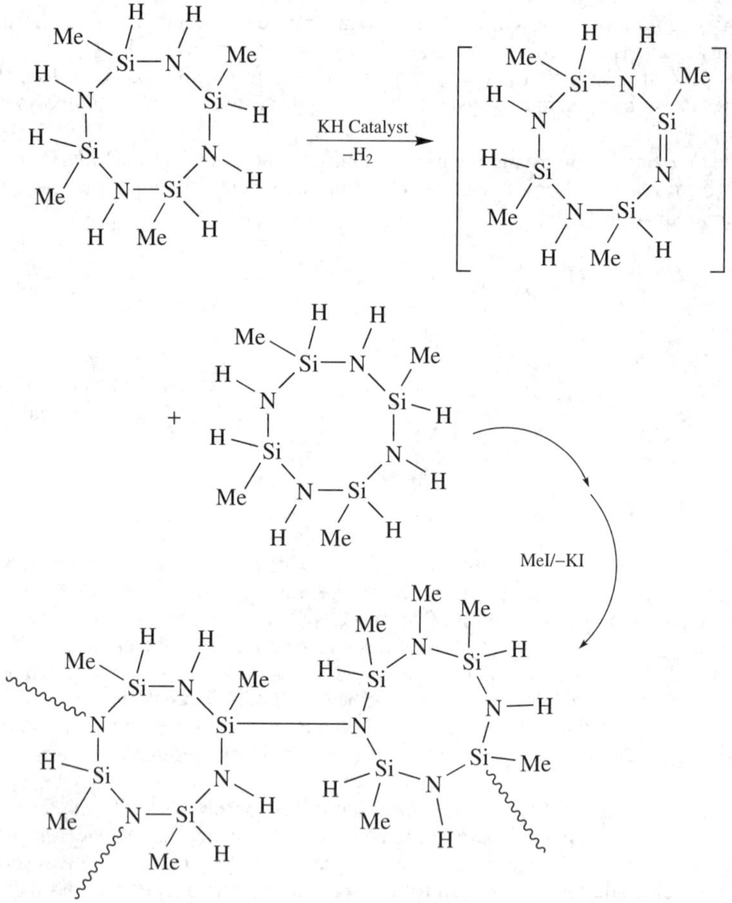

$$(13)$$

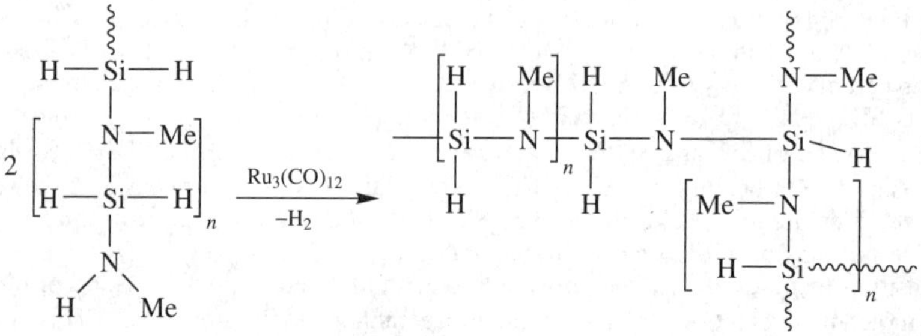

$$(14)$$

If $-[MeHSiNH]_x-$ is used in place of $-[H_2SiNMe]_x-$ in reaction 14, the resulting precursor is similar to the product of reaction 13. The solid state ^{29}Si NMR of this polymer heated to selected temperatures is shown in Figure 2[21]. The broad envelope of Si magnetic environments results from the series of SiN_xC_{4-x} structures, as seen previously[43].

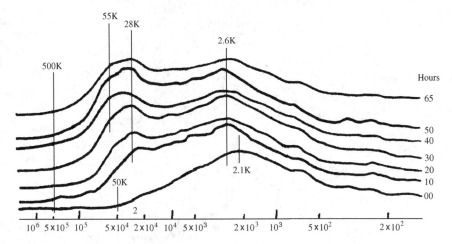

FIGURE 1. Gel permeation chromatogram of Ru-catalyzed chain extension of HMeN−[SiH$_2$NMe]$_x$−H

N-Methylpolysilazane is chemically identical to −[MeHSiNH]$_x$− (Si-methylpolysilazane), studied by many researchers[25,42], but isostructural. Following catalytic modification as in reaction 14, the chemical evolution of N-methylpolysilazane on pyrolysis was followed by ^{29}Si (Figure 3). A detailed analysis has already been published[21]. The pertinent points are that by 600 °C, the major portion of the Si centers experience a magnetic environment centered around −46 ppm, which shifts to −49 ppm on heating to 800 °C and then 1000 °C. Similar results are obtained for samples heated to 1400 °C. The 1000 °C ^{29}Si NMR spectrum can be simulated with a single line centered at −49 ppm with a 1110 Hz linewidth. This is typical of values (1600 Hz) reported previously for amorphous Si$_3$N$_4$[21]. Despite the fact that up to 25 wt% free carbon is present, it in no way influences the chemistry at Si, up to 1400 °C. Note that the 800 °C spectrum is similar to that seen for the Si(NHEt)$_4$ derived materials[41].

One further finding with the −[H$_2$NMe]$_x$− pyrolysis studies comes from Raman studies of the ceramic product. The NMR data show no traces of Si−C bonds, nor does the Raman spectrum. However, the Raman spectrum does show the presence of C−N bonds up to 1200 °C (the highest temperature studied). This suggests that the silicon nitride nanoparticles interact with the carbon matrix through C−N bonds. One might speculate that the interface looks somewhat like C$_3$N$_4$.

Thus, by shifting a methyl group from Si to N, the mechanism of transformation has been changed entirely. The products from pyrolysis of −[H$_2$NMe]$_x$− and −[MeHSiNH]$_x$− are quite different as clearly seen in Figure 4, which compares the ^{29}Si NMR spectra for both materials pyrolyzed to 1000 °C. This is proof that polymer architecture can strongly influence the type of ceramic material produced on pyrolysis.

Riedel and coworkers have also uncovered some extremely novel chemistry based on the use of carbodiimide chemistry, including a case where precursor architecture influences phase and compositional development[49]. The process begins with the very simple synthesis of silylcarbodiimides (equations 15 and 16). In the case where R = Me, the compound is a 16-atom heterocycle ($x = 4$). For R = H, a polymer is obtained that gives a 65 wt% ceramic yield at 1500 °C and a SiC$_{1.8}$N$_{1.3}$ (1000 °C) composition.

$$2Me_3SiCl + H_2NCN + 2py \longrightarrow Me_3SiN{=}C{=}NSiMe_3 + 2py{\cdot}HCl \qquad (15)$$

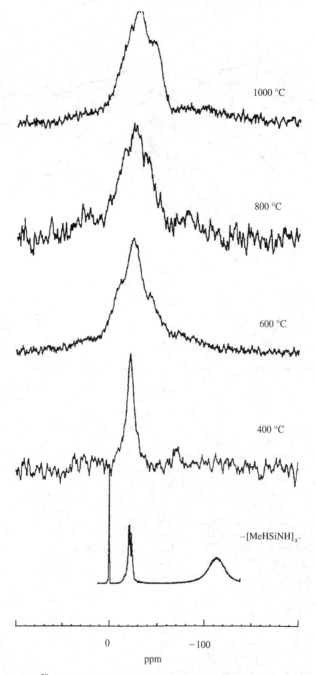

FIGURE 2. Solid state ^{29}Si MAS NMR of chain extended (65 h reaction time) $-[MeHSiNH_2]_x-$ heat treated (2 h/N$_2$) to selected temperatures. Note how the ^{29}Si peak broadens with the formation of various SiN$_x$C$_{4-x}$ species

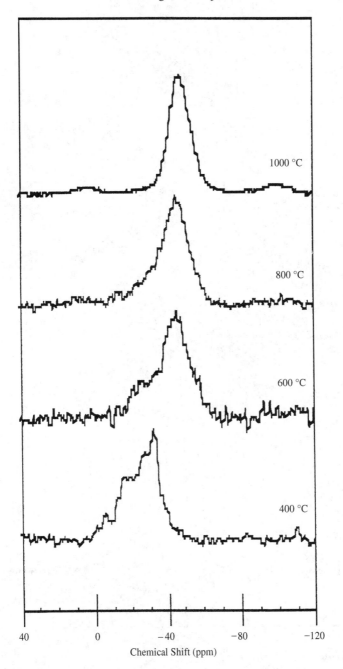

FIGURE 3. Solid state ^{29}Si MAS NMR of 65 h $-[H_2SiNMe]_x-$ heat treated (2 h/N$_2$) to selected temperatures. Note how the ^{29}Si peak becomes sharper at increasing temperatures with the formation of amorphous Si$_3$N$_4$

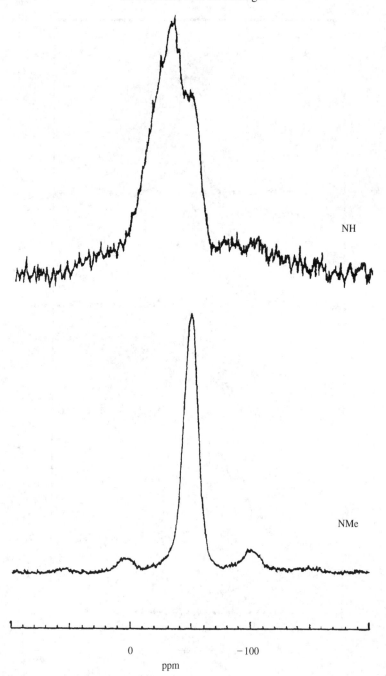

FIGURE 4. Comparison of $-[\text{MeHSiNH}_2]_x-$ and $-[\text{H}_2\text{SiNMe}]_x-$ heat treated at 1000 °C (2 h/N$_2$)

Another facet of the $Me_3SiN=C=NSiMe_3$ chemistry is that it can be reacted with chlorosilanes to obtain transparent gels (equation 17).

$$xMeRSiCl_2 + xH_2NCN + 2xpy \longrightarrow -[MeRSiN=C=N]_x- + 2xpy \cdot HCl \quad (16)$$
$$py = pyridine$$

$$2Me_3SiN=C=NSiMe_3 + SiCl_4 \xrightarrow{py\ catalyst} [Si(N=C=N)_2]_n + 4Me_3SiCl \uparrow \quad (17)$$

These gels are precursors to SiCN materials and offer opportunities for processing porous materials. However, perhaps the most important finding is the apparent discovery of a new phase in the Si–C–N ternary system. On heating the product from reaction 17 to 600 °C, an air-sensitive product is obtained that analyzes for SiC_2N_4. This product is crystalline by XRD (crystallizes at 400 °C), has a single ^{29}Si peak at -103 ppm and a single ^{13}C peak at 120 ppm in the MAS NMR spectra. Above 920 °C it decomposes with loss of cyanogen (C_2N_2) to form a colorless, air-stable material that analyzes for Si_2CN_2. This material contains a carbodiimide unit as indicated by the IR spectrum ($vN=C=N = 2170$ cm^{-1}) and the single ^{13}C peak at 116 ppm, with a ^{29}Si peak at -67 ppm in the MAS NMR spectra. The compound's crystal structure is similar to Si_2ON_2. On heating above 1000 °C, amorphous SiCN obtains, and then a mixture of SiC/Si_3N_4 at >1500 °C. This represents one of the clearest examples of how precursor design can provide access to totally new materials not available by standard ceramics processing methods.

To date, SiCN materials have been used to make coatings[44,50,51], monolithic and extruded shapes[52–55], for binder applications[44,56], fibers[45,57–61] and composite processing[44,52,55]. Recently, photopolymerizable precursors have been considered for rapid prototyping applications[62].

In general, the pyrolysis of polysilazanes with carbon groups on silicon leads to initial formation (400–1400 °C) of amorphous materials wherein silicon is present in a number of bonding environments of the general formula $Si_xN_yC_z$ ($x + y + z = 4$) (Si–H bonds not considered). Heating to higher temperatures in N_2 (1500–1700 °C), especially with sintering aids, causes crystallization of Si_3N_4/SiC nanocomposites[20,63–65]. Heating to the same temperatures in Ar drives the reaction toward the formation of SiC. In contrast, heating polysilazanes that contain only Si–N bonds can provide either Si_3N_4/Si or Si_3N_4/C nanocomposites depending on the polysilazane. At higher temperatures, in the absence of N_2, the Si_3N_4/C composites transform to SiC as Si_3N_4 decomposes. Surprisingly, heating shaped SiCN composites in air at temperatures of 1600 °C for periods of up to 60 h results in incorporation of only small amounts of oxygen (0.4 wt%) that passivate surfaces against further oxidation[54c].

C. Precursors to Silicon Oxynitride (Si_2ON_2)

Silicon oxynitride exhibits better resistance to thermal downshock and oxidation than silicon nitride, although it decomposes under similar conditions with loss of N_2[66]. Thus, it represents a practical target for precursor synthetic methods. Unfortunately, it is difficult to achieve the exact 2Si : O : 2N stoichiometry in a processable precursor and retain this ratio during pyrolysis. Hence, very little work has been directed toward developing Si_2ON_2 precursors[67–70]. Furthermore, the approaches used to date rely on processing with NH_3 to achieve the correct stoichiometry.

The earliest work in this area was that of Yu and Mah[67], and Seyferth and coworkers[68]. An approach related to reaction 13 is used, wherein $-[MeHSiO]_x-$ oligomers are grafted on before MeI is added. Alternately, the two oligomers are copolymerized. Based on reaction 13, a plausible copolymerization process is described in equation 18. Molecular weights (M_n) ranged from about 800 to 2700 Da with the graft-cooligomers giving somewhat higher average molecular weights.

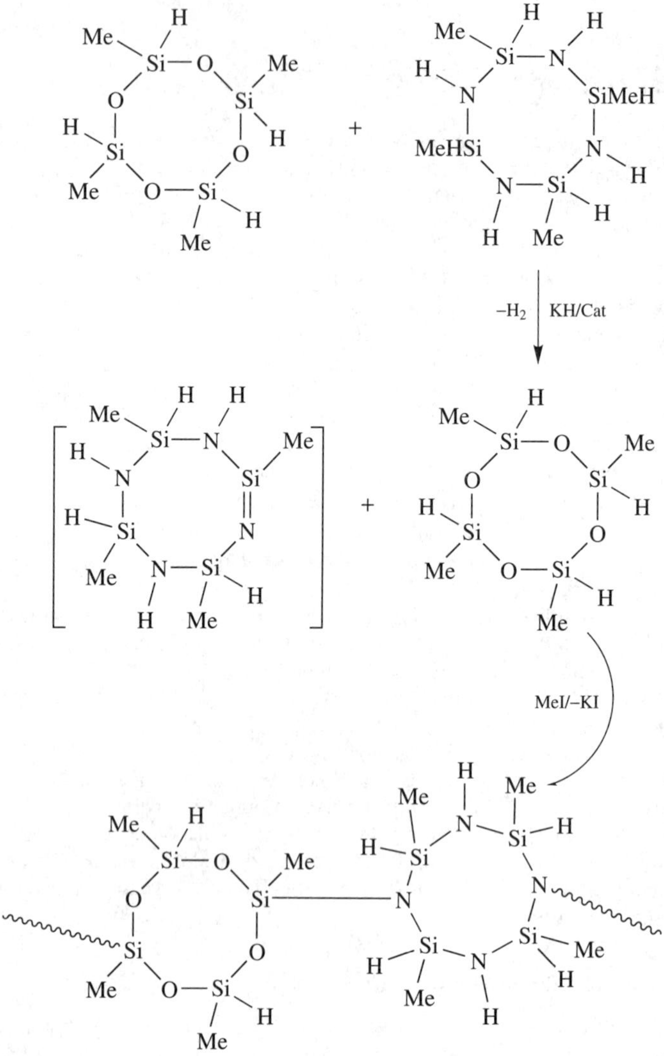

$$\text{(18)}$$

Recent work by Yu and coworkers suggests that Si–Si linkages may also form in reaction 18 as indicated by the presence of a ^{29}Si NMR signal at 0.3 ppm[71]. Furthermore, Yu and coworkers report that the hydridosiloxane reacts at a much slower rate than the silazane, leading to formation of block copolymer-like structures. It appears that most of the cyclosilazanes react to form oligomeric materials and then cyclosiloxanes react afterward. The end result would appear to be the same as the grafting reaction. However, Yu and Mah report that the grafted polymers behave somewhat differently from the cooligomers.

On heating the grafted and copolymerized oligomers to temperatures of 800 °C in NH_3, ceramic yields range from 78–86 wt%. At 800 °C the products are amorphous, but transform to crystalline Si_2ON_2 on heating to > 1300 °C. The 1 : 1 copolymerized oligomers gave essentially pure Si_2ON_2, whereas the graft 1 : 1 oligomers gave a material that was mostly Si_2ON_2 with some Si_3N_4. A 5 : 1 siloxane/silazane copolymer gave a material that contained both Si_2ON_2 and SiO_2.

Si_2ON_2 precursors have also been made via catalytic dehydrocoupling[69]. Linear polysiloxazanes prepared via reaction 19 gave materials with $M_n = 5–7$ KDa; however, the ceramic yields were poor.

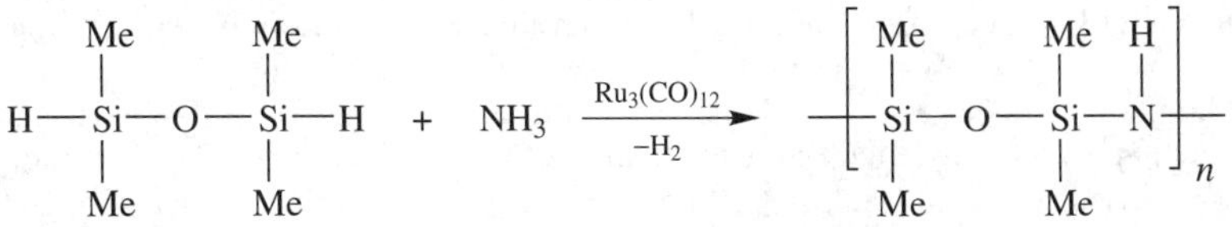

$$H-\underset{\underset{Me}{|}}{\overset{\overset{Me}{|}}{Si}}-O-\underset{\underset{Me}{|}}{\overset{\overset{Me}{|}}{Si}}-H \; + \; NH_3 \xrightarrow[-H_2]{Ru_3(CO)_{12}} \left[\begin{array}{c} \underset{\underset{Me}{|}}{\overset{\overset{Me}{|}}{Si}}-O-\underset{\underset{Me}{|}}{\overset{\overset{Me}{|}}{Si}}-\overset{\overset{H}{|}}{N} \end{array}\right]_n \quad (19)$$

By switching to a cyclic tetrameric hydridomethylsiloxane a processable, partially crosslinked liquid precursor could be obtained with $M_n = 1.2–1.5$ KDa (equation 20).

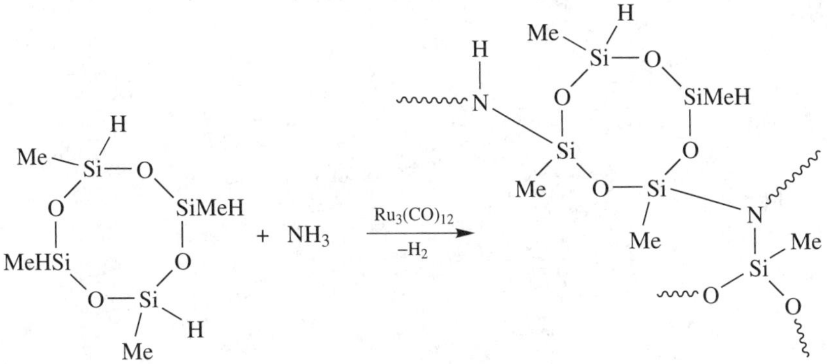

$$(20)$$

Pyrolysis of this type of polymer to 800 °C in N_2 provides ceramic yields of ca 75 wt%. In NH_3, the same material gave an 88 wt% ceramic yield at 800 °C. On heating to 1600 °C the ceramic yields drop to 64 wt% in N_2 and 78 wt% in NH_3. XRD studies of the 1600 °C materials heated in both atmospheres indicate that the primary crystalline phase is Si_2ON_2. This is despite the presence of some 12 wt% carbon for the sample heated in N_2. No carbon is detected for the sample heated in NH_3. This sample appears to be essentially phase pure Si_2ON_2.

Finally, Okamura and coworkers[70] have described the preparation of SiO_xN_y fibers from precursors. These 10–13 μm diameter fibers were prepared by pyrolysis of oxygen-cured polycarbosilane (see below) in an NH_3 atmosphere. These fibers, with densities of 2.3 g ml^{-1}, offer tensile strength of ca 1.8 GPa with elastic moduli of 200–220 GPa. Because the theoretical density of Si_2ON_2 is close to 2.8 g ml^{-1}, these values are not those expected of fully dense Si_2ON_2 fibers.

Because it is possible to convert amorphous SiO_2 directly to Si_2ON_2 using NH_3 at 900–1100 °C without recourse to precursors[66,72], the impetus for pursuing precursor routes to Si_2ON_2 is not high, except perhaps for processing thin films or fibers. However, the above studies are important as they provide the basis for processing more complex multicomponent oxynitrides, wherein Si_2ON_2 may be one of the pyrolysis products. The primary rationale for exploring the utility of multicomponent precursors is that mentioned above, wherein preventing high temperature crystallization allows fiber strength properties to be maintained. The following sections look at three- and four-component SiN-based materials.

D. Precursors to Si—Al—O—N Materials

Silicon–aluminum oxynitrides (SiAlONs) have been studied in great detail because they are: (1) easily made, high melting, glasses; (2) stable to reasonably high temperatures; (3) good sintering aids for silicon nitride and they (4) offer high hardness and (5) provide good-to-excellent mechanical properties[73-76]. Despite the importance of SiAlONs to the ceramics industry, very few groups have developed precursor routes to these materials[77-79].

Thus, Schmidt and coworkers[77] briefly described the synthesis of a Si—Al—O—N precursor by reacting a cyclovinyl silazane with an alkoxyaluminane (equation 21). The ratio of Si:Al was varied between 3 and 0.33. In all instances, the resulting products ranged from clear viscous liquids to slightly yellow solids. Fibers could be hand-drawn from the intermediate materials. Pyrolysis in NH_3 gave ceramic yields (1000 °C) ranging from 30 to 50 wt%. The resulting materials were reported to be SiAlON-like based on ^{27}Al solid state NMR. No other characterization was reported.

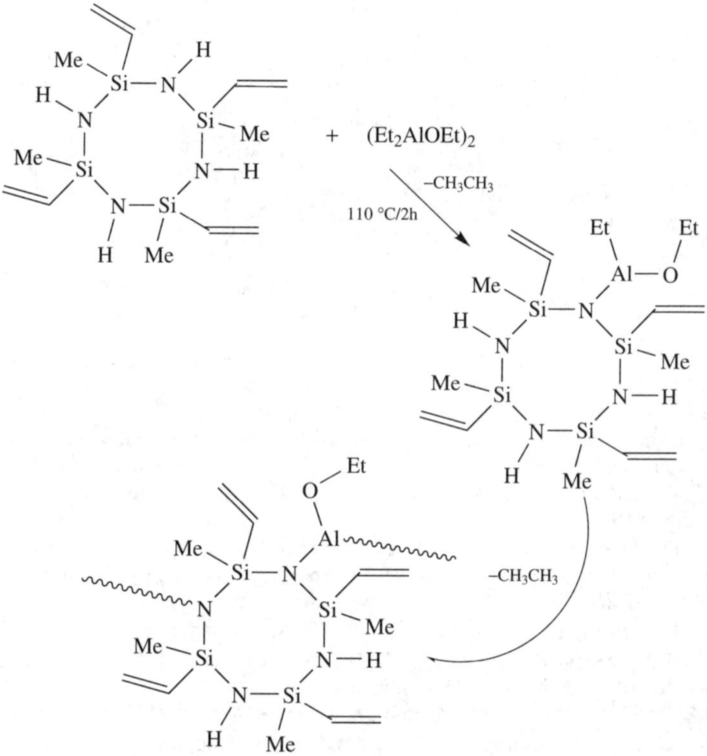

$$ (21) $$

The earliest successful efforts to produce SiAlON were those of Soraru and coworkers[78,79]. The approach was simply to react a commercial polycarbosilane, $-[MeHSiCH_2]_x[Me_2Si]_y-$, with $Al(OBu\text{-}t)_3$ in refluxing xylene for 1 h to obtain a clear solution. Removal of solvent gave a clear liquid that, on heating slowly to 300 °C, gave a 'polyaluminocarbosilane' which was then heated in flowing NH_3 at temperatures up to 1000 °C to form SiAlON. A typical chemical analysis for the precursor prior to nitridation in ammonia was $SiAl_{0.17}C_{2.37}O_{0.53}H_{5.84}$. Following nitridation at 1000 °C and then heating to 1500 °C, a typical chemical analysis was $SiAl_{0.16}O_{0.6}N_{1.48}$.

NMR characterization of the resulting clear liquid indicates that it is a blend rather than a copolymer. That is, no reaction appears to occur between $Al(OBu\text{-}t)_3$ and the polycarbosilane. However, the resulting blend provides materials with high enough viscosities to permit fibers to be spun and nitrided ($\geqslant 500$ °C) to SiAlON-like materials. Ceramic fibers of 12–30 μm diameter could be processed. The highest tensile strengths (1–1.8 GPa) were found for fibers with diameters of 10–15 μm after heating to 1000–1200 °C. Above 1200 °C, the fiber strengths drop as the material crystallizes, exhibiting a powder pattern in between that of β-Si_3N_4 and the crystalline oxynitride, β-SiAlON, with an average chemical composition of $Si_{2.5}Al_{0.5}O_{0.5}N_{3.5}$.

To delay crystallization, an additional element can be added, e.g. yttrium or magnesium[76]. Thus, Laine and coworkers[80] have shown that ammonolysis of a cordierite ($2MgO \cdot 2Al_2O_3 \cdot 5SiO_2$) precursor system provides access to MgSiAlON materials. The development of a cordierite precursor is described below in the sections on silicate precursors.

E. Precursors to Si−N−B−X Materials

The first report on SiNBC materials was published in the patent literature by Takamizawa and coworkers[81]. Based on this early work and arguments presented above, several research groups have sought to develop new materials wherein crystallization is inhibited, for high temperature fiber applications. These efforts have been primarily empirical in nature; however, most of them have targeted modifications of SiC or Si_3N_4, in part because precursors to SiC and Si_3N_4 have already been developed and in part because SiC and Si_3N_4 already offer good-to-excellent high temperature properties. Again, as noted above, there are three strategies for preparing these materials: one is through synthesis, another is through processing and a third is a combination of the first two approaches.

In the area of direct synthesis, only the Tonen group has explored the preparation of precursors to Si−B−N materials free from other elements[82]. They find that perhydropolysilazane (see above) will react with a variety of boron-containing compounds to produce SiNB precursors. Thus, reaction of perhydrosilazane with BH_3 leads to oligomers (equation 22). Pyrolysis under N_2 to 900 °C gives black ceramic materials in >95 wt% yield. On heating to 1700 °C, a typical chemical composition was found to be $SiB_{0.33}N_{1.52}C_{0.06}O_{0.06}$. XRD studies indicate that the 1700 °C material is amorphous; thus boron addition retards crystallization of Si_3N_4. Fibers (10 μm diameter) were spun from related systems; however, no details about the properties were provided. Su and coworkers[83] find that a Dow Corning HPZ[45,84] polymer with a $(HSi)_{0.33}(Me_3Si)_{0.17}(NH)_{0.33}N_{0.17}$ [$SiN_{1.04}C_{1.16}H_{4.69}$] composition will react with borazine according to the generic reaction shown in equation 23.

The above general structure is supported by both NMR and FTIR studies. No evidence was found for the formation of borazine oligomers via dehydrocoupling. Depending on the ratio of borazine to HPZ, the compositions and molecular weights of the resulting polymers varied. Thus, polymer compositions ranged from $SiB_{0.08}N_{0.84}C_{1.08}H_{4.58}$ to $SiB_{1.68}N_{2.59}C_{1.04}H_{6.25}$. The molecular weights of these polymers were $M_n = 23K$ Da and 5.5K Da with polydispersities of 5 and 10, respectively. They are soluble in most common organic solvents. The first polymer exhibited a 70 wt% ceramic yield (1400 °C/Ar) and a corresponding composition of $Si_{2.25}B_{0.2}N_{2.42}C$ whereas the second one gave a 76%

ceramic yield and a composition of $Si_{1.94}B_{2.57}N_{4.77}C$. Both pyrolysis products appeared nanocrystalline/amorphous by XRD. On heating to 1800 °C, the ceramic yield of the first polymer dropped to 51 wt% and the composition changed to $Si_{1.72}B_{0.18}N_{0.80}C$. At this temperature α- and β-Si_3N_4 and SiC crystallize. However, no crystallization is seen below this temperature. Thus, boron addition retards crystallization by some 300 °C. Furthermore, it inhibits decomposition of Si_3N_4.

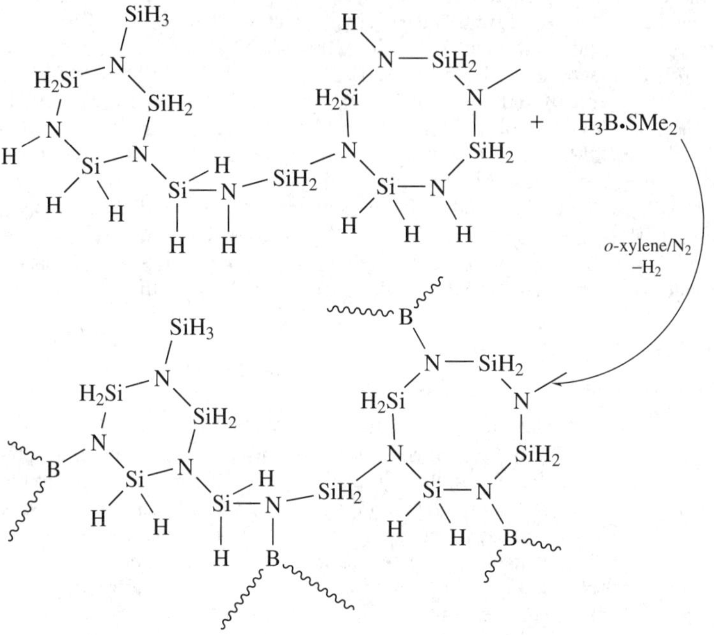

(22)

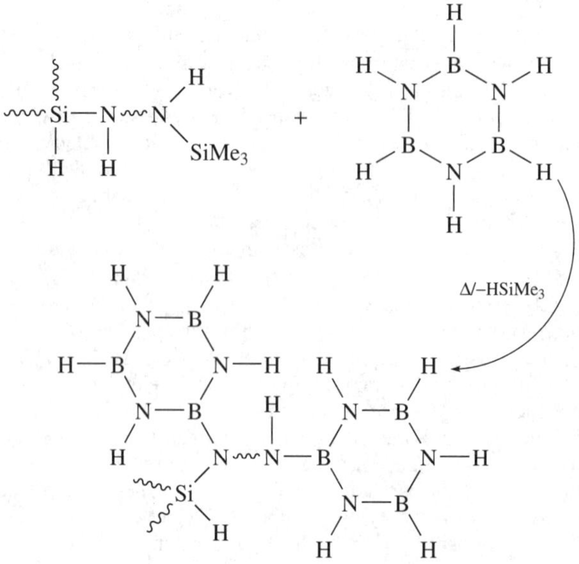

(23)

Riedel and coworkers have synthesized a series of Si−N−B−C precursors via reactions 24 and 25, as well as by hydroboration of the carbodiimide produced in reaction 17. The chemical composition of a typical reaction 24 product was found to be $Si_{2.84}BC_{9.05}N_{2.84}O_{0.64}H_{24.46}$. The polymer is soluble in common solvents and melts at $180\,°C^{49,85}$. This polymer provides a TGA (1000 °C/He) ceramic yield of ca 70 wt% and an apparent stoichiometry of $Si_3BC_4N_2$. A typical chemical composition for the reaction 25 product is $Si_{2.02}BC_{6.63}N_{2.76}O_{0.09}H_{19.84}$. This polymer is also soluble in common solvents and gives an amorphous material (ca 55 wt% ceramic yield, 1000 °C/He) on heating, with an approximate composition of Si_2BCN. Both materials remain amorphous on heating to higher temperatures. Furthermore, both materials are much more stable than related SiCN and Si_3N_4 materials, as illustrated in the high temperature TGA studies shown below. The 2−3% mass loss above 2000 °C for both compounds results from volatilization of oxygenate impurities introduced during polymer processing/handling.

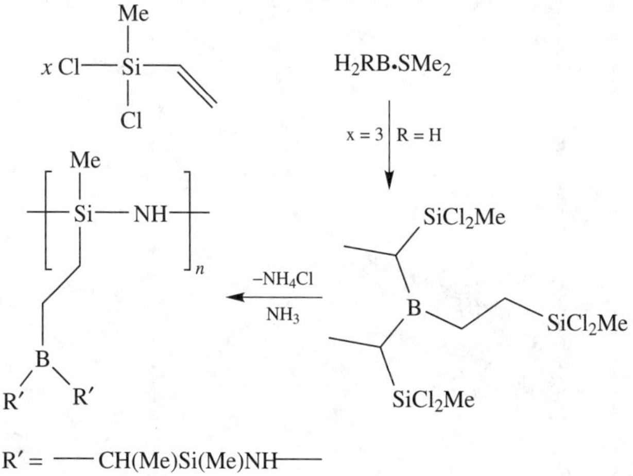

(24)

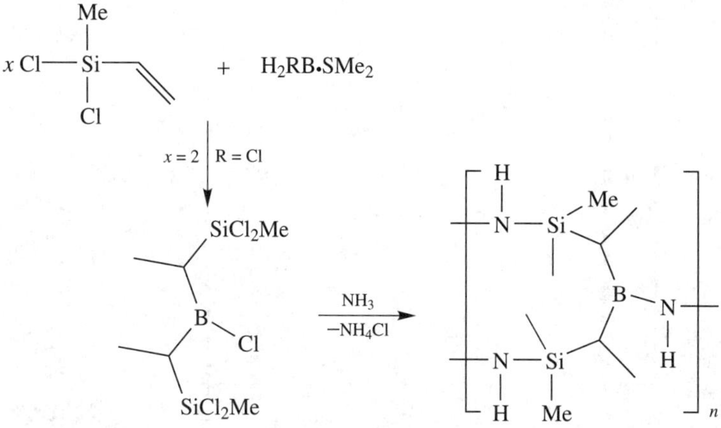

(25)

The stability of these two amorphous materials compared with Si_3N_4 and SiCN at high temperatures (see Figure 5) suggests that in fiber form they would offer superior properties. Preliminary studies suggest that Si−B−N−C fibers with diameters of 10−12 μm offer

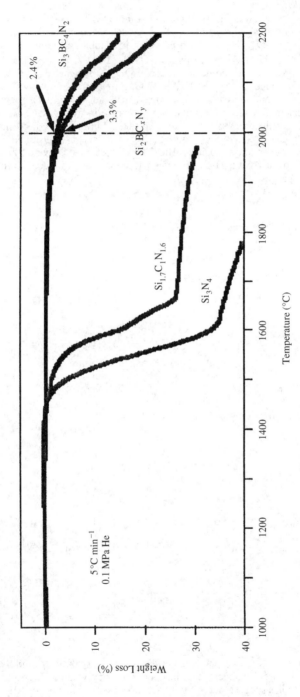

FIGURE 5. Comparison of the temperature stability (Ar) of various Si–N–C–B materials

tensile strengths of 3–4 GPa and elastic moduli of 200–210 GPa[6e]. Thus their mechanical properties are comparable with those of the known Tonen's silicon nitride fibers and Nicalon fibers, which degrade above 1400 °C[7,9].

Although Si−N−B−O materials containing trace amounts of nitrogen (2–3 wt%) have been prepared solely via a processing approach (ammonolysis of a borosilicate gel), complete conversion to Si−N−B was not achieved[86]. However, there are several literature examples of routes to Si−N−B materials that couple synthesis with a specific processing approach. These include early work on Si−N−B precursors by Seyferth and Plenio, who reacted the now ubiquitous −[MeHSiNH]$_x$− polysilazane with H$_3$B·SMe$_2$[87] as shown in equation 26.

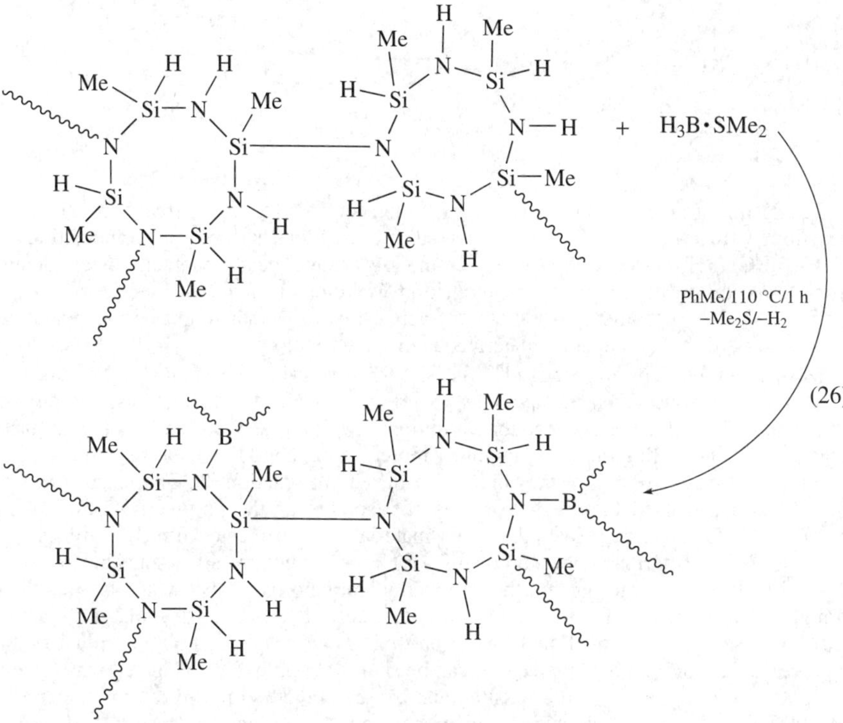

$$(26)$$

An oligomer made with a 4 : 1 ratio of silazane to borane had an M$_n$ ca 800 Da and a chemical composition of SiN$_{1.02}$B$_{0.24}$C$_{0.98}$H$_{3.65}$. This oligomer was soluble in common solvents and gave a 76 wt% ceramic yield at 1000 °C. Pyrolysis in NH$_3$ at 1000 °C gave a material with a composition of SiN$_{1.89}$B$_{0.31}$. By changing the ratio of the initial reactants, the Si:B:N ratio in the final ceramic product could be changed over a relatively wide range. Fibers could be hand-drawn from toluene solutions and pyrolyzed in ammonia to generate the corresponding ceramic fibers. No details of the phases formed at higher temperatures were presented, nor were efforts made to explore the direct conversion of these precursors to Si−N−B−C materials, which is unfortunate given the results of Riedel above, and Jansen and coworkers below.

Jansen and coworkers[88] described the synthesis of Si−N−B−C materials (equation 27). The ratio of the two reactants can be varied from 10 : 1 to 1 : 10 with a concomitant change in the composition of the ceramic product. For a typical reaction with a 1 : 1

ratio of reactants, the composition of the isolated polymer, following curing in NH_3/1 h/RT, was found to be $SiN_{4.37}BC_{3.68}H_{13.47}$. The TGA ceramic yield (1000 °C/He) was ca 65 wt%. Carbon can be removed by pyrolysis in NH_3 at 1000 °C/10–20 h. The resulting product is a pure Si–N–B material that does not crystallize on heating at 1500 °C for up to 72 h in N_2. The composition was close to that calculated for $Si_3B_3N_7$ or $Si_3N_4\cdot 3BN$. Ceramics produced in the Si–N–B system resist crystallization (1500 °C) even with B contents as low as 2 wt%.

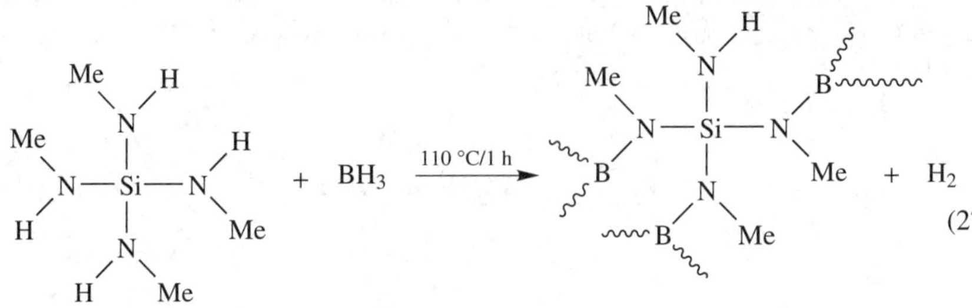

$$\text{(27)}$$

The same article describes the substitution of Al for B in reaction 27. For a reaction with a 1 : 2 Al:Si ratio, the resulting precursor's chemical composition was $SiAl_{0.59}BC_{3.03}H_{10.89}N_{2.85}$. Following ammonolysis and heat treatment under identical conditions, the resulting ceramic composition was close to that calculated for $Si_3Al_2N_6$ or $Si_3N_4\cdot 2AlN$. Although it was not considered, this precursor might be a candidate for processing AlN/SiC composite materials as discussed below.

Most recently, Jansen described[88] the simple synthesis of a spinnable precursor via the set of reactions given in equations 28–31. It appears that $(Me_3Si)_2NH$ reacts only once with $SiCl_4$ under mild reaction conditions to give the liquid chlorosilylsilazane of reaction 28. By heating this material under more forcing conditions, the second Me_3Si–N bond can be cleaved to form the chloroborosilazane in reaction 29. The product in reaction 30 is a proposed intermediate but is not isolated as the polymerization shown in reaction 31 proceeds coincidentally. This material can be reacted directly with NH_3 to form Si–N–B materials. However, the use of NH_3 does not provide a tractable precursor. Thus, $MeNH_2$ is used to prepare the polymethylaminoborosilazane, which is a distillable liquid and can be used for CVD studies. Alternately, by heating, deamination reactions akin to those shown in equation 10 are proposed to occur such that by controlled heating a processable oligomer is obtained that can be spun. If spinning is conducted with heating, the polymer is reported to self-crosslink and can be cured and pyrolyzed to temperatures of 2000 °C without decomposition and without crystallization. The resulting 10–20 μm diameter Si–N–B–C fibers are reported[88] to offer densities of ca 1.8 g ml^{-1} with tensile strengths of 4 GPa and elastic moduli of 400 GPa. Furthermore, the fibers are reported to be stable to 1500 °C in air for hundreds of hours. If these preliminary results prove correct, then these fibers represent a truly significant advance.

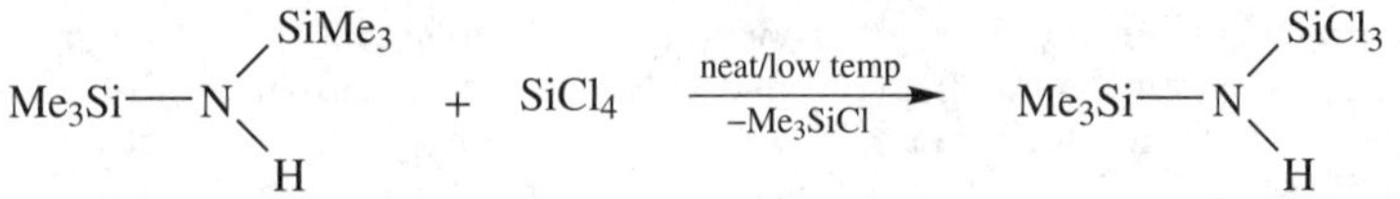

$$\text{(28)}$$

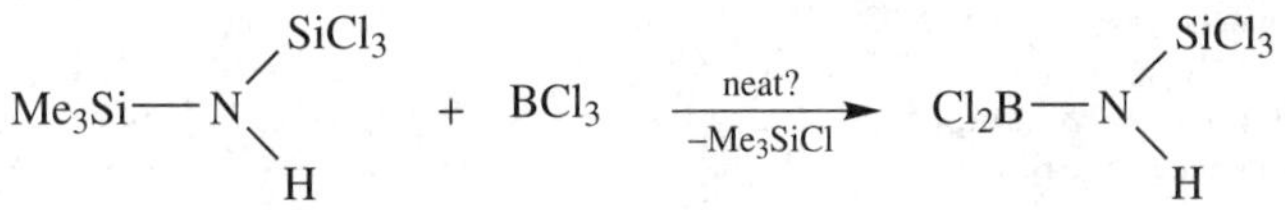

$$\text{(29)}$$

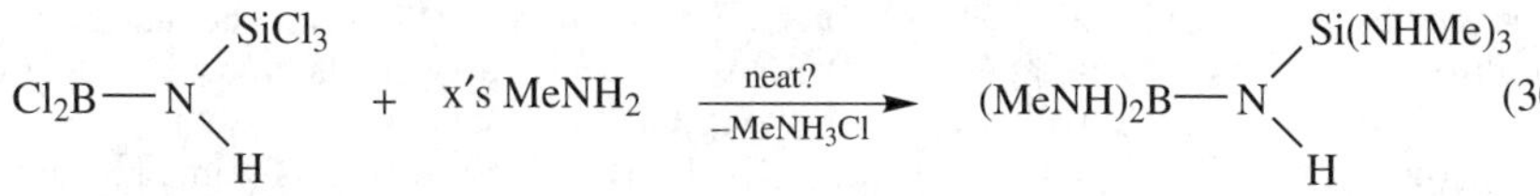

$$\text{Cl}_2\text{B}-\text{N}\begin{smallmatrix}\diagup\text{SiCl}_3\\ \diagdown\text{H}\end{smallmatrix} + \text{x's MeNH}_2 \xrightarrow[-\text{MeNH}_3\text{Cl}]{\text{neat?}} (\text{MeNH})_2\text{B}-\text{N}\begin{smallmatrix}\diagup\text{Si(NHMe)}_3\\ \diagdown\text{H}\end{smallmatrix} \quad (30)$$

$$(\text{MeNH})_2\text{B}-\text{N}\begin{smallmatrix}\diagup\text{Si(NHMe)}_3\\ \diagdown\text{H}\end{smallmatrix} \xrightarrow[-\text{MeNH}_2]{\text{heat}} \text{[B-N-Si-N network]} \quad (31)$$

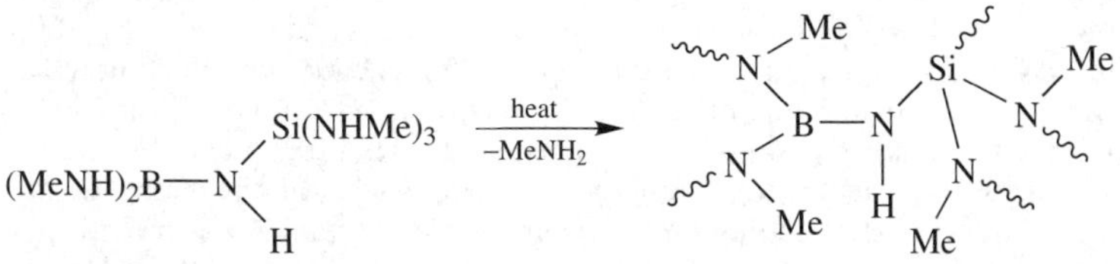

The Tonen group has also described a combined synthesis and processing approach to Si−N−B ceramics, but with considerable amounts of oxygen. Their work is based on the reaction of perhydridopolysilazane with trimethylborate [B(OMe)$_3$][89]. Thus, a pyridine solution of perhydrosilazane was mixed with 0.33 equivalents of B(OMe)$_3$ and heated at 120 °C in an autoclave for 3 h. Addition of dry o-xylene followed by distillative removal of solvent gave a white polymeric powder with M_n ca 2400 Da. A typical chemical analysis (SiB$_{0.36}$N$_{0.99}$O$_{0.42}$C$_{0.40}$H$_{4.86}$) shows the expected Si:B 3 : 1 ratio based on the reactant ratios. The given TGA ceramic yields were only for samples pyrolyzed in NH$_3$, which were $\leqslant$ 80 wt% at 1000 °C. A polymer sample heated in NH$_3$ (1200 °C/10 °C min^{-1}/10 min hold) provided an amorphous material with the following chemical analysis: SiB$_{0.38}$N$_{1.21}$O$_{0.38}$C$_{0.03}$H$_{0.68}$. Given that the oxygen content remains relatively constant from the polymer to the ceramic, and the observation by Riedel and coworkers that the weight losses observed in their Si−N−B−C ceramics occur because of oxygen impurities, it is likely that the Tonen ceramic product has a much lower use temperature. No information was provided about the processability of this material.

One other Si−N−B−X precursor type, where X = Ti, has been made by several groups[85,90,91]. The approach used has been to react Ti(NMe$_2$)$_4$ with a polysilazane as illustrated in equation 32. Note that the initial reaction with Ti(NMe$_2$)$_4$ can be exothermic and even violent if Si−H bonds are present. However, processable polymers are

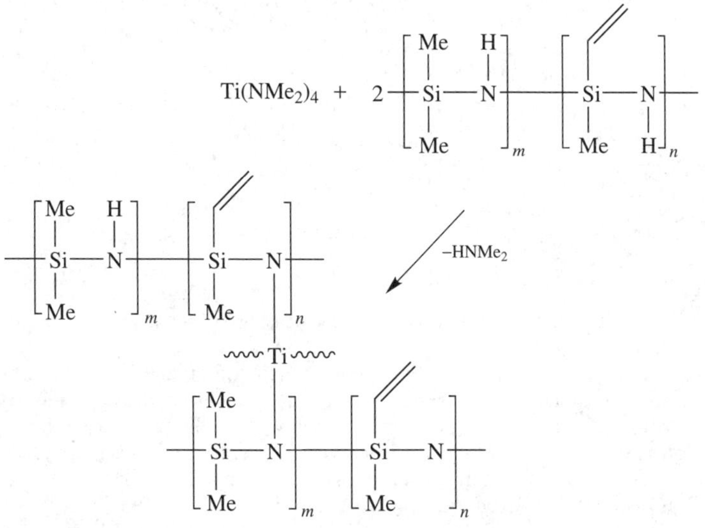

$$\text{Ti(NMe}_2)_4 + 2\left[\begin{smallmatrix}\text{Me} & \text{H}\\ | & |\\ \text{Si}-\text{N}\\ | \\ \text{Me}\end{smallmatrix}\right]_m \left[\begin{smallmatrix}\diagup\\ \text{Si}-\text{N}\\ | & |\\ \text{Me} & \text{H}\end{smallmatrix}\right]_n$$

$$\left[\begin{smallmatrix}\text{Me} & \text{H}\\ | & |\\ \text{Si}-\text{N}\\ |\\ \text{Me}\end{smallmatrix}\right]_m \left[\begin{smallmatrix}\diagup\\ \text{Si}-\text{N}\\ |\\ \text{Me}\end{smallmatrix}\right]_n \xrightarrow{-\text{HNMe}_2}$$

$$\sim\sim\sim\text{Ti}\sim\sim\sim$$

$$\left[\begin{smallmatrix}\text{Me}\\ |\\ \text{Si}-\text{N}\\ |\\ \text{Me}\end{smallmatrix}\right]_m \left[\begin{smallmatrix}\diagup\\ \text{Si}-\text{N}\\ |\\ \text{Me}\end{smallmatrix}\right]_n \quad (32)$$

only obtained on heating to ca 115 °C for several hours depending on whether the vinyl group or some other substituent (e.g. H) is used. Molecular weights were not reported. Only the precursors made with vinyl and hydrogen substituents gave reasonable ceramic yields (60–70 wt% at 800 °C) when the initial Si:Ti ratio was approximately 5 : 1. Thus, a composition for the 5 : 1 hydridopolysilazane/Ti(NMe$_2$)$_4$ reaction product was Si$_{4.55}$TiN$_{6.39}$C$_{11.48}$H$_{38.2}$. The composition of the ceramic product obtained at 1000 °C was Si$_{4.54}$TiN$_{5.84}$C$_{3.52}$H$_{0.21}$O$_{0.18}$. This material appears to be carbon rich and may be susceptible to oxidation, although the amounts of Ti incorporated can be adjusted by adjusting the initial ratios. Some work was done to demonstrate that these precursors could be used to infiltrate fiber preforms to make composite structures wherein the matrix was precursor derived. No information about high temperature phase formation was provided.

IV. PRECURSORS CONTAINING Si AND C

A. Precursors to SiC

Historically, one of the first routes to a processable SiC precursor was that reported by Yajima and coworkers[92–94] in 1975, wherein polydimethylsilane was processed (see Scheme 2) to produce SiC-containing ceramic fibers. This approach is still used to produce the only SiC precursor and SiC-containing fibers currently available commercially: Nicalon^{TM} fibers (Nippon Carbon, SiC$_{1.45}$O$_{0.36}$H$_{0.03}$)[7,92–94], Tyranno^{TM} fibers (Ube Industries, SiC$_{1.43}$O$_{0.46}$T$_{0.13}$)[95–97] and Mark I PCS (Shin-Etsu Co.) precursor polymer[6a]. As indicated by their respective compositional formulas, both Nicalon^{TM} and Tyranno^{TM} fibers are not phase pure SiC. Thus, their properties are inferior to those of phase pure SiC as shown in Table 1.

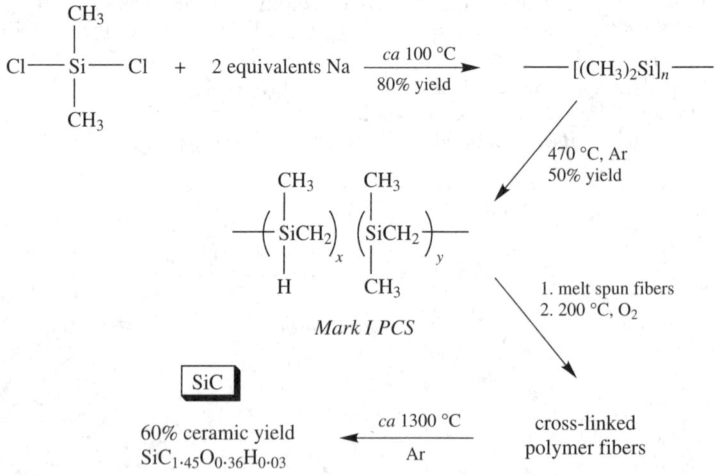

SCHEME 2. Original Yajima process for producing SiC$_{1.45}$O$_{0.36}$H$_{0.03}$ fibers

In addition to not offering properties expected for phase pure SiC, the original Yajima process suffers from other drawbacks that include a multistep precursor synthesis and the inability to self-cure. Finally, the presence of oxygen limits the upper use temperature for both Nicalon and Tyranno fibers to ca 1200 °C because above this temperature CO and SiO gases evolve, generating defects (large crystallites, pores and voids) that contribute to substantial decreases in mechanical properties.

TABLE 1. Comparison of properties of SiC-containing materials[7]

Type	Tensile strength (GPa)	Elastic moduli (GPa)
Nicalon^{TM}	2.0–2.5	<300
Tyranno^{TM}	3.0	>170
SiC whisker (single crystal)	8.0	580
Bulk SiC (hot pressed)	N/A	450

As a result of these disadvantages, tremendous efforts over the past 20 years have focused on developing chemistries/processes to improve or replace the Yajima process. At this point, as discussed below, several precursors can be synthesized and/or processed to produce phase pure SiC shapes with controlled microstructures that offer the exceptional properties expected of SiC. Thus, this area of precursor chemistry can now be considered to be mature. Consequently, only engineering and/or cost considerations now dictate which precursor systems are useful for a given application. Furthermore, because the general area of SiC precursors has been reviewed in detail[6,7], we focus here only on those precursors and processing methods that provide phase pure SiC.

To date, two precursor types have been identified that transform to nearly phase pure SiC. These are polymethylsilane (PMS, $-CH_3HSi-$) and polysilaethylene (equation 33), which are related by the fact that heating PMS >300 °C transforms it via the Kumada rearrangement[7] to polysilaethylene.

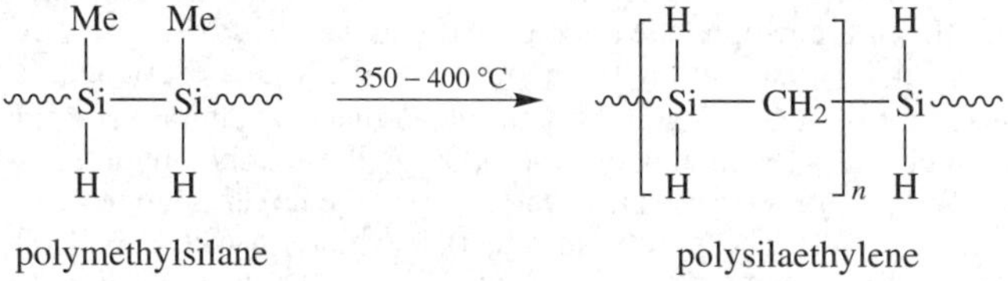

$$\begin{array}{c} \text{Me} \quad \text{Me} \\ | \quad\quad | \\ \sim\sim\text{Si}-\text{Si}\sim\sim \\ | \quad\quad | \\ \text{H} \quad\quad \text{H} \end{array} \xrightarrow{350-400\,°C} \begin{array}{c} \left[\begin{array}{c} \text{H} \\ | \\ \sim\sim\text{Si}-\text{CH}_2 \\ | \\ \text{H} \end{array}\right]_n \begin{array}{c} \text{H} \\ | \\ \text{Si}\sim\sim \\ | \\ \text{H} \end{array} \end{array} \quad (33)$$

polymethylsilane   polysilaethylene

Both PMS and polysilaethylene or polyperhydridocarbosilane have a 1 : 1 Si:C ratio and in principle are designed to generate phase pure SiC. In practice this is often not the case for a variety of reasons, as discussed below. Several related precursor systems and processing methods that provide essentially phase pure SiC are also discussed below to provide perspective.

Two general routes are used to synthesize PMS: Wurtz dehalocoupling of CH_3HSiCl_2 and transition-metal-catalyzed dehydrocoupling of methylsilane.

1. PMS via dehalocoupling[98,99]

Seyferth and coworkers[99] described dehalocoupling of CH_3HSiCl_2 with Na (Wurtz coupling), to synthesize PMS (equation 34). CH_3HSiCl_2 is added slowly to a mixture of Na sand in 7 : 1 hexane: THF with reflux under Ar for 20 h. The polymer product can be isolated in 60–70% yield as a viscous, hydrocarbon-soluble liquid that gives a negative Beilstein test for Cl.

^{1}H NMR characterization suggests a composition consistent with that shown, i.e. 20 mol% of the original Si—H bonds are consumed by Na and/or reactive silyl intermediates (silyl radicals) resulting in Si atoms bonded to three other Si atoms. ^{29}Si NMR data reveal the presence of $-SiH_2$ moieties in addition to the $-Si$ and $-SiH$ species suggested by ^{1}H NMR. The $-SiH_2$ groups are suggested to form by reaction of CH_3HSiCl_2

or $-CH_3HSiCl$ end groups with NaH (formed from Na and SiH) or by silyl radical processes.

$$n\ CH_3HSiCl_2 + 2.5\ \text{equivalents Na} \xrightarrow[\text{hexane/THF}]{\text{reflux, 20 h}} \left(\left[\begin{array}{c} CH_3 \\ | \\ -Si- \\ | \\ H \end{array}\right]_{0.7} \left[\begin{array}{c} CH_3 \\ | \\ -Si- \\ | \\ Si \\ / \backslash \\ CH_3 \end{array}\right]_{0.3}\right)_n + NaCl \quad (34)$$

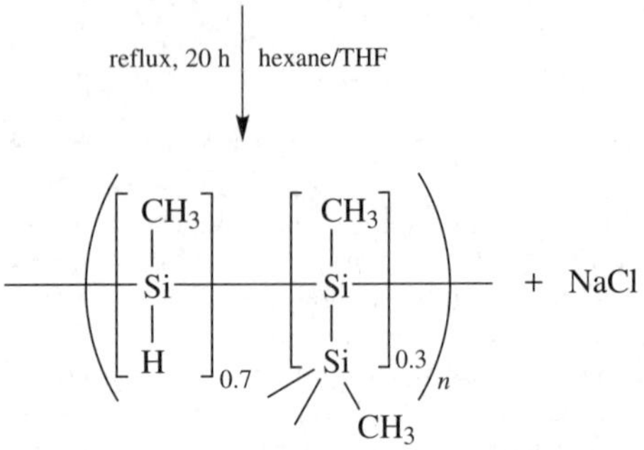

A molecular weight of 660 Da was determined cryoscopically (benzene); however, polydispersities were not reported. Pyrolysis of PMS ($10\,°C\,min^{-1}/950\,°C$) provides ceramic yields of 12–27%, with a $Si_{1.42}C_{1.00}H_{0.14}$ composition. Low MW oligomers present in PMS volatilize prior to pyrolysis, contributing to the low ceramic yields. The high Si contents are thought to arise from high temperature reactions ($>350\,°C$) of Si–H with SiCH$_3$ to form CH$_4$ (identified by TGA/FTIR studies) and Si–Si bonds.

To increase ceramic yields and carbon content during transformation of PMS to SiC, Seyferth and coworkers explored hydrosilylative crosslinking of PMS with $[CH_3(CH_2=CH)SiNH]_3$ using catalytic amounts of AIBN. For example, reactions using SiH:SiCH=CH$_2$ ratios $\geqslant 6$ in refluxing benzene provide quantitative yields of soluble precursor with a 68–77 wt% ceramic yield ($1000\,°C$). Presumably, hydrosilylation ties the volatile, low MW oligomers (MW values not given) to the larger oligomeric chains leading to higher ceramic yields. The pyrolyzed material exhibits a $(SiC)_{1.00}(Si_3N_4)_{0.033}C_{0.040}$ composition at $1000\,°C$, but at $1500\,°C$ only SiC was observed (XRD) to crystallize. Given the above work on SiCN, we presume that the $1000\,°C$ material is actually an SiCN composite (see work of Toreki and coworkers below).

These workers also explored transition-metal-promoted crosslinking of PMS as an approach to increase ceramic yields and SiC purity. Because transition metals, e.g. $Ru_3(CO)_{12}$, are known to catalyze redistribution reactions between Si–H and Si–Si bonds[100], efforts were made to modify PMS via a chain-extension process to generate higher M_n values. Thus, a set of metal carbonyls were used (1–2 mol%) to effect redistribution reactions that might lead to improved ceramic yields[99c]. For example, 1–2 wt% $Ru_3(CO)_{12}$ added to PMS followed by irradiation for 4 h (140 watts at ca 300 nm) provided a polymer with a 55% ceramic yield of a Si-rich material (amount of excess Si not reported). To increase carbon content Mark I [$(-MeSiHCH_2-)_n$] was combined with PMS in a 1 : 2 wt ratio and subjected to the $Ru_3(CO)_{12}$ catalyzed crosslinking. NMR analysis of the resultant polymer showed depletion of Si–H bonds, suggesting Si–Si bond redistribution reactions at the expense of the Si–H bonds. Molecular weights and polydispersities for the crosslinked PMS were not given. Pyrolysis of this polymer gave a 68% ceramic yield of high purity SiC, $Si_{0.99}C_{1.00}$. Although the polymer was reported to be moderately soluble in organic solvents, its processability was not discussed (Scheme 3).

Based on the extensive work of Harrod and coworkers[101], Seyferth and coworkers also examined metallocene (i.e. Cp_2ZrH_2) catalyzed dehydrogenative crosslinking of PMS as another method of increasing ceramic yields and carbon content during transformation

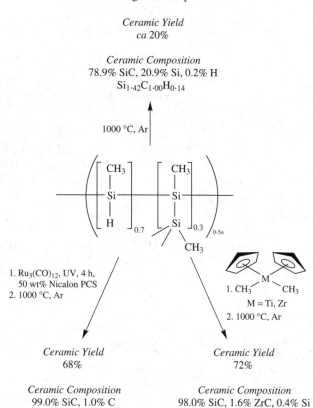

SCHEME 3. Modifying PMS to increase ceramic yields and SiC purity

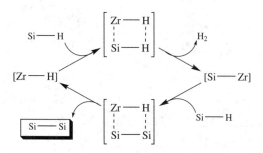

SCHEME 4. Dehydrocoupling mechanism to produce Si—Si bonds[102]

of PMS to SiC[99] (Scheme 4). The addition of 0.6 mol% Cp_2ZrH_2 to a hexane solution of PMS followed by reflux for 2 h resulted in loss of Si—H bonds, as determined by NMR, especially the SiH_2 groups as expected. Pyrolysis to 1500 °C gave ceramic yields of 70–80% with SiC purity as high as 98% (ZrC and Si were the primary side products in 1.6 and 0.4 wt%, respectively, $Si_{0.99}C_{1.00}Zr_{0.02}$). Unfortunately, fibers drawn from this

PMS required UV curing for 1.5 h prior to pyrolysis in order to retain the fiber shape. Additionally, the use of metallocene catalysts renders PMS highly oxygen-sensitive and pyrophoric.

In conclusion, Seyferth's work showed that PMS can be transformed from a poor (ceramic yields ca 20%, >20% excess Si) to an effective precursor (ceramic yields 60–80%, ca 1% excess Si or C) using relatively simple chemistry. The routes chosen provide the following advantages over the Yajima process: (1) a Wurtz coupling synthesis that provides a polymer with an inherent 1 : 1 Si:C ratio, (2) high ceramic yields, (3) some processability as fibers can be drawn and (4) essentially phase pure SiC following chemical modification. Drawbacks to this work include: (1) precursors that are not self-curing, (2) the use of pyrophoric catalysts (Cp_2ZrH_2) leading to pyrophoric PMS or (3) the use Ru catalysts, which may be too expensive for scale-up processes.

2. PMS by catalytic dehydrocoupling of MeSiH₃

PMS can also be synthesized by catalytic dehydrocoupling of methylsilane ($MeSiH_3$) using Cp_2MMe_2 (M = Ti, Zr) catalysts, as described by Harrod and coworkers, and Laine and coworkers[101,103,104] (equation 35). The reaction is run in a toluene/cyclohexene solvent system at 25–60 °C under ca 10 atm $MeSiH_3$ for 1–9 days using 0.2 mol% catalyst. The H_2 byproduct is consumed simultaneously by metal-catalyzed hydrogenation of cyclohexene to cyclohexane, to minimize pressure build-up and depolymerization. PMS can be obtained in >90% yield. The ^1H NMR spectrum suggests a partially branched structure based on the integration of the $\underline{CH}_3Si$:SiH ratio of 4 : 1. SEC indicates M_n of ca 1200–1300 Da (DP ca 30) with a PDI of ca 5–10. The M_n and PDI values were quite high relative to analogous dehydropolymerization of phenylsilane, which converts to both linear (DP ca 10) and cyclic structures (P = 5, 6), with no evidence of branching[105]. The higher M_n and PDI values for methylsilane polymerization were attributed to a lack of steric hindrance (methyl vs phenyl), resulting in a greater degree of branching, due to reaction with all three Si–H sites. Pyrolysis (1000 °C/1 h/10 °C min^{-1}/Ar) provides a ca 77 wt% ceramic yield of a material with a composition of $Si_{1.0}C_{0.9}H_{<0.2}O_{0.1}$ (6.0, 0.5 and 4.0 wt% excess Si, H and O respectively)[103,104]. The oxygen appears to arise from handling.

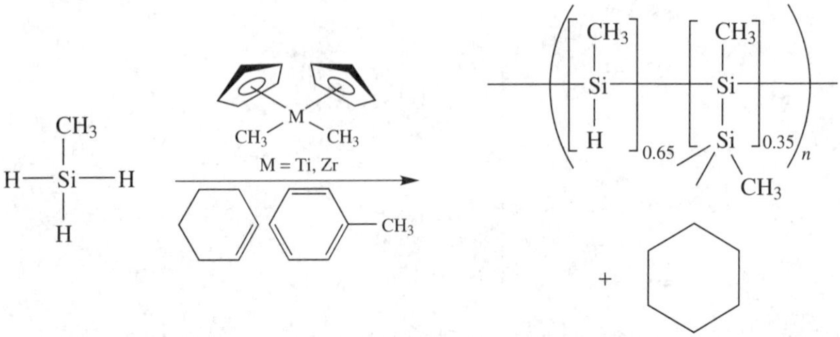

$$\underset{H}{\overset{CH_3}{\underset{|}{H-Si-H}}} \xrightarrow{\underset{\text{M = Ti, Zr}}{Cp_2MMe_2}} \left(\begin{array}{c}\begin{bmatrix}CH_3\\|\\Si\\|\\H\end{bmatrix}_{0.65} \begin{bmatrix}CH_3\\|\\Si\\|\\Si\\ \backslash\\CH_3\end{bmatrix}_{0.35}\end{array}\right)_n + \bigcirc \quad (35)$$

Figure 6 provides a series of ^{29}Si MAS NMRs taken of PMS samples heated to selected temperatures. The Kumada rearrangement occurs at 400 °C and essentially phase pure SiC (2–4 nm crystallites by TEM) crystallizes at 1000 °C. Holding the polymer for periods of 5–10 h at 800 or 900 °C also results in crystallization. Additional MAS^{29}Si NMR studies on samples pyrolyzed to 1400 °C showed no changes in peak position (from −15.7 ppm), suggesting the formation of well-defined β-SiC crystalline phases at 1000 °C.

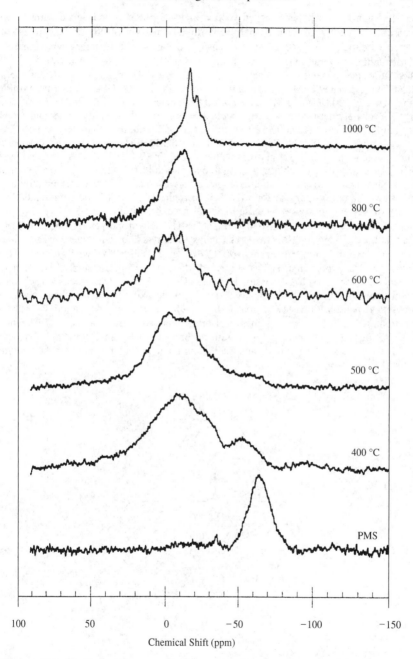

FIGURE 6. ^{29}Si MAS NMR spectra recording the pyrolytic transformation of polymethylsilane, $-[\text{MeSiH}]_n-$, heated to selected temperatures for 1 h. Note the Kumada rearrangement to polycarbosilane, $-[\text{H}_2\text{SiCH}_2]_x-$, that occurs at 400 °C. Crystalline β-SiC forms at 1000 °C

Fibers spun from a hyperbranched version of PMS (mPMS) and doped with 0.2 wt% boron are self-curing and can be pyrolyzed directly at rates of 20 °C min^{-1} to 1800 °C to give essentially phase pure, fully dense (>3.1 g ml^{-1} SiC fibers with controlled microstructures (crystallite sizes range from 2 nm at 1200 °C to 500 nm at 1800 °C)[104,106]. When subjected to bend stress analyses, these fibers exhibit bend strengths that average 3 GPa. The estimated elastic modulus for these fibers is >400 GPa. Furthermore, mPMS with M_n of ca 2000 Da and MW of ca 7000 Da can also be used (1) for joining ceramic parts, (2) for polymer infiltration and pyrolysis processing of ceramic powder and ceramic fiber reinforced matrices and (3) for coating other inorganic materials, e.g. carbon fibers.

The disadvantages of the PMS precursor are: (1) MeSiH$_3$ is costly and pyrophoric, forming potentially explosive mixtures with ambient air, (2) the Ti/Zr dehydropolymerization catalysts are also pyrophoric, leading to pyrophoric PMS and (3) metallocene-derived PMS is also highly pyrophoric. For example, a 20 wt% increase was observed after exposure of PMS samples to dry air for 5 h at RT. In contrast, the mPMS derivative mentioned above exhibits only a 3 wt% increase in air over a 5 h period.

Tanaka and coworkers[107] also describe MeSiH$_3$ dehydropolymerization to PMS in 68% yield, using Cp$_2$NdCH(SiMe$_3$)$_2$ as catalyst. The reaction, run under pressure (glass-lined autoclave, 8 kg cm^{-2}/benzene/>90 °C/2 d), gives solid PMS with H$_2$ and CH$_4$ as byproducts. The evolution of H$_2$ was expected whereas the appearance of CH$_4$ was not. A mechanism for Si–CH$_3$ bond scission leading to CH$_4$ formation was proposed (Scheme 5) wherein a methylneodymium complex forms. Both the [Nd]CH$_3$ and silylene species are proposed to react with Si–H to give CH$_4$, Nd–Si (active polymerization species) and branch sites within the growing chain. ^{1}H NMR analysis showed a SiCH$_3$:SiH ratio of ca 3 : 1, suggesting a linear polymer with no branching. Because β-methyl abstraction should result in loss of both Si–H and SiCH$_3$ simultaneously, a heavily branched polymer may still form even though the SiCH$_3$:SiH ratio remains 3 : 1.

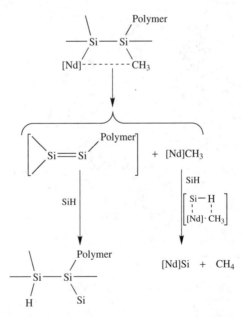

SCHEME 5. β-Methyl elimination during dehydropolymerization of MeSiH$_3$ using Cp$_2$NdCH(SiMe$_3$)$_2$

SEC analysis gave $M_n = 1470$ Da with a PDI = 5.00, very similar to the Ti and Zr values reported by Harrod. Pyrolysis to 900 °C in Ar provided ceramic yields of ca 74% with excess Si metal (amount not reported). To balance the excess Si, polyphenylsilane (MW = 1600 Da, 22 wt%) was blended with PMS to give phase pure β-SiC in 58% ceramic yield.

The disadvantages of this process, in addition to those described above for MeSiH$_3$, include (1) the availability of Nd catalysts for large-scale reactions and (2) the low (58%) ceramic yields.

3. PMS by dehydrocoupling of $CH_3SiH_2SiH_2CH_3$

Hengge and coworkers[108] studied Cp$_2$MMe$_2$ (M = Ti, Zr) catalyzed dehydrocoupling of CH$_3$SiH$_2$SiH$_2$CH$_3$. They report that Cp$_2$ZrMe$_2$ catalytically dehydropolymerizes neat, liquid CH$_3$SiH$_2$SiH$_2$CH$_3$ at room temperature in minutes to give an insoluble material with a general composition of H$-$[MeSiH)$_{0.58}$(MeSi)$_{0.42}$]$_n$$-$H (equation 36).

$$\text{CH}_3\text{-SiH}_2\text{-SiH}_2\text{-CH}_3 \xrightarrow{\text{Cp}_2\text{MMe}_2, \, M = \text{Ti, Zr}} \left(\left[\begin{array}{c} \text{CH}_3 \\ | \\ \text{Si} \\ | \\ \text{H} \end{array} \right]_{0.58} \left[\begin{array}{c} \text{CH}_3 \\ | \\ \text{Si} \\ | \\ \text{Si-CH}_3 \end{array} \right]_{0.42} \right)_n \quad (36)$$

During reaction, the liquid initially turns yellow, then orange as H$_2$ evolves vigorously and then gels are formed. The resultant crosslinked polymer is insoluble in common solvents, decomposes before melting and is pyrophoric. GC analysis of similar disilane systems reveals odd-numbered oligosilanes, suggesting Si$-$Si bond cleavage. A mechanism, termed β^*-bond elimination (Scheme 6), was proposed for this process.

$$\text{'Cp}_2\text{M'} + (\text{CH}_3)_3\text{SiSi}(\text{CH}_3)_2\text{H} \longrightarrow \text{Cp}_2\text{M}\begin{array}{c}\text{H} \\ \diagdown \\ \text{Si}(\text{CH}_3)_2\text{Si}(\text{CH}_3)_3\end{array}$$

$$\uparrow \, -:\text{Si}(\text{CH}_3)_2 \qquad\qquad \downarrow \beta\text{-elimination}$$

$$\text{Cp}_2\text{M}=\text{Si}(\text{CH}_3)_2 + \text{HSi}(\text{CH}_3)_3 \longleftarrow \begin{array}{c} \text{H}-\text{Si}(\text{CH}_3)_3 \\ | \qquad | \\ \text{Cp}_2\text{M}-\text{Si}(\text{CH}_3)_2 \end{array}$$

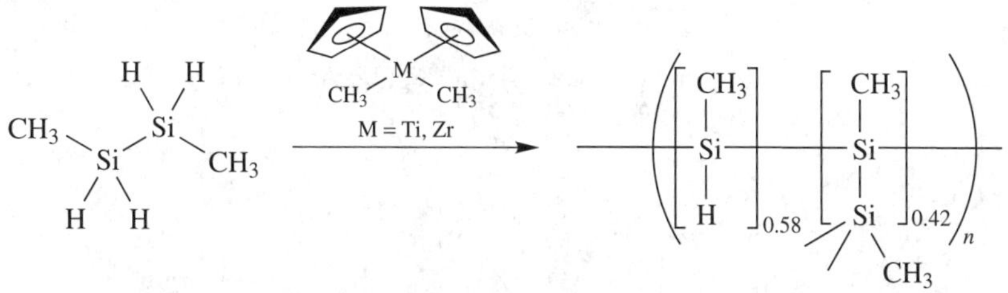

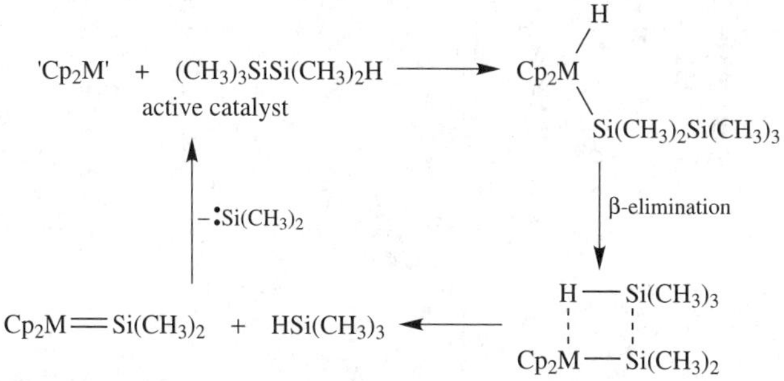

SCHEME 6. β^*-Bond elimination mechanism for the coupling of disilanes[109]

The authors suggest that polymeric products are derived from simultaneous dehydrocoupling and β^*-bond elimination mechanisms[110]. Presumably, the silylene species lost in the final step in the Scheme 6 cycle inserts into either Si$-$H or Si$-$Si bonds, creating odd-numbered oligosilanes. NMR studies of PMS reveal depletion of the Si$-$H bonds, suggesting crosslinking through Si$-$H sites, as found by Harrod and coworkers[101]. Pyrolysis provides ca 88% ceramic yields (1500 °C/Ar) of SiC.

The Hengge process appears attractive because the starting dimer (1) offers a 1 : 1 Si:C ratio, (2) is a liquid, rather than gaseous CH_3SiH, (3) polymerizes rapidly and because (4) the ceramic product appears to be phase pure SiC. The drawbacks are: (1) dimer requires $LiAlH_4$ reduction of $CH_3SiCl_2SiCl_2CH_3$ (from the direct process, Hengge obtained $CH_3SiH_2SiH_2CH_3$ from Wacker Chemie), (2) polymers without rheological utility and (3) pyrophoric polymers. It is likely that if a solvent were used, a soluble, processable precursor might be obtained. Unfortunately, it is very difficult to obtain $CH_3SiCl_2SiCl_2CH_3$ free from $(CH_3)_2SiClSiCl_2CH_3$ as the two compounds are difficult to separate.

$CH_3SiH_2SiH_2CH_3/(CH_3)_2SiHSiH_2CH_3$ mixtures can be copolymerized using Cp_2ZrMe_2 as shown above for $MeSiH_3$[111]. However, the resultant polymer gives low ceramic yields (50–60 wt% at 1000 °C) and is difficult to process. The Me_2SiH moiety appears to lead to early chain termination, resulting in low molecular weight products that volatilize on pyrolysis. Studies in this area are thwarted by the lack of a reliable source of pure $CH_3SiCl_2SiCl_2CH_3$ or its perhydrido analog.

4. PMS from the Bu_4PCl-catalyzed redistribution of chlorosilanes

Researchers at Dow Corning report synthesizing highly branched PMS by catalytic redistribution (1 wt% Bu_4PCl) of chlorodisilanes[112] (Scheme 7).

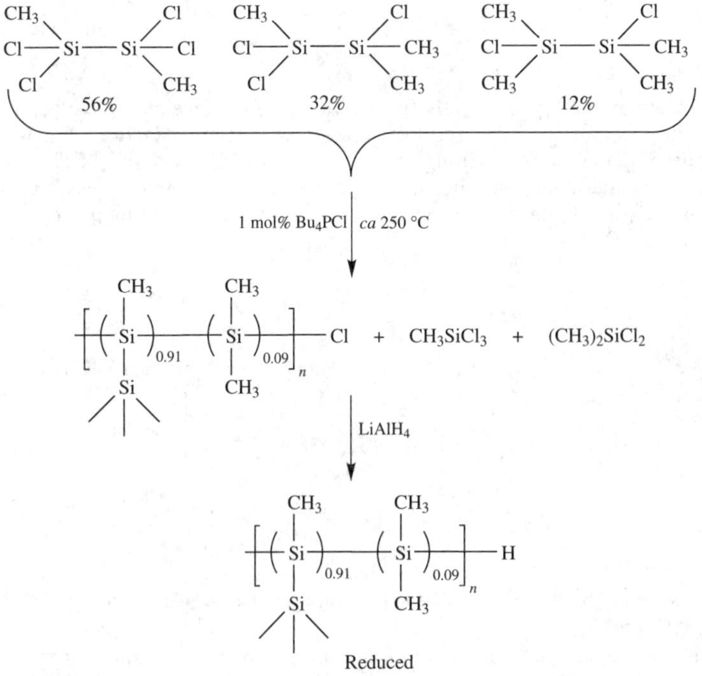

SCHEME 7. Catalytic redistribution of chlorodisilanes to give highly branched PMS[112]

This process has advantages in that the reaction is homogeneous and avoids the heterogeneous Na Wurtz coupling procedure, commonly used to polymerize chlorosilanes.

Furthermore, the disilanes used are byproducts of the 'direct process' used industrially to make methylchlorosilanes. The disilane monomer composition used was: 56.0% $CH_3SiCl_2SiCl_2CH_3$, 31.9% $(CH_3)_2SiClSiCl_2CH_3$, 12.5% $(CH_3)_2SiClSiCl(CH_3)_2$. Bu_4PCl (1 wt%) added to the mixture catalyzes redistribution of the disilanes. If the reaction mixture is heated slowly to 250 °C, a pyrophoric yellow glassy solid that is toluene-soluble forms in 15–20% yields. The remaining 80–85% of distillate consists of CH_3SiCl_3, $(CH_3)_2SiCl_2$ and $[(CH_3)_4Cl_2Si]_2$ (does not undergo redistribution), and small amounts of Si–H-containing species formed via a Kumada rearrangement (Scheme 8).

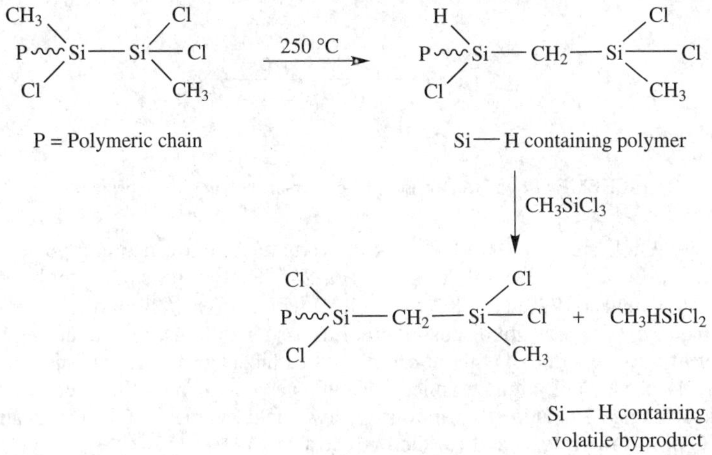

SCHEME 8. Kumada rearrangement of chlorosilanes[113]

LiAlH$_4$ reduction of the redistribution product gives a polymer with a composition of $-(MeSi)_{0.91}(Me_2Si)_{0.09}(H)_{0.4}-$. No NMR characterization was provided, although a M_n of ca 1130 Da was found by ebulliometry. Pyrolysis to 1560 °C gave 90% ceramic yields of 'essentially pure SiC', although no quantitative analyses were given. However, given the above polymer composition, the theoretical yield to SiC would be 88.4%, suggesting the presence of some excess carbon.

The advantages offered by this process are (1) a one-step synthesis to poly(methylchlorosilanes), (2) low cost dimers and (3) fiber processing that leads to phase pure SiC fibers. The primary drawbacks to the Dow-Corning process are (1) a low polymer yield (15–20%), (2) a multistep process and (3) the high cost of LiAlH$_4$ reduction of the chlorinated polymer.

5. Polycarbosilanes as precursors to SiC

Pillot and coworkers[113] reported the Wurtz polymerization of 2,4-dichloro-2,4-disilapentane (DCDP, **3**, to give a precursor to SiC (Scheme 9). DCDP was synthesized via Mg/Zn coupling of dichloromethane **1** with excess methyldichlorosilane **2** in ca 35% yield. DCDP was polymerized via Wurtz coupling, followed by LiAlH$_4$ reduction of residual Si–Cl, to poly(disilapentane), **4**, in 61% yield with $M_n = 1400$ Da and PDI = 3.1. No polymer yields were given for **4**.

^{1}H NMR spectra of **4** show depletion of Si–H bonds, as determined from an integration ratio (C$\underline{H}_3$+C$\underline{H}_2$)/Si–$\underline{H}$ of 5.5, higher than the expected value of 4.0. ^{29}Si NMR revealed three signals at −40, −36 and −6 assigned to $\underline{Si}Si_2C_2$, $-CH_2\underline{Si}(CH_3)H-Si$ and $\underline{Si}HC_3$

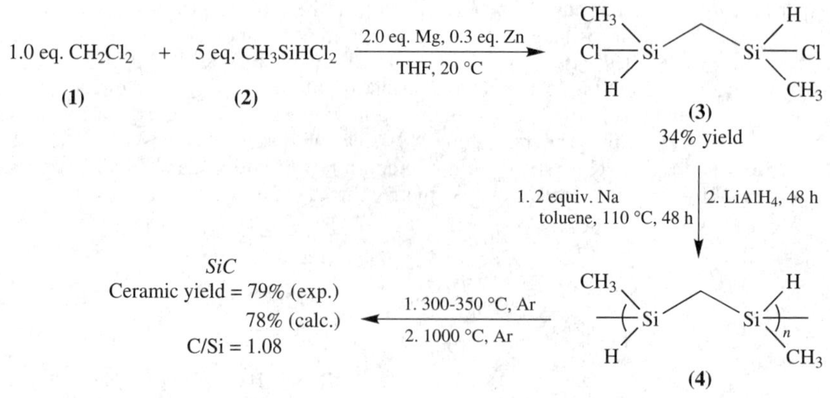

1.0 eq. CH$_2$Cl$_2$ + 5 eq. CH$_3$SiHCl$_2$ $\xrightarrow[\text{THF, 20 °C}]{\text{2.0 eq. Mg, 0.3 eq. Zn}}$ Cl—Si(CH$_3$)(H)—CH$_2$—Si(H)(CH$_3$)—Cl

(1) (2) (3) 34% yield

(3) → 1. 2 equiv. Na, toluene, 110 °C, 48 h | 2. LiAlH$_4$, 48 h → (4) [—Si(CH$_3$)(H)—CH$_2$—Si(H)(CH$_3$)—]$_n$

(4) $\xrightarrow[\text{2. 1000 °C, Ar}]{\text{1. 300-350 °C, Ar}}$ SiC
Ceramic yield = 79% (exp.)
78% (calc.)
C/Si = 1.08

SCHEME 9. Synthesis of the SiC precursor, poly(2,4-disilapentane)

respectively. On heating to 300–350 °C, **4** converts to the corresponding polycarbosilane (MW = 7650 Da, PDI = 3.0, softening point *ca* 245 °C). Pyrolysis of this polycarbosilane gave ceramic yields of 79% (vs 78% theoretical) with a 1.08 C∶Si ratio and 1.1 mol% O. The source of oxygen was not discussed, but it is assumed to incorporate during handling. Direct pyrolysis of polymer **4** without prior heat treatments gave <10 wt% ceramic yields.

Relatively phase pure, high ceramic yield SiC was obtained by this method; however, some disadvantages include (a) a multistep, low yield synthesis, (b) heat treatments to *ca* 350 °C prior to pyrolysis and (c) the high cost of LiAlH$_4$.

A related poly(disilapentane) precursor was developed by Corriu and coworkers[114] via metallocene catalyzed dehydropolymerization of 1,4-disilapentane (DSP) as depicted in Scheme 10. DSP was synthesized in 85% yield in two steps from the Pt catalyzed hydrosilylation of trichloro(vinyl)silane with MeSiHCl$_2$ followed by LiAlH$_4$ reduction. Reacting DSP with 0.5 mol% Cp$_2$TiMe$_2$ at room temperature for 48 h gave a polymer with M_n = 990 Da and PDI = 1.05. ^{1}H, ^{13}C and ^{29}Si NMR analyses indicate formation of a linear polymer with no reaction at the resulting SiH$_2$ sites. Reactions run for 72 h showed partial depletion of the SiH$_2$ groups giving branched, highly viscous liquids with M_n = 3260 Da and PDI = 10.1. At 50 °C, crosslinking occurred rapidly to give an insoluble material within 30 min.

Both low (M_n = 990 Da) and high (M_n = 3260 Da) molecular weight products, as described above, gave high ceramic yields (>73 wt% yield vs 78 wt% theoretical) of nearly phase-pure SiC (Si$_{1.01}$C$_{1.00}$) after pyrolysis. Conversion to crystalline β-SiC commenced at 1100 °C and was complete by 1400 °C, as determined by XRD. Ti was necessary during the pyrolytic transformation to SiC, as samples of poly(2,4-disilapentane) pyrolyzed without Ti gave 30% ceramic yields. Ti appears to catalyze crosslinking of volatile lower molecular weight species, which would otherwise volatilize on heating, into a more thermally stable crosslinked network, accounting for the high ceramic yields.

This method provides an attractive homogeneous route to SiC. The monomer can be synthesized in high yield from inexpensive starting materials and subsequent polymerizations were easily tuned by the reaction conditions. Thus, depending on the application, the resultant viscosity may easily be adjusted (i.e. fiber spinning). Drawbacks include (a) the use of LiAlH$_4$ (high cost and multistep) in the monomer synthesis, (b) use of pyrophoric Ti catalysts leading to pyrophoric poly(2,4-disilapentane) and (c) 2 : 3 Si∶C ratio in the precursor polymer, contributing to lower ceramic yields.

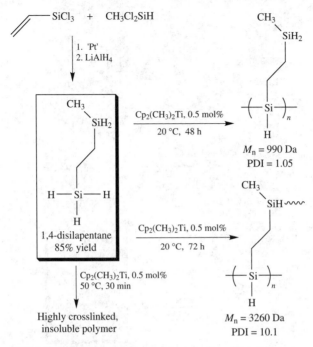

SCHEME 10. Ti-catalyzed dehydropolymerization of 2,4-disilapentane

6. Polysilaethylene (PSE) precursors

A brief patent by Smith[115] reports the Pt-catalyzed ring-opening polymerization of 1,3-disilacyclobutane in heptane or toluene to produce a linear PSE structure which, when pyrolyzed to 900 °C, gave SiC in 85% ceramic yield. The patent claims that the PSE can be processed and then pyrolyzed to give SiC fibers, films and molded shapes. However, neither the syntheses of 1,3-disilacyclobutane (from 1,1,3,3-tetrachloro-1,3-disilacyclobutane, Scheme 11) or PSE, nor ceramic product characterization (beyond brief IR claims) are provided.

Interrante and coworkers developed two polysilaethylene precursor systems, which upon pyrolysis gave nearly phase-pure SiC (95–99% SiC) in high ceramic yield (75–90%) (Scheme 11)[116]. The first procedure begins with a multistep synthesis of 1,1,3,3-tetrachloro-1,3-disilacyclobutane (monomer), **5**. Alternately, gas phase pyrolysis will convert **6** to **5**.

Pt-catalyzed ring-opening polymerization of monomer **5** gives poly(1,1-dichloro-1-silaethylene), **7**, which is reduced *in situ* with LiAlH$_4$. PSE, **8** is nearly linear with trace quantities of −SiHMe− branches (ca 1 per 200 chain Si atoms by NMR), which result from residual Cl$_2$Si(CH$_3$)$_2$ in the Cl$_3$SiMe starting material[116]. PSE molecular weights from SEC and NMR end group analysis gave M_n = 24 kDa and 11 kDa, respectively. The authors prefer NMR end group analysis rather than SEC (hydrodynamic volume relative to polystyrene) to estimate molecular weights.

As in traditional chain polymerization, the molecular weights are controlled by the catalyst concentration. Thus the more catalyst used, the lower the molecular weights. SEC indicates quite high MW (>130 kDa), that likely contributes to good rheological properties. PSE heated in the TGA under Ar to 1000 °C, undergoes smooth mass loss

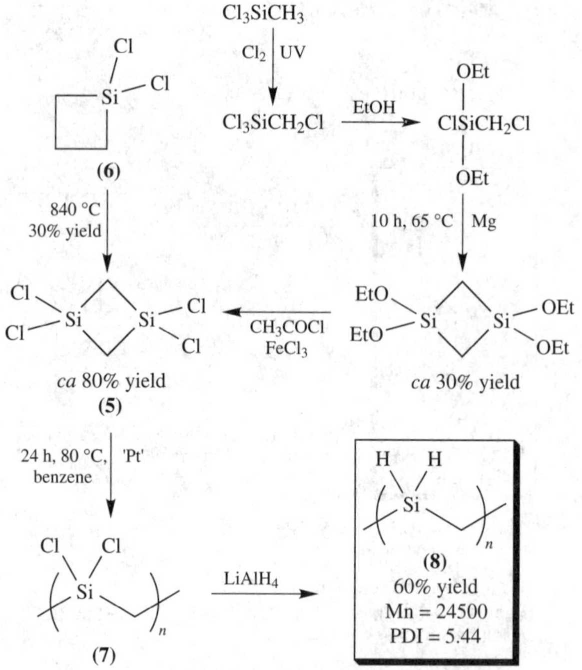

SCHEME 11. Synthesis of poly(silaethylene)[116]

between 100 and 600 °C, providing excellent ceramic yields of 87% (90.9% theoretical). The high ceramic yields are thought to result from the formation of silylene intermediates (=Si:) during pyrolysis, which crosslink the polymer prior to volatilization (self-curing). Samples heated to 1000 °C crystallized to β-SiC (by XRD) with average crystallite sizes of 2.5 nm.

Although their results were encouraging, the multistep, low yield monomer synthesis makes this route unlikely for the large-scale production of SiC. As a result, the Interrante group changed their synthetic tack. They recently reported the synthesis of a highly branched poly(silaethylene) **8** derived from Grignard-based polymerization of ClCH$_2$SiCl$_3$, followed by LiAlH$_4$ reduction[116] (Scheme 12).

The reaction, performed in refluxing diethyl ether (4 d), provides chlorinated polycarbosilane **7** in 50–60% yields with a typical composition of ca SiCl$_{1.70}$(OEt)$_{0.15}$Et$_{0.15}$CH$_2$, suggesting side reactions with the solvent. This highly branched intermediate was generally not isolated but was reduced in situ with LiAlH$_4$ to give SiH$_{1.85}$Et$_{0.15}$CH$_2$. The traces of the ethyl substituents appear to have minimal effect on the ceramic composition after pyrolysis (see below).

The product is an ether/hydrocarbon soluble, air stable (by TGA) yellow liquid at RT. Additionally, no changes in the NMR or IR spectra occur in samples exposed to air for days. IR and NMR data are consistent with a highly branched structure, as anticipated for tetra-functional ClCH$_2$SiCl$_3$. For example, ^{29}Si NMR indicates the presence of (CH$_2$)$_4$Si (1–7 ppm), (CH$_2$)$_3$SiH (−5 to −19 ppm), (CH$_2$)$_2$SiH$_2$ (−25 to −40 ppm) and CH$_2$SiH$_3$ (−50 to −75 ppm) moieties. SEC indicates M_n = 745 Da with a PDI of ca 7, also consistent with a highly branched structure.

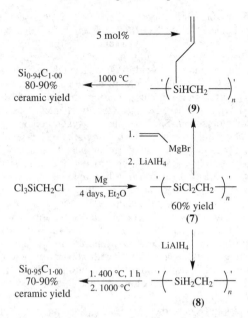

SCHEME 12. The synthesis of highly branched poly(silaethylene)

Pyrolysis of **8** in Ar to 1000 °C gave ceramic yields of only *ca* 50% vs 90.9% theoretical. The low ceramic yield results from vaporization of low MW fractions between 100 and 250 °C. However, ceramic yields of *ca* 80 wt% of nearly stoichiometric SiC (0.9, 0.7 and 3.5 wt% excess C, H and O, respectively) were obtained for partially crosslinked polymers, for example, on heating the polymer for 4 h in Ar between 200–400 °C, or on heating the polymer in the presence of a dehydrocoupling catalyst [$Cp_2M(CH_3)_2$, M = Zr, Ti] (as above) at 100 °C. Samples are amorphous after pyrolysis to 1000 °C and crystallize (crystallite size *ca* 30 nm) to β-SiC at 1600 °C with an additional weight loss of 10–15%. This weight loss was suggested to arise from loss of CO and SiO as the residual oxygen within the material is eliminated.

The addition of 4 mol% allyl or vinyl groups to **8** followed by $LiAlH_4$ reduction provided **9**, which crosslinks at 150 °C to give 80–90% amorphous SiC (5% wt excess C) on pyrolysis to 1000 °C.

This polymer precursor (1) requires relatively inexpensive starting materials, (2) is quite stable in air, (3) offers good processability for polymer infiltration processing of composites, (4) provides excellent SiC ceramic yields and (5) high purity with controllable microstructures. However, one important drawback is the use of costly $LiAlH_4$. This polymer is now available commercially (Starfire Inc., NY).

7. Phase pure SiC via processing

Many modifications to the existing Yajima polycarbosilanes (PCS) precursors have been reported in efforts to improve ceramic yields, phase purity and the mechanical properties[6,7]. Only a few of these actually provide phase pure SiC through processing efforts alone. For example, Dow Corning work reports that thermally stable, substantially dense polycrystalline SiC fibers (>2.9 g ml^{-1} vs 3.2 g ml^{-1} theoretical) can be processed from polydimethylsilane-derived PCS by adding boron during the processing step[117].

PCS was first melt spun into fibers at ca 300 °C and sequentially exposed to NO/diborane, or ammonia/BCl$_3$ or NO$_2$/BCl$_3$ gases (rather than O$_2$) at temperatures between 25–200 °C for periods of 4–24 h to render the fibers infusible. The residual N, O and excess carbon are eliminated during sintering as gaseous byproducts (e.g. SiO and CO) at 1400 °C resulting in pores and voids, that weakened the fiber. However, continued heating with B to >1600 °C results in smooth densification (decreasing porosity), and overall strengthening of the fiber. B aids sintering and promotes densification at high temperatures (>1400 °C) and is necessary to retain fiber integrity.

The final, near stoichiometric β-SiC fibers have oxygen contents of <0.1 wt%, when heated >1600 °C. They exhibit average tensile strengths of 2.6 GPa and elastic moduli >420 GPa. Nicalon fibers do not survive similar heat treatments. The advantages of these fibers are that they rely on an existing process and provide properties expected of fully dense, phase pure SiC. Dow Corning has announced that they will make these fibers available commercially in the near future. Although the properties found for these fibers are essentially those of bulk SiC, there are some drawbacks: (1) the precursor synthesis process is multistep, (2) the ceramic yields are low and (3) the spun fibers required multiple heat treatments at 100–300 °C for 4–24 h with BCl$_3$ or diborane.

As an aside, Jacobson and DeJonghe have reported that simply heating Nicalon or Tyranno fibers at temperatures of ca 1600 °C with sources of boron (e.g. boron metal, TiB$_2$ etc.) leads to sufficient incorporation of boron into the fibers such that they densify and stable, nearly phase pure, SiC fibers are obtained. No data are provided about fiber composition[118].

Toreki and coworkers have described processing SiC fibers (UF fibers) from a novel PCS with low oxygen content and better high-temperature stability than Nicalon fibers[119]. The PCS was synthesized by pressure pyrolysis of polydimethylsilane in an autoclave. Control of the molecular weight (MW) is the key to successfully producing PCS fibers by solution processing. For example, fibers derived from PCS with Mn <5000 Da melted before curing, while those derived from PCS >10000 Da were not soluble, thus not suitable for spinning fibers.

Thus, in the presence of spinning aids (polysilazane and polyisobutylene), PCS of Mn 5000–10000 Da can be dry spun and then pyrolyzed in N$_2$ to form SiC-containing fibers. The polymer is self-curing. The resultant fibers have 80% ceramic yields (950 °C, 20 °C per min), low oxygen contents (1.1–2.6 wt%) and mechanical properties similar to Nicalon (tensile strength of ca 3.0 GPa, ambient). The pyrolyzed fibers have excess carbon, and thus perform below expectations for phase pure SiC. However, Sacks and coworkers have recently[120] described near stoichiometric SiC fibers ($\leqslant$0.1 wt% O) with high tensile strength (ca 2.8 GPa), fine grain sizes (ca 0.05–0.2 µm), high densities (ca 3.1–3.2 g ml^{-1}) with small residual pore sizes ($\leqslant$0.1 µm). The synthesis of the polymer precursor for this process was not reported; however, it was stated that dopant additions were made. The fibers (designated UF-HM fibers) retained ca 92% of their initial strength (2.70 GPa) after heat treatments to 1800 °C, suggesting the dopants contained B or other sintering agents to prevent grain growth. Electron microprobe analysis (EMA) showed an average fiber composition of 68.5% Si, 31.5% C, $\leqslant$ 0.1% O (Si$_{0.93}$C$_{1.00}$).

Few fibers have the excellent mechanical properties reported for the UF-HM fibers; however, if the synthesis parallels the original one (see above), the expected drawbacks for the process include: (1) the use of an autoclave, not desirable for large-scale production and (2) low ceramic yields of SiC, given that the fiber is derived from a polymer precursor containing a 1 : 2 Si:C ratio.

As mentioned above, there are now several polymer systems that provide access to phase pure, fully dense SiC with the properties expected based on SiC materials made by traditional methods. This does not mean that no further development need take place in this area. There are still cost issues that limit the utility of many of these precursors.

Furthermore, not all applications will need or benefit from the availability of phase pure SiC. For example, in polymer infiltration and pyrolysis processing of composites, the reinforcing material frequently is oxidized at the surface. Thus, an SiC precursor that produces excess carbon may be required to ensure that the oxide surface layer is reduced off during processing so that good interfaces are obtained.

One continuing and unresolved issue is whether or not a crystalline fiber is more desirable than an amorphous fiber. Thus, efforts to make amorphous, SiC-based ceramics are still an area of interest as suggested by the following section.

B. Precursors to Si—C—B Materials

In work directed toward the development of phase and chemically pure SiC with controlled microstructures, boron was added to control oxygen content and improve sintering behavior. Unrelated efforts sought routes to Si—C—B materials because of their potential to provide lightweight, high-strength ceramic/amorphous materials. The earliest efforts in this area are those of Riccitiello and coworkers[121]. The objective of these studies was to make a material that would exhibit exceptional stability at high temperatures in oxidizing environments. It was proposed that Si—C—B materials would oxidize slowly at high temperatures to generate borosilicate glass coatings on the oxidizing surface, that would act as barriers to further oxidation.

Efforts to prepare Si—C—B materials focused on dehalocoupling mixtures of boron halides and Me_2SiCl_2 (equations 37 and 38).

$$BCl_3 + 1.5Me_2SiCl_2 + 6Na \xrightarrow{120\,°C/2-3\,d} -[B(Me_2Si)_3]_n- + NaCl \qquad (37)$$
$$50-60\%$$

$$MeBBr_2 + Me_2SiCl_2 + 6Na \xrightarrow{120\,°C/4-8\,h} -[MeB(Me_2Si)_2]_n- + NaCl/NaBr \qquad (38)$$
$$50-60\%$$

The stoichiometry used to synthesize the first polymer can in principle lead to a completely crosslinked and insoluble polymer; however, when xylenes were used as solvent, a 60% yield of soluble precursors were obtained. When octane was used as solvent, the soluble fraction was only 40–50%. The second polymer type, which in principle should give only linear materials, also gave 50–60% yields of soluble precursor.

A number of other polymers were made wherein the ratios of the two principle ingredients in both of the above reactions were varied. The resulting polymers differ very little. Efforts to incorporate vinyl groups attached to Si centers led to further crosslinking as the vinyl groups reacted under the dehalocoupling conditions.

FTIR analyses of the polymers revealed no surprises, i.e. the presence of Si—H and B—H bonds were minimal. The presence of possible Si—B bonds was inferred by the presence of three absorptions in the 620–690 cm^{-1} region of the polymer spectra. Theoretical calculations suggest that the Si—B band should appear at ca 650 cm^{-1}. An additional absorption at 1315 cm^{-1} was assigned to a B—C stretching vibration. UV absorption spectra contain no peaks in the 300–350 nm region that can be associated with long polysilane blocks, suggesting that the polymers consist of, at most, short Si—Si segments linked to boron. All of the polymers are air- and moisture-sensitive, oxidizing rapidly in air.

The chemical composition of the soluble fraction of the first polymer was $SiC_{1.96}B_{0.54}H_{5.14}$ and that of the second polymer was $SiC_{3.26}B_{0.93}H_{7.03}$. Oxygen contents were 3–4 wt%. Some of the xylene solvent appears to be incorporated in the polymers as evidenced by the presence of aromatic peaks in the proton NMRs. TGA of the first polymer shows a 65–70 wt% ceramic yield (1000 °C/N_2). The second polymer's ceramic yield was just slightly lower. Most of the polymers made had an $M_n = 700$–1500 Da range with the majority being at the low end of this molecular weight range. The polymers had softening points in the range of 50–180 °C, depending on boron content. Typically, polymers with less boron (lower crosslink density) softened and or melted at lower temperatures.

Pyrolysis of the first polymer (1100 °C per 1 h/Ar) gave a material with a $Si_{1.92}C_{2.53}B_{0.71}H_{0.08}$ composition. XRD analysis suggested the presence of crystalline SiC (polytype not specified) following further heating to 1300 °C (1 h/Ar). Solid state ^{13}C, ^{29}Si, and ^{11}B NMR of the 1000 and 1300 °C samples indicates that the carbon is present mostly as SiC with little if any B_4C. On heating in air (TGA), these high-temperature samples appear to exhibit excellent oxidation resistance with gains of only 1–2 wt% at 1000 °C, which the authors suggest indicates little free carbon. Note that phase pure SiC oxidizes to SiO_2 with a weight gain of 50%. Hence a combination of SiC and free carbon could also lead to weight gains of 1–2 wt%, thus care must be taken to ensure spectroscopically that following air oxidation the material remains substantially the same.

These same polymers, sometimes made with small amounts of isopropyl or phenylsilyl chlorides, can be melt-spun in air to produce green fibers with 10–17 μm diameters. UV/air curing for 18–24 h gave infusible fibers that could be converted to ceramic fibers by passing through (90 cm min^{-1}) a furnace held at 1100–1300 °C (30 cm hot zone). An alternate, low-oxygen cured green fiber was obtained by sealing green fibers in a quartz tube, introducing hydrazine vapor and irradiating *in situ* for the same times as above.

The tensile strengths of the oxygen-cured fibers heated to 1000–1300 °C averaged 1 GPa, with an elastic modulus of *ca* 150 GPa. The tensile strengths of the hydrazine-cured fibers were slightly higher at 1.2–1.6 GPa and modestly higher elastic moduli of up to 180 GPa. The oxygen contents of both fibers were still quite high (several percent) because they were spun in air.

Another approach to Si–C–B materials was explored by Riedel and coworkers wherein polymethylvinylsilanes were hydroborated (BH_3). Thus, polymethylvinylsilane was prepared by equation 39[122].

$$0.77 Me(CH_2\!\!=\!\!CH)SiCl_2 \; + \; 0.44 Me_2SiCl_2 \; + \; 0.39 Me_3SiCl \; + \; 3Na$$

$$\Big\downarrow \text{THF/toluene/100 °C}$$

$$NaCl \; + \; -\![Me(CH_2\!\!=\!\!CH)lSi]_{0.77}(Me_2Si)_{0.44}(Me_3Si)_{0.39}]_n- \tag{39}$$

This product is then treated with a borane derivative (equation 40).

$$-[(Me(CH_2\!\!=\!\!CH)lSi)_{0.77}(Me_2Si)_{0.44}(Me_3Si)_{0.39}]_n- \; + \; H_3B\!\cdot\!SMe_2$$

$$\Big\downarrow \text{toluene/RT to 110 °C}$$

$$Me_2S \; + \; -[Me(\!\!=\!\!BCH_2CH_2)Si)_{0.77}(Me_2Si)_{0.44}(Me_3Si)_{0.39}]_n- \tag{40}$$

Hydroboration reaction can also be performed (*ca* 65% yield) with $H_3B\!\cdot\!THF$. The M_n for the vinylsilane is *ca* 800 Da and goes up to *ca* 1220 Da on hydroboration. A typical chemical composition for the 1220 Da polymer is $SiC_{2.90}B_{0.29}H_{7.43}O_{0.06}$. FTIR analysis of the polymer shows a ν_{B-H} band at 2530 cm^{-1}, indicating that hydroboration is incomplete. A band at 2100 cm^{-1}, typical of ν_{Si-H}, suggests that some Si–Cl/B–H exchange occurs along with hydroboration. It also indicates that not all of the Si–Cl bonds are consumed during dehalocoupling. The ^{1}H NMR spectrum indicates a 1 : 6.5 vinyl-to-methyl ratio in the starting oligomer. Based on the initial stoichiometry, this should be closer to 1 : 3, thus some of the vinyl groups are probably consumed during dehalocoupling as seen by Riccitiello and coworkers.

In both cases, on heating to 200 °C for 1 h in argon, a yellow polymer powder is obtained. On further pyrolysis to 1000 °C (Ar) the polymers convert to amorphous Si–C–B materials. A typical found composition is $Si_{1.7}C_{3.20}B_{0.6}$. If this material is heated to higher temperatures, the composition changes slightly. Thus, heating at 2200 °C (1 h/Ar) leads to a composition of $Si_{1.8}C_{3.4}B_{0.5}$. Samples heated $\geqslant 1500$ °C start to crystallize as broad β-SiC peaks (25 nm average grain size) appear first. At temperatures $\geqslant 1700$ °C, SiC, graphitic carbon and B_4C powder patterns are discernible, with average grain sizes reaching ca 250 nm at ca 2000 °C. At this temperature, β-SiC converts to α-SiC. Thus, SiC/graphite/B_4C composite materials are produced as a consequence of phase segregation as the precursor-derived material is heated to higher temperatures. This work actually serves as the basis for the Si–N–B–C work from the Riedel group discussed above.

C. Precursors to Si–C–O Materials

Because it is possible to generate SiC from silsesquioxanes, $-[RO_{1.5}]_n-$, considerable early work on SiC syntheses focused on transforming siloxane and silsesquioxane precursors to SiC[123–126]. Unfortunately, the transformation process requires temperatures of >1300 °C to effect carbothermal reduction of the intermediate SiC_xO_{4-x} phase to generate CO and SiC[125,126]. Thus, interest in the use of silsesquioxanes as SiC precursors waned. However, these initial studies demonstrated that properly developed silsesquioxanes were very easy to process into a wide variety of shapes and the pyrolysis product, SiC_xO_{4-x} or black glass, offers reasonable mechanical properties[127–135]. Consequently, more recent work has targeted processing black glass. Much of the work in this area relies on sol-gel synthesis of processable silsesquioxanes and is therefore outside the objectives of this review; however, several pertinent references are provided.

Several chemical synthesis routes to black glass precursors have also been developed, as will be discussed below. In general, the properties of the materials produced by both methods are much the same. For example, Agaskar described the use of several novel polyhedral silsesquioxanes as precursors to microporous 'organolithic materials'. His approach was to synthesize functionalized polyhedral silsesquioxanes (equations 41 and 42)[136].

$$8Si(OEt)_4 + 8Me_4NOH \xrightarrow{RT} [Si_8O_{20}]^{8-}(^+NMe_4)_8 \quad (41)$$

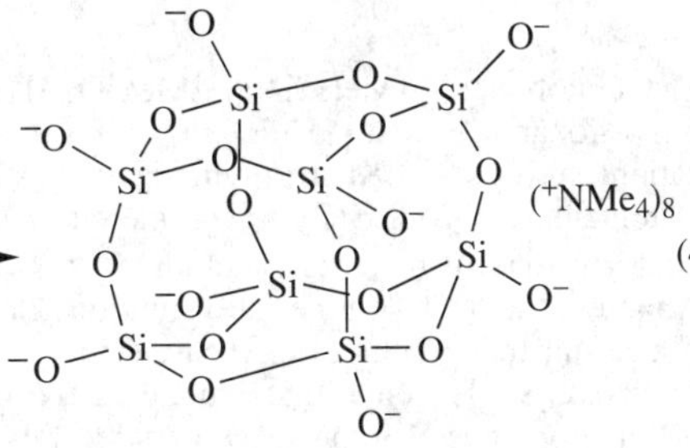

90$^+$% Yields

$[Me_4NOSiO_{1.5}]_8$ + excess RMe_2SiCl

(R = vinyl, CH_2Cl, etc.)

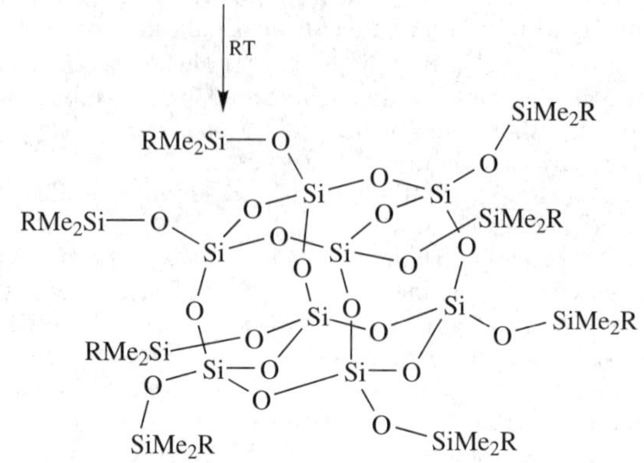

↓ RT

[octameric cube structure with eight $SiMe_2R$ groups attached via O to the Si_8O_{12} cage] (42)

The vinyl substituted cube readily hydrosilylates on reaction with (p-$HSiMe_2C_6H_4)_2O$ in the presence of platinum catalysts in THF. Pyrolysis in nitrogen to 800 °C (75 wt% ceramic yield) forms a black glass whose solid state ^{29}Si MAS NMR exhibits a variety of species consistent with the formation of SiC_xO_{4-x}. The NMR results are very similar to those discussed just below. Agaskar finds that if this material is leached with HF (48%/30 h), ca 40% of the mass is lost as a significant portion of the SiO_2 (actually Q_4) species are selectively removed with little change in the oxycarbide phase. The resulting material is microporous and exhibits a specific surface area of nearly 600 $m^2 g^{-1}$ and a pore volume of ca 0.2 ml g^{-1}. Microporous materials are of considerable interest for a variety of applications, as briefly discussed below.

An alternate, but less elegant route than that to polysilsesquioxane black glass precursors takes advantage of early work reported by Harrod and coworkers[137] on a novel disproportionation reaction (equation 43). When the same catalyst is applied to redistribute a polymethylhydridosiloxane, the result is a processable silsesquioxane polymer (equation 44)[138].

$$3MeHSi(OEt)_2 \xrightarrow{0.05 \text{ mol\% } Cp_2TiMe_2/20\,°C} MeSiH_3 + 2MeSi(OEt)_3 \quad (43)$$

$$\text{—[MeHSiO]}_x\text{—} \xrightarrow{0.05 \text{ mol\% } Cp_2TiMe_2/72 \text{ h}/20\,°C} 0.33x MeSiH_3 \\ + \\ \text{—[MeHSiO]}_{0.3}\text{[MeSiO}_{1.5}]_{0.7}\text{—} \quad (44)$$

The copolymer $-[MeHSiO]_{0.3}[MeSiO_{1.5}]_{0.7}-$ forms as a gel if neat methylhydridooligosiloxane, $Me_3Si-[MeHSiO]_x-H$ (M_n ca 2 kDa), is reacted with the catalyst. Gelation can be avoided if polymerization is conducted in toluene with a greater than 5 : 1 toluene to $-[MeHSiO]_x-$ volume ratio. The resulting polymer solution can be used to coat a variety of materials, including fibers[139].

Although the 5 : 1 volume ratio solutions can be used for coating applications, the gel-like material that results on solvent removal limits the copolymer's utility. Fortunately, the reactive Si—H bonds that remain in the copolymer can be used to further modify polymer properties. The presence of the catalyst in the final reaction solution provides a simple means of further modifying the original copolymer. Thus, after 72 h of reaction, the original royal blue toluene solutions of reaction 44 turn yellow-orange on addition of alcohols as the catalyst promotes alcoholysis of the remaining Si−H bonds (equation 45).

$$\text{—[MeHSiO]}_{0.3}\text{[MeSiO}_{1.5}\text{]}_{0.7}\text{—} \quad + \quad \text{ROH}$$

$$\downarrow \text{Cp}_2\text{TiMe}_2/20\,°\text{C} \quad\quad\quad (45)$$

$$\text{—[Me(RO)SiO]}_{0.3}\text{[MeSiO}_{1.5}\text{]}_{0.7}\text{—} \quad + \quad \text{H}_2$$

The blue color corresponds to a Ti(III) species, produced during reaction, which may or may not be the true catalyst. The yellow-orange color is typical of a Ti(IV) species. Alcoholysis creates more [MeSiO$_{1.5}$] groups, masked as alkoxy derivatives, [Me(RO)SiO]. If R is a long-chain alkyl group, more flexibility obtains and much of the gel character can be eliminated.

With MeOH, the reaction is rapid and vigorous H$_2$ evolution results. This polymer does not show much improvement in processability. However, the *n*-Bu derivative is completely tractable following solvent removal; however, it is expedient to redissolve the polymer in *n*-BuOH as gelation can still occur with time (days to weeks).

The original −[MeHSiO]$_{0.3}$[MeSiO$_{1.5}$]$_{0.7}$− and −[Me(RO)SiO]$_{0.3}$[MeSiO$_{1.5}$]$_{0.7}$− copolymers were characterized by ^{1}H, ^{13}C and by ^{29}Si NMR[138]. The starting oligomer, −[MeHSiO]$_n$−, is capped by Me$_3$SiO (M) and OH (DOH) groups. Thus, the actual composition of the 'Ti' catalyzed redistribution product is [MeSiO$_{1.5}$]$_{0.65}$ [MeHSiO]$_{0.28}$[MeSi(OH)O]$_{0.04}$[Me$_3$SiO]$_{0.03}$. The MeHSiO (D) ^{29}Si peak appears at −34.6 ppm. The ^{1}H MAS NMR spectrum also reveals the presence of Si−H groups at 4.5 ppm. The ^{13}C MAS spectrum shows two peaks at 1.1 and −2.9 ppm that can be assigned to the MeHSiO and MeSiO$_{1.5}$ (T) units, respectively. The ^{29}Si NMR spectrum also shows several peaks at −33 to −36 ppm and two peaks at −57.2 and −65.5 ppm, which are typical of cubic silsesquioxanes. These results, when coupled with the reproducible 30 : 70 [MeHSiO] : [MeSiO$_{1.5}$] ratio, suggest a polymer structure that consists of open cubes of T groups (MeSiO$_{1.5}$) bridged by one or two −MeHSiO− groups. The peak at −57.2 ppm is much smaller than the peak at −65.5 ppm. Thus, we assign this peak to the silicon vertices on the open edge of the cube and the −65.5 ppm peak to the remaining T group silicons in the cube. In the alkoxy derivatives, the −34.6 peak is replaced by a peak at *ca* −64 ppm. Note that silsesquioxane polymers are frequently described in the literature as ladder polymers when in fact they are best represented as discussed above. Frye and Klosowski addressed this misapprehension some 25 years ago[140].

Although the alkoxy derivatives are easier to process than −[MeHSiO]$_{0.3}$ [MeSiO$_{1.5}$]$_{0.7}$−, the pyrolysis behavior of both types of materials is much the same. Thus we will briefly look at the transformation processes that occur as −[MeHSiO]$_{0.3}$ [MeSiO$_{1.5}$]$_{0.7}$− decomposes to black glass. Pyrolysis of −[MeHSiO]$_{0.3}$[MeSiO$_{1.5}$]$_{0.7}$− in air leads to formation of pure silica (80–85% ceramic yield) at temperatures of 800–900 °C. In nitrogen, the ceramic yield is 70–75 wt% at 900 °C.

The simplest way to follow the polymer-to-ceramic transformation process is by solid state MAS ^{29}Si NMR spectroscopy. Thus, Figure 7 shows the MAS ^{29}Si NMR spectra of samples of copolymer heated to selected temperatures in N$_2$. Figure 8 provides an integration of the various species and how the relative amounts of each change on pyrolysis.

Heating the copolymer to 200 °C appears to have little effect on the polymer composition as seen in Figures 7 and 8. One might expect any Si−OH groups present to cocondense, react with Me$_3$SiO− or Si−H groups to form Si−O−Si bonds and release H$_2$O, Me$_3$Si−OH or H$_2$. However, the concentrations of these species and their diffusivities are too low for significant reaction to occur.

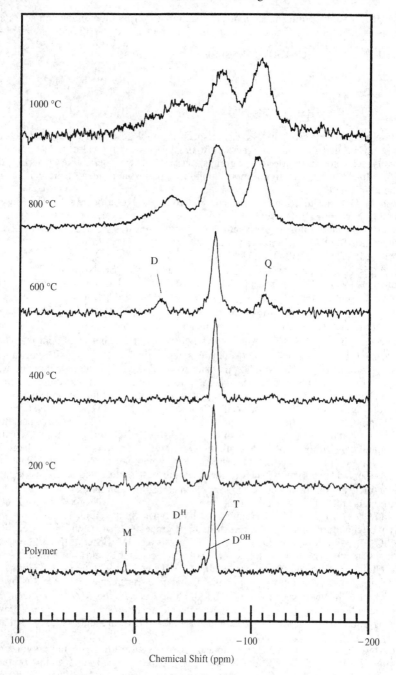

FIGURE 7. ^{29}Si MAS NMR spectra of samples of $-[\text{MeHSiO}]_{0.3}[\text{MeSiO}_{1.5}]_{0.7}-$ copolymer heated to selected temperatures in N_2. For definitions of abbreviations, see Figure 8

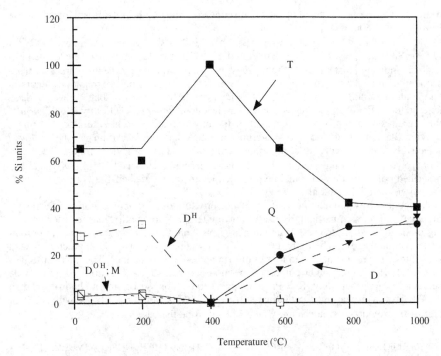

FIGURE 8. Integration of the various species on pyrolysis of $-[\text{MeHSiO}]_{0.3}[\text{MeSiO}_{1.5}]_{0.7}-$ to selected temperatures. M = $\text{Me}_3\text{SiO}-$, D = $\text{Me}_2\text{Si(O}-)_2$, D^H = $\text{MeHSi(O}-)_2$, D^{OH} = $\text{Me(OH)Si(O}-)_2$, T = $\text{MeSi(O}-)_3$, Q = $\text{Si(O}-)_4$

At 400 °C, the TGA (not shown) indicates a 20% mass loss which appears to result from depolymerization and redistribution of $-[\text{MeHSiO}]_x-$ segments. The ^{29}Si MAS and ^{1}H NMR spectra suggest that the 400 °C sample is now composed solely of $-[\text{MeSi(O)}_{1.5}]_x-$. However, the cross polarization NMR spectrum (not shown) reveals the presence of remnants of the $-[\text{MeHSiO}]_x-$ and $[\text{MeSi(OH)O}]-$ groups. The ^{13}C spectrum changes somewhat as the 1.1 ppm peak is reduced to a shoulder on the major peak which shifts from -2.9 to -3.5 ppm.

At 600 °C, a number of chemical changes occur in the polymer structure. For example, the number of T units decrease from *ca* 100% to 65% coincident with the appearance of D-type [MeSi(CH$_x$)O] units (14%) and Q (SiO$_4$, 20%) units (Figure 8). The relative amounts of each are close, suggesting that these species arise by direct redistribution of the T groups. ^{13}C NMR shows some changes as the peak attributable to Si$-$CH$_x$ species broadens and shifts to -4.2 ppm, in keeping with the formation of new D-type units. Groups containing Si$-$H or SiO$-$H functionality are no longer present.

On heating to 800 °C, all resemblance to a polymer structure disappears. All of the ^{29}Si peaks broaden greatly. The presence of Q units becomes significant as the number of T units continues to decrease (42%). The most important observation is the reappearance of DH units in the ^{29}Si spectrum and in the ^{1}H NMR spectrum (Si$-$H) as confirmed by FTIR studies[138]. These DH units result from carbothermal reduction by reaction of Si$-$O linkages with neighboring methyl groups. The 800 °C spectrum in Figure 7 is very similar to the ^{29}Si spectrum obtained by Agaskar on heating the polyhedral silsesquioxane above in nitrogen to 800 °C.

The ^{29}Si MAS NMR spectrum indicates that the 800 °C material is composed of SiO$_4$ (35%), SiO$_3$C (42%), SiO$_2$C$_2$ (<10%) and SiO$_2$CH (<20%) units. However, chemical analysis indicates that the Si:C ratio does not change during the heating process. Consequently, some free carbon appears to form. Thus, in principle, segregation occurs at temperatures $\geqslant$800 °C which has implications on the properties of black glass at these temperatures. For example, segregation could lead to a higher susceptibility to oxidation as the segregated carbon is no longer part of the oxycarbide network. Alternately, it might provide the opportunity to make microporous materials if the carbon can be removed selectively and uniformly. It is important to reiterate that the behavior of these precursors on pyrolysis is much the same as that of sol-gel derived materials[141,142]. Thus, the above described polymers may in principle provide the advantage of shelf-life not possible with the sol-gel derived materials.

Although silsesquioxane precursors do not provide the high-temperature properties of SiC, they still remain particularly useful for low-temperature applications, e.g. for providing oxidation and scratch-resistant coatings. Thus, adhesion and wettability are important issues as noted above. Therefore, efforts to improve the reactivity of these precursors with substrates can provide access to more uniform and adherent coatings. The $-$[MeHSiO]$_{0.3}$[MeSi(O)$_{1.5}$]$_{0.7}-$ copolymer could also be modified with acids (equation 46), to provide pendant SiO$_2$R groups (e.g. R = Pr) that appear to be more susceptible to hydrolysis than the OR groups. It was anticipated that surface hydroxyls would react faster with RCO$_2$Si groups to provide good adhesion, which proved to be the case[139].

$$-[\text{MeHSiO}]_{0.3}[\text{MeSi(O)}_{1.5}]_{0.7}- + \text{RCO}_2\text{H} \xrightarrow{\text{'Ti'}/20\,°\text{C}} \text{H}_2 + -[\text{Me(RCO}_2)\text{SiO}]_{0.3}[\text{MeSiO}_{1.5}]_{0.7}-$$

(46)

V. PRECURSORS TO Si–O CONTAINING MATERIALS

The development of processable organometallic and metalloorganic precursors to oxide materials that contain silicon has been very slow by comparison with development of precursors to nonoxide materials. This is chiefly a consequence of the exceptional success of sol-gel processing as a route to silicon containing ceramics[1,2,91]. However, sol-gel processing suffers some disadvantages that are difficult to overcome, especially in processing multimetallic materials. First, sol-gel processing relies on hydrolysis and condensation most commonly of Si(OEt)$_4$, TEOS. (Note that the methoxy derivative is poisonous.) However, when other metals are added, their hydrolysis and condensation rates are often orders of magnitude faster than that of TEOS. The end result is that atomic mixing, a primary motivation for using chemical processing, is lost. Alternately, the processing times required to retain atomic mixing become inordinately long. For example, the synthesis of atomically-mixed gel precursors to mullite (2SiO$_2$·3Al$_2$O$_3$) requires hydrolysis/condensation times as long as one month[14–17]. Another problem is that processable (spinnable, coatable etc.) gel intermediates often have limited shelf life, because continued condensation eventually results in a three-dimensional, highly-crosslinked covalent network. While this is often desirable (see latent reactivity above), it means that new gels must be made up each time they are used.

In precursor processing, shelf life is rarely a problem as noncovalent bonding interactions between molecules and solvent content define processability. Furthermore, atomic mixing and multimetallic compounds are readily accessible as will be amply demonstrated below. Thus, there is recent interest in developing oxide precursors for processing multimetallic oxides. Our discussion begins with a few simple SiO_2 precursors, but the more important precursors are the multimetallic precursors covered afterward.

A. SiO_2 Precursors

Although a number of SiO_2 precursors are used industrially, most are used for CVD of silica insulating layers in electronic device manufacture[1]. As such, they will not be discussed here. Many of the silicon oxycarbide precursors discussed above can be oxidized to SiO_2 and thus can be considered precursors. However, this is simply a processing approach. The chemistry remains the same as described above. Thus, further discussion is not warranted.

The generation of controlled porosity materials for membrane applications, catalysis, controlled release, filtration and insulation (both thermal and electronic) represents an area of intense interest wherein chemical processing is essential. Precursors to pure SiO_2 are receiving particular attention in these areas[1]. Perhaps the most recent and important use of silica precursors (on a cost basis) is spin-on silica films for electronics applications[143–147]. There are two objectives in this area: One is to develop spin-on coatings not based on sol-gel because of shelf life issues as above, and the second is to develop lower dielectric insulators for ultrahigh density integrated circuits[143,145,146]. We begin by addressing the second application.

As the distance between features in electronic circuitry becomes smaller (0.35 μm is the next target), even low dielectric insulating materials such as silica start to offer sufficiently strong capacitance that crosstalk can occur between features. Current, insulating silica (gap fill and interlevel) dielectrics are produced by spinning on sol-gel solutions made from $Si(OEt)_4$ followed by exposure to an oxygen plasma or through the use of methylsilsesquioxane precursors (see above). The resulting materials have dielectric constants in the range $k = 2.9–3.1$[143,145,146]. However, if one could use silsesquioxanes of the type **10** and **11**, then the open structure (pore size *ca* 3.3 Å) will give a material that is mechanically strong but with a lower dielectric constant. A number of groups have now studied the utility of preparing silica dielectrics with one or other of these materials and report dielectric constants as low as $k = 2.2$ after curing in an oxygen plasma at temperatures <400 °C[145]. If dielectric constants can be lowered to 2.2–2.4 without a sacrifice in mechanical properties, then these types of material will facilitate the production of the next generation of ultra-high, large scale integrated circuits.

Both $H_8Si_8O_{12}$ and the related polymer (resin) are made by hydrolysis of $HSiCl_3$ under water-starved conditions. One method that gives high yields of the polyhedral

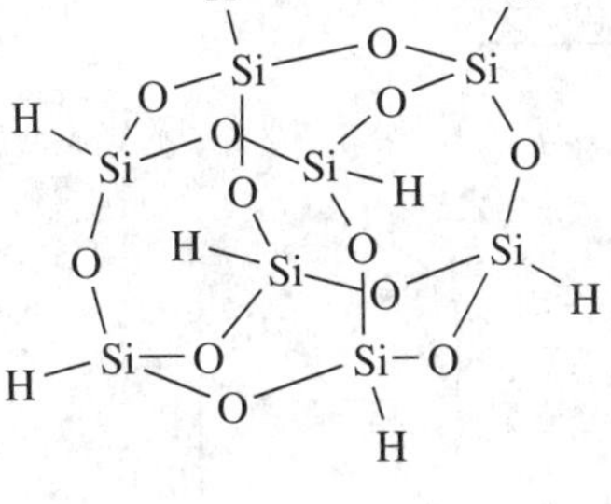

(10) T_8H_8

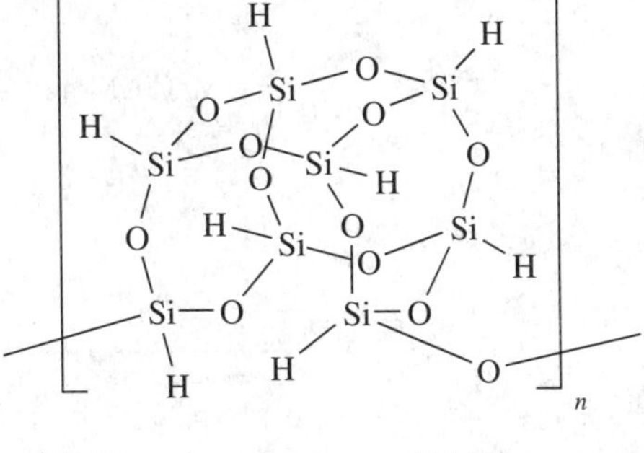

(11) Ring-opened polymer of T_8H_8

silsesquioxanes, developed by Agaskar, is shown in equation 47[148].

$$HSiCl_3 + (FeCl_3 \cdot H_2O) \longrightarrow H_{10}Si_{10}O_{15} + H_8Si_8O_{12} + resin \quad (47)$$
$$ 5\% \text{ yield} \quad 20\% \text{ yield} \quad ca\ 75\% \text{ yield}$$

All of the products dissolve in most organic solvents, although if the resin still retains Si—OH groups it can be expected to be less soluble. The resin is used for spin coating either as a toluene or heptane solution. Note that solvent volatility and ability to wet the coating substrate are important issues in determining whether or not a solvent and a precursor are useful. Another issue is the volatility of $H_8Si_8O_{12}$, which sublimes easily but has been used for both spin-on and CVD applications[149]. In air, $H_8Si_8O_{12}$ begins to decompose at ca 250 °C and converts to phase pure SiO_2 at approximately 400 °C, depending on the processing conditions[143]. One paper mentions a glass transition temperature at ca 250 °C, probably for the resin[145], which appears to be the basis of the Dow-Corning Flowable Oxide™ spin-on silica coating method[147]. At temperatures of 300 °C, following decomposition in oxygen, ^{29}Si MAS NMR spectroscopy suggests that the cubic structure (−83 ppm) of the starting material is retained in the resulting ceramic product. However, heating to 350 °C (0.5 h) converts the material to amorphous silica (−109 ppm). Thus, processing conditions are likely to be critical in achieving the desired drop in dielectric behavior.

Still another approach to developing spin-on silica precursors has been explored by Berry and Figge[146] and is based on unique β-haloethylsilsesquioxane resins. The key objective was to develop a processable silsesquioxane for spin-on glass applications that could be readily converted to pure silica at low-to-moderate temperatures. A critical step was to develop a 'mask' that provides processability but is easily removed without generating difficult-to-remove impurities. Hence, the mask must form an easily eliminated, volatile product on thermal or photochemical activation.

Based on these criteria, β-chloro- or β-bromoethyltrichlorosilanes are potentially attractive starting materials for this approach. First, because they are easily synthesized by equation 48 and then easily converted to the methoxy derivatives via equation 49. Finally, simple hydrolysis in methanol with a catalytic amount of HCl leads to a THF soluble gel (equation 50).

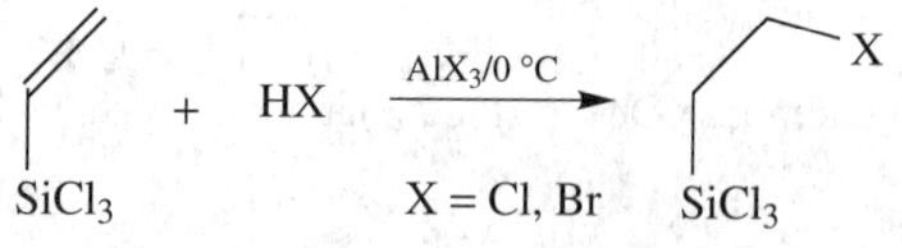

$$CH_2=CH-SiCl_3 + HX \xrightarrow{AlX_3/0\ °C} X-CH_2CH_2-SiCl_3 \quad X = Cl,\ Br \quad (48)$$

$$X-CH_2CH_2-SiCl_3 + MeC(OMe)_3 \xrightarrow[-3MeCl/40\ °C]{} X-CH_2CH_2-Si(OMe)_3 \quad (49)$$

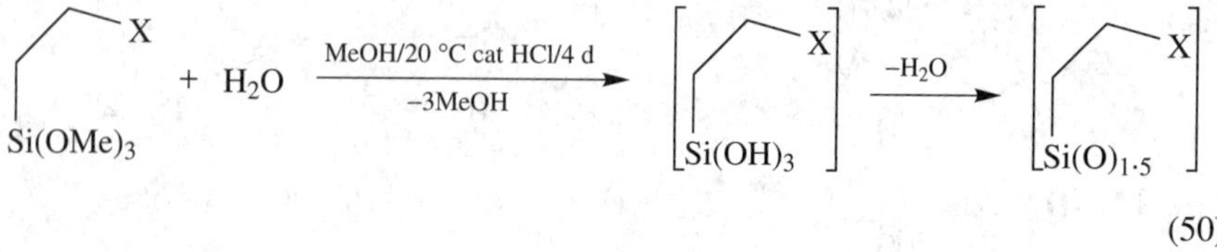

$$X-CH_2CH_2-Si(OMe)_3 + H_2O \xrightarrow[-3MeOH]{MeOH/20\ °C\ cat\ HCl/4\ d} [X-CH_2CH_2-Si(OH)_3] \xrightarrow{-H_2O} [X-CH_2CH_2-Si(O)_{1.5}]$$
$$\quad (50)$$

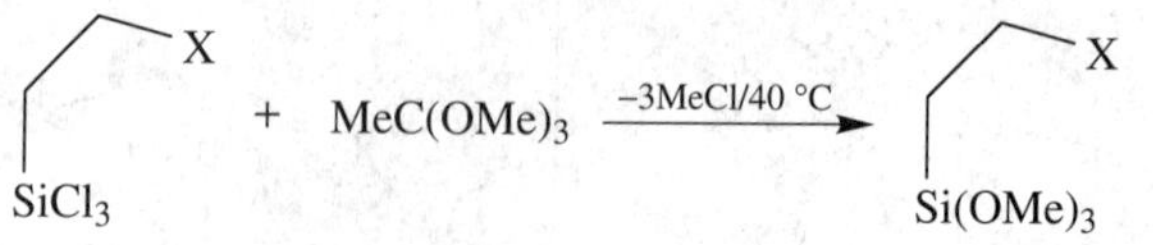

The resulting THF-soluble material is actually a low molecular weight oligomeric material that still retains some Si—OH groups. Molecular weight measurements indicate that the as-formed material consists of dimers, trimers and some high MW species; however, with time, continued condensation leads to materials with M_n ca 4k Da (after 70 d) that remain completely THF soluble.

The behavior of the β-chloro- and β-bromoethylsilsesquioxane on heating was studied by TGA. In a typical profile (10 °C per min/air/H$_2$O) mass loss onset for the β-chloro derivative occurs at 150 °C and coincides with detection of HCl and ethylene by mass spectrometry. The final ceramic yield was 52.5 wt% at 600 °C. A similar run in Ar gave a 63 wt% ceramic yield. The β-bromoethylsilsesquioxane exhibits, on heating, a nearly identical onset temperature with an 800 °C final ceramic yield of 38.5 wt% that reflects the higher mass of bromine. The primary decomposition process is thought to be as shown in equation 51. A second contributing mechanism is suggested to be that shown in equation 52.

$$\left[\begin{array}{c} \diagup X \\ | \\ Si(O)_{1.5} \end{array}\right] \xrightarrow[-CH_2=CH_2]{>150\,°C} \left[\begin{array}{c} X \\ | \\ Si(O)_{1.5} \end{array}\right] \xrightarrow[-HX]{+H_2O} \left[\begin{array}{c} H \\ \diagup \\ O \\ | \\ Si(O)_{1.5} \end{array}\right] \quad (51)$$

$$\begin{array}{c} H \\ \diagdown \\ O \\ \diagup \\ Si(O)_{1.5} \end{array} \quad \begin{array}{c} \diagup X \\ | \\ Si(O)_{1.5} \end{array} \longrightarrow \begin{array}{c} O_{1.5}Si-O \\ \diagdown \\ Si(O)_{1.5} \end{array} + HX + CH_2=CH_2 \quad (52)$$

On heating bulk samples of the β-chloro derivative to 1000 °C (1 h/air/H$_2$O), the resulting material is a black glass with a $C_{0.45}SiO_{1.95}$ composition. Thus, the original goal of producing phase pure, spun-on SiO$_2$ was not met. However, when thin spun-on glass films (100s nm) were photochemically decomposed with a 193-nm UV laser (in air or Ar), 200–300 nm films with negligible carbon contents (below detection limit) could be produced, albeit with exposure times of up to 8 h. In contrast, similar films heated to 450 °C for ca 8 h still contain as much as 8 wt% carbon. Of notable interest is the fact that chlorine or bromine contamination was rarely evident. The key problem was a partitioning of the decomposition reaction between loss of HCl and ethylene and formation of Si—carbon—Si bridges by other mechanisms. Some of these mechanisms are common to alkyl silsesquioxane decomposition processes as seen in the preparation of SiO$_x$C$_y$ materials, as discussed above.

Thus, high quality (high purity, uniform thickness, pore and defect-free) SiO$_2$ films for electronic applications remain a precursor and processing chemistry challenge. This challenge can be met, but the current solutions remain costly. Inexpensive precursors and processing methods remain viable targets for the materials chemist. One low-cost route to a processable SiO$_2$ precursor that may find use in some applications is discussed in the following paragraph.

A simple SiO$_2$ precursor can be made by direct reaction of SiO$_2$ with triethanolamine (equation 53)[72]. The resulting product, a silatrane glycol, **12**, forms quantitatively and can be recrystallized from hot ethylene glycol (EG) to give a high purity water-soluble precursor. **12** is an intermediate of considerable utility in the synthesis of aluminosilicate precursors (see below) but can also be used as a silica precursor. Thus, pyrolysis of a highly concentrated solution of **12** in ethylene glycol in a furnace preheated to 400–550 °C provides access to a carbon-contaminated porous material. Following oxidative removal of carbon at 450 °C, microporous silica can be made with specific surface areas (SSAs) as high as 570 m² g^{-1}[72]. These silicas are readily nitrided (NH$_3$) at low temperatures (700–800 °C) to produce SiON materials, as mentioned above.

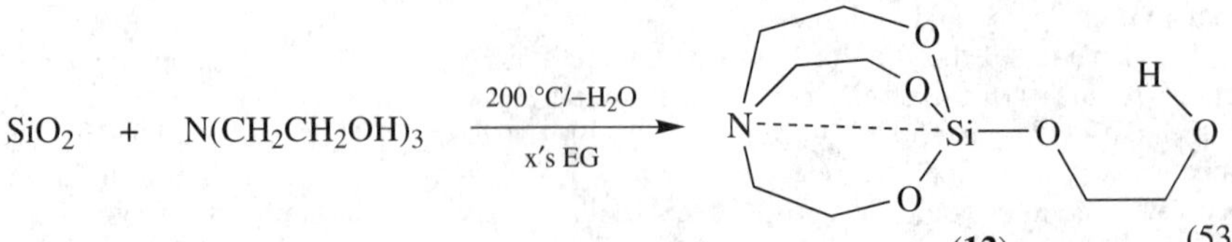

$$SiO_2 + N(CH_2CH_2OH)_3 \xrightarrow[\text{x's EG}]{200\,°C/-H_2O} \mathbf{(12)} \quad \mathbf{(53)}$$

One further comment on silatrane precursors. Frye and coworkers[149] report that several silatrane complexes, such as **12**, are soluble and quite stable in water. Thus, in principle, water-soluble silatranes can be used to minimize processing with organics and represent a 'green' or environmentally benign precursor (triethanolamine is used in cosmetics!).

B. Group I and II Silicate Precursors

Because group I and II silicates are readily available from commercial inorganic, high temperature reactions, there is no obvious need for precursors to these materials. Nonetheless, several precursors have been synthesized and their reactivity explored. A whole series of group I silicate precursors can be made via the general reaction shown in equation 54[150].

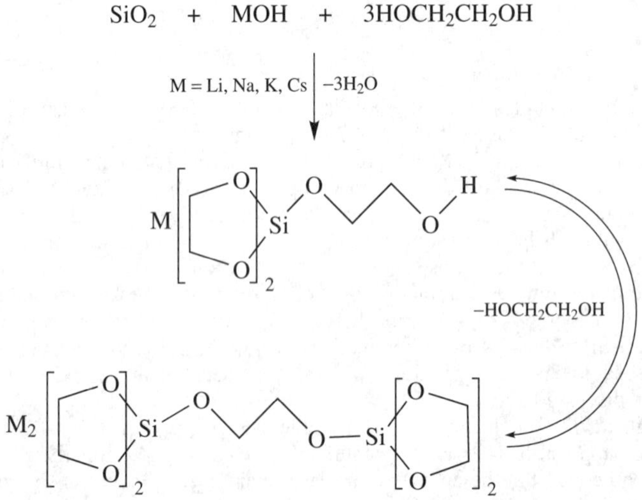

$$SiO_2 + MOH + 3HOCH_2CH_2OH$$

$$M = Li, Na, K, Cs \quad -3H_2O$$

$$\mathbf{(54)}$$

$$-HOCH_2CH_2OH$$

The yields of these compounds are nearly quantitative. Pyrolysis in air to 1000 °C gives the corresponding silicates quantitatively and usually phase pure[151]. Figure 9 shows the TGA of the monomeric lithium compound, LiSi(OCH$_2$CH$_2$O)$_2$OCH$_2$CH$_2$OH, on heating in air to 1000 °C. The ceramic yield is essentially that calculated for formation of phase pure Li$_2$O·2SiO$_2$, although trace amounts of the Li$_2$O·SiO$_2$ product are occasionally seen. BaO, CaO and (with difficulty) MgO can be made to work in reaction 55; however, if not done correctly, the reaction product is an unusual hexacoordinate complex[152].

These crystalline compounds are precursors to MO·2SiO$_2$ if the pentacoordinate precursor is made and to MO·SiO$_2$ if the hexacoordinated material is made[152]. On heating

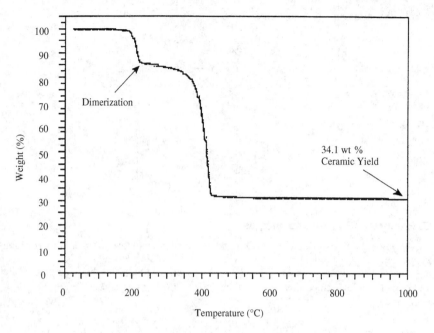

FIGURE 9. TGA of LiSi (OCH$_2$CH$_2$O)$_2$OCH$_2$CH$_2$OH on heating in air to 1000 °C

in air these materials transform with the coincident formation of the amorphous metal carbonates, MCO$_3$, even if oxygen is used instead of air. Thus, processing temperatures close to the melting or decomposition temperature of the carbonates are required to produce the phase pure material.

$$2SiO_2 \ + \ MO \ + \ x's \ HOCH_2CH_2OH$$
$$M = Mg, Ca, Sr, Ba$$

$$\downarrow -5 H_2O \quad 200 \ °C \tag{55}$$

Because these compounds are crystalline, they offer no useful rheological properties. However, exchange with high diols (1,4-diols or longer) leads to formation of polymeric materials that are soluble in alcoholic solvents and offer the potential for coatings and membrane applications[153]. These precursors can be used to produce aluminosilicates, as discussed in the following sections. Although it has not been attempted, it is conceivable that by exchange with transition metals, these materials may provide access to transition metal silicate precursors (equation 56).

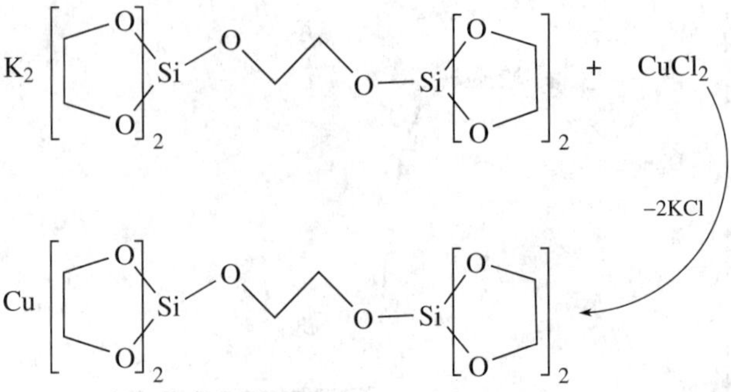

$$K_2\left[\begin{array}{c}O\\ \\ O\end{array}Si\begin{array}{c}O\\ \\ O\end{array}\right]_2 O-Si\left[\begin{array}{c}O\\ \\ O\end{array}\right]_2 + CuCl_2$$

$$\xrightarrow{-2KCl}$$

$$Cu\left[\begin{array}{c}O\\ \\ O\end{array}Si\begin{array}{c}O\\ \\ O\end{array}\right]_2 O-Si\left[\begin{array}{c}O\\ \\ O\end{array}\right]_2 \tag{56}$$

C. Precursors to Aluminosilicates

Although aluminosilicate double alkoxides have been available commercially for many years, their utility as direct precursors (via pyrolysis) to aluminosilicates has been poorly explored for the reasons stated above. One of the earliest studies in this area, from Tilley and coworkers, examined the use of compounds formed via equations 57 and 58[154].

$$Al_2Me_6 + 2(t\text{-}BuO)_3SiOH \xrightarrow[\text{toluene}]{-2MeH} (t\text{-}BuO)_3Si-O\cdots Al(Me)-O-Si(OBu\text{-}t)_3 \text{ (with Me-Al-Me bridge)} \tag{57}$$

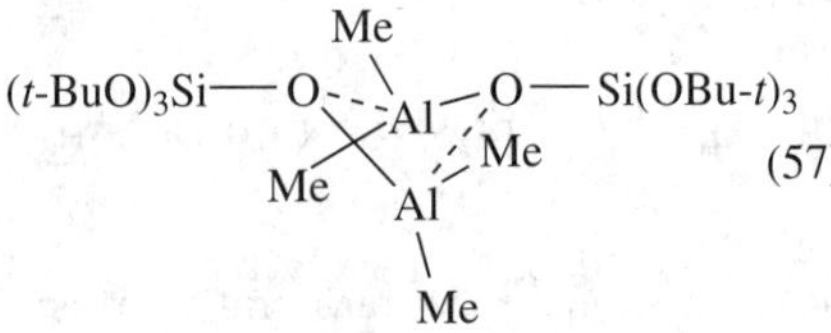

$$Al_2Si_2[O(t\text{-}Bu)O]_2Me_4 + t\text{-}BuOH \xrightarrow[\text{toluene}]{-2MeH} MeAl[OSi(OBu\text{-}t)_3](OBu\text{-}t) \tag{58}$$

These air- and moisture-sensitive compounds are soluble in most nonprotic organic solvents (presumably protic solvents will react). TGA studies on the first compound (O_2) show a rapid mass loss at 100–150 °C to give a ceramic yield of 30.9 wt% (at 1066 °C) vs an expected ceramic yield of 34.6 wt% for $Al_2O_3 \cdot 2SiO_2$. Note that this is not a stoichiometry known for aluminosilicate line compounds. On heating to 1000 °C, partial crystallization of mullite ($3Al_2O_3 \cdot 2SiO_2$) occurs, presumably with formation of amorphous silica. This dimer can also be thermally decomposed in refluxing toluene to generate materials with specific surface areas (SSAs) of 210 m² g⁻¹ but containing 4 wt% carbon. Heating to 800 °C/O_2 actually causes an increase in SSA to 280 m² g⁻¹ with a concomitant reduction in C content to 0.4 wt%.

TGA (O_2) of the second compound, $MeAl[OSi(OBu\text{-}t)_3](OBu\text{-}t)$, gives a very similar decomposition profile with a ceramic yield (1017 °C) of 29.5 wt% vs a theoretical value of 29.3 wt% for $Al_2O_3 \cdot 2SiO_2$. The decomposition process for both compounds is believed to occur by elimination of isobutene, methane and some water as discussed below for the Zr and Hf analogs.

A sample of $MeAl[OSi(OBu\text{-}t)_3](OBu\text{-}t)$ heated to 200 °C for 1 h was found to exhibit SSAs of 270 m² g⁻¹. Further heating to 800 °C for 1 h gave an SSA of 160 m² g⁻¹ as some sintering of the original nanocrystalline product occurred. At temperatures >1000 °C, mullite starts to crystallize. The carbon content of these materials was <0.1 wt%.

Most recently, Roesky and coworkers described several elegant molecules of the type shown in equations 59 that offer potential as aluminosilicate precursors[155].

$$Me_4Al_2[OSi(OBu\text{-}t)_2]_2 + 2t\text{-}BuOH \xrightarrow[\text{THF}]{-2i\text{-}BuH/H_2} Me_2Al_2[OSi(OBu\text{-}t)_2]_2(OBu\text{-}t)_2 \quad (59)$$

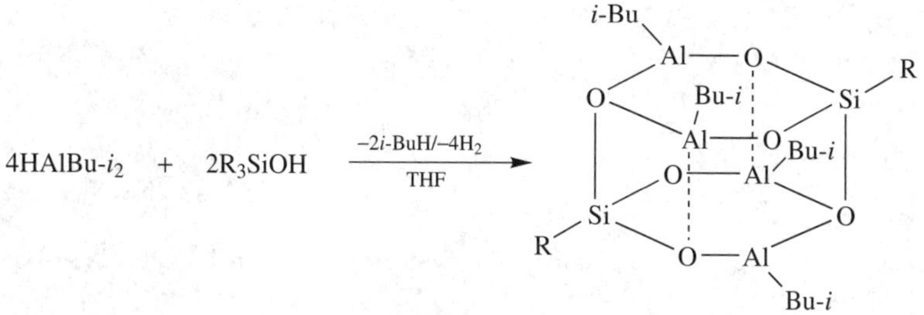

$$4HAlBu\text{-}i_2 + 2R_3SiOH \xrightarrow[\text{THF}]{-2i\text{-}BuH/-4H_2} \text{[cage structure]} \quad (60)$$

In principle these compounds offer access to materials with $Al_2O_3 \cdot SiO_2$ and $Al_2O_3 \cdot 2SiO_2$ stoichiometries. The latter stoichiometry is equivalent to the $Al[OSi(OBu\text{-}t)_3(OBu\text{-}t)]$ precursor. The major drawbacks with these materials are their air and moisture sensitivity, and the cost of the starting materials. Although the idealized stoichiometries of the above ceramics products are not those of crystalline aluminosilicates, amorphous aluminosilicate glasses are often important in optical applications or in scratch-resistant coatings. Furthermore, they may offer potential for CVD-type applications. There still remains considerable need for simple precursors to crystalline aluminosilicates, especially for structural applications. Dense, phase pure crystalline ceramic materials are desired for optimal mechanical properties, e.g. ceramic fibers for composite manufacture.

We recently described a general, low cost route to aluminosilicate precursors that permits control of precursor stoichiometry and processability[156-159]. This 'oxide one pot synthesis' (OOPS) approach basically evolved from several earlier studies, some of which are mentioned above. Thus, one can react any mixture of silica, $Al(OH)_3$ and a group I hydroxide, or group II oxide, in ethylene glycol with one equivalent of triethanolamine $[N(CH_2CH_2OH)_3]$ per metal. If this mixture is heated to the boiling point of ethylene glycol (ca 200 °C) such that byproduct water is removed by distillation, one forms a viscous, rheologically useful precursor to a ceramic with the stoichiometry of the initial ratio of oxides and hydroxides. To date, we have made precursors to mullite ($3Al_2O_3 \cdot 2SiO_2$), cordierite ($2MgO \cdot 2Al_2O_3 \cdot 5SiO_2$), spinel ($MgO \cdot Al_2O_3$), group I aluminosilicates including ($M_2O \cdot Al_2O_3 \cdot 2SiO_2$) where M = Li, Na and K, and the group II aluminosilicates ($MO \cdot Al_2O_3 \cdot 2SiO_2$) where M = Mg, Sr and Ba. In some instances, molecular complexes form as found for the spinel precursor **14**, as per equation 61.

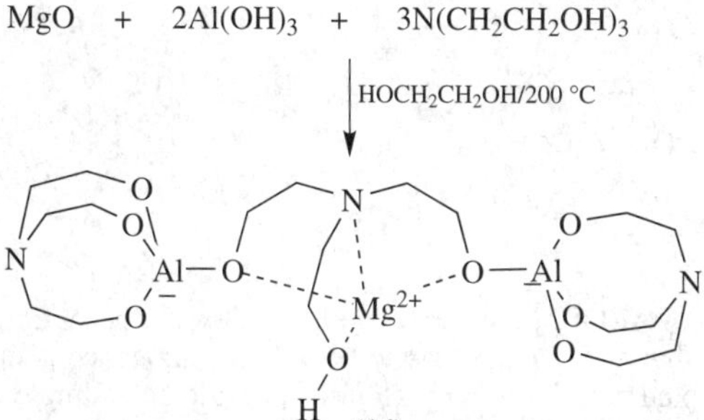

$$MgO + 2Al(OH)_3 + 3N(CH_2CH_2OH)_3 \xrightarrow{HOCH_2CH_2OH/200\,°C} \textbf{(14)} \quad (61)$$

In the case of the alkali aluminosilicate complexes, NMR, mass spectrometry TGA ceramic yield and ceramic product purity analyses support the formation of simple trimetallic precursors of the type **15**.

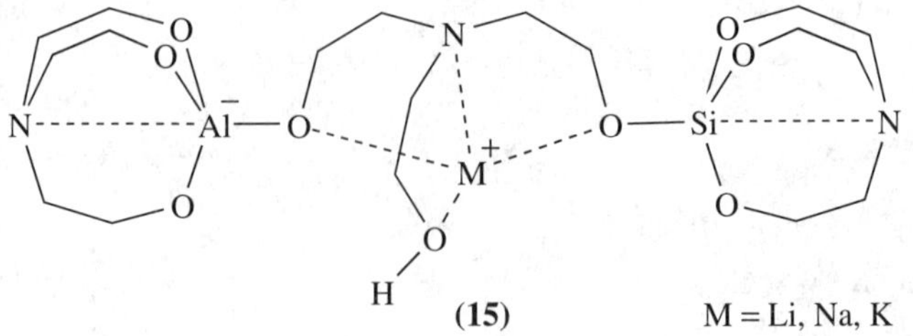

(15)   M = Li, Na, K

A related structure may form during the synthesis of group II analogs, as shown for the Ba complex **16**.

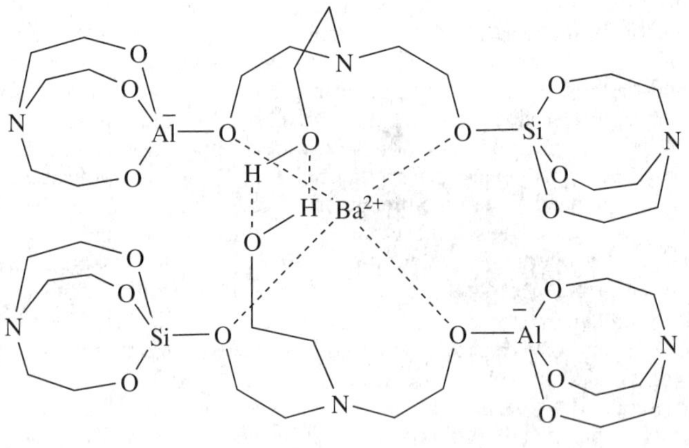

(16)

For other precursors, the polymer-like products appear to be simple, homogeneous mixtures of materials. Thus, the mullite precursor **17** likely consists of a 6 Al to 2 Si ratio.

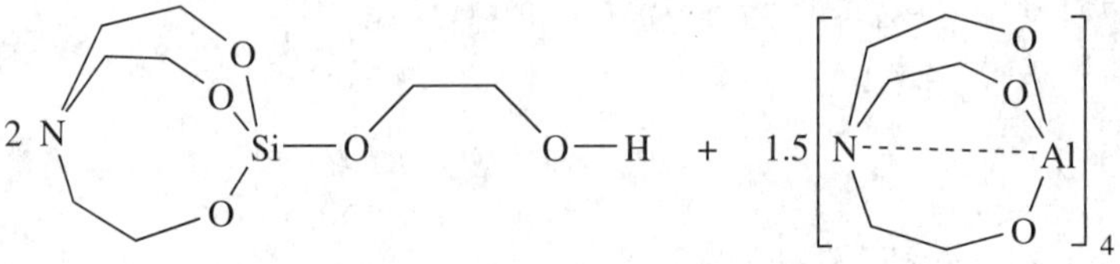

(17)

Note that reaction of $Al(OH)_3$ with triethanolamine produces a tetrameric alumatrane[159]. Despite the fact that this material is likely to be only a homogeneous mixture, it behaves as if it is atomically mixed. The following sections provide an example of how this mixture pyrolytically transforms to phase pure mullite. The purpose is to provide an example of how one establishes a processing window for conversion of a precursor to a desirable ceramic product. It also provides an example where instrumental analytical tools may fail to truly define what the actual ceramic products are. Thus, Figure 10 shows the TGA of the OOPS derived mullite precursor. The shown ceramic yield of only 23 wt% (vs ca 29 wt% theory) reflects the presence of residual ethylene glycol, which is difficult to remove even by vacuum drying, but its presence aids in providing desirable rheological properties.

Figure 11 shows the DTA of the precursor as it is heated in air (10 °C per min). Based on the TGA, most of the mass is actually lost before the first exotherm, maximum at ca 380 °C. Because this mass is associated with the decomposition of organic ligands, the first exotherm is only partly associated with oxidation of ligands as they decompose. This exotherm also appears to result partially from initial formation of an oxide network from the individual oxide fragments derived from the decomposition process. Indeed, a small exotherm is also seen in nitrogen which supports the contention that it is not indicative of oxidation. The second major exotherm centered at 610 °C arises as a result of char oxidation. Char forms as part of the ligand decomposition process (in air or nitrogen), which partitions between forming volatile fragments and crosslinked species. Although only a few wt% char material forms, it is actually the only serious drawback to using OOPS derived precursors. The final, very sharp exotherm at 990 °C is typical of crystallization of mullite and is often used as a primary indicator of mullite formation in atomically-mixed precursor systems.

The formation of mullite at ca 1000 °C is supported by the XRD data shown in Figure 12. Thus, based on DTA, TGA and XRD, one might decide that phase pure mullite without carbon contamination is produced by pyrolysis of precursor to 1000 °C. This then would represent a processing goal for making mullite materials. Also, it appears to be similar to the SiC precursor results, where nanocrystalline, phase pure SiC is obtained

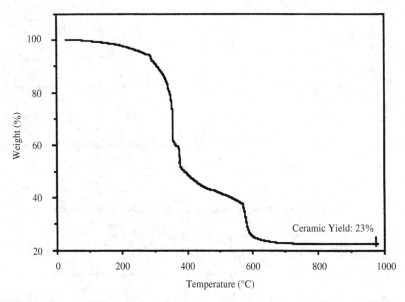

FIGURE 10. TGA of OOPS derived mullite precursor. The shown ceramic yield is only 23 wt% (vs. ca 29 wt% theory) owing to traces of excess ethylene glycol and triethanolamine retained in the sample to provide processability

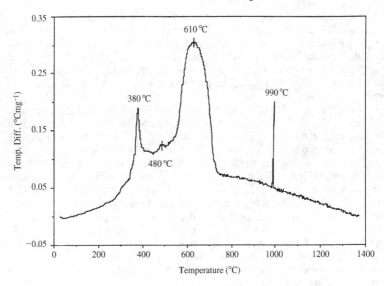

FIGURE 11. DTA (air/10 °C/min) of OOPS derived mullite precursor

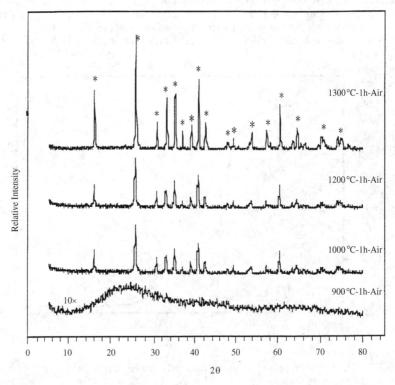

FIGURE 12. XRD of OOPS mullite precursor heated to selected temperatures (air/10 °C/min/1 h dwell)

at 1000–1200 °C. However, the diffuse reflectance IR (DRIFT) spectroscopy data tell another story as seen in Figure 13.

The DRIFT spectra show the typical broad peaks of an amorphous material up to approximately 1200 °C. Indeed, even at 1400 °C the sharp peaks indicative of a highly ordered, crystalline material are not really evident. These data suggest that the material is not particularly crystalline, possibly even up to 1400 °C. The solid state ^{29}Si MAS NMR data suggest the same thing as shown in Figure 14.

The spectra show an envelope of peaks resulting from nearly atomic mixing at the lowest temperatures. This envelope results from a statistical distribution of Al–O–Si linkages around tetrahedral silicon atoms, $Si(OAl)_{4-x}(OSi)_x$. Given that the Al:Si ratio is 3:1, then on average, each silicon will be surrounded by about 3 O–Al linkages — this is the maximum in intensity at temperatures below 950 °C. On moving above 950 °C, an unusual segregation occurs that leads to the formation of a poorly characterized, alumina rich, tetragonal mullite and nanoscale regions of pure silica. These react at higher temperatures to produce normal, orthorhombic mullite which is highly crystalline only at 1400 °C, as indicated by the sharp peaks finally seen in the NMR spectrum at this temperature. Thus, precursor processing may approach the correct material, but the final steps may require high temperature, post-pyrolysis heat treatments. The above material is presented in more detail elsewhere[157].

Only one paper that we are aware of explores a combined synthesis and processing route to aluminosilicates. Kemmitt and Milestone use precursors made by reaction of sodium hydroxide, boehmite $[Al(O)OH]_x$ and silica in ethylene glycol in a 4:3:1 ratio[160]. The precursor structures are related to those shown above. On removal of solvent (ethylene glycol) a glycolate precursor is obtained that contains a pentacoordinated

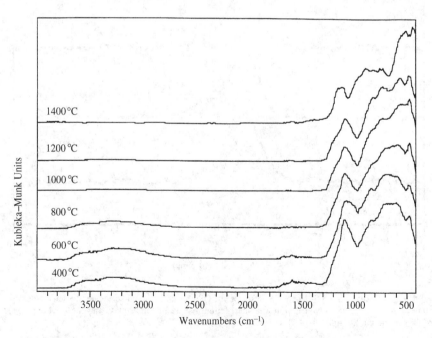

FIGURE 13. Diffuse reflectance infrared Fourier transform spectra of OOPS mullite precursor heated to selected temperatures (air/10 °C/min/1 h dwell)

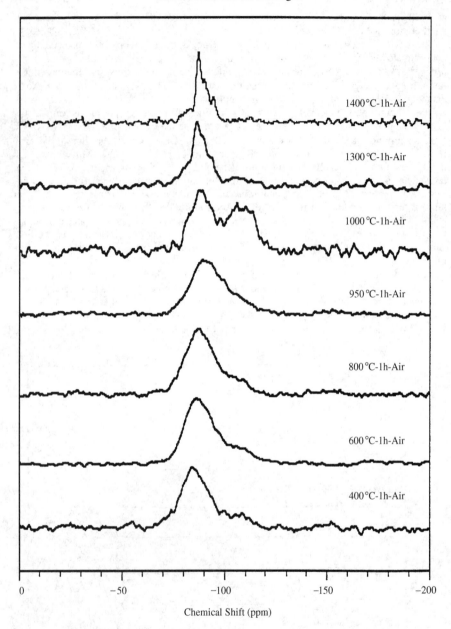

FIGURE 14. Solid state ^{29}Si MAS NMR spectra of OOPS mullite precursor heated to selected temperatures (air/10 °C/min/1 h dwell). Note the segregation that occurs at 1000 °C and the final appearance of crystalline mullite only at 1400 °C

silicon as determined by solid state MAS NMR spectroscopy (-103.8 ppm for ^{29}Si), and octahedral and tetrahedral Al (4.3 and 53 ppm for ^{27}Al) with the tetrahedral species being approximately three times as abundant. On pyrolysis of this precursor at 400, 500 or 950 °C the material transforms first to a mostly amorphous material at 400 °C, and then to a crystalline material with a composition $4Na_2O \cdot 3Al_2O_3 \cdot 2SiO_2$.

The unusual step occurs when the authors treat the 400 °C material with SO_2 for 4 h which converts all of the Na present to Na_2SO_4. The Na_2SO_4 is simply washed away with water and the resulting amorphous material then has the $3Al_2O_3 \cdot 2SiO_2$ composition which is that of mullite. Heating the amorphous material to 1200 °C for 16 h produces the desired mullite.

Finally, it is worth nothing that the mullite precursors shown above can be made at the kilogram scale in a matter of hours, whereas sol-gel derived materials can require up to one month to obtain atomically mixed gels. To date, no attempts have been made to form transition metal precursors via the OOPS or the Kemmitt and Milestone processes. However, the following section suggests that there are good reasons to explore this approach.

D. Precursors to Transition Metal Silicates

The earliest work in the development of simple molecular precursors to transition metal silicates is that of Hrncir and Skiles[161]. These workers report that the reaction of silanols with metal alkyls provides a low temperature, clean, quantitative route to pure Zr and Hf metallasiloxanes:

$$M(CH_2SiMe_3)_4 + 4R_3SiOH \xrightarrow{0-25\,°C} 4SiMe_4 + M(OSiR_3)_4 \quad (62)$$

$$M = Hf, Zr$$

The resulting moisture-sensitive compounds, $M(OSiR_3)_4$, are liquids when R = Et, and solids when the R groups are Me/t-Bu, Me/Ph or Ph. Those compounds with some R = Me groups are soluble in aromatic solvents, whereas those where R = Ph are insoluble.

On heating to 700 °C, all of the compounds decompose to white, insoluble products. The decomposition process was followed by IR and appears to begin with partial decomposition at 160 °C that retains most of the spectrum seen in the as-prepared product, with the exception of an additional Zr–O–Si stretching band at 970 cm^{-1}. At 350 °C, all resemblance to a molecular material is lost and broad, featureless spectra are observed which remain unchanged to 700 °C.

TGA analysis of the decomposition process for the Hf (R = Ph) compound suggests the stoichiometric conversion shown in equation 63.

$$Hf(OSiPh_3)_4 \xrightarrow{600\,°C/air/1\,h} HfSi_4O_{10} \quad (63)$$

ESCA indicates that only 2% carbon remains following pyrolysis; it also reveals a Si (2p) binding energy of 104.2 eV, which is higher than that found for silica (103.5 eV). This change in binding energy may be attributable to the influence of Hf–O–Si linkages. Unfortunately, substantiation via XRD was not possible as the material was X-ray amorphous.

Recent work by Terry and Tilley describes the preparation and pyrolytic behavior of the related $M[OSi(OBu-t)_3]_4$ (M = Ti, Zr, Hf) precursors. The $M[OSi(OBu-t)_3]_4$ compounds were prepared by reaction of the silanol with a metal amide[162,163] (equations 64 and 65).

$$Ti(NMe_2)_4 + 4(t\text{-BuO})_3SiOH \xrightarrow{110\,°C/PhMe} 4HNMe_2 + Ti[OSi(OBu-t)_3]_4 \quad (64)$$

$$\text{M(NEt}_2)_4 + 4t\text{-BuO}_3\text{SiOH} \xrightarrow{0-20\,°\text{C/pentane}} 4\text{HNEt}_2 + \text{M[OSi(OBu-}t)_3]_4 \quad (65)$$
$$(\text{M} = \text{Zr, Hf})$$

The Ti and Hf compounds are monomers, whereas the Zr complex was dimeric. As found by Hrncir and Skiles, the series of $\text{M[OSi(OBu-}t)_3]_4$ compounds are all moisture-sensitive. TGA studies indicate that the Ti complex decomposes cleanly at temperatures >240 °C and gives a ceramic yield of 25 wt% whereas the theoretical ceramic yield for TiSiO_4 should be 29.07 wt%. The primary gaseous thermolysis products were identified by mass spectroscopy to be isobutene and water. A likely pathway for decomposition appears to involve β-hydrogen elimination followed by condensation of the resulting Si–OH groups to generate isobutene, water and an oxide network as shown in equations 66 and 67.

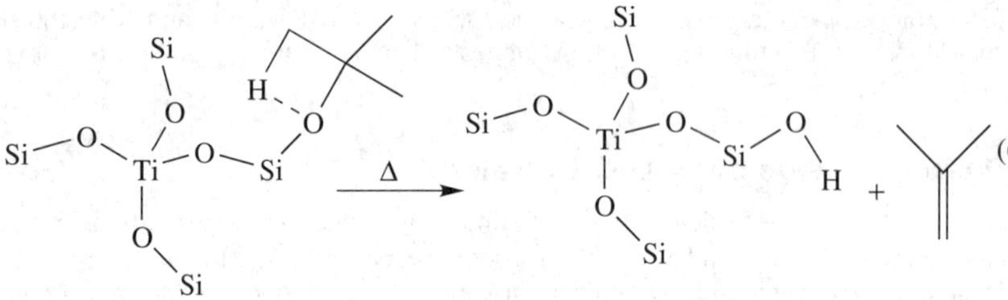

(66)

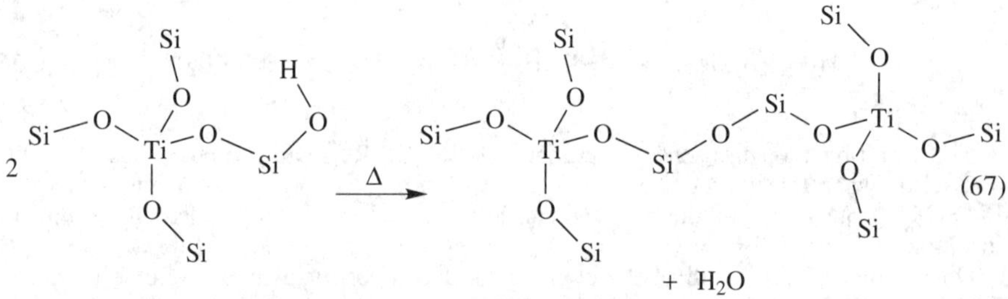

(67)

On heating to higher temperatures, no crystalline phases are observed until anatase crystallizes at 1000 °C. At 1400 °C, anatase, rutile and crystobalite are the only products. No single-phase material is obtained. The lack of correspondence between the TGA ceramic yield and the theoretical ceramic yield calculated for TiSiO_4 presages this problem. The exact reasons for the formation of a mixed-oxide phase are unknown at the moment, but they clearly contrast with the behavior of the Zr and Hf analogs.

The Zr and Hf compounds decompose cleanly at much lower temperatures (ca 140 °C) and give ceramic yields that are expected assuming MSiO_4 is the product. Heating to 400 °C generates amorphous, $\text{MSiO}_x(\text{OH})_y$ materials with surface areas of 500 m² g⁻¹. These materials are potentially interesting as catalysts or catalyst supports.

On heating to 1500 °C/6 h/Ar, the Zr material crystallizes to a mixture of monoclinic and tetragonal zirconia and crystobalite with loss of considerable original surface area (36 m² g⁻¹). The Hf material behaves similarly, although it partially crystallizes at 1000 °C to produce cubic or tetragonal hafnia. Cristobalite is only observed in materials heated to 1400 °C. Finally, thin films of the Zr and Hf derivatives could be cast from hydrocarbon solutions on quartz and then converted to thin films of the corresponding amorphous or ceramic materials.

A related study explores the use of $\text{OSi(OBu-}t)_3$ chemistry, to prepare manganese-doped zinc orthosilicates, e.g. Zn_2SiO_4 (willemite) which are used in electroluminescent devices. Two types of precursors were prepared, two molecular compounds and a set of oligomers (equations 68–70).

$$2Me_2Zn + 2HOSi(OBu\text{-}t)_3 \xrightarrow{-78\,°C/toluene} 2MeH + [MeZnOSi(OBu\text{-}t)_3]_2 \quad (68)$$

$$Me_2Zn + 2HOSi(OBu\text{-}t)_3 \xrightarrow{-78\,°C/toluene} 2MeH + Zn[OSi(OBu\text{-}t)_3]_2 \quad (69)$$

$$Me_2Zn + (HO)_2Si(OBu\text{-}t)_2 \xrightarrow{-40\,°C/toluene} 2MeH + [ZnOSi(OBu\text{-}t)_2O]_n \quad (70)$$

At set of toluene soluble oligomers of the type shown in equation 70 were made with various metal to silane diol ratios in order to achieve the appropriate Zn_2SiO_4 stoichiometry. The oligomer molecular weights varied from $M_n = ca$ 1.8 kDa to 11 kDa, as determined by NMR end group analysis.

Based on work done with the precursors synthesized in equations 57 and 58, these precursors were all assumed (and found) to decompose in a similar manner. The air- and moisture-sensitive dimer decomposes readily in the 100–250 °C range with a found ceramic yield of 22 wt%, which is close to the 24 wt% calculated for $Zn_2SiO_4 \cdot SiO_2$ by equation 71. Pyrolysis to 1100 °C for 1 h generated willemite (by XRD). No crystalline SiO_2 compound was observed coincidentally.

$$[MeZnOSi(OBu\text{-}t)_3]_2 \xrightarrow{>100\,°C} Zn_2SiO_4\cdot SiO_2 + 2HOSi(OBu\text{-}t)_3 + CH_2{=}CMe_2$$
$$+ H_2O + t\text{-BuOH} \quad (71)$$

As with the dimeric compound, the oligomers decompose in the 100–250 °C temperature range. However, unlike the dimer, $HOSi(OBu\text{-}t)_3$ is not observed as a pyrolysis product. One oligomer, with a chemical composition close to $C_8H_{18}O_4SiZn$, gave a 230 °C ceramic yield of 52.1 wt%, which is in good agreement with the 52.2 wt% calculated ceramic yield for $Zn_2SiO_4\cdot SiO_2$. However, mass loss continues with further heating. Thus, the 1110 °C/Ar ceramic yield is 48 wt% and at 1350 °C it is 42.5 wt%. The 1100 °C ceramic product composition was $(Zn_2SiO_4)_{0.37}(SiO_2)_{0.41}$ suggesting that some Zn is lost. This material contains <0.2 wt% carbon and <0.1 wt% H, thus the source of the ca 6 wt% change between 1100 and 1350 °C is not clear.

XRD analysis of the polymer pyrolysis products reveals formation of partially crystalline Zn_2SiO_4 on heating to $\geq 850\,°C/2$ h/Ar as expected based on an 800 °C exotherm seen in the DTA profile. SEMs of materials heated to 1350 °C indicate formation of a glassy material, although no XRD data are provided[162,163]. The authors point out that the melting points for cristobalite and α-Zn_2SiO_4 are 1713 °C and 1512 °C, respectively. Thus, there appears to be no explanation for the formation of the glassy phase. However, the almost 6 wt% mass change that occurs on heating to 1350 °C might result from loss of Zn metal (reduced by residual C/H) or loss of volatile SiO. More work needs to be done to understand this process.

Finally, polymer blends were prepared by mixing $[ZnOSi(OBu\text{-}t)_2O]_n$ with $Mn(CH_2SiMe_3)_2$ in diethyl ether in 1–15 wt% amounts in order to develop an electroluminescent precursor. Pyrolysis of these blends at 1000 °C/Ar produced materials which showed no evidence of segregated Mn containing phases at <3 wt% Mn. At higher concentrations ω-MnO_2 was observed to form. The resulting materials offer the desired photoluminescent properties, although more emission lines were observed than anticipated.

The absence of ZnO in any of the materials produced via these types of precursors suggests that, at least initially, atomic mixing is achieved. At higher temperatures, where segregation is seen, thermodynamics may play a role in driving phase segregation. The fact that exact stoichiometry precursors were not prepared may also strongly influence the final products and microstructures observed. Further precursor design studies are likely to sort out these problems.

VI. REFERENCES

1. (a) C. J. Brinker and G. Scherer, *Sol-Gel Science: The Physics and Chemistry of Sol-Gel Processing*, Academic Press, Boston, 1990.
   (b) D. Avnir, L. C. Klein, D. Levy, U. Schubert and A. B. Wojork, Chapter 40 of this book
2. *Journal of Sol-Gel Science and Technology*, Elsevier.
3. C. K. Narula, *Ceramic Precursor Technology and Its Applications*, M. Dekker, New York, 1995.
4. (a) K. Wynne, M. Zeldin and H. Allcock (Eds.), *Inorganic and Organometallic Polymers*, Am. Chem. Soc. Symp. Ser. Vol. 360, Am. Chem. Soc., Washington D.C., 1988.
   (b) K. Wynne, P. W. Nielson and H. Allock (Eds) *Inorganic and Organometallic Polymers II*, Am. Chem. Soc. Symp. Ser., Vol. 572, Am. Chem. Soc., Washington D.C. 1994.
5. (a) R. M. Laine (Ed.), *Transformation of Organometallics into Common and Exotic Materials: Design and Activation*, NATO ASI Ser. E: Appl. Sci.-No., Vol. 141, Kluwer Publ., Dordrecht, 1988.
   (b) R. M. Laine (Ed.), *Inorganic and Organometallic Polymers with Special Properties*, NATO ASI Ser. E: Appl. Sci.-No., Vol. 206, Kluwer Publ., Dordrecht, 1991.
   (c) J. F. Harrod and R. M. Laine (Eds.), *Applications of Organometallic Chemistry in the Preparation and Processing of Advanced Materials*, NATO ASI Ser. E: Appl. Sci.-No., Vol. 297, Kluwer Publ., Dordrecht, 1995.
6. (a) M. Birot, J-P. Pillot and J. Dunogues, *Chem. Rev.*, **95**, 1443 (1995).
   (b) K. J. Wynne and R. W. Rice, *Annu. Rev. Mater. Sci.*, **14**, 297 (1984).
   (c) R. R. Wills, R. A. Markle and S. P. Mukherjee, *Ceram. Bull.*, **62**, 904 (1983).
   (d) R. H. Baney and G. Chandra, in *Encyclopedia of Polymer Science and Engineering*, Vol 13, Wiley, New York, 1988, pp. 312–344.
   (e) J. Bill and F. Aldinger, *Adv. Mater.*, **7**, 775 (1995).
   (f) W. Toreki, *Polym. News*, **16**, 6 (1991).
   (g) G. Pouskouleli, *Ceram. Int.*, **15**, 213 (1989).
7. R. M. Laine and F. Babonneau, *Chem. Mater.*, **5**, 260 (1993).
8. J. A. Reed, *Principles of Ceramic Processing*, Second ed., Wiley Interscience, New York, 1995, p.57.
9. R. M. Laine, Z.-F. Zhang, K. W. Chew, M. Kannisto and C. Scotto, in *Ceramic Processing Science and Technology* (Eds. H. Hausner, G. Messing and S. Hirano), Am. Ceram. Soc., Westerville, OH, 1995, pp. 179–186.
10. C. D. Han, *Rheology in Polymer Processing*, Academic Press, New York, 1976.
11. A. Ziabicki, *Fundamentals of Fiber Formation*, Wiley, New York, 1976.
12. P. Calvert and M. J. Cima, *J. Am. Ceram. Soc.*, **73**, 575 (1990).
13. M. J. Cima, J. A. Lewis and A. D. Devoe, *J. Am. Ceram. Soc.*, **72**, 1192 (1989).
14. C. Gerardin, S. Sundaresan, J. Benziger and A. Navrotsky, *Chem. Mater.*, **6**, 160 (1994).
15. D. X. Li and W. J. Thomson, *J. Mater. Res.*, **6**, 819 (1991).
16. (a) J. C. Huling and G. Messing, *J. Non-Cryst. Solids*, **147**, 213 (1992).
    (b) J. C. Huling and G. L. Messing, *J. Am. Ceram. Soc.*, **72**, 1725 (1989).
17. P. Kansal, R. M. Laine and F. Babonneau, *J. Am. Ceram. Soc.*, **80**, 2597 (1997).
18. M. Scarlete, J. He, J. F. Harrod, and I. S. Butler, in *Applications of Organometallic Chemistry in the Preparation and Processing of Advanced Materials*, NATO ASI Ser. E: Appl. Sci.-Vol. 297 (Eds. J. F. Harrod and R. M. Laine), Kluwer Publ., Dordrecht, 1995, pp. 125–140 and references cited therein.
19. Z-F. Zhang, S. Scotto and R. M. Laine, *Ceram. Eng. Sci. Proc.*, **15**, 152 (1994).
20. F. Wakai, Y. Kodama, S. Sakaguchi, N. Murayama, K. Izaki and K. Niihara, *Nature* **344**, 421 (1990).
21. R. M. Laine, F. Babonneau, K. Y. Blohowiak, R. A. Kennish, J. A. Rahn, K. F. Waldner and G. J. Exarhos, *J. Am. Ceram. Soc.*, **78**, 137 (1995).
22. K. Su, E. E. Remsen, G. A. Zank and L. G. Sneddon, *Chem. Mater.*, **5**, 547 (1993).
23. (a) O. Funayama, M. Arai, Y. Tashiro, H. Aoki, T. Suzuki, K. Tamura, H. Kaya, H. Nishii and T. Isoda, *Nippon Seram. Kyokai, Gaku. Ron.*, **98** 104 (1990); *Chem. Abstr.* **111**, 218411s (1990).
    (b) H. Aoki, T. Suzuki, T. Katahata, M. Haino, G. Nishimura, H. Kaya, T. Isoda, Y. Tashiro, O. Funayama and M. Arai, European patent application 89302178.2; date of filing 03/03/89. Publication No. 0 332 357.
24. R. M. Laine, Y. Blum, D. Tse and R. Glaser, in *Inorganic and Organometallic Polymers*, Am. Chem. Soc. Symp. Ser. Vol. 360 (Eds. K. Wynne, M. Zeldin and H. Allcock), 1988, pp. 124–142.

25. (a) D. Seyferth, G. H. Wiseman and C. C. Prud'homme, *J. Am. Ceram. Soc.*, **66**, C-13 (1983).
    (b) D. Seyferth and G. H. Wiseman, in *Ultrastructure Processing of Ceramics and Composites* (Eds. L. L. Hench and D. R. Ulrich), Wiley, New York, 1984, p. 265.
    (c) D. Seyferth and G. H. Wiseman, in *Ultrastructure Processing of Ceramics and Composites II* (Eds. J. D. Mackenzie and D. R. Ulrich), Chap. 38, Wiley, New York, 1987.
    (d) D. Seyferth, in *Transformation of Organometallics into Common and Exotic Materials: Design and Activation*, NATO ASI Ser. E: Appl. Sci.-No. Vol. 141 (Ed. R. M. Laine), Kluwer Publ., Dordrecht, 1988, pp. 133-154 and references cited therein.
    (e) D. Seyferth, H. Plenio, W. S. Rees and K. Büchner. *Frontiers of Organosilicon Chemistry* (Eds. A. R. Bassindale and P. P. Gaspar), Royal Society of Chemistry, London, 1991, pp. 15-27.
26. N.S.C. K. Yive, R. Corriu, D. Leclercq, P. H. Mutin and A. Voiux, *New J. Chem.*, **15**, 85 (1991).
27. (a) G. T. Burns, T. P. Angelotti, L. F. Hanneman, G. Chandra and J. A. Moore, *J. Mater. Sci.*, **22**, 2609 (1987).
    (b) Y. Abe, T. Ozai, Y. Kumo, Y. Nagao and T. Misono, *J. Inorg. Organomet. Polym.*, **2**, 143 (1992).
28. (a) R. M. Laine, *Platinum Met. Rev.*, **32**, 64 (1988).
    (b) Y. D. Blum, K. B. Schwartz and R. M. Laine, *J. Mater. Sci.*, **24**, 1707 (1989).
    (c) K. A. Youngdahl, R. M. Laine, R. A. Kennish, T. R. Cronin and G. A. Balavoine, in *Better Ceramics Through Chemistry III.* Mat. Res. Symp. Proc. Vol. 121 (Eds. C. J. Brinker, D. E. Clark and D.R. Ulrich), 1988 pp. 489-495.
    (d) Y. D. Blum and R. M. Laine, *Organometallics*, **5**, 2801 (1986).
29. (a) H. Q. Liu and J. F. Harrod, *Organometallics*, **11**, 822 (1992).
    (b) J. He, H. Q. Liu, J. F. Harrod and R. Hynes, *Organometallics*, **13**, 336 (1994).
30. (a) E. Duguet, M. Schappacher and A. Soum, *Macromolecules*, **25**, 4835 (1992).
    (b) E. Duguet, M. Schappacher and A. Soum, WO 92/17527, PCT/FR92/0300 Oct. 15, 1992.
31. (a) C. R. Blanchard and S. T. Schwab, *J. Am. Ceram. Soc.*, **77**, 1729 (1994).
    (b) S. T. Schwab, R. C. Graef, Y. M. Pan and D. L. Davidson, NASA Conf. Publ., CP-3175 [Part 2], 721 (1992).
    (c) S. T. Schwab, C. R. Blanchard and R. C. Graef, *J. Mater. Sci.*, in press.
32. (a) T. Isoda, H. Kaya, H. Nishii, O. Funayama, T. Suzuki and Y. Tashiro, *J. Inorg. Organomet. Polym.*, **2**, 151 (1992).
    (b) M. Arai and T. Isoda, Japan Patent JP. 145903 (1985) *Chem. Abstr.*, **104**, 36340r (1985).
    (c) M. Arai, S. Sakurada, T. Isoda and T. Tomizawa, *Am. Chem. Soc., Polym. Div., Polym. Prepr.*, **27**, 407 (1987).
    (d) T. Isoda, in *Silicon Nitride Ceramics 2* (Eds. M. Mitomo and S. Somiya) U. Rokakuho 45 (1990).
    (e) Tonen technical data sheets, 9104 series.
33. G. T. Burns and G. Chandra, *J. Am. Ceram. Soc.*, **72**, 333 (1989).
34. T. Taki, M. Inui, K. Okamura and M. Sato, *J. Mater. Sci. Lett.*, **8**, 1119 (1989).
35. W. R. Schmidt, P. S. Marchetti, L. V. Interrante, W. J. Hurley, R. H. Lewis, R. H. Doremus and G. E. Maciel, *Chem. Mater.*, **4**, 937 (1992).
36. M. Peukert, T. Vaahs and M. Brück, *Adv. Mater.*, **2**, 398 (1990).
37. (a) S. Schaible, R. Riedel, E. Werner and U. Klingebiel, *Appl. Orgnaomet. Chem.*, **7**, 53 (1993).
    (b) R. Riedel and M. Seher, *J. Eur. Ceram. Soc.*, **7**, 21 (1991).
38. L. Maya, D. R. Cole and E. W. Hagaman, *J. Am. Ceram. Soc.*, **74**, 1686 (1991) and references cited therein.
39. (a) W. Verbeek, U. S. Patent No. 3, 853, 567 (Dec. 1974).
    (b) G. Winter, W. Verbeek and M. Mansmann, U. S. Patent No. 3,892,583 (July 1975).
40. (a) B. G. Penn, F. E. Ledbetter III and J. M. Clemons, *Ind. Eng. Chem., Process Des. Dev.*, **23**, 217 (1984).
    (b) B. G. Penn, J. G. Daniels, F. E. Ledbetter and J. M. Clemons, *Polym. Eng. Sci.*, **26**, 1191 (1986).
    (c) B. G. Penn, F. E. Ledbetter III, J. M. Clemons and J. G. Daniels, *J. Appl. Polym. Sci.*, **27**, 3751 (1982).
41. D. M. Narsavage, L. V. Interrante, P. S. Marchetti and G. E. Maciel, *Chem. Mater.*, **3**, 721 (1991).
42. D. Seyferth, C. Strohmann, N. R. Dando and A. J. Perrotta, *Chem. Mater.*, **7**, 2058 (1995).
43. N. R. Dando, A. J. Perrotta, C. Strohmann, R. M. Stewart and D. Seyferth, *Chem. Mater.*, **5**, 1624 (1993).

44. (a) Y. D. Blum, K. B. Schwartz, E. J. Crawford and R. D. Hamlin, in *Better Ceramics Through Chemistry III*, MRS Symp. Proc. Vol. 121 (Eds. C. J. Brinker, D. E. Clark and D. R. Ulrich), Mater. Res. Soc., Pittsburgh, 1988, pp. 565–570.
    (b) Y. D. Blum, S. M. Johnston and G. A. McDermott, in *Science, Technology, and Commercialization of Powder Synthesis and Shape Forming Processes* (Eds. J. J. Kingsley, C. H. Schilling and J. H. Adair), *Ceramic Trans.*, **62**, 67 (1996).
    (c) J. T. McGinn, Y. Blum, S. M. Johnson, M. I. Gusman and G. A. McDermott, in *Better Ceramics Through Chemistry VI*, MRS Symp. Proc. Vol. 346 (Eds. A. K. Cheetham, C. J. Brinker, M. L. Mecartney and C. Sanchez), Mater. Res. Soc., Pittsburgh, 1994, pp. 409–414.
45. (a) G. E. Legrow, T. F. Lim, J. Lipowitz and R. S. Reaoch, *Am. Ceram. Soc. Bull.*, **66**, 363 (1987).
    (b) J. Lipowitz, H. A. Freeman, R. T. Chen and E. R. Prack, *Adv. Ceram. Mater.*, **2**, 121 (1987).
    (c) G. E. Legrow, T. F. Lim, J. Lipowitz and R. S. Reaoch, *Am. Ceram. Soc. Bull.*, **66**, 363 (1987).
    (d) G. E. Legrow, T. F. Lim and J. Lipowitz, *J. Chim. Phys.*, **83**, 869 (1986).
46. F. Siriex, P. Goursat, A. Lecomte and A. Dauger, *Compos. Sci. Tech.*, **37**, 7 (1990).
47. (a) N. S. C. K. Yive, R. J. P. Corriu, D. Leclercq, P. H. Mutin and A. Voiux, *Chem. Mater.*, **4**, 141 (1992).
    (b) N. S. C. K. Yive, R. J. P. Corriu, D. Leclercq, P. H. Mutin and A. Vioux, *Chem. Mater.*, **4**, 1263 (1992).
48. T. Vaahs, Hoechst High Chem, Polysilazane ET 70, VT 50 Facts Sheets, AFE 23 25e, 1990.
49. (a) R. Riedel, A. Kienzle and M. Fuess, in *Applications of Organometalic Chemistry in the Preparation and Processing of Advanced Materials*, NATO ASI Ser. E: Appl. Sci.-Vol. 297 (Eds. J. F. Harrod and R. M. Laine) Kluwer Publ., Dordrecht, 1995, pp. 155–171 and references cited therein.
    (b) A. O. Gabriel and R. Riedel, poster PB107 at the XI International Symp. on Organosilicon Chemistry.
    (c) R. Riedel, A. Greiner, G. Miehe, W. Dressler, H. Fuss and J. Bill, submitted for publication.
50. S. J. Lenhart, Y. D. Blum and R. M. Laine, *J. Corrosion*, **45**, 503 (1989).
51. T. Sugama and N. Carciello, *Mater. Lett.*, **14**, 322 (1992).
52. J. M. Schwark and A. Lukacs, in *Inorganic and Organometallic Polymers II*, Am. Chem. Soc. Symp. Ser. Vol. 572, (Eds. K. Wynne, P. W. Nielson and H. Allcock), *Am. Chem. Soc.*, 1994, pp. 43–54.
53. B. C. Mutsuddy, in *High Tech Ceramics* (Ed. P. Vincenzini), Elsevier Sci. Publ., Amsterdam, 1987, pp. 571–589.
54. (a) R. Riedel, G. Passing, H. Schönfelder and R. J. Brook, *Nature*, **355**, 714 (1992).
    (b) R. Riedel, M. Seher and G. Becker, *J. Eur. Ceram. Soc.*, **5**, 113 (1989).
    (c) R. Riedel, H-J. Kleebe, H. Schönfelder and F. Aldinger, *Nature*, **374**, 526 (1995).
55. G. T. Burns, C. K. Saha, G. A. Zank and H. A. Freeman, *J. Mater. Sci.*, **27**, 2131 (1992).
56. M. F. Gonon, G. Fantozzi, M. Murat and J. P. Disson, *J. Eur. Ceram. Soc.*, **15**, 591 (1995).
57. V. S. R. Murthy and M. H. Lewis, *Curr. Sci.*, **62**, 744 (1992).
58. Y. Nakaido, Y. Otani, N. Kozakai and S. Otani, *Chem. Lett.*, 706 (1987).
59. H. Porte and J-J. Lebrun, European Pat. Application 0 202 176 A1 (Nov. 20, 1986).
60. D. Mocaer, R. Pailler, R. Naslain, C. Richard, J. P. Pillot, J. Dunogues, C. Darnez, M. Chambon and M. Lahaye, *J. Mater. Sci.*, **28**, 3049 (1993).
61. T. Vaahs, M. Brück and W. D. G. Böcker, *Adv. Mater.*, **4**, 224 (1992).
62. C. K. Saha, G. Zank and A. Ghosh, in *Manufacture of Ceramic Components, Ceramic Trans.*, **49**, 155 (1995).
63. A. Greiner, P. Kroll, R. Riedel and J. Bill, in *Adv. Ceram.-Matrix-Composites II, Ceramic Trans.*, **46**, 497 (1994).
64. N. Brodie, J-P. Majoral and J-P. Disson, *Inorg. Chem.*, **32**, 4646 (1993).
65. M. G. Salvetti, M. Pijolat, M. Soustelle and E. Chassagneux, *Sol. State Ion.*, **63–5**, 332 (1993).
66. R. van Weeren, E. A. Leone, S. Curran, L. C. Klein and S. C. Danforth, *J. Am. Ceram. Soc.*, **77**, 2699 (1994) and references cited therein.
67. (a) Y.-F. Yu and T.-I. Mah, in *Better Ceramics Through Chemistry II*, Mater. Res. Soc. Symp. Proc. Vol. 73 (Eds. C. J. Brinker, D. E. Clark and D. R. Ulrich), 1986, pp. 559–565.
    (b) Y.-F. Yu and T.-I. Mah, in *Ultrastructure Processing of Advanced Ceramics* (Eds. J. D. Mackenzie and D. R. Ulrich), Wiley, New York, 1988, pp. 773–781.

68. D. Seyferth, Y.-F. Yu and T. S. Targos, U. S. Patent 4,705,837 (Nov. 1987).
69. R. M. Laine, Y. D. Blum, R. D. Hamlin and A. Chow, in *Ultrastructure Processing of Advanced Ceramics* (Eds. D. D. Mackenzie and D. R. Ulrich), Wiley, New York, 1986, pp. 761–769.
70. K. Okamura, *Composites*, 107 (1988).
71. G-E. Yu, J. Parrick, M. Edirisinghe, D. Finch and B. Ralph, *J. Mater. Sci.*, **29**, 5569 (1994) and references cited therein.
72. C. R. Bickmore and R. M. Laine, *J. Am. Ceram. Soc.*, **79**, 2865 (1996) and references cited therein.
73. F. L. Riley (Ed.), *Progress in Nitrogen Ceramics*, NATO ASI Series E: Applied Sci. Vol. 65, 1983.
74. K. H. Jack, *J. Mater. Sci.*, **11**, 1135 (1976).
75. I-W. Chen, P. F. Becher, M. Mitomo, G. Petzow and T-S. Yen (Eds.), *Silicon Nitride Ceramics*, Mater. Res. Soc. Symp. Proc. Vol. 287, Pittsburgh, 1993.
76. D. R. Messier and R. P. Gleisner, U. S. Army Materials Technology Laboratory Publication MTL TR 92-6.
77. W. R. Schmidt, W. J. Hurley, R. H. Doremus, L. V. Interrante and P. S. Marchetti, in *Advanced Composite Materials, Ceram. Trans.*, **19**, 19 (1991).
78. G. D. Soraru, A. Ravagni, R. Campostrini and F. Babonneau, *J. Am. Ceram. Soc.*, **74**, 2220 (1991).
79. G. D. Soraru, M. Mercadini, R. D. Maschio, F. Taulelle and F. Babonneau, *J. Am. Ceram. Soc.*, **76**, 2595 (1993).
80. R. M. Laine, C. R. Bickmore, K. F. Waldner, B. L. Mueller and H. W. Estry, in *Silicon Nitride Ceramics* (Eds. I.-W. Chen, P. F. Becher, M. Mitomo, G. Petzow and T.-S. Yen), Mater. Res. Soc. Symp. Proc. Vol. 287, Mater. Res. Soc., Pittsburgh, 1993, pp. 251–256.
81. (a) M. Takamizawa, T. Kobayashi, A. Hayashida and Y. Takeda, U.S. Patent 4,550,151 (1985).
    (b) M. Takamizawa, T. Kobayashi, A. Hayashida and Y. Takeda, U.S. Patent 4,604, 367 (1986).
82. (a) O. Funayama, H. Nakhara, M. Okoda, M. Okumura and T. Isoda, *J. Mater. Sci.*, **30**, 410 (1995).
    (b) O. Funayama, Y. Tashiro, T. Kato and T. Isoda, Int. Symp. Organosilicon Chem. Directed Towards Mater. Sci., Sendai, Japan, Abstracts, 1990, pp. 95–96.
83. (a) K. Su, E. E. Remsen, G. A. Zank and L. G. Sneddon, *Chem. Mater.*, **5**, 547 (1993).
    (b) T. Wideman, K.Su, E. E. Remsen, G. A. Zank and L. G. Sneddon, *Chem. Mater.*, **7**, 2203 (1995).
    (c) T. Wideman, K. Su, E. E. Remsen, G. A. Zank and L. G. Sneddon, in *Mater. Res. Soc. Symp. Proc.* Vol. 410, Mater. Res. Soc., Pittsburgh, 1996, pp. 185–189.
84. (a) J. Lipowitz, *Am. Ceram. Soc. Ceram. Bull.*, **70**, 1888 (1991).
    (b) J. Lipowitz, *J. Inorg. Organomet. Polym.*, **1**, 277 (1991).
85. (a) R. Riedel and W. Dressler, *Ceram. Int.*, **22**, 233 (1996).
    (b) R. Riedel, A. Kienzle, W. Dressler, L. M. Ruwisch, J. Bill and F. Aldinger, *Nature*, **693**, 796 (1996).
86. Reference 1, pp. 655–662.
87. D. Seyferth and H. Plenio *J. Am. Ceram. Soc.*, **73**, 2131 (1990).
88. (a) J. Löffelholz and M. Jansen, *Adv. Mater.*, **7**, 289 (1995).
    (b) H.-P. Baldus, O. Wagner and M. Jansen in *Better Ceramics Through Chemistry VI*, (Eds. A. K. Cheetham, C. J. Brinker, M. L. McCartney and C. Sanchez) Mater. Res. Soc. Symp. Proc. Vol. 271, Pittsburgh, 1994, pp. 821–826.
    (c) H.-P. Baldus and G. Passing in *Mater. Res. Soc. Symp. Proc.* Vol. 346, Mater. Res. Soc., Pittsburgh, 1994 pp. 617–622.
    (d) M. Jansen, H.-P. Baldus and O. Wagner, Pat. Application DE 41 07 A1, disclosed Oct. 1992.
    (e) H.-P. Baldus, M. Jansen and O. Wagner, *Key Engineering Materials*, **89–91**, 75 (1994).
    (f) R. Riedel, A. Kienzle, G. Petzow M. Brück and T. Vaahs, Pat. Application DE 43 20 784 A1, disclosed May 1994.
89. O. Funayama, H. Nakahara, M. Okoda, M. Okumura and T. Isoda, *J. Mater. Sci.*, **30**, 410 (1995).
90. J. Bill, M. Friess, F. Aldinger and R. Riedel, in *Better Ceramics Through Chemistry VI*, (Eds. A. K. Cheetham, C. J. Brinker, M. L. McCartney and C. Sanchez) Mater. Res. Soc. Symp. Proc. Vol. 271, Pittsburgh, 1994, p. 839.
91. J. Hapke and G. Ziegler, *Adv. Mater.*, **7**, 380 (1995).

92. (a) S. Yajima, K. Okamura and J. Hayashi, *Chem. Lett.*, 1209 (1975).
    (b) S. Yajima, M. Omori, J. Hayashi, K. Okamura, T. Matsuzawa and C.-F. Liaw, *Chem. Lett.*, 551 (1976).
    (c) S. Yajima, H. Kayano, K. Okamura, M. Omori, J. Hayashi, T. Matsuzawa and K. Akutsu, *J. Am. Ceram. Soc.*, **55**, 1065 (1976).
93. (a) S. Yajima, K. Okamura, J. Hayashi and M. Omori, *J. Am. Ceram. Soc.*, **59**, 324 (1976).
    (b) S. Yajima, J. Hayashi, M. Omori and K. Okamura, *Nature*, **261**, 683 (1976).
    (c) S. Yajima, T. Shishido and H. Kayano, *Nature*, **273**, 525 (1978).
94. (a) S. Yajima, Y. Hasegawa, J. Hayashi and M. Iimura, *J. Mater. Sci.*, **13**, 2569 (1978).
    (b) Y. Hasegawa, M. Iimura and S. Yajima, *J. Mater. Sci.*, **15**, 720 (1980).
    (c) Y. Hasegawa and K. Okamura, *J. Mater. Sci.*, **18**, 3633 (1980).
95. T. Yamamura, T. Ishikawa, M. Shibuya and T. Hisayuki, *J. Mater. Sci.*, **23**, 2589 (1988).
96. Y. C. Song, Y. Hasegawa, S.-J. Yang and M. Sato, *J. Mater. Sci.*, **23**, 1911 (1988).
97. T. Yammura, *Am. Chem. Soc., Polym. Div., Polym. Prepr.*, **25**, 8 (1984).
98. K. A. Brown-Wensley and R. A. Sinclair, U. S. Patent 4,537,942 (1985).
99. (a) D. Seyferth, T. G. Wood, H. J. Tracy and J. L. Robison, *J. Am. Ceram. Soc.*, **75**, 1300 (1992).
    (b) D. Seyferth, H. J. Tracy and J. L. Robison, U. S. Patent No. 5,204,380 (1993).
    (c) D. Seyferth, C. A. Sobon and J. Borm, *New J. Chem.*, **14**, 545 (1990).
100. (a) F. Höfler, *Top. Curr. Chem.*, **50**, 129 (1974).
     (b) M. D. Curtis and P. S. Epstein. *Adv. Organomet. Chem.*, **19**, 213 (1981).
101. (a) Y. Mu and J. F. Harrod, in *Inorganic and Organometallic Oligomers and Polymers*, IUPAC 33rd Symp. on Macromol. (Eds. J. F. Harrod and R. M. Laine), Kluwer Publ., Dordrecht, 1991, p. 23.
     (b) J. F. Harrod, in *Inorganic and Organometallic Polymers with Special Properties* (Ed. R. M. Laine), NATO ASI Ser. E, Vol. 206, Kluwer Publ., Dordrecht, 1991, p. 87.
102. (a) T. D. Tilley, *Acc. Chem. Res.*, **26**, 22 (1993).
     (b) T. D. Tilley, H. G. Woo in *Inorganic and Organometallic Oligomers and Polymers*, IUPAC 33rd Symp. on Macromol. (Eds. J. F. Harrod and R. M. Laine), Kluwer Publ., Dordrecht, 1991, p. 3.
103. Z.-F. Zhang, F. Babonneau, R. M. Laine, Y. Mu, J. F. Harrod and J. A. Rahn, *J. Am. Ceram. Soc.*, **74**, 670 (1991).
104. (a) Z.-F. Zhang, S. Scotto and R. M. Laine, in *Covalent Ceramics II: Non-oxides*, Mater. Res. Soc. Symp. Proc. Vol. 327 (Ed. R. Gottschalk), Mater. Res. Soc., Pittsburgh, 1994, pp. 207–213.
     (b) Z.-F. Zhang, S. Scotto and R. M. Laine, in *Ceram. Eng. Sci. Proc.*, **15**, 152 (1994).
     (c) R. M. Laine, Z.-F. Zhang, K. W. Chew, M. Kannisto and C. Scotto, in *Ceramic Processing Science and Technology* (Eds. H. Hausner, G. Messing and S. Hirano), Am. Ceram. Soc., Westerville, OH, 1995, pp. 79–186.
105. J. P. Banovetz, R. M. Stein and R. M. Waymouth, *Organometallics* **10**, 3430 (1991).
106. Z.-F. Zhang, Ph.D. Thesis, University of Michigan, 1996.
107. T. Kobayashi, T. Sakakura, T. Hayashi, M. Yumura and M. Tanaka, *Chem. Lett.*, 1157 (1992).
108. E. Hengge, M. Weinberger and C. Jammegg, *J. Organomet. Chem.*, **410**, Cl (1991).
109. E. Hengge, *Organosilicon Chem. II*, **2**, 275 (1996).
110. E. Hengge and M. Weinberger, *J. Organometallic Chem.*, **433**, 21 (1992).
111. R. M. Laine and M. Kannisto, unpublished results.
112. (a) R. H. Baney, J. H. Gaul Jr and T. K. Hilty, *Organometallics*, **2**, 859 (1983).
     (b) R. H. Baney, U.S. Patent No. 4,310,482 (1982).
113. A. Tazi Hemida, J.-P. Pillot, M. Birot, J. Dunogues and R. Pailler, *J. Chem. Soc., Chem. Commun.*, 2337 (1994).
114. R. J. P. Corriu, M. Enders, S. Huille and J. J. E. Moreau, *Chem. Mater.*, **6**, 15 (1994).
115. T. L. Smith, U.S. Patent No. 4,631,179 (1986).
116. (a) L. V. Interrante and Q. H. Shen, *Macromolecules*, **29**, 5788 (1996).
     (b) L. V. Interrante, C. W. Whitmarsh, C.-Y. Yang and W. Sherwood, in *Silicon-Based Structural Ceramics*, Mat. Res. Soc. Symp. Proc. Vol. 365, Mater. Res. Soc., Pittsburgh, 1995, pp. 139–147.
117. (a) D. C. Deleeuw, J. Lipowitz and P. P. Lu, U.S. Patent No. 5,071,600 (1991).
     (b) J. Lipowitz, J. A. Rabe and G. A. Zank, *Ceram. Eng. Sci. Proc.*, **12**, 1819 (1991).
     (c) J. Lipowitz, T. Barnard, J. Bujalski, J. A. Rabe, G. A. Zank, A. Zangvil and Y. Xu, *Comp. Sci. Tech.*, **51**, 167 (1994).

118. C. P. Jacobson and L. C. DeJonghe, PCT WO 94/02430 (March, 1994).
119. W. Toreki, C. D. Batich, M. D. Sacks, M. Saleem, G. Choi and A. A. Morrone, *Comp. Sci. Tech.*, **51**, 145 (1994).
120. M. D. Sacks, G. W. Scheiffele, M. Saleem, G. A. Staab, A. A. Morrone and T. J. Williams in *Mater. Res. Soc. Symp. Proc.* Vol. **365**, 3 (1995).
121. (a) M.-T. S. Hsu, T. S. Chen and S. R. Riccitiello *J. Appl. Polym. Sci.*, **42**, 851 (1991).
  (b) S. R. Riccitiello, M. S. Hsu and T. S. Chen, *SAMPE Quarterly*, April, 1993, pp. 9-14.
122. R. Riedel, A. Kienzle, V. Szabo and J. Mayer, *J. Mater. Sci.*, **28**, 3931 (1993).
123. R. H. Baney, M. Itoh, A. Sakaibara and T. Suzuki, *Chem. Rev.*, **95**, 1409 (1995).
124. J. R. January, U. S. Pat. 4,472,510 (Sept. 1984).
125. D. A. White, S. M. Oleff, R. D. Boyer, P. A. Budinger and J. R. Fox, *Adv. Ceram. Mater.* **2**, 45 (1987).
126. D. A. White, S. M. Oleff and J. R. Fox, *Adv. Ceram. Mater.*, **2**, 53 (1987).
127. F. I. Hurwitz, L. Hyatt, J. Gorecki and L. D'Amore, *Ceram. Eng. Sci. Proc.*, **8**, 732 (1987).
128. F. Babonneau, K. Thorne and J. D. Mackenzie, *Chem. Mater.*, **1**, 554 (1989).
129. K. Kamiya, T. Yoko, K. Tanaka and M. Takeuchi, *J. Non-Cryst. Solids*, **121**, 182 (1990).
130. H. Ishida, R. Shick and F. Hurwitz, *Macromolecules*, **23**, 5279 (1990).
131. H. Ishida, R. Shick and F. Hurwitz, *J. Polym. Sci.: Part B: Polymer Physics*, **29** 1095 (1991).
132. F. I. Hurwitz, S. C. Farmer, F. M. Terepka and T. A. Leonhardt, *J. Mater. Sci.* **26** 1247 (1991).
133. H. Zhang and C. G. Pantano, *J. Am. Ceram. Soc.*, **73**, 958 (1990).
134. R. M. Renlund, S. Prochazka and R. H. Doremus, *J. Mater. Res.*, **6**, 2716, 2723 (1991).
135. F. I. Hurwitz, 24th International SAMPE Technical Conference, Oct. 20, 1992, pp. T950-961.
136. (a) P. A. Agaskar, *J. Am. Chem. Soc.*, **111**, 6859 (1989).
  (b) P. A. Agaskar, *J. Chem. Soc., Chem. Commun.*, 1024 (1992).
  (c) P. A. Agaskar, *Colloids and Surfaces*, **63** 131 (1992).
137. X. Xin, C. Aitken, J. F. Harrod and Y. Mu, *Can. J. Chem.*, **68**, 471 (1990).
138. (a) R. M. Laine, K. A. Youngdahl, F. Babonneau, J. F. Harrod, M. L. Hoppe and J. A. Rahn., *Chem. Mater.*, **2**, 464 (1990).
  (b) R. M. Laine, F. Babonneau, J. A. Rahn, Z.-F. Zhang and K. A. Youngdahl, in *Thirty-Seventh Sagamore Army Materials Research Conference Proceedings* (Ed.), D. J. Viechnicki), Publ. Dept. of the Army, 1991, pp. 159-169.
  (c) R. M. Laine, J. A. Rahn, K. A. Youngdahl and J. F. Harrod, in *Adv. Chem. Series; New Science in Transition Metal Catalyzed Reactions*, Vol. 230 (Eds. D. Slocum and W. R. Moser), 1992, pp. 553-565.
139. (a) M. Harris, T. Choudhary, D. Treadwell, R. M. Laine and L. Drzal *Mater. Sci. Eng. A*, **A195**, 223 (1995).
  (b) T. M. Choudhary, H. Ho, D. Treadwell, M. Harris, R. M. Laine and L. Drzal, *Mater. Sci. Eng. A*, **A195**, 237 (1995).
140. C. L. Frye and J. M. Klosowski *J. Am. Chem. Soc.*, **93**, 4599 (1971).
141. V. Belot, R. J. P. Corriu, D. Leclercq, P. H. Mutin and A. Vioux, *J. Non-Cryst. Solids*, **148**, 52 (1992).
142. L. Bois, J. Maquet, F. Babonneau, H. Mutin and D. Bahloul, *Chem. Mater.* **6**, 796 (1994).
143. M. D. Nyman, S. B. Desu and C. H. Peng *Chem. Mater.* **5**, 1636 (1993).
144. L. A. Haluska, K. W. Michael and L. Tarhay, U. S. Patent 4,756,977 (July, 1988).
145. S.-P. Jeng, K. Taylor, T. Seha, M.-C. Chang J. Fattaruso and R. H. Havemann, *Symp. on VLSI Tech.*, Digest of Tech. papers, Jap. Soc. Appl. Phys., Tokyo, Japan, 1995, pp. 61-62.
146. (a) L. K. Figge, Ph.D. dissertation, Univ. of Pennsylvania (Oct. 1996).
  (b) D. H. Berry and L. K. Figge, submitted.
147. See Dow Corning, Inc. fact sheets on Flowable Oxide™.
148. P. A. Agaskar, *Inorg. Chem.*, **30**, 2708 (1991).
149. C. L. Frye, G. A. Vincent and W. A. Finzel, *J. Am. Chem. Soc.*, **93**, 6805 (1971).
150. (a) R. M. Laine, K. Y. Blohowiak, T. R. Robinson, M. L. Hoppe, P. Nardi, J. Kampf and J. Uhm, *Nature*, **353**, 642 (1991).
  (b) K. Y. Blohowiak, D. R. Treadwell, B. L. Mueller, M. L. Hoppe, S. Jouppi, P. Kansal, K. W. Chew, C. L. S. Scotto, F. Babonneau, J. Kampf and R. M. Laine, *Chem. Mater.*, **6**, 2177 (1994).
151. P. Kansal and R. M. Laine, *J. Am. Ceram. Soc.*, **77**, 875 (1994).

152. (a) M. L. Hoppe, R. M. Laine, J. Kampf, M. S. Gordon and L. W. Burggraf, *Angew. Chem. Int. Ed. Engl*, **32**, 287 (1993).
  (b) P. Kansal and R. M. Laine, *J. Am. Ceram. Soc.*, **78**, 529 (1995).
153. R. M. Laine, in *Applications of Organometallic Chemistry in the Preparation and Processing of Advanced Materials*, NATO ASI Ser. E: Appl. Sci.-Vol. 297 (Eds. J. F. Harrod and R. M. Laine), Kluwer Publ., Dordrecht, 1995, pp. 69–78.
154. K. W. Terry, P. K. Ganzel and T. D. Tilley, *Chem. Mater.*, **4**, 1290 (1992).
155. M. L. Montero, I. Uson and H. W. Roesky, *Angew. Chem. Int. Ed. Engl.*, **33**, 2103 (1994).
156. R. Baranwal and R. M. Laine, *J. Am. Ceram. Soc.*, **80**, 1436 (1997).
157. R. M. Laine, D. R. Treadwell, B. L. Mueller, C. R. Bickmore, K. F. Waldner and T. Hinklin, *J. Chem. Mater.*, **6**, 1441 (1996).
158. R. Baranwal, A. Zika, B. L. Mueller and R. M. Laine, in *Electrorheological Fluids* (Eds. F. Filisko and K. Havelka), Plenum Press, New York, 1995, pp. 157–169.
159. K. Waldner, R. Laine, C. Bickmore, S. Dumrongvaraporn and S. Tayaniphan, *Chem. Mater.* **8**, 2850 (1996).
160. T. Kemmitt and N. B. Milestone, in *Silicon Containing Polymers* (Ed. R. G. Jones), Royal Soc. Chem., Cambridge, England, 1995, pp. 107–112.
161. D. C. Hrncir and G. D. Skiles, *J. Mater. Res.*, **3**, 410 (1988).
162. K. W. Terry and T. D. Tilley, *Chem. Mater.*, **3**, 1001 (1991).
163. K. Su, T. D. Tilley and M. J. Sailor, *J. Am. Chem. Soc.*, **118**, 3459 (1996).

CHAPTER **40**

Organo-silica sol–gel materials

DAVID AVNIR

Institute of Chemistry, The Hebrew University of Jerusalem, Jerusalem 91904, Israel

LISA C. KLEIN

Ceramics Department, Rutgers—The State University of New Jersey, Piscataway, NJ 08855-0909, USA

DAVID LEVY

Instituto de Ciencia de Materiales de Madrid, C.S.I.C., Cantoblanco, 28049 Madrid, Spain

ULRICH SCHUBERT

Institute for Inorganic Chemistry, the Technical University of Vienna, A-1060 Vienna, Austria

and

ANNA B. WOJCIK

Ceramics Department, Rutgers—The State University of New Jersey, Piscataway, NJ 08855-0909, USA

I. INTRODUCTION	2318
II. THE SILICA SOL–GEL PROCESS AND SILICA SOL–GEL MATERIALS	2319
III. DIRECT ENTRAPMENT OF ORGANIC AND BIOORGANIC MOLECULES IN SOL–GEL MATRICES	2320
A. General Aspects	2320
B. Applications in Analytical Chemistry: Sol–Gel Sensors	2321
C. Catalysis with Entrapped Organometallic Ion-pairs and Complexes	2322

The chemistry of organic silicon compounds, Vol. 2
Edited by Z. Rappoport and Y. Apeloig © 1998 John Wiley & Sons Ltd

D. Biochemistry Within Sol–Gel Matrices: The Entrapment of Enzymes
and Antibodies 2325
E. Electrochemistry with Organically Doped Sol–Gel Electrodes 2327
IV. COVALENT ENTRAPMENT OF ORGANIC FUNCTIONAL
GROUPS 2329
A. Trialkoxysilanes as Precursors for Functionalization 2329
B. Alkyl- and Aryl-substituted Materials 2331
C. Materials Substituted by Polymerizable Organic Groups 2333
1. Cocondensation with $E(OR')_n$ 2336
2. Cocondensation of different organically modified alkoxides 2336
3. Copolymerization with organic monomers 2337
D. Materials Substituted by Aminoalkyl or Phosphinoalkyl Groups 2339
E. Materials Substituted by Miscellaneous Groups 2341
V. ORGANIC/INORGANIC SOL–GEL COPOLYMERIZATIONS 2342
A. Physical Hybrids (Class I) 2342
B. Sequential Organic/Inorganic Interpenetrating Networks 2343
C. Simultaneous Organic/Inorganic Interpenetrating Networks 2344
D. Silsesquioxane-containing Hybrids (Class II) 2345
VI. PHOTOCHEMISTRY, PHOTOPHYSICS AND OPTICS OF DOPED
SOL–GEL MATERIALS 2346
A. Photoprobes for the Sol to Gel to Xerogel Transitions and for the Study
of the Properties of the Sol–Gel Cage 2346
B. Organic Photochemistry within Sol–Gel Matrices 2348
C. Sol–Gel Optics 2349
1. Photochromic sol–gel glasses 2349
2. Optical and photophysical properties of entrapped organic
molecules 2350
3. Materials substituted by chromophores 2353
VII. CONCLUDING REMARK 2354
VIII. ACKNOWLEDGMENTS 2354
IX. REFERENCES 2355

I. INTRODUCTION

This chapter describes a recent revolution in the chemistry and physics of materials. It has been founded on the interdisciplinary merging of organic chemistry and the world of ceramic materials, which traditionally have had little overlap between them. We hope that once the reader is exposed to the immense diversity of applications and unique properties of this novel family of materials, the justification of the word 'revolution' will become evident.

When one considers the unique properties of ceramics on the one hand, and the vastness of organic chemistry on the other, it seems almost inevitable that merging these domains of chemical research should lead to novel materials and to novel reaction configurations. We summarize in this chapter some representative research activities towards this goal, mainly since the early 80's which marked the emerging of this field from obscurity into a self-contained domain, with research activities in many academic and industrial laboratories[1–6]. The inherent difficulty in combining organic chemistry with glass and ceramic chemistry is the temperature parameter: organic compounds rarely survive temperatures higher than 200 °C (and biomolecules much less) but typical manipulation temperatures of ceramics and glasses can be well above 1000 °C. It is for this reason that over the millennia of the history of ceramics and glass production[7] one finds additives (e.g. colorants), only such that could withstand the high temperatures, namely inorganic salts and other oxides.

It has been known for quite some time[8] that inorganic oxides, with chemical compositions identical to glasses and ceramics, can be prepared at room temperatures, by polycondensation, precipitation and coagulation procedures of metal and semi-metal hydroxides or of their organic ethers (alkoxides) or esters[9]. Compared with the voluminous activity in the field of organic polymers, the study of formation of inorganic oxides by polycondensation remained for many years relatively low-key. The late 70's and early 80's witnessed a dramatic change: The recognition that better ceramic materials can be obtained through the detailed tailoring of the chemical aspects of the polycondensation[10] led to a world-wide burst of activity and to the birth of what became known as 'Sol–Gel Science'[9,10e,11]. Although sol–gel transitions are, of course, well documented in many polymerization and precipitation phenomena, the terms 'sol–gel materials', 'sol–gel process' etc. are used today in the connotation of the synthesis of inorganic materials from suitable monomers, passing indeed through sol, gel and xerogel (dry gel) stages.

It became clear that the sol–gel process is a promising candidate for closing the ceramics–organic gap, by virtue of the ambient temperatures employed up to the xerogel stage. This gap closure has three principle aspects, summarized in this chapter: the direct physical doping of ceramic materials with organic and bioorganic molecules; the functional modification of the ceramic matrix by copolymerization with organo-metal-alkoxides; and the formation of organo-ceramic composites by copolymerizations. As we shall see below, the applications of these novel materials cross chemistry from coast to coast, bridging in many instances traditional domains of chemistry, namely inorganic chemistry, organic chemistry, biochemistry, polymer chemistry, environmental chemistry, photochemistry, electrochemistry, optics, catalysis and more.

The main message of this chapter is that the chemistry and physics of organic molecules are carryable within ceramic matrices, allowing a plethora of novel and classical applications, and providing new insight on properties of both the organic component and of the matrix itself.

II. THE SILICA SOL–GEL PROCESS AND SILICA SOL–GEL MATERIALS

Most of the inorganic oxides, MO_n (where M is a metal or semi-metal, and n is not necessarily an integer), as well as many mixed oxides, have been prepared by the sol–gel process[9–11]. Many studies concentrated on SiO_2, the topic of this chapter, although for most applications described below, this need not be the optimal matrix, and one may also consider oxides such as TiO_2, ZrO_2, Al_2O_3 as well as their composites with silica such as SiO_2/ZrO_2.

The formation of a sol–gel porous material is through a hydrolysis–polycondensation reaction. An example is given in equation 1 with the methoxide of silicon (tetramethylorthosilicate, TMOS), but many other alkoxides, aryl oxides and acyl oxides can be used, as well as Si—N and Si—Cl compounds.

$$Si(OCH_3)_4 + H_2O \xrightarrow{H^+ \text{ or } OH^-} (SiO_mH_n)_p + CH_3OH \quad \text{(unbalanced)} \quad (1)$$

The hydrolysis introduces all of the oxygen into the silicon oxide matrix, i.e. all of the oxygen comes from the water molecules. The underlying details of the chemistry are extremely complex[12–14] and the detailed description of mechanistic studies is beyond the scope of the topic of this chapter, except for mechanistic studies which involve organic molecules and polymers—these are given in the various sections below. Here, some general aspects are summarized.

The values of m, n and p in the general silica equation are dictated by a host of parameters. These include water/silane ratio ($= r$), concentration of H^+ or OH^-, co-solvent, temperature, method and rate of drying, the existence of dopants and other additives, the size and shape of the product (thin film, monolith, powder) and even the chemical nature of the surface of the reaction container or thin film support. Typically, $(SiO_mH_n)_p$ are highly porous materials with pore size distributions which are centered around several tens of Angstroms, with surface areas usually in the range of several hundreds of $m^2 g^{-1}$, although surface areas as low as a few $m^2 g^{-1}$ have been recorded as well[13]. As for the values of m and n, the theoretical minimal composition is m close to 2 and n close to zero, for completely nonporous SiO_2. However, the surface of the porous product is densely coated with various types of silanols [$-SiOH$, $-Si(OH)_2$] moieties, so that the other hypothetical extreme (all Si is at the surface) is about SiO_3H_2. Adsorbed water molecules, which are always there, affect the n/m ratio as well. It is the fact that the silica has actually the SiO_mH_n composition (and not the commonly cited SiO_2) that is crucial for many of the special properties listed below. The value of p in the general formula can also be very high: since the monomer has four arms, the resulting polymer is extremely branched and a monolithic sol–gel block can in principle be one huge molecule with p approaching Avogradro's number.

The complexity of the polycondensation process has two levels. The first one is the competition between the hydrolysis rate of the Si—OR groups, forming Si—OH + ROH, and the rate of the polycondensation steps. The second one is that in the growing oligomers, each of the Si—OH and Si—OR groups has a distinctly different electronic and steric environment resulting in distinct hydrolysis and condensation rates for each of these groups. It is not surprising therefore that the control of the physical properties of porous sol–gel materials is still a matter of experimental exploration (for some typical mechanistic studies, see elsewhere[14]). It should be noted however, that the very ability to control it is an important advantageous property of this class of materials.

III. DIRECT ENTRAPMENT OF ORGANIC AND BIOORGANIC MOLECULES IN SOL–GEL MATRICES

A. General Aspects

The doping procedure is simple and straightforward: the host molecule is added to the polymerizing mixture. When the polycondensation is completed, the dopant molecules are entangled in the inorganic polymeric network (Figure 1).

Intensive studies[15] of the entrapment of organic molecules in sol–gel matrices evolved over the years into the following generalizations:

1. Most molecules can be entrapped in sol–gel matrices.
2. The entrapped molecule retains much of its characteristic physical properties.
3. The entrapped molecule retains much of its chemical properties.
4. The entrapped molecules are accessible to external reagents through the pore network: chemical reactions and interactions are possible.
5. Sol–gel materials such as monolithic silicas or silica films are transparent well into the UV: optical applications are possible.

Other properties which have contributed to the attractiveness and versatility of the sol–gel doping approach are the chemical, photochemical and electrochemical inertness as well as the thermal stability of the matrix; the ability to induce electrical conductivity[16]; the richness of ways to modify chemically the matrix and its surface as well as the above-mentioned controllability of matrix structural properties; the enhanced stability of the entrapped molecule[1,17]; the ability of employing the chromatographic properties of the matrix for enhanced selectivity and sensitivity of reactions with the dopant[4]; the simplicity of the entrapment procedure; the ability to obtain the doped sol–gel material in any desired shape (powders, monoliths, films, fibers); and the ability to miniaturize it[18,19].

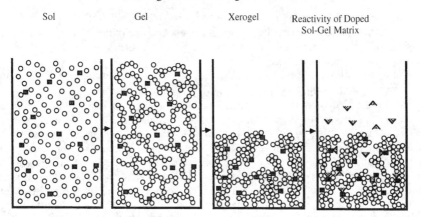

- ○ Sol Particles
- ■ Entrapped Species
- ▼ External Reactant

FIGURE 1. The sol–gel polymerization and entrapment process. (a) A sol of silica or other inorganic oxide particles (○) is prepared by polymerization of metal alkoxides in the presence of the desired dopant molecules, or of other monomers or polymers (■). (b) The sol turns into a gel within which the dopant molecules are entangled. Reactive monomers will form covalent bonds with the growing oxide network. It then dries and shrinks, forming a porous xerogel, (c), within which the dopant molecules are physically entrapped, or covalently attached. The pores are sufficiently narrow to render the matrix transparent, allowing optical applications. (d) The pore network allows external molecules (▲) to diffuse into the matrix and react with the trapped dopant (■). Sensing, catalysis, photochemistry and biocatalysis have been demonstrated

In the next sections we survey some of the applications of *reactive* organically doped sol–gel silicas. We return to optics and photophysical applications in Section VI, although historically, the first investigations were concentrated on optics[1,20] and it became evident only several years later that the entrapped molecule can also be involved in chemical reactions.

B. Applications in Analytical Chemistry: Sol–Gel Sensors

Our first family of applications of reactive doped sol–gel materials is the one that attracted most attention, namely in analytical chemistry[21]. Actually, the widest array of chemical reactions which has been carried out within sol–gel matrices was revealed while developing the doped sol–gel methodology for chemical sensing purposes[21–24]. They include proton-transfer reactions (such as pH indicators), redox reactions, complexations, ligand exchanges, enzymatic reactions and more. Common to these reactions is their ability to sense chemical changes in the environment of the entrapped molecule, revealing it through color-developing reactions between the entrapped reagent and an external diffusable chemical species or through emission of light. The chemical inertness and the optical transparency of SiO_2 sol–gel matrices have made the doping methodology quite attractive for sensor developments. Sensing reactions that have been demonstrated successfully with the sol–gel methodology include, for instance, many toxic metal cations, anions and various pollutants[21–24]. Another attractive feature for sensing purposes is that sol–gel materials can be made in various forms, as the analytical or optical method requires. Thus, one can prepare the sensing materials as a monolithic block[25], as a microscopic grain[18,19], as a thin film on a flat support[26], as an optical fiber coating[27,28] and in chromatographic capillary tubes[29].

High sensitivities are achievable with sol–gel sensors. An example is the sensing of Fe^{2+} [22], detected through its red-colored complex with three molecules of o-phenanthroline. This color evolution is far from being obvious: *three* entrapped phenanthroline molecules must complex with the penetrating Fe^{2+}, in order to form a color which is identical to the one formed in aqueous solution, as indeed is the case[23,30] (we return to this point below). The response curve of an e-*ortho*-phenanthroline (0.025%) SiO_2 porous glass to various concentrations of Fe^{2+} revealed an exceedingly high sensitivity with a detection limit of about 100 ppt (about two magnitudes of order better than conventional solution absorption spectroscopy)[23,30]. This very high sensitivity is due to a unique feature of the sol–gel matrix: being actually a high-surface area adsorbent, it *concentrates* by adsorption the impurity and then detects it. Facile detection of Fe^{+2} was also demonstrated by Lev and coworkers who developed the use of doped sol–gel particles as sensing chromatographic materials, packed within capillaries[29]. The idea is simple and convenient: the capillary is immersed in a given volume of solution containing the analyte; the solution is driven up by capillary forces, leaving behind a stain, the length of which is proportional to the concentration of the analyte. Besides Fe^{+2}, this approach was demonstrated for other analytes[29].

The development of sol–gel pH sensors attracted several laboratories, employing various indicators and various techniques based on both light absorption[29] and fluorescence emission[28,31]. Techniques included, as mentioned above, simple monolithic discs embedded with various routine indicators[22,25], thin films[27], various fluorescent fiber optics configurations[27,28], the capillaries[29,32] as well as prototype pH-meters constructed with some of these techniques[25,28]. For continuous, long-term, reversible applications leaching must be eliminated, and so covalent anchoring of trialkoxysilyl derivatives as described in Section IV below, of the pH indicators, may become desirable. A representative example is methyl-red, which was derivatized through its carboxylate moiety with trimethoxysilylpropylamine[19], shifting the pK_i from 5.2 to 3.0. Performance shifts in the case of copolymerization is a cost one should bear in mind. Another convenient method for eliminating leading is the co-entrapment of surfactants[25].

An important future aspect for the development of Useful applications of doped sol–gel sensors is miniaturization. One step towards that goal was undertaken[18] with the fluorescent pH-probe pyranine[33]. A pH-meter based on a 10-micron piece of doped sol–gel was constructed by incorporating it in a near-field optical microscopy configuration, which consists of metal-coated micropipette tips: a good, reversible titration curve was obtained. Further miniaturizing to *ca* 1 micron was achieved as well, with subsecond response time[18].

C. Catalysis with Entrapped Organometallic Ion-pairs and Complexes

The sol–gel approach has been used in catalysis in four main directions: preparation of inorganic oxide catalysts[34]; entrapment of metal microcrystallites[35]; polymerization of trialkoxysilyl derivatives of metal ligands[36,37]; and by direct, straightforward entrapment of organometallic catalysts in sol–gel matrices[38–40]. In this Section we concentrate on the latter and compare it with covalent attachment of trialkoxy derivatives, which is fully described in Section IV. As a representative example we consider the case of ion-pair catalysts[41,42] and, in particular, $RhCl_3$/Aliquat 336® and $RhCl_3/[Me_3N(CH_2)_3$

TABLE 1. Hydrogenation of several unsaturated compounds by the sol–gel entrapped ion-pair catalysts **1** and **2**[a]

Substrate	Catalyst	Products (yield in 1st run, %)
Styrene	1	ethylbenzene (100)
Styrene	2	ethylbenzene (48)
trans-Stilbene	1	bibenzyl (35)
trans-Stilbene	2	bibenzyl (64)
Benzalacetone	1	4-phenylbutan-2-one (62)
Benzalacetone	2	4-phenylbutan-2-one (72)
Phenylacetylene	1	styrene (72), ethylbenzene (20)
Phenylacetylene	2	styrene (18), ethylbenzene (3)
Benzene	1	cyclohexane (100)
Benzene	2	cyclohexane (68)
Acenaphthylene	1	acenaphthene (85)
Acenaphthylene	2	acenaphthene (54)
Naphthalene	1	tetralin (33), cis-decalin (66)
Naphthalene	2	tetralin (32), cis-decalin (66)
1-Naphthol	1	5-hydroxytetralin (24)
1-Naphthol	1	cis-decalin (12), trans-1-hydroxydecaline (37), trans-1-decalone (51)
1-Naphthol	2	1-hydroxytetralin (13), 5-hydroxytetralin (56)
1-Naphthol	2	cis-1-hydroxydecalin (52), cis-1-decalone (47)
Benzyl bromide	1	toluene (53)
Benzyl bromide	2	toluene (60)
(3-Bromopropyl) benzene	2	(3-bromopropyl)cyclohexane (50)
Nitrobenzene	1	aniline (34)
Nitrobenzene	2	aniline (50), aminocyclohexane (50)
Benzonitrile	1	benzylamine (98)
Benzonitrile	2	benzylamine (88)

[a]Typical procedure[39,40]: Entrapment of the catalysts was carried out by adding RhCl$_3$ and an equimolar amount of the appropriate quaternary ammonium salt to a water–methanol–tetramethoxysilane mixture in two steps. After drying and washing, the catalyst and the substrate were heated to 104 °C for 40 minutes. Upon reaction completion, the catalyst is simply filtered out and washed for re-use.

TABLE 2. Hydroformylation of some olefins by the sol–gel entrapped ion-pair catalysts **1** and **2**[a]

Substrate	Catalyst	Products (yield in 1st run, %)
Cyclopentene	1	c-C$_5$H$_9$CHO (95)
Cyclopentene	2	c-C$_5$H$_9$CHO (50)
Cyclohexene	1	c-C$_6$H$_{11}$CHO (94)
Cyclohexene	2	c-C$_6$H$_{11}$CHO (8)
1-Decene	2	Me(CH$_2$)$_9$CHO (41), Me(CH$_2$)$_7$CH(CHO)Me (41), Me(CH$_2$)$_6$CH(CHO)Et (8), Me(CH$_2$)$_5$CH(CHO)Pr (8)
Styrene	1	Ph(CH$_2$)$_2$CHO (68), PhCH(CHO)Me (31)
Styrene	2	Ph(CH$_2$)$_2$CHO (59), PhCH(CHO)Me (27)
1-Methylstyrene	1	PhCH(Me)CH$_2$CHO (12)
1-Methylstyrene	2	PhCH(Me)CH$_2$CHO (85)

[a] Reaction conditions: see footnote a to Table 1.

Si(OMe)$_3$]Cl, both entrapped in SiO$_2$ sol–gel matrices[39,40a]. While the former represents direct physical entrapment (catalyst **1**), the latter forms a covalent bond within the cage (catalyst **2**). These catalysts were used in numerous hydrogenation, hydroformylation and disproportionation reactions, and proved to be stable, leach-proof and recyclable. Occasionally, the catalytic efficiency dropped upon recycling, owing to pore blockage, but the activity could be restored by treatment with boiling water. The performances of the sol–gel-entrapped ion-pairs were compared with those of the homogeneous RhCl$_3$/Aliquat 336 catalyst and, in most cases, the immobilized catalysts proved superior to their homogeneous version. The results for many successful hydrogenation and hydroformylation reactions are collected in Tables 1 and 2. These immobilized ion-pair catalysts also proved to be stable, leach-proof and recyclable catalysts for disproportionation reactions of 1,3-cyclohexadiene and several other *vic*-dihydroarenes (Scheme 1 and Table 3). In these reactions, equimolar quantities of the respective tetrahydro and fully aromatic compounds were obtained. The entrapped catalysts, in most cases, proved to be more efficient and more selective than their homogeneous analogs. The reaction rates and conversions were shown to depend strongly on steric effects of substituents and on the bulkiness of the substrate skeleton. The recorded first-order kinetics in the substrates suggested that the mechanism involves stepwise addition of two molecules of the dihydroarenes to the rhodium nucleus, and that the addition of the first substrate molecule is rate-limiting.

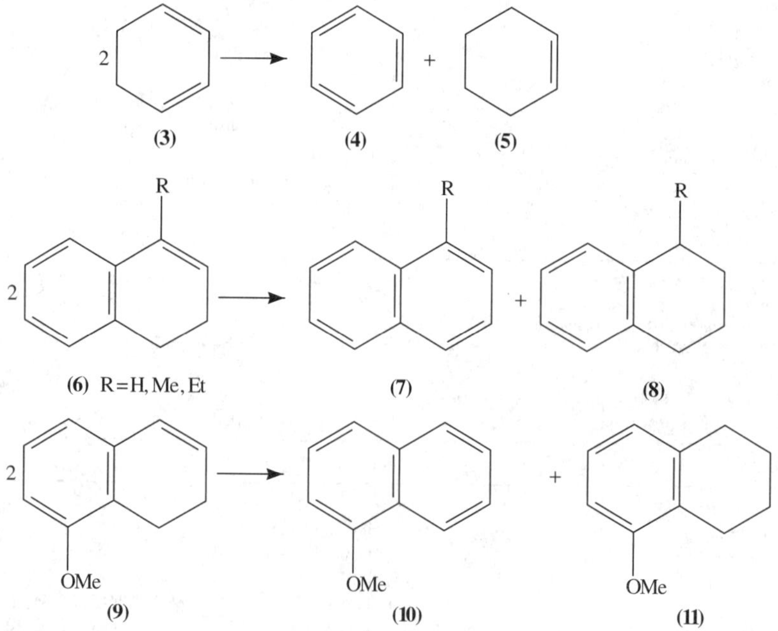

SCHEME 1

Finally, we mention the successful direct entrapment of the phosphinated complexes RuCl$_2$(PPh$_3$)$_3$, RhCl(PPh$_3$)$_3$, IrCl(CO)(PPh$_3$)$_2$ (see also Section IV), their water-soluble sulfonated analogs RuCl$_2$[Ph$_2$P(3-C$_6$H$_4$SO$_3$Na)]$_2$·4H$_2$O, RhCl[Ph$_2$P(3-C$_6$H$_4$SO$_3$Na)]$_3$·4H$_2$O, IrCl(CO)[Ph$_2$P(3-C$_6$H$_4$SO$_3$Na)]$_2$, and the dirhodium compounds *trans*-[Rh(CO)(PPh$_3$)(μ-1-pyrazole)]$_2$ and *trans*-[Rh(CO) (PPh$_3$)(μ-Cl)]$_2$, in silica sol–gel matrices. The catalysts proved to be active and efficient towards allylbenzene

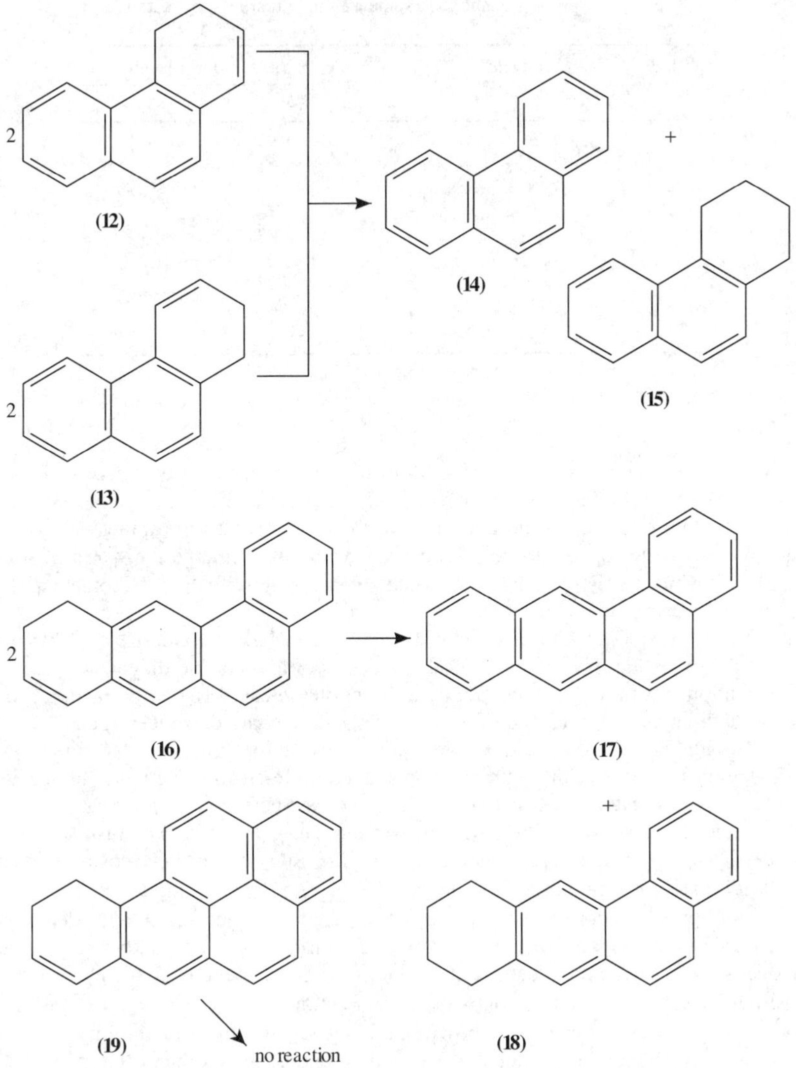

SCHEME 1. (*continued*)

isomerization[40b] (equation 2).

$$PhCH_2CH=CH_2 \xrightarrow{e\text{-}(RuCl_2)(PPh_3)_3} cis\text{- and } trans\text{-Ph}-CH=CHCH_3 \qquad (2)$$

D. Biochemistry Within Sol–Gel Matrices: The Entrapment of Enzymes and Antibodies

A priori there was no reason to believe that sol–gel encapsulation of enzymes will keep them alive. When one considers the ability of the alkoxide to react with the enzyme

TABLE 3. Disproportionation of some vic-dihydroarenes by catalysts **1** and **2**[a]

Substrate[b]	Products[b]	Yields after 40 min (%)	
		catalyst **1**	catalyst **2**
3	**4, 5**	97	96
6, R = H	**7**, R = H; **8**, R = H	83	99
6, R = Me	**7**, R = Me; **8**, R = Me	58	41
6, R = Et	**7**, R = Et; **8**, R = Et	16	17
9	**10, 11**	52	87
12	**14, 15**	28	73
13	**14, 15**	20	50
16	**17, 18**	8	23
19	no products	—	—

[a] See Reference 40a for experimental details; see Table 1.
[b] See Scheme 1 for substrates, reactions and products.

surface and active site, the pressure build-up on the protein as the matrix shrinks, the release of toxic alcohol during hydrolysis and the need to have the active site of the entrapped protein exposed to the pore-network, one appreciates the nontriviality of this remarkable aspect of doped sol–gel chemistry. The ability to merge the world of sol–gel ceramics with biochemistry[43], has motivated intensive activity[44,45d,e]. Some highlights are summarized next.

An intensively studied entrapped enzyme has been glucose oxidase[45a,b,46,47], which catalyzes the oxidation of D-glucose. That reaction is of appreciable diagnostic value, and this application is routinely carried out by utilizing the released H_2O_2 (a reaction product if oxygen is used as an oxidant) by forming a secondary colored product via a peroxidase catalyzed oxidation of aromatic aza compounds. Performing this procedure in a sol–gel matrix demonstrated the ability to carry out a multiple-enzyme network of reactions. Actually, enzymes that are hostile to each other and destructive in solution can be brought together in a harmless way when entrapped in the sol–gel matrix. For instance, trypsin autodigests itself completely in solution; by contrast, SiO_2-entrapped trypsin retained its full activity for several months, when incubated at pH 7.5[46]. Returning to glucose oxidase, a reversible optical glucose sensor was constructed by Shtelzer and Braun[48] employing the decoloration of the isoalloxazine moiety in the enzyme upon reductive complexation with glucose. Electrochemical detection of glucose with sol–gel enzymatic electrodes was carried out by several research groups (see also Section III.E)[49].

Some of the entrapped enzymes show remarkable enhancement in stability[43,50]. Thus, acid phosphatase, which in solution at 70 °C and pH 5.6 lived less than 0.1 min, had a two orders of magnitude jump in its half-life to 12 min in the entrapped form[50]. Since denaturation of enzymes involves folding–unfolding of the peptide chain, one may attribute the enhanced stability to the rigidity of the SiO_2 cage, which apparently restricts such motions. It has been demonstrated in a number of experiments that the geometric accessibility of the entrapped enzyme to incoming substrate molecules is perhaps the most important parameter which dictates its activity[43,50]. For instance, the efficiency of trypsin inhibitors, a set of benzoyl-arginine derivatives, was found to be linked to their size[46].

Many enzymes have been entrapped successfully in sol–gel matrices[44,45d–h]. The entrapment of parathion hydrolase[44,51], which is capable of detoxifying and detecting the pesticide Parathion, demonstrates the potential of the sol–gel methodology in environmental applications.

Of special interest are the entrapments of antibodies. Although still in its infancy, a number of promising recent reports point to the great potential of this direction. An example is that of monoclonal anti-atrazine mouse antibody which was entrapped successfully in SiO_2 sol–gel matrices, prepared from tetramethoxysilane by several methods, retaining its ability to bind free antigen from an aqueous solution[52]. Atrazine was selected as a model compound for that study, within the framework of the development of immunochemical-based methods for monitoring of pesticide residues and other organosynthetic environmental contaminants. Nanogram quantities of atrazine were applied on SiO_2-sol–gel columns doped with this antibody, and the amount of eluted antigen was determined using an Enzyme Linked ImmunoSorbent Assay (ELISA). The results showed that under appropriate sol–gel-forming conditions, the amounts of atrazine that were bound to the sol–gels were high, ranging between 60% to 91% of the amount applied to the column. The combination of the properties of the sol–gel matrix (e.g. stability, inertness, high porosity, high surface area and optical clarity) together with the selectivity and sensitivity of the antibodies bear a potential for further development of novel immunosensors which can be used for purification, concentration and monitoring of a variety of residues from different sources.

Indeed, more recent studies have shown the feasibility of this approach to yet another important group of pollutants, namely the nitroaromatics[58]. The nitroaromatic derivatives which are found most frequently as environmental contaminants are 2,4-dinitrotoluene and 2,6-dinitrotoluene used in plastics, dye and munitions manufacture, nitrophenols which are used as pesticides and 2,4,6-trinitrotoluene or 1,3,5-trinitrobenzene which are products of the munitions industry[53]. Large quantities of nitroaromatic compounds are currently manufactured all over the world, and their toxicity, mutagenicity and carcinogenicity are well established[53]. Accordingly, the need for extensive monitoring of nitroaromatic compounds in the environment (production effluents, toxic waste disposal sites, working places etc.) clearly exists. Nitroaromatic derivatives are also used in agriculture as insecticides, [e.g. parathion, (4,6-dinitro-*o*-cresol)], as herbicides (e.g. Ethalfluralin) and as fungicides (e.g. Quintozene). These compounds and their metabolites are often found as contaminants of food, soil and water. Towards this goal, the successful encapsulation of purified polyclonal IgG binding free antigen in aqueous solutions was reported[53]. The study was performed using a rabbit *anti*-2,4-dinitrobenzene (DNB) polyclonal antiserum and a 2,4-dinitrophenylhydrazine (DNPH) antigen which exhibits a high degree of cross-reactivity with the antiserum[54].

Additional reports include the covalent bonding of antibodies to functionalized sol–gel films[55], the entrapment of polyclonal fluorescein[56], the development of a sol–gel enzyme-linked immunosorbent assay (ELISA) test for antigenic parasitic protozoa[57] and an immunoassay for the detection of 1-nitropyrene[58].

To conclude this section we finally mention that various nonenzymatic proteins, whole cell extracts and whole intact cells were also entrapped successfully in silica sol–gel matrices[59], and these are reviewed elsewhere[44].

E. Electrochemistry with Organically Doped Sol–Gel Electrodes

The versatility of the doped sol–gel matrices was revealed also in the construction of electrochemical sensors. Tatsu and coworkers[60] constructed the first sol–gel glucose flow injection analyzer using an electrochemical sensor. The sensor was comprised of a silica sol–gel glucose oxidase powder which was attached to an oxygen electrode by a Nylon net and a cellulose membrane. Oxygen depletion was used for quantification of the converted glucose. The sol–gel methodology was also used employing redox-mediated electrodes. These sensors utilize an immobilized redox compound (e.g. ferrocene or hexacyanoferrate) which serves as an electron acceptor and affords oxygen-independent signal. The reduced mediator is regenerated by the anode and the exchanged Faradaic current is proportional to the amount of converted glucose. Audebert, Demaille and Sanchez constructed a ferrocene-mediated sol–gel biosensor by two successful methods[49a]: a two-stage silica sol–gel preparation procedure, and three gels prepared from commercial colloidal silica of various particle sizes. When this gel was kept in wet conditions, it maintained excellent activity and permitted good mobility of the analyte and the chemical mediator. The authors show that more than 80% of the glucose oxidase remained active in the gel, and that the Faradaic response of the electrode agrees well with theoretical calculations based on this activity.

Motivated by the need to produce stable bioactive silica films on conducting supports Lev and coworkers used two alternative approaches for the production of sol–gel derived amperometric biosensors[49b,61]. Thin films of glucose oxidase doped vanadium pentoxide (V_2O_5) were prepared from colloidal suspension. V_2O_5 doped with V^{4+} exhibits good electrical conductivity and adheres well to platinum and other conductive supports. This, and the ability to intercalate organic molecules[49b] made it suitable as a supporting matrix for active proteins. Sensing in this case is via the produced hydrogen peroxide which is electrooxidized on the metal support. A similar procedure was also used to prepare hydrogen peroxidase vanadium pentoxide biosensor[61].

Composite Carbon-silicate Electrodes (CCEs) were developed by Lev and coworkers[16,62–73]. The basic ingredients of the electrochemical sensors are graphite or carbon black powders dispersed in methyl-silicate sol–gel network. The carbon or graphite powder provides electric conductivity by an electron percolation mechanism and the porous silicate matrix contributes rigid and brittle construction. The hydrophobicity of the methyl modifiers prevents penetration of water to the electrode and thus only the outermost section of the electrode remains in contact with the electrolyte. These electrodes showed up to three orders improved Faradaic signal/background (capacitive) noise as compared to glassy carbon electrodes[63]. The electrode can be molded in diverse configurations including supported or unsupported thick films, rods, disks and even in the form of microelectrodes (approx. 20 micron in diameter)[62,63].

Incorporation of hydrophilic additives, such as polyethylene glycol or tetramethoxysilane in the sol–gel starting solution, endows a degree of hydrophilicity and a controlled section of the electrode can be wetted by the electrolyte. The wetted section is very stable and remains constant even after several weeks immersion in an electrolyte[65]. Graphite exhibits poor electrocatalytic activity but addition of trace metal or organometallic catalysts improves its electrocatalytic activity. CCEs containing organometallic catalysts such as cobalt phthalocyanine and cobalt porphyrin showed pronounced electrocatalytic activity toward the reduction of dioxygen[65,66]. Incorporation of the catalysts can be done by impregnation of the carbon powder with the organometallic catalysts prior to mixing with the sol–gel precursors or by adding the catalyst to the sol–gel precursors–carbon mixture. Inert metal ions (e.g. Pd, Pt) can also be added in ionic form to the sol–gel precursors and after electrode molding they can be reduced *in situ* by a flow of hydrogen gas through the porous electrodes[67,68]. Since the electrodes are porous, the inert metal and organometallic modified CCEs can act as gas electrodes such as oxygen, carbon monoxide and sulfur dioxide[67,68].

Since CCEs are molded at room temperature it is possible to incorporate enzymes into it as well[69–74]. The simplest form is comprised of glucose oxidase/methyl silicate (or silica)/graphite composite[16,69]. Dissolved oxygen served as electron acceptor and the resulting hydrogen peroxide is electrooxidized on the electrode. In order to lower the over-voltage that is required for hydrogen peroxide oxidation, palladium catalyst is added to the above construction. Electrodes for lactate (with lactate oxidase) and aminoacids (aminoacid oxidase) were prepared in a similar manner[64]. Another method to reduce the signal dependence on the level of oxygen was to incorporate a mediator into the sol–gel starting solution: Tetrathiafulvalene and glucose oxidase doped CCE was indeed found to show oxygen-independent response[69,70]. In another modification[71] ferrocenyl groups were covalently bonded to the glucose oxidase, which was encapsulated in the sol–gel matrix. Finally, a new type of porous, hybrid organic–inorganic material was synthesized and used for direct wiring of active enzymes[72,73]. The material was comprised of a dispersion of graphite powder and glucose oxidase incorporated in multifunctional, ferrocene-, amine- and methyl-modified silicate sol–gel backbone. Each component in this integrated construction accomplishes a specialized task: The graphite provides conductivity by percolation; the silicate provides highly crosslinked and rigid backbone, which is used to cage the redox enzyme; ferrocene is responsible for the signal transduction from the active center of the enzyme to the electron conductive surface; amine groups were incorporated for their high affinity to excess negative charges on the surface of glucose oxidase; and finally, the combination of methyl and amine groups allows control of the wetted electroactive section of the electrode. Recently, Wang and coworkers[74] showed that CCEs can be produced by thick film, ink jet technology. This technology enables highly reproducible mass production of biosensors.

IV. COVALENT ENTRAPMENT OF ORGANIC FUNCTIONAL GROUPS

A. Trialkoxysilanes as Precursors for Functionalization

The development of hybrid materials with covalent bonds between the organic and inorganic entities is based on molecular precursors of the general type $RSiY_3$, where R is an organic group bonded to the silicon atom via a Si−C bond. Compounds of the type R_2SiY_2 would serve the same purpose, but are hardly used for the development of sol–gel materials. We restrict ourselves to alkoxysilanes, $Y = OR'$, which are nearly exclusively employed in sol–gel chemistry. Hydrolysis and condensation of $RSi(OR')_3$ results in the formation of silsesquioxanes $RSiO_{3/2}$[75] in which the organic groups R are covalently bonded to the polysiloxane network. The materials can be inorganically 'diluted' by co-hydrolysis of $RSi(OR')_3$ (or mixtures of silanes with different organic groups R) with $Si(OR')_4$ in any ratio. The higher the amount of $Si(OR')_4$ in the starting mixture, the lower is the concentration of the organic groups in the final material and the more 'inorganic' are its properties. This blending of materials properties by using mixtures of different precursors is the basic idea behind the development of inorganic–organic hybrid materials. The sol–gel processing of mixtures of different alkoxides requires careful elaboration of the reaction conditions to avoid uncontrolled phase separations or to obtain a defined distribution of the different building blocks.

The group R can be almost any organic group if it is hydrolytically stable. It may serve several purposes, such as modifying the network structure, introducing organic functionalities into the inorganic network or providing reactive groups for organic crosslinking reactions.

When $Si(OR')_4$ is successively replaced by $RSi(OR')_3$, condensation sites are blocked and the average degree of crosslinking per silicon atom drops from 4 to 3. In reality the degree of crosslinking is somewhat smaller, due to residual Si−OR' and Si−OH groups in the final materials. Lowering the degree of crosslinking results in a modification of the microstructure and the physical properties associated with that (mechanical properties, for example). If this is the only goal of the system modification, simple organic groups, such as alkyl or phenyl groups, can be used as the substituents R. It should be kept in mind, however, that lowering the degree of crosslinking is an inevitable side-effect when organic groups are introduced for other purposes. For some applications it may therefore be necessary to balance the $RSi(OR')_3/Si(OR')_4$ ratio carefully.

The introduction of organo*functional* groups, i.e. the use of alkoxysilanes of the type $(R'O)_3Si-X-A$, where A is a functional organic group and X is a chemically inert spacer permanently linking Si and A, results in a more extensive chemical modification of the materials[76]. The properties of the organic functions A supplement the properties of the polysiloxane matrix formed by hydrolysis and condensation of the $Si(OR')_3$ and $Si(OR')_4$ units.

A variety of precursors of the type $(R'O)_3Si-X-A$ is commercially available or can easily be prepared. The spacer X is a $(CH_2)_n$ ($n = 2, 3$) chain in most cases. The preparation of such compounds has been reviewed elsewhere[77,78]. Therefore we restrict ourselves to demonstrating the more general routes by selected examples:

(i) *Hydrosilylation of alkenes or alkynes*, i.e. the rhenium catalyzed addition of Si—H bonds to double or triple bonds, e.g. equation 3[79].

(ii) *Substitution of organic groups*, e.g. equation 4[80].

(iii) *Addition to organic groups*, e.g. equations 5 and 6[81–84].

(iv) *Substitution of $(R'O)_3SiCl$ by Grignard or organolithium reagents*, e.g. equation 7[85].

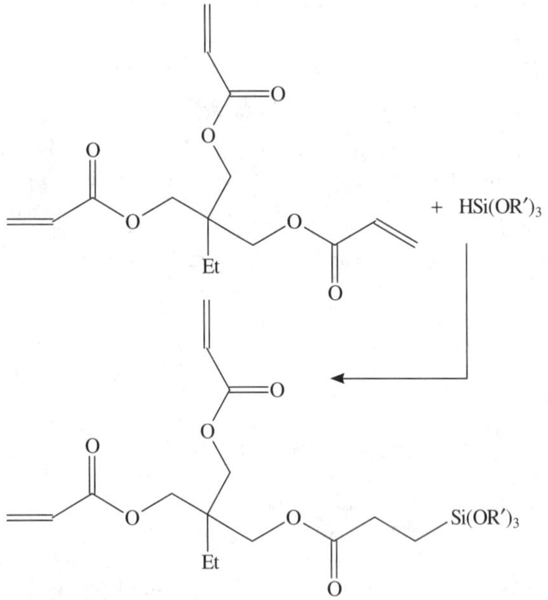

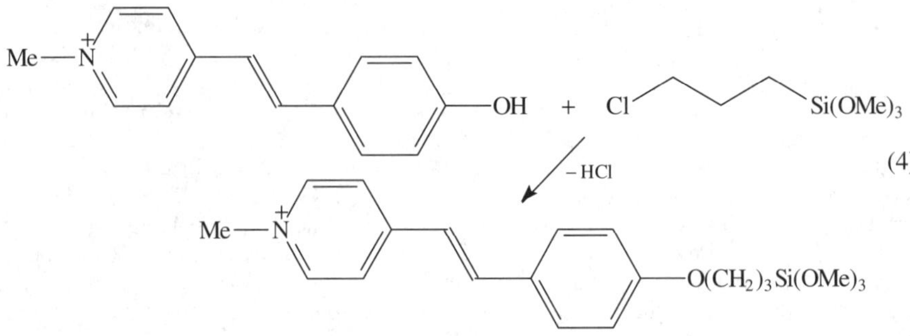

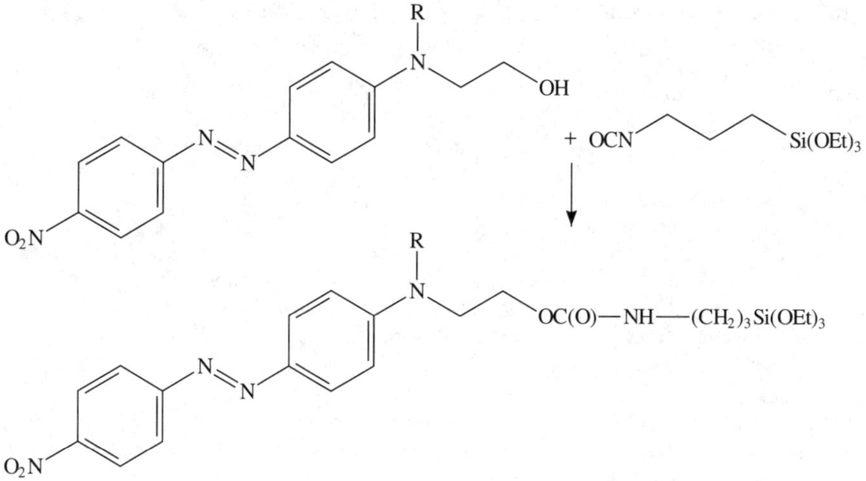

(5)

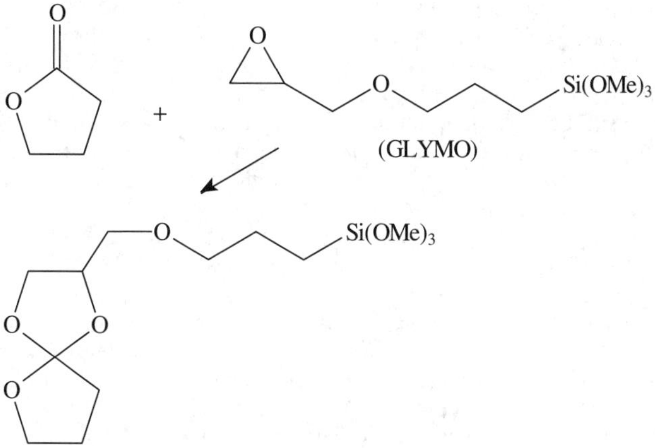

(6)

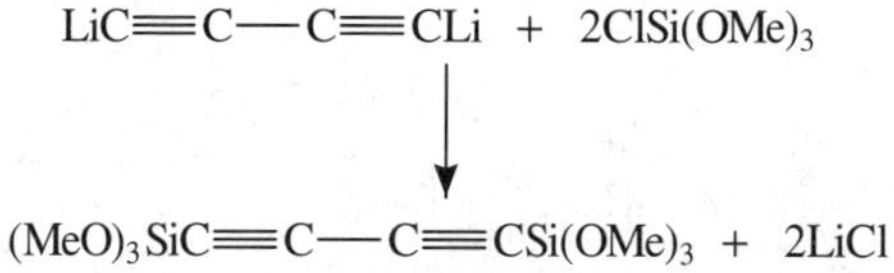

(7)

B. Alkyl- and Aryl-substituted Materials

Methyl- or phenyl-substituted silsesquioxanes obtained by hydrolysis and condensation of $RSi(OR')_3$ or $RSiCl_3$ (R = Me, Ph) were studied extensively by several groups[75]. MeSi— and PhSi— units are the classical components of silicones. The molecular weight and structure of methyl- or phenyl-substituted silsesquioxane polymers, and thus their physical properties (solubility, mechanical properties, etc.), depend very much on the reaction conditions. Since silicones are not the topic of this article, the interested reader is referred to the relevant literature.

In the context of sol–gel processing, $RSi(OR')_3/Si(OR')_4$ mixtures (R = simple aryl or alkyl groups) are interesting for several reasons:

(i) These groups allow an easy modification of the inorganic network (reduction of the average degree of crosslinking per silicon atom) without introducing too bulky or reactive organic groups.

(ii) Alkyl or aryl groups provide hydrophobicity to the resulting materials.

(iii) These mixtures are model systems for studying the sol–gel chemistry of $RSi(OR')_3/Si(OR')_4$ mixtures with less simple organic groups R.

Although alkyl groups already make sol–gel materials more hydrophobic than unmodified silica gels, fluorinated groups are even better for this purpose. The surface tension of 3,3,4,4,5,5,6,6,6-nonafluorohexylsilsesquioxane was found to be only one third of that of methylsilsesquioxane[86]. Addition of fluorinated alkoxysilanes of the type $(R'O)_3Si(CH_2)_n(CF_2)_mCF_3$ to a mixture of other precursors resulted in low energy coatings with antisoiling properties[87] or functional coatings on glass with increased hydrophobicity[88].

Considering the steric and electronic influence of different organic groups R on the hydrolysis and condensation rates of $RSi(OR')_3$ (relative to that of $Si(OR')_4$), it is obvious that in multicomponent precursor systems a statistical distribution of the molecular building blocks in the final materials is rather unlikely. This was, for example, demonstrated by ^{29}Si NMR spectra of gels obtained by cohydrolysis of $Me_2Si(OEt)_2$ and $Si(OEt)_4$ in a neutral or acidic medium[89,90]. Signals typical for hydrolyzed $Me_2Si(OEt)_2$ and for hydrolyzed $Si(OEt)_4$ were found, but also new signals assigned to links between both building blocks. Based on the NMR results structural models were proposed, depending on the $Me_2Si(OEt)_2/Si(OEt)_4$ ratio, x. For an excess of $Me_2Si(OEt)_2$ ($x \geqslant 2$), the tetrafunctional units derived from $Si(OEt)_4$ are rather isolated and serve to crosslink the polydimethylsiloxane chains derived from $Me_2Si(OEt)_2$. For $x < 2$, the spectra are characteristic for an interconnected network of $Si(-O-)_4$ and $MeSi(-O-)_3$ units. In any case, the $(Me_2SiO)_n$ chains appear to be rather long. $Ti(OR)_4$ and $Zr(OR)_4$ instead of $Si(OR)_4$ do not only act as crosslinking agents between the difunctional units, but also catalyze their condensation reaction[90].

The various reaction rates of different alkoxides are not just an obstacle on the way to homogeneous materials. They can be exploited for the deliberate engineering of some nanoheterogeneity (i.e. of the micro- and nanostructure) and thus provide additional possibilities for materials developments. A recent example to illustrate this point is the development of inorganic–organic hybrid aerogels from $RSi(OMe)_3/Si(OMe)_4$ mixtures (R = terminal alkyl and aryl, or certain functional organic groups)[91,92]. It is known that under basic conditions hydrolysis and condensation of $Si(OR')_4$ is faster than that of $RSi(OR')_3$. Therefore, the build-up of the gel network from $RSi(OMe)_3/Si(OMe)_4$ mixtures under these conditions is basically a two-step process. In the first stage, $SiO_x(OH)_y(OMe)_z$ clusters are formed by hydrolysis and condensation of $Si(OMe)_4$, while $RSi(OMe)_3$ basically is a co-solvent. The $RSi\equiv$ units condense at the surface of the primary clusters in the later stage of the reaction. Self-condensation of the $RSi-$ units does not play an important role under these conditions and with the proper $RSi(OMe)_3/Si(OMe)_4$ ratio. The latter depends on the nature of R and on the density of the gel. As a rule of thumb, 10–30 mol% $RSi(OMe)_3$ are sufficient to cover the inner surface of the aerogels, but do not affect their nanostructure from which the unique physical properties originate. The fate of the individual alkoxides was monitored by Raman spectroscopy, which provided clear evidence for the two-step process[92].

Since the $RSi\equiv$ units cap the $SiO_x(OH)_y(OMe)_z$ clusters formed during sol–gel processing of $Si(OMe)_4$, a hydrophobic inner surface is created, and the resulting aerogels are permanently stable against moisture (unmodified silica aerogels are immediately destroyed on contact with water). The uniform coverage of the inner aerogel surface by the organic groups also allows one to generate nanosized carbon structures by controlled pyrolysis of the organic groups. They partly cover the silica nanospheres from which the aerogel skeleton is composed, and result in an efficient infrared opacification necessary to improve the heat insulation properties of aerogels at high temperatures[93].

C. Materials Substituted by Polymerizable Organic Groups

Alkoxysilanes $(R'O)_3Si-X-A$, in which the group A can undergo organic polymerization reactions, are extremely interesting because they allow the preparation of inorganic–organic hybrid polymers. Polymerization (or any other crosslinking reaction) of the organic groups after sol–gel processing results in a dual network structure, where the inorganic and organic substructures are covalently linked to each other. Such polymers are mostly applied for coatings for different purposes, but recently also as bulk materials. They constitute the technically most advanced applications for inorganic–organic hybrid polymers.

The methacrylate substituted silane $(MeO)_3Si(CH_2)_3OC(O)C(CH_3)=CH_2$ (MEMO), vinyltrimethoxysilane $(MeO)_3SiCH=CH_2$ (VTMS) or the epoxy-substituted silane $(MeO)_3Si(CH_2)_3OCH_2(\overline{CHCH_2O})$, all being commercially available, are frequently used as precursors for this purpose.

Alkoxysilanes $(R'O)_3Si-X-A$ with different spacer groups X between the unsaturated organic group and the $Si(OR')_3$ moiety (different lengths and/or different chemical constitution) and/or with multiply unsaturated groups (to achieve a higher degree of organic crosslinking) allow the development of copolymers with tailored mechanical and optical properties. Such precursors were synthesized by using *one* double bond of a multiply unsaturated organic monomer to anchor the $Si(OR')_3$ group and retain the other double bond(s) for later polymerization reactions. An example is given in equation 3; many other multiply unsaturated organic compounds are available. Other coupling reactions can be utilized, such as SH addition of $(R'O)_3Si(CH_2)_3SH$ to $C=C$ bonds or coupling of $(EtO)_3Si(CH_2)_3NCO$ with OH-terminated bis(acrylates)[79,94,95]. Other reactive groups suitable for organic crosslinking by polymerization reactions can, of course, be linked analogously to $Si(OR')_3$ groups. For example, the spiro orthoester moiety shown in equation 6 is capable of ring-opening polymerization with only a very low shrinkage[84].

The influence of the steric and electronic properties of the group R on the hydrolysis and condensation kinetics of organically substituted alkoxysilanes $RSi(OR')_3$ was recently discussed in a review article[96]. It can be implied that functional and nonfunctional groups R of comparable size and electronic properties have the same influence if the organic function A in $(R'O)_3Si-X-A$ does not interact with the silicon atom or does not catalyze the reaction. Since polymerizable groups usually do not have basic or acidic properties, a catalytic effect on the sol–gel reactions cannot be expected. Only their steric bulk could influence the build-up and properties of the gel network. However, there are two other issues which are very important for the later polymerization of the organic groups: (i) are the functional groups retained during sol–gel processing? and (ii) where are the organic groups located in the primary hydrolysis products (which constitute the inorganic building blocks)? This is a particularly important question if mixtures of two or more different precursors are used, which is the case for most applications.

The hydrolysis and condensation reactions, and the development of the molar mass distribution of VTMS[97], GLYMO (equation 6)[98,99] and MEMO[98,100–103] was investigated by GPC (gas permeation chromatography), ^{29}Si NMR and FTIR. The uncatalyzed hydrolysis of both GLYMO and MEMO resulted in the slow formation of rather small oligomers[98,104]. The degree of oligomerization was influenced by the silane : water ratio.

Both basic and acidic catalysts accelerate the reaction rates and the growth of the oligomers, as expected. The influence of the kind of catalyst on the growth of oligomers from MEMO was studied by GPC[98]. With 1-methylimidazole, methacrylic acid or Zr(OPr-n)$_2$(OMe)$_2$ as catalysts, condensation essentially stopped at the stage of medium-sized oligomers ($M_p < 10^4$). Only with the strongly basic dimethylaminoethanol or with HF as the catalysts larger polymers were formed. In the dimethylaminoethanol catalyzed reaction, the polymer growth was not finished after 3 months[98,101]. The HCl-catalyzed hydrolysis of VTMS, MEMO or GLYMO also gave only medium-sized oligomers ($M_w = 1.0-1.5 \times 10^3$)[99,105].

The results of these studies can be summarized as follows. The growth mechanism of the siloxane oligomers is probably the same as proposed by Brown for the condensation of cyclohexyl- or phenylsilanetriol via cyclic tetramers and cubic octamers[106]. Condensation of the functionalized silanes stops at the stage of rather small oligomers unless *very* active catalysts are used. Even then, hydrolysis and condensation is not complete (residual Si—OH and Si—OR' groups) over rather long periods. This is probably due to steric effects of the rather bulky organic groups. Both the configuration and the molar mass distribution of the polymers depend on the solution pH.

Although random or ladder structures (Figure 2) have been proposed in some cases, the molecular weight distributions for the GLYMO and MEMO hydrolysates[98] indicate that octameric species play an important role in the formation of the polymers. Silsesquioxanes (RSiO$_{3/2}$)$_8$ (Figure 2) with various groups R have been synthesized[107] and can be considered model compounds for the hydrolysis intermediates.

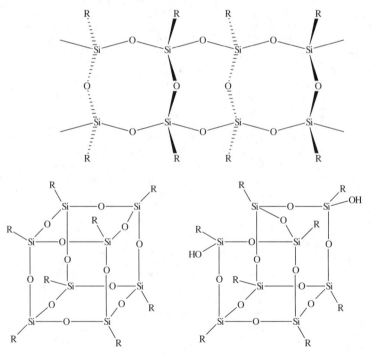

FIGURE 2. Structures of silsesquioxanes. Left: ladder structure; middle: 'double four-ring' (= cubic octamer); right: partial cage structure

In a mixture of two alkoxide precursors a mutual influence on the hydrolysis and condensation processes can be expected. How the molar mass develops in $Si(OR')_4/(R'O)_3Si-X-A$ mixtures is a question which was hardly investigated. For the MEMO/Si(OEt)$_4$ system this was studied by measuring the gel times (t_G)[108]. The addition of MEMO to an acidic mixture of Si(OEt)$_4$, ethanol and water resulted in a considerable increase of t_G. Under basic conditions the increase of t_G was less dramatic. The influence of MEMO is probably due to steric hindrance by the methacrylate group during formation of the gel network and the decrease of the average degree of crosslinking [partial replacement of $Si(-O-)_4$ by $RSi(-O-)_3$].

A major concern is the fate of the functional organic group A during sol–gel processing, particularly if it is not totally inert towards water. For example, hydrolytic ring opening of the epoxy group in GLYMO to give diol units is catalyzed by many basic or acidic compounds[109]. Since a catalyst has to be used for sol–gel processing to get real polymers (see above), its influence on the epoxy group is crucial. However, under neutral or slightly basic conditions the rate of the epoxide hydrolysis reaction is slow compared with the siloxane polycondensation reaction. The different reaction rates allow the preparation of hybrid materials by a two-step process, in which formation of the Si–O–Si network is carried out first and later the epoxy polymerization is induced[104,110]. For example, a highly flexible, nevertheless abrasion-resistant coating for polyethyleneterephthalate or polycarbonate was obtained from GLYMO, with 1-methylimidazole as the catalyst. 1-Methylimidazole acts as a sol–gel catalyst at room temperature, and in the second step of the materials synthesis as an initiator for the epoxide polymerization at a higher temperature[104,110]. Sol–gel processing of GLYMO can, of course, also be run in a way to deliberately produce diol units. Contact lenses[111] and antifogging coatings on glass[87] were developed by this approach. The preparation of hybrid polymers from the bifunctional molecular precursors $(R'O)_3Si-X-A$ is usually performed in the two-stage process described for GLYMO. The precursors are first reacted with water to form the inorganic structure to which the polymerizable organic groups A are attached. While gelation occurs, the materials can be molded or applied as coatings with conventional coating techniques (spray, dip, spin-on coating etc.). The organic groups are then polymerized or crosslinked to form the organic network. This results in a permanent hardening of the materials. Both reaction steps may occur more or less simultaneously, but the reaction in most cases is initiated by the addition of water and the catalyst, i.e. by starting the sol–gel process.

Another recent example also shows very elegantly this stepwise build-up of the inorganic and organic substructures. Transparent gels containing bridging 1,3-butadiyne units were obtained by sol–gel processing of $(MeO)_3SiC\equiv C-C\equiv CSi(OMe)_3$ (see equation 7). When the dried gels were heated in the solid state to about 200 °C, the highly reactive diyne units polymerized to give a more crosslinked inorganic–organic hybrid polymer[112].

There are several options by which the composition, structure and properties of the hybrid polymers can be varied:

(i) Cocondensation with other alkoxides: either with $Si(OR')_4$ or binary alkoxides of other elements $[E(OR')_n]$ and/or with other organically or organofunctionally substituted alkoxides.

(ii) Organic copolymerization with organic monomers, macromolecules or reactive substituents of other organofunctional alkoxides.

(iii) Choice of thermal or photochemical polymerization. Whether thermal or photochemical curing is more suitable depends on the chemical system and the kind of application. Photochemical curing is preferred for coatings of polymers with low thermal stability or for applications in which very rapid curing is essential.

With these options, there are many possibilities to modify organic–inorganic hybrid polymers chemically and thus to tailor their macroscopic properties. Not only the kind and connectivity of both the organic and inorganic building blocks can be varied (including combinations of different groupings in both parts), but also their ratio. Although sometimes difficult to control, multicomponent systems allow to some extent the fine-tuning of macroscopic materials properties, due to the specific functions and properties of each component.

1. Cocondensation with $E(OR')_n$

'Dilution' of an organofunctional alkoxysilane $(R'O)_3Si-X-A$ with an unsubstituted metal or semi-metal alkoxide $E(OR')_n$ results in materials with properties between those of the silsesquioxane $O_{3/2}Si-X-A$ and the oxide $EO_{n/2}$. A steady change of materials properties between the two extremes is in principle possible by varying the ratio of the two alkoxides. Employing another alkoxide than $Si(OR')_4$ additionally allows some variation of the chemical nature of the inorganic building blocks and of the materials properties originating from them.

The first commercial application of inorganic–organic hybrid polymers may illustrate this point. A scratch-resistant, transparent coating for optical polycarbonate lenses was developed from mixtures of GLYMO, $Si(OMe)_4$ and $E(OR)_n$ [$E(OR)_n$ = $Ti(OEt)_4$, $Zr(OPr)_4$ or $Al(OR)_3$, respectively][113]. The binary alkoxides have two functions: they build up the inorganic substructure of the final material, and the nonsilicon alkoxides $E(OR)_n$ already at room temperature catalyze the polymerization of the epoxide groups[109]. Since the rates for the reaction of the different alkoxides with water are very different, the way how the water (and the catalyst) is added to the precursor mixture is very crucial in obtaining homogeneous and transparent materials. The ratio of GLYMO, $Si(OMe)_4$ and $E(OR)_m$ is another parameter that influences both the rate of the polymerization reaction (and thus the viscosity increase and the processability) and the macroscopic properties of the coatings[113,114].

For the development of a scratch- and abrasion-resistant, corrosion-preventing coating system for brass, the system had to be modified. A better adhesion to the inorganic surface and also a reduction of the brittleness of the coating was achieved by sol–gel processing of a mixture of GLYMO, $PrSi(OMe)_3$ and $Al(OBu-s)_3$. Adhesion to the metal surface was very good even after weathering, due to the covalent bonding between the metal surface and the coating[115].

2. Cocondensation of different organically modified alkoxides.

The use of more than one organofunctional alkoxide precursor can serve different purposes:
  (i) to enable organic crosslinking by addition reactions without polymerization,
  (ii) to modulate the organic substructure, or
  (iii) to link different inorganic building blocks.

In the first example two different silanes were used, which allow different ways of organic crosslinking. A coating system for the mechanical protection of plastics, particularly polymethylmethacrylate, was obtained from $(MeO)_3SiCH=CH_2$ (VTMS) and $(MeO)_3Si(CH_2)_3SH$ (MPTMS). Two crosslinking modes of the organic groups are possible upon UV irradiation: the vinyl groups may polymerize to yield polymethylene links between silicon atoms, or the $-SH$ function may add to a vinyl group to give $-(CH_2)_2-S-(CH_2)_3-$ links. When VTMS and MPTMS were employed in a 9:1 molar ratio, vinyl polymerization prevailed, whereas with a 1:1 ratio the thiol addition reaction was dominating. In the latter case, the coatings were then cured in about half the time because this reaction is faster than vinyl polymerization. Owing to the different chain lengths of the alternative links, the flexibility of the coating can be varied by the ratio of the starting compounds[110].

Very versatile hybrid materials for applications in microsystem technologies were prepared from a mixture of MEMO, GLYMO, Si(OEt)$_4$, VTMS or diphenylsilanediol[116]. These materials were used, *inter alia*, for passivation and encapsulation of electrical components, as protective coatings for thin film capacitors[117] or as patternable coatings. The photolithographic structuring procedure illustrates the functions of the different precursors. The inorganic network is formed during sol–gel processing. The thus-obtained coating lacquer was applied to the substrates by standard techniques. The methacrylate groups of MEMO were photochemically polymerized by structuring UV irradiation (masks or laser writing). The nonexposed parts of the coating was then dissolved by a developer solution. Finally, the structured coating was thermally cured by polymerization of the epoxy groups of GLYMO[116].

Another example shows that two different inorganic building blocks, i.e. silicate and zirconate structures, can also be linked by organic polymer units. A scratch-resistant coating material for polycarbonate (for instance, compact disks)[101,110] for optical and microoptical applications[103] and transparent monoliths[118] was developed by copolymerizing the methacrylate units of (n-PrO)$_3$Zr(methacrylate) or (n-PrO)$_2$Zr(methacrylate)$_2$ and MEMO after sol–gel processing of the precursor mixture.

3. Copolymerization with organic monomers

When hybrid polymers are prepared from only (R′O)$_3$Si−X−A, the crosslinking of all available organic groups is often difficult, probably for steric reasons.

The dimension of the organic structures and the degree of organic crosslinking in inorganic–organic hybrid polymers can be increased by copolymerizing the reactive groups of the organofunctional alkoxide (R′O)$_3$Si−X−A with organic monomers or small oligomers. For example, photochemical polymerization of an aged gel, obtained from a 1:1 mixture of MEMO and Si(OEt)$_4$, left about 20% of the initial double bonds unreacted. However, when two equivalents of methylmethacrylate (MMA) were added, UV irradiation resulted in the complete polymerization of the methacrylate groups from both sources (MEMO and MMA)[102]. A mixture of MEMO, MeSi(OEt)$_3$ and CF$_3$CH$_2$CH$_2$Si(OMe)$_3$ was also copolymerized with fluorinated acrylic monomers and oligomers[119]. Other simple examples of this very useful approach include the copolymerization of hydrolyzed N-(3-trimethoxysilylpropyl)pyrrole with pyrrole[120] or GLYMO with cycloaliphatic diepoxide monomers[121].

The latter reaction provided a new type of Li$^+$ conducting polymer. MEMO and GLYMO were hydrolyzed and then ethylene glycol diglycidyl ether was added. Methylimidazole was used both as a catalyst for the sol–gel step and the copolymerization of the epoxide groups of GLYMO and the diglycidyl ether. Due to the presence of polyethylene oxide units, LiClO$_4$ is easily dissolved in the hybrid polymer. The obtained material is amorphous after curing and can be applied as coatings or in bulk[122].

A modification of the approach to facilitate crosslinking by adding organic monomers is to use pre-formed silicate clusters capped by reactive organic groups and to crosslink them by organic entities. Hybrid polymers with very defined structures are thus obtained[123]. As discussed above, spherosilicates (ROSiO$_{3/2}$)$_8$ or silsesquioxanes (RSiO$_{3/2}$)$_8$ (Figure 2) with various groups OR or R can be prepared. Examples for this approach include the coupling between (HMe$_2$Si−O−SiO$_{3/2}$)$_8$ and vinyl-substituted siloxanes or, *vice versa*, between (CH$_2$=CHSiMe$_2$−O−SiO$_{3/2}$)$_8$ and hydrogenosiloxanes by hydrosilylation[123,124]. Recently, heterocoupling between (HMe$_2$Si−O−SiO$_{3/2}$)$_8$ and allyl acetoacetate-modified aluminum and zirconium alkoxides was also reported[123].

The copolymerization of unsaturated organofunctional alkoxysilanes with organic monomers or the use of different organofunctional alkoxysilanes allows the tailoring of materials properties by variation of the organic polymer structures. In order to avoid the potential problems associated with precursor mixtures, but still to use this option, one was led to the development of the above-mentioned precursors by coupling multiply unsaturated organic monomers with $Si(OR')_3$-containing compounds

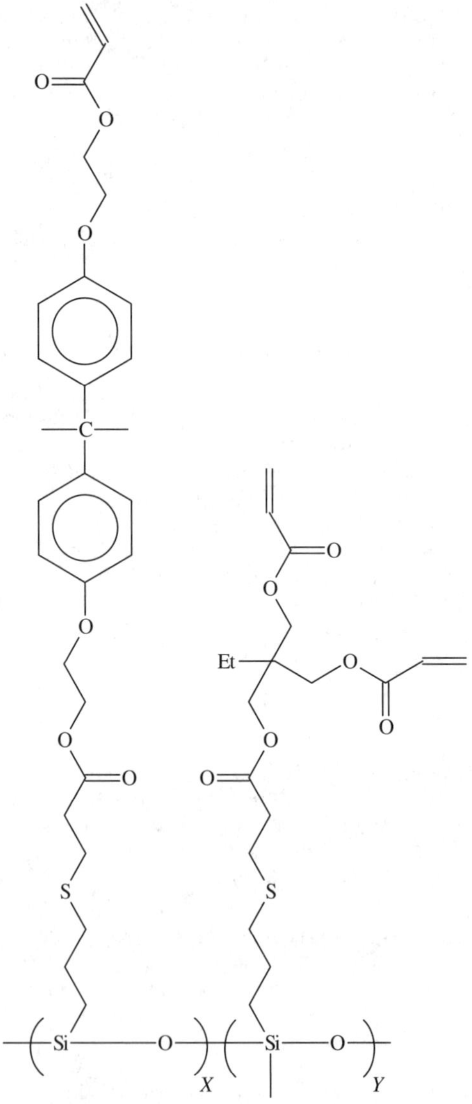

FIGURE 3. A schematic sketch of the polysiloxane used for rapidly curing coatings of optical glass fibers

(e.g. equation 3). Although they are single precursors, the structure of the organic spacer between the unsaturated group(s) and the $(R'O)_3Si$ group, as well as the number of unsaturated groups available for polymerization (i.e. the degree of organic crosslinking) is easily modulated[79,94,95]. It was shown that the Young's modulus of the photochemical crosslinked materials and the thermal expansion coefficient can be varied in a very wide range just by varying the organic structures[95]. Some other advantages of hybrid polymers prepared from these precursors, such as low shrinkage, high flexural strength and the low water adsorption, led to their application as dental filling materials (combined with some inorganic fillers) and for the production of optical lenses[95]. Lenses made from such precursors had comparable optical properties and densities to those made from purely organic polymers. However, their abrasion resistance was distinctly better[94].

Combination of different silanes allowed the development of fast curing primary and secondary coatings for optical glass fibers. UV curing is possible in less than 0.1 s, even on a technical scale, due to the high density of unsaturated organic groups (as shown schematically in Figure 3). The different modulus of elasticity and scratch resistance required for primary or secondary coatings was adjusted by the ratio of the two silanes ($x : y$ in Figure 3)[94].

D. Materials Substituted by Aminoalkyl or Phosphinoalkyl Groups

The most often used amino-substituted alkoxysilanes are $(R'O)_3Si(CH_2)_3NH_2$ (and the corresponding NMe_2 and NEt_2 derivatives) and the ethylene diamine derivative $(R'O)_3Si(CH_2)_3NH_2CHCH_2NH_2$ (DIAMO), which are commercially available. The phosphinoalkyl derivative $(R'O)_3Si(CH_2)_2PPh_2$ is easily prepared by addition of $HPPh_2$ to the vinylic double bond of VTMS.

Contrary to the above-discussed alkoxysilanes with unsaturated organic groups it can no longer be assumed that the organic functionalities will not interfere with the sol–gel chemistry of the $Si(OR')_3$ group during sol–gel processing. Both internal coordination of the basic amino group to the weakly Lewis-acidic silicon atom and strong hydrogen bonds between the Si−OH groups and the NH_2 group were debated for hydrolyzed aminopropyltrialkoxysilanes[125]. Due to these interactions, the rates of the uncatalyzed hydrolysis and condensation reactions of aminopropyltrialkoxysilanes are similar to those of alkyltrialkoxysilanes in the presence of a base catalyst[126,127].

Upon addition of water to $(EtO)_3Si(CH_2)_3NEt_2$, octameric species $[Et_2N(CH_2)_3SiO_{3/2}]_8$ were predominantly formed in methanolic solution according to gel permeation chromatographic studies[127]. A structure in which organofunctional groups are located at the outside of the polysiloxane building blocks was also postulated from a ^{29}Si and ^{31}P solid-state NMR study of a gel obtained from $(MeO)_3Si(CH_2)_6P(Ph)(CH_2)_2OMe$ and $Si(OEt)_4$ (1:2)[128].

There are four major functions of amino groups attached to organically modified silica gels:

(i) Amino groups provide hydrophilicity to the materials, due to their basic properties. For example, coatings with good long-term adhesion to glass, scratch and abrasion resistance, and stability against weathering and corrosive deamination were prepared from $PhSi(OEt)_3$, $Al(OBu-s)_3$ and MEMO (for photochemical curing) or GLYMO (for thermal curing). The polarity of the surface was varied by addition of tridecafluorooctyltrimethoxysilane (increased hydrophobicity) or DIAMO (increased hydrophilicity), respectively[88].

For the preparation of proton-conducting polymers, aminoalkyl-substituted siloxanes made by sol–gel processing of $(MeO)_3Si(CH_2)_3NR_2$ [for example, $NR_2 = NH_2$, $NH(CH_2)_2NH_2$, $NH(CH_2)_2NH(CH_2)_2NH_2$] were doped with CF_3SO_3H[129].

(ii) Amino groups may act as coupling sites for organic crosslinking reactions. For example, hydrolyzed GLYMO and $(EtO)_3Si(CH_2)_3NH_2$ react by formation of β-amino alcohol links[130]. Similarly, the reaction of $(EtO)_3Si(CH_2)_3NH_2$ with phthalic anhydride or pyromellitic dianhydride provided the hydrophilic function in a polymer with potential application for ultrafiltration and reverse osmosis membranes[131].

A sensor with a nonleachable pH indicator was prepared from a $(MeO)_3Si$-substituted derivative of methyl red [$(p$-$Me_2N)C_6H_4N{=}N(C_6H_4COOH$-$o)$] obtained by coupling of the COOH groups with $(MeO)_3Si(CH_2)_3NH_2$[132]. For the development of an optical immunosensor, a sol prepared from $(EtO)_3Si(CH_2)_3A$ (A = NHMe, NH_2 or SH) was coated on gold-coated glass surfaces. After activation of the organofunctional hybrid polymer films in aqueous buffers by bifunctional coupling agents, immunoglobulin was covalently anchored to the transducer surface with an optimal stability and biological activity[55].

(iii) Amino groups form weak adducts with SO_2 or CO_2. The incorporation of aminoalkyl groups into sol–gel materials can therefore be utilized for the development of gas sensors. For example, the sensitive layers of SO_2 gas sensor systems were synthesized from 7:3 mixtures of $(EtO)_3Si(CH_2)_3NR_2$ (R = Me, Et) and n-$PrSi(OEt)_3$[114,133,134]. Two gas-sensitive layers for different SO_2 concentration ranges were developed in another work: for low concentration (ca 2 ppm) based on $(EtO)_3Si(CH_2)_3NMe_2$, and for concentrations up to 6000 ppm based on $(EtO)_3Si(CH_2)_3NEt_2$[135]. Replacement of the NR_2 group in the starting alkoxysilane by NH_2 resulted in a material suitable for the detection of CO_2[134].

(iv) Amino groups, particularly the ethylenediamine entity, are capable of binding metal ions. Sol–gel processing of metal alkoxides in the presence of metal ions mostly does not result in a stable incorporation of the metal ions. They are leached from the resulting oxide materials by water or alcohols. This is prevented when amino-substituted alkoxides are employed. The ethylenediamine derivative $(EtO)_3Si(CH_2)_3NHCH_2CH_2NH_2$ proved to be very effective for anchoring Cd^{2+}, Co^{2+}, Cu^{2+}, Ni^{2+}, Pd^{2+} or Pt^{2+}, while Ag^+ was better coordinated by $(EtO)_3Si(CH_2)_3NH_2$ or $(EtO)_3Si(CH_2)_3CN$[136]. The substituted thiourea $(EtO)_3Si(CH_2)_3NHC(S)NHPh$ was prepared by reaction of $(EtO)_3Si(CH_2)_3NH_2$ with PhNCS and was used to coordinate Pd^{2+} ions[137]. When the organofunctional silane and a metal salt are mixed, metal (M) complexes of the type $[(R'O)_3Si(CH_2)_3A]_mM^{n+}$ are formed *in situ*. Coordination of the metal ions is retained during sol–gel processing[136].

The high dispersion of metal ions achieved by their coordination with $(EtO)_3Si(CH_2)_3A$ and by tethering the resulting metal complexes to the silicate matrix during sol–gel processing was used to develop a method of preparing nanosized metal or alloy particles in SiO_2 matrices. In the second step of the preparation procedure, metal oxide particles are formed by oxidation of the metal complex containing gels in air under carefully controlled conditions. Because of the high dispersion of the metal precursor, very small metal oxide particles with a very narrow size distribution are formed, which are then reduced to give the metal particles[136].

In a related approach, amino-substituted organofunctional alkoxysilanes were used to stabilize small CdS or metal clusters in sol–gel matrices[138].

The sol–gel method is also an attractive possibility to tether metal *complexes* to silica-type materials, mainly for the heterogenization of homogeneous metal complex catalysts. Work in this area was reviewed in other articles, where also a list of selected sol–gel heterogenized metal complexes is given[78,139]. For the tethering of catalytically active complexes, nearly any ligating group known from coordination chemistry can be used. In most examples, however, phosphanyl- or amino-substituted groups were employed. A typical approach for the preparation of the $Si(OR)_3$-containing complexes is shown in equation 8[140].

The (EtO)$_3$Si(CH$_2$)$_2$PPh$_2$-substituted complex is prepared as the corresponding complex (Ph$_3$P)$_2$Rh(CO)Cl, a very well-known homogeneous catalyst, i.e. PPh$_3$ is just replaced by another phosphine. The complexes [(R'O)$_3$Si—X—A]$_n$ML$_m$ (where ML$_m$ is a metal complex moiety) can be used in sol–gel chemistry as any other precursor with some functional group. In the known examples, the structures of the complexes are only insignificantly altered by the modification of the ligand and by sol–gel processing compared with the corresponding complexes with unmodified ligands. This means that the (porous) silica gel matrix surrounding the metal complexes does not interact significantly with the metal centers.

$$2\,(EtO)_3Si(CH_2)_2PPh_2 + Rh_2(CO)_4Cl$$

$$\downarrow \qquad\qquad\qquad\qquad (8)$$

$$2\,(EtO)_3Si(CH_2)_2PPh_2 \longrightarrow \underset{\underset{CO}{|}}{\overset{\overset{Cl}{|}}{Rh}} \longrightarrow PPh_2(CH_2)_2Si(OEt)_3$$

The chemistry of the complexes [(R'O)$_3$Si—X—A]$_n$ML$_m$ was mainly developed with regard to grafting catalytically active metal complexes on silica as a solid support[141]. The sol–gel approach for the heterogenization of such complexes has several advantages. Compared with other methods, a much higher metal loading can be achieved. The important difference is that in the sol–gel derived materials the catalytically active metal moieties are homogeneously distributed throughout the whole material. This has very beneficial effects on the catalyst stability. Since the structure (porosity, surface area, etc.) of the support can be tailored to some extent, the catalytically active centers are still accessible to the reactants. Therefore, the catalytic activity of the sol–gel catalysts is not inhibited, provided that the matrix structure is appropriate. This was shown, for example, by a comparison of the catalytic activity of sol–gel derived [O$_{3/2}$Si(CH$_2$)$_2$PPh$_2$]$_2$Rh(CO)Cl·SiO$_2$ with that of dissolved (Ph$_3$P)$_2$Rh(CO)Cl[142].

E. Materials Substituted by Miscellaneous Groups

In principle, (R'O)$_3$Si-containing groups can be anchored to almost any organic entity by one of the discussed methods. A selection of silanes is shown in Figure 4.

While considerable progress has been made in the preparation of ceramic membranes by sol–gel processing, the development of membranes from hybrid polymers is in its infancy (see also Section V). This is, nevertheless, a very promising area of development, because the possibility of forming mechanically stable membranes by inorganic polycondensation is implemented by the possibility to incorporate organic functions.

Membranes for gas filtration or nanofiltration were prepared from [(EtO)$_2$R'Si(CH$_2$)$_3$NHC(O)]$_2$C$_6$H$_4$-p (R' = Me or OEt) (Figure 4, bottom) by casting the gels obtained from these precursors on alumina supports[143]. Facilitated transport membranes were prepared by co-hydrolysis with the Si(OR')$_3$-substituted crown ether (Figure 4, top). The transport selectivity of K$^+$ over Li$^+$ was about 4 and only due to the presence of the crown ether moieties. Since the selectivity is in good agreement with the properties of benzo-15-crown-5, the character of the carrier was not changed by incorporation into the membrane[143].

The N,N-dioctadecylsuccinate-substituted silane (Figure 4, middle) was used for immobilized amphiphilic monolayers on glass plates. A covalently bonded monolayer of the silane amphiphile was immobilized on a porous glass plate by the Langmuir–Blodgett

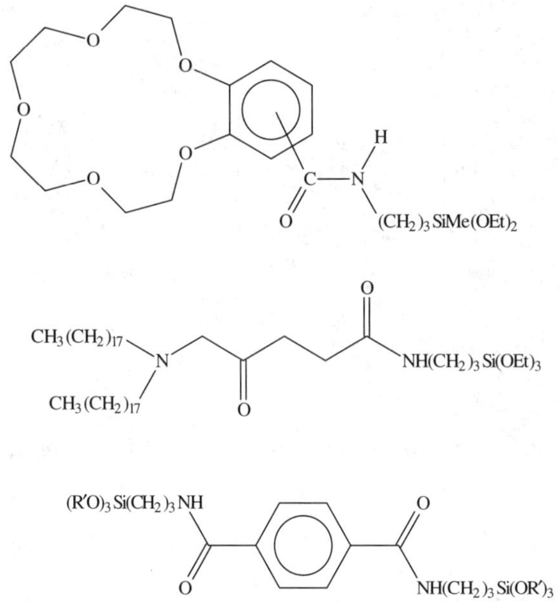

FIGURE 4. Di- and trialkoxysilyl derivatives with various functional organic groups

technique. The lipid monolayer acted as a gate membrane for permeation of ions and water-soluble fluorescent probes[144].

V. ORGANIC/INORGANIC SOL–GEL COPOLYMERIZATIONS

In a sol–gel process carried out with a precursor containing direct organic/inorganic links, it is possible to form homogeneous materials where all bonds are covalent bonds, C–C, Si–C and Si–O. This class of hybrid gels constitutes organic/inorganic polymers. Undoubtedly, the largest class of inorganic/organic gels is made up of hybrids linked via Si–C bonds that are stable and do not undergo hydrolysis under the sol–gel processing conditions[145].

As already described above, in the last decade sol–gel technology has become a popular way to create organic/inorganic hybrid materials[146]. Hybrid is a good generic term for an organic/inorganic material. Another name commonly applied to organic/inorganic gels is nanocomposite, to emphasize the nanometer level of mixing[147]. Nanocomposite usually refers to hybrids where organic and inorganic constituents are not covalently bonded. The forces acting between the constituents vary from weak to relatively strong. Sanchez and Ribot[148] divide hybrid gels into two classes: Class I corresponds to hybrids where organic molecules are blended into the inorganic network, whereas Class II includes hybrids, where inorganic and organic constituents are linked together via covalent or iono-covalent bonds. According to this definition, this Section deals with Class I hybrids, although some Class II hybrids are described at its end.

A. Physical Hybrids (Class I[149])

Physical hybrids containing silica and polymer are typically interpenetrating networks (IPNs). They can be subdivided into simultaneous or sequential IPNs. The terminology of

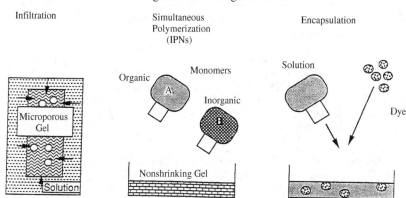

FIGURE 5. Schematics of a rigid porous silica xerogel infiltrated with monomer

interpenetrating polymer networks is borrowed from organic polymer technology[149]. In some cases, semisequential is used to reflect the linear structure of the organic polymer in which the silica gel forms.

Some of the earliest sol–gel hybrids were achieved by monomer infiltration into previously formed silica gel. Starting with a dried silica gel (xerogel), the porous shape is filled with monomer. The monomer is polymerized *in situ*. This is shown schematically in Figure 5. Interpenetration is achieved when the impregnating monomer polymerizes in the open pores of the rigid silica matrix.

In the polymer impregnated gels, some porosity typically remains. Although some copolymers of methyl methacrylate, butadiene and styrene have been used to impregnate silica, the best known system is still silica impregnated with polymethyl methacrylate (PMMA)[150–153]. While this type of hybrid was important at first, it has been surpassed by other methods of hybrid synthesis that are simpler, with fewer steps and shorter times.

B. Sequential Organic/Inorganic Interpenetrating Networks

Silica precursors, tetraethoxysilane and tetramethoxysilane (TEOS and TMOS) are able to solvate some organic polymers. This enables the silica precursor to polymerize in the environment of an organic polymer solution. The number of polymers that form solutions with sol–gel formulations is, however, limited. Some initially soluble polymers tend to precipitate during gelation when a change in solvent composition leads to phase separation.

For select polymers that are soluble, the method is very simple. The polymerization of the silica precursor occurs around the preformed polymer chains or domains. When the organic polymer is formed before the inorganic polymer, the result is a sequential organic/inorganic interpenetrating polymer network.

Generally, it is found that polymerization of TEOS in the presence of preformed polymers under acidic conditions generates small, well dispersed silica particles. Depending on the choice of polymer, highly transparent, lightweight materials can be obtained. Among the polymers used to form sequential IPNs, there are several capable of forming hydrogen bonds with hydroxylated SiO_2 particles[154]. Strong interactions between silanols, having the character of Bronsted acids, and specific groups on the polymer that are hydrogen acceptors are responsible for the high degree of two-phase mixing. Linear polymers with hydrogen acceptor groups, amide, carbonyl and carbinol have been used in semisequential organic/inorganic IPNs. These polymers are poly(N,N-dimethacrylamide), poly(2-methyl 2-oxazoline), poly(methyl methacrylate) (PMMA), poly(vinyl pyrrolidone) (PVP), poly(acrylic acid) (PAA) and its copolymers, poly(vinyl acetate) (PVAc) and poly(vinyl alcohol) (PVA)[154-158].

In detail, the properties of the hybrids are functions of the relative fractions of the polymer and TEOS in the sol–gel formulation, along with the chemistry and length of the polymer chain. When TEOS is polymerized in the presence of the polymer, the glass transition (T_g) of the polymer increases or becomes undetectable. The density of the hybrid does not follow a mixing rule for the inorganic and organic components, but is suppressed even with high inorganic contents.

C. Simultaneous Organic/Inorganic Interpenetrating Networks

While only a few organic polymers are soluble in sol–gel formulations, many organic monomers are soluble in TEOS. These monomers can be introduced directly into sol–gel formulations. Both ring-opening polymerizations and free-radical polymerizations can be carried out simultaneously with the hydrolysis–condensation of TEOS. The resulting hybrids have no covalent linkages between the organic and inorganic components. This simultaneous route captures insoluble organic polymers within a sol–gel inorganic network.

In the preparation of simultaneous networks, it is important to control both polymerizations rates. Systems with inorganic condensation rates much faster than the organic polymerization rates turn into brittle hybrids that shrink. The polymer content is low due to evaporation of unreacted monomer. Systems with fast organic polymerization rates usually show uncontrolled polymer precipitation leading to heterogeneous composites. In practice, the kinetics of polymerization are difficult to control, so the success of the simultaneous approach rests on the careful selection of the monomers and their composition.

The idea of simultaneous interpenetrating networks has been advanced by Novak and coworkers[159-162]. Several monomers of acrylate type (free-radical polymerization) or cyclic alkenes (ring-opening metathesis) have been used. They include acrylamide/N,N'-bisacrylamide monomer system, 2-hydroxyethyl or hydroxymethyl acrylate, and 7-oxanorbornene and its derivatives. In order to get nonshrinking hybrids, tetraacryloxyethoxysilane was used instead of TEOS. The acrylate monomer liberated due to hydrolysis acted as a cosolvent to solvate the polymerizing silica network. The result was transparent hybrids with no shrinkage. Ellsworth and Novak[159,160] also used soluble oligomers of silicic acid instead of TEOS in a condensation with (meth)acryl trialkoxysilane. This route involves a direct Si—C covalent bond to increase the silica content of the organic/inorganic hybrids and further reduce shrinkage, meaning the product is no longer simply a physical hybrid[163].

Wojcik and Klein[164] have studied the incorporation of silica with di- and triacrylate monomers, hexanediol diacrylate (HDDA) and glyceryl propoxy triacrylate (GPTA). These simultaneous IPNs are transparent hybrids with good mechanical strength. A drawback in these systems is the fact that the gelation times under acidic conditions are quite long, as long as weeks. The reason for the long gelation times may be the low concentration of alkoxysilyl groups and steric hindrance from organics in the system. However, the process can be accelerated by photocuring with UV or using a tri-monomer system. Related to the HDDA and GPTA/TEOS systems, Wojcik and Klein[164] tried the tri-monomer system TEOS-vinyltriethoxysilane (VTES)-hydroxyethyl methacrylate (HEMA). By introducing VTES, Si—C links result, meaning that this system, like the silicic acid/(meth)acryl trialkoxysilane system, is not strictly a physical hybrid[159,165].

Another example of a physical hybrid with a functionalized silica is the work of Hoebbel and coworkers[166], who used tetramethylammonium silicate ($[N(CH_3)_4]_8Si_8O_{20} \cdot 69H_2O$)

converted by an ion-exchange process into double-four ring (D4R) oligomers of silicic acid ($H_8Si_8O_{20}$). This oligomer was stabilized by reaction with functional siloxanes containing polymerizable groups like vinyl, allyl, hydrido- or chloromethyl. The functionalized derivative can be polymerized while preserving silicic acid cages in the structure.

Cellular structures are observed in poly(dimethylsiloxane) (PDMS)/silica gels when the sol–gel parameters (H_2O, HCl concentration and temperature) are adjusted to accelerate the gelation process[167]. PDMS, which contains direct Si—C bonds along its Si—O backbone, can co-condense with TEOS, as shown in equation 9.

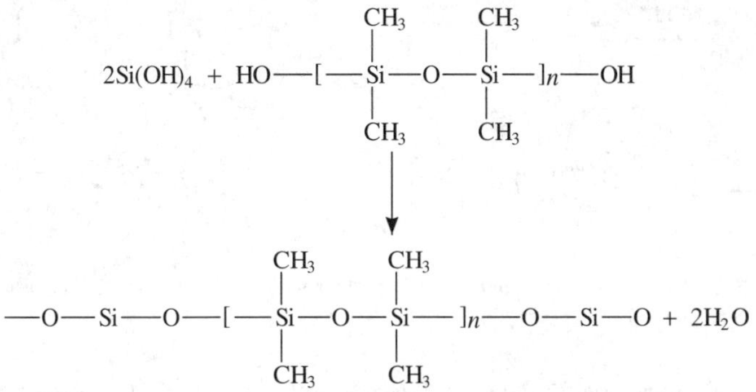

$$2Si(OH)_4 + HO-[-\underset{\underset{CH_3}{|}}{\overset{\overset{CH_3}{|}}{Si}}-O-\underset{\underset{CH_3}{|}}{\overset{\overset{CH_3}{|}}{Si}}-]_n-OH$$

$$\downarrow$$

$$-O-Si-O-[-\underset{\underset{CH_3}{|}}{\overset{\overset{CH_3}{|}}{Si}}-O-\underset{\underset{CH_3}{|}}{\overset{\overset{CH_3}{|}}{Si}}-]_n-O-Si-O + 2H_2O$$

(9)

Since this system contains covalent bonding, it represents a hybrid that is not strictly a Class I hybrid[167] although the co-condensation involves only silanol groups. Porosity remains for small concentrations of organic polymer in these PDMS/silica hybrids, although most hybrids are designed ideally to have low porosity. With high concentrations of the organic component, hybrids have low densities, indicating high free volumes characteristic of the organic polymers.

In a way similar to PDMS and TEOS, other organic polymers have been functionalized with alkoxysilyl groups and covalently bonded to silica. Polymers that have been coupled with silica are PTMO-based polyurethane oligomers[168,169], polyoxazolines[170], polyimide[168], poly(arylene ether ketone)[171], poly(arylene ether sulfone)[172], polystyrene[173], polyoxopropylene (PPO)[174], polyacrylonitrile[175], copolymers of methyl methacrylate and allyl methacrylate[176] and cyclophosphazenes[177].

Generally, it is claimed that PDMS/silica-type hybrids combine the flexibility and mechanical properties of the organic polymer chains with the hardness and stiffness of silica. In detail, the properties of the hybrids are governed by the chemistry, presence of side chains or pendant groups and length of polymeric chains. Especially important are the concentration and distribution of alkoxysilyl groups attached to the polymeric chain, as they directly affect the rigidity or flexibility of the system.

D. Silsesquioxane-containing Hybrids (Class II[148])

It is difficult to deal exclusively with Class I hybrids, because many of them fall into both categories to some degree. One type of Class II hybrid is discussed here. Organically modified silicas (Ormosils)[146], which have been treated more fully in Section IV, are formed from $R'Si(OR)_3$-type organosilanes where radical R has a network modifying effect and is used to introduce new functionalities into the inorganic network. Polymerization of $R'Si(OR)_3$ alone leads to formation of silsesquioxanes of general formula $(RSiO)_{1.5}$, i.e. a silicate framework where each Si is linked covalently to an organic radical R.

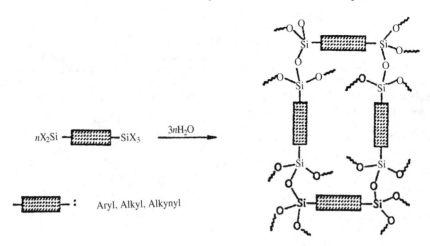

FIGURE 6. Schematics of polymerization of a monomer with an organic spacer between two silicon atoms (after Shea and coworkers[180])

Bifunctional silanes of formula $R'R''Si(OR)_2$ can also be used to prepare Ormosils. Upon polymerization, these organoalkoxysilanes form cyclic species or linear chains and are therefore not able to crosslink and form networks themselves. They can be used with tetrafunctional silanes (TEOS) to incorporate desirable functionality into the hybrid network. For example, hydrophobicity can be achieved by introducing a methyl, phenyl or fluoroalkyl chain as R[178,179]. There are also Ormosils where radical R forms a bridge between two Si atoms. A good example of this is arylene or alkylene-bridged polysilsesquioxanes synthesized from bis(triethoxysilyl) or bis(trichlorosilyl) aryl or alkyl monomers[180–182]. Various aliphatic or aromatic molecules have been used as the organic spacer between two silicon atoms and this is shown schematically in Figure 6.

In concluding this Section we note that, while attempts have been made to classify physical hybrids or interpenetrating networks, it is not always straightforward. A simple classification reflects the timing of polymerization — whether the silica and organic polymers react simultaneously or sequentially. Thermodynamic factors relating to solubility and miscibility and the kinetics of phase separation influence the outcome of the processing. The scale of mixing leads to the subdivision of the microstructure or the distribution of the components. The microstructure, in turn, influences a variety of physical, chemical and optical properties.

VI. PHOTOCHEMISTRY, PHOTOPHYSICS AND OPTICS OF DOPED SOL–GEL MATERIALS

A. Photoprobes for the Sol to Gel to Xerogel Transitions and for the Study of the Properties of the Sol–Gel Cage

Various spectroscopies were used for studying cage properties and their evolution in time along the monomer → oligomer → sol → gel → xerogel transition. The main tools have been UV-visible absorption and emission spectroscopies, and to a lesser extent other spectroscopies such as EPR[183]. The evolution of the sol–gel process and the evolution of various cage properties, was first followed with pyrene[184,185] (Py) because of its relatively long singlet life-time (*ca* 100 ns), its distinct vibronic peaks in the fluorescence spectrum, its ability to form an excited state dimer, i.e. the excimer, Py^*_2, and because of the sensitivity of these parameters to changes in environmental conditions. With this probe, the changing polarity and the changing structure around the probe, from adsorption onto the surface of the growing SiO_2 particles up to the complete isolation in microporous cages, were detected and monitored[185]. Many sol–gel studies have utilized that useful probe[186].

Some other photophysical probes which were used for following the sol–gel transitions include 7-azaindole[187], 1- and 2-naphthols[188], pyrene-3-carboxaldehyde[189], $ReCl(CO_3)$bipyridine (as a probe for cage rigidity)[190], thymol blue[191] and Rhodamine B[17b]. Photochromic compounds were also used for the investigation of the sol–gel xerogel transition in $Si(OCH_3)_4$[184,192,193] and the substituent's effects on the point of the change from normal to reversed photochromism were analyzed in terms of the Hammett relationship[192].

A pyrene derivative, 8-hydroxy-1,3,6-pyrenesulfonic acid (pyranine), was successfully used[33] for the detection of water during the early stages of the polycondensation of $Si(OCH_3)_4$. The idea of using this probe molecule is based on the fact that its excited state is a short-lived strong acid, capable of efficient proton transfer to neighboring water molecules, leaving behind the anion of pyranine in its excited state. Pyranine was also used by Pouxviel and coworkers for studying aluminosilicate sol–gels: here too, that probe proved to be a highly sensitive probe to changes in water content[194].

In addition to the spectroscopic follow-up of the sol–gel process, much research was devoted to the analysis of the cage properties at the final xerogel stage. Early detailed spectroscopic analysis of absorption and emission spectra of rhodamine 6G (R6G), a laser dye, in SiO_2 sol–gel[1] indicated that the polarity of the immediate environment of the dopant is high and due to hydroxyls, although less polar than water (as obtained also with Py as a probe). The ability of the matrix to isolate efficiently one molecule from the other, even at high concentration, first observed for that dopant[1], lead to an intensive study of doped sol–gel materials for a variety of optical uses, notably as dye laser components[195]. We know today, however, that this is not universal and that high concentrations may lead eventually to some aggregation. This was found for proteins[44] and for $Ru(bpy)_3^{+2}$[196]. Dopant aggregation may also be the way in which three doped o-phenanthroline ligands find their way to complex a single Fe^{2+} ion[23].

An indication as to the homogeneity of the entrapment comes from the analysis of the decay profile of the excited state in the final xerogel. First-order decays point to homogeneity of distribution, and were found, for instance, in $Ru(bpy)_3^{2+}/SiO_2-TiO_2$ films[20,196] and in biphenylcarboxylic acid/SiO_2[197]. The protective nature of the sol–gel cage has been clearly evident in these two cases and later on in many other studies. For instance, in the case of $Ru(bpy)_3^{2+}$, the observed life-time in solution at a concentration of 1×10^{-3} M is 0.3 μs, but at the same concentration in a sol–gel matrix, there is a sixfold increase to 1.9 μs[196]. Photostability to prolonged irradiation is another indication for the protective nature of the cage[1,17b]. For instance, the degradation half-life of 4-oxazine increased fourfold by entrapment, compared to solution[196]. One of the most striking manifestations of the special protective properties of the sol–gel caging has been the ability to obtain room-temperature phosphorescence (RTP), under regular atmospheric conditions, from a very wide range of organic molecules[197], many of which reveal that emission only at cryogenic temperatures and in vacuum. Examples include polycyclic hydrocarbons[184,197], dyes like eosin-y, bases like quinine and acids such as 4-biphenylcarboxylic acid.

The question of rigidity of the sol–gel cage[183,190] accompanied the research in this class of materials from its very beginning. Both the rigidity and polarity of the cage can be altered by changing the monomer, by the incorporation of suitable additives and by copolymerization. A nice example is the encapsulation of photochromic spiropyranes mentioned above[184,192,193]. When entrapped in pure SiO_2, the photochromism is observable only up to the wet-gel stage[192]. Beyond it, the cage becomes so rigid that the isomerization between the closed colorless spiro form and the open merocyanine colored zwitterion stops at the final dried xerogel stage. A straightforward method to relax the rigidity is to use a three-dentate monomer, rather than the normal tetraalkoxy monomer, thus reducing the degree of crosslinking. Indeed, entrapment of the spiropyranes in a sol–gel matrix prepared from ethyltriethoxysilane provided the needed flexibility for lasting photochromism. A marked effect of an additive on cage properties was observed in the case of entrapped pH-indicators[25]: addition of a surface active agent, such as cetyltrimethylammonium bromide, shifted the pK_i by 2 units to a more acidic range.

For practical applications, a delicate balance should be found between good isolation and protection on the one hand, and sufficient accessibility of the entrapped molecule to external reactants on the other. Evidence is accumulating that high acidity and low water/silane ratio do the trick, at least in some cases (for example in the entrapment of pyranine[18,19] as a fluorescent pH sensor and in the entrapment of Py[19,198] as a fluorescent sensor for oxygen). However, these need not be necessary conditions: the entrapped quaternary ammonium metal catalysts mentioned above do not leach out even though prepared under neutral conditions[38], and enzymes and antibodies do not leach out because of their sheer size[44].

The last cage property we would like to comment on is its symmetry. This property was studied by entrapment of Eu^{3+} (as its chloride, obtained *in situ* from Eu_2O_3/HCl) in SiO_2 sol–gel[199], and although it is an inorganic dopant, we are describing it here because that report led to many studies of Eu^{3+} doped sol–gel matrices, including both organic and inorganic ligands[200], as well as to many studies of doping with other transition-metal cations[201]. Symmetry in the chemical sciences is a rather complex issue[202], and the cage symmetry of sol–gel materials is no exception, especially since it involves a population of cages with varying degrees of symmetries (for the concept of degree of symmetry, see elsewhere[202]). Yet Eu^{3+}, which is known to be an excellent indicator for site-symmetry[200], allows at least an average picture of that property. The basis of this symmetry probe is that the lowering of local symmetry enhances radiative transitions in that cation, and particularly sensitive is the $^5D_0 \rightarrow {}^7F_2$ emission at 615 nm. In general, it was found that the symmetry around Eu^{3+} resembles the solvation shell found in a polar solvent, up to the nonheated, mature wet-gel stage. Upon heating to 100 °C, solvent leaves the gel and the fluorescence spectrum reveals this change by an overall increase in intensity and by a relative increase in the $^5D_0 \rightarrow {}^7F_2$ transition intensity. Thus, as the amount of O–H groups is reduced, the symmetry around the cation decreases.

B. Organic Photochemistry within Sol–Gel Matrices

The first family of reactions studied in detail were, as mentioned above, the photochromic isomerizations of Aberchrome 670[184] and of various spiropyranes[192,193]. The motivation was to develop novel materials for information processing and for other optics applications, and it serves as a good illustration of what the organically doped sol–gel methodology can offer: traditionally, photochromic glass preparations have been limited to a few dopants that can withstand the high temperatures of glass melting; by contrast, organic photochromism offers several tens of thousands of molecules with which one can tailor the shade of color change, the direction of the photochromism, the rate of change in color intensity, reversibility (optical gates) or unidirectionality (information recording). We return to it in the next section.

The ability to separate reactive species that otherwise annihilate each other is another feature which the sol-gel methodology allows by virtue of the entrapment-isolation. This was best demonstrated in a study aimed at storage of light energy in the form of a photogenerated ionic radical pair[203]. A notorious problem in this field of research has been the dissipation of the absorbed energy through fast back-reaction of the radical pair. An exceedingly long-lived—several hours!—charge-separated pair was obtained in SiO_2 sol-gel between the photogenerated pyrene cation (Py^+) and methyl-viologen cation (MV^+) (equation 10):

$$Py^* + MV^{+2} \rightleftharpoons Py^+ + MV^+ \tag{10}$$

This is a very reactive pair which, under free diffusing conditions, would immediately quench each other with the back-reaction. The back-reaction was inhibited by entrapment of both Py and MV^{+2} while allowing them to communicate chemically with a mobile charge carrier, N,N'-tetramethylene-2,2'-bipyridinium, which diffused freely in the aqueous solution within the porous network.

Photoinduced charge separations of this type can be used for subsequent useful reactions[204-206]. An example is the series of reactions in which H_3O^+ was reduced to H_2: A donor, $Ru(bpy)_3^{2+}$ [Ru(II)], and an acceptor, $Ir(bpy)_2(C,^3N')bpy)^{+3}$ [Ir(III)], were co-entrapped in an SiO_2 sol-gel matrix, and dimethoxybenzene (DMB) was used as the redox shuttler between the Ir and Ru (equations 11-13):

$$Ir(III)^* + DMB \longrightarrow Ir(II) + DMB^+ \tag{11}$$

$$DMB^+ + Ru(II) \longrightarrow Ru(III) + DMB \tag{12}$$

$$2H^+ + Ir(II) \longrightarrow Ir(IV) + H_2 \tag{13}$$

C. Sol-Gel Optics

1. Photochromic sol-gel glasses

In Section VI.B we have seen the use of photochromic materials for following the sol-gel process, and described the first successful preparation of organically doped photochromic materials. Here we comment further on this topic, from the point of view of the needs of optical materials. In addition to the entrapment in silica sol-gels, it is relevant in this context to mention also that aluminosilicate gels were reported as good host matrices for photochromic molecules[207-209]. These authors pointed out that thermal reversibility of the photochromism in silica gels was the main problem to become a thermally stable photochromic material. Improvements in the photochromic properties of the silica-doped gels have been performed both by modifications in the host matrix as well as by the use of different photochromic compounds. Modifications in the silica matrix by the use of trifunctional silane precursors[210,211] or by the impregnation of the porous material with an organic polymer[212] improved the photochromic response as well as the fading rate. Low levels of photofatigue[207] and a high color stability were thus achieved[210,211]. Improvements in this sense were also performed by the incorporation of new photochromic compounds such as diarylethene derivatives, which do not show thermal conversion below 140 °C. Other photochromic compounds which showed good thermal stability were reported[207,210,211], such as the spironaphtho-oxazines, both in pure silica matrix as in hybrid silica matrix. Thermal isomerization of azobenzenes in sol-gel glasses was described as dependent on the amount of water on the silica surface[212], which is in concordance with previous results found by other authors, where hybrid matrices improve the photochromic properties of the material[192,193,210,211,213,214].

The matrix itself must be photochemically stable, offering at the same time a favorable environment to the photochromic dye. Efforts to improve the photochromic properties of the material were conducted through the incorporation of several additives to the silica matrix. Materials with high photochromic intensity and low photofatigue were found using perfluoroalkoxysilane derivatives or an imidazole derivative[207]. Other types of photochromic materials are being studied and preliminary results are very promising[211].

In another application of doped sol-gel materials, it has been demonstrated that photochromic-doped sol-gel materials can be attached to optical fibers[215-217], although the properties of the light throughput may be modified. Simple fiberoptic/photochromic devices made of two optical fibers placed in a V-groove removable connector have been prepared. Once cured, these devices behave as optically addressed variable delay generators. The same devices can be used for preparing simple optical switches and routing systems. Silica gel-glasses obtained by the sol-gel process are chemically and optically very similar to the optical fiber itself. Therefore the Fresnel losses are drastically reduced, allowing the possibility of using doped sol-gel glasses for modifying the properties of the fiber light throughput. With these materials it is possible not only to make low-loss interconnections from fiber to bulk devices, but to build smart waveguides for applications traditionally associated with integrated optics (modulators, switches, wavelength filters) where the light can be optically processed.

The use of covalently bound photochromics via the use of trialkoxysilane derivatives of spiropyranes is mentioned below in Section VI.C.3.

2. Optical and photophysical properties of entrapped organic molecules

The development of novel optical devices requires suitable materials which have the required chemical and photochemical stability, with appropriate optical features, which can also be made into usable forms such as thin films and bulk pieces. Since the sol-gel process[9,10,218,219] offers a very attractive possibility with the ability to incorporate organic materials in inorganic matrices at low temperatures in the form of monolithic glasses or thin films, it has opened the way to many possible applications in optics[1,6c,190,192,193,196,220-222] and electrooptics[2,223,224].

From the point of view of optics applications, luminescent dyes (fluorescent and phosphorescent) were the organic compounds most frequently used for the preparation of doped sol-gel materials. The characteristics of these materials were studied extensively, from properties at the steady state (emission-excitation)[188] to properties of the dynamic state (lifetimes)[225], and to properties derived from anisotropy[226] and/or energy transfer[227]. Knowledge of the effects that the surface may have on the organic molecule is an important factor in understanding the spectroscopic properties of the entrapped molecule in its inorganic/organic environment.

Pure silica gels[228,229] or mixed titania-silica, alumina-silica, zirconia-silica gels[194,196,230,231] have been used as inorganic host for organic dopants. The use of mixtures is of interest because of the high Lewis acid character of the pure oxides of Ti, Al and Zr which may degrade the organic dye[232]. For this reason, the preparation of hybrid matrices has been intensively developed in recent years. Doped hybrid matrices[230] may be prepared by the impregnation of organic polymers or monomers which are polymerized inside the matrix[222-235], yielding matrices with a highly organic character, thus increasing the chemical compatibility between dye and surface. Other ways to prepare these hybrid materials consist of the use of the trifunctional silane precursors, yielding a silica matrix with nonfunctional ligand bonded to the silica surface[236,237] (see the next section). The presence of such ligands opens the possibility to bind covalently the organic dye which is incorporated to the silica matrix. Such systems may show enhancement in the stability of the organic dye[238].

Numerous lasing dyes have been incorporated within sol–gel, such as rhodamines[1,239], coumarins[240–242], oxazines[232,243] and cyanines[194,232]. A general feature of these dopants in the inorganic matrices is the high concentration which can be reached without self-quenching and/or self-absorption processes. Often observed is an increase in the fluorescence intensity of the dopant, ascribed to the greater rigidity and isolation of the dye in the silica host compared to solution. Laser dye doped sol–gel glasses for fabrication of narrow and broad band antireflective coatings, multilayered dielectric mirrors and polarizing thin films have been developed for the proposed[244] French megajoule neodymium-glass laser.

Lebeau and coworkers[232] have first reported on the optical properties of near-infrared (NIR) organic dyes (700–1000 nm)[245] when incorporated in silica-based gel. NIR dyes have led to applications well beyond optical recording technology: high-density optical data storage, thermal writing displays, laser printer and laser filter, infrared photography, medical applications and persistent (non) photochemical hole burning[246] as multiwavelength optical memory[247] are some examples. The dye employed for these purposes is the polymethine dye 1,1′,3,3,3′,3′-hexamethylindotricarbocyanine iodide (HITC) using 4-dimethylaminopyridine (DMAP) as a catalyst.

Recently, del Monte and Levy[248] reported on the spectral behavior and chemical stability of 1-oxazine and HITC doped gel-glasses. The fluorescence properties of HITC gel-glasses, which vary widely depending on conditions during gel formation, may be controlled with the chemical composition of the starting mixture. The entrapment of HITC in gel-glasses required dye protection from chemical degradation. This was achieved by using tetramethoxysilane (TMOS) in order to produce high-density matrices. Alternative strategies to protect HITC inside gel-glasses were achieved by using a modification of the techniques proposed by Pope and Mackenzie[222] and by Reisfeld and coworkers[249] in which the sol–gel matrix is blocked with organic polymers [by mixing dissolved polymethyl methacrylate (PMMA) in the starting solution]. This procedure results in transparent materials of excellent optical and mechanical quality. Chemical stabilization of HITC involves the detachment of the dopant from the surface silanols, which are the primary cause of the HITC degradation in gel-glasses. This isolation is apparently achieved by the use of the PMMA. The fluorescence of the resulting composite gel-glass showed no signal loss over 10 months[248].

The combination of organic dopants with oriented assemblies of organic compounds with high-order nonlinear polarization or organic substances of large third-order susceptibility, $\chi^{(3)}$[250,251], with inorganic sol–gel matrices, offers a greatly expanded capability for applications in nonlinear optics. These materials are essential for photonics in the coming 21st century and are thought to be applicable to optical switches, optical memories[252] and optical branches[253]. Semiconductor particles in a dielectric matrix have attracted much interest for nonlinear optics applications[254,255], and consequently, silica sol–gel materials were proposed as possible hosts due to their excellent physical performance[222].

Reisfeld[256] showed that nonlinear sol–gel glasses, and especially sol–gel thin films, form an interesting class of optoelectronic devices, and investigated the origin of strong nonlinearities on a millisecond to nanosecond scale of glasses doped with fluorescein, acridine and methyl orange.

Another unique aspect of the sol–gel process in this context is the synthesis of Ormocers (organically modified ceramics)[111]. Different organic–inorganic sol–gel precursors have been used for this purpose for the development of new glasses with second-order nonlinearities[257–261]. In these materials, organic chromophores are chemically bonded to the oxide backbone via the use of functionalized silicon alkoxide precursors (see also the next section). Chaput and coworkers[81] have reported the preparation of an organic–inorganic composite material in which the organic active molecule is covalently bound into the silica network, and demonstrated the ability of orientation of organic molecules via the photoisomerization of active azo molecules. For instance, azobenzenes are very useful molecules in the preparation of NLO (nonlinear optics) materials[262] when bonded to the silica surface. When this type of molecule is covalently attached to the surface by both ends, their thermal relaxation is decreased and a long-term stability of the nonlinear properties is obtained[263–265]. The covalent bond to the silicon atom must be introduced in order to avoid its rupture during the acid hydrolysis of the sol–gel process[81,266], which may cause a loss in the NLO properties[267].

Other types of organic molecules used as sol–gel dopants to form NLO materials are phthalocyanines, porphyrins and fullerenes. For the phthalocyanines entrapped in hybrid matrices prepared from trifunctional silane precursors[268,269], a very good stability was observed. Porphyrins doped in aluminosilicate gels showed a concentration-dependent nonlinear absorption behavior[270]. Fullerenes have recently incorporated into silica matrices of high porosity (aerogels)[271], also showing a concentration effect.

For electrooptics applications, the sol–gel process has been recently used[223,224,272] for the trapping of microdroplets of nematogenic organic compounds (i.e. liquid crystals, LCs) in a thin film of a silica-gel based matrix (GDLC, Gel-glass Dispersed Liquid Crystal). The compounds [a commercial nematic mixture (E7 from BDH Ltd)], composed of several n-alkyl cyanobiphenyl homologues and a cholesteric liquid-crystal dopant or two nematic liquid crystals, 4-pentyl-4,4'-biphenylcarbonitrile (K15, Aldrich) and ZLI-1221 (Merck, a well known commercial LC mixture for twisted nematic displays), were dissolved in the starting mixture; further hydrolysis and polycondensation reactions lead to a phase separation that results in the formation of pores which are filled with the organic

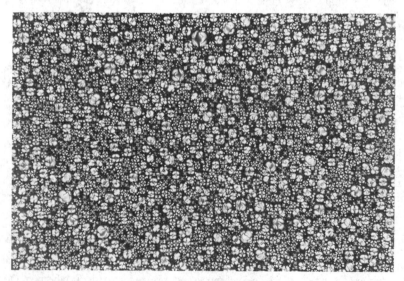

FIGURE 7. Optical microscope picture with crossed polarizers of the microdroplets showing a liquid-crystal state filling the pores of a GDLC thin film. The orientations in microdroplets can be observed as the result of the specific sol–gel processing (parallel polarizers gave Maltese crosses of LC microdroplets and a black image of the silica-gel substrate)

compound. If adequate processing of the preparation of the microdroplets is achieved, the organic compound recovers its nematic liquid-crystal state, its molecules being oriented in microdomains according to the pore inner-surface anchoring (organic group on the walls of the pores, e.g. $-CH_3$, $-C_2H_3$, $-C_6H_5$, etc.) and orienting properties (Figure 7). Introducing a two-functional aminosilane, 3-[diethoxymethylsilyl]propylamine, in the composition of matrix raw materials resulted in better matrix flexibility, higher conductivity and greater capacity for trapping LC^{273}. Once the phase separation in GDLC formation takes place, the LC microdroplet dispersion becomes related to the number and density of microdroplets within the silica-gel matrix. In this manner, the resulting properties of the LC microdroplets may allow the use of GDLC films as electrooptical devices. GDLC films scatter visible light according to the number of droplets and the relative refractive indices of the LC and the silica matrix[274]. Color may be included in the sol-gel matrix or in the liquid crystal itself. A comparative study of different dyes for GDLC color displays demonstrated that dyes may be either embedded in the sol-gel matrix or dissolved in the liquid-crystal microdroplets. In the first case, the liquid crystal scatters out the backlight when no voltage is applied, whereas a transparent colored pixel is obtained upon switching. In the second case, the dye is oriented by the liquid crystal, its absorption being modified by the liquid-crystal reorientation. This affects the display contrast, brightness and viewing angle[275,276].

3. Materials substituted by chromophores

The possibility to tailor the chemical and physical properties of sol-gel materials, their excellent optical quality, their low preparation temperatures and their easy deposition as coatings makes them ideally suited for optical applications. Mainly SiO_2 [prepared from $Si(OR')_4$], but also inorganic-organic hybrid polymers were used as a host for dyes. For example, the above-mentioned hybrid polymer obtained from the GLYMO and $Al(OBu-s)_3$ was shown to be a very interesting host for laser dyes. By doping with $YbCl_3$ and dihydroxyanthraquinone (DAQ) or Oxazine 170 (9-ethylamino-5-ethylimino-10-methyl-5H-benzo(a)phenoxazonium perchlorate), materials for optical data storage in the frequency domain were obtained[277].

Physical trapping dye molecules in sol-gel materials are reviewed elsewhere in this chapter. A comparatively recent approach is to have the dye molecules chemically *bonded* to the oxide matrix. This approach requires the chemical modification of the dye molecules by $Si(OR')_3$-containing groups, i.e. the preparation of compounds $(R'O)_3Si-X-A$, in which A is a chromophore, by the methods shown in equations 3-7. This is mostly not a trivial task, since the chromophore has to remain undisturbed.

Therefore, the chemical linking of the dye molecules is only worthwhile if it results in a substantially better performance of the resulting materials. For the reasons discussed below, this is mainly expected for nonlinear optics (NLO) dyes. However, other dye-containing materials were also prepared by this approach. For example, new transparent photochromic hybrid materials were prepared by sol-gel processing of $Si(OEt)_4$, silanol-terminated dimethylsiloxane oligomers and triethoxysilylpropyl-substituted spirobenzopyranes[278], and fluorescent glasses and films from pyrene derivatives covalently tethered to $RSiO_{3/2}$ (R = alkyl, phenyl)[279].

There are several potential benefits from chemically linking the dye molecules to the matrix compared to just trapping them:

(i) The dye concentration in the final material can be considerably increased. It was demonstrated for several types of $(R'O)_3Si$-functionalized NLO dyes that the chemical bonding to the oxide network allowed high dye concentrations without any crystallization occurring[81,82,250,257-260,280,281]. Coloration of glass by anthraquinone or naphthoquinone doped sols was not possible, because of their low solubility. This was not a problem when (EtO)$_3$ Si-substituted phenoxazinium and phenazinium dyes were used[282].

(ii) Leaching of the dye is prevented by its covalent bonding to the sol-gel matrix. For example, when Rhodamine 110 was only embedded in a sol-gel derived silica matrix, leaching was a severe problem. When the dye was covalently bonded to the matrix, no leaching was observed[283].

(iii) Glass-like sol-gel matrices appear to be more suitable hosts for NLO dyes than organic polymers, because they provide for better stability of the nonlinearity after alignment. Tethering the dye molecules to the matrix results in higher stabilities than physical doping.

The effect of binding the dye to the oxide network was studied in several cases[80,81,250,257-260,280-282] by preparing poled and cured films either by chemical binding of the Si(OR')$_3$-substituted dye to the sol-gel polymer or by incorporating the unmodified dye in the same matrix without chemical bonding. The former systems had a much better stability. This was, *inter alia*, attributed to restriction of the molecular motion of the segments, due to the bonding between the chromophore and the polysiloxane[260,280]. The relaxation behavior of NLO chromophores depends strongly on the rigidity of the matrix, which is influenced by the processing conditions and their composition[284,285]. Particularly the degree of crosslinking of the matrix appears to be critical. Preliminary attempts to get a higher degree of crosslinking (and thus a better optical response or stability) by using inorganic-organic hybrid polymers (organic *and* inorganic crosslinking) did not result in a real improvement[286]. It was also shown that the isomerization kinetics of azo dyes is different in sol-gel films and bulk materials, which was attributed to the larger free volume in the bulk material due to larger pores[287].

The presence of only one Si(OR')$_3$ unit per chromophore makes the formation of a highly crosslinked sol-gel network very difficult. The high mobility of the NLO dye in the free volume of the host network causes relaxation of the poling-induced order. Presently, the most promising approach to improve the stability of the second-order optical nonlinearity is therefore the use of multiple-substituted dyes, such as **20**[83].

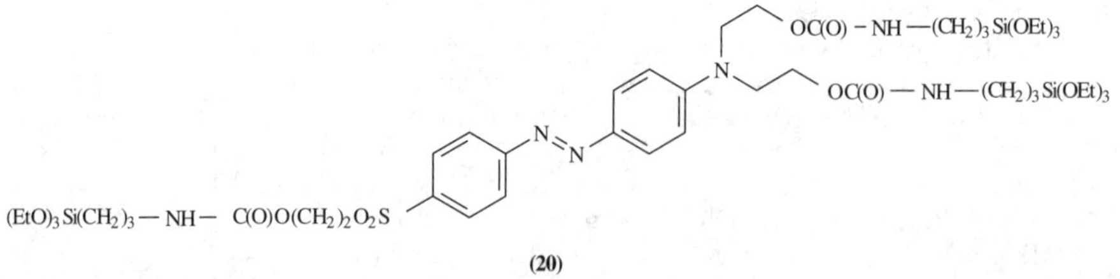

(EtO)$_3$Si(CH$_2$)$_3$—NH—C(O)O(CH$_2$)$_2$O$_2$S—... —N=N—... —N(CH$_2$CH$_2$OC(O)—NH—(CH$_2$)$_3$Si(OEt)$_3$)$_2$

(20)

VII. CONCLUDING REMARK

It is our hope that this chapter on a new family of materials, at the crossroads of organic and inorganic chemistry, will stimulate researchers and students to consider this new frontier of silicon materials chemistry for solving problems of their own scientific endeavors, for embarking on further basic science explorations and as a potential solution for the design of useful devices.

VIII. ACKNOWLEDGMENTS

D. A. thanks his collaborators R. Reisfeld, M. Altstein, J. Blum, S. Braun, D. Huppert, V. R. Kaufman, O. Lev, M. Ottolenghi, Y. Polevaya, A. Rosenfeld, C. Rotmann, J. Samuel, H. Sertchook, H. Schumann, A. Slama-Schwok and R. Zusman. D. A. also thanks the Volkswagen Foundation and the Israel Ministry for Science and Arts, for currently supporting the sol-gel research. L. C. K. thanks her collaborators, C. Beaudry, L. Laby and S. Yamazaki, for helpful discussions and greatly appreciates the support of the NJ Commission on Science and Technology. D. A. and L. C. K. acknowledge gratefully the joint support of the US-Israel Binational Science Foundation. D. L. thanks the CICYT for research grant MAT95-0040-C02-01 and his colleagues, C. J. Serna, J. M. Otón, M. Ocaña and M. L. Amo, A. Serrano, F. del Monte, E. Moreno, T. Prieto, G. Fiksman, I. R. Matías and P. Datta, and C. Alonso for his helpful technical assistance. U. S. acknowledges the assistance of N. Hüsing and A. Lorenz in preparing the chapter, and the Fonds zur Förderung der Wissenschaftlischen Forschung for current support of sol-gel research. D. A. is a member of the F. Haber Research Center for Molecular Dynamics and of the Farkas Center for Light Energy Conversion.

IX. REFERENCES

1. D. Avnir, D. Levy and R. Reisfeld, *J. Phys. Chem.*, **88**, 5956 (1984).
2. D. Levy, *J. Non-Cryst. Solids*, **147 & 148**, 621 (1992).
3. L. C. Klein, T. Bloxom and R. Woodman, *Coll. Surf.*, **63**, 173 (1992).
4. E.g., U. Schubert, *J. Chem. Soc. Dalton*, 3343 (1996).
5. J. I. Zink and B. Dunn, *J. Ceram. Soc. Jpn.*, **10**, 878 (1991).
6. (a) S. Sakka, *J. Sol-Gel Sci. Technol.*, **3**, 69 (1994).
   (b) L. L. Hench and J. K. West, *Chem. Rev.*, **90**, 33 (1990).
   (c) R. Reisfeld and C. K. Jorgensen, Editors, 'Chemistry, Spectroscopy and Applications of Sol-Gel Glasses', *Structure and Bonding*, **85**, (1996), Springer, Berlin, the whole volume.
7. A. Engle, *Reading in Glass History*, No. 23, Phoenix Publications, Jerusalem, 1991.
8. (a) T. Graham, *J. Chem. Soc.*, 618 (1864).
   (b) E. Hatschek (Ed.), *The Foundations of Colloid Chemistry*, Ernest Benn, London, 1925.
9. C. J. Brinker and G. W. Scherer, *Sol-Gel Science*, Academic Press, New York, 1990.
10. (a) J. Livage, M. Henry and C. Sanchez, *Prog. Solid State Chem.*, **18**, 259 (1988).
    (b) L. L. Hench and J. K. West, *Chem. Rev.*, **90**, 33 (1990).
    (c) H. Dislich, *J. Non-Cryst. Solids*, **73**, 599 (1985).
    (d) J. Zarzycki, *Heterogen. Chem. Rev.*, **1**, 243 (1994).
    (e) C. Sanchez, M. L. Mecartney, C. J. Brinker and A. Cheetham, (Eds.), *Better Ceramics Through Chemistry, VI, Mater. Res. Soc. Symp. Proc.*, **346** (1994); and other volumes in this series.
11. See current issues of the journal, *J. Sol-Gel Sci. Technol.*
12. (a) A. M. Buckley and M. Greenblatt, *J. Chem. Educ.*, **71**, 599 (1994).
    (b) V. R. Kaufman and D. Avnir, in *Better Ceramics Through Chemistry, II* (Eds. C. J. Brinker, D. E. Clarck and D. R. Ulrich), *Mater. Res. Soc. Symp. Proc.*, **73**, 145 (1986).
    (c) D. Avnir and V. R. Kaufman, *J. Non-Cryst. Solids*, **192**, 181 (1987).
13. Y. Polevaya, J. Samuel, M. Ottolenghi and D. Avnir, *J. Sol-Gel Sci. Technol.*, **5**, 65 (1995).
14. L. V. Ng and A. V. McCormick, *J. Phys. Chem.*, **100**, 12517 (1996).
15. D. Avnir, *Acc. Chem Res.*, **28**, 328 (1995).
16. M. Tsionsky, G. Gun, V. Glezer and O. Lev, *Anal. Chem.*, **66**, 1747 (1994).
17. (a) T. Fujii, A. Ishii and M. Anpo, *J. Photochem. Photobiol.*, **54A**, 231 (1990).
    (b) N. Negishi, M. Fujino, H. Yamashita, M. A. Fox and M. Anpo, *Langmuir*, **10**, 1772 (1994).
18. J. Samuel, A. Strinkovski, S. Shalom, K. Lieberman, M. Ottolenghi, D. Avnir and A. Lewis, *Mater. Lett.*, **21**, 431 (1994).
19. N. Aharonson, M. Altstein, G. Avidan, D. Avnir, A. Bronshtein, A. Lewis, K. Lieberman, M. Ottolenghi, Y. Polevaya, C. Rottman, J. Samuel, S. Shalom, A. Strinkovski and A. Turiansky, in *Better Ceramics Through Chemistry, VI* (Eds. C. Sanchez, M. L. Mecartney, C. J. Brinker and A. Cheetham), *Mater. Res. Soc. Symp. Proc.* **346**, 519 (1994).
20. T. Tani, A. Namikawa, K. Arai and A. Makishima, *J. Appl. Phys.*, **58**, 3559 (1985).
21. O. Lev, *Analusis*, **20**, 543 (1990).
22. R. Zusman, C. Rottman, M. Ottolenghi and D. Avnir, *J. Non-Cryst. Solids*, **122**, 107 (1990).
23. (a) B. Iosefzon-Kuyavskaya, I. Gigozin, M. Ottolenghi, D. Avnir and O. Lev, *J. Non-Cryst. Solids*, **147 & 148**, 808 (1992).

(b) O. Lev, B. Kuyavskaya, Y. Sacharov, C. Rottmann, A. Kuselman, D. Avnir, and M. Ottolenghi, in *Environmental Monitoring* (Ed. T. Vo-Dinh), *SPIE Proc. Ser.*, **1716**, 357 (1993).
24. Reviews: (a) O. S. Wolfbeis, R. Reisfeld and I. Oehme, *Structure and Bonding*, **85**, 52, (1996).
(b) O. Lev, M. Tsionski, L. Rabinovich, V. Glezer, S. Sampath, I. Pankratov and J. Gun, *Anal. Chem.*, **67**, 22A, (1995).
25. C. Rottman, M. Ottolenghi, R. Zusman, O. Lev, M. Smith, G. Gong, M. L. Kagan and D. Avnir, *Mater. Lett.*, **13**, 293 (1992).
26. P. Kiernan, C. McDonagh, B. D. MacCraith and K. Mongey *J. Sol-Gel Sci. Technol.* **2**, 513 (1994).
27. M. Kubeckova, M. Pospisilova and V. Matejec, *J. Sol-Gel Sci. Technol.*, **2**, 591 (1994).
28. K. T. V. Grattan, G. E. Badini, A. W. Palmer and A. C. C. Tseung, *Sens. Actuators A*, **25-27**, 483 (1991).
29. I. Kuselman, B. I. Kuyavskaya and O. Lev, *Anal. Chim. Acta*, **256**, 65 (1992).
30. O. Lev, B. Iosefson-Kuyavskaya, I. Gigozin, M. Ottolenghi and D. Avnir, *Fresenius J. Anal. Chem.*, **343**, 370 (1992).
31. V. Chernyak, R. Reisfeld, R. Gviski and D. Venezky, *Sens. Mater.*, **2**, 117 (1990).
32. M. Tsyonsky and O. Lev, *Analyst*, **118**, 557 (1993).
33. V. R. Kaufman, D. Avnir, P. Pines-Rojanski and D. Huppert, *J. Non-Cryst. Solids*, **99**, 379 (1988).
34. D. Bianchi, M. Lacross, J. M. Pajonk and J. Teichner, *J. Catal.*, **68**, 411 (1981).
35. T. Lopez, M. Villa and R. Gomez, *J. Phys. Chem.*, **95**, 1690 (1991).
36. J. R. Hardee, S. E. Tunney, J. Frye and J. K. Stille, *J. Polym. Sci.*, **28A**, 3669 (1991).
37. K. V. Parish, D. Habibi and V. Mohammadi, *J. Organomet. Chem.*, **369**, 17 (1989).
38. A. Rosenfeld, D. Avnir and J. Blum, *J. Chem. Soc., Chem. Commun.*, 583 (1993).
39. J. Blum, A. Rosenfeld, N. Polak, O. Israelson, H. Schumann and D. Avnir, *J. Mol. Catal.*, **107**, 217 (1996).
40. (a) A. Rosenfeld, J. Blum and D. Avnir, *J. Catal.*, **164**, 363 (1996).
(b) H. Sertchook, D. Avnir, J. Blum, F. Joo, A. Katho, H. Schumann, R. Weinmann and S. Wernik, *J. Mol. Catal. A*, **108**, 153 (1996).
41. I. Amer, V. Orshav and J. Blum, *J. Mol. Catal.*, **45**, 207 (1988).
42. J. Blum, H. Huminer and H. Alper, *J. Mol. Catal.*, **75**, 153 (1992).
43. S. Braun, S. Rappoport, R. Zusman, D. Avnir and M. Ottolenghi, *Mater. Lett.*, **10**, 1 (1990).
44. (a) *Biochemical Aspects of Sol-Gel Science and Technology*, D. Avnir, and S. Braun, Editors, Kluwer Academic Publishers, Boston, 1996.
(b) D. Avnir, S. Braun, O. Lev and M. Ottolenghi, *Chem. Mater.*, **6**, 1605 (1994).
45. (a) S. Shtelzer and S. Braun, *Biotechnol. Appl. Biochem.*, **19**, 293 (1994).
(b) S. A. Yamanaka, F. Nishida, L. M. Ellerby, C. R. Nishida, B. Dunn, J. S. Valentine and J. I. Zink, *Chem. Mater.*, **4**, 495 (1992).
(c) Y. P. Tatsu, K. Yamashita, M. Yamaguchi, S. Yamamura, H. Yamamoto and S. Yoshikawa, *Chem. Lett.*, 1615 (1992).
(d) B. C. Dave, B. Dunn, J. S. Valentine and J. I. Zink, *Anal. Chem.*, **66**, 1120A (1994).
(e) J. Livage, *C.R. Acad. Sci. Paris, II*, **322**, 417 (1996).
(f) M. T. Reetz, A. Zonta and A. Simpelkamp, *J. Biotechnol. Bioeng.*, **49**, 527 (1996).
(g) C. Dosoretz, R. Armon, J. Starosvetzky, and N. Rothschild, *J. Sol-Gel Sci. Tech.*, **7**, 7 (1996).
(h) J. Zuhlke, D. Knopp and R. Niessner, *Fresenius J. Anal. Chem.*, **352**, 654 (1995).
46. (a) S. Shtelzer, S. Rappoport, D. Avnir, M. Ottolenghi and S. Braun, *Biotechnol. Appl. Biochem.*, **15**, 227 (1992).
(b) S. Braun, S. Shtelzer, S. Rappoport, D. Avnir and M. Ottolenghi, *J. Non-Cryst. Solids*, **147 & 148**, 739 (1992).
47. U. Narang, P. N. Prasad, F. V. Bright, K. N. Ramanathan, K. Deepak, B. D. Malhotra, M. N. Kalamansan and S. Chandra, *Anal. Chem.*, **66**, 3139 (1994).
48. S. Shtelzer and S. Braun, *Biotechnol. Appl. Biochem.*, **19**, 293 (1994).
49. (a) P. Audebert, C. Demaille and C. Sanchez, *Chem. Mater.*, **5**, 911 (1993).
(b) V. Glezer and O. Lev, *J. Am. Chem. Soc.*, **115**, 2533 (1993).
50. S. Braun, S. Rappoport, S. Shtelzer, R. Zusman, S. Druckman, D. Avnir and M. Ottolenghi, in *Biotechnology: Bridging Research and Applications* (Eds. D. Kamely, A. M. Chakrabary and S. E. Kornguth), Kluwer, Boston, 1991, pp. 205-218.
51. C. Dozorets, R. Armon and N. Rothchild, *J. Sol-Gel Sci. Technol.*, **4**, 7 (1996).

52. A. Turniansky, D. Avnir, A. Bronshtein, N. Aharonson and M. Altstein, *J. Sol-Gel Sci. Technol.*, **7**, 135 (1996).
53. M. Sun, *Science*, **233**, 1143 (1986).
54. D. R. Nair and J. L. Schnoor, *Environ. Sci. Technol.*, **26**, 2298 (1992).
55. R. Collino, J. Therasse, P. Binder, F. Chaput, B.-P. Boilot and Y. Levy, *J. Sol-Gel Sci. Technol.*, **2**, 823 (1994).
56. R. Wang, U. Narang, P. N. Prasad and F. V. Bright, *Anal. Chem.*, **65**, 2671 (1993).
57. J. Livage, C. Roux and J. M. Da Cost, *J. Sol-Gel Sci. Technol.*, **4**, 45 (1996).
58. J. Zuhlke, D. Knopp and R. Niessner, *Fresenius J. Anal. Chem.*, **352**, 654 (1995).
59. P. L. Edmiston, C. L. Wambolt, M. K. Smith and S. S. Saavedra, *J. Colloid Interface Sci.*, **163**, 395 (1994).
60. Y. Tatsu, K. Yamashita, M. Yamaguchi, S. Yamamura, H. Yamamoto and S. Yoshikawa, *Chem. Lett.*, 1615 (1992).
61. V. Glezer, G. Gun, M. Tsionsky and O. Lev, *Abstracts of the 44th Int. Soc. Electrochem. Meeting*, Berlin, Germany, September 14, 1993.
62. G. Gun, M. Tsionsky and O. Lev, *Anal. Chim. Acta*, **294**, 261 (1994).
63. G. Gun, M. Tsionsky and O. Lev, in *Better Ceramics Through Chemistry, VI*, (Eds. C. Sanchez, M. L. Mecartney, C. J. Brinker and A. Cheetham), *Mater. Res. Soc. Symp. Proc.*, **346**, 1011 (1994).
64. S. Sampath and O. Lev, *Anal. Chem.*, **68**, 2015 (1996).
65. M. Tsionsky and O. Lev, *J. Electrochem. Soc.*, **142**, 2154 (1995).
66. M. Tsionsky and O. Lev, *Anal. Chem.*, **67**, 2409 (1995).
67. J. Gun, M. Tsionsky, L. Rabinovich, Y. Golan, I. Rubinstein and O. Lev, *J. Electroanal. Chem.*, **395**, 57 (1995).
68. L. Rabinovich, M. Tsionsky, V. Glezer, J. Gun and O. Lev, *J. Sol-Gel Sci. Technol.*, in press (1997).
69. I. Pankratov and O. Lev, *J. Electroanal. Chem.*, **393**, 35 (1995).
70. S. Sampath, I. Pankratov, J. Gun and O. Lev, *J. Sol-Gel Sci. Technol.*, **7**, 123 (1996).
71. S. Sampath and O. Lev, *Electroanalysis*, **8**, 1112 (1996).
72. J. Gun and O. Lev, *Anal. Lett.*, **29**, 1933 (1996).
73. J. Gun and O. Lev *Anal. Chim. Acta*, **330**, 95 (1996).
74. J. Wang, P. V. A. Pamidi and D. S. Park, *Anal. Chem.*, **68**, 2705 (1996).
75. Review: R. H. Baney, M. Itoh, A. Sakakibara and T. Suzuki, *Chem. Rev.*, **95**, 1409 (1995).
76. Review: U. Schubert, N. Hüsing and A. Lorenz, *Chem. Mater.*, **7**, 2010 (1995).
77. W. Noll, *Chemie und Technologie der Silicone*, Verlag Chemie, Weinheim, 1968.
78. U. Deschler, P. Kleinschmit and P. Panster, *Angew. Chem.*, **98**, 237 (1986); *Angew. Chem., Int. Ed. Engl.*, **25**, 236 (1986).
79. H. Wolter, W. Glaubitt and K. Rose, *Mater. Res. Soc. Proc.*, **271**, 719 (1992).
80. Y. Nosaka, N. Tohriiwa, T. Kobayashi and N. Fujii, *Chem. Mater.*, **5**, 930 (1993).
81. F. Chaput, D. Riehl, Y. Levy and J. P. Boilot, *Chem. Mater.*, **5**, 589 (1993); F. Chaput, J. P. Boilot, D. Riehl and Y. Levy, *J. Sol-Gel Sci. Technol.*, **2**, 779 (1994).
82. B. Lebeau, C. Guermeur and C. Sanchez, *Mater. Res. Soc. Symp. Proc.*, **346**, 315 (1994).
83. M. Ueda, H.-B. Kim, T. Ikeda and K. Ichimura, *J. Non-Cryst. Solids*, **163**, 125 (1993); Z. Yang, C. Xu, B. Wu, L. R. Dalton, S. Kalluri, W. H. Steier, Y. Shi and J. H. Bechtel, *Chem. Mater.*, **6**, 1899 (1994).
84. H. Wolter and W. Storch, *J. Sol-Gel Sci. Technol.*, **2**, 93 (1994).
85. J. L. Brefort, R. J. P. Corriu, P. Gerbier, C. Guerin, B. J. L. Henner, A. Jean, T. Kuhlmann, F. Garnier and A. Yassar, *Organometallics*, **11**, 2500 (1992).
86. H. Kobayashi, *Makromol. Chem.*, **194**, 2569 (1993).
87. R. Kasemann and H. Schmidt, *New J. Chem.*, **18**, 1117 (1994).
88. J. Kron, S. Amberg-Schwab and G. Schottner, *J. Sol-Gel Sci. Technol.*, **2**, 189 (1994).
89. F. Babonneau, K. Thorne and J. D. Mackenzie, *Chem. Mater.*, **1**, 554 (1989).
90. F. Babonneau, *Polyhedron*, **13**, 1123 (1994).
91. F. Schwertfeger, W. Glaubitt and U. Schubert, *J. Non-Cryst. Solids*, **145**, 85 (1992); F. Schwertfeger, N. Hüsing and U. Schubert, *J. Sol-Gel Sci. Technol.*, **2**, 103 (1994); F. Schwertfeger, A. Emmerling, J. Gross, U. Schubert and J. Fricke, in *Sol-Gel Processing and Applications* (Ed. Y. A. Attia), Plenum Press, New York, 1994, p. 351; U. Schubert, F. Schwertfeger, N. Hüsing and E. Seyfried, *Mater. Res. Soc. Symp. Proc.*, **346**, 151 (1994); N. Hüsing, F. Schwertfeger,

W. Tappert and U. Schubert, *J. Non-Cryst. Solids,* **186**, 37 (1995); N. Hüsing and U. Schubert, *J. Sol-Gel Sci. Technol.,* **8**, 807 (1997).
92. N. Hüsing, U. Schubert, B. Riegel and W. Kiefer, *Mater. Res. Soc. Symp. Proc.*, **435**, 339 (1996).
93. F. Schwertfeger and U. Schubert, *Chem. Mater.*, **7**, 1909 (1995).
94. K. Rose, H. Wolter and W. Glaubitt, *Mater. Res. Soc. Proc.*, **271**, 731 (1992).
95. H. Wolter, W. Storch and H. Ott, *Mater. Res. Soc. Proc.*, **346**, 143 (1994).
96. F. D. Osterholtz and E. R. Pohl, *J. Adhesion Sci. Technol.*, **6**, 127 (1992).
97. Y. Abe, K. Taguchi, H. Hatano, G. Takahiro, Y. Nagao and T. Misono, *J. Sol-Gel Sci. Technol.*, **2**, 131 (1994); F. Devreux, J. P. Boilot, F. Chaput and A. Lecomte, *Mater. Res. Soc. Symp. Proc.*, **180**, 211 (1990).
98. K. Piana and U. Schubert, *Chem. Mater.*, **6**, 1504 (1994).
99. J. P. Grey, *Adhesion (London)*, **12**, 106 (1988).
100. J. D. Miller, K.-P. Hoh and H. Ishida, *Polymer Compos.*, **5**, 18 (1984).
101. K. Greiwe, W. Glaubitt, S. Amberg-Schwab and K. Piana, *Mater. Res. Soc. Symp. Proc.*, **271**, 725 (1992).
102. L. Delattre, C. Dupuy and F. Babonneau, *J. Sol-Gel Sci. Technol.*, **2**, 185 (1994).
103. H. Krug and H. Schmidt, *New J. Chem.*, **18**, 1125 (1994).
104. R. Nass, E. Arpac, W. Glaubitt and H. Schmidt, *J. Non-Cryst. Solids*, **121**, 370 (1990).
105. N. Yamazaki, S. Nakahama, J. Goto, T. Nagawa and A. Hirao, *Contemp. Top. Polym. Sci.*, **4**, 105 (1984); Y. Abe, T. Namiki, K. Tsuchida, Y. Nagao and T. Misono, *J. Non-Cryst. Solids*, **147 & 148**, 47 (1992).
106. J. F. Brown and L. H. Vogt, *J. Am. Chem. Soc.*, **87**, 4313 (1965); J. F. Brown, *J. Am. Chem. Soc.*, **87**, 4317 (1965).
107. G. M. Voronkov and V. I. Lavrent'yev, *Top. Curr. Chem.*, **102**, 199 (1982); U. Dittmar, B. J. Hendan, U. Flörke and H. C. Marsmann, *J. Organomet. Chem.*, **489**, 185 (1995).
108. M. J. van Bommel, P. M. C. Ten Wolde and T. N. N. Bernards, *J. Sol-Gel Sci. Technol.*, **2**, 167 (1994).
109. G. Philipp and H. Schmidt, *J. Non-Cryst. Solids*, **82**, 31 (1986).
110. S. Amberg-Schwab, E. Arpac, W. Glaubitt, K. Rose, G. Schottner and U. Schubert, in *High Performance Films and Coatings* (Ed. P. Vincenzini), Elsevier, Amsterdam, 1991, p. 203.
111. H. Schmidt and G. Philipp, *J. Non-Cryst. Solids*, **63**, 283 (1984); H. Schmidt and G. Philipp, in *Glass: Current Issues* (Eds. A. F. Wright and J. Dupuy), Martinus Nijhoff Publ., Dordrecht, 1985, p. 580.
112. R. J. P. Corriu, J. J. E. Moreau, P. Thepot and M. Wong Chi Man, *Chem. Mater.*, **8**, 100 (1996).
113. H. Schmidt, B. Seiferling, G. Philipp and K. Deichmann, in *Ultrastructure Processing of Advanced Ceramics* (Eds. J. D. Mackenzie and D. R. Ulrich), Wiley, New York, 1988, p. 651.
114. H. Schmidt and B. Seiferling, *Mater. Res. Soc. Symp. Proc.*, **73**, 739 (1986).
115. K. Greiwe, *Farbe und lack*, **11**, 968 (1991).
116. M. Popall, J. Kappel, M. Pilz and J. Schulz, *Mater. Res. Soc. Symp. Proc.*, **264**, 353 (1992); M. Popall, J. Kappel and J. Schulz, in *Micro System Technologies '94* (Eds. H. Reichl and A. Heuberger), VDE-Verlag, Berlin, 1995.
117. M. Popall, J. Kappel, M. Pilz, J. Schulz and G. Feyder, *J. Sol-Gel Sci. Technol.*, **2**, 157 (1994).
118. R. Nass, H. Schmidt and E. Arpac, *SPIE, Sol-Gel Optics*, **1328**, 258 (1990).
119. N. Kaneko, Jap. Patent Kokai-H-5-586193 (1993); *Chem. Abstr.*, **119**, 73383 (1993).
120. C. Sanchez, B. Alonso, F. Chapusot, F. Ribot and P. Audebert, *J. Sol-Gel Sci. Technol.*, **2**, 161 (1994).
121. G. A. Sigel, R. C. Domszy and W. C. Welch, *Mates. Res. Soc. Symp. Proc.*, **346**, 135 (1994).
122. M. Popall and H. Durand, *Electrochim Acta*, **37**, 1593 (1992).
123. D. Hoebbel, K. Endres, T. Reinert and H. Schmidt, *Mater. Res. Soc. Symp. Proc.*, **346**, 863 (1994).
124. P. A. Agaskar, *J. Am. Chem. Soc.*, **111**, 6858 (1989); D. Hoebbel, I. Pitsch, D. Heidemann, H. Jancke and W. Hiller, *Z. Anorg. Allg. Chem.*, **583**, 133 (1990); D. Hoebbel, I. Pitsch and D. Heidemann, *Z. Anorg. Allg. Chem.*, **592**, 207 (1991); D. Hoebbel, K. Endres, T. Reinert and I. Pitsch, *J. Non-Cryst. Solids*, **176**, 179 (1994).
125. E. P. Plueddemann, in *Silylated Surfaces* (Eds. D. E. Leyden and W. T. Collins), Gordon & Breach, New York, 1980, p. 31; H. Ishida, C. Chiang and J. L. Koenig, *Polymer*, **23**, 251 (1982); H. Ishida, S. Naviroj, S. K. Tripathy, J. J. Fitzgerald and J. L. Koenig, *J. Polym. Sci.*, **20**, 701 (1982); K. C. Vrancken, P. van der Voort, I. Gillis-D'Hamers and E. F. Vansant, *J. Chem. Soc., Faraday Trans.*, **88**, 3197 (1992).

126. V. de Zea Bermudez, M. Armand, C. Poinsignon, L. Abello and J.-Y. Sanchez, *Electrochim. Acta*, **37**, 1603 (1992); W. Cao and A. J. Hunt, *Mater. Res. Soc. Symp. Proc.*, **346**, 631 (1994).
127. K. Piana and U. Schubert, *Chem. Mater.*, **7**, 1932 (1995).
128. E. Lindner, A. Bader and H. A. Mayer, *Inorg. Chem.*, **30**, 3783 (1991).
129. Y. Charbouillot, D. Ravaine, M. Armand and C. Poinsignon, *J. Non-Cryst. Solids*, **103**, 325 (1988); H. Schmidt, M. Popall, F. Rousseau, C. Poinsignon, M. Armand and J. Y. Sanchez, in *Proc. 2nd Int. Symp. on Polymer Electrolytes* (Ed. B. Scrosati), Elsevier, London, 1990, p. 325.
130. U. Posset, M. Lankers, W. Kiefer, H. Steins and G. Schottner, *Appl. Spectrosc.*, **47**, 1600 (1993).
131. M. P. Beslard, N. Hovnanian, A. Larbot, L. Cot, J. Sanz, I. Sobrados and M. Gregorkiewitz, *J. Am. Chem. Soc.*, **113**, 1982 (1991).
132. N. Aharonson, M. Altstein, G. Avidan, D. Avnir, A. Bronshtein, A. Lewis, K. Liberman, M. Ottolenghi, Y. Polevaya, C. Rottman, J. Samuel, S. Shalom, A. Strinkovski and A. Turniansky, *Mater. Res. Soc. Symp. Proc.*, **346**, 519 (1994).
133. H.-E. Endres, L. D. Mickle, C. Kösslinger, S. Drost and F. Hutter, *Sensors and Actuators B*, **6**, 285 (1992).
134. A. Brandenburg and R. Edelhäuser, *Sensors and Actuators B*, **11**, 361 (1993).
135. J. Lin, S. Möller and E. Obermeier, *Sensors and Actuators B*, **5**, 219 (1991).
136. B. Breitscheidel, J. Zieder and U. Schubert, *Chem. Mater.*, **3**, 559 (1991); U. Schubert, B. Breitscheidel, H. Buhler, C. Egger and W. Urbaniak, *Mater. Res. Soc. Symp. Proc.*, **271**, 621 (1992); U. Schubert, C. Görsmann, S. Tewinkel, A. Kaiser and T. Heinrich, *Mater. Res. Soc. Symp. Proc.*, **351**, 141 (1994); A. Kaiser, C. Görsmann and U. Schubert, *J. Sol-Gel Sci. Technol.*, **8**, 795 (1997); C. Görsmann and U. Schubert, *Mater. Res. Soc. Symp. Proc.*, **435**, 625 (1996).
137. C. Ferrari, G. Prediri and A. Tiripicchio, *Chem. Mater.*, **4**, 243 (1992).
138. L. Spanhel, E. Arpac and H. Schmidt, *J. Non-Cryst. Solids*, **147 & 148**, 657 (1992); T. Burkhart, M. Mennig, H. Schmidt and A. Licciulli, *Mater. Res. Soc. Symp. Proc.*, **346**, 779 (1994).
139. U. Schubert, *New. J. Chem.*, **18**, 1049 (1994).
140. K. G. Allum, R. D. Hancock, I. V. Horwell, S. McKenzie, R. C. Pikethly and P. J. Robinson, *J. Organomet. Chem.*, **107**, 393 (1976).
141. Leading references: F. R. Hartley, *Supported Metal Complexes*, Reidel, Dordrecht, 1985; Yu. I. Yermakov, B. N. Kuznetsov and V. A. Zakharov, *Catalysis by Supported Complexes*, Elsevier, Amsterdam, 1981.
142. U. Schubert, C. Egger, K. Rose and C. Alt, *J. Mol. Catal.*, **55**, 330 (1989).
143. C. Guizard and P. Lacan, *New J. Chem.*, **18**, 1097 (1994).
144. K. Arija and Y. Okahata, *J. Am. Chem. Soc.*, **111**, 5618 (1989).
145. J. E. Mark, *Heterog. Chem. Rev.*, **3**, 307 (1996).
146. H. Schmidt, in *Sol-Gel Optics* (Ed. L.C. Klein), Kluwer Academic Publ., Boston, 1994, pp. 451–481.
147. R. A. Roy and R. Roy, *Mater. Res. Bull.*, **19**, 169 (1984).
148. C. Sanchez and F. Ribot, *New J. Chem.*, **18**, 1007 (1994).
149. L. H. Sperling, *Interpenetrating Polymer Networks and Related Materials*, Plenum Press, New York, 1981.
150. E. J. A. Pope, M. Asami and J. D. Mackenzie, *J. Mater. Res.*, **4**, 1018 (1989).
151. L. C. Klein, in *Sol-Gel Optics* (Ed. L.C. Klein), Kluwer Academic Publ., Boston, 1994, p. 215.
152. B. Abramoff and L. C. Klein, in *Chemical Processing of Advanced Materials* (Eds. L. L. Hench and J. K. West), Wiley, New York, 1992, pp. 815–821.
153. B. Abramoff and L. C. Klein, *SPIE*, **1328**, 241 (1990).
154. C. J. T. Landry, B. K. Coltrain, J. A. Wesson, N. Zumbuluyadis and J. C. Lippart, *Polymer*, **33**, 1496 (1992).
155. C. J. T. Landry, B.K. Coltrain and B. K. Brady, *Polymer*, **33**, 1486 (1992).
156. A. B. Wojcik and L. C. Klein, *SPIE*, **2018**, 160 (1993).
157. C. L. Beaudry, L. C. Klein and R. A. McCauley, *J. Therm. Anal.*, **46**, 55 (1996).
158. C. L. Beaudry and L. C. Klein, in *Nanotechnology of Molecularly Designed Materials* (Eds. G.-M. Chow and K. E. Gonsalves), *Am. Chem. Soc. Symp.*, **622**, 1996, p. 382.
159. M. W. Ellsworth and B. M. Novak, *Polym. Prepr.*, **33**, 1088 (1992).
160. M. W. Ellsworth and B. M. Novak, *J. Am. Chem. Soc.*, **113**, 2756 (1991).
161. B. M. Novak, *Adv. Mater.*, **5**, 422 (1993).
162. B. M. Novak and C. Davies, *Macromolecules*, **24**, 5481 (1991).
163. A. B. Wojcik and L. C. Klein, *J. Sol-Gel Sci. Technol.*, **2**, 115 (1994).

164. A. B. Wojcik and L. C. Klein, *J. Sol-Gel Sci. Technol.*, **4**, 57 (1995).
165. A. B. Wojcik and L. C. Klein, *J. Sol-Gel Sci. Technol.*, **5**, 77 (1995).
166. D. Hoebbel and W. Wieker, *Z. Anorg. Allg. Chem.*, **384**, 43 (1991).
167. Y. Hu and J. D. Mackenzie, *MRS. Symp. Proc.*, **271**, 681 (1992); J. D. Mackenzie, Y. J. Chung and Y. Hu, *J. Non-Cryst. Solids*, **147 & 148**, 271 (1992); Y. Hoshino and J. D. Mackenzie, *J. Sol-Gel Sci. Technol.*, **5**, 83 (1995).
168. B. Wang, G. L. Wilkes, J. C. Hedrick, S. C. Liptak and J. E. McGrath, *Macromolecules*, **24**, 3449 (1991).
169. H. H. Huang, G. L. Wilkes and J. G. Carlson, *Polymer*, **30**, 2001 (1989).
170. Y. Chujo, E. Ihara, H. Ihara and T. Saegusa, *Macromolecules*, **22**, 2040 (1989).
171. J. Thompson, H. H. Fox, I. Gorodisher, G. Teowee, D. Calvert and D. R. Uhlmann, *MRS Symp. Proc.*, **180**, 987 (1990).
172. D. Xu, S. H. Wang and J. E. Mark, *MRS Symp. Proc.*, **180**, 445 (1990).
173. T. H. Mouray, S. M. Miller, J. A. Wesson, T. E. Long and L. W. Kelt, *Macromolecules*, **25**, 45 (1992).
174. S. Kohjiya, K. Ochial and S. Yamashita, *J. Non-Cryst. Solids*, **119**, 132 (1990).
175. Y. Wei, D. Yang and L. Tang, *Macromol. Chem., Rapid. Commun.*, **14**, 273 (1993).
176. Y. Wei, R. Bakthavatachalam and C. K. Whitecar, *Chem. Mater.*, **2**, 337 (1990).
177. M. Guglielmi, P. Colombo, G. Brusetin, G. Facchin and M. Gleria, *J. Sol-Gel Sci. Technol.*, **2**, 109 (1994).
178. H. Schmidt and M. Popall, *SPIE*, **1328**, 249 (1990).
179. H. Schmidt, H. Scholze and G. Tunker, *J. Non-Cryst. Solids*, **80**, 557 (1986).
180. K. J. Shea, D. A. Loy and O. W. Webster, *Chem. Mater.*, **1**, 572 (1989).
181. D. A. Loy, R. J. Buss, R. A. Assink, K. J. Shea and H. Oviatt, *Mater. Res. Soc. Symp. Ser.*, **346**, 825 (1994).
182. (a) K. M. Choi and K. J. Shea, *J. Sol-Gel Sci. Technol.*, **5**, 143 (1995).
     (b) D. A. Loy and K. J. Shea, *Chem Rev.*, **95**, 143 (1995).
183. (a) S. Ikoma, S. Takano, E. Nomoto and H. Yokoi, *J. Non-Cryst. Solids*, **113**, 130 (1989).
     (b) A. Shames, O. Lev and B. Iosefson-Kuyavskaya *J. Non-Cryst. Solids*, **175**, 14 (1994).
184. V. R. Kaufman, D. Levy and D. Avnir, *J. Non-Cryst. Solids*, **82**, 103 (1986).
185. V. R. Kaufman and D. Avnir, *Langmuir*, **2**, 717 (1986).
186. (a) For a review, see, M. Anpo, T. Fujii and N. Negishi, *Heterogen. Chem. Rev.*, **1**, 231 (1994); some other examples:
     (b) K. Matsui, M. Tominaga, Y. Arai, H. Satoh and M. Kyoto, *J. Non-Cryst. Solids*, **169**, 295 (1994).
     (c) C. J. Brinker, A. J. Hurd, P. R. Schank, G. C. Frye and C. S. Ashley, *J. Non-Cryst. Solids*, **147&148**, 424 (1992).
     (d) N. Negishi, T. Fujii and M. Anpo, *Langmuir*, **9**, 3320 (1993).
187. K. Matsui, T. Matsuzuka and H. Fujita, *J. Chem. Phys.*, **93**, 4991 (1989).
188. T. Fujii, T. Mabuchi and I. Mitsui, *Chem. Phys. Lett.*, **68**, 5 (1990).
189. K. Matsui and T. Nakazawa, *Bull. Chem. Soc. Jpn.*, **63**, 11 (1990).
190. J. McKiernan, J. C. Pouxviel, B. Dunn and J. I. Zink, *J. Phys. Chem.*, **93**, 2129 (1989).
191. T. Fujii and K. Toriumi, *J. Chem. Soc., Faraday Trans.*, **89**, 3437 (1993).
192. D. Levy and D. Avnir, *J. Phys. Chem.*, **92**, 4734 (1988).
193. D. Levy, S. Einhorn and D. Avnir, *J. Non-Cryst. Solids*, **113**, 137 (1989).
194. J. C. Pouxviel, B. Dunn and J. I. Zink, *J. Phys. Chem.*, **93**, 2134 (1989).
195. (a) *SPIE Proc. Series* on Sol-Gel Optics, (Ed. J. D. Mackenzie), every two years since 1990.
     (b) L. C. Klein (Ed.), *Sol-Gel Optics*, Kluwer, Boston, 1994.
196. D. Avnir, V. R. Kaufman and R. Reisfeld, *J. Non-Cryst. Solids*, **74**, 395 (1985).
197. D. Levy and D. Avnir, *J. Photochem. Photobiol. A: Chem.*, **57**, 41 (1991).
198. J. Samuel, Y. Polevaya, M. Ottolenghi and D. Avnir, *Chem. Mater.*, **6**, 1457 (1994).
199. D. Levy, R. Reisfeld and D. Avnir, *Chem. Phys. Lett.*, **109**, 593 (1984).
200. L. R. Matthews and E. T. Knobbe, *Chem. Mater.*, **5**, 1697 (1993) and references cited therein.
201. E.g.: S. S. Ostapenko, A. J. Neuhalfen and B. W. Wessels, *Mater. Sci. Forum*, **143-147**, 743 (1994).
202. (a) H. Zabrodsky, S. Peleg and D. Avnir, *J. Am. Chem. Soc.*, **114**, 7843 (1992).
     (b) H. Zabrodsky, S. Peleg and D. Avnir, *J. Am. Chem. Soc.*, **115**, 8278 (1993).
     (c) H. Zabrodsky and D. Avnir, *J. Am. Chem. Soc.*, **117**, 462 (1995).

203. A. Slama-Schwok, M. Ottolenghi and D. Avnir, *Nature*, **355**, 240(1992).
204. A. Slama-Schwok, D. Avnir and M. Ottolenghi, *J. Phys. Chem.*, **93**, 7544 (1989).
205. (a) A. Slama-Schwok, D. Avnir and M. Ottolenghi, *J. Am. Chem. Soc.*, **113**, 3984 (1991).
     (b) A. Slama-Schwok, D. Avnir and M. Ottolenghi, *Photochem. Photobiol.*, **54**, 525 (1991).
206. C.f. also: F. N. Castellano, T. A. Heimer, T. M. Tandhasetti and G. J. Meyer, *Chem. Mater.*, **6**, 1041 (1994).
207. M. Nogami and Y. Abe, *J. Mater. Sci.*, **30**, 5789 (1995).
208. D. Preston, J. Pouxviel, T. Novinson, W. C. Kaska, B. Dunn and J. I. Zink, *Phys. Chem.*, **94**, 4167 (1990).
209. M. Nogami and T. Sugiura, *J. Mater. Sci. Lett.*, **12**, 1544 (1993).
210. (a) M. Ueda, H.-B. Kim, T. Ikeda and K. Ichimura, *J. Non-Cryst. Solids*, **163**, 125 (1993).
     (b) M. Ueda, H.-B. Kim, T. Ikeda and K. Ichimura, *Chem. Mater.*, **4**, 1229 (1992).
211. L. Hou, M. Mennig and H. Schmidt, *SPIE Sol-Gel Optics III*, **2288**, 328 (1994).
212. E. J. A. Pope, *SPIE Sol-Gel Optics III*, **2288**, 410 (1994).
213. H. Nakazumi, R. Nagashiro, S. Matsumoto and K. Isagawa, *SPIE Sol-Gel Optics III*, **2288**, 402 (1994).
214. H. Nakazumi, K. Makita and R. Nagashiro, *J. Sol-Gel Sci. Technol.*, **8**, 901 (1997).
215. D. Levy, M. López-Amo, J. M. Otón, F. del Monte, P. Datta and I. Matías, in *Advanced Networks and Services, SPIE Proc.* (Ed. T. Russel Hsing), SPIE, Bellingham Washington 1995, p. 2449.
216. D. Yuan-Chieh, PhD Thesis, Rutgers, The State University of New-Jersey, 1993.
217. D. Levy, F. del Monte, J. M. Otón, I. Matias, P. Datta and M. Lopez-Amo, *J. Sol-Gel Sci. Technol.*, **8**, 931, (1997).
218. D. R. Ulrich, *J. Non-Cryst. Solids*, **100**, 174 (1988).
219. M. Yamane, S. Aso and T. Sakaino, *J. Mater. Sci.*, **13**, 865 (1978).
220. B. Dunn, E. Knobbe, J. M. MacKiernan, J. C. Pouxviel and J. I. Zink, *Mater. Res. Soc. Symp. Proc.*, **121**, 331 (1988).
221. T. Fujii, A. Ishii, H. Nagai, M. Niwano, N. Negishi and M. Anpo, *Chem. Express*, **4**, 1 (1989).
222. E. J. A. Pope and J. D. Mackenzie, *MRS Bulletin*, p. 29, March 17, 1987.
223. D. Levy, C. J. Serna and J. M. Otón, *Mater. Lett.*, **10**, 470 (1991).
224. J. M. Otón, A. Serrano, C. J. Serna and D. Levy, *Liq. Cryst.*, **10**, 733 (1991).
225. D. Levy, M. Ocaña and C. J. Serna, *Langmuir*, **10**, 2683 (1994).
226. U. P. Narang, R. Wang, P. N. Prasad and F. V. Bright, *J. Phys. Chem.*, **98**, 17 (1994).
227. S. C. Laperrière, J. W. Mullens, D. L'Espérance and E. L. Chronister, *Chem. Phys. Lett.*, **243**, 114 (1995).
228. Y. Kobayashi, Y. Imai and Y. Kurokawa, *J. Mater. Sci. Lett.*, **7**, 1148 (1988).
229. W. Nei, B. Dunn, C. Sanchez and P. Griesmar, *Mater. Res. Soc. Proc.*, **271**, 639 (1992).
230. Y. Haruvy, A. Heller and S. E. Webber, *Am. Chem. Soc. Symp. Ser.*, **499**, 405 (1992).
231. D. L'Esperance and E. L. Chronister, *Chem. Phys. Lett.*, **195**, 387 (1995).
232. B. Lebeau, N. Herlet, J. Livage and C. Sanchez, *Chem. Phys. Lett.*, **206**, 15 (1993).
233. M. D. Ranh and T. A. King, *SPIE Sol-Gel Optics III*, **2288**, 364 (1994).
234. R. Reisfeld, D. Brusilovsky, M. Eyal, E. Miron, Z. Burshtein and J. Ivri, *Chem. Phys. Lett.*, **160**, 43 (1989).
235. A. B. Wojcik and L. C. Klein, *SPIE Sol-Gel Optics III*, **2288**, 392 (1994).
236. M. Canva, A. Dubois, P. Georges, A. Brun, F. Chaput, A. Ranger and J. P. Boilot, *SPIE Sol-Gel Optics III*, **2288**, 298 (1994).
237. T. Surtwala, Z. Gardlund, J. M. Boulton, D. R. Uhlmann, J. Watson and N. Peyghambarian, *SPIE Sol-Gel Optics III*, **2288**, 310 (1994).
238. H. Nakazumi, S. Amano and K. Sakai, *SPIE Sol-Gel Optics III*, **2288**, 356 (1994).
239. T. Fujii and A. Ishii, *J. Photochem. Photobiol., A: Chem.*, **54**, 231 (1990).
240. B. Dunn and J. I. Zink, *J. Mater. Chem.*, **1**, 903 (1991).
241. E. Knobbe, B. Dunn, P. Fuqua and F. Nishida, *Appl. Opt.*, **29**, 2729 (1990).
242. L. A. M. K. Shun, L. O. Dennis and W. Kin Hung, *Appl. Opt.*, **34**, 3380 (1995).
243. R. Gvishi and R. Reisfeld, *J. Non-Cryst. Solids*, **128**, 69 (1991).
244. H. G. Floch, P. F. Belleville, J. J. Priotton, P. M. Pegon, C. S. Dijonneau and J. Guerain, *Am. Ceram. Soc., Bull.*, **74**, 60 (1995).
245. J. Fabian, H. Nakazumi and M. Matsuoka, *Chem. Rev.*, **92**, 1197 (1992).
246. H. Nakatsuka, K. Inouye, S. Uemura and R. Yano, *Chem. Phys. Lett.*, **171**, 245 (1990).

247. F. M. Schellenberg, W. Lenth and G. C. Bjorklund, *Appl. Opt.*, **25**, 3207 (1986).
248. F. del Monte and D. Levy, *Chem. Mater.*, **7**, 292 (1995).
249. R. Reisfeld, D. Brusilovsky, M. Eyal, E. Miron, Z. Burshtein and J. Ivri, *Chem. Phys. Lett.*, **160**, 43 (1989).
250. E. Toussaere, J. Zyss, P. Griesmar and C. Sanchez, *Nonlinear Optics*, **1**, 349 (1991).
251. M. Nakamura, H. Naso and K. Kamiya, *J. Non-Cryst. Solids*, **135**, 1 (1991).
252. M. Canva, G. Le Saux, P. Georges, A. Brun, F. Chaput and J. P. Boilot, *Opt. Lett.*, **17**, 218 (1992).
253. K. Kubodera and T. Kaino, in *Nonlinear Optics of Organic and Semiconductors* (Ed. T. Kobayashi), Springer, Berlin, 1989, p. 163.
254. R. K. Jain and R. C. Lind, *J. Opt. Soc. Am.*, **73**, 647 (1983).
255. S. S. Yao, C. Karaguleff, A. Gabel, R. Fortenberry, C. T. Seaton and G. I. Stegeman, *Appl. Phys. Lett.*, **46**, 801 (1985).
256. R. Reisfeld, *SPIE Sol–Gel Optics II*, **1758**, 546 (1992).
257. G. Pucetti, E. Toussaere, I. Ledoux, J. Zyss, P. Griesmar and C. Sanchez, *Am. Chem. Soc., Div. Polym. Chem.*, **32**, 61 (1991).
258. P. Griesmar, C. Sanchez, G. Pucetti, I. Ledoux and J. Zyss, *Mol. Eng.*, **1**, 205 (1991).
259. J. Kim, J. L. Plawsky, R. La Peruta and G. M. Korenowski, *Chem. Mater*, **4**, 249 (1992).
260. R. J. Jeng, Y. V. Chen, A. K. Jain, J. Kumar and S. K. Tripathy, *Chem. Mater.*, **4**, 972 (1992).
261. J. Livage, C. Schmutz, P. Griesmar, P. Barboux and C. Sanchez, *SPIE Sol–Gel Optics II*, **1758**, 274 (1992).
262. G. S. He, J. D. Bhawalkar, C. F. Zhao and P. N. Prasad, *Appl. Phys. Lett.*, **67**, 2433 (1995).
263. Z. Yang, C. Xu, B. Wu, L. R. Dalton, S. Kalluri, W. Steier, Y. Shi and J. H. Bechtel, *Chem. Mater.*, **6**, 1899 (1994).
264. B. Lebeau, C. Sanchez, S. Brasselet, J. Zyss, G. Froc and M. Dumont *New J. Chem.*, **20**, 13 (1996).
265. S. Kalluri, Y. Shi, W. Steier, Z. Yang, C. Xu, B. Wu and L. R. Dalton, *Appl. Phys. Lett.*, **65**, 2651 (1994).
266. D. Riehl, F. Chaput, A. Roustamian, Y. Lévy and J. P. Boilot, *Nonlinear Optics*, **8**, 141 (1994).
267. D. Riehl, F. Chaput, Y. Lévy, J. P. Boilot, F. Kajzar and P. A. Chollet, *Chem. Phys. Lett.*, **245**, 36 (1995).
268. C. A. Capozzi and A. B. Seddon, *SPIE Sol–Gel Optics III*, **2288**, 340 (1994).
269. G. J. Gall, T. A. King, S. N. Oliver, C. A. Capozzi, A. B. Seddon, C. A. S. Hill and A. E. Underhill, *SPIE Sol–Gel Optics III*, **2288**, 372 (1994).
270. X. Wang, L. M. Yates III and E. T. Knobe, *SPIE Sol–Gel Optics III*, **2288**, 264 (1994).
271. L. Zhu, Y. Li, J. Wang and J. Shen, *Chem. Phys. Lett.*, **239**, 393 (1995).
272. D. Levy, J. M. S. Pena and C. J. Serna, *J. Non-Cryst. Solids*, **147 & 148**, 646 (1992).
273. W.-P. Chang, W.-T. Whang and J.-C. Wong, *Jpn. J. Appl. Phys.*, **34**, 1888 (1995).
274. D. Levy, A. Serrano and J. M. Otón, *J. Sol–Gel Sci. Technol.*, **2**, 803 (1994).
275. D. Levy, X. Quintana, C. Rodrigo and J. M. Otón, *SPIE Sol–Gel Optics III*, **2288**, 60 (1994).
276. J. M. Otón, J. M. S. Pena, A. Serrano and D. Levy, *Appl. Phys. Lett.*, **66**, 2804 (1995).
277. G. Schottner, W. Grond, L. Kümmerl and D. Haarer, *J. Sol–Gel Sci. Technol.*, **2**, 657 (1994).
278. R. Nakao, N. Ueda, Y. Abe, T. Horii and H. Inoue, *Polym. Adv. Technol.*, **5**, 240 (1994).
279. R. C. Chambers, Y. Haruvy and M. A. Fox, *Chem. Mater.*, **6**, 1351 (1994).
280. S. K. Tripathy, J. Kumar, J. T. Chen, S. Marturunkalkul, R. J. Jung, L. Li and X. L. Jiang, *Mater, Res. Soc. Symp. Proc.*, **346**, 351 (1994).
281. C. Sanchez, P. Griesmar, E. Toussaere, G. Puccetti, I. Ledoux and J. Zyss, *Nonlinear Optics*, **2**, 245 (1992).
282. H. Nakazumi and S. Amano, *J. Chem. Soc., Chem. Commun.*, 1079 (1992).
283. I. J. M. Snijkers-Hendrickx, M. W. J. L. Oomen, N. Wittouck and F. De Schryver, *Proc. 1st Europ. Workshop on Hybrid Organic–Inorganic Materials* (Eds. C. Sanchez and F. Ribot), Paris, 1993, p. 237.
284. C. Sanchez and B. Lebeau, *Pure Appl. Opt.*, **5**, 689 (1996).
285. B. Lebeau, J. Maquet, C. Sanchez, E. Toussaere, R. Hierle and J. Zyss, *J. Chem. Mater.*, **4**, 1855 (1994).
286. L. Kador, R. Fischer, D. Haarer, R. Kasemann, S. Brück, H. Schmidt and H. Dürr, *Adv. Mater.*, **5**, 270 (1993).
287. M. Ueda, H.-B. Kim and K. Ichimura, *Chem. Mater.*, **6**, 1771 (1994).

CHAPTER 41

Chirality in bioorganosilicon chemistry

REINHOLD TACKE and STEPHAN A. WAGNER

Institut für Anorganische Chemie, Universität Würzburg, Am Hubland, D-97074 Würzburg, Germany

I. INTRODUCTION	2363
II. BIOLOGICAL RECOGNITION OF ENANTIOMERIC ORGANOSILICON DRUGS	2364
III. BIOCATALYSIS AS A METHOD FOR THE PREPARATION OF OPTICALLY ACTIVE ORGANOSILICON COMPOUNDS	2376
A. Introductory Remarks	2376
B. Reductions	2376
C. Hydrolyses	2384
D. Transesterifications	2388
E. Esterifications	2390
F. Oxidations	2394
IV. CONCLUDING REMARKS	2397
V. ACKNOWLEDGMENTS	2397
VI. REFERENCES	2398

I. INTRODUCTION

Bioorganosilicon chemistry represents a fascinating, rapidly expanding branch of organosilicon chemistry[1-9]. This area has been dominated by basic research for a long time; however, in recent years practical aspects are becoming of increasing importance. The development of new organosilicon drugs and agrochemicals and the application of biocatalysis in synthetic organosilicon chemistry are examples of this.

This short review deals with the topical subject 'chirality in bioorganosilicon chemistry'. Two different aspects will be discussed: (i) 'biological recognition of enantiomeric organosilicon drugs' and (ii) 'biocatalysis as a method for the preparation of optically active organosilicon compounds'. Most of the experimental material described in this article (which does not lay claim to completeness) comes from our own laboratory.

The chemistry of organic silicon compounds, Vol. 2
Edited by Z. Rappoport and Y. Apeloig © 1998 John Wiley & Sons Ltd

II. BIOLOGICAL RECOGNITION OF ENANTIOMERIC ORGANOSILICON DRUGS

During the past two decades, a variety of highly potent and receptor-selective silicon-based muscarinic antagonists have been developed[10–36]. The silanols sila-pridinol (**1**), sila-difenidol (**2**), sila-procyclidine (**3**), sila-trihexyphenidyl (**4**), hexahydro-sila-difenidol (**5**) and *p*-fluoro-hexahydro-sila-difenidol (**6**) are examples of this particular type of compound. The racemic mixtures of the silanols **5** (HHSiD) and **6** (*p*-F-HHSiD) are the most prominent drugs in this series of muscarinic antagonists. Both compounds are commercially available and are used worldwide as selective tools for the classification of muscarinic receptor subtypes. In the last few years, numerous biological data for HHSiD and *p*-F-HHSiD have been reported by many laboratories in more than 400 publications.

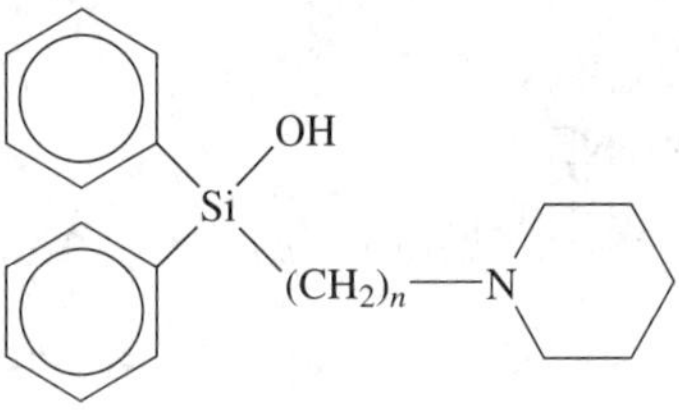

(**1**) *n* = 2
(**2**) *n* = 3

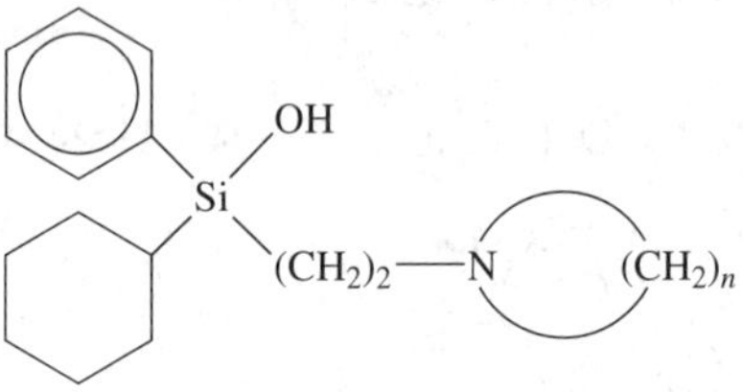

(**3**) *n* = 4
(**4**) *n* = 5

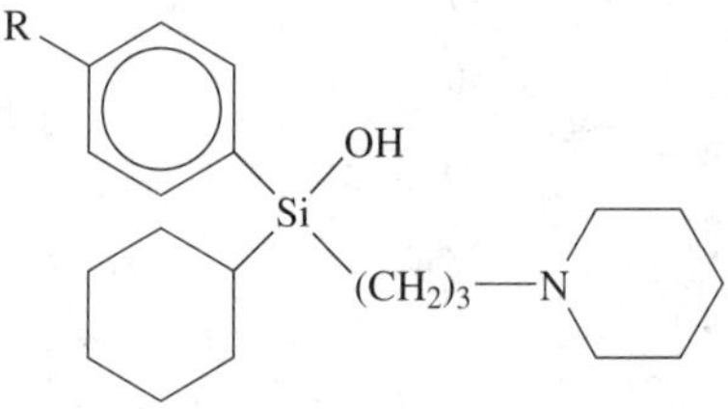

(**5**) R = H
(**6**) R = F

As can be seen from the formulas of **3–6**, these silanols are chiral compounds containing the silicon atom as the center of chirality. In order to study their stereoselectivity of antimuscarinic action, the enantiomers of sila-procyclidine (**3**) and of its methiodide sila-tricyclamol iodide (**7**) were prepared (Scheme 1)[20]. The optically active silanols (*R*)-**3** and (*S*)-**3** were obtained from the racemate *rac*-**3** (for syntheses of this compound[15,19], see Schemes 2 and 3) by a classical racemate resolution using (*L*)-(+)-tartaric acid and (*D*)-(−)-tartaric acid, respectively, as resolving agents (partial resolution by crystallization of diastereomeric salts and subsequent fractional crystallizations of the corresponding optically active free amine bases). The enantiomers (*R*)-**3** and (*S*)-**3** were isolated as crystalline solids with enantiomeric purities of >97% ee (NMR) and 99.7% ee (differential scanning calorimetry), respectively (in this context, see also Reference 32). The almost enantiomerically pure amines (*R*)-**3** and (*S*)-**3** were then transformed into the corresponding methiodides (*R*)-**7** and (*S*)-**7** by reaction with methyl iodide. The absolute configuration of the silanols (*R*)-**3**, (*S*)-**3**, (*R*)-**7** and (*S*)-**7** was established on the basis of a single-crystal X-ray diffraction analysis of (*R*)-sila-procyclidine [(*R*)-**3**][17]. The structure of the silanol (*R*)-**3** in the crystal is depicted in Figure 1.

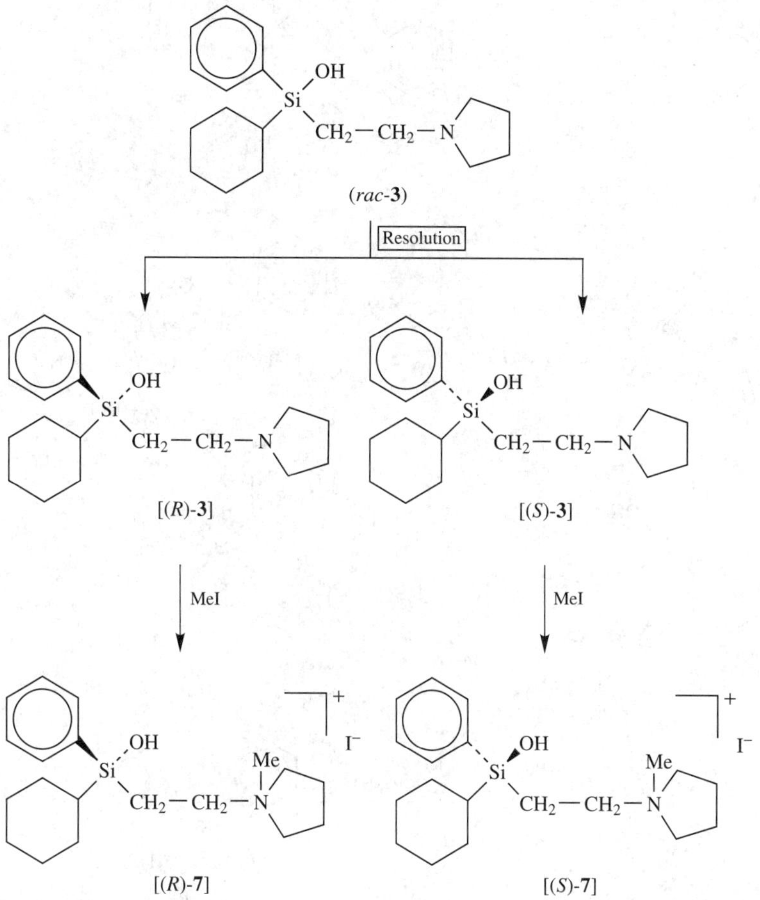

SCHEME 1

Compounds (*R*)-**3**, (*S*)-**3**, (*R*)-**7** and (*S*)-**7** are the first optically active silanols which were isolated as almost pure enantiomers. All compounds were found to be configurationally stable in the crystalline state and in inert organic solvents; however, in aqueous solution the optically active silanols racemize (**7** faster than **3**)[20]. This racemization can be described in terms of a nucleophilic attack of water (or OH$^-$) at the silicon atom which leads to substitution of the OH group bound to the silicon atom by the entering OH group.

The pure enantiomers of sila-procyclidine (**3**) and sila-tricyclamol iodide (**7**) were studied for their affinities for muscarinic M2 (guinea-pig atria) and M3 receptors (guinea-pig ileum) using carbachol as the agonist[20]. As the optically active silanols undergo a racemization under physiological conditions, a special experimental protocol was used for these studies (temperature, 32 °C; pH value, 7.4; equilibration time, 10 min). All compounds were found to exhibit an apparently competitive antagonism at muscarinic M2 and M3 receptors in these functional pharmacological experiments. The receptor affinities [pA_2

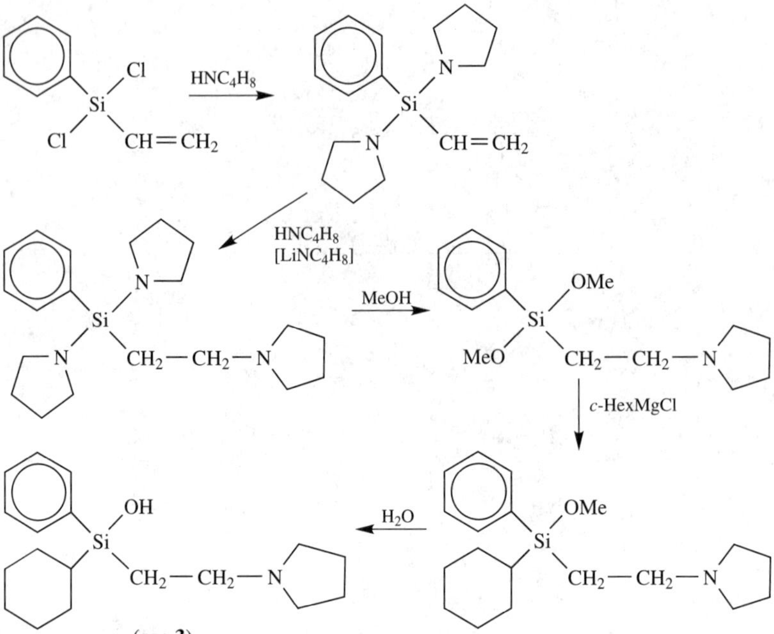

SCHEME 2

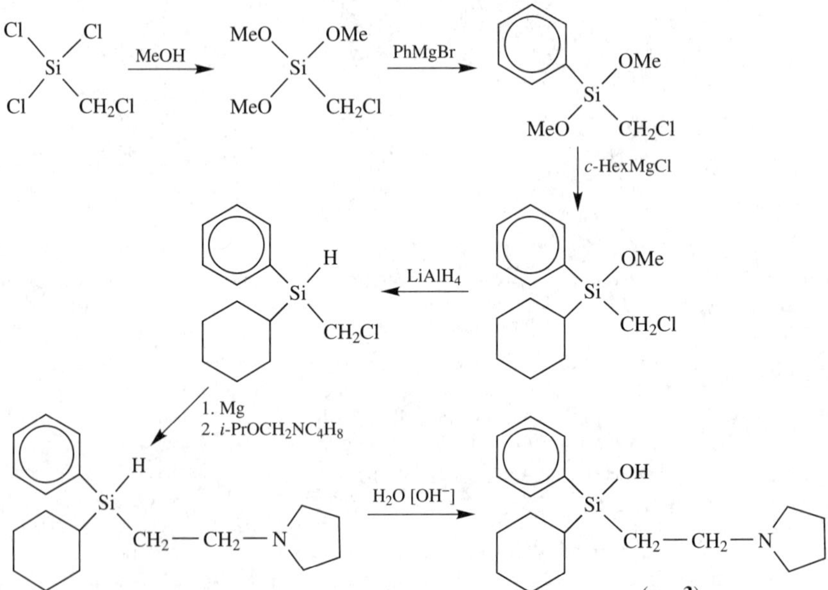

SCHEME 3

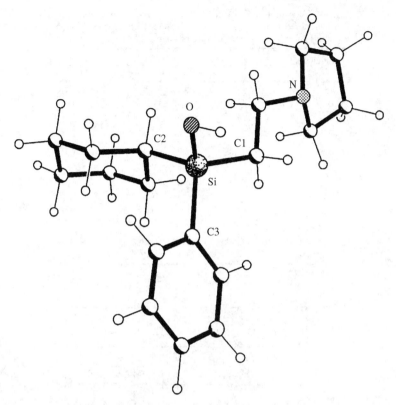

FIGURE 1. Structure of the silanol (R)-3 in the crystal. Selected bond distances (pm) and angles (deg): Si—C(1), 186.8(4); Si—C(2), 187.8(3); Si—C(3), 188.2(3); Si—O, 163.5(3); C(1)—Si—C(2), 109.9(2); C(1)—Si—C(3), 109.4(2); C(1)—Si—O, 111.3(1); C(2)—Si—C(3), 109.8(2); C(2)—Si—O, 107.1(1); C(3)—Si—O, 109.2(1). The molecules are connected by intermolecular O—H···N hydrogen bonds [O···N, 2.792(5) pm] to form infinite chains parallel to the c axis

values; $pA_2 = -\log K_D$ (K_D = dissociation constant of the drug–receptor complex)] obtained in these studies are given in Table 1 and Figure 2. From the kinetics of the racemization of the silanols, a maximum error of the pA_2 values by a factor of 2 can be estimated. The racemization kinetics were established by functional pharmacological experiments at muscarinic M3 receptors[20].

As can be seen from Figure 2, the (R)-enantiomers (eutomers) of the silanols **3** and **7** show a significantly higher affinity for muscarinic M2 and M3 receptors than the corresponding (S)-antipodes (distomers). To the best of our knowledge, this is the first example of a biological discrimination between enantiomeric silicon compounds, with the silicon atom as the center of chirality. The stereoselectivity indices SI [$SI = K_D(S)/K_D(R)$] for sila-procyclidine (**3**) are 1.8 (M2) and 4.1 (M3), respectively. For sila-tricyclamol iodide (**7**), SI values of 21 (M2) and 23 (M3) were found. Qualitatively analogous stereoselectivities were observed for the enantiomers of the related carbon analogues procyclidine (**8**) and tricyclamol iodide (**9**) which are configurationally stable under physiological conditions[37]. The silanols (R)-**3**, (S)-**3**, (R)-**7** and (S)-**7** exhibit a higher antimuscarinic

TABLE 1. Affinities (pA_2 values) of the (R)- and (S)-enantiomers of **3, 7** and **10–13** for muscarinic M1 (rabbit vas deferens), M2 (guinea-pig atria) and M3 receptors (guinea-pig ileum)

Compound	pA_2 value		
	M1	M2	M3
(R)-**3**	n.d.[a]	7.15 ± 0.04	8.26 ± 0.03
(S)-**3**	n.d.[a]	6.90 ± 0.05	7.65 ± 0.03
(R)-**7**	n.d.[a]	8.37 ± 0.05	8.72 ± 0.04
(S)-**7**	n.d.[a]	7.04 ± 0.04	7.36 ± 0.04
(R)-**10**	7.39 ± 0.05	6.76 ± 0.04	7.32 ± 0.03
(S)-**10**	6.53 ± 0.04	6.26 ± 0.03	6.15 ± 0.02
(R)-**11**	7.16 ± 0.06	6.24 ± 0.05	7.27 ± 0.06
(S)-**11**	6.23 ± 0.08	5.43 ± 0.05	6.25 ± 0.08
(R)-**12**	9.09 ± 0.06	8.21 ± 0.01	8.65 ± 0.05
(S)-**12**	7.74 ± 0.04	7.56 ± 0.04	7.36 ± 0.03
(R)-**13**	7.74 ± 0.03	7.15 ± 0.05	7.19 ± 0.04
(S)-**13**	7.15 ± 0.06	6.85 ± 0.03	6.54 ± 0.04

[a] Not determined.

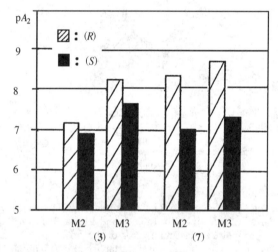

FIGURE 2. Affinities (pA_2 values) of the (R)- and (S)-enantiomers of **3** and **7** for muscarinic M2 (guinea-pig atria) and M3 receptors (guinea-pig ileum)

potency than the corresponding carbon analogues (R)-**8**, (S)-**8**, (R)-**9** and (S)-**9**; however, the stereoselectivity at the muscarinic receptors is more pronounced for the carbon compounds.

In order to overcome the experimental problems connected with the racemization of (R)-**3**, (S)-**3**, (R)-**7** and (S)-**7**, pharmacological stereoselectivity studies with the pure enantiomers of the related (hydroxymethyl)silanes **10–13** were carried out[34,38–40]. In

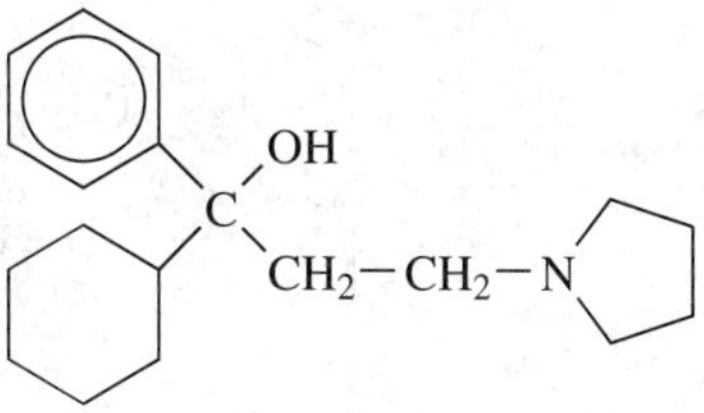

(8)

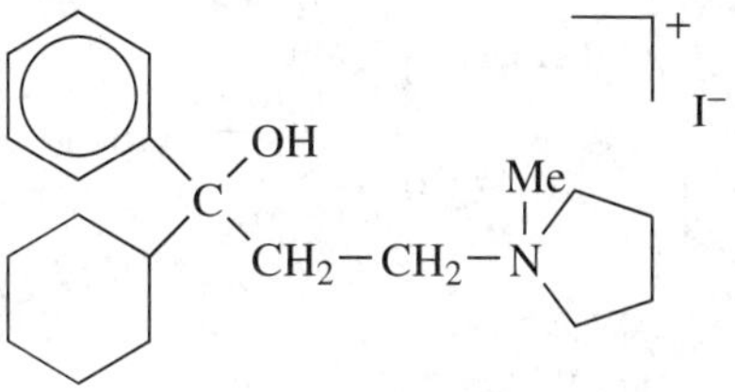

(9)

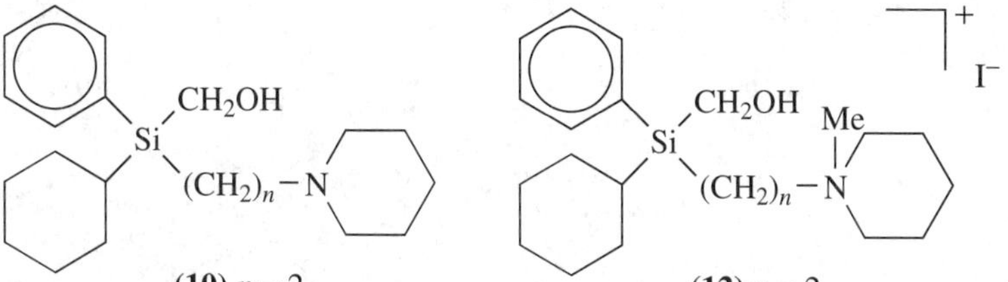

(10) n = 2
(11) n = 3

(12) n = 2
(13) n = 3

contrast to the silanols **3** and **7**, the (hydroxymethyl)silanes **10–13** are configurationally stable under physiological conditions.

The enantiomers of the silanes **10**[34] and **11**[39] were obtained from the corresponding racemic mixtures *rac*-**10** (for its synthesis see Scheme 4)[34] and *rac*-**11** (for its synthesis see Scheme 5)[39] by a classical racemate resolution using the enantiomers of *O,O'*-di-*p*-toluoyltartaric acid and 1,1'-binaphthyl-2,2'-diyl hydrogen phosphate, respectively, as resolving agents (for resolution by fractional crystallization of diastereomeric salts see Scheme 6)[34,39]. The silanes (*R*)-**10**, (*S*)-**10**, (*R*)-**11** and (*S*)-**11** were isolated as almost enantiomerically pure crystalline solids; their enantiomeric purities were determined by NMR experiments using chiral shift reagents[34,39]. Furthermore, the enantiomeric purities of (*R*)-**10** and (*S*)-**10** were established by liquid chromatography (HPLC) using chemically modified amylose as the chiral stationary phase[32]. The enantiomers of the ammonium salts **12** and **13** were obtained by reaction of the corresponding amines (*R*)-**10**, (*S*)-**10**, (*R*)-**11** and (*S*)-**11** with methyl iodide (Scheme 6). The absolute configurations of the silanes (*R*)-**10**, (*S*)-**10**, (*R*)-**12** and (*S*)-**12** were established on the basis of a single-crystal X-ray diffraction analysis of (*R*)-**12**[34]. The absolute configurations of the (*R*)- and (*S*)-enantiomers of the silanes **11** and **13** were determined by a crystal structure analysis of (*R*)-**11**·(*S*)-BNP·Me$_2$CO (BNP = 1,1'-binaphthyl-2,2'-diyl hydrogen phosphate)[39]. The

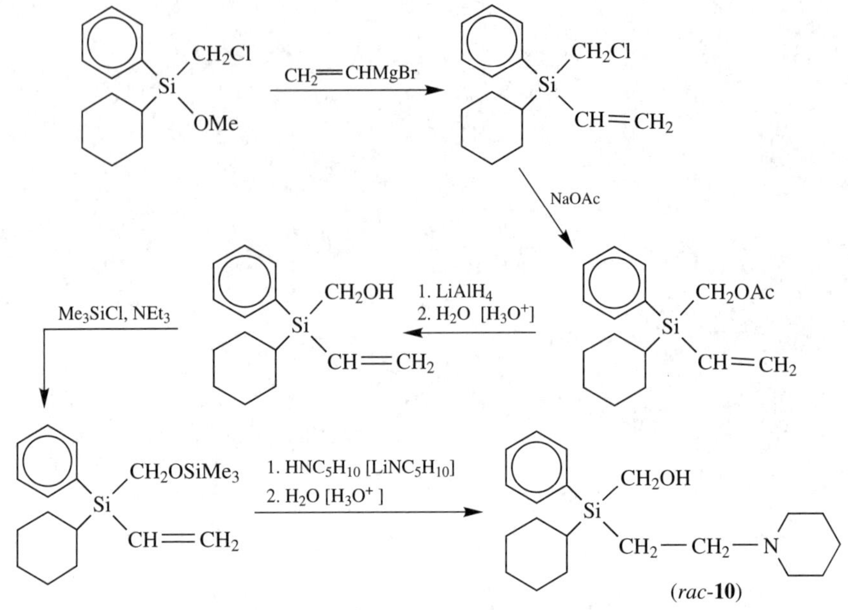

SCHEME 4

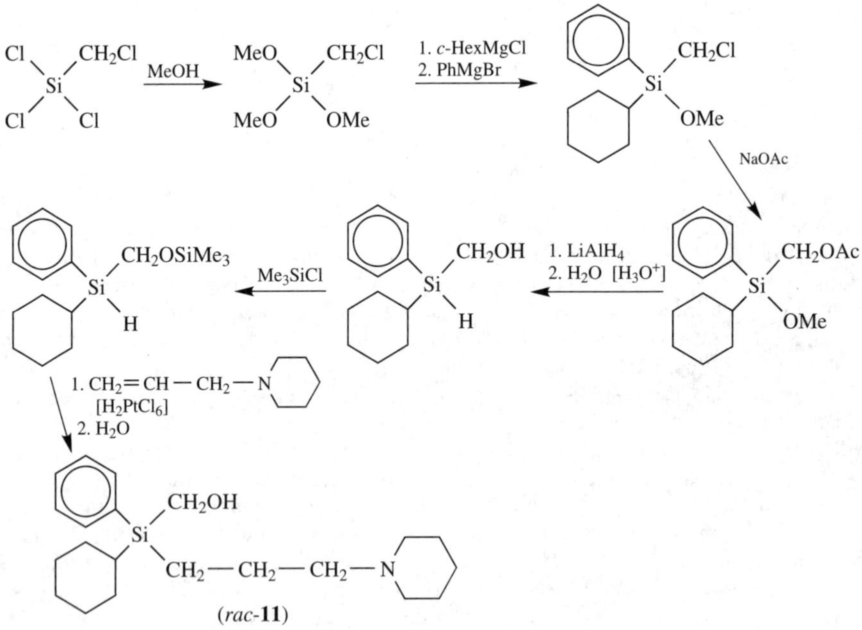

SCHEME 5

structures of the respective cations in the crystal of (R)-**12** and (R)-**11**·(S)-BNP·Me$_2$CO are depicted in Figures 3 and 4, respectively.

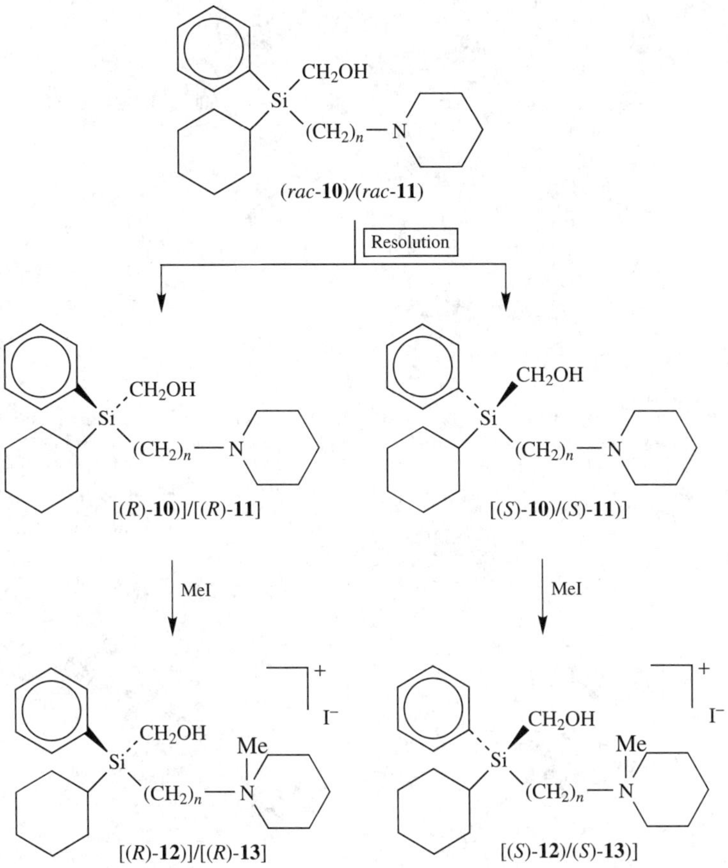

SCHEME 6

	10	11	12	13
n	2	3	2	3

The pure (R)- and (S)-enantiomers of the silanes **10–13** were studied for their affinities for muscarinic M1 (rabbit vas deferens), M2 (guinea-pig atria) and M3 receptors (guinea-pig ileum) in functional pharmacological experiments using 4-F-PyMcN$^+$ (= 1-[4-[[(4-fluorophenyl)carbamoyl]oxy]-2-butyn-1-yl]-1-methylpyrrolidinium tosylate) (M1) and arecaidine propargyl ester (M2, M3) as the agonist[34,38,40]. In addition, the affinities of these compounds for muscarinic M1 (human NB-OK 1 cells), M2 (rat heart), M3 (rat pancreas) and M4 receptors (rat striatum) were determined in radioligand binding studies using [^{3}H]-N-methylscopolamine as the radioligand[34,38,40]. All compounds investigated exhibited an apparently competitive antagonism at M1–M3 receptors in functional studies

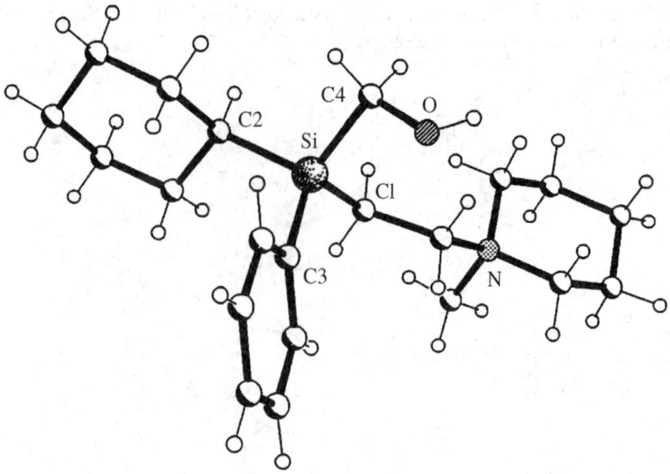

FIGURE 3. Structure of the cation of (*R*)-**12** in the crystal. Selected bond distances (pm) and angles (deg): Si−C(1), 188.8(2); Si−C(2), 188.5(2); Si−C(3), 187.7(2); Si−C(4), 188.2(2); C(1)−Si−C(2), 107.40(10); C(1)−Si−C(3), 111.47(10); C(1)−Si−C(4), 109.30(10); C(2)−Si−C(3), 110.57(9); C(2)−Si−C(4), 111.13(10); C(3)−Si−C(4), 106.99(11). The cation and anion of (*R*)-**12** are connected by an O−H···I hydrogen bond [O···I, 345.0(2) pm]

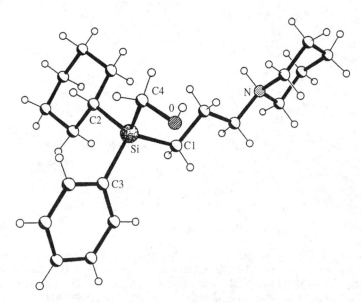

FIGURE 4. Structure of the cation of (*R*)-**11**·(*S*)-BNP·Me$_2$CO in the crystal. Selected bond distances (pm) and angles (deg): Si−C(1), 189(1); Si−C(2), 190.2(9); Si−C(3), 189(1); Si−C(4), 190(1); C(1)−Si−C(2), 112.7(5); C(1)−Si−C(3), 109.5(5); C(1)−Si−C(4), 112.7(5); C(2)−Si−C(3), 106.4(4); C(2)−Si−C(4), 108.6(5); C(3)−Si−C(4), 109.2(5)

and at M1–M4 receptors in binding experiments. The antimuscarinic potencies (pA_2 values) obtained in the functional studies are given in Table 1 as well as in Figure 5 (enantiomers of **10** and **12**) and Figure 6 (enantiomers of **11** and **13**). The corresponding binding affinities determined in the radioligand competition experiments correspond reasonably to these data.

As can be seen from Figures 5 and 6, the (R)-enantiomers (eutomers) of the silanes **10**–**13** show higher affinities for muscarinic M1–M3 receptors than their corresponding (S)-antipodes (distomers). Analogous results were obtained in the radioligand binding studies at M1–M4 receptors. These results again demonstrate that biological systems can

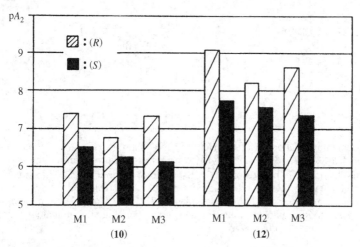

FIGURE 5. Affinities (pA_2 values) of the (R)- and (S)-enantiomers of **10** and **12** for muscarinic M1 (rabbit vas deferens), M2 (guinea-pig atria) and M3 receptors (guinea-pig ileum)

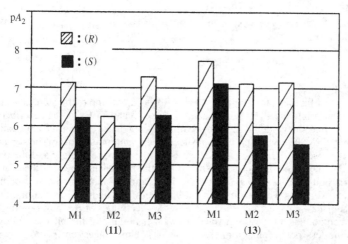

FIGURE 6. Affinities (pA_2 values) of the (R)- and (S)-enantiomers of **11** and **13** for muscarinic M1 (rabbit vas deferens), M2 (guinea-pig atria) and M3 receptors (guinea-pig ileum)

discriminate between the antipodes of chiral silicon compounds, with the silicon atom as the center of chirality. The highest stereoselectivity indices (functional studies) were observed for the silane **12** [$SI = 22$ (M1), 4.4 (M2), 20 (M3)].

Compounds **14–17** are derivatives of the corresponding silanes **10** and **12**. Functional pharmacological studies (M1–M3 receptors) and radioligand binding experiments (M1–M4 receptors) with the pure (R)- and (S)-enantiomers of these compounds revealed very similar pA_2 values as observed for the antipodes of the related analogues **10** and **12**[34]. The highest stereoselectivity indices (functional studies) were determined for **17** [$SI = 44$ (M1), 7.6 (M2), 20 (M3)].

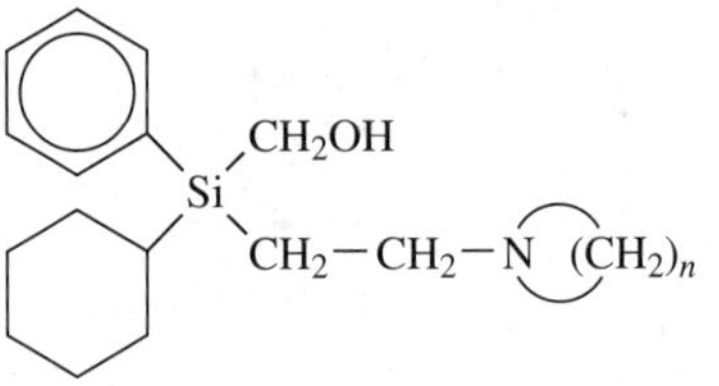

(**14**) $n = 4$
(**15**) $n = 6$

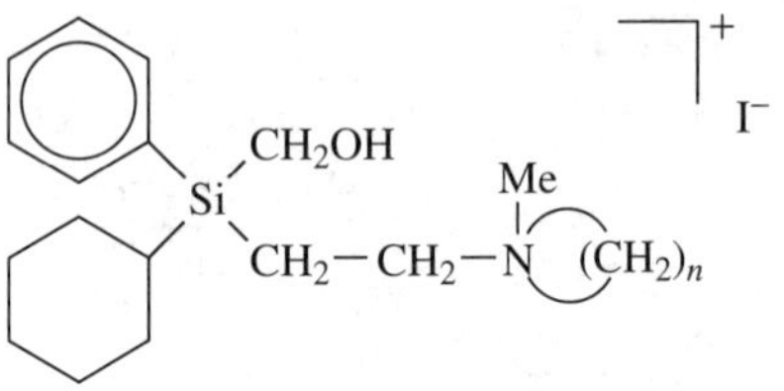

(**16**) $n = 4$
(**17**) $n = 6$

The germanium analogues of the (R)- and (S)-enantiomers of the silicon compounds **10** and **12** [(R)-**18**, (S)-**18**, (R)-**19**, (S)-**19**] were also studied for their antimuscarinic properties (studies on Si/Ge bioisosterism)[36]. By analogy with the parent silicon compounds, these germanes were also found to be configurationally stable under physiological conditions. The enantiomerically pure antipodes of **18** and **19** revealed similar pA_2 values and stereoselectivities as determined for their corresponding silicon analogues indicating a strongly pronounced Si/Ge bioisosterism.

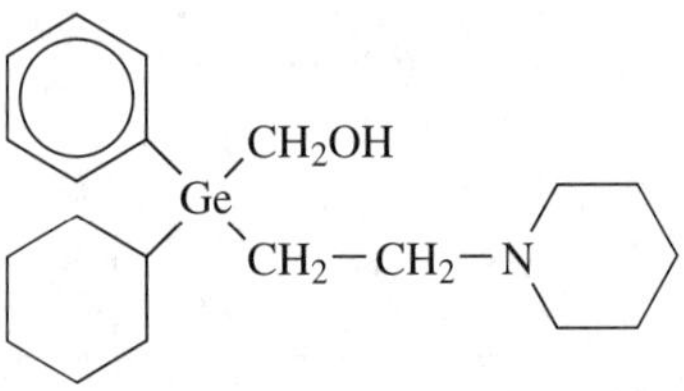

(**18**)

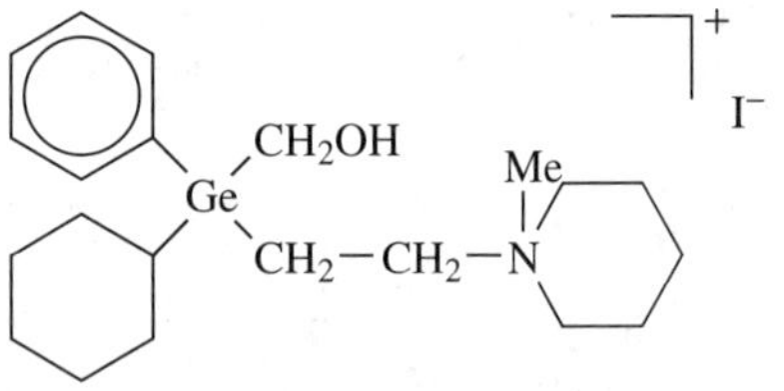

(**19**)

As shown in Figure 7 for sila-procyclidine (**3**) as an example, the stereoselective interaction of muscarinic receptors with the antipodes of **3** and **7–19** can be described in terms of a four-binding site model. According to this concept, these muscarinic antagonists might be recognized by the receptors with subsites for the ammonium group (protonated amino group in the case of **3**, **8**, **10**, **11**, **14**, **15** and **18**), the hydroxyl moiety and the phenyl and cyclohexyl group. Stereoselective interaction of the two enantiomers of these antagonists with muscarinic receptors is based on opposite binding of the phenyl and cyclohexyl ring to *site 1* and *site 2* as well as on identical binding of the hydroxyl and ammonium group to *site 3* and *site 4*, respectively. Thus, the stereoselectivities of receptor binding observed for these compounds are best explained by weaker binding of the phenyl and cyclohexyl group of the less potent enantiomers (distomers). This suggestion is strongly supported by the results obtained in pharmacological studies with the related achiral compounds **20–35**[36,38–40] (in this context, see also References 41–43).

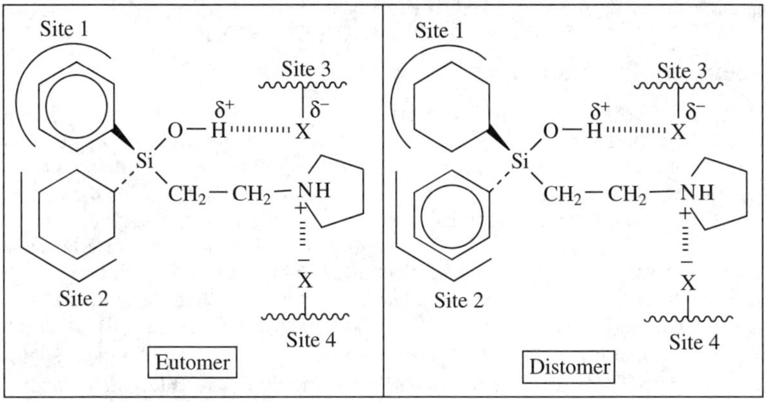

FIGURE 7. Four-binding-sites model for the stereoselective interaction of the (*R*)- and (*S*)-enantiomer of protonated sila-procyclidine (**3**) with four subsites of muscarinic receptors: *site 1* = phenyl-preferring hydrophobic subsite; *site 2* = cyclohexyl-preferring subsite; *site 3* = subsite for the hydroxyl group, probably forming a hydrogen bond (O—H···X); *site 4* = negatively charged subsite for the ammonium group (under physiological conditions, the pyrrolidino group of **3** is protonated). This model can also be used to describe the analogous stereoselective interaction of the antipodes of **7–19** with muscarinic receptors

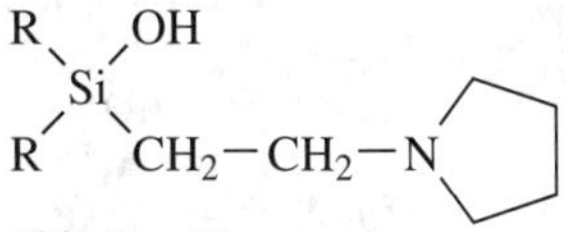

(**20**) R = Ph
(**21**) R = *c*-Hex

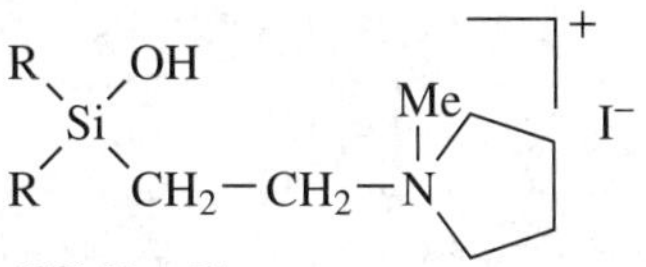

(**22**) R = Ph
(**23**) R = *c*-Hex

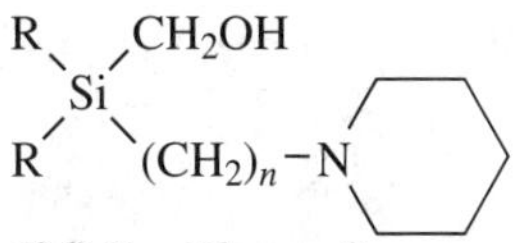

(**24**) R = Ph, *n* = 2
(**25**) R = *c*-Hex, *n* = 2
(**26**) R = Ph, *n* = 3
(**27**) R = *c*-Hex, *n* = 3

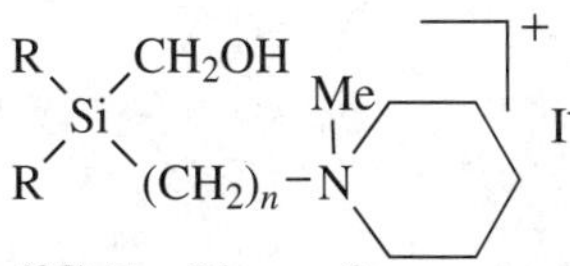

(**28**) R = Ph, *n* = 2
(**29**) R = *c*-Hex, *n* = 2
(**30**) R = Ph, *n* = 3
(**31**) R = *c*-Hex, *n* = 3

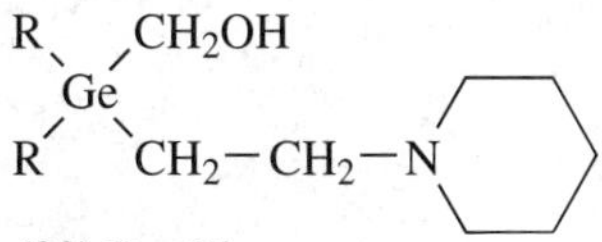

(**32**) R = Ph
(**33**) R = *c*-Hex

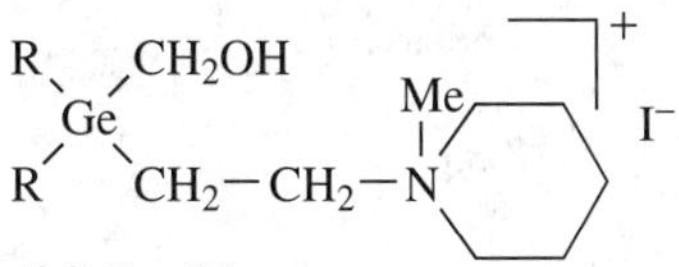

(**34**) R = Ph
(**35**) R = *c*-Hex

III. BIOCATALYSIS AS A METHOD FOR THE PREPARATION OF OPTICALLY ACTIVE ORGANOSILICON COMPOUNDS

A. Introductory Remarks

The results described in Section II clearly demonstrate that enantiomeric silanols and silanes may generally differ in their biological properties. Since biologically active organosilicon compounds have a great potential of application as agrochemicals, drugs and diagnostics, the development of appropriate preparative methods for the synthesis of enantiomerically pure silicon compounds, with the silicon atom as the center of chirality, is necessary (for reviews on optically active silicon compounds, see References 44 and 45). In most cases (see, for example, Section II) optically active silicon compounds have been obtained by (i) classical resolution of the respective racemic mixtures via fractional crystallization of appropriate diastereomeric derivatives and (ii) stereoselective chemical transformation of the resolved enantiomers. Asymmetric chemical syntheses[46] and chromatographic racemate resolutions[32], in principle, are alternative methods for the preparation of enantiomeric silicon compounds. As will be demonstrated in this section, stereoselective biotransformations represent a further challenge for the synthesis of optically active silicon compounds, with the silicon atom as the center of chirality.

B. Reductions

Since there are many examples of strongly pronounced bioisosteric relationships between analogous carbon and silicon compounds (in this context, see Reference 8), attempts have been made to apply well-known biotransformations of organic compounds to sila-analogous substrates. The first example of this is the enantioselective microbial reduction of the organosilicon compounds **36** and **38** with growing cells of the yeast *Kloeckera corticis* (ATCC 20109)[47]. By analogy with their corresponding carbon analogues (Si/C exchange), the prochiral silicon compounds **36** and **38** could be transformed into the optically active products (*S*)-**37** and (*S*)-**39**, respectively (Scheme 7). The enantiomeric purities of the isolated biotransformation products were moderate [(*S*)-**37**, 80% ee; (*S*)-**39**, 65% ee)].

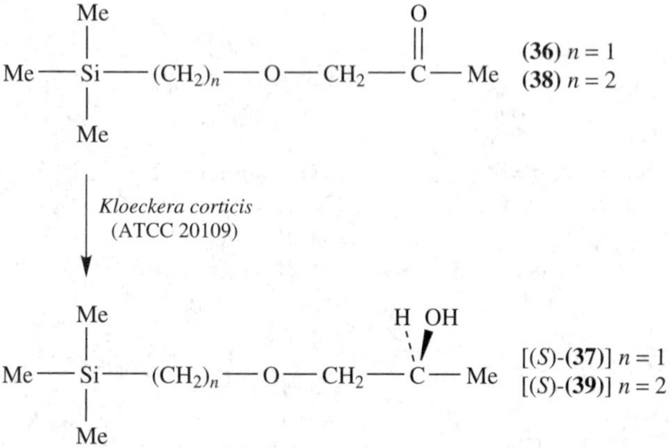

SCHEME 7

In the prochiral organosilicon substrates **36** and **38**, the carbonyl groups are relatively far away from the silicon atoms present in these molecules. However, by analogy with many organic ketones of the general formula type $R_3C-CO-CR_3$, the silaketones (acylsilanes) **40** and **42** (belonging to the formula type $R_3Si-CO-CR_3$) were also found to be accepted as substrates by ketone-reducing microorganisms and to be converted into the optically active biotransformation products (*R*)-**41** and (*R*)-**43**, respectively (Scheme 8).

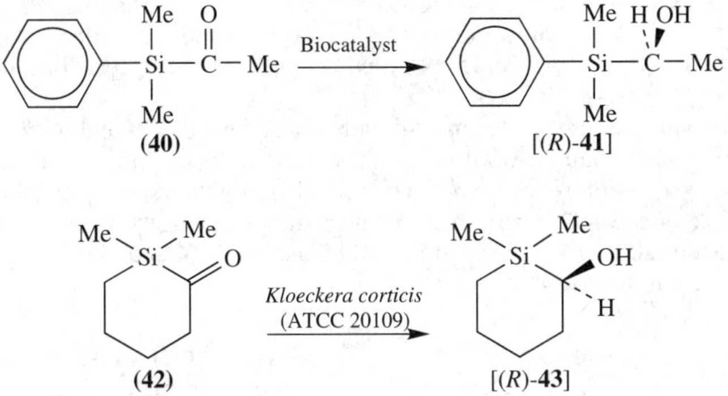

SCHEME 8

Thirty strains of microorganisms (bacteria, yeasts, fungi and green algae) were tested as resting free cells for their ability to transform the prochiral acetylsilane **40** into the corresponding (1-hydroxyethyl)silane (*R*)-**41**[48]. In these studies, some of these microorganisms were found to be good (or even excellent) biocatalysts for this conversion (Table 2). The biotransformation product (*R*)-**41** was obtained with enantiomeric purities of up to >95% ee. Immobilized cells (immobilization with calcium alginate) of the yeast *Trigonopsis variabilis* (DSM 70714), of the cyanobacterium *Synechococcus leopoliensis* (SAG 1402-1) and of the green alga *Chlorella fusca* (SAG 211/8b) can also be used for this particular type of bioconversion[49].

Studies with resting free cells of *Trigonopsis variabilis* (DSM 70714) demonstrated that the enantiomeric purity of (*R*)-**41** depends significantly on the substrate concentration

TABLE 2. Enantioselective microbial reduction of the acetylsilane **40** into the (1-hydroxyethyl)silane (*R*)-**41**

Microorganism	ee Value[a]
Acinetobacter calcoaceticus (ATCC 31012)	>95
Arthrobacter paraffinius (ATCC 15591)	85
Brevibacterium species (ATCC 21860)	90
Corynebacterium dioxydans (ATCC 21766)	>95
Candida albicans (ATCC 10231)	86
Candida humicola (DSM 70067)	90
Candida utilis (DSM 2361)	81
Kloeckera corticis (ATCC 20108)	80
Trigonopsis variabilis (DSM 70714)	86
Cunninghamella elegans (ATCC 26269)	94
Synechococcus leopoliensis (SAG 1402-1)	94
Chlamydomas reinhardii (Y-1)	85
Chlorella fusca (SAG 211/8b)	75

[a]Enantiomeric purity (% ee) of (*R*)-**41** as determined by HPLC using cellulose triacetate as a chiral stationary phase.

used[50]. At a concentration of 0.25 g of **40** per liter culture broth, the silane (R)-**41** was obtained on a preparative scale with an enantiomeric purity of 86% ee (yield ca 70%). An even enantiomerically pure (>99% ee) product could be prepared with resting cells of the yeast *Saccharomyces cerevisiae* (DHW S-3)[51]. This biotransformation was also performed on a preparative scale (yield 40%). Even plant cell suspension cultures of *Symphytum officinale* L. or *Ruta graveolens* L. were used for the enantioselective conversion of **40** into (R)-**41** on a preparative scale [enantiomeric purities: 81% ee (*Symphytum*), 60% ee (*Ruta*); yields: 15% (*Symphytum*), 9% (*Ruta*); reaction conditions not optimized][52].

Preliminary studies with the ketone **44** (carbon analogue of **40**) and with the germaketone **46** (germanium analogue of **40**) have shown that these compounds are also accepted as substrates by *Trigonopsis variabilis* (DSM 70714) to give the corresponding reduction products (S)-**45** and (R)-**47**, respectively[50]. Interestingly, the acetylsilane **40** was found to be reduced about 20 times faster than its carbon analogue **44** and about two times faster than its germanium analogue **46**.

(**44**) El = C
(**46**) El = Ge

[(S)-**45**] El = C
[(R)-**47**] El = Ge

Enantioselective reduction of the prochiral cyclic acylsilane **42** with growing cells of the yeast *Kloeckera corticis* (ATCC 20109) yielded the optically active reduction product (R)-**43** (Scheme 8)[53]. On a preparative scale, the 1-silacyclohexan-2-ol (R)-**43** was isolated in 60% yield with an enantiomeric purity of 92% ee. Repeated recrystallization of the biotransformation product from n-hexane raised the enantiomeric purity to 99% ee.

Based on the results obtained with the enantioselective microbial reduction of the achiral acylsilanes **40** and **42**, attempts were made to use this particular type of bioconversion for the synthesis of optically active silanes with the silicon atom as the center of chirality. The high preparative potential of this method was first demonstrated for the (R)-selective microbial reduction of racemic acetyl(t-butyl)methyl(phenyl)silane (rac-**48**; for the synthesis[54] of this compound, see Scheme 9) with resting free cells of the yeast *Trigonopsis variabilis* (DSM 70714) or with the bacterium *Corynebacterium dioxydans* (ATCC 21766)[55,56]. As shown in Scheme 10, these bioconversions yielded the diastereomeric optically active (1-hydroxyethyl)silanes (SiR,CR)-**49** and (SiS,CR)-**50** which were separated by column chromatography on silica gel. Under preparative conditions (10 g scale), compounds (SiR,CR)-**49** and (SiS,CR)-**50** were obtained with high enantiomeric purities [*Trigonopsis variabilis*: (SiR,CR)-**49**, 97% ee; (SiS,CR)-**50**, 96% ee. *Corynebacterium dioxydans*: (SiR,CR)-**49**, ⩾99% ee; (SiS,CR)-**50**, ⩾99% ee]. The products produced by *Trigonopsis variabilis* (DSM 70714) were obtained in good yields [(SiR,CR)-**49**, 74%; (SiS,CR)-**50**, 78%; yields relative to (R)-**48** and (S)-**48**, respectively, in the racemic substrate]. In the case of *Corynebacterium dioxydans* (ATCC 21766), the yields were significantly lower [(SiR,CR)-**49**, 20%; (SiS,CR)-**50**, 20%; reaction conditions not optimized]. Subsequent chemical oxidation (DMSO, DCC, pyridinium trifluoroacetate) of the diastereomeric (1-hydroxyethyl)silanes (SiR,CR)-**49** and (SiS,CR)-**50** gave the almost enantiomerically pure acetylsilanes (R)-**48** and (S)-**48**, respectively (Scheme 10). These examples clearly demonstrate that stereoselective biotransformations, as key steps in a

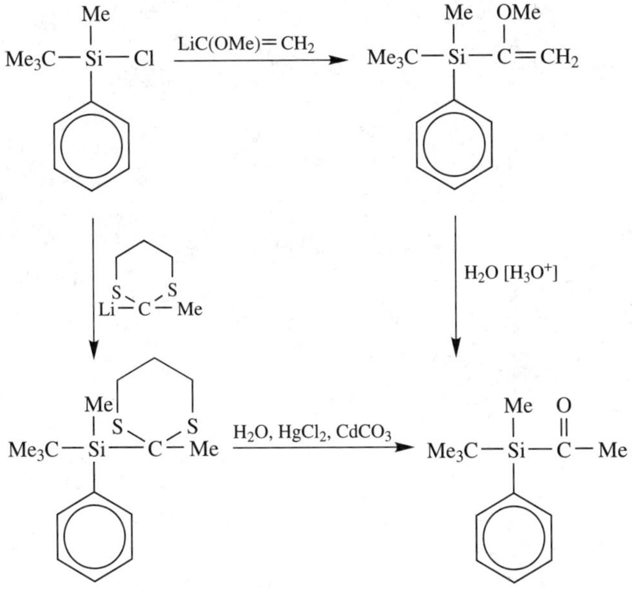

41. Chirality in bioorganosilicon chemistry

SCHEME 9

reaction sequence, can be used for the preparation of optically active silanes with the silicon atom as the center of chirality.

Recently, it has been demonstrated that the yeast *Saccharomyces cerevisiae* (DHW S-3) can also be used for the (*R*)-selective reduction of the acetylsilane *rac*-**48**. By analogy with the bioconversions illustrated in Scheme 10, incubation of *rac*-**48** with resting free cells of this microorganism yielded a 1:1 mixture of the corresponding diastereomeric (1-hydroxyethyl)silanes (Si*R*,C*R*)-**49** and (Si*S*,C*R*)-**50**[57]. Under preparative conditions, the biotransformation products were isolated in 43% yield (relative to *rac*-**48**). The enantiomeric purities of the silanes (Si*R*,C*R*)-**49** and (Si*S*,C*R*)-**50** were ⩾98% ee.

Very similar results were obtained with the germanium analogue of the acetylsilane *rac*-**48**, the acetylgermane *rac*-**51**; its microbial reduction with *Saccharomyces cerevisiae* (DHW S-3) yielded a 1:1 mixture of the corresponding diastereomeric (1-hydroxyethyl)germanes (Ge*R*,C*R*)-**52** and (Ge*S*,C*R*)-**53** (yield 60%; enantiomeric purity of both diastereomers ⩾98% ee)[57] which could also be separated by column chromatography on silica gel. Analogous (*R*)-selective microbial reductions (preparative scale) of the related acetylsilanes *rac*-**54** and *rac*-**57** with the yeast *Trigonopsis variabilis* (DSM 70714) have also been reported[7,58]. The bioconversion of the acetylsilane *rac*-**54** (resting free cells) yielded a 1:1 mixture of the (1-hydroxyethyl)silanes (Si*R*,C*R*)-**55** and (Si*S*,C*R*)-**56** (yield 72%; enantiomeric purity of both diastereomers *ca* 94% ee)[7]. The analogous biotransformation of the acetylsilane *rac*-**57** (immobilized resting cells; calcium alginate matrix) gave a 1:1 mixture of the (1-hydroxyethyl)silanes (Si*R*,C*R*)-**58** and (Si*S*,C*R*)-**59** (yield 99%; enantiomeric purity of both diastereomers >96% ee)[58].

By analogy with the stereoselective microbial reduction of the acetylsilane *rac*-**48** (Scheme 10) and its derivatives *rac*-**51**, *rac*-**54** and *rac*-**57**, the related acetyldisilane *rac*-**60** was also found to be reduced (*R*)-selectively with resting free cells of *Trigonopsis variabilis* (DSM 70714) to give a 1:1 mixture of the corresponding diastereomeric

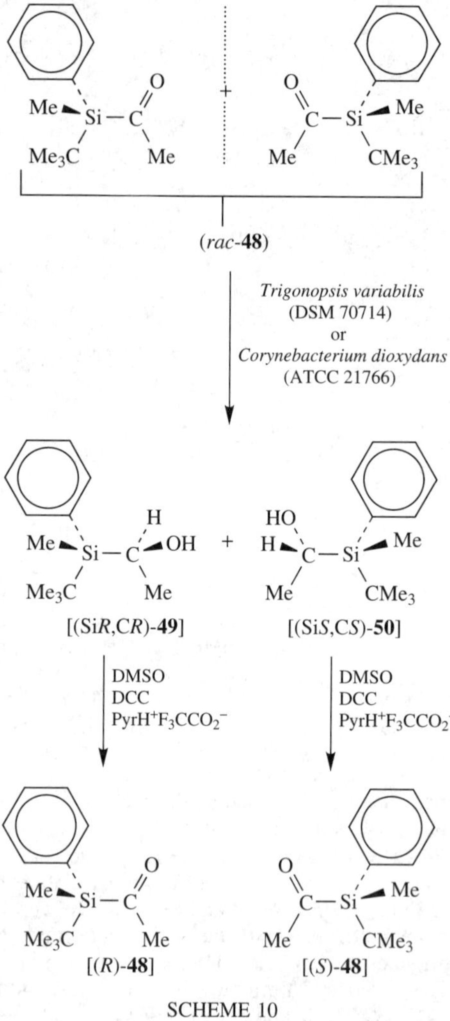

SCHEME 10

(1-hydroxyethyl)disilanes (SiR,CR)-**61** and (SiS,CR)-**62** (yield 75%; enantiomeric purity of both diastereomers $\geq$98% ee) (Scheme 11)[7,59]. Similar results were obtained with resting free cells of the yeast *Saccharomyces cerevisiae* (DHW S-3)[57]. These microbial transformations are especially remarkable since they could be realized without a noticeable degree of hydrolytic cleavage of the Si−Si bond. Compounds (SiR,CR)-**61** and (SiS,CR)-**62** are the first optically active disilanes obtained by stereoselective biotransformations.

Interestingly, the related acetyldigermane *rac*-**63** could be converted analogously into a 1:1 mixture of the diastereomeric (1-hydroxyethyl)digermanes (GeR,CR)-**64** and (GeS,CR)-**65** (yield 62%; enantiomeric purity of both diastereomers $\geq$98% ee)[57]. Obviously, the Ge−Ge bonds of the substrate and products are rather stable against hydrolytic cleavage under the bioconversion conditions used.

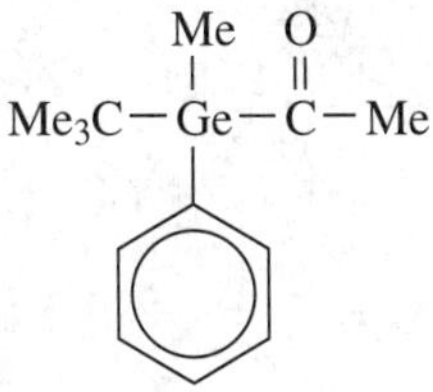

41. Chirality in bioorganosilicon chemistry

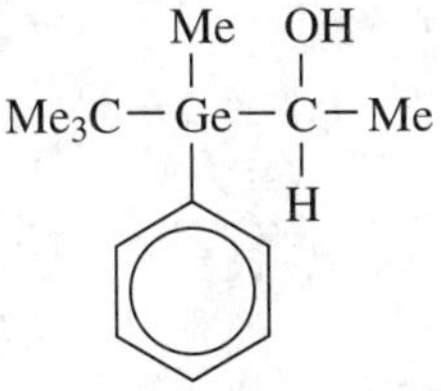

(51) Me$_3$C–Ge(Me)(Ph)–C(=O)–Me

(52/53) Me$_3$C–Ge(Me)(Ph)–C(OH)(H)–Me

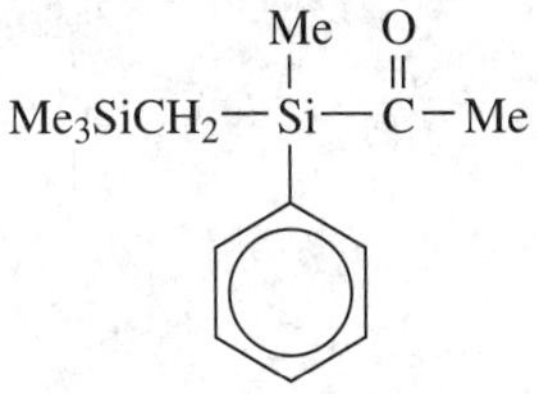

(54) Me$_3$SiCH$_2$–Si(Me)(Ph)–C(=O)–Me

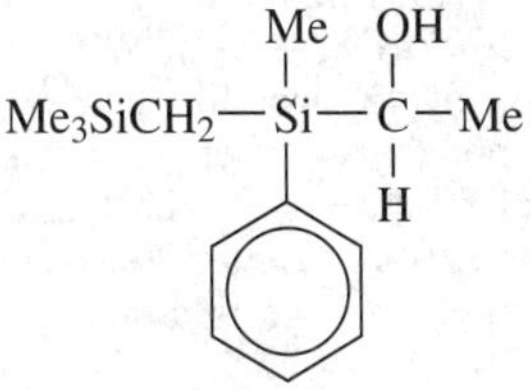

(55/56) Me$_3$SiCH$_2$–Si(Me)(Ph)–C(OH)(H)–Me

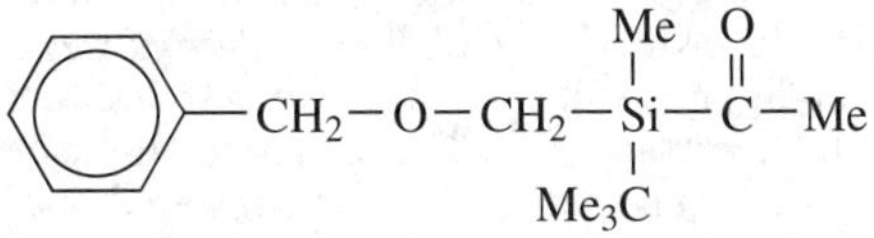

(57) Ph–CH$_2$–O–CH$_2$–Si(Me)(CMe$_3$)–C(=O)–Me

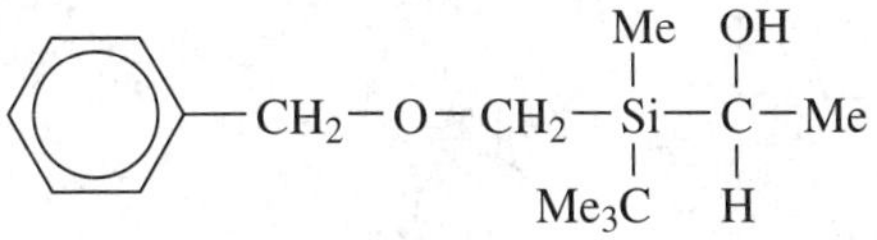

(58/59) Ph–CH$_2$–O–CH$_2$–Si(Me)(CMe$_3$)–C(OH)(H)–Me

(rac-**60**)

$\xrightarrow{\text{Trigonopsis variabilis (DSM 70714) or Saccharomyces cerevisiae (DHW S-3)}}$

[(Si*R*,C*R*)-**61**] + [(Si*S*,C*R*)-**62**]

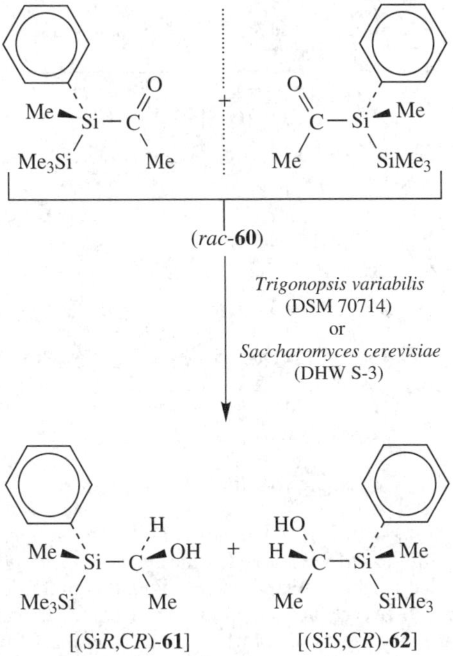

SCHEME 11

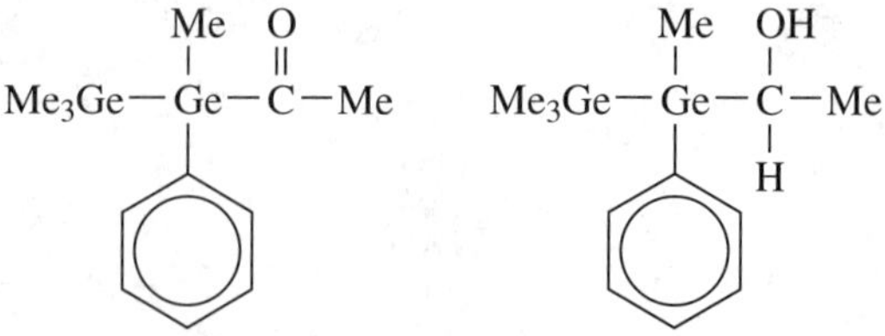

```
      Me  O                    Me  OH
      |   ||                   |   |
Me₃Ge—Ge—C—Me            Me₃Ge—Ge—C—Me
      |                        |   |
      Ph                       Ph  H

      (63)                     (64/65)
```

A quite different result was obtained when incubating the related acetyl(silyl)germane *rac*-**66** with resting free cells of *Saccharomyces cerevisiae* (DHW S-3). When using the same conditions as applied for the microbial reduction of the acetyldisilane *rac*-**60** and the acetyldigermane *rac*-**63**, cleavage of the Si—Ge bond was observed. As shown in Scheme 12, compound *rac*-**66** was converted into a 1:1 mixture of the diastereomeric (1-hydroxyethyl)hydridogermanes (Ge*S*,C*R*)-**67** and (Ge*R*,C*R*)-**68** (yield 50%; enantiomeric purity of both diastereomers ≥98% ee). The Si—Ge bond cleavage is probably a nonenzymatic reaction.

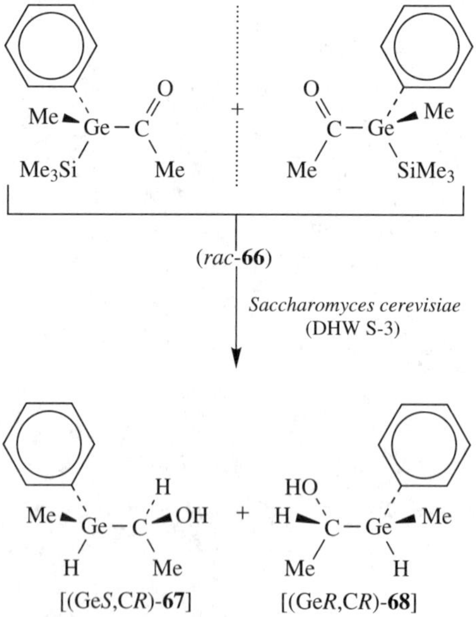

SCHEME 12

The microbial reductions of the acetylsilanes *rac*-**48**, *rac*-**54** and *rac*-**57**, of the acetylgermanes *rac*-**51** and *rac*-**66**, of the acetyldisilane *rac*-**60** and of the acetyldigermane *rac*-**63** represent (*R*)-selective bioconversions yielding 1:1 mixtures of the corresponding optically active diastereomeric reduction products. In contrast, incubation of the cyclic acylsilane *rac*-**69** with growing cells of the yeast *Kloeckera corticis* (ATCC 20109) gave a 1:1 mixture of the enantiomeric products (Si*S*,C*R*)-**70** and (Si*R*,C*S*)-**70** (yield ca 95%) (Scheme 13) [59,60]. Only small amounts of the corresponding diastereomers with the (Si*R*,C*R*)- and (Si*S*,C*S*)-configuration could be detected. Thus, this particular type of bioconversion can be regarded as a diastereoselective reduction (diastereomeric purity ca 90% de).

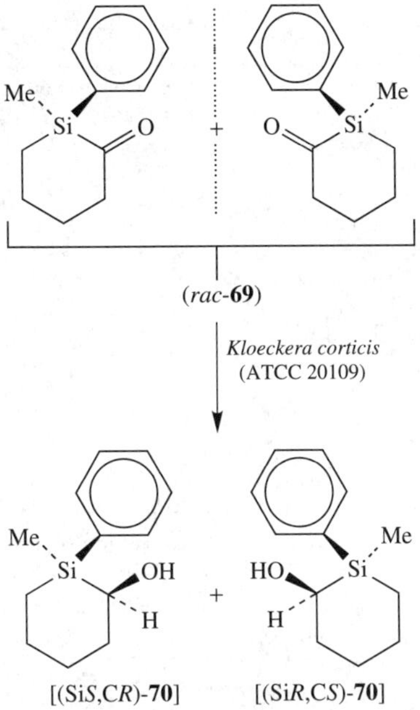

SCHEME 13

As the last example of stereoselective reductions of acylsilanes, the enantioselective microbial reduction of the benzoylsilane **71** shall be discussed. Nine from thirteen strains of yeasts or bacteria were found to convert the prochiral compound **71** enantioselectively into the corresponding optically active (1-hydroxybenzyl)silane (S)-**72**; the other microorganisms produced the corresponding antipode (R)-**72** (Scheme 14)[61]. The yeasts and bacteria used in these studies and the stereochemical results obtained are summarized in Table 3. Analogous investigations were carried out with the carbon (**73**) and germanium analogue (**75**) of the silane **71**. These compounds were also found to be reduced enantioselectively to yield the (R)- or (S)-enantiomers of the corresponding reduction products **74** and **76**, respectively (Table 3). Comparison of the conversion rates for the microbial transformations **71**→**72**, **73**→**74** and **75**→**76** demonstrated that the silicon compound **71** is most readily reduced, followed by its germanium analogue **75**. The lowest reaction rate was found for the carbon compound **73**. Analogous kinetics have already been reported for the microbial reduction of the acetylsilane **40** and its carbon (**44**) and germanium analogue (**46**)[50].

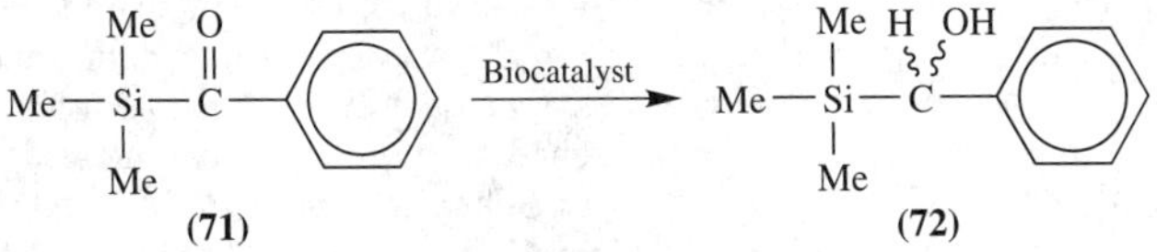

SCHEME 14

TABLE 3. Enantioselective microbial reduction of the benzoylsilane **71** and of its carbon (**73**) and germanium analogue (**75**) (formation of **72**, **74** and **76**)

Microorganism	Enantiomeric purity (% ee) and absolute configuration (in parentheses)		
	72	74	76
Saccharomyces cerevisiae	78 (*S*)	60 (*S*)	86 (*S*)
Torulopsis magnoliae (IFO 661)	94 (*S*)	75 (*S*)	96 (*S*)
Candida tropicalis (IFO 199)	98 (*S*)	93 (*S*)	99 (*S*)
Pichia miso (FERM P-404)	24 (*S*)	92 (*R*)	62 (*S*)
Saccharomyces rosei (IFO 428)	80 (*S*)	7 (*R*)	85 (*S*)
Saccharomyces (sp. H-1)	65 (*S*)	70 (*S*)	61 (*S*)
Rhodotorula rubra (IFO 889)	93 (*S*)	65 (*S*)	94 (*S*)
Nocardia erythropolis (IAM 12122)	25 (*R*)	50 (*R*)	19 (*S*)
Corynebacterium equi (IFO 3730)	65 (*R*)	88 (*R*)	57 (*R*)
Xanthomonas campestris (NIAS 1076)	17 (*S*)	57 (*R*)	56 (*S*)
Micrococcus luteus (M2-7)	12 (*R*)	80 (*R*)	31 (*S*)
Bacillus brevis (M4-5)	46 (*R*)	65 (*R*)	31 (*R*)
Baker's yeast	30 (*S*)	4 (*S*)	90 (*S*)

As can be seen from Table 3, there are also some significant stereochemical differences observed for the microbial reductions of the Si/C/Ge analogues **71**, **73** and **75**. In general, a gradual change from (*R*)-selectivity to (*S*)-selectivity was observed when going from the carbon compound **73** to its silicon (**71**) and germanium analogue (**75**). In some cases, this change even led to opposite absolute configurations of the analogous reduction products **72**, **74** and **76**. This interesting result can be probably explained by the different covalent radii of carbon, silicon and germanium.

$$\text{Me} - \underset{\underset{\text{Me}}{|}}{\overset{\overset{\text{Me}}{|}}{\text{El}}} - \overset{\overset{\text{O}}{\|}}{\text{C}} - \text{Ph} \qquad \text{Me} - \underset{\underset{\text{Me}}{|}}{\overset{\overset{\text{Me}}{|}}{\text{El}}} - \underset{\underset{\text{H}}{|}}{\overset{\overset{\text{OH}}{|}}{\text{C}}} - \text{Ph}$$

(**73**) El = C (**74**) El = C
(**75**) El = Ge (**76**) El = Ge

C. Hydrolyses

Stereoselective enzymatic hydrolyses of esters represent a further type of biotransformation that has been used for the synthesis of optically active organosilicon compounds. The first example of this particular type of bioconversion is illustrated in Scheme 15. Starting from the racemic (1-acetoxyethyl)silane *rac*-**77**, the optically active (1-hydroxyethyl)silane (*S*)-**41** was obtained by a kinetic racemate resolution using porcine liver esterase (*PLE*; E.C. 3.1.1.1) as the biocatalyst[7]. The silane (*S*)-**41** (isolated with an enantiomeric purity of 60% ee; bioconversion not optimized) is the antipode of compound (*R*)-**41** which was obtained by an enantioselective microbial reduction of the acetylsilane **40** (see Scheme 8).

The enantioselective hydrolysis of the racemic 2-acetoxy-1-silacyclohexane *rac*-**78** represents a further example of an enzymatic kinetic racemate resolution (Scheme 15). Hydrolysis of this compound in the presence of porcine liver esterase (*PLE*; E.C. 3.1.1.1) yielded the optically active 1-silacyclohexan-2-ol (*S*)-**43** which was isolated with an enantiomeric purity of 93% ee[7]. Similar results were obtained when using a crude lipase preparation from *Candida cylindracea* (*CCL*; E.C. 3.1.1.3) as the biocatalyst

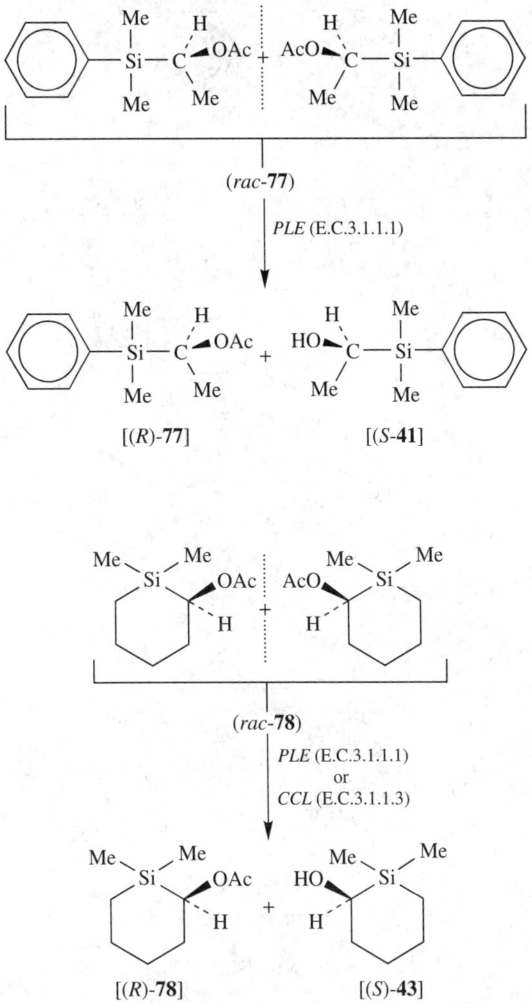

41. Chirality in bioorganosilicon chemistry

SCHEME 15

(Scheme 15)[62]. After terminating the reaction at a conversion of 38% (relative to total amount of substrate rac-**78**), the product (S)-**43** was separated from the nonreacted substrate by column chromatography on silica gel and isolated on a preparative scale in 71% yield (relative to total amount of converted rac-**78**) with an enantiomeric purity of 95% ee. Recrystallization led to an improvement of the enantiomeric purity by up to >98% ee. The biotransformation product (S)-**43** is the antipode of compound (R)-**43** which was obtained by enantioselective microbial reduction of the acylsilane **42** (see Scheme 8)[53]. The nonreacted substrate (R)-**78** was isolated in 81% yield (relative to total amount of nonconverted rac-**78**) with an enantiomeric purity of 57% ee. For further enantioselective enzymatic hydrolyses of racemic organosilicon esters, with the carbon atom as the center of chirality, see References 63 and 64.

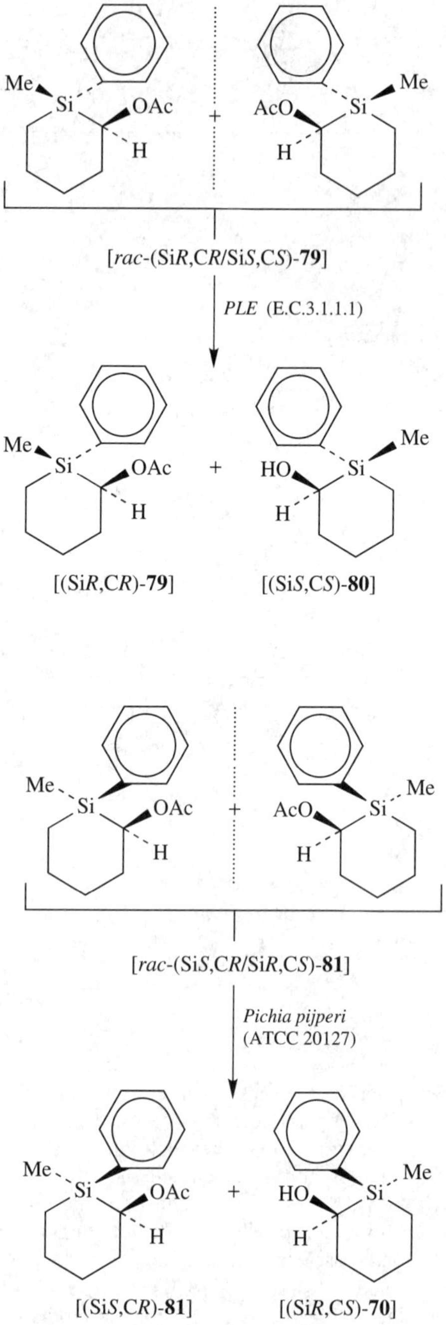
Me
Si
OAc
H
+
AcO
Si
Me
H
[rac-(SiR,CR/SiS,CS)-79]
PLE (E.C.3.1.1.1)
Me
Si
OAc
H
+
HO
Si
Me
H
[(SiR,CR)-79]
[(SiS,CS)-80]
Me
Si
OAc
H
+
AcO
Si
Me
H
[rac-(SiS,CR/SiR,CS)-81]
Pichia pijperi
(ATCC 20127)
Me
Si
OAc
H
+
HO
Si
Me
H
[(SiS,CR)-81]
[(SiR,CS)-70]

SCHEME 16

Enantioselective enzymatic ester hydrolyses have also been used for the preparation of optically active silicon compounds with the silicon atom as the center of chirality. An example of this is the kinetic resolution of the racemic 2-acetoxy-1-silacyclohexane rac-(SiR,CR/SiS,CS)-**79** with porcine liver esterase (*PLE*; E.C. 3.1.1.1) (Scheme 16)[65]. Under preparative conditions, the optically active 1-silacyclohexan-2-ol (SiS,CS)-**80** was obtained as an almost enantiomerically pure product (enantiomeric purity $\geqslant$96% ee) in ca 60% yield [relative to (SiS,CS)-**79** in the racemic substrate]. The biotransformation product could be easily separated from the nonhydrolyzed substrate by column chromatography on silica gel.

Intact microbial cells have also been used as biocatalyst for this particular type of bioconversion. Incubation of the racemic 2-acetoxy-1-silacyclohexane rac-(SiS,CR/SiR,CS)-**81** with growing cells of the yeast *Pichia pijperi* (ATCC 20127) yielded the optically active 1-silacyclohexan-2-ol (SiR,CS)-**70** (Scheme 16)[66,67]. Under preparative conditions, this biotransformation product was isolated as an almost enantiomerically pure compound (enantiomeric purity >96% ee) in ca 80% yield [relative to (SiR,CS)-**81** in the racemic substrate].

Enantioselective enzymatic ester hydrolyses of prochiral trimethylsilyl-substituted diesters of the malonate type have been applied for the synthesis of the related optically active monoesters[68]. As an example of this particular type of biotransformation, the enantioselective conversion of the diester **82** is illustrated in Scheme 17. Hydrolysis of compound **82** in phosphate buffer, catalyzed by porcine liver esterase (*PLE*; E.C. 3.1.1.1) or horse liver acetonic powder (*HLAP*), gave the optically active monoester **83** (absolute configuration not reported) in 86% and 49% yield, respectively. The enantiomeric purities

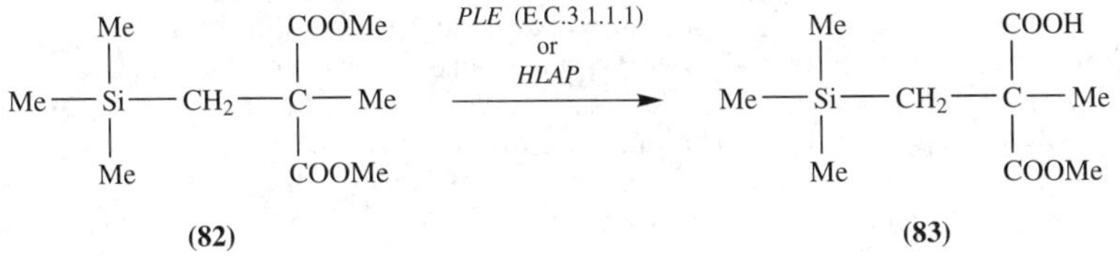

SCHEME 17

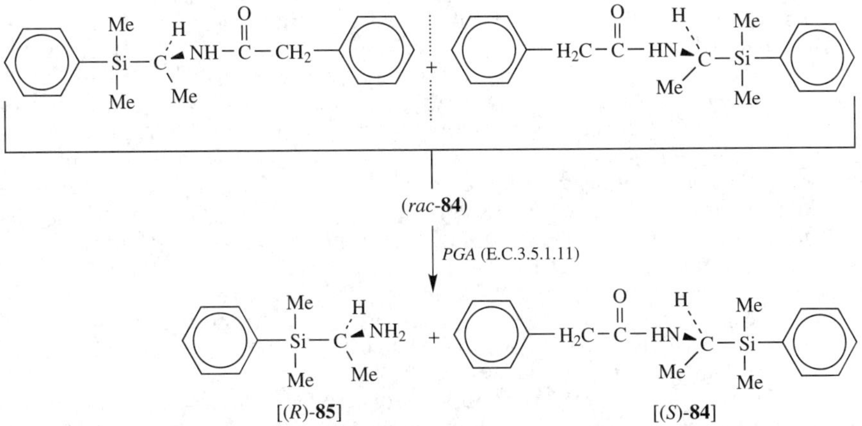

SCHEME 18

of the products were 88% ee (*PLE*) and 81% ee (*HLAP*). In the case of the *PLE*-catalyzed hydrolysis, addition of 50% DMSO to the reaction mixture led to an improvement of the yield (95%) and of the enantiomeric purity (98% ee). This particular type of biotransformation has not yet been applied for the synthesis of optically active silanes with the silicon atom as the center of chirality.

Enantioselective enzymatic amide hydrolyses can also be applied for the preparation of optically active organosilicon compounds. The first example of this is the kinetic resolution of the racemic [1-(phenylacetamido)ethyl]silane *rac*-**84** using immobilized penicillin G acylase (*PGA*; E.C. 3.5.1.11) from *Escherichia coli* as the biocatalyst (Scheme 18)[69]. (*R*)-selective hydrolysis of *rac*-**84** yielded the corresponding (1-aminoethyl)silane (*R*)-**85** which was obtained on a preparative scale in 40% yield (relative to *rac*-**84**). The enantiomeric purity of the biotransformation product was 92% ee. This method has not yet been used for the synthesis of optically active silicon compounds with the silicon atom as the center of chirality.

D. Transesterifications

Enantioselective enzymatic transesterifications have been used as a complementary method to enantioselective enzymatic ester hydrolyses. The first example of this particular type of biotransformation is the synthesis of the optically active 2-acetoxy-1-silacyclohexane (*S*)-**78** (Scheme 19). This compound was obtained by an enantioselective transesterification of the racemic 1-silacyclohexan-2-ol *rac*-**43** with triacetin (acetate source) in isooctane, catalyzed by a crude lipase preparation from *Candida cylindracea* (*CCL*; E.C. 3.1.1.3)[62]. After terminating the reaction at 52% conversion (relative to total amount of substrate *rac*-**43**), the product (*S*)-**78** was separated from the nonreacted substrate by column chromatography on silica gel and isolated in 92% yield (relative to total amount of converted *rac*-**43**) with an enantiomeric purity of 95% ee. The remaining 1-silacyclohexan-2-ol (*R*)-**43** was obtained in 76% yield (relative to total amount of nonconverted *rac*-**43**) with an enantiomeric purity of 96% ee. Repeated recrystallization of (*R*)-**43** led to an improvement of enantiomeric purity by up to >98% ee. Compound (*R*)-**43** has already earlier been prepared by an enantioselective microbial reduction of the 1-silacyclohexan-2-one **42** (see Scheme 8)[53]. The 1-silacyclohexan-2-ol (*R*)-**43** is the antipode of compound (*S*)-**43** which was obtained by a kinetic enzymatic resolution of the racemic 2-acetoxy-1-silacyclohexane *rac*-**78** (see Scheme 15)[62]. For further enantioselective enzymatic transesterifications of racemic organosilicon substrates, with a carbon atom as the center of chirality, see References 64 and 70–72.

Enantioselective enzymatic transesterifications have been successfully used for the synthesis of optically active silanes with the silicon atom as the center of chirality. As shown in Scheme 20, the prochiral bis(hydroxymethyl)silanes **86** and **88** were transformed into the corresponding chiral dextrorotatory isobutyrates (+)-**87** and (+)-**89**, respectively, using *Candida cylindracea* lipase (*CCL*; E.C. 3.1.1.3) as the biocatalyst[73]. For these bioconversions, methyl isobutyrate was used as solvent and acylation agent. When using acetoxime isobutyrate as the acylation agent and *Chromobacterium viscosum* lipase (*CVL*; E.C. 3.1.1.3) as the biocatalyst, the prochiral silanes **86** and **88** were transformed into the levorotatory isobutyrates (−)-**87** and (−)-**89**, respectively (Scheme 20)[73]. The latter conversions were carried out in diisopropyl ether (with **86**) and tetrahydrofuran (with **88**) as solvent. The yields and enantiomeric purities obtained with the transesterifications of **86** and **88** with *CCL* and *CVL* are summarized in Table 4. The absolute configurations of the biotransformation products are unknown.

A similar enantioselective enzymatic transesterification of the prochiral bis(hydroxymethyl)germane **90** (a germanium analogue of the silane **86**) has also been reported[74]. Transesterification of the germane **90** with vinyl acetate (serving as the acetate source and

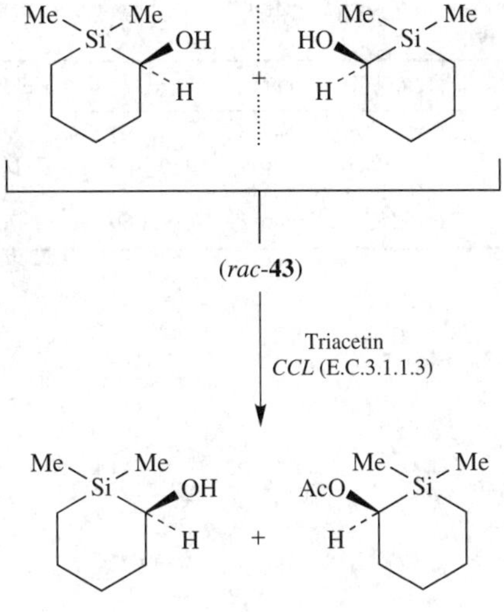

41. Chirality in bioorganosilicon chemistry

SCHEME 19

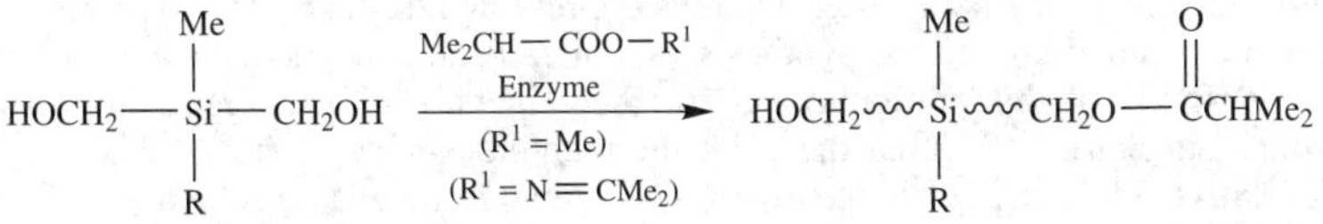

SCHEME 20

solvent), catalyzed by Eupergit-immobilized porcine liver esterase (*PLE*; E.C. 3.1.1.1), yielded the levorotatory germane (−)-**91** (absolute configuration unknown). Both the yield (57%) and the enantiomeric purity (50% ee) of the product were moderate.

Enantioselective enzymatic transesterification has also been used for a kinetic racemate resolution[75]. Starting from the racemic (hydroxymethyl)silane *rac*-**92** (analytical scale), transesterification with vinyl acetate in water-saturated 2,2,4-trimethylpentane, catalyzed by a commercial crude papain preparation (E.C. 3.4.22.2), yielded the corresponding optically active (acetoxymethyl)silane **93** (sign of optical rotation and absolute configuration not reported) (Scheme 21). The enantiomeric purity of the remaining dextrorotatory (hydroxymethyl)silane (+)-**92** was moderate (49% ee).

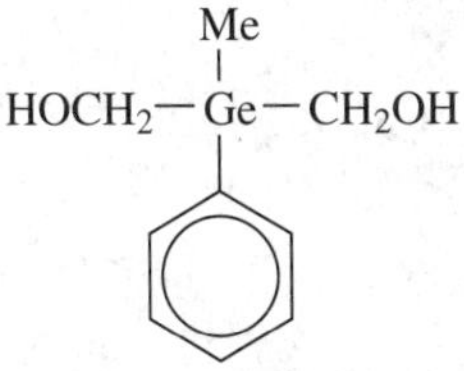

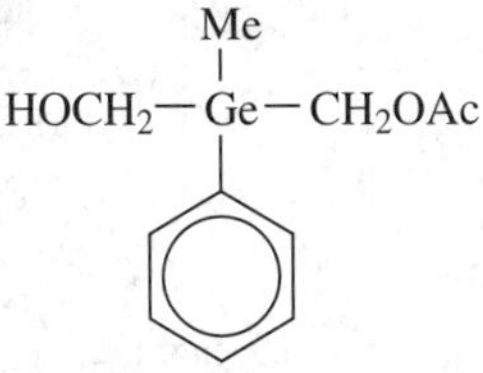

TABLE 4. Enantioselective transesterifications of the bis(hydroxymethyl)silanes **86** and **88** (formation of **87** and **89**)

Substrate	Enzyme	Acylation agent	Product	Yield (%)	ee Value (%)
86	CCL	Me$_2$CHCOOMe	(+)-**87**	80	70
88	CCL	Me$_2$CHCOOMe	(+)-**89**	63	75
86	CVL	Me$_2$CHCOON=CMe$_2$	(−)-**87**	50	70
88	CVL	Me$_2$CHCOON=CMe$_2$	(−)-**89**	70	76

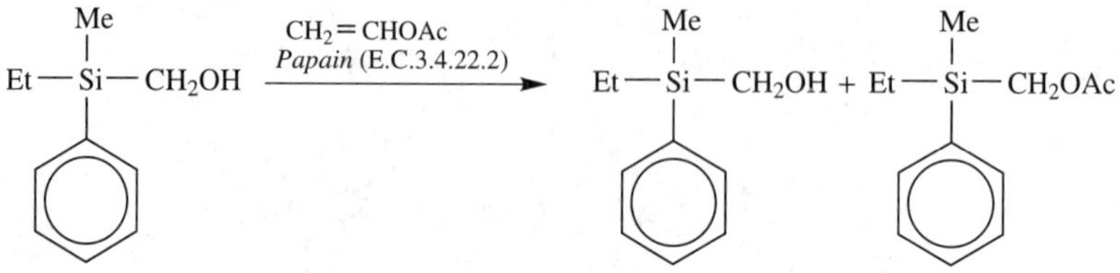

SCHEME 21

E. Esterifications

Enantioselective enzymatic esterifications represent a further type of biotransformation that has been used for the synthesis of optically active organosilicon compounds. The first example of this particular type of bioconversion (kinetic racemate resolution) is illustrated in Scheme 22. Starting from the racemic 1-silacyclohexan-2-ol *rac*-**43**, the optically active 5-phenylpentanoate (*S*)-**94** was prepared by enantioselective esterification with 5-phenylpentanoic acid using 2-methylheptane as solvent and crude *Candida cylindracea* lipase (CCL; E.C. 3.1.1.3) as biocatalyst[7]. The enantiomeric purity of (*S*)-**94** was ca 65% ee (bioconversion not optimized).

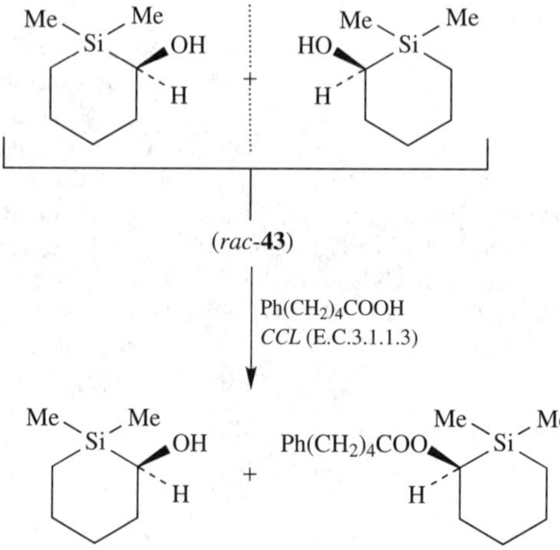

SCHEME 22

Enantioselective enzymatic esterifications of trimethylsilyl-substituted alcohols with racemic 2-(4-chlorophenoxy)propanoic acid in water-saturated benzene, catalyzed by the *Candida cylindracea* lipase OF 360 (*CCL* OF 360; E.C. 3.1.1.3) have been used to prepare (−)-2-(4-chlorophenoxy)propanoic acid[76,77]. As shown in Scheme 23, the (trimethylsilyl)alkanols **95**, **97** and **99** were converted enantioselectively into the corresponding (trimethylsilyl)alkyl (+)-2-(4-chlorophenoxy)propanoates **96**, **98** and **100**. The enantiomeric purity of the remaining (−)-2-(4-chlorophenoxy)propanoic acid was 95.8% ee (**95**), 76.1% ee (**97**) and 77.5% ee (**99**).

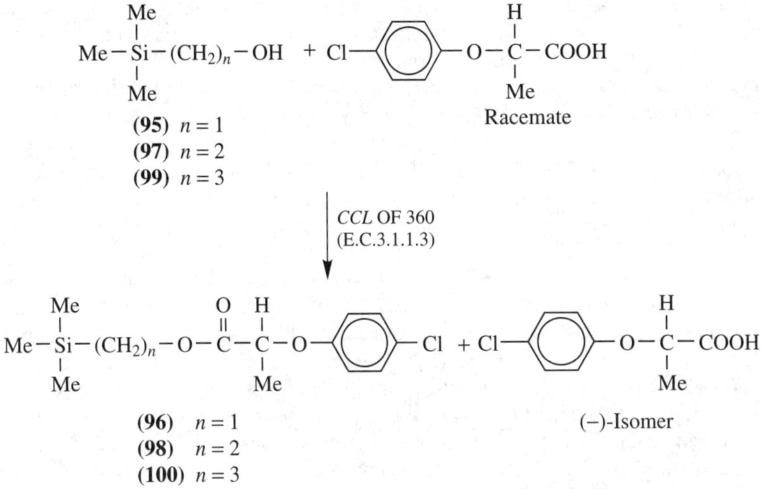

SCHEME 23

Analogous results were obtained for the enantioselective enzymatic esterifications of the related *t*-butyl-substituted alcohols **101** and **103** (carbon analogues of the silanes **95** and **97**, respectively). Reaction with racemic 2-(4-chlorophenoxy)propanoic acid in water-saturated benzene yielded the corresponding *t*-butylalkyl (+)-2-(4-chlorophenoxy)propanoates **102** and **104**, respectively[76,77]. The enantiomeric purity of the remaining (−)-2-(4-chlorophenoxy)propanoic acid was somewhat lower than that observed for the esterification of the analogous silicon compounds [91.1% ee (**101**), 71.6% ee (**103**)]. No esterification was observed for the Si/C analogues trimethylsilanol (Me_3SiOH) and *t*-butanol (Me_3COH).

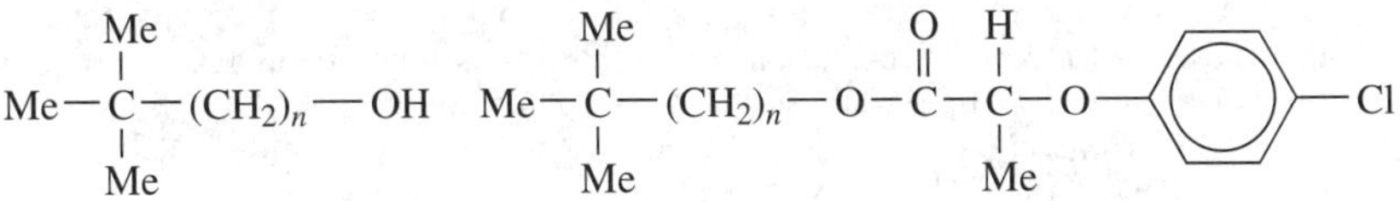

(**101**) $n = 1$
(**103**) $n = 2$

(**102**) $n = 1$
(**104**) $n = 2$

Interestingly, for the Si/C analogues **97** and **103** very similar reaction rates were observed, whereas the enzymatic esterifications of the Si/C analogues **95** and **101** are characterized by striking differences. The rate observed for the bioconversion of the silicon compound **95** was about ten times higher than that for the esterification of its carbon analogue **101**, indicating that *Candida cylindracea* lipase OF 360 (E.C. 3.1.1.3) clearly differentiates between silicon and carbon in this particular system. Analogous results were also obtained for lipases from other sources[76,77]. Both the different covalent radii and the different electronic properties of silicon and carbon may be responsible for this. Obviously, these differences do not affect the reaction rates for the Si/C analogues **97** and **103**; in these compounds, the longer ethylene group (instead of methylene) between the OH group and the silicon and carbon atom, respectively, prevents a significant enzymatic discrimination between silicon and carbon.

Enantioselective enzymatic esterifications of the racemic trimethylsilyl-substituted alcohols *rac*-**105**, *rac*-**107** and *rac*-**109** with 5-phenylpentanoic acid in water-saturated 2,2,4-trimethylpentane have been applied for efficient kinetic resolution of these compounds[78]. In systematic studies with five different Celite-adsorbed hydrolases, a particular enzyme was chosen for the bioconversion of each of the three isomers of (trimethylsilyl)propanol (Scheme 24). For the esterification of *rac*-**105**, cholesterol esterase (Type A from *Pseudomonas* sp.; E.C. 3.1.1.13) was used, yielding the alcohol (*S*)-**105** (93% ee) and the ester (*R*)-**106**. Esterification of *rac*-**107** with lipoprotein lipase (Typ A from *Pseudomonas* sp.; E.C. 3.1.1.34) gave the alcohol (*R*)-**107** (96% ee) and the ester (*S*)-**108**. Thus, the stereochemical course of these two conversions is analogous. For the esterification of *rac*-**109**, lipase Salken 100 (from *Rhizopus japonicus*; E.C. 3.1.1.3) was applied to give the alcohol (−)-**109** (95% ee; absolute configuration not reported) and the optically active ester **110** (sign of optical rotation and absolute configuration not reported). As can be seen from these data, the three isomers of (trimethylsilyl)propanol were found to undergo highly enantioselective esterifications. The conversions of the (2-hydroxyalkyl)silanes *rac*-**105** and *rac*-**109** are especially remarkable, since (2-hydroxyalkyl)silanes are known to undergo an Si−C bond cleavage (β-elimination) under the conditions of nonenzymatic acid- and base-catalyzed esterifications.

The carbon analogues (Si/C exchange) of the three (hydroxyalkyl)silanes *rac*-**105**, *rac*-**107** and *rac*-**109** (compounds *rac*-**111**, *rac*-**113** and *rac*-**115**) were also studied for an enzymatic kinetic resolution with hydrolases[78]. These experiments were performed under the same conditions as reported for the parent silicon compounds. By analogy with the (2-hydroxyalkyl)silanes *rac*-**105** and *rac*-**109**, the corresponding carbon analogues *rac*-**111** and *rac*-**115** were also found to undergo an enantioselective esterification with 5-phenylpentanoic acid. When using cholesterol esterase (Type A from *Pseudomonas* sp.; E.C. 3.1.1.13) (*rac*-**111**) and lipase Salken 100 (from *Rhizopus japonicus*; E.C. 3.1.1.3) (*rac*-**115**), the optically active esters (*S*)-**112** and (−)-**116** (absolute configuration not reported) were obtained with an enantiomeric purity of 88% ee and 45% ee, respectively. Thus, the enantioselectivies observed for the esterifications of the carbon compounds are lower than those for the kinetic resolutions of the corresponding silicon analogues. When using *Candida cylindracea* lipase OF 360 (E.C. 3.1.1.3) as the biocatalyst, the Si/C analogues *rac*-**105** and *rac*-**111** were esterified with inverse stereoselectivities, the remaining alcohols being (*R*)-**105** (8% ee) and (*S*)-**111** (15% ee), respectively. Interestingly, the carbon analogue of the (1-hydroxyalkyl)silane *rac*-**107** (compound *rac*-**113**) was found to be hardly esterified by all hydrolases studied (almost no formation of **114**)[78]. These results again demonstrate that enzymes can clearly differentiate between analogous silicon and carbon compounds.

Enzymatic kinetic resolutions via enantioselective esterifications have been successfully used for the preparation of optically active (hydroxymethyl)silanes, with the silicon atom

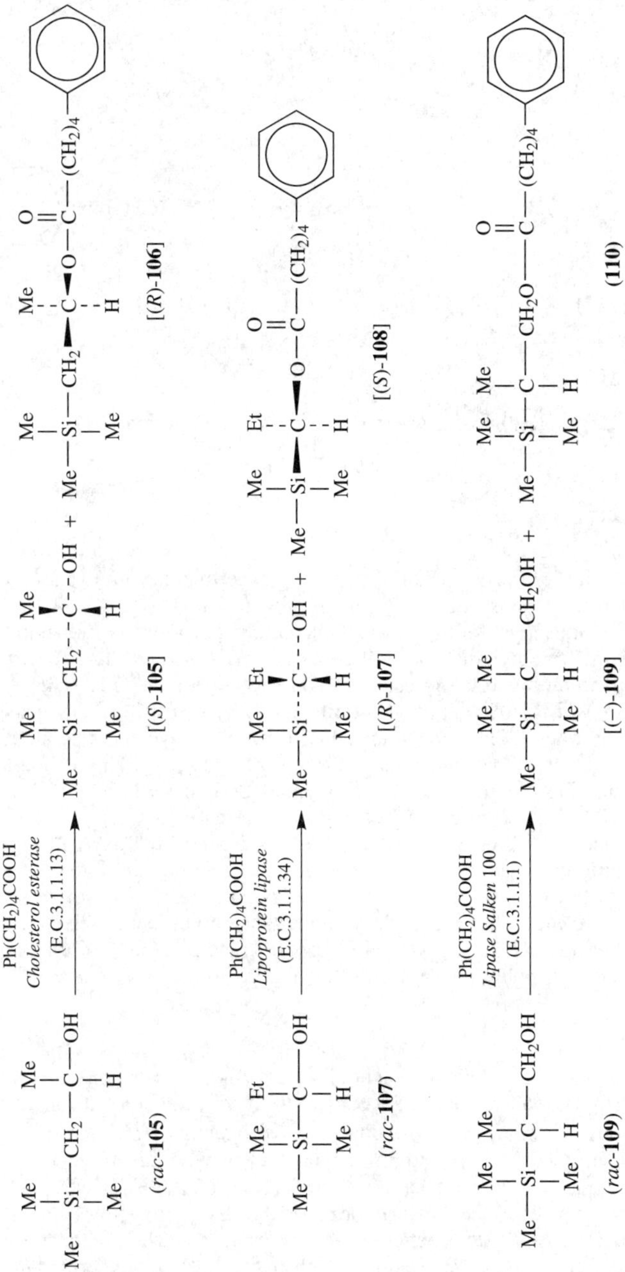

SCHEME 24

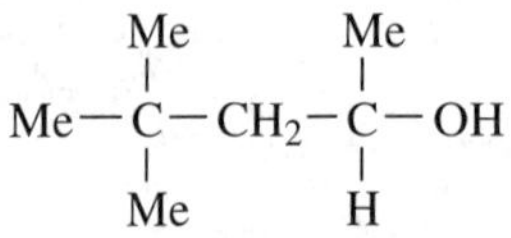

(111) Me−C(Me)(Me)−CH$_2$−C(H)(Me)−OH

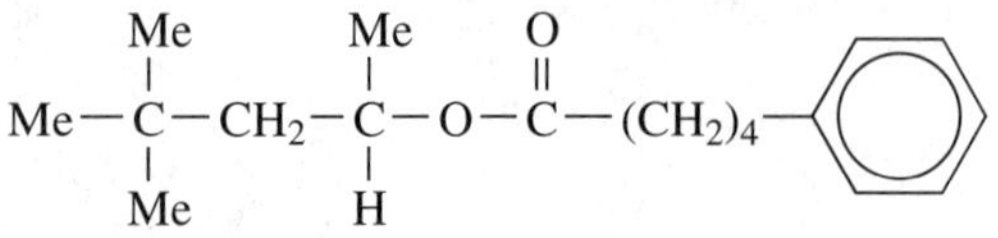

(112) Me−C(Me)(Me)−CH$_2$−C(H)(Me)−O−C(=O)−(CH$_2$)$_4$−C$_6$H$_5$

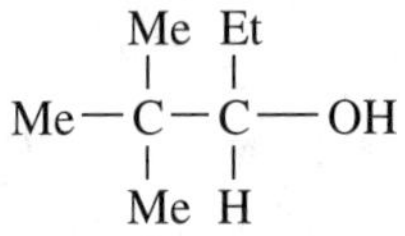

(113) Me−C(Me)(Me)−C(H)(Et)−OH

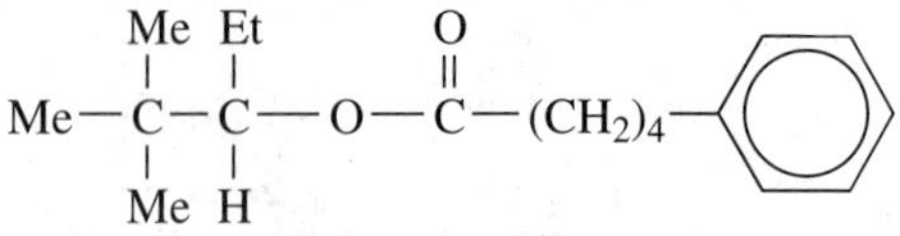

(114) Me−C(Me)(Me)−C(H)(Et)−O−C(=O)−(CH$_2$)$_4$−C$_6$H$_5$

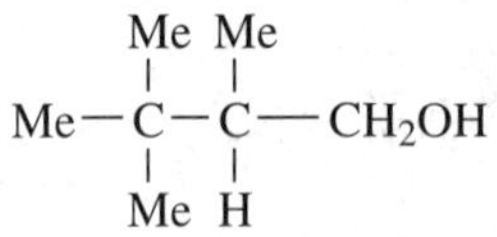

(115) Me−C(Me)(Me)−C(H)(Me)−CH$_2$OH

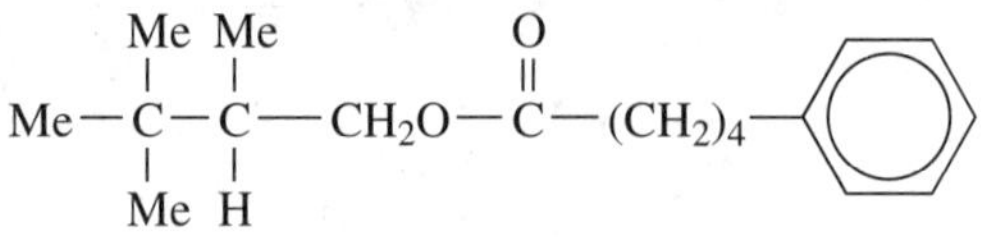

(116) Me−C(Me)(Me)−C(H)(Me)−CH$_2$O−C(=O)−(CH$_2$)$_4$−C$_6$H$_5$

as the center of chirality[75]. As a result of screening experiments with twenty kinds of commercially available hydrolases, a crude papain preparation (E.C. 3.4.22.2) was chosen as the most suitable biocatalyst for this particular synthetic purpose. As shown in Scheme 25, the racemic (hydroxymethyl)silanes *rac*-**92**, *rac*-**118**, *rac*-**120** and *rac*-**122** were converted into the corresponding levorotatory esters (−)-**117** (67% ee), (−)-**119** (64% ee), (−)-**121** (68% ee) and (−)-**123** (70% ee). The esterifications were carried out on a preparative scale in water-saturated 2,2,4-trimethylpentane using 5-phenylpentanoic acid as the acyl donor and Celite-adsorbed papain as the biocatalyst. In the sense of a kinetic racemate resolution, the remaining dextrorotatory (hydroxymethyl)silanes (+)-**92** (92% ee), (+)-**118** (93% ee), (+)-**120** (97% ee) and (+)-**122** (99% ee) were obtained with high enantiomeric purity. The optically active esters formed and the remaining (hydroxymethyl)silanes were isolated by column chromatography on silica gel. Data on the absolute configurations of these compounds have not been reported. The enzymatic resolutions of *rac*-**92**, *rac*-**118**, *rac*-**120** and *rac*-**122** are characterized by remarkably high enantioselectivities, demonstrating again the high synthetic potential of biocatalyzed reactions for the preparation of almost enantiomerically pure silanes, with the silicon atom as the center of chirality.

F. Oxidations

In order to assess the synthetic potential of enzymatic oxidations for organosilicon chemistry, the (hydroxyalkyl)silanes **95**, **97** and **99** have been studied for their oxidation (dehydrogenation) with horse liver alcohol dehydrogenase (*HLADH*; E.C. 1.1.1.1)[79]. For this purpose, these compounds were incubated with *HLADH* in a TRIS-HCl buffer/THF system in the presence of NAD$^+$. As monitored spectrophotometrically (increase of absorbance of the NADH formed), the (2-hydroxyethyl)silane **97** and the (3-hydroxypropyl)silane **99** were better substrates for *HLADH* than ethanol, whereas the related (hydroxymethyl)silane **95** was not a substrate under the experimental conditions used. Interestingly, the corresponding carbon analogue **101** was found to be accepted by *HLADH*. On the other hand, the (2-hydroxyethyl)silane **97** was found to be a better

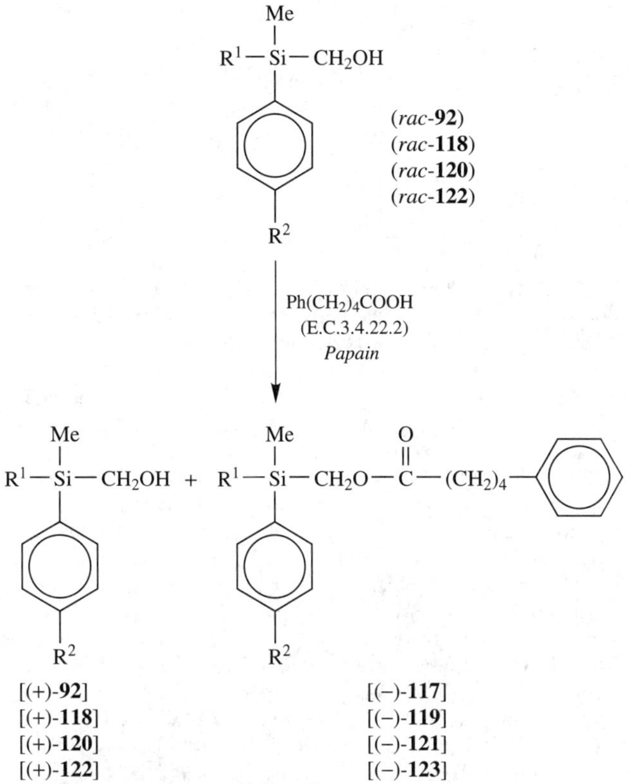

SCHEME 25

$R^1-\underset{\underset{C_6H_4-R^2}{|}}{\overset{\overset{Me}{|}}{Si}}-CH_2OH$

(rac-92)
(rac-118)
(rac-120)
(rac-122)

↓ Ph(CH₂)₄COOH
(E.C.3.4.22.2)
Papain

$R^1-\underset{\underset{C_6H_4-R^2}{|}}{\overset{\overset{Me}{|}}{Si}}-CH_2OH$ + $R^1-\underset{\underset{C_6H_4-R^2}{|}}{\overset{\overset{Me}{|}}{Si}}-CH_2O-\overset{\overset{O}{||}}{C}-(CH_2)_4-Ph$

[(+)-92]   [(−)-117]
[(+)-118]   [(−)-119]
[(+)-120]   [(−)-121]
[(+)-122]   [(−)-123]

	R¹	R²
92, 117	Et	H
118, 119	n-Pr	H
120, 121	Et	Me
122, 123	Et	F

substrate than its carbon analogue **103**. As demonstrated by kinetic studies at different temperatures (20–30 °C), replacement of the silicon atom in **97** by a carbon atom (cf **103**) leads to an increase of both the activation energy and the frequency factor. The different effects observed for the Si/C pairs **95/101** and **97/103** have been discussed in terms of the different electronic properties and covalent radii of silicon and carbon.

Another study on the enzymatic oxidation of the (hydroxyalkyl)silanes **95**, **97** and **99** [and related (hydroxyalkyl)silanes] with *HLADH* also demonstrated that these compounds (including **95**) are oxidized by this enzyme under the experimental conditions used[80]. Although the (hydroxymethyl)silane **95** was claimed to be not a substrate for *HLADH* in Reference 79, the order of activity of *HLADH* on the Si/C pairs **95/101** and **97/103** reported in References 79 and 80 is the same.

Based on these results, attempts have been made to apply enantioselective dehydrogenations with horse liver alcohol dehydrogenase (*HLADH*; E.C. 1.1.1.1), in the presence of NAD⁺, for kinetic resolution of the isomeric (hydroxyalkyl)silanes *rac*-**105**, *rac*-**107**

```
        Me                          Me                          Me
        |                           |                           |
Me —— El —— CH₂OH         Me —— El ——(CH₂)₂—— OH      Me —— Si ——(CH₂)₃—— OH
        |                           |                           |
        Me                          Me                          Me

   (95)  El = Si              (97)  El = Si                 (99)
   (101) El = C                (103) El = C
```

and rac-109[81]. The corresponding carbon analogues rac-111, rac-113 and rac-115 were included in these studies. The reactions were carried out in a water–n-hexane two-layer system with coenzyme regeneration (oxidation of NADH to NAD^+ coupled with L-glutamate dehydrogenase-catalyzed reductive amination of 2-oxoglutarate to L-glutamate). With the (2-hydroxyalkyl)silane rac-105, the dehydrogenation proceeded quickly and stopped at 50% conversion, whereas the carbon analogue rac-111 reacted significantly slower and the conversion reached over 50%. The Si/C analogues rac-109 and rac-115 were found to undergo dehydrogenation at high rates (the silicon compound reacting faster than its carbon analogue) and the conversions continued above 50% conversion. In contrast, the dehydrogenation of the (1-hydroxyalkyl)silane rac-107 and of its carbon analogue rac-113 was negligible under the reaction conditions used. Stereochemical studies (analysis of the remaining alcohols) demonstrated the enantioselective course of the dehydrogenations of rac-105, rac-109, rac-111 and rac-115. In the case of the Si/C analogues rac-105 and rac-111, the enzymatic kinetic resolution yielded the remaining alcohols (R)-105 and (R)-111 with enantiomeric purities of >99% ee and 85% ee, respectively (degree of conversion 50%). In the case of the Si/C analogues rac-109 and rac-115, the remaining levorotatory alcohols (−)-109 and (−)-115 (absolute configurations not reported) were obtained with enantiomeric purities of 70% ee and 59% ee, respectively (degree of conversion 70%). Thus, both the reaction rates and the stereoselectivities of the dehydrogenations of the silicon compounds were higher than those observed for their corresponding carbon analogues. In the two-layer system used for these biotransformations, accumulation of the oxidation products of the carbon compounds rac-111 and rac-115 was observed, while those obtained by oxidation of the silicon analogues rac-105 and rac-109 were not accumulated. This can be easily explained by hydrolytic Si−C bond cleavage of the resulting (1-acylalkyl)silanes yielding trimethylsilanol and the corresponding ketone and aldehyde, respectively. Absence of product inhibition derived from this hydrolytic decomposition may be one of the reasons for the higher reaction rates observed for the silicon compounds.

```
       Me      Me                  Me     Et                  Me     Me
       |       |                   |      |                   |      |
Me —— El —— CH₂—— C —— OH    Me —— El —— C —— OH       Me —— El —— C —— CH₂OH
       |       |                   |      |                   |      |
       Me      H                   Me     H                   Me     H

    (105) El = Si                (107) El = Si              (109) El = Si
    (111) El = C                 (113) El = C               (115) El = C
```

Due to the hydrolytic decomposition of (1-acylalkyl)silanes, enzymatic kinetic resolution of racemic (2-hydroxyalkyl)silanes, catalyzed by HLADH, can only be used for the preparation of one of the two enantiomers. In contrast, resolution of related carbon compounds (including subsequent chemical oxidation of the dehydrogenated bioconversion product), in principle, can yield both enantiomers.

IV. CONCLUDING REMARKS

Chirality has become an important subject in bioorganosilicon chemistry. As demonstrated by pharmacological studies with enantiomerically pure chiral silanols and silanes, with the silicon atom as the center of chirality, biological systems can discriminate between the two enantiomers of organosilicon drugs. In these investigations, significant differences in both pharmacological potency and pharmacological selectivity have been observed. Thus, for future developments of new chiral silicon-based agrochemicals, drugs and diagnostics, biological studies with the pure enantiomers of such compounds are necessary. For this reason, the availability of sufficient preparative methods for the synthesis of enantiomerically pure silicon compounds is of great importance.

On the other hand, it has been demonstrated that the synthetic potential of biological systems represent a challenge for the preparation of enantiomerically pure silicon compounds, with the silicon atom as the center of chirality. Stereoselective biotransformations with whole microbial cells or isolated enzymes have been demonstrated to be very useful for the synthesis of such compounds. Free microbial cells or enzymes and immobilized biocatalysts as well have been used for synthetic purposes. Preliminary studies have shown that plant cell suspension cultures may also be interesting biocatalysts. In general, the favorable characteristics of biotransformations (mild reaction conditions, chemoselectivity, regioselectivity, diastereoselectivity, enantioselectivity) offer a variety of synthetic applications in organosilicon chemistry. By taking advantage of the stereoselective course of enzyme-catalyzed reactions, the synthesis of new types of optically active silicon compounds, with the silicon atom as the center of chirality, is possible. As chemical reactions at the stereogenic silicon center of chiral silicon compounds often proceed with high stereoselectivity, optically active biotransformation products may serve as starting materials for the preparation of further, chemically modified optically active derivatives.

So far, only biocatalyzed reductions, oxidations, ester and amide hydrolyses, esterifications and transesterifications have been used for synthetic purposes in silicon chemistry. However, the application of many other reaction types also appears possible. Although the utilization of biocatalysis in organosilicon chemistry is generally restricted by the hydrolytic sensitivity of certain silicon-element bonds, this particular problem might by overcome, at least in part, (i) by optimizing the bioconversion conditions in aqueous solution (pH and temperature control) and (ii) by using biotransformations in organic media. Lipase-catalyzed conversions in organic solvents appear to be particularly promising in this respect.

Using the powerful methods of modern biotechnology, a scale-up of stereoselective biotransformations and thus the production of larger amounts of optically active organosilicon compounds, in principle, is possible. Development of organosilicon biotechnology, as the basis for large-scale applications of optically active organosilicon compounds in science and technology, represents a big challenge.

V. ACKNOWLEDGMENTS

R.T. wishes to express his sincere thanks to his coworkers and colleagues without whose contributions this article could not have been written; their names are cited in the references. In addition, financial support of our work by the Deutsche Forschungsgemeinschaft, the Volkswagenstiftung, the Fonds der Chemischen Industrie, the

Doktor Robert Pfleger-Stiftung and the Bayer AG (Leverkusen and Wuppertal-Elberfeld) is gratefully acknowledged.

VI. REFERENCES

1. R. J. Fessenden and J. S. Fessenden, *Adv. Drug Res.*, **4**, 95 (1967).
2. R. Tacke and U. Wannagat, *Top. Curr. Chem.*, **84**, 1 (1979).
3. M. G. Voronkov, *Top. Curr. Chem.*, **84**, 77 (1979).
4. R. J. Fessenden and J. S. Fessenden, *Adv. Organomet. Chem.*, **18**, 275 (1980).
5. R. Tacke and H. Zilch, *Endeavour, New Series*, **10**, 191 (1986).
6. R. Tacke and B. Becker, *Main Group Met. Chem.*, **10**, 169 (1987).
7. C. Syldatk, A. Stoffregen, A. Brans, K. Fritsche, H. Andree, F. Wagner, H. Hengelsberg, A. Tafel, F. Wuttke, H. Zilch and R. Tacke, in *Enzyme Engineering 9* (Eds. H. W. Blanch and A. M. Klibanov), *Ann. N. Y. Acad. Sci.*, Vol. 542, The New York Academy of Sciences, New York, 1988, pp. 330–338.
8. R. Tacke and H. Linoh, in *The Chemistry of Organic Silicon Compounds, Part 2* (Eds. S. Patai and Z. Rappoport), Wiley, Chichester, 1989, pp. 1143–1206, and references cited therein.
9. R. Tacke, S. Brakmann, M. Kropfgans, C. Strohmann, F. Wuttke, G. Lambrecht, E. Mutschler, P. Proksch, H.-M. Schiebel and L. Witte, in *Frontiers of Organosilicon Chemistry* (Eds. A. R. Bassindale and P. P. Gaspar), The Royal Society of Chemistry, Cambridge, 1991, pp. 218–228.
10. R. Tacke, M. Strecker, W. S. Sheldrick, E. Heeg, B. Berndt and K. M. Knapstein, *Z. Naturforsch.*, **34b**, 1279 (1979).
11. L. Steiling, R. Tacke and U. Wannagat, *Liebigs Ann. Chem.*, 1554 (1979).
12. R. Tacke, M. Strecker, W. S. Sheldrick, L. Ernst, E. Heeg, B. Berndt, C.-M. Knapstein and R. Niedner, *Chem. Ber.*, **113**, 1962 (1980).
13. R. Tacke, E. Zimonyi-Hegedüs, M. Strecker, E. Heeg, B. Berndt and R. Langner, *Arch. Pharm. (Weinheim)*, **313**, 515 (1980).
14. R. Tacke, H. Lange, W. S. Sheldrick, G. Lambrecht, U. Moser and E. Mutschler, *Z. Naturforsch.*, **38b**, 738 (1983).
15. R. Tacke, M. Strecker, G. Lambrecht, U. Moser and E. Mutschler, *Liebigs Ann. Chem.*, 922 (1983).
16. R. Tacke, M. Strecker, G. Lambrecht, U. Moser and E. Mutschler, *Arch. Pharm. (Weinheim)*, **317**, 207 (1984).
17. W. S. Sheldrick, H. Linoh, R. Tacke, G. Lambrecht, U. Moser and E. Mutschler, *J. Chem. Soc. Dalton Trans.*, 1743 (1985).
18. R. Tacke, H. Linoh, H. Zilch, J. Wess, U. Moser, E. Mutschler and G. Lambrecht, *Liebigs Ann. Chem.*, 2223 (1985).
19. R. Tacke, J. Pikies, H. Linoh, R. Rohr-Aehle and S. Gönne, *Liebigs Ann. Chem.*, 51 (1987).
20. R. Tacke, H. Linoh, L. Ernst, U. Moser, E. Mutschler, S. Sarge, H. K. Cammenga and G. Lambrecht, *Chem. Ber.*, **120**, 1229 (1987).
21. M. Waelbroeck, M. Tastenoy, J. Camus, J. Christophe, C. Strohmann, H. Linoh, H. Zilch, R. Tacke, E. Mutschler and G. Lambrecht, *Br. J. Pharmacol.*, **98**, 197 (1989).
22. G. Lambrecht, R. Feifel, M. Wagner-Röder, C. Strohmann, H. Zilch, R. Tacke, M. Waelbroeck, J. Christophe, H. Boddeke and E. Mutschler, *Eur. J. Pharmacol.*, **168**, 71 (1989).
23. R. Tacke, H. Linoh, K. Rafeiner, G. Lambrecht and E. Mutschler, *J. Organomet. Chem.*, **359**, 159 (1989).
24. G. Lambrecht, R. Feifel, U. Moser, M. Wagner-Röder, L. K. Choo, J. Camus, M. Tastenoy, M. Waelbroeck, C. Strohmann, R. Tacke, J. F. Rodrigues de Miranda, J. Christophe and E. Mutschler, *Trends Pharmacol. Sci. Suppl.*, **10**, 60 (1989).
25. M. Waelbroeck, J. Camus, M. Tastenoy, E. Mutschler, C. Strohmann, R. Tacke, G. Lambrecht and J. Christophe, *Eur. J. Pharmacol., Mol. Pharmacol. Sect.*, **206**, 95 (1991).
26. R. Tacke, K. Mahner, C. Strohmann, B. Forth, E. Mutschler, T. Friebe and G. Lambrecht, *J. Organomet. Chem.*, **417**, 339 (1991).
27. M. Waelbroeck, J. Camus, M. Tastenoy, G. Lambrecht, E. Mutschler, M. Kropfgans, J. Sperlich, F. Wiesenberger, R. Tacke and J. Christophe, *Br. J. Pharmacol.*, **109**, 360 (1993).
28. R. Tacke, J. Pikies, F. Wiesenberger, L. Ernst, D. Schomburg, M. Waelbroeck, J. Christophe, G. Lambrecht, J. Gross and E. Mutschler, *J. Organomet. Chem.*, **466**, 15 (1994).

29. M. Waelbroeck, J. Camus, M. Tastenoy, R. Feifel, E. Mutschler, R. Tacke, C. Strohmann, K. Rafeiner, J. F. Rodrigues de Miranda and G. Lambrecht, *Br. J. Pharmacol.*, **112**, 505 (1994).
30. R. Tacke, M. Kropfgans, A. Tafel, F. Wiesenberger, W. S. Sheldrick, E. Mutschler, H. Egerer, N. Rettenmayr, J. Gross, M. Waelbroeck and G. Lambrecht, *Z. Naturforsch.*, **49b**, 898 (1994).
31. R. Tacke, D. Terunuma, A. Tafel, M. Mühleisen, B. Forth, M. Waelbroeck, J. Gross, E. Mutschler, T. Friebe and G. Lambrecht, *J. Organomet. Chem.*, **501**, 145 (1995).
32. R. Tacke, D. Reichel, K. Günther and S. Merget, *Z. Naturforsch.*, **50b**, 568 (1995).
33. R. Tacke, B. Forth, M. Waelbroeck, J. Gross, E. Mutschler and G. Lambrecht, *J. Organomet. Chem.*, **505**, 73 (1995).
34. R. Tacke, D. Reichel, M. Kropfgans, P. G. Jones, E. Mutschler, J. Gross, X. Hou, M. Waelbroeck and G. Lambrecht, *Organometallics*, **14**, 251 (1995).
35. D. Reichel, R. Tacke, P. G. Jones, G. Lambrecht, J. Gross, E. Mutschler and M. Waelbroeck, in *Organosilicon Chemistry II—From Molecules to Materials* (Eds. N. Auner and J. Weis), VCH, Weinheim, 1996, pp. 231–236.
36. R. Tacke, D. Reichel, P. G. Jones, X. Hou, M. Waelbroeck, J. Gross, E. Mutschler and G. Lambrecht, *J. Organomet. Chem.*, **521**, 305 (1996).
37. R. Tacke, H. Linoh, D. Schomburg, L. Ernst, U. Moser, E. Mutschler and G. Lambrecht, *Liebigs Ann. Chem.*, 242 (1986).
38. E. Mutschler, H. A. Ensinger, J. Gross, A. Leis, K. Mendla, U. Moser, O. Pfaff, D. Reichel, K. Rühlmann, R. Tacke, M. Waelbroeck, J. Wehrle and G. Lambrecht, in *Perspectives in Receptor Research* (Eds. D. Giardinà, A. Piergentili and M. Migini), Elsevier Science Publishers B. V., Amsterdam, 1996, pp. 51–65.
39. M. Kropfgans, *Dissertation*, University of Karlsruhe, 1992; R. Tacke, M. Kropfgans, G. Mattern, J. Sperlich, G. Lambrecht, E. Mutschler and M. Waelbroeck, unpublished results.
40. H.-J. Egerer, *Dissertation*, University of Frankfurt, 1993; G. Lambrecht, E. Mutschler, H.-J. Egerer, R. Tacke and M. Waelbroeck, unpublished results.
41. M. Waelbroeck, J. Camus, M. Tastenoy, G. Lambrecht, E. Mutschler, R. Tacke and J. Christophe, *Eur. J. Pharmacol., Mol. Pharmacol. Sect.*, **189**, 135 (1990).
42. M. Waelbroeck, J. Camus, M. Tastenoy, E. Mutschler, C. Strohmann, R. Tacke, G. Lambrecht and J. Christophe, *Chirality*, **3**, 118 (1991).
43. M. Waelbroeck, J. Camus, M. Tastenoy, E. Mutschler, C. Strohmann, R. Tacke, L. Schjelderup, A. Aasen, G. Lambrecht and J. Christophe, *Eur. J. Pharmacol., Mol. Pharmacol. Sect.*, **227**, 33 (1992).
44. R. J. P. Corriu and C. Guérin, *Adv. Organomet. Chem.*, **20**, 265 (1982).
45. R. J. P. Corriu, C. Guérin and J. J. E. Moreau, *Top. Stereochem.*, **15**, 43 (1984).
46. T. Ohta, M. Ito, A. Tsuneto and H. Takaya, *J. Chem. Soc., Chem. Commun.*, 2525 (1994).
47. R. Tacke, H. Linoh, B. Stumpf, W.-R. Abraham, K. Kieslich and L. Ernst, *Z. Naturforsch.*, **38b**, 616 (1983).
48. C. Syldatk, A. Stoffregen, F. Wuttke and R. Tacke, *Biotechnol. Lett.*, **10**, 731 (1988).
49. C. Syldatk, J. Fooladi, A. Stoffregen, R. Tacke, F. Wagner and M. Wettern, in *Physiology of Immobilized Cells* (Eds. J. A. M. de Bont, J. Visser, B. Mattiasson and J. Tramper), Elsevier Science Publishers B. V., Amsterdam, 1990, pp. 377–385.
50. C. Syldatk, H. Andree, A. Stoffregen, F. Wagner, B. Stumpf, L. Ernst, H. Zilch and R. Tacke, *Appl. Microbiol. Biotechnol.*, **27**, 152 (1987).
51. L. Fischer, S. A. Wagner and R. Tacke, *Appl. Microbiol. Biotechnol.*, **42**, 671 (1995).
52. R. Tacke, S. A. Wagner, S. Brakmann, F. Wuttke, U. Eilert, L. Fischer and C. Syldatk, *J. Organomet. Chem.*, **458**, 13 (1993).
53. R. Tacke, H. Hengelsberg, H. Zilch and B. Stumpf, *J. Organomet. Chem.*, **379**, 211 (1989).
54. R. Tacke, K. Fritsche, A. Tafel and F. Wuttke, *J. Organomet. Chem.*, **388**, 47 (1990).
55. R. Tacke, S. Brakmann, F. Wuttke, J. Fooladi, C. Syldatk and D. Schomburg, *J. Organomet. Chem.*, **403**, 29 (1991).
56. R. Tacke, F. Wuttke and H. Henke, *J. Organomet. Chem.*, **424**, 273 (1992).
57. S. A. Wagner, S. Brakmann and R. Tacke, in *Organosilicon Chemistry II—From Molecules to Materials* (Eds. N. Auner and J. Weis), VCH, Weinheim, 1996, pp. 237–242.
58. F. Huber, S. Bratovanov, S. Bienz, C. Syldatk and M. Pietzsch, *Tetrahedron Asymmetry*, **7**, 69 (1996).
59. H. Hengelsberg, *Dissertation*, Technical University of Braunschweig, 1989; R. Tacke and H. Hengelsberg, unpublished results.

60. H. Zilch, *Dissertation*, Technical University of Braunschweig, 1986; R. Tacke, H. Zilch, B. Stumpf, L. Ernst and D. Schomburg, unpublished results.
61. Y. Yamazaki and H. Kobayashi, *Chemistry Express*, **8**, 97 (1993).
62. K. Fritsche, C. Syldatk, F. Wagner, H. Hengelsberg and R. Tacke, *Appl. Microbiol. Biotechnol.*, **31**, 107 (1989).
63. M. Shimizu, H. Kawanami and T. Fujisawa, *Chem. Lett.*, 107 (1992).
64. N. Belair, H. Deleuze, B. De Jeso and B. Maillard, *Main Group Met. Chem.*, **15**, 187 (1992).
65. R. Tacke, K. Fritsche, H. Hengelsberg, A. Tafel, F. Wuttke, H. Zilch, C. Syldatk, H. Andree, A. Stoffregen and F. Wagner, *VIIIth International Symposium on Organosilicon Chemistry, Abstracts*, St. Louis, 1987, p. 51.
66. R. Tacke, in *Organosilicon and Bioorganosilicon Chemistry: Structure, Bonding, Reactivity and Synthetic Application* (Ed. H. Sakurai), Ellis Horwood, Chichester, 1985, pp. 251–262.
67. C. Syldatk, H. Andree, F. Wagner, F. Wuttke, H. Zilch and R. Tacke, *4. DECHEMA Jahrestagung der Biotechnologen, Abstracts*, Frankfurt/Main, 1986, pp. 171–172.
68. B. De Jeso, N. Belair, H. Deleuze, M.-C. Rascle and B. Maillard, *Tetrahedron Lett.*, **31**, 653 (1990).
69. H. Hengelsberg, R. Tacke, K. Fritsche, C. Syldatk and F. Wagner, *J. Organomet. Chem.*, **415**, 39 (1991).
70. M. A. Sparks and J. S. Panek, *Tetrahedron Lett.*, **32**, 4085 (1991).
71. P. Grisenti, P. Ferraboschi, A. Manzocchi and E. Santaniello, *Tetrahedron*, **48**, 3827 (1992).
72. V. Fiandanese, O. Hassan, F. Naso and A. Scilimati, *Synlett*, 491 (1993).
73. A.-H. Djerourou and L. Blanco, *Tetrahedron Lett.*, **32**, 6325 (1991).
74. R. Tacke, S. A. Wagner and J. Sperlich, *Chem. Ber.*, **127**, 639 (1994).
75. T. Fukui, T. Kawamoto and A. Tanaka, *Tetrahedron Asymmetry*, **5**, 73 (1994).
76. A. Tanaka, T. Kawamoto and K. Sonomoto, in *Enzyme Engineering 10* (Eds. H. Okada, A. Tanaka and H. W. Blanch), *Ann. N. Y. Acad. Sci.*, Vol. 613, The New York Academy of Sciences, New York, 1990, pp. 702–706.
77. T. Kawamoto, K. Sonomoto and A. Tanaka, *J. Biotechnol.*, **18**, 85 (1992).
78. A. Uejima, T. Fukui, E. Fukusaki, T. Omata, T. Kawamoto, K. Sonomoto and A. Tanaka, *Appl. Microbiol. Biotechnol.*, **38**, 482 (1993).
79. M.-H. Zong, T. Fukui, T. Kawamoto and A. Tanaka, *Appl. Microbiol. Biotechnol.*, **36**, 40 (1991).
80. K. Sonomoto, H. Oiki and Y. Kato, *Enzyme Microbiol. Technol.*, **14**, 640 (1992).
81. T. Fukui, M.-H. Zong, T. Kawamoto and A. Tanaka, *Appl. Microbiol. Biotechnol.*, **37**, 209 (1992).

CHAPTER 42

Highly reactive small-ring monosilacycles and medium-ring oligosilacycles

WATARU ANDO and YOSHIO KABE

Department of Chemistry, University of Tsukuba, Tsukuba, Ibaraki 305, Japan

I. INTRODUCTION	2401
II. SILYL CARBENE-TO-SILENE REARRANGEMENT AND RING FORMATION	2403
A. Decomposition of Monosilanyl Diazo Compounds	2403
B. Decomposition of Oligosilanyl Diazo Compounds	2404
C. Oxasiletane and Oxasiletene	2406
D. Spectroscopic Studies of Silenes	2408
E. Decomposition of Bis(silyldiazomethyl)oligosilanes	2408
F. Silabenzene and 6-Silafulvenes	2412
III. YLIDE AND SMALL-RING FORMATIONS BY SILYLENE REACTIONS WITH HETEROATOMS AND CUMULENES	2414
A. Oxa(thia)siliranes and Sila(thio)carbonyl Ylides	2414
B. Siliranes and Alkylidenesiliranes	2418
IV. RING FORMATION BY WURTZ AND DIYNE COUPLING REACTIONS	2428
A. Heptasila[7]paracyclophane	2428
B. Bisalkylidenesiletanes	2429
V. RING FORMATION BY KIPPING AND GILMAN REACTIONS	2437
A. *trans*-Trisilacycloheptene and [3.1.1]Trisilapropellane	2437
B. Oligosilabridged Cyclic Acetylenes	2439
C. Oligosilabridged and Doubly Bridged Allenes and Bisallenes	2446
D. 1,2-Disilaacenaphthene and 9,10-Disilaanthracene Silyl Dianions	2452
VI. REFERENCES	2457

I. INTRODUCTION

Our research interest gravitated to organosilicon chemistry when the formation of a silicon carbon double bond (silene) intermediate was found in the decomposition of silyl

The chemistry of organic silicon compounds, Vol. 2
Edited by Z. Rappoport and Y. Apeloig © 1998 John Wiley & Sons Ltd

diazo compounds in 1973 (Section II.A, B and D). This process involves a 1,2-alkyl migration of a silyl carbene from a silicon to a carbene center which corresponds to a general process in carbene chemistry, i.e. 1,2-alkyl rearrangement of carbene to form alkenes. Although the first stable silene and disilene with bulky substituents were isolated by Brook and West, respectively, in 1979 and 1981, our reactive silene has proved to be a convenient source to a variety of functionalized silenes which are otherwise accessed with difficulty, such as silabenzene, silafulvene (Section II.F), silaacrylate, siladienone (Section I.C), 1,4-disilabutadiene and 1,5-disilapentadiene (Section I.E), respectively. Another aspect of this chemistry involves a ready assembly of new silicon ring systems, such as silaoxetane, silaoxetene (Section II.C), trisilabicyclo[1.1.1]pentane, trisilabicyclo[2.1.0]pentane, tetrasilabicyclo[3.1.0]hexane and tetrasilabicyclo[2.2.0]hexane (Section II.E) from the reactive silenes.

A silyl carbene can also undergo an intramolecular carbene insertion into the α-CH bond to form a silacyclopropane (silirane). The elusiveness of siliranes has long attracted our interest to small-ring compounds containing silicon atoms. At that time we suggested silirane as the intermediate in the decomposition of silyl diazo compounds. After isolation of the first stable silirane by Seyferth and of its unsaturated analogue (silirene) by Gaspar, Kumada and Sakurai, we started our study of silylene addition to ketones and thioketones. This research has not only culminated in the first isolation of oxa(thia)silirane but has also yielded a rich ylide chemistry (Section III.A). The silylene addition strategy to alkenes and cummulenes provided a variety of silicon ring systems such as silirane, alkylidenesilirane and bisalkylidenesilirane (Section III.B). In the latter case, the alkylidene moiety increased the reactivity of the silirane ring toward a variety of electrophiles and nucleophiles. Especially, transition metals promoted the Si—C bond cleavage to give silatrimethylenemethane complexes, i.e. silene complexes of trimethylenemethane analogue.

Meanwhile, the addition reaction of disilene has been reported by West's and Masamune's groups, respectively, to give a variety of three- and four-membered ring compounds containing Si—Si bonds. The structural analysis of disilacyclopropanes (disiliranes) indicates relatively short bonds and an almost planar geometry about each silicon atom, which is interpreted in terms of a π-complex of a disilene. This observation supported the conclusion that Si—Si σ-bonds in the small-ring systems have some analogous character and reactivity to C—C π-bonds. Similar structural and reactivity features of digermiranes were delineated by our group. In addition to digermiranes, the ring expansion reaction of disiliranes, disiletene and bisalkylidenedisiletane was found to be a successful tool for derivatization of silicon ring systems. Especially, this study was culminated in finding the transition metal promoted ring expansion of bisalkylidenesiletenes, which can be prepared by both intramolecular Wurtz and Diyne coupling reactions (Section IV.B).

Kumada's group developed the photochemical silylene extrusion and ring contraction reactions of cyclohexasilane in 1969 and this process has become very popular in organosilicon chemistry, e.g. in West's and Masamune's disilene syntheses of linear- and cyclotrisilanes. This process was adopted as another tool for ring contraction and derivatization of oligosilabridged cyclosilanes containing C—C unsaturated bonds (Section V.A). Fortunately, in the course of these studies, the presence of long Si—Si bonds was found to enable the direct preparation of strained ring systems by Kipping reaction (Section V.B). Our research along these lines culminated in the synthesis of *trans*-trisilacycloheptene, the first stable seven-membered cyclic *trans*-olefin, [7]pentasilacyclophane (Section IV.A), [3.1.1]trisilapropellane (Section V.A), oligosilabridged and doubly bridged allenes and bisallenes (Section V.C) and tetrasilacyclohexynes (Section V.B), the first stable six-membered cyclic acetylenes. West's and Masamune's disilenes which are discussed above are highly protected by bulky substituents and more functionalized disilenes constructed into an aromatic system have not been explored up to this point. In this context, it was found that silyl dianions such as 1,2-disilaacenaphthene and 9,10-disilaanthracene prepared by Gilman reaction act as synthetic equivalents of disilaaromatics in the presence of oxidizing reagents and in the reaction with α,ω-dichlorooligosilanes (Section V.D).

Hence this chapter provides analogy between silylene and carbene chemistry as a useful approach for construction of small-ring monosilacycles. Furthermore, intramolecular Wurtz, Diyne coupling, Kipping and Gilman reactions have been shown to be useful tools for construction of medium-ring oligosilacycles.

II. SILYL CARBENE-TO-SILENE REARRANGEMENT AND RING FORMATION

A. Decomposition of Monosilanyl Diazo Compounds

When ethyl trimethylsilyldiazoacetate (**1a**) is photolyzed with a high-pressure mercury lamp in alcohols, four products (**2–5**) are obtained. Products **2** and **3** may be rationalized in terms of insertion reaction of trimethylsilyl carboethoxycarbene and product **4** has been explained by a Wolf rearrangement.

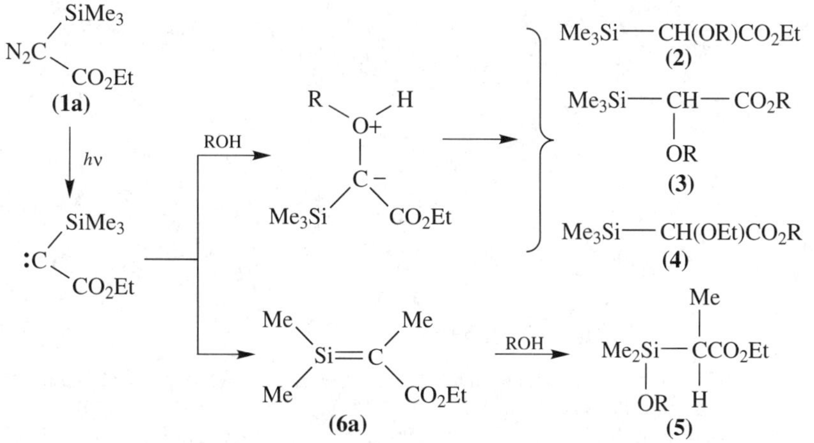

$$N_2C\begin{matrix}SiMe_3\\CO_2Et\end{matrix}\ (\mathbf{1a}) \xrightarrow{h\nu} :C\begin{matrix}SiMe_3\\CO_2Et\end{matrix}$$

$$\xrightarrow{ROH} Me_3Si\text{—}CH(OR)CO_2Et\ (\mathbf{2})$$
$$Me_3Si\text{—}CH\text{—}CO_2R,\ OR\ (\mathbf{3})$$
$$Me_3Si\text{—}CH(OEt)CO_2R \quad (1)$$
$$(\mathbf{4})$$

$$\begin{matrix}Me\\Me\end{matrix}Si=C\begin{matrix}Me\\CO_2Et\end{matrix}\ (\mathbf{6a}) \xrightarrow{ROH} Me_2Si\text{—}CCO_2Et,\ OR\ H\ |\ Me\ (\mathbf{5})$$

$$Me_3SiC(N_2)R + EtOH \longrightarrow Me_3Si\text{—}CHR\ |\ OEt\ +\ Me_2Si\text{—}CHR\ |\ Me\ |\ OEt \quad (2)$$

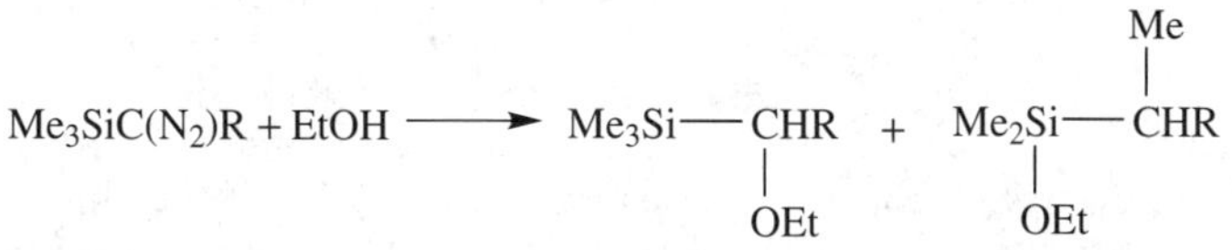

R = CO$_2$Et	**1a**		85%	15%
R = Ph	**7**		100	0
R = H	**8**		0	100

The formation of **5**, which involves a methyl migration, is interesting from a mechanistic point of view. It was rationalized by invoking the intermediacy of the silicon–carbon double bond species (silene) (**6a**) (equation 1)[1]. The extent of this silylcarbene-to-silene rearrangement is highly dependent on substituents. For example, phenyl trimethylsilyldiazomethane (**7**) gave only O–H insertion products, while trimethylsilyldiazomethane (**8**) afforded only methyl migration product (equation 2)[2]. The gas-phase pyrolysis of **1b** and **7** gave controversial results. When methyl trimethylsilyldiazoacetate (**1b**) was evaporated through a Pyrex tube maintained at 360 °C and the products were then treated with methanol, the Wolf rearrangement product (**9**) and methyl migration product (**10**) were obtained[3]. To our surprise the trapping of the pyrolysate in alcohols showed that both alkoxy groups in **10** are derived from the added alcohol. This observation led us to isolate ketene **11a** (R = Me) as the suspected intermediate. A major question that remained was the source of **11a** which could be formed by a dytropic rearrangement of ketene **12**, or by a rapid isomerization of **6b** (equation 3)[4].

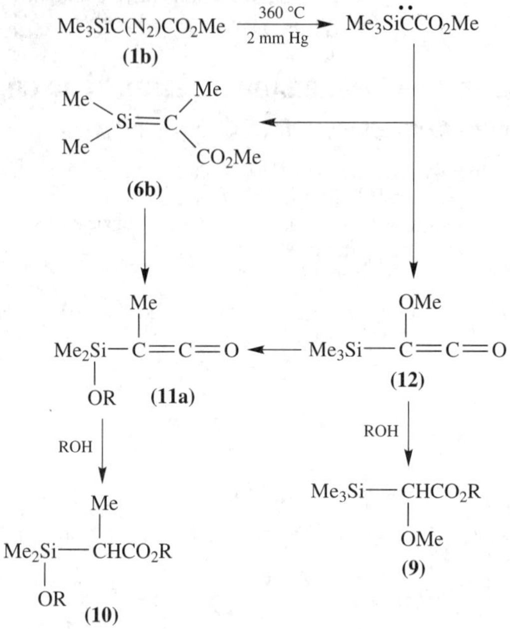

$$Me_3SiC(N_2)CO_2Me \xrightarrow[\text{2 mm Hg}]{360\,°C} Me_3Si\ddot{C}CO_2Me \quad (3)$$
(1b)

(with structures **6b**, **11a**, **12**, **10**, **9** shown in the scheme)

Pyrolysis of **7** gave four products (equation 4)[5]. Product **15** is easily rationalized as the dimerization product of the silyl carbene. Product **14** involves an abstraction of hydrogen by the silylcarbene and is a typical minor product of carbene reactions. A ^{13}C labeling experiment[6] was consistent with the proposal that **13** is formed as a result of a carbene–carbene rearrangement and also ruled out an earlier suggestion that a silirane was involved[7]. However, a silene is also formed in the reaction because, in the presence of an alcohol, a ketone and 2,3-dimethylbutadiene, alcohol insertion, Wittig type and Diels–Alder adducts were also isolated along with **13** and **14**[7,8]. The styrene is formed either by decomposition of the product of carbon–hydrogen insertion, 1,1-dimethyl-2-phenylsilirane or by dissociation of the silene to dimethylsilylene and methylphenylcarbene, which subsequently yield styrene.

B. Decomposition of Oligosilanyl Diazo Compounds

During the 1970s Brook's group had been studying the photochemistry of acylsilanes[9], which resulted in a 1,2-shift of the silyl group from carbon to oxygen forming a siloxycarbene. On replacing the silyl groups by disilanyl groups, it was found that the products formed included not only the 1,2-silyl shifted disiloxycarbene, but also a silene formed by a 1,3-shift of silyl group from silicon to oxygen[9]. Ultimately the use of a tris(trimethylsilyl)silyl group led to an exclusive 1,3-shift to afford the first stable silenes[10–12]. The good migrating ability of a silyl group induced us to pursue a project aimed at obtaining oligosilanyl silenes. An exclusive migration of a trimethylsilyl group to a carbene center was observed in both thermal and photochemical decomposition of oligosilylated diazomethanes (**16a,b** and **18**) (equation 5)[13]. The formation of the transient silenes **17** and **19** was deduced from trapping with alcohol and ketones. Photolysis of disilanyl diazoacetate (**1c**) in THF cleanly produced ketene **21** in quantitative yield (equation 6)[14]. Another interesting ketene **11b** was also obtained when **21** was treated with methanol. Support for the intermediacy of the silene (**6c**) in the reaction of **1c** was obtained by a reaction in methanol (**20**) and in cyclohexanone (**22**)[15]. It is now clear that silaacrylate (**6c**) isomerizes to alkoxyketene (**21**) by a spontaneous migration of the alkoxy group to silicon (see Section II.A).

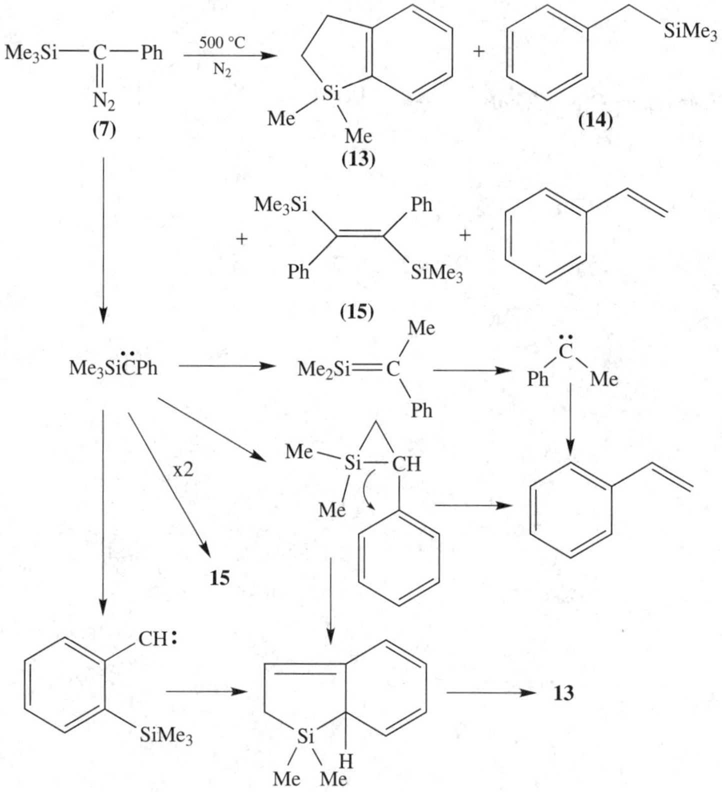

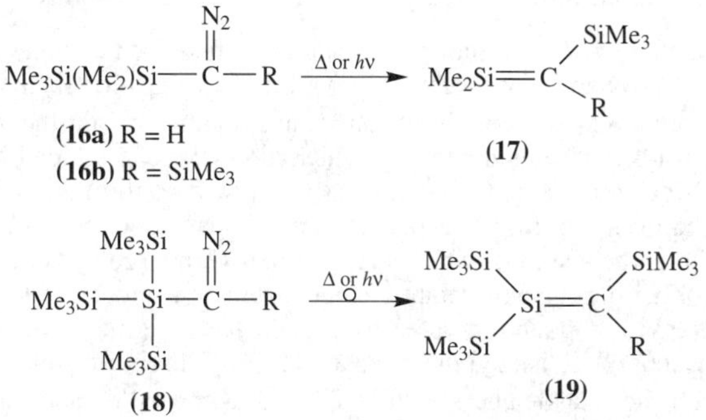

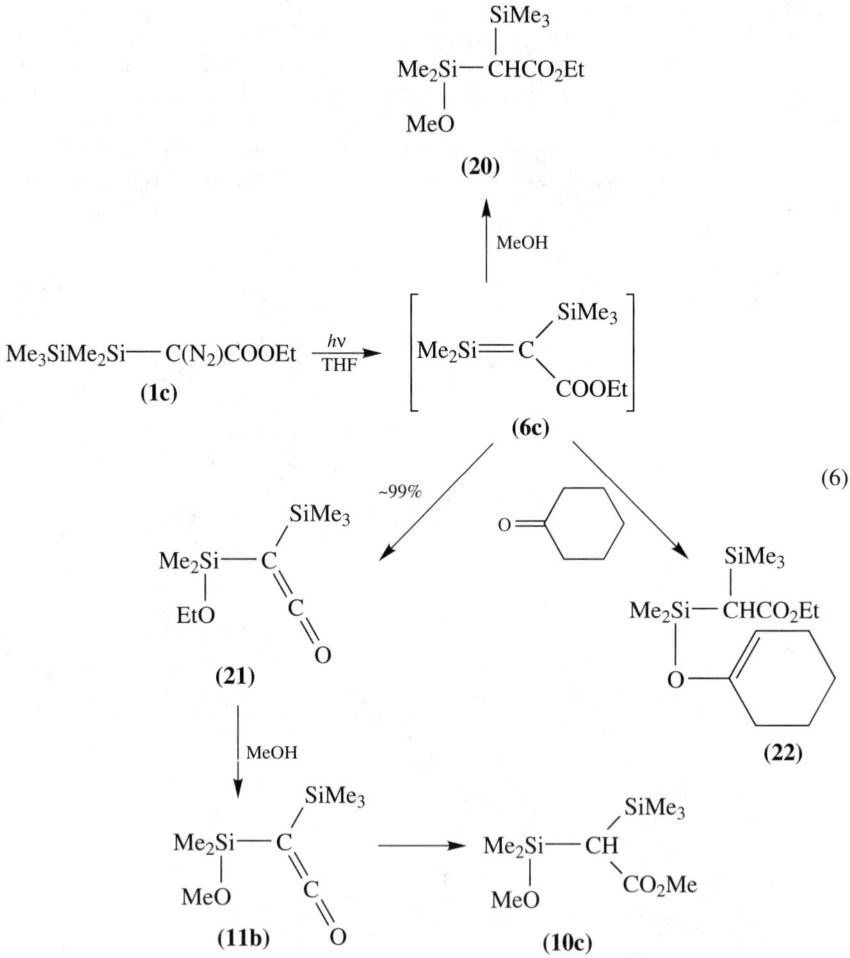

$$\text{Me}_3\text{SiMe}_2\text{Si}-\text{C}(\text{N}_2)\text{COOEt} \xrightarrow[\text{THF}]{h\nu} \left[\text{Me}_2\text{Si}=\text{C} \diagup^{\text{SiMe}_3}_{\text{COOEt}} \right]$$
(1c) → (6c)

With MeOH → (20): Me$_2$Si(OMe)–CH(SiMe$_3$)CO$_2$Et

~99% → (21): Me$_2$Si(OEt)–C(SiMe$_3$)=C=O

With cyclohexanone → (22): oxasiletane product

(21) + MeOH → (11b): Me$_2$Si(OMe)–C(SiMe$_3$)=C=O → (10c): Me$_2$Si(OMe)–CH(SiMe$_3$)CO$_2$Me

(6)

C. Oxasiletane and Oxasiletene

Pyrolysis of siletane had been the most reliable method for the generation of transient silene, and Gusel'nikov and coworkers confirmed the existence of such silenes by kinetic and chemical trapping experiments[16]. Barton[17] and Sommer[18] and their coworkers have found that co-pyrolysis of siletane with an aldehyde or ketone yielded a cyclic siloxane and an olefin. It is generally agreed that this reaction proceeds through cycloaddition of the silene to the carbonyl group to produce an unstable oxasiletane, which thermally decomposes to the olefin and a silanone, which then thermally undergo a cyclic oligomerization. In the hope of obtaining an oxasiletane, silene formation under mild conditions in the presence of carbonyl compounds was employed. In fact, thermal generation (180 °C) of silene **6c** from pentamethyldisilanyl diazoacetate (**1c**) in 7-norbornanone yields an isolable product for which the oxasiletane structure **23** was proposed[19]. Soon after, Barton and coworkers convincingly reinterpreted the structure as ketene acetal **24**[20]. If Barton is correct, the reaction should be classified as [4 + 2] cycloaddition of the silaacrylate **6c** (equation 7). Märkl, Wiberg and Brook demonstrated that stable silenes with ketones having aromatic rings undergo both [2 + 2] and [2 + 4] cycloadditions[21–23].

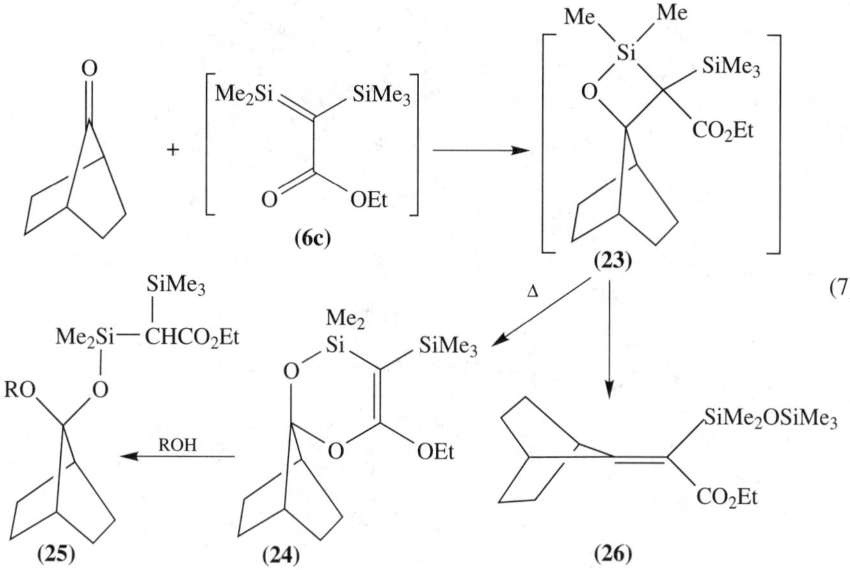

(7)

The formation of methanolysis product **25** (R = Me) was easily understood in terms of ketene acetal structure **24**. However, the isolated thermal product **26** is not directly derived from the ketene acetal **24**. It is not unreasonable to suggest that thermal transformation of the [2 + 4] adduct **24** into the [2 + 2] adduct **23** takes place, followed by intramolecular migration of the trimethylsilyl group to afford vinylsiloxane **26** (equation 7). The behavior of these stable oxasiletanes suggests that fragmentation to silanone and olefin is not a facile process. The formation of free silanones in reactions of this type has recently been questioned on the basis of calculations, which suggested that fragmentation of oxasiletane is too endothermic to occur at a significant rate under the pyrolytic conditions. Instead, Bachrach and Streitwieser proposed a bimolecular mechanism for the reactions[24].

Oxasiletene, the unsaturated analogue of oxasiletane, was first postulated by Seyferth and coworkers as a reactive intermediate in the reaction of 1,1-dimethyl-2,3-bis(trimethylsilyl)-1-silirene with dimethyl sulfoxide[25]. The photochemical generation of siladienone intermediate **28** from (pentamethyldisilanyl)diazomethyl 1-adamantyl ketone **27** led to an intramolecular [2 + 2] cyclization to yield the oxasiletene **29**, quantitatively (equation 8)[26]. Thermally, **29** is a very labile molecule. It is readily converted to **30** along with dimethylsilanone and the latter was trapped by using hexamethylcyclotrisiloxane as well as dimethyldimethoxysilane. More recently, we have studied the photochemical decomposition of a series of disilanyl diazomethyl ketones (**27**)[27,28]. We found that the products are highly dependent on the bulk of the substituents: oxasiletene (**29**) is formed when 1-adamantyl and *tert*-butyl groups are the substituents R, and head-to-tail cyclodimers **31** are formed when R is isopropyl or methyl. When R = Me, ketone **32** and bis(trimethyl) ketene are formed (equation 8).

$$
\begin{array}{c}
\text{R—C—C—SiMe}_2\text{SiMe}_3 \\
\text{|| ||} \\
\text{O } \text{N}_2 \\
(27)
\end{array}
\xrightarrow{h\nu}
\begin{array}{c}
\text{R—C—C—SiMe}_3 \\
\text{|| ||} \\
\text{O } \text{SiMe}_2 \\
(28)
\end{array}
\qquad (8)
$$

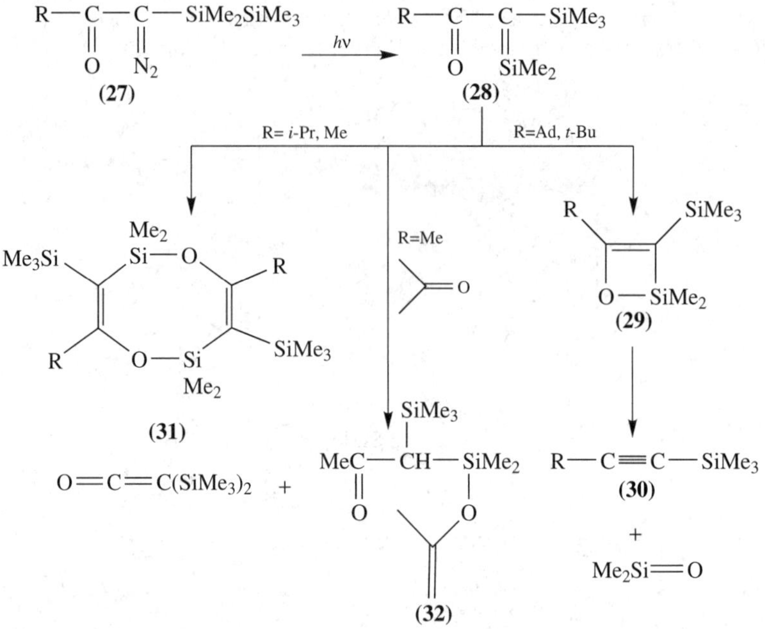

R = i-Pr, Me → (31)

R = Me → (32)

R = Ad, t-Bu → (29) → (30) + Me$_2$Si=O

(31): seven-membered ring with Me$_3$Si, Me$_2$Si—O, R, O—Si, Me$_2$, SiMe$_3$ substituents

O=C=C(SiMe$_3$)$_2$ +

(32): MeC—CH—SiMe$_3$/SiMe$_2$ cyclic structure with O

D. Spectroscopic Studies of Silenes

For comparative study, various silenes were isolated in 3-methylpentane matrix at 77 K by photolysis of the corresponding silyldiazo compounds and their ultraviolet spectra are measured. The results are summarized in Table 1. The introduction of trimethylsilyl group on carbon results in slight red shift compared with the parent silene (H$_2$Si=CH$_2$, 258 nm). As one might expect, considerable bathochromic shifts have been observed for conjugated silenes such as **6c, 6d** and **28**[29,30].

E. Decomposition of Bis(silyldiazomethyl)oligosilanes

Photolysis of **36a** with a high pressure mercury lamp in *tert*-butyl alcohol gave the adduct **37a** that incorporates two molecules of *tert*-butyl alcohol to give

TABLE 1. UV-absorption maxima of silenes

Silenes	λ_{max} (nm)	Precursor
Me$_2$Si=CHMe (**33**)	255	Me$_3$SiCHN$_2$ (**8**)
Me$_2$Si=CHSiMe$_3$ (**17a**)	265	Me$_3$SiMe$_2$SiCHN$_2$ (**16a**)
Me$_2$Si=C(Me)SiMe$_3$ (**35**)	274	(Me$_3$Si)$_2$CN$_2$ (**34**)
Me$_2$Si=C(SiMe$_3$)$_2$ (**17b**)	278	Me$_3$SiMe$_2$SiC(N$_2$)SiMe$_3$ (**16b**)
Me$_2$Si=C(Ph)CO$_2$Me (**6d**)	280	PhMe$_2$SiC(N$_2$)CO$_2$Me (**1d**)
Me$_2$Si=C(SiMe$_3$)COAd-1 (**28**)	284	Me$_3$SiMe$_2$SiC(N$_2$)COAd-1 (**27**)
Me$_2$Si=C(SiMe$_3$)CO$_2$Et (**6c**)	293	Me$_3$SiMe$_2$SiC(N$_2$)CO$_2$Et (**1c**)

38a[31]. The photolysis of **36a** without a trapping reagent gave the interesting trisilabicyclo[1.1.1]pentane **40a** in low yields. Photolysis of **36b** in cyclohexane gave **40b** whose structure was confirmed by X-ray analysis (Figure 1a) and also **43b**, which is derived from the oxygenation of trisilabicyclo[2.1.0]pentane **39b** (equations 9 and 10). Photolysis of **36b** in cyclohexane-d_{12} revealed the existence of **39b**, which is easily polymerized at room temperature but was trapped by *tert*-butyl alcohol to give **44b**. On the other hand, pyrolyses of **36a** and **36b** gave only **40a** and **40b** as almost a sole volatile product. 1,5-Disilapentadiene (**37a,b**) is thought to be a common intermediate of both **39b** and **40a,b**. Although transient silene is known to dimerize in a head-to-tail fashion, **39b** was rationalized to be also derived from a head-to-head [2 + 2] cycloaddition of **37b**. In another alternative, the formation of **39b** implies the bicycloazo compound (**42b**) as an intermediate formed via an intramolecular [2 + 3] cycloaddition of the diazosilene (**41b**). To corroborate this mechanism, a new bis(silyldiazomethyl)tetrasilane (**36c**) was employed in the photochemical reactions[32].

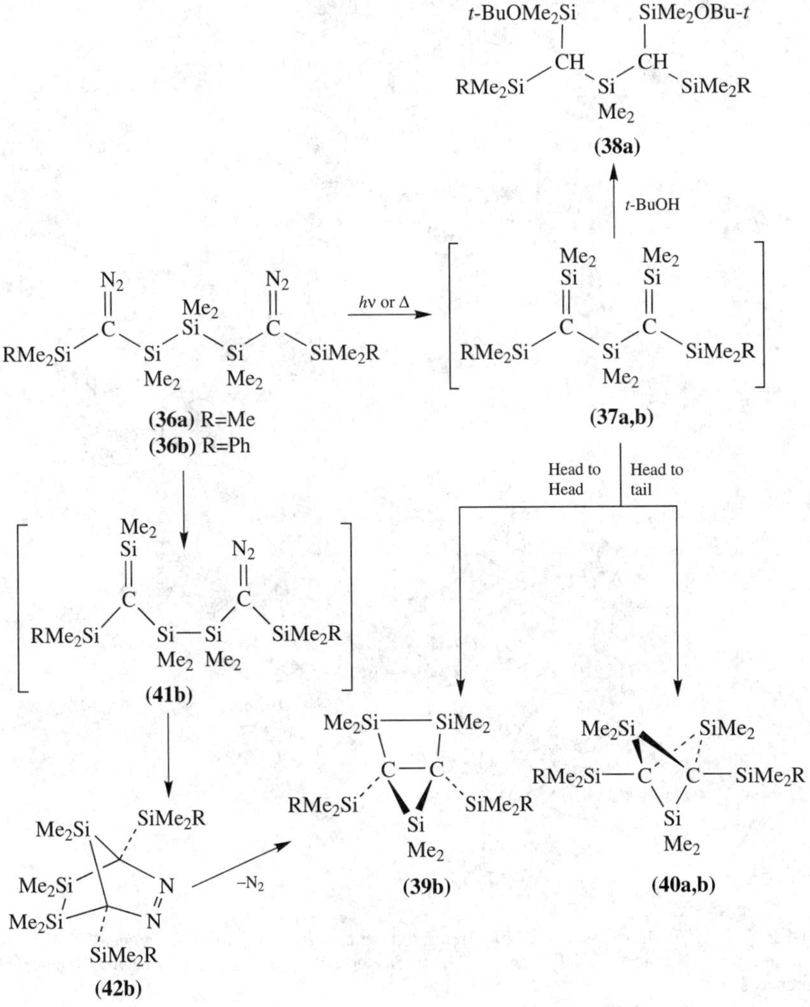

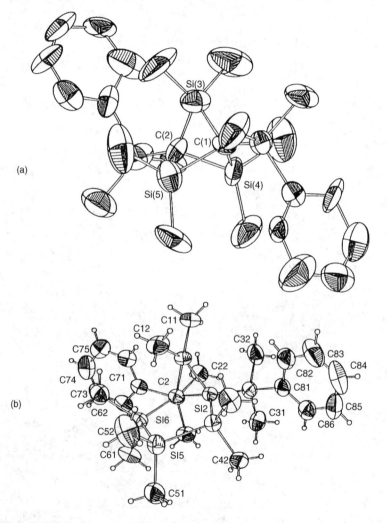

FIGURE 1. ORTEP drawings of (a) **40b** (reprinted with permission from Reference 31. Copyright 1991 American Chemical Society), and (b) **45** (reproduced by permission of Elsevier Science from Reference 32)

As long as the two silyldiazomethyl groups were linked with 1,1,2,2,3,3-hexamethyltrisilanylene unit (cf **36a,b**), the formation of product **39b** could be interpreted by both modes of reaction. However, if the number of dimethylsilanylene units was increased to four in **36c**, the two modes of reaction are expected to produce completely different ring frameworks. Actually, photolysis of **36c** under the same conditions and separation of the products by preparative TLC gave rise to tetrasilabicyclo[3.1.0]hexane (**45**) in 25% yield (equation 11)[32]. The three possible ring frameworks of tetrasilabicyclo[2.1.1]hexane (**40c**), tetrasilabicyclo[2.2.0]hexane (**39c**) and tetrasilabicyclo[3.1.0]hexane (**45**) require, respectively, four (4 : 4 : 2 : 2), three (4 : 4 : 4) and eight (2 : 2 : 2 : 2 : 1 : 1 : 1 : 1) resonances of SiMe groups in the ^{1}H-NMR spectra, based on their molecular symmetry. Indeed, compound **45** showed eight resonances of SiMe groups in the ^{1}H-NMR spectrum with four resonances in the ^{29}Si-NMR spectrum in which the most upfield shift at −54.1 ppm is characteristic of the silirane ring silicon. An X-ray analysis confirmed the structure of **45** (Figure 1b). Interestingly, compound **45** can be regarded as one of the most stable siliranes bearing methyl substituents on the silicon atom. Thus compound **45** is derived by nitrogen extrusion from the bicyclic azo compound (**42c**) and there is no need to invoke head-to-head [2 + 2] cycloaddition of 1,2,3,4-tetrasila-1,5-hexadiene (**37c**) to give tetrasila[2.2.0]hexane (**39c**), at least in the photochemical reaction (equation 11). Furthermore, flow pyrolysis of dilute solution of **36c** at 500 °C gave 9% of a compound **40c**, whose ^{1}H-NMR spectrum contains four resonances of SiMe group, which is consistent with tetrasila[2.1.1]hexane (**40c**). There was no indication that tetrasila[2.2.0]hexane (**39c**) was present among the pyrolytic products.

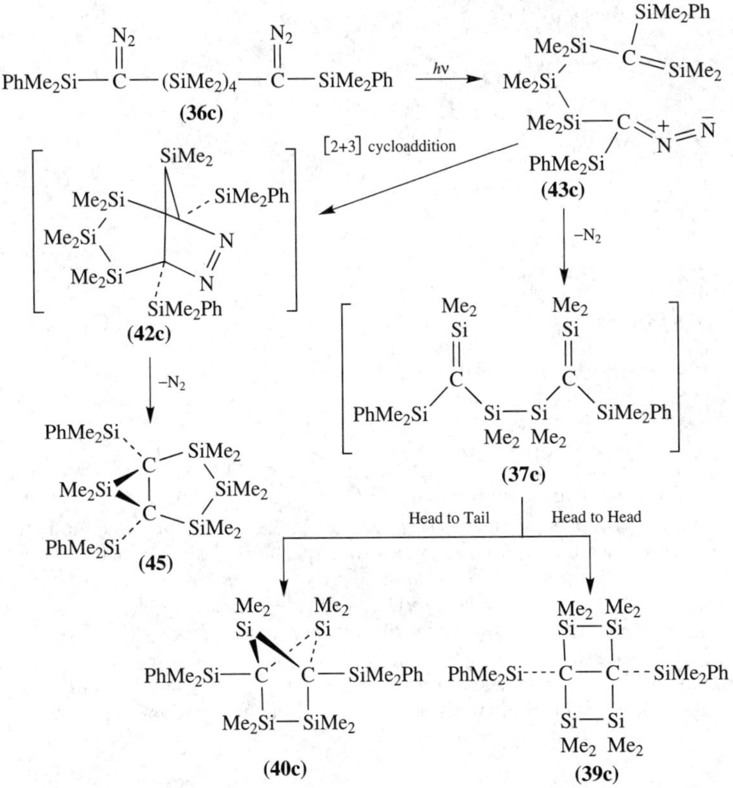

In conclusion, it is clear that the photochemical decomposition of bis(silyldiazomethyl) oligosilanes produces one silene moiety (**43c**) in the molecule, followed by a [2 + 3] silene–diazo cycloaddition, while thermal decomposition gives two silene moieties (**37c**) at the same molecule which undergo a [2+2] silene–silene cycloaddition in a head-to-tail (not head-to-head) fashion.

F. Silabenzene and 6-Silafulvenes

Both silabenzene (**46**) and 6-silafulvene (**47**) are attractive molecules in silicon chemistry. Thermally induced retro ene reactions to silabenzene and 6-silafulvene were explored by the groups of Barton and Sakurai[33−35]. One might expect that 2-silacyclohexadienylidene (**48**) or silacyclopentadienylcarbene (**49**) would give silabenzene (**46**) or 6-silafulvene (**47**) by a 1,2-migration of the substituent (equation 12). When a benzene solution of 1-diazo-2,2-dimethyl-3,4,5,6-tetraphenyl-2-silacyclohexane-3,5-diene (**50**) containing an excess of *tert*-butyl alcohol was refluxed in a sealed Pyrex tube at 100 °C, an adduct of *tert*-butyl alcohol (**51**) was formed in high yield (equation 13)[36]. The formation of **51** provides strong evidence for 6-silafulvene (**47a**) formed by migration of the dienyl group to the carbene center of **46a**. It seems reasonable that the dipolar form of the silafulvene would have enhanced importance due to the stability of the cyclopentadienyl anion. The evidence for the presence of silafulvene (**47a**) was further substantiated by a reaction with methanol and benzophenone to give adduct **52** which resembles **51**, and a Wittig type product **54**, (formed via **53**), respectively[37].

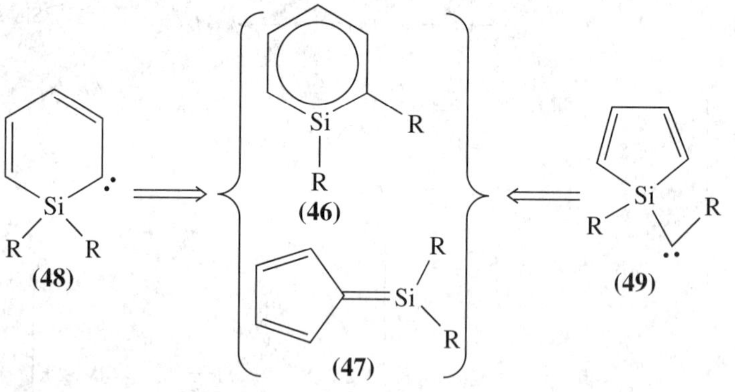

(12)

Photolysis of 1-methyl-2,3,4,5-tetraphenylsilacyclopentadienyldiazomethane (**55**) with an excess of methanol gave the diazirine **56** and the methanol adducts **59, 60** and **61** (equation 14)[38]. Products **59** and **60** appear to be derived from the methanolysis of silabenzene (**46b**), which could arise by a ring expansion of silacyclopentadienylmethylcarbene (**49b**). On the other hand, compound **61** is the evident product of methanolysis of 5-silafulvene (**57**) formed by a methyl shift in **49b**. Diazirine **56** was also found to produce silylcarbene under photochemical and thermal conditions and led to the same products **59,60** and **61**.

Further evidence for the generation of silabenzene (**46b**) and 5-silafulvene (**57**) as intermediates comes from their reaction with 2,3-dimethylbutadiene and benzophenone, which gave **58** and 1,1-diphenylpropene, respectively[37]. In an extension of this strategy Märkel and Schlosser succeeded later to stabilize silabenzene in solution at $-100\,^\circ\text{C}$[39].

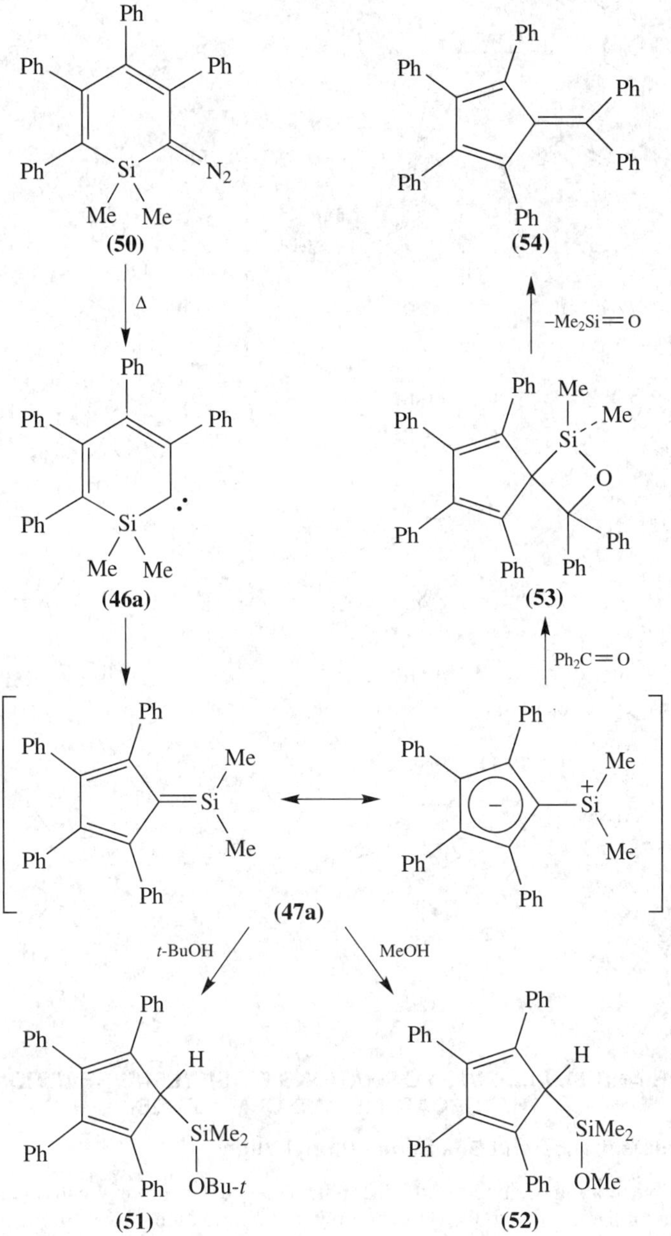

(13)

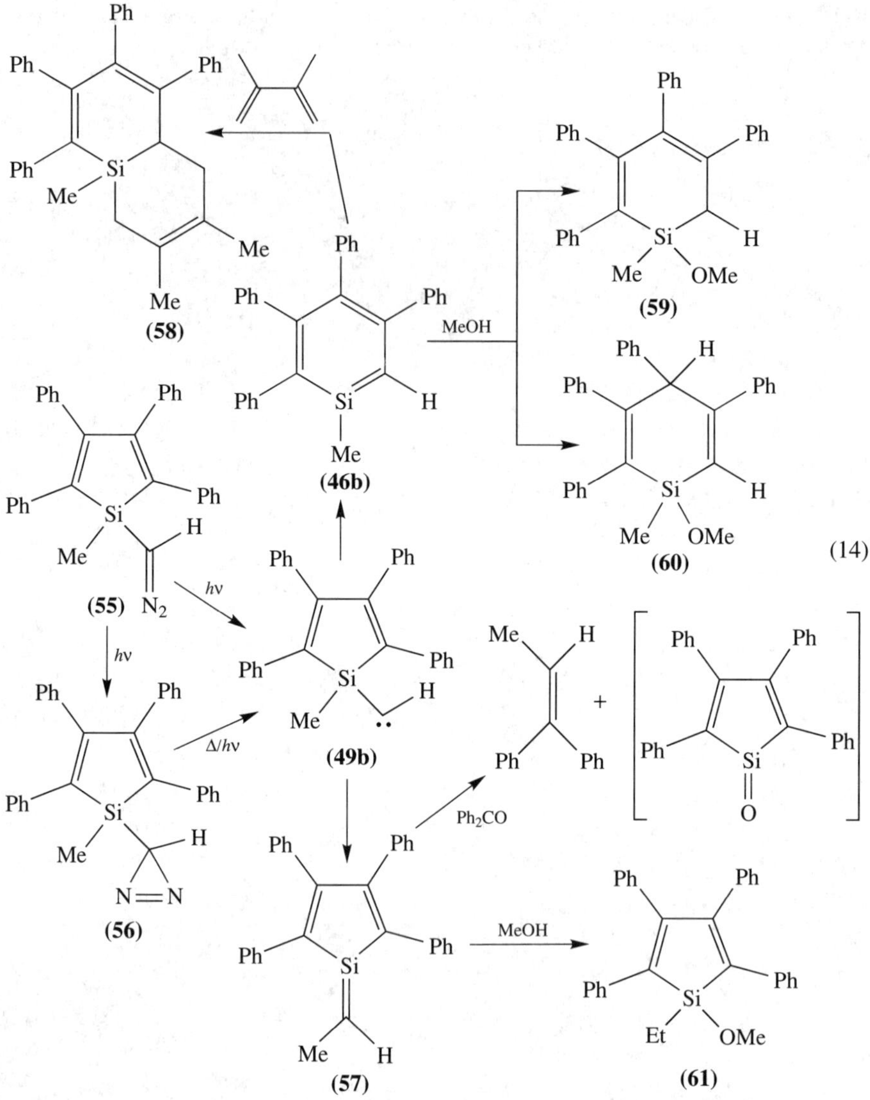

III. YLIDE AND SMALL-RING FORMATIONS BY SILYLENE REACTIONS WITH HETEROATOMS AND CUMULENES

A. Oxa(thia)siliranes and Sila(thio)carbonyl Ylides

Thermal generation of dimethylsilylene in the presence of carbonyl compounds leads to the addition of the silylene to the C=O bond with the formation of oxasilirane (e.g. **62** and **65**), followed by rearrangement[40]. Formation of tetraphenylethylene from benzophenone was attributed to carbenic fragmentation of the oxasilirane (**62**), whereas aliphatic ketones afford enol ether product **66** derived from oxasilirane intermediates (**65**). The formation of tetraphenylethylene by this mechanism is certainly not uniquely demanded by the data. Since the initial formation of ylide **63** leads to the same product (equation 15), oxasilirane **62** could thermally cleave the silicon–carbon bond to give 1,3-diradical followed by addition to benzene ring to give product **64**. Under milder conditions more direct evidence was found for the intermediacy of oxasiliranes. When 2-adamantanone (**67a**) and 7-norbornone(**67b**) were allowed to react with photochemically generated dimethylsilylene, the major products could be formulated as dimers (**69**) and insertion products (**70**) of oxasiliranes (**68**) (equation 16)[41]. Indeed **68** could be trapped with ethanol to give **71**. Dimethylsilylene reacts with α-diketones to yield 1,3-dioxa-2-silacyclopentane-4-ene derivatives (**72** and **73**) (equation 17)[42].

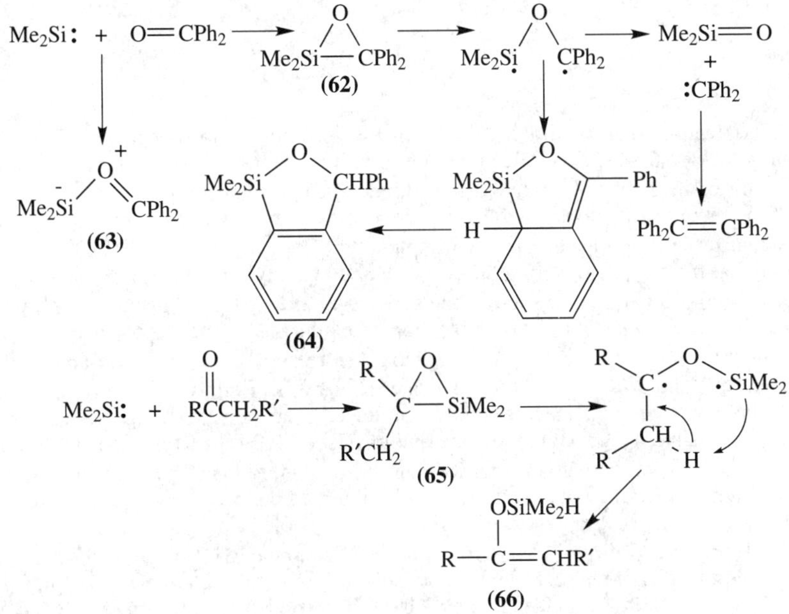

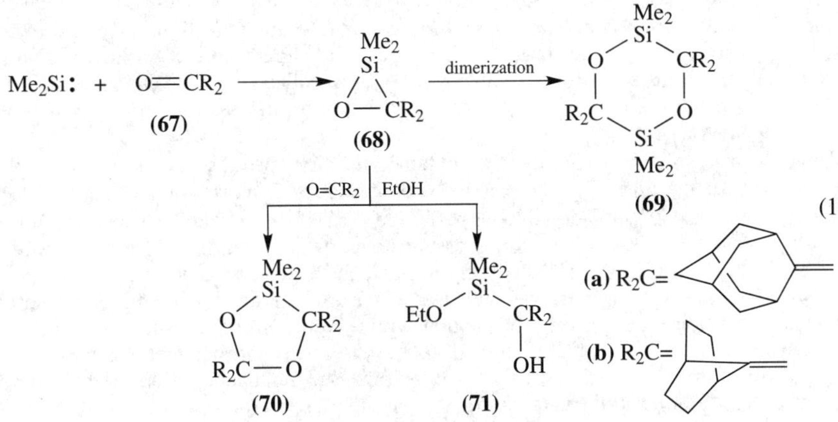

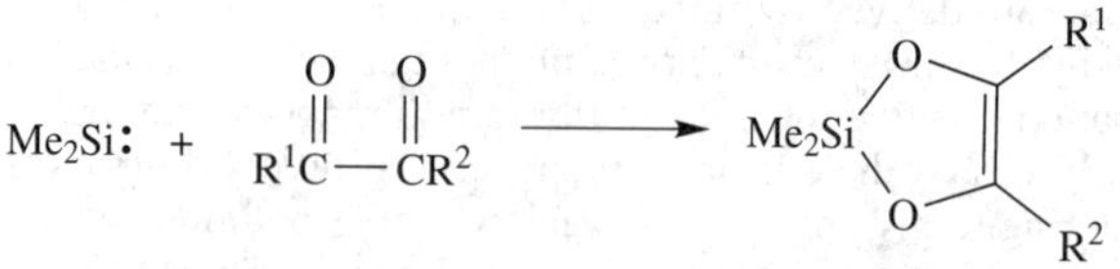

$Me_2Si:$ + $R^1\overset{O}{\overset{\|}{C}}-\overset{O}{\overset{\|}{C}}R^2$ ⟶ (72)

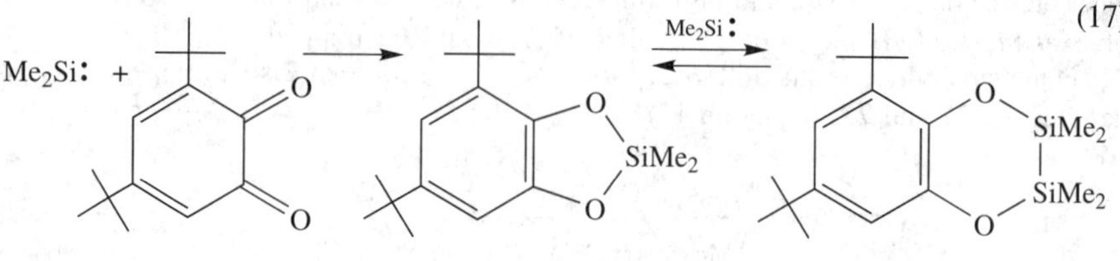

(73)

In 1980, Kumada and coworkers showed that when a bulky substituent, a mesityl group, was introduced onto the ring silicon atom of a silirene, atmospheric oxygen, moisture or alcohol surprisingly had no effect on the silirane at room temperature[43]. In the next year West and coworkers demonstrated that the dimesitylsilylene, isolated in matrix, dimerizes to a stable tetramesityldisilene at room temperature[44], while Masamune and coworkers reported that the employment of a bulky 2,6-dimethylphenyl group as substituents on silicon leads to the isolation of cyclotrisilane derivatives, the first of their kind[45]. From these studies, the simple concept of steric protection with bulky substituents provided significant progress in the small-ring chemistry of silicon. Guore and Barton[46], Tzeng and Weber[47] and Tortorelli and Jones[48] proposed the intermediacy of a sila-ylide in the reaction of dimethylsilylene with epoxides, which then produced silanone or rearranged. We observed that dimesitylsilylene generated by the photolysis of 2,2-dimesitylhexamethyltrisilane reacted with epoxides to give dimesitylsilanone-epoxide adducts (74) (equation 18)[49]. It seems that oxasilirane could be stabilized by the protection of a long silicon bond and by bulky substituents. Stable oxasilirane (75a) was isolated by the reaction of photochemically generated dimesitylsilylene with 1,1,3,3-tetramethyl-2-indanone (76a)[50]. The sulfur analogue, thiasilirane (75b), was also isolated from the reaction of dimesitylsilylene with 1,1,3,3-tetramethyl-2-indanethione (76b). Whereas adamantanone (67a) yielded with dimesitylsilylene 2 : 1 adduct (70a) instead of oxasilirane, adamantanethione (67c) gave a stable thiasilirane (75c) (equation 19)[51]. Oxasilirane (75a) and thiasiliranes (75b and 75c) were inert to oxygen and moisture and their structures were finally confirmed by X-ray crystal analyses (Figure 2). Belzner and coworkers succeeded later in isolating oxasilirane in the reaction with diarylsilylene, which is generated by thermolysis of cyclotrisilanes with adamantanone (67a)[52].

Using matrix isolation techniques, we obtained the first spectroscopic evidence for the intermediacy of the hitherto unknown silacarbonyl and silathiocarbonyl ylides in a low temperature matrix[53]. Irradiation of the oxasilirane (75a) in an isopentane/3-methylpentane (Ip/3-Mp) matrix at 77 K with a low-pressure mercury lamp led to the appearance of a new band at 610 nm in the UV-vis spectrum and the matrix became interestingly blue in color. This absorption band was stable at 77 K on prolonged standing. However, it immediately disappeared on brief irradiation with a Xenon lamp (λ >460 nm) or when the matrix was allowed to melt. This colored species was independently generated by the reaction of dimesitylsilylene with 1,1,3,3-tetramethyl-2-indanone (76a). Furthermore, irradiation of 2,2-dimesitylhexamethyltrisilane in the presence of 76a in IP/3-MP at 77 K with a low-pressure mercury lamp produced a colored species with a λ_{max} at 573 nm due to dimesitylsilylene[53]. The absorption bands at 610 and 420 nm increased in intensity in the dark at almost the same rate as the decrease of the band at 573 nm, showing that dimesitylsilylene reacts with **76a** or dimerizes to tetramesitylsilene (420 nm) in the matrix at 77 K. Irradiation of the red solution of the 610 nm species, with a wavelength of light greater than 460 nm or melting the matrix, resulted in the production of the oxasilirane (**75a**). On the basis of these results, it is quite reasonable to assume that the colored intermediate is the silacarbonyl ylide (**77a**). Similar spectroscopic evidence was gathered for the formation of silathiocarbonyl ylide (**77b**) (485 nm) from **75b** and **76b** (equation 20).

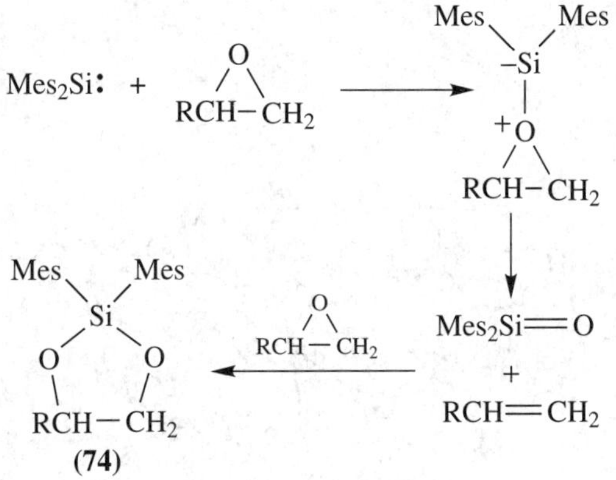

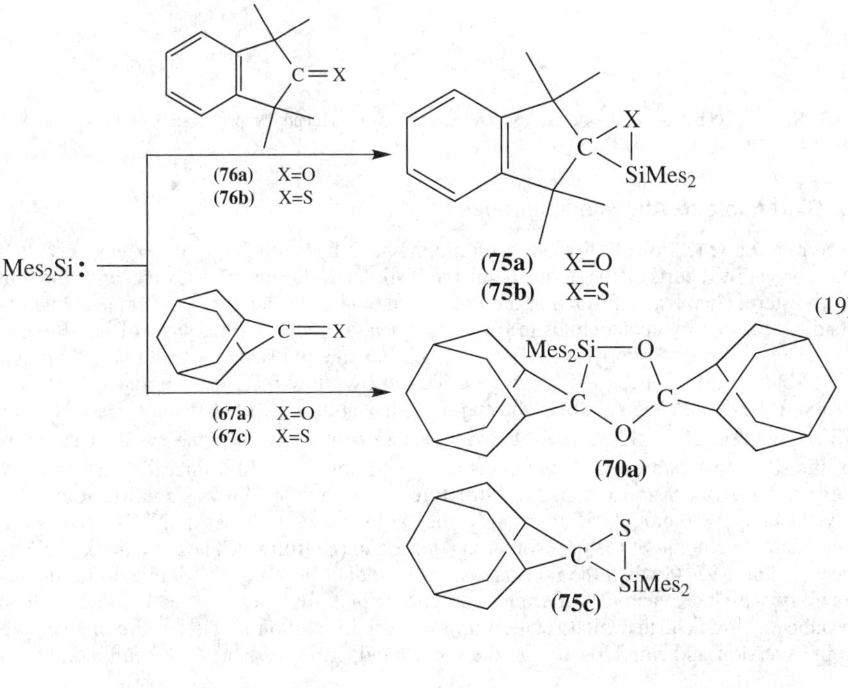

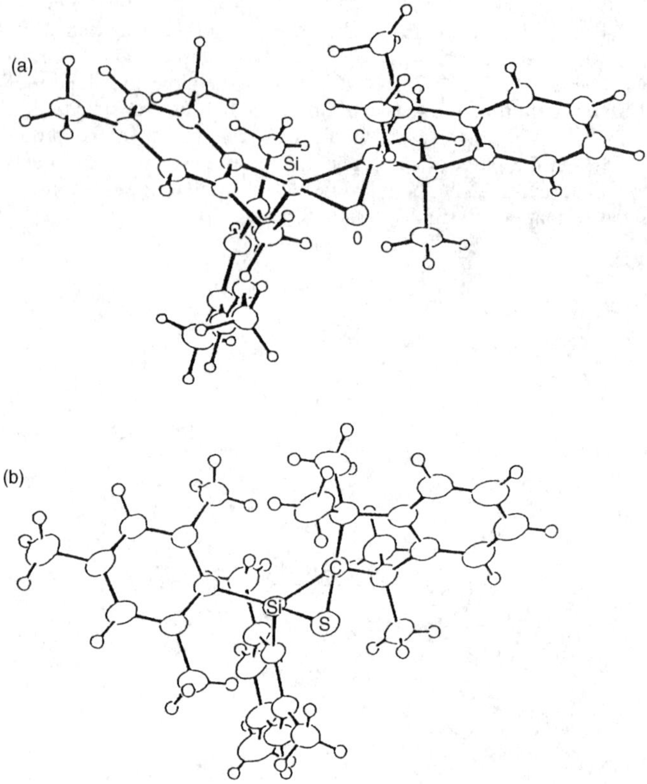

FIGURE 2. ORTEP drawings of (a) **75a** and (b) **75b** (reproduced by permission of Pergamon Press from References 50 and 51)

B. Siliranes and Alkylidenesiliranes

While the reactions of silylenes with olefins have been studied quite extensively, there have been few reports on the stereochemistry of silylene addition. The scarcity and ambiguity of the stereochemical information are due to the extreme instability of siliranes which are readily attacked by nucleophiles to give ring-opened products. Photolysis of 2,2-dimesityl or 2,2-bis(2,4,6-triisopropylphenyl)-1,1,1,3,3,3-hexamethyltrisilanes and a large excess of (Z)-2-butene in hexane at −5 °C was carried out by a low-pressure mercury lamp to give (Z)-silirane (**Z-78a** and **Z-78b**) as the major product and 1-7% of (E)-isomer (**E-78a** and **E-78b**) (equation 21)[54]. Siliranes **Z-78a,b** and **E-78a,b** are extremely stable and the structures of the (Z)-siliranes are easily distinguished by ^{1}H and ^{13}C NMR since **Z-78a** and **Z-78b** have nonequivalent aryl groups. The structure of products **Z-78b** was confirmed by X-ray crystal analysis (Figure 3a). Surprisingly, the photolysis of trisilanes in (E)-2-butene gave a considerable amount of the (Z)-isomer in addition to (E)-silirane. The most marked change occurred at −95 °C when the nonstereospecific adduct becomes the dominant product. In analogy with the reactivity of triplet carbene, triplet silylenes may add to the olefin to produce a diradical in which rotation around carbon–carbon single bond competes with spin inversion and ring closure. On the other hand, if rotation prevailed sufficiently over

42. Small-ring monosilacycles and medium-ring oligosilacycles

closure, one would expect that a stepwise addition of singlet silylene might occur.

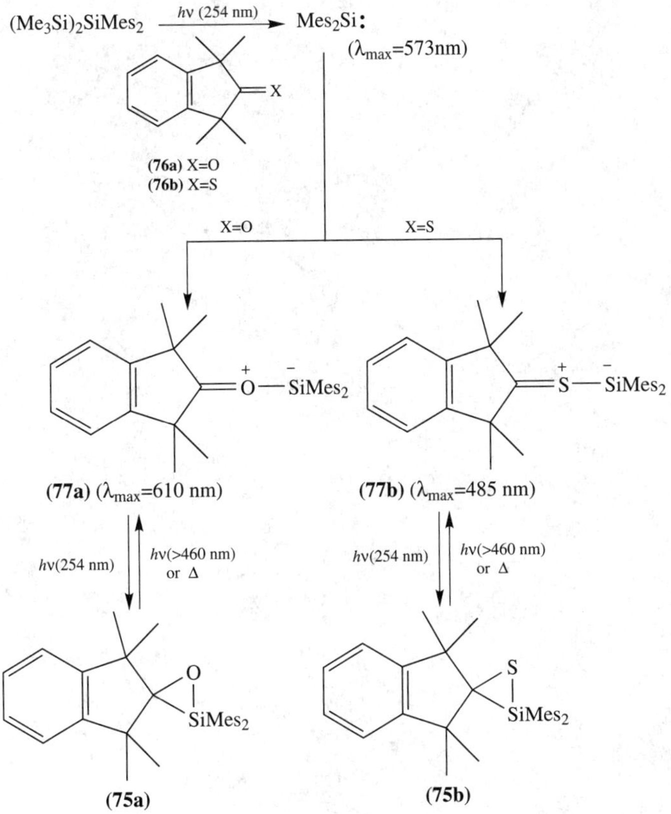

$(Me_3Si)_2SiMes_2 \xrightarrow{h\nu\ (254\ nm)} Mes_2Si:$ (λ_{max}=573 nm)

(76a) X=O
(76b) X=S

(77a) (λ_{max}=610 nm)
(77b) (λ_{max}=485 nm)

$h\nu(254\ nm)$ ⇅ $h\nu(>460\ nm)$ or Δ

(75a)
(75b)

(20)

$Ar-\underset{\underset{SiMe_3}{|}}{\overset{\overset{SiMe_3}{|}}{Si}}-Ar \xrightarrow{h\nu} Ar-\ddot{Si}-Ar$

(Z-78)   (E-78)

(a) Ar = Mes

(b) Ar = 2,4,6-i-Pr$_3$C$_6$H$_2$

	Z-78	E-78
(a)	30%	70%
(a)	99	1
(b)	41	59
(b)	93	7

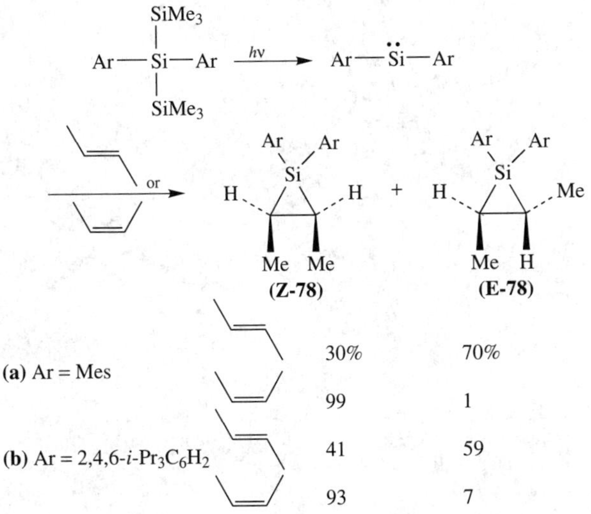

(21)

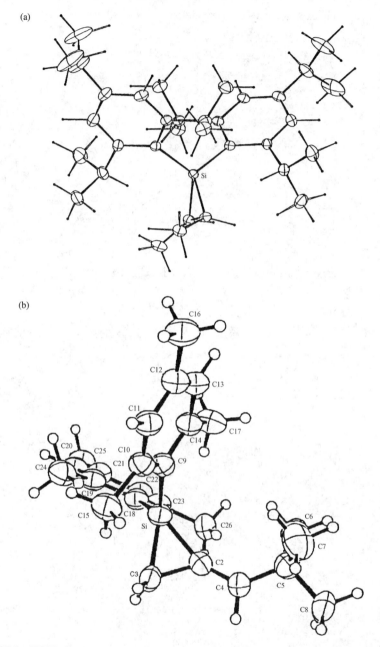

FIGURE 3. ORTEP drawings of (a) **Z-78b** (reprinted with permission from Reference 54. Copyright 1988. American Chemical Society), (b) **Z-79** (reproduced by permission of Pergamon Press from Reference 64) and (c) **91** (reprinted with permission from Reference 70. Copyright 1993 American Chemical Society)

(c)

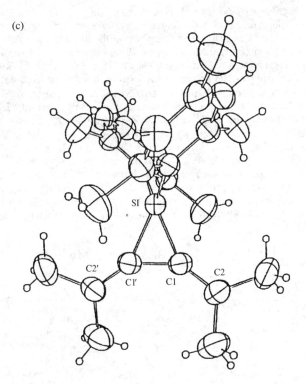

FIGURE 3. *(continued)*

In contrast, Conlin and coworkers reinvestigated our experiments and demonstrated that photogenerated dimesitylsilylene adds stereospecifically to other olefins with more than 98% retention[55,56]. They reinterpreted our nonstereospecific addition to (E)-2-butene on the basis that the (E)-2-butene used by us may have contained small amounts (< 5%) of the (Z)-isomer. Because (Z)-2-butene is significantly more reactive than the (E)-isomer, (E)-2-butene containing 2.3% (Z)-isomer actually gave 40% (Z)- to 60% (E)-isomer of the silirane adduct. More recently, Boudjouk and coworkers[57], and Gaspar and coworkers[58] reported that stereospecific addition by bulky silylenes also took place.

Methylenecyclopropane is known to have an extra strain of 11 kcal mol^{-1} (theory)[59] and 13 kcal mol^{-1} (experiment) over cyclopropane[60]. Alkylidenesiliranes (**E/Z-79** and **80**) were prepared by photolysis of 2,2-dimesitylhexamethyltrisilane in the presence of *tert*-butyl allenes[61–64]. Whereas these alkylidenesiliranes were stable in air at temperatures below 80 °C for at least one day, under photochemical conditions they rapidly decomposed. The photolysis of (isolated) **81** gave (E/Z)-3-silyl-1,3-hexadienes (**82**) which can be understood in terms of the intermediacy of a silatrimethylenemethane (**83**) (equation 22). The structure of **Z-79** was verified by X-ray crystal analysis (Figure 3b). Seyferth and coworkers found that aldehydes, ketones and olefins reacted with hexamethylsilirane and silirene with a two-atom insertion into the SiC rings to give silacyclopentanes and silacyclopentene[65–67]. The (E/Z)-alkylidenesilirane (**E/Z-**

79) showed high reactivity toward benzaldehyde in solvents other than methanol (equation 23). These reactions were initiated by nucleophilic attack on silicon atom, followed by Si—C bond cleavage. The selectivity in the direction of bond cleavage might depend on the (Z) and (E) regiochemistry of **79** and the nature of the nucleophile. The Si—vinyl bond cleavage for the formation of **84** and **88** is quite surprising since opening of alkylidenecyclopropane and its hetero analogue were found to proceed with the cleavage of C—C or C—hetero bond which is not bonded to the sp^2 center. A Si—C(sp^3) bond cleavage of **E-79** with simultaneous π-bond migration seems to afford **87**. Reaction with methanol gave **85** and **86**. The reaction of **Z-79** with oxygen-transfer reagents gave attractive 1,2-oxasiletane products **89** and **90** (equation 24)[62]. The different synthetic routes for 1,2-oxasiletane were mentioned in Section II.C. It is especially noteworthy that the formation of 1,2-oxasiletanes is regiospecific depending on the oxidants. Both the 1,2-oxasiletane products were thermally stable but were found to undergo thermal [2+2] cycloreversion reactions to produce silanone and allene (equation 24). More recently, Boudjouk and coworkers succeeded in showing that the reaction of di-*tert*-butylsilirane with elemental sulfur results in the formation of a stable thiasiletane via a single-atom insertion reaction[68,69].

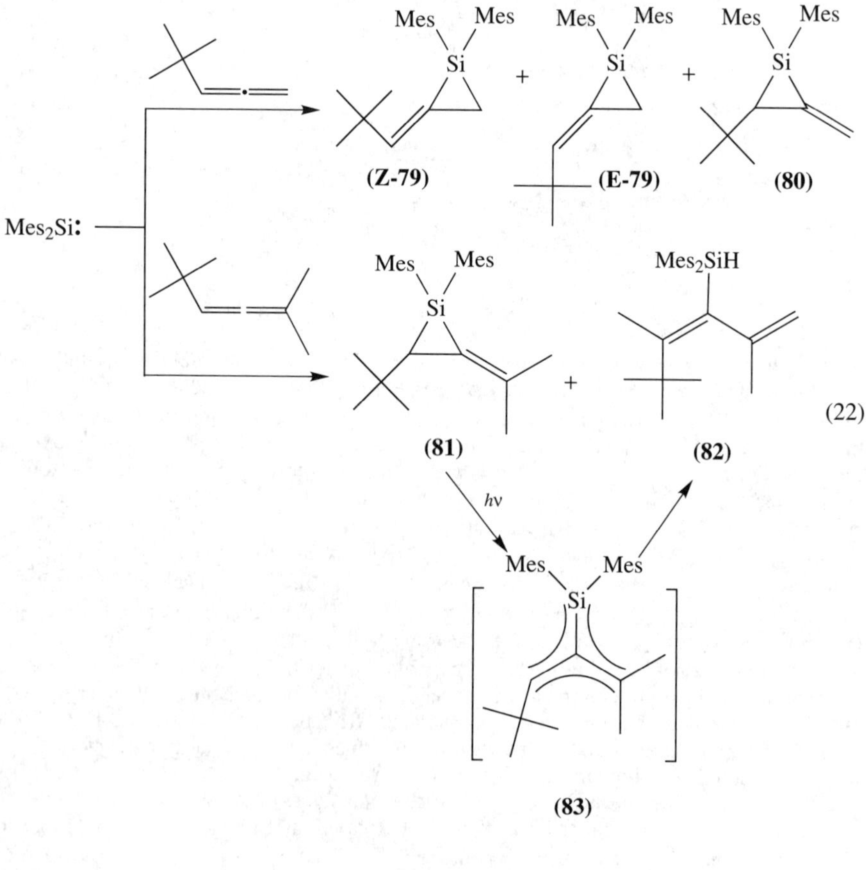

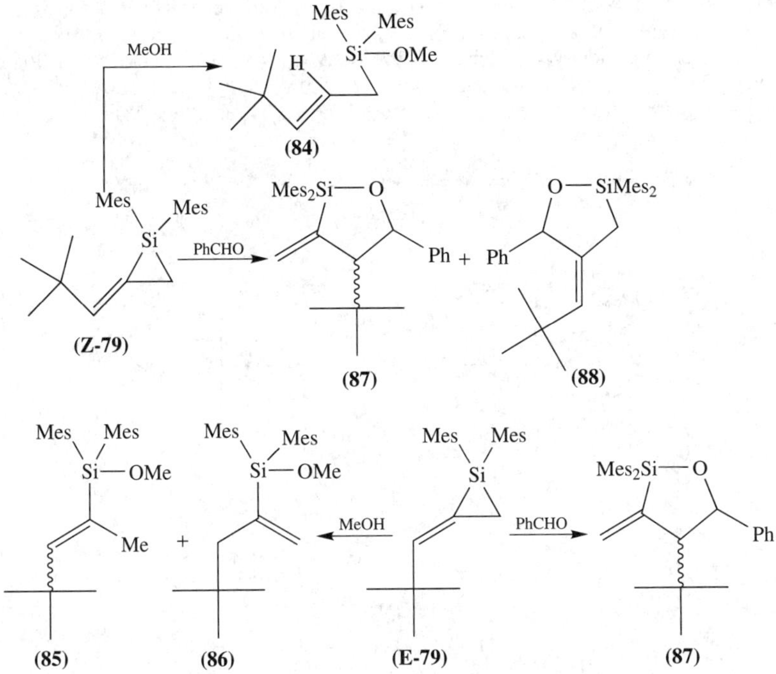

As a logical corollary, we have tried the reaction of dimesitylsilylene with 1,2,3-butatrienes, in the hope of an efficient route to the bisalkylidenesiliranes. In fact, the photolysis of 2,2-dimesitylhexamethyltrisilane and an excess of tetramethylbutatriene followed by methanolysis produced **94** together with trace amounts of **95** and **96** (equation 25)[70]. The structures **94** and **95** are consistent with the intermediacy of bisalkylidenesiliranes **91** and the allenic silirane **92**. Actually, in the absence of methanol, **91** can be isolated as a fairly stable compound, while low yields and hygroscopic instability precluded the isolation of **92**. In the photolysis of **91**, silatrimethylenemethane (**93**) is assumed to be initially formed. It then undergoes a 1,4-hydrogen shift to produce **96**.

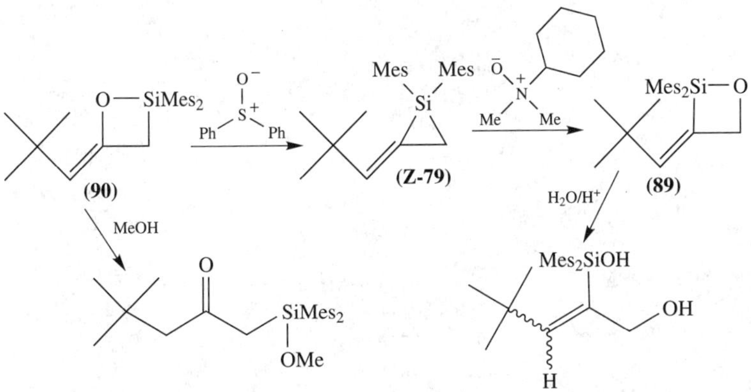

The structure of **91**, a new bisalkylidenesilirane (sila[3]radialene) system, was confirmed by X-ray structure analysis (Figure 3c). Because of the high kinetic stability of **91** due to steric protection, aldehydes such as benzaldehyde as well as ketones did not react. However, [2 + 4] cycloaddition chemistry of **91** with 4-methyl-1,2,4-triazoline-3,5-dione resulted in the formation of a new ring-fused silirene (**97**) (equation 26).

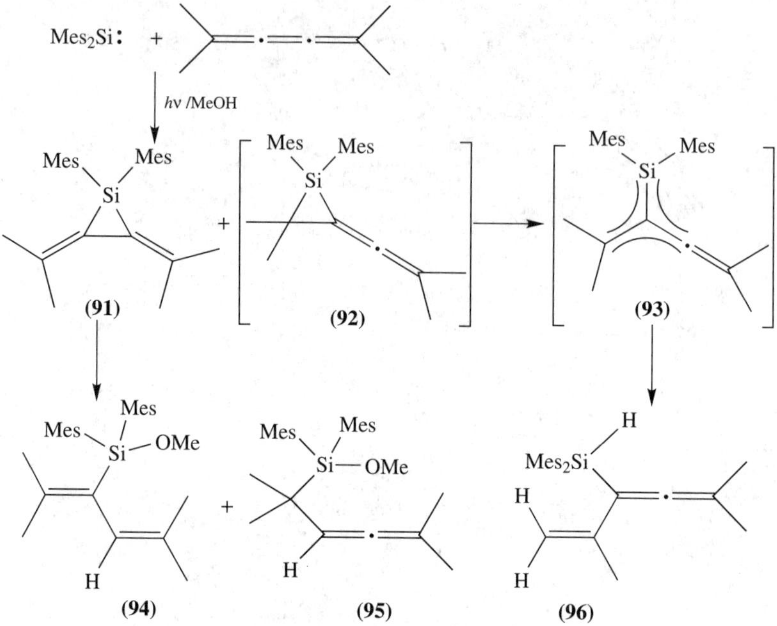

(25)

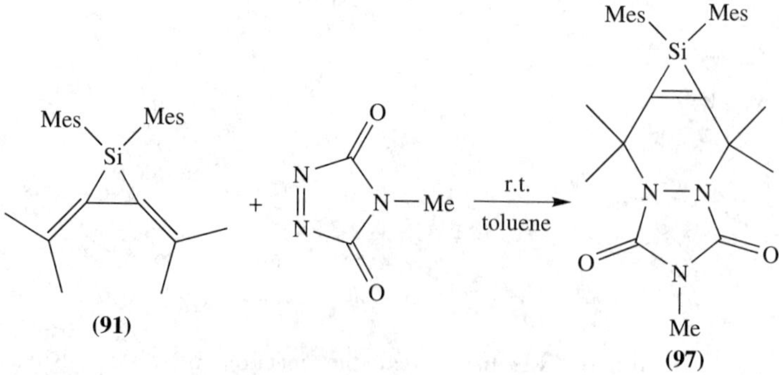

(26)

The reaction of methylenecyclopropanes with transition metal complexes is well known to promote a catalytic $\sigma-\pi$ cycloaddition reaction with unsaturated compounds, in which a trimethylenemethane complex might exist[71-76]. Recently, much interest has been focused on the interaction of strained silicon–carbon bonds with transition metal complexes. In particular, the reaction of siliranes with acetylene in the presence of transition metal catalysts was extensively investigated by Seyferth's and Ishikawa's groups[77-79]. In the course of our studies on alkylidenesilirane, we found that palladium catalyzed reaction of **Z-79** and **E-79** with unsaturated compounds displayed ring expansion reaction modes that depend on the (Z) and (E) regiochemistry of **79** as well as the nature of the unsaturated compounds[63]. In the reactions of **79** with electron-deficient olefins such as alkyl acetylenedicarboxylate and methyl methacrylate, insertion into Si—C bonds to form **98, 99** and **102** as well as silole formation (**101**) but no cyclization to **100** took place (equation 27). When electron-rich olefins or diphenylacetylene were used as reagent, no adduct with **Z-79** was formed; instead **103** and **104** were formed via elimination and subsequent addition of *tert*-butylallene (equation 28). In these reactions, no evidence for the existence of silatrimethylenemethane complex was obtained. Nevertheless, the formation of a hydrosilane (**82** and **96**) in the photolysis of alkylidenesiliranes does suggest the intermediacy of silatrimethylenemethane. In order to clarify this situation we tried the stoichiometric reaction of alkylidenesiliranes (**79** and **91**) with $Fe_2(CO)_9$. Silatrimethylenemethane (**E/Z-105** and **106**) was found to be attached to metal ligands in a η^4 manner[80,81]. The structures were confirmed by X-ray analysis (Figure 4). The alkylidenesiliranes retain their structural integrity in the complexation. The silatrimethylenemethane ligand is bound in a η^4-fashion, pyramidalized and staggered to the three carbonyl ligands in an analogous manner to that observed in the structurally characterized trimethylenemethane and η^4-heterotrimethylenemethane complexes[82–84]. The bond distances C(1)—Si, C(1)—C(2) and C(1)—C(3) are 1.840(8), 1.42(1), and 1.46(1) Å, respectively, values which are between C—Si or C—C single and double bonds. Especially notable is that the C(1)—Si distance is somewhat longer than the values found in the other silene complexes[85–87]: $Cp^*RuHPCy_3(\eta^2\text{-}CH_2=SiMe_2)$, 1.78–1.79 Å; $Cp^*IrPMe_3(\eta^2\text{-}CH_2=SiPh_2)$, 1.810 Å; $Cp_2W(\eta^2\text{-}CH_2=SiMe_2)$, 1.800 Å. The cross-conjugative interaction with the C—C double bond may be responsible for the slightly longer values of the C(1)—Si bond. When $Ru_3(CO)_{12}$ was used as the metal carbonyl, ruthenium analogues of (**E/Z-108**) were obtained along with an unexpected trinuclear ruthenium cluster (**107**), which demanded a unique metal bound CO-insertion into the silirane ring (equation 29)[81].

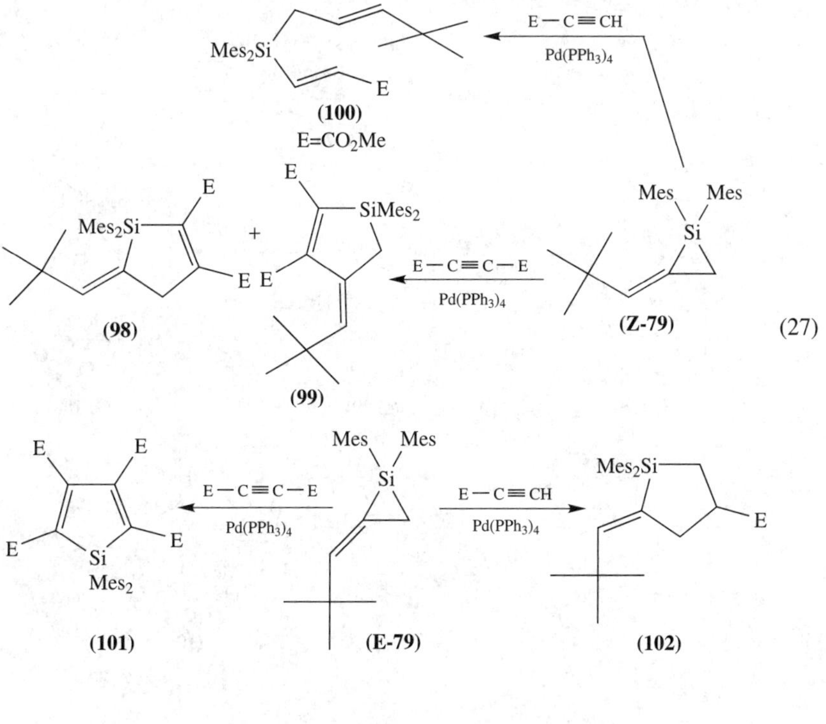

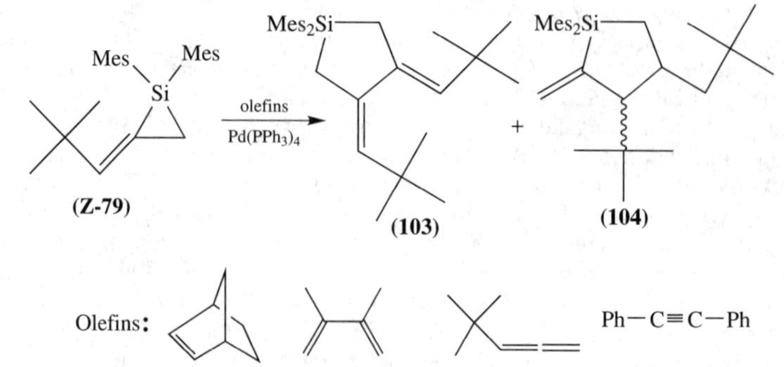
Mes
Mes
Si
olefins
Pd(PPh3)4
Mes2Si
Mes2Si
+
(Z-79)
(103)
(104)
Olefins:
Ph—C≡C—Ph

(28)

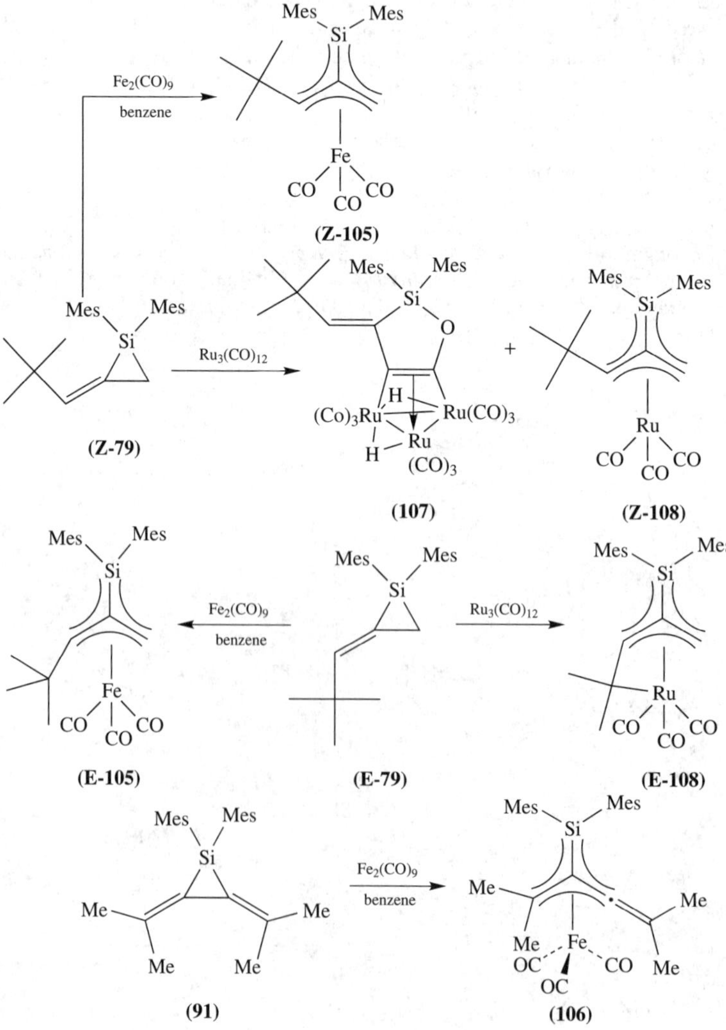
Fe2(CO)9
benzene
Mes
Mes
Si
Fe
CO
CO
CO
(Z-105)
Ru3(CO)12
O
H
H
(Co)3Ru
Ru(CO)3
Ru
(CO)3
+
Ru
(Z-79)
(107)
(Z-108)
(E-105)
(E-79)
(E-108)
Me
(91)
(106)
OC

(29)

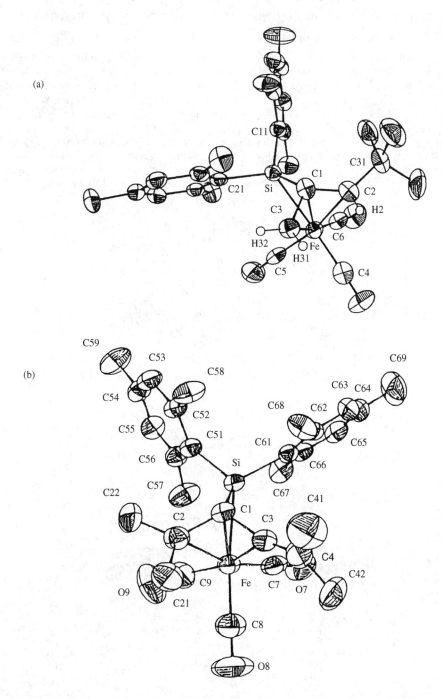

FIGURE 4. ORTEP drawings of (a) **Z-105** and (b) **106** (reprinted with permission from Reference 81. Copyright 1994 American Chemical Society)

IV. RING FORMATION BY WURTZ AND DIYNE COUPLING REACTIONS
A. Heptasila[7]paracyclophane

[n]Paracyclophanes have unique properties resulting from bending of the benzene ring. The smallest [n]paracyclophane so far isolated is the [6]isomer, first prepared by Jones and coworkers[88]. Recently Bickelhaupt, Tobe and coworkers succeeded in spectroscopic characterization of [5]paracyclophane[89], which is stable at low temperature in solution, but not isolable; more recently, a [4]paracyclophane system has been proposed as a reactive intermediate[90]. After many attempts, the Wurtz coupling of 1-chloro-6-*para*-chlorodimethylsilylphenyl-1,1,2,2,3,3,4,4,5,5,6,6-dodecamethylhexasilane **(109)** (prepared by equation 30) with sodium in refluxing toluene in the presence of [18]crown-6 produced heptasila[7]paracyclophane **(110)** in 1.6% yield, as colorless crystalline compound which is stable under atmospheric oxygen and moisture even when heated to its melting point (equation 31)[91].

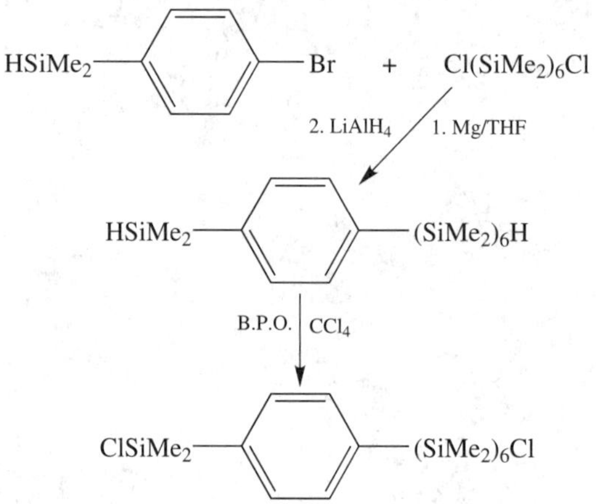

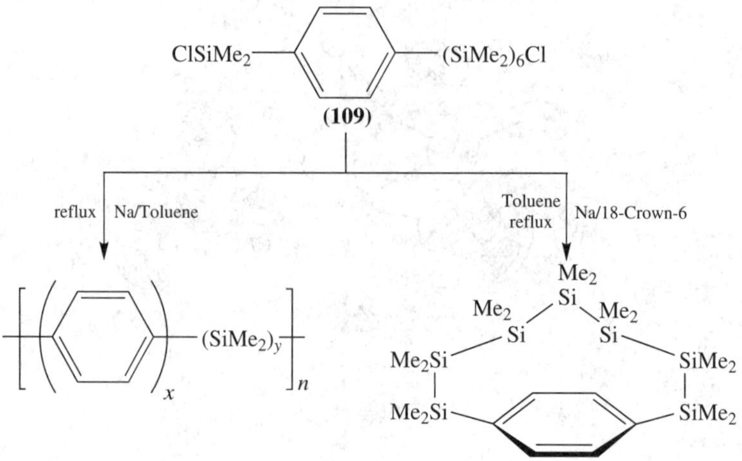

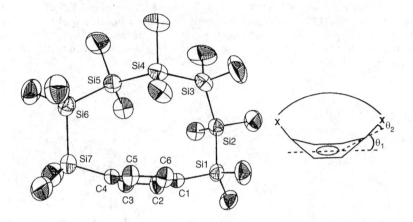

FIGURE 5. ORTEP drawing of **110** (reproduced by permission of VCH from Reference 91)

The presence of [18]crown-6 is essential for the formation of **110** and only polymeric substances were obtained in the absence of [18]crown-6 (equation 31). X-ray structural determination established unequivocally the molecular structure of **110** (Figure 5). Although the Si−Si bond lengths are normal and ranged between 2.338(5)–2.371(4) Å, there is an interesting alternation in the bond angles around the polysilane chain; the bond angles of C1−Si1−Si2 (105.5(3)°) and C4−Si7−Si6 (107.4(3)°) are contracted while those of Si1−Si2−Si3 (117.7(2)°), Si2−Si3−Si4 (126.5(2)°), Si3−Si4−Si5 (116.2(2)°), Si4−Si5−Si6 (115.6(2)°) and Si5−Si6−Si7 (113.7(2)°) are considerably expanded from the normal bond angles. The most interesting point in the structure of **110** is the deformation of the benzene ring from planarity, which is represented by the deviation angles θ_1 and θ_2. The deformation angles θ_1 of **110** are 6.6° and 4.5° and θ_2 are 6.5° and 9.0°, which are comparable with those observed for octamethyltetrasila[2,2]paracyclophane ($\theta_1 = 4.3°$, $\theta_2 = 15.0°$)[92] and its chromium complex ($\theta_1 = 2.5, 2.2°$, $\theta_2 = 10.0, 10.9°$)[93]. These θ values are smaller than those of [7]paracyclophane derivatives ($\theta_1 = 17°$, $\theta_2 = 23.5°$)[94] and those of [8]paracyclophane derivatives ($\theta_1 = 9°$, $\theta_2 = 15°$)[95]. These observations are best explained by the higher flexibility of the conformation in a polysilane chain compared with a methylene chain, involving an expansion of Si−Si−Si angles. Unfortunately, the low yield of **110** prevented further photochemical contraction to lower homologous [n]paracyclophanes (see Section V.B).

B. Bisalkylidenesiletanes

1,2-Dialkylidenecyclobutane possesses a large strain energy compared with cyclobutane (26.5 kcal mol^{-1})[96]. Tetramethyleneethane, or 2,2'-bis(allyl) diradical, has been postulated to be an intermediate in the thermal dimerization of allenes to form 1,2-dialkylidenecyclobutane. A small-ring system involving a silicon–silicon bond is interesting because of its expected high strain energy. Only a few examples concerning the formation of 1,2-disilacyclobutanes[97–100], 1,2-disilacyclobutenes[101–105], bisalkylidene-1,2-disilacyclobutane[106,107] and benzo-1,2-disilacyclobutene[108] have been reported.

Treatment of dilithiated tetramethylbutatriene with appropriate halosilane gave bissilanes that were chlorinated with PdCl$_2$/CCl$_4$. Wurtz coupling of the resulting bis(chlorosilane)s with sodium in toluene afforded the novel 3,4-dialkylidene-1,2-disilacyclobutanes (**111**), as oxygen- and moisture-sensitive liquids (equation 32). While **111a** is readily polymerized and must be stored as a dilute solution in order to avoid polymerization **111b** is stable at room temperature for several months, even in the neat form[109,110].

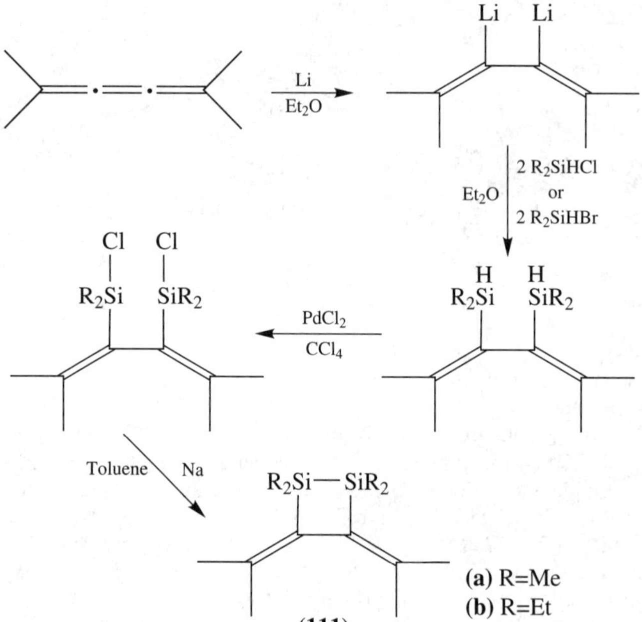

$$\text{(32)}$$

(a) R=Me
(b) R=Et

(**111**)

Transition metal catalyzed double silylation of C−C multiple bonds with disilanes is one of the most remarkable developments of organosilicon chemistry and numerous reports have dealt with this topic. Nevertheless, very little is known about the related disilane metathesis reaction in which, formally, a Si−Si bond is double silylated[111−117]. Diisopropylidenedisilacyclobutanes **111a** in the presence of a catalytic amount of Pd(PPh$_3$)$_4$ was cleanly converted into **112** in 93% yield (equation 33). When an equimolar mixture of **111a** and benzo derivatives **113** was treated with catalytic amounts of Pd(PPh$_3$)$_4$, the corresponding cross-metathesis products **115** were obtained, respectively, with high selectivity, in addition to minor amounts of homo-metathesis products **114**[110]. A favored formation of the cross metathesis product **116** was also observed in the conversion of a mixture of **113a** and **113b**, though with a lower degree of selectivity (equation 34). The structure of homo and cross metathesis products were confirmed by X-ray (Figure 6). Double silylation product (**117**) was obtained in the reaction of **111a** and excess alkynes in the presence of a catalytic amount of Pd(PPh$_3$)$_4$. In the case of diphenylacetylene, disilane metathesis and double silylation products were observed in the reaction, which showed moderate selectivity. The yield of **117** increased, if the ligands in catalyst were changed from the triphenylphosphines to benzonitrile ligands. The product distribution of disilane metathesis (**112**) and acetylene insertion (**117**) depends on the electrophilicity of the acetylenes, since in the case of dimethyl acetylenedicarboxylate mono-adduct **117** and a di-adduct **118** are obtained, and in the case of phenylacetylene only a mono-adduct **117**

42. Small-ring monosilacycles and medium-ring oligosilacycles

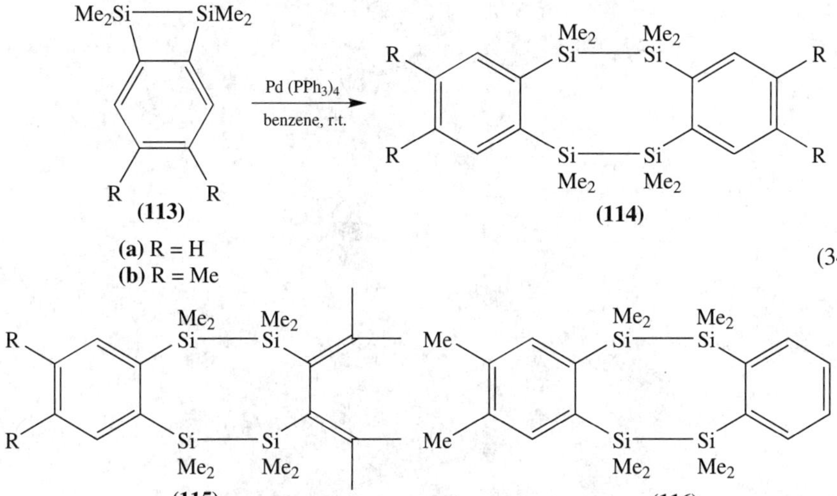

is obtained (equation 35).

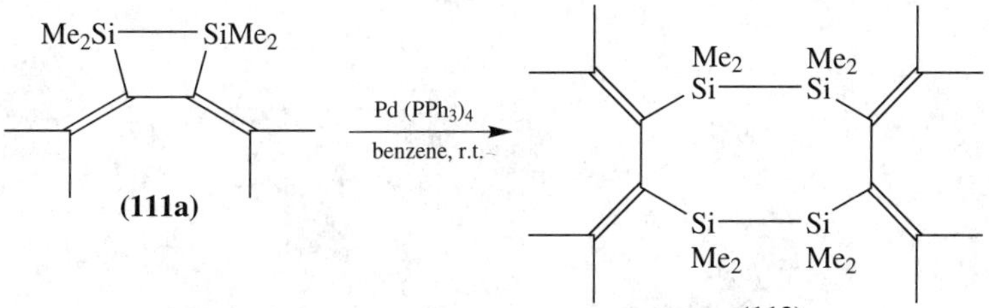

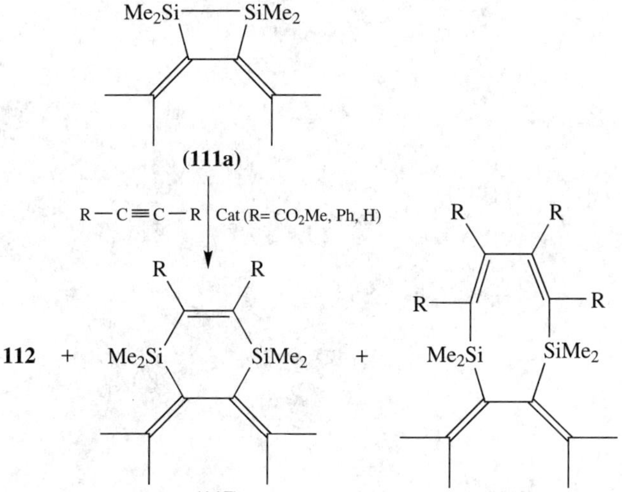

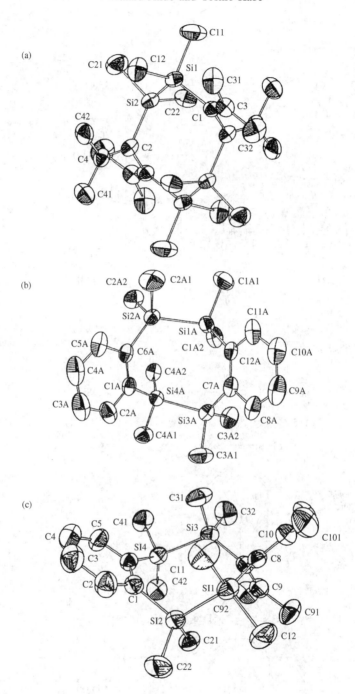

FIGURE 6. ORTEP drawings of (a) **112**, (b) **114a** and (c) **115a** (reprinted with permission from Reference 110. Copyright 1995 American Chemical Society)

2,3-Dimethyl-1,3-butadiene and benzaldehyde were unsuccessful in competing with a disilane metathesis reaction even if 5 equivalents of the unsaturated compound were employed in the presence of Pd(PPh$_3$)$_4$ catalyst at room temperature for 30 min, since only **112** was formed from **111a** in 74 and 80% yields, respectively. In contrast, Pt(CH$_2$=CH$_2$)(PPh$_3$)$_2$ catalyst (80 °C, 3 h) promoted a double silylation of **111a** to give **119** (41%), **120** (10%) and **121** (60%). Of special interest, 2,5-dimethyl-2,3,4-hexatriene moderately inhibited a disilane metathesis reaction to give **112** (15%) and 2,3,5,6-tetraisopropylidene-1,1,4,4-tetramethyl-1,4-disilacyclohexane (**122**) (68%) in the presence of Pd(PPh$_3$)$_4$. The latter was obtained as a 3 : 1 mixture of two conformational isomers, which do not interconvert at room temperature. *tert*-Butylallene also inhibited disilane metathesis to give 85% of **123** in the presence of Pd(PPh$_3$)$_4$. It is evident that the reaction involves a common intermediate with a double silylation reaction, i.e. bis(silyl)palladium(II) (**124**) (equation 36)[118,119]. Such an intermediate is followed by an insertion of unsaturated compounds. The selectivity of these pathways depends on the reactivity of the unsaturated compounds.

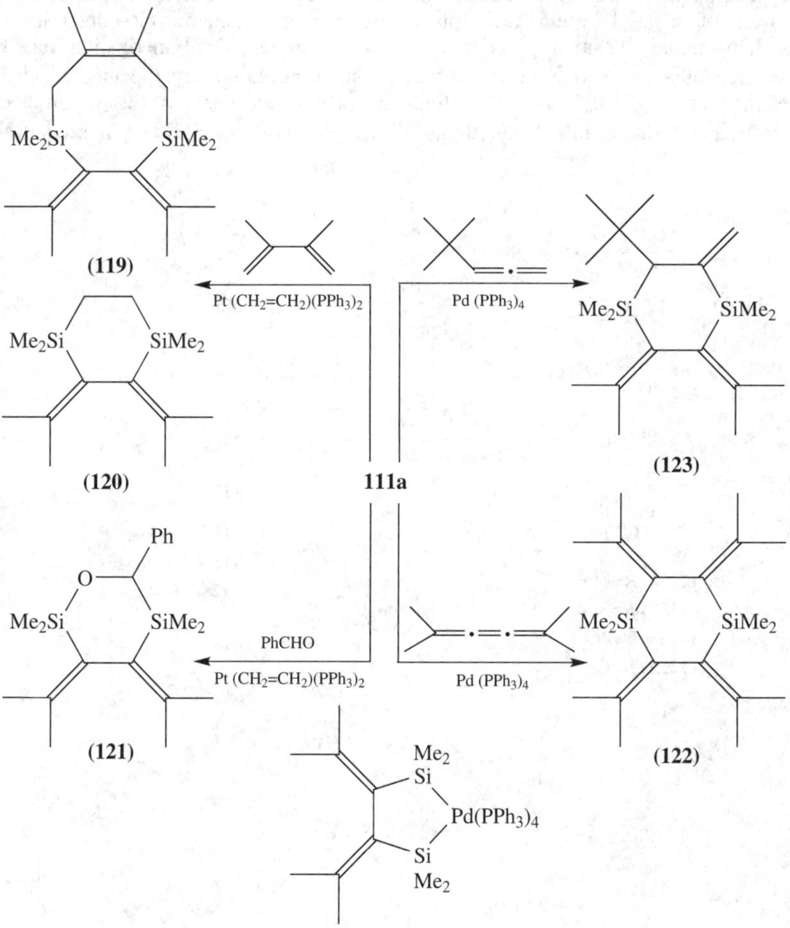

(36)

The group 4 metallocenes Cp_2M (M = Ti, Zr) induced bicyclization of terminally substituted α,ω-diyne is a powerful ring-forming method[120,121]. During the course of our study on bisalkylidene-1,2-disilacyclobutane, we have succeeded in the zirconocene-mediated synthesis of 1,2-disilacyclobutanes from 1,2-bis(ethoxyethynyl)disilane. Summarizing our attempts, we described a zirconocene-mediated small-ring formation which is dependent on the terminal substituents of the diyne. The small silacycles obtained in this route undergo unique spontaneous ring-opening polymerization (ROP).

Generation of a zirconocene transfer reagent $[Cp_2Zr(Bu\text{-}n)_2]$ (by addition of 2 equivalents of *n*-BuLi to Cp_2ZrCl_2 at $-78\,^{\circ}C$; Negishi's method) in the presence of 1,2-bis(1-propynyl)disilane **(125a)** resulted in the quantitative formation of a black air-sensitive polymer **126** (equation 37). Spectroscopic data of polymer **126** closely correspond to those for the model complex 2,5-bis(pentamethyldisilanyl)-3,4-dimethyl zirconacyclopentadiene **(127)**. Hydrolysis of **126** with CF_3CO_2H gave colorless polymer of high molecular weight (MW = 5780, MW/Mn = 1.8). These results are consistent with a regiospecific intermolecular coupling to afford 2,5-zirconacyclopentadiene rings in the polymer-backbone (equation 37). Once the polymer **126** was heated in THF, depolymerization followed by 1,2-migration of a silyl group took place to give a zirconabicyclo-containing 1,2-disilacyclopentene **(128a)** in moderate yield. A similar depolymerization has been applied recently to synthesize macrocyclic-containing zirconacyclopentadienes[122,123], while the same 1,2-silyl migration finds a recent precedent in the intramolecular zirconocene coupling of bis(phenylethynyl)dimethylmonosilane[124]. For trimethylsilyl and

(126) (128) (125) (129) (127)

(125a) R=Me
(125b) R=SiMe$_3$
(125c) R=Ph
(125d) R=OEt

(128a) R=Me
(128b) R=SiMe$_3$
(128c) R=Ph

(37)

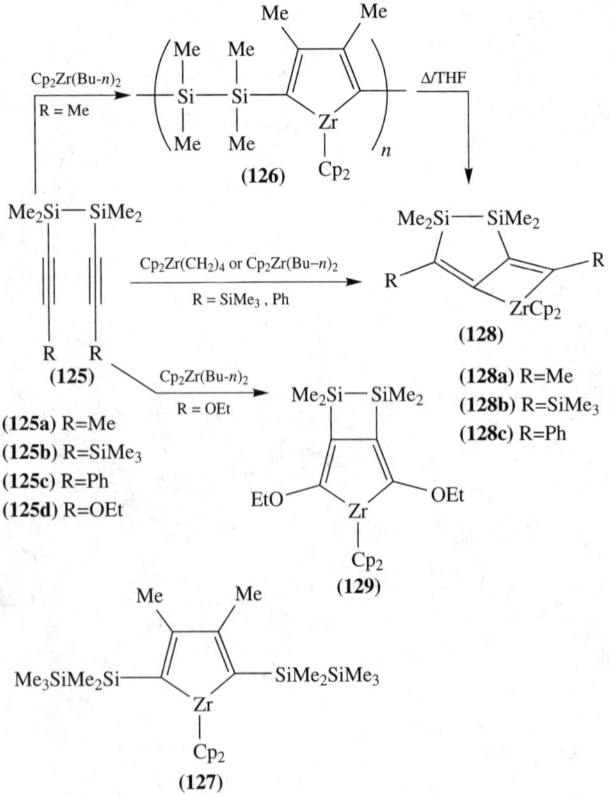

42. Small-ring monosilacycles and medium-ring oligosilacycles

phenyl-substituted bisalkynyldisilane (**125b,c**), the reaction with $Cp_2Zr(CH_2)_4$ at 40 °C and with $Cp_2Zr(Bu-n)_2$ at room temperature underwent direct intramolecular coupling to give 1,2-disilacyclopentene (**128b,c**) quantitatively (equation 37). When ethoxy groups which are strongly ligating to zirconium metal were employed as terminal substituents, the 1,2-migration was completely inhibited and the zirconabicycle **129**, which has a symmetrical 1,2-tetramethyldisilacyclobutane framework according to ^1H, ^{13}C and ^{29}Si NMR spectra, was obtained[125].

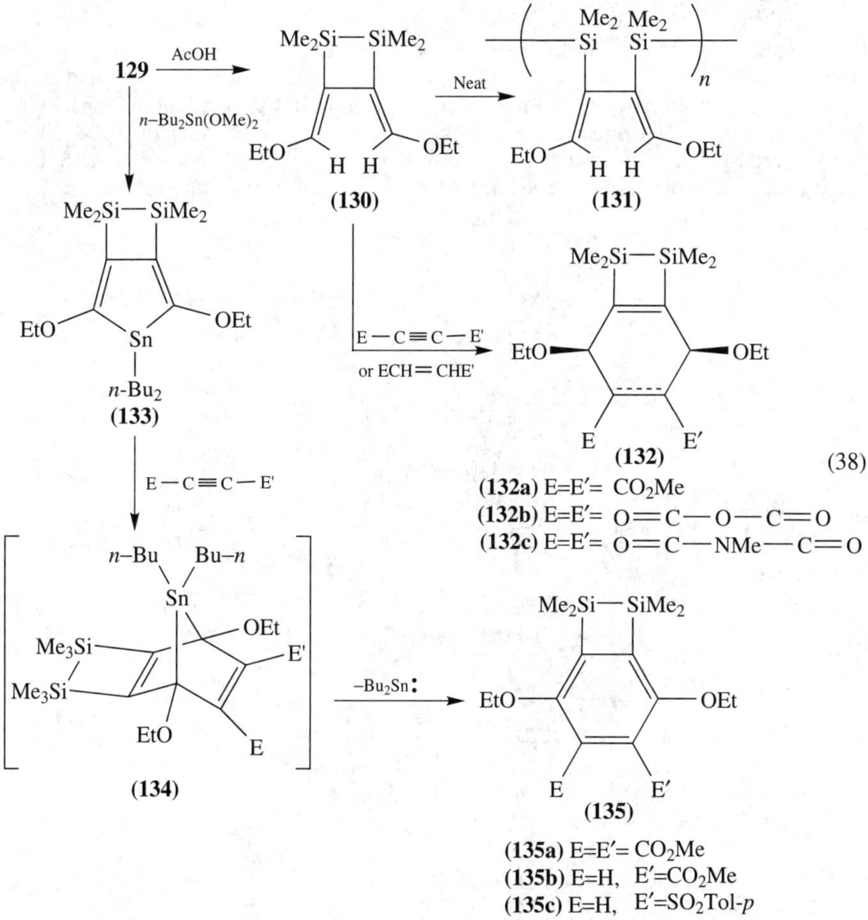

Further structural proof for **129** was demonstrated by the chemical derivatization illustrated in equation 38. First, hydrolysis of **129** with AcOH gave 3,4-bisalkylidene-1,2-disilacyclobutane **130** as an air-sensitive colorless oil in 78% yield. It is noteworthy that **130** showed a strong tendency to undergo a spontaneous ROP reaction and removal of the solvent is sufficient to initiate polymerization and to get a high molecular weight polymer (**131**; Mw = 1.8×10^5, Mw/Mn = 2.1). The relatively narrow line widths observed in both ^1H and ^{13}C NMR spectra of **131** strongly suggest that the polymer backbone of this material has a regular structure. As compound **130** also possesses 1,4-diethoxydiene moiety, Diels–Alder additions of activated dienophiles of **130** proceeded smoothly at room temperature to give adducts **132a, 132b** and **132c** in 32, 57 and 56% yields, respectively. Second, the conversion of readily prepared zirconacyclopentadienes into the corresponding main-group heterocycles by the Fagan–Nugent method has been shown to be extremely versatile[126]. Among various types of main-group electrophiles, treatment of **129** with one equivalent of dibutyltin dimethoxide[127] afforded quantitatively the expected stannole (**133**) based on ^{1}H and ^{13}C NMR spectra. Without isolation, stannole **133** was brought into contact with activated dienophiles. A Diels–Alder reaction readily occurred in moderate yields at room temperature due to activation by the strongly electron-donating ethoxy groups. Although the intermediate, 7-stannanorbornadiene adducts (**134**) could not be detected[128,129], the products obtained were 3,4-benzo-1,2-tetramethyldisilacyclobutanes (**135a-c**), which can be regarded as the dibutyl stannylene extrusion product from the intermediates (**134**). Compound **135a** was characterized by X-ray diffraction (Figure 7). This is important since the first 3,4-benzo-tetramethyldisilacyclobutane was synthesized in 1986[108], and so far no structural details could be obtained because of its polymerizability by a ROP reaction. The four-membered Si_2C_2 ring of compound **135a** is almost planar. The Si(1)–Si(2)–C(2) and Si(2)–Si(1)–C(1) angles have fairly acute values, such as 75.5° and 75.6°. Very similar values are observed for isolated 3,4-bisalkylidene-1,2-tetraphenyldisilacyclobutane (75.6° and 74.5°)[106] and 1,2-tetraaryldisilacyclobutene (75.4° and 74.7°)[104,105], respectively. Thus, the angle strain is responsible for the polymerizability of 1,2-tetramethyldisilacyclobutanes. Further inspection reveals that the conformational mobility of the two *ortho*-ethoxy groups of **135** could inhibit a new bond formation at the four-membered Si_2C_2 ring in a lateral direction. Actually **135a-c** were inert toward ROP reaction even at 180 °C. However, vertical insertion into the Si–Si bond could be accommodated, and indeed **135a-c** were still sensitive toward air oxidation.

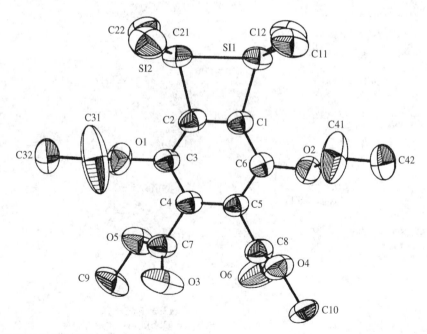

FIGURE 7. ORTEP drawing of **135a**

V. RING FORMATION BY KIPPING AND GILMAN REACTIONS
A. trans-Trisilacycloheptene and [3.1.1]Trisilapropellane

The first target molecules in these studies were considered to be the medium-ring trans-silacycloalkenes, which are a subject of considerable interest. trans-Cyclooctene is well known as the smallest trans-cycloalkene that is stable at room temperature. trans-Cycloheptene has been generated at room temperature and its existence was confirmed by chemical trapping with isobenzofuran or acidic methanol and also by ^{1}H NMR spectroscopy[130–133]. Even trans-cyclohexane has been generated[134]. However, isolation of trans-cycloheptenes has been unsuccessful, except for homoadamant-3-ene which is stabilized by a bulky substituent[135], because of the low thermal stability caused by the strained structure. Cis-1,2-diphenyl-4,4,5,5,6,6-hexamethyl-4,5,6-trisilacycloheptene (cis-**136**) was prepared by the reaction of 2,3-diphenyl-1,3-butadiene with Cl(SiMe$_2$)$_3$Cl in the presence of Mg in 23% yield (equation 39). When the benzene-d$_6$ solution of cis-**136** was irradiated with a low-pressure mercury lamp through a quartz NMR tube, ^{1}H NMR measurements showed that the resonance for cis-**136** decreased in intensity and new resonances (three singlets at −0.29, 0.37 and 0.39 ppm and an AB quartet at 1.76 and 3.22 ppm, $J = 14.0$ Hz) had appeared. After purification and crystallization this new compound, trans-**136**, was obtained in 13% yield[136]. The molecular structure of trans-**136** was determined by X-ray crystal diffraction (Figure 8). The most interesting structural feature is the twisting around the C=C double bond. The torsion angle was found to be 32.7°. Photoisomerization reaction of trans-**136** gave cis-**136**. In this isomerization, a photoequilibrium exists and the photostationary trans/cis ratio is 0.99. This value is higher than that obtained in the photosensitized isomerization of trans-cycloheptene.

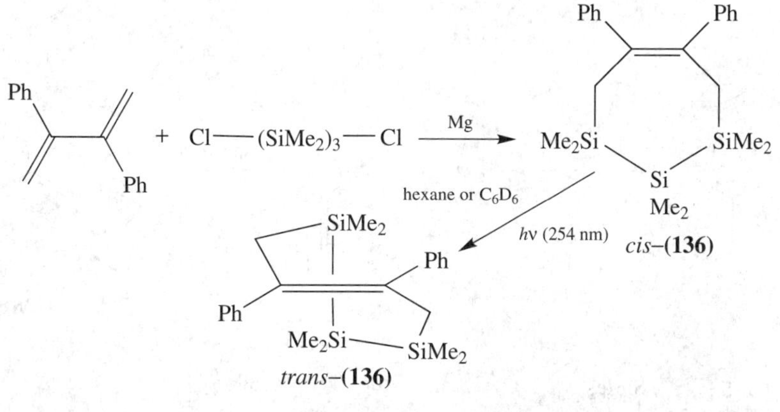

(39)

The next target molecule was a small-ring propellane such as [n.1.1]propellane, which have been of considerable interest because of the extremely high reactivity attributed to 'inverted tetrahedral' geometry at the bridgehead carbons[137,138]. The first [3.1.1]trisilapropellane (**138**) was prepared in 23% yield by the treatment of dilithium compound **137** with Cl(SiMe$_2$)$_3$Cl[139,140]. The air-sensitive compound **138** was also labile to acidic contamination such as silica gel in chromatography or HCl present in CHCl$_3$ and quantitatively rearranged to **140**. It is likely that the intermediate leading to **140** is the cation **139**, which is doubly stabilized by cyclopropyl and silyl substitution (equation 40). Cation **139** is probably attacked by chloride ion with concomitant rearrangement to the homoallyl skeleton of **140**. The ease of this reaction seems to be responsible for the liability of **138** toward acids.

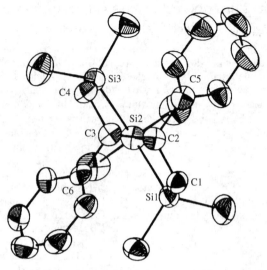

FIGURE 8. ORTEP drawing of *trans*-136 (reprinted with permission from Reference 136. Copyright 1991 American Chemical Society)

(40)

B. Oligosilabridged Cyclic Acetylenes

The last target molecule was the small silacycloalkyne ring, which is a subject of considerable interest because of bond-angle strain at the sp carbons[141]. In this series, various examples of silicon-containing cycloalkynes have been reported so far[142–148]. Cycloheptynes with their seven carbons are generally unstable at room temperature[149,150]. However, cycloheptynes which contain one sulfur or silicon atom in the skeleton are thermally stable[151–153]. This difference in the stability of medium-ring acetylenes is due to the longer C−S or C−Si bond lengths compared with the C−C bond. Generation of some six-membered cyclic acetylenes has been achieved either by matrix isolation studies or by trapping and oligomerization experiments. However, no example of isolation of a cyclohexyne has been reported. Treatment of Cl(SiMe$_2$)$_6$Cl with acetylene di-Grignard reagent under dilute conditions gave **141** as a stable colorless liquid in 46% yield. The photolysis of an hexane solution of **141** with a low-pressure mercury lamp in the presence of triethylsilane gave the lower homologous pentasilacycloheptyne (**142**) in 22% yield as colorless crystals (equation 41). This compound is stable to atmospheric oxygen and moisture[154]. The cycloheptyne (**142**) could also be obtained in 18% yield by thermolysis of **141**. The molecular structure of **142** could be established by X-ray diffraction (Figure 9). The Si−Si, Si−C single and C≡C triple bond lengths are normal, ranging between 2.34–2.35, 1.80–1.84 and 1.22 Å, respectively. However, there is an interesting alternation in the bond angle around the polysilane chain: the bond angles C(2)−Si(1)−Si(2) (100.4°) and C(1)−Si(5)−Si(4) (101.7°) are contracted, angles Si(1)−Si(2)−Si(3) (108.7°) and Si(3)−Si(4)−Si(5) (109.9°) are normal while the angle Si(2)−Si(3)−Si(4) (117.4°) is considerably expanded compared with the normal bond angles. The most interesting point in the structure of **142** is of course the bending of the acetylene bond from a linear geometry, as represented by the deviation angles θ_1 and θ_2. The bending angles θ_1 and θ_2 of **142** are 20.4 and 17.8°, respectively, values which are comparable with those observed in the nonsubstituted cyclooctyne ($\theta = 21.5°$) and cyclononyne (19.8°). Apparently, the structure of **142** reveals that the magnitude of the bending in acetylene decreases, but that the increase in the deformation of the polysilane chain is definitely noticeable. These observations are explained by the high flexibility of the conformation in the polysilane chain.

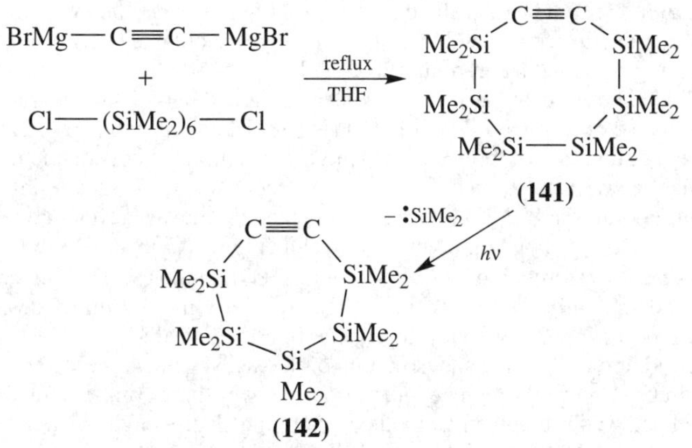

$$\text{BrMg—C≡C—MgBr} + \text{Cl—(SiMe}_2\text{)}_6\text{—Cl} \xrightarrow{\text{reflux THF}} \textbf{(141)} \xrightarrow[h\nu]{-:\text{SiMe}_2} \textbf{(142)} \tag{41}$$

The photoirradiation of **141** in the presence of six equivalents of triethylsilane provided the ring-contracted pentasilacycloheptyne **142** in 22% yield, together with a small amount of bicyclic compounds **143** and **144** in yields of 0.8 and 0.6%, respectively

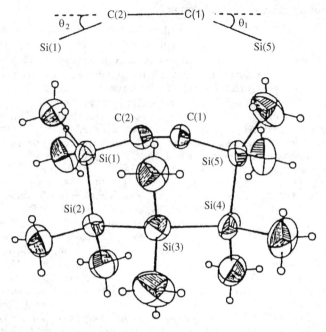

FIGURE 9. ORTEP drawing of **142** (reproduced by permission of Pergamon Press from Reference 154)

(equation 42)[155]. The structure of **144** was determined by X-ray crystallographic analysis. The bicyclic olefin **143** is considered to be formed by the oxidation of **145** after the addition of dimethylsilylene to **147**, which is generated in turn by the 1,2-silyl migration of **141**. The bicyclic olefin **144** is also considered to be formed by the oxidation of **146**, which is generated by the addition of dimethylsilylene to **148** after a 1,2-silyl migration of **142** (equation 43). When an acetone solution of **141** was irradiated with a low-pressure mercury lamp, bicyclic acetone adducts **149** and **150** were obtained in 4.4 and 1.8% yields, respectively, together with ring-reduced cycloheptyne **142** (21%) (equation 42). The formation of acetone adducts **149** and **150** can be explained by the acetone trapping process of the fused silacyclopropenes **147** and **148**. This photochemical behavior seems to indicate that the ring contraction of polysilacycloalkynes occurs via polysilabicyclo[n.1.0]alk-1(n + 2)-ene derivatives. A plausible reaction mechanism is shown in equation 43. However, the photolysis of 3,3,8,8-tetraphenyloctamethylhexasilacyclooctyne **151a** leads to formation of two types of ring-contracted cycloheptynes **153a** and **153b** in yields of 16 and 14%, respectively (equation 44)[156]. These results indicate that the ring contraction of polysilacycloalkynes proceeds via two different routes. One is a direct desilylation and another is a process involving silacyclopropene intermediate **152**.

Finally, irradiation of a hexane solution of hexasilacyclooctyne **151b** with a low-pressure mercury lamp through a quartz tube gave three ring-contracted products, dediethylsilylated **154a**, dedimethylsilylated **154b** and dediphenylsilylated **154c** (formed via the rearranged **155**) in yields of 10, 10 and 9%, respectively (equation 45). These yields of pentasilacycloheptynes **154a**, **154b** and **154c** are reasonable compared to those of pentasilacycloheptynes **153a** and **153b** formed by photoirradiation of hexasilacyclooctyne **151a**[156].

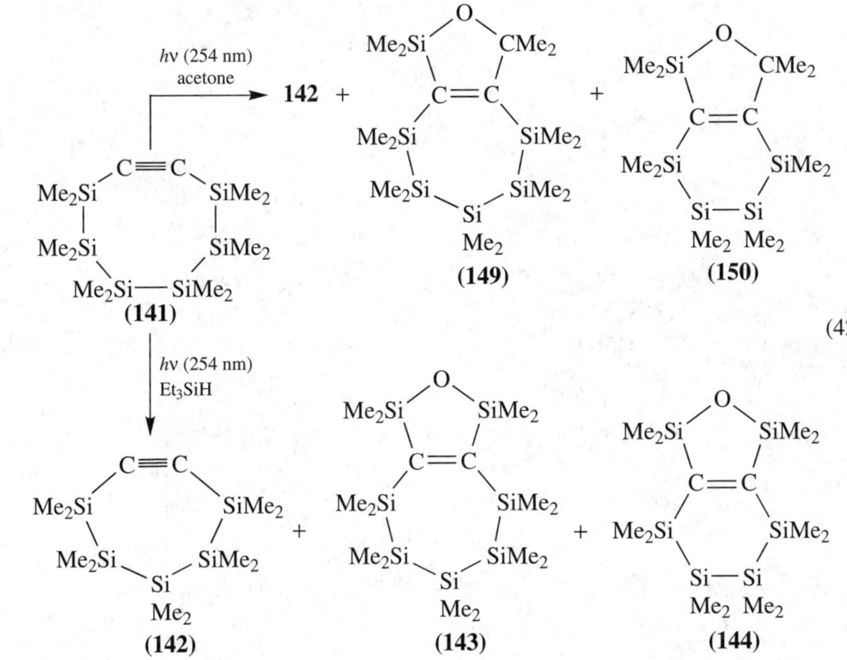

42. Small-ring monosilacycles and medium-ring oligosilacycles

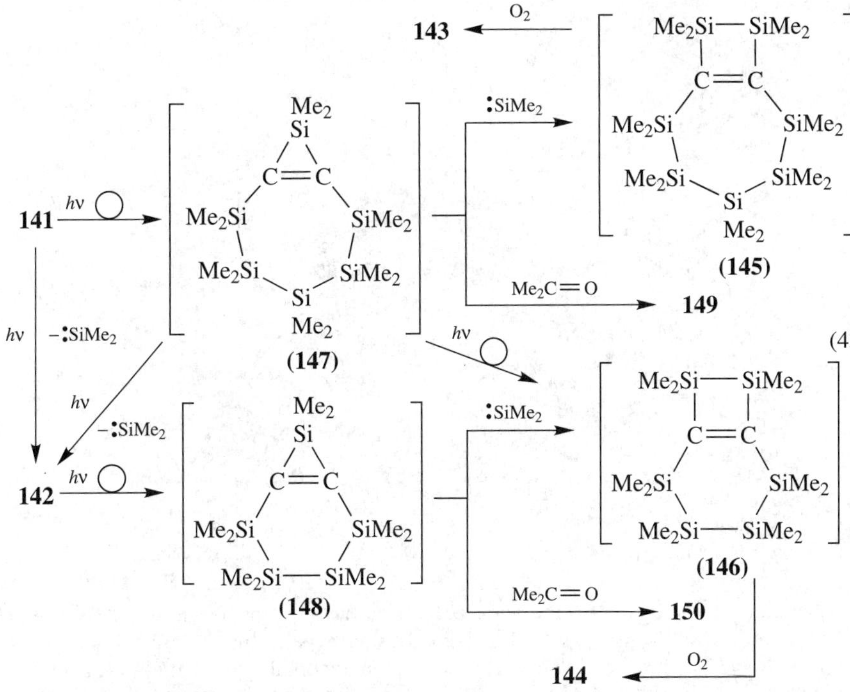

(42)

(43)

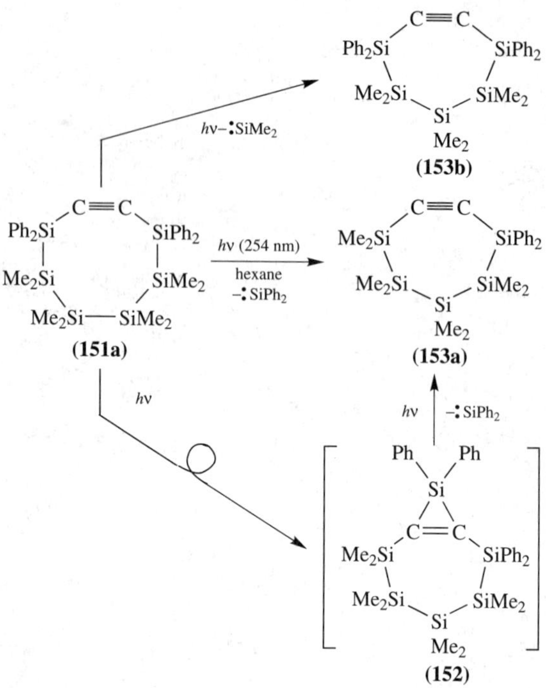

$$\text{(44)}$$

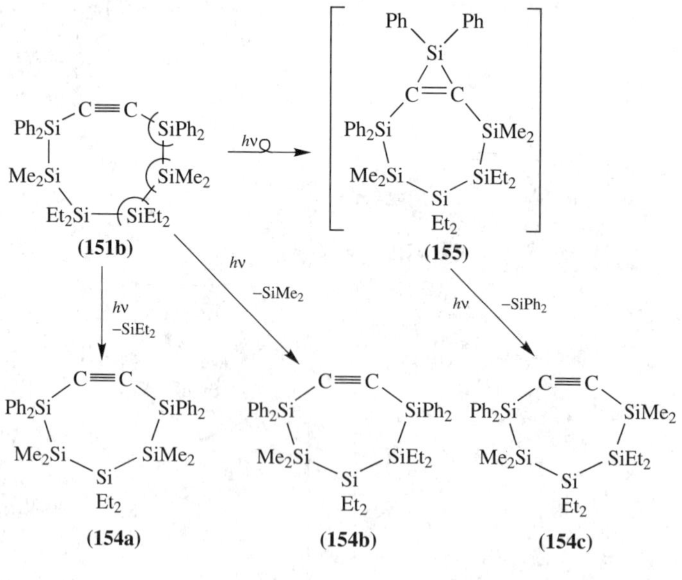

$$\text{(45)}$$

Generation of some small-ring cyclic acetylenes has been achieved either by matrix isolation studies or by trapping and oligomerization experiments[157–160]. However, no example of the isolation of cyclohexyne has been reported. While the six-membered cyclic acetylene **156a** was prepared for the first time by the ring-contracted photochemical reaction of pentasilacycloheptyne **142** in 20% yield, a later direct synthesis involved the reaction between the corresponding 1,4-dichlorotetrasilane with acetylene.

A di-Grignard reagent was also applied for the preparation of **156a** and **156b** in yields of 52% and 55%, respectively (equation 46)[155,161]. The spectroscopic analysis of **156a** indicates the strain at the sp carbons. Particularly, the ^{13}C-NMR signal of the acetylenic sp carbons of **156a** appears at 135.6 ppm, which is at a lower field than that observed in **141** and **142**. The UV absorption of **142** shows a maximum at 229 nm, which is at a shorter wavelength than that found for **141** (235 nm) because of the shortening of the silicon chain. However, the UV absorption maxima for **156a** appears at 237 nm, which is a bathochromic shift as compared with **142**. The X-ray crystallographic analysis of octaisopropyltetrasilacyclohexyne by Barton and coworkers shows the average bending angles of the sp carbons to be $31.4°$[162].

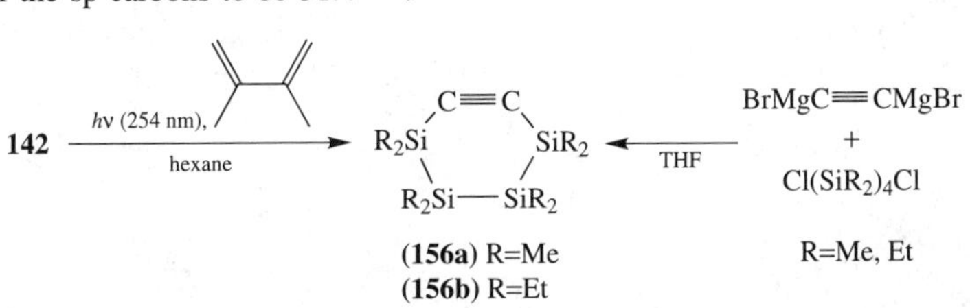

$$142 \xrightarrow[\text{hexane}]{h\nu \ (254 \text{ nm}),} \underset{\underset{R_2Si-SiR_2}{|\qquad\qquad |}}{R_2Si\diagup \overset{C\equiv C}{\diagdown} SiR_2} \xleftarrow{\text{THF}} \begin{array}{c} BrMgC\equiv CMgBr \\ + \\ Cl(SiR_2)_4Cl \end{array} \quad (46)$$

(**156a**) R=Me
(**156b**) R=Et

R=Me, Et

Tetrasilacyclohexyne **156a** was found to be thermally stable even in boiling hexane, but it slowly decomposed in boiling decane (174 °C; $t_{1/2} = 8$ h) to give a bicyclic compound **144**. However, tetrasilacyclohexynes **156a** and **156b** are unstable toward atmospheric moisture and give a complex mixture of products. Diels–Alder reaction of **156a** with 2,3-dimethyl-1,3-butadiene proceeds easily at room temperature to give the cycloadduct **157**. However, 1,3-diphenylisobenzofuran did not react with the tetrasilacyclohexyne **156a** because of a steric repulsion between the methyl groups of the latter with the phenyl groups of the former. Tetrasilacyclohexyne **156a** also reacted with phenyl azide and with diphenyldiazomethane to give the corresponding cycloadduct **158** and its rearranged product **159**, respectively (equation 47). Hexasilacyclooctyne **141** and pentasilacycloheptyne **142** did not react under the same conditions. The structure of **159** was also confirmed by X-ray diffraction.

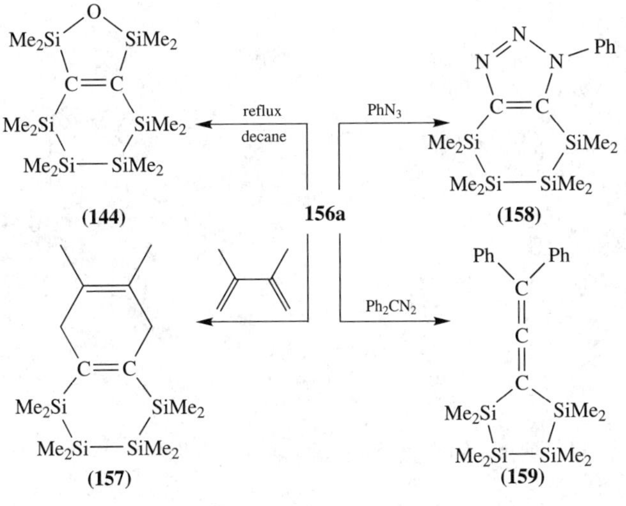

The reaction of hexasilacyclooctyne **141** with dimesitylsilylene generated from the photolysis of 2,2-dimesitylhexamethyltrisilane gave heptasilabicyclo[6.1.0]non-1(8)-ene **160** in 36% yield. The reaction of pentasilacycloheptyne **142** and tetrasilacyclohexyne **156a** give similarly the corresponding bicyclic compounds **161** and **162** in 37 and 15% yields, respectively (equation 48)[155]. The bicycic compounds **160** and **161** were purified by chromatography on silica gel. However, bicyclo[4.1.0]heptene **162** could not be isolated by silica gel chromatography because of its instability. Purification of **162** was achieved by gas chromatography. The structures of **160**, **161** and **162** were determined by spectroscopic analysis and the structure of **162** was further confirmed by X-ray diffraction (Figure 10).

$$\begin{array}{c} \text{Me}_2\text{Si} \diagup^{C \equiv C} \diagdown \text{SiMe}_2 \\ \diagdown (\text{SiMe}_2)_n \diagup \end{array} \xrightarrow[\text{cyclohexane}]{\text{Mes}_2\text{Si}(\text{SiMe}_3)_2, \, h\nu \, (254 \text{ nm})} \begin{array}{c} \text{Mes} \quad \text{Mes} \\ \diagdown \text{Si} \diagup \\ \text{Me}_2\text{Si} \diagup^{C = C} \diagdown \text{SiMe}_2 \\ \diagdown (\text{SiMe}_2)_n \diagup \end{array} \quad (48)$$

(**141**) n = 4
(**142**) n = 3
(**156a**) n = 2

(**160**) n = 4
(**161**) n = 3
(**162**) n = 2

Photochemical reactions of polysilabicyclo[n.1.0]alkenes **160** and **162** were examined. Irradiation of **160** in hexane with a high-pressure mercury lamp gave polysilacycloalkynes **141**, **163** and **164** in 23, 58 and 11% yields, respectively. Similarly, the polysilacycloalkynes **156a** and **165** were also formed by photoirradiation of **162** in 18 and 71% yields, respectively (equation 49). In these reactions, the formation of cyclooctynes **141** and **156a** can be explained by direct or stepwise elimination of dimesitylsilylene. The 1,2-silyl shift of the silacyclopropenes is probably an important step for ring enlargement. The

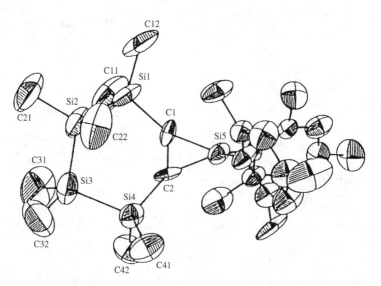

FIGURE 10. ORTEP drawing of **162** (reprinted with permission from Reference 155. Copyright 1993 American Chemical Society)

asymmetric cycloalkynes **163** and **165** can be considered to be formed by a 1,2-silyl migration of silacyclopropenes via 1,3-biradicals (**166**). The formation of **164** may involve the elimination of dimethylsilylene before or after 1,2-silyl migration via the 1,3-biradical **166**.

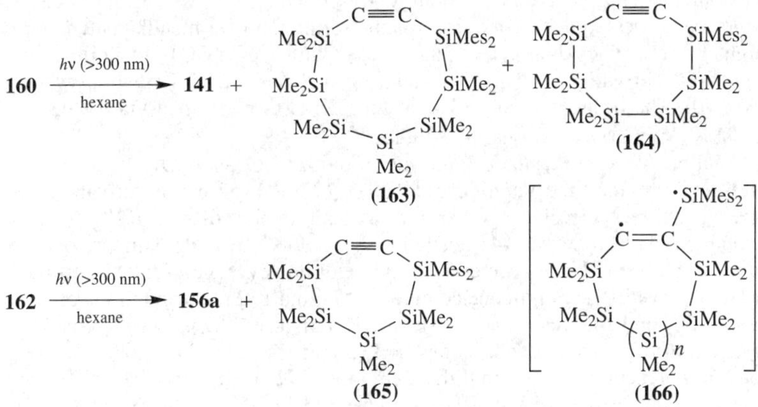

$$\mathbf{160} \xrightarrow[\text{hexane}]{h\nu\ (>300\ \text{nm})} \mathbf{141} + (\mathbf{163}) + (\mathbf{164}) \tag{49}$$

$$\mathbf{162} \xrightarrow[\text{hexane}]{h\nu\ (>300\ \text{nm})} \mathbf{156a} + (\mathbf{165}) \quad [(\mathbf{166})]$$

The rearrangement of alkynes into vinylidenes in the coordination sphere of a transition metal has been intensively investigated in recent years[163]. The 1,2-silyl migration of disilylacetylene to a vinylidene complex proceeds analogously to the 1,2-hydrogen shift. The corresponding manganese[164] and rhodium[165] vinylidene complexes are isolated. Vollhardt and coworkers reported that in the reaction of neat bis(trimethylsilyl)acetylene with CpCo(CO)$_2$, a tetrakis(trimethylsilyl)triene was formed in a low yield via a 1,2-silyl shift[166]. The reaction of the pentasilacycloheptyne **142** with CpCo(CO)$_2$ in boiling *n*-decane gave the metallapentene derivative **168** in 82% yield. The yellow crystals of **168** are unstable toward oxygen. The structure of the compound **168** was determined by spectroscopic methods and X-ray crystallographic diffraction (equation 50)[167].

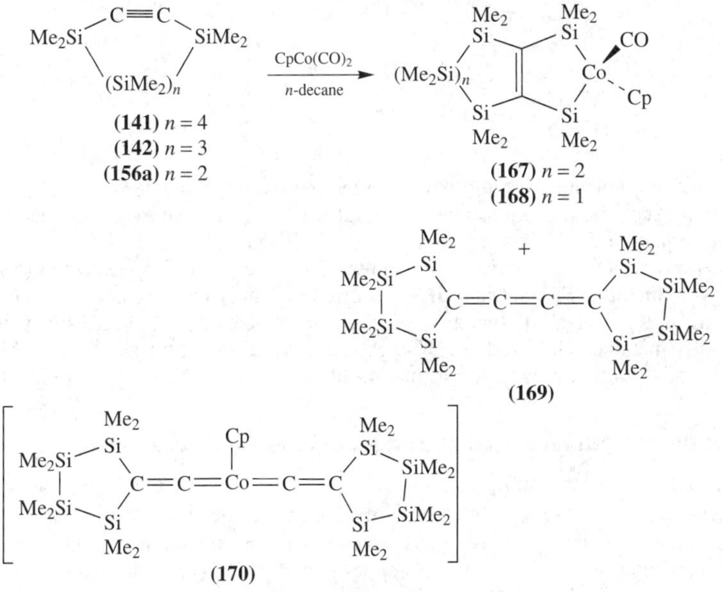

$$\begin{array}{c}(\mathbf{141})\ n = 4 \\ (\mathbf{142})\ n = 3 \\ (\mathbf{156a})\ n = 2\end{array} \xrightarrow[\text{n-decane}]{\text{CpCo(CO)}_2} \begin{array}{c}(\mathbf{167})\ n = 2 \\ (\mathbf{168})\ n = 1\end{array} + (\mathbf{169}) \tag{50}$$

[(**170**)]

The CpCo(CO)$_2$ also reacted readily with hexasilacyclooctyne **141** to give yellow stable crystals of metallapentene derivative **167** in 52% yield. The reaction of the acyclic acetylene with CpCo(CO)$_2$ under similar conditions did not give the metallapentene derivative and only an unidentified complex mixture was obtained. These results suggest that a ring structure is the requisite condition for the formation of a metallapentene derivative. Although the tetrasilacyclohexyne **156a** reacted with CpCo(CO)$_2$ in boiling hexane to give the triene derivative **169** in 52% yield, no metallapentene derivative was observed (equation 50). The triene derivative **169** is considered to be formed via the bisvinylidene complex **170**, similarly to Vollhardt's studies[166].

The reaction of the tetrasilacyclohexyne **156a** with CpMn(CO)$_3$ under photochemical conditions provided the vinylidene complex **172** and the triene derivative **169** in 46 and 11% yields, respectively, together with a small amount of trimer **173** (equation 51). When the hexasilacyclooctyne **141** reacted with CpMn(CO)$_3$ under similar conditions, the vinylidene complex **171** was formed in 36% yield; however, no triene and trimer derivatives were observed. The vinylidene complexes **171** and **172** regenerated the polysilacyclic acetylenes **156a** and **141**, respectively. The triene derivative **169** is considered to be formed by the reaction of the vinylidene complex **172** and the tetrasilacyclohexyne **156a**. The photochemical reaction of the vinylidene complex **172** or the tetrasilacyclohexyne **156a** gave the triene derivative **169** and the bisallene **173** in 30 and 18% yields, respectively. The structure was determined by spectroscopic and X-ray crystallographic analysis.

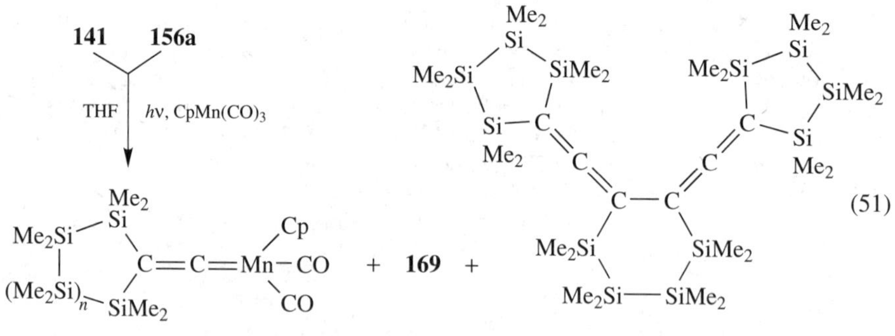

These results indicate that only when the tetrasilacyclohexyne **156a** reacts with vinylidene complexes, triene and bisallene derivatives formed. The reaction mechanism is shown in equation 52. A dissociation of carbonyl ligand from the vinylidene complex **172** generates an alkyne(vinylidene) complex **174**, which would be expected to form the metallacyclobutene **175**. The triene **169** is formed by reductive elimination reaction of **175** along with a 1,2-silyl shift. Insertion of a second molecule of the tetrasilacyclohexyne **156a** forms metallacyclohexadiene **176**. A reductive elimination reaction of **176** along with a 1,2-silyl shift results in the formation of the bisallene **173**.

C. Oligosilabridged and Doubly Bridged Allenes and Bisallenes

Cyclic allenes are the subject of considerable interest as strained unsaturated cyclic compounds[168–170]. *tert*-Butyl-substituted cycloocta-1,2-diene is known as the smallest isolable cyclic allene[170,171]. Seven- and six-membered cyclic allenes are also well known as important intermediates in organic synthesis, and many synthetic methods have been documented.

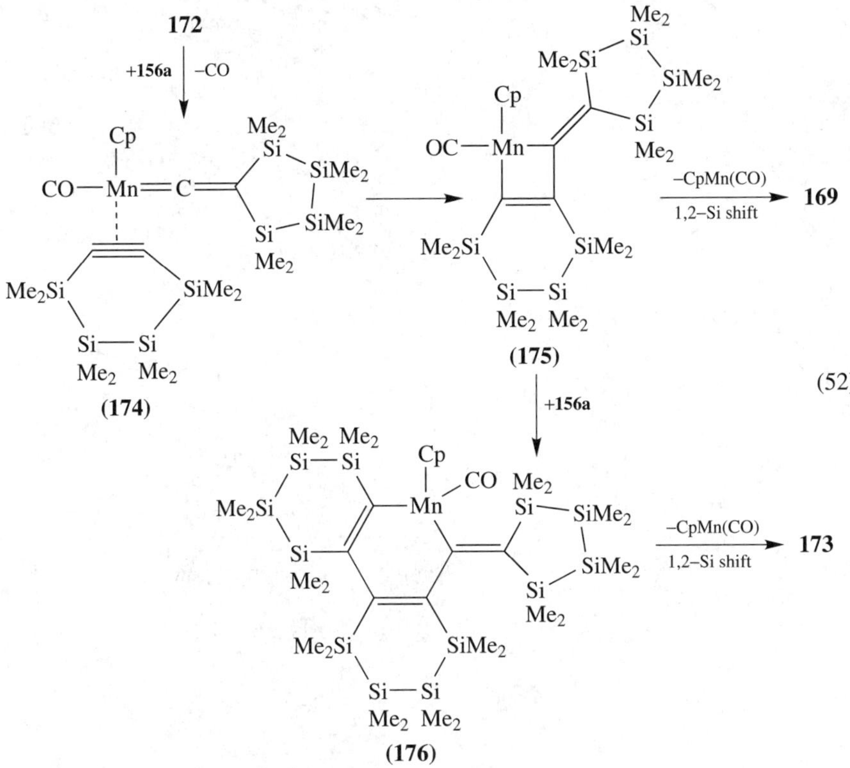

[Scheme showing reaction of **172** with +156a, −CO giving **174**, then forming intermediate (**175**), with −CpMn(CO) 1,2−Si shift to **169**; and via +156a forming (**176**), with −CpMn(CO) 1,2−Si shift to **173**]

(52)

Ph$_2$C$_3$ dianion, prepared from 1,3-diphenylpropyne and 2 equivalents of butyllithium, reacted with 1,4-dicholorohexasilane and 1,6-dichlorotetrasilane in THF/hexane to give tetrasilacyclohepta-1,2-diene **177** and hexasilacyclonona-1,2-diene **178** in 24 and 72% yields, respectively[172]. A reaction of Ph$_2$C$_3$ dianion with 1,3-dichlorohexamethyltrisilane gave mainly polymeric products and no 1 : 1 cycloadduct could be detected. The six-membered cyclic allene **179** was obtained in 11% yield from the reaction of 1,3-dichloro-2,2-diphenyltetramethyltrisilane with the Ph$_2$C$_3$ dianion (equation 53). The difference in the reactivities of the two types of dichlorotrisilanes might be attributed to the steric effects of two phenyl groups on a central silicon atom, which prevent an intermolecular reaction. This is the first example of isolated six- and seven-membered cyclic allenes[172]. The X-ray crystallographic analyses of **177** and **179** (Figure 11) showed that the slightly longer bond lengths of the skeletal C−Si and Si−Si bonds of **179** (1.94 and 2.38 Å on average, respectively) compared with those of **177** (1.91 and 2.34 Å) result in a greater release of strain caused by the allenic moiety. The most interesting point concerning the structures of **177** and **179** is the geometry of the allenic moiety. The bond angle C1−C2−C3 at the sp carbons of **179** was found to be highly strained (19°) from a linear geometry, while that of **177** is almost strain-free (6°). The dihedral angles around the allenic moieties of **177** and **179** are also influenced by the ring strain. The dihedral angle Si1−C1−C3−Si3 of **179** is strongly contracted (52.2°) from the normal vertical geometry, while that of **177** is nearly perpendicular (85.3°).

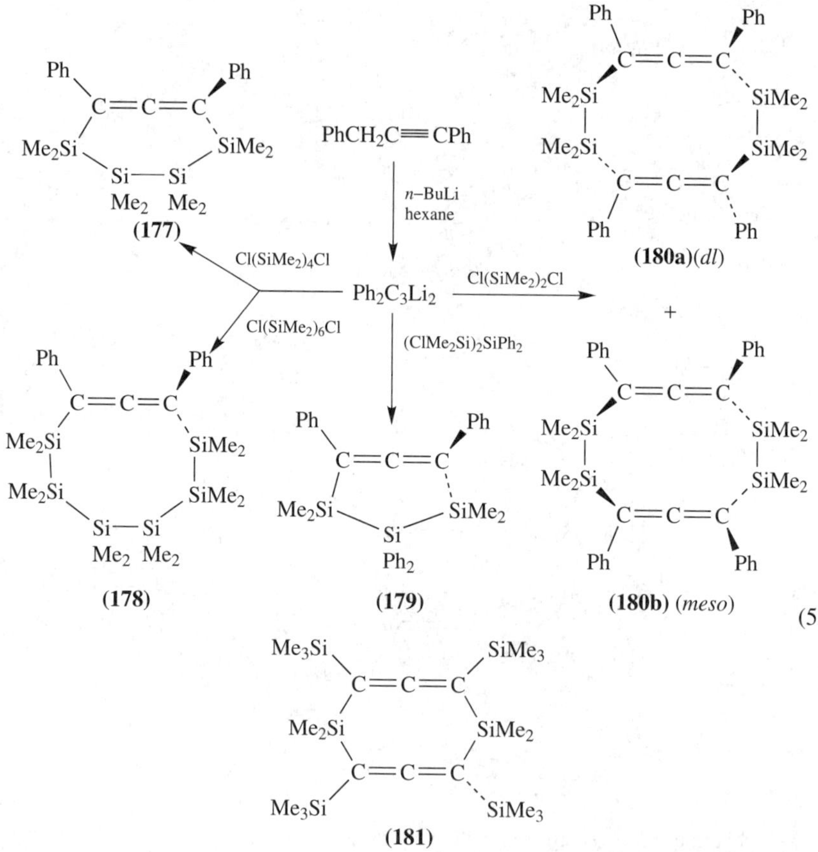

There have been attempts to synthesize dimeric allenes as well as cyclic bisallenes[173–178]. A ten-membered cyclic tetraene was synthesized by Skättebol for the first time as the smallest bisallene[173]. However, only the *meso* structure has been characterized, although the *dl* isomer does not appear to be seriously strained. The Ph_2C_3 dianion reacts readily with 1,2-dichlorotetramethyldisilane to give the two isomers of 1,2,6,7-tetrasilacyclodeca-3,4,8,9-tetraenes **180a** and **180b** (equation 53)[179]. The X-ray analysis of **180a** and **180b** shows that the average bent angle around the sp carbons C2 and C7 is 4° (Figure 12). The allene unit is slightly enlarged to 99–102° from the normal vertical geometry of allene. The eight-membered bisallene **181** was synthesized by the reaction of 1,3-dilithio-1,3-bis(trimethylsilyl)allene with dimethyldichlorosilane by Barton and coworkers[180]. Its X-ray diffraction shows that the allene unit is bent to 4.8° and twisted to 78.1°; the distance between the allenic central carbons is 2.859 Å.

There are few known examples of the bicyclic, doubly-bridged allenes as betweenallenes. The strained symmetrical betweenallenes can gain no strain relief by bending, since what is gained in one ring is lost in the other. Therefore, only twisting, as defined by a reduction in the normal allene dihedral angle of 90°, is available for strain relief. Nakazaki and coworkers synthesized [8.10]betweenallene, consisting of fused 11- and 13-membered rings, as the first example of a betweenallene[181].

42. Small-ring monosilacycles and medium-ring oligosilacycles

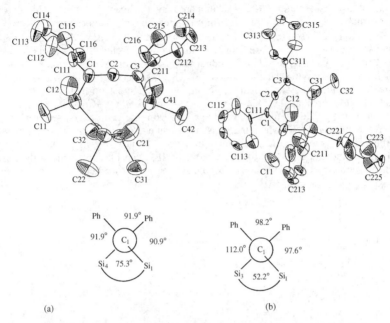

FIGURE 11. ORTEP drawings of (a) **177** and (b) **179** (reprinted with permission from Reference 172. Copyright 1993 American Chemical Society)

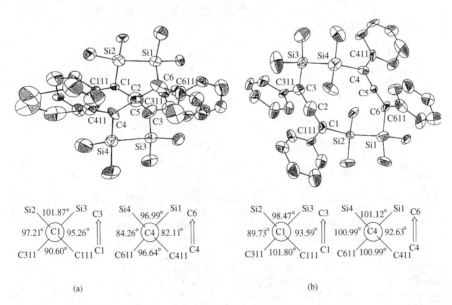

FIGURE 12. ORTEP drawings of (a) **180a** (*dl*) and (b) **180b** (*meso*) (reproduced by permission of The Chemical Society of Japan from Reference 179)

Octasila[4.4]betweenallene **182** was prepared by the reaction of hexachloropropene with 1,4-dichlorooctamethyltetrasilane in the presence of magnesium in 34% yield; the other expected isomer, i.e. **183**, could not be obtained[172]. The corresponding [3.3]betweenallene could not be isolated, and an unexpected 2 : 2 adduct **184** was isolated in 16% yield as a volatile product by the reaction of hexachloropropene with 1,3-dichlorohexamethyltrisilane (equation 54). The X-ray crystallographic analysis of **182** (Figure 13) showed a symmetrical conformation and almost normal bond lengths and angles. The torsional angles, Si1−C1−C3−Si4 and Si5−C1−C3−Si8, are 72.0° and 73.3°, respectively. This deformation of the allene from the vertical geometry is increased compared with that of **177**, which has the same-membered monocyclic system. The allenic sp carbon of **182** is almost linear, although in the case of **177** it is slightly bent. Therefore, the strain, as indicated by the dihedral angles of **181**, can be explained on the basis of the linear geometry of the allenic sp carbon. Barton and coworkers also reported on the synthesis of **182** (equation 54)[182,183].

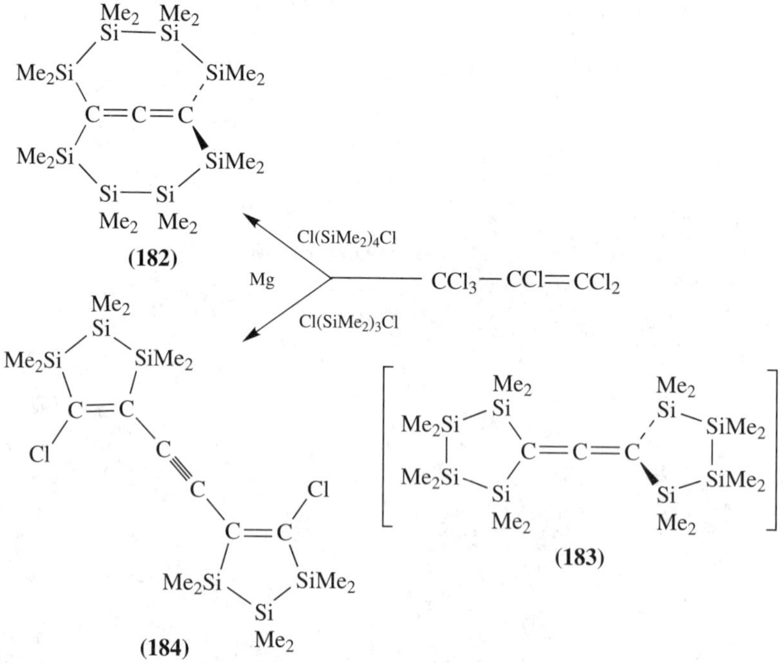

(54)

Bicyclo[n.1.0]alk-1(n + 3)-enes, as representatives of highly strained olefins, are of considerable interest. Since the photoisomerization of allene to cyclopropene and propyne was first reported by Chapman, allene photochemistry has received much attention[184]. Stierman and Johnson reported several photorearrangements of carbocyclic allenes giving the bicyclo[n.1.0] systems[185-189]. In a photochemical irradiation of hexasilacyclonona-1,2-diene **178** with a low-pressure mercury lamp, the corresponding hexasilabicyclo[6.1.0]non-1(9)-ene **185** was isolated in 59% yield; additionally, the ring-contracted allene **186** and pentasilabicyclo[5.1.0]oct-1(8)-ene **187** were obtained in 14% and 23% yields[190]. Irradiation of 5,7-diphenyl-1,2,3,4-tetrasilacyclohepta-5,6-diene **177** gave the tetrasilabicyclo[4.1.0]hept-1(7)-ene **188** and 2,3,4,7-tetrasilabicyclo[4.1.0]hept-5-ene **189** in 45% and 20% yields, respectively, probably via the vinylcarbene intermediate

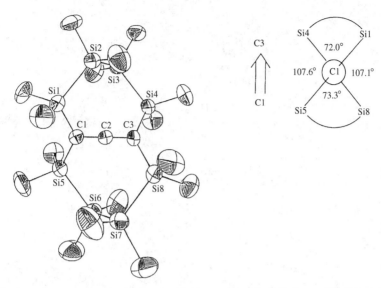

FIGURE 13. ORTEP drawing of **182** (reprinted with permission from Reference 172. Copyright 1993 American Chemical Society)

or via a concerted process (equations 55 and 56). The photolysis of a methanol solution

$$\text{(178)} \xrightarrow{h\nu\ (254\ \text{nm})}_{\text{cyclohexane}} \text{(185)} + \text{(186)} + \text{(187)} \quad (55)$$

$$\text{(177)} \xrightarrow{h\nu\ (254\ \text{nm})}_{\text{cyclohexane}} \text{(188)} + \text{(189)}$$

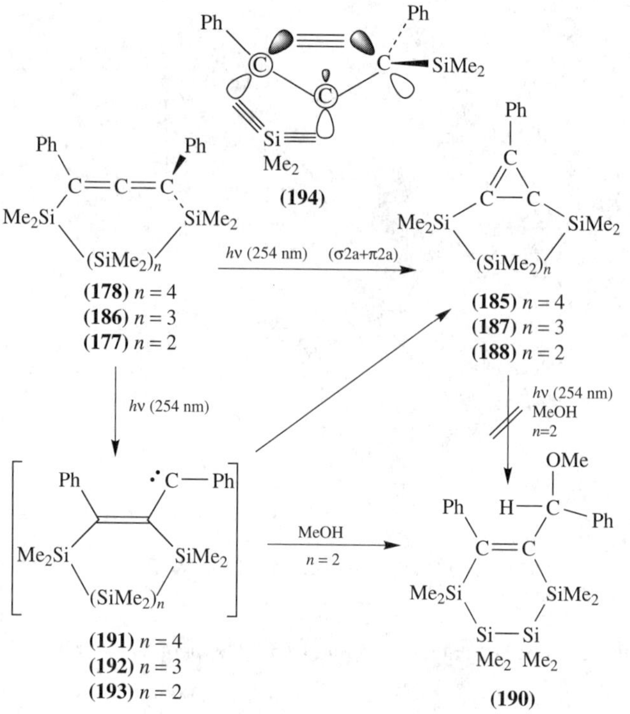

of tetrasilacyclohexa-1,2-diene **177** gave a methanol-incorporated product **190** in 54% yield, together with the bicyclo[4.1.0]heptene derivative **188** (8%). The formation of **190** can be explained by a methanol reaction of a vinylcarbene intermediate **193** generated by a 1,2-silyl shift. However, the photolysis of the bicyclo[4.1.0]heptene derivative **188** in methanol did not give a methanol-incorporated product **190**. No methanol-incorporated products **185**, **187** and **188** were obtained in the photolysis of a methanol solution of 1,2,3,4,5,6-hexasilacyclonona-7,8-diene **178**. It is likely that there are two reasons for the inability to observe the vinylcarbene intermediate. One is that the mechanism might be concerted through **194**; another may be the very short lifetime of the corresponding vinylcarbene intermediates **191** and **192**.

D. 1,2-Disilaacenaphthene and 9,10-Disilaanthracene Silyl Dianions

There has been a remarkable interest in the reactivity, structure and bonding of organolithium compounds in recent years. However, so far there is no report concerning the reduction of a silicon–silicon double bond to afford a 1,2-dilithiodisilane. One of the best candidates for the generation of a 1,2-dilithiodisilane might be 1,2-dichloro-1,2-diisopropyl-1,2-disilaacenaphthene (**195**). The particular advantage of **195** is that the two silicon atoms are in fixed positions and are separated by a single bond distance, and hence obviating the possibility of a double bond formation between them. We succeeded in the synthesis and characterization of the first 1,2-dilithiodisilane (**196**) derived by dilithiation of **195**. **195** and an excess lithium in THF were sonicated at room temperature under argon to afford a dark-blue solution (Gilman reaction)[191,192]. After removal of the unreacted lithium, addition of an excess of CH_3I to this solution afforded 1,2-dimethyl-1,2-diisopropyl-1,2-disilaacenaphthene in high yield. On treatment of this solution with CH_3OD, quantitative formation of 1,2-dideuterio-1,2-diisopropyl-1,2-disilaacenaphthene was observed. The formation of these products clearly reveals that the dichlorodisilane **195** is readily reduced by lithium to afford dilithium 1,2-disilaacenaphthendiide (**196**) (equation 57). The half-life of disappearance of **196** in THF is 13.5 h at 30 °C. The NMR measurement of the lithiated solution, however, showed the existence of a single product. Detailed NMR measurements (^{1}H, ^{13}C, H–H NOESY, H–C COSY, H–C COLOC and ^{13}C INVGATE NMR measurements) were carried out. The chemical shifts of the aromatic protons in **196** are shifted upfield in comparison with those of the corresponding disilanes. Upfield shifts of the ^{13}C NMR signals for the C-3-C-8 atoms in **196** are also observed. The most important factor affecting the upfield chemical shifts of protons and carbons in **196** is thought to be the presence of a negative charge on the silicon atom. The ^{29}Si NMR chemical shift for **196** was observed at -1.24 ppm, a large downfield shift compared with other silyl anions. Interestingly, the ^{29}Si resonance for **196**, however, is upfield shifted from 1-*tert*-butyl-2,3,4,5-tetraphenyl-1-silacyclopentadienide lithium. Based on the ^{1}H and ^{13}C NMR chemical shifts, the electronic structure of **196** is described in terms of the negative charge on the silicon atoms. In order to obtain further information on the structure of **196**, a low-temperature ^{29}Si NMR measurement was carried out in order to observe a scalar ^{29}Si–^{7}Li coupling. In the ^{29}Si NMR spectrum of **196**, a well-resolved quartet at low temperature was observed, indicative of the fact that **196** has at least a partial covalency at low temperature. In the presence of oxidizing reagents such as O_2 and Cd(II), the dilithium 1,2-disilaacenaphthendiide (**196**) further reacts with 2,3-dimethylbutadiene and 9,10-dimethylanthracene to afford cycloadducts **197** and **198** respectively. Consequently, **196** is synthetically equivalent to 1,2-disilaacenaphthene (**199**) (equation 57).

The chemistry of 9,10-dihydroanthracenes with silicon in position 9 and/or 10 has been considerably developed for some time by Jutzi[193,194] and Bickelhaupt and coworkers[195–197] and modified by Corey and coworkers[198,199].

A Wurtz coupling reaction of 9,10-dichloro-9,10-dimethyl-9,10-disilaanthracene (**200**) with sodium in toluene affords a bridged dimer (**202**) in 7% yield. The X-ray structure is given in Figure 14a. The reaction of a *cis/trans* mixture of 9,10-dihydro-9,10-dimethyl-9,10-disilaanthracene (**201**) with lithium in THF containing N,N,N',N'-tetramethylethylenediamine (TMEDA) also produced **202** in 61% yield, probably via the silicon-centered 9,10-disilaanthracene biradical or its equivalent intermediate (equation 58). A silicon–silicon bond cleavage of the dimer **202** by excess lithium or potassium produced 9,10-disilaanthracene dianion **203a**, or **203b** (Gilman reaction)[200,201]. The ^{29}Si NMR chemical shifts in THF-d_8 were observed at -45.4 and -42.8 ppm, respectively, a large upfield shift compared to that of other silyl anions [Ph_3SiM: -9.0 ppm (M = Li), -7.5 ppm (M = K); Ph_2MeSiM: -20.6 ppm (M = Li), -18.5 ppm (M = K)][202,203].

NMR studies have been most informative of the structure in solution, and particularly important evidence has recently been obtained from a low-temperature measurement of ^{29}Si–^{7}Li scalar coupling. In the ^{29}Si NMR spectrum for the dilithium compound (**203a**) at 168 K in THF-d_8, a quartet with ^1J [^{29}Si–^{7}Li] = 34 Hz was observed. This implies that the two silicon atoms in the molecule are magnetically equivalent and one lithium atom binds to the silicon atom at low temperature. The ^{7}Li NMR signal appears at 3.7 ppm as a singlet at 298 K. Dramatic change was observed for the low temperature NMR measurements which yielded two different singlets at 2.8 and 5.2 ppm; unfortunately, no satellite signals due to the ^{29}Si scalar coupling were observed. However, it is clear from the ^{29}Si NMR spectrum that **203a** has a partial covalent bond between ^{29}Si and ^{7}Li at low temperature. The result of ^{7}Li NMR studies would be compatible with different environments for the two lithium atoms in the molecule at low temperature. Thus, the observed NMR data are compatible with a structure that contains jointly bridged silicons by one lithium.

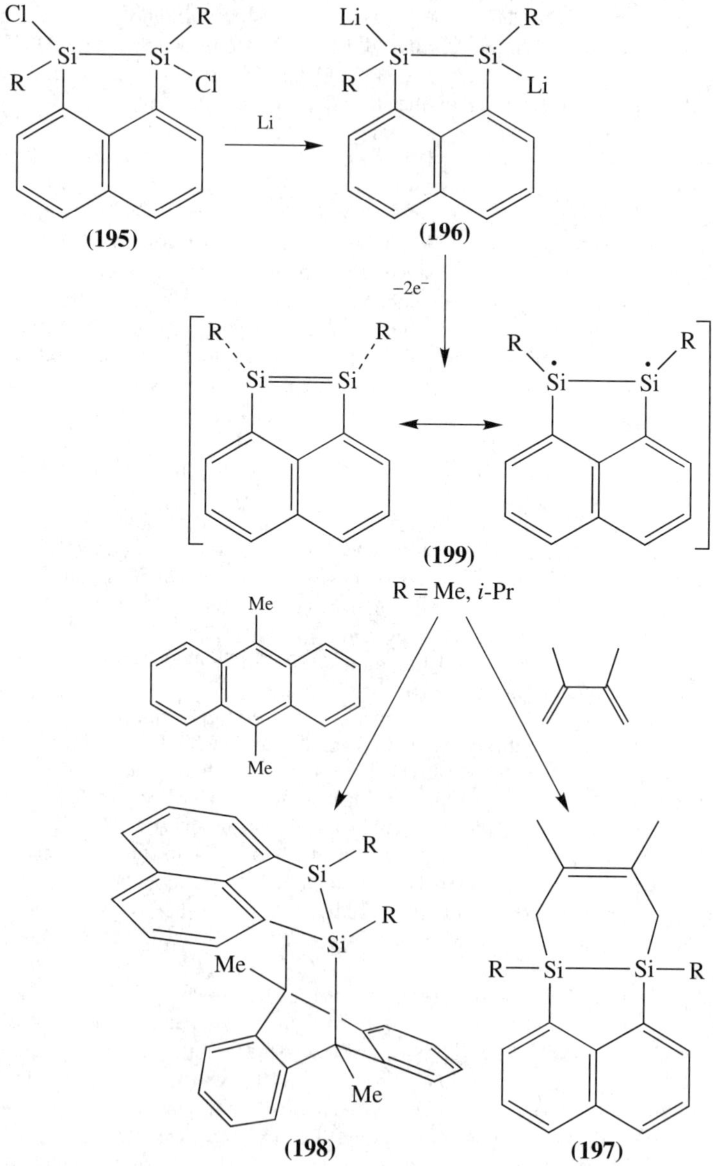

$$R = Me, i\text{-Pr} \tag{57}$$

During reactions of **202** with an equivalent amount of lithium or potassium, the reaction mixture became a yellow suspension which, when reacted with trimethylchlorosilane, gave

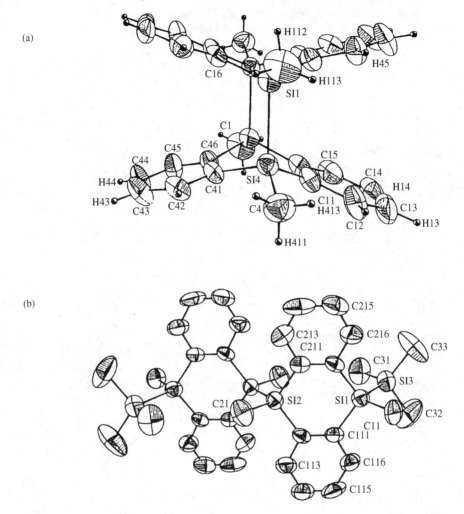

FIGURE 14. ORTEP drawings of (a) **202** and (b) **205** (reprinted with permission from Reference 200. Copyright 1995 American Chemical Society)

the opened dimer (**205**) via an intermediate **204** in 67% yield (equation 58). Since two methyl groups on the disilaanthracene unit of **205** were in a *cis* configuration (cf the X-ray structure in Figure 14b) the silicon–silicon bond cleavage of **202** with alkali metal proceeded with retention of the configuration around the silicon atom. The reactions of the dipotassium 9,10-dimethyl-9,10-disilaanthracene (**203b**) with trimethylchlorosilane gave the expected adduct (**206**) in 73% yield as one isomer. Compound **203b** reacts with dichlorodimethylsilane to give the expected dimethylsilyl adduct (**207**) bonded to the 9 and 10 positions of the disilaanthracene and its dimeric compound (**208**). Adducts **209** and **210** were obtained in good yields when the dipotassium salt **203b** reacted with 1,2-dichlorotetramethyldisilane and 1,3-dichlorohexamethyltrisilane, respectively, without forming dimeric products (equation 59). **203b** can be a valuable intermediate for the synthesis of a great variety of *cis*-9,10-disilaanthracene derivatives.

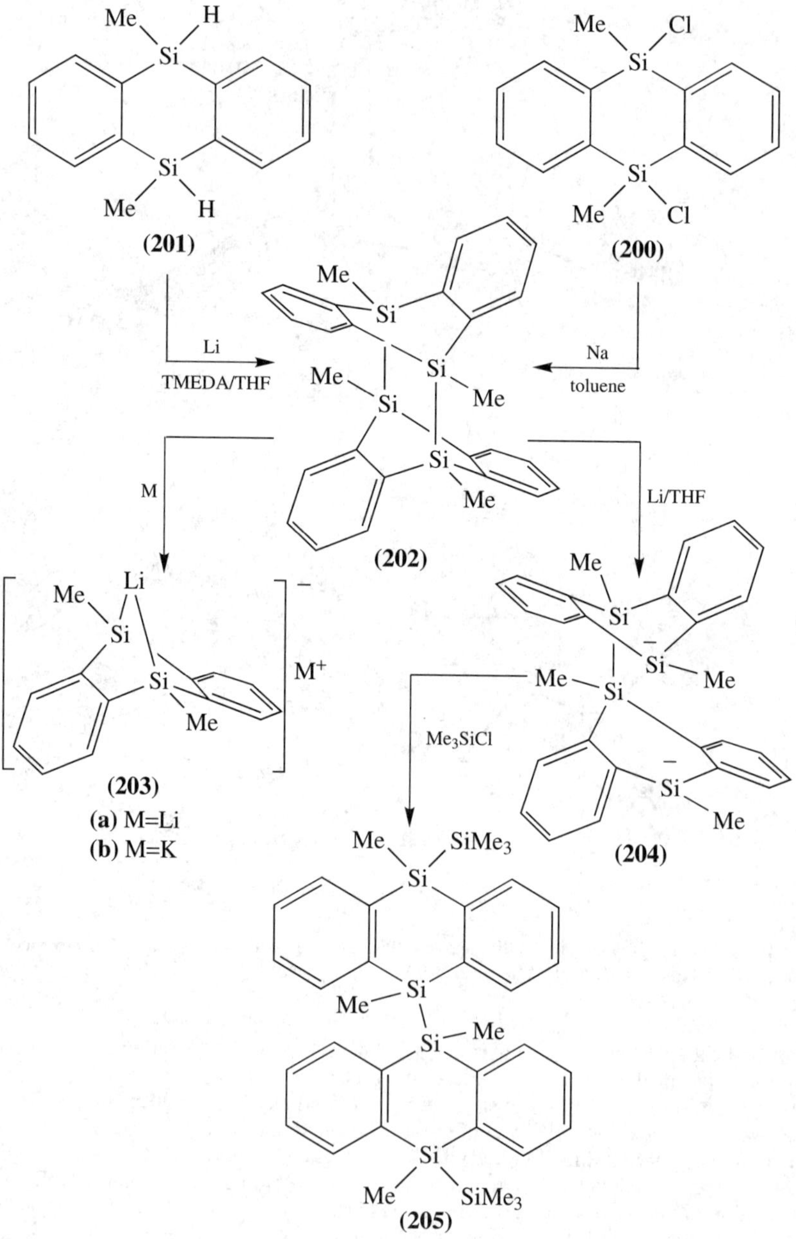

(58)

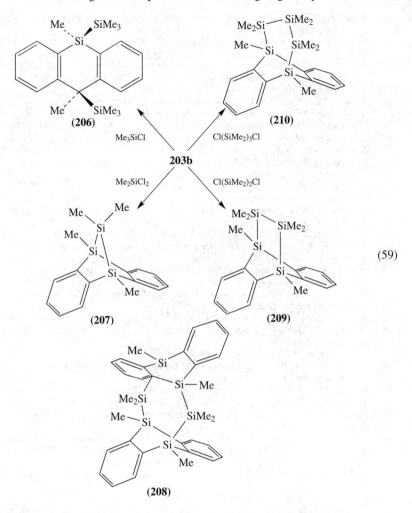

(59)

VI. REFERENCES

1. W. Ando, T. Hagiwara and T. Migita, *J. Am. Chem. Soc.*, **95**, 7518 (1973).
2. W. Ando, A. Sekiguchi and T. Migita, *Chem. Lett.*, 779 (1976).
3. W. Ando, A. Sekiguchi, T. Migita, S. Kammula, M. Green and M. Jones, Jr., *J. Am. Chem. Soc.*, **97**, 3818 (1975).
4. W. Ando, A. Sekiguchi, T. Hagiwara, T. Migita, V. Chowdhry, F. H. Westheimer, S. L. Kammula, M. Green and M. Jones, Jr., *J. Am. Chem. Soc.*, **101**, 6393 (1979).
5. W. Ando, A. Sekiguchi, T. Hagiwara and T. Migita, *J. Chem. Soc., Chem. Commun.*, 372 (1974).
6. T. J. Barton, J. A. Kilgour, R. R. Gallucci, A. J. Rothschild, J. Slutsky, A. D. Wolf and M. Jones, Jr., *J. Am. Chem. Soc.*, **97**, 657 (1975).
7. W. Ando, A. Sekiguchi, A. J. Rothschild, R. R. Gallucci, M. Jones, Jr., T. J. Barton and J. A. Kilgour, *J. Am. Chem. Soc.*, **99**, 6995 (1977).
8. W. Ando, A. Sekiguchi, J. Ogiwara and T. Migita, *J. Chem. Soc., Chem. Commun.*, 145 (1975).
9. A. G. Brook, J. W. Harris, J. Lennon and M. E. Sheikh, *J. Am. Chem. Soc.*, **101**, 83 (1979).

10. A. G. Brook, S. C. Nyburg, W. F. Reynolds, Y. C. Poon, Y.-M. Chang and J.-S. Lee, *J. Am. Chem. Soc.*, **101**, 6750 (1979).
11. A. G. Brook, F. Abdesaken, B. Gutekunst, G. Gutekunst and R. K. Kallury, *J. Chem. Soc., Chem. Commun.*, 191 (1981).
12. A. G. Brook, S. C. Nyburg, F. Abdesaken, B. Gutekunst, G. Gutekunst, R. Krishna, M. R. Kallury, Y. C. Poon, Y.-M. Chang and W. Wong-Ng, *J. Am. Chem. Soc.*, **104**, 5667 (1982).
13. A. Sekiguchi and W. Ando, *Organometallics*, **6**, 1857 (1987).
14. W. Ando, A. Sekiguchi and T. Sato, *J. Am. Chem. Soc.*, **103**, 5573 (1981).
15. A. Sekiguchi, T. Sato and W. Ando, *Organometallics*, **6**, 2337 (1987).
16. L. E. Gusel'nikov, N. S. Nametkin and V. M. Vdovin, *Acc. Chem. Res.*, **8**, 18 (1975).
17. T. J. Barton, G. Marquardt and J. A. Kilgour, *J. Organomet. Chem.*, **85**, 317 (1975).
18. C. M. Golino, R. D. Bush, P. On and L. H. Sommer, *J. Am. Chem. Soc.*, **97**, 1957 (1975).
19. A. Sekiguchi W. Ando and T. Sato, *J. Am. Chem. Soc.*, **104**, 6830 (1982).
20. T. J. Barton and G. P. Hussmann, *Organometallics*, **2**, 692 (1983).
21. G. Märkl and M. Horn, *Tetrahedron Lett.*, **24**, 1477 (1983).
22. N. Wiberg, *J. Organomet. Chem.*, **273**, 141 (1984).
23. A. G. Brook. W. J. Chatterton, J. F. Sawyer, D. W. Hughes and K. Vorspohl, *Organometallics*, **6**, 1246 (1987).
24. S. M. Bachrach and A. Streitwieser, Jr., *J. Am. Chem. Soc.*, **107**, 1186 (1985).
25. D. Seyferth, T. F. O. Lim and D. P. Duncan, *J. Am. Chem. Soc.*, **100**, 1626 (1978).
26. A. Sekiguchi and W. Ando, *J. Am. Chem. Soc.*, **106**, 1486 (1984).
27. G. Mass, K. Schneider and W. Ando, *J. Chem. Soc., Chem. Commun.*, 72 (1988).
28. K. Schneider, B. Daucher, A. Fronda and G. Mass, *Chem. Ber.*, **123**, 589 (1990).
29. A. Sekiguchi, W. Ando and K. Honda, *Tetrahedron Lett.*, **26**, 2337 (1985).
30. A. Sekiguchi and W. Ando, *Chem. Lett.*, 2025 (1986).
31. W. Ando, H. Yoshida, K. Kurishima and M. Sugiyama, *J. Am. Chem. Soc.*, **113**, 7790 (1991).
32. W. Ando, M. Sugiyama, T. Suzuki, C. Kato, Y. Arakawa and Y. Kabe, *J. Organomet. Chem.*, **499**, 99 (1995).
33. T. J. Barton and G. T. Burns, *J. Am. Chem. Soc.*, **100**, 5246 (1978).
34. T. J. Barton, G. T. Burns, E. V. Arnold and J. Clardy, *Tetrahedron Lett.*, **22**, 7 (1981).
35. Y. Nakadaira, H. Sakaba and H. Sakurai, *Chem. Lett.*, 1071 (1980).
36. A. Sekiguchi and W. Ando, *J. Am. Chem. Soc.*, **103**, 3579 (1981).
37. A. Sekiguchi, H. Tanikawa and W. Ando, *Organometallics*, **3**, 584 (1985).
38. W. Ando, H. Tanikawa and A. Sekiguchi, *Tetrahedron Lett.*, **24**, 4245 (1983).
39. G. Märkl and W. Schlosser, *Angew. Chem., Int. Ed. Engl.*, **27**, 963 (1988).
40. W. Ando, M. Ikeno and A. Sekiguchi, *J. Am. Chem. Soc.*, **99**, 6447 (1977).
41. W. Ando, M. Ikeno and A. Sekiguchi, *J. Am. Chem. Soc.*, **100**, 3613 (1978).
42. W. Ando and M. Ikeno *J. Chem. Soc., Chem. Commun.*, 655 (1979).
43. M. Ishikawa, K. Nishimura, H. Sugisawa and M. Kumada, *J. Organomet. Chem.*, **194**, 147 (1980).
44. R. West, M. J. Fink and J. Michl, *Science*, **214**, 1343 (1981).
45. T. Tsumuraya, S. A. Batcheller and S. Masamune, *Angew. Chem., Int. Ed. Engl.*, **30**, 902 (1991).
46. W. F. Guore and T. J. Barton, *J. Organomet. Chem.*, **199**, 33 (1980).
47. D. Tzeng and W. P. Weber, *J. Am. Chem. Soc.*, **102**, 1451 (1980).
48. V. J. Tortorelli and M. Jones, Jr, *J. Chem. Soc., Chem. Commun.*, 785 (1980).
49. W. Ando, M. Ikeno and Y. Hamada, *J. Chem. Soc., Chem. Commun.*, 621 (1981).
50. W. Ando, Y. Hamada, A. Sekiguchi and K. Ueno, *Tetrahedron Lett.*, **23**, 5323 (1982).
51. W. Ando, Y. Hamada, A. Sekiguchi and K. Ueno, *Tetrahedron Lett.*, **24**, 4033 (1983).
52. J. Belzner, H. Ihmels, L. Pauletto and M. Noltemeyer, *J. Org. Chem.*, **61**, 3315 (1996).
53. W. Ando, K. Hagiwara and A. Sekiguchi, *Organometallics*, **6**, 2270 (1987).
54. W. Ando, M. Fujita, H. Yoshida and A. Sekiguchi, *J. Am. Chem. Soc.*, **110**, 3310 (1988).
55. S. Zhang, P. E. Wagenseller and R. T. Conlin, *J. Am. Chem. Soc.*, **113**, 4278 (1991).
56. S. Zhang and R. T. Conlin, *J. Am. Chem. Soc.*, **113**, 4272 (1991).
57. P. Boudjouk, E. Blark and R. Kumarathasan, *Organometallics*, **10**, 2095 (1991).
58. D.-H. Pae, M. Xiao, M. Y. Chiang and P. P. Gaspar, *J. Am. Chem. Soc.*, **113**, 1281 (1991).
59. N. C. Baird and M. J. S. Dewar, *J. Am. Chem. Soc.*, **89**, 3966 (1967).
60. K. B. Wiberg and R. A. Fenoglio, *J. Am. Chem. Soc.*, **90**, 3395 (1968).
61. W. Ando and H. Saso, *Tetrahedron Lett.*, **27**, 5625 (1986).

42. Small-ring monosilacycles and medium-ring oligosilacycles

62. H. Saso, H. Yoshida and W. Ando, *Tetrahedron Lett.*, **29**, 4747 (1988).
63. H. Saso and W. Ando, *Chem. Lett.*, 1567 (1988).
64. H. Saso, W. Ando and K. Ueno, *Tetrahedron*, **45**, 1929 (1989).
65. D. Seyferth, D. P. Duncan, M. L. Shannon and E. W. Goldman, *Organometallics*, **3**, 574 (1984).
66. D. Seyferth, D. P. Duncan and M. L. Shannon, *Organometallics*, **3**, 579 (1984).
67. D. Seyferth, S. C. Vick and M. L. Shannon, *Organometallics*, **3**, 1897 (1984).
68. P. Boudjouk and U. Samaraweera, *Organometallics*, **9**, 2205 (1990).
69. P. Boudjouk, E. Black, R. Kumarathasan, U. Samaraweera, S. Castellino, J. P. Oliver and J. W. Kampf, *Organometallics*, **13**, 3715 (1994).
70. T. Yamamoto, Y. Kabe and W. Ando, *Organometallics*, **12**, 1996 (1993).
71. R. Noyori, T. Nishimura and H. Takaya, *J. Chem. Soc., Chem. Commun.*, 89 (1969).
72. R. Noyori, T. Odagi and H. Takaya, *J. Am. Chem. Soc.*, **92**, 5780 (1970).
73. R. Noyori, Y. Kumagai, I. Umeda and H. Takaya, *J. Am. Chem. Soc.*, **94**, 4018 (1972).
74. R. Noyori, M. Yamakawa and H. Takaya, *Tetrahedron Lett.*, 4823 (1978).
75. P. Binger and U. Schuchardt, *Angew. Chem., Int. Ed. Engl.*, **16**, 249 (1977).
76. P. Binger, Q.-H. Lüu and P. Wedemann, *Angew. Chem., Int. Ed. Engl.*, **24**, 316 (1985).
77. D. Seyferth, M. L. Shannon, S. C. Vick and T. F. O. Lim, *Organometallics*, **4**, 57 (1985).
78. M. Ishikawa, J. Oshita, Y. Ito and J. Iyoda, *J. Am. Chem. Soc.*, **108**, 7417 (1986).
79. J. Oshita, Y. Isomura and M. Ishikawa, *Organometallics*, **8**, 2050 (1989).
80. W. Ando, T. Yamamoto, H. Saso and Y. Kabe, *J. Am. Chem. Soc.*, **113**, 2791 (1991).
81. Y. Kabe, T. Yamamoto and W. Ando, *Organometallics*, **13**, 4606 (1994).
82. N. Choi, Y. Kabe and W. Ando, *Organometallics*, **11**, 1506 (1992).
83. J. F. Hartwig, R. A. Anderson and R. G. Bergman, *J. Am. Chem. Soc.*, **112**, 5670 (1990).
84. J. F. Hartwig, R. G. Bergman and R. A. Anderson, *Organometallics*, **10**, 3326 (1991).
85. B. K. Campion, R. H. Heyn and T. D. Tilley, *J. Am. Chem. Soc.*, **110**, 7558 (1988).
86. B. K. Campion, R. H. Heyn and T. D. Tilley, *J. Am. Chem. Soc.*, **112**, 4079 (1990).
87. T. S. Koloski, P. J. Carroll and D. H. Berry, *J. Am. Chem. Soc.*, **112**, 6405 (1990).
88. V. V. Kane, A. D. Wolf and and M. Jones Jr., *J. Am. Chem. Soc.*, **96**, 2643 (1974).
89. L. W. Jenneskens, F. J. J. de Kanter, P. A. Kraakman, L. A. M. Turkenburg, W. E. Koolhaas, W. H. de Wolf, F. Bickelhaupt, Y. Tobe, K. Kakiuchi and Y. Odaira, *J. Am. Chem. Soc.*, **107**, 3716 (1985).
90. T. Tsuji and S. Nishida, *J. Chem. Soc., Chem. Commun.*, 1189 (1987).
91. W. Ando, T. Tsumuraya and Y. Kabe, *Angew. Chem., Int. Ed. Engl.*, **29**, 778 (1990).
92. H. Sakurai, S. Hoshi, A. Kamiya, A. Hosomi and C. Kabuto, *Chem. Lett.*, 1781 (1986).
93. C. Elschenbroich, J. Hurley, W. Massa and G. Baum, *Angew. Chem., Int. Ed. Engl.*, **27**, 684 (1988).
94. N. L. Allinger, T. J. Walter and M. G. Newton, *J. Am. Chem. Soc.*, **96**, 4588 (1974).
95. M. G. Newton, T. J. Walter and N. L. Allinger, *J. Am. Chem. Soc.*, **95**, 5652 (1973).
96. P. Hemmersbach, M. Klessinger and P. Bruckmann, *J. Am. Chem. Soc.*, **100**, 6344 (1978).
97. D. Seyferth, E. W. Goldman and J. Escudie, *J. Organomet. Chem.*, **271**, 337 (1984).
98. L. E. Gusel'nikov and Y. P. Polyakov, in *Frontiers of Organosilicon Chemistry* (Eds. A. A. Bassindale and P. P. Gaspar), The Royal Society of Chemistry, 1991, pp. 50–61.
99. J. Ohshita, Y. Masaoka and M. Ishikawa, *Organometallics*, **10**, 3775 (1991).
100. D. Bravo-Zhivotovskii, V. Braude, A. Stanger, M. Kapon and Y. Apeloig, *Organometallics*, **11**, 2326 (1992).
101. W. H. Atwell and J. G. Uhlmann, *J. Organomet. Chem.*, **52**, C21 (1973).
102. H. Sakurai, T. Kobayashi and Y. Nakadaira, *J. Organomet. Chem.*, **162**, C43 (1978).
103. D. Seyferth, D. C. Annarelli and S. C. Vick, *J. Organomet. Chem.*, **272**, 123 (1984).
104. J. Belzner, H. Ihmels, B. O. Kneisel and R. Herbst-Irmer, *J. Chem. Soc., Chem. Commun.*, 1989 (1994).
105. K. Tamao, K. Nagata, M. Asahara, A. Kawachi, Y. Ito and M. Shiro, *J. Am. Chem. Soc.*, **117**, 11592 (1995).
106. M. Ishikawa, T. Fuchikami, T. Kumada, T. Higuchi and S. Miyamoto, *J. Am. Chem. Soc.*, **101**, 1348 (1979).
107. M. Ishikawa, K. Nishimura, H. Ochiai and M. Kumada, *J. Organomet. Chem.*, **236**, 7 (1982).
108. K. Shiina, *J. Organomet. Chem.*, **310**, C57 (1986).
109. T. Kusukawa, Y. Kabe and W. Ando, *Chem. Lett.*, 985 (1993).
110. T. Kusukawa, Y. Kabe, B. Nestler and W. Ando, *Organometallics*, **14**, 2556 (1995).

111. K. Tamao, T. Hayashi and M. Kumada, *J. Organomet. Chem.*, **114**, C19 (1976).
112. H. Sakurai, Y. Kamiyama and Y. Nakadaira, *J. Organomet. Chem.*, **131**, 147 (1977).
113. M. Suginome, H. Oike and Y. Ito, *Organometallics*, **13**, 4148 (1994).
114. M. Suginome, H. Oike and Y. Ito, *J. Am. Chem. Soc.*, **117**, 1665 (1995).
115. M. Suginome, H. Oike, P. H. Shuff and Y. Ito, *Organometallics*, **15**, 2170 (1996).
116. M. Suginome, H. Oike, S.-S. Park and Y. Ito, *Bull. Chem. Soc. Jpn.*, **69**, 289 (1996).
117. Y. Uchimaru, Y. Tanaka and M. Tanaka, *Chem. Lett.*, 164 (1995).
118. Y. Pan, J. T. Mauge and M. J. Fink, *Organometallics*, **11**, 3495 (1992).
119. M. Murakami, T. Yoshida and Y. Ito, *Organometallics*, **13**, 2900 (1994).
120. W. A. Nugent, D. L. Thorn and R. L. Harlow, *J. Am. Chem. Soc.*, **109**, 2788 (1987).
121. E. Negishi, J. S. Holmes, J. M. Tour, J. A. Miller, F. E. Cederbaum, D. R. Swanson and T. Takahashi, *J. Am. Chem. Soc.*, **111**, 3336 (1989).
122. S. S. H. Mao and T. D. Tilley, *J. Am. Chem. Soc.*, **117**, 5365 (1995).
123. S. S. H. Mao and T. D. Tilley, *J. Am. Chem. Soc.*, **117**, 7031 (1995).
124. T. Takahashi, Z. Xi, Y. Obora and N. Suzuki, *J. Am. Chem. Soc.*, **117**, 2665 (1995).
125. Y. Kabe, S. Kadoi, A. Sato and W. Ando, unpublished results.
126. P. J. Fagan, W. A. Nugent and J. C. Calabrese, *J. Am. Chem. Soc.*, **116**, 1880 (1994).
127. S. Kim and K. H. Kim, *Tetrahedron Lett.*, **36**, 3725 (1995).
128. C. Grugel, W. P. Neumann and M. Schriewer, *Angew. Chem., Int. Ed. Engl.*, **18**, 543 (1979).
129. K. Kuno, K. Kobayashi, M. Kawanishi, S. Kozima and T. Hitomi, *J. Organomet. Chem.*, **137**, 349 (1977).
130. E. J. Corey, F. A. Carey and R. A. Winter, *J. Am. Chem. Soc.*, **87**, 934 (1965).
131. P. J. Kroop, *J. Am. Chem. Soc.*, **91**, 5783 (1969).
132. Y. Inoue, S. Takamura and H. Sakurai, *J. Chem. Soc., Perkin Trans.* 2, 1635 (1977).
133. M. Squillacote, A. Bergman and J. D. Felippis, *Tetrahedron Lett.*, **30**, 6805 (1989).
134. J. L. Goodman, K. S. Peters, H. Misawa and R. A. Caldwell, *J. Am. Chem. Soc.*, **108**, 6803 (1986).
135. S. F. Sellers, T. C. Klebach, F. Hollowood and M. Jones, Jr., *J. Am. Chem. Soc.*, **104**, 5492 (1982).
136. T. Shimizu, K. Shimizu and W. Ando, *J. Am. Chem. Soc.*, **113**, 354 (1991).
137. K. B. Wiberg, *Chem. Rev.*, **89**, 975 (1989).
138. G. Szeimies, in *Strain and Its Implications in Organic Chemistry* (Eds. A. De Meijere and S. Blechert), Kluwer Academic, Dordrecht, 1989 p. 361.
139. W. Ando, Y. Igarashi, Y. Kabe and N. Tokitoh, *Tetrahedron Lett.*, **31**, 4185 (1990).
140. Y. Igarashi, Y. Kabe, T. Hagiwara and W. Ando, *Tetrahedron*, **48**, 89 (1992).
141. A. Krebs and J. Wilke, *Top. Curr. Chem.*, **109**, 189 (1983).
142. H. Sakurai, Y. Nakadaira, A. Hosomi, Y. Eriyama and C. Kabuto, *J. Am. Chem. Soc.*, **105**, 3359 (1983).
143. H. Sakurai, Y. Eriyama, A. Hosomi, Y. Nakadaira and C. Kabuto, *Chem. Lett.*, 595 (1984).
144. R. Bortolin, B. Parbhoo and S. S. D. Brown, *J. Chem. Soc., Chem. Commun.*, 1079 (1988).
145. M. Voronkov, O. Yarosh, G. Turkina, V. Vitkovskii and A. Albanov, *J. Organomet. Chem.*, **389**, 1 (1990).
146. T. Iwahara and R. West, *Chem. Lett.*, 545 (1991).
147. K. Sakamoto, M. Tumura and H. Sakurai, *Chem. Lett.*, 549 (1991).
148. M. Ishikawa, T. Hatano, Y. Hasegawa, T. Horio, A. Kunai, A. Miyai, T. Ishida, T. Tsukihara, T. Yamanaka, T. Koike and J. Shioya, *Organometallics*, **11**, 1604 (1992).
149. G. Wittig and J. Meske-Schüller, *Justus Liebigs Ann. Chem.*, **711**, 65 (1968).
150. A. Krebs and H. Kimling, *Angew. Chem., Int. Ed. Engl.*, **10**, 509 (1971).
151. S. F. Karaev and A. Krebs, *Tetrahedron Lett.*, 2853 (1973).
152. A. Krebs and H. Kimling, *Justus Liebigs Ann. Chem.*, 2074 (1974).
153. A. Krebs and G. Burgdöfer, *Tetrahedron Lett.*, 2063 (1973).
154. W. Ando, N. Nakayama, Y. Kabe and T. Shimizu, *Tetrahedron Lett.*, **31**, 3597 (1990).
155. F. Hojo, S. Sekigawa, N. Nakayama, T. Shimizu and W. Ando, *Organometallics*, **12**, 803 (1993).
156. S. Sekigawa, T. Shimizu and W. Ando, *Tetrahedron*, **49**, 6359 (1993).
157. P. Saxe and H. F. Schafer III, *J. Am. Chem. Soc.*, **102**, 3239 (1980).
158. G. Fitzgerald, P. Saxe and H. F. Schafer III, *J. Am. Chem. Soc.*, **105**, 690 (1983).
159. S. Olivella, M. A. Pericas, A. Riera and A. Sole, *J. Am. Chem. Soc.*, **108**, 6884 (1986).
160. S. Olivella, M. A. Pericas, A. Riera, F. Serratosa and A. Sole, *J. Am. Chem. Soc.*, **109**, 5600 (1987).

161. W. Ando, F. Hojo, S. Sekigawa, N. Nakayama and T. Shimizu, *Organometallics*, **11**, 1009 (1992).
162. Y. Pang, A. Schneider, T. J. Barton, M. S. Gordon and M. T. Carroll, *J. Am. Chem. Soc.*, **114**, 4920 (1992).
163. M. I. Bruce, *Chem. Rev.*, **91**, 197 (1991).
164. H. Sakurai, T. Fuji and K. Sakamoto, *Chem. Lett.*, 339 (1992).
165. D. Schnider and H. Werner, *Angew. Chem., Int. Ed. Engl.*, **30**, 700 (1991).
166. J. R. Fritch, K. P. C. Vollhardt, M. R. Thompson and V. W. Day, *J. Am. Chem. Soc.*, **101**, 2768 (1979).
167. F. Hojo, K. Fujiki and W. Ando, *Organometallics*, **15**, 3606 (1996).
168. W. Smadja, *Chem. Rev.*, **83**, 263 (1983).
169. D. J. Pasto, *Tetrahedron*, **40**, 2805 (1984).
170. R. P. Johnson, *Chem. Rev.*, **89**, 1111 (1989).
171. J. D. Price and R. P. Johnson, *Tetrahedron Lett.*, **27**, 4679 (1986).
172. T. Shimizu, F. Hojo and W. Ando, *J. Am. Chem. Soc.*, **115**, 3111 (1993).
173. L. Skättebol, *Tetrahedron Lett.*, 1967 (1961).
174. L. Skättebol and S. Solomon, *Org. Synth.*, **49**, 35 (1969).
175. P. J. Garratt, K. C. Nicolau and F. Sondheimer, *J. Am. Chem. Soc.*, **95**, 4582 (1973).
176. H. Irngartinger and H-U. Jäger, *Tetrahedron Lett.*, 3595 (1976).
177. W. R. Moore and H. R. Ward, *J. Org. Chem.*, **27**, 4179 (1962).
178. M. S. Baird and C. B. Reese, *Tetrahedron*, **32**, 2153 (1976).
179. F. Hojo, T. Shimizu and W. Ando, *Chem. Lett.*, 1171 (1993).
180. J. Lin, Y. Pang, V. G. Young, Jr. and T. J. Barton, *J. Am. Chem. Soc.*, **115**, 3794 (1993)
181. M. Nakazaki, K. Yamamoto, M. Maeda, O. Sato and T. Tsutsui, *J. Org. Chem.*, **47**, 1435 (1982).
182. S. A. Petrich, Y. Pang, V. G. Young, Jr. and T. J. Barton, *J. Am. Chem. Soc.*, **115**, 1591 (1993).
183. Y. Pang, S. A. Petrich, V. G. Young, Jr., M. S. Gordon and T. J. Barton, *J. Am. Chem. Soc.*, **115**, 2534 (1993).
184. O. L. Chapman, *Pure Appl. Chem.*, **40**, 511 (1974).
185. T. J. Stierman and R. P. Johnson, *J. Am. Chem. Soc.*, **105**, 2492 (1983).
186. T. J. Stierman and R. P. Johnson, *J. Am. Chem. Soc.*, **107**, 3971 (1985).
187. J. D. Price and R. P. Johnson, *J. Am. Chem. Soc.*, **107**, 2187 (1985).
188. T. J. Stierman, W. C. Shakespeare and R. P. Johnson, *J. Org. Chem.*, **55**, 1043 (1990).
189. J. P. Price and R. P. Johnson, *J. Org. Chem.*, **56**, 6372 (1991).
190. F. Hojo, T. Shimizu and W. Ando, *Organometallics*, **13**, 3402 (1994).
191. W. Ando, T. Wakahara, T. Akasaka and S. Nagase, *Organometallics*, **13**, 4683 (1994).
192. T. Wakahara, R. Kodama, T. Akasaka and W. Ando, *Bull. Chem. Soc. Jpn.*, **70**, 665 (1997).
193. P. Jutzi, *Chem. Ber.*, **104**, 1455 (1971).
194. P. Jutzi, *Angew. Chem., Int. Ed. Engl.*, **14**, 232 (1975).
195. Y. van den Winkel, B. L. M. van Baar, F. Bickelhaupt, W. Kulik, C. Sierakowski and G. Maier, *Chem. Ber.*, **124**, 185 (1991).
196. Y. van den Winkel, B. L. M. van Baar H., M. M. Bastiaans, F. Bickelhaupt, M. Schenkel and H. B. Stegmann *Tetrahedron*, **46**, 1009 (1990).
197. F. Bickelhaupt and G. L. van Mourik, *J. Organomet. Chem.*, **67**, 389 (1974).
198. W. Z. McCarthy, J. Y. Corey and E. R. Corey, *Organometallics*, **3**, 255 (1984).
199. J. Y. Corey and W. Z. McCarthy, *J. Organomet. Chem.*, **271**, 319 (1984).
200. W. Ando, K. Hatano and R. Urisaka, *Organometallics*, **14**, 3625 (1995).
201. K. Hatano, K. Morihashi, O. Kikuchi and W. Ando, *Chem. Lett.*, 293 (1997).
202. G. A. Olah and R. J. Hunadi, *J. Am. Chem. Soc.*, **102**, 6989 (1980).
203. U. Edlund and E. Buncel, *Prog. Phys. Org. Chem.*, **19**, 254 (1993).

CHAPTER **43**

Silylenes

PETER P. GASPAR

Department of Chemistry, Washington University, St. Louis, Missouri 63130-4899, USA
Fax: 314-935-4481;   e-mail: gaspar@wuchem.wustl.edu

and

ROBERT WEST

Department of Chemistry, University of Wisconsin at Madison, Wisconsin 53706, USA
Fax: 608-262-0381;   e-mail: west@chem.wisc.edu

I. INTRODUCTION	2464
II. GENERATION OF SILYLENES	2465
A. Thermolysis of Polysilanes	2465
B. Silicon Atom Reactions	2468
C. Photolysis of Silanes	2469
1. Photolysis of polysilanes	2469
2. Other photolyses generating silylenes	2473
D. Pyrolysis of Monosilanes	2475
E. Silylenes from Metal-Induced α-Eliminations	2481
F. Silylenes From Rearrangements	2484
III. REACTIONS OF SILYLENES	2485
A. Insertion	2485
B. Addition	2487
1. Mechanistic studies	2487
2. Recently discovered addition reactions	2491
a. Addition to n-donor bases	2491
b. Addition to diatomic molecules	2493
c. Addition to π-bonds	2493
d. Addition to dienes and heterodienes	2495
C. Silylene Dimerization and Disilene Dissociation	2496
1. Silylene dimerization	2496
2. Silylenes from disilenes	2498

The chemistry of organic silicon compounds, Vol. 2
Edited by Z. Rappoport and Y. Apeloig   © 1998 John Wiley & Sons Ltd

	D. Rearrangements and Other Reorganizations of Silylenes	2501
	E. Miscellaneous Reactions of Silylenes	2504
IV.	THEORETICAL CALCULATIONS	2505
	A. Singlet–Triplet Energy Gaps	2505
	B. Thermochemical Calculations	2507
	C. Theoretical Predictions of Silylene Spectra	2508
	D. Benchmark Calculations on Silylenes	2509
	E. Theoretical Treatments of Silylene Chemistry	2510
V.	SPECTROSCOPIC CHARACTERIZATION OF SILYLENES	2512
	A. Electronic Spectra	2512
	1. Free silylenes	2512
	2. Silylene–Lewis acid complexes	2518
	B. Other Spectroscopic Measurements	2521
	1. Silylene SiH_2	2521
	2. Other silylenes	2522
	a. HSiF	2522
	b. SiF_2	2523
	c. $SiCl_2$	2523
	d. $SiBr_2$, SiI_2	2523
	e. $SiMe_2$	2523
	f. $SiMes_2$	2523
	g. Si=O	2523
VI.	KINETICS OF SILYLENE REACTIONS	2524
VII.	SILYLENE–TRANSITION METAL COMPLEXES	2527
VIII.	SPECIAL TOPICS	2530
	A. Silylene Centers on Activated Silicon	2530
	B. Silylenes in the Direct Synthesis of Organosilicon Compounds	2533
	C. Stable Dicoordinate Silylenes	2534
	1. History	2534
	2. Synthesis and characterization	2535
	3. Bonding and 'aromaticity'	2537
	4. Chemical reactions	2539
	D. Silylene Intermediates in Chemical Vapor Deposition	2545
	E. Silylenes Stabilized by Intramolecular Coordination	2550
	F. Vinylidenesilylene and Other Double-Bonded Silylenes Z=Si:	2555
IX.	SPECULATIONS ON THE FUTURE OF SILYLENE CHEMISTRY	2557
X.	REFERENCES	2558

I. INTRODUCTION

It is now over a decade since the last attempts to survey all of silylene chemistry[1–5]. While new work continues to be published at a brisk pace, it is fair to state that the field has reached a stage of maturity in which silylenes are tools for answering questions as much as ends in themselves. That implies a belief that the broad outlines of silylene chemistry are known and can be applied. This is not to say that there are no surprises, and indeed there is considerable healthy controversy, extending even to what one means by 'silylene'.

The chapter table of contents contains subjects that were either unknown or merely distant hopes a decade ago, such as persistent silylenes, the dissociation of disilenes to silylenes and terminal silylene–transition metal complexes. The kinetics and spectroscopy of silylenes and theoretical treatments of silylene structure and reactivity have made such gigantic strides in the intervening years that they represent new vistas in our understanding.

Theoretical calculations of 'chemical accuracy' can now predict reaction pathways as well as structures and energies.

In what follows, emphasis will be placed on developments of the past decade, with sufficient (but brief) summaries of previous work and references to earlier reviews and seminal papers to allow this chapter to function as a stand-alone unit. We attempt to place silylene chemistry in a broader context and to interpret as well as describe important discoveries.

Reviews have been presented of inorganic silylenes[6], silylene chemistry in the gas phase[7] and sterically encumbered silylenes[8].

II. GENERATION OF SILYLENES

A. Thermolysis of Polysilanes

The pyrolysis of polysilanes played an important role in the discovery of silylene reactions[1-3,5] and is still widely used for the generation of silylenes in the gas phase. Many such reactions are concerted silylene extrusions in which a substituent migrates from the incipient divalent silicon atom during Si—Si bond cleavage (equation 1).

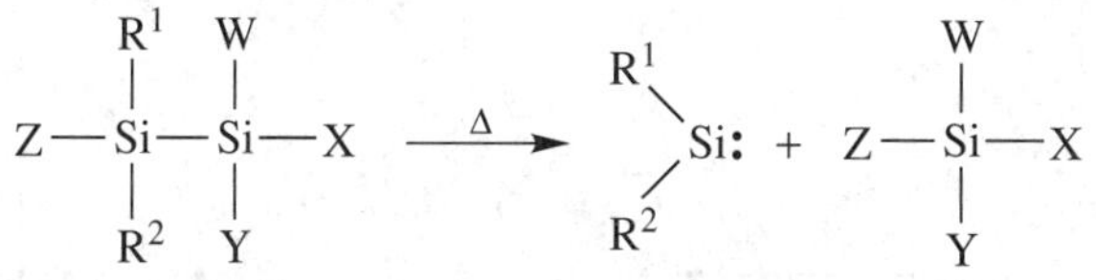

$$Z-\underset{\underset{R^2}{|}}{\overset{\overset{R^1}{|}}{Si}}-\underset{\underset{Y}{|}}{\overset{\overset{W}{|}}{Si}}-X \xrightarrow{\Delta} \underset{R^2}{\overset{R^1}{\diagdown}}Si: \ + \ Z-\underset{\underset{Y}{|}}{\overset{\overset{W}{|}}{Si}}-X \quad (1)$$

The migrating group Z can be hydrogen, halogen or alkoxy. It is not an overstatement that the 1966 discovery by Atwell and Weyenberg that the 1,2-shift of methoxy groups in disilanes takes place under relatively mild conditions ushered in the use of silylenes as synthetic reagents[9]. The ease of elimination, alkoxy > hydrogen > chlorine, has been measured quantitatively[10]. First-order rate constants for extrusion of Me_2Si at 350 °C from $Me_3SiSiMe_2X$ (X = OMe, H, Cl) were found to be 1.2×10^{-3}, 1.9×10^{-4} and 1.3×10^{-6} s^{-1}, respectively. The other groups can be virtually any substituent that migrates more slowly than Z^2. The symmetrical precursors shown in equations 2 and 3 have seen wide use in the generation of Me_2Si and H_2Si, respectively.

$$Me_2Si-\underset{\underset{OMe}{|}}{SiMe_2} \xrightarrow{\Delta} Me_2Si: \ + \ (MeO)_2SiMe_2 \quad (2)$$
(with MeO on left Si)

$$\underset{MeO}{\overset{}{|}}Me_2Si-SiMe_2\underset{OMe}{\overset{}{|}} \xrightarrow{\Delta} Me_2Si: + (MeO)_2SiMe_2 \quad (2)$$

$$H_3Si-SiH_3 \xrightarrow{\Delta} H_2Si: \ + \ SiH_4 \quad (3)$$

Trisilanes capable of the sequential pairwise loss of all four substituents from the central silicon can function as silicon atom synthons (see equation 4)[11]. Where comparison with reactions of atomic silicon has been possible, the product yield from the synthon has been higher.

The kinetics of the pyrolysis of disilane and substituted disilanes has received considerable attention in connection with the estimation of the heats of formation of silylenes. Martin, Ring and O'Neal reexamined the pyrolysis of Si_2H_6 (equation 3) in a static reactor and found both the high-pressure A-factor ($10^{15.6}$ to $10^{16.0}$ s^{-1}) and E_a (51.2 kcal mol^{-1}) to be higher than previously determined[12]. Walsh argued in 1988 that the A-factors for the disilanes $Me_nH_{3-n}SiSiH_3$ were all higher than the previous experimentally based estimates (with the exception of $Me_2HSiSiH_3$)[13]. Walsh's higher suggested activation parameters for MeH_2SiSiH_3 were very close to those subsequently found ($\log A = 15.6$ suggested versus 15.53 ± 0.28 found, $E_a = 51.7$ kcal mol^{-1} suggested versus 51.53 ± 0.65 kcal mol^{-1})[14]. Ring, O'Neal and coworkers found that eliminations of hydrogenated silylenes SiH_2 or H_3SiSiH proceed through 'looser' transition states with higher A-factors than do eliminations of alkylated silylenes such as MeHSi, Me_2Si, or $Me_2HSiSiH$[15]. Reexamination of the pyrolysis of $Me_3SiSiHMe_2$ and $Me_3SiSiMe_2SiHMe_2$ by Davidson and coworkers did, however, support earlier, lower values of the activation parameters[16].

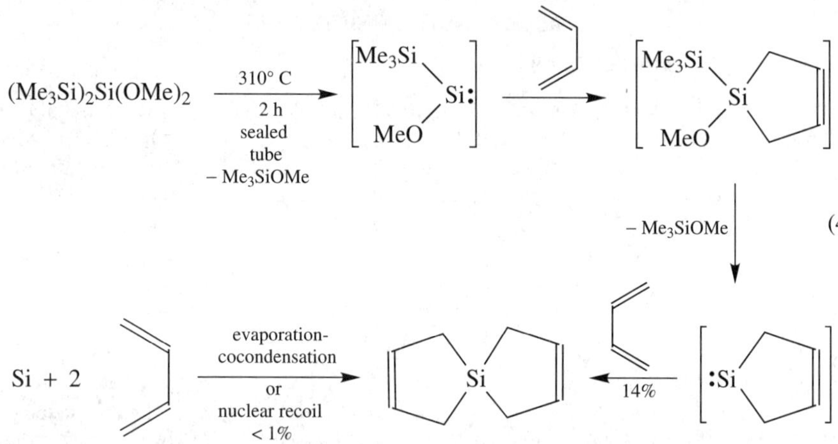

$$(Me_3Si)_2Si(OMe)_2 \xrightarrow[\substack{2\text{ h}\\ \text{sealed}\\ \text{tube}\\ -Me_3SiOMe}]{310°\text{ C}} \left[\begin{array}{c}Me_3Si\\ \diagdown\\ Si:\\ \diagup\\ MeO\end{array}\right] \longrightarrow \left[\begin{array}{c}Me_3Si\\ \diagdown\\ Si\\ \diagup\\ MeO\end{array}\right] \quad (4)$$

$$Si + 2 \xrightarrow[\substack{\text{or}\\ \text{nuclear recoil}\\ < 1\%}]{\text{evaporation-}\\ \text{cocondensation}} \left[\text{spirosilacycle}\right] \xleftarrow{14\%} [:Si\text{cycle}] \xleftarrow{-Me_3SiOMe}$$

Ab initio molecular orbital calculations have been carried out by Ignacio and Schlegel on the thermal decomposition of disilane and the fluorinated disilanes $Si_2H_nF_{6-n}$[17]. Both 1,1-elimination of H_2 or HF and silylene extrusion by migration of H and F atoms concerted with Si—Si bond cleavage were considered. The transition states for the extrusion reactions all involved movement of the migrating atom toward the empty p-orbital of the extruded silylene in the insertion which is the retro-extrusion (equation 5).

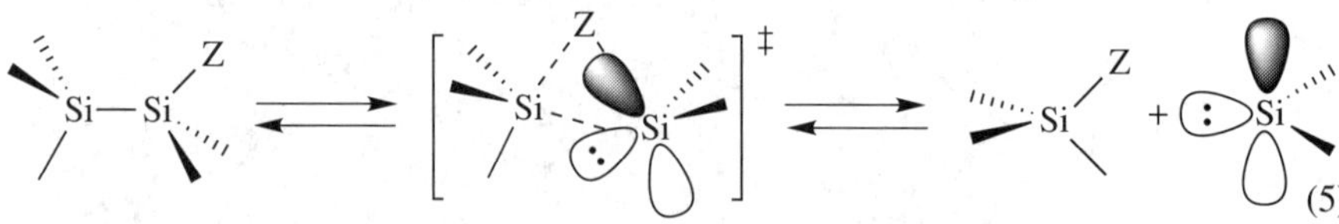

$$\text{Si—Si} \rightleftharpoons [\text{Si}\cdots\text{Si}]^{\ddagger} \rightleftharpoons \text{Si} + (:)\text{Si} \quad (5)$$

For all the fluorodisilanes considered, the barriers for the elimination of HF and the homolytic cleavage of an Si—Si bond were considerably higher than those for elimination of H_2 or the extrusion of a silylene.

Corriu and coworkers have suggested that pyrolysis of poly(silylethylene) $+SiH_2CH_2CH_2\frac{}{n}$ can lead to silylenes by two processes: (1) Loss of H_2 without chain cleavage; (2) extrusion of a silylene $(SiH_2CH_2CH_2)_n$SiH with cleavage of an Si—C bond and migration of an H-atom forming $CH_3CH_2(SiH_2CH_2CH_2\frac{}{n}$[18]. The first process leads to crosslinking via silylene insertion into Si—H bonds, while the second accounts for fragmentation of the polymer with the formation of $EtSiH_2$ chain ends.

Recently, flow thermolysis of disilanes has been used by Heinicke and coworkers to obtain the silylenes Me(Cl)Si, Me(MeO)Si and Me(Me$_2$N)Si, which were trapped with dienes to give silacyclopentenes[19]. The results with Me(Cl)Si are outlined in equation 6. With 1,3-butadiene and isoprene small amounts of the silacyclopent-2-ene were obtained along with the major product, the silacyclopent-3-ene isomer.

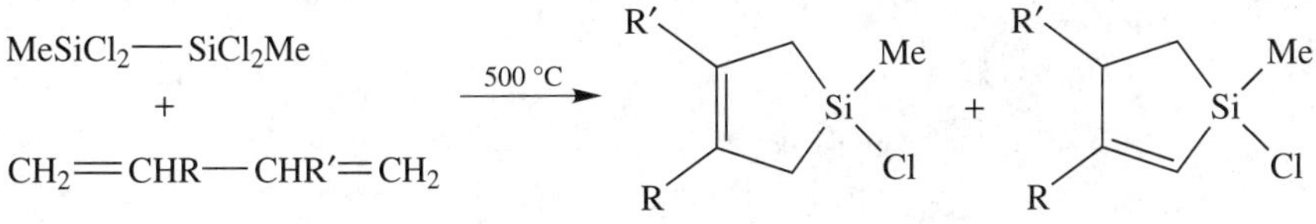

$$\text{MeSiCl}_2\text{—SiCl}_2\text{Me} + \text{CH}_2\text{=CHR—CHR'=CH}_2 \xrightarrow{500\,°C} \text{[silacyclopent-3-ene]} + \text{[silacyclopent-2-ene]}$$

(6)

R = R' = H	88 :	12
R = Me, R' = H	85 :	15
R = R' = Me	100 :	0

Similar results were obtained with Me(MeO)Si and Me(Me$_2$N)Si, consistent with a stepwise mechanism in which initial (2+1)cycloaddition is followed by rearrangement, as proposed earlier by Lei and coworkers[20]. Under the experimental conditions, the temperatures required for silylene generation were 400 °C for [MeSi(OMe)$_2$]$_2$, 420 °C for [MeSi(NMe$_2$)$_2$]$_2$ and 500 °C for [MeSiCl$_2$]$_2$[20].

The silylenes were also generated in the presence of heterodienes, including unsaturated aldehydes and ketones, diimines, diketones and iminoketones, to yield a variety of five-ring heterocycles containing Si—N and/or Si—O bonds[21]. Two examples are given in equations 7 and 8. With the α,β-unsaturated ketone in equation 7, the major product is the 3-ene isomer, but with the unsaturated imine in equation 8, both the 2-ene and 3-ene isomers are formed. Both reactions probably proceed through initial cycloaddition of the silylene to the C—heteroatom bond, as shown. Formation of the 2-ene isomer certainly suggests a three-membered ring intermediate.

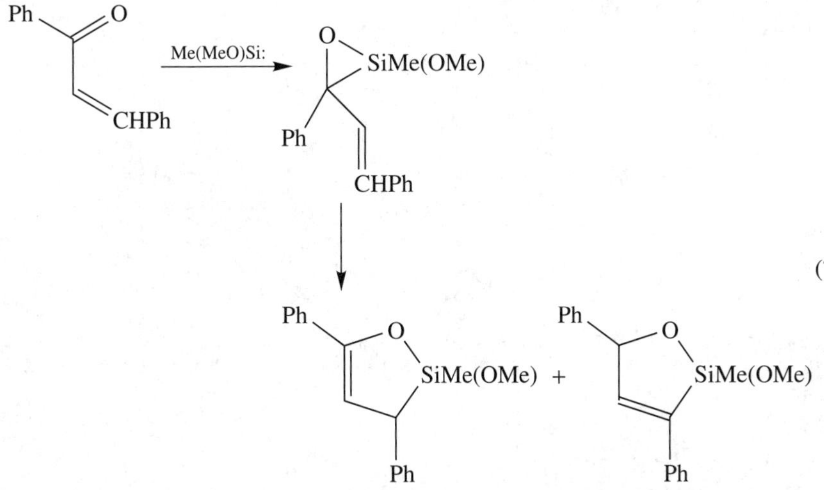

(7)

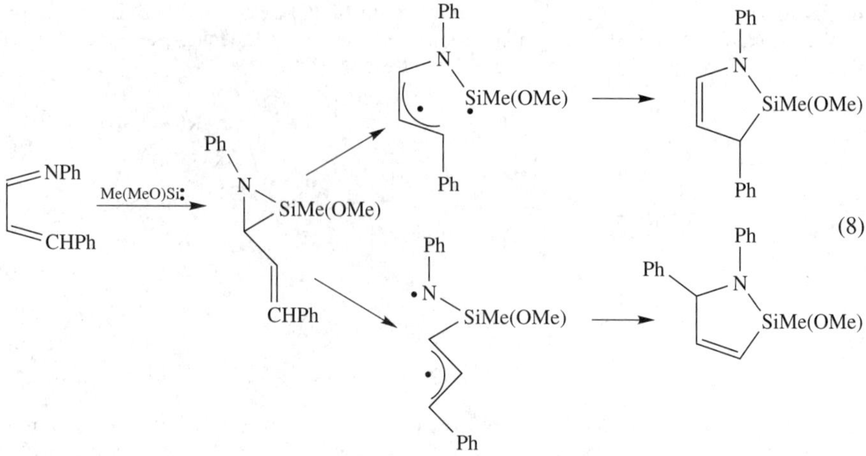

[Scheme showing reaction (8): butadiene-type NPh/CHPh compound reacting with Me(MeO)Si: to give intermediates with Ph-N-SiMe(OMe) groups, leading to cyclic products with SiMe(OMe)]

(8)

Intermolecular base-assisted disproportionation of disilanes also takes place, as was shown some years ago for Si_2Cl_6 and $MeSi_2Cl_5$[22]. These yielded branched oligosilanes upon treatment with Me_3N. Recently, the base-catalyzed disproportionation of the disilanes $MeSiCl_2SiCl_2Me$, $MeSiCl_2SiClMe_2$ and $Me_2SiClSiClMe_2$ has been investigated[23]. Catalysts employed were 1-methylimidazole, for homogeneous reactions, and a solid catalyst with $PO(NMe_2)_2$ groups attached to a silicate carrier for heterogeneous catalysis. A variety of branched oligomeric polysilanes was obtained. The reaction is thought to proceed via initial Lewis acid–Lewis base adduct formation, followed by a 1,2-chlorine shift to give a base-stabilized silylene which inserts into Si—Cl bonds to give the products. The preference for branched chains reflects the preference for insertion of the silylene into a central rather than a terminal Si—Cl bond. For example, see Scheme 1.

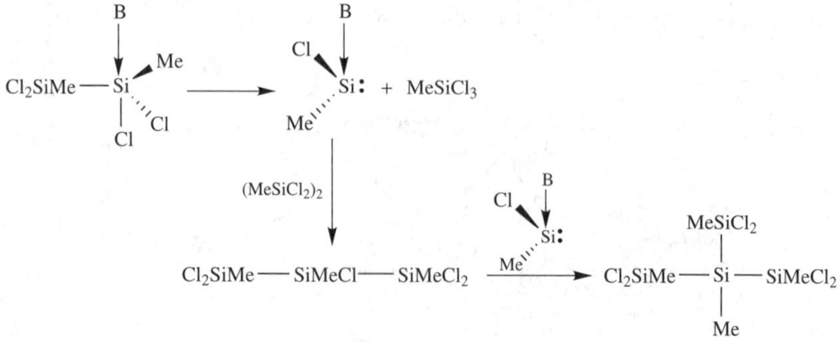

[Scheme 1: Cl_2SiMe—Si(B)(Me)(Cl) → Si:(B)(Cl)(Me) + $MeSiCl_3$; via $(MeSiCl_2)_2$ to Cl_2SiMe—SiMeCl—$SiMeCl_2$, then with Si:(B)(Cl)(Me) to Cl_2SiMe—Si($MeSiCl_2$)(Me)—$SiMeCl_2$]

SCHEME 1

Fragmentation of pentacoordinate di- or polysilanes, assisted by intramolecular coordination by chelating nitrogen atoms, can occur at very modest temperatures (see Section VIII.D).

B. Silicon Atom Reactions

Insertion, addition and abstraction reactions of free silicon atoms can lead to the formation of silylenes (equation 9)[1–5]. Silylene formation for reaction, spectroscopic and mechanistic studies has been initiated by high-temperature evaporation of bulk silicon and cocondensation of the resulting vapor (consisting largely of free silicon atoms) with substrates. Silicon atom insertion into H_2 delivered matrix-isolated H_2Si: for spectroscopic study[24]. Among the first mechanistic studies of silylenes were experiments employing free silicon atoms from the nuclear transformation $^{31}P(n,p)^{31}Si$[25–27].

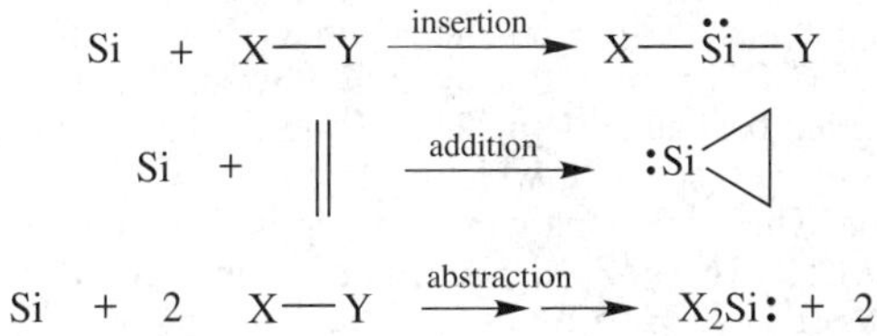

$$Si + X\!-\!Y \xrightarrow{\text{insertion}} X\!-\!\ddot{Si}\!-\!Y$$

$$Si + \| \xrightarrow{\text{addition}} :Si\!\triangleleft \qquad (9)$$

$$Si + 2\,X\!-\!Y \xrightarrow{\text{abstraction}} X_2Si: + 2\,Y\cdot$$

When a gaseous phosphorus-containing compound is irradiated with fast neutrons, high-energy silicon ions recoiling from this nuclear transformation break loose from the atoms to which their mother atoms were attached. Valence electrons are left behind as well, but neutralization and loss of energy occur before bond formation can occur. Thus the nuclear recoil method produces epithermal neutral silicon atoms that are radioactive ($t_{1/2} = 2.62$ h) and so carry a convenient label. It has been demonstrated that $^{31}SiH_2$ formed in recoil experiments is in its ground singlet state and at ambient temperature when it undergoes reaction[28].

C. Photolysis of Silanes

1. Photolysis of polysilanes

The generation of silylenes by extrusion from a chain or ring containing at least three adjacent Si atoms has long been known[2,3]. Polysilane high polymers undergo photolysis by two main processes, homolytic Si—Si bond cleavage to silyl radicals and silylene elimination[29]. Photolysis of poly(di-n-butylsilylene) and poly(di-n-hexlsilylene) with light of various wavelengths showed that silylenes are produced only with short-wavelength UV radiation ($\lambda < 300$ nm); only silyl radicals are formed at longer wavelengths[30] (equation 10).

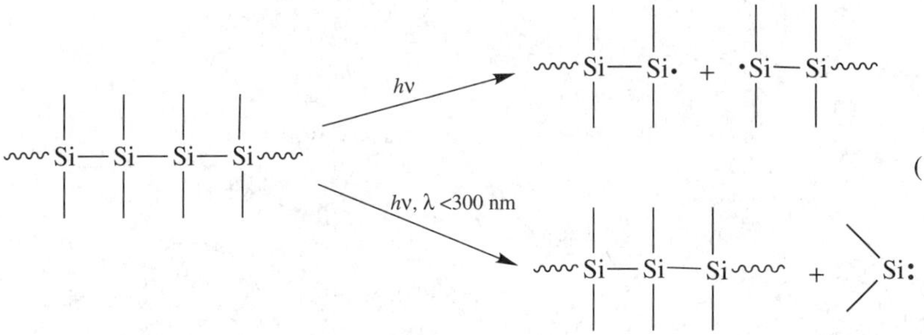

$$\sim\!\!Si\!-\!Si\!-\!Si\!-\!Si\!\!\sim \begin{array}{c} \xrightarrow{h\nu} \sim\!\!Si\!-\!Si\cdot + \cdot Si\!-\!Si\!\!\sim \\ \xrightarrow{h\nu,\ \lambda <300\ \text{nm}} \sim\!\!Si\!-\!Si\!-\!Si\!\!\sim + Si: \end{array} \qquad (10)$$

A recent flash photolytic study confirmed this unusual wavelength dependence for poly(phenylmethylsilylene). For this polymer also, irradiation at longer wavelengths (>300 nm) produced only silyl radicals. Light of higher energy yielded PhMeSi:[31], with $\lambda_{\text{max}} = 460$ nm, identified by trapping with triethylsilane. Similar results were obtained in the photolysis of network and branched polysilanes, $(Ph_2Si)_n(PhSi)_m$[32].

The photolysis of trisilanes, RR'Si(SiMe$_3$)$_2$, or of cyclosilanes (RR'Si)$_n$, is a well-established method for silylene generation[2,3]. Here we will describe only selected examples, concentrating especially on findings which bear on the mechanism of silylene elimination.

Photolysis of diphenylbis(trimethylsilyl)silane (**1**) produces diphenylsilylene, along with a significant amount of byproducts resulting from a 1,2-silyl shift to give a silatriene (Scheme 2). In an important study by Kira and coworkers[33] it was found that **1** undergoes nonresonant two-photon (NRTP) absorption upon laser irradiation at 532 nm, even though **1** shows no UV absorption beyond 300 nm. Thus **1** must belong to the class of molecules which have low-lying excited states to which NRTP is allowed, although single-photon absorption is forbidden. The most remarkable finding in this study was that the ratio of ethanol-trapped products, **2**:(**3A** + **3B**), depended on the absorption process. Irradiation of **1** at 266 nm gave yields of **2** and **3A** + **3B** of 53% and 30%, but NRTP at 532 nm produced these products in 80% and 13% yield, respectively. This unexpected selectivity provides an apparent exception to Kasha's rule[34], which states that rapid internal conversion to the lowest excited state takes place, so that photoproducts do not depend on the excitation mode. Such internal conversion must not happen for the NRTP excited state. It will be interesting to see if NRTP processes will provide product selectivity in other silane photolyses.

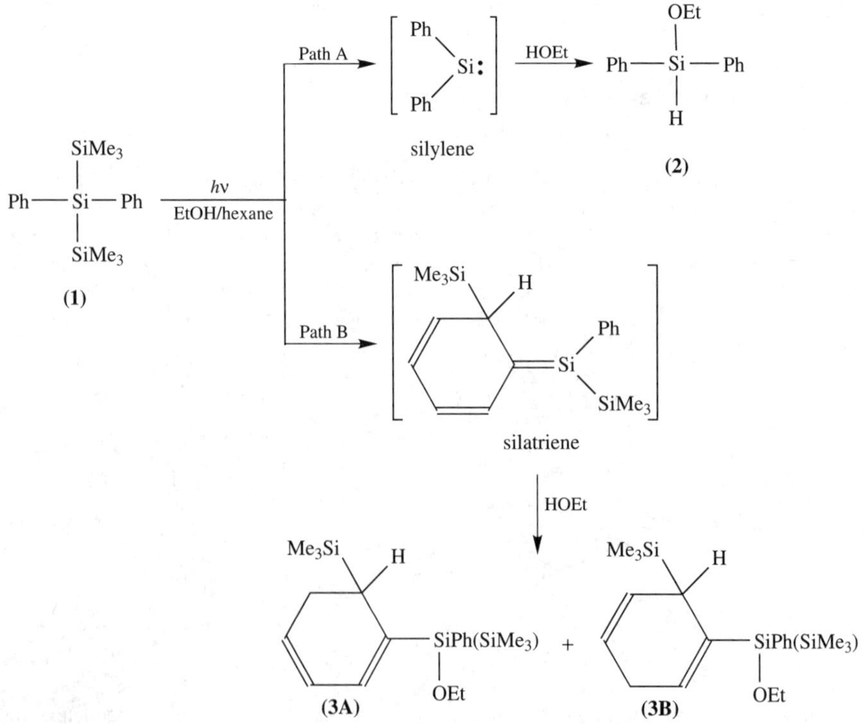

SCHEME 2

The photolysis of phenyltris(trimethylsilyl)silane in the presence of excess ethanol gave, as the first product, PhSi(H)OEt—SiMe$_3$. The amount of this product went through a maximum and then declined, being replaced by Me_3SiH and $PhSi(H)(OEt)_2$[35]. These results are best rationalized by sequential formation of two silylenes, as shown in equation 11.

$$PhSi(SiMe_3)_3 \xrightarrow[-(Me_3Si)_2]{h\nu} Ph(Me_3Si)Si: \xrightarrow{EtOH} Ph-\underset{\underset{OEt}{|}}{\overset{\overset{H}{|}}{Si}}-SiMe_3$$

$$\downarrow h\nu \; -Me_3SiH \qquad (11)$$

$$Ph-\underset{\underset{OEt}{|}}{\overset{\overset{H}{|}}{Si}}-OEt \xleftarrow{EtOH} Ph(OEt)Si:$$

Photolysis of several polysilanes in a molecular beam has been investigated[36]. For $PhMeSi(SiMe_3)_2$ and $PhSi(SiMe_3)_3$, a single photon resulted in loss of only one $SiMe_3$ group; capture of a second photon was necessary for cleavage of a second Si−Si bond and formation of the silylene. For the cyclic silane $(Me_2Si)_6$, on the other hand, a single photon sufficed to bring about elimination of $Me_2Si:$ and formation of $(Me_2Si)_5$. The implication is that the mechanism of the photolysis is quite different for these compounds. The result for $(Me_2Si)_6$ is consistent with, but does not require, a concerted mechanism for silylene extrusion. In any case, if the result is general for cyclosilanes, it may open up a new method for gas-phase generation of silylenes.

Several papers have appeared in which aryltris(trimethylsilyl)silanes have been employed as photoprecursors to trimethylsilyl-substituted silylenes, $Ar(Me_3Si)Si:$ [37,38]. In addition, Conlin and coworkers have photolyzed vinyltris(trimethylsilyl)silane at 254 nm, and shown by trapping experiments with 2-trimethylsiloxybutadiene that both silylene (**4**) and silene (**5**) are formed (equation 12)[39].

$$CH_2=CHSi(SiMe_3)_3 \begin{array}{c} \xrightarrow[-Me_6Si_2]{h\nu} CH_2=CH(Me_3Si)Si: \quad (\textbf{4}) \\ \xrightarrow[\sim SiMe_3]{h\nu} (Me_3Si)_2Si=CHCH_2SiMe_3 \\ (\textbf{5}) \end{array} \qquad (12)$$

Photolysis of cyclic polysilanes to give silylenes was first pioneered in the 1970s by Kumada and Ishikawa[40]. The classic example is dodecamethylcyclohexasilane, $(Me_2Si)_6$, which yields $Me_2Si:$ and the ring-contracted product, $(Me_2Si)_5$[41,42]. UV light of $\lambda = 254$ nm from a mercury lamp is commonly used. Photolysis of polysilanes continues to be an important method for generation of silylenes; the readily-available five- and six-membered rings are most often employed. An example is the photolysis of the rotane **6** to cyclic silylene **7** (equation 13)[43].

Photolysis of cyclosilane rings containing heteroatoms also leads to silylenes, and sometimes this is the preferred route. An example is shown in equation 14[44]. Likewise, hexaneopentyl trisilaoxetane undergoes photolysis to give the silylene and a disilaoxirane

(equation 15)[45].

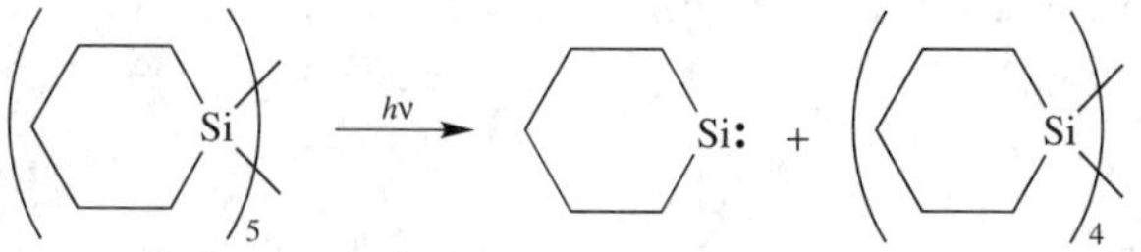

$$(6) \xrightarrow{h\nu} (7) \text{Si:} + \text{Si} \quad (13)$$

$$(t\text{-BuMeSi})_4\text{O} \xrightarrow{h\nu} t\text{-BuMeSi:} + (t\text{-BuMeSi})_3\text{O} \quad (14)$$

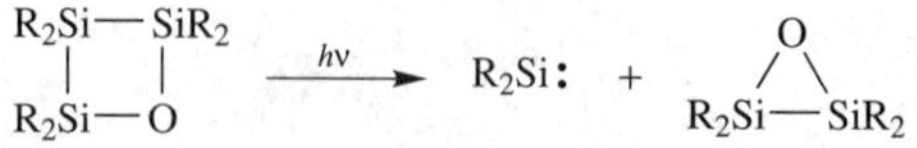

$$\begin{array}{c} R_2Si-SiR_2 \\ | \quad\quad | \\ R_2Si-O \end{array} \xrightarrow{h\nu} R_2Si: + \begin{array}{c} O \\ / \backslash \\ R_2Si-SiR_2 \end{array} \quad (15)$$

R = neopentyl

Shizuka and coworkers have shown that the photolysis of four-membered cyclosilanes depends on the ring conformation. Folded rings, such as $(i\text{-Pr}_2\text{Si})_4$ and $[\text{Me}(t\text{-Bu})\text{Si}]_4$, eliminate silylenes upon photolysis, but planar rings, $[(\text{Me}_3\text{Si})_2\text{Si}]_4$ and $[(\text{EtMe}_2\text{Si})_2]_4$, photolyze to two molecules of disilene, $R_2Si=SiR_2$[46].

In 1985[47], Weidenbruch and Pohl made the important discovery that hexa-*t*-butylcyclotrisilane photolyzes to yield both silylene and disilene (equation 16).

$$(t\text{-Bu}_2\text{Si})_3 \xrightarrow{h\nu} t\text{-Bu}_2\text{Si:} + t\text{-Bu}_2\text{Si}=\text{Si}(t\text{-Bu})_2 \quad (16)$$

This reaction, providing two highly reactive intermediates, has led to a remarkably rich and diverse reaction chemistry, too extensive for a complete treatment in this chapter. Readers are referred to Weidenbruch's comprehensive 1994 review of this area[8]; here we will describe only some developments stemming from this reaction since 1994.

Although the reaction of $t\text{-Bu}_2\text{Si:}$ with nitriles, isocyanides, carbonyl compounds and heterodienes had been studied earlier, the interaction with alkenes and dienes is new. Siliranes are obtained with alkenes, as predicted (equation 17)[48].

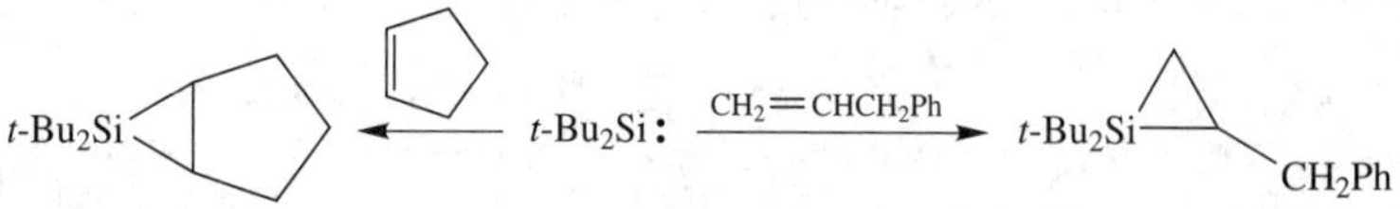

$$t\text{-Bu}_2\text{Si}\triangleleft \xleftarrow{} t\text{-Bu}_2\text{Si:} \xrightarrow{\text{CH}_2=\text{CHCH}_2\text{Ph}} t\text{-Bu}_2\text{Si}\triangleleft_{\text{CH}_2\text{Ph}} \quad (17)$$

Silylene and disilene adducts are obtained with dienes (equation 18). With thiophene, desulfurization takes place and the products are more complicated, as shown in equation 19[49].

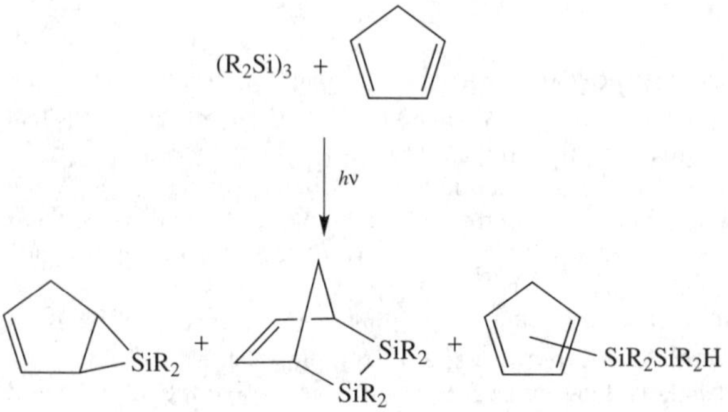

$$(R_2Si)_3 + \bigcirc \xrightarrow{h\nu} \text{—SiR}_2 + \text{SiR}_2/\text{SiR}_2 + \text{—SiR}_2\text{SiR}_2\text{H} \quad (18)$$

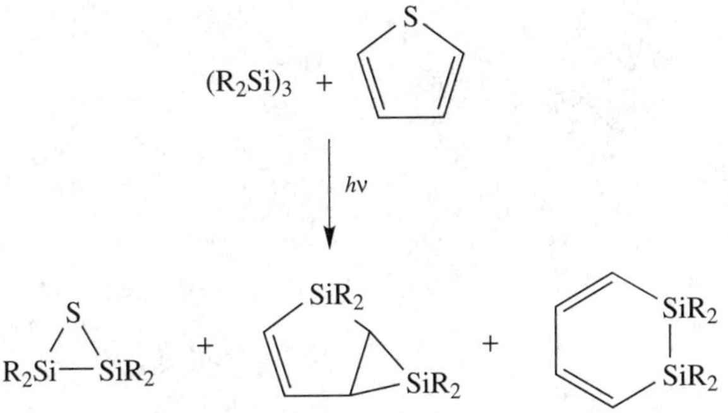

Complex products were also obtained from electron-poor dienes[50]. Photolysis of (t-Bu$_2$Si)$_3$ with tetrazine **8** gave a complex mixture of products, from which **9** and **10** could be isolated (equation 20). In **10**, one of the CF$_3$ groups has been completely degraded, with simultaneous rearrangement of the ring skeleton. The reaction with oxadiazole **11** also took an unexpected course (equation 21).

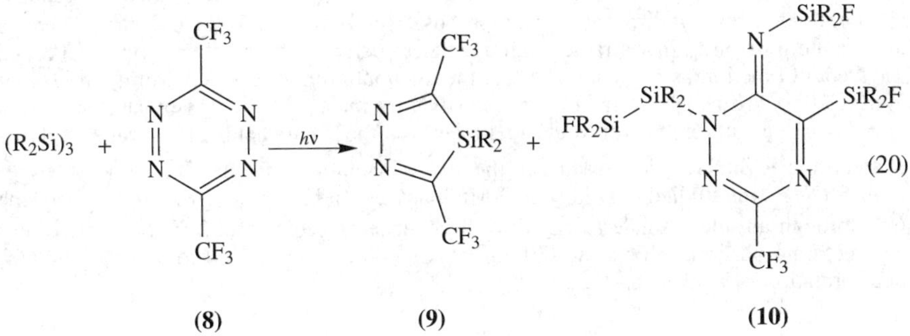

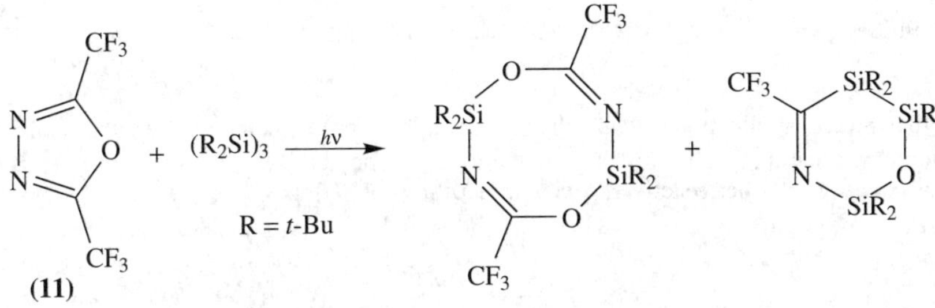

2. Other photolyses generating silylenes

Photochemical fragmentation of 7-silanorbornadienes is a convenient and frequently-used method for generation of silylenes. Nefedov and coworkers have studied this reaction for **12** using ^{1}H CIDNP to detect intermediate radicals[51]. The results indicate that the reaction proceeds through the reversible formation of a single biradical (**13**), which undergoes internal conversion to the triplet biradical. The latter eliminates dimethylsilylene irreversibly (Scheme 3). Similar conclusions were drawn for the germanium analog of **12**, which eliminates Me$_2$Ge: upon photolysis.

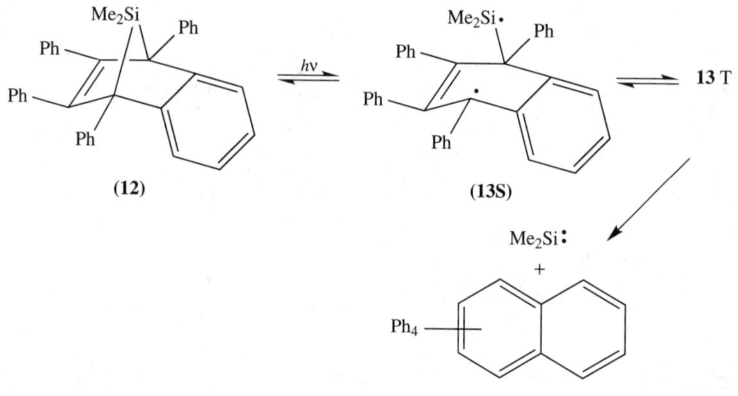

[Scheme 3 structures]

SCHEME 3

There is a difficulty with the mechanism of Scheme 3 in that the fragmentation of a triplet diradical should conserve spin, yet neither triplet Me_2Si: nor triplet tetraphenylnaphthalene have been detected. The diradical pathway for the photofragmentation of silanorbornadienes confirms an earlier proposal by Barton and coworkers[52]. There is always the possibility that diradical intermediates such as the singlet and triplet **13 S** and **13 T** could function as silylenoids. Thus, the assumption that products from pyrolysis of 7-silanorbornadienes are formed from free silylenes must be treated with caution.

A less-frequently employed route to silylenes is the photochemical cleavage of diazidosilanes, $RR'Si(N_3)_2$, first used for the matrix isolation of Me_2Si:[41]. Photolysis of t-$Bu_2Si(N_3)_2$ was studied in argon and 3-MP matrices, and found to give di-t-butylsilylene (**15**) through an intermediate assigned as diazosilane **14** (equation 22)[53]. Silylene **15** has an electronic absorption band at 480 nm; irradiation at 500 nm led to a hydrogen shift and formation of a silirane.

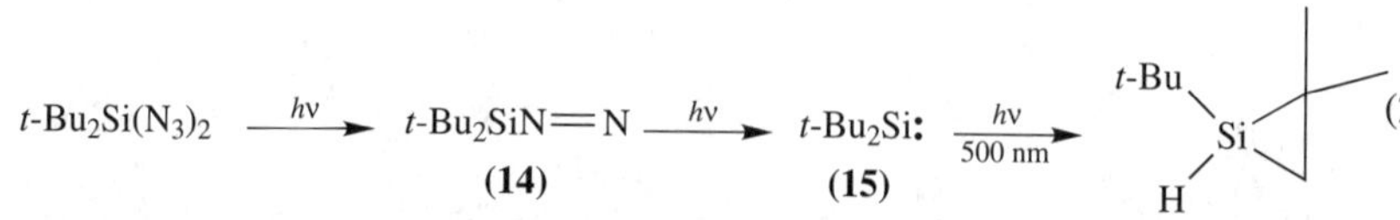

$$t\text{-}Bu_2Si(N_3)_2 \xrightarrow{h\nu} t\text{-}Bu_2SiN=N \xrightarrow{h\nu} t\text{-}Bu_2Si: \xrightarrow[500\text{ nm}]{h\nu} \underset{H}{\overset{t\text{-}Bu}{\text{Si}}}\triangleleft \quad (22)$$
$$\qquad\qquad\qquad\qquad\quad (14)\qquad\qquad\quad (15)$$

Other recent applications of the diazide photolysis route include the matrix isolation of some interesting cyclic silylenes (equations 23 and 24)[54,55]. Solution photolysis of diazides can give other reactive species in addition to silylenes.

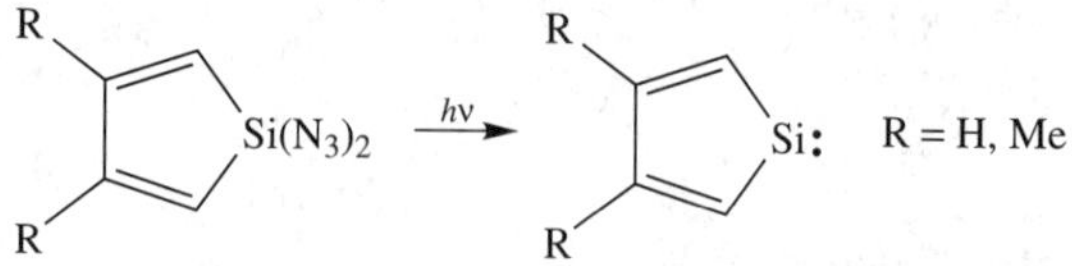

[Equation 23 structure] R = H, Me (23)

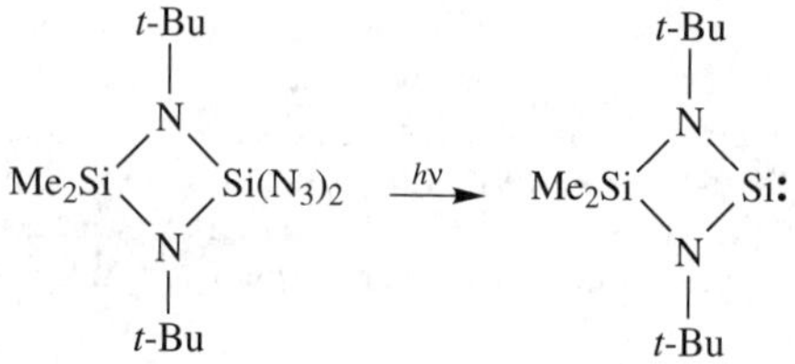

[Equation 24 structure] (24)

The use of siliranes as thermal sources of silylenes has long been known. At least in some cases, they can also yield silylenes photochemically. This has been demonstrated for di-t-butylsilylene[56] and di-1-adamantylsilylene (**16**)[57] as shown in equations 25 and 26.

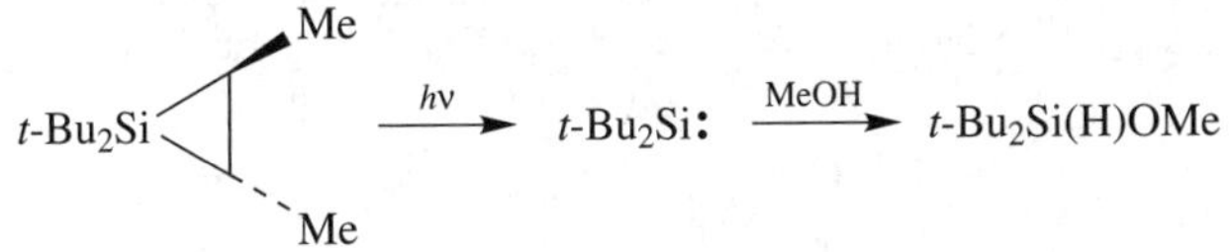

$$t\text{-Bu}_2\text{Si}\bigtriangleup\!\!\!\!\!{}^{Me}_{Me} \xrightarrow{h\nu} t\text{-Bu}_2\text{Si:} \xrightarrow{MeOH} t\text{-Bu}_2\text{Si(H)OMe} \quad (25)$$

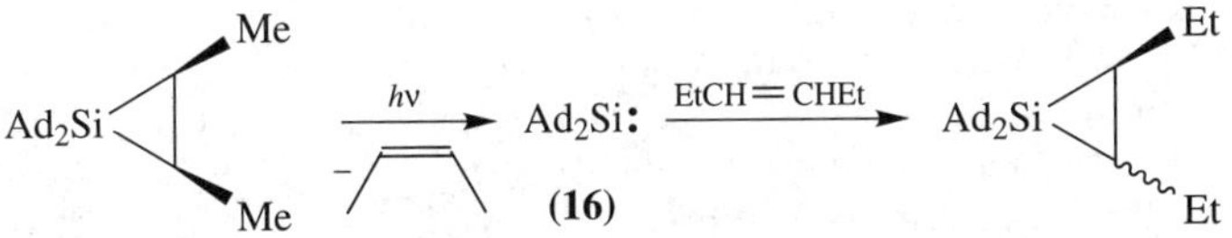

$$\text{Ad}_2\text{Si}\bigtriangleup\!\!\!\!\!{}^{Me}_{Me} \xrightarrow{h\nu} \text{Ad}_2\text{Si:} \xrightarrow{EtCH=CHEt} \text{Ad}_2\text{Si}\bigtriangleup\!\!\!\!\!{}^{Et}_{Et} \quad (26)$$

(**16**)

Most studies of silane photolysis have been conducted with radiation of wavelength 254 nm (high pressure mercury lamp) or lower energy. At higher energy, additional kinds of organosilanes may serve as silylene precursors. An initial foray in this direction has been reported by Steinmetz and Yu, who photolyzed silacyclopentene **17** with 214 nm radiation in methanol[58]. The predominant photoprocess was fragmentation to 1,3-butadiene and $Me_2Si:$. The silylene is partly trapped by methanol and partly reacts with the butadiene to give the vinylsilirane (**18**), which then undergoes stereoselective 1,4-addition of methanol (equation 27).

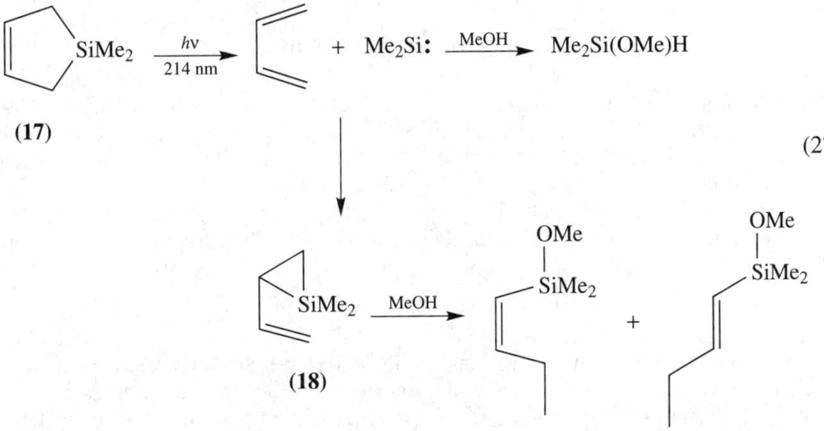

(**17**) $\xrightarrow[214 \text{ nm}]{h\nu}$ [butadiene] + $Me_2Si:$ $\xrightarrow{MeOH}$ $Me_2Si(OMe)H$

(**18**) $\xrightarrow{MeOH}$ [products with OMe–SiMe$_2$ groups]

(27)

What is the effect of extremely high energies on silylene precursors? The fragmentation of polysilanes by gamma rays has been investigated by Nakao and coworkers[59,60]. Irradiation of $(Me_2Si)_6$ in a mixture of benzene and ethanol produced the silylene trapping product, $Me_2Si(H)OEt$, in yields up to 65%. Similarly, $PhMeSi(SiMe_3)_2$ was exposed to gamma rays in benzene-methanol, yielding $PhMeSi(H)OMe$ as the major product. The reactions are thought to proceed through excitation of the benzene by the gamma rays, followed by collisional activation of the polysilane and loss of silylene.

D. Pyrolysis of Monosilanes

The simplest silylene SiH_2 can be formed by the homogeneous gas-phase elimination of H_2 from silane at temperatures above 380 °C (equation 28)[1-3]. This reaction has a long, controversial history and is of technological importance because of its contribution to the chemical vapor deposition of amorphous silicon from silane (*vide infra*). Pyrolysis of SiH_4 has seen little use in the study of silylene reactions, since it requires higher temperatures than the extrusion of SiH_2 from Si_2H_6 discussed in Section II.A (equation 3). The mechanism of silane pyrolysis has, however, been the subject of mechanistic studies for 60 years.

$$SiH_4 \xrightarrow{\Delta} H_2Si\colon + H_2 \quad (28)$$

Ring, O'Neal and coworkers have contributed much to the understanding of silane decomposition. In 1985 they reexamined the kinetics of silane loss and its inhibition in the presence of excess H_2 which is common under CVD conditions[61]. They proposed an explanation for the transition from the initial stage (<1% conversion) of SiH_4 pyrolysis in which the reaction is 3/2 order to the middle stage (3 to 30% conversion) in which the reaction is first order and more rapid. It was suggested that, in the middle stage, elimination of H_2 from SiH_4 occurs as both a wall reaction and a gas-phase process, the onset of surface initiation being promoted by the deposition of oligomeric decomposition products. It was suggested that disilane and trisilane concentrations reach steady state due to their reversible formation by SiH_2 insertion reactions, and the reversible extrusion of SiH_4 from trisilane (equations 29–31).

$$H_2Si\colon + SiH_4 \rightleftharpoons SiH_3SiH_3 \quad (29)$$

$$H_2Si\colon + SiH_3SiH_3 \rightleftharpoons SiH_3SiH_2SiH_3 \quad (30)$$

$$SiH_3SiH_2SiH_3 \rightleftharpoons SiH_3SiH\colon + SiH_4 \quad (31)$$

The dominant sink reaction for the removal of SiH_2 was believed to be 1,2-H_2-elimination from H_3SiSiH_3 forming $H_2Si=SiH_2$, which could undergo wall polymerization. In the final stage of silane pyrolysis (>30% conversion) reversal of H_2-elimination from SiH_4 could compete with other SiH_2 removal mechanisms, slowing the loss of SiH_4.

In 1991 Moffat, Jensen and Carr employed RRKM theory in the form of a non-linear regression analysis of experimental data to estimate the high-pressure Arrhenius parameters for elimination of H_2 from SiH_4 as $\log A = 15.79 \pm 0.5 \text{ s}^{-1}$, $E_a = 59.99 \pm 2.0 \text{ kcal mol}^{-1}$ and $\Delta H_f^\circ(SiH_2) = 65.5 \pm 1.0 \text{ kcal mol}^{-1}$[62].

In 1992 Ring and O'Neal once more analyzed the thermal decomposition of SiH_4[63]. That the rates of SiH_2 insertion reactions and H_3SiSiH to $H_2Si=SiH_2$ rearrangement were much faster than previously believed, necessitated revisions in the reactions postulated to act as silylene 'sinks'. In a revised view of the elimination of H_2 from SiH_3SiH_3, the exclusive pathway was 1,1-elimination forming $H_3Si\ddot{S}iH$. Adsorption and subsequent decomposition on the walls of higher silanes such as tri- and tetrasilanes were believed to be the major processes responsible for removal of the gaseous species. The modeling of silane thermal decomposition also indicated that all silylene Si—H insertion reactions have negative activation energies (see Section VI).

Also in 1992 Becerra and Walsh suggested that the increase in the rate of silane loss in the 3 to 30% conversion middle stage of silane pyrolysis is due to the chain process (equation 32) in which the chemically activated silylene insertion product gives rise to a silylsilylene that can consume another silane molecule[64].

$$SiH_4 \xrightarrow[-H_2]{\Delta} H_2Si\colon \xrightarrow{SiH_4} SiH_3SiH_3^* \xrightarrow[-H_2]{\Delta} SiH_3SiH\colon \xrightarrow{SiH_4} SiH_3SiH_2SiH_3 \quad (32)$$

Silylene SiH_2 was directly detected during the pyrolytic decomposition of silane (and ethylsilane $EtSiH_3$) for the first time by intracavity laser spectroscopy in 1988[65]. The high detection sensitivity of this experiment allowed the pyrolysis mechanism to be examined under conditions that optimized film growth rather than spectroscopic conditions.

The decomposition of methylsilane was studied in 1987 via high level *ab initio* molecular orbital calculations with correlation corrections to the MP4 level of perturbation theory[66]. 1,1-Elimination of H_2 forming CH_3SiH: was found to have the lowest activation energy, followed by elimination of methane forming H_2Si:. 1,2-H_2-elimination forming $CH_2=SiH_2$ was disfavored, as were homolytic bond cleavages.

Kinetic measurements in a shock tube on the pyrolysis of CH_3SiH_3[67,68], $EtSiH_3$[69], $CH_2=CHSiH_3$[70] and n-$PrSiH_3$[71] led to very similar activation parameters ($\log A$ ca 15, E_a 63–65 kcal mol^{-1}), and the primary dissociation processes were believed to be H_2-elimination, the 1,1-process being more important than the 1,2-process. In secondary decomposition processes $EtSiH$: and higher alkylsilylenes can decompose to form alkenes and H_2Si:. Study of the pyrolytic formation and decomposition of n-$BuSiH$: in a shock tube[72] and $EtSiH$: in a static system[73] to elucidate the mechanisms for secondary silylene decomposition led to the conclusion that intramolecular C–H insertions forming silacycles mediate the formation of H_2Si:, as shown in equation 33 for $EtSiH$:. It should be noted that 1,2-H_2-elimination is no longer believed to contribute to the primary pyrolysis.

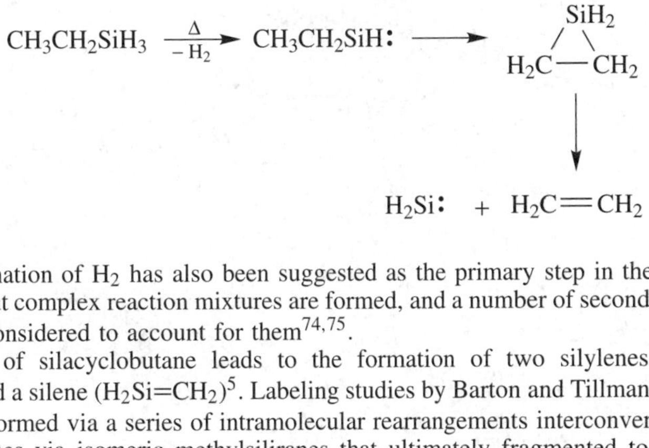

$$CH_3CH_2SiH_3 \xrightarrow[-H_2]{\Delta} CH_3CH_2SiH: \longrightarrow \underset{H_2C-CH_2}{SiH_2} \longrightarrow H_2Si: + H_2C=CH_2 \quad (33)$$

1,1-Elimination of H_2 has also been suggested as the primary step in the pyrolysis of Me_2SiH_2, but complex reaction mixtures are formed, and a number of secondary processes have been considered to account for them[74,75].

Pyrolysis of silacyclobutane leads to the formation of two silylenes (H_2Si: and $MeHSi$:) and a silene ($H_2Si=CH_2$)[5]. Labeling studies by Barton and Tillman revealed that H_2Si: was formed via a series of intramolecular rearrangements interconverting isomeric propylsilylenes via isomeric methylsiliranes that ultimately fragmented to silylene and propylene (equation 34)[76].

From studies of the reaction kinetics of butyl- and pentylsilylenes formed by ring opening of siliranes produced by addition of H_2Si to butenes and pentenes, generic high-pressure Arrhenius parameters could be deduced for silirane ring closure by intramolecular H–C insertion ($\log A = 12.3$, $E_a = 10.4$ kcal mol^{-1}), ring opening to a silylene ($\log A = 14.0$, $E_a = 64.3$ kcal mol^{-1}–silirane ring strain) and silylene extrusion ($\log A = 16.9$, $E_a = 75.7$ kcal mol^{-1}–silirane ring strain)[77].

Silirane pyrolysis has been employed as a mild route to silylenes since the pioneering work in the 1970s of Seyferth's group on hexamethylsilirane[2]. An important step forward was made by Boudjouk and coworkers when they found that bulky substituents (such as t-butyl) on silicon increased the stability of the silirane without preventing its thermolysis (equation 35)[56].

(35)

To explain the formation of the expected product of insertion into the H−Si bond of HSiEt$_3$ in similar yields from different precursors, it was suggested that t-Bu$_2$Si: also arises by ultrasound-promoted lithio-dehalogenation of t-Bu$_2$SiX$_2$ (X = Cl, Br, I); see also Section II.E. This would represent the first example of a metal-promoted α-elimination of dihalosilanes. In a related study of the chemistry of diadamantylsilylene Ad$_2$Si: the reactive species formed from Ad$_2$SiI$_2$ upon treatment with lithium under ultrasonic irradiation was compared with that formed upon pyrolysis of a silirane, as shown in equations 36 and 37[57].

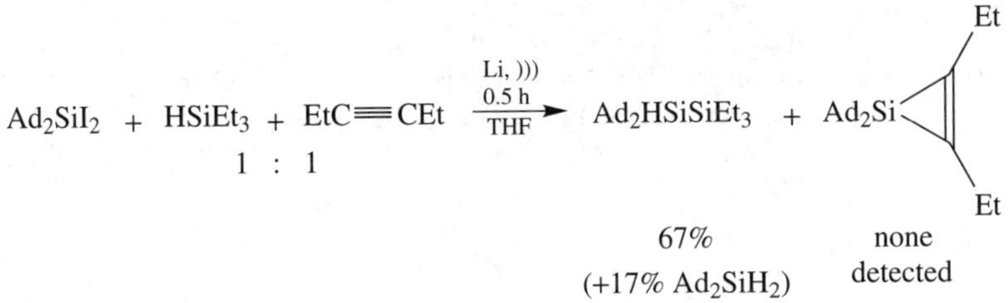

$$\text{Ad}_2\text{SiI}_2 + \text{HSiEt}_3 + \text{EtC} \equiv \text{CEt} \quad \xrightarrow[\text{THF}]{\text{Li,))), 0.5 h}} \quad \text{Ad}_2\text{HSiSiEt}_3 + \text{Ad}_2\text{Si}\underset{\text{Et}}{\overset{\text{Et}}{\triangleleft}} \quad (36)$$

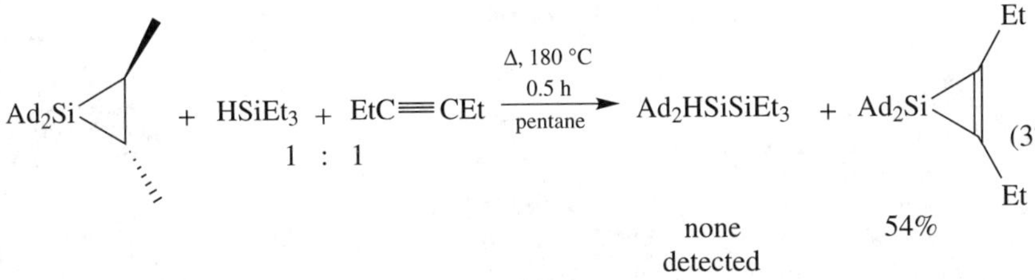

$$1 \; : \; 1$$

67% none
(+17% Ad$_2$SiH$_2$) detected

$$\text{Ad}_2\text{Si}\triangleleft + \text{HSiEt}_3 + \text{EtC} \equiv \text{CEt} \quad \xrightarrow[\text{pentane}]{\Delta, 180\,°\text{C}, 0.5\,\text{h}} \quad \text{Ad}_2\text{HSiSiEt}_3 + \text{Ad}_2\text{Si}\underset{\text{Et}}{\overset{\text{Et}}{\triangleleft}} \quad (37)$$

$$1 \; : \; 1$$

none 54%
detected

The inversion in selectivity between the insertion into an H−Si bond favored by the intermediate in the dehalogenation and the addition to a triple bond favored by the intermediate formed upon silirane pyrolysis suggests that different reactive intermediates are formed. Since free silylene is implicated in the silirane pyrolysis, the lithium-induced deiodination probably involves a silylenoid such as a complex of the silylene with LiI or with THF. For a further discussion of this possibility, see Section II.E.

The thermodynamic stability of unsubstituted silacyclopropane to fragmentation has been studied by *ab initio* quantum mechanical methods and the enthalpy of decomposition to H$_2$Si: + CH$_2$=CH$_2$ was predicted to be 44.8[78] and 43.2[79] kcal mol^{-1}. There is indirect experimental support for these theoretical estimates. When these values were employed in RRKM calculations on silirane decomposition, the pressure dependence of the bimolecular rate constant for addition of H$_2$Si: to ethylene could be accurately modeled[80].

Calculations indicate that silacyclopropenes might also be employed as silylene precursors in thermolysis reactions. The enthalpy of dissociation of unsubstituted 1-silacycloprop-2-ene is predicted to be 50.4 kcal mol^{-1}, only *ca* 7 kcal mol^{-1} greater than for silirane[81].

Extrusion of a silylene from a silirane or silirene is of course the inverse of silylene addition to alkenes or alkynes, respectively. The reversibility of most silylene reactions allows the inverse of addition to 1,3-dienes to also be employed as a silylene source. The first such reaction was reported by Chernyshev and coworkers, who found that transfer of SiCl$_2$ units from 1,1-dichlorosilacyclopent-3-enes was a unimolecular process and hence was likely to consist of silylene extrusion and readdition (equation 38)[82]. Dimethylsilylene extrusion has been found in the pyrolysis of silacyclopentenes and other products of

Me$_2$Si: addition[83,84].

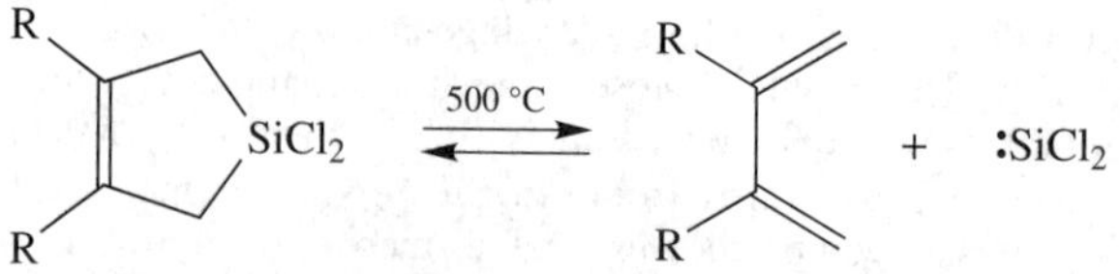

$$\text{(R-substituted silacyclopentene with SiCl}_2\text{)} \underset{}{\overset{500\,°C}{\rightleftarrows}} \text{(diene)} + :SiCl_2 \tag{38}$$

Product **19** in equation 39 is formed by trapping extruded Me$_2$Si:, but **20**, **21** and **22** are rearrangement products of silacyclopent-3-ene **23**. Their formation is in accord with stepwise retroaddition with silylene extrusion from a vinylsilirane that is an intermediate in a reversible addition mechanism. As expected from such a mechanism, **20**, **21** and **22** are coproducts with **23** in the addition of Me$_2$Si: to the piperylenes[20].

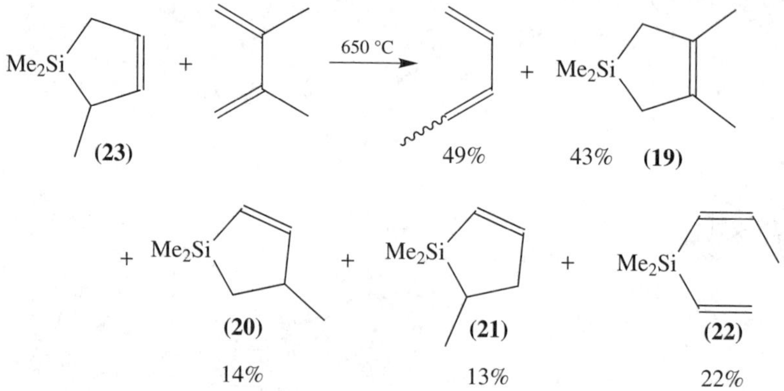

$$\text{Me}_2\text{Si (23)} + \text{(2,3-dimethylbutadiene)} \xrightarrow{650\,°C} \text{(diene, 49\%)} + \text{Me}_2\text{Si (19), 43\%}$$

$$+ \text{ Me}_2\text{Si (20)}\ 14\% + \text{ Me}_2\text{Si (21)}\ 13\% + \text{ Me}_2\text{Si (22)}\ 22\% \tag{39}$$

That product **22** in equation 39 is due to a retrohomo-ene reaction of a vinylsilirane intermediate was demonstrated by employing a vinyl(1-alkenyl)silirane as a starting material for silylene extrusion, as shown in equation 40[85]. As shown in equation 41, the first step in the extrusion of Me$_2$Si: from **23** is a homo-ene reaction forming a vinylsilirane that undergoes competing rearrangements to products **24** and **25** and extrusion of Me$_2$Si:. The silylene is trapped (reversibly) by isoprene as **26** that can rearrange to **27**.

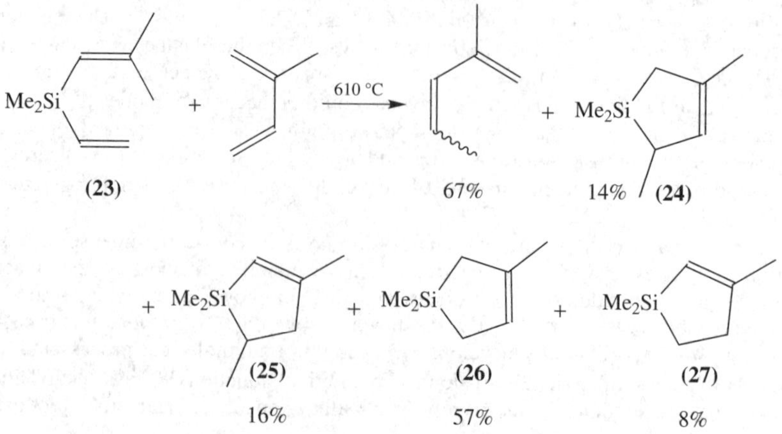

$$\text{Me}_2\text{Si (23)} + \text{(2,3-dimethylbutadiene)} \xrightarrow{610\,°C} \text{(diene, 67\%)} + \text{Me}_2\text{Si (24), 14\%}$$

$$+ \text{ Me}_2\text{Si (25)}\ 16\% + \text{ Me}_2\text{Si (26)}\ 57\% + \text{ Me}_2\text{Si (27)}\ 8\% \tag{40}$$

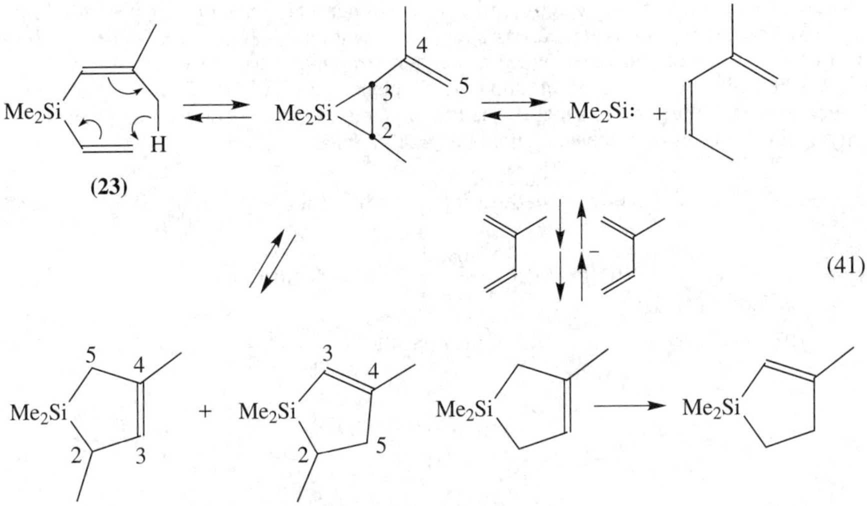

[Scheme showing structures (23), (41), (24), (25), (26), (27)]

E. Silylenes from Metal-Induced α-Eliminations

Some of the earliest evidence for organosilylenes arose from the gas-phase reactions of Me_2SiCl_2 with Na/K vapor, yielding $Me_2Si\!:\,$[86]. Beginning long before this, as early as the researches of Kipping, metal-induced elimination of halide from R_2SiX_2 in solution has been used to deliver R_2Si moieties in syntheses. Reductions with alkali metals are commonly employed for the synthesis of cyclic or linear polymeric polysilanes, for instance[87,88]. But these reactions do not necessarily involve free silylenes; they could equally well proceed through the intermediacy of *silylenoids*, R_2SiMX[89].

Carbenoids, α-heteroatom-substituted organometallic compounds which behave as masked carbenes, have been intensively investigated for more than thirty years, and have found important use as intermediates in organic synthesis[90]. A good example is CCl_3Li, which can give reactions like those of $Cl_2C\!:\,$. The silicon counterparts to the carbenoids, silylenoids, have often been suggested as reaction intermediates, especially in the synthesis of polysilanes from dichlorosilanes. Evidence for the existence of silylenoids was, however, lacking until the recent appearance of publications by the Kyoto University group of Tamao[91-93].

Earlier, these workers had found that electronegative atoms attached to silicon greatly stabilize silyl anions[94]. Thus the aminosilyllithium compounds $(Et_2N)Ph_2SiLi$, $(Et_2N)_2PhSiLi$ and $(Et_2N)PhMeSiLi$ can all be made from the corresponding chlorosilanes with lithium metal in THF at $0\,°C$, and are stable in solution at this temperature. These compounds behave simply as silyl anions, rather than silylenoids (they show no electrophilic behavior). The same is true for $(t\text{-BuO})_2PhSiLi$. *Mono*alkoxysilyllithium compounds, however, show both electrophilic and nucleophilic properties, and so can be accurately viewed as silylenoids.

These silylenoid species have been made by lithium–tin exchange from a Si–Sn precursor[96], or by reduction of chlorosilanes with lithium 1-(dimethylamino)naphthalenide (LDMAN) at $-78\,°C$ (equations 42 and 43)[97]. The resulting silyllithium species react as nucleophiles with Me_3SiCl to give the corresponding disilanes (equation 44). An important reaction showing electrophilic, hence silylenoid, behavior is that with n-BuLi, which displaces the alkoxy group as shown in equation 44. No free silylene is formed from the silylenoids, however, since no silylene trapping reactions were observed with Et_2MeSiH, diphenylacetylene or dimethylbutadiene. Also, when 12-crown-4-ether was added to $(t$-BuO$)Ph_2$SiLi to separate the lithium cation from the silicon, the molecule lost silylenoid character and behaved simply as a silyl anion.

$$(t\text{-BuO})Ph_2Si-SnMe_3 + n\text{-BuLi} \xrightarrow[-78\,°C]{THF} n\text{-BuSnMe}_3 + (t\text{-BuO})Ph_2SiLi \qquad (42)$$

$$(RO)Ph_2SiCl \xrightarrow[THF,\,-78\,°C]{LDMAN} (RO)Ph_2SiLi \qquad (43)$$

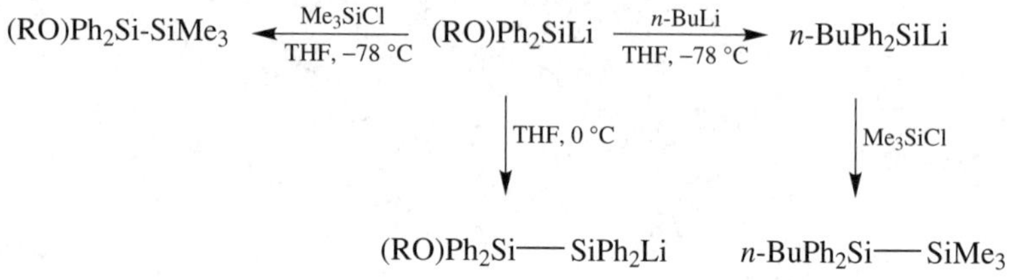

$$(RO)Ph_2Si\text{-}SiMe_3 \xleftarrow[THF,\,-78\,°C]{Me_3SiCl} (RO)Ph_2SiLi \xrightarrow[THF,\,-78\,°C]{n\text{-BuLi}} n\text{-BuPh}_2SiLi$$

$$\Big\downarrow THF,\,0\,°C \qquad\qquad\qquad \Big\downarrow Me_3SiCl \qquad (44)$$

$$(RO)Ph_2Si\text{—}SiPh_2Li \qquad n\text{-BuPh}_2Si\text{—}SiMe_3$$

$$R = t\text{-Bu},\, i\text{-Pr},\, Me$$

A significant finding in these studies was that at 0 °C in THF, the silylenoids undergo bimolecular condensation to the dimeric intermediates $ROPh_2Si-SiPh_2Li$, which can then be trapped by Me_3SiCl (equation 44). Dimerized species of this sort are quite likely to be intermediates in the coupling of dihalosilanes to cyclic and linear polysilanes.

As mentioned in Section II.D, an important publication bearing on the question of dichlorosilane–metal reactions appeared in 1988[56]. Boudjouk and coworkers treated $t\text{-Bu}_2SiX_2$, $X = Cl$, Br and I, with lithium and ultrasound activation in the presence of Et_3SiH. Good yields (ca 60%) of $Et_3Si-SiH(Bu\text{-}t)_2$ were obtained in all three cases. This is rather persuasive evidence in favor of silylene formation, since Si–H bond insertion is characteristic for silylenes and it is difficult to see how the disilane could arise from the reaction of Et_3SiH with a silylenoid[95].

When the same dehalogenations were carried out in the absence of a trapping agent, however, the products depended on the nature of the halogen. With $t\text{-Bu}_2SiBr_2$ and $t\text{-Bu}_2SiI_2$ the cyclotrisilane was obtained, but with $t\text{-Bu}_2SiCl_2$ a disilane and a cyclotetrasilane were produced (equations 45 and 46). It follows that all three of these reactions cannot have proceeded through the free silylene and, in fact, none may have.

$$t\text{-Bu}_2SiX_2 \xrightarrow[\text{ultrasound}]{Li} (t\text{-Bu}_2Si)_3 \qquad (45)$$
$$X = Br,\, I$$

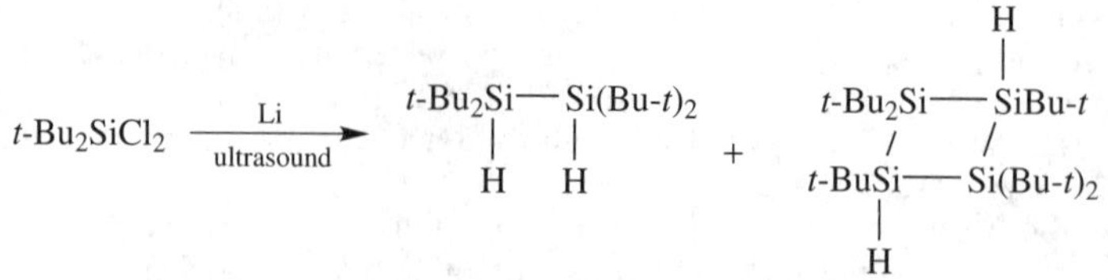

$$t\text{-Bu}_2SiCl_2 \xrightarrow[\text{ultrasound}]{Li} \begin{array}{c} t\text{-Bu}_2Si\text{—}Si(Bu\text{-}t)_2 \\ |\qquad\quad | \\ H\qquad\, H \end{array} + \begin{array}{c} \phantom{t\text{-Bu}_2Si}H \\ \phantom{t\text{-Bu}_2Si}| \\ t\text{-Bu}_2Si\text{—}SiBu\text{-}t \\ /\qquad\quad / \\ t\text{-BuSi}\text{—}Si(Bu\text{-}t)_2 \\ | \\ H \end{array} \qquad (46)$$

We will make a tentative suggestion as to how the results of Boudjouk's experiment might be explained. Possibly the halogen-containing silylenoids are in equilibrium with a very small amount of free silylene (equation 47).

$$R_2Si\begin{smallmatrix}Li\\X\end{smallmatrix} \rightleftharpoons Li^+X^- + R_2Si: \xrightarrow{Et_3SiH} Et_3Si-SiR_2H \quad (47)$$

If Et_3SiH is an efficient trapping agent for the silylene but does not react with the silylenoid, the reaction will proceed toward the right to give the disilane product. Other trapping agents may react instead with the silylenoid. When no trapping agent is present, silylenoid molecules would self-react, and the products might well depend on the nature of the halogen.

Another conceivable, but probably less likely, explanation of these experimental results[56] is that the triethylsilane somehow catalyzes the decomposition of silylenoid into free silylene, which can then insert into the Si—H bond. In any case, the intermediacy of silylenes in α-dehalogenations remains an open question, worthy of further study.

Different silylenoids may vary in their reactivity. When diiododiadamantylsilane was dehalogenated with lithium under ultrasonic activation in the presence of a mixture of Et_3SiH and 3-hexyne in THF, the silylenoid gave exclusively H—Si insertion product in high (67%) yield (equation 36)[57] Free Ad_2Si: formed by silirane photolysis (*vide supra* equation 26) gave products of both H—Si insertion and π-addition, addition being the favored reaction.

The formation of silacyclopentenes from dichlorosilanes, alkali metals and 1,3-dienes is sometimes taken as evidence for silylene formation. Lithium and ultrasound were used in one such study, in which R_2Si moieties were intercepted by 2,3-dimethylbutadiene (equation 48) albeit in low yield; the result was interpreted in terms of a silylene intermediate[96].

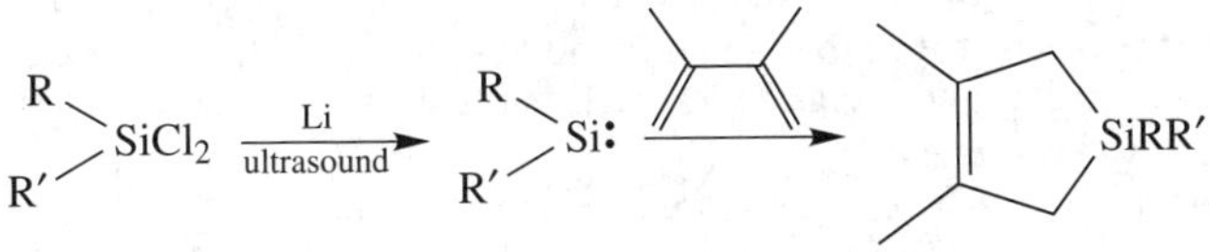

$$\begin{smallmatrix}R\\R'\end{smallmatrix}SiCl_2 \xrightarrow{Li, \text{ ultrasound}} \begin{smallmatrix}R\\R'\end{smallmatrix}Si: \longrightarrow \text{[silacyclopentene]} \quad (48)$$

A report by the Corriu group, however, shows that this is not necessarily the case (equation 49)[97]. When difluorosilane was treated with lithium metal in the presence of dimethylbutadiene, the silacyclopentene was obtained in good yield; but no silylene was intercepted if Et_3SiH was present as the trapping agent. The product silacyclopentene could result either from reaction of the diene with a silylenoid, or from reaction of the diene with lithium and trapping of the lithium compound by the difluorosilane. In earlier studies of the alkali metal induced reactions of dihalosilanes with 1,3-dienes, it was found that the diene, rather than the dihalosilane, reacts initially with the metal[2].

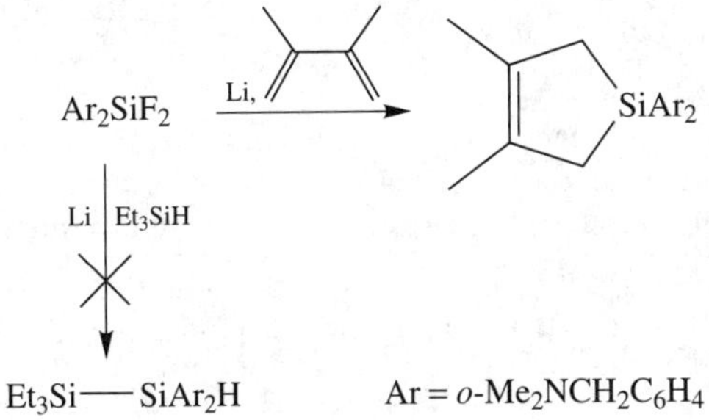

$$Ar_2SiF_2 \xrightarrow{Li,\ \diene} \text{[silacyclopentene SiAr}_2\text{]}$$

$$\downarrow Li\ |\ Et_3SiH$$

$$Et_3Si-SiAr_2H \qquad Ar = o\text{-}Me_2NCH_2C_6H_4 \quad (49)$$

On the other hand, there is no doubt whatsoever that stable, nitrogen-stabilized silylenes are produced from dihalosilanes and alkali metals[98]. The chemistry of these silylenes is treated in Section VIII.C.

F. Silylenes from Rearrangements

Rearrangements of siliranes to alkylsilylenes by ring-opening α-elimination from silicon were discussed in Section II.D (see equation 34). Similar mechanisms have been written for rearrangements of silirenes to alkenylsilylenes[5].

The rearrangements of silenes to silylenes via migration of an H-atom[5] remains controversial. Conlin has found that the temperature dependence of the product ratio and of the rate of decomposition of 1-methylsilacyclobutane in the gas phase is in accord with a mechanism in which 1-methylsilene MeHSi=CH$_2$ is formed as a trappable intermediate that rearranges to dimethylsilylene Me$_2$Si: (equation 50)[99]. Experimental evidence has been presented for the equilibration of CH$_3$HSi=CH$_2$ and (CH$_3$)$_2$Si: at the high temperatures at which they are generated in such experiments[100]. It has been pointed out that large differences in the rates of their addition reactions must be taken into account in the interpretation of trapping experiments[101].

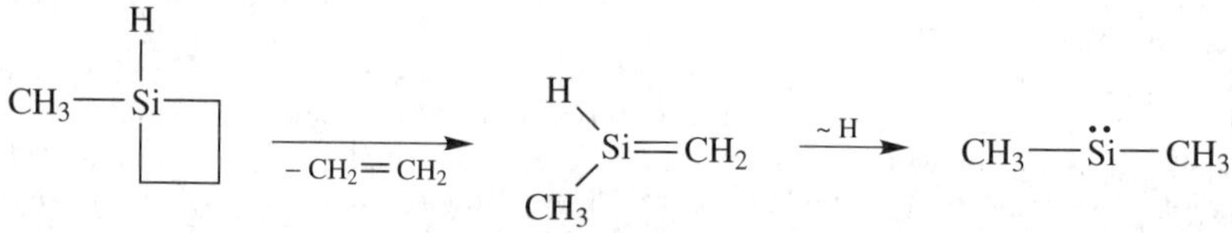

$$\text{CH}_3-\underset{\underset{\square}{|}}{\overset{\overset{H}{|}}{\text{Si}}} \xrightarrow{-\text{CH}_2=\text{CH}_2} \underset{\text{CH}_3}{\overset{H}{\text{Si}=\text{CH}_2}} \xrightarrow{\sim H} \text{CH}_3-\overset{..}{\text{Si}}-\text{CH}_3 \quad (50)$$

There seems to be agreement that the interconversion of MeHSi=CH$_2$ and Me$_2$Si: is nearly thermoneutral, with a barrier of *ca* 40 kcal mol$^{-1,102}$. MeHSi=CH$_2$ was found in higher concentration under conditions such that the two species could isomerize in a flowing-afterglow experiment[103].

Rearrangement of a silene to a silylene via migration of a Me$_3$Si group has been suggested as a step in the gas-phase silylene-to-silylene rearrangement Me$_3$Si–S̈i–Me → H–S̈i–CH$_2$SiMe$_3$[104]. Labeling experiments, however, have indicated that an alternative mechanism operates (equation 51)[105].

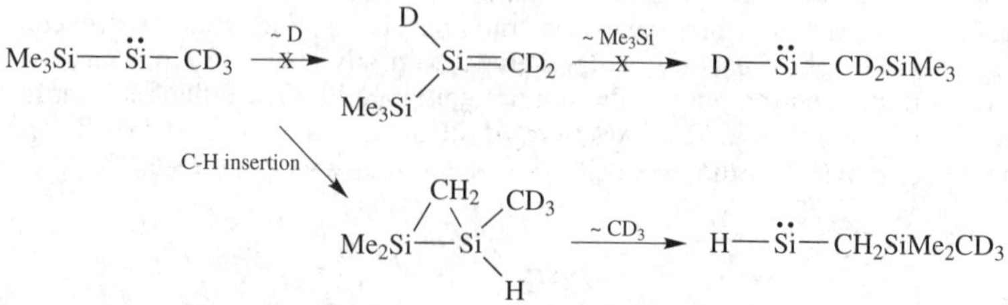

$$\text{Me}_3\text{Si}-\overset{..}{\text{Si}}-\text{CD}_3 \xrightarrow{\sim D} \underset{\text{Me}_3\text{Si}}{\overset{D}{\text{Si}=\text{CD}_2}} \xrightarrow{\sim \text{Me}_3\text{Si}} D-\overset{..}{\text{Si}}-\text{CD}_2\text{SiMe}_3$$

$$\downarrow \text{C-H insertion} \qquad\qquad (51)$$

$$\underset{\text{Me}_2\text{Si}-\underset{H}{\overset{\text{CH}_2\diagdown\text{CD}_3}{\text{Si}}}}{} \xrightarrow{\sim \text{CD}_3} H-\overset{..}{\text{Si}}-\text{CH}_2\text{SiMe}_2\text{CD}_3$$

Rearrangements of disilenes to α-silylsilylenes are, however, well established (equation 52)[5] and are involved in the exchange of substituents between a silylene center and an adjacent silicon, a process that has been called a transposition (equation 53)[106].

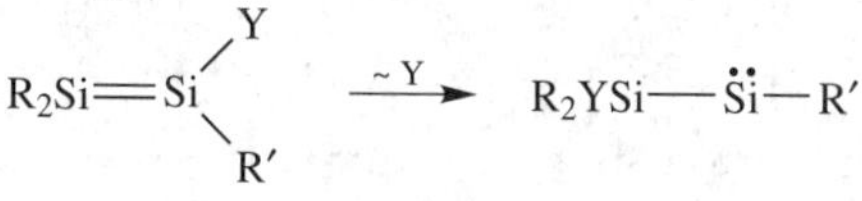

$$\text{R}_2\text{Si}=\text{Si}\diagup^Y_{\diagdown R'} \xrightarrow{\sim Y} \text{R}_2\text{YSi}-\overset{..}{\text{Si}}-R' \quad (52)$$

$$\text{Me}_3\text{Si}-\ddot{\text{Si}}-\text{H} \xrightarrow{\sim \text{Me}} \text{Me}_2\text{Si}=\text{Si}\begin{smallmatrix}\text{Me}\\ \diagup \\ \diagdown \\ \text{H}\end{smallmatrix} \xrightarrow{\sim \text{H}} \text{Me}_2\text{HSi}-\ddot{\text{Si}}-\text{Me} \quad (53)$$

Puranik and Fink found that the low-temperature photoinduced rearrangement of a silylene to a silacyclobutadiene is reversible in a thermal reaction at 160 K (equation 54)[107]. The silylene persists in fluid solution at 200 K!

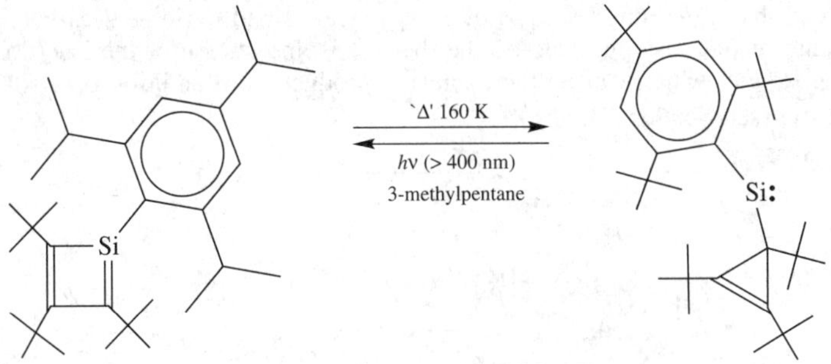

(54)

III. REACTIONS OF SILYLENES

A. Insertion

The first silylene reaction to be recognized, a half-century ago, was insertion of SiH_2 into silicon–hydrogen bonds[1]. The mechanism was found to be direct insertion[2]. Insertions of silylenes into a number of other bonds have been observed, including O–Si[1,2], N–Si[2], halogen–Si[2], strained O–C[2], Si–Si[3], H–O[3], H–N[3] and H–C (intramolecular only)[3]. There is both experimental and theoretical support for stepwise insertion of silylenes into O–Si, strained O–C and H–O bonds, initiated by the formation of silaylides by donation of a pair of electrons from an oxygen atom to form a bond to a divalent silicon atom (equation 55)[3,5]. Silylene insertion into a Cl–Si bond can follow a similar mechanism, or can result from halogen abstraction by the silylene followed by geminate recombination of the resulting radical pair[3,5].

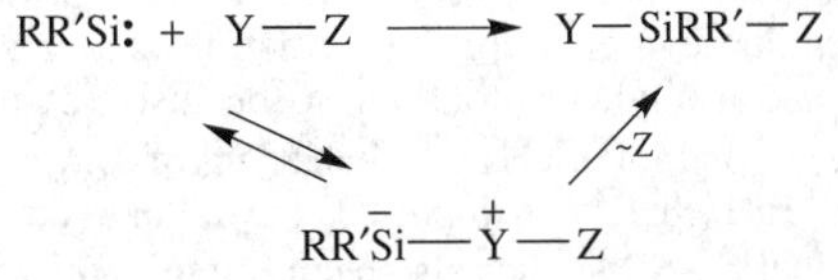

$$\text{RR}'\text{Si:} + \text{Y}-\text{Z} \longrightarrow \text{Y}-\text{SiRR}'-\text{Z}$$
$$\text{RR}'\overset{-}{\text{Si}}-\overset{+}{\text{Y}}-\text{Z} \quad (55)$$

Insertion of SiH_2 into H–H and H–C bonds, respectively, has been *assumed* in reports of rate constants from laser flash photolysis experiments for reactions of SiH_2 with H_2[108,109], D_2[109] and CH_4[108]. Since the rate of silylene insertion into the H–H bond was more rapid than previous estimates, interest was reawakened in the activation barrier[112-114]. The barrier for insertion of SiH_2 into H_2 was revised downward to *ca* 1 kcal mol^{-1}.

A number of Lewis base adducts of diorganosilylenes have been observed by electronic spectroscopy in frozen matrices (see Section V.A)[115]. The silylene-donor adducts of ethers, tertiary alcohols, tertiary amines and sulfides were found to revert to silylenes upon warming, but primary and secondary alcohol adducts underwent rearrangement to the H–O formal insertion products (equation 55).

The silaylides formed from coordination of silylenes to chlorocarbons are believed to undergo competitive rearrangement (to the formal Cl−C insertion product) and fragmentation (to the product of formal HCl abstraction by the silylene)[5]. In addition, these silaylides are believed to undergo dissociation into the radical pairs that would result from direct chlorine atom abstraction by the silylene[111].

A more complex reaction sequence may be responsible for the formation of the products of formal insertion of silylenes into the H−C bonds of carbaldimines[110,116]. As shown in equation 56, an initial Lewis base or π-bond adduct can undergo rearrangement to both of the products observed from reactions of diarylsilylenes and pyridine-2-carbaldimines. A common intermediate is suggested by the thermal isomerization of the insertion product to the cycloadduct which can be the exclusive product from addition of a silylene to a 2,2′-bipyridyl (see Section III.B)[117].

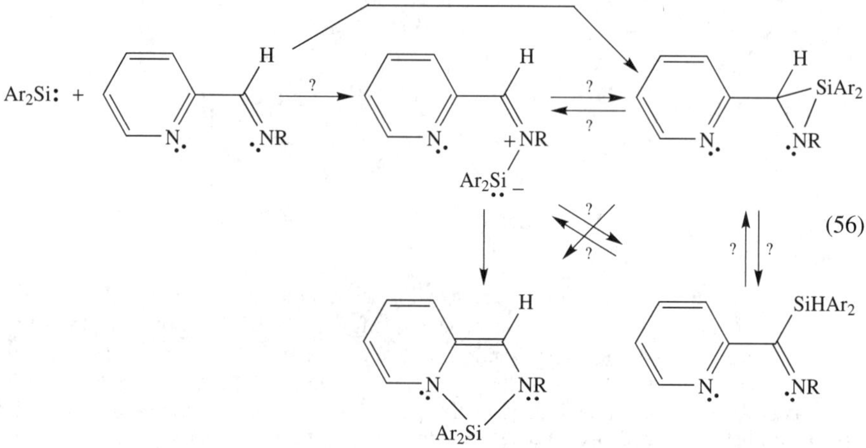

$$\text{(56)}$$

Insertion of bis(dialkylamino)silylenes into aromatic and benzylic H−C bonds has been suggested to explain the products formed from dehalogenation of the corresponding bis(dialkylamino)dihalosilanes by Na/K in solution[118,119]. Caution is advised regarding this mechanism, since previous investigations of the reductive dehalogenation of dihalosilanes in solution did *not* find evidence for the formation of free silylenes[2].

New insertion reactions of silylenes continue to be discovered. SiH_2 has been found to insert into an H−Ge bond of GeH_4[120]. Insertion of $SiMe_2$ into H−Ta bonds of $Cp_2TaH(PMe_3)$ and Cp_2TaH_3 has been reported, the latter reaction via a $Me_2Si(PMe_3)$ silylene–Lewis base adduct[121]. $SiMe_2$ has also been found to insert into the H−S bonds of trialkoxysilanethiols[122]. Apparent intermolecular H−C insertion by furyl(phenyl)silylene was observed to compete with addition to 2,3-dimethylbutadiene[123]. Formal insertion of two silylenes into a S=C double bond is believed to occur via rearrangement of a spiro di-adduct intermediate, as shown in equation 57[124].

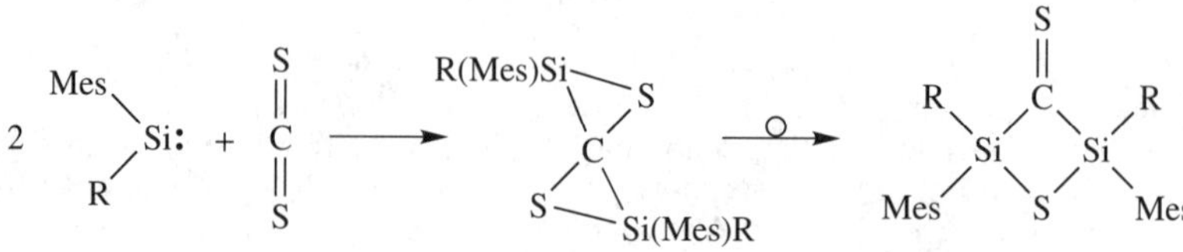

$$\text{(57)}$$

A bizarre silylene insertion into the I—I bond has been suggested as the initial step in the mechanism for the low-temperature (−90 °C) reaction of dihalosilylenes (SiF$_2$, SiCl$_2$, SiBr$_2$) with solutions of iodine in toluene, as shown in equation 58[125]. These reactions can be considered to involve electrophilic attack on an arene by SiX$_2$I$^+$[126].

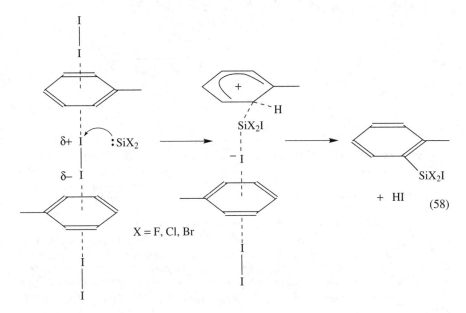

(58)

B. Addition

1. Mechanistic studies

Addition of silylenes to carbon–carbon π-bonds forming silirane and silirene intermediates had been inferred by the early 1960s, and by the middle 1970s such silylene addition products had been isolated (equation 59)[2]. The belief that addition of silylenes to olefins is a concerted process is based on the observation that additions to cis- and trans-olefins are stereospecific[3,5]. Since all reliably characterized silylenes have singlet ground electronic states, stereospecific addition is expected on the basis of the 'Skell rule' of carbene chemistry[127]. This pragmatically useful but theoretically shaky generalization can be stated as follows: since singlet carbenes (silylenes) *can* add to olefins in a single step, they will do so, and *E,Z*-stereochemistry of the olefin will be retained in the addition product; a triplet carbene(silylene) will add in a stepwise process including a triplet diradical intermediate in which stereochemistry can be scrambled by rotation about single bonds that competes with intersystem crossing and ring closure. A theoretical study of the reaction of SiH$_2$ with acetylene found that most observations can be accounted for by a model in which concerted formation of chemically activated silirene is followed by competition between dissociation, deactivation and rearrangement channels[128]. Experimental determinations of product yields from SiH$_2$ and ethylene were successfully modeled with a similar mechanism involving the formation of chemically activated silirane as the concerted primary step[129].

A singlet ground state was deduced for diadamantylsilylene when it was found to undergo stereospecific addition to *cis-* and *trans*-2-butene[57]. Stereospecific addition of cyclopropyl(phenyl)silylene has also been reported[130].

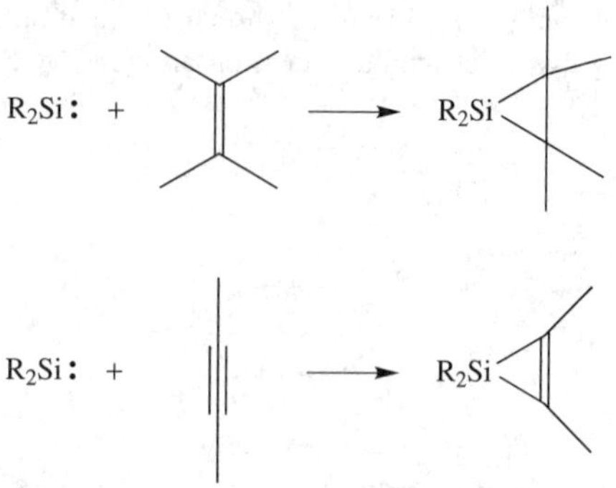

$$R_2Si: \ + \ \text{(alkene)} \longrightarrow R_2Si\text{(siliranes)} \qquad (59)$$

$$R_2Si: \ + \ \text{(alkyne)} \longrightarrow R_2Si\text{(silirene)}$$

When nonstereospecific addition of dimesityl- and bis(2,4,6-triisopropylphenyl)silylene to *trans*-2-butene, and to a lesser extent to *cis*-2-butene, was reported, a stepwise addition mechanism was suspected[131]. However, the apparent nonstereospecificity was caused by a *cis*-butene impurity in the *trans*-butene and by photoisomerization of the *cis*- and *trans*-siliranes[132].

Recently, stereospecific addition of bis(triisopropylsilyl)silylene (i-Pr$_3$Si)$_2$Si to *cis*-2-butene has been observed[133]. Since a triplet ground state has been predicted on the basis of density functional calculations[134], stereospecific addition could be due to a violation of the Skell rule. Alternatively, the Skell rule could be obeyed if the triplet ground state is siphoned off via more rapid reactions of the low-lying excited singlet (which, according to the calculations, lies only 1.7 kcal mol^{-1} above the triplet[134]) that may be in equilibrium with the triplet.

The observation of ^{1}H CIDNP signals during the addition of Me$_2$Si (and Me$_2$Ge) to a strained alkyne (equation 60), has been interpreted as indicating stepwise addition of excited triplet states of the divalent species to the triple bond[135]. One wonders, however, if there are not other viable explanations for the CIDNP signals, such as silylene(germylene) transfer from a diradical formed by homolysis of one of the bridging Si—C(Ge—C) bonds in the precursor (Scheme 3, Section II.C.2). There is evidence that silylene extrusion from 7-silanorbornadienes may be a stepwise process[2], and thus 'silylene products' may in some cases be formed via silylenoid intermediates rather than free silylenes.

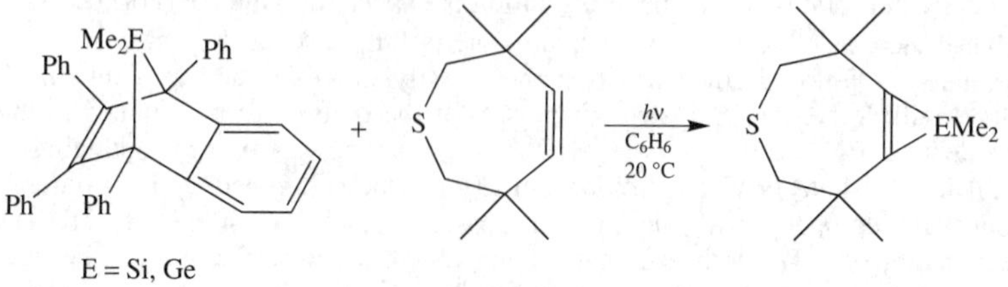

$$\text{precursor} \ + \ S\text{-cycloalkyne} \xrightarrow[\substack{C_6H_6 \\ 20\,°C}]{h\nu} S\text{-cycloalkyne-EMe}_2 \qquad (60)$$

E = Si, Ge

In the past decade there has been considerable mechanistic study of the addition of silylenes to 1,3-dienes. From the first report of the formation of 1-silacyclopent-3-enes as major products thirty years ago, a stepwise mechanism including vinylsilirane intermediates was proposed, but concerted 1,4-addition was also considered[1-5]. Vinylsilirane intermediates were clearly identified in trapping experiments by 1975[2], but questions have remained about the concertedness of both the primary addition and rearrangement steps of equation 61.

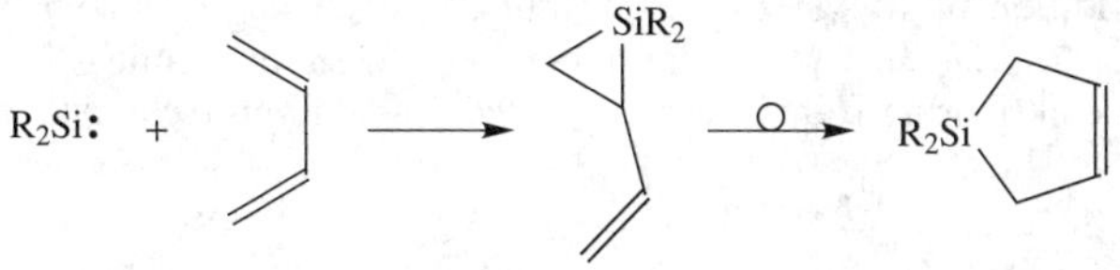

$$R_2Si: \quad + \quad \diagup\!\!\!\diagdown\!\!\!\diagup \quad \longrightarrow \quad \overset{SiR_2}{\diagup\!\!\!\diagdown}\!\!\!\diagup \quad \longrightarrow \quad R_2Si\diagdown\!\!\diagup \quad \quad (61)$$

The current view of the mechanism of addition of silylenes to 1,3-dienes is based on the following experimental observations:

(a) The dramatic differences (shown in Table 1) in the distribution of the products of addition of Me_2Si to the three stereoisomers of 2,4-hexadiene can be explained by formation of alkenylsilirane intermediates by concerted addition[20,136]. It can be seen that with substituents on the 1- and 4-positions of 1,3-butadiene, the 1-silacyclopent-3-enes that are formal 1,4-addition products are formed in low yield. The major product from the *cis, cis*-diene results from a retro-homo-ene reaction already depicted in the forward direction in Section II.D (equation 41). The 1,1,2,3-tetramethyl-1-silacyclopent-4-enes that are the major products from the *cis,trans*- and *trans,trans*-2,4-hexadienes are also rearrangement products of vinylsilirane intermediates, requiring the migration of C−C bonds, as shown in equation 41. The formation of both *cis*- and *trans*-stereoisomers of the silacyclopentenes suggested that the ring expansions of the alkenylsilirane intermediates were stepwise processes[11]. With phenyl substituents at the 1- and 4-positions of butadiene, facile C−C bond cleavage of the vinylsilirane intermediates leads to silacyclopent-4-enes as the exclusive addition products[137].

TABLE 1. Product yields (%, based on moles of Me_2Si generated) from addition of Me_2Si: to 2,4-hexadienes

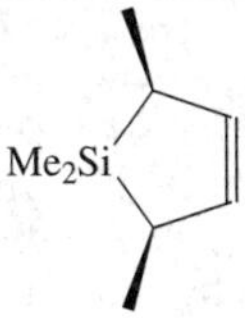

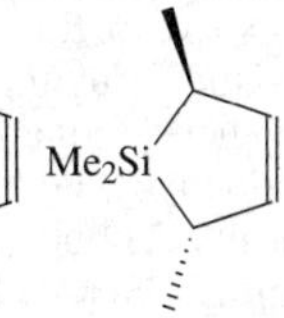

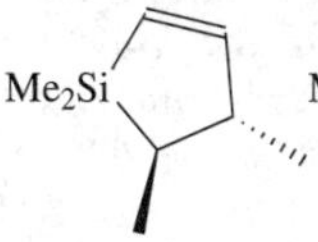

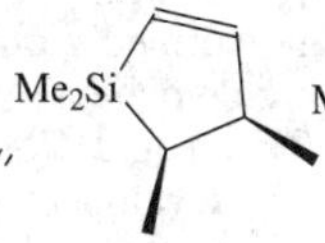

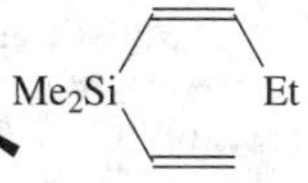

	1.2	1.3	12.7	8.3	50.5	
	7.5	1.8	27.3	18.4	20.4	
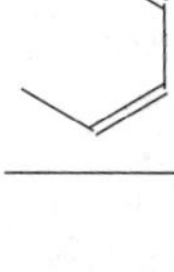	2.3	2.3	24.6	21.4	35.1	

(b) Use of 1,4-dideutero-1,3-butadiene allowed the demonstration that at least one-third of the addition of Me_2Si to 'unsubstituted' butadiene occurs via a vinylsilirane intermediate[138]. The labeling pattern in the absence of stereochemical information does not preclude a contribution from direct, concerted 1,4-addition (see equation 62).

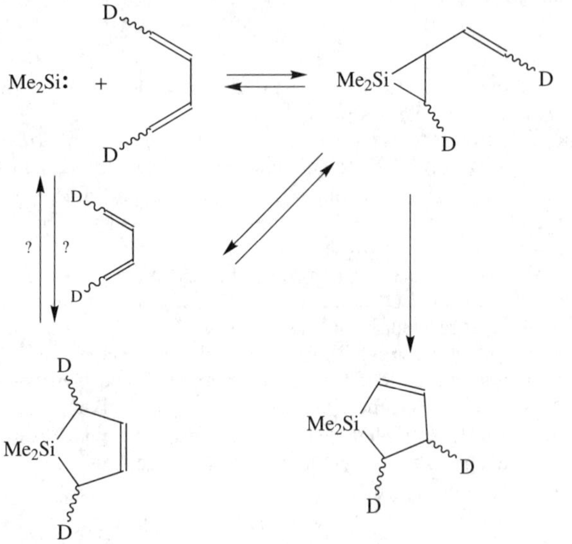

(62)

(c) Stereospecific 1,2-addition of the sterically shielded dimesitylsilylene to the stereoisomeric 2,4-hexadienes at room temperature in solution led to vinylsiliranes that were stable for months[139]. Longer irradiation led to the stereospecific conversion of the vinylsilirane formed from the *trans,trans*-isomer to the *cis*-2,3-dimethylsilirane that preserves the stereochemistry of the starting diene (equation 63). These important observations by Zhang and Conlin can be explained by concerted rearrangements of vinylsiliranes to 1-silacyclopent-3-enes, or by concerted photoreversion of vinylsiliranes to the silylene and diene from which they were formed, followed by concerted 1,4-addition of the silylene to the diene. Both possibilities were considered by Conlin[139].

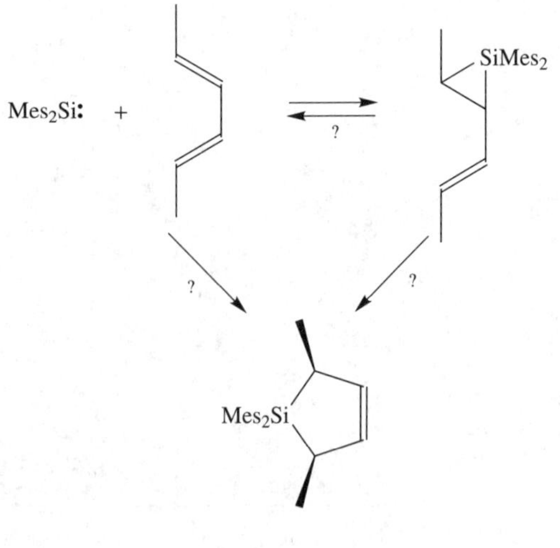

(63)

(d) The formation of 3,4-divinylsilacyclopentanes as 1 : 2-adducts from silylenes and unsubstituted butadiene not only provides examples of vinylsiliranes undergoing intermolecular reactions, but also lends support to the occurrence of direct 1,4-addition[140,141]. When silylenes (Ph_2Si:, PhMeSi:) are generated in solution in the presence of both butadiene and acetone, the products include vinylsiloxolanes incorporating a unit each of silylene, butadiene and acetone (equation 64).

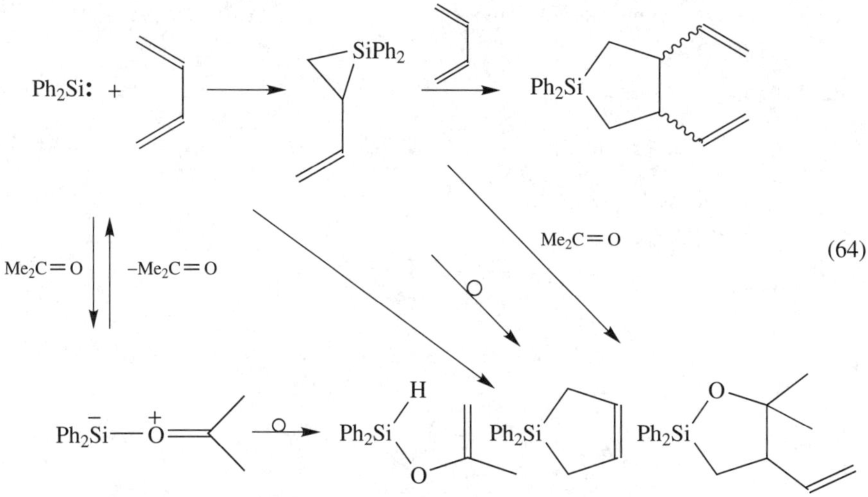

$$\text{(64)}$$

Steady-state kinetic analysis of a competition experiment led to the conclusion that the siloxolane is formed by reaction of a vinylsilirane intermediate with acetone, and that the vinylsilirane arises from addition of the free silylene to butadiene. Since silylenes are known to react more rapidly with acetone than with butadiene, the kinetic analysis further suggested that the carbonyl sila-ylide dissociates more rapidly than it rearranges to the silyl enol ether shown in equation 64[140].

Evidence for direct, 1,4-addition of Ph_2Si to butadiene comes from the observation that, at very low acetone concentrations, no divinylsilacyclopentane is formed, but the silacyclopent-3-ene is found. Since at the high (10 M) butadiene concentration employed in these experiments, reaction of the vinylsilirane with butadiene to form the divinylsilacyclopentane is more rapid than rearrangement to the silacyclopentene, there must be another route to the silacyclopentene, such as direct 1,4-addition[141].

2. Recently discovered addition reactions

a. Addition to n-donor bases. The first spectroscopic detection of silylene–ether complexes was achieved by annealing 3-methylpentane matrices doped with 2-methyltetrahydrofuran in which silylenes were photochemically generated at 77 K[142]. Trapping of the silylenes by the ether was shown to compete with silylene dimerization, and the silylene–ether complexes displayed a different reactivity from the free silylenes. Spectra of silylene complexes with N, O, P and S *n*-donor bases pyrrolidine, *N*-methylpyrrolidine, piperidine, tetrahydrofuran, tetrahydrothiophene, and Bu_3P have also been reported (see Section V.A.2)[113,143].

Sila-ylides from complexation of silylenes by carbonyl and thiocarbonyl compounds have been observed; their chemistry is discussed in Section V.A[144]. The first stable silylene–Lewis base complex was obtained in solution from an extremely hindered diarylsilylene and several isocyanides with bulky substituents as shown in equation 65[145]. Since the source of the silylene is the thermal dissociation of its highly congested disilene dimer, the possibility remains that the silylene complex was formed by addition of the isocyanide to the disilene rather than the free silylene. The free silylene route is quite plausible, since dimers of silaketenimines have been obtained from reactions of less crowded silylenes and isocyanides[146]. The spectroscopic properties of these silylene–isocyanide complexes are further discussed in Section III.C.2.

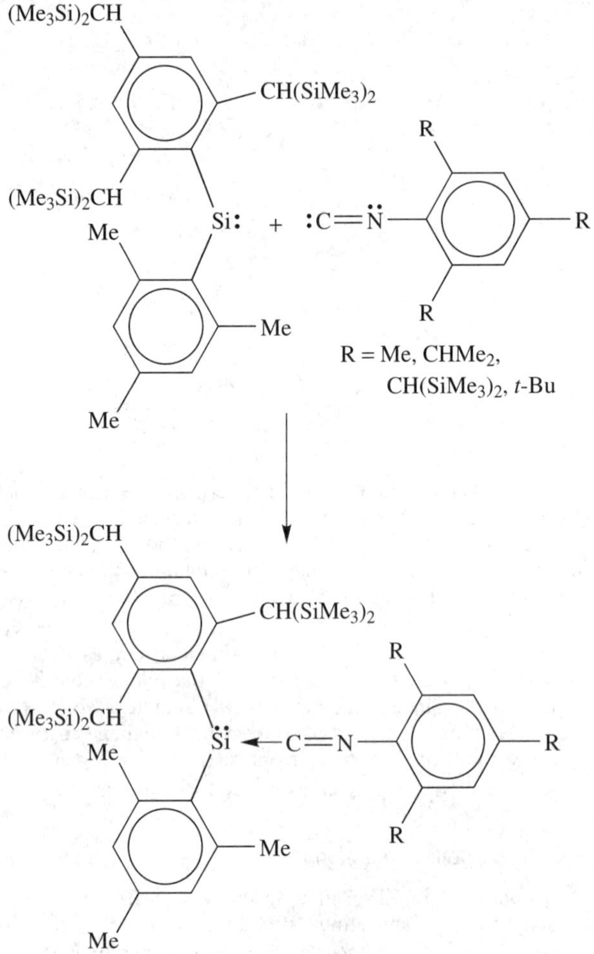

$$R = Me, CHMe_2, CH(SiMe_3)_2, t\text{-Bu} \tag{65}$$

Silylene–azide complexes are probable intermediates in the formation of silaimines from silylenes and covalent azides (equation 66)[147,148].

$$R_2Si: + N_3SiR'_3 \xrightarrow{-N_2} R_2Si=NSiR'_3 \tag{66}$$

Addition of a 'persistent silylene' (see Section V.III.C) to nickel carbonyl resulted in the first bis-silylene transition metal complex without Lewis base stabilization (equation 67)[149]. Other metal complexes of this 'stable silylene' are described in Section V.III.C.

$$2 \left[\begin{array}{c} t\text{-Bu} \\ | \\ \text{N}^+ \\ \| \quad \text{Si:}^- \\ \diagdown \text{N:} \\ | \\ t\text{-Bu} \end{array} \right] \longleftrightarrow \left[\begin{array}{c} t\text{-Bu} \\ | \\ \text{N:} \\ \diagup \quad \text{Si:} \\ \diagdown \text{N:} \\ | \\ t\text{-Bu} \end{array} \right] + \text{Ni(CO)}_4 \xrightarrow[-2\,\text{CO}]{} \left(\begin{array}{c} t\text{-Bu} \\ | \\ \text{N} \\ \diagup \quad \text{Si} \\ \diagdown \text{N} \\ | \\ t\text{-Bu} \end{array} \text{Ni(CO)}_2 \right)_2 \quad (67)$$

b. Addition to diatomic molecules. In a very difficult competition experiment it was found that SiH$_2$, formed by infrared multiphoton dissociation of SiH$_4$, reacts with nitric oxide NO with a rate constant similar to that for reaction of SiH$_2$ with SiH$_4$[150]. The products and the mechanism could not be determined.

Reaction of dimesitylsilylene with O$_2$ in an oxygen matrix at 16 K led to the formation of a silanone *O*-oxide Mes$_2$Si=O−O as deduced from the infrared spectra of isotopomers with the help of theoretical calculations[151,152]. However, isotopic labeling led to the assignment of dioxasilirane structures to the photoproducts from SiF$_2$ and SiCl$_2$ respectively with O$_2$ in an argon matrix at 10 K[153]. That the thermal reactions of SiF$_2$ with O$_2$ and H$_2$ are very slow ($k_2 < 2 \times 10^{-17}$ cm^3 molecule^{-1} s^{-1}) had been found earlier[154].

From the reactions of carbon monoxide in frozen matrices with Me$_2$Si[155] and several arylsilylenes Mes(R)Si (R = Mes, 2,6-diisopropylphenoxy, *t*-Bu)[156], adducts were formed (see Section V.A.2). Theoretical calculations led to consideration of both a linear silaketene and a pyramidal *n*-donor base complex structure for the Me$_2$Si(CO) adduct[249]. The observation that warming the carbon monoxide adducts of the arylsilylenes led to the formation of disilenes was interpreted as indicating the formation of a nonplanar complex that could dissociate as do other silylene−*n*-donor base complexes[156].

c. Addition to π-bonds. The repertoire of silylene π-bond additions, previously limited to the formation of siliranes and silirenes from alkenes and alkynes, respectively[2−5], has been vastly expanded in the past decade. The contributions of the Ando laboratory to the study of silylene additions to various olefins have been reviewed[157]. Addition of dimesitylsilylene to 1-*t*-butyl-3,3-dimethylallene led to the first stable alkylidenesilirane (equation 68). The first bis(alkylidene)silirane was obtained from addition of Mes$_2$Si to tetramethylbutatriene[158]. Azasilirenes from addition of silylenes to the C−N triple bond of nitriles undergo rapid dimerization[159,160], but isolable phosphasilirenes have been obtained from *t*-Bu$_2$Si: and phosphalkynes, and these were the first three-membered rings containing phosphorus−carbon double bonds (equation 69)[161].

$$\text{Mes}_2\text{Si:} + t\text{-BuCH}=\text{C}=\text{CMe}_2 \longrightarrow \begin{array}{c} \text{Mes} \quad \text{Mes} \\ \diagdown \diagup \\ \text{Si} \\ \diagup \quad \diagdown \\ t\text{-Bu} \quad \quad \text{Me} \\ \quad \quad \text{Me} \end{array} \quad (68)$$

$$t\text{-Bu}_2\text{Si:} + \text{RC}\equiv\text{P:} \longrightarrow \begin{array}{c} t\text{-Bu} \quad \text{Bu-}t \\ \diagdown \diagup \\ \text{Si} \\ \diagup \diagdown \\ \text{RC}=\text{P:} \end{array} \quad (69)$$

Addition of several sterically congested diarylsilylenes to carbon disulfide was presented in Section III.A. A diazasilirane was obtained from addition of Mes$_2$Si: to azobenzene, but the same product also resulted from nonsilylene pathways[162]. The sterically shielded and hence stable iminoboranes t-Bu—B=N—Bu-t and i-Pr$_2$N—B=N—Bu-t yielded stable three-membered rings upon addition of dimesitylsilylene Mes$_2$Si, but, with the less crowded dimethylsilylene Me$_2$Si, four- and six-membered rings were obtained[163].

There is a dramatic kinetic stabilization by bulky substituents of the three-membered ring products from silylene addition to π-bonds. Addition of t-Bu$_2$Si to ethylene led to the first silirane with no substituents on its ring carbon atoms as a distillable liquid![164]. Interestingly, 1,1-di($tert$-butyl)silirane does *not* undergo photochemical or thermal silylene extrusion, but instead polymerizes. A distillable silirane was also reported from addition of t-Bu$_2$Si to 2-methylstyrene[165].

The first addition of a silylene to benzene and naphthalene has been reported (equation 70)[166]. Product **28** can be rationalized by intramolecular insertion of the silylene into a nearby activated H—C bond. While products **29** and **30** may indeed be the result of stepwise addition of two silylenes to aromatic rings, one wonders whether direct additions of the disilene starting material, or a diradical formed from the disilene by homolysis of the π-bond, might have occurred. It is curious that no adducts containing a single silylene unit were found. Less congested silylenes have *not* been found to undergo addition to aromatic rings.

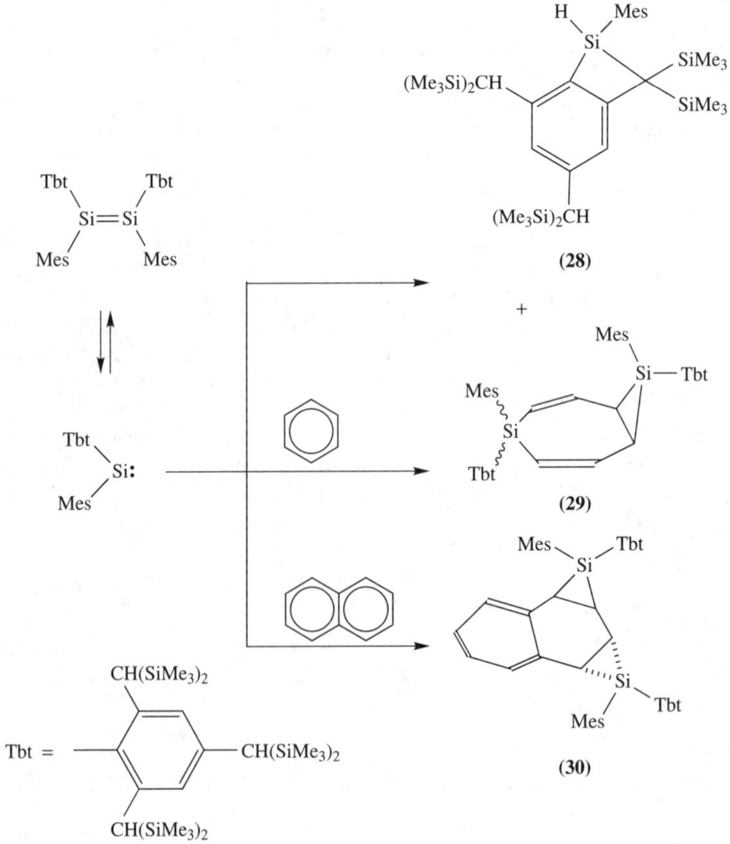

Silylene addition to 9,10-dimethylanthracene had previously been found to compete effectively with silylene dimerization (equation 71)[167]. Control experiments established that the silylene adduct did *not* undergo ring expansion to the disilene adduct.

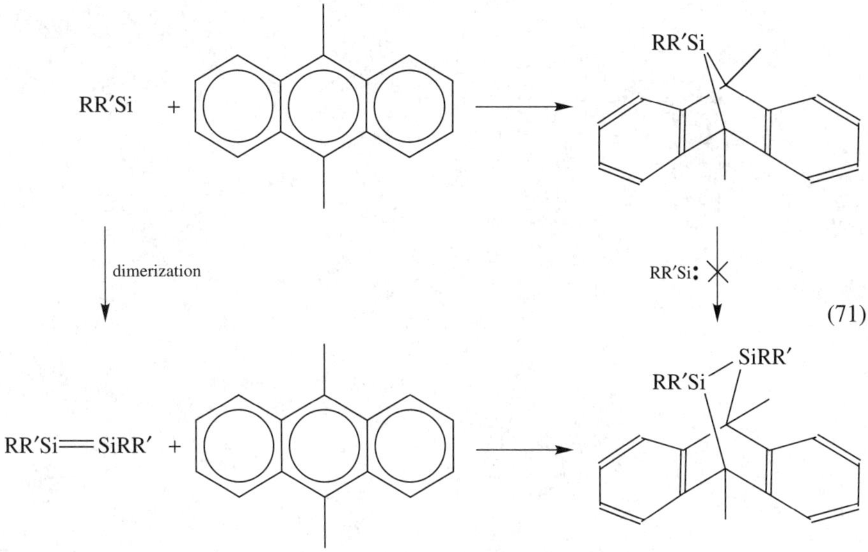

(71)

R = R' = Me; R = R' = Ph
R = Me, R' = Ph

d. Addition to dienes and heterodienes. Dimethoxysilylene, while unreactive toward mono-olefins, has been found to undergo addition to 2,3-dimethylbutadiene with formation of the formal 1,4-adduct, 1,1-dimethoxy-3,4-dimethylsilacyclopent-3-ene[168]. As mentioned in Section II.A, a series of thermally generated silylenes $\text{Me}\ddot{\text{S}}\text{iCl}$, $\text{Me}\ddot{\text{S}}\text{iOMe}$ and $\text{Me}\ddot{\text{S}}\text{iNMe}_2$ has been reacted with butadiene, isoprene and 2,3-dimethylbutadiene to yield mixtures of silacyclopent-2- and 3-enes[19]. Competition experiments indicated that the heterodienes (and heterotriene) shown in equation 72 react more rapidly than the simple dienes studied, and the structures of the heterocyclic products suggested that initial attack of the silylene was on the heteroatom or the hetero–olefinic bond[21,169].

The addition of $t\text{-Bu}_2\text{Si}$ to 1,4-diaza-1,3-butadienes competes with dimerization of the silylene only when the concentration of $t\text{-Bu}_2\text{Si}$ is low[170]. Subtle steric effects must also be responsible for the addition of $t\text{-Bu}_2\text{Si}$ to the *N*-cyclohexyl mono-imine of benzil, while only the silylene dimer undergoes addition under similar conditions in the presence of the *N*-methyl mono-imine[171]. It may be that $t\text{-Bu}_2\text{Si}$ and its dimer $t\text{-Bu}_2\text{Si}=\text{SiBu-}t_2$, both formed simultaneously upon photolysis of cyclo $(t\text{-Bu}_2\text{Si})_3$, are in equilibrium, and the steric effect is upon the (2+4) cycloaddition of the disilene.

Complex product mixtures have been obtained from addition of silylenes to 3,6-bis(trifluoromethyl)-1,2,4,5-tetrazine and 2,5-bis(trifluoromethyl)-1,3,4-oxadiazole[172].

Formal 1,4-addition of silylenes to 2,2'-bipyridyls leads to the formation of deep violet, highly air-sensitive adducts (equation 73)[173,174].

It is noteworthy that silylenes have been employed in the derivatization of fullerenes. Bis(2,6-diisopropylphenyl)silylene adds regiospecifically to a π-bond localized at a 6-ring,

6-ring juncture of C_{60} to give a single silirane derivative of C_{2v} symmetry[175]. The same silylene adds regioselectively to C_{70} to give two silirane adducts[176].

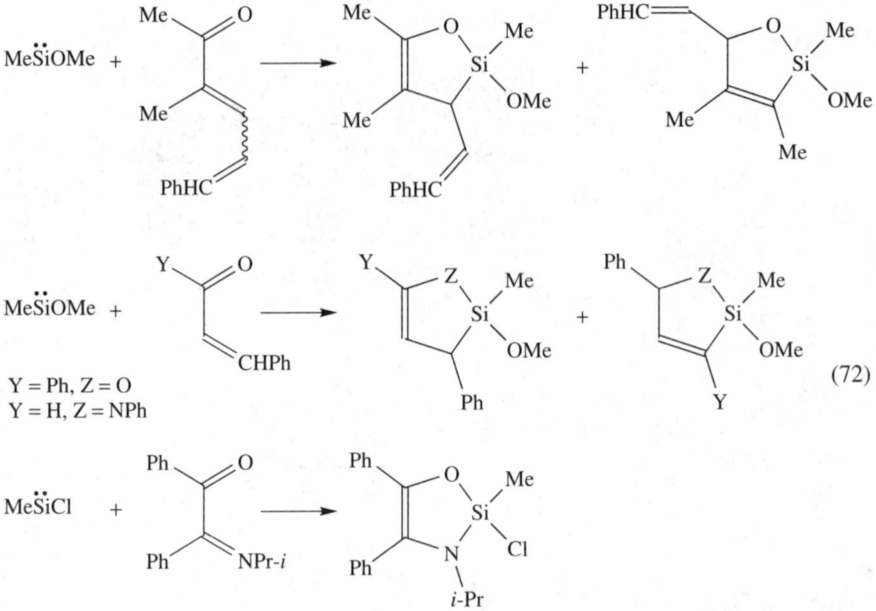

(72)

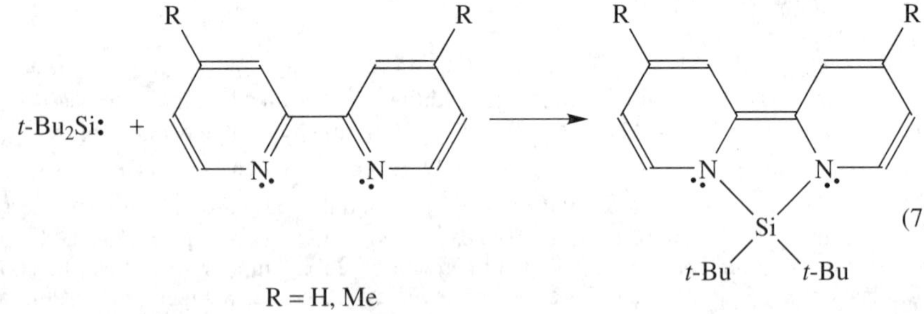

(73)

R = H, Me

C. Silylene Dimerization and Disilene Dissociation

1. Silylene dimerization

The intermolecular reaction of two silylene molecules to form a disilene was first proposed for gas-phase reactions[177]. The flow pyrolysis of $MeOSiMe_2-SiMe_2OMe$ at 600 °C, which produces Me_2Si, yielded 1,3-disilacyclobutanes as major products; these compounds are known rearrangement products of $Me_2Si=SiMe_2$, implying that the disilene was an intermediate. In a study by Sakurai and coworkers[178], silylenes generated from thermolysis of either silanorbornadienes or methoxydisilanes were trapped with anthracene yielding disilane-bridged products, as shown in Scheme 4. Anthracene is known to trap disilenes to give bridged products, so the results again suggest that dimerization of the silylenes has taken place.

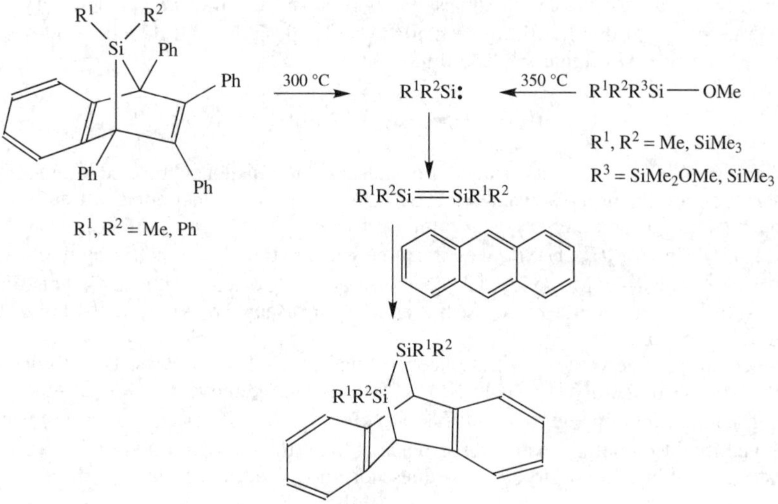

SCHEME 4

Direct evidence for dimerization of silylenes followed upon the isolation of silylenes in matrices, and it now appears that dimerization to the disilene is the usual route of deactivation for silylenes in solution if no trapping agent is present[179]. If the substituent groups are large enough, the resulting disilenes are thermally stable and can be isolated. Dimerization of silylenes, obtained by photolysis of trisilanes, led to the first isolation of a disilene[180] (equation 74) and is now the most general method for disilene synthesis. The 26 different stable disilenes which have been made by this route are listed in a review[181]. Although usually symmetrical disilenes, R'RSi=SiR'R, are obtained, cross-dimerization between two different silylenes has been employed to synthesize unsymmetrical disilenes (equation 75)[182].

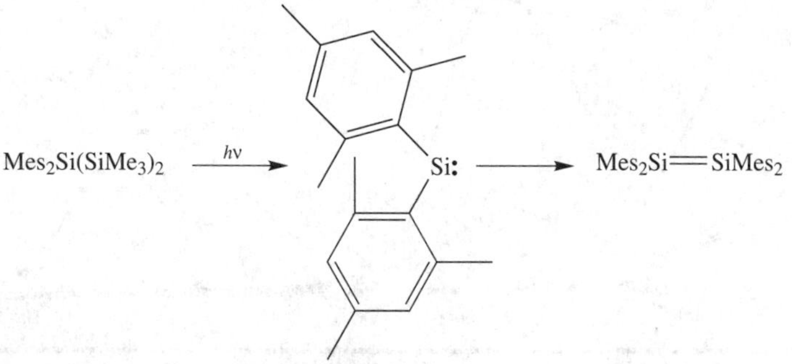

$$Mes_2Si(SiMe_3)_2 \xrightarrow{h\nu} Mes_2Si: \longrightarrow Mes_2Si=SiMes_2 \quad (74)$$

$$\begin{array}{c} Mes_2Si(SiMe_3)_2 \\ + \\ Xyl_2Si(SiMe_3)_2 \end{array} \xrightarrow{h\nu} \begin{array}{c} Mes_2Si: \\ + \\ Xyl_2Si: \end{array} \longrightarrow \begin{array}{c} Mes_2Si=SiMes_2 \\ + \\ Xyl_2Si=SiXyl_2 \\ + \\ Mes_2Si=SiXyl_2 \end{array} \quad (75)$$

Xyl = 2,6-Me$_2$C$_6$H$_3$–

Another route to stable disilenes is photolysis of hindered cyclotriarylsilanes (equation 76)[185]. In this reaction one disilene molecule is formed directly, and a second one is produced by eventual silylene dimerization.

$$(Ar_2Si)_3 \xrightarrow{h\nu} Ar_2Si=SiAr_2 + Ar_2Si: \qquad (76)$$

When smaller substituent groups are present, the disilenes formed from silylene dimerization are themselves unstable. They may undergo a second dimerization to a four-membered ring, or polymerize, or decompose by other pathways. An intermediate case is provided by Mes(Me$_3$CCH$_2$)Si:, which can be generated in 3-MP matrix by photolysis of the trisilane precursor, Mes(Me$_3$CCH$_2$)Si(SiMe$_3$)$_2$, at low temperatures[179]. This disilene persists in dilute solution up to 0 °C, but at 25 °C it disappears with a half-time of a few minutes (Figure 1).

Dimerization of silylenes has also been studied by flash photolysis. Dimethylsilylene, prepared by flash photolysis of (Me$_2$Si)$_6$ in cyclohexane, dimerizes to Me$_2$Si=SiMe$_2$ following second-order kinetics at the diffusion-controlled rate[183]. Disappearance of Mes$_2$Si: generated by flash photolysis also takes place at the diffusion-controlled limit[184], showing that steric shielding by mesityl groups does not impede dimerization.

2. Silylenes from disilenes

We have seen that silylenes normally undergo a dimerization to disilenes. The question then arises whether or not the reverse reaction takes place; that is, whether disilenes can separate at the Si=Si bond to produce silylenes.

There is fragmentary evidence for photolytic cleavage of disilenes to silylenes. For example, disilene **31** does not react with methanol, but under photolysis the methoxymonosilane product **32** is produced in good yield (equation 77)[185a]. A similar trapping reaction was reported earlier for Xyl$_2$Si=SiXyl$_2$ (Xyl = 2, 6-Me$_2$C$_6$H$_3$)[59]. These results suggest cleavage of the disilenes to silylenes, but of course other mechanisms could account for the products. In similar cases, photolytic cleavage of disilenes to

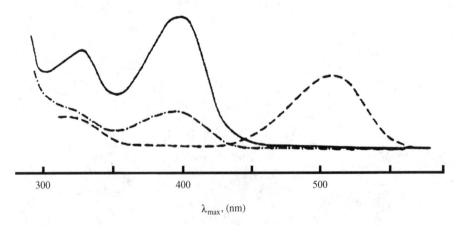

FIGURE 1. Mesityl(neopentyl)silylene and its dimerization: - - - - denotes electronic spectrum of Mes(Me$_3$CCH$_2$Si: in 3-MP glass at 77 K, ———— denotes spectrum after annealing at 120 K; the silylene has dimerized to the disilene, —·— ·— ·— denotes spectrum at 298 K, showing gradual disappearance of the disilene. Reproduced by permission from Reference 179

silylenes was not observed. For example, cophotolysis of a solution containing the two disilenes, $Xyl_2Si=SiXyl_2$ and $Mes_2Si=SiMes_2$, produced none of the mixed disilene, $Xyl_2Si=SiMes_2$[182].

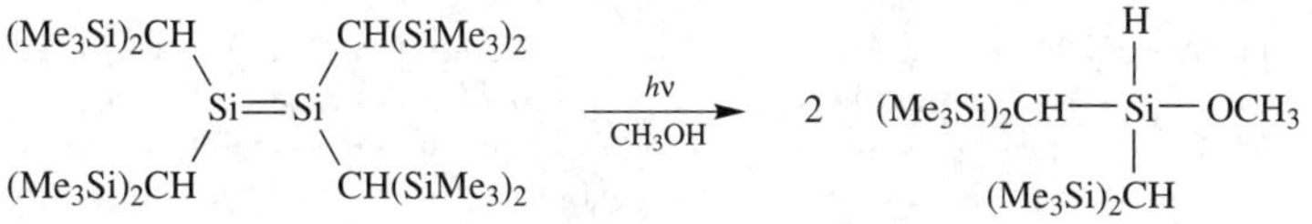

$$(Me_3Si)_2CH\diagdown_{Si=Si}\diagup^{CH(SiMe_3)_2} \xrightarrow[CH_3OH]{h\nu} 2\ (Me_3Si)_2CH-\underset{\underset{(Me_3Si)_2CH}{|}}{\overset{\overset{H}{|}}{Si}}-OCH_3 \quad (77)$$

(31) (32)

Apeloig and coworkers have trapped $(Me_3Si)_2Si$: from the photoinduced [2 + 2] cycloreversion to $(Me_3Si)_2Si=Si(SiMe_3)_2$ and 2,2′-biadamantylidene of the corresponding 1,2-disilacyclobutane[186]. Since laser flash photolysis experiments indicated that the disilene was the sole primary photoproduct, the silylene was believed to be formed by photodissociation of the disilene.

Until recently there was no evidence for thermally induced fragmentation of disilenes to silylenes. Early experiments with $Mes_2Si=SiMes_2$ showed that it did not react with the silylene trapping agent Et_3SiH on long heating at 80 °C. With Et_3SiH in toluene at 110 °C, the disilene slowly rearranged without cleaving to the disilene. Recent studies by Okazaki and coworkers have, however, suggested that thermal dissociation does take place for both the *cis* and *trans* isomer of the extremely hindered disilene **33**, under very mild conditions (70 °C)[166,187]. The resulting silylene **34**, was intercepted by the usual silylene trapping reagents, methanol, Et_3SiH and 2,3-dimethylbutadiene, to give the expected products (Scheme 5). Three-membered rings were obtained from **34** and unsaturated hydrocarbons, and trapping of **34** with sulfur and selenium led to interesting cyclic products. The most surprising reactions of **34** are those with naphthalene and benzene, illustrated in equation 70, Section III.B.2.C.

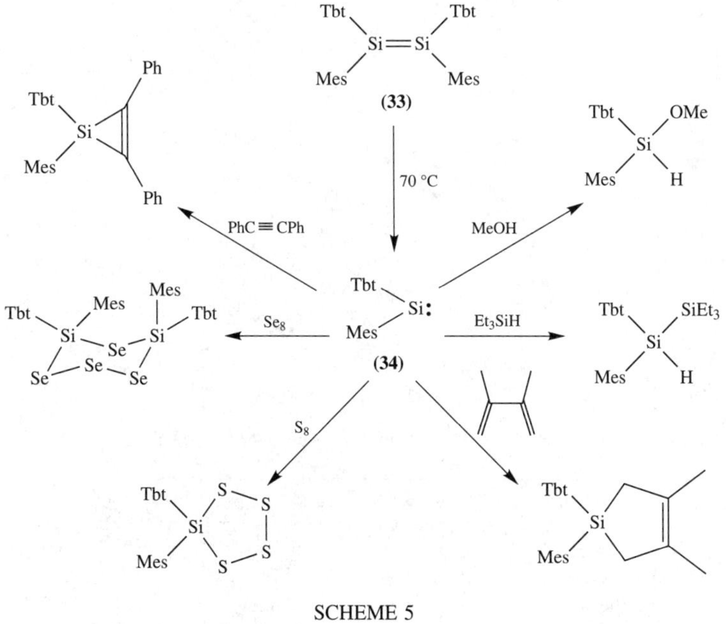

SCHEME 5

In a very recent report, Okazaki and coworkers describe the reaction of **34** with hindered isocyanides **35a–c**, leading to the first examples of stable silylene–Lewis base complexes **36a–c** (equation 78)[145]. The latter were isolated as blue-green to blue solids, stable up to 60 °C. Data for these complexes are listed in Table 2. The high-field ^{29}Si NMR resonances, and the very low values for the ^{29}Si–^{13}C coupling constants, are consistent with the acid–base complex formulation **36**, but incompatible with the alternative cumulene structure **37**. Crystal structures were not available, but quantum-mechanical calculations of the model complex Ph$_2$Si ← C=NPh suggested a strongly bent structure, as proposed for other silylene–Lewis base complexes (see Section V.A.1).

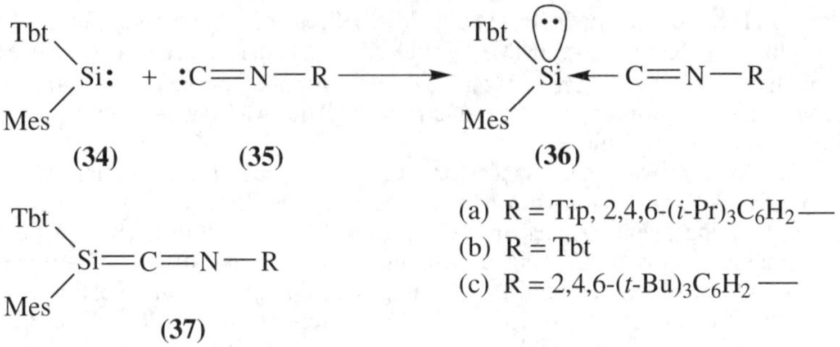

$$\text{Tbt}\diagdown\text{Si:} + \text{:C=N-R} \longrightarrow \text{Tbt}\diagdown\ddot{\text{Si}} \leftarrow \text{C=N-R} \qquad (78)$$
$$\text{Mes}\diagup \qquad\qquad\qquad \text{Mes}\diagup$$
$$(34) \qquad (35) \qquad\qquad (36)$$

(a) R = Tip, 2,4,6-(i-Pr)$_3$C$_6$H$_2$—
(b) R = Tbt
(c) R = 2,4,6-(t-Bu)$_3$C$_6$H$_2$—

$$\text{Tbt}\diagdown\text{Si=C=N-R}$$
$$\text{Mes}\diagup$$
$$(37)$$

Trapping reactions were carried out for **36a–c**, leading in most cases to silylene trapping products with liberation of the isocyanide. Thus, for instance, reaction with Et$_3$SiH yielded the disilane, along with **35a–c** (Scheme 6). Reaction of **36a–c** with methanol also led mainly to the silylene trapping product, but in one case the complex itself (**36a**) was trapped in small yield, producing the imine.

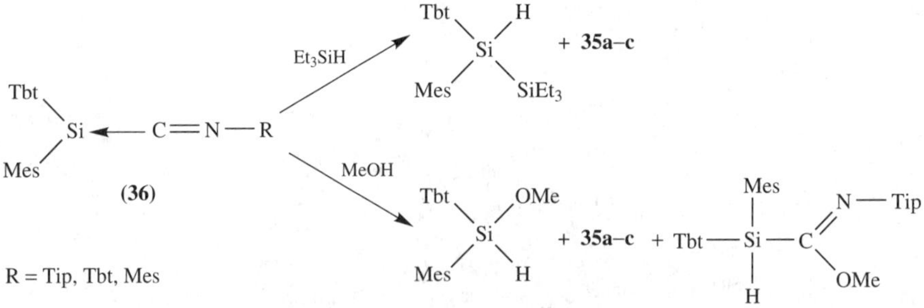

SCHEME 6

TABLE 2. Spectroscopic properties of silylene–isocyanide complexes

Compound	$\delta^{29}Si$–CN	δ Si–^{13}CN	$^1J_{\text{Si–CN}}$
36a	−53.6	209.2	38.6
36b	−57.4	196.6	22.1
36c	−48.6	178.5	1.0

Studies of the kinetics of *cis–trans* interconversion in disilene **33** indicate that it takes place by dissociation to the silylene and recombination, rather than by Si=Si bond rotation as is the case for other disilenes[187]. This isomerization occurs even at 50 °C.

A second example of thermal silylene cleavage has now been observed for the tetraalkyldisilene, $[(Me_3Si)_2CH]_2Si=Si[CH(SiMe_3)_2]_2$. No reaction occurs between this disilene and Et_3SiH at 80 °C, but at 120 °C the Si–H insertion product is formed (equation 79)[188]. These two examples suggest that the thermal fragmentation of disilenes into silylenes may be general for disilenes which do not undergo competing thermolytic reactions at lower temperatures.

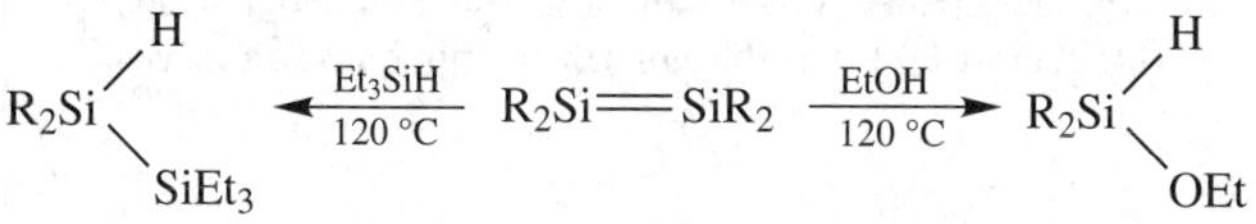

$$R_2Si\diagdown^{H}_{SiEt_3} \xleftarrow[120\,°C]{Et_3SiH} R_2Si=SiR_2 \xrightarrow[120\,°C]{EtOH} R_2Si\diagdown^{H}_{OEt} \quad (79)$$

$$R = (Me_3Si)_2CH$$

D. Rearrangements and Other Reorganizations of Silylenes

The pioneering studies by the Barton laboratory on the reactions of $Me_3Si\ddot{S}iMe$ revealed the richness of the intramolecular transformations of substituted silylenes[3,5]. In a 1978 paper that profoundly influenced the development of modern organosilicon chemistry, Wulff, Goure and Barton suggested that α-silylsilylenes can rearrange to β-silylsilylenes via unprecedented intramolecular H–C insertion by silylenes and alkyl as well as hydrogen shifts in the course of intramolecular silylene extrusions[189]. These steps are included in equation 80, which also contains Barton's finding that tetramethyldisilene $Me_2Si=SiMe_2$ can rearrange to $Me_3Si\ddot{S}iMe$ via a methyl shift. This is the inverse of the well-known rearrangement of carbenes to olefins. The lower stability of disilenes compared with olefins and of carbenes relative to silylenes inverts the thermodynamics of the silicon and carbon systems.

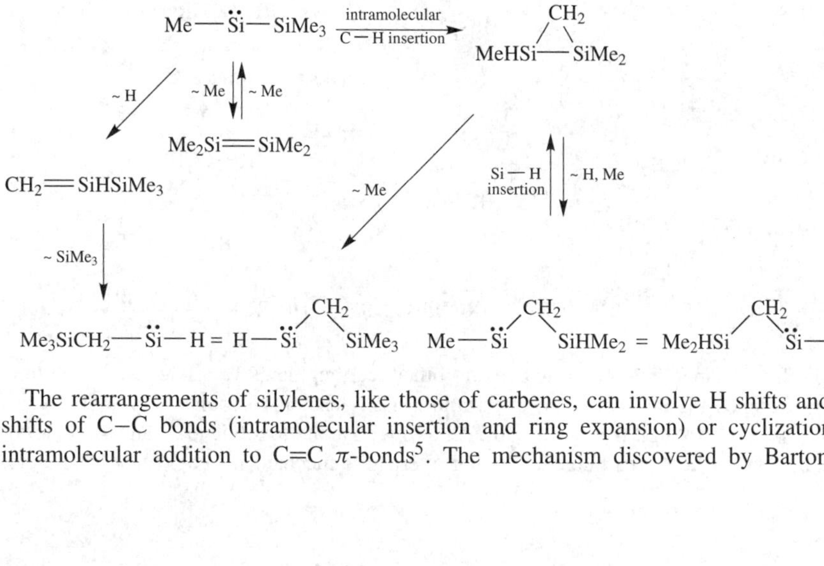

$$(80)$$

The rearrangements of silylenes, like those of carbenes, can involve H shifts and the shifts of C–C bonds (intramolecular insertion and ring expansion) or cyclization by intramolecular addition to C=C π-bonds[5]. The mechanism discovered by Barton for silylene-to-silylene rearrangements, silirane formation by H−C insertion, followed by ring cleavage via α-elimination, illustrates a major *difference* between silylenes and carbenes: silylene reactions are nearly always *reversible*, in contrast to carbene reactions.

Several questions about this reaction scheme did, however, remain unanswered until recently:

(a) Is the methyl shift that converts disilene $Me_2Si=SiMe_2$ to $Me_3Si\ddot{S}iMe$ reversible? Interconversions of silylenes and disilenes via silyl group shifts had been established[5,190], and a methyl shift was found to mediate the transposition of α- and β-substituents shown in equation 53[5,106]. The transposition deduced from trapping experiments (equation 53) and rendered nondegenerate and observable by use of a labeled silylene $CD_3\ddot{S}iSiMe_3$ (equation 81) finally established that the answer to this question is yes[105].

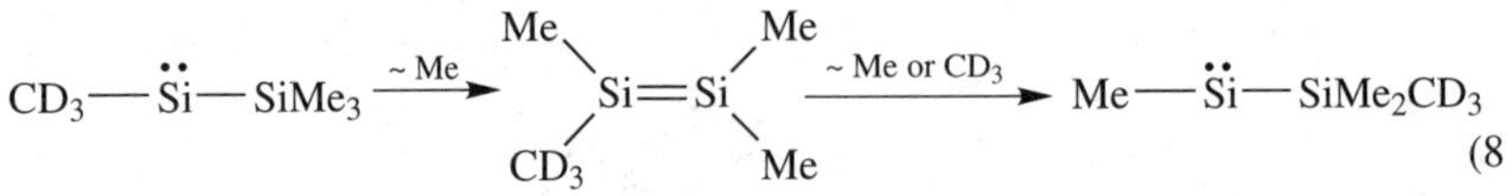

$$CD_3 - \ddot{Si} - SiMe_3 \xrightarrow{\sim Me} \underset{CD_3}{\overset{Me}{\diagdown}} Si = Si \underset{Me}{\overset{Me}{\diagup}} \xrightarrow{\sim Me \text{ or } CD_3} Me - \ddot{Si} - SiMe_2CD_3 \quad (81)$$

(b) Is ring opening of the disilirane in equation 80 with a concomitant methyl shift the major pathway leading to $H\ddot{S}iCH_2SiMe_3$? The alternative route shown in equation 80 from $Me\ddot{S}iSiMe_3$ to this β-silylsilylene via a hydrogen shift to a silene intermediate and a Me_3Si shift had been seriously considered[3,104]. Experiments with a labeled silylene $CD_3\ddot{S}iSiMe_3$ revealed that the silyl shift mechanism plays at most a minor role ($<5\%$)[105]. Discrepancies between quantum-mechanical calculations of this reaction and experiment have been resolved[191].

Ring opening of 2-ethylsilirane formed from addition of $H_2Si:$ to 1-butene led to the formation of *n*-butyl- and *sec*-butylsilylene[192]. It was deduced that *sec*-butylsilylene undergoes ring closure to 2,3-dimethylsilirane by intramolecular H−C insertion with an activation energy of 11.6 ± 2.6 kcal mol^{-1}. From this work a silylene-mediated polymerization of olefins can be envisioned (equation 82).

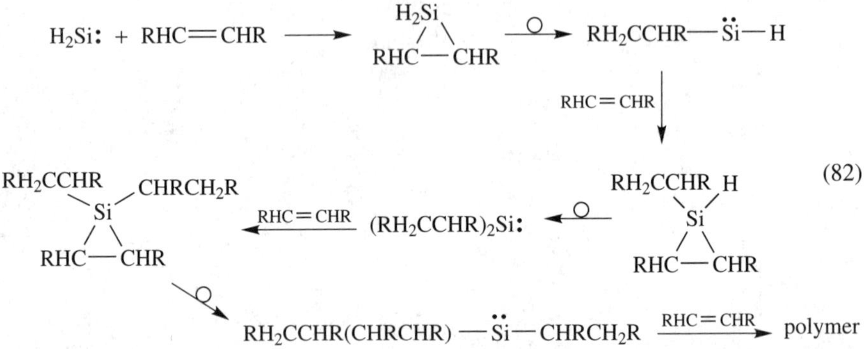

$$H_2Si: + RHC=CHR \longrightarrow \underset{RHC-CHR}{\overset{H_2Si}{\diagup \diagdown}} \xrightarrow{} RH_2CCHR - \ddot{Si} - H$$

$$\underset{RHC-CHR}{\overset{RH_2CCHR \diagup CHRCH_2R}{\diagdown Si \diagup}} \xleftarrow{RHC=CHR} (RH_2CCHR)_2Si: \xleftarrow{} \underset{RHC-CHR}{\overset{RH_2CCHR H}{\diagdown Si \diagup}} \quad (82)$$

$$\searrow RH_2CCHR(CHRCHR) - \ddot{Si} - CHRCH_2R \xrightarrow{RHC=CHR} \text{polymer}$$

The Barton laboratory has found many other rearrangments of silylenes[3,5] including strong evidence for the thermal isomerization of silylenes to silanones (equation 83)[193]. Results were also found that were consistent with silanone-to-silylene[194] and silylene-to-silanone-to-silylene[143] rearrangements, but alternative mechanisms could be written, e.g.

equation 84.

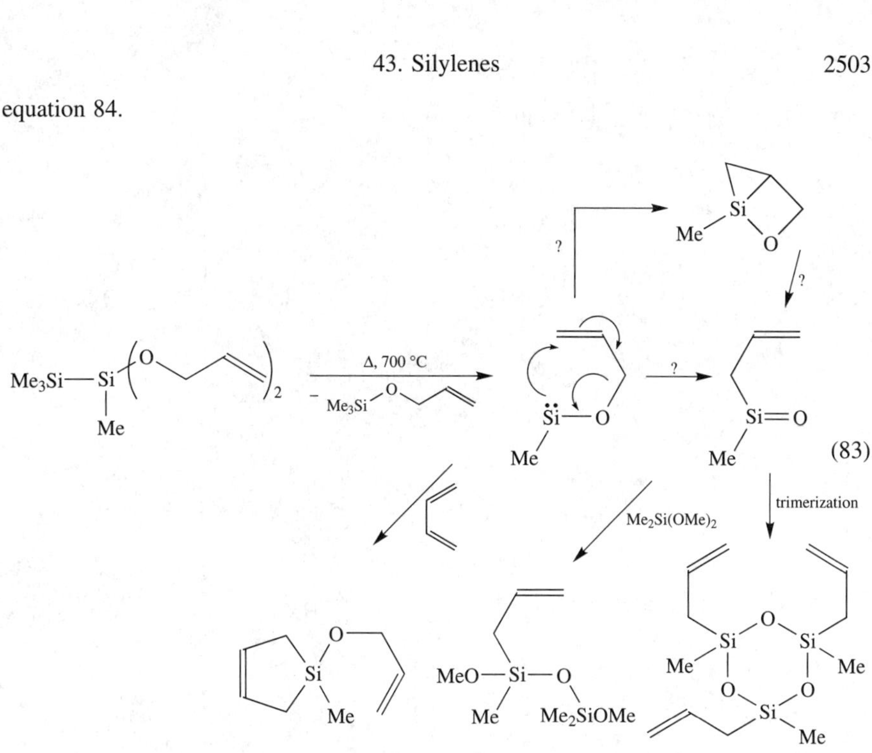

(83)

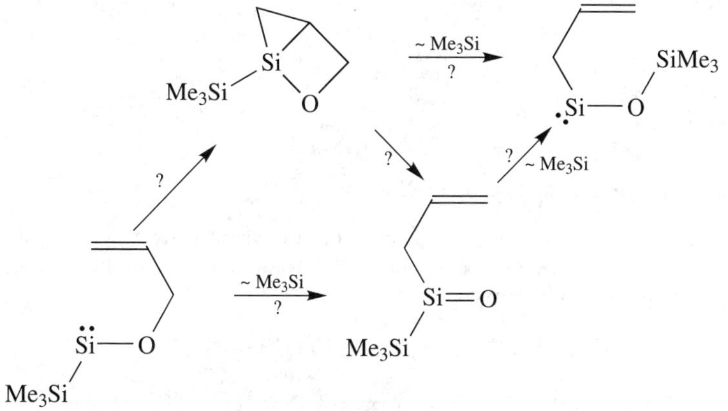

(84)

It should be noted that equation 84 is another example of a transposition in which α- and β-substituents on a silylene are exchanged.

The discovery by Fink, Puranik and Johnson of the rearrangement of a cyclopropenylsilylene to a silacyclobutadiene[195], an epic achievement, has been presented in the reverse direction in Section II.F (equation 54).

The rearrangement of 1-silacyclopent-3-en-1-ylidene to unsubstituted silole, first suggested in 1981[5], has been confirmed by the characterization of the silole dimer[196] and by the recording of UV-visible and IR spectra of the silylene, silole and 1-silacyclopenta-1,3-diene in matrix isolation experiments in which the isomeric silacyclopentadienes were interconverted photochemically (equation 85)[54,197].

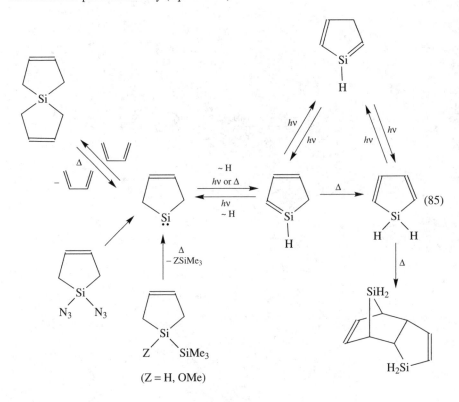

(85)

E. Miscellaneous Reactions of Silylenes

In a flowing afterglow experiment the reactions of pyrolytically generated Me_2Si: with the anions F^- and NH_2^- could be examined[103]. Addition (equation 86a) and deprotonation (equation 86b) were observed.

$$Me_2Si: + F^- \longrightarrow Me_2\ddot{S}iF^- \tag{86a}$$

$$Me_2Si: + NH_2^- \longrightarrow MeSiCH_2^- + NH_3 \tag{86b}$$

Dissociation of silylene into $^1D_2$ excited-state silicon atoms and H_2 was observed by excitation with a dye laser of $^1A_1$ ground-state H_2Si: to high bending vibrational levels of its $^1B_1$ singlet state[198]. From the vibrational level at which the dissociation channel opened, a heat of formation was established: ΔH_f° (H_2Si) = 65.4 ± 1.6 kcal mol^{-1}. Ground-state ^3P silicon atoms have also been observed upon laser excitation of H_2Si: from its $^1A_1$ state to $^1B_1$[199]. Extensive mixing of the $^1B_1$ state of H_2Si with other low-lying electronic states, including several triplet states, was believed to lead to predissociation to ^3P silicon atoms[199].

IV. THEORETICAL CALCULATIONS

The accessibility of ever larger and faster computers and the resulting increase in the experience of theorists and confidence in their results has led to a considerable number of calculations on silylenes, including their reaction paths. Reviews of these efforts have appeared[200,201]. In 1989 Apeloig identified the areas of silylene chemistry that had received most attention as: (a) the energy difference between the lowest triplet and singlet electronic states and the effects of substituents on ΔE_{ST}; (b) the electronic (UV-visible) spectra of silylenes; (c) pathways and transition states for insertion and addition reactions of silylenes; (d) energy differences between silylenes and their π-bonded isomers. A very useful interpretation of substituent effects on silylene structures and energetics in terms of the second-order Jahn–Teller effect has been given[202]. Papers published since 1988 are included here.

A. Singlet–Triplet Energy Gaps

One of the important differences between silylenes and carbenes is the contrast between the occurrence of both triplet and singlet ground states for carbenes and the universality of singlet ground states for silylenes. Conversion from the σ^2-configuration of the lowest singlet to the σ, π-configuration of the lowest triplet involves a one-electron energy increase due to the promotion of an unshared electron from an s-weighted orbital (for all nonlinear silylenes) to a pure p-orbital (equation 87). There is also a two-electron energy decrease as the repulsion between the lone pair electrons is relieved by their separation upon going from singlet to triplet. Kutzelnigg has pointed out that formation of triplet carbenes and their analogs ZR_2 (Z = C, Si, Ge, Sn) requires the promotion of an electron from the s^2p^2 configuration of the ground electronic state of the free atom to the sp^3 valence configuration that is associated with hybridization, while formation of singlet ZR_2 does not require this promotion[203]. Hence there should be an increase in the Z–R bond strength in going from singlet to triplet ZR_2. Kutzelnigg associates the occurrence of carbenes with triplet ground states with the gain in bond energy (for CR_2 but not SiR_2, GeR_2 or SnR_2) compensating for the promotion energy[203].

$$\underset{R}{\overset{R}{{\diagdown}}}Si{\diagup}\rightleftharpoons\underset{R}{\overset{R}{{\diagdown}}}Si{\diagup} \qquad (87)$$

Configuration interaction (CI) calculations based on generalized valence bond wave functions, called dissociation consistent CI(DCCI), were applied to the calculation of ΔE_{ST} for SiH_2, SiHF and SiF_2[204]. For SiH_2 and SiF_2 the theoretical values agreed well with (experiment): 21.5 (20.7) and 76.6 (76.2) kcal mol^{-1}, respectively. For SiHF the prediction was 41.3 kcal mol^{-1}. Stabilization by fluorine of the singlet state relative to the triplet state was ascribed to donation of fluorine pπ-electrons to the silicon p-orbital and an increase in the s-character of the nonbonding σ-orbital of silicon holding the lone pair due to the electronegativity of the fluorine substituent[204].

Similar calculations on SiHCl and $SiCl_2$ yielded ΔE_{ST} of 35.8 and 55.2 kcal mol^{-1}, respectively[205]. An interesting result was the difference in the contribution of the silicon p-orbital to the Si–Z bonds in the singlet and triplet states: for SiHCl, 73.6% p in Si–H and 62.3% p in Si–Cl of the singlet but 38.5% p in Si–H and 44.8% p in Si–Cl of the triplet. For $SiCl_2$ the silicon p-character of Si–Cl was predicted to be 61.1% in the singlet and 39.6% in the triplet. There is, of course, a corresponding opposite effect on the p-character of the σ-nonbonding orbital[205].

Apeloig has pointed out that the electronegativity of substituents at the divalent silicon atom of a silylene has opposite effects on singlet and triplet states, triplets being stabilized by less electronegative substituents while singlets are stabilized by more electronegative substituents[201]. It has been known since 1985 that extremely electropositive substituents like Li could lead to ground-state triplet silylenes[201], but these molecules are difficult to prepare.

Grev and coworkers have suggested that the effects of electronegativity can be combined with those of bulky groups to design a silylene whose ground state is a triplet[206]. The larger the bond angle at the divalent silicon atom, the greater is the p-character of the σ-nonbonding orbital, and thus the smaller is the one-electron promotion energy required to convert the σ^2-configuration that dominates the lowest singlet state to the σ,π-configuration that dominates the lowest triplet. The singlet state has both unshared electrons in the σ-orbital that increases in p-character, and hence energy, with an increase in bond angle. Therefore the σ^2 singlet state increases in energy more rapidly with an increase in bond angle than does the σ,π triplet, one of whose unshared electrons occupies a pure p-orbital whose energy does not vary with bond angle. The singlet and triplet potential curves cross[206].

Thus, while Me$_2$Si has a lowest singlet state with $\angle$C—Si—C = 98.8°, 25.1 kcal mol^{-1} below the triplet with $\angle$C—Si—C = 117.9°, for bond angles >140°, the triplet becomes the ground state. The implication is that for alkyl substituents with electronic effects similar to Me, it would be necessary to increase the bond angle by more than 40° in order to achieve a triplet ground state. This bond angle would be very difficult to achieve with even the bulkiest of substituents. However, for the less electronegative substituent SiH$_3$, there is a dramatic decrease in the predicted crossover angle beyond which the triplet is the ground state: 120° for (H$_3$Si)$_2$Si: versus 140° for (H$_3$C)$_2$Si:[206].

It has thus become an accepted strategy for the design of a ground-state triplet silylene to employ electropositive substituents like silyl groups SiR$_3$ that are sufficiently bulky that the $\angle$Si—Si—Si at which the singlet state has minimum energy is sufficiently large (>120°) that the triplet will be the ground state. The first of these 'designer silylenes' (i-Pr$_3$Si)$_2$Si: has been prepared, and preliminary results suggest that it has a triplet ground state[207] as predicted theoretically by Apeloig[134].

Calculations on a series of silylenes by Kalcher and Sax predicted that, relative to H, CH$_3$ is an electronegative substituent while SiH$_3$ and Li are electropositive[208]. The electron affinity of a silylene as well as its ΔE_{ST} were predicted to change with the electronegativity of the substituents. A decrease in ΔE_{ST} with decreasing electronegativity had been found previously, with LiHSi: and Li$_2$Si possessing triplet ground states. Unexpected and unexplained was the prediction that the electron affinities (EAs) of silylenes dramatically *increase* with silyl or Li substituents—to the point that not only the ground states but also the first-excited states of the substituted silylene anions are bound states. In the case of silyl substituents the stabilization may be the result of charge-induced dipole interactions due to the polarizability of the substituent, but the effect of Li substituents was attributed to delocalization[208].

Kalcher and Sax also examined the lowest singlet and triplet states of *cyclo*-(H$_2$Si)$_2$Si:, a three-membered cyclic silylene. The Si—Si—Si angles at the divalent silicon are constrained to be much smaller than those in the open-chain (H$_3$Si)$_2$Si for both the singlet (56.2° versus 95.9°) and the triplet (62.8° versus 125.4°) states. It is surprising that ΔE_{ST} is nearly the same in the cyclic (8.8 kcal mol^{-1}) and open-chain (8.3 kcal mol^{-1}) silylenes[208].

An interesting comparison has been made by Cremer and coworkers of calculations on the structures and energies of $^1A_1$ and $^3B_1$ H$_2$Si by two different high level computational methods, full valence active space SCF with multi-reference CI (CASSCF-MRCI), and density functional theory[209]. Both calculations predicted a ΔE_{ST} within 2 kcal mol^{-1} of the experimental estimate, but the CASSCF-MRCI calculations required several orders of magnitude more time than the density functional calculations (7,379 versus 50 s for $^1A_1$ and 13,873 versus 52 s for $^3B_1$ on a Cray supercomputer)! This large efficiency factor points to the growing importance of density functional calculations in dealing with the larger molecules of chemical interest. To predict ΔE_{ST} for $(i\text{-Pr}_3\text{Si})_2\text{Si}$ (-1.7 kcal mol^{-1}) required 170 *hours* of supercomputer time using density functional methods[210]! Apeloig and coworkers have calculated that $(t\text{-Bu}_3\text{Si})_2\text{Si}$ is also a ground-state triplet, with ΔE_{ST} of -7.1 kcal mol^{-1} at the BLYP/DZVP (ECP) level[210]. Such calculations would not be feasible by methods (such as CASSCF-MRCI) requiring several orders of magnitude more time.

B. Thermochemical Calculations

Whether methylsilylene CH$_3$S̈iH is more stable than its isomer silaethylene H$_2$C=SiH$_2$ has remained a subject of controversy and theoretical activity for two decades. Beauchamp, Goddard and coworkers summarized calculational results from 1978 to 1987 that predicted values for ΔE(CH$_3$S̈iH–H$_2$C=SiH$_2$) ranging from 11.6 to -6.8 kcal mol^{-1}[211]. The Beauchamp, Goddard value ($+11.6$ kcal mol^{-1}) obtained using a correlation consistent configuration interaction method based on generalized valence bond wavefunctions, which is in harmony with their experimental determination (see Section III.D), was immediately challenged by Grev, Schaefer, Gordon and coworkers[212]. Using much larger basis sets and higher level correlation treatments, these workers obtained ΔE(CH$_3$S̈iH–H$_2$C=SiH$_2$) = 3.6 kcal mol^{-1}. In 1990, Boatz and Gordon made a similar prediction (4.1 kcal mol^{-1}) using large basis sets and quadratically convergent configuration interaction[213]. The predicted ΔH_f for CH$_3$S̈iH, 50.6 kcal mol^{-1}, was similar to that predicted by Beauchamp, Goddard and coworkers (53 ± 4 kcal mol^{-1}). Boatz and Gordon had previously estimated ΔH_f(CH$_3$S̈iH) = 49.2 kcal mol^{-1} [and ΔH_f(Me$_2$Si) = 32–33 kcal mol^{-1}] from isodesmic reactions[214]. Silylsilylene H$_3$SiS̈iH (calculated ΔH_f = 72.8 kcal mol^{-1}) was predicted to be 7.9 kcal mol^{-1} less stable than disilene H$_2$Si=SiH$_2$ (ΔH_f = 64.9 kcal mol^{-1}. Trinquier predicted ΔE(H$_3$SiS̈iH–H$_2$Si=SiH$_2$) = 9.8 kcal mol^{-1}[215].

Relativistic effects on the structure and energy of the heavier group 14 dihydrides have been studied[216]. For H$_2$Si, inclusion of relativistic effects caused little change in the predicted values: $<0.05°$ in $\angle$ H–Si–H, <0.0005 Å in r(Si–H), <0.02 D in the dipole moment and <0.6 kcal mol^{-1} in energy.

Ho and Melius have employed Hartree–Fock SCF calculations with modest basis sets (6-31G*) with MP4(SDTQ)/6-31G** perturbation theory correlation corrections at the HF/6-31G* geometries and bond additivity corrections to predict revised ΔH_f(298 K) values (kcal mol^{-1}) for SiF$_2$ (-149.86 ± 4.00) and SiHF (-35.70 ± 2.85)[217]. Dissociation enthalpies for the formation of silylenes from SiF$_4$, SiF$_3$, SiHF$_3$, SiH$_2$F$_2$, SiH$_3$F, SiHF$_2$, SiH$_2$F and Si$_2$F$_6$ were also predicted[217].

Comparison of cyclic compounds **38** and **39** shown in equation 88 via correlated MP2/6-31G*//HF/6-31G* calculations indicated that the cyclic trisilylene **38** is more stable than the trisila-*s*-triazine structure **39** by 30 kcal mol^{-1} at the CCSD/6-31G*//MP2/ 6-31G* level of theory[218]. Similar calculations on 2-silapyridine-2-ylidene **40** and its silaimine isomer **41** (R = H) suggested that **40** might be a good candidate for a stable silylene, inert to dimerization[219]. **40** is calculated to be 8.8 kcal mol^{-1} more stable than **41**. Methyl substitution reverses the calculated order of energies, making **41** (R = Me) more stable than **40** (R = Me) by 4.7 kcal mol^{-1}.

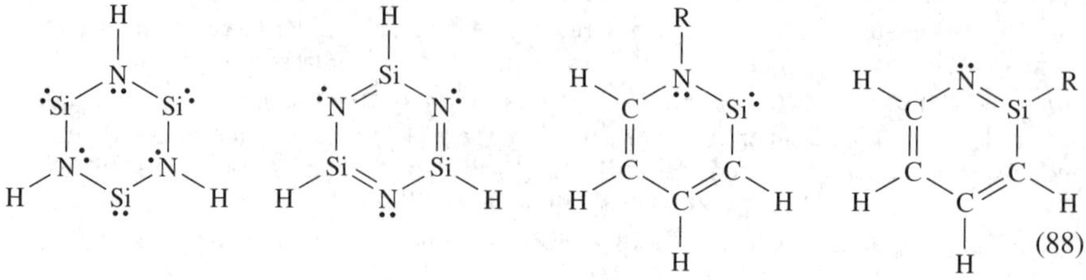

(38) (39) (40) (41) (88)

Despite the apparent difference in electron distribution suggested by the valence structures **38** and **39**, examination of homodesmic reactions leads to the prediction that both structures benefit from aromatic stabilization, the triazine more than the trisilylene. The greater strength of an N–H compared to a Si–H bond tips the scale in favor of the trisilylene structure[218].

Density functional calculations on the isomers of R$_2$Si$_2$ revealed that for R = H, Me, SiH$_3$, SiF$_3$, SiMe$_3$, SiPh$_3$ and Si(SiH$_3$)$_3$, the disilenylidene structure R$_2$Si=Si: is lower in energy than a *trans*-bent disilyne RSi≡SiR structure[220]. For R = H, the most stable structure is a doubly bridged 'butterfly'. Only for very bulky groups (*t*-Bu)$_3$Si and (2,6-Et$_2$C$_6$H$_3$)$_3$Si is the *trans*-bent RSi≡SiR more stable than R$_2$Si=Si:. For a discussion of vinylidenesilylenes and other double-bonded Z=Si: silylenes, see Section VIII.F.

C. Theoretical Predictions of Silylene Spectra

A diatomic silylene of astrophysical interest, silicon sulfide S=Si:, was treated by a three-stage calculation[221]. MOs from an HF-SCF treatment were employed in a complete active space SCF calculation whose optimized MOs were in turn used for a CI calculation that included as many as three million configurations! Comparison of various spectroscopic constants with those from experiment revealed fair agreement[221].

A coupled electron pair approach gave highly correlated wave functions for ground state $^1A_1$ and for first excited triplet $^3B_1$ SiH$_2$ from which theoretical rotational-vibrational spectra were calculated[222]. Calculated vibrational band origins agreed within 10–40 cm^{-1} with experimental values. Molecular geometries and vibrational frequencies, as well as force fields and mean-square amplitudes for SiF$_2$, SiCl$_2$ and SiBr$_2$ were calculated at the MP2/3-21G//HF/3-21G level in reasonable agreement with experiment[223].

Several calculations have aimed at interpreting the spectra of what has been called the first stable silylene (see Section VIII.C) shown in equation 89 as **42** (R = *t*-Bu). MP4/6-31G*//HF/6-31G* on **42** (R = H) suggested that the lowest energy peak in the photoelectron spectrum of **42** (R = *t*-Bu) corresponds to loss of an electron from a π-MO resulting from mixing of the Si 3p-orbital that would be the LUMO of a normal silylene with π_3 of a 1,4-diazabutadiene fragment[224]. This implies aromatic stabilization of **42** by N–Si π-bonding, also indicated by calculations on **43** and **44**. Density functional calculations led to similar assignments of the PES peaks, but the spherical distribution of electron density around the silicon atom of **42** (R = *t*-Bu) led to a description of the molecule as a silicon atom chelated to a diazabutadiene[225].

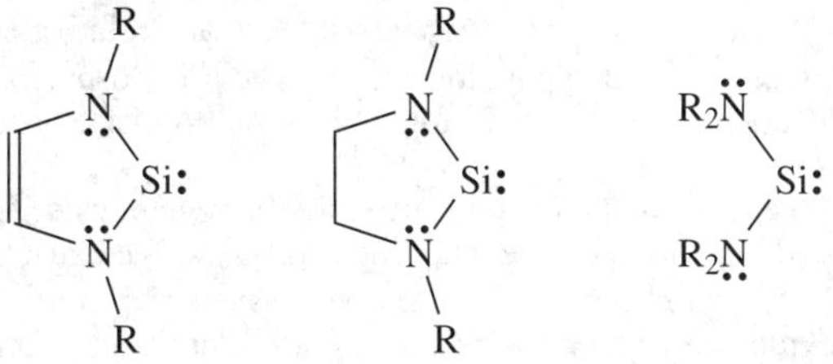

$$
\text{(42)} \quad \text{(43)} \quad \text{(44)} \tag{89}
$$

Calculations by Apeloig and coworkers on the electronic spectra of ethynyl- and vinylsilylenes have resolved earlier differences between theory and experiment and posed new questions[226]. The replacement of a methyl group of Me_2Si by a vinyl, ethynyl or aryl group causes a decrease in the transition energy, a red shift (see Section V.A.1). This is explained as due to excited-state stabilization by interaction of the singly occupied Si 3p-orbital with a π^* orbital of the substituent[226].

D. Benchmark Calculations on Silylenes

New advances in theoretical methodology have been tested by calculations on silylenes. The spin-coupled valence-bond theory is an *ab initio* technique that includes effects of electron correlation from the outset. Using small basis sets for Si (12s9p) and H (5s), accommodating the $(1s^2 2s^2 2p^6)$ core electrons in MOs from RHF calculations and expanding the spin-coupled valence orbitals in a basis of 30 MOs, compact wave functions were employed to calculate the lowest singlet and triplet states of SiH_2[227]. The $\Delta E_{ST} = 17.1$ kcal mol^{-1} value is in good agreement with the results of more extensive CI calculations with much larger basis sets, suggesting that modern valence-bond descriptions treat singlets and triplets even-handedly. The spin-coupled valence orbitals resembled the sp^2 and pure 3p orbitals employed in qualitative descriptions.

As a first application of a new analytical gradient method employing UHF reference functions, seven different methods for inclusion of correlation effects were employed to optimize the geometry and calculate the harmonic vibrational frequencies and dipole moments of the lowest open-shell states for three simple hydrides including $^3B_1$ SiH_2[228]. As the degree of correlation correction increased, results approached those from the best multiconfiguration SCF calculation.

When a new population analysis method for the calculation of atomic charges, the generalized atomic polar tensor (GAPT) approach, was applied to singlet SiH_2, the results were not changed by a correlation correction[229]. The insensitivity to the basis sets used (deduced from other examples), and the fact that the GAPT charges are obtained as a byproduct of the construction of the Hessian matrix in many electronic structure calculations, suggested that the GAPT analysis might replace the classical Mulliken population analysis.

Inclusion of connected triple excitations in the equation-of-motion coupled-cluster method for calculating excitation energies was also tested on SiH_2[230]. Excitation energies predicted for the lowest triplet state and the two lowest excited singlet states $^1B_1$ and $2\,^1A_1$ were within a kcal mol^{-1} of the predictions of the full CI calculation when triple excitations were included, but only for the $2\,^1A_1$ state was there marked improvement over the inclusion of only single and double excitations.

SiH_2 and SiF_2 were among several species used as 'guinea pigs' for new pseudospectral algorithms for electronic structure calculations[231,232]. Absolute energies agreed to within 0.25 kcal mol^{-1} with those from conventional basis sets, and relative energies were within 0.1 kcal mol^{-1}, with enhanced computation speeds. The pseudospectral numerical method is an alternative to all-integral methods for the calculation of Hartree–Fock molecular wave functions.

A perturbation-trajectory method for determining the dynamics of gas-surface collision processes was tested on the collision and subsequent surface reactions of SiH_2 on a Si(111) surface[233]. The predictions of an exact classical trajectory calculation[234] were confirmed: the sticking probabilities were unity at all temperatures, and it was found that surface SiH_2 can decompose by direct elimination of H_2 or by successive dissociation of Si–H bonds.

An efficient implementation of microcanonical classical variational transition state theory was applied to Si–H bond fission in SiH_2 and compared with trajectory calculations on the same potential surface[235].

E. Theoretical Treatments of Silylene Chemistry

Bond-order changes upon photoexcitation obtained from Sandorfy SCF-CI calculations have been employed to predict the ease of photoextrusion of a silylene from polysilane chains or rings[236]. Smaller bond-order changes with longer polysilane chains suggested that they are less efficient silylene sources. Similar conclusions were reached for branched and cyclic polysilanes. This approach focuses on σ to σ^* excitation without participation by Si 4s orbitals in descriptions of excited states[237].

Ab initio calculations suggested a bridged structure for the silver atom–SiO adduct[238]. The predicted structure was in better agreement with ESR observations than a previously proposed linear structure.

By theoretical prediction of its IR spectrum via *ab initio* SCF calculation, a structure of C_s symmetry with Si as the central atom was deduced for the photoproduct HClSi=S from Si=S and HCl in an argon matrix[239].

HF calculations have predicted two different transition states for the three-center concerted extrusion of SiH_2 from $EtSiH_3$[240]. The two transition states differ in energy by *ca* 10 kcal mol^{-1} and in their structure by the orientation of the SiH_2 group. It was suggested that contributions from concerted extrusion may account for the formation of some C_2H_6 and the wide range of rotational states for SiH_2 observed in IR multiphoton dissociation experiments. The reverse reaction, insertion of SiH_2 into a C–H bond of ethane (which, in principle, should have the same two transition states as the corresponding extrusion reaction) was examined by the same group with inclusion of electron correlation at the MP2/6-31G*//HF/6-31G* level[241]. The energy difference between the two transition states shrank to 7.8 kcal mol^{-1}, with a sizeable (22.3 kcal mol^{-1}) barrier for the favored transition state. By comparison with insertion reactions of several carbenes, it was concluded that barrier heights for insertion correlate with ΔE_{ST} rather than with frontier orbital interactions or steric effects. In such a model, the insertion barrier is due to an avoided crossing between a surface on which the singlet carbene (or silylene) is converted to an excited state of the insertion product and the surface on which the ground state of the insertion product lies. The higher the energy of the singlet silylene(carbene), the lower the relative energy (energy barrier) of the avoided crossing[241].

Calculations by Francisco examined decomposition reactions of ethylsilylene $EtSiH$[242]. Elimination to form SiH_2 and ethylene is computed to be the lowest energy pathway, with no barrier in the reverse direction. Rearrangement to $CH_3CH=SiH_2$ is predicted to have a 32 kcal mol^{-1} energy barrier, 4 kcal mol^{-1} higher than that for elimination of SiH_2. Elimination of an Si atom from EtSiH is predicted to have a barrier of 72 kcal mol^{-1}.

Transition states and their energies for insertion of SiH_2 into C−C, C−Si, Si−Si, C−H and Si−H bonds have been obtained at the MP4/6-31G(d)//HF/3-21G* level[243]. Since the barriers decreased in the order C−C>C−Si>Si−Si, steric interactions appeared to be a major factor in determining barrier heights. The barrier for insertion of SiH_2 into a C−C bond of cyclopropane (20 kcal mol^{-1}) was ca 40 kcal mol^{-1} less than that for ethane. Despite its low calculated barrier of <5 kcal mol^{-1}, insertion of SiH_2 into an unstrained Si−Si bond is unknown. The insertions into C−Si and C−C bonds have higher barriers[243] and are likewise unknown.

An important theoretical result has been the finding [at MP2/6-311 1G(d,p)] of a complex between SiH_2 and SiH_4 along the pathway for insertion into an H−Si bond[244]. This complex is 12.3 kcal mol^{-1} lower in energy than the separated reactants, but there is only a 1.6 kcal mol^{-1} barrier between the complex and the H_3SiSiH_3 insertion product. The experimentally observed negative activation energies for silylene insertion reactions (see Section VI) have been attributed to the intermediacy of such complexes, although other explanations are possible.

Ab initio study of the insertion of SiH_2 into the C−O and O−H bonds of methanol revealed that a complex is formed in which SiH_2 is coordinated to O[245]. Formation of the complex was found to be exothermic by 19.9 kcal mol^{-1} at MP2/6-31G*//HF/6-31G* with barriers for rearrangements to the O−H and C−O insertion products of 0.6 and 20.5 kcal mol^{-1}, respectively.

Insertion of SiH_2 into F_2 has been examined with a 6-31G* basis set and full geometry optimization at the MP2, CCD, CISD and QCISD levels[246]. Two transition states were found, differing by whether one of the F atoms approaches syn or anti to the Si lone pair. For anti approach there was no barrier found, while for syn approach a complex was found, whose energy is 2.5 kcal mol^{-1} below the reactants while a barrier of 0.5 kcal mol^{-1} separates the complex from the products.

Potential energy surfaces for the addition of SiH_2 to acetylene have been calculated at a very high ab initio level [QCISD/6-311G++(2df,2p)//MP2/6-31G(d,p), with corrections to the QCISD energies for triple excitations], and rate constants were calculated in the framework of the quantum-statistical RRK theory[128]. Negative activation energies for the formation of the silirene adduct were predicted, in accord with experimental observations, due to competition between stabilization and dissociation of vibrationally excited adduct. Rearrangement of silirene to HC≡CSiH$_3$ was predicted to occur via a 41.2 kcal mol^{-1} barrier, while a lower barrier of 36.8 kcal mol^{-1} was predicted for concerted rearrangement of silirene to vinylsilylene $H_2C=CHSiH$. The reverse rearrangement, ring closure of $H_2C=CHSiH$ to silirene, was calculated to have a 30.2 kcal mol^{-1} barrier[128].

Dimerization of silylenes XHSi, where X = NH_2, OH and F, has been studied theoretically by Apeloig and coworkers[247]. Alternative pathways for the dimerization of XHSi to the corresponding disilenes or bridged structures are shown in equation 90. For the three XHSi silylenes, dimeric bridged structures were found to be minima on the potential energy surface. According to MP4/6-311G**//6-31G** calculations, the bridged structure is less stable than the classical disilene structure by 10.0 and 3.2 kcal mol^{-1} for X = F and OH, respectively, but for X = NH_2 the bridged structure is more stable by 10 kcal mol^{-1}[247a,b].

Insertion of SiH_2 into H−C and H−Si bonds of CH_4 and SiH_4, respectively, and H-abstraction are among the model reactions for silicon carbide formation that were examined at the HF/6-31G* level[248].

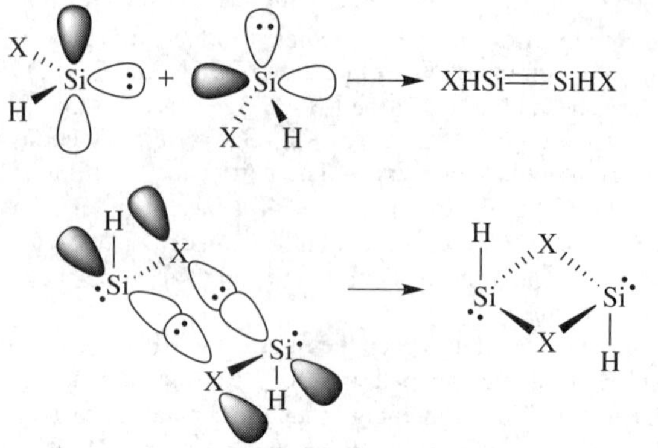

$$XHSi= SiHX \quad (90)$$

X = NH$_2$, OH, F

Theoretical investigation of the reaction between SiH$_2$ and CO found a weakly bound structure, pyramidal at Si, with a nonlinear array of the heavy atoms that corresponds to lone pair donation from the carbon and very little backbonding[249]. The addition is predicted to be exothermic by only 16 kcal mol^{-1}. Similar results have been reported, comparing the energies of planar and nonplanar adducts in which CO was coordinated to H$_2$Si at C and O, respectively[250].

The nature of the transition metal–silicon double bond has been examined by full optimized reaction space multiconfiguration SCF (FORS-MCSCF) calculations on MSiH$_2$$^+$ (M = Sc, Ti, V, Cr, Mn, Fe, Co and Ni) complexes[251]. It was found that the MSiH$_2$$^+$ complexes have significantly weaker bonds, as judged by the magnitude of the M−Si calculated force constants, than their MCH$_2$$^+$ counterparts. For the series Cr−EH$_2$$^+$ (E = C, Si, Ge and Sn) the silylene complex has a larger calculated force constant than the germylene and stannylene and thus was judged to represent a synthetically attainable target. At the SCF level the M−Si σ-bond was better described than the π-bond, the latter thus requiring a multiconfiguration wave function to include correlation effects for an accurate description of the bond. A qualitative description of the early transition metal complexes, e.g. ScSiH$_2$$^+$, is dominated by the double-bonded (M = E) and σ-ylide (M→E) resonance structures with the σ-dative/π-backbond (M⇌E) structure making a small contribution, while late transition metal silylenes, e.g. NiSiH$_2$$^+$, are dominated by (M⇌E) and to a lesser extent by (M⇌E) resonance structures[251].

Potential energy surfaces for dissociation of SiH$_2$ to SiH + H and Si + H$_2$, from various states up to 8 eV excitation energy, have been calculated by multiconfiguration SCF + multireference CI calculations[252].

V. SPECTROSCOPIC CHARACTERIZATION OF SILYLENES

A. Electronic Spectra

1. Free silylenes

The ultraviolet spectra of some inorganic silylenes, SiCl$_2$, SiF$_2$, SiH$_2$ and others, were measured in gas-phase experiments some years ago and are summarized in a 1992 review[253]. Studies of the electronic spectra of organosilylenes have appeared in recent years, following the discovery that photochemically generated silylenes could be isolated in argon matrices at temperatures near 10 K, or in hydrocarbon glasses at liquid nitrogen temperature[41]. Cyclosilanes or linear trisilanes are the most usual precursors, but other photochemical routes can also be used. The best studied of all silylenes in matrix conditions, dimethylsilylene, has been generated from no less than eight different precursors[254]. These are shown in Scheme 7, which illustrates the possible routes to silylenes in matrix.

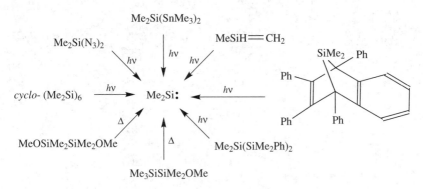

SCHEME 7

At liquid nitrogen temperature, 77 K, matrix isolation in hydrocarbons is successful for many silylenes because they are singlet species and so do not abstract hydrogen from C—H bonds, as would be expected for triplet molecules. The usual hydrocarbon is 3-methylpentane (3-MP) which forms a rigid glass at 77 K, but sometimes mixtures of hydrocarbons are used which are softer at this temperature, to allow some mobility of the silylene in the matrix. In a few cases, silylenes have also been identified from transient spectra obtained in flash photolysis experiments.

The electronic absorption of silylenes can be represented as a transition between the $^1A'$ singlet, S^0, with both electrons in a (mainly) 3s orbital on Si, to the $^1A''$ excited singlet, S^1, which has one electron in a 3p orbital perpendicular to the molecular plane (Figure 2). The energy of this transition usually places it in the visible region. Thus Me$_2$Si: has $\lambda_{max} = 453$ nm in 3-MP, with a fluorescence at 645 nm[254]. The large Stokes shift indicates a substantial difference in geometry in the ground and excited states. This is consistent with calculations on silylene H$_2$Si:, which predict the H—Si—H angle to be much larger in the S^1 excited state (120°) than in the S^0 ground state (93.2°)[255,256].

Table 3 presents a fairly complete list of electronic absorption maxima for the $S^0 \rightarrow S^1$ transition of silylenes isolated in matrices; data are also included for some silylenes identified in flash photolysis experiments[184,257–260]. The absorptions span a wide range, from 221 to *ca* 770 nm.

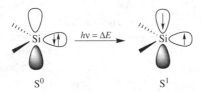

FIGURE 2. Electron distribution in ground and excited states of silylenes

TABLE 3. UV/vis absorption maxima for silylenes, RR'Si:

R	R'	λ_{max}	Reference
(A) In argon matrix, ca 10 K			
HC≡C	H	500	269
 (cyclopropenylidene)Si:		400–500	269
CH$_3$	C$_6$H$_5$	495	270,271
H	H	487	255,256
CH$_3$	H	480	270,271
CH$_3$	CH$_3$	460	254,270,271
CH$_3$	Cl	407	270,271
CH$_3$	OCH$_3$	355	168
H	NH$_2$	342	272
H	OCH$_3$	340	168
Cl	Cl	317	273
H$_2$C=C=Si:		310,325,340	269
 (cyclopropylidene)Si:		286	269
OPr-i	OPr-i	247	168
OCH$_3$	OCH$_3$	243	168
F	F	221 (gas phase)	274
(B) In 3-Methylpentane (3-MP), 77 K			
Me$_3$Si	Mes	760,776	37,275
Me$_3$Ge	Mes	730	37
Me$_3$Si	p-Tol	672	275
Me$_3$Si	o-Tol	662	275
Me$_3$Si	C≡CSiMe$_3$	660	37
Me$_3$Si	Ph	660	37,275
Me$_3$Ge	Ph	625	37
Me$_3$Si	CH=CH$_2$	619	275
Mes	Mes	577	172
Me$_3$Si	DEP^a	578	37
Me$_3$Si	Tip^b	568,570	37,275
Mes	C≡CPh	550	226
Mes	C≡CSiMe$_3$	545	226
Mes	Ph	530	179
Mes	1-Ad	526	142
Mes	C≡CH	524	226
Mes	t-Bu	505	142
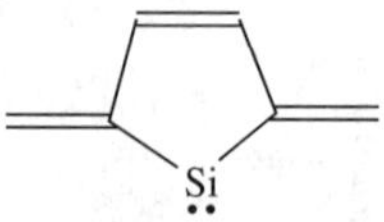 (2,5-dimethylenesilacyclopentanylidene)		505	265

TABLE 3. (continued)

R	R'	λ_{max}	Reference
Me₃Si	C≡CSiMe₃	500	37
Mes	Me₃CCH₂	500	226
Mes	i-Pr	500	179
Mes	CH₃	496	179
Ph	Ph	495	179
Ph	CH₃	490	179
Mes	Cl	487	179
t-Bu	t-Bu	480	53
Me₃Si	o-Me₂NCH₂C₆H₄	480	275
(methylenecyclopentyl-silylene)	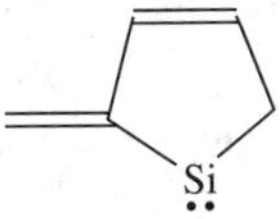	480	265
(cyclopentadienyl-silylene)		475	54, 197
CH₃	C≡CSiMe₃	473	226
CH₃	t-Bu	470	276
CH₃	Et	459	179
CH₃	CH₃	453	254
CH₃	p-CH₃OC₆H₄	450	179
Mes	TTBCP[c]	450	277
(cyclohexyl-silylene)		449	179
(cyclopentyl-silylene)	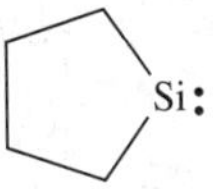	430	179
Mes	Tip[b]	430	275
Mes	OMes	425	278
Mes	NMe₂	405	179
Mes	N(SiMe₃)₂	404	179
Mes	OPh	400	278
Mes	OMe	390	278
(C) In cyclohexane solution (flash photolysis)			
Mes	Mes	580	150
(dibenzosilylene)	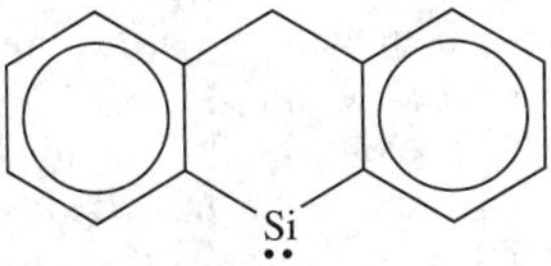	490	259

(continued overleaf)

TABLE 3. (*continued*)

R	R'	λ_{max}	Reference
	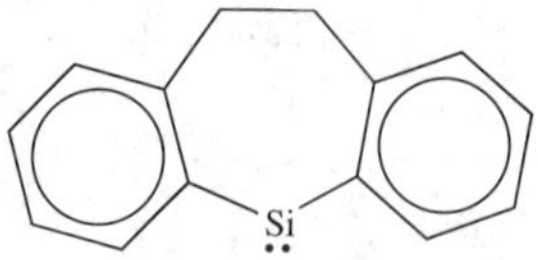 dibenzosilepine (Si)	485	259
	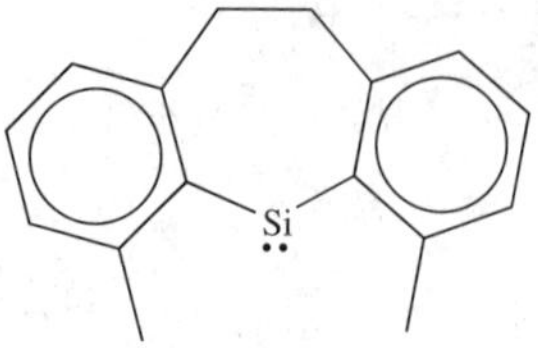 methyl-dibenzosilepine (Si)	485	259
Me	Me	465–470	257,258
Ph	Me	440	259
t-Bu	*t*-Bu	440	259

[a] DEP = 2,6-diethylphenyl,
[b] Tip = 2,4,6-triisopropylphenyl.
[c] TTBCP = tris(*t*-butyl)cyclopropyl.

Substituent effects on the electronic transition are now rather well understood, mainly due to theoretical calculations by Apeloig and coworkers[226,247,261]. π-Donor substituents, NR$_2$, OR or Cl, can donate electrons into the empty silicon 3p orbital in the ground state (Figure 3). The effect is to shift the p-orbital upward in energy, increasing the n–3p energy gap and causing a blue shift. Compare Mes(Me)Si:, 496 nm; Mes(Cl)Si:, 487 nm; Mes(MeO)Si:, 425 nm; Mes(Me$_2$N)Si:, 405 nm.

Electropositive, σ-donor substituents behave in opposite fashion. They donate electron density into the singly-occupied s-type orbital in the S^1 excited state, stabilizing it much more than the ground state, decreasing the energy gap and leading to a substantial red shift. Note the large effect on going from PhMeSi: (490 nm) to Me$_3$Ge(Ph)Si: (625 nm) and Me$_3$Si(Ph)Si: (660 nm). Finally, σ-acceptor substituents (F, CF$_3$) should withdraw electron density from the in-plane 3s-type orbital, stabilizing the ground state more than the excited state and producing a blue shift.

Aryl, vinyl and ethynyl substituents cause moderate bathochromic shifts compared with (CH$_3$)$_2$Si:. Calculations[226] suggest that this effect is best explained by excited-state stabilization resulting from mixing of the singly-occupied 3p orbital on silicon with the π^* orbitals of these sustituents, as originally suggested for aryl substituents (Figure 4)[179].

Steric effects on the excitation energy are also quite marked. As the R–Si–R angle increases with the introduction of bulky substituents, the S^0 state becomes more like the S^1 state geometrically, and the excitation energy is expected to decrease. Evidence for this is seen from the λ_{max} for Me$_2$Si: (453 nm), Et$_2$Si: (469 nm), and *t*-Bu$_2$Si: (480 nm). Decreasing the R–Si–R angle has the opposite effect, as seen in the series Et$_2$Si: (469 nm), *cyclo*-(CH$_2$)$_5$Si: (449 nm), *cyclo*-(CH$_2$)$_4$Si: (436 nm). Combination of aryl substitution with steric hindrance leads to large bathochromic shifts. For example, although PhMeSi:, MesMeSi: and Ph$_2$Si: all absorb near 490 nm, Mes$_2$Si: has λ_{max} = 577 nm.

Orientation of the substituent orbitals relative to the silicon 3p orbital is also significant. An example of this kind of conformational dependence has recently been reported by Kira

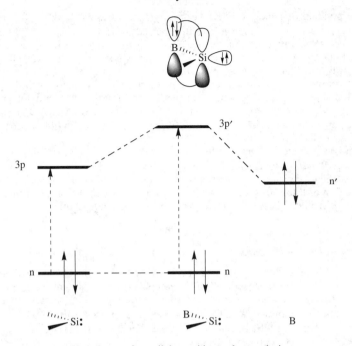

FIGURE 3. Qualitative orbital diagram for a silylene with a π-donor substituent

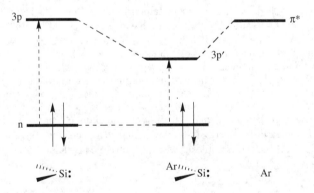

FIGURE 4. Orbital diagram for a silylene with a π-acceptor substituent

and coworkers[37]. In a series of aryl trimethylsilylsilylenes with increasing steric hindrance, λ_{max} (nm) increases from Ph(Me$_3$Si)Si: (660) to Mes(Me$_3$Si)Si: (790) but then decreases to 690 for 2,6-(C$_2$H$_5$)$_2$C$_6$H$_3$(Me$_3$Si)Si: and 2,4,6-(i-Pr)$_3$C$_6$H$_2$(Me$_3$Si)Si:. The decrease for the latter two silylenes was explained by rotation of the plane of the aromatic ring out of the plane containing the Si 3p orbital, reducing the conjugation.

Another example of an orientation effect was observed for the aryloxysilylenes Mes(OAr)Si:, where Ar = mesityl or 2,6-diisopropylphenyl, made by photolysis of the corresponding trisilanes at 77 K. These silylenes showed λ_{max} *ca* 430 nm in 3-MP glass but absorbed near 400 nm in a softer matrix, 3-MP/isopentane[142]. The difference was ascribed to a conformational change. In 3-MP the silylenes are believed to take a conformation similar to that of their precursors, in which the p lone pair on oxygen is twisted away from the vacant p-orbital on silicon, so that π-donation is limited. In the softer matrix the silylenes can relax into an equilibrium conformation with the p-orbitals on O and Si parallel, increasing conjugation and giving a further blue shift. Other silylenes with similar structure, Mes(OPh)Si: and Mes(OAlkyl)Si:, absorb near 400 nm in both matrices. For these less hindered silylenes, freezing-in of a nonequilibrium conformation does not take place.

The substituent and geometric effects on the electronic spectra of silylenes are well modeled by *ab initio* MO calculations, in all essential respects[226,247a,261]. For some substituents such as vinyl and ethynyl groups, the MO calculations are inaccurate because of contamination of the UHF states by higher-lying triplet states. Removal of this contamination by the method of spin projection gave results in good agreement with experimental trends[261].

2. Silylene–Lewis acid complexes

Since they have both a vacant low-energy orbital and a lone pair, silylenes might behave either as electron pair donors or acceptors. There is scant evidence for silylenes reacting as Lewis bases, but complexes of silylenes acting as Lewis acids are now well-established; these complexes can also be described as silaylides, $R_2Si^-\text{—}B^+$ [262]. Trinquier has calculated that even SiH_4 should form a weak complex with $H_2Si:$, in which a silane hydrogen binds to the p-orbital of the silylene[263].

Indirect evidence for such complexing was put forward in 1980, based on the observation that the selectivity of $(CH_3)_2Si:$ for reaction with pairs of alcohols was increased when the solvent was changed from hydrocarbon to diethyl ether to THF[264].

The first direct observation of silylene complexes was reported in 1987[142]. Four hindered silylenes, Mes(R)Si: where R = Mes, *t*-Bu, 1-adamantyl and 2,6-diisopropylphenyl, were generated by photoylsis of the corresponding trisilanes Mes(R)Si(SiMe₃)₂, in 3-MP and 2-methyltetrahydrofuran matrices. The long-wavelength band, falling from 577 to 425 nm for these silylenes in 3-MP, was replaced by an absorption near 350 nm for all four silylenes in 2-MeTHF, due to the formation of silylene complexes with the ether. In general, formation of silylene complexes with Lewis bases is signaled by a strong blue shift of the n–p absorption band. The base, by donating electrons into the 3p silicon orbital, greatly increases the energy of the S^1 excited state (equation 91). Large blue shifts upon formation of 2-MeTHF complexes were also observed by Kira, for vinylsilylenes[265].

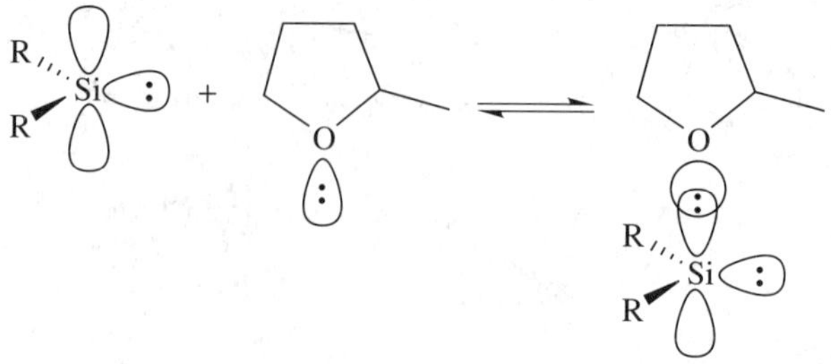

(91)

Later experiments were undertaken in mixed matrices containing 95% 3-MP and 5% Lewis base[113,143]. Irradiation of the trisilane precursors at 77 K gave initially the silylenes,

Mes$_2$Si:, Mes(t-Bu)Si:, Mes(Dip)Si: (Dip = 2,6-diisopropylphenyl) and Me$_2$Si:. When the matrix was annealed to allow some molecular motion, the absorption band for free silylene diminished and was replaced by a band at higher energy, assigned to the silylene complex. Representative data for absorption maxima of three silylenes with several bases are given in Table 4. The frequency shift upon complexation depends on the nature of both the base and the silylene and should be a measure of the strength of the interaction[113,143]. For the comparison of bases, the data for Me$_2$Si are most appropriate, since the silylene is unhindered. The frequency shifts increase in the order: phosphines<CO<sulfides<ethers<amines. This can be compared with the stability order for base–silylene complexes predicted from theoretical calculations: HCl<H$_2$S<H$_2$O<PH$_3$ <NH$_3$[266,267]. Agreement is satisfactory except for the phosphines, which are apparently much weaker bases than predicted according to frequency shifts.

Comparing different silylenes, it is interesting to note that the frequency shifts for the aryloxysilylene, Mes(ODip)Si:, are much smaller than for Me$_2$Si: or Mes$_2$Si:. This can be understood to result from π-donation by the aryloxy oxygen into the silicon 3p orbital, weakening the Lewis acidity[142].

In a similar study, dimesitylsilylene was generated photochemically from Mes$_2$Si(SiMe$_3$)$_2$ in a soft matrix, 4 : 1 isopentane:3-MP, in the presence of bases, at 77 K. The absorption frequencies in nm for the complexed silylene with various bases were: N-methylpyrrolidine, 324; pyrrolidine, 334; piperidine, 328; quinuclidine, 330; n-Bu$_3$P, 338; THF, 320; thiophene, 315. Agreement with the values in Table 4 is rather good[143].

Complexation of Me$_2$Si: by bases has also been observed in flash photolytic studies[257]. Photolysis of (Me$_2$Si)$_6$ in cyclohexane gave an absorption band at 465 nm assigned to Me$_2$Si:. This band shifted to 340 nm in CH$_3$CN, 310 nm in THF and 270 nm in Et$_3$N. Quenching experiments with various reactants showed that complexing reduced the reactivity of silylenes. Uncomplexed Me$_2$Si: reacted with alcohols, n-Pr$_3$SiH, THF and C=C and C≡C bonds essentially at the diffusion-controlled rate[257a], but reaction rates for silylene complexes were distinctly slower. Decreased reactivity of a complexed silylene has also been observed in solution. In 3-MP, Mes(1-Ad)Si: dimerizes immediately upon melting of the matrix, but in 2-MeTHF as solvent it gives a complex with 2-Me-THF which is stable up to ca −135 °C[142].

TABLE 4. Absorption bands (nm) for silylene complexes in 3MP[278]

Base	Me$_2$Si:	Mes$_2$Si:	Mes(ArO)Si:^{a}
none	453	577	395
n-Bu$_3$N	287	325	346
Et$_3$N	287	350	348
2-Me-THF	294	326	330
Et$_2$O	299	320	332
i-PrOH		322	321
t-Bu$_2$S	322	316	350
CO	345	354	328
n-Bu$_3$P	390	345	358
Et$_3$P		336	349

^{a}Ar = 2,6-diisopropylphenyl.

Alcohols are also bases toward silylenes. Complexes of alcohols with $Mes_2Si\colon$ and other hindered silylenes are observed spectroscopically when a 3-MP matrix containing the silylene and 5% of 2-propanol or 2-butanol is annealed[113]. Melting of the matrix results in rapid reaction of the complex to give the O—H insertion product (equation 92). These results imply that formation of an acid–base complex is the probable first step in the reaction of silylenes with alcohols and perhaps with many other reagents[259].

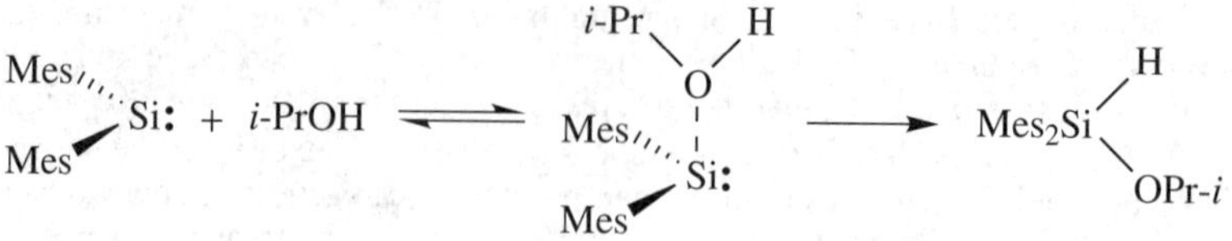

$$Mes_2Si\colon + i\text{-PrOH} \rightleftharpoons Mes_2Si\colon \cdot\cdot\cdot O(i\text{-Pr})H \longrightarrow Mes_2Si(H)(OPr\text{-}i) \quad (92)$$

As described in Section III.B.2.b, carbon monoxide forms complexes with silylenes which are relatively weak, losing CO easily to give the disilene[155,156]. Therefore, it is likely that they have the acid–base complex structure **45**, in accord with theoretical prediction[249], rather than the alternative silaketene structure **46**. However, in addition to the absorption bands near 350 nm (Table 4), a weak absorption near 600 nm was observed for matrices containing silylenes and CO. These long-wavelength bands may possibly be due to the silaketene isomers.

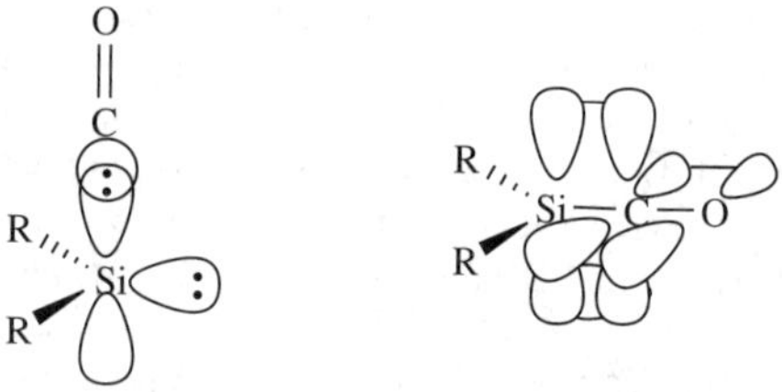

(45)   (46)

Carbonyl and thiocarbonyl complexes of Mes_2Si have been observed in a remarkable series of experiments by Ando and coworkers[144]. Photolysis of $Mes_2Si(SiMe_3)_2$ was carried out at 77 K in a 'soft' (3-MP/isopentane) matrix in the presence of tetramethyl-2-indanone or its sulfur analog. The free silylene was formed initially but it slowly reacted with the carbonyl (or thiocarbonyl) compound to give the complex, which can be formulated as a silacarbonyl ylide (Scheme 8). Further photolysis converted these ylide complexes to the silaoxirane or silathiirane, reversibly.

Another interesting compound which could be regarded as a silylene complex is the dioxygen adduct of dimesitylsilylene (mentioned in Section III.B.2.b) obtained by photogeneration of Mes_2Si in an oxygen matrix[151]. The infrared spectrum of the product showed a new band at 1084 cm^{-1}, assigned to Si—O stretching. Isotopic labeling with dioxygen was used to establish that the complex has an open structure, Mes_2Si—O—O (equation 93). The Si—O bonding is probably strong, so the product can perhaps equally well be described as a silanone-O-oxide, $Mes_2Si=O^+-O^-$. Dimethylsilylene appears to form a similar open-chain oxygen complex when photogenerated in an oxygen matrix[268]. $Cl_2Si\colon$ and $F_2Si\colon$, however, behave quite differently, forming dioxasiliranes (equation 94); see Section III.B.2.b[153].

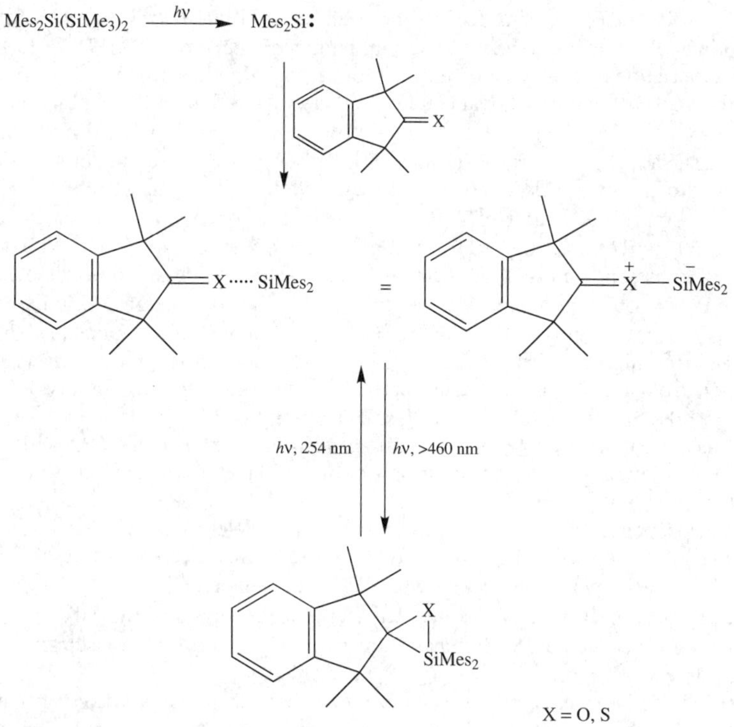

SCHEME 8

$$Mes_2Si: + O_2 \longrightarrow \underset{Mes_2Si\cdot}{O-O\cdot} \quad \text{or} \quad Mes_2Si\overset{+}{=}\overset{-}{O}-\overset{-}{O} \qquad (93)$$

$$SiX_2 + O_2 \longrightarrow X_2Si\overset{O}{\underset{O}{\diagdown|}} \qquad (94)$$

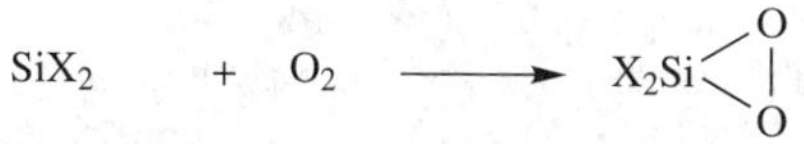

B. Other Spectroscopic Measurements

1. Silylene SiH$_2$

The importance of SiH_2 as an intermediate in the chemical vapor deposition (CVD) of silicon films from silanes (see Section VIII.D) has led to considerable effort in the development of sensitive spectroscopic methods for its detection *in situ*. The $^1B_1$ electronically excited state of SiH_2 reached by absorption is predissociated in higher vibrational levels, and thus laser-induced fluorescence (LIF) is difficult. The use of laser-spectroscopic techniques, including modified LIF and intracavity laser absorption spectroscopy (ILAS), has been reviewed[279,280]. ILAS has been used to detect SiH_2 in a parallel-plate RF-discharge system used in CVD of hydrogenated amorphous silicon[281]. A ring dye laser tuned to a single vibrational transition at *ca* 579 nm allowed the detection of SiH_2 during thermal decomposition of Si_2H_6[282].

SiH$_2$ was first detected using LIF by Inoue and Suzuki in 1984[283]. In 1986, SiH$_2$ was reported by LIF detection in the infrared multiphoton dissociation (IRMPD) of EtSiH$_3$[284]. The wide variations in the fluorescence lifetimes of individual rovibronic levels in the $^1B_1$ state of SiH$_2$ found by Thoman and coworkers was attributed to coupling of the $^1B_1$ (S_1), $^1A_1$ (S_0) and $^3B_1$ (T_1) states, manifested by predissociation of $^1B_1$ SiH$_2$ to Si(^{3}P) + H$_2$[285]. Dissociation of the $^1B_1$ SiH$_2$ formed in IRMPD of SiH$_4$ to singlet ($^1D_2$) as well as triplet silicon atoms was found by state-selective two-photon LIF from the Si atoms formed[286]. A potential barrier of 4.4 to 6.2 kcal mol^{-1} for the dissociation of $^3B_1$ (T_1) SiH$_2$ to Si(^{3}P) + H$_2$ was inferred from LIF in a supersonic jet experiment[287]. Rotational analysis of emission from SiD$_2$ in a similar experiment also suggested predissociation to Si(^{3}P) + D$_2$[288]. LIF experiments also led to an estimate of the energy distributions of the fragments from the photolysis of PhSiH$_3$ at 193 and 248 nm[289]. It was concluded that SiH$_2$ is formed from highly vibrationally excited ground-state PhSiH$_3$, produced via internal conversion from higher singlet states reached by photoexcitation. Photodissociation of PhSiH$_3$ to SiH$_2$ in a supersonic jet has also been monitored by LIF[290].

SiH$_2$ has been detected by LIF in RF plasmas in mixtures of SiH$_4$ and Ar, He, Xe and H$_2$, allowing the optimization of working parameters to maximize the silylene concentration[291,292].

The time dependence of the Si atom concentration determined by atomic resonance absorption spectroscopy during shock-wave-induced thermal decomposition of SiH$_4$ was in accord with a stepwise dissociation via SiH$_2$ intermediates[293].

LIF excitation spectra allowed a detailed vibrational analysis of the SiH$_2$ $^1A_1$ ground state yielding exact accurate vibrational frequencies and anharmonic constants[294]. A similar analysis was carried out on the $^1B_1$ (S_1) excited state[295].

Infrared diode laser absorption spectroscopy has been employed to detect ground-state SiH$_2$ and allowed the first observation of its high-resolution IR spectrum by Yamada and coworkers in 1989[296]. State-selective multiphoton IR excitation of SiH$_2$ has been examined computationally[297]. The resonance-enhanced multiphoton ionization (REMPI) spectrum of SiH$_2$ was observed for the first time by Robertson and Rossi and was employed to measure the SiH$_2$ sticking coefficient on a silicon surface[298].

Photoionization mass spectrometric studies, by Berkowitz and coworkers, of SiH$_2$ formed by successive abstraction of H atoms from SiH$_4$ by F atoms allowed determination of the ionization potentials of the two lowest states: $^1A_1(S_0) = 9.15 \pm 0.02$ or 9.02 ± 0.02 eV and $^3B_1(T_1) = 8.24_4 \pm 0.02_5$ eV[299]. This important series of experiments[300] yielded a singlet–triplet separation for SiH$_2$ of either 0.78 ± 0.03 or 0.91 ± 0.03 eV.

Synchrotron radiation of 115 to 170 nm has been used to dissociate SiH$_4$ in a pulsed supersonic free jet, and the abundance of SiH$_2$ was measured by quadrupole mass spectrometry using 11 V sub-ionization threshold electron-impact energy[301]. The possible detection of SiH$_2$ in the outer envelope of a stellar object has been reported[302].

An SiH$_2$–HF complex was detected by infrared spectroscopy following matrix isolation of the reaction products from SiH$_4$ and F$_2$[303].

2. Other silylenes

a. HSiF. HSiF was detected by LIF in 1983 by Lee and deNeufville[304]. Suzuki and coworkers were able to assign 1300 rotational transitions of HSiF by LIF using a dye laser tuned to the frequency of excitation (410–430 nm) from the zeroth vibrational energy level of the ground singlet state to that of the first excited singlet state[305]. The structure derived from the rotational constants agreed well with predictions of *ab initio* calculations. From a different LIF experiment it was concluded that ∠H−Si−F increases from 98° to 115° upon electronic excitation from S_0 to S_1[306]. Via an intermodulated fluorescence technique, magnetic interactions in HSiF were studied[307]. No hyperfine structure was observed, but rotational g-factors could be derived from the Zeeman effect. An IR spectrum of an HSiF−HF complex has been reported in a matrix isolation experiment[303].

b. SiF_2. The laser excitation spectrum of SiF_2 from pyrolysis of Si_2F_6 was reported by Stanton et al. in 1985, and the concentration of SiF_2 was monitored by LIF following excitation at 221.6 nm with a dye laser[308]. SiF_2 produced by reaction of F_2 and NF_3 with solid Si was detected by 320.25 to 321.50 nm multiphoton ionization and mass spectrometric detection[274]. SiF_2 has been detected by REMPI mass spectrometry, following reaction of $SiHF_3$ with F atoms and of heated silicon crystals with F_2 under single gas-surface reaction conditions[309].

c. $SiCl_2$. $SiCl_2$ was detected by LIF in the 310 to 410 nm region for the first time in 1986[310]. Fluorescence from $SiCl_2$ in this region had been observed previously upon IR multiphoton excitation of SiH_2Cl_2 and vacuum UV photolysis of SiH_2Cl_2 and of $SiHCl_3$[311,312]. High resolution LIF spectra of $SiCl_2$ have been observed in a molecular beam[313]. Analysis of the spectra revealed that ∠Cl−Si−Cl opens up by 22° upon excitation from the $^1A_1$ to the $^1B_1$ state (from 101° to 123°). LIF detection of $SiCl_2$ in a pyrolysis jet showed that pyrolysis of $SiHCl_3$ is a good source of $SiCl_2$ for spectroscopic studies[314].

The microwave spectrum of $SiCl_2$ formed by conproportionation of $SiCl_4$ and solid Si has been recorded[315]. The large asymmetry of the Cl nuclear quadrupole coupling tensor was accounted for by large π-electron back-donation from Cl to Si. Emission from the products of reaction of excited $^3P_{2,0}$ Ar atoms with $SiCl_4$ and with SiH_2Cl_2 in a DC discharge flow system has been attributed to phosphorescence from $^3B_1$ $SiCl_2$ on the basis of the vibrational spacings[316].

d. $SiBr_2$, SiI_2. The He(I) photoelectron spectra of $SiBr_2$ and SiI_2 have been presented and analyzed by Bock and coworkers and compared with those of SiF_2 and $SiCl_2$[317]. Relativistic effects must be included for $SiBr_2$.

e. $SiMe_2$. The IR spectrum of matrix-isolated $SiMe_2$ from photolysis of $Me_2Si(N_3)_2$ was recorded, and six IR transitions were assigned as in-plane or out-of-plane polarized[318].

f. $SiMes_2$. The IR spectrum of a complex between dimesitylsilylene $SiMes_2$ and O_2 (see Section III. B.2.b) was reported following reaction of $SiMes_2$ and O_2 in an oxygen matrix at 16 K[151]. Comparison of the experimental spectrum (and oxygen isotope shifts) with model calculations ruled out siladioxirane, silanone and dimeric structures and were in accord with a nonlinear silanone O-oxide structure $Mes_2Si=O-O$.

g. Si=O. $^{29}Si=O$ has been observed by absorption at 85.76 GHz in Sagittarius B2, the most massive and complex giant molecular cloud in the galaxy[319]. The observation was by interferometry employing a radio telescope. Earlier, $^{29}Si=O$ had been observed in a radio telescope by maser emission from various stellar sources[320]. In the laboratory, the K-shell absorption spectrum of Si in Si=O vapor has been measured in the 1830 to 1875 eV region by X-ray spectrometry[321]. The fine structure at the K-absorption edge resembled that of the PO molecule.

VI. KINETICS OF SILYLENE REACTIONS

Despite the short lifetimes of most silylenes, improvements in flash photolysis techniques for their generation and time-resolved spectroscopic detection methods in the past decade have made possible direct kinetic measurements on the reactions of silylenes. The purpose of these kinetic studies has been to elucidate the mechanisms of silylene reactions. While considerable work remains to be done, transition state structures and activation barriers are emerging from these experiments, and aspects of silylene insertion and addition mechanisms have been revealed that were not uncovered by product studies and were, indeed, unexpected.

There have been excellent reviews by some of the groups most active in this area, and only their most important conclusions are summarized here. Gas-phase kinetic studies on SiH_2 are discussed by Jasinski, Becerra and Walsh[322]. Since SiH_2 turns out to be one of the most reactive species known, laser flash photolysis of such precursors as $PhSiH_3$ and Si_2H_6 are employed to generate it for kinetic studies, with most kinetic measurements using laser resonance absorption on the nanosecond and microsecond time scales to detect it. Laser-induced fluorescence has also been employed for the detection of SiH_2. An immediate result of these kinetic studies was to dispel the belief based on relative rate measurements that the insertion of SiH_2 into H_2 has a positive activation energy variously estimated to be as high as 10 kcal mol^{-1}.

Becerra and Walsh have presented a clear exposition of the mechanistic problems addressed by kinetic studies on silylenes and the conclusions that have been drawn to date[323]. Absolute rate studies have begun to throw light on the questions: How do the structures and energies of silylenes affect their reactivity? What are the transition state structures? What is the balance between the electrophilic and nucleophilic character of the attacking silylene in the transition state? What roles do steric and electronic effects play in silylene reactions?

The small negative activation energy for reaction of $SiH_2 + D_2$ (-0.49 kcal mol^{-1}), its pressure dependence and the observed isotope effects, all suggested a transition state in which an electrophilic interaction of the H—H bonding electrons with the empty 3p orbital of singlet SiH_2 takes place. A similar electrophilically led reaction with a weakly bound complex formed between the attacking silylene and its reaction substrate was suggested by the observation of small negative activation energies for insertion of SiH_2 into an H—Si bond of SiH_4 and addition to ethylene and acetylene (see equations 95 and 96).

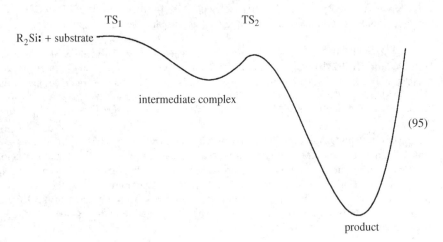

(95)

$$\text{Si} \overset{..}{\underset{R\ R}{\bigcirc}} + \underset{Z}{\overset{Y}{|}} \xrightarrow{TS_1} \underset{R\ R}{\overset{..}{\text{Si}}} \overset{Y}{\underset{Z}{\diagdown}} \xrightarrow{TS_2} \underset{R\ R}{\text{Si}} \overset{Y}{\diagup} \overset{Z}{\diagdown} \qquad (96)$$

$$\text{intermediate complex} \qquad \text{product}$$

In this model the reaction rate decreases with increasing temperature because dissociation of the intermediate complex to the reactants has a higher barrier (i.e. via TS_1), than does the rearrangement of the complex (i.e. via TS_2) to the product of the silylene reaction. Thus, as the temperature increases, product formation is disfavored relative to regeneration of the silylene from the intermediate complex.

An alternative explanation for the occurrence of negative activation energies without an intermediate complex has been presented for carbenes by Houk and coworkers[324]. It requires that the free energy of activation be dominated by the $T\Delta S^{\ddagger}$ term, and allows ΔH to decrease between the reactants and the transition state. This model is not discussed by Becerra and Walsh.

The Walsh group found that replacement of H by Me on the silylene lowers the rate of reaction with the H—Si bond of $SiH_{4-n}Me_n$ ($n = 0, 1, 2, 3$). While SiH_2 reacts at nearly the collision rate, $SiMe_2$ reacts more slowly by more than two orders of magnitude. Substitution of Me for H on the silane substrate *increases* the reaction rate on a per H—Si bond basis. These substituent effects are explained by Becerra and Walsh as arising from the stepwise nature of the insertion reaction (equation 95). They propose that the methyl groups do not affect the electrophilic stage, i.e. the formation of a complex through donation of the electrons of the H—Si bond to the silylene LUMO (step 1 in equation 96). Becerra and Walsh argue that methyl groups influence the nucleophilic stage of the reaction, i.e. the transformation of the complex to the insertion product. Methyl groups being more electronegative than hydrogen atoms, substitution of a hydrogen on the silylene by a methyl increases the barrier for electron donation by the silylene. Substitution of a hydrogen on the substrate by a more electronegative methyl group lowers the barrier for electron donation from the silylene in the conversion of the intermediate complex to insertion product.

Astonishingly, SiH_2 is about twice as reactive toward carbon–carbon π-bonds as singlet methylene CH_2, long held to be the most reactive molecule in all of organic chemistry. Becerra and Walsh point out that the greater reactivity of SiH_2 is in accord with an electrophilic reaction in which the initial interaction is dominated by the size of the acceptor orbital, 3p for SiH_2 and 2p for CH_2. Silylene reactions are much less exothermic than carbene reactions, and thus the secondary reactions of silylene adducts are rather different from those of carbenes.

An important result of the recent kinetic experiments, when combined with theoretical calculations, has been to provide a refined estimate of the heat of formation of SiH_2 $\Delta H_f = 65.3 \pm 0.5$ kcal mol^{-1}[325]. Walsh has reviewed the problems associated with assigning heats of formation to methylsilylenes[326].

Reactions of silylenes with lone-pair donors such as alcohols and ethers are also quite rapid (*ca* 10% as fast as addition to π-bonds). For such reactions, the silylene-donor complexes have been spectroscopically detected (Section V.A). Even when rearrangement to stable products is facile, as in the case of alcohols, the second, nucleophilic, step is believed to have a higher barrier than the electrophilic complex-forming step.

Nevertheless, the temperature dependence of the rate constant for the reaction of $SiMe_2$ with MeOH yielded a negative activation energy of -3.75 kcal mol^{-1}[323]. The silylene complexes act as silylenoids, and their reactions are slower by one to three orders of magnitude than those of the corresponding free silylenes.

Kinetic measurements on the gas-phase reactions of the halosilylenes SiF_2, SiHCl, $SiCl_2$ and $SiBr_2$ have been reviewed by Strausz and coworkers[327]. Donation of lone-pair electrons from halogen substituents to the divalent silicon atom would be expected to decrease reactivity, and indeed SiF_2 does not react at an appreciable rate with H_2 even at 1400 K, nor with molecular oxygen whose rates of reaction with $SiCl_2$ and $SiBr_2$ could be measured. Only with oxygen atoms has SiF_2 been found to react at near the collision rate, while reactions with F_2 and Cl_2 are two orders of magnitude slower. SiHCl reacts at appreciable rates with both SiH_4 and SiH_2Cl_2, presumably by H–Si insertion.

All the kinetic experiments discussed thus far have been in the gas phase, whose advantages include the possibility of studying the effects of energy release through pressure-dependence studies, and the absence of complications due to solvents, such as the rapid complexation of silylenes.

Kinetic studies on silylene reactions have also been carried out in solution. Shizuka, Ishikawa and coworkers generated $SiMe_2$ in methylcyclohexane solution by irradiation of *cyclo*-$(SiMe_2)_6$ and obtained rate constants for reaction with MeOH and Et_3SiH in good accord with gas-phase experiments[258]. Similar experiments in cyclohexane by Levin and coworkers gave similar rate constants for H–O and H–Si insertion reactions in the range 10^9 M^{-1} s^{-1} and rapid complexation of $SiMe_2$ with THF[257a]. These rate constants were 10 to 100 times those found by Gaspar and coworkers in experiments employing $(PhMe_2Si)_2SiMe_2$ as the precursor for $SiMe_2$ despite the observation of a similar transient absorption spectrum[260]. There was also close agreement in the Gaspar lab between isotope effects and some (but not all) reactivity ratios in static photolysis competition experiments with rate constant ratios in laser flash photolysis experiments[260].

In the same paper from the Gaspar lab the rates of reaction with a number of substrates were reported for the $\lambda_{max} = 440$ nm transient from the cyclohexane solution-phase 266 nm laser flash photolysis of $(Me_3Si)_2SiMePh$. The transient was believed, on the basis of chemical trapping experiments and a comparison of rate constant ratios, with static photolysis product ratios, to be the silylene SiMePh, despite the observed shift to the blue of the λ_{max} by *ca* 40 nm from the value attributed to SiMePh in a matrix isolation photolysis experiment with the same precursor[180]. While the work of Ishikawa and Kumada on the photolysis of $(Me_3Si)_2SiMePh$ had shown that a silene is formed by a 1,3-Me_3Si shift to the phenyl ring in competition with silylene extrusion[328], the reactivity profile of the 440 nm transient in the laser flash experiments was inconsistent with a silene, an excited state of the precursor, or a silyl radical. Although the laser photolysis rate constant ratio for reaction of the 440 nm transient with EtOH and Et_3SiH was 10^3, the relative reactivity deduced from static photolysis product ratios was only 3. This was rationalized by pointing out that the reversible formation of a complex would consume a silylene upon reaction with alcohol, but the product yield would depend on the relative rates of dissociation and rearrangement of the complex. Hence the rate constant ratios and the product ratios might depend on different processes. A similar low ratio of reactivity toward EtOH and Et_3SiH for SiMePh had been found by Hawari *et al.* in product studies employing a different precursor[329].

Nevertheless, the rate constants for bimolecular decay of the 440 nm transient found for addition ($k_2 = ca$ 10^5-10^6 M^{-1} s^{-1}) and H–Si insertion ($k_2 = 10^4$-10^5 M^{-1} s^{-1}) now seem too low to have been due to a silylene. Conlin and coworkers have found that dimesitylsilylene adds to π-bonds with $k_2 = 10^6 - 10^7$ M^{-1}s^{-1} and inserts into an H—Si bond of Et$_3$SiH with $k_2 = 8 \times 10^7$ M^{-1}s^{-1}[184]. It must therefore be advised with some regret that the rate constants for disappearance of other silylene precursors reported by Gaspar and coworkers[260] should be treated with suspicion until they are independently confirmed.

A number of rate constants for individual silylene reactions and the references to their original reports can be found in the major reviews cited here[322,323,327]. One must echo Becerra and Walsh[323] in stating that there remains much to be done in silylene kinetic studies. They have suggested intermediate complexes as a likely explanation for the activation parameters observed, but definite proof of the occurrence of these complexes along the reaction coordinates of most insertion and addition reactions remains to be accomplished. If electrophilic and nucleophilic stages of silylene reactions are indeed separated by energy barriers, then their individual kinetic characterization will be a formidable task. It would be marvelous if reaction kinetics could be developed into reaction dynamics by the study of state- and orientation-selected reactants in beam studies. Becerra and Walsh argue that silylenes, because of their lone-pair acceptor and donor properties, offer a rare example of experimentally accessible acid–base chemistry in the gas phase whose insights can be extended to other types of reactants. Since most silylene reactions occur in solution, more reliable ways to identify the transients produced in solution experiments, such as time-resolved laser-Raman spectroscopy, need to be employed.

VII. SILYLENE–TRANSITION METAL COMPLEXES

The past ten years have seen extensive development of the chemistry of silylene–transition metal complexes. Since this topic has been thoroughly treated in several recent reviews by Zybill and coworkers[330–332] and by Tilley[333,334], who have carried out much of the fundamental work, this active area will only be summarized here.

The most general method for synthesis of silylene–metal complexes is the reaction of a transition metal dianion with a dichlorosilane, in the presence of a donor molecule D to stabilize the resulting silylene complex (equation 97).

$$R_2SiCl_2 + Na_2[M(CO)_n] \xrightarrow{D} R_2Si\overset{D}{=\!=\!=}M(CO)_n + 2\,NaCl \qquad (97)$$

The metal, M, can be Fe, Ru, Os (with $n = 4$) or Cr ($n = 5$). Substituents on silicon can vary widely, including aryl, alkyl, alkoxy and Et$_2$N groups. With a slight modification, the same reaction is useful for silylene–tungsten complexes (equation 98).

$$R_2SiCl_2 + Na_2W_2(CO)_{10} \xrightarrow{HMPA} R_2Si\overset{HMPA}{=\!=\!=}W(CO)_5 \qquad (98)$$

When the silicon can be intramolecularly coordinated by a nitrogen or phosphorus atom, no added base is necessary. An example is shown in equation 99.

Silylene complexes with two intramolecular coordinating 'arms' can participate in dynamic behavior, involving a 'flip-flop' alternating coordination, illustrated in equation 100[335,336].

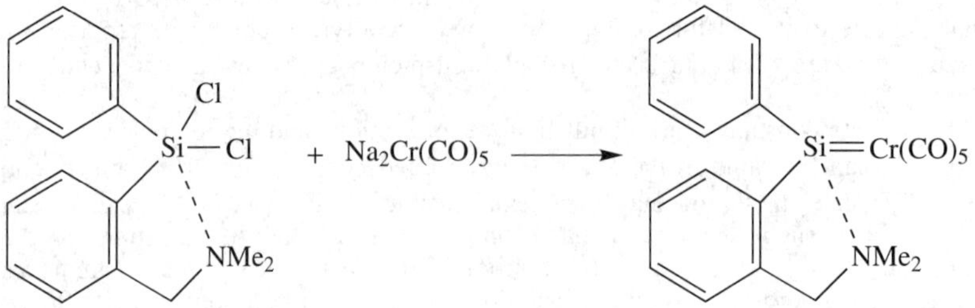

$$Ph(o-Me_2NCH_2C_6H_4)SiCl_2 + Na_2Cr(CO)_5 \longrightarrow Ph(o-Me_2NCH_2C_6H_4)Si=Cr(CO)_5 \quad (99)$$

$$(o-Me_2NCH_2C_6H_4)_2SiCl_2 \rightleftharpoons (o-Me_2NCH_2C_6H_4)_2SiCl_2 \quad (100)$$

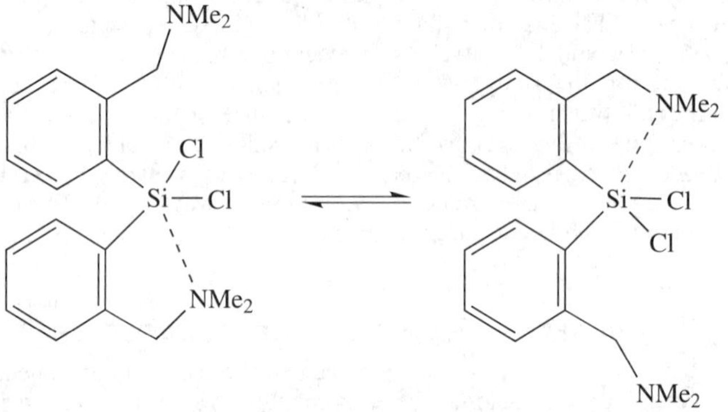

In some cases, insertion of coordinately unsaturated metals into Si—H bonds has been used for the synthesis of silylene complexes, as in the reaction of equation 101.

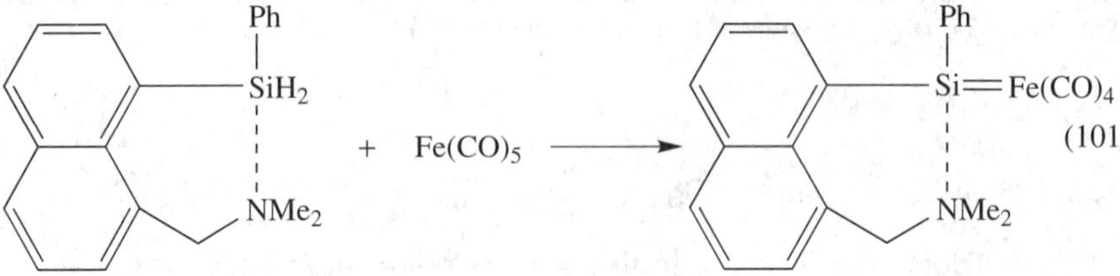

$$\text{(8-Me}_2\text{NCH}_2\text{-naphth-1-yl)PhSiH}_2 + Fe(CO)_5 \longrightarrow \text{(8-Me}_2\text{NCH}_2\text{-naphth-1-yl)PhSi}=Fe(CO)_4 \quad (101)$$

The displacement of triflate ion from silicon is important as a route to cationic silylene complexes, like the one shown in equation 102.

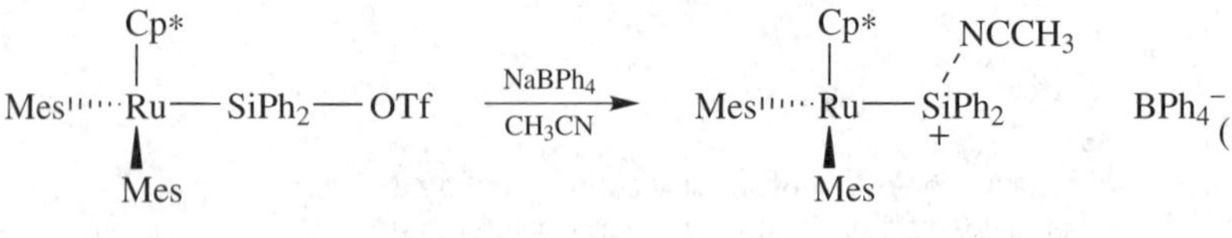

$$Cp^*(Mes)_2Ru\text{-}SiPh_2\text{-}OTf \xrightarrow[CH_3CN]{NaBPh_4} [Cp^*(Mes)_2Ru\text{-}SiPh_2(NCCH_3)]^+ BPh_4^- \quad (102)$$

Silylene–metal complexes, with tricoordinate silicon, free of coordinating donors, have been difficult to synthesize. Only a few examples were known until recently; one example is the osmium complex $Cp^*(Me_3P)_2RuSi=Os(CO)_4$ synthesized by Tilley and coworkers[337]. Now, with the availability of stable silylenes, tricoordinate silylene complexes are becoming much more easily available. These compounds are discussed in Section VIII.C.

The nature of the bonding in silylene–metal complexes, as compared with the better known metal–carbene complexes, is a question of considerable interest. MO calculations on $H_2Si=Mo(CO)_5$ indicate that the Si–Mo bond consists of a σ-donor and π-backbond component, like the carbon–metal complexes. The π-component is, however, weaker than for metal carbenes[251]. Infrared C=O frequencies for the base-free silylene metal complexes support this model. Theoretical considerations of the bonding in silylene–metal complexes are treated more fully in Section IV.E.

Bridged silylene complexes are the subject of a recent comprehensive review by Ogino and Tobita[338]. These complexes can be classified into three types: A, B and C (Scheme 9). In type A complexes there is no metal–metal bonding, the silicon is essentially tetravalent, and the bonding is similar to that in mononuclear metal–silyl complexes. In type C complexes, the bonding is best described as η^2-coordination of the Si–H bond to the metal, or alternatively as a metal–hydrogen–silicon 3-center 2-electron bond.

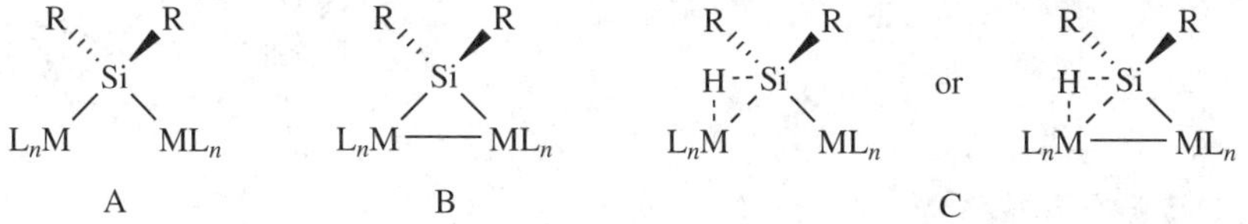

SCHEME 9

It is only in type B complexes that the silicon exhibits silylene character. In these molecules the M–Si–M angle is usually very acute, from 58° to 77°, and the silicon atoms show large downfield shifts in the ^{29}Si NMR. The bonding is best described in terms of a model originally suggested by Triplett and Curtis for germanium complexes[339], in which interaction takes place between a lone pair on the silylene fragment and the metal–metal bond. The silylene lone-pair orbital overlaps in σ-fashion with a symmetry-adapted combination of metal orbitals, and the vacant silylene p-orbital takes part in π-overlap with another such combination (Scheme 10). This orbital interaction is similar to, but slightly different from, that suggested for alkylidene-bridged metal complexes[340].

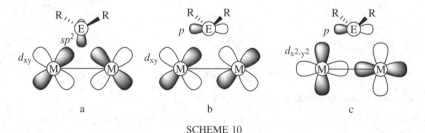

SCHEME 10

Many of the bridged silylene complexes show complex fluxional behavior or slow isomerization between different isomeric forms. An example is the iron–silylene complex shown in Scheme 11, in which the proposed mechanism is illustrated. A full discussion appears in Ogino and Tobita's review[338].

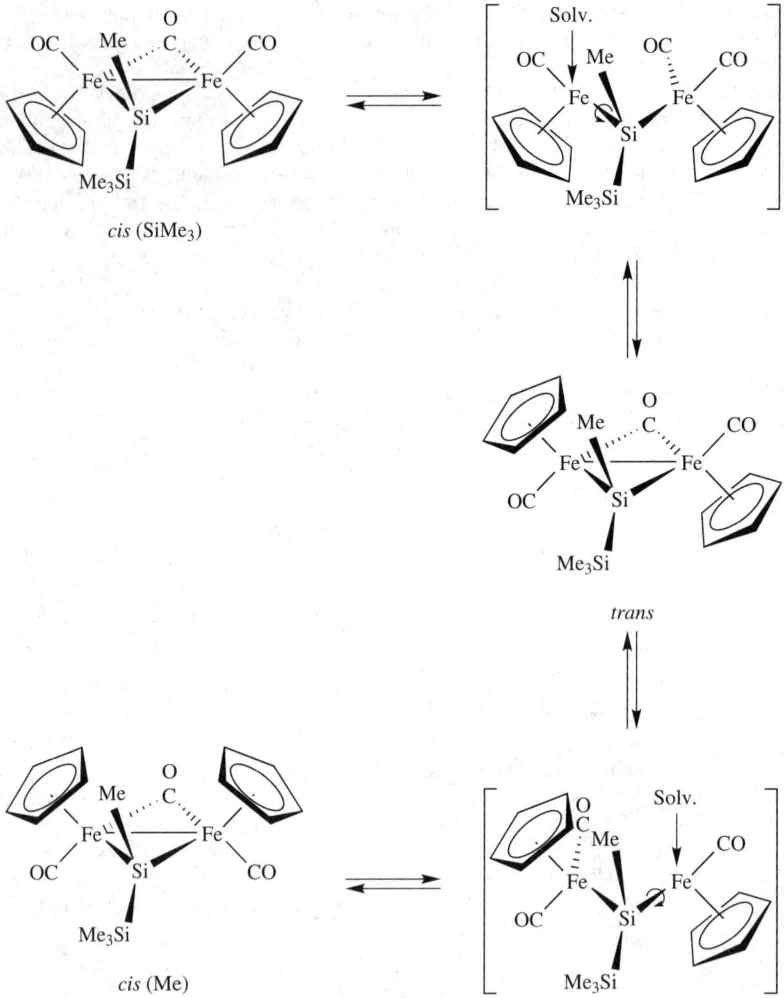

SCHEME 11

VIII. SPECIAL TOPICS

A. Silylene Centers on Activated Silicon

When finely divided silica is treated with methanol and then pyrolyzed, it becomes activated toward chemisorption of various gases[341–343]. Recent careful spectroscopic studies by Radzig and coworkers[344–347] and by Razskazovskii et al.[348,349] establish beyond reasonable doubt that the principal reactive sites are divalent silicons, $(\equiv Si-O)_2Si\colon$, 'silylene centers'[350], which participate in a rich chemistry.

A typical procedure for the preparation of reactive silica involves annealing of the sample in oxygen at 1220 K, then treatment with methanol vapor at 720 K followed by pyrolysis at 1220 K with pumping. Reactive silica may also be prepared by thorough grinding of SiO_2[351]. Like other silylenes, the silylene centers on silica are ground state singlets. They have a strong absorption based at 240 nm ($S_0 \rightarrow S_1$), a fluorescent emission at 290 nm ($S_1 \rightarrow S_0$, lifetime 5.6 ns) and a phosphorescence at 460 nm ($T_1 \rightarrow S_0$, lifetime 10.2 ns)[344,345]. The excitation spectrum is identical to the absorption spectrum. These properties are similar to those of gaseous $F_2Si:$.

Some of the silylene-like reactions which take place at the silylene centers are displayed in Scheme 12. The divalent silicon inserts into the H—H bond of dihydrogen, the S—H and O—H bonds of H_2S and H_2O, and the C—S bond of Me_2S[348].

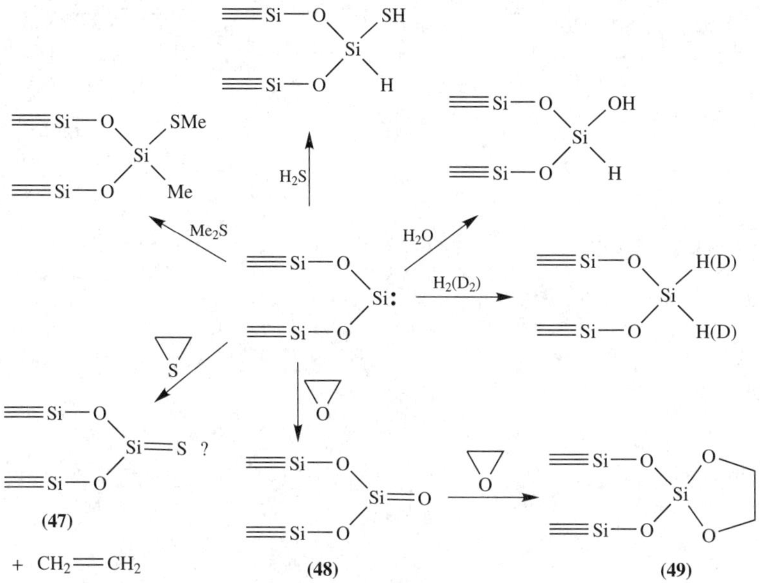

SCHEME 12

Reaction with ethylene sulfide yields ethylene and, possibly, a silanethione (**47**). With ethylene oxide, the first step is abstraction of an oxygen atom to yield the silanone (**48**), which adds to a second ethylene oxide molecule to give the siladioxolane (**49**). The products were identified mainly by infrared spectroscopy; in several cases the reaction kinetics were studied and activation energies determined[348].

Reaction of the silylene centers with dioxygen, or with N_2O at high temperature, produces the siladioxirane (**50**) (Scheme 13). The silylene center itself does not bind carbon monoxide, but the siladioxirane reacts with CO to yield the cyclic carbonate **51**. At 650 K, **51** loses CO_2 reversibly to give the silanone **48**[346]. The latter reacts reversibly with N_2O at 300–450 K to give **52**, but at higher temperature an irreversible reaction takes place to regenerate siladioxirane **50**.

The silylene center does not react with ethylene. In this respect it is similar to $Cl_2Si:$, which reacts with ethylene at a significant rate only at high temperatures (793 K). Ethylene, however, quenches the silylene fluorescence, as does CO[348]. In both cases a charge-transfer complex may be involved. (CO forms weak complexes with matrix-isolated silylenes; see Section V.A.2)

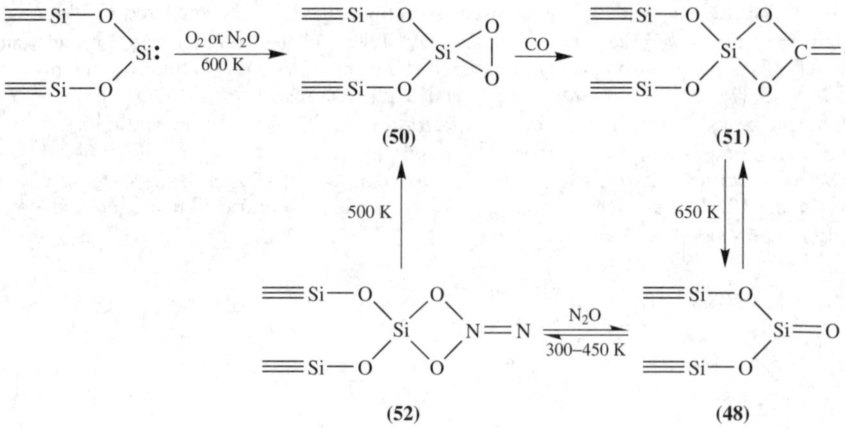

SCHEME 13

The reaction of silylene centers with dihydrogen has been studied in some detail[346]. It appears to proceed by a free-radical chain mechanism, initiated by radical sites on the silica surface which abstract hydrogen from H_2. The H atoms react with the silylene center to give a free radical, and a reaction chain can then ensue (equations 103–105). The hydrogen addition to silylene centers is reversible; the hydrogen is completely removed at temperatures near 1000 K.

$$—Si^{\bullet} + H_2 \longrightarrow —Si—H + H^{\bullet} \quad (103)$$

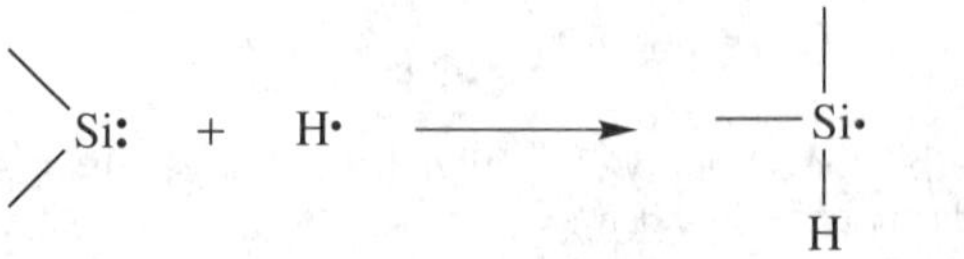

$$>Si: + H^{\bullet} \longrightarrow —Si^{\bullet}—H \quad (104)$$

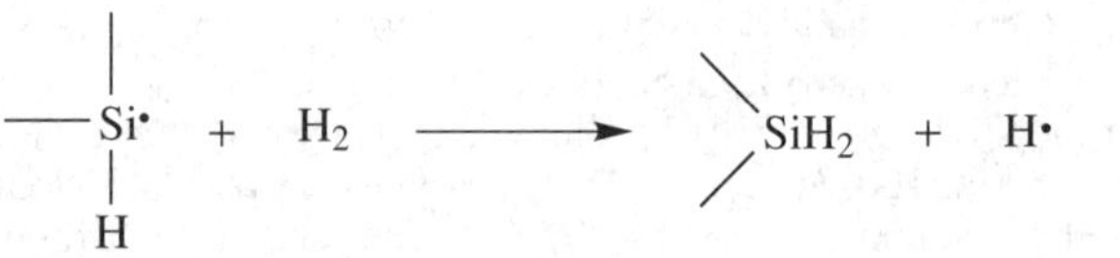

$$—Si^{\bullet}—H + H_2 \longrightarrow >SiH_2 + H^{\bullet} \quad (105)$$

Quite recently, the reaction of photoexcited (singlet or triplet) silylene centers with dihydrogen has also been studied[346]. These excited states abstract hydrogen atoms from H_2 directly. For the excited singlet state, the process takes place with no discernable barrier.

The remarkable progress which has been made in the study of silylene centers on silica reflects the recent advances in surface spectroscopy. The silylene centers, $(\equiv Si-O)_2Si:$, represent a strongly stabilized form of divalent silicon. The reactions which they undergo may have important implications for the reactivity of silylenes generally.

B. Silylenes in the Direct Synthesis of Organosilicon Compounds

The 'direct reaction' of methyl chloride with silicon metal is the foundation stone of the worldwide silicone industry[352]. In corporate laboratories over the years, the reaction has been carefully engineered to provide the maximum amount of the desired product, dimethyldichlorosilane. Despite the industrial importance of the direct reaction, and the great amount of research devoted to it, its mechanism is still obscure[353]. Recently, however, a model has been suggested in which silicon in silylene form provides the crucial intermediate.

A careful study of the direct reaction was carried out by Clarke and Davidson[354], who interpreted their results in terms of surface-bound 'silylenoid' species, MeClSi: (surf) and Cl_2Si: (surf), which were thought to react with methyl chloride to give the principal products, Me_2SiCl_2 and $MeSiCl_3$. Consistent with this model, addition of butadiene to the gas stream led to formation of silacyclopent-3-enes (equation 106). Small amounts of gaseous MeClSi: and Cl_2Si: were also believed to be present, but were probably not the key intermediates in the direct process.

$$\begin{array}{c}X\\ \diagdown\\ Si: \text{(surf)} \\ \diagup\\ Cl\end{array} + CH_2{=}CH{-}CH{=}CH_2 \longrightarrow \begin{array}{c}\diagup X\\ Si\\ \diagdown Cl\end{array} \qquad (106)$$

$$X = Me, Cl$$

Copper is an essential catalyst in the direct reaction, usually added in the form of Cu_3Si. Lewis and coworkers, in a review of the direct reaction[355], have suggested that a divalent silicon atom, perhaps attached to two copper atoms, reacts with methyl chloride to insert into the C−Cl bond, a typical silylene reaction (equation 107). The resulting intermediate (53) could then react with a second methyl chloride molecule by insertion into the C−Cl bond to yield Me_2SiCl_2. The other major products of the direct reaction, namely $MeSiCl_3$, $MeSiHCl_2$, $HSiCl_3$ and Me_3SiCl, can be accounted for by assuming the intermediacy of similar silylene-like species. In another book chapter, Ono and coworkers have presented kinetic evidence consistent with the silylene mechanism for the direct reaction[356]. At this point the evidence for silylene intermediates in the methyl chloride−silicon reaction is suggestive but not truly compelling.

$$\begin{array}{c}\quad\ddot{\ }\\ \ \ Si\\ \diagup\ \diagdown\\ Cu\quad Cu\end{array} + CH_3Cl \longrightarrow \begin{array}{c}H_3C\ \diagdown\ \diagup Cl\\ Si\\ \diagup\ \diagdown\\ Cu\quad Cu\end{array} \xrightarrow{CH_3Cl} (CH_3)_2SiCl_2 \qquad (107)$$

$$(53)$$

Silylene intermediates have also been proposed for the reaction of alcohols with silicon[357]. When phenol is reacted with silicon in the presence of CuCl as a catalyst, the major product $(PhO)_3SiH$ is obtained in 94% selectivity. This result is explained by insertion of a silylene $:Si(OPh)_2$ into the O−H bond of phenol. Addition of ethylene to the reaction produces 5.8% of $EtSi(H)(OPh)_2$, which can be accounted for by the reactions of Scheme 14[358].

Similarly, reaction of silicon, methanol and ethylene in an autoclave at 433 K produces ethylmethoxysilanes, $EtSi(H)(OMe)_2$ and $EtSi(OMe)_3$, as well as $HSi(OMe)_3$ and $Si(OMe)_4$. And, although 2-propanol does not react with silicon, a mixture of methanol and 2-propanol reacted with silicon to yield $i\text{-}PrOSi(H)(OMe)_2$[359]. Taken together, these results strongly suggest the intermediacy of silylene-like species in these reactions.

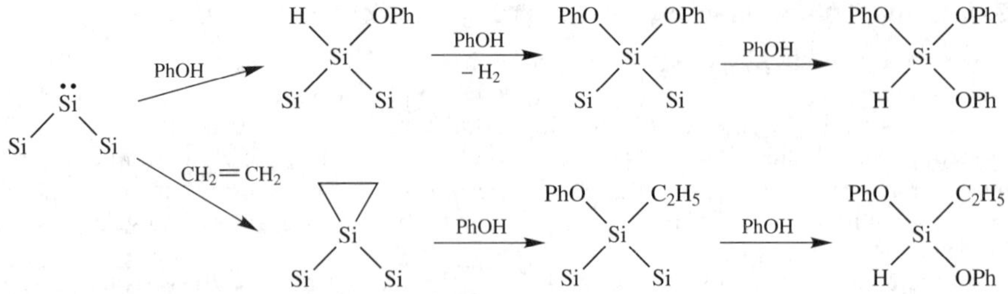

SCHEME 14

C. Stable Dicoordinate Silylenes

1. History

A major advance in silylene chemistry over the past five years has been the synthesis of persistent dicoordinate silicon compounds: 'stable silylenes'.

Thermally stable dicoordinate silicon compounds have been known since 1986, when Jutzi and coworkers reported the remarkable π-complex, $(Me_5C_5)_2Si$[360]. The chemistry of this compound, developed over the past decade, makes a fascinating story[361]. Other compounds of dicoordinate silicon include the four-coordinate phosphorous compound, $Si[(Me_2P)_2C(SiMe_3)]_2$[362]. These compounds, with coordination numbers at silicon greater than 2, behave rather differently from silylenes in their reactions and will not be covered here.

Dicoordinate lead and tin compounds have been known since the early days of chemistry. The study of dicoordinate germylenes began with Lappert's and Harris's pioneering work on the synthesis and reactions of the dinitrogen-substituted germylene, $[(Me_3Si)_2N]_2Ge$ (**54**)[363]. The chemistry of this and analogous germylenes has been well explored[364]. Compound **54** is a monomer both in solution and as a solid, but the isostructural carbon-substituted germylene, $[(Me_3Si)_2CH]_2Ge$ (**55**), although monomeric in solution, dimerizes to the digermene in the solid state[365]. The difference probably reflects stabilization of monomeric **54** by electron donation from nitrogen. Steric shielding by the bulky $(Me_3Si)_2N$ and $(Me_3Si)_2CH$ substituents, however, is undoubtedly important in stabilizing both **54** and **55**.

Can silylenes be stabilized in monomeric form simply by steric shielding? No final answer can be given, but the results so far are not very promising. The design of a group sufficiently bulky to discourage intermolecular reactions and sufficiently inert to prevent loss through intramolecular processes is a formidable problem. The silicon counterpart to **55** exists in the dimeric disilene form $[(Me_3Si)_2CH]_2Si=Si[CH(SiMe_3)_2]_2$ (**56**), although, as mentioned in Section. III.C.2, it gives silylene trapping products at 120 °C, suggesting that partial dissociation to silylene takes place at this temperature. The silicon analog to **54** is unknown, but a silylene bearing a $(Me_3Si)_2N$ and a mesityl substituent has been prepared in 3-MP at 77 K. Upon melting of the matrix, immediate dimerization of the silylene took place to give the disilene, $(Me_3Si)_2N(Mes)Si=Si(Mes)N(SiMe_3)_2$[179]. Perhaps the best example to date of stabilization by steric hindrance is the silylene **57a** of Puranik and Fink[107]. This species has a limited stability in solution, at least up to 200 K, but could not be isolated at room temperature.

A related silylene, the four-membered ring **57b** prepared by photolysis of the corresponding diazide, was reported by Veith and coworkers to decompose at very low temperatures[55].

A carbene, fully stable at room temperature, **58a**, was isolated in 1991 by Arduengo and coworkers, and since that time several other carbenes have been prepared, all having similar structure (**58a–f**)[366]. The synthesis of these carbenes left silicon as the last element with no dicoordinate, divalent compound stable at ordinary temperature. The stability of carbenes **58a–f** suggested that silicon-containing rings with similar structure might be good candidates for isolation at room temperature.

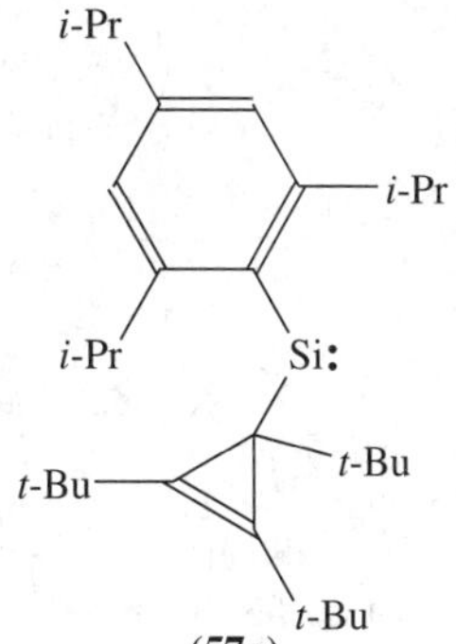

(**57a**)

(**57b**)

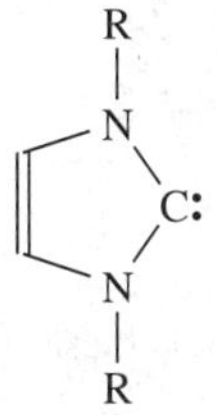

(**58a**) R = 1-adamantyl
(**58b**) R = *t*-butyl
(**58c**) R = *p*-tolyl
(**58d**) R = *p*-chlorophenyl
(**58e**) R = mesityl

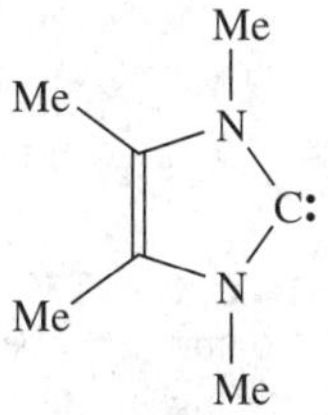

(**58f**)

The field of stable silylenes commenced with the synthesis of compound **59**, isostructural with carbene **58b**, in 1994[367]. This development was promptly followed by reports of four additional examples: **60**[98], the saturated analog to **59**; and the benzene-[368] and pyridine-[369] fused bicyclic silylenes **61**, **62** and **63**.

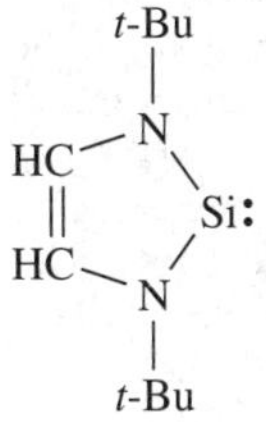

(**59**)

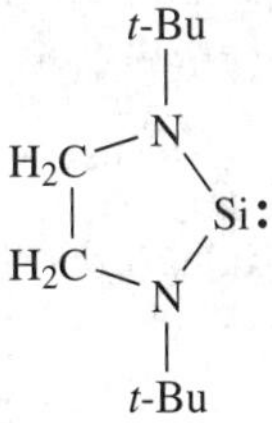

(**60**)

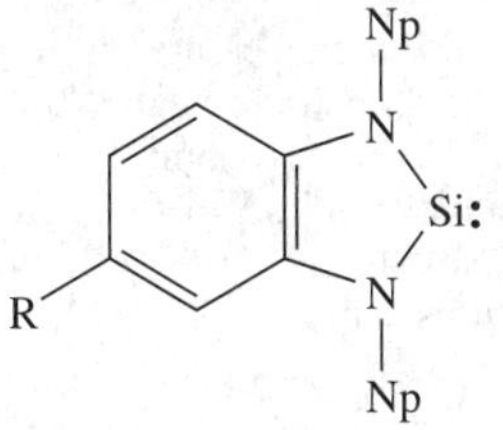

(**61**) R = H
(**62**) R = Me

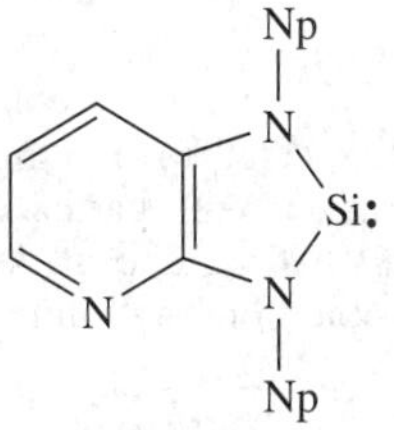

(**63**)

Np = neopentyl

2. Synthesis and characterization

Silylene **59** was obtained starting from glyoxal and *t*-butylamine, according to Scheme 15[367]. The final step, dehalogenation of the dichloride, was carried out using molten potassium metal in refluxing THF, and these rather vigorous conditions have become standard for the synthesis of stable silylenes of this type.

Silylenes **59** and **60** are colorless solids, while the benzannelated compounds **61** and **62** are yellow. Compound **59** possesses truly surprising thermal stability. It was purified by distillation at 90 °C (0.1 torr) and survived unchanged after heating in toluene at 150 °C for

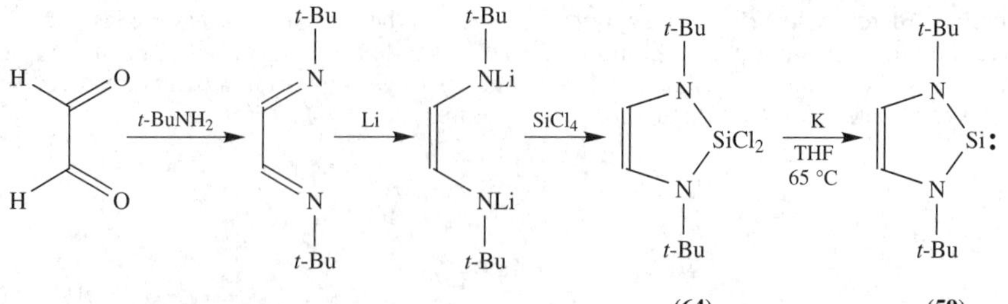

SCHEME 15

4 months. Solid **59** decomposes only at its melting point, *ca* 220 °C. Compounds **61** and **62** also appear to be quite stable to heating; for example, **62** was sublimed from a 155 °C bath during purification[368]. The saturated-ring compound **60**, on the other hand, is much less stable toward heating and there are indications that it slowly dimerizes. Apeloig and Müller[247b] predicted on the basis of *ab initio* calculations that **60** dimerizes to a μ^2-, nitrogen bridged compound as shown in equation 90, and not to the corresponding disilene.

In their ^{29}Si NMR spectra, the silylenes all show rather deshielded silicon atoms, consistent with theoretical calculations. Values recorded, in ppm, are **59**, 78.3; **60**, 119; **61**, 96.92; **62**, 97.22; **63**, 95.1.

X-ray crystal structures have been reported for **60** and **61**, and the structure of **59** was obtained from a gas-phase electron diffraction study. Structures for **59** and **60** are shown in Figures 5 and 6. The five-membered ring in **59** is nearly planar, but in **60** it is puckered as shown in the figure. Significant bond lengths and angles are given in Table 5.

As expected, the intraring C—C distances increase from **59** to **61** to **60**, as the C—C bond order is reduced from 2 to 1. Also predictable is the larger N—C distance in **60**, where the nitrogen is bonded to a tetrahedral carbon atom, than in **59** and **61**, where the bonding is N—C sp^2. The N—Si bond distances are slightly longer than for normal single bonds from nitrogen to silicon, which average near 170 pm. However, bonds to *divalent* silicon are expected to be longer, by about 8 pm. The N—Si distances in **59–61** are therefore consistent with partial double bonding in these molecules. Particularly interesting is the small but significant decrease in the N—Si distance, going from **59** and **61** to **60**. The possible bonding implications will be discussed in the following section.

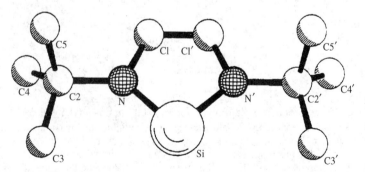

FIGURE 5. Structure of unsaturated silylene **59** from electron diffraction. Reprinted with permission from Reference 367. Copyright (1994) American Chemical Society

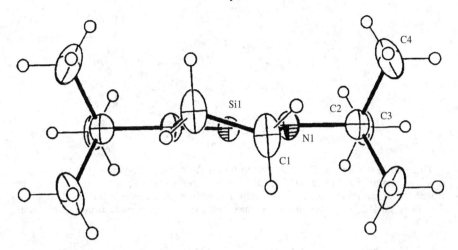

FIGURE 6. Structure of saturated silylene **60** from X-ray diffraction. Permission of The Royal Society of Chemistry from Reference 224.

TABLE 5. Selected intraring bond lengths (pm) and angles (deg) for silylenes

Silylene	Si–N	N–C	C–C	N–Si–N	Reference
59	175	140	135	90.5	367
60	171	150	152	92.0	224
61	175	138	142	88.2	370

3. Bonding and 'aromaticity'

The properties of the stable silylenes raise interesting questions of chemical bonding, many of which apply also to the carbenes, **58a–f**. First, why are these compounds so stable compared to ordinary silylenes (carbenes)? Steric hindrance may be a factor, but it is unlikely to be a major one; silylenes far more hindered than **59–63** dimerize at low temperatures, as was described in Section III.C[181].

Electron donation from nitrogen to silicon is undoubtedly important in stabilizing all of these molecules. Partial double-bonding between N and Si will tend to occupy the vacant p-orbital on silicon, which is the usual site of electrophilic reactivity of silylenes. This stabilization is shown in resonance drawings in Scheme 16. Greater π-donation from N to Si would be expected for **60** than for **59** or **61**, since in **60** the nitrogens are more basic. The Si–N bond lengths are consistent with this model, as mentioned above.

Structures **B** and **C** in Scheme 16 are ylidic, and objection is sometimes raised to the description of **59–63** as silylenes and **58a–f** as carbenes, on the grounds that they are really ylides. Indeed, it is likely that all these compounds have significant ylidic character (although a contrary view has been put forward for stable carbenes, see below). But this is really a question of degree, since stabilization of the vacant p-orbital on silicon, by π-electron donation or hyperconjugation, probably takes place to some extent for all silylenes except SiH_2. π-Donation to silicon probably explains the relatively low reactivity of SiF_2, for example. The nitrogen atoms in **59–63** serve as particularly good π-donors, and electrophilic reactivity is therefore reduced more strongly to the point where these compounds can be isolated.

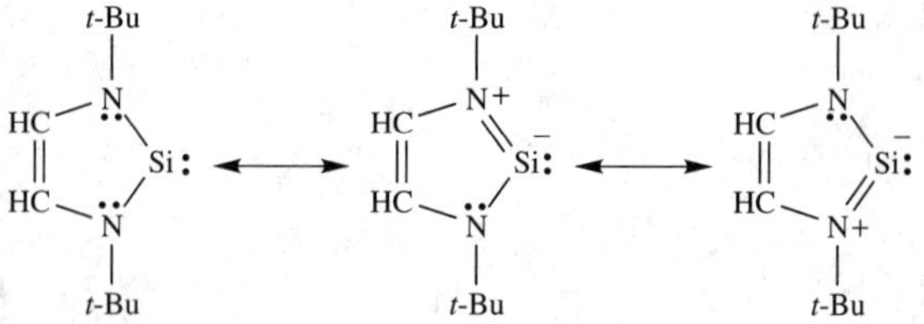

SCHEME 16

The greater stability of **59** compared to **60** also requires explanation. The ring is more nearly planar in **59** than in **60**, which may marginally increase steric shielding of the silicon atom, but this seems an insufficient rationalization. Instead, it has been suggested that **59** is stabilized by aromatic resonance, since it forms a 6 π-electron cyclic system, **59a**[98,367].

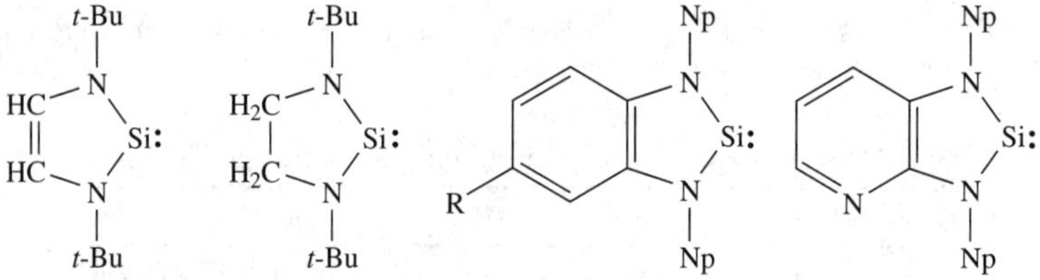

(59)   (60)   (61) R = H   (63)
              (62) R = Me

Np = neopentyl

Possible evidence for aromatic delocalization in **59** could come from proton NMR data. The chemical shifts of the vinyl protons in the model compounds **64** and **65** lie at 5.73 and 6.00 ppm. In **59** the vinyl proton resonance is shifted downfield to 6.75 ppm, consistent with deshielding due to an aromatic ring current.

To explore this possibility further, quantum chemical calculations were carried out for the isodesmic reactions of the model compounds **66** and **67** with dihydrogen to give the corresponding dihydrides (Scheme 17). These calculations showed that the reaction of **66** with H_2 is about 14 kcal mol^{-1} less exothermic than that of **67**. This difference may reflect aromatic resonance energy in the unsaturated molecule **66**, reducing the enthalpy of the hydrogenation reaction.

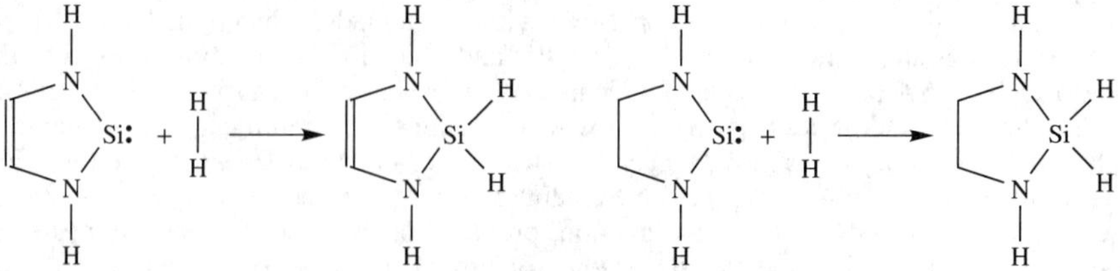

(66)   (67)

SCHEME 17

The He(I) and He(II) photoelectron spectra have been determined for **59** and **60** and for the unsaturated dihydride **65**; in both **59** and **65**, the lowest energy ionization was assigned to an orbital of the π system (C−C bonding, C−N antibonding)[224]. In **65**, this first ionization was observed at 6.56 eV, but for **59** it was significantly higher, 7.13 eV. Evidently the Si 3p orbital participates in the HOMO, stabilizing this ring π-orbital, just as expected if aromatic resonance is present. The ionization energies of the lone pair electrons in **59** and **60** were not very different, 8.21 and 8.11 eV, indicating that hydrogenation of the C=C bond in **59** does not greatly change the nature of the lone pair. A very similar study of photoelectron spectra has recently been reported for **61** and its dihydride, **68**[370]. The highest-lying molecular orbital was stabilized by 0.41 eV in **61** compared with **68**, again indicating mixing of the out-of-plane 3p orbital on silicon with the ring π-orbital, as expected for aromatic resonance. Calculations of isodesmic reactions carried out on model compounds for **61** were also consistent with aromatic stabilization in the silylene, although this is believed to be reduced by considerable ring strain.

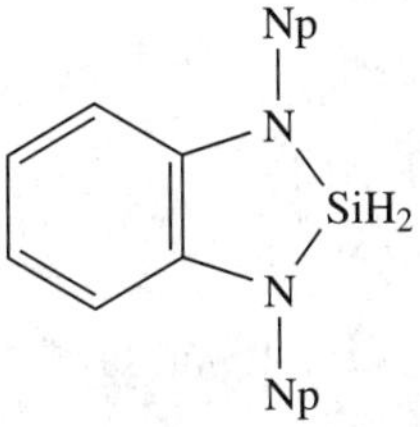

(**68**)

All these indications are consistent with some degree of aromatic delocalization in the unsaturated silylenes, **59** and **61**–**63**. This conclusion is supported by two recent theoretical papers, one based on calculations of hydrogenation enthalpies, magnetic anisotropy and a careful analysis of the electron density[371], and another quite complete treatment of energetics, structure, ionization energies and magnetic properties[372]. Both conclude that aromatic delocalization is present in silylenes like **59**, although it is less pronounced than in the isostructural carbenes. A distinctly different model, also based on photoelectron spectroscopy and theoretical considerations, has, however, been suggested by Arduengo et al.[225,373]. In this treatment the bonding in **59** is viewed in terms of σ-donation from a diimine to a neutral silicon atom **59b**, and the theoretical and experimental evidence is interpreted as showing that aromatic delocalization is not significant in carbenes, **58a**–**f**. The question of aromatic stabilization in these silylenes and carbenes is not fully closed; further studies are needed.

4. Chemical reactions

To date, chemical properties have been reported mainly for **59** and **61**. Occupancy of the silicon 3p orbital by electrons from nitrogen greatly reduces the electrophilicity of these silylenes. This, together with probable aromatic stabilization, significantly mutes the behavior of **59** and **61** as silylenes. For example, **59** does not insert into Si−H bonds, or react with alkynes such as PhC≡CPh[367]. Moreover, **59** shows no Lewis acidic behavior, even toward bases as strong as pyridine.

The stable silylenes react, however, with water and alcohols, by straightforward insertion into O−H bonds. Silylenes **59**, **61** and **62** all react with methyl iodide, inserting into the C−I bond. Examples of these reactions are shown for **59** in Scheme 18. Compounds **59** and **61** also react with elemental chalcogens. For **61** with S_8, Se_8 or Te [370], and for **59** with

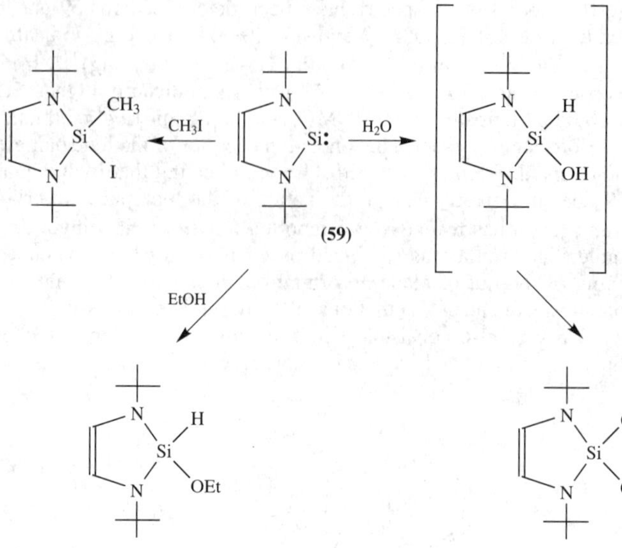

SCHEME 18

Se[374], the products are four-membered rings, as illustrated for **61** in equation 108. These products may arise from dimerization of a silicon–chalcogen doubly bonded intermediate.

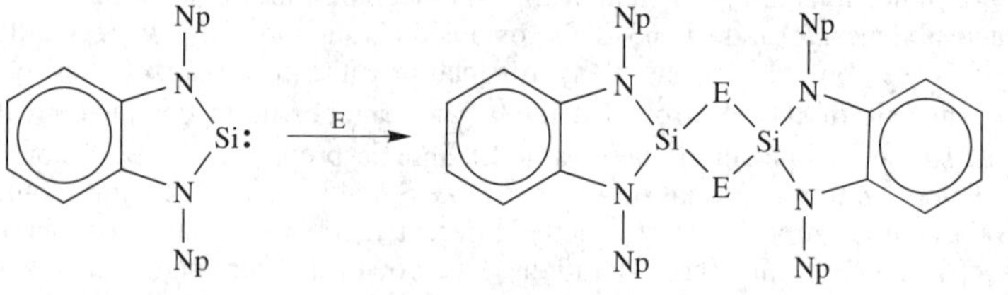

(108)

Np = neopentyl   E = S, Se, Te

The reaction of **59** with sulfur is somewhat more complicated (Scheme 19). At low temperatures, an intermediate is observed with a ^{29}Si NMR chemical shift of +122.6 ppm; this deshielded value is consistent with the silanethione, **69**. Warming of the solution produces the 4-ring dimer, **70**. But when the reaction is carried out at room temperature this intermediate is not observed, and the major product is the trisilicon compound **71** in which a molecule of diimine has been lost from the central silicon[374]. The structure of **71** is somewhat similar to that of solid silicon disulfide.

In a number of ways the reactions of stable silylenes resemble those of phosphines, R_3P, to which they are isolobal analogs. Examples are provided by the reactions of **59** with covalent azides. Phosphines are known to react with azides to give phosphineimines, $Ph_3P=NR$. In similar fashion, **59** reacted with triphenylmethyl azide in THF to give the silanimine **72** as its THF complex (equation 109)[148]. This reaction provides a new method for synthesizing compounds containing Si=N double bonds, which have previously been made by salt elimination reactions[375].

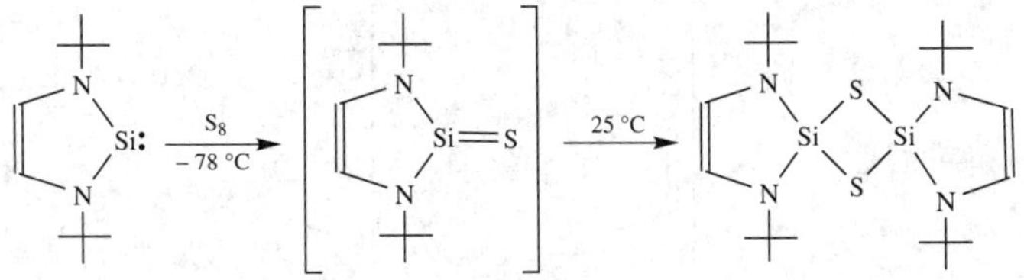

SCHEME 19

(59) (69) (70)
$\delta^{29}Si$ +122.6 −36.3

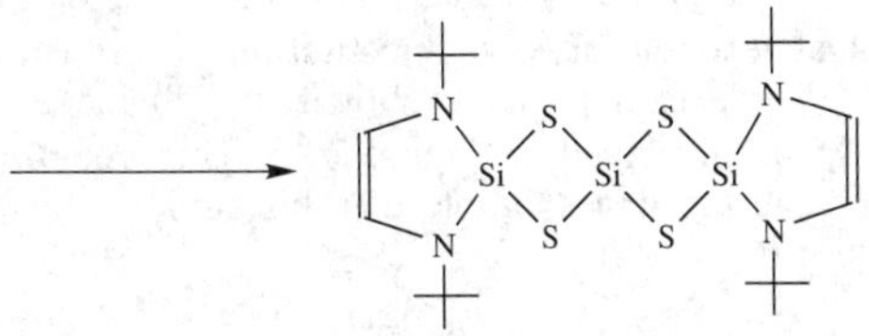

(71)
−38.1, −44.3

$$\text{(59)} + N_3CPh_3 \xrightarrow{THF} \text{(72)} \quad (109)$$

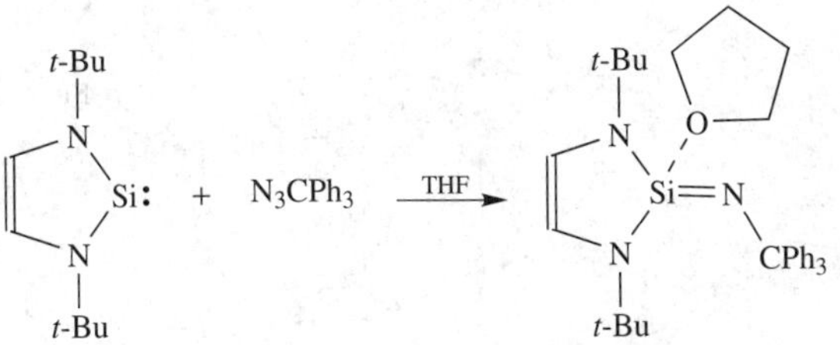

Reactions of **59** with less hindered azides can be more complex. With trimethylsilyl azide, the isolated product was the azidosilane **73**. The formation of **73** can be rationalized, however, as proceeding through the initial formation of a silanimine, followed by addition of Me$_3$Si—N$_3$ across the Si=N double bond, as shown in equation 110. When **59** was treated with 1-adamantyl azide, the product was the silatetrazoline **74**. Once again this product may have resulted from initial formation of a silanimine, followed in this case by 2 + 3 cycloaddition of the azide to the Si=N double bond (equation 111)[98,148].

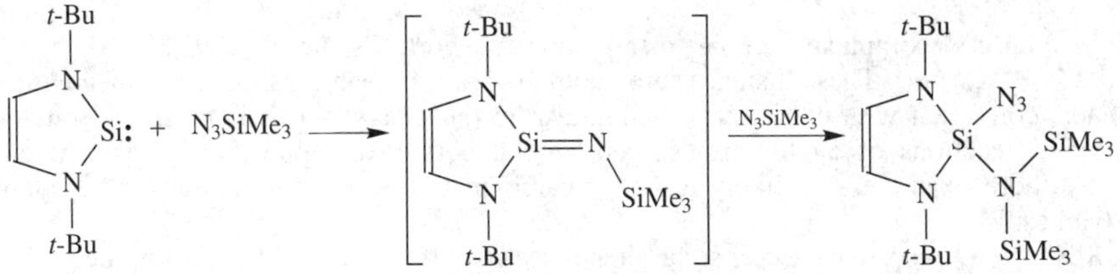

(110)

(59) (73)

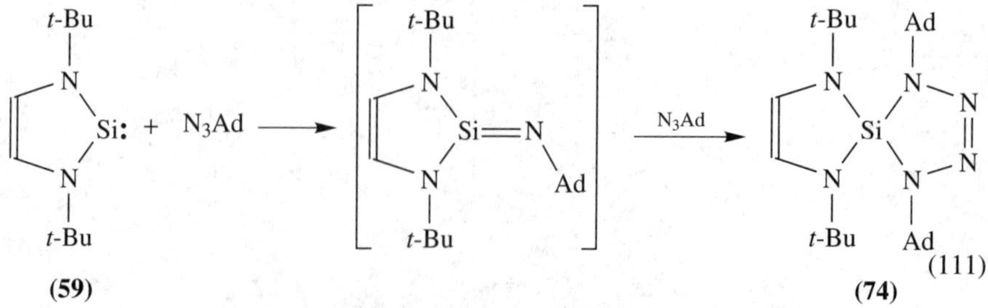

Silylene **59** also behaves somewhat like a phosphine in its interactions with metal carbonyls[98,149,376]. Typical reactions involve substitution of silylene for CO, to give a silylene–metal complex. Three examples are shown in Scheme 20, and the structure of the nickel complex **75** is displayed in Figure 7[149]. This complex is both the first silylene–nickel complex, and the first example of a bis-silylene–metal complex free of stabilization by Lewis base donors.

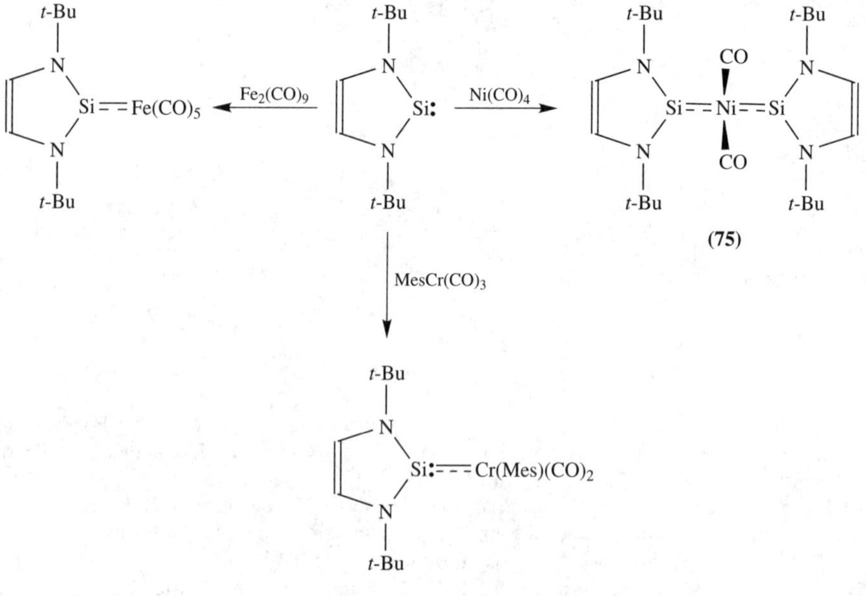

SCHEME 20

In all three complexes the Si–metal bond is short (Si–Fe = 219, Si–Ni = 221, Si–Cr = 228 pm). These distances are about 10 pm less than for typical Si–metal single bonds, consistent with partial back-bonding from the metal d-orbitals to the p-orbital on silicon. Reactions of **59** to form silylene–metal complexes appear to be general, and many more examples are likely to be prepared in the future, for this and other stable silylenes.

Silylene **59** also reacts with some dienes to give 1 + 4 cycloaddition products. With *trans,trans*-1,4-diphenylbutadiene, stereospecific addition takes place to give the *cis*-product, as expected for a singlet silylene (equation 112). A second example is the reaction

FIGURE 7. Structure of silylene–nickel carbonyl complex, **75**. Reproduced by permission of The Royal Society of Chemistry from Reference 149.

of **59** with the diimine used in its synthesis, illustrated in equation 113; the product is the spiro compound **76**[98,376].

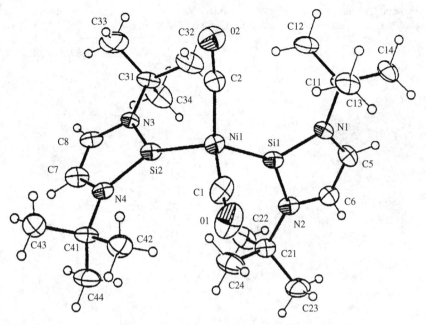

Recently, a number of interesting products have been obtained from the reaction of **61** with various compounds containing C–O or C–N multiple bonds[377]. Several such reactions are illustrated in Scheme 21. The behavior of **59** as a Lewis base has been demonstrated in a reaction with $(C_6F_5)_3B$ to give the acid–base adduct, which slowly transfers a C_6F_5 group to silicon as shown in Scheme 22[378].

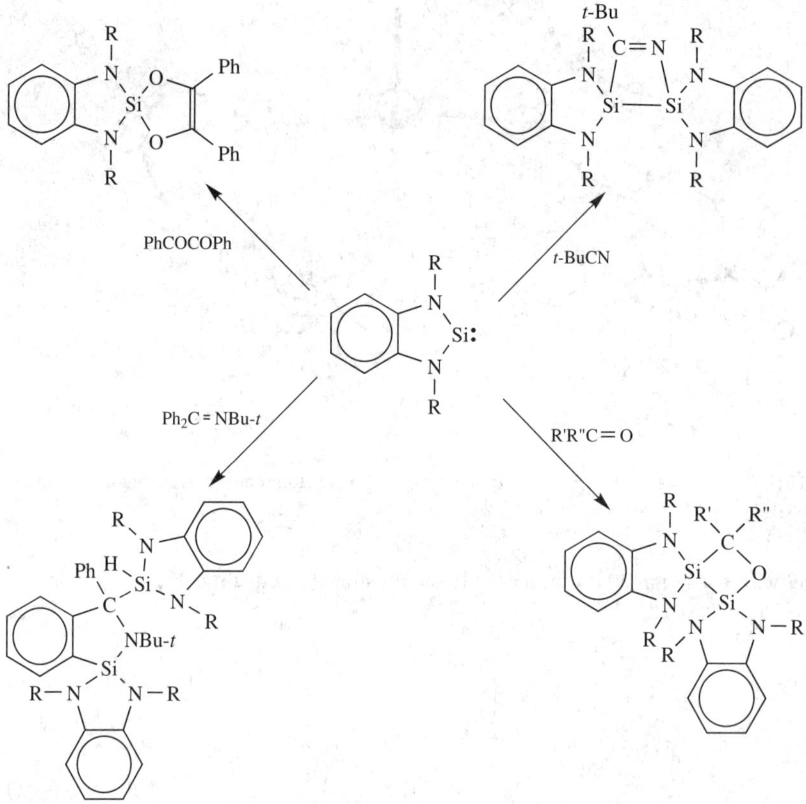

SCHEME 21

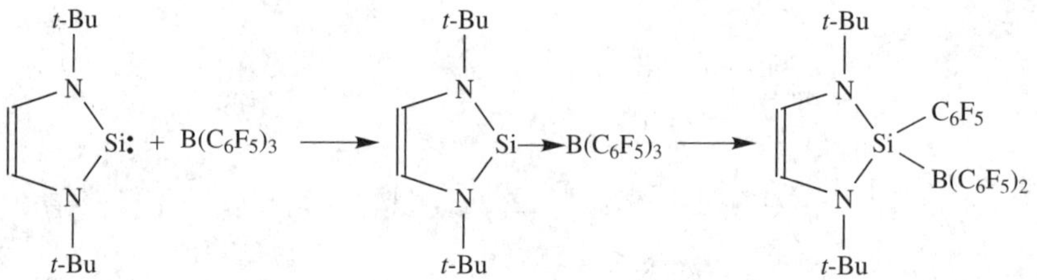

$R = CH_2C(CH_3)_3$

SCHEME 22

43. Silylenes

We see that although the stable silylenes lack some reaction pathways common for transient silylenes, they participate in many reactions previously unknown for silylenes. A number of these lead to products with structures of considerable interest. Further developments in the chemistry of stable silylenes are to be expected.

D. Silylene Intermediates in Chemical Vapor Deposition

The formation of thin solid films of silicon by the decomposition of a gaseous precursor such as SiH_4 plays a vital role in many areas of modern technology, including the manufacture of microelectronic devices and solar energy collectors. Depending on the conditions of temperature, pressure and on the precursor, the product film can range from amorphous hydrogenated Si to single-crystal epitaxial Si. Jasinski and Gates have provided a succinct introduction to the mechanisms of film formation from silicon hydrides[379]. The decomposition of silane (equation 114) is exothermic by 8.2 kcal mol^{-1}, but energy to overcome significant kinetic barriers must be provided by heat, light or plasma excitation to achieve useful rates of film growth.

$$SiH_4(g) \longrightarrow Si(s) + 2H_2(g) \tag{114}$$

$$SiH_4 \longrightarrow H_2Si\colon + H_2 \tag{115}$$

When SiH_4 is decomposed at pressures >0.1 torr and temperatures >900 K, the formation of SiH_2 in the gas phase is the dominant primary process (equation 115), and SiH_2 has been detected by laser absorption spectroscopy during silane pyrolysis[380]. SiH_2 has also been detected by laser-induced fluorescence at 0.2 torr in a silane plasma similar to those employed in plasma-enhanced CVD (PECVD) of amorphous hydrogenated silicon for solar cells[381]. Initially formed SiH_2 can undergo reversible insertion into an Si−H bond of SiH_4 and the resulting Si_2H_6 can form other reactive species that can lead to higher silane oligomers. The higher silanes react much more efficiently at surfaces than does parent SiH_4. Various species formed in the gas phase, such as SiH_2 and $H_3SiSiH\colon$, can react at surfaces if they are not consumed in gas-phase reactions, but only $H_3Si^\bullet$ is sufficiently unreactive toward saturated silanes to have a good chance of reaching the surface where film growth occurs. Even in low pressure CVD (LPCVD) at 0.1 to 1 torr, it has been estimated that SiH_2 reaching the surface can make a contribution to film growth only ca 5% that of SiH_4[382].

It has been suggested that when phosphorus-doped polycrystalline silicon is produced by LPCVD of SiH_4–PH_3 mixtures, gaseous SiH_2 that reaches the surface becomes the dominant contributor to film formation[383]. It seems, however, that the quality of the films deposited by LPCVD from SiH_4 increases if gas-phase decomposition, and hence SiH_2 formation, *decreases*[384].

PECVD can be carried out at temperatures sufficiently low (<500 K) that surface decomposition of SiH_4 plays a minor role. In the plasma, high-energy electrons are formed whose impact on silanes dissociates them into fragments, including SiH_2 and SiH_3. Orlicki and coworkers have carried out an extensive modeling study of a glow-discharge reactor operated at ca 1 torr and 475–525 K[385]. It was predicted that for film growth at the cathode (where the growth rate is higher) the contribution from SiH_2 reaching the surface can be slightly higher than that of SiH_3, or be significantly less, depending on the electron energy distribution function and on the branching ratio for formation of SiH_2 and SiH_3 from SiH_4 under electron impact. It has been suggested that, under PECVD conditions, reactions of silylenes with their silane precursors lead to the formation of higher silanes that are converted into gas-phase silicon clusters, which are detrimental to the growth of high quality films[386]. Recent modeling studies have revised sharply downward the ratio of SiH_2 to SiH_3 (from 5 : 1 to 1 : 10) produced upon electron impact on SiH_4 in PECVD reactors[387].

SiH_2 is also believed to contribute to film growth when vacuum ultraviolet radiation is employed to decompose silane, thus avoiding both high temperatures and plasma damage[388]. These conditions produce SiH, SiH_2 and H atoms as primary dissociation products in unknown ratios[388].

There is a range of mechanisms that can lead to Si film growth from SiH_4[379]. The competition between gas-phase chemistry and surface chemistry depends on the pressures and temperatures employed. At very low pressures (10^{-5}–10^{-2} torr) gas-phase collisions are rare, and unimolecular decomposition is slower than dissociative adsorption on the growing Si surface. Decomposition of surface SiH_3 and SiH_2 groups forms surface H atoms whose desorption in the form of H_2 can be the rate-controlling process in film growth, since desorption frees unsaturated surface sites ('dangling bonds') for further adsorption of SiH_4 from the gas phase. It has been argued from theoretical studies that desorption of H_2 occurs predominantly from a surface SiH_2 group[389]. Surface SiH is more robust, but also dissociates into another surface H and a surface Si atom that is incorporated into the lattice of the growing film[379].

It has been suggested that the difference between the lower temperatures required for PECVD (300–400 °C) and those required for thermal CVD from SiH_4 (500–700 °C) is due to the formation of surface unsaturation by silylene rearrangements $-Si-SiH_2- \rightarrow -SiH=SiH-$ in adsorbed polysilane chains[390]. The Si=Si double bonds thus formed are able to capture SiH_3 radicals formed in the plasma, but are unreactive toward saturated silanes. In thermal CVD processes SiH_3 radicals are not important, and in this model saturated silanes are captured by surface silylene H–Si insertions that can occur only at temperatures high enough to produce neighboring silylene sites without nearby hydrogens that can participate in rearrangements.

Surface SiH_3, SiH_2 and SiH have all been detected by infrared reflection absorption spectroscopy under conditions of plasma-enhanced CVD using a glow discharge in SiH_4 at 0.02 torr at 250 °C[391]. As the surface temperature is increased, there is a shift from surface SiH_3 to SiH_2 to SiH, also found with SiD_4 as the precursor[392]. A new Fourier-transform IR attenuated total reflectance technique has revealed the dominance of adsorbed SiH_2 in the early stages of plasma CVD from SiH_4[393].

There has been considerable controversy regarding the nature of the gas-phase reactive species that contribute to film growth formed in plasma-assisted CVD from SiH_4. Strong arguments have been presented in favor of SiH_2 formed by electron-impact induced fragmentation of SiH_4[394]. Insertion reactions of SiH_2 convert SiH_4 to Si_2H_6 and Si_2H_6 to Si_3H_8. The resulting higher silanes have much higher sticking coefficients to growing surfaces and thus contribute significantly to the formation of the hydrogenated amorphous Si films that are the desired products under these conditions.

If SiH_2 survives to reach a growing film surface, insertion into surface Si–H bonds can act as a primary step for film formation. It has been reported that the sticking coefficient for vibrationally excited SiH_2 formed by multiphoton IR laser dissociation (IRMPD) of n-$BuSiH_3$ is significantly higher (0.5 vs 0.1) than for cold SiH_2[395,396]. Use of resonance-enhanced multiphoton ionization to detect SiH_2 allowed hot and cold species to be distinguished.

Use of IRMPD to dissociate precursor silanes can initiate rather complex processes. The mechanism presented in equation 116 can explain the formation of both ground-state singlet SiH_2 and excited-state singlet ^1D Si atoms[396]. This mechanism is analogous to the phenylcarbene rearrangement[397] and has *not* been found for vibrationally unexcited phenylsilylenes formed in simple pyrolysis or photolysis experiments.

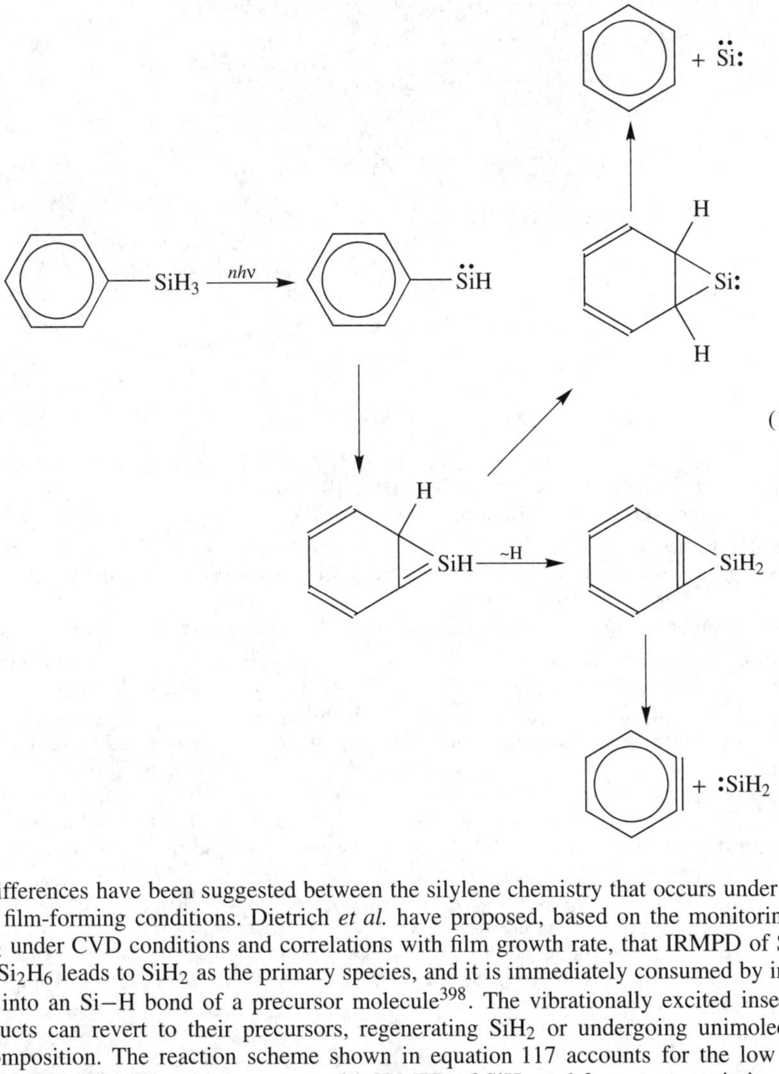

(116)

Differences have been suggested between the silylene chemistry that occurs under various film-forming conditions. Dietrich et al. have proposed, based on the monitoring of SiH_2 under CVD conditions and correlations with film growth rate, that IRMPD of SiH_4 and Si_2H_6 leads to SiH_2 as the primary species, and it is immediately consumed by insertion into an Si—H bond of a precursor molecule[398]. The vibrationally excited insertion products can revert to their precursors, regenerating SiH_2 or undergoing unimolecular decomposition. The reaction scheme shown in equation 117 accounts for the low efficiency of the film deposition process with IRMPD of SiH_4 and for a monotonic increase with SiH_4 concentration, a higher efficiency with a maximum with variation of precursor concentration upon IRMPD of Si_2H_6, and a yet higher efficiency but a monotonic decrease with concentration upon 193-nm laser photolysis of Si_2H_6.

In this mechanism the silylenes $H_2Si\colon$ and $H_3Si\ddot{S}iH$ are consumed before they reach the growing film surface. Disilene $H_2Si=SiH_2$ is postulated to be the species that can reach the growing surface and stick. The direct formation of silylsilylene $H_3Si\ddot{S}iH$ that can rearrange to $H_2Si=SiH_2$ by UV photolysis of Si_2H_6 leads to film formation with a high efficiency that diminishes due to the siphoning off of $H_3Si\ddot{S}iH$ at increasing precursor concentrations.

$$SiH_4 \xrightarrow[10.6\ \mu m]{nh\nu} :SiH_2 + H_2$$

$$Si_2H_6 \xrightarrow[10.6\ \mu m]{nh\nu} :SiH_2 + SiH_4$$

$$Si_2H_6 \xrightarrow[193\ nm]{h\nu} H_3Si-\ddot{S}iH + H_2$$

$$:SiH_2 + SiH_4 \rightleftharpoons Si_2H_6^* \longrightarrow H_3Si-\ddot{S}iH + H_2$$

$$:SiH_2 + Si_2H_6 \rightleftharpoons Si_3H_8^* \longrightarrow H_3Si-\ddot{S}iH + SiH_4$$

$$H_3Si-\ddot{S}iH \longrightarrow H_2Si=SiH_2$$

$$H_3Si-\ddot{S}iH + SiH_4 \longrightarrow Si_3H_8$$

$$H_3Si-\ddot{S}iH + Si_2H_6 \longrightarrow Si_4H_{10} \tag{117}$$

The tetrasilane insertion product from $H_3Si\ddot{S}iH$ and Si_2H_6 is less effectively incorporated into the growing film than is $H_2Si=SiH_2$. IRMPD of both SiH_4 and Si_2H_6 produce SiH_2. The higher efficiency of film formation from IRMPD of Si_2H_6 over SiH_4 is believed to be due to the greater ease of formation of $SiH_3\ddot{S}iH$ (and hence $H_2Si=SiH_2$) by unimolecular decomposition of Si_3H_8 (formed from SiH_2 and Si_2H_6) than from unimolecular decomposition of Si_2H_6 (formed from SiH_2 and SiH_4).

Detailed models continue to be devised for CVD reactors for the deposition of silicon from SiH_4. Recently SiH_2, Si_2H_6, $H_3Si\ddot{S}iH$, $H_2Si=SiH_2$ have all been included as potential participants in film formation in a LPCVD reactor model, with SiH_2 and Si_2H_6 estimated as contributing up to 20%[399]. A significant problem of many such models is that these mathematical models include large numbers of rate constants and their temperature dependences that have been estimated or measured by indirect means. The test of such a model is often a comparison of predicted and measured growth rates, but, like a reaction mechanism, such models can be disproved but never proved. One must therefore retain a certain degree of skepticism while appreciating the effort and ingenuity that has gone into developing them.

The deposition of silicon films from disilane Si_2H_6 occurs at lower temperatures than from SiH_4, thus permitting the use of a wider range of substrates. Since the formation of SiH_2 from Si_2H_6 (equation 118) occurs at lower temperatures than from SiH_4 (equation 115), it has been suggested that SiH_2 makes a greater contribution to film growth from Si_2H_6[400].

$$H_3Si-SiH_3 \xrightarrow{\Delta} H_2Si: + SiH_4 \quad \text{(major)}$$

$$\xrightarrow{\Delta} H_3Si-\ddot{S}iH + H_2 \quad \text{(minor)} \tag{118}$$

Since the reactive sticking coefficient of Si_2H_6 is known to be much higher than that for SiH_4[379], the differences between Si_2H_6 and SiH_4 as CVD precursors may extend beyond differences in ease of SiH_2 formation. Success has been claimed for models for film growth from Si_2H_6 in which both SiH_2 and higher silanes from SiH_2 gas-phase insertions reach the growing surface and react[401].

An attempt has been reported to detect reactive intermediates in the gas-phase chemistry in thermal CVD from Si_2H_6, from 300 to 1000 K and 1 to 10 torr[402]. The reaction mixture was continuously sampled in the form of a supersonic molecular beam analyzed by multiphoton ionization MPI or resonance enhanced MPI (REMPI) and mass spectrometry. Species of mass 2, 28, 32 and 60 were detected, with maximum intensities at *ca* 700 K (a commonly employed CVD temperature) corresponding to H_2, Si, SiH_4 and Si_2H_4. The last of these could be due to $H_3Si\ddot{S}iH$, a minor product of disilane pyrolysis (equation 118), or its rearrangement product $H_2Si=SiH_2$. SiH_4 is a major product of Si_2H_6, together with SiH_2, whose expected detection was unsuccessful. SiH and SiH_3 were also not detected. The mass of 28 was believed to be due to Si atoms formed by secondary decomposition of $H_3Si\ddot{S}iH$ (major source) and SiH_2 (minor source).

Even lower temperatures can be employed in photochemical CVD from Si_2H_6, employing vacuum ultraviolet radiation to decompose the precursor. An elaborate and convincing model has been proposed for 147 and 185 nm radiation[403]. At 147 nm, much higher quality films can be grown from Si_2H_6 than from SiH_4. While SiH_2 is believed to be the major primary reactive species from both precursors (83% from SiH_4 and 61% from Si_2H_6), photoinduced homolysis of both precursors to silyl radicals and H atoms also occurs at these low wavelengths. Since the insertion of SiH_2 into H−Si bonds of Si_2H_6 is believed to be an order of magnitude more rapid than the insertion into SiH_4, the gas-phase concentration of SiH_2 is much lower in Si_2H_6. Due to their lower reactivity, the concentration of SiH_3 and Si_2H_5 radicals (formed mostly in secondary reactions) is much higher than that of SiH_2 despite their comparable or lower initial production rates (equation 119). In this model it is the SiH_3 and Si_2H_5 radicals that make the major contribution to the formation of high quality films.

$$SiH_4 \xrightarrow{h\nu,\ 147\ nm} H_2Si\colon + 2H^\bullet \qquad 83\%$$

$$\longrightarrow H_3Si^\bullet + H^\bullet \qquad 17\%$$

$$H_3Si-SiH_3 \xrightarrow{h\nu,\ 147\ nm} H_2Si\colon + H_3Si^\bullet + H^\bullet \qquad 61\%$$

$$\longrightarrow Si_2H_4 + 2H^\bullet \qquad 21\%$$

$$\longrightarrow H_3Si-SiH_2^\bullet + H^\bullet \qquad 18\% \qquad (119)$$

It is believed that the smaller contribution to film formation from SiH_2 under the conditions of photo-CVD with Si_2H_6 precursor leads to better quality films, because interconnections of surface sites created from SiH_2 are slower than the interconnection of surface sites created from mono-radicals.

SiH_2 is believed to play an important role in the deposition of semi-insulating silicon films that are nonstoichiometric sub-oxides of silicon. When large ratios of N_2O and Si_2H_6 are subjected to LPCVD conditions, it has been suggested that Si and SiO are codeposited, with SiH_2 contributing to Si deposition[404]. It has been suggested that the species believed to be responsible for SiO deposition, the elusive parent silanone $H_2Si=O$, is formed from SiH_2 (equation 120)[405].

$$H_2Si\colon + N_2O \longrightarrow H_2Si=O + N_2 \qquad (120)$$

SiH_2 is believed to play a role in the silicon doping of Ga−As films grown under MOCVD conditions. The efficiency of Si incorporation is much higher with Si_2H_6 as the silicon source than with SiH_4[406]. In a model for the doping process it was assumed that SiH_2 is formed by gas-phase pyrolysis of Si_2H_6 and inserts into an As−H bond of AsH_3 to form H_3SiAsH_2, which reacts at the growing film surface.

Known addition reactions of SiH_2 with ethylene and acetylene are believed to contribute to the carbon deposition process in the growth of SiC films from mixtures of Si_2H_6 and C_2H_4 or C_2H_2 at 873 to 1273 K under LPCVD conditions[407]. It is believed that SiF_2 can act as both the source of Si and as a halogen scavenger in the growth of titanium silicide films by CVD from gaseous mixtures of SiF_2 and $TiCl_4$[408].

E. Silylenes Stabilized by Intramolecular Coordination

In 1992 Belzner prepared the cyclotrisilane **77** by magnesium coupling of the pentacoordinate dichlorosilane **78** (equation 121). **77** reacted with silylene trapping agents under truly mild conditions, at 50 °C, to give three equivalents of trapping products[409]. The reactions with benzil and diphenylacetylene[410] are shown as examples in equations 122 and 123.

(121)

(77)

(78) Ar = *o*-$Me_2NCH_2C_6H_4$

(122)

(77)

Ar = *o*-$Me_2NCH_2C_6H_4$

(123)

(77)

Ar = *o*-$Me_2NCH_2C_6H_4$

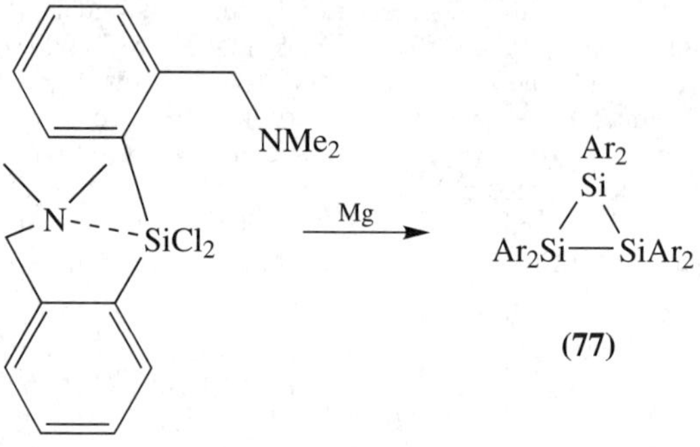

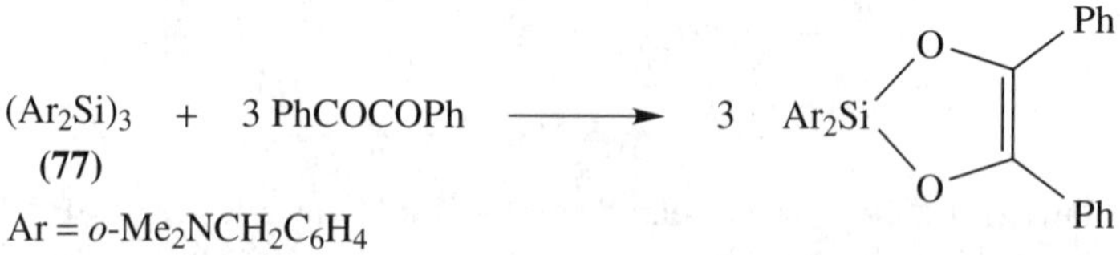

Many further reactions of **77** have been investigated, leading to a rather extensive chemistry of the base-stabilized silylene, **79**. In Scheme 23 the structure of **79** is shown as tetracoordinated, but could also be tricoordinate. Intramolecularly base-stabilized silylenes can also be depicted as silaylides, for instance **79a**.

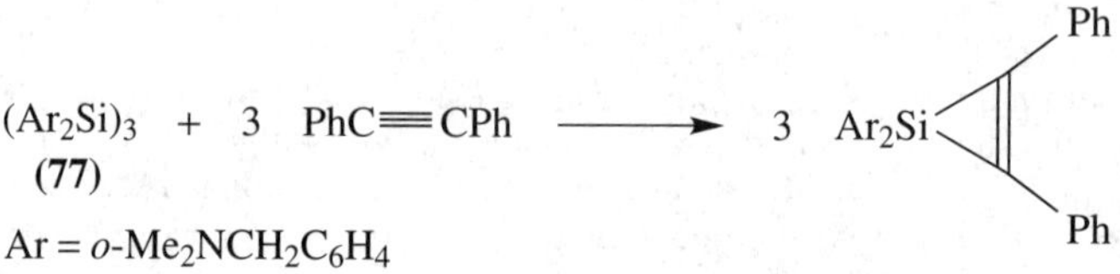

In an especially significant paper Belzner reports the reaction of **77** with the terminal olefins 1-pentene and 1-hexene[411]. The products, siliranes **80a,b**, were found to be in

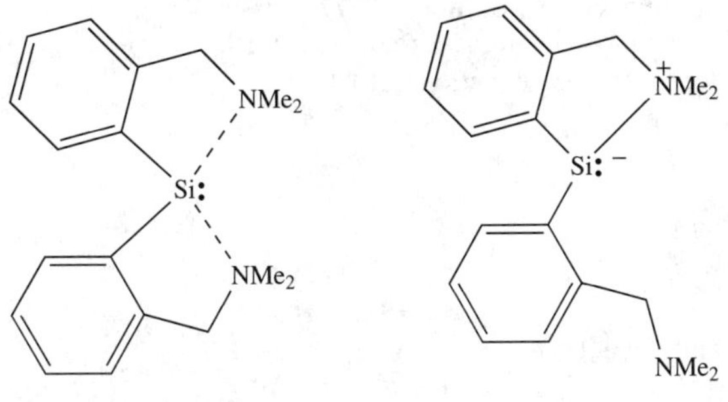

SCHEME 23

thermal equilibrium with the cyclotrisilane and the starting olefin (equation 124).

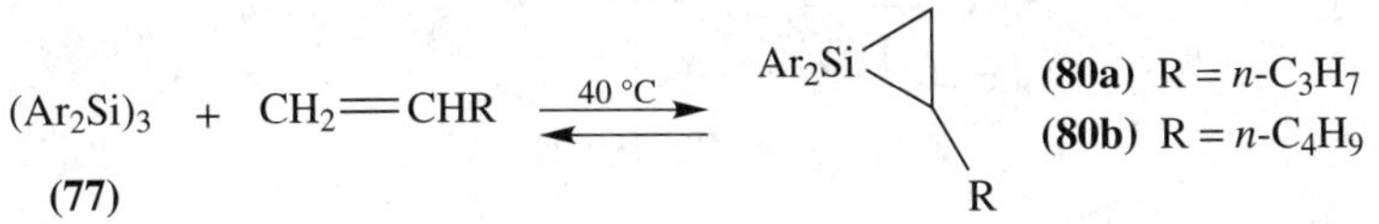

$$(Ar_2Si)_3 + CH_2=CHR \xrightleftharpoons{40\,°C} Ar_2Si\text{-cyclopropane-}R \quad \begin{array}{l}(80a)\ R = n\text{-}C_3H_7 \\ (80b)\ R = n\text{-}C_4H_9\end{array} \quad (124)$$
(77)

The first step of the retro-reaction involves loss of silylene **79**, which could be trapped with 1-pentyne to give the known silirene **81** (equation 125). In the absence of a trapping agent, **79** recondenses to **77**, probably by first dimerizing to the disilene $Ar_2Si=SiAr_2$ followed by 2+1 cycloaddition to give **77** (equation 126). From the principle of microscopic reversibility, the fact that silylene is formed in the retro-reaction leads to the conclusion that **79** must also be an intermediate in the cycloaddition reaction.

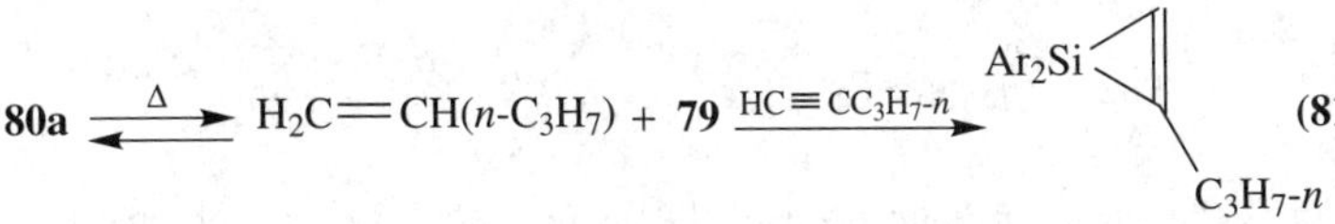

$$\textbf{80a} \xrightarrow{\Delta} H_2C=CH(n\text{-}C_3H_7) + \textbf{79} \xrightarrow{HC\equiv CC_3H_7\text{-}n} Ar_2Si\text{-silirene-}C_3H_7\text{-}n \quad (\textbf{81}) \quad (125)$$

$$3Ar_2Si: \rightleftharpoons Ar_2Si: + Ar_2Si=SiAr_2 \rightleftharpoons (Ar_2Si)_3 \quad (126)$$
$\qquad\quad$ (**79**) $\hspace{4.5cm}$ (**77**)

$$Ar = o\text{-}Me_2NCH_2C_6H_4$$

This is the first time that an equilibrium between a silylene, a disilene and a cyclotrisilane has been demonstrated, but a similar photoequilibrium may exist for $(t\text{-}Bu_2Si)_3$, as shown by Weidenbruch[8].

Cyclotrisilane **77** reacts with *strained* (but not unstrained) internal olefins to give stable siliranes; an example is the reaction with norbornene (equation 127).

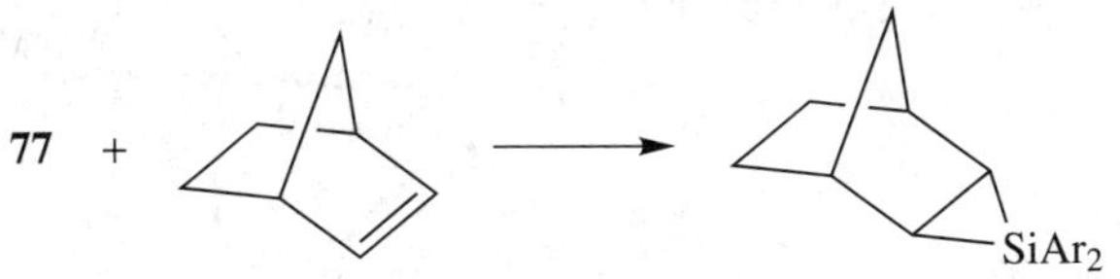

$$\textbf{77} + \text{norbornene} \longrightarrow \text{norbornene-SiAr}_2 \quad (127)$$

$$Ar = o\text{-}Me_2NCH_2C_6H_4$$

Other interesting and useful reactions of **77** are shown in equations 128–131. With pivalonitrile, a four-membered ring is formed, presumably through intermediate formation of the azasilacyclopropene **82** (equation 128)[412].

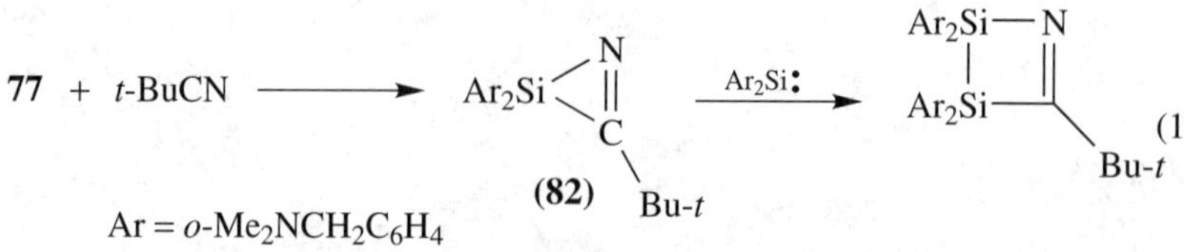

$$77 + t\text{-BuCN} \longrightarrow Ar_2Si\underset{(82)}{\overset{N}{\underset{C}{\diagdown\!\!\diagup}}}\!\!\underset{Bu\text{-}t}{} \xrightarrow{Ar_2Si:} \underset{Ar_2Si}{\overset{Ar_2Si-N}{\underset{\diagdown}{|\|}}}\!\!\underset{Bu\text{-}t}{} \quad (128)$$

Ar = o-Me$_2$NCH$_2$C$_6$H$_4$

The silylene **79** abstracts oxygen or sulfur from isocyanates and isothiocyanates[413]. Cyclic siloxanes are the products with isocyanates (equation 129); these probably arise by polymerization of the silanone, Ar$_2$Si=O (**83**).

$$77 + \text{RNCO} \xrightarrow{70\,°\text{C}} [\underset{(83)}{Ar_2Si=O}] \longrightarrow \underset{n=2,3}{(Ar_2SiO)_n}$$

Ar = o-Me$_2$NCH$_2$C$_6$H$_4$ (129)

When (Me$_2$SiO)$_3$ (D$_3$) was present, the diarylsilanone (**83**) was intercepted as the cyclotetrasiloxane (equation 130); this is a known reaction for silanone capture.

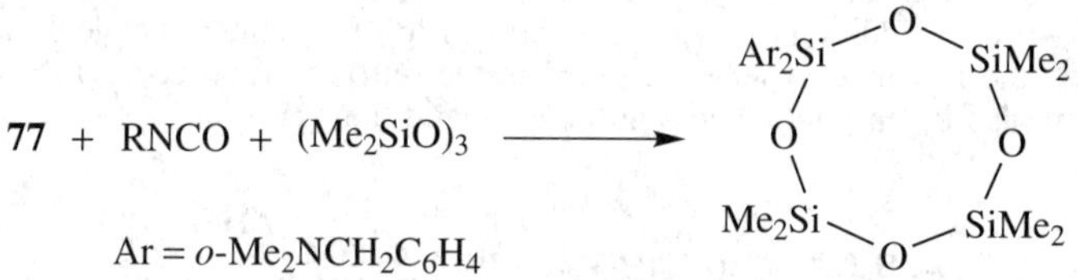

$$77 + \text{RNCO} + (Me_2SiO)_3 \longrightarrow \begin{array}{c} Ar_2Si\!-\!O\!-\!SiMe_2 \\ || \\ OO \\ || \\ Me_2Si\!-\!O\!-\!SiMe_2 \end{array} \quad (130)$$

Ar = o-Me$_2$NCH$_2$C$_6$H$_4$

With isothiocyanates, the 'silanethione' (**84**) is obtained, without oligomerization. The ^{29}Si NMR of **84**, at −21.0 ppm, shows that the silicon atom is highly shielded, consistent with a pentacoordinate structure as shown in equation 131. The chemical shift can be compared with that for singly N-coordinated silanethiones, +22 to +41 ppm, and with that for a true tricoordinate silanethione, +166.6 ppm.

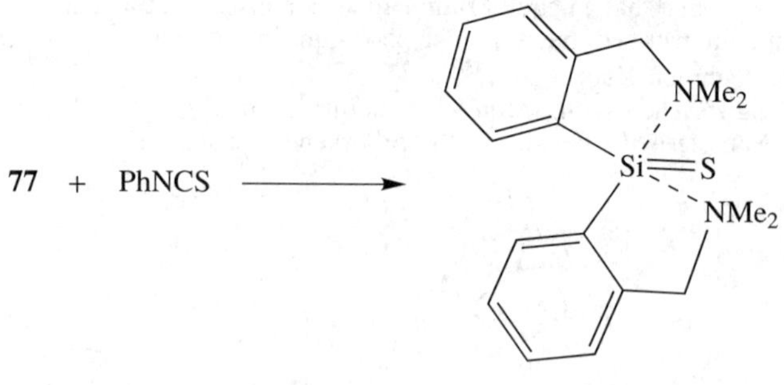

77 + PhNCS ⟶

(**84**) (131)

Various complex products were obtained in reactions of **77** with ketones[414]. With benzophenone, for example, the rearranged compound **85** was obtained, while fluorenone reacted in 2 : 1 ratio to give **86**, both shown in Scheme 24[415]. Reaction pathways leading to both products were suggested. For additional examples, the original papers should be consulted.

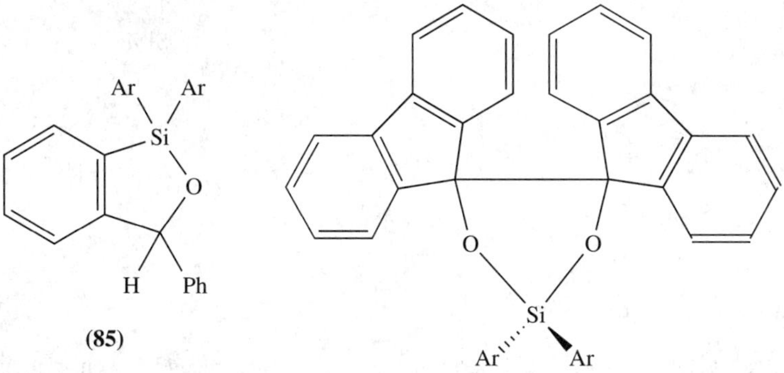

Ar = o-Me$_2$NCH$_2$C$_6$H$_4$

SCHEME 24

The same hypercoordinate silylene (**79**) was generated by the Corriu group in dehalogenation reactions; the best results were with the difluoride Ar$_2$SiF$_2$ (Ar = o-Me$_2$NCH$_2$Ph) and lithium metal or lithium naphthalene[97]. The silylene was trapped by 1,4-addition to 2,3-dimethylbutadiene. Similar defluorination reactions were used to obtain silylenes **87–89** shown in Scheme 25, all trapped with dimethylbutadiene.

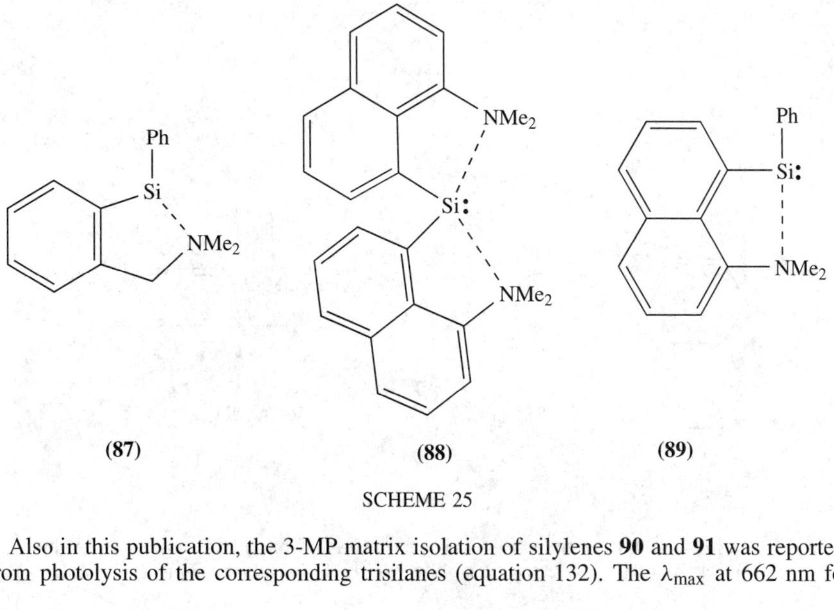

SCHEME 25

Also in this publication, the 3-MP matrix isolation of silylenes **90** and **91** was reported, from photolysis of the corresponding trisilanes (equation 132). The λ_{max} at 662 nm for

91 shifted to 478 nm in **90**, consistent with strong N→Si electron donation in the latter.

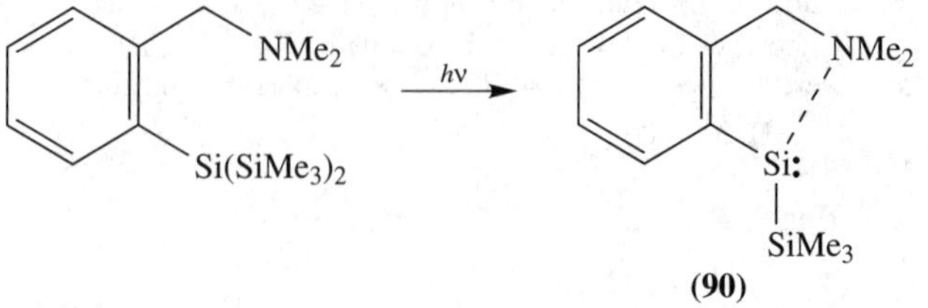

(132)

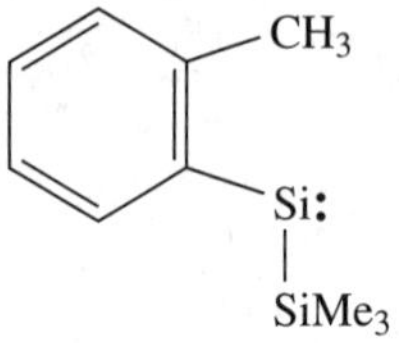

(90)

(91)

Remarkably facile thermal generation of a silylene has recently been demonstrated for the pentacoordinate alkoxydisilane **92**[416]. Intramolecular N→Si coordination in **92** was shown by X-ray crystallography. This disilane underwent thermolysis at 110 °C in toluene, or 90 °C in DMF, producing the silylene **93**, which was trapped with 2,3-dimethylbutadiene and diphenylacetylene (Scheme 26). The 2 : 1 adduct with diphenylacetylene was shown to have *one* nitrogen intramolecularly coordinated to silicon, even though the silicon atom lacks any electronegative substituents.

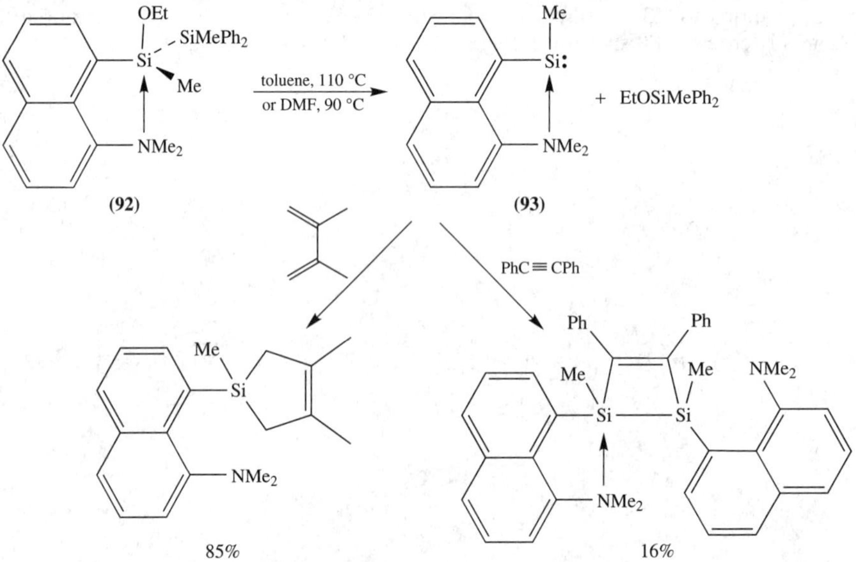

SCHEME 26

In a related investigation, the pseudo-pentacoordinated naphthylsilane (**94**), in the presence of a nickel catalyst, was found to react with diphenylacetylene to give the silole

95[417]. Presumably, this reaction proceeds through intermediate formation of the silylene **96**, stabilized by intramolecular coordination from nitrogen (equation 133).

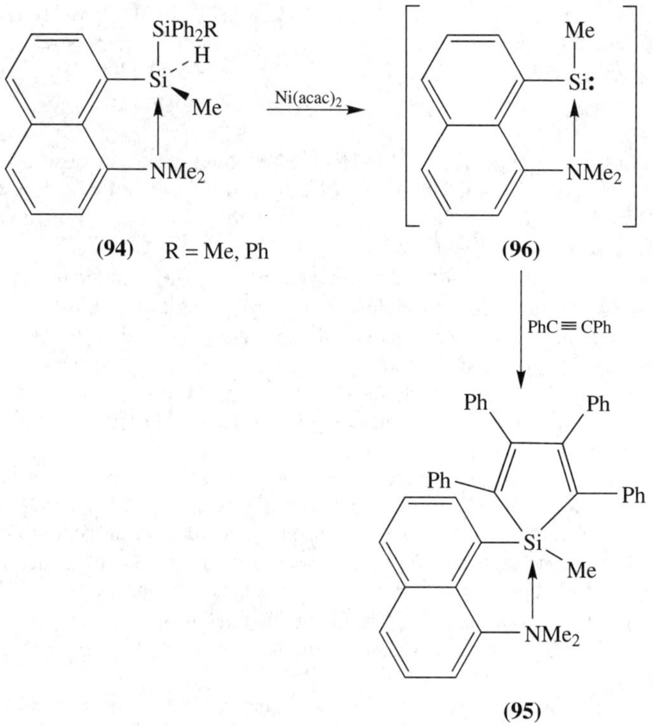

(133)

The research on silylenes with intramolecular coordination shows beyond doubt that such electron donation from nitrogen greatly stabilizes the silylenes. A persistent silylene stabilized in this manner has not yet appeared, but it seems possible that one might be made if just the right substituents are employed.

F. Vinylidenesilylene and other Double-bonded Silylenes Z=Si:

Silylenes with double bonds to silicon Z=Si: have long interested theorists (see Section IV.B for density functional calculations on disilenylidenes $R_2Si=Si$:) but experimental data concerning them are scarce. Theoretical calculations up to 1989 have been reviewed by Apeloig[201], himself an active investigator in the field.

In 1995 Apeloig and Albrecht carried out high level *ab initio* calculations on the effects of substituents on the energy barriers separating silanitriles RSi≡N: and silaisocyanides R—N=Si:[418]. For most substituents, i.e. R = H, Li, BeH, BH_2, CH_3, SiH_3, PH_2 and SH, the isocyanide was predicted to be more stable than the silanitrile by 22 to 76 kcal mol^{-1}, and the effects of substituents correlated well with R—Si *vs* R—N bond strengths. Only for R = F and OH was RSi≡N predicted to be more stable than RN=Si:. At the QCISD(T)/6-311G(2df,p)//QCISD/6-31G* level the calculated energy differences and barriers (kcal mol^{-1}) between RN=Si: and RSi≡N: were: R = H (65.2, 11.4), OH (−6.5, 35.8), F (−26.6, 49.3).

In 1997 Apeloig and Karni published similar computational studies on the effects of substituents on the potential energy surface for RHSiC (R = H, CH$_3$, SiH$_3$, F, OH)[419]. In all cases except R = F, the energy calculated for the vinylidenesilylene RHC=Si: was lower than that for the silacetylene RSi≡CH. At MP4SDTQ/6-31G(d,p)//MP2/6-31G(d,p) the energy differences and the barriers (kcal mol^{-1}) were: R = H (36.1, 9.0), CH$_3$ (25.8, 17.5), SiH$_3$ (33.0, 3.2), OH (4.3, 25.0), F(−6.4, 22.5). Thus, both silaisocyanides R−N=Si:[418] and silaacetylenes RHC=Si:[419] appear to be viable targets for synthesis.

Experimental information is limited to H$_2$C=Si:, whose electronic absorption spectrum was observed by Leclercq and Dubois in 1979 as a well resolved band system between 3400 and 3100 Å following a flash discharge through CH$_3$SiH$_3$[420]. For the $^1B_2$ excited state, an HCH bending frequency of 1110 cm^{-1} and a Si−C stretching frequency of 700 cm^{-1} were deduced from the vibrational analysis. The rotational analysis was in accord with C_{2v} symmetry for the molecule and a $^1A_1$ singlet ground state.

In 1988, Damrauer and coworkers found the gas-phase acidity of H$_2$C=Si: to lie between that of CH$_3$SH and acetophenone, making H$_2$C=Si: a stronger acid than acetylene[421]. It was necessary for these workers to assume the HCSi$^-$ structure for the ion that they studied in tandem flowing afterglow selected ion flow tube proton-transfer experiments, as well as the H$_2$C=Si: structure for the product.

In neutralization−reionization mass spectrometric experiments on CH$_2$Si$^{+\bullet}$ formed by electron-impact dissociative ionization of ClCH$_2$SiH$_3$, Srinivas, Sülzle and Schwarz found evidence for the formation of a viable neutral molecule whose fragmentation pattern and collisional activation mass spectrum were in accord with a H$_2$C=Si: structure[422]. These authors suggested that their experiments supported electron-capture by CH$_2$Si$^{+\bullet}$ as a mechanism for the formation of H$_2$C=Si: in interstellar space. Various models have predicted that H$_2$C=Si: is one of the most abundant forms of silicon in dense interstellar clouds[423].

In 1995, Sherrill and Schaefer published new quantum mechanical calculations beginning at the restricted Hartree−Fock level with basis sets as large as TZ2pf and employing the coupled clusters method to the CCSD(T) level for correlation corrections[424]. Their work was stimulated by still unpublished photoelectron spectroscopic studies of H$_2$CSi$^-$ and D$_2$CSi$^-$ by Bengali and Leopold[425]. The equilibrium geometry predicted by Sherrill and Schaefer for the ground state of H$_2$C=Si: was similar to those of earlier calculations, but the geometrical parameters, particularly the Si−C bond distance, were found to vary noticeably with basis set and correlation method. At the highest level of treatment r(C−Si) = 1.720 Å, r(C−H) = 1.089 Å and ∠H−C−H = 113.4°[425].

There have been several other recent theoretical studies. In 1995, Nguyen, Sengupta and Vanquickenborne carried out high level calculations on the H$_2$CSi potential surface to study the effects of tunneling on the rearrangement of trans-HC≡SiH to H$_2$C=Si:[426]. With a 6-3111++G(3df,3pd) basis set and geometry optimization including configuration interaction to the QCISD(T) level, the geometry predicted for H$_2$C=Si: was r(C−Si) = 1.726 Å, r(C−H) = 1.095 Å and ∠H−C−H = 118.4°. The calculated heats of formation were ΔH_f^o(H$_2$C=Si:) = 84.5 ± 2 kcal mol^{-1} and ΔH_f^o(HC≡SiH) = 118.3 ± 2 kcal mol^{-1}[426]. The 1996 calculations on H$_2$CSi isomers by Stegmann and Frenking also employed correlated wave functions in geometry optimizations[427]. At CCSD(T)/TZ2P, the predicted structure of H$_2$C=Si was r(C−Si) = 1.721 Å, r(C−H) = 1.087 Å and ∠H−C−H = 118.4°. The energy difference between H$_2$C=Si: and trans-HC≡SiH predicted by Stegmann and Frenking, 34.1 kcal mol^{-1}, is almost exactly that calculated by Nguyen and coworkers[426]. In 1994, with a smaller 6-31G(d,p) basis set, Schoeller and Strutwolf had calculated $r(C-Si) = 1.701$ Å, $r(C-H) = 1.082$ Å and $\angle H-C-H = 120.7°$ as geometrical parameters for $H_2C=Si$:[428].

The equilibrium rotational constants and the theoretical harmonic frequencies calculated by Sherrill and Schaefer[424] were in good agreement with the experimental ground-state rotational constants obtained by Leclercq and Dubois[420] and in reasonable accord with the experimental fundamental vibrational frequencies obtained by Bengali and Leopold[425].

In 1996, Izuha, Yamamoto and Saito observed the microwave spectrum of $H_2C=Si$: produced via a glow-discharge plasma in a gaseous mixture of SiH_4 and CO[429]. The rotational constants deduced were very close to those of Leclercq and Dubois[420]. The vibrational frequency of the CH_2 rocking mode was estimated to be 331 ± 5 cm^{-1}[429], rather close to the 305 cm^{-1} predicted by Sherrill and Schaefer[424] but higher than the ca 265 cm^{-1} found by Bengali and Leopold[425].

In 1997, Clouthier et al. presented laser-induced fluorescence spectra of $H_2C=Si$: produced by an electric discharge through dilute Me_4Si-Ar mixtures[430]. The rotational analysis yielded geometric parameters in good agreement with theoretical predictions (given in brackets) employing TZ(2df,2pd) basis sets and correlation corrections at the CISD level: $r(C-Si) = 1.706(5)$ [1.704] Å, $r(C-H) = 1.099(3)$ [1.083] Å and $\angle H-C-H = 114.4(2)$ [113.4]°[430]. The emission spectrum observed by Clouthier et al. agreed with the absorption spectrum reported by Leclercq and Dubois[420].

Thus, experiment and theory agree that $H_2C=Si$: is a planar C_{2v} molecule with C–H single and C=Si double bonds, and that it is the global minimum on the H_2SiC ground-state potential surface. Its chemistry, other than C–H ionization, is totally unexplored, but we predict that clean routes to its generation as well as that of substituted vinylidenesilylenes RR'C=Si will be developed in the near future that will allow the potentially useful and surely interesting chemistry of these molecules to be explored.

IX. SPECULATIONS ON THE FUTURE OF SILYLENE CHEMISTRY

In writing this chapter, the authors have been much impressed by the tremendous strides made since the silylenes were last comprehensively reviewed, more than a decade ago. Here we shall attempt to look ahead, and predict some of the discoveries concerning silylenes which may take place during the next ten years, foolhardy as this exercise of imagination may seem. We ask then: where is silylene chemistry likely to advance in the future?

An area obviously ready for expansion is the chemistry of persistent or 'stable' silylenes. Only four such compounds are now known, and their status as 'true' silylenes in the sense of containing a divalent silicon atom remains a subject of discussion, but dozens can readily be imagined. The persistent silylenes known today are stabilized by electron donation from a pair of nitrogen atoms. It seems likely that ways will be found to prepare persistent silylenes stabilized by other electron donors, oxygen, sulfur or phosphorus. The nitrogen atoms in known stable silylenes are directly attached to silicon, but this may not be necessary; persistent silylenes stabilized by intramolecular chelation may be synthesized. Moreover, the chemistry of stable silylenes is just beginning to be explored, and appears to be sufficiently different from that of transient silylenes. Especially, we look for an entire new series of base-free transition metal complexes of the stable silylenes stretching across the periodic table. The quest for silylenes stabilized solely by steric hindrance to attack at the silylene center has not been abandoned and may succeed in the near future. To find groups of sufficient size is no problem; the difficulty has been to make them sufficiently inert to avoid consuming the divalent silicon atom in an intramolecular reaction. It is confidently predicted that the challenge of generating ground-state triplet silylenes and studying their chemistry will finally be met in the immediate future.

Methods for the synthesis of transient silylenes are now well-developed, so that it is risky to predict the discovery of entirely new methods for the generation of silylenes, as desirable as these might be. But many questions remain about the details of silylene-generating reactions, and these may yield to mechanistic investigations in the coming years. For instance, the precise pathways for silylene extrusion reactions, both thermal and photochemical, are still obscure. Here, it seems likely that theoretical calculations, as much as experiment, will aid our understanding of these phenomena. This is not only a matter of intellectual curiosity. Silylene-generating reactions are often accompanied by undesired side-reactions, and thus deeper understanding is desired to enhance our control of silylene generation and thus increase its efficiency. The lack of silylene precursors as versatile and generally useful as diazo compounds are, for carbene generation, has materially limited the applications of silylene chemistry in organic and main-group synthesis.

The reactions of transient silylenes are so rapid that most of the limited mechanistic information that has been obtained over the past quarter-century has been through indirect means. Direct measurements of silylene reaction rates by kinetic spectroscopy in the past decade have yielded important new insights. One can predict with some confidence an explosion of mechanistic studies of silylenes employing fast spectroscopies capable of providing more structural information than traditional electronic absorption and emission techniques. The nearly universal reversibility of silylene reactions remains to be fully exploited through kinetic studies of retro-reactions. The mechanisms of most silylene reactions remain to be fully elucidated, and this task will increase in urgency as silylenes see more use in synthesis.

After more than fifty years of industrial use, the 'direct reaction' of methyl chloride with silicon, which underlies the entire silicone industry, is not understood. Promising recent experiments on this process are likely to be continued, and should at least settle the question of the nature of the intermediates. Are silylenes important in the process, either in the gas phase or at the silicon surface?

Also obscure is the nature of the reaction of dihalosilanes with active metals. Further experiments should show under what circumstances, if any, such dehalogenation reactions lead to silylenes, or to silylenoids, another area of likely future activity. In a related area, isolable silylenoids are quite new, and further development of their chemistry can be anticipated.

In the past few years, theoretical calculations have become increasingly important in guiding our understanding of silylenes, and indeed of organosilicon chemistry generally. The partnership between theory and experiment is likely to become even stronger and more important in the years ahead. The development of new computational techniques such as density functional theory, and the availability of larger and ever-faster computers have made chemically accurate calculations on large molecules and the mapping of potential surfaces for complex reactions an everyday reality. The computational experiment is an important component in the study of silylenes.

Last of all the authors continue to expect the unexpected. That is one of the reasons silylene chemistry is still such fun.

X. REFERENCES

1. P. P. Gaspar and B. Jerosch Herold, in *Carbene Chemistry*, 2nd ed. (Ed. W. Kirmse), Academic Press, New York, 1971, pp. 504–550.
2. P. P. Gaspar, in *Reactive Intermediates*, Vol. 1 (Eds. M. Jones, Jr. and R. A. Moss), Wiley, New York, 1978, pp. 229–277.

3. P. P. Gaspar, in *Reactive Intermediates*, Vol. 2 (Eds. M. Jones, Jr. and R. A. Moss), Wiley, New York, 1981, pp. 335–385.
4. Y.-N. Tang, in *Reactive Intermediates*, Vol. 2 (Ed. R. A. Abramovitch), Plenum Press, New York, 1982, pp. 297–366.
5. P. P. Gaspar, in *Reactive Intermediates*, Vol. 3 (Eds. M Jones, Jr. and R. A. Moss), Wiley, New York, 1985, pp. 333–427.
6. C.-S. Liu and T.-H. Hwang, *Adv. Inorg. Chem. Radiochem.*, **29**, 1 (1985).
7. E. A. Chernyshev and N. G. Komalenkova, *Russ. Chem. Rev.*, **59**, 531 (1990).
8. M. Weidenbruch, *Coord. Chem. Rev.*, **130**, 275 (1994).
9. W. H. Atwell and D. R. Weyenberg, *J. Organometal. Chem.*, **5**, 594 (1966); *Angew. Chem., Int. Ed. Engl.*, **8**, 469 (1969).
10. I. M. T. Davidson, K. J. Hughes and S. Ijadi-Maghsoodi, *Organometallics*, **6**, 639 (1986).
11. T. J. Barton, S. A. Burns, P. P. Gaspar and Y.-S. Chen, *Synth. React. Inorg. Met.-Org. Chem.*, **13**, 881 (1983).
12. J. G. Martin, M. A. Ring and H. E. O'Neal, *Int. J. Chem. Kinet.*, **19**, 715 (1987).
13. R. Walsh, *Organometallics*, **7**, 75 (1988).
14. M. E. Harris, M. A. Ring and H. E. O'Neal, *Organometallics*, **11**, 983 (1992).
15. K. E. Nares, M. E. Harris, M. A. Ring and H. E. O'Neal, *Organometallics*, **8**, 1964 (1989).
16. M. P. Clarke, I. M. T. Davidson and M. P. Dillon, *J. Chem. Soc., Chem. Commun.*, 1251 (1988).
17. E. W. Ignacio and H. B. Schlegel, *J. Phys. Chem.*, **96**, 1758 (1992).
18. R. J. P. Corriu, D. Leclercq, P. H. Mutin, J.-M. Planeix and A. Vioux, *Organometallics*, **12**, 454 (1993).
19. J. Heinicke, B. Gerhus and S. Meinel, *J. Organometal. Chem.*, **474**, 71 (1994).
20. D. Lei, R.-J. Hwang and P. P. Gaspar, *J. Organometal. Chem.*, **271**, 1 (1984).
21. J. Heinicke and B. Gerhus, *J. Organometal. Chem.*, **423**, 13 (1993).
22. A. Kaczmarczyk and G. Urry, *J. Am. Chem. Soc.*, **82**, 751 (1960); R. F. Trandell and G. Urry, *J. Inorg. Nucl. Chem.*, **40**, 1305 (1978).
23. U. Herzog, R. Richter, E. Brendler and G. Roewer, *J. Organometal. Chem.*, **507**, 221 (1996).
24. L. Fredin, R. H. Hauge, Z. H. Kafafi and J. L. Margrave, *J. Chem. Phys.*, **82**, 3542 (1985).
25. P. P. Gaspar, in *Organosilicon and Bioorganosilicon Chemistry* (Ed. H. Sakurai), Horwood, Chichester, 1985, pp. 87–98.
26. P. P. Gaspar, *Radiochim. Acta*, **43**, 89 (1988).
27. P. P. Gaspar, in *Handbook of Hot Atom Chemistry* (Eds. J.-P. Adloff, P. P. Gaspar, M. Imamura, A. G. Maddock, T. Matsuura, H. Sano and K. Yoshihara), Kodansha, Tokyo and VCH, Weinheim, 1992, pp. 85–104.
28. P. P. Gaspar, S. Konieczny and S. H. Mo, *J. Am. Chem. Soc.*, **106**, 424 (1984).
29. P. Trefonas, R. D. Miller and R. West, *J. Am. Chem. Soc.*, **107**, 2737 (1985).
30. T. Karatsu, R. D. Miller, R. Sooriyakumaran and J. Michl, *J. Am. Chem. Soc.*, **111**, 1140 (1989).
31. A. Watanabe and M. Matsuda, *Macromolecules*, **25**, 484 (1992).
32. A. Watanabe, H. Miike, Y. Tsutsumi and M. Matsuda, *Macromolecules*, **26**, 2111 (1993).
33. M. Kira, T. Miyazawa, S. Koshihara, Y. Segawa and H. Sakurai, *Chem. Lett.*, 3 (1995).
34. M. Kasha, *Discuss. Faraday Soc.*, **9**, 14 (1950).
35. K. Oka and R. Nakao, *Res. Chem. Intermed.*, **13**, 143 (1990).
36. Y. Huang, M. Sulkes and M. J. Fink, *J. Organomet. Chem.*, **499**, 1 (1995).
37. M. Kira, T. Maruyama and H. Sakurai, *Chem. Lett.*, 1345 (1993).
38. G. Bott, P. Marshall, P. E. Wagensteller, Y. Wang and R. T. Conlin, *J. Organomet. Chem.*, **489**, 11 (1995).
39. S. Zhang, M. B. Ezhova and R. T. Conlin, *Organometallics*, **14**, 1471 (1995).
40. M. Ishikawa and M. Kumada, *Adv. Organomet. Chem.*, **19**, 51 (1981).
41. T. J. Drahnak, R. West and J. Michl, *J. Am. Chem. Soc.*, **101**, 5427 (1979).
42. H. Vancik, G. Raabe, M. J. Michalczyk, R. West and J. Michl, *J. Am. Chem. Soc.*, **107**, 4097 (1985).
43. C. W. Carlson, K. J. Haller, X.-H. Zhang and R. West, *J. Am. Chem. Soc.*, **106**, 5521 (1989).
44. B. J. Helmer and R. West, *Organometallics*, **1**, 1458 (1982).
45. H. Watanabe, E. Tabei, M. Goto and Y. Nagai, *J. Chem. Soc., Chem. Commun.*, 522 (1987).
46. H. Shizuka, K. Murata, Y. Arai, K. Tonokura, H. Tanaka, H. Matsumoto, Y. Nagai, G. Gillette and R. West, *J. Chem. Soc., Faraday Trans.*, **85**, 2369 (1989).
47. M. Weidenbruch and S. Pohl, *J. Organomet. Chem.*, **282**, 305 (1985).

48. E. Kroke, S. Willms, M. Weidenbruch, S. Pohl and H. Marsmann, *Tetrahedron Lett.*, **37**, 3675 (1996).
49. E. Kroke, M. Weidenbruch, W. Saak, S. Pohl and H. Marsmann, *Organometallics*, **14**, 5695 (1995).
50. M. Weidenbruch, P. Will, K. Peters, H. G. von Schnering and H. Marsmann, *J. Organomet. Chem.*, **521**, 355 (1996).
51. S. P. Kolesnikov, M. D. Egorov, A. M. Galminas, M. B. Ezhova, O. M. Nefedov, T. V. Leshina, M. B. Taraban, A. I. Kruppa and V. I. Maryasova, *J. Organomet. Chem.*, **391**, C1 (1990).
52. T. J. Barton, W. F. Goure, J. L. Witiak and W. D. Wulff, *J. Organomet. Chem.*, **225**, 87 (1982).
53. K. M. Welsh, J. Michl and R. West, *J. Am. Chem. Soc.*, **110**, 6689 (1988).
54. V. N. Khabashesku, V. Balaji, S. E. Boganov, O. M. Nefedov and J. Michl, *J. Am. Chem. Soc.*, **116**, 320 (1994).
55. M. Veith, E. Werle, R. Lisowsky, R. Koppe and H. Schnoeckel, *Chem. Ber.*, **125**, 1375 (1992).
56. P. Boudjouk, V. Samaraweera, R. Sooriyakumaran, J. Chrusciel and K. R. Anderson, *Angew. Chem., Int. Ed. Engl.*, **27**, 1355 (1988).
57. D. H. Pae, M. Xiao, M. Y. Chiang and P. P. Gaspar, *J. Am. Chem. Soc.*, **113**, 1281 (1991).
58. M. G. Steinmetz and C. Yu, *Organometallics*, **11**, 2686 (1992).
59. (a) K. Oka, T. Dohmaru, Y. Nagai and R. Nakao, *J. Chem. Soc., Chem. Commun.*, 552 (1987).
    (b) K. Oka, R. Nakao, Y. Nagata and T. Dohmaru, *J. Chem. Soc., Perkin Trans. 2*, 337 (1987).
60. R. Nakao, K. Oka, S. Irie, T. Dohmaru, Y. Abe and T. Horii, *J. Chem. Soc., Perkin Trans. 2*, 755 (1991).
61. R. T. White, R. L. Espino-Rios, D. S. Rogers, M. A. Ring and H. E. O'Neal, *Int. J. Chem. Kinet.*, **17**, 1029 (1985).
62. H. K. Moffat, K. F. Jensen and R. W. Carr, *J. Phys. Chem.*, **95**, 145 (1991).
63. M. A. Ring and H. E. O'Neal, *J. Phys. Chem.*, **96**, 10848 (1992).
64. R. Becerra and R. Walsh, *J. Phys. Chem.*, **96**, 10856 (1992).
65. J. J. O'Brien and G. H. Atkinson, *J. Phys. Chem.*, **92**, 5782 (1988).
66. M. S. Gordon and T. N. Truong, *Chem. Phys. Lett.*, **142**, 110 (1987).
67. I. M. T. Davidson and M. A. Ring, *J. Chem. Soc., Faraday Trans. 1*, **76**, 1520 (1980).
68. B. A. Sawrey, H. E. O'Neal, M. A. Ring and D. Coffey, *Int. J. Chem. Kinet.*, **16**, 31 (1984).
69. S. F. Rickborn, M. A. Ring and H. E. O'Neal, *Int. J. Chem. Kinet.*, **16**, 1371 (1984).
70. S. F. Rickborn, M. A. Ring, H. E. O'Neal and D. Coffey, *Int. J. Chem. Kinet.*, **16**, 289 (1984).
71. B. A. Sawrey, H. E. O'Neal, M. A. Ring and D. Coffey, *Int. J. Chem. Kinet.*, **16**, 801 (1984).
72. B. A. Sawrey, H. E. O'Neal and M. A. Ring, *Organometallics*, **6**, 720 (1987).
73. R. E. Jardine, H. E. O'Neal, M. A. Ring and M. E. Beatie, *J. Phys. Chem.*, **99**, 12507 (1995).
74. S. F. Rickborn, D. S. Rogers, M. A. Ring and H. E. O'Neal, *J. Phys. Chem.*, **90**, 408 (1986).
75. H. E. O'Neal and M. A. Ring, *Organometallics*, **7**, 1017 (1988).
76. T. J. Barton and N. Tillman, *J. Am. Chem. Soc.*, **109**, 6711 (1987).
77. A. P. Dickinson, H. E. O'Neal and M. A. Ring, *Organometallics*, **10**, 3513 (1991).
78. J. A. Boatz and M. S. Gordon, *J. Phys. Chem.*, **93**, 3025 (1989).
79. D. A. Horner, R. S. Grev and H. F. Schaefer III, *J. Am. Chem. Soc.*, **114**, 2093 (1992).
80. N. Al-Rubaiey, H. M. Frey, B. P. Mason, C. McMahon and R. Walsh, *Chem. Phys. Lett.*, **204**, 301 (1993).
81. J. A. Boatz, M. S. Gordon and L. R. Sita, *J. Phys. Chem.*, **94**, 5488 (1990).
82. E. A. Chernyshev, N. G. Komalenkova and S. A. Bashkirova, *J. Organometal. Chem.*, **271**, 129 (1984) and references cited therein.
83. D. Lei and P. P. Gaspar, *Organometallics*, **4**, 1471 (1985).
84. P. P. Gaspar and D. Lei, *Organometallics*, **5**, 1276 (1986).
85. D. Lei, *A Mechanistic Study of the Addition of Silylenes to 1,3-Dienes in Both the Forward and Reverse Directions*, Doctoral Dissertation, Washington University, St. Louis, August 1988.
86. P. J. Skell and E. J. Goldstein, *J. Am. Chem. Soc.*, **86**, 1442 (1964).
87. R. West, in *The Chemistry of Organic Silicon Compounds* (Eds. S. Patai and Z. Rappoport), Chap. 19, Wiley, Chichester, 1989.
88. R. West, in *Comprehensive Organometallic Chemistry II* (Eds. E. W. Abel, F. G. A. Stone and G. Wilkinson), Vol. 2, Chap. 3, Pergamon Press, Oxford, 1995.
89. The familiar Wurtz synthesis of polysilanes from dichlorosilanes and alkali metal evidently does not involve free silylenes. See R. G. Jones, R. E. Benfield, R. H. Cragg, A. C. Swain and S. J. Webb, *Macromolecules*, **26**, 4878 (1993).

90. Reviews: G. Köbrich, *Angew. Chem., Int. Ed. Engl.*, **11**, 473 (1972); A. Maercker, *Angew. Chem., Int. Ed. Engl.*, **32**, 1023 (1993).
91. K. Tamao and A. Kawachi, *Angew. Chem., Int. Ed. Engl.*, **34**, 818 (1995).
92. K. Tamao and A. Kawachi, *Organometallics*, **14**, 3108 (1995).
93. A. Kawachi and K. Tamao, *Organometallics*, **15**, 4653 (1996).
94. Review: K. Tamao and A. Kawachi, *Adv. Organomet. Chem.*, **38**, 1 (1995).
95. Recently, similar experiments have been carried out with Mes_2SiCl_2, t-BuMesSiCl$_2$ and t-BuPhSiCl$_2$. The Si—H insertion product was obtained in all of these cases, although the yields reported were smaller than with t-Bu$_2$SiCl$_2$. E. P. Black, Ph.D. Thesis, North Dakota State University, 1995.
96. J. Grobe and T. Scherholt, in *Organosilicon Chemistry II* (Eds. N. Auner and J. Weis), VCH, Weinheim, 1996, pp. 317–319.
97. R. Corriu, G. Lanneau, C. Priou, F. Soulairol, N. Auner, R. Probst, R. Conlin and C. Tan, *J. Organomet. Chem.*, **466**, 55 (1994).
98. R. West and M. Denk, *Pure Appl. Chem.*, **68**, 785 (1996).
99. R. T. Conlin and Y.-W. Kwak, *J. Am. Chem. Soc.*, **108**, 834 (1986).
100. I. M. T. Davidson, S. Ijadi-Maghsoodi, T. J. Barton and N. Tillman, *J. Chem. Soc., Chem. Commun.*, 478 (1984).
101. R. Walsh, *J. Chem. Soc., Chem. Commun.*, 1415 (1982).
102. H. F. Schaefer III, *Acc. Chem. Res.*, **15**, 283 (1982).
103. R. Damrauer, C. H. DePuy, I. M. T. Davidson and K. J. Hughes, *Organometallics*, **5**, 2054 (1986).
104. I. M. T. Davidson and R. J. Scampton, *J. Organometal. Chem.*, **271**, 249 (1984).
105. M. E. Lee, M. A. North and P. P. Gaspar, *Phosphorus, Sulfur, and Silicon*, **56**, 203 (1991).
106. B.-H. Boo and P. P. Gaspar, *Organometallics*, **5**, 698 (1986).
107. D. B. Puranik and M. J. Fink, *J. Am. Chem. Soc.*, **111**, 5951 (1989).
108. G. Inoue and M. Suzuki, *Chem. Phys. Lett.*, **122**, 361 (1985).
109. J. M. Jasinski, *J. Phys. Chem.*, **90**, 555 (1986).
110. M. Weidenbruch, H. Piel, A. Lesch, K. Peters and H. G. von Schnering, *J. Organometal. Chem.*, **454**, 35 (1993).
111. K. Oka and R. Nakao, *J. Organometal. Chem.*, **390**, 7 (1990).
112. L. M. Raff, I. Noorbatcha and D. L. Thompson, *J. Chem. Phys.*, **85**, 3623 (1986).
113. J. M. Jasinski, *J. Chem. Phys.*, **86**, 3057 (1986).
114. L. M. Raff and D. L. Thompson, *J. Chem. Phys.*, **86**, 3059 (1986).
115. G. R. Gillette, G. H. Noren and R. West, *Organometallics*, **8**, 487 (1989).
116. M. Weidenbruch, H. Piel, K. Peters and H. G. von Schnering, *Organometallics*, **13**, 3990 (1994).
117. M. Weidenbruch, H. Piel, K. Peters and H. G. von Schnering, *Organometallics*, **12**, 2881 (1993).
118. F. Huppmann, W. Maringgele, T. Kottke and A. Meller, *J. Organometal. Chem.*, **434**, 35 (1992).
119. F. Huppmann, M. Noltemeyer and A. Meller, *J. Organometal. Chem.*, **483**, 217 (1994).
120. P. Zhu, M. Piserchio and F. W. Lampe, *J. Phys. Chem.*, **89**, 5344 (1985).
121. D. H. Berry and Q. Jiang, *J. Am. Chem. Soc.*, **109**, 6210 (1987).
122. S. Konieczny, K. Wrzesich and W. Wojnowski, *J. Organometal. Chem.*, **446**, 73 (1994).
123. S. Wu, G. Wu, F. Tao and Z. Lin, *Kexue Tongbao*, **31**, 1365 (1986).
124. N. Tokitoh, H. Suzuki and R. Okazaki, *J. Chem. Soc., Chem. Commun.*, 125 (1996).
125. P. L. Timms, *Chem. Soc. Rev.*, **25**, 93 (1996).
126. S. B. Church, C. G. Davies, M. Lümen, P. A. Mounier, G. Saint and P. L. Timms, *J. Chem. Soc., Dalton Trans.*, 227 (1996).
127. P. P. Gaspar and G. S. Hammond, in *Carbenes*, Vol. II (Eds. R. A. Moss and M. Jones, Jr.), Wiley, New York, 1975, pp. 207–362.
128. M. T. Nguyen, D. Sengupta and L. G. Vanquickenborne, *Chem. Phys. Lett.*, **240**, 513 (1995).
129. D. S. Rogers, K. L. Walker, M. A. Ring and H. E. O'Neal, *Organometallics*, **6**, 2313 (1987).
130. S.-H. Wu, F. Xiao and Y. Li, *Sci. China (Ser B)*, **33**, 129 (1990).
131. W. Ando, M. Fujita, H. Yoshida and A. Sekiguchi, *J. Am. Chem. Soc.*, **110**, 3310 (1988).
132. S. Zhang, P. E. Wagenseller and R. T. Conlin, *J. Am. Chem. Soc.*, **113**, 4278 (1991).
133. W. R. Winchester and P. P. Gaspar, unpublished results presented at the 30th Symposium on Organosilicon Chemistry, London, Ontario, May 30–31, 1997, Abstract A-16.
134. Y. Apeloig, unpublished results presented at the 30th Symposium on Organosilicon Chemistry, London, Ontario, May 30–31, 1997, Abstract C-4.

135. M. P. Egorov, M. B. Ezhova, S. P. Kolesnikov, O. M. Nefedov, M. B. Taraban, A. I. Kruppa and T. V. Leshina, *Mendeleev Commun.*, **4**, 143 (1991).
136. D. Lei and P. P. Gaspar, *J. Chem. Soc., Chem. Commun.*, 1149 (1985).
137. (a) H. Sakurai, Y. Kobayashi and Y. Nakadaira, *Chem. Lett.*, 1197 (1983).
    (b) H. Appler and W. P. Neumann, *J. Organometal. Chem.*, **314**, 261 (1986).
138. D. Lei and P. P. Gaspar, *Res. Chem. Intermed.*, **12**, 103 (1989).
139. S. Zhang and R. T. Conlin, *J. Am. Chem. Soc.*, **113**, 4272 (1991).
140. K. L. Bobbitt and P. P. Gaspar, *J. Organometal. Chem.*, **499**, 17 (1995).
141. K. L. Bobbitt, *Photochemical Generation of Germylenes and Silylenes: Mechanism of Germylene and Silylene Addition to 1,3-Dienes*, Doctoral Dissertation, Washington University, St. Louis, December 1990.
142. G. R. Gillette, G. H. Noren and R. West, *Organometallics*, **6**, 2617 (1987).
143. W. Ando, A. Sekiguchi, K. Hagiwara, A. Sakakibara and H. Yoshida, *Organometallics*, **7**, 558 (1988).
144. W. Ando, K. Hagiwara and A. Sekiguchi, *Organometallics*, **6**, 2270 (1987).
145. N. Takeda, H. Suzuki, N. Tokitoh, R. Okazaki and S. Nagase, *J. Am. Chem. Soc.*, **119**, 1456 (1997).
146. M. Weidenbruch, B. Brandt-Roth, S. Pohl and W. Saak, *Angew. Chem., Int. Ed. Engl.*, **29**, 90 (1990); *Polyhedron*, **10**, 1147 (1991).
147. M. Weidenbruch, B. Brand-Roth, S. Pohl and W. Saak, *J. Organometal. Chem.*, **379**, 217 (1989).
148. M. Denk, R. K. Hayashi and R. West, *J. Am. Chem. Soc.*, **116**, 10813 (1994).
149. M. Denk, R. K. Hayashi and R. West, *J. Chem. Soc., Chem. Commun.*, 33 (1994).
150. T. Dohmaru and F. W. Lampe, *J. Photochem. Photobiol., A*, **41**, 275 (1988).
151. T. Akasaka, S. Nagase, A. Yabe and W. Ando, *J. Am. Chem. Soc.*, **110**, 6270 (1988).
152. S. Nagase, T. Kudo, T. Akasaka and W. Ando, *Chem. Phys. Lett.*, **163**, 23 (1989).
153. A. Patyk, W. Sander, J. Gauss and D. Cremer, *Chem. Ber.*, **123**, 89 (1990).
154. A. C. Stanton, A. Freedman, J. Wormhoudt and P. P. Gaspar, *Chem. Phys. Lett.*, **122**, 190 (1985).
155. C. A. Arrington, J. T. Petty, S. E. Payne and W. C. K. Haskins, *J. Am. Chem. Soc.*, **110**, 6240 (1988).
156. M.-A. Pearsall and R. West, *J. Am. Chem. Soc.*, **110**, 7228 (1988).
157. W. Ando, *Pure Appl. Chem.*, **67**, 805 (1995).
158. T. Yamamoto, Y. Kabe and W. Ando, *Organometallics*, **12**, 1996 (1993).
159. M. Weidenbruch, A. Schäfer, K. Peters and H. G. von Schnering, *J. Organometal. Chem.*, **314**, 25 (1986).
160. M. Weidenbruch, B. Flintjer, S. Pohl and W. Saak, *Angew. Chem., Int. Ed., Engl.*, **28**, 95 (1989).
161. A. Schäfer, M. Weidenbruch, W. Saak and S. Pohl, *Angew. Chem., Int. Ed., Engl.*, **26**, 776 (1987).
162. A. Sakakibara, Y. Kabe, T. Shimizu and W. Ando, *J. Chem. Soc., Chem. Commun.*, 43 (1991).
163. P. Paetzold, D. Hahnfeld, U. Englert, W. Wojnowski, B. Dreczewski, Z. Pawelec and L. Walz, *Chem. Ber.*, **125**, 1073 (1992).
164. P. Boudjouk, E. Black and R. Kumarathasan, *Organometallics*, **10**, 2095 (1991).
165. M. Weidenbruch, E. Kroke, H. Marsmann, S. Pohl and W. Saak, *J. Chem. Soc., Chem. Commun.*, 1233 (1994).
166. H. Suzuki, N. Tokitoh and R. Okazaki, *J. Am. Chem. Soc.*, **116**, 11572 (1994).
167. A. Sekiguchi and R. West, *Organometallics*, **5**, 1911 (1986).
168. G. Maier, H. P. Reisenauer, K. Schöttler and U. Wessolek-Kraus, *J. Organometal. Chem.*, **366**, 25 (1989).
169. J. Heinicke and B. Gerhus, *Heteroatom Chem.*, **6**, 461 (1995).
170. M. Weidenbruch, A. Lesch and K. Peters, *J. Organometal. Chem.*, **407**, 31 (1991).
171. K. Peters, E. M. Peters, H. G. von Schnering, H. Piel and M. Weidenbruch, *Z. Kristallogr.*, **209**, 611 (1994).
172. M. Weidenbruch, P. Will, K. Peters, H. G. von Schnering and H. Marsmann, *J. Organometal. Chem.*, **521**, 355 (1996).
173. M. Weidenbruch, A. Schäfer and H. Marsmann, *J. Organometal. Chem.*, **354**, C12 (1988).
174. M. Weidenbruch, A. Lesch and H. Marsmann, *J. Organometal. Chem.*, **385**, C47 (1990).
175. T. Akasaka and W. Ando, *J. Am. Chem. Soc.*, **115**, 1605 (1993).
176. T. Akasaka, E. Mitsuhida, W. Ando, K. Kobayashi and S. Nagase, *J. Chem. Soc., Chem. Commun.*, 1529 (1995).
177. R. T. Conlin and P. P. Gaspar, *J. Am. Chem. Soc.*, **98**, 868 (1976).

178. Y. Nakadaira, T. Kobayashi, T. Otsuka and H. Sakurai, *J. Am. Chem. Soc.*, **101**, 486 (1979).
179. M. J. Michalczyk, M. J. Fink, D. J. DeYoung, C. W. Carlson, K. M. Welsh and R. West, *Silicon, Germanium, Tin and Lead Compounds*, **9**, 75 (1986).
180. R. West, M. J. Fink and J. Michl, *Science (Washington, D.C.)*, **214**, 1343 (1981).
181. R. Okazaki and R. West, *Adv. Organomet. Chem.*, **39**, 231 (1996).
182. H. B. Yokelson, D. A. Siegel, A. J. Millevolte, J. Maxka and R. West, *Organometallics*, **9**, 1005 (1990).
183. M. Yamaji, K. Hamanishi, T. Takahashi and H. Shizuka, *J. Photochem. Photobiol., A*, **81**, 1 (1994).
184. R. T. Conlin, J. C. Netto-Ferreira, S. Zhang and J. C. Scaiano, *Organometallics*, **9**, 1332 (1990).
185. (a) S. Masamune, Y. Eriyama and T. Kawase, *Angew. Chem., Int. Ed. Engl.*, **26**, 584 (1989).
    (b) S. Masamune, S. Murakami, H. Tobita and D. J. Williams, *J. Am. Chem. Soc.*, **105**, 7776 (1983).
186. Y. Apeloig, D. Bravo-Zhivotovskii, I. Zharov, V. Panov, W. J. Leigh and G. W. Sluggett, *J. Am. Chem. Soc.*, **120**, 1398 (1998).
187. H. Suzuki, N. Tokitoh and R. Okazaki, *Bull. Chem. Soc. Jpn.*, **68**, 2471 (1995).
188. H. Spitzner and R. West, unpublished results.
189. W. D. Wulff, W. F. Goure and T. J. Barton, *J. Am. Chem. Soc.*, **100**, 6236 (1978).
190. H. Sakurai, Y. Nakadaira and H. Sakaba, *Organometallics*, **2**, 1484 (1983).
191. I. M. T. Davidson and G. H. Morgan, *Organometallics*, **12**, 289 (1993).
192. A. P. Dickenson, K. E. Nares, M. A. Ring and H. E. O'Neal, *Organometallics*, **6**, 2596 (1987).
193. L. Linder, A. Revis and T. J. Barton, *J. Am. Chem. Soc.*, **108**, 2742 (1986).
194. T. J. Barton and G. P. Hussmann, *J. Am. Chem. Soc.*, **107**, 7581 (1985).
195. M. J. Fink, D. B. Puranik and P. Johnson, *J. Am. Chem. Soc.*, **110**, 1315 (1988).
196. D. Lei, Y.-S. Chen, B. H. Boo, J. Frueh, D. L. Svoboda and P. P. Gaspar, *Organometallics*, **11**, 559 (1992).
197. V. N. Khabashesku, V. Balaji, S. E. Boganov, S. A. Bashkirova, P. M. Matveichev, E. A. Chernyshev, O. M. Nefedov and J. Michl, *Mendeleev Commun.*, 38 (1992).
198. C. M. Van Zoeren, J. W. Thoman, Jr., J. I. Steinfeld and M. W. Rainbird, *J. Phys. Chem.*, **92**, 9 (1988).
199. R. I. McKay, I. S. Uichanco, A. J. Bradley, J. R. Holdsworth, J. S. Francisco, J. I. Steinfeld and A. E. W. Knight, *J. Chem. Phys.*, **95**, 1688 (1991).
200. K. K. Baldridge, J. A. Boatz, S. Koseki and M. S. Gordon, *Annu. Rev. Phys. Chem.*, **38**, 211 (1987).
201. Y. Apeloig, in *The Chemistry of Organic Silicon Compounds*, Vol. 1 (Eds. S. Patai and Z. Rappoport), Wiley, New York, 1989, p. 57.
202. R. S. Grev, *Adv. Organometal. Chem.*, **33**, 125 (1991).
203. W. Kutzelnigg, *Angew. Chem., Int. Ed. Engl.*, **23**, 272 (1984).
204. S. K. Shin, W. A. Goddard III and J. L. Beauchamp, *J. Chem. Phys.*, **93**, 4986 (1990).
205. S. K. Shin, W. A. Goddard III and J. L. Beauchamp, *J. Phys. Chem.*, **94**, 6963 (1990).
206. R. S. Grev, H. F. Schaefer III and P. P. Gaspar, *J. Am. Chem. Soc.*, **113**, 5638 (1991).
207. P. P. Gaspar, A. M. Beatty, T. Chen, T. Haile, W. T. Klooster, T. F. Koetzle, D. Lei, T. S. Lin and W. R. Winchester, unpublished results presented at the 30th Symposium on Organosilicon Chemistry, London, Ontario, May 30–31, 1997, Abstract A-16.
208. J. Kalcher and A. F. Sax, *J. Mol. Struct. (Theochem)*, **253**, 287 (1992).
209. C. J. Cremer, F. J. Dulles, J. W. Storer and S. E. Worthington, *Chem. Phys. Lett.*, **218**, 387 (1994).
210. Y. Apeloig, Private communication.
211. S. K. Shin, K. K. Irikura, J. L. Beauchamp and W. A. Goddard, III, *J. Am. Chem. Soc.*, **110**, 24 (1988).
212. R. S. Grev, G. E. Scuseria, A. C. Scheiner, H. F. Schaefer, III and M. S. Gordon, *J. Am. Chem. Soc.*, **110**, 7337 (1988).
213. J. A. Boatz and M. S. Gordon, *J. Phys. Chem.*, **94**, 7331 (1990).
214. M. S. Gordon and J. A. Boatz, *Organometallics*, **8**, 1978 (1989).
215. G. Trinquier, *J. Am. Chem. Soc.*, **112**, 2130 (1990).
216. K. G. Dyall, *J. Chem. Phys.*, **96**, 1210 (1992).
217. P. Ho and C. F. Melius, *J. Phys. Chem.*, **94**, 5120 (1990).
218. L. Nyulászi, T. Kárpáti and T. Veszprémi, *J. Am. Chem. Soc.*, **116**, 7239 (1994).

219. T. Veszprémi, L. Nyulászi and T. Kárpáti, *J. Phys. Chem.*, **100**, 6262 (1996).
220. K. Kobayashi and S. Nagase, *Organometallics*, **16**, 2489 (1997).
221. S.-z. Li, D. Moncrieff, J.-g Zhao and F. B. Brown, *Chem. Phys. Lett.*, **151**, 403 (1988).
222. W. Gabriel, P. Rosmus, K. Yamashita and K. Morokuma, *Chem. Phys.*, **174**, 45 (1993).
223. M. Spoliti, F. Ramondo, L. Bencivenni, P. Kolandaivel and R. Kumaresann, *J. Mol. Struct. (Theochem)*, **283**, 73 (1993).
224. M. Denk, J. C. Green, N. Metzler and M. Wagner, *J. Chem. Soc., Dalton Trans.*, 2405 (1994).
225. A. J. Arduengo, III, H. Bock, H. Chen, M. Denk, D. A. Dixon, J. C. Green, W. A. Herrmann, N. L. Jones, M. Wagner and R. West, *J. Am. Chem. Soc.*, **116**, 6641 (1994).
226. Y. Apeloig, M. Karni, R. West and K. Welsh, *J. Am. Chem. Soc.*, **116**, 9719 (1994).
227. S. C. Wright, D. L. Cooper, M. Sironi, M. Raimondi and J. Garratt, *J. Chem. Soc., Perkin Trans. 2*, 369 (1990).
228. J. D. Watts, G. W. Trucks and R. J. Bartlett, *Chem. Phys. Lett.*, **164**, 502 (1989).
229. J. Cioslowski, T. Hamilton, G. Scuseria, B. A. Hess, Jr., J. Hu, L. J. Schaad and M. Dupuis, *J. Am. Chem. Soc.*, **112**, 4183 (1990).
230. J. D. Watts and R. J. Bartlett, *J. Chem. Phys.*, **101**, 3073 (1994).
231. J.-M. Langlois, R. P. Muller, T. R. Coley, W. A. Goddard III, M. N. Ringnalda, Y. Won and R. A. Friesner, *J. Chem. Phys.*, **92**, 7488 (1990).
232. B. H. Greeley, T. V. Russo, D. T. Mainz, J.-M Langlois, W. A. Goddard III, R. E. Donnelly, Jr. and M. N. Ringnalda, *J. Chem. Phys.*, **101**, 4028 (1994).
233. M. Jezercak, P. M. Agrawal, D. L. Thompson and L. M. Raff, *J. Chem. Phys.*, **90**, 3363 (1989).
234. P. M. Agrawal, D. L. Thompson and L. M. Raff, *Surf. Sci.*, **195**, 283 (1988).
235. H. W. Schranz, L. M. Raff and D. L. Thompson, *J. Chem. Phys.*, **94**, 4219 (1991).
236. S. Bloński, A. Herman and S. Konieczny, *Spectrochim. Acta*, **45A**, 747 (1989).
237. A. Herman, *Chem. Phys.*, **122**, 53 (1988).
238. J. S. Tse, *J. Chem. Soc., Chem. Commun.*, 1179 (1990).
239. R. Köppe and H. Schnöckel, *Z. Anorg. Allg. Chem.*, **607**, 41 (1992).
240. C. Gonzales, H. B. Schlegel and J. S. Francisco, *Mol. Phys.*, **66**, 859 (1989).
241. R. D. Bach, M.-D. Su, E. Aldebagh, J. L. Andrés and H. B. Schlegel, *J. Am. Chem. Soc.*, **115**, 10237 (1993).
242. J. S. Francisco, *Mol. Phys.*, **73**, 235 (1991).
243. D. R. Gano, M. S. Gordon and J. A. Boatz, *J. Am. Chem. Soc.*, **113**, 6711 (1991).
244. R. Becerra, H. M. Frey, B. P. Mason, R. Walsh and M. S. Gordon, *J. Am. Chem. Soc.*, **114**, 2751 (1992).
245. S. Y. Lee and B. H. Boo, *J. Mol. Struct.*, 79 (1996).
246. P. Y. Ayala and H. B. Schlegel, *Chem. Phys. Lett.*, **225**, 410 (1994).
247. For a review see:
    (a) Y. Apeloig, M. Karni and T. Müller, in *Organosilicon Chemsitry II. From Molecules to Materials*, (Eds. N. Auner and J. Weis), VCH, Weinheim, 1996, pp. 263–288).
    (b) Y. Apeloig and T. Müller, *J. Am. Chem. Soc.*, **117**, 5363 (1995).
    (c) Y. Apeloig, in *Heteroatom Chemistry: ICHAC-2*. (Ed. E. Block), VCH, New York, 1990, pp. 27–46.
    (d) Maxka and Y. Apeloig, *J. Chem. Soc., Chem. Commun.*, 737 (1990).
248. A. Tachibana, Y. Kurosaki, K. Yamaguchi and T. Yamabe, *J. Phys. Chem.*, **95**, 6849 (1991).
249. T. P. Hamilton and H. F. Schaefer III, *J. Chem. Phys.*, **90**, 1031 (1989).
250. I. I. Zhakharov and G. M. Zhidomirov, *React. Kinet. Catal. Lett.*, **41**, 59 (1990).
251. T. R. Cundari and M. S. Gordon, *J. Phys. Chem.*, **96**, 631 (1992).
252. C. Winter and P. Millie, *Chem. Phys.*, **174**, 177 (1993).
253. For a review see O. M. Nefedov, M. P. Egorov, A. I. Ioffe, C. G. Menchikov, P. S. Zuev, V. L. Minikin, B. V. Simkin and M. N. Glukhovtsev, *Pure Appl. Chem.*, **64**, 265 (1992).
254. H. Vancik, G. Raabe, M. J. Michalczyk, R. West and J. Michl, *J. Am. Chem. Soc.*, **107**, 4097 (1985).
255. I. Dubois, *Can. J. Phys.*, **46**, 2485 (1968).
256. D. E. Milligan and M. E. Jacox, *J. Chem. Phys.*, **52**, 2594 (1970).
257. (a) G. Levin, P. K. Das, C. Bilgren and C. L. Lee, *Organometallics*, **8**, 1206 (1989).
    (b) G. Levin, P. K. Das and C. L. Lee, *Organometallics*, **7**, 1231 (1988).
258. H. Shizuka, H. Tanaka, K. Tonokura, K. Murata, H. Hiratsuka, J. Ohshita and M. Ishikawa, *Chem. Phys. Lett.*, **143**, 225 (1988).

259. M. Yamaji, K. Hamanishi, T. Takahashi and H. Shizuka, *J. Photochem. Photobiol., A*, **81**, 1 (1994).
260. P. P. Gaspar, D. Holten, S. Konieczny and J. Y. Corey, *Acc. Chem. Res.*, **20**, 329 (1987).
261. Y. Apeloig and M. Karni, *J. Chem. Soc., Chem. Commun.*, 1048 (1985).
262. An early inorganic example of a silylene complex is $HSiOH-OH_2$, obtained by the reaction of silicon atoms with water. See Z. K. Ismail, R. M. Hauge, L. Fredin, J. W. Kauffman and J. L. Margrave, *J. Chem. Phys.*, **77**, 1617 (1982).
263. G. Trinquier, *J. Chem. Soc., Faraday Trans.*, **89**, 775 (1993).
264. K. P. Steele and W. P. Weber, *J. Am. Chem. Soc.*, **102**, 6095 (1980).
265. M. Kira, T. Maruyama and H. Sakurai, *Tetrahedron Lett.*, **33**, 243 (1992).
266. K. Raghavachari, J. Chandrashekar, M. S. Gordon and K. J. Dykema, *J. Am. Chem. Soc.*, **106**, 5853 (1984).
267. For a theoretical study of amine-silylene donor-acceptor bonding, see R. T. Conlin, D. Laasko and P. Marshall, *Organometallics*, **13**, 838 (1994).
268. T. Akasaka, W. Ando, S. Nagase and A. Yabe, *Nippon Kagaku Kaishi*, 1440 (1989).
269. G. Maier, H. D. Reisenauer and H. Pacl, *Angew. Chem., Int. Ed. Engl.*, **33**, 1248 (1994).
270. H. P. Reisenauer, G. Mihm and G. Maier, *Angew. Chem., Int. Ed. Engl.*, **21**, 854 (1982).
271. G. Maier, G. Mihm, H. P. Reisenauer and D. L. Littman, *Chem. Ber.*, **117**, 2369 (1984).
272. G. Maier, S. Glatthaar and H. P. Reisenauer, *Chem. Ber.*, **122**, 2403 (1989).
273. B. D. Ruzsicska, A. Jodhan, I. Safarik, O. P. Strausz and T. N. Bell, *Chem. Phys. Lett.*, **113**, 67 (1985).
274. J. A. Dagata, D. W. Squire, C. S. Dulcey, D. S. Y. Hsu and M. C. Lin, *Chem. Phys. Lett.*, **134**, 151 (1987).
275. S. G. Bott, P. Marshall, P. E. Wagenseller, Y. Wang and R. T. Conlin, *J. Organometal. Chem.*, **499**, 11 (1995).
276. B. D. Helmer and R. West, *Organometallics*, **1**, 1463 (1982).
277. M. J. Fink and D. B. Puranik, *Organometallics*, **6**, 1809 (1987).
278. G. R. Gillette, G. Noren and R. West, *Organometallics*, **9**, 2925 (1990).
279. T. Goto, *Mater. Res. Soc. Symp. Proc.*, **297**, 3 (1993).
280. G. H. Atkinson and J. J. O'Brien, *Prepr. Pap., Am. Chem. Soc., Div. Fuel Chem.*, **34**, 483 (1989).
281. K. Tachibana, T. Shirafuji and Y. Matsui, *Jpn. J. Appl. Phys.*, **31**, 2588 (1992).
282. M. W. Markus and P. Roth, *J. Quantum. Spectrosc. Radiat. Transfer*, **52**, 783 (1994).
283. G. Inoue and M. Suzuki, *Chem. Phys. Lett.*, **105**, 641 (1984).
284. D. M. Rayner, R. P. Steer, P. A. Hackett, C. L. Wilson and P. John, *Chem. Phys. Lett.*, **123**, 449 (1986).
285. J. W. Thoman, Jr., J. I. Steinfeld, R. I. McKay and A. E. W. Knight, *J. Chem. Phys.*, **86**, 5909 (1987).
286. E. Borsella and R. Pantoni, *Chem. Phys. Lett.*, **150**, 544 (1988).
287. M. Fukushima, S. Mayama and K. Obi, *J. Chem. Phys.*, **96**, 44 (1992).
288. M. Fukushima and K. Obi, *J. Chem. Phys.*, **100**, 6221 (1994).
289. H. Ishikawa and O. Kajimoto, *J. Phys. Chem.*, **98**, 122 (1994).
290. K. Obi, M. Fukushima and K. Saito, *Appl. Surf. Sci.*, **79/80**, 465 (1994).
291. A. Kono, N. Koike, K. Okuda and T. Goto, *Jpn. J. Appl. Phys.*, **32**, L543 (1993).
292. A. Kono, N. Koike, H. Nomura and T. Goto, *Jpn. J. Appl. Phys.*, **34**, 307 (1995).
293. H. J. Mick, Y. N. Smirnov and P. Roth, *Ber. Bunsenges. Phys. Chem.*, **97**, 793 (1993).
294. H. Ishikawa and O. Kajimoto, *J. Mol. Spectrosc.*, **150**, 610 (1991).
295. H. Ishikawa and O. Kajimoto, *J. Mol. Spectrosc.*, **160**, 1 (1993).
296. C. Yamada, H. Kanamori, E. Hirota, N. Nishiwaki, N. Itabashi, K. Kato and T. Goto, *J. Chem. Phys.*, **91**, 4582 (1989).
297. W. Gabriel and P. Rosmus, *J. Phys. Chem.*, **97**, 12644 (1993).
298. R. M. Robertson and M. J. Rossi, *J. Chem. Phys.*, **91**, 5037 (1989).
299. J. Berkowitz, J. P. Greene, H. Cho and B. Ruscic, *J. Chem. Phys.*, **86**, 1235 (1987).
300. J. Berkowitz, *Acc. Chem. Res.*, **22**, 413 (1989).
301. P. Fons, T. Motooka, K. Awazu and H. Onuki, *Appl. Surf. Sci.*, **79**, 476 (1994).
302. B. E. Turner, *Astrochemistry of Cosmic Phenomena* (Ed. P. D. Singh), I. A. U., Amsterdam, 1992, pp. 181-186.
303. T. C. McInnis and L. Andrews, *J. Phys. Chem.*, **96**, 5276 (1992).
304. H. U. Lee and J. P. deNeufville, *Chem. Phys. Lett.*, **99**, 394 (1983).

305. T. Suzuki, K. Hakuta, S. Saito and E. Hirota, *J. Chem. Phys.*, **82**, 3580 (1985).
306. R. N. Dixon and N. G. Wright, *Chem. Phys. Lett.*, **117**, 280 (1985).
307. T. Suzuki and E. Hirota, *J. Chem. Phys.*, **85**, 5541 (1986).
308. A. C. Stanton, A. Freedman, J. Wormhoudt and P. P. Gaspar, *Chem. Phys. Lett.*, **122**, 190 (1985).
309. J. S. Horwitz, C. S. Dulcey and M. C. Lin, *Chem. Phys. Lett.*, **150**, 165 (1988).
310. M. Suzuki, N. Washida and G. Inoue, *Chem. Phys. Lett.*, **131**, 24 (1986).
311. R. C. Sausa and A. M. Ronn, *Chem. Phys.*, **96**, 183 (1985).
312. N. Washida, Y. Matsumi, T. Hayashi, T. Ibuki, A. Hiraya and K. Shobatake, *J. Chem. Phys.*, **83**, 2769 (1985).
313. J. T. Hougen, *J. Mol. Spectrosc.*, **138**, 251 (1989).
314. J. Carolczak and D. J. Clouthier, *Chem. Phys. Lett.*, **201**, 409 (1993).
315. M. Tanimoto, H. Takeo, C. Matsumura, M. Fujitake and E. Hirota, *J. Chem. Phys.*, **91**, 2102 (1989).
316. H. Sekiya, Y. Nishimura and M. Tsuji, *Chem. Phys. Lett.*, **176**, 477 (1991).
317. H. Bock, M. Kremer, M. Dolg and H.-W. Preuss, *Angew. Chem., Int. Ed. Engl.*, **30**, 1186 (1991).
318. G. Raabe, H. Vancik, R. West and J. Michl, *J. Am. Chem. Soc.*, **108**, 671 (1986).
319. Y. Peng, S. N. Vogel and J. E. Carlstrom, *Astrophys. J.*, **455**, 223 (1995).
320. J. Chernicharo, V. Bujarrabul and R. Lucas, *Astron. Astrophys.*, **249**, L27 (1991).
321. E. Bouisset, J. M. Esteva, R. C. Karnatak, J. P. Connersde, A. M. Flank and P. Lagarde, *J. Phys. B*, **34**, 1609 (1991).
322. J. M. Jasinski, R. Becerra and R. Walsh, *Chem. Rev.*, **95**, 1203 (1995).
323. R. Becerra and R. Walsh, in *Research in Chemical Kinetics* (Eds. R. G. Compton and G. Hancock), Vol. 3, Elsevier Science, Amsterdam, 1995, p. 263.
324. K. N. Houk, N. G. Rondan and J. Mareda, *Tetrahedron*, **41**, 1555 (1985).
325. R. Becerra, H. M. Frey, B. P. Mason, R. Walsh and M. S. Gordon, *J. Chem. Soc., Faraday Trans.*, **91**, 2723 (1995).
326. R. Walsh, *Organometallics*, **8**, 1973 (1989).
327. I. Safarik, V. Sandhu, E. M. Lown, O. P. Strausz and T. N. Bell, *Res. Chem. Intermed.*, **14**, 105 (1990).
328. M. Ishikawa, K.-L. Nakagawa, R. Enokida and M. Kumada, *J. Organomet. Chem.*, **201**, 151 (1980).
329. J. A. Hawari, M. Lesage, D. Griller and W. P. Weber., *Organometallics*, **6**, 880 (1987).
330. C. Zybill, H. Handwerker and H. Friedrich, *Adv. Organometal. Chem.*, **36**, 229 (1994).
331. C. Zybill and C. Liu, *Synlett.*, 687 (1995).
332. C. Zybill, *Top. Curr. Chem.*, 160 (1991).
333. T. D. Tilley, in *The Silicon–Heteroatom Bond* (Eds. S. Patai and Z. Rappoport), Wiley, New York, 1991, p. 309.
334. T. D. Tilley, in *The Chemistry of Organic Silicon Compounds* (Eds. S. Patai and Z. Rappoport), Chap. 24, Wiley, Chichester, 1989, p. 1415.
335. T. Carré, G. Cerrau, C. Chuit, R. J. P. Corriu and C. Réyé, *Angew. Chem., Int. Ed. Engl.*, **28**, 489 (1989).
336. H. Handwerker, C. Leis, R. Probsty, P. Bissinger, A. Grohmann, P. Kiprof, E. Herdtweck, J. Bümel, N. Auner and C. Zybill, *Organometallics*, **12**, 2162 (1993).
337. S. O. Grumbine, T. D. Tilley and A. L. Rheingold, *J. Am. Chem. Soc.*, **115**, 358 (1993).
338. N. Ogino and H. Tobita, *Adv. Organometal. Chem.*, **42**, 223 (1998).
339. K. Triplett and M. D. Curtis, *J. Am. Chem. Soc.*, **97**, 5747 (1975).
340. W. A. Herrmann, *Adv. Organomet. Chem.*, **20**, 159 (1982).
341. C. Morterra and M. J. D. Low, *J. Phys. Chem.* **73**, 321, 327 (1989).
342. C. Morterra and M. J. D. Low, *Ann. N. Y. Acad. Sci.*, **220**, 173 (1973).
343. M. J. D. Low, Y. E. Rhodes and P. O. Orphanos, *J. Catal.*, **40**, 236 (1975).
344. V. A. Radzig, *Colloid Surf: A: Physicochem. Eng. Aspects*, **74**, 91 (1993).
345. V. A. Radzig, E. G. Baskis and V. A. Karobeu, *Kinet. Catal.*, **36**, 142, 568 (1995).
346. V. A. Radzig, *Kinet. Catal.*, **37**, 310, 418 (1996).
347. A. A. Bobyshev and V. A. Radzig, *Khim. Fiz.*, **7**, 950 (1988).
348. Yu. V. Razskazovskii, M. V. Roginskaya and M. Ya. Mel'nikov, *J. Organometal. Chem.*, **486**, 249 (1995).
349. M. V. Roginskaya, Yu. V. Razskazovskii and M. Ya. Mel'nikov, *Kinet. Catal.*, **33**, 521 (1992).
350. Free-radical sites are also formed by breaking of Si—O bonds, but evidently these are far less abundant than silylene centers.

351. V. A. Radzig and A. V. Bystrikov, *Kinet. Catal.*, **19**, 713 (1978).
352. E. G. Rochow, *Silicon and Silicones*, Springer, Berlin, 1987.
353. M. P. Clarke, *J. Organomet. Chem.*, **376**, 165 (1989).
354. M. P. Clarke and I. M. T. Davidson, *J. Organomet. Chem.*, **408**, 149 (1991).
355. K. M. Lewis, D. McLeod, B. Kanner, J. L. Falconer and T. Frank, in *Studies in Organic Chemistry 49. Catalyzed Direct Reactions at Silicon* (Eds. K. M. Lewis and D. S. Rethwisch), Elsevier, Amsterdam, 1993, pp. 333–340.
356. Y. Ono, M. Okano, N. Watanabe and E. Suzuki, in *Silicon for the Chemical Industry II* (Eds. H. A. Oye, H. M. Rong, L. Nygard, G. Schuessler and J. K. Tuset), Tapir Forlag, Trøndheim, Norway, 1994, pp. 185–186.
357. M. Okamoto, E. Suzuki and Y. Ono, *J. Catal.*, **145**, 537 (1994).
358. M. Okamoto, N. Watanabe, E. Suzuki and Y. Ono, *J. Organometal. Chem.*, **515**, 51 (1996).
359. M. Okamoto, N. Watanabe, E. Suzuki and Y. Ono, *J. Organometal. Chem.*, **489**, C12 (1995).
360. (a) P. Jutzi, D. Kanne and C. Krüger, *Angew. Chem., Int. Ed. Engl.*, **25**, 164 (1986).
    (b) P. Jutzi, U. Holtmann, D. Kanne, C. Krüger, R. Blom, R. Gleiter and I. Hyla-Krispin, *Chem. Ber.*, **122**, 1629 (1989).
361. (a) P. Jutzi, in *Organosilicon Chemistry: From Molecules to Materials* (Eds. N. Auner and J. Weis), VCH, Weinheim, 1994, pp. 87–92.
    (b) P. Jutzi, D. Eikenberg, B. Neumann and H. G. Stammler, *Organometallics*, **15**, 3659 (1998) and references cited therein.
    (c) M. Tacke, C. H. Klein, D. J. Stafkens, A. Oskam, P. Jutzi and E. A. Bunte, *Z. Anorg. Allg. Chem.*, **619**, 865 (1993).
362. H. M. Karsch, U. Keller, S. Gamper and G. Müller, *Angew. Chem., Int. Ed. Engl.*, **29**, 295 (1990).
363. D. H. Harris and M. F. Lappert, *J. Chem. Soc., Chem. Commun.*, 895 (1974).
364. M. S. J. Gynane, D. H. Harris, M. E. Lappert, P. P. Power, P. Riviere and M. Riviere-Baudet, *J. Chem. Soc., Dalton Trans.*, 2004 (1977).
365. D. H. Harris, M. F. Lappert, J. B. Pedley and G. J. Sharp, *J. Chem. Soc., Dalton Trans.*, 945 (1976).
366. A. J. Arduengo III, H. V. Rasika Dias, R. L. Harlow and M. Kline, *J. Am. Chem. Soc.*, **114**, 5330 (1992).
367. M. Denk, R. Lennon, R. Hayashi, R. West, A. V. Belyakov, H. P. Verne, A. Haaland, M. Wagner and N. Metzler, *J. Am. Chem. Soc.*, **116**, 2691 (1994).
368. B. Gerhus, M. F. Lappert, J. Heinicke, R. Boese and D. Blaser, *J. Chem. Soc., Chem. Commun.*, 11931 (1996).
369. J. Heinicke and A. Oprea, Abstracts of Papers, 11th International Symposium on Organosilicon Chemistry, Montpellier, France, Sept. 1–6, 1996, p. PA 28.
370. B. Gerhus, P. B. Hitchcock, M. F. Lappert, J. Heinicke, R. Boese and D. Bläser, *J. Organomet. Chem.*, **521**, 211 (1996).
371. C. Heinemann, T. Müller, Y. Apeloig and H. Schwartz, *J. Am. Chem. Soc.*, **118**, 2023 (1996).
372. (a) C. Boehme and G. Frenking, *J. Am. Chem. Soc.*, **118**, 2039 (1996).
    (b) C. Heinemann, W. A. Hermann and W. Thiel, *J. Organomet. Chem.*, **475**, 73 (1994).
373. (a) A. J. Arduengo, D. A. Dixon, R. L. Harlow, K. K. Kumashiro, C. Lee, W. P. Power and K. Zilm, *J. Am. Chem. Soc.*, **116**, 6361 (1994).
    (b) A. J. Arduengo, H. V. Rasika Dias, D. A. Dixon, R. L. Harlow, W. T. Klooster and T. F. Koetzle, *J. Am. Chem. Soc.*, **116**, 6812 (1994).
374. M. Haaf and R. West, unpublished investigation.
375. I. Hemme and U. Klingebiel, *Adv. Organomet. Chem.*, **39**, 159 (1996).
376. M. Denk and R. West, unpublished investigations.
377. B. Gerhus, P. B. Hitchcock and M. E. Lappert, Abstracts of Papers, 11th International Symposium on Organosilicon Chemistry, Montpellier, France, Sept. 1–6, 1996, p. OB 2.
378. N. Metzler and M. Denk, *J. Chem. Soc., Chem. Commun.*, 2657 (1996).
379. J. M. Jasinski and S. M. Gates, *Acc. Chem. Res.*, **24**, 9 (1991).
380. G. H. Atkinson and J. J. O'Brien, *J. Phys. Chem.*, **92**, 5782 (1988).
381. M. Heintze and G. H. Bauer, *J. Phys. D*, **28**, 2470 (1995).
382. S. Bismo, P. Duverneuil, L. Pibouleau, S. Domenech and J. P. Couderc, *Chem. Eng. Sci.*, **47**, 2921 (1992).
383. C. Azzaro, P. Duverneuil and J.-P. Couderc, *J. Electrochem. Soc.*, **139**, 305. (1992).
384. C. Cobianu, P. Cosimin, R. Plugaru, D. Descalu and J. Holloman, *Electrochem. Soc. Proc.*, **96**, 300 (1996).

385. D. Orlicki, V. Hlavacek and H. J. Viljoen, *J. Mater. Res.*, **7**, 2160 (1992).
386. S. Veprek, K. Schopper, O. Ambocher, W. Rieger and M. G. J. Veprek-Heijman, *J. Electrochem. Soc.*, **140**, 1935 (1993).
387. L. Layeillon, P. Duverneuil, J. P. Couderc and B. Despax, *Plasma Sources Sci. Technol.*, **3**, 61 (1994); L. Layeillon, A. Dollet, J. P. Couderc and B. Despax, *Plasma Sources Sci. Technol.*, **3**, 72 (1994).
388. Y. Sawado, T. Akiyama, T. Ueno, K. Kamisako, K. Kuroiwa and Y. Tarui, *Jpn. J. Appl. Phys.*, **33**, 950 (1994).
389. C. J. Wu, I. V. Ionova and E. A. Carter, *Surf. Sci.*, **295**, 64 (1993).
390. S. Oikawa, S. Ohtsuka and M. Tsuda, *Appl. Surf. Sci.*, **60/61**, 29 (1992).
391. Y. Toyoshima, K. Arai, A. Matsuda and K. Tanaka, *Appl. Phys. Lett.*, **56**, 1540 (1990); **57**, 1028 (1990).
392. Y. Toyoshima, A. Matsuda and K. Arai, *J. Noncrystal. Solids*, **164–166**, 103 (1993).
393. S. Miyazaki, H. Shin, Y. Miyoshi and M. Hirose, *Jpn. J. Appl. Phys.*, **34**, 787 (1995).
394. S. Veprek and M. G. J. Veprek-Heijman, *Appl. Phys. Lett.*, **56**, 1766 (1990).
395. R. M. Robertson and M. J. Rossi, *Appl. Phys. Lett.*, **54**, 185 (1989).
396. J. I. Steinfeld, *Spectrochim. Acta*, **46A**, 589 (1990).
397. P. P. Gaspar, J.-P. Hsu, S. Chari and M. Jones, Jr., *Tetrahedron*, **41**, 1479 (1985).
398. T. R. Dietrich, S. Chiussi, M. Marek, A. Roth and F. J. Comes, *J. Phys. Chem.*, **95**, 9302 (1991).
399. W. L. M. Weerts, M. H. J. M. de Croon and G. B. Marin, *Chem. Eng. Sci.*, **51**, 2109 (1996).
400. E. Scheid, J. J. Pedroviejo, P. Duverneuil, M. Gueye, J. Samitier, A. El Hassani and D. Bielle-Daspet, *Mater. Sci. Eng.*, **B17**, 72 (1993).
401. R. J. Bogaert, T. W. F. Russell, M. T. Klein, R. E. Rocheleau and R. N. Baron, *J. Electrochem. Soc.*, **136**, 2960 (1989).
402. J. E. Johannes and J. G. Ekerdt, *J. Electrochem. Soc.*, **141**, 2135 (1994).
403. T. Shirafuji, S. Nakajima, Y. F. Wang, T. Genji and K. Tachibana, *Jpn. J. Appl. Phys.*, **32**, 1546 (1993).
404. E. Dehan, J. J. Pedroviejo, E. Scheid and J. R. Morante, *Jpn. J. Appl. Phys.*, **34**, 4666 (1995).
405. J. D. Chapple-Sokol, C. J. Goiunta and R. G. Gordon, *J. Electrochem. Soc.*, **136**, 2993 (1989).
406. H. K. Moffat, T. F. Kuech, K. F. Jensen and P.-J. Wang, *J. Crystal Growth*, **93**, 594 (1988).
407. S. Tanaka, *J. Am. Ceram. Soc.*, **73**, 3046 (1990).
408. C. C. Chen, J. L. Yu, C. Y. Lee, C. S. Liu and H. T. Chiu, *J. Mater. Chem.*, **2**, 983 (1992).
409. J. Belzner, *J. Organomet. Chem.*, **430**, C51 (1992).
410. J. Belzner and H. Ihmels, *Tetrahedron Lett.*, 6541 (1993).
411. J. Belzner, H. Ihmels, B. O. Kneisel, R. O. Gould and R. Herbst-Irmer, *Organometallics*, **14**, 305 (1995).
412. J. Belzner, H. Ihmels and M. Noltemeyer, *Tetrahedron Lett.*, **36**, 8187 (1995).
413. J. Belzner, H. Ihmels, B. O. Knebel and R. Herbst-Irmer, *Chem. Ber.*, **129**, 125 (1996).
414. J. Belzner, H. Ihmels, L. Pauletto and M. Noltemeyer, *J. Org. Chem.*, **61**, 3315 (1995).
415. J. Belzner, D. Schaer, B. O. Kneisel and R. Herbst-Irmer, *Organometallics*, **14**, 1840 (1995).
416. K. Tamao, K. Nagata, M. Asahara, A. Kawachi, Y. Ito and M. Shiro, *J. Am. Chem. Soc.*, **117**, 11592 (1995).
417. K. Tamao, M. Asahara and A. Kawachi, *J. Organomet. Chem.*, **521**, 325 (1996).
418. Y. Apeloig and K. Albrecht, *J. Am. Chem. Soc.*, **117**, 7263 (1995).
419. Y. Apeloig and M. Karni, *Organometallics*, **16**, 310 (1997).
420. H. Leclercq and I. Dubois, *J. Mol. Spectr. Sc.*, **76**, 39 (1979).
421. R. Damrauer, C. H. DePuy, S. E. Barlow and S. Gronert, *J. Am. Chem. Soc.*, **110**, 2005 (1988).
422. R. Srinivas, D. Sülzle and H. Schwarz, *J. Am. Chem. Soc.*, **113**, 52 (1993).
423. E. Herbst, T. J. Millar, S. Wlodek and D. K. Bohme, *Astron. Astrophys.*, **222**, 205 (1989).
424. C. D. Sherrill and H. F. Schaefer III, *J. Phys. Chem.*, **99**, 1949 (1995).
425. A. A. Bengali and D. G. Leopold, to appear; A. A. Bengali, PhD Thesis, University of Minnesota, 1992.
426. M. T. Nguyen, D. Sengupta and L. G. Vanquickenborne, *Chem. Phys. Lett.*, **244**, 83 (1985).
427. R. Stegmann and G. Frenking, *J. Comput. Chem.*, **17**, 781 (1996).
428. W. W. Schoeller and J. Strutwolf, *J. Mol. Struct.*, **305**, 127 (1994).
429. M. Izuha, S. Yamamoto and S. Saito, *J. Chem. Phys.*, **105**, 4923 (1996).
430. W. W. Harper, E. A. Ferrall, R. K. Hilliard, S. M. Stogner, R. S. Grev and D. J. Clouthier, *J. Am. Chem. Soc.*, **119**, 8361 (1997).

Author index

This author index is designed to enable the reader to locate an author's name and work with the aid of the reference numbers appearing in the text. The page numbers are printed in normal type in ascending numerical order, followed by the reference numbers in parentheses. The numbers in *italics* refer to the pages on which the references are actually listed.

Aamodt, L.C. 588(197), *594*
Aasen, A. 2374(43), *2399*
Abboud, J.L.M. 1477(269), *1533*
Abboud, K.A. 2079, 2092, 2093(126), *2125*
Abdali, A. 1860(355), *1868*
Abdel Halim, S.H. 1374, 1376, 1377(90, 91), 1380, 1381(91), 1398, 1400(171), 1401(90, 91, 171), 1403(171), 1434, 1435(241), *1440, 1442, 1444*
Abdelqader, W. 1700, 1701(57), *1786*
Abdesaken, F. 3(4), *95*, 250(305–307), *264*, 859(3), 880(3, 87, 91), 881(3, 87), 977, 978, 981(3), 985, 986, 992, 996(3, 91), *1054–1056*, 1064, 1068(1), 1080(47), *1101, 1102*, 1145(15), 1146(22), *1181*, 1601(24, 25), 1602, 1604(31), 1609(24, 31), 1645(24), *1661*, 2404(11, 12), *2458*
Abdol Latif, L. 2110, 2120(223), *2127*
Abdulrahman, M. 1697(36), *1786*
Abe, H. 559(30), *590*
Abe, T. 1601, 1602(12), 1614(94, 95), 1620(12, 115), 1623(129), 1651, 1652(12), *1660, 1662, 1663*
Abe, Y. 2251, 2253(27b), *2311*, 2333(97), 2334(105), 2349, 2350(207), 2353(278), *2358, 2361, 2362*, 2475(60), *2560*
Abel, E.W. 2130, 2139(1), 2140(1, 52), 2141, 2143, 2144(52), *2170, 2172*
Abele, E. 1740(171), *1789*
Abello, L. 2339(126), *2358*
Abicht, H.P. 1743(219), *1790*
Abiko, A. 1671(38), *1683*
Abraham, W.-R. 2376(47), *2399*

Abram, T.S. 661(136), 667(143), 669(146), *700, 701*
Abramoff, B. 2343(152, 153), *2359*
Abramovich, R. 1019(318, 319), *1061*
Abrash, S.A. 1318(66), *1334*
Abriel, W. 2152, 2153(121, 122), *2173*
Abronin, I.A. 240(258), *263*, 580, 584(167), *593*
Abstreiter, G. 2194(77), *2215*
Abu-Eid, M. 1700(55), *1786*
Abu-Freih, A. 363(32a), *426*, 482(239), *493*, 611, 613(40), 634(87), *698, 699*
Acemoglu, M. 466(180), *492*
Achiba, Y. 1120, 1121(133), *1139*, 1302(203), *1310*, 1942(19), *1960*
Achiwa, K. 1754, 1755(245), *1790*
Ackerman, J.H. 465(177), *492*
Ackermann, K. 243(268), *263*, 2027(203), *2036*
Adachi, K. 1590(88), *1596*
Adaci, T. 2075, 2091(115), *2125*
Adam, A. 310(134), *353*
Adam, W. 449(121, 122), 467(185), *491, 492*, 1841(255), 1842(256), 1846(278), 1858(341), *1866–1868*
Adamovich, S.N. 1457, 1458, 1467(121), 1488(310), 1491(321), 1494(335), 1496(344), 1497(349, 350), 1499(354, 355), 1513(121), *1528, 1534, 1535*
Adams, B.R. 278(28), *351*
Adams, H. 1807(85, 87), *1863*
Adams, J. 436, 479(63), *489*

Adams, R.D. 2069, 2071(78), 2075(106–108), 2091(106, 107, 184), *2124–2126*
Adamson, C.L. 1405(173), *1443*
Adcock, W. 361(17), 365(40), 380(70, 73, 75), 381(76), *426, 427*, 641(102), 642(102, 106, 108), 643(106, 108), 644(108), *700*
Adiwidjaja, G. 434, 445(26), *489*, 1679(110), *1684*, 1872(18), *1892*
Adjé, N. 1671(33), *1683*
Adley, A.D. 236(254), *263*
Adlington, M.G. 1740(178), *1789*
Adlington, R.M. 1675(80), *1684*
Afshari, M. 1807(83), *1863*
Agars, R.F. 2226(53), *2242*
Agarunov, M.J. 164(76), *179*
Agaskar, P.A. 1923(102), 1924(103, 104), 1925(111), *1928*, 2289(136a–c), 2296(148), *2315*, 2337(124), *2358*
Ager, D.J. 415, 418(182), *429*, 456(143), *491*, 1608(69), 1626(134), *1662, 1663*, 1794(21), *1862*
Aggarwal, E.H. 1899(35), *1926*
Aggarwal, V.K. 401(142), *428*
Agolini, F. 1603(38, 41), 1605(38), *1661*
Agosta, W.C. 772(151, 152), 773(153), *778*
Agrawal, P.M. 2510(233, 234), *2564*
Agrofoglio, L. 435, 470(58), *489*
Agwaramgbo, E.L.O. 419(198), *429*, 1831(200), *1865*
Aharonson, N. 2320, 2322(19), 2327(19, 52), 2340(132), 2348(19), *2355, 2356, 2359*
Ahlberg, P. 683(155), *701*
Ahlrichs, R. 762(131), *777*, 1341, 1349–1351, 1357, 1361, 1368(30b), *1438*, 1878, 1886(50), *1893*
Ahmed, K.J. 437, 484(75), *490*
Ahmed, M. 174, 175(138), *180*
Ahmed, M.E. 425(215), *430*, 1679(108), *1684*
Aihara, H. 1826(173), 1856(327), *1865, 1868*
Aitken, C. 2042, 2045(13), *2122*, 2290(137), *2315*
Aitken, C.T. 2047(26), *2123*
Aiube, Z.H. 577(156), *593*
Aizenberg, M. 1074(29), *1102*, 1872(21), *1892*, 2102(198, 199), 2108(198, 199, 219, 220), *2127*
Aizpurua, J.M. 423(209), *430*, 446(113), 460(162), 476(225), *490, 491, 493*, 1226(115), *1232*
Akagi, K. 1325(133), *1335*
Akasaka, K. 277(23), *350*
Akasaka, T. 200(102), *259*, 804(77), *824*, 1169(178), *1185*, 1301(198, 199), 1302(202–204), *1310*, 1897(29–31), 1899(37), 1900(31), 1901(29–31), *1926*, 1931(11), 1934(12), 1936(13), 1940(14–17), 1942(18, 19), *1960*, 2452(191, 192), *2461*,
2493(151, 152), 2496(175, 176), 2520(151, 268), 2523(151), *2562, 2565*
Akasaki, Y. 1091(58c), *1103*
Akerman, E. 436, 482(67, 68), *490*
Åkermark, B. 760(123), *777*
Akhrem, A.A. 1996(108b), *2033*
Akhrem, I.S. 1825(162), 1838(226), *1865, 1866*
Akiba, E. 345–347(153), *354*
Akiba, K.-y. 1850(299, 300), *1867*
Akita, M. 1418(218a), *1444*, 1797(46), *1862*, 2080(137–141), *2125*
Akiyama, T. 435, 477(59), *489*, 1849(297), 1852(314, 315), *1867, 1868*, 2546(388), *2568*
Akkerman, O.S. 1151(49), *1182*
Akovali, G. 2224(42b), *2241*
Aksamentova, T.N. 1455, 1467(112), *1528*
Akutsu, K. 2272(92c), *2314*
Albach, R.W. 2136, 2137, 2147(31), *2171*
Albanov, A. 2439(145), *2460*
Albanov, A.I. 1374, 1376(92a), 1390(128–131), 1391(140), 1392(129, 131), 1393(151, 153), 1394(151), 1395(153), 1396(151), 1398, 1404(153), *1440–1442*, 1450(40), 1454(105, 108, 109), 1455(108, 109), 1467(105, 108, 109), 1490(317), 1495(109, 340), 1500(109), 1501(369), *1526, 1528, 1534–1536*, 1691(13), *1786*
Alberti, A. 1552, 1553, 1555(50), 1558(64), *1578*
Albertsen, J. 1884(69), *1893*, 1896, 1904(3), *1925*
Albini, A. 1290(172), *1309*
Albinsson, B. 1314(34), *1333*
Albrecht, J. 1593(112), *1597*
Albrecht, K. 707, 709(18), *775*, 1049(368), *1062*, 1128, 1129(228), *1141*, 2555(418), *2568*
Albright, T.A. 77, 79(152a), *100*
Alcaraz, G. 761(130), 763(136), *777*
Alcaraz, J.M. 1998(121), *2034*
Aldebagh, E. 2510(241), *2564*
Ald Elhafez, F.A. 1651(218), *1665*
Alder, R.W. 1466, 1467(175), *1530*
Aldinger, F. 203(112), *259*, 2246, 2250, 2253, 2254(6e), 2261(54c), 2267(85b), 2269(6e), 2271(85b, 90), 2273, 2285(6e), *2310, 2312–2314*
Aleksandrov, Y.A. 435, 456(50), *489*
Alekseev, N.V. 183(14), *257*, 284(51), *351*, 1340, 1351, 1381(1), *1437*, 1449(20), 1450(34, 35), 1460(20), 1461(34, 35), 1463, 1466(20), 1473(20), 1474(20, 244), 1475(20, 244, 253), 1508, 1509(391), 1523(20), *1525, 1532, 1537*
Alexander, E.K. 209(138), *260*

Author index

Alexander, R.S. 1266(92), *1307*
Alexeev, N.V. 245(284), *264*, 586(193), *594*
Alezeev, N.V. 240(257), *263*
Al'fonsov, V.A. 1870(6), *1891*
Alford, J.M. 1120(125), *1138*
Al-Ghamdi, A.M.S. 2224(41c), *2241*
Al-Gurashi, M.A.M.R. 454(136), *491*, 577(145), *592*
Alhassan, M.I. 397(132), *428*
Ali, S. 1979(57f), *2032*
Alias, S.B. 1925(110), *1928*
Aliev, A.E. 333(146), *353*
Aliev, M.A. 1996(108c), *2033*
Aliev, Z.G. 2152(116), *2173*
Alijah, A.A. 1107(12), *1136*
Al-Inaid, S.S. 2131(12), *2171*
Al-Juaid, S.S. 196(74), 203(111), 222(191, 194), 245(288), *258, 259, 261, 264*
Alla, M.A. 1474, 1476–1478(263), *1532*
Alla, M.J. 1381(107), *1440*
Allaf, A.W. 1929, 1930(1a), *1960*
Al-Laham, M.A. 1119(112), *1138*
Allcock, H. 2246(4a, 4b), *2310*
Allcock, H.A. 2237(118), *2244*
Allen, L.C. 123(24f), *151*, 835(41), *854*, 1157(84), *1183*, 1911(64, 65), *1927*
Allen, M.J. 1120, 1121(129), *1138*
Allen, S.R. 1723(128), *1788*
Allen, W.D. 66–68(119), *99*
Allen, W.N. 607(27, 28), 608(27), *698*
Allendorf, M.D. 155(22, 23, 25), 159–161(22), 164(23, 25), 166(25), 167, 168(22), 169(114), 170(25), 171–174(22), 175(22, 25), *178, 180*, 901(148), *1057*
Allgeier, A.M. 198(84), *259*
Allinger, N.L. 2429(94, 95), *2459*
Allison, C. 996(284), *1060*
Allison, J.S. 2054(40), *2123*
Allred, A.L. 356(4), *426*, 806(85), *824*, 1207(50), 1211(63, 64, 67), 1213(72), 1214(77), *1230, 1231*, 1312, 1313(14), *1332*, 2193(69), *2215*
Allum, K.G. 2340(140), *2359*
Al-Mansour, A.I. 454(136), *491*, 577(158), *593*
Almennigen, A. 116(46), *118*
Almenningen, A. 2181(19), *2214*
Almlöf, J. 72(145), *100*
Almond, G.G. 315, 317, 318(139), *353*
Almond, M.J. 1144(9), 1162(116), *1181, 1183*
Almstead, N.G. 1802(70), *1863*
Al-Nasr, A.A. 203(111), *259*
Al-Nasr, A.K.A. 222(191), 245(288), *261, 264*
Alonso, B. 1758(257), *1791*, 2337(120), *2358*
Alper, H. 1682(130), *1685*, 1719(112), 1743(215), 1767(285), 1776(327), 1778(333), *1788, 1790–1792*, 1840(248), *1866*, 2322(42), *2356*
Al-Rubaiey, N. 161(59), 163(59, 75), *178, 179*, 2479(80), *2560*
Alste, J. 361(17), *426*
Al-Subu, M. 1700(55), *1786*
Alt, C. 2341(142), *2359*
Alt, H. 1212, 1213(70), *1231*, 1322(99), *1334*
Alt, M. 434, 447(28), *489*, 752, 753, 755(114), 756(114, 117), 757(118, 120, 122), 758(122), 760(120), *777*, 1276(131), *1308*
Altstein, M. 2320, 2322(19), 2327(19, 52), 2340(132), 2348(19), *2355, 2356, 2359*
Alverez-García, L.J. 1797(47), *1862*
Alyev, I.Y. 1188(4), *1229*
Amano, S. 2351(238), 2354(282), *2361, 2362*
Amberg, W. 476(219), *493*
Amberger, K. 1743(224), *1790*
Amberg-Schwab, S. 2332(88), 2333, 2334(101), 2335(110), 2337(101, 110), 2339(88), *2357, 2358*
Ambocher, O. 2546(386), *2568*
Ambrus, L. 1470(208), *1531*
Améduri, B. 1769(288), *1791*
Amer, I. 2322(41), *2356*
Ammon, H.L. 1294(183), *1309*
Amor, F. 2151, 2152(117), 2159(149), *2173, 2174*
Amos, R.D. 66, 68(120a), *99*, 1125(173), *1139*, 1163(133), *1184*
Amosova, S.V. 1501(368, 369), *1536*
Ananthan, S. 1834(211), *1866*
Ancelle, J. 847(71), *855*
Ancens, G. 1362(53a), 1368(56), *1438*
Ancens, G.A. 1362(54a, 54b), *1438*
Ancillotti, M. 1608(68), *1662*
Ancker, T.van den 1377, 1378, 1380(97a), *1440*
Andersen, N.H. 1499(353), *1535*
Andersh, B. 1879(55), *1893*
Anderson, C.L. 1638(169), 1639(183, 184), *1664*
Anderson, D.A. 479(233), *493*
Anderson, D.G. 416(186), *429*, 1875(35), *1892*
Anderson, D.W.N. 296(82), *352*
Anderson, H.L. 1958(28), *1960*
Anderson, J.S. 1162(110), *1183*
Anderson, K.R. 1218(95), *1231*, 1238(21), *1306*, 1331(192), *1336*, 2475, 2477, 2482, 2483(56), *2560*
Anderson, L.R. 1120(123), *1138*
Anderson, O.P. 578(160), *593*
Anderson, P.P. 2223(30b), *2241*
Anderson, R. 704, 739(6), *775*, 1675(83), *1684*
Anderson, R.A. 2425(83, 84), *2459*
Anderson, S.B. 704, 739(6), *775*, 1675(83), *1684*
Anderson, T.J. 784, 786, 787(22), *822*

Andersson, C. 1840(252), *1866*
Andersson, P.G. 472(198), *492*
Ando, H. 1291(175), *1309*
Ando, I. 293(77), 294, 295(78), 296(81), 302(78), 329, 330(144), *352, 353*
Ando, K. 1726, 1727(133), *1788*
Ando, W. 20(51a, 51b), *96*, 186(39), 194(63, 64, 68), 200(102), 205(117), *258, 259*, 277(23, 24), 278(25–27), 329(26), *350, 351*, 709(19), 713(29–31, 33, 34), 714(29, 31, 36a, 36b), 715(34), 716(39, 40), 718(29, 30), 719(42, 43), 720(31, 42), 722(36a, 36b, 45, 46, 48, 49), 724, 726(49, 50), 727(40), 735(69), 737(69–71), 741(33, 69, 70), 743(33, 48–50, 85, 86), 744(48, 49, 85, 86), 748(33, 39, 40, 98), 749(40, 85, 86), 750(39, 40, 85, 86, 98), 751(101), 752(33, 113, 115), 753(33, 113), 755(116), *775–777*, 791(42), 804(77), *822, 824*, 904(164), 905(165), 906(168, 169), 908(169), 931(169, 195), 969(165), 973(165, 168), 992(164, 165, 277), 996(164, 165), 1030(344), 1051(375), *1057, 1058, 1060–1062*, 1075(32c), 1080(43b, 43c), *1102*, 1146(19), 1158(89), 1169(170, 178), *1181, 1183, 1185*, 1238(23), 1242(29), 1276(125, 126, 129, 130), 1278(135, 136), 1282(141), 1299(195), 1301(198–201), 1302(202), *1306, 1308, 1310*, 1331(187–189), *1336*, 1601, 1604, 1612, 1621(15), *1660*, 1881(59), 1883(68), 1887(59, 93), *1893, 1894*, 1896(4, 5), 1897(6, 12–15, 28–31), 1899(37), 1900(31), 1901(29–31, 39), 1906(4, 5), 1911(12–15, 75), 1913(13), 1914(15), 1915(14, 15, 75), 1916(13, 14, 75, 84), 1918(88), 1923(6, 88, 91), *1925–1927*, 1931(11), 1934(12), 1936(13), 1940(14–16), 1942, 1944(20), 1945(21), 1947(21, 22), 1951(23, 24), 1955(25), 1959(30), *1960*, 1968, 1969, 1976(24a, 24b), 1977(24a), 1996(24a, 24b), 2012(168a, 168b), *2030, 2035*, 2089, 2092, 2094(181, 182), *2126*, 2403(1–3), 2404(4, 5, 7, 8), 2405(13–15), 2406(19), 2407(26, 27), 2408(29, 30), 2409, 2410(31, 32), 2411(32), 2412(36–38), 2413(37), 2414(40), 2415(41, 42), 2416(49–51, 53), 2417(53), 2418(50, 51, 54), 2420(54, 64, 70), 2421(61–64), 2422(62), 2423(70), 2425(63, 80–82), 2427(81), 2428, 2429(91), 2430(109, 110), 2432(110), 2435(125), 2437(136, 139, 140), 2438(136), 2439(154), 2440(154–156), 2443(155, 161), 2444(155), 2445(167), 2447(172), 2448(179), 2449(172, 179), 2450(172, 190), 2451(172), 2452(191, 192), 2453(200, 201), 2455(200), *2457–2461*, 2488(131), 2491(143, 144), 2493(151, 152,

157, 158), 2494(162), 2496(175, 176), 2518, 2519(143), 2520(144, 151, 268), 2523(151), *2561, 2562, 2565*
Andree, H. 2363(7), 2378(50), 2379, 2380(7), 2383(50), 2385(7), 2387(65, 67), *2398–2400*
Andreetti, G.D. (231), *262*
Andreoni, W. 1121(140), *1139*
Andrés, J.L. 2510(241), *2564*
Andrew, E.R. 309(116), *353*
Andrews, L. 992, 995(275), *1060*, 1068(18), *1101*, 1127(207), *1140*, 1161(101–103), *1183*, 2522, 2523(303), *2565*
Andrews, P. 520(67), *553*
Andrews, S.W. 1821(148), *1864*
Andriamisaka, J.D. 1983(72), *2032*
Andriamizaka, J. 1051(380), *1062*
Andrianarion, M. 1051(382), *1062*
Andrianarison, M. 211(146), 213(150), 234(241), *260, 263*
Andrianov, K.A. 1583(17), *1595*
Anema, S.G. 2107(217), *2127*
Anet, F.A.L. 7(27), *96*
Ang, H.G. 2075, 2091(109, 110), *2125*
Angelaud, R. 1681(121), *1685*
Angelini, G. 561(52), *590*, 1109(33), *1136*
Angelini, J. 385(97), *428*
Angelotti, T.P. 2251, 2253(27a), *2311*
Angermund, K. 1918(85), *1927*
Anh, N.T. 497, 498(10), *510*, 1639(181), *1664*, 1996(108e), *2033*
Anhaus, J.A. 1840(249), *1866*
Anicich, V.G. 1109(52), *1137*
Aniszfeld, R. 520, 524, 540(70), *553*, 563, 583(82), *591*
Anklekar, T.V. 434, 440(16), *489*
Annarelli, D.C. 727(57), *776*, 2429(103), *2459*
Annelin, R.B. 2231(81c), *2242*
Annet, F.A.L. 69, 72(129), *100*
Anpo, M. 2320(17a, 17b), 2347(17b, 186a, 186d), 2350(221), *2355, 2360, 2361*
Anthony, J. 186(33), *257*
Antic, D. 2027(204), *2036*
Antiñolo, A. 2055–2057(44), *2123*
Antipin, A.Yu. 1418, 1422, 1423(216), *1444*
Antipin, M.Y. 1244(34), *1306*
Antipin, M.Yu. 194(66), 209(134), 211(142), 219(173), 226(206), 237(245), 243(270), 254(335), *258, 260–263, 265*, 1395(155), 1396(158, 163), *1442*, 2002(139), *2034*
Antipin, Yu. 1397(166, 167), *1442*
Antipova, B.A. 196(71), *258*
Antoniotti, P. 560(39), *590*, 1120(115), *1138*
Antonova, N.D. 2133(20), *2171*
Anwar, S. 1740(181), *1789*
Anwari, F. 11, 33, 42, 43(41a), *96*, 814(106), *825*

Aoki, H. 115(34), *118*, 2250, 2252, 2253(23a, 23b), *2310, 2311*
Aoki, M. 22, 24, 25(56), *97*, 122, 123(21a), *151*, 1211(66), *1231*
Aoki, S. 2226(52), *2242*
Aoyagi, M. 1201, 1203(33), *1230*
Aoyama, T. 704, 739(5), 741, 742(82), 743(87, 88), *775, 776*, 1612(88), *1662*, 1675(83, 85), 1676(86–88), *1684*
Apasov, E.T. 1377(98), *1440*
Apeloig, Y. 3(1b, 1c), 4(1b, 10b), 14, 16(47), 18, 20(48), 21(55), 29(1c, 65), 37(10b, 85), 39, 40(10b), 44(10b, 93b), 47(93b), 48(10b, 93b), 49(85, 93b), 52(98d), 53(65, 98d, 100a, 102), 55(100a, 102), 56, 57(100a), 59(98d, 102), 60(98d, 100a, 102, 107), 61(100a), 62(98d, 100a), 63(102, 110, 111), 64, 66(102, 110), *94–99*, 250(317–319), 251(318), *265*, 345, 347–349(154), *354*, 362(27), 363(27, 32a–c), *426*, 437(79), 482(238, 239), 485(79), *490, 493*, 515(8–10), 517(51, 54), 518, 520(9), 521–523, 527, 529(73), 532(99a), 538(9, 10), 542, 544(73), *551, 553, 554*, 559(11), 560(51), 575(11), 576(140–142), 577(11), 578(162), 584(189), *590, 592–594*, 596(1), 597(2, 6), 598(6, 8), 599(10), 601(11, 13, 14), 603(1, 6), 604(6, 10), 606(22, 23), 607(14), 611(8, 39, 40), 612(8, 39), 613(39, 40), 615(6, 22, 23, 44), 630(11), 634(87), 647(11, 13, 119), 648(120, 121), 652(127), 685(156), *697–701*, 705(12), 707(12, 18), 709(18), 746, 770(90b), *775, 776*, 780(1g), 781(5), 794(1g), 795(69a, 69b), 796(69c), 800(69a–c), 801(69b), 818, 819(116), *821, 823, 825*, 833(35), 835(35, 37, 43), 842(53, 54), 843(54), 853(88), *854, 855*, 860(11, 12), 866(39), 884–886(39, 111), 901(11, 12, 147), 910(111), 917(39), 919(11, 12), 924, 925, 928(39), 957(39, 111), 958(241), 966(246a), 981(11, 12, 111, 260), 982(246a), 983(111, 246a), 984(260, 266), 985(111, 270), 987(111, 260), 989(260), 992, 996(111), 1001(246a), 1008(11, 12, 297c), 1009(12, 297c), 1010(297c), 1049(368), *1054–1057, 1059, 1060, 1062*, 1123(156), 1128, 1129(228), 1130(236), 1131(236, 240), *1139, 1141*, 1157(83), 1164(137), 1166(149, 156, 157), 1169(137, 171, 176), 1176(137, 176), 1177(171), 1178(137, 171), 1180(137), *1183–1185*, 1330(176), 1331(175, 176, 195), *1336*, 1373(74), *1439*, 1465, 1467(171), *1530*, 1544(26), *1578*, 1603, 1605(39), *1661*, 2023(196), 2024, 2026(198), *2036*, 2122(258), *2128*, 2166(192), *2175*, 2182(30), *2214*, 2429(100), *2459*, 2488(134), 2499(186), 2505(201), 2506(134, 201), 2509(226), 2511(247a–d), 2514, 2515(226), 2516(226, 247a–d, 261), 2518(226, 247a, 261), 2536(247b), 2539(371), 2555(201, 418), 2556(419), *2561, 2563–2565, 2567, 2568*
Apodaca, R. 1670(31), *1683*
Appel, A. 129(36), *151*
Appel, R. 213(152), *260*
Appell, J. 1128(216, 217), *1140, 1141*
Apperley, D.C. 333(146), *353*
Appler, H. 2001(133), *2034*, 2489(137b), *2562*
Arad, D. 363(32b, 32c), *426*, 532(99a), *554*, 601, 630, 647(11), *698*, 1603, 1605(39), *1661*
Arai, H. 1710(86, 87), 1725(87), *1787*
Arai, K. 2321, 2347(20), *2355*, 2546(391, 392), *2568*
Arai, M. 90(170c), *101*, 1614(93), 1649(210), *1662, 1665*, 2250(23a, 23b), 2252, 2253(23a, 23b, 32b, 32c), *2310, 2311*
Arai, T. 2116(241), *2127*
Arai, Y. 1157(75), *1182*, 1268(95), *1307*, 2347(186b), *2360*, 2472(46), *2559*
Arakawa, Y. 735, 737, 741(69), *776*, 906, 973(168), *1057*, 1278(136), *1308*, 2409–2411(32), *2458*
Araki, K. 438(92), *490*
Araki, T. 1849(291), *1867*
Archibald, R.S. 248(301), *264*, 348(162), *354*, 833(33), *854*, 1266(91), 1283(148), 1298(194), *1307, 1308, 1310*
Arcoleo, J.P. 425(217), *430*
Arduengo, A.J. 763(133a), *777*, 2539(373a, 373b), *2567*
Arduengo, A.J.III 52(95a–c, 95e, 95f, 95h, 96), 53(96), 54(95e, 95f, 95h, 96), 61(96), *98*, 520(66), *553*, 710(26), *775*, 2509(225), 2534(363), 2539(225), *2564, 2567*
Arefev, O.A. 1996(108a), *2033*
Arens, J.F. 1621(123), *1663*
Arias, J. 516(35c), *552*
Arif, A.M. 241(267), *263*, 846(70), *855*, 1154(62), *1182*, 2069, 2071(77), *2124*
Ariga, T. 364(33), *426*, 603(18), *698*
Arija, K. 2342(144), *2359*
Arimoto, M. 1193(14a), *1230*, 1828(187, 188), *1865*
Arita, N. 1372, 1373(73), *1439*
Ariza-Castolo, A. 1979(57d), *2031*
Armand, M. 2339(126, 129), *2358, 2359*
Armbrust, R. 763(137), *778*
Armentrout, P.B. 167(95), 168(101), 171, 172(95), 173(101, 137), *179, 180*, 532(100), *554*, 559(28), *590*, 1107(11, 13), 1108(23–29), 1110(54–57), 1113(67), 1118(24, 90–92), 1123(90, 91), *1136–1138*

Armitage, D.A. 1896(1, 2), 1903, 1904(1, 41b), *1925, 1926*, 2130, 2138(3), *2170*
Armon, R. 2326(51), *2356*
Arnason, I. 10, 22, 70(37), *96*, 1110(62, 63), *1137*
Arndt, J.L. 2231(81c), *2242*
Arnett, E.M. 436, 481(65), *489*, 842(49), *855*
Arnold, E.V. 1972, 1997, 1999, 2002(35c), *2031*, 2155(140), *2174*, 2412(34), *2458*
Arnold, F.P. 37, 42, 44, 47(86), *98*, 571, 574(129, 130), *592*, 1879(53), *1893*, 2021, 2028(192), *2036*, 2087, 2090, 2094(164), 2118, 2120, 2121(254), *2126, 2128*
Arnold, J. 1869(3), 1875(31), 1888(96–98), 1889(99, 100, 102), 1890(31, 96, 106, 107, 109–112), 1891(97, 98, 113, 114), *1891, 1892, 1894*
Aroca, P. 312, 314, 316, 317(137), *353*
Arpac, E. 2333(104), 2335(104, 110), 2337(110, 118), 2340(138), *2358, 2359*
Arrignton, C.A. 1331(198), *1336*
Arrington, C.A. 901(154), *1057*, 1075(33), *1102*, 1168(165), *1185*, 2493, 2520(155), *2562*
Arshadi, M. 522, 527, 528, 532, 535, 547(83b), *554*, 564, 569, 570(90), *591*
Artaki, I. 290(75), *352*
Artamkina, O.B. 1390(130), 1392(143), 1393(148), 1396(162), *1441, 1442*
Arthur, N.L. 163, 174, 175(71), *179*, 1125(185), *1140*
Arvanaghi, M. 1740(175), *1789*
Arya, P. 1064, 1069, 1082–1084, 1086, 1099(9), *1101*, 1387, 1388(114, 117), *1440, 1441*, 1545(28–31), 1557(62), 1564(28, 76), *1578, 1579*
Asahara, M. 1390(123a), 1434, 1435(243), *1441, 1444*, 2429, 2436(105), *2459*, 2554(416), 2555(417), *2568*
Asami, M. 2343(150), *2359*
Asano, K. 1242(29), *1306*, 1881, 1887(59), *1893*, 1897, 1911(13, 14), 1913(13), 1915(14), 1916(13, 14, 84), 1923(91), *1925, 1927*
Asao, N. 1672(45), *1683*, 1794(25), 1856(330), *1862, 1868*
Asaoka, M. 459(155, 156), 462(169), *491, 492*
Asayama, M. 482(237), *493*
Ashe, A.J.III 741(81), *776*, 1148, 1149(30), *1182*, 1970(28), 1980(60a, 60b), 1986(60a, 77), 2004(60b), 2012(60a, 77), *2031, 2032*
Ashida, P. 310(130), *353*
Ashley, C.S. 2347(186c), *2360*
Ashraf, A. 2236(114), *2244*
Ashwell, M. 1620(120), *1663*
Asirvatham, V.S. 2080, 2092(130, 132), *2125*
Askew, E. 2151, 2152(113), *2173*

Aslam, , M. 1677(97), *1684*
Aso, S. 2350(219), *2361*
Aso, Y. 194(67), *258*, 1828(190), *1865*
Assink, R.A. 2346(181), *2360*
Ast, T. 1134(259–262), *1142*
Astapov, B. 302(101), *352*
Astruc, D. 2137, 2159(32), *2171*
Atagi, L.M. 1830(198), *1865*
Atkins, R.M. 1162(119), *1183*
Atkinson, G.H. 2477(65), 2521(280), 2545(380), *2560, 2565, 2567*
Atkinson, R.S. 1843(268, 269), *1867*
Atrinkovski, A. 2340(132), *2359*
Attar-Bashi, M.T. 1459, 1467, 1488(131), *1529*
Attinà, M. 562(69), *591*
Atwell, W.H. 1312(10, 11), 1321(88), 1322(94, 95), *1332, 1334*, 1972(34), 1976(52), 1980(34), 1998, 1999, 2001(52), 2003, 2005, 2008, 2009(34), 2010, 2013(164), *2031, 2035*, 2429(101), *2459*, 2465(9), *2559*
Atwood, D.A. 217(170), 245(283), *261, 264*
Atwood, V.O. 245(283), *264*
Atzkern, H. 2159(151, 152, 154), 2160, 2162(151), *2174*
Aubert, C. 452(129), *491*
Audebert, P. 2326, 2328(49a), 2337(120), *2356, 2358*
Audran, G. 1849(288, 289), 1856(331), *1867, 1868*
Aue, D.H. 1470(219), *1531*
Aue, W.P. 310(132), *353*
Auer, H. 125, 126, 128, 129(32b), *151*
Augelli-Szafran, C.E. 1828(186), *1865*
Auner, N. 3(2a, 3f), *95*, 169(112), *179*, 543–545, 547(110), *555*, 570, 573(125), 581(181), *592, 593*, 696(170), *701*, 860(17), 863(26, 27a, 27b), 866, 867(40), 872(57), 878(80), 879(81), 913(178, 179), 915(199), 931(80), 932, 938(199), 945(81, 205–211), 946(81, 205, 209–224), 948(225–234), 949(178, 179, 233), 988, 990(27a, 27b), 1007, 1008(232), 1016(217, 312), *1054, 1055, 1057–1059, 1061*, 1064, 1068(11), 1075(35c), *1101, 1102*, 1123(162), *1139*, 1374, 1375(80–82), 1377(81), 1379(80–82), 1380(81), 1386(112, 113), 1387(113, 120), 1389(112, 113, 120), 1390(123b), 1430, 1434(81, 82), *1439–1441*, 1965(9b), *2030*, 2059(51, 52), 2060(52), 2066, 2067(51, 52), 2069, 2071, 2084(76), 2085(158), 2089(76), 2090(158), 2093(76), 2094(76, 158), *2123, 2124, 2126*, 2481, 2483(97), 2528(336), 2553(97), *2561, 2566*
Auranagzeb, M. 361(17), *426*
Aust, J. 2075(108), *2125*
Avakyan, V.G. 981, 985, 992(254), *1059*
Avdyukhina, N.A. 240(258), *263*

Avidan, G. 2320, 2322, 2327(19), 2340(132), 2348(19), *2355, 2359*
Avila, D.V. 1542(14), *1577*
Avnir, D. 2318(1), 2319(12b, 12c, 13), 2320(1, 13, 15, 18, 19), 2321(1, 22, 23a, 23b), 2322(18, 19, 22, 23a, 23b, 25, 30, 33, 38, 39, 40a, 40b), 2323(39, 40a, 40b), 2325(40b), 2326(40a, 43, 44a, 44b, 46a, 46b, 50), 2327(19, 44a, 44b, 52), 2340(132), 2346(184, 185), 2347(1, 33, 44a, 44b, 184, 185, 192, 193, 196, 197), 2348(18, 19, 25, 44a, 44b, 184, 192, 193, 198, 199, 202a–c), 2349(192, 193, 203, 204, 205a, 205b), 2350(1, 192, 193, 196), 2351(1), *2355, 2356, 2359–2361*
Avtomonov, E.V. 1450, 1456, 1457, 1461, 1496(37), *1525*
Awakyam, W.G. 1910(57), *1927*
Awakyan, W.G. 229(212), *262*, 1913, 1914(77), *1927*
Awasthi, A.K. 645(110, 111), *700*
Awazu, K. 2522(301), *2565*
Ayala, P.Y. 2511(246), *2564*
Aylett, B.J. 36(76b), *97*, 2054(40), 2086(162), *2123, 2126*
Ayoko, A. 577(150), *593*
Ayoko, G.A. 203(111), *259*, 577(145), *592*
Azpeleta, S. 1129(235), *1141*
Azzari, E. 1844(270), *1867*
Azzaro, C. 2545(383), *2567*
Azzaro, M. 1110(61a), *1137*

Baar, B.L.M.van 10(35), *96*, 1151(46, 48), *1182*, 2453(195, 196), *2461*
Babaian, E.A. 222(187), *261*
Babin, P. 385(94), 386(102), *427, 428*, 2226(50), *2242*
Babisch, J.-A.H. 1241(25b), *1306*
Babkin, Yu.A. 1201(35), *1230*
Baboneau, F. 2249(17), 2294(17, 142), *2310, 2315*
Babonneau, F. 226(205), *262*, 2232(92), *2243*, 2246, 2248, 2249(7), 2250(7, 21), 2253, 2255–2257(21), 2264(78, 79), 2269, 2272, 2273(7), 2276(103), 2285(7), 2289(128), 2290, 2291, 2293(138a, 138b), 2298(150b), *2310, 2313–2316*, 2332(89, 90), 2333, 2337(102), *2357, 2358*
Baboul, A. 1053(394), *1062*
Babu, R. 2224(39), *2241*
Babudri, F. 1819(141, 142), *1864*
Baceiredo, A. 761(127–130), 762(131), 763(127, 129, 135, 136), *777*
Bach, R.D. 2510(241), *2564*
Bach, T. 446(112), *490*, 562(66), *591*, 1297(187), *1309*, 2141(60), *2172*
Bachi, M.D. 1576(100), *1579*

Bachmann, W.E. 125(31b), *151*
Bachrach, S. 1080(44), *1102*
Bachrach, S.M. 959(242), *1059*, 1963(2d), 1997, 1998(2d, 115), *2030, 2034*, 2407(24), *2458*
Bäcklund, S.J. 1613(90), *1662*
Bäckvall, J.-E. 1828(184), *1865*
Badaoui, E. 1546(33), *1578*
Bader, A. 2339(128), *2359*
Bader, R.F.W. 528(88a, 88b), *554*, 1911(67), *1927*
Bader, R.W.F. 61, 82(109a–c), *99*
Badini, G.E. 2322(28), *2356*
Badirova, G.T. 1996(108c), *2033*
Baek, E.K. 1250(50), *1306*
Baffy, J.J. 60(107), *99*
Baggott, J.E. 1123(163), *1139*
Bahloul, D. 2294(142), *2315*
Bahr, S.R. 284(50), *351*, 566, 570, 572(101), *592*, 1097, 1098(73b), *1103*, 1289(168), *1309*, 1879(57), 1887(57, 91), *1893, 1894*, 1897(9), *1925*
Bai, H. 863(23), *1054*, 1240(25a), *1306*
Baier, M. 1318(48, 51), *1333*
Bailar, J.C. 1414, 1424(195a), *1443*
Bailey, D.L. 364(36), *426*, 1606, 1628(46), *1661*
Bailey, S.M. 159, 160(54), *178*
Bailleaux, S. 3(7a), *95*
Bailleux, S. 864, 981, 982(31–33), *1054*
Bain, A.D. 275, 276(21), *350*
Bain, S. 1022(331), *1061*
Baines, J.E. 437, 483(72), *490*
Baines, K.M. 3(3a), *95*, 247(297), *264*, 487(263), *494*, 578(159), *593*, 713(35), 775, 795, 799(62c), *823*, 845(65, 66), 855, 880(89, 90), 881(89, 90, 96), 913(90, 96, 180), 914(90), 917(90, 180), 922, 923(89), 924(89, 90), 985, 986, 992(90), 996(89), 1004(90), *1056, 1057*, 1064(7b), *1101*, 1145(14), 1155(66, 67), *1181, 1182*, 1234(4), 1238(24), 1273, 1276, 1285(109), *1305, 1306, 1308*, 1601, 1604(26), 1609(26, 72), 1645(26, 72, 201), *1661, 1662, 1664*
Bains, K.M. 827, 828, 844(6), *854*, 1064, 1068(4c), *1101*
Baird, C.A. 209(131), *260*
Baird, M.C. 2043, 2051(16), *2122*
Baird, M.S. 746(91), 747(92), *776*, 2448(178), *2461*
Baird, N.C. 2421(59), *2458*
Bakhareva, E.V. 1491(323), *1534*
Bakhtiar, R. 362(24), *426*, 560(43, 49, 50), 590, 601, 602, 606(16), *698*, 707(16), *775*, 1111(64, 65a), 1113(66), 1131(239, 241), *1137, 1141*, 2088(178–180), *2126*
Bakker, R.H. 1797(49), *1863*

Bakowies, D. 72(144), *100*
Bakthavatachalam, R. 2345(176), *2360*
Balaban, A.T. 120(2), *150*
Balaishis, D. 2153, 2154(134), *2173*
Balaji, R. 2234(104b), *2243*
Balaji, V. 901, 902, 993, 995, 996(159), 1019(321), 1021(326), 1044(321, 326), 1045, 1047(321), *1057, 1061*, 1128(225), *1141*, 1170(179), *1185*, 1283(156), 1289(169), *1309*, 1313, 1314(33), 1316(35, 43), 1317(43), 1320(82), *1333, 1334*, 1967, 1986, 1996, 1997(20a, 20b), 2013(172a), *2030, 2035*, 2474(54), 2504, 2515(54, 197), *2560, 2563*
Balakhchi, G.K. 1466(179), 1467(179, 188a), 1468(188a, 196), 1470, 1471(196), 1476–1478(188a), *1530*
Balanov, A. 1576(100), *1579*
Balasubramanian, K. 1166(152), *1184*
Balasubramanian, R. 1998, 1999(125), 2005–2007(155), 2010(163), 2011(155), *2034, 2035*
Balavoine, G. 1743(202), *1789*
Balavoine, G.A. 2251, 2253, 2255(28c), *2311*
Baldridge, K. 1151(42a, 45), *1182*
Baldridge, K.K. 11(41b, 43a), 12(43a), 13–16(41b), 18, 19(50a), 21(41b, 50a), 32(50a), 96, 860, 983(10a), *1054*, 2505(200), *2563*
Baldus, H.-P. 2269, 2270(88b–e), *2313*
Baldwin, J.C. 164, 166(77), *179*
Baldwin, J.E. 1091(58d), *1103*, 1675(80), *1684*
Baldwin, K.J. 110(11), *118*
Balegroune, F. 2072, 2090(91), *2124*
Balint-Kurti, G.G. 1107(15), *1136*
Ball, G.E. 1889(100), *1894*
Ball, J.L. 3(6), *95*, 250(312), *264*, 859, 981, 984(5), 987(5, 271), 988, 989(271), 993, 995, 996(5), 1000(5, 271), 1004(271), *1054, 1060*, 1145(16), *1181*, 1285(159), *1309*
Ballestri, M. 1542(9), 1544(21), 1545(21, 27), 1546(21), 1549(21, 41a), 1553(21, 52, 54), 1558(64), 1572(21, 94, 95), 1573(21), *1577–1579*, 1703, 1722(68), *1787*, 1831(203), *1865*, 1875(32), *1892*
Bally, T. 131(38), *151*, 1153(56), *1182*, 2182(34), *2214*
Balm, S.P. 1929, 1930(1a), *1960*
Balooch, M. 1120, 1121(129), *1138*
Balter, S. 1887(86), 1890(104), *1893, 1894*
Balzarini, J. 1550(44), *1578*
Banasiak, D. 1148(31), *1182*
Banasink, D. 10(32a), *96*
Banciu, M. 120(2), *150*
Band, E.I. 1879(54a), *1893*
Bandini, E. 435, 456(55), 459(157), *489, 491*
Bandodakar, B.S. 487(267), *494*
Banerji, P. 1782(338), *1792*

Baney, R.H. (27), *1102*, 1923(93), *1928*, 2246, 2250, 2253, 2254, 2273(6d), 2280(112a, 112b), 2285(6d), 2289(123), *2310, 2314, 2315*, 2329, 2331(75), *2357*
Banger, K.K. 1878(47), *1892*
Banik, G.M. 1606(52b), *1661*
Banisch, J.-A. 864, 932, 934(34), *1054*
Banisch, J.-A.H. 849, 852(84), *855*
Bank, S. 544(111c, 111d), *555*
Banks, K.E. 1329, 1330(170), *1336*
Bankwitz, U. 4(10b), 37(10b, 84), 38(84), 39, 40, 44, 48(10b), *95, 97*, 818, 819(116), *825*, 981, 984, 987, 989(260), *1059*, 1972, 1988, 1998, 2023(40), *2031*
Bankwitz, V. 2023(196), *2036*
Bannikova, O.B. 243(274), *264*, 1374, 1377(94), 1390(132–137), 1391(140), 1392(132, 135–137, 146, 147), 1393(147, 149), 1396(160, 164, 165), 1398(133, 137, 147, 149), 1399(160, 164, 165), 1400, 1401(137), 1404(147), 1405(133), 1418, 1420(210, 212), 1423(212), *1440–1442, 1444*
Banno, H. 1619(114), *1663*
Banovetz, J.P. 2276(105), *2314*
Baran, J.R.Jr. 820(119), *825*
Baranwal, R. 2301(156, 158), *2316*
Barashkova, N.V. 1220(96), *1231*
Barbero, A. 1819, 1842(140), *1864*
Barboux, P. 2351(261), *2362*
Barchi, J.J.Jr. 1550(47), *1578*
Barczynski, P. 8(29), *96*
Bardin, V.V. 1341, 1344(23), *1437*, 1838(226), *1866*
Barens, G.H. 1699(47), *1786*
Barie, W.P.Jr. 437, 483(74), *490*
Baris, H. 2227(69), *2242*
Barley, S.H. 2225(48), *2242*
Barlow, S. 2150, 2153, 2154(110), 2162(175), *2173, 2174*
Barlow, S.E. 1010(298d), *1060*, 2556(421), *2568*
Barluenga, J. 1797(47), *1862*
Barnard, T. 2285(117c), *2315*
Bar-Ner, N. 1576(100), *1579*
Barnes, G.H. 1487(305), *1534*
Barnhart, R.W. 1756(251), *1790*
Barnighausen, H. 205(116), *259*
Baron, R.N. 2548(401), *2568*
Barrau, J. 52(94c), *98*, 1064(4f), *1101*, 1882(65), *1893*
Barrett, A.G.M. 416(187), *429*, 435, 456(52, 53), *489*
Barris, G.C. 2107(218), *2127*
Barron, A.R. 1878(45), *1892*
Barrow, M.J. 116(47), *118*, 196(75), *258*, 1463, 1467(162), *1529*

Bart, J.C. 1923(90), *1927*
Bart, J.C.J. 229(222), *262*
Bartelmehs, K.L. 109(8, 10), *117*
Barthelat, J.-C. 63(112a), *99*, 1163(126), *1184*
Bartik, T. 1721(123), *1788*
Bartlett, P.D. 544(111c, 111d), *555*, 563(72), *591*
Bartlett, R.A. 207(119), *259*, 485(248), *493*, 793(50), *823*, 1043(360), *1062*
Bartlett, R.J. 68(122b), *99*, 2509(228, 230), *2564*
Bartók, M. 435, 479(60), *489*, 1733(144), 1738(160), *1788, 1789*
Bartoletti, L. 1628, 1656(140), *1663*
Barton, D.H.R. 1682(127, 128), *1685*
Barton, T.J. 10(32a–c, 33a), *96*, 169(113), *179*, 186(38), *258*, 709, 714(20), 716(40, 41), 722(47), 727(40, 54), 741(54), 748–750(40), 767(145), 768(146), 769(147), 770(148, 149), 772(149), *775, 776, 778*, 860(15, 16b), 862(16b), 917(15), 994, 995(283), 1022(331), *1054, 1060, 1061*, 1071(22, 24), 1075(32a), 1095(70, 71), *1102, 1103*, 1148(31, 32), 1149(33), *1182*, 1303(208), *1310*, 1452, 1453, 1458(52), *1526*, 1767(278), *1791*, 1860(350), *1868*, 1964(5), 1965(5, 9a, 11), 1967(5), 1972(35c, 36), 1980(61a, 62a), 1984(5), 1994(105a, 105b), 1995(106), 1997(5, 35c, 106), 1999(35c), 2001(132), 2002(35c, 142), 2003(145, 146), 2016(176a), *2030–2035*, 2155(140), *2174*, 2404(6), 2406(17), 2407(20), 2412(33, 34), 2416(46), 2443(162), 2448(180), 2450(182, 183), *2457, 2458, 2461*, 2465(11), 2474(52), 2477(76), 2484(100), 2489(11), 2501(189), 2502(193, 194), *2559–2561, 2563*
Baruah, J.B. 1692, 1720(17), *1786*, 1872(20), *1892*
Baruda, H. 1875(34), *1892*
Baryshok, P. 237(245), *263*
Baryshok, V.P. 154, 156(11, 12), 159(11), 160, 161, 164(11, 56), 165(11), 166(12, 56), *177, 178*, 1373(75c), 1381(107, 108), *1439, 1440*, 1449(16), 1452(56, 57, 59), 1453(62), 1454(56, 70, 75–79, 81, 93, 100, 105), 1455(59, 76, 114), 1457(62, 120), 1459(134), 1460(120, 155), 1462(151, 155), 1467(81, 93, 100, 105, 114, 120, 134, 151, 155), 1469(200–202), 1470(212, 216, 217), 1471(216, 217, 221, 223, 226, 230), 1472(216, 217, 221, 230, 233a, 233b, 234), 1473(235, 238), 1474(235, 242, 243, 263, 266), 1475(59, 62, 212, 242, 253, 255), 1476(57, 238, 243, 259, 261, 263), 1477(243, 259, 263, 264, 266), 1478(263, 266), 1479(57, 242, 274, 276–278),
1480(276, 277), 1481(288), 1483(288, 293), 1484(59, 294), 1485(62, 299), 1487(308, 309), 1488, 1489(16), 1490(309), 1491(16), 1493(329), 1494(335, 336), 1503(155), 1506, 1510(261), 1511(401), *1525–1535, 1537*
Basak, A. 412(174), *429*
Basch, H. 21(54a), *97*, 586(190), *594*
Basenko, S.V. 1507(383), 1508(383, 384, 391), 1509(383, 391), *1536, 1537*
Bashkirova, S.A. 1967, 1986(20a), 1989, 1994(89a), 1996, 1997(20a), *2030, 2033*, 2479(82), 2504, 2515(197), *2560, 2563*
Baskir, E.G. 1069, 1070(19), 1071(23), 1072(19, 23), 1083(23), *1102*, 1161(104), *1183*
Baskis, E.G. 2530, 2531(345), *2566*
Bassindale, A.R. 268(2), 284(52), 287(65), 288(52), *350–352*, 356(1), 361(15, 19), 416(189), 420(200), 421(202), 424(212), 425(216), *426, 429, 430*, 435(51), 456(51, 144), *489, 491*, 495(1), 500(27), 508(57), 509(59–61, 63–65), *510, 511*, 564(87), 569(118, 121), 577(144), *591, 592*, 828(13), *854*, 1371(63), 1374, 1376(93), 1378(103, 104), 1382(103), 1390(104, 138), 1392(144), 1398(138), 1401, 1402(104, 138), 1403(138), 1404(144), 1409(183), 1420(144), *1439–1443*, 1603, 1605(38), 1614(91), 1638(171), 1646(205), *1661, 1662, 1664, 1665*, 1667(8), 1680(114), *1682, 1684*, 1795, 1797(33), *1862*
Basso, N. 576(138, 139), *592*
Bassoul, P. 1512(404), *1537*
Bastiaans, H. 1052(387), *1062*
Bastiaans, H.M.M. 1064, 1068(6b), *1101*
Bastiaans, M.M. 2453(196), *2461*
Bastianns, H. 1051(379), *1062*
Bastiansen, O. 116(46), *118*
Bastl, Z. 1006, 1008(295), *1060*, 1237(17), *1306*
Basu, S. 862(21), *1054*
Basu, S.C. 66(115b), *99*
Batcheller, S.A. 123, 125, 130(23), *151*, 812(99a), *824*, 827, 828(11), 829(21), *854*, 1064(4g), *1101*, 2183, 2205(43), *2214*, 2416(45), *2458*
Batchelor, R.J. 487(264), *494*
Bates, D.K. 2005(154), *2034*
Bates, F.S. 2234(105a), *2243*
Bates, R.B. 1997(112a), *2033*
Bates, T. 874, 934(68), *1055*
Bates, T.F. 846(69, 70), *855*, 1154(62, 63), *1182*
Batich, C.D. 2286(119), *2315*
Bats, J.W. 219(178), *261*, 1374, 1375, 1377, 1378(87), *1439*

Battiste, M.A. 466(182), *492*
Batyeva, E.S. 1870(6), *1891*
Bau, R. 197(79), *259*, 514(3a, 3b), 520(3a, 3b, 70, 71), 521(3a, 3b), 522(3a), 524(3b, 70, 71), 525(71), 529, 530(90), 532(3a), 540(70), *551, 553, 554*, 561(59), 563(82), 567, 570(103), 572(59, 103), 583(59, 82, 103, 185), 585(59, 185), 586(103, 185), 587(185), 588(59, 185, 196), *591–594*, 654(132), *700*, 1675(82), *1684*, 2069, 2071(79), *2124*
Baudler, M. 213(153), 215(157), *260*
Baudrillard, V. 269(6), *350*
Bauer, B. 1798(50), *1863*
Bauer, C. 1107(16), *1136*
Bauer, G. 2177(4c), 2181(18), *2213, 2214*
Bauer, G.H. 2545(381), *2567*
Bauer, I. 747, 748(94), *776*
Bauer, S.H. 133(39, 40), *151*, 1899(35), *1926*
Bäuerle, P. 584(184), *593*
Baughman, G.A. 617(54), *698*
Baukov, Yu.A. 1463, 1464, 1467, 1476(165), *1530*
Baukov, Yu.I. 243(271–273, 277), 245(286), *263, 264*, 500, 501(26), *510*, 1374, 1376(92a, 92b), 1390(128–130, 139), 1391(140), 1392(129, 139, 143, 145a, 145b), 1393(148), 1395(139, 156, 157), 1396(156–159, 162, 163), 1397(139, 156, 166, 167), 1398(139), 1399(156), 1400, 1401, 1403(139), 1404(172), 1434, 1435(240), *1440–1442, 1444*, 1454, 1467(106), 1501(363), *1528, 1536*
Baukov, Yu.L. 587(195), *594*
Baukov, Yu.T. 226(206, 207), *262*
Baum, G. 215(156), *260*, 406(154), *429*, 547(112a), *555*, 1848(287), *1867*, 1887(88), *1893*, 2152, 2153(121, 122), *2173*, 2429(93), *2459*
Baum, M. 2106, 2109(210), *2127*
Baum, M.W. 748, 750(97), *777*
Bauman, D.L. 507(49), *511*
Baumann, W. 2051, 2052(34), *2123*
Baumegger, A. 186(43), *258*, 795–798(57), *823*, 880(94), 881(94, 95), 882(95), 914(94), 915, 924, 932, 954, 956(95), 967(94), 978, 985, 986(95), *1056*, 1272, 1273(108), 1274(112, 122), 1276(112), *1308*
Baumgärtner, R.O.W. 11, 13, 21(40), *96*, 995(281), *1060*, 1150(38), *1182*
Bauml, E. 397(134), *428*
Bausch, J. 540(105), *554*, 563, 567–569(83), *591*
Bausch, M.J. 363(30), *426*, 631, 632(85), *699*
Bax, A. 310(131), *353*

Bayston, D.J. 1807(87–89), *1863*
Bazant, V. 1600(3), *1660*
Beach, D.L. 1705(78, 79), 1714(79), *1787*
Beagley, B. 234(237), *263*
Beatie, M.E. 161(57), *178*, 2477(73), *2560*
Beauchamp, A.L. 1280(138), *1308*
Beauchamp, J.L. 34, 35(72a, 72b), *97*, 171(125), 172(125, 131), 173–175(131), *180*, 520(68), *553*, 558(5), 559(29), 560(29, 41, 42), *590*, 901(150), *1057*, 1108(19), 1132(244, 245), *1136, 1141*
Beaudry, C.L. 2344(157, 158), *2359*
Becerra, R. 154(6), 157, 158(46), 161, 163(60), 167(6), 171(6, 117), 172(117, 133), 173(134, 135), 174(134), 175(135, 140), *177, 178, 180*, 901(149), *1057*, 2476(64), 2511(244), 2524(322, 323), 2525(325), 2526(323), 2527(322, 323), *2560, 2564, 2566*
Becher, P.F. 2264(75), *2313*
Bechgaard, K. 2160(172), 2162(172, 173), *2174*
Bechstein, O. 245(278), *264*
Becht, J. 238(253), *263*, 334, 335(148), *354*, 1341(28, 29, 30b), 1349(28, 29, 30b, 31a, 31b), 1350, 1351(30b), 1357(28, 29, 30b, 31a, 31b), 1358(28), 1359(28, 29), 1360(28), 1361(28, 30b), 1362(28), 1366(29), 1368(30b), *1437, 1438*
Bechtel, J.H. 2330(83), 2352(263), 2354(83), *2357, 2362*
Becker, B. 219(184, 185), 229(218, 220), 232(184, 185, 225–228, 230), *261, 262*, 1370, 1371(59), *1439*, 1989(91), *2033*, 2363(6), *2398*
Becker, G. 215(155), 217(167), *260, 261*, 780, 781(2e), 796(66b, 74), 799(65, 66b), 800(66b), 802(74), *821, 823, 824*, 1890(105, 107, 108), *1894*, 2261(54b), *2312*
Becker, J.Y. 1207(53), 1208(53–55), *1230*
Becker, S. 205(118), *259*
Beckers, H. 193(61), *258*
Beckwith, A.L.J. 1542(14), *1577*
Bedard, T.C. 1697(40), *1786*, 2047(25), *2123*
Beek, D.A.V.Jr. 806(85), *824*
Beese, R. 454(133), *491*
Beest, B.W.H.van 115(36), *118*
Beetz, A.G. 241(261), *263*
Beevor, R.G. 1723(128), *1788*
Befurt, R. 241(260), *263*
Begemann, C. 1053(398), *1062*
Beggiato, G. 1210(60), *1231*
Begum, M.K. 1669(22), *1682*
Behbehani, H. 1925(109), *1928*
Behm, J. 913(179), 948(233), 949(179, 233), *1057, 1059*
Behn, J. 52(97), *98*

Behnam, B.A. 487(263), *494*, 795, 799(62c), *823*, 845(66), *855*, 880, 881, 913, 914, 917, 924, 985, 986, 992, 1004(90), *1056*
Behnke, J.S. 1244(35), *1306*
Behrens, S. 1887(90), *1894*
Belair, N. 2387(64, 68), 2388(64), *2400*
Beletskaya, J.P. 2155(141), *2174*
Belgardt, T. 1034, 1038(353), *1061*
Belin, C. 1374, 1375, 1377, 1378, 1382, 1383, 1385, 1424(77), *1439*
Bell, D. 456(145), *491*
Bell, K.L. 1816(129, 131), *1864*
Bell, L.G. 2054(41), *2123*, 2150(106), *2173*
Bell, T.N. 2514(273), 2526, 2527(327), *2565, 2566*
Bellama, J.M. 1459, 1467(132), 1473–1475(236), 1493(132), *1529, 1532*
Bellamy, F. 1547(37), *1578*
Bellassoued, M. 423(210), *430*
Belleville, P.F. 2351(244), *2361*
Belot, V. 2231(90a, 90b, 91), *2243*, 2294(141), *2315*
Belousova, L.I. 1390(132, 133), 1391(140), 1392(132), 1398, 1405(133), *1441*, 1454, 1467(107), *1528*
Belova, V.V. 1875(36), *1892*
Belsky, V.K. 2155(141), *2174*
Beltzner, J. 46(90b), *98*, 1087(55), *1102*
Belu, A.M. 2219(15a–c), 2225(46), *2240, 2241*
Belyaeva, V.V. 1454, 1467(105), 1470(216), 1471(216, 221, 223), 1472(216, 221, 234), 1479(277, 278), 1480(277), *1528, 1531, 1533*
Belyakov, A.V. 52–55, 57–59(98a), *98*, 279, 280(38), *351*, 1144(11), *1181*, 2023(195), *2036*, 2535–2539(367), *2567*
Belyakov, S. 1451(48b), *1526*
Belyavsky, A.B. 1714(95), *1787*
Belzner, J. 541(107), 551(117b), *554, 555*, 581(180), *593*, 790, 792(41), 796(41, 75), 798(41), 802, 803(75), *822, 824*, 870(52), *1055*, 1075(35a, 35b, 36, 37), 1086, 1090(37), *1102*, 1264(83), *1307*, 1377, 1379, 1380(96), 1411, 1412(188), *1440, 1443*, 1910, 1911(60e), *1927*, 2183, 2185(41), 2187(48), *2214*, 2416(52), 2429, 2436(104), *2458, 2459*, 2550(409–411), 2552(412, 413), 2553(414, 415), *2568*
Belznerand, J. 1973(47), *2031*
Bencivenni, L. 2508(223), *2564*
Bender, H. 1051(384, 385), *1062*
Bender, H.R.G. 213(147), 252(331), *260, 265*
Bender, S. 727(58), *776*
Bendikov, M. 18, 20(48), *96*, 648(121), *700*, 795, 800(69a, 69b), 801(69b), *823*, 884–886, 910, 957, 981, 983, 985, 987, 992, 996(111), *1056*, 2122(258), *2128*

Benesi, A. 283(46), *351*, 514, 520–522(3a), 529, 530(90, 91), 532(3a), *551, 554*, 567, 569, 570(106), 583, 585(106, 185), 586(185), 587(106, 185), 588(106, 185, 196), *592–594*
Benesi, R. 654(132), *700*
Benfield, R.E. 2481(89), *2560*
Bengali, A.A. 1010(298b), *1060*, 2556, 2557(425), *2568*
Benham, B.A. 1601, 1604, 1609, 1645(26), *1661*
Ben Hamida, N. 1882(65), *1893*
Benin, V.A. 564(92), *591*, 1410, 1412(187), *1443*
Benkeser, R.A. 789(28f), 813(103, 105b), *822, 825*, 1722(125), *1788*
Bennetau, B. 385(94), 386(101, 102), 411(169), *427–429*, 2226(50), *2242*
Bennett, F. 434, 444(23–25), *489*
Benno, R.H. 1897(10), *1925*
Benson, S.W. 6(22), *95*, 155(29–31), 156(29, 31), 157(37–45), 158(39, 42–44), 159(44), 160(41, 44), 161(44), 176(29–31), *178*
Bent, H.A. 300(95), *352*, 1330(171–173), *1336*
Bentham, J.E. 296(82), *352*, 2131(7), *2171*
Bentley, T.W. 637, 639, 640(98), *699*
Bentz, P. 1705, 1707, 1709(74), *1787*
Bentz, P.O. 1707(83), *1787*
Benziger, J. 2249, 2294(14), *2310*
Ben-Zvi, N. 1459, 1467(132), 1473–1475(236), 1493(132), *1529, 1532*
Beresis, R. 1852(316), 1853(318), *1868*
Berg, J.H.N.van den 1585(37), *1595*
Berg, W.van den 2152, 2153(123), *2173*
Bergens, S.H. 1756(250, 251, 253), 1757(250), *1790*
Berger, A. 310(132), *353*
Berger, D. 1832(206), *1866*
Berger, S. 168(105, 109), 169(105), *179*, 381(81), *427*, 540(103), *554*, 1109(37), *1136*
Bergerat, B. 2159(152), *2174*
Bergerat, P. 2159(154, 155), *2174*
Berglund, M. 1840(252), *1866*
Bergman, A. 2437(133), *2460*
Bergman, R.G. 1709(84, 85), *1787*, 2106(208), *2127*, 2425(83, 84), *2459*
Bergmeier, S.C. 1822(149), *1864*
Bergsträsser, U. 763(137), *778*
Berk, R. 1585(35), *1595*
Berk, S.C. 1671(42), *1683*, 1741(184), *1789*
Berkessel, A. 1742(185), *1789*
Berkowitz, J. 155(27), 167(27, 94), 171, 172(127), 175(94), 177(27), *178–180*, 517(56), *553*, 1106(2, 6), 1107(6), *1136*, 1166(154), *1184*, 2522(299, 300), *2565*
Berlekamp, U.H. 1434, 1435(239), *1444*

Berlin, A. 1543(18), *1577*
Berliner, E. 382(84), *427*
Bernadez, L. 1120, 1121(129), *1138*
Bernal, I. 198(83), *259*
Bernardi, A. 441(102), *490*
Bernardi, F. 6(20), *95*, 919, 921(187, 188), *1058*, 1320(81), *1334*, 1602(29), *1661*
Bernards, T.N.N. 2335(108), *2358*
Berndt, A. 685(156), *701*
Berndt, B. 2364(10, 12, 13), *2398*
Bernstein, S. 435, 456(49), *489*
Berrier, A.L. 645, 647, 661(113), *700*, 1602, 1604, 1655(30), *1661*
Berry, D.H. 163(74), *179*, 229(221), 250(314), 262, *264*, 437, 484(76), *490*, 1032, 1040(348–350), 1042, 1043(349, 350), 1044, 1045, 1050(349), *1061*, 2049(30–33), 2051(30, 31), 2052(30), 2054(42), 2055(32, 42, 43), 2056(42, 43, 46), 2057(42), 2062(63–65), 2064(71, 72), 2065(72), 2066(63, 65, 72), 2068(63, 65), 2069(65, 72), 2080, 2092(129), 2105(64), *2123–2125*, 2295, 2296(146b), *2315*, 2425(87), *2459*, 2486(121), *2561*
Berryhill, S.R. 2142(62, 64, 65), *2172*
Bertók, B. 1850, 1851(304), *1867*
Bertram, M. 234(236), *263*
Bertrand, G. 709(24), 710(26), 761(127–130), 762(131, 132), 763(127, 129, 134a, 134b, 135–137), 765(140), *775*, *777*, *778*, 847(71), *855*, 946(217), 1005, 1006(294), 1016(217), *1058, 1060*, 1069, 1070(20, 21), *1102*, 1982(69), *2032*
Bertrand, J. 1998, 1999(124a), *2034*
Bertsch, K. 668(144), *701*
Bertz, S.H. 1668(12), *1682*
Beruda, H. 241(263), *263*, 2159(154), *2174*
Berwe, H. 1897(11), *1925*
Berwin, H.J. 358, 359(9), *426*, 616(50–52), *698*, 838(45), *854*
Besenyei, I. 1968, 1986, 1996–1998(20c), *2030*
Beslard, M.P. 2340(131), *2359*
Beslin, P. 1870(8), *1891*
Bestmann, H.J. 1601, 1602(13, 23), 1603(42), 1604(13), 1610(23), 1612(13), 1642(23), 1643(13), *1660, 1661*
Besztercey, G. 2220(19a), *2240*
Béteille, J.P. 1964, 1965, 1967(6), 1970(27), 1981–1983(67c), 1984(6, 74), 1985(75), 1996(67c, 75), 1997(6), 1998(67c, 74), 1999(6, 67c), 2007(75, 159), *2030, 2032, 2035*
Bethune, D.S. 72, 74(146), *100*, 149(61), *152*
Betowski, L.D. 1372(68), *1439*
Betsuyaku, T. 1304(209), *1310*

Bettenhausen, M. 1887(90), *1894*
Beurskens, P.T. 2152, 2153(123), *2173*
Beyer, W. 385(96), *427*
Beynon, J.H. 1134(259–262), *1142*
Beyou, E. 2226(50), *2242*
Bhacca, N.S. 497(12), *510*
Bhalfacharyya, K. 1323(116b), *1335*
Bhamidipati, R.S. 1679(107), 1681(121), *1684, 1685*, 1854(321), *1868*
Bhandari, A. 1659(225), *1665*
Bhatia, Q. 2234(104b), *2243*
Bhawalkar, J.D. 2352(262), *2362*
Bhowmick, A.K. 2239(127), 2240(127, 129, 130), *2244*
Bi, C. 1763(266–268), 1764(268, 269), *1791*
Bi, D. 2225(44), 2231(89), *2241, 2243*
Bianchi, D. 2322(34), *2356*
Bicerano, J. 184, 250(28), *257*
Bickelhaupt, F. 10(35), *96*, 1051(378, 379), 1052(387), *1062*, 1064, 1068(6a, 6b), *1101*, 1151(46, 48, 49), *1182*, 1388(122), *1441*, 2453(195–197), *2461*
Bickmore, C. 2301, 2302(159), *2316*
Bickmore, C.R. 2263(72), 2265(80), 2297, 2298(72), 2301, 2305(157), *2313, 2316*
Biedrzicka, Z. 304(106), *353*
Bielle-Daspet, D. 2548(400), *2568*
Bielmeier, S.R. 987–989, 1000, 1004(271), *1060*
Bienlein, F. 1377, 1378, 1380(97b), 1434, 1436(250), *1440, 1445*
Bienz, S. 1638(175, 176), 1649(211), *1664, 1665*, 2379(58), *2399*
Bierbaum, V.M. 1109, 1115(38), *1137*, 1642(193), *1664*
Bieri, J.H. 186(31), *257*
Biffar, W. 795, 799(63c), 800(70a), *823*
Bigelow, R.W. 1316(36, 37), *1333*
Bihatsi, I. 1481, 1509, 1510(287), *1533*
Bihátsi, L. 1450(33, 42, 43), 1452(55), 1453(33, 63), 1454(55), 1460(154), 1461(137), 1462(136, 149, 154), 1463(149), 1467(136, 137, 149, 154), 1471(33), 1472(229), 1477, 1478(137), 1481(282), 1491(154), *1525, 1526, 1529, 1531, 1533*
Bikelhaupt, F. 2428(89), *2459*
Bilgren, C. 2513, 2516, 2519, 2526(257a), *2564*
Bill, E. 2159(150), *2174*
Bill, J. 2246, 2250, 2253, 2254(6e), 2257(49c), 2261(63), 2267(49c, 85b), 2269(6e), 2271(85b, 90), 2273, 2285(6e), *2310, 2312–2314*
Billeb, G. 1969, 1976, 1977(26b), *2030*
Biltueva, I.S. 888, 990(122), *1056*
Binder, D. 215(159), *260*

Binder, P. 2327(55), *2356*
Bindl, J. 2227(68), *2242*
Binger, P. 2424(75, 76), *2459*
Binkley, J.S. 155(18, 19), 159(19), 171(18, 19, 128, 129), *178, 180*, 1107(9), *1136*
Biran, C. 1220(97), 1222(101, 102), 1223(103–105), 1224(106), 1225(109), 1229(126), *1231, 1232*, 1609, 1635(75), *1662*
Bird, C.W. 9(31c), *96*
Bird, P.H. 236(254), *263*
Birgele, I. 296, 299, 300(94), 301(94, 96), 304, 306(107), *352, 353*, 1474(241, 245), 1479(245), *1532*
Birgele, I.S. 1449, 1460(6), 1470(6, 207), 1474(240), 1475(250), 1481(6), 1506(6, 207, 250), 1510(6, 250), *1525, 1531, 1532*
Birke, P. 1587(58), *1596*
Birkofer, L. 1450(45), 1501(364), *1526, 1536*, 1897(16), 1903(41a), *1926*
Birot, M. 2246, 2250, 2251, 2253, 2254, 2272, 2273(6a), 2281(113), 2285(6a), *2310, 2314*
Bischoff, R. 2220(19c), 2222(22b), *2241*
Bismo, S. 2545(382), *2567*
Bissinger, P. 207(126, 127), *260*, 1386, 1387, 1389(113), 1434, 1436(250), *1440, 1445*, 2059, 2060, 2066, 2067(52), *2123*, 2528(336), *2566*
Biton, R. 363(32a), *426*, 482(239), *493*, 611, 613(40), *698*
Bjarnason, A. 10, 22, 70(37), *96*, 1110(62, 63), *1137*
Bjorklund, G.C. 2351(247), *2362*
Black, E. 1238(22), 1283(151), *1306, 1309*, 1331(193), *1336*, 1883(67), *1893*, 2422(69), *2459*, 2494(164), *2562*
Black, E.P. 2482(95), *2561*
Black, J.R. 577(158), *593*
Blackman, C.S. 1878(47), *1892*
Blair, I.A. 434, 448(35), *489*
Blake, A.J. 464(175), *492*
Blake, J.F. 184, 222(22), *257*, 482(241), *493*
Blakeman, P. 52, 55, 56, 58, 61, 62(99c), *98*
Blanchard, C.R. 2252(31a, 31c), *2311*
Blanco, F.J. 1860(352), *1868*
Blanco, L. 2389(73), *2400*
Blark, E. 2421(57), *2458*
Bläser, D. 52(99a, 99b), 62(99b), *98*, 884–886, 910, 957, 981, 983, 985, 987, 992, 996(111), 1051(382), *1056, 1062*, 2535, 2536(368), 2537, 2539(370), *2567*
Blechert, S. 1294(182), *1309*
Bleiber, A. 184, 221(20), *257*
Bleidelis, J. 1506(379), 1508(390), 1510(396), *1536, 1537*
Bleidelis, J.J. 1362(54a, 54b), *1438*
Bleidelis, Ya.Ya. 1460(155), 1461(140, 143), 1462(155), 1467(140, 143, 155), 1503(155), *1529*
Bliefert, C. 487(269), *494*
Blinka, Th. 269(8), *350*
Blintjer, B. 245(287), *264*
Blitz, M.A. 1123(163), *1139*
Block, E. 872(56), *1055*, 1677(97), *1684*, 1873(23), *1892*
Blohowiak, K.Y. 226(205), *262*, 1353, 1368, 1369(36), *1438*, 2250, 2253, 2255–2257(21), 2298(150a, 150b), *2310, 2315, 2316*
Blom, R. 77–79(154b), *100*, 1434(232b), *1444*, 2163–2166(182), *2175*, 2534(360b), *2567*
Blondeau, J.P. 1118(99), *1138*
Błoński, S. 2510(236), *2564*
Blount, J.F. 131(38), *151*, 2182(34), *2214*
Bloxom, T. 2318(3), *2355*
Bluestein, B.A. 1700(52), *1786*
Blum, J. 2322(38, 39, 40a, 40b, 41, 42), 2323(39, 40a, 40b), 2325(40b), 2326(40a), *2356*
Blum, Y. 2251–2253(24), *2311*
Blum, Y.D. 2251(28b, 28d), 2253, 2255(28b, 28d, 44a–c), 2261(44a–c, 50, 69), 2263(69), *2311–2313*
Blümel, J. 1386, 1387, 1389(113), *1440*, 2059, 2060, 2066, 2067(52, 53), *2123*, 2159(155), *2174*
Blumenkopf, T.A. 1794(14), *1862*
Blumich, B. 310(123, 126, 127), *353*
Blustin, P.H. 11, 12(39), *96*
Boag, N.M. 1260(74), *1307*
Boatz, J.A. 123(24e, 24g), *151*, 159–161(52), 162(66), 163(52, 67, 72), 172, 173(132, 144), 175(132), *178–180*, 860, 983(10a), *1054*, 1166(151), *1184*, 2479(78, 81), 2505(200), 2511(243), *2560, 2563, 2564*
Bobbitt, K.L. 866(37), 916(183), 971(37), 993, 996(183), *1054, 1058*, 1241(26), 1248(45), 1283(154), *1306, 1309*, 2491(140, 141), *2562*
Boberski, W.G. 1207(50), *1230*, 1312, 1313(14), *1332*
Bobyshev, A.A. 2530(347), *2566*
Bocarsly, A.B. 2153–2155(132), *2173*
Boche, G. 520(67), *553*, 811(97b), *824*
Bochenska, W. 229(216), *262*
Bochkarev, L.N. 2042, 2045(13), *2122*
Bochkarev, V.N. 1481(285, 292), 1482(292), *1533*
Bock, C.W. 6(21b, 21c), 11(43b), 12, 32(43b, 45a, 45b), *95, 96*
Bock, H. 10(32c, 33c), 12(44), 52(96, 97), 53, 54, 61(96), *96, 98*, 207(128), 219(178), *260, 261*, 484(246), 485(250, 254), *493*, 864, 992, 996(30), 1018, 1021(315), *1054, 1061*,

Bock, H. (cont.) 1106, 1109(1b), 1123(159), 1128(220), *1136, 1139, 1141*, 1150(37), 1159(92), *1182, 1183*, 1212, 1213(70), *1231*, 1314(21–23, 26), 1320(80), 1322(99, 103), *1332–1334*, 1603, 1605(38), *1661*, 2131(15), 2132(15, 16), 2165(16), *2171*, 2181(23), 2182(29), *2214*, 2509(225), 2523(317), 2539(225), *2564, 2566*
Bock, W. 540(103), *554*
Böcker, W.D.G. 2261(61), *2312*
Boddeke, H. 2364(22), *2398*
Bodenhausen, G. 310(124), *353*
Bodensieck, U. 2085, 2089, 2094(157), *2126*
Bodenstedt, H. 1991(98), *2033*
Boeckman, R.K.Jr. 417(196), *429*
Boeffel, C. 310(134), *353*
Boehme, C. 53–55, 57, 58, 61, 62(103), *98*, 2539(372a), *2567*
Boer, F.P. 228(210), *262*, 1357(42), *1438*, 1448(3), 1456(115), 1462(3, 115), 1465(3), 1467(115), 1510(397), *1524, 1528, 1537*
Boese, R. 52(99a, 99b), 62(99b), *98*, 215(158), 216(160), 252(326), *260, 265*, 648(121), *700*, 747, 748(94, 95), *776*, 884–886, 910, 957, 981, 983, 985, 987, 992, 996(111), 1043(357), 1051(382), *1056, 1062*, 1884(72), *1893*, 1979(57b, 57c, 57e, 57g, 58b), *2031, 2032*, 2535, 2536(368), 2537, 2539(370), *2567*
Bogaert, R.J. 2548(401), *2568*
Boganov, S.E. 901, 902, 993, 995, 996(159), *1057*, 1170(179), *1185*, 1283(156), *1309*, 1967(20a, 20b), 1968(20c), 1986, 1996, 1997(20a–c), 1998(20c), 2013(172a), 2027(204), *2030, 2035, 2036*, 2474(54), 2504, 2515(54, 197), *2560, 2563*
Bogen, S. 1566(81a, 81b, 82), 1567(82), *1579*
Bogey, M. 3(7a, 7b), *95*, 864, 981, 982(31–33), 1024, 1046(333), *1054, 1061*, 1128(222), *1141*, 1157(86, 87), *1183*
Bögge, H. 833(34), *854*, 1099(75), *1103*, 2166(191), 2167(194), *2175*
Boggs, J.E. 184, 192(17), *257*, 1468(195), *1530*
Bogillo, V.I. 2228, 2229(76), *2242*
Bogoradovsky, E.T. 304(107, 109), 306(107), *353*
Böhme, D.K. 3(7c), 36(75), 68(121), *95, 97, 99*, 161, 175(58), *178*, 597, 601(3, 4), 607(3), 608(4), 697, 1028(339, 340), 1029(341), *1061*, 1109(43–47, 50), 1117(44, 45, 50, 84, 86, 88), 1118(93–95), 1122(43–46, 153), 1123(157, 158), 1125(153, 178, 179, 183), 1126(50, 194, 200, 205), 1129(47), *1137–1140* 1372(68), *1439*, 2556(423), *2568*

Boileau, S. 1769(290), *1791*, 2218(11a–d), 2226(50), *2240, 2242*
Boilot, B.-P. 2327(55), *2356*
Boilot, J.P. 2330(81), 2333(97), 2350(236), 2351(252), 2352(81, 266, 267), 2354(81), *2357, 2358, 2361, 2362*
Bois, C. 2075, 2091(112), *2125*
Bois, L. 2294(142), *2315*
Boisen, M.B.Jr. 104(2), 106, 107(6), 108(7), 109(8–10), 110(17), 111(20), 115(2, 35, 40), 116(41), 117(2), *117, 118*
Bokerman, G.N. 1593(114), *1597*
Bokii, N.G. 1897(17), *1926*
Boldi, A.M. 186(33), *257*
Boldskul, I.E. 211(142), *260*
Boldyrev, A.I. 1127(211), *1140*, 1372, 1373(70), *1439*
Bolestova, G.I. 1794(11), *1862*
Boley, T. 2151(118), *2173*
Bolm, C. 1671(37), *1683*
Bols, M. 1674(67b), *1683*, 1794, 1819, 1853, 1854(28), *1862*
Bolts, J.M. 2153–2155(132), *2173*
Bolvin, H. 1157(86), *1183*
Bommel, M.J.van 2335(108), *2358*
Bonaccorsi, R. 528, 535(87b–d), *554*
Bonafoux, D. 1223(105), 1229(126), *1232*
Bonasai, P.J. 1875, 1890(31), *1892*
Bonasia, P.J. 1888(95), 1890(106, 110, 111), *1894*
Bondybey, V.E. 1156(74), *1182*
Bone, R.G.A. 90(170g), *101*
Bongini, A. 435, 456(55), *489*
Bonini, B.F. 473(207), *492*, 1610, 1630(79, 80), 1638(174), 1644, 1649, 1650(196), 1658(224), *1662, 1664, 1665*, 1830(197), *1865*
Bonny, A. 2132(19), *2171*
Boo, B.H. 167, 171, 172(95), *179*, 532(100), *554*, 559(28), *590*, 1107(13), 1118, 1123(90, 91), *1136, 1138*, 1965(18), 1966, 1967, 1997–1999(19), *2030*, 2484, 2502(106), 2503(196), 2511(245), *2561, 2563, 2564*
Boo, D.W. 530(94), *554*
Boom, J.H.van 475(216), *493*, 1681(124), *1685*
Boon, W.H. 1991, 1992(96b), *2033*
Boorman, P.M. 1885(77), *1893*
Boos, H. 1323(113, 115), *1335*
Boot, C.E. 2152, 2153(123), *2173*
Borbaruah, M. 500(27), 508(57), *510, 511*, 1390(138), 1392(144), 1398, 1401–1403(138), 1404, 1420(144), *1441, 1442*
Borcic, S. 622(71), *699*
Bordeau, M. 1217(84), 1220(97), 1222(101, 102), 1223(103–105), 1224(106), 1225(109), 1226(115), 1229(126), *1231, 1232*

Borden, W.T. 90(170a), *101*, 168, 169(103), *179*, 380(74), *427*, 645(112), *700*, 748, 750(99), *777*, 1998(117), *2034*
Bordwell, F.G. 382(83), *427*, 849(76–78), *855*
Borgdorff, J. 2107(214), *2127*
Borisova, J.V. 2155(141), *2174*
Borisova, L. 1451(48a, 48b), 1459, 1467(135), *1526, 1529*
Borkowsky, S.L. 1639(180), *1664*
Börlin, K. 1106(7), *1136*
Borm, J. 2273–2275(99c), *2314*
Borman, S. 559(16), *590*
Bormann, H. 203(109), *259*
Born, D. 747, 748(93, 94), *776*
Börner, U. 784, 787(23), *822*
Borrmann, H. 1034(352), *1061*
Borsella, E. 2522(286), *2565*
Bortolin, H. 2439(144), *2460*
Bortolin, R. 186(45), *258*
Bos, H.J.T. 1601, 1624(22b), *1660*
Bosch, E. 1576(100), *1579*
Bosman, W.P. 2152, 2153(123), *2173*
Bosnich, B. 1756(250–253), 1757(250), *1790*, 1801(65), *1863*
Bosold, F. 811(97b), *824*
Botoshansky, M. 437, 485(79), *490*
Bott, G. 2471(38), *2559*
Bott, R.W. 383(86), 388(106), *427, 428*
Bott, S.G. 1169, 1177(172), *1185*, 2514, 2515(275), *2565*
Bottcher, P. 1890(105), *1894*
Botteghi, C. 1743(218), *1790*
Bottoni, A. 6(20), *95*, 919, 921(187, 188), *1058*
Bouchoule, A. 1118(99), *1138*
Boudin, A. 1415(201b), *1443*, 1972(38), *2031*
Boudjouk, P. 4(10c), 11, 33(41a), 37(10c, 79, 80), 38(79, 80), 39(10c), 40(79), 42, 43(41a), 46(79), *95–97*, 284(50), *351*, 566, 570, 572(101), *592*, 813(104), 814(106, 109, 110a), 815(111), 818(115), *825*, 984, 987, 989(265), *1059*, 1097(73a, 73b), 1098(73b, 74), *1103*, 1218(95), *1231*, 1238(21, 22), 1247(40), 1283(151), 1289(168), *1306, 1309*, 1319(67), 1325(125), 1331(192, 193), 1332(201), *1334–1336*, 1699(51), 1700(54), *1786*, 1879(57), 1883(66, 67), 1887(57, 91, 92), *1893, 1894*, 1897(9), 1904(45), 1907(53), 1918, 1923(86), *1925–1927*, 2019(186), 2020(186, 188), 2021(189), 2022, 2023(194), 2024(197), 2025(199), *2035, 2036*, 2421(57), 2422(68, 69), *2458, 2459*, 2475, 2477, 2482, 2483(56), 2494(164), *2560, 2562*
Boufendi, L. 1118(99), *1138*
Bouffard, F.A. 1607, 1634(58), *1661*
Bouisset, E. 2523(321), *2566*

Boulton, J.M. 2350(237), *2361*
Bouma, W.J. 1125(180), *1140*
Bouman, T.D. 516(33), *552*
Bouquet, P. 1547(37), *1578*
Bourgeois, P. 385(94), *427*, 1614(92, 96), 1618(108), *1662*
Bourne, S.A. 222(186, 188, 190), *261*
Bourque, R.A. 1147(25), *1181*, 1646(204), *1665*
Boutevin, B. 1769(288, 289), *1791*
Bower, J.E. 1120(121, 127), *1138*
Bowers, M.T. 1470(219), *1531*
Bowie, J.H. 284, 288(52), *351*, 1109(40, 42), 1133(250, 251), 1134(253), *1137, 1141*, 1372(66), *1439*
Bowling, R.A. 10(32c), *96*
Bowry, V.W. 1542(14), *1577*
Bo Yang 2227(57), *2242*
Boyd, D.B. 515(23), *552*
Boyd, P.D.W. 283(46), *351*, 529, 530(91), *554*, 567, 569, 570, 583, 585, 587, 588(106), *592*
Boyd, R.J. 499(23), *510*
Boyer, J. 505(35), *511*, 1064, 1069, 1082–1084, 1086, 1099(9), *1101*, 1387, 1388(114, 117), *1440, 1441*, 1738(161, 162, 164), *1789*
Boyer, R.D. 2289(125), *2315*
Boyer, R.J. 1487(306), *1534*
Boyer-Elma, K. 1418, 1420, 1423, 1427(214), *1444*
Boyle, F.T. 1670(32), *1683*
Boyle, T.J. 2134(26), *2171*
Bradaric, C. 864(34, 35a, 35b), 932, 934(34), 992(35a, 35b), *1054*
Bradaric, C.J. 849(82–84), 852(82–84, 87), *855*, 1241(25b), *1306*, 1332(202, 203), *1336, 1337*
Braddock-Wilking, J. 900(145a), *1057*, 1257(65), *1307*, 1325(147), *1335*, 1431(230), *1444*
Bradley, A.J. 2504(199), *2563*
Bradley, C. 1214(77), *1231*
Bradley, M. 1675(82), *1684*
Bradshaw, J.D. 186(32, 34, 35), *257*
Brady, B.K. 2344(155), *2359*
Brady, K.A. 1723(127), *1788*
Braga, A.L. 1801(63), *1863*
Brakmann, S. 1649(208, 209a), *1665*, 2363(9), 2378(52, 55), 2379, 2382(57), *2398, 2399*
Brammer, L. 1431(230), *1444*
Brandenburg, A. 2340(134), *2359*
Brandl, P. 1743(196, 197, 199), 1748(196), *1789*
Brandow, C.G. 66–68(119), *99*
Brand-Roth, B. 1025(334), *1061*
Brandsma, L. 1601(22b), 1621(123), 1624(22b), *1660, 1663*, 1991(98), *2033*

Brandt, M.S. 2196(78a), *2215*
Brandt-Roth, B. 2492(146, 147), *2562*
Brans, A. 2363, 2379, 2380, 2385(7), *2398*
Brard, L. 1697, 1698, 1709, 1710, 1755(43), *1786*, 2039–2041(5), *2122*, 2153(126), *2173*
Brasselet, S. 2352(264), *2362*
Bratovanov, S. 2379(58), *2399*
Bratt, K. 435, 455(47), *489*
Braude, V. 634(87), 685(156), *699, 701*, 866, 884–886, 917, 924, 925, 928, 957(39), 966, 982, 983, 1001(246b), *1055, 1059*, 2429(100), *2459*
Brauer, D.J. 193(61), *258*
Brauer, H. 1969, 1976, 1977(26a, 26b), *2030*
Brauman, J.I. 168(105, 109), 169(105), *179*, 381(80, 81), 382(80), *427*, 1109(33, 37), *1136*
Braun, K. 1918(89), *1927*
Braun, R. 2223(27c), *2241*
Braun, S. 661(136), 669(146), *700, 701*, 2326(43, 44a, 44b, 45a, 46a, 46b, 48, 50), 2327, 2347, 2348(44a, 44b), *2356*
Braunschweig, H. 209(133), *260*
Braunstein, P. 2057(50), 2072(84–95), 2073(96, 100, 101), 2074(87, 102–105), 2085, 2089(157), 2090(84, 85, 87, 88, 90, 91, 94, 105), 2091(86, 89, 94, 96, 100, 102), 2094(157), 2110(102), *2123–2126*
Bravo-Zhivotovskii, D. 795, 800(69a, 69b), 801(69b), *823*, 866(39), 884–886(39, 111), 910(111), 917, 924, 925, 928(39), 957(39, 111), 958(241), 981, 983, 985, 987, 992, 996(111), (191), *1055, 1056, 1058, 1059*, 2122(258), *2128*, 2429(100), *2459*
Bravo-Zhivotovskii, D.A. 888, 990(122), *1056*, 2044, 2051(19), *2123*
Bravo-Zhivotovskii, O. 437, 485(79), *490*
Bravo-Zhivotskii, D. 2499(186), *2563*
Braye, E.H. 1971(32), *2031*
Breault, G.A. 1854(319), *1868*
Brefort, J.L. 508(55), *511*, 781(9), *822*, 1356, 1357(39), *1438*, 2016(176c), 2017(179), *2035*, 2330(85), *2357*
Breidung, J. 864, 981, 982(31, 33), *1054*
Breitscheidel, B. 2340(136), *2359*
Breker, J. 1878(45), *1892*
Breliere, C. 287(64), 541(106b), *554*, 581(179), *593*, *352*, 1374, 1375(79), 1377(79, 99), 1378(99), 1380(79), 1409, 1411(185a), 1412(185a, 191, 194), 1414(191), 1418(213, 217), 1420, 1421(213), 1422(194), 1423(194, 213, 217), 1424(217), 1427(194, 217), 1428(217), 1430(227, 228), 1431, 1432, 1434(228), *1439, 1440, 1443, 1444*
Brendler, E. 2468(23), *2559*

Brengal, G.P. 405(153), *429*
Brengel, G.P. 478(229), *493*, 1848(285), *1867*
Breslow, R. 1998(122b), *2034*
Breton, S. 1110(61a), *1137*
Bretschneider-Hurley, A. 2159(159), *2174*
Brett, A.M. 6(21b, 21c), *95*
Breuckmann, R. 1903(40b), *1926*
Brewer, J.C. 1280(138), *1308*
Brey, W.S. 301(98), *352*
Brian, C. 1217(84), *1231*
Brianse, N. 2053(36), *2123*
Brickhouse, M.D. 1116(74, 75), *1137*
Bridson, J.N. 1620(118), *1663*
Brigaud, T. 1641, 1652(188), *1664*
Bright, F.V. 2326(47), 2327(56), 2350(226), *2356, 2357, 2361*
Brinker, C.J. 2246(1), 2269(86), 2294, 2295(1), *2310, 2313*, 2319(9, 10e), 2347(186c), 2350(9, 10e), *2355, 2360*
Brinkman, E.A. 168(109), *179*, 381(81), *427*, 1109(37), *1136*
Brintzinger, H.H. 2153(128, 129), 2158(148), *2173, 2174*
Brisdon, A.K. 1878(47), *1892*
Brisdon, B.J. 1925(109), *1928*
Brisse, F. 229(223), *262*
Brisset, H. 1189, 1190(9), *1230*
Brittain, J. 1678(100), *1684*, 1874(26), *1892*
Britten, J. 2043, 2051(16), *2122*
Britten, J.F. 275, 276(21), *350*, 580(168), *593*
Brittingham, K.A. 2115(234), *2127*
Britton, D. 196(76), *258*
Brix, Th. 163, 174, 175(71), *179*
Brodie, N. 2261(64), *2312*
Brodskaya, E.I. 752, 760(110), 777, 1381(108), *1440*, 1454(79, 105), 1467(105), 1469(204), 1470(216, 217), 1471(216, 217, 221–223, 226), 1472(216, 217, 221, 233a, 233b, 234), 1479(277, 278), 1480(277), 1484(294), 1485(299), 1507–1509(383), *1527, 1528, 1531, 1533, 1534, 1536*
Broka, K. 1199(31), 1200(32), *1230*
Broka, K.A. 1480, 1481(280, 281), *1533*
Brondani, D. 1341, 1347(24), *1437*
Bronnimann, C.E. 311, 312(135), *353*
Bronshtein, A. 2320, 2322(19), 2327(19, 52), 2340(132), 2348(19), *2355, 2356, 2359*
Brood, A.G. 1601(24–26), 1604(26), 1609, 1645(24, 26), *1661*
Brook, A.G. 3(3a, 3b, 4), *95*, 196(74), 250(305–307), *258, 264*, 416(186), 425(216), *429, 430*, 487(263), *494*, 578(159), *593*, 713(35), 743, 750(84), *775, 776*, 789, 790(28b), 795(57, 62c, 63b), 796–798(57), 799(62c, 63b), *822, 823*, 827, 828(6), 844(6, 56), 845(65, 66), *854, 855*, 859(3), 865, 866(38), 880(3, 38, 86, 87,

Author index

89–94), 881(3, 38, 86, 87, 89, 90, 93–96), 882(95), 883(100), 885, 890(116), 913(38, 90, 96, 180), 914(90, 93, 94, 181), 915(95), 917(90, 180), 919(86), 922(38, 86, 89), 923(89), 924(86, 89, 90, 92, 95), 928(86), 932(95), 937, 938(116), 939(202), 949(181), 950(181, 235), 951(236, 237, 238b, 239), 953(240), 954(95, 240), 955(240), 956(86, 95), 959–961(235), 962, 963(244), 964(244, 245), 965(245), 966(244, 245), 967(94, 247), 968, 969(248), 977(3), 978(3, 95), 981(3), 985(3, 38, 90, 91, 93, 95), 986(3, 38, 90, 91, 93, 95, 181), 992(3, 90, 91), 996(3, 89, 91), 1004(90, 181), *1054–1059*, 1064, 1068(1, 4c, 8a), 1080(45, 47), *1101, 1102*, 1145(14, 15), 1146(21, 22), 1147(24), 1155(66, 67), *1181, 1182*, 1234(2, 4), 1236(12), 1238, 1249, 1270(2), 1271(105, 106), 1272(108), 1273(2, 108, 109), 1274(111–122), 1276(2, 109, 112, 123, 124, 127), 1282(144), 1285(109), *1305, 1308*, 1312(1, 2), 1332(2), *1332*, 1600(2, 4–6), 1601(5, 6, 17), 1602(5, 31), 1603(38), 1604(5, 17, 31), 1605(5, 38), 1606(5, 56), 1608(2, 60), 1609(17, 31, 72), 1611(4, 5, 83), 1612(4), 1614(91), 1631, 1632(5), 1633(5, 153, 157, 158), 1638(171), 1639(5, 182), 1640(5), 1642(157, 158), 1645(17, 72, 201, 202), 1646(5, 203, 205), 1647(5, 60), *1660–1665*, 1881, 1887(62), *1893*, 2404(9–12), 2407(23), *2457, 2458*

Brook, G.A. 860, 868, 934(9), *1054*
Brook, M.A. 3(3b), *95*, 275, 276(21), *350*, 363, 368(29), 369(46), 372(57), 391(115), 392(116, 117), 396(131), *426–428*, 438(95), *490*, 580(169), *593*, 604(19), 605, 606(19, 21), 616(46), 626(78, 79), 629(21), 630(46), *698, 699*, 860, 868, 934(9), *1054*, 1064, 1068(8a), *1101*, 1282(144), *1308*, 1771(301), *1791*
Brook, R.J. 2261(54a), *2312*
Brookhart, M. 1705, 1707, 1708(72), *1787*
Brough, L.F. 2178(6), 2181(32), *2214*
Brough, P.A. 416(184), *429*
Brown, C.A. 72, 74(146), *100*, 149(61), *152*
Brown, D.P. 234(237), *263*
Brown, H.C. 361(12), *426*, 618(61), 635(89c), 647(118), 677(151, 152), *699–701*
Brown, I.D. 106, 110(4), *117*
Brown, J.F. 1923(96), 1925(108), *1928*, 2334(106), *2358*
Brown, R.S. 358, 359(9), *426*, 616(51), *698*
Brown, S.S.D. 186(45), *258*, 509(65), *511*, 1299(196), *1310*, 1374, 1376(93), *1440*, 2439(144), *2460*
Brownfain, D.S. 1606, 1611(47), *1661*

Brownlee, R.T.C. 361(16), *426*
Brown-Wensley, K.A. 1693(25), *1786*, 2273(98), *2314*
Bruce, M.I. 2445(163), *2461*
Brück, M. 2253(36), 2261(61), 2269, 2270(88f), *2311–2313*
Brück, S. 2354(286), *2362*
Bruckmann, P. 2429(96), *2459*
Brückmann, R. 751(103), 752(103–106), 760(103), *777*
Brückner, R. 435(43), 453(43, 130), *489, 491*, 1849(290), *1867*, 1871(10), *1891*
Brumme, J. 1587(58), *1596*
Brummond, K.M. 402(145), *428*, 1795(34), *1862*
Brun, A. 2350(236), 2351(252), *2361, 2362*
Brun, M.-C. 2083(147, 148), *2126*
Bruna, P.J. 1067(17), *1101*, 1107(14), *1136*
Brunet, J.C. 1981(64), 1998(64, 124a, 124b), 1999(124a, 124b), 2001(64, 124b), *2032, 2034*
Brunner, H. 1688, 1733(1), 1743(1, 189, 195–201, 203–205, 218, 224), 1746(195, 201, 203), 1747(203), 1748(196), *1785, 1789, 1790*
Bruno, G. 2165(185), *2175*
Brusetin, G. 2345(177), *2360*
Brusilovsky, D. 2350(234), 2351(249), *2361, 2362*
Bryan, J.C. 2065, 2066(73, 74), *2124*
Bryant, G.L.Jr. 190(48), *258*
Bubaraju, J. 842(50), *855*
Buch, R.R. 2229(79), *2242*
Buchanan, A.C.III 2193(72), *2215*
Bucher, G. 704, 709(7), *775*
Buchholz, H.A. 521(74, 77), 522, 529(77), 530(74), 532–534(77), *553*, 568–570(109), 584(109, 188), 585(109), *592, 593*, 652(129, 130), *700*, 1225(110–112), *1232*
Buchholz, U. 2238(125), *2244*
Büchner, K. 2251, 2253, 2255, 2257(25e), *2311*
Buchner, M. 1387, 1389(121a, 121b), *1441*
Büchner, W. 1189, 1191(7), *1229*
Buchner, W. 2061, 2066–2068(60), *2123*
Buchwald, S.L. 1671(42), *1683*, 1741(184), 1743, 1746, 1747(225), 1749(229), *1789, 1790*, 1992(99, 100), *2033*, 2042(14), 2044(18), 2051(14, 18), 2052(18), *2122*
Buck, C.E. 1606, 1628(46), *1661*
Buck, H.M. 1473(237), *1532*
Buckle, M.J.C. 408(164), 413(175), *429*
Buckley, A.M. 2319(12a), *2355*
Buckley, L.J. 1767(280), *1791*
Buckwalter, B.L. 1397, 1398, 1404(170), *1442*

Buda, A.B. 288, 289(68), *352*
Budinger, P.A. 2289(125), *2315*
Budzichowski, T.A. 1923(94, 100), 1924(105), *1928*
Buechner, W. 1582(10), *1594*
Buell, G.L. 1603, 1609(28), *1661*
Buell, G.R. 435, 456(49), *489*
Buenker, R.J. 1107(14), 1128(226), *1136, 1141*
Bueno, A.B. 470(194), *492*
Buffy, J.J. 345, 347–349(154), *354*, 985(270), *1059*
Bugaeva, S.B. 1377(98), *1440*
Bugerenko, E.F. 1458, 1467(127), *1529*
Bühl, M. 7(27), 69(129, 132, 133), 72(129), 74(132, 133), 77(132), *96, 100*, 516(35b), 517(38), *552*
Buhler, H. 2318, 2320(4), 2340(136), *2355, 2359*
Buijink, J.-K. 1431(231), *1444*
Bujalski, J. 2285(117c), *2315*
Bujarrabul, V. 2523(320), *2566*
Bukowinski, M.S.T. 115(33), 116(41), *118*
Buller, B. 1986, 2011(76b), *2032*
Bulliard, M. 1555(58c), *1578*
Bullock, W.J. 167, 169, 170(92), *179*
Bümel, J. 2528(336), *2566*
Buncel, E. 789(13d), 790(36), 792(46), 793(36, 46, 47b, 51a, 51b), 794(47b, 51b), *822, 823*, 2453(203), *2461*
Bunge, A. 66(118c), *99*
Bunker, P.R. 704(9), *775*, 1166(155), *1184*
Bunnelle, W.H. 435, 455(48), *489*
Bunte, E.-A. 571, 574(131), *592*
Bunte, E.A. 1434(232a), *1444*, 2166(189, 190), 2167(189, 190, 193), 2168(189, 196), 2169(189, 198), 2170(189, 200), *2175*, 2534(361c), *2567*
Bunz, U. 1910, 1911(60e), *1927*
Buono, G. 1743, 1746(223), *1790*
Buravtseva, E. (36), *1054*
Burdasov, E.N. 163(69), *179*
Burdett, J.K. 77, 79(152a), 82(157c), *100*
Burdorf, H. 2152, 2153(122), *2173*
Burdulu, F.Y. 2142(64), *2172*
Bures, E.J. 447(115), 454(134), *490, 491*
Burford, C. 422(204), *429*
Burgdöfer, G. 2439(153), *2460*
Bürger, H. 183(10), *257*, 780(2d), 794(2d, 55), *821, 823*, 864, 981, 982(31–33), *1054*
Burger, H. 193(61), *258*, 1237(17), *1306*
Burger, K. 1377, 1378, 1380(97b), *1440*
Burger, P. 2106(208), *2127*
Burgers, P.C. 601, 607(14), *698*, 1131(240), *1141*
Burgess-Henry, J. 1676(90), *1684*
Burggrae, L.W. 1372(64), *1439*
Burggraf, L.W. 228(211), 262, 505(36, 37), 508(54), *511*, 1109, 1126, 1130(35), *1136*, 1372, 1373(71b–d, 72a, 72b), *1439*, 1467(190), *1530*, 2298(152a), *2316*
Burği, H.B. 110(14), *118*, 420(201), *429*, 1341(19), 1397, 1401(19, 168), *1437, 1442*, 1463, 1467(164), *1530*
Bürgy, H. 110(14), *118*
Burk, M.J. 1750(231), *1790*
Burk, R.M. 396(129), *428*, 1817(133), *1864*
Burke, S.D. 1823(153–155), *1864*
Burkey, D.J. 209(138), *260*
Burkhard, C.A. 2177, 2193(2), *2213*
Burkhart, D.J.M. 113(23), *118*
Burkhart, T. 2340(138), *2359*
Burlakov, V.V. 2051, 2052(34), *2123*
Burmiester, M.J. 2233(97), *2243*
Burns, C.J. 205(115), *259*, 2065, 2066(73, 74), *2124*
Burns, E.G. 1963, 1986, 1998(1a), *2030*
Burns, G.T. 10(32b, 32c, 33a), *96*, 994, 995(283), *1060*, 1148(32), 1149(33), *1182*, 1964(5), 1965(5, 11), 1967, 1984(5), 1995(106), 1997(5, 106), 1998(119), *2030, 2033, 2034*, 2155(140), *2174*, 2232(96), *2243*, 2251(27a), 2253(27a, 33), 2261(55), *2311, 2312*, 2412(33, 34), *2458*
Burns, M.R. 1635(162), *1664*
Burns, S.A. 1965(9a), *2030*, 2465, 2489(11), *2559*
Bursey, M.M. 561(57), *591*, 1122(145), *1139*, 1312, 1313(17), *1332*
Burshtein, Z. 2350(234), 2351(249), *2361, 2362*
Burstinghaus, R. 1608(65), *1662*
Burton, D.J. 1828(193), *1865*
Burton, G.W. 1557(62), *1578*
Busch, R. 1903(40b), *1926*
Busch, T. 123(24c), *151*, 1053(396, 397), *1062*
Bush, L.W. 1211(67), *1231*
Bush, R.D. 1080(43a), *1102*, 2406(18), *2458*
Busi, F. 1610, 1630(79), *1662*
Buskas, T. 1872(16), *1892*
Buske, G.R. 2005(154), *2034*
Buss, J.H. 155, 156, 176(29), *178*
Buss, R.J. 2346(181), *2360*
Butakov, K.A. 2152(116), *2173*
Butcher, R.J. 449(124), *491*
Buterakos, L.A. 109(9), *117*
Butler, I.S. 200(96), *259*, 2179(15), *2214*, 2249(18), *2310*
Butler, J.R. 2153, 2154(133), *2173*
Butler, W.M. 1986, 2012(77), *2032*, 2054(41), *2123*
Buttle, L.A. 1566(80), *1579*
Büttner, H. 1156(70), *1182*
Buttrus, N.H. 577(151), *593*

Butts, M.D. 1709(84), *1787*
Buxton, S.R. 746(91), *776*
Buynak, J.D. 222(192), *261*, 457(149), *491*, 1638(167, 170, 173), *1664*
Buzek, P. 521–523, 527, 529, 542, 544(73), *553*, 567(107), 584(189), *592, 594*, 597, 599, 603–605(7), 652(127), 690(7), *697, 700*
Bye, T.S. 1606, 1628(46), *1661*
Bystrikov, A.V. 2531(351), *2566*

Cabeza, J.A. 2075(111–114), 2091(111, 112), *2125*
Cacace, F. 561(62, 64), 562(62, 68, 69), *591*, 608(30), *698*
Cai, G. 437, 446(85, 86), *490*
Cai, S. 2224(42a), *2241*
Caille, J.-C. 1842(258), *1867*
Calabrese, J. 4, 37, 39, 40, 44, 48(10b), *95*, 818, 819(116), *825*, 981, 984, 987, 989(260), *1059*, 2023(196), *2036*
Calabrese, J.C. 1963, 1986(1a, 1b), 1988(1b), 1998(1a, 1b), *2030*, 2112, 2120(227), *2127*, 2436(126), *2460*
Calabrese, S.C. 2209(99), *2216*
Calas, R. 386(98–100), 411(169), *428, 429*, 1592(108), *1597*, 1609(75), 1614(96), 1635(75), *1662*, 1833(209), *1866*
Calbo, D.A. 84, 87, 88(165a), *101*
Caldwell, R.A. 2437(134), *2460*
Calestani, G. (231), *262*
Calvert, D. 2345(171), *2360*
Calvert, P. 2247(12), *2310*
Calzaferri, G. 110(14), *118*, 2107, 2109(213), *2127*
Camail, M. 457(148), *491*
Camaioni, N. 1210(60), *1231*
Cambie, R.C. 1651(216), *1665*
Cameron, R.A. 1585(40, 41), *1595*
Cammenga, H.K. 2364, 2365, 2367(20), *2398*
Campana, C.F. 247(296), 248(302), *264*, 348(156), *354*
Campbell, B.M. 2155(139), *2174*
Campbell, D.J. 198(84), *259*
Campbell, V.A. 1875(35), *1892*
Campiom, B.K. 2080, 2082, 2092, 2093(135), *2125*
Campion, B.K. 250(315), *265*, 789(28c), *822*, 1606(49), *1661*, 2039(2), 2079(125), 2082(125, 142), 2092, 2093(125), *2122, 2125*, 2425(85, 86), *2459*
Campostrini, R. 2264(78), *2313*
Camus, J. 2364(21, 24, 25, 27, 29), 2374(41–43), *2398, 2399*
Canadell, E. 1996(108e), *2033*
Candela, G.A. 2162(173), *2174*

Caneiro, W.de M. 687(158), *701*
Cannady, J.P. 1593(114), *1597*
Cano, A. 2157–2159(146), *2174*
Canva, M. 2350(236), 2351(252), *2361, 2362*
Cao, D.H. 2137(35), 2143(75), *2171, 2172*
Cao, W. 2339(126), *2358*
Cao, Y. 530(93), *554*, 1107(17), *1136*
Capdevila, A. 434, 447(31), *489*
Capka, M. 1701(60–62), 1702(61), *1787*
Caporiccio, G. 1769(289), *1791*
Caporusso, A.M. 1696, 1717(35), *1786*
Capozzi, C.A. 2352(268, 269), *2362*
Capperucci, A. 1601, 1602, 1610(10), 1625(130), 1626(132), 1628(137–140), 1635(10), 1650(214), 1655(137), 1656(137–140), 1658(132), *1660, 1663, 1665*, 1677(98), *1684*, 1796(38), 1830(197), *1862, 1865*, 1871(14), 1877(40, 41), *1892*
Carberry, E. 1213(71), *1231*, 1314(29, 30), *1333*, 2177(3b), 2178(7), 2193(70), *2213–2215*
Carciello, N. 2261(51), *2312*
Carda, M. 437, 438(84), *490*
Cardi, N. 1545(27), *1578*
Cardona, M. 2196(78a), *2215*
Cardoso, A.M. 2143(80), *2172*
Care, F. 287(64), *352*
Carey, F.A. 2437(130), *2460*
Carey, R.N. 1211(65), *1231*, 1322(100), *1334*
Cargioli, J.D. 307(110, 111), *353*, 1341, 1343, 1409, 1422(21a), *1437*
Carini, D.J. 412(173, 174), *429*
Carlson, C.W. 1157(77), 1169, 1176–1180(169), *1183, 1185*, 1283(147), *1308*, 1329–1331(163), *1336*, 2181(33), *2214*, 2471(43), 2497, 2498, 2514–2516, 2534(179), *2559, 2563*
Carlson, J.G. 2345(169), *2360*
Carlstrom, J.E. 2523(319), *2566*
Carneiro, J.W.d.M. 567(107), *592*
Carneiro, T.W.de M. 530, 532(95b), *554*
Carolczak, J. 2523(314), *2566*
Carpenter, I.W. 163(75), *179*
Carpenter, J.C. 2230(80), *2242*
Carr, R.W. 171(120–122), *180*, 2476(62), *2560*
Carre, F. 541(106b, 106c), *554*, 581(179), *593*, 1064, 1069, 1082–1084, 1086, 1099(9), *1101*, 1374, 1375(77, 79), 1377(77, 79, 99), 1378(77, 99), 1380(79), 1382, 1383, 1385(77), 1387, 1388(114), 1409, 1411(185a), 1412(185a, 191, 192, 194), 1414(191, 192), 1417(192, 205, 206, 207a), 1418(192, 205, 206, 207a, 215), 1422, 1423(194, 206, 215), 1424(77, 215), 1427(194), 1428(206, 215), 1429(215), 1430(228, 229), 1431(228–230), 1432(228,

Carre, F. (*cont.*) 1434(228, 229), *1439, 1440, 1443, 1444*, 1450, 1464, 1465(44), *1526*, 1972, 1981(37b), *2031*
Carre, F.H. 503(33), *511*, 1341, 1347(24), 1374, 1375, 1377(78), 1418, 1420, 1423, 1427(214), *1437, 1439, 1444*
Carré, T. 2528(335), *2566*
Carreño, M.C. 470(194), *492*
Carrillo, F. 2055–2057(44), *2123*
Carrol, P.J. 250(314), *264*
Carroll, K.M. 1582, 1583(11), *1594*
Carroll, M.R. 186(38), *258*
Carroll, M.T. 500(25), 508(54), *510, 511*, 1372, 1373(71a), *1439*, 1467(190), *1530*, 1911, 1913(73, 74a), *1927*, 2443(162), *2461*
Carroll, P.J. 229(221), *262*, 437, 484(76), *490*, 1032, 1040(348–350), 1042, 1043(349, 350), 1044, 1045, 1050(349), *1061*, 2049(30–33), 2051(30, 31), 2052(30), 2054(42), 2055(32, 42, 43), 2056(42, 43, 46), 2057(42), 2062(63, 65), 2064(71, 72), 2065(72), 2066(63, 65, 72), 2068(63, 65), 2069(65, 72), 2080, 2092(129), *2123–2125*, 2425(87), *2459*
Carsten, D.H. 1672(48), *1683*
Carter, E.A. 705, 706(11), *775*, 2546(389), *2568*
Carter, M.B. 1743, 1746, 1747(225), *1790*
Carter, P. 544(111b), *555*
Cartledge, F.K. 441(99), *490*, 497(12, 13), 507(47), *510, 511*, 611, 612(37), *698*, 2179(8), *2214*
Cartledge, F.R. 362(22), *426*
Cary, D.R. 1889(99, 100), 1891(114), *1894*
Casalbore-Miceli, G. 1210(60), *1231*
Casanova, J. 662(138), *700*, 1225(110), *1232*
Casares, A.M. 406(155), *429*
Casellato, U. 2053(36), *2123*
Casey, C.P. 2134(28), *2171*
Casida, M.E. 517(44c), *553*, 658(134b), *700*
Castaño, A.M. 1828(184), *1865*
Castedo, L. 1855(326), *1868*
Castellano, F.N. 2349(206), *2361*
Castellino, S. 4, 37, 39(10c), *95*, 818(115), *825*, 984, 987, 989(265), *1059*, 1883(67), *1893*, 2022, 2023(194), *2036*, 2422(69), *2459*
Castillo-Ramirez, J. 2142(67), *2172*, 2213(110), *2216*
Castro, P. 396(130), *428*
Cattoz, R. 1583(19), *1595*
Caulton, K.G. 2079(127, 128), 2093(127), *2125*
Cavalieri, J.D. 147, 148(58a, 58b), *152*, 345, 347(154), 348(154, 155), 349(154), *354*, 985(270), *1059*
Cavé, A. 1675(74), *1684*

Cave, N.G. 2235(108), *2243*
Cavelier, F. 1674(68), *1683*
Cavezzan, J. 2007(158b), *2035*, 2231(82), *2242*
Cazacu, M. 2223(24), *2241*
Cazaux, F. 2226(56), *2242*
Cederbaum, F.E. 1986(80c), *2032*, 2434(121), *2460*
Cederbaum, L.S. 1135(267, 268), *1142*
Cella, J.A. 1341, 1343, 1409, 1422(21a), *1437*, 2230(80), *2242*
Cerfontain, H. 1797(49), *1863*
Čermák, J. 487(269), *494*
Cerrau, G. 2528(335), *2566*
Cerreta, F. 1677(98), *1684*, 1877(40), *1892*
Cervantes, F. 437, 485(77), *490*, 1602, 1604, 1605(32), *1661*
Cervantes, J. 2083(147), *2126*
Cervantes, L. 2142(67), *2172*
Cervantes-Lee, F. 437, 485(78), *490*, 1237(18b), *1306*, 2083(147), *2126*, 2143(72), 2155(135), *2172, 2173*, 2213(110), *2216*
Cerveau, G. 1367, 1369(55), 1415(201b), 1417, 1418(205, 206), 1422, 1423, 1428(206), *1438, 1443*, 1450(39, 44), 1457(39), 1464, 1465(44), 1480(279), 1488, 1489, 1492, 1496(311), 1498, 1499(39), *1526, 1533, 1534*, 1972(38), *2031*
Cha, H.T. 1669(23), *1682*
Chaban, G.M. 1467(187), *1530*
Chabot, B.M. 1682(127, 128), *1685*
Chadha, R.K. 1879(53), *1893*, 2086, 2090(161), *2126*
Chadwick, K.M. 1593(110, 111), *1597*
Chaffe, K.P. 2233(97), *2243*
Chai, W. 1481, 1483, 1506(286), *1533*
Chakoumakos, B.C. 110(12), *118*
Chalk, A.J. 1704(69), *1787*
Chamberlin, A.R. 407(162), *429*
Chambers, R.C. 2353(279), *2362*
Chambon, M. 2261(60), *2312*
Chan, K.S. 436, 479(62), *489*
Chan, T.H. 393(120–122), 394(123), 422(206), *428, 430*, 433(4, 5), 434, 442(19), *488, 489*, 1736, 1737(155), *1788*, 1794(7, 24, 29), 1797(45), 1801(24), 1808(95), 1860(24, 29), *1862, 1863*
Chan, T.-Y. 1674(67a), *1683*, 1854(319), *1868*
Chan, Y.-M. 844(56), *855*
Chance, J.M. 288, 289(68), *352*, 1692, 1718, 1719, 1733(20), *1786*
Chandra, G. 1694, 1695(31), *1786*, 2246, 2250(6d), 2251(27a), 2253(6d, 27a, 33), 2254, 2273, 2285(6d), *2310, 2311*
Chandra, S. 2326(47), *2356*
Chandrasekaran, A. 1406(174a, 174b), *1443*
Chandrasekhar, J. 11, 13(40), 19, 20(50b),

21(40, 50b), 36(76a), *96, 97*, 250(319), *265*, 363, 365(28), *426*, 597, 598, 603, 604(5), *697*, 705, 707(12), *775*, 781(5), *821*, 835(37), *854*, 901(147), *1057*, 1125(182), *1140*, 1151(44), *1182*, 1331(182, 183), *1336*
Chandrasekhar, V. 226(203), *262*, 284(60), *351*, 1352, 1353, 1356, 1357(34), (43), *1438*
Chandrashekar, J. 1166(149, 157), *1184*, 2519(266), *2565*
Chandrashekhar, J. 1123(156), *1139*
Chandresekhar, J. 601(12), *698*
Chang, B. 2075, 2091(109, 110), *2125*
Chang, C.-C. 709, 714(20), *775*, 901, 917, 981, 992(157), *1057*
Chang, I. 2153, 2162(124), *2173*
Chang, L.S. 1989(86), *2032*, 2111, 2120, 2121(226), *2127*
Chang, M.-C. 2295(145), *2315*
Chang, S.-Y. 1644(200), *1664*
Chang, W.-P. 2353(273), *2362*
Chang, Y.-M. 250(306), *264*, 859(3), 865, 866(38), 880, 881(3, 38), 913, 922(38), 977, 978, 981(3), 985, 986(3, 38), 992, 996(3), *1054*, 1080(47), *1102*, 1146(22), *1181*, 1601, 1609, 1645(24), *1661*, 2404(10, 12), *2458*
Channareddy Sreelatha 1354–1357(37a, 37b), *1438*
Chao, D. 1764(272), *1791*
Chao, L.C.F. 580(168), *593*
Chapeaurouge, A. 1638(175, 176), *1664*
Chapleur, Y. 1547(37), *1578*
Chapman, O.L. 10(32b), *96*, 709, 714(20), *775*, 901, 917, 981, 992(157), 994, 995(283), *1057, 1060*, 1149(33), *1182*, 2450(184), *2461*
Chapman, S.E. 434, 448, 449(36), *489*
Chapple-Sokol, J.D. 2549(405), *2568*
Chapusot, F. 2337(120), *2358*
Chaput, F. 2327(55), 2330(81), 2333(97), 2350(236), 2351(252), 2352(81, 266, 267), 2354(81), *2356–2358, 2361, 2362*
Charbouillot, Y. 2339(129), *2359*
Chari, S. 2546(397), *2568*
Charton, M. 486(259), *493*
Chase, D.B. 196(77), *259*, 1356, 1357(41), *1438*
Chase, K.J. 1743, 1745, 1748(213), *1790*
Chase, M.W. 156, 158, 159, 173, 177(35), *178*
Chaseand, D.B. 284(58), *351*
Chassagneux, E. 2261(65), *2313*
Chatani, N. 1616(102), *1662*, 1714(100), 1715(101), 1771(309, 310), 1772(312, 313, 315, 316), 1773, 1774(317), 1783, 1784(346), *1787, 1792*, 1838(232, 233, 235–237, 239–241), *1866*, 1978(55), *2031*, 2110(224), *2127*

Chatgilialoglu, C. 155, 167(28), 168(28, 106, 107), *178, 179*, 475(214), *493*, 827(2b), *854*, 1074(28), *1102*, 1539(4), 1540(4–7), 1542(8–13, 16), 1543(17–19, 24), 1544(16, 20, 21), 1545(21, 27), 1546(16, 21, 35a), 1547(7), 1549(4, 7, 21, 41a), 1550(48), 1552(50), 1553(21, 50–52, 54), 1555(50, 51), 1556(51), 1557(60, 61), 1558(64), 1559(16), 1561(61), 1567(83), 1570(16, 88), 1572(21, 92, 94, 95), 1573(21, 92), *1577–1579*, 1703, 1722(66–68), 1723(66), *1787*, 1831(203), *1865*, 1875(32), *1892*
Chatterton, W. 1645(202), *1664*
Chatterton, W.J. 487(263), *494*, 795, 799(62c), *823*, 845(66), *855*, 880, 881, 913, 914, 917, 924(90), 950(235), 953–955(240), 959–961(235), 962–964, 966(244), 967(247), 968, 969(248), 985, 986, 992, 1004(90), *1056, 1059*, 1080(45), *1102*, 1273(109), 1274(111, 113, 114, 119, 121), 1276, 1285(109), *1308*, 1601, 1604, 1609, 1645(26), *1661*, 2407(23), *2458*
Chaubon, M. 909(175), *1057*
Chaudhari, M.A. 1903(43), *1926*
Chaudhry, S.C. 243(275), 245(280–282), *264*, 1374, 1376, 1377, 1380, 1381, 1401(89), 1429(224, 225), *1440, 1444*
Chauhan, B.P.S. 1387, 1389(118–120), *1441*, 2060, 2067(56, 57), 2069(76), 2071(57, 76), 2084(56, 57, 76), 2089(76), 2093(57, 76), 2094(76), *2123, 2124*
Chauhan, M. 541(106d), *554*, 564(93, 94), *591*, 1409(185b), 1410(186b), 1411, 1412(185b, 186b), 1434(244), 1435(186b, 244), *1443, 1444*
Chauret, D.C. 438, 476(91), *490*
Chauviere, G. 498(21), *510*
Chedekel, M.R. 709, 714, 727, 741, 743(21), *775*
Cheeseman, J.R. 517(44a), *553*
Cheetham, A. 2319, 2350(10e), *2355*
Chelikowsky, J.R. 115(37, 39), 116(37), *118*, 1121(138, 139), *1139*
Chelius, E.C. 366(42), *426*, 620, 621(64), 622(64, 69), 623(64), *699*
Chelucci, G. 1743(218), *1790*
Chemg, C.-D. 1644(195), *1664*
Chen, , H. 485(248), *493*
Chen, B.-L. 434, 442(18), *489*, 1835(215), *1866*, 1875(37), *1892*
Chen, B.-Q. 1477(265), *1533*
Chen, C. 1470, 1483(210), *1531*
Chen, C.C. 2550(408), *2568*
Chen, C.N. 434, 443(20), *489*
Chen, E. 479(233), *493*
Chen, G. 226(201), *262*
Chen, H. 52–54, 61(96), *98*, 505(41), *511*,

Chen, H. (cont.) 1043(360), *1062*, 1810(103), *1863*, 2134, 2147(25), *2171*, 2509, 2539(225), *2564*
Chen, I.W. 2264(75), *2313*
Chen, J. 981, 982(257, 258), *1059*
Chen, J.T. 2354(280), *2362*
Chen, K.-L. 1834(211), *1866*
Chen, L.-F. 434, 447(32), *489*
Chen, M.W. 1769(293), *1791*
Chen, P.-C. 1467, 1474, 1475(183), *1530*
Chen, Q. 898(142), *1057*, 1242(30), *1306*
Chen, R.-M. 1854(324), *1868*
Chen, R.T. 2253, 2261, 2265(45b), *2312*
Chen, S.-H. 1548(39), *1578*
Chen, T.S. 2287(121a, 121b), *2315*
Chen, W. 2095, 2108(186), *2126*
Chen, X. 434, 439(12, 13), *488*, 2234, 2235(106), 2236(111, 112), *2243*
Chen, Y. 1550(45), *1578*, 1767(279), *1791*
Chen, Y.-S. 1965(16c), 1966, 1967, 1997–1999(19), *2030*, 2465, 2489(11), 2503(196), *2559, 2563*
Chen, Y.-T. 1049(366), *1062*
Chen, Y.V. 2351, 2354(260), *2362*
Chen, Z. 475(217), *493*, 1743(226), *1790*
Cheng Zhang, L. 1434, 1435(242), *1444*
Chenier, J.H.B. 1162(117), *1183*
Cherest, M. 1639(181), *1664*
Cherkasov, R.A. 1497(346, 347), *1535*
Chernega, A.N. 209(134), 211(142), 219(173), *260, 261*
Chernicharo, J. 2523(320), *2566*
Chernikova, N.Yu. 1396(158), *1442*
Chernov, N.F. 237(248), *263*, 1454(80–82, 89), 1455(112, 113), 1458(123), 1460(157), 1467(80–82, 89, 112, 113, 123, 157), 1493(331), 1503(372), *1527–1529, 1535, 1536*
Chernyak, V. 2322(31), *2356*
Chernyavskii, A.I. 1501(363), *1536*
Chernyshev, A.E. 1481, 1482(292), *1533*
Chernyshev, E.A. 1582(8), 1583(17), *1594, 1595*, 1967, 1986(20a), 1989(88, 89a, 89b, 90a, 90b), 1994(89a), 1996, 1997(20a), *2030, 2033*, 2465(7), 2479(82), 2504, 2515(197), *2559, 2560, 2563*
Chernyshov, A.E. 1481(285), *1533*
Chernyshov, E.A. 1458, 1467(127), *1529*
Chertkov, V.A. 304(105), *353*
Chew, K.W. 226(205), *262*, 2247–2250, 2269(9), 2276, 2278(104c), 2298(150b), *2310, 2314, 2316*
Chey, J. 229(221), *262*
Chey, J.C. 2064(71), *2124*
Chiang, C. 2339(125), *2358*
Chiang, C.-M. 1127(210), *1140*
Chiang, M. 900(145b), *1057*

Chiang, M.Y. 194(65), *258*, 900(145a), *1057*, 1166(159), *1184*, 1283(150), *1309*, 1325(147), 1331(194), *1335, 1336*, 2421(58), *2458*, 2475, 2479, 2483, 2487(57), *2560*
Chickos, J.S. 163(70), *179*, 861–863, 926(20), *1054*
Chicote, M.T. 1986, 2012(78a–c), *2032*
Chieh, P.C. 1605, 1606(43), *1661*
Chien, P.C. 196(69), *258*
Chihi, A. 1331(181), *1336*
Childress, T.E. 1590(86), 1591(90), *1596, 1597*
Chimichi, S. 1601, 1602(23), 1603(34), 1610, 1642(23), *1660, 1661*
Chin, E. 413(177), *429*
Chinnery, D. 1298(194), *1310*
Chipanina, N.N. 1458, 1467(123), *1528*
Chistovalova, N.M. 1825(162), 1838(226), *1865, 1866*
Chiu, H.T. 2550(408), *2568*
Chiu, P. 885, 890, 937, 938(116), *1056*, 1274(118), *1308*
Chiu, P.K. 1997(114a), *2033*
Chiussi, S. 2547(398), *2568*
Chive, A. 1024, 1046(333), *1061*
Chivers, T. 1214(74), *1231*
Chmielewski, D. 1700, 1701(57), *1786*
Cho, H. 171, 172(127), *180*, 1106, 1107(6), *1136*, 1166(154), *1184*, 2522(299), *2565*
Cho, I.-S. 474(209), *492*
Cho, S. 2153, 2162(124), *2173*
Cho, S.G. 362(25), *426*, 561(53), *591*, 602, 603(17), *698*, 1132(242), *1141*
Choi, G. 2286(119), *2315*
Choi, H.-K. 520, 524, 540(70), *553*, 563, 583(82), *591*
Choi, H.-S. 1871(15), *1892*
Choi, J.-H. 530(93), *554*, 1107(17), *1136*
Choi, K.M. 2346(182a), *2360*
Choi, N. 1242(29), *1306*, 1716(103), *1787*, 1881(59), 1883(68), 1887(59), *1893*, 1896(4, 5), 1897(6, 13, 14), 1906(4, 5), 1911(13, 14, 75), 1913(13), 1915(14, 75), 1916(13, 14, 75, 84), 1923(6, 91), *1925, 1927*, 2425(82), *2459*
Choi, S.-B. 37(83), *97*, 818(114), *825*, 2018, 2022(183), *2035*
Choi, Y.S. 1294(183), *1309*
Chojnowski, J. 508(58), *511*, 558(3), 559(3, 17), 564(86), 575(136, 137), *589–592*, 2218(1, 2), 2221(21a, 21b), 2223(23, 32a, 32b), 2228(21b, 77), 2229(78), *2240–2242*
Chollet, P.A. 2352(267), *2362*
Chonan, Y. 2002(141), *2034*
Chong, D.P. 1049(366), *1062*
Chong, J.M. 438, 476(91), *490*
Choo, K.Y. 1319(69), *1334*

Choo, L.K. 2364(24), *2398*
Chopra, S.K. 287(63), *352*
Choquette, D.M. 437, 484(75), *490*
Chosa, J.-i. 1257(67), *1307*
Chou, S.-S.P. 1843(262), *1867*
Choudhary, T.M. 2290, 2294(139a), *2315*
Choudhury, A. 441(104), *490*
Choudhury, M. 1323(116b), *1335*
Chow, A. 2261, 2263(69), *2313*
Chow, Y.M. 196(76), *258*
Chowdhry, V. 713, 741, 743, 748, 752, 753(33), *775*, 2404(4), *2457*
Chowdhury, A.K. 1115(73), *1137*
Christensen, B.G. 1607, 1634(58), *1661*
Christensen, J.W. 1711(91), *1787*
Christophe, J. 219(180), 222(195), *261*, 2364(21, 22, 24, 25, 27, 28), 2374(41–43), *2398, 2399*
Christou, V. 1875(31), 1888(97), 1890(31, 112), 1891(97), *1892, 1894*
Chronister, E.L. 2350(227, 231), *2361*
Chrostowskasenio, A. 1051, 1052(381), *1062*
Chrusciel, J. 1238(21), *1306*, 1331(192), *1336*, 2187(52), *2215*, 2475, 2477, 2482, 2483(56), *2560*
Chu, C.K. 1550(45), *1578*
Chu, H.K. 508(52), *511*, 1714(93), *1787*, 2223(26), *2241*
Chuang, C. 981, 982(257), *1059*
Chuang, S. 311, 312(135), *353*
Chuhan, J. 1675(80), *1684*
Chuiko, A.A. 2228, 2229(76), *2242*
Chuit, C. 495(3), *510*, 519(64), 541(64, 106a, 106c, 106d), *553, 554*, 564(91, 93, 94), *591*, 1340, 1343, 1351, 1354(6), 1367, 1369(55), 1375, 1378, 1382(6), 1409(185b, 186a), 1410(186b), 1411(185b, 186a, 186b), 1412(6, 185b, 186a, 186b, 192), 1414(192), 1415(6, 201b), 1417(192, 205, 206, 207a, 207b), 1418(192, 205, 206, 207a, 207b, 215), 1420, 1421(6), 1422, 1423(206, 215), 1424(215), 1428(206, 215), 1429(6, 215), 1434(244), 1435(186a, 186b, 244), *1437, 1438, 1443, 1444*, 1449(25), 1450(39, 44), 1457(39), 1464(25, 44), 1465(44), 1480(279), 1488, 1489, 1492, 1496(311), 1498, 1499(39), *1525, 1526, 1533, 1534*, 1738(163), *1789*, 1972(38), *2031*, 2528(335), *2566*
Chujo, Y. 1768(287), 1771(296), *1791*, 2345(170), *2360*
Chuklanova, E.B. 240(258), *263*
Chukovskaya, E.C. 1714(95), *1787*
Chung, C.K. 1574(97), *1579*
Chung, D.-I. 1758(255), *1790*
Chung, G. 1503(374), *1536*
Chung, T.-M. 1460, 1467, 1503(159), *1529*

Chung, Y.J. 2345(167), *2360*
Chung, Y.K. 1460, 1467(159–161), 1493(330), 1503(159–161, 330, 373–375), 1504(330), 1505(330, 376, 377), *1529, 1535, 1536*
Churakov, A.V. 1450, 1456, 1457, 1461, 1496(37), *1525*, 2144, 2145(86), *2172*
Church, S.B. 2487(126), *2561*
Church, S.R. 578(163), *593*
Churchill, M.R. 272(17, 19), *350*
Churney, K.L. 159, 160(54), *178*
Chuvashov, D.D. 1470(216, 217), 1471(216, 217, 222), 1472(216, 217), *1531*
Chvalovsk'y, V. 487(269), *494*
Chvalovsky, V. 296(92), *352*, 617(55), *698*, 1452, 1454(56), *1526*, 1600(3), 1604(35), *1660, 1661*
Chyall, L.J. 1116(74, 75), *1137*
Cibura, G. 397(134), *428*
Ciliberto, E. 2165(185), *2175*
Cima, M.J. 2247(12, 13), *2310*
Cimiraglia, R. 528, 535(87b), *554*
Ciolowski, J. 149(62), *152*
Ciommer, B. 607(24), *698*
Ciorba, V. 120(2), *150*
Cioslowski, J. 52, 54(95d), *98*, 2509(229), *2564*
Cioslowsky, J. 73(149, 150), 74(149), *100*
Cirillo, P.F. 433(8), 435, 454(46), 459(153), *488, 489, 491*, 1638(172), 1639(179), 1650(212), *1664, 1665*, 1804(80), 1845(271, 272), *1863, 1867*
Ciro, S.M. 280(42), *351*, 532, 547(101), *554*, 562, 565, 567–570, 581, 583–585(67), *591*, 1408(181), *1443*
Ciruelos, S. 2151, 2152(111), *2173*
Clabo, D.A. 22(57), *97*
Clabo, D.A.Jr. 123(27a), *151*
Claessens, H.A. 343–346(152), *354*
Claggett, A.R. 784, 786, 787(22), *822*
Clardy, J. 1972(35c, 36), 1997, 1999, 2002(35c), *2031*, 2155(140), *2174*, 2412(34), *2458*
Clark, C.I. 365(40), *426*, 642–644(108), *700*
Clark, G.M. 1904, 1906(47), *1926*
Clark, G.R. 2076, 2091(118), *2125*
Clark, K.B. 168(107), *179*, 1544–1546, 1549, 1553, 1572, 1573(21), *1577*, 1703, 1722(68), *1787*
Clark, R.J.H. 2143(80), *2172*
Clark, T. 515(21), *552*, 781(6), *821*
Clarke, M.P. 1589(73), *1596*, 1964, 1965, 1967, 1984, 1997, 1999(6), *2030*, 2466(16), 2533(353, 354), *2559, 2567*
Clarson, S. 2219(14), *2240*
Clarson, S.J. 2224(41a–c), 2235(108), 2236(114), *2241, 2243, 2244*
Claudio, M.R. 459(154), *491*

Clauss, J. 310(134), *353*
Cleary, B.P. 2102(200), *2127*
Clegg, W. 470(193), *492*, 1620(120), *1663*, 1840(249), *1866*
Clement, A. 751(102), *777*
Clemons, J.M. 2253, 2254(40a–c), *2311*
Clevenger, G.L. 2142(64), *2172*
Clinet, J.C. 1601, 1624(22d), *1660*, 1743(202), *1789*
Clinton, N.A. 358, 359(9), *426*, 616(51), *698*
Clos, N. 1718(106, 107), 1723(106), 1733(140), 1774(322, 323), 1776(107, 323), *1787, 1788, 1792*
Clouthier, D.J. 2523(314), 2557(430), *2566, 2568*
Coan, P.S. 2079, 2093(127), *2125*
Coates, R.M. 464(171), *492*
Cobianu, C. 2545(384), *2567*
Cochran, B.B. 1774(318, 319), *1792*
Cockayne, E. 69(130), *100*
Coe, D.M. 1809(96), 1810(104), *1863*
Coffey, D. 2477(68, 70, 71), *2560*
Cohen, N. 6(22), *95*
Cohen, S.C. 1988(85a–c), 1989(87), 2004(85c), *2032, 2033*
Cohen, T. 456(142), *491*, 1614(98), *1662*, 1669(17), *1682*
Coker, E.N. 1416, 1417(202), *1443*
Colberg, J.C. 435, 451(39), *489*, 1837(222), *1866*
Cole, D.R. 2253(38), *2311*
Cole, S.J. 68(122b), *99*
Colegrove, B.T. 66, 67(118b), *99*
Cole-Hamilton, D.J. 1771(298), *1791*
Coleman, B. 11, 12(42), *96*, 296(91), *352*, 748, 750(97), *777*, 1148, 1149(29), *1182*, 1325(124), *1335*
Coles, H.J. 2227(59), *2242*
Coley, T.R. 2510(231), *2564*
Colgrove, B.T. 1163(131), *1184*
Collin, J. 1671(41), *1683*
Collingwood, S.P. 1840(249), *1866*
Collino, R. 2327(55), *2356*
Collins, R. 2194(77), *2215*
Collins, R.L. 2162(173), *2174*
Collins, S. 123(26a), *151*
Collins, W.T. 1923(98), *1928*
Collman, J.P. 77, 81(153), *100*
Colombo, P. 2345(177), *2360*
Colomer, E. 789(28g, 30), 790(28g), *822*, 1367, 1369(55), *1438*, 1480(279), *1533*, 1963(4), 1972, 1981(37b), 1996(4), 1998(119), 2019, 2026(4), *2030, 2031, 2034*, 2072, 2074, 2090(87), *2124*
Colquhoun, H.M. 36(76b), *97*, 2054(40), *2123*

Coltrain, B.K. 2343(154), 2344(154, 155), *2359*
Coltrin, M.E. 155(18, 19), 159(19), 171(18, 19), *178*
Colvin, E.W. 402(143), *428*, 780(1k), *821*, 1236(8), *1305*, 1496(341), *1535*, 1631(151, 152), 1651(152), *1663*, 1667(1, 3–5), *1682*, 1794(1, 5), 1849(295), *1862, 1867*
Colvin, M.E. 83, 84(160), *100*, 184, 250(28), *257*
Colwell, S.M. 517(44d, 44e), *553*
Comes, F.J. 2547(398), *2568*
Comes-Franchini, M. 1610, 1630(80), 1644, 1649, 1650(196), *1662, 1664*
Comina, P.J. 1680(115, 116), *1684*
Comins, D.L. 790(31), *822*
Compton, R.N. 1135(266), *1142*
Condom, R. 435, 470(58), *489*
Conlin, R. 871, 940(54), 972(54, 250), 973, 974(54), *1055, 1059*, 1075(35c), *1102*, 1390(123b), *1441*, 2481, 2483, 2553(97), *2561*
Conlin, R.H. 1860(351), *1868*
Conlin, R.T. 163(70), *179*, 861(20), 862, 863(20, 22), 866(37), 880, 881(88), 901(22), 922, 924(88), 926(20, 88), 940(88), 971(37), 992(88), *1054, 1055*, 1165(145), 1169, 1177(172), *1184, 1185*, 1241(26), 1248(45–47), 1266(93), 1274(110), *1306–1308*, 1329, 1330(170), 1331(190, 191), *1336*, 2421(55, 56), *2458*, 2471(38, 39), 2484(99), 2488(132), 2490(139), 2496(177), 2498, 2513(184), 2514, 2515(275), 2519(267), 2527(184), *2559, 2561–2563, 2565*
Conlin, S. 916, 993, 996(183), *1058*
Connell, J.W. 2231(89), *2243*
Connell, R.D. 760(123), *777*
Connersde, J.P. 2523(321), *2566*
Connolly, J.W. 727(55, 56), 728(56), 729(59), 730(56), 739, 740(55), 743(56), *776*
Conrad, N.D. 748, 750(97), *777*
Consalvo, D. 1120, 1121(128), *1138*
Constant, V.A. 1120(124), *1138*
Constantieux, T. 460(162), *491*, 1226(115, 116), *1232*
Conticello, V.P. 2153(126), *2173*
Contreras, R. 1512(403), *1537*
Conway, J. 416(185), *429*
Coogan, M.P. 1843(269), *1867*
Cook, M.A. 362(23), *426*, 612(36), 618(60), *698, 699*
Cooke, F. 422(204), *429*, 1497(348), *1535*
Cooke, J.A. 1238(24), *1306*
Coolidge, M.B. 168, 169(103), *179*
Coope, J. 380(73), *427*, 635, 637(95), 641(102), 642(102, 106), 643(106), *699, 700*

Cooper, B.E. 1196–1198(21), *1230*
Cooper, D.L. 1163(132), *1184*, 2509(227), *2564*
Cooper, J.C. 2076, 2091(120), *2125*
Coote, S.J. 1671(43), *1683*
Coqueret, X. 1768(286), *1791*, 2226(56), 2227(58), *2242*
Corderman, R.R. 559, 560(29), *590*
Cordes, A.W. 2151, 2152(113), *2173*
Cordfunke, E.P.H. 164, 165(83, 84), *179*
Cordonnier, M. 1157(87), *1183*
Corey, E.J. 434, 447(29), 475(217), *489, 493*, 1606(57), *1661*, 2437(130), *2460*
Corey, E.R. 1667(6), *1682*, 2453(198), *2461*
Corey, J.Y. 503(32), *511*, 518(60a, 60b), *553*, 563(74), *591*, 860, 899, 901(14), *1054*, 1236(14), *1305*, 1383(111), 1409(184), *1440, 1443*, 1667(6), *1682*, 1697(38, 40), 1708(38), *1786*, 1972, 1981(37a, 37b), 1989(86), *2031, 2032*, 2047(25), *2123*, 2453(198, 199), *2461*, 2513, 2526, 2527(260), *2565*
Cormier, J.F. 434, 447(32), *489*
Cornet, H. 421(203), *429*
Cornett, B.J. 1319(69), *1334*
Correa-Duran, F. 1211(64), *1231*
Corrigan, J.F. 1890(104), *1894*
Corriu, R. 497(14), *510*, 581(179), *593*, 1052(388), *1062*, 1064, 1069(9), 1075(35c), 1082–1084, 1086, 1099(9), *1101, 1102*, 1387, 1388(114, 116), *1440, 1441*, 1612(87), *1662*, 2016(176c), 2017(179–181), 2018(181, 182), 2021(190), *2035, 2036*, 2251, 2253, 2255(26), *2311*, 2481, 2483, 2553(97), *2561*
Corriu, R.J.P. 229(218), *262*, 284(51, 52), 287(64), 288(52), *351, 352*, 495(3), 496(5, 6), 497(5, 6, 9), 498(20, 21), 503(32, 33), 505(35), 507(44–46, 48, 50), 508(48, 51, 55, 56), 509(62, 66), *510, 511*, 519(64), 541(64, 106a–d), *553*, 554, 559, 563(8), 564(91, 93, 94), 565(8), *590, 591*, 781(8a, 8b, 9), 782, 784(8a, 8b), 789(8a, 28g, 30), 790(8b, 28g), 814(8b), *821, 822*, 828(14), *854*, 1086, 1090(54), *1102*, 1215(78–80), *1231*, 1340(2, 5, 6), 1341(24), 1343(6), 1347(24), 1351(2, 5, 6), 1354(5, 6), 1356, 1357(39), 1367, 1369(55), 1370(59–62), 1371(59–61), 1374(76–79), 1375(6, 76–79), 1376(76), 1377(77–79, 99), 1378(6, 77, 99), 1380(79), 1381(106), 1382(2, 5, 6, 76, 77, 109), 1383(76, 77, 110, 111), 1385(76, 77), 1386(76), 1387(115, 117–120), 1388(115, 117), 1389(118–120), 1390(123b), 1409(2, 185a, 185b, 186a), 1410(186b), 1411(185a, 185b, 186a, 186b), 1412(6, 185a, 185b,

186a, 186b, 191, 192, 194), 1414(191, 192), 1415(6, 201b), 1417(192, 205, 206, 207a, 207b), 1418(192, 205, 206, 207a, 207b), 213–215, 217), 1420(6, 213, 214), 1421(6, 213), 1422(194, 206, 215), 1423(194, 206, 213–215, 217), 1424(77, 215, 217), 1427(194, 214, 217), 1428(206, 215, 217), 1429(6, 215), 1430(227–229), 1431(228, 229), 1432(228), 1434(228, 229, 244), 1435(186a, 186b, 244), *1437–1441, 1443, 1444*, 1449(23, 25, 31), 1450(39, 44), 1457(39), 1464(25, 44), 1465(44), 1477(23), 1480(279), 1487(306), 1488, 1489, 1492, 1496(311), 1498, 1499(39), *1525, 1526, 1533, 1534*, 1633(155), *1663*, 1738(161–165), 1751(232, 233, 235, 236), *1789, 1790*, 1824(157), *1864*, 1963(4), 1972(37a, 37b, 38), 1981(37a, 37b), 1989(91), 1996(4), 1998(119), 2007(158a), 2019, 2026(4), *2030, 2031, 2033, 2034*, 2060, 2067(56, 57), 2069(76), 2071(57, 76), 2084(56, 57, 76), 2089(76), 2093(57, 76), 2094(76), *2123, 2124*, 2142(65), *2172*, 2231(90a, 90b, 91), *2243*, 2253(47a, 47b), 2282(114), 2294(141), *2312, 2314, 2315*, 2330(85), 2335(112), *2357, 2358*, 2376(44, 45), *2399*, 2466(18), 2528(335), *2559, 2566*
Cortopassi, J.E. 2069, 2071(78), 2075(106–108), 2091(106, 107, 184), *2124–2126*
Cosimin, P. 2545(384), *2567*
Cossio, F.P. 423(209), *430*
Cossy, J. 1573(96a), *1579*
Cot, L. 2340(131), *2359*
Cote, B. 2042, 2045(13), *2122*
Cottis, S.G. 1976, 1998, 1999, 2001(52), *2031*
Cotton, J.D. 487(266), *494*
Cotts, P.M. 1318(47, 48, 51), *1333*
Couderc, J.-P. 2545(382, 383), 2546(387), *2567, 2568*
Couret, C. 879, 987(82), 1051(380), *1055, 1062*, 1983(72), *2032*
Courter, J.H. 1550(46), *1578*
Coutant, R.W. 1989(93), *2033*
Couture, A. 421(203), *429*
Cowan, D.O. 2159(156), 2160(172), 2162(172, 173), *2174*
Coward, J.K. 1635(162), *1664*
Cowley, A.F. 846(70), *855*, 1154(62), *1182*
Cowley, A.H. 211(143), 217(170), 245(283), 247(293, 294), *260, 261, 264*, 793(49b), *823*, 2131(10), 2141, 2149(56), *2171, 2172*
Cowley, H. 518(61), *553*
Cox, J.D. 154, 165(7), *177*
Crabtree, J.D. 1454, 1467(85), *1527*
Crabtree, R.H. 1720(113, 115, 116), 1721(113), 1723(113,

Crabtree, R.H. (*cont.*) 115, 116), 1732(113, 115), *1788*, 2070(81, 82), 2071(82), 2099, 2108(192), *2124, 2126*
Cradock, S. 1017(314), *1061*, 1128(221), *1141*, 1159(95), *1183*, 1459, 1467(130), *1529*, 1996(108d), *2033*, 2131(13, 14), *2171*
Cragg, R.H. 296(89), *352*, 2481(89), *2560*
Craig, D. 462(167), *492*
Cram, D.J. 1651(218), *1665*
Cramer, C.J. 505(38), *511*
Cramers, C.A. 343–346(152), *354*
Cramp, M.C. 1563(74a), *1579*
Crawford, E.J. 1735(147), *1788*, 2253, 2255, 2261(44a), *2312*
Crawford, R.J. 544(111d), *555*
Crawley, J.E. 474(210), *492*
Creasy, W.R. 1117(87), 1124(171), *1138, 1139*
Creegan, K.M. 1120(121, 122), *1138*
Cremer, D. 113(27), 114(30), 115(27, 30), *118*, 122(19), *150*, 281, 282, 296(43), *351*, 522, 527, 528, 532(83a, 83b), 533, 534(83a), 535(83a, 83b), 536, 538, 540(83a), 541, 542(108), 547(83a, 83b, 108, 113), 548, 549(113), *554, 555*, 564(90, 95), 567(105), 568(108, 110), 569(90, 105, 114), 570(90, 110, 114), 584(105, 108, 114), *591, 592*, 747, 748(94), 776, 1075(34), *1102*, 1146, 1147(23), 1162, 1168(108), 1181(201), *1181, 1183, 1185*, 1277(132), *1308*, 1434, 1435(245), *1444*, 1606(50), *1661*, 2493, 2520(153), *2562*
Cremer, E.J. 122(19), *150*
Crespo, F.G. 1129(232), *1141*
Crespo, R. 72(141, 142), 73(141), 75–77(151), *100*
Crestioni, M.E. 608(30, 31), *698*
Crestoni, M.E. 362(21), *426*, 561(62), 562(62, 68, 70), *591*
Crich, D. 1539, 1540(2a), 1571(90), *1577, 1579*
Crimmins, M.T. 1296(186), *1309*, 1849(292), *1867*
Crisp, G.T. 487(266), *494*
Critchlow, S.C. 205(114), *259*, 1830(198), *1865*
Crivello, J.V. 1763(266–268), 1764(268–271), *1791*, 2225(44), 2227(57), 2231(89), 2241–2243
Crocco, G.L. 2142(69), *2172*
Crocker, L.S. 1910, 1911(60b), *1927*
Croix, C.le 1603, 1609(28), *1661*
Cronin, T.R. 2251, 2253, 2255(28c), *2311*
Croon, M.H.J.M.de 2548(399), *2568*
Cross, R.P. 2223(26), *2241*
Crossen, D.I. 2223(26), *2241*
Crouch, R.D. 1674(65), *1683*
Crowe, E.A. 456(145), *491*

Crowe, W.E. 1840(250), *1866*
Crudden, C.M. 473(201, 202), *492*, 1840(248), 1861(360), *1866, 1868*
Cruikshank, F.R. 155, 176(30), *178*
Crump, R.A.N.C. 476(218), *493*
Csizmadia, I.G. 1603, 1605(38), *1661*
Csonka, G. 1467(182), 1470(206), 1508(385, 388), *1530, 1531, 1536, 1537*
Csonka, G.I. 1466(180), 1467(180, 191, 192), 1468(180, 191–193), 1472(191, 192), *1530*
Cuadrado, I. 1758(257), *1791*
Cuadrado, P. 1860(352), *1868*
Cuadrodo, P. 1819, 1842(140), *1864*
Cuenca, J. 2144, 2150, 2151(84), *2172*
Cuenca, T. 2151, 2152(111), 2157–2159(146), *2173, 2174*
Cukanova, D. 1237(16), *1305*
Cullen, W.R. 1701, 1703(65), 1743(220), 1755(65), *1787, 1790*, 2153, 2154(133), *2173*
Cummingham, A.F.Jr. 2145, 2146(97), *2173*
Cundari, T.R. 21, 24, 30(53a), *96*, 1110(58), *1137*, 2049(29), 2116(246), 2118(29), *2123, 2128*, 2512, 2529(251), *2564*
Cunico, R.F. 440(98), *490*, 1627, 1656(135), *1663*, 1833(210), *1866*
Curl, R.F. 69(124a, 124b), 70(134a), *99, 100*, 1929(1b), *1960*
Curphey, T.J. 1875(39), *1892*
Curran, D.P. 1539, 1540(3a–c), 1545(32), 1555(58a), 1558(64), 1559(67a), 1572(94, 95), *1577–1579*, 1644(199), *1664*, 1831(203), *1865*, 1875(32), *1892*
Curran, S. 2261, 2263(66), *2313*
Curtis, M.D. 1211(63), *1231*, 2006, 2007(157), *2034*, 2054(41), *2123*, 2150(106), *2173*, 2274(100b), *2314*, 2529(339), *2566*
Curtiss, L.A. 34(74a–d, 74f), *97*, 164(80), 167(97, 98), 168(102), 171(97), 173, 175(102), *179*, 184, 221, 223(21), *257*, 526(85), *554*, 1107(8, 10), 1119(111), 1126(191), *1136, 1138, 1140*
Curzon, E.H. 269(4), *350*
Cushner, M.C. 518(61), *553*
Cuthbertson, A.F. 2131(9), *2171*
Cutler, A.R. 1735(147–149), *1788*, 2102, 2109(203), *2127*
Cypryk, M. 508(58), *511*, 564(86), *591*, 805(81), *824*, 2221(21a–c), 2222(22a), 2223(32a, 32b), 2228(21b), *2241*
Czakoova, M. 1701(60–62), 1702(61), *1787*
Czaputa, R. 2204(89), *2215*
Czekay, G. 560(51), *590*, 601, 647(13), *698*
Czermak, J. 617(55), *698*

Dabbagh, G. 69(130), *100*
Dabbousi, B.O. 1890(106, 111), *1894*

Dabestani, S. 2152, 2153(112), *2173*
Dabisch, T. 123(24b, 24c), *151*
Dabosi, G. 507(44, 45), 509(62), *511*, 1215(78–80), *1231*
Dacis, W. 829(21), *854*
Da Cost, J.M. 2327(57), *2357*
Dagani, M. 380(72), *427*, 641(101), *700*
Dagata, J.A. 2514, 2523(274), *2565*
Dahan, F. 763(136), *777*
Dahl, L.F. 1909, 1910(56), *1927*
Dahlhaus, J. 2145, 2146(95), 2150(109), *2172, 2173*
Dahlhoff, W.V. 1418(208), *1444*
Dahn, H. 1603(42), *1661*
Dahn, J.R. 2232(94), *2243*
Dai, J. 1508(386, 387), *1536*
Dai, L.-X. 1668(14), *1682*
Dailey, W.P. 1910, 1911(60b, 60d), *1927*
Dallaire, C. 275, 276(21), *350*, 616, 630(46), *698*
Dalton, J.C. 1147(25), *1181*, 1646(204), *1665*
Dalton, L.R. 2330(83), 2352(263, 265), 2354(83), *2357, 2362*
Daly, J.J. 229(222), *262*, 1923(90), *1927*
Daly, M.J. 1845(275, 276), *1867*
Damewood, J.R.Jr. 42, 43, 46, 47, 50(87), *98*, 200(97), *259*, 814(107), *825*
Damm, W. 1555(58a, 58b), *1578*
Dammel, R. 1018, 1021(315), *1061*, 1128(220), *1141*, 1159(92), *1183*
D'Amore, L. 2289(127), *2315*
Damrauer, A.R. 2556(421), *2568*
Damrauer, B. 1434(236), *1444*
Damrauer, N.H. 2064–2066, 2069(72), *2124*
Damrauer, R. 284(52, 54), 288(52), *351*, 487(268), *494*, 518(59), *553*, 1010(298d), 1026(337, 338), 1027(337), 1028, 1050(338), *1060, 1061*, 1106(3, 4), 1109(3, 4, 42), 1126(204), 1133(252), 1134(253), *1136, 1137, 1140, 1141*, 1340, 1341, 1343(11a, 11b), 1372(64, 66, 67a, 67b), *1437, 1439*, 1642(193), *1664*, 2484, 2504(103), *2561*
Damrauer, R.J. 1372, 1373(71b), *1439*
Danahey, S.E. 284(54), *351*, 487(268), *494*, 1340, 1341, 1343(11a, 11b), *1437*
Dance, I. 1869(2), *1891*
Dando, N.R. 2253, 2255(42, 43), 2256(43), 2257(42), *2312*
Daneshrad, A. 1452, 1471, 1479(54), 1484(298), *1526, 1534*
Danforth, S.C. 2261, 2263(66), *2313*
Dang, H.-S. 1576(101), *1579*, 1874(28), *1892*
Danheiser, R.L. 405(152), 412(172–174), *428, 429*, 1601, 1602, 1604(20), 1608(62), 1620(116), 1622(20), 1628(20, 116, 141), 1630(141, 148), 1653(116), 1655(62, 116, 141), 1656(141), *1660, 1661, 1663*, 1847(284), 1850, 1851(303, 304), *1867*
D'Aniello, F. 1803(75), *1863*
Daniels, J.G. 2253, 2254(40b, 40c), *2311*
Daniels, L.M. 226(200), *261*, 485(257), *493*, 1460, 1462, 1465, 1467, 1472(150), 1517(410, 411), 1518(410, 412), 1521(411), 1522(410–412), *1529, 1537*
Daniels, R.G. 739–741(74), *776*
Danilova, T.F. 154, 156(11, 14–16), 159(11), 160, 161(11, 56), 162, 163(14), 164(11, 15, 56), 165(11, 15), 166(16, 56), 174(14), *177, 178*, 1469(200, 201), *1530*
Danishefsky, S. 1804(79), *1863*
Dannappel, O. 1341, 1349(29, 30a, 30b), 1350(30a, 30b), 1351(30b), 1357(29, 30a, 30b, 52), 1359(29), 1361(30a, 30b), 1364(52), 1365(30a), 1366(29), 1368(30a, 30b), *1437, 1438*
Danovich, D. 576(140), *592*, 685(156), *701*, 2182(30), *2214*
Danovitch, D. 5, 6, 24(15a), *95*
Daran, J.C. 1992(101, 102), 1993(103), *2033*
Darden, T. 515(18a–c), *551*
Darling, C.L. 164(81), 168(100), 170(81), 173(100), 175(81), *179*, 1126(190), *1140*
Darnez, C. 2261(60), *2312*
Daroszewski, J. 1875(32), *1892*
Dartiguenave, M. 1280(137, 138), *1308*, 1987, 1998(81), 2021(191), *2032, 2036*
Dartiguenave, Y. 1280(137, 138), *1308*, 1987, 1998(81), 2021(191), *2032, 2036*
Dartmann, M. 945(208), *1058*
Das, P.K. 2513, 2516, 2519(257a, 257b), 2526(257a), *2564*
Das, S.K. 226(200), *261*, 1460, 1462, 1465, 1467, 1472(150), *1529*
Dat, Y. 1051, 1052(381), *1062*
Dathe, C. 1494(334), *1535*
Datta, P. 2350(215, 217), *2361*
DattaGupta, A. 1674(69), *1683*
Dau, R.O. 238(252), *263*
Daub, G.W. 396(127), *428*
Dauben, H.J.Jr. 7(24), *96*
Daubert, B.F. 1500(357), *1536*
Daucher, B. 222(193), *261*, 713, 719(32), 730, 731(61), 735(67), 750, 752, 754(32), *775, 776*, 931(196), *1058*, 1276(128), *1308*, 2407(28), *2458*
Dauger, A. 2253(46), *2312*
Daunheimer, S.A. 2227(60), *2242*
Dauvarte, A. 1501(366), *1536*
Dave, B.C. 2326(45d), *2356*
Daves, G.D.Jr. 1828(182), *1865*
David, A.P. 1740(181), 1741(182, 183), *1789*
David, L.D. 580(174), *593*
Davidson, D.L. 2252(31b), *2311*

Davidson, E.D. 635, 636(92), *699*
Davidson, E.R. 90(170a), *101*
Davidson, F. 196(77, 78), *259*, 284(58), *351*, 1356, 1357(41), *1438*
Davidson, I.M. 169(113), *179*
Davidson, I.M.T. 860, 861(18), 862(18, 21), 866, 867(40), 872(57), 992(18), *1054, 1055*, 1069, 1070(21), 1080(46), *1102*, 1266(88), *1307*, 1319(73), *1334*, 1964(6), 1965(6, 9a, 9b), 1967, 1984, 1997, 1999(6), *2030*, 2465(10), 2466(16), 2477(67), 2484(100, 103, 104), 2502(104, 191), 2504(103), 2533(354), *2559–2561, 2563, 2567*
Davies, A.G. 1074(26), *1102*, 2137, 2138(37), 2165(186), *2171, 2175*
Davies, C. 2344(162), *2359*
Davies, C.A. 156, 158, 159, 173, 177(35), *178*
Davies, C.G. 578(163), *593*, 2487(126), *2561*
Davies, D.E. 1670(32), *1683*
Davies, I.W. 1838(238), *1866*
Davies, P.B. 559(26, 27), *590*
Davies, S.G. 482(240), *493*, 1842(259), *1867*
Davis, A.P. 434, 440(15), *489*, 1799(56, 57), *1863*
Davis, D.D. 619(62), *699*, 780(1h), *821*
Davis, H.T. 2234(105a), *2243*
Davis, J.K. 1670(32), *1683*
Davis, L.P. 505(36, 37), 508(54), *511*, 1109, 1126, 1130(35), *1136*, 1372(64, 71b–d, 72a, 72b), 1373(71b–d, 72a, 72b), *1439*, 1467(190), *1530*
Davis, M.E. 798(59), *823*
Davis, N.R. 1606(56), *1661*
Davis, P.D. 1147(25), *1181*, 1646(204), *1665*
Davis, W.M. 2042(14), 2044(18), 2051(14, 18), 2052(18), *2122*
Davoust, D. 269(6), *350*
Day, M.C.Jr. 183(5), *257*
Day, P. 2159, 2162(157), *2174*
Day, R.O. 226(203), *262*, 284(55, 56, 60), *351*, 1340(12, 13a, 13b, 14), 1341(12, 13a, 13b, 14, 20a, 20b), 1342(13a, 13b, 20a, 20b), 1343, 1347(13a, 13b), 1352(33, 34), 1353(34), 1354, 1355(37a, 37b), 1356, 1357(34, 37a, 37b), 1406(174a, 174b, 175a, 175b), 1407(176a, 176b), (43–45), *1437, 1438, 1443*
Day, V.W. 1923(99, 102), *1928*, 2151(114), 2152(112, 114), 2153(112), *2173*, 2445, 2446(166), *2461*
Dean, C.E. 1080(46), *1102*
Deaton, D.N. 1823(153), *1864*
Debad, J.D. 487(264), *494*
Debaerdemaeker, T. 245(280, 282), *264*, 1429(224), *1444*
DeBruyn, J.H.N. 1585(37), *1595*
DeCian, A. 2072(86, 88, 89, 95), 2074(104), 2090(88), 2091(86, 89), *2124, 2125*
Deck, P.A. 2137(35), *2171*
Deck, W. 2085, 2089, 2094(157), *2126*
Decken, A. 580(168, 169), *593*
Decker, G.T. 2227(70), *2242*
Declercq, D. 1318(51), *1333*, 2294(141), *2315*
DeClercq, E. 1550(44), *1578*
Declerq, J.P. 200(95), *259*
DeCooker, M.G.R.T. 1585(37), *1595*
Decouzon, M. 1110(60, 61a), *1137*
Decrem, M. 1050(371), *1062*
Dee, G.T. 2234(103a–e), *2243*
Deepak, K. 2326(47), *2356*
Deev, L.E. 1494(336), *1535*
Deffieux, A. 580(171), *593*
Deffieux, D. 1223(104, 105), 1224(108), 1225(109, 111, 112), 1226(115), *1232, 1675*(81, 82), *1684*
DeFrees, D.J. 690(162), *701*
De Gala, S.R. 2070, 2071(82), *2124*
Degen, B. 1584(23, 28, 33, 34), 1586(45, 46), 1588(28), *1595*
Degl'Innocenti, A. 1600(1), 1601, 1602(10, 13, 23), 1604(13), 1608(68), 1610(10, 23, 76), 1611(86), 1612(13), 1625(130), 1626(132), 1628(136–140), 1632, 1633(1), 1635(10), 1642(23, 192), 1643(13), 1650(214), 1655(137), 1656(136–140), 1658(132), *1660, 1662–1665*, 1677(98), *1684*, 1796(38), 1830(197), *1862, 1865*, 1871(14), 1877(40, 41), *1892*
Degueil, M. 1875(32), *1892*
DeGunzbourg, A. 2219(16), *2240*
Dehan, E. 2549(404), *2568*
Dehnen, S. 1878, 1886(50), 1887(86), *1893*
Dehnert, U. 46(90b), *98*, 790, 792(41), 796(41, 75), 798(41), 802, 803(75), *822, 824*
Dehnicke, K. 254(336, 337), *265*, 1885(77), *1893*
Deichmann, K. 2336(113), *2358*
Deiters, J.A. 284(56), *351*, 497(11), 502(29), 508(53), *510, 511*, 1340(14), 1341(14, 20a, 20b), 1342(20a, 20b), 1434(237), *1437, 1444*
De Jeso, B. 2387(64, 68), 2388(64), *2400*
DeJong, B.H.W.S. 113(23), *118*
DeJonghe, L.C. 2286(118), *2315*
Dekina, T.A. 1493(331), *1535*
DeKock, R.L. 1127(212), *1140*
Delaloge, F. 1671(34), *1683*
Delanghe, P.H.M. 453(131, 132), 467(188), *491, 492*
Delattre, L. 2333, 2337(102), *2358*
Delcroix, B. 3(7b), *95*
DeLeeuw, B.J. 515(7a), *551*
Deleeuw, D.C. 2285(117a), *2315*
Deleris, G. 1592(108), *1597*

Deleuze, H. 2387(64, 68), 2388(64), *2400*
Delft, F.L.van 475(216), *493*, 1681(124), *1685*
Delogu, G. 1743(200), *1789*
DeLong, R. 1582(3), *1594*
Delpon-Lacaze, G. 879, 987(82), *1055*
DeLucca, G. 621, 627(82), *699*
Demaille, C. 2326, 2328(49a), *2356*
Demaison, J. 864, 981, 982(33), *1054*
Dembech, P. 1601, 1602(10, 13), 1604(13), 1608(68), 1610(10), 1612(13), 1635(10), 1643(13), *1660, 1662*
Dement'ev, V.V. 196(71), *258*, 2155(135), 2160(168, 171), *2173, 2174*
Demey, N. 1981, 1998, 2001(64), *2032*
Demidov, M.P. 1461, 1467(142, 145), *1529*
Demidova, N.V. 163(69), *179*
Dempeier, W. 245(281), *264*
Demuynck, C. 3(7a), *95*, 1128(222), *1141*, 1157(86, 87), *1183*
De Neef, A. 951, 952(238a), *1059*
Denenmark, D. 1576(100), *1579*
Dener, J.M. 1835(217), *1866*
Denes, F. 560(45, 46), *590*
deNeufville, J.P. 2522(304), *2565*
DeNinno, M. 1804(79), *1863*
Denis, J.-M. 1022, 1025(332), *1061*
Denk, M. 52(96, 97, 98a–d), 53(96, 98a–d), 54(96, 98a), 55(98a, 98b), 56(98b), 57(98a, 98b), 58(98a), 59(98a, 98d), 60(98d), 61(96, 98b), 62(98c, 98d), 63(98b), *98*, 252(328), *265*, 279, 280(38), *351*, 1026(335), *1061*, 1144(11), 1166(156), *1181, 1184*, 2023(195), *2036*, 2118–2121(255), *2128*, 2484(98), 2492(148), 2493(149), 2508(224), 2509(225), 2535(98, 367), 2536(367), 2537(224, 367), 2538(98, 367), 2539(224, 225, 367), 2540(148), 2541(98, 148), 2542, 2543(98, 149, 376), 2544(378), *2561, 2562, 2564, 2567*
Denmark, S.E. 1711, 1712(92), *1787*, 1802(70), 1809(96), 1810(104), *1863*
Dennis, L.O. 2351(242), *2361*
D'Enrico, J.J. 2150(106), *2173*
Depmeier, W. 1429(225), *1444*
De Poorter, B. 301(99), *352*
Deppisch, B. 215(159), 243(275), *260, 264*, 1374, 1376, 1377, 1380, 1381, 1401(89), *1440*
De Proft, F. 72–74(148), *100*
DePuy, C.E. 284, 288(52), *351*
DePuy, C.H. 518(59), *553*, 1010(298d), *1060*, 1109(38, 42), 1115(38), 1126(193, 204), 1134(253), *1137, 1140, 1141*, 1372(66), *1439*, 1642(193), *1664*, 2484, 2504(103), 2556(421), *2561, 2568*
Deriglazov, N.M. 1471, 1472(221), 1485(299), *1531, 1534*

Deriglazova, E.S. 1485(299), *1534*
Derouiche, Y. 196(74), *258*
Dervan, P.B. 782, 801(13c), *822*
De Saxce, A. 508(56), *511*
Descalu, D. 2545(384), *2567*
Deschler, U. 2330, 2340(78), *2357*
De Schryver, F. 2354(283), *2362*
De Schryver, F.C. 1318(51), *1333*
Desclaux, J.P. 122(17), *150*
DeShong, P. 434(33, 34), 448(33, 34, 117), *489, 491*, 1677(95), *1684*
DeSimon, J.M. 2219(15a–c), *2240*
DeSimone, J.M. 2225(46), *2241*
Desimone, J.M. 2236(115), *2244*
Desmond, R. 1814(116), *1864*
Despax, B. 2546(387), *2568*
Desper, J.M. 248(301), *264*, 833(33), *854*, 1266(91), *1307*
Dessy, R.E. 1214(74), *1231*, 2011(165), *2035*
Destombes, J. 3(7a), *95*
Destombes, J.-L. 1157(86, 87), *1183*
Destombes, J.L. 1128(222), *1141*
Desu, S.B. 2295(143), *2315*
Deubelly, B. 1434, 1436(249, 250), *1445*
Deubzer, B. 2223(27c), *2241*
Deutsch, P.W. 168, 173, 175(102), *179*, 1119(111), *1138*
Dev, S. 2143(81), *2172*
Devasagayaraj, A. 1668(16), *1682*
Deveson, A.C. 1887(90), *1894*
Devine, R.A.B. 2231(84), *2243*
Devoe, A.D. 2247(13), *2310*
Devora, G.A. 1715, 1716(102), *1787*
Devore, D.D. 580(175), *593*
Devreux, F. 2333(97), *2358*
deVries, M. 72, 74(146), *100*
Devylder, N. 2017(180, 181), 2018(181), *2035*
Dewald, H.-P. 1875, 1885(33), *1892*
Dewan, J.C. 272(18), *350*
Dewar, M.J.S. 11(38), *96*, 183(6), *257*, 357(6), *426*, 1149(34), *1182*, 2421(59), *2458*
Dexheimer, E.M. 1603, 1609(28), 1617, 1620(103), *1661, 1662*
Deydier, E. 1280(138), *1308*
De Young, D.J. 247, 248(291), *264*, 839(47), *855*,, 870(52), *1055*, 1169, 1176–1180(169), *1185*, 1264(83), 1283(147), 1288(163), *1308, 1309*, 1329–1331(163), *1336*, 1897(22), *1926*, 2497, 2498, 2514–2516, 2534(179), *2563*
DeYoung, L. 1677(94), *1684*
Dhanya, S. 863(24), *1054*, 1237(18a), *1306*
Dhaul, A.K. 1593(110, 111), *1597*
Dhimane, A.-L. 1566(81b), *1579*

Dhurjati, M.S.K. 1248(46), *1306*
Dias, A.R. 168(108), *179*
Dias, H.V.R. 485(248), *493*, 763(133a), 777, 793(49a, 50), 794, 799(49a), *823*, 1043(360), *1062*
Diaz, A. 1207(51), *1230*, 2155(135), *2173*
Diaz, A.F. 2160(164, 168), *2174*
Diaz, M. 516(35c), *552*
Dick, S. 873, 943(67), *1055*
Dickenson, A.P. 2502(192), *2563*
Dickhaut, J. 1542(13), 1555(58a–c), *1577, 1578*
Dickinson, A.P. 2477(77), *2560*
Diebold, J. 2158(148), *2174*
Diederich, F. 186(33, 36, 37), *257*, 1930, 1934(3), 1958(28), 1959(3), *1960*
Dielkus, S. 219(176), *261*
Diephouse, T.R. 1980(60a), 1986, 2012(60a, 77), *2032*
Diercksen, G.H.F. 68(122a), *99*
Dietrich, R. 1587(55, 56), *1596*
Dietrich, R.J. 1586(51), *1596*
Dietrich, T.R. 2547(398), *2568*
Dietze, P.E. 498(16–19), *510*
Diez, A. 1815(126), *1864*
Di Grandi, M.J. 413(178), *429*, 467(184), *492*
Dijonneau, C.S. 2351(244), *2361*
Dikic, B. 252(329), *265*
Dillon, M.P. 2466(16), *2559*
Dilthey, W. 1415(196a), *1443*
Diminnie, J.B. 2051, 2052, 2056, 2057, 2064(35), *2123*
Ding, L. 167(91, 93, 99), 168(93), *179*, 1125(188), *1140*
Dinnocenzo, J.P. 437, 486(80), *490*
Diogo, H.P. 168(108), *179*
Dioumaev, V.K. 571(127), 574(127, 134, 135), *592*, 2047, 2052(24), *2123*
Dippel, K. 209(140), 235(238), 241(265), *260, 263*
Diré, S. 2232(92), *2243*
Dirk, J.N. 1459, 1467, 1493(132), *1529*
Dirnens, V. 1451(51), 1494(338), *1526, 1535*
Disapio, A.J. 2227(67), *2242*
Disch, R.L. 26(63), *97*
Dislich, H. 2319, 2350(10c), *2355*
Disson, J.-P. 2261(56, 64), *2312*
Ditchfield, R. 6(21a), 59(105a), *95, 98*, 516(34), 517(36, 55), *552, 553*
Dittman, W.R. 1740(178), *1789*
Dittmar, U. 1923(92), *1928*, 2334(107), *2358*
Dix, N.D. 2070, 2071(80), *2124*
Dixon, A.D. 763(133a), *777*
Dixon, B.R. 405(152), 412(172), *428, 429*, 1850, 1851(303, 304), *1867*

Dixon, D.A. 52(95e, 95f, 95h, 96), 53(96), 54(95e, 95f, 95h, 96), 61(96), *98*, 156(33), *178*, 184(29), 196(77, 78), 238(251), 254(29), *257, 259, 263*, 284(58), *351*, 710(26), *775*, 1127(208), *1140*, 1356, 1357(41), *1438*, 2509(225), 2539(225, 373a, 373b), *2564, 2567*
Dixon, N.J. 456(145), *491*
Dixon, R.N. 2523(306), *2565*
Djerourou, A.-H. 2389(73), *2400*
Djuric, S. 780(1j), *821*, 1794(10), *1862*
Djurovich, P.I. 2080, 2092(129), 2104, 2108, 2109(205), *2125, 2127*
Dmitriev, P.I. 1491(323), *1534*
Do, J.Y. 1567, 1574(84), *1579*
Do, S. 2205, 2206(90), *2215*
Do, Y. 461(165), *492*, 1462, 1467(153), 1490(315), 1491(325), *1529, 1534, 1535*, 1856(329), *1868*
Dobbs, D.A. 2137(35), 2141, 2144, 2145(54), *2171, 2172*
Dobbs, K.D. 710(26), *775*
Dodd, K. 2063, 2066(66), *2124*
Doddrell, D. 449(123), *491*
Dogaev, O.B. 1463, 1467(163), *1529*
Doherty, N.M. 205(114), *259*
Dohmaru, T. 2475(59a, 59b, 60), 2493, 2515(150), *2560, 2562*
Doi, N. 810(94), *824*
Doi, S. 125, 130(33), 134(42, 43), 137, 143, 146(46), 147(56), *151, 152*
Dolbier, W.R.Jr. 1542(15), *1577*
Dolenko, E.V. 1875(36), *1892*
Dolenko, G.N. 1470, 1475(212), *1531*, 1875(36), *1892*
Dolg, M. 2523(317), *2566*
Dolgopolov, N.N. 860, 862(16a), *1054*
Dolgushina, G.S. 1458, 1467(122), *1528*
Dölle, A. 1803(76), *1863*
Dollet, A. 2546(387), *2568*
Domenech, S. 2545(382), *2567*
Domszy, R.C. 2337(121), *2358*
Donahue, P.E. 190(48), *258*, 1717(104), *1787*
Doncaster, A.M. 154, 156, 159(4), 164(79), 169(111), *177, 179*
Dong Ho Pae 194(65), *258*
Donnelly, R.E.Jr. 2510(232), *2564*
Donnelly, S.J. 1740(177), *1789*
Donovan, D.J. 690(165), *701*
Donovan, R.J. 1733(140), 1774(323), 1776(323, 324), 1778(331), 1779(324), 1780(336), 1781(337), 1782(336–338), *1788, 1792*
Donovan, R.L. 1718(106, 107), 1723(106), 1776(107), *1787*
Doonquah, K.A. 1831(200), *1865*
Doremus, R.H. 2253(35), 2264(77), 2289(134),

2311, 2313, 2315
Dorfman, E. 364, 376(35), *426*
Dorfmann, E. 610, 616, 635(34), *698*
Dorn, H. 72, 74(146), *100*
Dorn, H.C. 149(61), *152*
Dorn, S.B. 2230(80), *2242*
Dorofeev, Y.I. 2231(86), *2243*
Dosaj, V.D. 1584(31), *1595*
Dostrovsky, I. 362(26), *426*, 612(38), *698*
Dou, D. 209(132), 216(162), *260*
Dou, S. 1463, 1464, 1467(167), *1530*
Douce, L. 2072, 2090(90), *2124*
Douglas, T. 1998(120), *2034*
Douglas, W. 2018(182), *2035*
Douglas, W.E. 1412, 1414(191), 1418, 1420, 1423, 1427(214), *1443, 1444*
Doumaux, H.A. 2234(105a), *2243*
Dousmanis, G. 588(197), *594*
Doussot, P. 1641, 1652(188), *1664*
Dow, A.W. 741(79, 80), *776*
Dowd, P. 1570(89), *1579*
Downard, K.M. 1133(251), *1141*
Downey, J.R. 156, 158, 159, 173, 177(35), *178*
Downing, J.W. 1312(5), 1313(33), 1314(33, 34), 1316(38), 1317(5), 1318(5, 38), *1332, 1333*
Downs, A.J. 1144(9), 1162(116), *1181, 1183*
Downs, J.W. 104(2), 113, 114(29), 115(2, 29), 117(2), *117, 118*
Downs, R.T. 109(10), 110(17), 115(35), *117, 118*
Doyle, M.J. 374(59), *427*
Doyle, M.M. 435(40), 451(40, 127), *489, 491*, 1840(247), *1866*
Doyle, M.P. 481(236), *493*, 711, 718(28a), 775, 1715, 1716(102), 1719, 1732(111), 1740(177), 1776(325, 326), *1787–1789, 1792*
Dozorets, C. 2326(51), *2356*
Drahnak, T.J. 901(155), *1057*, 1144, 1167(10), 1169(10, 175), *1181, 1185*, 1329, 1330(159), *1336*, 2471, 2474, 2513(41), *2559*
Drahnak, T.S. 2209(99), *2216*
Dransfeld, A. 7, 9, 42, 44, 45, 47, 69(26), 82, 83(158), 90(26), *96, 100*
Draxl, K. 6(22), *95*
Drechsler, L.E. 2224(41a), *2241*
Dreczewski, B. 229(215, 219), *262*, 2494(163), *2562*
Dreiding, A.S. 186(31), *257*
Dreier, K. 1585(35), *1595*
Dreihaupl, K.-H. 207(127), *260*
Dreizler, R.M. 515, 516(14c), *551*
Dresselhaus, G. 69(126), *99*

Dresselhaus, M.S. 69(126), *99*
Dressler, W. 2257(49c), 2267(49c, 85a, 85b), 2271(85a, 85b), *2312, 2313*
Drew, G.M. 381(76), *427*
Drewello, T. 560(51), *590*, 601(13, 14), 607(14), 647(13), *698*, 1130(236), 1131(236, 240), *1141*
Drieß, M. 520(65b), *553*
Driess, M. 3(3e), *95*, 213(148), 216(161), 252(332), *260, 265*, 1046(377), 1047(361), 1051(376, 377), 1052(389, 390, 392), 1053(376, 377, 390, 391, 393), *1062*, 1064(6c, 7a, 7c), 1068(6c), 1077(39), 1080(7a), 1081(7a, 48), *1101, 1102*
Drone, F.J. 1980, 2004(60b), *2032*
Drost, S. 2340(133), *2359*
Druckman, S. 2326(50), *2356*
Drzal, L. 2290, 2294(139a, 139b), *2315*
D'Souza, M.J. 645(111), *700*
Dubac, J. 37(77a, 77b), *97*, 401(141), *428*, 497(13), *510*, 847(71), *855*, 1857(339), *1868*, 1963(3), 1964(6, 8), 1965(3, 6, 10a, 10b, 12–14), 1967(6, 10a, 10b), 1968(13, 21, 22, 23a, 23b), 1970(27), 1971(30), 1973(23b), 1976(53), 1980(3, 62b), 1981(10a, 10b, 12, 21, 65, 67a–c), 1982(12, 14, 22, 67a–c, 70, 71), 1983(14, 22, 67c), 1984(6, 14, 22, 74), 1985(75), 1987(81), 1992(101, 102), 1996(3, 14, 21, 67a–c, 75, 107, 110), 1997(6), 1998(3, 10a, 10b, 67a, 67c, 74, 81, 110, 118, 119, 123), 1999(6, 10a, 10b, 12, 21, 67c, 107, 123), 2001(123), 2003(12), 2007(12, 75, 158b), 2008, 2009(110), 2011(167), 2021(191), 2022(193), *2030–2036*
Dubaq, J. 580(166), *593*
Dubinskaya, E.I. 1490(317), *1534*
Dubois, A. 2350(236), *2361*
Dubois, I. 2513, 2514(255), 2556, 2557(420), *2564, 2568*
Dubois, L.H. 1127(210), *1140*
Duboudin, F. 390(109), *428*
Dubrous, F. 1586(49), *1595*
Duchamp, J.C. 147, 148(58a, 58b), *152*, 345, 347(154), 348(154, 155), 349(154), *354*, 985(270), *1059*
Duckett, S.B. 1707, 1708(82), *1787*, 2102(201), *2127*
Duda, A. 2218(3), *2240*
Duesler, E.N. 209(132), 216(162), *260*
Duff, J.M. 416(186), *429*, 1147(24), *1181*, 1271(105, 106), *1308*, 1606(56), 1646(203), *1661, 1665*
Duffaut, N. 1592(108), *1597*, 1609(75), 1614(92, 96), 1635(75), *1662*
Duffi, D.N. 2150(106), *2173*
Duffy, N.W. 2107(214), *2127*

Dufour, C. 1573(96b), *1579*
Dufour, P. 1981–1983(67c), 1987(81), 1993(103), 1996(67c), 1998(67c, 81), 1999(67c), 2021(191), *2032, 2033, 2036*
Duggan, P.J. 380(73), *427*, 641, 642(102), *700*
Duguet, E. 2251(30a, 30b), *2311*
Dujuric, S. 1631, 1632(150), *1663*
Dulcey, C.S. 2514(274), 2523(274, 309), *2565, 2566*
Dumont, M. 2352(264), *2362*
Dumont, W. 416(188), *429*, 1546(33), *1578*
Dumrongvaraporn, S. 2301, 2302(159), *2316*
Dunbar, R.C. 1108(21), 1109(48), 1121(48, 143), *1136, 1137, 1139*
Duncan, D.P. 1973, 1974(45a, 45b), *2031*, 2407(25), 2421(65, 66), *2458, 2459*
Duncan, T.M. 350(164), *354*
Dunitz, J. 1341, 1397, 1401(19), *1437*
Dunitz, J.D. 420(201), *429*, 1397, 1401(169), *1442*, 1463, 1467(164), *1530*
Dunlap, N.K. 1797(43, 44), *1862*
Dunn, B. 2318(5), 2326(45b, 45d), 2347(190, 194), 2349(208), 2350(190, 194, 220, 229), 2351(194, 240, 241), *2355, 2356, 2360, 2361*
Dunn, G.E. 565(96), *591*
Dunoguès, J. 447(116), *491*, 1217(84), 1220(97), 1222(101, 102), 1223(103–105), 1224(106), 1225(109), 1226(115), 1229(126), *1231, 1232*, 1794, 1795, 1797, 1801, 1811, 1821(19), 1833(209), *1862, 1866*, 2226(50), *2242*
Dunogues, J. 385(94), 386(98, 99, 101, 102), 392, 398(118), 411(168, 169), 412(118), *427–429*, 1592(108), *1597*, 1609(75), 1614(92, 96), 1635(75), *1662*, 1861(359), *1868*, 2246, 2250, 2251, 2253, 2254(6a), 2261(60), 2272, 2273(6a), 2281(113), 2285(6a), *2310, 2312, 2314*
Dunster, M.O. 2130, 2139, 2140(1), *2170*
Dupray, L. 2225(46), *2241*
Dupuis, M. 2509(229), *2564*
Dupuy, C. 2333, 2337(102), *2358*
Durand, H. 2337(122), *2358*
Durand, P. 457(148), *491*
Durham, B. 2151, 2152(113), *2173*
Dürr, H. 2354(286), *2362*
Durup, J. 1128(216, 217), *1140, 1141*
Dutt, N.K. 1414, 1424(195b), *1443*
Duval, P.B. 2147(99), *2173*
Duverneuil, P. 2545(382, 383), 2546(387), 2548(400), *2567, 2568*
Duxbury, G. 1107(12), *1136*
Dvorak, L. 1583(21), *1595*
Dvornic, P.R. 1764(273, 274), *1791*
Dwight, D.W. 2236(115), *2244*

D'yakov, V.M. 237(244), *263*, 1449(5, 13), 1452(56, 57), 1454(56, 70–76, 80, 84, 86–88, 90, 91), 1455(76), 1457(13), 1458(122, 127), 1460(5, 13), 1461(140, 143, 145), 1465, 1466(13), 1467(80, 84, 86–88, 90, 91, 122, 127, 140, 143, 145), 1469(201), 1470(5, 13), 1471–1473(13), 1475(249, 252), 1476(57), 1479(57, 249, 276), 1480(249, 276), 1481, 1483, 1484(13), 1485(299), 1487(13), 1492(327), 1500(71, 72, 74, 88, 327, 358–360), 1501(327, 368), 1506, 1510(13), 1511(401), 1523(13, 249), *1525–1530, 1532–1537*
Dyke, J.M. 992, 996(276), *1060*
Dykema, K.J. 252(330), *265*, 1331(183), *1336*, 2519(266), *2565*
Dymchenko, V.I. 1454(77), *1527*
Dyszlewski, A.D. 744(89), *776*

Eaborn, C. 203(111), 222(191, 194), 245(288), *259, 261, 264*, 361(19), 362(23), 383(85, 86), 384(89, 90), 388(105, 106), *426–428*, 437(72), 454(136), 483(72), *490, 491*, 498(15), *510*, 540(104), *554*, 559(12), 563(12, 84), 577(12, 145–156, 158), *590–593*, 612(36), 618(60), *698, 699*, 1074(25), *1102*, 1212(68), *1231*, 1452(54), 1459(131), 1462(156a), 1467(131, 156a), 1471, 1479(54), 1484(298), 1488(131), 1491(324), *1526, 1529, 1534, 1535*, 2110(223), 2115(233), 2120(223), *2127*, 2131(12), *2171*
Eaton, P.E. 642, 645(109), *700*
Ebata, K. 133(41), *151*, 234(242), 248(303), *263, 264*, 288(66, 67), 289(67), 348(163), *352, 354*, 484(242–244), 488(275, 276), *493, 494*, 796, 802(76), *824*, 829, 832, 835(27), *854*, 1157(81), *1183*, 1238(19), 1244(33), *1306*, 1407, 1408(177), *1443*, 2139, 2141, 2147(43), *2171*
Ebeling, S. 1854(322), *1868*
Ebenhoch, J. 296(93), *352*
Eberhart, S.T. 1879(54a), *1893*
Eberlein, T.H. 1091(58g), *1103*
Ebina, Y. 1826, 1827(176), *1865*
Ebraheen, K.A.K. 293(76), *352*
Ebsworth, E.A.V. 116(47), *118*, 356(2), *426*, 1165(148), *1184*, 1459(130), 1463(162), 1467(130, 162), 1470(215), *1529, 1531*, 2131(10, 14), *2171*
Eckberg, R.P. 2226(53), *2242*
Eckelman, W.C. 901(160), *1057*, 1965(16a), *2030*
Eckert, H. 311, 313(136), *353*, 1086, 1091(53), *1102*, 1406(175a, 175b), *1443*, 1870(5), *1891*
Eddy, V.J. 1690(12), *1786*
Edelhäuser, R. 2340(134), *2359*

Edelmann, E.T. 1431(231), *1444*
Edelstein, N. 2137(33), *2171*
Edema, J.J. 219(181), *261*
Edema, J.J.H. 1373(75e), *1439*
Edirisinghe, M. 2262(71), *2313*
Edlund, U. 522, 527, 528, 532, 535, 547(83b), *554*, 564, 569, 570(90), *591*, 789(13d), 790(36), 793(36, 47b, 51a, 51b), 794(47b, 51b), *822*, *823*, 2453(203), *2461*
Edmiston, P.L. 2327(59), *2357*
Edwards, A.J. 2095, 2108(186), *2126*
Effenberger, F. 584(184), *593*
Effinger, G. 217(163), *260*
Efimov, Yu.T. 580, 584(167), *593*
Efremova, G.G. 1455, 1467(114), 1501(368), *1528*, *1536*
Efrimov, V.G. 1470–1472(217), *1531*
Egawa, Y. 1314(27), *1333*
Egenolf, H. 1156, 1158(73a), 1164(144), 1175, 1179(73a, 144), *1182*, *1184*
Egerer, H. 2364(30), *2399*
Egerer, H.-J. 2368, 2371, 2374(40), *2399*
Egger, C. 2340(136), 2341(142), *2359*
Egger, K.W. 872(58), *1055*
Egorochkin, A.N. 304(109), *353*
Egorov, M.D. 2473(51), *2560*
Egorov, M.P. 52(94a), *98*, 194(66), *258*, 1244(32, 34), *1306*, 2001(137), 2002(139), *2034*, 2488(135), 2512(253), *2562*, *2564*
Eguchi, M. 1776(324), 1779(324, 335), *1792*
Eguchi, S. 467(186), *492*, 1798(53), 1811(110), *1863*, *1864*
Ehlinger, E. 422(204, 205), *429*, *430*
Ehrenson, S. 361(16), *426*
Eibl, M. 279(30), *351*, 2083(146), *2125*, 2191(64), 2209(100–102), 2210(100, 101), *2215*, *2216*
Eichhofer, A. 1887(90), *1894*
Eichinger, P.C.H. 1133(250), *1141*
Eidenschink, R. 498(15), *510*
Eidensshink, R. 1484(298), *1534*
Eikema Hommes, N.J.R.van 551(114), *555*
Eikenberg, D. 1079, 1088, 1089(41b), *1102*, 1434(232a), *1444*, 1879, 1880(58), *1893*, 2166(189), 2167(189, 195), 2168(189, 195–197), 2169, 2170(189), *2175*, 2534(361b), *2567*
Eiki, I. 2116(239, 240), *2127*
Eilbracht, P. 2162(173), *2174*
Eilert, U. 1649(208), *1665*, 2378(52), *2399*
Einhorn, S. 2347–2350(193), *2360*
Einstein, F.W.B. 487(264), *494*
Eisch, J.J. 772(150), *778*, 1986(76a), 2006(156), 2011(76a), *2032*, *2034*
Eisen, M.S. 2053(38), *2123*
Eisenbarth, P. 748(96), *777*
Eisenberg, C. 434, 438(11), *488*

Eisenberg, R. 2100(193–195), 2102(200), 2108(193), *2127*
Eisenberger, F. 222(195), *261*
Eisenhart, E.K. 466(183), *492*, 1601, 1653(21), *1660*
Eisenstein, O. 82(157c), *100*, 1639(181), *1664*, 1996(108e), *2033*
Eisenthal, K.B. 1325(121), *1335*
Ekerdt, J.G. 2548(402), *2568*
Eklund, P.C. 69(126), *99*
Ekouya, A. 1614(96), *1662*
Elagina, O.V. 1989(89b), *2033*
El Anba, F. 1982(69), *2032*
Elattar, A. 1584(24), *1595*
Elattar, A.A. 1587(53), *1596*
Elemes, Y. 1858(342), *1868*
Elhanine, M. 1026, 1045, 1049, 1050(336), *1061*, 1128(223), *1141*
El Hassani, A. 2548(400), *2568*
Eliason, R. 622(71), *699*
Eliassen, G.A. 186(42), *258*
Eliasson, B. 790, 793(36), *822*
Eliel, E.L. 434(12, 13), 439(12, 13, 96), *488*, *490*
Elin, V.P. 1470, 1475(212), *1531*, 1875(36), *1892*
Eliseev, V.N. 1470(214), *1531*
Elkind, J.L. 1118, 1123(90), *1138*
Elkins, T.M. 793(49b), *823*
Eller, K. 1110, 1115(53), *1137*
Ellerby, L.M. 2326(45b), *2356*
Ellern, A. 504(34), *511*, 1374, 1377, 1380, 1383–1386(95a), *1440*
Ellinger, Y. 1050(370), *1062*, 1129(234), *1141*
Ellington, J.C.Jr. 436, 481(65), *489*
Ellis, R.J. 416(189), 420(200), 421(202), *429*, 435(51), 456(51, 144), *489*, *491*
Ellison, G.B. 155, 167, 177(27), *178*, 517(56), *553*, 794(54), *823*
Ellsworth, E.L. 477(226), *493*, 784(21), *822*
Ellsworth, M.W. 2344(159, 160), *2359*
El-Nahas, A.M. 518, 519(63), *553*
El Omar, F. 1188(1), 1189(6), *1229*
Elschenbroich, C. 2159(159), *2174*, 2429(93), *2459*
Elseikh, M. 1019(322), *1061*
Elser, V. 69(130), *100*
El-Shahat, M.F. 1434, 1435(241), *1444*
El-Shall, M.S. 1049(367), *1062*, 1128(227), *1141*
El Sheikh, M. 880, 881, 919, 922, 924, 928, 956(86), *1055*, 1146(21), *1181*, 1601, 1609, 1645(24), *1661*
Elsner, F.H. 1606(51), *1661*
Elyusufi, A. 1861(359), *1868*
Emberger, E. 780(2c), *821*
Emblidge, R.W. 367(43), *426*, 622(70), *699*

Emel'yanov, I.S. 1471(224), *1531*
Emerson, D.W. 1608(59), *1661*
Emig, N. 762(132), *777*
Emilsson, T. 981, 982(257, 258), *1059*
Emmer, G. 1553(55), *1578*
Emmerling, A. 2332(91), *2357*
Emokpae, T.A. 388(105), *428*
Emziane, M. 1672(46), *1683*
Enda, J. 1601, 1602(18), 1636(18, 166), *1660, 1664*
Enders, D. 472(200), 473(205), *492*
Enders, M. 2282(114), *2314*
Endo, K.-I. 2088(175), *2126*
Endo, Y. 1021, 1045, 1046, 1049(325), *1061*
Endres, H.-E. 2340(133), *2359*
Endres, K. 2337(123, 124), *2358*
Enev, V. 1649(211), *1665*
Engdahl, C. 683(155), *701*
Engel, R.P. 774(155), *778*
Engeler, M.P. 2053(39), *2123*
Engelhardt, G. 269(5), *350*, 1474(242), 1475(242, 255), 1479(242), *1532*, 1897(20), *1926*
Engelhardt, L.M. 2140(46), *2171*
Engle, A. 2318(7), *2355*
Englehardt, L.M. 210(141), *260*
Engler, T.A. 414(181), *429*
Englert, U. 2494(163), *2562*
Enjabal, C. 1674(68), *1683*
Ennan, A.A. 1340, 1412(9b), *1437*
Enokida, R. 2526(328), *2566*
Ensinger, H.A. 2368, 2371, 2374(38), *2399*
Ensinger, M.W. 376(65, 66), 377(68), 380(73), *427*, 635(90, 91, 93–95), 636(90, 91), 637(91, 93–95), 641, 642(102), *699, 700*
Ensslin, W. 1314(21, 22), *1332*
Epiotis, N.D. 1911(66), *1927*
Epishev, V.I. 1996(108a), *2033*
Epstein, P.S. 2150(106), *2173*, 2274(100b), *2314*
Erata, T. 277(24), *350*, 1301(200), *1310*, 1887(93), *1894*, 1897(6, 15), 1911, 1914, 1915(15), 1923(6), *1925*, 1945, 1947(21), *1960*
Erb, W. 784(8c), *822*
Erchak, N. 1362(53a, 53b), 1368(56), *1438*
Erchak, N.P. 1362(54a, 54b), *1438*, 1451, 1461(49), 1467(185), 1481, 1483(289, 290), 1487(49), 1496(342, 343), *1526, 1530, 1533, 1535*
Eriks, K. 434, 441(17), *489*
Eriksson, L.A. 516(31), *552*
Eriksson, M. 1668(13), *1682*
Eriyama, Y. 186(41), *258*, 348(161), *354*, 1157(82), *1183*, 2439(142, 143), *2460*, 2498(185a), *2563*

Ermikov, A.F. 1470–1472(216), *1531*
Ernst, L. 219(180), 222(195), *261*, 2364(12, 20, 28), 2365(20), 2367(20, 37), 2376(47), 2378(50), 2383(50, 60), *2398, 2399*
Ernst, M.C. 68, 69(123), *99*
Ernst, R.R. 310(124), *353*
Ernst, T.D. 1674(57), *1683*
Ervin, K.M. 1116(82), *1138*
Escudié, J. 52(94c), *98*, 909(175), 1051(380), *1057, 1062*, 1064(4f), *1101*, 1983(72), *2032*, 2429(97), *2459*
Espino-Rios, R.L. 2476(61), *2560*
Esser, L. 2153(127), *2173*
Esterbauer, J. 2225(49), *2242*
Esteruelas, M.A. 1720(118, 119), 1721(120–122), 1723(120), 1724(118–122), 1725(119), 1732(118, 119), (129), *1788*, 2095(186), 2096(187), 2097(188, 189), 2098(190, 191), 2108(186, 190, 191), *2126*
Esteva, J.M. 2523(321), *2566*
Estry, H.W. 2265(80), *2313*
Etchepare, J. 111(21), *118*
Eujen, R. 780(2d), 794(2d, 55), *821, 823*
Evans, A.G. 793(48), *823*, 1211(62), *1231*
Evans, D.A. 455(140), *491*
Evans, D.F. 285–287(62), *352*, 1355–1357(38), 1416(202, 203), 1417(202–204), *1438, 1443*, 2234(105a), *2243*
Evans, F.J. 1606, 1628(46), *1661*
Evans, J.S.O. 2162(175), *2174*
Evans, P.A. 1565(78a, 78b), *1579*, 1673(55, 56), *1683*
Evans, W.H. 159, 160(54), *178*
Evans, W.J. 2134(26), *2171*
Evans-Stoeckli, H. 2154(136), *2174*
Evenson, K.M. 704(9), *775*
Ewing, G.D. 1212(69), *1231*
Ewing, V. 116(46), *118*
Exarhos, G.J. 2250, 2253, 2255–2257(21), *2310*
Exner, O. 361(13), *426*
Exner, R. 232(229), *262*
Eyal, M. 2350(234), 2351(249), *2361, 2362*
Ezhova, M. 1248(46, 47), *1306*
Ezhova, M.B. 1244(32), 1266(93), *1306, 1307*, 2001(137), 2002(139), *2034*, 2471(39), 2473(51), 2488(135), *2559, 2560, 2562*
Ezhova, T.M. 196(71), *258*

Faber, D. 648(121), *700*
Faber, T. 746, 770(90b), *776*
Fabian, J. 90(170d), *101*, 2351(245), *2362*
Fabré, A. 1573(96b), *1579*
Fabry, L. 1021, 1044(326), *1061*, 1128(225), *1141*, 1289(169), *1309*, 1325(133), *1335*
Facchin, G. 2345(177), *2360*

Fagan, P.J. 1316(44), *1333*, 1963(1a, 1b), 1986(1a, 1b, 79), 1988(1b), 1998(1a, 1b), *2030, 2032*, 2436(126), *2460*
Faggi, C. 1601, 1602, 1610(10), 1625(130), 1635(10), *1660, 1663*, 1844(270), *1867*
Fagnoni, M. 1290(172), *1309*
Fajarado, M. 2055–2057(44), *2123*
Fajgar, R. 864, 981, 982(31–33), 1006, 1008, 1009(296a), *1054, 1060*, 1237(17), *1306*
Falconer, J.L. 1588(65–71), 1589(65, 66), *1596*, 2533(355), *2567*
Falgueirettes, J. 196(73), *258*
Falk, J. 1606(49), *1661*
Faller, J.W. 1743, 1745, 1748(213), *1790*, 2099, 2108(192), *2126*
Famil-Ghiria, J. 175(141), *180*
Famili, A. 1986(80b), *2032*
Fan, K. 184, 192(17), *257*, 1050(372), *1062*
Fan, M. 1763(266, 267), 1764(271), *1791*
Fang, J. 2224(42a), *2241*
Fang, L. 1454, 1467(95), *1527*
Fang, Yu.W. 1481(283), *1533*
Fanta, A. 870(52), *1055*, 1298(194), *1310*, 1412, 1414, 1417, 1418(192), *1443*
Fanta, A.D. 1264(83), *1307*
Fantozzi, G. 2261(56), *2312*
Fanwick, P.E. 2043, 2051(15), *2122*
Farid, S. 437, 486(80), *490*
Fariña, F. 478(230), *493*
Farina, V. 1548(39), *1578*
Farinola, G.M. 1828(192), *1865*
Farkas, R. 1470(208), *1531*
Farmer, S.C. 2232(95), *2243*, 2289(132), *2315*
Farnan, I. 325–328(143), *353*
Farnham, W.B. 196(77), 238(251), *259, 263*, 284(57), *351*, 505(39), *511*, 1353(35b), 1356(41), 1357(35b, 41), 1358(35b), 1430(226), *1438, 1444*
Farnham, W.G. 284(58), *351*
Farona, M.F. 1986(80b), *2032*
Farooq, O. 569(115), *592*
Farrar, J.M. 1109(49), *1137*
Farrenq, R. 1026, 1045, 1049, 1050(336), *1061*, 1128(223), *1141*
Fatome, M. 1453, 1492(67), 1524(422), *1526, 1537*
Fattah, F.A. 454(136), *491*
Fattaruso, J. 2295(145), *2315*
Faul, M. 1019, 1048(320), *1061*
Faure, T. 2072(91, 93, 95), 2085, 2089(157), 2090(91), 2094(157), *2124, 2126*
Faust, R. 1958(28), *1960*
Faustov, V.I. 1968, 1986, 1996–1998(20c), *2030*
Favaretto, L. 1210(60), *1231*, 1625(130), *1663*
Favart, J.-F. 1189, 1190(9), *1230*
Favier, J.-C. 2219(16), *2240*

Fayssoux, J. 497(12), *510*
Fearon, G. 3(3f), *95*
Feaster, E. 1750(231), *1790*
Fedorova, E.A. 783(18), *822*
Fedorova, E.O. 1454, 1467(82), *1527*
Fedot'ev, B.V. 752, 760(110), *777*
Fedot'eva, I.B. 752, 760(110), *777*
Fedotov, N.S. 1454, 1467(94), *1527*
Fehér, F. 780, 794(4a), *821*, 1314(22), *1332*
Feher, F.J. 1923(94, 100), 1924(105, 106), *1928*
Fehér, M. 213(149), *260*, 780, 794(4a), *821*
Fehsenfeld, F.C. 1128(216, 217), *1140, 1141*
Feifel, R. 2364(22, 24, 29), *2398, 2399*
Feixas, J. 434, 447(31), *489*
Feldman, K.S. 186(34, 35), *257*
Feldner, K. 1584(23, 33, 34), 1586(46), 1592(106, 107), *1595, 1597*
Felföldi, K. 435, 479(60), *489*, 1733(144), 1738(160), *1788, 1789*
Felippis, J.D. 2437(133), *2460*
Felix, G. 386(98, 99), *428*
Felix, R.A. 1824, 1828(160), *1865*
Felkin, H. 1639(181), *1664*
Fellenberger, K.J. 635(89a), *699*
Feng, J. 90(170f), *101*
Feng, X.M. 1671(36), *1683*
Feng, Y. 167(88), *179*, 270(14–16), *350*, 518(57), *553*
Fenoglio, R.A. 2421(60), *2458*
Fenske, D. 215(156), *260*, 1743(214), *1790*, 1878(48–50), 1885(77), 1886(50, 80, 82, 84, 85), 1887(86–88, 90), 1889(80, 103), 1890(104), *1892–1894*
Fensterbank, L. 446(114), *490*, 1566(81b), *1579*, 1854(320), *1868*
Fenton, A. 1069, 1070(21), *1102*
Fenton, A.M. 860–862, 992(18), *1054*
Fenton, G. 434, 444(24, 25), *489*
Féraud, M. 1801(67), *1863*
Férézou, J.-P. 1671(34), *1683*
Ferguson, G. 219(175), *261*
Ferguson, S.P. 1593(109), *1597*
Ferhati, A. 1110(61b), *1137*
Fernandez, M.-J. 1705, 1707, 1709(74), 1720, 1732(117), *1787, 1788*
Fernandez-Vincent, C. 1512(403), *1537*
Ferraboschi, P. 2388(71), *2400*
Ferrall, E.A. 2557(430), *2568*
Ferrante, P.F. 1126(201), *1140*
Ferrante, R.F. 1127(213), *1140*, 1179(195), *1185*
Ferrara, M.C. 1830(197), *1865*
Ferrari, A.M. 184, 221(20), *257*
Ferrari, C. 2340(137), *2359*

Ferreri, C. 1540(7), 1542(10, 16), 1544, 1546(16), 1547, 1549(7), 1553(52, 54), 1559(16), 1570(16, 88), 1572(95), *1577–1579*, 1831(203), *1865*
Ferrieri, R.A. 1965(17), *2030*
Feshin, V.P. 1470, 1475(212), 1479, 1480(276), *1531, 1533*
Fessenden, J.S. 2363(1, 4), *2398*
Fessenden, R.J. 380(72), *427*, 641(101), *700*, 2363(1, 4), *2398*
Fevig, T.L. 1539, 1540(3b), *1577*
Fey, O. 2078, 2091, 2095(121, 122), *2125*
Feyder, G. 2337(117), *2358*
Feyereisen, M. 1126(206), *1140*
Fiandanese, V. 1819(141, 142), 1828(192), *1864, 1865*, 2388(72), *2400*
Fiander, H. 1557(62), *1578*
Fichtner, H. 1589(74), *1596*
Fickes, G.N. 1318(47), *1333*
Field, L. 567, 569, 572(104), *592*
Field, L.D. 645, 647, 661(113), *700*, 1602, 1604, 1655(30), *1661*
Fieldhouse, S.A. 1639(182), *1664*
Figadère, B. 1675(74), *1684*
Figge, L.K. 2062, 2066, 2068, 2069(65), *2124*, 2295, 2296(146a, 146b), *2315*
Filimonova, L.V. 2235(109), *2243*
Filimonova, N.P. 1989(88), *2033*
Finch, D. 2262(71), *2313*
Finckh, W. 2154(137), *2174*
Findea, H.O. 2159(158), *2174*
Findlay, R.H. 1996(108d), *2033*, 2131(13), *2171*
Finestone, A.B. 1450(32), *1525*
Finger, Ch. 876, 917, 931(78), *1055*
Finger, C.M. 2205(92), *2215*
Finger, C.M.M. 88(166), *101*, 125, 126, 128, 129(32a, 32b), *151*
Finger, L.W. 106, 107(6), *117*
Fink, D.M. 1601, 1602, 1604(20), 1608(62), 1622(20), 1628(20, 141), 1630(141), 1655(62, 141), 1656(141), *1660, 1661, 1663*
Fink, G. 2153(129), *2173*
Fink, M.J. 3(5), 83–86(161), *95, 100*, 247, 248(291), *264*, 348(157, 160), *354*, 795, 799(62a, 62b), *823*, 839(46, 47), 847(72–74), *854, 855*, 903(161–163), 904(163), 916(183), 922(163, 189, 190), 932, 933, 937(163), 993, 996(161, 163), *1057, 1058*, 1064, 1068(2), *1101*, 1153(54, 55), 1154(55, 64, 65), 1155(64, 65), 1169, 1176–1180(169), *1182, 1185*, 1241(26), 1283(145, 147, 155), 1288(163), *1306, 1308, 1309*, 1329(162, 163, 165), 1330(162, 163), 1331(163), *1336*, 1897(21–23, 25), 1898(25), *1926*, 2111(226), 2112(227), 2113(228), 2120(226–228), 2121(226), *2127*, 2416(44), 2433(118), *2458, 2460*, 2471(36), 2485(107), 2497(179, 180), 2498(179), 2503(195), 2514(179), 2515(179, 277), 2516(179), 2526(180), 2534(107, 179), *2559, 2561, 2563, 2565*
Finke, R.G. 77, 81(153), *100*
Finzel, R.B. 366(41), *426*, 620–622(65, 66), 637(66), *699*
Finzel, W.A. 1453, 1454, 1456, 1458, 1459, 1461, 1465, 1472, 1483, 1486, 1491, 1492, 1506, 1509, 1511, 1513(69), *1526*, 2296, 2298(149), *2315*
Fioravanti, S. 460(159), *491*, 1843(267), *1867*
Fiorenza, M. 1601, 1602(13, 23), 1604(13), 1610(23), 1612(13), 1633(156), 1642(23, 192), 1643(13), *1660, 1664*
Fiori, C. 2231(84), *2243*
Fiorin, G.L. 1801(63), *1863*
Firgo, H. 1214(76), *1231*, 2179(13), *2214*
Fisch, H. 1743(204), *1789*
Fischer, A.B. 2153(132), 2154(130, 132), 2155(132), *2173*
Fischer, D. 2153(128), *2173*
Fischer, E.O. 190(54), *258*, 2163(178, 180), *2174, 2175*
Fischer, E.R. 168(101), 173(101, 137), *179, 180*
Fischer, G. 252(322), *265*, 873(60, 61), 876(72–76), 917, 931(72, 73), 937, 938(201), 940(72–76), 941(60, 203, 204), 969(60, 61), 1011(302), *1055, 1058, 1060*, 1064, 1068(5a), *1101*
Fischer, J. 2072(86, 88, 89, 95), 2074(104), 2090(88), 2091(86, 89), *2124, 2125*
Fischer, L. 1649(208, 209b), *1665*, 2378(51, 52), *2399*
Fischer, R. 2354(286), *2362*
Fish, P.V. 1861(358), *1868*
Fisher, E.R. 1107(11), 1108(25–28), 1118(92), *1136, 1138*
Fisher, K. 1869(2), *1891*
Fisher, R.A. 2042, 2051(14), *2122*
Fisher, S. 416(184), *429*
Fisk, T.E. 611, 615(45), *698*
Fitzgerald, G. 68(122b), *99*, 2442(158), *2460*
Fitzgerald, J.J. 2339(125), *2358*
Fitzmaurice, N. 451(128), *491*
Fitzpatrick, N.J. 1019, 1048(320), *1061*
Fitzpatrick, R.J. 270(13), *350*
Fjeldberg, T. 29(66b), *97*
Flament, J.-P. 5, 6, 24, 25, 87(15c), *95*
Flank, A.M. 2231(90a), *2243*, 2523(321), *2566*
Flann, C.J. 1817(134), *1864*
Fleischer, C.A. 2239(126), 2240(132), *2244*
Fleischer, E.B. 142, 143(49), *152*
Fleischer, H. 1434, 1436(247), *1445*, 1878(49), 1886(84, 85), *1892, 1893*

Fleischer, U. 60(106a), *99*, 281(44), *351*, 516(32a, 32b), 532(32b), *552*
Fleischmann, E.D. 73, 74(149), *100*, 149(62), *152*
Fleisher, M. 1508(390), *1537*
Fleming, I. 381(77), 389(108), 392(118), 393(120), 398(118, 135), 407(161, 163), 408(164), 409, 410(165), 412(118), 413(175), 422(207, 208), *427–430*, 433(6), 447(116), 455(138, 139), 460(160, 161), 472(139), 473(206), 476(218, 220, 223), *488, 491–493*, 780(1i), 790(38), *821, 822*, 827(3), *854*, 1609, 1618(74), 1631, 1632(149), 1633(154), 1639(149), *1662, 1663*, 1667(9), 1680(120), *1682, 1685*, 1794(7, 8, 19), 1795(19, 35), 1797, 1801, 1811(19), 1819(140), 1821(19), 1842(140), *1862, 1864*, 1997(111), *2033*, 2130, 2138(4), *2170*
Fleming, S.A. 1298(193), *1310*
Flemming, S. 1668(15), *1682*
Fleury, E. 2227(69), *2242*
Fleury, G.I. 1471, 1472(228), *1531*
Flick, K.E. 1888(95), *1894*
Flintjer, B. 198(91, 92), 205(113), *259*, 1077(38), *1102*, 2493(160), *2562*
Flippin, L.A. 1109, 1115(38), *1137*
Floch, H.G. 2351(244), *2361*
Flock, J.W. 1582, 1583(11, 12), *1594*
Floerke, U. 1923(92), *1928*
Flood, T.C. 741(79, 80), *776*
Floquet, N. 1588, 1589(65, 66), *1596*
Florensova, O.N. 1452(56), 1454(56, 73, 81, 89), 1458(122), 1467(81, 89, 122), 1503(372), *1526–1528, 1536*
Flores, J.C. 2151, 2152(111), *2173*
Flores, J.R. 1050(369), *1062*, 1123(166), 1124(167), 1129(230–232, 235), *1139, 1141*
Florio, S. 481(234), *493*
Floris, F. 528, 535(87e), *554*
Flörke, H. 434, 445(26), *489*, 1679(109, 110), *1684*
Flörke, U. 2334(107), *2358*
Flott, H. 194(62), *258*
Flower, K.R. 2076, 2091(118), *2125*
Flowers, M.C. 859, 860(1, 2), 862(2), 917, 981(1, 2), *1054*, 1144(7, 8), *1181*
Flygare, J.A. 416(187), *429*, 435, 456(52, 53), *489*
Flygare, W.H. 6(23a), *96*, 1475(257), *1532*
Flynn, J.J. 228(210), *262*, 1357(42), *1438*
Fochi, M. 473(207), *492*
Foerster, C. 498(19), *510*
Foitzik, N. 404(151), *428*, 547(112b), *555*, 1850(306), *1867*
Fokas, D. 1563(73), *1579*
Fokin, V.N. 240(256), *263*

Foley, K.M. 813(105b), *825*
Foltynowicz, Z. 1840(251), *1866*
Fons, P. 559(30), *590*, 2522(301), *2565*
Fontana, F. 1558(65), *1578*
Fooladi, J. 1649(209a), *1665*, 2377(49), *2399*
Forbes, D.C. 1711, 1712(92), *1787*
Ford, R.R. 487(263), *494*, 795, 799(62c), *823*, 845(66), *855*, 880, 881, 913, 914, 917, 924(90), 953–955(240), 985, 986, 992, 1004(90), *1056, 1059*, 1273(109), 1274(111), 1276, 1285(109), *1308*, 1601, 1604, 1609, 1645(26), *1661*
Forgacs, G. 1465, 1467, 1472(170), *1530*
Fornarini, S. 362(21), *426*, 561(62, 63), 562(62, 63, 68–70), *591*, 608(29–31), *698*
Forster, W.R. 1872(18), *1892*
Forstner, J.A. 1897, 1918(8), *1925*
Forsyth, C.M. 2040(10), *2122*
Forsyth, D.A. 647(116), 677(153), 687(158), *700, 701*
Forsyth, G.A. 1875(35), *1892*
Fortenberry, R. 2351(255), *2362*
Forth, B. 2364(26, 31, 33), *2398, 2399*
Fortuniak, W. 575(136, 137), *592*, 2223(32a, 32b), *2241*
Forwald, A.G. 1586(47), *1595*
Fossum, E. 2187(49, 50, 52, 53), *2215*
Fostiropoulos, F. 69(125), *99*
Fostiropoulos, K. 1930(1c, 1d), *1960*
Fouch, R.A. 506(42), *511*
Foucher, D.A. 2150(110), 2153(110, 125, 134), 2154(110, 134, 137), 2160(166), 2162(125), *2173, 2174*
Fournier, P. 1128(216, 217), *1140, 1141*
Fournier, R. 515, 516(14b), *551*
Fowler, P.W. 5, 31, 32, 42(18), *95*
Fox, A. 68(121), *99*, 1109(43), 1117(86), 1122(43), 1123(158), 1126(205), *1137–1140*
Fox, H.H. 2345(171), *2360*
Fox, J.R. 2289(125, 126), *2315*
Fox, M.A. 1189(5), *1229*, 2320, 2347(17b), 2353(279), *2355, 2362*
Fracchiolla, D.A. 1782(338, 341), *1792*
Fracois, P. 5, 44, 68(17a), *95*
Fragala, J.L. 2165(185), *2175*
Frampton, C.S. 275, 276(21), *350*
Franchini, M.C. 473(207), *492*
Franciotti, M. 402(146), *428*, 1816, 1819(128), *1864*
Francisco, J.S. 155, 159, 160, 168, 173, 175(26), *178*, 1106(5), *1136*, 2504(199), 2510(240, 242), *2563, 2564*
Franck, X. 1675(74), *1684*
Franck-Neumann, M. 1860(354, 355), *1868*
Franco, R.J. 2075(112, 114), 2091(112), *2125*
Frank, D. 311, 313(136), *353*
Frank, T. 2533(355), *2567*

Frank, T.C. 1588(68, 70), *1596*
Franke, R. 516(32a), *552*
Franken, S. 1918(89), *1927*
Frankenau, A. 1885(77), *1893*
Frankhauser, P. 216(161), *260*
Frankiss, S.G. 1165(148), *1184*
Fraser, A.R. 236(254), *263*
Fredin, L. 1127(209), *1140*, 1167(161), 1175(161, 188), 1176(188), 1180(196), 1181(200), *1184*, *1185*, 2469(24), 2518(262), *2559, 2565*
Freeburger, M.E. 435, 456(49), *489*
Freeburne, S.K. 1586(44), *1595*
Freedman, A. 2493(154), 2523(308), *2562, 2566*
Freedman, H.H. 2004(147–150), *2034*
Freedman, R. 1606(57), *1661*
Freeman, F. 1875(38), *1892*
Freeman, H.A. 2253(45b), 2261(45b, 55), 2265(45b), *2312*
Freeman, J.M. 234(237), *263*
Freeman, P. 8, 9, 42, 43, 47, 90(31a), *96*
Freeman, P.K. 814(108b), *825*
Freeman, R. 269(7), *350*
Freeman, W.P. 4(10a, 10d), 37(10a, 10d, 81, 82, 86), 38(10a, 10d), 39(82), 40(10a, 10d, 82), 42, 44(86), 46(10a), 47(10d, 86), *95, 97, 98*, 816(112a), 817(112b), 819(112a, 112b), 820(112a), *825*, 981, 984(261, 262), *1059*, 1606, 1610(55), *1661*, 2019(187), 2021(192), 2024(187), 2026(187, 200), 2027(202), 2028(187, 192), 2029(210), *2035, 2036*, 2045, 2051(20), 2082, 2092, 2093(143), *2123, 2125*
Frei, B. 1602(33), 1611, 1630(85), 1647(33), 1649(207), *1661, 1662, 1665*
Freidlina, R.K. 1714(95), *1787*
Freiser, B.S. 1134(256), *1141*
Freitag, S. 203(108), *259*
Frejd, T. 403(148), *428*
French, L.G. 1852(317), *1868*
Frenck, H.J. 2145, 2146(95), *2172*
Frenking, G. 53–55, 57, 58, 61, 62(103), 66–68(118a), *98, 99*, 283(45), *351*, 527–529, 533(86), 538(86, 102a, 102b), 539(86), 540(86, 103), 547(86), *554*, 568, 571, 573(112), *592*, 1008, 1009(297b), *1060*, 1108(31), 1125(172), 1126(202), *1136, 1139, 1140*, 1163(130), *1184*, 2539(372a), 2556(427), *2567, 2568*
Freudenberger, J.H. 272(17), *350*
Freund, R. 1314(22), *1332*
Frey, H. 1758(258, 259), *1791*
Frey, H.M. 171(117, 118), 172(117, 133), 175(140), *180*, 901(149), *1057*, 1123(163), *1139*, 2479(80), 2511(244), 2525(325), *2560, 2564, 2566*

Frey, J. 437, 485(79), *490*
Frey, V. 1585(42), *1595*
Fricke, J. 2332(91), *2357*
Friebe, T. 2364(26, 31), *2398, 2399*
Friedrich, D. 1546(34), *1578*
Friedrich, H. 2057(48), *2123*, 2527(330), *2566*
Friedrich, H.B. 1584(22), *1595*
Friedrich, P. 190(54), *258*
Friese, C. 740(76), 766(142), 767(76, 142, 143), *776, 778*
Friesner, R.A. 2510(231), *2564*
Friess, M. 2271(90), *2314*
Frisch, M.J. 171(128, 129), *180*, 515, 516(17), 517(44a), *551, 553*, 1107(9), *1136*, 1331(182), *1336*
Fritch, J.R. 2445, 2446(166), *2461*
Fritchie, C.J. 1897(10), *1925*
Fritsche, K. 1608, 1617(64), *1661*, 2363(7), 2378(54), 2379, 2380(7), 2385(7, 62), 2387(65), 2388(62, 69), *2398–2400*
Fritz, G. 215(156), 241(260–262), *260, 263*
Fritz, H.P. 2132(17), *2171*
Fritz, M. 190(56), *258*, 2159(152, 154, 155), *2174*
Froc, G. 2352(264), *2362*
Frohn, H.J. 1341, 1344(23), *1437*
Frolov, Yu.L. 237(244), *263*, 1393, 1404(150), *1442*, 1458(123), 1461(145), 1467(123, 145, 184), 1469(184), 1471(224, 226), 1472(233a), 1483(293), 1485(299), 1500(358), *1528–1531, 1533, 1534, 1536*
Frolow, F. 241(266), *263*
Fronczek, F.R. 198(82), *259*
Fronda, A. 222(193), *261*, 713, 719(32), 720(44), 735(66–68), 737(72), 750, 752, 754(32), *775, 776*, 906(166), 931(196), *1057, 1058*, 1276(128, 131), *1308*, 2407(28), *2458*
Fronda, M. 906(167), *1057*
Fronzoni, G. 517(40a), *552*
Frosch, W. 1387, 1389(121b), *1441*
Frueh, J. 1966, 1967, 1997–1999(19), *2030*, 2503(196), *2563*
Frunze, T.M. 196(71), *258*
Frurip, D.J. 156, 158, 159, 173, 177(35), *178*
Fry, A.J. 1224(107), *1232*
Fry, C. 345, 347–349(154), *354*
Fry, Ch. 985(270), *1059*
Fry, J.L. 1740(178), *1789*
Frye, C.L. 508(52), *511*, 1352(32b, 32c), *1438*, 1448, 1450(1), 1452(58a, 58b), 1453(1, 69), 1454, 1456, 1458, 1459, 1461, 1465(69), 1470(1), 1472, 1483(69), 1484(58a, 58b), 1485(58b), 1486, 1491, 1492, 1506, 1509(69), 1511(69, 402), 1512(402), 1513(69, 402), *1524, 1526, 1537*, 1714(93), *1787*, 1923(98), *1928*,

2230(81a), 2231(81b), *2242*, 2291(140), 2296, 2298(149), *2315*
Frye, G.C. 2347(186c), *2360*
Frye, J. 2322(36), *2356*
Frye, S.V. 434, 439(12, 13), *488*
Fryer, P.F. 729(59), *776*
Fryzuk, M.D. 2101(196), 2102(197), 2108(196, 197), 2109(197), *2127*, 2147(99), *2173*
Fu, P.-W. 1697, 1698, 1709, 1710, 1755(43), *1786*
Fuchigami, T. 1190, 1192(10a, 11, 12), 1193(14b), 1196(12), 1197(12, 24, 27, 28), 1200(28), 1215(81), *1230, 1231*
Fuchikami, T. 852(85), *855*, 891(124), 917(184), 998(184, 287), 1002(184), *1056, 1058, 1060*, 1251, 1254, 1256(51), *1306*, 1325(127–132), *1335*, 1705, 1709, 1714(80), *1787*, 1973, 1974(50a), *2031*, 2429, 2436(106), *2459*
Fuchs, H.D. 2196(78a), *2215*
Fuchs, P.L. 1877(44), *1892*
Fuess, M. 2257, 2267(49a), *2312*
Fuji, T. 2445(164), *2461*
Fujii, K. 434, 445(27), *489*
Fujii, N. 459(156), *491*, 1651(215), *1665*, 2330, 2354(80), *2357*
Fujii, O. 2231(83), *2242*
Fujii, T. 310(130), *353*, 1822(150), *1864*, 2320(17a), 2347(186a, 186d, 188, 191), 2350(188, 221), 2351(239), *2355, 2360, 2361*
Fujiki, K. 2445(167), *2461*
Fujimoto, H. 506(43), *511*, 1372, 1373(73), *1439*
Fujimura, Y. 90(170c), *101*
Fujino, M. 137(44a–c), 138, 140(44b), 142(44a–c), 143(44a), 146(44b), 147(44a), *151*, 1318(53), *1333*, 2320, 2347(17b), *2355*
Fujio, M. 610, 624(33), *698*
Fujisawa, M. 1656(223), *1665*
Fujisawa, T. 1843(263), *1867*, 2387(63), *2400*
Fujise, Y. 2002(141), *2034*
Fujita, E. 1193(14a), *1230*, 1828(187, 188), 1860(356), *1865, 1868*
Fujita, H. 2347(187), *2360*
Fujita, M. 194(63), *258*, 435, 451(41), *489*, 780(1c), *821*, 1292(176), *1309*, 1331(189), *1336*, 1739(166–170), 1740(174, 179), *1789*, 2418, 2420(54), *2458*, 2488(131), *2561*
Fujitake, M. 2523(315), *2566*
Fujitsuka, H. 465(176), *492*
Fujiwara, T. 1291(175), *1309*, 1801(64), *1863*
Fujiwara, Y. 1641(189), 1651(215), *1664, 1665*
Fujiyama, R. 371(55), *427*, 617(56), *698*
Fukawa, N. 248(299), *264*, 279(35), 348(159), *351, 354*

Fukin, G.K. 2041(12), *2122*
Fukuchi, M. 1318(52), *1333*
Fukuda, Y. 279(32), *351*, 1269(103), *1308*
Fukui, K. 559(24), *590*, 998(287), *1060*, 1312(20), 1325(133), *1332, 1335*
Fukui, T. 2389(75), 2392(78), 2394(75, 78, 79), 2396(79, 81), *2400*
Fukumoto, K. 1548(38), 1561(71a, 71b), *1578, 1579*
Fukumoto, Y. 1772(315), 1783, 1784(346), *1792*
Fukunaga, T. 1998(122a), *2034*
Fukusaki, E. 2392, 2394(78), *2400*
Fukushima, M. 1304(216), *1310*, 2015, 2016(173), *2035*, 2522(287, 288, 290), *2565*
Fukute, Y. 1641(190), *1664*
Fukuyo, E. 1752(240), *1790*
Fukuzawa, S. 1743, 1747(210), *1790*
Fukuzumi, S. 1292(176), *1309*, 1550(42), *1578*, 1740(179), *1789*
Fuller, E.W. 2232(94), *2243*
Fülöp, V. 226(199), *261*, 1454(78), 1461(139), 1463(166), 1467(139, 166), 1508(385), 1510(166), *1527, 1529, 1530, 1536*
Funada, Y. 829(25), *854*
Funahashi, H. 1221(99, 100), *1231*
Funasaka, H. 1940(17), 1942(18, 19), *1960*
Funasaki, H. 1302(203, 204), *1310*
Funayama, O. 2250(23a, 23b), 2252, 2253(23a, 23b, 32a), 2265(82a, 82b), 2271(89), *2310, 2311, 2313, 2314*
Fundamenskii, V.S. 237(246, 247), *263*
Funicella, M. 1877(40, 41), *1892*
Funk, R.L. 402(145), *428*, 1795(34), *1862*
Fuqua, P. 2351(241), *2361*
Furin, G.G. 1824(158), *1865*, 2143(74), *2172*
Fürstner, A. 784, 789, 790(20), *822*, 1635(164), *1664*
Furuhashi, K. 1821(146), *1864*
Furukawa, K. 137(44a–c), 138, 140(44b), 142(44a–c), 143(44a, 51), 146(44b), 147(44a), *151, 152*, 2028(208), *2036*
Furumori, K. 2017(177a, 177c), *2035*
Fusik, P. 241(262), *263*
Fuss, H. 2257, 2267(49c), *2312*
Fuss, M. 675(149), 691(168), 693(169), *701*
Fuzukawa, S. 1743, 1747, 1749(211), *1790*
Fyfe, C.A. 270(14–16), 312, 314, 316, 317(137), *350, 353*

Gabe, E.J. (233), *262*
Gabel, A. 2351(255), *2362*
Gabelica, V. 369, 397(49), *427*, 627–630(81), *699*

Gabor, A.H. 1771(299), *1791*
Gabriel, A.O. 403(147), *428*, 2257, 2267(49b), *2312*
Gabriel, J. 811(97a), *824*
Gabriel, W. 2522(297), *2565*
Gabrielli, R. 561, 562(62), *591*
Gadea, F.X. 1163(126), *1184*
Gadre, S.R. 82(157a), *100*
Gadzhiev, S.N. 164(76), *179*
Gagne, M.R. 2153(126), *2173*
Gaines, G.L. 1583(13, 14), *1594*
Gajda, C. 1860(353), *1868*
Gal, J.-F. 559(22), 560(39), *590*, 1110(60), 1120(114, 115), *1137, 1138*
Gal, M. 1450, 1453(33), 1470(208), 1471(33), 1472(229), *1525, 1531*
Galasso, V. 517(40a, 40b), *552*
Galemmo, R.A.Jr. 1842(258), *1867*
Galigne, J.L. 196(73), *258*
Gall, G.J. 2352(269), *2362*
Gallagher, J.F. 219(175), *261*
Gallagher, P. 2234(104a), *2243*
Gallaher, K.L. 133(39), *151*
Gallaher, T.N. 2115(234), *2127*
Gallardo, T. 1802(69), *1863*
Galle, J.E. 772(150), *778*, 2006(156), *2034*
Gallot, Y. 2238(121), *2244*
Gallucci, J.C. 460(163), *491*
Gallucci, R.R. 716, 727, 748–750(40), *775*, 2404(6, 7), *2457*
Galm, W. 2107, 2109(212), *2127*
Gal'minas, A.L. 194(66), *258*
Gal'minas, A.M. 1244(32, 34), *1306*
Galminas, A.M. 2001(137), *2034*, 2473(51), *2560*
Galzaferri, G. 219(182), *261*
Gammie, L. 1319(70), *1334*
Gamper, S. 186(43), 217(165), *258, 260*, 581(181), *593*, 1374, 1375, 1379(80), 1386, 1389(112), *1439, 1440*, 2059, 2066, 2067(51), 2085, 2090, 2094(160), *2123, 2126*, 2534(362), *2567*
Ganboa, I. 423(209), *430*
Gangloff, A.R. 760(123), *777*
Gangodawila, H. 381(76), *427*
Gani, P. 1877(43), *1892*
Gano, D.R. 171(129), *180*, 2511(243), *2564*
Gantzel, P.K. 37, 42, 44, 47(86), *98*, 2021, 2028(192), *2036*
Ganz, K.-T. 1746(227), *1790*
Ganzel, P.K. 2300(154), *2316*
Gao, J. 515(19a, 19b), *552*
Gao, Q. 1548(39), *1578*
Gar, T.K. 245(284), *264*, 586(193), *594*, 1450(47), *1526*

Garant, R.J. 226(200), *261*, 1460, 1462, 1465, 1467, 1472(150), *1529*
Garavelas, A. 435, 455(47), *489*
Garbaccio, R.M. 1847(283), *1867*
Garber, L.J. 1565(78b), *1579*
Garbisch, E.W.Jr. 1998(122c), *2034*
Garcia, J.M. 423(209), *430*
García-Granda, S. 2075(113, 114), *2125*
Gardella, J.A. 2234, 2235(106), *2243*
Gardella, J.A.Jr. 2236(111–113), *2243*
Garden, S.J. 1542(14), *1577*
Gardlund, Z. 2350(237), *2361*
Gareau, Y. 1678(100), *1684*, 1874(26, 27), *1892*
Garegg, P.J. 1872(16), *1892*
Garnier, F. 2017(179), *2035*, 2330(85), *2357*
Garone, E. 184, 221(20), *257*
Garratt, J. 2509(227), *2564*
Garratt, P.J. 5, 7(13b), *95*, 741(77), *776*, 2448(175), *2461*
Garreau, R. 1189, 1191(7), *1229*
Garrity, P.M. 577(158), *593*
Garst, M.E. 1825(167), *1865*
Gasanova, L.V. 1585(36), *1595*
Gasking, D.I. 1601, 1602, 1604, 1624, 1651(11), *1660*
Gaspar, P.D. 827(1), *854*
Gaspar, P.P. 52(94b), *98*, 194(65), *258*, 860, 899(14), 900(145a, 145b), 901(14, 160), *1054, 1057*, 1145(13), 1157(85), 1165(145), 1166(85, 159), 1169(85), 1170(13), 1176(85), *1181, 1183, 1184*, 1236(14), 1257(65), 1283(150, 154), *1305, 1307, 1309*, 1319(69), 1320, 1321(78), 1325(147), 1331(194, 196), *1334–1336*, 1667(6, 8), *1682*, 1965(16a–c, 18), 1966, 1967, 1997–1999(19), *2030*, 2421(58), *2458*, 2464(1–3, 5), 2465(1–3, 5, 11), 2467(20), 2468(1–3, 5), 2469(2, 3, 25–28), 2470(2, 3), 2475(1–3, 57), 2477(2, 5), 2479(57), 2480(20, 83, 84), 2483(2, 57), 2484(5, 105, 106), 2485(1–3, 5), 2486(2), 2487(2, 3, 5, 57, 127), 2488(1–3, 5, 133), 2489(11, 20, 136), 2490(138), 2491(140), 2493(2, 3, 5, 154), 2496(177), 2501(3, 5), 2502(3, 5, 105, 106), 2503(5, 196), 2513(260), 2523(308), 2526, 2527(260), 2546(397), *2558–2563, 2565, 2566, 2568*
Gaspard-Ilhoughmane, H. 580(166), *593*
Gasparis-Ebeling, T. 800(70a), *823*
Gasper, P.P. 268(2), *350*
Gasper-Galvin, L.D. 1584(22), *1595*
Gassman, P.G. 2027(201), *2036*, 2137(34, 35), *2171*
Gates, S.M. 2545, 2546(379), *2567*
Gattow, G. 1875, 1885(33), *1892*

Gaukhman, A. 1481, 1483(289–291), 1514, 1523(291), *1533*
Gaukhman, A.P. 1481(284), *1533*
Gaul, B. 485(255), *493*
Gaul, J.H.Jr. 2280(112a), *2314*
Gaul, J.M. 813(105b), *825*
Gauss, J. 59(105b), *98*, 122(19), *150*, 516(27), 517(27, 38, 42a, 42b, 43a–e), 552, 1075(34), *1102*, 1162, 1168(108), 1181(201), *1183*, *1185*, 2493, 2520(153), *2562*
Gautheron, B. 1992(101, 102), 1993(103), *2033*
Gavrilova, G.A. 1455(113), 1458(123), 1467(113, 123), 1523(419), *1528, 1537*
Gay, S.C. 1175, 1176(186), *1185*
Geanangel, R.A. 198(83), *259*
Gee, J.R. 922(189), *1058*, 1154, 1155(65), *1182*
Geen, G.R. 456(145), *491*
Geerlings, P. 72–74(148), *100*
Gehrhus, B. 52(99a–c), 55, 56, 58(99c), 60(107), 61(99c), 62(99b, 99c), *98, 99*
Geib, S.J. 566(100), *591*
Geisinger, K.L. 110(16), 111, 112(16, 22), 113(26), *118*
Geißler, H. 2232(93), *2243*
Gelachvili, G. 1026, 1045, 1049, 1050(336), *1061*
Gelius, R. 1988(84), *2032*
Gel'mbol'dt, V.O. 1340, 1412(9b), *1437*
Gelsomini, N. 1844(270), *1867*
Genchel', V.G. 163(69), *179*
Genderen, M.H.P.van 1473(237), *1532*
Geng, B. 1638(173), *1664*
Geniès, E.M. 1188(1), 1189(6), *1229*
Genji, T. 2549(403), *2568*
Gennari, C. 441(102), *490*
Gentle, T.M. 1171(181), *1185*
Gentzkow, W.V. 2226(52), *2242*
George, C. 184, 221(18), *257*
George, C.F. 2076, 2091(120), *2125*
George, M.V. 788–790(28a), *822*, 836(44a), *854*, 1998, 1999(125), 2005–2007(155), 2010(163), 2011(155), *2034, 2035*
George, M.W. 1260(74), *1307*
George, P. 6(21b, 21c), 11(43b), 12, 32(43b, 45a, 45b), *95, 96*
George, T.F. 70(135), *100*
Georges, P. 2350(236), 2351(252), *2361, 2362*
Gerardin, C. 2249, 2294(14), *2310*
Gerbier, P. 2016(176c), 2017(179), *2035*, 2330(85), *2357*
Gergö, E. 1470(206), *1531*
Gerhus, B. 2467(19, 21), 2495(19, 21, 169), 2535, 2536(368), 2537, 2539(370), 2544(377), *2559, 2562, 2567*

Geribaldi, S. 1110(60, 61a), *1137*
Gerlach, C.P. 1888, 1891(97, 98), *1894*, 2053(39), *2123*
Germain, G. 200(95), *259*
German, M.I. 1471(225), *1531*
Gerov, V.V. 1764(273, 274), *1791*
Gerstmann, S. 203(109, 110), *259*
Gerval, J. 1609(75), 1614(96), 1635(75), *1662*
Gesing, E. 186(31), *257*
Gevorgyan, V. 1451(48a, 48b), 1459, 1467(135), *1526, 1529*
Gewald, R. 1808(93), *1863*
Ghitti, G. 2235(107), *2243*
Ghorai, B.K. 434, 449(37), *489*
Ghosh, A. 2261(62), *2312*
Ghosh, U. 1609, 1618(74), *1662*
Giannozi, P. 1121(140), *1139*
Giansiracusa, J.J. 647(118), 677(151), *700, 701*
Giardello, M.A. 2153(126), *2173*
Giardini-Guidoni, A. 1120, 1121(128), *1138*
Gibbons, W.M. 1766(276, 277), *1791*
Gibbs, G.V. 104(2), 106(5, 6), 107(6), 108(7), 109(8–10), 110(5, 11, 12, 16, 17), 111(5, 16, 18, 20, 22), 112(5, 16, 22), 113(26), 115(2, 5, 32, 35, 40), 116(41), 117(2, 5), *117, 118*
Gibby, M.G. 309(114), *353*
Gibson, F.S. 1565(77), *1579*
Gibson, V.C. 1840(249), *1866*, 1878(46), *1892*
Giering, W.P. 434, 441(17), *489*
Giese, B. 1539, 1540(2b), 1542(13), 1544–1546, 1549(21), 1553(21, 51), 1555(51, 57, 58a–c), 1556(51), 1557(60, 61), 1561(61), 1572, 1573(21, 92), *1577–1579*, 1703, 1722(66, 68), 1723(66), *1787*
Gigozin, I. 2321(23a), 2322(23a, 30), *2355, 2356*
Gil, S. 408(164), *429*
Gilbert, A. 2225(48), *2242*
Gilbert, B.A. 414(180), *429*, 1294(184), *1309*
Gilbert, S. 2073(97–99), 2091(97), *2124*
Gilday, J.P. 460(163), *491*
Gilette, G.R. 1169(177), *1185*
Gilges, H. 2064(69), *2124*
Gillan, E.G. 1878(45), *1892*
Gillette, G. 1157(75), *1182*, 1268(95), *1307*, 2183(40), *2214*, 2472(46), *2559*
Gillette, G.R. 323, 324(141), *353*, 761(128, 129), 763(129), 777, 833(32), *854*, 1006, 1008(296c), *1060*, 1169, 1180(168), *1185*, 1266(90), *1307*, 1331(184, 185), *1336*, 1887(94), *1894*, 1897, 1908(26), *1926*, 2001(131b), *2034*, 2485(115), 2491, 2514(142), 2515(278), 2518(142), 2519(142, 278), *2561, 2562, 2565*
Gilliam, W. 1582(2), *1594*
Gilliam, W.F. 1589(77), *1596*

Gilloir, F. 452(129), *491*, 1680(118), *1684*
Gillot, B. 1587(63), *1596*
Gilman, H. 565(96), *591*, 789(29), 790(33, 40), 791, 792(40), 795(63a, 68, 71a, 79), 796(71a, 71b, 72, 73), 799(63a), 800(68, 71a), 801(72), 802(73), 804(79), 806(73), (110b), *822–825*, 836(44a), *854*, 1312(10, 11), 1321(88), 1322(94–96), *1332, 1334*, 1609(71), *1662*, 1972(34), 1976(52), 1980(34), 1988(83a–d), 1998, 1999, 2001(52), 2003, 2005, 2008, 2009(34), *2031, 2032*, 2177(3a), *2213*
Gilman, H.J. 788(28a), 789, 790(28a, 28b), *822*
Gilman, J.W. 2233(97), *2243*
Gimarc, B.M. 20, 22–28, 84, 87, 88(52), *96*
Gimisis, T. 1550(48), 1567(83), *1578, 1579*
Gimmy, M. 434, 447(28), *489*, 752, 753, 755, 756(114), *777*
Gindelberger, D.E. 1888(95), 1889(102), 1890(106, 107), 1891(113), *1894*
Ginkel, R.van 1835(216), *1866*
Giolando, D.M. 1904(46, 47), 1906(47), 1909, 1910(46), *1926*
Girard, L. 580(169), *593*
Girija, K. 1292(178), *1309*
Girolami, G.S. 2076(116, 117), 2091(117), *2125*
Giwa-Agboreirele, P.A. 2227(61), *2242*
Gladiali, S. 1743(200), *1789*
Gladyshev, E.N. 783(17, 18), *822*
Gladysz, J.A. 241(267), *263*, 2069, 2071(77), *2124*
Glarum, S.H. 69(130), *100*
Glasdysz, J.A. 2142(69), *2172*
Glaser, D.M. 2226(54), *2242*
Glaser, R. 499(24), *510*, 2251–2253(24), *2311*
Glaser, R.H. 290(73), *352*
Glass, E. 1213(71), *1231*
Glass, G.E. 1314(30), *1333*, 2193(70), *2215*
Glassgold, A.E. 1124(168), *1139*
Glatthaar, J. 1021(328–330), 1045, 1046, 1049(328), *1061*, 1128(224), *1141*, 1145(12, 18), 1158(18), 1159(12, 18, 96), 1160(12), 1178(96), *1181, 1183*
Glatthaar, S. 2514(272), *2565*
Glatthaur, J. 1156, 1158, 1175, 1179(73b), *1182*
Glaubitt, W. 2330(79), 2332(91), 2333(79, 94, 101, 104), 2334(101), 2335(104, 110), 2337(101, 110), 2339(79, 94), *2357, 2358*
Glavee, G.N. 2063, 2066(66), *2124*
Gleason, R.W. 405(152), 412(172), *428, 429*, 1850, 1851(303), *1867*
Gleisner, R.P. 2264, 2265(76), *2313*
Gleiter, R. 77–79(154b), *100*, 1434(232b), *1444*, 2163–2166(182), *2175*, 2534(360b), *2567*
Gleixner, R. 948(231), *1059*
Gleria, M. 2345(177), *2360*
Glezer, V. 1200(32), *1230*, 2320(16), 2321(24b), 2326(49b), 2328(16, 49b, 61, 68), *2355–2357*
Glezer, V.T. 1480, 1481(280, 281), *1533*
Glick, M.D. 784(22), 785(26), 786, 787(22), *822*
Glidewell, C. 183(9), *257*, 381, 382(79), *427*, 2131(9), *2171*
Glinnemann, J. 115, 116(37), *118*
Glinski, R.J. 1127(208), *1140*
Glosik, J. 1118(96), *1138*
Glover, D.E. 1211(64), *1231*
Glover, S.O. 2226(55), *2242*
Glukhikh, N.G. 1475(252), *1532*
Glukhikh, V.I. 1474(242), 1475(242, 249, 252), 1479(242, 249, 274), 1480, 1523(249), *1532, 1533*
Glukhovtsev, M. 33, 34, 50, 51(71), *97*
Glukhovtsev, M.N. 5, 6(13a, 14), 7, 8, 32(13a), 52(94a), 69, 77(13a), 90(170h), *95, 98, 101*, 601(12), *698*, 2512(253), *2564*
Glynn, S.G. 1374, 1376(93), *1440*
Glynn, S.J. 509(65), *511*
Gobbi, A. 538(102a, 102b), *554*, 1108(31), *1136*
Gobin, P. 2220(18), *2240*
Göcke, H.J. 1083, 1096(50), *1102*, 1163(123), *1184*
Goda, K.-I. 1755(247), 1756(247, 248), *1790*, 1827(177, 179), *1865*
Goddard, J.D. 705(14), *775*
Goddard, R. 784, 787, 788(25), *822*, 1970, 1974(29b), *2031*
Goddard, W.A. 901(150), *1057*
Goddard, W.A.III 167, 171(96), 172(131), 173(96, 131), 174(131), 175(96, 131), *179, 180*, 705, 706(11), *775*, 2510(231, 232), *2564*
Godde, G. 1583(19), *1595*
Godleski, S.A. 517(53, 54), *553*
Goebel, D.W.Jr. 784, 787(24), *822*
Goerlich, J.R. 52(95c), *98*
Goesmann, H. 254(336), *265*, 547(112b), *555*, 1743(214), *1790*, 1850(306), *1867*
Goetze, B. 948, 1007, 1008(232), *1059*
Goetz-Schatowitz, P.-R. 1228(124), *1232*
Goh, B.J. 473(202), *492*
Goh, J.B. 453(131, 132), *491*
Goikhman, R. 1074(29), *1102*, 1872(21), *1892*, 2108(219, 220), *2127*
Goiunta, C.J. 2549(405), *2568*
Golan, Y. 2328(67), *2357*
Goldberg, D.E. 29(66b), *97*

Goldberg, D.R. 1840(250), *1866*
Goldberg, G.M. 364, 376(35), *426*, 610, 616, 635(34), *698*, 1606, 1628(46), *1661*
Goldberg, N. 1029, 1049, 1050(342), *1061*, 1118(89), 1122, 1125(151), 1126(195–199), 1127(195, 196), 1128(197), 1129, 1130(195, 198, 199), 1135(195, 199), *1138–1140*
Goldberg, Y. 1682(130), *1685*, 1719(112), 1740(171, 172), 1743(215), 1767(285), *1788–1791*
Goldberg, Y.S. 1717(105), *1787*
Golden, D.M. 155, 176(30), 177(142), *178, 180*
Goldfuss, B. 8, 9(31a), 42(31a, 88, 89), 43(31a, 88, 89, 93a), 44(88, 93a), 45(88), 46(88, 89), 47(31a, 88, 89, 93a), 48(88, 93a), 50(89), 90(31a), *96, 98*, 814(108a, 108b), 815(108a), 817–819(113), *825*, 984, 988(267), *1059*, 2013(172b), 2019(184b, 184c), 2026(184b), *2035*
Goldman, E.W. 2421(65), 2429(97), *2459*
Goldstein, E.J. 727, 739(53), *776*, 2481(86), *2560*
Gole, J.L. 184, 254(29), *257*, 1127(208), *1140*
Golensnikov, S.P. 194(66), *258*
Golik, J. 1548(39), *1578*
Golino, C.M. 496(8), *510*, 1080(43a), *1102*, 1247(40), *1306*, 1325(125), *1335*, 2406(18), *2458*
Golovina, N.A. 254(335), *265*
Gomberg, M. 125(31a, 31b), *151*
Gomei, M. 1120, 1121(132), *1139*
Goméz, R. 2151, 2152(111), *2173*, 2322(35), *2356*
Goméz-Sal, P. 2144, 2150, 2151(84), 2157, 2158(146), 2159(146, 149), *2172, 2174*
Goméz-Sol, P. 2151(111, 117), 2152(111, 115, 117), *2173*
Gömöry, A. 1968, 1986, 1996–1998(20c), *2030*
Gong, G. 2322, 2348(25), *2356*
Gong, L. 782, 791, 795(11c), *822*
Gong, L.Z. 1671(36), *1683*
Gong, Y. 363(30), *426*, 631, 632(85), *699*
Gönne, S. 2364(19), *2398*
Gonon, M.F. 2261(56), *2312*
Gonzales, C. 2510(240), *2564*
González, A.M. 1819, 1842(140), 1860(352), *1864, 1868*
González, F. 437, 438(84), *490*
González, J.M. 1797(47), *1862*
Good, A.M. 1144(5), *1181*
Goodfellow, C.L. 482(240), *493*
Goodman, J.L. 437, 486(80), *490*, 2437(134), *2460*
Goodman, J.M. 441(102), *490*
Gorath, G. 1798(50), *1863*

Gordon, B.III 1997(112a), *2033*
Gordon, M. 171, 172(117), *180*, 872(55), *1055*
Gordon, M.S. 11(41a, 41b, 43a), 12(43a), 13(41b, 46a), 14–16(41b), 18, 19(50a), 21(41b, 50a, 53a, 53b), 22, 23(53b), 24(53a), 29(53b), 30(53a, 53b), 32(50a), 33, 42, 43(41a), 66, 68(120b), 83(159, 161), 84(161, 162), 85(161), 86(161, 162), *96, 99, 100*, 123(21e, 24e, 24g), *151*, 155(26), 159, 160(26, 52), 161(52), 162(66), 163(52, 67, 72), 168(26), 171(129), 172(132, 144), 173(26, 132, 144), 174(139), 175(26, 132), *178–180*, 184(18, 27), 186(38), 221(18), 228(211), 250(27), 251(320), 252(330), *257, 258, 262, 265*, 296(86, 88), *352*, 500(25), 505(36, 37), 508(54), *510, 511*, 515(11), *551*, 704–706(8), 707(15, 16), 770, 772(149), *775, 778*, 814(106), *825*, 829(23), *854*, 860(10a, 16b), 862(16b), 901(151), 983(10a), 997(151), 1046(373), 1048(365), *1054, 1057, 1062*, 1067(15), *1101*, 1106(5), 1109(35), 1110(58), 1111(65b), 1126, 1130(35), *1136, 1137*, 1150(40), 1151(42a, 45), 1153(53, 54), 1159(98), 1166(150, 151), 1172(184), *1182–1185*, 1241(27), *1306*, 1331(183), *1336*, 1372(64, 71a–d, 72a, 72b), 1373(71a–d, 72a, 72b), *1439*, 1467(190), 1468, 1469, 1477, 1511, 1521(194), *1530*, 1860(350), *1868*, 1911, 1913(73, 74a), *1927*, 2049(29), 2116(246), 2118(29), *2123, 2128*, 2298(152a), *2316*, 2443(162), 2450(183), *2461*, 2477(66), 2479(78, 81), 2505(200), 2511(243, 244), 2512(251), 2519(266), 2525(325), 2529(251), *2560, 2563–2566*
Gordon, R.D. 792, 793(46), *823*
Gordon, R.G. 2549(405), *2568*
Gordon-Wylie, S.W. 2187(53), *2215*
Gore, W.F. 1075(32a), *1102*
Gorecki, J. 2289(127), *2315*
Gorelova, M.M. 2235(109), 2236(110a, 110b), *2243*
Görls, H. 2051, 2052(34), *2123*
Gornowicz, G.A. 2227(70), *2242*
Gorodisher, I. 2345(171), *2360*
Görs, S. 576(139), *592*
Gorshkov, A.G. 1459, 1467(134), *1529*
Gorsich, R.D. (110b), *825*, 1988(83a, 83b, 83d), *2032*
Görsmann, C. 2340(136), *2359*
Gorton, J.E. 2162(174), *2174*
Gostevskii, B.A. 243(274), *264*, 1374, 1377(94), 1390(132–134, 136, 137), 1392(132, 136, 137, 146, 147), 1393(147, 149), 1396(160, 164, 165), 1398(133, 137, 147, 149), 1399(160, 164, 165), 1400, 1401(137), 1404(147), 1405(133), 1418(210,

Gostevskii, B.A. (*cont.*) 212, 216), 1420(210, 212), 1422(216), 1423(212, 216), *1440–1442, 1444*
Gostevsky, B.A. 1824(158), *1865*, 2143(74), *2172*
Goto, J. 2334(105), *2358*
Goto, M. 137(44c, 45), 139, 140(45), 142(44c), 143(45, 52c), 144, 146(45), *151, 152*, 200(93, 100), 248(299), *259, 264*, 279(31–33, 35, 36), 348(159), *351, 354*, 1090(56), 1092(63), *1102, 1103*, 1264(85), 1270(104), *1307, 1308*, 1884(70), *1893*, 1904, 1907(48), 1924(107), *1926, 1928*, 2114(230), *2127*, 2183(42), 2206(94), 2207(96), 2208(98), *2214–2216*, 2472(45), *2559*
Goto, T. 810(93b), *824*, 2521(279), 2522(291, 292, 296), *2565*
Gottaut, I. 690(166c), *701*
Gottsman, E.E. 1980(61a), *2032*
Gould, I.R. 437, 486(80), *490*
Gould, R.O. 1075(35b), *1102*, 2187(48), *2214*, 2550(411), *2568*
Goumri, A. 167, 168(89, 93), *179*
Goure, W.F. 2001(132), *2034*, 2474(52), 2501(189), *2560, 2563*
Goursat, P. 2253(46), *2312*
Govedarica, M.N. 1764(274), *1791*
Grabek, P.J. 2226(54), *2242*
Grabovskaya, Zh.E. 1467(186, 187), *1530*
Grabowski, J.J. 1109, 1115(38), 1116(80, 83), *1137, 1138*
Grabowski, Z.R. 1323(117), *1335*
Gradock, S. 1470(215), *1531*
Graef, R.C. 2252(31b, 31c), *2311*
Graf, D.D. 2143(75), *2172*
Graf, E. 1743(200), *1789*
Graf, R. 404(151), 406(154), *428, 429*, 547(112a, 112b), *555*, 1848(287), 1850(305, 306, 308), *1867*
Grafen, K. 1450(45), *1526*
Graham, I.M. 1771(298), *1791*
Graham, P. 315, 317, 318(139), *353*
Graham, T. 2319, 2346(8a), *2355*
Graham, W.R.M. 68(122a), *99*, 1164(140–142), 1175(185–187), 1176(186, 187), *1184, 1185*
Grandclaudon, P. 421(203), *429*
Grant, B.E. 1705, 1707, 1708(72), *1787*
Grape, W. 1592(106, 107), *1597*
Graschy, S. 1218(94), *1231*, 2179(11, 12), 2201(84), *2214, 2215*
Grasso, R.P. 1766(276, 277), *1791*
Grattan, K.T.V. 2322(28), *2356*
Gray, C.E. 780(1h), *821*
Gray, H.B. 1280(138), *1308*
Gray, R.C. 1470(209), *1531*
Graziani, R. 2053(36), *2123*

Greasley, P.M. 383(86), *427*
Greault, E.A. 1674(67a), *1683*
Greaves, J. 1369, 1370(58b), *1439*
Greeley, B.H. 2510(232), *2564*
Green, A.J. 374, 375(64), *427*
Green, J.C. 52(96, 98b, 99c), 53(96, 98b), 54(96), 55, 56(98b, 99c), 57(98b), 58(99c), 61(96, 98b, 99c), 62(99c), 63(98b), *98*, 2508(224), 2509(225), 2537(224), 2539(224, 225), *2564*
Green, J.R. 1860(353), *1868*
Green, L.G. 158(47, 48), *178*
Green, M. 713, 741, 743, 748(33), 752(33, 115), 753(33), *775, 777*, 1260(74), *1307*, 1723(128), *1788*, 2403(3), 2404(4), *2457*
Greenberg, A. 120(1), *150*, 1449, 1461, 1463(17), 1467, 1471(189), 1513(17), *1525, 1530*
Greenblatt, M. 2319(12a), *2355*
Greene, J.P. 171, 172(127), *180*, 1106, 1107(6), *1136*, 1166(154), *1184*, 2522(299), *2565*
Greene, T.W. 1674(62), *1683*
Greengard, L. 515, 516(16), *551*
Greenwood, P.F. 1120, 1121(131), *1139*
Gregg, B.T. 1735(148, 149), *1788*, 2102, 2109(203), *2127*
Greil, P. 2227, 2233(73), *2242*
Grein, F. 1067(17), *1101*
Greiner, A. 2257(49c), 2261(63), 2267(49c), *2312*
Greisinger, D. 2142(63), *2172*
Greiwe, K. 2333, 2334(101), 2336(115), 2337(101), *2358*
Grelier, S. 1226(115), *1232*
Grellmann, K.H. 1323(117), *1335*
Gressinger, A. 1886, 1889(80), *1893*
Greszta, D. 812(101b), *824*
Grev, R. 29(66a), *97*
Grev, R.S. 13(46a), 68(122c), *96*, *99*, 158(49), 163(73), 167(49), 171(49, 116), 172(49), 174(139), *178–180*, 247(297), *264*, 515(7a), *551*, 835(42), *854*, 860(10b), 901(151), 919, 920(186), 983(10b), 997(151), *1054, 1057, 1058*, 1157(85), 1166(85, 150), 1169, 1176(85), *1183, 1184*, 1331(196), *1336*, 2479(79), 2557(430), *2560, 2568*
Grevels, F.-W. 1700, 1701(57), *1786*
Grey, J.P. 2333, 2334(99), *2358*
Griedel, B.D. 1809(96), 1810(104), *1863*
Grierson, J.R. 1550(46), *1578*
Griesmar, P. 2350(229), 2351(250, 257, 258, 261), 2354(250, 257, 258, 281), *2361, 2362*
Griffin, J.B. 1108(29), *1136*
Griffin, R.G. 310(132), *353*
Griffith, J.R. 1767(280), *1791*

Grignon-Dubois, M. 301(97), *352*, 411(168), *429*
Grigoras, S. 184, 221, 222(19), *257*
Grigorieva, L. 1227(118), *1232*
Grigorieva, N.Y. 456(146), *491*
Griller, D. 168(106–108), *179*, 1320, 1321(78), *1334*, 1540(5, 6), 1543(24), 1544(20, 21), 1545(21, 28), 1546, 1549, 1553(21), 1564(28), 1572, 1573(21), *1577, 1578*, 1703, 1722(68), *1787*, 2526(329), *2566*
Grimme, W. 1903(40b), *1926*
Grimmer, A.R. 334(147), *353*
Grimshire, M.J. 1843(268), *1867*
Grinberga, S. 1481, 1483, 1514, 1523(291), *1533*
Grisdale, P.J. 357(6), *426*
Grisenti, P. 2388(71), *2400*
Grissinger, A. 1878(48), *1892*
Griswold, R. 2226(53), *2242*
Grob, C.A. 376(67), *427*, 637, 638(96, 97), *699*
Grobe, J. 193(60), *258*, 860(17), 863(26, 27b), 945(208), 946(223), 988, 990(27b), *1054, 1058*, 1123(162), *1139*, 1524(420), *1537*, 2481, 2483(96), *2561*
Grogger, C. 2179(11), *2214*
Grogorkiewitz, M. 2340(131), *2359*
Groh, B.L. 768(146), 769(147), *778*, 1095(70), *1103*, 1994(105a), *2033*
Grohmann, A. 1386, 1387, 1389(113), *1440*, 2059, 2060, 2066, 2067(52), *2123*, 2528(336), *2566*
Grohmann, I. 1589(74), *1596*
Grond, W. 2353(277), *2362*
Grondey, H. 270(14–16), *350*
Gronert, S. 499(24), *510*, 1010(298d), *1060*, 1126(204), *1140*, 2556(421), *2568*
Groot, H.J.M.de 310(132), *353*
Gross, E.K. 515, 516(14c), *551*
Gross, G. 1021, 1044(326), *1061*, 1128(225), *1141*, 1158(91), 1159(93), *1183*, 1289(169), *1309*
Gross, J. 219(180), 222(195), *261*, 2332(91), *2357*, 2364(28, 30, 31, 33–36), 2368(34, 38), 2369(34), 2371(34, 38), 2374(34, 36, 38), *2398, 2399*
Gross, L.W. 2001(133), *2034*
Gross, T. 1611(84), *1662*
Grosser, M. 1178(193), *1185*
Grosskopf, D. 219(176), *261*, 1011(307), 1034, 1038(353), 1043, 1044, 1048(307), *1061*
Grossman, R.B. 1325(122), *1335*
Grover, R. 559(22), 560(39), *590*, 1120(114, 115), *1138*
Gruber, K. 2191(66), 2201(84), *2215*
Grubert, H. 2163(180), *2175*
Grudenberg, D.W.von 254(336, 337), *265*

Grugel, C. 1999(126), *2034*, 2436(128), *2460*
Grumbine, S.D. 566(100), 571, 574(128–130), *591, 592*, 1879(53), *1893*, 2082(142), 2086(161, 163), 2087(164), 2090(161, 163, 164, 183), 2093(163), 2094(163, 164, 183), 2118, 2120, 2121(254), *2125, 2126, 2128*
Grumbine, S.O. 2529(337), *2566*
Gründel, M. 637, 638(97), *699*
Grundy, K.N. 1454, 1467(85), *1527*
Grüner, N.E. 68(122a), *99*
Grunhagen, A. 211(144), *260*
Grützmacher, H. 520(65a, 65b), *553*, 761, 763(127), *777*, 2019, 2026(184a), *2035*
Grutzner, J.B. 449(123), *491*
Gspaltl, P. 796(83b, 83c), 805(83c), 806, 807(86), *824*, 2179(17), 2191(65, 66), 2192(68), 2194(74), 2199(81), 2200(81, 82), 2201(83, 84), 2203(74), 2210(106), 2213(106, 107), *2214–2216*
Gu, T.-Y.Y. 1331(179), *1336*
Gu, Y.G. 450(125), *491*
Guarini, A. 1543(17), *1577*
Guba, G.Ya. 2228, 2229(76), *2242*
Gubanova, L.I. 237(244, 246, 247), *263*, 1340(9a), 1393, 1395, 1398(152), 1412(9a), *1437, 1442*, 1454(71, 84), 1458(123), 1461(142), 1467(84, 123, 142), 1475(252), 1500(71), *1527–1529, 1532*
Gudat, D. 213(152), *260*, 485(257), *493*, 1514, 1515(409), 1517(409–411), 1518(410), 1519, 1520(414), 1521(409, 411, 414), 1522(409–411), *1537*, 2145, 2146(96), *2172*
Gudovich, L.V. 1471(220), *1531*
Guedj, R. 435, 470(58), *489*
Guelachvili, G. 1128(223), *1141*
Guenot, P. 1016(313), 1022(332), 1023(313), 1025(332), *1061*
Guenschel, H. 1587(58), *1596*
Guerain, J. 2351(244), *2361*
Guérin, C. 781(8a, 8b, 9), 782, 784(8a, 8b), 789(8a), 790, 814(8b), *821, 822*, 1972, 1981(37a, 37b), 2007(158a), 2016(176c), 2017(179–181), 2018(181), 2021(190), *2031, 2034–2036*, 2376(44, 45), *2399*
Guerin, C. 37(77b), *97*, 284, 288(52), *351*, 496(5, 6), 497(5, 6, 9, 14), 508(55), 509(66), *510, 511*, 828(14), *854*, 1356, 1357(39), 1370(59–62), 1371(59–61), *1438, 1439*, 1633(155), *1663*, 2330(85), *2357*
Guermeur, C. 2330, 2354(82), *2357*
Guerra, M. 168(107), *179*
Guerrero, A. 434, 447(31), *489*
Guerrini, A. 168(107), *179*, 460(162), *491*, 1543(17), *1577*
Guest, M.F. 1150(39), *1182*
Gueye, M. 2548(400), *2568*
Guglielmi, M. 2345(177), *2360*

Guida-Pietrasanta, F. 1769(289), *1791*
Guidry, R.M. 1467(190), *1530*
Guillemin, J.-C. 3(7b), *95*
Guimon, C. 184, 203(25), *257*, 1018, 1042, 1046, 1047(316), *1061*, 1968, 1982–1984(22), *2030*
Guise, L.E. 1296(186), *1309*, 1849(292), *1867*
Guizard, C. 2341(143), *2359*
Gulinski, J. 1688(6), 1692(18), 1735(150), *1785, 1786, 1788*
Gullinane, J.A. 1390, 1392(127), *1441*
Gun, G. 2320(16), 2328(16, 61–63, 67, 68, 72, 73), 2329(72, 73), *2355, 2357*
Gun, J. 2321(24b), *2356*
Gundu Rao, C. 635(89c), *699*
Gung, B.W. 484(247), *493*, 506(42), *511*
Gunji, T. 1771(297), *1791*
Gunn, S.R. 158(47, 48), *178*
Günther, K. 2364, 2369, 2376(32), *2399*
Guo, H. 1769(291), *1791*, 1839(242, 243), *1866*
Guo, Li 186(32), *257*
Guo, Y. 1116(80), *1137*
Guore, W.F. 2416(46), *2458*
Gupta, H.K. 580(169), *593*
Gupta, Y. 805(81), *824*
Guram, A.S. 1830(196), *1865*
Gurkova, S.N. 240(257), 245(284), *263, 264*, 586(193), *594*, 1450, 1461(34, 35), *1525*
Gusarova, N.K. 1501(368), *1536*
Gusel'nikov, L. (36), *1054*
Gusel'nikov, L.E. 154, 156, 162(14), 163(14, 69), 174(14), *177, 179*, 715(37), *775*, 844(55), *855*, 859(1, 2), 860(1, 2, 16a), 861(19), 862(2, 16a, 19), 917(1, 2), 981(1, 2, 254), 985, 992(254), 1006, 1008(295), *1054, 1059, 1060*, 1064, 1068(4a), *1101*, 1144(6–8), *1181*, 1237(15), *1305*, 1913, 1914(77), *1927*, 2406(16), 2429(98), *2458, 2459*
Guselnikow, L.E. 229(212), *262*, 1910(57), *1927*
Gusev, A.I. 240(258), 245(284), *263, 264*, 586(193), *594*, 1450, 1461(34, 35), *1525*
Gusev, A.J. 240(257), *263*
Gusman, M.I. 2253, 2255, 2261(44c), *2312*
Gust, D. 518(60b), *553*
Gustavson, W.A. 2005(152, 153), *2034*
Gutekunst, B. 3(4), *95*, 250(305, 306), *264*, 859(3), 880, 881(3, 87), 977, 978, 981, 985, 986, 992, 996(3, 91), *1054, 1055*, 1064, 1068(1), 1080(47), *1101, 1102*, 1145(15), 1146(22), *1181*, 1601, 1609, 1645(24), *1661*, 2404(11, 12), *2458*
Gutekunst, G. 3(4), *95*, 250(305–307), *264*, 795, 799(63b), *823*, 859(3), 880(3, 87, 91), 881(3, 87), 977, 978, 981(3), 985, 986, 992, 996(3, 91), *1054–1056*, 1064, 1068(1), 1080(47), *1101, 1102*, 1145(15), 1146(22), *1181*, 1601(24), 1602, 1604(31), 1609(24, 31), 1645(24), *1661*, 2404(11, 12), *2458*
Guth, M. 1147(27), *1182*
Gutiérrez, A. 1743, 1746, 1747(225), *1790*
Gutman, D. 155(27), 167(27, 85–89), 168(89), 169(114), 177(27), *178–180*, 517(56), 518(57, 58), *553*
Gutmann, V. 538(84), *554*
Gutowsky, H.S. 981, 982(257, 258), *1059*
Gutowsky, R. 982(263), *1059*
Guy, A. 1189(7, 9), 1190(9), 1191(7), *1229, 1230*
Guyot, B. 402(144), *428*, 1802(71), *1863*
Guziec, J.C. 1907(52a), *1926*
Gvishi, R. 2351(243), *2361*
Gviski, R. 2322(31), *2356*
Gyname, M.S.J. 2534(364), *2567*
Gyobu, S. 805(82), *824*

Haaf, K. 1128(229), *1141*
Haaf, M. 60(107), *99*, 2540(374), *2567*
Haaima, G. 462(168), *492*
Haaland, A. 29(66b), 52–55, 57–59(98a), *97, 98*, 279, 280(38), *351*, 1144(11), *1181*, 2535–2539(367), *2567*
Haan, J.W.de 343–346(152), *354*
Haar, J.P.Jr. 1799(55), *1863*
Haarer, D. 2353(277), 2354(286), *2362*
Haas, A. 198(87), *259*, 1885(73), *1893*, 1897(11), *1925*
Haas, B.-M. 530(93), *554*, 1107(17), *1136*
Habaue, S. 443(109, 111), *490*, 1635(161), *1664*
Habben, C. 229(217), *262*
Habibi, D. 2322(37), *2356*
Habimana, J. 2223(32a), *2241*
Haces, A. 1550(47), *1578*
Hache, B. 1874(27), *1892*
Hacker, H. 2226(51), *2242*
Hackett, P.A. 2522(284), *2565*
Hada, M. 1808(91), *1863*, 2056(47), *2123*
Haddon, R.C. 8(28), 69(127, 128, 130, 131a–c), *96, 99, 100*
Hadi, M.A. 626(78), *699*
Hagaman, E.W. 2253(38), *2311*
Hagemeyer, A. 310(129), *353*
Hagen, A.P. 2143(76), *2172*
Hagen, G. 399(136, 137), *428*, 441(101), *490*, 576(138), *592*, 625(75, 76), 626, 627(76), *699*
Hagen, K. 161(63), *178*
Hagen, V. 1972(35a), 2012(169), *2031, 2035*
Hagger, R. 2223(27a–c, 28), *2241*
Hagiwara, K. 1169(170), *1185*, 1331(187, 188), *1336*, 2416, 2417(53), *2458*,

Author index

2491(143, 144), 2518, 2519(143), 2520(144), *2562*
Hagiwara, T. 476(222), *493*, 713, 741, 743(33), 748(33, 98), 750(98), 752, 753(33, 113), 755(116), *775, 777*, 1692, 1738(16), *1786*, 2403(1), 2404(4, 5), 2437(140), *2457, 2460*
Hague, D.N. 1321(89), *1334*
Hague, M.S. 1636, 1652(165), *1664*
Hahn, F. 1374, 1375, 1377, 1379, 1380, 1430, 1434(81), *1439*
Hahn, F.E. 228(208), *262*, 1415, 1430(201a), *1443*, 2040(7), *2122*
Hahn, J. 1341, 1349, 1350, 1357, 1361, 1365, 1368(30a), *1438*
Hahnfeld, D. 2494(163), *2562*
Haino, M. 2250, 2252, 2253(23b), *2311*
Hairston, T.J. 559(7), *590*
Hajdasz, D. 370(50), *427*, 601, 604(15), *698*
Hajdasz, D.J. 1109(32), *1136*, 1372, 1373(65), *1439*
Hajduk, P.J. 981, 982(257, 258), *1059*
Hakuta, K. 2522(305), *2565*
Halbach, R. 1555(58c), *1578*
Hale, M.R. 476(224), *493*
Halevi, E.A. 1316(41), *1333*
Halim, S.H.A. 243(276), *264*
Hall, D.I. 1107(12, 16), *1136*
Hall, J.A. 1448, 1450, 1453, 1470(1), *1524*
Hall, M.H. 616(52), *698*
Hallberg, A. 1825(161, 168), 1828(182), 1838(227–229), *1865, 1866*
Haller, I. 559(23), *590*, 1120(113), *1138*
Haller, K.J. 200(97), *259*, 348(157, 160), *354*, 1329, 1330(162), *1336*, 1897(23, 25), 1898(25), *1926*, 2134(28), *2171*, 2471(43), *2559*
Halley, K.A. 376(65), *427*, 635–637(91), *699*
Hallgren, J.E. 1690(12), *1786*
Halloran, D.J. 2227(60, 62, 63), *2242*
Halm, R.L. 1584(29–32), 1585(38, 39), 1586(44), 1593(110, 111), *1595, 1597*
Halow, I. 159, 160(54), *178*
Halterman, R.L. 1743(226), *1790*
Haltermann, R.L. 2138(40), *2171*
Haltiwanger, R.C. 219(177), *261*
Haluska, L.A. 2295(144), *2315*
Hamada, Y. 1075(32c), *1102*, 1313, 1317, 1320(45), *1333*, 2416(49–51), 2418(50, 51), *2458*
Hamanishi, K. 2498(183), 2513, 2515, 2516, 2520(259), *2563, 2564*
Hamann, J. 200(101), *259*
Hamashima, N. 2087(166, 167), 2090, 2094(166), *2126*
Hamer, G.K. 361(15), *426*
Hamid, M.A. 793(48), *823*
Hamilton, T. 2509(229), *2564*

Hamilton, T.P. 1168(167), *1185*, 2493, 2512, 2520(249), *2564*
Hamish, D.F. 1603(37), *1661*
Hamlin, R.D. 2253, 2255(44a), 2261(44a, 69), 2263(69), *2312, 2313*
Hammerschmidt, F. 1550(44), *1578*
Hammerum, S. 1125(181), *1140*
Hammett, L.P. 359(10), *426*
Hammond, G. 1930(5), *1960*
Hammond, G.B. 1836(221), *1866*
Hammond, G.S. 2487(127), *2561*
Hampel, B. 2134(27), 2140(45), *2171*
Hampel, F. 43, 44, 47, 48(93a), *98*, 817–819(113), *825*, 984, 988(267), *1059*, 2019(184c), *2035*
Han, B.-H. 1218(95), *1231*
Han, C.D. 2247(10), *2310*
Han, J.S. 461(165), *492*, 1590(80), *1596*, 1856(329), *1868*, 1871(12, 13), *1891*, 2085, 2090, 2094(159), *2126*
Han, N.F. 1701, 1703, 1755(65), *1787*
Han, Y.-K. 1842(257), *1867*
Hanack, M. 517(47), *553*, 669(145), *701*
Hanaoka, M. 441(105), *490*
Hancock, R.D. 2340(140), *2359*
Handwerker, H. 1386(113), 1387, 1389(113, 120), *1440, 1441*, 2057(48), 2059(52, 53), 2060(52–55), 2066, 2067(52, 53, 55), 2069, 2071(76), 2084(55, 76), 2085(160), 2089(55, 76), 2090(160), 2093(76), 2094(55, 76, 160), *2123, 2124, 2126*, 2527(330), 2528(336), *2566*
Handy, N.C. 5(12), 66, 68(120a), *95, 99*, 515(24), 516(29), 517(44d, 44e), *552, 553*, 1125(173), *1139*, 1163(133), *1184*
Hangazheev, S.H. 1490, 1494(319), *1534*
Hankin, J.A. 1026, 1028, 1050(338), *1061*, 1106, 1109(4), *1136*, 1372(67b), 1434(236), *1439, 1444*
Hanna, P.K. 1735(147, 148), *1788*
Hanneman, L.F. 2251, 2253(27a), *2311*
Hannington, J.P. 2227(59), *2242*
Hanquet, B. 1992(102), *2033*
Hansch, C. 360, 361(11), *426*
Hansen, A.E. 516(33), *552*
Hansen, D. 1746(227), *1790*
Hansen, K.B. 1672(49), *1683*
Hanshow, D. 2159(158), *2174*
Hanson, E.M. 739, 741(75), 767(75, 144), *776, 778*
Hanson, J. 870(53), *1055*, 1094(69), *1103*, 1264(84), *1307*, 1882(63), *1893*
Hanstein, W. 358, 359(9), *426*, 616(49–51), *698*, 838(45), *854*
Hanus, M. 1050(370), *1062*, 1129(234), *1141*
Hanusa, T.P. 209(138), *260*
Hanzawa, Y. 131(38), *151*, 2182(34), *2214*

Hapke, J. 2271, 2294(91), *2314*
Happ, B. 1721(123), *1788*
Happer, D.A.R. 577(158), *593*
Hara, M. 810(93a), *824*
Hara, Y. 2226(52), *2242*
Harada, J. 279(36), *351*
Harada, M. 72, 74, 75(147), *100*, 1590(89), *1596*
Harada, Y. 1832(204), *1865*
Harata, O. 1973(48a), *2031*
Haraya, A. 2523(312), *2566*
Harbison, G.S. 310(121, 122), *353*
Hardee, J.R. 2322(36), *2356*
Harding, M. 464(175), *492*
Harding, M.M. 116(47), *118*, 1463, 1467(162), *1529*
Hargittai, I. 1465, 1467, 1472(170), *1530*
Harland, J. 284(55), *351*
Harland, J.J. 226(203), 238(252), *262, 263*, 284(60), *351*, 1340, 1341(12), 1352(33, 34, 34), 1353, 1356, 1357(34, 34), (44, 45), *1437, 1438*
Harlow, R.L. 52(95a, 95b, 95e, 95f), 54(95e, 95f), *98*, 520(66), *553*, 763(133a), *777*, 1986(80a), *2032*, 2434(120), *2460*, 2535(366), 2539(373a, 373b), *2567*
Harmony, M.D. 586(191), *594*
Harms, K. 520(67), *553*
Harns, R.K. 1897(19), *1926*
Harper, W.W. 2557(430), *2568*
Harrah, L.A. 192(57), *258*, 1318(46), *1333*
Harrell, R.L.Jr. 795, 796, 800(71a), *823*
Harris, D.H. 2534(363–365), *2567*
Harris, J. 1646(205), *1665*
Harris, J.M. 1644(197), *1664*
Harris, J.W. 880, 881, 919, 922, 924, 928, 956(86), *1055*, 1146(21), *1181*, 1601, 1609(24), 1614(91), 1645(24), *1661, 1662*, 2404(9), *2457*
Harris, K.D.M. 333(146), *353*
Harris, M. 2290, 2294(139a, 139b), *2315*
Harris, M.E. 2466(14, 15), *2559*
Harris, R.K. 269(4), 307(112), 315(139), 317(139, 140), 318(139), 319–322(140), *350, 353*, 533(118), *555*, 1474, 1475(246), *1532*
Harrison, K.A. 1557(62), *1578*
Harrison, R.J. 184, 221, 223(21), *257*
Harrison, R.W. 1605(44), *1661*
Harrod, J.F. 200(96), 207(120), 219(174), *259–261*, 571(127), 574(127, 134, 135), *592*, 1697(39, 41, 42), 1704(69), 1714(94), *1786, 1787*, 2042(13), 2043(16, 17), 2045(13), 2047(24, 26), 2051(16), 2052(24), *2122, 2123*, 2179(15), *2214*, 2246(5c), 2249(18), 2251(29a, 29b), 2274(101a, 101b), 2276(101a, 101b, 103), 2279(101a, 101b), 2290(137, 138a, 138c), 2291, 2293(138a, 138c), *2310, 2311, 2314, 2315*
Harrowven, D.C. 1669(18), *1682*
Hart, D.J. 1563(74b), *1579*, 1835(217), *1866*
Hartley, F.R. 2341(141), *2359*
Hartman, J.R. 175(141), *180*
Hartmann, H.-M. 217(167), *261*, 780, 781(2e), 796, 799(65), 802(74), *821, 823, 824*
Hartog, M.den 1671(34), *1683*
Hartwig, J.F. 2425(83, 84), *2459*
Haruvy, Y. 2350(230), 2353(279), *2361, 2362*
Hasebe, H. 807, 808(87c), *824*, 883(98), 888, 889, 986, 988, 990(121), *1056*
Hasegawa, M. 1559(68), *1579*, 1798(51), *1863*, 2043, 2051(15), *2122*
Hasegawa, T. 1669(25), *1682*
Hasegawa, Y. 186(40), *258*, 2017(178), *2035*, 2272(94a–c, 96), *2314*, 2439(148), *2460*
Hasenhindl, A. 331–333(145), *353*
Häser, M. 72(145), *100*
Hashida, I. 581(176), *593*, 1292(177), *1309*
Hashiguchi, D. 1587(55, 56), *1596*
Hashiguchi, D.H. 1586(51), *1596*
Hashimoto, M. 269, 273, 275(12), *350*, 2207(97), *2216*
Hashimoto, T. 384(87, 88), 385(92), *427*
Hashimoto, Y. 1798(51), *1863*
Hashizume, S. 441(105), *490*
Haskins, W.C.K. 1168(165), *1185*, 2493, 2520(155), *2562*
Hass, A. 1918(85), *1927*
Hass, B.A. 6(21d), *95*
Hass, D. 245(278), *264*
Hassan, O. 2388(72), *2400*
Hässelbarth, A. 1114(71a), 1115(72), *1137*
Hasselbring, R. 203(106), 209(130), *259, 260*
Hässig, R. 811(97a), *824*
Hassler, C. 1555(58c), *1578*
Hassler, K. 790(35), *822*, 2178(5b), 2187(54), 2189(54, 56, 60, 62), 2204(60), *2213, 2215*
Hassner, A. 611, 614(42), *698*, 1601, 1602, 1604(11), 1613, 1622(89), 1624, 1651(11), *1660, 1662*
Hastie, J.W. 1162(111), 1181(198), *1183, 1185*
Haszeldine, R.N. 741, 742(78), *776*
Hatakenaka, K. 1293(181), *1309*
Hatanaka, Y. 804(80), *824*, 1755(247), 1756(247, 248), *1790*, 1825(169–172), 1826(174, 176), 1827(176–179), *1865*
Hatano, H. 2333(97), *2358*
Hatano, K. 791(42), *822*, 2453(200, 201), 2455(200), *2461*
Hatano, T. 186(40), *258*, 2017(178), *2035*, 2439(148), *2460*
Hataya, K. 1872(19), *1892*
Hatayama, Y. 1771(308), *1792*

Hatschek, E. 2319, 2346(8b), *2355*
Hattori, T. 457(147), *491*
Hauenstein, P. 2073(101), *2124*
Haufe, G. 1885(76), *1893*
Hauge, R.H. 1127(209), *1140*, 1162(111), 1167(161), 1175(161, 188), 1176(188), 1180(196), 1181(198–200), *1183–1185*, 2469(24), *2559*
Hauge, R.M. 2518(262), *2565*
Haugen, G.R. 155, 176(30), *178*
Haumann, T. 648(121), *700*
Hauschildt, G.L. 1511–1513(402), *1537*
Havemann, R.H. 2295(145), *2315*
Haves, A.J. 2131, 2133(11), *2171*
Havlas, Z. 219(178), *261*
Haward, M.T. 1260(74), *1307*
Hawari, J.A. 168(106), *179*, 1320, 1321(78), *1334*, 1540(5), *1577*, 2526(329), *2566*
Hayakawa, F. 371(53, 56, 54), *427*, 621, 623, 624(73), 632, 633(86a), 634(86a, 88a), 635(88a), *699*
Hayakawa, K. 1716(103), *1787*
Hayama, T. 417(194), *429*, 1861(357), *1868*
Hayase, S. 2016(176b), *2035*
Hayashi, C. 1743, 1746(208), *1789*
Hayashi, H. 1257(66), *1307*
Hayashi, J. 2272(92a–c, 93a, 93b, 94a), *2314*
Hayashi, K. 1558(66a, 66b), *1579*
Hayashi, M. 867(44), 870(44, 51), *1055*, 1671(35, 40), 1672(47), *1683*
Hayashi, R. 52–55, 57–59(98a), *98*, 279, 280(38), *351*, 1144(11), 1166(156), *1181*, *1184*, 2023(195), *2036*, 2535–2539(367), *2567*
Hayashi, R.K. 52, 53, 59, 60, 62(98d), *98*, 200(103), 252(328), *259*, *265*, 2118–2121(255), *2128*, 2492(148), 2493(149), 2540, 2541(148), 2542, 2543(149), *2562*
Hayashi, S. 345–347(153), *354*
Hayashi, T. 340–342(151), *354*, 407(157–159), *429*, 1340, 1341(17a–c, 18), 1342, 1343(18), 1344, 1345(17a–c), 1346(17a), 1347(17c, 18), 1348(17c), *1437*, 1692(22), 1743, 1746(208), 1751(234), 1752(238–241), 1753(241, 242), 1754(242–244), 1755, 1756(246), *1786*, *1789*, *1790*, 2110(222), 2116(238), 2120(222), *2127*, 2159(162), *2174*, 2278(107), *2314*, 2430(111), *2460*, 2523(312), *2566*
Hayashi, Y. 1628(144), *1663*
Hayashida, A. 2265(81a, 81b), *2313*
Hayashida, H. 1800(60), *1863*
Hayaya, K. 2107, 2109(211), *2127*
Hayden, G.W. 1316(39, 40), *1333*
Hayes, R.N. 1133(251), *1141*
Haynes, A. 1260(74), *1307*
Hays, M.K. 2100, 2108(193), *2127*

He, C. 226(201), *262*
He, G.S. 2352(262), *2362*
He, J. 207(120), 219(174), *260, 261*, 2249(18), 2251(29b), *2310, 2311*
He, M. 2234(105a), *2243*
He, X.W. 2223(25), *2241*
Heacock, D.J. 517(53, 54), *553*
Hearn, M.P. 2220(20), *2241*
Heath, J.R. 69(124a–c), *99*, 1929(1b), *1960*
Heathcock, C. 1601, 1602(9), *1660*
Heathcock, C.H. 440(97), *490*, 1642(191), *1664*
Heaton, S.N. 1299(196), *1310*
Heberle, J. 1674(63), *1683*
Heck, J. 2152, 2153(120–123), *2173*
Heckel, M. 1377, 1378, 1380(97b), 1434, 1436(250), *1440, 1445*, 1720, 1724(114), *1788*, 2085, 2090, 2094(158), 2106, 2109(209), *2126, 2127*
Hedberg, K. 72, 74(146), *100*, 116(46), *118*, 149(61), *152*, 1910, 1911(60c), *1927*
Hedberg, L. 72, 74(146), *100*, 149(61), *152*, 1910, 1911(60c), *1927*
Hedgecock, C.J.R. 1842(259), *1867*
Hedrick, J.C. 2345(168), *2360*
Heeg, E. 2364(10, 12, 13), *2398*
Heeg, M.J. 2141(54), 2143(75), 2144, 2145(54), *2172*
Heerding, D.A. 396(127), *428*
Hegarty, A.F. 710(25a), *775*
Hegarty, S.C. 1741(182, 183), *1789*, 1799(56), *1863*
Hegedus, L.S. 77, 81(153), *100*
Hehre, W.J. 6(21a), *95*, 407(162), *429*, 515(22), 517(55), 518(62), *552, 553*, 620(63), 642(103, 105), *699, 700*
Heidemann, D. 2337(124), *2358*
Heider, G.L. 798(58), *823*
Heikenwalder, C. 1374, 1375, 1379(80), *1439*
Heikenwälder, C.-R. 945(210), 946(210, 218–221, 224), *1058*
Heil, B. 1689(8), *1785*
Heilbronner, E. 1910, 1911(60d), *1927*
Heilemann, W. 238(251), *263*
Heiliger, L. 520, 524(70), 540(70, 105), *553, 554*, 563(81–83), 567(81, 83), 568(83), 569(81, 83), 583(82), *591*
Heilinger, L. 1408(182), *1443*
Heilman, S.M. 1670(29), *1683*
Heimann, P. 2232(95), *2243*
Heimer, T.A. 2349(206), *2361*
Heine, A. 198(81, 85), *259*, 799(64), 800(67a, 67b), *823*, 884–886, 937, 938(103), *1056*
Heinemann, C. 52(95g), 53(100a), 54(95g), 55(100a, 104), 56, 57, 60, 61(100a), 62(100a, 104), 65(104), *98*, 1134(258), *1142*, 2539(371, 372b), *2567*

Heinicke, J. 52(99a–c), 55, 56, 58, 61(99c), 62(99b, 99c), *98*, 2467(19, 21), 2495(19, 21, 169), 2535(368, 369), 2536(368), 2537, 2539(370), *2559, 2562, 2567*
Heinis, T. 1106(7), *1136*
Heinrich, L. 1352, 1412(32a), *1438*
Heinrich, T. 2340(136), *2359*
Heintze, M. 2545(381), *2567*
Heinze, U. 1587(54, 59), *1596*
Helary, G. 2224(38), *2241*
Helber, M.J. 1823(154), *1864*
Helfer, A.P. 1965(16c), *2030*
Helm, A.van der 2224(41a), *2241*
Helmchen, G. 1746(227), *1790*
Helmer, B.D. 2515(276), *2565*
Helmer, B.J. 269(8), *350*, 508(56), *511*, 1157(76), *1183*, 1268(96), *1307*, 1329(161), *1336*, 1381(106), *1440*, 2471(44), *2559*
Helquist, P. 760(123), *777*
Hembree, D.M.Jr. 2232(95), *2243*
Hemery, P. 2219(16), *2240*
Hemme, I. 3(3d), *95*, 1010(299b), *1060*, 1064, 1068(5d), *1101*, 2541(375), *2567*
Hemmersbach, P. 2429(96), *2459*
Hench, L.L. 2318(6b), 2319, 2350(10b), *2355*
Hencher, J.L. 784, 787(24), *822*
Hencken, G. 780, 781(3), *821*
Hencsei, P. 183(11), 226(199), *257, 261*, 1449(14, 15, 18, 19), 1450(33, 42, 43), 1452(55), 1453(33, 63), 1454(55, 78, 93), 1460(14, 15, 18, 19, 136, 154), 1461(137–139), 1462(42, 149, 152, 154), 1463(15, 18, 19, 149, 166), 1466(180), 1467(93, 136–139, 149, 152, 154, 166, 180, 182, 191, 192), 1468(180, 191–193), 1470(206, 208), 1471(33, 229, 231), 1472(191, 192, 229, 231), 1474, 1475(256), 1477(154), 1478(137), 1481(282, 287, 288), 1483(288), 1490(19), 1491(154), 1503(152), 1508(385, 388, 389), 1509(287, 395), 1510(166, 287), *1525–1527, 1529–1533, 1536, 1537*
Hendan, B.J. 1923(92), *1928*, 2334(107), *2358*
Hendrickson, N. 2160(170), *2174*
Heneghan, M. 1859(348), *1868*
Hengelsberg, H. 196(72), *258*, 2363(7), 2378(53), 2379(7), 2380(7, 59), 2383(59), 2385(7, 62), 2387(53, 65), 2388(53, 62, 69), *2398–2400*
Hengge, E. 186(43), *258*, 269–272(11), 279(30), *350, 351*, 782(12b), 796(74, 83a–c, 84), 802(74), 805(83c, 84), 806(84, 86), 807(86), *822, 824*, 1214(75, 76), 1218(94), *1231*, 1267(94), *1307*, 2047(27), 2083(145, 146), *2123, 2125*, 2177(4a–d), 2178(5b, 5d), 2179(10–14, 17), 2181(5d, 18–20, 24, 31), 2182, 2187(5d), 2189(56), 2191(63–66), 2192(67, 68), 2193(73), 2194(74, 76), 2197(5d, 79, 80), 2199(81), 2200(81, 82), 2201(83–85), 2203(74, 86), 2204(5d, 87, 89), 2205(5d), 2209(5d, 100–102), 2210(73, 100, 101, 103, 104, 106), 2213(106–108), *2213–2216*, 2279(108–110), *2314*
Henke, H. 196(72), *258*, 2378(56), *2399*
Henkel, G. 1524(420), *1537*
Henner, B. 781(9), *822*, 1370, 1371(59, 60), *1439*, 1972, 1981(37a), 2016(176c), 2017(179–181), 2018(181), 2021(190), *2031, 2035, 2036*
Henner, B.J.L. 229(218), *262*, 498(20, 21), 508(55), 509(66), *510, 511*, 1356, 1357(39), 1370(61, 62), 1371(61), *1438, 1439*, 1972, 1981(37b), 1989(91), *2031, 2033*, 2330(85), *2357*
Henner, M. 559, 563, 565(8), *590*
Henner-Leard, M. 507(46), 508(51), *511*
Henrichs, C. 1743, 1746(195), *1789*
Henriquez, R. 2137, 2138(37), *2171*
Henry, C. 391(115), 392(116, 117), 396(131), *428*, 438(95), *490*
Henry, G.K. 1981, 1983, 1998(68), *2032*
Henry, M. 2319, 2350(10a), *2355*
Henry, S.S. 1671(44), *1683*
Hensen, K. 520, 524, 525(69), *553*, 564(88), 581(178), 583, 586(88), *591, 593*, 1374, 1375(85, 87, 88), 1377(85, 87), 1378(87), 1380(88), 1434, 1436(247), *1439, 1440, 1445*
Herberhold, M. 203(109, 110), 209(136), *259, 260*
Herberich, G.E. 1986, 2011(76b), *2032*
Herbst, E. 1109, 1117(50), 1126(50, 200), *1137, 1140*, 2556(423), *2568*
Herbst-Irmer, R. 198(81), 219(176), *259, 261*, 541(107), 554, 581(180), *593*, 799(64), 800(67b), *823*, 1075(35b, 37), 1086(37), 1087(55), 1090(37), *1102*, 1411, 1412(188), *1443*, 2183, 2185(41), 2187(48), *2214*, 2429, 2436(104), *2459*, 2550(411), 2552(413), 2553(415), *2568*
Hercules, D.M. 1470(209), *1531*
Herdtreck, E. 2069, 2071, 2084, 2089, 2093, 2094(76), *2124*
Herdtweck, E. 186(43), *258*, 581(181), *593*, 945(206, 209, 211), 946(209, 211, 214, 222), 948(228, 229), *1058, 1059*, 1374, 1375(80, 81), 1377(81), 1379(80, 81), 1380(81), 1386(112, 113), 1387(113), 1389(112, 113), 1430, 1434(81), *1439, 1440*, 2059(51, 52), 2060(52), 2066, 2067(51, 52), 2085, 2090, 2094(158), *2123, 2126*, 2135(30), 2136, 2137(31), 2147(31, 98), *2171, 2173*, 2528(336), *2566*
Herlet, N. 2350, 2351(232), *2361*

Herman, A. 229(215), *262*, 2510(236, 237), *2564*
Hermann, C. 1189(5), *1229*
Hermann, U. 2203(86), *2215*
Hermann, W.H. 2539(372b), *2567*
Hermans, B. 1801(62), *1863*
Hermeling, D. 1206(47), *1230*
Hernandez, C. 1261(76), *1307*, 2083(152), *2126*
Herold, B.J. 2464, 2465, 2468, 2475, 2485, 2488(1), *2558*
Herrero, J. 1721, 1723, 1724(120), *1788*
Herrmann, W.A. 52(96, 97), 53, 54(96), 55(104), 61(96), 62, 65(104), *98*, 2509(225), 2529(340), 2539(225), *2564, 2566*
Herron, J.T. 6(22), *95*
Herrschaft, B. 948, 1007, 1008(232), *1059*
Hersh, K.A. 1415, 1420(198), *1443*
Hershberger, D.W. 2137(34), *2171*
Hertel, D. 1450, 1456, 1457, 1461, 1496(37), *1525*
Hertkorn, N. 190(56), *258*
Hertler, W.R. 196(77, 78), *259*, 284(58), *351*, 1356, 1357(41), *1438*
Herz, J.E. 2223(25), *2241*
Herzfeld, J. 310(132), *353*
Herzig, C. 2225(49), *2242*
Herzig, C.J. 2117(253), *2128*
Herzog, A. 209(130), *260*
Herzog, F. 2143(79), *2172*
Herzog, U. 2194(75), *2215*, 2468(23), *2559*
Hess, B.A. 1019, 1044, 1045, 1047(321), *1061*
Hess, B.A.Jr. 2509(229), *2564*
Hesse, M. 578(159), *593*, 953–955(240), 1011(304), *1059, 1060*, 1064, 1068(5c), *1101*, 1274(111), *1308*
Hessner, B. 1986, 2011(76b), *2032*
Hetflej, J. 617(55), *698*
Hetflejs, J. 1700(53), *1786*
Heucheroth, R.O. 473(208), *492*
Hevesi, L. 1801(62), *1863*
Hewkin, C.T. 1620(119, 120), *1663*
Hey, E. 217(166, 169), *261*
Heydenreich, M. 1885(76), *1893*
Heydt, H. 765(140), *778*
Hey-Hawkins, E. 217(168), *261*
Heyn, R.H. 250(315), *265*, 566(100), *591*, 789(28c), 791(44b), 800(70b), *822, 823*, 2039(2), 2046(21), 2049(28), 2051, 2052(21, 28), 2079(125, 128), 2080(135), 2082(125, 135, 142), 2092, 2093(125, 135), 2113, 2114, 2120(229), *2122, 2123, 2125, 2127*, 2425(85, 86), *2459*
Hiberty, P.C. 5, 6, 24(15a–c), 25, 87(15b, 15c), *95*, 499(22), *510*
Hicks, F.A. 1671(42), *1683*
Hidaka, A. 1771(302–304), *1791, 1792*

Hidaka, T. 20(51a), *96*
Hideki, A. 1611, 1617(82), *1662*
Hiemstra, H. 1815(122–125), 1835(216), *1864, 1866*
Hierle, R. 2354(285), *2362*
Hiermeier, J. 190(56), *258*, 2155(142), 2159(142, 151, 152, 154, 155), 2160, 2162(151), *2174*
Higashimura, T. 1861(361), *1868*
Higgins, D. 407(163), *429*, 1795(35), *1862*
Higgins, T.B. 198(84), *259*
High, K.G. 1715, 1716(102), *1787*
Highsmith, T.K. 1997(112a), *2033*
Higichi, T. 998, 1001(288), *1060*
Hign, K.G. 1719, 1732(111), *1788*
Higuchi, J. 517(49), *553*
Higuchi, K. 120(3), 137, 139, 140(45), 142(3), 143(3, 45, 52a, 52b), 144(45), 145(54), 146(3, 45), *150–152*, 810(93a, 93b), *824*, 1670(30), *1683*, 1740(176, 180), *1789*, 1924(107), *1928*, 2206(94), 2207(96), *2215*
Higuchi, N. 1710(86, 87), 1725(87), *1787*
Higuchi, T. 917, 998, 1002(184), *1058*, 1064(3), *1101*, 1973(48b, 48c), *2031*, 2429, 2436(106), *2459*
Hikage, S. 1885(78), *1893*
Hikida, T. 2110, 2120(222), *2127*
Hilal, H.S. 1692(21), 1700(55), *1786*
Hildebrandt, R.L. 163(67), *179*
Hildenbrand, T. 1890(107, 108), *1894*
Hilderbrandt, R.I. 1465, 1467, 1472, 1478(169), *1530*
Hill, C.A.S. 2352(269), *2362*
Hill, D.K. 434, 448(34), *489*
Hill, F.C. 111(20), *118*
Hill, J.E. 1735(151), *1788*
Hill, J.H.M. 476(218), *493*
Hill, J.M. 435, 456(52), *489*
Hill, M.G. 2016(174), *2035*
Hill, R.J. 110(12), *118*
Hill, R.M. 2234(105a), *2243*
Hiller, W. 209(137), 217(163), *260*, 1720, 1724(114), *1788*, 2085, 2090, 2094(158), 2106, 2109(209), *2126, 2127*, 2337(124), *2358*
Hilliard, R.K. 2557(430), *2568*
Hillyard, R.W. 1390(124, 125), 1396, 1399(124), *1441*
Hilty, T.K. 1735(152), *1788*, 2280(112a), *2314*
Himeshima, Y. 630(84), *699*
Hindahl, K. 2062, 2078, 2091(62), *2123*
Hinkln, T. 2301, 2305(157), *2316*
Hino, T. 569(113, 120, 122), 570, 571(113), 572(120), *592*, 1814(121), 1821(145), *1864*
Hinton, J.F. 517(37), *552*
Hinzen, B. 1555(58c), *1578*
Hioki, R. 1640(186), *1664*

Hiramatsu, K. 1973, 1974(44), *2031*
Hirao, A. 457(147), *491*, 1638(168), *1664*, 2334(105), *2358*
Hirao, T. 1669(25), *1682*, 1822(150), 1830(195), *1864, 1865*
Hirashima, T. 1226(114), *1232*
Hirata, H. 829(26), *854*
Hiratsuka, H. 1151(47), *1182*, 1246(38), *1306*, 1322(110), *1334*, 2513, 2516, 2526(258), *2564*
Hirle, B. 192(58), *258*, 2072(83, 86), 2074(102), 2091(83, 86, 102), 2110(102), *2124, 2125*, 2318, 2320(4), *2355*
Hirohara, Y. 1848(286), *1867*
Hirose, M. 1983, 2001(73), *2032*, 2546(393), *2568*
Hirose, T. 1767(282), *1791*
Hirota, E. 1021, 1045, 1046, 1049(325), *1061*, 2522(296, 305), 2523(307, 315), *2565, 2566*
Hirotsu, K. 998, 1001(288), *1060*, 1064(3), *1101*, 1973(48b, 48c), *2031*
Hirotsu, M. 1452, 1453, 1473(61), *1526*
Hirsch, A. 1930(4), *1960*
Hirsch, G. 1107(14), *1136*
Hirsl-Starcevic, S. 642(105), *700*
Hirst, D.M. 1107(15, 16), *1136*
Hisayuki, T. 2272(95), *2314*
Hitchcock, P. 203(111), *259*
Hitchcock, P.B. 29(66b), 52, 62(99b), *97, 98*, 196(74), 217(166), 222(191, 194), 245(288), *258, 261, 264*, 577(151), *593*, 1694, 1695(31, 32), *1786*, 2537, 2539(370), 2544(377), *2567*
Hitchcock, P.P. 2131(12), *2171*
Hitchcock, S.A. 1561(70), *1579*
Hite, G.E. 70, 71(139b), *100*
Hitomi, T. 2001(136), *2034*, 2436(129), *2460*
Hitze, R. 1918(85), *1927*
Hiyama, T. 435, 451(41), *489*, 783(14–16), 789(15), 795(16), 801(15), 804(80), *822, 824*, 1688(5), 1726, 1727(133), 1731(138, 139), 1739(166–170), 1740(174), 1755(247), 1756(247, 248), *1785, 1788–1790*, 1825(169–172), 1826(174, 176), 1827(176–179), 1861(362), *1865, 1868*
Hiyama, T.J. 780(1c), *821*
Hjortkjaer, J. 1701(62), *1787*
Hlasta, D.J. 465(177), *492*
Hlavacek, V. 2545(385), *2568*
Ho, H. 2290, 2294(139b), *2315*
Ho, P. 155(18–21, 24, 25), 156(20), 159(19), 164(24, 25), 165(24), 166(21, 25), 167, 168(20), 170(21, 24, 25), 171(18–20), 173(20), 175(20, 25), *178*, 1048(363), *1062*, 1129(233), *1141*
Ho, T. 2236(112, 113), *2243*
Ho, T.-L. 1811(109), *1864*

Ho, Y. 1109(32), *1136*, 1372, 1373(65), *1439*
Hoa, H.A. 1189, 1191(7), *1229*
Hobbs, J.L. 437, 484(75), *490*
Hobbs, R.H. 173(136), *180*
Hochmuth, W. 129(36), *151*
Hochstrasser, R.M. 1316(44), 1318(60–63, 66), *1333, 1334*
Hockemayer, J. 1283(149), *1309*
Hodge, P. 2218(4), *2240*
Hodgson, D.M. 1680(115, 116), *1684*
Hoebbel, D. 2337(123, 124), 2344(166), *2358, 2360*
Hoekman, S.K. 716(41), *775*
Hoff, S. 1621(123), *1663*
Hoffman, D.K. 727, 741(54), *776*
Hoffman, J.J. 2224(40), *2241*
Hoffmann, C. 1836(220), *1866*
Hoffmann, D. 884(110), 885(110, 115, 118), 889(110), 925(110, 118), 926(118), 928(110), *1056*
Hoffmann, H.M.R. 1857(337), *1868*
Hoffmann, J.M.Jr. 1998(122b), *2034*
Hoffmann, M.R. 710(25b), *775*
Hoffmann, R. 435(43), 453(43, 130), *489, 491*, 620(63), 642(103, 104), *699, 700*, 1130(238), *1141*, 1871(10), *1891*, 1911(62), *1927*, 1997(113a), *2033*
Hoffmann, U. 1555(58c), *1578*
Höfle, G. 1550(43), *1578*
Höfler, F. 2274(100a), *2314*
Hofman, W. 2141(61), *2172*
Hofmann, J. 2141, 2147, 2149(58), *2172*
Hofmann, M. 517(38), *552*
Hofmann, P. 1720, 1724(114), *1788*, 2106, 2109(209), 2117(251), *2127, 2128*
Hogen-Esch, T.E. 2224(34), *2241*
Hoh, K.-P. 2333(100), *2358*
Hohenberg, P. 516(30a, 30c), *552*
Hohenwarter, T.E. 2226(52), *2242*
Hojo, F. 194(68), *258*, 278(26, 27), 329(26), *351*, 2440(155), 2443(155, 161), 2444(155), 2445(167), 2447(172), 2448(179), 2449(172, 179), 2450(172, 190), 2451(172), *2460, 2461*
Hojo, M. 400(138), *428*, 1826(173), 1848(286), 1856(327), *1865, 1867, 1868*
Holcomb, R.C. 466(181), *492*
Holdsworth, J.R. 2504(199), *2563*
Hollemann, A.F. 1583(18), *1595*
Hollenstein, S. 558(6), *590*
Hollfelder, H. 190(54), *258*
Hollinger, G. 2231(84), *2243*
Hollinworth, G.J. 1561(72), *1579*
Hollis, T.K. 1801(65), *1863*
Holloman, J. 2545(384), *2567*
Hollowood, F. 2437(135), *2460*
Holme, T.A. 1172(184), *1185*

Holmes, J. 1354–1357(37a, 37b), *1438*
Holmes, J.L. 1122(149, 152), 1125(152), *1139*
Holmes, J.M. 226(203), *262*, 284(56, 60), *351*, 795, 800(68), *823*, 1340(13b), 1341, 1342(13b, 20a, 20b), 1343, 1347(13b), 1352, 1353, 1356, 1357(34), (43–45), *1437, 1438*
Holmes, J.S. 2434(121), *2460*
Holmes, R.R. 226(203), 238(252), *262, 263*, 284(52, 55, 56, 60), 288(52), *351*, 495(2), 497(11), 502(29), 508(53), *510, 511*, 1340(4a, 4b, 12, 13a, 13b, 14), 1341(4a, 4b, 12, 13a, 13b, 14, 20a, 20b), 1342(13a, 13b, 20a, 20b), 1343, 1347(13a, 13b), 1351(4a, 4b), 1352(33, 34), 1353(4a, 4b, 34), 1354, 1355(4a, 4b, 37a, 37b), 1356, 1357(34, 37a, 37b), 1378(4a, 4b), 1406(174a, 174b, 175a, 175b), 1407(176a, 176b), 1424(4a, 4b), 1434(237), (43–45), *1437, 1438, 1443, 1444*
Holmes, T.A. 84, 86(162), *100*
Holmes-Smith, R.D. 2132(19), *2171*
Holst, A. 1872(18), *1892*
Holt, A. 365(38), *426*, 617(57), 618(57, 59), 620(59), *699*, 1897(19), *1926*
Holtan, R.C. 1601, 1602, 1604, 1621, 1624, 1625(14), 1639(180), 1658(14), *1660, 1664*
Holten, D. 860, 899, 901(14), *1054*, 1236(14), *1305*, 2513, 2526, 2527(260), *2565*
Holtmann, U. 77–79(154b), *100*, 833(34), *854*, 1434(232b), *1444*, 2163–2165(182), 2166(182, 189–191), 2167(189, 190), 2168–2170(189), *2175*, 2534(360b), *2567*
Holtschmidt, N. 782(12b), *822*
Holzangel, C.M. 2088(178), *2126*
Holznagel, C.M. 362(24), *426*, 560(43, 49, 50), *590*, 601, 602, 606(16), *698*, 1111(65a), 1131(239, 241), *1137, 1141*
Homan, G.R. 2226(55), *2242*
Homme, C.P. 1583(20), *1595*
Hommes, N.J.R.v.E. 781(7), *821*
Hommes, N.v.E. 5(19), 7(19, 26), 9(26), 22–24, 31, 32(19), 42, 44, 45, 47, 69, 90(26), *95, 96*
Honda, A. 2115, 2120, 2121(232), *2127*
Honda, K. 1158(89), *1183*, 2408(29), *2458*
Honda, M. 1641(190), *1664*
Honda, Y. 2207(95), *2215*
Hondo, T. 390(113), 391(114), *428*, 1833(207, 208), *1866*
Honea, E.C. 70(134b), *100*, 1120(126), *1138*
Honegger, E. 1910, 1911(60d), *1927*
Honeyman, C.H. 2150, 2153, 2154(110), 2160(166), *2173, 2174*
Hong, B. 1189(5), *1229*
Hong, E. 461(165), *492*, 1856(329), *1868*
Hong, J.-H. 4(10c), 37(10c, 79, 80, 83), 38(79, 80), 39(10c), 40, 46(79), *95, 97*, 814(109, 110a), 815(111), 818(114, 115), *825*, 984, 987, 989(265), *1059*, 1981(66), 2018(183), 2019(185, 186), 2020(186, 188), 2021(189), 2022(183, 194), 2023(194), 2024(197), 2025(199), *2032, 2035, 2036*
Hong, L. 1767(279), *1791*
Hong, M.-S. 226(198), *261*
Hong, P. 2064–2066, 2069(72), *2124*
Hong, Q. 1606(53), *1661*
Hong, R.-F. 2017(177c), *2035*
Honjou, N. 997(286), *1060*
Hoogenstraaten, W. 466(178), *492*
Hooz, J. 1620(118), *1663*
Hop, C.E.C.A. 560(45), 561(54), *590, 591*
Hopfinger, A.J. 184, 221, 223(21), *257*
Hopkinson, A.C. 161, 175(58), *178*, 381, 382(78), *427*, 517(52), 553, 597, 601(3, 4), 607(3), 608(4), *697*, 1029(341), *1061*, 1118(93–95), 1126(205), *1138, 1140*
Hoppe, D. 752–755(112), *777*
Hoppe, M.L. 226(205), 228(211), *262*, 1353, 1368, 1369(36), *1438*, 2290, 2291, 2293(138a), 2298(150a, 150b, 152a), *2315, 2316*
Horata, K. 2016(175d), *2035*
Horchler, K. 296(90), 302, 304(102), *352*, 1979(58a, 58b), *2032*
Hori, K. 679(154), *701*
Horie, O. 1125(185), *1140*
Horiguchi, Y. 438(89, 93), 465(89), *490*
Horihata, M. 1743(190), *1789*
Horii, T. 2353(278), *2362*, 2475(60), *2560*
Horio, T. 186(40), *258*, 998(289, 290), 1004(289), 1005(289, 290), *1060*, 1769(294), 1770(294, 295), *1791*, 2017(178), *2035*, 2439(148), *2460*
Horiuchi, Y. 1620(121), *1663*, 1680(117), *1684*, 1807(82), *1863*
Horn, H.-G. 1869(1), 1879(56), *1891, 1893*
Horn, K. 2117(250), *2128*
Horn, K.A. 1325(122), *1335*, 1794, 1824(31), *1862*
Horn, M. 961, 964, 966(243), *1059*, 2407(21), *2458*
Horner, D.A. 163(73), *179*, 2479(79), *2560*
Hornung, G. 1114(71b), *1137*
Horowitz, D.S. 167, 171, 173, 175(96), *179*
Hortelano, E.R. 434, 439(12, 13), *488*
Horwell, I.V. 2340(140), *2359*
Horwitz, J.S. 2523(309), *2566*
Hoshi, S. 277(22), *350*, 2159(160), *2174*, 2429(92), *2459*
Hoshi, T. 832, 838(30), *854*, 1341(22, 26), 1343(22), 1347(26), 1434(235), *1437, 1444*
Hoshino, M. 1250(48), *1306*

Hoshino, Y. 120, 142, 143, 146(3), *150*, 2345(167), *2360*
Hoskin, D.H. 739, 740(73), *776*
Hosomi, A. 186(41), *258*, 277(22), *350*, 390(113), 391(114), 400(138), 404(149), *428*, 1493(328), 1497(352), *1535*, 1794(16), 1800(60), 1808(94), 1809(102), 1826(173), 1833(207, 208), 1848(286), 1856(327), *1862, 1863, 1865–1868*, 2159(160), *2174*, 2429(92), 2439(142, 143), *2459, 2460*
Hossain, M.A. 1897(18), *1926*
Hostetler, M.J. 1709(84, 85), *1787*
Hota, N.K. 1986(76a), 2001(135), 2011(76a), *2032, 2034*
Hoteiya, K. 1254(63), *1307*
Hotoda, K. 1201, 1203(33), *1230*
Hou, L. 2349, 2350(211), *2361*
Hou, X. 2364(34, 36), 2368, 2369, 2371(34), 2374(34, 36), *2399*
Hou, X.-L. 1668(14), *1682*
Hougen, J.T. 2523(313), *2566*
Houk, K.N. 9(31b), *96*, 521(78), *554*, 559(14), *590*, 2525(324), *2566*
House, H.O. 419(197), *429*, 838(44b), *854*
Hoveyda, A.H. 476(224), *493*
Hovnanian, N. 2340(131), *2359*
Howard, J.A. 1162(117), 1163(125), *1183, 1184*
Howard, J.A.K. 2070, 2071(80), *2124*
Howard, W.A. 922(189), *1058*, 1154, 1155(65), *1182*
Howarth, O.W. 269(4), *350*
Howe, L. 1341, 1342(20a, 20b), *1437*
Howes, A.J. 193(59), *258*, 2145(88), *2172*
Howson, B. 1619(113), *1663*
Hoy, A.R. 234(236), *263*
Hoyt, L.K. 2051, 2052, 2056, 2057, 2064(35), *2123*
Hrncir, D.C. 222(187), *261*, 2307(161), *2316*
Hrovat, D. 380(74), *427*
Hrovat, D.A. 645(112), *700*, 1998(117), *2034*
Hrusàk, J. 21(55), 36(75), *97*, 1029, 1049, 1050(342), *1061*, 1118(89), 1125(179, 183), 1126(197, 198), 1128(197), 1129, 1130(198), *1138, 1140*
Hrzhanovskaya, Y.N. 2041(12), *2122*
Hsiao, J.-S. 1313, 1317, 1320(45), *1333*
Hsiao, Y.-L. 1769(292), *1791*
Hsu, D.S.Y. 2514, 2523(274), *2565*
Hsu, G.J.-H. 1617, 1620(103), *1662*
Hsu, J.-P. 2546(397), *2568*
Hsu, M.-T.S. 2287(121a, 121b), *2315*
Hu, C. 1461, 1467(146), *1529*
Hu, C.-H. 531(98), 532(98, 99b), *554*
Hu, H.S.W. 1767(280), *1791*
Hu, J. 1116(76, 79), *1137*, 2509(229), *2564*
Hu, N.X. 1828(190), *1865*

Hu, S.S. 962, 963(244), 964(244, 245), 965(245), 966(244, 245), *1059*, 1274(114, 115), *1308*, 1325(153), *1335*
Hu, S.-W. 1481(283), *1533*
Hu, S.-Z. 226(198), *261*
Hu, W. 2238(121), *2244*
Hu, W.H. 1671(36), *1683*
Hu, Y. 1605(45), *1661*, 2345(167), *2360*
Hua, R. 2080(137–139), *2125*
Hua, Z.Q. 560(45, 46), *590*
Huang, H. 290(70–73), *352*, 2236(115), *2244*
Huang, H.H. 2345(169), *2360*
Huang, J. 1605(45), *1661*
Huang, L.-M. 1313, 1317, 1320(45), *1333*
Huang, M.-S. 226(198), *261*
Huang, S. 1548(39), *1578*
Huang, W. 1652(220, 222), *1665*
Huang, W.-T. 1454, 1467(95), *1527*
Huang, Y. 1283(145), *1308*, 2471(36), *2559*
Hua Qin Liu 207(120), *260*
Hübel, K.W. 1971(32), *2031*
Huber, B. 1340(15), *1437*
Huber, C. 1743(198), *1789*
Huber, F. 2379(58), *2399*
Huber, H. 1910, 1911(60d), *1927*
Huber, J. 2226(51), *2242*
Huber, P. 1649(211), *1665*
Huch, V. 205(118), *259*, 1178(193), *1185*
Hückel, E. 4(11), *95*
Hudalla, C. 311, 313(136), *353*
Hudeszek, P. 2159(152, 154, 155), *2174*
Hudrlik, A.M. 413(177), 419(198), 424(211), 425(215, 217), *429, 430*, 456(141), *491*, 782, 791, 795(11a, 11b), *822*, 1679(106–108), *1684*, 1831(200), *1865*
Hudrlik, P.F. 413(177), 419(198), 424(211), 425(215, 217), *429, 430*, 456(141), *491*, 618(58), *699*, 782, 791, 795(11a, 11b), *822*, 1679(106–108), *1684*
Hudson, C.M. 1196(19, 20), *1230*, 1823(152), *1864*
Huff, M. 222(187), *261*
Huffaker, H.B. 1860(351), *1868*
Huffman, D.R. 69(125), *99*, 1930(1c, 1d), *1960*
Huffman, J.C. 205(115), *259*, 280(41), *351*, 377(68), *427*, 514, 520–522, 529, 532, 533(2a), *551*, 581, 583, 584(182), *593*, 635, 637(94), 652(125), *699, 700*, 1992(99), *2033*, 2079(128), *2125*
Hughes, D.W. 950, 959–961(235), *1059*, 1080(45), *1102*, 1274(113), *1308*, 2407(23), *2458*
Hughes, E.D. 362(26), *426*, 612(38), *698*, 1633(159), *1664*
Hughes, K.J. 169(113), *179*, 2465(10), 2484, 2504(103), *2559, 2561*

Hughey, M.R. 362, 363(27), *426*, 598, 611, 612(8), *697*
Huhmann, J. 1431(230), *1444*
Huhn, K.G. 2227(68), *2242*
Hui, R.C. 1618(107), *1662*
Huille, S. 2282(114), *2314*
Huisgen, R. 1849(290), *1867*
Huling, J.C. 2249, 2294(16a, 16b), *2310*
Hull, K. 406(155), *429*, 1814(116), *1864*
Hulme, C. 1673(55), *1683*
Hultsch, K.C. 2153, 2154(134), *2173*
Huminer, H. 2322(42), *2356*
Humphries, M. 2227(65), *2242*
Hunadi, R.J. 793(47a), *823*, 2453(202), *2461*
Hundai, R.J. 46(90a), *98*
Hünig, S. 459(158), *491*
Hunt, A.J. 2339(126), *2358*
Hunt, M.O. 2219(15a–c), *2240*
Hunter, G. 2132(19), *2171*
Huppert, D. 2322, 2347(33), *2356*
Huppmann, F. 207(122), *260*, 2486(118, 119), *2561*
Hurd, A.J. 2347(186c), *2360*
Hurd, D.T. 1589(76), *1596*
Hurd, T. 457(149), *491*, 1638(167), *1664*
Hurdrlik, P.F. 1831(200), *1865*
Hurlburt, P.K. 578(160), *593*
Hurley, J. 2159(159), *2174*, 2429(93), *2459*
Hurley, W.J. 2253(35), 2264(77), *2311, 2313*
Hursthouse, M. 193(59), *258*, 2145(88), *2172*
Hursthouse, M.B. 2131, 2133(11), 2155, 2158, 2159(143), *2171, 2174*
Hursthouse, M.D. 1897(18), *1926*
Hurwitz, F. 2289(130, 131), *2315*
Hurwitz, F.I. 2232(95), *2243*, 2289(127, 132, 135), *2315*
Huryn, D.M. 1797(43), *1862*
Husain, D. 66(115b), *99*
Husain, D.J. 66(115a), *99*
Hüsing, N. 2330(76), 2332(91, 92), *2357*
Hussain, H.H. 747(92), *776*
Hussain, S. 772(151), *778*
Hussmann, G. 1071(22), *1102*
Hussmann, G.P. 722(47), 767(145), *776, 778*, 2407(20), *2458*, 2502(194), *2563*
Husson, H.-P. 1818(136), *1864*
Hutchings, G.J. 1589(75), *1596*
Hutchins, R.R. 456(142), *491*
Hutter, D.H. 6(23a), *96*
Hutter, F. 2340(133), *2359*
Huttner, G. 190(54), 193(60), *258*
Huvenne, J.P. 1471, 1472(228), *1531*
Hwang, H.S. 1981(66), *2032*
Hwang, R.-J. 901(160), *1057*, 1965(16a, 16b), *2030*, 2467, 2480, 2489(20), *2559*
Hwang, S.S. 2237(117), *2244*
Hwang, T.-H. 2465(6), *2559*

Hwang-Park, H.-S. 873, 943(67), *1055*
Hwangpark, H.S. 909(173, 174), *1057*
Hwu, J.R. 414(179, 180), *429*, 433(1–3), 434(18, 20), 435(38), 438(2), 442(18), 443(20), 449(38, 124), 464(173, 174), 479(232, 233), 480(3), *488, 489, 491–493*, 1294(184), *1309*, 1678(101), 1682(125), *1684, 1685*, 1834(211), 1835(215), *1866*, 1875(37), 1877(43), *1892*
Hyatt, L. 2289(127), *2315*
Hyde, J.F. 1582(3), *1594*
Hyla-Krispin, I. 77–79(154b), *100*, 2534(360b), *2567*
Hyla-Kryspin, I. 1434(232b), *1444*
Hyla-Kryspin, J. 2163–2166(182), *2175*
Hynes, R. 207(120), 219(174), *260, 261*, 2251(29b), *2311*

Ialongo, G. 1550(48), *1578*
Ibrahim, M.R. 365, 368, 370(39), *426*, 599, 603, 604, 618, 622, 623(9), *697*
Ibuki, T. 2523(312), *2566*
Ichikawa, H. 1215(81, 82), 1217(83), 1218–1220(91), *1231*
Ichikawa, K. 1203, 1212, 1213(40), *1230*
Ichimura, K. 2330(83), 2349(210a, 210b), 2354(83, 287), *2357, 2361, 2362*
Ichinake, M. 589(198), *594*
Ichinohe, M. 190(52), *258*, 551(116), *555*
Ichinose, M. 1322(93), *1334*
Ida, H. 1125, 1133(184), *1140*
Ida, M. 145(54, 55), 146(55), *152*, 2206(93), 2207(93, 96), *2215*
Ida, T. 1783, 1784(346), *1792*
Igarashi, Y. 2437(139, 140), *2460*
Igau, A. 761(127, 129), 763(127, 129, 135), *777*
Ignacio, E.W. 156, 167, 171, 173(32), *178*, 184, 234, 239(23), *257*, 2466(17), *2559*
Ignatenko, M.A. 1450, 1461(34, 35), *1525*
Ignat'ev, I.S. 1471, 1472(230), *1531*
Ignatovich, L.M. 1481, 1483(289, 290), *1533*
Ignatovich, M.M. 1451, 1455(50), *1526*
Ignatyev, I.S. 560(48), 581(177), *590, 593*, 1120, 1121(137), *1139*
Igonin, V. 958(241), *1059*
Igonin, V.A. 190(50), *258*, 2044, 2051(19), *2123*
Ihara, E. 2345(170), *2360*
Ihara, H. 2345(170), *2360*
Ihara, M. 1561(71a, 71b), *1579*
Ihmels, H. 1075(35a, 35b, 37), 1086, 1090(37), *1102*, 1973(47), *2031*, 2183, 2185(41), 2187(48), *2214*, 2416(52), 2429, 2436(104), *2458, 2459*, 2550(410, 411), 2552(412, 413), 2553(414), *2568*
Iida, Y. 2219(12), *2240*

Iijima, S. 1493(328), *1535*, 2160(169), *2174*
Iimura, M. 2272(94a, 94b), *2314*
Ijadi-Maghsoodi, S. 169(113), *179*, 770, 772(149), *778*, 866, 867(40), 872(57), *1055*, 1303(208), *1310*, 1860(350), *1868*, 1965(9a, 9b), 2016(176a), *2030, 2035*, 2465(10), 2484(100), *2559, 2561*
Ijadi-Magsoodi, S. 1767(278), *1791*
Ikeda, I. 1669(25), *1682*, 1830(195), *1865*
Ikeda, M. 387(104), *428*
Ikeda, N. 436, 480(64), *489*
Ikeda, S. 1616(102), *1662*, 2150, 2151(105), *2173*
Ikeda, T. 1771(309), 1772(312, 313), 1773, 1774(317), *1792*, 2330(83), 2349(210a, 210b), 2354(83), *2357, 2361*
Ikeda, Y. 1828(191), *1865*
Ikenaga, K. 364(33), *426*, 603(18), *698*, 1825(163–166), 1838(231), *1865, 1866*, 1973, 1974(44), *2031*
Ikeno, M. 1030(344), *1061*, 1075(32c), 1080(43c), *1102*, 2414(40), 2415(41, 42), 2416(49), *2458*
Ikoma, S. 2346, 2347(183a), *2360*
Ilcewicz, H. 304(108), *353*
Iloughmane, H. 1965(10a, 10b, 12–14), 1967(10a, 10b), 1968(13, 21, 22), 1981(10a, 10b, 12, 21, 67b, 67c), 1982(12, 14, 22, 67b, 67c, 70, 71), 1983(14, 22, 67c), 1984(14, 22), 1996(14, 21, 67b, 67c, 110), 1998(10a, 10b, 67c, 110, 119), 1999(10a, 10b, 12, 21, 67c), 2003, 2007(12), 2008, 2009(110), *2030, 2032–2034*
Ilsley, W.H. 784(22), 785(26, 27), 786(22, 27), 787(22), *822*
Imai, Y. 2218(10), *2240*, 2350(228), *2361*
Imaizumi, S. 1733(142), *1788*
Imakoma, T. 2231(83), *2242*
Imakubo, T. 578(164), *593*, 1873(25), *1892*
Imamura, A. 1318(55), *1333*
Imanieh, H. 416(185), *429*
Imazu, S. 1846(279), 1856(333), *1867, 1868*
Imbenotte, M. 1471, 1472(227, 228), 1485(300), *1531, 1534*
Imhof, R. 219(182), *261*
Imi, K. 1838(238), *1866*
Imma, H. 1796, 1858(37), *1862*
Imme, S. 784, 787(23), *822*
Imori, T. 2016(175c), *2035*, 2049, 2051, 2052(28), *2123*
Imrie, C. 1550(41b), *1578*
Inada, T. 234(242), *263*, 288, 289(67), *352*, 1407, 1408(177), *1443*
Inami, H. 475(211), *492*
Inamoto, N. 486(260, 261), *493, 494*, 867(43), *1055*, 1064(3), *1101*
Inazu, T. 371(56), *427*, 437, 483(70), *490*

Indolese, A. 434, 443(21), *489*
Indriksons, A. 2204(88), *2215*
Ingallina, P. 1718(106, 107), 1723(106), 1774(322, 323), 1776(107, 323, 324), 1779(324), *1787, 1792*
Ingebridtson, D.N. 2229(79), *2242*
Ingold, C.K. 1633(159), *1664*
Ingold, K.U. 1074(28), *1102*, 1542(8, 14), *1577*
Inhof, R. 2107, 2109(213), *2127*
Inokuchi, H. 1318(53), *1333*
Inokuchi, T. 1205(43), *1230*
Inomata, K. 458(150), *491*
Inomata, S. 2088(175), *2126*
Inose, J. 810(93b), *824*
Inoue, G. 2485(108), 2522(283), 2523(310), *2561, 2565, 2566*
Inoue, H. 1291(174, 175), 1293(179), *1309*, 1778(332), *1792*, 2218(9), *2240*, 2353(278), *2362*
Inoue, T. 1616(100), *1662*, 1671(35), *1683*
Inoue, Y. 2437(132), *2460*
Inouye, K. 2351(246), *2362*
Inshhakova, V.T. 384(91), *427*
Interrante, L.V. 2253(35, 41), 2254, 2257(41), 2264(77), 2283, 2284(116a, 116b), *2311–2314*
Inui, M. 2253(34), *2311*
Inui, N. 1756(249), *1790*
Ioffe, A. 576(140), *592*
Ioffe, A.I. 2512(253), *2564*
Ionkin, A. 883(100), *1056*, 1276(124), *1308*
Ionova, I.V. 2546(389), *2568*
Iosefson-Kuyavskaya, B. 2322(30), 2346, 2347(183b), *2356, 2360*
Iosefzon-Kuyavskaya, B. 2321, 2322(23a), *2355*
Iovel, I.G. 1717(105), *1787*
Iraqi, A. 1771(298), *1791*
Iraqi, M. 1029, 1049, 1050(342), *1061*, 1123(165), 1126(195–199), 1127(195, 196), 1128(197), 1129, 1130(195, 198, 199), 1135(195, 199), *1139, 1140*
Ireland, R.E. 1606, 1611(48), *1661*
Irie, S. 2475(60), *2560*
Iriguchi, J. 1616(99), *1662*
Irikura, K.K. 172–175(131), *180*, 901(150), *1057*
Irngartinger, H. 2448(176), *2461*
Isaac, M.B. 434, 447(32), *489*
Isaacs, L. 1930, 1934, 1959(3), *1960*
Isagawa, K. 2349(213), *2361*
Isbell, T.A. 435, 455(48), *489*
Iseard, B.S. 177(143), *180*
Ishibashi, H. 387(104), *428*, 1782(339), *1792*
Ishibashi, T. 1692(15), 1733(15, 145), 1734, 1737(145), *1786, 1788*

Ishichi, Y. 1617, 1620, 1624(105), *1662*
Ishida, H. 2289(130, 131), *2315*, 2333(100), 2339(125), *2358*
Ishida, T. 186(40), *258*, 2017(178), *2035*, 2439(148), *2460*
Ishifune, M. 1216, 1217(87), *1231*, 1811(107), *1864*
Ishiguro, M. 436, 480(64), *489*, 1283(146), *1308*, 1328(156), 1332(199), *1336*
Ishihara, H. 1885(79), *1893*
Ishihara, K. 440(97), *490*
Ishii, A. 2320(17a), 2350(221), 2351(239), *2355, 2361*
Ishii, N. 1840(244, 245), *1866*
Ishii, T. 1897, 1923(6), *1925*
Ishikawa, H. 1323(118), *1335*, 2150, 2151(105), *2173*, 2522(289, 294, 295), *2565*
Ishikawa, K. 435, 477(59), *489*, 925, 998, 1000, 1002(192), *1058*, 1852(314, 315), *1868*
Ishikawa, M. 186(40), 194(67), *258*, 798(60a, 60b), 807, 808(87a–c), (110c), *823–825*, 852(85), *855*, 867(44–46), 868(46), 870(44, 47–51), 883(98, 99), 884(112, 113), 885(45, 112, 113), 888(112, 113, 120, 121, 123), 889(121), 891(124–127), 892(128–131), 895(132–134), 917(184, 185), 922–924(112), 925(129), 928(112, 193, 194), 935(133), 936(134), 937, 938(113), 957(112), 958(120), 960–962(99), 970(134), 974(45, 46, 133), 975(134), 976(251), 986, 988, 990(121), 998(184, 287–291), 1001(185, 288), 1002(184), 1004(289), 1005(289, 290), *1055, 1056, 1058–1060*, 1208–1210(57), 1216, 1217(85, 86), 1218(86, 89, 90), 1219(89, 90), 1227(120), *1230–1232*, 1248(43), 1251(51), 1252(53–56), 1253(57, 58), 1254, 1256(51), 1258(68), 1259(69, 70), 1283(146), 1303(206), *1306–1308*, *1310*, 1319(71, 72, 74, 75), 1322(106–110), 1325(72, 120, 126–141), 1328(72, 155, 156), 1332(199), *1334–1336*, 1769(294), 1770(294, 295), *1791*, 1973(42, 43, 48a–c, 49, 50a, 50b), 1974(50a, 50b), 1990(94, 95a, 95b), 2007–2009(95a), 2016(175a, 175d), 2017(43, 177a–c, 178), *2031, 2033, 2035*, 2416(43), 2424(78, 79), 2429(99, 106, 107), 2436(106), 2439(148), *2458–2460*, 2471(40), 2513, 2516(258), 2526(258, 328), *2559, 2564, 2566*
Ishikawa, T. 1215(81, 82), 1217(83), 1218–1220(91), *1231*, 1830(195), *1865*, 2272(95), *2314*
Ishimaru, K. 1850(299, 300), *1867*
Ishiwata, T. 1218(93), *1231*

Ishiyama, J. 1733(142), *1788*
Ishuii, M. 1991(97), *2033*
Ismail, Z.K. 1127(209), *1140*, 1180(196), 1181(200), *1185*, 2518(262), *2565*
Ismial, I.M.K. 2233(97), *2243*
Isoda, T. 2250(23a, 23b), 2252, 2253(23a, 23b, 32a–d), 2265(82b), 2271(89), *2310, 2311, 2313, 2314*
Isoe, S. 469(189), *492*, 1190, 1192(13), 1193(16–18), 1194(17), 1196(22, 23), 1197(22, 23, 25), 1198, 1199(22), 1201(34, 36, 37), 1202(25, 36, 38), 1203(25, 39, 41), 1204(37), *1230*, 1617, 1620, 1624(105), 1650(213), *1662, 1665*, 1831(199), 1843(261), *1865, 1867*
Isomura, Y. 998(291), *1060*, 2424(79), *2459*
Israëli, Y. 2231(82), *2242*
Israelson, O. 2322, 2323(39), *2356*
Itabashi, N. 2522(296), *2565*
Itani, A. 1207(48), *1230*
Itenberg, A.M. 616(47), *698*
Itenberg, I.M. 364(34), *426*
Ito, F. 1816(131), *1864*
Ito, H. 407(157, 159), *429*, 488(273), *494*
Ito, K. 839(48), *855*, 1638(168), *1664*, 1764(272), *1791*
Ito, M. 1674(61), *1683*, 1751(237), *1790*, 2376(46), *2399*
Ito, O. 1254(63), *1307*, 1930(9, 10), *1960*
Ito, S. 2002(141), *2034*, 2088(175), *2126*
Ito, T. 475(211), 488(273), *492, 494*
Ito, Y. 198(88), 207(124), *259, 260*, 340–342(151), *354*, 434, 445(27), *489*, 809(88–90), 810(92), *824*, 1014, 1033, 1034(311), *1061*, 1304(216), *1310*, 1340, 1341(17a–c, 18), 1342, 1343(18), 1344, 1345(17a–c), 1346(17a), 1347(17c, 18), 1348(17c), 1390(123a), *1437, 1441*, 1611, 1617(82), *1662*, 1710(86, 87, 89), 1711(89, 90), 1725(87), 1743(206, 207), 1745–1747(207), 1748(206), 1755(246), 1756(246, 249), 1783(345), *1787, 1789, 1790, 1792*, 1828(183), 1843(264), 1854(323), *1865, 1867, 1868*, 1973(49), 1978(54a, 54b), 2015(54b, 173), 2016(173), *2031, 2035*, 2116(236, 237), 2117(248, 249), 2120(236, 237, 248), *2127, 2128*, 2227(64), *2242*, 2424(78), 2429(105), 2430(113–116), 2433(119), 2436(105), *2459, 2460*, 2554(416), *2568*
Itoh, K. 742(83), *776*, 1688(1), 1692(15), 1733(1, 15, 145, 146), 1734, 1737(145), 1743(1, 190–194), 1745(191), 1746(191–193), 1747(191, 192), 1748(191), *1785, 1786, 1788, 1789*, 1957(26), 1958(27), *1960*
Itoh, M. 1201(34), *1230*, 1296(185), *1309*, 1650(213), *1665*, 1689(7), *1785*, 1923(93),

Itoh, M. (cont.) *1928*, 2289(123), *2315*, 2329, 2331(75), *2357*
Itoh, Y. 742(83), *776*
Iton, L.E. 184, 221, 223(21), *257*
Itsuno, S. 1638(168), *1664*, 1764(272), *1791*
Iturburu, M. 446(113), 476(225), *490, 493*
Ivanov, A.G. 2218(5), *2240*
Ivanov, A.P. 1996(108b), *2033*
Ivanov, P.V. 2218(8), *2240*
Ivanova, Z.G. 1450(40), *1526*
Ivri, J. 2350(234), 2351(249), *2361, 2362*
Iwahara, T. 1767(282), *1791*, 2016(176b), *2035*, 2439(146), *2460*
Iwama, N. 284(61), *352*, 420(199), *429*
Iwamija, H. 1474–1477(254), *1532*
Iwamiya, J.H. 337, 338(150), *354*
Iwamoto, M. 437, 452(88), *490*
Iwamoto, T. 290–292(74), 345, 347–349(154), *352, 354*, 985(270), *1059*, 1304(210), *1310*, 1332(207), *1337*
Iwamoto, Y. 741, 742(82), *776*
Iwamura, A. 642(103), *700*
Iwamura, H. 1197(26), *1230*
Iwasa, S. 1573(96b), *1579*
Iwasaki, F. 1293(181), *1309*
Iwasaki, H. 1850(300), *1867*
Iwasaki, T. 1669(28), *1683*, 1801(64), *1863*
Iwasawa, N. 437, 452(88), *490*
Iwata, K. 1689(7), *1785*
Iwata, S. 1050(372), *1062*, 1120, 1121(132), *1139*
Iyengar, D.R. 2237(117), *2244*
Iyer, V.S. 381(76), *427*
Iyoda, J. (110c), *825*, 998, 1001(288), *1060*, 1973(49), 1990(94), *2031, 2033*, 2424(78), *2459*
Izaki, K. 2250, 2261(20), *2310*
Izawa, T. 469(189), *492*
Izawa, Y. 761(124, 125), *777*
Izuha, M. 66, 67(117c), *99*, 1009, 1010(298a), *1060*, 1163(135), *1184*, 2557(429), *2568*
Izuka, Y. 810(93b), *824*
Izumi, Y. 1670(30), *1683*, 1740(176, 180), 1774(320), *1789, 1792*
Izumizawa, T. 2028(208), *2036*

Jaballas, J. 516(35c), *552*
Jack, K.H. 2264(74), *2313*
Jackson, J.E. 123(24f), *151*, 1911(64, 65), *1927*
Jackson, P. 860–862, 992(18), *1054*
Jackson, R.A. 1212(68), *1231*
Jackson, R.F.W. 470(193), *492*, 1620(119, 120), *1663*
Jackson, W.P. 1682(126), *1685*
Jackson, W.R. 435(40), 451(40, 127, 128), *489, 491*, 1840(247), *1866*
Jacobi, P.A. 437, 446(85, 86), *490*
Jacobi, V. 752–755(112), *777*
Jacobine, A.F. 2226(54), *2242*
Jacobsen, E.N. 1672(48–54), *1683*
Jacobsen, H. 184, 247(26), *257*, 829(22), *854*, 1110(59), *1137*, 2116(247), *2128*
Jacobsen, H.J. 301(98), *352*
Jacobson, C.P. 2286(118), *2315*
Jacobson, D.B. 362(24), *426*, 560(43, 47, 49), *590*, 601, 602, 606(16), *698*, 707(16), *775*, 1111(64, 65a), 1113(66), 1131(239, 241), *1137, 1141*, 2088(178–180), *2126*
Jacobson, H. 28(64), *97*
Jacobson, R.A. 226(200), *261*, 1460, 1462, 1465, 1467, 1472(150), *1529*
Jacock, H.M.III 619(62), *699*
Jacox, M.E. 1167(160), 1181(197), *1184, 1185*, 2513, 2514(256), *2564*
Jada, S.S. 580(172, 173), *593*
Jäger, H.-U. 2448(176), *2461*
Jagirdar, B.R. 243(269), *263*, 2063(66–68), 2066(66, 68), *2124*
Jain, A.K. 2351, 2354(260), *2362*
Jain, N.F. 1804(80), 1847(283), 1850(310), 1851(310, 311), *1863, 1867, 1868*
Jain, R.K. 2351(254), *2362*
Jakoubková, M. 1006, 1008, 1009(296a), *1060*
Jalbert, C. 2234(104a, 104b), *2243*
Jamaji, M. 2498(183), *2563*
James, B.R. 1692(18), *1786*
Jammegg, C. 2179(12), *2214*, 2279(108), *2314*
Jammegg, Ch. 1218(94), *1231*
Jamrodgiewicz, Z. 1482(295), 1484(295–297), *1533, 1534*
Jamrogiewicz, Z. 1454, 1467(97), *1527*
Janaik, C. 1453, 1462, 1467, 1483, 1491(68), *1526*
Janakiraman, M.N. 226(200), *261*, 1460, 1462, 1465, 1467, 1472(150), *1529*
Jancke, H. 2337(124), *2358*
Janda, K.D. 2150(104), *2173*
Janes, R.R. 2144(82), *2172*
Jang, B.Z. 2231(88), *2243*
Janiak, C. 2138(39), *2171*
Jankovska, I.S. 1506(380), *1536*
Jankowski, P. 424(213, 214), *430*, 1679(111), *1684*
Janochek, R. 1267(94), *1307*
Janoschek, R. 22(58, 60), 23(60), 24(58), 25, 26(58, 60), 66–68(117b), *97, 99*, 123(21b, 21c, 24a), *151*, 186(43), *258*, 1047(361), *1062*, 1125(175), *1139*, 1151(43), 1163–1165(128), 1166(158), *1182, 1184*, 1286(162), *1309*, 1316(41), *1333*, 1911(68), *1927*, 2178(5d), 2181(5d, 25, 31), 2182, 2187, 2197, 2204, 2205, 2209(5d), *2214*
Janowski, P. 1835(214), *1866*

Jansen, I. 2220(17b), *2240*
Jansen, M. 203(107), *259*, 2269, 2270(88a, 88b, 88d, 88e), *2313*
January, J.R. 2289(124), *2315*
Janzen, E.G. 2010, 2013(164), *2035*
Jaouen, G. 580(168), *593*
Jaquet, R. 252(333), *265*
Jardine, R.E. 161(57), *178*, 2477(73), *2560*
Jarek, R.L. 34, 36(73c), *97*
Jarrold, M. 1109, 1120(51), *1137*
Jarrold, M.F. 70(134b), *100*, 1119(100), 1120(120–122, 124, 126, 127), *1138*
Jarvie, A.W. 795(79), 796, 801(72), 804(79), *824*
Jarvie, A.W.P. 364(37), 365(38), *426*, 617(57), 618(57, 59), 620(59), *699*
Jarvis, R.F. 1593(109), *1597*
Jasian, P. 21(54b), *97*
Jasinski, J.M. 154, 167, 171(6), *177*, 2485(109, 113), 2491, 2518–2520(113), 2524, 2527(322), 2545, 2546(379), *2561, 2566, 2567*
Jaspars, M. 1799(57), *1863*
Jasperse, C.P. 1539, 1540(3b), *1577*
Jean, A. 2016(176c), 2017(179–181), 2018(181), *2035*, 2330(85), *2357*
Jeannin, Y. 2075, 2091(112), *2125*
Jeffrey, D. 1771(298), *1791*
Jeh, J.-Y. 1850(307), *1867*
Jelinek, T. 197(79), *259*, 514, 520, 521, 524(3b), *551*, 567, 570, 572, 583, 586(103), *592*
Jelski, D.A. 70(135), *100*
Jemmis, E.D. 33(70), 47, 77(91), 82(157a, 157b), 90(170b), *97, 98, 100, 101*, 551(115), *555*
Jeng, R.J. 2351, 2354(260), *2362*
Jeng, S.-P. 2295(145), *2315*
Jenkner, P.K. 279(30), *351*, 796(83b, 84), 805, 806(84), *824*, 2191(63), 2204(89), 2213(107), *2215, 2216*
Jenneskens, L.W. 2428(89), *2459*
Jennings, N.A. 1107(15), *1136*
Jensen, B.L. 1852(317), *1868*
Jensen, J.H. 1467(190), *1530*
Jensen, K.F. 171(120–122), *180*, 2476(62), 2549(406), *2560, 2568*
Jensen, P. 1049(366), *1062*
Jeong, E. 1493, 1503–1505(330), *1535*
Jeong, J.H. 1503(373, 374), *1536*
Jeong, J.U. 1610, 1630(77), *1662*
Jerkunica, J. 616(52), *698*
Jerome, B. 1211(62), *1231*
Jérôme, R. 2235(107), *2243*
Jesús, E.de 2151(117), 2152(115, 117), 2159(149), *2173, 2174*
Jevons, W. 1021(323), *1061*

Jezercak, M. 2510(233), *2564*
Ji, J. 1563(75), *1579*
Jiaang, W.-J. 1644(198, 199), *1664*
Jiang, B. 1621(122), 1652(221), *1663, 1665*
Jiang, J. 287(65), *352*, 509(65), *511*, 1371(63), 1374, 1376(93), *1439, 1440*
Jiang, L.-J. 1477(265), *1533*
Jiang, Q. 2054(42), 2055(42, 43), 2056(42, 43, 46), 2057(42), *2123*, 2486(121), *2561*
Jiang, X.L. 2354(280), *2362*
Jiang, Y.Z. 1671(36), *1683*
Jiao, H. 5(16, 19), 6(16, 23b), 7(16, 19, 26, 27), 8(31a), 9(16, 26, 31a), 10(16), 22–24, 31, 32(19), 42(26, 31a), 43(31a), 44, 45(26), 46(16), 47(26, 31a, 92), 69(26, 129), 72(129), 90(26, 31a), *95, 96, 98, 100*, 515(19), *552*, 601(12), *698*, 859, 990, 994, 995(6), *1054*
Jiao, H.J. 814(108b), (108c), *825*
Jin, F. 1621(122), 1652(220–222), *1663, 1665*
Jin, M.J. 1743(217), *1790*
Joanteguy, S. 1051, 1052(381), *1062*
Jodhan, A. 2514(273), *2565*
Jödicke, K. 446(112), *490*
Joe, G.H. 1567, 1574(84), *1579*
Joffe, A.I. 52(94a), *98*
Jogun, K.H. 584(184), *593*
Johannes, J.E. 2548(402), *2568*
Johansson, A. 1668(13), *1682*
John, P. 2522(284), *2565*
Johnels, D. 522, 527, 528, 532, 535, 547(83b), *554*, 564, 569, 570(90), *591*
Johnson, A.E. 1454, 1467(85), *1527*
Johnson, B.F.G. 1426(223), *1444*
Johnson, C.A. 515(20), *552*
Johnson, C.E. 1109(33), *1136*
Johnson, D.W. 1553(53), *1578*
Johnson, F. 1971(31), *2031*
Johnson, J.S. 798(58), *823*
Johnson, L. 222(186), *261*
Johnson, L.M. 1018, 1045(317), *1061*
Johnson, M.D. 508(52), *511*
Johnson, M.J.A. 473(202), *492*
Johnson, M.P. 795, 799(62b), *823*, 847(72), 855, 903(161), 922(190), 993, 996(161), *1057, 1058*, 1154, 1155(64), *1182*, 1283(155), *1309*, 2111, 2120, 2121(226), *2127*
Johnson, M.S. 530(93), *554*, 1107(17), *1136*
Johnson, P. 2503(195), *2563*
Johnson, R.D. 72, 74(146), *100*, 149(61), *152*
Johnson, R.G. 1593(110, 111), *1597*
Johnson, R.P. 2446(170, 171), 2450(185–189), *2461*
Johnson, S.E. 284(56), *351*, 1340(13a, 13b, 14), 1341(13a, 13b, 14, 20a, 20b),

Johnson, S.E. (*cont.*) 1342(13a, 13b, 20a, 20b), 1343, 1347(13a, 13b), *1437*
Johnson, T.J. 2079, 2093(127), *2125*
Johnston, L.J. 649(123), *700*, 1301(197), *1310*
Johnston, S.M. 2253, 2255, 2261(44b, 44c), *2312*
Jolly, B.S. 1377, 1378, 1380(97a), *1440*
Joly, H.A. 1162(117), *1183*
Joly, M. 497(13), *510*
Jonas, J. 290(75), *352*
Jonas, V. 540(103), *554*
Jondi, W. 1692(21), *1786*
Jones, G.R. 409, 410(165), *429*, 1680(119), *1685*
Jones, J. 1474, 1475(246), *1532*
Jones, J.P. 362(22), *426*, 611, 612(37), *698*
Jones, K. 2083(148), *2126*
Jones, K.L. 577(147–149), *593*, 1260(75), *1307*, 2083(151), *2126*
Jones, L.L. 1312(15), *1332*
Jones, M.Jr. 11, 12(42), *96*, 704, 712(2), 713(33), 716, 727(40), 741, 743(33), 748(33, 40, 97), 749(40), 750(40, 97), 752(33, 115), 753(33), *774, 775, 777*, 1148, 1149(29), *1182*, 1325(124), *1335*, 2403(3), 2404(4, 6, 7), 2416(184), 2428(88), 2437(135), *2457–2460*, 2546(397), *2568*
Jones, M.T. 1115(73), *1137*
Jones, N.L. 52–54, 61(96), *98*, 2509, 2539(225), *2564*
Jones, P. 867(41), 876(70), *1055*
Jones, P.A. 2233(97), *2243*
Jones, P.F. 743, 750(84), *776*, 1276(127), *1308*, 1606(56), 1638(171), *1661, 1664*
Jones, P.G. 374(60), *427*, 547(112c), *555*, 1357, 1360, 1362, 1364, 1366(49), 1368(57b), *1438, 1439*, 1850(298, 301), *1867*, 1986, 2012(78c), *2032*, 2364(34–36), 2368, 2369, 2371(34), 2374(34, 36), *2399*
Jones, P.R. 846(67–70), *855*, 874(68), 878, 931(79), 934(68), 945(79), 990(273), *1055, 1060*, 1154(62, 63), *1182*, 1241(28), 1242(31), *1306*, 2155(139), *2174*
Jones, R. 123(26a, 26b), *151*, 1163(125), *1184*
Jones, R.A. 217(170), *261*, 793(49b), *823*
Jones, R.G. 2481(89), *2560*
Jones, W.M. 2079, 2092, 2093(126), *2125*
Joo, B.J. 1574(97), *1579*
Joo, F. 2322, 2323, 2325(40b), *2356*
Joo, K.-S. 250(313), *264*, 879, 932, 940, 981(84), *1055*
Joo, W.-C. 37(83), *97*, 818(114), *825*, 1981(66), 2018(183), 2019(185), 2022(183), *2032, 2035*
Jordan, K.D. 2001(138a, 138b), *2034*
Jorgensen, C.K. 2318, 2350(6c), *2355*

Jorgensen, W.J. 597, 598(5), 599(9), 603, 604(5, 9), 618, 622, 623(9), *697*
Jorgensen, W.L. 184, 222(22), *257*, 363(28), 365(28, 39), 368, 370(39), *426*, 482(241), *493*
Josland, G.D. 992, 996(276), *1060*
Jost, R. 683(155), *701*
Jouanne, J.von 1374, 1375, 1380(88), *1440*
Jouany, C. 1163(126), *1184*
Joubert, O. 2231(84), *2243*
Jouikov, V. 1225(113), 1227(118, 119), *1232*
Jouppi, S. 226(205), *262*, 2298(150b), *2316*
Journet, M. 1566(81a), *1579*
Joyner, H.H. 1875(39), *1892*
Joyner, R.W. 1589(75), *1596*
Judeika, I. 1452, 1453, 1458(52), *1526*
Judeika, I.A. 1505, 1506(378), *1536*
Jueschke, R. 391(115), 392(116), *428*
Jug, K. 9(30), 88(168), *96, 101*
Julg, A. 5, 44, 68(17a), *95*
Juliano, P.C. 307(110, 111), *353*
Julius, M. 1697, 1698, 1709, 1710(45), *1786*
Jun, C.-H. 1720, 1721, 1723, 1732(113), *1788*
Jung, D.M. 611, 615(45), *698*
Jung, I. 867(41, 42), *1055*
Jung, I.N. 401(140), *428*, 461(165), *492*, 874, 875(69), 876(70), 934(69), *1055*, 1241(28), 1242(31), *1306*, 1460, 1467, 1503(159), *1529*, 1589(78), 1590(78–80), *1596*, 1856(328, 329), *1868*
Jung, J. 1171(182, 183), 1172(182), *1185*
Jung, M.E. 709, 714(20), *775*, 901, 917, 981, 992(157), *1057*
Jung, R.J. 2354(280), *2362*
Jung, W.-H. 1871(15), *1892*
Jungen, M. 1106(7), *1136*
Jungermann, H. 203(107), *259*
Junk, P.C. 210(141), *260*, 2140(46), *2171*
Jun Liao 1455, 1464, 1467, 1481, 1483, 1486(110), *1528*
Jurkschat, K. 1510, 1511(398–400), *1537*
Just, U. 2232(93), *2243*
Jutzi, P. 10(34), 77–79(154a, 154b), *96, 100*, 190(49), 193(59), *258*, 571, 574(131), *592*, (110d), *825*, 833(34), *854*, 1079, 1088, 1089(41a, 41b), 1099(75), *1102, 1103*, 1144(2), 1152(50), *1181, 1182*, 1434(232a, 232b, 239), 1435(239), *1444*, 1879, 1880(58), *1893*, 1997(113c), 2007(160, 161), 2008(160), 2011(160, 161), *2033, 2035*, 2130(5, 6), 2131(11, 12), 2132(5), 2133(5, 11, 22), 2134(25, 27), 2138(6), 2139(44), 2140(45, 47, 48, 51), 2141(53, 56), 2143(77–79), 2144(77, 83, 85), 2145(85, 87, 88, 95), 2146(95), 2147(25), 2148(22, 101), 2149(56), 2150(108, 109), 2151(108), 2155(143, 144), 2156(144, 145),

2158(143), 2159(143, 150), 2163(6, 181, 182), 2164(182), 2165(6, 182, 185, 187), 2166(182, 188–191), 2167(189, 190, 193–195), 2168(188, 189, 195–197), 2169(189, 198), 2170(189, 200, 201), *2170–2175*, 2453(193, 194), *2461*, 2534(360a, 360b, 361a–c), *2567*
Juvet, M. 1994(105b), *2033*

Käb, H. 2061(60), 2062(62), 2066–2068(60), 2078, 2091(62), *2123*
Kabalka, G.W. 1639(184), *1664*
Kabbara, J. 412(171), *429*, 1668(15), *1682*, 1811(108), 1836(220), *1864, 1866*
Kabe, Y. 123(26a, 26b), *151*, 186(39), 205(117), *258, 259*, 269, 273, 275(12), 277(24), 278(25), *350, 351*, 735, 737, 741(69), *776*, 906, 973(168), 1051(375), *1057, 1062*, 1278(136), 1301(200, 201), *1308, 1310*, 1601, 1604, 1612, 1621(15), *1660*, 1881, 1887(59), *1893*, 1897(6, 12, 30), 1901(30), 1910(61), 1911(12, 61), 1918(88), 1923(6, 88, 91), *1925–1927*, 1945(21), 1947(21, 22), 1959(30), *1960*, 1968, 1969, 1976, 1996(24b), *2030*, 2089, 2092, 2094(181, 182), *2126*, 2207(95, 97), *2215, 2216*, 2409–2411(32), 2420, 2423(70), 2425(80–82), 2427(81), 2428, 2429(91), 2430(109, 110), 2432(110), 2435(125), 2437(139, 140), 2439, 2440(154), *2458–2460*, 2493(158), 2494(162), *2562*
Kabeta, K. 870(53), *1055*, 1094(69), *1103*, 1264(84), *1307*, 1882(63), *1893*
Kabuto, C. 120(4), 130(37), 133(4, 37, 41), 134(37), 137, 140, 143(47), 146(37, 47), *150–152*, 186(41), 190(52), 234(242), 248(303), *258, 263, 264*, 284(59), 288(66, 67), 289(67), 348(163), *351, 352, 354*, 484(242, 243), 488(274–276), *493, 494*, 796(56, 61, 76), 797(56), 798(56, 61), 802(76), *823, 824*, 829, 832, 835(27), *854*, 1157(81), *1183*, 1244(33), 1268(101), 1304(210), *1306, 1307, 1310*, 1332(207), *1337*, 1341, 1347(26), 1407, 1408(177), 1434, 1435(242), *1437, 1443, 1444*, 1548(38), *1578*, 2009(162), *2035*, 2139, 2141, 2147(43), 2159(160), *2171, 2174*, 2205(91), *2215*, 2429(92), 2439(142, 143), *2459, 2460*
Kabuto, K. 1238(19), *1306*
Kaczmarczyk, A. 2468(22), *2559*
Kadina, M.A. 1220(96), *1231*
Kadoi, S. 2435(125), *2460*
Kador, L. 2354(286), *2362*
Kadowaki, T. 1897(6), 1918(88), 1923(6, 88, 91), *1925, 1927*

Kaeriyama, K. 1189(8), *1229*
Kaesz, H.D. 2073(100, 101), 2091(100), *2124*
Kafafi, Z.H. 1167(161), 1175(161, 188), 1176(188), *1184, 1185*, 2469(24), *2559*
Kaftory, M. 437, 485(79), *490*
Kagan, M.L. 2322, 2348(25), *2356*
Kageyama, H. 1885(79), 1889(101), *1893, 1894*
Kageyama, M. 810(93b), 812(98), *824*
Kahl, W. 2150, 2152(102), *2173*
Kahn, O. 2159(152, 154, 155), *2174*
Kahr, B. 288, 289(68), *352*
Kai, Y. 2145(91), *2172*
Kaim, W. 1314(26), *1333*, 2131, 2132(15), *2171*, 2182(29), *2214*
Kaino, T. 2351(253), *2362*
Kaiser, A. 2340(136), *2359*
Kaiser, H.J. 1584(33), *1595*
Kajikawa, Y. 1771(310), 1772(313), *1792*
Kajimoto, O. 2522(289, 294, 295), *2565*
Kajzar, F. 2352(267), *2362*
Kakiuchi, F. 1714(100), 1715(101), *1787*, 1838(232, 233, 235–237, 239–241), *1866*
Kakiuchi, K. 2428(89), *2459*
Kako, M. 200(102), *259*, 1293(181), *1309*, 1897(29–31), 1899(37), 1900(31), 1901(29–31), *1926*, 1965, 1996, 1998(15), *2030*
Kakui, T. 1412(190), *1443*, 1824, 1828(159), *1865*
Kakuma, S. 1293(181), *1309*
Kalamansan, M.N. 2326(47), *2356*
Kalbalka, G.W. 385(95), *427*
Kalcher, J. 22, 23, 25, 26(60), 68, 69(123), 87(163a, 163b), *97, 99, 100*, 123(24a, 27b, 27c), *151*, 159, 167, 168, 171, 173(50), *178*
Kaldor, S.W. 455(140), *491*
Kalikhman, I. 504(34), *511*, 1418, 1420, 1422(211a, 211b), 1423(211a, 211b, 221, 222), 1424(211a, 211b, 220, 222), 1425(211a, 211b, 220–222), 1426(211a, 211b, 222), 1427(221, 222), *1444*
Kalikhman, I.D. 243(274), *264*, 752, 760(110), 777, 888, 990(122), *1056*, 1374, 1377(94, 95a, 95b), 1379(105), 1380, 1383–1386(95a, 95b), 1390(132, 132), 1391(140), 1392(132, 137, 146, 147), 1393(147, 149), 1395(105), 1396(105, 160, 164, 165), 1398(137, 147, 149), 1399(105, 160, 164, 165), 1400, 1401(105, 137), 1404(147), 1418(210, 212, 216), 1420(210, 212), 1422(216), 1423(212, 216), *1440–1442, 1444*, 1476(262), *1532*
Kalikhman, L.D. 1390(133–136), 1392(135, 136), 1398, 1405(133), *1441*
Kalinin, A.B. 1377(98), *1440*
Kalinovski, I.J. 167, 168(89), *179*
Kalinowski, H.-O. 747, 748(95), *776*

Kalluri, S. 2330(83), 2352(265), 2354(83), 2357, 2362
Kallury, M.R. 250(306), 264, 2404(12), 2458
Kallury, R.K. 3(4), 95, 250(305), 264, 1064, 1068(1), 1101, 1145(15), 1181, 2404(11), 2458
Kallury, R.K.M.R. 859(3), 880(3, 87, 92), 881(3, 87), 924(92), 977, 978, 981, 985, 986, 992, 996(3), 1054–1056, 1080(47), 1102, 1146(22), 1181, 1601, 1609, 1645(24), 1661
Kalman, A. 1508(385), 1536
Kaluri, S. 2352(263), 2362
Kamata, T. 1591(93), 1597
Kamatani, A. 1838(232, 233, 235, 237), 1866
Kamatani, H. 137, 140, 143, 146(47), 152, 2205(91), 2215
Kambe, N. 1616(100), 1662
Kamenska-Trela, K. 304(106, 108), 353
Kamiga, A. 2159(160), 2174
Kamisako, K. 2546(388), 2568
Kamitori, S. 1973(48b, 48c), 2031
Kamiya, A. 277(22), 350, 2429(92), 2459
Kamiya, K. 2289(129), 2315, 2351(251), 2362
Kamiyama, Y. 1973(46), 2031, 2430(112), 2460
Kamlet, M.J. 1477(267, 269), 1533
Kammula, S. 752(115), 777, 2403(3), 2404(4), 2457
Kammula, S.L. 713, 741, 743, 748, 752, 753(33), 775
Kampf, J. 226(205), 228(211), 262, 1353, 1368, 1369(36), 1438, 2298(150a, 150b, 152a), 2315, 2316
Kampf, J.W. 1883(67), 1893, 2422(69), 2459
Kamphuis, J. 1815(122), 1864
Kamyshova, A.A. 1494(337), 1535
Kan, H.-C. 254(337), 265
Kanabus-Kaminska, J.M. 168(106, 107), 179, 1540(5), 1577
Kanakubo, O. 1203, 1212, 1213(40), 1230
Kanamori, H. 2522(296), 2565
Kanatani, R. 417(193), 429
Kanaya, D. 2016(175b), 2035
Kanda, T. 1689(9), 1786, 1889(101), 1894
Kandror, I.I. 1374, 1376(92a), 1440
Kane, K.M. 1418(218b), 1444
Kane, V.V. 2428(88), 2459
Kanehisa, N. 2145(91), 2172
Kaneko, H. 1226(117), 1232
Kaneko, N. 2337(119), 2358
Kaneko, T. 805(82), 824
Kaneko, Y. 1852(313), 1868
Kanellakopulus, B. 2159(152), 2174
Kanemitsu, Y. 143(52a–d), 152
Kaneta, N. 1796(39), 1862

Kanetani, F. 892(128), 1056, 1252(54), 1306
Kang, C.H. 610, 624(33), 698
Kang, E. 1758(255, 256), 1790
Kang, J. 1610, 1630(77), 1662
Kang, K.K. 1669(20), 1682
Kang, K.-T. 867(43), 1055, 1819(138), 1837(224), 1864, 1866
Kang, S.H. 434, 444(22), 489
Kang, T.S. 1574(97), 1579
Kang, Y.K. 1505(376), 1536
Kania, L. 280(39), 351, 559(18), 563(18, 77, 80), 564(18, 77), 565(18), 566(77)567(18, 77), 569(77), 590, 591, 1108(18), 1136, 1408(180), 1443, 2223(33), 2241
Kanishita, A. 1602(27), 1661
Kanj, A. 1992(101, 102), 1993(103), 2033
Kanne, D. 77–79(154a, 154b), 100, 193(59), 258, 1434(232b), 1444, 2131, 2133(11), 2163(181, 182), 2164–2166(182), 2171, 2175, 2534(360a, 360b), 2567
Kanner, B. 1582, 1584(1), 1585(40, 41), 1589(1, 72), 1592(104, 105), 1594–1597, 2533(355), 2567
Kannisto, M. 2247–2250, 2269(9), 2276, 2278(104c), 2280(111), 2310, 2314
Kanouchi, S. 2009(162), 2035
Kansal, P. 226(205), 262, 2249, 2294(17), 2298(150b, 151, 152b), 2310, 2316
Kantak, U.N. 1903(44), 1926
Kanter, F.J.J.de 1434(234), 1444, 2428(89), 2459
Kanyha, P.J. 301(98), 352
Kapocsi, I. 1733(144), 1788
Kapon, M. 866, 884–886, 917, 924, 925, 928, 957(39), 1055, 2429(100), 2459
Kapoor, R.N. 437, 485(78), 490, 2083(149), 2126, 2143(72), 2172
Kapp, J. 24, 29, 30(62), 33(70), 97, 515(6), 518(6, 63), 519(63), 520, 522, 526–529, 532, 533, 535, 537, 542, 543(6), 551(115), 551, 553, 555, 558, 559, 573, 577, 584, 588(4), 589, 652(131), 700
Kappel, J. 2337(116, 117), 2358
Käppler, K. 435(57), 458(57, 151), 489, 491, 2232(93), 2243
Kaptein, B. 1815(122), 1864
Kapura, A.A. 1497(346, 347), 1535
Kapustin, V.Yu. 580, 584(167), 593
Karabelas, K. 1825(168), 1838(227–229), 1865, 1866
Karaev, S.F. 2439(151), 2460
Karaghiosoff, K. 215(156), 260
Karaguleff, C. 2351(255), 2362
Karampatses, P. 1012(308, 309), 1033(308), 1034, 1035, 1038(351), 1051(308, 351), 1061
Karatsu, T. 1266(86, 87), 1307, 1312, 1317,

1318(5), 1320(84), 1321(84–86), *1332,
1334, 2469(30), *2559*
Kargin, Yu.M. 1201(35), *1230*
Karl, A. (110d), *825*, 2007(160, 161),
2008(160), 2011(160, 161), *2035*
Karl, R.R. 133(39, 40), *151*
Karle, I.L. 110(13), *118*, 224(196), *261*
Karle, J.M. 110(13), *118*, 224(196), *261*
Karlov, S.S. 1450, 1456, 1457, 1461, 1496(37),
1525
Karnatak, R.C. 2523(321), *2566*
Karni, M. 3(1c), 14, 16(47), 29(1c, 65),
52(98d), 53(65, 98d, 102), 55(102), 59,
60(98d, 102), 62(98d), 63, 64, 66(102),
94, 96–98, 250(317–319), 251(318),
265, 515(8), *551*, 560(51), *590*, 601(13),
647(13, 119), *698, 700*, 705, 707(12),
775, 835(37, 43), *854*, 860(12), 901(12,
147), 919(12), 966(246a), 981(12), 982,
983, 1001(246a), 1008, 1009(12, 297c),
1010(297c), *1054, 1057, 1059, 1060*,
1123(156), 1130, 1131(236), *1139, 1141*,
1157(83), 1164(137), 1166(156), 1169(137,
171, 176), 1176(137, 176), 1177(171),
1178(137, 171), 1180(137), *1183–
1185*, 1330(176), 1331(175, 176), *1336*,
2509(226), 2511(247a), 2514, 2515(226),
2516, 2518(226, 247a, 261), 2556(419),
2564, 2565, 2568
Karni, M.J. 1166(149), *1184*
Karobeu, V.A. 2530, 2531(345), *2566*
Kárpáti, T. 30–32(68, 69), 53(68), 64(68, 69),
65, 66(69), *97*
Karplus, M. 1475, 1477(248), *1532*
Karrass, S. 1113(68, 70), *1137*
Karsai, E.B. 1450, 1453, 1471(33), 1481,
1509, 1510(287), *1525, 1533*
Karsch, H.H. 190(55), 217(164, 165),
258, 260, 1377, 1378, 1380(97b), 1434,
1436(248–250), *1440, 1445*
Karsch, H.M. 2534(362), *2567*
Karstedt, B. 1689, 1694(10), *1786*
Karthikeyan, M. 1292(178), *1309*
Kartsev, G.N. 1467(186, 187), *1530*
Kasai, N. 2145(91), *2172*
Kasal, A. 487(269), *494*
Kasemann, R. 2332, 2335(87), 2354(286),
2357, 2362
Kasemura, T. 2236, 2237(116), *2244*
Kasha, M. 2470, 2499(34), *2559*
Kashaev, A.A. 237(244, 246, 247), *263*, 1461,
1467(142), *1529*
Kashik, T.V. 1479, 1480(277), 1501(369),
1533, 1536
Kashikami, S. 1216, 1217(87), *1231*
Kashimura, S. 1218(92), *1231*, 1811(107),
1864

Kaska, W.C. 2349(208), *2361*
Kass, D.A.F. 1782(340), *1792*
Kass, S.R. 518(59), *553*
Kassim, A.M. 1831(200), *1865*
Kaszynski, P. 1019, 1044, 1045, 1047(321),
1061
Katagiri, T. 438(90), *490*
Katahara, T. 2250, 2252, 2253(23b), *2311*
Katari, J.E.B. 1888(95), *1894*
Katho, A. 2322, 2323, 2325(40b), *2356*
Kato, C. 735, 737, 741(69), *776*, 906,
973(168), *1057*, 1278(136), *1308*, 2409–
2411(32), *2458*
Kato, H. 1418(218a), *1444*
Kato, K. 1312(20), *1332*, 1846(280), *1867*,
2522(296), *2565*
Kato, M. 200(93), 248(299), *259, 264*, 279(29,
31, 35), 348(159), *351, 354*, 1157(79,
80), *1183*, 1614(97), *1662*, 2183(38, 42),
2214
Kato, R. 279(32), *351*, 1269(102, 103), *1308*
Kato, S. 1689(9), 1742(186), 1771(308), *1786,
1789, 1792*, 1885(79), 1889(101), *1893,
1894*
Kato, T. 1302(203, 204), *1310*, 1940(17),
1942(18, 19), *1960*, 2265(82b), *2313*
Kato, Y. 2395, 2396(80), *2400*
Katoh, S. 1221(98), *1231*
Katritzky, A.R. 8(29), *96*, 361(18), *426*,
1606(53), *1661*
Katsumata, K. 1885(74), *1893*
Katsuta, S. 1676(87), *1684*
Katz, T.J. 274(20), *350*
Katzer, G. 68, 69(123), *99*
Kauffman, J.W. 1180(196), *1185*, 2518(262),
2565
Kauffmann, J.W. 1127(209), *1140*
Kaufman, V.R. 2319(12b, 12c), 2322(33),
2346(184, 185), 2347(33, 184, 185, 196),
2348(184), 2350(196), *2355, 2356, 2360*
Kaufmann, F. 2159(156), *2174*
Kaufmann, F.-P. 370(51), *427*, 646(114),
670(114, 147, 148), 676, 677(114),
679(154), 685(156), *700, 701*
Kaufmann, S. 310(128), *353*
Kaupp, G. 2062, 2078, 2091(62), *2123*
Kaupp, M. 125(30), *151*
Kaur, G. 1801(66), *1863*
Kautsky, H. 2196(78b), *2215*
Kawa, H. 820(119), *825*
Kawabata, N. 1221(99, 100), *1231*
Kawabata, Y. 1318(64, 65), *1333*
Kawachi, A. 37(78), 46(90b), *97, 98*, 198(88),
207(124), *259, 260*, 780(1b), 790, 791(45),
794(52), 797(45), 809(88–91), 810(92, 94,
95), 811(45, 95, 96), 812(45), *821, 823,
824*, 827(4), *854*, 1390(123a), 1434,

Kawachi, A. (*cont.*) 1435(243), *1441, 1444*, 2119(256), *2128*, 2429, 2436(105), *2459*, 2481(91–94), 2554(416), 2555(417), *2561, 2568*
Kawada, M. 1608(66), *1662*
Kawada, Y. 1675(75), *1684*
Kawahara, S.-I. 434, 447(30), *489*
Kawahara, Y. 1885(79), *1893*
Kawai, T. 1897, 1901(29), *1926*
Kawakami, H. 998(287), *1060*, 1973, 1974(50b), *2031*
Kawakami, S. 1228(122), *1232*
Kawakami, T. 1208–1210(57), 1216, 1217(85, 86), 1218(86, 89, 90), 1219(89, 90), *1230, 1231*
Kawamoto, K. 1705(76, 77), 1714(96–99), 1771(307), *1787, 1792*, 1804(78), *1863*
Kawamoto, T. 2389(75), 2390, 2391(76, 77), 2392(76–78), 2394(75, 78, 79), 2396(79, 81), *2400*
Kawamura, T. 482(237), *493*, 1889(101), *1894*
Kawanami, H. 2387(63), *2400*
Kawanami, S. 2116, 2120(237), *2127*
Kawanishi, M. 2436(129), *2460*
Kawano, Y. 2088(168–170), 2090, 2094(168, 169), *2126*
Kawasaki, Y. 1771(308), *1792*
Kawase, T. 348(161), *354*, 1157(82), *1183*, 1910, 1911(61), *1927*, 2207(95), *2215*, 2498(185a), *2563*
Kawashima, T. 284(61), *352*, 420(199), *429*, 437, 487(81), *490*
Kawasini, M. 2001(136), *2034*
Kawata, K.-i. 1831(202), *1865*
Kawauchi, S. 559(20, 24), *590*, 870(50), *1055*
Kaya, H. 2250(23a, 23b), 2252, 2253(23a, 23b, 32a), *2310, 2311*
Kaya, K. 1120, 1121(132), *1139*
Kayano, H. 2272(92c, 93c), *2314*
Kaye, A.D. 474(210), *492*
Kazmierski, K. 2223(32a, 32b), *2241*
Kazmirski, K. 2223(23), *2241*
Kazoura, S.A. 1069, 1070(20), *1102*
Kazumi, N. 2080(140), *2125*
Kealy, T.J. 2163(177), *2174*
Kean, E.S. 1314(31), *1333*, 2193(71), *2215*
Keay, B.A. 447(115), 454(133, 134), 473(203), *490–492*, 1674(70, 71), *1683*
Kebe, Y. 194(64), *258*
Keck, M. 228(208), *262*, 1415, 1430(201a), *1443*
Keder, N.L. 1086, 1091(53), *1102*, 1870(5), *1891*, 2104, 2108, 2109(205), *2127*
Kee, I.S. 1675(72), *1683*
Kee, T.P. 209(131), *260*, 477(227), *493*, 1878(46), *1892*
Keen, J.D. 981, 982(257, 258), *1059*

Keheyan, Y. 561(52), *590*
Kehr, G. 1979(56d, 57b, 57f, 57g), *2031, 2032*
Keider, F. 2232(93), *2243*
Keijzer, A.H.J.F.de 1434(234), *1444*
Keiko, V.V. 1466, 1467(176, 179), 1475, 1479, 1480(249), 1485(302), 1523(249), *1530, 1532, 1534*, 1691(13), *1786*
Keinan, E. 1738(156), *1788*, 1826(175), *1865*
Keiner, P. 1871(10), *1891*
Keith, T.A. 517(44a), *553*
Keller, H. 763(138), 765(139), *778*
Keller, P.A. 2070, 2071(80), *2124*
Keller, U. 217(165), *260*, 1434, 1436(249, 250), *1445*, 2534(362), *2567*
Kelling, H. 1507, 1508(381), *1536*
Kellog, R.M. 219(181), *261*
Kellogg, R.M. 1373(75e), *1439*
Kelly, B.J. 1843(268), *1867*
Kelly, D.P. 647(118), 677(151), *700, 701*
Kelly, M.J. 466(183), *492*, 1601(14, 21), 1602, 1604, 1621, 1624, 1625(14), 1653(21), 1658(14), *1660*
Kelt, L.W. 2345(173), *2360*
Kemme, A. 1362(53a), *1438*, 1451(51), 1494(338), 1506(379), 1508(390), 1510(396), *1526, 1535–1537*
Kemme, A.A. 1362(54a, 54b), *1438*, 1449(6), 1460(6, 155), 1461(140, 143), 1462(155), 1467(140, 143, 155), 1470, 1481(6), 1503(155), 1506, 1510(6), *1525, 1529*
Kemmitt, T. 2305(160), *2316*
Kempe, R. 884, 885, 889, 925, 928(110), *1056*, 2051, 2052(34), *2123*
Kendrick, T.C. 2227(72), *2242*
Kenmoto, N. 1270(104), *1308*
Kenmotu, N. 279(33), *351*
Kennedy, V.O. 2105, 2109(207), *2127*
Kennish, R.A. 2250(21), 2251(28c), 2253, 2255(21, 28c), 2256, 2257(21), *2310, 2311*
Kephart, S.E. 505(41), *511*, 1810(103), *1863*
Kerschl, S. 296(90), 302, 304(102), *352*
Kerst, C. 849, 852(84), *855*, 864, 932, 934(34), 993, 996, 999, 1001(278), *1054, 1060*, 1241(25b), *1306*
Kersten, H.-J. 1886(81), *1893*
Kerzina, Z.A. 1069, 1070(19), 1071(23), 1072(19, 23), 1083(23), *1102*, 1161(104–106), *1183*
Keskar, N.R. 115(39), *118*
Kester, K.B. 1588(68, 70), *1596*
Kesti, M.R. 1697(36, 37), 1708, 1709(37), *1786*
Ketelaar, J.A.A. 1422(219), *1444*
Ketelson, H.A. 1771(301), *1791*
Ketvirtis, A.E. 161, 175(58), *178*, 597, 601(3, 4), 607(3), 608(4), *697*, 1118(93–95), *1138*
Kevill, D.N. 577(143), *592*, 645(111), *700*

Keweloh, N. 1011, 1044, 1048(305), *1060*
Kewley, R. 234(234), *262*
Keyaniyan, S. 520, 524, 540(70), *553*, 563, 583(82), *591*
Khabashesku, V.N. 901, 902(159), 981, 992(255), 993(159), 995(159, 255), 996(159), *1057, 1059*, 1069, 1070(19), 1071(23), 1072(19, 23), 1083(23), *1102*, 1161(104–106), 1170(179), *1183, 1185*, 1283(156), *1309*, 1967(20a, 20b), 1968(20c), 1986, 1996, 1997(20a–c), 1998(20c), 2013(172a), 2027(204), *2030, 2035, 2036*, 2474(54), 2504, 2515(54, 197), *2560, 2563*
Khalaf, S. 1692(21), 1700(55), *1786*
Khalifa, R.B. 2218(11c), *2240*
Khan, I.M. 1767(281), *1791*
Khan, K.M. 448(118), *491*
Khan, S. 2073(100, 101), 2091(100), *2124*
Khan, S.D. 407(162), *429*
Khananashvili, L.M. 1585(36), *1595*, 2218(5), *2240*
Khanbabaee, K. 782, 791, 795, 796(11d), *822*
Khara, E. 1771(296), *1791*
Kharasch, M.S. 1915(83), *1927*
Khasnis, D. 222(192), *261*, 1638(170), *1664*
Khetani, V. 406(155), *429*, 1814(118), *1864*
Khoudary, K.P. 433, 480(3), *488*
Khouzami, F. 2137(33), *2171*
Khramtsova, S.Yu. 1496(344), *1535*
Khromova, N.Yu. 1450(47), *1526*
Khudobin, Yu.I. 154, 156(11, 15, 16), 159–161(11), 164, 165(11, 15), 166(16), *177*
Khvalovskii, V. 436, 479(61), *489*
Kickel, B.L. 168, 173(101), *179*, 1108(27–29), 1110(54–57), 1118(92), *1136–1138*
Kido, K. 1885(79), *1893*
Kido, M. 928(194), *1058*, 1325(140), *1335*
Kiefer, W. 2061(58–60), 2066(59, 60), 2067(60), 2068(58–60), 2078, 2091, 2095(121), *2123, 2125*, 2332(92), 2339(130), *2357, 2359*
Kienzle, A. 203(112), *259*, 2257(49a), 2267(49a, 85b), 2269, 2270(88f), 2271(85b), 2288(122), *2312, 2313, 2315*
Kiernan, P. 2322(26), *2356*
Kiesgen, F. 1341, 1349–1351, 1357, 1361, 1368(30b), *1438*
Kiesler, R.P. 2152, 2153(112), *2173*
Kieslich, K. 2376(47), *2399*
Kikuchi, K. 1302(203), *1310*, 1942(19), *1960*
Kikuchi, M. 928(193, 194), *1058*, 1325(140), *1335*
Kikuchi, O. 20(51a), *96*, 794(53), *823*, 2453(201), *2461*
Kikukawa, K. 364(33), *426*, 603(18), *698*, 1825(163–166), 1838(231), *1865, 1866*

Kilburn, J.D. 460(160), *491*, 1569(86), *1579*
Kilesso, M.V. 1500(356), *1536*
Kilesso, V.M. 1475, 1477(251), *1532*
Kilgour, J.A. 716, 727, 748–750(40), *775*, 860, 917(15), *1054*, 2404(6, 7), 2406(17), *2457, 2458*
Killian, L. 1979(57a), 1986, 2011(76c), *2031, 2032*
Killpack, M.O. 790(31), *822*
Kim, C. 1758(255, 256), *1790*
Kim, C.-H. 37(83), *97*
Kim, C.H. 2083(144), 2085, 2090, 2094(159), *2125, 2126*
Kim, C.-K. 1012(308), 1014(310), 1033(308), 1034(310, 351), 1035(351), 1038(310, 351), 1051(308, 310, 351), *1061*
Kim, D. 1676(91), *1684*
Kim, D.U. 1294(183), *1309*
Kim, D.Y. 1828(189), *1865*
Kim, H.-B. 2330(83), 2349(210a, 210b), 2354(83, 287), *2357, 2361, 2362*
Kim, J. 1490(315), *1534*, 2351, 2354(259), *2362*
Kim, J.D. 1875(38), *1892*
Kim, J.-H. 2019(185), *2035*
Kim, J.P. 1583(15), 1586(50), *1594, 1595*
Kim, K. 1460, 1467, 1503(161), *1529*, 2153, 2162(124), *2173*
Kim, K.D. 454(135), *491*
Kim, K.H. 2436(127), *2460*
Kim, K.S. 1610, 1630(77), *1662*
Kim, M.W. 1462, 1467(153), 1490(315), *1529, 1534*
Kim, S. 1460(160), 1462(153), 1467(153, 160), 1503(160), *1529*, 1567, 1574(84, 85), *1579*, 1675(72), 1676(91), *1683, 1684*, 1834(213), *1866*, 2436(127), *2460*
Kim, S.H. 1803(74), *1863*
Kim, S.-I. 1589, 1590(78), *1596*
Kim, S.-J. 530(95a), 531(96), 532(95a), *554*
Kim, S.S. 1837(224), *1866*
Kim, W.-G. 1764(270), *1791*
Kim, W.G. 2227(57), *2242*
Kim, Y. 1503(373, 374), *1536*
Kim, Y.G. 1676(91), *1684*
Kim, Y.H. 1803(74), *1863*
Kimata, Y. 1210(58, 59), *1231*
Kimber, B.J. 307(112), *353*, 1897(19), *1926*
Kimijima, K. 1257(66), *1307*
Kimling, H. 2439(150, 152), *2460*
Kimoto, H. 1798(51), *1863*
Kimpe, N.de 1651(217), *1665*
Kimura, A. 1120, 1121(133), *1139*
Kimura, E. 1957(26), 1958(27), *1960*
Kimura, F. 1689(9), *1786*
Kimura, M. 1197(26), 1229(128), *1230, 1232*

Kindermann, M. 52, 55, 56, 58, 61, 62(99c), 98
Kindon, N.D. 409, 410(165), *429*, 476(223), *493*
King, A. 484(247), *493*
King, G.K. 1109, 1115(38), *1137*
King, H.E.Jr. 115, 116(37), *118*
King, K. 167, 169, 170(92), *179*
King, K.G. 182(7), *257*
King, R.A. 18, 21, 22, 32(49), *96*
King, T.A. 2350(233), 2352(269), *2361, 2362*
King, W.A. 2053(37), *2123*
Kingma, A.J. 1872(18), *1892*
Kin Hung, W. 2351(242), *2361*
Kinney, D.R. 311, 312(135), *353*
Kinney, J.B. 2154(130), *2173*
Kinoshita, C. 371(53), *427*, 632, 633(86a, 86b), 634(86a, 86b, 88a, 88b), 635(88a, 88b), *699*
Kinoshita, I. 122(16), 142, 143(50), *150, 152*, 1910, 1911(60g, 60h), *1927*
Kinting, A. 1743(219), *1790*
Kippenhan, R.C. 2003(145), *2034*
Kipping, F.S. 1897(7a, 7b), *1925*, 2177(1), *2213*
Kiprof, P. 913, 949(178), *1057*, 1386, 1387, 1389(113), *1440*, 2059, 2060, 2066, 2067(52), *2123*, 2528(336), *2566*
Kira, H. 1852(313), *1868*
Kira, M. 248(303), *264*, 284(59), 345, 347(154), 348(154, 163), 349(154), *351, 354*, 569(113, 120, 122), 570, 571(113), 572(120), *592*, 782(13a), 796, 802(76), *822, 824*, 834(36), 842(51), 845(63), 847, 850(75), *854, 855*, 885(114), 887(114, 119), 897(138), 898(143), 899(144), 911(119, 177), 922, 934, 957(119), 985(270), 987, 989(114), *1056, 1057, 1059*, 1157(81), 1169(173, 174), 1176(173), 1178(191), *1183, 1185*, 1235(7), 1236(10, 11), 1248(44), 1256(44, 64), 1262(82), 1283(157), 1290(170), 1304(210), *1305–1307, 1309, 1310*, 1314(25, 26), 1322(93, 101, 102, 104, 105, 112), 1323(118), 1324(105, 112, 119), 1325(105, 148–150), 1326, 1327(105), 1328(157, 158), 1329(157, 158, 166–169), 1330(166, 168, 169, 174), 1331(166, 167), 1332(200, 207), *1333–1337*, 1341(22, 26), 1343(22), 1347(26), 1434(235, 242), 1435(242), *1437, 1444*, 1808(90–93), 1814(121), 1821(145), *1863, 1864*, 1989(92), *2033*, 2137, 2138(36), *2171*, 2182(29), *2214*, 2470(33), 2471(37), 2514, 2515(37, 265), 2517(37), 2518(265), *2559, 2565*
Kirby, A.J. 374(59, 60), *427*
Kirby, G.W. 1091(58e), *1103*

Kirino, M. 1849(297), *1867*
Kirmse, W. 377(69), *427*, 471(197), *492*, 637(98), 639(98–100), 640(98, 99), 641(100), *699*, 704, 712, 728, 730(1), *774*
Kirpichenko, S.V. 1298(190), *1309*, 1449, 1457, 1460, 1465, 1466, 1470–1473, 1481, 1483, 1484, 1487, 1506, 1510, 1523(13), *1525*, 1691(13), *1786*
Kirsanov, A.V. 1500(360), *1536*
Kirschning, A. 434, 445(26), *489*, 1679(110), *1684*, 1796(41), 1821(144), *1862, 1864*
Kirste, B. 2182(28), *2214*
Kise, N. 1206(46), 1221(98), 1226(117), 1228(123), *1230–1232*
Kishan Reddy, Ch. 1668(16), *1682*
Kishi, N. 394(125), *428*, 1796(37), 1858(37, 343), 1859(343), *1862, 1868*
Kishi, R. 1120, 1121(132), *1139*
Kishida, H. 137, 142(44c), *151*, 200(100), *259*
Kisin, A.V. 1475(253), 1497(351), *1532, 1535*, 2132(18), 2133(20), *2171*
Kiso, K. 895(132), *1056*
Kiso, Y. 1252(56), *1306*, 1325(137), *1335*, 1699(49), *1786*
Kitagawa, H. 761(124), *777*
Kitahara, T. 810(93a), *824*
Kitayama, H. 438(90), *490*
Kitayama, K. 1752(240), 1754(244), *1790*
Kitchen, D.B. 123(24f), *151*, 1911(65), *1927*
Kitching, W. 381(76), *427*, 449(123), *491*, 1214(74), *1231*
Kitching, W.J. 407(160), *429*
Kitson, F.G. 196(78), *259*
Kiviskik, R. 1600, 1601(6), *1660*
Kiyooka, S.-i. 1852(313), *1868*
Kizil, M. 1569(87), *1579*
Klabunde, K.J. 243(269), *263*, 2063(66–68), 2066(66, 68), 2080, 2092(130–132), *2124, 2125*
Klabunovskii, E.I. 1743(222), *1790*
Klair, S.S. 433(7), *488*
Klanberg, F. 503(31), *511*, 1340(16), *1437*
Klar, E. 1584(24), 1586(51), 1587(56), *1595, 1596*
Klarner, F.G. 1903(40b), *1926*
Kläui, W. 2133(23), *2171*
Klaus, U. 2028(205), *2036*
Klebach, T.C. 2437(135), *2460*
Klebe, G. 183(12), 245(279), *257, 264*, 520, 524, 525(69), *553*, 564, 583, 586(88), *591*, 1374, 1375(83, 85–88), 1377(83, 85–87), 1378(86, 87), 1380(88), 1381, 1382(83), 1418, 1420, 1422, 1423(209), *1439, 1440, 1444*
Kleebe, H.-J. 2261(54c), *2312*
Kleewein, A. 2197(80), *2215*
Kleimeier, J. 190(49), *258*, 2148(101), *2173*

Klein, C. 2169(198), *2175*
Klein, C.H. 2534(361c), *2567*
Klein, D.J. 70, 71(139b), *100*
Klein, E. 1051(384, 386), *1062*
Klein, H.F. 1280(137), *1308*
Klein, L.C. 2261, 2263(66), *2313*, 2318(3), 2343(151–153), 2344(156–158, 163–165), 2347(195b), 2350(235), *2355, 2359–2361*
Klein, M.T. 2548(401), *2568*
Kleinschmit, P. 2330, 2340(78), *2357*
Kleman, B. 68(122a), *99*, 1164(138), *1184*
Klemann, L.P. 1991, 1992(96a, 96b), *2033*
Klemenko, S. 1603, 1605(38), *1661*
Klemperer, W.G. 1923(99, 102), *1928*
Klessinger, M. 2429(96), *2459*
Kletsko, F.P. 1454, 1467(86, 90), *1527*
Klimenko, N.M. 1467(186, 187), *1530*
Kline, M. 52(95a, 95b), *98*, 520(66), *553*, 763(133a), *777*, 2535(366), *2567*
Klingan, F.R. 52(97), *98*
Klingebdil, U. 3(3d), *95*
Klingebiel, U. 203(105), 207(129), 209(140), 211(145, 146), 213(150), 219(176), 234(241, 243), 235(238–240), 241(264, 265), 252(325–327), *259–261, 263, 265*, 915(182), 1010(299b), 1011(304–307), 1012(306), 1034(306, 353), 1036(354), 1037(355), 1038(353, 354), 1040(355), 1043(306, 307, 354, 357–359), 1044(305–307, 354, 359), 1048(305, 307, 359), 1051(382), *1057, 1060–1062*, 1064, 1068(5c, 5d), *1101*, 2253(37a), *2311*, 2541(375), *2567*
Klingensmith, K.A. 901(154), *1057*, 1316, 1318(38), 1331(198), *1333, 1336*
Klinger, R.J. 1188(3), *1229*
Klingner, E. 196(72), *258*
Klinkhammer, K. 795, 800, 801(69b), *823*, 2122(258), *2128*, 2203(86), *2215*
Klinkhammer, K.W. 796, 799, 800(66a, 66b), *823*, 1890(105, 107, 108), *1894*
Kloos, S. 813(104), *825*, 1699(51), *1786*
Klooster, W.T. 52, 54(95f), *98*, 2539(373b), *2567*
Klopper, W. 530, 532(95c), *554*
Klosin, J. 2079, 2092, 2093(126), *2125*
Klosowski, J.M. 2291(140), *2315*
Kloster-Jensen, E. 186(42), *258*
Klug, P. 449(122), *491*
Klumpp, G.W. 1434(234), *1444*, 1601, 1624(22c), 1628(143), *1660, 1663*
Klyba, L.V. 1523(419), *1537*
Klyuchnikov, V.A. 154, 156(11–16, 17a), 159(11), 160, 161(11, 56), 162, 163(14), 164(11, 15, 56), 165(11, 15), 166(12, 13, 16, 56), 174(14), *177, 178*, 1469(199–203), *1530, 1531*

Knapp, S. 1565(77), *1579*
Knapstein, C.-M. 2364(12), *2398*
Knapstein, K.M. 2364(10), *2398*
Knebel, B.O. 2552(413), *2568*
Kneisel, B.O. 541(107), *554*, 581(180), *593*, 1075(35b, 37), 1086(37), 1087(55), 1090(37), *1102*, 1411, 1412(188), *1443*, 2183, 2185(41), 2187(48), *2214*, 2429, 2436(104), *2459*, 2550(411), 2553(415), *2568*
Knieren, B. 304(106), *353*
Knight, A.E.W. 2504(199), 2522(285), *2563, 2565*
Knight, C.T.G. 270(13), *350*
Knight, D.W. 434, 444(23–25), 473(204), *489, 492*
Knobbe, E. 2350(220), 2351(241), *2361*
Knobbe, E.T. 2348(200), *2360*
Knobe, E.T. 2352(270), *2362*
Knobler, C.B. 186(33, 36), *257*
Knochel, P. 434, 438(11), *488*, 1668(16), *1682*, 1837(225), *1866*
Knölker, H.-J. 404(151), 406(154), *428, 429*, 547(112a–c), *555*, 1848(287), 1850(298, 301, 305, 306, 308, 309), *1867, 1868*
Knopp, 'D. 2327(58), *2357*
Knorr, M. 2057(50), 2072(84–95), 2073(96, 97, 99–101), 2074(87, 102–105), 2085, 2089(157), 2090(84, 85, 87, 88, 90, 91, 94, 105), 2091(86, 89, 94, 96, 97, 100, 102), 2094(157), 2110(102), *2123–2126*
Knunyants, I.L. 1188(4), *1229*
Knuppel, P.C. 211(143), *260*
Ko, J. 2083(144), *2125*
Kobayashi, H. 630(84), *699*, 1411(189), *1443*, 2087, 2090, 2092, 2093(165), *2126*, 2234(105b), *2243*, 2332(86), *2357*, 2383(61), *2400*
Kobayashi, J. 184, 250(28), *257*
Kobayashi, K. 30(67), 70(137, 138), 71(137), 72–74(143), 77(137), *97, 100*, 121, 122(15), 123(15, 22), 124(29), 125, 129, 131(34), 136(29), 140, 142(48), 148(15), 149(15, 60a–d), *150–152*, 277(23), *350*, 1296(185), 1301(198, 199), 1302(202–204), *1309, 1310*, 1557(63), *1578*, 1739(168), 1783(345), *1789, 1792*, 1828(183), 1854(323), *1865, 1868*, 1931(11), 1934(12), 1936(13), 1940(14–17), 1942(18, 19), *1960*, 1978(54a), 2001(136), *2031, 2034*, 2436(129), *2460*, 2496(176), *2562*
Kobayashi, M. 810(93a), *824*, 1689(7), *1785*
Kobayashi, N. 1120, 1121(133), *1139*
Kobayashi, S. 459(152), *491*, 510(67), *511*, 1610, 1630(78), 1639(178), *1662, 1664*, 1809(97–99), *1863*, 1871(12), *1891*

Kobayashi, T. 829(16, 17), *854*, 1205(43), *1230*, 1999(128), 2001(130a), *2034*, 2083(149), *2126*, 2265(81a, 81b), 2278(107), *2313, 2314*, 2330, 2354(80), *2357*, 2429(102), *2459*, 2496(178), *2563*
Kobayashi, Y. 1319(76, 77), 1320(77), *1334*, 1680(113), *1684*, 1699, 1735(48), *1786*, 1847(282), *1867*, 2350(228), *2361*, 2489(137a), *2562*
Koberstein, J.T. 2234(104a–c), 2238(121), 2239(126), 2240(132), *2243, 2244*
Köbrich, G. 2481(90), *2561*
Koch, E.-W. 669(145), *701*
Koch, H.J. 2145(89), *2172*
Koch, W. 567(107), *592*, 687(158), 690(162), *701*, 1126(193, 196), 1127(196), *1140*
Kochi, J.K. 1188(3), 1207(48), *1229, 1230*
Kochina, T.A. 559(9), 581(9, 177), *590, 593*
Kocienski, P. 1670(32), *1683*
Kocks, P. 2231(81e), *2242*
Koda, S. 1125(187), *1140*
Kodama, R. 2452(192), *2461*
Kodama, Y. 2250, 2261(20), *2310*
Koe, J.R. 2088(174), *2126*
Koegler, G. 331–333(145), *353*
Koegler, W. 1587(58), *1596*
Koelle, U. 2137(33), *2171*
Koellner, G. 1878(47), *1892*
Koenig, J.L. 2339(125, 125), *2358*
Koenig, K.E. 392(119), *428*, 1824, 1828(160), *1865*
Koert, U. 1539(1), *1577*
Koerwitz, F.L. 1836(221), *1866*
Koetzle, T.F. 52, 54(95f), *98*, 2539(373b), *2567*
Koft, E.R. 1797(42), *1862*
Koga, G. 1675(73), *1683*
Koga, N. 2116(244, 245), *2127, 2128*
Kogure, T. 1735(153), 1736, 1737(154), *1788*
Kohane, J.P. 1586(44), *1595*
Kohjiya, S. 2345(174), *2360*
Kohl, B.E. 1588(67), *1596*
Kohl, F.X. 2141, 2149(56), *2172*
Kohl, G.S. 2227(61), *2242*
Kohler, C. 209(136), *260*
Köhler, F.H. 190(56), *258*, 2155(142), 2159(142, 151, 152, 154, 155), 2160, 2162(151), *2174*
Köhler, H.J. 295(80), *352*, 705(13), *775*
Kohmura, K. 1672(47), *1683*
Kohn, W. 516(30a–c), *552*
Köhne, F. 1923(101), *1928*
Kohra, S. 1808(94), 1826(173), *1863, 1865*
Koie, K. 1197(26), *1230*
Koike, H. 120, 142, 143, 146(3), *150*
Koike, N. 2522(291, 292), *2565*
Koike, T. 186(40), *258*, 2017(177c, 178), *2035*, 2439(148), *2460*
Koike, Y. 810(93a), *824*
Koizumi, T. 794(53), *823*, 1190, 1192(10a, 11, 12), 1196(12), 1197(12, 24), 1215(81, 82), *1230, 1231*, 1699, 1735(48), *1786*
Kojima, M. 1218–1220(91), *1231*, 1452, 1453, 1473(61), *1526*
Kok, G.B. 380(70), 381(76), *427*
Kök, T.R. 1918(89), *1927*
Kokko, B.J. 1771(300), *1791*
Kokotailo, G.T. 270(14–16), *350*
Kolaczowski, S.T. 2224(35a), *2241*
Kolandaivel, P. 2508(223), *2564*
Kolani, B. 781, 782, 784, 790, 814(8b), *821*, 1972, 1981(37a, 37b), *2031*
Kolb, H.C. 467, 474(187), *492*
Kolbasov, V.I. 384(91), *427*
Kolbert, A.C. 310(132), *353*
Kolbuszewski, M. 1134(263), *1142*
Kolc, J. 709, 714(20), *775*, 901, 917, 981, 992(157), *1057*
Kole, S. 2240(130), *2244*
Kolenbrander, K.D. 560(32, 33), *590*, 1119(102), *1138*
Kolesnikov, S.P. 1244(32, 34), *1306*, 1312(16), *1332*, 2001(137), *2034*, 2473(51), 2488(135), *2560, 2562*
Kollar, L. 1689(8), *1785*
Kollegger, G. 1635(164), *1664*
Kolodyazhnyi, Yu.V. 1470(214), *1531*
Kolodziej, G. 2226(51), *2242*
Kolodziejski, W. 203(108), *259*
Kolomiets, T.A. 2218(7), *2240*
Kolonits, M. 1465, 1467, 1472(170), *1530*, 2181(21), *2214*
Koloski, T.S. 250(314), *264*, 437, 484(76), *490*, 2062(63, 64), 2066, 2068(63), 2105(64), *2124*, 2425(87), *2459*
Kolosova, N.D. 2155(141), *2174*
Komalenkova, N.G. 1381(108), *1440*, 1989(89a, 89b, 90a, 90b), 1994(89a), *2033*, 2465(7), 2479(82), *2559, 2560*
Komarov, V.G. 1490(317), *1534*
Komatsu, K. 1959(29), *1960*
Komatu, C. 2236, 2237(116), *2244*
Komori, E.-i. 1826(173), *1865*
Kondo, F. 782(13b), *822*
Kondo, K. 1669(28), *1683*, 2056(47), *2123*
Kondo, M. 143(52d), *152*, 761(125), *777*, 1743(190–192), 1745(191), 1746, 1747(191, 192), 1748(191), *1789*, 2150, 2151(105), *2173*
Kondo, T. 2150, 2151(105), *2173*
Kong, K.Y. 1274(116), *1308*
Kong, S. 466(182), *492*
Kong, Y. 2083(144), *2125*

Kongo, Y. 1157(80), *1183*
Konieczny, S. 860, 899, 901(14), *1054*, 1236(14), *1305*, 1965(16c), *2030*, 2469(28), 2486(122), 2510(236), 2513, 2526, 2527(260), *2559, 2561, 2564, 2565*
König, B. 1091(58b), *1103*
König, H. 1280(137), *1308*
Konings, R.J.M. 164, 165(84), *179*
Konishi, M. 407(157, 158), *429*
Konitz, A. 232(230), *262*
Konnert, J. 196(76), *258*
Kono, A. 2522(291, 292), *2565*
Kono, K. 1838(231), *1866*
Konradsson, P. 1872(16), *1892*
Koob, R.D. 66, 68(120b), *99*, 707(15), *775*, 1319(67, 68), *1334*
Koolhaas, W.E. 2428(89), *2459*
Koot, W.-J. 1835(216), *1866*
Köpf, H. 844(62), *855*, 873(61, 62), 879(85), 881(97), 910(62), 912(97), 932, 937(62), 940(97), 941, 942(62), 969(61, 62), *1055, 1056*, 2150, 2152(102), *2173*
Kopkov, V.I. 1500(356), *1536*
Köppe, R. 53(101), *98*, 1083(50, 51), 1096(50, 51, 72), *1102, 1103*, 1162(114, 115, 118, 120), 1163(120, 123, 124, 126), 1178(192), *1183–1185*, 2474, 2534(55), 2510(239), *2560, 2564*
Koppel, I.A. 1477(268), *1533*
Kopping, B. 1544–1546, 1549(21), 1553(21, 51), 1555, 1556(51), 1557(60, 61), 1561(61), 1572, 1573(21, 92), *1577–1579*, 1703, 1722(66, 68), 1723(66), *1787*
Kopylov, V.M. 580, 584(167), *593*, 1450(46), 1454(103, 104), 1461(46), 1464, 1467(103, 104), *1526, 1528*, 2218(5), *2240*
Kopylova, L.I. 436, 479(61), *489*, 1692(19), *1786*
Korchagina, A.N. 154, 156(11, 12, 14), 159(11), 160, 161(11, 56), 162, 163(14), 164(11, 56), 165(11), 166(12, 56), 174(14), *177, 178*, 1469(200, 202, 201), *1530, 1531*
Korda, A. 1776(324), 1779(324, 335), *1792*
Koreeda, M. 463(170), *492*, 1675(76), *1684*
Korenowski, G.M. 2351, 2354(259), *2362*
Korkin, A. 33(70, 71), 34, 50, 51(71), *97*, 1048(364), *1062*
Korkin, A.A. 90–92(169), *101*, 551(115), *555*
Kornath, A. (232), *262*, 1875(29, 30), *1892*
Korneva, S.P. 783(17), *822*
Korolev, V.A. 981, 992, 995(255), *1059*
Korolev, V.K. 1125(177), *1139*
Korotaeva, I.M. 1450(38), 1490(317), *1526, 1534*
Kort, M.de 475(216), *493*, 1681(124), *1685*
Kort, M.E. 1823(155), *1864*
Kortan, A.R. 69(130), *100*

Korte, W.D. 496(7), *510*
Koschinsky, R. 397(134), *428*
Koseki, S. 90(170c), *101*, 860, 983(10a), 1048(365), *1054, 1062*, 2505(200), *2563*
Köser, H.G. 748, 750(99), *777*
Koshi, M. 1125(186), *1140*
Koshihara, S. 1318(65), 1328, 1329(157), *1333, 1336*, 2470(33), *2559*
Kosima, S. 2001(136), *2034*
Kositsyna, E.I. 1454, 1467(102), *1528*
Kosse, P. 2051, 2052(34), *2123*
Kösslinger, C. 2340(133), *2359*
Kost, D. 504(34), *511*, 1067(16), *1101*, 1374, 1377, 1380, 1383–1386(95a, 95b), 1418, 1420, 1422(211a, 211b), 1423(211a, 211b, 221, 222), 1424(211a, 211b, 220, 222), 1425(211a, 211b, 220–222), 1426(211a, 211b, 222), 1427(221, 222), *1440, 1444*, 1476(262), *1532*
Köster, A.M. 9(30), *96*
Köster, R. 1979(56c), *2031*
Koster, R. 216(160), *260*, 1884(72), *1893*
Kosugi, M. 1831(202), *1865*
Kotani, J. 805(82), *824*
Kotani, S. 2075, 2091(115), *2125*
Kotov, V.V. 2056(45), *2123*
Kottke, T. 211(144), 235(238, 239), *260, 263*, 1423, 1425, 1427(221), *1444*, 2486(118), *2561*
Kougo, Y. 279(31), *351*, 2183(38, 39, 42), *2214*
Koumaglo, K. 393(121), 394(123), *428*, 434, 442(19), *489*, 1797(45), *1862*
Kouzai, H. 1861(361), *1868*
Kovacs, I. 215(156), 226(199), *260, 261*, 1452(55), 1454(55, 93), 1461(138, 139), 1462(149, 152), 1463(149, 166), 1467(93, 138, 139, 149, 152, 166), 1481(287, 288), 1483(288), 1503(152), 1508(385, 388), 1509(287, 395), 1510(166, 287), *1526, 1527, 1529, 1530, 1533, 1536, 1537*
Kovacz, I. 1450, 1453, 1471(33), *1525*
Kovalchuk, A.V. 2231(85), *2243*
Kovar, D. 917, 998, 1002(184), *1058*, 2177(4a, 4b, 4d), 2181(19), *2213, 2214*
Kovyazin, V.A. 580, 584(167), *593*
Kovyazina, T.G. 1450(46), 1454(103, 104), 1461(46), 1464, 1467(103, 104), *1526, 1528*
Kowalewski, J. 296(81), *352*
Koyakumaru, K. 1674(61), *1683*
Koyama, K. 1847(282), *1867*
Kozakai, N. 2261(58), *2312*
Kozawa, S.M. 1550(46), *1578*
Kozima, S. 1986(76a), 2006(156), 2011(76a), *2032, 2034*, 2436(129), *2460*
Kozlova, G.N. 1585(36), *1595*
Kozlova, G.V. 1458, 1467(122), *1528*

Kozyrev, A.K. 1467(185), *1530*
Kozyreva, O.B. 1523(419), *1537*
Kpoton, A. 503(32), 505(35), *511*, 1374, 1375(76, 77, 79), 1376(76), 1377(77, 79), 1378(77), 1380(79), 1382(76, 77), 1383(76, 77, 111), 1385(76, 77), 1386(76), 1424(77), *1439, 1440*
Kraakman, P.A. 2428(89), *2459*
Kraatz, H.-B. 1885(77), *1893*, 2108(219), *2127*
Krafft, G.A. 1091(58f), *1103*, 1678(99), *1684*
Kraft, G. 243(268), *263*
Krahé, E. 487(269), *494*
Kraitchman, J. 588(197), *594*
Kraka, E. 601(12), *698*
Krakar, E. 113, 115(27), *118*
Krallmann, R. 190(49), *258*, 2133(21), 2148(101), 2150, 2151(108), 2156(145), *2171, 2173, 2174*
Kramarova, E.P. 243(271–273), 245(286), *263, 264*, 500, 501(26), *510*, 587(195), *594*, 1390(128–130), 1391(140), 1392(129, 143, 145a, 145b), 1393(148), 1395(156, 157), 1396(156–158, 162), 1397(156, 166, 167), 1399(156), 1404(172), 1434, 1435(240), *1441, 1442, 1444*, 1454, 1467(106), *1528*
Kramer, E.J. 2237(117), *2244*
Kramer, G.J. 115(36, 38), 116(38), *118*
Kramer, J.B. 1797(49), *1863*
Kramer, K. 198(92), *259*, 791(44a), *823*
Kramerova, E.P. 226(206), *262*
Krancher, M. 780, 794(4a), *821*
Kranenburg, M. 1835(216), *1866*
Krapivin, A.M. 1475, 1477(251), *1532*
Kräschmer, W. 69(125), *99*, 1930(1c, 1d), *1960*
Krasnoperov, L. 167, 168(89), *179*
Krasnoperov, L.N. 169(114), *180*, 518(58), *553*
Krasnov, V. 1225(113), *1232*
Krasnova, T.L. 1989(88), *2033*
Kratel, G. 1587(62), *1596*
Kratky, C. 2181(20), 2204(89), *2214, 2215*
Krato, B. 2145(88), *2172*
Kratt, A. 1153(59), *1182*
Kraus, G.A. 1879(55), *1893*
Kraus, M. 21(54a, 54b), *97*
Krause, J. 1970, 1974(29b), *2031*
Krause, N. 690(167), *701*
Krautscheid, H. 1878(49), 1886(82, 84, 85), 1887(86–88), *1892, 1893*
Krebs, A. 2439(141, 150–153), *2460*
Krebs, B. 945(208), *1058*, 1524(420), *1537*, 1910(59), 1913, 1914(78, 79), *1927*
Krebs, F. 222(193), *261*, 730(62), 732(63), 735(67), *776*
Kreeger, R.L. 709, 714(21), 715(38), 727, 741(21, 38), 743(21), *775*

Kreil, C.L. 10(32b), *96*, 994, 995(283), *1060*, 1149(33), *1182*
Kreissl, F.R. 190(54), *258*
Kreiter, C.G. 2132(17), *2171*
Kremer, M. 2523(317), *2566*
Krempner, C. 884(104, 107, 108, 110), 885(104, 107, 108, 110, 117), 889(107, 108, 110, 117), 917(104, 107, 108), 922(107), 925(107, 108, 110, 117), 926(107, 108, 117), 928(110), 934(200), 937(104, 107, 117), 938(104, 117, 200), 939, 940(108), 957(104, 107), *1056, 1058*
Krempp, M. 1026(337, 338), 1027(337), 1028, 1050(338), *1061*, 1133(252), *1141*
Krepski, L. 1670(29), *1683*
Kresge, A.J. 369(47–49), 397(47, 49), *427*, 611, 614(43), 626(80), 627(80, 81), 628–630(81, 83), *698, 699*
Kretzschmar, U. 1583(16), *1595*
Kreutzer, K.A. 2042(14), 2044(18), 2051(14, 18), 2052(18), *2122*
Kreuzfeld, H.J. 1743(219), *1790*
Kriebisch, K.A. 2152, 2153(120, 121), *2173*
Krief, A. 416(188), *429*, 1546(33), *1578*
Krishna, A. 1292(178), *1309*
Krishna, R. 250(306), *264*, 2404(12), *2458*
Krishnamurti, R. 1667(10), 1675(77), *1682, 1684*
Krivinos, S. 504(34), *511*
Krivitskii, V.V. 1470(213), *1531*
Krivonos, S. 1374, 1377, 1380, 1383–1386(95a), 1423(221, 222), 1424(222), 1425(221, 222), 1426(222), 1427(221, 222)*1440, 1444*
Kriz, G.S. 376(65), *427*, 635, 636(90, 91), 637(91), *699*
Krogh-Jespersen, K. 36(76a), *97*, 247(295), *264*, 1000(293), *1060*
Krogh-Jespersen, M. 1123(156), *1139*
Krogh-Jespersen, M.-B. 250(319), *265*, 705, 707(12), *775*, 781(5), *821*, 901(147), *1057*, 1166(149, 157), *1184*
Krogh-Jesperson, M.-B. 835(37), *854*
Krogsrud, S. 2160(170), *2174*
Krohn, K. 782, 791, 795, 796(11d), *822*
Kroke, E. 1268(98), 1283(152), 1304(213), *1307, 1309, 1310*, 1881(61), *1893*, 2184(45), 2185(46, 47), *2214*, 2472(48, 49), 2494(165), *2560, 2562*
Kroll, P. 2261(63), *2312*
Kron, J. 2332, 2339(88), *2357*
Kroop, P.J. 2437(131), *2460*
Kropfgans, M. 1341, 1349, 1357, 1359, 1366(29), *1437*, 2363(9), 2364(27, 30, 34), 2368, 2369(34, 39), 2371(34), 2374(34, 39), *2398, 2399*

Kroto, H.W. 69(124a–c), 70, 71(139a), 99, 100, 1150(39), 1182, 1929(1a, 1b), 1930(1a), 1960
Krotz, A. 1746(227), 1790
Krstic, A.R. 380(73), 427, 641, 642(102), 700
Krug, H. 2333, 2337(103), 2358
Krüger, C. 77–79(154a, 154b), 100, 784, 787, 788(25), 822, 2140(47), 2163(181, 182), 2164–2166(182), 2171, 2175, 2534(360a, 360b), 2567
Kruger, C. 198(87), 259, 1434(232b), 1444, 1918(85), 1927
Kruglaya, O.A. 752, 760(110), 777
Kruithof, K.J.H. 1601, 1624(22c), 1628(143), 1660, 1663
Krukonis, V. 2234(104a), 2239(126), 2243, 2244
Krul'ko, D.P. 1456, 1467, 1497, 1501(119), 1528
Krumpe, K.E. 1236(13), 1305
Kruppa, A.I. 2473(51), 2488(135), 2560, 2562
Krynitz, U. 207(128), 260, 484(246), 493
Ku, J. 479(233), 493
Ku, W.-S. 1678(102), 1684, 1877(44), 1892
Kuan, C.P. 440(98), 490
Kuan, Y. 374(63, 64), 375(64), 427
Kubas, G.J. 2065, 2066(73, 74), 2124
Kubeckova, M. 2322(27), 2356
Kubicki, J.D. 113(24), 118
Kubiniok, S. 1885(73), 1893
Kubo, Y. 1291(175), 1309, 1849(291), 1867
Kubodera, H. 1314(27), 1333
Kubodera, K. 2351(253), 2362
Kucera, H.W. 1608, 1647(60), 1661
Kuck, V.J. 1930(5), 1960
Kudiakov, N.M. 1393, 1394, 1396(151), 1442
Kudo, K. 1798(51), 1863
Kudo, T. 22(56, 59), 23(59), 24, 25(56), 30(67), 72(59), 81(156), 97, 100, 120(13), 121(13, 15), 122(13, 15, 21a), 123(13, 15, 21a, 21d, 24d, 25b), 129, 131, 133(13), 148, 149(15), 150, 151, 252(334), 265, 839(48), 855, 901(152), 1057, 1065(13, 14), 1066, 1067(13), 1082(49), 1083(13, 14), 1092(62), 1101–1103, 1898, 1899(33), 1911(69–71), 1926, 1927, 2493(152), 2562
Kudyakov, N.M. 1454(72, 74, 108, 109), 1455, 1467(108, 109), 1495(109, 340), 1496(344), 1500(72, 74, 109, 359), 1527, 1528, 1535, 1536
Kuech, T.F. 2549(406), 2568
Kuehr, H. 780, 781(3), 821
Küerzinger, A. 1743(203, 218), 1746, 1747(203), 1789, 1790
Kugita, T. 1762(263), 1791
Kuhler, K. 710(25b), 775
Kuhlmann, B. 563(80), 591

Kuhlmann, T. 2017(179), 2035, 2330(85), 2357
Kuhn, M. 2141(53), 2143(78, 79), 2144(85), 2145(85, 89), 2172
Kühnel, E. 1034(352), 1061
Kuhs, W. 243(276), 264, 1357, 1359, 1360, 1362, 1364(46), 1374, 1376, 1377(90, 91), 1380, 1381(91), 1401(90, 91), 1438, 1440
Kuhs, W.F. 334(149), 354, 1349, 1357(31a), 1438
Kukkola, P.J. 505(40), 511
Kulicke, K.J. 1555(57), 1572, 1573(92), 1578, 1579
Kulik, W. 10(35), 96, 1151(46, 48), 1182, 2453(195), 2461
Kulisch, W. 2145, 2146(95), 2172
Kulkarni, A.K. 424(211), 430
Kum, M. 925, 998, 1000, 1002(192), 1058
Kumada, M. 407(157–159), 417(193), 429, 798(60a, 60b), (110c), 823, 825, 852(85), 855, 891(124, 125), 917(184), 998(184, 287, 288), 1001(288), 1002(184), 1056, 1058, 1060, 1248(43), 1251, 1254, 1256(51), 1261(79), 1283(146), 1306–1308, 1312, 1313(12), 1319(71, 72, 74, 75), 1321(87), 1322(90–92, 106–109), 1325(72, 127–134), 1328(72, 155, 156), 1332(199), 1332, 1334–1336, 1412(190), 1418(218a), 1443, 1444, 1699(49), 1743(188), 1751(234), 1786, 1789, 1790, 1797(46), 1824, 1828(159), 1862, 1865, 1973(41a, 41b, 48a, 50a, 50b), 1974(50a, 50b), 1990(94, 95a, 95b), 2007–2009(95a), 2031, 2033, 2150, 2151(105), 2173, 2177(3c), 2213, 2416(43), 2429(106, 107), 2430(111), 2436(106), 2458–2460, 2471(40), 2526(328), 2559, 2566
Kumagai, M. 1699(50), 1718, 1723(108), 1735(50), 1786, 1788
Kumagai, Y. 2424(73), 2459
Kumano, K. 1983(73), 1999(127), 2001(73), 2032, 2034
Kumar, A. 863(24, 25), 864, 932, 934, 982(25), 1054, 1237(18a), 1306
Kumar, J. 2351(260), 2354(260, 280), 2362
Kumar, M. 1453, 1462, 1467, 1483, 1491(68), 1526
Kumara, S.K.C. 284(60), 351
Kumara Swamy, K.C. 1352, 1353(34), 1354, 1355(37a, 37b), 1356, 1357(34, 37a, 37b), 1438
Kumarathasan, R. 578(159), 593, 951(239), 967(247), 968, 969(248), 1059, 1238(22), 1274(119–121), 1283(151), 1306, 1308, 1309, 1645(202), 1664, 1881(62), 1883(67), 1887(62), 1893, 2421(57), 2422(69), 2458, 2459, 2494(164), 2562
Kumaresann, R. 2508(223), 2564
Kumareswaran, R. 1674(59), 1683

Kumashiro, K.K. 52, 54(95e), *98*
Kumashito, K.K. 2539(373a), *2567*
Kumler, P.L. 2234, 2235(106), *2243*
Kummer, D. 228(209), 243(275, 276), 245(280–282), *262, 264*, 1374, 1376, 1377(89–91), 1380, 1381(89, 91), 1398, 1400(171), 1401(89–91, 171), 1403(171), 1429(224, 225), 1434, 1435(241), *1440, 1442, 1444*
Kümmerl, L. 2353(277), *2362*
Kumo, Y. 2251, 2253(27b), *2311*
Kumoyama, H. 1590(88), *1596*
Kunai, A. 186(40), 194(67), *258*, 870(51), 891(127), 892(130), 928(193, 194), 998(289, 290), 1004(289), 1005(289, 290), *1055, 1056, 1058, 1060*, 1208–1210(57), 1216, 1217(85, 86), 1218(86, 89, 90), 1219(89, 90), 1227(120), *1230–1232*, 1252(55), *1306*, 1325(140), *1335*, 1769, 1770(294), *1791*, 1973(43), 2016(175d), 2017(43, 178), *2031, 2035*, 2439(148), *2460*
Kunat, D. 1471(226), *1531*
Kundai, A. 1770(295), *1791*
Kundler, S. 1979(57c–e), *2031, 2032*
Kunimi, N. 1228(123), *1232*
Kun Kong, Y. 951(236), *1059*
Kunô, K. 2001(136), *2034*, 2436(129), *2460*
Kunz, H. 1814(119, 120), *1864*
Künzer, H. 1575(98), *1579*
Kuo, C.M. 2224(41a, 41b), *2241*
Kuo, H.-L. 1843(262), *1867*
Kupce, E. 296(84, 85), 302(100, 101, 103, 104), 303(104), 304(103, 104), 305(104), *352, 353*, 1373(75d), *1439*, 1458, 1467(124), 1474(247), 1475(124, 247), 1479(124, 272), 1508(247), 1509(124, 247), 1521, 1523(418), *1528, 1532, 1533, 1537*
Kupcè, E.L. 1449, 1460, 1470, 1481, 1506, 1510(6), *1525*
Kupche, E.L. 1459, 1467(133), 1474(240), 1478, 1479(270), *1529, 1532, 1533*
Kupfer, S. 2220(17b), *2240*
Kurakaka, T. 1911(70), *1927*
Kurarathasan, R. 1331(193), *1336*
Kurata, H. 1831(202), *1865*
Kure, S. 1771(296), *1791*
Kurihara, A. 1559(68), *1579*
Kurishima, K. 737, 741(70), *776*, 906, 908, 931(169), *1057*, 1278(135), *1308*, 2409, 2410(31), *2458*
Kurita, A. 1412(190), *1443*
Kurita, J. 1991(97), *2033*
Kuriyama, A. 1210(58, 59), *1231*
Kurjata, J. 2228(77), 2229(78), *2242*
Kurka, B. 1583(21), *1595*

Kuroboshi, M. 1755, 1756(247), *1790*, 1826(174), *1865*
Kuroda, M. 269, 273, 275(12), *350*, 2207(95, 97), *2215, 2216*
Kuroiwa, K. 2546(388), *2568*
Kurokawa, Y. 2350(228), *2361*
Kurosaki, Y. 892(131), 895(132), *1056*, 1252(56), 1253(57), *1306, 1307*, 1325(135, 137), *1335*, 2512(248), *2564*
Kurosawa, H. 1771(308, 309), *1792*, 2110(224), *2127*
Kurreck, H. 2182(28), *2214*
Kusakawa, T. 277(24), 278(25), *350, 351*, 1301(200), *1310*
Kusama, H. 1628(144), *1663*
Kusche, A. 1871(10), *1891*
Kuselman, A. 2321, 2322(23b), *2356*
Kuselman, I. 2322(29), *2356*
Kushner, M.J. 1118(98), *1138*
Kushner, M.J.J. 1122(154), *1139*
Kusomoto, T. 1688(5), *1785*
Kusuda, S. 476(222), *493*
Kusukawa, T. 1301(201), *1310*, 1942, 1944(20), 1945(21), 1947(21, 22), 1951(23, 24), 1955(25), *1960*, 2430(109, 110), 2432(110), *2459*
Kusumoto, T. 1726, 1727(133), *1788*
Kutlubaev, R.G. 1467(185), *1530*
Kutoglu, V.A. 1909, 1910(54, 55), *1927*
Kutyrev, G.A. 1497(346, 347), *1535*
Kutzelnigg, W. 60(106a–c), *99*, 122(18), *150*, 182(8), 252(333), *257*, 265, 281(44), *351*, 516(32a–c), 517(41), 530(95c), 532(32b, 32c, 95c), *552, 554*, 829(24), *854*
Kuwabara, H. 1157(80), *1183*, 2183(38), *2214*
Kuwajima, I. 435(44), 438(89, 93), 449(120), 453(44), 465(89), 469(191, 192), *489–492*, 1601, 1602(12, 18), 1614(93–95, 97), 1620(12, 115), 1623(129), 1636(18, 166), 1649(210), 1651(12), 1652(12, 219), *1660, 1662–1665*, 1859(344, 345), *1868*
Kuwano, R. 1743(206, 207), 1745–1747(207), 1748(206), *1789*
Kuyavskaya, B. 2321, 2322(23b), *2356*
Kuyavskaya, B.I. 2322(29), *2356*
Kuzenetsov, V.A. 304(109), *353*
Kuzmenko, L.P. 1485(302), *1534*
Kuzmich, D. 1563(74b), *1579*
Kuzmina, L.G. 2056(45), *2123*
Kuznetsov, B.N. 2341(141), *2359*
Kuznetsov, I.G. 1452, 1454(56), *1526*
Kuznetsova, E.E. 1450(41), 1454(81), 1458(122), 1467(81, 122), 1494(336), *1526–1528, 1535*
Kuznetsova, G.A. 1452(59), 1453(62), 1454(78, 79, 105), 1455(59), 1457(62, 120), 1459(134), 1460(120), 1467(105, 120,

134), 1475(59, 62, 249), 1479, 1480(249), 1484(59), 1485(62), 1501(368), 1523(249), *1526–1529, 1532, 1536*
Kuznetsov, V.A. 1870(6), *1891*
Kvintovics, P. 1721(123), *1788*
Kwak, K.H. 1674(60), *1683*
Kwak, Y.-W. 862, 863, 901(22), *1054*, 1860(351), *1868*, 2484(99), *2561*
Kwart, H. 182(7), *257*
Kwiatkowski, R. 1701, 1702(63), *1787*
Kwik, W.L. 2075, 2091(109, 110), *2125*
Kwon, H. 1616(101), *1662*
Kwon, J.H. 1250(48), *1306*
Kyba, E. 1019(318, 319), *1061*
Kyoto, M. 2347(186b), *2360*
Kyushin, S. 137, 139, 140(45), 143(45, 52a–d), 144(45), 145(54), 146(45), *151, 152*, 1290(171), 1304(209), *1309, 1310*, 2188(55), 2206(94), 2207(96), 2208(98), *2215, 2216*

Laaksonen, A. 296(81), *352*
Laaksonen, R.T. 1120(125), *1138*
Laali, K. 569(115), *592*
Laasko, D. 2519(267), *2565*
Labrouillère, M. 580(166), *593*
Lacan, P. 2341(143), *2359*
Lachmann, J. 1434, 1436(249), *1445*, 2084, 2090(156), *2126*
Lacoste, J. 2231(82), *2242*
LaCote, E. 1566(81b), *1579*
Lacour, J. 1673(55), *1683*
Lacross, M. 2322(34), *2356*
Ladika, M. 362, 363(27), *426*, 598, 611, 612(8), *697*
Laet, R.C.de 1610, 1630(79), *1662*
Lagarde, P. 2523(321), *2566*
Lagow, R.J. 819(117), 820(119), *825*
Laguerre, M. 386(99), 411(168), *428, 429*
Laguzzi, G. 561(52), *590*
Lahaye, M. 2261(60), *2312*
Lahoz, F.J. 2095(186), 2097(188), 2098(190, 191), 2108(186, 190, 191), *2126*
Lahtin, V.G. 296, 299–301(94), *352*
Lai, Z.-G. 1467, 1474, 1475(183), *1530*
Laidig, K.E. 690(164a), *701*
Laine, R.M. 226(205), 228(211), *262*, 1353, 1368, 1369(36), *1438*, 2246(5a–c, 7), 2247(9), 2248(7, 9), 2249(7, 9, 17), 2250(7, 9, 19, 21), 2251(24, 28a–d), 2252(24), 2253(21, 24, 28a–d), 2255(21, 28a–d), 2256, 2257(21), 2261(50, 69), 2263(69, 72), 2265(80), 2269(7, 9), 2272, 2273(7), 2276(103, 104a–c), 2278(104a–c), 2280(111), 2285(7), 2290(138a–c, 139a, 139b), 2291, 2293(138a–c), 2294(17, 139a, 139b), 2297(72), 2298(72, 150a, 150b, 151,

152a, 152b), 2299(153), 2301(156–159), 2302(159), 2305(157), *2310–2316*
Laity, J.L. 7(24), *96*
Lallemand, J.-Y. 1671(34), *1683*
Lamb, G.W. 222(192), *261*, 1638(170), *1664*
Lamb, L. 1930(1c), *1960*
Lambecht, G. 219(180), *261*
Lambert, J. 2223(33), *2241*
Lambert, J.B. 198(84), *259*, 280(39–42), 283(47, 48), *351*, 366(41, 42), 367(43, 44), 373(58), 380(71), *426, 427*, 514(2a, 2b, 4, 5), 520(2a, 2b), 521(2a, 2b, 75), 522(2a, 75, 79, 80a, 80b, 81), 529(2a), 532(2a, 2b, 5, 101), 533(2a, 2b), 536(4, 5), 540(79, 80a, 80b, 81), 542(4), 543(5), 547(2b, 101), *551, 553, 554*, 559(10, 18), 562(67), 563(18, 75–80), 564(18, 75–78, 85), 565(18, 67, 75), 566(75–77), 567(18, 67, 77, 102), 568(67, 75, 102, 111), 569(67, 77, 102), 570(67, 102, 123, 124, 126), 572(124), 581(67, 123, 182), 583(67, 182), 584(67, 182, 187), 585(67, 187), *590–593*, 616, 618(48), 620(64–68), 621(64–67), 622(64–70), 623(64), 637(66), 652(125, 126), 662(137), *698–700*, 780(1f), 790, 794(37), *821, 822*, 1108(18), *1136*, 1408(180, 181), *1443*
Lambrecht, G. 222(195), *261*, 2363(9), 2364(14–18, 20–31, 33–36), 2365(20), 2367(20, 37), 2368(34, 38–40), 2369(34, 39), 2371(34, 38, 40), 2374(34, 36, 38–43), *2398, 2399*
Lambrecht, W. 1585(35), *1595*
Lammertsma, K. 1134(255), *1141*
Lampe, F.W. 560(44), *590*, 607(25–28), 608(26, 27), *698*, 1109(41), 1120(119), 1124(169), *1137–1139*, 2486(120), 2493, 2515(150), *2561, 2562*
Lampe, J. 1601, 1602(9), *1660*
Lan, A.J.Y. 473(208), *492*
Landa, L.M. 160, 161, 164, 166(56), *178*, 1469(200), *1530*
Landais, Y. 409, 410(165), *429*, 1576(102), *1579*, 1680(119), 1681(121), *1685*
Landgrebe, J.A. 757(121), *777*
Landor, S. 999(292), *1060*
Landry, C.J.T. 2343(154), 2344(154, 155), *2359*
Lane, R.D. 296(89), *352*
Lane, T.H. 184, 221, 222(19), *257*
Laneau, G. 1075(35c), *1102*
Laneau, G.F. 1086, 1090(54), *1102*
Lanfranchi, M. 2055–2057(44), 2072, 2090(84, 90), *2123, 2124*
Lang, H. 1374, 1375(84), 1387, 1389(121a, 121b), *1439, 1441*, 2107(215), *2127*
Lange, H. 2364(14), *2398*
Lange, L. 1885(73), *1893*

Lange, L.D. 1697(40), *1786*, 2047(25), *2123*
Lange, U.E.W. 1749(229), *1790*
Langhoff, S.R. 704(9), *775*
Langkopf, E. 1794, 1795, 1801(32), *1862*
Langlois, J.-M. 2510(231, 232), *2564*
Langner, R. 2364(13), *2398*
Lankat, R. 2061(58, 59), 2062(61), 2066(59), 2068(58, 59), *2123*
Lankers, M. 2339(130), *2359*
Lanneau, G. 1052(388), *1062*, 1064, 1069, 1082–1084, 1086, 1099(9), *1101*, 1387, 1388(114, 116), *1440, 1441*, 2481, 2483, 2553(97), *2561*
Lanneau, G.F. 503(33), *511*, 1374, 1375, 1377(78), 1382(109), 1387(115, 117–120), 1388(115, 117), 1389(118–120), 1390(123b), *1439–1441*, 2060, 2067(56, 57), 2069(76), 2071(57, 76), 2084(56, 57, 76), 2089(76), 2093(57, 76), 2094(76), *2123, 2124*
Lapasset, J. 1064, 1069, 1082–1084, 1086, 1099(9), *1101*, 1387, 1388(114), 1430–1432, 1434(228), *1440, 1444*
Laperrière, S.C. 2350(227), *2361*
La Peruta, R. 2351, 2354(259), *2362*
Lapidot, H. 1587(61), *1596*
Lapointe, R.E. 580(175), *593*
Laporte, F. 1963(2c), *2030*
Laporterie, A. 1857(339), *1868*, 1963(3), 1964(8), 1965(3, 10a, 10b, 12–14), 1967(10a, 10b), 1968(13, 21, 22, 23a, 23b), 1970(27), 1971(30), 1973(23b), 1976(53), 1980(3, 62a), 1981(10a, 10b, 12, 21, 63, 65, 67a–c), 1982(12, 14, 22, 67a–c, 70, 71), 1983(14, 22, 67c, 72), 1984(14, 22, 74), 1985(75), 1996(3, 14, 21, 67a–c, 75, 107, 110), 1998(3, 10a, 10b, 67a, 67c, 74, 110, 118, 119, 123), 1999(10a, 10b, 12, 21, 67c, 107, 123), 2001(123), 2003(12), 2007(12, 75, 158b), 2008, 2009(110), 2011(167), *2030–2035*
Lapouyade, P. 1614(92, 96), *1662*
Lappert, M. 52(99a–c), 55, 56, 58, 61(99c), 62(99b, 99c), *98*
Lappert, M.E. 2534(364), 2544(377), *2567*
Lappert, M.F. 29(66b), 60(107), *97, 99*, 164, 166(77), *179*, 217(166), *261*, 1377, 1378, 1380(97a), *1440*, 1694(31, 32), 1695(31–33), *1786*, 2534(363, 365), 2535, 2536(368), 2537, 2539(370), *2567*
Lapsina, A. 1373(75d), *1439*, 1452, 1453(52), 1458(52, 124), 1467(124), 1474(247), 1475(124, 247), 1479(124), 1501(366), 1508(247), 1509(124, 247), 1514(408b), 1521(418), 1522(408b), 1523(418), *1526, 1528, 1532, 1536, 1537*
Lapsina, A.F. 1508(390), *1537*

Lapsinya, A.F. 1475(250), 1481(284), 1506(250), 1507(382), 1510(250), *1532, 1533, 1536*
Laquerre, M. 301(97), *352*
Larbot, A. 2340(131), *2359*
Larcher, F. 507(48, 50), 508(48), *511*
Largo-Cabrerizo, A. 1123(166), *1139*
Largo-Cabrerizo, J. 1050(369), *1062*, 1123(166), 1124(167), 1129(230–232), *1139, 1141*
Larin, M.F. 1390(128, 129), 1392(129), 1393, 1395, 1398(152, 153), 1404(153), *1441, 1442*
Larnerd, J.M. 1585(40, 41), *1595*
Laroze, G. 1586(48), 1587(63), *1595, 1596*
Larrow, J.F. 1672(50), *1683*
Larsen, K. 757(121), *777*
Larson, G.L. 459(154), *491*
Larsson, R. 1840(252), *1866*
Lartiges, S. 1890(105), *1894*
Lasaga, A.C. 115(32, 35), *118*
Lasch, J.G. 2141, 2149(56), *2172*
Laschat, S. 1814(119, 120), *1864*
Lask, G.M. 1156(74), *1182*
Laskovenko, N.N. 2218(7), *2240*
Lassacher, P. 880, 881, 914, 985, 986(93), *1056*, 1276(123), *1308*, 2083(145), *2125*, 2193(73), 2204(87), 2210(73), 2213(108), *2215, 2216*
Latif, L. 487(266), *494*
Lau, J.C.-Y. 420(200), *429*, 456(144), *491*, 509(61, 63, 64), *511*
Lau, P.W.K. 393(122), 422(206), *428, 430*
Lau, W.W.Y. 1763(265), *1791*
Laube, T. 558(6), *590*
Lauble, St. 241(260), *263*
Laupert, R. 862(21), *1054*
Laure, C. 1118(99), *1138*
Laurent, C. 1997(112b, 112c), *2033*
Laurent, D.St. 1546(34), *1578*
Laurie, V.W. 234(235), *262*
Lautens, M. 453(131, 132), 467(188), 472(199), 473(201, 202), *491, 492*, 1669(27), *1683*, 1861(360), *1868*
Lautenschlager, H. 1697, 1698(44), *1786*
Lautenschlager, H.J. 1691(14), *1786*, 2227(68), *2242*
Laval, J.D. 1524(422), *1537*
Lavrent'yev, V.I. 1923(97), *1928*, 2334(107), *2358*
Lawrence, F.T. 860–862(18), 866, 867(40), 992(18), *1054, 1055*, 1080(46), *1102*
Lawrence, N.J. 455, 472(139), 476(220), *491, 493*
Lawton, R.G. 544(111a), *555*
Lay, U. 2107(215), *2127*
Lay, W.-J. 226(198), *261*

Layeillon, L. 2546(387), *2568*
Layh, M. 1890(108), *1894*
Layton, T.W.J. 1719, 1732(111), *1788*
Lazarescu, S. 2223(24), *2241*
Lazarev, A.N. 1471, 1472(230), *1531*
Lazareva, N.F. 237(245), *263*, 1454(79, 93, 100, 105), 1455(114), 1467(93, 100, 105, 114), 1479(278), 1481, 1483(288), 1493(329), 1523(419), *1527, 1528, 1533, 1535, 1537*
Lazraq, M. 1524(422), *1537*
Lazzeretti, P. 295–297(79), *352*
Léandri, G. 1807(83), 1849(288, 289), 1856(331), *1863, 1867, 1868*
Leavitt, F.C. 1971(31), *2031*
Lebeau, B. 2330(82), 2350, 2351(232), 2352(264), 2354(82, 284, 285), *2357, 2361, 2362*
Le Bideau, F. 452(129), *491*
Lebrun, J.-J. 2261(59), *2312*
Lecea, B. 423(209), *430*
Leclercq, D. 2231(90a, 90b, 91), *2243*, 2251(26), 2253(26, 47a, 47b), 2255(26), *2311, 2312*, 2466(18), *2559*
Leclercq, H. 2556, 2557(420), *2568*
Lecomte, A. 2253(46), *2312*, 2333(97), *2358*
Ledbetter, F.E.III 2253, 2254(40a–c), *2311*
Ledoux, I. 2351(257, 258), 2354(257, 258, 281), *2362*
Lee, A.M. 517(44d, 44e), *553*
Lee, B.W. 401(140), *428*, 1589, 1590(78), *1596*, 1856(328), *1868*
Lee, C. 52, 54(95e), *98*, 2539(373a), *2567*
Lee, C.L. 2513, 2516, 2519(257a, 257b), 2526(257a), *2564*
Lee, C.S. 1763(265), *1791*
Lee, C.Y. 2550(408), *2568*
Lee, D. 1503(374), *1536*, 1680(120), *1685*
Lee, D.H.T. 2234(104c), *2243*
Lee, E. 1544(25), 1574(97), *1578, 1579*
Lee, F.L. (233), *262*
Lee, G.H. 1590(79), *1596*
Lee, H. 516(35c), *552*, 1812(113), *1864*
Lee, H.F. 2236(111), *2243*
Lee, H.U. 2522(304), *2565*
Lee, I.S. 1505(377), *1536*
Lee, I.Y. 1834(213), *1866*
Lee, J. 1116(83), *1138*
Lee, J.C. 1819(138), 1837(224), *1864, 1866*
Lee, J.G. 1669(20, 23), 1674(60), *1682, 1683*
Lee, J.H. 1491(325), *1535*, 1610, 1630(77), *1662*
Lee, J.-S. 844(56), *855*, 865, 866, 880, 881, 913, 922, 985, 986(38), *1054*, 1460, 1467(161), 1503(161, 375), *1529, 1536*, 1601, 1609, 1645(24), *1661*, 2404(10), *2458*
Lee, J.S. 449(124), *491*

Lee, K. 1828(189), *1865*
Lee, K.E. 241(267), *263*, 2069, 2071(77), *2124*, 2142(69), *2172*
Lee, K.-J. 485(256), *493*
Lee, M. 867(41, 42), *1055*
Lee, M.E. 846(67, 68), *855*, 874, 875(69), 876(70), 934(69), 990(273), *1055, 1060*, 1241(28), 1242(31), *1306*, 2083(144), 2085, 2090, 2094(159), *2125, 2126*, 2484, 2502(105), *2561*
Lee, M.Y. 1875(38), *1892*
Lee, S.B. 434, 444(22), *489*
Lee, S.K. 2107(217), *2127*
Lee, S.-L. 70(138), *100*, 149(60d), *152*
Lee, -S.-S. 1871(15), *1892*
Lee, S.S. 1493(330), 1503(330, 375), 1504(330), 1505(330, 377), *1535, 1536*
Lee, S.T. 1250(48–50), *1306*
Lee, S.W. 1877(44), *1892*
Lee, S.Y. 1752, 1753(241), *1790*, 2511(245), *2564*
Lee, T.J. 66(118c), 78, 79(155), *99, 100*, 2164(184), *2175*
Lee, T.V. 1822(151), *1864*
Lee, T.-Y. 1505(376), *1536*
Lee, W.K. 1872(17), *1892*
Lee, Y.A. 1460, 1467(159), 1503(159, 373, 374), *1529, 1536*
Lee, Y.-G. 1850(299, 300), *1867*
Lee, Y.-J. 1294(183), *1309*
Lee, Y.R. 1297(188), *1309*
Lee, Y.T. 530(94), *554*
Leeuwen, P.W.N.M.van 1758, 1759(254), *1790*
Lefevre, V. 1024, 1046(333), 1051, 1052(381), *1061, 1062*, 1878(51), *1893*
Leffers, W. 2134(25, 27), 2140(45, 48, 51), 2147(25), *2171, 2172*
Lefour, J.-M. 5, 6, 24, 25, 87(15b, 15c), *95*
Léger, M.-P. 1220(97), *1231*
Léger-Lambert, M.-P. 1217(84), 1222(102), 1223(103), 1225(109), *1231, 1232*
Legrand, C. 1993(103), *2033*
Legrand, P. 1471, 1472(227, 228), 1485(300), *1531, 1534*
Le Grow, G.E. 1514(407), *1537*, 1600, 1601(6), *1660*
Legrow, G.E. 2253, 2261, 2265(45a, 45c, 45d), *2312*
Legzdins, P. 487(264), *494*
Lehman, M.F. 1397, 1398, 1404(170), *1442*
Lehmann, R.G. 2230(81a), 2231(81b–d), *2242*
Lehner, E.A. 1771(299), *1791*
Lehnert, R. 2223(29a–d), *2241*
Lei, D. 1966, 1967, 1997–1999(19), *2030*, 2467(20), 2480(20, 83–85), 2489(20, 136), 2490(138), 2503(196), *2559, 2560, 2562, 2563*

Leichliter, R.P. 1590(86), *1596*
Leichtweis, I. 209(130), *260*
Leigh, G.J. 558(1), *589*
Leigh, W. 864(34, 35a, 35b), 932, 934(34), 969(249), 992(35a, 35b), 993, 996, 999, 1001(278), *1054, 1059, 1060*
Leigh, W.J. 849(80–84), 852(82–84, 87), *855*, 897(135–137), 898(139, 140), 932, 934(198), 969(136, 137, 140), 977(140), 994, 996(137), *1056–1058*, 1241(25b), 1248(46), 1253(59–62), 1254(59–61), 1256(62), 1303(207), *1306, 1307, 1310*, 1325(142–146), 1332(202, 203), *1335–1337*, 2499(186), *2563*
Leighton, J.L. 1672(48, 49, 51), *1683*
Leir, C.M. 2224(40), *2241*
Leis, A. 2368, 2371, 2374(38), *2399*
Leis, C. 581(181), *593*, 1386(112, 113), 1387(113), 1389(112, 113), *1440*, 2059(51, 52), 2060(52, 55), 2066, 2067(51, 52, 55), 2084(55, 156), 2085(160), 2089(55), 2090(156, 160), 2094(55, 160), *2123, 2126*, 2528(336), *2566*
Leising, G. 2197(80), *2215*
Leitão, M.L.P. 161, 162, 164, 171–173(64), *178*
Leiter, K. 1135(264), *1142*
Lejon, T. 793(47b, 51a), 794(47b), *823*
Lellouche, I. 1743(202), *1789*
Lemaire, M. 1189, 1191(7), *1229*
Lembke, F.R. 2145(92), *2172*
Lembke, R.R. 1179(195), *1185*
Lemenovskii, D.A. 2056(45), *2123*
Lemke, F.R. 1418(218b), *1444*, 2078(123), 2079, 2093(123, 124), *2125*
Lenhart, S.J. 2261(50), *2312*
Lenk, T.J. 2234(104c), *2243*
Lennartz, H.W. 1903(40b), *1926*
Lennon, J. 880, 881, 919, 922, 924, 928, 956(86), *1055*, 1146(21), *1181*, 1601, 1609, 1645(24), *1661*, 2404(9), *2457*
Lennon, J.M. 1638(171), *1664*
Lennon, R. 52–55, 57–59(98a), *98*, 279, 280(38), *351*, 1144(11), *1181*, 2023(195), *2036*, 2535–2539(367), *2567*
Le Noble, W.J. 642, 643(107), *700*
Le Nocher, A.-M. 1870(7), *1891*
Lenoir, D. 517(47), *553*, 616(53), 688(160a), *698, 701*
Lenth, W. 2351(247), *2362*
Lenthe, J.H.van 113(23), *118*
Lentzner, H.L. 2162(174), *2174*
Leo, A. 360, 361(11), *426*
León-Colón, G. 1837(223), *1866*
Leone, E.A. 2261, 2263(66), *2313*
Leonhardt, T.A. 2289(132), *2315*
Leopold, D.G. 704(10), *775*, 1010(298b),
1060, 1116(81), *1138*, 2556, 2557(425), *2568*
Leopold, E.J. 464(174), *492*
Leporterie, A. 37(77a), *97*
Lerner, H.-W. 873, 943(67), *1055*
Le Roux, C. 580(166), *593*
LeRoux, C. 401(141), *428*
Leroux, Y. 1601, 1624(22a), *1660*
Leroy, G. 159, 161, 162(51), *178*
Lesage, M. 1540(6), 1543(24), 1544(20), 1545(28–30), 1564(28), *1577, 1578*, 2526(329), *2566*
Le Saux, G. 2351(252), *2362*
Lesch, A. 1268(99), 1283(153), *1307, 1309*, 2486(110), 2495(170, 174), *2561, 2562*
Leshina, T.V. 2473(51), 2488(135), *2560, 2562*
Leslie, C.P. 409, 410(165), *429*
L'Espérance, D. 2350(227, 231), *2361*
Lestel, L. 2218(11a, 11d), *2240*
Leszczynski, J. 33(70), 90(169, 170e, 170f), 91, 92(169), *97, 101*, 551(115), *555*
Lett, R.M. 396(128), *428*
Letulle, M. 1016(313), 1022(332), 1023(313), 1025(332), *1061*
Leung-Toung, R. 782, 791, 795(11c), *822*
Lev, O. 2320(16), 2321(21, 23a, 23b, 24b), 2322(23a, 23b, 25, 29, 30, 32), 2326(44b, 49b), 2327(44b), 2328(16, 49b, 61–73), 2329(69–73), 2346(183b), 2347(44b, 183b), 2348(25, 44b), *2355–2357, 2360*
Le Vanda, C. 2159(156), 2160(172), 2162(172, 173), *2174*
Levick, A.P. 1107(12), *1136*
Levin, G. 2513, 2516, 2519(257a, 257b), 2526(257a), *2564*
Levin, V.Yu. 2236(110a, 110b), *2243*
Levroy, G. 1050(371), *1062*
Levy, A. 1989(93), *2033*
Levy, C.J. 2110, 2120(221), *2127*
Levy, D. 2318(1, 2), 2320, 2321(1), 2346(184), 2347(1, 184, 192, 193, 197), 2348(184, 192, 193, 199), 2349(192, 193), 2350(1, 2, 192, 193, 215, 217, 223–225), 2351(1, 248), 2352(223, 224, 272), 2353(274–276), *2355, 2360–2362*
Levy, G.C. 307(110, 111), *353*
Lévy, Y. 2352(266, 267), *2362*
Levy, Y. 2327(55), 2330, 2352, 2354(81), *2356, 2357*
Lew, C.S.Q. 649(123), 651(124), *700*, 1301(197), *1310*
Lewis, A. 2320, 2322(18, 19), 2327(19), 2340(132), 2348(18, 19), *2355, 2359*
Lewis, C.M. 1765(275), *1791*
Lewis, F.D. 1695(34), *1786*
Lewis, J.A. 2247(13), *2310*
Lewis, K.B. 2237(120), *2244*

Author index

Lewis, K.M. 1582, 1584(1), 1585(40, 41), 1589(1, 72), 1592(104, 105), *1594–1597*, 2533(355), *2567*
Lewis, L.N. 190(48), *258*, 1693(23, 28–30), 1694(30), 1717(104), *1786, 1787*
Lewis, M.H. 2261(57), *2312*
Lewis, N. 1693(23, 28), *1786*
Lewis, R.A. 992, 996(276), *1060*
Lewis, R.H. 2253(35), *2311*
Ley, S.V. 467, 474(187), *492*
Leyh-Nihaut, B. 1128(218), *1141*
Lheureux, M. 1963, 1996(4), 1998(119), 2019, 2026(4), *2030, 2034*
Lhoste, P. 1672(46), *1683*
Li, C.S. 1467, 1474, 1475(183), *1530*
Li, D. 622(69), *699*
Li, D.X. 2249, 2294(15), *2310*
Li, G.S. 813(105b), *825*
Li, H. 200(96), *259*, 2179(15), *2214*
Li, J. 1454, 1467(83), 1489(313, 314), 1492(83), *1527, 1534*, 1802(69), *1863*
Li, J.S. 1808(95), *1863*
Li, K.S. 1574(97), *1579*
Li, L. 1293(180), *1309*, 1328(154), *1335*, 2051, 2052, 2056, 2057, 2064(35), *2123*, 2354(280), *2362*
Li, N. 1782(339), *1792*
Li, S. 1470, 1483(210), *1531*
Li, X. 368(45), *426*, 560(40), 561(61), *590, 591*, 605, 606, 629(20), *698*
Li, X.-Y. 521(74, 77), 522, 529(77), 530(74), 532–534(77), *553*, 563, 567(81), 568(109), 569(81, 109, 119), 570(109), 571, 572(119), 578(161), 584(109, 188), 585(109), *591–593*, 652(129, 130), *700*, 1674(58), *1683*
Li, X.Y. 1408(182), *1443*
Li, Y. 1697, 1698, 1709, 1710, 1755(43), *1786*, 2352(271), *2362*, 2488(130), *2561*
Li, Z. 1671(36), *1683*, 1781, 1782(337), *1792*
Liable-Sands, L.M. 4, 37, 38, 40, 46(10a), *95*, 817, 819(112b), *825*, 981, 984(262), *1059*, 2029(210), *2036*
Lian-Fang Shen 1455, 1464, 1467, 1486(111), *1528*
Liang, C. 835(41), *854*, 1157(84), *1183*
Liang, F.-S. 1811(109), *1864*
Liang, G. 616(53), *698*
Liang, H.-C. 1872(22), *1892*
Liao, C.X. 1767(283), 1769(293), *1791*
Liao, R. 1454, 1467(83), 1489(313, 314), 1492(83), *1527, 1534*
Liaw, B.R. 1294(184), *1309*
Liaw, C.-F. 2272(92b), *2314*
Libbers, R. 219(181), *261*, 1373(75e), *1439*
Liberman, K. 2340(132), *2359*
Libert, L.I. 1454(96), 1458, 1460(125), 1467(96, 125), 1473(239), 1501(96), 1506,
1508(239), 1509(393, 394), 1510(239), *1527, 1528, 1532, 1537*
Licciardi, G.F. 160, 162, 171–173(55), *178*
Licciulli, A. 2340(138), *2359*
Licht, E. 1584(28, 34), 1586(45), 1588(28), *1595*
Lichtenhan, J.D. 205(114), *259*, 2233(97), *2243*
Lichtenwalter, G.D. 790(33), *822*
Lickiss, P. 2131(12), *2171*
Lickiss, P.D. 196(74), 198(82), 222(194), *258, 259, 261*, 454(136), 487(263), *491, 494*, 559, 563, 565, 567(13), 577(145–149, 151–153), *590, 592, 593*, 780(1a), 790(34), 795(34, 62c), 799(62c), 800(34), *821–823*, 827(5), 845(65, 66), *854, 855*, 880(90), 881, 913(90, 96), 914, 917, 924, 985, 986, 992, 1004(90), *1056*, 1273, 1276, 1285(109), *1308*, 1601, 1604, 1609(26), 1645(26, 201), *1661, 1664*
Lide, D.R. 65(113b), *99*
Liebau, F. 110, 111(15), *118*
Lieberman, K. 2320, 2322(18, 19), 2327(19), 2348(18, 19), *2355*
Liebhafsky, H.A. 1582(5), *1594*
Liebman, J.F. 120(1), *150*
Liehr, W. 565(97), *591*
Lien, M.H. 381, 382(78), *427*, 517(52), *553*, 1029(341), *1061*, 1126(205), *1140*
Liepins, E. 296(84, 85, 94), 299, 300(94), 301(94, 96), 302(100, 101), 304, 306(107), *352, 353*, 1373(75d), *1439*, 1454(98), 1458(124), 1467(98, 124), 1474(241, 245, 247), 1475(124, 247), 1479(124, 245), 1508(247), 1509(124, 247), 1521, 1523(418), *1527, 1528, 1532, 1537*
Liepins, E.E. 1362(54b), *1438*, 1449, 1460, 1470, 1481(6), 1505(378), 1506(6, 378), 1510(6), *1525, 1536*
Liepinsh, E. 1390, 1398, 1405(133), *1441*
Liepin'sh, E.E. 1459, 1467(133), 1470(207), 1474(240), 1475(250), 1479(275), 1506(207, 250), 1510(250), 1514, 1519, 1521, 1522(408a), *1529, 1531–1533, 1537*
Lieske, H. 1583(16), 1585(35), 1587(54, 58–60), 1589(74), *1595, 1596*
Lifshitz, C. 1132(243), *1141*
Lightfoot, P.D. 1123(163), *1139*
Lilla, G. 561(52), *590*
Lim, K.P. 560(44), *590*, 1120(119), 1124(169), *1138, 1139*
Lim, T.F. 2253, 2261, 2265(45a, 45c, 45d), *2312*
Lim, T.F.O. 878, 931, 945(79), *1055*, 1331(177), *1336*, 1973, 1974(45b, 45d), 1976(45d), *2031*, 2407(25), 2424(77), *2458, 2459*

Lim, Y. 1872(17), *1892*
Limberg, C. 1885(73), *1893*
Limburg, W. 1633, 1642(158), *1664*
Limburg, W.W. 1236(12), *1305*
Lin, C.-Y. 1108(21), *1136*
Lin, J. 770, 772(149), *778*, 1719, 1732(111), 1788, 1860(350), *1868*, 2340(135), *2359*, 2448(180), *2461*
Lin, J.-M. 1454, 1467(95), 1471(232), *1527, 1531*
Lin, L.C. 1294(184), *1309*, 1875(37), *1892*
Lin, L.T. 846(68), *855*
Lin, M.C. 2514(274), 2523(274, 309), *2565, 2566*
Lin, N.-H. 1816(130), *1864*
Lin, P.-Y. 1678(102), *1684*, 1877(44), *1892*
Lin, S.-H. 437, 485(78), *490*
Lin, W. 2076(116, 117), 2091(117), *2125*
Lin, X. 1454, 1467(83), 1489(313, 314), 1492(83), *1527, 1534*
Lin, Y. 1121(142), *1139*
Lin, Z. 2486(123), *2561*
Lind, R.C. 2351(254), *2362*
Lindeman, S.V. 2044, 2051(19), *2123*
Linden, A. 186(31), *257*, 1649(211), *1665*
Linden, J.G.M.van der 2152, 2153(123), *2173*
Linder, L. 727, 741(54), *776*, 2502(193), *2563*
Linderman, R.J. 434, 440(16), *489*, 1606, 1611, 1628(54), *1661*
Lindinger, W. 1118(96), *1138*
Lindner, E. 2339(128), *2359*
Lindsley, C. 1611(84), 1659(225), *1662, 1665*
Lindsley, C.W. 1669(18), *1682*
Lineberger, W.C. 704(10), *775*, 1116(79, 81, 82), *1137, 1138*
Lingelbach, P. 1153(60), *1182*
Lingelser, J.P. 2238(121), *2244*
Link, M. 873(63, 64), 881, 913(64), 937, 938(63, 201), 943, 969(63, 64), *1055, 1058*
Linoh, H. 2363(8), 2364(17–21, 23), 2365(20), 2367(20, 37), 2376(47), *2398, 2399*
Linstrumelle, G. 1601, 1624(22d), *1660*
Linti, G. 209(132), *260*
Linton, R.W. 2219(15a–c), 2225(46), *2240, 2241*
Lioa-Rong Shen 1455, 1464, 1467, 1486(111), *1528*
Lion, C. 1453, 1492(67), 1524(422), *1526, 1537*
Lipkowitz, K.B. 515(23), *552*
Lipowitz, J. 361(14), *426*, 1479, 1480(273), *1533*, 2253, 2261(45a–d), 2265(45a–d, 84a, 84b), 2285(117a–c), *2312, 2313, 2315*
Lippart, J.C. 2343, 2344(154), *2359*
Lippert, E. 1323(113, 115), *1335*
Lippert, E.Z. 1323(114), *1335*
Lippert, F. 2151(118), *2173*
Lippmaa, E. 296(92), *352*, 1475(255), *1532*, 1604(35), *1661*, 1897(20), *1926*
Lippmaa, E.T. 1381(107), *1440*, 1474, 1476, 1477(260, 263), 1478(263), 1510(260), *1532*
Lipshutz, B. 1659(225), *1665*
Lipshutz, B.H. 477(226), *493*, 784(21), *822*, 1611(84), *1662*, 1669(18), 1676(90), *1682, 1684*
Liptak, S.C. 2345(168), *2360*
Liptuga, N.I. 1500(360), *1536*
Lischka, H. 295(80), *352*, 705(13), *775*
Lisovski, A. 2053(38), *2123*
Lisowsky, R. 53(101), *98*, 1178(192), *1185*, 2474, 2534(55), *2560*
Liston, D.J. 197(79), *259*, 514, 520, 521, 524(3b), *551*, 567, 570, 572, 583, 586(103), *592*
Litchin, N. 563(72), *591*
Litscher, G. 1214(75), *1231*, 2179(10, 14), *2214*
Littman, D. 1329(160), *1336*
Littman, D.L. 2514(271), *2565*
Littmann, D. 981, 992(256), 1019(321), 1021(326), 1044(321, 326), 1045, 1047(321), *1059, 1061*, 1123(161), 1128(225), *1139, 1141*, 1156(72), 1167–1169(162), *1182, 1184*, 1289(169), *1309*
Liu, B. 690(162), *701*
Liu, C. 450(125), *491*, 2057(49), *2123*, 2527(331), *2566*
Liu, C.-S. 2465(6), *2559*
Liu, C.S. 2550(408), *2568*
Liu, H. 1669(17), *1682*
Liu, H.Q. 2043(17), *2122*, 2251(29a, 29b), *2311*
Liu, K.T. 677(152), *701*
Liu, Q. 1605(45), *1661*, 1828(193), *1865*
Liü, Q.-H. 2424(76), *2459*
Liu, R. 531(97), *554*
Liu, T. 70(134a), *100*
Liu, X. 367(44), *426*, 620–622(67), *699*
Liu, Y. 69(124b, 124c), *99*, 864, 981, 982(31, 32), *1054*
Livage, J. 2319(10a), 2326(45e), 2327(57), 2350(10a, 232), 2351(232, 261), *2355–2357, 2361, 2362*
Livant, P.D. 485(256), *493*
Livantsova, L.I. 1456, 1467(119), 1497(119, 351), 1501(119), *1528, 1535*
Llamazares, A. 2075(111–113), 2091(111, 112), *2125*
Llewellyn, G. 637, 639, 640(98), *699*
Lloyd, C.D. 422(207), *430*
Lo, D.H. 11(38), *96*, 183(6), *257*, 1149(34), *1182*
Lo, P.Y. 1694, 1695(31), 1725(131, 132), *1786, 1788*

Lobkovski, E.B. 240(256), *263*
Lobreyer, T. 780(2f, 4b), 781(7), 794(4b), *821*
Lochead, A.W. 1091(58e), *1103*
Lochrie, I.S.T. 1843(269), *1867*
Loewenstamm, W. 1415(196b), *1443*
Löffelholz, J. 2269, 2270(88a), *2313*
Loft, M.S. 435, 454(45), *489*
Loh, K.L. 1965(17), *2030*
Loh, T.-P. 469(190), *492*, 1804(78), 1856(334), *1863, 1868*
Lohde, A. 1887(90), *1894*
Lohray, B.B. 473(205), *492*
Lohrenz, J.C.W. 811(97b), *824*
London, F. 7(25), *96*
Long, L. 559(19), *590*
Long, T.E. 2345(173), *2360*
Loon, J.-D.van 186(37), *257*
Loos, M. 1878(48), *1892*
López, A.M. 1721, 1724(122), *1788*
Lopez, C. 423(209), *430*, 435, 456(54), *489*
López, C.J. 1845(273), *1867*
Lopez, R.C.G. 1091(58d), *1103*
Lopez, T. 2322(35), *2356*
López-Amo, M. 2350(215, 217), *2361*
Lopez-Mras, A. 226(204), 238(253), *262, 263*, 324, 325(142), 334(148, 149), 335(148), *353, 354*, 1341(28, 29), 1349(28, 29, 31b), 1357(28, 29, 31b, 46, 47, 49), 1358(28), 1359(28, 29, 46, 47), 1360(28, 46, 47, 49), 1361(28), 1362(28, 46, 47, 49), 1363(47), 1364(46, 49), 1366(29, 49), *1437, 1438*
Lopirev, V.A. 1390, 1392(136), *1441*
Lorberth, J. 1450(37), 1456(37, 119), 1457(37), 1461(37, 141), 1467(119, 141), 1496(37), 1497(119, 141), 1501(119), *1525, 1528, 1529*
Lorbeth, L.J. 226(202), *262*
Lorena, A. 2233(99), *2243*
Lorenz, A. 2330(76), *2357*
Lorenz, I.-P. 217(163), *260*
Lorenz, K. 1758(258, 259), *1791*
Loreto, M.A. 460(159), *491*, 1843(266, 267), *1867*
Lorquet, J.C. 1128(218), *1141*
Lorz, P. 2145, 2146(93), *2172*
Losada, J. 1758(257), *1791*
Lottes, H.C. 757(121), *777*
Lotts, K.D. 739, 740(73), *776*
Lou, X.-L. 2065, 2066(73), *2124*
Lough, A. 2154(137), *2174*
Lough, A.J. 219(175), *261*, 578(159), *593*, 795–798(57), *823*, 880(93, 94), 881(93–95), 882(95), 883(100), 885, 890(116), 914(93, 94), 915, 924, 932(95), 937, 938(116), 951(239), 954, 956(95), 962, 963(244), 964(244, 245), 965(245), 966(244, 245), 967(94), 978(95), 985, 986(93, 95), *1056*, *1059*, 1272, 1273(108), 1274(112, 114, 115, 118, 120), 1276(112, 123, 124), *1308*, 1881, 1887(62), *1893*, 2111, 2120, 2121(225), *2127*, 2150(110), 2153(110, 134), 2154(110, 134, 138), 2162(176), *2173, 2174*
Loustau Cazalet, C. 1547(37), *1578*
Louwrier, S. 1835(216), *1866*
Low, M.J.D. 2530(341–343), *2566*
Lowe, J.A. 709, 714(20), *775*, 901, 917, 981, 992(157), *1057*
Lowery, D. 1814(118), *1864*
Lowinger, T.B. 1565(79), *1579*
Lown, E.M. 2526, 2527(327), *2566*
Loy, D.A. 1341, 1347, 1367(25), 1369(25, 58a), *1437, 1439*, 2346(180, 181, 182b), *2360*
Loza, M. 2099, 2108(192), *2126*
Loza, M.L. 2070, 2071(82), *2124*
Lu, J. 1767(284), *1791*
Lu, K. 1454(99, 101), 1456(116, 117), 1463, 1464(167), 1467(99, 101, 116, 117, 167), 1470(211), 1481, 1483(286), 1500, 1501(99), 1506(286), *1528, 1530, 1531, 1533*
Lu, L. 1828(193), *1865*
Lu, P.P. 2285(117a), *2315*
Lu, R.S. 2069, 2071(79), *2124*
Lu, S.-P. 1640(185), *1664*
Lu, X. 1606(52b), *1661*, 1767(279), *1791*
Lu, Z.-R. 1464, 1467, 1486(168), *1530*
Lucarini, M. 1542(10, 16), 1544, 1546(16), 1549(41a), 1559, 1570(16), *1577, 1578*
Lucas, C.R. (233), *262*
Lucas, D.J. 164(80), *179*, 1126(191), *1140*
Lucas, R. 2523(320), *2566*
Luckenbach, R. 502(30), *511*
Lüder, W. 1323(113, 115), *1335*
Luderer, C. 884, 885, 889, 925–927(109), *1056*
Luderer, F. 884, 885, 925, 937, 938(105), *1056*
Ludvig, M.M. 820(119), *825*
Luh, T.-Y. 1794(26), 1796, 1798, 1818, 1820(40), 1835(218), 1847(281), 1854(26, 324), *1862, 1866–1868*
Lukacs, A. 2261(52), *2312*
Lukasiak, J. 1454, 1467(97), 1482(295), 1484(295–297), *1527, 1533, 1534*
Luke, B.T. 171(128), *180*, 250(319), *265*, 532(99a), *554*, 705, 707(12), *775*, 781(5), *821*, 835(37), *854*, 901(147), *1057*, 1107(9), 1123(156), *1136, 1139*, 1166(149, 157), *1184*
Lukevics, E. 296(84, 85, 94), 299, 300(94), 301(94, 96), 302(100, 101, 103, 104), 303(104), 304(103, 104, 107), 305(104), 306(107), *352, 353*, 485(252), *493*, 780(1e), *821*, 1199(31), 1200(32), *1230*, 1373(75d), 1434(233), *1439, 1444*, 1449(7a, 7b, 29),

Lukevics, E. (*cont.*) 1451(48a, 48b, 49–51), 1452, 1453(52), 1455(50), 1458(52, 124, 125), 1459(135), 1460(7a, 7b, 29, 125), 1461(29, 49), 1462, 1463, 1465(29), 1467(124, 125, 135), 1473(239), 1474(241, 245, 247), 1475(124, 247), 1476(261), 1479(124, 245), 1480(281), 1481(281, 289–291), 1483(289–291), 1487(49), 1492(326), 1494(338), 1496(342, 343), 1501(366), 1505(326, 378), 1506(239, 261, 326, 378–380), 1508(239, 247, 390), 1509(124, 247, 393, 394), 1510(239, 261, 396), 1514(291), 1521(417, 418), 1523(291, 418), 1524(421), *1525, 1526, 1528, 1529, 1532, 1533, 1535–1537*, 1717(105), 1727, 1730(134), 1740(171, 172), *1787–1789*
Lukevics, E.J. 1362(54b), *1438*, 1454, 1467(96, 98), 1501(96), *1527*
Lukevics, E.Ya. 1449(4a–c), *1525*
Lukevits, E. 1459(133), 1467(133, 185), 1470(207), 1474(240), 1475(250), 1478(270), 1479(270, 272, 275), 1506(207, 250), 1507(382), 1510(250), 1514(408a, 408b), 1519, 1521(408a), 1522(408a, 408b), *1529–1533, 1536, 1537*
Lukevits, E.J. 1501(365), *1536*
Lukevits, E.Ya. 1481(284), *1533*
Lukina, Yu.A. 1454(72, 74), 1461, 1467(145), 1473(235, 238), 1474(235), 1476(238), 1492(327), 1500(72, 74, 327, 358–360), 1501(327), *1527, 1529, 1532, 1535, 1536*
Lümen, R. 578(163), *593*, 2487(126), *2561*
Lunazzi, L. 1602(29), *1661*
Luo, B.-S. 1464, 1467, 1486(168), *1530*
Luo, X. 1501(367), *1536*, 2065, 2066(74), *2124*
Luo, Y. 1481, 1483, 1506(286), *1533*
Luo, Y.-R. 157(37–45), 158(39, 42–44), 159(44), 160(41, 44), 161(44), *178*
Lusztyk, J. 1542(14), *1577*, 1875(32), *1892*, 2165(186), *2175*
Lutsenko, I.F. 1501(363), *1536*
Luttke, W. 304(106), *353*
Luzin, A.P. 1475, 1477(251), *1532*
Lycka, A. 1237(16), *1305*
Lynch, M.-J. 462(168), *492*
Lyons, J.E. 1571(91), *1579*

Ma, D. 1679(107), *1684*
Ma, E.C.-L. 1965(16c), *2030*
Ma, R. 559(19), *590*
Ma, Z. 770, 772(149), *778*, 1303(208), *1310*, 1860(350), *1868*
Maas, G. 222(193), *261*, 434(28), 435(42), 447(28), 452(42), *489*, 704(4), 711(28b), 713(32), 718(28b), 719(32, 43), 720(44), 730(62), 732(63), 735(66–68), 750(32), 751(103), 752(32, 103–109, 114), 753(114), 754(32), 755, 756(4, 114), 757(118, 120, 122), 758(122), 760(103, 120), 765(141), *774–778*, 906(166, 167), 931(195, 196), *1057, 1058*, 1276(128, 130, 131), *1308*
Maas, W.P.M. 1135(265), *1142*
Maass, G. 1181(199), *1185*
Mabuchi, T. 2347, 2350(188), *2360*
Macciantelli, D. 1558(64), *1578*
MacCraith, B.D. 2322(26), *2356*
MacDiarmid, A.G. 1159(94), *1183*
Macdonald, T.L. 1622(127), *1663*
MacDougall, P.J. 538(102b), *554*
Macharashvili, A.A. 226(207), 243(270–274, 277), 245(286), *262–264*, 500, 501(26), *510*, 587(195), *594*, 1374, 1376(92b), 1392(143), 1393, 1394(151), 1395(156, 157), 1396(151, 156–162, 164, 165), 1397(156, 161), 1399(156, 160, 161, 164, 165), *1440–1442*, 1462, 1467(151), 1514, 1522(408b), *1529, 1537*
Machiguchi, Y. 1675(73), *1683*
Machinek, R. 304(106), *353*
Machnik, D. 1778(331), *1792*
Maciejewski, H. 479(231), *493*, 1700(56), *1786*
Maciel, G.E. 310(118, 131), 311, 312(135), 337, 338(150), *353, 354*, 1474–1477(254), *1532*, 2253(35, 41), 2254, 2257(41), *2311, 2312*
MacInnes, I. 1644(197), *1664*
Mackay, K.M. 2107(216–218), *2127*
Mackenzie, J.D. 290–292(74), *352*, 2289(128), *2315*, 2343(150), 2345(167), 2347(195a), 2350, 2351(222), *2359–2361*
MacKiernan, J.M. 2350(220), *2361*
MacMahon, T.B. 1110(61b), *1137*
MacMillan, A. 939(202), *1058*
Macomber, D.W. 2137(34), *2171*
MacRae, D.M. 1236(12), *1305*
Maddock, S.M. 2093(185), *2126*
Maddox, M.L. 1843(260), *1867*
Made, A.W.van der 1758, 1759(254), *1790*
Maeda, K. 1797(46), *1862*
Maeda, M. 2448(181), *2461*
Maeda, S. 798(60b), *823*
Maekawa, T. 1201, 1202(36), 1203(39), *1230*, 1617, 1620, 1624(105), *1662*
Maercker, A. 1991(98), *2033*, 2481(90), *2561*
Maerker, C. 7, 9, 42, 44, 45, 47, 69, 90(26), 96, 515(6), 516(35a), 518, 520, 522, 526–529, 532, 533, 535, 537, 542(6), 543(6, 110), 544, 545, 547(110), 551(114), *551, 552, 555*, 558, 559(4), 570(125), 573(4, 125), 577, 584, 588(4), *589, 592*, 652(131), 696(170), *700, 701*
Magi, M. 296(92), 334(147), *352, 353*,

1604(35), *1661*, 1897(20), *1926*
Magnette, J.E. 1881(60), *1893*
Magnus, F. 1497(348), *1535*
Magnus, P. 422(204, 205), *429, 430*, 1673(55), *1683*
Magnus, P.D. 780(1j), *821*, 1631, 1632(150), *1663*, 1794(10), *1862*
Magnusson, E. 1465, 1467(172), *1530*
Magrini, K.A. 1588(67), *1596*
Magriotis, P.A. 454(135), *491*
Mague, J.T. 2113, 2120(228), *2127*
Magull, J. 1878(48, 49), 1886(84, 85), *1892, 1893*
Mah, T.-I. 2261(67a, 67b), *2313*
Mahaffy, P.G. 982(263), *1059*
Mahayan, M.P. 2010(163), *2035*
Mahmood, K. 448(118), *491*
Mahmoud, F.M.S. 563(84), *591*
Mahmoud, S. 1970(28), *2031*
Mahner, K. 2364(26), *2398*
Mahon, M.F. 1925(109), *1928*
Maichle-Mossmer, C. 254(337), *265*
Maier, G. 3(2d), 10(33b–d, 34–36), 12(44), 17, 18, 20(36), 66(117a, 117b), 67(117b), 68(117a, 117b), *95, 96, 99*, 128(35), *151*, 648(122), *700*, 747, 748(93–95), 757(119), *776, 777*, 864(28–30), 901(29, 158), 977(28), 981(29, 158, 256), 985(29), 992(28–30, 158, 256), 994(282), 995(29, 279–282), 996(28–30), 997, 998(285b), 1021(328–330), 1045, 1046, 1049(328), *1054, 1057, 1059–1061*, 1123(159–161), 1125(174, 175), 1128(224), *1139, 1141*, 1145(12, 17, 18), 1146(20), 1149(35, 36), 1150(37, 38, 41), 1151(46, 48), 1152(50), 1153(57, 59), 1156(73a, 73b), 1158(18, 73a, 73b), 1159(12, 18, 96), 1160(12), 1162(107), 1163(127–129), 1164(127–129, 144), 1165(128, 147), 1167, 1168(162), 1169(107, 162), 1170(107, 180), 1171, 1172(182), 1175(73a, 73b, 144), 1178(96), 1179(73a, 73b, 144), 1180(107), *1181–1185*, 1189(5), *1229*, 1286(162), *1309*, 1329(160), 1331(197), *1336*, 2453(195), *2461*, 2495(168), 2514(168, 269–272), *2562, 2565*
Maillard, B. 1644(197), *1664*, 1875(32), *1892*, 2387(64, 68), 2388(64), *2400*
Mainz, D.T. 2510(232), *2564*
Mainz, V.V. 1923(99), *1928*
Maio, F. 1463, 1464, 1467(167), *1530*
Mairanovski, S.G. 1220(96), *1231*
Maitlis, P. 1705(74, 75), 1707, 1709(74), *1787*
Maitlis, P.M. 1707(83), *1787*
Majetich, , G. 1794(17, 18), *1862*
Majetich, G. 406(155), *429*, 1803(72), 1814(116, 118), *1863, 1864*

Majoral, J.-P. 2261(64), *2312*
Mak, C.C. 436, 479(62), *489*
Makarov, V.M. 2041(12), *2122*
Makarova, L.I. 2235(109), 2236(110a, 110b), *2243*
Makarova, N.N. 254(335), *265*
Makhija, A.V. 69(130), *100*
Makino, S. 1771(304), *1792*
Makino, T. 1656(223), *1665*
Makioka, Y. 1641(189), 1651(215), *1664, 1665*
Makishima, A. 2321, 2347(20), *2355*
Makita, K. 2349(214), 2350(230), *2361*
Malacria, M. 452(129), 475(213), *491, 493*, 1554(56), 1566(81a, 81b, 82), 1567(82), *1578, 1579*, 1680(118), *1684*
Malany, S. 622(70), *699*
Malar, E.J.P. 88, 89(167), *101*
Malek, J.R. 1592(101), *1597*
Malhotra, B.D. 2326(47), *2356*
Malik, K.M.A. 1897(18), *1926*
Malik, S. 448(118), *491*
Malisch, W. 2061(58–60), 2062(61, 62), 2066(59, 60), 2067(60), 2068(58–60), 2078, 2091(62, 121, 122), 2095(121, 122), *2123, 2125*, 2142(63), 2145, 2146(93), *2172*
Malisza, K.L. 580(168), *593*
Malkin, V.G. 5, 7, 22–24, 31, 32(19), *95*, 516(31), 517(44c), *552, 553*, 658(134a, 134b), *700*
Malkina, O. 5, 7, 22–24, 31, 32(19), *95*, 655(133), 658(134a, 134b), *700*
Malkina, O.L. 516(31), 517(44c), *552, 553*
Malkova, T.I. 1454, 1467(81), *1527*
Mallela, S.P. 198(83), *259*
Maloisel, J.-L. 1872(16), *1892*
Malony, S. 367(43), *426*
Malrieu, J.-P. 29(66c, 66e), 63(112c), *97, 99*, 835(39, 40), *854*, 984(264), *1059*
Maltsev, A.K. 981, 992, 995(255), *1059*, 1069, 1070(19), 1071(23), 1072(19, 23), 1083(23), *1102*, 1161(104, 105), *1183*
Malusare, M.G. 1669(26), *1682*
Mamedov, D.H. 2224(42b), *2241*
Mamzano, B.R. 1720, 1732(117), *1788*
Man, M.W.C. 541(106b), *554*, 581(179), *593*
Mancini, D. 2226(54), *2242*
Mancuso, A.J. 1606, 1611(47), *1661*
Mandai, T. 1608(66), *1662*
Mandich, M.L. 560(31–33, 36, 37), *590*, 1119(100–105, 110), *1138*
Mandt, J. 1910(59), 1913, 1914(79), *1927*
Manju, K. 1801(66), *1863*
Mann, A. 402(146), *428*, 1803(75), 1816, 1819(128), *1863, 1864*
Mann, B.E. 533(118), *555*, 1707(83), *1787*
Mann, I.S. 456(145), *491*
Mann, K.R. 2016(174), *2035*

Manners, I. 2111, 2120, 2121(225), *2127*, 2150(110), 2153(110, 125, 134), 2154(110, 134, 137, 138), 2159(163), 2160(166, 167), 2162(125, 163, 176), *2173, 2174*
Manning, J. 283(46), *351*, 529, 530(91), *554*, 567, 569, 570, 583, 585, 587, 588(106), *592*
Manolopoulos, D.E. 70, 71(139c), *100*
Mansmann, M. 2253(39b), *2311*
Manson, G.A. 1124(168), *1139*
Mantell, G. 1915(83), *1927*
Mantione, R. 1601, 1624(22a), *1660*
Mantz, R.A. 2233(97), *2243*
Manue, G. 37(77a), *97*
Manuel, G. 1005, 1006(294), *1060*, 1069, 1070(20, 21), *1102*, 1963, 1965(3), 1968(21, 22, 23a, 23b), 1971(30), 1973(23b), 1976(53), 1980(3), 1981(21, 65, 67a–c), 1982(22, 67a–c, 69, 70), 1983(22, 67c, 72), 1984(22), 1996(3, 21, 67a–c, 107), 1998(3, 67a, 67c, 118, 123), 1999(21, 67c, 107, 123), 2001(123), 2011(167), *2030–2035*
Manuel, T.A. 1971(31), *2031*
Manz, B. 752(109), *777*
Manzanero, A. 2144, 2150, 2151(84), *2172*
Manzocchi, A. 2388(71), *2400*
Mao, G. 1771(299), *1791*
Mao, S.S.H. 2434(122, 123), *2460*
Mao, S.S.M. 2147(99), *2173*
Mao, Z. 1735(149), *1788*
Maquet, J. 2294(142), *2315*, 2354(285), *2362*
Marais, C. 222(186), *261*
Marchese, G. 1819(141), *1864*
Marchetti, P.S. 2253(35, 41), 2254, 2257(41), 2264(77), *2311–2313*
Marcinec, B. 1688(6), *1785*
Marciniec, B. 479(231), *493*, 1700(56), 1735(150), *1786, 1788*, 1840(251), *1866*
Marco, J.A. 437, 438(84), *490*
Marcus, L. 207(129), *260*, 1011, 1043, 1044, 1048(307), *1061*
Marcus, M. 2223(24), *2241*
Marczak, S. 424(213), *430*
Mardanov, M.A. 1996(108c), *2033*
Mardones, M.A. 217(170), *261*
Mareda, J. 2525(324), *2566*
Marek, M. 2547(398), *2568*
Marel, G.A.van der 475(216), *493*, 1681(124), *1685*
Marenkova, L.I. 154, 156(16, 17a), 166(16), *177, 178*
Margaillan, A. 457(148), *491*
Margaretha, P. 772(152), *778*
Margaria, T. 1584, 1588(28), *1595*
Margolin, Z. 849(77, 78), *855*
Margrave, J.L. 1127(209), *1140*, 1162(111), 1167(161), 1175(161, 188), 1176(188), 1180(196), 1181(198–200), *1183–1185*, 2469(24), 2518(262), *2559, 2565*
Maria, P.-C. 559(22), 560(39), *590*, 1120(114, 115), *1138*
Mariano, P.S. 396(130), *428*, 473(208), 474(209), 492, 1193(15), *1230*
Maricq, M.M. 310(120, 132), *353*
Marie, P. 2238(123), *2244*
Marin, G.B. 2548(399), *2568*
Marinetti-Mignani, A. 2001(134), *2034*
Maringgele, W. 2486(118), *2561*
Marino, P.S. 1294(183), *1309*
Marion, P. 1870(8), *1891*
Mark, J.E. 2228(75), *2242*, 2342(145), 2345(172), *2359, 2360*
Mark, T.D. 1135(264), *1142*
Märkel, G. 3(9), *95*, 2407(21), *2458*
Märkl, G. 724(51), 725(52), *776*, 908(170), 961, 964, 966(243), 987, 990(274), *1057, 1059, 1060*, 1152(51, 52), 1153(52), *1182*, 1282(139, 140), *1308*, 2413(39), *2458*
Markle, R.A. 2246, 2250, 2253, 2254, 2273, 2285(6c), *2310*
Markó, I.E. 1807(84–89), *1863*
Marks, T.B. 1697, 1698, 1709, 1710, 1755(43), *1786*
Marks, T.J. 487(265), *494*, 2039–2041(5), 2053(37), *2122, 2123*, 2153(126), *2173*
Markus, M.W. 2521(282), *2565*
Maroldo, S.G. 2005(154), *2034*
Maroshina, M.Yu. 1875(36), *1892*
Marquand, C. 734(65), *776*, 1165(146), *1184*, 1277(133), *1308*
Marquardt, G. 860, 917(15), *1054*, 2406(17), *2458*
Márquez, A. 2116(242, 243), *2127*
Marquez, V.E. 1550(47), *1578*
Marsch, M. 520(67), *553*
Marshall, P. 167(89–91, 93, 99), 168(89, 90, 93), 169(115), *179, 180*, 1047(362), *1062*, 1125(188), *1140*, 1169, 1177(172), *1185*, 2471(38), 2514, 2515(275), 2519(267), *2559, 2565*
Marshall, W.J. 52(95c), *98*
Marsmann, H. 804(78), *824*, 1268(98), 1283(152), 1304(213), *1307, 1309*, 1310, 1356, 1357, 1422, 1434(40), *1438*, 1881(61), *1893*, 2184(45), 2185(46, 47), 2214, 2472(48, 49), 2473(50), 2494(165), 2495(172–174), 2514(172), *2560, 2562*
Marsmann, H.C. 268(3), *350*, 780, 794(2d), *821*, 1474, 1475(256), *1532*, 1923(92), *1928*, 2334(107), *2358*
Martelli, G. 435, 456(55), 459(157), *489, 491*
Martin, A. 2151, 2152(117), *2173*
Martin, G.F. 1390, 1392(126, 127), *1441*
Martin, G.I. 1467, 1474, 1475(183), *1530*
Martin, J.C. 284(57), 287(63), *351, 352*,

505(39), *511*, 564(92), *591*, 1353(35a, 35b), 1357, 1358(35b), 1410, 1412(187), *1438, 1443*
Martin, J.G. 171(119), *180*, 2465(12), *2559*
Martin, M. 1903(40a), *1926*
Martin, R. 193(60), *258*
Martineau, M. 507(44, 45), 509(62), *511*, 1215(78–80), *1231*
Martineau, P.M. 559(27), *590*
Martínez, L.E. 1672(48, 53), *1683*
Martinez Urreaga, J. 2233(99), *2243*
Martinho Simões, J.A. 168(107, 108), *179*
Martino, R.L. 515(20), *552*
Martins, J.L. 115, 116(37), *118*
Marturunkalkul, S. 2354(280), *2362*
Maruca, R. 2002(140), *2034*
Maruki, I. 240(259), *263*, 829(27), 830, 831(28), 832(27, 28), 835(27), 836, 839–842(28), *854*, 1244(33), *1306*
Marumo, K. 1676(88), *1684*
Marumoto, S. 435, 453(44), *489*, 1859(344, 345), *1868*
Maruoka, K. 442(108), *490*
Maruoko, K. 1619(114), *1663*
Maruyama, S. 1120(123), *1138*
Maruyama, T. 248(303), *264*, 348(163), *354*, 796, 802(76), *824*, 847(75), 849(79), 850(75), *855*, 898(143), *1057*, 1157(81), 1169(173, 174), 1176(173), *1183, 1185*, 1248, 1256(44), 1283(157), *1306, 1309*, 1329(166–169), 1330(166, 168, 169), 1331(166, 167), *1336*, 2471(37), 2514, 2515(37, 265), 2517(37), 2518(265), *2559, 2565*
Marwin, L.H. 2164(183), *2175*
Maryasova, V.I. 2473(51), *2560*
Marzabadi, M.R. 1823(152), *1864*
Marzien, J. 2235(107), *2243*
Masaki, H. 1849(294), *1867*
Masaki, Y. 1871(11), *1891*
Masamune, S. 123(23, 26a, 26b), 125, 130(23), 131(38), *151*, 248(298), *264*, 269, 273, 275(12), 348(161), *350, 354*, 461(164), *492*, 812(99a), *824*, 827, 828(11), 829(21), *854*, 1064(4g), *1101*, 1153(56), 1157(82), *1182, 1183*, 2182(34, 36), 2183, 2205(43), 2207(95, 97), *2214–2216*, 2416(45), *2458*, 2498(185a, 185b), *2563*
Masaoka, S. 807, 808(87a–c), *824*, 888(120, 121, 123), 889(121), 958(120), 986, 988, 990(121), *1056*
Masaoka, Y. 807, 808(87a, 87c), *824*, 883(98), 884, 885(112, 113), 888(112, 113, 120, 121), 889(121), 922–924, 928(112), 937, 938(113), 957(112), 958(120), 986, 988, 990(121), *1056*, 2429(99), *2459*
Mascareñas, J.L. 1855(326), *1868*

Maschio, R.D. 2264(79), *2313*
Masiero, S. 1638(174), 1658(224), *1664, 1665*
Maslen, E.N. 113(28), *118*
Masnuk, M. 1835(214), *1866*
Masnyk, M. 424(213), *430*
Maso, G.N. 245(278), *264*
Mason, B.P. 171(117), 172(117, 133), 175(140), *180*, 901(149), *1057*, 2479(80), 2511(244), 2525(325), *2560, 2564, 2566*
Mason, R. 790(39), *822*
Mass, G. 2407(27, 28), *2458*
Massa, W. 215(155), 226(202), 254(337), *260, 262, 265*, 1461, 1467, 1497(141), *1529*, 2159(159), *2174*, 2429(93), *2459*
Masse, C.E. 410(166), *429*, 1794(30), *1862*
Masse, J. 497(14), *510*, 1612(87), *1662*
Massey, A.G. 1988(85a–c), 1989(87), 2004(85c), *2032, 2033*
Masterson, M. 2226(54), *2242*
Mastryukov, V.S. 2181(21), *2214*
Masuda, H. 1189(8), *1229*, 1638(168), *1664*
Masuda, T. 1861(361), *1868*
Masummarra, G. 8(29), *96*
Masumoto, Y. 143(52a, 52b), *152*
Masure, M. 2220(18, 19a–c), *2240, 2241*
Mata-Mata, J.L. 2083(147), *2126*
Matano, Y. 1828(185), *1865*
Matejec, V. 2322(27), *2356*
Matern, E. 241(261, 262), *263*
Mathey, F. 1963(2a, 2c), 1996(109), 1997(114b), 1998(2a, 121), *2030, 2033, 2034*
Mathias, L.J. 1765(275), *1791*
Mathur, R. 283(46), *351*, 529, 530(91), *554*, 567, 569, 570, 583, 585, 587, 588(106), *592*
Matías, I. 2350(215, 217), *2361*
Matias, M.C. 2233(99), *2243*
Matsubara, S. 1197(26), *1230*
Matsubara, T. 1075(30), *1102*
Matsuda, A. 2546(391, 392), *2568*
Matsuda, I. 1774(320), 1777(329, 330), 1778(332–334), 1779(334), 1782(339), *1792*
Matsuda, M. 1254(63), *1307*, 2469(31, 32), *2559*
Matsuda, T. 1825(163–166), 1838(231), *1865, 1866*
Matsuguchi, A. 2017(177c), *2035*
Matsuhashi, H. 1755, 1756(247), *1790*, 1826(174), *1865*
Matsuhashi, Y. 1090(56), 1093(66a), *1102, 1103*, 1884(70), *1893*, 1904, 1907(48, 49), *1926*
Matsui, H. 1125(186), *1140*
Matsui, K. 998, 1001(288), *1060*, 2347(186b, 187, 189), *2360*
Matsui, M. 115(34), *118*
Matsui, S. 883, 960–962(99), *1056*

Matsui, Y. 2521(281), *2565*
Matsukawa, S. 1856(332), *1868*
Matsumi, Y. 2523(312), *2566*
Matsumoto, H. 120(3), 137, 139, 140(45), 142(3), 143(3, 45, 52a–d), 144(45), 145(53–55), 146(3, 45, 55), *150–152*, 487(270), *494*, 1157(75), *1182*, 1211(66), *1231*, 1268(95), 1304(209, 211, 212), *1307, 1310*, 1918(87), 1924(107), 1925(110), *1927, 1928*, 2188(55), 2206(93, 94), 2207(93, 96), 2208(98), *2215, 2216*, 2472(46), *2559*
Matsumoto, K. 1601, 1602, 1620, 1651(12), 1652(12, 219), *1660, 1665*, 1810, 1830(105, 106), *1864*
Matsumoto, M. 137, 142(44c), *151*, 200(100), 259, 1849(294), *1867*
Matsumoto, N. 137(44a–c), 138, 140(44b), 142(44a–c), 143(44a, 51), 146(44b), 147(44a), *151, 152*, 1318(52, 54, 57), *1333*
Matsumoto, S. 1302(205), *1310*, 1973, 1974(44), *2031*, 2349(213), *2361*
Matsumoto, T. 143(52a), *152*, 1090(56), 1093(65, 68a, 68b), *1102, 1103*, 1884(70), *1893*, 1904, 1907(48–50), *1926*
Matsumoto, Y. 1755, 1756(246), *1790*
Matsumura, C. 2523(315), *2566*
Matsumura, Y. 1221(98), *1231*
Matsunaga, N. 21(53a, 53b), 22, 23(53b), 24(53a), 29(53b), 30(53a, 53b), *96*, 123(21e), *151*
Matsunaga, S. 1197(25), 1201(34, 36), 1202(25, 36, 38), 1203(25, 41), *1230*, 1617, 1620, 1624(105), 1650(213), *1662, 1665*
Matsuo, Y. 194(67), *258*
Matsuoka, M. 2351(245), *2362*
Matsutani, T. 1636(166), *1664*
Matsuura, S. 1749(228), *1790*
Matsuyama, M. 2002, 2009(144), *2034*
Matsuzak, Y. 1304(216), *1310*
Matsuzaki, Y. 2015, 2016(173), *2035*
Matsuzawa, S. 917, 1001(185), *1058*, 1973(48b, 48c), *2031*
Matsuzawa, T. 2272(92b, 92c), *2314*
Matsuzuka, T. 2347(187), *2360*
Mattern, G. 215(159), 226(204), 243(275, 276), 245(281), *260, 262, 264*, 334(149), *354*, 1349(31a), 1357(31a, 46, 48), 1359, 1360(46), 1362(46, 48), 1364(46), 1374, 1376, 1377(89–91), 1380, 1381(89, 91), 1401(89–91), *1438, 1440*, 2368, 2369, 2374(39), *2399*
Mattes, S.L. 437, 486(80), *490*
Matthews, E.W. 196(78), *259*
Matthews, L.R. 2348(200), *2360*
Matthies, D. 1854(322), *1868*
Mattii, D. 1803(75), *1863*
Matur, D. 1134(257), *1141*

Matveichev, P.M. 1967, 1986, 1996, 1997(20a), *2030*, 2504, 2515(197), *2563*
Matyjaszewski, K. 798(59), 805(81), *823, 824*, 2187(49–53), 2189(61), *2215*
Matyjazewski, K. 812(101b), *824*
Matz, J.R. 456(142), *491*, 1614(98), *1662*
Matzner, E. 648(121), *700*
Mauge, J.T. 2433(118), *2460*
Mauriello, G. 1877(41), *1892*
Maurino, V. 2231(87), *2243*
Mauris, R.J. 1600, 1611, 1612(4), *1660*
Mautner, K. 1679(105), *1684*
Maxka, J. 63(111), *99*, 833, 835(35), *854*, 1313, 1317, 1320(45), *1333*, 2166(192), *2175*, 2497, 2499(182), 2511, 2516(247d), *2563, 2564*
Maxma, J. 1288(165), *1309*
Maxwell, R. 311, 313(136), *353*
Maya, L. 2253(38), *2311*
Mayama, S. 2522(287), *2565*
Mayer, B. 2001(133), *2034*, 2073(101), 2121(257), *2124, 2128*, 2142(70), *2172*
Mayer, D. 704, 755, 756(4), 757(118), *774, 777*
Mayer, H.A. 2339(128), *2359*
Mayer, J. 2288(122), *2315*
Mayer, J.M. 1830(198), *1865*
Mayer, S. 765(141), *778*
Mayer, T.M. 607(25, 26), 608(26), *698*
Mayer-Posner, F.J. 1758(258), *1791*
Maynard, G.D. 1628, 1630(145), *1663*
Maynard, T.L. 1877(42), *1892*
Mayr, H. 397(134), 399(136, 137), 403(147), *428*, 441(101), *490*, 576(138, 139), *592*, 624(74), 625(74–76), 626(76, 77), 627(76), *699*, 1798(50), *1863*
Mazeika, I. 1481, 1483(289–291), 1506(380), 1514, 1523(291), *1533, 1536*
Mazeika, I.B. 1449, 1460, 1470, 1481(6), 1505(378), 1506(6, 378), 1510(6), *1525, 1536*
Mazeiku, I.B. 1453(65), *1526*
Mazerolles, P. 497(13), *510*, 847(71), *855*, 1005, 1006(294), *1060*, 1965, 1967(10a, 10b), 1968(21), 1981(10a, 10b, 21, 67a), 1982(67a), 1996(21, 67a), 1997(112b, 112c), 1998(10a, 10b, 67a), 1999(10a, 10b, 21), 2007(158b), *2030, 2032, 2033, 2035*
Mazhar, M. 505(35), *511*, 1383(110), *1440*
Mazheika, I.B. 1470(207), 1481(284), 1506(207), 1514, 1519, 1521, 1522(408a), *1531, 1533, 1537*
Mazur, D.J. 1091(58g), *1103*
Mazzanti, G. 473(207), *492*, 1610, 1630(79, 80), 1638(174), 1644, 1649, 1650(196), 1658(224), *1662, 1664, 1665*, 1830(197), *1865*

Mazzone, L. 1828(192), *1865*
McBride, B.J. 1825(167), *1865*
McCamley, A. 470(193), *492*
McCarthy, D. 1854(322), *1868*
McCarthy, W.Z. 2453(198, 199), *2461*
McCauley, R.A. 2344(157), *2359*
McClelland, R.A. 649(123), 651(124), *700*, 1301(197), *1310*
McClenaghnan, J. 885, 890, 937, 938(116), *1056*, 1274(118), *1308*
McClure, C.K. 455(137), *491*, 1091(58g), *1103*
McConnell, J.A. 522, 540(79, 80a), *554*, 563(76–78, 80), 564(76–78, 85), 566(76, 77), 567, 569(77), *591*
McConnell, W.W. 2150(104), *2173*
McCord, D.J. 1369, 1370(58b), *1439*
McCormack, J.J. 1550(47), *1578*
McCormick, A.V. 2319, 2320(14), *2355*
McCullagh, J.V. 1782(342), 1783(343, 344), *1792*
McCune, J.A. 2105, 2109(207), *2127*
McDermott, G.A. 2253, 2255, 2261(44b, 44c), *2312*
McDermott, J.X. 1903(42), *1926*
McDonagh, C. 2322(26), *2356*
McDonald, J.R. 1117(87), *1138*
McDonald, R.A. 156, 158, 159, 173, 177(35), *178*
McDonald, R.N. 1115(73), *1137*
McElvany, S.W. 1124(171), *1139*
McGarry, P.F. 880, 881, 922, 924, 926, 940, 992(88), *1055*, 1274(110), *1308*
McGarvey, G. 390(111, 112), 394(111), *428*
McGeachin, S.G. 1144(5), *1181*
McGibbon, G.A. 363, 368(29), 369(46), *426*, 604(19), 605, 606(19, 21), 629(21), *698*
McGinn, J.T. 2253, 2255, 2261(44c), *2312*
McGinn, M.A. 710(25a), *775*
McGlinchey, M.J. 580(168, 169), *593*
McGrath, J.E. 1768(287), *1791*, 2224(37), 2236(115), *2241*, 2244, 2345(168), *2360*
McGrath, K.J. 2076, 2091(120), *2125*
McGrath, T.D. 488(272), *494*
McHenry, B.M. 456(142), *491*
McInnis, T.C. 2522, 2523(303), *2565*
McIntosh, M.C. 1817(135), *1864*
McKay, R.A. 310(133), *353*
McKay, R.I. 2504(199), 2522(285), *2563, 2565*
McKean, D.C. 184, 192(17), *257*
McKee, M.L. 485(256), *493*, 690(163), *701*
McKellar, A.R.W. 704(9), *775*
McKelvey, J.M. 517(53, 54), *553*
McKenna, J.M. 1842(259), *1867*
McKenzie, S. 2340(140), *2359*
McKiernan, J. 2347, 2350(190), *2360*

McKillop, K.L. 323, 324(141), *353*, 1897, 1908(26), *1926*
McKinley, A.J. 1266(86, 87), *1307*, 1312, 1317, 1318(5), 1321(85, 86), *1332, 1334*
McKinney, P.M. 234(234), *262*
McKinnie, B.G. 497(12), 507(47), *510, 511*
McLafferty, F.W. 1122(147, 148), 1125(148), 1128(218), *1139, 1141*
McLean, A.D. 1166(152), *1184*
McLeod, D. 1589(72), *1596*, 2533(355), *2567*
McLeod, D.Jr. 68(122a), *99*, 1164, 1175(139), *1184*
McLick, J. 496(8), *510*
McMahon, C. 2479(80), *2560*
McMahon, T. 996(284), *1060*
McMahon, T.B. 1108(22), *1136*
McMillan, D.F. 177(142), *180*
McOsker, C.C. 1740(177), *1789*
McPhail, A.T. 1390, 1396, 1399(124), *1441*
McPherson, G.L. 922(189), *1058*, 1154, 1155(65), *1182*
McRae, D. 425(216), *430*
McVie, J. 2223(23), *2241*
McWilliams, P.C.M. 1316(40), *1333*
Meagher, E.P. 111(18), *118*
Mealli, C. 1603(34), *1661*
Mecartney, M.L. 2319, 2350(10e), *2355*
Meenu 1434, 1436(246), *1444*
Meersche, M.van 1510, 1511(399), *1537*
Meese-Marktschaffel, J.A. 2153(127), *2173*
Meese-Marktscheffel, A. 2040(7), *2122*
Meggers, E. 1294(182), *1309*
Meguro, M. 1672(45), *1683*
Mehdi, A. 541(106a, 106c), *554*, 564(91, 94), *591*, 1409, 1411(186a), 1412(186a, 192), 1414(192), 1417(192, 207a, 207b), 1418(192, 207a, 207b, 215), 1422–1424, 1428, 1429(215), 1430, 1431(229), 1434(229, 244), 1435(186a, 244), *1443, 1444*
Mehlhorn, A. 90(170d), *101*
Mehner, T. 1127(215), *1140*, 1162(113, 116, 118), *1183*
Mehring, M. 310(119), *353*
Mehrotra, S.K. 820(119), *825*, 2131(10), *2171*
Mehta, R. 2102(200), *2127*
Mehta, V.D. 1086, 1090(54), *1102*, 1387, 1388(115), *1441*
Meichel, E. 1374, 1375(84), *1439*
Meier, C. 1720, 1724(114), *1788*, 2106, 2109(209), *2127*
Meier, G. 1585(35), 1587(54, 58–60), *1595, 1596*
Meier, H. 751(100), *777*
Meier, M. 1128(229), *1141*
Meier-Brocks, F. 1987(82), *2032*

Meijere, A.de 648(121), *700*, 746, 770(90b), 776, 1236(9), *1305*
Meinel, S. 2467, 2495(19), *2559*
Meinke, P.T. 1091(58f), *1103*, 1678(99), *1684*
Meisl, M. 1316(41), *1333*
Mekhalfia, A. 1807(84–87), *1863*
Mekhtiev, S.D. 1996(108c), *2033*
Mele, A. 1120, 1121(128), *1138*
Melius, C.F. 155(18–25), 156(20), 159(19, 22), 160, 161(22), 164(23–25), 165(24), 166(21, 25), 167, 168(20, 22), 169(114), 170(21, 24, 25), 171(18–20, 22), 172(22), 173(20, 22), 174(22), 175(20, 22, 25), *178, 180*, 901(148), 1048(363), *1057, 1062*, 1129(233), *1141*
Mella, M. 1290(172), *1309*
Meller, A. 207(122), 229(217), *260, 262*, 2486(118, 119), *2561*
Mellinghoff, H. 2152, 2153(120), *2173*
Melnick, J.P. 484(247), *493*
Mel'nikov, M.Ya. 2530(348, 349), 2531(348), *2566*
Melter, M. 1374, 1375(84), *1439*
Mena, G.H. 2145(90), *2172*
Menchikov, C.G. 2512(253), *2564*
Menchkov, C.G. 52(94a), *98*
Mendel, M. 9(31b), *96*
Mendicino, F.D. 1590(85, 85), *1596*
Mendla, K. 2368, 2371, 2374(38), *2399*
Mengele, W. 2158(148), *2174*
Meng-Yan, Y. 161, 162, 164, 171–173(64), *178*
Mennig, M. 2340(138), 2349, 2350(211), *2359, 2361*
Menon, M. 71, 72(140), *100*
Menu, M.J. 1280(137, 138), *1308*
Menzel, H. 741(79, 80), *776*
Mercadini, M. 2264(79), *2313*
Mercier, F. 1963(2c), *2030*
Mergardt, B. 1819(139), *1864*
Merget, S. 2364, 2369, 2376(32), *2399*
Merian, M. 111(21), *118*
Merin-Aharoni, O. 576(140), 578(162), *592, 593*
Merritt, J. 845(64), *855*
Merzweiler, K. 1886(81, 83), *1893*
Meske-Schüller, J. 2439(149), *2460*
Messier, D.R. 2264, 2265(76), *2313*
Messing, G.L. 2249, 2294(16a, 16b), *2310*
Messmer, R.P. 1120(136), *1139*
Metail, V. 1051, 1052(381), *1062*
Metham, T.N. 2115(233), *2127*
Metxler, N. 279, 280(38), *351*
Metzler, N. 52, 53(98a, 9b), 54(98a), 55(98a, 98b), 56(98b), 57(98a, 98b), 58, 59(98a), 61, 63(98b), *98*, 1144(11), *1181*, 2023(195), *2036*, 2508(224), 2535, 2536(367),

2537(224, 367), 2538(367), 2539(224, 367), 2544(378), *2564, 2567*
Metzner, P. 1870(7), *1891*
Meudt, A. 66–68(117b), *99*, 1125(175), *1139*, 1156–1158(69), 1163–1165(128), 1166, 1167(69), 1171(182), 1172(69, 182), 1174(69), *1182, 1184, 1185*, 1286(162), *1309*
Meunier, P. 37(77b), *97*, 1992(101, 102), 1993(103), 2022(193), *2033, 2036*
Meunier-Piret, J. 1510, 1511(399), *1537*
Meuret, J. 219(178), *261*, 484(246), 485(250, 254), *493*
Mews, R. 238(251), *263*
Meyer, A.J.H.M. 113(23), *118*
Meyer, G. 1434, 1437(251), *1445*
Meyer, G.J. 2349(206), *2361*
Meyer, J. 789(28e), *822*
Meyer, M. 10(34), *96*, 211(145), *260*, 1152(50), *1182*
Meyer, T.J. 2225(46), *2241*
Meyerhoffer, W.J. 561(57), *591*, 1122(145), *1139*
Meyers, A.I. 405(153), *429*, 478(229), *493*, 1671(44), *1683*, 1848(285), *1867*
Meyring, M. 1910(59), 1913, 1914(79), *1927*
Mi, A.Q. 1671(36), *1683*
Mi, Z.-Y. 1692(22), *1786*
Miaga, M.Ja. 1381(107), *1440*
Miao, G. 1668(12), *1682*
Michael, K.W. 247, 248(291), *264*, 417(192), *429*, 2295(144), *2315*
Michalczyk, M.J. 348(157, 160), *354*, 829, 833(19), *854*, 901(153), *1057*, 1167, 1168(163), 1169(163, 169), 1176–1180(169), *1185*, 1283(147), 1284(158), 1288(164), *1308, 1309*, 1329, 1330(162, 163), 1331(163), *1336*, 1897(24, 25), 1898(25), *1926*, 2112, 2120(227), *2127*, 2471(42), 2497, 2498(179), 2513(254), 2514, 2515(179, 254), 2516, 2534(179), *2559, 2563, 2564*
Michalik, M. 1507, 1508(381), *1536*, 1743(214), *1790*
Michalska, Z.M. 1701, 1702(63, 64), *1787*
Michalski, J. 564(86), *591*
Michalski, M. 508(58), *511*
Michel, D. 269(5), *350*
Michels, H.H. 173(136), *180*
Michl, J. 3(1a, 5), 63(1a), *94, 95*, 122(20b, 20d), *150*, 247(292), *264*, 348(157, 160), *354*, 485(253), *493*, 713(35), *775*, 827, 828(7, 8), 829(8, 19), 832(8), 833(19), 839(46, 47), 844(7, 8), *854, 855*, 860(7, 8), 901(8, 153–156, 159), 902(159), 919, 940, 959, 977(8), 981, 991(7, 8), 992(156),

993(159), 995(156, 159), 996(159), 1006, 1008(296b), 1010(8), 1019(321), 1021(326), 1030(345), 1044(321, 326), 1045(321, 345), 1047(321), 1048(345), *1054, 1057, 1060, 1061*, 1064(2, 4b, 4e), 1068(2, 4b), 1071(4b, 4e), 1075(4e, 33), 1077, 1080(4e), *1101, 1102*, 1128(225), *1141*, 1144(1, 10), 1145(1), 1158(1, 90, 91), 1159(93), 1160(1), 1167(10, 163), 1168(163, 164), 1169(10, 163, 175), 1170(164, 179), *1181, 1183, 1185*, 1234(5, 6), 1246(39), 1266(86–88), 1283(147, 156), 1284(158), 1286(161), 1288(164), 1289(169), *1305–1309*, 1312(4, 5), 1313(33, 45), 1314(32–34), 1316(35, 38, 42, 43), 1317(4, 5, 42, 43, 45), 1318(5, 38, 49, 50), 1319(73), 1320(42, 45, 82, 84), 1321(84–86), 1329(159, 162, 163, 165), 1330(159, 162, 163), 1331(163, 198), *1332–1334, 1336*, 1609, 1645(73), *1662*, 1897(21–25), 1898(25), 1907(52b), *1926*, 1930(7), *1960*, 1964(7), 1967, 1986, 1996, 1997(20a, 20b), 2001(7, 131a), 2013(172a), 2027(204), *2030, 2034–2036*, 2159(161), *2174*, 2416(44), *2458*, 2469(30), 2471(41, 42), 2474(41, 53, 54), 2497(180), 2504(54, 197), 2513(41, 254), 2514(254), 2515(53, 54, 197, 254), 2523(318), 2526(180), *2559, 2560, 2563, 2564, 2566*
Mick, H.J. 171(123, 124), 172(124), *180*, 2522(293), *2565*
Mickle, L.D. 2340(133), *2359*
Middleton, D.S. 1858(340), *1868*
Middleton, W.J. 1091(58a), *1103*
Miehe, G. 2257, 2267(49c), *2312*
Mieling, I. 2155, 2156(144), 2157(147), *2174*
Miertus, S. 528, 535(87a), *554*
Migata, T. 1831(202), *1865*
Miginiac, L. 402(144), *428*, 1802(71), *1863*
Miginiac, P. 1820(143), *1864*
Migita, T. 713(33, 34), 715(34), 716(39), 741, 743(33), 748(33, 39, 98), 750(39, 98), 752(33, 113, 115), 753(33, 113), 755(116), *775, 777*, 1080(43b), *1102*, 2403(1–3), 2404(4, 5, 8), *2457*
Miguel, A.V.de 2159(149), *2174*
Miguel, D. 1815(126), *1864*
Mihm, G. 10(33b), *96*, 864(28–30), 901(29, 158), 977(28), 981(29, 158, 256), 985(29), 992(28–30, 158, 256), 995(29, 279–281), 996(28–30), *1054, 1057, 1059, 1060*, 1145(17), 1146(20), 1149(35, 36), 1150(38), 1167–1169(162), *1181, 1182, 1184*, 1329(160), 1331(197), *1336*, 2514(270, 271), *2565*
Miike, H. 2469(32), *2559*
Mikami, K. 389(107), 394(124, 125), *428*, 437, 449(87), 469(190), *490, 492*, 1302(205),
1310, 1796(37), 1804(78), 1856(332, 334), 1857(335), 1858(37, 343), 1859(343), *1862, 1863, 1868*
Mikami, N. 1323(118), 1324(119), *1335*
Mikhailova, A.N. 302–305(104), *353*
Miklos, P. 1481, 1509, 1510(287), *1533*
Mikulcik, P. 909(174), *1057*
Mile, B. 1162(117), *1183*
Mileshkevich, V.P. 213(172), *261*
Milestone, N.B. 2305(160), *2316*
Milius, W. 203(110), 209(136), *259, 260*, 1979(57e), *2032*
Millan, A. 1705(74, 75), 1707, 1709(74), *1787*
Millar, D.M. 1923(99), *1928*
Millar, T.J. 1109, 1117, 1126(50), *1137*, 2556(423), *2568*
Miller, A.E.S. 704(10), *775*, 1116(81), *1138*
Miller, D.M. 113(25), *118*
Miller, D.P. 517(36), *552*
Miller, J.A. 1601, 1602, 1604, 1613(8), 1626, 1642, 1655(133), *1660, 1663*, 2434(121), *2460*
Miller, J.D. 2333(100), *2358*
Miller, J.R. 2231(81d), *2242*
Miller, L.L. 1205(44), *1230*, 2016(174), *2035*
Miller, M. 1771(297), *1791*
Miller, R.B. 390(110–112), 394(111), *428*
Miller, R.D. 1207(51), *1230*, 1266(86, 87), *1307*, 1312(4, 5), 1316(38), 1317(4, 5), 1318(5, 38, 47–51, 64), 1320(83, 84), 1321(84–86), *1332–1334*, 1930(7), *1960*, 2469(29, 30), *2559*
Miller, S.L. 588(197), *594*
Miller, S.M. 2345(173), *2360*
Millevolte, A. 348(162), *354*
Millevolte, A.J. 248(301), *264*, 278(28), *351*, 833(33), *854*, 1266(91), 1288(165), *1307, 1309*, 2497, 2499(182), *2563*
Millie, P. 2512(252), *2564*
Milligan, D.E. 1167(160), 1181(197), *1184, 1185*, 2513, 2514(256), *2564*
Mills, I.M. 234(236), *263*
Mills, R.J. 385(93), 386(103), *427, 428*, 476(221), 484(245), *493*
Milstein, D. 241(266), *263*, 1074(29), *1102*, 1872(21), *1892*, 2102(198, 199), 2108(198, 199, 219, 220), *2127*
Mimura, K. 2150, 2151(105), *2173*
Mimura, N. 1590(83), *1596*
Minami, N. 1614(94, 95), 1620(115), 1623(129), *1662, 1663*
Minami, T. 1731(138, 139), *1788*
Minarik, M. 1237(16), *1305*
Minato, A. 1252(54), 1253(57), *1306, 1307*, 1325(135), *1335*
Minero, C. 2231(87), *2243*
Mingxin Fang 2225(44), *2241*

Ming Yu, J. 2218(11c, 11d), *2240*
Minikin, V.L. 2512(253), *2564*
Minisci, F. 1558(65), *1578*
Minkin, V.I. 5–8, 32(13a), 52(94a), 69, 77(13a), *95, 98*
Minkwitz, R. (232), *262*, 1875(29, 30), *1892*
Minot, C. 497, 498(10), *510*
Mintmire, J.W. 1318(56), *1333*
Mintz, E.A. 487(265), *494*
Minuth, K.-P. 1586(45), *1595*
Miracle, G.E. 3(6), *95*, 250(312), *264*, 859(5), 981, 984(5, 259), 987(5, 259, 271), 988(271), 989(259, 271), 993, 995, 996(5), 1000(5, 259, 271), 1002, 1003(259), 1004(259, 271), *1054, 1059, 1060*, 1145(16), *1181*, 1285(159), *1309*
Miranda, E.I. 1606(52a), 1638(169), *1661, 1664*
Mirecki, J. 1700(56), *1786*
Miron, E. 2350(234), 2351(249), *2361, 2362*
Mironenko, E.V. 154, 156, 164, 165(15), *177*
Mironov, B.F. 1582(8), *1594*
Mironov, V.A. 1996(108b), *2033*
Mironov, V.F. 1450(36, 47), *1525, 1526*
Mirskov, R.G. 1373(75c), *1439*, 1449(16), 1457, 1458, 1467(121), 1474, 1477, 1478(266), 1487(309), 1488(16, 310), 1489(16), 1490(309, 316, 319, 320), 1491(16, 321), 1494(319), 1496(344), 1497(349, 350), 1499(354, 355), 1507(383), 1508(383, 384, 391), 1509(383, 391), 1513(121), *1525, 1528, 1533–1537*
Mirzel, N.W. 1875(34), *1892*
Misawa, H. 2437(134), *2460*
Mishima, M. 364(33), *426*, 603(18), 610, 624(33), *698*
Mishima, S. 1205(43), *1230*
Mislow, K. 288, 289(68), *352*, 518(60b), *553*, 1408(178), *1443*
Mison, P. 616(53), *698*
Misono, T. 2251, 2253(27b), *2311*, 2333(97), 2334(105), *2358*
Misra, R.N. 425(217), *430*, 1636, 1652(165), *1664*, 1733(141), *1788*
Mitani, M. 1847(282), *1867*
Mitchell, G. 761(126), *777*
Mitchell, G.P. 1888, 1890(96), *1894*, 2105, 2109(206), *2127*
Mitchell, G.R. 2225(48), *2242*
Mitchell, T.D. 307(110, 111), *353*
Mitchener, J.C. 1705(73), *1787*
Mitomo, M. 2264(75), *2313*
Mitra, M. 746(91), *776*
Mitro, V. 197(79), *259*, 514, 520, 521, 524(3b), *551*, 567, 570, 572, 583, 586(103), *592*
Mitsuhida, E. 1302(202), *1310*, 1934(12), 1940(14, 16), *1960*, 2496(176), *2562*
Mitsui, I. 2347, 2350(188), *2360*
Mittal, J.P. 863(24, 25), 864, 932, 934, 982(25), *1054*, 1237(18a), *1306*
Mittendorf, J. 1815(125), *1864*
Mitter, F.K. 796(83a), *824*, 2192(67), *2215*
Mituoka, T. 1641(190), *1664*
Mitzel, N.W. 207(121, 125–127), *260*, 586(192), *594*
Miura, H. 123(22), *151*
Miura, K. 390(113), 391(114), *428*, 1555, 1556(59), 1572(93), *1578, 1579*, 1833(207, 208), 1856(327), *1866, 1868*
Miura, T. 1871(11), *1891*
Miwa, K. 1675(85), 1676(86), *1684*
Mix, A. 1434, 1435(239), *1444*
Miyagawa, J. 1801(64), *1863*
Miyai, A. 186(40), *258*, 2017(178), *2035*, 2439(148), *2460*
Miyaji, H. 1630(146, 147), 1653(146), *1663*
Miyamoto, S. 917, 998, 1002(184), *1058*, 2429, 2436(106), *2459*
Miyamoto, Y. 1671(35), *1683*
Miyashita, K. 1640(186), *1664*
Miyazaki, K. 1638(168), *1664*
Miyazaki, S. 2546(393), *2568*
Miyazawa, M. 470(196), *492*
Miyazawa, T. 897(138), *1057*, 1235(7), *1305*, 1322(105), 1323(118), 1324(105, 119), 1325–1327(105), 1328, 1329(157, 158), *1334–1336*, 2470(33), *2559*
Miyazuma, T. 1290(170), *1309*
Miyoshi, N. 1325(133), *1335*
Miyoshi, Y. 2546(393), *2568*
Mizhiritskii, M. 1576(100), *1579*
Mizhiritskii, M.D. 1677(96), *1684*
Mizuno, K. 581(176), *593*, 1257(67), 1290(173), 1291(174), 1292(177), 1293(179), 1298(191, 192), *1307, 1309, 1310*, 1849(293), *1867*
Mizushima, T. 1940(15), *1960*
Mizutani, T. 2104, 2108, 2109(204), *2127*
Mlynar, J. 1583(21), *1595*
Mlynarski, P. 515, 516(14b), *551*
Mneimne, O. 1778(331), *1792*
Mo, S.-H. 1965(16c), *2030*, 2469(28), *2559*
Mo, Y.K. 1108(20), *1136*
Moc, J. 1111(65b), *1137*
Mocaer, D. 2261(60), *2312*
Mochida, K. 564(89), *591*, 1203(40), 1207(48, 49, 52), 1210, 1211(61), 1212, 1213(40), 1214(73), 1229(127), *1230–1232*, 1257(66), *1307*, 1930(10), *1960*
Mock, S. 2073, 2091(97), *2124*
Modi, D.P. 1673(56), *1683*
Modi, P. 391(115), *428*
Modi, S. 222(192), *261*, 1638(170), *1664*

Moeller, K.D. 1196(19, 20), *1230*, 1823(152), *1864*
Moeller, M. 331–333(145), *353*
Moerek, R. 1497(348), *1535*
Moerlein, S.M. 385(96), *427*
Moffat, H.K. 171(121, 122), *180*, 2476(62), 2549(406), *2560, 2568*
Mohammadi, V. 2322(37), *2356*
Mohan, M. 2187(49), *2215*
Mohr, P. 1800(61), 1841(253), *1863, 1866*
Möhrke, A. 1079, 1088, 1089(41a, 41b), 1099(75), *1102, 1103*, 2131(12), 2166(189), 2167(189, 194, 195), 2168(189, 195, 196), 2169(189), 2170(189, 201), *2171, 2175*
Mohrke, A. 1434(232a), *1444*, 1879, 1880(58), *1893*
Moiseenkov, A.M. 456(146), *491*
Mokhi, M. 1860(354), *1868*
Mokhlesur Rahman, A.F.M. 1743(221), *1790*
Molander, G.A. 1643(194), *1664*, 1679(105), 1681(123), *1684, 1685*, 1697, 1698, 1709, 1710(45), 1721(124), *1786, 1788*, 1799(55), 1821(148), *1863, 1864*, 2039(4), *2122*
Molenberg, A. 2219(13a, 13b), *2240*
Moll, F. 1323(113), *1335*
Moller, M. 2219(13a, 13b), *2240*
Möller, S. 2078, 2091, 2095(121, 122), *2125*, 2340(135), *2359*
Molloy, K.C. 1925(109), *1928*
Molnár, A. 435, 479(60), *489*, 1738(160), *1789*
Monger, S.J. 448(119), *491*
Mongey, K. 2322(26), *2356*
Mont, W.-W.du 1885(73), 1887(89), *1893, 1894*
Monte, F.del 2350(215, 217), 2351(248), *2361, 2362*
Monteil, F. 1778(333), *1792*
Monteith, M. 402(143), *428*, 1849(295), *1867*
Montero, M.L. 2301(155), *2316*
Montgomery, L.K. 982(263), *1059*
Monti, H. 1801(67), 1807(83), 1849(288, 289), 1856(331), *1863, 1867, 1868*
Monti, J.-P. 1849(288, 289), 1856(331), *1867, 1868*
Montle, J.F. 1591(100), *1597*
Moody, L.G. 1590(86), *1596*
Moody, M.A. 1669(22), *1682*
Mooiweer, H.H. 1815(123, 124), *1864*
Moolenaar, M.J. 1835(216), *1866*
Moore, H.W. 1799(54), *1863*
Moore, J.A. 2251, 2253(27a), *2311*
Moore, M.H. 1299(196), *1310*
Moore, W.R. 2448(177), *2461*
Moorhouse, S. 2140, 2141(52), 2143(52, 80), 2144(52), *2172*
Morales, A.R. 2240(132), *2244*
Moran, M. 1758(257), *1791*

Morante, J.R. 2549(404), *2568*
Mordini, A. 402(146), *428*, 460(162), *491*, 1626(132), 1628(137–140), 1655(137), 1656(137–140), 1658(132), *1663*, 1677(98), *1684*, 1796(38), 1816, 1819(128), *1862, 1864*, 1871(14), 1877(40), *1892*
More, S. 1304(216), *1310*
Moreau, J.J.E. 284, 288(52), *351*, 497(9), *510*, 828(14), *854*, 1341, 1347(24), *1437*, 1751(232, 233, 235, 236), *1790*, 2007(158a), *2034*, 2282(114), *2314*, 2335(112), *2358*, 2376(45), *2399*
Moreau, M. 2220(18, 19a, 19c), *2240, 2241*
Morehouse, E.L. 1509(392), *1537*
Moretto, H. 2131(14), *2171*
Morgan, G.H. 2502(191), *2563*
Morgan, I.T. 409, 410(165), 422(208), *429, 430*
Morgunova, M.M. 1470(214), *1531*
Mori, I. 783(14), *822*
Mori, M. 1796(39), *1862*
Mori, S. 561(55), *591*, 1125(184), 1133(184, 247, 249), *1140, 1141*, 1959(29), *1960*, 2015, 2016(173), *2035*
Mori, Y. 1322(110), *1334*
Moriarty, R.M. 645(110, 111), *700*
Morihashi, K. 20(51a), *96*, 794(53), *823*, 2453(201), *2461*
Morikawa, I. 1755, 1756(246), *1790*
Morimoto, T. 1754, 1755(245), *1790*, 1978(55), *2031*
Morimoto, Y. 888(123), *1056*
Morino, S. 1883(68), *1893*, 1896, 1906(5), *1925*
Morita, K. 290–292(74), *352*
Morita, Y. 443(110), *490*, 1203(39), *1230*
Moritami, N. 2002, 2009(144), *2034*
Moritani, T. 1675(84), *1684*
Moritomo, Y. 1318(64), *1333*
Morley, C.P. 2140(47), *2171*
Morokuma, K. 515(13), *551*, 2116(244, 245), *2127, 2128*
Moro-oka, Y. 2080(133, 137–140), 2092(133), *2125*
Morosin, B. 192(57), *258*
Morris, A. 992, 996(276), *1060*
Morris, G.A. 269(7), *350*
Morris, P.J. 1322(96), *1334*
Morrison, D.S. 611, 615(45), *698*
Morrison, H. 1298(189), *1309*
Morrison, J.A. 819(117), *825*
Morrison, W.H. 1618(109), *1662*, 2160(170), *2174*
Morrone, A.A. 2286(119, 120), *2315*
Morse, P.M. 111, 114(19), *118*
Mort, S.P. 1107(15), *1136*
Morterra, C. 2530(341, 342), *2566*

Morvai, L. 1470(208), *1531*
Moser, U. 2364(14–18, 20, 24), 2365(20), 2367(20, 37), 2368, 2371, 2374(38), *2398, 2399*
Moskovich, R.Ya. 1454, 1467(98), *1527*
Moskovits, M. 1899(38), *1926*
Moss, R.A. 704, 712(2), *774*
Motherwell, W.B. 1539, 1540(2a), 1566(80), *1577, 1579*, 1669(19), *1682*
Motooka, T. 559(30), *590*, 2522(301), *2565*
Motoyama, J. 2160(169), *2174*
Motoyama, Y. 1983(73), 1999(127), 2001(73), *2032, 2034*
Motsarev, G.V. 384(91), *427*
Motz, D. 2141(57), *2172*
Mounier, P.A. 578(163), *593*, 2487(126), *2561*
Mouray, T.H. 2345(173), *2360*
Mourik, G.L.van 2453(197), *2461*
Mourik, T.von 516(32a), *552*
Mouriño 1855(326), *1868*
Mowlem, T.J. 435, 454(45), *489*
Mozzhukhin, A.O. 237(245), *263*, 1396(163), 1397(166, 167), 1418, 1422, 1423(216), *1442, 1444*
Mozzhurkhin, A.O. 226(206), *262*
Mu, Y. 2042(13), 2043(16), 2045(13), 2051(16), *2122*, 2274(101a), 2276(101a, 103), 2279(101a), 2290(137), *2314, 2315*
Muchowski, J.M. 1843(260), *1867*
Muehlhofer, E. 780(2c), *821*
Mueller, B.L. 226(205), *262*, 2265(80), 2298(150b), 2301(157, 158), 2305(157), *2313, 2316*
Mueller, T. 2023(196), 2024, 2026(198), *2036*
Mueller-Westerhoff, U.T. 2162(173), *2174*
Muetterties, E.L. 503(31), *511*, 1171(181), *1185*, 1340(16), *1437*, 1897, 1918(8), *1925*
Mugge, C. 1510, 1511(398), *1537*
Muguruma, Y. 1669(25), *1682*
Muhkerjee, B. 434, 449(37), *489*
Mühleisen, M. 236(255), *263*, 1341, 1349(29, 30a), 1350(30a), 1357(29, 30a, 47, 50, 51), 1359(29, 47), 1360(47, 50, 51), 1361(30a), 1362(47, 50), 1363(47), 1364(51), 1365(30a, 51), 1366(29), 1368(30a, 57a, 57b), 1412, 1415(193), *1437–1439, 1443*, 2364(31), *2399*
Mühlhaupt, R. 2153(128, 129), *2173*
Mui, J.Y.P. 1586(52), *1596*
Mui, L.C.M. 1651(216), *1665*
Muire, J.B. 1470(215), *1531*
Muiry, I.B. 1459, 1467(130), *1529*
Mujsce, A.M. 69(130), *100*
Mukai, C. 441(105), *490*
Mukaiyama, T. 1846(280), *1867*, 1871(12, 13), *1891*

Mukherjee, S.P. 2246, 2250, 2253, 2254, 2273, 2285(6c), *2310*
Mülhaupt, R. 1758(258, 259), *1791*, 2238(125), *2244*
Mullens, J.W. 2350(227), *2361*
Müller, A. 833(34), *854*, 1099(75), *1103*, 2166(191), 2167(194), *2175*
Muller, A.J. 69(130), *100*
Müller, B. 646, 647(115), 655(115, 133), 659, 660(115, 135), 675(149), *700, 701*
Muller, B. 1671(34), *1683*
Müller, E.G. 1909, 1910(56), *1927*
Müller, F. 540(103), *554*
Müller, G. 183(13), 250(308–311), 252(323, 324), *257, 264, 265*, 859(4), 873(59), 876(4), 879(83), 909(174), 932(83, 197), 940(83), 941(59), 944(83, 197), 981(4, 83, 197), 985, 986(268), 988(83), 991(4), 1011(301), 1021(327), 1033(301), 1037(59), 1042, 1043(327), *1054, 1055, 1057–1061*, 1064, 1068(5b), *1101*, 1158(88), *1183*, 1374, 1375, 1379(80), *1439*, 2084, 2090(156), *2126*, 2140(51), 2155, 2159(142), *2172, 2174*, 2534(362), *2567*
Muller, G. 190(56), 203(104), 216(160), 217(164, 165), *258–260*, 296(93), *352*, 844(57, 59), 845, 847(59), *855*, 1340(15), 1434, 1436(249, 250), *1437, 1445*
Müller, H. 784(8c), *822*
Müller, J. 2027(203), *2036*
Muller, M. 1872(18), 1886(82), *1892, 1893*
Müller, M.A. 1318(59), *1333*
Muller, N. 517(48), *553*
Müller, P. 1671(37), *1683*
Müller, R. 1340, 1351(8a, 8b), 1352(32a), 1412(8a, 8b, 32a), *1437, 1438*, 1493(332), 1494(332–334), *1535*, 1923(101), *1928*, 1972(35b), *2031*
Muller, R.P. 2510(231), *2564*
Müller, T. 3(2a), 4(10b), 37(10b, 85), 39, 40(10b), 44(10b, 93b), 47(93b), 48(10b, 93b), 49(85, 93b), 53, 55(100a, 102), 56, 57(100a), 59(102), 60(100a, 102, 107), 61, 62(100a), 63, 64, 66(102, 110), *95, 97–99*, 283(48), *351*, 514(5), 521–523, 527, 529(73), 532, 536(5), 542(73), 543(5, 110), 544(73, 110), 545, 547(110), *551, 553, 555*, 570(125), 573(125, 133), 581(178), 584(189), *592–594*, 599, 604(10), 606, 615(22), 648(120, 121), 652(127), 663(141), 696(170), *698, 700, 701*, 746, 770(90b), 776, 818, 819(116), *825*, 863(27b), 981(260), 984(260, 266), 985(270), 987(260), 988(27b), 989(260), 990(27b), 997(285a), *1054, 1059, 1060*, 1169, 1177, 1178(171), *1185*, 1460, 1462, 1467, 1477(154), 1481(282, 287),

1491(154), 1509, 1510(287), *1529, 1533,*
2511, 2516(247a, 247b), 2518(247a),
2536(247b), 2539(371), *2564, 2567*
Muller, T. 345, 347–349(154), *354*
Muller, U. 1878(47), *1892*
Mulliken, R.S. 517(48), *553,* 1021(324), *1061,*
1159(97), *1183*
Munasinghe, V.R.N. 462(167), *492*
Münch, A. 799(65), *823*
Mundt, O. 780, 781(2e), *821*
Munechika, T. 371(55), *427,* 617(56), *698*
Munschauer, R. 435, 452(42), *489,* 752(107, 108), *777*
Munson, B. 1121(142), *1139*
Munsted, R. 1635(163), *1664*
Munster, I. 1871(10), *1891*
Murahashi, S. 1750(230), *1790*
Murai, S. 1091(58h), *1103,* 1616(99, 102), 1618(110, 111), *1662, 1663,* 1678(99), *1684,* 1705(77), 1714(96–100), 1715(101), 1771(302–304, 306, 308–310), 1772(306, 311–316), 1773, 1774(317), 1783, 1784(346), *1787, 1791, 1792,* 1838(232–237, 239–241), *1866,* 1978(55), *2031,* 2110(224), *2127*
Murai, T. 1689(9), 1742(186), 1771(308), *1786, 1789, 1792,* 1885(79), 1889(101), *1893, 1894*
Murai, Y. 1811(107), *1864*
Murakami, C. 1826(173), 1856(327), *1865, 1868*
Murakami, M. 1611, 1617(82), *1662,* 1716(103), *1787,* 2116, 2120(236, 237), *2127,* 2433(119), *2460*
Murakami, S. 131(38), *151,* 248(298), *264,* 2182(34, 36), *2214,* 2498(185b), *2563*
Muraki, K. 1221(99, 100), *1231*
Muraoka, T. 810(93b), 812(98), *824,* 2182(27), *2214*
Murase, H. 1218(92), *1231*
Murashov, V.V. 90–92(169), *101*
Murat, M. 2261(56), *2312*
Murata, K. 1157(75), *1182,* 1268(95), *1307,* 2472(46), 2513, 2516, 2526(258), *2559, 2564*
Murata, M. 1412(190), *1443*
Murata, S. 1749(228), *1790*
Murata, T. 1190, 1192(13), 1193(16), 1197(25), 1201(36), 1202(25, 36), 1203(25), *1230,* 1831(199), *1865*
Murata, Y. 1959(29), *1960*
Murayama, N. 2250, 2261(20), *2310*
Murphy, D. 219(175), *261*
Murphy, J.A. 1569(87), *1579*
Murphy, P.J. 1727–1730(135), *1788,* 1841(254), *1866*
Murphy, V.J. 2162(175), *2174*

Murray, K.K. 704(10), *775*
Murray, R.C. 272(18), *350*
Murray, R.W. 1622(126), *1663*
Murrell, J.N. 1150(39), *1182*
Murthy, S. 34, 35(72a, 72b), *97,* 560(41, 42), *590,* 1132(244, 245), *1141*
Murthy, V.S.R. 2261(57), *2312*
Murugavel, R. 2218(6), *2240*
Musaev, D.G. 2116(245), *2128*
Musaev, M.R. 1996(108c), *2033*
Musaki, Y. 1797(48), *1862*
Musashi, Y. 2116(241), *2127*
Musher, J.I. 1378(100), *1440,* 1465, 1467(173), *1530*
Muskulus, B. 1903(40b), *1926*
Mussell, R.D. 580(175), *593*
Mutchler, E. 219(180), *261*
Mutin, H. 2294(142), *2315*
Mutin, P.H. 2231(90a, 90b, 91), *2243,* 2251(26), 2253(26, 47a, 47b), 2255(26), 2294(141), *2311, 2312, 2315,* 2466(18), *2559*
Muto, T. 1978(55), *2031*
Mutschler, E. 222(195), *261,* 2363(9), 2364(14–18, 20–31, 33–36), 2365(20), 2367(20, 37), 2368(34, 38–40), 2369(34, 39), 2371(34, 38, 40), 2374(34, 36, 38–43), *2398, 2399*
Mutsuddy, B.C. 2261(53), *2312*
Muzart, J. 1674(64), *1683*
Myagi, M.Ya. 1474, 1476, 1477(260, 263), 1478(263), 1479(271), 1510(260), *1532, 1533*
Mychajlowskij, W. 393(122), *428*
Myers, A.G. 505(40, 41), *511,* 1810(103), *1863*
Myhre, P.C. 690(164b), *701*
Mynott, R. 1153(58), *1182*
Myoshi, A. 1125(186), *1140*
Myrick, M. 2075(108), *2125*

Naasz, B.M. 1593(109), *1597*
Nadjo, L. 2153–2155(132), *2173*
Nadler, E. 363(32c), *426*
Nadler, E.B. 485(258), *493,* 782, 795, 796(10b), *822,* 1603, 1605(39), *1661*
Naef, R. 1843(260), *1867*
Nagafuji, A. 1641(189), *1664*
Nagahara, K. 1560(69), *1579*
Nagai, H. 2350(221), *2361*
Nagai, Y. 120, 142, 143, 146(3), *150,* 200(93), 248(299), *259, 264,* 279(29, 31, 35), 348(159), *351, 354,* 810(93a, 93b), 812(98), *824,* 1157(75, 79, 80), *1182, 1183,* 1211(66), *1231,* 1268(95), *1307,* 1699(50), 1718, 1723(108), 1733(143), 1735(50, 153), 1738(158), *1786, 1788,* 1897, 1901(29),

Nagai, Y. (cont.) *1926*, 2182(27), 2183(38, 39, 42), *2214*, 2472(45, 46), 2475(59a), *2559, 2560*
Nagami, K. 441(105), *490*
Nagamoto, I. 1770(295), *1791*
Nagandrappa, G. 413(177), *429*
Nagano, Y. 34, 35(72a, 72b), *97*, 1132(244, 245), *1141*
Nagao, Y. 2251, 2253(27b), *2311*, 2333(97), 2334(105), *2358*
Nagase, N. 30(67), *97*
Nagase, S. 3, 10, 11, 13, 16, 17(8), 22(56, 59, 61a, 61b), 23(59, 61a, 61b), 24, 25(56), 28, 29(61a), 32(8), 69(61a, 61b), 70(61a, 61b, 137, 138), 71(137), 72(59, 61a, 61b, 143), 73, 74(61a, 61b, 143), 77(61a, 61b, 137), 81(156), 84, 87, 88(164), *95, 97, 100, 101*, 120(5, 8, 10, 13, 14), 121(5, 10, 13–15), 122(5, 13–15, 18, 21a), 123(5, 13, 15, 21a, 21d, 22, 24d, 25a, 25b, 28), 124(8, 10, 29), 125(10, 28, 34), 127, 128(10), 129, 131(13, 34), 133(13, 14), 134(42), 136(8, 29), 140(10, 48), 142(10, 14, 48), 143(14), 147(57), 148(10, 15), 149(8, 10, 15, 60a–d), *150–152*, 200(102), 252(334), *259*, 265, 277(23), *350*, 515(12), *551*, 804(77), *824*, 839(48), *855*, 859(6), 901(152), 990(6, 272), 994, 995(6), *1054, 1057, 1060*, 1064(10), 1065(13, 14), 1066, 1067(13), 1068(10), 1082(49), 1083(10, 13, 14), 1090, 1091(10), 1092(10, 62, 63), *1101–1103*, 1163(122), 1169(178), *1184, 1185*, 1301(198, 199), 1302(202–204), *1310*, 1879(52), *1893*, 1897(29, 31), 1898, 1899(33), 1900(31), 1901(29, 31), 1911(69–71, 74b), 1913(74b), *1926, 1927*, 1931(11), 1934(12), 1936(13), 1940(14–17), 1942(18, 19), *1960*, 2452(191), *2461*, 2492(145), 2493(151, 152), 2496(176), 2500(145), 2520(151, 268), 2523(151), *2562, 2565*
Nagashima, H. 1692(15), 1733(15, 145, 146), 1734, 1737(145), 1777(330), 1778(332), *1786, 1788, 1792*, 1957(26), 1958(27), *1960*
Nagashima, M. 124, 136(29), *151*
Nagashiro, R. 2349(213, 214), 2350(230), *2361*
Nagata, K. 1390(123a), *1441*, 2429, 2436(105), *2459*, 2554(416), *2568*
Nagata, R. 1682(129), *1685*
Nagata, Y. 2475(59b), *2560*
Nagawa, T. 2334(105), *2358*
Nägele, W. 1323(113), *1335*
Nagendrappa, G. 487(267), *494*, 782, 791, 795(11b), *822*
Nagy, G. 1721(123), *1788*
Nagy, J. 1460(136, 158), 1462(147, 148), 1467(136, 147, 148, 158), *1529*

Naik, P.D. 863, 864, 932, 934, 982(25), *1054*
Nair, D.R. 2327(54), *2356*
Naito, I. 1602(27), *1661*
Naitou, H. 998, 1005(290), *1060*
Naitou, K. 1641(190), *1664*
Najim, S.T. 577(146, 151–153), *592, 593*
Naka, A. 867(44, 46), 868(46), 870(44, 47–51), 883, 960–962(99), 974(46), *1055, 1056*
Nakada, M. 459(152), *491*, 1610, 1630(78), 1639(177, 178), *1662, 1664*
Nakadaira, Y. 186(41), *258*, 488(273–276), *494*, 829(16, 17), *854*, 1247(41, 42), 1290(171), 1293(181), *1306, 1309*, 1319(76, 77), 1320(77), 1325(148), *1334, 1335*, 1965(15), 1972(39), 1973(46), 1981(39), 1996, 1998(15), 1999(128, 129), 2001(130a–c), 2002(39, 143), 2009(162), 2011(166), *2030, 2031, 2034, 2035*, 2412(35), 2429(102), 2430(112), 2439(142, 143), *2458–2460*, 2489(137a), 2496(178), 2502(190), *2562, 2563*
Nakagawa, K.-I. 1328(156), 1332(199), *1336*
Nakagawa, K.-L. 2526(328), *2566*
Nakagawa, N. 1283(146), *1308*
Nakagawa, Y. 198(88), 207(124), *259, 260*, 1014, 1033, 1034(311), *1061*, 1197(27, 28), 1200(28), *1230*, 1710(87–89), 1711(89, 90), 1725(87), *1787*, 1978, 2015(54b), *2031*
Nakahama, S. 457(147), *491*, 1638(168), *1664*, 1906, 1907(51), *1926*, 2334(105), *2358*
Nakahara, H. 2271(89), *2314*
Nakahiro, H. 1226(114), *1232*
Nakai, S. 472(200), *492*
Nakai, T. 310(130), *353*, 389(107), 394(124, 125), *428*, 469(190), *492*, 1796(37), 1804(78), 1856(334), 1857(335), 1858(37, 343), 1859(343), *1862, 1863, 1868*
Nakaido, Y. 2261(58), *2312*
Nakajima, A. 1120, 1121(132), *1139*
Nakajima, I. 2231(83), *2242*
Nakajima, K. 707(17), *775*, 1957(26), 1958(27), *1960*
Nakajima, M. 829(26), *854*
Nakajima, S. 2549(403), *2568*
Nakajima, T. 90(170c), *101*, 1091(58h), *1103*, 1630(146, 147), 1640(186), 1641(190), 1653(146), *1663, 1664*, 1678(99), *1684*, 1710(86), *1787*
Nakajo, E. 1843(264), *1867*
Nakamura, A. 1972, 1981, 2002(39), *2031*, 2145(91), *2172*
Nakamura, E. 746(90a), *776*, 1667(11), *1682*
Nakamura, H. 1640, 1655(187), *1664*
Nakamura, K. 1303(206), *1310*, 1325(141), *1335*
Nakamura, M. 2351(251), *2362*
Nakamura, S. 1610, 1630(78), *1662*

Nakamura, T. 438(93), 449(120), 469(191, 192), 486(260), *490–493*, 1743(190, 191), 1745–1748(191), *1789*
Nakamura, Y. 1832(204), *1865*, 2107, 2109(211), *2127*
Nakanishi, K. 1257(67), 1290(173), 1298(191, 192), *1307, 1309, 1310*, 1849(293), *1867*
Nakano, H. 860, 862(16b), *1054*
Nakano, K. 1201, 1204(37), *1230*
Nakano, M. 84, 87, 88(164), *101*, 120–122(13), 123(13, 25a, 28), 125(28), 129, 131, 133(13), *150, 151*, 1638(168), *1664*, 1829(194), *1865*
Nakano, T. 1738(158), *1788*
Nakao, N. 1120, 1121(132), *1139*
Nakao, R. 2353(278), *2362*, 2471(35), 2475(59a, 59b, 60), 2486(111), *2559–2561*
Nakaoka, A. 1733, 1734, 1737(145), *1788*
Nakash, M. 437, 485(79), *490*, 795, 800(69a), *823*, 842(53, 54), 843(54), *855*, 884–886, 910, 957, 981, 983, 985, 987, 992, 996(111), *1056*, 1544(26), *1578*
Nakatani, J. 1640, 1655(187), *1664*
Nakatani, K. 469(189), *492*
Nakatani, S. 1843(261), *1867*
Nakatsuji, H. 2056(47), *2123*
Nakatsuka, H. 2351(246), *2362*
Nakayama, M. 1836(219), *1866*
Nakayama, N. 186(39), 194(68), *258*, 278, 329(26), *351*, 2439(154), 2440(154, 155), 2443(155, 161), 2444(155), *2460, 2461*
Nakayama, O. 1756(249), *1790*
Nakayama, Y. 1296(185), *1309*, 1608(66), *1662*
Nakazaki, M. 2448(181), *2461*
Nakazawa, T. 2347(189), *2360*
Nakazumi, H. 2349(213, 214), 2351(238, 245), 2354(282), *2361, 2362*
Nakhara, H. 2265(82a), *2313*
Nakos, S.T. 2226(54), *2242*
Namavari, M. 163(70), *179*, 861–863(20), 871(54), 916(183), 926(20), 940(54), 972(54, 250), 973, 974(54), 993, 996(183), *1054, 1055, 1058, 1059*, 1241(26), *1306*
Nametkin, N.S. 163(69), *179*, 715(37), *775*, 844(55), *855*, 860, 862(16a), 981, 985, 992(254), *1054, 1059*, 1064, 1068(4a), *1101*, 1144(6), *1181*, 2406(16), *2458*
Namikawa, A. 2321, 2347(20), *2355*
Namiki, T. 2334(105), *2358*
Nanami, H. 1740(176, 180), *1789*
Nando, G.B. 2239(127, 128), 2240(127, 129, 131), *2244*
Nandy, S.K. 434, 449(37), *489*
Nanjo, M. 796(56, 61), 797(56), 798(56, 61), *823*
Naoi, Y. 120, 142, 143, 146(3), *150*

Narang, U. 2326(47), 2327(56), *2356, 2357*
Narang, U.P. 2350(226), *2361*
Narasaka, K. 1625(131), 1628(144), *1663*
Nardi, P. 1353, 1368, 1369(36), *1438*, 2298(150a), *2315*
Nares, K.E. 2466(15), 2502(192), *2559, 2563*
Narjes, F. 1796(41), 1821(144), *1862, 1864*
Narsavage, D.M. 2253, 2254, 2257(41), *2312*
Narske, R. 1828(193), *1865*
Narula, C.K. 2246(3), *2310*
Narula, D.K. 216(162), *260*
Narula, S.P. 1434, 1436(246), *1444*, 1453, 1462, 1467, 1483, 1491(68), *1526*
Nasaka, N. 1973, 1974(44), *2031*
Nasim, M. 226(202), *262*, 1456(118, 119), 1457, 1460(118), 1461(141), 1467(118, 119, 141), 1494(337), 1496(345), 1497(119, 141, 351), 1501(119), *1528, 1529, 1535*
Naslain, R. 2261(60), *2312*
Naso, F. 1819(141, 142), 1828(192), *1864, 1865*, 2388(72), *2400*
Naso, H. 2351(251), *2362*
Nass, R. 2333, 2335(104), 2337(118), *2358*
Nassimbeni, L.R. 222(186, 188, 190), *261*
Nate, K. 891(126), *1056*, 2016(175a), *2035*
Näther, C. 207(128), *260*, 484(246), 485(250), *493*
Nativi, C. 410(167), *429*, 1849(296), *1867*
Natsume, H. 1821(146), *1864*
Nava, D.F. 167(86), *179*
Navarro, P. 1875(32), *1892*
Naviroj, S. 2339(125), *2358*
Navrotsky, A. 110–112(16), *118*, 2249, 2294(14), *2310*
Naylor, D.M. 580(171), *593*
Nayyar, N.K. 1450, 1464, 1465(44), 1488, 1489, 1492, 1496(311), *1526, 1534*
Nedalec, J.-Y. 2137, 2138(37), *2171*
Nedolya, N.A. 1454(77), *1527*
Neely, J.D. 1594(116, 117), *1597*
Nefedov, A.O. 1715, 1716(102), *1787*
Nefedov, O.M. 52(94a), *98*, 194(66), *258*, 901, 902(159), 981, 992(255), 993(159), 995(159, 255), 996(159), *1057, 1059*, 1069, 1070(159), 1071(23), 1072(19, 23), 1083(23), *1102*, 1161(104–106), 1170(179), *1183, 1185*, 1244(32, 34), 1283(156), *1306, 1309*, 1312(16), *1332*, 1967(20a, 20b), 1968(20c), 1986, 1996, 1997(20a–c), 1998(20c), 2001(137), 2013(172a), 2027(204), *2030, 2034–2036*, 2473(51), 2474(54), 2488(135), 2504(54, 197), 2512(253), 2515(54, 197), *2560, 2562–2564*
Nefedov, V.D. 559, 581(9), *590*
Negishi, E. 1986(80c), *2032*, 2043, 2051(15), *2122*, 2434(121), *2460*

Negishi, N. 2320(17b), 2347(17b, 186a, 186d), 2350(221), *2355, 2360, 2361*
Negishi, Y. 1120, 1121(133), *1139*
Negrebetsky, Vad.V. 1404(172), *1442*
Negrebetsky, Vit.V. 1404(172), *1442*
Negrebetsky, V.V. 1434, 1435(240), *1444*
Negrebezkii, V.V. 1392(145b), *1442*
Neh, H. 1668(15), *1682*
Nei, W. 2350(229), *2361*
Neithamer, D.R. 580(175), *593*
Nelson, A.J. 1972(36), 2003(145), *2031, 2034*
Nelson, G.O. 2150(104), 2151(114), 2152(114, 119), 2153(119), *2173*
Nelson, J. 2002(142), *2034*
Nelson, J.M. 2160(166), *2174*
Nelson, T.D. 1674(65), *1683*
Nesloney, C.L. 1719, 1732(111), *1788*
Nesmeyanov, A.N. 1714(95), *1787*
Nestler, B. 277(24), *350*, 1301(200), *1310*, 1945, 1947(21), *1960*, 2430, 2432(110), *2459*
Nestunovich, V.A. 436, 479(61), *489*
Netto-Ferreira, J.C. 2498, 2513, 2527(184), *2563*
Neudert, J. 648(122), *700*, 757(119), *777*
Neuhalfen, A.J. 2348(201), *2360*
Neuman, A. 1671(34), *1683*
Neumann, B. 190(49), *258*, 1079, 1088, 1089(41b), *1102*, 1434(232a, 239), 1435(239), *1444*, 1879, 1880(58), *1893*, 2148(101), 2150, 2151(107, 108), 2155(143, 144), 2156(107, 144, 145), 2158, 2159(143), 2166(190), 2167(190, 195), 2168(195–197), *2173–2175*, 2534(361b), *2567*
Neumann, R. 1593(112), *1597*
Neumann, W.P. 52(94c), *98*, 1969(25, 26a, 26b), 1970(25), 1976(25, 26a, 26b), 1977(26a, 26b), 1999(126), 2001(133), *2030, 2034*, 2436(128), *2460*, 2489(137b), *2562*
Neuret, J. 207(128), *260*
Neuy, A. 372(57), *427*, 626(78, 79), *699*
Neville, A.G. 1074(26), *1102*
New, D.G. 1823(152), *1864*
Newlands, M.J. (233), *262*
Newman, T.H. 1214(77), *1231*
Newton, M.D. 1910(60b), 1911(60b, 63), *1927*
Newton, M.G. 2429(94, 95), *2459*
Newton, S.Q. 113(25), *118*
Newton, W.E. 1590–1592(82), *1596*
Ng, L.V. 2319, 2320(14), *2355*
Ng, S. 1474, 1475(246), *1532*
Nguyen, K.A. 1911, 1913(73, 74a), *1927*
Nguyen, M. 1053(395), *1062*
Nguyen, M.T. 161, 163(61, 62), *178*, 710(25a), 775, 951, 952(238a), 1008, 1009(297a), 1019, 1048(320), 1050(371), *1059–1062*, 2155(135), 2160(168), *2173, 2174*, 2487, 2511(128), 2556(426), *2561, 2568*
Nguyen-Dang, T.T. 1911(67), *1927*
Ni, H. 891(126), *1056*
Ni, Y. 2153, 2154(134), *2173*
Nibbering, N.M.M. 1135(265), *1142*
Niburg, S.C. 1080(47), *1102*
Nicholas, J.B. 184, 221, 223(21), *257*
Nicholas, J.N. 1126(206), *1140*
Nichols, P.J. 1681(123), *1685*, 2039(4), *2122*
Nicholson, B.K. 2107(216–218), *2127*
Nickisch, K. 1668(15), *1682*
Nicol, P. 2220(19b), *2241*
Nicolaides, A. 34(73a, 73b), 35(73a), 36(73a, 73b), 37(73a), *97*, 1132(246), *1141*
Nicolau, K.C. 2448(175), *2461*
Nicoll, J.S. 109(10), *117*
Niecke, E. 209(135), 213(147, 152), 252(331), *260, 265*, 1051(384–386), *1062*, 2145, 2146(96), *2172*
Niedner, R. 2364(12), *2398*
Nieger, M. 209(135), 213(147, 151, 152), 252(331), *260, 265*, 762(131), *777*, 1051(384–386), *1062*, 1851(312), *1868*, 2145, 2146(96), *2172*
Nieh, H.-C. 1644(195), *1664*
Nielsen, C.J. 224(196), *261*
Nielsen, R.B. 1992(100), *2033*
Nielson, D.G. 1391(141), *1441*
Nielson, P.W. 2246(4b), *2310*
Nies, J.D. 1473–1475(236), *1532*
Niessner, R. 2327(58), *2357*
Nifant'ev, J.E. 2152(116), *2173*
Nihm, G. 10(33d), *96*
Nihonyanagi, M. 1733(143), *1788*
Niibo, Y. 1608(67), *1662*
Niihara, K. 2250, 2261(20), *2310*
Niiranen, J.T. 169(114), *180*, 518(58), *553*
Nikiforova, T.I. 1501(369), *1536*
Nikitin, E.V. 1201(35), *1230*
Nikonov, G.I. 2056(45), *2123*
Nikson, D.F. 219(173), *261*
Nile, T.A. 1692, 1718, 1719(20), 1723(127), 1733(20), 1735(151), *1786, 1788*
Nilsson, M. 1668(13), *1682*
Nilsson, Y. 452(129), *491*
Nimlos, M.R. 794(54), *823*
Nishi, K. 389(107), *428*
Nishibayashi, Y. 1743(209–211), 1745(209), 1747(209–211), 1748(209), 1749(211), *1790*
Nishida, C.R. 2326(45b), *2356*
Nishida, F. 2326(45b), 2351(241), *2356, 2361*
Nishida, R. 1216, 1217(87), *1231*
Nishida, S. 2428(90), *2459*
Nishide, K. 1861(362), *1868*
Nishidie, H. 1803(72), *1863*
Nishigaki, S. 741, 742(82), *776*

Nishiguchi, I. 1226(114), *1232*
Nishihara, H. 2236, 2237(116), *2244*
Nishii, H. 2250(23a), 2252, 2253(23a, 32a), *2310, 2311*
Nishimura, G. 2250, 2252, 2253(23b), *2311*
Nishimura, H. 1771(310), *1792*
Nishimura, K. 917, 998, 1002(184), *1058*, 1325(134), *1335*, 1990(95a, 95b), 2007–2009(95a), *2033*, 2416(43), 2429(107), *2458, 2459*
Nishimura, M. 925, 998, 1000, 1002(192), *1058*
Nishimura, T. 1871(12), *1891*, 2424(71), *2459*
Nishimura, Y. 892(129), 895(133, 134), 925(129), 935(133), 936, 970(134), 974(133), 975(134), *1056*, 1259(69, 70), *1307*, 1325(136, 139), *1335*, 2523(316), *2566*
Nishio, K. 510(67), *511*, 1809(97–99), *1863*
Nishitani, K. 1832(204), *1865*
Nishiwaki, K. 1190, 1192(10b), *1230*
Nishiwaki, N. 2522(296), *2565*
Nishiyama, H. 742(83), 776, 1688, 1733(1), 1743(1, 190–194), 1745(191), 1746(191–193), 1747(191, 192), 1748(191), *1785, 1789*
Nishiyama, K. 1151(47), *1182*, 1246(38), *1306*, 1548(40), *1578*
Nishiyama, T. 1291(174), 1293(179), *1309*
Nitta, S. 1728, 1729(136, 137), *1788*, 1860(349), *1868*
Niwa, H. 976(251), *1059*, 1303(206), *1310*, 1325(141), *1335*
Niwano, M. 2350(221), *2361*
Nix, M. 1374, 1375, 1377(85), *1439*
Nizamov, I.S. 1870(6), *1891*
Niznik, G.E. 1618(109), *1662*
Nogami, K. 1715(101), *1787*
Nogami, M. 2349(207, 209), 2350(207), *2361*
Noheda, P. 478(230), *493*, 1756(250, 251, 253), 1757(250), *1790*
Noirot, M.D. 578(160), *593*
Nojori, R. 709(19), *775*
Nolan, S.P. 2040(10), *2122*
Noll, J.E. 1500(357), *1536*
Noll, W. 117(48), *118*, 1582(7), *1594*, 2330(77), *2357*
Noltemeyer, M. 222(189), 235(239, 240), *261, 263*, 1011(305, 307), 1034(353), 1036(354), 1038(353, 354), 1043(307, 354), 1044(305, 307, 354), 1048(305, 307), *1060, 1061*, 1431(231), *1444*, 2061, 2066, 2068(59), *2123*, 2416(52), *2458*, 2486(119), 2552(412), 2553(414), *2561, 2568*
Noltmeyer, M. 203(106), 207(122, 129), 209(130), 229(217), 232(225), 238(251), *259, 260, 262, 263*, 648(121), *700*
Nomoto, E. 2346, 2347(183a), *2360*
Nomura, H. 2522(292), *2565*
Nomura, R. 1228(123), *1232*
Nomura, T. 2009(162), *2035*
Nomura, Y. 417(194), *429*, 1861(357), *1868*
Nonaka, T. 1190, 1192(10a, 11, 12), 1196(12), 1197(12, 24, 27), 1198, 1200(29), 1215(81, 82), 1217(83), 1218(91, 93), 1219, 1220(91), *1230, 1231*
Noorbatcha, I. 2485(112), *2561*
Nora, G. 2183(40), *2214*
Norbeck, D.W. 1606, 1611(48), *1661*
Noren, G. 833(32), *854*, 1169, 1180(168), *1185*, 1266(90), *1307*, 2515, 2519(278), *2565*
Noren, G.H. 1169(177), *1185*, 1331(184, 185), *1336*, 2485(115), 2491, 2514, 2518, 2519(142), *2561, 2562*
Norman, A.D. 219(177), *261*
Norman, N.C. 247(294), *264*, 1723(128), *1788*, 2141, 2149(56), *2172*
Normant, J.F. 1837(225), *1866*
Norris, P.E. 66(115a), *99*
North, M. 1671(39), *1683*
North, M.A. 2484, 2502(105), *2561*
Norton, J.R. 77, 81(153), *100*
Nosaka, Y. 2330, 2354(80), *2357*
Nosova, V.M. 1450(47), *1526*
Nöth, H. 129(36), *151*, 209(132), 216(162), *260*, 795, 799(63c), 800(70a), *823*
Nott, A.P. 473(204), *492*
Nouiri, M. 1769(288), *1791*
Noura, S. 1550(42), *1578*
Novak, B.M. 2344(159–162), *2359*
Novikova, E.O. 1454, 1467(89), *1527*
Novinson, T. 2349(208), *2361*
Nowick, J.S. 1620, 1628(116), 1630(148), 1653, 1655(116), *1663*
Noyori, R. 443(110), *490*, 1688(2), *1785*, 2424(71–74), *2459*
Nozaki, H. 783(14), *822*, 1608(67), *1662*
Nozaki, M. 810(93b), *824*
Nugent, W.A. 1672(53), *1683*, 1963(1b), 1986(1b, 79, 80a), 1988, 1998(1b), *2030, 2032*, 2434(120), 2436(126), *2460*
Nundnberg, W. 1915(83), *1927*
Nunn, C.M. 211(143), *260*, 793(49b), *823*
Nürnberg, O. (129), *1788*, 2096(187), *2126*
Nuttall, R.L. 159, 160(54), *178*
Nyburg, S.C. 250(306, 307), *264*, 844(56), *855*, 859(3), 865, 866(38), 880, 881(3, 38), 913, 922(38), 977, 978, 981(3), 985, 986(3, 38), 992, 996(3), *1054*, 1146(22), *1181*, 1601, 1609, 1645(24), *1661*, 2404(10, 12), *2458*

Nyguyen, V.Q. 561(54), *591*
Nyman, M.D. 2295(143), *2315*
Nyulászi, L. 30–32(68, 69), 53(68), 64(68, 69), 65(69, 114), 66(69), *97, 99*, 1449, 1460, 1463, 1490(19), *1525*

Oba, M. 1151(47), *1182*, 1246(38), *1306*, 1548(40), *1578*
Oba, S. 1965, 1996, 1998(15), *2030*
Obata, K. 829(26), *854*
Obayashi, M. 783(14–16), 789(15), 795(16), 801(15), *822*, 1861(362), *1868*
Ober, C.K. 2237(117), *2244*
Oberdorfer, R. 209(135), *260*
Oberhammer, H. 780, 794(4b), *821*
Obermann, U. 1743(201, 205, 218), 1746(201), *1789, 1790*
Obermeier, E. 2340(135), *2359*
Obermeyer, A. 203(112), *259*
Obi, K. 2522(287, 288, 290), *2565*
Obora, Y. 482(237), *493*, 2434(124), *2460*
O'Brien, D.H. 559(7), *590*
O'Brien, J.J. 2477(65), 2521(280), 2545(380), *2560, 2565, 2567*
O'Brien, M.K. 1766(276, 277), *1791*
O'Brien, S.C. 69(124a–c), *99*, 1929(1b), *1960*
Obuchi, H. 1322(106, 109), *1334*
Obuchi, K. 459(155), *491*
Ocaña, M. 2350(225), *2361*
Ochiai, H. 925, 998, 1000, 1002(192), *1058*, 2429(107), *2459*
Ochiai, M. 1193(14a), *1230*, 1797(48), 1828(187, 188), 1860(356), *1862, 1865, 1868*
Ochial, K. 2345(174), *2360*
O'Connell, B. 284(54), *351*, 487(268), *494*, 1340, 1341, 1343(11b), *1437*
O'Connor, G. 1858(340), *1868*
O'Connor, J.M. 2134(28), *2171*
O'Connor, N.J. 269(4), *350*
Oda, M. 1325(133, 134), *1335*
Oda, T. 1689(9), *1786*
Odagi, T. 2424(72), *2459*
Odaira, Y. 2428(89), *2459*
Oddershede, J. 68(122a), *99*, 517(39a, 39b), *552*
Odell, K.J. 1462, 1467(156a), *1529*
O'Dell, R. 1341, 1347(27), *1437*
Oehlert, W. 213(153), *260*
Oehlschlager, A.C. 791(43a–d), *823*
Oehme, H. 198(85), *259*, 884(101–110), 885(101–110, 115, 117, 118), 886(103), 889(106–110, 117), 917(104, 107, 108), 922(107), 925(105–110, 117, 118), 926(107–109, 117, 118), 927(109), 928(110), 937(103–105, 107, 117), 938(103–105, 117), 939, 940(108), 957(104, 107), *1056*
Oehme, I. 2321(24a), *2356*
Oeler, J. 780(2f, 4b), 794(4b), *821*
Oezkar, S. 1700, 1701(57), *1786*
Ofori-Okai, G. 1873(23), *1892*
Ogasawara, J. 845(63), *855*, 885(114), 887(114, 119), 911(119, 177), 922, 934, 957(119), 987, 989(114), *1056, 1057*
Ogata, K. 1808(94), *1863*
Ogata, T. 2011(166), *2035*
Ogawa, A. 1091(58h), *1103*, 1559(68), *1579*, 1678(99), *1684*
Ogawa, K. 279(36, 37), *351*, 487(262), *494*, 812(99b), *824*
Ogawa, M. 2116(241), *2127*
Ogden, J.S. 1162(110), *1183*
Ogilvie, J.F. 1017(314), *1061*, 1128(221), *1141*, 1159(95), *1183*
Ogilvy, A.E. 1593(114), *1597*
Ogino, H. 1260(72, 73), *1307*, 1332(204), *1337*, 1411(189), *1443*, 2069, 2071(75), 2087(165–167), 2088(168–177), 2090(165, 166, 168, 169), 2092, 2093(165), 2094(166, 168, 169), 2105(171–173), *2124, 2126*
Ogino, K. 1133(247), *1141*
Ogino, N. 2529(338), *2566*
Ogiso, A. 1774(320), 1777(329), *1792*
Ogiwara, J. 716, 748, 750(39), *775*, 2404(8), *2457*
Ogoshi, H. 2104, 2108, 2109(204), *2127*
Ogoshi, S. 2110(224), *2127*
Ogune, N. 1671(35), *1683*
Oguni, N. 1671(40), 1672(47), *1683*
Ogura, F. 194(67), *258*, 1828(190), *1865*
Oh, A.-S. 1460, 1467, 1503(160), *1529*
Oh, D.H. 1834(213), *1866*
Oh, D.Y. 1828(189), *1865*
O'Hair, R.A. 1126(204), *1140*
O'Hair, R.A.J. 1026, 1027(337), *1061*, 1134(253), *1141*
Ohanessian, G. 5, 6, 24, 25, 87(15b, 15c), *95*, 499(22), *510*
Ohannesian, G. 1110(61b), *1137*
Ohannesian, L. 1740(175), *1789*
Ohara, Y. 1731(138, 139), *1788*
O'Hare, D. 2150, 2153, 2154(110), 2162(175), *2173, 2174*
Ohashi, H. (110c), *825*
Ohashi, M. 1290(171), *1309*
Ohba, T. 2226(52), *2242*
Ohe, K. 1616(102), 1618(110, 111), *1662, 1663*, 1743(209, 211), 1745(209), 1747(209, 211), 1748(209), 1749(211), 1772(312, 313), *1790, 1792*, 2110(224), *2127*
Ohff, A. 2051, 2052(34), *2123*
Ohga, K. 1193(15), *1230*

Ohi, F. 1283(146), *1308*, 1328(156), 1332(199), *1336*
Ohira, S. 1675(84), *1684*
Ohkata, K. 1850(299, 300), *1867*
Ohkubo, K. 1951(23), *1960*
Ohl, K. 1585(35), 1587(58), *1595, 1596*
Öhler, E. 1550(44), *1578*
Öhme, H. 813(102), *824*
Ohmizu, H. 1206(46), 1228(122), *1230, 1232*
Ohmura, H. 1755, 1756(247), *1790*
Ohnishi, Y. 1091(58c), *1103*
Ohno, A. 1091(58c), *1103*
Ohno, M. 459(152), 467(186), *491, 492*, 1610, 1630(78), 1639(177, 178), *1662, 1664*, 1798(53), 1811(110), *1863, 1864*
Ohno, S. 2028(209), *2036*
Ohno, T. 1226(114), *1232*, 1871(12), *1891*
Ohsaki, H. 892(131), 895(132), *1056*, 1252(56), 1253(57, 58), *1306, 1307*, 1325(135, 137, 138), *1335*
Ohsako, Y. 1318(63), *1333*
Ohshiro, Y. 1822(150), *1864*
Ohshita, J. 194(67), *258*, 870(48), 883(98, 99), 884, 885(112, 113), 888(112, 113, 120, 121, 123), 889(121), 892(131), 895(132), 922–924, 928(112), 937, 938(113), 957(112), 958(120), 960–962(99), 976(251), 986, 988, 990(121), 998(291), *1055, 1056, 1059, 1060*, 1252(56), 1253(57, 58), 1303(206), *1306, 1307, 1310*, 1325(135, 137, 138, 141), *1335*, 1973(49), 2016(175b), 2017(177a, 177c), *2031, 2035*, 2429(99), *2459*, 2513, 2516, 2526(258), *2564*
Ohsumi, K. 400(138), *428*
Ohta, T. 1674(61), *1683*, 1751(237), *1790*, 2376(46), *2399*
Ohtaki, T. 1051(375), *1062*
Ohtsuka, S. 2546(390), *2568*
Oikawa, S. 2546(390), *2568*
Oike, H. 2117(248, 249), 2120(248), *2128*, 2430(113–116), *2460*
Oiki, H. 2395, 2396(80), *2400*
Ojima, I. 1688(3), 1699(50), 1700(3), 1705(3, 80), 1709(80), 1714(3, 80), 1717(3), 1718(3, 106–108), 1722(3), 1723(3, 106, 108), 1733(3, 140, 143), 1735(3, 50, 153), 1736(3, 154), 1737(154), 1742(3), 1743(3, 187, 188), 1751, 1752, 1772(3), 1774(321–323), 1776(107, 323, 324, 328), 1777(328), 1778(331, 334), 1779(324, 334, 335), 1780(336), 1781(328, 337), 1782(336–338, 340–342), 1783(343, 344), *1785–1789, 1792*
Oka, K. 1771(297), *1791*, 2471(35), 2475(59a, 59b, 60), 2486(111), *2559–2561*
Okabe, K. 279(32), *351*, 1264(85), 1269(102, 103), *1307, 1308*
Okabe, M. 438(94), *490*

Okada, A. 782(13a), *822*
Okada, F. 561(55, 56), *591*, 1125(184), 1133(184, 247–249), *1140, 1141*
Okada, J. 1910, 1911(61), *1927*
Okada, K. 1810, 1830(106), *1864*
Okada, T. 1754, 1755(245), *1790*
Okahata, Y. 2342(144), *2359*
Okai, K. 1675(84), *1684*
Okajima, S. 390(113), 391(114), *428*, 1833(207, 208), *1866*
Okamoto, M. 1590(83), 1591(91, 94–96, 98), *1596, 1597*, 2533(357–359), *2567*
Okamoto, T. 1206(45), *1230*, 2145(91), *2172*
Okamoto, Y. 361(12), *426*
Okamura, K. 2253(34), 2261, 2263(70), 2272(92a–c, 93a, 93b, 94c), *2311, 2313, 2314*
Okano, K. 1601, 1602, 1604, 1622, 1628(20), *1660*
Okano, M. 564(89), *591*, 1207(49, 52), 1210, 1211(61), 1214(73), 1229(127), *1230–1232*, 1591(97), *1597*, 2533(356), *2567*
Okasaki, R. 990(272), *1060*
Okawa, T. 200(93), *259*, 279(29, 31), *351*, 1157(79, 80), *1183*, 2183(38, 42), *2214*, 2220(17a), 2225(45), *2240, 2241*
Okazaki, K. 1322(110), 1325(120), *1334, 1335*
Okazaki, R. 3(2c, 3c, 8), 10, 11, 13, 16, 17, 32(8), *95*, 122(20e), *151*, 248(304), *264*, 279(36, 37), 284(61), 350(165), *351, 352, 354*, 420(199), *429*, 437(81), 486(261), 487(81, 262), *490, 494*, 578(164), *593*, 812(99b), *824*, 827, 828(12), *854*, 859(6), 867(43), 985(269), 990, 994, 995(6), *1054, 1055, 1059*, 1064, 1068(8b, 10), 1083(10), 1090(10, 56), 1091(10, 57, 59a, 59b, 60, 61), 1092(10, 59a, 59b, 60, 61, 63), 1093(59a, 59b, 60, 61, 64, 65, 66a, 66b, 67, 68a, 68b), 1100(76), *1101–1103*, 1163(122), *1184*, 1282(143), 1288(166), 1289(167), *1308, 1309*, 1332(206), *1337*, 1677(93), *1684*, 1873(25), 1879(52), 1882(64), 1884(70, 71), *1892, 1893*, 1897(27), 1904, 1907(48–50), *1926*, 2486(124), 2492(145), 2494(166), 2497(181), 2499(166, 187), 2500(145), 2501(187), 2537(181), *2561–2563*
Okazaki, S. 867(44, 46), 868(46), 870(44, 47, 49–51), 974(46), *1055*
O'Keefe, A. 1117(87), *1138*
O'Keefe, M. 116(42, 44), *118*
Okinoshima, H. 1075(31), *1102*, 1325(152), *1335*, 1973(41a, 41b), *2031*
Okoda, M. 2265(82a), 2271(89), *2313, 2314*
Oksinoid, D.W. 2132(18), *2171*
Oku, T. 2080(137–139), *2125*

Okuda, J. 272(18), *350*, 2133(24), 2135(30), 2136, 2137(31), 2138(38), 2147(31, 98, 100), *2171, 2173*
Okuda, K. 2522(291), *2565*
Okui, S. 1203, 1212, 1213(40), *1230*
Okumoto, H. 1834(212), *1866*
Okumura, M. 530(93), *554*, 1107(17), *1136*, 2265(82a), 2271(89), *2313, 2314*
Okumura, Y. 1590(84, 84, 87), 1591(99), *1596, 1597*
Okutsu, T. 1151(47), *1182*, 1246(38), *1306*
Olah, G.A. 46(90a), *98*, 401(139), *428*, 517(45), 520(70), 521(74, 77), 522(77), 524(70), 529(77, 89), 530(74, 89), 532(77), 533(77, 89), 534(77), 540(70, 105), *553, 554*, 562(66), 563(81–83), 567(81, 83, 104), 568(83, 109), 569(81, 83, 104, 109, 115, 117, 119), 570(109, 117), 571(119), 572(104, 119), 578(161), 583(82), 584(109, 117, 188), 585(109), 588(117), *591–593*, 616(53), 645(113), 647(113, 116, 117), 652(129, 130), 661(113), 662(138), 677(150, 153), 688(159, 160b), 690(165), *698, 700, 701*, 793(47a), *823*, 1108(20), *1136*, 1224(108), 1225(110–112), *1232*, 1408(182), *1443*, 1602, 1604, 1655(30), *1661*, 1667(10), 1669(24), 1674(57, 58), 1675(81, 82), *1682–1684*, 1740(175), *1789*, 1818(137), *1864*, 1870(9), *1891*, 2141(60), *2172*, 2224(34), *2241*, 2453(202), *2461*
Olbrich, G. 829(18), *854*
Oldfield, E. 270(13), *350*
Oleff, S.M. 2289(125, 126), *2315*
Oleneva, G.I. 500, 501(26), *510*, 587(195), *594*, 1390(130), 1393(148), 1395(156, 157), 1396(156–158), 1397, 1399(156), *1441, 1442*
Olenova, G.I. 243(271–273), 245(286), *263, 264*
Oliván, M. 1720, 1724(118, 119), 1725(119), 1732(118, 119), (129), *1788*, 2095(186), 2096(187), 2097(188, 189), 2108(186), *2126*
Olivella, S. 2442(159, 160), *2460*
Oliver, C. 1886(85), *1893*
Oliver, J.P. 782(12a), 784(22, 24), 785(26, 27), 786(22, 27), 787(22, 24), 822, 1883(67), *1893*, 2422(69), *2459*
Oliver, S.N. 2352(269), *2362*
Olivucci, M. 919, 921(187, 188), *1058*, 1320(81), *1334*
Ollivier, J. 435, 457(56), *489*
Ollmann, G.W. 1228(125), *1232*
Olmstead, M. 207(119), *259*
Olmstead, M.M. 485(248), *493*, 793, 794, 799(49a), *823*, 1043(360), *1062*, 2134, 2147(25), *2171*
Olmsted, W.N. 849(77, 78), *855*

Olofson, R.A. 739, 740(73), *776*
Olson, R.E. 1601(14, 21), 1602, 1604(14), 1608(61), 1621, 1624, 1625(14), 1632, 1639(61), 1653(21), 1658(14), *1660, 1661*
Olsson, L. 281, 282, 296(43), *351*, 522, 527, 528, 532–536, 538, 540, 547(83a), *554*, 567(105), 568(108), 569(105, 114), 570(114), 584(105, 108, 114), *592*
Olsson, T. 1668(13), *1682*
Olstikova, L.L. 1298(190), *1309*
Oltmanns, K.K. 1593(109), *1597*
Omata, T. 2392, 2394(78), *2400*
Omel'chaiko, S.I. 2218(7), *2240*
Omori, M. 2272(92b, 92c, 93a, 93b), *2314*
Omoto, K. 506(43), *511*
On, P. 1080(43a), *1102*, 2406(18), *2458*
Onak, T. 516(35c), *552*
Onaka, M. 1670(30), *1683*, 1740(176, 180), *1789*
Onan, K.D. 1390, 1396, 1399(124), *1441*
Oñate, E. 2097(188), 2098, 2108(190, 191), *2126*
O'Neal, H.E. 155(30), 160(55), 161(57), 162(55), 171(55, 119), 172, 173(55), 175(141), 176(30), *178, 180*, 1156, 1167(68), *1182*, 2465(12), 2466(14, 15), 2476(61, 63), 2477(68–75, 77), 2487(129), 2502(192), *2559–2561, 2563*
Onishi, H. 1689(9), *1786*
Onmura, H. 2236, 2237(116), *2244*
Ono, Y. 1590(83), 1591(91–98), *1596, 1597*, 2533(356–359), *2567*
Onodera, T. 310(130), *353*
Onopchenko, A. 1693(24), 1705(78, 79), 1714(79), *1786, 1787*
Onozawa, S. 1698, 1779(46), *1786*
Onsager, L. 1467, 1468, 1476, 1477(188b), *1530*
Onuki, H. 2522(301), *2565*
Onyszchuk, M. 236(254), *263*
Oomen, M.W.J.L. 2354(283), *2362*
Opel, A. 811(97b), *824*
Operti, L. 559(21, 22), 560(38, 39), *590*, 1120(114–118), *1138*
Opitz, K. 1910, 1911(60e), *1927*
Oppenstein, A. 1109(41), *1137*
Oppolzer, W. 476(221), *493*
Oprea, A. 2535(369), *2567*
Orfanopoulos, M. 1740(178), *1789*, 1858(342), *1868*
Organ, H.M. 1823(155), *1864*
Orgel, L.E. 1603(40), *1661*
Oribe, T. 1699, 1735(48), *1786*
Orita, A. 1618(110, 111), *1662, 1663*
Oriyama, T. 1675(73, 75), *1683, 1684*
Orler, B. 290(70–72), *352*
Orlicki, D. 2545(385), *2568*

Orlinkov, A. 1669(24), *1682*
Orlov, G.I. 1458, 1467(127), *1529*
Orlova, N.A. 1396(163), *1442*
Oro, L.A. 1720(117–119), 1721(120–122), 1723(120), 1724(118–122), 1725(119), 1732(117–119), (129), *1788*, 2095(186), 2096(187), 2097(188, 189), 2098(190, 191), 2108(186, 190, 191), *2126*
Orpen, A.G. 1723(128), *1788*
Orphanos, P.O. 2530(343), *2566*
Orshav, V. 2322(41), *2356*
Ortí, E. 72, 73(141), *100*
Orvlova, T.M. 186(46, 47), *258*
Osajima, E. 436, 482(66), *490*, 611, 613, 614(41), 632, 633(86a, 86b), 634(86a, 86b, 88a, 88b), 635(88a, 88b), *698, 699*
Osajina, E. 371(53), *427*
Osaka, M. 1591(94), *1597*
Osakada, K. 1692, 1720(17), *1786*, 1872(19, 20), *1892*, 2107, 2109(211), *2127*
Osawa, E. 72, 74, 75(147), *100*
Osawa, S. 72, 74, 75(147), *100*
Osborn, H.M.I. 1619(113), *1663*
Osborne, A.G. 2153(131), 2154(131, 136), *2173, 2174*
Oschmann, W. 1986, 2011(76b), *2032*
Oshibe, Y. 2236, 2237(116), *2244*
Oshima, K. 462(166), *492*, 1555, 1556(59), 1572(93), *1578, 1579*, 1620(121), *1663*, 1680(117), *1684*, 1797(48), 1807(82), 1810, 1830(105, 106), *1862–1864*
Oshita, J. 807, 808(87a–c), *824*, 2424(78, 79), *2459*
Osijama, E. 363(31), *426*
Osipov, O.A. 1470(214), *1531*
Oskam, A. 2169(198), *2175*, 2534(361c), *2567*
Ossola, F. 2053(36), *2123*
Ostapenko, S.S. 2348(201), *2360*
Ostaszewski, B. 1701, 1702(63, 64), *1787*
Osterholtz, F.D. 2333(96), *2358*
Ostrander, R.L. 37(81), *97*, 1759(262), *1791*, 2026(200), *2036*
Otani, S. 2261(58), *2312*
Otani, Y. 2261(58), *2312*
Otera, J. 1292(176), *1309*, 1608(66, 67), *1662*, 1740(179), *1789*
Otero, A. 2055–2057(44), *2123*
Othani, H. 810(92), *824*
Otón, J.M. 2350(215, 217, 223, 224), 2352(223, 224), 2353(274–276), *2361, 2362*
Otsubo, T. 194(67), *258*, 1828(190), *1865*
Otsuda, T. 2001(130a), *2034*
Otsuji, Y. 581(176), *593*, 1257(67), 1290(173), 1291(174), 1292(177), 1298(191, 192), *1307, 1309, 1310*, 1849(293), *1867*
Otsuka, T. 829(17), *854*, 1247(41, 42), *1306*, 2496(178), *2563*

Ott, H. 2333, 2339(95), *2358*
Otten, U. 1873(24), *1892*
Otter, J.C. 1405(173), *1443*
Otto, M. 1456, 1467, 1497, 1501(119), *1528*
Ottolenghi, M. 2319(13), 2320(13, 18, 19), 2321(22, 23a, 23b), 2322(18, 19, 22, 23a, 23b, 25), 2326(43, 44b, 46a, 46b, 50), 2327(19, 44b), 2340(132), 2347(44b), 2348(18, 19, 25, 44b, 198), *2355, 2356, 2359, 2360*
Ottolengi, M. 2322(30), *2356*
Ottoson, C.-H. 1146, 1147(23), *1181*, 1277(132), *1308*, 1606(50), *1661*
Ottosson, C.-H. 522(83a, 80b), 527, 528, 532–536, 538(83a), 540(83a, 80b), 541, 542(108), 547(83a, 108, 113), 548, 549(113), *554, 555*, 564(90, 95), 567(105), 568(110), 569(90, 105), 570(90, 110), 584(105), *591, 592*, 1434, 1435(245), *1444*
Ottosson, H. 569, 570, 584(114), *592*
Ouahab, L. 2075, 2091(111), *2125*
Oussaid, A. 580(166), *593*
Ovchinnikov, Y. 958(241), *1059*
Ovchinnikov, Yu.E. 186(44, 46, 47), 190(50), 196(71), 226(207), 237(245, 248, 249), 243(277), *258, 262–264*, 1374, 1376(92b), 1392(143), 1396(159, 162), *1440–1442*, 1450(46), 1454(103, 104), 1455(113), 1460(157), 1461(46), 1464(103, 104), 1467(103, 104, 113, 157), *1526, 1528, 1529*
Ovchinnikova, Z.A. 1475(253), *1532*
Ovechinnikov, Y.E. 2044, 2051(19), *2123*
Overman, L.E. 396(126–130), *428*, 1794(14), 1816(129–132), 1817(133, 134), 1821(147), 1832(205, 206), *1862, 1864–1866*
Oviatt, H. 2346(181), *2360*
Owen, M.J. 2234(100–102, 105b), *2243*
Owen, P.W. 773(154), *778*
Owen, W.J. 1196–1198(21), *1230*
Ozai, T. 2251, 2253(27b), *2311*
Ozaki, S. 435, 477(59), *489*, 1852(314, 315), *1868*
Ozanne, N. 423(210), *430*
Ozawa, F. 2110(222), 2116(238), 2120(222), *2127*
Ozin, G.A. 1899(38), *1926*

Pacansky, J. 997(286), *1060*
Pachaly, B. 1585(42, 43), 1587(57), 1593(113), 1594(115), *1595–1597*
Pachinger, W. 476(221), *493*
Pacl, H. 66(117a, 117b), 67(117b), 68(117a, 117b), *99*, 997, 998(285b), *1060*, 1125(174, 175), *1139*, 1163, 1164(127–129, 136), 1165(128, 136, 147), 1171, 1172(182),

Pacl, H. (*cont.*) *1184, 1185*, 1286(162), *1309*, 2514(269), *2565*
Paddon-Row, M.N. 2001(138a, 138b), *2034*
Padwa, A. 466(179), *492*, 732(64), 744(89), 776, 1236(13), *1305*
Pae, D. 867(41), 900(145b), *1055, 1057*
Pae, D.-H. 2421(58), *2458*
Pae, D.H. 1166(159), *1184*, 1241(28), 1283(150), *1306, 1309*, 1331(194), *1336*, 2475, 2479, 2483, 2487(57), *2560*
Paek, C. 2083(144), *2125*
Paetzold, P. 209(133), *260*, 2494(163), *2562*
Page, E.M. 161(63), *178*
Page, P.C.B. 433(7), *488*, 1601, 1602, 1604, 1608(7), 1620(7, 117), 1622(7, 124, 125), 1633(154), 1634, 1642, 1643(160), 1653(7, 125), *1660, 1663, 1664*
Pai, Y.M. 307, 308(113), *353*
Paidorova, I. 517(39a, 39b), *552*
Pailler, R. 2261(60), 2281(113), *2312, 2314*
Paine, R.T. 209(132), 216(162), *260*
Pajdowska, M. 304(108), *353*
Pajonk, J.M. 2322(34), *2356*
Pakusch, J. 1128(229), *1141*
Pal, I.P. 1456, 1457, 1460, 1467(118), *1528*
Pala, P. 528, 535(87c), *554*
Palacios, F. 2145(90), *2172*
Palavit, G. 1471, 1472(227, 228), 1485(300), *1531, 1534*
Palazotto, M.C. 2153–2155(132), *2173*
Palio, G. 410(167), *429*
Palm, V.A. 1477(268), *1533*
Palma, P. 2044, 2051, 2052(18), *2122*
Palmer, A.W. 2322(28), *2356*
Palmer, M.H. 1996(108d), *2033*, 2131(13), *2171*
Palmer, R. 243(269), *263*, 2063, 2066(68), *2124*
Palomo, C. 423(209), *430*, 446(113), 460(162), 476(225), *490, 491, 493*, 1226(115), *1232*
Palovich, M. 1644(199), *1664*
Pamidi, , P.V.A. 2328, 2329(74), *2357*
Pan, H.-Q. 1542(15), *1577*
Pan, J. 1763(265), *1791*
Pan, L.-R. 1812(111, 112), *1864*
Pan, W.D. 1671(36), *1683*
Pan, Y. 804(78), *824*, 2025(199), *2036*, 2113, 2120(228), *2127*, 2433(118), *2460*
Pan, Y.M. 2252(31b), *2311*
Pancrazi, A. 1671(34), *1683*
Pande, L.M. 1456, 1457, 1460, 1467(118), *1528*
Pandey, G. 1292(178), *1309*, 1885(75), *1893*
Panek, J.S. 410(166), *429*, 433(8), 434(17), 435(46), 437(82), 441(17), 454(46), 459(153), 477(228), 487(82), *488–491, 493*, 1638(172), 1639(179), 1650(212), *1664*,
1665, 1794(23, 30), 1800(58, 59), 1804(80), 1806(81), 1845(271, 272), 1847(283), 1850(310), 1851(310, 311), 1852(316), 1853(318), 1859(346, 347), *1862, 1863, 1867, 1868*, 2388(70), *2400*
Pang, Y. 186(38), *258*, 1767(278), *1791*, 2443(162), 2448(180), 2450(182, 183), *2461*
Paniez, P. 2231(84), *2243*
Panke, G. 411(170), *429*
Pankratov, I. 2321(24b), 2328, 2329(69), *2356, 2357*
Panne, K.H. 1237(18b), *1306*
Pannek, J.-B. 404(151), *428*, 547(112c), *555*, 1850(298, 301), *1867*
Pannel, K.H. 1602, 1604, 1605(32), *1661*
Pannell, K. 2213(110), *2216*
Pannell, K.H. 437, 485(77, 78), *490*, 1260(71, 75), 1261(76–78), 1262(80, 81), *1307*, 1332(205), *1337*, 2083(147–155), 2087(150, 155), *2126*, 2142(66, 67), 2143(71, 72), 2155(135), 2160(164, 165, 168, 171), *2172–2174*, 2213(109), *2216*
Panov, V. 2499(186), *2563*
Pansare, S.V. 1669(26), *1682*
Panster, P. 2330, 2340(78), *2357*
Pantano, C.G. 2289(133), *2315*
Pantel, R. 2231(84), *2243*
Pantoni, R. 2522(286), *2565*
Panunzio, M. 435, 456(55), 459(157), *489, 491*
Panyachotipun, C. 441(103), *490*
Papai, I. 515, 516(14b), *551*
Papousek, D. 1049(366), *1062*
Paquette, L.A. 149(59), *152*, 460(163), *491*, 621, 627(82), *699*, 739–741(74), 776, 1212(69), *1231*, 1546(34), *1578*, 1628, 1630(145), *1663*, 1794(13), 1842(257, 258), *1862, 1867*
Parbhoo, B. 186(45), *258*, 2227(72), *2242*, 2439(144), *2460*
Paredes, M.C. 478(230), *493*
Parent, D.C. 1120(130), 1121(130, 141), *1138, 1139*
Parisel, O. 1050(370), *1062*, 1129(234), *1141*
Parish, K.V. 2322(37), *2356*
Park, C.M. 1544(25), *1578*
Park, D.S. 2328, 2329(74), *2357*
Park, E. 1758(255, 256), *1790*
Park, J. 2153(124), 2159(156), 2162(124), *2173, 2174*
Park, S.-B. 742(83), *776*, 1743, 1746(193), *1789*
Park, S.-S. 2430(116), *2460*
Párkányi, L. 1237(18b), *1306*, 1460(154, 158), 1461(137, 138), 1462(147–149, 152, 154), 1463(149, 166), 1467(137, 138, 147–149, 152, 154, 158, 166), 1477(154), 1478(137,

1491(154), 1503(152), 1508(388, 389), 1510(166), *1529, 1530, 1537*
Parkanyi, L. 183(11), 200(95), 226(199), *257, 259, 261*, 437, 485(78), *490*, 1449(14, 15, 19), 1450(42), 1454(78), 1460(14, 15, 19), 1462(42), 1463(15, 19), 1490(19), 1509(395), *1525–1527, 1537*, 2155(135), *2173*
Parker, D. 476(218), *493*
Parker, D.J. 509(65), *511*, 1374, 1376(93), *1440*
Parker, J.K. 374(60), *427*
Parker, K.A. 1563(73), *1579*
Parker, P.J. 2137, 2138(37), *2171*
Parker, V.B. 159, 160(54), *178*
Parker, W. 1469(205), *1531*
Parlar, H. 2231(81e), *2242*
Parnes, Z.N. 1794(11), *1862*
Parr, J. 1416(202), 1417(202, 204), *1443*
Parr, R.G. 515, 516(14d), *551*
Parrick, J. 2262(71), *2313*
Parry, D.M. 448(119), *491*
Parshall, G.W. 2039(3), *2122*
Parsons, P.J. 1563(74a), *1579*
Parvez, M. 1885(77), *1893*
Paryk, A. 709, 713, 714, 763, 764(22), *775*
Pascual-Ahuir, J.L. 528, 535(87d), *554*
Pasero, M. 1696, 1717(35), *1786*
Pasman, P. 2142(68), *2172*
Pasquerello, A. 69(131a, 131b), *100*
Passamonti, U. 1610, 1630(80), *1662*
Passing, G. 2261(54a), 2269, 2270(88c), *2312, 2313*
Passler, T. 876(77, 78), 913(77), 917, 931(78), *1055*
Pasto, D.J. 2446(169), *2461*
Patai, S. 514(1), *551*, 1609, 1645(73), *1662*, 1794(6), *1862*, 2130, 2135, 2138(2), *2170*
Patalinghug, W.C. 210(141), *260*
Patel, A.M. 2220(20), *2241*
Patel, H.V. 1682(125), *1685*, 1834(211), *1866*
Patel, S.K. 398(135), *428*
Paterson, I. 441(102), *490*
Pattenden, G. 474(210), *492*, 1561(70, 72), 1575(99), *1579*
Patterson, C.H. 1120(136), *1139*
Patyk, A. 978(253), *1059*, 1075(34), *1102*, 1147, 1148(28), 1162, 1168(108), 1181(201), *1182, 1183, 1185*, 1286(160), *1309*, 2493, 2520(153), *2562*
Patz, M. 626(77), *699*
Pätzold, U. 2194(75, 76), 2197(80), *2215*
Pau, C.F. 407(162), *429*
Pauer, F. 203(108), 235(238), *259, 263*
Paul, G.C. 169(113), *179*, 770(148), *778*, 1095(71), *1103*
Paul, M. 207(125), *260*, 2059(53), 2060(53, 54), 2066, 2067(53), *2123*, 2159(154), *2174*
Pauletto, L. 2416(52), *2458*, 2553(414), *2568*
Pauling, L. 104(3), 117(49), *117, 118*, 157(36), *178*, 183, 228, 250(3), *257*, 356(3), 357(7), *426*, 521, 522(72), *553*, 584, 588(186), *593*, 652(128), *700*
Pauncz, R. 1166(156), *1184*
Paunz, R. 52, 53, 59, 60, 62(98d), *98*
Pauson, P.L. 77, 81(153), *100*, 2163(177), *2174*
Paust, J. 635(89a), *699*
Pavanaja, U.B. 863, 864, 932, 934, 982(25), *1054*
Pavlov, K.V. 1450(36), *1525*
Pavlov, V.A. 1743(222), *1790*
Pavlova, T.O. 1875(36), *1892*
Pawelec, Z. 2494(163), *2562*
Pawlenko, S. 1794(4), *1862*
Payne, J.S. 238(252), *263*, 284(55, 56), *351*, 1340, 1341(12, 13b), 1342, 1343, 1347(13b), *1437*
Payne, S.E. 1168(165), *1185*, 2493, 2520(155), *2562*
Payzant, J.D. 1372(68), *1439*
Pazdernik, L. 229(223), *262*, 435, 456(49), *489*
Paz-Sandoval, M.A. 1512(403), *1537*
Peach, M.E. 563(73), *591*
Pearce, A. 389(108), *428*
Pearce, R. 1212(68), *1231*
Pearsall, M.-A. 1168, 1169(166), *1185*, 1331(186), *1336*, 2493, 2520(156), *2562*
Pebler, J. 2159(159), *2174*
Péchy, P. 1603(42), *1661*
Peddle, G.J. 828(15), *854*
Peddle, G.J.D. 1144(5), *1181*, 1600(4), 1601, 1604(16), 1611, 1612(4), 1618(16), *1660*
Pedersen, L. 515(18a–c), *551*
Pedersen, S.F. 272(19), *350*
Pederson, L.A. 707(16), *775*
Pederson, S. 1828(193), *1865*
Pedley, J.B. 154, 159, 163(9), 164(9, 77), 165(9), 166(77), 177(143), *177, 179, 180*, 2534(365), *2567*
Pedrielli, P. 1542, 1544, 1546, 1559, 1570(16), *1577*
Pedroviejo, J.J. 2548(400), 2549(404), *2568*
Pedulli, G.F. 1542, 1544, 1546(16), 1549(41a), 1559, 1570(16), *1577, 1578*
Peel, J.B. 1470, 1471(218), *1531*
Pegon, P.M. 2351(244), *2361*
Pehk, T. 1475(255), *1532*
Peleg, S. 2348(202a, 202b), *2361*
Pelizzetti, E. 2231(87), *2243*
Pellacani, L. 460(159), *491*, 1843(267), *1867*
Pellerin, B. 1022, 1025(332), *1061*
Pelletier, R. 1804(80), *1863*

Pellinghelli, M.A. 2055–2057(44), *2123*
Pellissier, H. 1798(52), 1801(68), *1863*
Pellmann, A. 804(78), *824*
Pelton, R.H. 1771(301), *1791*
Pena, J.M.S. 2352(272), 2353(276), *2362*
Penezek, S. 2218(3), *2240*
Peng, C.H. 2295(143), *2315*
Peng, Ch. 1468(195), *1530*
Peng, Y. 2523(319), *2566*
Peng-Fei, F. 2039–2041(5), *2122*
Penkett, C.S. 1563(74a), *1579*
Penmasta, R. 645(110, 111), *700*
Penn, B.G. 2253, 2254(40a–c), *2311*
Penneau, J.-F. 2016(174), *2035*
Penner, G.H. 362(20), *426*
Penny, D.A. 1091(58g), *1103*
Penzenstadler, E. 948(225–228), *1058*
Pepekin, K.L. 1469(202, 203), *1531*
Pepikin, V.I. 154, 156(11–13), 159–161, 164, 165(11), 166(12, 13), *177*
Perchonock, C. 1998(122b), *2034*
Percy, J.M. 374(59), *427*
Pereyre, M. 475(215), *493*, 1616(101), *1662*, 1994(104), *2033*
Pérez, A.J. 2151(117), 2152(115, 117), *2173*
Perez, D. 1738(156), *1788*
Periasamy, M. 647(118), 677(151, 152), *700, 701*
Pericas, M.A. 2442(159, 160), *2460*
Perich, J.W. 1547(36), *1578*
Perichon, J. 1222(101), *1231*
Perlmutter, P. 435(40, 47), 451(40, 127, 128), 455(47), *489, 491*, 1840(247), *1866*
Perozzi, E.F. 1353(35a), *1438*
Perrin, H.M. 1147(26), *1182*
Perriot, L. 1997, 1998(115), *2034*
Perrot, M. 1064, 1069, 1082–1084, 1086, 1099(9), *1101*, 1387, 1388(114, 117), *1440, 1441*
Perrotta, A.J. 2253, 2255(42, 43), 2256(43), 2257(42), *2312*
Perrotta, E. 1849(296), *1867*
Pertsin, A.J. 2235(109), 2236(110a, 110b), *2243*
Pertz, S.V. 2226(55), *2242*
Perutz, R.N. 1299(196), *1310*, 1707, 1708(82), *1787*, 2102(201, 202), 2109(202), *2127*
Perutz, S. 2237(117), *2244*
Perz, R. 1487(306), *1534*, 1738(161–165), *1789*, 1824(157), *1864*
Pestana, D.C. 437, 484(76), *490*, 2055, 2056(43), 2062(63, 64), 2066, 2068(63), 2105(64), *2123, 2124*
Pestunovicch, V.A. 1474(260), 1475(258), 1476(259–261), 1477(259, 260), 1506(261), 1510(260, 261), *1532*
Pestunovich, A.E. 1374, 1376(92a), 1390, 1392(131), *1440, 1441*, 1454, 1467(102), 1501(369), *1528, 1536*
Pestunovich, S.V. 243(277), *264*, 1374, 1376(92a, 92b), *1440*
Pestunovich, V.A. 243(271, 272, 274), *263, 264*, 501(28), *510*, 1341, 1343(21b), 1373(75c), 1374, 1377(94), 1378(101, 102), 1381(107, 108), 1390(128–132, 134–137, 139), 1391(140), 1392(129, 131, 132, 135–137, 139, 146, 147), 1393(147, 150–153), 1394(151), 1395(139, 152, 153), 1396(151, 160, 164, 165), 1397(139), 1398(137, 139, 147, 152, 153), 1399(160, 164, 165), 1400, 1401(137, 139), 1403(139), 1404(147, 150, 153), 1418(210, 212, 216), 1420(210, 212), 1422(216), 1423(212, 216), *1437, 1439–1442, 1444*, 1449(12, 16, 28), 1452(57), 1460, 1462(155), 1463(28, 163, 165), 1464(165), 1465(12, 174), 1466(12, 28, 176–179), 1467(12, 155, 163, 165, 174, 176–179, 181, 188a), 1468(188a, 195), 1469(12, 198), 1470(12, 28), 1471(12), 1472(198), 1473(12, 235, 238), 1474(235, 242, 243, 263, 266), 1475(242, 251, 255), 1476(28, 57, 165, 188a, 238, 243, 263), 1477(174, 188a, 243, 251, 263, 264, 266), 1478(28, 174, 188a, 263, 266), 1479(57, 242, 271, 274), 1483(293), 1485(301, 302), 1487(308, 309), 1488, 1489(16), 1490(309, 316, 319, 320), 1491(16, 321–323), 1494(319), 1503(155), 1506, 1508(174), 1510(12, 174), 1521(417), 1523(12, 28, 174, 419), *1525, 1526, 1529, 1530, 1532–1534, 1537*
Petakhov, V.A. 1312(16), *1332*
Petasis, N.A. 1640(185), *1664*
Peters, E.-M. 219(179, 183–185), 229(214–216, 219, 220), 232(184, 185, 226–228), 241(260–262), *261–263*
Peters, E.M. 2495(171), *2562*
Peters, J. 1910(59), 1913, 1914(78, 79), *1927*
Peters, K. 190(53), 194(62), 198(90–92), 200(99), 219(179, 183–185), 229(213–216, 218–220), 232(184, 185, 224, 226–228), 241(260–262), 245(287), *258, 259, 261–264*, 783(19b), 791(44a), *822, 823*, 1268(99), 1283(153), *1307, 1309*, 1910(58), 1913, 1914(76), *1927*, 1989(91), *2033*, 2182, 2183(37), *2214*, 2473(50), 2486(110, 116, 117), 2493(159), 2495(170–172), 2514(172), *2560–2562*
Peters, K.S. 2437(134), *2460*
Peters, M.A. 2225(46), *2241*
Petersen, J.L. 1418(218b), *1444*, 1909, 1910(56), *1927*
Peterson, D. 618(58), *699*
Peterson, D.J. 415, 417(183), *429*, 788–

790(28a), 796, 801(72), *822, 824*, 836(44a), *854*
Peterson, M.A. 1677(94), *1684*, 1743(217), *1790*
Peterson, W.R.Jr. 2141(59), *2172*
Petitjean, S. 2235(107), *2243*
Petrich, S.A. 2450(182, 183), *2461*
Petrie, M.A. 213, 215(154), *260*
Petrosyan, V.S. 226(202), *262*, 1461, 1467(141), 1496(345), 1497(141, 351), *1529, 1535*
Petrov, A. 1582(8), *1594*, 1996(108a), *2033*
Petrova, E.I. 1989(88), *2033*
Petrova, R.G. 1714(95), *1787*
Petty, J.T. 1168(165), *1185*, 2493, 2520(155), *2562*
Petuchov, L.P. 1390, 1392(135), *1441*
Petukhov, L.P. 1373(75c), 1381(107), *1439, 1440*, 1449(16), 1474(263, 266), 1476(263), 1477, 1478(263, 266), 1483(293), 1485(301), 1487(308, 309), 1488(16, 310), 1489(16), 1490(309, 316, 319, 320), 1491(16, 321–323), 1494(319, 335), *1525, 1532–1535*
Petukhov, V.A. 1471(220), *1531*
Petzow, G. 2264(75), 2269, 2270(88f), *2313*
Peukert, M. 2253(36), *2311*
Peukert, S. 409, 410(165), *429*
Peyerimhoff, S.D. 1107(14), 1128(226), *1136, 1141*
Peyghambarian, N. 2350(237), *2361*
Peynircioglu, N.B. 1700, 1701(57), *1786*
Peyronel, G. 1899(36a, 36b), *1926*
Pfab, W. 2163(178), *2174*
Pfaff, O. 2368, 2371, 2374(38), *2399*
Pfister-Guillouzo, G. 184, 203(25), *257*, 1018, 1042, 1046, 1047(316), 1051, 1052(381), *1061, 1062*, 1968, 1982–1984(22), *2030*
Pflug, J.L. 198(84), *259*
Pham, E.K. 198(89), *259*
Phan, A. 457(149), *491*, 1638(167), *1664*
Philip, D. 1930, 1934, 1959(3), *1960*
Philipp, G. 290(69), *352*, 2335(109, 111), 2336(109, 113), 2351(111), *2358*
Phillips, J.C. 115(31), *118*, 1121(138, 139), *1139*
Phtaki, T. 205(117), *259*
Piana, H. 2072, 2090(85), *2124*
Piana, K. 2333, 2334(98, 101), 2337(101), 2339(127), *2358*
Pianese, G. 1558(65), *1578*
Pibouleau, L. 2545(382), *2567*
Picard, J.-P. 460(162), *491*, 844(56), *855*, 1226(115, 116), *1232*, 1601, 1609(24), 1614(96), 1645(24), *1661, 1662*, 1861(359), *1868*

Pickardt, J. 784, 787(23), *822*
Pickel, P. 520, 524, 525(69), *553*, 564(88), 581(178), 583, 586(88), *591, 593*
Pickett, J.B. 2010, 2013(164), *2035*
Picotin, G. 1820(143), *1864*
Pidcock, A. 1462, 1467(156a), *1529*, 2115(233), *2127*
Pidcock, A.P. 2110, 2120(223), *2127*
Pidum, U. 283(45), *351*
Pidun, U. 527–529, 533, 538–540, 547(86), *554*, 568, 571, 573(112), *592*
Piecha, G. 2226(51), *2242*
Piel, H. 1268(99), 1283(153), *1307, 1309*, 2486(110, 116, 117), 2495(171), *2561, 2562*
Pieper, U. 211(144, 145), 234(243), 235(239), 252(327), *260, 263, 265*, 1043, 1044, 1048(359), *1062*
Pierce, A.B. 1584(29), *1595*
Pierce, J. 1611(83), *1662*
Pierce, O.R. 437, 483(73), *490*, 2228(74), *2242*
Pierce, R.A. 878, 931, 945(79), *1055*
Pierre, I. 1144(3), *1181*
Pietro, W.J. 518(62), *553*
Pietropaolo, D. 1210(60), *1231*
Pietrusza, E.W. 1722(126), *1788*
Pietzsch, M. 2379(58), *2399*
Pihter, E. 2194, 2203(74), *2215*
Pijolat, M. 2261(65), *2313*
Pike, R.A. 813(105a), *825*
Pike, R.D. 1505(376), *1536*
Pike, R.M. 1415, 1420(200), *1443*
Pike, S. 1642(192), *1664*
Pikethly, R.C. 2340(140), *2359*
Pikies, J. 219(179, 180), 222(195), *261*, 2364(19, 28), *2398*
Pikl, R. 2061(58, 59), 2066(59), 2068(58, 59), 2078, 2091, 2095(121), *2123, 2125*
Pilcher, A.S. 434, 448(33, 34), *489*
Pilcher, G. 154(7), 161, 162, 164(64), 165(7), 171–173(64), *177, 178*
Pilling, M.J. 167(88), *179*, 518(57, 58), *553*
Pillot, J.-P. 411(169), *429*, 1833(209), *1866*, 2246, 2250, 2251, 2253, 2254(6a), 2261(60), 2272, 2273(6a), 2281(113), 2285(6a), *2310, 2312, 2314*
Pilz, M. 2337(116, 117), *2358*
Pinard, E. 1546(34), *1578*
Pinas de Piedade, M.E. 168(108), *179*
Pinelli, R.F. 361(18), *426*
Pines, A. 309(114), *353*
Pines-Rojanski, P. 2322, 2347(33), *2356*
Pinkas, J. 485(255), *493*, 1520(415), *1537*
Pinna, L. 1743(200), *1789*
Pinon, M. 437, 485(78), *490*
Pinsker, O.A. 456(146), *491*
Pinter, E. 2201(85), 2209(102), *2215, 2216*

Piper, U. 1043(358), *1062*
Piqueras, M.C. 72(141, 142), 73(141), 75–77(151), *100*
Pirazzini, G. 1611(86), *1662*
Pirrung, M.C. 1297(188), *1309*
Pirzard, S. 2143(75), *2172*
Pisano, D. 8(29), *96*
Piserchio, M. 2486(120), *2561*
Pitrolo, R.L. 1590(86), *1596*
Pitsch, I. 2337(124), *2358*
Pitt, C.G. 358(8), *426*, 1211(65), *1231*, 1312(15, 17), 1313(17), 1322(97, 98, 100, 103), *1332, 1334*
Pittman, C.U.Jr. 685(157), *701*
Pius, K. 1125(182), *1140*
Plagne, J.L. 1583(19), 1587(63), *1595, 1596*
Plain, A. 1118(99), *1138*
Planalp, R.P. 437, 484(75), *490*
Planchenault, D. 1576(102), *1579*
Planeix, J.-M. 2466(18), *2559*
Plant, C. 1467, 1471(189), *1530*
Plantier-Royon, R. 1628(142), *1663*
Platz, M.S. 1147(26), *1182*
Plavac, N. 880, 985, 986, 992, 996(91), *1056*, 1602, 1604, 1609(31), *1661*
Plawsky, J.L. 2351, 2354(259), *2362*
Ple, G. 269(6), *350*
Plenio, H. 2150(103), *2173*, 2251, 2253, 2255, 2257(25e), 2269(87), *2311, 2313*
Plitt, H.S. 1127(215), *1140*, 1162(113), *1183*, 1313(33), 1314(32, 33), 1316, 1317(43), *1333*
Plitzko, C. 1434, 1437(251), *1445*
Pluddemann, E.P. 2233(98), *2243*
Plueddemann, E.P. 2339(125), *2358*
Plugaru, R. 2545(384), *2567*
Plunkett, S.J. 434, 440(15), *489*
Pluta, C. 1970(29a, 29b), 1974(29b), *2031*
Pock, R. 624, 625(74), *699*
Poggi, G. 1210(60), *1231*, 1312, 1317, 1318(5), *1332*
Pohl, E.R. 2333(96), *2358*
Pohl, R.L. 2011(165), *2035*
Pohl, S. 200(94, 101), 205(113), *259*, 578(165), 584(183), *593*, 804(78), *824*, 1025(334), *1061*, 1077(38), *1102*, 1268(98), 1283(152), 1304(213), *1307, 1309, 1310*, 1881(61), 1885(73), *1893*, 1976(51), *2031*, 2134(27), 2140(45), *2171*, 2179(9), 2184(45), 2185(46, 47), *2214*, 2472(47–49), 2492(146, 147), 2493(160, 161), 2494(165), *2559, 2560, 2562*
Pohlmann, S. 915(182), *1057*
Poinsignon, C. 2339(126, 129), *2358, 2359*
Poirier, M. 287(64), *352*, 503(32), 505(35), 508(56), *511*, 1374, 1375(76, 77, 79), 1376(79), 1377(77, 79, 99), 1378(77, 99), 1380(79), 1381(106), 1382(76, 77), 1383(76, 77, 110, 111), 1385(76, 77), 1386(76), 1412(191, 194), 1414(191), 1422, 1423(194), 1424(77), 1427(194), *1439, 1440, 1443*, 1738(164), *1789*
Pokorna, D. 1237(17), *1306*
Pokrovska, N. 1494(338), *1535*
Pola, I. 436, 479(61), *489*
Pola, J. 861, 862(19), 863(24, 25), 864(25, 31–33), 932, 934(25), 981(31–33), 982(25, 31–33), 1006, 1008(295, 296a), 1009(296a), *1054, 1060*, 1237(15–17, 18a), *1305, 1306*
Polak, N. 2322, 2323(39), *2356*
Polborn, K. 88(166), *101*, 125, 126, 128, 129(32a, 32b), *151*, 250(313), *264*, 876(77, 78), 879(84), 913(77), 917, 931(78), 932, 940, 981(84), *1055*, 2205(92), *2215*
Poleschner, H. 1885(76), *1893*
Polevaya, Y. 2319(13), 2320(13, 19), 2322, 2327(19), 2340(132), 2348(19, 198), *2355, 2359, 2360*
Poliakoff, M. 1260(74), *1307*
Polishchuk, A.P. 254(335), *265*
Politzer, P. 515, 516(14a), *551*
Polizzi, C. 1696, 1717(35), *1786*
Poll, W. 1890(105), *1894*
Polla, M. 403(148), *428*
Pollack, S. 642(105), *700*
Pollitte, J.L. 2051, 2052, 2056, 2057, 2064(35), *2123*
Polmanteer, K.E. 2228(74), *2242*
Polowin, J. 2043, 2051(16), *2122*
Polsterer, J.P. 1550(44), *1578*
Polt, R. 1677(94), *1684*
Polyakov, A.V. 194(66), *258*
Polyakov, Y.P. 2429(98), *2459*
Polzer, T. 2061, 2066–2068(60), *2123*
Pompeo, M.P. 2075, 2091(106, 107), *2125*
Ponomarenko, V.A. 1220(96), *1231*, 1582(8), *1594*
Ponomarev, A.N. 2231(85), *2243*
Pons, P. 1220(97), 1222(101, 102), 1224(106), *1231, 1232*
Poon, H.S. 1669(18), *1682*
Poon, Y.C. 250(306), *264*, 844(56), *855*, 859(3), 865, 866(38), 880(3, 38, 92), 881(3, 38), 913, 922(38), 924(92), 977, 978, 981(3), 985, 986(3, 38), 992, 996(3), *1054, 1056*, 1080(47), *1102*, 1146(22), *1181*, 1601, 1609, 1645(24), *1661*, 2404(10, 12), *2458*
Popall, M. 2337(116, 117, 122), 2339(129), 2346(178), *2358–2360*
Pope, E.J.A. 2343(150), 2349(212), 2350, 2351(222), *2359, 2361*
Popelis, J. 1451(48b, 49, 51), 1461, 1487(49), 1494(338), *1526, 1535*

Popkova, V. 946(222), *1058*
Pople, J.A. 6(21a), 34(74a, 74b), *95, 97*, 164(80), 167(97, 98), 168(102), 171(97, 128), 173, 175(102), *179, 180*, 250(319), *265*, 515(22), 517(55), 532(99a), *552–554*, 597(2), 606(23), 615(23, 44), 620(63), *697–699*, 705, 707(12), *775*, 781(5), *821*, 835(37), *854*, 901(147), *1057*, 1107(8–10), 1119(111), 1123(156), 1126(191), *1136, 1138–1140*, 1150(40), 1166(149, 157), *1182, 1184*, 1475, 1477(248), *1532*
Popowski, E. 576(139), *592*, 1507(381), 1508(381, 389), *1536, 1537*
Porchia, M. 2053(36), *2123*
Pornet, J. 402(144), *428*, 1802(71), 1809(100, 101), *1863*
Pörschke, K.R. 1970(29a, 29b), 1974(29b), *2031*
Porte, H. 2261(59), *2312*
Portella, C. 1628(142), 1641, 1652(188), *1663, 1664*
Porter, C.J. 1134(259–261), *1142*
Porter, J.R. 1822(151), *1864*
Porz, C. 213(152), *260*
Porzel, A. 458(151), *491*
Porzel, K. 435, 458(57), *489*
Pöschl, U. 2187(54), 2189(54, 60, 62), 2204(60), *2215*
Pospisilova, M. 2322(27), *2356*
Posset, U. 2061(58–60), 2066(59, 60), 2067(60), 2068(58–60), 2078, 2091, 2095(121), *2123, 2125*, 2339(130), *2359*
Potochnik, S.J. 1588(71), *1596*
Potzinger, P. 163(71), 174, 175(71, 138), *179, 180*, 829(18), *854*, 862(21), *1054*, 1125(185), *1140*
Pouliquen, L. 2227(58), *2242*
Poulos, A. 1553(53), *1578*
Pouskouleli, G. 2246, 2250, 2253, 2254, 2273, 2285(6g), *2310*
Pouxviel, J. 2349(208), *2361*
Pouxviel, J.C. 2347(190, 194), 2350(190, 194, 220), 2351(194), *2360, 2361*
Powel, D.R. 1094(69), *1103*
Powell, D. 870(53), 1031, 1043, 1045(347), *1055, 1061*
Powell, D.J. 348(158), *354*
Powell, D.R. 3(6), 37(84, 85), 38(84), 49(85), *95, 97*, 248(300), 250(312), *264*, 323, 324(141), 348(162), *353, 354*, 829(20), 842(50, 52), *854, 855*, 859(5), 981, 984(5, 259), 987(5, 259, 271), 988(271), 989(259, 271), 993, 995, 996(5), 1000(5, 259, 271), 1002, 1003(259), 1004(259, 271), *1054, 1059, 1060*, 1145(16), *1181*, 1264(84), 1266(89, 92), 1283(148), 1285(159), *1307–1309*, 1881(60), 1882(63), 1887(94), *1893,*
1894, 1897, 1908(26), *1926*, 1972, 1988, 1998, 2023(40), 2024, 2026(198), *2031, 2036*
Power, M.D. 770, 772(149), *778*, 1860(350), *1868*
Power, P.P. 207(119), 213, 215(154), *259, 260*, 485(248), *493*, 793(49a, 50), 794, 799(49a), *823*, 1043(360), *1062*, 2134, 2147(25), *2171*, 2534(364), *2567*
Power, W.P. 52, 54(95e), *98*, 2539(373a), *2567*
Pozhidaev, Yu.N. 1501(369), *1536*
Prack, E.R. 2253, 2261, 2265(45b), *2312*
Prakash, C. 434, 448(35), *489*
Prakash, G.K.S. 520(70), 521(74, 77), 522(77), 524(70), 529(77, 89), 530(74, 89), 532(77), 533(77, 89), 534(77), 540(70, 105), *553, 554*, 562(66), 563(81–83), 567(81, 83), 568(83, 109), 569(81, 83, 109, 117, 119), 570(109, 117), 571, 572(119), 578(161), 583(82), 584(109, 117, 188), 585(109), 588(117), *591–593*, 616(53), 645, 647(113), 652(129, 130), 661(113), 662(138), 688(159, 160b), 690(165), *698, 700, 701*, 1224(108), 1225(110–112), *1232*, 1408(182), *1443*, 1870(9), *1891*, 2141(60), *2172*, 2224(34), *2241*
Prakash, G.S. 401(139), *428*
Prakasha, T.K. 1406(174a, 174b, 175a, 175b), *1443*
Prangé, T. 1671(34), *1683*
Prasad, P.N. 2326(47), 2327(56), 2350(226), 2352(262), *2356, 2357, 2361, 2362*
Pratt, L.M. 1767(281), *1791*
Pratt, N.E. 1809(96), *1863*
Predieri, G. 2340(137), *2359*
Prein, M. 467(185), *492*
Preiner, G. 873(60–63), 876(72–74), 910(62), 917, 931(72, 73), 932(62), 937(62, 63), 938(63), 940(72–74), 941(60, 62), 942(62), 943(63), 969(60–63), 1012(309), 1034, 1035, 1038(351), 1039(356), 1051(351), *1055, 1061, 1062*, 1077(40), 1079(42), *1102*
Prescott, P.I. 1923(96), *1928*
Presilla-Márquez, J.D. 68(122a), *99*, 1164(142), 1175(185–187), 1176(186, 187), *1184, 1185*
Preston, D. 2349(208), *2361*
Preuss, F. 232(229), *262*
Preuss, H.-W. 2523(317), *2566*
Preuss, R. 1128(226), *1141*
Preut, H. (232), *262*, 1875(29, 30), *1892*
Prewitt, C.T. 110(11), *118*
Prewo, R. 186(31), *257*
Pribytkova, J.M. 2133(20), *2171*
Price, G.J. 2220(20), *2241*
Price, J.D. 2446(171), 2450(187), *2461*
Price, J.P. 2450(189), *2461*

Prichtl, F. 1841(255), *1866*
Prigge, H. 1323(113), *1335*
Prignano, A.L. 1693(27), *1786*
Prince, R.H. 1321(89), *1334*
Principe, L.M. 2005(152, 153), *2034*
Prinzbach, H. 1543(23), *1577*
Priotton, J.J. 2351(244), *2361*
Priou, C. 1052(388), *1062*, 1064, 1069(9), 1075(35c), 1082–1084, 1086, 1099(9), *1101, 1102*, 1387(114, 116, 120), 1388(114, 116), 1389(120), 1390(123b), *1440, 1441*, 2069, 2071, 2084, 2089, 2093, 2094(76), *2124*, 2481, 2483, 2553(97), *2561*
Pritchard, T.N. 317, 319–322(140), *353*
Pritzkow, H. 213(148), 216(161), 252(332), *260, 265*, 781(7), *821*, 1052(390, 392), 1053(390, 391, 393), *1062*, 1064, 1080(7a), 1081(7a, 48), *1101, 1102*, 2107, 2109(212), *2127*
Prizkow, H. 1077(39), *1102*
Probert, G.D. 1671(43), *1683*
Probst, M. 1879(56), *1893*
Probst, R. 581(181), *593*, 1075(35c), *1102*, 1374, 1375(80, 81), 1377(81), 1379(80, 81), 1380(81), 1386(112, 113), 1387(113), 1389(112, 113), 1390(123b), 1430, 1434(81), *1439–1441*, 2059(51, 52), 2060(52), 2066, 2067(51, 52), *2123*, 2481, 2483, 2553(97), *2561*
Probsty, R. 2528(336), *2566*
Prochazka, S. 2289(134), *2315*
Prock, A. 434, 441(17), *489*
Procopio, L.J. 1032, 1040(348–350), 1042, 1043(349, 350), 1044, 1045, 1050(349), *1061*, 2049(30–33), 2051(30, 31), 2052(30), 2055(32), *2123*
Procter, G. 1727–1730(135), *1788*, 1841(254), 1845(274–277), *1866, 1867*
Proctor, C.J. 1134(259–261), *1142*
Proctor, G. 1859(348), *1868*
Prodnuk, S. 1126(204), *1140*
Prokopyev, V.Yu. 1457, 1458, 1467(121), 1497(349), 1513(121), *1528, 1535*
Proksch, P. 2363(9), *2398*
Prout, T.R. 219(177), *261*
Prudent, N. 1639(181), *1664*
Prud'Homme, C. 1584(25–27), *1595*
Prud'homme, C.C. 2251–2253, 2255, 2257(25a), *2311*
Prunet, J. 1671(34), *1683*
Prüsse, T. 1114(71a), 1115(72), *1137*
Pryde, D.C. 1671(44), *1683*
Przyjemska, K. 232(224), *262*
Puccetti, G. 2354(281), *2362*
Pucetti, G. 2351, 2354(257, 258), *2362*
Puddephatt, R.J. 2110, 2120(221), *2127*
Pudelski, J.K. 2150(110), 2153(110, 125, 134), 2154(110, 134), 2162(125), *2173*
Pudova, O. 296(84), *352*, 780(1e), *821*
Pudova, O.A. 1434(233), *1444*, 1449(7a, 7b, 29), 1451(49), 1460(7a, 7b, 29), 1461(29, 49), 1462, 1463, 1465(29), 1487(49), *1525, 1526*, 1727, 1730(134), *1788*
Pudovik, A.N. 1497(346, 347), *1535*, 1870(6), *1891*
Pudowa, O. 485(252), *493*
Puff, H. 1918(89), *1927*
Pukhnarevich, V.B. 436, 479(61), *489*, 1458, 1467(122), *1528*, 1692(19), *1786*
Pulay, P. 517(37, 44b), *552, 553*
Pulido, F.J. 1819, 1842(140), 1860(352), *1864, 1868*
Puranic, D.B. 1283(155), *1309*
Puranik, D. 1153, 1154(55), *1182*
Puranik, D.B. 795, 799(62a, 62b), *823*, 847(72, 73), *855*, 903(161, 162), 922(190), 993, 996(161), *1057, 1058*, 1154, 1155(64), *1182*, 2485(107), 2503(195), 2515(277), 2534(107), *2561, 2563, 2565*
Purs, N.E. 2145, 2146(94), *2172*
Pushechkina, T.A. 1454, 1467(81, 102), *1527, 1528*
Puyear, S. 517(44b), *553*
Pyne, S.G. 252(329), *265*
Pyun, C. 1610, 1630(77), *1662*
Pyykkö, P. 793(51a), *823*

Quaiser, S. 1225(110), *1232*
Quayle, P. 416(185), *429*
Quimbita, G.E. 566(100), *591*
Quintana, X. 2353(275), *2362*
Quintard, J.P. 475(215), *493*, 1994(104), *2033*
Quittmann, W. 1501(364), *1536*

Raab, G. 417(190), *429*
Raabe, C. 1907(52b), *1926*
Raabe, G. 3, 63(1a), *94*, 122(20b, 20d), *150*, 485(253), *493*, 713(35), *775*, 827, 828(7, 8), 829, 832(8), 844(7, 8), *854*, 860(7, 8), 901(8, 153, 156), 919, 940, 959, 977(8), 981, 991(7, 8), 992, 995(156), 1010(8), *1054, 1057*, 1064(4b, 4e), 1068(4b), 1071(4b, 4e), 1075, 1077, 1080(4e), *1101*, 1144, 1145, 1158, 1160(1), 1167(163), 1168(163, 164), 1169(163), 1170(164), *1181, 1185*, 1234(5, 6), 1284(158), *1305, 1309*, 1609, 1645(73), *1662*, 1964, 2001(7), *2030*, 2471(42), 2513–2515(254), 2523(318), *2559, 2564, 2566*
Raabe, G.R. 247(292), *264*
Raban, M. 1418, 1420, 1422–1426(211b), *1444*
Rabe, J.A. 2285(117b, 117c), *2315*

Rabezzana, R. 559(21, 22), 560(39), *590*, 1120(114, 115), *1138*
Rabinovich, L. 2321(24b), 2328(67, 68), *2356, 2357*
Rabinovitch, I.B. 154, 165, 166(10), *177*
Rabrenovic, M. 1134(262), *1142*
Rachlin, V.I. 1373(75c), *1439*
Radacki, K. 232(230), *262*
Radecky, A. 1454, 1467(97), *1527*
Radom, L. 6(21a), 34(73a, 73b, 74e, 74f), 35(73a), 36(73a, 73b), 37(73a), *95, 97*, 515(22), 517(55), *552, 553*, 620(63), *699*, 1125(180), 1132(246), *1140, 1141*
Radonovich, L.J. 243(269), *263*, 2063, 2066(66, 68), *2124*
Radu, N.S. 2040(8, 9), 2041(11), 2053(39), *2122, 2123*
Radzig, V.A. 2530(344–347), 2531(344–346, 351), 2532(346), *2566*
Radziszewski, J.G. 1019(321), 1021(326), 1044(321, 326), 1045, 1047(321), *1061*, 1128(225), *1141*, 1289(169), *1309*
Rafeiner, K. 2364(23, 29), *2398, 2399*
Raff, L.M. 2485(112, 114), 2510(233–235), *2561, 2564*
Raghavachari, K. 34(74a–d), 69(130), *97, 100*, 167(98), 168, 173, 175(102), *179*, 560(34, 35), *590*, 1119(106–109, 111, 112), 1120(134, 135), 1127(214), *1138–1140*, 1331(182, 183), *1336*, 1898(34), *1926*, 2519(266), *2565*
Rahim, M.M. 437, 484(75), *490*
Rahimian, K. 1923(94), *1928*
Rahm, A. 475(215), *493*, 1994(104), *2033*
Rahn, J.A. 2250, 2253, 2255–2257(21), 2276(103), 2290, 2291, 2293(138a–c), *2310, 2314, 2315*
Rai, A.K. 217(166), *261*
Raimondi, M. 2509(227), *2564*
Raimondi, S. 1843(267), *1867*
Rainbird, M.W. 2504(198), *2563*
Rainbird, N. 171, 172(126), *180*
Raithby, P.R. 374(59), *427*
Rajarison, F. 385(94), 386(102), *427, 428*
Rajca, A. 821(120), *825*
Rajca, G. 197(80), *259*
Rajkumar, A.B. 813(104), *825*, 1699(51), 1700(54), *1786*
Rakas, M.A. 2226(54), *2242*
Rakhlin, V.I. 1449(16), 1457, 1458, 1467(121), 1474, 1477, 1478(266), 1487(309), 1488(16, 310), 1489(16), 1490(309, 316, 319, 320), 1491(16, 321, 322), 1494(319, 335), 1496(344), 1497(349, 350), 1499(354, 355), 1513(121), *1525, 1528, 1533–1535*
Raksit, A.B. 1117(84), *1138*
Ralle, B. 194(62), *258*
Ralph, B. 2262(71), *2313*
Ramadan, N.A. 784, 787, 788(25), *822*
Ramaiah, P. 1669(24), 1675(77), *1682, 1684*
Ramanathan, K.N. 2326(47), *2356*
Rammo, A. 1434(238), *1444*
Ramondo, F. 2508(223), *2564*
Ramsden, C.A. 11(38), *96*, 183(6), *257*, 1149(34), *1182*
Ramsey, B.G. 1312(15, 18), 1320(79), *1332, 1334*, 1603, 1605(38), *1661*
Ramsey, T.M. 1743(226), *1790*
Ranaivonjatovo, H. 909(175), 1051(384), *1057, 1062*
Randolph, C.L. 1705(71), *1787*
Rane, A.M. 1845(273), *1867*
Rangappa, K.S. 520(67), *553*
Ranger, A. 2350(236), *2361*
Ranh, M.D. 2350(233), *2361*
Rankers, R. 1084(52), *1102*
Rankin, D.W.H. 296(82), *352*, 475(212), *492*, 1875(35), *1892*, 2131(7, 10, 14), *2171*
Ranzuvaev, G.A. 304(109), *353*
Rao, K.S.S.P. 1885(75), *1893*
Rapp, U. 1758(258), *1791*
Rappoli, B.J. 2076, 2091(120), *2125*
Rappoport, S. 2326(43, 46a, 46b, 50), *2356*
Rappoport, Z. 363(32b, 32c), *426*, 437(79), 485(79, 258), *490, 493*, 514(1), *551*, 782, 795, 796(10b), *822*, 1603, 1605(39), 1609, 1645(73), *1661, 1662*, 1794(6), *1862*, 2130, 2135, 2138(2), *2170*
Rascle, M.-C. 2387(68), *2400*
Rasika Dias, H.V. 52(95b, 95f), 54(95f), *98*, 2535(366), 2539(373b), *2567*
Rasmussen, J.K. 1670(29), *1683*
Rasol, G. 2224(34), *2241*
Rastah, C.L. 2140(46), *2171*
Raston, C.L. 210(141), 217(169), *260, 261*, 1377, 1378, 1380(97a), *1440*
Rasul, G. 521(74, 77), 522(77), 529(77, 89), 530(74, 89), 532(77), 533(77, 89), 534(77), 540(105), *553, 554*, 563, 567(83), 568(83, 109), 569(83, 109, 117, 119), 570(109, 117), 571, 572(119), 584(109, 117, 188), 585(109), 588(117), *591–593*, 652(129, 130), 662(138), *700*
Ratcliffe, R.W. 1607, 1634(58), *1661*
Rathmann, H. 863, 988, 990(27b), *1054*
Rathousky, J. 1600(3), *1660*
Ratner, B.D. 2237(120), *2244*
Ratovski, G.V. 1470–1472(217), *1531*
Ratovskii, G.V. 1471(222), *1531*
Ratsimihety, A. 1769(289), *1791*
Raubo, P. 1619(112), *1663*, 1670(32), 1679(112), *1683, 1684*
Rauchfuss, T.B. 1904(46, 47), 1906(47), 1909, 1910(46), *1926*

Rauhut, G. 517(44b), *553*
Rausch, M.D. 1991, 1992(96a, 96b), *2033*, 2141(55), *2172*
Rautschek, H. 2223(29a, 29b), *2241*
Ravagni, A. 2264(78), *2313*
Ravaine, D. 2339(129), *2359*
Ravindranathan, M. 635(89c), *699*
Rawal, V.H. 1573(96b), *1579*
Ray, P. 1414, 1424(195c), *1443*
Ray, U. 1120(120), *1138*
Raymond, K.N. 228(208), *262*, 1415, 1430(201a), *1443*
Raymond, M.K. 1313, 1314(33), *1333*
Rayner, D.M. 2522(284), *2565*
Razskazovskii, Yu.V. 2530(348, 349), 2531(348), *2566*
Razuvaev, G.A. 783(17, 18), *822*
Razzano, J.S. 2223(30b), *2241*
Reaoch, R.S. 2253, 2261, 2265(45a, 45c), *2312*
Réau, R. 762(132), 763(137), *777, 778*
Reber, G. 190(56), 203(104), 250(311), 252(323, 324), *258, 259, 264, 265*, 879, 932, 940, 944, 981, 988(83), 1011(301), 1021(327), 1033(301), 1042, 1043(327), *1055, 1060, 1061*, 1064, 1068(5b), *1101*, 1158(88), *1183*
Recatto, C.A. 2112, 2120(227), *2127*
Reddy, J.P. 414(181), *429*
Reddy, M.L.N. 1989(87), *2033*
Reddy, M.V.R. 1674(59), *1683*
Reddy, N.L. 476(218), *493*
Reddy, V.P. 662(138), 688(159), *700, 701*
Redfern, P.C. 34(74f), *97*
Reding, M.T. 1749(229), *1790*
Redondo, P. 1129(235), *1141*
Reed, A.E. 61(108), *99*, 526(85), *554*, 819, 820(118), *825*, 1109(34), *1136*, 1372, 1373(69), *1439*
Reed, C.A. 197(79), *259*, 283(46), *351*, 514(3a, 3b), 520(3a, 3b, 71), 521(3a, 3b, 76), 522(3a, 76), 524(3b, 71), 525(71), 529, 530(90, 91), 532(3a), *551, 553, 554*, 561(59), 567(103, 106), 569(106), 570(103, 106), 572(59, 103), 583(59, 103, 106, 185), 585(59, 106, 185), 586(103, 185), 587(106, 185), 588(59, 106, 185, 196), *591–594*, 654(132), *700*
Reed, D.E. 577(155), *593*
Reed, J.A. 2247(8), *2310*
Reed, R. 709(24), 761(130), *775, 777*
Reed, R.W. 283(46), *351*, 529, 530(91), *554*, 567, 569, 570, 583, 585, 587, 588(106), *592*
Reents, W.D. 1119(100–105, 110), *1138*
Reents, W.D.Jr. 560(31–33, 36, 37), *590*
Rees, C.W. 761(126), *777*

Rees, N.H. 793(48), *823*, 1211(62), *1231*
Rees, W.S. 2251, 2253, 2255, 2257(25e), *2311*
Reese, C.B. 2448(178), *2461*
Reets, M.T. 11, 13, 21(40), *96*
Reetz, M.T. 1130(237), *1141*
Reginato, G. 1626(132), 1628(136–140), 1650(214), 1655(137), 1656(136–140), 1658(132), *1663, 1665*, 1677(98), *1684*, 1796(38), *1862*, 1871(14), 1877(40), *1892*
Regitz, M. 711(27), 748(96), 763(133b, 137), 765(139, 140), *775, 777, 778*, 1153(58), *1182*, 1963(2b), 1983(72), *2030, 2032*
Reich, , P. 1471(226), *1531*
Reich, H.J. 466(183), *492*, 1601(14, 21), 1602, 1604(14), 1608(61, 70), 1621, 1624, 1625(14), 1632(61), 1639(61, 180), 1653(21), 1658(14), *1660–1662, 1664*
Reich, P. 1472(233a), *1531*
Reichel, C.L. 1705(70), *1787*
Reichel, D. 2364(32, 34–36), 2368(34, 38), 2369(32, 34), 2371(34, 38), 2374(34, 36, 38), 2376(32), *2399*
Reichenbach, T. 390(110), *428*
Reid, S.T. 1234(3), *1305*
Reide, J. 586(192), *594*
Reikhsfel'd, V.O. 1677(96), *1684*
Reimann, B. 1125(185), *1140*
Reiner, J. 1771(305), *1792*
Reinert, T. 2337(123, 124), *2358*
Reinhard, G. 192(58), *258*, 2072(83, 86), 2073(100, 101), 2074(102), 2091(83, 86, 100, 102), 2110(102), *2124, 2125*
Reinke, H. 884(105, 107–109), 885(105, 107–109, 115, 117, 118), 889(107–109, 117), 917(107, 108), 922(107), 925(105, 107–109, 117, 118), 926(107–109, 117, 118), 927(109), 937(105, 107, 117), 938(105, 117), 939, 940(108), 957(107), *1056*
Reinmann, B. 829(18), *854*
Reinsch, E.-A. 1166(153), *1184*
Reisenauer, H. 1021(329), *1061*
Reisenauer, H.D. 2514(269–272), *2565*
Reisenauer, H.P. 10(33b, 33d, 34, 35), 66(117a, 117b), 67(117b), 68(117a, 117b), *96, 99*, 864(28, 29), 901(29, 158), 977(28), 981(29, 158, 256), 985(29), 992(28, 29, 158, 256), 994(282), 995(29, 279–282), 996(28, 29), 997, 998(285b), *1054, 1057, 1059, 1060*, 1123(160, 161), 1125(174, 175), *1139*, 1145(17), 1146(20), 1149(35, 36), 1150(38), 1152(50), 1156, 1158(73a), 1159(96), 1162(107), 1163(127–129), 1164(127–129, 144), 1165(128, 147), 1167, 1168(162), 1169(107, 162), 1170(107, 180), 1171, 1172(182), 1175(73a, 144), 1178(96), 1179(73a, 144), 1180(107), *1181–1185*,

1286(162), *1309*, 1331(197), *1336*, 2495, 2514(168), *2562*
Reisenayer, H.P. 1150(41), *1182*
Reisenhauer, H.P. 1329(160), *1336*
Reisfeld, R. 2318(1, 6c), 2320(1), 2321(1, 24a), 2322(31), 2347(1, 196), 2348(199), 2350(1, 6c, 196, 234), 2351(1, 243, 249, 256), *2355, 2356, 2360–2362*
Reisgys, M. 213(148), *260*
Reising, J. 2061(59), 2062(61), 2066, 2068(59), *2123*
Reisinger, F. 584(184), *593*
Reiter, B. 790(35), *822*
Reizig, K. 215(158), *260*
Rell, S. 252(332), *265*, 1052(390), 1053(390, 391), *1062*, 1081(48), *1102*
Remez, I.M. 1494(336), *1535*
Remington, R.B. 66–68(118a), *99*, 1125(172), *1139*, 1163(130), *1184*
Remsen, E.E. 2250, 2253(22), 2265(83a–c), *2310, 2313*
Ren, Y. 1669(27), *1683*
Renhowe, P.A. 1821(147), 1832(206), *1864, 1866*
Renlund, R.M. 2289(134), *2315*
Renning, J. 1144(4), *1181*
Rentsch, S.F. 2227(67), *2242*
Ren-Xi Zhuo 1455, 1464, 1467(110, 111), 1481, 1483(110), 1486(110, 111), *1528*
Renzoni, G.E. 748, 750(99), *777*
Repinec, S.T. 1318(66), *1334*
Reshetova, E.V. 1825(162), *1865*
Résibois, B. 1998, 1999(124a, 124b), 2001(124b), *2034*
Retch, W.H. 1721(124), *1788*
Rethwisch, D.G. 1584(22), 1586(50), *1595*
Rettenmayr, N. 2364(30), *2399*
Rettig, S.J. 196(70), *258*, 2101(196), 2102(197), 2108(196, 197), 2109(197), *2127*, 2147(99), 2153, 2154(133), *2173*
Rettig, W. 1323(116a), *1335*
Reuter, D.C. 477(226), *493*, 784(21), *822*
Reuter, K.A. 560(47), *590*
Revelle, L.K. 872(56), *1055*
Revis, A. 169(113), *179*, 1735(152), *1788*, 2502(193), *2563*
Reyé, C. 541(106a, 106d), *554*, 564(91, 93, 94), *591*, 1480(279), *1533*, 2528(335), *2566*
Reye, C. 495(3), *510*, 519, 541(64), *553*, 1340, 1343, 1351, 1354(6), 1367, 1369(55), 1375, 1378, 1382(6), 1409(185b, 186a), 1410(186b), 1411(185b, 186a, 186b), 1412(6, 185b, 186a, 186b, 192), 1414(192), 1415(6, 201b), 1417(192, 205, 206, 207a, 207b), 1418(192, 205, 206, 207a, 207b, 215), 1420, 1421(6), 1422, 1423(206, 215), 1424(215), 1428(206, 215), 1429(6, 215), 1430, 1431(229), 1434(229, 244), 1435(186a, 186b, 244), *1437, 1438, 1443, 1444*, 1449(25), 1450(39, 44), 1457(39), 1464(25, 44), 1465(44), 1487(306), 1488, 1489, 1492, 1496(311), 1498, 1499(39), *1525, 1526, 1534*, 1738(161–165), *1789*, 1972(38), *2031*
Reynolds, W.F. 361(15), *426*, 844(56), *855*, 865, 866, 880, 881, 913, 922, 985, 986(38), *1054*, 1601, 1609, 1645(24), *1661*, 2404(10), *2458*
Reys, C. 1824(157), *1864*
Rhee, W.Z. 2005(151–153), *2034*
Rheingold, A.L. 4(10a, 10d), 37(10a, 10d, 81, 82, 86), 38(10a, 10d), 39(82), 40(10a, 10d, 82), 42, 44(86), 46(10a), 47(10d, 86), *95, 97, 98*, 250(315), *265*, 566(100), 571, 574(129, 130), *591, 592*, 816(112a), 817(112b), 819(112a, 112b), 820(112a), *825*, 981, 984(261, 262), *1059*, 1405(173), *1443*, 1759(262), 1762(263), *1791*, 1879(53), *1893*, 1904, 1909, 1910(46), *1926*, 2019(187), 2021(192), 2024(187), 2026(187, 200), 2027(202), 2028(187, 192), 2029(210), *2035, 2036*, 2040(8, 9), 2049, 2051, 2052(28), 2053(39), 2079(125), 2082(125, 143), 2086(163), 2087(164), 2090(163, 164, 183), 2092(125, 143), 2093(125, 143, 163), 2094(163, 164, 183), 2105, 2109(206), 2118, 2120, 2121(254), *2122, 2123, 2125–2128*, 2153, 2162(125), *2173*, 2529(337), *2566*
Rhodes, Y.E. 2530(343), *2566*
Ribo, J. 751(102), *777*
Ribot, F. 2337(120), 2342, 2345(148), *2358, 2359*
Ricard, L. 1963(2c), *2030*
Ricci, A. 460(162), 473(207), *491, 492*, 1600(1), 1601(10, 13, 23), 1602(10, 13, 23, 29), 1604(13), 1608(68), 1610(10, 23, 80), 1611(86), 1612(13), 1625(130), 1626(132), 1628(136–139), 1630(80), 1632(1), 1633(1, 156), 1635(10), 1642(23, 192), 1643(13), 1644, 1649(196), 1650(196, 214), 1655(137), 1656(136–139), 1658(132), *1660–1665*, 1677(98), *1684*, 1796(38), 1830(197), 1849(296), *1862, 1865, 1867*, 1871(14), 1877(40), *1892*
Riccitiello, S.R. 2287(121a, 121b), *2315*
Rice, D.A. 161(63), *178*
Rice, J.E. 78, 79(155), *100*, 2164(184), *2175*
Rice, R.W. 2246, 2247, 2250, 2253, 2254, 2273, 2285(6b), *2310*
Rich, J.D. 1246(39), *1306*
Richard, C. 2261(60), *2312*
Richardson, W.H. 160, 162, 171–173(55), *178*
Richter, A. 1670(32), *1683*

Richter, M. 449(121), *491*, 1842(256), 1858(341), *1866, 1868*
Richter, M.J. 1841(255), 1846(278), *1866, 1867*
Richter, R. 190(55), *258*, 1434, 1436(248, 250), *1445*, 2468(23), *2559*
Rickard, C.E.F. 2076, 2091(118, 119), 2093(185), *2125, 2126*
Rickborn, S.F. 2477(69, 70, 74), *2560*
Ridder, A. 219(181), *261*, 1373(75e), *1439*
Ridge, D.P. 1110(63), 1121(142), *1137, 1139*
Rieber, N. 752(111, 112), 753, 754(112), 755(111, 112), *777*
Riede, J. 203(104), 207(121, 125, 127), 250(309, 311), 252(324), *259, 260, 264, 265*, 859(4), 873(59), 876(4), 879(83), 932(83, 197), 940(83), 941(59), 944(83, 197), 981(4, 83, 197), 988(83), 991(4), 1021(327), 1037(59), 1042, 1043(327), *1054, 1055, 1058, 1061*, 1720, 1724(114), *1788*, 2060(54), 2106, 2109(209), *2123, 2127*
Riedel, R. 203(112), *259*, 2253(37a, 37b), 2257(49a–c), 2261(54a–c, 63), 2267(49a–c, 85a, 85b), 2269, 2270(88f), 2271(85a, 85b, 90), 2288(122), *2311–2315*
Riegel, B. 2332(92), *2357*
Rieger, B. 2153(128), *2173*
Rieger, W. 2546(386), *2568*
Riehl, D. 2330(81), 2352(81, 266, 267), 2354(81), *2357, 2362*
Riemers, J.N. 2232(94), *2243*
Riera, A. 2442(159, 160), *2460*
Riera, V. 2075(111–114), 2091(111, 112), *2125*
Ries, W. 2142(63), *2172*
Riess, G. 2238(122), *2244*
Rietropaolo, D. 1633(156), *1664*
Rietzel, M. 203(106), *259*
Riffel, H. 799(65), *823*
Riffle, J.S. 2224(39), *2241*
Riga, J. 2235(107), *2243*
Rihm, G. 1123(159–161), *1139*
Riley, F.L. 2264(73), *2313*
Riley, P.E. 518(61), *553*
Rima, F.G. 1524(422), *1537*
Rima, G. 1453, 1492(67), *1526*
Ring, M.A. 160(55), 161(57), 162(55), 171(55, 119), 172, 173(55), 175(141), *178, 180*, 780(2a, 2b), 781(2b), *821*, 1156, 1167(68), *1182*, 2465(12), 2466(14, 15), 2476(61, 63), 2477(67–75, 77), 2487(129), 2502(192), *2559–2561, 2563*
Ringe, K. 412(171), *429*, 1814(117), *1864*
Ringnalda, M.N. 2510(231, 232), *2564*
Ripoll, J. 1051, 1052(381), *1062*
Ripoll, J.-L. 1016(313), 1022(332), 1023(313), 1024(333), 1025(332), 1046(333), *1061*, 1878(51), *1893*
Rithner, C. 1848(285), *1867*
Ritscher, J.S. 1590(86), 1591(90), *1596, 1597*
Rittby, C.M.L. 1175(186, 189), 1176(186, 190), *1185*
Ritter, A. 784, 787, 788(25), *822*
Ritter, D.M. 780(2a, 2b), 781(2b), *821*
Ritter, W. 1135(264), *1142*
Ritzer, A. 1582(11, 12), 1583(11–14), 1587(61), 1594(116–118), *1594, 1596, 1597*
Rivera, I. 1638(169), *1664*
Rivest, R. 229(223), *262*
Riviére, P. 29(66e), *97*, 2534(364), *2567*
Riviere-Baudet, M. 2534(364), *2567*
Rizvi, S.Q.A. 361(17), *426*
Roan, B.L. 1838(230), *1866*
Roark, D.N. 828(15), *854*, 1144(5), *1181*
Robb, M.A. 919, 921(187, 188), *1058*, 1320(81), *1334*
Robbins, J.L. 2137(33), *2171*
Roberge, R. 1319(70), *1334*
Roberts, B.P. 1576(101), *1579*, 1874(28), *1892*
Roberts, J.D. 690(165), *701*
Roberts, J.R. 1247(40), *1306*, 1325(125), *1335*
Roberts, L.R. 1669(19), *1682*
Roberts, R.R. 425(215), *430*, 1679(108), *1684*, 1831(200), *1865*
Roberts, S.M. 448(119), 474(210), *491, 492*
Robertson, G.B. 374(62), *427*
Robertson, J. 1858(340), *1868*
Robertson, R.M. 2522(298), 2546(395), *2565, 2568*
Robichard, A.T. 396(126), *428*
Robichaud, A.L. 1816(132), *1864*
Robiette, A.G. 234(234), *262*
Robin, M.V. 2159, 2162(157), *2174*
Robinson, B.H. 2107(214), *2127*
Robinson, J.L. 2273, 2275(99a, 99b), *2314*
Robinson, N.P. 1801(65), *1863*
Robinson, P.J. 2340(140), *2359*
Robinson, R. 1897(7a), *1925*
Robinson, T.R. 1353, 1368, 1369(36), *1438*, 2298(150a), *2315*
Robinson, W.T. 222(189), *261*
Robl, J.A. 479(232, 233), *493*, 1543(22), *1577*
Rocheleau, R.E. 2548(401), *2568*
Rochow, E.G. 356(4), *426*, 563(71), *591*, 1582(2), 1588(64), 1589(77), 1590(81, 82), 1591(64, 82), 1592(82, 102), *1594, 1596, 1597*, 2533(352), *2567*
Roddick, D.M. 791(44b), *823*
Roden, B.A. 1546(34), *1578*
Roden, F.S. 1822(151), *1864*
Rodger, A. 1426(223), *1444*
Rodgers, A.S. 155, 176(30), *178*

Rodrigez, C.F. 1029(341), *1061*
Rodrigo, C. 2353(275), *2362*
Rodrigues de Miranda, J.F. 2364(24, 29), *2398, 2399*
Rodriguez, C. 1126(205), *1140*
Rodríguez, L. 2098, 2108(190, 191), *2126*
Rodríguez, S. 437, 438(84), *490*
Roe, M.B. 1673(55), *1683*
Roenijk, K.F. 171(120), *180*
Roesch, L. 1587(62), *1596*
Roesky, H.W. 203(106), 209(130), 211(144), 222(189), *259–261*, 1873(24), *1892*, 2145(89), *2172*, 2218(6), *2240*, 2301(155), *2316*
Roewer, G. 2194(75), *2215*, 2468(23), *2559*
Rogachevskii, V.L. 1989(89b), *2033*
Roger, R. 1391(141), *1441*
Rogers, C.W. 993, 996, 999, 1001(278), *1060*
Rogers, D.S. 2476(61), 2477(74), 2487(129), *2560, 2561*
Rogers, L.M. 1878(45), *1892*
Rogers, R.D. 1878(45), *1892*
Rogerson, P.F. 1312, 1313(17), *1332*
Rogido, R.J. 2003(146), *2034*
Roginskaya, M.V. 2530(348, 349), 2531(348), *2566*
Rogler, W. 2226(52), *2242*
Rohlfing, C.M. 1120(135), *1139*
Rohr-Aehle, R. 2364(19), *2398*
Röll, W. 2158(148), *2174*
Romakhin, A.S. 1201(35), *1230*
Roman, V.K. 1450(40), *1526*
Romanelli, M.N. 577(152–154), *593*
Romanenko, L.S. 1454, 1467(105), 1479, 1480(276), *1528, 1533*
Romanenko, V.D. 211(142), *260*
Romming, C. 186(42), *258*
Rona, R.J. 425(217), *430*
Roncali, J. 1189(7, 9), 1190(9), 1191(7), *1229, 1230*
Rondan, N.G. 2525(324), *2566*
Rong, G. 559(19), *590*
Rong, H.M. 1586(47), 1589(75), *1595, 1596*
Rong, X.X. 1542(15), *1577*
Ronn, A.M. 2523(311), *2566*
Roore, C.E. 183(4), *257*
Roos, E.C. 1815(122), *1864*
Roovers, J. 1759(260, 261), *1791*
Roper, W.R. 2076, 2091(118, 119), 2093(185), *2125, 2126*
Roques, C. 1996, 1998, 2008, 2009(110), *2033*
Rosa-Fauzza, L. 1644, 1649, 1650(196), *1664*
Rosch, J. 2238(125), *2244*
Rösch, L. 784(8c, 23), 787(23), *822*
Rose, K. 2330(79), 2333(79, 94), 2335, 2337(110), 2339(79, 94), 2341(142), *2357–2359*

Roseman, J.D. 1565(78a, 78b), *1579*
Rosenberg, H. 2141(55), *2172*
Rosenberg, L. 2101(196), 2102(197), 2108(196, 197), 2109(197), *2127*
Rosenberg, V.R. 384(91), *427*
Rosenblum, M. 2163(179), *2174*
Rosenfeld, A. 2322(38, 39, 40a), 2323(39, 40a), 2326(40a), *2356*
Rosenheim, A. 1415(196b), *1443*
Rosenstock, H.M. 6(22), *95*
Rosenthal, M.R. 529(92), *554*
Rosenthal, S. 433(7), *488*, 1601, 1602, 1604, 1608(7), 1620(7, 117), 1622(7, 124, 125), 1633(154), 1634, 1642, 1643(160), 1653(7, 125), *1660, 1663, 1664*
Rosenthal, U. 2051, 2052(34), *2123*
Roser, H.G. 1903(40b), *1926*
Rosmus, P. 10(33c), 12(44), *96*, 864, 992, 996(30), *1054*, 1107(16), 1123(159), *1136, 1139*, 2522(297), *2565*
Rospenk, M. 304(108), *353*
Ross, J.L. 1878(45), *1892*
Rosseinsky, M.J. 69(130), *100*
Rossetto, G. 2053(36), *2123*
Rossi, M.J. 2522(298), 2546(395), *2565, 2568*
Rossini, G. 1601, 1602, 1610, 1642(23), *1660*
Rossini, S. 1542(12), *1577*
Rossiter, B.E. 1668(12), *1682*
Rotella, D.P. 470(195), *492*
Roth, A. 2547(398), *2568*
Roth, G.P. 1676(90), *1684*
Roth, L.M. 1134(256), *1141*
Roth, P. 171(123, 124), 172(124), *180*, 2521(282), 2522(293), *2565*
Roth, W.R. 1903(40a, 40b), *1926*, 1998(116), *2034*
Rothchild, N. 2326(51), *2356*
Rothlisberger, U. 1121(140), *1139*
Rothschild, A.J. 716, 727, 748–750(40), *775*, 2404(6, 7), *2457*
Rotkiewicz, K. 1323(117), *1335*
Rottman, C. 2320(19), 2321(22, 23b), 2322(19, 22, 23b, 25), 2327(19), 2340(132), 2348(19, 25), *2355, 2356, 2359*
Rouillard, M. 1110(60), *1137*
Rousseau, F. 2339(129), *2359*
Rousset, C.J. 2043, 2051(15), *2122*
Roustamian, A. 2352(266), *2362*
Routledge, A. 462(168), *492*
Roux, C. 2327(57), *2357*
Roy, R. 2342(147), *2359*
Roy, R.A. 2342(147), *2359*
Roy, S. 2239(127, 128), 2240(127, 131), *2244*
Royer, J. 1818(136), *1864*
Royo, B. 2157–2159(146), *2174*
Royo, G. 287(64), *352*, 503(32), 505(35), 507(48, 50), 508(48, 56), *511*, 1374,

Royo, G. (*cont.*) 1375(76, 77, 79), 1376(76), 1377(77, 79, 99), 1378(77, 99), 1380(79), 1381(106), 1382(76, 77), 1383(76, 77, 110, 111), 1385(76, 77), 1386(76), 1412(191, 194), 1414(191), 1418(213, 217), 1420, 1421(213), 1422(194), 1423(194, 213, 217), 1424(77, 217), 1427(194, 217), 1428(217), 1430(227, 228), 1431, 1432, 1434(228), *1439, 1440, 1443, 1444*
Royo, P. 2144(84), 2145(90), 2150(84), 2151(84, 111, 117), 2152(111, 115, 117), 2157, 2158(146), 2159(146, 149), *2172–2174*
Rozell, J.M. 1261(77), *1307*, 2142(66), 2160(164, 165), *2172, 2174*, 2213(109), *2216*
Rozell, J.M.Jr. 2144(82), 2155(139), *2172, 2174*
Rozhkov, I.N. 1188(4), *1229*
Rozite, S. 1481, 1483(289, 290), *1533*
Rozsondai, B. 2181(21), *2214*
Ruano, J.L.G. 470(194), *492*
Rub, C. 2121(257), *2128*
Ruben, D.J. 310(132), *353*
Rubenstahl, T. 254(336), *265*
Rubin, Y. 186(36), *257*, 1958(28), *1960*
Rubina, K. 1740(172), *1789*
Rubinstein, I. 2328(67), *2357*
Rubinsztajn, S. 1762(264), *1791*, 2221(21a, 21b), 2223(30a, 30b), 2228(21b, 77), 2229(78), *2241, 2242*
Rubiralta, M. 1815(126), *1864*
Rüchardt, C. 1128(229), *1141*
Rücker, C. 433(9), *488*, 1674(66), *1683*
Rudinger, C. 241(263), *263*
Ruehl, K.E. 798(59), *823*, 2189(61), *2215*
Ruehlmann, K. 2220(17b), *2240*
Ruffolo, R. 580(169), *593*, 993, 996, 999, 1001(278), *1060*
Ruggeri, R. 1091(58g), *1103*
Ruhlandt-Senge, K. 207(119), *259*, 793, 794, 799(49a), *823*, 1878(47), *1892*
Rühlman, K. 1972(35a), *2031*
Rühlmann, K. 435(57), 458(57, 151), *489, 491*, 2012(169), *2035*, 2368, 2371, 2374(38), *2399*
Ruhs, M. 2159(152), *2174*
Ruiz, J. 1707(83), *1787*
Rulkens, R. 2153, 2154(134), 2162(176), *2173, 2174*
Rumbo, A. 1855(326), *1868*
Ruppert, I. 1675(78), *1684*
Ruppert, K. 485(250, 254), *493*
Ruscic, B. 167(94), 171, 172(127), 175(94), *179, 180*, 1106, 1107(6), *1136*, 1166(154), *1184*, 2522(299), *2565*
Rusek, J.J. 1608, 1632, 1639(61), *1661*

Russell, J.J. 167(85–87), *179*
Russell, T.P. 1318(48), *1333*
Russell, T.W.F. 2548(401), *2568*
Russo, P.J. 2143(76), *2172*
Russo, T.V. 2510(232), *2564*
Ruther, M. 1813(114), *1864*
Rutkowske, R.D. 376(66), *427*, 635, 637(93), *699*
Rutledge, P.S. 1651(216), *1665*
Ruwisch, L.M. 2267, 2271(85b), *2313*
Ruzsicska, B.D. 2514(273), *2565*
Ryabova, V. 1368(56), *1438*
Ryan, C.M. 1390(125, 126), 1392(126), *1441*
Rybakov, V.B. 245(278), *264*
Rybczynski, P.J. 1677(95), *1684*
Rychnovsky, S.D. 1678(103), *1684*
Rylance, J. 154, 159, 163–165(9), *177*
Ryu, I. 1559(67a, 67b, 68), 1560(69), *1579*, 1616(99, 100), *1662*, 1678(99), *1684*
Rzaev, Z.M.O. 2224(42b), *2241*

Saak, W. 200(94, 101), 205(113), *259*, 578(165), *593*, 804(78), *824*, 1025(334), *1061*, 1077(38), *1102*, 1268(98), 1283(152), 1304(213), *1307, 1309, 1310*, 1881(61), 1885(73), *1893*, 2134(27), 2140(45), *2171*, 2179(9), 2184(45), 2185(46, 47), *2214*, 2472(49), 2492(146, 147), 2493(160, 161), 2494(165), *2560, 2562*
Saam, J.C. 2228(74), *2242*
Saavedra, S.S. 2327(59), *2357*
Saba, S. 772(152), *778*
Sabata, A. 2235(108), 2236(114), *2243, 2244*
Sabin, J.R. 68(122a), *99*
Sabourin, E.T. 1693(24), 1705(78, 79), 1714(79), *1786, 1787*
Saburi, M. 2043, 2051(15), *2122*
Sacharov, Y. 2321, 2322(23b), *2356*
Sacks, M.D. 2286(119, 120), *2315*
Sada, K. 458(150), *491*
Sadahiro, T. 1100(76), *1103*
Saegusa, J. 805(82), *824*
Saegusa, T. 1771(296), *1791*, 2345(170), *2360*
Saeki, T. 1238(20), *1306*
Safa, K.D. 845(65), *855*, 881, 913(96), *1056*
Safarik, I. 1319(70), *1334*, 2514(273), 2526, 2527(327), *2565, 2566*
Safir, A.L. 2104, 2108, 2109(205), *2127*
Sagai, Y. 2139, 2141, 2147(43), *2171*
Saha, C.K. 2261(55, 62), *2312*
Saheki, Y. 891(126), *1056*
Saigo, K. 1798(51), *1863*
Saika, A. 310(130), *353*
Sailor, M.J. 2307, 2309(163), *2316*
Saini, R.D. 863(24, 25), 864, 932, 934, 982(25), *1054*
Saini, R.K. 1237(18a), *1306*

Saint, G. 578(163), *593*, 2487(126), *2561*
Saint-Roch, B. 1051(380), *1062*
Saito, H. 1925(110), *1928*
Saito, I. 1682(129), *1685*
Saito, K. 2522(290), *2565*
Saito, M. 487(270), *494*, 1091, 1092(61), 1093(61, 66b), *1103*, 1809(102), *1863*
Saito, S. 66, 67(117c), *99*, 1009, 1010(298a), 1021, 1045, 1046, 1049(325), *1060, 1061*, 1163(135), *1184*, 2522(305), 2557(429), *2565, 2568*
Saito, Y. 1075(30), *1102*
Sakaba, H. 2001(130b, 130c), *2034*, 2412(35), *2458*, 2502(190), *2563*
Sakaguchi, H. 1743(190), *1789*
Sakaguchi, K. 1193(17, 18), 1194(17), *1230*
Sakaguchi, M. 1830(195), *1865*
Sakaguchi, S. 2250, 2261(20), *2310*
Sakai, K. 2351(238), *2361*
Sakaibara, A. 2289(123), *2315*
Sakaino, T. 2350(219), *2361*
Sakaki, J. 1743, 1746(212), *1790*
Sakaki, S. 2116(239–241), *2127*
Sakakibara, A. 1331(188), *1336*, 1923(93), *1928*, 2329, 2331(75), *2357*, 2491(143), 2494(162), 2518, 2519(143), *2562*
Sakakibara, J. 1692(15), 1733(15, 145), 1734, 1737(145), 1777(330), 1778(332), *1786, 1788, 1792*
Sakakibara, M. 1885(74), *1893*
Sakakura, T. 1697(44), 1698(44, 46), 1779(46), *1786*, 2278(107), *2314*
Sakamoto, H. 867(45), 870(47, 49), 885(45), 891(127), 892(128–130), 895(133, 134), 925(129), 928(193), 935(133), 936, 970(134), 974(45, 133), 975(134), *1055, 1056, 1058*, 1252(53–55), 1258(68), 1259(69, 70), *1306, 1307*, 1325(136, 139), *1335*
Sakamoto, K. 488(276), *494*, 829(25, 26), 831, 832(29), 834(36), 836, 837(29), 842(51), 845(63), *854, 855*, 885(114), 887(114, 119), 911(119, 177), 922, 934, 957(119), 987, 989(114), *1056, 1057*, 1238(20), 1262(82), *1306, 1307*, 1325(149, 150), 1330(174), *1335, 1336*, 1989(92), *2033*, 2439(147), 2445(164), *2460, 2461*
Sakamoto, S. 798(60a, 60b), *823*
Sakane, T. 1742(186), *1789*
Sakashita, H. 387(104), *428*
Sakata, Y. 384(88), *427*
Sakeyeni, H. 2489(137a), *2562*
Sakigawa, S. 278, 329(26), *351*
Sakka, S. 2318(6a), *2355*
Sako, K. 1640, 1655(187), *1664*
Sakurada, S. 2252, 2253(32c), *2311*
Sakurai, H. 3(2b), *95*, 120(4, 6, 7, 9), 125(33), 130(33, 37), 132(9), 133(4, 37, 41), 134(37, 42, 43), 137(46, 47), 140(9, 47), 143(46, 47), 146(37, 46, 47), 147(56), *150–152*, 186(41), 190(52), 234(242), 240(259), 248(303), *258, 263, 264*, 277(22), 284(59), 288(66, 67), 289(67), 348(163), *350–352, 354*, 404(149), *428*, 484(242–244), 488(273–276), *493, 494*, 569(113, 120, 122), 570, 571(113), 572(120), *592*, 782(13a, 13b), 796(56, 61, 76), 797(56), 798(56, 61), 802(76), *822–824*, 827(2a), 829(16, 17, 25–27), 830(28), 831(28, 29), 832(27–30), 834(36), 835(27), 836(28, 29), 837(29), 838(30), 839–841(28), 842(28, 51), 845(63), 847, 850(75), *854, 855*, 887(119), 897(138), 898(143), 899(144), 911, 922, 934, 957(119), *1056, 1057*, 1064, 1068(12), *1101*, 1157(81), 1169(173, 174), 1176(173), 1178(191), *1183, 1185*, 1236(10, 11), 1244(33), 1247(41, 42), 1248(44), 1251(52), 1256(44), 1262(82), 1268(100, 101), 1283(157), *1305–1307, 1309*, 1314(25), 1319(76, 77), 1320(77), 1321(87), 1322(90–93), 101, 104, 105, 112), 1324(105, 112, 119), 1325(105, 123, 148–151), 1326, 1327(105), 1328(157, 158), 1329(157, 158, 166–169), 1330(166, 168, 169), 1331(166, 167), 1332(202, 200), *1333, 1334, 1336*, 1341(22, 26), 1343(22), 1347(26), 1407, 1408(177), 1434, 1435(242), *1437, 1443, 1444*, 1458, 1467(126), 1493(328), 1497(352), *1528, 1535*, 1794(3, 9, 12, 20), 1808(20, 90–93), 1809(102), 1811(20), 1814(121), 1821(145), *1862–1864*, 1972(39), 1973(46), 1981(39), 1989(92), 1999(128, 129), 2001(130a–c), 2002(39, 141, 143), 2009(162), 2011(166), *2031, 2033–2035*, 2137, 2138(36), 2139, 2141, 2147(43), 2159(160), *2171, 2174*, 2188(55), 2205(90, 91), 2206(90), 2208(98), *2215, 2216*, 2412(35), 2429(92, 102), 2430(112), 2437(132), 2439(142, 143, 147), 2445(164), *2458–2461*, 2470(33), 2471(37), 2496(178), 2502(190), 2514, 2515(37, 265), 2517(37), 2518(265), *2559, 2563, 2565*
Sakurai, K. 1238(19, 20), *1306*
Sakurai, T. 1208–1210(57), *1230*
Salaheev, G. 1227(119), *1232*
Salahub, D.R. 515(14b), 516(14b, 31), 517(44c), *551–553*, 658(134a, 134b), *700*
Salaün, J. 435, 457(56), *489*
Saleem, M. 2286(119, 120), *2315*
Saleh, S. 434, 448(35), *489*
Salem, L. 620(63), *699*
Sallé, L. 1573(96a), *1579*
Salomon, K.E. 168, 169(105), *179*
Salter, D.M. 2076, 2091(118, 119), *2125*

Salvadori, P. 1696, 1717(35), *1786*
Salvati, L. 2234(104b), *2243*
Salvetti, M.G. 2261(65), *2313*
Salvi, G.D. 1695(34), *1786*
Salzmann, T.N. 1607, 1634(58), *1661*
Samantaray, B.K. 2240(129), *2244*
Samaraweera, U. 1238(21), *1306*, 1331(192), *1336*, 1883(66), *1893*, 2422(68, 69), *2459*
Samaraweera, V. 2475, 2477, 2482, 2483(56), *2560*
Samitier, J. 2548(400), *2568*
Sammes, M.P. 1997(114a), *2033*
Samoson, A.V. 1474, 1476, 1477(260), 1479(271), 1510(260), *1532, 1533*
Samour, C.M. 1452–1454, 1492(53), *1526*
Sampath, S. 2321(24b), 2328(64, 70, 71), 2329(70, 71), *2356, 2357*
Samreth, S. 1547(37), *1578*
Samson, C. 1545, 1564(28), *1578*
Samsonova, G.A. 1454(72, 74), 1500(72, 74, 359, 360), *1527, 1536*
Samuel, E. 2042(13), 2043(16), 2045(13), 2047(26), 2051(16), *2122, 2123*
Samuel, J. 2319(13), 2320(13, 18, 19), 2322(18, 19), 2327(19), 2340(132), 2348(18, 19, 198), *2355, 2359, 2360*
Sana, M. 159, 161, 162(51), *178*, 1050(371), *1062*
Sánchez, A. 1819, 1842(140), *1864*
Sanchez, C. 2319(10a, 10e), 2326, 2328(49a), 2330(82), 2337(120), 2342, 2345(148), 2350(10a, 10e, 229, 232), 2351(232, 250, 257, 258, 261), 2352(264), 2354(82, 250, 257, 258, 281, 284, 285), *2355–2359, 2361, 2362*
Sanchez, J.-Y. 2339(126, 129), *2358, 2359*
Sanchez, M. 763(137), *778*
Sanchez, R. 1261(78), *1307*, 2083(154), *2126*
Sandel, V.R. 2004(150), 2005(154), *2034*
Sander, W. 704(7), 709(7, 22, 23), 713(22, 23), 714(22), 727(23), 734(65), 763, 764(22), *775, 776*, 978(252, 253), *1059*, 1075(34), *1102*, 1125(176), *1139*, 1146(23), 1147(23, 28), 1148(28), 1162(108), 1164(143), 1165(146), 1168(108), 1181(201), *1181–1185*, 1272(107), 1277(132, 133), 1278(134), 1286(160), *1308, 1309*, 1606(50), *1661*, 2493, 2520(153), *2562*
Sanderson, R.T. 183(2), *257*
Sandford, G. 521, 530(74), *553*, 584(188), *593*, 652(129), *700*, 1669(24), *1682*
Sandhu, V. 2526, 2527(327), *2566*
Sandorfy, C. 1312, 1314(19), 1319(70), *1332, 1334*
Sands, J.E. 2177(1), *2213*
Sanemitsu, K. 1226(114), *1232*
Sanji, T. 832, 838(30), *854*

Sano, H. 1677(92), *1684*, 2160(169), *2174*
Sano, T. 1771(309), *1792*
Santagostino, M. 1569(86), *1579*
Santaniello, E. 2388(71), *2400*
Santelli, M. 1798(52), 1801(68), *1863*
Santen, R.A. 115(36), *118*
Santen, R.A.van 115, 116(38), *118*
Santiago, B. 435, 456(54), *489*, 1843(265), *1867*
Santra, R.A. 2239(127, 128), 2240(127, 129–131), *2244*
Sanz, J. 2340(131), *2359*
Sanz, J.F. 2116(242, 243), *2127*
Sapozhnikov, Yu.M. 1472(233b), *1531*
Saraev, V.V. 1454, 1467(93), 1481, 1483(288), *1527, 1533*
Sard, H. 1847(284), *1867*
Sarge, S. 2364, 2365, 2367(20), *2398*
Sarina, T.V. 211(142), *260*
Sarkar, A.K. 409, 410(165), 422(208), *429, 430*, 455(138), *491*
Sarkar, T. 780(1j), *821*, 1631, 1632(150), *1663*, 1794(10), *1862*
Sarkar, T.K. 434, 449(37), *489*, 1499(353), *1535*, 1794(22), *1862*
Sarre, P.J. 1107(12, 16), *1136*
Saruyama, T. 2226(52), *2242*
Sasaki, A. 438, 465(89), *490*
Sasaki, R. 1885(78), *1893*
Saso, H. 1238(23), *1306*, 2089, 2092, 2094(181), *2126*, 2420(64), 2421(61–64), 2422(62), 2425(63, 80), *2458, 2459*
Sassaman, M.B. 1870(9), *1891*
Sassmannshausen, J. 219, 232(185), *261*
Sasvari, K. 200(95), *259*
Satgé, J. 909(175), 1051(380), *1057, 1062*, 1064(4f), *1101*
Satge, J. 52(94c), *98*, 1453, 1492(67), 1524(422), *1526, 1537*, 1882(65), *1893*, 1907(52c), *1926*
Satici, H. 439(96), *490*
Sato, A. 2435(125), *2460*
Sato, F. 475(211), *492*, 1557(63), *1578*, 1680(113), *1684*, 1796(36), 1831(201), 1842(36), *1862, 1865*
Sato, J. 442(108), *490*
Sato, K. 284(59), *351*, 1808(90, 91, 93), *1863*
Sato, M. 1796(36), 1831(201), 1842(36), *1862, 1865*, 2253(34), 2272(96), *2311, 2314*
Sato, N. 279(32), *351*, 1269(102, 103), *1308*, 1840(244–246), *1866*, 1881, 1887(59), *1893*
Sato, O. 2448(181), *2461*
Sato, R. 2009(162), *2035*
Sato, S. 1774(320), 1777(329), *1792*
Sato, T. 714(36b), 722(36b, 45, 46), *775, 776*, 905, 969, 973, 992, 996(165), *1057*, 1211(66), *1231*, 1276(129), *1308*, 1601,

1602(12), 1614(93–95, 97), 1620(12), 1649(210), 1651, 1652(12), *1660, 1662, 1665*, 1838(239), 1849(294), *1866, 1867*, 1897, 1901(29), *1926*, 2002, 2009(144), *2034*, 2405(14, 15), 2406(19), *2458*
Sato, Y. 1322(107, 108), *1334*
Satoh, H. 2347(186b), *2360*
Satoh, S. 1210(58, 59), *1231*
Sattelberger, A.P. 205(115), *259*
Sattler, E. 217(168), *261*
Sau, A.C. 1352(33), (44), *1438*
Sauer, B.B. 2234(103a–e), *2243*
Sauer, G. 1575(98), *1579*
Sauer, J. 2133, 2148(22), *2171*
Sauer, S.P.A. 517(39a, 39b), *552*
Saunders, J. 1109(49), *1137*
Saunders, M. 7(27), 69, 72(129), *96, 100,* 690(161, 164a, 167), *701*
Sausa, R.C. 2523(311), *2566*
Sauvet, G. 2224(38), *2241*
Sawada, H. 1836(219), *1866*
Sawada, Y. 506(43), *511*
Sawado, Y. 2546(388), *2568*
Sawahata, M. 783, 789, 801(15), *822*
Sawaki, Y. 1197(26), 1229(128), *1230, 1232*
Sawamura, M. 1743(206, 207), 1745–1747(207), 1748(206), *1789*
Sawitzki, G. 228(209), *262*
Sawlewicz, P. 376(67), *427*, 637, 638(96, 97), *699*
Sawrey, B.A. 2477(68, 71, 72), *2560*
Sawyer, J.F. 487(263), *494*, 795, 799(62c), *823*, 845(66), *855*, 880, 881, 913, 914, 917, 924(90), 950(235), 951(236, 237), 959–961(235), 985, 986, 992, 1004(90), *1056, 1059*, 1080(45), *1102*, 1274(113, 116, 117), *1308*, 1601, 1604, 1609(26), 1645(26, 201), *1661, 1664*, 2407(23), *2458*
Sax, A. 123(21b, 21c), *151*, 1151(43), *1182*
Sax, A.F. 22(58, 60), 23(60), 24(58), 25, 26(58, 60), 68, 69(123), 87(163a, 163b), *97, 99, 100*, 123(24a, 27b, 27c), *151*, 159, 167, 168, 171, 173(50), *178*
Saxce, A.de 1374–1376(76), 1381(106), 1382, 1383, 1385, 1386(76), *1439, 1440*
Saxe, P. 2442(157, 158), *2460*
Saxena, A.K. 487(263), *494*, 795, 799(62c), *823*, 845(66), *855*, 880, 881, 913, 914, 917, 924(90), 951(236, 237), 964–966(245), 985, 986, 992, 1004(90), *1056, 1059*, 1273(109), 1274(115–117), 1276, 1285(109), *1308*, 1456, 1457, 1460, 1467(118), *1528*, 1601, 1604, 1609, 1645(26), *1661*
Sayer, B.G. 580(168), *593*
Saykally, R.J. 704(9), *775*
Scafato, P. 1877(40, 41), *1892*
Scaiano, J.C. 880, 881, 922, 924, 926, 940,

992(88), *1055*, 1074(28), *1102*, 1542(8), *1577*, 2498, 2513, 2527(184), *2563*
Scampton, R.J. 2484, 2502(104), *2561*
Scarlete, M. 2249(18), *2310*
Schaad, L.J. 6(21d), *95*, 2509(229), *2564*
Schaaf, T.F. 782(12a), 785(26), *822*
Schabacher, A.W. 1877(42), *1892*
Schade, C. 518, 519(63), *553*
Schaefer, A. 190(53), *258*
Schaefer, H.F. 247(297), 250(316), *264, 265*, 532(99b), *554*
Schaefer, H.F.III 5(12), 13(46a), 18, 21(49), 22(49, 57), 29(66a), 32(49), 66(118a–c, 119), 67(118a, 118b, 119), 68(118a, 119, 122c), 83(160), 84(160, 165a), 87(165a, 165b), 88(165a), *95–97, 99–101*, 158(49), 163(73), 167, 171, 172(49), 174(139), *178–180*, 184, 250(28), *257*, 515(7a, 7b, 24), 530(95a), 531(96, 98), 532(95a, 98), *551, 552, 554*, 705(14), *775*, 901(146, 151), 919, 920(186), 997(151), *1057, 1058*, 1120, 1121(137), 1125(172), 1126(202, 203), 1127(212), *1139, 1140*, 1157(85), 1163(130, 131), 1166(85, 150), 1168(167), 1169, 1176(85), *1183–1185*, 1331(196), *1336*, 2479(79), 2484(102), 2493, 2512, 2520(249), 2556, 2557(424), *2560, 2561, 2564, 2568*
Schaefer, J. 309(115, 117), *353*
Schaefer, T. 362(20), *426*
Schaeffer, C.D.Jr. 1397, 1398, 1404(170), *1442*
Schaeffer, R. 219(177), *261*
Schaer, D. 2553(415), *2568*
Schäfer, A. 517(43a), *552*, 578(165), 584(183), *593*, 1084(52), *1102*, 2182, 2183(37), *2214*, 2493(159, 161), 2495(173), *2562*
Schafer, A. 200(99), *259*, 1878, 1886(50), *1893*
Schafer, H. 215(159), *260*
Schafer, H.F.III 123(27a), *151*, 2442(157, 158), *2460*
Schäfer, H.J. 1204(42), 1206(47), *1230*
Schafer, J. 310(133), *353*
Schafer, L. 113(25), *118*
Schäfer, M. 459(158), *491*
Schäfer, T. 945(208), *1058*
Schaible, S. 2253(37a), *2311*
Schakel, M. 1434(234), *1444*
Schamp, N. 1651(217), *1665*
Schank, P.R. 2347(186c), *2360*
Schappacher, M. 2251(30a, 30b), *2311*
Schär, D. 541(107), *554*, 581(180), *593*, 1087(55), *1102*, 1377, 1379, 1380(96), 1411, 1412(188), *1440, 1443*
Scharma, H.K. 1332(205), *1337*
Schauer, S.N. 1135(266), *1142*
Schaumann, E. 434, 445(26), *489*, 740(76),

Schaumann, E. (*cont.*) 766(142), 767(76, 142), *776, 778*, 1679(109, 110), *1684*, 1796(41), 1819(139), 1821(144), *1862, 1864*
Schaumann, W. 1872(18), *1892*
Schaus, S.E. 1672(50, 52), *1683*
Scheid, E. 2548(400), 2549(404), *2568*
Scheiffele, G.W. 2286(120), *2315*
Scheim, U. 435(57), 458(57, 151), *489, 491*
Scheiner, A.C. 174(139), *180*, 901, 997(151), *1057*, 1166(150), *1184*
Schellenberg, F.M. 2351(247), *2362*
Scheller, M.E. 1602(33), 1611, 1630(85), 1647(33), 1649(207), *1661, 1662, 1665*
Scheller, M.K. 1135(267, 268), *1142*
Schenk, A. 1585(35), 1587(59, 60), *1595, 1596*
Schenkel, A. 2143(73), *2172*
Schenkel, M. 2453(196), *2461*
Schepp, N.P. 649(123), *700*, 1301(197), *1310*
Scherbaum, F. 1340(15), *1437*
Scherer, G. 2246(1), 2269(86), 2294, 2295(1), *2310, 2313*
Scherer, G.W. 2319, 2350(9), *2355*
Scherer, P. 407(156), *429*
Scherer, W. 52(97), *98*
Scherholt, T. 2481, 2483(96), *2561*
Schiavelli, M.D. 362, 363(27), *426*, 598(8), 611(8, 45), 612(8), 615(45), *697, 698*
Schick, A. 1153(61), *1182*
Schickmann, H. 2223(29a, 29c, 29d), *2241*
Schiebel, H.-M. 2363(9), *2398*
Schieda, O. 876(72–74), 917, 931(72, 73), 940(72–74), *1055*
Schiemann, K. 1803(76, 77), *1863*
Schiemenz, B. 1387, 1389(121b), *1441*
Schier, A. 190(55), 207(121, 125), *258, 260*, 1875(34), *1892*
Schieser, M. 1431(230), *1444*
Schiesser, C.H. 365(40), *426*, 642–644(108), *700*, 1547(35b), 1571(91), *1578, 1579*
Schilf, W. 522, 540(80a, 80b), *554*, 563(75–77), 564(75–77, 85), 565(75), 566(75–77), 567(77), 568(75), 569(77), *591*
Schilling, B.E.R. 29(66b), *97*
Schimeczek, M. 520(67), *553*
Schimpf, R. 1827(180, 181), *1865*
Schinabeck, A. 1593(113), *1597*
Schindler, M. 60(106a, 106c), *99*, 281(44), *351*, 516, 532(32b, 32c), *552*
Schinzer, D. 404(150), 411(170), 412(171), *428, 429*, 442(107), *490*, 1601, 1602(19), 1642(191), 1649(19), *1660, 1664*, 1667(7), *1682*, 1794(15, 32), 1795, 1801(32), 1811(108), 1814(117), 1836(220), *1862, 1864, 1866*
Schionato, A. 1743(218), *1790*
Schiøtt, B. 1743, 1746, 1747(225), *1790*
Schjelderup, L. 2374(43), *2399*

Schklower, W.E. 229(212), *262*
Schklower, W.E.v. 1910(57), 1913, 1914(77), *1927*
Schlachter, R. 1156(74), *1182*
Schlegel, H. 1053(394), *1062*
Schlegel, H.B. 11, 12(42), *96*, 155(26), 156(32, 34), 159, 160(26), 164(81), 167(32), 168(26, 100), 170(81), 171(32), 173(26, 32, 100), 175(26, 81), *178, 179*, 184, 234(23, 24), 239(23), *257*, 1106(5), 1126(190), *1136, 1140*, 1148, 1149(29), *1182*, 2466(17), 2510(240, 241), 2511(246), *2559, 2564*
Schlenk, W. 1144(4), *1181*
Schleyer, P.v.R. 3(8), 5(12, 16, 19), 6(16, 23b), 7(16, 19, 26, 27), 8(31a), 9(16, 26, 31a), 10(8, 16), 11, 13(8, 40), 16, 17(8), 19, 20(50b), 21(40, 50b), 22, 23(19), 24(19, 62), 29, 30(62), 31(19), 32(8, 19), 33(70, 71), 34(71), 36(76a), 42(26, 31a, 88, 89), 43(31a, 88, 89, 93a), 44(26, 88, 93a), 45(26, 88), 46(16, 88, 89), 47(26, 31a, 88, 89, 91, 92, 93a), 48(88, 93a), 50(71, 89), 51(71), 69(26, 129), 72(129), 77(91), 82, 83(158), 90(26, 31a, 169, 170b, 170h), 91, 92(169), *95–98, 100, 101*, 123(24a), 125(30), *151*, 250(319), 251(321), *265*, 283(49), 296(87), *351, 352*, 515(6, 22, 24), 516(35a, 35b), 517(38, 45, 51), 518(6, 63), 519(63), 520(6), 521(73), 522(6, 73), 523(73), 526(6), 527(6, 73), 528(6), 529(6, 73), 530(95a, 95b), 531(96, 98), 532(6, 95a, 95b, 98, 99a), 533(6, 119), 535, 537(6), 542(6, 73), 543(6, 110), 544(73, 110), 545, 547(110), 551(114, 115, 117a), *551–555*, 558, 559(4), 567(107), 570(125), 572(132), 573(4, 125), 577(4), 584(4, 189), 588(4), *589, 592, 594*, 597(2), 601(12), 606(23), 615(23, 44), 616(53), 620(63), 635(89a, 89b), 652(127, 131), 687(158), 690(164a), 696(170), *697–701*, 705, 707(12), 775, 781(5–7), 814(108a, 108b), 815(108a), 817, 818(113), 819(113, 118), 820(118), 821(120), (108c), *821, 825*, 827(5), 835(37), *854*, 859(6), 901(147), 984, 988(267), 990, 994, 995(6), 1046(374), *1054, 1057, 1059, 1062*, 1067(16), *1101*, 1109(34), 1123(156), 1134(255), *1136, 1139, 1141*, 1151(44), 1166(149, 157), *1182, 1184*, 1372, 1373(69), *1439*, 1911(68), *1927*, 2013(172b), 2019(184b, 184c), 2026(184b), *2035*
Schlich, K. 1675(78), *1684*
Schlosser, Th. 209(137), *260*
Schlosser, W. 3(9), *95*, 724(51), 725(52), *776*, 908(170), 987, 990(274), *1057, 1060*, 1152(51), *1182*, 1282(139, 140), *1308*, 2413(39), *2458*

Schlotz, G. 215(157), *260*
Schlüter, A.D. 1910, 1911(60e), *1927*
Schlüter, E. 2141, 2149(56), *2172*
Schlüter, M. 69(131a, 131b), *100*
Schmalz, T.G. 70, 71(139b), *100*
Schmid, G.H. 544(111d), *555*
Schmidbaur, H. 207(121, 125–127), 209(137), 241(263), *260, 263*, 296(93), *352*, 586(192), *594*, 1340(15), *1437*, 1875(34), *1892*
Schmidt, A.H. 1490(318), *1534*
Schmidt, C. 310(126, 127), *353*
Schmidt, H. 290(69), *352*, 2332(87), 2333(103, 104), 2335(87, 104, 109, 111), 2336(109, 113, 114), 2337(103, 118, 123), 2339(129), 2340(114, 138), 2342, 2345(146), 2346(178, 179), 2349, 2350(211), 2351(111), 2354(286), *2357–2362*
Schmidt, H.-G. 209(130), *260*
Schmidt, J. 1510, 1511(398), *1537*
Schmidt, M. 1318(59), *1333*
Schmidt, M.U. 1720, 1724(114), *1788*, 2106, 2109(209), *2127*
Schmidt, M.W. 21, 24, 30(53a), 84, 86(162), *96, 100*, 829(23), *854*, 1048(365), *1062*, 1067(15), *1101*, 1172(184), *1185*, 1468(194), 1469(194, 197), 1477, 1511, 1521(194), *1530*
Schmidt, R.E. 215(155), *260*
Schmidt, T. 1738(157), *1788*
Schmidt, W.R. 2253(35), 2264(77), *2311, 2313*
Schmidt-Amelunxen, M. 129(36), *151*
Schmidt-Bäse, D. 203(105), 209(140), 235(240), 241(264, 265), 252(325), *259, 260, 263, 265*, 1011, 1012, 1034(306), 1037, 1040(355), 1043, 1044(306), *1060, 1061*
Schmidt-Rohr, K. 310(129, 134), *353*
Schmitt, R.J. 1109, 1115(38), *1137*
Schmitz, R.F. 1434(234), *1444*, 1628(143), *1663*
Schmitzer, S. 2061(58, 60), 2062(61, 62), 2066, 2067(60), 2068(58, 60), 2078, 2091(62), *2123*
Schmock, F. 1878(47), *1892*
Schmutz, C. 2351(261), *2362*
Schmutzler, R. 1878(45), *1892*
Schneemeyer, L.F. 69(130), *100*
Schneggenburger, L.A. 1771(299), *1791*, 2237(117), *2244*
Schneider, A. 186(38), *258*, 2443(162), *2461*
Schneider, D. 2106, 2109(210), *2127*
Schneider, H. 1557(62), *1578*
Schneider, J.J. 2063, 2066(66), *2124*
Schneider, K. 713(32), 719(32, 43), 720(44), 750(32), 751(103), 752(32, 103), 754(32), 760(103), 765(141), 775–778, 931(195, 196), *1058*, 1276(128, 130, 131), *1308*, 2407(27, 28), *2458*
Schneider, M. 2145(87), *2172*
Schneider, O. 2223(27c), *2241*
Schneider, U. 1341, 1349–1351, 1357, 1361, 1368(30b), *1438*
Schnering, H.G.v. 791(44a), *823*, 1913, 1914(76), *1927*
Schnering, H.G.von 190(53), 194(62), 198(91, 92), 200(99), 219(179, 183–185), 228(209), 229(213–216, 218–220), 232(184, 185, 224, 226–228), 241(260–262), 245(287), *258, 259, 261–264*, 1283(153), *1309*, 1910(58), *1927*, 1989(91), *2033*, 2182, 2183(37), *2214*, 2473(50), 2486(110, 116, 117), 2493(159), 2495(171, 172), 2514(172), *2560–2562*
Schnider, D. 2445(165), *2461*
Schnöckel, H. 53(101), *98*, 1083(50, 51), 1096(50, 51, 72), *1102, 1103*, 1127(215), *1140*, 1162(109, 112–116, 118, 120), 1163(120, 121, 123, 124, 126), 1178(192), *1183–1185*, 2510(239), *2564*
Schnoeckel, H. 2474, 2534(55), *2560*
Schnoor, J.L. 2327(54), *2356*
Schnuck, S. 1127(215), *1140*
Schnute, M.E. 1116(75), *1137*, 1810(104), *1863*
Schoeller, W. 1053(396–398), *1062*
Schoeller, W.W. 123(24b, 24c), *151*, 2557(428), *2568*
Schoepe, G. 1587(55), *1596*
Schofield, C.J. 1675(80), *1684*
Schöllkopf, K. 584(184), *593*
Schöllkopf, U. 635(89a), *699*, 752(111, 112), 753, 754(112), 755(111, 112), *777*
Scholten, A.B. 343–346(152), *354*
Scholz, B.P. 1903(40b), *1926*
Scholz, G. 2027(203), *2036*
Scholze, H. 2346(179), *2360*
Schomburg, D. 219(180), 222(195), 238(250), *261, 263*, 1340, 1341(10a, 10b), *1437*, 1649(209a), *1665*, 2364(28), 2367(37), 2383(60), *2398, 2399*
Schomburg, D.J. 284(53), *351*
Schönberg, A. 1091(58b), *1103*
Schönfelder, H. 2261(54a, 54c), *2312*
Schonk, R.M. 1797(49), *1863*
Schopper, K. 2546(386), *2568*
Schore, N.E. 2139(41, 42), 2153(41), *2171*
Schottland, E. 437, 485(79), *490*
Schöttler, K. 10, 17, 18, 20(36), *96*, 994, 995(167), *1060*, 1150(41), 1162, 1169(107), 1170(107, 180), 1180(107), *1182, 1183, 1185*, 2495, 2514(168), *2562*
Schottner, G. 2332(88), 2335, 2337(110), 2339(88, 130), 2353(277), *2357–2359, 2362*

Schöwälder, K.H. 584(184), *593*
Schraml, J. 269(10), 296(92), *350, 352*, 487(269), *494*, 1475, 1477(251), *1532*, 1604(35), *1661*
Schrank, F. 269–272(11), *350*, 796, 802(74), *824*, 2209, 2210(100), *2216*
Schranz, H.W. 2510(235), *2564*
Schreiber, S.L. 184, 222(22), *257*, 482(241), *493*
Schreiner, P.R. 530(95a), 531(96, 98), 532(95a, 98), *554*
Schreiner, S. 2115(234), *2127*
Schreyeck, G. 2238(123), *2244*
Schriewer, M. 1980(61b), 1999(61b, 126), *2032, 2034*, 2436(128), *2460*
Schriver, G.W. 83–86(161), *100*, 1153(54), *1182*
Schröder, C. 772(152), *778*
Schröder, D. 1114(71b), 1125(183), 1134(258), *1137, 1140, 1142*
Schröder, F. 1010(299a), *1060*
Schröder, H. 946(223), *1058*
Schroder, S. 13(46b), *96*
Schroeder, M.A. 1700(58, 59), 1705(59), *1786, 1787*
Schubert, U. 192(58), 243(268), 245(285), *258, 263, 264*, 586(194), *594*, 789(28d, 28e), *822*, 1701(61, 62), 1702(61), *1787*, 2027(203), *2036*, 2064(69, 70), 2072(83–86), 2073(97–101), 2074(102), 2090(84, 85), 2091(83, 86, 97, 100, 102), 2110(102), 2117(252), 2121(257), *2124, 2125, 2128*, 2142(70), 2143(73), *2172*, 2318, 2320(4), 2330(76), 2332(91, 92), 2333(93, 98), 2334(98), 2335, 2337(110), 2339(127), 2340(136, 139), 2341(142), *2355, 2357–2359*
Schuchardt, U. 2424(75), *2459*
Schuh, W. 1918(89), *1927*
Schulman, J.M. 26(63), *97*, 1911(63), *1927*
Schultz, A.G. 1812(113), *1864*
Schultz, R.H. 1113(67), *1137*
Schultz, W.J.Jr. 780(1f), *821*
Schulz, D.J. 1561(72), 1575(99), *1579*
Schulz, J. 2337(116, 117), *2358*
Schulz, S. 1878(45), *1892*, 2145(89), *2172*
Schulz, W. 2223(33), *2241*
Schulz, W.J. 563(76, 78, 79), 564(76, 78), 566(76), *591*
Schulz, W.J. 522, 540(79, 80a, 81), *554*
Schulz, W.J.Jr. 559(10), 564(85), *590, 591*
Schulze, M. 1584(23, 28), 1586(45, 46), 1588(28), *1595*
Schumacher, R. 403(147), *428*
Schumann, H. 2040(7), *2122*, 2138(39), 2153(127), *2171, 2173*, 2322, 2323(39, 40b), 2325(40b), *2356*

Schummer, D. 1550(43), *1578*
Schunck, S. 1162(113), *1183*
Schünke, C. 1813(115), *1864*
Schurz, K. 203(104), 252(322–324), *259, 265*, 783(19a), *822*, 873(59, 60), 876, 940(75, 76), 941(59, 60), 969(60), 1011(301–303), 1021(327), 1025(303), 1033(300, 301, 303), 1034(300, 303, 351, 352), 1035(300, 303, 351), 1036(300), 1037(59), 1038(300, 303, 351), 1039(356), 1040(303), 1042(327), 1043(303, 327), 1044(300, 303), 1045(303), 1051(351), *1055*, 1060–1062, 1064, 1068(5a, 5b), 1077(40), *1101, 1102*, 1158(88), *1183*
Schuster, H. 198(90), 215(156), *259, 260*, 783(19b), *822*, 1051(383), *1062*
Schuster, H.G. 2181(20, 21), *2214*
Schuster, I.I. 1408(178), *1443*
Schwab, S.T. 2252(31a–c), *2311*
Schwark, J.M. 2261(52), *2312*
Schwartz, D. 1888(95), *1894*
Schwartz, E. 1091(58g), *1103*
Schwartz, H. 2539(371), *2567*
Schwartz, K.B. 2251(28b), 2253, 2255(28b, 44a), 2261(44a), *2311, 2312*
Schwartz, N.V. 1633, 1642(157), *1664*
Schwartz, W. 217(167), *261*
Schwarz, H. 3(7c), 36(75), 53, 55–57, 60–62(100a), 66, 68(116), *95*, 97–99, 558, 559(2), 560(51), *589, 590*, 601(13, 14), 607(14, 24), 647(13), 698, 1010(298c), 1029, 1049, 1050(342), *1060, 1061*, 1106, 1109(1a), 1110(53), 1113(68, 70), 1114(71a, 71b), 1115(53, 72), 1118(89), 1122(146, 151), 1123(155, 157, 164, 165), 1124(170), 1125(151, 179, 183), 1126(193–199), 1127(195, 196), 1128(197, 219), 1129(195, 198, 199), 1130(195, 198, 199, 236), 1131(236, 240), 1134(254, 255, 258), 1135(195, 199), *1136–1142*, 1163(134), *1184*, 1408(179), *1443*, 2556(422), *2568*
Schwarz, W. 197(80), *259*, 796, 799, 800(66a, 66b), *823*, 1890(107), *1894*
Schwebke, G.L. 796, 802, 806(73), *824*, 1312(10), 1321(88), *1332, 1334*, 2177(3a), *2213*
Schweitzer, G.W. 1487(305), *1534*
Schweizer, W.B. 1743, 1746(212), *1790*
Schwertfeger, F. 2332(91), 2333(93), *2357*
Schwindeman, J. 1497(348), *1535*
Sciano, J.C. 1274(110), *1308*
Scibiorek, M. 2228(77), 2229(78), *2242*
Scilimati, A. 2388(72), *2400*
Scollary, G.R. 1462, 1467(156a, 156b), *1529*
Scott, D.L. 741, 742(78), *776*
Scott, D.W. 1923(95), *1928*
Scott, F.P.A. 1695(33), *1786*

Scott, R.S.III 2143(71), *2172*
Scotto, C. 2247–2250, 2269(9), *2310*
Scotto, C.L.S. 226(205), *262*, 2298(150b), *2316*
Scotto, S. 2250(19), 2276, 2278(104a–c), *2310, 2314*
Scriven, L.E. 2234(105a), *2243*
Scrocco, E. 528, 535(87a), *554*
Scrock, R.R. 272(17–19), *350*
Scuseria, G. 2509(229), *2564*
Scuseria, G.E. 13(46a), 72(145), *96, 100*, 174(139), *180*, 515, 516(17), *551*, 901, 997(151), *1057*, 1166(150), *1184*
Seakins, P.W. 167(88), *179*, 518(57, 58), *553*
Seanina, Z. 1151(42b), *1182*
Sears, T.J. 704(9), *775*, 1166(155), *1184*
Seaton, C.T. 2351(255), *2362*
Sebald, A. 238(253), *263*, 324, 325(142), 334(149), *353, 354*, 2164(183), *2175*
Sebastian, R. 362(20), *426*
Sebestyen, A. 1471, 1472(231), *1531*
Seconi, A. 168(107), *179*
Seconi, G. 460(162), *491*, 1543(17), *1577*, 1601(10, 13), 1602(10, 13, 29), 1604(13), 1608(68), 1610(10), 1611(86), 1612(13), 1625(130), 1628(136), 1635(10), 1642(192), 1643(13), 1656(136), *1660–1664*, 1875(32), *1892*
Seddon, A.B. 2352(268, 269), *2362*
Sedrati, M. 1860(354, 355), *1868*
Seebach, D. 476(219), *493*, 811(97a), *824*, 1606(57), 1608(65), *1661, 1662*, 1743, 1746(212), *1790*
Seebald, S. 2142(70), *2172*
Seelbach, W. 2061, 2066, 2068(59), 2078, 2091, 2095(122), *2123, 2125*, 2145, 2146(93), *2172*
Seeler, K. 641(101), *700*
Seeley, A.J. 1163(125), *1184*
Seetula, J.A. 167(85–88), *179*, 518(57), *553*
Segawa, K. 1743(209–211), 1745(209), 1747(209–211), 1748(209), 1749(211), *1790*
Segawa, Y. 1328, 1329(157), *1336*, 2470(33), *2559*
Segi, M. 1091(58h), *1103*, 1630(146, 147), 1640(186), 1641(190), 1653(146), *1663, 1664*, 1678(99), *1684*
Seguin, K.J. 1244(35), *1306*
Seha, T. 2295(145), *2315*
Seher, M. 2253(37b), 2261(54b), *2311, 2312*
Seibold-Blankenstein, I. 1323(113), *1335*
Seidel, G. 216(160), *260*, 1884(72), *1893*, 1979(56c), *2031*
Seidenschwarz, C. 945(205, 209, 211), 946(205, 209, 211, 214, 216, 222), *1058*
Seidl, E.T. 919, 920(186), *1058*
Seidl, H. 1212, 1213(70), *1231*

Seiferling, B. 2336(113, 114), 2340(114), *2358*
Seifert, J. 243(275), *264*, 1374, 1376, 1377, 1380, 1381, 1401(89), *1440*
Seiler, P. 186(37), *257*, 1910, 1911(60f), *1927*
Seiles, K. 380(72), *427*
Seio, K. 1550(49), *1578*
Seip, H.M. 2181(18), *2214*
Seishi, T. 1548(38), *1578*
Seitz, F. 1705(81), *1787*
Seitz, W.A. 70, 71(139b), *100*
Sekegudi, A. 2139, 2141, 2147(43), *2171*
Seki, K. 1318(53), *1333*
Seki, M. 384(87), *427*, 1669(28), *1683*
Seki, S. 2088, 2105(172), *2126*
Seki, Y. 1705(76, 77), 1714(96–99), 1715(101), 1771(302–304, 307), 1772(314), *1787, 1791, 1792*
Sekigawa, S. 194(68), *258*, 2440(155, 156), 2443(155, 161), 2444(155), *2460, 2461*
Sekiguchi, A. 120(4, 6, 7, 9), 125(33), 130(33, 37), 132(9), 133(4, 37, 41), 134(37, 42, 43), 137(46, 47), 140(9, 47), 143(46, 47), 146(37, 46, 47), 147(56), *150–152*, 190(52), 194(63), 240(259), *258, 263*, 484(242–244), *493*, 551(116), *555*, 589(198), *594*, 713(29–31, 33, 34), 714(29, 31, 36a, 36b), 715(34), 716(39, 40), 718(29, 30), 719(42), 720(31, 42), 722(36a, 36b, 45, 46, 48, 49), 724, 726(49, 50), 727(40), 741(33), 743(33, 48–50, 85, 86), 744(48, 49, 85, 86), 748(33, 39, 40, 98), 749(40, 85, 86), 750(39, 40, 85, 86, 98), 752(33, 115), 753(33), *775–777*, 796(56, 61), 797(56), 798(56, 61), *823*, 829(27), 830, 831(28), 832(27, 28), 835(27), 836, 839–842(28), *854*, 904(164), 905, 969, 973(165), 992(164, 165, 277), 996(164, 165), 1006, 1008(296b, 296c), *1057, 1060*, 1080(43b, 43c), *1102*, 1146(19), 1158(89), 1169(170), *1181, 1183, 1185*, 1238(19), 1244(33), 1268(100, 101), 1276(125, 126, 129), 1282(141), *1306–1308*, 1331(187–189), *1336*, 1601, 1604, 1612, 1621(15), *1660*, 2001(131a, 131b), 2012(168a, 168b), *2034, 2035*, 2205(90, 91), 2206(90), *2215*, 2403(2, 3), 2404(4, 5, 7, 8), 2405(13–15), 2406(19), 2407(26), 2408(29, 30), 2412(36–38), 2413(37), 2414(40), 2415(41), 2416(50, 51, 53), 2417(53), 2418(50, 51, 54), 2420(54), *2457, 2458*, 2488(131), 2491(143, 144), 2495(167), 2518, 2519(143), 2520(144), *2561, 2562*
Sekiguti, A. 288(66), *352*
Sekimoto, K. 1808(93), *1863*
Sekine, M. 434, 447(30), *489*, 1550(49), *1578*
Sekine, S. 1838(232, 233, 235), *1866*
Sekiya, H. 2523(316), *2566*

Selbin, J. 183(5), *257*
Selby, C.E. 2235(108), 2236(114), *2243, 2244*
Selegue, J.P. 2143(81), *2172*
Selenina, M. 1587(54), 1589(74), *1596*
Seligson, A.L. 1888(96), 1890(96, 109), *1894*
Selin, T.G. 790(32), *822*
Sellers, S.F. 2437(135), *2460*
Seltzman, H.H. 1669(22), *1682*
Semeneko, K.N. 240(256), *263*
Seminario, J.M. 515, 516(14a), *551*
Semlyen, J.A. 2218(4), *2240*
Semmelhack, M.F. 1733(141), *1788*
Semmelmann, M. 1887(88), *1893*
Semmler, K. 1910, 1911(60e), *1927*
Sen, D. 1903(44), *1926*
Sen, P.K. 1322(95), *1334*
Senda, Y. 1733(142), *1788*
Sengupta, D. 161, 163(61, 62), *178*, 1008, 1009(297a), *1060*, 2487, 2511(128), 2556(426), *2561, 2568*
Senn, M. 1555(58c), *1578*
Sentenac-Roumanou, H. 1453, 1492(67), *1526*
Sentenas-Roumanou, H. 1524(422), *1537*
Senzlober, M. 864, 981, 982(31–33), *1054*, 1237(17), *1306*
Seo, Y. 2153, 2162(124), *2173*
Serbetcioglu, S. 2224(35a, 35b), *2241*
Serbny, H.-G. 2223(29a, 29b), *2241*
Sereda, S.V. 1244(34), *1306*
Serein-Spirau, F. 1223(103), *1232*
Sergeev, V.N. 243(277), *264*, 1374, 1376(92a, 92b), *1440*
Sergeyev, N.M. 304(105), *353*
Serna, C.J. 2350(223–225), 2352(223, 224, 272), *2361, 2362*
Serpone, N. 1415, 1420(198), *1443*
Serrano, A. 2350, 2352(224), 2353(274, 276), *2361, 2362*
Serrano, F. 1915(82), *1927*
Serrano, R. 2145(90), *2172*
Serratosa, F. 2442(160), *2460*
Sertchook, H. 2322, 2323, 2325(40b), *2356*
Servis, K.L. 307, 308(113), *353*
Seshadri, T. 228(209), *262*
Seth, P.P. 1822(149), *1864*
Seth, S. 1771(298), *1791*
Setsu, F. 1561(71a, 71b), *1579*
Seufert, A. 2143(77), 2144(77, 83), *2172*
Sevenich, D.M. 1584(22), *1595*
Severson, R.G. 789(28f), *822*
Sewald, N. 945(205, 211), 946(205, 211, 212, 214, 222), *1058*
Seyferth, D. 417(190, 191), *429*, 727(57), 739(75), 741(75, 79, 80), 767(75, 144), *776, 778*, 1331(177), *1336*, 1618(106, 107), *1662*, 1759(262), 1762(263), *1791*, 1809(100, 101), *1863*, 1973, 1974(45a–d), 1976(45d), 1995(45c), *2031*, 2251(25a–e), 2252(25a), 2253, 2255(25a–e, 42, 43), 2256(43), 2257(25a–e, 42), 2261(68), 2269(87), 2273(99a–c), 2274(99c), 2275(99a–c), *2311–2314*, 2407(25), 2421(65–67), 2424(77), 2429(97, 103), *2458, 2459*
Seyfried, E. 2332(91), *2357*
Seymour, C.M.- 2160(164), *2174*
Sgarabotto, P. (231), *262*
Shäfer, A. 1976(51), *2031*
Shaffer, S.A. 561(54), *591*
Shafiee, F. 200(97), *259*
Shafrin, E.G. 2237(119), *2244*
Shagun, V.A. 1466(177, 178), 1467(177, 178, 181), 1469, 1472(198), *1530*
Shah, P.M. 448(118), *491*
Shah, S.K. 1608(70), *1662*
Shaik, S. 576(140), *592*
Shaik, S.S. 5, 6, 24(15a–c), 25, 87(15b, 15c), *95*, 499(22), *510*
Shainyan, B.A. 1484(294), *1533*
Shakespeare, W.C. 2450(188), *2461*
Shakkour, E. 1207, 1208(53), *1230*
Shalom, S. 2320, 2322(18, 19), 2327(19), 2340(132), 2348(18, 19), *2355, 2359*
Shaltout, R. 1431(230), *1444*
Sham, L.J. 516(30b, 30c), *552*
Shambayati, S. 184, 222(22), *257*, 482(241), *493*
Shames, A. 2346, 2347(183b), *2360*
Shamshin, L.N. 1989(90a, 90b), *2033*
Shanamuganthan, K. 1852(317), *1868*
Shankar, R. 1434, 1436(246), *1444*, 1453, 1462, 1467, 1483, 1491(68), *1526*
Shankaran, K. 1823(154), *1864*
Shanklin, M.S. 481(236), *493*, 1776(325, 326), *1792*
Shannon, M.L. 1973, 1974(45b–d), 1976(45d), 1995(45c), *2031*, 2421(65–67), 2424(77), *2459*
Shannon, P.J. 1766(276, 277), *1791*
Shannon, R.D. 106, 110(4), *117*
Shapiro, A.B. 1491(323), *1534*
Shapley, P.A. 1872(22), *1892*
Sharapov, V.A. 245(284), *264*, 586(193), *594*
Sharifova, S.M. 1996(108c), *2033*
Sharma, H. 1262(80, 81), *1307*, 2083(148, 150, 152, 153, 155), 2087(150, 155), *2126*, 2155(135), *2173*
Sharma, H.K. 437, 485(77), *490*, 1237(18b), 1260(71), 1261(76), *1306, 1307*, 1602, 1604, 1605(32), *1661*
Sharma, S. 791(43a–c), *823*, 2083(147, 148), *2126*, 2143(72), *2172*
Sharmow, B. 2142(62), *2172*
Sharp, G.J. 2534(365), *2567*

Sharp, J.T. 464(175), *492*
Sharp, R.R. 1148, 1149(30), *1182*
Shashkov, A.S. 1500(356), *1536*
Shatz, V.D. 1449, 1460, 1470, 1481, 1506, 1510(6), *1525*
Shaw, A. 1878(46), *1892*
Shaw, R. 155, 176(30), *178*
Shay, W.R. 1776, 1779(324), 1780(336), 1782(336, 342), *1792*
Shchembelov, G.A. 2131(8), *2171*
Shchepinov, S.A. 1989(90a), *2033*
Shchukina, L.V. 1450(38, 41), *1526*
Shea, K.J. 1341, 1347, 1367(25), 1369(25, 58a, 58b), 1370(58b), *1437, 1439*, 1854, 1860(325), *1868*, 2346(180, 181, 182a, 182b), *2360*
Shechter, H. 709, 714(21), 715(38), 727, 741(21, 38), 743(21), *775*
Sheehan, J.P. 219(175), *261*
Shehepinov, S.A. 1989(88), *2033*
Sheikh, M.E. 2404(9), *2457*
Sheiner, A.C. 13(46a), *96*
Shekhani, M.S. 448(118), *491*
Sheldon, J.C. 284, 288(52), *351*, 1109(42), 1134(253), *1137, 1141*, 1372(66), *1439*
Sheldrake, G.N. 1091(58e), *1103*
Sheldrick, G.M. 198(81, 85), 213(150), 229(217), 234(241), 235(238), *259, 260, 262, 263*, 799(64), *823*, 884–886, 937, 938(103), 1011, 1044, 1048(305), *1056, 1060*, 1986, 2012(78c), *2032*, 2145(89), *2172*
Sheldrick, W.S. 183, 198, 218, 224, 229, 234(15), 238(253), 247(15), *257, 263*, 324, 325(142), 334, 335(148), *353, 354*, 485(251), 488(271), *493, 494*, 724(51), *776*, 780(1d), *821*, 1086, 1091(53), *1102*, 1282(139), *1308*, 1340(3), 1341(3, 30b), 1349(30b, 31b), 1350(30b), 1351(3, 30b), 1354(3), 1357(30b, 31b, 47), 1359, 1360(47), 1361(30b), 1362, 1363(47), 1368(30b), *1437, 1438*, 1449(30), *1525*, 1914(80), *1927*, 2134(29), *2171*, 2364(10, 12, 14, 17, 30), *2398, 2399*
Sheludyakov, V.D. 240(258), *263*, 296, 299–301(94), 302–305(104), *352, 353*, 1454, 1467(94), *1527*
Shen, J. 2352(271), *2362*
Shen, L.-R. 1464, 1467, 1486(168), *1530*
Shen, M. 532(99b), *554*
Shen, Q. 1465, 1467, 1472, 1478(169), *1530*
Shen, Q.H. 2283, 2284(116a), *2314*
Shepherd, B.D. 247(296), 248(300, 302), *264*, 348(156), *354*, 829(20), *854*, 1208, 1214(56), *1230*, 1266(89), *1307*
Shepherd, R.A. 68(122a), *99*, 1164(140–142), *1184*

Sherbine, J.P. 456(142), *491*
Sheridan, J.B. 2111, 2120, 2121(225), *2127*, 2154(138), *2174*
Sherrill, C.D. 66–68(119), *99*, 2556, 2557(424), *2568*
Sherwood, W. 2283, 2284(116b), *2314*
Shestakov, E.E. 1475(253), *1532*
Shestunovich, V.A. 245(286), *264*, 587(195), *594*
Shevchenko, S.G. 888, 990(122), *1056*, 1467, 1469(184), 1470(212), 1471(226, 230), 1472(230, 233a), 1475(212), 1485(299), *1530, 1531, 1534*
Shi, Y. 2330(83), 2352(263, 265), 2354(83), *2357, 2362*
Shi, Z. 499(23), *510*
Shiao, M.-J. 1678(102), *1684*, 1877(44), *1892*
Shiao, S.-S. 434(18, 20), 442(18), 443(20), *489*, 1835(215), *1866*
Shiba, T. 20(51a), *96*
Shibasaki, M. 1796(39), *1862*
Shibata, F. 1846(279), *1867*
Shibata, K. 1677(93), *1684*
Shibata, T. 1625(131), *1663*
Shibley, J.L. 200(103), *259*
Shibuya, M. 2272(95), *2314*
Shick, R. 2289(130, 131), *2315*
Shida, T. 1314(27), *1333*
Shields, A.F. 1550(46), *1578*
Shien, M.-Q. 1208(55), *1230*
Shigihara, A. 707(17), *775*
Shih, C. 1646(206), *1665*
Shiina, K. 2429, 2436(108), *2459*
Shike, A. 1942, 1944(20), *1960*
Shilin, S.V. 1454, 1467(89), *1527*
Shim, S.C. 1250(48–50), *1306*
Shima, I. 1064(3), *1101*
Shima, K. 459(156), *491*
Shimada, K. 1885(78), *1893*
Shimada, S. 2115, 2120, 2121(232), *2127*
Shimada, Y. 832, 838(30), *854*
Shimanskaya, M.V. 1740(171), *1789*
Shimazaki, M. 470(196), *492*
Shimizu, H. 704–706(8), 770, 772(149), *775, 778*, 1860(350), *1868*
Shimizu, K. 1299(195), *1310*, 2437, 2438(136), *2460*
Shimizu, M. 437, 449(87), *490*, 1857(335), *1868*, 2387(63), *2400*
Shimizu, N. 363(31), 370(52), 371(53, 54, 56), *426, 427*, 436(66, 69), 437(70), 482(66), 483(69, 70), *490*, 611, 613, 614(41), 621, 623, 624(72, 73), 632, 633(86a, 86b), 634(86a, 86b, 88a, 88b), 635(88a, 88b), *698, 699*, 1846(279), 1856(333), *1867, 1868*
Shimizu, S. 467(186), *492*

Shimizu, T. 186(39), 194(68), *258*, 278(26, 27), 329(26), *351*, 1299(195), *1310*, 2437, 2438(136), 2439(154), 2440(154–156), 2443(155, 161), 2444(155), 2447(172), 2448(179), 2449(172, 179), 2450(172, 190), 2451(172), *2460, 2461*, 2494(162), *2562*
Shimizutani, T. 1415(199), *1443*
Shimoi, M. 1260(72), *1307*, 2087(166), 2088(168), 2090, 2094(166, 168), *2126*
Shimon, L.J.W. 1074(29), *1102*
Shimshock, S.J. 434(34), 448(34, 117), *489, 491*
Shin, H. 2546(393), *2568*
Shin, H.C. 1490(315), *1534*
Shin, S.K. 34, 36(73c), *97*, 171(125), 172(125, 131), 173–175(131), *180*, 520(68), *553*, 558(5), 559, 560(29), *590*, 901(150), *1057*, 1108(19), *1136*
Shiner, V.J.Jr. 376(65, 66), 377(68), 380(73), *427*, 635(90–95), 636(90–92), 637(91, 93–95), 641(102), 642(102, 106), 643(106), *699, 700*
Shinimoto, R. 1981, 1983, 1998(68), *2032*
Shinoda, I. 1733(142), *1788*
Shinokubo, H. 462(166), *492*
Shioiri, T. 704, 739(5), 741, 742(82), 743(87, 88), *775, 776*, 1612(88), *1662*, 1675(83, 85), 1676(86–88), *1684*
Shiomi, Y. 1852(313), *1868*
Shiota, T. 1750(230), *1790*
Shioya, J. 186(40), *258*, 2013(171), 2017(177c, 178), *2035*, 2439(148), *2460*
Shioyama, H. 145(54, 55), 146(55), *152*, 1918(87), *1927*, 2206(93), 2207(93, 96), *2215*
Shiozaki, M. 1978, 2015(54b), *2031*
Shiozawa, M. 1590(84, 84, 87), 1591(99), *1596, 1597*
Shipov, A.G. 226(206, 207), 243(273), *262, 263*, 500, 501(26), *510*, 1390(130), 1391(140), 1392(143, 145a, 145b), 1393(148), 1395(156), 1396(156, 158, 159, 162, 163), 1397(156, 166, 167), 1399(156), 1404(172), 1434, 1435(240), *1441, 1442, 1444*, 1454, 1467(106), *1528*
Shippey, M.A. 782, 801(13c), *822*
Shipton, M.R. 456(145), *491*
Shirafuji, T. 2521(281), 2549(403), *2565, 2568*
Shirai, J. 1743, 1748(206), *1789*
Shirakawa, K. 437, 438(83), *490*
Shirchin, B. 1475(252), *1532*
Shiro, M. 190(51), *258*, 340–342(151), *354*, 1304(215), *1310*, 1340, 1341(17b, 18), 1342, 1343(18), 1344, 1345(17b), 1347(18), 1390(123a), *1437, 1441*, 1979, 2013, 2014(59a), *2032*, 2429, 2436(105), *2459*, 2554(416), *2568*

Shirohata, A. 1458, 1467(126), *1528*
Shiromura, H. 1120, 1121(133), *1139*
Shiryaev, V.I. 1312(16), *1332*
Shishido, T. 2272(93c), *2314*
Shiue, C.-Y. 385(97), *428*
Shizuka, H. 998(287), *1060*, 1157(75), *1182*, 1268(95), *1307*, 1322(106–111), 1325(120), *1334, 1335*, 2472(46), 2498(183), 2513(258, 259), 2515(259), 2516(258, 259), 2520(259), 2526(258), *2559, 2563, 2564*
Shklover, V.E. 186(44, 47), 190(50), 196(71), 224(197), 237(244), 243(270–274), 245(286), *258, 261, 263, 264*, 500, 501(26), *510*, 587(195), *594*, 1393, 1394(151), 1395(154, 156, 157), 1396(151, 156–158, 160, 164, 165), 1397(156), 1399(156, 160, 164, 165), *1442*, 1449(21, 22), 1450(46), 1454(103, 104), 1460(21, 22), 1461(46, 142, 144, 145), 1462(151), 1463(21, 22), 1464(103, 104), 1467(103, 104, 142, 144, 145, 151), 1514, 1522(408b), *1525, 1526, 1528, 1529, 1537*
Shkol'nik, O.V. 2218(5), *2240*
Shobatake, K. 2523(312), *2566*
Shoda, H. 449(120), 469(191), *491, 492*
Shoemaker, H.E. 1815(122), *1864*
Shohda, M. 1561(71a, 71b), *1579*
Shomburg, D. 1352(32d), *1438*
Shono, T. 1206(46), 1216, 1217(87, 88), 1218(88, 92), 1219(88), 1221(98), 1228(121–123), *1230–1232*, 1811(107), *1864*
Short, R.P. 461(164), *492*
Shorygin, P.P. 1312(16), *1332*
Shtefan, O.V. 580, 584(167), *593*
Shtelzer, S. 2326(45a, 46a, 46b, 48, 50), *2356*
Shterenberg, B.Z. 1381(107, 108), *1440*, 1454, 1467(102), 1474(243, 260, 263, 266), 1476(243, 259–261, 263), 1477(243, 259, 260, 263, 264, 266), 1478(263, 266), 1479(271), 1485(301), 1487(309), 1490(309, 316, 319, 320), 1491(321), 1494(319, 335), 1497(349), 1506(261), 1510(260, 261), *1528, 1532–1535*
Shuff, P.H. 2430(115), *2460*
Shukla, P. 1318(48, 51), *1333*
Shull, B.K. 1675(76), *1684*
Shumm, R.H. 159, 160(54), *178*
Shun, L.A.M.K. 2351(242), *2361*
Shurki, A. 5, 6, 24(15a), *95*
Shuvaev, A.T. 1470(213, 214), *1531*
Shvets, G.N. 154, 156(13–16), 162, 163(14), 164, 165(15), 166(13, 16), 174(14), *177*, 1469(201, 203), *1530, 1531*
Shvetz, G.N. 1469(199), *1530*
Shymanska, M. 1740(172), *1789*
Shymanska, M.V. 1717(105), *1787*

Sibao, R.K. 1086, 1091(53), *1102*, 1870(5), *1891*
Sibi, M.P. 1563(75), *1579*, 1711(91), *1787*
Sibille, S. 1222(101), *1231*
Siddiqui, M.R.H. 1589(75), *1596*
Sidorkin, V.F. 501(28), *510*, 1341, 1343(21b), 1378(101, 102), *1437, 1440*, 1449(12, 28), 1463(28, 163), 1465(12, 174), 1466(12, 28, 176–179), 1467(12, 163, 174, 176–179, 181, 188a), 1468(188a, 195, 196), 1469(12, 198), 1470(12, 28, 196), 1471(12, 196), 1472(198), 1473(12), 1475(255, 258), 1476(28, 188a), 1477(174, 188a), 1478(28, 174, 188a), 1506, 1508(174), 1510(12, 174), 1523(12, 28, 174), *1525, 1529, 1530, 1532*
Sidorov, V.I. 388(105), *428*
Sieber, S. 567(107), *592*
Siebert, W. 216(161), *260*
Sieburth, S.M. 446(114), *490*, 1854(320), *1868*
Sieburth, S.McN. 382(82), *427*
Siedem, C.S. 1643(194), *1664*
Siefert, E.E. 1965(17), *2030*
Siegel, C. 442(106), *490*
Siegel, D.A. 1288(165), *1309*, 2497, 2499(182), *2563*
Siegl, H. 2189(60), 2191(66), 2201(84), 2204(60), 2210(103), *2215, 2216*
Siegle, H. 2083(146), *2125*
Siehl, H.-U. 370(51), *427*, 517(47), 521–523, 527, 529, 542, 544(73), *553*, 584(189), *594*, 599, 604(10), 647(119), 648(120), 652(127), 655(133), 659, 660(135), 663(139, 140, 142), 669(145), 670(147), 675(149), 676, 677(140), 679(154), 685(156), 688(160a), 690(161, 166a, 166b, 166d), 691(168), 693(169), *698, 700, 701*
Siekhaus, W.J. 1120, 1121(129), *1138*
Siemeling, U. 2150, 2151(107, 108), 2155(143), 2156(107), 2158(143), 2159(143, 150, 153), *2173, 2174*
Sierakowski, C. 10(35), *96*, 1151(46, 48), *1182*, 2453(195), *2461*
Siewek, B. 1878(47), *1892*
Sigalov, M.V. 1479, 1480(276), 1492(327), 1499(354), 1500, 1501(327), *1533, 1535*
Sigel, G.A. 2337(121), *2358*
Sigwalt, P. 580(170, 171), *593*, 2220(18, 19a–c), 2222(22a, 22b), *2240, 2241*
Silks, L.A.III 1823(155), *1864*
Silla, E. 528, 535(87d), *554*
Silveira, C.C. 1801(63), *1863*
Silverman, R.B. 1606(52b), *1661*
Silverman, S.B. 1740(178), *1789*
Silwinski, S. 1923(101), *1928*
Sim, E.S.H. 2075, 2091(110), *2125*
Simard, M. 1280(138), *1308*
Simchen, G. 1674(63), 1676(89), *1683, 1684*

Simionescu, M. 2223(24), *2241*
Simkin, B.V. 2512(253), *2564*
Simkin, B.Y. 52(94a), *98*
Simkin, B.Ya. 5–8, 32, 69, 77(13a), *95*
Simmons, H.E. 1998(122a), *2034*
Simon, A. 198(90), 203(112), *259*, 783(19b), *822*, 1034(352), *1061*
Simon, G. 1584(25, 26), *1595*
Simon, J. 1512(404, 405), *1537*
Simon, K. 1460(158), 1462(147, 148), 1467(147, 148, 158), *1529*
Simon, R. 284(54), *351*, 487(268), *494*, 1340, 1341, 1343(11b), *1437*
Simons, J. 1127(211), *1140*, 1372, 1373(70), *1439*
Simons, R.S. 2079, 2093(124), 2105, 2109(207), *2125, 2127*, 2145(92), *2172*
Simpson, J. 2107(214), *2127*
Simpson, T. 1266(88), *1307*, 1319(73), *1334*
Sinai-Zingde, G. 2224(39), *2241*
Sinclair, R.A. 2273(98), *2314*
Singer, L. 1415(196b), *1443*
Singer, R.A. 373(58), *427*
Singer, R.D. 791(43d), *823*
Singewald, E.T. 198(84), *259*
Singh, J. 417(195), *429*
Singh, J.D. 1743, 1747(210, 211), 1749(211), *1790*
Singh, M. 1622(126), *1663*
Singh, R. 1674(69), *1683*
Singh, V.K. 1674(69), *1683*
Sini, G. 499(22), *510*
Sinnott, M.V. 361(18), *426*
Sinotova, E.N. 559, 581(9), *590*
Sinou, D. 1672(46), *1683*
Siriex, F. 2253(46), *2312*
Sironi, M. 2509(227), *2564*
Sita, L.R. 120(11), 122(16), 142, 143(50), *150, 152*, 163(72), *179*, 1910, 1911(60g, 60h), *1927*, 2479(81), *2560*
Sivavec, T.M. 274(20), *350*
Sizoi, V.F. 1838(226), *1866*
Skalsky, V. 1118(96), *1138*
Skancke, P.N. 184, 234(24), *257*
Skättebol, L. 2447(173), 2448(173, 174), *2461*
Skell, P.J. 2481(86), *2560*
Skell, P.S. 727, 739(53), 773(154), 774(155), *776, 778*
Skelton, B.W. 210(141), 217(169), 252(329), *260, 261, 265*, 1377, 1378, 1380(97a), *1440*
Skiles, G.D. 2307(161), *2316*
Sklenak, S. 18, 20(48), *96*
Skobeleva, S.E. 304(109), *353*
Skobridis, K. 222(186, 188, 190), *261*
Skoda, L. 915(182), *1057*
Skoda-Foldes, R. 1689(8), *1785*
Skoglund, M. 709, 714, 727, 741, 743(21), *775*

Skrydstrup, T. 1674(67b), *1683*, 1794, 1819, 1853, 1854(28), *1862*
Skurat, V.E. 2231(86), *2243*
Sladek, A. 209(137), *260*, 1377, 1378, 1380(97b), *1440*
Slama-Schwok, A. 2349(203, 204, 205a, 205b), *2361*
Slanana, Z. 149(60d), *152*
Slanina, Z. 70(138), *100*, 1151(43), *1182*
Slater, J.C. 116(45), *118*, 183(1), *257*, 516(28), *552*
Slawin, A.M.Z. 285–287(62), *352*, 467, 474(187), *492*, 1355–1357(38), *1438*
Sleiksa, I. 1199(31), *1230*
Sliwinski, W.F. 635(89a, 89b), *699*
Sluggett, G.W. 849(80–82), 852(82), *855*, 864(35a), 897(135–137), 898(139, 140), 932, 934(198), 969(136, 137, 140), 977(140), 992(35a), 994, 996(137), *1054, 1056–1058*, 1248(46), 1253(59–62), 1254(59–61), 1256(62), *1306, 1307*, 1325(142–146), 1332(202), *1335, 1336*, 2499(186), *2563*
Slutsky, J. 2404(6), *2457*
Smadja, W. 475(213), *493*, 1554(56), *1578*, 2446(168), *2461*
Small, J.H. 1341, 1347, 1367(25), 1369(25, 58a, 58b), 1370(58b), *1437, 1439*
Smalley, R.E. 69(124a–c), 70(134a), *99, 100*, 1120(123, 125), *1138*, 1929(1b), *1960*
Smart, B.A. 1547(35b), *1578*
Smart, B.E. 1542(15), *1577*
Smart, J.C. 2137(33), *2171*
Smart, R.T. 806(85), *824*, 1213(72), *1231*, 2193(69), *2215*
Smerz, A.K. 1841(255), *1866*
Smetarkine, L. 111(21), *118*
Smirnov, V.N. 171(123, 124), 172(124), *180*
Smirnov, Y.N. 2522(293), *2565*
Smit, C. 1051(378), *1062*
Smit, C.N. 1064, 1068(6a), *1101*, 1388(122), *1441*
Smith, A.B.III 1797(42–44), *1862*
Smith, B.A. 1318(48), *1333*
Smith, B.J. 34(74e, 74f), *97*
Smith, C.L. 795(63a, 68), 796(71b), 799(63a), 800(68), *823, 824*, 1322(95), *1334*
Smith, C.M. 780(1a), 790, 795, 800(34), *821, 822*
Smith, D. 1117(85), *1138*
Smith, D.C. 1877(44), *1892*
Smith, D.E. 1211(64), *1231*, 2237(118), *2244*
Smith, D.M. 559(26, 27), *590*
Smith, E.G. 317, 319–322(140), *353*
Smith, J.V. 116(43), *118*
Smith, M. 2322, 2348(25), *2356*
Smith, M.K. 2327(59), *2357*
Smith, P.A. 413(176), *429*
Smith, S.D. 2235(108), 2236(114, 115), *2243, 2244*
Smith, S.G. 1502(371), *1536*
Smith, T.L. 2283(115), *2314*
Smith, W.D. 1397, 1398, 1404(170), *1442*
Smith, W.H. 205(115), *259*
Smith, Z. 2181(18, 19), *2214*
Smithers, R. 392, 398, 412(118), *428*, 447(116), *491*, 1794, 1795, 1797, 1801, 1811, 1821(19), *1862*
Smitrovich, J.H. 1681(122), *1685*
Smits, J.M.M. 2152, 2153(123), *2173*
Sneddon, L.G. 2250, 2253(22), 2265(83a–c), *2310, 2313*
Snegova, A.D. 1220(96), *1231*
Snider, B.B. 1850(302), 1857(336), *1867, 1868*
Snieckus, V. 385(93), 386(103), 416(184), 427–429, 484(245), *493*, 1803(73), *1863*
Snijkers-Hendrickx, I.J.M. 2354(283), *2362*
Snow, A.W. 1767(280), *1791*
Snow, J.T. 248(298), *264*
Snyder, D.C. 1669(21), *1682*
Snyder, J.P. 1668(12), *1682*
Snyder, L.C. 1127(214), *1140*, 1898(34), *1926*
Sobczyk, L. 304(108), *353*
Sobon, C.A. 2273–2275(99c), *2314*
Sobrados, I. 2340(131), *2359*
Soderquist, J. 1613, 1622(89), *1662*
Soderquist, J.A. 435(39, 54), 451(39), 456(54), 459(154), *489, 491*, 611, 614(42), *698*, 1601, 1602, 1604(11), 1606(52a), 1617(103, 104), 1620(103), 1624(11), 1638(169), 1639(183, 184), 1642(193), 1651(11), *1660–1662, 1664*, 1837(222, 223), 1843(265), 1845(273), *1866, 1867*
Soga, T. 1871(13), *1891*
Sohn, H. 4(10b), 37(10b, 84, 85), 38(84), 39, 40, 44, 48(10b), 49(85), *95, 97*, 818, 819(116), *825*, 845(64), *855*, 981(260), 984(260, 266), 987, 989(260), *1059*, 1972, 1988, 1998(40), 2023(40, 196), 2024, 2026(198), *2031, 2036*
Sohoni, G.B. 2228(75), *2242*
Sokolova, E.V. 154, 156, 162, 163, 174(14), *177*
Sokol'skaya, I.B. 580, 584(167), *593*
Sole, A. 2442(159, 160), *2460*
Soleilhavoup, M. 762(131), *777*
Sollenböhmer, F. 377(69), *427*, 471(197), *492*, 637(98), 639(98–100), 640(98, 99), 641(100), *699*
Sollot, G.P. 2141(59), *2172*
Sollradl, H. 1601(25), *1661*
Solomennikova, I.I. 1476(261), 1481(284), 1492(326), 1505(326, 378), 1506(261, 326, 378–380), 1510(261), 1514, 1519(408a),

1521(408a, 417), 1522(408a), 1524(421), *1532, 1533, 1535–1537*
Solomennikova, I.S. 1475, 1506, 1510(250), *1532*
Solomon, S. 2448(174), *2461*
Solouki, B. 10(32c, 33c), 12(44), 52(97), *96, 98*, 1106, 1109(1b), 1123(159), *1136, 1139*, 1150(37), *1182*, 1314(23), *1333*, 2181(23), *2214*
Somasivi, N.L.D. 1591(100), *1597*
Somers, J.J. 382(82), *427*
Sommazzi, A. 1545(27), 1570(88), *1578, 1579*
Sommer, J. 683(155), 688(160b), *701*
Sommer, L. 1019(322), *1061*
Sommer, L.H. 364(35, 36), 376(35), *426*, 437, 483(71), *490*, 496(4, 7, 8), 497, 498(4), 507(49), *510, 511*, 610, 616(34, 35), 617(54), 635(34, 35), *698*, 790(39), *822*, 1080(43a), *1102*, 1247(40), *1306*, 1325(125), 1332(201), *1335, 1336*, 1606, 1628(46), *1661*, 1722(126), *1788*, 2406(18), *2458*
Sommerfeld, T. 1135(267, 268), *1142*
Sommerlade, R. 2231(81e), *2242*
Son, D.Y. 1759(262), *1791*
Son, H. 44, 47–49(93b), *98*
Son, H.-E. 37(83), *97*, 818(114), *825*, 2018(183), 2019(185), 2022(183), *2035*
Sondheimer, F. 2448(175), *2461*
Song, Y.C. 2272(96), *2314*
Sonnenberger, D.C. 487(265), *494*
Sonoda, M. 1838(232, 233, 235–237), *1866*
Sonoda, N. 1091(58h), *1103*, 1559(67a, 67b, 68), 1560(69), *1579*, 1616(99, 100), *1662*, 1678(99), *1684*, 1705(77), 1714(96–99), 1771(302–304, 306, 308, 309), 1772(306, 311, 314), *1787, 1791, 1792*
Sonoda, T. 630(84), *699*
Sonogashira, K. 2075, 2091(115), *2125*
Sonomoto, K. 2390, 2391(76, 77), 2392(76–78), 2394(78), 2395, 2396(80), *2400*
Sooriyakakumaran, R. 1238(21), *1306*
Sooriyakumaran, R. 1266(86), *1307*, 1312, 1317(5), 1318(5, 49), 1320(84), 1321(84, 85), 1331(192), *1332–1334, 1336*, 2469(30), 2475, 2477, 2482, 2483(56), *2559, 2560*
Soos, Z.G. 1316(39, 40), *1333*
Soraru, G.D. 2264(78, 79), *2313*
Sorensen, O.W. 269(9), *350*
Sorley, S.D. 485(256), *493*
Sorokin, M.S. 154, 156, 166(13), *177*, 1393, 1395, 1398, 1404(153), *1442*, 1454(80, 86–88, 90–92), 1461(144), 1467(80, 86–88, 90–92, 144), 1469(199, 203), 1471(225), 1472(234), 1473, 1474(235), 1495(339), 1497(346, 347), 1500(88, 92, 361, 362),

1502(92, 361, 370), *1527, 1529–1532, 1535, 1536*
Sorokon, M.S. 237(249), *263*
Soulairol, F. 1075(35c), *1102*, 1390(123b), *1441*, 2481, 2483, 2553(97), *2561*
Soulie, C. 1512(404, 405), *1537*
Soulouki, B. 864, 992, 996(30), *1054*
Soum, A. 2251(30a, 30b), *2311*
Soustelle, M. 2261(65), *2313*
Soysa, H.S.D. 1075(31), *1102*
Spackman, M.A. 113(26, 28), *118*
Spagnolo, P. 1877(40, 41), *1892*
Spalding, T.R. 219(175), *261*
Spaltenstein, E. 2042(14), 2044(18), 2051(14, 18), 2052(18), *2122*
Spangler, C.W. 1997(113b), *2033*
Spanhel, L. 2340(138), *2359*
Spaniol, T.P. 209(133), *260*
Spanka, S. 740, 767(76), *776*
Sparks, M.A. 437, 487(82), *490*, 1859(346, 347), *1868*, 2388(70), *2400*
Spear, R.J. 647(116), 677(153), *700, 701*
Speckamp, W.N. 1815(122–125), 1835(216), *1864, 1866*
Spehr, M. 1885(73), *1893*
Speier, J.L. 1500(357), *1536*, 1592(101), *1597*, 1688(4), 1699(47), 1722(4), *1785, 1786*
Speier, O.L. 417(192), *429*
Spek, A.L. 219(181), *261*
Spencer, B. 2137(33), *2171*
Spencer, C.M. 1707(83), *1787*
Spencer, J.L. 1727–1730(135), *1788*, 2070, 2071(80), *2124*
Speranza, M. 385(97), *428*, 561(65), *591*
Sperlich, G. 1341, 1349, 1357(28, 29), 1358(28), 1359(28, 29), 1360–1362(28), 1366(29), *1437*
Sperlich, J. 226(204), *262*, 334(149), *354*, 1357(46, 48), 1359, 1360(46), 1362(46, 48), 1364(46), *1438*, 2364(27), 2368, 2369, 2374(39), 2389(74), *2398–2400*
Sperling, L.H. 2342, 2343(149), *2359*
Spialter, L. 435, 456(49), *489*, 1372, 1373(72c), *1439*, 1617, 1620(103), *1662*
Spielberger, A. 279(30), *351*, 796(83b, 83c), 805(83c), *824*, 2191(65, 66), 2199(81), 2200(81, 82), 2201(83), 2210(106), 2213(106, 107), *2215, 2216*
Spiess, H.W. 310(121–123, 125–129, 134), *353*
Spindler, K. 1885(76), *1893*
Spinu, M. 2224(37), *2241*
Spitzner, H. 2501(188), *2563*
Splendore, M. 560(38), *590*, 1120(116–118), *1138*
Spoliti, M. 2508(223), *2564*
Sprecher, R.F. 1998(122c), *2034*
Spunta, G. 435, 456(55), 459(157), *489, 491*

Squillacote, M. 2437(133), *2460*
Squire, D.W. 2514, 2523(274), *2565*
Squires, R. 370(50), *427*, 601, 604(15), *698*
Squires, R.R. 505(38), *511*, 1109(32, 36), 1116(74–79), *1136, 1137*, 1372, 1373(65), *1439*
Sreelatha, C. 1341, 1342(20a, 20b), *1437*
Srinivas, G.N. 33(70), *97*, 551(115), *555*
Srinivas, R. 3(7c), 36(75), 66, 68(116), *95, 97, 99*, 1010(298c), *1060*, 1123(155, 157, 164), 1124(170), 1125(179, 183), 1126(193, 194), *1139, 1140*, 1163(134), *1184*, 2556(422), *2568*
Srinivasan, S. 1406(174a), *1443*
Srivastava, I.H. 82(157a), *100*
Staab, A.J. 1854, 1860(325), *1868*
Staab, G.A. 2286(120), *2315*
Stadelmann, B. 2083(145, 146), *2125*, 2193(73), 2210(73, 103, 106), 2213(106, 108), *2215, 2216*
Stader, C. 296(90), 302, 304(102), *352*
Stadnichuk, M.D. 1794, 1854(27), *1862*
Staemmler, V. 252(333), *265*
Stafkens, D.J. 2534(361c), *2567*
Stahl, A. 1851(312), *1868*
Stahl, M. 283(45), *351*, 527–529, 533, 538–540, 547(86), *554*, 568, 571, 573(112), *592*
Stährfeldt, T. 2072(95), 2073(101), 2074(103), *2124, 2125*
Staley, R.H. 2154(130), *2173*
Stalke, D. 198(81, 85), 203(108), 209(139), 211(145, 146), 213(150), 234(241, 243), 235(238, 239), 252(327), *259, 260, 263, 265*, 796(75), 799(64), 800(67a, 67b), 802, 803(75), *823, 824*, 884–886, 937, 938(103), 1011(305), 1043(358, 359), 1044, 1048(305, 359), *1056, 1060, 1062*, 1423, 1425, 1427(221), *1444*, 2145(89), *2172*
St-Amant, A. 515, 516(14b), *551*
Stamatis, N. 685(156), *701*
Stamatovic, A. 1135(264), *1142*
Stammler, H.-G. 190(49), *258*, 1079, 1088, 1089(41b), *1102*, 1434, 1435(239), *1444*, 1879, 1880(58), *1893*, 2148(101), 2155(143, 144), 2156(144, 145), 2158, 2159(143), 2166(190), 2167(190, 195), 2168(195–197), *2173–2175*
Stammler, H.G. 1434(232a), *1444*, 2150, 2151(107, 108), 2156(107), *2173*, 2534(361b), *2567*
Stanczyk, W.A. 558(3), 559(3, 17), 575(136), 577(146), *589, 590, 592*, 1074(25), *1102*
Standen, S.P. 470(193), *492*
Stang, P.J. 362, 363(27), *426*, 598(8), 611(8, 45), 612(8), 615(45), *697, 698*
Stanger, A. 362, 363(27), *426*, 482(238), *493*, 560(51), 576(142), *590, 592*, 597(6), 598(6,

8), 601(13), 603, 604(6), 611, 612(8, 39), 613(39), 615(6), 647(13), *697, 698*, 866, 884–886, 917, 924, 925, 928, 957(39), *1055*, 1130, 1131(236), *1141*, 2429(100), *2459*
Stannett, V. 580(170), *593*
Stannett, V.T. 580(171), *593*
Stansfield, R.F.D. 2027(203), *2036*
Stanton, A.C. 2493(154), 2523(308), *2562, 2566*
Stanton, J.F. 517(43b–d), *552*
Staral, J.S. 690(165), *701*
Starke, W.A. 2227(71), *2242*
Starodubtsev, E.S. 1585(36), *1595*
Statakis, M. 1858(342), *1868*
Stebbins, J.F. 325–328(143), *353*
Steck, A. 2159, 2160, 2162(151), *2174*
Steck, J.-C. 1889(103), *1894*
Steckhan, E. 1294(182), *1309*, 1851(312), *1868*
Steele, K.P. 1331(178), *1336*, 2518(264), *2565*
Steele, W.V. 159, 160(53), 163, 174(68), *178, 179*, 1469(205), *1531*
Steer, R.P. 2522(284), *2565*
Stegeman, G.I. 2351(255), *2362*
Stegmann, H.B. 2453(196), *2461*
Stegmann, R. 1008, 1009(297b), *1060*, 2556(427), *2568*
Steib, C. 245(285), *264*, 586(194), *594*
Steidel, M. 232(229), *262*
Steier, W. 2352(263, 265), *2362*
Steier, W.H. 2330, 2354(83), *2357*
Steigelmann, O. 190(56), *258*, 1434, 1436(249), *1445*
Steiling, L. 2364(11), *2398*
Stein, R.M. 2276(105), *2314*
Steinberger, H.-U. 543–545, 547(110), *555*, 570, 573(125), *592*, 696(170), *701*
Steiner, A. 209(139), *260*
Steiner, B.W. 6(22), *95*
Steiner, E. 5, 31, 32, 42(18), *95*
Steinfeld, J.I. 171, 172(126), *180*, 2504(198, 199), 2522(285), 2546(396), *2563, 2565, 2568*
Steinhauser, N. 2238(125), *2244*
Steinmetz, M. 860(13), 863(23), 872(55), 898(141, 142), 978(13), *1054, 1055, 1057*
Steinmetz, M.G. 433(10), *488*, 852(86), *855*, 1234(1), 1240(25a), 1241(27), 1242(30), 1244(35–37), 1293(180), *1305, 1306, 1309*, 1312(3), 1328(154), 1332(3), *1332, 1335*, 1725(130), *1788*, 2475(58), *2560*
Steinmeyer, R.D. 1593(111), *1597*
Steins, H. 2339(130), *2359*
Stejskal, E.O. 309(115, 117), 310(133), *353*
Stepanenko, B.N. 1500(356), *1536*

Stepanov, B.I. 1471(225), *1531*
Stern, C. 2085, 2089, 2094(157), *2126*
Stern, C.L. 198(84), *259*, 280(41), *351*, 514, 520–522, 529, 532, 533(2a), *551*, 581, 583, 584(182), *593*, 652(125), *700*
Steudel, R. 1884(69), *1893*, 1896, 1904(3), *1925*
Steudel, W. 790–792(40), *822*
Stevens, R.C. 520, 524, 540(70), *553*, 563, 583(82), *591*
Stevens, W.J. 21(54a, 54b), *97*
Stevenson, T. 476(221), *493*
Stevenson, W.H.III 284(57), *351*, 505(39), *511*, 1353, 1357, 1358(35b), *1438*
Steward, O.W. 437, 483(73), *490*, 798(58), *823*
Stewart, R.F. 1208(55), *1230*
Stewart, R.M. 2253, 2255, 2256(43), *2312*
Stezowski, J.J. 584(184), *593*
Stibbs, W.G. 1064(7b), *1101*
Stierman, T.J. 2450(185, 186, 188), *2461*
Still, W.C. 782(10a), *822*, 1622(127), *1663*
Stille, J.K. 2322(36), *2356*
Stirling, W. 1469(205), *1531*
Stixrude, L. 115(33), *118*
Stobart, S.R. 2132(19), *2171*
Stober, M.R. 417(192), *429*
Stöckigt, D. 1118(89), *1138*
Stocklin, G. 385(96), *427*
Stoenescu, I. 37, 38(80), *97*, 815(111), *825*, 2020(188), *2035*
Stoffer, J.O. 1591(100), *1597*
Stoffregen, A. 2363(7), 2377(48, 49), 2378(50), 2379, 2380(7), 2383(50), 2385(7), 2387(65), *2398–2400*
Stogner, S.M. 2557(430), *2568*
Stohrer, W.-D. 584(184), *593*, 1911(62), *1927*
Stone, F.G.A. 1903(43), *1926*
Stone, J. 605, 606, 629(21), *698*
Stone, J.A. 368(45), 369(46), *426*, 560(40), 561(58, 60, 61), 566(98), *590*, *591*, 605, 606(20), 609(32), 629(20), *698*, 1122(144), *1139*
Stone, J.M. 566(98), *591*, 1122(144), *1139*
Storch, W. 800(70a), *823*, 2330(84), 2333(84, 95), 2339(95), *2357, 2358*
Stork, G. 417(195), *429*, 1674(67a), *1683*, 1854(319), *1868*
Stout, P.D. 251(321), *265*, 296(87), *352*, 1046(374), *1062*
Stout, T. 509(59–61), *511*, 564(87), 569(121), *591, 592*, 1409(183), *1443*
Stradiņš, J. 1480, 1481(281), *1533*
Stradins, J. 1199(31), 1200(32), *1230*
Stradyn', Ya.P. 1480, 1481(280), *1533*
Strahle, J. 254(337), *265*
Strain, M.C. 515, 516(17), *551*

Strampfer, M. 2072(88, 90, 92, 94), 2090(88, 90, 94), 2091(94), *2124*
Strand, M.R. 586(191), *594*
Stranges, D. 1120, 1121(128), *1138*
Straus, D.A. 2082(142), 2086, 2090, 2093, 2094(163), *2125, 2126*
Strausberger, H. 1585(42), *1595*
Strauss, D.A. 566(100), 571, 574(128), *591, 592*
Strauss, S.H. 559(15), 578(160), 587(15), *590, 593*
Straussberger, H. 1587(57), 1594(115), *1596, 1597*
Strausz, O.P. 751(102), *777*, 1319(70), *1334*, 2514(273), 2526, 2527(327), *2565, 2566*
Streckel, W. 1587(57), 1594(115), *1596, 1597*
Strecker, M. 2364(10, 12, 13, 15, 16), *2398*
Street, S.D.A. 416(185), *429*
Streitwieser, A. 499(24), *510*, 821(120), *825*, 959(242), *1059*, 1911(72), *1927*
Streitwieser, A.Jr. 1080(44), *1102*, 2407(24), *2458*
Strekowski, L. 466(182), *492*
Strelenko, Yu.A. 1474, 1475(244), *1532*
Strelkov, S.A. 2181(21), *2214*
Streu, R.D. 1452, 1484(58a, 58b), 1485(58b), *1526*
Strickland, J.B. 222(192), *261*, 457(149), *491*, 1638(167, 170, 173), *1664*
Strickland, M.S.S. 1823(155), *1864*
Strinkovski, A. 2320, 2322(18, 19), 2327(19), 2348(18, 19), *2355*
Stroh, A. 1587(62), *1596*
Strohmann, C. 334(149), *354*, 1357(46, 48), 1359, 1360(46), 1362(46, 48), 1364(46), *1438*, 2253, 2255(42, 43), 2256(43), 2257(42), *2312*, 2363(9), 2364(21, 22, 24–26, 29), 2374(42, 43), *2398, 2399*
Strouse, C.E. 2073, 2091(100), *2124*
Struchkov, Y.T. 958(241), *1059*, 1244(34), *1306*, 2041(12), *2122*
Struchkov, Yu.T. 186(44, 46, 47), 190(50), 194(66), 196(71), 209(134), 211(142), 219(173), 224(197), 226(206, 207), 229(212), 237(244, 245, 248, 249), 243(270–274, 277), 245(286), 254(335), *258, 260–265*, 500, 501(26), *510*, 587(195), *594*, 1374, 1376(92b), 1392(143), 1393, 1394(151), 1395(154, 156, 157), 1396(151, 156–160, 162–165), 1397(156, 166, 167), 1399(156, 160, 164, 165), 1418, 1422, 1423(216), *1440–1442, 1444*, 1449(21, 22), 1450(46), 1454(103, 104), 1455(113), 1460(21, 22, 157), 1461(46, 142, 144, 145), 1462(151), 1463(21, 22), 1464(103, 104), 1467(103, 104, 113, 142, 144, 145, 151,

Struchkov, Yu.T. (*cont.*) 157), 1514, 1522(408b), *1525, 1526, 1528, 1529, 1537*, 2002(139), *2034*
Struchkov, Yu.T.Z. 1897(17), *1926*
Struehkov, Y.T. 2044, 2051(19), *2123*
Strukovich, R. 780(1e), *821*
Struth, G. 1228(124), *1232*
Strutschkow, Y.T. 1910(57), 1913, 1914(77), *1927*
Strutwolf, J. 1053(398), *1062*, 2557(428), *2568*
Strzelec, K. 1701, 1702(63, 64), *1787*
Stuart, J.O. 2235(108), 2236(114), *2243, 2244*
Stucchi, E. 1628(137, 139), 1655(137), 1656(137, 139), *1663*, 1796(38), *1862*
Stufkens, D.J. 2169(198), *2175*
Stüger, H. 2083(145), *2125*, 2181(24, 25, 31), 2193(73), 2204(87), 2210(73), 2213(108), *2214–2216*
Stuhl, O. 1897(16), 1903(41a), *1926*
Stuhlmeier, M. 1236(9), *1305*
Stults, J.S. 1091(58g), *1103*
Stumpf, B. 2376(47), 2378(50, 53), 2383(50, 60), 2387, 2388(53), *2399*
Stumpf, T. 1434, 1436(247), *1445*
Sturkovich, R. 485(252), *493*, 1449, 1460(7a, 7b), *1525*
Sturkovich, R.Y. 1727, 1730(134), *1788*
Sturla, J.-M. 1110(60), *1137*
Stutzmann, M. 2196(78a), *2215*
Su, K. 2250, 2253(22), 2265(83a–c), 2307, 2309(163), *2310, 2313, 2316*
Su, M.-D. 66, 68(120a), 99, 156(34), *178*, 1125(173), *1139*, 1163(133), *1184*, 2510(241), *2564*
Su, T.M. 635(89b), *699*
Süb, J. 1979(56c, 56d), *2031*
Subbaswamy, K.R. 71, 72(140), *100*
Subramanian, G. 82(157a, 157b, 158), 83(158), *100*
Subramanian, L.R. 517(47), *553*
Subrt, J. 1237(17), *1306*
Suda, K. 1198, 1200(30), 1201, 1203(33), 1204(30), *1230*, 1608(63), *1661*
Sue, R.E. 210(141), *260*
Suehiro, I. 438, 465(89), *490*
Suel, J.J. 2142(68), *2172*
Suga, S. 1091(58h), *1103*, 1630(146, 147), 1653(146), *1663*, 1678(99), *1684*
Sugama, T. 2261(51), *2312*
Sugawara, M. 2116(238), *2127*
Sugawara, S. 1675(73), *1683*
Sugaya, T. 891(124), *1056*, 1325(127), *1335*
Sugi, S. 1883(68), *1893*, 1896, 1906(4, 5), *1925*
Sugihara, Y. 895, 936, 970, 975(134), *1056*, 1259(69), *1307*, 1325(136), *1335*
Sugimoto, F. 1640(186), *1664*

Sugimoto, T. 1749(228), *1790*
Suginome, H. 1296(185), *1309*
Suginome, M. 434, 445(27), *489*, 2430(113–116), *2460*
Sugisawa, H. 998(287, 288), 1001(288), *1060*, 1973(48a, 50b), 1974(50b), 1990(95a, 95b), 2007–2009(95a), *2031, 2033*, 2416(43), *2458*
Sugita, N. 1959(29), *1960*
Sugiura, M. 451(126), *491*, 1840(245), *1866*
Sugiura, T. 2349(209), *2361*
Sugiyama, H. 897(138), 899(144), *1057*, 1228(122), *1232*, 1322, 1324(105, 112), 1325(105, 148), 1326, 1327(105), *1334, 1335*
Sugiyama, M. 735(69), 737, 741(69, 70), *776*, 906(168, 169), 908, 931(169), 973(168), *1057*, 1278(135, 136), *1308*, 2409, 2410(31, 32), 2411(32), *2458*
Suh, E.B. 515(20), *552*
Suh, H. 1491(325), *1535*
Suhr, Y. 1606, 1611, 1628(54), *1661*
Suk, M.-Y. 401(140), *428*, 1590(79), *1596*, 1856(328), *1868*
Sukata, K. 1678(104), *1684*
Sulkes, M. 1283(145), *1308*, 2471(36), *2559*
Sullentrup, R. 198(87), *259*
Sullivan, K.J. 1397, 1398, 1404(170), *1442*
Sullivan, M.J. 310(118), *353*
Sullivan, S.A. 1109, 1115(38), *1137*
Sulmon, P. 1651(217), *1665*
Sülzle, D. 3(7c), 36(75), 66, 68(116), *95, 97, 99*, 1010(298c), *1060*, 1118(89), 1123(155, 164), 1124(170), 1125(179), 1126(193, 194, 204), 1128(219), *1138–1141*, 1163(134), *1184*, 2556(422), *2568*
Sum, V. 209(131), *260*, 477(227), *493*
Sumitani, M. 1325(120), *1335*
Sumiya, R. 1710(86), *1787*
Sun, C.-M. 1843(262), *1867*
Sun, G.-R. 809(91), *824*, 2119(256), *2128*
Sun, H. 1998(117), *2034*
Sun, H.-n. 568(111), *592*
Sun, J. 2069, 2071(79), *2124*
Sun, M. 2327(53), *2356*
Sun, R.-C. 438(94), *490*
Sun, S. 1545(32), 1555(58a), *1578*
Sun, S.T. 1766(276, 277), *1791*
Sun, X.-Y. 1477(265), *1533*
Sun, Y.-P. 1313(45), 1316(42), 1317(42, 45), 1318(49, 50), 1320(42, 45), *1333*
Sundar, S. 2139(42), *2171*
Sundaresan, S. 2249, 2294(14), *2310*
Sunderlin, L.S. 1116(74), *1137*
Sundermeyer, W. 780(2f, 4b), 781(7), 794(4b), *821*
Sundholm, D. 517(43a), *552*

Sundius, T. 560(48), *590*
Sung, D.-D. 1758(255), *1790*
Sunginome, M. 2117(248, 249), 2120(248), *2128*
Sünkel, K. 2141(57, 58), 2147, 2149(58), *2172*
Sunko, D.E. 622(71), 642(105), *699, 700*
Surtwala, T. 2350(237), *2361*
Surya Prakash, G.K. 1602, 1604, 1655(30), *1661*, 1667(10), 1669(24), 1674(58), 1675(77, 79, 81, 82), *1682–1684*, 1818(137), *1864*
Susfalk, R. 1611(84), *1662*
Suslova, E.N. 1298(190), *1309*
Süss-Fink, G. 1771(305), *1792*
Susumu, K. 1639(177), *1664*
Sutej, K. 1571(91), *1579*
Sutowardoyo, K.I. 1672(46), *1683*
Suzuke, T. 1716(103), *1787*
Suzuki, E. 1590(83), 1591(91–98), *1596, 1597*, 2533(356–359), *2567*
Suzuki, H. 248(304), *264*, 279(32, 33, 36, 37), *351*, 487(262), *494*, 812(99b), *824*, 832(31), *854*, 1064, 1068, 1083(10), 1090(10, 56), 1091(10, 57), 1092(10, 63), *1101–1103*, 1163(122), *1184*, 1210(58, 59), *1231*, 1264(85), 1269(102, 103), 1270(104), 1288(166), 1289(167), *1307–1309*, 1332(206), *1337*, 1828(185), *1865*, 1879(52), 1882(64), 1884(70, 71), *1893*, 1897(27), 1904, 1907(48, 49), *1926*, 2080, 2092(133, 134, 136), 2093(134, 136), *2125*, 2486(124), 2492(145), 2494(166), 2499(166, 187), 2500(145), 2501(187), *2561–2563*
Suzuki, K. 143(52a, 52b, 52d), *152*, 470(196), *492*, 1959(30), *1960*
Suzuki, M. 443(110), *490*, 805(82), *824*, 2485(108), 2522(283), 2523(310), *2561, 2565, 2566*
Suzuki, N. 2043, 2051(15), *2122*, 2434(124), *2460*
Suzuki, S. 1610(81), *1662*, 1771(308), *1792*
Suzuki, T. 735, 737, 741(69), *776*, 906, 973(168), *1057*, 1278(136), 1302(203), *1308, 1310*, 1725(131, 132), *1788*, 1923(93), *1928*, 1942(18, 19), *1960*, 2225(45), *2241*, 2250(23a, 23b), 2252, 2253(23a, 23b, 32a), 2289(123), *2310, 2311, 2315*, 2329, 2331(75), *2357*, 2409–2411(32), *2458*, 2522(305), 2523(307), *2565, 2566*
Suzuki, Y. 2231(83), *2242*
Svara, J. 1998(121), *2034*
Svejda, S.A. 1743(216, 217), *1790*
Svoboda, D.L. 1966, 1967, 1997–1999(19), *2030*, 2503(196), *2563*
Svoboda, P. 1700(53), *1786*
Swain, A.C. 2481(89), *2560*

Swamy, K.C.K. 226(203), *262*
Swansiger, W.A. 435, 456(49), *489*
Swanson, D.R. 2434(121), *2460*
Sweeney, J.B. 1619(113), *1663*
Sweigardt, D.A. 1505(376), *1536*
Swenton, J.S. 1646(206), *1665*
Swern, D. 1606, 1611(47), *1661*
Sy, K.G. 190(48), *258*, 1717(104), *1787*
Sykes, D. 113(24), *118*
Syldatk, C. 1649(208, 209a), *1665*, 2363(7), 2377(48, 49), 2378(50, 52, 55), 2379(7, 58), 2380(7), 2383(50), 2385(7, 62), 2387(65, 67), 2388(62, 69), *2398–2400*
Sylvain, M. 1126(205), *1140*
Sypniewski, G. 2005(154), *2034*
Systermans, A. 1016, 1023(313), 1024, 1046(333), *1061*
Syverud, A.N. 156, 158, 159, 173, 177(35), *178*
Szabó, K.J. 547–549(113), *555*
Szabo, V. 2288(122), *2315*
Szafran, M. 8(29), *96*
Szakacs, L. 1470(208), *1531*
Szalay, E. 1450, 1453, 1471(33), *1525*
Szczepanski, S.W. 1601, 1602, 1604, 1622, 1628(20), *1660*
Sze, S.M. 1118(97), *1138*
Szeimies, G. 1910, 1911(60e), *1927*, 2437(138), *2460*
Szeverenyi, N.M. 310(131), *353*
Szieberth, D. 65(114), *99*
Szöllösy, A. 1450, 1453, 1462, 1463, 1467(149), 1471(33), *1525, 1529*

Taba, K.M. 1418(208), *1444*
Tabei, E. 561(55), *591*, 1125(184), 1133(184, 247, 249), *1140, 1141*, 2472(45), *2559*
Tabei, T. 1738(159), *1788*
Taber, D.F. 1681(121), *1685*, 1854(321), *1868*
Tabohashi, T. (110c), *825*, 1990(94, 95a), 2007–2009(95a), *2033*
Tabuchi, T. 867, 885, 974(45), *1055*
Tachibana, A. 559(20, 24), *590*, 807, 808(87a, 87c), *824*, 870(50), 888(120, 121), 889(121), 892(131), 895(132), 958(120), 986, 988, 990(121), *1055, 1056*, 1325(135, 137), *1335*, 2512(248), *2564*
Tachibana, H. 137, 142(44c), *151*, 200(100), *259*, 1318(64, 65), *1333*
Tachibana, K. 2521(281), 2549(403), *2565, 2568*
Tachibani, A. 1252(56), *1306*
Tachikawa, M. 1716(103), *1787*
Tacke, M. 2168(199), 2169(198, 199), *2175*, 2534(361c), *2567*

Tacke, R. 196(72), 219(180), 222(195), 226(204), 236(255), 238(253), *258, 261–263*, 324, 325(142), 334(148, 149), 335(148), *353, 354*, 1341(28, 29, 30a, 30b), 1349(28, 29, 30a, 30b, 31a, 31b), 1350(30a, 30b), 1351(30b), 1357(28, 29, 30a, 30b, 31a, 31b, 46–52), 1358(28), 1359(28, 29, 46, 47), 1360(28, 46, 47, 49–51), 1361(28, 30a, 30b), 1362(28, 46–50), 1363(47), 1364(46, 49, 51, 52), 1365(30a, 51), 1366(29, 49), 1368(30a, 30b, 57a, 57b), 1412, 1415(193), *1437–1439, 1443*, 1608, 1617(64), 1649(208, 209a, 209b), *1661, 1665*, 2363(2, 5–9), 2364(10–36), 2365(20), 2367(20, 37), 2368(34, 38–40), 2369(32, 34, 39), 2371(34, 38, 40), 2374(34, 36, 38–43), 2376(32, 47), 2377(48, 49), 2378(50–56), 2379(7, 57), 2380(7, 59), 2382(57), 2383(50, 59, 60), 2385(7, 62), 2387(53, 65–67), 2388(53, 62, 69), 2389(74), *2398–2400*

Tada, T. 1109(39), *1137*

Taddei, M. 402(146), 410(167), *428, 429*, 1633(156), *1664*, 1803(75), 1816, 1819(128), 1844(270), 1849(296), *1863, 1864, 1867*

Tae, J.S. 1574(97), *1579*

Taege, R. 1125(185), *1140*

Tafel, A. 1608, 1617(64), *1661*, 2363(7), 2364(30, 31), 2378(54), 2379, 2380, 2385(7), 2387(65), *2398–2400*

Taft, R.W. 360(11), 361(11, 16), *426*, 1477(267, 269), *1533*

Taggart, D.L. 1212(69), *1231*

Taguchi, K. 2333(97), *2358*

Tajima, S. 561(55, 56), *591*, 707(17), 775, 1125(184), 1133(184, 247–249), *1140, 1141*

Tajima, Y. 1323(118), *1335*

Takabe, K. 437(83), 438(83, 90), *490*

Takacs, J.M. 1816(127), *1864*

Takahara, Y. 1887(93), *1894*, 1897, 1911, 1914, 1915(15), *1925*

Takahashi, H. 1325(137), *1335*, 1771(297), *1791*

Takahashi, K. 1252(56), *1306*, 1325(137), *1335*, 1731(138, 139), *1788*

Takahashi, M. 911(177), *1057*, 1412(190), 1434(235), *1443, 1444*, 1884(70), *1893*, 1904, 1907(49), *1926*

Takahashi, N. 1293(179), *1309*

Takahashi, S. 2236, 2237(116), *2244*

Takahashi, T. 1302(203, 204), *1310*, 1850, 1851(304), *1867*, 1940(17), 1942(18, 19), *1960*, 2043, 2051(15), *2122*, 2434(121, 124), *2460*, 2498(183), 2513, 2515, 2516, 2520(259), *2563, 2564*

Takahashi, Y. 895(132), *1056*

Takahiro, G. 2333(97), *2358*

Takaki, K. 895, 936, 970, 975(134), *1056*, 1259(69), *1307*, 1325(136), *1335*, 1641(189), 1651(215), *1664, 1665*

Takamizawa, M. 2265(81a, 81b), *2313*

Takamura, S. 2437(132), *2460*

Takanami, T. 1198, 1200(30), 1201, 1203(33), 1204(30), *1230*, 1608(63), *1661*

Takano, S. 2346, 2347(183a), *2360*

Takao, T. 2080, 2092(133, 134, 136), 2093(134, 136), *2125*

Takaoka, T. 1319(71), *1334*

Takaya, H. 1674(61), *1683*, 1751(237), *1790*, 2376(46), *2399*, 2424(71–74), *2459*

Takayama, T. 294, 295, 302(78), 329, 330(144), *352, 353*

Takeda, K. 1318(52–54, 57, 58), *1333*, 1640, 1655(187), 1656(223), *1664, 1665*

Takeda, M. 1215(81, 82), *1231*

Takeda, N. 1091, 1092(59a, 59b), 1093(59a, 59b, 64), *1103*, 2492, 2500(145), *2562*

Takeda, T. 1296(185), *1309*, 1801(64), *1863*

Takeda, Y. 2265(81a, 81b), *2313*

Takei, H. 459(155, 156), 462(169), *491, 492*

Takeishi, Y. 1885(78), *1893*

Takenaka, K. 457(147), *491*

Takenoshita, H. 1871(13), *1891*

Takenouchi, K. 462(169), *492*

Takeo, H. 2523(315), *2566*

Takeshita, K. 1714(96–99), *1787*

Takesue, N. 436(69), 437(70), 483(69, 70), *490*

Takeuchi, C.S. 1782(341), *1792*

Takeuchi, K. 248(299), *264*, 348(159), *354*

Takeuchi, M. 2289(129), *2315*

Takeuchi, R. 451(126), *491*, 1689(7), 1718(109, 110), 1728, 1729(136, 137), *1785, 1788*, 1840(244–246), 1860(349), *1866, 1868*

Takeuchi, T. 559(25), *590*, 884, 885, 888(113), 928(194), 937, 938(113), *1056, 1058*, 1325(140), *1335*, 2069, 2071(75), *2124*

Takeuchi, Y. 417(194), *429*, 1861(357), *1868*

Taki, T. 1236(11), *1305*, 1332(200), *1336*, 2253(34), *2311*

Takiguchi, T. 1198, 1200(29), *1230*, 1325(148), *1335*

Takikawa, Y. 1885(78), *1893*

Takimoto, N. 1959(29), *1960*

Takizawa, N. 1611, 1617(82), *1662*

Takuwa, T. 1120, 1121(132), *1139*

Tal, Y. 1911(67), *1927*

Talbi, M. 1769(288), *1791*

Tamai, T. 581(176), *593*, 1292(177), *1309*

Tamaka, N. 2011(166), *2035*

Tamao, K. 37(78), 46(90b), *97, 98*, 190(51), 198(88), 207(124), *258–260*, 340–342(151), *354*, 780(1b), 790, 791(45), 794(52), 797(45), 809(88–91), 810(92, 94, 95),

811(45, 95, 96), 812(45), *821, 823, 824,*
827(4), *854,* 1014, 1033, 1034(311), *1061,*
1304(214–216), *1310,* 1312, 1313(12),
1332, 1340, 1341(17a–c, 18), 1342,
1343(18), 1344, 1345(17a–c), 1346(17a),
1347(17c, 18), 1348(17c), 1372, 1373(73),
1390(123a), 1412(190), 1418(218a), 1434,
1435(243), *1437, 1439, 1441, 1443, 1444,*
1699(49), 1710(86–89), 1711(89, 90),
1725(87), 1756(249), 1783(345), *1786,
1787, 1790, 1792,* 1797(46), 1824(159),
1828(159, 183), 1843(264), 1854(323),
1862, 1865, 1867, 1868, 1978(54a, 54b),
1979(59a, 59b), 2013(59a, 59b, 170),
2014(59a), 2015(54b, 173), 2016(173),
2028(206–209), *2031, 2032, 2035,
2036,* 2119(256), *2128,* 2177(3c), *2213,*
2429(105), 2430(111), 2436(105), *2459,
2460,* 2481(91–94), 2554(416), 2555(417),
2561, 2568
Tamas, J. 1968, 1986, 1996–1998(20c),
2030
Tamatsu, T. 1986(80c), *2032*
Tamazaki, Y. 2383(61), *2400*
Tamelen, E.E.van 464(172, 173), *492*
Tamura, K. 2250, 2252, 2253(23a), *2310*
Tamura, M. 1671(40), *1683*
Tan, C. 1075(35c), *1102,* 1390(123b), *1441,*
2481, 2483, 2553(97), *2561*
Tan, R. 1031, 1043, 1045(347), *1061*
Tan, R.P.-K. 1887(94), *1894*
Tanabe, H. 1133(247), *1141*
Tanaka, A. 2389(75), 2390, 2391(76, 77),
2392(76–78), 2394(75, 78, 79), 2396(79,
81), *2400*
Tanaka, H. 1157(75), *1182,* 2472(46), 2513,
2516, 2526(258), *2559, 2564*
Tanaka, K. 279(33), *351,* 1270(104),
1308, 1372(68), *1439,* 2289(129), *2315,*
2546(391), *2568*
Tanaka, M. 559(25), *590,* 1151(47), *1182,*
1246(38), *1306,* 1325(120), *1335,* 1691(14),
1692(22), 1697(44), 1698(44, 46), 1779(46),
1786, 1838(230), *1866,* 2080, 2092(133,
134, 136), 2093(134, 136), 2107, 2109(211),
2114(230, 231), 2115(232, 235), 2117(231),
2120(231, 232), 2121(231, 232, 235), *2125,
2127,* 2278(107), *2314,* 2430(117), *2460*
Tanaka, S. 2550(407), *2568*
Tanaka, Y. 810(92), *824,* 1714(100), *1787,*
1796(36), 1831(201), 1838(232, 233, 235,
239), 1842(36), *1862, 1865, 1866,* 2115,
2121(235), *2127,* 2430(117), *2460*
Tanake, H. 1268(95), *1307*
Tandhasetti, T.M. 2349(206), *2361*
Tandura, S.N. 183(14), *257,* 1340, 1351(1),
1381(1, 107, 108), *1437, 1440,* 1449(20),

1450(38), 1452(57, 59), 1454(80, 90),
1455(59), 1460, 1463, 1466(20), 1467(80,
90), 1473(20, 235, 238), 1474(20, 235,
242–244, 260), 1475(20, 59, 242, 244,
253, 255), 1476(57, 238, 243, 259–261),
1477(243, 259, 260, 264), 1479(57, 242,
274), 1484(59), 1487(308), 1500(358),
1506(261), 1508, 1509(391), 1510(260,
261), 1521(417), 1523(20), *1525–1527,
1532–1534, 1536, 1537*
Tandura, St.N. 284(51), *351*
Tandura, T.A. 1458, 1467(127), *1529*
Tang, B.Z. 2154(137), 2160(166), *2174*
Tang, K. 1873(23), *1892*
Tang, K.-H. 1644(198), *1664*
Tang, L. 2345(175), *2360*
Tang, Y.-N. 1965(17), *2030,* 2464, 2468, 2488,
2493(4), *2559*
Tani, T. 2321, 2347(20), *2355*
Taniguchi, E. 1849(291), *1867*
Taniguchi, M. 1620(121), *1663,* 1680(117),
1684
Taniguchi, N. 1561(71a, 71b), *1579*
Taniguchi, Y. 1641(189), 1651(215), *1664,
1665*
Tanikawa, H. 722(48, 49), 724, 726(49),
743, 744(48, 49), *776,* 1282(141), *1308,*
2012(168a, 168b), *2035,* 2412(37, 38),
2413(37), *2458*
Taniki, Y. 1189(8), *1229*
Tanimori, S. 1852(313), *1868*
Tanimoto, M. 2523(315), *2566*
Tanino, K. 449(120), 469(191, 192), *491, 492*
Tanizawa, T. 2075, 2091(115), *2125*
Tanke, R.S. 1720, 1723, 1732(115), *1788*
Tannenbaum, S. 154(8), *177*
Tanouchi, N. 1718(109, 110), *1788*
Tao, F. 2486(123), *2561*
Tappert, W. 2332(91), *2357*
Tapsak, M.A. 1839(242, 243), *1866*
Taraban, M.B. 2473(51), 2488(135), *2560,
2562*
Tarada, M. 1857(335), *1868*
Tardella, P.A. 460(159), *491,* 1843(266, 267),
1867
Targos, T.S. 2261(68), *2313*
Tarhay, L. 2295(144), *2315*
Tartakovskii, B.A. 1377(98), *1440*
Tarui, Y. 2546(388), *2568*
Tarunin, B.I. 435, 456(50), *489*
Tasaka, M. 1452, 1453, 1473(61), *1526*
Tasaka, S. 1322(104), *1334*
Tasch, S. 2197(80), *2215*
Tashiro, Y. 2250(23a, 23b), 2252, 2253(23a,
23b, 32a), 2265(82b), *2310, 2311, 2313*
Tastenoy, M. 2364(21, 24, 25, 27, 29),
2374(41–43), *2398, 2399*

Tatebe, K. 1692(15), 1733(15, 145, 146), 1734, 1737(145), *1786, 1788*
Tatevskii, V.M. 154, 156(17b), *178*
Tatsu, Y. 2327(60), *2357*
Tatsu, Y.P. 2326(45c), *2356*
Taulelle, F. 2264(79), *2313*
Taya, K. 1692, 1738(16), *1786*
Tayaniphan, S. 2301, 2302(159), *2316*
Taylor, A.L. 2070, 2071(80), *2124*
Taylor, B.F. 1707(83), *1787*
Taylor, K. 2295(145), *2315*
Taylor, N.J. 386(103), *428*, 484(245), *493*
Taylor, P.G. 284, 288(52), *351*, 356(1), 416(189), 420(200), 421(202), 424(212), *426, 429, 430*, 435(51), 456(51, 144), *489, 491*, 495(1), 509(61, 63–65), *510, 511*, 577(144), 592, 828(13), *854*, 1374, 1376(93), 1378, 1382(103), *1440*, 1680(114), *1684*, 1795, 1797(33), *1862*
Taylor, R. 388(105, 106), *428*, 434, 448, 449(36), *489*, 2223(32a, 32b), *2241*
Taylor, R.B. 2232(96), *2243*
Tazi Hemida, A. 2281(113), *2314*
Tebbe, K.-F. 213(149, 153), 215(157), *260*
Tebbe, M. 760(123), *777*
Teclé, B. 785, 786(27), *822*
Teghil, R. 1120, 1121(128), *1138*
Teichner, J. 2322(34), *2356*
Tejeda, J. 762(132), *777*
Tel'noi, V.I. 154, 165, 166(10), *177*
Tempkin, O. 829(21), *854*
Temsamani, D.R. 159, 161, 162(51), *178*
Tench, R.J. 1120, 1121(129), *1138*
Ten Wolde, P.M.C. 2335(108), *2358*
Teowee, G. 2345(171), *2360*
Teppen, B.J. 113(25), *118*
Teramae, H. 22, 23, 72(59), *97*, 123(21d), 143(51), *151, 152*, 835(38), *854*, 1314(34), 1318(54, 55, 58), *1333*
Teramura, D.H. 366(41), *426*, 620–622, 637(66), *699*
Terao, T. 310(130), *353*
Terasaki, H. 1957(26), 1958(27), *1960*
Terent'ev, A.B. 1494(337), *1535*
Terepka, F.M. 2289(132), *2315*
Terets, M.I. 2228, 2229(76), *2242*
Terlouw, J.K. 363, 368(29), *426*, 604–606(19), 698, 1122(146), *1139*
Terrett, N.K. 407(161), *429*
Terry, K.W. 2300(154), 2307, 2309(162), *2316*
Terry, L.W. 1236(13), *1305*
Terui, Y. 288(66), *352*, 484(244), *493*
Terunuma, D. 1983(73), 1999(127), 2001(73), *2032, 2034*, 2364(31), *2399*
Tessier, C.A. 186(32), 200(103), *257, 259*, 2105, 2109(207), *2127*
Tewinkel, S. 2340(136), *2359*

Teyssié, D. 1769(290), *1791*, 2218(11c, 11d), 2226(50), *2240, 2242*
Teyssié, Ph. 2235(107), *2243*
Thaetner, R. 1585(35), 1587(58), *1595, 1596*
Thanedar, S. 1986(80b), *2032*
The, U. 2145, 2146(94), *2172*
Theopold, K.H. 1998(120), *2034*
Thepot, P. 2335(112), *2358*
Therasse, J. 2327(55), *2356*
Thewalt, M. 2194(77), *2215*
Thewalt, U. 245(280, 282), *264*, 1429(224), *1444*
Thiel, F.A. 69(130), *100*
Thiel, W. 7(27), 13(46b), 52, 54(95g), 55, 62, 65(104), 69(129, 132, 133), 72(129, 144), 74(132, 133), 77(132), *96, 98, 100*, 864, 981, 982(31, 33), *1054*, 2539(372b), *2567*
Thiery, D. 219(179), *261*
Thobie-Gautier, C. 1189, 1190(9), *1230*
Thölmann, D. 1108(22), *1136*
Thom, K.-L. 200(94), *259*, 2179(9), *2214*
Thoma, A. 1156(74), *1182*
Thoman, J.W. 171, 172(126), *180*
Thoman, J.W.Jr. 2504(198), 2522(285), *2563, 2565*
Thomas, A.B. 563(71), *591*
Thomas, A.P. 455(138), *491*
Thomas, D.R. 2227(59), *2242*
Thomas, K.M. 29(66b), *97*
Thomas, R.C. 416(184), *429*
Thomas, S.E. 1659(226), *1665*, 1667(2), *1682*
Thompson, A.S. 1832(205), *1865*
Thompson, D.F. 1845(275), *1867*
Thompson, D.L. 2485(112, 114), 2510(233–235), *2561, 2564*
Thompson, D.P. 1097(73a, 73b), 1098(73b, 74), *1103*, 1266(87), 1289(168), *1307, 1309*, 1321(86), 1334, 1887(91), *1894*, 1904(45), *1926*, 2223(30b), *2241*
Thompson, H.R. 2151, 2152(114), *2173*
Thompson, J. 365(38), *426*, 617(57), 618(57, 59), 620(59), *699*, 2345(171), *2360*
Thompson, M.L. 219(177), *261*
Thompson, M.R. 2445, 2446(166), *2461*
Thomson, C. 381, 382(79), *427*
Thomson, W.J. 2249, 2294(15), *2310*
Thorn, D.L. 1986(80a), *2032*, 2434(120), *2460*
Thorne, A.J. 29(66b), *97*
Thorne, J.R.G. 1316(44), 1318(63, 66), 1325(122), *1333–1335*
Thorne, K. 2289(128), *2315*
Thornton, E.R. 441(103, 104), 442(106), *490*
Thornton-Pett, M. 209(131), *260*
Thum, G. 2142(63), 2145, 2146(93), *2172*
Thuring, J.-W.J. 1610, 1630(79), *1662*
Tibbals, F.A. 222(187), *261*

Tidwell, T.T. 782, 791, 795(11c), *822*
Tietze, L.F. 1803(76, 77), 1813(114, 115), 1827(180, 181), *1863–1865*
Tikku, V.K. 2240(131), *2244*
Tikurs, S. 1256(64), *1307*
Tiley, T.D. 789(28c), *822*
Tilgner, A. 1318(60–62), *1333*
Tillack, A. 1743(214), *1790*, 2051, 2052(34), *2123*
Tilley, D. 4(10a, 10d), 37(10a, 10d, 81, 82, 86), 38(10a, 10d), 39(82), 40(10a, 10d, 82), 42, 44(86), 46(10a), 47(10d, 86), *95, 97, 98*
Tilley, T.D. 250(315), *265*, 566(100), 571, 574(128–130), *591, 592*, 791(44b), 800(70b), 816(112a), 817(112b), 819(112a, 112b), 820(112a), *823, 825*, 981, 984(261, 262), *1059*, 1606(49, 51, 55), 1610(55), *1661*, 1879(53), *1893*, 2016(175c), 2019(187), 2021(192), 2024(187), 2026(187, 200), 2027(202), 2028(187, 192), 2029(210), *2035, 2036*, 2038(1a, 1b), 2039(2, 6), 2040(8, 9), 2041(11), 2042(1a, 1b), 2045(20), 2046(21–23), 2047(22, 23), 2049(28), 2051(20, 21, 28), 2052(21, 28), 2053(39), 2079(125), 2080(135), 2082(125, 135, 142, 143), 2086(161, 163), 2087(164), 2090(161, 163, 164, 183), 2092(125, 135, 143), 2093(125, 135, 143, 163), 2094(163, 164, 183), 2105, 2109(206), 2113, 2114(229), 2118(254), 2120(229, 254), 2121(254), *2122, 2123, 2125–2128*, 2210(105), *2216*, 2275(102a, 102b), 2300(154), 2307, 2309(162, 163), *2314, 2316*, 2425(85, 86), 2434(122, 123), *2459, 2460*, 2527(333, 334), 2529(337), *2566*
Tillman, N. 2477(76), 2484(100), *2560, 2561*
Timm, M.J. 437, 484(75), *490*
Timms, P.L. 578(163), *593*, 1162(117, 119), 1163(125), *1183, 1184*, 2487(125, 126), *2561*
Timms, R.E. 1513(406), *1537*
Timofeeva, T.V. 254(335), *265*, 2044, 2051(19), *2123*
Timokhin, V. 1543(18), *1577*
Timonen, R.S. 167(86), *179*
Timosheva, N.V. 1406(174b), *1443*
Tinao-Wooldridge, L.V. 1196(19), *1230*
Tipker, J. 466(178), *492*
Tipping, A.E. 741, 742(78), *776*
Tiripicchio, A. 2072(84, 87, 90, 94), 2073(96), 2074(87, 105), 2090(84, 87, 90, 94, 105), 2091(94, 96), *2124, 2125*, 2340(137), *2359*
Tiripicchio-Camellini, M. 2072(87), 2073(96), 2074, 2090(87), 2091(96), *2124*
Tischler, M. 2194(77), *2215*
Tittel, F.K. 69(124b), *99*

Tittle, F.K. 70(134a), *100*
Tkach, V.S. 1692(19), *1786*
Tláskal, J. 1006, 1008, 1009(296a), *1060*, 1237(16), *1305*
Tobe, Y. 2428(89), *2459*
Tobin, J.B. 369(47, 48), 397(47), *427*, 611, 614(43), 626, 627(80), 628–630(83), *698, 699*
Tobita, H. 248(298), *264*, 488(273, 274), *494*, 1260(72, 73), *1307*, 1332(204), *1337*, 1999(129), *2034*, 2069, 2071(75), 2088(168–170, 173–177), 2090, 2094(168, 169), 2105(173), *2124, 2126*, 2182(36), *2214*, 2498(185b), 2529(338), *2563, 2566*
Tobita, S. 561(55, 56), *591*, 707(17), *775*, 1125(184), 1133(184, 247–249), *1140, 1141*
Tobukuro, K. 2227(70), *2242*
Tochibana, A. 1253(57), *1307*
Toda, A. 1210, 1211(61), *1231*
Todani, T. 1291(175), *1309*
Todd, W.P. 437, 486(80), *490*
Tofani, D. 1843(266), *1867*
Togni, A. 2159(162), *2174*
Togo, H. 1558(66a, 66b), *1579*
Tohma, T. 1756(249), *1790*
Tohriiwa, N. 2330, 2354(80), *2357*
Tokach, S.K. 1319(68), *1334*
Tokareva, L.I. 1454, 1467(81), *1527*
Tokitoh, D.M. 1904, 1907(48–50), *1926*
Tokitoh, N. 3(2c, 8), 10, 11, 13, 16, 17, 32(8), *95*, 248(304), *264*, 279(36, 37), *351*, 487(262), *494*, 578(164), *593*, 812(99b), *824*, 859(6), 990(6, 272), 994, 995(6), *1054, 1060*, 1064, 1068, 1083(10), 1090(10, 56), 1091(10, 57, 59a, 59b, 60, 61), 1092(10, 59a, 59b, 60, 61, 63), 1093(59a, 59b, 60, 61, 64, 65, 66a, 66b, 67, 68a, 68b), 1100(76), *1101–1103*, 1163(122), *1184*, 1288(166), 1289(167), *1309*, 1332(206), *1337*, 1677(93), *1684*, 1873(25), 1879(52), 1882(64), 1884(70, 71), *1892, 1893*, 1897(27), *1926*, 2437(139), *2460*, 2486(124), 2492(145), 2494(166), 2499(166, 187), 2500(145), 2501(187), *2561–2563*
Tokoroyama, T. 1812(111, 112), *1864*
Tokuda, Y. 137, 142(44c), *151*
Tokunaga, Y. 1561(71b), *1579*
Tokura, Y. 200(100), *259*, 1318(64, 65), *1333*
Tokuyama, H. 746(90a), *776*
Tokuyama, T. 559(30), *590*
Tolan, J.W. 1148, 1149(30), *1182*
Tolman, C.A. 441(100), *490*
Tolosa, J.I. 1720(118), 1721(122), 1724(118, 122), 1732(118), *1788*
Toltl, N. 969(249), *1059*
Tomao, K. 417(193), *429*

Tomàs, F. 72(141, 142), 73(141), 75–77(151), *100*
Tomasi, J. 528, 535(87a–e), *554*
Tomic, M. 622(71), *699*
Tomietto, M. 1163(125), *1184*
Tominaga, K. 1322(91), *1334*
Tominaga, M. 2347(186b), *2360*
Tominaga, T. 1808(94), *1863*
Tominaga, Y. 1800(60), 1826(173), *1863, 1865*
Tomioka, H. 704(3), 761(124, 125), *774, 777*
Tomita, K. 1848(286), 1856(327), *1867, 1868*
Tomizawa, T. 2252, 2253(32c), *2311*
Tomoda, S. 279(36), *351*, 417(194), *429*, 1861(357), *1868*
Tomsons, P. 301(96), *352*, 1474(241), *1532*
Tonachini, G. 1602(29), *1661*
Tonner, D.S. 1108(22), *1136*
Tonokura, K. 1157(75), *1182*, 1268(95), *1307*, 2472(46), 2513, 2516, 2526(258), *2559, 2564*
Top, S. 580(168), *593*
Toporowski, P.M. 1759(260), *1791*
Topsom, R.D. 357(5), 361(18), *426*
Toreki, W. 2246, 2250, 2253, 2254, 2273, 2285(6f), 2286(119), *2310, 2315*
Toren, E.C.Jr. 1211(65), *1231*, 1322(100), *1334*
Torii, S. 1205(43), 1206(45), *1230*
Toritani, K. 1838(231), *1866*
Toriumi, K. 488(273), *494*, 2347(191), *2360*
Törnroos, K.W. 110(14), *118*, 219(182), *261*, 2107, 2109(213), *2127*
Töröek, B. 1738(160), *1789*
Török, B. 435, 479(60), *489*
Torres, M. 751(102), *777*
Torto, I. 184, 192(17), *257*
Tortorelli, V.J. 2416(48), *2458*
Toru, T. 476(222), *493*, 1885(74), *1893*
Toryashinova, D.D. 1470–1472(216, 217), *1531*
Toryashinova, D.-S.D. 1484(294), *1533*
Toryashinova, T.D. 1507(383), 1508(383, 384), 1509(383), *1536*
Toscano, J.P. 288, 289(68), *352*
Toshida, H. 194(63), *258*
Toshida, J. 1412(190), *1443*
Toskas, G. 2220(19a), *2240*
Tossell, J.A. 111(18), *118*, 295–297(79), *352*
Totyl, N.P. 1303(207), *1310*
Toupet, L. 1801(68), *1863*
Tour, J.M. 2434(121), *2460*
Toussaere, E. 2351(250, 257), 2354(250, 257, 285), *2362*
Touster, J. 1224(107), *1232*
Townes, C.H. 588(197), *594*
Towns, E. 1705(75), *1787*

Toyoda, E. 1208–1210(57), 1216, 1217(85, 86), 1218(86, 89, 90), 1219(89, 90), 1227(120), *1230–1232*, 1769(294), 1770(294, 295), *1791*, 1973(43), 2016(175d), 2017(43), *2031, 2035*
Toyoshima, Y. 2546(391, 392), *2568*
Toyota, M. 1548(38), *1578*
Trabi, M. 806, 807(86), *824*, 2192(68), *2215*
Trachtman, M. 6(21b, 21c), 11(43b), 12, 32(43b, 45a, 45b), *95, 96*
Tracy, H.J. 2273, 2275(99a, 99b), *2314*
Traedwell, D.R. 226(205), *262*
Traetteberg, M. 116(46), *118*
Trandell, R.F. 2468(22), *2559*
Tran Qui, D. 245(279), *264*, 1418, 1420, 1422, 1423(209), *1444*
Trautwein, A.V. 2159(150), *2174*
Traven', V.F. 1314(24), *1333*, 1471(225), *1531*
Traylor, T.G. 358, 359(9), *426*, 616(49–52), *698*, 838(45), *854*, 1322(93), *1334*
Traynor, S.G. 1212(69), *1231*
Treadwell, D. 2290, 2294(139a, 139b), *2315*
Treadwell, D.R. 2298(150b), 2301, 2305(157), *2316*
Trefonas, P. 2469(29), *2559*
Trefonas, P.III 1320(83), *1334*
Trehan, S. 1801(66), *1863*
Treofonas, P.T.III 1318(47), *1333*
Tretner, C. 2189(59), *2215*
Treutler, O. 762(131), *777*
Treverton, J.A. 164, 166(77), *179*
Triki, S. 2075, 2091(111), *2125*
Trinquier, G. 29(66c–e), 63(66d, 112a–c), *97, 99*, 763(135), *777*, 835(39, 40), *854*, 984(264), *1059*, 1126(192), *1140*, 2518(263), *2565*
Tripathy, S.K. 2339(125), 2351(260), 2354(260, 280), *2358, 2362*
Triplett, K. 2529(339), *2566*
Trofimenko, S. 245(283), *264*
Trofimov, B.A. 436, 479(61), *489*, 1501(368), *1536*
Trofimova, I.V. 1583(17), *1595*
Trofimova, O.M. 237(248), *263*, 1455(112, 113), 1460(157), 1467(112, 113, 157), *1528, 1529*
Trogler, W.C. 1693(26, 27), *1786*
Troisi, L. 481(234), *493*
Trojanov, S.I. 245(278), *264*
Troll, C. 2158(148), *2174*
Troll, T. 1228(125), *1232*
Trommer, M. 709, 713(22, 23), 714(22), 727(23), 734(65), 763, 764(22), *775, 776*, 978(252, 253), 981, 984, 987, 989, 1000, 1002–1004(259), *1059*, 1146(23), 1147(23,

28), 1148(28), 1165(146), *1181, 1182, 1184*, 1272(107), 1277(132, 133), 1278(134), 1286(160), *1308, 1309*, 1606(50), *1661*
Trommsdorf, H.P. 1318(60–62), *1333*
Trost, B.M. 434, 443(21), 466(180, 181), *489, 492*, 1826(175), 1838(238), *1865, 1866*
Trotter, J. 196(69, 70), *258*, 1605(43, 44), 1606(43), *1661*
Troullier, N. 115, 116(37), *118*
Trout, N.A. 380(75), *427*, 642, 643(106), *700*
Trucks, G.W. 34(74a, 74b), *97*, 167(98), *179*, 517(44a), *553*, 2509(228), *2564*
Truong, N.T. 1159(98), *1183*
Truong, P.N. 829(23), *854*, 1067(15), *1101*
Truong, T.N. 251(320), 252(330), *265*, 296(86), *352*, 1046(373), *1062*, 2477(66), *2560*
Trus, B.L. 515(20), *552*
Tsai, C.-Y. 1776, 1777, 1781(328), *1792*, 1843(262), *1867*
Tsai, W.M. 2142(66), *2172*
Tsai, Y.-M. 1601, 1602, 1604, 1622, 1628(20), 1644(195, 198–200), *1660, 1664*
Tsang, W. 164, 170, 175(82), *179*, 1126(189), *1140*
Tsay, S.-C. 433, 480(3), *488*, 1678(101), *1684*, 1875(37), 1877(43), *1892*
Tse, D. 2251–2253(24), *2311*
Tse, J.S. 1163(125), *1184*, 2510(238), *2564*
Tsetlina, E.O. 436, 479(61), *489*, 1450(38), *1526*
Tseung, A.C.C. 2322(28), *2356*
Tsionski, M. 2321(24b), *2356*
Tsionsky, M. 2320(16), 2328(16, 61–63, 65–68), *2355, 2357*
Tsomaya, N.I. 1585(36), *1595*
Tsotinis, A. 741(77), *776*
Tsuchida, K. 2334(105), *2358*
Tsuchihashi, G.-I. 470(196), *492*
Tsuchiya, T. 1203(40), 1207(48), 1212, 1213(40), *1230*, 1991(97), *2033*
Tsuda, M. 2546(390), *2568*
Tsudo, S. 1560(69), *1579*
Tsuji, J. 1610(81), *1662*, 1824(156), 1834(212), *1864, 1866*
Tsuji, M. 2523(316), *2566*
Tsuji, T. 2428(90), *2459*
Tsuji, Y. 482(237), *493*, 675(149), *701*
Tsujishima, H. 1201, 1204(37), *1230*
Tsukada, T. 115(34), *118*
Tsukamoto, M. 551(116), *555*, 589(198), *594*
Tsukazaki, M. 1803(73), *1863*
Tsukihara, T. 186(40), *258*, 895(132), 928(194), 998, 1005(290), *1056, 1058, 1060*, 1252(56), *1306*, 1325(137, 140), *1335*, 2017(178), *2035*, 2439(148), *2460*

Tsumaki, H. 1030(344), *1061*
Tsumura, M. 1767(282), *1791*
Tsumuraya, T. 123, 125, 130(23), *151*, 812(99a), *824*, 827, 828(11), 829(21), *854*, 1064(4g), *1101*, 1968, 1969, 1976(24a, 24b), 1977(24a), 1996(24a, 24b), *2030*, 2183, 2205(43), *2214*, 2416(45), 2428, 2429(91), *2458, 2459*
Tsuneto, A. 1751(237), *1790*, 2376(46), *2399*
Tsuneyuki, S. 115(34), *118*
Tsuno, Y. 363(31), 364(33), 370(52), 371(53, 54, 56), *426, 427*, 436(66, 69), 482(66), 483(69), *490*, 603(18), 610(33), 611, 613, 614(41), 621, 623(72, 73), 624(33, 72, 73), 632, 633(86a, 86b), 634(86a, 86b, 88a, 88b), 635(88a, 88b), *698, 699*, 1846(279), 1856(333), *1867, 1868*
Tsunoi, S. 1559(68), *1579*
Tsutsui, S. 834(36), 842(51), *854, 855*, 1330(174), *1336*
Tsutsui, T. 2448(181), *2461*
Tsutsumi, T. 436, 483(69), *490*
Tsutsumi, Y. 2469(32), *2559*
Tsvetnitskaya, S.I. 154, 156, 166(16), *177*
Tsybulya, G.F. 1486(303), 1487(307), *1534*
Tsyonsky, M. 2322(32), *2356*
Tubbesing, U. 1053(398), *1062*
Tuladhar, S.M. 645(110, 111), *700*
Tumey, M.L. 709, 714(20), *775*
Tumura, M. 2439(147), *2460*
Tunker, G. 2346(179), *2360*
Tunney, S.E. 2322(36), *2356*
Turchaninov, V.K. 1483(293), *1533*
Turecek, F. 561(54), *591*, 1122(150), *1139*
Turetskaya, R.A. 1583(17), *1595*
Turiansky, A. 2320, 2322, 2327, 2348(19), *2355*
Turkenburg, L.A.M. 2428(89), *2459*
Turkina, G. 2439(145), *2460*
Turkina, G.Yu. 186(44, 46, 47), 190(50), *258*
Turley, J.W. 1357(42), *1438*, 1448(3), 1456(115), 1462(3, 115), 1465(3), 1467(115), 1510(397), *1524, 1528, 1537*
Turner, B.E. 2522(302), *2565*
Turner, J.J. 1260(74), *1307*
Turniansky, A. 2327(52), 2340(132), *2356, 2359*
Turtle, R. 509(65), *511*, 1374, 1376(93), *1440*
Tustin, G.J. 1659(226), *1665*
Tycko, R. 69(130), *100*
Tyutyulkov, N. 90(170d), *101*
Tzachach, A. 1510, 1511(398–400), *1537*
Tzamarioudaki, M. 1776, 1777(328), 1778, 1779(334), 1781(328, 337), 1782(337), *1792*
Tzeng, D. 1075(32b), *1102*, 1331(180), *1336*, 2416(47), *2458*

U, J.S. 1819(138), *1864*
Uang, S. 1638(173), *1664*
Uchida, M. 2028(208), *2036*
Uchida, S. 1264(85), *1307*
Uchida, T. 1314(25), *1333*, 1824, 1828(159), *1865*
Uchimaru, Y. 1691(14), *1786*, 2430(117), *2460*
Udayakumar, B. 872(55), *1055*
Udayakumar, B.S. 1241(27), 1244(35), *1306*, 1725(130), *1788*
Udodov, Yu.N. 1472(233b), *1531*
Udre, V.E. 1501(365), *1536*
Ueda, M. 2330(83), 2349(210a, 210b), 2354(83, 287), *2357, 2361, 2362*
Ueda, N. 2353(278), *2362*
Ueda, T. 1227(120), *1232*
Uejima, A. 2392, 2394(78), *2400*
Ueki, Y. 998(287), *1060*, 1322(108), *1334*
Uemoto, N. 1226(117), *1232*
Uemura, S. 1743(209–211), 1745(209), 1747(209–211), 1748(209), 1749(211), *1790*, 2351(246), *2362*
Ueno, K. 1238(23), 1260(72, 73), *1306, 1307*, 1332(204), *1337*, 1411(189), *1443*, 2087(165–167), 2088(171–173, 175, 177), 2090(165, 166), 2092, 2093(165), 2094(166), 2105(171–173), *2126*, 2416, 2418(50, 51), 2420, 2421(64), *2458, 2459*
Ueno, M. 1849(294), *1867*
Ueno, N. 1206(45), *1230*
Ueno, T. 2546(388), *2568*
Ueno, Y. 476(222), *493*, 1885(74), *1893*
Uesaka, T. 2104, 2108, 2109(204), *2127*
Ugliengo, P. 184, 221(20), *257*
Ugozzoli, F. 2072(94), 2074(105), 2090(94, 105), 2091(94), *2124, 2125*
Uguen, D. 1671(33), *1683*
Uh, D.S. 1462, 1467(153), 1491(325), *1529, 1535*
Uhl, G. 215(155), *260*
Uhl, W. 1890(108), *1894*
Uhlig, F. 806, 807(86), *824*, 2192(68), 2193(73), 2201(85), 2203(86), 2209(102), 2210(73), *2215, 2216*
Uhlig, W. 542(109), *555*, 1879, 1883(54b), *1893*, 2179(16), 2189(57–59), *2214, 2215*
Uhlmann, D.R. 2345(171), 2350(237), *2360, 2361*
Uhlmann, J.G. 2429(101), *2459*
Uhm, J. 1353, 1368, 1369(36), *1438*, 2298(150a), *2315*
Uichanco, I.S. 2504(199), *2563*
Ukaji, Y. 458(150), *491*, 1843(263), *1867*
Ukita, T. 1860(356), *1868*
Ukuda, Y. 1269(102), *1308*
Uliva, P. 1626, 1658(132), *1663*
Ulivi, P. 1871(14), *1892*

Ulrich, D.R. 2350(218), *2361*
Umeda, I. 2424(73), *2459*
Umemura, M. 561(55), *591*, 1133(249), *1141*
Umeno, M. 1699(49), *1786*
Umezawa, M. 1215(81, 82), 1217(83), 1218(91, 93), 1219, 1220(91), *1231*
Umstead-Daggett, J. 402(145), *428*, 1795(34), *1862*
Un, D.S. 1490(315), *1534*
Underhill, A.E. 2352(269), *2362*
Underiner, G.E. 1030(346), 1031(347), 1043(346, 347), 1044(346, 347), 1045(346, 347), *1061*
Unkefer, C.J. 2065, 2066(73, 74), *2124*
Unno, M. 145(53–55), 146(55), *152*, 486(261), 487(270), *494*, 1304(211, 212), *1310*, 1918(87), 1925(110), *1927, 1928*, 2206(93), 2207(93, 96), *2215*
Uno, Y. 1676(88), *1684*
Untiedt, S. 1236(9), *1305*
Uozumi, Y. 1743, 1746(208), 1752(238–241), 1753(241, 242), 1754(242–244), *1789, 1790*
Upadhyaya, H.P. 863, 864, 932, 934, 982(25), *1054*
Urabe, H. 475(211), *492*, 1557(63), *1578*
Urano, Y. 459(152), *491*, 1639(177, 178), *1664*
Uratani, Y. 2011(166), *2035*
Urazovskii, J.F. 2152(116), *2173*
Urbaniak, W. 2340(136), *2359*
Urbanova, M. 1006, 1008(295), *1060*, 1237(17), *1306*
Urchegui, R. 446(113), 476(225), *490, 493*
Urdaneta-Pérez, M.J. 790, 794(37), *822*
Uriarte, R.J. 1693(29), *1786*
Uribe, G. 1512(403), *1537*
Urisaka, R. 791(42), *822*, 2453, 2455(200), *2461*
Urry, G. 589(199), *594*, 727, 739, 740(55), *776*, 2468(22), *2559*
Urtane, I. 1458, 1467, 1475, 1479, 1509(124), 1510(396), *1528, 1537*
Urtane, I.P. 1459, 1467(133), 1476, 1506, 1510(261), *1529, 1532*
Urtane, L.P. 1487(307), *1534*
Ushakov, S.N. 364(34), *426*, 616(47), *698*
Ushio, J. 515, 516(14b), *551*
Uski, V.A. 2152, 2153(112), *2173*
Uson, I. 2301(155), *2316*
Uson, R. 1986, 2012(78a–c), *2032*
Ustinov, M.V. 2044, 2051(19), *2123*
Ustynyuk, Yu.A. 2131(8), 2132(18), 2133(20), *2171*
Utimoto, K. 462(166), *492*, 1555, 1556(59), 1572(93), *1578, 1579*, 1620(121), *1663*, 1680(117), *1684*, 1807(82), 1810, 1830(105, 106), *1863, 1864*
Uto, H. 143(52a, 52b), *152*

Utsuno, S. 1452, 1453, 1473(61), *1526*
Uyehara, T. 1849(294), *1867*

Vaahs, T. 2253(36, 48), 2261(61), 2269, 2270(88f), *2311–2313*
Vacek, G. 18, 21, 22, 32(49), 66, 67(118b), *96, 99*, 1163(131), *1184*
Vaden, A.K. 611, 615(45), *698*
Vaglio, G.-A. 559(21, 22), 560(38, 39), *590*, 1120(114–118), *1138*
Vakul'skaya, T.I. 1454, 1467(93), 1481(288), 1483(288, 293), 1487(308), *1527, 1533, 1534*
Valentine, J.S. 2326(45b, 45d), *2356*
Valero, C. 1721, 1724(121), *1788*
Valle, L.D. 435, 451(39), *489*
Valpey, R.S. 1842(258), *1867*
Van Alesony, C. 72–74(148), *100*
Van Beek, D.A.Jr. 1213(72), *1231*, 2193(69), *2215*
Vancik, A. 1284(158), *1309*
Vancik, H. 901(153, 156), 992, 995(156), *1057*, 1167(163), 1168(163, 164), 1169(163), 1170(164), *1185*, 2471(42), 2513–2515(254), 2523(318), *2559, 2564, 2566*
Van Den Berghe, E.V. 569(116), *592*
Van den Kerk, S.M. 5, 44, 68(17b), *95*
Van den Winkel, Y. 1266(91), *1307*
Van Der Kelen, G.P. 569(116), *592*
Van der Maelen, J.F. 2075(113, 114), *2125*
Vandersar, T.S.D. 1633, 1642(158), *1664*
Vanderwielen, A.J. 1156, 1167(68), *1182*
Van de Weghe, P. 1671(41), *1683*
Van Dine, G.W. 635(89a), *699*
Van Dyke, M.E. 2219(14), *2240*
Vankar, Y.D. 1674(59), *1683*
Vankeer, A. 1053(395), *1062*
Vannoorenberghe, Y. 1743, 1746(223), *1790*
Van Ooij, W.J. 2235(108), *2243*
Vanquickborne, L.G. 951, 952(238a), *1059*
Vanquickenborne, L. 1053(395), *1062*
Vanquickenborne, L.G. 161, 163(62), *178*, 1008, 1009(297a), 1050(371), *1060, 1062*, 2487, 2511(128), 2556(426), *2561, 2568*
Van Seggan, D.M. 578(160), *593*
Vansweevelt, H. 951, 952(238a), *1059*
Van Tiel, M. 2107(216), *2127*
VanVliet, D.S. 401(139), *428*, 1818(137), *1864*
VanZee, R.J. 1126(201), 1127(213), *1140*
Van Zoeren, C.M. 171, 172(126), *180*, 2504(198), *2563*
Varaprath, s. 2230(81a), 2231(81b, 81c), *2242*
Varezhkin, Y.M. 302–305(104), *353*
Varezhkin, Yu.M. 240(257), *263*
Varie, D.L. 1091(58g), *1103*

Varma, R.S. 385(95), *427*
Vasilets, V.N. 2231(85), *2243*
Vasisht, S. 909(171), *1057*
Vatsa, R.K. 863(24, 25), 864, 932, 934, 982(25), *1054*, 1237(18a), *1306*
Vázquez de Migel, A. 2151, 2152(117), *2173*
Vdovin, V.M. 844(55), *855*, 1144(6), *1181*, 2406(16), *2458*
Vdovin, V.N. 163(69), *179*
Vecchi, D. 1553(52), 1572(94), *1578, 1579*
Vedejs, E. 455(137), *491*, 1091(58g), *1103*, 1677(92), *1684*
Vega, A.J. 315(138), *353*
Veith, M. 53(101), *98*, 205(116, 118), *259*, 1178(192–194), *1185*, 1320(80), *1334*, 1434(238), *1444*, 2474, 2534(55), *2560*
Ven, L.J.M.von de 343–346(152), *354*
Venanzi, C.A. 1467, 1471(189), *1530*
Vencel, J. 1459, 1467, 1488(131), *1529*
Vencl, J. 617(55), *698*
Venezky, D. 2322(31), *2356*
Venkatachalam, T.K. 790(36), 792(46), 793(36, 46, 47b, 51a, 51b), 794(47b, 51b), *822, 823*
Venkatashalam, T.K. 789(13d), *822*
Vennall, G.P. 401(142), *428*
Venneri, P. 1248(46), *1306*
Venturini, A. 6(20), *95*, 1320(81), *1334*
Venturino, A. 919, 921(187, 188), *1058*
Veprek, S. 2546(386, 394), *2568*
Veprek-Heijman, M.G.J. 2546(386, 394), *2568*
Verbeek, W. 2253(39a, 39b), *2311*
Verbist, J. 2235(107), *2243*
Verdagnuer, X. 1741(184), *1789*
Verdaguer, X. 1749(229), *1790*
Verkade, J.G. 198(86), 207(123), 226(200), *259–261*, 481(235), 485(255, 257), *493*, 1373(75a), *1439*, 1449(26, 27), 1460(27, 150), 1462, 1465(150), 1466(27), 1467, 1472(150), 1489(26, 27, 312), 1513(27), 1514(26, 27, 312, 409), 1515(312, 409), 1516(312), 1517(409–411), 1518(410, 412, 413), 1519(312, 413, 414), 1520(414–416), 1521(312, 409, 411, 414), 1522(312, 409–413, 416), 1523(312), 1524(27), *1525, 1529, 1534, 1537*
Verlhac, J.B. 1616(101), *1662*
Verloop, A. 466(178), *492*
Verne, H.P. 52–55, 57–59(98a), *98*, 279, 280(38), *351*, 1144(11), *1181*, 2535–2539(367), *2567*
Verne, H.R. 2023(195), *2036*
Vernet, J.L. 457(148), *491*
Veszprémi, T. 30–32(68, 69), 52(99c), 53(68), 55, 56, 58, 61, 62(99c), 64(68, 69), 65(69, 114), 66(69), *97–99*, 842(51), *855*, 911(177), *1057*
Vetter, A.H. 1742(185), *1789*

Vhandra, R.K. 1453, 1462, 1467, 1483, 1491(68), *1526*
Vicari, R. 437, 485(77), *490*, 1602, 1604, 1605(32), *1661*
Vicente, J. 1986, 2012(78a–c), *2032*
Vick, S.C. 1973, 1974(45a–d), 1976(45d), 1995(45c), *2031*, 2421(67), 2424(77), 2429(103), *2459*
Vidal, E. 1778, 1779(334), *1792*
Vidal, J.P. 196(73), *258*
Vidal, L. 1818(136), *1864*
Viktorov, N.A. 1450(36), *1525*
Viljoen, H.J. 2545(385), *2568*
Vilkov, L.V. 2181(21), *2214*
Villa, M. 2322(35), *2356*
Villarroya, E. 2072, 2091(89), *2124*
Vincent, A.M. 2227(66), *2242*
Vincent, C.A. 1771(298), *1791*
Vincent, G.A. 1453, 1454, 1456, 1458, 1459, 1461, 1465, 1472, 1483, 1486, 1491, 1492, 1506, 1509(69), 1511(69, 402), 1512(402), 1513(69, 402), *1526, 1537*, 2296, 2298(149), *2315*
Vincent, J.M. 2227(60, 61), *2242*
Vincenti, S.P. 437, 485(77), *490*, 1602, 1604, 1605(32), *1661*
Vincenti, S.R. 2143(71), *2172*
Vioux, A. 2231(90a, 90b, 91), *2243*, 2253(47a, 47b), 2294(141), *2312, 2315*, 2466(18), *2559*
Vis, M.G.M.van der 164, 165(83, 84), *179*
Visger, D.C. 463(170), *492*
Visser, R.G. 1601, 1624(22b), *1660*
Vitins, P. 872(58), *1055*
Vitkovskii, V. 2439(145), *2460*
Vitkovskii, V.Y. 888, 990(122), *1056*
Vitkovskii, V.Yu. 186(44), *258*, 1452(57), 1459, 1467(134), 1476, 1479(57), 1481(285), 1508(384, 391), 1509(391), *1526, 1529, 1533, 1536, 1537*
Vittal, J.J. 2110, 2120(221), *2127*
Vitulli, G. 1696, 1717(35), *1786*
Vjater, A. 1451(48b), *1526*
Vladimirov, V.V. 501(28), *510*
Vlasova, N.N. 1390, 1392(131), *1441*, 1454, 1467(86, 90, 102, 107), 1501(369), *1527, 1528, 1536*, 1875(36), *1892*
Voaden, M. 416(185), *429*
Vogel, G.E. 1448, 1450, 1453, 1470(1), *1524*
Vogel, M. 2141(55), *2172*
Vogel, P. 517(46), *553*
Vogel, S.N. 2523(319), *2566*
Vogelaar, G.C. 1586(47), *1595*
Vogelbacher, U.-J. 1153(58), *1182*
Vogeleisen, F. 1671(33), *1683*
Vogt, L.H. 1923(96), 1925(108), *1928*, 2334(106), *2358*

Vogt, T. 2070, 2071(80), *2124*
Vogt, V.-D. 310(122), *353*
Voight, A. 2218(6), *2240*
Voiux, A. 2251, 2253, 2255(26), *2311*
Volbach, W. 1675(78), *1684*
Volfson, P.G. 1486(303), *1534*
Volkov, I.O. 2235(109), *2243*
Volkova, L.I. 1390, 1392(136), *1441*
Volkova, V. (36), *1054*
Volkova, V.V. 154, 156, 162, 163, 174(14), *177*, 981, 985, 992(254), *1059*
Vollano, J.F. 1352(33), *1438*
Vollbrecht, S. 234(243), 252(327), *263, 265*, 1037, 1040(355), 1043(358), *1061, 1062*
Vollhardt, K.P.C. 2445, 2446(166), *2461*
Volnina, E. (36), *1054*
Volnina, E.A. 163(69), *179*, 861, 862(19), *1054*, 1237(15), *1305*
Voloboev, A.A. 1743(222), *1790*
Volpe, P. 559(21, 22), 560(38, 39), *590*, 1120(114–118), *1138*
Vol'pin, M.E. 1825(162), 1838(226), *1865, 1866*
Volvina, E.A. 1006, 1008(295), *1060*
Volz, D. 648(122), *700*, 757(119), *777*
Von-Lampe, F. 334(147), *353*
Von Schnering, H.G. 1268(99), *1307*
Voorhoeve, R.J.H. 1582(9), *1594*
Voort, P.van der 2339(125), *2358*
Vorob'era, N.S. 1996(108a), *2033*
Voronkov, G.M. 2334(107), *2358*
Voronkov, M. 2439(145), *2460*
Voronkov, M.G. 154, 156(11–16, 17a), 159(11), 160, 161(11, 56), 162, 163(14), 164(11, 15, 56), 165(11, 15), 166(12, 13, 16, 56), 174(14), *177, 178*, 183(14), 186(44, 46, 47), 190(50), 213(172), 237(244–249), 243(274), *257, 258, 261, 263, 264*, 284(51), *351*, 436, 479(61), 489, 501(28), *510*, 888, 990(122), *1056*, 1298(190), *1309*, 1340(1, 9a), 1341, 1343(21b), 1351(1), 1373(75b, 75c), 1374(92a, 94), 1376(92a), 1377(94), 1378(101, 102), 1381(1, 107, 108), 1390(128–131, 135–137, 139), 1391(140), 1392(129, 131, 135–137, 139, 146, 147), 1393(147, 150–153), 1394(151), 1395(139, 152–154), 1396(151, 164, 165), 1397(139), 1398(137, 139, 147, 152, 153), 1399(164, 165), 1400, 1401(137, 139), 1403(139), 1404(147, 150, 153), 1412(9a), 1418(210, 212, 216), 1420(210, 212), 1422(216), 1423(212, 216), *1437, 1439–1442, 1444*, 1448(2), 1449(2, 4a–c, 5, 8–13, 16, 20–22, 24, 28), 1450(38, 40, 41, 46), 1452(56, 57, 59), 1453(62, 64–66), 1454(56, 70–82, 84, 86–93, 96, 100, 102–105, 107–109), 1455(59, 76, 108, 109, 112–

Author index

114), 1457(13, 62, 120, 121), 1458(121–123, 125), 1459(128, 129, 134), 1460(5, 13, 20–22, 120, 125, 155, 157), 1461(46, 140, 142–145), 1462(151, 155), 1463(20–22, 28, 163, 165), 1464(24, 103, 104, 165), 1465(2, 12, 13, 174), 1466(12, 13, 20, 24, 28, 176–179), 1467(12, 80–82, 84, 86–93, 96, 100, 102–105, 107–109, 112–114, 120–123, 125, 128, 129, 134, 140, 142–145, 151, 155, 157, 163, 165, 174, 176–179, 181, 184, 188a), 1468(188a), 1469(12, 184, 199–204), 1470(5, 12, 13, 28, 213, 216, 217), 1471(12, 13, 216, 217, 220–226), 1472(2, 13, 216, 217, 221, 233a, 233b, 234), 1473(2, 12, 13, 20, 235, 238, 239), 1474(20, 235, 242–244, 260, 263, 266), 1475(20, 59, 62, 242, 244, 249, 252, 253, 255, 258), 1476(28, 57, 165, 188a, 238, 243, 259–261, 263), 1477(174, 188a, 243, 259, 260, 263, 264, 266), 1478(28, 174, 188a, 263, 266), 1479(57, 242, 249, 271, 274, 276–278), 1480(249, 276, 277), 1481(13, 285, 288), 1483(13, 288, 293), 1484(13, 24, 59, 294), 1485(62, 299, 301, 302), 1486(129, 303, 304), 1487(13, 307–309), 1488(16, 129, 310), 1489(16), 1490(309, 316, 317, 319, 320), 1491(16, 321–323), 1492(327), 1493(329, 331), 1494(319, 335, 336), 1495(109, 339, 340), 1496(344), 1497(346, 347, 349, 350), 1499(354, 355), 1500(71, 72, 74, 88, 92, 109, 327, 358–362), 1501(96, 327, 368, 369), 1502(92, 361, 370), 1503(155, 372), 1506(13, 174, 239, 261), 1507(383), 1508(174, 239, 383, 384, 391), 1509(383, 391, 393, 394), 1510(12, 13, 174, 239, 260, 261), 1511(401), 1513(2, 121), 1521(417), 1523(12, 13, 20, 28, 174, 249, 419), *1524–1537*, 1691(13), 1692(19), *1786*, 1875(36), *1892*, 1923(97), *1928*, 2363(3), *2398*
Voropaeva, T.I. 1794, 1854(27), *1862*
Vorspohl, K. 950(235), 953–955(240), 959–961(235), *1059*, 1080(45), *1102*, 1274(111, 113), *1308*, 2407(23), *2458*
Voss, J. 1228(124), *1232*
Voulgarakis, N. 1524(420), *1537*
Vrancken, K.C. 2339(125), *2358*
Vreven, T. 1320(81), *1334*
Vries, M.de 149(61), *152*
Vugmeister, E.K. 1500(359), *1536*
Vulpetti, A. 441(102), *490*
Vyazankin, N.S. 752, 760(110), 777, 783(17, 18), *822*, 888, 990(122), *1056*, 1390(132–134, 136), 1392(132, 136), 1398, 1405(133), *1441*, 1824(158), *1865*, 2143(74), *2172*
Vyazankina, O.A. 1390(132–134, 136), 1392(132, 136), 1398, 1405(133), *1441*, 1824(158), *1865*, 2143(74), *2172*

Wachholz, S. 2232(93), *2243*
Wachter, U. 2145, 2146(93), *2172*
Wachtler, U. 2062, 2078, 2091(62), *2123*
Wada, F. 1825(163, 165), 1838(231), *1865, 1866*
Wada, H. 2088(177), *2126*
Wada, T. 417(190, 191), *429*, 434, 447(30), *489*
Waddell, S.T. 1910, 1911(60b), *1927*
Waddington, T.C. 563(73), *591*
Wadepohl, H. 2107, 2109(212), *2127*
Wadsworth, C.L. 2182(26, 27), *2214*
Waelbroeck, M. 219(180), 222(195), *261*, 2364(21, 22, 24, 25, 27–31, 33–36), 2368(34, 38–40), 2369(34, 39), 2371(34, 38, 40), 2374(34, 36, 38–43), *2398, 2399*
Wagenseller, P.E. 1169, 1177(172), *1185*, 1331(191), *1336*, 2421(55), *2458*, 2488(132), 2514, 2515(275), *2561, 2565*
Wagensteller, P.E. 2471(38), *2559*
Wagman, D.D. 159, 160(54), *178*
Wagner, C. 945, 946(210), 948(234), *1058, 1059*, 2085, 2090, 2094(158), *2126*
Wagner, F. 2363(7), 2377(49), 2378(50), 2379, 2380(7), 2383(50), 2385(7, 62), 2387(65, 67), 2388(62, 69), *2398–2400*
Wagner, F.E. 2136, 2137, 2147(31), *2171*
Wagner, G. 250(308–311), *264*, 844(57–60), 845, 847(59), *855*, 859(4), 873(61, 62, 65, 66), 876(4, 65, 66), 879(83), 910(62), 917(66), 932(62, 83, 197), 937(62), 940(83), 941, 942(62), 944(83, 197), 969(61, 62), 981(4, 83, 197), 985, 986(268), 988(83), 991(4), *1054, 1055, 1058, 1059*, 1079(42), *1102*, 1584(23, 34), 1586(45, 46), *1595*
Wagner, H. 789(28d), *822*
Wagner, H.Gg. 174, 175(138), *180*
Wagner, I. 1885(73), 1887(89), *1893, 1894*
Wagner, M. 52(96, 97, 98a, 98b), 53(96, 98a, 98b), 54(96, 98a), 55(98a, 98b), 56(98b), 57(98a, 98b), 58, 59(98a), 61(96, 98b), 63(98b), *98*, 279, 280(38), *351*, 1144(11), *1181*, 2023(195), *2036*, 2238(124), *2244*, 2508(224), 2509(225), 2535, 2536(367), 2537(224, 367), 2538(367), 2539(224, 225, 367), *2564, 2567*
Wagner, O. 2269, 2270(88b, 88d, 88e), *2313*
Wagner, S. 876(78), 909(172), 910(176), 917, 931(78), 941(176, 203, 204), 942, 953(176), *1055, 1057, 1058*
Wagner, S.A. 1649(208, 209b), *1665*, 2378(51, 52), 2379, 2382(57), 2389(74), *2399, 2400*
Wagner-Röder, M. 2364(22, 24), *2398*
Wahl, F. 1543(23), *1577*
Waizumi, N. 438(93), *490*
Wakabayashi, H. 394(124), *428*

Wakahara, T. 804(77), *824*, 2452(191, 192), *2461*
Wakai, F. 2250, 2261(20), *2310*
Wakasa, M. 1257(66), *1307*
Wakatsuki, Y. 1829(194), *1865*
Wakita, K. 3, 10, 11, 13, 16, 17, 32(8), *95*, 859(6), 990(6, 272), 994, 995(6), *1054, 1060*
Walawalker, M.G. 2218(6), *2240*
Walborsky, H.M. 1618(109), *1662*
Waldner, K. 2301, 2302(159), *2316*
Waldner, K.F. 2250, 2253, 2255–2257(21), 2265(80), 2301, 2305(157), *2310, 2313, 2316*
Wales, D.J. 90(170g), *101*
Walker, F.H. 1910, 1911(60a, 60b), *1927*
Walker, K.L. 2487(129), *2561*
Walkinshaw, M.D. 2131(10), *2171*
Walkow, W. 1585(35), 1587(58–60), 1589(74), *1595, 1596*
Wallace, E.M. 435, 456(52), *489*
Wallace, E.N.K. 2220(20), *2241*
Wallis, J.M. 2140(47), *2171*
Wallraff, G.M. 1266(86, 87), *1307*, 1312, 1317(5), 1318(5, 48, 50, 51), 1321(85, 86), *1332–1334*
Walsh, J. 1108(30), *1136*
Walsh, K.M. 1283(147), *1308*
Walsh, R. 65(113a), *99*, 154(1–6, 1–17), 155(1, 30, 1, 18–31), 156(1, 3, 4, 1, 3, 4, 11–17, 20, 29, 31–35), 157(46, 36–46), 158(46, 35, 39, 42–44, 46–49), 159(1, 4, 52, 1, 4, 9, 11, 19, 22, 26, 35, 44, 50–54), 160(1, 52, 1, 11, 22, 26, 41, 44, 52–56), 161(1, 52, 59, 60, 63–65, 1, 11, 22, 44, 51, 52, 56–65), 162(1, 64, 65, 1, 14, 51, 55, 64–66), 163(52, 59, 60, 70, 75, 9, 14, 52, 59–62, 67–75), 164(1, 64, 78, 79, 1, 9, 11, 15, 23–25, 56, 64, 76–84), 165(1, 1, 7, 9–11, 15, 24, 83, 84), 166(1, 2, 5, 1, 2, 5, 10, 12, 13, 16, 21, 25, 56, 77), 167(1, 5, 6, 92, 1, 5, 6, 20, 22, 27, 28, 32, 49, 50, 85–99), 168(2, 5, 110, 2, 5, 20, 22, 26, 28, 50, 89, 90, 93, 100–110), 169(1, 2, 92, 1, 2, 92, 103, 105, 111, 112), 170(1, 92, 1, 21, 24, 25, 81, 82, 92), 171(1, 5, 6, 64, 1, 5, 6, 18–20, 22, 32, 49, 50, 55, 64, 95–97, 117, 118), 172(64, 65, 22, 49, 55, 64, 65, 95, 117, 130, 133), 173(3, 64, 65, 3, 20, 22, 26, 32, 35, 50, 55, 64, 65, 96, 100–102, 130, 134, 135), 174(1, 3, 5, 78, 110, 1, 3, 5, 14, 22, 68, 71, 78, 110, 134), 175(1, 1, 20, 22, 25, 26, 58, 71, 81, 82, 94, 96, 102, 135, 140), 176(30, 29–31), 177(1, 3, 1, 3, 27, 35), (0), *177–180*, 566(99), *591*, 829(18), *854*, 861–863(20), 901(149), 926(20), *1054, 1057*, 1123(163), *1139*, 1236(9), *1305*, 2466(13),

2476(64), 2479(80), 2484(101), 2511(244), 2524(322, 323), 2525(325, 326), 2526(323), 2527(322, 323), *2559–2561, 2564, 2566*
Walsingham, R.W. 1601, 1604, 1618(16), *1660*
Walter, M.W. 1675(80), *1684*
Walter, O. 1387, 1389(121a), *1441*
Walter, S. 203(105), 235(239, 240), 252(325), *259, 263, 265*, 1011, 1012, 1034(306), 1036, 1038(354), 1043, 1044(306, 354, 359), 1048(359), *1060–1062*
Walter, T.J. 2429(94, 95), *2459*
Waltermire, R.E. 434(34), 448(34, 117), *489, 491*
Walters, A. 3(7a, 7b), *95*, 1024, 1046(333), *1061*, 1128(222), *1141*
Walton, D.R.M. 361(19), 362(23), 397(133), *426, 428*, 498(15), *510*, 612(36), 618(60), 698, 699, 1452(54), 1459, 1467(131), 1471, 1479(54), 1484(298), 1488(131), *1526, 1529, 1534*, 1611(86), 1642(192), *1662, 1664*
Walton, J.C. 1644(197), *1664*
Waltz, L. 232(228), *262*
Waltz, M. 2145, 2146(96), *2172*
Walz, L. 2494(163), *2562*
Walzer, J.F. 2016(175c), *2035*, 2046, 2047(22), *2123*
Wambolt, C.L. 2327(59), *2357*
Wan, J. 198(86), *259*
Wan, T.S.M. 1959(29), *1960*
Wan, Y. 207(123), *260*, 1489, 1514–1516, 1519(312), 1520(416), 1521(312), 1522(312, 416), 1523(312), *1534, 1537*
Wang, B. 2345(168), *2360*
Wang, C.-J. 1843(262), *1867*
Wang, C.R.C. 560(31), *590*, 1119(110), *1138*
Wang, D. 433(4, 5), *488*, 1470(210, 211, 218), 1471(218), 1483(210), *1531*, 1794(24, 29), 1801(24), 1808(95), 1860(24, 29), *1862, 1863*
Wang, D.-K. 1668(14), *1682*
Wang, G. 366(41), *426*, 1481, 1483, 1506(286), *1533*, 1671(38), *1683*, 1769(291), *1791*, 1839(243), *1866*
Wang, G.-T. 620(66, 68), 621(66), 622(66, 68, 69), 637(66), *699*
Wang, J. 1454(83), 1463, 1464(167), 1467(83, 167), 1489(313, 314), 1492(83), *1527, 1530, 1534*, 2328, 2329(74), 2352(271), *2357, 2362*
Wang, J.T. 1314(28), *1333*
Wang, K.K. 450(125), *491*
Wang, L.J. 2213(109), *2216*
Wang, N. 433, 438(2), 479(233), *488, 493*
Wang, O. 2224(34), *2241*
Wang, P. 436, 479(63), *489*, 821(120), *825*
Wang, P.-J. 2549(406), *2568*

Wang, Q. 401(139), *428*, 509(66), *511*, 569, 571, 572(119), 578(161), *592, 593*, 1370(59–62), 1371(59–61), *1439*, 1674(58), *1683*, 1818(137), *1864*, 2021(190), *2036*
Wang, R. 2327(56), 2350(226), *2357, 2361*
Wang, S. 1461, 1467(146), *1529*
Wang, S.H. 2345(172), *2360*
Wang, T. 1454, 1467(101), *1528*
Wang, W.-D. 2100(194, 195), *2127*
Wang, X. 1756(251, 252), *1790*, 1875(38), *1892*, 2352(270), *2362*
Wang, Y. 515, 516(14d), *551*, 1169, 1177(172), *1185*, 1329, 1330(170), *1336*, 1930(8), *1960*, 2471(38), 2514, 2515(275), *2559, 2565*
Wang, Y.C. 133(39, 40), *151*
Wang, Y.F. 2549(403), *2568*
Wannagat, U. 565(97), *591*, 1635(163), *1664*, 2363(2), 2364(11), *2398*
Wannamaker, M.W. 466(179), *492*, 744(89), 776, 1236(13), *1305*
Wanzl, G. 1850(309), *1868*
Ward, A.H. 2227(67), *2242*
Ward, B.J. 2225(47), *2242*
Ward, H.R. 2448(177), *2461*
Ward, R.A. 1845(274, 275, 277), *1867*
Ward, S.C. 1298(193), *1310*
Ward, W.J. 1582(11, 12), 1583(11–14), *1594*
Warhurst, N.J.W. 1694, 1695(32), *1786*
Warren, E.S. 1563(74a), *1579*
Warrick, E.L. 1582(6), *1594*, 2228(74), *2242*
Warrick, G.L. 1593(109), *1597*
Washida, N. 2523(310, 312), *2566*
Waszczak, J.V. 69(130), *100*
Watanabe, A. 1254(63), *1307*, 1897, 1923(6), *1925*, 1930(9, 10), *1960*, 2469(31, 32), *2559*
Watanabe, D. 1728, 1729(136, 137), *1788*, 1860(349), *1868*
Watanabe, H. 200(93), 248(299), *259, 264*, 279(29, 31–33, 35), 348(159), *351, 354*, 810(93a, 93b), 812(98), *824*, 891(126), 928(193), *1056, 1058*, 1157(79, 80), *1183*, 1211(66), *1231*, 1264(85), 1269(102, 103), 1270(104), *1307, 1308*, 1796, 1842(36), *1862*, 1885(78), *1893*, 2182(27), 2183(38, 39, 42), *2214*, 2472(45), *2559*
Watanabe, J. 1198, 1200, 1204(30), *1230*, 1608(63), *1661*
Watanabe, K. 891(127), 892(130), *1056*, 1252(55), *1306*
Watanabe, M. 2137, 2138(36), *2171*
Watanabe, N. 1120, 1121(133), *1139*, 1591(96, 97), *1597*, 1796(39), *1862*, 2533(356, 358, 359), *2567*
Watanabe, S. 370(52), 371(54, 56), *427*, 621, 623, 624(72, 73), 699, 1750(230), *1790*,

1911, 1915(75), 1916(75, 84), 1923(91), *1927*
Watanabe, Y. 1885(74), *1893*
Waters, A. 2130, 2139, 2140(1), *2170*
Waterson, D. 476(218), *493*
Watkinson, M.D. 1771(298), *1791*
Watson, B.T. 1992(99), *2033*
Watson, J. 2350(237), *2361*
Watson, M. 1557(62), *1578*
Watson, P.L. 2039(3), *2122*
Watt, I. 1469(205), *1531*
Watts, I.M. 171(118), *180*
Watts, J.D. 2509(228, 230), *2564*
Watts, R.J. 2104, 2108, 2109(205), *2127*
Watts, W.E. 661(136), 667(143), 669(146), *700, 701*, 2162(174), *2174*
Waugaman, M. 1767(281), *1791*
Waugh, F. 397(133), *428*
Waugh, J.S. 309(114), 310(120, 132), *353*
Waugh, M.A. 782, 791, 795(11a, 11b), *822*
Wawsonek, S. 1392(142), *1441*
Waymouth, R. 2153(128), *2173*
Waymouth, R.M. 1697(36, 37), 1708, 1709(37), 1769(292), *1786, 1791*, 2276(105), *2314*
Wayner, D.D.M. 1545(29, 31), 1564(76), *1578, 1579*
Weavers, R.T. 462(168), *492*
Webb, G.A. 296(81), *352*
Webb, G.G. 690(164b), *701*
Webb, H.M. 1470(219), *1531*
Webb, S.J. 2481(89), *2560*
Webb, S.W. 1594(116, 117), *1597*
Webb, W.A. 293(76, 77), *352*
Weber, B. 2159(154), *2174*
Weber, D. 229(213), *262*, 1910(58), 1913, 1914(76), *1927*
Weber, E. 222(186, 188, 190), *261*
Weber, G. 1587(63), *1596*
Weber, J. 2196(78a), *2215*
Weber, L. 215(158), *260*
Weber, M.E. 1108(23, 24), 1118(24), *1136*
Weber, V. 1576(102), *1579*
Weber, W.P. 307, 308(113), *353*, 392(119), *428*, 1069, 1070(20), 1075(31, 32b), *1102*, 1320, 1321(78), 1325(152, 153), 1331(178–181), *1334–1336*, 1740(173), 1767(283, 284), 1769(291, 293), *1789, 1791*, 1794(2), 1824, 1828(160), 1839(242, 243), *1862, 1865, 1866*, 1981, 1983, 1998(68), *2032*, 2416(47), *2458*, 2518(264), 2526(329), *2565, 2566*
Weber-Roth, S. 1553(55), *1578*
Webster, D.E. 384(89, 90), *427*
Webster, J.A. 1699(47), *1786*
Webster, O.W. 2346(180), *2360*
Wecker, U. 763(136), *777*

Wedemann, P. 2424(76), *2459*
Wedepohl, K.H. 103(1), *117*
Weeren, R.van 2261, 2263(66), *2313*
Weerts, W.L.M. 2548(399), *2568*
Wefing, S. 310(125, 126, 128), *353*
Wegner, C. 1803(77), *1863*
Wegner, G. 1318(59), *1333*, 1768(286), *1791*
Wegner, P.A. 2152, 2153(112), *2173*
Wehrle, J. 2368, 2371, 2374(38), *2399*
Wei, Y. 2345(175, 176), *2360*
Wei, Z.Y. 1808(95), *1863*
Weidenbruch, M. 190(53), 194(62), 198(91, 92), 200(94, 99, 101), 205(113), 245(287), *258, 259, 264*, 578(165), 584(183), *593*, 791(44a), 804(78), 812(100b), *823, 824*, 1025(334), *1061*, 1077(38), 1084(52), *1102*, 1268(97–99), 1282(97, 142), 1283(97, 142, 152, 153), 1304(213), *1307–1310*, 1881(61), *1893*, 1976(51), *2031*, 2179(9), 2182, 2183(35, 37), 2184(35, 44, 45), 2185(46, 47), *2214*, 2465(8), 2472(8, 47–49), 2473(50), 2486(110, 116, 117), 2492(146, 147), 2493(159–161), 2494(165), 2495(170–174), 2514(172), 2551(8), *2559–2562*
Weidlein, J. 197(80), *259*
Weidmann, H. 784, 789, 790(20), *822*, 1635(164), *1664*
Weidner, J.J. 1816(127), *1864*
Weil, D.A. 1113(69), *1137*
Weiler, L. 1565(79), *1579*
Weinberger, M. 2047(27), *2123*, 2279(108, 110), *2314*
Weiner, B. 90(170f), *101*
Weingarten, M.D. 732(64), *776*
Weingartner, A. 1016(312), *1061*
Weingartner, A.W. 946(217), 948(229), 1016(217), *1058, 1059*
Weinhold, F. 61(108), *99*, 526(85), *554*
Weinkauf, A. 1850(301), *1867*
Weinmann, M. 1374, 1375(84), 1387, 1389(121a, 121b), *1439, 1441*
Weinmann, R. 2322, 2323, 2325(40b), *2356*
Weinreb, C.K. 186(34, 35), *257*
Weinreb, S.M. 1817(135), *1864*
Weinstein, R.M. 1618(106), *1662*, 1809(100), *1863*
Weinstock, R.B. 61(108), *99*
Weis, H. 813(102), *824*
Weis, J. 3(3f), *95*, 2223(27a, 27b, 28), *2241*
Weis, U. 2061, 2066–2068(60), *2123*
Weisbeck, M. 1969, 1976, 1977(26b), *2030*
Weisgerber, S. 1886(85), *1893*
Weiske, T. 66, 68(116), *99*, 1124(170), 1128(219), *1139, 1141*, 1163(134), *1184*
Weisman, F. 90(170e), *101*
Weiss, E. 780, 781(3), *821*, 1987(82), *2032*

Weisse, L. 1886(83), *1893*
Weissensteiner, W. 1408(178), *1443*
Wekel, H-.U. 2078, 2091, 2095(121), *2125*
Welch, A.J. 488(272), *494*
Welch, J.T. 438(92), *490*
Welch, W.C. 2337(121), *2358*
Weller, F. 254(336), *265*
Weller, K.J. 1924(106), *1928*
Welsh, K. 1164, 1169, 1176, 1178, 1180(137), *1184*, 1331(175), *1336*, 2509, 2514–2516, 2518(226), *2564*
Welsh, K.M. 1029(343), 1030(345), 1045(343, 345), 1048(345), *1061*, 1160(99, 100), 1169(169), 1170(100), 1176–1180(169), *1183, 1185*, 1246(39), 1286(161), *1306, 1309*, 1329–1331(163), *1336*, 2474(53), 2497, 2498, 2514(179), 2515(53, 179), 2516, 2534(179), *2560, 2563*
Weltner, W. 1126(201), 1127(213), *1140*
Weltner, W.Jr. 68(122a), *99*, 1164, 1175(139), 1179(195), *1184, 1185*
Wendler, C. 884, 885, 889, 925(106), *1056*
Wendt, H.-D. 2223(29a–d), *2241*
Weng, N.S. 2110, 2120(223), *2127*
Weng, W.-W. 1796, 1798, 1818, 1820(40), 1835(218), 1847(281), 1854(324), *1862, 1866–1868*
Wenthold, P.G. 1116(76–79), *1137*
Wenz, G. 1318(59), *1333*
Werle, E. 53(101), *98*, 1178(192), *1185*, 2474, 2534(55), *2560*
Werle, T. 222(193), *261*, 729(60), 735(67), 757(118), *776, 777*
Wermuth, U. 1878(45), *1892*
Werner, E. 2253(37a), *2311*
Werner, H. (129), *1788*, 2096(187), 2106, 2109(210), *2126, 2127*, 2133(23), 2141(61), 2151(118), *2171–2173*, 2445(165), *2461*
Werner, H.-J. 1166(153), *1184*
Wernik, S. 2322, 2323, 2325(40b), *2356*
Wess, J. 2364(18), *2398*
Wessels, B.W. 2348(201), *2360*
Wessely, H.-J. 914, 949, 950, 986, 1004(181), *1057*
Wessolek-Kraus, U. 1156(71), 1162, 1169, 1170, 1180(107), *1182, 1183*, 2495, 2514(168), *2562*
Wesson, J.A. 2343, 2344(154), 2345(173), *2359, 2360*
West, C.T. 1740(177), *1789*
West, J.K. 2318(6b), 2319, 2350(10b), *2355*
West, R. 3(3c, 5, 6), 4(10b), 37(10b, 84, 85), 38(84), 39, 40(10b), 44(10b, 93b), 47(93b), 48(10b, 93b), 49(85, 93b), 52, 53(96, 98a, 98c, 98d), 54(96, 98a), 55, 57, 58(98a), 59(98a, 98d), 60(98d, 107), 61(96), 62(98c, 98d), *95, 97–99*, 122(20a,

20c, 20e), 147(56, 58a, 58b), 148(58a, 58b), *150–152*, 198(89), 200(97, 98, 103), 247(289–291, 296), 248(291, 300–302), 250(312), 252(328), *259, 264, 265*, 269(8), 278(28), 279(34, 38), 280(38), 323, 324(141), 345, 347(154), 348(154–158, 160, 162), 349(154), 350(165), *350, 351, 353, 354*, 508(56), *511*, 790(32), 812(100a, 101a), 818, 819(116), *822, 824, 825*, 827, 828(9, 10, 12), 829(19, 20), 833(19, 32, 33), 839(46, 47), 842(50, 52), 845(64), *854, 855*, 859(5), 870(52, 53), 901(153–156), 981(5, 259, 260), 984(5, 259, 260, 266), 985(269, 270), 987(5, 259, 260, 271), 988(271), 989(259, 260, 271), 992(156), 993(5), 995(5, 156), 996(5), 1000(5, 259, 271), 1002, 1003(259), 1004(259, 271), 1006, 1008(296b, 296c), 1018(317), 1026(335), 1029(343), 1030(345, 346), 1031(347), 1043(346, 347), 1044(346), 1045(317, 343, 345–347), 1048(345), *1054, 1055, 1057, 1059–1061*, 1064(2, 4d, 8b), 1068(2, 8b), 1075(33), 1094(69), (27), *1101–1103*, 1144(10, 11), 1145(16), 1157(75–77), 1158(90, 91), 1159(93), 1160(99, 100), 1164(137), 1166(156), 1167(10, 163), 1168(163, 164, 166), 1169(10, 137, 163, 166, 168, 169, 175, 177), 1170(100, 164), 1176(137, 169), 1177(169), 1178(137, 169), 1179(169), 1180(137, 168, 169), *1181–1185*, 1208(54–56), 1213(71), 1214(56), *1230, 1231*, 1246(39), 1264(83, 84), 1266(89–92), 1268(95, 96), 1282(143), 1283(147, 148), 1284(158), 1285(159), 1286(161), 1288(163–165), 1298(194), *1306–1310*, 1312(7–9, 13), 1313(45), 1314(24, 26, 29–31), 1317(7–9, 45), 1318(47), 1320(45, 83), 1329(159, 161–165), 1330(159, 162, 163), 1331(163, 175, 184–186, 198), *1332–1334, 1336*, 1381(106), 1409(184), 1415, 1420(197), *1440, 1443*, 1603(36, 37), 1622(128), *1661, 1663*, 1771(297), *1791*, 1881(60), 1882(63), 1887(94), *1893, 1894*, 1897(21–26), 1898(25, 32), 1908(26), *1926*, 1930(6, 8), *1960*, 1972, 1988, 1998(40), 2001(131a, 131b, 134), 2016(176b), 2023(40, 195, 196), 2024, 2026(198), *2031, 2034–2036*, 2118–2121(255), *2128*, 2177(3b), 2178(5a, 5c, 6, 7), 2181(22, 32, 33), 2182(22, 26–29), 2183(40), 2193(70–72), 2197(22), 2204(88), 2205(22), 2209(99), *2213–2216*, 2416(44), 2439(146), *2458, 2460*, 2469(29), 2471(41–44), 2472(46), 2474(41, 53), 2481(87, 88), 2484(98), 2485(115), 2491(142), 2492(148), 2493(149, 156), 2495(167), 2497(179–182), 2498(179), 2499(182), 2501(188), 2509(225, 226), 2513(41, 254), 2514(142, 179, 226, 254), 2515(53, 179, 226, 254, 276, 278), 2516(179, 226), 2518(142, 226), 2519(142, 278), 2520(156), 2523(318), 2526(180), 2534(179), 2535(98, 367), 2536(367), 2537(181, 367), 2538(98, 367), 2539(225, 367), 2540(148, 374), 2541(98, 148), 2542, 2543(98, 149, 376), *2559–2567*

West, R.I. 1563(74a), *1579*
Westall, S. 2224(36), *2241*
Westerhausen, M. 209(132), 217(171), *260, 261*, 1890(107), *1894*
Westerlund, C. 1825(161), *1865*
Westerman, P.W. 647(117), 677(150), *700, 701*
Westermann, H. 213(151, 152), *260*
Westermann, J. 1668(15), *1682*
Westheimer, F.H. 713, 741, 743, 748, 752, 753(33), *775*, 2404(4), *2457*
Westman, G. 435, 455(47), *489*
Westrup, J. 169(112), *179*
Wetmore, P.A. 2239(126), *2244*
Wetter, H. 407(156), *429*
Wetterich, F. 1555(58a, 58b), *1578*
Wettern, M. 2377(49), *2399*
Wetzel, D.M. 168, 169(105), *179*, 381, 382(80), *427*
Wetzel, J.M. 414(179), *429*, 433(1), 435(38), 449(38, 124), *488, 489, 491*
Weyenberg, D.R. 1972, 1980, 2003, 2005, 2008, 2009(34), *2031*, 2465(9), *2559*
Whang, D. 1460, 1467, 1503(161), *1529*
Whang, W.-T. 2353(273), *2362*
Whangbo, M.-H. 77, 79(152a), *100*
Whelan, J. 1756(250, 251, 253), 1757(250), *1790*, 1801(65), *1863*
Whipple, W.L. 466(183), *492*
Whitaker, A.F. 2231(88), *2243*
Whitby, R.J. 1671(43), *1683*
White, A.H. 210(141), 217(169), 252(329), *260, 261, 265*, 1377, 1378, 1380(97a), *1440*, 2140(46), *2171*
White, D.A. 2289(125, 126), *2315*
White, J.B. 1802(69), *1863*
White, J.J. 1997(112a), *2033*
White, J.M. 374(61–64), 375(64), 377, 381(61), *427*
White, J.W. 2227(72), *2242*
White, R.T. 2476(61), *2560*
White, W.R. 1147(26), *1182*
Whitecar, C.K. 2345(176), *2360*
Whiteley, R.H. 2153(131), 2154(131, 136), *2173, 2174*
Whitely, D.L. 1593(109), *1597*
Whitenack, A.A. 1325(122), *1335*
Whitesell, J.K. 1670(31), *1683*, 1857(338), *1868*
Whitesides, G.M. 1903(42), *1926*

Whitham, C.J. 1107(12), *1136*
Whitham, G.H. 1601, 1602, 1604, 1624, 1651(11), *1660*
Whithnall, R. 1161(101–103), *1183*
Whiting, M.C. 2163(179), *2174*
Whitman, D. 2005(154), *2034*
Whitmarsh, C.W. 2283, 2284(116b), *2314*
Whitmore, F.C. 364(35, 36), 376(35), *426*, 610, 616, 635(34, 35), *698*, 1606, 1628(46), *1661*, 1722(126), *1788*
Whitney, J.F. 1430(226), *1444*
Whittaker, S.M. 577(157), *593*
Wiberg, K.B. 120(12), *150*, 522(82), *554*, 690(164a), *701*, 1910, 1911(60a, 60b, 60d), *1927*, 2421(60), 2437(137), *2458*, *2460*
Wiberg, N. 88(166), *101*, 125, 126, 128(32a, 32b), 129(32a, 32b, 36), *151*, 198(90), 203(104), 215(156), 250(308–311, 313), 252(322–324), *259*, *260*, *264*, *265*, 485(249), *493*, 783(19a, 19b), *822*, 844(57–62), 845(59, 61), 847(59), 848(61), *855*, 859(4), 873(59–67), 876(4, 65, 66, 71–78), 879(83–85), 881(64, 97), 909(171–174), 910(62, 176), 912(97), 913(64, 77), 917(66, 72, 73, 78), 931(72, 73, 78), 932(62, 83, 84, 197), 937(62, 63, 201), 938(63, 201), 940(72–76, 83, 84, 97), 941(59, 60, 62, 176, 203, 204), 942(62, 176), 943(63, 64, 67), 944(83, 197), 953(176), 969(60–64), 981(4, 83, 84, 197), 985, 986(268), 988(83), 991(4), 1011(301–303), 1012(308, 309), 1014(310), 1021(327), 1025(303), 1033(300, 301, 303, 308), 1034(300, 303, 310, 351, 352), 1035(300, 303, 351), 1036(300), 1037(59), 1038(300, 303, 310, 351), 1039(356), 1040(303), 1042(327), 1043(303, 327), 1044(300, 303), 1045(303), 1051(308, 310, 351, 383), *1054–1062*, 1064, 1068(5a, 5b), 1077(40), 1079(42), *1101*, *1102*, 1158(88), *1183*, 1320(80), *1334*, 2205(92), *2215*, 2407(22), *2458*
Wicha, J. 424(213, 214), *430*, 1619(112), *1663*, 1679(111, 112), *1684*, 1835(214), *1866*
Wickenheiser, E.B. 1743(220), *1790*
Wickham, G. 407(160), *429*
Widdowson, D.A. 435, 454(45), *489*
Widdowson, K.L. 505(40), *511*
Wideman, T. 2265(83b, 83c), *2313*
Wiechert, R. 1575(98), *1579*
Wiegand, G. 1228(124), *1232*
Wieker, W. 2344(166), *2360*
Wiemer, D.F. 1836(221), *1866*
Wierlacher, S. 704, 709(7), *775*
Wierschke, C. 597, 598, 603, 604(5), *697*
Wierschke, S.G. 184, 222(22), *257*, 363, 365(28), *426*, 482(241), *493*

Wiesenberger, F. 219(180), 226(204), *261*, *262*, 2364(27, 28, 30), *2398*, *2399*
Wilante, C. 159, 161, 162(51), *178*
Wilbur, D.S. 385(97), *428*
Wilczek, L. 575(137), *592*
Wild, S.B. 1743(221), *1790*
Wilde, R.G. 1091(58g), *1103*
Wilding, O.K. 1584(29–32), *1595*
Wilhelm, D. 2226(51), *2242*
Wiliams, E.A. 268(1), *350*
Wilke, C.J. 193(61), *258*
Wilke, J. 2439(141), *2460*
Wilkes, G.L. 2236(115), *2244*, 2345(168, 169), *2360*
Wilkesg, G.L. 290(70–73), *352*
Wilkins, C.L. 1113(69), *1137*
Wilkinson, D.L. 2060, 2066, 2067, 2084, 2089, 2094(55), *2123*
Wilkinson, G. 2163(179), *2174*
Will, P. 2473(50), 2495, 2514(172), *2560*, *2562*
Willard, A.K. 1824, 1828(160), *1865*
Willcott, M.R. 564(91), *591*, 1410, 1412(187), *1443*
Willeke, C. 520(67), *553*
Willett, G.D. 1120, 1121(131), *1139*
Willey, P.R. 456(142), *491*, 2225(47), *2242*
Willhite, D.L. 1372, 1373(72c), *1439*
Williams, C.J. 123(26a, 26b), *151*
Williams, D. 222(192), *261*, 1638(170), *1664*
Williams, D.A. 2225(47), *2242*
Williams, D.J. 248(298), *264*, 285–287(62), *352*, 467, 474(187), *492*, 1355–1357(38), *1438*, 2498(185b), *2563*
Williams, E.A. 1341, 1343, 1409, 1422(21a), *1437*
Williams, F. 1314(28), *1333*
Williams, I.D. 1723(128), *1788*
Williams, J.P. 1546(34), *1578*
Williams, P. 1135(266), *1142*
Williams, R.V. 1634, 1642, 1643(160), *1664*
Williams, S.A. 1316(44), *1333*
Williams, T.J. 2286(120), *2315*
Willing, D.N. 1689, 1694(11), *1786*
Willis, C.J. 2001(135), *2034*
Willms, S. 1304(213), *1310*, 2185(46), *2214*, 2472(48), *2560*
Willnecker, J. 789(28e), *822*
Willoughby, C.A. 2044, 2051, 2052(18), *2122*
Wills, M. 1667(5), *1682*
Wills, R.R. 2246, 2250, 2253, 2254, 2273, 2285(6c), *2310*
Wilson, A.A. 2227(66), *2242*
Wilson, A.M. 2232(94), *2243*
Wilson, C.L. 2522(284), *2565*
Wilson, D. 2145, 2146(93), *2172*
Wilson, G. 1299(196), *1310*

Wilson, J.D. 7(24), *96*
Wilson, M.A. 1120, 1121(131), *1139*
Wilson, M.E. 1903(42), *1926*
Wilson, N.S. 1674(70, 71), *1683*
Wilson, S. 284(57), *351*, 505(39), *511*, 1353, 1357, 1358(35b), *1438*
Wilson, S.R. 413(178), *429*, 467(184), *492*, 1636, 1652(165), *1664*, 1828(186), *1865*, 1904, 1909, 1910(46), *1926*, 2076(116, 117), 2091(117), *2125*
Wimmer, P. 1743(205), *1789*
Winans, R.E. 184, 221, 223(21), *257*
Wincel, H. 1122(153), 1125(153, 178), *1139, 1140*
Winchester, W.R. 2488(133), *2561*
Wind, M. 348(158), *354*, 842(52), *855*
Windmüller, B. 2106, 2109(210), *2127*
Windus, T.L. 500(25), 505(36, 37), *510, 511*, 1109, 1126, 1130(35), *1136*, 1372, 1373(71a, 71d, 72a, 72b), *1439*, 1468, 1469, 1477, 1511, 1521(194), *1530*
Wineland, J.D. 1586(44), *1595*
Winkel, Y.van den 10(35), *96*, 348(162), *354*, 1051(379), 1052(387), *1062*, 1064, 1068(6b), *1101*, 1151(46, 48, 49), *1182*, 1283(148), *1308*, 2453(195, 196), *2461*
Winkel, Y.van der 248(301), *264*
Winkel, Y.v.d. 833(33), *854*
Winkel, Y.von den 1266(92), *1307*
Winkelhofer, G. 1316(41), *1333*
Winkhofer, N. 222(189), *261*
Winkler, H.S.J. 796, 801(72), *824*
Winkler, U. 1052(390, 392), 1053(390), *1062*, 1081(48), *1102*
Winstein, S. 544(111b), *555*, 1502(371), *1536*
Winter, C. 2512(252), *2564*
Winter, C.H. 2027(201), *2036*, 2137(35), 2141(54), 2143(75), 2144, 2145(54), *2171, 2172*
Winter, G. 2253(39b), *2311*
Winter, R.A. 2437(130), *2460*
Winter, S.B.D. 790(38), *822*, 1680(120), *1685*
Wirsam, B. 517(50), *553*
Wiseman, G.H. 2251(25a–c), 2252(25a), 2253, 2255, 2257(25a–c), *2311*
Witanovsky, M. 1474, 1475, 1479(242), *1532*
Withers, G.P. 425(217), *430*
Withnall, R. 992, 995(275), *1060*, 1068(18), *1101*, 1127(207), *1140*
Witiak, J.L. 2001(132), *2034*, 2474(52), *2560*
Witt, E. 1434, 1436(248), *1445*
Witte, F. 2363(9), *2398*
Wittel, K. 1320(80), *1334*
Wittenberg, D. 1609(71), *1662*, 1988(83c), *2032*
Wittenberger, S. 1091(58g), *1103*

Wittig, G. 2439(149), *2460*
Wittouck, N. 2354(283), *2362*
Witzgall, K. 1601, 1602, 1604, 1612, 1643(13), *1660*
Wlochowicz, A. 1701, 1702(63), *1787*
Wlodawer, A. 515(18c), *551*
Wlodek, S. 68(121), *99*, 1029(341), *1061*, 1109(43, 50), 1117(50, 86, 88), 1122(43, 153), 1123(158), 1125(153, 178), 1126(50, 200, 205), *1137–1140*, 2556(423), *2568*
Wo, W.K.L. 1128(229), *1141*
Wocadlo, S. 226(202), *262*, 1461, 1467, 1497(141), *1529*, 2159(159), *2174*
Woerpel, K.A. 1681(122), *1685*
Woggon, U. 1887(90), *1894*
Wojcik, A.B. 2344(156, 163–165), 2350(235), *2359–2361*
Wojnowska, M. 232(225), *262*
Wojnowski, M. 219(183), *261*
Wojnowski, W. 219(179, 183–185), 229(213–216, 218–220), 232(184, 185, 224–228, 230), *261, 262*, 1913, 1914(76), *1927*, 1989(91), *2033*, 2486(122), 2494(163), *2561, 2562*
Wojtyniak, A.C.M. 561(58, 60), *591*, 609(32), *698*
Wokaun, A. 310(124), *353*
Wolcott, J.M. 497(13), 507(47), *510, 511*
Wolf, A. 2107, 2109(212), *2127*
Wolf, A.D. 2404(6), 2428(88), *2457, 2459*
Wolf, A.P. 385(97), *428*
Wolf, B.A. 2238(124), *2244*
Wolf, G. 2156(145), *2174*
Wolf, M.A. 484(247), *493*
Wolf, R. 747, 748(94, 95), *776*
Wolf, W.H.de 2428(89), *2459*
Wolfbeis, O.S. 2321(24a), *2356*
Wolff, A. 946(212, 213, 215), *1058*
Wolff, S. 438(94), *490*, 772(152), 773(153), *778*
Wolinski, K. 517(37, 44b), *552, 553*
Wolkowa, W.W. 229(212), *262*, 1910(57), 1913, 1914(77), *1927*
Wolter, H. 2330(79, 84), 2333(79, 84, 94, 95), 2339(79, 94, 95), *2357, 2358*
Won, Y. 2510(231), *2564*
Wong, C.-L. 168(104), *179*
Wong, C.Y. 285–287(62), *352*, 1340, 1351, 1354(7), 1355–1357(38), 1412, 1415(7), 1416(203), 1417(203, 204), 1420(7), *1437, 1438, 1443*
Wong, J.-C. 2353(273), *2362*
Wong, K.-T. 1794, 1854(26), *1862*
Wong, L.-J. 1261(77), *1307*
Wong, S.S. 2001(138b), *2034*
Wong Chi Man, M. 1341, 1347(24), 1409, 1411(185a), 1412(185a, 191), 1414(191),

Wong Chi Man, M. (cont.) 1430–1432, 1434(228), *1437, 1443, 1444,* 2335(112), *2358*
Wong Chi Man, W.W.C. 508(55), *511,* 1356, 1357(39), 1418, 1420, 1421, 1423(213), *1438, 1444,* 1972, 1981(37a, 37b), *2031*
Wong-Ng, W. 250(306, 307), *264,* 859, 880, 881, 977, 978, 981, 985, 986, 992, 996(3), *1054,* 1080(47), *1102,* 1146(22), *1181,* 1601, 1609, 1645(24), *1661,* 2404(12), *2458*
Woning, J. 481(235), *493,* 1518(412, 413), 1519(413), 1522(412, 413), *1537*
Woo, H. 1606, 1610(55), *1661*
Woo, H.-G. 1606(51), *1661,* 2016(175c), *2035,* 2045(20), 2046(21, 22), 2047(22), 2051(20, 21), 2052(21), *2123,* 2275(102b), *2314*
Wood, B.R. 2218(4), *2240*
Wood, G.L. 209(132), 216(162), *260*
Wood, I.T. 1965(9a), *2030*
Wood, M.D. 1897(19), *1926*
Wood, M.R. 1669(18), *1682*
Wood, T.G. 2273, 2275(99a), *2314*
Woodgate, P.D. 1651(216), *1665*
Woodman, R. 2318(3), *2355*
Woods, J.G. 2226(54), *2242*
Woodward, R.B. 1130(238), *1141,* 1997(113a), *2033,* 2163(179), *2174*
Woollins, J.D. 285–287(62), *352,* 1340, 1351, 1354(7), 1355–1357(38), 1412, 1415, 1420(7), *1437, 1438*
Wörle, B. 243(268), *263,* 2027(203), *2036*
Worley, S.D. 1207(48), *1230*
Wormhoudt, J. 2493(154), 2523(308), *2562, 2566*
Wörth, J. 1543(23), *1577*
Woynar, H. 1074(28), *1102*
Wrackmeyer, B. 203(109, 110), 209(136), 216(160), *259,* 260, 296(83, 90), 302, 304(102), *352,* 800(70a), *823,* 1979(56a–d, 57a–g, 58a, 58b), 1986, 2011(56a, 76c), 2028(205), *2031, 2032, 2036,* 2164(183), *2175*
Wright, A. 1622(128), *1663*
Wright, A.P. 1592(101), *1597*
Wright, C.D.III 1212(69), *1231*
Wright, J.S. 1134(263), *1142*
Wright, L.J. 2076, 2091(118, 119), 2093(185), *2125, 2126*
Wright, M.E. 1743(216, 217), 1774(318, 319), *1790, 1792,* 2150(104), 2151(114), 2152(114, 119), 2153(119), *2173*
Wright, N.G. 2523(306), *2565*
Wright, S.C. 2509(227), *2564*
Wrighton, M.S. 1700(58, 59), 1705(59, 70, 71, 73, 81), *1786, 1787,* 2153(132), 2154(130,

132), 2155(132), *2173*
Wrobel, D. 2231(81e), *2242*
Wrzesich, K. 2486(122), *2561*
Wu, B. 2330(83), 2352(263, 265), 2354(83), *2357, 2362*
Wu, C.J. 2546(389), *2568*
Wu, G. 1449(17), 1454(99), 1456(116, 117), 1461, 1463(17), 1467(99, 116, 117), 1470(210, 211), 1481(286), 1483(210, 286), 1500, 1501(99), 1506(286), 1508(386), 1513(17), *1525, 1528, 1531, 1533, 1536,* 2486(123), *2561*
Wu, G.-L. 1467, 1474, 1475(183), 1477(265), 1481(283), *1530, 1533*
Wu, H.-J. 1467, 1474, 1475(183), 1477(265), *1530, 1533*
Wu, J. 1675(76), *1684*
Wu, M.H. 1672(54), *1683*
Wu, M.-J. 1850(307), *1867*
Wu, S. 2486(123), *2561*
Wu, S.-H. 2488(130), *2561*
Wu, T.C. 789(29), *822*
Wu, Y. 1454(99), 1456(117), 1463, 1464(167), 1467(99, 117, 167), 1470(211), 1500, 1501(99), 1508(386, 387), *1528, 1530, 1531, 1536*
Wu, Y.-D. 168(104), *179*
Wu, Z.C. 70(135), *100*
Wu, Z.-Z. 1298(189), *1309*
Wuang, D. 2153, 2162(124), *2173*
Wulff, W.D. 1071(22), *1102,* 1972, 1997, 1999(35c), 2001(132), 2002(35c), *2031, 2034,* 2474(52), 2501(189), *2560, 2563*
Wüllen, C.v. 517(41), *552*
Wüllen, C.van 7(27), *96,* 516(32a), *552*
Wuller, S.P. 1888(96), 1890(96, 112), *1894*
Wurfel, B.E. 1156(74), *1182*
Wustrack, R. 198(85), *259,* 884, 885(101–103), 886, 937, 938(103), *1056*
Wuts, P.G.M. 1674(62), *1683*
Wüttke, F. 1608, 1617(64), 1649(208, 209a), *1661, 1665,* 2363(7, 9), 2377(48), 2378(52, 54–56), 2379, 2380, 2385(7), 2387(65, 67), *2398–2400*
Wynberg, H. 1608(59), *1661*
Wynne, K. 2246(4a, 4b), *2310*
Wynne, K.J. 2236(112, 113), *2243,* 2246, 2247, 2250, 2253, 2254, 2273, 2285(6b), *2310*

Xi, Z. 2434(124), *2460*
Xia, H. 1799(54), *1863*
Xiao, F. 2488(130), *2561*
Xiao, M. 194(65), *258,* 900(145b), *1057,* 1166(159), *1184,* 1283(150), *1309,* 1331(194), *1336,* 2421(58), *2458,* 2475, 2479, 2483, 2487(57), *2560*

Xiao-Dong Zhang 1455, 1464, 1467, 1486(111), *1528*
Xie, M. 642, 643(107), *700*
Xie, Q. 1454, 1467(83), 1489(313, 314), 1492(83), *1527, 1534*
Xie, Y. 1126(203), *1140*
Xie, Z. 197(79), *259*, 283(46), *351*, 514(3a, 3b), 520(3a, 3b, 71), 521(3a, 3b, 76), 522(3a, 76), 524(3b, 71), 525(71), 529, 530(90, 91), 532(3a), *551, 553, 554*, 561(59), 567(103, 106), 569(106), 570(103, 106), 572(59, 103), 583(59, 103, 106, 185), 585(59, 106, 185), 586(103, 185), 587(106, 185), 588(59, 106, 185, 196), *591–594*, 654(132), *700*
Xin, S. 1714(94), *1787*
Xin, X. 2290(137), *2315*
Xu, C. 2330(83), 2352(263, 265), 2354(83), *2357, 2362*
Xu, D. 2345(172), *2360*
Xu, M.-S. 1481(283), *1533*
Xu, S.L. 1799(54), *1863*
Xu, Y. 424(212), *430*, 498(18, 19), *510*, 1621(122), 1652(220–222), *1663, 1665*, 1680(114), *1684*, 2232(96), *2243*, 2285(117c), *2315*
Xue, L. 1828(193), *1865*
Xue, Z. 2051, 2052, 2056, 2057, 2064(35), *2123*

Yabe, A. 1169(178), *1185*, 2493(151), 2520(151, 268), 2523(151), *2562, 2565*
Yabushita, S. 84, 86(162), *100*, 1172(184), *1185*
Yajima, S. 2272(92a–c, 93a–c, 94a, 94b), *2314*
Yakeshita, K. 1705(76, 77), *1787*
Yakushigawa, Y. 1959(30), *1960*
Yamabe, S. 465(176), *492*
Yamabe, T. 559(20, 24), *590*, 807, 808(87a, 87c), *824*, 870(50), 888(120, 121), 889(121), 892(131), 895(132), 958(120), 986, 988, 990(121), 998(287), *1055, 1056, 1060*, 1252(56), 1253(57), 1303(206), 1304(216), *1306, 1307, 1310*, 1318(55), 1325(133, 135, 137, 141), *1333, 1335*, 1973, 1974(50b), 2015, 2016(173), *2031, 2035*, 2512(248), *2564*
Yamada, C. 2522(296), *2565*
Yamada, M. 1871(13), *1891*
Yamada, S. 1592(103), *1597*, 2225(45), *2241*
Yamada, T. 746(90a), *776*
Yamada, W. 438(90), *490*
Yamagishi, H. 1229(128), *1232*
Yamaguchi, H. 1828(188), *1865*, 2208(98), *2216*
Yamaguchi, K. 1640, 1655(187), *1664*, 1906, 1907(51), *1926*, 2512(248), *2564*

Yamaguchi, M. 897(138), *1057*, 1322, 1324–1327(105), *1334*, 1608(66), *1662*, 2326(45c), 2327(60), *2356, 2357*
Yamaguchi, S. 190(51), *258*, 1304(215, 216), *1310*, 1743, 1746(192, 193), 1747(192), *1789*, 1978(54b), 1979(59a, 59b), 2013(59a, 59b, 170), 2014(59a), 2015(54b, 173), 2016(173), 2028(206–209), *2031, 2032, 2035, 2036*
Yamaguchi, T. 1906, 1907(51), *1926*
Yamaguchi, Y. 2013(171), *2035*
Yamaji, M. 2513, 2515, 2516, 2520(259), *2564*
Yamaji, T. 1796, 1842(36), *1862*
Yamakado, Y. 436, 480(64), *489*
Yamakawa, K. 1832(204), *1865*
Yamakawa, M. 709(19), *775*, 2424(74), *2459*
Yamamori, H. 1322(90, 92), *1334*
Yamamoto, A. 436, 483(69), *490*
Yamamoto, H. 434(14), 436(64), 439(14), 440(97), 442(108), 443(109, 111), 480(64), *489, 490*, 1412(190), *1443*, 1619(114), 1635(161), *1663, 1664*, 1824(159), 1828(159, 191), *1865*, 2326(45c), 2327(60), *2356, 2357*
Yamamoto, I. 1772(314), *1792*
Yamamoto, J.H. 2069, 2071(78), *2124*
Yamamoto, K. 1193(14b), 1197, 1200(28), 1203, 1212, 1213(40), *1230*, 1302(203, 204), *1310*, 1591(94, 98), *1597*, 1610(81), *1662*, 1738(159), 1743(188), 1751(234), *1788–1790*, 1940(17), 1942(18, 19), *1960*, 1973(41a, 41b), *2031*, 2150, 2151(105), *2173*, 2448(181), *2461*
Yamamoto, M. 559(25), *590*
Yamamoto, S. 66, 67(117c), *99*, 1009, 1010(298a), *1060*, 1163(135), *1184*, 2557(429), *2568*
Yamamoto, T. 1872(19, 20), *1892*, 2089, 2092, 2094(181, 182), 2107, 2109(211), *2126, 2127*, 2236, 2237(116), *2244*, 2420, 2423(70), 2425(80, 81), 2427(81), *2459*, 2493(158), *2562*
Yamamoto, Y. 434, 445(27), *489*, 1672(45), *1683*, 1692(16, 17), 1720(17), 1738(16), *1786*, 1794(25), 1798(53), 1811(110), 1829(194), 1838(240), 1856(330), *1862–1866, 1868*
Yamamura, S. 2326(45c), 2327(60), *2356, 2357*
Yamamura, T. 2272(95), *2314*
Yamanaka, S.A. 2326(45b), *2356*
Yamanaka, T. 186(40), *258*, 2016(175b), 2017(177a, 177c, 178), *2035*, 2439(148), *2460*
Yamane, M. 2350(219), *2361*
Yamasaki, Y. 1771(308), *1792*
Yamashita, F. 1756(248), *1790*

Yamashita, H. 1838(230), *1866*, 2114(230, 231), 2115(235), 2117, 2120(231), 2121(231, 235), *2127*, 2320, 2347(17b), *2355*
Yamashita, K. 2326(45c), 2327(60), *2356, 2357*
Yamashita, N. 437, 487(81), *490*
Yamashita, O. 2207(95), *2215*
Yamashita, S. 2345(174), *2360*
Yamauchi, M. 1838(241), *1866*
Yamazaki, H. 1560(69), *1579*, 1692, 1738(16), *1786*, 1829(194), *1865*
Yamazaki, N. 1906, 1907(51), *1926*, 2334(105), *2358*
Yamazaki, S. 465(176), *492*
Yammamoto, Y. 194(64), *258*
Yammura, T. 2272(97), *2314*
Yan, L. 1481, 1483, 1506(286), *1533*
Yan, Y.M. 1558(65), *1578*
Yanagi, K. 1752(240), *1790*
Yanagi, T. 1808(94), *1863*
Yanagisawa, A. 434, 439(14), 443(109, 111), *489, 490*, 1635(161), *1664*
Yanaka, M. 1755, 1756(247), *1790*
Yang, B. 1764(270), *1791*
Yang, C.-Y. 2283, 2284(116b), *2314*
Yang, D. 2345(175), *2360*
Yang, D.B. 2224(43), *2241*
Yang, G.K. 2069, 2071(79), *2124*
Yang, M. 477(228), *493*, 1800(58, 59), 1806(81), *1863*
Yang, S.-Y. 2272(96), *2314*
Yang, Y. 226(201), *262*, 1452(60), *1526*
Yang, Z. 1606(53), *1661*, 2330(83), 2352(263, 265), 2354(83), *2357, 2362*
Yang, Z.-X. 2018(182), *2035*
Yankovskaya, I.S. 1454, 1467(98), 1505, 1506(378), 1514, 1519, 1521, 1522(408a), *1527, 1536, 1537*
Yannoni, C.S. 690(164b), *701*
Yano, R. 2351(246), *2362*
Yano, T. 559(20), *590*, 807, 808(87a, 87c), *824*, 888(120, 121), 889(121), 958(120), 986, 988, 990(121), *1056*
Yanovsky, A.I. 2041(12), *2122*
Yanwu, L. 2039–2041(5), *2122*
Yao, Q. 1571(90), *1579*
Yao, S.S. 2351(255), *2362*
Yao, Z. 2080, 2092(130–132), *2125*
Yap, G.P.A. 37, 39, 40(82), *97*, 816, 819, 820(112a), *825*, 981, 984(261), *1059*, 1762(263), *1791*, 2019, 2024, 2026, 2028(187), *2035*, 2105, 2109(206), *2127*, 2153, 2162(125), *2173*
Yap, T.K. 515(20), *552*
Yarosh, O. 2439(145), *2460*
Yarosh, O.G. 186(44, 46, 47), 190(50), *258*, 1450(38, 40, 41), 1472(234), 1474(244), 1475(244, 249), 1479, 1480(249, 277), 1523(249), *1526, 1531–1533*
Yassar, A. 2017(179), *2035*, 2330(85), *2357*
Yasuda, H. 2145(91), *2172*
Yasue, K. 434, 439(14), *489*
Yasuhara, S. 371(56), *427*, 436(69), 437(70), 483(69, 70), *490*
Yasui, E. 1771(308), *1792*
Yasui, M. 1293(181), *1309*
Yasuike, S. 1991(97), *2033*
Yasunaga, E. 1592(103), *1597*
Yasusa, T. 1852(314), *1868*
Yatabe, K. 1675(73, 75), *1683, 1684*
Yatabe, M. 1705, 1709, 1714(80), *1787*
Yatabe, T. 130, 133(37), 134(37, 43), 137, 140, 143(47), 146(37, 47), *151, 152*, 1268(101), *1307*, 2205(90, 91), 2206(90), *2215*
Yates, B.F. 84(165a), 87(165a, 165b), 88(165a), *101*, 1125(180), 1127(212), *1140*
Yates, K. 1603(38, 41), 1605(38), *1661*
Yates, L.M.III 2352(270), *2362*
Yatsumoto, H. 1412(190), *1443*
Yau, L. 1601, 1604, 1609, 1645(17), *1660*
Yavari, I. 690(165), *701*
Ye, F. 1501(367), *1536*
Yeakey, E. 1392(142), *1441*
Yee, N.K.N. 464(171), *492*
Yeh, M.-H. 727, 741(54), *776*
Yen, T.-S. 2264(75), *2313*
Yeon, S.H. 401(140), *428*, 461(165), *492*, 1589(78), 1590(78–80), *1596*, 1856(328, 329), *1868*
Yep, G.L. 1875(37), *1892*
Yermakov, Yu.I. 2341(141), *2359*
Yet, L. 1681(121), *1685*, 1854(321), *1868*
Yi, K.Y. 434, 447(29), *489*
Yilgor, I. 2234(104a, 104b), *2243*
Yilmaz, S. 1588(65, 66, 69), 1589(65, 66), *1596*
Yimenu, T. 413(177), *429*, 782, 791, 795(11b), *822*
Yin, C. 226(201), *262*, 1452(60), *1526*
Yin, J. 2079, 2092, 2093(126), *2125*
Yive, N.S.C.K. 2251(26), 2253(26, 47a, 47b), 2255(26), *2311, 2312*
Yoda, H. 437(83), 438(83, 90), *490*
Yoder, C.H. 1390(124–127), 1392(126, 127), 1396(124), 1397, 1398(170), 1399(124), 1404(170), 1405(173), *1441–1443*
Yoffe, S.L. 1377(98), *1440*
Yokelson, H.B. 278(28), *351*, 1288(165), *1309*, 2497, 2499(182), *2563*
Yoko, T. 2289(129), *2315*
Yokoi, H. 2346, 2347(183a), *2360*
Yokoo, K. 1832(204), *1865*

Yokota, S. 1640(186), *1664*
Yokota, T. 145(53), *152*, 1304(212), *1310*
Yokota, Y. 559(25), *590*
Yokoyama, M. 1207(48), *1230*, 1548(38), 1558(66a, 66b), *1578*, *1579*
Yokoyama, Y. 1797(43, 44), *1862*
Yokozawa, T. 1821(146), *1864*
Yonenitsu, T. 1602(27), *1661*
Yonezawa, K. 782(13a), *822*
Yonezawa, T. 1312(20), *1332*
Yoo, B. 867(41, 42), *1055*, 1875(32), *1892*
Yoo, B.R. 401(140), *428*, 874, 875(69), 876(70), 934(69), *1055*, 1241(28), 1242(31), *1306*, 1856(328), *1868*
Yoon, J.-Y. 1834(213), *1866*
Yoon, U.C. 1294(183), *1309*
York, D.M. 515(18a–c), *551*
York, G. 2236(115), *2244*
Yoshiaki, T. 2080(140), *2125*
Yoshida, A. 1843(263), *1867*
Yoshida, H. 737, 741(70), *776*, 906, 908, 931(169), *1057*, 1236(10), 1278(135), *1305*, *1308*, 1331(188, 189), *1336*, 1881(59), 1887(59, 93), *1893*, *1894*, 1897, 1911(12, 15), 1914, 1915(15), 1923(91), *1925*, *1927*, 2409, 2410(31), 2418, 2420(54), 2421, 2422(62), *2458*, *2459*, 2488(131), 2491, 2518, 2519(143), *2561*, *2562*
Yoshida, J. 1188(2), 1190, 1192(10b, 13), 1193(16–18), 1194(17), 1196(22, 23), 1197(22, 23, 25), 1198, 1199(22), 1201(34, 36, 37), 1202(2, 25, 36, 38), 1203(2, 25, 39, 41), 1204(37), 1221(99, 100), 1226(117), *1229–1232*, 1617, 1620, 1624(105), 1650(213), *1662*, *1665*, 1824, 1828(159), *1865*
Yoshida, J.-i. 1831(199), 1843(261), *1865*, *1867*
Yoshida, K. 1325(148), *1335*
Yoshida, M. 831, 832, 836, 837(29), *854*
Yoshida, N. 559(20), *590*
Yoshida, S. 2080, 2092, 2093(136), *2125*
Yoshida, T. 2075, 2091(115), 2116, 2120(236, 237), *2125*, *2127*, 2433(119), *2460*
Yoshidain, M. 829(25), *854*
Yoshido, N. 559(24), *590*
Yoshifuji, M. 486(260), *493*, 1064(3), *1101*
Yoshihara, K. 1325(120), *1335*
Yoshii, E. 1640, 1655(187), 1656(223), *1664*, *1665*, 1699, 1735(48), *1786*
Yoshii, T. 1303(206), *1310*, 1325(141), *1335*
Yoshikawa, E. 1856(330), *1868*
Yoshikawa, S. 2326(45c), 2327(60), *2356*, *2357*
Yoshikawa, Y. 1415(199), *1443*, 1452, 1453, 1473(61), *1526*

Yoshimine, M. 997(286), *1060*
Yoshimune, M. 1828(185), *1865*
Yoshimura, R. 1109(39), *1137*
Yoshinaga, M.K. 2219(12), *2240*
Yoshinori, Y. 1590(89), *1596*
Yoshioka, Y. 705(14), *775*
Yoshitake, M. 2226(52), *2242*
Yoshizumi, K. 812(98), *824*
Young, C.S. 2142(69), *2172*
Young, D. 381(76), *427*
Young, D.J. 361(19), *426*
Young, J.C. 284(51), *351*, 495(3), *510*, 519, 541(64), *553*, 1340(2, 6), 1343(6), 1351(2, 6), 1354(6), 1374(76, 77, 79), 1375(6, 76, 77, 79), 1376(76), 1377(77, 79), 1378(6, 77), 1380(79), 1382(2, 6, 76, 77), 1383, 1385(76, 77), 1386(76), 1409(2), 1412, 1415, 1420, 1421(6), 1424(77), 1429(6), *1437*, *1439*, 1449(25, 31), 1464(25), *1525*
Young, R.A. 560(45, 46), *590*
Young, V.G.Jr. 2448(180), 2450(182, 183), *2461*
Youngdahl, K.A. 2251, 2253, 2255(28c), 2290, 2291, 2293(138a–c), *2311*, *2315*
Youngs, W.J. 186(32, 34, 35), *257*, 2079, 2093(124), *2125*, 2145(92), *2172*
Youssaere, E. 2354(281), *2362*
Yu, C. 898(141), *1057*, 1244(36, 37), 1293(180), *1306*, *1309*, 1328(154), *1335*, 2475(58), *2560*
Yu, C.-M. 1871(15), *1892*
Yu, G.-E. 2262(71), *2313*
Yu, J.L. 2550(408), *2568*
Yu, J.M. 1769(290), *1791*
Yu, S. 473(203), *492*
Yu, X. 2224(42a), *2241*
Yu, Y.-F. 2261(67a, 67b, 68), *2313*
Yu, Z. 503(33), *511*, 1374, 1375, 1377(78), 1382(109), *1439*, *1440*
Yuan, C.-H. 1930(8), *1960*
Yuan, W.-J. 167, 168(89, 93), *179*
Yuan-Chieh, D. 2350(216), *2361*
Yudin, A.K. 1224(108), *1232*, 1675(81, 82), *1684*
Yumura, M. 2278(107), *2314*
Yun, J.S. 1544(25), *1578*
Yun, S.S. 1697(39), *1786*
Yuntila, L.O. 783(18), *822*
Yushmanova, T.G. 1390, 1392(136), *1441*
Yuuki, M. 1849(294), *1867*
Yuzefovich, M. 795, 800(69a, 69b), 801(69b), *823*, 884–886, 910, 957, 981, 983, 985, 987, 992, 996(111), *1056*, 2122(258), *2128*
Yuzhelevskii, Yu.A. 213(172), *261*
Yuzuriha, Y. 998(289, 290), 1004(289), 1005(289, 290), *1060*

Zablocki, J. 396(128), *428*
Zaborovskiy, A. 1543(18), *1577*
Zabrodsky, H. 2348(202a–c), *2361*
Zacharia, M.R. 1126(189), *1140*
Zachariah, M.R. 155(25), 164(25, 82), 166(25), 170, 175(25, 82), *178, 179*
Zagorevskii, D.V. 1122, 1125(152), *1139*
Zahouily, M. 475(213), *493*, 1554(56), *1578*
Zahurak, S.M. 69(130), *100*
Zaitseva, G.S. 1450(37), 1456(37, 119), 1457(37), 1461(37, 141), 1467(119, 141), 1494(337), 1496(37, 345), 1497(119, 141, 351), 1501(119, 363), *1525, 1528, 1529, 1535, 1536*
Zakharov, L.N. 2041(12), *2122*
Zakharov, V.A. 2341(141), *2359*
Zakharova, G.N. 1897(17), *1926*
Zakouril, P. 1118(96), *1138*
Zamaev, I.A. 186(44, 47), 190(50), *258*
Zamble, D.B. 2154(137), *2174*
Zamboni, M. 1550(48), *1578*
Zandi, K.S. 1854, 1860(325), *1868*
Zanella, P. 2053(36), *2123*
Zangvil, A. 2232(96), *2243*, 2285(117c), *2315*
Zani, P. 473(207), *492*, 1610, 1630(79, 80), 1638(174), 1644, 1649, 1650(196), 1658(224), *1662, 1664, 1665*, 1830(197), *1865*
Zank, G. 2261(62), *2312*
Zank, G.A. 2232(96), *2243*, 2250, 2253(22), 2261(55), 2265(83a–c), 2285(117b, 117c), *2310, 2312, 2313, 2315*
Zapf, F. 2226(52), *2242*
Zapp, R.H. 1585(38, 39), *1595*
Zarate, E.A. 2105, 2109(207), *2127*
Zarkin, A.A. 2133(20), *2171*
Zarzycki, J. 2319, 2350(10d), *2355*
Zaslonko, I.S. 171, 172(124), *180*
Zasyadko, O.A. 1471(224), *1531*
Zavgorodny, V.S. 304(109), *353*
Zavgovodny, V.S. 304, 306(107), *353*
Zavistoski, J.G. 1971, 2009(33), *2031*
Zav'yalov, V.I. 1144(6), *1181*
Zbiral, E. 1550(44), *1578*
Zea Bermudez, V.de 2339(126), *2358*
Zechel, D.L. 2153(125, 134), 2154(134), 2162(125), *2173*
Zechmann, A. 2083(146), *2125*, 2193(73), 2200(82), 2210(73, 104), *2215, 2216*
Zegarski, B.R. 1127(210), *1140*
Zehnder, M. 1553, 1555, 1556(51), *1578*, 1703, 1722, 1723(66), *1787*
Zeidler, G. 2226(52), *2242*
Zeigler, J.M. 1312, 1317(6), 1318(46, 60–63, 66), *1332–1334*

Zeigler, R. 311, 312(135), *353*
Zeitseva, G.S. 226(202), *262*
Zelbst, E.A. 237(244, 246, 247), *263*, 1461, 1467(142), *1529*
Zelcans, G. 1373(75d), *1439*
Zelchan, G. 1474, 1475(247), 1506(379), 1508, 1509(247), 1521, 1523(418), *1532, 1536, 1537*
Zelchan, G.I. 1449(4a–c), 1466, 1467(176–178), 1473(235, 238), 1474(235, 242), 1475(242, 250, 255), 1476(238, 261), 1479(242, 275), 1480(280), 1481(280, 284, 291), 1483(291), 1486(303, 304), 1487(307), 1506(250, 261), 1507(382), 1510(250, 261), 1514(291, 408a), 1519, 1521, 1522(408a), 1523(291), *1525, 1530, 1532–1534, 1536, 1537*
Zelchans, G. 1452, 1453(52), 1458(52, 124), 1467(124), 1471(220), 1474(245), 1475(124), 1479(124, 245), 1480, 1481(281), 1492(326), 1501(366), 1505(326), 1506(326, 380), 1509(124), 1510(396), 1514, 1522(408b), *1526, 1528, 1531–1533, 1535–1537*
Zelchans, G.I. 1453(64–66), 1459, 1467(128, 129, 133), 1486, 1488(129), 1505, 1506(378), 1508(390), 1521(417), 1524(421), *1526, 1529, 1536, 1537*
Zeldin, M. 2228(77), *2242*, 2246(4a), *2310*
Zeller, K.-P. 751(100), *777*
Zellers, E.T. 413(177), *429*
Zellner, K. 217(164), *260*
Zemlyanov, A.P. 1470(213, 214), *1531*
Zemlyansky, N.N. 2155(141), *2174*
Zeng, Q. 1776(324), 1779(324, 335), *1792*
Zengerley, T. 564(88), 581(178), 583, 586(88), *591, 593*
Zengerly, T. 520, 524, 525(69), *553*
Zerda, T.W. 290(75), *352*
Zerner, M.C. 90(170e, 170f), *101*
Zhakharov, I.I. 2512(250), *2564*
Zhang, C. 566(100), *591*, 1627, 1656(135), *1663*
Zhang, C.H. 453(131, 132), 473(201, 202), *491, 492*, 1861(360), *1868*
Zhang, D. 1470(211), *1531*
Zhang, H. 222(192), *261*, 1638(170), *1664*, 2224(34), *2241*, 2289(133), *2315*
Zhang, J. 1508(386, 387), *1536*
Zhang, Q. 69(124b, 124c), *99*, 1850(302), *1867*
Zhang, Q.L. 70(134a), *100*
Zhang, R.T. 916, 993, 996(183), *1058*
Zhang, S. 280(39–42), *351*, 382(83), *427*, 514, 520(2a, 2b), 521(2a, 2b, 75), 522(2a, 75), 529(2a), 532(2a, 2b, 101), 533(2a, 2b), 547(2b, 101), *551, 553, 554*, 559(18),

562(67), 563, 564(18), 565(18, 67), 567(18, 67, 102), 568–570(67, 102), 581, 583(67, 182), 584(67, 182, 187), 585(67, 187), *590–593*, 652(125, 126), *700*, 880, 881, 922, 924, 926, 940, 992(88), *1055*, 1108(18), *1136*, 1241(26), 1266(93), 1274(110), *1306–1308*, 1331(190, 191), *1336*, 1408(180, 181), *1443*, 2421(55, 56), *2458*, 2471(39), 2488(132), 2490(139), 2498, 2513, 2527(184), *2559*, *2561–2563*
Zhang, W. 369(46), *426*, 605, 606, 629(21), *698*, 1570(89), *1579*
Zhang, X. 770, 772(149), *778*, 1860(350), *1868*
Zhang, X.-H. 2471(43), *2559*
Zhang, X.-M. 382(83), 396(130), *427, 428*
Zhang, Y. 312, 314, 316, 317(137), *353*, 1803(72), *1863*
Zhang, Z. 2083(154), *2126*
Zhang, Z.-F. 2247–2249(9), 2250(9, 19), 2269(9), 2276(103, 104a–c), 2278(104a–c, 106), 2290, 2291, 2293(138b), *2310, 2314, 2315*
Zhang, Z.G. 108(7), *117*
Zhang, Z.J. 1840(250), *1866*
Zhang, Z.-R. 1208(54), *1230*
Zhao, B. 416(184), *429*
Zhao, C.F. 2352(262), *2362*
Zhao, M. 20, 22–28, 84, 87, 88(52), *96*
Zhao, Y. 283(47, 48), *351*, 514(4, 5), 532(5), 536(4, 5), 542(4), 543(5), *551*, 570(123, 124), 572(124), 581(123), *592*, 662(137), *700*
Zharov, I. 2499(186), *2563*
Zhau, X.-X. 2141, 2144, 2145(54), *2172*
Zhdanov, A.A. 2235(109), *2243*
Zheng, G.Z. 1736, 1737(155), *1788*
Zheng, M. 1550(46), *1578*
Zheng, Z. 1261(78), *1307*
Zheng-Rong Lu 1455, 1464, 1467(110, 111), 1481, 1483(110), 1486(110, 111), *1528*
Zhidkova, T.I. 1875(36), *1892*
Zhidomirov, G.M. 2512(250), *2564*
Zhong, H.-Z. 2236(113), *2243*
Zhorov, E.Y. 1743(222), *1790*
Zhou, C. 559(19), *590*
Zhou, J.P. 642, 645(109), *700*
Zhou, J.-Q. 1776(327), *1792*
Zhou, L.-L. 1759(260, 261), *1791*
Zhou, Q. 1981, 1983, 1998(68), *2032*
Zhou, X.J. 531(97), *554*
Zhou, Z. 506(42), *511*
Zhu, J. 1782(340), *1792*
Zhu, J.-C. 1467, 1474, 1475(183), 1477(265), *1530, 1533*
Zhu, L. 2352(271), *2362*
Zhu, P. 2486(120), *2561*
Zhu, X.-H. 1697(38, 40), 1708(38), *1786*, 2047(25), *2123*
Zhun, V.I. 240(258), *263*
Zhuo, R. 1501(367), *1536*
Zhuo, R.-X. 1464, 1467, 1486(168), *1530*
Ziabicki, A. 2247(11), *2310*
Ziche, W. 913(178, 179), 945(206, 211), 946(211, 212, 222, 224), 948(233, 234), 949(178, 179, 233), 997(285a), 1016(312), *1057–1061*
Zicke, W. 3(2a), *95*
Zidermane, A. 1501(366), *1536*
Zieder, J. 2340(136), *2359*
Ziegler, G. 2271, 2294(91), *2314*
Ziegler, J.M. 2160(165), *2174*
Ziegler, S.S. 1006, 1008(296b), *1060*
Ziegler, T. 28(64), *97*, 184, 247(26), *257*, 515, 516(15), *551*, 829(22), *854*, 1110(59), *1137*, 2116(247), *2128*
Ziembinski, R. 2154(137), *2174*
Ziemelis, M.J. 2227(71), *2242*
Ziemer, B. 245(278), *264*
Zigler, S.S. 1018(317), 1029(343), 1045(317, 343), *1061*, 1158(90, 91), 1160(99), *1183*, 2001(131a), *2034*
Zika, A. 2301(158), *2316*
Zilch, H. 2363(5, 7), 2364(18, 21, 22), 2378(50, 53), 2379, 2380(7), 2383(50, 60), 2385(7), 2387(53, 65, 67), 2388(53), *2398–2400*
Ziller, J.W. 272(17, 19), *350*, 2134(26), *2171*
Ziller, Z.W. 1923(94), *1928*
Zilm, K. 2539(373a), *2567*
Zilm, K.W. 52, 54(95e), *98*, 147, 148(58a, 58b), *152*, 345, 347(154), 348(154, 155), 349(154), *354*, 985(270), *1059*
Zimmermann, R. 1583(16), 1589(74), *1595, 1596*
Zimonyi-Hegedüs, E. 2364(13), *2398*
Zink, J.I. 2318(5), 2326(45b, 45d), 2347(190, 194), 2349(208), 2350(190, 194, 220), 2351(194, 240), *2355, 2356, 2360, 2361*
Zinn, A. 2073, 2091(100), *2124*
Zipin, H.S. 229(221), *262*, 2064(71), *2124*
Zitsmane, I.A. 1479(275), *1533*
Zlota, A.A. 241(266), *263*
Zolnai, L. 193(60), *258*
Zombeck, A. 2227(66), *2242*
Zong, M.-H. 2394(79), 2396(79, 81), *2400*
Zsolnai, L. 2107(215), *2127*
Zsombok, G. 1470(206), *1531*
Zsombok, Gy. 1460, 1467(136), *1529*
Zubieta, J. 1873(23), *1892*

Zuckerman, J.J. 1582(4), 1592(102), *1594, 1597*, 1971(33), 2005(151–153), 2009(33), *2031, 2034*
Zuev, P.S. 52(94a), *98*, 2512(253), *2564*
Zuhlke, J. 2327(58), *2357*
Zulauf, P. 811(97b), *824*
Zumbuluyadis, N. 2343, 2344(154), *2359*
Zummack, W. 601, 607(14), *698*, 1131(240), *1141*
Zundel, T. 2218(11d), *2240*
Zurmühlen, F. 1963(2b), *2030*
Zusman, R. 2321(22), 2322(22, 25), 2326(43, 50), 2348(25), *2355, 2356*
Zwanenburg, B. 473(207), *492*, 1610, 1630(79), *1662*
Zwecker, J. 1412(194), 1418(213, 217), 1420, 1421(213), 1422(194), 1423(194, 213, 217), 1424(217), 1427(194, 217), 1428(217), *1443, 1444*
Zweifel, G. 1601, 1602, 1604(8), 1613(8, 90), 1626, 1642, 1655(133), *1660, 1662, 1663*
Zybill, C. 70(136), *100*, 1386(112, 113), 1387(113), 1389(112, 113), *1440*, 2057(48), 2059(51–53), 2060(52–55), 2066, 2067(51–53, 55), 2084(55, 156), 2085(160), 2089(55), 2090(156, 160), 2094(55, 160), *2123, 2126*, 2527(330–332), 2528(336), *2566*
Zybill, C.E. 2057(49), *2123*
Zybill, L. 581(181), *593*
Zyss, J. 2351(250, 257, 258), 2354(250, 257, 258, 281, 285), *2362*

Index compiled by K. Raven

Subject index

Ab initio calculations,
  for acylsilanes 1605
  for disilyl-substituted model cation 687
  for fluorosilicates 1350
  for geometrical distortions 616
  for hydride transfer reactions 604
  for isodesmic reactions 597, 603
  for oligosilanes 1314, 1316
  for silaformamide ion 1028
  for silaisocyanides 1021
  for silanimines 1019
  for silanorbornyl cations 696, 697
  for silatranes 1467–1469
  for silenes 982, 997
    complexes of 990
    cycloadditions of 951, 959
    dimerization of 919
  for silicenium ions 515, 516
  for δ-silicon effect in norbornyl system 644
  for silole dianions 989
  for silylarenium ions 652–654
  for α-silyl cations 597, 598, 647
  for β-silyl cations 598–601, 607, 622, 656, 658, 659
  for silylcyclobutonium ions 692, 693, 695
  for silylenes 997, 1169, 1331, 2510–2512
  for ^{29}Si NMR parameters 293–296
  for solvolysis 613
  for spirosilicates 1364
  for tricarbasilatranes 1511
  for tris(trimethylsilyl)cyclopropenium cation 648, 649
  potential energy diagrams for 602, 608, 609
Absorption spectroscopy, two-photon 1316, 1318
Acetals, hydrolysis of 1606–1608
Acetamidosilanes, enzymatic hydrolysis of 2387, 2388
(Acetoxyethyl)silanes, enzymatic hydrolysis of 2384, 2385
2-Acetoxy-1-silacyclohexanes, enzymatic hydrolysis of 2385–2387
Acetylferrocene, structure of 1605
Activation parameters, for dimerization of silenes 926
Acyl anions 1642, 1643
Acylation, steric effects of silyl groups on 438
Acyldigermanes, microbial reduction of 2380, 2382
Acyldisilanes,
  microbial reduction of 2382
  photolysis of 1272, 1273
Acyl(disilanyl)carbenes 719
Acylgermanes—*see also* Acyldigermanes, Acyl(silyl)germanes
  IR spectra of 1601
  microbial reduction of 2379–2382
  PE spectra of 1605
Acyl halides, reduction of, using (TMS)$_3$SiH 1545
Acylmetallic species 1606
  silylation of 1618, 1619
1-Acyloxysilatranes, synthesis of 1487, 1489
Acylpolysilanes,
  as silene precursors 962, 966, 967
  formation of 889
  isomerization of 880–883
  photolysis of 913, 1273–1276
  α,β-unsaturated—*see also* α,β-Unsaturated acylpolysilanes
Acyl radicals 1570, 1571
Acylsilanes—*see also* Acyldisilanes, Acylpolysilanes, Acyltris(trimethylsilyl)silanes, α-Cyclopropylacylsilanes, α-Epoxyacylsilanes, α-Haloacylsilanes, α-Ketoacylsilanes, Polysilylacylsilanes, Thioacylsilanes
  as acyl anion precursors 1642, 1643
  biotransformations of 1649
  carbonyl group in, n–π^* transitions of 1645

Acylsilanes (cont.)
  chiral—see Chiral acylsilanes
  cyclization of 1643–1645
  IR spectra of 1600, 1601
  NMR spectra of 1601–1604
  nucleophilic additions to 1632–1642
  oxidation of, anodic 1203, 1204, 1650
  PE spectra of 1605
  photolysis of 1270–1276
    mechanism of 1271
  reactions of 1649–1651
    photochemical 1645–1649
  reduction of,
    cathodic 1212, 1213
    microbial 2377–2383, 2385, 2387
  structure of 195, 196, 1605, 1606
  synthesis of 1198, 1199
    by coupling, Pd-catalysed 1610, 1611
    by hydroboration–oxidation 1613, 1614
    by hydrolysis of acetals 1606–1608
    by oxidation reactions 1611–1613
    from acylmetallic species 1618, 1619
    from enol ethers 1614–1617
    from silylmetallic species 1608–1610
    from silyloxycarbenes 1614
  α,β-unsaturated—see α,β-Unsaturated acylsilanes
  UV/visible spectra of 1603–1605
Acyl(silyl)carbenes,
  formation of 769
  rearrangement of 712, 718–722, 751
  spin state of 751
Acyl(silyl)germanes, microbial reduction of 2382
Acyltris(trimethylsilyl)silanes, synthesis of 1609
Adamantanes—see Bromoadamantanes
Adamantyl derivatives, solvolysis of 482
  δ-silicon effect in 642, 643
Addition reactions,
  reactivity of Si-containing reagents in 475–477
  steric effects of silyl groups on,
    comparison of 438–441
    comparison with H atom and R groups 459, 460
Additivity rules 155, 156
Agostic interactions 2050, 2051, 2079
Albite, NMR spectra of 326
Alcohols,
  as substrates in Direct Process 1590, 1591
  deoxygenation of, using $(TMS)_3SiH$ 1547, 1548
Aldehydes, reactions with $(TMS)_3Si$ radical 1574
Aldimines—see Metalloaldimines
Aldol–Peterson reaction 1626, 1627

Aldol reactions 1641, 1649, 1656, 1810
  steric effects of silyl groups on,
    comparison of 441, 442
    comparison with H atom and R groups 460
Alkaloids, synthesis of 1561–1563, 1569
Alkanals, silylformylation of 1774–1777
Alkenes—see also Arsaalkenes, Bicyclo[n.1.0]-1(n+3)-enes, Dienes, Germaalkenes, Phosphaalkenes, Silaalkenes
  hydrosilylation of 1552–1555, 1688–1716
    by Group IV metallocenes and lanthanides 1697–1701, 1708–1710
    by Group VIII transition metal catalysts 1688–1692
    by immobilized catalysts 1701–1703
    by metallic and colloidal metal catalysts 1693–1696
    by radical initiators 1703, 1704
    intramolecular 1710–1714
    mechanism of 1704–1710
    other reactions associated with 1714–1716
  prochiral—see Prochiral alkenes
  protonation of, β-silicon effect in 628, 629
  silylation of, electrochemical 1227
β-Alkenyloxyenones, reactions with $(TMS)_3Si$ radical 1573
Alkenylsilanes—see also Cyclopropenylsilanes, Hexenyl(trimethyl)silanes
  structure of 189–191
1-Alkenylsilatranes,
  reactions of 1494–1499
  synthesis of 1450, 1456, 1457
(Alkenylsilyl)carbenes, reactions of 729–732
Alkenyl(silyl)carbenes—see also 1-Cycloalkenyl(silyl)carbenes, (Trimethylsilyl)vinylcarbene
  reactions of 746, 747
  rearrangement of 744–746
Alkoxycarbonyl(disilanyl)carbenes,
  rearrangement of 718, 721
Alkoxycarbonyl(silyl)carbenes,
  reactions of 754–760
  rearrangement of 718
  synthesis of 752, 753
Alkoxydisilenes,
  formation of 831, 833
  reactions with alcohols 836–838
Alkoxysilanes—see also Dialkoxysilanes, Trialkoxysilanes
  reactions of 1084
  structure of 221, 223
  substituted by polymerizable organic groups 2333–2342
    cocondensation of 2336, 2337
    copolymerization of 2337–2339

Subject index

1-Alkoxysilatranes,
  reactions of 1484, 1491, 1492
  synthesis of 1453
Alkoxysilylenes,
  matrix isolation of 1179, 1180
  reactions of 2467, 2495
Alkylation reactions, steric effects of silyl
    groups on,
  comparison of 442, 443
  comparison with H atom and R groups 460
(Alkyldimethylsilyl)carbenes, reactions of 728
Alkylidenesiliranes 2420–2423—*see also*
    Bisalkylidenesiliranes
Alkyl radicals 1561
Alkyl selenides, reduction of using $(TMS)_3SiH$
    1545, 1546
Alkylsilanes—*see also* Halomethylsilanes,
    Hexaalkyldisilanes
  anodic oxidation of 1188, 1189
  cyclic—*see* Cyclic alkylsilanes
  structure of 192–194
  thermochemistry of 159–161
1-Alkylsilatranes—*see also* 1-
    (Haloalkyl)silatranes, 1-
    (Mercaptoalkyl)silatranes
  reactions of 1492–1494
  synthesis of 1450, 1456
Alkyl(silyl)carbenes, reactions of,
    intramolecular 743, 744
Alkyl silyl ethers,
  synthetic applications of 1674, 1675
  thermochemistry of 164, 165
Alkynes—*see also* Bis(trimethylsilyl)ethynes,
    Silynes
  hydrosilylation of 1555, 1556, 1716–1733
    by platinum catalysts 1717
    by radical initiators 1722, 1723
    by rhodium catalysts 1718–1720
    intramolecular 1725
    mechanism of 1723, 1724
    other reactions associated with 1732,
      1733
  protonation of, β-silicon effect in 628–630
  reactions with $(TMS)_3Si$ radical 1575
  silylation of, electrochemical 1227
  silylformylation of 1774, 1775
Alkynols, hydrosilylation of 1727–1730
Alkynyldisilanes, photolysis of 1249–1251
Alkynylsilanes—*see also* Alkynyldisilanes,
    Ethynylsilanes
  hydroboration–oxidation of 1613, 1614
  matrix isolation of 1165, 1166, 1175
  reactions of, effect of R_3Si on 397, 398
  structure of 185–189
  thermochemistry of 161
1-Alkynylsilatranes, synthesis of 1450, 1491
(Alkynylsilyl)carbenes, reactions of 732, 733

Alkynylsilylenes—*see also* Butadiynyl-
    silylenes, Dialkynylsilylenes, Ethynyl-
    silylene
  matrix isolation of 1163, 1164, 1171–1174
  photolysis of 1174
Allenes—*see* Cyclic allenes, Silaallenes, Silyl-
    oxyallenes
Allenyl ethers, as acylsilane precursors 1624,
    1625
Allenylsilanes, reactions of 1656
  effect of R_3Si on 412, 413
Allylation—*see also* Photoallylation
  radical 1558, 1572
  steric effects of silyl groups on 443
Allylboration, steric effects of silyl groups on
  461
Allyl chloride, as substrate in Direct Process
  1589, 1590
(Allyldimethylsilyl)carbene, reactions of 730
Allyloxysilanes, thermolysis of 1071–1073
Allylsilanes—*see also* Allyloxysilanes, Allyl-
    trimesitylsilane
  addition of electrophiles to 1832, 1834
  cyanothianation of 1830
  cycloadditions of 1847–1853
  halodesilylation of 1796, 1797
  isomerization of 1236
  oxidation of 1840, 1841, 1843–1846
    anodic 1190, 1192–1196
  protodesilylation of 1795, 1796
  protonation of 542
  reactions of,
    coupling 1290–1292, 1826–1828, 1836
    effect of R_3Si on 398–411
    metathesis 1840
    photochemical 1236, 1290–1292
    radical 1834
    β-silicon effect in 625
    with acetals 1798–1801
    with acyl halides 1818, 1819
    with alkenes 1822–1824
    with alkyl halides 1820, 1821
    with carbonyls 1801–1811
    with enones 1811–1814
    with iminium ions 1814–1818
    with tropylium ion 1820
  rearrangement of 1856, 1857
  sulphonation of 1797
  transmetallation of 1828, 1829
1-Allylsilatranes, reactions of 1494, 1496,
    1497
Allylsilylation, steric effects of silyl groups on
  461
Allyl silyl ethers,
  as acylsilane precursors 1622, 1623
  cycloadditions of 1298
  ring closure of 1296

Allyltrimesitylsilane, protonation of 542
Aluminosilicates, precursors to 2300–2307
Aminoalkylsiloxanes 2339
Aminocarbonyl(silyl)carbenes, reactions of 760, 761
Aminodisilenes,
  formation of 832, 833
  reactions with alcohols 836, 838
1-(3′-Aminopropyl)silatranes, reactions of 1501
Aminosilanes—*see also* Diaminosilanes, Silylamines, Tri(dialkylamino)silanes
  anodic oxidation of 1200
  thermochemistry of 166
Aminosilylenes—*see also* Diaminosilylenes
  cyclic—*see* Cyclic aminosilylenes
  matrix isolation of 1178
  photoformation of 1158
  photoisomerization of 1159
  reactions of 2467, 2495
Amphibole 103
Anchimeric effect, in solvolysis 619
1,6-Anhydro-D-glucose, dideoxygenation of, using (TMS)$_3$SiH 1547, 1548
Anions, silicon-containing 1026–1028—*see also* Sila anions, Silenolate anions, Silirenyl anion, Silole anions, Siloyl anions, Silyl anions
  rearrangement of 1133, 1134
  theoretical studies of 1050
  thermochemistry of 1109
α-Anisyl-β-silylvinyl cations, NMR spectra of 679–685
Annelation reactions, of acylsilanes 1656
Anomeric effects, in silatranes 1469
Antenna chromophores, for carbonyl photoreduction 1298
Antiaromaticity 4
  of silacyclobutadienes 83–88
  of silacyclopentadienyl cations 89–93
  of silirenyl anion 88, 89
Apeloig–Ishikawa–Oehme-type silenes,
  cycloadditions of 957, 958
  dimerization of 884, 917, 919, 928, 1002
  synthesis of 884–890
Apeloig-type silenes,
  NMR spectra of 985
  UV spectra of 996
1-Aralkylsilatranes,
  reactions of 1492–1494
  synthesis of 1450
Arbuzov reaction 1628, 1653
Arenium complexes, σ-bonded 585
Aromaticity 4
  computational criteria for 5–10
    energetic criterion 6
    geometric criterion 5, 6

interrelations between 8–10
    magnetic criterion 6, 7
  of silaaromatics, charged 32–51
  of silabenzenoids 10–32
  of silafullerenes 69, 70
  of silylenes 52–69, 2537–2539
  three-dimensional 77–83
Arrhenius parameters,
  for decomposition of silacyclobutanes 861, 862
  for retro-ene fragmentation of diallylsilanes 872
Arsaalkenes, synthesis of 1081
Arsanilidenesilanes, cycloadditions of 1080–1082
Arsasilenes 1053, 1100
1-Aryl-2-azasilatranes 1523
Arylbis(trimethylsilyl)silanes, photolysis of 2470, 2471
Aryldisilanes—*see also* Dinaphthyldisilanes, Haloaryldisilanes
  CT studies of 1322–1325, 1328
  dual fluorescence phenomena of 1322–1325
  emission studies of, time-resolved 1325
  PE spectra of 1322
  photolysis of 891, 1251–1259, 1303
    mechanism of 1325–1328
    solvent effects on 1327
  photophysics of 1321–1325
1-Aryloxysilatranes,
  hydrolysis of 1484
  synthesis of 1453
Arylsiladienes, photoformation of 1252
Arylsilanes—*see also* Arylbis(trimethylsilyl)silanes, Aryldisilanes, Benzylsilanes, Poly(phenylsilane)s, Triphenylsilanes
  dπ–pπ effect in 1211
  oxidation of, anodic 1189, 1190
  polymerization of 574
  reduction of, cathodic 1211, 1212
  structure of 191, 192
1-Arylsilatranes,
  reactions of 1492–1494
  synthesis of 1450
Aryl(silyl)carbenes—*see also* Phenyl(trimethylsilyl)carbene
  carbene/carbene rearrangement of 749
  formation of silenes from 750
  reactions of, intermolecular 750
  spin state of 749
  synthesis of 748, 749
Arylthiosilanes, reactions of 578, 580
Aryltrisilanes—*see also* 2,2-Diaryltrisilanes
  photolysis of 1328, 1329
Asymmetry parameters 319
Auner–Jones-type silenes, dimerization of 917

1-Aza-2-silacyclobutanes, cycloreversion of 912
Azasilacyclopropenylidenes, matrix isolation of 1179
Azasilatranes—*see* 1-Aryl-2-azasilatranes, Triazasilatranes
Azasilatranones 1523
Azides,
　cyclization of, radical 1569
　reactions with (TMS)$_3$Si radical 1574, 1575
Azidosilanes—*see also* Diazidosilanes, Triazidosilanes
　irradiation of 1158, 1179
Azo dyes 2354

Backbiting reactions, of silicenium ions 580
Barton–McCombie reaction, deoxygenation of, using (TMS)$_3$SiH 1547
Bayer, Direct Process and 1584, 1586, 1592
Bent's rule 300, 304, 1330, 2078
Benzene, stability of 122
Benzocyclobutenols, synthesis of 1646, 1647
Benzodisilacyclobutanes,
　photolysis of 1258, 1259
　synthesis of 2435, 2436
　structure of 2436
Benzodisilacyclobutenes, reactions with C$_{60}$ 1944, 1945
Benzometalloles—*see also* Dibenzometalloles
　synthesis of 1991–1995
Benzoselenophenes, synthesis of 1571
Benzosilacyclobutenes—*see also* Benzodisilacyclobutenes
　formation of 1003, 1026
　photolysis of 867, 869, 870
　polymerization of 866
Benzosilatranes—*see also* Tribenzosilatranes
　synthesis of 1511–1513
Benzoylsilanes,
　IR spectra of 1600, 1601
　NMR spectra of 1602–1604
　nucleophilic additions to 1632, 1640
　photolysis of 1272
　reduction of, microbial 2383, 2384
　synthesis of 1608
　UV/visible spectra of 1604
Benzvalene, stability of 122
Benzylic *p*-toluenesulphonates, solvolysis of 482
Benzylsilanes,
　oxidation of, anodic 1190, 1191
　photocoupling of 1292
　photoisomerization of 1236
Betweenallenes 2448–2451
Bicyclo[*n*.1.0]alk-1(*n*+3)-enes 2450, 2451
Bicyclobutonium ions 688–690

α-silicon effect in 690–693
γ-silicon effect in 693–695
Bicyclohexasilanes,
　isomerism in 123
　oxidation of 2208
Bicyclo[2.2.2]octyl derivatives, solvolysis of, δ-silicon effect in 641, 642
Bicyclosilachalcogenides 1910–1916
Bicyclo[1.1.0]tetrasilane, isomerism in 123
Bikitaite 116
Biocatalysis 2376–2397
　of esterifications 2390–2394
　of hydrolyses 2384–2388
　of oxidations 2394–2397
　of reductions 2376–2384
　of transesterifications 2388–2390
Bioisosterism, Si/Ge 2374
Biphenyls, silyl-bridged, photolysis of 1263
Biradicals—*see also* Diradicals
　formation of,
　　in cleavage of disilacyclobutenes 868, 870
　　in photolysis of silabicyclooctanes 875, 876
Bis(alkylidene)disilacyclobutanes 2429
　reactions with C$_{60}$ 1945–1947
Bisalkylidenesiletanes 2429–2436
Bisalkylidenesiliranes 2423, 2424
Bis(allyl)disiloxanes, photolysis of 1298
Bis(catecholate)s, pentacoordinate, NMR spectra of 285–287
Bis-catecholato complexes 1417, 1418
Bis(cyclohexasilanyl)benzenes, structure of 2201, 2202
Bis(diazomethyl)silanes, carbene reactions of 732–739
Bis(hydroxymethyl)germanes, enzymatic transesterification of 2389, 2390
Bis(hydroxymethyl)silanes, enzymatic transesterification of 2388–2390
Bis(pentamethyldisilanyl)butadiyne, photlysis of 1250
Bis(pentamethyldisilyl)naphthalene, photolysis of 1253
1,3-Bis(silanols), synthesis of 578
Bis-silenes,
　head-to-tail dimers of 906, 907
　photoformation of 1280
Bis(siliconate)s,
　NMR spectra of 339, 341–343
　X-ray studies of 339–341
Bis(silyl) complexes,
　with group 8 transition metals 2079–2082
　with group 9 transition metals 2095–2107
　with group 10 transition metals 2110–2117
Bis(silyldiazomethyl)oligosilanes, decomposition of 2408–2412

1-Bis(silyl)methyl-2-bis(silyl)ethenyl cations,
    NMR spectra of 685–687
Bis(silyl)selenides 1885–1887
1,4-Bis(trimethylsilyl)butadiyne, hydro-
    silylation of 1726, 1727
Bis(trimethylsilyl)ethyne, hydrosilylation of
    1725, 1726
Bis(trimethylsilyl) peroxide, synthesis of 1682
Bis(trimethylsilyl)thioketenes, pyrolysis of
    1095, 1096
α,α'-Bis(trimethylsilyl)xylenes, anodic
    oxidation of 1190
Bond angles, effect of silyl groups on 484,
    485
Bond cleavage reactions, O—O, reactivity of
    Si-containing reagents in 481
Bond dissociation energies,
    Si—H 166–169
    Si—Y 169–171
Bond energies, π 1066, 1067
Bonding model, 3c4e 1465, 1466, 1506, 1523
Bond lengths, effect of silyl groups on 484,
    485
Bonds, π, shortening of 1066, 1093
Bond strengths, π, of Si=X 1067
Bond-stretch isomerism 123, 125–127
Brassinosteroids, synthesis of 1652
Bridged ion structures 600, 601, 604, 605, 607,
    618
    partial 610, 624
Bromoadamantanes, solvolysis of, γ-silicon
    effect in 638, 639
Brønsted acids, as catalysts 1799, 1815
Brookhart–Grant mechanism, for alkene
    hydrosilylation 1707, 1708
Brook rearrangement 1632–1636, 1639–1642,
    1651–1654, 1656, 1658
Brook-type silenes,
    cycloadditions of 949–956
    with carbonyls 959–968
    with imines 968, 969
    dimerization of 917, 919, 1002
    IR spectra of 995
    NMR spectra of 985
    reactions of 890
        with alcohols 932
        with alkoxysilanes 937
    rearrangement of, photochemical 913, 914
    structure of 983
    synthesis of 880, 881
    UV spectra of 996
Brosylates, solvolysis of,
    γ-silicon effect in 638
    δ-silicon effect in 642, 643
Buckminsterfullerene, hydrosilylation of 1771
Butadienes—see Polybutadienes, Silabuta-
    dienes

Butadiynes—see Bis(pentamethyldisilanyl)
    butadiyne, 1,4-Bis(trimethyl-
    silyl)butadiyne, Disilabutadiynes
Butadiynylsilylene, matrix isolation of 1172,
    1173
Butterfly molecules 1158
Butyrolactones, reactions with alkyl halides
    1557, 1558
C_{60} adducts, NMR spectra of 277, 278
Cadmium, as promoter in Direct Process 1590
Cambridge Structural Database (CSD) 184
Cancrinite 116
Cannizaro reaction 1649
2-Carba-3-oxahomosilatranes,
    silyl analogues of 1524
    synthesis of 1500
Carbasilatranes—see also 2-Carba-3-oxahomo-
    silatranes, Tricarbasilatranes
    mass spectra of 1510
    NMR spectra of 1510
    synthesis of 1509
    X-ray studies of 1510
Carbenes—see also Dicarbenes, Disilanyl-
    carbenes, Phosphinocarbenes, Siloxy-
    carbenes, Silylcarbenes, α-Thioketo-
    carbenes
    matrix isolation of 709
    silicenium-substituted 707
    singlet 704–706
    stable 709, 710
    triplet 704–706
Carbenium ions, α-silyl-substituted,
    rearrangement of 1131, 1132
Carbocyclization, carbonylative 1783–1785
1,3-Carbon shifts 898
Carbonylation reactions, radical 1559
Carbonylbis(trimethylsilane),
    as acyl anion precursor 1643
    colour of 1605
    synthesis of 1608, 1612
Carbonyl compounds, hydrosilylation of 1733–
    1741
    intramolecular 1740, 1741
    mechanism of 1737
Carboxy(trimethylsilyl)carbene 707–709
Cations, silicon-containing 35, 1028, 1029,
    1134, 1135—see also Silabenzyl cation,
    Silacyclopentadienyl cations, Silacyclo-
    propenium cations, Silacyclopropylmethyl
    cations, Silanorbornadienyl cations,
    Silanorbornyl cations, Silylaryl cations,
    Silyl carbocations, Silylene cations
    formation of 1122–1130
    ion–molecule reactions of 1117–1122
    rearrangement of 1130–1133
    theoretical studies of 1050

thermochemistry of 1106–1108
C—C bonds, formation of, mediated by
  (TMS)$_3$SiH,
  intermolecular 1557–1561
  intramolecular 1561–1572
Ceramic precursors,
  containing Si and C 2272–2294
  containing Si and N 2251–2272
  containing Si and O 2294–2310
  criteria for 2247–2250
Chalcogenides—see Silachalcogenides
Chalcogenoacetals—see Homoallylchalcogenoacetals
Chalk–Harrod mechanism, for alkene hydrosilylation 1704, 1705, 1709, 1710
Charge-transfer studies, in aryldisilanes 1322–
  1325
Chemical shielding tensors, calculated
  296
CHFPT 295
Chiral acylsilanes,
  reactions with organometallics 1638
  rearrangement of 1633–1635
  synthesis of 1619
Chiral drugs, biological recognition of 2364–
  2375
Chiral (hydroxymethyl)silanes 2368–2375
Chiral silanols 2364–2368
Chlorobenzene, as substrate in Direct Process
  1589
3-Chloropropyltrimethoxysilane, synthesis of
  1692
Chlorosilanes—see also Chlorotrimethylsilane
  rearrangement of 2281
  structure of 239–243
Chlorotrimethylsilane, synthetic applications of
  1667–1670
Claisen rearrangement 1619, 1859
Clustering reactions, of silicon-containing ions
  560, 1118–1121
Coalescence phenomena,
  in fluorosilicates 1347
  in Si—N complexes 1382–1386
Cocondensation reactions 1175, 1176, 1179,
  1180
Coesite 109, 113–115
Coil-to-rod transitions 1318
Collisional activation mass spectrometry 1123–
  1125, 1127, 1128, 1130, 1131
Colloidal metal catalysts, for hydrosilylation of
  alkenes 1693–1695
Composite Carbon-silicate Electrodes (CCEs)
  2328, 2329
Condensation reactions, reactivity of Si-
  containing reagents in 477
Conductance measurements, of silicenium ions
  562–565

Conformation, effect of silyl groups on 484,
  485
Conjugate addition reactions 1637
Conjugation,
  σ,
    in oligosilanes 1316
    in polysilanes 1312
  σ–π, in aryldisilanes 1322
Connectivity experiments 315
Continuum model 528
Copper, use of, in Direct Process 1583, 1584,
  1586–1589, 1591–1593
Copper halides, use of, in Direct Process 1586,
  1587, 1590, 1591
Copper oxides, use of, in Direct Process 1585,
  1587
Copper–silicon alloys, use of, in Direct Process
  1586–1589, 1591
COSY 269
Coumarins, as lasing dyes in sol–gel matrices
  2351
Coupling reactions 1669
  electrochemical 1228
  of allylsilanes 1290–1292, 1826–1828,
    1836, 1837
  of halosilanes 1214–1218
  of vinylsilanes 1228, 1824–1826, 1836,
    1837
  Pd-catalysed, in synthesis of acylsilanes
    1610, 1611, 1628, 1656
  photochemical 1290–1294
Cox–Yates plots 630
Cristobalite 109, 112, 115, 116
Cross-polarization 309, 311–313, 315, 319–
  321, 343
Cryoscopic measurements, of silicenium ions
  565
C—Si bonds—see Si—C bonds
Cubane, strain energy of 120
Cubyl derivatives, solvolysis of, δ-silicon effect
  in 645
Cumyl cations, NMR spectra of 646
Cuprates,
  in synthesis of acylsilanes 1609, 1618,
    1627, 1628, 1630
  reactions with acylsilanes 1656
Cyanines, as lasing dyes in sol–gel matrices
  2351
Cyclic alkylsilanes, electrosynthesis of
  1225
Cyclic allenes 2446–2452
Cyclic aminosilylenes 64–66
Cyclic disilanes, photolysis of 1262–1265,
  1303
Cyclic disilenes, photolysis of 1304
Cyclic disilthianes 1883
Cyclic selenides 1887, 1888

Cyclic silanes—*see also* Cyclic alkylsilanes, Cyclic disilanes, Cyclopolysilanes
  nucleophilic cleavage of 804, 805
  reductive cleavage of 801–804
Cyclic silaselenanes—*see also* Cyclodisiladiselenanes, Cyclotrisilatriselenanes
  reactions of 1907, 1908
  synthesis of 1903, 1904
  thermolysis/photolysis of 1098
Cyclic silathianes,
  reactions of 1907, 1908
  structure of 1908–1910
  synthesis of 1903, 1904
Cyclic silenes—*see also* Cyclic disilenes
  photoformation of 1242, 1248
Cyclic silthianes 1879–1881—*see also* Cyclic disilthianes
Cyclic silylenes,
  complexes of 2057
  matrix isolation of 2474
  rearrangement of 901–903
Cyclization–fragmentation–transannulation–ring expansion, radical 1575
Cyclization reactions—*see also* Macrocyclization reactions, Photocyclization reactions, Silylcarbocyclization reactions
  radical 1561–1573
  stereoselective 1567
  steric effects of silyl groups on,
    comparison of 443–445
    comparison with H atom and R groups 462–465, 473, 474
  tandem 1563
Cycloaddition reactions,
  of metalloles 1998–2003
  reactivity of Si-containing reagents in 477, 478
  steric effects of silyl groups on,
    comparison of 445, 446
    comparison with H atom and R groups 465–467
[2+1]Cycloaddition reactions, of Brook-type silenes 949–952
[2+2]Cycloaddition reactions,
  of Apeloig–Ishikawa–Oehme-type silenes 957
  of Brook-type silenes 953–955
    with carbonyls 959, 964
    with imines 968
  of miscellaneous silenes 970–972
  of neopentylsilenes 945, 946, 948
  of silaacrylates 969, 970
  of Wiberg-type silenes 941, 943
[2+2+2]Cycloaddition reactions, of neopentylsilenes 945, 946, 948, 949
[2+3]Cycloaddition reactions,
  of miscellaneous silenes 973
  of Wiberg-type silenes 942, 943
[2+4]Cycloaddition reactions,
  of Apeloig–Ishikawa–Oehme-type silenes 957
  of Brook-type silenes 954
    with carbonyls 960–962, 964, 965
    with imines 968
  of miscellaneous silenes 973, 974
  of neopentylsilenes 945, 946
  of Wiberg-type silenes 941, 943
[6+2]Cycloaddition reactions, of neopentylsilenes 945, 946
Cycloalkenyl(silyl)carbenes 773
  reactions of 747, 748
Cycloalkyl(silyl)carbenes, reactions of, intramolecular 743, 744
Cyclobutanes—*see also* Silacyclobutanes
  strain energy of 121
Cyclobutanones, synthesis of 1641
Cyclobutenols—*see* Benzocyclobutenols
Cyclodisiladiselenanes, formation of 1098
Cyclodisilagermanes, photolysis of 1264
Cyclodisilazanes, synthesis of 1016, 1018, 1033
Cyclodisiloxanes 1897–1899
  NMR spectra of 278
Cyclohexasilanes—*see also* Bicyclohexasilanes, Dodecamethylcyclohexasilanes, Tricyclohexasilanes
  conformation of 2180
  photolysis of 1261, 1262, 1267
  reactions of 2189–2196, 2200, 2201
  synthesis of 2178, 2180
  with silanyl side-chains 2193–2196
Cyclohexasilanylpotassiums 2192, 2193
Cyclopentadienes—*see also* Heterocyclopentadienes, Silacyclopentadienes
  multisilylation of 2146–2149
Cyclopentadienyl complexes—*see also* Silacyclopentadienyl complexes
  multisilylation of 2146, 2147, 2149
  silicon-bridged 2149–2163
Cyclopentadienyl radicals, orbital degeneracy in 2137, 2138
Cyclopentadienylsilanes—*see also* Silylcyclopentadienes
  dynamic behaviour of 2131
  GED studies of 2131
  rearrangement of 2132, 2133
  X-ray studies of 2131
Cyclopentadienyl–silicon π-bonds 2163–2170
Cyclopentadienyl–silicon σ-bonds,
  basic features of 2130–2138
  cleavage of 2143–2146
  synthesis of 2138–2143
Cyclopentasilanes,
  cleavage of, reductive 802

NMR spectra of 279
  photoformation of 1261, 1262
  reactions of 2188–2190
Cyclopentenes, ring expansion of 1656
Cyclopentenols—*see* Metallacyclopentenols
Cyclopolysilanes—*see also* Cyclohexasilanes, Cyclopentasilanes, Cyclotetrasilanes, Cyclotrisilanes, Perphenylcyclopolysilanes, Tetracyclooctasilanes
  oxidation of, anodic 1207, 1208
  photolysis of 1267–1270, 1304, 2470–2472
  properties of, general 2180–2182
  reduction of, cathodic 1213
  structure of 198–202
  synthesis of 2178–2180
  transition-metal containing 2209–2213
  UV spectra of 2181
Cyclopropanation,
  of alkenes 741, 742
  steric effects of silyl groups on 467–469
Cyclopropane, strain energy of 121
  substituent effects on 124
Cyclopropenium cations—*see* Silacyclopropenium cations, Tris(trimethylsilyl)cyclopropenium cation
Cyclopropenylsilanes—*see also* Silylcyclopropenes
  photolysis of 1153
Cyclopropenylsilylenes, photorearrangement of 1154
α-Cyclopropylacylsilanes, synthesis of 1630, 1631, 1656, 1657
Cycloreversion reactions, in synthesis of silanimines 1012–1018
Cyclosilachalcogenides 1896, 1897—*see also* Bicyclosilachalcogenides, Monocyclosilachalcogenides, Tricyclosilachalcogenides
Cyclosiloxanes 1897—*see also* Cyclodisiloxanes, Cyclotrisiloxanes, Polycyclosiloxanes
  formation of 1076, 1077
    pyrolytic 1071, 1072
  polymerization of,
    anionic 2218–2220
    cationic 2220, 2221
  structure of 223, 224
Cyclotetragermanes,
  reactions with C_{60} 1947–1951
  strain energy of 121
Cyclotetrasilanes—*see also* Bicyclo[1.1.0]tetrasilanes
  as disilene photolytic precursors 1157
  geometry of 2181
  photoformation of 1261, 1267
  photolysis of 1267, 1268, 2472
  polymerization of, ring-opening 2187

reactions of 2187, 2188
  with C_{60} 1947–1951
ring contraction of, photolytic 2182, 2183
strain energy of 121
structure of 123, 2188
synthesis of 2184, 2185
UV spectra of 2188
Cyclotetrastannanes, strain energy of 121
Cyclotrigermanes, strain energy of 121
Cyclotrisilanes,
  cleavage of, reductive 802
  NMR spectra of 279, 2183
  photolysis of 1077, 1283, 1304, 2184–2186, 2472, 2473
  reactions of 2183–2187
  stability of 2180
  strain energy of 121
    substituent effects on 124
  structure of 123, 2183
  synthesis of 2182, 2183
  thermolysis of 1075–1077, 2184, 2187
Cyclotrisilatriselenanes, formation of 1098
Cyclotrisiloxanes 1897–1899
Cyclotristannanes, strain energy of 121

Dative bonds 210, 211, 225–227
Decamethylsilicocene,
  reactions of 2166–2169
  structure and bonding in 2163–2166
  synthesis of 2163, 2164
Decomposition reactions, steric effects of silyl groups on 446, 447
Decoupling, high-power 309
Dehydration reactions, steric effects of silyl groups on 469
Dehydrocoupling, transition-metal catalysed 2255
Dehydrofukinone, synthesis of 1653, 1654
Dendrimers, silicon-containing, synthesis of 1758–1763
Density functional theory 515, 516
Deoxygenation, radical 1547–1552
Deprotection, selective, of silyl ethers 1674
Deprotonation reactions, steric effects of silyl groups on 447
DEPT 269, 302
Desilylation reactions 1795–1797, 1811, 1812
  fluorine-induced 1115–1117
  steric effects of silyl groups on 447, 448
Deuterium isotope effects,
  on protonation of alkynes 630
  on solvolysis 618, 619, 622, 623, 635
Dewar benzene, stability of 122
Dewar disilabenzene 19
Dewar hexasilabenzene 25
Dewar silabenzenes 904

Diadamantylsilylene,
  formation of 2475, 2479
  reactions of 2487
Dialkoxysilanes 2346
  synthesis of 1633
Dialkylsilenes, reactions of 1606
Dialkylsilylenes—see also Diadamantylsilylene
  formation of 2472, 2474, 2475, 2479, 2480
  IR spectra of 2523
  matrix isolation of 1167–1169, 1329
  reactions of 2472, 2485–2487, 2489, 2490, 2504
Dialkynylsilylenes, matrix isolation of 1172, 1173
Diaminosilanes, reactions of 1086
Diaminosilylenes, reactions of 834, 835
Dianions, silicon-containing 1135
Diarylsilacyclobutanes, photolysis of 1332
Diarylsilylenes—see also Dimesitylsilylene
  formation of 2470
  reactions of 2486, 2490, 2491, 2493
2,2-Diaryltrisilanes, reactions of,
  with C_{60} 1931–1934
  with C_{70} 1934–1936
Diazasiletanes, reactions of 1012, 1013
Diazides, photolysis of 1170, 1178
Diazidosilanes—see also Methyldiazidosilane
  cleavage of, photochemical 2474
Diazirines, photolysis of 1277, 1278, 1286
Diazo compounds, silanyl—see also Bis(diazomethyl)silanes, Diazosilanes
  decomposition of 2403–2406
Diazosilanes, matrix isolation of 1160
Dibenzometalloles, synthesis of 1988–1990
Dicarbenes 733–735—see also Disilanediyl-1,2-dicarbenes
Dicarbonyl coupling reaction 1669
Dications, silicon-containing 1134, 1135
Di(dimethyldichlorosilane), synthesis of 1582–1589, 1592–1594
Diels–Alder reactions 1636, 1656, 1658, 1853–1855
  of silapyrans 1070
Dienes—see also Butadienes, Cyclopentadienes, Siladienes, Silyldienes
  hydrosilylation of 1698, 1699
  photocatalytic 1700, 1701
  reactions with silylenes 2488–2491, 2495
1,3-Dienes, prochiral 1755, 1756
1,6-Dienes, reactions with (TMS)$_3$Si radical 1573
Dienol ethers, as acylsilane precursors 1624
Difenidols 2364
Dihalodisilenes, formation of 833
Dihydrofurans, synthesis of 1644
Dihydro-1,3-oxathiins, synthesis of 1658

Dihydropyrans, synthesis of 1644
9,10-Dihydrosilaanthracenes, photolysis of 1246
Dimesitylsilylene, addition reactions of 2490, 2493
Dimetallofullerenes, photoadditions of 1302
Dimethyl ether, as substrate in Direct Process 1592
Dinaphthyldisilanes, photolysis of 1253, 1254
Dioxadisiletanes, formation of 1081, 1082
Dioxasiletanes—see also Dioxadisiletanes
  retro-cycloadditions of 1080, 1081
Dioxolanation, selective, of ketones 1676
2,5-Diphenylmetalloles, synthesis of 1972, 1981
Diphosphenes 1064
Dipolar-dephasing experiments 311, 312
Diradicals 892—see also Biradicals
  formation of,
    in silene cycloadditions 950
    in silene dimerizations 919–930
    in silene oxidations 978
1,4-Diradicals, formation of 919
  in cleavage of disilacyclobutanes 867
  photochemical 1241, 1274, 1275
Direct Process,
  effect of Cu catalyst preparation on 1587
  effect of intermetallics in contact mass on 1587, 1588
  effect of oxygen in silicon metal used on 1586
  effect of promoters on 1582–1585
  effect of silicon size on 1586
  effect of surface area on 1585, 1586
  recovery and use of by-products from 1592–1594
  surface fundamentals and mechanistic studies of 1588, 1589
  synthesis of catalyst, silicon and promoters for 1586, 1587
  with alcohols as substrate 1590, 1591
  with dimethyl ether as substrate 1592
  with methyl chloride as substrate 1582–1589
  with other organic halides as substrate 1589, 1590
Diselenadisiletanes, formation of 1099
1,2-Disilaacenaphthendiides 2453
1,2-Disilaacenaphthenes 2452–2454
9,10-Disilaanthracene silyl dianions 2453–2457
Disilabenzenes 10—see also Dewar disilabenzenes
  matrix isolation of 1150, 1151
  photoformation of 1246, 1247, 1251
  theoretical aspects of 17–20
Disilabenzvalene 20

Disilabicycloalkanes—*see* Disilabicyclopentanes, Polyselenadisilabicycloalkanes, Polythiadisilabicycloalkanes
Disilabicyclopentanes—*see* Tetrathiadisilabicyclopentanes, Trioxadisilabicyclopentanes, Trithiadisilabicyclopentanes
Disilabutadiynes, photolysis of 1250
Disilachalcogenides, synthesis of 1904
Disilacyclobutadienes 86
Disilacyclobutanes—*see also* Benzodisilacyclobutanes, Bis(alkylidene)disilacyclobutanes
   cycloreversion of 864–870
   synthesis of 884, 1645
1,2-Disilacyclobutanes, synthesis of, in dimerization of silenes 917, 919–930, 1273, 1274, 1278–1280
1,3-Disilacyclobutanes 899, 900
   photolysis of 1241
   synthesis of, in dimerization of silenes 714, 917, 928, 1278, 1279
1,4-Disilacycloheptenes, photolysis of 1244, 1245
1,4-Disilacyclohexadienes, photoformation of 1250, 1251
Disilacyclopropanes—*see also* Thiadisilacyclopropanes
   photoreactions with fullerenes 1301, 1302
1,2-Disila-3,8-dioxacyclooctynes, photolysis of 1303
Disiladioxetanes, X-ray–NMR studies of 322–324
Disiladithietanes, formation of 1084
Disilaindacenes 2155
Disilaindanes, formation of 915, 916
Disilametallacycles,
   reactions of 1906, 1907
   synthesis of 1906
Disilanediyl-1,2-dicarbenes 735
Disilane radical cations 580
Disilanes—*see also* Acyldisilanes, Alkynyldisilanes, Aryldisilanes, Hexaalkyldisilanes, Styryldisilanes, Vinyldisilanes
   cleavage of,
      nucleophilic 782–784, 788–790, 792
      reductive 784, 788, 790, 792
   cyclic—*see* Cyclic disilanes
   disproportionation of, base-assisted 2468
   formation of, cathodic 1214–1216
   photolysis of 891–900, 1247–1265
      sensitized 1257
   pyrolysis of 2465, 2466
   structure of 197, 198
   thermochemistry of 161, 162
   thermolysis of 1170–1173
   transition-metal substituted 1260–1263
Disilanylbenzenes, photolysis of 891, 892

Disilanyl bridges 2150, 2155
(Disilanyl)carbenes 706, 707—*see also* Acyl(disilanyl)carbenes, Alkoxycarbonyl(disilanyl)carbenes
Disilanylene–3,4-diethynylsilole conjugated polymers 2016, 2017
Disilanylnaphthalenes, photolysis of 892, 894–896
Disilaoxiranes, ring opening of, photochemical 1269, 1270
1,5-Disilapentadiene 2409
Disilathianes,
   synthesis of 1904, 1905
   UV spectra of 1915
Disilathietanes 1881, 1882
   photolysis of 1264
Disilazanes—*see also* Cyclodisilazanes, 1,3-Divinyldisilazanes
   coupling constants for 302–304
Disilene complexes 2057
Disilenes 1064, 1068—*see also* Alkoxydisilenes, Aminodisilenes, Dihalodisilenes, Phenyldisilenes, Tetrasilyldisilenes, Trimethylsilylmethyldisilenes
   addition of alcohols to 830, 831
      diastereochemistry of 839–842
      mechanism for 842–844
      regiochemistry of 835–839
   cleavage of 2498–2501
   cyclic—*see* Cyclic disilenes
   cycloadditions of 958, 1094
   electronic spectra of 832
   E/Z isomerism in 828, 829, 839
   formation of 812, 833–835
      photochemical 829–832, 1263, 1264, 1268, 1269, 1282, 1283
   matrix isolation of 1156, 1157
   NMR spectra of 278, 279, 985
      solid-state 345, 347–350
   photolysis of 1288, 1289, 1332
   reduction of, cathodic 1214
   structure of 247–250
   thermochemistry of 175
   UV spectra of 829, 830
Disiletanes—*see* Dioxadisiletanes, Diselenadisiletanes, Thiadisiletanes
Disilicic acid, potential energy surface for Si—O—Si skeleton of 109–113
Disiliranes—*see also* Oxadisiliranes
   oxidation of 1899–1901
   reactions of,
      with C_{60} 1936–1939
      with C_{70} 1940
      with metallofullerenes 1302, 1940–1943
Disiloxanes—*see also* Bis(allyl)disiloxanes, Cyclodisiloxanes, 1,3-Divinyldisiloxanes

Disiloxanes (cont.)
　structure of 222–225
Disiloxetanes, photolysis of 870, 871, 1264
Disilthianes 1875–1887
　cyclic—see Cyclic disilthianes
Disilylcarbenes, photoformation of 1276, 1277
Disilyl ethers—see Hydroquinone disilyl ethers
Disilylethylenes, coupling constants for 296, 299–301
Disilynes, formation of 1008
Distomers 2367, 2373–2375
Dithiadigermetanes, synthesis of 1905
1,4-Dithia-2,3-disilametallacyclopentanes,
　reactions of 1906, 1907
　synthesis of 1906
Dithiadisiletanes, formation of 1088
1,3-Dithianes—see also 2-Silyl-1,3-dithianes
　hydrolysis of 1606, 1628
　reduction of, using (TMS)$_3$SiH 1545
Dithiasiletanes, formation of 1093
Dithioacetals, hydrolysis of 1606
1,3-Dithiolanes, reduction of, using (TMS)$_3$SiH 1545
1,3-Divinyldisilazanes, Rh complexes of, photolysis of 1299, 1300
1,3-Divinyldisiloxanes, Rh complexes of, photolysis of 1299, 1300
Divinyl ether, hydrosilylation of 1691
Divinylsilacyclopentanes, formation of 2491
Dodecamethylcyclohexasilane,
　photolysis of 2471, 2472
　synthesis of 2178
Double-quantum coherence spectroscopy 269
Dow Corning, Direct Process and 1582, 1584–1586, 1593
Drugs, enantiomeric, biological recognition of 2364–2375
Duckett–Perutz two-silicon cycle mechanism, for alkene hydrosilylation 1706–1708
Dyotropic rearrangements 881, 1130
　photolytic, of silenes 1285, 1288

Electrochemical sensors 2327–2329
Electronic spectroscopy,
　of silanethiones 1067, 1068, 1091, 1092
　of silicenium ions 565
Electronic transitions, σ–σ*,
　in oligosilanes 1312
　in polysilanes 1318
Electron spin resonance spectroscopy,
　of oligosilanes 1314
　of polysilanes 1321
Electron transfer reactions, photoinduced 1290–1294
Electrophilic substitution, of C—Si bonds 1794–1824

Elimination reactions, steric effects of silyl groups on 448, 449
ELIZA test 2327
Elkem, Direct Process and 1586
Enamides, hydrosilylation of 1689
Enamines, silylcarbonylation of 1772
Enediynes, synthesis of 1811
Ene reactions 1856–1858—see also Retro-ene reactions
　of silenes 922, 923, 926, 941, 944–946, 948, 953–955, 967, 974–977
　steric effects of silyl groups on,
　　comparison of 449
　　comparison with H atom and R groups 469, 470
Enol ethers—see also Silyl enol ethers
　as acylsilane precursors 1614–1617, 1623–1627
　electrophilic halogenation of 1619, 1620
Enthalpies of formation,
　group increment contributions to 176
　of alkylsilanes 159–161
　of alkynylsilanes 161
　of π-bonded silicon species 174–176
　of halosilanes 156, 164, 177
　of methylsilylamines 166
　of organosilanes, O-containing 164, 165
　of radicals 167–169
　of silacycloalkanes 162, 163
　of silicenium ions 559
　of silicon hydrides 158, 159, 177
　of silylenes 171–174
　of vinylsilanes 161
Enthalpy/electronegativity correlations 157, 158
Enthalpy of activation 1351
Entropy of activation 1343
Enynes, hydroboration–oxidation of 1622
Enzymatic reactions, in sol–gel matrices 2321
Epoxidation, steric effects of silyl groups on,
　comparison of 449, 450
　comparison with H atom and R groups 470
α-Epoxyacylsilanes, synthesis of 1630
α,β-Epoxysilanes, synthetic applications of 1679, 1680
Esterification, enzymatic 2390–2394
Ethyl chloride, as substrate in Direct Process 1589
Ethynylsilanes, coupling constants for 304–306
Ethynylsilylene, matrix isolation of 1163, 1164
Ethynylstannanes, coupling constants for 304
Eutomers 2367, 2373, 2375

FA-SIFT technique 1026, 1027
Feldspar 103
Felkin–Anh transition states 1639
anti-Felkin selectivity 1804, 1805

Ferrocenes—*see also* Acetylferrocene, Persilaferrocene, Trimethylsilylferrocenes
    silicon-bridged 2159–2163
Ferrocenophanes, silicon-bridged 2153–2155
α-Ferrocenyl-β-silylvinyl cations, NMR spectra of 667–669
1-Ferrocenyl-2-triisopropylsilylethyl cation, NMR spectra of 660, 661
Flip-flop coordination 2051
Fluorenes—*see* 9-Metallafluorenes, Silafluorenes
Fluorenyl cations 631, 632, 649, 650
Fluorine, bridged 1434
    in fluorosilicates 1344–1349
p-Fluorohexahydrosila-difenidol 2364
Fluorosilanes,
    ^{29}Si chemical shifts of, calculated 294, 295
    structure of 192–194, 234–236
Fluorosilicates,
    hexacoordinate 1412
        NMR spectra of 1413–1415
    pentacoordinate 1340–1351
        crown ether complexes of 1340–1342
        NMR spectra of 1342, 1343, 1347, 1350
        zwitterionic 1349–1351
Formyl anions, synthons of 1198
Formylsilanes,
    photolysis of 1271, 1272
    synthesis of 1606, 1610
Free energy of activation, effect of silyl groups on 485–487
Friedel–Crafts reaction 1820, 1860
Frontier molecular orbital (FMO) theory 1331
Fullerenes—*see also* Buckminsterfullerene, Dimetallofullerenes, Silafullerenes
    organosilicon derivatives of 1929–1960
    photoadditions of 1301, 1302
Furans—*see* Dihydrofurans, Tetrahydrofurans

Gas-phase ion techniques,
    afterglow ionization 601, 602
    CAD 607
    FTMS 601, 602
    high-pressure MS 605, 607, 608
    ion/molecule reactions 602
    NR-MS 604
    pulsed ICR 603
    tandem MS 607, 608
General Electric, Direct Process and 1582, 1583, 1587, 1594
Germaalkenes—*see also* Germenes
    as germanimine precursors 1051
    formation of 717
Germacyclopentadienide anions 2021, 2022
η^5-Germacyclopentadienyl transition metal complexes 2026, 2027

Germacyclopentanes—*see* Tetrasilagermacyclopentanes
Germacyclopropanes—*see* Siladigermacyclopropanes
Germadihydrotetrazoles 1051
Germadihydrotriazoles 1051
Germaindenide anions 2022
Germametallacycles 1905
Germanes—*see* Acylgermanes, Bis(hydroxymethyl)germanes, Cyclodisilagermanes, Polygermanes
Germaneselenones, synthesis of 1093, 1094
Germanethiones,
    electronic spectra of 1092
    synthesis of 1093, 1094
Germanimines 1051
Germanium cations 589
δ-Germanium effect 642
Germaprismanes—*see* Pergerma[*n*]prismanes
Germasilathietanes 1882
Germatrisilacyclobutanes,
    NMR spectra of 279
    photolysis of 1269
Germenes—*see also* Germaalkenes, Silagermenes
    synthesis of 909
Germetanes—*see* Dithiadigermetanes
Germoles,
    anions of 2024–2026, 2028, 2029
    synthesis of 1968–1970, 2027, 2028
Germylenes, photoformation of 1237, 1264, 2473
Germyllithium reagents, reactions with C_{60} 1955
GIAO methods 516, 658
GIAO-MP2 method 567, 692, 693, 695
GIAO-SCF method 692
Gitonic dication intermediates 672
Glass transition 327
Glucose oxidase 2326
GLYMO 2333–2337
Grignard reagents,
    reactions with acylsilanes 1635–1638, 1651, 1652
    silyl 782, 784
Grunwald–Winstein effect 633

Halides, reduction of, using (TMS)$_3$SiH 1543, 1544
α-Haloacylsilanes,
    reactions of 1651–1653
    synthesis of 1619–1621
1-(Haloalkyl)silatranes, reactions of 1500
Haloaryldisilanes, photolysis of 1256, 1257
Halomethylsilanes—*see also* (Trihalomethyl)trialkylsilanes
    cathodic reduction of 1218–1220

Halosilanes—see also Chlorosilanes, Iodotrimethylsilane
  reactions of 1457, 1458
  reduction of, cathodic 1214–1218
  thermochemistry of 164
1-Halosilatranes,
  reactions of 1488–1491
  solvolytic stability of 1483
  synthesis of 1458, 1486–1488
Halo(silyl)carbenes 767
Halosilylenes,
  formation of 2480
  matrix isolation of 1180, 1181
  reactions of 2467, 2487, 2495
  spectra of 2522, 2523
Hammond postulate 610
Heck reaction 1824–1827, 1837, 1838
Hemithioacetals, hydrolysis of 1606, 1608
Heptacoordinate silicon complexes 1430–1433
Heptasilanes—see also Oxabicycloheptasilanes
  emission properties of 1316
Heptasila[7]paracyclophane 2428, 2429
Heteroaromatic compounds, alkylation of 1558
1-(Heteroarylalkyl)silatranes, synthesis of 1455, 1456
1-Heteroarylsilatranes, synthesis of 1450, 1459
Heterocyclopentadienes, transmetallation of 1986–1988
Hetero-Diels–Alder reaction 1658
Heterosilabenzenoids, theoretical aspects of 30, 31
Hexaalkyldisilanes, photolysis of 1319
Hexacoordinate silicon compounds, NMR spectra of 334
Hexacoordinate silicon complexes,
  ionic 1415–1418
  neutral,
    intermolecular 1429, 1430
    intramolecular 1418–1429
Hexadecasilanes, spectra of 1317
Hexahalocarboranes, as counterions for silicenium ions 283
Hexahydrosila-difenidol 2364
Hexakis(fluorodimethylsilyl)benzene, NMR spectra of 288, 289
Hexakis(trimethylsilyl)benzene, photolysis of 1237
Hexasilabenzenes—see also Dewar hexasilabenzene
  higher congeners of 29, 30
  theoretical aspects of 22–29
Hexasilabenzvalene 25
Hexasilabicyclo[6.1.0]non-1(9)-ene 2450, 2451
Hexasilacyclonona-1, 2-diene 2447
Hexasilacyclooctyne 2446
Hexasilanes—see also Cyclohexasilanes
  absorption/emission spectra of 1316

Hexasilaprismanes 120, 2205, 2206
  absorption spectra of 133, 134
  NMR spectra of 146–148
  photoreactions of 134–136, 1268, 1269
  photosensitivity of 2206
  strain energy of, substituent effects on 124
  structure of 123, 133
  synthesis of 129, 130
HexasilylDewar benzene, photoformation of 1238
Hexenyl(trimethyl)silanes, photoreactions of 1290
HMBC spectroscopy, ^{29}Si–^{1}H 1947
Homoallylchalcogenoacetals, synthesis of 1801
Homoallyl ethers, synthesis of 1798, 1803, 1804, 1807
Homoallylic alcohols, synthesis of 1801–1804, 1808, 1809
Homodesmotic reactions 538, 539
Homoenolates 1637
Homohyperconjugation 642, 695
3-Homosilatranes,
  properties of 1506
  synthesis of 1505, 1506
Horner–Emmons reaction 1628, 1653
Hybridization 122, 123
Hydridosilicates 1370–1373
  gas-phase studies of 1372
  NMR spectra of 1371
  theoretical studies of 1372, 1373
Hydridosilyl anions, synthesis of 780, 781
Hydroboration, steric effects of silyl groups on, comparison of 450, 451
  comparison with H atom and R groups 472, 473
Hydrocoupling, cathodic 1228
Hydroethoxycarbonylation, steric effects of silyl groups on 451
Hydroformylation,
  catalysis by sol–gel entrapped ion-pair catalysts 2323
  steric effects of silyl groups on 451
Hydrogenation, catalysis by sol–gel entrapped ion-pair catalysts 2323
Hydrogen bonding, in silatranes 1465
1,2-Hydrogen migrations 901, 902
1,5-Hydrogen migrations, in rearrangements of silenes 914
Hydrolysis, enzymatic 2384–2388
Hydroquinone disilyl ethers, anodic oxidation of 1205, 1206
Hydrosilanes,
  oxidation of, anodic 1208–1210
  photolysis of, with ozone 1068
  polymerization of, electrochemical 1210
  reactions of,
    with CO_2 1068, 1069

with elemental sulphur 1084, 1085
reduction of, cathodic 1229
1-Hydrosilatranes,
  reactions of 1486–1488
  synthesis of 1450, 1457, 1459
Hydrosilylation 1552–1556, 2330, 2337
  asymmetric 1743–1758
    intramolecular 1750, 1751, 1756–1758
  modification of polymers by 1768–1771
  of alkenes 1552–1555, 1688–1716
    intramolecular 1710–1714
    mechanism of 1704–1710
  of alkynes 1555, 1556, 1716–1733
    intramolecular 1725
    mechanism of 1723, 1724
  of carbonyls 1733–1741
    intramolecular 1740, 1741
    mechanism of 1737
  of imines 1742
  of nitriles 1742
  reactivity of Si-containing reagents in 479
  steric effects of silyl groups on 451, 452
  syntheses of Si-containing dendrimers and polymers using 1758–1771
Hydrosilylation–allylation 1758
Hydrosilylation–cross-coupling 1731
Hydrosilylation–ethenylation 1759, 1762
Hydrosilylation–ethynylation 1762
Hydrosilylation–isomerization, of sec-propargyl alcohols 1729, 1730
Hydrosilylation–polymerization 1763–1768
(Hydroxyalkyl)silanes—see also Bis(hydroxymethyl)silanes, Chiral (hydroxymethyl)silanes
  enzymatic oxidation of 2394–2397
α-Hydroxysilanes,
  desilylation of 1636
  oxidation of 1611
  synthesis of 1638, 1649
1-Hydroxysilatranes, reactions of 1491
Hydroxysilylenes, matrix formation of 1180
Hydrozirconation 1659
Hyperconjugation 605, 606, 615, 616, 619, 634—see also Homohyperconjugation
  antiperiplanar geometry in 618
  cosine-squared dependency of 620, 627
  double 642, 644
  geometrical consequences of 599, 600
  in 1-bis(silyl)methyl-2-bis(silyl)ethenyl cations 686, 687
  in cyclopropenium ions 648, 649
  in dienyl cations 654, 666
  in α-ferrocenyl-β-silylethyl cations 660, 661
  in β-silylallyl cations 662, 663
  in β-silylvinyl cations 669, 671, 675, 677, 682–685
  in stabilization of silicenium ions 576

isodesmic equation for 598
orientation-dependent 598
rotational barrier in 618
β-silicon 584, 585
Hypercoordination 616, 688–697

IGLO method 281, 516, 567, 653, 658, 686
Imidazoles, as sol–gel catalysts 2335
Imine N-oxides, prochiral 1750
Imines—see also Aldimines, Germanimines, Silaimines, Silanimines
  hydrosilylation of 1742
  prochiral 1749
Iminosilanes 1064—see also Silanimines
Imino(silyl)carbenes, rearrangement of 761
INADEQUATE spectroscopy 269–274
  $^{13}C–^{13}C$ 1940
Indenes—see 1-Metallaindenes, Silaindenes
INEPT 269, 302
INEPT INADEQUATE spectroscopy 272–274
Infrared spectroscopy,
  of acylsilanes 1600, 1601
  of silaisonitriles 1021, 1046
  of silanimines 1044, 1045
  of silatranes 1471, 1472
  of silatrane-3, 7, 10-triones 1509
  of silenes 991–995
  of silicenium ions 559
  of triazasilatranes 1521
Insertion reactions, reactivity of Si-containing reagents in 479
Intersystem crossing 1320, 1327
Inversion recovery method 308, 329, 330
Iodotrimethylsilane 1667
Ion cyclotron resonance mass spectrometry 558, 560, 1109
  of silenes 996, 997
Ion–molecule reactions 1109–1122
Iridium catalysts,
  for dehydrogenative silylation 1732, 1733
  for hydrosilylation 1720, 1721, 1732
  for silylcarbonylation 1772
Iron catalysts, for dehydrogenative silylation 1714, 1715
Isocyanides, reduction of, using $(TMS)_3SiH$ 1549
Isocyanosilylenes, matrix isolation of 1179
Isodesmic reactions 649
Isomerization—see also Photoisomerization
  steric effects of silyl groups on 452
1-Isothiocyanatosilatranes,
  methanolysis of 1483, 1491
  structure of 1460
Isotope effects—see also Deuterium isotope effects
  in solvolysis 618, 619, 622, 623, 633, 635
Isotopic enrichment, in ^{29}Si NMR 269

Isotropization 331
Itsuno reagent 1638

Kaolinite, NMR spectra of 343, 346, 347
Karplus–Pople equation 1603
Karstedt's catalyst 1689, 1690, 1693–1695, 1710–1714, 1725, 1726, 1758, 1760, 1764, 1765, 1771
Kesti–Waymouth's olefin-first mechanism, for alkene hydrosilylation 1708, 1709
Ketene silyl acetals, photolysis of 1293
α-Ketoacylsilanes,
  reactions of 1653
  synthesis of 1621, 1622
α-Ketoesters, synthesis of 1653, 1654
Ketones,
  hydrosilylation of 1555
  prochiral—*see* Prochiral ketones
β-Ketosilanes, synthesis of 1640
Kinetic stabilization, in multiple bonds to Si 1064
Kumada rearrangement 2281

Ladder C method 1313, 1314, 1316
Ladder H method 1316
Langmuir–Blodgett technique 2341, 2342
Lanthanides, as hydrosilylation catalysts 1697, 1698
LArSR equation 610
Laser flash photolysis, in synthesis of silenes 897
Laser-induced fluorescence spectroscopy, of silylenes 2521–2523
Lewis acids,
  as catalysts,
    in reactions of acylsilanes 1638, 1639, 1655, 1658
    in reactions of allylsilanes 1801–1810, 1812–1816, 1821–1823, 1832, 1834, 1850
    in synthesis of acylsilanes 1620, 1625
  complexes with silylenes 2518–2521
Lewis bases, complexes with silylenes 2485, 2492, 2500
Lewis base stabilization 1387–1390
Lewis mechanism, for colloid-catalysed hydrosilylation 1694
Ligand site exchange,
  in fluorosilicates 1343
  in Si—N complexes 1382–1387
  in Si—O complexes 1403–1405
Liquid crystals, in sol–gel matrices 2352
Lithiation, steric effects of silyl groups on 473
LORG methods 516
Luminescent dyes 2350

Macrocyclization reactions 1560, 1561
Magadiite, NMR spectra of 315, 317, 318
Magic angle spinning (MAS) 309–313, 317, 319–321, 343
Manganese complexes, as hydrosilylation catalysts 1700
Mass spectrometry,
  MIKE 561
  neutralization–reionization 1029, 1106, 1122-1127, 1130
  of silatranes 1481–1483, 1506, 1510, 1523
  of silatranones 1509
  of silicenium ions 559–561
  selected-ion flow tube (SIFT) methods in 1109
Matrix isolation 1144, 1145
  of carbenes 709
  of diazosilanes 1160
  of disilenes 1156, 1157
  of silaantiaromatics 1152–1156
  of silaaromatics 1148–1152
  of silacarbonates 1161
  of silacarboxylic esters 1161, 1162
  of silacyclopropenes 1165, 1166, 1175
  of 1-silacyclopropenylidene 1163–1165, 1173–1175
  of silacyclopropyne 1163–1165
  of silaisonitriles 1158–1160
  of silanediimines 1160
  of silanethiones 1096, 1097, 1163
  of silanimines 1158–1160
  of silanitriles 1158–1160
  of silanones 1161
  of silenes 714, 1145–1148
  of silicon oxides 1162
  of silylenes 1163–1181, 1329, 2474
  of silynes 1148
McCoy reaction 1653
MCS reaction,
  effect of Cu catalyst preparation on 1587
  effect of intermetallics in contact mass on 1587, 1588
  effect of oxygen in silicon metal used on 1586
  effect of promoters on 1582–1585
  effect of silicon size on 1586
  effect of surface area on 1585, 1586
  recovery and use of by-products from 1592–1594
  surface fundamentals and mechanistic studies of 1588, 1589
  synthesis of catalyst, silicon and promoters for 1586, 1587
MEMO 2333–2337
1-(Mercaptoalkyl)silatranes, reactions of 1503
(Mesityl)$_3$Si$^+$ ion 572, 573, 589

Subject index

α-Mesityl-β-silylvinyl cations, NMR
  spectra of 669–676
Mesylates, solvolysis of, δ-silicon effect in
  641, 642
Metallacyclopentenols, esters of, thermolysis of
  1982–1984
9-Metallafluorenes, synthesis of 1988–1990
1-Metallaindenes, synthesis of 1991–1995
7-Metallanorbornadienes, reactions with
  alkynes 1976, 1977
Metalloaldimines, silylation of 1618
Metalloles—*see also* Benzometalloles, 2,5-
  Diphenylmetalloles
  cycloadditions of 1998–2003
  isomerization of 1996, 1997
  oxidation of 2009, 2010
  reactions of,
    with acids 2005, 2006
    with bases 2006, 2007
    with halogens 2003–2005
    with organometallics 2007–2009
  reduction of 2010, 2011
  ring expansion of 2012, 2013
  stability of 1996, 1998
  synthesis of 1963–1988
  tautomerism in 1997, 1998
  transmetallation of 2011, 2012
Metasilicates, NMR spectra of 326
Methyl chloride, as substrate in Direct Process
  1582–1589, 1592–1594
(η^5-Methylcyclopentadienyl)silanium cation 35
Methyldiazidosilane, as silaethylene photolytic
  precursor 1145
1,3-Methyl migrations, silicon-to-silicon 1285
(Methylsilyl)carbene 706
Methylsilylene, matrix isolation of 1167
M_6H_6, valence isomers of, relative stability of
  122
Mica 103
Microwave spectroscopy, of silenes 864, 982
Migrations—*see also* Hydrogen migrations,
  1,3-Methyl migrations, Silyl migrations,
  Stevens migration
  1,2-H/alkyl, in gas-phase studies of α-silyl
    cations 601
  steric effects of silyl groups on 452–454
Migratory aptitudes,
  effect of silyl groups on 487
  of alkyl groups 602
Millimeter wave spectroscopy 1024, 1046
MOCVD processes 2145
Molecular modelling, of bonded interactions of
  crystalline silica 103–117
Monocyclosilachalcogenides 1897–1910
Monosilabenzenes, theoretical aspects of 11–16
Monosilacycles 2401–2408, 2412–2427
Monosilacyclobutadienes 83–86

Monosilacyclopropenium cation 33
Monosilanes,
  oligolithiated 819–821
  pyrolysis of 2475–2481
Mosher's acid, synthesis of 1682
Mukaiyama reaction 1649
Mullite, formation of 2303–2307
Murai reaction 1838, 1839
Muscarinic antagonists 2364
Muscarinic receptors 2364, 2365, 2367–2369,
  2371, 2373–2375

Naphthalenes—*see also* Bis(pentamethyldisilyl)naphthalenes, Disilanylnaphthalenes, Silanaphthalenes
  silyl-bridged, photolysis of 1244
Nematogenic compounds 2352
Neopentylsilenes 913
  cycloadditions of 945–949
  dimerization of 931, 945, 946, 948, 949
  formation of 874, 875, 878, 879
  NMR spectra of 985, 991
  reactions with alcohols 934
Neutralization–reionization mass spectrometry
  1029, 1106, 1122–1127, 1130
NIR organic dyes 2351
Nitrenium ions 520
Nitriles—*see also* Silanitriles
  hydrosilylation of 1742
Nitrone formation, reactivity of Si-containing
  reagents in 479, 480
NLO materials 2352
Noble gases, complexes with silicenium ions
  527, 535, 537
Nonresonant two-photon (NRTP) method 1329
Norbornadienes—*see* 7-Metallanorbornadienes,
  Silanorbornadienes
Norbornenes, hydrosilylation of 1690
Norbornyl derivatives, solvolysis of,
  γ-silicon effect in 639
  δ-silicon effect in 643, 644
Norbornyl–norpinyl rearrangement 640
Norrish Type II cleavage reactions, of acylsilanes 1647
Nuclear magnetic resonance spectroscopy,
  ^{11}B 567
  ^{13}C 566, 991, 1355, 1366, 1397, 1401,
    1474, 1475, 1602, 1603
  ^{35}Cl 566–568
  dynamic 1364, 1365
  effect of silyl groups on 487
  ^{19}F 567, 1343, 1347, 1350, 1479
  ^{1}H 566, 1473, 1474, 1601, 1602
  ^{15}N 1476–1479
  ^{17}O 1479
  of acylsilanes 1601–1604
  of arsasilenes 1053

Nuclear magnetic resonance spectroscopy
(cont.)
  of bicyclosilachalcogenides 1915
  of cyclotrisilanes 2183
  of fluorosilicates,
    hexacoordinate 1413–1415
    pentacoordinate 1342, 1343, 1347, 1350
  of hydridosilicates 1371
  of hypervalent Si compounds 284–289,
    334–343
  of phosphasilenes 1051, 1052
  of silaamidides 1030, 1031, 1043
  of silanethiones 1086, 1087
  of silanimines 1043, 1044
  of silanorbornyl cations 696, 697
  of silapolyhedranes 146–148
  of silatranes 1473–1479, 1506, 1510, 1514,
    1521
  of silatranones 1508, 1509
  of silenes 985–991
  of silicates, organically modified 290–293
  of silicenium ions 280–284, 532–538, 565–
    575, 584
  of α-silyl cations 645–648
  of β-silyl cations 655–663
  of silylcyclobutonium ions 691, 692, 694,
    695
  of Si—N complexes 1379–1382, 1384, 1385
  of Si—O complexes 1397–1403
  of spirosilicates 1355–1357, 1362–1366,
    1369, 1370
  ^{29}Si 146–148, 268–350, 532–538, 567–575,
    985–991, 1340, 1342, 1343, 1355–
    1357, 1362–1366, 1369–1371, 1379–
    1382, 1384, 1385, 1397–1403, 1413–
    1415, 1475, 1476, 2183
    solid-state 309–350, 584
Nuclear Overhauser effect (NOE) 269
Nucleosides, deoxygenation of, using
  (TMS)$_3$SiH 1550–1552

Octacoordinate silicon complexes 1430–1433
Octasilacubanes 120, 2205–2207
  absorption spectra of 142–144
  halogenation of 2207
  NMR spectra of 146–148
  photolysis of 1268, 1304
  reactivity of 143, 145, 146
  stereoisomerism in 145, 146
  strain energy of, substituent effects on 124
  structure of 123, 137–142
  synthesis of 136, 137
Ojima–Crabtree mechanism, for alkyne hydrosilylation 1723, 1724
Ojima–Kogure mechanism, for carbonyl hydrosilylation 1737

Olefination, reactivity of Si-containing reagents
  in 480
Olefin-to-carbene isomerization 769–772
Oligosilacycles 2408–2412, 2428–2457
Oligosilanes—see also Bis(silyl-
  diazomethyl)oligosilanes
  cleavage of,
    nucleophilic 794, 795, 801
    reductive 795, 796
  α,ω-dilithiated, synthesis of 801
  dimers of 1317
  ESR spectra of 1314
  excited state nature of 1312–1317
  photolysis of 1168–1170, 1178–1180
    mechanism of 1319, 1320
Oligosiloles, synthesis of 2013–2015
Olivine 103
Optoelectronic devices 2351
Organocuprate conjugate additions, role of
  TMSCl in 1667, 1668
Organometallic reagents, reactions of,
  with acylsilanes 1635–1638, 1651, 1652
  with metalloles 2007–2009
1-Organyloxysilatranes, synthesis of 1486
Organylsilanes, reactions of 1458
1-Organylsilatranes,
  hydrolysis of 1484
  synthesis of 1491
1-(Organylthioalkyl)silatranes, reactions of
  1502
1-Organylthiocyanatosilatranes, synthesis of
  1453
1-Organylthiosilatranes, synthesis of 1453,
  1490, 1492
Ormocers 2351
Ormosils 2345, 2346
  NMR spectra of 290–293
Orthosilicates, NMR spectra of 326
Osmium catalysts,
  for dehydrogenative silylation 1714, 1715
  for hydrosilylation 1720, 1721, 1730
Osmylation, steric effects of silyl groups on
  454, 455
Oxabicycloheptasilanes, formation of 2208
Oxadisiliranes,
  oxidation of 1899–1901
  stereochemistry of 1901–1903
  reactions with C$_{60}$ 1939, 1940
Oxahexasilanorbornanes, structure of 2191
6-Oxa-3-silabicyclo[3.1.0]hexanes, pyrolysis of
  1069, 1070
1-Oxa-2-silacyclobut-3-enes, formation of 931
Oxasiletanes 2406, 2407—see also
  Dioxasiletanes
Oxasiletanides, NMR spectra of 284, 285
Oxasiletenes 2406, 2407
Oxasiliranes 2414–2416, 2418

Oxathiins—see Dihydro-1,3-oxathiins
Oxathioacetals, hydrolysis of 1606
1,3-Oxathiolanes, reduction of, using (TMS)$_3$SiH 1545
1,3-Oxathiolanones, reduction of, using (TMS)$_3$SiH 1545
Oxatrisilacyclobutanes, photolysis of 1269, 1270
Oxazine 170 2353
Oxazines, as lasing dyes in sol–gel matrices 2351
Oxetanes—see also Silaoxetanes
  synthesis of 1849, 1850
Oxidation—see also Photooxidation
  anodic 1188–1211
  enzymatic 2394–2397
  steric effects of silyl groups on 473
Oxidation potentials,
  of acylsilanes 1204
  of allylsilanes 1192
  of aminosilanes 1200
  of benzylsilanes 1191
  of disilenes 1209
  of Group 14 metals 1207
  of polysilanes 1207
  of silyl ethers 1202
  of tetraalkylsilanes 1188
  of trimethylsiloxyarenes 1205
Oxiranes, as α-haloacylsilane precursors 1620, 1621
Oxocarbenium ions, as intermediates 1799
3-Oxo-1,4-dienes, reactions with (TMS)$_3$Si radical 1575
Oxonium ions—see also Silaoxonium ions
  as intermediates 1807
Oxymercuration, steric effects of silyl groups on 455
Ozonolysis, steric effects of silyl groups on 455, 456

Pagodane structure, of Si$_{20}$H$_{20}$ 148, 149
Palladium catalysts, polyamide-supported, for hydrosilylation 1702
Pechiney, Direct Process and 1586
Pentacoordinate silicon adducts, NMR spectra of 287
Pentacoordinate silicon compounds, NMR spectra of 334, 335
Pentacoordinate silicon complexes,
  cationic 1408–1412
  F—Si 1407, 1408
  N—Si,
    ligand exchange in 1382–1387
    NMR spectra of 1379–1382, 1384, 1385
    stereodynamics of 1382–1387
    synthesis of 1373–1377
    X-ray studies of 1377–1379

O—Si 1390–1405
  ligand exchange in 1403–1405
  NMR spectra of 1397–1403
  synthesis of 1390–1395
  X-ray studies of 1395–1397
S—Si 1406, 1407
Pentamethylcyclopentadienyl ligands 2131
Pentasilabicyclo[5.1.0]oct-1(8)-ene 2450, 2451
Pentasilacycloheptyne 2439–2441
Pentasilacyclopentadienyl anion 50, 51
Peracid oxidation, in synthesis of acylsilanes 1628, 1629
Pergerma[n]prismanes, strain energies of 121, 122
Perhydropolysilazanes 2252
Permethylpolysilanes, anodic oxidation of 1207
Perphenylcyclopolysilanes,
  reactions of 2189
  synthesis of 2178
Per-rhenates 1628
Persiladodecahedrane 148, 149
Persilaferrocene 81
Persila[n]prismanes,
  strain energies of 121, 122
  structure of 148, 149
Perstanna[n]prismanes, strain energies of 121, 122
Peterson elimination 1617
Peterson olefination, steric effects of silyl groups on 456
Peterson reaction 417–421—see also Sila-Peterson reaction
  variations of 421–423
Phenazasilane, anodic oxidation of 1210, 1211
Phenothiasilane, anodic oxidation of 1210, 1211
Phenyldisilenes,
  formation of 830, 831
  reactions with alcohols 830, 831, 835, 840–843
  UV spectra of 830
α-Phenyl-β-silylvinyl cations, NMR spectra of 676–679
Phenyl(trimethylsilyl)carbene, reactions of 727
Pheromones, synthesis of 1639, 1640
pH indicators 2322, 2340
Phosphaalkenes, formation of 1081, 1082
Phosphasilenes 1051–1053, 1064, 1068, 1100
  cycloadditions of 1081, 1082
Phosphinocarbenes 709, 710
Phosphino(silyl)carbenes 710, 711, 761–765
Phosphonium ylids, oxidation of 1612, 1613
Phosphonylation, steric effects of silyl groups on 456, 457
Phosphoryl(silyl)carbenes 765, 766
Photoallylation 1290–1292
Photochromic materials 2348–2350

Photocyclization reactions 1296, 1298, 1299
Photoelectron spectroscopy 1046
  of acylsilanes 1605
  of aryldisilanes 1322
  of silatranes 1470, 1471
  of silenes 992, 996
Photoenolization 1645
Photoionization mass spectrometry 1107
Photoisomerization,
  of divinylsiloxane metal complexes 1299, 1300
  of transition-metal substituted cyclohexasilanes 1261
  of trisilacycloheptenes 1298–1300
Photooxidation, of octasilacubanes 1304
Pinacolone, UV spectra of 1603
β-Pinene, reactions with $(TMS)_3Si$ radical 1572
Platinum catalysts,
  for dehydrogenative silylation 1716
  for hydrosilylation,
    of alkenes 1688–1692, 1702
    of alkynes 1717, 1727, 1728, 1730–1732
    of polymers 1768–1771
  highly dispersed metallic 1696
  photoactivated 1695, 1696
Plumbanethiones, synthesis of 1093, 1094
Polybutadienes, hydrosilylation of 1771
Polycarbosilanes, electrosynthesis of 1218–1220
Polycyclosiloxanes 1923–1925
Poly(dibutylsilane), photolysis of 1321
Polydimethylsiloxanes,
  degradation in soil 2229–2231
  depolymerization of, thermal 2228, 2229
  oxidation of 2231
  rearrangement of, acid-catalysed 2228
Poly(ferrocenylsilane)s, synthesis of 2146
Polygermanes—*see also* Cyclotetragermanes, Cyclotrigermanes
  fullerene-doped, photoconductivity of 1930, 1931
Polyhedranes—*see also* Silapolyhedranes
  strain energies of 120, 121
Polymerization,
  co-catalysts for 580
  steric effects of silyl groups on 457
Polymers,
  modification of,
    by hydrosilylation 1768–1771
    by silicenium ions 560
  Si-containing, synthesis of 1763–1768
  siloxene-like 2197
  thermoset nonlinear optical (NLO) 1765, 1766
Polymethylaminoborosilazanes 2270
Polymethylhydridosiloxanes 2290

Polymethylsilane (PMS), synthesis of,
  via dehalocoupling 2273–2276
  via dehydrocoupling 2276–2280
  via redistribution of chlorosilanes 2280, 2281
Poly(phenylacetylene), synthesis of 1767
Poly(phenylsilane)s, hydrosilylation of 1769
Polyselenadisilabicycloalkanes,
  NMR spectra of 1915
  structure of 1913–1915
  synthesis of 1912, 1913
Polysilabicyclo[n.1.0]alkenes 2444, 2445
Poly(1-silacyclobutane) graft polymers 1767
Polysilaethylene (PSE), precursors to 2283–2285
Polysilane cages 2205–2209
Polysilane high polymers,
  excited state nature of 1317, 1318
  segment model for 1317, 1318
  worm-like model for 1318
Polysilanes—*see also* Acylpolysilanes, Heptasilanes, Hexadecasilanes, Hexasilanes, Pentasilanes, Permethylpolysilanes, Polycarbosilanes, Poly(ferrocenylsilane)s, Polymethylsilane, Poly(phenylsilane)s, Poly(siloxysilane)s, Tetrasilanes, Trisilanes
  cyclic—*see* Cyclopolysilanes
  ESR spectra of 1321
  formation of, cathodic 1215–1218
  fullerene-doped, photoconductivity of 1930, 1931
  γ-irradiation of 2475
  ladder 1217
  oxidation of, anodic 1207
  photolysis of 1265–1267, 1303, 2469–2473
    mechanism of 1319–1321
  polycyclic,
    annelated 2204, 2205
    linearly connected 2197, 2199–2204
    synthesis of 2203, 2204
  reduction of, cathodic 1212–1214
  thermolysis of 2465–2468
Polysiloxane macrocycles, synthesis of 1768
Polysiloxanes 2329, 2338, 2354—*see also* Polydimethylsiloxanes, Polymethylhydridosiloxanes
  degradation of 2227–2233
  NMR spectra of 306–309
    solid-state 315–317, 329–332
  surface activity of 2234–2240
    blend compatibilization of copolymers 2238–2240
    end-group effect on 2234
    siloxane migration to solid–air interfaces 2234–2237
  synthesis of 2218–2226

Subject index

Poly(siloxysilane)s, starburst, synthesis of 1762, 1763
Polysilylacylsilanes, reactions of 577, 578
  with organometallics 888, 889
Polysilylation, electrochemical 1223
Polysilylcarbinols, deprotonation of 889
Polysilyldiazomethanes, as silene precursors 904
Poly(silylethylene), pyrolysis of 2466
Polythiadisilabicycloalkanes—*see also* Tetrathiadisilabicyclopentanes, Trithiadisilabicyclopentanes
  NMR spectra of 1915
  reactions of 1916
  structure of 1913–1915
  synthesis of 1911–1913
  UV spectra of 1915, 1916
Powder patterns 350
Prismanes—*see also* Germaprismanes, Silaprismanes, Stannaprismanes
  stability of 122
  strain energy of 120
Prochiral alkenes, asymmetric hydrosilylation of 1752–1755
Prochiral 1,3-dienes, asymmetric hydrosilylation of 1755, 1756
Prochiral imine *N*-oxides, asymmetric hydrosilylation of 1750
Prochiral imines, asymmetric hydrosilylation of 1749
Prochiral ketones, asymmetric hydrosilylation of 1743–1748
  intramolecular 1750, 1751
Propargylic alcohols, hydrosilylation of 1727, 1728
Propargylsilanes, reactions of, effect of R_3Si on 411, 412
Prostaglandins, synthesis of 1635
Proton affinities, of $H_2Si(C)=X$ 1066, 1067
Pseudopotential methods, for calculation of energy levels 1320
Pseudorotation 1633–1635
Pyrans—*see* Dihydropyrans, Silapyrans, Tetrahydropyrans
Pyroglutamates, synthesis of 1576
Pyroxene 103

Quartz 103, 109, 112, 115, 116

Racemate resolution 2364, 2365, 2368, 2369
Radical reactions,
  chain 1540, 1541
  consecutive,
    initiation by atom (or group) abstraction 1557–1572
    initiation by radical addition to unsaturated bonds 1572–1577

Radicals,
  enthalpies of formation of 167–169
  photoformation of 1237, 1241, 1244, 1274, 1275
[1,5]-Radical translocation 1567
Raman spectroscopy,
  of silanethiones 1090
  of silatranes 1471, 1472
Reaction rates, effect of silyl groups on 485–487
Rearrangements,
  dyotropic 881, 1130, 1285, 1288
  haptotropic 1099
  norbornyl–norpinyl 640
  silene–silene 881, 882, 911–914
    photochemical 1285
  silene–silylene 910, 911, 2484, 2485
  silylcarbene–silene 712–726, 735, 737–739, 743, 744, 2402–2414
  silylcarbene–silylketene 712, 718, 720
  silylene–silene 1331
  silylsilylene–disilene 1320, 1321
  steric effects of silyl groups on,
    comparison of 452–454
    comparison with H atom and R groups 462, 474
Reduction,
  cathodic 1211–1229
  microbial 2376–2384
  steric effects of silyl groups on,
    comparison of 457
    comparison with H atom and R groups 470–474
Reduction potentials,
  of acylsilanes 1213
  of arylsilanes 1211
  of cyclooctatetraenes 1212
  of disilenes 1214
Reformatsky reaction 1635
Relaxation times,
  for polysiloxanes 306–309
  for silicate glasses 325–329
Retro-ene reactions, in synthesis of silanimines 1022–1024
Rhodamines, as lasing dyes in sol–gel matrices 2351
Rhodium catalysts,
  for dehydrogenative silylation 1714–1716
  for hydrosilylation,
    asymmetric 1743–1749
    of alkenes 1688–1690, 1692, 1701–1703
    of alkynes 1718–1720, 1727–1730, 1732
    of carbonyls 1733–1738
    of epoxyalkenes 1763, 1764
    of polymers 1769, 1770
  for silylcarbocyclization 1780–1784
  for silylcarbonylation 1772–1779

Ruppert's reagent 1675
Ruthenium catalysts,
  for dehydrogenative silylation 1714, 1715
  for hydrosilylation 1702
    asymmetric 1750
    of alkynes 1720, 1721
Rydberg states 1316, 1320

Sacrificial anodes 1214, 1222, 1224, 1229
Sakurai–Hosomi reaction 1801, 1802, 1811, 1850, 1856
Sandorfy C method 1312–1314
SASS 310
Schwartz reagent 1659
SCM, Direct Process and 1586
Seitz–Wrighton mechanism, for alkene hydrosilylation 1705–1707
Selective population transfer, in ^{29}Si NMR 269
Selenacyclohexanes—see Trisilatriselenacyclohexanes
Selenides—see Bis(silyl)selenides, Cyclic selenides
Selenoacetals, reduction of, using (TMS)$_3$SiH 1546
Selenoaldehydes, synthesis of 1093, 1094
Selenoesters, reduction of, using (TMS)$_3$SiH 1546
Selenophenes—see Benzoselenophenes
Selenosilanes 1884, 1885
  synthetic applications of 1677–1679
Selones—see Silaneselones, Stannaneselones
Semiempirical calculations,
  for acylsilanes 1605
  for oligosilanes 1316
  for ^{29}Si NMR parameters 293–296
(+)-Sesbanimide, synthesis of 1638
Sesquiterpenes, synthesis of 1653
Sharpless oxidation 1845
Shielding anisotropy 319
Shielding tensors 318, 319, 322
Si$_{60}$,
  endohedral complexes of 72–75
  structure of 70–72
Si$_{70}$ 75–77
SiAlON materials, precursors to 2264, 2265
Si—Br bonds, geometry of 245, 247
SiC,
  phase-pure 2286, 2287
  precursors to 2272–2287
SiCB materials, precursors to 2287–2289
Si—C bonds,
  cleavage of 1451, 1459, 1492, 1493, 1500, 1829–1831
    oxidative 1680, 1681
    promoted by organometallics 1824–1829
    protolytic 1457

  electrophilic substitution of 1794–1824
  geometry of 185–197
Si—C(aryl) bonds, geometry of 191, 192
Si—C(sp) bonds, geometry of 185–189
Si—C(sp^2) bonds, geometry of 189–191
Si—C(sp^3) bonds, geometry of 192–194
Si=C bonds, geometry of 250, 251
Si—Cl bonds, geometry of 239–246
SiCN, precursors to 2253–2261
SiCO materials, precursors to 2289–2294
Si—F bonds, geometry of 233–239
SIFT technique 1028
1,5-Sigmatropic rearrangements, in cyclopentadienylsilanes 2132
Si—I bonds, geometry of 245, 247
Silaacenaphthendiides—see 1,2-Disilaacenaphthendiides
Silaacenaphthenes—see 1,2-Disilaacenaphthenes
Silaacrylates 905, 2405
  cycloadditions of 969
  dimerization of 931
Silaacyl zirconium complexes, reactions of 1606
Silaalkenes—see also Silacycloalkenes, Siladienes, Silaethylenes, Silenes
  formation of 717
  thermochemistry of 174, 175
Silaallenes,
  complexes of 2057
  dimerization of 917, 919
  NMR spectra of 987, 989, 991
  photoformation of 1175, 1249, 1250
  photolysis of 1285, 1286
  reactivity of 1000–1008
  stability of 977
  structure of 251, 984
  synthesis of 998–1000
  theoretical studies of 997, 998
  UV spectra of 992, 993, 996
Silaamidides 1030–1032
Sila anions—see also Silacyclopentadienide anions, Silacyclopentadienyl anions, Silafluorenide anions
  rearrangement of 1133, 1134
Silaanthracenes—see also 9,10-Dihydrosilaanthracenes
  matrix isolation of 1151
  photoformation of 1246
Silaantiaromatics 83–93
  charged 88–93
  matrix isolation of 1152–1156
Silaaromatics—see also Silaanthracenes, Silabenzenes, Silabenzenoids, Silanaphthalenes, Silatoluenes
  charged 32–51

Subject index

matrix isolation of 1148–1152
1-Sila-3-azacyclobutanes, synthesis of 951
Silaazetidines,
  cycloreversion of 1014, 1015
  formation of 968, 969
  thermolysis of 1016, 1018
Silaaziridines, formation of 951
Silabenzenes 10, 2412, 2413—*see also* Dewar silabenzenes, Disilabenzenes, Hexasilabenzenes, Monosilabenzenes, Trisilabenzenes
  formation of 722–726, 908, 909, 1071
  photochemical 1282
  IR spectra of 995
  matrix isolation of 1148–1152
  NMR spectra of 987, 990
  theoretical aspects of 11–13
  UV spectra of 994
Silabenzenoids—*see also* Heterosilabenzenoids, Silaaromatics
  experimental background to 10
  theoretical aspects of 11–32
Silabenzvalenes—*see* Disilabenzvalene, Hexasilabenzvalene
Silabenzyl cation 34
closo-Silaboranes 81–83
Silabutadienes 902, 903—*see also* Silacyclobutadienes
  cycloadditions of 973, 974
  IR spectra of 993, 995
  rearrangement of 916
  synthesis of 860, 861, 870–872, 933
  UV spectra of 993
Silabutatrienes, dimerization of, head-to-tail 1002
Silacarbonates, matrix isolation of 1161
Silacarbonyl ylides 2416, 2417
Silacarboxylic esters, matrix isolation of 1161, 1162
Silacarbyne complexes 2057
Silachalcogenides—*see* Cyclosilachalcogenides, Disilachalcogenides, Silaselenides, Silasesquichalcogenides
Silacycloalkanes—*see also* Disilabicycloalkanes, Silacyclobutanes, Silacycloheptanes, Silacyclohexanes, Silacyclopentanes, Silacyclopropanes
  thermochemistry of 162, 163
Silacycloalkenes—*see* Polysilabicyclo[*n*.1.0]alkenes, Silacyclobutenes, Silacycloheptenes, Silacyclohexenes, Silacyclooctenes, Silacyclopentenes, Silacyclopropenes
Silacycloalkynes—*see also* Pentasilacycloheptyne, Silacyclooctynes, Silacyclopropyne, Tetrasilacyclohexynes

  photoformation of 1286, 1288
Silacyclobutadienes 83–88—*see also* Disilacyclobutadienes, Monosilacyclobutadienes, Tetrasilacyclobutadiene
  dimerization of 1155
  head-to-head 922, 923
  matrix isolation of 1153–1155
  photoformation of 1283, 1284
  reactions of,
    with alcohols 932, 933
    with alkoxysilanes 937
Silacyclobutanes—*see also* 1-Aza-2-silacyclobutanes, Diarylsilacyclobutanes, Disilacyclobutanes, 1-Sila-3-azacyclobutanes, Siletanes, Trisilacyclobutanes
  cycloreversion of 860–866
  formation of 947, 955
  photoreactions of 1237, 1240, 1241
    with fullerenes 1301, 1302
  pyrolysis of 982, 1080, 2477
  thermolysis of 844
Silacyclobutenes—*see also* Benzosilacyclobutenes, 1-Oxa-2-silacyclobut-3-enes, Tetrasilacyclobutenes
  photolysis of 916, 933, 1241
  pyrolysis of 870–872
  synthesis of 881, 882, 910, 956
Silacyclobutenylidene, matrix isolation of 1171
Silacycloheptanes—*see* Trisilacycloheptanes
Silacycloheptenes—*see* 1,4-Disilacycloheptenes, Tetrasilabicyclo[4.1.0]hept-1(7)-ene, Trisilacycloheptenes
Silacyclohexanes—*see* 2-Acetoxy-1-silacyclohexanes, 6-Oxa-3-silabicyclo[3.1.0]hexanes, Tetrasilabicyclohexanes, Trisilatriselenacyclohexanes
Silacyclohexanols, enzymatic transesterification of 2388, 2389
Silacyclohexenes—*see* 1-Sila-2-oxacyclohex-3-enes
Silacyclooctenes—*see* Pentasilabicyclo[5.1.0]oct-1(8)-ene
Silacyclooctynes—*see* 1,2-Disila-3,8-dioxacyclooctynes, Hexasilacyclooctyne
Silacyclopentadienes 1170
  light-emitting polymers of 1304, 1305
  photoformation of 1283
Silacyclopentadienide anions 2019–2021
Silacyclopentadienyl anions 37–51
Silacyclopentadienyl cations 89–93
Silacyclopentadienyl complexes, with transition metals 2026, 2027, 2057, 2082, 2083
Silacyclopentanes—*see* Disilabicyclopentanes, Divinylsilacyclopentanes, 1-Thia-2-silacyclopentanes, Trisilabicyclopentanes, Trithiadisilacyclopentanes

Silacyclopentenes,
  formation of 2467, 2488–2490
  photolysis of 1245, 1246
  pyrolysis of, gas-phase 1984–1986
Silacyclopentenylidenes, rearrangement of 2503
Silacyclopropanes—*see also* Disilacyclopropanes, Siladigermacyclopropane, Trisilacyclopropanes
  bridged 600, 601
  photoformation of 1286, 1288
  photolysis of 1238, 1239
  pyrolysis of 2479
  structure of 194, 195
Silacyclopropanimines, formation of 951
Silacyclopropenes,
  as silaallene precursors 998, 999
  matrix isolation of 1165, 1166, 1175
  photoformation of 1249–1251
  photolysis of 1238
  pyrolysis of 2479
  reactions of 1004
  stability of 997
  structure of 194, 195
Silacyclopropenium cations—*see* Monosilacyclopropenium cation, Trisilacyclopropenium cation
Silacyclopropenylidenes 66–68—*see also* Azasilapropenylidenes
  matrix isolation of 1163–1165, 1173–1175
  photoformation of 1286, 1288
  stability of 997
Silacyclopropylmethyl cations 543
Silacyclopropyne, matrix isolation of 1163–1165
Silacyclopropynylidene 66–68
Siladienes—*see also* Arylsiladienes, Silabutadienes, Silaheptadienes, Silahexadienes, Silanonadienes, Silapentadienes, Silapropadienes
  photoformation of 1236, 1237, 1241, 1242
  UV spectra of 996
Siladienones 2407, 2408
Sila-difenidol 2364
Siladigermacyclopropane, photolysis of 1238, 1239
Siladihydrotetrazoles, synthesis of 1013
Siladioxetanes—*see also* Disiladioxetanes
  formation of 978, 979
Siladioxiranes, photoformation of 1286
Siladithietanes 1882, 1883—*see also* Disiladithietanes
Silaethylenes—*see also* Polysilaethylene
  formation of 1071
Silafluorenes, photoformation of 1263
Silafluorenide anions 2019–2021
Silaformamide ion 1028

Silafullerenes 69–77
Silafulvenes 2130, 2412–2414—*see also* Silatriafulvenes
  formation of 722–724
  photochemical 1282
Silagermenes, photoformation of 1238, 1239
Silaguanidinium ion 528, 529, 538–540
Silaheptadienes—*see* Tetrasilacyclohepta-1,2-diene
Silaheterocycles, formation of 2168
  from acylsilenes 719, 720
  from silylcarbenes 727, 730, 732
  from silylene-dicarbenes 734
Silahexadienes—*see* 1,4-Disilacyclohexadienes, Trisilacyclohexadienes
Silahydrotriazoles, synthesis of 1012, 1013
Silaimines, complexes of 2057
Silaindacenes—*see* Disilaindacenes
Silaindanes—*see also* Disilaindanes
  formation of 914
    photochemical 1273, 1274, 1276, 1285
Silaindenes, formation of 1004, 1005
Silaisonitriles,
  IR spectra of 1021, 1046
  matrix isolation of 1158–1160
  synthesis of 1017, 1018, 1021, 1024, 1026
  theoretical studies of 1049
  UV spectra of 1021
Silametallacycles 1904, 1905—*see also* Disilametallacycles
Silanaphthalenes 10
  formation of 726
  NMR spectra of 987, 990
  Raman spectra of 994, 995
  structure of 984
  theoretical aspects of 16, 17
  UV spectra of 994
Silanediimines 1029, 1030
  matrix isolation of 1160
  theoretical studies of 1046, 1048
Silanediyl complexes 2057
Silanephosphimines, structure of 252–254
Silaneselones 1064
  formation of 1097–1100
Silanetellones 1097
Silanethiols 1874, 1875—*see also* Triphenylsilanethiol
Silanethiones 1064, 1068, 1101, 1878, 1879
  electronic spectra of 1067, 1068, 1091, 1092
  formation of 1083
    from diaminosilanes 1086
    from disilenes 1094, 1095
    from hydrosilanes 1084–1086
    from Si-containing cyclic polysulphides 1090–1094
    from silylenes 1086–1089
    from thioketenes 1095, 1096

matrix isolation of 1096, 1097, 1163
NMR spectra of 1086, 1087
Raman spectra of 1090
reactivity of 1091
structure of 252, 254, 1084, 1086, 1090, 1091
theoretical studies of 1065–1068
UV/visible spectra of 1090
Silanimines 1068, 1100—*see also* Iminosilanes, Silanediimines
  complexes of 1032, 1033, 1040–1042, 2049
  cycloadditions of 1035–1040
  dimerization of 1036
  donor addition to 1033, 1034
  ene reactions of 1035
  insertion reactions of 1034, 1035
  IR spectra of 1044, 1045
  matrix isolation of 1158–1160
  NMR spectra of 1043, 1044
  photolysis of 1158
  structure of 251–253, 1042, 1043
  synthesis of 1010–1020, 1022–1026
  theoretical studies of 1046–1048
  UV spectra of 1044, 1045
Silanitriles,
  formation of 1021, 1022
  matrix isolation of 1158–1160
  theoretical studies of 1049
Silanium ions 1107, 1108
Silanolates, elimination from α-hydroxysilanes 884
Silanol complexes,
  with group 9 transition metals 2107–2109
  with group 10 transition metals 2110
Silanol ions 520
Silanols 2320—*see also* 1,3-Bis(silanol)s
  chiral—*see* Chiral silanols
  protonated 580
  reactions of 1074, 1075
  structure of 221, 222
Silanonadienes—*see* Hexasilacyclonona-1,2-diene
Silanones,
  as intermediates 1038
  dimerization of 1081
  formation of 977–979, 1022
    from allyloxysilanes 1071–1073
    from arsanilidenesilanes 1080–1082
    from cyclic silyl ethers 1069–1071
    from hydrosilanes 1068, 1069
    from phosphasilenes 1081, 1082
    from silanethiones 1082, 1083
    from silanols 1074, 1075
    from silenes 1080
    from silylenes 1075–1080, 2502, 2503
  photochemical 1264, 1265
  matrix isolation of 1161

structure of 252, 254
theoretical studies of 1065–1067
thermochemistry of 175, 176
Silanorbornadienes,
  photofragmentation of 2473, 2474
  photolysis of 1293
  photorerrangement of 1293
Silanorbornadienyl cations 543
Silanorbornanes—*see* Oxahexasilanorbornanes
Silanorbornenes, photolysis of 1242, 1244
Silanorbornyl cations 543, 544, 573, 695–697
  NMR spectra of 696, 697
1-Sila-2-oxacyclohex-3-enes, formation of 966
Silaoxetanes 716, 719–721—*see also* Siladioxetanes
  formation of 727, 728
Silaoxiranes—*see* Disilaoxiranes
Silaoxonium ions 586
Silaparacyclophanes—*see* Heptasila[7]paracyclophane, Tetrasila[2.2]paracyclophane
Silapentadienes—*see* 1,5-Disilapentadiene, Silacyclopentadienes
Sila-Peterson reaction 884–890, 937, 957, 983
Silapolychalcogenides, synthesis of 1904
Silapolyhedranes—*see also* Silatetrahedranes
  NMR spectra of 146–148
  synthesis of 125, 128, 129, 136, 137
  theoretical studies of 120–125
Sila-pridinol 2364
Silaprismanes—*see* Persila[*n*]prismanes, Trisilaprismanes
Sila-procyclidine 2364–2368, 2374, 2375
Silapropadienes, dimerization of 1002
Silapropellanes—*see* [3.1.1]Trisilapropellane
Silapropenylidenes—*see* Trisilapropenylidene
Sila-Pummerer rearrangement 1608
Silapyrans,
  Diels–Alder reaction of 1070
  thermolysis of 1072, 1073
Silaselenanes, cyclic—*see* Cyclic silaselenanes
Silaseleniranes, synthesis of 951, 952
Silaselenones, photoformation of 1289
Silasesquichalcogenides 1896, 1897
  adamantane-type 1916–1920, 1923
  bis-nor-adamantane-type 1917, 1921–1923
  bis-nor-double-decker-type 1917
  double-decker-type 1916–1920, 1923
  nor-adamantane-type 1916, 1917, 1921–1923
  nor-double-decker-type 1917
Silasesquioxanes, X-ray–NMR studies of 324, 325
Silasesquioxides 1916
Silastyrene, formation of 867, 870
Silatetrahedranes—*see also* Tetrasilatetrahedranes
  matrix isolation of, theoretical 1153

Silatetrathialanes 1884
Silatetrathialenes 1883, 1884
Silathianes—*see* Cyclic silathianes, Disilathianes
Silathietanes 1882, 1883—*see also* Disilathietanes, Germasilathietanes, Siladithietanes
Silathiiranes 1881
  synthesis of 951, 952
Silathiocarbonyl ylides 2416, 2417
Silathiones, photoformation of 1264, 1265
Silatoluenes 10
  formation of 1071
  IR spectra of 995
  matrix isolation of 1149
  UV spectra of 994
Silatrane-4-carboxylic acids,
  structure of 1464
    zwitterionic 1485, 1486
  synthesis of 1455
Silatrane-3, 7-diones, synthesis of 1507
Silatrane hydrochlorides, synthesis of 1485
Silatranes—*see also* 1-Acyloxysilatranes, 1-Alkenylsilatranes, 1-Alkoxysilatranes, 1-Alkylsilatranes, 1-Alkynylsilatranes, 1-Allylsilatranes, 1-(3'-Aminopropyl)silatranes, 1-Aralkylsilatranes, 1-Aryloxysilatranes, 1-Arylsilatranes, Azasilatranes, Benzosilatranes, Carbasilatranes, 1-Halosilatranes, 1-(Heteroarylalkyl)silatranes, 1-Heteroarylsilatranes, 3-Homosilatranes, 1-Hydrosilatranes, 1-Hydroxysilatranes, 1-Isothiocyanatosilatranes, 1-Organylsilatranes, Trithiasilatranes, 1-Vinylsilatranes
  basicity of oxygen atoms in 1465, 1472
  bonding model for 1465, 1466
  carbofunctional 1454
    reactions of 1500–1503
  C-substituted 1452–1454
    conformation of 1463
    solvolytic stability of 1484
  dipole moments for 1469, 1470
  geometry of 183
  hydrolysis of 1483–1486
    mechanism of 1484
  mass spectra of 1481–1483
  molecular structure of 1460–1465
  NMR spectra of 336–338, 1473–1479
  oxidation of, anodic 1200, 1480, 1481
  PE spectra of 1470, 1471
  quantum-mechanical studies for 1466–1469
  synthesis of 1449
    from organylhalosilanes 1457, 1458
    from organyltriacetoxysilanes 1456, 1457
    from organyltrialkoxysilanes 1450–1456
    from organyltris(dialkylamino)silanes 1458
  thermochemistry of 1469
  UV spectra of 1471
  vibrational spectra of 1471, 1472
Silatrane-3, 7, 10-triones,
  IR spectra of 1509
  synthesis of 1507, 1508
Silatranones—*see also* Azasilatranones
  mass spectra of 1509
  NMR spectra of 1508, 1509
  reactions of 1508
  synthesis of 1507
  X-ray studies of 1508
2-Silatranylacetaldehyde, synthesis of 1501
Silatranyl complexes 1503–1505
Silatranyl group, electronic effects of 1479, 1480
Silatranyl oxonium salts, structure of 1460
Silatriafulvenes 887, 888
  NMR spectra of 989
  rearrangement of 910, 911
Sila-tricyclamol iodide 2364, 2365, 2367, 2368, 2374, 2375
Silatrienes 895, 897
  cycloadditions of 969, 970
  dimerization of 928, 930
  ene reactions of 977, 1252, 1254
  methanol adducts of 1256, 1257
  photoformation of 1251–1254, 1256
  trapping of 1252, 1253, 1255
  UV spectra of 994, 996
Sila-trihexyphenidyl 2364
Silatrimethylenemethane 2425–2427
  complexes with transition metals 2057, 2089–2092
Silatropylium cation 34–36
Silavinylidenes, theoretical studies of 1008–1010
Sila-Wagner–Meerwein rearrangement 547
Sila-Wittig reaction 1081
Silaylides 2485, 2486, 2491
Silazanes—*see also* Disilazanes, Perhydropolysilazanes, Polymethylaminoborosilazanes
  coupling constants for 302–304
Silenes 859, 860, 1064, 1068—*see also* Apeloig–Ishikawa–Oehme-type silenes, Apeloig-type silenes, Arsasilenes, Auner–Jones-type silenes, Bis-silenes, Brook-type silenes, Dialkylsilenes, Disilenes, Neopentylsilenes, Phosphasilenes, Silaalkenes, Siloxysilenes, α-Silyloxysilenes, Wiberg-type silenes
  addition of alcohols to 845
    diastereochemistry of 845–853
  adducts of 980

NMR spectra of 988
  with amines 844, 845
  as reaction intermediates 2403–2406
  as silylcarbene rearrangement products 712–726
  complexes of 2057, 2083–2088
  cyclic—*see* Cyclic silenes
  cycloadditions of 940–974, 1080, 1280
    with carbonyls 958–970
  dimerization of 916–931, 939
    head-to-head 881, 888, 917, 919–930
    head-to-tail 881, 884, 886, 900, 917–919, 926, 931
  ene reactions of 974–977
  E/Z isomers of 1276
  ICR spectra of 996, 997
  IR spectra of 991–995
  long-lived 1274
  matrix isolation of 714, 1145–1148
  microwave spectra of 864
  NMR spectra of 985–991
  nucleophilic additions of 932–940
  oxidation of 977–980, 1080, 1081
  PE spectra of 996
  photoformation of 1237, 1241, 1245, 1247, 1248, 1251, 1257, 1262, 1264–1266, 1272, 1273, 1276–1278
  photoisomerization of 1167, 1168
  photolysis of 1273–1275, 1284, 1285
  reactions of,
    in argon matrices 1147, 1148
    with alcohols 1256, 1257
    with alkoxysilanes 846, 847
  rearrangement of 881, 882, 910–916, 1285, 2484, 2485
  structure of 250, 251, 981–985
    calculated 982–985
  synthesis of 844
    by cycloreversion reactions 860–876
    by donor cleavage 879
    by isomerization of acylpolysilanes 880–883
    by photolysis of disilanes 891–900
    by rearrangement of silylenes and carbenes 900–909
    by salt elimination 873, 876–879
    by sila-Peterson reaction 884–890
  transformations of 714
  UV spectra of 992–994, 996, 2408
Silene–silene rearrangements 881, 882, 911–914
  photochemical 1285
Silene–silylene rearrangements 910, 911, 2484, 2485
Silenolate anions, NMR spectra of 988, 990, 991
Silenolates 888

cycloadditions of 957, 958
synthesis of 807, 808
Silepines, synthesis of 892, 893
Siletanes—*see also* Bisalkylidenesiletanes, Diazasiletanes, Disiletanes, Dithiasiletanes, Oxasiletanes, Silacyclobutanes
  thermochemistry of 162, 163
Siletanides—*see* Oxasiletanides
Siletenes—*see* Oxasiletenes
Silica gels 2350
  as solid supports 311–313, 315
  NMR spectra of 343
Silicates 103–117—*see also* Aluminosilicates, Hydridosilicates, Metasilicates, Orthosilicates, Spirosilicates, Tetrasilicates
  bond length–bond strength variations in 104–109
  electron density distribution of skeletal Si—O—Si unit in 113–115
  precursors of 2298–2300, 2307–2310
  structures for 115, 116
Silica-W 116
Silicenium ions 620, 652, 1087, 1408, 1411—*see also* Trimesitylsilicenium ion, Tris-(9-borylbicyclo[3.3.1]nonyl)silicenium ion
  application in cationic polymerization 551
  as reaction intermediates 575–581
  bridged 577, 624
  caged 550, 551
  clustering reactions of 560
  computational methods for,
    ab initio 515, 516
    DFT-based 516
    magnetic property 516, 517
  π-donor-stabilized 538–547
  electronic spectra of 565
  electrophilicity of 558
  exchange reactions of 560
  free 519, 520, 558, 559
  gas-phase studies of 559–562, 1108
  hydride affinities of 558, 560
  ion–molecule reactions of 1121, 1122
  IR spectra of 559
  mass spectra of 559–561
  NMR spectra of 280–284, 565–575
  organoboryl-substituted 547–549
  radiative association reactions of 1121
  rearrangement of 1130–1133
  solid-state studies of 581–588
  solution studies of 562–581
  stability of, compared to silyl cations 597, 607
  thermochemistry of 559, 560, 1108
  thermodynamic stability of, calculated 517–519

Silicenium ions (cont.)
  trialkyl-substituted 520–532
    complexes of 524–532, 608–610
    unimolecular decomposition of 707
Silicocenes 77–80—see also Decamethyl-
    silicocene
  reactions of 574
    with CO_2 1079, 1080
    with COS 1088
    with CS_2 1089
    with isothiocyanates 1088
    with n-Bu_3P 1099
Silicon, activating and directive effects of 356–425
Siliconates—see also Bis(siliconate)s
  pentacoordinate, NMR spectra of 284
  spirobicyclic 1436
Silicon atoms, free 2504
  reactions of 2468, 2469
Silicon carbide, synthesis of 559, 2511, 2512
Silicon–carbon clusters, matrix isolation of 1175, 1176
Silicon cations, atomic, reactions with neutral molecules 1117, 1118
Silicon clusters 149, 150
Silicon complexes,
  heptacoordinate 1430–1433
  hexacoordinate 1415–1430
  octacoordinate 1430–1433
  pentacoordinate 1373–1387, 1390–1412
Silicon connection, temporary 1674
Silicon-containing reagents, reactivity of,
  in addition 475–477
  in condensation 477
  in cycloaddition 477, 478
  in hydrosilylation 479
  in insertion 479
  in nitrone formation 479, 480
  in olefination 480
  in O—O bond cleavage 481
  in silylation 481
  in silylformylation 481
  in silylstannation 482
  in substitution 475–477
α-Silicon effect 610–616
  in bicyclobutonium ions 690–693
  in ethers 1201, 1202
β-Silicon effect 616–635, 706, 707
γ-Silicon effect 635–641
  in bicyclobutonium ions 693–695
  stereochemical requirements of 635
δ-Silicon effect 641–645
Silicones, polymerization of 2218–2224
Silicon hydride plasmas 560
Silicon hydrides,
  NMR chemical shifts of, calculated 293, 294
  thermochemistry of 158, 159, 1106–1109
Silicon nitride, synthesis of 559
Silicon oxide diradical 2523
  adducts of 2510
Silicon oxides, matrix isolation of 1162
Silicon oxynitride, precursors to 2261–2264
Silicon sulphide diradical 2508, 2510
Silicon tethers 1196
Siliranes 907, 2418–2421—see also Alkyl-
    idenesiliranes, Disiliranes, Oxasiliranes,
    Thiasiliranes
  as intermediates in silylene reactions 2501, 2502
  photolysis of 1283, 2475, 2477
  thermochemistry of 163
Siliranethiones—see Thiasiliranethiones
Silirenes,
  formation of 734
  in silylene addition reactions 2487
  photochemical 1244
  thermochemistry of 163
Silirenyl anion 88, 89
Silolanes—see Tetrathiasilolanes
Silole anions 37–42, 2028, 2029
Silole dianions 37–42, 2022–2024, 2028, 2029
  higher congeners of 50
  lithiated complexes of 47–49
  NMR spectra of 989
  structure of 984
Silole polymers 1304, 1305
  involving only Si of silole ring 2017, 2018
Silole–pyrrole co-oligomers, synthesis of 2028
Siloles—see also Oligosiloles, 2,3,4,5-Tetra-
    methylsiloles
  as silylene rearrangement products 901–903
  metallated,
    aromaticity of 814, 817
    NMR spectra of 814–818
    structure of 818, 819
    synthesis of 814
  synthesis of 1963–1968
Silole–thiophene copolymers, synthesis of 1978, 2015, 2016
Siloxane resins, thermal degradation of 2231–2233
Siloxanes—see also Aminoalkylsiloxanes,
    Cyclosiloxanes, Disiloxanes, Poly-
    siloxanes
  formation of 977
  NMR spectra of 306–309
    solid-state 315–317
  organofunctional,
    applications of 2226, 2227
    synthesis of 2224–2226
  polycondensation of,
    anionic 2224
    cationic 2221–2224

reactions of 1084
thermochemistry of 164, 165
Siloxene,
  photoluminescence of 2196
  structure of 2197, 2198
Siloxetanes—*see also* Disiloxetanes
  arsenic-containing 1081, 1082
  as reaction intermediates 1070
  formation of 959–964, 967, 969, 970
    photochemical 1303
  isomerization of 960
Siloxycarbenes, photoformation of 1270, 1271
Siloxysilanes—*see* Poly(siloxysilane)s
Siloxysilenes,
  cycloadditions to 953–956, 1645
  dimers of 881
  reactions with organometallics 890
  synthesis of 880–882
Siloyl anions 42–47, 2020
  higher congeners of 50
  silyl-substituted 47
Silsesquioxanes 1923–1925, 2289, 2290, 2331–2334
Silthianes—*see* Cyclic silthianes, Disilthianes
Silyconium ions 1408
Silylacetylides, reactions with C_{60} 1958, 1959
Silyl acyloin reaction 1614, 1615
1-Silylalkylamides, synthesis of 1689
1-Silylalkylureas, synthesis of 1689, 1690
α-Silylallyl cations, *ab initio* calculations for 597
β-Silylallyl cations, NMR spectra of 662, 663
O-Silylallylic alcohols, intramolecular hydrosilylation of 1710
N-Silylallylic amines, intramolecular hydrosilylation of 1710–1712
β-(Silyl)allylic ethers, rearrangement of 1858, 1859
Silylamines—*see* Aminosilanes
Silylammonium cations 564
Silyl anions—*see also* Hydridosilyl anions, α-Silyl carbanions
  alkoxy-substituted 810–812
  alkyl-substituted 781–788
  amino-substituted 808–810
  aryl-substituted,
    ion-pairing of 793
    resonance effects in 793
    structure of 793, 794
    synthesis of 788–792
  functionalized 807–813
  halogen-substituted 812, 813
  silyl-substituted 794–807
  thermochemistry of 1109
Silylarenium ions, structure of 652–654
Silylaryl cations, *ab initio* calculations for 601

Silylation—*see also* Allylsilylation, Hydrosilylation
  dehydrogenative 1714–1716, 1772
  double 1691
  electrophilic aromatic 561
  electroreductive 1221–1229
  reactivity of Si-containing reagents in 481
4-Silylazetidin-2-ones, anodic oxidation of 1201
Silylbromonium zwitterion 655
Silylcarbamates, photoreactions of 1294
α-Silylcarbamates, anodic oxidation of 1199, 1200
α-Silyl carbanions,
  effect of R_3Si on 381, 382
  formation of 415–417
Silylcarbenes 914, 915—*see also* Acyl(silyl)carbenes, (Alkenylsilyl)carbenes, Alkenyl(silyl)carbenes, Alkoxycarbonyl(silyl)carbenes, (Alkyldimethylsilyl)carbenes, (Alkynylsilyl)carbenes, (Allyldimethylsilyl)carbene, Aminocarbonyl(silyl)carbenes, Aryl(silyl)carbenes, Carboxy(trimethylsilyl)carbene, Cycloalkyl(silyl)carbenes, Disilylcarbenes, Halo(silyl)carbenes, Imino(silyl)carbenes, (Methylsilyl)carbene, Phosphino(silyl)carbenes, Phosphoryl(silyl)carbenes, Silylketocarbenes, (Silylmethyl)carbene, (Trialkylsilyl)carbenes, (Trimethoxysilyl)carbene, (Trimethylgermyl)(trimethylsilyl)carbene
  C—H insertion reactions of 1276, 1278, 1285
    intramolecular 727–729
  electronic structure of 704–711
  formation of 711, 712, 881
    from 2-lithio-2-silyloxiranes 772
    from (silyl)alkynones and alkenes 772, 773
    from 2-silylfurans 767–769
    from silylketenes 769, 770
    from vinylsilanes 770–772
    photochemical 1273, 1276, 1278, 1285
  general reactivity of 711, 712
  geometry of 704–711
  isolable 520
  metal complexes of 742
  1,2-methyl shifts in 1280
  reactions of,
    intramolecular 727–732
    not involving the silyl group 739–767
  rearrangement of 712–726, 735–739, 743–746, 904, 906
    photochemical 1276, 1277, 1285

Silylcarbenes (cont.)
    singlet–triplet gap in 705–707
    spin state of 712, 748
    sulphur-substituted 766, 767
    thermochemistry of 901
Silylcarbene–silene rearrangement 712–726,
    735, 737–739, 743, 744, 2402–2414
Silylcarbene–silylketene rearrangement 712,
    718, 720
α-Silyl carbocations—see also
        α-Silylallyl cations, α-Silylethyl cations,
        α-Silylmethyl cation, α-Silylvinyl cations
    ab initio calculations for 597, 598, 647
    effect of R$_3$Si on 362–364
    gas-phase studies of 601–604
    in solvolysis 610–616
    NMR spectra of 645–648
β-Silyl carbocations—see also
        β-Silylallyl cations, β-Silyldienyl cations,
        β-Silylethyl cations, β-Silylvinyl cations
    ab initio calculations for 598–601, 607, 622,
        656, 658, 659
    electron demand of 623
    formation of 413, 414
    gas-phase studies of 604–610
    in solvolysis 616–635
    NMR spectra of 655–663
    UV spectra of 649, 650
γ-Silyl carbocations 635–641
    formation of 414, 415
    percaudal interactions in 636, 637
δ-Silyl carbocations 641–645
    percaudal interactions in 642
Silylcarbocyclization 1779–1785
    cascade 1782, 1783
Silylcarbonylation 1771–1779
    of enamines 1772, 1773
Silyl complexes,
    with actinides 2053
    with copper 2121
    with group 3 transition metals 2038, 2039
    with group 4 transition metals 2042–2052
    with group 5 transition metals 2054–2057
    with group 6 transition metals 2061–2069
    with group 7 transition metals 2069–2071
    with group 8 transition metals 2071–2095
        bimetallic 2072–2074
    with group 9 transition metals 2095–2107
    with group 10 transition metals 2110–2117
    with lanthanides 2039–2041
    with mercury 2121, 2122
Silylcuprates,
    structure of 800
    synthesis of 784, 791
Silyl cyanides, structure of 196
α-Silylcyclobutonium ions 690–693
    NMR spectra of 691, 692

γ-Silylcyclobutonium ions 693–695
    NMR spectra of 694, 695
β-Silylcyclohexadienyl cations 651, 652
    X-ray studies of 652–655
β-Silylcyclohexanones, photolysis of
    1294, 1295
Silylcyclooctatetraenes, cathodic reduction of
    1212
Silylcyclopentadienes—see also Cyclopenta-
        dienylsilanes
    desilylation of 2140, 2141, 2143
    metallation of 2140
    synthesis of 2139
Silylcyclopentadienide anions 2130
    X-ray studies of 2134–2136
Silylcyclopentadienyl metal complexes 2130
    synthesis of 2140–2143
Silylcyclopentadienyl radicals 2130
Silylcyclopropenes—see also Cyclopropenyl-
        silanes
    photoformation of 1249
    photorearrangement of 1236
Silyldiazoalkanes—see also Silyldiazomethanes
    photolysis of 1276–1282, 1286, 1287
(Silyl)diazo compounds,
    as acylsilane precursors 1612, 1613
    catalytic decomposition of 732
    photolysis of 709, 713–716, 731, 732
    pyrolysis of 715, 716, 727
    thermolysis of 731, 732
Silyldiazomethanes, as silene photolytic
    precursors 1146
Silyldiazonium compounds, reactions with C$_{60}$
    1959, 1960
Silyldienes,
    anodic methoxylation of 1193
    cycloadditions of 1846, 1847
β-Silyldienyl cations—see also
        β-Silylcyclohexadienyl cations
    NMR spectra of 663–667
Silyldihydrotetrazoles, formation of 913
2-Silyl-1,3-dithianes, anodic oxidation of 1198,
    1199
Silylene–base complexes 1331
Silylene cations, thermochemistry of 1107
Silylene centres, on activated silicon 2530–
    2532
Silylene complexes 575, 2527–2530
    cyclic—see Cyclic silylene complexes
    with group 6 transition metals 2058–2061,
        2066–2069
    with group 7 transition metals 2069
    with group 8 transition metals 2083–2088
    with group 9 transition metals 2095
    with group 10 transition metals 2117–2121
    with Lewis acids 2518–2521
    with Lewis bases 2485, 2492, 2500

Subject index

Silylene-dicarbenes 733, 734
Silylenes—*see also* Alkoxysilylenes, Alkynylsilylenes, Aminosilylenes, Cyclopropenylsilylenes, Dialkylsilylenes, Diarylsilylenes, Halosilylenes, Hydroxysilylenes, Isocyanosilylenes, Silylsilylenes, Trisilylenes, Vinylidenesilylenes, Vinylsilylene
   Arduengo-type 52–64
   aromaticity of 53–62
   as disilabenzene isomers 19, 20
   as intermediates, in chemical vapour deposition 2545–2550
   as trapping agents 1001
   cyclic—*see* Cyclic silylenes
   cycloadditions to silenes 949, 950
   1,3,2-diazasilol-2-ylidene 52–64
   dimerization of 63, 64
      calculations for 2511
   electronic spectra of 2512–2521
      substituent effects on 1330
   electronic structure of 1329–1331
   energy calculation for 707
   formation of 898, 899, 901–903, 1004
      by pyrolysis of monosilanes 2475–2481
      by silicon atom reactions 2468, 2469
      by thermolysis of polysilanes 2465–2468
      from metal-induced α-eliminations 2481–2484
      from rearrangements 2484, 2485
      photochemical 1236, 1238, 1244, 1245, 1251, 1260, 1261, 1263–1270, 1282–1284, 1319–1321, 2469–2475
   future of chemistry of 2557, 2558
   in direct synthesis of organosilicon compounds 2533, 2534
   IR spectra of 2523
   laser-induced fluorescence spectra of 2521–2523
   matrix isolation of 1163–1181
   oxidation of 1075–1080
   PE spectra of 2523
   reactions of 578, 1025, 2504
      addition 2487–2496
      comparison with carbene reactions 2502
      dimerization 2496–2498
      insertion 2485–2487
      kinetics of 2524–2527
      photochemical 1286–1288, 1301
      with isothiocyanates 1086, 1087
   rearrangement of 62, 63, 2501–2504
   selenation of 1100
   spectral properties of,
      calculated 2508, 2509
      substituent effects on 1169
   stabilized by intramolecular coordination 2550–2555
   stable dicoordinate 2534–2545
      aromaticity of 2538, 2539
      bonding in 2537, 2538
      reactions of 2539–2545
      structure of 2536, 2537
      synthesis of 2535
   theoretical studies of 997, 2505–2512
   thermochemistry of 171–174
      calculated 2507, 2508
   triplet 1331
Silylene–silene photorearrangement 1331
Silylenoids 2481–2483
Silyl enol borinates, bromination of 1620
Silyl enol ethers,
   as acylsilane precursors 1614–1616, 1623, 1624
   oxidation of, anodic 1204, 1205
   photocyclization of 1292
   photocycloadditions of 1295–1297
   synthesis of 1640, 1644, 1645
      electrochemical 1229
α-Silylepoxides, reactions of, effect of R_3Si on 423–425
Silyl ethers—*see also* Alkyl silyl ethers, Allyl silyl ethers, Disilyl ethers, Silyl enol ethers, α-Silyl ethers, Vinyl silyl ethers
   solvolysis of 482, 483
   thermochemistry of 164, 165
α-Silyl ethers, anodic oxidation of 1201–1203
α-Silylethyl cations, *ab initio* calculations for 597
β-Silylethyl cations,
   ab initio calculations for 598, 601
   gas-phase studies of 604
   NMR spectra of 655, 656, 658–662
Silylethylenes—*see also* Disilylethylenes, Poly(silylethylene)
   coupling constants for 296, 298–301
Silylformylation 1774–1779
   reactivity of Si-containing reagents in 481
Silyl Grignard reagents,
   reactions with C_{60} 1957, 1958
   synthesis of 782, 784
γ-Silyl groups, stabilizing effect of 635
Silylidene complexes, Schrock-type 2049
Silyl isocyanides, structure of 196, 197
Silylium complexes, with zirconium 2047
Silylium ions 1436—*see also* Silicenium ions
Silyl ketene acetals, photoreactions of 1293, 1302
Silylketenes 754
   formation of 712, 715, 718, 720, 722, 1649
Silyl-ketocarbenes 718, 719
Silyl ketones—*see also* β-Trimethylsilyl ketones
   IR spectra of 1601

α-Silyl ketones, synthesis of 1576, 1577
Silyllithium compounds,
 configurational stability of 794
 reactions with C_{60} 1951–1957
 transmetallation of 784
Silylmagnesium compounds, synthesis of 782
Silylmercury compounds, transmetalation of 781, 795, 796, 805
Silylmetallic species, reactions of 1608–1610
(Silylmethyl)carbene 706, 707
α-Silylmethyl cation, *ab initio* calculations for 597
N-Silylmethylphthalimides, photolysis of 1294
Silyl migration pathway, for alkene hydrosilylation 1705–1708
Silyl migrations,
 in silylcyclopentadienyl metal complexes 2142, 2143
 nitrogen-to-carbon 1618, 1619
1,2-Silyl migrations 895, 1616, 1651, 2132
 photolytic,
  in acylsilanes 1270
  in disilanes 1247–1250, 1260
  in silylcarbenes 1280
  in silylcyclopropenes 1236
1,3-Silyl migrations 880, 891, 895, 897–899, 927, 1645
 in silene cycloadditions 974
 photolytic 1236
  in acylpolysilanes 1273, 1275, 1276
  in allylsilanes 1332
  in disilanes 1247, 1249, 1250, 1252, 1253, 1256, 1259, 1272, 1303
  in silanorbornenes 1242
1,4-Silyl migrations, photolytic 1253
1,8-Silyl migrations 892, 894
 photolytic 1253
Silylnitrium ions, NMR spectra of 284
α-Silyloxyalkyl hydroperoxides, reactions of 1682
Silyloxyallenes, synthesis of 1653
Silyloxycarbenes,
 formation of 1646, 1647
 insertion reactions of 1646–1648
 rearrangement of 1614
α-Silyloxysilenes, synthesis of 1645
Silyl perchlorates, ionization of 563, 565
Silylphosphines, structure of 211–215
Silylpotassium compounds, transmetallation of 784
Silyl protecting groups, for alcohols 1297
Silyl radicals 891, 892, 898, 900
 ionization potentials for 558
 long-lived 1266
 photoformation of,
  from acylsilanes 1271, 1272, 1274
  from disilanes 1247, 1248, 1251, 1253–1255, 1257, 1262, 1325
  from polysilanes 1264, 1266, 1267, 1269, 1270
  from silacyclobutanes 863, 864, 1258, 1259
Silyl shifts—*see* Silyl migrations
Silylsilylenes—*see also* Trimethylsilylsilylenes
 as disilene isomers 1156, 1320, 1321
Silylstannanes, transmetallation of 810, 811
Silylstannation, reactivity of Si-containing reagents in 482
β-Silylstyrenes, reactions of, β-silicon effect in 626
Silyltriazides, photolysis of 1289
α-Silylvinyl cations,
 ab initio calculations for 597
 gas-phase studies of 604
 in solvolysis 615, 616
β-Silylvinyl cations—*see also* α-Anisyl-β-silylvinyl cations, α-Mesityl-β-silylvinylcations, α-Phenyl-β-silylvinyl cations, α-Tolyl-β-silylvinyl cations
 ab initio calculations for 599–601
 gas-phase studies of 604, 605, 607
 NMR spectra of 663–687
β-Silylvinylsulphones, desulphonylation of 1860, 1861
Silylyne complexes, with group 8 transition metals 2086
Silynes—*see also* Disilynes
 theoretical studies of 1008–1010, 1148
Simmons–Smith reaction 1847
 steric effects of silyl groups on 457, 458
Si_3N_4, precursors to 2252, 2253
Si—N bonds, geometry of 198, 200, 203–211
 in silatranes 1460, 1467–1469, 1476–1478
Si=N bonds, geometry of 251–253
SiNBX materials, precursors to 2265–2272
Singlet–triplet energy gaps, in silylenes 2505–2507
SiN_2 isomers, matrix isolation of 1179
SiO_2, precursors to 2295–2298
Si—O bonds,
 cleavage of 1488–1490
 geometry of 213, 218–228
Si=O bonds,
 formation of 1068–1083
 geometry of 252, 254
 theoretical studies of 1065–1067
Si—O—Si skeleton,
 bond critical point properties of electron density distribution of 113–115
 potential energy surface for, in disilicic acid 109–113
Si—P bonds, geometry of 211–218

Si=P bonds, geometry of 252–254
Si—S bonds, geometry of 228–233
Si=S bonds,
  formation of 1083–1097
  geometry of 252, 254
  theoretical studies of 1065–1068
Si=Se bonds, formation of 1097–1100
Si—Si bonds,
  electroformation of 2179
  geometry of 197–202
  homolytic cleavage of 1319, 1320
Si=Si bonds, geometry of 247–250
Si–Si coupling, cathodic, of halosilanes 1214–1218
Sodalite 116
Sol–gel cage 2346–2348
Sol–gel glasses, photochromic 2349, 2350
Sol–gel hybrids 2342–2346
  physical 2342, 2343
  sequential interpenetrating networks 2343, 2344
  silsesquioxane-containing 2345, 2346
  simultaneous interpenetrating networks 2344, 2345
Sol–gel materials,
  dyes as dopants in 2350, 2351
  formation of 2319, 2320
Sol–gel matrices,
  covalent entrapment of organic functional groups in 2329–2342
  direct entrapment of molecules in 2320–2329
  enzymes and antibodies 2325–2327
  organometallic catalysts 2322–2325
  photochemistry within 2348, 2349
Sol–gel optics 2349–2354
Sol–gel process 2319, 2320
  photoprobes for 2346–2348
Sol–gel sensors 2321, 2322
Solvolysis,
  additivity of substituent effects 619
  anchimeric effect in 619
  β-germanium effect in 625, 626, 630
  δ-germanium effect in 642
  ground-state effects in 612, 613, 615
  isotope effects in 618, 619, 622, 623, 633, 635
  kinetic effect of disilanyl group in 632–635
  leaving group effects in 613
  mechanisms for 612, 614, 616–618, 621–624, 630, 633
  neighbouring group participation in 612, 619
  nucleophilic solvent assistance in 612
  rates of 612, 613
  α-silicon effect in 610–616
  β-silicon effect in 616–635

  dihedral angle dependence of 620
  γ-silicon effect in 635–641
  δ-silicon effect in 641–645
  solvent effects in 618, 621
  steric effects in 612–615
  β-tin effect in 625, 626, 630
  δ-tin effect in 642, 645
  transition state of 616
SOS-DFPT method 567
SOS-DFT method 695
Speier's catalyst 1691, 1726, 1727
Spinning sideband 310
Spironucleosides, synthesis of 1567
Spirosilicates 1351–1370
  bridged 1368
  NMR spectra of 1355–1357, 1362–1366, 1369, 1370
  polynuclear 1367–1370
  X-ray studies of 1355, 1362, 1368
  zwitterionic 1357–1367, 1435
λ^5-Spirosilicate zwitterions, X-ray–NMR studies of 324, 325
Stability, effect of silyl groups on 485–487
Stannaindenide anions 2022
Stannanes—see Cyclotetrastannanes, Cyclotristannanes, Ethynylstannanes, Silylstannanes
Stannaneselones, synthesis of 1093, 1094
Stannanethiones,
  electronic spectra of 1092
  synthesis of 1093, 1094
Stannaprismanes—see Perstanna[n]prismanes
Stannenes, synthesis of 909
Stannoles, synthesis of 1970, 1971
Stephens–Castro coupling 1836, 1837
Steric effects, of silyl groups,
  comparison of 433–438
  on acylation 438
  on addition 438–441
  on aldol condensation 441, 442
  on alkylation 442, 443
  on allylation 443
  on cyclization 443–445
  on cycloaddition 445, 446
  on decomposition 446, 447
  on deprotonation 447
  on desilylation 447, 448
  on elimination 448, 449
  on ene reaction 449
  on epoxidation 449, 450
  on hydroboration 450, 451
  on hydroethoxycarbonylation 451
  on hydroformylation 451
  on hydrosilylation 451, 452
  on isomerization 452
  on migration and rearrangement 452–454

Steric effects, of silyl groups (*cont.*),
   on osmylation 454, 455
   on oxymercuration 455
   on ozonolysis 455, 456
   on Peterson olefination 456
   on phosphonylation 456, 457
   on polymerization 457
   on reduction 457
   on Simmons–Smith reaction 457, 458
   on substitution 458
  comparison with H atom and R groups,
   on addition 459, 460
   on aldol condensation 460
   on alkylation 460
   on allylboration 461
   on allylsilylation 461
   on carbenoid rearrangement 462
   on cyclization 462–465, 473, 474
   on cycloaddition 465–467
   on cyclopropanation 467–469
   on dehydration 469
   on ene reaction 469, 470
   on epoxidation 470
   on hydroboration 472, 473
   on lithiation 473
   on oxidation 473
   on rearrangement 474
   on reduction 470–474
   on substitution 474, 475
Steric protection, in multiple bonds to Si 1064, 1101
Stevens migration 1033
Stishovite 115
Stokes shifts 1316
Styrenes—*see also* β-Silylstyrenes
  hydrosilylation of 1691, 1692
Styryl cations 603
  NMR spectra of 646
Styryldisilanes, *p*-cyano-substituted, photolysis of 1328
Substitution at silicon,
  geometry of intermediate/transition state 498–503
  kinetics of 498
  nucleophile-catalysed 506–510
  pseudorotation at pentacoordinate Si 503–506
  racemization during 506–510
  stereochemistry of 496
    effect of nucleophile and leaving group 496, 497
    effect of substrate structure 497, 498
Substitution reactions,
  electrophilic aromatic, selectivity of 562
  reactivity of Si-containing reagents in 475–477

  steric effects of silyl groups on,
    comparison of 458
    comparison with H atom and R groups 474, 475
Sulphides, anodic alkoxylation of 1197, 1198
α-Sulphinyl carbanions, reactions with acylsilanes 1658
Sulphonate esters, solvolysis of, γ-silicon effect in 637
Sulphur ylids 1653
  reactions with acylsilanes 1640
Suzuki–Miyaura coupling 1828, 1836, 1837
Swern oxidation 1622, 1628
Switching-angle sample spinning—*see* SASS
Synchronous hydride transfer (SHT) 576

Taft constants 625
Tamao oxidation 1850
Tellurosilanes 1889, 1890
Tessier–Young dimers 2113
Tetracyclooctasilanes 2207
  NMR spectra of 272–275
Tetragermatetrahedrane, synthesis of 129
Tetrahedranes—*see also* Silatetrahedranes, Tetragermatetrahedrane
  strain energy of 120, 121
Tetrahydrofurans, synthesis of 1806
Tetrahydropyrans, synthesis of 1800, 1807–1809
Tetrakis(trimethylsilyl)silane, NMR spectra of 333
2,3,4,5-Tetramethylsiloles, synthesis of 1972
Tetrasilabicyclo[4.1.0]hept-1(7)-ene 2450, 2451
Tetrasilabicyclohexanes 2410, 2411
  photoformation of 1280
Tetrasilacyclobutadiene 87, 88
  dication of 34
Tetrasilacyclobutenes, photolysis of 1304
Tetrasilacyclobutene–tetrasilabicyclo-[1.1.0]butane photoisomerization 1332
1,2,6,7-Tetrasilacyclodeca-3,4,8,9-tetraene 2448, 2449
Tetrasilacyclohepta-1,2-diene 2447
Tetrasilacyclohexynes 2443, 2444, 2446
  NMR spectra of 278
Tetrasilagermacyclopentanes, photolysis of 1270
Tetrasilanes—*see also* Cyclotetrasilanes
  molecular orbitals of 1314, 1315
  nucleophilic cleavage of 799
  spectra of 1314
Tetrasila[2.2]paracyclophane, NMR spectra of 277
Tetrasilatetrahedranes 120, 2205
  as reaction intermediates 125, 126
  geometry of 123, 124
  isomerism in 125–127

NMR spectra of 146
reactions of 128
strain energy of, substituent effects on 124
structure of 123, 128, 129
synthesis of 128
Tetrasilicates, NMR spectra of 326–328
Tetrasilyldisilenes, matrix isolation of 1157
Tetrathiadisilabicyclopentanes, photolysis of 1242
Tetrathiasilolanes, reactions with Ph_3P 1090, 1092, 1093
Tetrazoles—see Germadihydrotetrazoles, Siladihydrotetrazoles
Thermochroism, in peralkylpolysilanes 1314, 1318
Thiadisilacyclopropanes 1881
Thiadisiletanes 1882—see also Dithiadisiletanes
photolysis of 1094
1-Thia-2-silacyclopentanes 1883
Thiasiliranes 2416, 2418
Thiasiliranethiones, formation of 1089
1,3-Thiazolidines, reduction of, using $(TMS)_3SiH$ 1545
Thienamycin, synthesis of 1607
Thioacetals—see Dithioacetals, Hemithioacetals, Oxathioacetals
Thioacylsilanes, synthesis of 1650, 1651
Thioaldehydes, synthesis of 1093, 1094
Thioketenes—see Bis(trimethylsilyl)thioketenes
α-Thioketocarbenes 1095, 1096
Thiolanes—see 1,3-Oxathiolanes
Thiolanones—see 1,3-Oxathiolanones
α-Thiomethylsilanes, anodic oxidation of 1197, 1198
Thiones—see Plumbanethiones, Silanethiones, Silathiones, Stannanethiones
Thionocarbonates, as radical precursors 1569
Thionoesters, reduction of, using $(TMS)_3SiH$ 1549
Thiophenes—see α-Trimethylsilylthiophenes
Thiosilanes 1870–1874—see also Arylthiosilanes
synthetic applications of 1677, 1678
δ-Tin effect 642, 645
Titanium catalysts, for hydrosilylation, asymmetric 1749
Toluene, trimethylsilylation of 562
α-Tolyl-β-silylvinyl cations, NMR spectra of 676–678
Transesterification, enzymatic 2388–2390
Transetherification 1633
Transition-metal ions, reactions with silicon-containing molecules 1110–1115
Transition-metal–silene complexes 1111, 1112
Transition-metal–silicon bond, thermochemistry of 1110

Transition-metal–silylene complexes 1111–1113
Trialkoxysilanes,
as precursors for functionalization 2329–2331
reactions of 1450–1457
Trialkylsilanes, hydrogen donor abilities of 1541, 1542
Trialkylsilyl azides, synthetic applications of 1672–1674
(Trialkylsilyl)carbenes—see also (Trimethylsilyl)carbene
reactions of 727
Trialkylsilyl cyanides, synthetic applications of 1670–1672
Trialkylsilyl groups,
activating and directive effects of 356–425
donor/acceptor qualities of 2138
Trialkylsilyl halides, synthetic applications of 1667–1670
Trialkylsilyl halocarboranes 583, 585, 586
silicenium ion/halonium ion character of 530, 587, 588
Trialkylsilyl trifluoromethanesulphonates, synthetic applications of 1676, 1677
Triazasilatranes,
basicity of 1518
IR spectra of 1521
mass spectra of 1523
NMR spectra of 1521
reactions of 1515–1520
salts of 1518, 1519
synthesis of 1514, 1515
X-ray studies of 1522
Triazasilatranyl group, substituent effects of 1523
Triazidosilanes, as sila-isonitrile/-nitrile photolytic precursors 1159
Triazoles—see Germadihydrotriazoles, Silahydrotriazoles
Tribenzosilatranes,
NMR spectra of 1514
reactions of 1513
synthesis of 1511–1513
X-ray studies of 1513
Tricarbasilatranes,
synthesis of 1510
X-ray studies of 1511
Tricyclohexasilanes 2207
NMR spectra of 272–275
Tricyclosilachalcogenides 1916–1923
Tridymite 109, 112, 116
(Trifluoroacetyl)triphenylsilane, synthesis of 1621
Trigermanium ions 551
(Trihalomethyl)trialkylsilanes, synthetic applications of 1675

Trimesitylsilicenium ion 542, 543
(Trimethoxysilyl)carbene 709
  reactions of 727, 743
(Trimethylgermyl)(trimethylsilyl)carbene 716, 717
1,2-Trimethylsiloxy shifts, in photolysis of silenes 1273, 1276, 1285
Trimethylsilylation, electrochemical 1223–1225
(Trimethylsilyl)carbene 704, 709, 715
  formation of 739–741
  reactions of 727, 740–742, 774
β-Trimethylsilylcarboxylic acids, anodic oxidation of 1206
Trimethylsilyl cyanohydrin, enantioselective formation of 1671
Trimethylsilyldiazomethane, synthetic applications of 1675, 1676
Trimethylsilylferrocenes,
  cyclic voltammetry of 2136, 2137
  X-ray studies of 2135
β-Trimethylsilyl ketones 1636, 1637
Trimethylsilylmethyldisilenes, reactions with alcohols 838, 839
Trimethylsilyl phosphites, anodic oxidation of 1201
N-Trimethylsilylpyrrole, anodic oxidation of 1210, 1211
1,3-Trimethylsilyl shifts 880
(Trimethylsilyl)silanes—*see* Arylbis(trimethylsilyl)silanes, Tetrakis(trimethylsilyl)silane, Tris(trimethylsilyl)silanes,
Trimethylsilylsilylenes, matrix isolation of 1176, 1177
Trimethylsilyl substituents, electronic effects of, in cyclopentadienyl π-complexes 2137
α-Trimethylsilylthiophenes, anodic oxidation of 1189, 1190
(Trimethylsilyl)vinylcarbene 745, 746
Tri(methyltrichlorosilane), synthesis of 1582–1585, 1587, 1592–1594
Triorganyl halides, solvolysis of 483
Triorganyl hydrides, solvolysis of 483
Trioxadisilabicyclopentanes 1911
Triphenylsilanes—*see also* (Trifluoroacetyl)triphenylsilane
  reactions of 578
Triphenylsilanethiol, reactions of 1678
Triquinanes, synthesis of 1561, 1566
Tris-(9-borylbicyclo[3.3.1]nonyl)silicenium ion 547–549
Tris-catecholato complexes 1416
Tris(dialkylamino)silanes, reactions of 1458
Trisilabenzenes 10
  theoretical aspects of 21, 22
Trisilabicyclopentanes 2409, 2410
Trisilacyclobutanes—*see also* Germatrisilacyclobutanes, Oxatrisilacyclobutanes
  formation of 958
Trisilacycloheptanes, photolysis of 1319, 1320
Trisilacycloheptenes 2437, 2438
  photoreactions of 1298–1300
Trisilacyclohexadienes 2447
  NMR spectra of 278
Trisilacyclopropanes,
  photoformation of 1269
  photolysis of 1268, 1269
Trisilacyclopropenium cation 33
Trisilanes—*see also* Aryltrisilanes, Cyclotrisilanes, Tri(methyltrichlorosilane)
  cleavage of,
    nucleophilic 783, 797
    reductive 798
  photolysis of 2470
  pyrolysis of 2466
  transition-metal substituted, photolysis of 1261
Trisilanyl bridges 2155
Trisilaprismanes, as trisilabenzene isomers 22
[3.1.1]Trisilapropellane 2437, 2438
Trisilapropenylidene 68, 69
Trisilatriselenacyclohexanes, photolysis of 1289
Trisilylenes, as trisilabenzene isomers 22
Trisilyloxonium ions 571, 572
Tris(trimethylsilyl)cyclopropenium cation, X-ray studies of 648, 649
Tris(trimethylsilyl)silanes—*see also* Vinyltris(trimethylsilyl)silane
  as hydrosilylating agent 1552–1556
  as reducing agent 1540, 1543–1552
  autoxidation of 1542, 1543
  reactions of 1911, 1912
Tris(trimethylsilyl)silyl selenide derivatives 1888, 1889
Tris(trimethylsilyl)silyl tellurol derivatives 1890
Trithiadisilabicyclopentanes 1911
Trithiadisilacyclopentanes 1910
Trithiasilatranes, synthesis of 1524
Trityl salts 518
Troger's base, rhodium(III) complex of, as hydrosilylation catalyst 1719

Ultraviolet spectroscopy,
  of bicyclosilachalcogenides 1915, 1916
  of cyclopolysilanes 2181, 2188
  of disilenes 829, 830
  of silaisonitriles 1021
  of silanediimines 1030
  of silanimines 1044, 1045
  of silatranes 1471
  of silenes 996
  of β-silyl cations 649, 650

of silylenes 2512–2516
Ultraviolet/visible spectroscopy,
  of acylsilanes 1603–1605
  of silanethiones 1090
Umpolung, oxidizable, of allylsilanes 1193
Union Carbide, Direct Process and 1585
α,β-Unsaturated acylpolysilanes,
  photolysis of 1276
α,β-Unsaturated acylsilanes,
  iron–carbonyl complexes of 1659
  reactions of 1653–1659
  synthesis of 1622–1629

Vibrational frequencies, of $H_2Si=X$ 1066
Vinylcarbene–cyclopropene isomerization 744–746
Vinyl chloride, as substrate in Direct Process 1589
Vinyldisilanes, photolysis of 1248, 1249, 1259, 1260, 1293
Vinyl ethers, hydrolysis of, β-silicon effect in 626, 627
Vinylidenesilylene 2555–2557
  matrix isolation of 1164
Vinyloxysilanes, photoformation of 1272
Vinylsilanes—*see also* Vinyldisilanes, Vinyltris(trimethylsilyl)silane
  addition of electrophiles to 1832–1834
  as chain-transfer agents 1861
  carbonylation of 1839, 1840
  coupling of 1824–1826, 1836, 1837
  cathodic 1228
  cycloadditions of 1847–1849, 1853–1855
  halodesilylation of 1796, 1797
  hydrogenation of 1861
  hydrosilylation of 1693
  oxidation of 1841–1843, 1846
  protodesilylation of 1795, 1796
  reactions of,
    effect of R_3Si on 388–397
    ene 1858
    radical 1835, 1836
    with acyl halides 1819
    with imines 1816, 1817
  thermochemistry of 161
  transmetallation of 1828, 1829
1-Vinylsilatranes, reactions of 1494–1499
Vinylsilylenes,
  matrix isolation of 1165
  synthesis of 2471
Vinyl silyl ethers, photocycloadditions of 1295, 1296
Vinyltris(trimethylsilyl)silane, photolysis of 1266, 2471
N-Vinylureas, hydrosilylation of 1689
Vitamin A, synthesis of 1635

Wacker, Direct Process and 1585, 1587
Wadeite 111
Wagner–Meerwein rearrangement 641
Wheland intermediates 561
Wiberg-type silenes,
  adducts of 879, 932
  cycloadditions of 940–944, 953
  dimerization of 917, 931
  NMR spectra of 985, 991
  rearrangement of 911–913
  structure of 983
  synthesis of 873, 876, 878
Wideline separation experiment (WISE) 310
Wittig reactions 1622, 1623, 1639, 1655, 1858, 1859
Wolff rearrangement 751, 753
  of silylcarbenes 712, 718, 720, 761
Woodward–Hoffmann rules 1320
Wurtz reactions 730, 740, 2281, 2428–2430

Xerogel 2321
X-ray–NMR studies 275–277, 280, 315, 317–325, 339–341
X-ray structure determination 183, 184
  effect of silyl groups on 487, 488
  for acylsilanes 1605, 1606
  for arsasilenes 1053
  for benzosilatranes 1513
  for bis(germanimine)s 1051
  for carbasilatranes 1510, 1511
  for cyclopentadienylsilanes 2131
  for phosphasilenes 1052
  for Si—Br bonds 245, 247
  for Si—C bonds 185–197
  for Si=C bonds 250, 251
  for Si—Cl bonds 239–246
  for Si—F bonds 233–239
  for Si—I bonds 245, 247
  for silaallenes 1000
  for silanethiones 1090, 1091
  for silanimines 1042, 1043
  for silatranes 1460–1465
  for silatranones 1508
  for silenes 983, 984
  for siliconates, pentacoordinate 284
  for silole dianions 989
  for β-silylcyclohexadienyl cations 652–655
  for silylcyclopentadienide anions 2134–2136
  for Si—N bonds 198, 200, 203–211
  for Si=N bonds 251–253
  for Si—N complexes 1377–1379
  for Si—O bonds 213, 218–228
  for Si=O bonds 252, 254
  for Si—O complexes 1395–1397
  for Si—P bonds 211–218
  for Si=P bonds 252–254
  for Si—S bonds 228–233

X-ray structure determination (*cont.*)
  for Si=S bonds 252, 254
  for Si—Si bonds 197–202
  for Si=Si bonds 247–250
  for spirosilicates 1355, 1362, 1368
  for triazasilatranes 1522
  for trimethylsilylferrocenes 2135
  for tris(trimethylsilyl)cyclopropenium cation 648, 649
  summary of values obtained 254–257

Xylenes—*see* α,α'-Bis(trimethylsilyl)xylenes

Yttrium catalysts, for hydrosilylation of alkynes 1721, 1722

Zinc,
  as inhibitor in Direct Process 1590
  as promoter in Direct Process 1582–1589
Zinc oxide, use in Direct Process 1585

Index compiled by P. Raven